Henry Gray appears in the foreground, third from the left.
(Reproduced by permission of the Governors of St. George's Hospital.)

Anatomy of the Human Body

Anatomy
of the Human Body

by Henry Gray, F.R.S.

Late Fellow of the Royal College of Surgeons;
Lecturer on Anatomy at St. George's Hospital Medical School,
London

Thirtieth American Edition

EDITED BY

Carmine D. Clemente, A.B., M.S., Ph.D.

Professor of Anatomy and Director of the Brain Research Institute
University of California at Los Angeles School of Medicine
AND
Professor of Surgery (Anatomy), Charles R. Drew Postgraduate
Medical School, Los Angeles, California

Williams & Wilkins
A WAVERLY COMPANY

BALTIMORE • PHILADELPHIA • LONDON • PARIS • BANGKOK
HONG KONG • MUNICH • SYDNEY • TOKYO • WROCLAW

A Lea & Febiger Book

Williams & Wilkins
351 West Camden Street
Baltimore, Maryland 21201-2436 USA

Rose Tree Corporate Center
1400 North Providence Road
Building II, Suite 5025
Media, Pennsylvania 19063-2043 USA

Executive Editor - R. Kenneth Bussy
Project Editor - Frances M. Klass
Production Manager - Thomas J. Colaiezzi

GRAY'S ANATOMY
Translations:
29th Edition—
  Japanese Edition by Hirokawa Publishing Co.,
    Tokyo, Japan—1981
  Portuguese Edition by Editora Guanabara Koogan,
    Rio de Janeiro, Brazil—1977
  Spanish Edition by Salvat Editores, S.A., Barcelona,
    Spain—1976

Library of Congress Cataloging in Publication Data
Gray, Henry, 1825–1861.
  Anatomy of the human body.

  Includes bibliographies and index.
  1. Anatomy, Human.   I. Clemente, Carmine D.
  II. Title.
[DNLM: 1. Anatomy.   QS 4 G779an]
QM23.2.G73 1984   611   84-5741
ISBN 0-8121-0644-X

Print Number: 5

DEDICATION

My work on this 30th Edition of Gray's Anatomy
is gratefully dedicated to my father and mother,
ERMANNO CLEMENTE AND CAROLINE CLEMENTE, who
with great wisdom, taught me in my youth the
everlasting value of education,
and to
DR. WILLIAM FREDERICK WINDLE, my teacher,
who taught me to understand a way of life so
effectively stated in the following passage from
Robert Frost:

. . . My object in living is to unite
My avocation and my vocation,
As my two eyes make one in sight.
Only when love and need are one,
And the work is play for mortal stakes,
Is the deed ever truly done,
For Heaven and the future's sakes.

C.D.C.

American Editions of Gray's Anatomy

DATE	EDITION	EDITOR
June 1859	*First American Edition*	Dr. R. J. Dunglison
February 1862	*Second American Edition*	Dr. R. J. Dunglison
May 1870	*New Third American from Fifth English Edition*	Dr. R. J. Dunglison
July 1878	*New American from the Eighth English Edition*	Dr. R. J. Dunglison
August 1883	*New American from the Tenth English Edition*	Dr. R. J. Dunglison
September 1887	*New American from the Eleventh English Edition*	Dr. W. W. Keen
September 1893	*New American from the Thirteenth English Edition*	(Dr. W. W. Keen; changes as in 1887)
September 1896	*Fourteenth Edition*	Drs. Gallaudet, Brockway and McMurrich
October 1901	*Fifteenth Edition*	Drs. Gallaudet, Brockway and McMurrich
October 1905	*Sixteenth Edition*	Dr. J. C. DaCosta
September 1908	*Seventeenth Edition*	Drs. DaCosta and Spitzka
October 1910	*Eighteenth Edition*	Dr. E. A. Spitzka
July 1913	*Nineteenth Edition*	Dr. E. A. Spitzka
October 1913	*New American from Eighteenth English Edition*	Dr. R. Howden
September 1918	*Twentieth Edition*	Dr. W. H. Lewis
August 1924	*Twenty-first Edition*	Dr. W. H. Lewis
August 1930	*Twenty-second Edition*	Dr. W. H. Lewis
July 1936	*Twenty-third Edition*	Dr. W. H. Lewis
May 1942	*Twenty-fourth Edition*	Dr. W. H. Lewis
August 1948	*Twenty-fifth Edition*	Dr. C. M. Goss
July 1954	*Twenty-sixth Edition*	Dr. C. M. Goss
August 1959	*Twenty-seventh Edition*	Dr. C. M. Goss
August 1966	*Twenty-eighth Edition*	Dr. C. M. Goss
January 1973	*Twenty-ninth Edition*	Dr. C. M. Goss
October 1984	*Thirtieth Edition*	Dr. C. D. Clemente

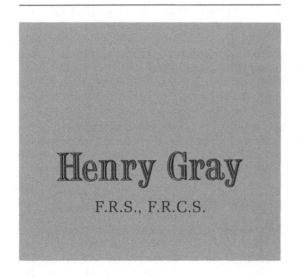

Henry Gray

F.R.S., F.R.C.S.

THE ORIGINATOR of this now famous book, which has reached its one hundred and twenty-fifth year of continuous publication in America, was born in Windsor, England, in 1825. A letter written to Lea & Febiger, by Henry Gray's great nephew, F. Lawrence Gray, has supplied this date as well as some interesting biographical information.

Henry's father, William Gray, was born in 1787. He is reported to have been an only child. Whether for protection or from special interest, he was placed in the royal household where he was brought up. In 1811, before the Regency, and while he was in his twenties, he was Deputy Treasurer to the Household of the Prince of Wales. On the accession of George IV as Regent in 1820, William Gray became the King's private messenger and continued in this capacity, with William IV, until his death at the age of forty-seven. He married Ann Walker in 1817 and during George IV's reign lived at Windsor Castle in accommodations provided for him and his wife. When William IV acceded to the throne in 1830, the Grays moved to London in order to be near Buckingham Palace, and either immediately or soon after that No. 8 Wilton Street became the family residence.

Henry Gray had two brothers and one sister. His younger brother, Robert, trained to become a naval surgeon. He died at sea in his twenty-second year. Henry's sister, younger than he, apparently never was married. His older brother, Thomas William, adopted the legal profession, and was Attor-

ney to the Queen's Bench in 1841, and, in 1846, Solicitor of the High Court of Chancery. He had a large family—seven sons and three daughters. His third son, Charles, the nephew whom Henry treated for smallpox in 1861, lived to the ripe age of fifty-three. His fifth son, Frederick, moved to South Africa, married in 1881 and had six children, one of whom was the F. Lawrence Gray mentioned above.

After his father's death in 1834, Henry Gray continued living at 8 Wilton Street with his mother, brothers and sister. This is undoubtedly the address from which he later carried on his short practice and the place where he wrote his famous book. He and his brothers were educated at one of the large London schools, possibly Westminster, Charterhouse, or St. Paul's. He was engaged to be married at the time of his death.

Henry Gray's signature appears on the pupil's book at St. George's Hospital, London, as a "perpetual student" entering on the 6th of May, 1845. Four years later, in 1849, his name appears in the Proceedings of the Royal Society with the M.R.C.S. after it, the approximate equivalent of our M.D. While still a student in 1848, he won the triennial prize of the Royal College of Surgeons for an essay entitled "The Origin, Connections and Distribution of the Nerves to the Human Eye and Its Appendages, Illustrated by Comparative Dissections of the Eye in Other Vertebrate Animals." He was appointed for the customary year as house surgeon to St. George's Hospital in 1850. Successively thereafter he held the posts of demonstrator of anatomy, curator of the museum, and lecturer on anatomy at St. George's Hospital. In 1861 he was surgeon to St. James Infirmary and was a candidate for the post of assistant surgeon to St. George's Hospital and would certainly have been elected had he not died from confluent smallpox which he contracted from his nephew, Charles, whom he was treating. He was buried at Highgate Cemetery in June 1861. Sir Benjamin Brodie, president of the Royal Society, wrote, "I am much grieved about poor Gray. His death, just as he was on the point of obtaining the reward of his labors, is a sad event indeed . . . Gray is a great loss to the hospital and school. Who is there to take his place?"

During his lifetime Henry Gray received outstanding recognition for his original investigations. That they have received so little mention since his death is as surprising as the lack of information about his life. The study of the eye which won him the Royal College of Surgeons prize was expanded into an embryological work, "On the development of the retina and optic nerve, and of the membranous labyrinth and auditory nerve," published in the Philosophical Transactions of the Royal Society in 1850. It contains the earliest description of the histogenesis of the retina. Two years later the Transactions contained another article, "On the development of the ductless glands in the chick." This must have stimulated Gray's interest in the spleen, which he classed as a ductless gland, because he obtained an allotment of funds for further study from the annual grant placed at the disposal of the Royal Society by Parliament for the promotion of science. The result was a monograph of 380 pages, "On the Structure and Use of the Spleen," which won him the triennial Astley Cooper Prize of £300 (about $1,500.00) in 1853. It was published by J. W. Parker and Son in 1854, but appears now to be rare. Numerous "first observations" recorded in this book have escaped notice by all subsequent authors writing about the spleen. Gray described, among other things, the origin of the spleen from the dorsal mesogastrium ten years before Müller, who is usually given credit for the discovery

As a result of his ability and accomplishment, Henry Gray was made a Fellow of the Royal Society at the very young age of twenty-five. Besides anatomy, his interests also included pathologic and clinical investigation. In 1853 he had a paper in the Medico-Chirurgical Proceedings entitled, "An Account of a Dissection of an Ovarian Cyst Which Contained Brain," and in 1856 a more extensive treatise entitled, "On Myeloid and Myelo-Cystic Tumours of Bone; Their Structure, Pathology, and Mode of Diagnosis."

His crowning achievement, however, and the one which is the source of his lasting fame is the publication, *Anatomy, Descriptive and Surgical,* now widely known as *Gray's Anatomy.*

125 Years of Gray's Anatomy

The first edition of *Gray's Anatomy* was published in London by J. W. Parker and Son in 1858 and in June of 1859 in Philadelphia

by Blanchard and Lea, who purchased the American rights for the book. The American edition was identical with the English, except that many typographic errors had been corrected, the index considerably improved, and the binding made more rugged. It contained xxxii + 754 pages and 363 figures. The drawings were the work of Dr. H. Van Dyke Carter, of whom Gray writes in his preface, "The Author gratefully acknowledges the great services he has derived, in the execution of this work, from the assistance of his friend, Dr. H. V. Carter, late Demonstrator of Anatomy at St. George's Hospital. All the drawings, from which the engravings were made, were executed by him. In the majority of cases, they have been copied from, or corrected by, recent dissections, made jointly by the Author and Dr. Carter."

Blanchard and Lea obtained the services of Dr. R. J. Dunglison to edit the first and the next four American editions. Dunglison corrected the typographic and other small errors and improved the index but made few alterations or additions in the text. Almost no adaptation was required because medical education in this country and in England were still much alike.

Henry Gray had just finished preparing the second edition before his untimely death. This was published in 1860 in England and in 1862 in this country under the editorship of Dunglison. The next edition published in America was a "New American from the 5th English Edition." This appeared in 1870 and was bound in either cloth or sheep. Another pause, and there appeared the "New American from the 8th English" in 1878, and "The New American from the 10th English" in 1883. Dunglison was again the editor, but a new editor, W. W. Keen, revised extensively the section of topographic anatomy at the end of the book. The following edition, a "New American from the Eleventh English," published in 1887, was edited and thoroughly revised by W. W. Keen. Color was used for the first time in the chapters on blood vessels and nerves. The "New American from the Thirteenth English" appeared in 1893, apparently edited by Keen and others.

In 1896 the reference to the English edition was dropped and we have the fourteenth edition with the editors Bern B. Gallaudet, F. J. Brockway, and J. P. McMurrich. J. C.

DaCosta, editor of the sixteenth edition in 1905, expanded the book, introducing much new material. E. A. Spitzka assisted DaCosta with the seventeenth edition in 1908 and edited the next two alone in 1910 and 1913. The publishers also tried a "New American from the Eighteenth English Edition" in 1913, but its sale was much smaller than that of the American edition of that year. In 1918, W. H. Lewis began his editorship, giving the book a scholarly treatment, reducing its length, improving the sections on embryology, and generally giving a more straightforward treatment of the various chapters. For his last edition, the twenty-fourth, published in 1942, he had the assistance of six associate editors.

Beginning with the twentieth edition in 1918, new editions appeared at regular intervals of six years; their dates and the names of the editors are listed on page vi. Dr. Charles Mayo Goss became responsible for the twenty-fifth through the twenty-ninth American Editions, these being spaced at five- to seven-year intervals. The twenty-seventh Edition was published in 1959 in order to have its publication fall on the 100th year.

Since the beginning of this century, the American and English editions have tended to drift apart. Even from an early date, new American editions were less frequent than the English, and if the American editions had been numbered independently from the beginning, the present edition would be the twenty-fifth, whereas the current English edition is the thirty-sixth. Somewhat more of the imprint of Henry Gray has been preserved in this American edition, in the illustrations especially, as it still has nearly 200 drawings based on the original illustrations by Carter. This treatise, offered initially by a young English anatomist and surgeon, is the oldest continuously revised textbook still currently in use by the medical professions. After one hundred and twenty-five years, this enduring recognition is a tribute that, perhaps, even Henry Gray may not have conceived would happen.

Preface

When I was asked to consider the preparation of this 30th American Edition of *Gray's Anatomy,* I spent an afternoon with Dr. Charles Mayo Goss at his home near Washington, D.C. Dr. Goss, who was affectionately known as "Chiz" by his many friends, had edited the five previous editions and had for the preceding three decades masterfully succeeded in retaining both the effectiveness and popularity of the book to students and practitioners of the medical, dental and other allied health professions. Dr. Goss warmly encouraged me to assume the responsibility for this 30th edition and recommended that I attempt to retain the classical tradition and style of the book, even if I planned, as he thought was necessary, a major revision of the text and figures. Chiz's twinkling smile should have alerted me to the enormous task that I was considering. Sadly, Dr. Goss's death in 1981 came about two years after I commenced serious work on this project, and I never had a further opportunity to benefit from his wisdom.

Eleven years have elapsed since the 29th American Edition appeared in 1973. Recognizing the special regard held for this book by generations of American anatomists, I felt that it was of special importance for me to understand the views held by many of my esteemed colleagues in the United States and Canada about the current usefulness of the book and how it might be altered or improved. Many informal discussions almost invariably pointed to the lasting value of having available for student and profes-sional use a comprehensive anatomic trea-tise that is organized by systems. Many valu-able suggestions from these conversations have been incorporated in this edition and my appreciation is extended to my many friends for their interest and their patience with my many questions.

Nearly two thirds of the chapters in this edition have been thoroughly revised and the remainder have been edited. The chapter on Developmental and Gross Anatomy of the Central Nervous System has virtually been rewritten as have extensive sections in at least ten other chapters. Many hundreds of new paragraphs and pages have replaced those of the 29th edition and, thus, numerous subjects have received some expanded attention and updating. To list them all here would not be appropriate, but a few of these are: certain aspects of the new cellular biology and embryology; the physical features and properties of bone and its formation; the development, classification and movement of joints; the form, attachment and actions of muscles, as well as the structure of skeletal and cardiac muscle; the vascular supply to the central nervous system; and the development of veins and lymphatics. Certainly, due to time limitations, many subjects did not get the attention they deserved, and thus, much still needs to be done in subsequent editions. Because the value of this book is essentially its information on the gross anatomy of the human body, it was not the intention of the editor to offer the reader comprehensive treatments in the fields of embryology, microscopic anatomy, physiology, and neuroscience. Excellent texts written by scholars expert in these subjects already exist, and to recount detailed features of their disciplines here is unnecessary and would have been presumptuous.

I have attempted to retain the classical style of the text, long considered one of the most elegant examples of descriptive scientific literature. In the extensively revised chapters, however, few paragraphs remain totally intact. I have tried to match the new with the old, to simplify some of the more complex sentence structure in order to clarify meaning, and to maintain a uniformity of prose. I hope that at least to some extent this effort has been successful.

A nearly complete change has been introduced in the headings within the chapters, yet the chapter subjects remain the same.

Using color print for various headings and subheadings will make it easier to find pertinent information on structures that in the past had to be identified by bold print within the paragraphs. In this respect, the student who must study and the practitioner who more quickly seeks information may both find the format of this edition beneficial. Additionally, the entire text has been reset and hundreds of more modern references and others that have proved valuable over time have been added to the reference lists at the end of each chapter. Many citations, in addition to those used in the text, have been made available for readers who require more knowledge than this book attempts to offer.

A significant number of new figures, never before published, have been introduced into this edition, along with numerous others borrowed from various sources. Many of the latter have come from original journal articles. The labels on most of the figures have been replaced with type that is easier to read. Certain older, less effective figures have either been eliminated or replaced by drawings that illustrate more clearly the subject matter. I believe the publishers should be commended for having done an excellent job in abiding by the principle to modernize the information and format and yet retain the classical look of this oldest (now 125 years), continuously used textbook in biomedical science.

Grateful acknowledgment must go to the many friends and colleagues who have contributed to this new edition, and my heartfelt appreciation is extended to so many authors and publishers who have allowed the reproduction of figures from their works or copyrighted properties. Especially mentioned, however, should be Professors Helmut Ferner and Jochen Staubesand, editors of Benninghoff and Goerttler's *Lehrbuch der Anatomie des Menschen,* published by Urban and Schwarzenberg in Munich, from which a considerable number of figures were taken. Along the way, I have benefited greatly from the advice, suggestions, or contributions on one subject or another from Professors Leslie B. Arey, Malcolm B. Carpenter, James D. Collins, Jack Davies, George E. Erikson, Fakhry G. Girgis, W. Henry Hollinshead, Lawrence Kruger, Lee V. Leak, Raymond J. Last, David S. Maxwell, Keith L. Moore, Charles R. Noback, Keith R. Porter, Nabil N. Rizk, Charles H. Sawyer, William F. Windle, Luciano Zamboni, and the late Elizabeth C. Crosby and Jan Langman.

Many others, as well, have contributed greatly to the work of this revision. Among those who deserve special mention and much praise are Jill Penkhus, who created much of the new artwork for this edition; Dr. Charles F. Bridgman, who allowed the use of several of his drawings made of brain dissections prepared by the late Professor A. T. Rasmussen; Thomas Colaiezzi, at Lea & Febiger, who handled production; Howard N. King, who designed this new edition; Anne Cope who prepared the index; Dr. Caroline Belz Caloyeras, whose dedicated assistance with typing and editing was absolutely essential for this book to have appeared; and Dorothy Di Rienzi and, earlier, Mary Mansor of Lea & Febiger, who copy-edited all of the chapters. My appreciation also is extended to John F. Spahr and George Mundorff of Lea & Febiger who suggested that I undertake the work on this edition and then patiently waited for its completion.

Most importantly, I wish to offer my gratitude to Julie, my wife, who through the years has made her special contributions to whatever work I have undertaken. Her heartening and unfailing encouragement to me and her belief in the fundamental objectives of education have been my inspiration and have lightened the thousands of hours spent on this book.

CARMINE D. CLEMENTE
Los Angeles, California

Contents

11. Development and Gross Anatomy of the Central Nervous System

12. The Peripheral Nervous System

13. Organs of the Special Senses

14. The Integument

15. The Respiratory System

16. The Digestive System

17. The Urogenital System

1

Introduction

In ancient Greece, at the time of Hippocrates (460 B.C.), the word anatomy, Ἀνατομή, meant a dissection, from τομή, a cutting, and the prefix ἀνά, meaning up. Today anatomy is still closely associated in our minds with the dissection of a human cadaver, but the term was extended very early to include the whole field of knowledge dealing with the structure of living things.

The history of anatomy is one of the most exciting sagas in the ascent of human civilization. Since the subject deals with the structure of the body, it has lured some of the greatest minds not only to contemplate the nature of man, but to discover first hand by the method of dissection how the human being is put together. However, religious and other social pressures and taboos over the centuries made dissection of the human cadaver frequently a dangerous and secret activity. The study of the structure of the human body was the first of the medical sciences, and from it have derived all the rest. Even before human dissection was practiced by Herophilos and Erasistratos (300–250 B.C.), Aristotle (384–323 B.C.) dissected animals, wrote a treatise on their anatomy, and laid the groundwork for a scientific study of their form and structure.

Today it is considered that the greatest contributor to the subject matter of human anatomy was a young Flemish physician from Brussels who worked in Padua, Italy

Fig. 1-1. Andreas Vesalius (1514–1564). This figure was copied from the second edition of Vesalius' *De Humani Corporis Fabrica*, published in 1555. Having also appeared in the first edition (1543), it illustrates Vesalius during his late twenties and is still the only authenticated portrait of him known to exist.

during the height of the Renaissance. In 1542, at the age of 28, Andreas Vesalius, breaking boldly from the millennia and the prevailing traditions of Galen, published his great treatise *De Humani Corporis Fabrica* (Fig. 1-1). His work was so impressive and accurate that his books became the principal texts used by students of medicine for the next 200 years. Vesalius had substituted observation by dissection for authoritarianism, thereby grounding his findings on a scientific basis. Anatomy has therefore become the branch of knowledge concerned with **structure** or **morphology.** Our fund of anatomical information has been increased greatly during the last three or four centuries by the study of minute structure with the microscope, by following the development of the embryo, and by the addition of new and refined technical methods. The whole field of anatomy has become very large, and as a result a number of subdivisions of the subject have been recognized and named, usually to correspond with a specialized interest or avenue of approach.

Gross anatomy is the study of morphology, by means of dissection, with the unaided eye or with low magnification such as that provided by a hand lens. The study of the structure of organs is **organology.** The study of the tissues is **histology.**

Microscopic anatomy is the study of structure with the aid of a microscope, and since it deals largely with the tissues, it is often called **histology.** Microscopic anatomy has another branch, the study of the cell, or **cytology.** An enormous increase in research into the finer structure of the cell has taken place in recent years, due to the adaptation of the **electron microscope** to the observation of ultrathin sections of the tissues.

Embryology or **developmental anatomy** deals with the growth and differentiation of the organism from the single-celled ovum to birth. These processes of development and maturation are also called *ontogeny.* It is impossible to understand clearly all the structures found in the adult body without some knowledge of embryology. Since human embryos, particularly the younger stages, are difficult to obtain for study, a large part of our knowledge has been gained from animals, that is, from **comparative embryology.** The second chapter in this book is devoted to a brief summary of **human embryology.** The development of each system is outlined in the appropriate chapters and sections.

Comparative anatomy refers to the study of the structure of all living creatures, in contradistinction to human anatomy which refers only to man. A comparison of all the known animal forms, both living and fossil, indicates that they can be arranged in a scale which begins with the simplest forms and progresses through various gradations of complexity and specialization to the most highly evolved forms. The unfolding of a particular race or species is called *phylogeny.* Many of the earlier stages of development in man and other higher animals resemble the adult stages of animals lower in the scale; hence it has been said that *ontogeny repeats phylogeny.* The study of comparative anatomy has contributed vitally to the understanding of human anatomy and physiology.

Terms of Position and Direction (Fig. 1-2)

THE ANATOMICAL POSITION. The traditional anatomical position, which has been arbitrarily accepted, places the body in the erect posture with the feet together, the arms hanging at the sides, and the thumbs pointing away from the body. This position is used in giving topographic relationships, especially those which are medial and lateral in reference to the limbs. The muscle actions and motions at the joints are also given with reference to this position unless it is stated otherwise. It is particularly important that the anatomical position be remembered in descriptions using anterior and posterior, or the more general and ambiguous terms such as up, down, over, under, below, and above.

The **median plane** is a vertical plane that *divides the body into right and left halves;* it passes approximately through the sagittal suture of the skull, and therefore any plane parallel to it is called a **sagittal plane. Frontal planes** are also vertical, but they course at right angles to the sagittal planes. Since one frontal plane passes approximately through the coronal suture of the skull, they are also called **coronal planes. Transverse** or **horizontal planes** cut across the body at right angles to both sagittal and frontal planes.

Ventral and **dorsal** refer to the front and back of the body, and are synonymous with **anterior** and **posterior.** The use of these terms in courses of comparative anatomy may lead to some confusion for the student of human anatomy, since man's posture is erect. In the hands, **palmar** or **volar** is substituted for ventral or anterior. The sole of the foot is called **plantar,** while **dorsal** is retained for the top or opposite surface of both the hand and foot.

Cranial and **caudal** indicate the head and tail respectively, and are synonymous with **superior** and **inferior** when human subjects are in the anatomical position. In comparative anatomy the latter terms might be confused with dorsal and ventral when referring to four-footed beasts.

Median refers to structures along the midline or median sagittal plane of the trunk or limb.

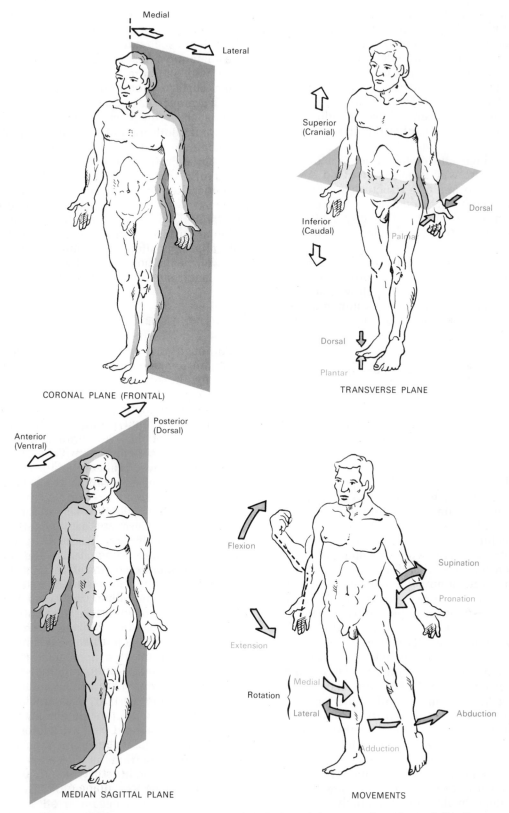

Medial

Lateral

CORONAL PLANE (FRONTAL)

Superior (Cranial)

Inferior (Caudal)

Dorsal

Palmar

Dorsal

Plantar

TRANSVERSE PLANE

Posterior (Dorsal)

Anterior (Ventral)

MEDIAN SAGITTAL PLANE

Flexion

Extension

Rotation { Medial Lateral

Supination

Pronation

Abduction

Adduction

MOVEMENTS

FIG. 1-2. Diagrams illustrating the planes of the body and the terms of position and direction.

Medial and **lateral** refer to structures nearer or farther away from the median plane. It should be noted that the Latin word *"medius"* means middle, not medial. The Latin word for medial is *"medialis."*

Superficial and **deep** indicate the relative depth from the surface, and are preferable to over and under, and above or below, since the latter may be confused with superior and inferior.

Proximal and **distal** indicate a direction toward or away from the attached end of a limb, the origin of a structure, or the center of the body. Proximal refers to a structure closer to the center of the body, while distal refers to a structure farther from the body center.

External and **internal** are most commonly used for describing the body wall, or the walls of cavities and hollow viscera. At times these terms may be synonymous with superficial and deep, or even with lateral and medial.

Terms for Movements at Joints (Fig. 1-2)

Flexion is that action which approximates two parts of the body connected at a joint. Flexion of the forearm at the elbow joint approximates the forearm to the arm; flexion of the calf at the knee joint approximates the calf to the posterior thigh. **Extension** is the act of straightening a limb, thereby diminishing the angle formed by flexion. Thus, straightening the flexed forearm at the elbow joint is an example of extension. Flexion and extension are, therefore, antagonistic actions. **Abduction** is the movement of a limb or other structure *away* from the middle line. Elevating an upper limb from the anatomical position at the side of the trunk to any point more lateral is an example of abduction. The opposite action, i.e., returning the abducted limb *toward* the midline is called **adduction.** The movements of flexion, extension, abduction, and adduction may be sequentially performed in a manner by which the end of a limb describes a circle or a cone; this is called **circumduction.**

Pronation is that movement of the forearm by which the palm of the hand is rotated to a backward position when the upper extremity hangs by the side of the body in the anatomical position. The pronated hand will rest palm down on the surface of the table. **Supination** is the opposite, i.e., the rotation of the hand so that the palm faces forward in the anatomical position. The supinated hand rests with its palmar surface upward on a table surface. **Rotation** is a revolving or turning movement of a limb or other structure, such as the back or head, around its long axis. Both **medial** and **lateral rotation** are commonly observed at the shoulder and hip joints. **Protraction** is the forward movement of a body part such as the jaw or shoulder. **Retraction** is the opposite, i.e., the backward movement of these structures. **Eversion** is the outward rotation of the sole (plantar surface) of the foot so that it faces laterally. Oppositely, **inversion** is the inward rotation of the sole of the foot resulting in the plantar surface facing medially.

Nomenclature

Although the majority of the common anatomical names are derived from those used by the ancient Greeks, considerable disagreement and confusion have arisen because of conflicting loyalties to teachers, schools, or national traditions. Shortly before the end of the 19th century, however, the need for a comprehensive system of nomenclature was realized, and a commission of eminent authorities from various countries was organized for this purpose. The system devised by this commission was adopted by the German Anatomical Society in 1895 at its meeting in Basel, Switzerland, and it has since been called the **Basle Nomina Anatomica** or the **BNA** (His, 1895).

The **BNA** gives the names in Latin, and many nations, including the United States, and Great Britain, adopted it, translating the terms into their own language whenever desirable. The French anatomists, however, did not entertain this nomenclature with much enthusiasm, preferring their own quite ancient tradition. The British anatomists gradually found certain inaccuracies and inconsistencies with their time-honored tradition which favored human anatomy related to the arbitrary "anatomical position." They broke away from the BNA and adopted their own revision in 1933 which they called the **Birmingham Revision** or **BR.** Even the German anatomists became restless under the restrictions of the BNA adoption of the "anatomical position" and preferred an attempt to establish a scientific language applicable to the comparative anatomy of vertebrates as well as to human anatomy. In 1937 the Anatomische Gesellschaft met in Jena and adopted a revision that is known as the **Jena Nomina Anatomica** or **INA.** The anatomists of the United States remained faithful to the BNA.

At the Fifth International Congress of Anatomists held at Oxford in 1950 a committee was appointed to attempt a new revision of anatomical names. This International Nomenclature Committee met at Paris in July 1955, and approved the **Paris Nomina Anatomica** or **PNA.**

This book utilizes the terms designated in the **PNA** except in a few instances where it has been abandoned deliberately in the interests of clarity. Since Latin is less familiar to students today than in the past, names are usually given in English with the official Latin name at times italicized in parentheses. Other recognized names are also added in parentheses.

Anatomical eponyms, that is, designations by the names of persons, are popular among clinicians and specialists. Many eponyms have been added to the other names in parentheses in order to expose the student to terms commonly used in clinical medicine.

Approaches to the Study of Anatomy

Human anatomy may be presented from at least three points of view: (1) **descriptive** or **systemic anatomy,** (2) **regional** or **topographical anatomy,** and (3) **applied** or **practical anatomy.** This book is primarily concerned with *systemic anatomy,* but *regional* and *applied anatomy* are introduced when their use facilitates communication.

SYSTEMIC ANATOMY. The body is composed of a number of systems whose parts are related by physiological as well as anatomical considerations. Each system is composed of several organs or tissues whose combined functions serve the body uniquely. Thus, although the anatomy of the oral cavity is remarkably different from that of the stomach or duodenum, these organs are presented in the same section because they all contribute to the functions of the digestive system. Although the study of

anatomy is concerned primarily with morphology, the knowledge of structure becomes more understandable and of practical value if the close association between structure and function is continually considered.

The systems of the body are as follows:

The **skeletal system,** the study of which is called *osteology,* consists of bones and cartilage. Its function is to support and protect the soft parts of the body.

The **system of joints** or **articulations** (*arthrology*) allows for the movement of bones while they are held together by strong fibrous bands, the ligaments.

The **muscular system** (*myology*) forms the fleshy parts of the body and is responsible for the purposeful movements of the bones at the joints.

The **vascular system** (*angiology*) includes the **circulatory system** consisting of the heart and blood vessels and the **lymphatic system** which transports the lymph and tissue fluids.

The **nervous system** (*neurology*) includes the **central nervous system,** composed of the brain and spinal cord, the **peripheral nervous system,** consisting of nerves and ganglia, and the **sense organs,** such as the eye and ear. The nervous system controls and coordinates all the other organs and structures, and relates the individual to his environment.

The **integumentary system** forms a protective covering over the entire body and is composed of the skin, hair and nails.

The **alimentary system** is composed of the food passages and the associated organs and glands of digestion.

The **respiratory system** supplies the body with oxygen and eliminates the carbon dioxide which results from the metabolic processes in the cells. It is composed of the air passages and lungs.

The **urogenital system** includes the kidneys and urinary passages and the reproductive organs in both sexes.

The **endocrine system** includes the thyroid, suprarenal, pituitary, and other ductless glands which control certain bodily functions by secreting hormones into the blood stream.

Splanchnology is the term that applies to the study of all the internal organs, especially those in the thorax and abdomen.

An account of the anatomy of any system would be incomplete without consideration of the microscopic anatomy and embryology of the organs in that system. Sections dealing with these fields are appropriately located in the text.

TOPOGRAPHICAL OR REGIONAL ANATOMY. **Regional anatomy** deals with the structural relationships within the various parts of the body, e.g., the pectoral region or axilla. Students in the laboratory usually approach the subject of anatomy regionally because they dissect by regions rather than by systems. Although the acquisition and organization of anatomical knowledge are easier for beginners if followed by systems, students in the medical fields must continually be on the alert to learn the relationship of the various parts to each other and to the surface of the body, because the final purpose of their study is to visualize these relationships in living subjects. The recent introduction of new radiologic scanning techniques in the clinics has reemphasized the importance of human dissection and of learning anatomical relationships. In addition to the dissection of the body, the study of topographical anatomy is carried out by the study of **surface anatomy, cross sectional anatomy,** and **radiographic** or **x-ray anatomy.**

Applied anatomy has a number of subdivisions, and is concerned with the practical application of anatomical knowledge in some clinical field or specialty. **Surgical anatomy, pathological anatomy,** and **radiological anatomy** are probably the largest fields.

Physical anthropology is closely related to human anatomy because it relies heavily on anatomical studies and measurements of man and other primates. It is more broadly a subdivision of human biology with particular emphasis on racial development, evolution, genetics, and paleontology.

References

(References are listed not only to those articles and books cited in the text, but to others as well which are considered to contain valuable resource information for the student who desires it.)

Barclay, J. 1803. *A New Anatomical Nomenclature Related to the Terms which are Expressive of Position and Aspect in the Animal System.* Ross and Blackwood, Edinburgh, 182 pp.

Donath, T. 1969. *Anatomical Dictionary with Nomenclature and Explanatory Notes.* English Translation edited by G. N. C. Crawford. Pergamon Press, Oxford, 634 pp.

His, W. 1895. *Die anatomische Nomenclature. Nomina Anatomica.* Verzeichniss der von der anatomischen Gesellschaft auf ihrer IX. Versammlung in Basel angenommenen Namen. Archiv fuer Anatomie und Physiologie, anatomische Abteilung, Supplementband. 1895. Viet & Comp., Leipzig, 180 pp.

Knese, K. H. 1957. *Nomina Anatomica,* 5th edition. Georg Thieme Verlag, Stuttgart, 155 pp.

Lassek, A. M. 1958. *Human Dissection, Its Drama and Struggle.* Charles C Thomas, Springfield, Ill., 310 pp.

Nomina Anatomica. 1965. Third edition with index. Revised by the International Anatomical Nomenclature Committee and adopted at the Eighth International Congress of Anatomists. Wiesbaden. Excerpta Medical Foundation, Amsterdam, 164 pp.

Nomina Embryologica. 1970. Compiled by the Embryology Subcommittee of the I.A.N.C. and adopted at the Ninth International Congress of Anatomists, Leningrad 1970. 70 pp.

Nomina Histologica. 1970. Compiled by the Histology Subcommittee of the I.A.N.C. and adopted at the Ninth International Congress of Anatomists, Leningrad 1970, 111 pp.

O'Malley, C. D. 1964. *Andreas Vesalius of Brussels 1514-1564.* University of California Press, Berkeley, 480 pp.

Rufus of Ephesus. Second Century. *On Names of the Parts of the Human Body.* In: Oeuvres de Rufus d'Ephèse by Daremberg and Ruelle, 1879. Paris, pp. 133-185.

Triepel, H. 1957. *Die Anatomischen Namen.* Ihre Ableitung und Aussprache. 25th edition by Robert Herrlinger. J. F. Bergmann, München, 82 pp.

Skinner, H. A. 1963. *The Origin of Medical Terms.* 2nd ed. The Williams & Wilkins Company, Baltimore, 437 pp.

2

Cytology and Embryology

The Cell

The human body is composed of structural units of protoplasm, called **cells,** which together with the intercellular material—especially plentiful in the bones, cartilages, tendons, and ligaments—form the tissues and organs. The individual cells are microscopic in size and occur in many gradations of shape and composition corresponding to their diversified functions (Figs. 2-1; 2-4). Every human being begins life as a single cell, the fertilized ovum. By the process of cell division there soon results a great increase in the number of cells. Additionally, by the processes of differentiation and metabolism, the tissues assume their varied functional states. All cells are composed of two parts: the outer soft, jelly-like **cytoplasm** surrounded by a limiting **plasma membrane,** and an inner **nucleus** composed of nucleoplasm which is surrounded by a **nuclear envelope.** During the interval between periodic divisions of the cell, the nucleus is in a resting stage; during periods of division, the nucleus undergoes a series of changes called mitosis.

The **cytoplasm** of a cell appears homogenous when seen with the light microscope, but in reality is very complex in structure. Within it are embedded many intracellular bodies and organelles (Fig. 2-1). Although most of these structures are recognizable with the light microscope, their finer details have only been seen through the higher magnifications of the electron microscope. Today it is appreciated that most cytoplasmic organelles are composed of delicate lipid-protein membranes. The outer surface of the cell, the **cell membrane (plasma membrane** or **plasmalemma),** is frequently studded with microvilli or cilia or other surface specializations. Generally considered to be composed of two layers of protein between which rests a molecular layer of lipid, the cell membrane, or **unit membrane,** as it is now frequently called, appears trilaminar in structure. Functionally, the membrane is a very important structure because it controls the exchange of materials between the cell and its environment by osmosis, phagocytosis, pinocytosis, secretion, and other phenomena. A specialized region of cytoplasm called the **central body** or **centrosome** with its tubular **centrioles,** not always visible with the light microscope, usually lies at one side of the nucleus near the central part of the cell. It serves as a dynamic center with other intracellular structures such as the Golgi complex arranged around it, and it is especially important during cell division, as described on page 12. The **mitochondria,** visible in living cells seen with the phase contrast microscope, appear in the form of threads, rodlets, and granules. They are likely to be destroyed by the action of lipid solvents during the preparation of routinely preserved and stained microscope slides. Their internal organization has been revealed by the electron microscope (Figs. 2-1; 2-2). It has been demonstrated that mitochondria are composed of two membranes. One, a continuous outer limiting membrane, surrounds a second inner membrane whose surface area is enlarged by a series of folds called **cristae** (Palade, 1952, 1953). By their isolation with differential centrifugation, mitochondria have been shown to contain important enzymes necessary for the metabolic processes of oxidative phosphoryla-

7

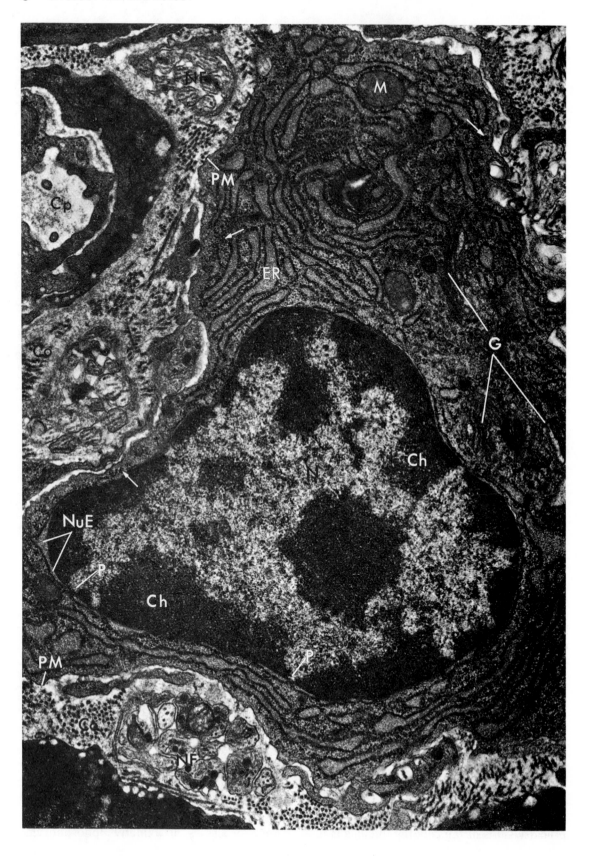

tion. The **Golgi apparatus** (also called **Golgi bodies** or **complex;** Fig. 2-1) can be demonstrated in nearly all cells by special methods of preparation, but is seldom identifiable in living cells. With the electron microscope it has been shown to include large and small vacuoles, granules, and membranes associated with vesicles or clefts (Dalton and Felix, 1954). The Golgi apparatus appears to be importantly related to secretory functions in many cell types. Protein synthesized in the endoplasmic reticulum appears to move into the Golgi apparatus where it is "packaged" with a membrane and then discharged as a **vesicle.** The Golgi apparatus also seems to play a role in the secretion of mucus, hormones, lysosomes, and various intracellular granules. Electron micrographic studies have also shown that within the cell's cytoplasm can be found irregular networks of branching tubules with membranous walls which have been called the **endoplasmic reticulum** (Figs. 2-1; 2-4). Two types of endoplasmic reticulum have been identified, a **rough** or granular type and a **smooth** or agranular type. The rough endoplasmic reticulum contains particles called **ribosomes,** which are composed of basophilic **ribonucleic acid (RNA)** and protein, and either adhere to the membrane of the endoplasmic reticulum or lie free between the membranes. These structures appear to be importantly involved in protein synthesis. At times a number of these ribosomes are seen together in groups or clusters and, as such, are called **polyribosomes.** These granules, about 150 angstrom units (Å) in diameter (see white arrows, Fig. 2-1) may also be scattered throughout the cell (Palade, 1955). The agranular (smooth) endoplasmic reticulum does not contain basophilic ribosomes attached to its membranes and, therefore, usually is acidophilic in its staining property. It is found in large amounts in muscle cells as the sarcoplasmic

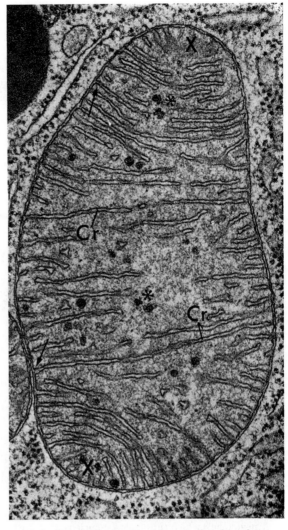

FIG. 2-2. Longitudinal section of a mitochondrion from a secretory cell of the pancreas. 64,000 ×. Arrows indicate where the inner of the two limiting membranes is continuous with the membrane forming the cristae. (From Porter and Bonneville, *Fine Structure of Cells and Tissues,* Lea & Febiger, 1973.)

   Cr—Cristae

   X—Sites where mitochondrion may be growing or enlarging

   *—Dense granules

   Arrows—Inner limiting membrane-cristae sites of continuity

FIG. 2-1. Electron micrograph showing a plasma cell and illustrating a number of cellular organelles. 29,000 ×. (From Porter and Bonneville, *Fine Structure of Cells and Tissues,* Lea & Febiger, 1973.)

Co—Collagen	N—Nucleus
Ch—Chromatin	NF—Nerve fiber
Cp—Capillary	NuE—Nuclear envelope
ER—Endoplasmic reticulum	P—Pore
G—Golgi apparatus	White arrows—Free ribosomes
M—Mitochondrium	

NUCLEAR PORE

A.

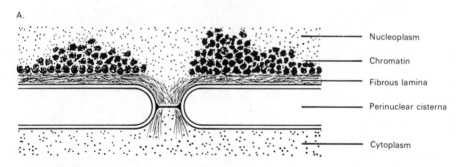

— Nucleoplasm

— Chromatin

— Fibrous lamina

— Perinuclear cisterna

— Cytoplasm

B.

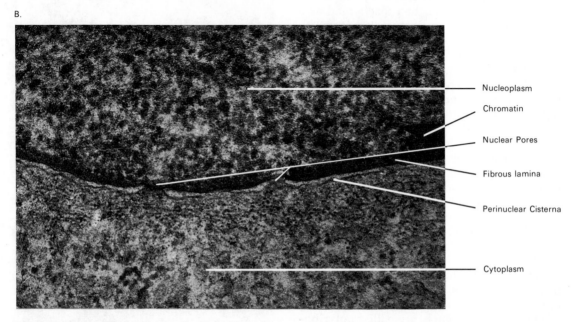

— Nucleoplasm

— Chromatin

— Nuclear Pores

— Fibrous lamina

— Perinuclear Cisterna

— Cytoplasm

FIG. 2-3. Pores in the nuclear envelope of cells. *A*, Diagrammatic representation of a nuclear pore in the envelope ("membrane") which surrounds the nucleus. Observe that the pore is traversed by a thin septum and by the fibrous lamina of filamentous material. *B*, An electron micrograph of the nuclear envelope showing at least two such pores. (From Fawcett, Amer. J. Anat., 1966.)

reticulum and in the cells of certain glands. Its function appears to be related to cholesterol and its derivative steroids as well as to certain cellular ionic mechanisms. Another cytoplasmic organelle is the **lysosome.** These electron dense bodies are also surrounded by membranes, and they contain certain hydrolytic enzymes that are capable of breaking down and "digesting" proteins and carbohydrates. Lysosomes also release their enzymes when a cell dies, and autolysis of the cell results. Other cellular organelles include microtubules, fibrils, and fibrillar material. The cytoplasm of the cell also contains many **cytoplasmic inclusions,** among which are granules of metabolic products such as glycogen, secretion granules, vacu-

oles, lipid inclusions, and inert or ingested material.

The **nucleus** of a cell in the intermitotic or resting stage is a spherical or elongated body surrounded by a thin **nuclear envelope** (membrane), which is interrupted by openings called **nuclear pores** (Fig. 2-3, *A* and *B*). It appears that through these pores certain substances may be interchanged between the **nucleoplasm** and the cytoplasm. The nucleoplasm contains several semisolid, basophilic staining bodies called nucleoli, which are variable in size in different cells. The nucleus also contains the genetic substance, **desoxyribonucleic acid (DNA),** which in combination with the histone proteins forms a threadlike material called

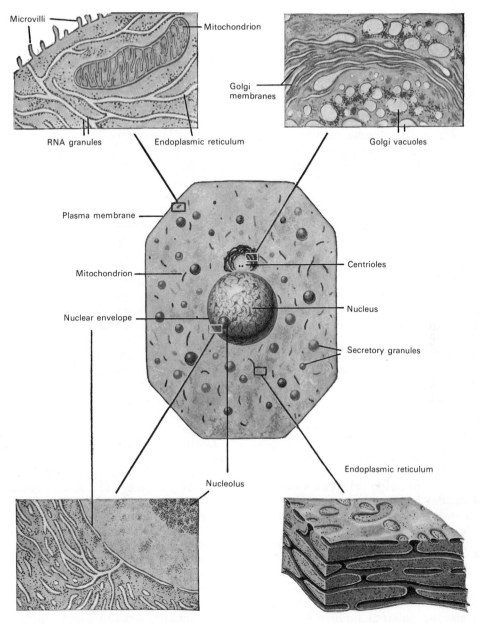

FIG. 2-4. Schematic representation of a cell as seen with the light microscope and enlargements showing the fine structure of some of the cell constituents as revealed by electron microscopy. The endoplasmic reticulum is shown in three dimensions, other structures are in section. (From Copenhaver, *Bailey's Textbook of Histology*, Williams & Wilkins Co., 1964.)

chromatin. These strands may be brought out by staining the nuclei with various dyes, the one most commonly used being hematoxylin. The chromatin material is invisible in living resting nuclei except possibly with phase microscopy, but during cell division it is aggregated first into threads and then into refractive bodies called **chromosomes.** The process is called **mitosis** from the Greek word mitos (μίτος) meaning thread. The chromosomes of the germ cells carry the factors for the genetic or hereditary constitution of the individual.

CELL DIVISION

Although the division of a cell by mitosis is a continuous process, it is the custom to identify four different stages: prophase, metaphase, anaphase, and telophase (Fig. 2-5). In preparation for the prophase, the **centrosome** divides into two parts which become oriented at opposite poles of the nucleus. During the **prophase,** the chromatin, which has been scattered throughout the nucleus, condenses into **chromomeres** which are scattered along a thread or **chromonema.** The threads are individual entities composed especially of desoxyribonucleic acid (DNA); they become condensed into more compact bodies, the **chromosomes.** During the **metaphase** the chromosomes become oriented in a plane, the **equatorial plate,** located midway between the two centrosomes, with a clear area, the **spindle,** leading toward each centrosome. At this time each chromosome has doubled. The two parts are named **chromatids** and are attached to each other at a particular site, the **centromere.** In late metaphase and early anaphase, each centromere divides. During **anaphase** the two chromatids of each chromosome separate, pulled apart by a spindle fiber leading to one or the other of the two centrosomes. The two sets of chromatids, now called daughter chromosomes, are grouped around the two centrosomes, which move to opposite ends of the cell. In **telophase** the chromosomes form a compact mass, lose their individuality, and disperse into the chromatin of the intermitotic nucleus.

The division of living cells cannot be observed in the intact body, even with high magnifications of the microscope. Harrison (1910), however, discovered that embryonic cells can survive and will divide on glass slides if they are placed in an appropriate medium with aseptic precautions. Since then many observations have been made using this method, known as *tissue culture,* on a wide range of different animal forms, including tissues from human fetuses, normal adults, and tumors. The following description (by W. H. Lewis) of cell division is based on observations of connective tissue cells or fibroblasts in cultures of tissues from chick embryos; descriptions of the process in other cells, such as those in the whitefish blastula, are virtually the same (Fig. 2-6, A to E).

MITOSIS IN TISSUE CULTURE. An hour or two before division begins, the centrosome material separates into two parts and is placed at opposite poles of the nucleus. During **prophase** (Figs. 2-5; 2-6, B), the chromatin is formed into threads and then into chro-

mosomes. The cell retracts its processes and tends to become spherical, with the nucleus centrally placed and mitochrondria or other intracellular bodies arranged around it. Near the end of prophase, which takes about an hour, the nuclear membrane disappears and the chromosomes are released and move into the equatorial plane.

During **metaphase** (Figs. 2-5; 2-6, C), which lasts about six minutes, the chromosomes oscillate back and forth in short paths in the equatorial region, as if they were pulled first toward one pole and then toward the other by strings or fibers. Although these fibers are not visible in living fibroblasts, they are visible in many cell types, especially after fixation. Together with the chromosomes these fibers make up the *mitotic spindle.* Each chromosome splits longitudinally into two equal and similar daughter chromosomes.

During **anaphase** (Figs. 2-5; 2-6, D and E), which lasts about three minutes, one set of daughter chromosomes moves to each pole of the spindle, or is pulled there by contraction of the spindle fibers, and is clumped into a compact mass. As the chromosomes approach the poles, the cell becomes elongated and a groove encircles it at the equator.

During **telophase** (Figs. 2-5; 2-6, F), which lasts about three minutes, the groove sinks farther into the cell, gradually pinching it into two daughter cells, which remain connected for a short time by a stalk. Each daughter cell receives approximately one-half the mitochondria, but the division of fat droplets and other intracellular bodies appears more haphazard. The chromosomes lose their identity in the compact mass of the daughter nuclei. The daughter nuclei slowly increase in size, as clear areas split the compact mass into granules, and a nuclear envelope forms. As the clear areas increase, the chromosomal granules become invisible except for those that become the nucleoli. It takes several hours for the nucleus to attain the final resting or intermitotic condition.

AMITOSIS. Direct division of the cell, or amitosis, has been described but probably does not occur under normal conditions. Degenerating cells may show **nuclear fragmentation** without division of the cytoplasm or the centrosome.

Binucleate cells are not uncommon, especially in certain organs, and are formed either by mitotic division of the nucleus with-

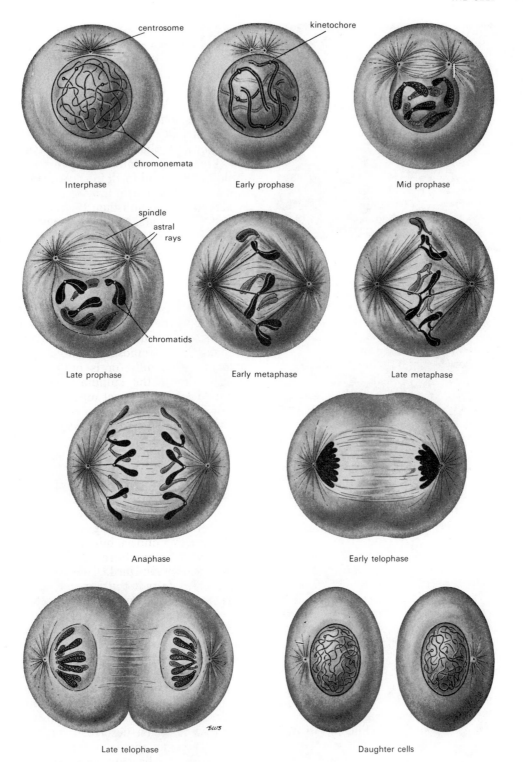

FIG. 2-5. Diagrammatic representation of various stages in mitosis. (From Winchester, *Genetics, A Survey of the Principles of Heredity,* Houghton Mifflin, 1972.)

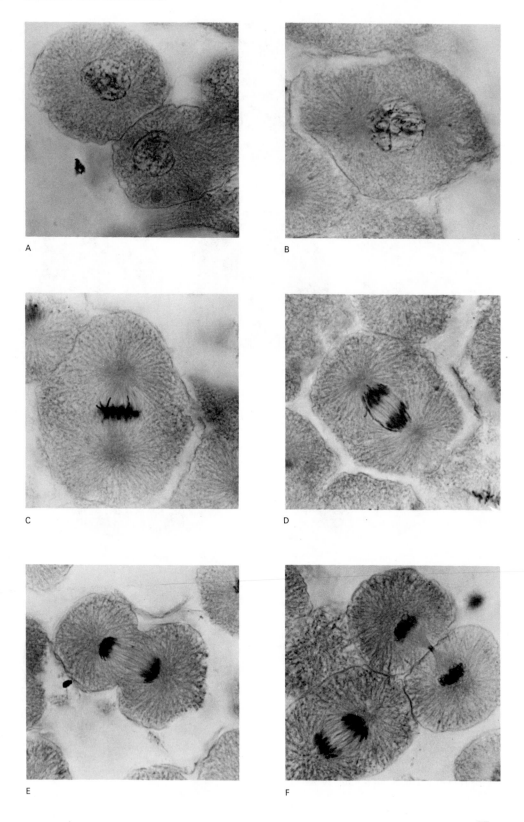

out cleavage of the cytoplasm, or by some process not clearly understood such as fragmentation.

Chromosomes

In the last three decades, dramatic advances in electron microscopy, biochemical genetics, and cytogenetics have enhanced our understanding of the mechanisms involved in cell division and of the fundamental biochemistry of genetic material. Concepts have emerged as to the nature of the control exerted by the nucleus over the activities of the cytoplasmic organelles. Additionally, this knowledge has even extended into an appreciation of how and where these processes can go wrong, thereby elucidating better than ever before the mechanics of genetic mutation and the hereditary nature of certain diseases. The interdisciplinary fields of cellular and molecular biology are oriented toward closing the gap between the structure observed by the anatomist using the electron microscope and the molecular chains so beautifully unravelled and reconstructed by the biological and physical chemists.

Chromosomes are the unique biological structures responsible for the continuity of an animal species through its generations. Further, through their genetic material, chromosomes dictate the proper functional activities of every living animal cell. In each animal species there is a characteristic number of chromosomes. In man there are 23 pairs or 46 chromosomes. This number, i.e., 46 chromosomes, is known as the **diploid** number. In the nucleus of the human sperm or ovum the chromosome number is exactly half, or 23, and this is known as the **haploid** number. When the male gamete fertilizes the female gamete, the diploid number of 46 is attained. Thus, each pair of chromosomes consists of a maternal and a paternal chromosome. The random distribution of maternal and paternal chromosomes is such that in the germ cells, during maturation, over 16,000,000 possible combinations can result in the daughter cells. Of the 23 pairs in man, 22 are called **autosomes** and the remaining pair are the **sex chromosomes** which can be identified under the microscope by their size and shape. The sex pair in all the cells of a male organism consists of a large X and a small Y chromosome (Fig. 2-8). The sex pair of a female contains two X but no Y chromosomes (Fig. 2-7).

STRUCTURE OF CHROMOSOMES. At metaphase just prior to cell division, each chromosome can be observed to consist of two identical strands of chromatin called **chromatids.** The chromatids are connected at a constriction in the chromosome called the **centromere** or **kinetochore,** and each will become a separate chromosome when the centromere divides. The location of the centromere allows for a descriptive morphologic classification of chromosomes (Fig. 2-9). When the centromere is located in the middle of the chromosome, the chromosome is referred to as **metacentric.** As such its **arms** are of equal length. A chromosome in which the centromere is situated somewhere between its midpoint and one of its ends is called a **submetacentric** chromosome. In such chromosomes one arm is somewhat longer than the other. When the centromere is located very close to the end of a chromosome, the term **acrocentric** is used in its description, and it follows that such a chromosome has one very short arm and one that is considerably longer. The short arm of a chromosome is called "p," for "petite," while the long arm is designated by the letter "q" (Fraser and Nora, 1975).

FIG. 2-6. Stages of mitosis in whitefish blastula.

A, Interphase: This is a "resting" stage during which the chromatin material appears dispersed, although it is thought that the chromosomes continue to maintain their individual integrity.

B, Prophase: Chromatin commences to aggregate into discrete threads. Asters develop around the divided centriole at each end of the cell, and the nuclear envelope disintegrates.

C, Metaphase: The chromosomes move into a horizontal equatorial plane. Each chromosome is attached to a spindle fiber at its centromere and begins to split into two chromatids.

D, Anaphase: Each pair of chromatids abruptly becomes pulled apart to form daughter chromosomes which move toward the poles of the spindle.

E, Late Anaphase: The spindle continues to lengthen and a constriction of the cell body begins to pinch the cytoplasm into two cells.

F, Telophase: Separation is virtually complete. The daughter cells, however, are still connected by a fine cytoplasmic stalk which eventually is broken. (From *Genes in Action,* The Upjohn Co., 1967.)

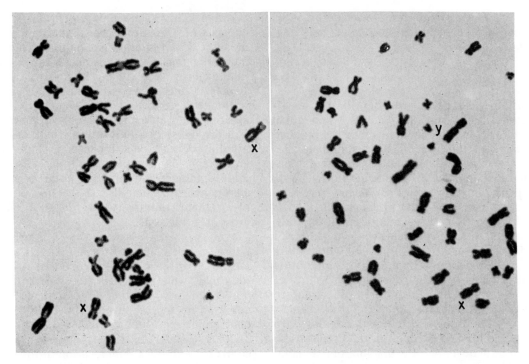

FIG. 2-7 (female) and FIG. 2-8 (male). Colchicine metaphase of human cells grown in vitro, showing chromosomes. (From Tijo and Puck, Proc. Nat. Acad. Sci. (U.S.A.), 1958.)

HUMAN KARYOTYPE OR IDIOGRAM. In the study of human genetics, living cells from the blood or other tissues are prepared in a tissue culture that allows them to proliferate and divide by mitosis. When individual cells are found in the appropriate stage of division, that is, the metaphase, colchicine is added to the culture, and one of the cells is flattened out on a glass slide in order to separate the chromosomes (Figs. 2-7; 2-8). When examined under the microscope, the individual chromosomes can be identified and mapped or arranged in a diagram called a **karyotype** (Figs. 2-10; 2-11). At one time the terms **karyotype** and **idiogram** were used interchangeably, but now the word karyotype is used for a systematized arrangement to typify the chromosomes of an individual according to a standard classification. The term idiogram is now used to mean a diagrammatic representation of a karyotype that may be based on many observations and measurements (Roberts, 1963). Thus, an individual's karyotype may be compared with normal chromosomes, and if abnormalities are identified, they may point to the diagnosis of some hereditary aberration in the individual.

A standard system has now been adopted for the purpose of classifying the chromosomes composing a human karyotype. This classification was determined at international conferences held in Denver (1960) and in London (1963).

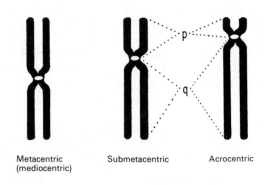

Metacentric (mediocentric)   Submetacentric   Acrocentric

FIG. 2-9. The nomenclature of chromosomes. The centromere (site of attachment) of the chromosome lies either in the middle (metacentric), off center (submetacentric), or near the end (acrocentric). "p" designates the short arm, while "q" is the large arm of a chromosome. (From Fraser and Nora, *Genetics in Man*, Lea & Febiger, 1975.)

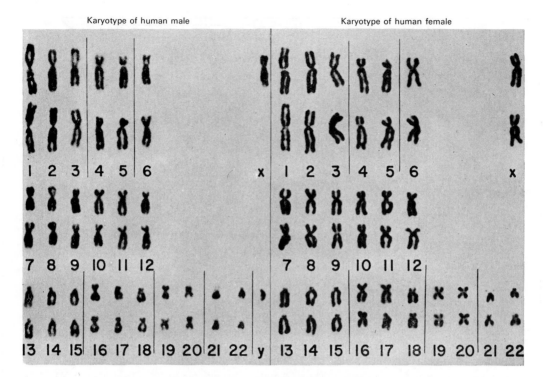

Karyotype of human male   Karyotype of human female

FIGS. 2-10 and 2-11.   Karyotypes of human chromosomes. (From Tijo and Puck, Proc. Nat. Acad. Sci. (U.S.A.), 1958.)

GROUP 1–3   Large chromosomes with centromeres approximately in the middle. The three chromosomes can be readily distinguished from each other by their size and centromere position.

GROUP 4–5   Large chromosomes with centromeres in the submetacentric position. Although these two chromosomes are more difficult to distinguish from each other, chromosome 4 is slightly larger.

GROUP 6–12 AND THE X CHROMOSOME   Medium-sized chromosomes with centromeres in the submetacentric position. Chromosomes 6, 7, 8, 11, and X are more metacentric than are 9, 10, and 12. The X chromosome most resembles 6. This group presents the major difficulty in identification of individual chromosomes.

GROUP 13–15   Medium-sized chromosomes with centromeres in the acrocentric position. Chromosomes 13 and 14 have satellites on the short arms.

GROUP 16–18   Rather short submetacentric chromosomes. Chromosome 16 is more metacentric.

GROUP 19–20   Short metacentric chromosomes.

GROUP 21–22 AND THE Y CHROMOSOME   Very short acrocentric chromosomes. Chromosomes 21 and 22 may have a satellite on their short arm.

CHEMISTRY OF CHROMOSOMES AND CHROMATIN.   Investigations into the biochemical nature of the genetic substance have resulted in one of the more significant and exciting advances in the history of biological science. It has now been accepted that chromosomes consist principally of desoxyribose nucleic acid (DNA) embedded within proteins which are largely histones in type. Based on observations made by Wilkins with the techniques of x-ray diffraction crystallography, a breakthrough occurred with the publication of a paper by Watson and Crick in 1953 in which it was proposed that the DNA molecule consists of two strands of nucleotide molecules wound together in the form of a double helix (Fig. 2-12, *A*). Additionally, the *in vitro* synthesis of biologically active or infective viral DNA by Goulian, Kornberg, and Sinsheimer in 1967 elucidated further the nature of the nuclear substance which was first separated from the nuclei of cells by Friedrich Miescher almost a century earlier in 1871 and called **nuclein.** In later years nuclein was renamed nucleic acid.

DNA is composed of a linear assemblage of molecules of desoxyribonucleotides. Each

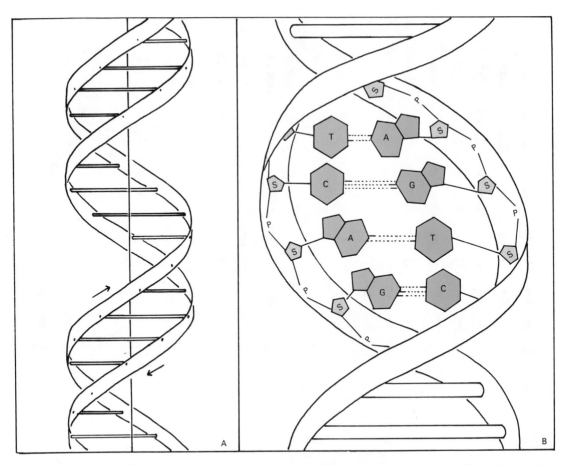

FIG. 2-12. *A*, Diagram of a part of the double-stranded DNA molecule as proposed in the double helix model by Watson and Crick. The horizontal rods represent the pairs of bases which connect the two strands. The vertical line is the longitudinal axis around which the coils of the double helix twist. *B*, A diagram showing the single-ringed pyrimidine bases, cytosine (C) and thymine (T) along with the double-ringed purine bases, adenine (A) and guanine (G) which connect the two strands of the DNA molecule. The nitrogenous bases link with the sugar (S), desoxyribose, which in turn binds with phosphate groups (P) in an alternating pattern along DNA strands.

molecule of the nucleotide consists of either a large double-ring purine base (adenine or guanine) or a smaller single-ring pyrimidine base (cytosine or thymine), plus the sugar, desoxyribose, and a phosphate group (Fig.2-12, *B*). The sugar of one molecule is attached to the phosphate of the next in each of the two strands. The critical aspect of the contribution by Watson and Crick was the manner in which the two strands of desoxyribose and phosphate are interconnected. Knowing that in DNA the amount of adenine always equals the amount of thymine and that the amount of guanine always equals the amount of cytosine, and realizing that their model had to give DNA the property of self-replication, they proposed that wherever a two-ring adenine base is located on one strand, it is attached to a one-ring thymine base on the other. Additionally, they proposed that a guanine base on one strand is always connected to a cytosine base on the other strand. (Fig. 2-12, *B*). The bases were then proposed to be held together loosely by hydrogen atoms. In the authors' own words:

The novel feature of the structure is the manner in which the two chains are held together by the purine and pyrimidine bases. The planes of the bases are perpendicular to the fibre axis. They are joined together in pairs, a single base from one chain being hydrogen-bonded to a single base from the other chain, so that the two lie side by side. . . . One of the pair must be a purine and the other a pyrimidine for bonding to occur. . . . If it is assumed that the bases only occur in the structure in the most plausible tautomeric forms (that is, with the keto rather than the enol configurations), it is found that only specific pairs of bases can bond together. These pairs are: adenine (purine) with thymine (pyrimidine) and guanine (purine) with cytosine (pyrimidine).

In other words, if an adenine forms one member of a pair, on either chain, then on these assumptions the other member must be thymine; similarly for guanine and cytosine. The sequence of bases on a single chain does not appear to be restricted in any way. However, if only specific pairs of bases can be formed, it follows that if the sequence of bases on one chain is given, then the sequence on the other chain is automatically determined.

It has been found experimentally that the ratio of the amounts of adenine to thymine, and the ratio of guanine to cytosine, are always very close to unity for deoxyribose nucleic acid. . . .

It has not escaped our notice that the specific pairing we have postulated immediately suggests a possible copying mechanism for the genetic material (Watson and Crick, 1953).

REPLICATION OF DNA. The simplicity of the Watson-Crick model for the structure of DNA was also combined with a unique property, i.e., the molecular structure at any part of one strand dictated specifically the molecular structure of the other. With a separation of the hydrogen bonds, each of the two strands could become the molds for the **replication** of DNA because of the necessity of the divided strand to bind exactly as before with a specific sequence of bases. In this manner the model accounted for the requirement necessitated by genes, that they have the ability to duplicate themselves exactly. The process of separation of the two DNA strands at the site of hydrogen bonding therefore would allow each of the strands to act as a template against which to synthesize a new companion strand (Fig. 2-13). In this manner the original double helix could produce two exact replicas of itself, chemically identical in every respect, following the separation of the chromosomes at cell division.

MECHANISMS OF GENE ACTION: TRANSCRIP-TION. Although chromosomes show little visible internal structure when viewed by conventional microscopes, it is now appreciated that the DNA component of chromosomes includes submicroscopic structures, which are spaced linearly along the DNA thread and are called **genes.** Genes can be considered the centers for the control of particular hereditary traits. The nature of the genes' interactions with their environments (the cytoplasm of the cells) establishes the hereditary peculiarities or characteristics of an individual. It is called that person's **genotype.** Anatomically a gene is a section of the DNA molecule and is functionally involved in the determination of the sequential order of the amino acids, which form a single polypeptide chain of a protein.

The process by which the arrangement of the linear chemical groupings on the DNA molecule is communicated to the cytoplasm for the formation of proteins required for a cell's functions is called **transcription,** and the chemical language which has to be translated to understand this communication is called the **genetic code.** The number of possible differences in the sequential arrangement of the nucleotide bases provides DNA with an exceedingly high number of possible combinations, a property which the genetic material must have in order to account for the many different genes.

Another nucleic acid, **ribonucleic acid** or **RNA,** is present in larger quantities than DNA in cells known to be particularly active in protein synthesis. Unlike DNA, ribonucleic acid is found not only in the nucleus of the cell, but in the cytoplasm as well. It has been shown that the RNA found in the cytoplasm of cells is synthesized in the nucleus and moves from the nucleus to the cytoplasm. Although quite similar in chemical structure, RNA differs from DNA in three ways. RNA contains the sugar, ribose, instead of desoxyribose; generally RNA is a single chain while DNA is double-stranded; and RNA contains the pyrimidine base uracil instead of thymine.

In the process of transcription the double-stranded DNA separates as it would in DNA replication. Alongside of one of the two DNA strands (the master strand), however, nucleotides containing ribose (ribonucleotides) align in proper sequence to the DNA nucleotides. The RNA, now containing the coded sequence, separates from the DNA, leaves the nucleus through the pores in the nuclear envelope, and moves into the cytoplasm. As such it is called the **messenger RNA (mRNA).** The synthesis of protein actually takes place in the ribosomes of the cytoplasm, and the mRNA becomes attached to

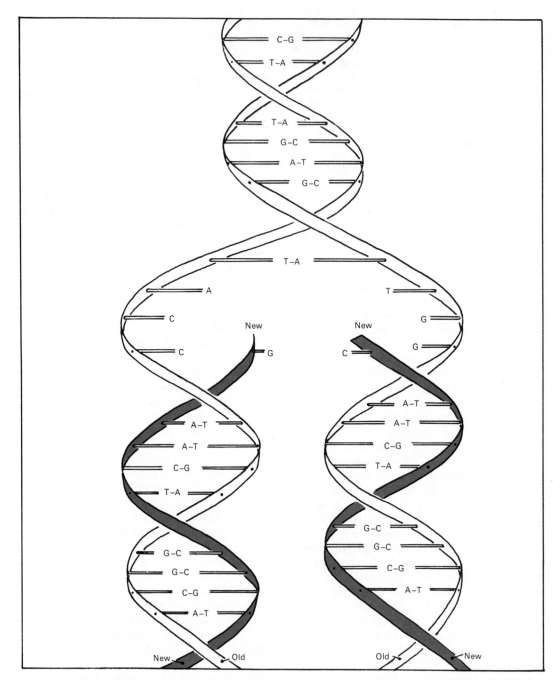

Fig. 2-13. A diagram illustrating the replication of the DNA molecule. The two strands of the old molecule separate and act as templates for the newly formed strands. In this manner the sequence of bases is copied exactly, thereby replicating the original molecule.

the ribosomes. In the eventual formation of a protein, two other types of RNA are involved. These are **ribosomal RNA (rRNA),** a normal structural component of the ribosomes, and **transfer RNA (tRNA),** which functions as the carrier of amino acids from the rest of the cell's cytoplasm to the site of protein synthesis in the ribosomes. Thus, the genetically coded information along the DNA strand (gene) for the production of a specific polypeptide becomes transferred to cellular elements in the cytoplasm, which not only are capable of recognizing the chemical code, but also are capable of producing the protein required.

GERM CELLS. For an individual to begin life and develop into an embryo, it is necessary for specialized cells, the germ cells from each sex, to be combined at fertilization. The female germ cell is the ovum; the male germ cell is the spermatozoon.

THE SPERMATOZOON

The spermatozoa or male germ cells are produced in the testis and are present in the seminal fluid in enormous numbers. Each mature human spermatozoon is a highly specialized cell resembling a free-swimming protozoan and possessing a prominent head and a long tail (Figs. 2-14; 2-15). The length of the entire spermatozoon is 52 to 62 μ. The tail is subdivided into a middle piece, principal piece, and end piece. The **head** is a somewhat flattened oval body 4 or 5 μ long and 2 or 3 μ thick. The head is composed of a condensed nucleus, the anterior two thirds of which is covered by an **acrosomal cap** with an **acrosome** at the tip. Running the entire length of the tail is an **axial core** composed of filaments whose ultrastructure appears identical with that of a cilium. In the center of the axial core are two single filaments surrounded by nine uniformly spaced double filaments. The cylindrical **middle piece** surrounds the axial structures from the head to the ring-shaped annulus. It is 5 to 7 μ long, 1 μ thick, and contains a single layer of spirally arranged thread-like mitochondria. The **principal piece** is 45 μ long and $\frac{1}{2}$ μ thick and is composed of a circular arrangement of dense fibers. In the **end piece** the axial filaments are covered only by the **plasma membrane** covering the entire spermatozoon. The axial filaments are assumed to be the contractile elements that provide the undulating motion of the tail. The development of the spermatozoa within the seminiferous tubules can be considered as a threefold process: (1) **spermatocytogenesis,** during which the primary germ cells replace themselves and produce primary spermatocytes; (2) **meiosis,** that phase during which the spermatocytes become spermatids, each containing the haploid number of chromosomes; and (3) **spermiogenesis,** during which the immature spermatids become transformed into mature sperm.

SPERMATOCYTOGENESIS. Among the epithelial cells lining the seminiferous tubules of the testis are found the primary male germ cells which are called the **spermatogonia.** By the process of mitosis these cells proliferate. Some of the daughter cells form more spermatogonia, while others migrate toward the lumen and enlarge into the **primary spermatocytes.** The primary spermatocytes then begin their process of **maturation** by dividing, at the first maturation division, into two secondary spermatocytes. Each **secondary spermatocyte** in the second maturation division divides into two spermatids. Thus, each primary spermatocyte provides four spermatids, all of which develop into motile spermatozoa (Fig. 2-16).

MEIOSIS. One of the most important aspects of germ cell maturation is the distribution of chromosomes and their constituent carriers of heredity, the genes. This takes place by a special type of cell division called **meiosis.** During the prophase of a usual somatic mitosis each threadlike chromosome initiates a longitudinal splitting. In the **first maturation division** of meiosis in the primary spermatocyte, however, the individual chromosomes do not carry the splitting to completion, and the two chromosomes of each pair except the sex chromosomes come into close union by a process known as **synapsis** or **conjugation.** The resulting chromosomal configuration is known as a **tetrad** from its quadruplicate appearance (Painter, 1923). The individual chromosomes of the synaptic pair, without splitting, separate from each other, and one chromosome of each pair goes to each daughter cell. As a result the number of chromosomes in each daughter cell is **reduced** to one half, or 23, in the secondary spermatocytes, one of which contains an X chromosome and the other a Y chromosome. In the **secondary maturation division** the chromosomes of each secondary spermatocyte now complete their splitting, and each divides into two spermatids. The result is four viable spermatids, each containing 23 chromosomes, the haploid number. Two of the spermatids contain X chromosomes while the other two contain Y chromosomes.

SPERMIOGENESIS. The differentiation of the spermatid into the mature sperm begins at one pole of the nucleus with the formation of a fluid-filled **acrosomal vesicle** from the Golgi complex. The acrosomal vesicle ad-

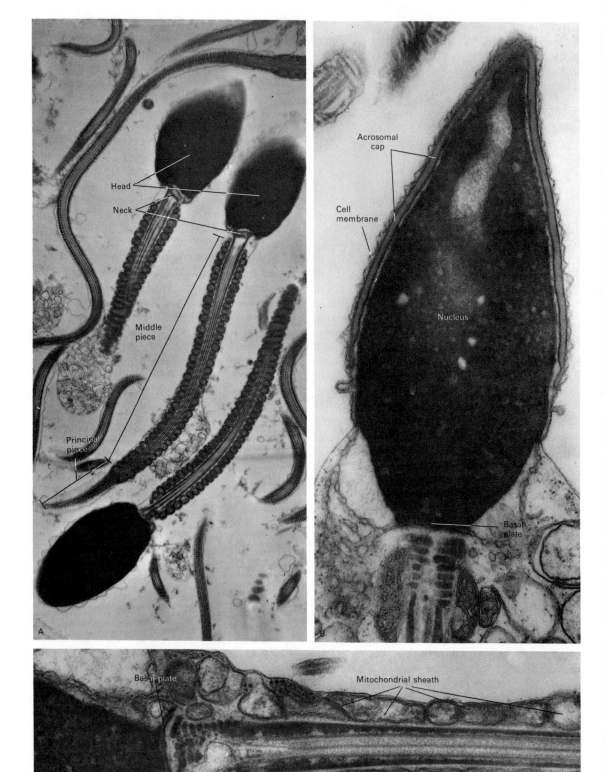

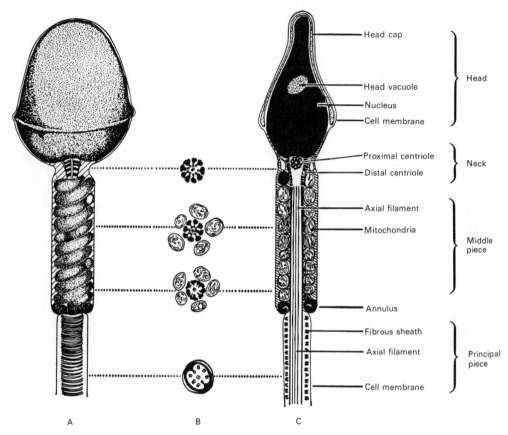

FIG. 2-15. Diagram of the structure of the human spermatozoon based on studies with the electron microscope. Most of the principal piece and end piece have been omitted. *A*, Depicted as though the cell membrane and cytoplasmic matrix had been removed. *B*, Diagrammatic cross sections showing the arrangement of the axial fibrils and their relation to adjacent structures. *C*, Diagram of a longitudinal section passing through the center of *A*. (From Fawcett, Internat. Rev. Cytol., 1958.)

heres to the outer aspect of the nuclear membrane, and this becomes the future anterior tip of the sperm (Fawcett and Hollenberg, 1963). The acrosomal vesicle then loses its fluid and fits over the nucleus at the anterior end as a head cap with the acrosome at the tip. The centrioles migrate toward the opposite pole of the nucleus where a slender **flagellum** is forming. One of the centrioles forms the base of the flagellum which then elongates and develops its axial filaments. The other centriole forms a ring or annulus which migrates away from the nucleus to the caudal end of the middle piece, and the spiral mitochondrial sheath occupies the rest of

the middle piece. The residual cytoplasm is finally cast off, leaving the plasma membrane covering the entire mature spermatozoon.

THE OVUM

The **human ovum** is a large cell, about 0.14 mm in diameter, and contains the visible internal structures found in most cells (Fig. 2-17). The nucleus with nucleolus and chromatin is known as the **germinal vesicle.** The cytoplasm contains a centrosome, mitochondria, granules, fat droplets, Golgi com-

FIG. 2-14. *A*, Spermatozoa in the seminal fluid of the monkey, Macaca mulatta. 7,500 ×. *B*, Head of a human spermatozoon. 37,500 ×. *C*, Neck and middle piece of a human spermatozoon. Note the sheath of irregularly arranged mitochondria. 36,500 ×. (From Zamboni, Zemjanis and Stefanini, Anat. Rec., 1971.)

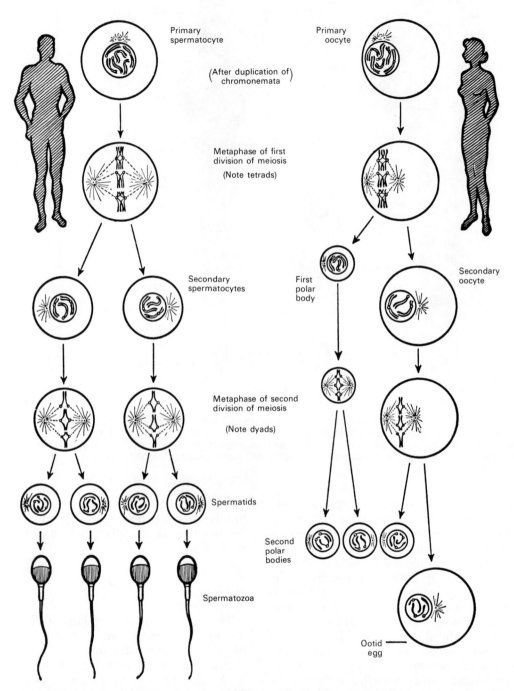

Spermatogenesis—Oogenesis

Primary spermatocyte

(After duplication of chromonemata)

Metaphase of first division of meiosis

(Note tetrads)

Secondary spermatocytes

Metaphase of second division of meiosis

(Note dyads)

Spermatids

Spermatozoa

Primary oocyte

First polar body

Secondary oocyte

Second polar bodies

Ootid — egg

FIG. 2-16. Diagrammatic representation of gametogenesis. Compare the mechanism involved in oogenesis and spermatogenesis, particularly with reference to their differences. These simplified diagrams only indicate six of the forty-six chromosomes in the human being. (From Winchester, *Genetics, A Survey of the Principles of Heredity,* Houghton Mifflin Co., 1972.)

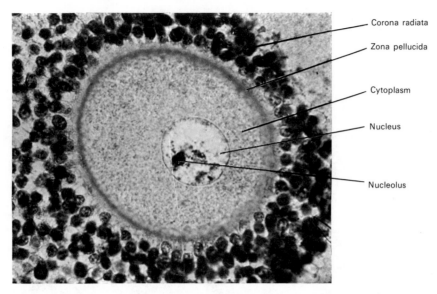

— Corona radiata

— Zona pellucida

— Cytoplasm

— Nucleus

— Nucleolus

Fɪɢ. 2-17. Mature human ovum. (Carnegie Institute Collection.)

plex, and yolk material. The ovum or **vitellus** is enclosed in a tough transparent membrane about 12 μ in thickness, named the **zona pellucida.** With the ovary, the mature ovum is contained in a spherical vesicle about 1 cm in diameter called an **ovarian** or **graafian follicle** (Fig. 2-18). The ovum is held at one side of the follicle by a mass of follicular cells named the **cumulus oophorus;** the follicular cells immediately around the zona pellucida are radially arranged and make up the **co-**rona radiata (Fig. 2-17). While in the ovary and at the time of ovulation, the ovum completely fills the cavity of the zona, but it shrinks shortly after ovulation and a perivitelline space filled with clear fluid is developed.

Ovulation occurs when the mature graafian follicle ruptures through the outer wall of the ovary and the ovum is discharged into the peritoneal cavity. Normally it does not remain free in the peritoneal cavity because

Tᴀʙʟᴇ 2-1. Comparative Size of Living One-Cell Tubal Eggs After the Formation of the Perivitelline Space (Lewis and Wright, 1935).

Animal	Eggs	Author	ODZ	IDZ	DV	Volume*	Surface*
Mouse	26	Lewis and Wright (1935)	113.0	87.8	71.6	192000	16100
Guinea-pig	1	Squier (1932)	121.3	96.8	84.3	314000	22300
Macaque I	1	Corner (1923)		109.0*	86.0	333000	23400
Macaque two-cell	1	Lewis and Hartman (1933)	150.0	125.0	103.0*	562000	
Macaque IV	1	Allen (1928)	178.5	138.5*	104.0	589000	34000
Human	1	Allen *et al.* (1930)	176.0*	139.0*	104.0*	589000	34000
Human (abnormal)	1	Lewis (1931)	148.0	136.0			
Rabbit	2	Lewis and Gregory (1929)	174.0	126.0*			
Rabbit	9		174.2	126.5	111.0	718000	38700
Pig	4	Heuser and Streeter	160.0	130.0	111.0	718000	38700
Cow (nonfertile)	1	Hartman, Lewis,	170.0	143.0	120.0	907000	45300
Cow two-cell	1	Miller and Swett (1931)	162.5	135.0		740000	
Dog	3	Hartman	172.0	141.0	120.0	907000	45300
Sheep	6	Clark	178.0*	150.0*	133.0*	1232000	55700

Volume and surface area of ovum in cubic and square microns estimated by authors. ODZ: outside diameter of zona. IDZ: inside diameter of zona. DV: Diameter of vitellus (ovum) in microns.
*Determined or estimated from illustrations and data in articles quoted.

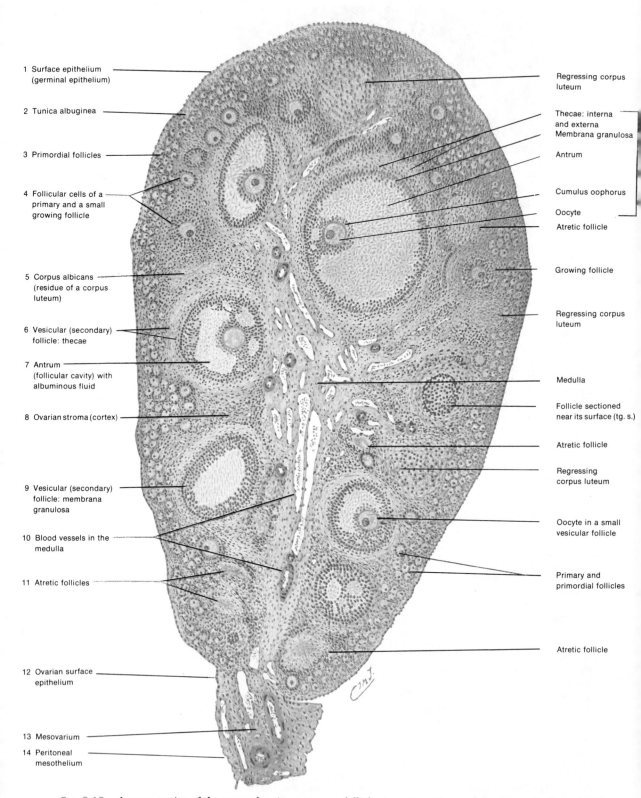

1 Surface epithelium
  (germinal epithelium)

2 Tunica albuginea

3 Primordial follicles

4 Follicular cells of a
  primary and a small
  growing follicle

5 Corpus albicans
  (residue of a corpus
  luteum)

6 Vesicular (secondary)
  follicle: thecae

7 Antrum
  (follicular cavity) with
  albuminous fluid

8 Ovarian stroma (cortex)

9 Vesicular (secondary)
  follicle: membrana
  granulosa

10 Blood vessels in the
   medulla

11 Atretic follicles

12 Ovarian surface
   epithelium

13 Mesovarium

14 Peritoneal
   mesothelium

Regressing corpus
luteum

Thecae: interna
and externa
Membrana granulosa

Antrum

Cumulus oophorus

Oocyte

Atretic follicle

Growing follicle

Regressing corpus
luteum

Medulla

Follicle sectioned
near its surface (tg. s.)

Atretic follicle

Regressing
corpus luteum

Oocyte in a small
vesicular follicle

Primary and
primordial follicles

Atretic follicle

FIG. 2-18.　A cross section of the ovary showing numerous follicles in various stages of development. Stained with hematoxylin and eosin. 20 ×. (From diFiore, *An Atlas of Human Histology*, Lea & Febiger, 1981.)

the fimbriated end of the uterine tube covers the point of rupture, and the ovum is immediately swept into the lumen of the tube, probably by ciliary action (Rock and Hertig, 1944). Although such a close relationship between the fimbriae and the ovary has been observed in many mammals, in the human female the ciliated epithelial cells of the oviducts increase significantly in height at ovulation, thereby increasing the assurance of the ovum's entrance into the uterine tube.

In the sexually adult ovulating woman, an ovarian follicle matures and ruptures from the ovary in a cyclical manner at intervals of about 28 days. Although there may be some variation in the exact date of ovulation during the menstrual cycle, typically the mature ovum is discharged 14 days prior to the expected date of commencement of the next menstruation. With ovulation there frequently is a small amount of hemorrhage accompanied at times by slight pain. The basal body temperature then rises 0.5°F, and in carefully recorded and charted studies, the temperature has been shown to remain elevated for about the next 14 days. The temperature lowers again during the next menstruation and remains down on the days of the new cycle prior to the next ovulation. It is presumed that the right and left ovaries alternate in the production of newly ruptured ova.

MATURATION OF THE OVUM. Before an ovum can be fertilized, it must reach a particular stage of maturation, which begins just prior to ovulation. During the **first maturation division,** the ovum, now known as the **primary oocyte,** divides into one large cell, the **secondary oocyte,** and one small cell which is known as the first **polar body** or **polocyte** (N.E. *polocytum*) (Fig. 2-16). The secondary oocyte quickly initiates the **second maturation division** by forming a mitotic spindle, but it is arrested in the metaphase stage until after it escapes from the follicle and is penetrated by a spermatozoon. During the second maturation division, which can be completed normally only if the ovum is fertilized, a large **mature ootid** and a small **second polocyte** are formed (Figs. 2-16; 2-19). The first polocyte may or may not divide into two smaller polocytes.

MEIOSIS. When the ovum or primary oocyte undergoes the first maturation division, there is a reduction in the number of chromosomes in the same manner as in spermatogenesis (Fig. 2-16). The secondary oocyte and the first polar body both contain 22 autosomes and one sex chromosome. Since it is a female cell, the primary oocyte has two X chromosomes and the secondary oocyte also must have an X chromosome. When it undergoes its second maturation division, therefore, there must be only an X chromosome in the ootid and the second polocyte. It will be seen from this that every ovum from a ruptured follicle can provide only one fertilizable ootid and that it will have an X chromosome. Since there are equal numbers of spermatozoa with X chromosomes and with Y chromosomes in the ejaculate, there is an equal opportunity for the fertilized ovum to contain two X or an X and a Y chromosome and the genetic equality of the sexes is assured.

Fertilization of the Ovum

Ovulation takes place in a woman with regular menstrual cycles approximately 10 to 14 days after the onset of the preceding menstrual flow or approximately 14 days before the expected date of commencement of the next menstruation. The released ovum immediately begins its course through the uterine tube. It must be met by sperm in the ampulla of the tube within approximately 6 to 12 hours or it will begin to show signs of degeneration and no longer be fertilizable.

Fertilization takes place when a spermatozoon enters the ovum (Fig. 2-19). Normally only one sperm takes part in the process; its entrance causes the peripheral layer of the ovum to change into the vitelline membrane, which prevents the entrance of additional sperm. It is thought that the spermatozoon undergoes important changes which allow it to achieve and penetrate the ovum. The acrosome of the sperm becomes perforated, and this appears to allow certain enzymes to escape, thereby creating a path for the sperm through the zona pellucida to the ovum cell membrane. The plasma membranes of the two gametes seem to fuse, and as the sperm head enters the ovum, the latter completes its second meiotic division and the second polar body is formed. Once the spermatozoon has penetrated the ovum, the tail portion separates, and the head and middle piece expand into the **male pronucleus** and the **centrosome.** The nucleus of the ovum, known as the **female pronucleus,** and the

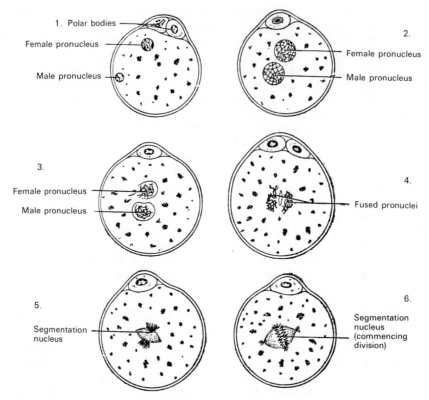

1. Polar bodies

Female pronucleus

Male pronucleus

2.

Female pronucleus

Male pronucleus

3.

Female pronucleus

Male pronucleus

4.

Fused pronuclei

5.

Segmentation nucleus

6.

Segmentation nucleus (commencing division)

FIG. 2-19. The process of fertilization in the ovum of a mouse. (After Sobotta, *Atlas und Lehrbuch der Histologie und mikroskopischen Anatomie des Menschen,* Lehmann, 1911.)

male pronucleus migrate toward each other, and, breaking down into their constituent chromosomes near the center of the cell, are combined into the new **segmentation nucleus.** The mitotic division of this nucleus and the cleavage of the cell produce the **two-cell stage** which is the beginning of the embryological development of the individual.

Fertilization restores the diploid number of chromosomes, since each of the gametes contains the haploid number and their junction achieves genetic completeness. Additionally, fertilization also results in the determination of the sex of the newly formed zygote. Since the ovum is an X-bearing gamete, fertilization by an X-bearing sperm results in an XX-bearing female zygote. In contrast, penetration and fertilization of the ovum by a Y sperm causes the formation of an XY-bearing male zygote.

Development of the Fertilized Ovum

Division of the fertilized ovum into the two-cell stage probably occurs 30 to 36 hours after fertilization. Hertig et al. (1954) have recovered a two-cell human embryo surrounded by the zona pellucida and containing two polar bodies from a uterine tube. These first two cells, known as **blastomeres** (Figs. 2-20, B; 2-21), soon undergo division, and the process of **cleavage** or **segmentation** continues until a cluster of daughter cells, the **morula,** has been produced. During segmentation the ovum does not enlarge; the morula is about the same size as the single-celled ovum. The exact timing is not known, but probably two or three days are required for these five or six cleavage divisions, and while they are taking place the ovum makes its passage through the uterine tube.

MORULA. Although there is no visible difference between the early daughter cells, a segregation of materials and potencies takes place during cleavage, which first becomes evident in the morula (Fig. 2-20, E). By the fourth day two separate parts of the morula become recognizable, an outer layer of cells known as the **trophoblast,** and an inner cluster of cells, known as the **inner cell mass** or **embryoblast.** Together they form the **blastocyst.**

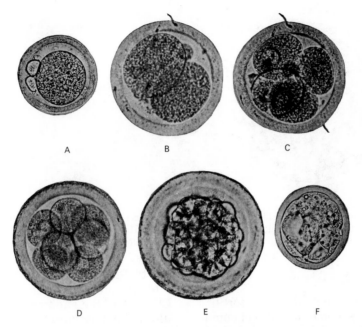

FIG. 2-20. Photographs of living eggs. 200 ✕. *A,* Fertile one-cell mouse egg showing the zona pellucida, perivitelline space, ovum, first and second polar bodies (From Lewis and Wright, Carneg. Inst., Contr. Embryol., 1935). *B,* Two-cell monkey egg. *C,* Four-cell stage of same egg, extra sperm in zona (From Lewis and Hartman, Carneg. Inst., Contr. Embryol., 1933). *D,* Eight-cell rabbit egg. *E,* Rabbit morula (From Lewis and Gregory, Science, 1929). *F,* Early blastocyst, 76-hour mouse egg with a small amount of fluid.

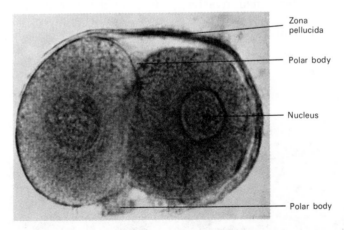

Zona pellucida

Polar body

Nucleus

Polar body

FIG. 2-21. A two-celled human ovum recovered from the fallopian tube. Observe the prominent and closely adhering zona pellucida. Note one polar body between the two blastomeres at the upper margin of the cleavage plane and the other polar body below. 550 ✕. (From Hertig, Rock, Adams and Mulligan, Carneg. Inst., Contr. Embryol., 1954.)

The **blastocyst** develops by the passage of fluid from the uterine cavity into the interior of the morula (Fig. 2-22). The trophoblast enlarges to form a vesicle, the center of which is a fluid-filled cavity called the **blastocyst cavity** or **blastocele.** The cavity is surrounded by the trophoblast cells which multiply, spread out, and become flattened against the zona pellucida. This latter structure is thereby stretched into a thin membrane and disappears. The inner cell mass remains as a relatively small cluster of cells attached to the inner surface of the trophoblast at a region which is known as the **embryonic pole.** The blastocyst stage is reached soon after the ovum enters the uterus, where it remains free in the uterine cavity for an estimated three to five days.

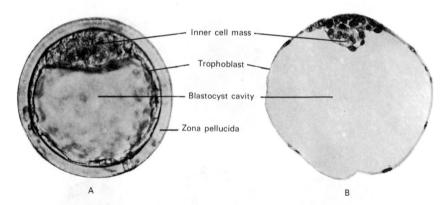

Fig. 2-22.   *A*, Blastocyst, 92-hour rabbit egg. Trophoblast and inner cell mass. 200 ×. (From Lewis and Gregory, Science, 1929.) *B*, Blastocyst, nine-day monkey egg. Stained section. The zona has disappeared, the trophoblast is thin, and the inner cell mass small. 200 ×. (From Carnegie Institute Collection.)

Implantation

On the sixth or seventh day after fertilization, the blastocyst comes into direct contact with the uterine mucosa or endometrium and adheres to it. Since the zona pellucida has disappeared by this time, the trophoblast cells at the embryonic pole work their way into the endometrium by digesting the uterine tissue. By the eighth to tenth day the blastocyst has become a thickened invasive mass buried in the mucosa (Figs. 2-22; 2-23; 2-25). A thin wall of trophoblastic cells op-posite the embryonic pole still protrudes into the uterine cavity (Figs. 2-23; 2-25). Changes in the endometrial wall and in the uterine glands following ovulation prepare the uterus for the reception of the blastocyst. Usually implantation occurs on the posterior wall of the uterus near the fundus.

By the twelfth day the embryo has become completely buried and the site of implantation in the endometrium has been closed over by the uterine epithelium (Figs. 2-24; 2-26). The trophoblast has formed a spongy mass which has broken into the walls of

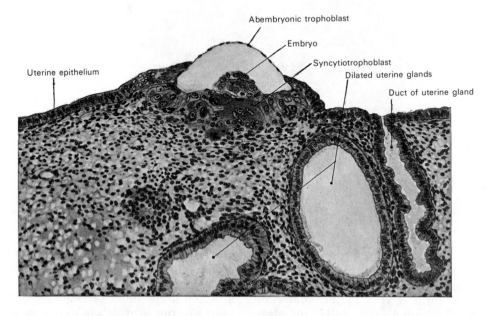

Fig. 2-23.   A human blastocyst (Carnegie 8020), fertilization age 7 to 7½ days, in process of embedding in the uterine mucosa. In the actual specimen the abembryonic trophoblast had collapsed on the inner cell mass, but for the purposes of clarity, it is shown projecting into the uterine cavity. Drawing from a photomicrograph. 150 ×. (From Rock and Hertig, Amer. J. Obstet. Gynec., 1942 and 1944.)

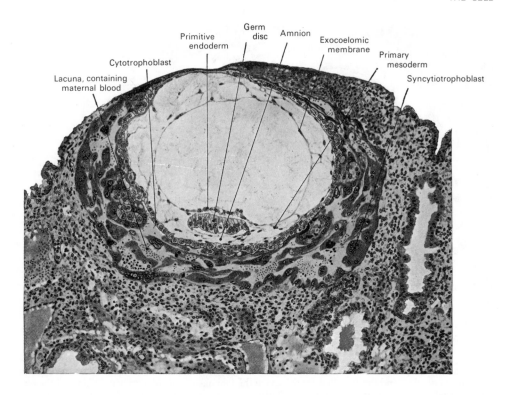

FIG. 2-24. A human blastocyst (Carnegie 7700), fertilization age 12 to 12½ days, embedded in the stratum compactum of the endometrium. Drawing from a photomicrograph. 105 ×. (From Hertig and Rock, Carneg. Inst., Contr. Embryol., 1941.)

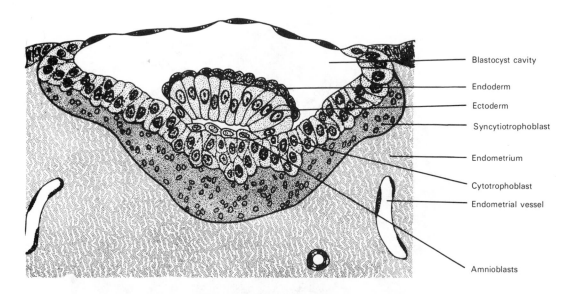

FIG. 2-25. Diagrammatic representation of a 7½-day human blastocyst in the process of becoming embedded in the endometrium of the uterine wall. Compare with Fig. 2-23. (Redrawn after Langman, *Medical Embryology: Human Development, Normal and Abnormal,* Williams & Wilkins Co., 1969, and Rock and Hertig, Amer. J. Obstet. Gynec., 1942 and 1944.)

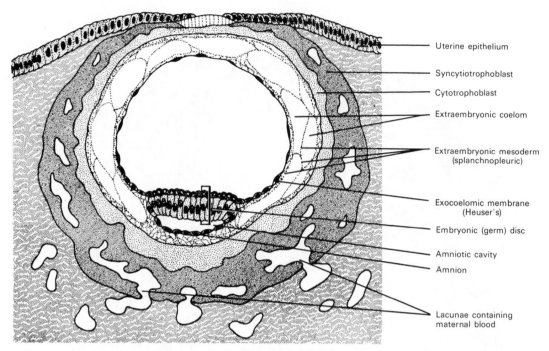

Uterine epithelium

Syncytiotrophoblast

Cytotrophoblast

Extraembryonic coelom

Extraembryonic mesoderm (splanchnopleuric)

Exocoelomic membrane (Heuser's)

Embryonic (germ) disc

Amniotic cavity

Amnion

Lacunae containing maternal blood

FIG. 2-26. Schematic representation of a twelve-day blastocyst in the endometrial wall of the uterus. Observe the developing embryonic disc and the fact that by this stage the lacunae of the syncytiotrophoblast of the embryo are filled with maternal blood. Compare with Fig. 2-24. (Redrawn after Langman, *Medical Embryology: Human Development, Normal and Abnormal,* Williams & Wilkins Co., 1969, and Hertig and Rock, Carneg. Inst., Contr. Embryol., 1941.)

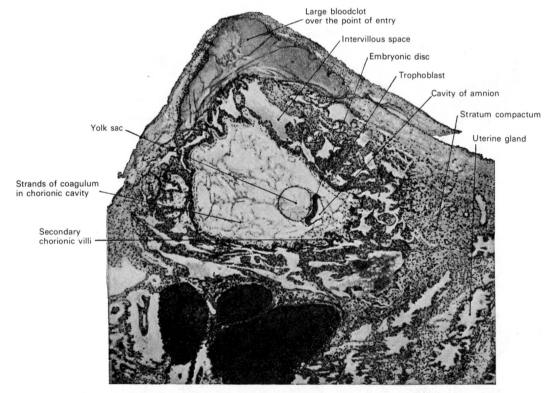

Large bloodclot over the point of entry

Intervillous space

Embryonic disc

Trophoblast

Cavity of amnion

Stratum compactum

Uterine gland

Yolk sac

Strands of coagulum in chorionic cavity

Secondary chorionic villi

FIG. 2-27. A human ovum embedded in the stratum compactum. Estimated age—13½ days. 35 ×. (Hertig and Rock, Amer. J. Obstet. Gynec., 1944.)

some of the maternal vessels, and the strands of cells are bathed in maternal blood. This multinucleated noncellular cytoplasmic mass formed by the trophoblast is called the **syncytiotrophoblast** (Figs. 2-23 to 2-26). On the embryonic side of the syncytiotrophoblast, a cellular trophoblastic layer called the **cytotrophoblast** also forms (Figs. 2-24 to 2-28). The trophoblast continues to grow rapidly and later combines with mesoderm to form the **chorion,** the extraembryonic membrane which protects the embryo and makes contact with the maternal blood for the absorption of oxygen and food substances and for the elimination of waste.

GERM LAYERS. While the blastocyst is becoming embedded in the uterine mucosa, the inner cell mass proliferates rapidly, and within its cluster two groups of cells become distinguishable. A thicker plate of large cells is called the **outer germ layer** or **ectoderm**

(germ disc) (Figs. 2-24 to 2-28). The single layer sheet of cells protruding into the blastocyst cavity is the **inner germ layer** or **endoderm.** Between the invading trophoblast and the adjacent embryonic ectoderm there is a coalescence of small spaces which combine to form a cavity.

The cells surrounding the cavity within the ectoderm stretch out into a thin-roofed vesicle known as the **amnion;** the cavity becomes the **amniotic cavity.** Simultaneously other cells in continuity with the endoderm separate from the inner aspect of the cytotrophoblast, giving rise to the **exocoelomic membrane** (of Heuser) (Figs. 2-24; 2-26). In this manner the **primitive yolk sac** is formed and its cavity is known as the **yolk sac cavity.**

The parts of the ectoderm and endoderm which are adjacent to each other remain in close contact, sharing in the formation of a

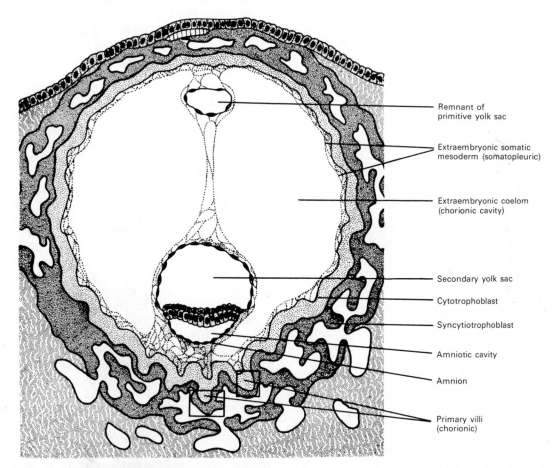

Remnant of primitive yolk sac

Extraembryonic somatic mesoderm (somatopleuric)

Extraembryonic coelom (chorionic cavity)

Secondary yolk sac

Cytotrophoblast

Syncytiotrophoblast

Amniotic cavity

Amnion

Primary villi (chorionic)

FIG. 2-28. Diagrammatic representation of a 13-day human embryo which has become completely implanted into the uterine wall. Primary chorionic villi have developed as well as the secondary yolk sac. (Redrawn after Langman, *Medical Embryology: Human Development, Normal and Abnormal,* Williams & Wilkins Co., 1969, and Hertig and Rock, Amer. J. Obstet. Gynec., 1944.)

thickened plate named the **embryonic** or **germ disc** (Figs. 2-23 to 2-28), from which the embryo proper will be developed. This two-layered or bilaminated disc consists of the columnar cells of the ectoderm, which lie adjacent to the amniotic cavity, and the flatter cells of the developing endoderm, which at first lie adjacent to the blastocyst cavity and then lie adjacent to the cavity of the primitive yolk sac. At about this time, scattered cells migrate outward from the border of the germ disc, forming a loose network of cellular processes which stretch across the blastocyst cavity and occupy the space between the trophoblast and the yolk sac, helping to hold the latter in place. This is the first representation of the **middle germ layer** or **mesoderm,** and since it occupies the blastocyst cavity rather than the germ disc, it is called the **extraembryonic mesoderm** (Fig. 2-26).

EXTRAEMBRYONIC COELOM.   The first representation of the body cavity or coelom in the human embryo occurs very early, at a time when the amniotic cavity and yolk sac are still small and the germ disc has just been established (Fig. 2-26). The extraembryonic mesoderm, just beginning to form, spreads

out in a layer over the amnion and yolk sac, and around the inner surface of the trophoblast, as well as filling in the cavity of the blastocyst with a loose meshwork (Figs. 2-26; 2-27, labeled chorionic cavity). Between the layers of mesoderm covering the embryo and lining the blastocyst cavity, the loosely arranged mesodermal cells separate from each other and leave a space, the **extraembryonic coelom** (Figs. 2-26; 2-28; 2-31). The body cavity of the embryo proper is formed somewhat later and is at first independent of this extraembryonic coelom, but during the period when the yolk sac is being constricted at the body stalk, the two cavities do become confluent for a short time.

Development of the Embryo

EMBRYONIC DISC.   As the thickened layers of the germ disc grow out into a flat oval plate, they begin the formation of the embryo proper. The narrow end of the disc is attached at the **body stalk** and represents the caudal end of the embryo (Figs. 2-32; 2-33). The median line or axis in the caudal part of the embryonic disc is marked by the **primitive groove** (Figs. 2-29; 2-30,D; 2-33). Lying along the bottom of the primitive groove is an elongated mass of rapidly proliferating cells known as the **primitive streak** (Figs. 2-30, D; 2-33). The cells formed by this rapid proliferation spread out laterally between the ectoderm and endoderm as the **definitive mesoderm** of the embryo **(intraembryonic mesoderm),** part of which migrates forward as far as the cranial portion of the embryo. In this manner the third of the three primary germ layers, i.e., the mesoderm, forms between the other two (Fig. 2-30, B to D), thereby transforming the bilaminar disc into a three-layered or **trilaminated disc** (Fig. 2-30, A to D). Between the cranial end of the primitive streak and this anterior growth of mesoderm, the endoderm is thickened into a plate known as the **prochordal plate.** The endoderm and ectoderm remain in close contact at this point without the intervention of mesoderm, and the bilaminar area later becomes the buccopharyngeal membrane of the future stomodeum.

THE NOTOCHORD.   At the cranial end of the primitive streak there appears a knot of ectodermal cells known as the **primitive (Henson's) node** (Figs. 2-29; 2-30, B), from which a

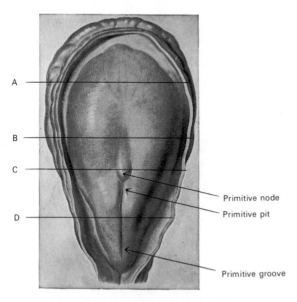

Primitive node
Primitive pit

Primitive groove

FIG. 2-29.   Embryonic disc of 16-day-old human embryo surrounded by cut edge of amnion. Body stalk cut off at posterior end. Primitive groove overlies primitive streak. Primitive node at anterior end of primitive groove. A, B, C, and D are section levels illustrated in Fig. 2-30. A, Prochordal region. B, Notochordal region. C, Primitive node region. D, Primitive streak region. 50 ×. (Heuser, Carneg. Inst., Contr. Embryol., 1932.)

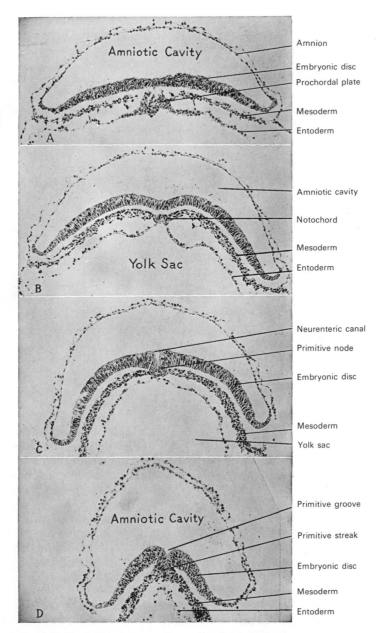

Amnion

Embryonic disc

Prochordal plate

Mesoderm

Entoderm

Amniotic cavity

Notochord

Mesoderm

Entoderm

Neurenteric canal

Primitive node

Embryonic disc

Mesoderm

Yolk sac

Primitive groove

Primitive streak

Embryonic disc

Mesoderm

Entoderm

Fig. 2-30. Transverse sections through presomite embryo shown in Fig. 2-29. *A*, Prochordal region. *B*, Notochordal region. *C*, Primitive node region. *D*, Primitive streak region. 100 ×. (Heuser, Carneg. Inst., Contr. Embryol., 1932.)

strand of cells grows cranialward between the ectoderm and endoderm in the midline until it is blocked at the prochordal plate. This is the **notochordal process,** and it delineates the primitive axis of the developing embryo. Since the primitive streak must grow by a caudalward migration when the neural groove lengthens, the further growth of the notochordal process occurs by caudalward migration of the primitive node. During this migration the notochordal cells form

a flattened strand or plate which displaces the endoderm in the midline for a short time. The plate soon becomes a groove and then a tube, the **notochordal canal.** Cells bounding the floor of the notochordal canal separate to form openings so that the notochordal canal now communicates freely with the yolk sac. More caudally a temporary communication between the yolk sac and the amniotic cavity is established. This communicating channel is called the **neurenteric canal** and ex-

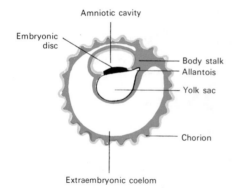

FIG. 2-31.　Diagram illustrating early formation of allantois and differentiation of body stalk.

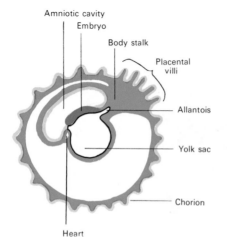

FIG. 2-32.　Diagram showing later stage of allantoic development with commencing constriction of the yolk sac.

tends by means of the **primitive pit** (Fig. 2-29) through both ectoderm and endoderm at the primitive node. The primitive pit is generally considered homologous with the **blastopore** of lower vertebrates, a structure that is importantly involved in the processes of gastrulation in these animals. Certain early notochordal structures are transient, lasting only a short time before the somites are formed, but the notochord remains in the embryo as a solid rod of cells lying between the neural groove and the endoderm in the midline as a structural guiding axis for the growing vertebral column (Fig. 2-34). Only remnants of it persist in the adult.

EARLY GROWTH OF THE EMBRYO

NEURAL TUBE.　In the middle of the third week (about the 18th day), the broad region

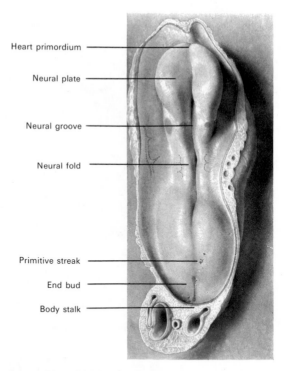

FIG. 2-33.　Dorsal view human embryo 1.38 mm. in length, medullary groove open, about 18 days old. 52.5 ×. (Ingalls, Carneg. Inst., Contr. Embryol., 1920.)

of ectoderm in front of the primitive streak of the embryonic disc becomes thickened and elevated into ridges on both sides of the midline, forming the **neural folds,** the first evidence of the nervous system (Fig. 2-35, *A* and *B*). As the primitive streak migrates caudalward, the neural folds become elongated (Fig. 2-33). Intervening between the neural folds is a longitudinal invagination called the **neural groove.** At about the end of the third week the crests of the neural folds rise into ridges, which gradually curve over the groove until they join, forming the **neural tube** (Figs. 2-35, *C* and *D;* 2-36). The closing of the neural groove occurs first along the early somites in the region of the future hindbrain (Fig. 2-36) and progresses both cranialward and caudalward. In the middle of the fourth week the cranial opening (anterior neuropore) of the neural groove finally closes at the future forebrain (Fig. 2-37). The caudal part of the neural tube is also temporarily open and is called the **caudal** or **posterior neuropore** (Fig. 2-37). It normally closes on the 26th or 27th day, about two days after the closure of the cranial opening.

　While the neural folds are closing, neuroectodermal cells that lie along the dorsal edges of the neural plates on both sides fuse

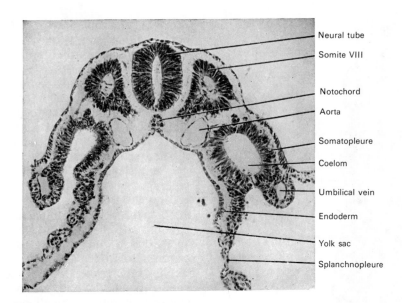

Neural tube
Somite VIII

Notochord
Aorta

Somatopleure

Coelom

Umbilical vein

Endoderm

Yolk sac

Splanchnopleure

FIG. 2-34. Transverse section through somite VIII (primitive segments) of the 14-somite human embryo shown in Fig. 2-37. 150 ×. (Heuser, Carneg. Inst., Contr. Embryol., 1930.)

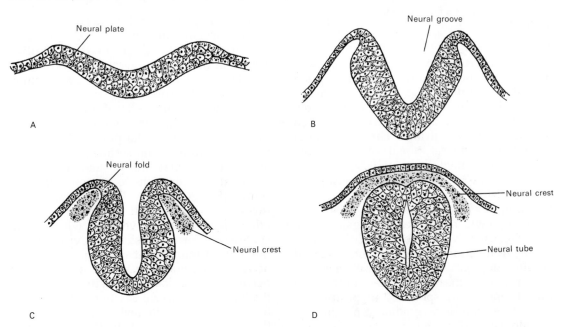

Neural plate

Neural groove

A

B

Neural fold

Neural crest

Neural crest

Neural tube

C

D

FIG. 2-35. Successive stages in the development of the neural tube. *A*, The neural plate develops from neuroectoderm, which probably is induced into differentiation by the underlying notochord and paraxial mesoderm. *B*, The neural plate invaginates to form a neural groove between the elevated neural folds on each side. *C*, The neural folds then converge, and at about the end of the third week begin to fuse to form the neural tube. *D*, As the neural tube fuses, certain neuroectodermal cells that do not participate in the formation of the neural tube develop into the neural crest and eventually form the sensory and autonomic ganglia, as well as other structures.

to form the **neural crest** or **ganglionic ridge** (Fig. 2-35, *C, D*). These cells then migrate lateralward and ventralward beside the neural tube, eventually forming certain ganglionic nerve and sheath cell components of

the cranial, spinal, and autonomic nerves (see Chap. 11 for more details).

The cranial end of the neural tube becomes expanded into a large vesicle with three subdivisions which correspond to the

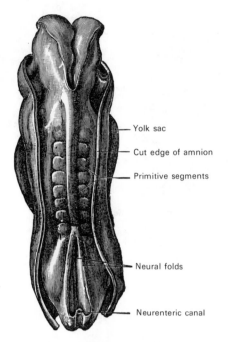

Yolk sac

Cut edge of amnion

Primitive segments

Neural folds

Neurenteric canal

FIG. 2-36. Dorsum of human embryo, 2.11 mm. in length. (After Eternod, Actes helv. Sci. nat. Zurich, 1896.)

future **forebrain** (*prosencephalon*), **midbrain** (*mesencephalon*), and **hindbrain** (*rhombencephalon*) (Fig. 2-45). The walls develop into the nervous tissue and neuroglia of the brain, and the cavity is modified into the cerebral ventricles. The more caudal portion of the neural tube develops into the spinal cord or medulla spinalis.

PRIMITIVE GUT. The embryo increases rapidly in size once the neural tube is established in the fourth week, but it does so by the overgrowth of the central region of the embryonic disc, while the margin of the disc, or line of junction between the embryo and amnion, stops growing and later even gradually becomes narrower. The constriction thus formed, corresponding in position to the future umbilicus, gradually pinches off the dorsal part of the yolk sac included in the disc, so that it becomes enclosed in the embryo proper as the **primitive gut** (Fig. 2-38).

Because of the rapid elongation of the neural tube, the embryo grows more rapidly in length than in width. This results in the cranial and caudal ends pushing out beyond the margin of the embryonic disc, and bending ventrally into **head** and **tail** folds (Figs. 2-37; 2-38). At the cranial end the forebrain grows beyond the **buccopharyngeal membrane** curving over the developing heart.

The diverticulum of the primitive gut, which occupies the head, is known as the **foregut.** Its most cranial extremity becomes the **buccopharyngeal membrane,** which marks the location of the future opening of the mouth, the **stomodeum** (Fig. 2-38). The caudal end of the embryo is at first connected to the chorion by a band of mesoderm called the body stalk, but with the formation of the tail fold, the body stalk is moved into the position of the **umbilicus,** and a diverticulum of the yolk sac called the **hindgut** is formed. The continuation of the hindgut caudally is a dilated cavity called the **cloaca.** For a time the opening into the yolk sac between the foregut and the hindgut remains wide, but the communication is gradually reduced to a slender tube known as the **vitelline duct** or **yolk stalk** and the yolk sac itself, being rudimentary, remains a small pear-shaped sac.

Mesoderm and Somites

At the primitive streak stage (16 to 17 days) intraembryonic mesodermal cells migrate forward and laterally from the primitive streak on both sides, thereby separating the ectoderm and endoderm and transforming the bilaminated disc into a trilaminated one (Fig. 2-39). Certain cells of the mesoderm which migrate forward join across the midline in the most cranial portion of the embryo and form a horseshoe-shaped plate around the cranial ends of the neural folds. At about the time of the formation of the first somite (Fig. 2-40), the cells in the interior of this newly formed plate separate from each other, leaving small vesicles which soon coalesce to form a U-shaped cavity. This is the earliest representative of the **embryonic body cavity** or **coelom,** which is destined to become the **pericardial cavity** (containing the heart), the **pleural cavity** (containing the lungs), and the **peritoneal cavity** (containing the abdominal viscera). In the cardiogenic area the mesoderm of the outer coelomic wall becomes the **parietal pericardium** and, combined with the overlying ectoderm, is known as the **somatopleure.** The mesodermal cells of the inner wall of the coelom will form the **myocardium** and **epicardium** of the **heart** and, combined with the endoderm, are called the **splanchnopleure** (Fig. 2-34).

The median portion of the pericardial coelom expands greatly to accommodate the rapidly enlarging heart. As the rim of the embryonic shield is contracted in pinching

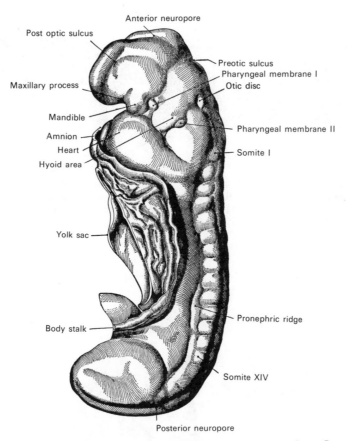

FIG. 2-37. Lateral view of a 14-somite human embryo. 50 ×. (Heuser, Carneg. Inst., Contr. Embryol., 1930.)

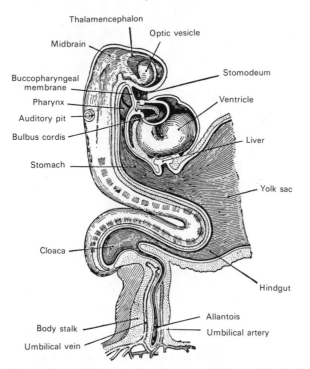

FIG. 2-38. Human embryo about 15 days old. Brain and heart represented from right side. Digestive tube and yolk sac in median section. (After His, *Anatomie menschlicher Embryonen,* F. C. W. Vogel, 1880–1885.)

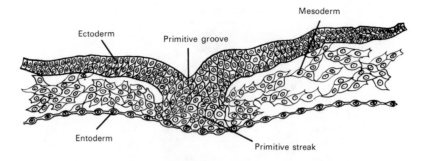

FIG. 2-39. Cross-sectional diagram through an embryo of about 16 days. At this time mesenchymal cells migrate laterally from the primitive streak between the ectoderm and endoderm (entoderm). This mesenchyme forms the (intraembryonic) mesoderm, thereby changing the bilaminated disc into a three-layered one.

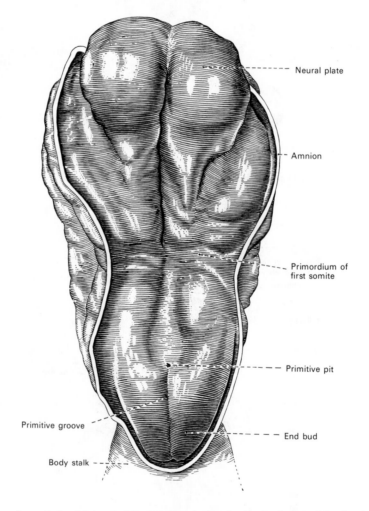

FIG. 2-40. Human embryo at about the end of the third week showing the beginning of the development of somites. The crown-rump length at this stage is about 2 mm. and subsequent somites develop in a craniocaudal sequence behind the first pair. (From Blechschmidt, *The Stages of Human Development Before Birth*, S. Karger, 1961.)

off the foregut, the heart and pericardial cavity protrude caudalward from the cranial ends of the neural folds, gradually becoming ventral to them with the foregut in between (see Figs. 7-1 to 7-4). Since the heart is the fastest growing part of the embryo at this time, it forms a prominent cardiac swelling between the yolk sac and the head fold of the embryo (Figs. 2-37; 2-38).

VASCULAR MESENCHYME. In the middle of the third week (15 to 16 days) and as the primitive pericardial cavity begins to expand, scattered cells separate from the mesoderm of the pericardial splanchnopleure and gather in the interval between the mesoderm and endoderm (Fig. 2-34). These cells are called **mesenchyme** (*mesenchyma*) or **angioblasts** and represent the primordium of the endothelial lining of the vascular system. They proliferate rapidly and in this cardiac region form a tubular sac, the **primitive endocardium** of the heart. A strand of cells grows out from the endocardium on either side of the cranial end of the foregut to become the primordium of the **first pair of aortic arches.** The latter join similar strands of cells lying along the notochord which are destined to become the **dorsal aortae** (Figs. 7-1 to 7-4).

At about this time also mesenchymal cells appear between the endoderm of the yolk sac and its covering of mesoderm. They proliferate rapidly, some gathering into groups to form aggregated masses and cords, which are called **blood islands;** others form strands that acquire a lumen and become the **yolk sac capillaries.** The capillaries join into larger vessels, which become the **vitelline** or **omphalomesenteric veins** leading to the heart. Similarly, the capillary systems in the **chorion** develop and become the **umbilical vessels** (Fig. 2-41).

BLOOD.   The cells forming the blood islands of the vascular mesenchyme of the yolk sac are at first in compact groups. The fluid within the lumen of the capillaries, the **primitive blood plasma,** gradually separates the individual cells of the group into free cells, which are swept into the circulation when the final circuit is established. These are the **primitive hemocytoblasts,** and most of them rapidly acquire hemoglobin in their cytoplasm to become the **primitive erythroblasts** and **primitive erythrocytes** or red blood cells.

It has been theorized that the vascular endothelium, or perhaps even connective tissue cells of the later embryo, may revert

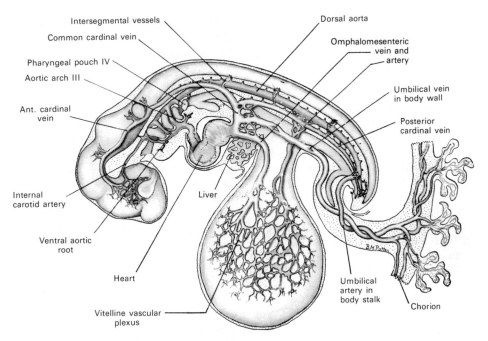

FIG. 2-41.   Semischematic diagram shows basic vascular plan of human embryo at end of first month. For the sake of simplicity, the paired vessels are shown only on side toward observer. (Patten, *Human Embryology,* McGraw-Hill, 1968.)

into hemocytoblasts and produce hemoglobin-containing blood cells. Another theoretical point of view, however, is that tissue as highly specialized as hemoglobin-containing cells must logically come from a specific primordium. The mechanism for this would be the seeding of undifferentiated descendents of the primitive blood island hemocytoblasts in various tissues such as the liver, spleen, and bone marrow. In favor of this view are experiments with amphibian embryos in which the entire blood island was removed surgically. The resulting embryos grew into free-swimming larvae with complete vascular systems including hearts and livers but no erythrocytes. The larvae were kept alive for as long as 32 days with an increased oxygen supply. The circulation of clear plasma pumped by the heart could be detected by occasional passage of particles and non-erythrocytic cells in the capillaries of the gills (Federici, 1926; Goss, 1928).

The **beginning of circulation** occurs at about the eight-somite stage, during the early part of the fourth week (22nd day) in the human embryo. By this time the vascular lumen has become continuous from the endocardium, through the first aortic arches, aortae, yolk sac capillaries, and vitelline veins. The myocardium has become a muscular tube contracting with its own regular intrinsic rhythm. The cells within the blood islands become circulating hemocytoblasts and are swept through the vascular lumen by the primitive blood plasma and propelled by the pumping of the heart.

The changes that the heart undergoes during its development from the fusion of two symmetrically straight tubes into its more complex four-chambered structure are described in Chapter 7. This precocious development of the heart and the commencement of the circulation is necessary for the survival of the embryo because of the deficiency in stores of nutriment within the ovum. Since the circulation becomes established early in the fourth week, the circulatory system becomes the first functional system in the embryo. The growth of the umbilical vessels in the body stalk and the development of the chorionic villi soon establish the interrelationship between the embryonic and maternal circulations.

SOMITES. As the neural folds increase in size, the mesoderm along both sides of the notochord and neural folds expands into a column of cells, the **paraxial mesoderm** (Fig.

2-34). Early in the third week the cells of these columns become organized into paired blocks called **primitive body segments** or **somites,** which lie just under the ectoderm, where they are visible in the intact embryo and can be counted (Figs. 2-36; 2-37). From the somites are derived most of the axial skeleton, its associated musculature, and the skin dermis. This primitive segmentation of the embryonic mesoderm is retained as the muscular and skeletal segments of the mature organism. During development each individual segment also achieves its segmental nerve from the adjacently developing neural tube. The first pair of somites can be recognized on about the 20th day and differentiates in the future occipital region; the separation of new somites progresses in a caudal direction until eventually there are 42 to 44 pairs. The number of somites in each part of the human embryo are 4 occipital, 8 cervical, 12 thoracic, 5 lumbar, 5 sacral, and 8 or 10 coccygeal. The first occipital and last few coccygeal somites regress and disappear. The other occipital somites become incorporated into structures of the head. Each of the remaining somites differentiates into three parts, a sclerotome, a myotome, and a dermatome (Fig. 2-42, D).

At first the somite contains a central cavity, called the **myocele,** surrounded by cells in a tightly packed epithelioid arrangement (Fig. 2-42, C). The ventromedial aspect of the somite soon breaks down into irregular-shaped cells which migrate toward the notochord giving rise to the **sclerotome.** These cells multiply rapidly, surround the notochord, and extend upward beside the neural tube to become the primordium of the vertebral column. The cells remaining in the dorsal and ventral parts of the somite next migrate into the region adjacent to the sclerotome, become elongated into myoblasts, and constitute the **myotome** from which the segmentally arranged musculature is developed. The remaining lateral part of the somite also breaks down; its cells migrate closely beneath the ectoderm as the **dermatome,** which provides the future dermis (Fig. 2-42).

While the somites develop from the medial portion of the paraxial mesoderm, the lateral portion spreads out into two thin sheets, one lying against the ectoderm, the other against the endoderm (Fig. 2-34). The space between the two sheets is part of the embryonic coelom (Fig. 2-42). The ectoderm,

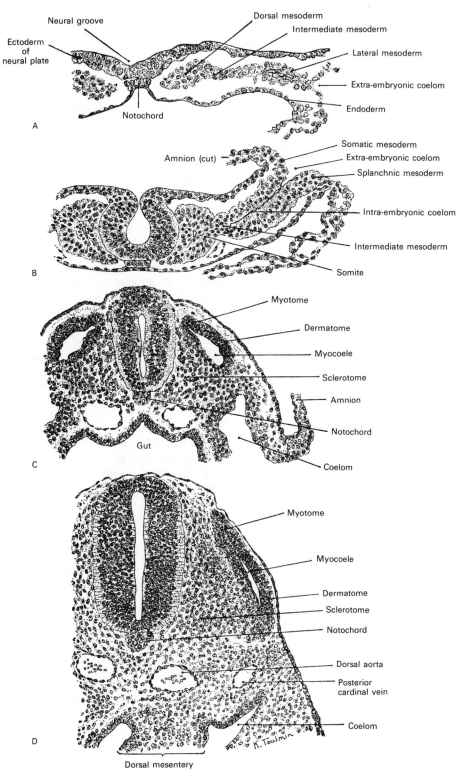

Fig. 2-42.   Drawings from transverse sections of pig embryos of various ages to show formation and early differentiation of somites. 150 ×. *A*, Beginning of somite formation. *B*, Seven-somite embryo. *C*, Sixteen-somite embryo. *D*, Thirty-somite embryo. (Specimens from the Carnegie Institute Collection; figure reproduced from Crouch, *Functional Human Anatomy*, Lea & Febiger, 1972.)

with its layer of somatic mesoderm, is known as the **somatopleure** and forms the skin, muscles, bones, and fasciae of the body wall. The endoderm, with its layer of splanchnic mesoderm, forms the **splanchnopleure** from which develop the serous membranes of the future body cavities (Fig. 2-34). At the junction between the somites and the lateral mesoderm a special strand of cells appears which is known as the **intermediate cell mass** or **nephrotome.** It is the source of the future genitourinary system.

Development of the Body Cavities

At the time of the commencement of somite formation, a cavity appears within the cardiogenic mesoderm at the cranial end of the embryonic disc. This cavity is the first indication of the **intraembryonic coelom** (Fig. 2-42, B). The coelom begins with the formation of scattered vesicles, which coalesce into a horseshoe-shaped cavity that is continuous across the midline in front of the forebrain, but is at first closed caudally in the region of the early somites. This cavity is the pericardial portion of the primitive coelomic cavity. The two lateral ends of the U-shaped cavity progress caudally, and in a short time become confluent with the extraembryonic coelom on each side of the body of the embryo. As segmentation of the me-

dial portion of the paraxial mesoderm into somite formation continues, clefts occur within the mesoderm of the lateral plate resulting in the formation of the splanchnic and somatic mesoderm layers (Figs. 2-42, B; 2-43, A). The coelomic cavity forms between these layers. The splanchnic mesoderm with the underlying endoderm forms the intraembryonic splanchnopleure, while the somatic mesoderm with its overlying ectoderm forms the intraembryonic somatopleure (Fig. 2-43, B).

More rostrally as the pericardial cavity moves into its position ventral to the foregut by the formation of the head fold, a thick plate of splanchnic mesoderm comes to lie between the enlarged heart and the constricted portion of the yolk sac. This plate is known as the **septum transversum.** Through it the vitelline veins from the yolk sac reach the heart, and into it grows the diverticulum from the gut which forms the liver (Fig. 2-41). The more caudal parts of the coelom remain as narrow extensions on each side of the foregut in the region of the somites. This narrowed portion of the coelom dorsal to the septum transversum is called the **pleural canal** because a little later diverticula from the foregut which develop into the lungs push out this canal and expand it into the pleural cavities (Fig. 2-44).

At this early stage, the coelomic cavity is continuous from the pericardial cavity

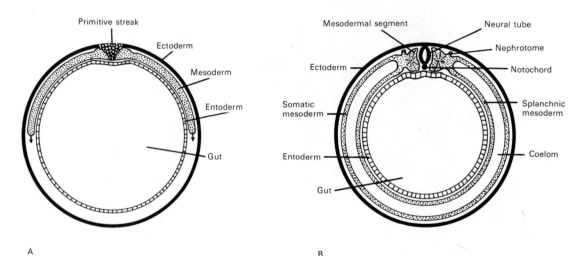

A   B

FIG. 2-43. Diagrammatic representations of cross sections through developing embryos to illustrate the migration of mesoderm laterally from the region of the primitive streak between the ectoderm and endoderm (entoderm). *A,* Through the primitive streak at about 16 days. *B,* Through the neural tube at a somewhat later stage of development, perhaps in the fourth week. (Modified from Arey, *Developmental Anatomy: A Textbook and Laboratory Manual of Embryology,* W. B. Saunders Co., 1974.)

around the heart, through the pleural canals around the lung buds, into the peritoneal cavity in the abdomen. Later the coelomic cavity is divided into the pericardial, pleural, and peritoneal cavities by the formation of two septa. One septum begins as a fold of tissue on each side from the cranial border of the septum transversum and the lateral body wall; it protrudes into the pleural canal cranial to the lung bud and is known as the **pleuropericardial fold.** Within this fold are found the **common cardinal vein** and the **phrenic nerve.** The fold continues to grow outward until it completely closes the canal as a septum, separating the pericardial cavity from the pleural cavities (Fig. 2-44).

The second fold is more caudally placed and appears at the lower part of the septum transversum in the region where the common cardinal veins (*ducts of Cuvier*) open into the sinus venosus at the base of the heart. This fold, known as the **pleuroperitoneal fold,** is also paired, and it grows dorsally from the septum transversum. As the growing lung buds expand the thoracic cavities at both sides of the heart, the pleuroperitoneal fold contributes an important part of the tissue that forms the diaphragm. Eventually the diaphragm closes off the coelom and separates the pleural cavities from the peritoneal cavity.

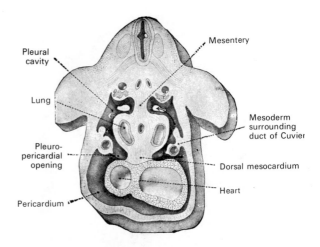

Fig. 2-44. Figure obtained by combining several successive sections of a human embryo of about the fourth week. The upper arrow is in the pleuroperitoneal opening, the lower in the pleuropericardial. (After Kollman, *Handatlas der Entwicklungsgeschichte des Menschen*, G. Fischer, 1907.)

midline of the neck. In all, six arches make their appearance, but of these only the first four are visible externally. The fifth and sixth branchial arches are more rudimentary and less well defined and can be seen only on the inner surface of the pharynx. The first arch is named the **mandibular,** and the sec-

The Branchial Region

PHARYNGEAL POUCHES AND BRANCHIAL OR VISCERAL ARCHES. In the lateral wall of the anterior foregut or pharynx there develops during the 4th week a series of alternating depressions and ridges visible on each side of the embryo. The depressions are called the **pharyngeal pouches,** of which five pairs appear (Fig. 2-41); each of the upper four pouches is prolonged into a dorsal and a ventral diverticulum. On the external surface of the embryo, corresponding indentations of the ectoderm occur, forming what are known as the **branchial** or **outer pharyngeal grooves** (Fig. 2-45, shown but not labeled). The grooves separate a series of rounded bars or arches, the **branchial** or **visceral arches,** in which thickening of the mesoderm takes place (Figs. 2-45; 2-47). The dorsal ends of the branchial arches are attached to the sides of the head, while the ventral extremities ultimately meet at the

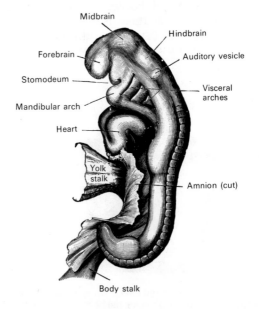

Fig. 2-45. Embryo of the fourth week showing pharyngeal grooves between visceral arches. (His, *Anatomie menschlicher Embryonen*, F. C. W. Vogel, 1880–1885.)

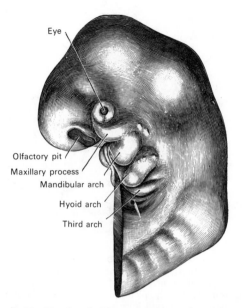

Eye

Olfactory pit

Maxillary process

Mandibular arch

Hyoid arch

Third arch

FIG. 2-46. Head end of human embryo, about the end of the fourth week. (From model by Peter, *Atlas der Entwicklung der Nase und des Gaumens beim Menschen*, G. Fischer, 1913.)

ond the **hyoid**; the others do not have distinctive names and are referred to numerically (Fig. 2-46). The mandibular and hyoid arches are the first ones to appear and are recognizable in the 14-somite stage (Fig. 2-37). In each arch a cartilaginous bar, consisting of right and left halves, is developed, and with each of these there is one of the primitive aortic arterial arches. Additionally, each arch contains a primordium which will de-

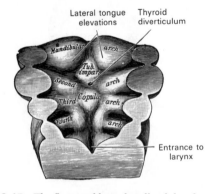

Lateral tongue elevations   Thyroid diverticulum

Mandibular   arch

Tub. impar

Second   arch

Copula

Third   arch

Fourth   arch

Entrance to larynx

FIG. 2-47. The floor and lateral walls of the pharynx of a four-week embryo showing the pharyngeal pouches as depressions between the pairs of branchial arches. (Head end of same embryo shown in Fig. 2-46.)

velop into muscle tissue, and receives nerves on each side which grow into the arch from the developing nervous system.

The **mandibular arch** lies between the first branchial groove and the stomodeum (Fig. 2-45). This first branchial arch is associated with the development of the face, and from it develop the lower lip, mandible, muscles of mastication, and anterior part of the tongue. Its cartilaginous bar is formed by what are known as the right and left **Meckel's cartilages** (Figs. 2-48; 4-14). The dorsal end of each cartilage, related to the developing ear, becomes ossified and forms two bones of the middle ear, the **incus** and the **malleus.** The ventral ends of each cartilage meet in the region of the symphysis menti, and whereas most of the cartilage disappears, the portion immediately adjacent to the malleus is replaced by fibrous membrane and becomes the **sphenomandibular ligament.** From the connective tissue covering the remainder of the cartilage, the greater part of the **mandible** is ossified. The triangular-shaped **maxillary process** grows forward on each side from the dorsal ends of the mandibular arch to form the cheek and lateral part of the upper lip. In addition to the muscles of mastication, the mylohyoid muscle, the anterior belly of the digastric muscle, and the tensors tympani and palati are derived from the mandibular arch. The nerve of the first arch is the **mandibular division of the trigeminal nerve.**

The **second** or **hyoid arch** assists in forming the side and front of the neck. From its cartilage are developed the styloid process, stylohyoid ligament, and lesser cornu of the hyoid bone. The stapes probably arises in the upper part of this arch. The muscles of facial expression along with the stapedius muscle, stylohyoid muscle, and posterior belly of the digastric muscle develop from the hyoid arch. The nerve of the second arch is the seventh cranial or **facial nerve.**

The cartilage of the **third arch** gives origin to the greater cornu of the hyoid bone. The ventral ends of the second and third arches unite with those of the opposite side, and form a transverse band, from which the body of the hyoid bone and the posterior part of the tongue are developed. The stylopharyngeus muscle and the upper pharyngeal musculature are derived from the third arch. The **glossopharyngeal** or ninth cranial

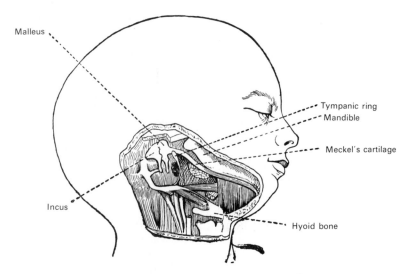

FIG. 2-48. Head and neck of a human embryo 18 weeks old, with Meckel's cartilage and hyoid bar exposed. (After Kölliker, *Grundriss der Entwicklungsgeschichte des Menschen und der Hoheren Thiere,* von Wilhelm Engelmann, 1880.)

nerve is associated with the third branchial arch.

The ventral portions of the cartilages of the **fourth arch** unite to form the thyroid cartilage, while from the **fifth** and **sixth arches** develop the arytenoid, corniculate, cuneiform, and cricoid cartilages of the larynx. The musculature of the pharynx and larynx also forms from these more caudal arches, and the superior and recurrent laryngeal branches of the **vagus nerve** are associated with them.

The mandibular and hyoid arches grow more rapidly than those behind them, with the result that the latter become, to a certain extent, telescoped within the former, and a deep ectodermal depression, the **sinus cervicalis,** is formed on either side of the neck. This sinus is ultimately obliterated by the fusion of its walls; if this fails to happen, a **branchial cyst** results, or if the overgrowing mandibular and hyoid arches fail to fuse with the fifth arch below, a **branchial fistula** will develop.

From the first **branchial groove** the concha of the ear and external acoustic meatus are developed, while around the groove there appear, on the mandibular and hyoid arches, a number of swellings from which the auricula or pinna of the external ear is formed (Fig. 2-57). No traces of the second, third, and fourth branchial grooves persist.

The **first pharyngeal pouch** is prolonged dorsally to form the auditory tube and the tympanic cavity; the closing membrane between the mandibular and hyoid arches is invaded by mesoderm, and forms the tympanic membrane. The inner part of the **second pharyngeal pouch** is named the **sinus tonsillaris;** in it the palatine tonsil is developed, above which a trace of the sinus persists as the supratonsillar fossa. The fossa of Rosenmüller or lateral recess of the pharynx is regarded by some as a persistent part of the second pharyngeal pouch, but it is probably developed as a secondary formation. From the **third pharyngeal pouch** the thymus arises as an endodermal diverticulum on either side, and from the fourth pouches small diverticula project and become incorporated with the thymus, but in man these diverticula probably never form true thymus tissue. The parathyroid glands also arise as diverticula from the third and fourth pouches. From the fifth pouches the **ultimobranchial bodies** originate and are enveloped by the lateral prolongations of the median thyroid rudiment; they do not, however, form true thyroid tissue, nor are any traces of them found in the human adult (see Development of Thyroid Gland, Chapter 18).

NOSE AND FACE. During the third week two sites of thickened ectoderm, the **olfactory areas,** appear immediately under the forebrain in the anterior wall of the stomodeum, one on either side of a region termed the **frontonasal process** (Fig. 2-49). By the upgrowth of the surrounding parts these areas are converted into shallow depressions, the **olfactory pits,** which indent the frontonasal process and divide it into a **medial** and two **lateral nasal processes** (Fig. 2-50). The rounded lateral angles of the medial process constitute the **globular processes.** The olfactory pits form the rudiments of the nasal cavities, and the epithelium which lines the nasal cavities (except that of the inferior meatuses) is derived from the ectodermal lining of the olfactory pits. The globular processes are prolonged backward as plates, which are termed the **nasal laminae.** These laminae at first are some distance apart, but they gradually approximate and ultimately fuse to form the nasal septum. More superficially, the globular processes themselves meet at the midline, and form the **premaxillary bones,** as well as the **philtrum** or central part of the upper lip (Fig. 2-53). Between the globular processes, the depressed region of the medial nasal process forms the lower part of the nasal septum or **columella.** Above this is seen a prominent angle, which becomes the future apex of the nose (Figs. 2-50; 2-51), and still higher a flat area, the future bridge of the nose. The lateral nasal processes form the alae of the external nose.

Continuous with the dorsal end of the mandibular arch and growing forward from its cranial border is the triangular **maxillary process,** the ventral extremity of which is separated from the mandibular arch by a V-shaped notch (Fig. 2-49). The maxillary process forms the lateral wall and floor of the orbit, and within it are ossified the zygomatic bone and the greater part of the maxilla. As development ensues, the maxillary process meets with the lateral nasal process, but is separated from it for a time by a groove, the **nasolacrimal sulcus,** which extends from the furrow encircling the eyeball to the olfactory pit. Ultimately the maxillary processes fuse with the lateral nasal and globular processes to form the lateral parts of the upper lip and the posterior boundaries of the nares (Figs. 2-52; 2-53). From the third to the fifth month the nares are filled by masses of epithelium, which then break down and disappear, thereby forming the permanent openings. Additionally, the maxillary process gives rise to the lower portion of the lateral wall of the nasal cavity. The roof of the nasal cavity and the remaining parts of its lateral wall (the ethmoidal labyrinth, the inferior nasal concha), as well as the lateral cartilage and the lateral crus of the alar cartilage of the external nose, are developed in the lateral nasal process.

The **primitive palate** develops by the fusion of the maxillary and nasal processes in the roof of the stomodeum (Fig. 2-54). The olfactory pits (also called **nasal pits**), which will become the nasal cavities, extend backward above the primitive palate. The posterior end of each pit is closed by an epithelial membrane, the **bucconasal (oronasal) membrane,** formed by the apposition of the nasal and stomodeal epithelium. As these mem-

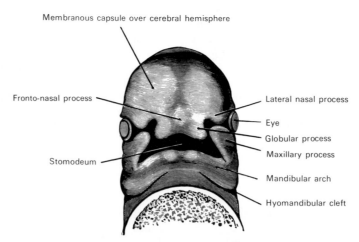

FIG. 2-49.   Facial region of a human embryo about 29 days old. (After His, *Anatomie menschlicher Embryonen,* F. C. W. Vogel, 1880–1885.)

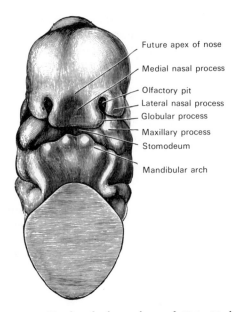

Future apex of nose
Medial nasal process
Olfactory pit
Lateral nasal process
Globular process
Maxillary process
Stomodeum
Mandibular arch

Future apex of nose
Medial nasal process
Olfactory pit
Lateral nasal process
Globular process
Maxillary process
Roof of pharynx
Hypophyseal diverticulum
Dorsal wall of pharynx

FIG. 2-50. Head end of a embryo of 30 to 31 days. (From model by Peter, *Atlas der Entwicklung der Nase und des Gaumens beim Menschen*, G. Fischer, 1913.)

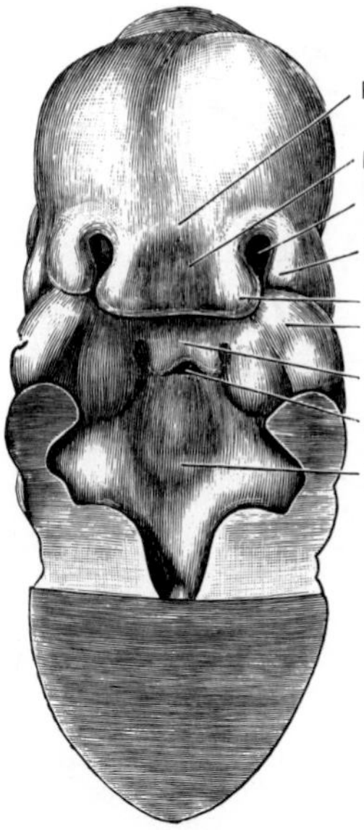

FIG. 2-51. Same embryo as shown in Fig. 2-50 with front wall of pharynx removed.

branes rupture, the **primitive choanae** or openings between the olfactory pits and the stomodeum are established.

The floor of the nasal cavity develops from a pair of shelf-like **lateral palatine processes,** which extend medialward from the maxillary processes (Figs. 2-54 to 2-56); these coalesce in the midline to constitute the entire palate, except for a small anterior portion, which is formed by the premaxillary bones. Two apertures persist for a time between the palatine processes and the premaxillae. In lower animals these persist as permanent channels while in the adult human palate the apertures are represented by the **incisive foramen** anteriorly in the hard palate. The union of the parts that form the palate commences anteriorly, the premaxillary and palatine processes joining in the ninth week, whereas the region of the future hard palate is completed by the tenth, and that of the soft palate by the end of the third month. By this time the **permanent choanae** are formed and are situated at the junction of the nasal cavities and the pharynx, a considerable distance posterior to the primitive choanae.

Failure in the union of the palatine processes with each other and with the nasal septum occurs about once in 2500 births (Moore, 1973). The resulting malformation is called **cleft palate** and may manifest itself with varying degrees of severity. A mild deformity may be simply a small furrow in the soft palate, while in more severe cases, a large midline fissure may extend throughout the length of both the hard and soft palate. Perhaps the most important congenital deformity of the face is **cleft lip** (hare lip). This results through a failure in the fusion of the maxillary process with the corresponding

FIG. 2-52. Head of a human embryo of about eight weeks in which the nose and mouth are formed. (His, *Anatomie menschlicher Embryonen*, F. C. W. Vogel, 1880–1885.)

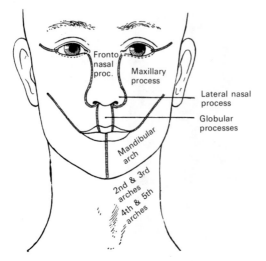

FIG. 2-53. Diagram showing the regions of the adult face and neck related to the frontonasal process and the branchial arches.

extends from the inferior aspect of the ethmoid plate of the chondrocranium. The anterior part of this cartilaginous plate persists as the septal cartilage of the nose and the medial crus of the alar cartilage, but the posterior and superior parts are replaced by the vomer and perpendicular plate of the ethmoid. On each anteroinferior surface of the nasal septum, the ectoderm is invaginated to form a small blind pouch or diverticulum. These extend posteriorly for 3 to 5 mm, and each is supported by a tiny curved plate of cartilage. These pouches are called the **vomeronasal organs** (of Jacobson) and in man they are rudimentary (Fig. 2-54). In certain other mammals, however, they apparently serve as accessory organs of olfaction, since their nerve filaments join the olfactory nerve.

The Limbs

The limbs first appear in the fourth week as small elevations or buds at the side of the trunk. Unsegmented somatic mesoderm pushes into the limb buds and multiplies by division of its cells into closely packed cellular masses. During the sixth week, the forelimbs have grown anteriorly and lie lateral to the developing heart (Fig. 2-57). As the seventh and eighth weeks progress, the em-

globular process (medial nasal process). Occurring about once in 900 births, it may be either unilateral or bilateral.

The nasal cavity becomes divided by the vertical **nasal septum,** which extends downward and backward from the medial nasal process and nasal laminae, and unites below with the transversely oriented palatine processes. Into this septum a plate of cartilage

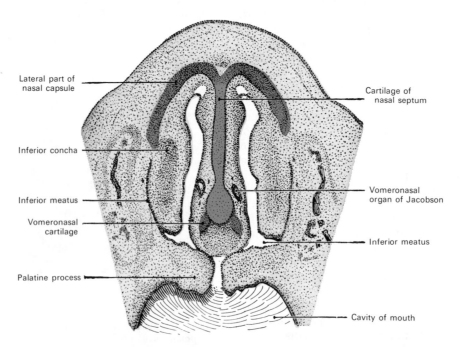

FIG. 2-54. Frontal section of nasal cavities of a human embryo 28 mm. long. (Kollman, *Handatlas der Entwicklungsgeschichte des Menschen,* G. Fischer, 1907.)

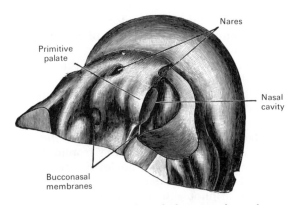

FIG. 2-55. Primitive palate of a human embryo of 37 to 38 days. On the left side of the specimen the lateral wall of the nasal cavity has been removed. (From a model by Peter, *Atlas der Entwicklung der Nase und des Gaumens beim Menschen,* G. Fischer, 1913.)

bryo becomes distinctly human in its form and the developing hands approximate the nose and chin (Fig. 2-58). The intrinsic muscles of the limbs differentiate *in situ*. The upper limb, developing in the caudal neck region, receives its innervation from the fourth cervical to the second thoracic spinal nerves. The lower limb bud develops in the lumbosacral region and receives nerves from the twelfth thoracic through the fourth sacral segments before the limbs migrate caudally.

The axial part of the mesoderm of the limb bud becomes condensed and converted into its cartilaginous skeleton. By its ossification the bones of the limb are formed. The three chief divisions of each limb can already be identified by the end of the sixth week. Furrows separate the upper limb into arm, forearm, and hand, and the lower limb

into thigh, leg, and foot (Fig. 2-57). The limbs are at first directed *inferiorly* nearly parallel to the long axis of the trunk, and each presents two surfaces and two borders. Of the surfaces, one, the future **flexor** surface of the limb, is initially directed ventrally; the other, the **extensor** surface, dorsally; one border, the **preaxial,** is oriented toward the cranial end of the embryo, and the other, the **postaxial,** toward the caudal end. The lateral epicondyle of the humerus, the radius, and the thumb lie along the preaxial border of the upper limb; and the medial epicondyle of the femur, the tibia, and the great toe lie along the preaxial border of the lower limb.

The limbs next undergo a rotation through an angle of 90 degrees around their long axes, the rotation being effected almost entirely at the limb girdles. Flexion increases at the elbow and knee, with the elbow pointing caudally and the knee cranially (Fig. 2-58). The upper limb rotates laterally and as a consequence its preaxial (radial) border is directed lateralward while the ventral (flexor) surface of the forelimb is turned cranially or anteriorly. The opposite (medial) rotation occurs in the lower limb. In the hindlimb, rotation directs the preaxial (tibial) border medially while the ventral (flexor) surface is oriented posteriorly.

FETAL MEMBRANES AND PLACENTA

The fetal membranes which protect and nourish the human embryo and which allow the elimination of metabolic materials are the **chorion** (which develops from the **trophoblast**), **amnion, yolk sac,** and **allantois.** The

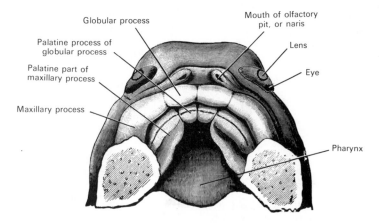

FIG. 2-56. The roof of the mouth of a human embryo, aged about $2\frac{1}{2}$ months, showing the mode of formation of the palate. (His, *Anatomie menschlicher Embryonen,* F. C. W. Vogel, 1880–1885.)

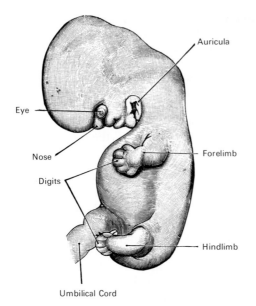

FIG. 2-57. Embryo of about six weeks. (His, *Anatomie menschlicher Embryonen,* F. C. W. Vogel, 1880–1885.)

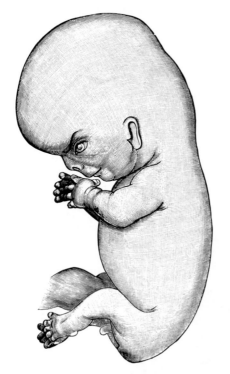

FIG. 2-58. Human embryo about $8\frac{1}{2}$ weeks old. (His, *Anatomie menschlicher Embryonen,* F. C. W. Vogel, 1880–1885.)

organ of metabolic transfer between the mother and fetus is called the **placenta** and includes a fetal portion derived from the chorion and a maternal part from the endometrial wall of the uterus.

In the early blastocyst stage, the trophoblast consists of a single layer of cells (Figs. 2-22 to 2-25). As the blastocyst becomes embedded in the endometrium of the uterus, two layers are formed in the trophoblast: The outer layer, rich in nuclei but with no evident cell boundaries, is called the **syncytiotrophoblast;** the inner layer (of Langhans) is cellular and is called the **cytotrophoblast** (Figs. 2-25 to 2-28; 2-59; 2-60). The trophoblast invades the tissue and vessels of the uterine wall and becomes converted into a thick sponge-like mass of strands consisting of cores of cytotrophoblast and outer coverings of syncytiotrophoblast. The walls of the uterine vessels lying in the path of the trophoblast become eroded, and maternal blood seeps into the spaces between the strands of trophoblast. These spaces are now called the **intervillous spaces** (Figs. 2-59; 2-60).

Chorion

The outermost fetal membrane is known as the **chorion.** Its structure includes the strands of trophoblast that have become anchored to the uterine tissue and others that protrude into the intervillous space as primary villi (Figs. 2-59; 2-61). Its internal cavity (chorionic cavity or extraembryonic coelom) at first contains only fluid and loose strands of extraembryonic mesoderm. Later the amnion expands so rapidly that it encroaches upon the chorionic cavity until it comes into contact with the inner wall of the chorion, thereby obliterating its cavity (Figs. 2-63; 2-64). The chorionic tissue, however, continues to expand throughout pregnancy to accommodate the growing fetus and serves as the outer barrier between it and the maternal tissue across which physiological exchange must be maintained.

CHORIONIC VILLI. The **primary villi** of the chorion develop from the solid strands of trophoblast which extend into the uterine wall. The villi consist of an inner cluster of proliferating cytotrophoblast cells covered with an irregular layer of syncytiotrophoblast (Fig. 2-59). The primary villi are converted

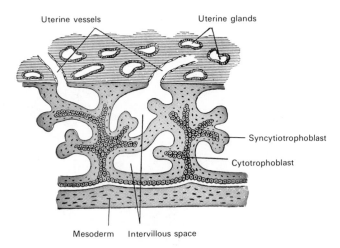

Uterine vessels   Uterine glands

Syncytiotrophoblast

Cytotrophoblast

Mesoderm   Intervillous space

FIG. 2-59.   Primary chorionic villi. Diagrammatic. (Modified from Bryce *et al.*, 1908.)

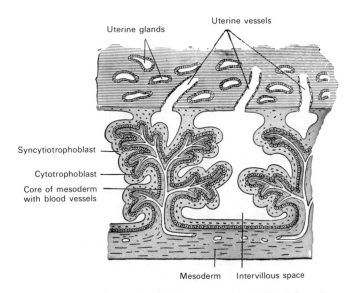

Uterine glands   Uterine vessels

Syncytiotrophoblast

Cytotrophoblast

Core of mesoderm
with blood vessels

Mesoderm   Intervillous space

FIG. 2-60.   Secondary or true chorionic villi. Diagrammatic. (Modified from Bryce *et al.*, 1908.)

into the **secondary** or **true chorionic villi** by an ingrowth into each primary villus of a central core of mesoderm. Capillaries begin to differentiate within this mesodermal core and soon connect with vascular sprouts from the umbilical vessels which have grown into the chorionic mesoderm by way of the body stalk. As the chorionic villi continue to grow and ramify, the ingrowth of mesoderm and the proliferation of the umbilical vessels keep pace, establishing the structural substrate for the exchange of nutrients and metabolites between the fetal and maternal circulation (Figs. 2-60; 2-64; 2-65).

Until the end of about the second month,

the villi sprout from the entire outer surface of the chorion and are more or less uniform in size. After this time the villi nearest the body stalk of the embryo (Figs. 2-62; 2-65) grow more elaborate and complicated than the rest, becoming the fetal portion of the placenta. This is called the **chorion frondosum** (leafy chorion) and the portion of the endometrium to which it is attached is known as the **decidua basalis.** The villi in the rest or nonplacental chorion gradually atrophy and by the end of the fourth month scarcely a trace of them is left. The resulting smooth surface over this part of the chorion accounts for its name, the **chorion laeve** (Fig. 2-64).

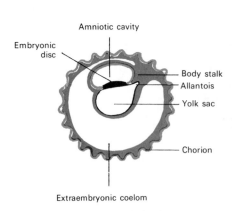

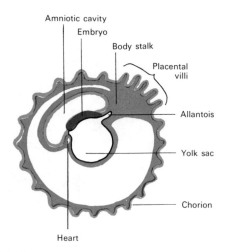

Fig. 2-61. Diagram illustrating early formation of allantois and differentiation of body stalk.

Fig. 2-62. Diagram showing later stage of allantoic development with commencing constriction of the yolk sac.

MATERNAL TISSUES. Important morphologic changes occur in the endometrial wall of the uterus in preparation for the possible attachment of a developing blastocyst. It is said to enter the **progravid phase.** These changes may also be considered premenstrual because the superficial layer of the endometrial wall is lost during the menstrual period in the event that a fertilized ovum does not become implanted. If the implantation of an ovum occurs, however, the thickness and vascularity of the endometrium are greatly increased. During this **gravid phase,** the uterine glands are elongated with funnel-shaped orifices at the surface, and their deeper portions are dilated into irregular, tortuous spaces.

The trophoblastic cells of the implanted blastocyst chorion secrete a hormone called **chorionic gonadotropin,** which sustains the life of the **corpus luteum.** The resulting corpus luteum of pregnancy continues to secrete another hormone, **progesterone.** As a result of these hormonal influences, the menstrual period does not occur, and the endometrial wall of the uterus continues to develop into a favorable site for gestation. Since large amounts of chorionic gonadotropin are secreted at this time, its excess is excreted in the urine of the pregnant woman. This forms the basis of the urine pregnancy tests, one of which is the Friedman modification of the Ascheim-Zondek test. In this test 4 ml of urine are injected intravenously twice daily for two days into a

virgin, mature, female rabbit. The presence of chorionic gonadrotropin, and hence a positive test for pregnancy, causes the rabbit to ovulate resulting in the presence of fresh corpora lutea or hemorrhagic corpora detectable in the ovaries of the rabbit.

The changes in the uterine wall are well advanced by the second month of pregnancy and the tissue between the uterine glands has become crowded with enlarged, glycogen-rich, connective tissue cells known as decidual cells. The endometrium consists of the following strata (Fig. 2-66): (1) the **stratum compactum,** next to the free surface, is traversed by the necks of the uterine glands and contains a rather compact layer of interglandular tissue; (2) the **stratum spongiosum** contains the tortuous and dilated glands and only a small amount of interglandular tissue; (3) the **stratum basalis,** or boundary layer next to the uterine muscular wall, contains the deepest parts of the uterine glands. At each menstrual flow and at the termination of pregnancy, the stratum compactum and stratum spongiosum are sloughed. Together these two strata of the endometrium are known as the **pars functionalis** because they are cyclically undergoing a changing histological structure in reaction to the hormonal secretions of the ovary. The stratum basalis remains relatively unchanged and serves as the source for the regenerating tissue for the functional layer.

The **decidua** (Latin *deciduus* = falling off) is the name given to the pars functionalis of

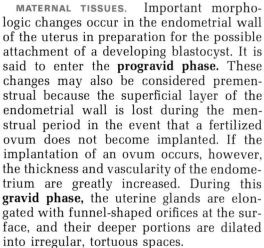

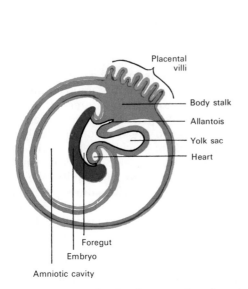

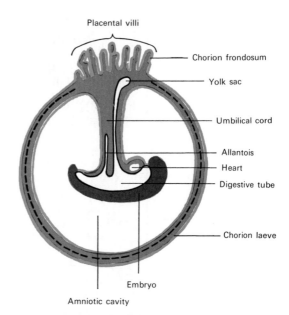

FIG. 2-63. Diagram showing the expansion of amnion and constriction of the yolk sac.

FIG. 2-64. Diagram illustrating a later stage in the development of the umbilical cord.

the endometrium during pregnancy. The various parts of the decidua are given special names according to their relationship to the chorion of the embryo and fetus. That part beneath the chorion frondosum is the **decidua basalis.** It is the thickest and deepest part of the endometrium between the embryo and the muscular layer of the uterus.

(Since it is also the part of the decidua involved in forming the placenta, it is sometimes called the **decidua placentalis.**) The portion of the decidua that is thin, continuous with the decidua basalis, and closes over the rest of the embryo after it has burrowed into the uterine tissue is the **decidua capsularis.** Finally, the part that lines the rest of

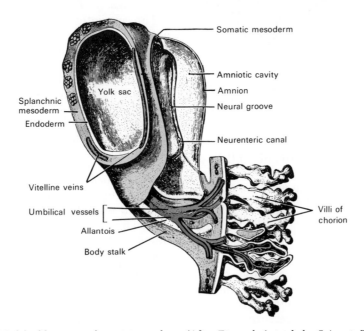

FIG. 2-65. Model of human embryo 1.3 mm. long. (After Eternod, Actes helv. Sci. nat. Zurich, 1896.)

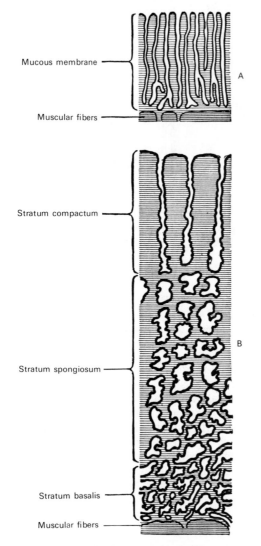

Mucous membrane

Muscular fibers

A

Stratum compactum

Stratum spongiosum

B

Stratum basalis

Muscular fibers

FIG. 2-66. Diagrammatic sections of the uterine mucous membrane. A, The nonpregnant uterus. B, The pregnant uterus, showing the thickened mucous membrane and the altered condition of the uterine glands. (Kundrat and Englemann, Stricker's Med. Jahrb., 1873.)

the uterus is the **decidua parietalis** or the **decidua vera** (Fig. 2-67).

During the growth of the embryo, the decidua capsularis is stretched but not broken. Gradually, by the third month the cavity of the uterus is filled, and the decidua capsularis comes into contact with the decidua parietalis (vera). By the fifth month the decidua capsularis has practically disappeared, and during the succeeding months the decidua parietalis also undergoes atrophy from the increased pressure. By this time also the glands of the stratum compactum are obliterated, and in the spongiosum they become compressed into slit-like fissures with degenerated epithelium. In the stratum basalis, the glandular epithelium is retained.

The Amnion

The human **amnion** is the inner fetal membrane that surrounds the embryo and contains the amniotic fluid *in utero*. It arises in the ectodermal part of the inner cell mass and lies next to the trophoblast during the second week of development. Fluid is secreted within the cluster of cells, causing the formation of the **amniotic cavity** (Figs. 2-26 to 2-28; 2-61 to 2-64). Thus, during the second and third weeks the amniotic cavity is covered by a layer of epithelial cells called **amnioblasts,** which are attached to the margins of the embryonic disc, while the floor of the cavity is formed by the columnar ectodermal cells of the disc (Fig. 2-26). As the amniotic cavity increases in size, the amnioblasts become flat and the outer surface is separated from the cytotrophoblast by a layer of extraembryonic mesoderm. About the fourth week, amniotic fluid accumulates and expands the amnion. By the end of about the second month, the amnion comes into contact with the inner wall of the chorionic sac, thereby obliterating the primitive blastocyst cavity. The amnion also grows around the body stalk and yolk sac and thus forms the covering of the umbilical cord (Figs. 2-61 to 2-64).

The amniotic fluid increases in quantity up to the sixth or seventh month of pregnancy, after which it diminishes somewhat; at the end of pregnancy it amounts to about 1 liter. The initial fluid in the amniotic cavity is probably secreted by the amnioblasts. As development continues, amniotic fluid is derived by filtration from the maternal circulation. Fetal urine also is normally found in the fluid.

The fetus is not only surrounded by the fluid but also swallows it, and the fluid not only is found in the fetal gastrointestinal tract but probably in the respiratory tract as well. The amniotic fluid allows the fetus free movement, and the buoyancy of the amniotic sac protects it from mechanical injury during the later stages of pregnancy.

The Yolk Sac

The primary yolk sac probably commences development from the cluster of endodermal cells in the inner cell mass as

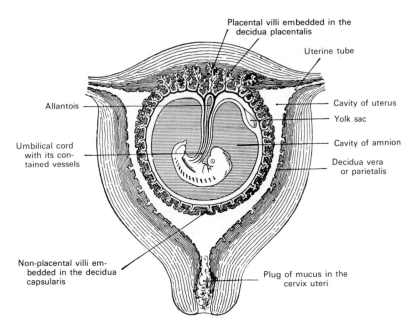

Placental villi embedded in the decidua placentalis

Uterine tube

Allantois

Cavity of uterus

Yolk sac

Cavity of amnion

Umbilical cord with its con- tained vessels

Decidua vera or parietalis

Non-placental villi em- bedded in the decidua capsularis

Plug of mucus in the cervix uteri

FIG. 2-67. Sectional plan of the gravid uterus in the third and fourth month. (After Wagner.)

early as the second week (Figs. 2-61 to 2-64). As it expands into a vesicle, two parts become distinguishable. The part lying against the embryonic disc becomes the future primitive gut of the embryo; the thinner expansion growing out into the cavity of the chorion is the more definitive yolk sac. The extraembryonic mesoderm on the outer surface of the endodermal yolk sac becomes a compact layer which differentiates quite early. During the presomite and early somite stages, the cells of this mesodermal layer proliferate rapidly, producing aggregates of cells that are known as **blood islands** and give the yolk sac a lumpy appearance (Fig.

2-68). The outermost cells of the blood islands gradually become flattened into primitive endothelium; the inner cells become separated from each other by a small amount of primitive blood plasma, and most begin to elaborate hemoglobin in their cytoplasm. The blood vessels of the yolk sac remain as the **vitelline plexus,** fed by the **vitelline** or **omphalomesenteric artery** and drained by the **vitelline** or **omphalomesenteric vein** (Fig. 2-41). In the human embryo, after it has supplied the blood cells for the beginning of circulation, the yolk sac appears to have little further function and undergoes regression (Fig. 2-69).

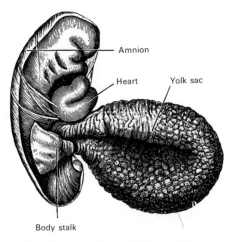

Amnion

Heart

Yolk sac

Body stalk

FIG. 2-68. Human embryo of 2.6 mm. (His, *Anatomie menschlicher Embryonen,* F. C. W. Vogel, 1880–1885.)

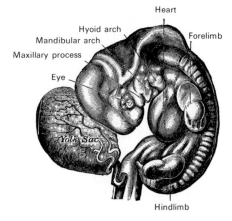

Heart

Hyoid arch

Mandibular arch

Forelimb

Maxillary process

Eye

Yolk Sac

Hindlimb

FIG. 2-69. Human embryo from 31 to 34 days. (His, *Anatomie menschlicher Embryonen,* F. C. W. Vogel, 1880–1885.)

At the end of the fourth week the yolk sac presents the appearance of a small pear-shaped vesicle (umbilical vesicle) and opens into the digestive tube by the long narrow **yolk stalk.** As the amnion spreads around the body stalk and over the inner surface of the chorion, the proximal part of the yolk stalk becomes enclosed in the umbilical cord. The distal part extends to the placenta and ends in the yolk sac vesicle, which lies between amnion and chorion, either on the placenta or a short distance from it (Fig. 2-67). The vesicle can be seen in the **afterbirth** as a small oval body whose diameter varies from 1 to 5 mm. As a rule the yolk stalk undergoes complete obliteration during the seventh week, but occasionally it persists within the embryo. In about 2% of adult bodies a patent omphalomesenteric duct (yolk stalk) is found as a diverticulum from the small intestine, **Meckel's diverticulum.** When present it is situated about 1 meter proximal to the ileocolic junction and may be attached by a fibrous cord to the abdominal wall at the umbilicus.

The Allantois

The human **allantois** (Figs. 2-38; 2-61 to 2-65; 2-67) arises as a tubular diverticulum from the caudal wall of the yolk sac on about the 16th day of development. When the hindgut is developed the allantois retains an opening into its terminal part, the cloaca. The allantois grows out into the body stalk as a diverticulum and with it are carried the **allantoic vessels,** which eventually form the **umbilical vessels** and their branches in the chorionic villi of the placenta. In reptiles, birds, and many mammals the allantois expands into a large sac, acquires an elaborate blood supply, and plays an important metabolic role as well as serving as a reservoir for excretions. In man and other primates, however, the allantois itself remains rudimentary and of minor significance, but its blood vessels become functionally important as the umbilical vessels. The allantoic diverticulum thus extends from the fetal hindgut into the umbilical cord as a narrow canal called the **urachus.** In the fully developed infant this canal becomes obliterated. This results in the formation of a fibrous cord, the **median umbilical ligament,** which extends from the fundus of the urinary bladder to the umbilicus.

Body Stalk and Umbilical Cord

The **body stalk** or **connecting stalk** develops between the embryo and the chorion, and along this stalk the embryonic blood vessels communicate with those of the chorion. Proximally, the body stalk contains the allantoic duct, whereas the umbilical vessels travel throughout its length. Subsequently the body stalk becomes a part of the **umbilical cord.**

The inner cell mass of the blastocyst during the first week is directly in contact with the trophoblast. After the development of the embryonic disc, chorionic mesoderm anchors the embryo to the inner wall of the chorion. As the embryo and the chorion grow, this mesodermal attachment becomes elongated into the body stalk (Figs. 2-62; 2-65), which extends from the posterior end of the embryo to the chorion. When the tail fold is formed, the attachment moves to the midventral region of the embryo (Figs. 2-62 to 2-64), becoming the umbilical cord.

The umbilical blood vessels grow beside the allantois into the body stalk and ramify as the chorionic vessels. These channels provide the vascular pathways to and from the embryo and the chorion. With further development of the embryo and the expansion of the amnion, the body stalk and yolk sac become enclosed in the formation of the umbilical cord. The mesoderm of the body stalk blends with that of the yolk sac and its elongated yolk stalk, being converted into a gelatinous tissue known as **Wharton's jelly.**

The umbilical cord at this stage consists of an outer covering of epithelial cells derived from the amnion and an inner mass of mesodermally derived gelatinous tissue. Within this soft tissue are the vitelline and allantoic ducts and the vitelline and umbilical (allantoic) vessels. The vitelline vessels, together with the right umbilical vein, undergo atrophy and disappear. Thus, the cord at birth contains a pair of umbilical arteries and one (the left) umbilical vein. It attains a length of about 50 cm and a diameter of about 1.5 cm (Fig. 2-70).

THE PLACENTA

The **placenta** is the highly specialized organ which provides the fetus with nutrition and oxygen and removes the excretory

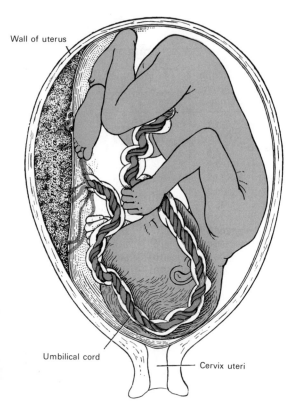

Wall of uterus

Umbilical cord

Cervix uteri

FIG. 2-70.   Fetus in utero, between fifth and sixth months.

products of fetal metabolism. It is a transient organ through which fetal and maternal circulations, although closely associated functionally, remain physically separate. By means of the placenta the fetus is structurally attached to the uterine wall. It has both a fetal and a maternal portion.

The **fetal portion** of the placenta consists of the villi of the chorion frondosum. The elaborate system of villi are suspended in the intervillous space, and are bathed in maternal blood. In this manner the surface area of the trophoblastic epithelium is enormously enlarged, thereby enhancing the physiological exchange between the two circulations. Branches of the umbilical arteries enter each chorionic villus and end in capillary plexuses, which in turn are drained by tributaries of the umbilical vein. Within a villus, the endothelium of the blood vessels is surrounded by a thin layer of gelatinous mesodermal connective tissue and the trophoblast. During the first half of pregnancy the trophoblast consists of two layers: (1) the deeper stratum next to the connective tissue is the cellular **cytotrophoblast** or

Langhans' layer, and (2) the superficial layer in contact with the maternal blood of the intervillous space is the **syncytiotrophoblast.** The cells of the cytotrophoblast form a germinal layer of rapidly dividing cells, which then coalesce with the overlying syncytiotrophoblast. After about the fifth month this cell division appears to stop, and thereafter the finger-like villi are lined by only a single stratum, that of the syncytiotrophoblast.

The **maternal portion** of the placenta is formed by the decidua basalis, which lies adjacent to the chorion frondosum. Changes in the pars functionalis of the endometrial wall involve the conversion of the greater portion of the stratum compactum and the stratum spongiosum into a **basal plate** and **placental septa** (Fig. 2-71), through which the uteroplacental branches of the uterine arteries and veins pass to and from the intervillous space. The endothelial lining of these vessels ceases at the intervillous space, where the tunica media and intima become replaced by syncytiotrophoblast. Portions of the stratum compactum persist in the form of septa and extend from the basal plate through the thickness of the placenta, subdividing it into as many as 30 **lobules** or **cotyledons.** The septa project into the intervillous space and each cotyledon, considered a placental unit, contains several branching chorial trunks from which the individual free villi bathe in the surrounding pool of maternal blood (Figs. 2-60; 2-71). Fetal blood passes through the vessels of the placental villi, yet the two currents do not intermingle, being separated from each other by the delicate walls of the villi. Nevertheless, through the walls of the villi the fetal blood is able to absorb oxygen and nutritive materials from the maternal blood, and to give up its waste products to the latter. The purified blood is then returned to the body of the fetus by the umbilical vein. The term **placental barrier** has been used to refer not only to the nature of the transfer of some substances between the two circulations, but also to the blockage of others. The cellular biology of the mother-fetal relationship is also of considerable interest, especially in light of the fact that the successful fetal chorionic graft onto the maternal endometrium represents the functional union of cells of two different genotypes.

The placental attachment usually is located near the fundus of the uterus, and

TABLE 2-2. Menstrual Age with Mean Sitting Height and Weight of Fetus. (Streeter, 1921)

Menstrual age, weeks	Sitting height at end of week, mm	Increment in height		Formalin weight, grams	Increment in weight	
		mm	%		grams	%
8	23	. . .	. . .	1.1		
9	31	8	26.0	2.7	1.6	59.3
10	40	9	22.5	4.6	1.9	41.3
11	50	10	20.0	7.9	3.3	41.8
12	61	11	18.0	14.2	6.3	44.4
13	74	13	17.6	26.0	11.8	45.4
14	87	13	15.0	45.0	19.0	42.2
15	101	14	14.0	72.0	27.0	37.5
16	116	15	13.0	108.0	36.0	33.3
17	130	14	10.8	150.0	42.0	28.0
18	142	12	8.4	198.0	48.0	24.2
19	153	11	7.2	253.0	55.0	21.7
20	164	11	6.7	316.0	63.0	20.0
21	175	11	6.3	385.0	69.0	18.0
22	186	11	6.0	460.0	75.0	16.3
23	197	11	5.6	542.0	82.0	15.0
24	208	11	5.3	630.0	88.0	14.0
25	218	10	4.6	723.0	93.0	13.0
26	228	10	4.4	823.0	100.0	12.2
27	238	10	4.2	830.0	107.0	11.5
28	247	9	3.6	1045.0	115.0	11.0
29	256	9	3.5	1174.0	129.0	11.0
30	265	9	3.4	1323.0	149.0	11.3
31	274	9	3.3	1492.0	169.0	11.3
32	283	9	3.1	1680.0	188.0	11.2
33	293	10	3.4	1876.0	196.0	10.4
34	302	9	3.0	2074.0	198.0	9.5
35	311	9	3.0	2274.0	200.0	8.8
36	321	10	3.1	2478.0	204.0	8.2
37	331	10	3.0	2690.0	212.0	8.0
38	341	10	3.0	2914.0	224.0	7.7
39	352	11	3.1	3150.0	236.0	7.5
40	362	10	2.8	3405.0	255.0	7.5

more frequently on the posterior than on the anterior wall. It may be situated at a lower position, however, and in rare cases, its site is close to the internal os or opening into the cervix. When this occurs, the internal os may be occluded, giving rise to the condition known as **placenta previa.** In these instances the placenta may become a barrier to a normal delivery.

SEPARATION OF THE PLACENTA. After the child is delivered, the placenta and fetal membranes (chorion laeve and amnion) are expelled from the uterus as the **afterbirth.** Not only the decidua basalis, which underlies the chorion frondosum, but also the contiguous decidua parietalis and decidua capsularis separate from the uterine wall. The detachment occurs through the stratum spongiosum, and necessarily causes rupture of the uterine vessels. The vessels are tightly compressed and closed, however, by the firm contraction of the uterine muscular fibers, and thus, postpartum hemorrhage normally is limited. During the postpartum period, the epithelial lining of the uterus is restored by the proliferation and extension of the epithelium, which lines the remaining portions of the uterine glands in the stratum basalis of the endometrium.

The expelled placenta is a flattened discoid mass which weighs between 450 and 600 gm and attains a diameter of 15 to 20 cm. Its average thickness is about 3 cm, but this diminishes rapidly toward the circumfer-

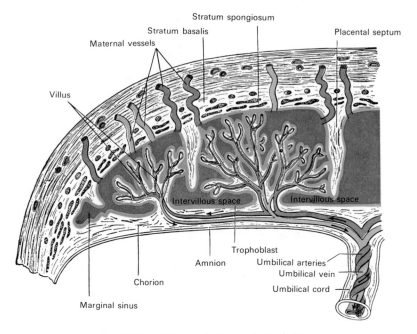

Stratum spongiosum

Stratum basalis

Maternal vessels

Placental septum

Villus

Intervillous space   Intervillous space

Trophoblast

Amnion   Umbilical arteries

Umbilical vein

Chorion

Umbilical cord

Marginal sinus

FIG. 2-71.   Scheme of placental circulation.

ence of the disc, which is continuous with the ruptured amnion and chorion laeve. Its uterine surface is rough and uneven and is divided by a series of fissures into lobules or cotyledons. The fissures contain the remains of the endometrial septa, most of which end in irregular or pointed processes. Some septa, especially those near the edge of the placental disc, pass through its thickness and are attached to the chorion. The fetal surface of the placenta is smooth and glistening, being closely invested by the amnion. Seen through the latter, the chorion presents a mottled appearance, consisting of gray, purple, or yellowish areas. The **umbilical cord** is usually attached near the center of this surface, but may connect closer to the margin of the placental disc. In some cases the umbilical cord is inserted into the chorion laeve, peripheral to the placental disc. This so-called **velamentous insertion** can result in a more precarious placental vascular pattern.

From the attachment of the umbilical cord the larger branches of the umbilical vessels radiate under the amnion, the veins being deeper and larger than the arteries. The remains of the yolk stalk and yolk sac may also be observed beneath the amnion, close to the cord, the former as an attenuated thread, the latter as a minute sac.

Growth of the Embryo and Fetus

FIRST WEEK.   The human ovum is usually fertilized in the upper end of the uterine tube by sperm which have passed from the vagina, through the uterus to the oviduct. The fertilized ovum or zygote commences to divide, descends in the oviduct, and segments into about eight cells before it passes into the uterus by the end of the third day. In the uterus it continues to segment and develops into a blastocyst with a trophoblast and an inner cell mass. By the end of the first week the blastocyst has become superficially attached to the uterine mucosa.

SECOND WEEK.   The blastocyst enlarges, loses its zona pellucida and becomes implanted more securely in the uterine wall. The trophoblast, enclosing the blastocyst cavity, develops into an actively invading outer syncytiotrophoblast and inner cytotrophoblast and forms primitive chorionic villi into which mesoderm and then blood vessels grow. From the inner cell mass develops a bilaminated embryonic disc, the amnion, and the yolk sac. The primitive streak differentiates toward the end of the second week, and mesoderm and notochord are formed.

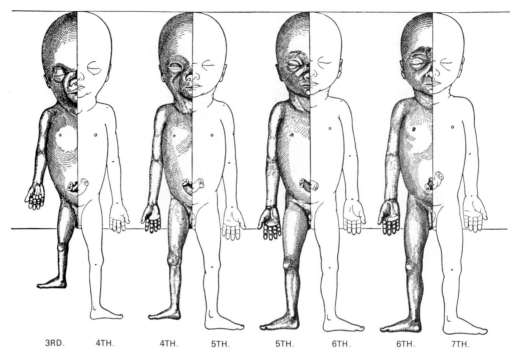

3RD. 4TH. 4TH. 5TH. 5TH. 6TH. 6TH. 7TH.

FIG. 2-72. Body proportions during fetal growth and in adult life based upon average measurements and reduced to the same sitting height. With advancing growth the head, particularly its cranial part, becomes relatively smaller and the extremities in general relatively longer. The upper limb reaches its maximum prenatal length during the fifth month, but increases again in relation to the trunk height after birth. The tremendous development of the lower limb, typical for man, does not become evident until postnatal growth. The forearm grows faster than the upper arm and the leg faster than the thigh. The width between the hips increases at a more rapid rate than the width between the shoulders. In fetuses the shoulders are relatively higher above the suprasternal notch, the nipples relatively higher on the chest, and the umbilicus relatively lower on the abdominal wall than in adults. With advance in growth the breadth of the head decreases in relation to the head length, the ears become relatively larger, the eyes move closer together, the nose increases in height in relation to the face height and decreases in width in relation to the face width. (After Schultz, A. H.)

THIRD WEEK. During the first part of the third week the neural folds appear (Figs. 2-33; 2-35, C), the allantoic duct begins to develop, the yolk sac enlarges and blood vessels begin to form on the yolk sac and allantois and in the chorion. Before the end of the week the neural folds begin to unite (Figs. 2-35, D; 2-40) and the neurenteric canal opens. The primitive segments begin to form and the developing heart is represented by a pair of longitudinally oriented tubes. The changes during this week occur with great rapidity.

FOURTH WEEK. During the fourth week (Figs. 2-41; 2-45; 2-46) the neural folds close, the primitive segments increase in number, and the branchial arches appear. The connection of the yolk sac with the embryo becomes considerably narrowed, and the embryo assumes a more humanoid form. The limb buds appear and the heart increases greatly in size, producing a prominent bulge in the branchial region.

FIFTH WEEK. The extensive development of the head region during the fifth week and its increase in size are due largely to the growing brain. The branchial arches undergo profound changes and partly disappear. The superficial nose, eye, and ear rudiments become prominent, and the limbs begin to show regional identity. The upper limbs develop more quickly than the lower limbs.

SIXTH WEEK. The curvature of the embryo's trunk is further diminished, but the cervical flexure of the head over the developing heart is more pronounced. The branchial grooves—except the first—have disappeared, and the rudiments of the fingers and toes can be recognized (Fig. 2-57).

SEVENTH AND EIGHTH WEEKS. The flexure of the head is gradually reduced and the

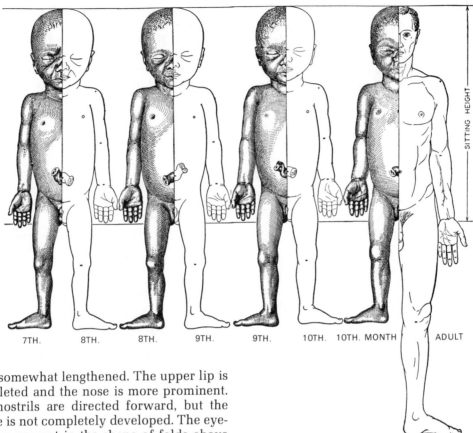

7TH.　8TH.　8TH.　9TH.　9TH.　10TH.　10TH. MONTH　ADULT

SITTING HEIGHT

neck somewhat lengthened. The upper lip is completed and the nose is more prominent. The nostrils are directed forward, but the palate is not completely developed. The eyelids are present in the shape of folds above and below the eye, and the different parts of the auricula are distinguishable. By the end of the second month the fetus measures about 28 to 30 mm in length (Fig. 2-58).

THIRD MONTH. The head is extended and the neck is lengthened. The eyelids meet and fuse, remaining closed until the end of the sixth month. The limbs are well developed and nails appear on the digits. The external generative organs are so differentiated by the end of the third month that it is possible to distinguish the sex. By 13 weeks the length of the fetus is about 7 cm, but if the legs are included it measures from 9 to 10 cm.

FOURTH MONTH. The loop of gut which projected into the umbilical cord during the sixth week **(umbilical herniation)** is withdrawn into the fetal abdomen. The hair begins to make its appearance. There is a rapid increase in general body size so that by the end of the fourth month the fetus is from 12 to 13 cm in length, but if the legs are included it measures from 16 to 20 cm.

FIFTH MONTH. It is during this month that the first movements of the fetus are usually observed. The eruption of hair on the head commences, and the **vernix caseosa,** a fatty, sebaceous material that covers the skin, begins to be deposited. By the end of this month the total length of the fetus, including the legs, is 25 to 27 cm.

SIXTH MONTH. The body is covered by fine hairs (*lanugo*) and the deposit of vernix caseosa is considerable. The papillae of the skin are developed, but the skin is quite wrinkled. The free border of the nail projects from the corium of the dermis. Measured from vertex to heel, the total length of the fetus at the end of the sixth month is from 30 to 32 cm.

SEVENTH MONTH. The pupillary membrane atrophies and the eyelids open. The testes descend with the vaginal sac of the peritoneum. From vertex to heel the total length at the end of the seventh month is 35 to 36 cm. The weight is between three and four pounds.

EIGHTH MONTH. The skin assumes a pink color (even the skin of fetuses of dark-skinned races) and is now entirely coated with vernix caseosa. The lanugo begins to disappear. Subcutaneous fat has been developed to a considerable extent, and the fetus loses its wrinkles and assumes a plump appearance. The total length, i.e., from head to heel, at the end of the eighth month is about 40 cm, and fetal weight varies between $4\frac{1}{2}$ and $5\frac{1}{2}$ pounds.

NINTH MONTH. The lanugo has largely disappeared from the trunk. The umbilicus is almost in the middle of the body, and the testes are usually in the scrotum. At full term, the fetus weighs from $6\frac{1}{2}$ to 8 pounds, and measures from head to heel about 50 cm.

References

(References are listed not only of those articles and books cited in the text, but of others as well which are considered to contain valuable resource information for the student who desires it.)

Histology and Cytology

Bargmann, W. 1964. *Histologie und Mikroscopische Anatomie des Menschen.* Georg Thieme Verlag, Stuttgart, 856 pp.

Barrnett, R. J., and G. E. Palade. 1957. Histochemical demonstration of the sites of activity of dehydrogenase systems with the electron microscope. J. Biophys. Biochem. Cytol., 3:577–588.

Beams, H. W., and R. G. Kessel. 1969. The Golgi apparatus: structure and function. Internat. Rev. Cytol., 22:209–276.

Behnke, O., and A. Forer. 1967. Evidence for four classes of microtubules in individual cells. J. Cell Sci., 2:169–192.

Bloom, W., and D. W. Fawcett. 1975. *A Textbook of Histology.* 10th ed. W. B. Saunders Co., Philadelphia, 1033 pp.

Christensen, A. K., and D. W. Fawcett. 1961. The normal fine structure of opossum testicular interstitial cells. J. Biophys. Biochem. Cytol., 9:653–670.

Copenhaver, W. M., and R. P. Bunge. 1971. *Bailey's Textbook of Histology.* 16th ed. The Williams & Wilkins Co., Baltimore, 745 pp.

Dalton, A. J., and M. D. Felix. 1954. Cytologic and cytochemical characteristics of the Golgi substance of epithelial cells of the epididymis—in situ, in homogenates and after isolation. Amer. J. Anat., 94:171–207.

De Duve, C. 1963. The lysosome. Sci. Amer., 208:64–72.

De Duve, C., and R. Wattiaux. 1966. Functions of lysosomes. Ann. Rev. Physiol., 28:435–492.

Di Fiore, M. S. 1981. *An Atlas of Human Histology.* 5th ed. Lea & Febiger, Philadelphia, 252 pp.

Farquar, M. G., and G. E. Palade. 1963. Junctional complexes in various epithelia. J. Cell Biol., 17:375–412.

Fawcett, D. W. 1961. Cilia and flagella. In *The Cell: Biochemistry, Physiology, Morphology.* (Brachet, J., and A. E. Mirsky, eds.) Academic Press, New York, Vol. 2, pp. 217–297.

Fawcett, D. W. 1966. *The Cell: Its Organelles and Inclusions. An Atlas of Fine Structure.* W. B. Saunders Co., Philadelphia, 448 pp.

Fawcett, D. W. 1966. On the occurrence of a fibrous lamina on the inner aspect of the nuclear envelope in certain cells of vertebrates. Amer. J. Anat., 119:129–146.

Finerty, J. C., and E. V. Cowdry. 1960. *A Textbook of Histology.* 5th ed. Lea & Febiger, Philadelphia, 573 pp.

Fox, C. F. 1972. The structure of cell membranes. Sci. Amer., 226:31–38.

Freeman, J. A. 1964. *Cellular Fine Structure, An Introductory Student Text and Atlas.* The McGraw-Hill Book Co., New York, 198 pp.

Gahan, P. B. 1967. Histochemistry of lysosomes. Internat. Rev. Cytol., 21:1–63.

Greep, R. O. 1973. *Histology.* 3rd ed. The McGraw-Hill Book Co., New York, 1044 pp.

Hackenbrock, C. 1968. Ultrastructural basis for metabolically linked mechanical activity in mitochondria. J. Cell Biol., 37:345–369.

Ham, A. W. 1974. *Histology.* 7th ed. J. B. Lippincott Co., Philadelphia, 1066 pp.

Harrison, R. G. 1910. The development of peripheral nerve fibers in altered surroundings. Roux's Arch. f. Entwicklungsmechanik d. Org., 30:15–31.

Hendler, R. W. 1971. Biological membrane ultrastructure. Physiol. Rev., 51:66–97.

Ito, S. 1961. The endoplasmic reticulum of gastric parietal cells. J. Biophys. Biochem. Cytol., 11:333–347.

Kelly, D. 1966. Fine structure of desmosomes, hemidesmosomes and an adepidermal globular layer in developing newt epidermis. J. Cell Biol., 28:51–72.

Matthews, J. L., and J. H. Martin. 1971. *Atlas of Human Histology and Ultrastructure.* Lea & Febiger, Philadelphia, 382 pp.

Moreland, J. E. 1962. Electron microscopic studies of mitochondria in cardiac and skeletal muscle from hibernated ground squirrels. Anat. Rec., 142:155–167.

Neutra, M., and C. P. Le Blond. 1969. The Golgi apparatus. Sci. Amer., 220:100–107.

Northcote, D. H. 1973. The Golgi complex. In *Cell Biology in Medicine.* (Bittar, E. E., ed.) John Wiley & Sons, New York, pp. 197–214.

Novikoff, A. B., and E. Holtzman. 1970. *Cells and Organelles.* Holt, Rinehart & Winston, New York, 337 pp.

Novikoff, P. M., A. B. Novikoff, N. Quintana, and J. J. Hauw. 1971. Golgi apparatus, GERL, and lysosomes of neurons in rat dorsal root ganglia, studied by thick section and thin section cytochemistry. J. Cell Biol., 50:859–886.

Palade, G. E. 1952. The fine structure of mitochondria. Anat. Rec., 114:427–452.

Palade, G. E. 1953. An electron microscope study of the mitochondrial structure. J. Histochem. Cytochem., 1:188–211.

Palade, G. E. 1955. A small particulate component of the

cytoplasm. J. Biophys. Biochem. Cytol., 1:59–68.

Palade, G. E. 1955. Studies on the endoplasmic reticulum. II. Simple dispositions in cells in situ. J. Biophys. Biochem. Cytol., 1:567–582.

Palade, G. E., and K. R. Porter. 1954. Studies on the endoplasmic reticulum. I. Its identification in cells in situ. J. Exp. Med., 100:641–656.

Palay, S. L. 1958. Frontiers in Cytology. Yale University Press, New Haven, 529 pp.

Porter, K. R. 1953. Observations on a submicroscopic basophilic component of cytoplasm. J. Exp. Med., 97:727–750.

Porter, K. R. 1966. Cytoplasmic microtubules and their functions. In Principles of Biomolecular Organization. (Wolstenholme, G. E. W., and M. O'Conner, eds.) Ciba Foundation Symposium. Churchill, London, pp. 308–345.

Porter, K. R., and M. A. Bonneville. 1973. Fine Structure of Cells and Tissues. 4th ed. Lea & Febiger, Philadelphia, 204 pp.

Racker, E. 1968. The membrane of the mitochondrion. Sci. Amer., 218:32–39.

Revel, J. P., and S. Ito. 1967. The surface components of cells. In The Specificity of Cell Surfaces. (Davis, B. D., and L. Warren, eds.) Prentice-Hall, Englewood Cliffs, New Jersey, pp. 211–234.

Sabatini, D. D., Y. Tashiro, and G. E. Palade. 1966. On the attachment of ribosomes to microsomal membranes. J. Molec. Biol., 19:503–524.

Sobotta, J. 1911. Atlas und Lehrbuch der Histologie und mikroskopischen Anatomie des Menschen. Lehmann, München, 307 pp.

Stöhr, P., W. Von Möllendorf, and K. Goerttler. 1959. Lehrbuch der Histologie und der mikroskopischen Anatomie des Menschen. Fischer, Jena, 560 pp.

Warner, J. R., P. Knopf, and A. Rich. 1963. A multiple ribosomal structure in protein synthesis. Proc. Nat. Acad. Sci. (U.S.A.), 49:122–129.

Weissmann, G. 1965. Lysosomes. New Eng. J. Med., 273:1084–1090, 1143–1149.

White, E. H. 1964. Cell Physiology and Biochemistry. Prentice-Hall, Inc., Englewood Cliffs, New Jersey, 120 pp.

Windle, W. F. 1976. Textbook of Histology. 5th ed. McGraw-Hill Book Co., New York, 561 pp.

Cell Division and Genetics

Barr, M. L., and E. G. Bertram. 1949. A morphological distinction between neurones of the male and female, and the behaviour of the nucleolar satellite during accelerated nucleoprotein synthesis. Nature, 163:676–677.

Bartalos, M., and T. A. Baramki. 1967. Medical Cytogenetics. The Williams & Wilkins Co., Baltimore, 419 pp.

Ford, C. E. 1961. Human cytogenetics. Brit. Med. Bull., 17:179–183.

Ford, C. E., and J. L. Hamerton. 1956. The chromosomes of man. Nature, 178:1020–1023.

Ford, C. E., P. A. Jacobs, and L. G. Lajtha. 1958. Human somatic chromosomes. Nature, 181:1565–1568.

Fraser, F. C. 1970. The genetics of cleft lip and palate. Amer. J. Human Genet., 22:336–352.

Fraser, F. C. and J. J. Nora. 1975. Genetics of Man. Lea & Febiger, Philadelphia, 270 pp.

Genes in Action. 1967. The Upjohn Company, Kalamazoo, Michigan, 25 pp.

Goulian, M., A. Kornberg, and R. L. Sinscheimer. 1967. Enzymatic synthesis of DNA. Proc. Nat. Acad. Sci. (U.S.A.), 58:2321–2328.

Kindred, J. E. 1963. The chromosomes of the ovary of the human fetus. Anat. Rec., 147:295–311.

Le Blond, C. P., and B. E. Walker. 1956. Renewal of cell populations. Physiol. Rev., 36:255–276.

Levitan, M., and A. Montagu. 1971. Textbook of Human Genetics. Oxford University Press, New York, 922 pp.

Lewis, W. H. 1947. Interphase (resting) nuclei, chromosomal vesicles and amitosis. Anat. Rec., 97:433–445.

Lewis, W. H., and C. G. Hartman. 1933. Early cleavage stages of the egg of the monkey (Macacus rhesus). Carneg. Inst., Contr. Embryol., 24:187–201.

Lewis, W. H., and M. R. Lewis. 1917. The duration of the various phases of mitosis in the mesenchyme cells of tissue cultures. Anat. Rec., 13:359–367.

Mazia, D. 1974. The cell cycle. Sci. Amer., 230:54–64.

Mc Kusick, V. A. 1964. Human Genetics. Prentice-Hall, Inc., Englewood Cliffs, N.J., 148 pp.

Mc Kusick, V. 1971. The mapping of human chromosomes. Sci. Amer., 224:104–113.

Miescher, F. 1871. Ueber die chemische Zusammensetzung der Eiterzellen. Hoppe-Seyler's Medicinisch-chemische Untersuchungen. A. Hirschwald, Berlin IV. pp. 441.

Moore, K. L. 1966. The Sex Chromatin. W.B. Saunders Co., Philadelphia, 474 pp.

Osborne, R. H. and F. V. De George. 1959. Genetic Basis of Morphological Variation, an Evaluation and Application of the Twin Study Method. Harvard University Press, Cambridge, 204 pp.

Ris, H. 1955. Cell division. In Analysis of Development. (Willier, B. H., P. A. Weiss, and V. Hamburger, eds.) W. B. Saunders Co., Philadelphia, pp. 91–125.

Ris, H. 1961. Ultrastructural and molecular organization of genetic systems. Canad. J. Genet. Cytol., 3:95–120.

Thompson, J. S., and M. W. Thompson. 1973. Genetics in Medicine. 2nd ed. W. B. Saunders Co., Philadelphia, 400 pp.

Tijo, J. H., and T. T. Puck. 1958. The somatic chromosomes of man. Proc. Nat. Acad. Sci. (U.S.A.), 44:1229–1236.

Watson, J. D. 1976. The Molecular Biology of the Gene. 3rd ed. W. A. Benjamin, Inc., Menlo Park, California, 739 pp.

Watson, J. D., and F. A. C. Crick. 1953. Genetical implications of the structure of deoxyribonucleic acid. Nature, 171:964–967.

Watson, J. D., and F. A. C. Crick. 1953. Molecular structure of nucleic acid. Nature, 171:737–738.

Winchester, A. M. 1972. Genetics, a Survey of the Principles of Heredity. 4th ed. Houghton Mifflin Co., Boston.

Ovum, Sperm and Fertilization

Allen, E. 1928. An unfertilized tubal ovum from Macacus rhesus. Anat. Rec., 37:351–356.

Allen, E., J. P. Pratt, Q. U. Newell, and L. J. Bland. 1928. Recovery of human ova from the uterine tubes. J.A.M.A., 91:1018–1020.

Allen, R. C. 1959. The moment of fertilization. Sci. Amer., 201:124–134.

Austin, C. R., and M. W. H. Bishop. 1958. Some features of the acrosome and the perforatorium in mammalian spermatozoa. Proc. Roy. Soc. Lond., B, 149:234–240.

Bedford, J. M. 1972. An electron microscopic study of

sperm penetration into the rabbit egg after natural mating. Amer. J. Anat., *133*:213–253.

Blandau, R. J., and D. L. Odor. 1949. The total number of spermatozoa reaching various segments of the reproductive tract in the female albino rat at intervals after insemination. Anat. Rec., *103*:93–109.

Chang, M. C. 1955. The maturation of rabbit oocytes in culture and their maturation, activation, fertilization, and subsequent development in the fallopian tubes. J. Exp. Zool., *128*:379–406.

Chiquoine, A. D. 1960. The development of the zona pellucida of the mammalian ovum. Amer. J. Anat., *106*:149–169.

Clark, R. T. 1934. Studies on the physiology of reproduction in the sheep. II. The cleavage stages of the ovum. Anat. Rec., *60*:135–160.

Corner, G. W. 1923. Ovulation and menstruation in *Macacus rhesus*. Carneg. Inst., Contr. Embryol., *15*:73–101.

Everett, J. W. 1956. The time of release of ovulating hormone from the rat hypophysis. Endocrinology, *59*:580–585.

Everett, J. W., and C. H. Sawyer. 1949. A neural timing factor in the mechanism by which progesterone advances ovulation in the cyclic rat. Endocrinology, *45*:581–595.

Everett, N. B. 1943. Observational and experimental evidences relating to the origin and differentiation of the definitive germ cells in mice. J. Exp. Zool., *92*:49–92.

Fawcett, D. W. 1958. The structure of the mammalian spermatozoon. Internat. Rev. Cytol., *7*:195–234.

Fawcett, D. W., and R. D. Hollenberg. 1963. Changes in the acrosome of guinea pig spermatozoa during passage through the epididymis. Zeitsch. f. Zellforsch., *60*:276–292.

Hartman, C. G. 1929. How large is the mammalian egg? A review. Quart. Rev. Biol., *4*:373–388.

Hartman, C. G., and G. W. Corner. 1941. The first maturation division of the macaque ovum. Carneg. Inst., Contr. Embryol., *29*:1–7.

Hartman, C. G., W. H. Lewis, F. W. Miller, and W. W. Swett. 1931. First findings of tubal ova in the cow together with notes on oestrus. Anat. Rec., *48*:267–275.

Heller, C. G., and Y. Clermont. 1963. Spermatogenesis in man: An estimate of its duration. Science, *140*:184–186.

Lewis, W. H. 1931. A human tubal egg, unfertilized. Bull. Johns Hopkins Hosp., *48*:368–372.

Lewis, W. H., and P. W. Gregory. 1929. Cinematographs of living developing rabbit eggs. Science, *69*:226–229.

Lewis, W. H., and C. G. Hartman. 1933. Early cleavage stages of egg in the monkey (*Macacus rhesus*). Carneg. Inst., Contr. Embryol., *24*:187–201.

Lewis, W. H., and C. G. Hartman. 1941. Tubal ova of the rhesus monkey. Carneg. Inst., Contr. Embryol., *29*:7–15.

Lewis, W. H., and E. S. Wright. 1935. On the early development of the mouse egg. Carneg. Inst., Contr. Embryol., *25*:113–144.

Painter, T. S. 1923. Studies in mammalian spermatogenesis. II. The spermatogenesis of man. J. Exp. Zool., *37*:291–336.

Purshottam, N., and G. Pincus. 1961. In vitro cultivation of mammalian eggs. Anat. Rec., *140*:51–55.

Raven, C. P. 1961. *Oogenesis: The Storage of Developmental Information.* Pergamon Press, Inc., New York, 274 pp.

Rock, J., and A. T. Hertig. 1944. Information regarding the time of human ovulation derived from a study of 3 unfertilized and 11 fertilized ova. Amer. J. Obstet. Gynec., *47*:343–356.

Shettles, L. B. 1957. The living human ovum. Obstet. Gynec., *10*:359–365.

Speck, G. 1959. The determination of the time of ovulation. Obstet. Gynec. Survey, *14*:798–818.

Squier, R. R. 1932. The living egg and early stages of its development in the guinea pig. Carneg. Inst., Contr. Embryol., *23*:223–250.

Zamboni, L., R. Zemjanis, and M. Stefanini. 1971. The fine structure of monkey and human spermatozoa. Anat. Rec., *169*:129–154.

Implantation, Fetal Membranes, and Placenta

Amoroso, E. C. 1962. Histology of the placenta. Brit. Med. Bull., *17*:81–90.

Bartelmez, G. W. 1957. Cyclic changes in the endometrium of the Rhesus monkey (*Macaca mulatta*). Carneg. Inst., Contr. Embryol., *34*:101–144.

Benirschke, K. 1961. Examination of the placenta. Obstet. Gynec., *18*:309–333.

Benirschke, K. 1972. Implantation, placental development, uteroplacental blood flow. In *Principles and Management of Human Reproduction.* (Reid, D. E., K. J. Ryan, and K. Benirschke, eds.) W. B. Saunders Co., Philadelphia, pp. 179–196.

Benirschke, K., and S. G. Driscoll. 1967. *The Pathology of the Human Placenta.* Springer-Verlag, New York, 512 pp.

Boyd, J. D., and W. J. Hamilton. 1970. *The Human Placenta.* Heffer, Cambridge, England, 365 pp.

Bryce, T. H., J. H. Teacher, and J. M. M. Kerr. 1908. *Contributions to the Study of the Imbedding of the Human Ovum.* J. Maclehose & Sons, Glasgow, 93 pp.

Dallenback-Hellweg, G., and G. Nette. 1964. Morphological and histochemical observations on trophoblast and decidua of the basal plate on the human placenta at term. Amer. J. Anat., *115*:309–326.

Eckstein, P. 1959. *Implantation of Ova.* (Memoirs, Society for Endocrinology, no. 6) Cambridge, University Press, 97 pp.

Freda, V. J. 1962. Placental transfer of antibodies in man. Amer. J. Obstet. Gynec., *84*:1756–1777.

Hagerman, D. D., and C. A. Villee. 1960. Transport functions of the placenta. Physiol. Rev., *40*:313–330.

Hamilton, W. J., and J. D. Boyd. 1960. Development of the human placenta in the first three months of gestation. J. Anat. (Lond.), *94*:297–328.

Kundrat, H., and Englemann, G. J. 1873. Untersuchungen über die Uterusschleimhaut. Stricker's Med. Jahrb. pp. 135–177.

Larsen, J. F., and J. Davies. 1962. The paraplacental chorion and accessory fetal membranes of the rabbit. Histology and electron microscopy. Anat. Rec., *143*:27–45.

Plenti, A. 1957. The origin of amniotic fluid. In *Transactions of 4th Conference on Gestation.* Josiah Macy Jr. Foundation, New York, pp. 71–114.

Ramsey, E. M. 1954. Venous drainage of the placenta of the Rhesus monkey (*Macaca mulatta*). Carneg. Inst., Contr. Embryol., *35*:151–173.

Ramsey, E. M. 1955. Vascular patterns in the endometrium and the placenta. Angiology, *6*:321–338.

Ramsey, E. M. 1959. Vascular adaptations of the uterus to pregnancy. Ann. N. Y. Acad. Sci., *75*:726–745.

Ramsey, E. M. 1965. The placenta and fetal membranes. In *Obstetrics* (Greenhill, J. P., ed.) 13th ed. W. B. Saunders Co., Philadelphia, pp. 101–136.

Sternberg, J. 1962. Placental transfers: modern methods of study. Amer. J. Obstet. Gynec., *84*:1731–1748.

Torpin, R. 1969. *The Human Placenta; Its Shape, Form, Origin and Development*. Charles C Thomas, Springfield, Ill., 190 pp.

Villee, C. A. 1960. *The Placenta and Fetal Membranes*. The Williams & Wilkins Co., Baltimore, 404 pp.

Wislocki, G. B., and H. S. Bennett. 1943. The histology and cytology of the human and monkey placenta, with special reference to the trophoblast. Amer. J. Anat., *73*:335–450.

Wislocki, G. B., and G. L. Streeter. 1938. On the placentation of the Macaque (*Macaca mulatta*) from the time of implantation until the formation of the definitive placenta. Carneg. Inst., Contr. Embryol., *27*:1–66.

Embryology and Human Embryos

Albers, G. D. 1963. Branchial anomalies. J.A.M.A., *183*:399–409.

Arey, L. B. 1938. The history of the first somite in human embryos. Carneg. Inst., Contr. Embryol., *27*:233–270.

Arey, L. B. 1974. *Developmental Anatomy: A Textbook and Laboratory Manual of Embryology*. Rev. 7th ed. W. B. Saunders Co., Philadelphia, 695 pp.

Barry, A. 1951. The aortic arch derivatives in the human adult. Anat. Rec., *111*:221–238.

Blechschmidt, E. 1961. *The Stages of Human Development Before Birth*. W. B. Saunders Co., Philadelphia, 640 pp.

Blechschmidt, E. 1963. *The Human Embryo. Documentations on Kinetic Anatomy*. Schattauer-Verlag, Stuttgart, 105 pp.

Bridges, J. B., and W. R. M. Morton. 1964. Multiple anomalies in a human fetus associated with absence of one umbilical artery. Anat. Rec., *148*:103–109.

Corner, G. W. 1929. A well-preserved human embryo of 10 somites. Carneg. Inst., Contr. Embryol., *20*:81–102.

Crelin, E. S. 1969. *Anatomy of the Newborn: An Atlas*. Lea & Febiger, Philadelphia, 256 pp.

Crouch, J. E. 1972. *Functional Human Anatomy*. Lea & Febiger, Philadelphia, 649 pp.

Davies, J. 1963. *Human Developmental Anatomy*. Ronald Press, New York, 298 pp.

Eternod, A. C. F. 1896. Sur un oeuf humain de 16.3 mm avec embryon de 2.1 mm. Actes helv. Sci. nat. Zurich, pp. 164–170.

Federici, H. 1926. Recherches experimentales sur les potentialités de l'ilot sanguin chez l'embryon de *Rana fusca*. Arch. de Biol., *36*:466–487.

Gasser, R. J. 1967. The development of the facial muscles in man. Amer. J. Anat., *120*:357–376.

George, W. C. 1942. A presomite human embryo with chorda canal and prochordal plate. Carneg. Inst., Contr. Embryol., *30*:1–8.

Gitlin, G. 1968. Mode of union of right and left coelomic channels during development of the peritoneal cavity in the human embryo. Acta Anat., *71*:45–52.

Goss, C. M. 1928. Experimental removal of the blood islands of Amblystoma punctatum embryos. Anat. Rec., *38*:12.

Gruenwald, P. 1966. Growth of the human fetus. I. Normal growth and its variations. Amer. J. Obstet. Gynec., *94*:1112–1119.

Hertig, A. T. 1935. Angiogenesis in the early human chorion and in the primary placenta of the macaque monkey. Carneg. Inst., Contr. Embryol., *25*:37–82.

Hertig, A. T., and J. Rock. 1941. Two human ova of the pre-villous stage, having an ovulation age of about-eleven and twelve days respectively. Carneg. Inst., Contr. Embryol., *29*:127–156.

Hertig, A. T., and J. Rock. 1944. On the development of the early human ovum with special reference to the trophoblast of the previllous stage: A description of 7 normal and 5 pathological ova. Amer. J. Obstet. Gynec., *47*:149–184.

Hertig, A. T. and J. Rock. 1945. Two human ova of the pre-villous stage, having a developmental age of about seven and nine days respectively. Carneg. Inst., Contr. Embryol., *31*:65–84.

Hertig, A. T., J. Rock and E. C. Adams. 1956. A description of 34 human ova within the first 17 days of development. Amer. J. Anat., *98*:435–493.

Hertig, A. T., J. Rock, E. C. Adams, and W. J. Mulligan. 1954. On the preimplantation stages of the human ovum: A description of four normal and four abnormal specimens ranging from the second to the fifth day of development. Carneg. Inst., Contr. Embryol., *35*:199–220.

Heuser, C. H. 1930. A human embryo with fourteen pairs of somites. Carneg. Inst., Contr. Embryol., *22*:135–154.

Heuser, C. H. 1932. A presomite human embryo with a definite chorda canal. Carneg. Inst., Contr. Embryol., *23*:251–267.

Heuser, C. H., J. Rock, and A. T. Hertig. 1945. Two human embryos showing early stages of the definitive yolk sac. Carneg. Inst., Contr. Embryol., *31*:85–99.

Heuser, C. H., and G. L. Streeter. 1929. Early stages in the development of pig embryos from the period of initial cleavage to the time of the appearance of limb buds. Carneg. Inst., Contr. Embryol., *20*:1–30.

His, W. 1880–1885. *Anatomie menschlicher Embryonen*. F. C. W. Vogel, Leipzig, 3 vols. & atlas.

Ingalls, N. W. 1920. A human embryo at the beginning of segmentation, with special reference to the vascular system. Carneg. Inst., Contr. Embryol., *11*:61–90.

Kölliker, A. 1880. *Grundriss der Entwicklungsgeschichte des Menschen und der Hoheren Thiere*. Von Wilhelm Engelmann Verlag, Leipzig, 418 pp.

Kollman, J. 1907. *Handatlas der Entwicklungsgeschichte des Menschen*. Gustav Fischer, Jena, 2 vols.

Langman, J. 1969. *Medical Embryology: Human Development, Normal and Abnormal*. 3rd ed. The Williams & Wilkins Co., Baltimore, 421 pp.

Malcolm, R. B., and R. E. Benson. 1940. Branchial cysts. Surgery, *7*:187–203.

Mall, F. P. 1918. On the age of human embryos. Amer. J. Anat., *23*:397–422.

Moore, K. L. 1973. *The Developing Human; Clinically Oriented Embryology*. W. B. Saunders Co., Philadelphia, 374 pp.

O'Rahilly, R. 1963. The early development of the otic vesicle in staged human embryos. J. Embryol. Exp. Morph., *11*:741–755.

O'Rahilly, R. 1966. The early development of the eye in staged human embryos. Carneg. Inst., Contr. Embryol., *38*:1–42.

Patten, B. M. 1968. *Human Embryology*. 3rd ed. The Blakiston Division, McGraw-Hill Book Co., New York, 651 pp.

Payne, F. 1925. General description of a seven somite

human embryo. Carneg. Inst., Contr. Embryol., *16*:115–124.

Peter, K. 1913. *Atlas der Entwicklung der Nase und des Gaumens beim Menschen.* G. Fischer, Jena, 130 pp.

Rock, J., and A. T. Hertig. 1942. Some aspects of early human development. Amer. J. Obstet. Gynec., *44*:973–983.

Rudnick, D. 1958. *Embryonic Nutrition.* The University of Chicago Press, Chicago, 113 pp.

Shaner, R. F. 1945. A human embryo of two to three pairs of somites. Canad. J. Res. E., *23*:235–243.

Smith, W. N. A. 1963. The site of action of trypan blue in cardiac teratogenesis. Anat. Rec., *147*:507–523.

Stark, R. B. 1961. Embryology, pathogenesis and classification of cleft lip and cleft palate. In *Congenital Anomalies of the Face and Associated Structures* (Pruzansky, S., ed.) Charles C Thomas, Springfield, Ill., pp. 68–84.

Streeter, G. L. 1921. Weight, sitting height, head size, foot length and menstrual age of the human embryo. Carneg. Inst., Contr. Embryol., *11*:143–170.

Streeter, G. L. 1942. Developmental horizons in human embryos: Description of age group XI, 13 to 20 somites, and age group XII, 21 to 29 somites. Carneg. Inst., Contr. Embryol., *30*:211–246.

Warbrick, J. G. 1960. The early development of the nasal cavity and upper lip in the human embryo. J. Anat., *94*:351–362.

West, C. M. 1930. Description of a human embryo of eight somites. Carneg. Inst., Contr. Embryol., *21*:25–36.

Wilson, J. G., and J. Warkany. 1965. *Teratology Principles and Techniques.* The University of Chicago Press, Chicago, 279 pp.

3

Surface and Topographical Anatomy

Variability is one of the few biological constants. The human body, despite fundamental similarities in structure, shows remarkable variation among individuals in its anatomy. Quite apart from specific variations that might exist as anomalies in one or another system, such as differences in vascular or neural patterns, it has been recognized that differences in general body types and stature might be categorized. The term **hypersthenic** has been used to describe the short, stout, stocky individual in contrast to **hyposthenic,** which refers to taller, more slender subjects. Naturally, the relationships of the internal organs of these differing physiques reflect the body build and become a matter for consideration in radiographic examination and interpretation.

The Head

The shape of the cranium has also been used by physical anthropologists as a general basis for describing skull types. In the **dolichocephalic** (long headed) skull, the maximum skull width is less than 75% of the maximum length. When the maximum width is 80% or more of the length, the term **brachycephalic** (short or broad) skull is used. **Mesaticephalic** denotes a skull in which the width is between 75% and 80% of the length.

SCALP. The covering of the cranial part of the head is the **scalp,** a structure composed of the following layers: (1) skin; (2) dense subcutaneous connective tissue; (3) the occipitofrontal muscle, including the galea aponeurotica; (4) subaponeurotic loose connective tissue; and (5) pericranium or periosteum of the bones (Fig. 3-1). These layers are of rather uniform thickness, so that the contours of the scalp conform largely to those of the underlying frontal, parietal, occipital, and temporal bones.

The outermost three layers of the scalp are closely adherent to each other and glide together when manipulated over the periosteum and bone, as if they were a single layer. This mobility can result in separation of the outer three layers of the scalp from the periosteum through the loose connective tissue layer as a result of accidents that cause shearing injuries.

BONY CRANIUM. Clearly visible landmarks are the **parietal** and **frontal eminences,** and the opening of the **external auditory meatus** of the ear. In addition, certain prominences can be identified by palpation. In the occipital region or back of the head, the **external occipital protuberance** or **inion** is a bony prominence in the midline, at the junction of the head and neck (Fig. 3-2). The **superior nuchal line** is a slight, upward curving ridge that extends laterally from the protuberance to the mastoid process of the temporal bone. Certain muscles in the neck attach to the superior nuchal line, and above it the cranium is covered only by scalp. The **mastoid process** projects downward and forward from behind the ear. Its anterior border lies immediately behind the concha of the external ear, and its rounded apex lies inferiorly at a level with the ear lobe (Fig. 3-2).

Some recognizable bony points are used in surgery and in anthropological measurements and therefore are given special names; other bony landmarks, not recognizable by observation or palpation, can be located by the use of identifiable structures as points of reference.

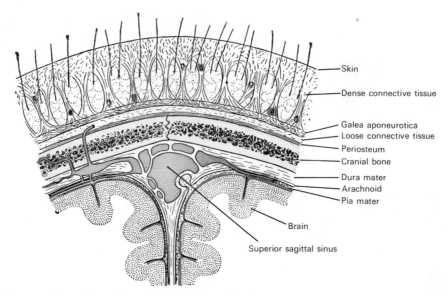

FIG. 3-1. Diagrammatic section of the scalp, the skull, and the brain. Observe the important loose connective tissue layer lying between the galea aponeurotica and the periosteum. It is through this layer that shearing accidents can separate the overlying layers of the scalp from the subjacent periosteum and bone.

Auricular Point. The center of the opening of the external acoustic meatus (Fig. 3-2).

Preauricular Point. At the root of the zygomatic arch immediately in front of the external acoustic meatus (Fig. 3-2).

Asterion. A point 4 cm behind and 12 mm above the auricular point. It marks the meeting of the lambdoidal, occipitomastoid, and parietomastoid sutures (Fig. 3-2).

Zygomatic Tubercle. A prominence on

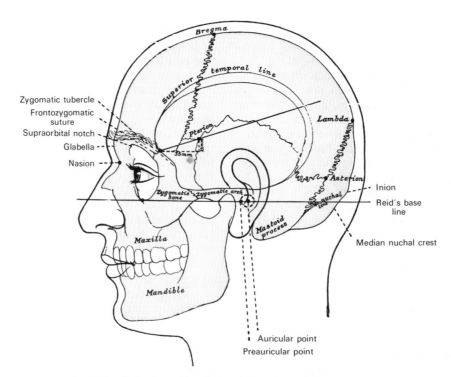

FIG. 3-2. Side view of head, showing surface relations of bones.

the posterior margin of the zygomatic bone at the level of the lateral palpebral commissure of the eye (Fig. 3-2).

Frontozygomatic Suture. A slight depression on the posterior margin of the zygomatic bone, 1 cm above the zygomatic tubercle. It is the line of union between the zygomatic process of the frontal bone and the frontal process of the zygomatic bone (Fig. 3-2).

Pterion. A point 35 mm behind and 12 mm above the level of the frontozygomatic suture. It marks the point where the great wing of the sphenoid meets the sphenoidal angle of the parietal bone (Fig. 3-2).

Inion. The external occipital protuberance. It is palpable in the midline at the back of the skull where the posterior neck muscles attach (Figs. 3-2; 3-6).

Lambda. The point where the lambdoidal and sagittal sutures meet, about 6.5 cm above the inion. It marks the junction of the occipital and the two parietal bones (Figs. 3-2; 3-6).

Bregma. The junction site of the coronal and sagittal sutures. It can be located at the intersection of a line drawn vertically upward from the preauricular point and the midline at the top of the head (Fig. 3-2).

Lambdoidal Suture. This articulation between the parietal and occipital bones can be located by the upper two-thirds of a line from the mastoid process to the lambda.

Sagittal Suture. The midline junction of the parietal bones, between the lambda and the bregma.

Obelion. The point on the sagittal suture between the parietal emissary foramina; a flattened area at the top of the head.

Coronal Suture. This junction of the frontal and parietal bones can be approximated by a line from the bregma to the middle of the zygomatic arch.

Glabella. A flattened triangular area on the forehead between the two superciliary ridges.

Nasion. A slight depression at the root of the nose marking the frontonasal suture (Figs. 3-2; 3-6).

Reid's base line. This line, which is used in cranial topography (Figs. 3-2; 3-6), passes through the inferior margin of the orbit to the auricular point, and then extends back to the center of the occipital bone.

The fontanelles or soft spots in the skull of a newborn infant correspond to the above-mentioned points and sutures as follows: (1) the **anterior fontanelle,** the largest, is at the bregma; (2) the **posterior fontanelle** is at the lambda; (3) the **lateral** or **sphenoidal fontanelle** is at the pterion; and (4) the **mastoid fontanelle** is at the asterion.

FACE. Facial features are the principal structures used for human visual recognition. Adult facial contours develop with the growth of the facial skeleton and most rapidly between the childhood years and the early post-adolescent years. Thus, the main contours of the face are governed by bony landmarks, most of which are evident to the eye but require palpation for establishment of their details.

The **superciliary ridges** are horizontal arches that extend across the forehead on each side above the eyebrow. They are part of the frontal bone above the eye, and each is marked by a depression, the supraorbital notch or foramen (Fig. 3-2).

The rim of the orbital cavity is called the **orbital margin,** and it is formed by the frontal, zygomatic, and maxillary bones.

In front of the ear, the **zygomatic arch,** formed by the temporal process of the zygomatic bone and the zygomatic processes of the temporal and frontal bones, can be palpated throughout its length (Fig. 3-2).

The **superior temporal line** is a curved ridge along the lateral aspect of the parietal bone, and it extends between the zygomatic process of the frontal bone and the region behind the ear (Fig. 3-2).

The **mandible** can be recognized and palpated throughout most of its extent. Its features include the prominence of the chin or **mental protuberance,** the **angle of the jaw,** the **alveolar portion** containing the teeth, and the **condyle,** which articulates with the temporal bone.

The **temporomandibular joint** can easily be manipulated when the mouth is opened wide; this draws the mandible forward and a depression can be felt between the condyle of the mandible and a fossa in the temporal bone, just in front of the tragus of the ear.

MUSCLES AND SOFT PARTS. The **masseter muscle** produces the fullness of the posterior part of the cheek, between the angle of the mandible and the zygomatic arch. Its posterior border is masked by the substance of the **parotid gland,** lying over the muscle and between it and the ear. The important duct of the parotid gland (Stensen's duct) courses

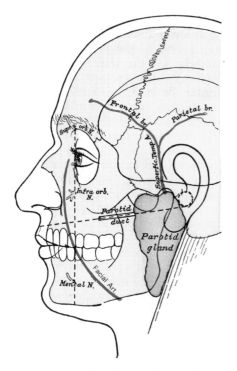

FIG. 3-3. Outline of side of face, showing chief surface markings as well as the superficial temporal and facial (maxillary) arteries.

transversely across the face, superficial to the masseter muscle and approximately 2 cm below the zygomatic arch (Fig. 3-3).

The anterior border of the masseter muscle may be palpated readily when the muscle is contracted by clenching the teeth; in front of it the fullness of the cheek is produced by the **buccopharyngeal fat pad** and the **facial muscles.**

The **temporalis muscle** occupies the temporal fossa, extending from the hollow behind the zygomatic arch to the superior temporal line (Fig. 3-5). The thick temporal fascia, covering the muscle, makes the superior border of the zygomatic arch difficult to palpate.

ARTERIES (Fig. 3-3). The pulsations of the **superficial temporal artery** may be felt on the surface of the temples just above the zygomatic arch in front of the ear. In older individuals its frontal branch frequently is visible making a serpentine course across the surface of the temple. The **facial artery** also may be felt as it crosses the margin of the mandible at the anterior border of the masseter muscle; it has a tortuous course from this point upward across the face on

the lateral side of the nose toward the medial angle of the eye.

VEINS. The **facial vein** crosses the margin of the mandible with the facial artery. It then takes a relatively straight course to the angle between the eye and nose where it is usually called the **angular vein.**

NERVES (Fig. 3-3). The **supraorbital nerve** emerges from the skull through the supraorbital notch (or foramen) and then courses superiorly over the superciliary ridge to innervate the anterior scalp. The **infraorbital nerve** becomes subcutaneous at the infraorbital foramen, just below the orbit. The **mental nerve** emerges from the mandible above and lateral to the mental protuberance. The **facial nerve** divides within the substance of the parotid gland, and its branches spread across the face like a fan.

LYMPH NODES. Usually lymph nodes are not palpable in the head, but if enlarged, the following nodes may be felt: posterior auricular, parotid, occipital, buccal, and submandibular.

THE EYE. The **palpebral fissure** is the elliptical opening between the two eyelids. At the extremities of the fissure, the lids are united by the **medial** and **lateral commissures.** At the medial commissure is a reddish elevation called the caruncula lacrimalis. Just lateral is a crescenteric fold of conjunctiva, the **plica semilunaris,** and two small openings in the eyelids, the **puncta lacrimalia.** The **nasolacrimal duct** runs from the puncta to an opening in the inferior meatus in the lateral wall of the nasal cavity. Its direction is illustrated in Figure 3-4. The **lacrimal sac,** at the top of the duct, lies behind the medial palpebral ligament. The latter may be felt at the medial commissure if the eyelids are drawn laterally to tighten the skin. The inner surface of the eyelids is lined by the smooth and glistening epithelial layer, the **palpebral conjunctiva.** At the margin of each eyelid (**fornix**), the conjunctiva doubles back over the eyeball as the **bulbar conjunctiva.**

The colored portion of the eye is observable because the pigmented layer of the retina extends behind the **iris** and because the eyeball is covered anteriorly by a transparent sheet of tissue, the **cornea.** The central aperture in the iris is the **pupil** through which the **lens,** the **retinal vessels,** and the **optic disc** can be seen with an ophthalmoscope.

THE EAR. The **pinna** of the external ear is a skin-covered cartilaginous structure also

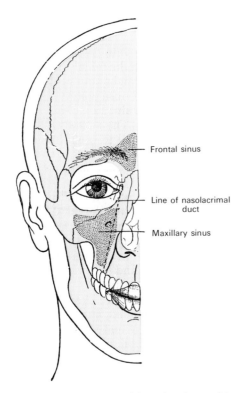

Frontal sinus

Line of nasolacrimal
duct

Maxillary sinus

Fig. 3-4. Outline of bones of face showing position of the frontal and maxillary air sinuses and the line of direction of the nasolacrimal duct.

called **auricula,** and it is marked by various prominences and fossae. The external opening of the ear leads into a canal, the **external acoustic meatus,** which can be more fully exposed by drawing the **tragus** forward. The orifice is guarded by crisp hairs and contains a coating of wax, the secretion of the **ceruminous glands.** The interior of the meatus can be examined through an otoscope more easily if the auricula is drawn upward, backward, and slightly outward in order to straighten the slight curvature at the junction of the cartilaginous and bony portions of the wall. At the internal end of the meatus is the **tympanic membrane** on which are seen certain structures and markings when viewed with an otoscope. The **suprameatal triangle** (Figs. 3-5; 3-6) is an important landmark for internal structures such as the **tympanic antrum, facial nerve,** and **transverse sinus.**

THE BRAIN.   The surface outline of certain brain structures of considerable practical importance can be established by external cranial landmarks and measurements (Figs. 3-6; 3-7). The **longitudinal fissure** separates

the two hemispheres at the midline and corresponds to the median line of the scalp between the nasion and the inion. The **lateral sulcus** (Sylvian) can be located by a point, termed the Sylvian point, which practically corresponds to the pterion, 3.5 cm behind and 12 mm above the level of the frontozygomatic suture. The **central sulcus** (of Rolando), which separates the cortical somatosensory centers in the precentral gyrus from the cortical motor centers of the postcentral gyrus, descends in an anterolateral direction from a projected point on the scalp 1.25 cm behind the center of the median line joining the nasion and inion. The position of the **lateral ventricle** may be shown on a ventriculogram obtained after air has been injected into it.

The branches of the **internal carotid artery** may be visualized on x-ray examination performed after injection of an opaque medium (see arteriogram, Fig. 8-31). The **middle meningeal artery** supplies the **dura mater** and courses on the inner aspect of the skull at the level of the middle of the superior border of the zygomatic arch; its anterior branch may be reached through a trephined opening slightly anterior to and below the pterion, while the posterior branch can be found through an opening about 2.5 cm above the external auditory meatus (Fig. 3-6). The **transverse dural sinus** may be approximated by a line through the asterion, and curving downward toward the mastoid process (Fig. 3-6).

PARANASAL AIR SINUSES.   The sinuses in the head (Fig. 3-4) are mucous membrane-lined cavities within bones of the skull that vary greatly in size, shape, and position. The **frontal sinus** occupies a region in the frontal bone deep to the medial part of the superciliary ridge. The **maxillary sinus** lies in the body of the maxilla, between the orbit, nasal cavity, and upper teeth. Other sinuses are also found in the ethmoid and sphenoid bones, as well as in the mastoid portion of the temporal bone.

The Neck

BONES.   The relationships between the cartilaginous and bony structures in the neck can be used to practical advantage in identifying the location of soft structures that may be of clinical importance. Although

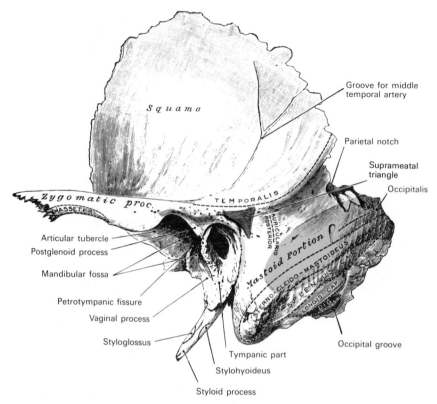

Fig. 3-5. Left temporal bone showing surface markings for the transverse sinus (blue) and facial nerve (yellow). Also observe the suprameatal triangle (red) lying above the external auditory meatus at the posterior end of the zygomatic process. An instrument can readily be pushed through this triangular region into the underlying tympanic antrum.

the **vertebral column** is deeply placed in the neck, certain bony landmarks can be identified by palpation.

The **transverse process of the atlas** can be felt about 1 cm below and in front of the mastoid process.

The **anterior tubercle** on the **transverse process** of the **sixth cervical vertebra** may be felt deep in the neck at the anterior border of the sternocleidomastoid muscle; it has been named the **carotid** or **Chassaignac's tubercle,** and is the point of preference for compressing the common carotid artery to stop bleeding.

The **spinous process** of the **seventh cervical vertebra (vertebra prominens)** can be distinguished by palpation, since it is somewhat longer than the spinous processes of the upper cervical vertebrae, and since a band of fibroelastic tissue called the **ligamentum nuchae** partially cushions the spinous processes of the upper cervical vertebrae.

The **hyoid bone** may be felt in the receding angle between the chin and anterior part of the neck. Its vertebral level is approximately at the disc between the third and fourth cervical vertebrae. The **greater horn (greater cornu)** of the hyoid, which extends backward on a level with the angle of the mandible, can best be felt when the bone is pushed laterally toward the side to be palpated.

LARYNX AND TRACHEA. The **laryngeal prominence** of the **thyroid cartilage** (commonly called the Adam's apple) is visible at the midline, 1 or 2 cm below the hyoid bone (Fig. 3-8). It constitutes the anterior cartilaginous protection of the **vocal folds** of the **larynx,** and if projected posteriorly, the thyroid cartilage spans the fourth and fifth cervical vertebrae. The upper margin of the thyroid cartilage is attached to the hyoid bone by the **thyrohyoid membrane,** and its lower margin to the **cricoid cartilage** by the **cricothyroid ligament.** The level of the vocal folds corresponds to the middle of the anterior border of the thyroid cartilage.

The **cricoid cartilage** lies at the level of the sixth cervical vertebra (Fig. 3-8). Below it the

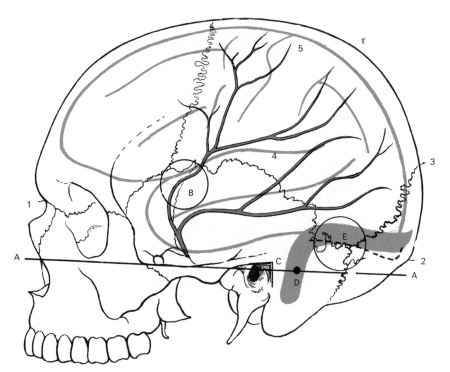

FIG. 3-6.   Relations of the brain and middle meningeal artery to the surface of the skull. *1*, Nasion. *2*, Inion. *3*, Lambda. *4*, Lateral cerebral fissure. *5*, Central sulcus. *AA*, Reid's base line. *B*, Point for trephining the anterior branch of the middle meningeal artery. *C*, Suprameatal triangle. *D*, Sigmoid bend of the transverse sinus. *E*, Point for trephining over the straight portion of the transverse sinus, exposing dura mater of both cerebrum and cerebellum. Outline of cerebral hemisphere indicated in blue; course of middle meningeal artery in red.

trachea can be felt, but only in thin subjects can the individual cartilaginous rings be distinguished. As a rule there are seven or eight rings above the **jugular notch** of the sternum, and the isthmus of the **thyroid gland** covers the second, third, and fourth rings. The trachea measures about 10 cm in length before it bifurcates into the primary bronchi. Of these 10 cm, 5 extend in the neck above the suprasternal notch, while 5 extend in the thorax below the suprasternal notch.

MUSCLES.   The **sternocleidomastoid muscle** is the most prominent muscle in the neck. Its entire extent, from the mastoid process to the sternal and clavicular heads (Fig. 3-8), is clearly visible in most persons. The **trapezius** muscle forms the superior curvature of the shoulder, sloping upward toward the posterior neck and back of the head. The **platysma** is a thin superficial sheet of muscle overlying the anterior neck. It wrinkles the skin of the neck when contracted and can be detected when thrown into prominence by expressions of horror or sudden and intense muscular efforts requiring deep inspiration.

TRIANGLES OF THE NECK.   The lack of prominent topographical features and the large number of important nerves and blood vessels contained within the neck have made it customary to subdivide this region into smaller areas. These areas, triangular in shape and related to some of the more superficial muscles, are known as the triangles of the neck. The side of the neck presents a quadrangular shape limited *above* by the lower border of the **mandible,** *below* by the **clavicle,** *in front* by the **midline,** and *behind* by the **anterior border of the trapezius muscle.** This is subdivided obliquely into two large triangles by the **sternocleidomastoid** muscle. The triangle in front of this muscle is called the **anterior triangle;** the one behind it, the **posterior triangle.**

Anterior Triangle.   The anterior triangle is bounded anteriorly by the **midline,** and posteriorly by the anterior border of the **sternocleidomastoid muscle;** its apex is below at the sternum, and its base is formed by the lower margin of the **body of the mandible** and an extension of this line to the mastoid

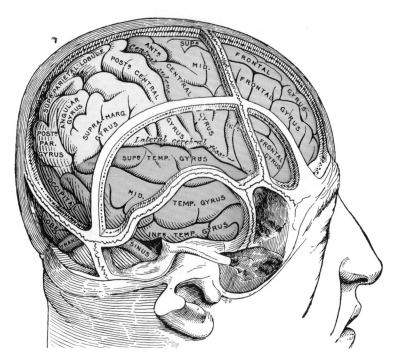

Fig. 3-7. Drawing of a cast to illustrate the relationships of the brain with the overlying skull. Note that the convolutions of the cerebral cortex can be described as comprising the frontal (blue), temporal (green), parietal (yellow), and occipital (red) lobes.

process. This larger triangle is subdivided, by the **digastric muscle** above and the superior belly of the **omohyoid muscle** below, into four smaller triangles: **muscular**, **carotid**, **submandibular**, and **submental** (Fig. 3-9).

The **muscular triangle** is bounded *in front* by the midline of the neck; *behind* by the

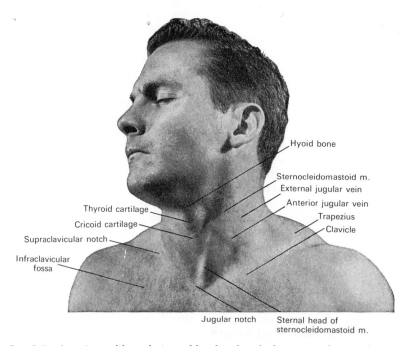

Fig. 3-8. Anterior and lateral view of head and neck showing surface markings.

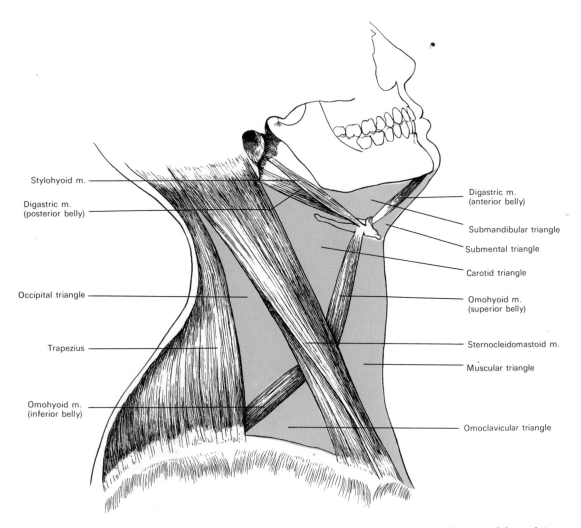

Stylohyoid m.

Digastric m.
(posterior belly)

Occipital triangle

Trapezius

Omohyoid m.
(inferior belly)

Digastric m.
(anterior belly)

Submandibular triangle

Submental triangle

Carotid triangle

Omohyoid m.
(superior belly)

Sternocleidomastoid m.

Muscular triangle

Omoclavicular triangle

Fɪɢ. 3-9.   The triangles of the neck. The sternocleidomastoid muscle divides the anterolateral aspect of the neck into large *anterior* and *posterior triangles* (not labeled). The *posterior triangle* is further subdivided by the inferior belly of the omohyoid into the *occipital and omoclavicular triangles.* The *anterior triangle* is crossed by the superior belly of the omohyoid, thereby forming the *carotid and muscular triangles.* Above the hyoid bone are located the *submandibular and submental triangles.*

superior belly of the omohyoid above, and the anterior margin of the lower part of the sternocleidomastoid below. In this triangle, under cover of the skin, subcutaneous fascia, platysma, and deep fascia, are the **sternohyoid** and **sternothyroid** muscles, the **isthmus of the thyroid gland,** the **larynx,** and the **trachea.** Although not within the triangle as defined above, the following structures can be approached surgically through the triangle by displacing the sternocleidomastoid laterally: the lower part of the common carotid artery in the carotid sheath with the internal jugular vein and vagus nerve, the ansa cervicalis, the sympathetic trunk, recurrent laryngeal nerve, and the esophagus.

The **carotid triangle** is bounded *behind* by the sternocleidomastoid muscle; *below* by the superior belly of the omohyoid muscle; and *above,* by the stylohyoid muscle and the posterior belly of the digastric muscle (Fig. 3-9). It is covered by the skin, subcutaneous fascia, platysma, and deep fascia, layers that contain the ramifications of the cutaneous cervical nerves and the cervical branch of the facial nerve. The floor or deepest part of the triangle is formed by the thyrohyoid, hyoglossus, and middle and inferior pharyngeal constrictor muscles. In the triangle, especially if it is enlarged by displacing the sternocleidomastoid backward, is the upper part of the **common carotid artery,** *which* *bifurcates at the level of the upper border of the thyroid cartilage into the* **internal and external carotid arteries.** The internal carotid artery lies posterior and somewhat lateral to the external carotid and *gives off no branches in the neck.* The branches of the external carotid artery in the carotid triangle are the superior thyroid, the lingual, the facial, the occipital, and the ascending pharyngeal. Enclosed with the arteries in a fascial membrane, the **carotid sheath,** are the **internal jugular vein** and the **vagus nerve.** The tributaries of the vein correspond with the branches of the external carotid: superior thyroid, lingual, facial, ascending pharyngeal, and sometimes the occipital. Superficial to the carotid sheath is the ansa cervicalis, and deep to it or embedded in its substance is the **sympathetic trunk.** The **hypoglossal nerve** crosses both the internal and external carotids in the upper part of the triangle, curving around the origin of the occipital artery. Between the external carotid and the pharynx, just below the hyoid bone, is the internal branch of the **superior laryngeal nerve;** the external branch is slightly lower. The **accessory nerve** may cross the uppermost corner of the triangle.

Between the hyoid bone and the superior border of the thyroid cartilage at the level of the carotid bifurcation is a slight dilatation of the common carotid artery, the **carotid sinus.** This is an organ of sensory receptors, innervated by the glossopharyngeal nerve, which is capable of sensing blood pressure changes. Also located at the carotid bifurcation is the **carotid body.** Innervated by fibers of both the glossopharyngeal and vagus nerves, the carotid body contains chemoreceptors capable of responding to undue changes in carbon dioxide or oxygen in the blood.

The **submandibular triangle,** located immediately below the body of the mandible, is bounded *above* by the lower border of the body of the mandible and an extension of this line to the mastoid process; *below* by the posterior belly of the digastric muscle and the stylohyoid muscle; and *in front* by the anterior belly of the digastric muscle (Fig. 3-9). It is covered by skin, subcutaneous fascia, platysma, and deep fascia, within which ramify motor branches of the facial nerve (to platysma) and sensory filaments from the cervical cutaneous nerves. The muscular floor of the triangle is formed by the mylohyoid and hyoglossus muscles. The stylomandibular ligament at the angle of the mandible divides the triangular area into an anterior and a posterior part. The *anterior part* contains the **submandibular gland,** which is crossed by the anterior facial vein and facial artery. Deep to the gland are the submental artery and the mylohyoid artery and nerve. The *posterior part* of this triangle is largely occupied by the **parotid gland.** By displacing the muscles slightly, certain deeper structures may be included. In its superior course, the external carotid artery is crossed by the facial nerve. The vessel then becomes embedded within the parotid gland, and gives off its posterior auricular, superficial temporal, and maxillary branches. At this level the internal carotid artery, internal jugular vein, and vagus nerve are more deeply situated in the neck and are separated from the external carotid by the styloglossus and stylopharyngeus muscles and the glossopharyngeal nerve.

The **submental triangle** is bounded *posteriorly* by the anterior belly of the digastric muscle, *anteriorly* by the midline, and *inferiorly* by the body of the hyoid bone (Fig. 3-9). Its floor is formed by the **mylohyoid muscle,** and the triangle contains several lymph nodes which drain the chin and lower lip along with the anterior tongue and floor of the mouth. Several smaller veins drain into the anterior jugular vein.

Posterior Triangle. The posterior triangle is bounded *in front* by the **sternocleidomastoid muscle** and *behind* by the anterior margin of the **trapezius;** its base is the middle third of the **clavicle,** and its apex is at the occipital bone. About 2.5 cm above the clavicle, it is crossed obliquely by the inferior belly of the omohyoid, dividing it into an upper or **occipital** and a lower or **omoclavicular** (supraclavicular or subclavian) triangle.

The **occipital triangle** (Fig. 3-9) is bounded *in front* by the sternocleidomastoid; *behind* by the trapezius; and *below* by the omohyoid. Its muscular floor is formed by the splenius capitis, the levator scapulae, and the scalenus medius and posterior, and it is covered by the skin, subcutaneous fascia, superficial fascia, and deep fascia. About midway along the posterior border of the sternocleidomastoid muscle, the **spinal accessory nerve** emerges from beneath the muscle to cross the triangle obliquely in an inferolateral direction on its path to innervate the trapezius muscle. Somewhat lower, the cervical cutaneous and supraclavicular nerves also emerge from under the posterior border of the sternocleidomastoid. In the lowest part of the triangle, the transverse cervical vessels along with the uppermost part of the brachial plexus may be seen.

The **omoclavicular triangle** (Fig. 3-9) is bounded *above* by the inferior belly of the omohyoid muscle; *below* by the clavicle; and *in front* by the sternocleidomastoid muscle. Its floor is formed by the first rib and the first digitation of the serratus anterior muscle. To some extent the size of the omoclavicular triangle depends on the breadth of the attachment of the sternocleidomastoid and trapezius to the clavicle and the position of the omohyoid. Some of the structures listed below might be reached by displacement of the muscles, since the triangle is increased in size by drawing the shoulder downward to lower the clavicle. The triangular area is covered by skin, subcutaneous fascia, platysma, and the deep fascia that contains the supraclavicular nerves. Medially, within the triangle, the third part of the **subclavian artery** emerges from behind the scalenus anterior and curves down behind the clavicle. Sometimes the arch of the artery rises by as much as 4 cm above the clavicle, and this is the most commonly chosen place for ligature. The **subclavian vein** usually rests behind the clavicle, but may be found partially within the triangle. The suprascapular vessels cross the lower part and the superficial cervical vessels the upper part of the triangle. The **external jugular vein** enters the anterior part of the triangle from its position on the surface of the sternocleidomastoid muscle and empties into the subclavian vein; its tributaries, the transverse scapular and transverse cervical veins, form a plexus superficial to the subclavian artery. The **brachial plexus** of nerves crosses the lateral part of the triangle, coming into close relationship with the subclavian artery in its descent behind the clavicle to the axilla. A portion of the plexus courses above the subclavian artery, while another part courses behind the vessel.

ARTERIES. The position of the **common** and **external carotid arteries** can be approximated by a line drawn from the upper part of the sternal end of the clavicle to a point midway between the tip of the mastoid process and the angle of the mandible. Above the carotid bifurcation at the upper border of the thyroid cartilage, this line overlies both internal and external carotids until the latter arches backward slightly toward the external acoustic meatus. The *main branches of the external carotid* originate near the tip of the greater horn of the hyoid bone. The arching course of the *subclavian artery* upward from the sternoclavicular joint to the middle of the clavicle can be observed in Figure 3-10.

VEINS. The **internal jugular vein** is parallel and slightly lateral to the internal carotid artery. The **external jugular** runs from the angle of the mandible to the middle of the clavicle.

NERVES. The **facial nerve** emerges from the stylomastoid foramen about 2.5 cm deep to the surface of the skin, opposite the anterior border of the mastoid process. The **accessory nerve** courses quite superficially as it crosses the posterior triangle. Emerging from under the posterior border of the sternocleidomastoid muscle, it achieves the anterior border of the trapezius muscle. Also appearing from under the posterior border of the sternocleidomastoid in this same region are the following **cutaneous nerves:** the *lesser occipital,* the *greater auricular,* the *anterior cervical cutaneous,* and the *supraclavicular* nerves (Fig. 3-10). The **phrenic nerve** is formed from cervical roots at the level of the middle of the thyroid cartilage. It then passes inferiorly to the thorax beneath and about half way between the two borders of the sternocleidomastoid. In its course it lies in front of the scalenus anterior muscle. The upper part of the **brachial plexus** can be located by a line drawn from the cricoid cartilage to the middle of the clavicle. The **vagus nerve** and **sympathetic trunk** run parallel and deep to the internal carotid artery.

LYMPH NODES. Lymph nodes may be felt between the ramus of the mandible and the sternocleidomastoid muscle, along the course of the internal jugular vein, and along both the anterior and posterior borders of this muscle.

The Back

BONES. The furrow down the middle of the back lies over the tips of the *spinous processes of the vertebrae.* The upper cervical vertebrae cannot be felt, but the seventh, or **vertebra prominens,** can be distinguished easily and lies just above the even more

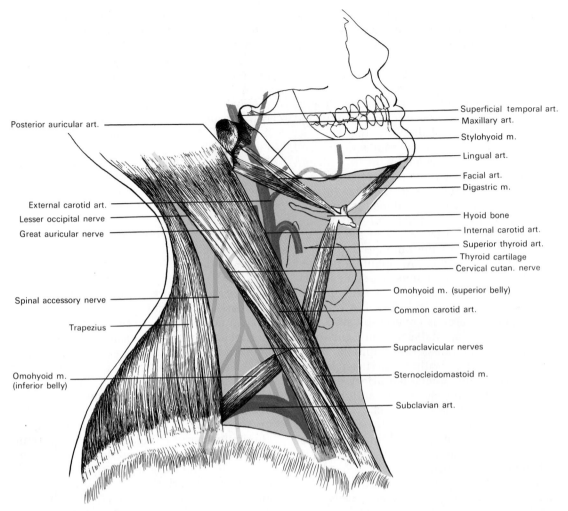

FIG. 3-10. The anterolateral aspect of the neck showing the course of principal arteries, cutaneous nerves, and the spinal accessory nerve.

prominent *first thoracic*. Other vertebral spinous processes can be identified by counting from these. The posterior surface of the **scapula** is marked by a prominent obliquely crossing plate of bone called the *spine of the scapula*. This lies on a level with the third thoracic spinous process, while the inferior angle of the scapula is at the level of the seventh thoracic spinous process. The highest point of the **crest of the ilium** is on a level with the fourth lumbar spinous process (Fig. 3-11), and the **posterior superior iliac spine** is level with that of the second sacral.

SPINAL CORD. The **spinal cord** extends downward to the level of the spinous process of the second lumbar vertebra in the adult body (Fig. 3-12), but as far as the fourth in an infant. The **subarachnoid space,** containing the spinal fluid, extends downward to the third sacral vertebra. Because the sub-

stance of the spinal cord ends at the second lumbar level, the dural and arachnoid sacs below that level containing the lumbar and sacral nerve roots **(cauda equina)** surrounded by spinal fluid are used to advantage for the induction of spinal anesthesia. Injections of anesthetics, for example, through the fourth lumbar interspace may be performed in the adult without much danger of striking the spinal cord.

MUSCLES. The **muscles** of the back are large and their surface contours usually can be identified with ease in a living subject (Fig. 3-33).

The Thorax

Several prominent muscles that extend either into the scapula or into the upper extremity cover the bony thoracic cage. Anteri-

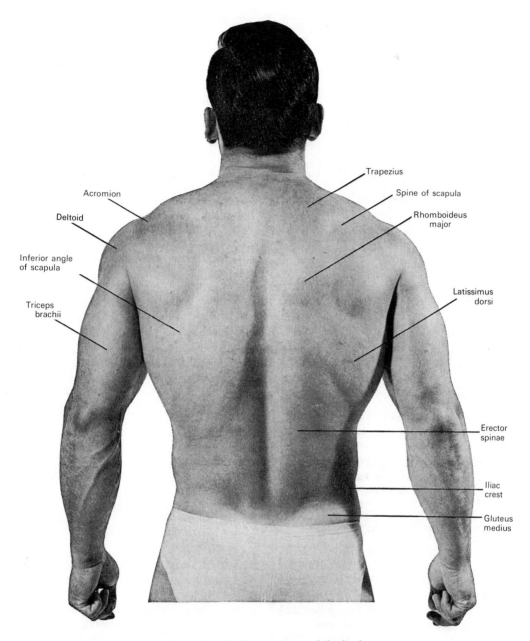

Acromion

Deltoid

Inferior angle
of scapula

Triceps
brachii

Trapezius

Spine of scapula

Rhomboideus
major

Latissimus
dorsi

Erector
spinae

Iliac
crest

Gluteus
medius

FIG. 3-11.   Surface anatomy of the back.

orly, these include the pectoralis major and the serratus anterior, along with the upper portions of the abdominal muscles (Figs. 3-13; 3-18). Posteriorly the superficial back muscles (trapezius and latissimus dorsi) overlie the deeper muscles of the shoulder and back (Fig. 3-11). Each **axilla,** or hollow of the armpit (Fig. 3-37), is limited by two fleshy folds: the **anterior axillary fold** is the prominence caused by the pectoralis major, while the **posterior axillary fold** is the prominence of the latissimus dorsi. A **triangle of auscul-**

tation can be identified on the posterior thoracic wall bounded *above* by the trapezius, *below* by the latissimus dorsi, and *laterally* by the medial border of the scapula. Drawing the scapula forward by folding the arms across the chest will reveal the sixth and seventh ribs subcutaneously at this site, and allow for auscultation of the thorax.

The size of the **breast** or **mamma** is subject to great variation. In most adult nulliparous females it extends vertically from the second to the sixth ribs, and transversely from the

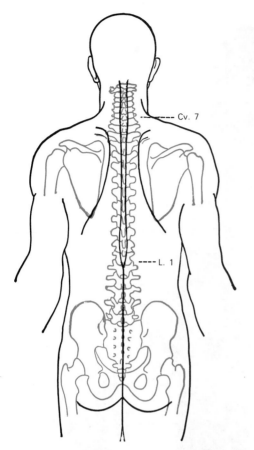

Cv. 7

L. 1

Fig. 3-12. Diagram showing the relationships of the spinal cord to the dorsal surface of the trunk. Observe that the substance of the spinal cord reaches to the upper lumbar levels of the spinal column. Below this the spinal canal contains spinal nerves and it can be punctured to achieve the cerebrospinal fluid compartment.

about 5 cm below the jugular notch. This ridge, called the **sternal angle** or **angle of Louis,** is opposite the sternochondral junction of the second rib (Fig. 3-18). At the lower end of the body of the sternum is the **infrasternal notch** in the midline between the sternal attachments of the seventh costal cartilages of both sides. In the triangular depression, called the epigastric fossa, below the notch, the **xiphoid process** of the sternum can be felt.

Additional bony relationships of these sternal structures can be identified with the vertebrae of the spinal column. If projected posteriorly, the *jugular notch* lies at the lower level of the second thoracic vertebra, the *sternal angle* is at the level of the lower part of the fourth thoracic vertebra, and the junction between the xiphoid process and the body of the sternum (*xiphisternal junction*) projects backward to the body of the tenth thoracic vertebra.

The **ribs** can usually be felt at the sternochondral junction in front, on the sides of the thorax, and in back as far as their angles, although they are mostly covered by muscles. The first rib is difficult to palpate because it is deeply situated and partly hidden by the clavicle. The **second rib** is the one most reliably identified because of its attachment at the sternal angle, and from this point the other ribs can be counted. The lower boundary of the thorax is formed by the xiphoid process, the cartilages of the seventh, eighth, ninth, and tenth ribs, and the ends of the eleventh and twelfth cartilages (Figs. 3-17; 3-19).

LINES FOR ORIENTATION (Fig. 3-21). Certain arbitrarily accepted lines offer some value for clinical orientation and descriptive purposes. The **midsternal line** is the midline of the body over the sternum. The **midclavicular** or **mammary line** is a vertical line, parallel with the midsternal, through a point midway between the center of the jugular notch and the tip of the acromion or point of the shoulder. The **lateral sternal line** is a vertical line along the sternal margin over the sternochondral junctions. The **anterior** and **posterior axillary lines** are vertical lines drawn along the corresponding axillary folds, while the **midaxillary line,** halfway between them, passes through the apex of the axilla. On the back, the **scapular line** is drawn vertically through the inferior angle of the scapula (Fig. 3-11).

sternum to the midaxillary line. In males the **mammary papilla** or **nipple** is situated in the fourth intercostal space (Fig. 3-14), while in females the position of the nipple is so variable that it is much less useful as a surface landmark for intrathoracic structures.

BONY LANDMARKS. The **sternum, ribs, scapula,** and **clavicle** can be seen in many individuals and can be felt in all but the very muscular or obese subjects. The upper border of the sternum is marked by the **jugular** (or suprasternal) **notch,** between the sternal heads of the two sternocleidomastoid muscles (Fig. 3-8). In the midline the sternum is subcutaneous, leaving the sternal furrow between the origins of the two pectoralis major muscles. The junction between the two parts of the sternum, the **manubrium** above and the **body** (or gladiolus) below, is marked by a well-defined transverse ridge

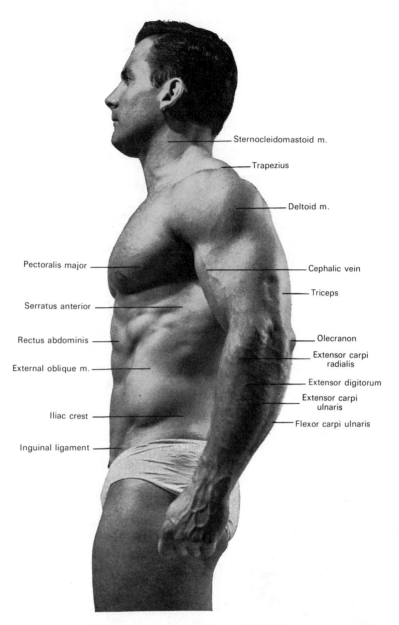

Sternocleidomastoid m.

Trapezius

Deltoid m.

Pectoralis major

Cephalic vein

Triceps

Serratus anterior

Rectus abdominis

Olecranon

Extensor carpi
radialis

External oblique m.

Extensor digitorum

Extensor carpi
ulnaris

Iliac crest

Flexor carpi ulnaris

Inguinal ligament

Fig. 3-13.   Surface anatomy of left side.

THE LUNGS (Figs. 3-14, 3-15). The **apex of the lung** lies behind the medial third of the clavicle and extends superiorly into the neck from 1 to 5 cm, but usually about 2.5 cm. The **anterior border** of the right lung approaches the midsternal line, that of the left does likewise as far as the fourth costal cartilage, where it deviates laterally because of the **cardiac notch.** The **inferior border** of each lung at expiration follows a curving line downward from the sixth sternocostal junction anteriorly to the spinous process of the tenth thoracic vertebra posteriorly. This line crosses the midclavicular line at the sixth and the midaxillary line at the eighth rib. The **posterior border** of each lung is parallel with the midline (Fig. 3-15), 2 to 3 cm from it, and extends from the spinous process of the seventh cervical vertebra to that of the tenth thoracic vertebra.

The reflections of the **pleura** correspond in general with the surface of the lungs, but are more extensive. At expiration the **parietal pleurae** of the two sides of the thorax ap-

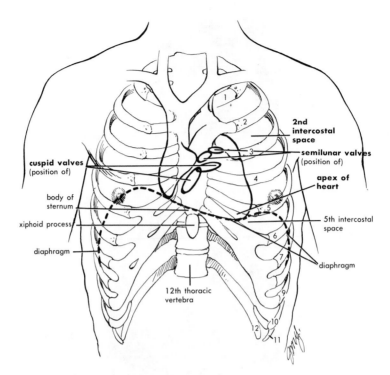

FIG. 3-14. Anterior view of the thorax to show the position of the heart in relationship to the ribs, sternum, and diaphragm, and the position of the heart valves. (From Crouch, J.E.; *Functional Human Anatomy.* Lea & Febiger, 1978.)

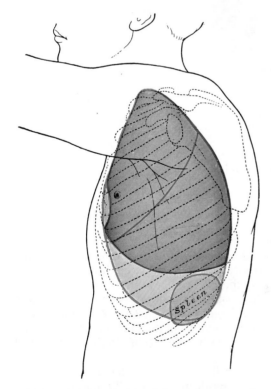

FIG. 3-15. Left lateral view of thorax showing surface markings for bones, lungs (purple), pleura (blue), and spleen (green).

proximate each other in the midline anteriorly, opposite the second or fourth costal cartilages. The most medial 1 or 2 cm of the pleural cavity anteriorly are unoccupied by lung, but consist of a slit-like cavity forming a potential space between the two layers of pleura known as the **costomediastinal sinus.** The inferior limit of the pleura extends 2 to 5 cm below that of the lung at expiration, leaving the **phrenicocostal sinus** also unoccupied by lung. This site is frequently used for the introduction of a needle into the pleural cavity in order to aspirate fluid from the chest. The inferior limit of the thorax is the **diaphragm** (Fig. 3-16), and its attachment to the thoracic wall along the costal margin is 2 to 3 cm below the inferior limit of the pleura.

THE HEART. The **apex of the heart** is directed inferolaterally in the **middle mediastinum** of the thorax, and its pulsation may be felt in the fifth intercostal space just below the nipple or about 9 cm to the left of the midsternal line. A projection of the heart outline on the anterior chest wall can be approximated in the following manner: the **superior margin** is marked by a horizontal line at the third sternochondral attachment;

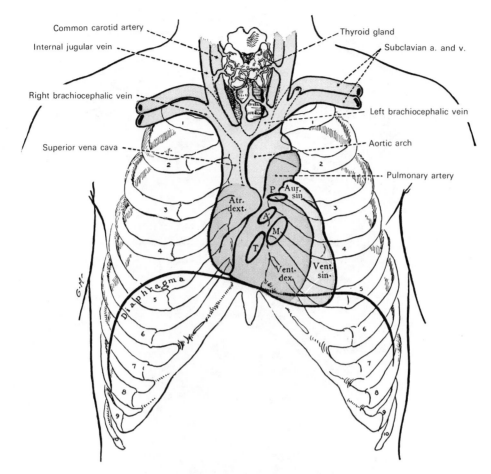

FIG. 3-16. The heart and cardiac valves projected on the anterior chest wall, showing their relationship to the ribs, sternum, and diaphragm. (Eycleshymer and Jones, 1925.)

the **right margin** corresponds to a vertical line drawn 2.5 cm lateral to the right sternal margin; the **inferior** or **diaphragmatic margin** is marked by a line sloping slightly downward and to the left at the xiphisternal junction; and the **left margin** angles upward from the apex (located 9 cm to the left in the fifth intercostal space) to the second intercostal space 2.5 cm from the left sternal margin. The superior margin is also the **base of the heart** and marks the beginning of the great vessels. The position of the chambers, sulci, and valves is shown in Figure 3-16.

It should be noted that a portion of the heart is not covered by lung and lies directly beneath the bony thorax (Figs. 3-14; 3-17), while the remainder is overlapped by lung tissue. Upon percussion of the chest over that region of the heart not covered by lung, a less resonant tone is heard and this zone has been called the area of **superficial car-**diac dullness. The chest area over the remainder of the heart covered by lung is called the area of **deep cardiac dullness.** A circle 5 cm in diameter around a point midway between the male nipple (fourth interspace) and the end of the sternum usually outlines the area of superficial dullness.

The **internal thoracic vessels** course vertically 1 cm lateral to the sternal margin and extend inferiorly from the subclavian vessels above the first rib to the sixth cartilage. The **intercostal vessels** and **nerves** generally lie along the inferior borders of the ribs in their intercostal spaces in the anterior wall and tend to lie deep to the ribs posteriorly.

The Abdomen

MUSCLES. The contours of the abdominal wall are established largely by the muscles, with modifications brought about by the

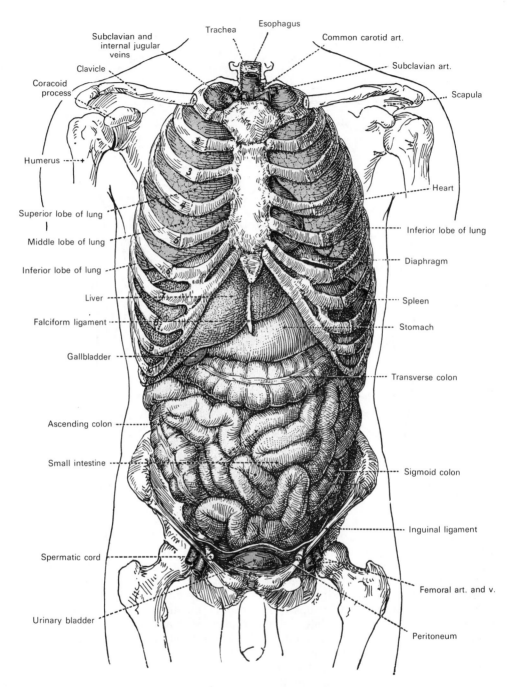

FIG. 3-17.   Thoracic and abdominal viscera shown in their normal relationship to the skeleton. Anterior view. (Eycleshymer and Jones, 1925.)

accumulation of adipose tissue in the subcutaneous layers. A groove in the midline called the **linea alba** separates the medial margin of the **rectus abdominis** of one side from that of the other. It extends from the infrasternal notch above to the pubic symphysis below, and its fibrous structure is formed by the decussating fibers of the apo-

neuroses of the underlying flat abdominal muscles (Figs. 3-18; 3-28). The **umbilicus** (or navel) interrupts the linea alba about midway along its extent. This depressed scar is the site of attachment of the umbilical cord of the fetus. In most individuals it lies at the level of the intervertebral disc between the third and fourth lumbar vertebrae. The **linea**

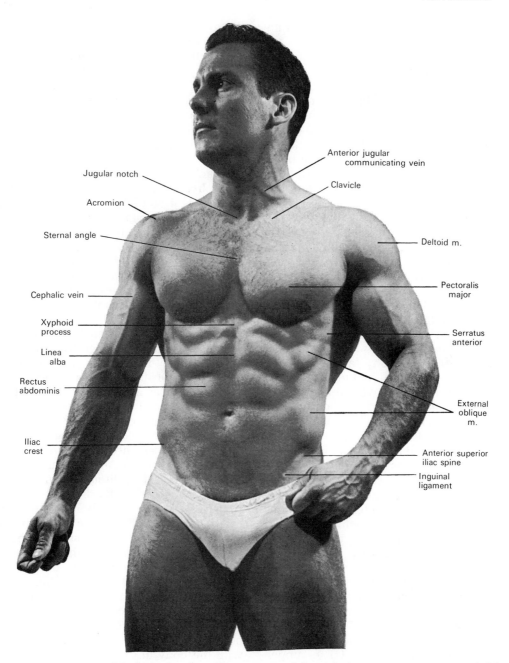

Anterior jugular
communicating vein

Jugular notch

Clavicle

Acromion

Sternal angle

Deltoid m.

Cephalic vein

Pectoralis
major

Xyphoid
process

Serratus
anterior

Linea
alba

Rectus
abdominis

External
oblique
m.

Iliac
crest

Anterior superior
iliac spine

Inguinal
ligament

FIG. 3-18. Anterior view of the body showing surface contours of the neck, thorax, upper extremity, and abdomen.

semilunaris is a curved, vertically oriented groove located on both sides of the anterior abdominal wall at the lateral margin of the rectus abdominis muscles (Fig. 3-28). It is not as clearly marked as the linea alba except in muscular subjects, but becomes more evident when the rectus abdominis is in action. At the level of the umbilicus, the linea semilunaris lies about 7 cm from the midline, curving medially as it nears the pubis.

Lateral to the rectus, the **external oblique muscle of the abdomen** (Fig. 3-18) becomes the most superficial muscle; it interdigitates with the **serratus anterior** above, and posteriorly it is partly overlapped by the **latissimus dorsi.** A separation between the external oblique and the latissimus at their attachment to the crest of the ilium is known as the **lumbar triangle** (of Petit). This triangle is bounded *medially* by the latissimus dorsi, *laterally* by the external oblique, and *inferiorly* by the crest of the ilium. The floor of the triangle is formed by the internal oblique muscle.

At the lower border of the external oblique, the **inguinal ligament** lies in the floor of the groove of the groin. The ligament is formed by the thickened inferior border of the aponeurosis of the external oblique. The surface of the rectus abdominis muscle, especially in a muscular subject, has three transverse furrows caused by the **tendinous inscriptions:** one is a little below the xiphoid process, one is at the umbilicus, and the third is between these two.

The **superficial inguinal ring** (Fig. 3-28), an opening in the aponeurosis of the external oblique, is situated 1 cm above and lateral to the pubic tubercle; the **deep inguinal ring** lies 1 to 2 cm above the middle of the inguinal ligament. The position of the **inguinal canal,** which contains the **spermatic cord** in the male and the **round ligament of the uterus** in the female, is indicated by a line joining these two rings.

BONY LANDMARKS. Bony landmarks that help define the abdomen include: the lower border of the thorax above, which has already been described, and the bony pelvis below (Figs. 3-19; 3-20). The **crest of the ilium** may be identified by the prominent **tubercle** on its outer lip which can be seen or palpated. The **anterior superior iliac spine** may be felt in the groin at the lateral end of the inguinal ligament (Figs. 3-17; 3-18), and the **pubic tubercle** may be felt at the medial end of the ligament, a point which may also be identified as the inferior attachment of the rectus abdominis muscle.

ABDOMINAL PLANES AND REGIONS. For convenience of description and reference, the abdomen can be divided into nine regions by four imaginary planes, two horizontal and two sagittal (Fig. 3-21). Indicated by lines on the surface of the abdominal wall, the two transverse lines are (1) an upper transverse, the **transpyloric,** situated halfway between the jugular notch and the upper border of the symphysis pubis; this line cuts through the pylorus of the stomach, the tips of the ninth costal cartilages, and the lower border of the first lumbar vertebra; (2) a lower transverse line termed the **transtubercular,** which passes across the iliac tubercles and cuts the body of the fifth lumbar vertebra. By means of these horizontal planes the abdomen is divided into three zones. These are further subdivided into

three regions each by two sagittal planes, which are indicated on the surface by **right** and **left lateral lines** drawn vertically through points halfway between the anterior superior iliac spines and the midline. The middle region of the upper zone is called the **epigastric,** and the two lateral regions the **right** and **left hypochondriac.** The central region of the middle zone is the **umbilical,** and the two lateral regions the **right** and **left lumbar.** The middle region of the lower zone is the **hypogastric** or **pubic,** and the lateral are the **right** and **left iliac** or **inguinal.**

Another more simplified but quite useful subdivision of the abdomen consists of two planes at right angles to each other, one corresponding to the midsagittal plane of the body, the other to a transverse plane through the umbilicus. The resulting four portions are known as **quadrants,** the upper and lower, right and left.

VISCERA. Under normal conditions the various portions of the digestive tube cannot be identified by simple palpation. The greater part of the liver lies under cover of the ribs and cartilages, especially in the supine position, but during a deep inspiration it may be pushed out below the costal margin on the right side and felt (Figs. 3-17; 3-22). Other viscera can only be palpated in emaciated subjects with lax abdominal walls, or if they are the seat of disease or tumors.

Stomach (Figs. 3-17; 3-22; 3-23). The shape of the stomach changes throughout the day and is affected by the particular phase of digestion, by the state of the surrounding viscera, and by the amount and character of its contents. Its position also varies with that of the body, so that it is impossible to indicate its position on the body surface with any degree of accuracy. The reference points and measurements given below refer to a moderately filled stomach with the body in the supine position.

The **cardiac orifice** lies deep to the seventh left costal cartilage about 2.5 cm from the left sternal border; it corresponds to the level of the tenth thoracic vertebra. The **pyloric orifice** is along the transpyloric line about 1 cm to the right of the midline, or alternately 5 cm below the seventh right sternocostal articulation; it is at the level of the first lumbar vertebra.

Duodenum (Fig. 3-24). The duodenum consists of four parts, **superior, descending, horizontal,** and **ascending.** It is approxi-

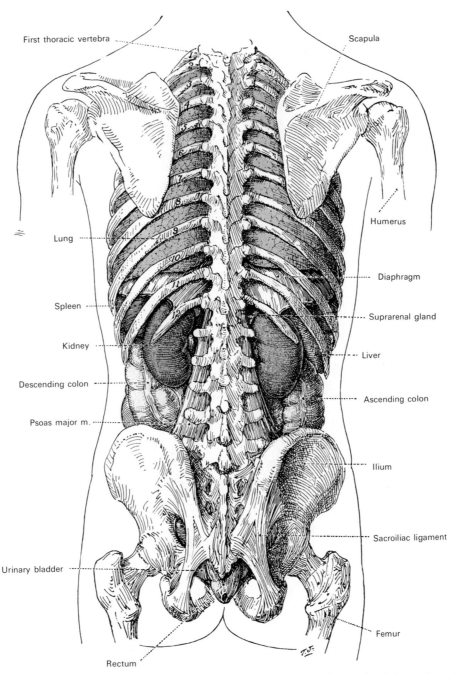

First thoracic vertebra

Scapula

Lung

Humerus

Diaphragm

Spleen

Suprarenal gland

Kidney

Liver

Descending colon

Ascending colon

Psoas major m.

Ilium

Sacroiliac ligament

Urinary bladder

Femur

Rectum

FIG. 3-19. Thoracic and abdominal viscera shown in their normal relationship to the skeleton. Posterior view. (Eycleshymer and Jones, 1925.)

mately 25 cm in length, shaped in the form of a "C," and is the commencement of the small intestine. The *superior part* passes transversely to the right and somewhat posteriorly. It extends from the pylorus nearly to the right lateral line. The *descending part,* situated medial to the right lateral line,

courses inferiorly from the transpyloric line (L1) to a point midway between the transpyloric and transtubercular lines (between L3 and L4). The *horizontal part* runs transversely, but with a slight upward slope from the end of the descending part to the left of the midline. The *ascending part* is

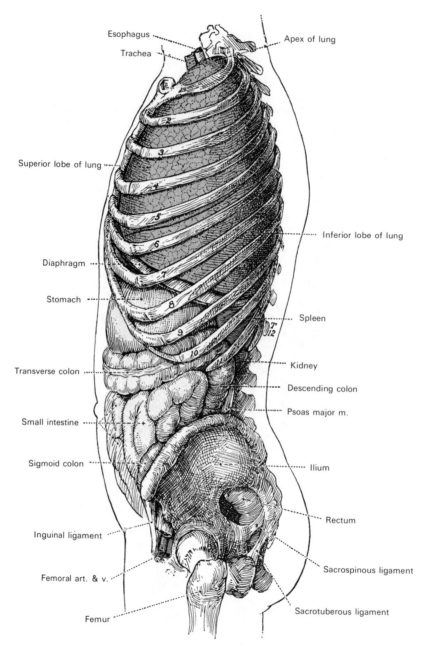

Esophagus
Trachea
Apex of lung
Superior lobe of lung
Inferior lobe of lung
Diaphragm
Stomach
Spleen
Transverse colon
Kidney
Descending colon
Psoas major m.
Small intestine
Sigmoid colon
Ilium
Rectum
Inguinal ligament
Sacrospinous ligament
Femoral art. & v.
Sacrotuberous ligament
Femur

Fig. 3-20. Thoracic and abdominal viscera shown in their normal relationship to the skeleton, from the left side. (Eycleshymer and Jones, 1925.)

vertical, reaches the transpyloric line, and ends at the duodenojejunal flexure, about 2.5 cm to the left of the midline.

Jejunum and Ileum. The coils of small intestine, comprising the **jejunum** and **ileum,** occupy most of the abdomen. Usually the greater part of the jejunum is situated on the left side of the abdomen, while the coils of the ileum are located on the right side and partially within the pelvis. The small intestine terminates at the **ileocolic junction,** which is slightly below and medial to the intersection of the right lateral and transtubercular lines (Fig. 3-21).

Cecum and Vermiform Appendix. The **cecum** is in the right iliac and hypogastric regions. Its position varies with its degree of distention, but a line drawn from the right

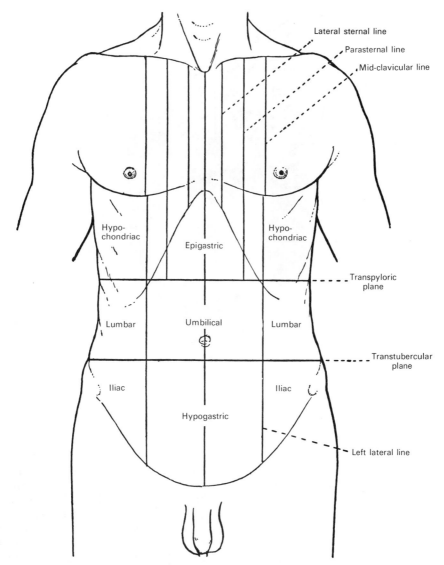

Fig. 3-21. Surface lines on the anterior aspect of the thorax and abdomen. Observe the arbitrary subdivision of the abdomen into nine zones by two transverse planes (transpyloric and transtubercular) and two vertical planes (right and left lateral lines).

anterior superior iliac spine to the upper margin of the symphysis pubis will mark approximately the middle of its lower border. The position of the base of the **appendix** is indicated by a point on the lateral line, level with the anterior superior iliac spine (Fig. 3-22).

Ascending Colon (Figs. 3-17; 3-19; 3-22; 3-26). The ascending colon, in contact with the posterior wall of the abdomen, passes upward from the right iliac region through the right lumbar region, lateral to the right lateral line. It bends abruptly at the **right colic flexure,** which is situated at the inter-

section of the subcostal and right lateral lines. The right costal margin, the right lobe of the liver, and the gallbladder overlie the right colic flexure.

Transverse Colon (Figs. 3-17; 3-20; 3-22; 3-23). The transverse colon is the most mobile part of the large intestine and it crosses the abdomen in the epigastric region. Its lower border is located on a level slightly above the umbilicus, while its upper border lies just below the greater curvature of the stomach (Fig. 3-17).

Descending Colon (Figs. 3-19; 3-20; 3-26). Under cover of the left costal margin and in

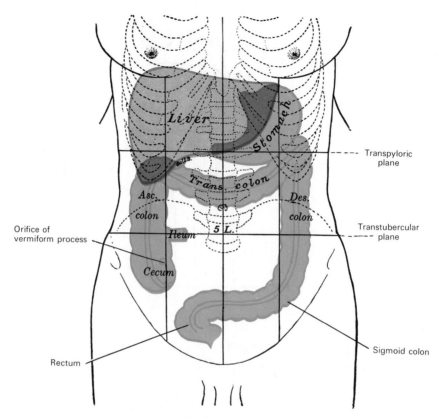

FIG. 3-22. Anterior aspect of abdomen showing the surface projections for the liver, stomach, and large intestine.

relationship to the spleen, the transverse colon turns sharply downward and becomes the descending colon at the **left colic flexure.** This flexure is situated approximately at the intersection of the left lateral and transpyloric lines. The descending colon courses inferiorly through the left lumbar region, lateral to the left lateral line, as far as the iliac crest.

Sigmoid Colon (Figs. 3-17; 3-20). The iliac and pelvic portions of the large intestine form the **sigmoid colon.** The iliac portion is contained in the left iliac fossa, and continues as the pelvic colon at the left lateral line. The sigmoid colon is freely moveable since it has a mesentery, and it becomes the **rectum** anterior to the sacrum in the midline (Figs. 3-19; 3-20; 3-23; 3-27).

Liver (Fig. 3-17; 3-22). The **liver** is located in the right hypochondrium, extending across the midline toward the left into the upper part of the epigastric region. Although the liver moves up and down with respiration, the upper limit of its right lobe, in the midline, is at the level of the junction between the body of the sternum and the xiph-oid process. On the right side the superior border of the liver carries upward as far as the fifth costal cartilage in the midclavicular line. On the left side the liver extends across the midline as far as 7 or 8 cm. Its upper margin is somewhat lower than on the right, reaching the sixth costal cartilage. The inferior extent of the liver can be indicated by a line drawn 1 cm below the lower margin of the thorax on the right side as far as the ninth costal cartilage, then obliquely upward to the eighth left costal cartilage, crossing the midline just above the transpyloric plane. Finally, with a slight left convexity, the inferior border reaches the end of the line, indicating the upper limit at about the fifth left interspace.

Gallbladder. The **gallbladder** is attached to the inferior (visceral) surface of the liver (Figs. 3-17; 3-23). Its *fundus* is directed inferiorly and projects just beyond the lower border of the right lobe of the liver at the ninth right costal cartilage close to the lateral margin of the rectus abdominis.

Pancreas (Fig. 3-24). The **pancreas** lies in front of the second lumbar vertebra. Its *head*

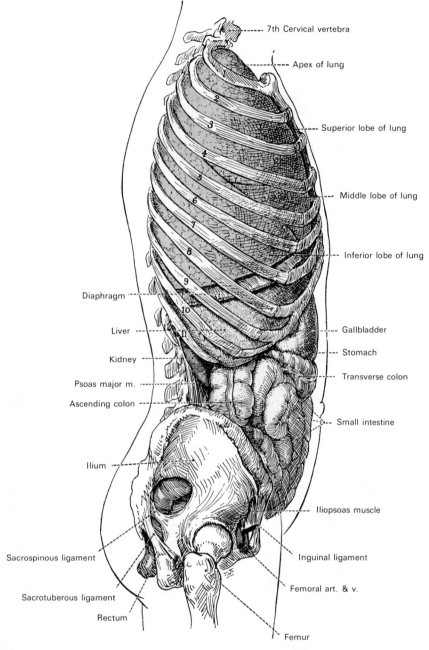

----- 7th Cervical vertebra

---- Apex of lung

---- Superior lobe of lung

---- Middle lobe of lung

---- Inferior lobe of lung

Diaphragm ----

Liver ----

Kidney ----

Psoas major m. ----

Ascending colon ----

--- Gallbladder

--- Stomach

--- Transverse colon

--- Small intestine

Ilium ----

--- Iliopsoas muscle

Sacrospinous ligament ----

Inguinal ligament

Sacrotuberous ligament ----

Femoral art. & v.

Rectum ----

Femur

FIG. 3-23. Thoracic and abdominal viscera shown in their normal relationship to the skeleton from the right side. (Eycleshymer and Jones, 1925.)

occupies the curve of the duodenum and is therefore indicated by the same lines as that viscus; its *neck* corresponds to the pylorus. Its *body* extends along the transpyloric line, the bulk of it lying above this line; the *tail* is in the left hypochondriac region, extending slightly to the left of the lateral line and above the transpyloric line.

Spleen (Figs. 3-15; 3-17; 3-19; 3-20; 3-25). The long axis of the spleen corresponds to that of the tenth rib on the left side. It is situated between the upper border of the ninth and the lower border of the eleventh ribs. It is separated from the ribs by the diaphragm. Its medial end is about 4 cm to the left of the midline of the back. Its lateral extent is the

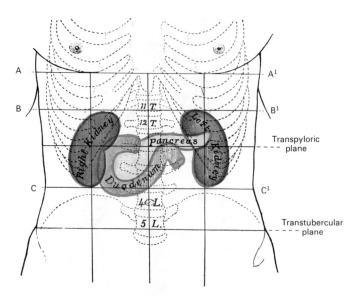

Fig. 3-24. Front of abdomen, showing surface projections for duodenum, pancreas, and kidneys. *AA'*, Plane through the xiphisternal junction. *BB'*, Plane midway between *AA'* and transpyloric plane. *CC'*, Plane midway between transpyloric and transtubercular planes. The vertical lines are the left and right lateral lines between which is the midline.

left midaxillary line at the ninth intercostal space.

Lying obliquely in the left hypochondrium, the superior border of the spleen at times can extend more medially into the epigastrium. The inferior pole of the spleen is generally described as extending to the level of the first lumbar vertebra in the cadaver; however, in the erect, living person, it usually extends significantly lower, to the third lumbar vertebra.

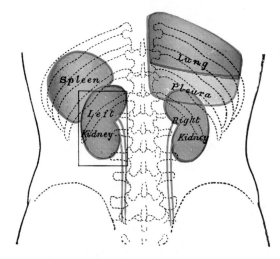

Fig. 3-25. Back of lumbar region, showing surface projections for kidneys, ureters, and spleen. The lower portions of the lung and pleura are shown on the right side.

Kidneys (Figs. 3-19; 3-20; 3-23; 3-24 to 3-27). The **kidneys** are located in the posterior abdomen on each side of the spinal column. The right kidney usually lies about 1 cm lower than the left because of the presence of the liver on the right side, but for practical purposes similar surface markings may be taken for each.

Projected anteriorly, the *upper pole of each kidney* lies along a line midway between the plane of the xiphisternal junction and the transpyloric plane. The upper pole of each kidney is found about 5 cm from the midline (see Plane BB' in Fig. 3-24). The *lower poles* are situated along a plane midway between the transpyloric and intertubercular planes, 7 cm from the midline (see Plane CC' in Fig. 3-24). The *hilum* of each kidney lies along the transpyloric plane, 5 cm from the midline. Around these three points a bean-shaped figure 4 to 5 cm broad is drawn, two thirds of which lies medial to the lateral line. To indicate the position of the kidney from the back, a parallelogram may be used. *Two vertical lines are drawn, the first 2.5 cm, the second 9.5 cm from the midline; the parallelogram is completed by two horizontal lines drawn at the level of the tip of the spinous process of the eleventh thoracic and the lower border of the spinous process of the third lumbar vertebra.* The hilum is 5 cm from the midline at the level of the spinous process of the first lumbar vertebra.

Ureters. The **ureters** are flattened tubes, 28 to 34 cm in length, which extend vertically, one on each side, from the kidneys to the bladder (Figs. 3-26; 3-27). Projected on the anterior aspect of the abdomen, the line of the ureter courses from the hilum of the kidney to the pubic tubercle. Their vertical course can be visualized posteriorly extending from the hilum of the kidney and passing practically across the posterior superior iliac spine (Fig. 3-26).

Vessels (Fig. 3-28). The course of the **inferior epigastric artery** can be approximated to the umbilicus. This line also indicates the lateral boundary of **Hesselbach's triangle,** the other boundaries of this triangle being the lateral border of the rectus abdominis muscle and the medial half of the inguinal ligament. This triangle is of importance in the classification of **inguinal hernias.** There are two types of inguinal hernias, **indirect** (lateral, oblique, or congenital) and **direct** (medial or acquired). An *indirect inguinal hernia* is generally congenital, and the herniation commences lateral to the inferior epigastric artery. The path of the herniation follows the course of the spermatic cord or round ligament of the uterus through the inguinal canal, and hence it does not penetrate through Hesselbach's triangle. A *direct inguinal hernia* is usually due to some underlying weakness or defect in the lower anterior abdominal wall. The herniation occurs medial to the inferior epigastric artery and, therefore, commences through Hesselbach's triangle.

The **abdominal aorta** begins in the midline about 4 cm above the transpyloric line and

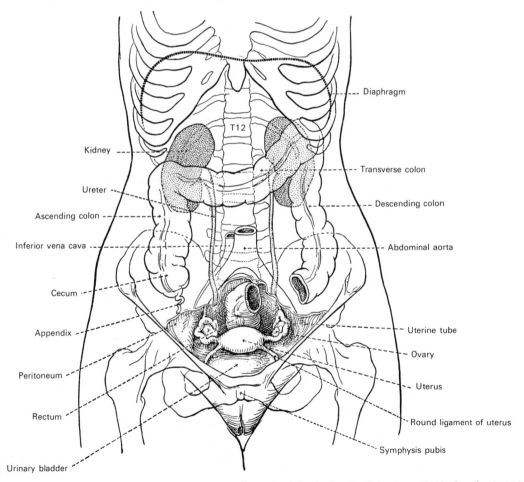

FIG. 3-26. Projection showing the average position of certain abdominal and pelvic viscera in the female. Anterior view. (Eycleshymer and Jones, 1925.)

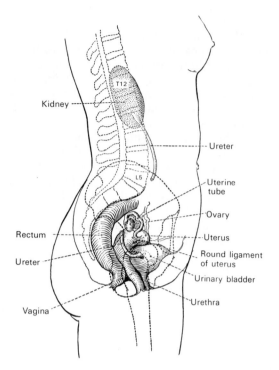

FIG. 3-27. Projection showing the average position of the female pelvic organs. Lateral view. (Eycleshymer and Jones, 1925.)

extends to a point 2 cm below and to the left of the umbilicus (AA', Fig. 3-28). The point of bifurcation of the abdominal aorta corresponds to the level of the fourth lumbar vertebra. A line drawn from that site to a point midway indicates the course of the common iliac artery and one of its branches, the external iliac.

Of the larger branches of the abdominal aorta, the **celiac artery** is 4 cm above the transpyloric line. The **superior mesenteric artery** is 2 cm above and the **renal arteries** are 2 cm below the same line. The **inferior mesenteric artery** is 4 cm above the bifurcation of the abdominal aorta (Fig. 3-28).

Nerves. The skin of the anterior abdominal wall is innervated by anterior and lateral cutaneous branches of spinal nerves **T7** through **L1**. These course around the trunk in a manner that could be represented by lines continuing those of the bony ribs. The termination of the **seventh thoracic nerve** is at the level of the xiphoid process; the **tenth** reaches the vicinity of the umbilicus; the

twelfth ends about midway between the umbilicus and the upper border of the symphysis pubis. The **first lumbar nerve** courses parallel to the thoracic nerves. Its **iliohypogastric** branch becomes cutaneous above the subcutaneous inguinal ring, while its **ilioinguinal** branch becomes cutaneous by penetrating through the ring.

The Perineum

A line drawn transversely between the ischial tuberosities divides the **perineum** into a **posterior** or **rectal triangle** and an **anterior** or **urogenital triangle** (see Fig. 6-38). This line passes through the **central tendinous point** of the perineum, which is situated in the female about 2.5 cm anterior to the center of the anal aperture, or, in the male, midway between the anus and the reflection of the skin onto the scrotum.

RECTAL EXAMINATION (Fig. 3-29). A finger inserted through the anal orifice is compressed by the **external anal sphincter,** passes into the region of the **internal anal sphincter,** and more deeply encounters the resistance of the **puborectalis;** beyond this it may reach the lowest of the **transverse rectal folds.** Anteriorly, the **urethral bulb** and the **membranous urethra** are first identified, and then about 4 cm deep to the anal orifice in the male the **prostate** is felt. Beyond the prostate, the **seminal vesicles,** if enlarged, and the **fundus** of the **bladder,** when distended, are palpable. On either side of the anus is the **ischiorectal fossa.** Posterior to the anus can be felt the **anococcygeal body,** the pelvic surfaces of the **coccyx** and lower end of the **sacrum,** and the **sacrospinous ligaments.**

Additionally, rectal examination in the female allows the posterior wall and **fornix** of the **vagina,** and the **cervix** and **body** of the **uterus** to be felt in front, while somewhat laterally the **ovaries** can just be reached.

SUPERFICIAL MALE UROGENITAL ORGANS. The **penis** and the **testes** within their scrotal sac constitute the male superficial urogenital organs. The penis consists of two **corpora cavernosa penis,** which can be followed backward to their **crura** attached to the sides of the pubic arch, and the **corpus spongiosum penis,** the middle portion, which is tra-

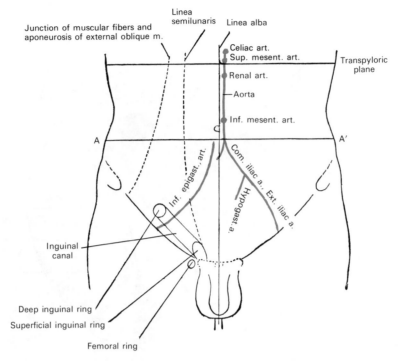

Linea semilunaris

Junction of muscular fibers and aponeurosis of external oblique m.

Linea alba

Celiac art.
Sup. mesent. art.

Transpyloric plane

Renal art.

Aorta

Inf. mesent. art.

Inf. epigast. art.

Com. iliac a., Ext. iliac a.

Hypogast. a.

Inguinal canal

Deep inguinal ring

Superficial inguinal ring

Femoral ring

FIG. 3-28.   The front of the abdomen showing the surface projection of the arteries and the inguinal canal. The line *AA′* is drawn at the level of the highest points of the iliac crests. Observe that the region bounded by the inguinal ligament, the linea semilunaris (lateral border of the rectus abdominis muscle), and the inferior epigastric artery is the triangle of Hesselbach.

versed by the urethra. The distal portion of the corpus spongiosum penis is the **glans penis,** which is covered by the **prepuce.** The external urethral orifice can be examined at the tip of the glans penis, and the course of the urethra can be traced along the undersurface of the penis to the **bulb,** which is situated immediately in front of the central point of the perineum. The **testis** can be palpated through the wall of the **scrotum** on either side. It lies toward the back of the scrotum, and along its posterior border the **epididymis** can be felt. Passing upward along the medial side of the epididymis is the **spermatic cord,** which is palpable as far as the superficial inguinal ring.

SUPERFICIAL FEMALE UROGENITAL ORGANS (Figs. 3-26; 3-27).   The female urogenital structures that are observable superficially include the **labia majora,** the **labia minora,** and the **clitoris.** The labia majora are two prominent and rounded folds of skin covering fatty and areolar tissue which extend from a site anterior to the anus to the **mons pubis,** a soft mound of tissue in front of the symphysis pubis. Between the labia majora are two smaller and more narrow longitudinal folds, the labia minora. In the **pudendal cleft** between the labia minora are the openings of the **vagina** and **urethra.** In the virgin the vaginal opening is partly closed by a thin fold of mucous membrane, the **hymen.** After coitus the remains of the hymen are represented by small rounded elevations called the carunculae hymenales. Between the hymen and the frenulum of the labia minora is the **fossa navicularis,** while in the grooves between the hymen and the labia minora, the small openings of the **greater vestibular** (*Bartholin's*) **glands** can be seen. These glands, when enlarged, can therefore be felt on either side of the posterior part of the vaginal orifice. The clitoris, composed of erectile tissue, is situated in the midline anterior to the urethral orifice. It is palpable at the anterior ends of the labia minora and is an organ homologous to the male penis.

VAGINAL EXAMINATION (Fig. 3-30).   With the

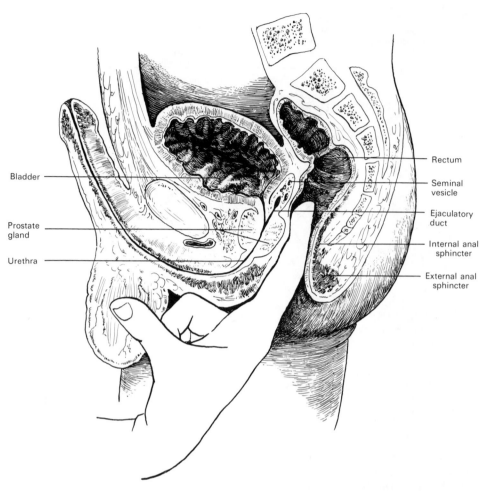

Bladder

Prostate
gland

Urethra

Rectum

Seminal
vesicle

Ejaculatory
duct

Internal anal
sphincter

External anal
sphincter

Fig. 3-29.   The digital rectal examination. With the gloved index finger the anal canal in both sexes can be examined for tumors or other pelvic problems for a distance of about three inches. In the male the prostate gland and seminal vesicle can be palpated anteriorly. In the female the posterior wall of the vagina and the uterine cervix may be explored and by moving the finger laterally, the uterine tubes and ovaries may be felt.

examining finger inserted into the vagina the following structures can be palpated through its posterior wall from below upward: the **anal canal,** the **rectum,** and the **rectouterine excavation.** Projecting into the roof of the vagina is the vaginal portion of the uterine cervix with its external uterine orifice. In front of and behind the cervix the anterior and posterior **vaginal fornices** can be examined. With a finger still in the vagina and with the other hand exerting pressure on the abdominal wall, the entire **cervix** and **body of the uterus** as well as the **uterine tubes** and **ovaries** can be palpated. If a speculum is introduced into the vagina, its walls, the vaginal portion of the cervix, and the external uterine orifice can all be visually examined.

The external urethral orifice lies in front of the vaginal opening, and the angular gap in which the orifice is situated between the two converging labia minora is termed the **vestibule.** The urethral canal in the female is easily dilatable and can be explored with the finger. About 2.5 cm in front of the external orifice of the urethra are the **glans** and **prepuce of the clitoris,** and still farther forward is the **mons pubis.**

The Upper Limb

BONES. The **clavicle** can be felt throughout its entire length (Figs. 3-17; 3-18). It is a curved bone and its extremities are enlarged both medially, where it articulates with the

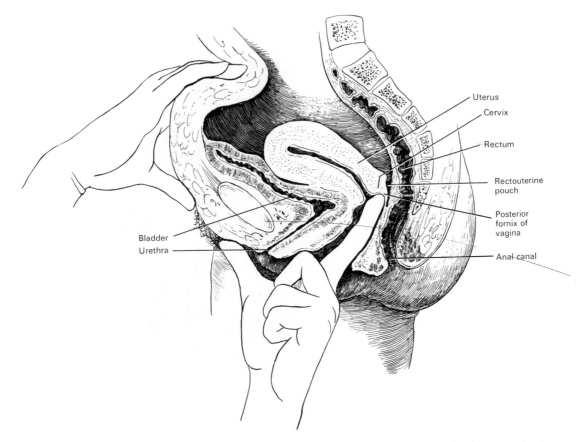

Fɪɢ. 3-30. The digital vaginal examination. Digital palpation of the vaginal walls also allows for direct examination of the uterine cervix, as well as palpation of the uterus, bladder, and urethra anteriorly and the uterine tubes and ovaries laterally.

sternum, and laterally, where it articulates with the **acromion** of the **scapula.** The clavicle is fractured more frequently than any other bone in the body.

The only parts of the **scapula** (Figs. 3-19; 3-31) that are truly subcutaneous are the **spine** and **acromion,** but the **coracoid process,** the **vertebral border,** the **inferior angle,** and to a lesser extent the **axillary border** can also be readily palpated. The acromion and spine are recognizable surface landmarks throughout their entire extent, forming with the clavicle the arch of the shoulder. The acromion forms the point of the shoulder; it joins the clavicle at an acute angle—the acromial angle—slightly medial to, and behind the tip of the acromion. The scapular spine can be felt as a distinct ridge, marked on the surface as an oblique depression which becomes less distinct and ends in a slight dimple a little lateral to the spinous processes of the vertebrae (Figs. 3-11, 3-31). The spine of the scapula helps to define the

muscle-filled **infraspinatus** and **supraspinatus fossae.** Below the spine the contour of the infraspinatus muscle can readily be inspected, whereas above the spine the supraspinatus muscle is covered smoothly by the trapezius (Fig. 3-31). The coracoid process is situated about 2 cm below the junction of the middle and lateral thirds of the clavicle. It is covered by the anterior border of the deltoid, and thus lies a little lateral to the infraclavicular fossa or depression which marks an interval bounded by the pectoralis major, the deltoid, and the clavicle called the **deltopectoral triangle.** Through this triangle the superficially coursing **cephalic vein** penetrates the deep fascia to enter the **axillary vein** just below the level of the clavicle.

The **humerus** (Figs. 3-32, 3-33, 3-34) is almost entirely surrounded by muscles, and the only parts that are strictly subcutaneous are small portions of the **medial** and **lateral epicondyles.** Additionally, however, the **tubercles** and a part of the **head** of the hu-

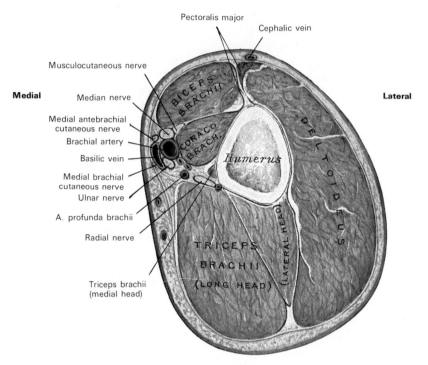

FIG. 3-31. Cross-section, viewed from above, through the right arm at a level immediately below the proximal one-third of the humerus.

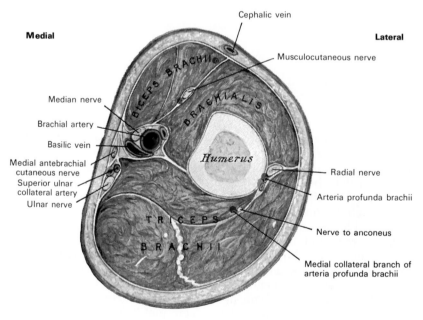

FIG. 3-32. Cross section, viewed from above, through the right arm, a little below the middle of the shaft of the humerus.

Extensor digitorum

Ext. carpi ulnaris

Ext. carpi radialis

Vertebra
prominens

Trapezius

Lateral epicondyle

Biceps
brachii

Triceps, lat.
head

Acromion

Deltoid m.

Spine of scapula

Infraspinatus

Inferior angle of
scapula

Latissimus dorsi

Erector spinae

FIG. 3-33. Posterior view of the back and upper extremities demonstrating surface muscle contours.

merus can be felt under the skin and muscles by which they are covered. The greater **tubercle** forms the most prominent bony point of the shoulder, extending even beyond the acromion. This tubercle, covered by the deltoid muscle, forms the roundness of the lateral aspect of the shoulder. On the distal end of the humerus the medial and lateral epicondyles can be palpated on either side of the elbow joint. Of these, the medial epicondyle is the more prominent, but the **medial supracondylar ridge** passing upward from it is much less marked than the lateral, and as a rule is not palpable.

The most prominent part of the **ulna,** the **olecranon,** can easily be identified behind the elbow joint. The prominent dorsal border of the ulna can be felt along its whole length in the forearm (Figs. 3-35; 3-36), and

the **styloid process** forms a prominent tubercle at the distal end of the ulna. It is continuous above with the dorsal border and ends below in a blunt apex at the level of the wrist joint.

Below the lateral epicondyle of the humerus a portion of the head of the **radius** is palpable. When the forearm is extended, a depression can be observed in the posterolateral aspect of the elbow joint. Rotation of the head of the radius can be palpated in this depression while the subject pronates and supinates the forearm. The upper half of the body of the radius is obscured by muscles (Figs. 3-35; 3-36); the lower half, though not subcutaneous, can be traced downward to the lateral aspect of the wrist joint. Here, a conical projection of bone, the **styloid process** of the radius, can be observed to project

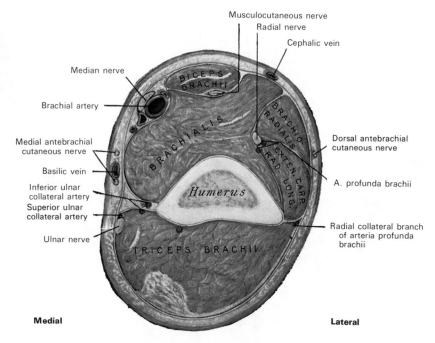

FIG. 3-34. Cross section, viewed from above, through the right arm 2 cm. proximal to the medial epicondyle of the humerus.

5 to 6 mm distal to the styloid process of the ulna.

The radius and ulna articulate distally with the **carpal bones** at the wrist joint. On the front of the wrist are two subcutaneous eminences: one, on the radial side, the larger and flatter, is produced by the adjacent tu-

bercle of the **scaphoid bone** and ridge of the **trapezoid bone;** the other, on the ulnar side, is formed by the **pisiform bone.** The rest of the palmar surface of the carpal bones is covered by tendons and the transverse carpal ligament, and is entirely concealed. An exception is the hamulus (or hook) of the **ha-**

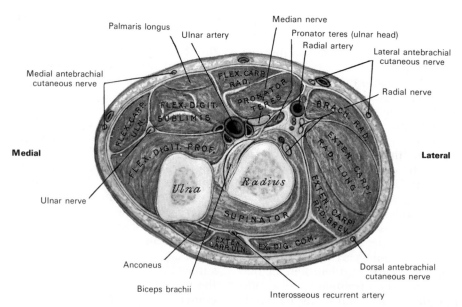

FIG. 3-35. Cross section, viewed from above, through the right forearm at the level of the radial (bicipital) tuberosity.

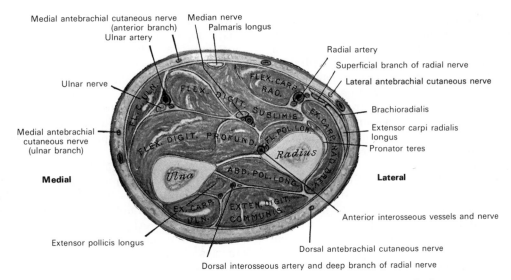

FIG. 3-36.   Cross section, viewed from above, through the middle of the right forearm.

mate bone on the ulnar side of the wrist, which, however, is also somewhat difficult to define. On the dorsal surface of the carpus only the **triquetral bone** can be clearly recognized by palpation.

Distal to the wrist the dorsal surfaces of the **metacarpal bones,** covered by the extensor tendons, are visible only in very thin hands. The dorsal surface of the fifth metacarpal bone is, however, subcutaneous throughout almost its entire length. The heads of the metacarpal bones are their distal extremities, and they can be plainly seen and felt, rounded in contour and standing out under the skin when the fist is clenched, forming the prominences of the knuckles. The head of the third metacarpal is the most prominent.

The enlarged ends of the **phalanges** can be easily felt. When the digits are bent the proximal ends of the phalanges form prominences. In the joints between the first and second phalanges, these prominences are slightly hollow, but they are flattened and square-shaped between the second and third.

ARTICULATIONS.   The **sternoclavicular joint** is subcutaneous, and its position can be identified by the enlarged sternal extremity of the clavicle, lateral to the cord-like sternal head of the sternocleidomastoid. Upon relaxing this muscle, a depression between the end of the clavicle and the sternum can be felt, defining the position of the joint. The sternoclavicular joint is the only articulation between the upper extremity and the trunk.

Posterior to this joint on the right side is found the pleural covering of the right lung along with the right brachiocephalic artery and vein. On the left side, in addition to the pleura, the left brachiocephalic vein, the left common carotid artery, and the thoracic duct underlie this joint.

The position of the **acromioclavicular joint** can generally be ascertained by determining the slightly enlarged acromial end of the clavicle, which usually projects somewhat above the level of the acromion, thereby overriding it. Sometimes this enlargement is considerable, forming a rounded eminence.

The **shoulder joint** is deeply seated and cannot be palpated. It should be appreciated, however, that the **glenoid cavity of the scapula,** into which the spherical **head of the humerus** fits, is exceedingly shallow. This, along with the looseness of the articular capsule, allows for the elegant freedom of movement at that joint. At the same time, however, *the shoulder joint is weakly constructed.* Although it receives some protection from tendons, muscles, and the coracoid and acromial processes and their associated ligaments, *dislocations occur more frequently at this joint than at any other.* These usually are directed initially inferiorly to assume a *subglenoid* position, but then secondarily the head of the humerus may be directed *anteriorly,* beneath the coracoid process of the clavicle, or *posteriorly,* beneath the acromion or spine of the scapula.

The **elbow joint** consists of two articulations: the **humeroulnar,** in which the **trochlea** of the humerus fits into the **trochlear notch** of the ulna, and the **humeroradial,** in which the **capitulum** of the humerus articulates with the **head** of the radius. Behind the curved trochlear notch of the ulna is the thick and prominent **olecranon.** When the forearm is flexed, the olecranon becomes the pointed bony eminence of the elbow. When the forearm is extended, the olecranon fits snugly between the epicondyles of the humerus, forming a continuous line between forearm and arm. If the forearm is slightly flexed, a curved crease or fold with its convexity downward is seen in front of the elbow, extending from one epicondyle to the other; the elbow joint is slightly distal to the center of the fold. The position of the humeroradial joint can be ascertained by feeling for a slight groove or depression between the head of the radius and the capitulum of the humerus, at the back of the elbow joint.

In addition to their articulation with the humerus, the radius and ulna articulate with each other proximally, below the elbow joint, along their shafts by way of the interosseous membrane, and distally, just above the carpal bones. The position of the **proximal radioulnar joint** is marked on the surface at the back of the elbow by a slight depression, which indicates the position of the head of the radius. The site of the **distal radioulnar joint** can be defined by feeling for the slight groove at the back of the wrist between the prominent head of the ulna and the lower end of the radius, when the forearm is pronated.

Of the three transverse skin furrows on the front of the wrist, the middle corresponds fairly accurately with the **wrist joints,** while the most distal indicates the position of the **midcarpal articulation.**

The **metacarpophalangeal** and **interphalangeal joints** are readily available for surface examination. The former are situated just distal to the prominences of the knuckles; the latter are sufficiently indicated by the furrows on the palmar and the wrinkles on the dorsal surfaces of the fingers.

MUSCLES. The large muscles that arise on the trunk and insert on the shoulder girdle form the superficial layers of muscle of the thorax, both on the chest and on the back. The **pectoralis major** (Figs. 3-18; 3-37), **latis-**simus dorsi (Fig. 3-31), and **trapezius** (Figs. 3-11; 3-13; 3-31) muscles are easily visible in most subjects. The **serratus anterior** (Figs. 3-13; 3-37) is mostly concealed by the scapula, but its serrations, interdigitating with the external oblique on the anterolateral aspect of the chest, are usually visible in muscular or thin subjects (Fig. 3-37). The **pectoralis minor** is completely hidden by the pectoralis major.

The **deltoid** (Figs. 3-11; 3-13; 3-31) and the **teres major** (Fig. 3-37) are prominent on the shoulder. The **biceps** (Fig. 3-37), **coracobrachialis,** and **brachialis** are easily distinguishable on the anterior, and the **triceps** (Fig. 3-37) on the posterior aspect of the arm. The muscles of the forearm form two groups: a *medial group,* consisting of the flexors and the pronators arising from the medial epicondyle, and a *lateral group,* comprised of the extensors and supinators arising from the lateral epicondyle. Many tendons of the individual forearm muscles can be identified at the wrist (Fig. 3-37).

The **antecubital fossa,** in front of the elbow joint, is triangular in shape. Its boundaries are a lateral group forearm muscle, the **brachioradialis,** a medial group muscle, the **pronator teres,** and a line joining the medial and lateral epicondyles above. The brachial and supinator muscles form the floor of the fossa, and within it the **brachial artery** and **median nerve** may be palpated. Through the fossa courses the **median cubital vein,** making it readily accessible clinically for intravenous injections or blood withdrawal.

The **synovial tendon sheaths** of the palm and dorsum of the hand cannot be identified except as they follow the various tendons across the wrist and distally in the fingers.

ARTERIES. Above the middle of the clavicle the pulsation of the **subclavian artery** can be detected by pressing downward, backward, and medialward against the first rib. The pulsation of the **axillary artery** as it crosses the second rib can be felt below the middle of the clavicle just medial to the coracoid process. The course of the artery can be followed along the lateral wall of the axilla to the medial border of the coracobrachial muscle. The **brachial artery** can be recognized along practically its entire extent, coursing down the medial margin of the biceps. The **radial artery** becomes superficial over the lower end of the radius, between the styloid process and flexor carpi

Flexor digitorum

Palmaris longus

Biceps

Sternocleidomastoid m.

Triceps, medial head

Triceps, lateral head

Pectoralis major

Teres major

Subscapularis

Serratus anterior

External oblique m.

FIG. 3-37. Anterior view of head and chest and the upper extremities demonstrating surface muscular contours.

radialis. It is used clinically for observations of the pulse (Figs. 3-38; 3-40). The **ulnar artery** also becomes more superficial in its course down the medial aspect of the distal forearm, and at the wrist, the ulnar nerve lies just medial to the artery. Both the ulnar and the radial arteries enter the hand (Fig. 3-40).

In the hand the **superficial palmar arch** (Fig. 3-40), derived mainly from the ulnar artery, can be indicated by a line starting from the radial side of the pisiform bone and curving distalward and lateralward as far as the base of the thumb, with its convexity toward the fingers. The summit of the arch is usually on a level with the ulnar border of the outstretched thumb. The **deep palmar arch** (Fig. 3-40), derived mainly from the radial artery, is practically transverse, and is situated deeper in the hand and somewhat more distal to the superficial palmar arch.

VEINS. The superficial veins of the upper limb are easily rendered visible by compressing the proximal trunks. The **basilic vein** forms from the venous network on the ulnar side of the dorsal hand and then passes up the *medial* aspect of the forearm and arm

to join the more deeply coursing **brachial vein** to form the **axillary vein.** The **cephalic vein** (Fig. 3-13) commences on the radial aspect of the dorsal hand, courses up the *lateral* aspect of the forearm and arm, and passes in the groove between the deltoid and pectoralis major muscles to enter the axillary vein. In front of the elbow, the **median cubital vein** communicates between the cephalic and basilic veins.

NERVES (Figs. 3-38; 3-39). The uppermost trunks of the **brachial plexus** are palpable for a short distance above the clavicle as they emerge from under the lateral border of the sternocleidomastoid; the larger nerves derived from the plexus can be rolled under the finger against the lateral axillary wall, but cannot be identified with certainty at this site.

The **axillary nerve** (Fig. 3-39) emerges from the posterior cord of the brachial plexus. It then courses around the shaft of the humerus under cover of the deltoid muscle.

The **musculocutaneous nerve** (Fig. 3-32) lies deep within the musculature of the

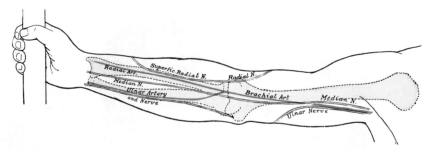

FIG. 3-38.   Anterior aspect of right upper extremity, showing surface projection of bones, arteries, and nerves.

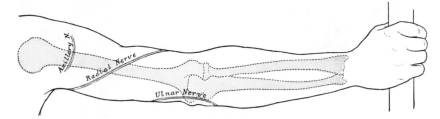

FIG. 3-39.   Posterior aspect of right upper extremity showing surface projection of bones and nerves.

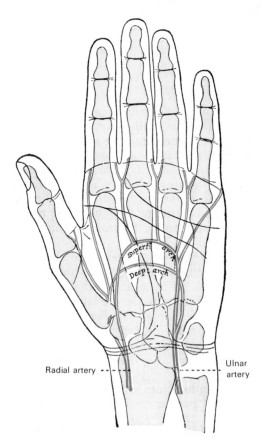

FIG. 3-40.   Palm of left hand, showing position of skin creases and bones, and surface projection for the palmar arterial arches.

flexor compartment in the arm, but becomes superficial about 7.5 cm above the cubital fossa, lateral to the tendon of the biceps brachii.

The line of the **median nerve** (Figs. 3-32; 3-33; 3-38) in the arm is practically the same as that for the brachial artery and may even be injured when pressure is applied to the artery. Anterior to the elbow, the nerve lies medial to the artery in the center of the cubital fossa. Descending the forearm under cover of the anterior forearm muscles, it reaches the wrist lateral to the palmaris longus, and enters the palm of the hand by passing deep to the flexor retinaculum within the carpal tunnel.

The **ulnar nerve** (Figs. 3-34 to 3-36; 3-38) also follows the course of the brachial artery in the upper arm, then diverges and descends in the groove behind the medial epicondyle over the elbow, where it lies just beneath the skin. In the upper forearm it descends along the ulnar side under the flexor carpi ulnaris, while more distally in the forearm it courses with the ulnar artery, with which it enters the ulnar aspect of the palm of the hand superficial to the flexor retinaculum.

The **radial nerve** (Figs. 3-32; 3-33; 3-34; 3-35; 3-38; 3-39) derives from the posterior cord of the brachial plexus and achieves the posterior compartment of the arm by coursing deeply between the medial and long heads of the triceps brachii. Here it is accompanied by the profunda brachii artery along the spiral groove of the humerus. Because of this relationship with the bone, it is liable to injury in fractures of the humerus or may be severely compressed by pressure. Supplying the extensor compartment muscles in the forearm, the radial nerve eventually achieves the dorsum of the hand, which it helps to innervate with sensory fibers.

The Lower Limb

BONES. The **pelvis** or **hip bones** are largely covered with muscle, so that only at a few sites do they lie immediately beneath the skin surface. The highest part of the pelvis on each side is the thick **iliac crest.** This can be easily palpated, except in very obese individuals, and can be traced anteriorly to the **anterior superior iliac spine,** and posteriorly to the **posterior superior iliac spine.** The latter bony spine is indicated by a slight depression. On the outer lip of the crest, about 5 cm behind the anterior superior spine, is the prominent **iliac tubercle.** In thin subjects the **pubic tubercle,** located near the superior lateral border of the symphysis pubis, is very apparent, but in the obese it is obscured by the pubic fat. It can, however, be detected by following the tendon of origin of the adductor longus superiorly. Additionally, the inguinal ligament can be followed from the anterior superior iliac spine to the pubic tubercle. Another part of the bony pelvis that is accessible to touch is the **ischial tuberosity,** situated beneath the gluteus maximus. When the hip is flexed, this tuberosity can easily be felt, since it is then not covered by muscle.

The **femur** (Figs. 3-41 to 3-43) is enveloped by muscles, so that the only palpable parts are the lateral surface of the **greater trochanter** and the lower expanded end of the bone. The greater trochanter can be felt on the upper lateral aspect of the thigh. It is generally indicated by a depression, owing to the thickness of the gluteus medius and gluteus minimus which project above it. The greater trochanter becomes a more prominent protuberance if the thigh is flexed and crossed over the opposite one. At the lower end of the femur the **lateral condyle** is more easily felt than the **medial.** Both **epicondyles** can

be readily identified, and at the summit of the medial epicondyle the sharp **adductor tubercle** can be recognized without difficulty.

The anterior surface of the **patella** is subcutaneous (Fig. 3-44). This sesamoid bone lies within the tendon of the quadriceps femoris muscle, and when the muscle is relaxed, the patella can be moved from side to side. Upon flexion of the leg the patella recedes between the condyles of the femur, but upon extension it lies anterior to the condyles. By its position, the patella helps to protect the knee joint from injury, strengthening the structure of the quadriceps tendon as well.

A considerable portion of the **tibia** or shin bone is subcutaneous (Figs. 3-46; 3-47). At the upper end the **tibial condyles** can be felt just below the knee joint. In front of the upper

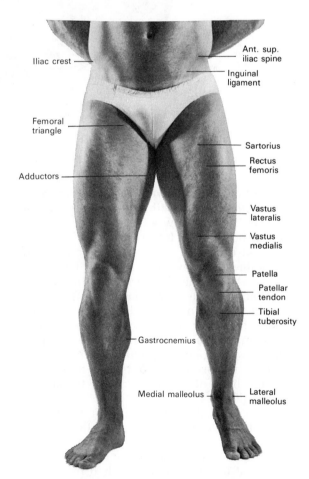

Iliac crest

Ant. sup. iliac spine

Inguinal ligament

Femoral triangle

Sartorius

Rectus femoris

Adductors

Vastus lateralis

Vastus medialis

Patella

Patellar tendon

Tibial tuberosity

Gastrocnemius

Medial malleolus

Lateral malleolus

FIG. 3-41. Anterior view of lower limbs and abdomen demonstrating surface muscular contours.

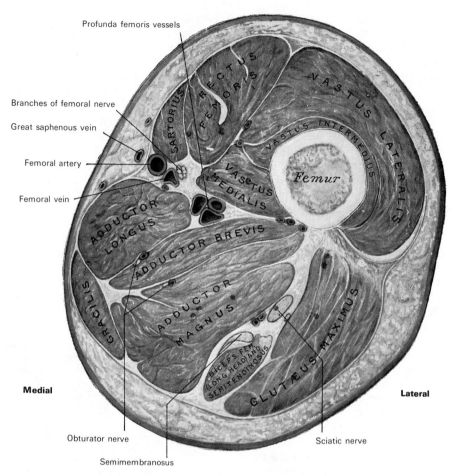

Profunda femoris vessels

Branches of femoral nerve

Great saphenous vein

Femoral artery

Femoral vein

Medial

Lateral

Obturator nerve

Sciatic nerve

Semimembranosus

FIG. 3-42. Cross section through the right thigh, viewed from above, at the level of the apex of the femoral triangle.

end of the bone, between the condyles, is an oval eminence, the **tibial tuberosity,** which is continuous below with the anterior crest of the bone. At the lower end of the bone, the **medial malleolus** (Fig. 3-44) forms a broad prominence, situated at a higher level and somewhat farther forward than the **lateral malleolus** of the **fibula** (Fig. 3-44). The only subcutaneous parts of the fibula are the **head,** the lower part of the **body,** and the **lateral malleolus.** The lateral malleolus is a narrow elongated prominence, from which the lower third or half of the lateral surface of the body of the bone can be traced upward (Fig. 3-44).

It is difficult to distinguish the individual **tarsal bones** on the dorsum of the foot by palpation alone. An exception is the **head** of the **talus,** which forms a rounded projection in front of the ankle joint when the foot is forcibly extended. The whole dorsal surface of the foot has a smooth convex outline, the

summit of which is the ridge formed by the head of the talus, the **navicular,** the **intermediate cuneiform,** and the **second metatarsal** bone. On the medial side of the foot the medial process of the **tuberosity** of the **calcaneus** and the ridge separating the posterior from the medial surface of the bone are distinguishable; in front of this, and below the medial malleolus, is the **sustentaculum tali.** The **tuberosity** of the **navicular** is palpable about 2.5 to 3 cm in front of the medial malleolus.

Farther forward, the ridge formed by the base of the short and thick **first metatarsal bone** can be felt. The head of the first metatarsal is large, and its prominence on the plantar aspect of the foot is called the ball of the large toe. Beneath the base of the first phalanx is the medial **sesamoid bone.** On the lateral side of the foot, in the region of the heel, the most posterior bony point is the lateral process of the tuberosity of the calca-

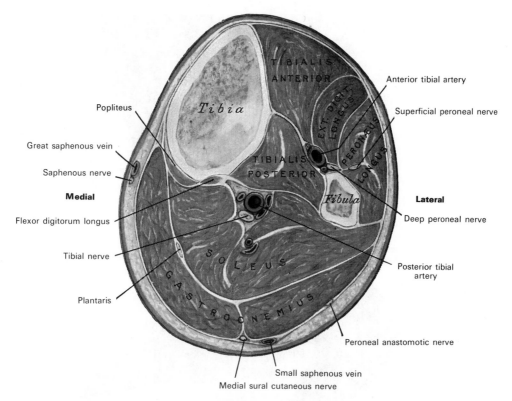

Popliteus

Great saphenous vein

Saphenous nerve

Medial

Flexor digitorum longus

Tibial nerve

Plantaris

T I B I A L I S
A N T E R I O R

Tibia

EXT. DIGIT. LONGUS

TIBIALIS
POSTERIOR

PERONEUS LONGUS

Fibula

S O L E U S

G A S T R O C N E M I U S

Anterior tibial artery

Superficial peroneal nerve

Lateral

Deep peroneal nerve

Posterior tibial artery

Peroneal anastomotic nerve

Small saphenous vein
Medial sural cutaneous nerve

FIG. 3-46. Cross section, viewed from above, through the right leg, 9 cm. distal to the knee joint. Observe the relationship of the tibial nerve with the posterior tibial artery and the deep peroneal nerve with the anterior tibial artery. Note also the superficial peroneal nerve in the lateral compartment.

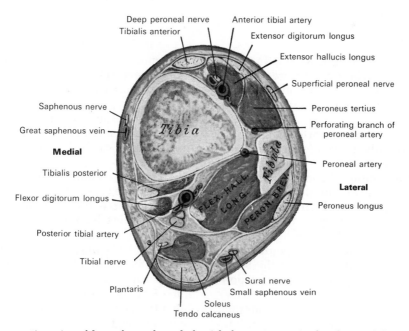

Deep peroneal nerve
Tibialis anterior

Saphenous nerve

Great saphenous vein

Medial

Tibialis posterior

Flexor digitorum longus

Posterior tibial artery

Tibial nerve

Plantaris

Tibia

FLEX. HALL. LONG.

Fibula

PERON. BREV.

Anterior tibial artery

Extensor digitorum longus

Extensor hallucis longus

Superficial peroneal nerve

Peroneus tertius

Perforating branch of
peroneal artery

Peroneal artery

Lateral

Peroneus longus

Sural nerve
Small saphenous vein

Soleus
Tendo calcaneus

FIG. 3-47. Cross section, viewed from above, through the right leg, 6 cm. proximal to the tip of the medial malleolus.

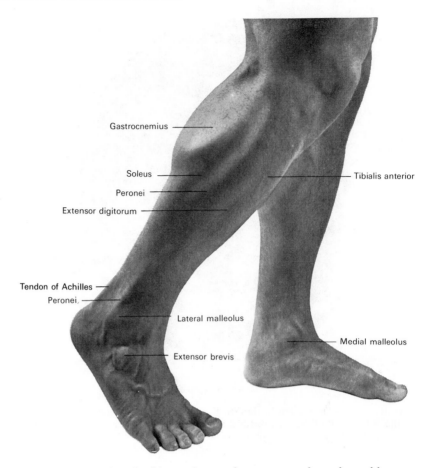

Gastrocnemius

Soleus

Peronei

Extensor digitorum

Tibialis anterior

Tendon of Achilles

Peronei

Lateral malleolus

Medial malleolus

Extensor brevis

FIG. 3-48. Right side of leg to show surface contours of muscles and bones.

gastrocnemius muscle. In this fossa the **popliteal artery** and **sciatic nerve** are covered only by subcutaneous tissue and fat, and through it the **small saphenous vein** joins the **popliteal vein.**

ARTERIES. The **femoral artery** enters the anterior thigh midway between the anterior superior iliac spine and the pubic tubercle, just below the inguinal ligament. Its pulsations can be felt at this point, but they become increasingly difficult to detect as the artery courses inferiorly toward and into the adductor canal. When the leg is flexed at the knee joint, the pulsation of the **popliteal artery,** which is the continuation of the femoral artery, can easily be detected in the popliteal fossa behind the joint.

On the lower part of the front of the tibia, the **anterior tibial artery** becomes superficial (Fig. 3-47) and can be traced over the ankle joint into the dorsal foot as the **dorsalis pedis artery.** This latter vessel can be followed to the proximal end of the first intermetatarsal space. On the back of the leg, the pulsation of the **posterior tibial artery** becomes detectable near the lower end of the tibia, and is easily felt behind the medial malleolus.

VEINS. By compressing the proximal trunks, the venous arch on the dorsum of the foot, together with the **great** and **small saphenous veins** leading from it, is rendered visible.

NERVES. The only nerve of the lower limb that can be located by palpation is the **common peroneal** as it winds around the lateral side of the neck of the fibula. Because of its subcutaneous location and its position adjacent to bone, this nerve is quite vulnerable to injury at this site.

References

(References are listed not only to those articles and books cited in the text, but to others as well which are considered to contain valuable resource information for the student who desires it.)

Anson, B. J. 1966. *Morris' Human Anatomy. A Complete Systemic Treatise.* 12th ed. McGraw-Hill Book Co., New York, 1623 pp.

Bargmann, W., H. Leonhardt, and G. Töndury. 1968. *Lehrbuch und Atlas der Anatomie des Menschen (von) Rauber (und) Kopsch.* 20th ed. Georg Thieme Verlag, Stuttgart, 3 vols.

Basmajian, J. V. 1975. *Grant's Method of Anatomy.* 9th ed. The Williams and Wilkins Co., Baltimore, 654 pp.

Bassett, D. L. 1952–63. *A Stereoscopic Atlas of Human Anatomy.* Sawyer's Inc., Portland, and The C. V. Mosby Co., St. Louis, 23 vols.

Clemente, C. D. 1981. *Anatomy, An Atlas of the Human Body,* 2nd ed., Urban & Schwarzenberg, Baltimore, 387 pp.

Crouch, J. E. 1978. *Functional Human Anatomy.* 3rd ed. Lea & Febiger, Philadelphia, 663 pp.

Eycleshymer, A. C., and T. Jones. 1925. *Hand-Atlas of Clinical Anatomy.* 2nd ed. Lea & Febiger, Philadelphia, 425 pp.

Ferner, H., and J. Staubesand. 1975. *Benninghoff und Goerttler's Lehrbuch der Anatomie des Menschen.* 11th ed. Urban and Schwarzenberg, Munich, 3 vols.

Figge, F. H. J., and W. J. Hild. 1974. *Sobotta's Atlas of Human Anatomy.* 9th English ed. Urban and Schwarzenberg, Munich, 3 vols.

Gardner, E. D., D. J. Gray, and R. O'Rahilly. 1975. *Anatomy, A Regional Study of Human Structure.* 4th ed. W. B. Saunders Co., Philadelphia, 821 pp.

Grant, J. C. B. 1972. *Atlas of Anatomy.* 6th ed. The Williams and Wilkins Co., Baltimore, no pagination, 655 figures.

Hamilton, W. J., G. Simon, and S. G. Hamilton. 1971. *Surface and Radiological Anatomy.* 5th ed. The Williams and Wilkins Co., Baltimore, 397 pp.

Healey, Jr., J. E., and W. D. Seybold. 1969. *A Synopsis of Clinical Anatomy.* W. B. Saunders Co., Philadelphia, 324 pp.

Hollinshead, W. H. 1971. *Anatomy for Surgeons.* 2nd ed. Harper and Row, Hagerstown, Md., 3 vols.

Kaplan, E. B. 1965. *Functional and Surgical Anatomy of the Hand.* 2nd ed. J. B. Lippincott, Philadelphia, 488 pp.

Kiss, F., and J. Szentagothai. 1964. *Atlas of Human Anatomy.* 17th ed. Macmillan Publishing Co., Inc., New York, 3 vols.

Lachman, E. 1971. *Case Studies in Anatomy.* 2nd ed. Oxford University Press, London, 342 pp.

Last, R. J. 1972. *Anatomy, Regional and Applied.* 5th ed. Churchill-Livingstone, Edinburgh, 925 pp.

Lockhart, R. D., G. F. Hamilton, and F. W. Fyfe. 1969. *Anatomy of the Human Body.* 3rd ed. J. B. Lippincott, Philadelphia, 705 pp.

Lopez-Altunez, L. 1971. *Atlas of Human Anatomy.* Translated by H. Monsen, W. B. Saunders Co., Philadelphia, 366 pp.

Pernkopf, E. 1963. *Atlas of Topographical and Applied Human Anatomy.* Edited by H. Ferner. Translated by H. Monsen, W. B. Saunders Co., Philadelphia, 2 vols.

Pansky, B., and E. L. House. 1975. *Review of Gross Anatomy.* 3rd ed. Macmillan Publishing Co., Inc., New York, 508 pp.

Rasch, P. J., and R. K. Burke. 1978. *Kinesiology and Applied Anatomy: The Science of Human Movement.* 6th ed. Lea & Febiger, Philadelphia, 496 pp.

Rehman, I., and N. Hiatt. 1965. *Descriptive Atlas of Surgical Anatomy.* McGraw-Hill, New York, 180 pp.

Romanes, G. J. 1972. *Cunningham's Textbook of Anatomy.* 11th ed. Oxford University Press, London, 996 pp.

Rouviere, H. 1962. *Anatomie Humaine, Descriptive et Topographique.* 9th ed. Rev. by G. Cordier and A. Delmas, Masson, Paris, 3 vols.

Snell, R. S. 1973. *Clinical Anatomy for Medical Students.* Little, Brown and Co., Boston, 909 pp.

Thorek, P. 1962. *Anatomy in Surgery.* 2nd ed. J. B. Lippincott, Philadelphia, 904 pp.

Töndury, G. 1970. *Angewandte und Topographische Anatomie.* 4th ed. Georg Thieme Verlag, Stuttgart, 636 pp.

Warfel, J. H. 1984. *The Head, Neck and Trunk.* 5th ed. Lea & Febiger, Philadelphia, 128 pp.

Warfel, J. H. 1984. *The Extremities.* 5th ed. Lea & Febiger, Philadelphia, 124 pp.

Wolf-Heidegger, G. 1972. *Atlas of Systemic Human Anatomy.* Hafner, New York, 3 vols. in one, 635 pp.

Woodburne, R. T. 1973. *Essentials of Human Anatomy.* 5th ed. Oxford University Press, New York, 629 pp.

Osteology

The study of bones is called **osteology.** Bones provide a framework for the body, and when assembled in their proper position, they form the **skeleton** (Fig. 4-1). In the living body, bones are frequently held in position by strong fibrous bands, the **ligaments,** which form articulations or **joints.** Attached to bones across the joints are most of the **muscles** of the body, and muscular contraction results in the movement of bones and their associated body parts. In addition to providing internal support for the body, the skeleton protects the vital organs of the head and chest by forming the cranial and thoracic cavities. Because of its large mineral content, bone is a hard unyielding tissue and is ideally suited for the mechanical functions of support, protection, and movement. In life it is also a dynamic substance and is importantly involved with other tissues and organs in the metabolism of calcium. The interior of bones contains the **bone marrow,** which consists of the blood-forming, hemopoietic tissue.

In the skeleton of the adult there are 206 distinct bones, as follows:

Axial Skeleton	Vertebral column	26	
	Skull	28	
	Hyoid bone	1	
	Ribs and sternum	25	
			80
Appendicular Skeleton	Upper extremities	64	
	Lower extremities	62	
			126
Total			206

The patellae are included in this enumeration, but the smaller sesamoid bones are not; also, the sternum is considered as one bone. Bones can be classified into four general types according to their shape and form. **Long bones,** such as the humerus and femur, are found in the limbs (Fig. 4-1), whereas **short bones,** such as those in the hand and foot, are more compact and subjected to more limited movement. Other bones, illustrated by the parietal bones of the skull or the scapulae (Fig. 4-1), are **flat bones,** while **irregular bones** form the vertebral column (Fig. 4-1), as well as other structures.

Ossification and Skeletal Development

The skeleton, like some other structures, is derived from **mesoderm,** which migrates from the primitive streak during the formation of the three primary germ layers. Its cells multiply, undergo certain changes, move into appropriate regions, and form cellular aggregates or sheets which can be recognized as the **membranous skeleton.** The cells of these primordia, except in certain parts of the cranium, soon elaborate an intercellular substance that converts the tissue into cartilage. After the **cartilaginous skeleton** is well established, centers of ossification appear which gradually spread to form the **bony skeleton.** In some of the cranial bones ossification centers arise in the membranous skeleton without the intervention of cartilage.

OSSIFICATION

The term **ossification** refers to the formation of bone. In the development of certain bones, such as those forming the roof and sides of the skull, bone formation occurs directly within clusters of primitive mesenchyme cells surrounded by a more definitive connective tissue membrane. In other bones, such as those in the limbs, the formation of bone occurs within rods of pre-existing cartilage. Hence, two forms of ossification are described: **intramembranous** and **endochondral.** The latter type is sometimes called intracartilaginous.

INTRAMEMBRANOUS OSSIFICATION. In the case of bones that are developed in membrane, no cartilaginous mold precedes the

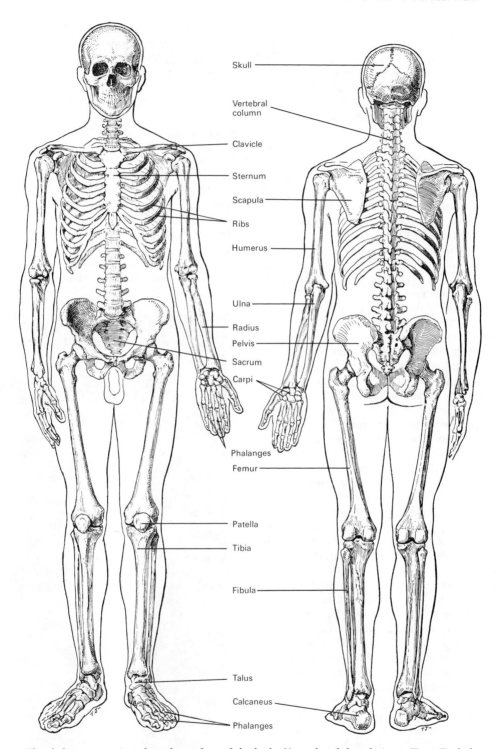

Skull
Vertebral column
Clavicle
Sternum
Scapula
Ribs
Humerus
Ulna
Radius
Pelvis
Sacrum
Carpi
Phalanges
Femur
Patella
Tibia
Fibula
Talus
Calcaneus
Phalanges

FIG. 4-1. The skeleton as projected on the surface of the body. Ventral and dorsal views. (From Eycleshymer and Jones.)

appearance of bony tissue. The connective tissue membrane, which surrounds the centers of ossification, persists and ultimately forms the **periosteum;** it is composed of fibers and granular cells in a matrix. The peripheral portion is more fibrous, while in the interior the primitive mesenchymal cells differentiate into **osteoblasts;** the whole tissue is richly supplied with blood vessels. At the outset of the process of bone formation, a little network of eosinophilic extracellular spicules is noticed forming between the branching vascular elements. At their growth points, these rays consist of a network of fine clear fibers and osteoblast cells with an intervening ground substance. The fibers are termed **osteogenic fibers,** and are made up of fine collagen fibrils differing little from those of other connective tissue. The membrane soon assumes a dark and granular appearance from the deposition of calcium-containing crystals (hydroxyapatite) in the fibers and in the intervening matrix. As the bony trabeculae enlarge, some of the osteoblasts become surrounded by the newly deposited calcified material. With the deposition of mineral salts, the newly formed osteoid tissue again assumes a more transparent appearance, and the fibrils are no longer so distinctly seen. The involved osteoblasts form the corpuscles of the future bone, the spaces in which they are enclosed constituting the **lacunae.** As the osteogenic fibers grow toward the periphery, they continue to ossify and give rise to fresh bone spicules. Thus a network of woven bone is formed, the meshes of which contain blood vessels and a delicate connective tissue crowded with osteoblasts. The bony trabeculae thicken by the addition of fresh layers of bone formed by the osteoblasts on their surface. Subsequently, successive layers of bony tissue are deposited under the periosteum and around the larger vascular channels, which become the Haversian canals as the bone increases in thickness. The entrapped osteoblasts of developing bone become the **osteocytes** of mature bone, and these cells may well be importantly involved in the release of calcium from bone to blood.

ENDOCHONDRAL OSSIFICATION. Ossification in long bones of the extremities, in portions of the basal bones of the skull, and in the vertebrae, the sternum, the ribs, and the pelvis occurs in masses of pre-existing developed hyaline cartilage, which present a model form of the fully ossified bone. In a long bone, which may be taken as an example (Fig. 4-2), the process commences in the center of the **diaphysis,** or **shaft,** and proceeds toward the extremities, called the **epiphyses,** which for some time remain cartilaginous. Subsequently a similar ossification process commences in one or more sites of the epiphyses and gradually extends through them. The epiphyses do not, however, become joined to the shaft of the bone by bony tissue until growth has ceased; between the diaphysis and either epiphysis a layer of cartilaginous tissue termed the **epiphyseal cartilage** persists for a definite period.

The *first stage* in the ossification of cartilage takes place at the **center of ossification,**

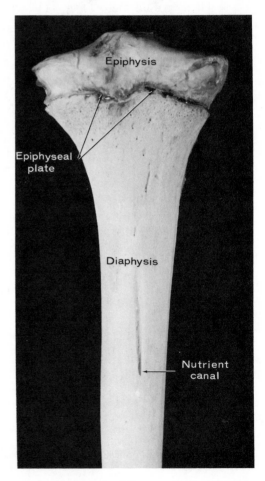

FIG. 4-2. Photograph of the upper half of the tibia of young girl, showing the proximal bony epiphysis, the cartilaginous epiphyseal plate, and the shaft or diaphysis. (From Bloom and Fawcett, *A Textbook of Histology,* W. B. Saunders Co., 1975.)

where cartilage cells enlarge and arrange themselves in rows (Fig. 4-3). The matrix in which they are embedded increases in quantity, so that the cells become separated from each other. Deposits of calcium-containing crystals now accumulate in this matrix, between the rows of cells, so that the cartilage cells become separated from each other by longitudinal columns of calcified matrix which appear granular and opaque. At some sites the matrix between two cartilage cells of the same row also becomes calcified, and transverse bars of calcified substance stretch across from one calcified column to another. Thus, there are longitudinal groups of cartilage cells enclosed in oblong cavities, the walls of which consist of calcified matrix and effectively cut off all nutrition from the cells. In consequence, the cells atrophy, leaving spaces called the **primary areolae.**

At the same time that this process occurs in the center of the solid bar of cartilage, certain changes are taking place on its surface. This is covered by a highly vascular membrane, the **perichondrium,** entirely similar to the embryonic connective tissue already described as constituting the basis of membrane bone. On its inner surface, that is, on the surface in contact with the cartilage, are gathered rows of osteoblasts. These cells form a thin layer of bony tissue, called a periosteal bone collar, between the perichondrium and the cartilage by the *intramembranous* mode of ossification already described. There are, then, in this first ossification stage of cartilaginous bone, two processes going on simultaneously: in the center of the cartilage, the formation of a number of oblong spaces, surrounded by calcified matrix and containing the withered cartilage cells, and on the surface of the cartilage, the formation of a layer of true membranous bone.

The *second stage* of intracartilaginous ossification (Fig. 4-3) occurs when blood vessels perforate the periosteal bone collar to invade the deeper osteogenic regions of the cartilage. Additionally, **osteoclasts,** or bone destroyers, make their appearance. These giant cells seem associated with the absorption of osteogenic matrix, possibly by their secretion of hydrolytic enzymes. Osteoclasts can be observed occupying minute chambers called **Howship's lacunae,** which, by erosion, they may have created themselves on the bone surface. There is evidence that osteoclasts are derived from the fusion of

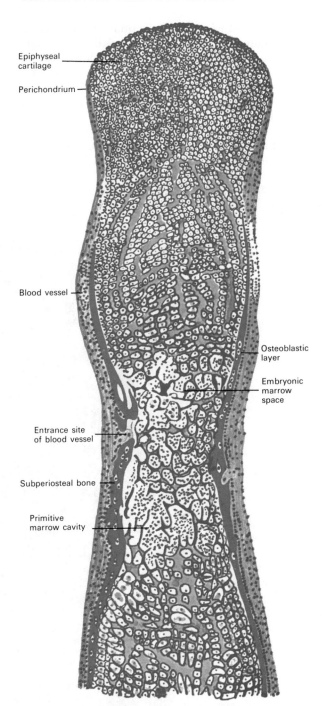

FIG. 4-3. Osteogenesis. Subperiosteal stage of ossification in an embryonic long bone. 20 ×. (Redrawn from Sobotta.)

mononuclear osteogenic cells, yet some may come from mononuclear cells in the blood such as monocytes. Regardless of their origin, as they come into contact with the calcified walls of the primary areolae, osteoclasts cause a fusion of the original cavities, which creates larger spaces termed the secondary areolae or **medullary spaces.** These secondary spaces become filled with embryonic marrow, osteoblasts, and vessels, derived from the osteogenic layer of the periosteum (Fig. 4-4).

Traced thus far has been the formation of enlarged medullary spaces, the perforated walls of which are still formed by calcified cartilage matrix, containing an **embryonic marrow** derived from the osteogenic layer of periosteum, and consisting of blood vessels

and osteoblasts. At this time the walls of these secondary areolae are of only inconsiderable thickness, but they become thickened by the deposition of layers of true bone on their surface. This process takes place in the following manner: some osteoblasts in the embryonic marrow, after undergoing rapid division, arrange themselves in an epithelioid manner on the surface of the wall of the space. This layer of osteoblasts forms a bony stratum, and thus the wall of the space becomes gradually covered with a layer of true osseous substance in which some of the

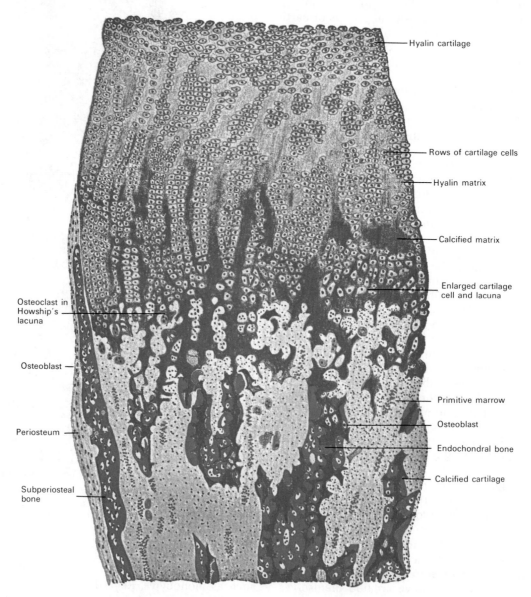

FIG. 4-4.   Endochondral bone formation proceeding from the center of the shaft of a long bone toward one of the two extremities. 40 ×. (Redrawn from Sobotta.)

bone-forming cells are included as bone corpuscles. The next step in the process consists in the removal of these primary bone spicules by the osteoclasts. One of these giant cells may be found lying in a Howship's lacuna at the free end of each spicule. The removal of the primary spicules goes on *pari passu* with the formation of permanent bone by the periosteum, and in this way the medullary cavity of the body of the bone is formed.

Commencing in the center of the diaphysis of the cartilaginous bony model, this series of changes gradually proceeds toward the extremities of the bone, so that in a specimen of ossifying bone all the aforementioned changes may be seen at different sites, from true bone at the center of the shaft to hyaline cartilage at the extremities.

While the ossification of the cartilaginous body is extending toward the articular ends, the cartilage immediately in advance of the osseous tissue continues to grow until the length of the adult bone is reached.

During the period of growth the articular end, or epiphysis, remains for some time entirely cartilaginous, then a bony center appears within it, initiating at this site, as well, the process of intracartilaginous ossification. This is a **secondary center of ossification.** The epiphysis remains distinct from the ossified diaphysis by a narrow cartilaginous zone called the **epiphyseal plate.** This, too, ultimately ossifies, and the histologic distinction between diaphysis and epiphysis is obliterated, the bone assuming its completed form and shape. Similar events also occur in bony processes that ossify separately, such as the trochanters of the femur. The bones continue to grow until the body has acquired its full stature, increasing in length by ossification toward the epiphyses and in circumference by the deposition of new bone from the deeper inner layer of the periosteum. At the same time bone resorption takes place from within, thereby enlarging the medullary marrow cavities. A review of the stages in the development of a typical long bone is presented in the diagrams of Figure 4-5.

Permanent bone, when first formed by the periosteum, is spongy or **cancellous** in structure. Later, as osteoblasts contained in its spaces become arranged in the concentric layers characteristic of the **Haversian systems,** the bone assumes the more solid structure of **compact** bone.

The number of ossification centers varies in different bones. In most of the short bones, ossification commences at a single central site, and proceeds toward the surfaces. In long bones there is a central point of ossification for the shaft or diaphysis, and one or more for each extremity, or epiphysis. That for the diaphysis is the first to appear. The times of union of the epiphyses with the diaphysis vary inversely with the dates at which their ossifications began (with the exception of the fibula). This process also influences the direction of growth of the nutrient arteries of the bones. Thus, the nutrient arteries of the bones of the arm and forearm are directed toward the elbow, since the epiphyses at this joint become united to the diaphyses before those at the opposite extremities at the shoulder and wrist. In the lower limb, on the other hand, the nutrient arteries are directed away from the knee: that is, upward in the femur and downward in the tibia and fibula. In these bones it is observed that the upper epiphysis of the femur, and the lower epiphyses of the tibia and fibula unite first with the diaphyses. Where there is only one epiphysis, the nutrient artery is directed toward the opposite end of the bone; for example, toward the acromial end of the clavicle, toward the distal ends of the metacarpal bone of the thumb and the metatarsal bone of the great toe, and toward the proximal ends of the other metacarpal and metatarsal bones.

SKELETAL DEVELOPMENT

VERTEBRAL COLUMN. The central axis of the embryo around which the vertebrae develop is the notochord. This rod of cells lies in the median line between the neural tube and the primitive gut, extending from the level of the midbrain to the end of the tail. The somites in the paraxial mesoderm on either side of the neural tube give rise to three primordia, the **sclerotome, myotome,** and **dermotome.** The cells of the sclerotome eventually form the vertebrae, as well as other structures. During the fourth embryonic week these cells multiply rapidly and migrate toward the notochord, forming a mesenchymal layer on either side. The **segmental primordia** of the individual vertebrae are still recognizable, *but soon each sclerotome differentiates into a densely arranged cau-*

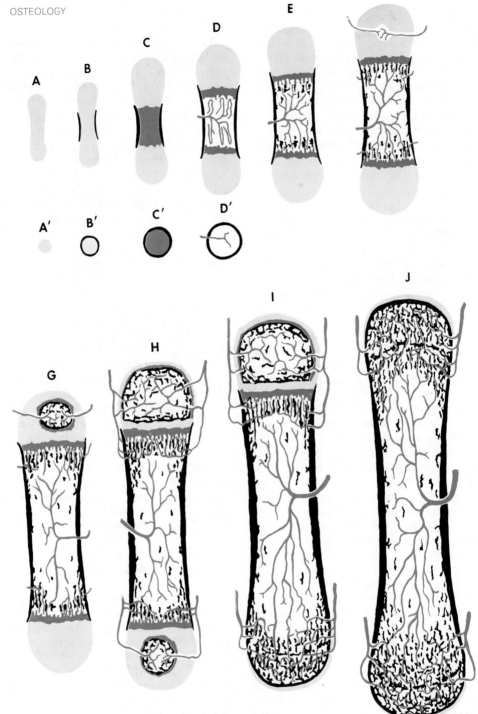

FIG. 4-5. Diagram of the development of a typical long bone as shown in longitudinal sections (*A* to *J*) and in cross sections *A'*, *B'*, *C'*, and *D'* through the centers of *A*, *B*, *C*, and *D*. Pale blue, cartilage; purple, calcified cartilage; black, bone; red, arteries. *A*, Cartilage model. *B*, Periosteal bone collar appears before any calcification of cartilage. *C*, Cartilage begins to calcify. *D*, Vascular mesenchyme enters the calcified cartilage matrix and divides it into two zones of ossification, *E*. *F*, Blood vessels and mesenchyme enter upper epiphyseal cartilage and the epiphyseal ossification center develops in it, *G*. A similar ossification center develops in the lower epiphyseal cartilage, *H*. As the bone ceases to grow in length, the lower epiphyseal plate disappears first, *I*, and then the upper epiphyseal plate, *J*. The bone marrow cavity then becomes continuous throughout the length of the bone, and the blood vessels of the diaphysis, metaphyses, and epiphyses intercommunicate. (From Bloom and Fawcett, *A Textbook of Histology,* W. B. Saunders Co., 1975).

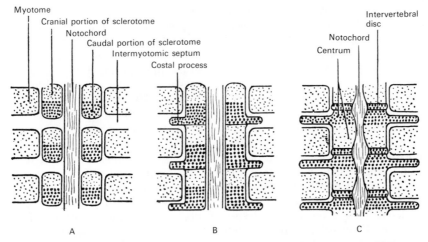

FIG. 4-6. Diagrams showing the manner in which each vertebral centrum is developed from portions of two adjacent somites or body segments. *A,* The segmental structure of the sclerotome differentiates into a more dense caudal portion and a more loosely arranged cranial portion around the notochord. *B,* The lateral and somewhat cranial migration of some of the densely arranged cells forms a costal process laterally and the primordium of the intervertebral disc cranially as seen in *C. C,* Note the disc formed by densely arranged cells and the centrum of the vertebrae formed by parts of two adjacent sclerotomes.

dal portion and loosely arranged cranial portion (Fig. 4-6,*A*). The ventral portion of the densely arranged cells then migrates laterally and cranially (Fig. 4-6,*B*). Upon reaching the level of the center of their original somite, these cells spread back to the midline around the notochord. The **dense portion** becomes the future **intervertebral disc,** and the **loosely arranged portion** and the adjacent dense portion left in close juxtaposition combine to become the future **vertebral body** or **centrum** (Fig. 4-6,*C*). Similarly, the dorsal portion of the sclerotome around the neural tube forms the future **vertebral arch** from its loosely arranged cells, and ligaments from its dense portion. Since the primordia of each vertebra come from two

sclerotomes, they are intersegmental in origin, in contrast to the intervertebral discs, which are segmentally derived. The most ventral extension of the sclerotome forms the future **costal process** from its looser portion and ligaments from the dense portion (Figs. 4-6,*B* and *C;* 4-7,*A*).

This **mesenchymal stage** is succeeded by that of the **cartilaginous vertebral column** (Fig. 4-7). In the sixth week two cartilaginous centers make their appearance, one on each side of the notochord; these extend around the notochord and form the *body* of the cartilaginous vertebra. A second pair of cartilaginous foci appear more dorsally in the mesenchyme surrounding the neural tube to form the cartilaginous *vertebral arch,* and a

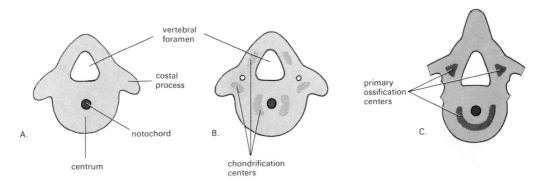

FIG. 4-7. Diagrams illustrating the stages of vertebral development. *A,* Precartilaginous (mesenchymal) vertebra at five weeks. *B,* Chondrification centers in a mesenchymal vertebra at six weeks. *C,* Primary ossification centers in a cartilaginous vertebra at seven weeks. (From Moore, K. L., *The Developing Human,* W. B. Saunders Co., 1973.)

separate cartilaginous center appears for each *costal process*. Somewhat after the second month the cartilaginous arch has fused with the body, and in the fourth month the two halves of the arch are joined on the dorsal aspect of the neural tube. The *spinous process* develops from the junction of the two halves of the neural or vertebral arch. The *transverse process* grows out from the vertebral arch dorsal to the costal process.

In the more cranial cervical vertebrae a band of mesodermal tissue connects vertebral ends of the costal processes, spanning the ventral aspects of the centra and the intervertebral discs. This is termed the **hypochordal bar** or **arch**; in all except the first vertebra, it is transitory and disappears by fusing with the discs. In the atlas, however, the entire bar persists and undergoes chondrification; it develops into the ventral arch of the atlas, while the cartilage representing the body of that bone forms the **dens** or odontoid process, which then fuses with the body of the second cervical vertebra, the **axis.**

The portions of the notochord that are surrounded by the bodies of the vertebrae atrophy, and ultimately disappear, while those that lie in the centers of the intervertebral discs undergo enlargement, and persist throughout life as part of the central **nucleus pulposus** of the discs.

THE RIBS. The **ribs** are formed from the ventral or costal processes of the primitive vertebrae which extend between the muscle plates (Fig. 4-6,*B* and *C*). In the *thoracic region* of the vertebral column the costal processes grow lateralward to form a series of precartilaginous arches, the **primitive costal arches.** As already described, the transverse process grows out dorsal to the vertebral end of each arch. Initially the transverse process is connected to the costal process by continuous mesoderm, but the interconnecting mesoderm cells become differentiated later to form the costotransverse ligament. Between the costal process and the lateral tip of the transverse process, the costotransverse joint is formed by absorption. The costal process becomes separated from the vertebral centrum by the development of the costocentral joint. In the *cervical vertebrae* (Fig. 4-8) the transverse process forms the posterior boundary of the foramen transversarium, while the costal process, corresponding to the head and neck of a thoracic rib, fuses with the body of the vertebra and forms the anterolateral boundary of the foramen. The distal portions of the primitive costal arches remain undeveloped. Occasionally the arch of the seventh cervical vertebra undergoes greater development, forming an accessory **cervical rib** complete with costovertebral joints. A cervical rib (which may even be bilateral) may compress the trunks of the brachial plexus and the subclavian vessels, resulting in symptoms in the upper extremity. In the *lumbar region* (Fig. 4-8) the distal portions of the primitive costal arches fail to develop, while the proximal portions fuse with the true transverse processes to form the "transverse processes" of descriptive anatomy. Occasionally movable ribs may develop in connection with the first lumbar vertebra. In the *sacral region* costal processes are developed only in connection with the upper three or four vertebrae; the processes of adjacent sacral segments fuse

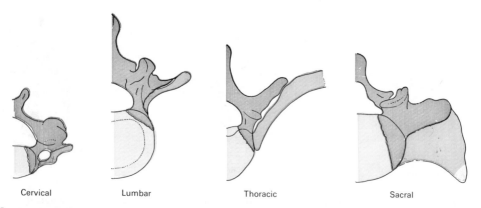

Cervical   Lumbar   Thoracic   Sacral

FIG. 4-8. Diagrams showing the portions of the adult vertebrae derived respectively from the bodies (yellow), vertebral arches (red), and costal processes (blue) of the embryonic vertebrae.

with one another to form the lateral masses of the sacrum. The *coccygeal vertebrae* are devoid of costal processes.

THE STERNUM. The **sternum** arises from a pair of mesenchymal bands in the ventrolateral body wall. Although the sternum develops separately from the ribs, on each side the ventral ends of the upper six or seven ribs become joined to these longitudinal bars of mesenchyme, which by this time are called **sternal plates.** These plates become cartilaginous and migrate ventrally toward the midline. The sternal plates of the two sides then fuse in a craniocaudal direction in the midline to form the manubrium and body of the sternum. The xiphoid process is formed by a caudal extension of the sternal plates.

THE SKULL. The bones of the skull, like those elsewhere in the body, are derived from mesenchyme. The cranial vault grows around the developing brain and is therefore referred to as the **neurocranium.** Additionally, other bones in the head are derived from the branchial arches and the term **viscerocranium** is applied to these.

The mesenchymal membranes that form the neurocranium in some regions become transformed into bone by endochondral ossification and in other sites by way of intramembranous ossification. Thus, reference is frequently made to the **cartilaginous neurocranium** or **chondrocranium,** which includes the bones at the base of the skull, and the **membranous neurocranium,** which includes certain flat bones of the skull.

The first observable indications of the developing skull are found in the basal occipital and basal sphenoid regions and around the auditory vesicles. Condensation of the primordial mesodermal cells from which the skull will be formed gradually extends from these basal areas completely around the brain until the latter becomes enclosed by a **membranous cranium,** also called the **blastemal skull** or **desmocranium.** The membranous cranium is incomplete in the regions where large nerves and vessels pass into or out of the cranium.

Cartilaginous Neurocranium. Before the membranous cranium is even complete, chondrification begins to show in the basilar part of the developing occipital bone (basioccipital). It should be appreciated that chondrification centers develop in relation to the notochord, which extends into the developing cranium from spinal levels (Fig.

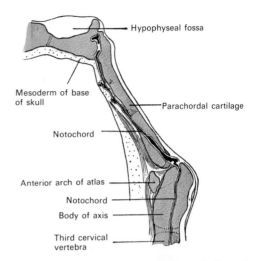

FIG. 4-9. Sagittal section through the cephalic end of the developing skeleton showing that the notochord extends cranially as high as the basisphenoid. Note that above the atlas the rostral and caudal parts of the notochord become surrounded by the parachordal cartilage, while its middle portion lies between the parachordal cartilage and the wall of the pharynx. (From Gaupp.)

4-9). Two centers appear, one on each side of the notochord near its entrance into the condensed mesoderm of the occipital blastema. Chondrification gradually spreads from these centers—medially around the notochord **(parachordal cartilages),** laterally about the roots of the hypoglossal nerve, and cranialward to unite with the spreading cartilaginous center of the developing basal sphenoid bone **(hypophyseal cartilages).** This process forms an elongated **basal plate** of cartilage extending from the foramen magnum caudally to the cranial end of the sphenoid. It continues into the blastema of the ethmoid region, which later also becomes chondrified **(trabeculae cranii).** Other smaller cartilages lateral to the hypophyseal cartilages unite with the main continuous cartilaginous mass. The **ala orbitalis (orbitosphenoid)** forms the lesser wing of the sphenoid bone, and the **ala temporalis (alisphenoid)** forms the greater wing of the sphenoid bone (Fig. 4-10). When the auditory capsules begin to chondrify, they are quite widely separated from the basal plate. By the time the embryo is 20 mm in length the cochlear portion of the auditory or otic capsule **(periotic capsule)** is fused to the widened basal plate, and the jugular foramen has become separated from the foramen lacerum (Fig. 4-11). From the lateral region

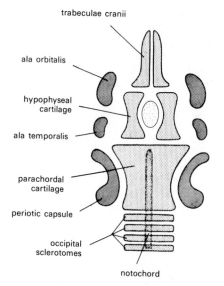

trabeculae cranii

ala orbitalis

hypophyseal cartilage

ala temporalis

parachordal cartilage

periotic capsule

occipital sclerotomes

notochord

Fig. 4-10. Diagram showing the different components that play a role in the formation of the base of the skull or chondrocranium. (From Langman, Williams and Wilkins, 1975.)

of the occipital cartilage a broad thin plate of cartilage **(tectum posterius** or **nuchal plate)** extends around the caudal region of the brain in a complete ring (Fig. 4-12), forming the primitive foramen magnum. The complete chondrocranium or cartilaginous neurocranium is shown in Figure 4-12.

Although the chondrocranium forms only a small part of the future ossified skull, it does give rise to most of the base of the skull. Ossification centers then develop in the cartilage and give rise to all of the **occipital bone** (except the upper part of its squama), to the petrous and mastoid portions of the **temporal bone,** to the **sphenoid bone** (except its medial pterygoid plates and part of the temporal wings), and to the **ethmoid bone.**

Membranous Neurocranium. The membranous neurocranium consists of those bones of the cranial vault and face which ossify directly in the mesoderm of the membranous cranium. They comprise the upper part of the squama of the occipital bone (interparietal), the squamae and tympanic parts of the temporal bones, the parietal bones, the frontal and vomer bones, the medial pterygoid plates of the sphenoid bone, and the bones of the face. Some membranous bones remain as distinct bones throughout life, e.g., parietal and frontal, while others join with bones of the chondrocranium, e.g., the squama of the occipital, squamae of temporals, and medial pterygoid plates.

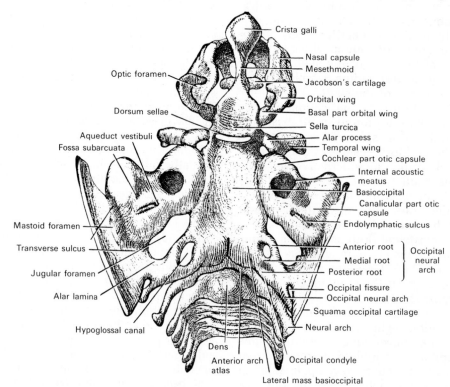

Fig. 4-11. Cartilaginous skull of 21-mm human embryo. Age of embryo is estimated to be six weeks.

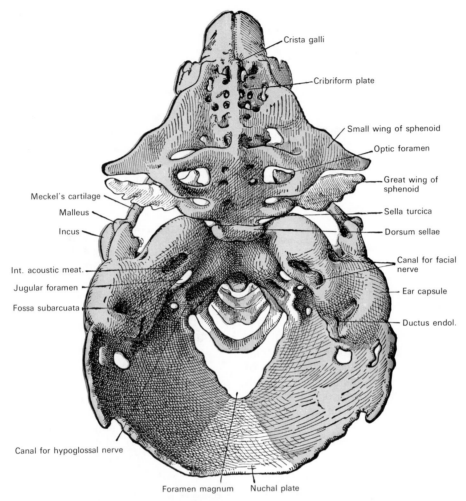

Crista galli

Cribriform plate

Small wing of sphenoid

Optic foramen

Great wing of sphenoid

Meckel's cartilage

Malleus

Incus

Sella turcica

Dorsum sellae

Canal for facial nerve

Int. acoustic meat.

Jugular foramen

Fossa subarcuata

Ear capsule

Ductus endol.

Canal for hypoglossal nerve

Foramen magnum   Nuchal plate

Fig. 4-12. Model of the chondrocranium of a human embryo 8 cm long. Its age might be estimated to be 11 weeks. The developing intramembranous bones are not shown. (From Gaupp.)

Cartilaginous Viscerocranium (Figs. 4-13; 4-14). Each developing branchial arch has a mesodermal core of mesenchyme covered by an ectodermal exterior and an ectodermal interior. In addition to musculature and blood vessels, the branchial mesoderm develops a skeletal primordium which subsequently chondrifies (Fig. 4-13,A). The dorsal end of the cartilage of the *first branchial arch* **(Meckel's cartilage)** forms the malleus and also probably the incus of the middle ear, while its intermediate portion forms the anterior malleolar and sphenomandibular ligaments (Fig. 4-13,B). The ventral portion of Meckel's cartilage largely regresses and becomes surrounded by the mesenchyme of the mandible, which then develops by intramembranous ossification. The cartilage of the *second branchial arch* is called **Reichert's**

cartilage. Its dorsal end, directed toward the developing ear, separates to form the stapes of the middle ear. Continuing ventrally it gives rise to the styloid process of the temporal bone (Fig. 4-14), the stylomandibular ligament, and the lesser horn (and probably the upper part of the body) of the hyoid bone (Fig. 4-13,B). The cartilage of the *third branchial arch* gives rise to the greater horn and the lower part of the body of the hyoid bone. The cartilages of the fourth, fifth, and sixth branchial arches do not ossify but develop as the laryngeal cartilages (Fig. 4-13,B).

Membranous Viscerocranium. The remaining skeletal derivatives of the branchial arches develop by intramembranous ossification. From the maxillary process of the first branchial arch, intramembranous ossification of mesenchyme results in the forma-

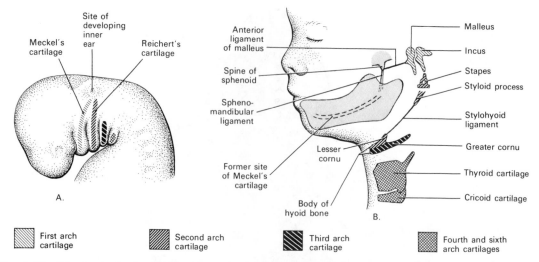

First arch cartilage | Second arch cartilage | Third arch cartilage | Fourth and sixth arch cartilages

FIG. 4-13. The developing branchial arch cartilage. *A,* Schematic lateral view of the head and neck region of a four-week embryo, illustrating the location of the branchial arch cartilages. *B,* Similar view of a 24-week fetus illustrating the derivations of the branchial arch cartilages. (From Moore, K. L., *The Developing Human,* W. B. Saunders Co., 1973.)

tion of the premaxilla, maxilla, and the squama and zygomatic process of the temporal bone. From the mandibular process of the first branchial arch, the mandible develops around the ventral portion of Meckel's cartilage by intramembranous ossification as described (Fig. 4-14). The other branchial arches do not develop bone by intramembranous ossification.

Axial Skeleton

The axial skeleton consists of the **vertebral column,** its cranial extension, the **skull,** and the **thoracic cage,** comprised of the sternum and ribs, to which the vertebral column is articulated. In contrast, the term appendicular skeleton relates to the bones of the extremities. This central osseous axis of the

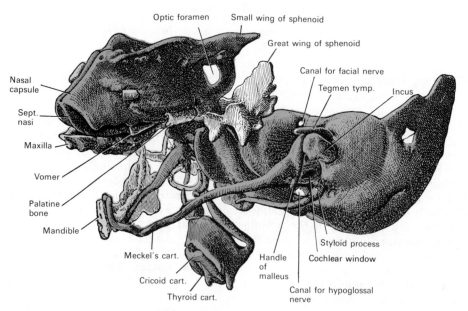

FIG. 4-14. Model of the chondrocranium of a human embryo, 8 cm in length. Viewed from the left side, this is the same model illustrated in Fig. 4-12. Note that certain of the membranous bones of the right side are shown in yellow. (From Gaupp.)

body, although flexible in its structure, affords stability and support to the viscera of the head, trunk, and abdomen and allows for the effective locomotor functions of the appended extremities.

VERTEBRAL COLUMN

The **vertebral column** or **backbone** (*columna vertebralis; spinal column*) (Fig. 4-15) is formed by a series of bones called vertebrae. The 33 vertebrae are grouped under the names cervical, thoracic, lumbar, sacral, and coccygeal, according to the regions they occupy. There are seven in the cervical region, twelve in the thoracic, five in the lumbar, five in the sacral, and four in the coccygeal.

The cervical, thoracic, and lumbar vertebrae remain distinct throughout life, and are known as **true** or **movable vertebrae.** Those of the sacral and coccygeal regions are termed **false** or **fixed vertebrae,** because they unite or fuse with one another to form two adult bones: five form the sacrum, and four form the terminal bone or coccyx. Between any two adjacent moveable vertebrae (except between the first and second cervical vertebrae) rests a tough, but elastic, fibrocartilaginous intervertebral disc.These intervertebral cushions contribute to the stability of the spinal column because they are strongly bound to the vertebrae, yet they allow considerable movement between the adjoining bones. Furthermore, they allow the spinal column to absorb significant shocks such as those received in jumping from heights or from certain types of exercises and conditioning procedures.

A typical vertebra consists of two essential parts: a ventral segment, the **body,** and a dorsal part, the **vertebral arch** (*arcus vertebralis; neural arch*), which encloses the vertebral foramen. Since the bodies of the vertebrae are articulated by the intervertebral discs, they form a strong pillar for the support of the head and trunk, and the vertebral foramina form a longitudinal tube, the vertebral canal, which encloses the spinal cord.

Vertebral Column as a Whole

The average length of the vertebral column in the male is about 71 cm. Of this length the cervical part measures 12.5 cm,

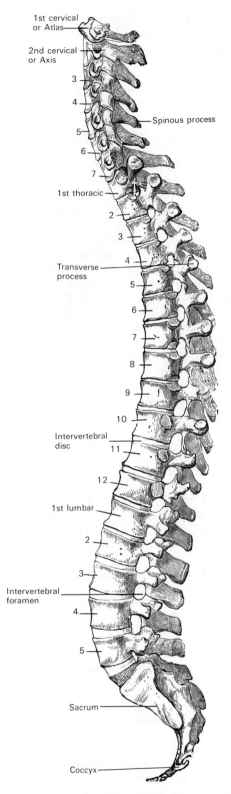

FIG. 4-15. Lateral view of the vertebral column. Observe the cervical, thoracic, lumbar, and pelvic curvatures.

the thoracic about 28 cm, the lumbar 18 cm, and the sacrum and coccyx 12.5 cm. The female column averages about 61 cm in length.

CURVATURES. A lateral view (Fig. 4-15) of the vertebral column presents several curves, which correspond with the different regions of the column and are called cervical, thoracic, lumbar, or pelvic. The **cervical curve,** convex ventrally, from the apex of the odontoid process of the axis to the middle of the second thoracic vertebra, is the least marked of all the curves. The **thoracic curve,** concave ventrally, begins at the middle of the second and ends at the middle of the twelfth thoracic vertebra. Its most prominent point dorsally corresponds to the spinous process of the seventh thoracic vertebra. The **lumbar curve** is more marked in the female than in the male; it begins at the middle of the last thoracic vertebra and ends at the sacrovertebral angle. It is convex ventrally, the convexity of the more caudal three vertebrae being much greater than that of the more cranial two. The **pelvic curve** begins at the sacrovertebral articulation, and ends at the point of the coccyx; its concavity is directed caudally and ventrally. The thoracic and pelvic curves are termed **primary curves,** because they alone are present during fetal life. The cervical and lumbar are **compensatory** or **secondary curves,** and are developed after birth, the former when the child is able to hold up its head (at three or four months) and to sit upright (at nine months), the latter when the child begins to walk (at twelve to eighteen months).

The vertebral column also has a slight **lateral curvature,** the convexity of which is directed toward the right side in right-handed people and toward the left side in individuals who are left-handed. Because of the correlated handedness of the lateral curvature, it is thought by some to be produced by long-continued muscular action. Others, however, believe this curvature to be produced by the aortic arch and upper part of the descending thoracic aorta, a view that is supported by the fact that in cases where the viscera are transposed and the aorta is on the right side, the convexity of the curve is directed to the left side.

SURFACES. *Ventral Surface.* Consistent with an increasing weight-bearing function from above downward, the width of the bodies of the vertebrae increases from the second cervical to the first thoracic, diminishes slightly in the next three, then again progressively increases from the fifth thoracic vertebra as far as the sacrovertebral angle. From this point there is a rapid diminution to the apex of the coccyx.

Dorsal Surface. The median line of the dorsal surface of the vertebral column is occupied by the **spinous processes.** In the cervical region (with the exception of the second and seventh vertebrae), these are short and nearly horizontal, with bifid extremities. In the upper part of the thoracic region they are directed obliquely downward; in the mid-thoracic region they are almost vertical; and in the lower thoracic and lumbar regions they are nearly horizontal. The spinous processes are separated by considerable intervals in the lumbar region and by narrower intervals in the neck, but are closely approximated in the middle of the thoracic region. Occasionally one of these processes deviates a little from the median line, a fact to be remembered in practice, as irregularities of this sort are attendant also in fractures or dislocations of the vertebral column. Important clinical landmarks are the protruding spinous processes of the seventh cervical vertebra, which is called the **vertebra prominens,** and that of the first thoracic vertebra, which is virtually as prominent. With the upper extremity in the anatomical position, the inferior angle of the scapula lies at the level of the seventh thoracic spinous process, while the superior border of the iliac crest lies at the plane of the spinous process of the fourth lumbar vertebra.

Lateral to the spinous processes on both sides are longitudinal **vertebral grooves** containing the deep back muscles. In the cervical and lumbar regions these grooves are shallow and they are formed by the laminae of the dorsal arches. In the thoracic region the grooves are deep and broad, and they are formed not only by the laminae but by the transverse processes as well. Lateral to the laminae are the superior and inferior articular processes, and still more laterally the transverse processes. In the thoracic region, the transverse processes are on a plane considerably dorsal to that of the transverse processes in the cervical and lumbar regions. In the cervical region, the transverse processes are placed ventral to the articular processes, lateral to the pedicles, and between the intervertebral foramina. In the thoracic region they are dorsal to the pedicles, intervertebral foramina, and articular processes.

In the lumbar region they are ventral to the articular processes, but dorsal to the intervertebral foramina.

Lateral Surfaces (Fig. 4-15).   The lateral surfaces are separated from the dorsal surface by the articular processes in the cervical and lumbar regions, and by the transverse processes in the thoracic region. The ventrolateral aspect of the vertebral column is formed by the sides of the bodies of the vertebrae, marked in the thoracic region by the facets for articulation with the heads of the ribs. More dorsally the **intervertebral foramina** are formed by the juxtaposition of the vertebral notches. These foramina are oval in shape, smallest in the cervical and cranial part of the thoracic regions, and gradually increase in size to the last lumbar vertebra. They transmit the spinal nerves and vessels and are situated between the transverse processes in the cervical region, and ventral to them in the thoracic and lumbar regions.

VERTEBRAL CANAL.   The vertebral canal, which contains the spinal cord, follows the different curves of the column; it is large and triangular in those parts of the column that enjoy the greatest freedom of movement, such as the cervical and lumbar regions; it is small and rounded in the thoracic region, where motion is more limited.

General Characteristics of a Vertebra

With the exception of the first and second cervical, the true or movable vertebrae have in common certain characteristics that may be studied in a typical vertebra from the middle of the thoracic region (Fig. 4-16). Characteristically a mid-thoracic vertebra consists of a **body** or **centrum** on its anterior aspect and a **vertebral arch** on its posterior aspect surrounding the **vertebral foramen,** within which lies the spinal cord. The vertebral arch is formed by two **pedicles,** which attach to the vertebral body, and a roof, formed by two **laminae** between which projects the **spinous process** dorsally. Additionally, two superior and two inferior intervertebral **articular processes** stud the pedicles and, dorsolaterally, project the **transverse processes** which, in the thoracic region, afford attachments on each side for the ribs, as do the **costal fovea** on the vertebral body.

The **body** (*corpus vertebra; centrum*) is the largest part of a vertebra, and is nearly cylindrical in shape. On its cranial and caudal surfaces attach the intervertebral discs. These surfaces are rough in the central areas, but a smooth and slightly elevated rim forms their circumference. The surface compact bone ventrally presents a few small apertures for the passage of nutrient blood vessels. The dorsal surface, which faces the vertebral canal, has one or more large, irregular apertures (*foramen vasculare*) for the passage of the basivertebral veins (Fig. 4-17).

The **vertebral arch** (*arcus vertebralis*), which lies posterior to the vertebral body, consists of a pair of pedicles and a pair of laminae which together support seven processes: four articular, two transverse, and one spinous.

The **pedicles** (*pediculi arci vertebrae*) are two short, thick processes that project dorsally, one on either side, from the cranial

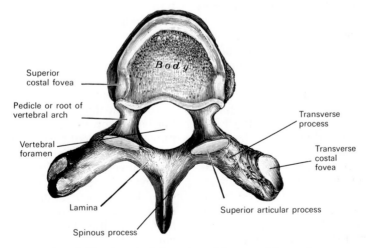

Superior costal fovea

Pedicle or root of vertebral arch

Vertebral foramen

Lamina

Spinous process

Body

Transverse process

Transverse costal fovea

Superior articular process

FIG. 4-16.   A typical thoracic vertebra, cranial aspect.

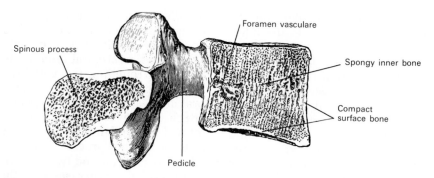

FIG. 4-17. Sagittal section of a lumbar vertebra showing the foramen vasculare for the passage of the basivertebral vein.

part of the vertebral body at the junction of its dorsal and lateral surfaces (Fig. 4-16). The pedicles are interposed between the transverse processes and the body. They are constricted in the middle, forming the concave **superior** and **inferior vertebral notches** which, when combined in two articulated adjacent vertebrae, form the **intervertebral foramina.**

The **laminae** (*laminae arci vertebrae*) are two broad plates directed dorsally and medially from the pedicles (Fig. 4-16). They fuse at the midline to complete the dorsal part of the arch surrounding the vertebral foramen and they provide a base for the spinous process. Their superior borders and the caudal parts of their ventral surfaces are rough for the attachment of the ligamenta flava.

PROCESSES. The **transverse processes** (*processus transversi*) project laterally on each side from the point where the laminae join the pedicles and between the superior and inferior articular processes (Fig. 4-16). They serve for the attachment of muscles and ligaments.

The **spinous process** (*processus spinosus*) is directed dorsally from the junction of the laminae, and serves for the attachment of muscles and ligaments (Fig. 4-16).

The **articular processes** (*processus articulares*), two superior and two inferior, spring from the vertebral arch at the junctions of the pedicles and laminae. The superior processes project cranially, but their articular surfaces are directed dorsally. The inferior processes are projected caudally and their articular surfaces face ventrally. The articular surfaces are coated with hyaline cartilage and, meeting with those of adjoining vertebrae, form synovial joints.

STRUCTURE OF A VERTEBRA (Fig. 4-17). The body is composed of cancellous or spongy bone, covered by a thin coating of compact bone, whereas the arches and processes have a thick covering of compact bone. The interior of the body is traversed by one or two large canals for the basivertebral veins, which converge toward a single large, irregular aperture (*foramen vasculare*) at the dorsal part of the body. The thin bony lamellae of the cancellous tissue are more pronounced in lines perpendicular to the upper and lower surfaces and develop in response to greater pressure from this direction. The dorsal vertebral arch and the processes projecting from it have thick coverings of compact tissue.

Cervical Vertebrae

The **cervical vertebrae** (*vertebrae cervicales*; Figs. 4-12; 4-13) are the smallest of the true vertebrae, and can readily be distinguished from those of the thoracic or lumbar regions by the presence of a foramen in each transverse process. The first, second, and seventh present exceptional features and will be described separately; the following characteristics are common to the remaining four.

The body is small, oval, and broader transversely than dorsoventrally. The ventral and dorsal surfaces are flat and of equal height; the ventral surface is positioned at a slightly lower level than the dorsal and its inferior border is prolonged downward, overlapping the subjacent vertebra in the articulated column (Fig. 4-15). The cranial surface of the cervical vertebral body (Fig. 4-18) is concave transversely, and presents a projecting lip on each lateral aspect; the caudal surface is convex from side to side, and presents shallow concavities laterally which receive the corresponding projecting lips of the subja-

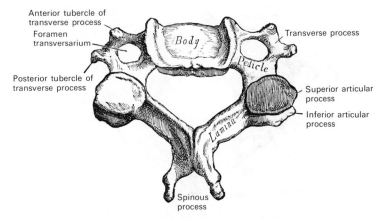

FIG. 4-18.   A typical cervical vertebra, cranial view.

cent vertebra. The interlocking nature of the adjacent cervical vertebrae increases the stability of the intervertebral joints. The **pedicles** are attached to the body midway between its cranial and caudal surfaces, so that the superior **vertebral notch** is as deep as the inferior, yet at the same time it is narrower. The **laminae** are narrow and long, and the superior border is thinner than the inferior. The **vertebral foramen** is large and triangular in shape, rather than round. The **spinous process** is short and bifid, the two projections often being of unequal size (Fig. 4-18). The superior and inferior **articular processes** are fused to form an articular pillar (Fig. 4-19), which is situated on the lateral aspect of the junction of the pedicle and lamina. The articular facets are flat and oval: the superior are directed dorsally, cranially, and slightly medially, while the inferior face ventrally, caudally, and slightly laterally.

The **transverse processes** (Fig. 4-18) are each pierced by a **foramen transversarium,** which, in the first six vertebrae, *transmits the vertebral artery and vein, accompanied by a plexus of sympathetic nerves.* Each transverse process consists of a ventral and a dorsal part. The portion ventral to the foramen is the homologue of the rib, and is therefore named the *costal process* or *costal element* (see Fig. 4-8). It arises from the side of the vertebral body and ends as the **anterior tubercle.** The dorsal part may be considered the "true" transverse process, homologous to that seen in a thoracic vertebra. It springs from the vertebral arch dorsal to the foramen and is directed ventrally and laterally, ending in a flattened vertical prominence, the **posterior tubercle.** The bar of bone joining the two portions of the transverse process lateral to the foramen exhibits a deep sulcus on its cranial surface for the passage of the corresponding spinal nerve (Fig. 4-19). The costal element of a cervical vertebra includes not only the portion that springs from the side of the vertebral body, but also the anterior and posterior tubercles and the bar of bone that connects them (Fig. 4-8).

The **first cervical vertebra** (Fig. 4-20) is named the **atlas** after a Titan in Greek mythology because it supports the globe of the head. Its chief peculiarity is that it has no vertebral body, and this is because during

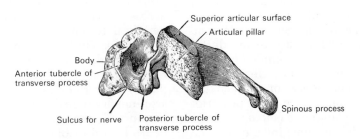

FIG. 4-19.   Side view of a typical cervical vertebra.

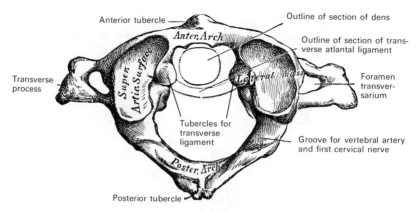

FIG. 4-20.   First cervical vertebra, or atlas.

development its centrum fused with that of the second vertebra to form an osseous pivot, the dens (odontoid process) of the axis, which projects superiorly through the atlas and around which the atlas turns. The dens is held in place by the transverse atlantal ligament (see Fig. 4-20). Its other peculiarities are that it has no spinous process, is ring-like, and consists of an anterior and a posterior arch and two lateral masses.

The **anterior arch** forms about one-fifth of the ring; the center of its ventral surface presents the **anterior tubercle** for attachment of the anterior longitudinal ligament and the longus colli muscles. The dorsal or internal surface of the anterior arch is marked by a smooth, oval or circular facet (*fovea dentis*) for articulation with the dens of the axis. The cranial border gives attachment to the anterior atlanto-occipital membrane which stretches to the occipital bone; along the caudal border of the arch attaches the anterior atlantoaxial ligament which connects it with the axis.

The **posterior arch** forms about two-fifths of the circumference of the ring (Fig. 4-20); it ends dorsally in the **posterior tubercle,** which is the rudiment of a spinous process and gives origin to the two recti capitis posteriores minores. The diminutive size of the spinous process obviates interference with the movements between the atlas and the skull. The cranial surface of the dorsal part of the arch presents a rounded edge for the attachment of the posterior atlanto-occipital membrane. Posterior to each lateral mass, the cranial surface of the posterior arch is marked by a groove (*sulcus arteriae vertebralis*), which is sometimes converted into a foramen by a delicate bony spicule

growing from the posterior end of the superior articular process. *It serves for the transmission of the vertebral artery,* which, after ascending through the foramen in the transverse process, winds around the lateral mass in a dorsal and medial direction. Within the groove is also found the suboccipital (first spinal) nerve. Along the caudal surface of the posterior arch, dorsal to the articular facets, are two shallow grooves, the **inferior vertebral notches,** which contribute to the formation of the intervertebral foramina through which pass the second cervical pair of spinal nerves. The inferior border gives attachment to the posterior atlantoaxial ligament.

The **lateral masses** are the most bulky and solid parts of the atlas and support the weight of the head. Each has a superior and an inferior articular facet. The **superior facets** are large, oval, and concave, and converge ventrally. They face cranialward, medialward, and dorsalward, each forming a cup for the corresponding condyle of the occipital bone. At times the superior facets are partially subdivided by indentations that encroach upon their margins. Just inferior to the medial margin of each superior facet is a small tubercle for the attachment of the transverse atlantal ligament, which stretches across the ring of the atlas and divides the vertebral foramen into two unequal parts (Fig. 4-20). The ventral or smaller part receives the dens of the axis, the dorsal transmits the spinal cord and its membranes. This part of the vertebral canal is of considerable size, much greater than is required for the accommodation of the spinal cord, and hence lateral displacement of the atlas may occur as a result of trauma without compression of the cord. The **inferior articular facets** are circular in form, flattened or slightly convex, and directed caudalward and medialward. They articulate with the axis. The **transverse processes** of the atlas project laterally from the lateral masses, and serve for the attachment of muscles that assist in rotating the head. They are long, their anterior and posterior tubercles are fused into a single mass, and the foramen transversarium is slanted dorsalward (Fig. 4-20).

The **second cervical vertebra** (Figs. 4-21; 4-22) is named the **axis** or **epistropheus** because it forms the pivot upon which the first vertebra, carrying the head, rotates. The axis is the strongest and thickest of the cervical vertebrae. Its most distinctive characteristic is the cranial extension of the body into a strong, tooth-like bony process, the **dens** (*odontoid process*). The ventral surface of the body is marked by a median ridge for the attachment of the longus colli muscle and is prolonged inferiorly into a lip, which overlaps the third vertebra when articulated.

The **dens** (Figs. 4-21; 4-22), which measures 12 to 15 mm in length, exhibits a slight constriction or *neck* where it joins the body. Within a shallow groove on the dorsal surface of the neck, which frequently extends onto its lateral surfaces, rests the transverse atlantal ligament. On its ventral surface is an oval or nearly circular facet for articulation with the facet on the anterior arch of the

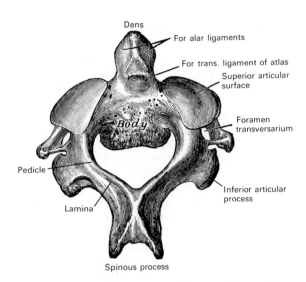

FIG. 4-21.   Second cervical vertebra or axis, cranial aspect.

atlas. The *apex* of the dens is pointed and gives attachment to the apical dental ligament. Near its junction with the vertebral body, the sides of the dens have rough impressions for the attachment of the alar ligaments, which connect it to the occipital bone. The internal structure of the dens is more compact than that of the body.

The **pedicles** are broad and strong, especially ventrally, where they coalesce with the sides of the vertebral body and the root

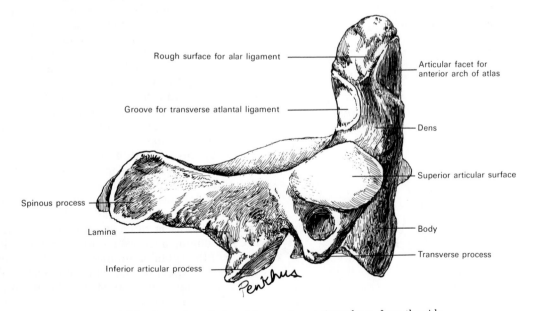

FIG. 4-22.   Second cervical vertebra, axis or epistropheus, from the side.

of the dens. They are covered cranially by the superior articular surfaces. The **laminae** are thick and strong, and the **vertebral foramen** is large, but smaller than that of the atlas. The small **transverse processes** end in a single tubercle, perforated by the **foramen transversarium,** which is oriented in an obliquely lateral direction. The **superior articular surfaces,** which are oval, slightly convex, and directed cranially and laterally, are supported by the body, pedicles, and transverse processes. They articulate with the inferior articular facets of the atlas. The **inferior articular surfaces** are directed inferiorly and ventrally in a manner similar to that of the other cervical vertebrae (Fig. 4-22). The superior **vertebral notches** are shallow and lie dorsal to the superior articular processes; the inferior lie ventral to the inferior articular processes, as in the other cervical vertebrae. The **spinous process** is large, strong, deeply channeled on its caudal surface, and presents a bifid, tuberculated extremity.

The **seventh cervical vertebra** (Fig. 4-23) is distinguished by its characteristic long and prominent spinous process. This spine, along with that of the first thoracic vertebra, protrudes dorsally beyond the upper six cervical spines in the articulated skeleton, and since it is the most cranial spinous process that is palpable, it is a useful landmark in

identifying and counting the other spines. The seventh cervical vertebra is named, therefore, the **vertebra prominens.** Its spinous process is thick, resembling that of a thoracic vertebra; it is nearly horizontal in direction, not bifurcated, and terminates in a tubercle to which the caudal end of the ligamentum nuchae is attached. The transverse processes are of considerable size, their dorsal components are large and prominent, their ventral, small and faintly marked. The ventral component is derived from the costal element or process (see Fig. 4-8) and may develop into an anomalous cervical rib. The **foramen transversarium** may be as large as that in the other cervical vertebrae, but more usually it is smaller on one or both sides; at times it is double and occasionally it is absent. On the left side it may give passage to the vertebral artery. More frequently, however, the vertebral vein traverses it on both sides. The usual arrangement is for both artery and vein to pass ventral to the transverse process of the seventh cervical vertebra and not through its foramen.

Thoracic Vertebrae

The **thoracic vertebrae** (*vertebrae thoracicae;* Fig. 4-24) are intermediate in size between the cervical and lumbar, providing a gradual transition from the small cervical cranially to the large lumbar vertebrae caudally. This enlargement of successively more caudal vertebrae may be correlated with the increased weight-bearing requirements of these lower spinal column bones. The thoracic vertebrae are distinguished by the presence of costal facets on the sides of their bodies for articulation with the heads of the ribs, and other articular facets on the transverse processes of the upper ten thoracic vertebrae for articulation with the tubercles of the ribs.

The **bodies** of vertebrae in the mid-thoracic region are heart-shaped, and as broad anteroposteriorly as the transverse diameter. They are flat and slightly thicker dorsally. In fact the ventral length of a thoracic vertebral body measures as much as 2 mm less than its dorsal length, accounting for the thoracic curve of the spinal column (see Fig. 4-15). On the lateral aspect of each thoracic vertebral body are found two **costal demifacets,** one superior near the root of the pedicle, the

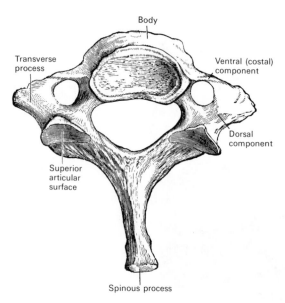

FIG. 4-23. Seventh cervical vertebra, viewed from above.

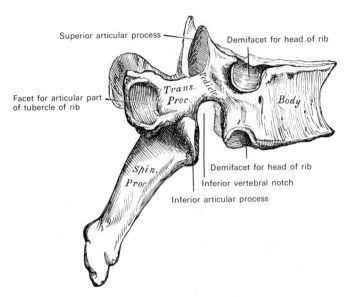

Superior articular process

Demifacet for head of rib

Facet for articular part
of tubercle of rib

*Trans.
Proc.*

Pedicle

Body

*Spin.
Proc.*

Demifacet for head of rib

Inferior vertebral notch

Inferior articular process

FIG. 4-24. A typical thoracic vertebra, right lateral view.

other inferior, ventral to the inferior vertebral notch. In life these articular surfaces are covered with cartilage, and when the vertebrae are articulated in sequence, each two adjacent demifacets form oval surfaces for the reception of the heads of the ribs.

The **pedicles** (Fig. 4-24) are directed dorsally and slightly cranialward from the transverse processes, and the inferior vertebral notches are large and extend more deeply than in any other region of the vertebral column. The **laminae** are broad, thick, and imbricated—that is to say, they overlap those of subjacent vertebrae like tiles on a roof. The **vertebral foramen** is small and circular, and accommodates the narrowed thoracic spinal cord. The **spinous processes** are long, triangular in coronal section, directed obliquely caudalward, and end in tuberculated extremities. The fifth to the eighth thoracic spines are more oblique and overlapping than those of the upper and lower thoracic vertebrae.

The **superior articular processes** (Fig. 4-24) are thin plates of bone that project superiorly from the junctions of the pedicles and laminae. Their articular facets are practically flat, being directed dorsally and a little laterally. The **inferior articular processes** are fused to a considerable extent with the laminae, and project but slightly beyond the inferior borders of the laminae. Their facets are directed ventrally and a little medially

and inferiorly to receive the superior articular processes of subjacent vertebrae. The **transverse processes** arise on each side from the dorsal vertebral arch at the junction of the lamina and the pedicle. These processes are long, thick, and strong and are directed obliquely dorsally and laterally. Each transverse process ends in a clubbed extremity, on the ventral aspect of which is a small, concave surface for articulation with the tubercle of a rib (Fig. 4-24).

The first, ninth, tenth, eleventh, and twelfth thoracic vertebrae present certain peculiarities, and will be considered separately (Fig. 4-25).

The **first thoracic vertebra** has an **entire articular facet** for the head of the first rib and a **demifacet** for the cranial half of the head of the second rib, on each side of the vertebral body. Its centrum is broad transversely like that of a cervical vertebra, which emphasizes the transitional structure of the first thoracic vertebra. The superior articular surfaces are directed cranially and dorsally; the spinous process is thick, long, and almost horizontal. The transverse processes are also long, and each contains an articular facet for the tubercles of the first ribs. The superior vertebral notches are deeper than those of the other thoracic vertebrae.

The **ninth thoracic vertebra,** although otherwise the same as other thoracic vertebrae, may have no demifacets caudally on its

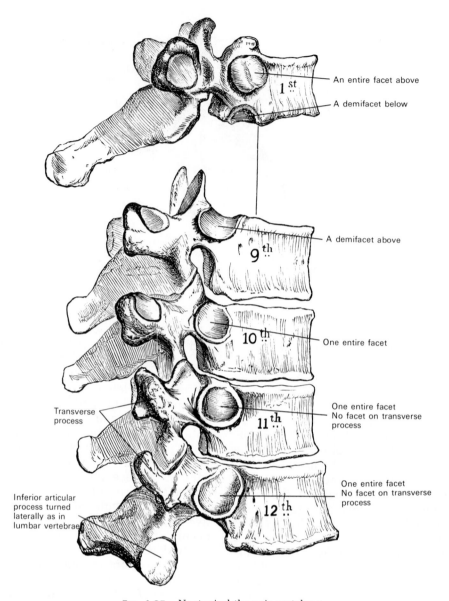

An entire facet above

A demifacet below

A demifacet above

One entire facet

One entire facet
No facet on transverse
process

Transverse
process

One entire facet
No facet on transverse
process

Inferior articular
process turned
laterally as in
lumbar vertebrae

FIG. 4-25. Nontypical thoracic vertebrae.

vertebral body (Fig. 4-25). In some instances, however, two demifacets have been found on each side of the vertebral centrum; when this occurs the tenth vertebral body has demifacets only on its cranial portion.

The **tenth thoracic vertebra** has (except as just mentioned) a single entire articular facet on each side, which also extends partly onto the lateral surface of the pedicle (Fig. 4-25).

The **eleventh thoracic vertebra** has a body that approaches the form and size of the lumbar vertebrae. The articular facets for the heads of the ribs are large and placed chiefly on the pedicles. The pedicles on the

eleventh and twelfth thoracic vertebrae are the thickest and strongest of any in the thoracic portion of the spinal column. The spinous process is short, and nearly horizontal in direction. The transverse processes are very short, tuberculated at their extremities, but have no costal articular facets, since the eleventh pair of ribs articulates only on the lateral aspect of the pedicles.

The **twelfth thoracic vertebra** (Fig. 4-25) has a number of characteristics in common with the eleventh. It may, however, be distinguished from it because (1) its inferior articular surfaces are convex and directed laterally, like those of the lumbar vertebrae, (2) in general form of the body, laminae, and spinous process, it resembles the lumbar vertebrae, and (3) each transverse process is subdivided into three elevations, the **superior, inferior,** and **lateral tubercles.** The superior and inferior tubercles correspond to the mammillary and accessory processes of the lumbar vertebrae. Less marked but similar elevations are identifiable on the transverse processes of the tenth and eleventh thoracic vertebrae.

Lumbar Vertebrae

The **lumbar vertebrae** (*vertebrae lumbales;* Figs. 4-26; 4-27) are the largest of the true or moveable vertebrae, a feature that allows them to be distinguished from other vertebrae. Additionally, lumbar vertebrae have no articular facets on the sides of their centra as are found in thoracic vertebrae, and they have no foramina in their transverse processes as are found in cervical vertebrae. The foramen vasculare for the basivertebral vein is larger than in the other vertebrae.

The **body** of a lumbar vertebra is large, wider transversely than dorsoventrally, and a little thicker ventrally than dorsally (Fig. 4-26). It is flat or slightly concave superiorly and inferiorly. The sides of the centrum are concave, so that a finger can easily encircle the circumference of the vertebral body within its grooved concavity. Strong but short **pedicles** unite on the cranial part of the body, leaving deep inferior **vertebral notches** (Fig. 4-26). The **laminae** are broad, short, and strong. The **vertebral foramen** is

triangular, larger than in the thoracic, but smaller than in the cervical region. The **spinous process** is thick, broad, and somewhat quadrilateral in shape and ends in a rough, uneven border, which is thickest caudally where it is occasionally notched. The superior and inferior **articular processes** are well defined, projecting from the junctions of pedicles and laminae (Fig. 4-27). The facets on the superior processes are concave and face dorsally and medially; those on the inferior are convex, and are directed ventrally and laterally. The superior facets are wider apart than the inferior; in the articulated column the inferior articular processes are embraced by the superior processes of the subjacent vertebra. The **transverse processes** are long, slender, and horizontal in the cranial three lumbar vertebrae, but incline a little dorsally in the caudal two. In the upper three vertebrae the transverse processes arise from the junctions of the pedicles and laminae, but in the lower two they are set farther ventrally and spring from the pedicles and dorsal parts of the bodies. Situated ventral to the articular processes instead of dorsal as in the thoracic vertebrae, the transverse processes can be considered homologous with the ribs.

Two additional tubercles found on lumbar vertebrae bear description. On the posterior borders of the large superior articular process are rounded enlargements called the **mammillary processes,** and arising from the dorsal part of the bases of the transverse processes are small rough elevations called the **accessory processes** (Fig. 4-27).

The **fifth lumbar vertebra** (Fig. 4-28) is

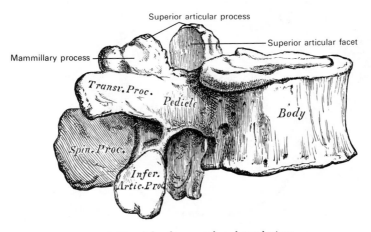

Fɪɢ. 4-26. A lumbar vertebra, lateral view.

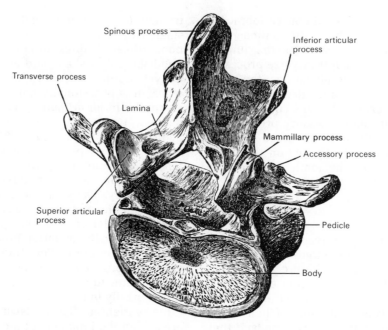

FIG. 4-27.   A lumbar vertebra, craniodorsal aspect viewed from left.

massive in its structure and its body is much deeper ventrally than dorsally, which accords with the prominent *sacrovertebral angle* formed at the lumbosacral articulation. The spinous process is smaller than that of other lumbar vertebrae, but the transverse processes are bulky, short, and thick and arise from the body as well as from the pedicles. The inferior articular processes, which attach to the sacrum, are more widely separated in the fifth lumbar vertebra than in other lumbar vertebrae; the distance approaches or even exceeds that which separates the superior articular processes.

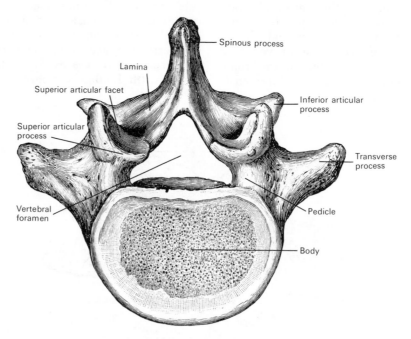

FIG. 4-28.   Fifth lumbar vertebra, cranial aspect.

VARIATIONS. The last lumbar vertebra is subject to certain defects described as *bifid* and *separate neural arches,* the latter occurring three times as frequently as the former. Both defects result from an incomplete union of the bony vertebra and cause weakness in the spinal column: the bifid arch by impairing ligamentous attachments, the separate arch through loss of bony anchorage of the column to its base.

Sacral and Coccygeal Vertebrae

At an early period of life the sacral and coccygeal vertebrae consist of nine separate segments which, in the adult, are united to form two bones, five entering into the formation of the **sacrum,** four into that of the **coccyx.**

THE SACRUM (os sacrum)

The **sacrum** is a large, triangular bone, which is situated in the lower part of the vertebral column and forms the dorsal bony wall of the true pelvis (see Fig. 4-191). The broad, superior end of the bone, considered its **base,** articulates with the last lumbar vertebra, while its more narrow apex articu-

lates inferiorly with the coccyx. Inserted like a wedge between the two hip bones, its base projects ventrally and forms the prominent **sacrovertebral angle** when articulated with the last lumbar vertebra. Its central part is projected dorsally (see Fig. 4-190). It encases the sacral canal within which course the sacral nerves.

The **pelvic surface** (*facies pelvina;* Fig. 4-29) is concave, giving increased capacity to the pelvic cavity. Its median part is crossed by four **transverse ridges,** the positions of which correspond to the original planes of separation between the five segments of the fetal bone (see Fig. 4-44). The four ridges represent the intervertebral discs of the immature sacrum which then ossified and fused. Intervening between the ridges are the **bodies** of the sacral vertebrae. The body of the first segment is large, and in form resembles that of a lumbar vertebra. Succeeding ones diminish in size, are flat and curve with the form of the whole sacrum. Four pairs of rounded **pelvic sacral foramina** are located at the ends of the transverse ridges. These foramina communicate with the sacral canal by means of intervertebral foram-

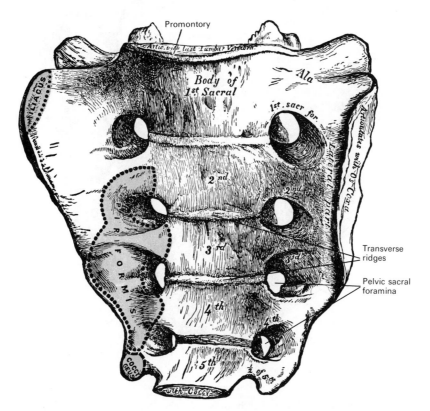

FIG. 4-29. Sacrum, pelvic surface.

ina. Through them course the ventral divisions of the first four sacral nerves along with the lateral sacral arteries and veins.

Lateral to the pelvic sacral foramina are the **lateral parts of the sacrum,** which consist of five separate segments at an early period of life, but blend with the bodies and with each other during maturation. Four broad, shallow grooves which lodge the ventral divisions of the sacral nerves extend laterally from the foramina. These are separated by prominent ridges of bone which give origin to the piriformis muscle. Inserting onto the lateral aspect of the most inferior sacral segment is the coccygeus muscle.

The **dorsal surface** (*facies dorsalis*) (Fig. 4-30) is convex and narrower than the pelvic surface. In its midline is located the **median sacral crest.** On this elevated ridge are mounted three or four tubercles, the rudimentary **spinous processes** of the upper three or four sacral segments (Fig. 4-31). On both sides of the median sacral crest, shallow **sacral grooves** are formed by the united **laminae** of the corresponding vertebrae, and give origin to the multifidus. Overlying the multifidus and taking origin in a U-shaped manner around the multifidus is the erector spinae muscle. Some aponeurotic fibers of origin for the latissimus dorsi muscle also arise from the median sacral crest. The laminae of the fifth sacral vertebra fail to meet in the midline dorsally, leaving an enlarged aperture into the sacral canal, the **sacral hiatus** (Fig. 4-31). On the lateral aspect of the sacral groove is a linear series of tubercles produced by the fusion of the **articular processes,** which together form the indistinct **intermediate crests** (Fig. 4-30). The **articular processes** of the first sacral vertebra are large and oval in shape. Their facets are concave, face dorsomedially, and articulate with the facets on the inferior processes of the fifth lumbar vertebra. The tubercles that represent the inferior articular processes of the fifth sacral vertebra are prolonged down-

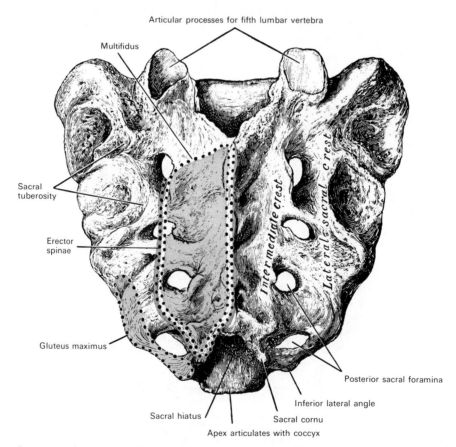

Articular processes for fifth lumbar vertebra

Multifidus

Sacral tuberosity

Erector spinae

Gluteus maximus

Intermediate crest

Lateral sacral crest

Posterior sacral foramina

Inferior lateral angle

Sacral cornu

Sacral hiatus

Apex articulates with coccyx

FIG. 4-30. The sacrum, dorsal aspect. Observe that the origin of the multifidus muscle is surrounded by the U-shaped origin of the erector spinae (sacrospinalis) muscle.

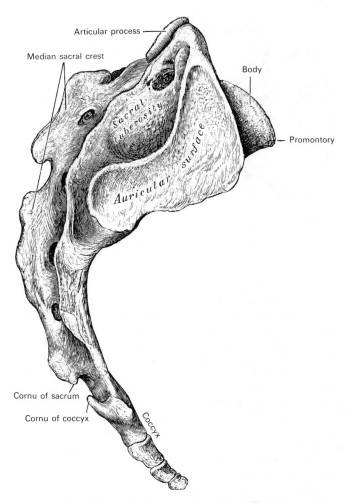

Articular process —

Median sacral crest —

Body

Sacral tuberosity

Promontory

Auricular surface

Cornu of sacrum —

Cornu of coccyx —

Coccyx

Fɪɢ. 4-31. Lateral aspect of sacrum and coccyx.

ward as rounded processes named the **sacral cornua,** which are connected to the cornua of the coccyx (Fig. 4-30). Lateral to the intermediate crest are the four **posterior sacral foramina** (Fig. 4-30). Smaller in size and less regular in form than the ventral foramina, they transmit the dorsal divisions of the sacral nerves. Lateral to the dorsal foramina, a series of tubercles represents the **transverse processes** of the sacral vertebrae and forms the **lateral crests** of the sacrum (Fig. 4-31). The tubercles of the first and second sacral segments receive the attachment of the horizontal parts of the posterior sacroiliac ligaments; those of the third vertebra give attachment to the oblique fasciculi of the dorsal sacroiliac ligaments; and those of the fourth and fifth to the sacrotuberous ligaments.

The lateral surface (Fig. 4-31) of the sac-

rum is broad and irregular cranially, narrow caudally, and formed by the union of elements comparable to the costal and transverse processes seen in other vertebrae (Fig. 4-32). The upper half of this surface presents a large ear-shaped articular facet, the **auricular surface,** which is covered with cartilage in the fresh state for articulation with the ilium. Dorsal to it is a rough surface, the **sacral tuberosity** (Fig. 4-31), on which are three deep and uneven impressions for the attachment of the posterior sacroiliac ligament. The caudal half of the lateral surface is thin and ends in a projection called the **inferior lateral angle** (Fig. 4-30); medial to this angle is a notch that is converted by the transverse process of the first coccygeal segment into a foramen for transmission of the ventral division of the fifth sacral nerve. The thin caudal half of the lateral surface gives attachment

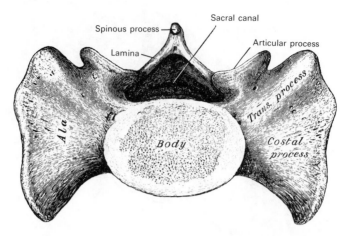

FIG. 4-32. Base of sacrum. This superior end of the sacrum articulates with the fifth lumbar vertebra.

to the sacrotuberous and sacrospinous ligaments, to some fibers of the gluteus maximus muscle, and to the coccygeus muscle.

The **base of the sacrum** (*basis ossis sacri;* Fig. 4-32), which is its cranial surface, is broad and expanded laterally. In its center is the upper surface of the **body** of the first sacral vertebra. The ventral border of the body projects into the pelvis as the **promontory.** The articulation between the body of the first sacral segment and the body of the fifth lumbar vertebra is interposed by an intervertebral disc. The orifice of the **sacral canal** lies dorsal to this articulation. As is seen throughout the vertebral column, the sacral canal is completed dorsally by the laminae and spinous process. The superior articular processes on either side are directed dorsomedially, as are the superior articular processes of a lumbar vertebra. They are attached to the body of the first sacral vertebra and to the wings, or **alae,** by short thick pedicles. On the cranial surfaces of each pedicle is a vertebral notch that forms the caudal part of the intervertebral foramen between the last lumbar and the first sacral vertebrae. On either side of the body, a large triangular surface called the ala supports the psoas major muscle and the lumbosacral nerve trunk, and receives the attachment of a few fibers of the iliacus muscle. The dorsal quarter of the ala represents the transverse process, and its ventral three-quarters the costal process of the first sacral segment (Fig. 4-32).

The **apex** (*apex ossis sacri*) is the caudal extremity of the sacrum and presents an oval facet for articulation with the coccyx.

The **sacral canal** (*canalis sacralis;* Figs. 4-26; 4-27), the vertebral canal in the sacrum, is incomplete inferiorly due to the nondevelopment of the laminae and spinous processes of the last one or two segments. The resultant widened aperture into the caudal end of the sacral canal, the **sacral hiatus** (Fig. 4-30), is used by anesthesiologists for the insertion of a flexible needle to produce caudal analgesia. The canal lodges the sacral nerves of the lower spinal cord's cauda equina, and its walls are perforated by the dorsal and ventral sacral foramina through which these nerves pass.

STRUCTURE. The sacrum consists of cancellous bone enveloped by a thin layer of compact bone. In a sagittal section through the center of the sacrum (Fig. 4-33), the bodies appear united at their circumferences by bone, but in the center are wide intervals which, in the intact body, are filled by the intervertebral discs. This union may be more completely ossified between the caudal segments than between the cranial segments in some sacra.

ARTICULATIONS. The sacrum articulates with four bones: the last lumbar vertebra cranially, the coccyx caudally, and the hip bone on each side.

DIFFERENCES IN THE SACRUM OF THE MALE AND THE FEMALE. The wider female pelvic cavity is partially explained by the fact that the female sacrum is shorter and wider than the male. The cranial half of the sacrum is nearly straight while the caudal half presents a curvature. In women the caudal half forms a greater angle with the cranial half. The female sacrum being more oblique and dorsally directed renders the sacrovertebral angle more prominent. In the male the curvature of the sacrum is more evenly distributed over the whole length of the bone.

VARIATION. The sacrum may consist of the remnants of six segments, in which case a part of the

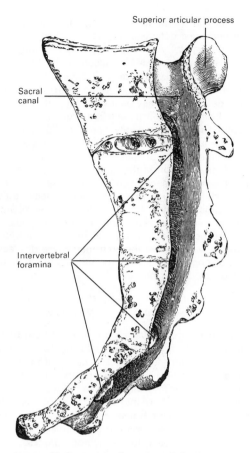

FIG. 4-33. Median sagittal section of the sacrum.

last lumbar or first coccygeal vertebra is included (sacralization). More occasionally the number is reduced to four. The bodies of the first and second may fail to unite. Sometimes the uppermost transverse tubercles are not joined to the rest of the ala on one or both sides, or the sacral canal may be open throughout a considerable part of its length in consequence of the imperfect development of the laminae and spinous processes. The sacrum also varies considerably with respect to its degree of curvature.

THE COCCYX (os coccygis)

Usually formed of four rudimentary vertebrae, the coccyx (Figs. 4-34; 4-35) may consist of three or as many as five segments. This small triangular bone articulates superiorly with the apex of the sacrum, and its inferior tip is the lowest part of the spinal column. Rudimentary bodies and articular and transverse processes can be identified in the first three segments, but coccygeal vertebrae are destitute of pedicles, laminae, and spinous processes. The first segment is the largest, and, resembling the last sacral vertebra, it

often exists as a separate bone. The lowest three segments diminish in size, and are usually fused with one another. Of these, the last segment is a mere module of bone.

The **pelvic surface** is marked with three transverse grooves, which indicate the junctions of the different segments. It gives attachment to the ventral sacrococcygeal ligament, and to its lateral margins are inserted muscle fibers of the coccygeus and levator ani muscles. Additionally, it is in contact with a part of the rectum. The **dorsal surface** has a row of tubercles on both sides; these are the rudimentary articular processes. The large first pair of tubercles is called the **coccygeal cornua**; it articulates with the cornua of the sacrum, and on each side completes the intervertebral foramen for the transmission of the dorsal division of the fifth sacral nerve.

On the **lateral borders** of the coccyx a series of small eminences represents the transverse processes. The uppermost pair is the largest, and on each side the transverse proc-

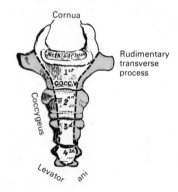

FIG. 4-34. Coccyx, pelvic surface.

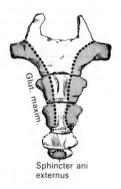

FIG. 4-35. Coccyx, dorsal surface.

ess often joins the thin lateral edge of the sacrum, thus completing the foramen for the transmission of the ventral division of the fifth sacral nerve. The narrow borders of the coccyx receive the attachments of the sacrotuberous and sacrospinous ligaments laterally, the coccygeus muscle and the iliococcygeal portion of the levator ani muscle ventrally, and the origin of the gluteus maximus muscle dorsally. The oval facet on the surface of the base articulates with the sacrum. The rounded apex may be bifid, or deflected to one side, and has the tendon of the sphincter ani externus attached to it.

Variations and Anomalies of the Vertebral Column

NUMBER OF VERTEBRAE.   Variation in the total number of vertebrae cranial to the coccyx is less common than variations in the normal number within the different regions. The number of segments in the coccyx is the most variable, the sacral next, then the thorax, and finally variation in the number of cervical vertebrae the least common. The number of vertebrae in one region may be increased at the expense of the number in a neighboring region however. Variations in the lumbosacral and cervicothoracic region are rather frequent and usually can be traced to anomalies in the development of the costal processes (Fig. 4-8). The costal processes of the twelfth thoracic vertebra may be wanting, in which case there will be eleven thoracic and six lumbar vertebrae. Conversely, the costal element of the first lumbar may develop into a rib, resulting in thirteen thoracic and four lumbar vertebrae. The costal element of the last lumbar may unite with the first sacral, a condition known as *sacralization of the lumbar vertebra,* resulting in four lumbar and six sacral segments. True sacralization does not take place, however, unless the fifth lumbar enters into the articulation with the ilium. Again, the first sacral segment may fail to unite with the other sacrals, resulting in six lumbar and four sacral segments. There may be a secondary sacral promontory.

CERVICAL RIB.   Occasionally a cervical rib condition develops in which the costal element of the seventh cervical is elongated into a separate rib-like structure. In rare instances, such a rib may be long enough to articulate with the sternum. This anomaly may cause interference with blood flow in the subclavian artery or paralysis in the upper extremity due to pressure on the brachial plexus.

ASYMMETRY.   One side of a vertebra may develop defectively to a greater or lesser degree. When this occurs on one side of the centrum or body, the resulting condition is called *congenital scoliosis* or lateral curvature. In other situations an overgrowth or failure of the costal process may occur on one side,

especially in the lumbosacral region, or there may exist one-sided defects in the pedicles and laminae.

SPINA BIFIDA.   At times laminae fail to unite and form a spinous process. Not uncommonly it involves only one or two vertebrae and causes few symptoms (*spina bifida occulta*). Depending on the size of the defect, there may be a small fat pad or lipoma in its place, but larger defects may contain a *meningocele* or a *myelomeningocele* produced by fluid and the protrusion of tissues from the spinal canal. The most common site of spina bifida is the lumbar region, less common is the cervical region. In some instances the absence of spinous processes may extend distally through the entire sacrum, leaving the sacral canal exposed. Other parts of the neural arch may have similar defects. The union of the pedicles with the body may fail on one or both sides; the resulting weakness of the arch may cause displacement of one vertebra over the other, the condition named *spondylolisthesis*.

Ossification of the Vertebral Column

Each cartilaginous vertebra is ossified from **three primary centers** (Fig. 4-36), two for the vertebral arch and one for the body. Ossification of the vertebral arches begins in the upper cervical vertebrae during the seventh or eighth week of fetal life, and gradually extends to more caudal vertebrae. The ossific granules first appear at the sites from which the transverse processes eventually project. Ossification spreads dorsally to the spinous process, ventrally into the pedicles, and laterally into the transverse and articular processes. Ossification of the vertebral bodies begins about the eighth week in the lower thoracic region, and subsequently extends cranially and caudally along the column. The initial ossification center for the body does not give rise to the entire centrum of the adult vertebra, since the dorsolateral portions are ossified by extensions from the vertebral arch. During the first few years of life the body of a vertebra, therefore, shows two lines of union of the neural arch with the centrum. These sites of junction are called **neurocentral synchondroses** (Fig. 4-37). In the thoracic region, the facets for the heads of the ribs lie dorsal to the neurocentral synchondroses and are ossified from the centers for the vertebral arch.

At birth the vertebra consists of three parts, the body and the two halves of the vertebral arch joined by cartilage. During the first year the halves of the arch unite dorsally, union taking place first in the lumbar region and then extending cranially through the thoracic and cervical regions. About the third year the bodies of the upper cervical vertebrae are joined to the arches on each side, but this union in the caudal lumbar vertebrae is not completed until the sixth year. During the years before puberty these primary ossification centers gradually increase in size, while the cranial and caudal surfaces of the bodies and the ends of the transverse and spinous

THREE PRIMARY CENTERS

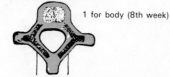

1 for body (8th week)

1 for each side of vertebral arch (7th week)

FIG. 4-36.

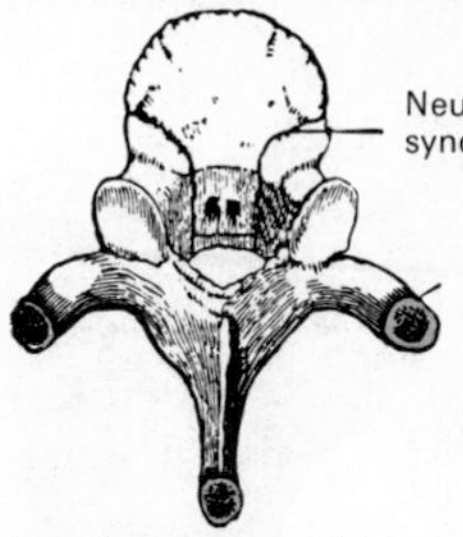

FIVE SECONDARY CENTERS

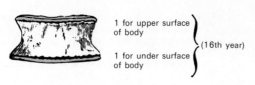

Neurocentral synchondrosis

1 for each transverse process (16th year)

1 for spinous process (16th year)

FIG. 4-37.

1 for upper surface of body

1 for under surface of body

(16th year)

FIG. 4-38.

FIGS. 4-36 to 4-38. Ossification of a vertebra. Fig. 4-36. The three primary ossification centers are indicated. Fig. 4-37. Three secondary centers, to which are added two other secondary centers on the upper and lower surfaces of the vertebral body (Fig. 4-38).

processes remain cartilaginous. About the sixteenth year (Fig. 4-37), **five secondary centers** appear, one for the tip of each transverse process, one for the extremity of the spinous process, one for the cranial and one for the caudal surface of the body (Fig. 4-38). The secondary centers fuse with the rest of the bone about the age of twenty-five.

The mode of ossification just described occurs in vertebrae of the spinal column from the third through the sixth cervical and the twelve thoracic segments. Exceptions to this ossification pattern, however, occur in the first, second, and seventh cervical vertebrae, and in the lumbar, sacral, and coccygeal vertebrae.

ATLAS.   The atlas is usually ossified from three centers (Fig. 4-39). Two of these usually appear in the lateral masses during the seventh week of fetal life. At birth, the dorsal portions of bone are separated from one another by a narrow interval filled

THREE OSSIFICATION CENTERS

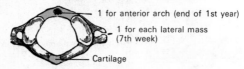

1 for anterior arch (end of 1st year)

1 for each lateral mass (7th week)

Cartilage

FIG. 4-39. Ossification centers in the atlas (first cervical vertebra).

with cartilage. Between the third and fourth years the posterior arch becomes united either directly or through the development of a separate ossification center in the cartilage. The anterior arch consists of cartilage at birth; a separate center appears toward the end of the first year after birth and eventually joins the lateral masses between the sixth and eighth years. The lines of this union extend across the ventral portions of the superior articular facets (Fig. 4-39). Occasionally there is no separate center for the ventral arch, or it may ossify from two centers, one on each side of the midline.

AXIS OR EPISTROPHEUS.   The axis ossifies from five primary and two secondary centers (Fig. 4-40). The body and arch are ossified in the same manner as in the other vertebrae, i.e., one primary center for the body, and two for the vertebral arch. The centers for the arch appear about the seventh or eighth week of fetal life, that for the body about the fourth or fifth month. The **dens** (being the representation of the body of the atlas) commences ossification during the sixth fetal month from two laterally placed primary centers in its base. These two centers join before birth and form a conical mass of bone which has a cleft superiorly. The interval between the sides of the cleft is filled by a wedge-shaped piece of cartilage, thereby forming the summit of the process. This cartilaginous apex of the dens has a separate secondary ossification center which appears during the second year and joins the main part of the process about the twelfth year. The base of the dens is separated from the vertebral body by a cartilaginous disc, which gradually becomes ossified at its circumference, but remains cartilaginous in its center until advanced age. In this cartilage, rudiments of the caudal epiphyseal lamella of the atlas and the cranial epiphyseal lamella of the axis may sometimes be found. Additionally, there is a secondary center for a thin epiphyseal plate on the caudal surface of the body of the bone which ossifies at about the seventeenth year.

SEVENTH CERVICAL VERTEBRA.   In addition to the ossification centers described for typical cervical and thoracic vertebrae, the ventral or costal part of the transverse processes of the seventh cervical vertebra is sometimes ossified from separate centers which appear about the sixth month of fetal life. These generally join the body and dorsal part of the transverse process between the fifth and sixth years. They may continue to grow laterally as cervical ribs.

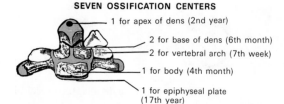

SEVEN OSSIFICATION CENTERS

1 for apex of dens (2nd year)

2 for base of dens (6th month)

2 for vertebral arch (7th week)

1 for body (4th month)

1 for epiphyseal plate (17th year)

FIG. 4-40. Ossification centers in the axis (second cervical vertebra).

LUMBAR VERTEBRAE. Each lumbar vertebra (Fig. 4-41) has two additional ossification centers for the mammillary processes, as well as those described for typical cervical and thoracic vertebrae. Similar to those in the transverse and spinous processes, the mammillary processes ossify at about the seventeenth year. The transverse process of the first lumbar is sometimes developed separately, and may remain permanently apart from the rest of the bone, thus forming a rare peculiarity, a lumbar rib.

THE SACRUM (Figs. 4-42 to 4-45). The body of each sacral vertebra is ossified from a primary center and two epiphyseal plates, while each vertebral arch becomes ossified from two centers. Cranial and lateral to the pelvic foramina, six or more additional ossification centers develop, two (one on each side) for each of the upper three or four sacral vertebrae. These represent the costal elements of the sacral segments (Figs. 4-42; 4-43). On the cranial and caudal surfaces of each sacral vertebral body, ossification centers for the epiphyseal plates develop (Fig. 4-44). Further, on each lateral surface two epiphyseal plates are formed (Figs. 4-44; 4-45): one for the auricular surface, and another for the remaining lower part of the thin lateral edge of the bone.

About the eighth or ninth week of fetal life, ossification of the central part of the body of the first sacral vertebra commences; similarly, ossification rapidly follows in the second and third. Ossification does not commence in the bodies of the last two sacral segments until the fifth to eighth month of fetal life. Between the sixth and eighth months ossification of the vertebral arches occurs, and at about the same time the costal centers make their appearance laterally. The vertebral arches join with the

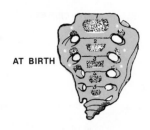

AT BIRTH

FIG. 4-42. The sacrum at birth shows ossification not only in the five centra, but also in the costal elements on each side of the upper three or four sacral vertebrae (marked by asterisks).

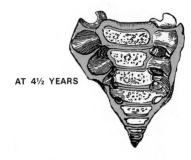

AT 4½ YEARS

FIG. 4-43. By 4½ years, the ossified lateral sacral processes have joined the centra, but the lateral epiphyseal plates and those between the bodies are still cartilaginous.

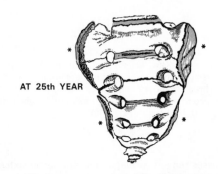

AT 25th YEAR

FIG. 4-44. By the twenty-fifth year, ossification centers have appeared not only in the intervertebral epiphyseal plates, but in those bordering the lateral margins of the sacrum (marked by asterisks).

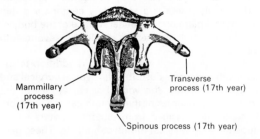

Mammillary process (17th year)

Transverse process (17th year)

Spinous process (17th year)

FIG. 4-41. Ossification centers in lumbar vertebrae include an additional two secondary centers for the mammillary processes.

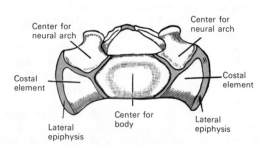

Center for neural arch

Center for neural arch

Costal element

Costal element

Center for body

Lateral epiphysis

Lateral epiphysis

FIG. 4-45. Base of the sacrum in a child of four to five years (viewed from above).

bodies in the more caudal vertebrae as early as the second year, whereas this does not occur in the more cranial until the fifth or sixth year. About the sixteenth year the epiphyseal plates for the superior and inferior surfaces of the bodies are formed; and between the eighteenth and twentieth years, those for the lateral surfaces make their appearance. During early life the bodies of the sacral vertebrae are separated from each other by intervertebral discs, but by about the eighteenth year the last two segments become united by bone, and the process of bony union gradually extends cranially. Between the twenty-fifth and thirtieth years of life all the sacral segments are united intervertebrally.

THE COCCYX. The coccyx is ossified from four primary centers, one for each segment. These centers make their appearance soon after birth in the first segment and between the fifth and tenth years in the second segment. Ossification of the third segment commences before puberty (between the tenth and fifteenth years), and after puberty in the fourth segment (between the fourteenth and twentieth years). As age advances, the segments unite intervertebrally. The union between the first and second segments is frequently delayed until after the age of twenty-five or thirty. At a late period of life, especially in women, the coccyx often fuses with the sacrum.

THE THORAX

The skeleton of the **thorax** or **chest** (Figs. 4-46 to 4-48) forms an osseocartilaginous cage which protects the principal organs of respiration and circulation. It extends inferiorly over the upper part of the abdomen, thereby covering portions of certain abdominal organs. It is conical in shape, so that its superior or cervical inlet is of lesser diameter than its inferior abdominal end. In transverse section it appears kidney-shaped because of the projection of the vertebral bodies into the cavity. The thoracic cage is formed by the ribs with their costal cartilages attached anteriorly to the sternum (Fig. 4-46) and posteriorly to the thoracic vertebrae of the spinal column (Fig. 4-48).

BOUNDARIES. The **dorsal surface** is formed by the twelve thoracic vertebrae and the dorsal parts of the twelve ribs. The **ventral surface** is formed by the sternum and costal cartilages. The **lateral surfaces** are formed by the ribs, separated from each other by the eleven intercostal spaces within which are located the intercostal muscles and membranes.

The **superior opening** of the thoracic cage is broader from side to side than dorsoven-trally. It is formed by the first thoracic vertebra, the cranial margin of the sternum, and the first rib on each side. Its dorsoventral diameter is about 5 cm, and its transverse diameter about 10 cm. The plane of the aperture slopes caudally from the spinal column toward the sternum, since the upper border of the sternum lies 3 to 5 cm below the upper border of the body of the first thoracic vertebra. Thus, the thoracic inlet faces ventralward as well as cranialward (Figs. 4-46; 4-48).

The **inferior opening** or thoracic outlet is formed dorsally by the twelfth thoracic vertebra, laterally by the eleventh and twelfth ribs, and ventrally by the cartilages of the tenth, ninth, eighth, and seventh ribs, which slope on each side to form the **infrasternal angle** at the xiphoid process. This caudal opening, also wider transversely than dorsoventrally, inclines obliquely caudalward and dorsalward. It is closed by the diaphragm, which forms both the floor of the thorax and the roof of the abdomen.

The thorax of the female differs from that of the male as follows: (1) its capacity is less; (2) the sternum is shorter; (3) the cranial margin of the sternum is on a level with the caudal part of the body of the third thoracic vertebra, whereas in the male it is on a level with the caudal part of the body of the second; and (4) the upper ribs are more movable, permitting a greater expansion of the cranial part of the thorax.

The Sternum

The **sternum** (Figs. 4-49 to 4-53) is an elongated, flat bone, which forms the middle portion of the ventral wall of the thorax. Its cranial end supports the clavicles, and its lateral margins articulate with the cartilages of the first seven pairs of ribs. The adult sternum consists of three parts, the **manubrium,** the **body,** and the **xiphoid process.** In early life, the sternum consists of six segments. The first and sixth remain as separate parts, the manubrium and xiphoid process respectively, while the intermediate four segments or **sternebrae** fuse to become the sternal body. The sternum is broad superiorly where the first ribs attach and in the region of the body where the fourth and fifth ribs attach. It narrows at the junction of the manubrium with the body, where the second ribs articulate, and once again at its inferior extremity (Fig. 4-49). In its natural position, the sternum is directed inferiorly, the caudal

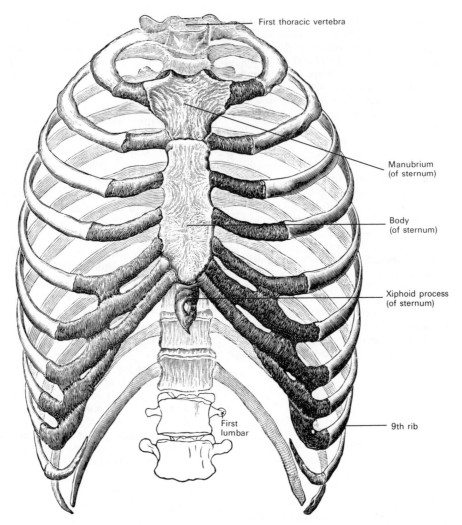

First thoracic vertebra

Manubrium
(of sternum)

Body
(of sternum)

Xiphoid process
(of sternum)

First
lumbar

9th rib

FIG. 4-46. The ventral aspect of the thoracic cage. Observe that all 12 pairs of ribs articulate with their respective thoracic vertebrae; however, only the first seven pairs attach directly to the sternum. The eighth and ninth ribs attach to the cartilage of the seventh rib on each side, while the tenth, eleventh and twelfth pairs have no anterior articulation. (From Spalteholz.)

end of the bone being inclined somewhat anteriorly and away from the vertebral column. Its average length in the adult is about 17 cm, being rather longer in males than in females. Projected posteriorly onto the vertebral column, the manubrium lies at the level of the third and fourth thoracic vertebrae. Its articulation with the sternal body is level with the intervertebral disc between the fourth and fifth vertebrae. The body of the sternum lies anterior to the fifth to tenth thoracic vertebrae. A line from the xiphoid process projects back to the eleventh thoracic vertebra.

The **manubrium** (*manubrium sterni*) is quadrangular in shape, broad and thick

cranially, but more narrow caudally at its junction with the body.

Its **ventral surface** is smooth, and on each side affords attachment to the pectoralis major and sternocleidomastoideus muscles (Fig. 4-49). The ridges limiting the attachments of these muscles are at times quite distinct. Its **dorsal surface,** concave and smooth, affords attachment on each side to the sternohyoid and sternothyroid muscles (Fig. 4-50).

The **cranial border** of the manubrium is thick and presents at its center the **jugular** (suprasternal) **notch,** on each side of which, at the **clavicular notch,** is an oval articular surface for the sternal end of the clavicle.

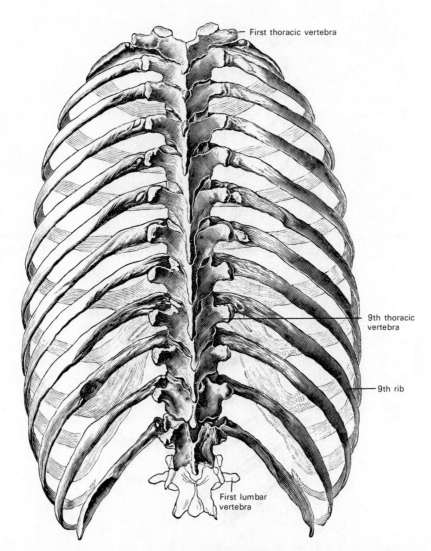

First thoracic vertebra

9th thoracic vertebra

9th rib

First lumbar vertebra

FIG. 4-47. The dorsal aspect of the thoracic cage. Note that each rib articulates with its respective thoracic vertebra, and upon encircling the trunk in an inferolateral direction, articulates anteriorly at a site lower than its vertebral commencement. Thus, any respective intercostal space lies higher dorsally than ventrally. (From Spalteholz.)

The **caudal border,** oval and rough, is surfaced by a thin layer of cartilage for articulation with the body. The upper aspect of the **lateral borders** is marked by a depression for the first costal cartilage. Found caudally is a small facet that forms part of the notch for the reception of the costal cartilage of the second rib.

The **body** (*corpus sterni*) of the sternum is considerably longer, narrower, and thinner than the manubrium. Its **ventral surface** is nearly flat and is marked by three transverse ridges which cross the bone opposite the third, fourth, and fifth costal articular depressions. These ridges define the sites of ossification of the original four sternebrae.

This anterior surface affords attachment on each side to the sternal origin of the pectoralis major, the line of origin of this muscle continuing down from the manubrium. At the junction of the third and fourth sternebrae of the body, a sternal foramen of varying size and form is occasionally seen. The slightly concave **dorsal** or **internal surface** is also marked by three transverse lines, less distinct, however, than the ventral; from its caudal part, the transversus thoracis muscle takes origin on each side (Fig. 4-50).

The **cranial border** of the sternal body is oval and articulates with the manubrium, the junction of the two forming the **sternal angle** (*angle of Louis*). The **caudal border** is

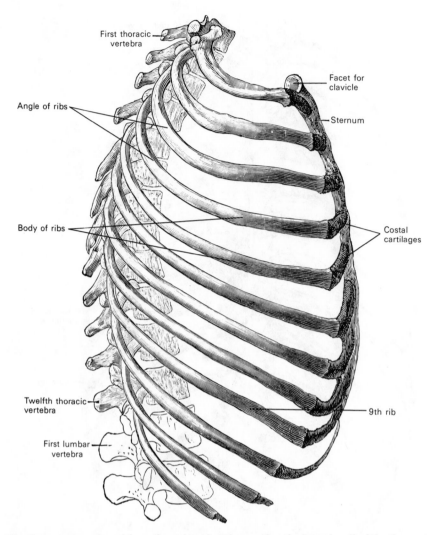

First thoracic vertebra

Facet for clavicle

Angle of ribs

Sternum

Costal cartilages

Body of ribs

Twelfth thoracic vertebra

9th rib

First lumbar vertebra

FIG. 4-48. The thoracic cage viewed from the right side. Observe that the lateral wall of the thorax is supported by the bodies of the ribs. Note that the ribs initially course in an inferolateral direction from the vertebral column and then inferomedially to achieve the costal cartilages and the sternum. (From Spalteholz.)

narrow, and articulates with the xiphoid process. Each **lateral border** (Fig. 4-51), at its cranial end, has a small facet which, with a similar facet on the manubrium, forms a cavity for the cartilage of the second rib. Caudal to this are found four angular depressions which receive the cartilages of the third, fourth, fifth, and sixth ribs. At its caudal end the sternal body has a small facet, which combines with a corresponding one on the xiphoid process to form a notch for the cartilage of the seventh rib. These articular depressions are separated by a series of curved interarticular intervals, which diminish in length caudally, and correspond to the ventral extremities of the intercostal spaces. Most of the costal cartilages extend-

ing from the true ribs articulate with the sternum at the lines of junction of its primitive component sternebrae segments. This is frequently seen even more clearly in many lower animals, in which the sternebrae remain distinct longer than in man.

The **xiphoid process** (*processus xiphoideus*) is the smallest of the three parts of the sternum. It is thin and elongated, cartilaginous in youth, but more or less ossified proximally in the adult. Its **ventral surface** affords attachment on each side to the ventral costoxiphoid ligament and to a small part of the rectus abdominis muscle. Onto the **dorsal surface** are attached the dorsal costoxiphoid ligament and some of the fibers of the diaphragm and the transversus thoracis

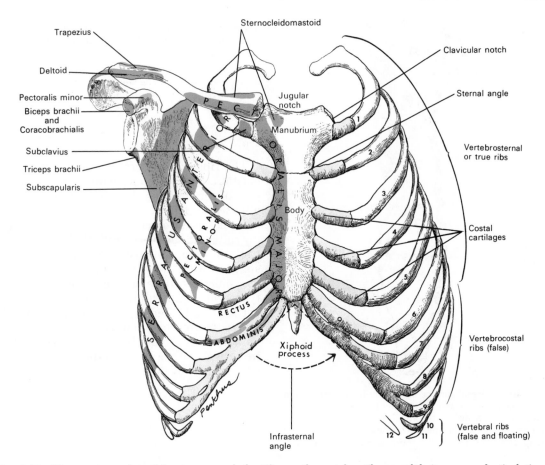

FIG. 4-49. The anterior surface of the sternum and ribs. Observe the costal cartilages and their manner of articulation with the sternum. Suprasternally note the jugular notch on the superior border of the manubrium, lateral to which articulate the clavicles. The infrasternal angle is formed by the midline articulation of the costal margins of the two sides. Muscle origins are in red, insertions in blue.

muscles. The **lateral borders** give attachment to some fibers of the aponeuroses of the abdominal muscles.

The xiphoid process articulates with the sternal body, and at each superior angle presents a facet for part of the cartilage of the seventh rib. The pointed caudal extremity gives attachment to the linea alba. The xiphoid process varies much in form; it may be broad and thin, pointed, bifid, perforated, curved, or deflected considerably to one side or the other.

STRUCTURE. The sternum is composed of highly vascular cancellous bone, covered by a thin, compact layer which is thickest in the manubrium between the articular facets for the clavicles. When necessary, the sternum is utilized by hematologists for *sternal puncture,* in which a needle of large bore is thrust through the thin subcutaneous cortical bone and a specimen of red marrow aspirated.

OSSIFICATION. The sternum early appears as two cartilaginous bars, situated one on each side of the median plane. Each of these two sternal plates is connected with the cartilages of the upper nine ribs of its own side. The two bars then fuse along the midline to form the cartilaginous sternum, which is subsequently ossified from six centers: one for the manubrium, four for the body, and one for the xiphoid process (Fig. 4-52). The ossification centers appear in the intervals between the articular depressions for the costal cartilages, in the following order: in the manubrium and first piece of the body during the sixth prenatal month; in the second and third sternebrae of the body during the seventh month of fetal life; in the fourth sternebrae of the body during the first year after birth; and in the xiphoid process between the fifth and eighteenth years. As a rule, ossification commences in the more superior parts of the segments and proceeds caudalward. Occasionally two additional small episternal centers may develop on each side of the jugular notch. These are probably vestiges of the episternal bone of the monotremes and lizards.

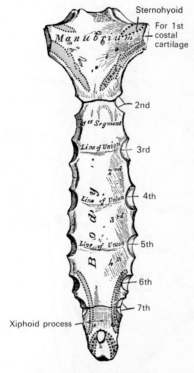

FIG. 4-50. Dorsal surface of the sternum. Observe the transverse lines of union between the segments of the sternal body. Note also the attachments of the sternohyoid and sternothyroid muscles on the manubrium and the diaphragm and transverse thoracis muscles inferiorly.

Multiple ossification centers within the sternal segments are frequently encountered, the number and position of which vary. Thus, manubrium may develop from two, three, or even six centers. When two are present, they are generally situated one superior to the other, the cranial being the larger. The second piece seldom has more than one center, while the third, fourth, and fifth pieces are often formed from two laterally placed centers, the irregular union of which explains the rare occurrence of a sternal foramen. The imperfect union of the longitudinally oriented cartilaginous bars may result in a vertical fissure constituting the malformation known as **fissura sterni.** This condition is seen more frequently lower in the sternum, but may occur even in the manubrium. If extensive, pulsations of the heart may be readily observed through the skin of the anterior chest wall.

Union of the four segments of the sternal body commences at puberty and proceeds cranially from the lower segments (Fig. 4-53). By the age of twenty-five the segments of the sternal body are all united. The xiphoid process may become ossified to the sternal body before the age of thirty, but this occurs more frequently after forty; at times the xiphisternal junction remains ununited in old age. In advanced life the manubrium may occasionally be joined to the body by bone. When this happens,

however, the bony tissue is usually superficial, and the central portion of the intervening cartilage remains unossified.

ARTICULATIONS.   The sternum articulates on each side with the clavicle and first seven costal cartilages.

The Ribs

The **ribs** (*costae*) are flattened, narrowed, elastic arches of bone, which form a large part of the thoracic skeleton. There are usually twelve pairs. The first seven pairs, called **true** or **vertebrosternal ribs,** are connected dorsally with the vertebral column, and ventrally by means of costal cartilages, with the sternum (Figs. 4-46 to 4-49). The remaining five pairs are false ribs and consist of two types: the eighth, ninth, and tenth have their cartilages attached to the cartilage of the rib above **(vertebrochondral)**, while the eleventh and twelfth are free at their ventral extremities and termed **floating** or **vertebral ribs,** since they do not attach to the sternum.

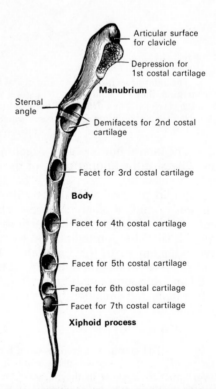

FIG. 4-51. Lateral border of the sternum. Observe that the facets for the second costal cartilages lie laterally at the junction of the manubrium and sternal body, while the facets for the seventh cartilages lie laterally at the xiphisternal junction.

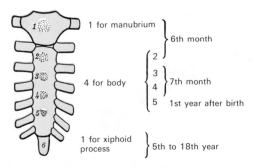

1 for manubrium } 6th month

4 for body { 2 \
3 \
4 } 7th month \
5 1st year after birth

1 for xiphoid process } 5th to 18th year

Fig. 4-52. Ossification centers for the sternum and their times of appearance. Multiple centers are frequently seen in certain of the segments.

The ribs vary somewhat in their shape and direction, the upper ones being less oblique, and curving at a more acute angle than the lower. The obliquity reaches its maximum at the ninth rib, and gradually decreases again from that rib to the twelfth. The ribs are arranged in a serial manner below one another, each being separated from the next by intercostal spaces. The length of each space around the thorax corresponds to that of the adjacent ribs and their cartilages. The breadth of the intercostal spaces is greater ventrally than dorsally and greater between the more superior ribs than between the lower ones. The ribs increase in length from the first to the seventh, then diminish to the twelfth.

CHARACTERISTICS OF THE RIBS (Figs. 4-54; 4-55). A rib from the middle of the series may be used to study the common characteristics of these bones. Each rib has two extremities, a dorsal or vertebral, and a ventral or sternal. The intervening portion is called the body or shaft.

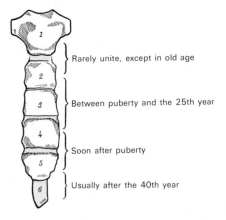

Rarely unite, except in old age

Between puberty and the 25th year

Soon after puberty

Usually after the 40th year

Fig. 4-53. Times when ossified union occurs between the six segments of the sternum.

The **dorsal** or **vertebral extremity** presents for examination a head, neck, and tubercle.

The **head** is marked by two articular facets which are separated by a horizontal **interarticular crest** (Fig. 4-55). The more inferior of the two facets is usually the larger and articulates onto the centrum of the segmentally corresponding vertebra. The cranial facet is smaller and articulates with the next superior vertebral body. The crest is attached to the intervertebral disc by the interarticular ligament.

The **neck** (Fig. 4-55) is the flattened portion of the rib which extends lateralward from the head. It is about 2.5 cm long, and is placed ventral to the transverse process of the more caudal of the two vertebrae with which the rib head articulates. Its ventral surface is flat and smooth; its dorsal surface is roughened for the attachment of the costotransverse ligament (of the neck of the rib) and perforated by numerous foramina. Of its two borders the cranial is roughest and presents a sharp crest (*crista colli costae*) for the attachment of the anterior costotransverse ligament. The caudal border is rounded and is continuous laterally with the ridge of the costal groove. The **tubercle** is an eminence on the posterior surface of the rib at the junction of the neck and body, and consists of an articular and a nonarticular portion (Fig. 4-55). The articular portion, the more caudal and medial of the two, presents a small, oval facet for articulation with the end of the transverse process of the numerically corresponding vertebra (the more caudal of the two vertebrae to which the head is connected). The nonarticular portion is a rough elevation, and affords attachment to the lateral costotransverse ligament (ligament of the tubercle of the rib). The tubercle is much more prominent in the upper than in the lower ribs.

The **body** or **shaft** of a rib is thin and flat and presents an external and an internal surface along with two borders, a superior and an inferior. About six centimeters beyond the tubercle, the **external surface** is marked by a prominent line which gives attachment to a tendon of the iliocostalis muscle. At this site the shaft of the rib turns ventrally at a point, the **angle**. The rib is bent in two directions, and at the same time twisted on its long axis. If placed upon its inferior border, the portion dorsal to the angle is bent medialward and at the same time tilted

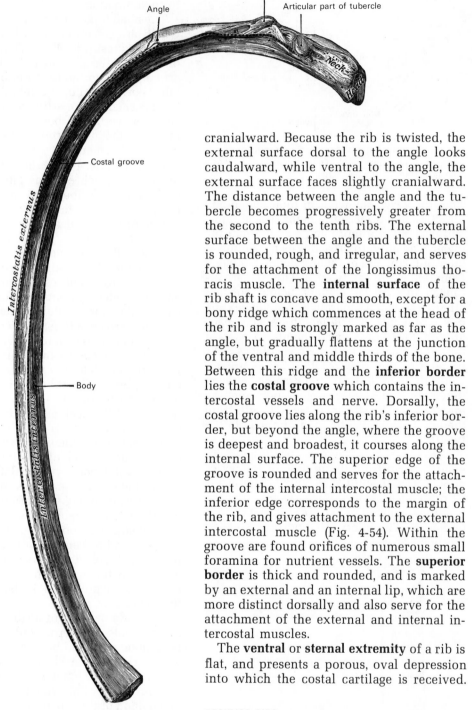

FIG. 4-54.   The inferior aspect of a typical central rib from the left side. Note that the principal parts of the bone include the head, neck, tubercle (which possesses an articular eminence and a nonarticular eminence), and the curved body or shaft. The costal groove lodges the intercostal vessels and nerve.

cranialward. Because the rib is twisted, the external surface dorsal to the angle looks caudalward, while ventral to the angle, the external surface faces slightly cranialward. The distance between the angle and the tubercle becomes progressively greater from the second to the tenth ribs. The external surface between the angle and the tubercle is rounded, rough, and irregular, and serves for the attachment of the longissimus thoracis muscle. The **internal surface** of the rib shaft is concave and smooth, except for a bony ridge which commences at the head of the rib and is strongly marked as far as the angle, but gradually flattens at the junction of the ventral and middle thirds of the bone. Between this ridge and the **inferior border** lies the **costal groove** which contains the intercostal vessels and nerve. Dorsally, the costal groove lies along the rib's inferior border, but beyond the angle, where the groove is deepest and broadest, it courses along the internal surface. The superior edge of the groove is rounded and serves for the attachment of the internal intercostal muscle; the inferior edge corresponds to the margin of the rib, and gives attachment to the external intercostal muscle (Fig. 4-54). Within the groove are found orifices of numerous small foramina for nutrient vessels. The **superior border** is thick and rounded, and is marked by an external and an internal lip, which are more distinct dorsally and also serve for the attachment of the external and internal intercostal muscles.

The **ventral** or **sternal extremity** of a rib is flat, and presents a porous, oval depression into which the costal cartilage is received.

PECULIAR RIBS

The first, second, tenth, eleventh, and twelfth ribs present certain variations from the common characteristics described above, and require special consideration.

The **first rib** (Fig. 4-56) is the most curved

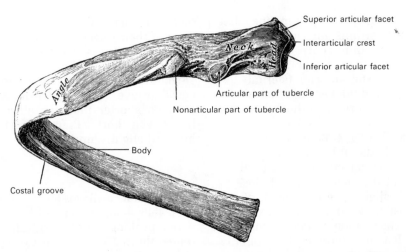

FIG. 4-55. A typical central rib from the left side viewed from behind. Observe that the vertebral extremity at the head end of the rib contains a large inferior facet for the numerically corresponding vertebra and a smaller facet for the adjacent vertebra above. Between them is the interarticular crest which attaches to the intervertebral disc.

and usually the shortest of all the ribs. It is broad and flat, its surfaces face cranially and caudally, and its borders internally and externally. The rounded head is small, and possesses only a single articular facet by which it articulates with the body of the first thoracic vertebra. The neck is narrow and also rounded, while the tubercle, thick and prominent, is placed on the outer border. There is no angle, but at the tubercle the rib is slightly bent, presenting a cranial convexity. The **superior surface** of the body is

marked by two shallow grooves, separated from each other by a slight ridge prolonged internally as the **scalene tubercle,** onto which attaches the scalenus anterior muscle. *The ventral groove transmits the subclavian vein; the dorsal groove, the subclavian artery and the inferior trunk of the brachial plexus* (Fig. 4-56). Beyond the dorsal groove is a rough area for the insertion of the scalenus medius. The **caudal surface** is smooth, and has no costal groove. The **outer border** is convex, thick, and rounded, and at

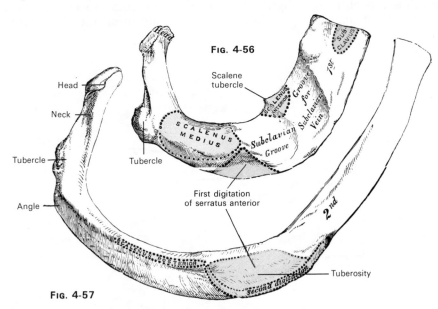

FIG. 4-56. The right first rib viewed from above. Observe the large tubercle below the rib head and the scalene tubercle accentuating the shaft.

FIG. 4-57. The right second rib viewed from above. Note the small tubercle, the slight angle nearby, and the tuberosity for serratus anterior.

its dorsal part gives origin to the first digitation of the serratus anterior. The inner border is concave, thin, sharp, and marked near its center by the scalene tubercle already described. The anterior (sternal) extremity is larger and thicker than that of any other rib and gives origin to the subclavius muscle.

The **second rib** (Fig. 4-57) is considerably longer than the first, but has a similar curvature. The nonarticular portion of the tubercle is small and the angle, situated close to the tubercle, is slight. The body shaft of the rib is not twisted, so that both ends touch any plane surface upon which it may be laid. The shaft is not flattened horizontally like that of the first rib, but proceeding ventrally from the tubercle, there is a cranial convexity, which is smaller than that seen in the first rib. Near the middle of its **outer surface** is the roughened **tuberosity of the second rib,** on which arises part of the first and the whole of the second digitation of the serratus anterior. Extending dorsally from the tuberosity is the insertion of the scalenus posterior. The internal surface is smooth and concave, and on its posterior part there is a short costal groove.

The **tenth rib** (Fig. 4-58) usually has only a single articular facet on its head for articulation with the body of the tenth thoracic vertebra.

The **eleventh** and **twelfth ribs** (Figs. 4-59; 4-60) each have a large single articular facet

on the rib head for articulation with the centrum of their respective vertebra. These two ribs have no necks or tubercles, and are narrow or pointed at their anterior ends. The eleventh has a slight angle and a shallow costal groove, but the twelfth has neither. It is much shorter than the eleventh, and at times even shorter than the first. Its head is inclined slightly upward.

COSTAL CARTILAGES

The **costal cartilages** (*cartilagines costales*; Figs. 4-46; 4-49) are bars of hyaline cartilage which prolong the ribs ventrally, and which render the chest wall a more elastic nature in order to accommodate, among other things, the movements of respiration. The first seven pairs are connected with the sternum; the next three each articulate with the lower border of the cartilage of the preceding rib; the last two have pointed anterior extremities which end in the musculature of the abdominal wall. They increase in length from the first to the seventh, then gradually decrease to the twelfth. Their breadth diminishes from the first to the last, as do the intercostal spaces between them. The first two are the same breadth throughout their extent, while the third, fourth, and fifth are broader at their lateral attachments to the ribs than at their sternal ends. The sixth, seventh, and eighth are enlarged where their margins are in contact. The cartilages also

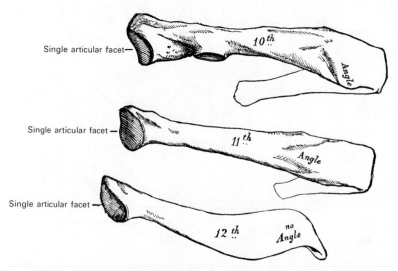

Single articular facet

10th

Angle

Single articular facet

11th

Angle

Single articular facet

12th

no Angle

FIGS. 4-58 to 4-60. Peculiar or atypical ribs. Each has but a single facet on its rib head. The eleventh and twelfth have no neck or tubercle, and the twelfth has no angle or costal groove.

vary in their orientation; the first three are horizontal, the others angular. Each costal cartilage presents two surfaces, two borders, and two extremities.

The **ventral surface** of the first cartilage gives attachment to the costoclavicular ligament and the subclavius muscle. The sternal ends of the first six or seven cartilages give origin to the pectoralis major (Fig. 4-49). The lower costal cartilages are covered by, and give partial attachment to, some of the flat muscles of the abdomen. The **dorsal surface** of the first cartilage gives attachment to the sternothyroid muscle, those of the third to the sixth inclusive to the transversus thoracis, and the last six or seven to the transversus abdominis and diaphragm.

The **superior borders** are concave while the **inferior borders** are convex. They afford attachment to the internal intercostal muscles along with the external intercostal membranes. Generally, the upper aspect of the fifth and sixth cartilages also give origin to some fibers of the pectoralis major. The inferior borders of the sixth, seventh, eighth, and ninth cartilages present heel-like projections at their points of greatest convexity. These projections bear smooth oblong facets which articulate with facets on slight projections from the superior borders of the seventh, eighth, ninth, and tenth cartilages, respectively.

EXTREMITIES. The lateral end of each cartilage is continuous with the osseous tissue of its corresponding rib. The medial end of the first is continuous with the sternum, while the rounded medial ends of the six succeeding ones are received into shallow concavities on the lateral margins of the sternum (Fig. 4-51). The medial ends of the eighth, ninth, and tenth costal cartilages are more pointed, and each is connected with the cartilage immediately above. Those of the eleventh and twelfth are pointed and free. In old age the costal cartilages are prone to lose some of their elasticity and undergo superficial ossification.

STRUCTURE. The ribs consist of highly vascular cancellous bone enclosed in a thin layer of compact bone.

OSSIFICATION. Cartilaginous models of the ribs are seen in the eight-week-old fetus. At about this time a primary ossification center develops near the angle of the rib shaft. Generally this is seen first in the sixth and seventh ribs. Three additional epiphyseal centers, one for the head and one each for the articular and nonarticular parts of the tubercle, develop postnatally. The epiphyses for the head and tubercle make their appearance between the sixteenth and twentieth years, and are united to the body about the twenty-fifth year. The first rib has only three ossification centers, a primary center for the shaft and one secondary center for the rib head and only one for the tubercle. The eleventh and twelfth ribs each have only two centers, those for the tubercles being wanting.

VARIATIONS AND CLINICAL INTEREST. **Cervical ribs** derived from the costal process of the seventh cervical vertebra (Fig. 4-8) are not infrequent (about 1%) and are important clinically because they may give rise to neural and vascular symptoms. The cervical rib may be a mere epiphysis articulating only with the transverse process of the vertebra, but more commonly it consists of a defined head, neck, and tubercle, with or without a body. It extends laterally into the posterior triangle of the neck where it may terminate freely or join the first thoracic rib, the first costal cartilage, or even the sternum. It may vary much in shape, size, direction, and mobility. If a cervical rib reaches far enough ventrally, the roots or trunks of the brachial plexus and the subclavian artery and vein cross over it and are apt to be compressed. Pressure on the artery may obstruct the circulation while pressure on the nerves, especially the lower trunk of the brachial plexus (eighth cervical and first thoracic), may cause paralysis, pain, and paresthesia.

The thorax is frequently altered in shape by certain diseases. **Rickets** causes the ends of the ribs to become enlarged where they join the costal cartilages, giving rise to the so-called "**rachitic rosary,**" which in mild cases is only found on the internal surface of the thorax. Lateral to these bead-like enlargements, the softened ribs are depressed, leaving a groove along each side of the sternum, which in turn is forced outward by the bending of the ribs, thereby increasing the dorsoventral diameter of the chest. The ribs so affected are the second to the eighth. Those below are prevented from being depressed by the underlying abdominal organs. Since liver and spleen enlargement frequently occurs in rickets, the abdomen becomes distended and the lower ribs may be pushed anteriorly, causing a transverse groove just above the costal arch. This deformity or forward projection of the sternum, often asymmetrical, is known as **pigeon breast,** and may be taken as evidence of active or old rickets except in cases of primary spinal curvature.

An opposite condition known as **funnel breast** occurs in some rachitic children or adults, and also in others who give no further evidence of having had rickets. The lower part of the sternum is deeply depressed backward, producing an oval hollow in the lower sternal and upper epigastric regions. This does not appear to produce disturbances in vital functions.

The flattened **phthisical chest,** sometimes seen in pulmonary tuberculosis, is long and narrow, presenting a great obliquity of the ribs and projection of the scapulae. Contrastingly, the **barrel chest** of pulmonary emphysema is enlarged in all its diameters, and presents on section an almost circular outline.

In severe cases of **scoliosis,** or lateral curvature of the vertebral column, the thorax becomes much distorted. Due to rotation of the vertebral bodies, the ribs opposite the convexity of the dorsal curve become extremely convex behind. This distortion results in a bulging of the ribs posteriorly and a flattening anteriorly, so that the two ends of the same rib are almost parallel. Coincidentally, the ribs on the opposite side, i.e., on the concavity of the curve, are sunken and depressed behind, and bulging and convex in front.

THE SKULL

The **skull,** or bony framework of the head, presents by far the most complex skeletal anatomy of the body. Lodging and protecting the vitally required brain with its equally important vasculature, the human head, as in other vertebrates, contains organs housing the special senses of sight, hearing, balance, taste, and smell, as well as structures in the mouth and nasal cavity related to the specialized life-maintaining mechanisms of feeding, respiration, and communication. The fact that food is ingested into the oral cavity, and that this entrance to the digestive tract is located conveniently close to visual, auditory, and olfactory receptors, and to the brain itself, is probably more than coincidental.

The bones of the head may be divided into two types: (1) those of the **cranium,** which enclose the brain and consist of eight bones, and (2) those of the **skeleton of the face,** numbering fourteen bones. For the most part the bones are either flat or irregular and, except for the mandible, are joined to each other by immovable articulations called **sutures.** Prior to describing the characteristics of each separate bone, important features of the anatomy of the skull as a whole will be outlined. The exterior of the skull may be viewed from above **(norma verticalis);** from the side **(norma lateralis);** toward the face **(norma frontalis);** from the back **(norma occipitalis);** and from below **(norma basalis).**

Exterior of the Skull

NORMA VERTICALIS

The skull cap or **calvaria,** viewed from above, varies greatly in different skulls. In some skulls its shape is more or less oval, in others more nearly circular. It is traversed by three sutures: (1) the **coronal suture,** nearly transverse in direction, between the frontal and parietal bones; (2) the **sagittal suture,** placed in the midline, between the two parietal bones, and (3) the upper part of the **lambdoidal suture,** separating the parietals and the occipital. The point of junction of the sagittal and coronal sutures is named the **bregma,** and the junction of the sagittal and lambdoid sutures is called the **lambda.** In the adult skull, these two sites indicate the positions of the anterior and posterior fontanelles, or the so-called soft spots in the skull of an infant. On each side of the sagittal suture are the **parietal eminences** where the smooth, rounded surface of the parietal bones is maximally convex. Lateral to the sagittal suture and anterior to the lambda may be found, on one or both sides, the **parietal foramen,** through which courses a small emissary vein. A craniometric term, **obelion,** is applied to that point on the sagittal suture at the level of the parietal foramina, while the **vertex,** near the middle of the sagittal suture, is the topmost point of the skull. Anteriorly, the frontal eminences form the two prominences of the forehead. Below these is a smooth prominence above the root of the nose, the **glabella,** on each side of which are the two **superciliary arches.** Immediately above the glabella, in a small percentage of skulls, a frontal suture persists and extends along the midline to the bregma.

NORMA LATERALIS (Fig. 4-61)

Both cranial and facial bones of the skull can be inspected from the lateral aspect. Visible are the frontal, parietal, occipital, temporal, and zygomatic bones, as well as the great wing of the sphenoid, the orbital lamina of the ethmoid, and the lacrimal; nasal, and maxillary bones. The mandible also can be seen, but its description here is omitted. The following sutures can be observed from a lateral view: **temporozygomatic, frontozy-**

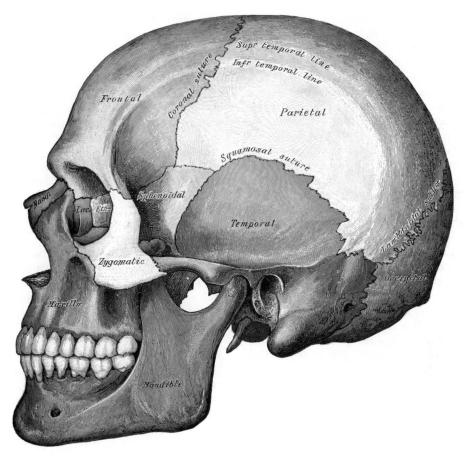

FIG. 4-61. Side view of the skull. (Norma lateralis.)

gomatic, **squamosal, zygomaticomaxillary,** and surrounding the great wing of the sphenoid—the **sphenofrontal, sphenoparietal,** and **sphenosquamosal sutures.** Additionally, several sutures on the medial aspect of the orbit between the lacrimal, ethmoid, maxillary, and frontal bones can be seen, as well as those sutures that extend laterally from above. Near the posterior base of the skull is the **occipitomastoid suture** joining the occipital and temporal bones. The sphenoparietal suture varies in length in different skulls, and is absent when the frontal bone articulates with the temporal squama. The point corresponding with the posterior end of the sphenoparietal suture is named the **pterion.** It is situated about 4 cm above the zygomatic arch and about 3 cm posterior to the frontozygomatic suture. The pterion overlies the anterior division of the middle meningeal artery.

The **squamosal suture** arches posteriorly from the pterion and lies between the temporal squama and the inferior border of the parietal. This suture is continuous posteriorly with the short, nearly horizontal **parietomastoid suture,** which unites the mastoid process of the temporal with the parietal bone at its mastoid angle.

Extending across the cranium are the coronal and lambdoidal sutures, of which the latter is continuous with the occipitomastoid suture. The mastoid foramen, which transmits an emissary vein, lies in or near the occipitomastoid suture. The meeting point of the parietomastoid, occipitomastoid, and lambdoidal sutures is known as the **asterion.** Arching across the side of the cranium are the temporal lines (Fig. 4-61), which mark the superior limit of the temporal fossa.

The **temporal fossa** (*fossa temporalis*) is a region on the lateral aspect of the skull occupied by the temporalis muscle which arises from the cranium as far superiorly as the **temporal lines** (Fig. 4-61). Extending from the zygomatic process of the frontal bone anteriorly, these two lines cross the frontal and parietal bones. Behind, they then curve anteriorly to become continuous with the supramastoid crest and the posterior root of the zygomatic arch. The craniometric term **stephanion** refers to the point where the **superior temporal line** cuts the coronal suture. The **inferior temporal line** lies somewhat below the superior. The temporal fossa is bounded *anteriorly* by the frontal and zygomatic bones. *Inferiorly,* it is separated from the infratemporal fossa by the **infratemporal crest** on the great wing of the sphenoid (Fig. 4-62), and by a ridge, continuous with this crest, which is carried posteriorly across the temporal squama to the anterior root of the zygomatic process. *Laterally* the temporal fossa is limited by the zygomatic arch. The fossa communicates with the orbital cavity through the **inferior orbital fissure** and, inferiorly, the temporal fossa is directly continuous with the **infratemporal fossa.** The floor of the fossa is deeply concave anteriorly and convex behind, and is formed by the zygomatic, frontal, parietal, sphenoid, and temporal bones. The temporal fossa contains the temporalis muscle and its vessels and nerves, together with the zygomaticotemporal nerve. The muscle descends to its insertion on the coronoid process of the mandible medial to the zygomatic arch.

The **zygomatic arch** is formed by the zygomatic process of the temporal bone and the temporal process of the zygomatic bone, the two being united by an oblique suture. The arch can easily be felt, extending anteriorly from the external auditory meatus. At its junctions with the squamous part of the temporal bone, the zygomatic process broadens into two bony ridges or roots. An **anterior root** is directed medially to the mandibular fossa, where it expands to form the articular tubercle. The **posterior root** expands posteriorly above the external acoustic meatus and is continuous with the supramastoid crest. The superior border of the zygomatic arch gives attachment to the temporal fascia, while the inferior border and medial surface give origin to the masseter muscle.

Between the posterior root of the zygomatic arch and the mastoid process is the elliptical orifice of the **external acoustic meatus** (Fig. 4-62). The opening is principally bounded by the tympanic part of the tempo-

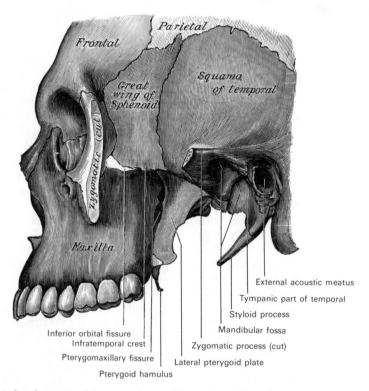

FIG. 4-62.   Left infratemporal fossa as observed from the left lateral aspect with the zygoma cut.

ral bone, although its superior margin is formed by the squamous part of the temporal bone. The cartilaginous segment of the external auditory canal is fixed to this roughened outer bony margin. The small triangular area between the posterior root of the zygomatic arch and the posterosuperior part of the orifice is termed the **suprameatal triangle,** on the anterior border of which is sometimes seen a small spinous process, the **suprameatal spine.** Deep to this triangular area lies the mastoid antrum of the middle ear. Between the tympanic part of the temporal bone and the articular tubercle lies the **mandibular fossa,** which is divided into two parts by the **petrotympanic fissure.** The anterior and larger part of the fossa receives the condyle of the mandible, while the posterior part is nonarticular and sometimes lodges a portion of the parotid gland. The delicate and pointed **styloid process** extends downward and forward for a distance of 2.5 to 3.0 cm from the undersurface of the temporal bone inferior to the external acoustic meatus. It gives attachment to the styloglossus, stylohyoid, and stylopharyngeus muscles and to the stylohyoid and stylomandibular ligaments. Posterior to the external acoustic meatus is the **mastoid process,** to the outer surface of which are inserted the sternocleidomastoideus, splenius capitis, and longissimus capitis. The posterior belly of the digastric muscle arises from the mastoid notch, which grooves the medial aspect of the process.

The **infratemporal fossa** (*fossa infratemporalis;* Fig. 4-62) is an irregularly shaped cavity situated below and medial to the zygomatic arch. It is bounded *in front* by the infratemporal surface of the maxilla; *behind* by the articular tubercle of the temporal bone and the spine of the sphenoid bone; *above* by the great wing of the sphenoid and by the undersurface of the temporal squama; *below* by the alveolar border of the maxilla; *medially* by the lateral pterygoid plate of the sphenoid; and *laterally* by the zygomatic arch. It contains part of the temporalis and the medial and lateral pterygoid muscles, the maxillary vessels, the mandibular and maxillary nerves, and the pterygoid plexus of veins. The foramen ovale and foramen spinosum open into it from above, while the alveolar canals open on its anterior wall. At its superior and medial part are two fissures which meet at right angles, the horizontal

limb being named the **inferior orbital fissure,** and the vertical limb the **pterygomaxillary fissure.**

The **inferior orbital fissure** (*fissura orbitalis inferior*) opens into the posterolateral part of the orbit (Figs. 4-61, 4-62). It is bounded *above* by the inferior border of the orbital surface of the great wing of the sphenoid; *below* by the lateral border of the orbital surface of the maxilla and the orbital process of the palatine bone; *laterally* by a small part of the zygomatic bone; and *medially* it joins the ptergyomaxillary fissure at right angles. Through the inferior orbital fissure, the orbit communicates with the temporal, infratemporal, and pterygopalatine fossae. It transmits the maxillary nerve, its zygomatic branch, ascending branches from the pterygopalatine ganglion, the infraorbital vessels, and the clinically important vein that connects the inferior ophthalmic vein with the pterygoid venous plexus.

The **pterygomaxillary fissure** is oriented vertically and descends at right angles from the medial end of the inferior orbital fissure (Fig. 4-61). It is a triangular interval, formed by the divergence of the maxilla from the pterygoid process of the sphenoid. It connects the infratemporal fossa with the **pterygopalatine fossa,** and transmits the terminal part of the maxillary artery and veins.

The **pterygopalatine fossa** (*fossa pterygopalatina;* Fig. 4-63) is a small cone-shaped or pyramidal region at the junction of the inferior orbital and pterygomaxillary fissures. It lies just beneath the apex of the orbit and is bounded *above* by the inferior surface of the body of the sphenoid bone and the orbital process of the palatine bone; *in front* by the infratemporal surface of the maxilla; *behind* by the base of the pterygoid process and the lower part of the anterior surface of the sphenoid bone; and *medially* by the perpendicular plate of the palatine bone. This fossa communicates with the orbit through the **inferior orbital fissure,** with the nasal cavity by the **sphenopalatine foramen,** and with the infratemporal fossa by way of the **pterygomaxillary fissure.** Additionally, the pterygopalatine fossa communicates with the cranial cavity by way of the **pterygoid canal,** which extends to the foramen lacerum, and with the roof of the oral cavity through the **pterygopalatine canal.** Of these five communicating foramina, the inferior orbital fissure opens anteriorly into the fossa while the

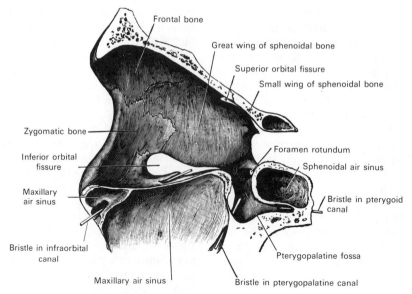

Frontal bone

Great wing of sphenoidal bone

Superior orbital fissure

Small wing of sphenoidal bone

Zygomatic bone

Inferior orbital fissure

Maxillary air sinus

Foramen rotundum

Sphenoidal air sinus

Bristle in pterygoid canal

Bristle in infraorbital canal

Pterygopalatine fossa

Maxillary air sinus

Bristle in pterygopalatine canal

FIG. 4-63. The pterygopalatine fossa and the lateral wall of the right orbit.

foramen rotundum and pterygoid canal are posterior. On the medial wall is the spheno-palatine foramen and below is found the pterygopalatine canal. The fossa contains the maxillary nerve, the pterygopalatine ganglion, and the terminal part of the maxillary artery.

NORMA FRONTALIS (Fig. 4-64)

The facial aspect of the skull is outlined by the frontal bone superiorly, below which are the nasal bones, maxilla, and body of the mandible medially, and the zygomatic bones and rami of the mandible laterally. The **frontal eminences** stand out more or less prominently above the **superciliary arches,** which merge into one another at the **glabella.** Below the glabella is the **frontonasal suture,** the midpoint of which is termed the **nasion.** Lateral to the frontonasal suture and at the medial margin of the orbit, the frontal bone articulates with the frontal process of the maxilla and with the lacrimal. Below the superciliary arches, the superior part of the orbital margin is thin and prominent in its lateral two-thirds, but rounded and more massive in its medial third. At the junction of these two portions is the **supraorbital foramen** or **notch** for the supraorbital nerve and vessels. The **supraorbital margin** ends laterally in the **zygomatic process** of the frontal bone (Fig. 4-65), which articulates with the zygomatic bone. Below the fronto-

nasal suture is the bridge of the nose, formed by the two nasal bones supported posteriorly in the midline by the **perpendicular plate** of the ethmoid, and laterally by the **frontal processes** of the maxillae, which form the lower and medial part of the margin of each orbit. The **bony anterior nasal aperture** is bounded by the sharp margins of the nasal and maxillary bones. The **lateral** and **alar cartilages** of the nose are attached to the lateral nasal margins, while the inferior margins are thicker and end in the **anterior nasal spine** medially. The bony septum that separates the nasal cavity in the midline shows a large triangular deficiency in the bone anteriorly. This is filled by **septal cartilage** in the intact body. On the lateral wall of each nasal cavity the anterior part of the **inferior nasal concha** is visible.

Below and on each side of the pear-shaped aperture of the nose are the **maxillae,** which articulate in the midline and form much of the anterior bony substructure of the face. The anterior surface of the maxilla is perforated near the inferior margin of the orbit by the **infraorbital foramen** for the passage of the infraorbital nerve and vessels. Inferiorly, the **alveolar process** of each maxilla contains the osseous sockets for the upper teeth. The vertical ridges formed by the roots of the teeth are clearly visible, and that of the canine tooth, the **canine eminence,** is the most prominent.

The zygomatic bone forms the promi-

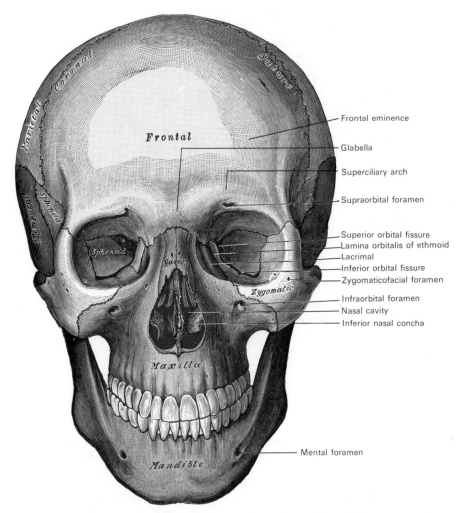

Frontal eminence
Glabella
Superciliary arch
Supraorbital foramen
Superior orbital fissure
Lamina orbitalis of ethmoid
Lacrimal
Inferior orbital fissure
Zygomaticofacial foramen
Infraorbital foramen
Nasal cavity
Inferior nasal concha
Mental foramen

FIG. 4-64. The skull from the front. (Norma frontalis.)

nence of the cheek, the inferior and lateral portion of the orbital cavity, and the anterior part of the zygomatic arch. On the face it articulates medially with the maxilla, superiorly with the zygomatic process of the frontal bone, and laterally with the zygomatic process of the temporal bone. The zygomatic bone is perforated by the zygomaticofacial foramen for the passage of the zygomaticofacial nerve.

The maxillae rest on the **mandible,** or lower jaw, which extends from a characteristically human bony landmark, the chin or **mental protuberance** on the mandibular **body,** to the mandibular fossae laterally, where the mandibular **rami** articulate with the temporal bones. Lateral to the mental protuberance are the **mental tubercles.** The superior or alveolar border contains the sockets for the lower teeth. Below the incisor teeth is the incisive fossa, and inferior to the second premolar tooth is the **mental foramen,** which transmits the mental nerve and vessels. An **oblique line** courses posteriorly from the mental tubercle and becomes continuous with the anterior border of the ramus. The posterior border of the ramus is directed downward and forward from the **condyle** to the **angle.** The latter frequently is more or less everted.

THE ORBITS (orbitae; Fig. 4-65)

The orbits are conical cavities which contain the eyeballs and their associated structures and are so placed that their medial walls are approximately parallel with each other and with the midline, but their lateral

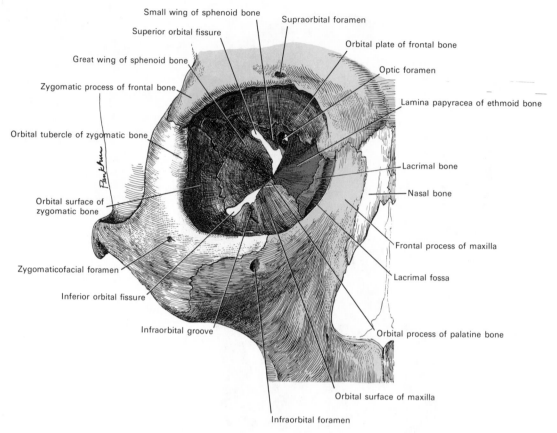

Small wing of sphenoid bone

Supraorbital foramen

Superior orbital fissure

Orbital plate of frontal bone

Great wing of sphenoid bone

Optic foramen

Zygomatic process of frontal bone

Lamina papyracea of ethmoid bone

Orbital tubercle of zygomatic bone

Lacrimal bone

Nasal bone

Orbital surface of zygomatic bone

Frontal process of maxilla

Zygomaticofacial foramen

Lacrimal fossa

Inferior orbital fissure

Infraorbital groove

Orbital process of palatine bone

Orbital surface of maxilla

Infraorbital foramen

FIG. 4-65. The anterior aspect of the right orbital cavity. Note the superior orbital fissure, inferior orbital fissure, and optic foramen within the orbital cavity, and the supraorbital, infraorbital, and zygomaticofacial foramina opening onto the surface.

walls are widely divergent. Thus, the longitudinal axes of the two orbits diverge widely anteriorly, but if prolonged posteriorly, they would cross dorsal to the body of the sphenoid bone near the center of the base of the skull. It is customary to describe the orbits in terms of a roof, a floor, a medial wall, a lateral wall, a base, and an apex.

The **roof** of the orbit (see Fig. 4-191), which separates the orbital contents from the overlying frontal lobes of the brain, is formed by the thin **orbital plate** of the frontal bone and the **lesser wing** of the sphenoid bone. The orbital surface of this plate presents medially the **trochlear fovea** for the attachment of the cartilaginous pulley of the obliquus superior oculi laterally. Anterolaterally is located the deep **lacrimal fossa** for the orbital part of the lacrimal gland. The posterior end of the orbital roof is marked by the **sphenofrontal suture.**

The **floor** of the orbit (Fig. 4-66) is formed chiefly by the orbital surface of the maxilla, but also by the orbital process of the zygomatic bone, and to a small extent by the orbital process of the palatine. Near the orbital margin medially is the lacrimal sulcus for the **nasolacrimal canal,** and just lateral to this a small depression for the origin of the obliquus inferior oculi. Located laterally is the maxillozygomatic suture, and at its posterior part is the suture between the maxilla and the orbital process of the palatine (palatomaxillary suture). Near the middle of the floor is the **infraorbital groove,** which leads anteriorly into the **infraorbital canal** and transmits the infraorbital nerve and vessels to the anterior aspect of the face.

The **medial wall** (Fig. 4-67) is delicate and oriented nearly vertically. It is formed by the frontal process of the maxilla, the lacrimal bone, the lamina orbitalis of the ethmoid, and a small part of the body of the sphenoid. Anteriorly, the **lacrimal fossa** is a hollow region between the lacrimal bone and the frontal process of the maxilla. It lodges the

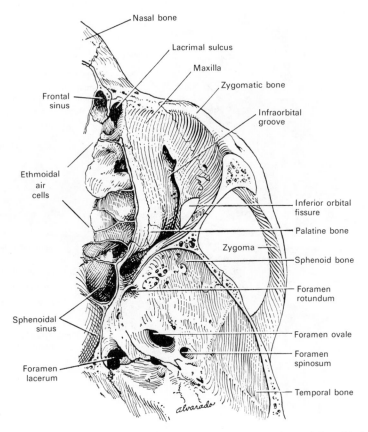

Nasal bone

Lacrimal sulcus

Maxilla

Zygomatic bone

Frontal
sinus

Infraorbital
groove

Ethmoidal
air
cells

Inferior orbital
fissure

Palatine bone

Zygoma

Sphenoid bone

Foramen
rotundum

Sphenoidal
sinus

Foramen ovale

Foramen
spinosum

Foramen
lacerum

Temporal bone

alvarado

FIG. 4-66. Horizontal section through the right orbit demonstrating the bony floor of the orbital cavity from above. Observe the infraorbital groove and the longitudinally oriented suture between the maxilla and the zygomatic bone.

lacrimal sac and is limited behind by the **posterior lacrimal crest,** from which the lacrimal part of the orbicularis oculi arises. The medial wall contains three vertical sutures: the **lacrimomaxillary, lacrimoethmoidal,** and **sphenoethmoidal.** At the junction of the medial wall and the roof are the **frontomaxillary, frontolacrimal, frontoethmoidal,** and **sphenofrontal sutures.** The point of junction of the anterior border of the lacrimal with the frontal is named the **dacryon.** In the frontoethmoidal suture are the **anterior** and **posterior ethmoidal foramina,** the former transmitting the anterior ethmoidal vessels and nerve, and the latter the posterior ethmoidal vessels and nerve. The nerves are branches of the nasociliary while the vessels are branches of the ophthalmic.

The **lateral wall** of the orbit (Fig. 4-63) is directed forward and medially and is formed by the orbital process of the zygomatic bone and the orbital surface of the great wing of the sphenoid. These two bones

articulate at the sphenozygomatic suture, which terminates at the anterior end of the **inferior orbital fissure.** Between the roof and the lateral wall, near the apex of the orbit, is the **superior orbital fissure.** Communicating with the cranial cavity, this fissure transmits the oculomotor, trochlear, ophthalmic division of the trigeminal, and abducent nerves to the orbit. Also entering the orbit through the superior orbital fissure are sympathetic nerve filaments from the cavernous plexus as well as the orbital branches of the middle meningeal artery. Leaving through the fissure for the cranial cavity are the superior ophthalmic vein and the recurrent branch from the lacrimal artery to the dura mater. The lateral wall and the orbital floor are separated posteriorly by the inferior orbital fissure.

The **base** of the orbit is its anterior end; it is formed by the orbital margin, and has already been described. The **apex** is situated at the back of the orbit and corresponds to

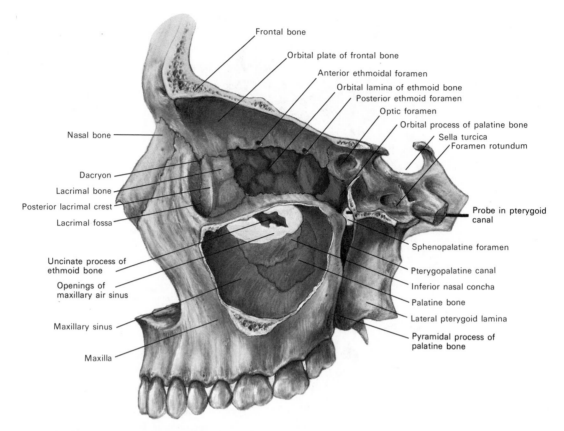

FIG. 4-67. Medial wall of left orbit. In this dissection the lateral wall of the left orbit has been removed in order to expose the medial wall, and the left maxillary sinus has been opened from the lateral aspect. Note that the medial wall of the orbit is formed by the maxilla, lacrimal, ethmoid, and, most posteriorly, the sphenoid bone. The ethmoid bone is shown in red, the palatine bone in blue.

the **optic foramen,** a short, cylindrical canal, which transmits the optic nerve and ophthalmic artery.

Ten openings communicate with each orbit: the optic foramen, superior and inferior orbital fissures, supraorbital foramen, infraorbital canal, anterior and posterior ethmoidal foramina, zygomaticofacial and zygomaticotemporal foramina, and the canal for the nasolacrimal duct.

NORMA OCCIPITALIS

When viewed from behind the cranium presents a circular, dome-like shape, except at its base, which is flattened. In the midline is observed the posterior continuation of the **sagittal suture,** which extends between the parietal bones to the deeply serrated **lambdoidal suture.** The latter lies between the parietal bones and the occipital and contin-

ues laterally into the **parietomastoid** and **occipitomastoid sutures.** Frequently, one or more sutural bones, such as an **interparietal (inca) bone,** may be interposed between the parietal and occipital bones. Near the middle of the occipital squama is the **external occipital protuberance,** and extending laterally from it on each side is the **superior nuchal line.** Also seen is the more faintly marked **highest nuchal line,** which extends laterally about 1 cm above the superior. If the skull is slightly elevated, the **inferior nuchal line** comes into view coursing laterally below the superior line. The **inion** is a craniometric term applying to the midline point on the external occipital protuberance. Descending from the inion to the **foramen magnum** is a bony ridge, the **external occipital crest,** which affords attachment to the ligamentum nuchae. Laterally and somewhat anteriorly on each side can be seen the mastoid process of the temporal bone, and near the occipitomastoid suture is the mastoid foramen through which passes the mastoid emissary vein.

NORMA BASALIS (Figs. 4-68; 4-69)

With the mandible removed, the inferior surface of the base of the skull is formed by the palatine processes of the **maxillae,** the horizontal plates of the **palatine** bones, the **vomer,** the pterygoid processes, the inferior surfaces of the great wings, the spinous processes and part of the body of the **sphenoid,** the inferior surfaces of the squamae and petrous portions of the **temporals,** and the inferior surface of the **occipital** bone.

The *anterior part of the normal basalis* projects downward beyond the level of the rest of the basal surface and is formed principally by the **hard palate.** The bony palate is bounded anteriorly and laterally by the alveolar arch containing the sixteen teeth of the maxillae. Just behind the incisor teeth is found the **incisive fossa.** Within the fossa are two **lateral incisive foramina** (foramina of Stenson) which continue as the incisive canals and transmit the terminal branches of the greater palatine vessels and the nasopalatine nerves from the nasal cavity. Occasionally two other **median incisive foramina** (foramina of Scarpa) are found. When present they transmit the nasopalatine nerves. The vault of the hard palate is marked by depressions for the palatine glands. It is traversed by a cruciate suture formed by the junction of the two palatine processes of the maxillae and the two horizontal plates of the palatine bones of which it is composed. At either posterior angle of the hard palate is the **greater palatine foramen,** for the transmission of the greater palatine vessels and nerve, which descend in the greater palatine canal from the pterygopalatine fossa. Posterior to the foramen is the pyramidal process of the palatine bone, perforated by one or more **lesser palatine foramina** through which course the lesser palatine vessels and nerve to the soft palate. At this site also is observed a transverse ridge for the attachment of the tendinous expansion of the tensor veli palatini muscle. Projecting backward from the center of the posterior border of the hard palate is the **posterior nasal spine,** on which attaches the musculus uvulae.

The *middle part of the norma basalis*

commences just behind the hard palate and extends to the level of the anterior border of the foramen magnum. Just behind the hard palate are the **choanae** or posterior openings into the nasal cavities. They are separated by the vomer, and each is bounded above by the body of the sphenoid, below by the horizontal part of the palatine bone, and laterally by the medial pterygoid plate of the sphenoid. At the superior border of the vomer, the expanded alae of this bone receive between them the rostrum of the sphenoid. The **medial pterygoid plate** is long and narrow; on the lateral side of its base is the **scaphoid fossa,** for the origin of the tensor veli palatini muscle, and at its lower extremity the **pterygoid hamulus,** around which the tendon of this muscle turns. The **lateral pterygoid plate** is broad; its lateral surface forms the medial boundary of the infratemporal fossa, and affords attachment to the pterygoideus lateralis muscle.

Posterior to the vomer is the basilar portion of the occipital bone, presenting near its center the **pharyngeal tubercle** for the attachment of the fibrous raphe of the pharynx, with depressions on each side for the insertions of the rectus capitis anterior and longus capitis. At the base of the lateral pterygoid plate is the **foramen ovale,** through which passes the mandibular nerve and the accessory meningeal artery. Lateral to this is the prominent **sphenoidal spine,** which gives attachment to the sphenomandibular ligament and the tensor veli palatini. Posterior and somewhat lateral to the foramen ovale is the **foramen spinosum,** which transmits the middle meningeal vessels and a small meningeal branch of the mandibular nerve. Lateral to the sphenoidal spine is the **mandibular fossa,** which is divided into two parts by the **petrotympanic fissure.** The *anterior portion*—concave, smooth, and bounded in front by the articular tubercle— articulates with the condyle of the mandible; the *posterior portion* is rough and bounded behind by the tympanic part of the temporal bone.

Observable in a dried skull at the base of the medial pterygoid plate is an aperture, irregular in shape and variable in size, named the **foramen lacerum.** It is not a complete foramen in the intact body, because its inferior part is covered over by a fibrocartilaginous plate, across the superior (inner or cerebral) surface of which passes the internal carotid artery. The foramen lacerum is bounded in front by the great wing of the sphenoid, behind by the apex of the petrous portion of the temporal bone, and medially by the body of the sphenoid and the basilar portion of the occipital bone. The inferior surface of the petrous temporal bone is pierced by the round opening of the **carotid canal.** The internal carotid artery, coursing within the canal, immediately takes a right-angle turn to reach the side of the foramen lacerum. At the upper end of the foramen lacerum the internal carotid artery enters the cranial cavity, taking an S-shaped course to achieve the undersurface of the brain. Lateral to the foramen lacerum is a groove, the **sulcus tubae auditivae,** between the petrous part of the temporal and the great wing of the sphenoid. This sulcus lodges the cartilaginous part of the **auditory tube** which is continuous with the bony part within the temporal bone. At the bottom of this sulcus is a narrow cleft, the **petrosphenoidal fissure.** Near the apex of the petrous portion of the temporal bone, the quadrilateral rough surface affords attachment to the levator veli palatini, lateral to which is the orifice or entrance of the carotid canal.

The *posterior part of the norma basalis* is formed principally by the occipital bone; however, laterally are the roughened, bony **mastoid processes** of the temporal bones. On the medial side of each mastoid process is the **mastoid notch** for the posterior belly of the digastricus, and medial to the notch is the occipital groove for the occipital artery. Medial and slightly anterior to the mastoid processes and emerging from between the laminae of the **vaginal process** of the tympanic part of the temporal bone is the **styloid process,** at the base of which is the **stylomastoid foramen.** Through this foramen the facial nerve exits in its course toward the side of the face, and the stylomastoid artery enters to supply the tympanic cavity of the middle ear. Medial to the styloid process and posterior to the carotid canal is the **jugular foramen,** a large aperture, formed between the jugular fossa of the petrous portion of the temporal bone anteriorly, and the occipital bone posteriorly. It is generally larger on the right than on the left side, and may be subdivided into three compartments. The *anterior compartment* transmits the inferior petrosal sinus; the *intermediate,* the glossopharyngeal, vagus, and accessory nerves; the

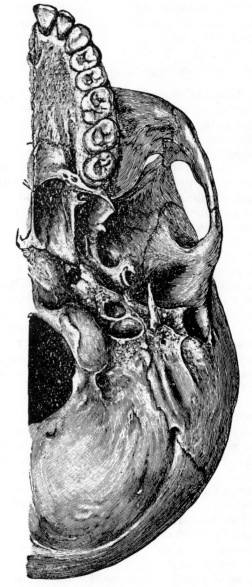

FIG. 4-68. The external surface of the left half of the base of the skull. (Norma basalis.)

posterior, the sigmoid sinus which leads to the internal jugular vein, and some meningeal branches from the occipital and ascending pharyngeal arteries. Extending anteriorly from the jugular foramen to the foramen lacerum is the **petro-occipital fissure,** occupied in the intact body by a plate of cartilage.

Posterior to the basilar portion of the occipital bone is the **foramen magnum,** bounded laterally by the **occipital condyles,** on the medial surfaces of which attach the alar ligaments. These condyles are oval in

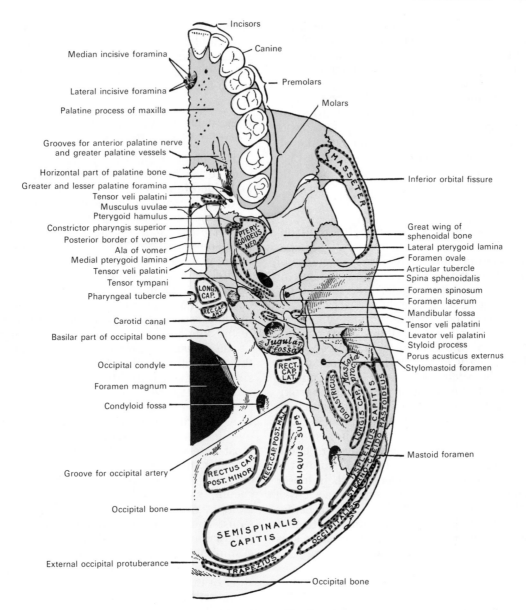

Fig. 4-69. Key to Figure 4-68.

shape and project inferiorly to articulate with the superior articular surfaces overlying the lateral masses of the atlas. Lateral to each condyle is the **jugular process,** to which attaches the rectus capitis lateralis muscle and the lateral atlanto-occipital ligament, which strengthens the atlanto-occipital articular capsules laterally. The foramen magnum transmits the medulla oblongata and its membranes, the accessory nerves, the vertebral arteries, the anterior and posterior spinal arteries, and the ligaments connecting the occipital bone with the axis. The mid-

points on the anterior and posterior margins of the foramen are respectively termed the **basion** and the **opisthion.**

The **hypoglossal canal** courses forward and laterally from a foramen on the inner aspect of the occipital bone within the cranium just above the foramen magnum to an opening that perforates the occipital bone externally at the lateral part of the base of the condyle. It transmits the hypoglossal nerve and a branch of the posterior meningeal artery. Posterior to each condyle is the **condyloid fossa,** perforated on one or both sides by the **condyloid canal,** for the transmission of a vein from the sigmoid sinus to the vertebral veins in the upper cervical region. Posterior to the foramen magnum is the **external occipital crest,** ending superiorly at the **external occipital protuberance.** On either side are the **superior** and **inferior nuchal lines** which, along with the surfaces of bone between them, are rough for the attachment of a number of muscles (see Figs. 4-69; 4-85).

Interior of the Skull

The **cranial cavity** is formed by the frontal, ethmoid, sphenoid, temporal, parietal, and occipital bones. The internal surface of the floor of the cranial cavity, i.e., the base of the skull, is quite irregular in shape, but the inner surfaces of the bones forming the walls and roof of the cranial vault are smoother and encase the rounded hemispheres of the cerebrum and cerebellum. The capacity of the normal cranial cavity varies between 1300 and 1500 cc, and the weight of the brain in grams is approximately the same as the figure for cranial capacity expressed in cubic centimeters. The cranial cavity is lined by two fibrous layers or membranes: the periosteal layer, or **endocranium,** which is closely adherent to the bone, and the **dura mater,** which is looser and completely covers the brain in a sac-like fashion. The interior of the skull will be described in two sections: the inner surface of the skullcap or **calvaria** and the internal surface of the base of the skull.

INNER SURFACE OF THE SKULLCAP

Formed principally by the frontal and parietal bones, the skullcap, or **calvaria,** also includes the upper parts of the squamae of the temporal and occipital bones laterally and posteriorly. Its inner surface is marked by depressions for the convolutions of the cerebrum, and by numerous furrows for the branches of the meningeal vessels. Along the midline is a longitudinal groove, narrow where it commences at the **frontal crest,** but broader posteriorly. It lodges the superior sagittal sinus, and its margins afford attachment to the falx cerebri. On each side of this groove are several **granular foveolae** or pits for the arachnoid granulations, and posteriorly, the openings of the **parietal foramina** when these are present. The coronal and lambdoidal sutures can be seen crossing the inner surface of the calvaria transversely while the sagittal suture lies in the median plane between the parietal bones.

INTERNAL OR SUPERIOR SURFACE OF THE BASE OF THE SKULL (Fig. 4-70)

The superior surface of the base of the skull forms the **floor of the cranial cavity** and is divided into three fossae, called the anterior, middle, and posterior cranial fossae.

ANTERIOR CRANIAL FOSSA (*fossa cranii anterior*). The floor of the anterior cranial fossa, on which rest the frontal lobes of the brain, is formed by the *orbital plates* of the frontal, and *cribriform plate* of the ethmoid, and the *small wings* and *anterior part of the body* of the sphenoid. It is limited behind by the posterior borders of the small wings of the sphenoid and by the anterior margin of the chiasmatic groove. It is traversed by the frontoethmoidal, sphenoethmoidal, and sphenofrontal sutures. Its lateral portions or **orbital plates** are convex and form the bony roof of the orbital cavities. The middle portion, formed primarily by the ethmoid bone, overlies the nasal cavities. In the midline most anteriorly is the frontal crest (of the frontal bone) on which attaches the **falx cerebri.** Just behind this crest is the **foramen cecum,** which usually transmits a small vein from the nasal cavity to the superior sagittal sinus. The upward continuation of the perpendicular plate of the ethmoid bone forms a midline bony crest, the **crista galli,** to which also attaches the falx cerebri. On both sides of the crista galli, the olfactory grooves of the **cribriform plate** support the olfactory bulbs. These grooves are perforated by foramina for the transmission of the olfactory nerves and, more rostrally, by a slit for the nasociliary nerve (Fig. 4-70). In the lateral wall of the olfactory groove are the internal openings of the **anterior** and **posterior ethmoidal foramina.** The anterior transmits the anterior ethmoidal vessels and the nasociliary nerve; the posterior ethmoidal foramen opens more caudally under cover of the projecting lamina of the sphenoid, and transmits the posterior ethmoidal vessels and nerve. Behind the ethmoid bone are the **ethmoidal spine** of the sphenoid and the anterior margin of the **chiasmatic groove.** This groove courses laterally on each side to the

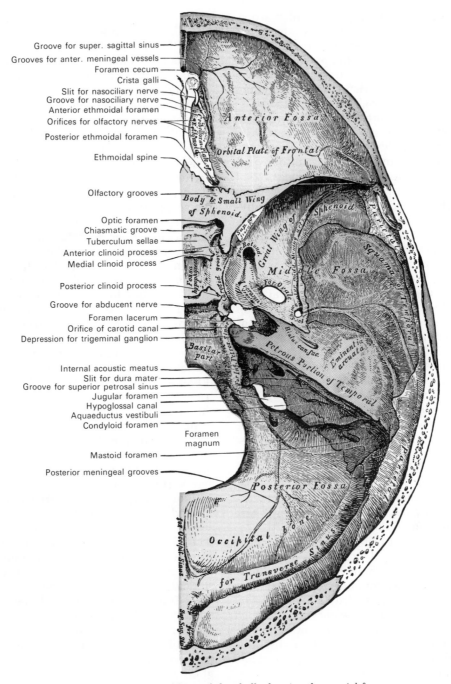

Groove for super. sagittal sinus
Grooves for anter. meningeal vessels
Foramen cecum
Crista galli
Slit for nasociliary nerve
Groove for nasociliary nerve
Anterior ethmoidal foramen
Orifices for olfactory nerves
Posterior ethmoidal foramen
Ethmoidal spine
Olfactory grooves
Optic foramen
Chiasmatic groove
Tuberculum sellae
Anterior clinoid process
Medial clinoid process
Posterior clinoid process
Groove for abducent nerve
Foramen lacerum
Orifice of carotid canal
Depression for trigeminal ganglion
Internal acoustic meatus
Slit for dura mater
Groove for superior petrosal sinus
Jugular foramen
Hypoglossal canal
Aquaeductus vestibuli
Condyloid foramen
Foramen magnum
Mastoid foramen
Posterior meningeal grooves

Anterior Fossa
Orbital Plate of Frontal
Body & Small Wing of Sphenoid
Great Wing of Sphenoid
Middle Fossa
Squama of Temporal
Fossa hypophyseos
Petrous Portion of Temporal
Eminentia arcuata
Basilar part
Posterior Fossa
Occipital bone
for Transverse Sinus

FIG. 4-70. Interior of base of the skull, showing the cranial fossae.

upper margin of the **optic foramen.** Through each optic foramen courses an optic nerve and ophthalmic artery.

   MIDDLE CRANIAL FOSSA (*fossa cranii media*). The lateral parts of the middle cranial fossa are of considerable depth, and they support the temporal lobes of the brain.

More medially rests the hypothalamus and midbrain. On each side the middle fossa is bounded *anteriorly* by the posterior margins of the small wings of the sphenoid, the anterior clinoid processes, and the ridge forming the anterior margin of the chiasmatic groove; *posteriorly,* by the superior angles of the petrous portion of the temporals and the dorsum sellae; *laterally* by the temporal squamae, sphenoidal angles of the parietals, and great wings of the sphenoid. The middle fossa on each side is traversed by the **squamosal, sphenoparietal, sphenosquamosal** and **sphenopetrosal sutures.**

In the anterior part of the median region are the **chiasmatic groove** and **tuberculum sellae.** On each side the chiasmatic groove ends at the optic foramen, posterior to which extends the **anterior clinoid process** that gives attachment to the tentorium cerebelli. Behind the tuberculum sellae, the **hypophyseal fossa** forms a deep depression in the **sella turcica** and lodges the hypophysis. A tubercle, the **medial clinoid process,** may be present on each side of the anterior wall of the sella turcica. The sella turcica is bounded posteriorly by a quadrilateral plate of bone, the **dorsum sellae,** the free angles of which are surmounted by the **posterior clinoid processes.** These afford attachment to the tentorium cerebelli, and below each is a notch for the abducent nerve. On each side of the sella turcica is the **carotid groove,** which is broad, shallow, and curved. The groove lodges the internal carotid artery and extends from the foramen lacerum to the medial side of the anterior clinoid process, where it is sometimes converted into a **caroticoclinoid foramen** by the union of the anterior with the medial clinoid process. The carotid groove also contains the cavernous sinus and may be bounded posterolaterally by the **lingula** of the sphenoid.

The bony lateral surface of the middle cranial fossa is traversed by furrows for the anterior and posterior branches of the **middle meningeal vessels.** Commencing near the **foramen spinosum,** the anterior furrow runs to the sphenoidal angle of the parietal, where it is sometimes converted into a bony canal; the posterior runs lateralward across the temporal squama, passing onto the parietal bone near the middle of its inferior border. The following apertures are also to be seen. Rostrally, the **superior orbital fissure** is bounded above by the small wing, below by the great wing, and medially by the body of the sphenoid; it is usually completed laterally by the orbital plate of the frontal bone. The fissure transmits: (1) *to the orbital cavity*— the oculomotor, the trochlear, the ophthalmic division of the trigeminal, and the abducent nerves, sympathetic nerve fibers

from the cavernous plexus, and the orbital branch of the middle meningeal artery; and (2) *to the cranial cavity* from the orbit—the ophthalmic veins and a recurrent branch of the lacrimal artery to the dura mater. Posterior to the medial end of the superior orbital fissure is the **foramen rotundum** through which passes the maxillary nerve. Posterolaterally is the larger **foramen ovale** through which courses the mandibular nerve, the accessory meningeal artery, and the lesser petrosal nerve. The **foramen of Vesalius,** often absent, opens at the lateral aspect of the scaphoid fossa inferiorly, and transmits a small vein. Lateral and posterior to the foramen ovale is the **foramen spinosum,** which transmits the middle meningeal vessels and a recurrent branch of the mandibular nerve which supplies the dura mater. Medial to the foramen ovale the **foramen lacerum** may be seen in a dried skull, but in the fresh state the inferior part of this aperture is occupied by a layer of fibrocartilage. The internal carotid artery, surrounded by a plexus of sympathetic nerves, passes across the superior part of the foramen but does not traverse its whole extent. The small nerve of the pterygoid canal and a small meningeal branch from the ascending pharyngeal artery alone pierce the layer of fibrocartilage and are thus the only structures to pass through the foramen lacerum. On the anterior surface of the petrous portion of the temporal bone is seen the **arcuate eminence** formed by the projection of the anterior semicircular canal. Also on this anterior surface of the petrous temporal bone is found a groove called the **hiatus for the greater petrosal nerve.** This nerve, coming from the geniculate ganglion of the facial nerve, courses along the groove and then traverses the foramen lacerum to join the deep petrosal nerve. A bit more lateral is another somewhat smaller **hiatus for the lesser petrosal nerve,** which is bound for the otic ganglion and leaves the cranial cavity through the foramen ovale or its own foramen. Behind the foramen lacerum can be seen the **depression for the trigeminal ganglion.**

POSTERIOR CRANIAL FOSSA (*fossa cranii posterior*). The posterior cranial fossa is the largest and deepest of the three. Within it lies the pons and medulla oblongata medially, over which expands the lobes of the cerebellum. The bony floor of this fossa is formed by the dorsum sellae and clivus of the sphenoid, the occipital bone, the petrous and mastoid portions of the temporals, and the mastoid angles of the parietal bones. It is crossed by the occipitomastoid, parieto-mastoid, and spheno-occipital sutures. The posterior fossa is separated from the middle fossa in and near the midline by the dorsum sellae of the sphenoid and on each side by the grooved superior margin of the petrous portion of the temporal bone. This margin gives attachment to the tentorium cerebelli, and lodges in its groove the superior petrosal sinus. In the center of the posterior fossa is the **foramen magnum,** and on each of its lateral sides is the **hypoglossal canal,** through which courses the hypoglossal nerve and a meningeal branch from the ascending pharyngeal artery. Anterior to the foramen magnum, the basilar portion of the occipital and the posterior part of the body of the sphenoid form a grooved surface which supports the medulla oblongata and pons. In the young skull these bones are joined by a synchondrosis. This grooved basilar region is separated from the petrous temporal by the **petro-occipital fissure,** which is occupied in life by a thin plate of cartilage. The fissure is continuous with the **jugular foramen,** and its margins are grooved for the **inferior petrosal sinus.**

The **jugular foramen** is situated between the lateral part of the occipital and the petrous part of the temporal. The *anterior portion* of this foramen transmits the inferior petrosal sinus; the *posterior portion,* the sigmoid sinus and some meningeal branches from the occipital and ascending pharyngeal arteries; and the *intermediate portion,* the glossopharyngeal, vagus, and accessory nerves. Superior to the jugular foramen is the **internal acoustic meatus,** for the facial and vestibulocochlear nerves and internal auditory artery. A narrow fissure is located behind and lateral to the meatus, which contains the external opening of the **aqueduct of the vestibule** (*aqueductus vestibuli*). In the aqueduct is located the endolymphatic duct. Behind the foramen magnum are the **inferior occipital fossae,** which support the hemispheres of the cerebellum. These are separated from one another by the **internal occipital crest,** which serves for the attachment of the falx cerebelli and lodges the occipital sinus. The posterior cranial fossae are bounded posterolaterally by S-shaped **grooves for the sigmoid sinuses**

which lead to the jugular foramen. Posteriorly these grooves are continuous with those of the **transverse sinuses.** These grooves accentuate the inner surface of the occipital bone, the mastoid angle of the parietal, the mastoid portion of the temporal, and the jugular process of the occipital, and end at the jugular foramen on each side. Where this sinus grooves the mastoid portion of the temporal, the orifice of the **mastoid foramen** may be seen, and just prior to its termination, the **condyloid canal** opens into it. Both of these orifices transmit emissary veins, but neither opening is constant.

NASAL CAVITY (*cavum nasi;* Figs. 4-64; 4-71 to 4-73). The **nasal cavities** are two irregularly shaped spaces which extend from the base of the cranium to the roof of the mouth. They are separated from each other in the midline by a thin vertical **septum.** They open on the face through a single pear-shaped **anterior nasal aperture,** since the anterior cartilaginous part of the septum is lacking in a dried skull. Posteriorly, two openings, the **choanae,** communicate with the nasal part of the pharynx. The nasal cavities are much narrower superiorly than inferiorly, and they communicate with the **frontal, maxillary, ethmoidal,** and **sphenoidal sinuses.** Each cavity is bounded by a roof, a floor, a medial wall, and a lateral wall.

The **roof** (Figs. 4-71; 4-72) is horizontal centrally, but slopes downward in front and behind. Its middle is 2 to 3 mm wide and is formed by the **cribriform plate** of the ethmoid bone, through which pass nearly 20 foramina for the filaments of the olfactory nerves. Anteriorly, the roof is formed by the nasal bone and the spine of the frontal, and posteriorly, by the body of the sphenoid, the ala of the vomer, and the sphenoidal process of the palatine bone. On the superior and posterior part of the roof is the **opening of the sphenoidal sinus.**

The bony **floor** (Fig. 4-22) is smooth and is formed by the **palatine process** of the maxilla and the **horizontal part** of the palatine bone. These represent the bones of the hard palate, which separate the nasal cavity from the oral cavity. Near the anterior end is the opening of the **incisive canal,** through which pass the nasopalatine nerves and branches of the greater palatine arteries.

The **medial wall** (*septum nasi;* Fig. 4-71) is the osseous **nasal septum** which separates

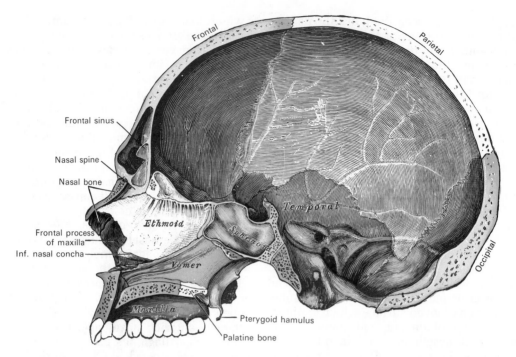

Frontal

Parietal

Frontal sinus

Nasal spine

Nasal bone

Temporal

Ethmoid

Frontal process
of maxilla

Sphenoid

Inf. nasal concha

Vomer

Occipital

Maxilla

Pterygoid hamulus

Palatine bone

FIG. 4-71. Sagittal section of skull showing the osseous nasal septum formed principally by the perpendicular plate of the ethmoid bone and the vomer anteriorly. Behind and above the septum are seen the bones forming the inner surface of the cranial cavity.

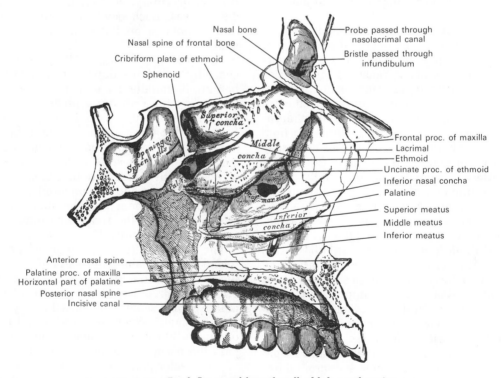

Nasal bone

Nasal spine of frontal bone

Cribriform plate of ethmoid

Sphenoid

Probe passed through
nasolacrimal canal

Bristle passed through
infundibulum

Superior
concha

Middle
concha

Opening of
Sphen. cells

Frontal proc. of maxilla

Lacrimal

Ethmoid

Uncinate proc. of ethmoid

Inferior nasal concha

Palatine

Sphenopalat.

op. max.sinus

Superior meatus

Middle meatus

Inferior
concha

Inferior meatus

Anterior nasal spine

Palatine proc. of maxilla

Horizontal part of palatine

Posterior nasal spine

Incisive canal

FIG. 4-72. Roof, floor, and lateral wall of left nasal cavity.

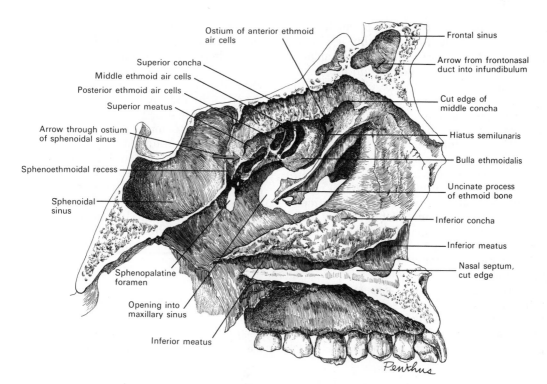

Ostium of anterior ethmoid
air cells

Superior concha

Middle ethmoid air cells

Posterior ethmoid air cells

Superior meatus

Arrow through ostium
of sphenoidal sinus

Sphenoethmoidal recess

Sphenoidal
sinus

Sphenopalatine
foramen

Opening into
maxillary sinus

Inferior meatus

Frontal sinus

Arrow from frontonasal
duct into infundibulum

Cut edge of
middle concha

Hiatus semilunaris

Bulla ethmoidalis

Uncinate process
of ethmoid bone

Inferior concha

Inferior meatus

Nasal septum,
cut edge

FIG. 4-73. Lateral wall of the nasal cavity after the removal of the middle concha to expose the structures and orifices of the middle meatus. Arrows indicate openings from the frontal and sphenoid sinuses.

the two nasal cavities. It extends from the roof to the floor and is frequently deflected to one side or the other, more often to the left than to the right. Although cartilaginous in front, the septum is formed primarily by the **perpendicular plate of the ethmoid bone** and the **vomer.** Posteriorly, these bones articulate with the sphenoid, while anteriorly the crest of the nasal bones and the spine of the frontal also contribute to the septum.

The **lateral wall** (Figs. 4-72; 4-73) appears irregular, but in fact is a rather flat surface upon which are projected three curved laminae or **conchae.** This wall is formed, *anteriorly,* by the frontal process of the maxilla and by the lacrimal bone; *in the middle,* by the ethmoid, maxilla, and inferior nasal concha; *posteriorly,* by the vertical plate of the palatine bone, and the medial pterygoid plate of the sphenoid. On this wall are three curved anteroposterior passages, termed the superior, middle, and inferior meatuses of the nose (Fig. 4-72). The **superior meatus** is the shortest of the three and occupies the middle third of the lateral wall between the superior and middle nasal conchae. The **posterior ethmoidal cells** open into the superior

meatus, while the **sphenopalatine foramen,** an opening in the perpendicular plate of the palatine bone which serves as a means of communication between the pterygopalatine fossa and the nasal cavity, lies just behind the meatus. The orifice of the sphenoidal sinus drains into the **sphenoethmoidal recess,** which lies in the most superior and posterior part of the nasal cavity, formed by the angle of junction of the sphenoid and ethmoid bones. The **middle meatus** is situated between the middle and inferior conchae (Fig. 4-72), and it courses anteroposteriorly above the superior border of the inferior concha along its entire extent. The lateral wall of this meatus can be satisfactorily studied only after the removal of the middle concha (Fig. 4-73). A curved fissure, the **hiatus semilunaris,** lies along this meatus between the **uncinate process** of the ethmoid bone and the **bulla ethmoidalis.** The **middle ethmoidal cells** are contained within the bulla and open on or near it. The middle meatus communicates through the hiatus semilunaris with a curved passage termed the **infundibulum,** into which the **anterior ethmoidal cells** open. The **frontonasal duct** leading from the frontal sinus opens into the anterior end of the infundibulum or directly into the anterior part of the middle meatus. Below the bulla ethmoidalis and hidden by the uncinate process of the ethmoid is the **opening** of the **maxillary sinus,** and an **accessory opening** frequently is present above the posterior part of the inferior nasal concha. The **inferior meatus** is the space between the inferior concha and the floor of the nasal cavity. It extends almost the entire length of the lateral wall of the nasal cavity, and presents anteriorly the **inferior orifice** of the **nasolacrimal canal.** Observe the locations of the frontal maxillary and sphenoid sinuses in Figure 4-74.

The **anterior nasal aperture** (Fig. 4-64) is a pear-shaped opening, the long axis of which is vertical. It is bounded *superiorly* by the inferior borders of the nasal bones; *laterally* by the thin, sharp margins which separate the anterior from the nasal surfaces of the maxillae; and *inferiorly* by the same borders,

where they curve medialward to join at the anterior nasal spine. Attached to the borders of the bony anterior nasal aperture are the external cartilages of the nose. These provide the principal surface features and shape of the external nose and open by way of two elliptical orifices, the **nares.**

The two **posterior nasal apertures,** or choanae, are bounded *superiorly* by the body of the sphenoid and ala of the vomer; *inferiorly,* by the posterior border of the horizontal part of the palatine bone, which actually forms the posterior part of the hard palate; and *laterally,* by the medial pterygoid plate. The choanae are separated from each other in the midline by the posterior border of the vomer.

Differences in the Skull Due to Age

At birth the skull is large in proportion to the other parts of the skeleton, but the facial portion of the cranium in the newborn is small, and equals only

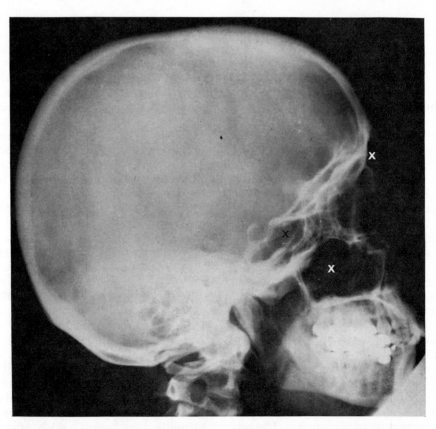

Fig. 4-74. Lateral view roentgenogram of the adult skull. Crosses are placed in the frontal, maxillary, and sphenoidal sinuses. The hypophyseal fossa can be identified directly behind the sphenoid sinus. The dense white region below and behind the fossa is due to the petrous part of the temporal bone. (Courtesy of Doctors Moreland and Arndt.)

about one-eighth of the bulk of the skull compared to one-half in the adult. The smallness of the **face at birth** is due to the rudimentary condition of the maxillae and mandible, the noneruption of the teeth, and the small size of the maxillary air sinuses and nasal cavities. At birth the nasal cavities lie almost entirely between the orbits, and the lower border of the anterior nasal aperture is only a little below the level of the orbital floor. In contrast, the **cranial cavity** is large to accommodate the large neonatal brain. The frontal and parietal eminences are prominent, and the skull is maximally wide at the level of the parietal tubers (Fig. 4-75). On the other hand, the glabella, superciliary arches, and mastoid processes are not even developed at birth. Ossification of the skull bones is not completed, and many, e.g., the occipital, temporals, sphenoid, frontal, and mandible, consist of more than one piece. Unossified membranous intervals, termed **fontanelles,** are seen at the angles of the parietal bones. There are six fontanelles in number: two, an anterior and a posterior, are situated in the midline; two others are found on each side, an anterolateral (sphenoidal) and a posterolateral (mastoid).

The **anterior** or **bregmatic fontanelle** (*fonticulus anterior;* Fig. 4-75) is the largest, and is located at the junction of the sagittal, coronal, and frontal sutures. It is diamond-shaped and measures about 4 cm in its anteroposterior and 2.5 cm in its transverse diameter. The **posterior fontanelle** is triangular in form and is situated at the junction of the sagittal and lambdoidal sutures. The **sphenoidal and mastoid fontanelles** (Fig. 4-76) are small, irregular in shape, and correspond respectively to the sphenoidal and mastoid angles of the parietal bones. The posterior and lateral fontanelles become closed within a month or two after birth, but the anterior does not completely close until about the middle of the second year.

Postnatal growth of the face and enlargement of the jaws are temporally correlated with the eruption of the deciduous teeth. These changes are still more marked after the second dentition. The skull grows rapidly from birth to the seventh year, by which time the foramen magnum and petrous parts of the temporal bones have reached their full size and the orbital cavities are only a little smaller than those of the adult. Growth of the skull is slow from the seventh year until the approach of puberty, when a second period of rapid skull enlargement occurs. This results in an increase of skull size in all directions, but it is especially marked in the frontal and facial regions, where it is associated with the development of the air sinuses.

The time of closure of the individual cranial sutures varies between individuals, but generally begins between the twentieth and thirtieth years. Usually the process commences in the sagittal and sphenofrontal sutures, but closure progresses very slowly and may not be complete in all the sutures until old age. The most striking feature of the old skull is the diminution in the size of the maxillae and

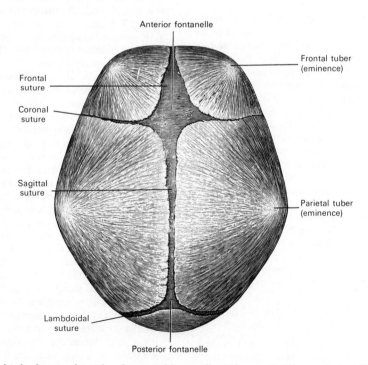

Anterior fontanelle

Frontal tuber (eminence)

Frontal suture

Coronal suture

Sagittal suture

Parietal tuber (eminence)

Lambdoidal suture

Posterior fontanelle

FIG. 4-75.   Skull at birth, showing frontal and occipital fontanelles. Observe that the greatest width of the skull at this stage is at the level of the parietal tubers.

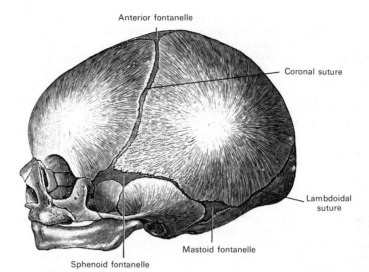

Anterior fontanelle

Coronal suture

Lambdoidal suture

Mastoid fontanelle

Sphenoid fontanelle

FIG. 4-76. Skull at birth, showing sphenoidal and mastoid fontanelles. Observe the small facial bones and the great difference in size between the facial and cranial vault bones in the neonatal skull.

mandible after the loss of the teeth and the absorption of the alveolar processes. This is associated with a marked reduction in the vertical (occlusal) measurement of the face and with an alteration in the condylar angles of the mandible (Fig. 4-83).

Sex Differences in the Skull

Until the age of puberty there is little difference between the skull of the two sexes. The skull of an adult female is, as a rule, lighter and smaller, and its cranial capacity about 10% less than that of the male. Its walls are thinner and its muscular ridges less strongly marked. The glabella, superciliary arches, and mastoid processes are less prominent in females, and the corresponding air sinuses are smaller or may be rudimentary. The upper margin of the orbit is sharp, the forehead vertical, the frontal and parietal eminences prominent, and the vault somewhat flattened. The contour of the face is more rounded, the facial bones are smoother, and the maxillae and mandible and their contained teeth smaller. More of the morphological characteristics seen during the pre-puberty years are retained in the skull of the adult female than in that of the adult male. A well-marked male or female skull can easily be recognized as such, but in some cases the respective characteristics are so indistinct that the determination of the sex may be difficult or impossible.

Fractures of the Skull

Consistent with the primary function of the skull, which is protection of the brain, is the fact that those portions of the skull that are most exposed to external violence are thicker than those that are shielded

from injury by overlying muscles. The skull-cap is thick and dense, whereas the temporal squamae, protected by the temporal muscles, and the inferior occipital fossae, shielded by the muscles at the back of the neck, are thin and fragile. The calvaria average about 5 to 6 mm in thickness, while the temporal squama is only 1 to 2 mm thick. The thickness of the skull at the glabella is slightly over 1 cm and at the external occipital protuberance between 1.5 and 2.0 cm. The skull is further protected by its elasticity, its rounded shape, and the fact that its structure contains a number of secondary elastic arches, each made up of a single bone.

The manner in which vibrations are transmitted through the bones of the skull is of importance as a protective mechanism, especially as far as the skull base is concerned. The vibrations resulting from a blow on the head are transmitted in a uniform manner in all directions to the top and sides of the skull, but at the base, due to the varying thickness and density of the bones, this is not so. In this latter region there are special buttresses that serve to carry the vibrations in certain definite directions. The laterally projecting ridge that is formed by the lesser and greater wings of the sphenoid bone and separates the anterior from the middle cranial fossae and, more posteriorly, the ridge that separates the middle from the posterior fossa form such buttresses. If any violence is applied to the cranial vault, the vibrations would be carried along these buttresses to the sella turcica, where they meet and form a center of resistance. Additionally, the subarachnoid cavity is somewhat dilated at this site and forms a cushion of fluid, helping to dissipate the vibrations. In like manner, when violence is applied to the base of the skull, as in falls when a person lands on his feet, the vibrations are carried backward through the occipital crest, and forward through the basilar part of the occipital bone and body of the sphenoid up to the vault of the skull.

The calvaria and sides of the cranial cavity are formed by an inner layer of compact bone between which is the spongy diploic layer containing the vascular channels. Fractures involving the outer compact layer are relatively easy to identify; however, a blow on the head may be transmitted at times to the inner compact layer, resulting in a fracture of this layer with bony spicules and splinters without evidence of fracture to the outer compact layer.

Basal skull fractures of the **anterior cranial fossa** may involve the delicate cribriform plate of the ethmoid bone and/or the very thin orbital plate of the frontal bone. Such injury can cause lacerations of the dura mater with subsequent discharge of cerebrospinal fluid through the nose (cerebrospinal rhinorrhea) and injury to the olfactory nerves resulting in the loss of the sense of smell. Fractures of the orbital plate may result in hemorrhage into the orbital cavity, evidenced by subconjunctival discoloration and protrusion of the eyeball (exophthalmos).

Fractures of the bones of the **middle cranial fossa** are the most common basal skull fractures, probably because this region is weakened by the many foramina and canals. The petrous portion of the temporal bone is further weakened by containing the cavities of the inner ear. Fractures of the sphenoid bone frequently injure the meningeal vessels and may injure the internal carotid artery as well as the cavernous sinus. Injury might also be expected to the third, fourth, and sixth cranial nerves. Fractures in the middle fossa frequently involve the tegmen tympani (roof of the tympanic cavity). In these situations, if the tympanic membrane is ruptured, blood and cerebrospinal fluid can escape through the external auditory meatus. Fractures of the petrous portion of the temporal bone may injure the facial and vestibulocochlear nerves.

Fractures of the **posterior cranial fossa** may be exceedingly dangerous because cerebrospinal fluid and blood cannot escape, and even a seemingly small fracture might be fatal because of the pressure that develops on the vital centers in the brain stem. Fractures of the body of the sphenoid bone may paralyze the sixth cranial nerve, whereas fractures of the occipital bone may injure the hypoglossal nerve.

Malformations of the Skull

One of the most common developmental malformations of the facial bones is **cleft palate.** The cleft may involve only the uvula or it may extend anteriorly through the soft palate and involve part or even all of the hard palate. In the latter instance, the cleft would extend as far forward as the incisive foramen. In the severest forms, the cleft extends through the alveolar process and continues between the premaxillary bone and the rest of the maxilla, that is between the lateral incisor and canine teeth. In some instances, the cleft courses between the central and lateral incisor teeth, lending some support to the belief that the premaxillary bone develops from two centers and not one. The cleft may affect one or both sides. If it is bilateral, the central part is frequently displaced forward and remains united to the septum of the nose, the deficiency in the bony palate being complicated with a cleft lip.

Individual Bones of the Skull

THE MANDIBLE

The **mandible** (Figs. 4-77; 4-78) is the largest and strongest of the facial bones and extends from the chin in the midline to the two mandibular fossae laterally. It contains the lower teeth and consists of a horizontal portion, the **body,** and two perpendicular portions, the **rami,** which join the body nearly at right angles.

BODY OF THE MANDIBLE. The **body** (*corpus mandibulae*) is curved somewhat like a horseshoe, and has two *surfaces* and two *borders.*

The **external surface** of the body (Figs. 4-64; 4-77) is marked in the median line by a faint ridge, indicating the **symphysis menti** or line of fusion of the two separate halves of the mandible as seen in the fetus. This ridge divides inferiorly to enclose a triangular eminence, the **mental protuberance,** the base of which is depressed in the center but raised on each side to form the **mental tubercle.** On either side of the symphysis is a depression, the **incisive fossa,** which gives origin to the mentalis and a small portion of the orbicularis oris. Inferior to the second premolar tooth on each side is the **mental foramen,** for the passage of the mental vessels and nerve. Coursing posteriorly and upward from each mental tubercle is a faint ridge, the **oblique line,** which becomes continuous with the anterior border of the ramus. At its anterior end attaches the depressor labii inferioris and depressor anguli oris muscles. The platysma inserts near the inferior border (Fig. 4-77).

The **internal surface** of the body (Fig. 4-78) is concave and anteriorly presents two small bony projections on each side. Near the inferior part of the symphysis on each side is a laterally placed spine, termed the **mental spine,** which gives origin to the genioglossus. Immediately below these is a second pair of spines, or more frequently a median ridge or

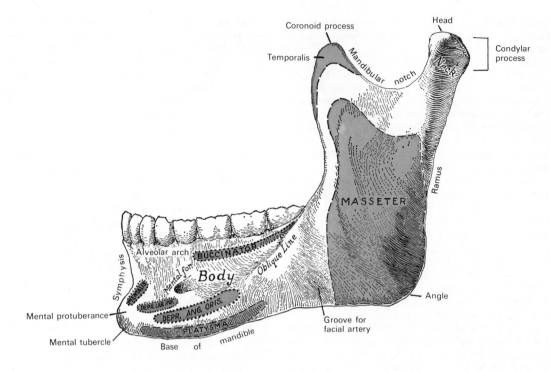

FIG. 4-77.   The left half of the mandible. Lateral aspect.

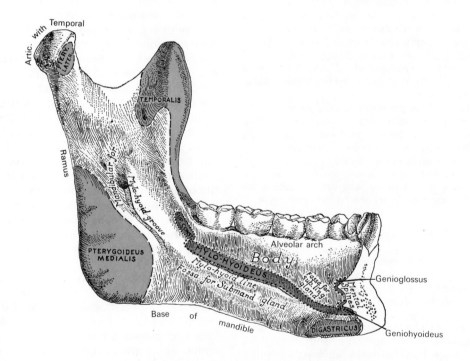

FIG. 4-78.   The left half of the mandible. Medial aspect.

impression, for the origin of the geniohyoid muscles. On each side of the midline near the inferior margin of the symphysis is an oval depression for the attachment of the anterior belly of the digastric muscle. Extending posteriorly from the mental spine on each side of the symphysis is the **mylohyoid line,** which gives origin to the mylohyoid muscle. The posterior part of this line, near the alveolar margin, behind the third molar tooth, gives attachment to a small part of the superior pharyngeal constrictor, and to the pterygomandibular raphe. Superior to the anterior part of this line is the smooth, triangular **sublingual fossa** within which the sublingual gland rests, and inferior to the posterior part is the oval **submandibular fossa** which supports the submandibular gland.

The **superior border** or **alveolar part** of the mandibular body is hollowed into sockets for the reception of the sixteen teeth. The sockets, or **alveoli,** arranged sequentially in a curved bar of bone, the **alveolar arch,** vary in depth and size according to the teeth they contain. The sockets are separated by bony partitions, the **interalveolar septa.** The buccinator is attached to the outer lip of the superior border as far forward as the first molar tooth.

The **inferior border** is called the **base of the mandible,** and it is rounded, thicker anteriorly than behind, and at the point where it joins the lower border of the ramus there may be a shallow groove for the facial artery. The base of the mandible ends posteriorly at the mandibular angle, diagonal to the socket of the third molar tooth. From this point the ramus ascends in a nearly vertical manner.

RAMUS OF THE MANDIBLE. The **ramus** (*ramus mandibulae*) is quadrilateral in shape and has two *surfaces,* four *borders,* and two *processes.*

The **lateral surface** of the ramus (Fig. 4-77) is flat and marked by oblique ridges in its lower portion where the masseter muscle inserts. The **medial surface** (Fig. 4-72) presents near its middle the oblique **mandibular foramen** for the entrance of the inferior alveolar vessels and nerve into the **mandibular canal.** The margin of the mandibular foramen presents a prominent ridge, surmounted by a sharp spine anteriorly, the **lingula mandibulae,** to which attaches the sphenomandibular ligament. The **mylohyoid groove**

runs obliquely downward and forward from the foramen and lodges the mylohyoid vessels and nerve. Posterior to this groove is a roughened surface extending to the mandibular angle for the insertion of the medial pterygoid muscle. From the mandibular foramen and within the substance of the ramus, the **mandibular canal** courses obliquely downward and forward. After reaching the mandibular body, the canal continues horizontally forward, and is placed under the alveoli, communicating with them by small openings. At the level of the second premolar tooth, the canal splits, forming the **mental canal,** which courses upward and laterally to reach the **mental foramen,** and the incisive canal, which continues toward the symphysis, giving off two small channels to the incisor teeth. In the posterior two-thirds of the bone the mandibular canal is situated near the internal surface of the mandible, whereas in the anterior third the canal is nearer its external surface. Through the canal courses the inferior alveolar vessels and nerve, from which branches are distributed to the lower teeth.

The **inferior border** of the ramus is thick and straight and joins the posterior border at the **angle of the mandible.** The mandibular angle, which measures about 120 degrees, may be either a bit inverted or everted and serves for attachment on its outer aspect of the masseter and on its inner aspect of the medial pterygoid muscle. Between these two muscles, the stylomandibular ligament attaches to the angle. The **posterior border** is thick, smooth, and rounded. It extends from the condylar process above to the mandibular angle below and is covered by the parotid gland. The **anterior border** is continuous above with the **coronoid process** and below with the oblique line. The **superior border** is thin, and is surmounted by the coronoid process anteriorly and the condylar process posteriorly, separated by the mandibular notch.

The **coronoid process** (*processus coronoideus;* Fig. 4-77) is a thin, triangular eminence, and it varies in shape and size. Its anterior border is convex and is continuous with the anterior border of the ramus; its posterior border is concave and forms the anterior boundary of the mandibular incisure (notch). Along its entire medial surface inserts the temporalis muscle, the insertion extending along the entire rim of the process and over onto the lateral surface. The medial surface is also marked by a bony ridge, which begins near the apex of the process and runs downward and forward to the inner side of the last molar tooth.

The **condylar process** (*processus condylaris*) is thicker than the coronoid, and consists of two portions: the articular portion, called the **head** (*condyle*), and the constricted portion that supports it, the **neck.** The head presents an oval surface covered with fibrocartilage for articulation with the temporal bone at the mandibular fossa. Interposed between the two articular surfaces is the articular disc of the temporomandibular joint. In a living subject the head can be manipulated about 1 cm anterior to the tragus. With the jaw closed it lies totally within the mandibular fossa. When the jaw opens, the head and disc glide forward within the joint capsule, leaving a small depression in front of the tragus. At the lateral extremity of the head is a small tubercle for the attachment of the lateral (temporomandibular) ligament. The **neck** is flattened and strengthened by ridges that descend from the anterior and lateral parts of the head. Its posterior surface is convex; its anterior presents a depression, the **pterygoid fovea,** for the insertion of the lateral pterygoid muscle. On the medial surface of the neck just posterior to the insertion of the lateral pterygoid is a blunt crest called the dental trajectory or the **mandibular crest.**

The **mandibular incisure** or **notch,** separating the two processes, is a deep semilunar depression, and is crossed by the masseteric vessels and nerve.

OSSIFICATION. The mandible is ossified in the fibrous membrane covering the outer surface of the right and left **Meckel's cartilages.** These are the cartilaginous bars of the first branchial arch **(mandibular arch).** Their proximal or cranial ends are connected with the ear capsules, and their distal extremities are joined at the symphysis menti by mesodermal tissue. From the proximal end of each cartilage, the *malleus* and *incus,* two of the bones of the middle ear, are developed. The next succeeding portion, as far as the lingula, is replaced by fibrous tissue, which persists to form the *sphenomandibular ligament.* Between the lingula and the canine tooth the cartilage disappears; the remaining portion adjacent to the incisor teeth becomes ossified and incorporated into this part of the mandible.

Ossification takes place in the membrane covering the outer surface of the ventral end of Meckel's carti-

lage (Figs. 4-79 to 4-82). Each half of the mandible is formed from a single center, which appears near the mental foramen at about the sixth week of fetal life. By the tenth week the portion of Meckel's cartilage that lies adjacent to the incisor teeth is surrounded and invaded by the membranous bone. Somewhat later, accessory cartilaginous nuclei make their appearance—as a wedge-shaped nucleus in the condylar process which extends downward through the ramus (Fig. 4-82), as a small strip along the anterior border of the coronoid process (Fig. 4-82), and as smaller cartilaginous nuclei in the front part of both alveolar walls and along the front of the lower border of the bone (Fig. 4-83). These accessory nuclei possess no separate ossific centers, but are invaded by the surrounding membranous bone and undergo absorption. The inner alveolar border, usually described as arising from a separate ossific center (splenial center), is formed in the human mandible by an ingrowth from the main mass of the bone. At birth the bone consists of two parts, united anteriorly by the fibrous symphysis menti which ossifies during the first year.

ARTICULATIONS. The mandible articulates with the two temporal bones.

CHANGES PRODUCED BY AGE. **At birth** (Fig. 4-83, *A*) the body of the bone is a mere shell, containing the sockets of the two incisors, the canine, and the two deciduous molar teeth, imperfectly partitioned from one another. The mandibular canal is large and runs near the lower border of the bone, and the mental foramen opens beneath the socket of the first

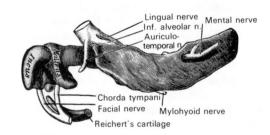

FIG. 4-79. Mandible of human embryo 24 mm long. Outer aspect. Cartilage is in blue, bone in yellow. (From model by Low.)

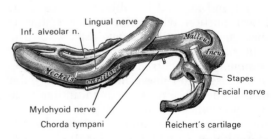

FIG. 4-80. Mandible of human embryo 24 mm long. Inner aspect. Cartilage is in blue, bone in yellow. (From model by Low.)

deciduous molar tooth. The mandibular angle is obtuse (175 degrees), and the condylar process is nearly in line with the body. The coronoid process is comparatively large and projects above the level of the head of the mandible.

After birth (Fig. 4-83, *B*) the two segments of the bone become joined from below upward at the symphysis in the first year (Fig. 4-83, *A*), but a trace of separation may be visible in the beginning of the second year near the alveolar margin. The entire length of the mandibular body becomes elongated, but more especially posterior to the mental foramen, which provides space for the three additional teeth developed in this part. The depth of the body increases due to enlargement of the alveolar part. This allows room for the roots of the teeth, and results in the thickening of the portion below the oblique line, which enables the jaw to withstand the powerful action of the masticatory muscles. The alveolar portion, however, becomes the deeper of the two parts, and consequently the largest portion of the mandibular body lies above the oblique line. The mandibular canal, after *second dentition* (Fig. 4-83, *C*), is situated just above the level of the mylohyoid line and the mental foramen occupies the more adult position. The mandibular angle diminishes due to the separation of the jaws by the teeth, and by the fourth year it is 140 degrees.

In the adult (Fig. 4-83, *D*) the alveolar portion and the portion below the oblique line are usually of equal depth. The mental foramen opens midway between the upper and lower borders of the bone, and the mandibular canal runs nearly parallel with the mylohyoid line. The ramus is almost vertical in direction, and the mandibular angle measures about 120 degrees.

In old age (Fig. 4-83, *E*) the mandible becomes greatly reduced in size because the alveolar process is absorbed after the loss of the teeth, leaving the chief bulk of the bone below the oblique line. The mandibular canal with the mental foramen opening from it becomes positioned close to the alveolar border. The angle widens once more, measuring about 140 degrees, and the neck of the condylar process becomes oriented slightly backward.

HYOID BONE

The **hyoid bone** (*os hyoideum*; Fig. 4-84), named from its resemblance to the Greek letter *U*, is suspended from the tips of the styloid processes of the temporal bones by the stylohyoid ligaments. It is located in the anterior part of the neck between the chin and the larynx, and is composed of five por-

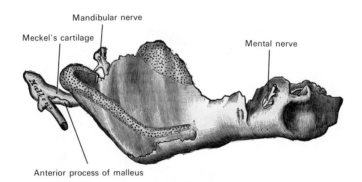

Fig. 4-81. Mandible of human embryo 95 mm long. Outer aspect. Nuclei of accessory cartilages stippled. Cartilage is in blue, bone in yellow. (From model by Low.)

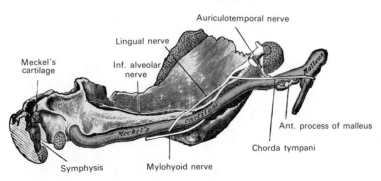

Fig. 4-82. Mandible of human embryo 95 mm long showing symphysis. Inner aspect. Nuclei of accessory cartilages stippled. Cartilage is in blue, bone in yellow. (From model by Low.)

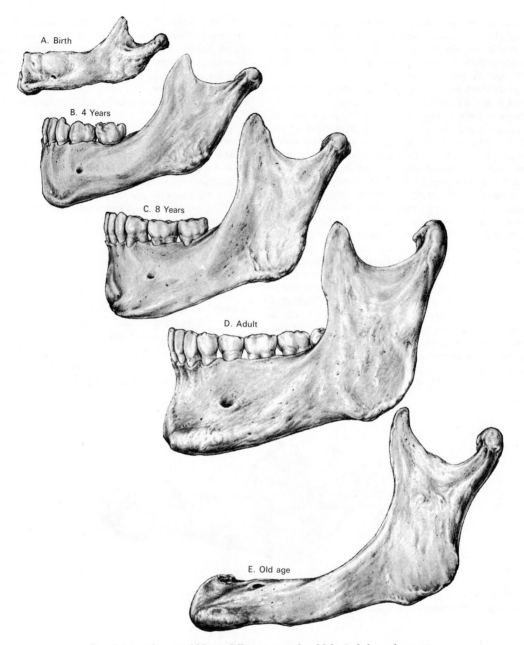

A. Birth

B. 4 Years

C. 8 Years

D. Adult

E. Old age

FIG. 4-83. The mandible at different periods of life. Left lateral aspect.

tions: a body, two greater cornua, and two lesser cornua.

The **body** (*corpus ossis hyoidei*) or central part is quadrilateral in shape and placed transversely across the midline of the neck. Its **ventral surface** (Fig. 4-84) is convex and is directed forward and somewhat cranially. Its upper half is crossed by a well-marked transverse ridge, and in many cases a vertical median ridge divides it into two lateral halves. The geniohyoid muscle is inserted into the greater part of the ventral surface, both above and below the transverse ridge. A portion of the origin of the hyoglossus muscle encroaches upon the lateral margin of the geniohyoid attachment. Below the transverse ridge are inserted the mylohyoid, sternohyoid, and omohyoid muscles. The **dorsal surface** is smooth and concave, and is directed somewhat caudally, being sepa-

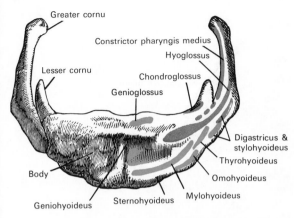

Greater cornu

Constrictor pharyngis medius

Hyoglossus

Lesser cornu

Chondroglossus

Genioglossus

Digastricus & stylohyoideus

Thyrohyoideus

Body

Omohyoideus

Mylohyoideus

Geniohyoideus   Sternohyoideus

FIG. 4-84. Hyoid bone. Anterior surface viewed from above.

rated from the epiglottis by the thyrohyoid membrane, an intervening bursa, and a quantity of loose areolar tissue. The **superior border** is rounded and gives attachment to the thyrohyoid membrane and some aponeurotic fibers of the genioglossus. The **inferior border** receives the insertion of the sternohyoid muscle medially, the omohyoid laterally, and occasionally the medial portion of the thyrohyoid. It also gives attachment to the levator glandulae thyroideae, when this muscle is present. In early life the lateral borders of the body are connected by cartilage to the greater cornua, but after middle life this union is osseous.

The two **greater cornua** (*cornua majora*) project dorsally from the lateral border of the body (Fig. 4-84). Each cornu, or horn, is flattened and becomes more slender posteriorly, ending in a tubercle to which is fixed the lateral thyrohyoid ligament. The greater cornua can be palpated above the lateral aspects of the thyroid cartilage and their cranial surfaces are rough near the lateral borders for muscular attachments. The most extensive attachments are the origins of the hyoglossus and constrictor pharyngis medius which stretch along the whole length of the horn. The digastric and stylohyoid muscles have small insertions near the junction of the horn with the body. The digastric is attached by a small connective tissue loop through which its tendon courses. The thyrohyoid membrane is attached along the medial border of each horn and the thyrohyoid muscle is attached anteriorly along the lateral border.

The two **lesser horns** (*cornua minora*) are

small, conical eminences, each attached by its base to the angle of junction between the body and the greater horn (Fig. 4-84). The lesser horn is connected to the body of the hyoid by fibrous tissue, and occasionally to the greater horn by a distinct synovial joint, which may either persist throughout life, or become ankylosed.

The lesser horn is situated in line with the transverse ridge on the body and appears to be in morphological continuation to it. The stylohyoid ligament is attached to the apex of the lesser horn and the chondroglossus arises from the medial side of its base.

DEVELOPMENT AND OSSIFICATION. The greater horns and much of the body of the hyoid are derived embryologically from the third visceral arch cartilage. A portion of the body, the lesser horns, and the stylohyoid ligaments to which they are attached are derived from the second or hyoid arch.

The hyoid is ossified from six centers: two for the body, and one for each horn. Ossification commences in the greater horns toward the end of fetal life, in the body shortly afterward, and in the lesser cornua during the first or second year after birth.

OCCIPITAL BONE

The **occipital bone** (*os occipitale*), which forms most of the posterior aspect of the skull as well as part of its base, is trapezoid in shape and internally concave in a cup-like manner (Figs. 4-85; 4-86). It is pierced by a large oval aperture, the **foramen magnum,** through which the cranial cavity communicates with the vertebral canal. The curved, expanded plate posterior to the foramen magnum is named the **squama;** the thick, somewhat quadrilateral portion anterior to the foramen is called the **basilar part,** and on each side of the foramen is the **lateral portion.**

THE SQUAMA. The **squama** (*squama occipitalis*), situated above and bordering the posterior part of the foramen magnum, is curved and rounded from above downward and from side to side. The **external surface** is convex and, midway between the summit of the bone and the foramen magnum, presents a bony prominence, the **external occipital protuberance** (Fig. 4-85). Extending lateralward from this on each side are two curved lines, one a little above the other. The more superior, often faintly marked, is named the

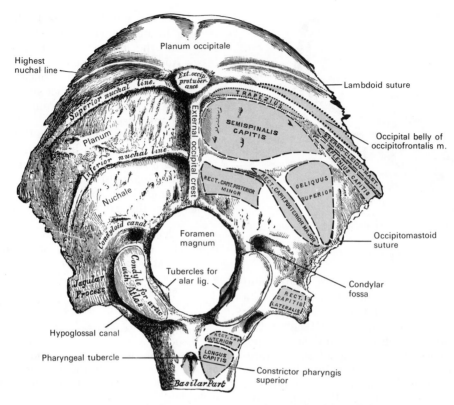

FIG. 4-85.   Occipital bone. Outer surface as visualized from below.

highest nuchal line, and to it the galea aponeurotica is attached. The lower is termed the **superior nuchal line.** That part of the squama that lies above the highest nuchal line is named the **planum occipitale,** and is covered by the occipital belly of the occipitofrontalis muscle; that below, termed the **planum nuchale,** is rough and irregular for the attachment of several muscles. From the external occipital protuberance a ridge or crest, **external occipital crest,** which is also called the **median nuchal line,** runs to the foramen magnum and affords attachment to the ligamentum nuchae. Midway along the crest on each side, the **inferior nuchal line** courses laterally from the crest across the planum nuchale, dividing it approximately in half. The muscles attached to the external surface of the squama include the origins of the occipital belly of occipitofrontalis and trapezius along the superior nuchal line and the insertions of the sternocleidomastoid and the splenius capitis laterally also along this line. Into the surface between the superior and inferior nuchal lines are inserted the semispinalis capitis and

obliquus capitis superior. Along the inferior nuchal line and below it are inserted the recti capitis posteriores major and minor. The posterior atlanto-occipital membrane is attached around the posterolateral part of the foramen magnum, just outside its margin.

The **internal surface** of the occipital bone is deeply concave and divided into four fossae by bony ridges which cross each other and along which course venous sinuses. The two superior fossae are triangular and lodge the occipital lobes of the cerebrum, while the inferior two are quadrilateral and accommodate the hemispheres of the cerebellum (Fig. 4-86). At the point of intersection of the bony ridges is the **eminentia cruciata,** on the peak of which is the **internal occipital protuberance.** From this protuberance one ridge courses superiorly to the upper angle of the bone, and on one side of it (generally the right) is a deep groove, the **sagittal sulcus,** which lodges the posterior part of the *superior sagittal sinus.* Along the margins of this sulcus the falx cerebri is attached. The ridge coursing inferiorly from the internal

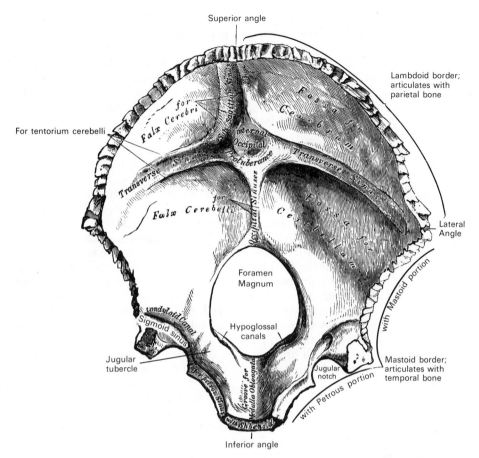

FIG. 4-86.   Occipital bone. Inner surface as visualized from above.

occipital protuberance is named the **internal occipital crest,** and it bifurcates near the foramen magnum, giving attachment to the falx cerebelli. In the attached margin of this falx is the *occipital sinus,* which is sometimes duplicated. Along the internal occipital crest a small depression, called the vermian fossa and occupied by the vermis of the cerebellum, is sometimes distinguishable. Transverse grooves extend on each side from the internal occipital protuberance to the lateral angles of the bone. They accommodate the *transverse sinuses,* and the margins of the grooves give attachment to the tentorium cerebelli. The groove on the right side is usually larger than that on the left and is continuous with that for the superior sagittal sinus. Exceptions to this description exist, however, since the left may be larger than the right or the two may be almost equal in size. The point of junction between the superior sagittal and transverse sinuses

is named the *confluence of the sinuses,* and its position is indicated by a depression situated on one side or the other of the protuberance.

LATERAL PARTS.   The **lateral parts** (*pars lateralis*) of the occipital bone are situated at the sides of the foramen magnum. These regions of the bone are sometimes called the condylar parts because on their **inferior surfaces** are the **occipital condyles** for articulation with the superior facets of the atlas (Fig. 4-85). The condyles are oval in shape, with their anterior extremities encroaching on the basilar portion of the bone, and their posterior extremities diverging toward the middle of the foramen magnum. The articular surface of the condyle is convex, and it is directed laterally and downward. To its margin is attached the capsule of the atlanto-occipital articulation, and on the medial side of each condyle is a rough impression or tubercle for the alar ligament. At the base of the condyle the bone is tunnelled by a short canal, the **hypoglossal canal,** through which exits the hypoglossal or twelfth cranial nerve from the cranial cavity. Additionally, a meningeal branch of the ascending pharyngeal artery enters the cranial cavity through this canal, which actually may be partially or completely divided into two canals by a spicule of bone. Posterior to each condyle is a depression, the **condyloid fossa,** into which the posterior margin of the superior facet of the atlas is received when the head is extended backward. The floor of this fossa is sometimes perforated by the **condyloid canal,** through which an emissary vein passes. Extending laterally from the posterior half of the condyle is a quadrilateral plate of bone, the **jugular process.** Anteriorly this process is indented by the **jugular notch** (Fig. 4-85), which, in the articulated skull, forms the posterior part of the **jugular foramen.** The jugular notch may also be divided into two indentations by a bony spicule, the **intrajugular process,** which projects laterally above the hypoglossal canal. The inferior surface of the jugular process is roughened and affords insertion to the rectus capitis lateralis muscle and attachment of the lateral atlanto-occipital ligament. From this surface an eminence, the **paramastoid process,** sometimes projects downward and may be of sufficient length to reach and articulate with the transverse process of the atlas. Laterally the jugular process also presents a roughened triangular area, which is joined to the jugular surface of the temporal bone by a plate of cartilage that tends to ossify after the age of twenty-five.

The **superior surface** of the lateral part presents an oval eminence, the **jugular tubercle,** which overlies the hypoglossal canal and is frequently crossed by an oblique groove for the glossopharyngeal, vagus, and accessory nerves. On the superior surface of the jugular process is a deep groove that curves medially and anteriorly and is continuous with the jugular notch. This groove lodges the continuation of the transverse sinus, the terminal portion of which is called the *sigmoid sinus.* Close to the medial margin of the groove opens the orifice of the condyloid canal.

BASILAR PART. The **basilar part** (*pars basilaris*) of the occipital bone is virtually quadrilateral in shape and extends anteriorly and superiorly from the foramen mag-

num (Fig. 4-85). In the young skull the basilar part is joined to the body of the sphenoid by a plate of cartilage (Fig. 4-87). By the twenty-fifth year, however, this cartilaginous plate is ossified, and the occipital and sphenoid bones become fused at this site.

On its **inferior surface,** about 1 cm anterior to the foramen magnum, is the **pharyngeal tubercle,** which gives attachment to the fibrous raphe of the pharynx. On each side of the midline, the longus capitis and rectus capitis anterior muscles are inserted, and immediately anterior to the foramen magnum the anterior atlanto-occipital membrane is attached.

The **superior surface** presents a broad, shallow groove upon which rests the lower pons and medulla oblongata of the brainstem. Near the margin of the foramen magnum is attached the tectorial membrane as well as the apical dental ligament. On the lateral margins of this surface are faint grooves for the *inferior petrosal sinuses.*

The **foramen magnum** is a large oval aperture, wider behind than in front, with its long diameter anteroposterior. It transmits the medulla oblongata and its membranes, the accessory nerves, the vertebral arteries, the anterior and posterior spinal arteries, the tectorial membrane, and alar ligaments.

The **superior angle** of the occipital bone (Fig. 4-86) articulates with the occipital an-

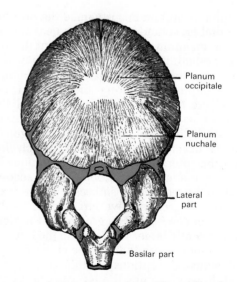

FIG. 4-87. Occipital bone at birth. Observe that the neonatal occipital bone consists of four ossified parts (squama, two lateral parts and the basilar part) interconnected by cartilaginous strips.

gles of the parietal bones and, in the fetal skull, corresponds in position with the posterior fontanelle (Fig. 4-75). The **inferior angle** is fused with the body of the sphenoid. The **lateral angles,** located at the extremities of the grooves for the transverse sinuses, are received into the interval between the mastoid angle of the parietal and the mastoid part of the temporal bone.

The **lambdoid borders** (*margo lambdoideus*) extend from the superior to the lateral angles: they are deeply serrated for articulation with the occipital borders of the parietals, and form by this union the lambdoidal suture. The **mastoid borders** (*margo mastoideus*) extend from the lateral angles to the inferior angle. The posterior half of each mastoid border articulates with the mastoid portion of the corresponding temporal, while the anterior half articulates with the petrous part of the same bone. Between these two parts of the mastoid border is the jugular process, the notch on the anterior surface of which forms the posterior part of the jugular foramen.

STRUCTURE. The occipital, like the other cranial bones, consists of two compact lamellae, called the outer and inner tables, between which is the cancellous tissue or diploë. The bone is especially thick at the ridges, protuberances, condyles, and anterior region of the basilar part. In the inferior fossae, within which rests the cerebellum, the bone is thin, semitransparent, and destitute of diploë.

OSSIFICATION (Fig. 4-87). The portion of the squama above the nuchal lines, the *planum occipitale,* is developed in membrane, and may remain separate throughout life, in which case it constitutes the *interparietal bone.* The rest of the bone is developed in cartilage. Ossification commences in the planum occipitale from two centers which appear near the midline at about the eighth week of fetal life. Two other centers appear at some distance from the midline at about the third fetal month. The *planum nuchale* of the squama, located below the highest nuchal lines, is ossified from two centers, which appear in the cartilage at about the seventh week of fetal life and soon unite to form a single center. Union of the upper and lower portions of the squama takes place in the third month of fetal life. Each of the *lateral parts* begins to ossify from a single center also during the eighth week of fetal life. The *basilar portion* is ossified from either one or two centers, which appear at about the sixth week of prenatal life.

At birth, the occipital bone consists of four ossified pieces—the squama, two lateral parts, and the basilar part—united by cartilaginous strips (Fig.

4-87). The lines of union between the portions of the squama are also still evident. About the fourth year the squama and the two lateral portions unite, and by about the sixth year the bone consists of a single piece. Between the eighteenth and twenty-fifth years the synchondrosis between the occipital and sphenoid becomes united, forming a single bone.

ARTICULATIONS. The occipital bone articulates with six other bones: the two parietals, the two temporals, the sphenoid, and the atlas.

PARIETAL BONE

The **parietal bones** (*os parietale*), by their union, form the sides and roof of the cranium. Each bone is irregularly quadrilateral in shape and presents two surfaces, four borders, and four angles.

SURFACES. The **external surface** (Fig. 4-88) is convex, smooth, and marked near the center by the mound-like **parietal eminence** (*tuber parietale*), an elevation of even greater convexity and the site at which ossification commences in the fetus. Crossing the middle of the bone in an arched direction are two curved lines, the **superior** and **inferior temporal lines.** The former gives attachment to the temporal fascia, and the latter indicates the upper limit of the muscular origin of the temporalis. Above these lines the bone is covered by the galea aponeurotica of the scalp, while below the lines the parietal bone forms the medial boundary of the temporal fossa and also lends origin to the temporalis. Close to the upper or sagittal border posteriorly is the **parietal foramen,** which transmits a vein to the superior sagittal sinus and, sometimes, a small branch of the occipital artery. The foramen is not constantly present, and its size varies considerably.

The **internal surface** (Fig. 4-89) is concave and presents depressions corresponding to the cerebral convolutions, and numerous furrows for the ramifications of the middle meningeal vessels. These vascular furrows course upward and backward from the anteroinferior sphenoidal angle and from the central and posterior part of the squamous border. Along the superior edge of the bone is a shallow groove, which combines with that of the opposite parietal to form the midline **sulcus for the superior sagittal sinus.** The edges of the sulcus afford attachment to the falx cerebri. Near the groove are several depressions or pits, **granular foveolae,** best marked in the skulls of old persons, which

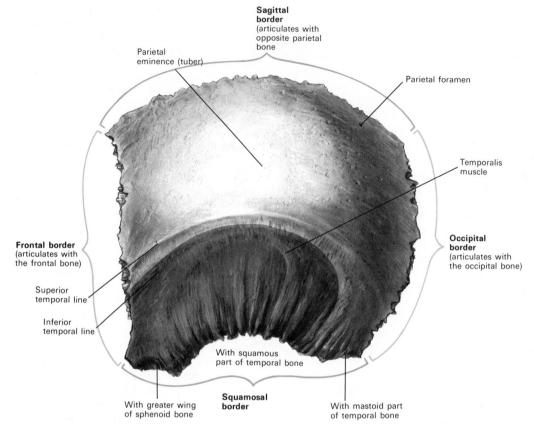

Sagittal border (articulates with opposite parietal bone

Parietal eminence (tuber)

Parietal foramen

Temporalis muscle

Occipital border (articulates with the occipital bone)

Frontal border (articulates with the frontal bone)

Superior temporal line

Inferior temporal line

With squamous part of temporal bone

With greater wing of sphenoid bone

Squamosal border

With mastoid part of temporal bone

FIG. 4-88. Left parietal bone, external surface. Observe the serrated borders of the parietal bone. These contribute to the formation of the coronal suture in front, the sagittal suture above, the lambdoidal suture behind, and the squamosal suture below. Additionally, the squamosal border of the parietal bone contributes to the sphenoparietal suture anteriorly and the parietomastoid suture posteriorly.

contain the **arachnoid granulations** (*Pacchionian bodies*). In the groove is the internal orifice of the parietal foramen.

BORDERS. The **sagittal border** is the longest and thickest and its deeply serrated edge articulates with the same border of the opposite parietal bone, thereby forming the sagittal suture. The **squamosal border** is the inferior border of the parietal bone and it is divided into three parts: of these, the *anterior* is thin, bevelled externally, and pointed, and is overlapped by the tip of the great wing of the sphenoid; the *middle* portion is arched and also bevelled externally, and is overlapped by the squama of the temporal bone; the *posterior part* is thick and serrated for articulation with the mastoid portion of the temporal bone. The **frontal border** is deeply serrated and is externally bevelled above and internally bevelled below. This border articulates with the frontal bone, forming one-half of the coronal suture. The

occipital border is deeply denticulated and forms one-half of the lambdoidal suture by articulating with the occipital bone.

ANGLES. The **frontal angle** is practically a right angle. It is the anterosuperior angle of the bone and corresponds with the site of junction of the sagittal and coronal sutures, the **bregma.** In the fetal skull and for about a year and a half after birth, this region is not ossified and is called the **anterior fontanelle** (Figs. 4-75; 4-76). The **sphenoidal angle,** located anteroinferiorly on the parietal bone, is thin and acute, and is received into the interval between the frontal bone and the great wing of the sphenoid. On the side of the skull in the vicinity of the sphenoidal angle there are four bones in proximity: the parietal, sphenoid, temporal, and frontal. This site is called the **pterion.** The inner surface of the bone near the sphenoidal angle is marked by a deep groove, which sometimes forms a canal for the frontal branch of the

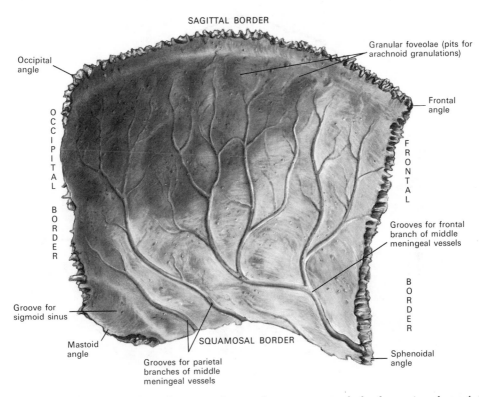

SAGITTAL BORDER

Occipital
angle

Granular foveolae (pits for
arachnoid granulations)

Frontal
angle

O
C
C
I
P
I
T
A
L

B
O
R
D
E
R

F
R
O
N
T
A
L

B
O
R
D
E
R

Grooves for frontal
branch of middle
meningeal vessels

Groove for
sigmoid sinus

Mastoid
angle

SQUAMOSAL BORDER

Sphenoidal
angle

Grooves for parietal
branches of middle
meningeal vessels

FIG. 4-89.   Left parietal bone, internal surface. Note the vascular grooves not only for the meningeal vessels but also for the superior sagittal sinus and the sigmoid sinus. It is at the mastoid angle that the transverse sinus becomes the sigmoid sinus.

middle meningeal artery. The **occipital angle** is located posterosuperiorly on the bone and is more rounded than the previous two. It corresponds with the intersection of the sagittal and lambdoidal sutures, the **lambda.** In the newborn skull this site is yet to ossify and is called the **posterior fontanelle** (Fig. 4-75). The thick and truncated **mastoid angle** is located posteroinferiorly. It articulates with the occipital bone and with the mastoid portion of the temporal. On its inner surface is found a broad, shallow groove which lodges the transition site of the *transverse sinus,* beyond which it is known as the *sigmoid sinus.* The point where this angle meets with the mastoid part of the temporal bone is named the *asterion.*

OSSIFICATION.   The parietal bone ossifies intramembranously from two centers, one above the other, which appear at the parietal eminence about the eighth week of fetal life. The two centers fuse during the third month and ossification gradually extends in a radial manner toward the margins of the bone. The angles are consequently the last sites to be ossified and it is here that the fontanelles exist. Occasionally the parietal bone is divided into two parts by an anteroposterior suture caused by a failure of the two primary ossification sites to fuse.

ARTICULATIONS.   The parietal usually articulates with five bones: the opposite parietal, the occipital, frontal, temporal, and sphenoid. At times the parietal bone does not articulate with the sphenoid. In these instances the frontal bone reaches posteriorly as far as the squama of the temporal without the greater wing of the sphenoid achieving the squamosal border of the parietal bone.

FRONTAL BONE

The **frontal bone** (*os frontale*) encases the cranial cavity anteriorly and is shaped as an externally convex shell. It consists of two portions: the **squama,** which is vertical and forms the osseous convexity of the forehead, and the **orbital portion,** which is horizontal and enters into the formation of the roofs of the orbital and nasal cavities.

SQUAMA OR VERTICAL PART (*squama frontalis*).   The **external surface** (Fig. 4-90) of the squama of the frontal bone is convex and in the midline usually retains evidence of the **frontal** or **metopic suture.** In infancy this

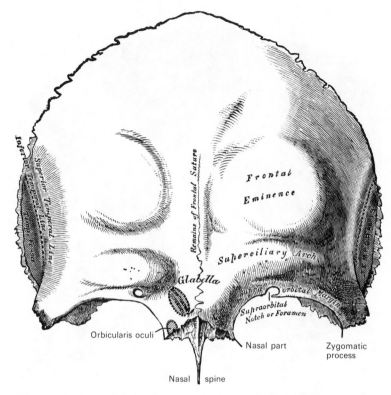

FIG. 4-90. Frontal bone. Outer surface as viewed from anterior aspect.

suture separates the two lateral halves of the bone, a condition that may persist throughout life. On each side of this suture, about 3 cm superior to the supraorbital margin, is a rounded elevation, the **frontal eminence** (*tuber frontale*). These eminences vary in size, are occasionally asymmetrical, and are especially prominent in young skulls. Inferior to the frontal eminences, and separated from them by a shallow groove, are the **superciliary arches,** which are prominent medially and joined to one another by a smooth elevation named the **glabella.** These curved arches are larger in males than in females, and to some extent their degree of prominence depends on the size of the frontal air sinuses. Beneath each superciliary arch is a curved and prominent margin, the **supraorbital margin,** which forms the boundary of the orbit and separates the squama from the orbital portion of the bone. The lateral part of this margin is sharp and prominent, affording protection to the eye, and its medial part is rounded. Along the margin, a third of the distance from its medial end, is the **supraorbital notch** or **foramen,** which transmits the supraorbital vessels and nerve. A small aperture in the notch transmits a vein from the diploë to join the supraorbital vein. The supraorbital margin ends laterally in the **zygomatic process,** which is strong and prominent and articulates with the zygomatic bone. Coursing upward and backward from the zygomatic process is the **temporal line,** which then divides into upper and lower temporal lines, continuous in the articulated skull with the corresponding lines on the parietal bone.

Between the supraorbital margins the squama projects inferiorly to a level even farther than the zygomatic process. This portion of the frontal bone is the **nasal part,** and it presents a rough, uneven interval, the **nasal notch,** which articulates with the nasal bone, the frontal process of the maxilla, and the lacrimal. The center of the frontonasal suture is a craniometric point called the **nasion.** From the center of the notch, the nasal part projects inferiorly beneath the nasal bones and the frontal processes of the maxillae to support the bridge of the nose. The nasal part ends inferiorly as the sharp **nasal spine,** on each side of which is a small grooved nasal surface that enters into the formation of the roof of the corresponding nasal cavity (Fig. 4-91). The spine forms a small portion of the nasal septum (Figs. 4-71; 4-72) and articulates anteriorly with the crest of the nasal bones and with the perpendicular plate of the ethmoid posteriorly.

The **internal surface** (Fig. 4-91) of the squama is concave and presents in the midline a vertical groove, the **sulcus for the superior sagittal sinus,** the edges of which unite below to form a ridge, the **frontal crest.** The sulcus lodges the anterior part of the superior sagittal sinus, while its margins and the crest afford attachment to the falx cerebri. The crest ends in a small notch, which is converted into the **foramen cecum** by articulation with the ethmoid bone. This foramen varies in size and is frequently closed; when open, it transmits a vein from the roof of the nasal cavity to the superior sagittal sinus. On each side of the midline the bone presents depressions for convolutions of the frontal lobes of the brain, and numerous small pits,

the **granular foveolae,** for the arachnoid granulations.

The **parietal border** of the squama of the frontal bone is thick, strongly serrated, and bevelled. It articulates posteriorly and laterally with the parietal bones. This border is continued below into a triangular, rough surface which articulates with the great wing of the sphenoid.

ORBITAL OR HORIZONTAL PART (*pars orbitalis*). The orbital portion of the frontal bone consists of two smooth **orbital plates** which form most of the roofs of the orbits. They are separated from one another by a gap in the midline, the **ethmoidal notch.** The concave *inferior surface* (Fig. 4-91) presents a shallow depression laterally, the **fossa for the lacrimal gland.** Medially, near the nasal part, is a depression, the **fovea trochlearis,** or occasionally a small trochlear spine, for the attachment of the cartilaginous pulley of the superior oblique muscle of the eye. The *superior* (or intracranial) *surface* is convex

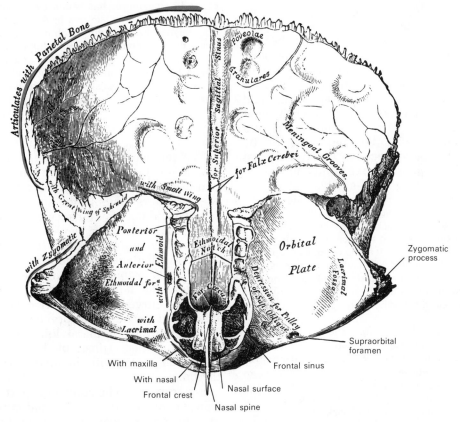

FIG. 4-91. Frontal bone. Inner surface as viewed from its inferior aspect. Observe that on the left side of the figure is accentuated the lines of articulation of the frontal bone. These are with the parietal, zygomatic, and greater wing of the sphenoid bone laterally and the small wing of the sphenoid, ethmoid, lacrimal, maxillary, and nasal bones medially.

and marked by depressions for gyri on the orbital surface of the frontal lobes of the brain, and by faint grooves for the meningeal branches of the ethmoidal vessels. The **posterior borders of the orbital plates** are thin and serrated, and articulate with the small wings of the sphenoid.

The **ethmoidal notch** that separates the two orbital plates is filled, in the articulated skull, by the cribriform plate of the ethmoid. The margins of the notch are indented by several air cells which, when united with corresponding cells on the superior surface of the ethmoid, form the ethmoidal air cells. Two grooves cross these edges transversely and are converted, by articulation with the ethmoid bone, into the **anterior and posterior ethmoidal canals,** which open on the medial wall of the orbit and transmit the anterior and posterior ethmoidal vessels and nerves.

The **frontal air sinuses** open on both sides of the frontal spine and anterior to the ethmoid air cells and ethmoid notch. The sinuses are two irregularly shaped hollow cavities that extend for a variable distance between the two tables of the skull. They are separated from one another by a thin bony septum, which often deviates to one side or the other, with the result that the sinuses are rarely symmetrical. They vary in size and are larger in men than in women. The frontal sinuses are lined by mucous membrane, and communicate with the nasal cavity by means of a passage called the **frontonasal duct,** which opens into the middle meatus of the corresponding nasal cavity.

STRUCTURE. The squama and the zygomatic processes are very thick, consisting of cancellous bone contained between two compact laminae. The cancellous bone is absent in the regions occupied by the frontal air sinuses. The orbital portion is thin, translucent, and composed entirely of compact bone, explaining the facility with which instruments can penetrate the cranium through this part of the orbit.

OSSIFICATION (Fig. 4-92). The frontal bone is ossified in membrane from two primary centers. These appear toward the end of the second month of fetal life, one above each superciliary arch. From each of these centers ossification extends upward to form the corresponding half of the squama, inferiorly to form the nasal part, and backward to form the orbital plate. At birth the bone consists of two pieces, separated by the **frontal** or **metopic suture,** which is usually fused by the eighth year except at its lower part. Occasionally it persists throughout

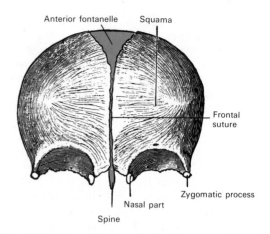

FIG. 4-92. Frontal bone at birth. The neonatal frontal bone consists of two halves between which is found the frontal or metopic suture.

life. During childhood and after the metopic suture has united (between the tenth and twelfth years), secondary ossification centers appear at the lower end of the suture to form the nasal spine of the frontal bone. It is generally maintained that development of the frontal sinuses has begun by the end of the first year. The frontal sinuses have developed considerably by the seventh or eighth year, but do not attain their maximum size until adulthood.

ARTICULATIONS. The frontal bone articulates with twelve bones: the sphenoid, the ethmoid, the two parietals, the two nasals, the two maxillae, the two lacrimals, and the two zygomatics (Fig. 4-91).

TEMPORAL BONE

The **temporal bone** (*os temporale;* Figs. 4-93 to 4-97) is situated at the lateral aspect and base of the skull. It contains the peripheral receptors for both the special sense of hearing and the maintenance of equilibrium or vestibular function. Developmentally, the temporal bone forms from a fusion of three parts: the flattened **squamous** portion; the hard **petromastoid** part, which encases the membranous labyrinth and its receptors; and the curved, ring-like **tympanic** part, interposed between the previous two (Fig. 4-98). Additionally, the **styloid process,** derived from the cartilage of the second visceral arch, unites with the inferior aspect of the bone.

THE SQUAMA. The squamous part is thin, translucent, and scale-like, and forms the anterior and superior part of the bone. Its **temporal** or **external surface** (Fig. 4-93) is smooth and convex, forms part of the temporal fossa, and affords attachment to the

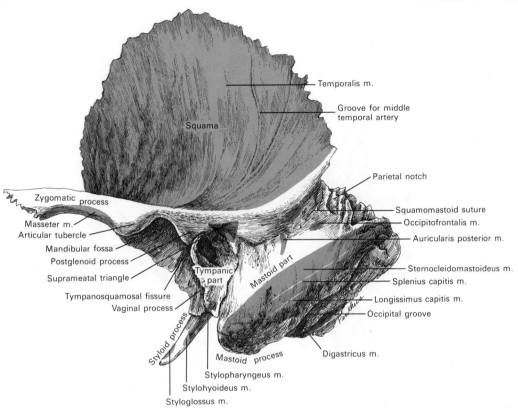

FIG. 4-93. The left temporal bone. External aspect.

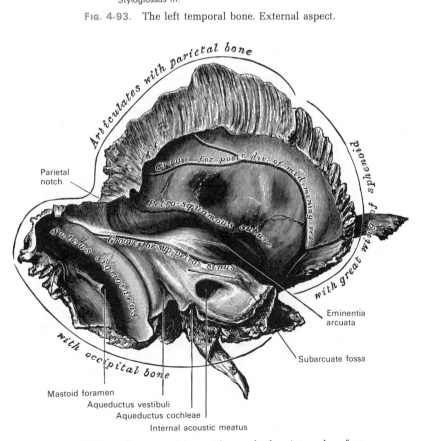

FIG. 4-94. Left temporal bone. The cerebral or internal surface.

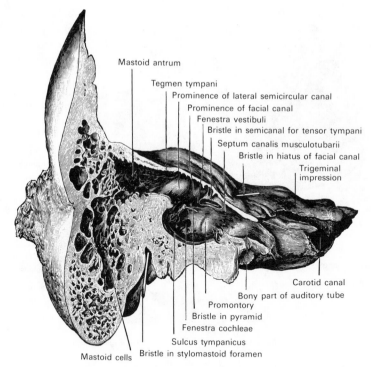

Mastoid antrum

Tegmen tympani

Prominence of lateral semicircular canal

Prominence of facial canal

Fenestra vestibuli

Bristle in semicanal for tensor tympani

Septum canalis musculotubarii

Bristle in hiatus of facial canal

Trigeminal impression

Carotid canal

Bony part of auditory tube

Promontory

Bristle in pyramid

Fenestra cochleae

Sulcus tympanicus

Bristle in stylomastoid foramen

Mastoid cells

FIG. 4-95. Coronal section of right temporal bone which has been opened to display the mastoid antrum and cells as well as the roof and medial wall of the tympanic cavity.

temporalis muscle. The curved **temporal line** limits the origin of the temporalis muscle. The boundary between the squama and the mastoid process of the bone lies about 1 cm below this line. Superior to the external acoustic meatus, the external surface presents a vertically coursing **groove for the middle temporal artery.**

The **zygomatic process** is a narrow arch of bone that projects forward from the inferior part of the squama. Its thin superior border is sharp and serves for the attachment of the temporal fascia; its inferior border is somewhat thicker and arched, and affords attachment to some fibers of the masseter muscle. The lateral surface is convex and lies just beneath the skin, while the medial surface is concave and also allows origin to the masseter muscle. The anterior end is deeply serrated and articulates with the zygomatic bone. The posterior end arises from the squama by two roots. The **posterior root,** a prolongation of the upper border, courses above the opening of the external acoustic meatus and is continuous with the temporal line. The **anterior root** is continuous with the inferior border. It is short, broad, and strong, and is directed medialward, ending in a

rounded eminence, the **articular tubercle.** This tubercle forms the anterior boundary of the **mandibular fossa.** In the intact body the tubercle is covered with cartilage and the fossa receives the condyle of the mandible to form the temporomandibular joint. Between the posterior wall of the external acoustic meatus and the posterior root of the zygomatic process is an area called the **supra-meatal triangle,** through which an instrument may be pushed into the tympanic antrum.

The **mandibular fossa** (glenoid fossa) is bounded anteriorly by the articular tubercle, and posteriorly by the bony rim surrounding the external acoustic meatus. The fossa is ovoid in shape and divided into two parts by a narrow slit, the **tympanosquamosal fissure,** which, when traced medially, becomes continuous with the **petrotympanic fissure.** The anterior part of the fossa, formed by the squama, is smooth and covered with cartilage for articulation with the condyle of the mandible. Behind this part of the fossa is a small conical eminence, the **postglenoid process.** This process is representative of a prominent tubercle, seen in some mammals, which projects behind the condyle of the

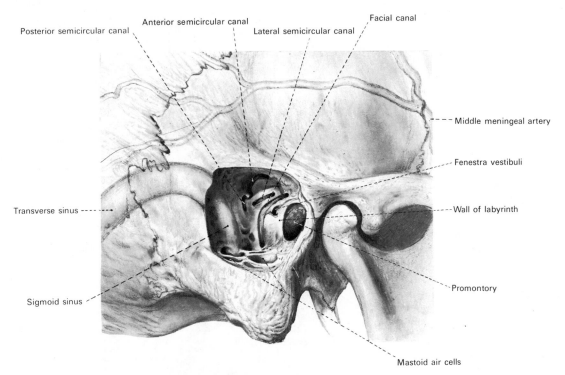

Posterior semicircular canal

Anterior semicircular canal

Lateral semicircular canal

Facial canal

Middle meningeal artery

Fenestra vestibuli

Transverse sinus

Wall of labyrinth

Promontory

Sigmoid sinus

Mastoid air cells

FIG. 4-96. The right temporal bone has been opened to reveal the medial wall of the tympanic cavity. Observe that the facial canal and semicircular canals have also been opened. Note that the course of the facial canal in the petrous temporal bone is at first directed laterally (toward the middle ear), then posteriorly, and finally inferiorly. (From Benninghoff and Goerttler, *Lehrbuch der Anatomie des Menschen,* Urban and Schwarzenberg, 1975.)

mandible and prevents its backward displacement. The *posterior part* of the mandibular fossa is formed by the tympanic part of the bone and is nonarticular. The petrotympanic fissure leads into the middle ear or tympanic cavity, where it lodges the anterior ligament of the malleus and transmits the anterior tympanic branch of the maxillary artery. The chorda tympani nerve passes through a canal **(anterior canaliculus of chorda tympani),** which is separated from the anterior edge of the petrotympanic fissure by a thin scale of bone and situated on the lateral side of the auditory tube in the angle between the squama and the petrous portion of the temporal bone.

The **cerebral** or **internal surface** of the squama (Fig. 4-94) presents depressions corresponding to the convolutions of the temporal lobe of the brain, and grooves for the branches of the middle meningeal vessels. The **superior border** is thin, bevelled at the expense of the internal table, and overlaps the squamous border of the parietal bone, forming with it the squamosal suture. Posteriorly, the superior border forms an angle,

the **parietal notch,** with the mastoid portion of the bone. The **antero-inferior border** is thick, serrated, and bevelled at the expense of the inner table above and of the outer below. It articulates with the great wing of the sphenoid.

PETROMASTOID PART. The petromastoid part of the temporal bone consists of a **mastoid portion** and a **petrous portion,** each of which is best described separately.

Mastoid Portion. The **mastoid portion** forms the posterior part of the bone. Its **external surface** (Fig. 4-93) is rough and gives attachment to the occipital belly of the occipitofrontalis muscle, as well as to the auricularis posterior muscle. It is perforated by numerous foramina. One of these, the **mastoid foramen,** is larger, and is often situated near the posterior border. It transmits a vein to the sigmoid sinus (Fig. 4-94) and a small branch of the occipital artery to the dura mater. The position and size of this foramen are variable. It may be absent or situated in the occipital bone, or even in the suture between the temporal and the occipital bones. Inferiorly, the mastoid portion is con-

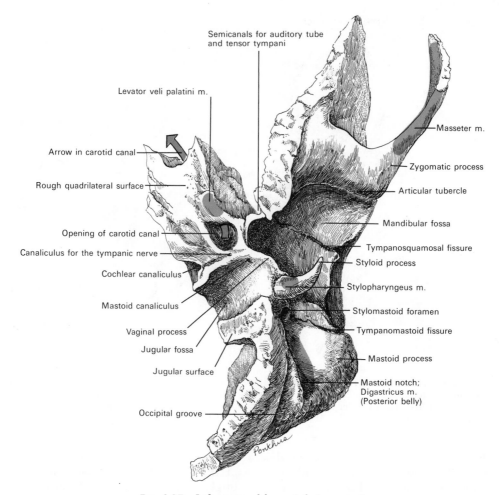

Semicanals for auditory tube
and tensor tympani

Levator veli palatini m.

Masseter m.

Arrow in carotid canal

Zygomatic process

Rough quadrilateral surface

Articular tubercle

Mandibular fossa

Opening of carotid canal

Tympanosquamosal fissure

Canaliculus for the tympanic nerve

Styloid process

Cochlear canaliculus

Stylopharyngeus m.

Mastoid canaliculus

Stylomastoid foramen

Vaginal process

Tympanomastoid fissure

Jugular fossa

Mastoid process

Jugular surface

Mastoid notch;
Digastricus m.
(Posterior belly)

Occipital groove

FIG. 4-97. Left temporal bone. Inferior aspect.

tinued as a conical projection, the **mastoid process.** Although its size and form may vary, this process serves for the insertion of the sternocleidomastoid, splenius capitis, and longissimus capitis muscles. On the medial side of the process is a deep groove, the **mastoid notch,** which gives attachment to the posterior belly of the digastric muscle. Medial to the notch is a shallow furrow, the **occipital groove,** which lodges the occipital artery.

The **internal surface** (Fig. 4-94) of the mastoid portion presents a deep, curved groove, the **sigmoid sulcus,** which lodges the sigmoid sinus. Behind the sulcus is frequently seen the opening of the mastoid foramen. The groove for the sigmoid sinus is separated from the innermost of the mastoid air cells by a very thin lamina of bone, and even this may be partly deficient.

The **superior border** of the mastoid portion is broad and serrated, for articulation with the mastoid angle of the parietal bone. The **posterior border,** also serrated, articulates with the inferior border of the occipital bone between the lateral angle and jugular process. As observable from an external view (Fig. 4-93), the mastoid portion is fused above and anteriorly with the descending process of the squama. Below and posteriorly it enters into the formation of the posterior wall of the tympanic cavity, and extends nearly to the external acoustic meatus.

A section of the mastoid process (Fig. 4-95) shows it to be hollowed out into a number of spaces, the **mastoid cells.** These exhibit a great variability, both in size and number. In the superior part of the mastoid process they are large and irregular, and contain air. Toward the inferior part they diminish in size, while those at the apex of the process are frequently quite small and contain marrow. Occasionally the mastoid cells are entirely absent, and the process is then solid throughout. Additionally, a large irregular cavity called the **mastoid antrum** is situated at the superior and anterior part of the process (Fig. 4-95). It can be distinguished from the mastoid cells, although it communicates with them. The tympanic antrum also contains air, and it is lined by a prolongation of the mucous membrane of the tympanic cavity with which it communicates as well. The mastoid antrum is bounded above by a thin plate of bone, the **tegmen tympani,** which separates it from the middle fossa of the cranial cavity. Lateral to the antrum is the temporal squama, while medially is the **lateral semicircular canal** of the internal ear which projects into its cavity. It opens anteriorly into that portion of the tympanic cavity known as the **attic** or **epitympanic recess.** The mastoid antrum is a cavity of some considerable size at birth. The mastoid air cells may be regarded as diverticula from the antrum, and begin to appear at or before birth. By the fifth year they are well marked, but their development is not completed until puberty.

Petrous Portion. The **petrous portion** or **pyramid** is wedged between the sphenoid and occipital bones at the base of the skull (see Fig. 4-70). Enclosed within it is the labyrinth containing its receptors for the senses of hearing and equilibration. It presents for examination a base, an apex, three surfaces, and two margins.

Base and Apex. The **base** is fused with the internal surfaces of the squamous and mastoid portions.

The **apex,** rough and uneven, is received into the angular interval between the posterior border of the great wing of the sphenoid and the basilar part of the occipital bone. It presents the anterior or internal orifice of the carotid canal and forms the posterolateral boundary of the foramen lacerum.

Surfaces. The **anterior surface** helps to form the base of the skull at the posterior part of the middle cranial fossa. It is continuous with the internal or cerebral surface of the squamous portion, to which it is united by the petrosquamous suture, the remains of which are distinct even late in life. It is marked by depressions for the convolutions of the brain, and presents six points for examination: (1) near the center is the **arcuate eminence** (Fig. 4-94), which indicates the site of the anterior semicircular canal; (2) a shallow depression is found slightly lateral and anterior to the eminence, and indicates the position of the tympanic cavity (Fig. 4-95) and its thin, bony roof, the **tegmen tympani;** (3) a shallow groove, sometimes double, leads into the **hiatus** of the **facial canal** (Fig. 4-95), for the passage of the greater petrosal nerve and the petrosal branch of the middle meningeal artery; (4) lateral to the hiatus is occasionally seen a smaller opening for the passage of the lesser petrosal nerve; (5) near the apex of the bone is found the termination of the **carotid canal** (Fig. 4-95), the wall of which at this site is deficient; and (6) just behind the apex and above the carotid canal is the shallow **trigeminal impression** for the trigeminal (*semilunar*) ganglion (Fig. 4-70).

The **posterior surface** of the petrous portion forms the anterior part of the posterior cranial fossa (Fig. 4-70). Near the center is the large orifice of the **internal acoustic meatus** (Fig. 4-94) leading into the **internal acoustic canal.** The margins of the meatus are smooth and rounded, and the canal measures about 1 cm in length. It transmits the facial and vestibulocochlear nerves as well as the nervus intermedius, and the internal auditory branch of the basilar artery. Posterior to the internal acoustic meatus is a small slit, almost hidden by a thin plate of bone, that leads to a canal, the **aqueduct of the vestibule,** which transmits the endolymphatic duct together with a small artery and vein. Above and between these two openings is an irregular depression which lodges a process of the dura mater and transmits a small vein. In the infant this depression is represented by a large fossa, the **subarcuate fossa,** which extends backward as a blind tunnel under the anterior semicircular canal.

The **facial canal** is a bony passage within the petrous portion of the temporal bone through which the facial nerve passes. The canal commences at the internal acoustic meatus and ends at the stylomastoid foramen. In its course the canal is initially directed *laterally* toward the medial wall of the tympanic cavity with the cochlea lying in front and the semicircular canals behind. Then the canal suddenly courses *posteriorly* (Fig. 4-96), lying above the **fenestra vestibuli** and the **promontory.** With the semicircular canals still lying above and behind, the facial canal then arches *downward* toward its exit at the stylomastoid foramen.

The **inferior surface** of the petrous portion of the temporal bone (Fig. 4-97) is rough and irregular, and forms part of the external surface of the base of the skull. It presents a number of features which may be examined: (1) near the apex is a rough quadrilateral surface, partly for the attachment of the levator veli palatini and the cartilaginous portion of the auditory tube, and partly for connection with the basilar part of the occipital bone through the intervention of some dense fibrous tissue; (2) posterior to this is the large circular aperture of the **carotid canal,** which transmits the internal carotid artery and the carotid plexus of sympathetic nerve fibers into the cranium; (3) medial and posterior to the carotid opening and anterior to the jugular fossa is a triangular depression at the apex of which is a small opening, the **cochlear canaliculus;** it lodges a tubular prolongation of the dura mater establishing a communication between the perilymphatic space and the subarachnoid space and it transmits a vein from the cochlea to the internal jugular; (4) posterior to these openings is a deep depression of varying depth and size, the **jugular fossa,** containing the bulb of the internal jugular vein; (5) in the bony ridge between the carotid canal and the jugular fossa is the small **canaliculus for the tympanic nerve** through which passes the tympanic branch of the glossopharyngeal nerve; (6) in the lateral part of the jugular fossa is the **mastoid canaliculus** for the entrance into the petrous portion of the temporal bone of the auricular branch of the vagus nerve; (7) behind the jugular fossa is a quadrilateral area, the **jugular surface,** which is covered with cartilage in the intact body, and which articulates with the jugular process of the occipital bone; (8) extending posteriorly from the carotid canal is the **vaginal process,** a sheath-like plate of bone, which divides behind into two laminae: the lateral lamina is continuous with the tympanic part of the bone; the medial, with the lateral margin of the jugular surface; (9) between these laminae is a sharp spine, the **styloid process;** (10) between the styloid and mastoid processes is the **stylomastoid foramen,** which is the termination of the facial canal and transmits the facial nerve and stylomastoid artery; (11) situated between the tympanic portion of the temporal bone and the mastoid process is the **tympanomastoid fissure,** through which the auricular branch of the vagus nerve usually exits from the temporal bone.

Margins. The **superior margin** of the petrous portion is grooved for the superior petrosal sinus (Fig. 4-94), and gives attachment to the tentorium cerebelli. At its medial extremity is a notch, in which the roots of the trigeminal nerve lie.

The **posterior margin** is marked on its medial half by a sulcus which forms, with a corresponding sulcus on the occipital bone, the channel for the inferior petrosal sinus. The lateral half of the posterior margin presents an excavation, the jugular fossa, which, with the jugular notch on the occipital bone, forms the **jugular foramen.** An eminence occasionally projects from the center of the fossa, dividing the foramen into two.

The petrosquamous suture (Fig. 4-94), which is quite distinct in early life, unites the petrous and squamosal parts. This border of the petrous bone might be considered the **anterior margin,** and here are located two canals, one above the other, separated by a thin plate of bone, the **septum canalis musculotubarii** (Fig. 4-95). Both canals lead into the tympanic cavity. The upper one **(semicanalis m. tensoris tympani)** transmits the tensor tympani muscle; the lower one **(semicanalis tubae auditivae)** forms the bony part of the auditory tube.

The tympanic cavity, auditory ossicles, and internal ear are described with the organ of hearing in the chapter on the Organs of the Senses.

TYMPANIC PART. The **tympanic part** of the temporal bone is a curved osseous plate lying inferior to the squama and anterior to the mastoid process (Fig. 4-93). Internally, the tympanic part is fused with the petrous portion, and appears in the angle between it and the squama, where it lies inferior and lateral to the orifice of the auditory tube. Posteriorly, it blends with the squama and mastoid portion and forms the anterior boundary of the tympanomastoid fissure. The **posterior surface** of the tympanic part is concave, and forms the anterior wall, the floor, and part of the posterior wall of the bony **external acoustic meatus.** Medially, this surface presents a narrow furrow, the **tympanic sulcus,** for the attachment of the tympanic membrane. The **anterior surface** of the tympanic part is quadrilateral and slightly concave. It constitutes the posterior boundary of the mandibular fossa and frequently is in contact with the retromandibular part of the parotid gland. Its **lateral border** is largely free and rough, but does give attachment to the cartilaginous part of the external acoustic meatus. Its **superior border** fuses laterally with the postglenoid process, while medially it bounds the tympanosquamosal fissure. The medial part of the **inferior border** is thin and sharp, while more laterally this border splits to enclose the root of the **styloid process.** It is called the **vaginal process** or **sheath of the styloid process.** The central portion of the tympanic part is thin, and in a considerable percentage of skulls presents a small foramen (of Huschke) which represents a nonossified site.

The **external acoustic meatus** is nearly 2 cm long and is directed inward and slightly forward. At the same time it forms a slight curve, so that the floor of the canal is convex superiorly. Its anterior wall, floor, and inferior part of the posterior wall are formed by the tympanic portion, while the roof and superior part of the posterior wall are formed by the squama. The inner end of the meatus is closed in the intact body by the *tympanic membrane,* thereby separating the tympanic cavity of the middle ear from the external ear. The superior limit of its outer orifice is bounded above by the posterior root of the zygomatic process, immediately below which is sometimes seen the small **suprameatal spine,** situated at the upper and posterior part of the orifice.

The **styloid process** of the temporal bone (Figs. 4-93; 4-97) is slender and pointed, and varies in length from 1 to as much as 4 cm. It projects downward and forward, from the inferior surface of the temporal bone immediately in front of the stylomastoid foramen. Its proximal or **basal part** (*tympanohyal*) is ensheathed by the vaginal process of the tympanic portion, while its distal or **projecting part** (*stylohyal*) gives attachment to the stylohyoid and stylomandibular ligaments, and to the styloglossus, stylohyoid, and stylopharyngeus muscles. The stylohyoid ligament, which may be more or less ossified, extends from the apex of the process to the lesser horn of the hyoid bone. The stylomandibular ligament, which is fibrous in structure, courses from its attachment on the styloid process below the origin of the styloglossus to the angle of the mandible.

OSSIFICATION. The temporal bone is ossified from at least eight centers, exclusive of those for the ossicles of the middle ear and the labyrinthine structures of the internal ear. The centers include one for the squama and zygomatic process, one for the tympanic part, four for the petrous and mastoid parts, and two for the styloid process.

At birth (Fig. 4-98) the temporal bone consists of three principal parts: (1) The **squama** is ossified in membrane from a single nucleus, which appears near the *root of the zygomatic process* about the second month. (2) The **petromastoid part** is developed from at least four centers, which make their appearance in the cartilaginous **otic** (ear) **capsule** at the end of the fifth month and during the sixth prenatal month. The centers are named according to their position relative to the embryonic otic capsule. One (pro-otic) appears in the *region of the arcuate eminence* and spreads anterior and superior to the internal acoustic meatus, extending to the apex of the bone. It forms part of the cochlea, vestibule, superior semicircular canal, and medial wall of the tympanic cavity. A second (opisthotic) appears at the *promontory on the medial wall of the tympanic cavity* and surrounds the cochlear window. It forms the floor of the tympanic cavity and vestibule, surrounds the carotid canal, invests the lateral and lower part of the cochlea, and spreads medially below the internal acoustic meatus. A third (pterotic) appears *over the lateral semicircular canal* and spreads to the tegmen tympani, roofing in the tympanic cavity and antrum. The fourth (epiotic) appears *near the posterior semicircular canal* and extends to form the mastoid process. (3) At birth the **tympanic part** is an incomplete ring, in the concavity of which is a groove, the tympanic sulcus, for the attachment of the circumference of the tympanic membrane. This ring ossifies in membrane from a single center which appears during the ninth or tenth fetal week.

The **styloid process** is developed from the proximal part of the cartilage of the second visceral or hyoid arch. It ossifies by means of two centers: one for the proximal part, the *tympanohyal,* appears before birth; the other, comprising the rest of the process, is named the *stylohyal,* and does not appear until the second year after birth.

The tympanic ring unites with the squama shortly before birth. The petromastoid part and squama join during the first year, and at about the same time, the tympanohyal portion of the styloid process unites with the inferior surface of the temporal bone (Figs. 4-99; 4-100). The stylohyal portion of the styloid process does not unite with the rest of the bone until after puberty, and in some skulls never at all.

After birth the principal changes in the temporal bone apart from increase in size are: (1) The tym-

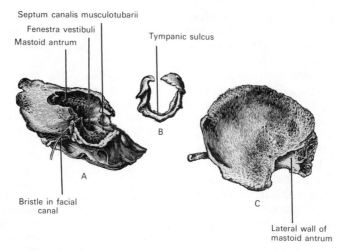

Septum canalis musculotubarii
Fenestra vestibuli
Mastoid antrum
Tympanic sulcus
Bristle in facial canal
Lateral wall of mastoid antrum

FIG. 4-98.   The three principal parts of the temporal bone at birth. *A*, Outer surface of petromastoid part. *B*, Inner surface of tympanic ring. *C*, Inner surface of squama.

panic ring extends laterally and backward to form the tympanic part of the bone. This growth does not take place at an equal rate all around the circumference of the ring, but occurs most rapidly on its anterior and posterior portions. These outgrowths meet and blend, and thus, for a time, there exists in the floor of the meatus an opening called the *foramen of Huschke*. This foramen is usually closed about the fifth year, but may persist throughout life. (2) The mandibular fossa is at first extremely shallow, and faces laterally as well as inferiorly. It becomes deeper and is ultimately directed inferiorly. Its change in direction is the result of growth phenomena. The part of the squama that forms the fossa lies at first inferior to the level of the zygomatic process. As the base of the skull increases in width, however, this lower part of the squama is directed horizontally inward to contribute to the bony middle fossa of the skull, and its surfaces therefore come to face superiorly and inferiorly. The attached portion of the zygomatic process also becomes everted and projects like a shelf at right angles to the squama. (3) The mastoid portion is at first quite flat, and the stylomastoid foramen and rudimentary styloid process lie immediately behind the tympanic ring. With the development of the air cells the lateral part of the mastoid portion grows inferiorly and anteriorly to form the mastoid process. The styloid process and stylomastoid foramen now come to lie on the inferior surface. The descent of the foramen is necessarily accompanied by a corresponding lengthening of the facial canal. (4) The downward and forward growth of the mastoid process also pushes forward the tympanic part, so that the portion of it that formed the original floor of the meatus at birth (and contained the foramen of Huschke) is ultimately found in the

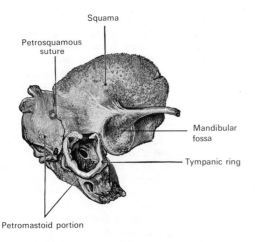

Squama
Petrosquamous suture
Mandibular fossa
Tympanic ring
Petromastoid portion

FIG. 4-99.   Lateral aspect of the temporal bone at birth. Note the shallow, laterally oriented mandibular fossa.

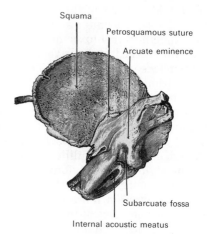

Squama
Petrosquamous suture
Arcuate eminence
Subarcuate fossa
Internal acoustic meatus

FIG. 4-100.   Medial aspect of the temporal bone at birth. Note the widely open subarcuate fossa.

anterior wall. (5) On the posterior aspect of the petrous portion, the **subarcuate fossa,** which is quite pronounced and widely open at birth (Fig. 4-100), gradually becomes filled and almost obliterated.

ARTICULATIONS. The temporal articulates with five bones: occipital, parietal, sphenoid, mandible, and zygomatic.

SPHENOID BONE

The **sphenoid bone** (*os sphenoidale;* Figs. 4-101 to 4-103) is situated at the base of the skull anterior to the temporal bones and basilar part of the occipital. Its shape somewhat resembles a bat with its wings extended.

The sphenoid bone consists of a median portion or **body,** two **great wings,** two **small wings** extending outward from the sides of the body, and two **pterygoid processes,** which project from the inferior surface of the bone.

THE BODY. The **body** (*corpus sphenoidale*), more or less cubical in shape, is hollow and forms two large cavities, the **sphenoidal air sinuses,** which are separated from each other by a thin septum.

The **superior** or **intracranial surface** of the body (Fig. 4-101) presents the prominent **ethmoidal spine** rostrally, for articulation with the cribriform plate of the ethmoid bone. Posterior to this is a smooth surface called the **jugum sphenoidale,** which is slightly

raised in the midline and grooved on each side for the olfactory tracts of the brain. This surface is bounded posteriorly by a ridge that forms the anterior border of the narrow, transversely oriented **chiasmatic groove** (*optic groove*), on which the optic chiasma lies. On each side the groove leads to the **optic canal,** which transmits the optic nerve and ophthalmic artery into the orbital cavity. Posterior to the chiasmatic groove is an elevation, the **tuberculum sellae.** Still more posteriorly is located a hollow depression, the **sella turcica,** the deepest part of which lodges the hypophysis cerebri and is known as the **hypophysial fossa.** The anterior boundary of the sella turcica is completed by two small eminences, one on each side, called the **middle clinoid processes.** The posterior boundary is formed by a square plate of bone, the **dorsum sellae,** which ends at its superior angles in two tubercles, the **posterior clinoid processes.** These may vary considerably in size and form in different skulls. The posterior clinoid processes deepen the sella turcica and give attachment to the tentorium cerebelli. On either side of the dorsum sellae is a notch along which passes the abducent nerve. Inferior to the notch is located the **petrosal process,** which articulates with the apex of the petrous portion of the temporal bone and forms the medial boundary of the foramen lacerum.

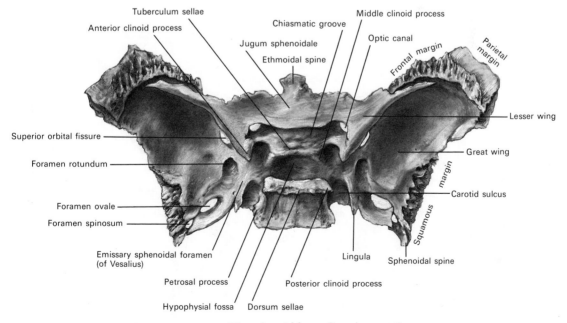

FIG. 4-101. The sphenoid bone. Superior aspect.

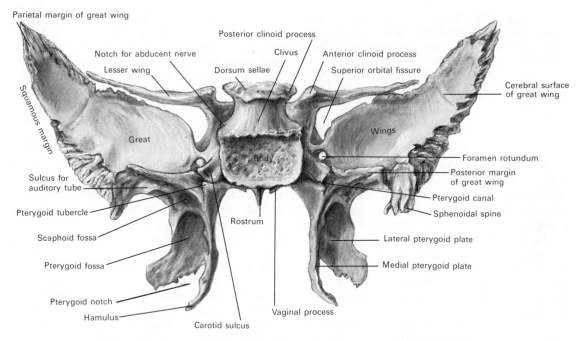

Parietal margin of great wing
Posterior clinoid process
Notch for abducent nerve
Clivus
Lesser wing
Anterior clinoid process
Dorsum sellae
Superior orbital fissure
Squamous margin
Cerebral surface of great wing
Great
Wings
Body
Foramen rotundum
Posterior margin of great wing
Sulcus for auditory tube
Pterygoid canal
Pterygoid tubercle
Sphenoidal spine
Rostrum
Scaphoid fossa
Lateral pterygoid plate
Pterygoid fossa
Medial pterygoid plate
Pterygoid notch
Hamulus
Vaginal process
Carotid sulcus

FIG. 4-102. Sphenoid bone. Posterior aspect, viewed from above.

Posterior to the dorsum sellae is a shallow depression, the **clivus,** which slopes posteriorly and is continuous with a similar groove on the basilar portion of the occipital bone. It supports the pons and the basilar artery.

The **lateral surfaces** of the body are united with the great wings and the medial pterygoid plates. Above the attachment of each

great wing is a broad groove, curved something like the italic letter *f,* named the **carotid sulcus.** It lodges the internal carotid artery, the cavernous sinus, and their related autonomic nerve fibers. Along the posterior part of the lateral margin of this groove, in the angle between the body and the great wing, is a ridge of bone called the **lingula.**

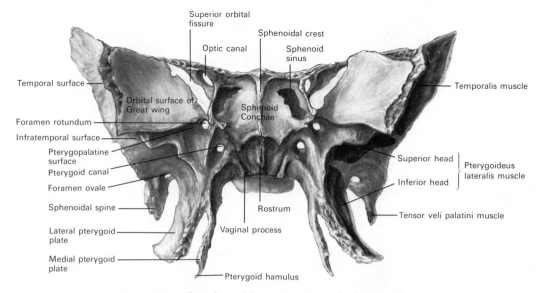

Superior orbital fissure
Sphenoidal crest
Optic canal
Sphenoid sinus
Temporal surface
Temporalis muscle
Orbital surface of Great wing
Sphenoid Conchae
Foramen rotundum
Infratemporal surface
Pterygopalatine surface
Superior head
Pterygoideus lateralis muscle
Pterygoid canal
Inferior head
Foramen ovale
Sphenoidal spine
Rostrum
Tensor veli palatini muscle
Lateral pterygoid plate
Vaginal process
Medial pterygoid plate
Pterygoid hamulus

FIG. 4-103. The sphenoid bone. Anterior and inferior surfaces.

The **posterior surface** (Fig. 4-102) is quadrilateral in shape and joined, during infancy and adolescence, to the basilar part of the occipital bone by a disc of hyaline cartilage forming the sphenoccipital synchondrosis. This becomes ossified between the eighteenth and twenty-fifth years.

The **anterior surface** of the body (Fig. 4-103) forms the posterior wall of the nasal cavity and presents, in the midline, the **sphenoidal crest,** which articulates with the perpendicular plate of the ethmoid bone to form part of the septum of the nose. On each side of the crest is a rounded opening leading into the corresponding **sphenoidal air sinus.** These sinuses are two large irregular cavities within the body of the sphenoid bone. They are seldom symmetrical and vary considerably in form and size. Often they are partially subdivided by irregular bony laminae, and occasionally extend into the basilar part of the occipital bone almost as far as the foramen magnum. Partially closed in front by two thin, curved plates of bone, the **sphenoidal conchae,** the sinuses communicate anteriorly through rounded openings with the sphenoethmoidal recess of the nasal cavity. The lateral margin of the anterior surface is serrated and articulates with the orbital lamina of the ethmoid bone, completing the posterior ethmoidal cells. The inferior margin articulates with the orbital process of the palatine bone, and the superior margin with the orbital plate of the frontal bone.

The **inferior surface** of the body of the sphenoid bone also forms part of the posterior wall of the nasal cavity. It presents, in the midline, a triangular spine, the **sphenoidal rostrum,** which is continuous with the sphenoidal crest of the anterior surface and is received into a deep fissure between the alae of the vomer. On each side of the rostrum is a projecting lamina, the **vaginal process,** directed medially from the base of the medial pterygoid plate.

GREAT WINGS.   The **great wings** of the sphenoid bone are two strong processes that arise from the sides of the body and curve upward and laterally. The posterior part of each wing projects as a triangular process, which fits into the angle between the squama and the petrous portion of the temporal bone and presents at its apex an inferiorly directed process, the **sphenoidal spine** (Fig. 4-101).

The **superior** or **cerebral surface** of each great wing (Figs. 4-70; 4-101) forms part of the middle cranial fossa. It is deeply concave and presents depressions for the convolutions of the temporal lobe of the brain. At its anterior and medial part is a circular aperture, the **foramen rotundum,** through which passes the maxillary nerve. Posterior and lateral to this is the **foramen ovale,** for the transmission of the mandibular nerve, the accessory meningeal artery, and sometimes the lesser petrosal nerve. The lesser petrosal nerve sometimes passes through a special canal (**canaliculus innominatus**) situated medial to the foramen spinosum. Medial to the foramen ovale, a small aperture, the **emissary sphenoidal foramen** (of Vesalius), is frequently seen. It opens inferiorly near the scaphoid fossa, and transmits a small vein from the cavernous sinus to the pterygoid plexus. In the posterior angle near the sphenoidal spine is the **foramen spinosum,** which transmits the middle meningeal vessels and a recurrent branch from the mandibular nerve.

The **temporal surface** (*lateral surface;* Fig. 4-62) is convex, and divided by a transverse ridge, the **infratemporal crest,** into **temporal** and **infratemporal portions.** The superior or temporal portion gives attachment to the temporalis muscle, while the inferior or infratemporal, smaller in size, enters into the formation of the **infratemporal fossa** and, together with the infratemporal crest, affords attachment to the pterygoideus lateralis muscle. It is pierced by the **foramen ovale** and **foramen spinosum,** and at its posterior part is the **sphenoidal spine,** which frequently is grooved on its medial surface for the chorda tympani nerve. To the sphenoidal spine are attached the sphenomandibular ligament and the tensor veli palatini muscle. Medial to the anterior extremity of the infratemporal crest is a triangular process that serves to increase the attachment of the lateral pterygoid muscle. A ridge extends downward and medialward from this process on the anterior part of the lateral pterygoid plate. This ridge defines the anterior limit of the infratemporal surface and, in the articulated skull, the posterior boundary of the **pterygomaxillary fissure.**

The **orbital surface** of the great wing is smooth and quadrilateral (Fig. 4-103) and forms the posterior part of the lateral wall of the orbit. Its serrated *superior margin* articulates with the orbital plate of the frontal bone. Its rounded *inferior margin* forms the posterolateral boundary of the **inferior orbital fissure.** Its sharp *medial margin* forms the inferior boundary of the **superior orbital fissure** (Fig. 4-102). Projecting from about its center is a little tubercle that gives attachment to part of the fibrous tendinous ring from which arise the rectus muscles of the eyeball. At this site on the ring arises the inferior head of the rectus lateralis oculi muscle. At the upper part of this medial margin is a notch for the transmission of a recurrent branch of the lacrimal artery. Its *lateral margin* is serrated and articulates with the zygomatic bone. Inferior to the medial end of the superior orbital fissure is a grooved surface, which forms the posterior wall of the **pterygopalatine fossa** and is pierced by the **foramen rotundum.**

The **posterior margin of the great wing** (Figs. 4-101; 4-102), which extends from the body to the sphenoidal spine, is irregular in shape. Medially near the body it forms the anterior wall of the **foramen lacerum.** It also presents the posterior aperture of the **pterygoid canal** (Fig. 4-102), through which passes the corresponding nerve and artery. Its lateral half articulates, by means of a cartilaginous joint, with the petrous portion of the temporal bone, forming the **sphenopetrosal synchondrosis.** Between the two bones, on the inferior surface of the skull, is a furrow, the **sulcus tubae auditivae,** which lodges the cartilaginous part of the auditory tube. Anterior to the sphenoidal spine, the concave, serrated **squamous margin,** bevelled at the expense of the inner table inferiorly, and of the outer table superiorly, articulates with the temporal squama. At the tip of the great wing is a triangular portion, the **parietal margin** (Fig. 4-102), bevelled at the expense of the internal surface, for articulation with the sphenoidal angle of the parietal bone. On the lateral surface of the skull, this site is named the **pterion.** Medial to this is a triangular, serrated surface, the **frontal margin** for articulation with the frontal bone (Fig. 4-103). This surface is continuous medially with the sharp edge, which forms the inferior boundary of the superior orbital fissure, and laterally with the **zygomatic margin** for articulation with the zygomatic bone (Fig. 4-103).

LESSER WINGS. The **lesser wings** of the sphenoid bone are two thin triangular plates, which arise from the superior and anterior parts of the sphenoidal body and, projecting laterally, end in sharp points (Fig. 4-101).

The **superior surface** is flat and supports part of the frontal lobe of the brain. The **inferior surface** forms the posterior part of the roof of the orbit and the superior boundary of the **superior orbital fissure.** The **anterior border** is serrated for articulation with the orbital plate of the frontal bone. The **posterior border,** smooth and rounded, is received into the lateral fissure of the brain; the medial end of this border forms the **anterior clinoid process** and gives attachment to the tentorium cerebelli. The anterior clinoid process is sometimes united to the middle clinoid process by a spicule of bone, and when this happens the termination of the groove for the internal carotid artery is converted into the **caroticoclinoid foramen.** The small wing is connected to the body by two roots: a thin, flat superior root and a thicker, triangular inferior root. Between the two roots is the **optic canal** for the transmission of the optic nerve and ophthalmic artery.

A **superior orbital fissure** (Fig. 4-103) is triangular in shape and leads from the cavity of the cranium into that of the orbit. It is bounded *medially* by the sphenoidal body, *superiorly* by the small wing, *inferiorly* by the medial margin of the orbital surface of the great wing, and completed *laterally* by the frontal bone between the great and lesser wings. Into the orbit, the fissure transmits the oculomotor, trochlear, and abducent nerves, the three branches of the ophthalmic division of the trigeminal nerve, some sympathetic nerve filaments from the internal carotid and cavernous plexuses, and the orbital branch of the middle meningeal artery. Into the cranial cavity from the orbit, the fissure transmits a recurrent branch from the lacrimal artery to the dura mater and the ophthalmic veins.

PTERYGOID PROCESSES. The **pterygoid processes,** one on each side of the sphenoid bone, project downward perpendicularly from the inferior region where the body and great wing unite. Each pterygoid process consists of a **medial** and **lateral pterygoid plate,** the superior parts of which are fused anteriorly. The **pterygopalatine groove,** a vertical sulcus, descends anterior to the line of fusion. The plates are separated inferiorly by an angular cleft, the **pterygoid fissure,** the rough margins of which articulate with the pyramidal process of the palatine bone. The two plates diverge posteriorly and enclose between them a V-shaped fossa, the **pterygoid fossa** (Fig. 4-102), which contains the medial pterygoid and tensor veli palatini muscles. Superior to this fossa is a small, oval depression, the **scaphoid fossa,** which gives origin to part of the tensor veli palatini. The anterior surface of the pterygoid process is broad and triangular near its root, where it forms the posterior wall of the pterygopalatine fossa. The anterior orifice of the pterygoid canal opens at this site (Fig. 4-103).

The **lateral pterygoid plate** is broad, thin, and everted. Its *lateral surface* forms part of the medial wall of the infratemporal fossa and gives origin to the lateral pterygoid muscle. Its *medial surface* forms part of the pterygoid fossa and gives attachment to the medial pterygoid.

The **medial pterygoid plate** is narrower and longer than the lateral. It curves laterally at its inferior extremity into a hook-like process, the **pterygoid hamulus,** around which the tendon of the tensor veli palatini muscle glides. The *lateral surface* of this plate forms part of the pterygoid fossa, and the belly of the tensor veli palatini lies adjacent to it. The *medial surface* of the medial pterygoid plate constitutes the lateral boundary of the choana or posterior aperture of the corresponding nasal cavity. Superiorly the medial plate is prolonged onto the inferior surface of the body as a thin lamina, named the **vaginal process** (Fig. 4-103), which articulates anteriorly with the sphenoidal process of the palatine and posteriorly with the ala of the vomer. The angular prominence between the posterior margin of the vaginal process and the medial border of the scaphoid fossa is named the **pterygoid tubercle,** and immediately superior to this is the posterior opening of the **pterygoid canal** (Fig. 4-102). On the inferior surface of the vaginal process is a furrow, which is converted into the **palatovaginal canal** by the sphenoidal process of the palatine bone. This canal transmits the pharyngeal branch of the maxillary artery and the pharyngeal nerve from the pterygopalatine ganglion. The pharyngobasilar fascia is attached to the entire length of the posterior margin of the medial pterygoid plate, and the constrictor pharyngis superior muscle takes origin from its inferior third. Projecting dorsally from near the middle of the posterior edge of this plate is an angular process, the **processus tubarius,** which supports the pharyngeal end of the auditory tube. The anterior margin of the medial pterygoid plate articulates with the posterior border of the perpendicular plate of the palatine bone.

The **pterygoid canal** (*Vidian canal*) courses anteroposteriorly within the part of the body of the sphenoid to which each pterygoid process is attached. When the sphenoidal sinus is extensive, the canal is covered by a ridge of bone in the floor of the sinus. The posterior orifice opens into the foramen lacerum (Fig. 4-102), the anterior into the pterygopalatine fossa (Fig. 4-103). The canal measures 1 to 2 mm in diameter and transmits the nerve of the pterygoid canal (*Vidian nerve*) and a small branch of the pterygopalatine portion of the maxillary artery called the artery of the pterygoid canal.

The **sphenoidal conchae** (Fig. 4-103) are two thin, curved plates, situated at the anterior part of the body of the sphenoid. An aperture of variable size exists in the anterior wall of each, and through this the **sphenoidal sinus** opens into the sphenoethmoidal recess of the nasal cavity. The inferior surface of these plates is convex and forms part of the roof of the nasal cavity.

OSSIFICATION. The greater part of the sphenoid bone is first formed in cartilage. Until the seventh or eighth month of fetal life the body of the sphenoid may be considered to consist of two parts. One part, anterior to the tuberculum sellae, is named the **presphenoid** and is continuous with the lesser wings. The other, comprising the sella turcica and dorsum sellae, is called the **postsphenoid** and has attached to it the great wings and pterygoid processes. There are fourteen ossification centers in all, six for the presphenoid and eight for the postsphenoid.

Presphenoid Ossification. About the ninth week of fetal life an ossific center appears for each of the lesser wings (orbitosphenoids) just lateral to the optic canal. Shortly afterward two nuclei appear in the presphenoid part of the body. The sphenoidal conchae are each developed from ossification centers, which make their appearance about the fifth prenatal month. At birth the conchae consist of small triangular laminae, and it is not until the third year that they become hollow and cone-shaped. About the fourth year the conchae join the labyrinths of the ethmoid, and between the ninth and twelfth years their union with the sphenoid becomes ossified.

Postsphenoid Ossification. The first ossific nuclei are those for the great wings (ali-sphenoids). One makes its appearance at the base of each wing between the foramen rotundum and foramen ovale about the eighth week. These centers ossify only the more proximal parts of the great wings, while the orbital plate and the lateral part of the great wings, as well as the lateral pterygoid plates, are ossified in membrane. Two centers for the postsphenoid part of the body appear soon after those of the great wings. These develop on each side of the sella turcica and fuse about the middle of fetal life. Each medial pterygoid plate (with the exception of its hamulus) is ossified in membrane, and its center probably appears about the ninth or tenth week. The hamulus becomes chondrified during the third month, and almost at once undergoes ossification. The medial and lateral pterygoid plates join about the sixth month. During the fourth month a center appears for each lingula and speedily joins the rest of the bone.

The presphenoid fuses to the postsphenoid portion of the body about the eighth month, and at birth the bone consists of three parts (Fig. 4-104): a central, which includes the body and small wings, and two lateral, each consisting of a great wing and pterygoid process. In the first year after birth the great wings and body unite in the region of the pterygoid canal. The lesser wings extend inward above the anterior part of the body and, meeting in the midline, form an elevated smooth surface, termed the *jugum sphenoidale.* By the twenty-fifth year the sphenoid and occipital bones are completely fused.

The *craniopharyngeal canal,* through which the hypophyseal diverticulum is connected with buccal ectoderm in fetal life, occasionally persists between the presphenoid and postsphenoid parts of the sphenoid bone. Epithelial lined cysts (craniopharyngeomas) may develop along the course of a persisting craniopharyngeal canal.

The *sphenoid sinuses* are present as minute cavities in the body of the sphenoid at the time of birth, but do not attain their full size until after puberty.

INTRINSIC LIGAMENTS.   The more important of these are: the **pterygospinous,** stretching between the sphenoidal spine and the superior part of the lateral pterygoid plate; the **interclinoid,** a fibrous process joining the anterior to the posterior clinoid process; and the **caroticoclinoid,** connecting the anterior to the middle clinoid process. These ligaments occasionally ossify.

ARTICULATIONS.   Centrally positioned, the sphenoid bone articulates with all the other bones of the cerebral cranium. Totally, it articulates with twelve bones: four single—the vomer, ethmoid, frontal, and occipital; and four paired—the parietal, temporal, zygomatic, and palatine. It also sometimes articulates with the tuberosity of the maxilla.

ETHMOID BONE

The **ethmoid bone** (*os ethmoidale*; Figs. 4-105 to 4-108) is cuboidal in shape and exceedingly light and spongy. It is situated at the anterior part of the base of the cranium, between the two orbits, and forms part of the medial wall of each orbit as well as part of the nasal septum and much of the roof and lateral walls of the nasal cavity. The ethmoid consists of four parts: a horizontal or **cribriform plate,** forming part of the base of the cranium; a **perpendicular plate,** constituting part of the nasal septum; and two lateral masses called **ethmoidal labyrinths.**

The **cribriform plate** (Fig. 4-105) is received into the ethmoidal notch of the frontal bone and forms most of the roof of the nasal cavities. Projecting superiorly into the cranial fossa from the midline of this plate is a thick, smooth triangular process, the **crista galli,** so called from its resemblance to a cock's comb. The long thin posterior border of the crista galli serves for the attachment of the falx cerebri. The anterior border of the cribriform plate is short and thick, and it presents two small projecting **alae,** which are received into corresponding depressions in the frontal bone, thereby completing the **foramen cecum.** Its sides are smooth and sometimes bulge due to the presence of a small air sinus in the interior. On each side of the crista galli, the cribriform plate is nar-

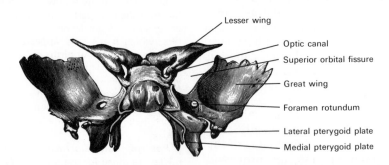

FIG. 4-104.   Sphenoid bone at birth. Posterior aspect.

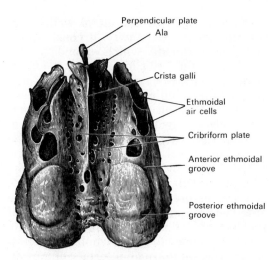

Perpendicular plate
Ala
Crista galli
Ethmoidal air cells
Cribriform plate
Anterior ethmoidal groove
Posterior ethmoidal groove

FIG. 4-105.   Ethmoid bone from above.

row and deeply grooved; it supports the olfactory bulb and is perforated by the **olfactory foramina** for the passage of the olfactory nerves. At the anterior part of the cribriform plate, on each side of the crista galli, is a small fissure that is occupied by a process of dura mater. Lateral to this fissure is a notch or foramen, which transmits the nasociliary nerve; from this notch a groove extends backward to the anterior ethmoid foramen.

The **perpendicular plate** (Figs. 4-106; 4-107) is a thin, flattened lamina, which descends from the cribriform plate and assists in forming the septum of the nose. The *ante-*

rior border articulates with the spine of the frontal bone and the crest of the nasal bones. The *posterior border* articulates with the sphenoidal crest above and the vomer below. The *inferior border* is thicker than the posterior, and serves for the attachment of the septal cartilage of the nose. The surfaces of the plate are smooth except superiorly, where numerous grooves and canals lead from the medial foramina in the cribriform plate and lodge filaments of the olfactory nerves.

The **labyrinth** or **lateral mass** (Fig. 4-107) consists of a number of thin-walled cellular cavities which are arranged in three groups, the **anterior, middle,** and **posterior ethmoidal air cells.** These are interposed between two vertical plates of bone. The lateral plate forms part of the orbit; the medial, part of the corresponding nasal cavity. In the disarticulated bone many of the cells appear open, but in the articulated skull they are completely closed except where their ostia open into the nasal cavity.

The **superior surface of the labyrinth** (Fig. 4-105) presents a number of incompletely closed cells, the walls of which are completed, in the articulated skull, by the edges of the ethmoidal notch of the frontal bone. Crossing this surface on each side are two grooves, which are converted into canals by articulation with the frontal bone. These are the **anterior** and **posterior ethmoidal canals,** which open on the inner wall of the orbit. The **posterior surface of the labyrinth** (Fig.

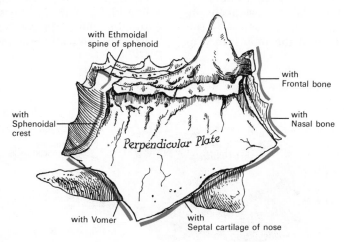

with Ethmoidal spine of sphenoid
with Frontal bone
with Sphenoidal crest
with Nasal bone
Perpendicular Plate
with Vomer
with Septal cartilage of nose

FIG. 4-106.   Perpendicular plate of ethmoid bone, exposed by removing the right labyrinth. Lines of articulation are shown along the borders.

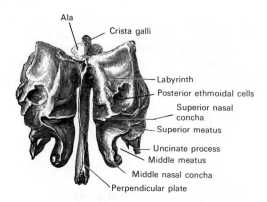

FIG. 4-107. Ethmoid bone. Posterior aspect.

4-107) presents large irregular cellular cavities, which are closed by articulation with the sphenoidal conchae and orbital processes of the palatine bones. The **lateral surface of the labyrinth** (Fig. 4-108) is formed by a thin, smooth, oblong plate, the **lamina orbitalis,** which covers the middle and posterior ethmoidal cells and forms a large part of the medial wall of the orbit. This plate articulates *above* with the orbital plate of the frontal bone, *below* with the maxilla and orbital process of the palatine, *anteriorly* with the lacrimal, and *posteriorly* with the sphenoid.

Anterior to the lamina orbitalis are some air cells, which are overlapped and completed by the lacrimal bone and the frontal process of the maxilla. A thin, curved, bony lamina, the **uncinate process,** projects inferiorly and posteriorly from this part of the labyrinth (Figs. 4-107; 4-108). It forms part of the medial wall of the maxillary sinus and articulates with the ethmoidal process of the inferior nasal concha.

The **medial surface of the labyrinth** (Fig.

4-109) forms part of the lateral wall of the corresponding nasal cavity. It consists of a thin, rough lamella, which extends downward from the cribriform plate and ends below in a free, scroll-shaped, convoluted margin, the **middle nasal concha.** Superiorly, the medial surface is marked by numerous grooves lodging branches of the olfactory nerves which are distributed to the uppermost nasal mucous membrane. The posterior part of the medial surface of the labyrinth is subdivided by a narrow anteroposterior fissure, the **superior meatus** of the nose, which separates the **superior nasal concha** from the middle concha. The middle concha extends along the whole anteroposterior extent of the medial surface of the labyrinth, and its margin is free and thick. The superior concha extends only about half as far anteriorly as the middle concha, and its free margin is less thick. Posteriorly, the superior meatus extends upward as a deep groove under cover of the superior concha, and the *posterior ethmoidal air cells* open into it. The **middle meatus** of the nose is an extensive narrow channel extending anteroposteriorly and upward under cover of the middle concha. Continuing upward and forward from the middle meatus is a deep curved passage, the **infundibulum.** The *anterior ethmoidal cells* open into the infundibulum, and in about half the skulls the infundibulum continues upward as the **frontonasal duct** into the **frontal sinus.** The **ostium** of the **maxillary sinus** opens into the bottom of the middle part of the infundibulum. The lateral wall of the middle meatus contains a rounded swelling called the **bulla ethmoidalis.** The *middle ethmoid air cells* lie within the bulla and open above it. The portion of the wall of the meatus posterior to

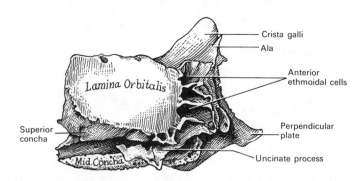

FIG. 4-108. Ethmoid bone from the right lateral aspect.

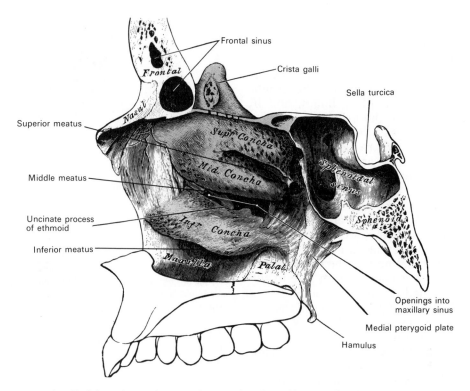

FIG. 4-109.   Lateral wall of the right nasal cavity, showing the ethmoid bone (red) and the inferior nasal concha (blue).

the infundibulum is lacking in bone and may fail to be closed by mucous membrane in the intact body, thus providing an **accessory opening** into the **maxillary sinus.**

OSSIFICATION.   The ethmoid is ossified in the cartilage of the nasal capsule by three centers: one for the perpendicular plate and one for each labyrinth.

The labyrinths are developed first as granules of bone that make their appearance between the fourth and fifth months of fetal life in the region of the lamina orbitalis, which then extends into the conchae. At birth, the ethmoid bone consists of two partially ossified labyrinths, which are small and poorly developed. The remainder of the bone is still cartilaginous. During the first postnatal year the perpendicular plate and crista galli begin to ossify from a single center, fusing with the labyrinths at about the beginning of the second year. The cribriform plate is ossified partly from the perpendicular plate and partly from the labyrinths. The ethmoidal cells commence forming before birth, and become well defined by the third year. They form by bone resorption, even as the ethmoid bone grows, and come to occupy the two labyrinths.

ARTICULATIONS.   The ethmoid bone articulates with thirteen other bones. Two of these, the frontal and sphenoid, are of the cranium. The remaining

eleven are of the face, and include the two nasals, two maxillae, two lacrimals, two palatines, two inferior nasal conchae, and the vomer.

SUTURAL BONES

In addition to the usual centers of ossification of the cranium, other centers may occur in the course of the sutures, especially those related to the parietal bones. These may give rise to irregular, isolated bones, termed **sutural** or **Wormian bones.** They occur most frequently in the course of the lambdoid suture, but are occasionally seen at the fontanelles, especially the posterior. A **pterion ossicle,** or **epipteric bone,** sometimes exists between the sphenoidal angle of the parietal bone and the great wing of the sphenoid. Sutural bones have a tendency to be more or less symmetrical on the two sides of the skull, and vary much in size. Their number is generally limited to two or three, but more than a hundred have been found in the skull of an adult hydrocephalic subject. Sutural bones are rarely seen in lines of union between facial bones.

NASAL BONES

The **nasal bones** (Figs. 4-110 to 4-112) are two small oblong bones which vary in form and size in different individuals. They are placed side by side between the frontal processes of the maxillary bones, and by their junction, form the bridge of the nose (Fig. 4-64). The nasal bones are wider and thinner inferiorly, where they join the nasal cartilages, than they are superiorly. Each nasal bone presents two surfaces and four borders.

The **internal surface** (Fig. 4-110) is concave from side to side and is traversed by a longitudinal groove, the **ethmoidal sulcus,** for the passage of the external nasal branch of the anterior ethmoidal nerve. The **external surface** (Fig. 4-111) is convex from side to side and is perforated at about its center by one or two foramina for the transmission of small external nasal veins. It is covered by the procerus and nasalis muscles.

The **superior border** is narrow, thick, and serrated for articulation with the nasal notch of the frontal bone. The **inferior border** is thin and gives attachment to the lateral cartilage of the nose. Near its middle is a small notch which marks the end of the ethmoidal sulcus lodging the external nasal nerve. The **lateral border** is serrated and articulates with the frontal process of the maxilla. The **medial border** is thicker above than below and articulates with the opposite nasal bone. This border is prolonged posteriorly as a vertical crest which forms a small part of the nasal septum. This crest articulates with the nasal spine of the frontal bone, the perpendicular plate of the ethmoid, and the cartilage of the nasal septum.

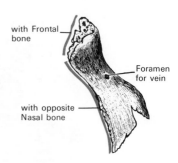

Fig. 4-111.   Right nasal bone. External surface. Borders of articulation indicated in green.

OSSIFICATION.   Each nasal bone is ossified from one center which appears during the eighth or ninth week of fetal life in the membrane overlying the anterior part of the cartilaginous nasal capsule. The nasal bones at birth are nearly square in shape, but as facial development ensues, they become nearly three times as long as they are wide.

ARTICULATIONS.   Each nasal bone articulates with four bones: two of the cranium—the frontal and ethmoid, and two of the face—the opposite nasal and the maxilla.

MAXILLARY BONE

The **maxillae** (*upper jaw*) are the largest bones of the face, excepting the mandible, and by their union form the whole of the upper jaw. Each assists in forming the boundaries of *four cavities:* the roof of the mouth, the floor and lateral wall of the nasal cavity, the floor of the orbit, and the maxillary sinus. The maxillae support the upper teeth and enter into the formation of *two fossae,* the infratemporal and pterygopalatine, and *two fissures,* the inferior orbital and pterygomaxillary. Each bone consists of a **body** and four processes—**zygomatic, frontal, alveolar,** and **palatine.**

BODY OF THE MAXILLA.   The **body** is somewhat pyramidal in shape and is hollowed by a large cavity, the **maxillary sinus.** It has four surfaces—an anterior or facial, a posterior or infratemporal, a superior or orbital, and a medial or nasal.

The **anterior** or **facial surface** (Fig. 4-112) is directed forward and laterally, and presents inferiorly a series of eminences corresponding to the positions of the roots of the teeth. Superior to those of the incisor teeth is a depression, the **incisive fossa** (Fig. 4-113), which gives origin to the depressor septi muscle. To the alveolar border below the

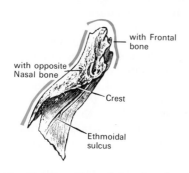

Fig. 4-110.   Right nasal bone. Internal surface. Borders of articulation indicated in green.

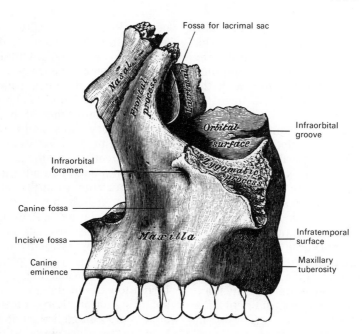

FIG. 4-112. Lateral aspect of the maxilla showing the articulations of the nasal and lacrimal bones with the maxilla.

fossa a slip of the orbicularis oris is attached, and above and a little lateral to it, the nasalis muscle arises. Lateral to the incisive fossa is another depression, the **canine fossa.** It is larger and deeper than the incisive fossa, and is separated from it by a vertical ridge, the **canine eminence,** corresponding to the

socket of the canine tooth. The canine fossa gives origin to the levator anguli oris. Above the canine fossa is the **infraorbital foramen,** which is the anterior orifice of the infraorbital canal and transmits the infraorbital vessels and nerve. Above the foramen is the sharp, inferior margin of the orbit, which

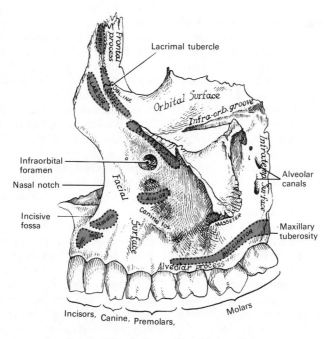

FIG. 4-113. Left maxilla. Outer surface.

affords attachment to the levator labii superioris muscle. Medially, the anterior surface is limited by a deep concavity, the **nasal notch,** which ends below in a pointed process. This latter process joins with the one from the opposite maxilla to form the **anterior nasal spine.**

The **posterior** or **infratemporal surface** (Fig. 4-113) is convex, directed laterally and posteriorly, and forms the anterior wall of the infratemporal fossa. It is separated from the anterior surface of the maxilla by the zygomatic process and by a strong ridge, extending upward from the socket of the first molar tooth. It is pierced about its center by the apertures of two or three **alveolar canals** which transmit the posterior superior alveolar vessels and nerves. At the inferior part of this surface is a rounded eminence, the **maxillary tuberosity,** especially prominent after the growth of the wisdom tooth. The tuberosity is rough on its medial side for articulation with the pyramidal process of the palatine bone. It gives origin to a few fibers of the medial pterygoid muscle, and in some instances it articulates with the lateral pterygoid plate of the sphenoid. Immediately above this is a smooth surface, which forms the anterior boundary of the pterygopalatine fossa and presents a laterally directed groove for the maxillary nerve. This groove is continuous with the infraorbital groove on the orbital surface.

The **superior** or **orbital surface** (Figs. 4-65; 4-113) is smooth and triangular and forms the greater part of the floor of the orbit. It is bounded *medially* by an irregular margin, the **lacrimal notch,** behind which it articulates with the lacrimal bone, the lamina orbitalis of the ethmoid bone, and the orbital process of the palatine bone. It is bounded *posteriorly* by a smooth rounded edge, which forms the anterior margin of the inferior orbital fissure. At times it articulates at its lateral extremity with the orbital surface of the great wing of the sphenoid. **Anteriorly,** the orbital surface is limited by a small portion of the circumference of the orbit, continuous medially with the frontal process and laterally with the zygomatic process. Near the middle of the posterior part of the orbital surface is the **infraorbital groove,** for the passage of the infraorbital vessels and nerve. The groove begins near the middle of the posterior border (Fig. 4-112), where it is continuous with the groove for the maxillary

nerve. Passing rostrally, the infraorbital groove ends in a canal which then subdivides into two canals. One of these, the **infraorbital canal,** opens just below the margin of the orbit at the infraorbital foramen. The other canal, which is smaller, courses downward in the substance of the anterior wall of the maxillary sinus, and transmits the anterior superior alveolar vessels and nerve to the front teeth. From the posterior part of the infraorbital canal, an additional small canal sometimes courses in the lateral wall of the sinus, conveying the middle alveolar nerve to the premolar teeth. Anteromedially on the orbital surface, just lateral to the lacrimal groove, is a depression from which arises the inferior oblique muscle of the eyeball.

The **nasal surface** (Fig. 4-114) assists in the formation of the lateral wall of the corresponding nasal cavity and presents posterosuperiorly a large, irregular opening leading into the **maxillary sinus.** At the superior border of this aperture (also called the **maxillary hiatus**) are some incomplete air cells which, in the articulated skull, are closed in by the ethmoid and lacrimal bones. Below the aperture of the maxillary sinus is a smooth concavity which forms part of the inferior meatus of the nasal cavity, and posterior to it is a rough surface for articulation with the perpendicular part of the palatine bone. A groove commencing near the middle of the posterior border and running obliquely downward and forward is converted into the **greater palatine canal** by articulation with the palatine bone. Rostral to the opening of the sinus is the **lacrimal groove,** which is converted into the **nasolacrimal canal** by the lacrimal bone and inferior nasal concha. This canal opens into the inferior meatus of the nose and transmits the nasolacrimal duct. More anteriorly the bone is marked by an oblique ridge, the **conchal crest,** for articulation with the inferior nasal concha. The shallow concavity superior to this ridge forms part of the atrium of the middle meatus, and that inferior to it, part of the inferior meatus.

MAXILLARY SINUS. The **maxillary sinus** (Figs. 4-114; 4-115) is a large pyramidal cavity within the body of the maxilla. Its walls are thin and correspond to the nasal, orbital, anterior, and infratemporal surfaces of the body of the bone. Its **apex** is oriented laterally and corresponds to the zygomatic process. The **base** of the maxillary sinus is

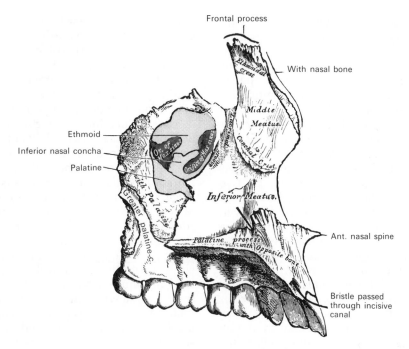

Frontal process

With nasal bone

Ethmoid
Inferior nasal concha
Palatine

Ant. nasal spine

Bristle passed
through incisive
canal

FIG. 4-114.   Left maxilla. Nasal surface.

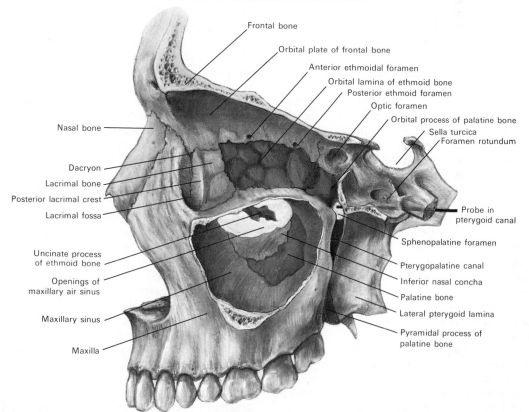

Frontal bone
Orbital plate of frontal bone
Anterior ethmoidal foramen
Orbital lamina of ethmoid bone
Posterior ethmoid foramen
Optic foramen
Orbital process of palatine bone
Sella turcica
Foramen rotundum

Nasal bone

Dacryon
Lacrimal bone
Posterior lacrimal crest
Lacrimal fossa

Probe in
pterygoid canal

Sphenopalatine foramen

Uncinate process
of ethmoid bone
Openings of
maxillary air sinus

Pterygopalatine canal
Inferior nasal concha
Palatine bone
Lateral pterygoid lamina

Maxillary sinus

Maxilla

Pyramidal process of
palatine bone

FIG. 4-115.   The infratemporal surface of the maxilla has been opened from the lateral side to reveal the maxillary sinus. This exposure also demonstrates the nasal (medial) wall of the sinus and the three bony structures that partially cover it: perpendicular plate of the palatine bone, the inferior nasal concha, and the uncinate process of the ethmoid bone.

bounded by the *nasal wall,* which is directed medially and presents, in the disarticulated bone, a large, irregular aperture, communicating with the nasal cavity. In the articulated skull this aperture is partly closed by the following bones: the uncinate process of the ethmoid, the maxillary process of the inferior nasal concha, the perpendicular plate of the palatine, and a small part of the lacrimal (Figs. 4-72; 4-73; 4-114; 4-115). In the dried skull the sinus communicates with the middle meatus, generally by two small apertures between the above-mentioned bones. In the intact body, however, usually one small opening exists near the upper part of the cavity, since the other is closed by mucous membrane. On the posterior wall of the maxillary sinus are ridges formed by the **alveolar canals,** which transmit the posterior superior alveolar vessels and nerves to the molar teeth.

The *floor* of the sinus is formed by the alveolar process of the maxilla. When the sinus is average in size, its floor is on a level with the floor of the nose; if the sinus is large, it extends below this level. Projecting into the floor of the sinus are several conical processes corresponding to the roots of the teeth. The bone separating the roots of the teeth from the sinus is quite thin. The roots of the upper molar teeth occasionally extend into the sinus, but diseased roots of any of the maxillary teeth can communicate with the sinus. The infraorbital canal usually projects into the cavity as a well-marked ridge extending from the *roof* to the *anterior wall.* The size of the sinus cavity varies in different skulls, and even on the two sides of the same skull.

PROCESSES OF THE MAXILLA.   Of the four bony processes of the maxilla, two—the **zygomatic** and **frontal**—belong to the upper part of the bone, and two—the **alveolar** and **palatine**—belong to the lower part.

The **zygomatic process** (Fig. 4-112) is a rough triangular eminence, situated at the angle of separation of the anterior, infratemporal, and orbital surfaces. *Above* it is rough and serrated for articulation with the zygomatic bone. *Below* it presents a prominent arched border which marks the division between the anterior and infratemporal surfaces. *In front* it forms part of the anterior surface of the maxilla, while *behind* it is concave and forms part of the infratemporal fossa.

The **frontal process** (Figs. 4-112; 4-113) is a strong plate which projects upward and backward and forms part of the lateral boundary of the nose. Its *lateral surface* is smooth and gives attachment to the levator labii superioris alaeque nasi, the orbicularis oculi, and the medial palpebral ligament. Its *medial surface* forms a portion of the lateral wall of the nasal cavity. At its upper part is a rough, uneven area which articulates with the ethmoid bone and closes the anterior ethmoidal cells. Inferior to this is an oblique ridge, the **ethmoidal crest,** the posterior end of which articulates with the middle nasal concha and forms the upper limit of the atrium of the middle meatus, while the anterior part, which is termed the **agger nasi,** is an elevation on the lateral wall of the nasal cavity. The *superior border* articulates with the frontal bone while the *anterior border* articulates with the nasal bone. The *posterior border* is thick and hollowed into a groove, which is continuous below with the lacrimal groove on the nasal surface of the body. The lateral margin of the groove is named the **anterior lacrimal crest.** At its junction with the orbital surface is the small **lacrimal tubercle,** which serves as a guide to the position of the lacrimal sac. The posterior border of the frontal process articulates with the lacrimal bone to form the **lacrimal fossa** for the lodgement of the lacrimal sac.

The **alveolar process** (Figs. 4-113; 4-114) is the thickest and most spongy part of the maxilla. Its arched shape is broader behind than anteriorly, and it is excavated by eight deep cavities for the reception of the teeth. These sockets vary in size and depth according to the teeth they contain. The socket for the canine tooth is the deepest, while those for the molars are widest and subdivided by bony septa into three smaller sockets, each receiving a root. The incisors and the second upper premolar generally have one root and therefore their sockets are single, but the upper first premolar frequently has a bifid root and therefore a two-cavity socket. The buccinator arises from the outer surface of the alveolar process, as far anteriorly as the first molar tooth. When the maxillae are articulated with each other, their alveolar processes together form the **alveolar arch,** and the center of the anterior margin of this arch is named the **alveolar point.**

The **palatine process** (Figs. 4-114; 4-116) is horizontal and projects medially from the

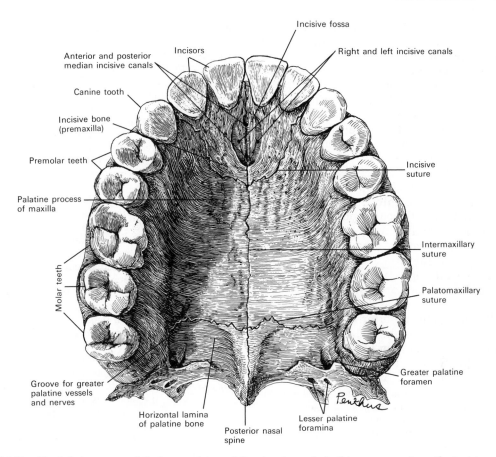

FIG. 4-116. The inferior aspect of the bony palate and the alveolar arch. In this young specimen the incisive suture was well delineated. Note that the skeleton of the hard palate consists of the palatine processes of the maxillae and the horizontal laminae (plates) of the palatine bones.

nasal surface of the bone. It is thick and strong and forms a considerable part of the floor of the nose and roof of the mouth. Its *inferior surface* (Fig. 4-116) is concave, rough, and uneven, and forms, with the palatine process of the opposite bone, the anterior three-fourths of the **hard palate.** It is perforated by numerous foramina for the passage of nutrient vessels and presents little depressions for the palatine glands. At the posterior part of the lateral border near the last molar tooth are grooves for the transmission of the greater palatine vessels and nerve. When the two maxillae are articulated, an oblong depression, the **incisive fossa,** is seen in the midline immediately behind the incisor teeth. In this fossa the orifices of two lateral canals, named the **right** and **left incisive canals** (or *foramina of Stenson*), are visible. Through each passes a terminal branch of the greater palatine artery into the nasal cavity from the oral cavity

and the nasopalatine nerve into the oral cavity from the nasal cavity. Occasionally two additional canals are present in the midline. They are termed the **anterior** and **posterior median incisive canals** (or the *foramina of Scarpa*) and, when present, transmit branches of the nasopalatine nerves, the left passing through the anterior and the right through the posterior canal. On the inferior surface of the palatine process, a delicate linear **incisive suture,** best seen in young skulls, may sometimes be noticed extending laterally and rostrally on both sides of the incisive fossa to the interval between the lateral incisor and the canine tooth. The small part anterior to this suture constitutes the **incisive bone** or **premaxilla,** which in most vertebrates is identifiable as an independent bone. It includes the whole thickness of the alveolus, the corresponding part of the floor of the nose, and the anterior nasal spine, and contains the sockets of the incisor teeth.

The classic view that the human incisive bone is quite comparable to the premaxilla of other vertebrates has been questioned on the basis of certain ossification studies.

The *superior surface* of the palatine process is smooth and concave from side to side and forms a large part of the floor of the nasal cavity (Fig. 4-114). The lateral border of the palatine process is fused with the nasal surface of the body of the maxilla above and the alveolar process below. The *medial border* is thicker anteriorly than behind, and is raised into a ridge, the **nasal crest,** which, with the corresponding ridge of the opposite maxilla, forms a groove for the reception of the vomer. The anterior part of this ridge is raised considerably and is named the **incisor crest.** This is prolonged forward into a sharp process which, together with the same process from the opposite maxilla, forms the **anterior nasal spine.** The *posterior border* of the palatine process is serrated for articulation with the horizontal part of the palatine bone.

OSSIFICATION. Several descriptions of the ossification in membrane of the maxilla have been offered over the years and there is still some uncertainty on the subject. The commonly read and classically described mode of ossification goes back to, among others, that of Franklin Mall (1906), and was reasserted by Chase (1942). These authors claimed that the maxilla ossifies from only two centers, one for the maxilla proper, which appears above the canine fossa, and one for the premaxilla, which is located in front of the developing incisor teeth. These two centers appear between the sixth and eighth week of fetal life and unite at the beginning of the third month. Others (Kölliker, 1885; Fawcett, 1911; Woo, 1949) described at least two centers of ossification in the premaxilla, one along the incisor tooth germ and the other along the medial wall of the paraseptal cartilage, thereby claiming that the maxillary bone as a whole develops from at least three ossification centers. Finally, some investigators (Wood et al., 1967, 1969) feel that separate maxillary and premaxillary ossification centers do not develop, and that the entire maxilla ossifies from a single primary ossification lamina that extends from the region of the body of the maxilla.

Controversy on the premaxillary bone actually dates to the appearance in 1543 of Vesalius' great work, *De Humani Corporis Fabrica,* in which Vesalius repudiated the views of Galen who had described the human premaxilla as a separate bone, the condition Galen had encountered in his dissections of monkeys, dogs, and swine. Vesalius denied for the first time the existence of the premaxillary suture line on the facial aspect of the human skull, which Galen had described, even if the palatal suture line could at times be seen. In a very complete review of this subject, Ashley-Montegue (1935) stated ''that *in primates, with the exception of man,* the premaxilla may be observed upon the facial aspect of the skull suturally quite distinctly separated from the maxilla in almost all sub-adult as well as in many adult individuals. *In late fetal and postnatal man,* no trace of a separation of this bone from the maxilla upon the face is observed, although the palatine portion of the premaxilla may be distinctly observed in the majority of infant crania, and may be encountered in some 26.0 percent of adult crania.''

ARTICULATIONS. The maxilla articulates with nine bones: *two of the cranium,* the frontal and ethmoid, and *seven of the face,* the nasal, zygomatic, lacrimal, inferior nasal concha, palatine, vomer, and the opposite maxilla. Sometimes it articulates with the orbital surface and sometimes with the lateral pterygoid plate of the sphenoid.

CHANGES WITH AGE. *At birth* the transverse and anteroposterior diameters of the bone are each greater than the vertical (Figs. 4-117; 4-118). The frontal process is well marked and the body of the bone consists of little more than the alveolar process with the teeth sockets reaching almost to the floor of the orbit. The maxillary sinus is represented by a furrow on the lateral wall of the nasal cavity. *In the adult* the vertical diameter is the greatest, owing to the development of the alveolar process and the increase in size of the sinus. *In old age* the maxilla reverts in some measure to the infantile condition. Its height is diminished, and after loss of the teeth the alveolar process is absorbed, and the lower part of the bone is contracted and reduced in thickness.

PALATINE BONE

The **palatine bone** is situated in the posterior part of each nasal cavity between the maxilla and the pterygoid process of the sphenoid (Fig. 4-119). Each palatine bone assists in forming the posterior part of the hard palate, part of the floor and lateral wall of the nasal cavity, and the floor of the orbit. It enters into the formation of the **pterygopalatine, pterygoid,** and **infratemporal fos-**

FIG. 4-117.　Anterior surface of maxilla at birth.

FIG. 4-118.   Inferior surface of maxilla at birth.

sae, and the **inferior orbital fissure.** The palatine bone somewhat resembles the letter L, and consists of a **horizontal** and a **perpendicular plate.** It also presents three prominent processes, the **pyramidal,** which is directed backward and laterally from the junction of the two parts, and the **orbital** and **sphenoidal processes,** which extend from the vertical part and are separated by the **sphenopalatine notch.**

HORIZONTAL PLATE.   The **horizontal plate** resembles somewhat the palatine process of the maxilla, and it is quadrilateral in shape, presenting two surfaces and four borders (Figs. 4-120; 4-121).

The **nasal** or **superior surface** is smooth and concave from side to side, and forms the posterior part of the floor of the nasal cavity. The **palatine** or **inferior surface** is more rough and forms, with the corresponding surface of the opposite bone, the posterior

one-fourth of the **hard palate** (Fig. 4-116). Near its posterior margin may be seen a transverse ridge, called the **palatine crest,** to which attaches a part of the aponeurosis of the tensor veli palatini muscle.

The **anterior border** is serrated, and articulates with the palatine process of the maxilla. The **posterior border** is concave, free, and serves for the attachment of the palatine aponeurosis, which forms the tendinous core of the soft palate. Its medial end is sharp and pointed and, when united with that of the opposite bone, forms a projecting process, the **posterior nasal spine** for the attachment of the musculus uvulae. The **lateral border** is united to the lower margin of the perpendicular plate and is grooved by the lower end of the greater palatine canal. The **medial border** is the thickest and is serrated for articulation with the corresponding border of the opposite palatine bone. Its superior edge is raised into a ridge, which, united with the opposing ridge, forms the **nasal crest** for articulation with the posterior part of the lower edge of the vomer.

PERPENDICULAR PLATE.   The **perpendicular plate** is thin and long, and presents two surfaces and four borders (Figs. 4-120; 4-121).

The inferior part of the **medial** or **nasal surface** exhibits a broad, shallow depression, which forms part of the inferior meatus of the nasal cavity. Superior to this is a prominent horizontal ridge, the **conchal**

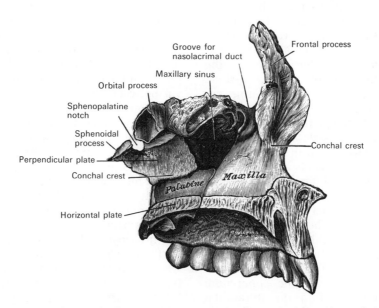

FIG. 4-119.   Articulation of left palatine bone with maxilla. Viewed from the medial aspect.

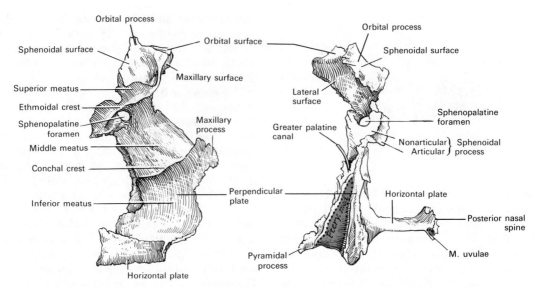

FIG. 4-120. Left palatine bone. Nasal aspect. Enlarged.

FIG. 4-121. Left palatine bone. Posterior aspect. Enlarged.

crest, which articulates with the inferior nasal concha. Still higher is a second broad, shallow depression, which forms part of the middle meatus and is limited by the **ethmoidal crest,** for articulation with the middle nasal concha. Above the ethmoidal crest is a narrower and deeper horizontal groove, which forms part of the superior meatus.

The **lateral** or **maxillary surface** is rough and irregular throughout the greater part of its extent for articulation with the nasal surface of the maxilla. Its posterosuperior part is smooth where it helps form the medial wall of the pterygopalatine fossa. Its anterior portion is also smooth, and here it forms the posterior part of the medial wall of the maxillary sinus. On the posterior part of the maxillary surface is the deep, vertical **greater palatine groove,** which upon articulation with the maxilla becomes the greater palatine canal. Through the canal courses the greater palatine vessels and nerve.

The **anterior border** is thin and irregular. Opposite the conchal crest is a pointed, projecting lamina, the **maxillary process,** which articulates with the inferior nasal concha and helps form the medial wall of the maxillary sinus (Fig. 4-115). The **posterior border** presents a deep groove, the edges of which are serrated for articulation with the medial pterygoid plate of the sphenoid. This border is continuous above with the sphenoidal process and expands into the **pyramidal process.** The **superior border** is marked by

the **orbital process** in front and the **sphenoidal process** behind. These processes are separated by the **sphenopalatine notch,** which is converted into the **sphenopalatine foramen** by articulation with the inferior surface of the body of the sphenoid. In the articulated skull this foramen leads from the pterygopalatine fossa into the posterior part of the nasal cavity and transmits the sphenopalatine vessels and the posterior superior nasal nerves. The **inferior border** is fused with the lateral edge of the horizontal plate. Immediately anterior to the pyramidal process is the lower end of the greater palatine groove.

PYRAMIDAL PROCESS.   The **pyramidal process** (Fig. 4-121) projects laterally from the junction of the horizontal and perpendicular plates and is received into the angular interval between the inferior extremities of the pterygoid plates. On its **posterior surface** is a smooth, grooved triangular area limited on each side by a rough articular furrow. The furrows articulate with the pterygoid plates, while the grooved intermediate area completes the lower part of the pterygoid fossa and gives origin to a few fibers of the medial pterygoid muscle. The anterior part of the **lateral surface** is rough, for articulation with the tuberosity of the maxilla. Its posterior part consists of a smooth triangular area which appears, in the articulated skull, between the tuberosity of the maxilla and the inferior part of the lateral pterygoid plate, and completes the infratemporal fossa. On the **base** of the pyramidal process, close to its union with the horizontal plate, are the **lesser palatine foramina** for the transmission of the lesser palatine nerves and vessels.

ORBITAL PROCESS.   The **orbital process** (Figs. 4-119 to 4-121) is directed superiorly and laterally from the perpendicular plate, to which it is connected by a constricted neck. It encloses an air sinus and presents five surfaces, three of which are articular and two nonarticular. The articular surfaces are: (1) the **anterior** or **maxillary,** oblong in form, and rough for articulation with the maxilla; (2) the **posterior** or **sphenoidal,** directed backward, upward, and medially and presenting the opening of the air sinus that usually communicates with the sphenoidal sinus (the margins of the opening are serrated for articulation with the sphenoidal concha); (3) the **medial** or **ethmoidal,** directed forward and articulating with the labyrinth of the ethmoid. In some cases the air sinus opens on this surface of the bone and then communicates with the posterior ethmoidal cells. More rarely it opens on both the ethmoidal and sphenoidal surfaces, and then communicates with the posterior ethmoidal cells and the sphenoidal sinus. The nonarticular surfaces are: (1) the **superior** or **orbital,** which is triangular in shape, is directed upward and laterally, and forms the posterior part of the floor of the orbit; and (2) the **lateral,** which is oblong in form and separated from the orbital surface by a rounded border that enters into the formation of the inferior orbital fissure.

SPHENOIDAL PROCESS.   The **sphenoidal process** (Fig. 4-119) is a thin, compressed plate, much smaller than the orbital, and directed superiorly and medially. It presents three surfaces and three borders. The **superior surface** articulates with the root of the medial pterygoid plate and the inferior surface of the sphenoidal concha. It presents a groove that contributes to the formation of the palatovaginal canal. The **medial surface** is concave and forms a small part of the lateral wall of the nasal cavity. The **lateral surface** is divided into an articular portion, which is rough, for articulation with the medial pterygoid plate, and a nonarticular portion, which is smooth and forms part of the pterygopalatine fossa. The **medial border** reaches as far as the ala of the vomer for articulation. The **anterior border** forms the posterior boundary of the sphenopalatine

notch. The serrated **posterior border** articulates with the vaginal process of the medial pterygoid plate.

The **orbital** and **sphenoidal processes** are separated from one another by the **sphenopalatine notch** (Fig. 4-119), which becomes the sphenopalatine foramen by the inferior surface of the sphenoid bone. Sometimes the two processes are united above and form between them a complete sphenopalatine foramen (Fig. 4-120), or the notch may be crossed by one or more spicules of bone, giving rise to two or more foramina.

OSSIFICATION.   The palatine bone is ossified in membrane from a single center, which makes its appearance about the eighth week of fetal life at the angle of junction between the horizontal and perpendicular plates of the bone. From this point ossification spreads medialward to the horizontal plate and downward into the pyramidal process. By the tenth week ossification has spread upward into the perpendicular plate and its orbital and sphenoidal processes. At the time of birth the height of the perpendicular plate is about equal to the transverse width of the horizontal plate. As the maxilla grows and the nasal cavities increase in depth, the perpendicular plate in the adult eventually measures about twice as much as the horizontal plate.

ARTICULATIONS.   The palatine bone articulates with six other bones: the sphenoid, ethmoid, maxilla, inferior nasal concha, vomer, and the opposite palatine.

INFERIOR NASAL CONCHA

The **inferior nasal conchae** are a slender pair of bones, each of which extends horizontally along the lateral wall of the nasal cavity (Figs. 4-72; 4-73; 4-109). Each inferior concha consists of a lamina of spongy bone, curled upon itself like a scroll, and it is so located as to separate the middle meatus in the nasal cavity from the inferior meatus. It has two surfaces, two borders, and two extremities.

The **medial surface** (Fig. 4-122) is convex, perforated by numerous apertures, and traversed by longitudinal grooves for the lodgment of vessels. The **lateral surface** is concave (Fig. 4-123), faces the lateral wall of the nasal cavity, and forms part of the inferior meatus. Its **superior border** is thin and irregular, and articulates with four bones: the maxillary, lacrimal, ethmoid, and palatine bones. This border may be divided into three portions: anterior, middle, and posterior. The *anterior* articulates with the

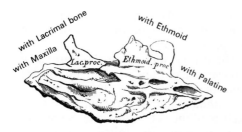

F ig. 4-122. Right inferior nasal concha. Medial surface.

conchal crest of the maxilla; the *posterior* with the conchal crest of the palatine. The *middle portion* presents three well-marked processes, which vary much in size and form. Of these, the **lacrimal process** is small and pointed and is situated at the junction of the anterior one-fourth with the posterior three-fourths of the bone. By its apex, it articulates with the descending process of the lacrimal bone, and by its margins, with the nasolacrimal groove on the medial surface of the maxilla just behind the frontal process. Thus, it assists in forming the canal for the nasolacrimal duct. Posterior to the lacrimal process is a broad, thin plate, the **ethmoidal process,** which ascends to articulate with the uncinate process of the ethmoid. Extending inferiorly and laterally from the ethmoidal process is a thin lamina, the **maxillary process,** which articulates with the maxilla and forms a part of the medial wall of the maxillary sinus. The **inferior border** is free, thick, and spongy in structure, more especially in the middle of the bone. Both **extremities** are more or less pointed, the posterior being the more tapering.

OSSIFICATION. The inferior nasal concha is ossified from a single center, which appears about the fifth month of fetal life in the lateral wall of the cartilaginous nasal capsule.

ARTICULATIONS. The inferior nasal concha articulates with four bones. From anterior to posterior these are the maxillary, lacrimal, ethmoid, and palatine (Fig. 4-122).

F ig. 4-123. Right inferior nasal concha. Lateral surface.

THE VOMER

The **vomer** is situated in the median plane, but its anterior portion is frequently deviated to one side or the other. It is thin, unpaired, flat, and irregularly quadrilateral, and forms the posterior and inferior part of the **nasal septum** (Fig. 4-124). It has two surfaces and four borders. The **surfaces** are marked by small furrows for blood vessels, and on each is the **nasopalatine groove,** which courses obliquely downward and forward and lodges the nasopalatine nerve and vessels (Fig. 4-125). The **superior border,** the thickest, presents a deep furrow, bounded on each side by a horizontal projecting **ala** of bone. The furrow receives the rostrum of the sphenoid, while the margins of the alae articulate with the vaginal processes of the medial pterygoid plates of the sphenoid behind, and with the sphenoidal processes of the palatine bones in front. The **inferior border** articulates with the nasal crest formed by the maxillae and palatine bones. The **anterior border** is the longest, and it slopes downward and forward. Its superior half is fused with the perpendicular plate of the ethmoid, while its inferior half is grooved to receive the inferior margin of the septal cartilage of the nose. The **posterior border** is free and concave, and separates the choanae, or posterior nasal apertures. It is thick and bifid above, and thin below.

OSSIFICATION. At an early period the septum of the nose consists of a cartilaginous plate. The posterior superior part of this cartilage is ossified to form the perpendicular plate of the ethmoid. Its anterior inferior portion persists as the septal cartilage, while the vomer is ossified in the connective tissue membrane covering its posterior inferior part. Two ossification centers, one on each side of the midline, appear about the eighth week of fetal life in this part of the membrane, and hence the vomer consists primarily of two lamellae. At about the third month these unite below the cartilaginous plate, and thus a deep groove is formed in which the septal cartilage of the nose becomes lodged. As growth proceeds, the union of the ossified lamellae extends anteriorly and superiorly, and at the same time the intervening plate of cartilage undergoes absorption. By puberty the lamellae are almost completely united to form a median plate, but evidence of the bilaminar origin of the bone is seen in the everted alae of its upper border and the groove on its anterior margin.

ARTICULATIONS. The vomer articulates with six bones: two of the cranium—the sphenoid and ethmoid; and four of the face—the two maxillae and the two palatine bones. It also articulates with the septal cartilage of the nose.

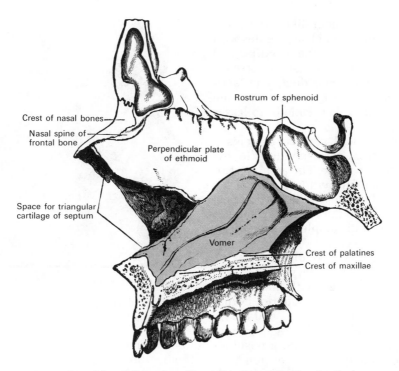

Rostrum of sphenoid

Crest of nasal bones

Nasal spine of frontal bone

Perpendicular plate of ethmoid

Space for triangular cartilage of septum

Vomer

Crest of palatines

Crest of maxillae

FIG. 4-124.   The anterior portion of the skull has been bisected in the midline leaving the bony nasal septum which separates the two nasal cavities. Observe that the septum is a thin sheet of bone formed almost entirely by the vomer and the perpendicular plate of the ethmoid, although small contributions to the septum are made by the rostrum of the sphenoid posteriorly and the nasal spine of the frontal bone anteriorly. Note the articular surfaces of the vomer with the sphenoid, ethmoid, maxillae, and palatine bones. Additionally, it articulates with the septal cartilage.

LACRIMAL BONE

The **lacrimal bone,** the smallest and most fragile bone of the face, is situated at the anterior part of the medial wall of the orbit (Fig. 4-112). It is quadrilateral in shape and presents two surfaces and four borders.

The **lateral** or **orbital surface** (Fig. 4-126) is divided by a vertical ridge, the **posterior lacrimal crest,** into two parts. Anterior to the crest is a longitudinal groove, the **lacrimal**

sulcus, which unites with the frontal process of the maxilla to form the **fossa of the lacrimal sac.** The upper part of this fossa lodges the lacrimal sac, and the lower part, the nasolacrimal duct. Posterior to the crest, the lateral surface of the lacrimal bone is smooth and forms part of the medial wall of the orbit. The crest, with a part of the orbital surface immediately behind it, gives origin to the lacrimal part of the orbicularis oculi muscle. The posterior lacrimal crest ends

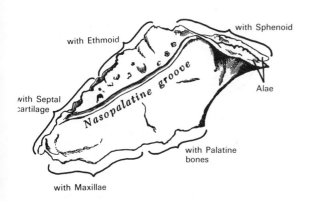

with Ethmoid

with Sphenoid

with Septal cartilage

Nasopalatine groove

Alae

with Palatine bones

with Maxillae

FIG. 4-125.   The vomer. Left lateral aspect.

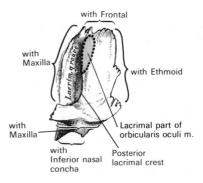

with Frontal

with Maxilla

Lacrim. groove

with Ethmoid

with Maxilla

with Inferior nasal concha

Lacrimal part of orbicularis oculi m.

Posterior lacrimal crest

FIG. 4-126.   The left lacrimal bone. The lateral (or orbital) aspect.

inferiorly in a small, hook-like projection, the **lacrimal hamulus,** which articulates with the lacrimal tubercle of the maxilla, and completes the upper orifice of the osseous canal for the nasolacrimal duct. The lacrimal hamulus sometimes exists as a separate small bone, and it is then called the **lesser lacrimal bone.**

The **medial** or **nasal surface** presents a longitudinal furrow, corresponding to the crest on the lateral surface. The area anterior and inferior to this furrow forms part of the middle meatus of the nose. The portion posterior and superior to the furrow articulates with the ethmoid and completes some of the anterior ethmoidal cells.

The **anterior border** articulates with the frontal process of the maxilla, while the **posterior border** articulates with the lamina orbitalis of the ethmoid. The **superior border** articulates with the frontal bone, and the **inferior border** is divided by the lower edge of the posterior lacrimal crest into a *posterior part,* which articulates with the orbital plate of the maxilla, and an *anterior part,* which is prolonged downward as a descending process. The latter articulates with the lacrimal process of the inferior nasal concha and assists in forming the canal for the nasolacrimal duct.

OSSIFICATION. The lacrimal bone is ossified from a single center, which appears about the twelfth week of prenatal life in the membrane covering the cartilaginous nasal capsule. Ossification of the hamulus from a separate center has been reported.

ARTICULATIONS. The lacrimal bone articulates with four bones: two of the cranium—the frontal and ethmoid, and two of the facial skeleton—the maxilla and the inferior nasal concha.

ZYGOMATIC BONE

The **zygomatic bone** (Figs. 4-127 to 4-129) is nearly quadrangular and is situated in the upper lateral aspect of the face. It forms the prominence of the cheek, part of the lateral wall and floor of the orbit, and part of the bony walls of the temporal and infratemporal fossae. It presents lateral, orbital, and temporal surfaces; frontal and temporal processes; and five borders.

The **lateral surface** (Figs. 4-127; 4-128) is convex, and it is perforated toward the orbital border by a small aperture, the **zygomaticofacial foramen,** through which passes

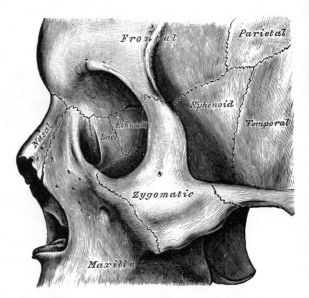

FIG. 4-127. Left zygomatic bone *in situ.*

the zygomaticofacial nerve and vessels. Below this foramen is a slight elevation, which gives origin to the zygomaticus minor and the levator labii superioris muscles.

The **temporal surface** (Fig. 4-129) is directed medially and posteriorly. It is concave and presents medially a rough, triangular area for articulation with the maxilla. Laterally, its smooth, concave surface extends superiorly behind the frontal process to form the bony anterior boundary of the temporal fossa. More posteriorly the temporal surface of the zygomatic bone also forms

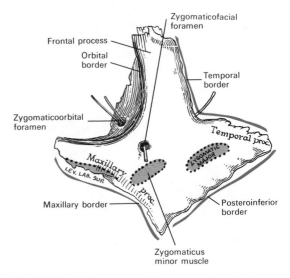

FIG. 4-128. Left zygomatic bone. Lateral surface.

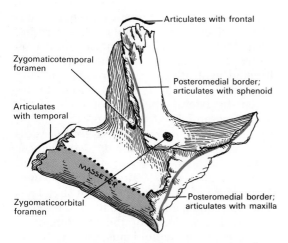

FIG. 4-129. Left zygomatic bone, temporal surface.
Note that the posteromedial border articulates with two
bones, the sphenoid and the maxilla.

part of the boundary of the infratemporal
fossa. Near the center of this surface is the
zygomaticotemporal foramen for the transmission of the zygomaticotemporal nerve.

The **orbital surface** forms, by its junction
with the orbital surface of the maxilla and
with the great wing of the sphenoid, part of
the floor and lateral wall of the orbit. Usually it presents the orifices of two canals, the
zygomatico-orbital foramina, which communicate with the zygomaticotemporal and
zygomaticofacial foramina.

The **frontal process** (Fig. 4-128) is thick
and serrated, and articulates superiorly with
the zygomatic process of the frontal bone
and posteriorly with the great wing of the
sphenoid and the orbital surface of the maxilla (Fig. 4-129). On its orbital surface, just
within the orbital margin and about 1 cm
below the zygomaticofrontal suture is the **tuberculum marginale.** Onto this tubercle, located laterally in the orbit, are attached: (1)
the lateral cheek ligament, (2) the lateral palpebral ligament, (3) the suspensory ligament
of the eye, and (4) a lateral extension of the
aponeurosis of the levator palpebral superioris muscle.

The **temporal process** is blunt and its
oblique serrated margin helps to form the
zygomatic arch by articulation with the zygomatic process of the temporal bone.

In the older nomenclature, reference is
also made to the *maxillary process* of the
zygomatic bone (Fig. 4-128), but the more
recent *Nomina Anatomica* does not recog-
nize this as an officially named structure.

The term simply refers to the rough, triangular inferior extension of the bone, which articulates with the maxilla.

The **orbital border** (anterosuperior) is
smooth and rounded (Fig. 4-128) and forms
approximately one-third of the bony circumference of the orbit. The **maxillary border** (anteroinferior) is rough and, in apposition with the maxilla, helps to form the
prominence of the cheek (Fig. 4-127). Close
to the orbit, this border gives origin to the
levator labii superioris muscle. The **temporal border** (posterosuperior) extends from
the frontozygomatic suture to the temporozygomatic suture (Fig. 4-128). It is curved
and smooth and gives attachment to the temporal fascia. The **posteroinferior border** allows attachment of the masseter muscle (Fig.
4-129). On the medial or temporal aspect of
the zygomatic bone can be observed the
posteromedial border, which is serrated and
articulates with the great wing of the sphenoid bone above and the maxilla below (Fig.
4-129). Between these two articular surfaces,
this border also presents a short, concave,
nonarticular part, which forms the lateral
boundary for the inferior orbital fissure.
Occasionally this nonarticular part is absent,
and the fissure is then completed by the
junction of the maxilla and sphenoid, or by
the interposition of a small sutural bone in
the angular interval between them.

OSSIFICATION. The zygomatic bone ossifies in
membrane from a single center, which appears just
beneath and to the lateral aspect of the orbit at about
the eighth week of prenatal life. After birth the bone
is sometimes divided by a horizontal suture into an
upper large and a lower smaller division, and in
some animals the zygomatic bone is composed of
two parts, consistent with this human variation.

ARTICULATIONS. The zygomatic bone articulates
with four other bones: the frontal, sphenoidal, temporal, and maxillary.

Appendicular Skeleton

The bones by which the upper and lower
limbs are attached to the trunk constitute
respectively the shoulder and pelvic girdles.
The **shoulder girdle** or girdle of the superior
limbs is formed by the scapulae and clavicles. Ventrally it is completed by the cranial
end of the sternum, with which the medial
end of both clavicles articulates. Dorsally the
skeleton of the shoulder girdle is incomplete

and the scapulae are connected to the trunk by muscles only. The **pelvic girdle** or girdle of the inferior limbs is formed by the hip bones, which articulate with each other at the symphysis pubis. The pelvic girdle is completed posteriorly by the sacrum, thereby forming a complete ring, massive and comparatively rigid, in marked contrast to the light and mobile shoulder girdle.

BONES OF THE SHOULDER GIRDLE AND UPPER LIMB

THE CLAVICLE

The **clavicle** (Figs. 4-130; 4-131), commonly called the collar bone, forms the ventral portion of the shoulder girdle. It is a long, curved bone, which lies subcutaneously across the root of the neck. It is directed nearly horizontally toward the tip of the shoulder, immediately above the first rib. It articulates medially with the manubrium sterni and laterally with the acromion of the scapula.

In the female the clavicle is generally shorter, thinner, less curved, and smoother than in the male. In those persons who perform considerable manual labor it becomes thicker and more curved, and its ridges for muscular attachment are prominently marked.

The clavicle acts especially as a fulcrum to enable the muscles to give lateral motion to the arm. Accordingly, it is absent in those animals whose forelimbs are used only for progression, but is present for the most part in animals whose anterior extremities are clawed and used for prehension. In some animals, however, such as many of the carnivora, it is only a rudimentary bone which is suspended among the muscles, not articulating with either the scapula or the sternum.

The human clavicle presents a double curvature. At the sternal end the convexity is directed ventrally, at the acromial end the convexity is dorsal. It is convenient to describe the lateral one-third of the clavicle and its **acromial extremity** and the medial two-thirds with its **sternal extremity.** Additionally, the clavicle presents two surfaces and two borders.

LATERAL ONE-THIRD.   The **cranial** or **superior surface** of the lateral third of the clavicle is flat, rough, and marked toward the borders by impressions for the attachment of the trapezius and deltoid muscles (Fig. 4-130). Between these impressions the strip of bone is smooth and subcutaneous. The **caudal** or **inferior surface** at the dorsal border near the flat end presents a rough eminence, the **conoid tubercle** (Fig. 4-131), for the attachment of the conoid ligament. An oblique ridge, the **trapezoid line,** runs ventrally and laterally from this tubercle and affords attachment to the trapezoid ligament. The **dorsal border** of the lateral third is convex, rough, and thicker than the ventral border, and receives the insertion of the trapezius muscle. The **ventral border** is concave, thin, and rough, and gives origin to the deltoid muscle. The distal extremity or acromial end of the lateral one-third of the clavicle articulates with the acromion of the scapula and will be described below.

MEDIAL TWO-THIRDS.   The medial two-thirds of the clavicle is rounded or prismatic and curved with a ventral convexity. Medially, its **ventral border** presents an elliptical surface for the attachment of the clavicular head of the pectoralis major muscle. The ventral border and free surface of the bone between the attachments of the pectoralis

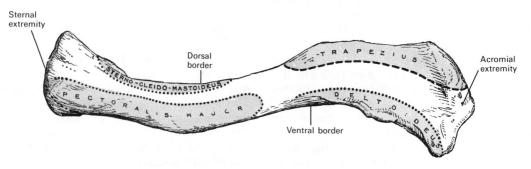

Sternal extremity

Dorsal border

Ventral border

Acromial extremity

FIG. 4-130.   Left clavicle. Superior surface.

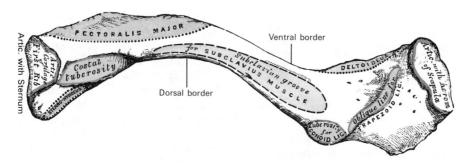

FIG. 4-131. Left clavicle. Inferior surface. Attachment of articular capsules indicated by solid blue lines.

major and deltoid is smooth. The superior surface of the medial two-thirds is smooth and rounded laterally, but becomes more roughened medially for the attachment of the sternocleidomastoid muscle. The ventral aspect of the middle portion of the clavicle is smooth, convex, and nearly subcutaneous, being covered only by skin and platysma muscle fibers. The **dorsal border,** from the conoid tubercle to the costal tuberosity, forms the posterior boundary of the groove for the subclavius muscle, and gives attachment to a layer of cervical fascia which envelops the omohyoid muscle. The surface of the bone along the dorsal border is smooth and is oriented toward the root of the neck. It is concave mediolaterally, and lies in relation to the course of the transverse scapular vessels. This aspect of the clavicle, at the junction of the curves of the bone, is also adjacent to the brachial plexus of nerves and the subclavian vessels. It gives attachment, near the sternal extremity, to part of the sternohyoid muscle, and presents, near the middle, an oblique foramen directed lateralward, which transmits the chief nutrient artery of the bone. On the **inferior surface** of the medial two-thirds of the clavicle, near its sternal extremity, is a broad roughened area, the **costal tuberosity,** which is rather more than 2 cm in length and to which attaches the costoclavicular ligament. The rest of this surface is occupied by a groove, which gives attachment to the subclavius muscle. The clavipectoral fascia, which splits to enclose the subclavius muscle, is attached to the margins of the groove.

STERNAL EXTREMITY. The **sternal extremity** of the clavicle is directed medialward and presents an **articular facet,** which articulates with the clavicular notch of the *manubrium sterni* through the intervention of an **articular disc.** The inferior part of the articular facet continues as a triangular surface for articulation with the *cartilage of the first rib.* The circumference of the articular surface is roughened for the attachment of numerous ligaments. The sternal end of the clavicle projects above the level of the manubrium sternum and is covered by the sternal end of the sternocleidomastoid muscle. It is easily palpable and frequently can be seen subcutaneously in the root of the neck.

ACROMIAL EXTREMITY. The **acromial extremity** presents a small flat oval surface directed obliquely downward for articulation with the *acromion of the scapula.* The circumference of the articular facet is rough, especially above, for the attachment of the acromioclavicular ligaments. The acromial extremity of the clavicle projects somewhat above the acromion of the scapula, and the acromioclavicular joint is palpable about 3 cm medial to the lateral border of the acromion.

STRUCTURE. The clavicle consists of cancellous bone enveloped by a compact layer, which is much thicker in the intermediate part than at the extremities of the bone.

OSSIFICATION. The clavicle begins to ossify before any other bone in the body. Ossification commences from three centers: *two primary centers,* a medial and a lateral, for the clavicular shaft, which appear during the fifth or sixth week of fetal life, and a *secondary center* for the sternal end, which appears about the eighteenth or twentieth year and unites with the rest of the bone about the twenty-fifth year (Fig. 4-132).

FRACTURES OF THE CLAVICLE. The clavicle is fractured more frequently than any other bone in the body. This usually is caused by a fall that results in an inordinate force applied to the bone from the hand or the shoulder. The usual location of the fracture is at the junction of the lateral and intermediate thirds of the bone. The lateral fragment of bone then

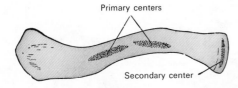

Primary centers

Secondary center

FIG. 4-132.   Diagram showing the three centers of ossification of the clavicle.

becomes displaced downward by contractions of the pectoralis major and teres major muscles, and by the weight of the arm. The medial fragment is tilted upward by contractions of the sternocleidomastoid and trapezius muscles.

THE SCAPULA

The **scapula** or shoulder blade is a large, flat, triangular bone located on the dorsal aspect of the thorax, extending between the second and seventh ribs. It articulates with the clavicle and humerus laterally to form the shoulder joint and is attached to the vertebral column medially by means of muscles. It has two **surfaces** (costal and dorsal) and is marked by three prominent **bony processes:** the spine of the scapula, the acromion, and the coracoid process. Additionally, it has three **borders** (superior, lateral or axillary, and medial or vertebral), and three **angles** (inferior, superior, and lateral).

SURFACES.   The **costal surface** (Fig. 4-133) presents a broad concavity, the **subscapular fossa.** The medial two-thirds of the fossa are marked by several oblique ridges which course laterally and superiorly. These ridges give attachment to the tendinous origin and the surfaces between them to the fleshy fibers of the subscapularis muscle. The lateral third of the fossa is smooth and also covered by the fibers of this muscle. The fossa is separated from the two extremities of the medial or vertebral border by smooth triangular areas at both the medial and inferior angles. The interval between these two angles is marked by a narrow ridge, which at times is deficient. The triangular areas at the angles and this intervening ridge receive the insertion of the serratus anterior muscle.

The **dorsal surface** (Fig. 4-134) is arched from above downward and is subdivided into two unequal parts by a prominent plate of bone, the **spine of the scapula.** The smaller portion above the spine is called the **supraspinatous fossa,** and that below it the **infraspinatous fossa.**

The **supraspinatous fossa** is a smooth concavity, which is broader at its vertebral than at its humeral end. Its medial two-thirds give origin to the supraspinatus muscle.

The **infraspinatus fossa** is about three times as large as the supraspinatus fossa. It has a shallow concavity toward the cranial part of its vertebral margin, a prominent convexity at its center, and a deep groove near the axillary border. The medial two-thirds of the fossa give origin to the infraspinatus muscle and the lateral third is covered by this muscle.

The dorsal surface of the scapula is marked near the axillary border by an elevated ridge, which runs from the lower part of the glenoid cavity, downward and backward to the medial or vertebral border, about 2.5 cm above the inferior angle. The ridge serves for the attachment of a fibrous septum which separates the infraspinatus muscle from the teres major and teres minor muscles. The surface between the ridge and the axillary border is narrow in the cranial two-thirds of its extent and is crossed near its center by a groove for the passage of the circumflex scapular vessels. This ridge affords attachment to the teres minor above, while its caudal third presents a broader, somewhat triangular surface, which gives origin to the teres major. The latissimus dorsi muscle glides over this lower region, and frequently a few of its muscle fibers arise at the inferior angle of the scapula. An oblique line from the axillary border to the elevated ridge has attached to it a fibrous septum which separates the teres muscles from each other (Fig. 4-134).

BONY PROCESSES.   The scapula possesses three important bony processes: the spine of the scapula, the acromion, and the coracoid process.

The **spine of the scapula** (Fig. 4-134) is a strong, triangular bony projection from the dorsal surface of the scapula, and it separates the supraspinatous and infraspinatous fossae. It begins at the medial or vertebral border as a smooth, triangular zone over which the tendon of insertion of the trapezius glides, and gradually becoming more elevated, it ends in the acromion, which overhangs the shoulder joint. The spine is flattened from above downward and the apex of its triangular shape is directed toward the medial border of the scapula. The spine presents two surfaces and three bor-

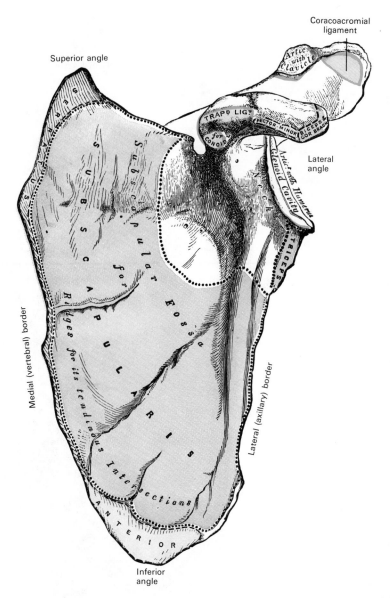

FIG. 4-133. Left scapula, costal surface. Attachment of articular capsules indicated by solid blue lines.

ders of its own. Its **superior surface** is concave, and it assists in forming the supraspinatous fossa, giving origin to part of the supraspinatus muscle. Its **inferior surface** forms part of the infraspinatous fossa, gives origin to a portion of the infraspinatus muscle, and presents near its center the orifice of a nutrient canal. Of its three borders, the longitudinal **anterior border** is the ridge of bone by which the spine is attached to the dorsal scapular surface. The **dorsal border,** or **crest of the spine,** is broad, and presents two lips with an intervening rough interval.

The trapezius is attached to the superior lip, and a rough tubercle is generally seen on that portion of the spine that receives the tendon of insertion of the caudal part of this muscle. The deltoid muscle is attached to the whole length of the caudal lip. The interval between the lips is subcutaneous and partly covered by the tendinous fibers of these muscles. The **lateral border** or **base** is the shortest of the three borders. It is slightly concave and its thick, rounded edge is continuous with the undersurface of the acromion and the neck of the scapula. It forms

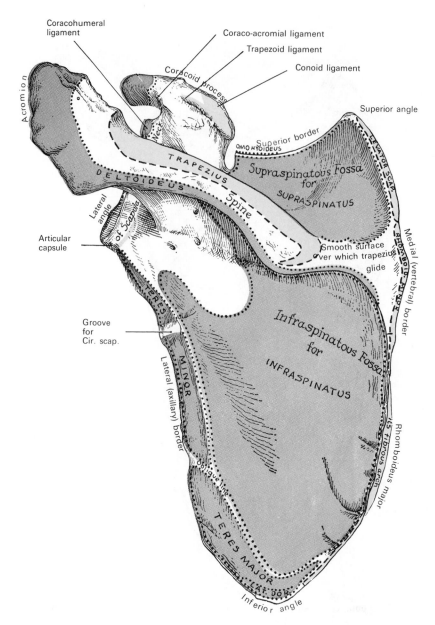

FIG. 4-134.   Left scapula. Dorsal surface.

the medial boundary of the **great scapular notch,** which serves to connect the supraspinatus and infraspinatous fossae.

The **acromion** is a flattened, lateral prolongation of the spine of the scapula, which forms the summit of the shoulder and overhangs the glenoid cavity. Its **superior surface** is convex and rough, and affords attachment to some fibers of the deltoid muscle, while the rest of its extent is subcutaneous. Its **inferior surface** is smooth and concave. Its **lateral border** is thick and irregular, and presents three or four tubercles for the tendinous origins of the deltoid muscle. Its **medial border,** shorter than the lateral, is concave, gives attachment to a portion of the trapezius, and presents about its center a small, oval **surface for articulation** with the acromial end of the clavicle. Its **apex,** which corresponds to the point of meeting of these two borders, is thin and has attached to it the coracoacromial ligament. Inferiorly, where the lateral border of the acromion becomes continuous with the lower border of the crest of the spine of the scapula, is located the **acromial angle** (Fig. 4-135), which can be palpated subcutaneously and therefore used as a landmark.

The **coracoid process** (Figs. 4-134; 4-135) is a thick, curved bony projection attached by a broad base to the upper part of the neck of the scapula. From its base it is initially directed upward and medially, then it becomes somewhat thinner and changes direction to project upward and laterally. With the arm by the side, the tip of the coracoid process is oriented anteriorly and can be palpated by applying deep pressure through

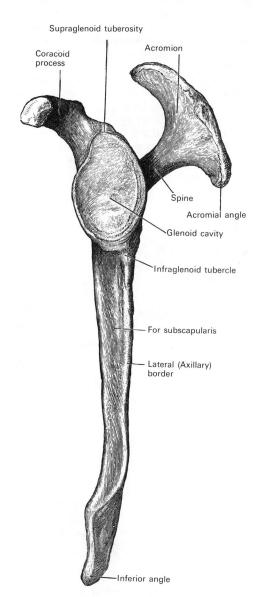

Supraglenoid tuberosity

Coracoid process

Acromion

Spine

Acromial angle

Glenoid cavity

Infraglenoid tubercle

For subscapularis

Lateral (Axillary) border

Inferior angle

FIG. 4-135. Left scapula. Lateral view.

the anterior part of the deltoid muscle about 2.5 mm below the lateral part of the clavicle on the lateral side of the infraclavicular fossa. Its concave surface faces laterally and is smooth to accommodate the passage of the tendon of the subscapularis muscle. The distal portion is horizontal, and its convex outer surface is rough and irregular for the attachment of the pectoralis minor muscle. Its medial and lateral borders are rough. The former gives attachment to the pectoralis minor and the latter to the coracoacromial ligament. The **apex** is embraced by the conjoined tendon of origin of the coracobrachialis and short head of the biceps brachii and gives attachment to the clavipectoral fascia. On the medial part of the root of the coracoid process is a rough impression for the attachment of the conoid ligament. Coursing from it obliquely to the convex surface of the horizontal portion is an elevated ridge for the attachment of the trapezoid ligament.

BORDERS. The **superior border** of the scapula is the shortest and thinnest of the three scapular borders. It is concave, and extends from the medial angle to the base of the coracoid process (Fig. 4-134). At its lateral part is a semicircular notch, the **scapular notch,** formed partly by the base of the coracoid process. This notch is converted into a foramen by the superior transverse ligament (which is at times ossified), and serves for the passage of the suprascapular nerve. The adjacent part of the superior border affords attachment to the omohyoid muscle.

The **lateral (axillary) border** is the thickest of the three borders. It begins above at the lower margin of the glenoid cavity and inclines obliquely downward and backward to the inferior angle. Immediately below the glenoid cavity is a rough impression, the **infraglenoid tubercle,** about 2.5 cm in length, which gives origin to the long head of the triceps brachii muscle. The inferior third is thin and sharp, and serves for the attachment of a few fibers of the teres major behind and the subscapularis muscle in front (Figs. 4-133; 4-134).

The **medial (vertebral) border** is the longest of the three borders, and extends from the superior to the inferior angle. It is arched and intermediate in thickness between the superior and lateral borders, and the portion of it above the spine forms an obtuse angle with the part below. This border presents an anterior and a posterior lip, and an intermediate narrow area. The **anterior lip** affords attachment to the serratus anterior muscle (Fig. 4-133), while the **posterior lip** gives attachment to the supraspinatus muscle above the spine and the infraspinatus muscle below (Fig. 4-134). To the area between the two lips is inserted the levator scapulae muscle above the triangular surface at the commencement of the spine. The rhomboid minor muscle inserts on the edge of that surface and rhomboid major below it (Fig. 4-134). The latter is attached by means of a fibrous arch, connected above to the lower part of the triangular surface at the base of the spine and below to the lower part of the border.

ANGLES.   The **superior angle** (Fig. 4-134), formed by the junction of the superior and medial (or vertebral) borders, is thin, smooth, and rounded, and gives attachment to a few fibers of the levator scapulae muscle. The **inferior angle** (Figs. 4-134; 4-135) is formed by the union of the medial (vertebral) and lateral (axillary) borders. It is thick and rough, and its dorsal surface affords attachment to the teres major and frequently to a few fibers of the latissimus dorsi. The **lateral angle** (Fig. 4-134) is the thickest part of the bone, and the adjacent broadened process is sometimes called the **head of the scapula.** This is connected with the rest of the bone by a slightly constricted **neck.** It is this region of the scapula that enters into the formation of the shoulder joint, and it is hollowed out to form the **glenoid cavity** (Fig. 4-135). The surface of this shallow cavity faces laterally and, in the intact body, is covered with articular cartilage for reception of the head of the humerus. Its margins or lips are covered by a fibrocartilaginous ring, the **glenoid labrum,** which broadens and deepens the cavity. At its apex, near the base of the coracoid process, is a slight elevation, the **supraglenoid tubercle,** to which the long head of the biceps brachii muscle is attached.

STRUCTURE.   The head, processes, and thickened parts of the scapula contain cancellous bone; the rest consists of a thin layer of compact tissue. The central part of the supraspinatous and infraspinatous fossae are usually so thin as to be translucent. Occasionally the bone is even partially deficient at these sites, and the adjacent muscles are separated only by fibrous tissue.

OSSIFICATION.   The scapula is ossified from at least seven centers: one for the body, two for the coracoid process, two for the acromion, one for the medial or vertebral border, and one for the inferior angle (Fig. 4-136).

Ossification of the body of the scapulae begins about the second month of fetal life, by the formation of an irregular quadrilateral plate of bone near the scapular neck adjacent to the glenoid cavity. This plate is extended to form the chief part of the bone, the spine growing up from its dorsal surface about the third month. At birth, a large part of the scapula is osseous, but the glenoid cavity, the coracoid process, the acromion, the vertebral border, and the inferior angle are cartilaginous. During the first year after birth an ossification center appears in the middle of the coracoid process which as a rule becomes joined with the rest of the bone about the fifteenth year. Between the fourteenth and twentieth years, ossification of the remaining parts of the scapula takes place in quick succession. During this post-puberty period, the following structures ossify in order: the root of the coracoid process, the base of the acromion, the inferior angle and contiguous part of the medial border, the extremity of the acromion, and the remainder of the medial border.

The base of the acromion is formed by an extension from the spine of the scapula. The two separate

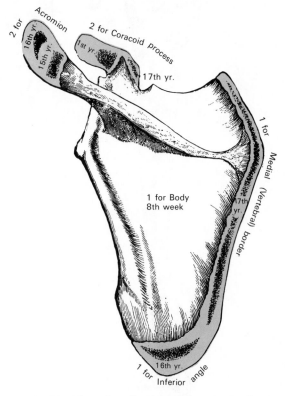

FIG. 4-136.   Plan of ossification of the scapula. From seven centers.

ossification centers of the acromion unite and then join with the extension from the spine. The upper third of the glenoid cavity is ossified from a separate center (subcoracoid), which makes its appearance between the tenth and eleventh years and joins between the sixteenth and eighteenth. Further, an epiphyseal plate appears for the lower part of the glenoid cavity, while the tip of the coracoid process frequently presents a separate nucleus. These various epiphyses are joined to the bone by the twenty-fifth year. Failure of bony union between the acromion and spine sometimes occurs, the junction being effected by fibrous tissue or by an imperfect articulation. In some cases of supposed fracture of the acromion with ligamentous union, it is probable that the detached segment was never united to the rest of the bone.

THE HUMERUS

The **humerus** (Figs. 4-137 to 4-146) is the longest and largest bone of the upper limb. It extends from the shoulder joint above, where it articulates with the scapula, to the elbow joint below, where it articulates with the radius and ulna of the forearm. The humerus consists of a **body** or **shaft** and enlarged **proximal** and **distal extremities.** The proximal extremity includes a rounded **head,** a neck, and two tubercles, the **greater** and **lesser tubercles.** The distal extremity is condylar in form and includes the rounded **capitulum,** the grooved **trochlear,** and two **epicondyles.**

PROXIMAL EXTREMITY. The **head** (Fig. 4-137) of the humerus is nearly hemispherical in form. With the upper extremity by the

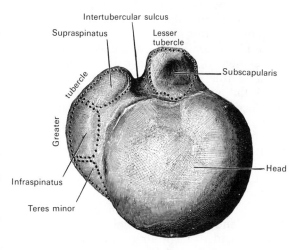

FIG. 4-137.   The upper end of the left humerus. Superior aspect.

side of the body, it is directed superiorly, medially and slightly dorsally, and its smooth surface, covered with cartilage, articulates with the glenoid cavity of the scapula (Fig. 4-138). The circumference of the articular surface of the head of the humerus is slightly constricted and is termed the **anatomical neck,** in contradistinction to a constriction below—a tubercle which is called the **surgical neck** and is frequently the site of fracture. Fracture of the anatomical neck rarely occurs.

The **anatomical neck** (Figs. 4-139; 4-141; 4-142) is obliquely directed, forming an obtuse angle with the body. It is well marked in the lower half of its circumference, and in the upper half, it is represented by a narrow groove which separates the humeral head from the tubercles. The circumference of the anatomical neck affords attachment to the articular capsule of the shoulder joint, and it is perforated by numerous vascular foramina.

The **greater tubercle** (Figs. 4-137 to 4-139; 4-141; 4-142) is situated lateral to the head of the humerus and lesser tubercle. Its upper surface is rounded and marked by three flat impressions: the highest of these gives insertion to the supraspinatus muscle, the middle to the infraspinatus, and the lowest to the teres minor. The insertion of the latter muscle extends about an inch below the greater tubercle along the humeral shaft. The lateral surface of the greater tubercle is convex and rough, and merges distally into the lateral surface of the body.

The **lesser tubercle** (Figs. 4-139; 4-141), although smaller than the greater tubercle, is more prominent. It is situated anteriorly on the bone, adjacent to the anatomical neck. Its anterior surface serves for the insertion of the tendon of the subscapularis muscle.

The tubercles are separated from each other by a deep groove, the **intertubercular groove (bicipital groove),** which lodges the tendon of the long head of the biceps brachii muscle and transmits a branch of the anterior humeral circumflex artery to the shoulder joint. The groove courses obliquely downward and ends nearly one-third of the way along the humeral shaft. In the intact body its upper part is coated with a thin layer of cartilage and enclosed by a prolongation of the synovial membrane of the shoulder joint. Its lower portion becomes shallow and receives the insertion of the

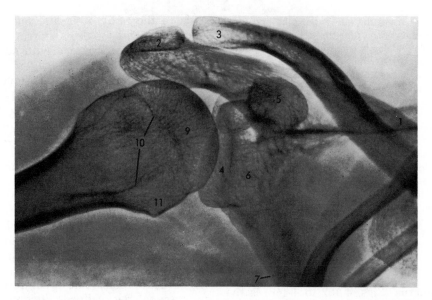

FIG. 4-138. Roentgenogram of the right shoulder joint with the upper arm abducted and medially rotated. (From Benninghoff and Goerttler, 1975.)

Key to Numerals

1. Spine of scapula	6. Neck of scapula
2. Acromion	7. Lateral border of scapula
3. Acromial end of clavicle	9. Head of humerus
4. Glenoid cavity	10. Greater tubercle
5. Coracoid process	11. Lesser tubercle

tendon of the latissimus dorsi muscle. Its lips are called, respectively, the **crests** of the **greater** and **lesser tubercles (bicipital ridges).** Just below the two tubercles, the humerus becomes constricted to form the **surgical neck** of the humerus.

BODY OR SHAFT. The **body** or **shaft** of the humerus is nearly cylindrical in the upper half of its extent. It is prismatic and flattened distally, and has two borders and three surfaces (Figs. 4-139 to 4-142).

The **lateral border** (Figs. 4-139; 4-140) runs from the dorsal part of the greater tubercle to the lateral epicondyle and separates the anterolateral surface from the posterior surface. Its proximal half is rounded and indistinctly marked, serving for the attachment of part of the insertion of the teres minor, and the origin of the lateral head of the triceps brachii. Its center is traversed by a broad but shallow oblique depression, the **sulcus** or **groove for the radial nerve** (Figs. 4-140; 4-142). Its distal part forms a prominent, rough margin, the **lateral supracondylar ridge,** which presents an *anterior lip* for the origin of the brachioradialis above and extensor carpi radialis longus below, a *posterior lip* for the medial head of the triceps brachii,

and an *intermediate ridge* for the attachment of the lateral intermuscular septum.

The **medial border** (Figs. 4-139; 4-140) extends from the lesser tubercle to the medial epicondyle. Its proximal third consists of a prominent ridge, the **crest of the lesser tubercle,** which receives the insertion of the tendon of the teres major. About its center is a rough impression for the insertion of the coracobrachialis, and just distal to this is the entrance of the nutrient canal. Sometimes there is a second nutrient canal at the commencement of the radial sulcus. The distal third of this border is raised into a slight ridge, the **medial supracondylar ridge,** which becomes very prominent distally. It presents an *anterior lip* for the origin of the brachialis muscle, a *posterior lip* for the medial head of the triceps brachii, and an *intermediate ridge* for the attachment of the medial intermuscular septum.

The anterior aspect of the humerus is divided longitudinally into an **anterolateral surface** and an **anteromedial surface** by an oblique ridge beginning laterally at the greater tubercle and ending near the medial epicondyle.

The **anterolateral surface** receives the in-

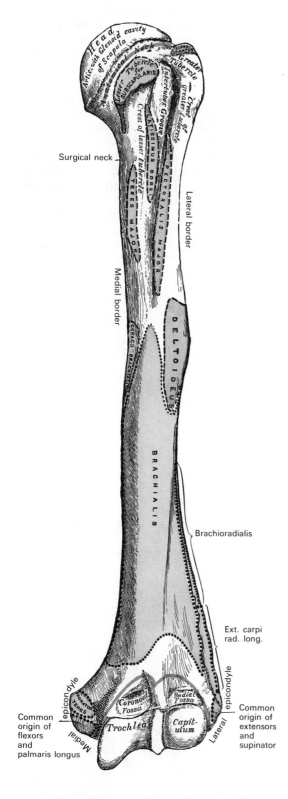

FIG. 4-139. Left humerus. Ventral view. Articular capsule in blue line.

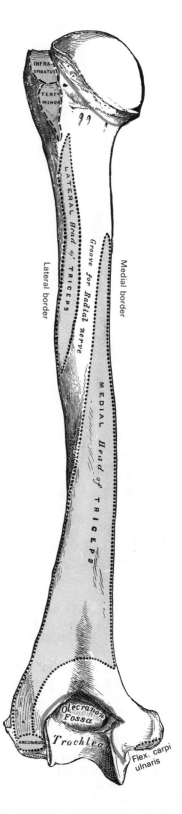

FIG. 4-140. Left humerus. Dorsal view. Articular capsule in blue line.

Head of humerus
Anatomical neck
Crest of the lesser tubercle
Greater tubercle
Lesser tubercle
Intertubercular sulcus
Surgical neck
Crest of the greater tubercle
Coronoid fossa
Radial fossa
Capitulum
Trochlea

FIG. 4-141. The left humerus viewed from the anterior aspect. (From Benninghoff and Goerttler, 1975.)

sertion of the pectoralis major muscle along the distal part of the crest of the greater tubercle. Lateral and distal to this is an oblong area onto which inserts the deltoid muscle. At this site a rough triangular elevation, the **deltoid tuberosity**, marks the anterolateral surface. Distally there is a broad area, slightly concave, which gives origin to part of the brachialis. Beyond the deltoid tuberosity is the radial sulcus, which is directed obliquely distalward, and marks the path of the radial nerve and profunda artery.

The **anteromedial surface** proximally receives the tendon of insertion of the latissimus dorsi in the intertubercular groove. Somewhat distal and medial to this, near the

medial border, is the insertion of the teres major. At about the middle of the shaft is a rough impression for the coracobrachialis muscle. The distal part of the surface is flat and smooth, giving origin to the brachialis muscle.

The **dorsal surface** appears somewhat twisted, so that its upper part is directed a little medially, while the lower part is directed posteriorly and a little laterally. It is almost completely covered by the lateral and medial heads of the triceps brachii, the former arising proximal and the latter distal to the radial sulcus.

DISTAL EXTREMITY. The **distal extremity** is flat and ends inferiorly in a broad, articular surface, generally called a **condyle,** which

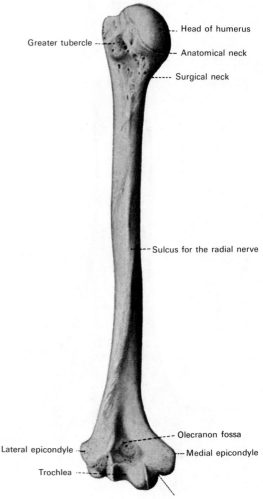

Greater tubercle
Head of humerus
Anatomical neck
Surgical neck
Sulcus for the radial nerve
Olecranon fossa
Lateral epicondyle
Medial epicondyle
Trochlea
Sulcus for the ulnar nerve

FIG. 4-142. The left humerus viewed from the posterior aspect. (From Benninghoff and Goerttler, 1975.)

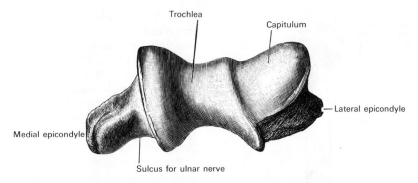

Trochlea

Capitulum

Lateral epicondyle

Medial epicondyle

Sulcus for ulnar nerve

Fig. 4-143. The lower end of the left humerus. Inferior aspect.

is divided into two parts by a slight ridge. The lateral portion of this surface consists of a smooth, rounded eminence named the **capitulum** of the humerus (Figs. 4-139; 4-141; 4-143). It articulates with the fovea on the head of the radius. On the medial side of this eminence is a shallow groove, in which is received the medial margin of the head of the radius. Above the anterior part of the capitulum is a slight depression, the **radial fossa** (Fig. 4-139), which receives the anterior border of the head of the radius when the forearm is flexed. The medial portion of the articular surface is named the **trochlea** (Figs. 4-139; 4-141; 4-143), which occupies the anterior, inferior, and posterior surfaces of

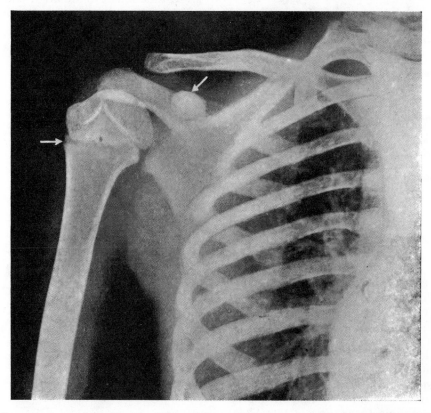

Fig. 4-144. Roentgenogram of shoulder of a child aged six years. The upper arrow indicates the coracoid process; the lower arrow indicates the epiphyseal line. Note that the upper end of the diaphysis is conical and projects into the center of the epiphysis. The centers for the head of the humerus and the tuberosities have fused to form a single epiphysis. Observe also the compact nature of the bone along the surface of the humeral shaft in contrast to the inner cancellous tissue.

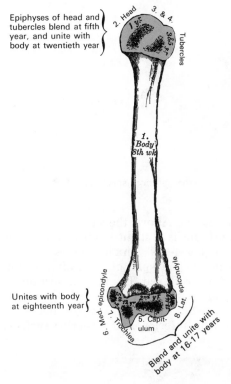

Epiphyses of head and tubercles blend at fifth year, and unite with body at twentieth year

2. Head   3. & 4.   Tubercles

1. y.

3. & 3. y.

1. Body 8th wk

Unites with body at eighteenth year

9. Med. epicondyle   8. Lat. epicondyle

7. Trochlea   5. Capit- ulum

Blend and unite with body at 16–17 years

Fig. 4-145.   Plan of ossification of the humerus.

Sites on Bone	Time of Ossification or Union
1. Body	8th week (ossification center)
2. Head	Tubercles and head blend
3. & 4. Tubercles	at 5th year, unite with body at 20th year
5. Capitulum	16th–17th years
6. Med. Epicondyle	18th year
7. Trochlea	16th–17th years
8. Lat. Epicondyle	16th–17th years

(5, 6, 7, 8: unite with body)

the condyle. The trochlea presents a deep groove between two well-marked borders. Its lateral border is separated from the capitulum, which articulates with the margin of the head of the radius, by a fine groove. The medial border is thicker and more prominent, and projects farther distally than the lateral. The grooved portion of the articular surface fits accurately within the trochlear notch of the ulna, and it is broader and deeper on the dorsal than on the ventral aspect of the bone. Above the anterior part of the trochlea is a small depression, the **coronoid fossa** (Figs. 4-139; 4-141), which receives the coronoid process of the ulna during flexion of the forearm. Above the posterior part of the trochlea is a deep triangular depression, the **olecranon fossa**

(Figs. 4-140; 4-142), into which the summit of the olecranon is received when the forearm is extended. These fossae are separated from one another by a thin, transparent lamina of bone, which is sometimes perforated to produce a **supratrochlear foramen.** They are lined in the intact body by the synovial membrane of the elbow joint, and their margins afford attachment to the anterior and posterior ligaments of this articulation.

Projecting from both the lateral and medial aspects of the articular surface of the condyle are the **medial** and **lateral epicondyles** (Fig. 4-142). These nonarticular processes allow for the attachment of many muscles.

The **lateral epicondyle** is a small, tuberculated and curved eminence giving attachment to the radial collateral ligament of the elbow joint, and to the common tendon of origin of the supinator and extensor muscles of the forearm. The **medial epicondyle,** larger and more prominent than the lateral, gives attachment to the ulnar collateral ligament of the elbow joint, to the pronator

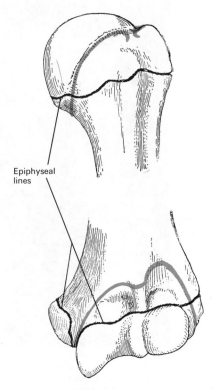

Epiphyseal lines

Fig. 4-146.   Epiphyseal lines of the humerus in a young adult. Ventral aspect. The lines of attachment of the articular capsules are in blue.

teres, and to the common tendon of origin of the flexor muscles of the forearm. The ulnar nerve runs in a groove on the dorsum of the medial epicondyle. The epicondyles are continuous superiorly with the **supracondylar ridges.**

STRUCTURE.   The extremities of the humerus consist of cancellous tissue, covered with a thin, compact layer (Fig. 4-144). The body or shaft is composed of a cylinder of compact tissue, thicker at the center than toward the extremities, and it contains a large medullary canal which extends along its whole length.

OSSIFICATION (Figs. 4-145; 4-146).   The humerus is ossified from eight centers, one for each of the following parts—the body, the head, the greater tubercle, the lesser tubercle, the capitulum, and the trochlea—and one for each epicondyle. The center for the body appears near the middle of the bone in the eighth week of fetal life, and soon extends toward the extremities. At birth the humerus is ossified in nearly its whole length, only the extremities remaining cartilaginous. During the first year, but in some individuals before birth, ossification commences in the head of the bone. During the third year the center for the greater tubercle, and during the fifth that for the lesser tubercle, make their appearance. By the sixth year (Fig. 4-144) the centers for the head and tubercles have joined, thus forming a single large epiphysis, which fuses with the body about the twentieth year. The conical shape of the proximal end of the diaphysis, where the epiphyseal cap fits over it as shown by the epiphyseal line in Figure 4-144, is an adult condition. In the fetus and newborn, the end of the diaphysis is flat. The conical shape is established during the first year and gradually reaches its adult height by the twelfth year. The distal end of the humerus is ossified as follows: at the end of the second year ossification begins in the capitulum, and extends medially to form the chief part of the articular end of the bone; the center for the medial part of the trochlea appears about the age of ten. Ossification begins in the medial epicondyle about the fifth year, and in the lateral about the twelfth or thirteenth year. About the sixteenth or seventeenth year, the lateral epicondyle and both portions of the articulating surface, having already joined, unite with the body, and at about the eighteenth year the medial epicondyle becomes joined to it.

VARIATIONS.   A small, hook-shaped process of bone, the **supracondylar process,** varying from 2 to 20 mm in length, is a common variation found projecting from the ventromedial surface of the body of the humerus 5 cm cranial to the medial epicondyle. Curved distally and ventrally, the pointed end of a well-developed supracondylar process may be connected to the medial epicondyle by a fibrous band that gives origin to a portion of the pronator teres. Through the arch completed by this fibrous band, the median nerve and brachial artery may pass. Sometimes the nerve alone is transmitted through it, or the nerve may be accompanied by the ulnar artery, in cases of high division of the brachial artery. A well-marked groove is frequently found dorsal to the process, along which the nerve and artery are lodged. This arch is the homologue of the **supracondyloid foramen** found in many animals, in which it probably serves to protect the nerve and artery from compression during the contraction of the muscles in this region.

FRACTURES OF THE HUMERUS.   Fractures of the humerus are encountered frequently and may occur at the proximal end, at the shaft, and at the distal end (Thorek, Healey and Seybold, 1966). Proximally *fractures at the anatomical neck* are rare, but they may occur in older people, and usually result from the impact of a fall on the shoulder. These fractures are intracapsular, because the articular capsule of the shoulder joint attaches to the humerus below the anatomical neck. *Fractures of the surgical neck* are more common and frequently occur from a direct impact or from falls on the elbow with the arm abducted. The distal fragment in this fracture (shaft of the humerus) is adducted or pulled medially by the attached pectoralis major, teres major, and latissimus dorsi, whereas the proximal fragment is abducted by the supraspinatus muscle. The type of injury that would cause fracture at the surgical neck in older individuals may cause a *separation of the upper humeral epiphysis* from the head of the humerus in children and adolescents prior to the time of permanent union of the epiphysis. Separation or *fracture of the greater tubercle* can result from a direct impact or from the pull of the attached supraspinatus muscle. The greater tubercle in these instances is usually pulled superiorly and posteriorly. *Fractures of the lesser tubercle* are more rare.

Fractures of the humeral shaft are usually the result of direct violence and generally break the bone transversely. If the fracture occurs *above* the insertion of the deltoid muscle, the upper fragment of the humerus is adducted or pulled medially by the pectoralis major, latissimus dorsi, and teres major. The lower fragment is displaced laterally by the deltoid. If the fracture occurs *below* the insertion of the deltoid, the upper fragment of the humerus becomes displaced laterally by the deltoid and supraspinatus muscles, while the lower fragment is pulled medially and upward by the triceps, biceps, and coracobrachialis muscles. Fractures of the humeral shaft endanger the radial nerve which, along with the profunda brachii artery, courses in the sulcus for the radial nerve around the body of the bone.

Distally, fractures of the humerus may involve the shaft of the humerus above the condyle. Thus, a *supracondylar fracture* occurs frequently in children who sustain a fall on the outstretched hand. The shaft of the humerus is thin just above the elbow joint, and the break results in the distal fragment of the humerus frequently being displaced posteriorly and upward (extension type). At times the distal fragment is displaced anteriorly (flexion type). Supracondylar fractures may injure the medial nerve and brachial artery anteriorly or the radial nerve laterally. Fractures may also occur at the level of the condyle. The break in these *condylar fractures* may extend into the elbow joint and even injure the ulnar nerve which courses behind the joint.

THE ULNA

The **ulna** (Figs. 4-147; 4-148; 4-154 to 4-156) is a long prismatic bone that occupies the medial or little finger side of the forearm and lies parallel with the radius. It is divisible into a **body** and **two extremities.** Its proximal extremity is of great thickness and strength. It is hook-like in shape and forms a large part of the elbow joint. The bone diminishes in size from above downward, so

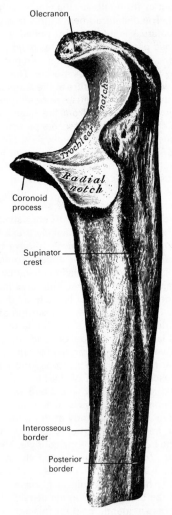

FIG. 4-147. Proximal extremity of the left ulna. Lateral aspect.

that its distal extremity is quite small, and is excluded from the wrist joint by the interposition of an articular disc.

PROXIMAL EXTREMITY. The **proximal extremity** (Fig. 4-147) presents two curved processes, the **olecranon** and **coronoid processes,** and two concave, articular cavities, the **trochlear** and **radial notches.** The latter articulate, respectively, with the humerus and radius.

The **olecranon** (Figs. 4-147; 4-148; 4-156) is a large, thick, curved eminence situated on the superior and posterior part of the ulna, and forming the point of the elbow. It is bent forward at the summit so as to present a prominent lip which is received into the olecranon fossa of the humerus when the forearm is extended. The **base** of the olecranon is constricted where it joins the body of the ulna, and this is the narrowest part of the upper end of the ulna. Its **posterior surface** is triangular and smooth and can readily be palpated through the skin. It is covered by a bursa. The **superior surface** of the olecranon is quadrilateral in shape and is marked behind by a rough impression for the insertion of the triceps brachii, and in front, near the margin, by a slight transverse groove for the attachment of part of the posterior ligament of the elbow joint. Its **anterior surface** is smooth and concave, and forms the upper part of the trochlear notch. To the anterior aspect of the superior surface of the olecranon is attached the capsule of the elbow joint, while medially the olecranon affords attachment to the oblique and posterior parts of the ulnar collateral ligament. Also attached medially is the ulnar head of the flexor carpi ulnaris muscle. The lateral surface of the olecranon affords attachment to the anconeus muscle.

The **coronoid process** (Figs. 4-147; 4-148) is a triangular eminence that projects by means of its **base** from the proximal and anterior part of the body of the bone, just below the olecranon. Its **apex** is pointed, slightly curved upward, and during flexion of the forearm is received into the coronoid fossa of the humerus. The **superior surface** of the coronoid process is smooth and concave, and forms the lower part of the trochlear notch. Its **inferior surface** is concave and rough, and at its junction with the front of the body is a rough eminence, the **tuberosity of the ulna,** which gives insertion to the brachialis muscle and the ligamentous oblique cord of the radius. Its **lateral surface**

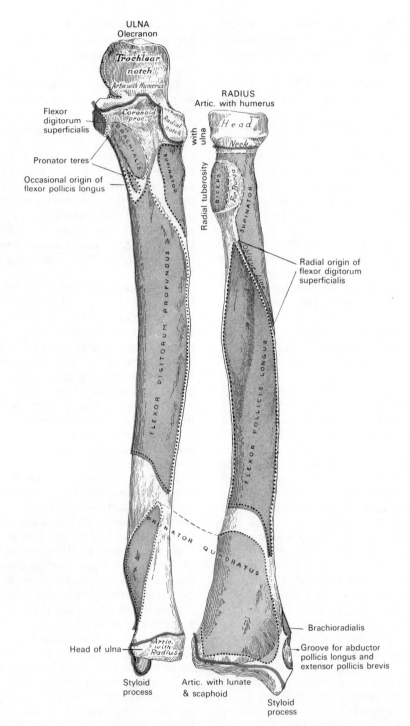

FIG. 4-148. The bones of the left forearm. Anterior aspect. Attachment of articular capsules outlined in blue.

presents a narrow, oblong, articular depression, the **radial notch.** Its **medial surface,** by its prominent, free margin, serves for the attachment of the anterior and oblique parts of the ulnar collateral ligament. At the front part of the medial surface is a small rounded

eminence for the origin of the humeroulnar head of the flexor digitorum superficialis. Behind the eminence descends a ridge from which arises the uppermost fibers of the flexor digitorum profundus, along with the ulnar head of the pronator teres. A small ulnar head of the flexor pollicis longus may arise from the distal part of the coronoid process by a rounded bundle of muscular fibers.

The **trochlear notch** (Figs. 4-147; 4-148) is a large depression which is semilunar in shape and formed by the coronoid process and the olecranon. It serves for articulation with the trochlea of the humerus. The borders of the notch are indented near its middle, thereby subdividing the articular surface into a proximal olecranon part and a distal coronoid part. Additionally, the notch is divided by a longitudinal ridge into a larger medial and a smaller lateral part.

The **radial notch** (Figs. 4-147; 4-148) is a narrow, oblong, articular depression on the lateral side of the coronoid process. It receives the circumferential articular surface of the head of the radius. The anterior and posterior margins of its concave surface afford attachment of the annular ligament. Radiographically, the proximal extremity of the ulna can be seen in Figures 4-151 to 4-153.

BODY OR SHAFT.   The **body** or **shaft** (Figs. 4-148; 4-154 to 4-156) is prismatic, or three-sided, in its upper two-thirds, but then becomes rounded or cylindrical distally. It is curved so as to be convex dorsally and laterally. Its central part is straight, its distal part a little concave lateralward. It tapers gradually and presents *anterior, posterior, and interosseous borders* as well as *anterior, posterior, and medial surfaces.*

The **anterior border** begins above at the prominent medial angle of the coronoid process and ends below in front of the styloid process. Its proximal part is well defined, and its middle portion is smooth and rounded, giving origin to the flexor digitorum profundus muscle. The distal one-fourth of the anterior surface is called the **pronator ridge,** which serves for the origin of the pronator quadratus. This border separates the anterior from the medial surface.

The **posterior border** begins above at the apex of the triangular subcutaneous surface of the olecranon, and ends below at the back of the styloid process. It is well marked in the proximal three-fourths, and gives attachment to an aponeurosis which affords a common origin to the flexor carpi ulnaris, the extensor carpi ulnaris, and the flexor digitorum profundus. Its distal one-fourth is smooth and rounded. This border separates the medial from the posterior surface.

The **interosseous border** (Figs. 4-147; 4-154; 4-155) is actually the lateral margin of the ulna. The union proximally of two lines that converge from the extremities of the radial notch form a sharp, prominent crest along the middle two-fourths of this border. In the upper part the converging lines enclose between them a triangular space for the origin of part of the supinator muscle. One of these lines descends from the medial margin of the radial notch and is called the **supinator crest** (Fig. 4-147). Thus, the proximal part of the interosseous border is sharp, while its distal one-fourth is smooth and rounded. The interosseous border gives attachment to the interosseous membrane and separates the anterior from the posterior surface.

The **anterior surface** (Figs. 4-148; 4-154) is concave in its upper three-fourths. This surface is much broader proximally where it gives origin to the flexor digitorum profundus. Its distal fourth, covered by the pronator quadratus, is separated from the remaining portion by an oblique ridge, which marks the extent of origin of the pronator quadratus (Fig. 4-148). At the junction of the proximal with the middle third of the bone is the nutrient canal, which is directed upward and contains a branch of the anterior interosseous artery.

The **posterior surface** (Figs. 4-155; 4-156) of the ulna is broad and concave above, convex and somewhat narrower in the middle, and narrow, smooth, and rounded distally. This surface is marked by an oblique ridge which commences proximally at the dorsal end of the radial notch and courses distally. The triangular surface proximal to this ridge receives the insertion of the anconeus, while the proximal part of the ridge affords attachment to the supinator. Distally, the surface is subdivided into two parts by another longitudinal ridge, sometimes called the **perpendicular line:** the *medial part* is smooth and covered by the extensor carpi ulnaris; the *lateral part,* wider and rougher, gives origin to the supinator, the abductor pollicis longus, the extensor pollicis longus, and the extensor indicis (Fig. 4-156).

The **medial surface** of the ulna is broad and concave proximally, but narrow and convex distally. Its proximal three-fourths give origin to the flexor digitorum profundus; its distal one-fourth is subcutaneous.

DISTAL EXTREMITY.   The **distal extremity** of the ulna (Figs. 4-148; 4-154 to 4-156) is smaller than the proximal extremity and it presents two eminences. Of the two, the lateral is larger, and it is a rounded, articular eminence, termed the **head of the ulna.** The medial eminence is the narrower, more projecting, nonarticular **styloid process.** The head of the ulna articulates with the proximal surface of a triangular articular disc which separates it from the wrist joint. The remaining portion of the ulnar head is received into the ulnar notch of the radius. The styloid process projects from the medial and posterior part of the bone. It descends a little lower than the head of the ulna, and its rounded end affords attachment to the ulnar collateral ligament of the wrist joint. The head is separated from the styloid process by a depression on which attaches the apex of the triangular articular disc. Posteriorly, these two eminences are separated by a shallow groove for the tendon of the extensor carpi ulnaris.

STRUCTURE. The long, narrow medullary cavity is enclosed in a strong wall of compact tissue, which is thickest along the interosseous border and dorsal surface. At the extremities the compact layer becomes thinner and is continued onto the back of the olecranon as a plate of close spongy bone with parallel lamellae. From the inner surface of this plate and the compact layer distal to it, trabeculae arch toward the olecranon and coronoid, and cross other trabeculae passing posteriorly over the medullary cavity from the proximal part of the shaft distal to the coronoid process. Distal to the coronoid process there is a small area of compact bone from which trabeculae curve proximally to end oblique to the surface of the trochlear notch, which is coated with a thin layer of compact bone. The trabeculae at the distal end have more longitudinal direction.

OSSIFICATION   (Figs. 4-149; 4-150). The ulna is ossified from three centers: one each for the body or shaft, the distal extremity, and the proximal portion of the olecranon. Ossification begins near the middle of the shaft about the eighth week of fetal life and soon extends through the greater part of the bone. At birth the distal extremity and the greater part of the olecranon are cartilaginous. Between the fifth and sixth years, a center appears in the middle of the ulnar head and soon extends into the styloid process. About the tenth year, a center appears in the olecranon near its extremity. The major portion of the olecranon that is being ossified develops from a proximal extension of the ossification center in the shaft. The proximal epiphysis joins the body about the sixteenth year, while the distal extremity unites with the shaft about the twentieth year.

ARTICULATIONS.   The ulna articulates with the humerus and radius. Although it is customary to exclude the ulna from the wrist joint because of the

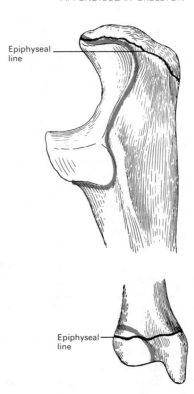

FIG. 4-149. Epiphyseal lines of ulna in a young adult. Lateral aspect. The lines of attachment of the articular capsules are in blue.

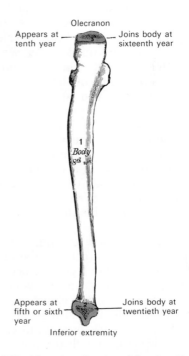

FIG. 4-150. Plan of ossification of the ulna. From three centers.

intervention of the articular disc, it should be appreciated that it contributes to this joint in much the same way as the clavicle does to the sternoclavicular and the mandible to the temporomandibular joints.

THE RADIUS

The **radius** (Figs. 4-148; 4-154 to 4-156) lies parallel with the ulna and is on the lateral, or thumb, side of the forearm. It is shorter than the ulna, and although both ends are expanded, the proximal end is smaller than the distal. The radius participates in forming the elbow joint, and it is the chief proximal bone in the formation of the wrist joint. It presents a body or shaft and two extremities.

PROXIMAL EXTREMITY. The **proximal extremity** presents a head, neck, and tuberosity. The **head** is cylindrical, and on its proximal surface is a shallow cup or **fovea** for articulation with the capitulum of the humerus. The **articular circumference** of the head is smooth. It is broad medially where it articulates with the radial notch of the ulna, narrow in the rest of its extent where it is embraced by the annular ligament. The head is supported on a round, smooth, and constricted portion called the **neck,** on the back of which is a slight ridge for the insertion of part of the supinator. Distal to the medial part of the neck is an oval eminence, the **radial tuberosity.** Its surface is divided into a posterior, rough portion, for the insertion of the tendon of the biceps brachii muscle, and an anterior, smooth portion, on which a bursa is interposed between the tendon and the bone. Radiographs of the proximal extremity of the radius can be seen in Figures 4-151 to 4-153.

BODY OR SHAFT. The **body** or **shaft** is round or cylindrical above where it joins the neck, but becomes triangular or prismatic in shape more distally. It also gradually increases in size from above downward. The radius is gently curved, so that it is convex laterally, and it presents three borders and three surfaces.

The **anterior border** extends from the lower part of the tuberosity to the anterior part of the base of the styloid process, and separates the anterior from the lateral surface. Its proximal third is elevated to form the **anterior oblique line of the radius** (Fig. 4-148), which gives origin to the flexor digitorum superficialis and flexor pollicis longus muscles. The surface proximal to the

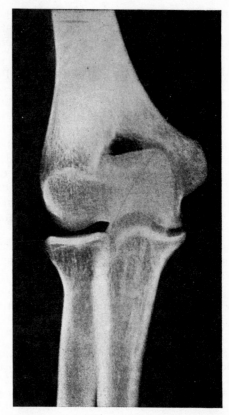

FIG. 4-151.   Adult elbow. Frontal view. The shadow of the olecranon extends upward to the olecranon fossa and obscures the outline of the trochlea. The gap between the humerus and the bones of the forearm is occupied by the articular cartilage of the bones concerned.

oblique line receives the insertion of part of the supinator muscle. The middle portion of the anterior border is rounded and less distinct, but the distal one-fourth receives the insertion of the pronator quadratus, and attachment of the dorsal carpal ligament. It ends in a small tubercle, into which the tendon of the brachioradialis muscle is inserted.

The **posterior border** begins proximally at the posterior aspect of the neck and ends distally at the posterior part of the base of the styloid process. It separates the posterior from the lateral surface. This border is rounded and indistinct near the extremities, but well marked in the middle third of the bone.

The **interosseus border** is rounded and indistinct where it begins above at the radial tuberosity. It becomes sharp and prominent as it descends, and at its distal part divides

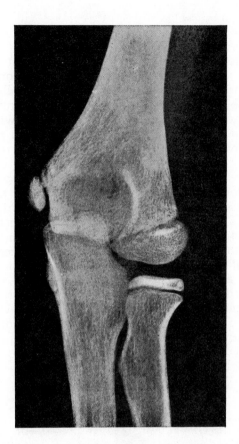

FIG. 4-152.   Elbow of a child aged 11 years. Frontal view. The upper epiphysis of the radius, the epiphysis for the medial epicondyle, and the center for the capitulum and the lateral part of the trochlea can be recognized.

into two ridges which are continued to the anterior and posterior margins of the ulnar notch. The triangular surface between the ridges receives the insertion of part of the pronator quadratus muscle. This border separates the anterior from the posterior surface of the bone, and gives attachment to the interosseous membrane, which interconnects the opposed borders of the radius and ulna.

The **anterior surface** (Figs. 4-148; 4-154) is concave in its proximal three-fourths, and gives origin to the flexor pollicis longus. It is broad and flat in its distal one-fourth, and receives the insertion of the pronator quadratus. A prominent ridge limits the insertion of the pronator quadratus distally, and between this and the inferior edge of the radius is a triangular rough surface for the attachment of the palmar radiocarpal ligament. At the level of the junction of the upper and middle third of the anterior surface is the nutrient foramen which transmits a branch of the anterior interosseous artery to the bone.

The **posterior surface** of the radius (Figs. 4-155; 4-156) is convex and smooth in the proximal third of its extent, and covered by the supinator. Its middle third is broad, slightly concave, and gives origin to the abductor pollicis longus and extensor pollicis brevis muscles. Its distal third is broad, convex, and covered by the tendons of the muscles that subsequently run in grooves along

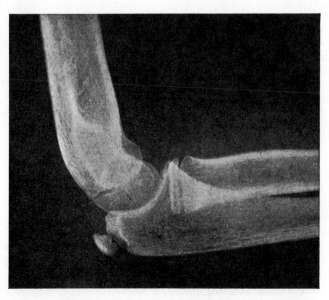

FIG. 4-153.   Elbow of a child aged 10 years. Lateral view. The upper epiphysis of the radius, the olecranon epiphysis, and the center for the capitulum and the lateral part of the trochlea can be observed.

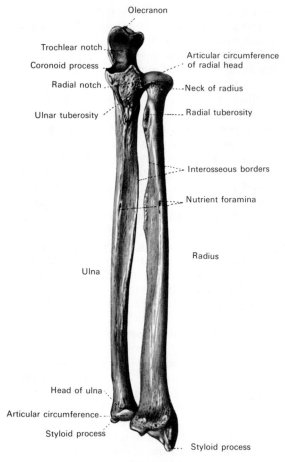

FIG. 4-154. Left forearm bones. Anterior aspect. (From Benninghoff and Goerttler, 1975.)

the lateral is somewhat triangular in shape and it articulates with the scaphoid bone, while the medial is more quadrangular in shape and it articulates with the lunate bone. The articular surface for the ulna is called the **ulnar notch of the radius.** It is narrow, smooth, and concave, and articulates with the head of the ulna to form the inferior radioulnar joint.

The two articular sites on the distal extremity of the radius are separated by a prominent ridge, to which is attached the base of the triangular articular disc which lies between the wrist joint and the distal radioulnar articulation. The distal extremity also presents three nonarticular surfaces: anterior, posterior, and lateral. The palmar or **anterior surface,** rough and irregular, affords attachment to the palmar radiocarpal ligament. The **posterior surface** is convex and affords attachment to the dorsal radiocarpal ligament. Near its middle, this surface

the dorsal aspect of the distal extremity. These are described below.

The **lateral surface** is convex throughout its entire extent. Its upper third receives the insertion of the supinator. About its center is a rough ridge for the insertion of the pronator teres muscle. Its lower part is narrow and covered by the tendons of the abductor pollicis longus and extensor pollicis brevis muscles.

DISTAL EXTREMITY. The **distal extremity** of the radius (Figs. 4-148; 4-154; 4-156) is quadrilateral in shape and is the broadest portion of the bone. It is provided with two articular surfaces—one on the distal end of the bone articulates with the carpal bones at the wrist joint; the other is directed medially for articulation with the ulna. The **carpal articular surface** (Fig. 4-157) is concave and smooth, and it is divided by a slight anteroposterior ridge into lateral and medial parts. Of these,

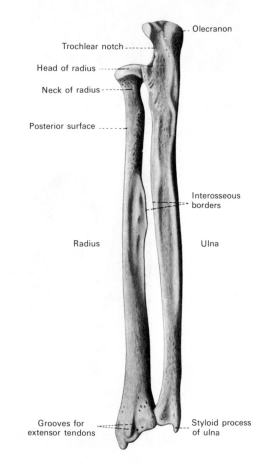

FIG. 4-155. Left forearm bones. Posterior aspect. (From Benninghoff and Goerttler, 1975.)

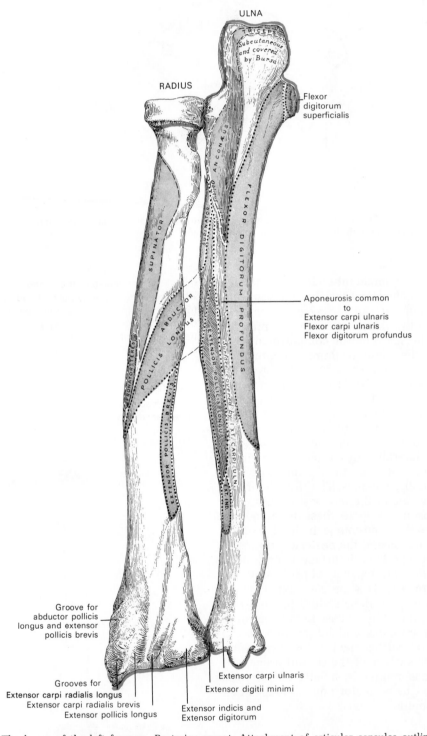

ULNA

RADIUS

Flexor
digitorum
superficialis

Aponeurosis common
to
Extensor carpi ulnaris
Flexor carpi ulnaris
Flexor digitorum profundus

Groove for
abductor pollicis
longus and extensor
pollicis brevis

Grooves for
Extensor carpi radialis longus
Extensor carpi radialis brevis
Extensor pollicis longus

Extensor indicis and
Extensor digitorum

Extensor carpi ulnaris

Extensor digitii minimi

Fɪɢ. 4-156.   The bones of the left forearm. Posterior aspect. Attachment of articular capsules outlined in blue.

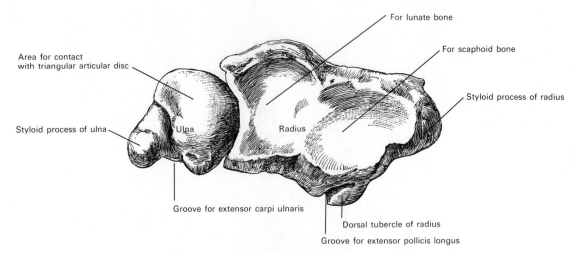

FIG. 4-157. The distal ends of the left radius and ulna showing the carpal articular surface of the radius and the area on the ulna for contact with the articular disc. Inferior aspect.

is marked by a **dorsal tubercle** (Fig. 4-157), to which is attached a part of this extensor retinaculum. Further, this surface contains three grooves within which lie tendons of muscles of the extensor forearm compartment. Beginning at the lateral or thumb side, the first groove is broad but shallow, and subdivided into two parts by a slight ridge; the lateral part transmits the tendon of the extensor carpi radialis longus, the medial transmits the tendon of the extensor carpi radialis brevis. The second groove is deep but narrow and bounded laterally by a sharply defined oblique ridge; it transmits the tendon of the extensor pollicis longus. The third is broad, for the passage of the tendons of the extensor digitorum. Deep to these tendons and adjacent to the bone course the tendon of the extensor indicis and the posterior interosseous nerve. The **lateral surface** is prolonged distally into a strong, conical projection, the **styloid process.** This process extends distally beyond the ulnar styloid process and gives attachment by its base to the tendon of the brachioradialis, and by its apex to the radial collateral ligament of the wrist joint. The lateral surface of the styloid process of the radius is marked by a flat groove for the tendons of the abductor pollicis longus and extensor pollicis brevis.

STRUCTURE. The long narrow medullary cavity of the radius is enclosed in a strong wall of compact tissue, which is thickest along the interosseous border and thinnest at the extremities except over the cup-shaped articular surface of the head, where it is

thick. The trabeculae of the spongy tissue are somewhat arched at the proximal end and pass proximally from the compact layer of the shaft to the fovea of the head of the radius. They are crossed by other trabeculae which are oriented parallel to the surface of the fovea. The arrangement at the distal end is somewhat similar.

OSSIFICATION (Figs. 4-158; 4-159). The radius is ossified from three centers: one for the body and one for each extremity. That for the shaft makes its appearance near the center of the bone during the

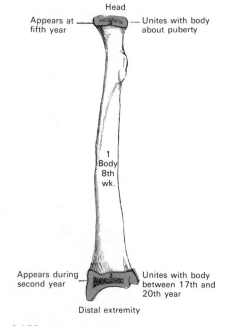

FIG. 4-158. Plan of ossification of the radius. From three centers.

those of the distal row, in the same order, are named the **trapezium, trapezoid, capitate,** and **hamate.**

SURFACES OF THE CARPAL BONES.   With the exception of the pisiform bone which lies on the palmar surface of the triquetral, each carpal bone has six surfaces: palmar and dorsal, proximal and distal, medial and lateral. The **palmar** (volar) and **dorsal surfaces** are roughened for ligamentous attachment, the dorsal surfaces generally being broader than the palmar, except for the scaphoid and lunate. The **proximal** and **distal surfaces** are articular, the proximal generally being convex and the distal concave. The **medial** (ulnar) and **lateral** (radial) **surfaces** are also

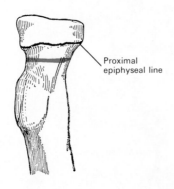

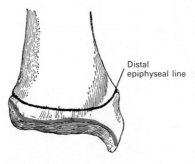

FIG. 4-159.   Epiphyseal lines of radius in a young adult. Anterior aspect. The line of attachment of the articular capsules of the wrist joint is in blue.

eighth week of fetal life. During the second year, ossification commences in the distal end, and at about the fifth year, in the proximal end. The proximal epiphysis fuses with the body at the age of fifteen to eighteen years, while the distal epiphysis fuses between the seventeenth and twentieth year. An additional center sometimes found in the radial tuberosity appears about the fourteenth or fifteenth year.

ARTICULATIONS.   The radius articulates with the humerus, ulna, scaphoid, and lunate bones.

The Hand

The skeleton of the hand (Figs. 4-160; 4-161) is subdivided into three segments: the **carpal** or wrist bones, the **metacarpal** bones of the palm, and the **phalanges** or bones of the digits.

CARPAL BONES

The **carpal bones** are eight in number and they are arranged in two rows of four bones each. Those of the proximal row, from the radial to the ulnar side, are named the **scaphoid, lunate, triquetral,** and **pisiform;**

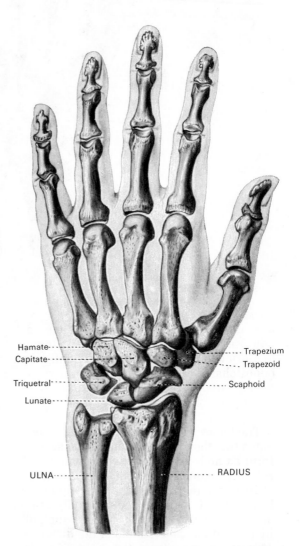

FIG. 4-160.   Outline of the left hand, showing the skeleton from the dorsal aspect. (From Benninghoff and Goerttler, 1975.)

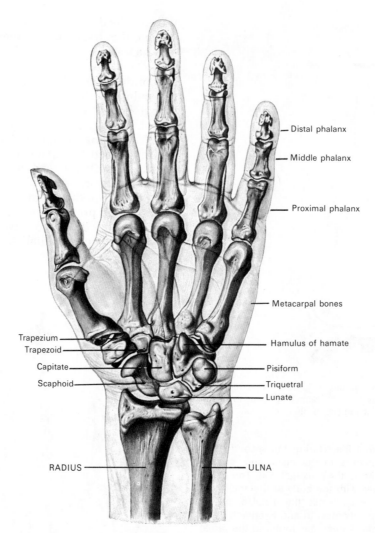

Distal phalanx

Middle phalanx

Proximal phalanx

Metacarpal bones

Trapezium
Trapezoid
Capitate
Scaphoid

Hamulus of hamate

Pisiform

Triquetral

Lunate

RADIUS

ULNA

FIG. 4-161. Outline of the left hand, showing the skeleton from the palmar aspect. (From Benninghoff and Goerttler, 1975.)

articular and the means by which the carpal bones articulate with each other, allowing the formation of the two rows of contiguous bones characteristic of the human carpus.

BONES OF THE PROXIMAL ROW

The **bones of the proximal row** are so oriented that their proximal and distal surfaces form an arch, the convexity of which is directed proximally for articulation with the radius and the articular disc capping the lower end of the ulna. The concavity of this arch, directed distally, is filled principally by the capitate and hamate of the distal row of carpal bones (Figs. 4-160; 4-161). The dorsal surface of the bones in the proximal row

is slightly convex and uneven, and it is covered by ligaments and tendons. The volar surface is concave, contributing to the **carpal groove** which is filled by structures entering the palm of the hand from the forearm. The scaphoid, lunate, triquetral, and pisiform bones will be described in order.

SCAPHOID BONE. The **scaphoid bone** (Fig. 4-162) is the largest bone of the proximal row. Its long axis extends obliquely distally and laterally and it is situated at the radial side of the carpus. The **proximal surface** is convex, smooth, and triangular, and articulates with the radius. The **distal surface** is also smooth, convex, and triangular, and is divided by a slight ridge into two parts, the lateral articulating with the trape-

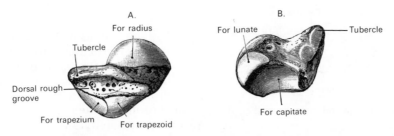

FIG. 4-162.   The left scaphoid bone. *A*, From the dorsal aspect. *B*, From the palmar aspect.

zium, the medial with the trapezoid. On the **dorsal surface** is a narrow, rough groove, which courses the entire length of the bone and serves for the attachment of ligaments. The **palmar surface** is concave and elevated at its lateral part into a rounded projection, the **tubercle,** to which is attached the flexor retinaculum and sometimes the origin of a few fibers of the abductor pollicis brevis. The tubercle can be felt through the skin on the anterior aspect of the extended wrist at the base of the thenar eminence and in line with the radial aspect of the middle finger. The **lateral surface** is rough and narrow, and gives attachment to the radial collateral ligament of the wrist. The **medial surface** presents two articular facets. Of these, the proximal is smaller, flattened, and of semilunar form, and articulates with the lunate bone, while the distal is concave, forming with the lunate a concavity for the head of the capitate bone.

Articulations.   The scaphoid articulates with five bones: the radius proximally, trapezium and trapezoid distally, and capitate and lunate medially.

Fracture of the Scaphoid.   The scaphoid is the most commonly fractured bone in the wrist. A fall upon the radially deviated hand may result in a fracture across the middle of the bone at its narrowest site.

LUNATE BONE.   The **lunate bone** (Fig. 4-163) may be distinguished by its deep concavity and crescentic outline. It is situated in the center of the proximal row of carpal bones, between the scaphoid and triquetral. The **proximal surface,** convex and smooth, articulates with the radius. The **distal surface** is deeply concave and articulates with the head of the capitate. This surface also articulates with the edge of the hamate bone by a long, narrow facet which is separated from the capitate articular surface by a curved

ridge. The **dorsal** and **palmar surfaces** are both roughened for the attachment of ligaments. The **lateral surface** presents a narrow, flattened, semilunar facet for articulation with the scaphoid. The **medial surface** is marked by a smooth, quadrilateral facet for articulation with the triquetral bone.

Articulations.   The lunate articulates with five bones: the radius proximally, capitate and hamate distally, scaphoid laterally, and triquetral medially.

Dislocation of the Lunate.   The lunate bone is the carpal bone that is most frequently dislocated. It usually results from a fall upon the hyperextended hand in which all the other bones in the wrist are carried backward by the force of the fall except the lunate and part of the scaphoid.

TRIQUETRAL BONE.   The **triquetral bone** (Fig. 4-164) may be distinguished by its pyramidal shape, and by an oval isolated facet for articulation with the pisiform bone. It is situated at the proximal and ulnar side of the wrist. The **proximal surface** presents a medial, rough, nonarticular portion, and a lateral convex articular portion which articulates with the triangular articular disc of the distal radioulnar joint of the wrist. The **distal surface,** directed lateralward, is concave, sinuously curved, and smooth for articulation with the hamate. The **dorsal surface** is rough for the attachment of ligaments. The

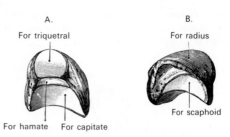

FIG. 4-163.   The left lunate bone. *A*, From the inferomedial aspect. *B*, From the superolateral aspect.

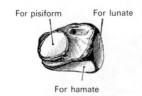

For pisiform   For lunate

For hamate

FIG. 4-164. The left triquetral bone. From the palmar aspect.

palmar surface presents, on its medial part, an oval facet for articulation with the pisiform, while its lateral part is rough for ligamentous attachment. The **lateral surface,** which forms the base of the pyramid, is marked by a flat, quadrilateral facet for articulation with the lunate. The roughened **medial surface** is the pointed summit of the pyramid, and to it attaches the ulnar collateral ligament of the wrist.

Articulations. The triquetral articulates with three bones: the lunate laterally, the pisiform anteriorly, and the hamate distally, and with the triangular articular disc, which separates it from the lower end of the ulna proximally.

PISIFORM BONE. The **pisiform bone** (Fig. 4-165) is the smallest of the carpal bones and can be identified by its spheroidal shape, which is flat on the one surface that serves as its only articular facet. It is situated on a plane anterior to the other carpal bones, at the base of the hypothenar eminence on the medial side of the wrist. Its **dorsal surface** presents a smooth, **oval facet** for articulation with the triquetral bone. This facet approaches the proximal but not the distal border of the bone. The **palmar surface** is rounded and rough, and gives attachment to the flexor retinaculum and to the flexor carpi ulnaris and abductor digiti minimi muscles. The **lateral** and **medial surfaces** are also rough, and the lateral surface usually presents a shallow groove which lies in relation to the ulnar artery.

Articulation. The pisiform articulates with one bone, the triquetral. It does not participate in the formation of the radiocarpal joint.

For triquetral

FIG. 4-165. The left pisiform bone. From the dorsal aspect.

BONES OF THE DISTAL ROW

The four **bones of the distal row** (Figs. 4-160; 4-161) articulate proximally with the scaphoid, lunate, and triquetral, distally with the five metacarpal bones and with each other. Their arrangement is such that they form a gentle convex arch dorsally, but a deep concavity, the carpal groove, on the palmar surface. They are straighter in their alignment across the wrist than the proximal row, especially at their distal articulations with the metacarpal bones. From radial to ulnar sides, they include the trapezium, trapezoid, capitate, and hamate.

THE TRAPEZIUM. The **trapezium** (Fig. 4-166) is quite irregular in its shape, but may be distinguished by a deep groove and a distinct ridge or tubercle on its palmar surface. The bone is situated at the radial side of the wrist between the scaphoid and first metacarpal. The **proximal surface** is smooth and contains a shallow facet which articulates with the scaphoid bone. The **distal surface** is oval and saddle-shaped, and articulates with the base of the first metacarpal bone. The **dorsal surface** is rough and rests in relationship to the overlying radial artery. The **palmar surface** is narrow and rough. At its proximal part is a deep groove that transmits the tendon of the flexor carpi radialis. It is bounded laterally by an oblique, ridge-like **tubercle.** This surface gives origin to the opponens pollicis, abductor pollicis brevis, and flexor pollicis brevis muscles, and affords attachment to the flexor retinaculum. The **lateral surface** is broad and rough for the attachment of ligaments. The **medial surface** presents two facets: a large, concave, proximal one, which articulates with the trapezoid, and a small, oval, distal one, which articulates with the base of the second metacarpal.

Articulations. The trapezium articulates with four bones: the scaphoid proximally, the first metacarpal distally, and the trapezoid and second metacarpal medially.

TRAPEZOID BONE. The **trapezoid bone** (Fig. 4-167) is the smallest bone in the distal row. It may be identified by its wedge shape, the broad end of the wedge constituting the dorsal surface, and the narrow end forming the palmar surface of the bone. Additionally, the trapezoid has four articular facets touching each other, and these are separated by

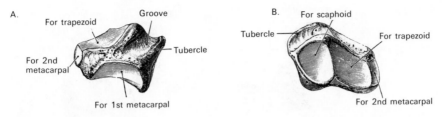

FIG. 4-166.   The left trapezium. *A,* From the palmar aspect. *B,* From the superomedial aspect.

sharp edges. The **proximal surface** articulates with the scaphoid. The **distal surface** is subdivided by an elevated ridge into two unequal facets, which articulate with the base or proximal end of the second metacarpal bone. The large **dorsal** and smaller **palmar surfaces** are rough for the attachment of ligaments. The **lateral surface,** convex and smooth, articulates with the trapezium. The **medial surface** is concave and smooth anteriorly for articulation with the capitate bone, but rough dorsally for the attachment of an interosseous ligament.

Articulations.   The trapezoid articulates with four bones: the scaphoid proximally, second metacarpal distally, trapezium laterally, and capitate medially.

CAPITATE BONE.   The **capitate bone** (Fig. 4-168) is the largest of the carpal bones and occupies the center of the wrist. It presents a rounded portion or **head,** which is received into the concavity formed by the scaphoid and lunate bones, a constricted portion or **neck,** and a **body.** The **proximal surface** is round and smooth and articulates with the lunate. The **distal surface** is divided by two ridges into three facets, which articulate with the bases of the second, third, and fourth metacarpal bones, that for the third being the largest. The **dorsal surface** is broad and rough. The **palmar surface** is narrow, rounded, and rough for the attachment of ligaments and a part of the oblique head of

the adductor pollicis muscle. The **lateral surface** articulates with the trapezoid by a small facet at its anterior inferior angle, behind which is a rough depression for the attachment of an interosseous ligament. Proximal to this is a deep, rough groove, forming part of the neck, and serving for the attachment of ligaments. Proximally, the lateral surface is bounded by a smooth, convex surface for articulation with the scaphoid. The **medial surface** articulates with the hamate by a smooth, concave, oblong facet, which occupies its proximal and more dorsal part. The palmar portion of the medial surface is rough for the attachment of an interosseous ligament.

Articulations.   The capitate articulates with seven bones: the scaphoid and lunate proximally; the second, third, and fourth metacarpals distally; the trapezoid on the radial side, and the hamate on the ulnar side.

Clinical Note.   The large size and central location of the capitate bone places its head in a position to transmit the force of a fall on the outstretched hand to the radius through the scaphoid and lunate bones. Thus, it is through the capitate bone that the most common injuries to the carpal bones occur, i.e., fractures of the scaphoid and dislocations of the lunate.

HAMATE BONE.   The **hamate bone** (Fig. 4-169) is wedge-shaped and may readily be distinguished by the hook-like process that projects from its palmar surface. This bone is situated most medially along the distal row of the carpus, and the base of the wedge is directed distally and rests on the fourth and fifth metacarpal bones. The **proximal surface** is narrow, convex, and smooth, and is the apex of the wedge, articulating with the lunate. The **distal surface** articulates with the fourth and fifth metacarpal bones by concave facets that are separated by a ridge. The **dorsal surface** is triangular and rough for ligamentous attachment. The **pal-**

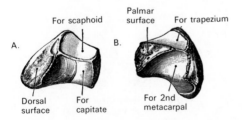

FIG. 4-167.   The left trapezoid bone. *A,* From the medial aspect. *B,* From the inferolateral aspect.

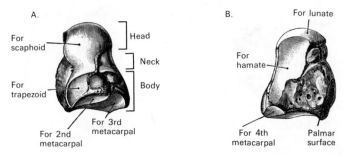

FIG. 4-168.　The left capitate bone. *A*, From the lateral aspect. *B*, From the medial aspect.

mar surface presents a curved hook-like process, the **hamulus,** at its distal and ulnar side. The hamulus can be felt through the skin at the base of the hypothenar eminence in line with the ulnar border of the ring finger. To the tip of the hamulus is attached the flexor retinaculum and the flexor carpi ulnaris, and to its medial surface, the flexor digitorum brevis and opponens digiti minimi. Its lateral side is grooved for the passage of the flexor tendons into the palm of the hand. It is one of the four eminences on the front of the carpus to which the flexor retinaculum of the wrist is attached; the others being the pisiform medially, the oblique ridge of the trapezium, and the tubercle of the scaphoid laterally. The **ulnar surface** of the hamate articulates with the triquetral bone by an oblong facet. The **radial surface** articulates with the capitate by its proximal and more posterior part, the remaining portion being rough for the attachment of ligaments.

Articulations. The hamate articulates with five bones: the lunate proximally, the fourth and fifth metacarpals distally, the triquetral medially, and the capitate laterally.

METACARPAL BONES

COMMON CHARACTERISTICS. The **metacarpus** (Figs. 4-160; 4-161; 4-170 to 4-176) consists of five slender bones which are numbered 1 to 5 from the radial or thumb side to the ulnar or little finger side. Consisting of a body or shaft and two extremities, the metacarpal bones might be described as long bones in miniature. Each extends from the carpal bones to the proximal phalanx of the respective finger.

They can be palpated throughout their entire extent on the dorsum of the hand, and they terminate distally at the proximal row of knuckles. On the palmar side, they are difficult to palpate proximally since they course beneath the flexor tendons, but at a site 2 cm proximal to the roots of the fingers, they can easily be felt on the palmar surface. This site on the palm of the hand lies opposite the knuckle on the dorsum of the hand. Each metacarpal consists of a **body,** a **base,** and a **head.**

The body is curved longitudinally so as to be slightly convex on the dorsal surface, the concavity on the palmar side containing muscles. The metacarpal body presents

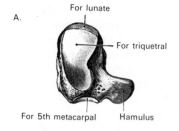

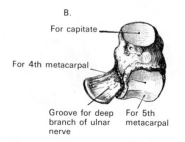

FIG. 4-169.　The left hamate bone. *A*, From the medial aspect. *B*, From the inferolateral aspect.

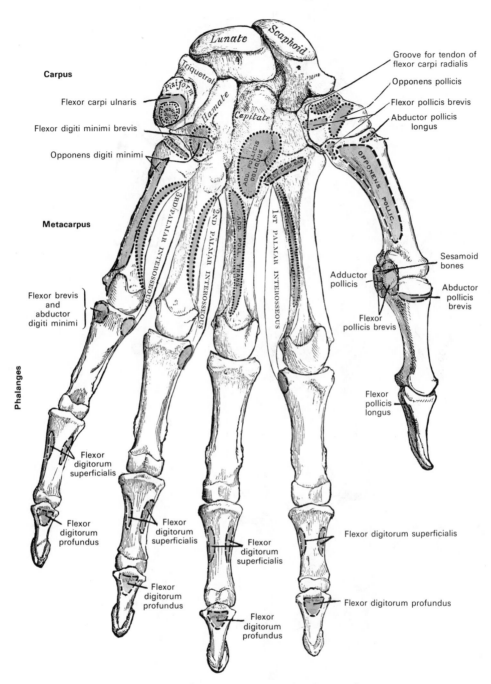

Carpus

Flexor carpi ulnaris

Flexor digiti minimi brevis

Opponens digiti minimi

Metacarpus

Flexor brevis and abductor digiti minimi

Phalanges

Flexor digitorum superficialis

Flexor digitorum profundus

Flexor digitorum superficialis

Flexor digitorum profundus

Flexor digitorum superficialis

Flexor digitorum profundus

Lunate

Scaphoid

Triquetral

Pisiform

Hamate

Capitate

Groove for tendon of flexor carpi radialis

Opponens pollicis

Flexor pollicis brevis

Abductor pollicis longus

3RD PALMAR INTEROSSEOUS

2ND PALMAR INTEROSSEOUS

1ST PALMAR INTEROSSEOUS

ADD. POLLICIS OBLIQUUS

ADD. POLL. TRANS.

FLEX. CARP. RAD.

OPPONENS POLLIC.

Sesamoid bones

Abductor pollicis brevis

Adductor pollicis

Flexor pollicis brevis

Flexor pollicis longus

Flexor digitorum superficialis

Flexor digitorum profundus

FIG. 4-170.   Bones of the left hand. Palmar surface.

three surfaces: medial, lateral, and dorsal. The **medial** and **lateral surfaces** are concave for the attachment of the interosseous muscles, and these surfaces are separated from one another by a prominent palmar ridge. The **dorsal surface** presents in its distal two-thirds a smooth, triangular, flattened area, which is covered in the intact body by the tendons of the extensor muscles. This surface is bounded by two lines which commence as small tubercles on each side of the distal extremity. These lines converge some distance proximal to the center of the bone to form a ridge that runs along the rest of the dorsal surface to the proximal extremity. This ridge separates two sloping surfaces for

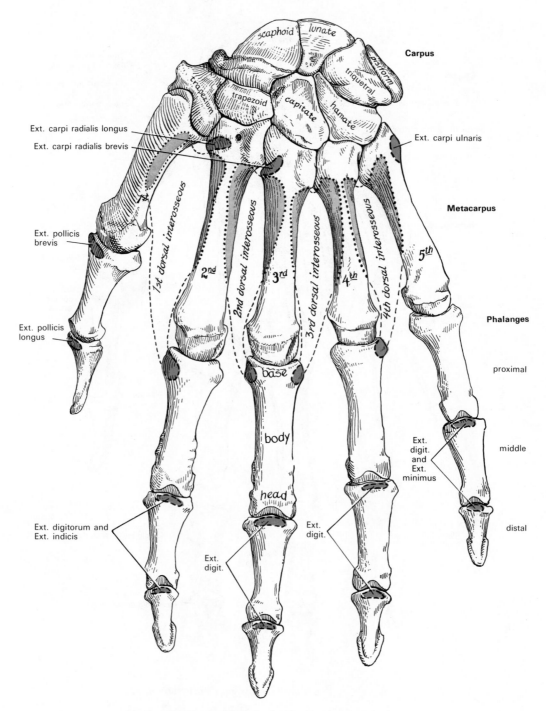

FIG. 4-171. Bones of the left hand. Dorsal surface.

the attachment of the dorsal interosseous muscles. To the tubercles on the distal extremities are attached the collateral ligaments of the metacarpophalangeal joints.

The **base** or **proximal extremity** is the expanded carpal end of the bone and is cuboidal in shape but broader dorsally. It articulates proximally with the carpus, and on each side with the adjacent metacarpal bones. An exception is the base of the first metacarpal, which does not articulate with the second. The dorsal and palmar surfaces

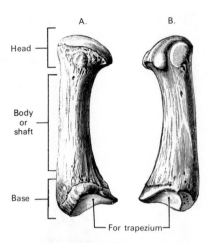

FIG. 4-172. The left first metacarpal. *A*, Lateral or radial aspect. *B*, Medial or ulnar aspect.

of the metacarpal bases are rough for the attachment of ligaments.

The **head** or **distal extremity** articulates with the proximal phalanx on each finger. It presents a smooth, oblong, convex surface flattened from side to side, which extends farther proximally on the palmar than on the dorsal aspect. On each side of the head is a **tubercle,** and between the tubercles on the palmar side, a hollow fossa for the attachment of the collateral ligament of the metacarpophalangeal joint. The dorsal surface is broad and flat and supports the extensor tendons. The palmar surface is grooved in the midline for the passage of the flexor tendons and it is marked on each side by an articular eminence continuous with the terminal articular surface.

FIRST METACARPAL BONE. The **first metacarpal bone** (to the thumb) is shorter and stouter, and diverges to a greater degree from the carpus than the other metacarpals (Fig. 4-172). It is rotated about its long axis so that its flexor surface is directed medially toward the radial side of the second metacarpal. The **body** is flattened and broad, and its dorsal surface is slightly convex. Its palmar surface is concave and divided by a blunt ridge into a larger lateral part, which gives rise to the opponens pollicis muscle (Fig. 4-170), and a smaller medial part, which gives origin to the lateral head of the first dorsal interosseous muscle (Fig. 4-171). The **base** or proximal extremity presents a saddle-shaped surface for articulation with the trapezium (Fig. 4-172). It has no facets on its sides, but on its lateral side is a small tuber-

cle for the insertion of the abductor pollicis longus. The **head** is rounded and less convex than those of the other metacarpal bones, and is broader from side to side, than from palmar to dorsal surface. On its palmar surface of the head are two articular eminences, of which the lateral is larger, for the two sesamoid bones in the tendons of the two heads of the flexor pollicis brevis.

SECOND METACARPAL BONE. The **second metacarpal bone** (Fig. 4-173) is the longest, and its **base** the largest of all the metacarpal bones. The medial aspect of its base is prolonged proximally, thereby forming a prominent groove. On the base are four articular facets, three on the proximal surface and one, for articulation with the third metacarpal bone, on the ulnar side. Of the facets on the proximal surface the intermediate is the largest and is for articulation with the trapezoid bone. The facet on the lateral aspect of the base is flat and oval for articulation with the trapezium, while the one on the medial aspect is long and narrow for articulation with the capitate.

The extensor carpi radialis longus is inserted on the dorsal surface and the flexor carpi radialis on the palmar surface of the base (Figs. 4-171; 4-172). Along the **body** of the second metacarpal bone originate three

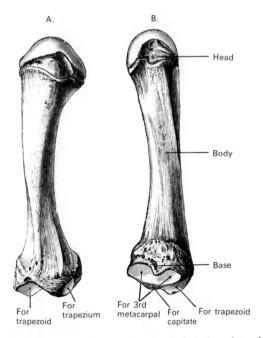

FIG. 4-173. The left second metacarpal. *A*, Dorsolateral aspect. *B*, Medial aspect.

interosseous muscles, two dorsal and one palmar. The convex, rounded **head** articulates with the proximal phalanx of the index finger.

THIRD METACARPAL BONE.   The **third metacarpal bone** (Fig. 4-174) is a little smaller than the second. The dorsal aspect of its **base** presents on its radial side a pyramidal eminence, the **styloid process,** which extends proximally behind the capitate bone. Immediately distal to this is a rough surface on the dorsal aspect of the base for the insertion of the extensor carpi radialis brevis muscle. The base articulates by means of a concave facet with a single carpal bone, the capitate. On the radial side is a smooth, concave facet for articulation with the second metacarpal, and on the ulnar side two small oval facets for the fourth metacarpal. The dorsal surface of the **body** gives origin to the second and third dorsal interosseous muscles, while along the palmar surface arises the transverse head of the adductor pollicis muscle. The **head** of the bone articulates with the proximal phalanx of the middle finger.

FOURTH METACARPAL BONE.   The **fourth metacarpal bone** (Fig. 4-175) is shorter and more slender than the third. The **base** is

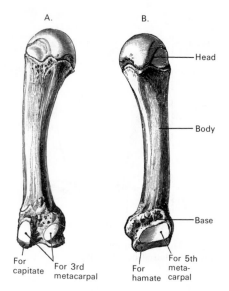

FIG. 4-175.   The left fourth metacarpal. A, Lateral aspect. B, Medial aspect.

small and quadrilateral, and its proximal surface presents two facets, the larger of which is oriented medially for articulation with the hamate, while the smaller one is lateral for the capitate. On the lateral aspect of the head are two oval facets for articulation with the third metacarpal, and on the medial side is found a single concave facet for the fifth metacarpal. The thin **body** gives attachment to two dorsal interosseous muscles and one palmar interosseous. The **head** articulates with the proximal phalanx of the ring finger.

FIFTH METACARPAL BONE.   The **fifth metacarpal bone** (Fig. 4-176) presents on its **base** a concave facet on its proximal surface that articulates with the hamate, and another on its lateral side that articulates with the fourth metacarpal. Unlike all the other metacarpal bones, the medial or ulnar side of the base is nonarticular, but does present a prominent **tubercle** for the insertion of the tendon of the extensor carpi ulnaris. The dorsal surface of the **body** is divided by an oblique ridge, lateral to which arises the fourth dorsal interosseous muscle (Fig. 4-171), and medial to which course the extensor tendons of the little finger. The palmar surface of the body gives origin laterally to the third palmar interosseous muscle and insertion medially to the opponens digiti minimi muscle (Fig. 4-170).

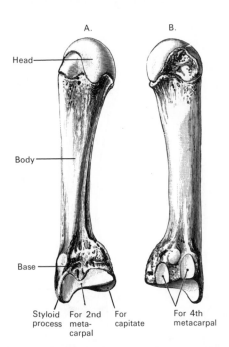

FIG. 4-174.   The left third metacarpal. A, Lateral aspect. B, Medial aspect.

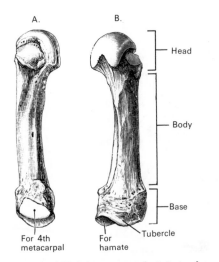

A.   B.

— Head

— Body

— Base

Tubercle

For 4th metacarpal   For hamate

FIG. 4-176. The fifth left metacarpal. *A*, Lateral aspect. *B*, Medial aspect.

ARTICULATIONS. Besides their phalangeal articulations, the metacarpal bones articulate as follows: the **first** with the trapezium; the **second** with the trapezium, trapezoid, capitate, and third metacarpal; the **third** with the capitate, and the second and fourth metacarpals; the **fourth** with the capitate, hamate, and third and fifth metacarpals; and the **fifth** with the hamate and fourth metacarpal.

PHALANGES OF THE HAND

The **phalanges** of the hand (Figs. 4-160; 4-161; 4-170; 4-171) are fourteen in number, three for each finger, and two for the thumb. The phalanges of the four fingers contain three bones called **proximal, middle,** and **distal.** The two of the thumb are called proximal and distal. Each phalanx consists of a **body** and two extremities. The proximal extremity is called the **base** and the distal extremity, the **head,** in a manner similar to the metacarpal bones. The **body** tapers distally and is convex dorsally, and its sides are marked by rough ridges which give attachment to the fibrous sheaths of the flexor tendons.

The **bases** of the bones of the **proximal row of phalanges** present oval, concave articular surfaces, which are broader from side to side and articulate with the rounded, smooth heads of the metacarpal bones. The **heads** of the proximal row of phalanges are smaller than the bases, and each ends in two condyles separated by a shallow groove. The articular surface extends farther on the palmar than on the dorsal surface.

The bases of each of the bones in the **middle row of phalanges** present two small concave facets separated by an elevated median ridge, allowing them to articulate snugly and conform to the grooved heads of the proximal phalanx. The heads of the phalanges of the middle row are pulley-shaped and end, as do those in the proximal row, in two condyles between which is a groove (Fig. 4-180).

The **distal row of phalanges** are convex on their dorsal and flat on their palmar surfaces, and they can be recognized by their small size. Their bases are identical in shape to those in the middle row, but their distal extremities present a roughened, horseshoe-shaped elevation on the palmar surface, called the **ungual tuberosity,** which supports the fingernail and the sensitive pulp of the fingertips.

MUSCLE INSERTIONS. The edges of the palmar surface of the middle phalangeal bases in the four ulnar fingers receive the insertion of the *flexor digitorum superficialis* tendons. Onto the palmar surface of the distal phalangeal bases are inserted the tendons of the *flexor digitorum profundus.* The *interosseous muscles* insert onto the bases of the proximal phalanges. Onto the medial aspect of the phalangeal base of the little finger insert the *flexor digiti minimi brevis* and the *abductor digiti minimi.* The base of the proximal phalanx of the thumb receives the insertions of the *flexor* and *abductor pollicis brevis* as well as the *adductor pollicis.* The *extensor digitorum* tendons insert onto the dorsal aspect of the bases of the middle and distal phalanges. Onto the dorsum of the base of the middle phalanx of the index finger is inserted the *extensor indicis* tendon. The dorsum of the proximal phalangeal base of the thumb receives the insertion of the *extensor pollicis brevis,* while onto the distal phalanx is inserted the *extensor pollicis longus* (Figs. 4-170; 4-171).

ARTICULATIONS. In the four fingers the proximal phalanges articulate with the middle phalanges and with the metacarpals. The middle phalanges articulate with both the proximal and distal phalanges, while the distal phalanges articulate only with the middle. In the thumb, which has only two phalanges, the first phalanx articulates by its base with the metacarpal bone and by its head with the ungual phalanx.

OSSIFICATION OF THE BONES OF THE HAND. The **carpal bones** (Figs. 4-178; 4-179) are each ossified from a single center, and although individual variation is observed, ossification generally proceeds in the following order (Fig. 4-177): in the *capitate* and *hamate,* during the first year after birth, the former preceding the latter; in the *triquetral,* during the

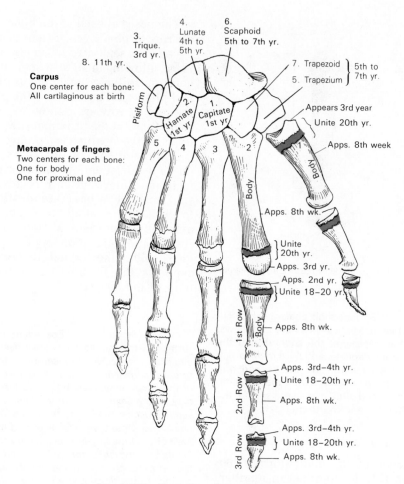

FIG. 4-177. Plan of ossification of the hand.

third year; in the *lunate,* during the fourth year in females and the fifth year in males; in the *trapezium, trapezoid,* and *scaphoid,* during the fifth year in females and during the sixth and seventh years in males; and in the *pisiform* during the ninth year in females and the eleventh year in males. Ossification is usually bilaterally symmetrical.

The **metacarpal bones** are each ossified from two centers: one for the body and one for the distal extremity (the heads) of each of the second, third, fourth, and fifth bones. One of the two ossification centers of the first metacarpal bone occurs in the body, but in contrast to the other metacarpals, the second ossification center occurs at the base. The first metacarpal bone is therefore ossified in the same manner as the phalanges, and this has led some anatomists to regard the thumb as being made up of three phalanges, and not of a metacarpal bone and two phalanges. Ossification commences in the middle of the body about the eighth or ninth week of fetal life. The centers for the second and third metacarpals are the first to appear, and that for the first metacarpal, the last. At about the third year the distal extremities (heads) of the metacarpals of the fin-

gers and the base of the metacarpal of the thumb begin to ossify. They unite with the bodies between the eighteenth and twentieth year.

The **phalanges** are each ossified from two centers: one for the body and one for the base. Ossification begins in the body of the phalanges at about the eighth week of fetal life. Ossification of the base commences in the proximal phalanges during the second year, and a year later in middle and distal phalanges. The two centers become united in each row before the twentieth year.

In the distal phalanges the centers for the bodies appear at the heads (distal extremity) of the phalanges, instead of at the middle of the bodies, as in other phalanges. Moreover, of all the bones of the hand, the distal phalanges are the first to ossify.

VARIATIONS. Occasionally an additional bone, the *os centrale,* is found on the dorsum of the wrist, lying between the scaphoid, trapezoid, and capitate. During the second month of fetal life it is represented by a small cartilaginous nodule, which usually fuses with the cartilaginous scaphoid. Sometimes the **styloid process** of the third metacarpal is detached and forms an additional bone.

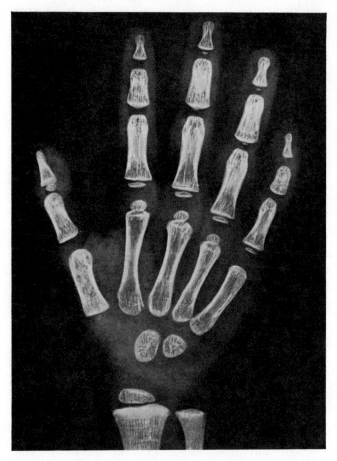

FIG. 4-178. Hand and wrist of a child aged 2½ years. The capitate and hamate bones are in process of ossification, but the other carpal bones are still cartilaginous. The center for the head of the ulna has not yet appeared, but the center for the lower epiphysis of the radius is present. Note the condition of the metacarpal bones and phalanges.

BONES OF THE LOWER LIMB

HIP BONE

The **hip bone,** also called the innominate bone (Figs. 4-181 to 4-184), is large, flat, and irregularly shaped. Its middle is constricted, and it is expanded both above and below. Articulating with the opposite hip bone anteriorly in the midline, and with the sacrum posteriorly, it forms the greater part of the bony walls of the pelvic cavity. Also, the two hip bones together form the **pelvic girdle** from which extends the skeleton of the lower extremities. Thus, in position, it is interposed between the vertebral column and the femur, thereby connecting the trunk to the lower limb. Each hip bone consists of three parts, the **ilium, ischium,** and **pubis,** which are separated from each other by cartilage in young subjects, but are fused in the adult. The union of the three parts takes place in and around a large cup-shaped articular cavity, the **acetabulum,** which is situated near the middle of the lateral surface of the bone (Fig. 4-181). The **ilium** is the superior, broad, and expanded portion which extends cranialward from the acetabulum. It forms the upper two-fifths of the acetabulum and acts as the weight-bearing part of the bone. The **ischium** is the most inferior and strongest portion. It proceeds downward from the acetabulum and expands into the large **ischial tuberosity.** Curving ventrally, the ischium then forms with the pubis a large aperture, the **obturator foramen.** The **pubis** extends medialward from the acetabulum and articulates in the midline at the **symphysis pubis** with the bone of the opposite side. The anterior aspect of the pubic bones forms the front of the pelvis and supports the overlying external genital organs.

THE ILIUM.   The **ilium** (Figs. 4-182 to 4-184) is divisible into two parts, the smaller **body**

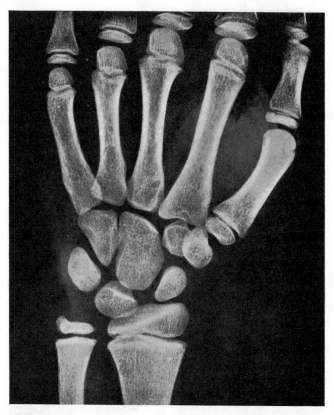

FIG. 4-179. Hand and wrist of a child aged 11 years. All the centers of ossification are present except that for the pisiform bone. Note how the first metacarpal differs from the other metacarpal bones.

and the expanded upper portion called the **ala.** The junction of these two parts is indicated on the internal surface by the **arcuate line** (Fig. 4-183) and on the external surface by the margin of the acetabulum.

The **body** of the ilium enters into the formation of the acetabulum, of which it forms rather less than two-fifths. Its *external surface* (Fig. 4-182) is partly articular, partly nonarticular. The articular segment forms a portion of the **lunate surface** of the acetabulum, i.e., the walls of the acetabular cup. The nonarticular portion contributes to the roughened floor of the cavity, or the **acetabular fossa.** The *internal surface* (Fig. 4-183) of the body is part of the wall of the lesser pelvis and gives origin to some fibers of the obturator internus muscle. It is continuous below with the pelvic surfaces of the ischium and pubis, only a faint line indicating the place of union.

The **ala** or wing is the large expanded portion which bounds the greater pelvis laterally. It presents for examination two surfaces—a gluteal and a sacropelvic, and a crest, and two borders—an anterior and a posterior.

The **gluteal surface** is oriented laterally and posteriorly behind, and laterally and somewhat inferiorly in front. It is smooth and convex ventrally, but deeply concave dorsally. The gluteal surface is bounded superiorly by the **iliac crest,** inferiorly by the superior border of the acetabulum, and in front and behind by the anterior and posterior borders of the ala. This surface is crossed in an arched direction by three ridges—the posterior, anterior, and inferior gluteal lines. The **posterior gluteal line,** the shortest of the three, begins above on the outer lip of the iliac crest about 5 cm in front of its posterior extremity. The upper part of this line is distinctly marked, but as it passes downward toward its termination at the **greater sciatic notch,** it becomes more obscure and difficult to define (Figs. 4-182; 4-184). Dorsal to this line is a narrow semilunar surface which is rough and gives origin to a small portion of the gluteus maximus. From a smoother region below this origin of

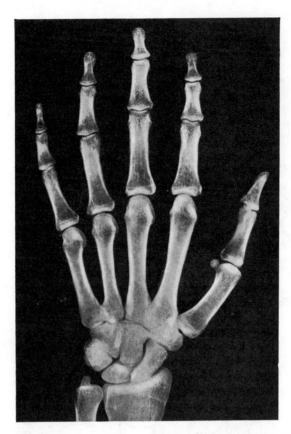

FIG. 4-180. Radiograph of the adult wrist and hand. Note the sesamoid bone in the thumb in relation to the metacarpophalangeal joint.

the gluteus maximus attach fibers of the sacrotuberous ligament as well as the iliac head of the piriformis muscle. The **anterior gluteal line** is the longest of the three. It begins at the outer lip of the iliac crest, about

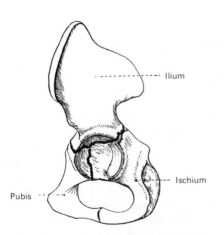

FIG. 4-181. The hip bone of a 14-year-old child viewed from the lateral aspect. (From Benninghoff and Goerttler, 1975.)

4 cm from its ventral extremity, and taking a curved direction downward and backward, ends at the greater sciatic notch (Figs. 4-182 to 4-184). Near the middle of this line a nutrient foramen is often seen. The space between the anterior and posterior gluteal lines and the iliac crest is concave, and gives origin to the gluteus medius. The **inferior gluteal line,** the least distinct of the three, begins anteriorly at the border of the ilium somewhat above the anterior inferior iliac spine. It curves backward and downward to end near the upper part of the greater sciatic notch. The surface of bone included between the anterior and inferior gluteal lines gives origin to the gluteus minimus. Between the inferior gluteal line and the upper part of the acetabulum is a rough, shallow groove, from which the reflected head of the rectus femoris muscle arises.

The **sacropelvic surface** of the ala is oriented medially and posteriorly toward the pelvis and sacrum. It is the internal surface, and is bounded above by the iliac crest, below by the **arcuate line,** and in front and behind by

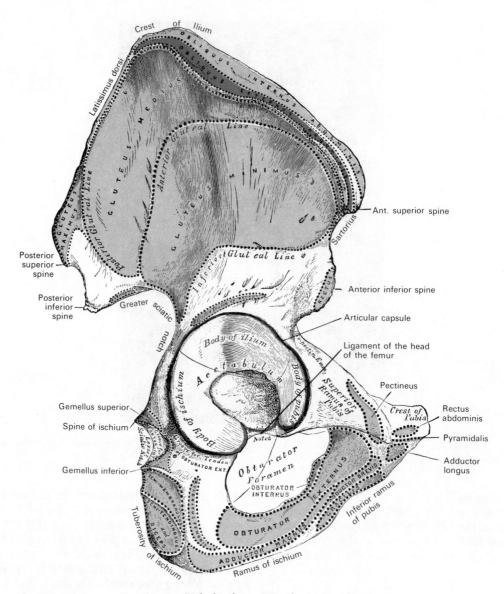

FIG. 4-182.   Right hip bone. Lateral or external aspect.

the anterior and posterior borders of the ilium (Fig. 4-183). It presents a large, smooth, concave surface, called the **iliac fossa,** from which arises the iliacus muscle and which is perforated at its inner part by a nutrient canal. Behind the iliac fossa is a rough surface, divided into two portions, an anterior and posterior. The anterior part or **auricular surface,** so-called from its resemblance in shape to the ear, is coated with cartilage, and articulates with a similar surface on the side of the sacrum. The posterior portion, known as the **iliac tuberosity,** is elevated and rough, for the attachment of the dorsal sacroiliac ligament and for the origins of the erector

spinae muscle. Anterior and below the auricular surface is the **preauricular sulcus,** better marked in the female than in the male. To this rough groove attach fibers of the ventral sacroiliac ligament.

The **iliac crest** is convex and like an arch in its general outline, but it is sinuously curved. To it attach the broad, flat muscles of the anterior abdominal wall, certain muscles of the lower back, and others of the lower extremity. The crest is thinner at the center than at the extremities, and ends in the **anterior** and **posterior superior iliac spines.** The surface of the crest is broad; it is divided into **external** and **internal lips,** and an **intermedi-**

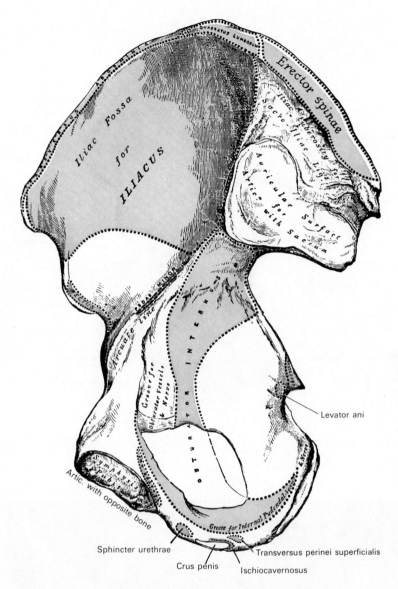

Levator ani

Sphincter urethrae

Crus penis

Transversus perinei superficialis

Ischiocavernosus

FIG. 4-183.   Right hip bone. Medial or internal aspect.

ate line. About 5 cm dorsal to the anterior superior iliac spine there is a prominent **tubercle** on the outer lip. To the external lip are attached the fascia lata along with the tensor fasciae lata muscle, the external oblique muscle, and more posteriorly the latissimus dorsi muscle. Along the intermediate line arises the internal oblique, while to the internal lip attach the transversus abdominis, quadratus lumborum, and erector spinae muscles, the lumbodorsal fascia and the iliacus muscle and iliac fascia (Figs. 4-182; 4-183).

The **anterior border** of the ala is concave. It presents two projections, separated by a

notch. Of these, the more cranial, situated at the junction of the crest and ventral border, is called the **anterior superior iliac spine** (Fig. 4-184). Its outer border gives attachment to the fascia lata and the tensor fasciae latae, its inner border to the iliacus; its extremity affords attachment to the inguinal ligament and the origin of the sartorius. To a notch inferior to the spine extends the origin of the sartorius, and across this notch passes the lateral femoral cutaneous nerve. Below this notch is the **anterior inferior iliac spine** (Fig. 4-185), which ends near the anterior lip of the acetabulum. This lower spine gives attachment above to the straight tendon of the

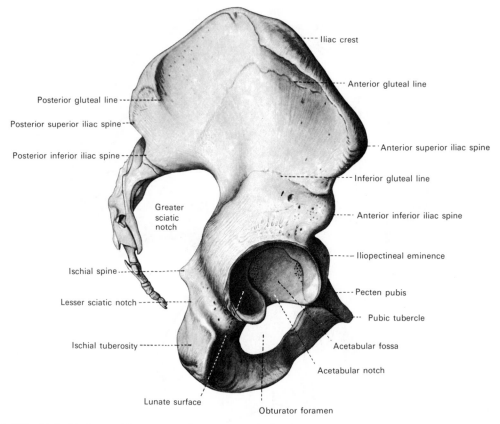

Iliac crest

Anterior gluteal line

Posterior gluteal line

Posterior superior iliac spine

Posterior inferior iliac spine

Anterior superior iliac spine

Inferior gluteal line

Greater sciatic notch

Anterior inferior iliac spine

Iliopectineal eminence

Ischial spine

Lesser sciatic notch

Pecten pubis

Pubic tubercle

Ischial tuberosity

Acetabular fossa

Acetabular notch

Lunate surface

Obturator foramen

FIG. 4-184. Right hip bone with sacrum and coccyx. External or lateral surface. (From Benninghoff and Goerttler, 1975.)

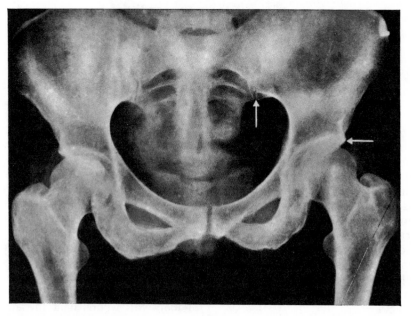

FIG. 4-185. Roentgenogram of adult pelvis. The upper arrow indicates the line of the sacroiliac joint; the lower arrow points to the anterior inferior iliac spine.

rectus femoris muscle and below to the ilio-femoral ligament of the hip joint. Medial to the anterior inferior spine is a broad, shallow groove, over which the iliacus and psoas major muscles pass. This groove is bounded medially by an eminence, the **iliopectineal eminence** (Fig. 4-182), which marks the point of union of the ilium and pubis.

The **posterior border** of the ala is shorter than the anterior and also presents two projections, the **posterior superior iliac spine** and the **posterior inferior iliac spine,** which are separated by a notch. The upper spine serves for the attachment of the oblique portion of the dorsal sacroiliac ligaments and the multifidus. The lower spine corresponds with the dorsal extremity of the auricular surface. Inferior to the posterior inferior spine is the deeply indented **greater sciatic notch.**

THE ISCHIUM. The **ischium** (Figs. 4-182 to 4-184) forms the inferior and posterior part of the hip bone. It is L-shaped, interposed between the ilium above and the pubis in front, and is divided into a **body** and a **ramus.**

The *upper portion* of the **body** of the ischium enters into and constitutes a little more than two-fifths of the acetabulum. Its external surface forms part of the lunate surface of the acetabulum and a portion of the acetabular fossa. The internal surface of the upper part of the ischial body forms a portion of the wall of the lesser pelvis, and gives origin to some fibers of the obturator internus muscle. From the posterior border there extends a thin and pointed triangular eminence, the **ischial spine,** which is more or less elongated in different subjects. The outer surface of the spine gives attachment to the gemellus superior, while to its inner surface attach the coccygeus and levator ani muscles, and the pelvic fascia. The sacrospinous ligament is attached to the pointed extremity of the ischial spine. Superior to the spine is a large notch, the **greater sciatic notch,** which becomes converted into the **greater sciatic foramen** by the sacrospinous ligament. Through this foramen pass a number of structures from within the pelvis into the gluteal region. It transmits the piriformis muscle, the superior and inferior gluteal vessels and nerves, the sciatic and posterior femoral cutaneous nerves, the internal pudendal vessels, the pudendal nerve, and the nerves to the obturator internus and

quadratus femoris muscles. Of these, the superior gluteal vessels and nerve pass out above the piriformis, the other structures below it. Also inferior to the spine is a smaller notch, the **lesser sciatic notch.** It is smooth, coated with cartilage, and presents two or three ridges corresponding to the subdivisions of the tendon of the obturator internus, which glides over it. It is converted into the **lesser sciatic foramen** by the sacro-tuberous and sacrospinous ligaments. Through the foramen course the tendon of the obturator internus, the nerve that supplies that muscle, and the internal pudendal vessels and pudendal nerve. Having emerged from the pelvis through the greater sciatic foramen, the internal pudendal vessels and the pudendal nerve cross over the ischial spine and go through the lesser sciatic foramen to enter the perineum between the obturator internus muscle and its fascia.

The *lower portion* of the **body** of the ischium (formerly called the superior or descending ramus) projects caudally and dorsally from the acetabulum. It presents three surfaces: external, internal, and dorsal. The **external surface** is quadrilateral and, in the articulated hip, faces anterolaterally and inferiorly toward the thigh. It is bounded *above* by a groove that lodges the tendon of the obturator externus. *Below,* this surface becomes continuous with the ramus of the ischium. *In front,* it is bounded by the posterior margin of the obturator foramen, while *behind,* a prominent margin separates it from the internal or dorsal surface. Anterior to this margin the external surface gives origin to the quadratus femoris and to some of the fibers of origin of the obturator externus. The lower part of the surface gives origin to part of the adductor magnus. The **internal surface** forms part of the bony wall of the lesser pelvis. *In front,* it is bounded by the posterior margin of the obturator foramen. *Below,* it is bounded by a sharp ridge which gives attachment to the falciform process of the sacrotuberous ligament, and more anteriorly gives origin to the transversus perinei and ischiocavernosus. Posteriorly, a large swelling, the **ischial tuberosity,** is divided into two portions: a lower, rough, somewhat triangular part, and an upper, smooth, quadrilateral portion. The *lower portion* is subdivided by a prominent vertical ridge into two parts: the larger, lateral area gives attachment to the adductor magnus, the medial affords attachment to the sacrotuberous ligament, and supports the body in the sitting position. The *upper portion* allows attachment of the hamstring muscles. An oblique ridge subdivides this portion into a proximal area, from which the semimembranosus arises, and a distal area for the head of the biceps femoris and the semitendinosus.

The **ramus** of the ischium is the thin, flattened part of the ischium, which ascends from the lower part of the body and joins the inferior ramus of the pubis, the junction being indicated in the adult by a raised line. The combined rami are sometimes called the **ischiopubic ramus.** The ischial ramus presents an **external** (anterior) **surface** oriented anteroinferiorly toward the thigh, and an **internal** (posterior) **surface** oriented posterosuperiorly toward the perineum and pelvis. The external surface is uneven, and gives origin to the obturator externus and some of the fibers of the adductor magnus, while the internal surface forms part of the anterior wall of the pelvis. The **medial border** of the ischial ramus is thick, rough, slightly everted, and, together with the medial border of the opposite ischium, forms the bony rim of the inferior outlet of the pelvis. The border presents an outer and an inner ridge along with an intervening space. The ridges are continuous with similar ones along the inferior ramus of the pubis. To the outer ridge is attached the deep layer of the superficial perineal fascia (fascia of Colles), and to the inner ridge, the inferior fascia of the urogenital diaphragm. If the ridges are traced downward, they are found to join with each other just dorsal to the origin of the transversus perinei profundus muscle. The transversus perinei is attached to the intervening space, and ventral to this, a portion of the crus penis or crus clitoris and the ischiocavernosus muscles are attached. The **lateral border** of the ischial ramus is thin and sharp, and forms part of the medial margin of the obturator foramen.

THE PUBIS.   The **pubis** forms the anterior part of the hip bone. It articulates with the opposite pubis to form the symphysis pubis. It is divisible into a **body,** a **superior ramus,** and an **inferior ramus** (Fig. 4-182).

The **body** of the pubis forms one-fifth of the acetabulum, contributing by its **external surface** both to the lunate surface and the acetabular fossa. Its **internal surface** enters into the formation of the wall of the lesser pelvis and gives origin to a portion of the obturator internus.

The **superior ramus** extends inferomedially from the body to the median plane, where it articulates with the superior ramus of the opposite side. It is conveniently described in two portions, a flat, **medial portion** and a narrow, prismoid **lateral portion.**

The **medial portion** of the superior ramus is somewhat quadrilateral and presents for examination two *surfaces*—medial and lateral—and three *borders*—superior, medial, and lateral. The **external surface** is rough and directed inferolaterally, and serves for the origin of certain muscles. The adductor longus muscle arises from the upper and medial angle immediately below the crest. More inferiorly, the obturator externus and the adductor brevis muscles, and the proximal part of the gracilis take origin. The **internal surface** is smooth and forms part of the anterior wall of the pelvis. It gives origin to the levator ani and obturator internus muscles, and gives attachment to the puboprostatic ligaments and to a few muscular fibers prolonged from the bladder. The **superior border** presents a prominent elevation, the **pubic tubercle,** which projects ventrally. To it are attached the inferior crus of the superficial inguinal ring and the inguinal ligament. Passing laterally from the pubic tubercle is a well-defined ridge forming a part of the **pecten pubis** (or pectineal line), which marks the brim of the lesser pelvis. To it are attached a portion of the inguinal falx (the conjoined tendon of the internal oblique and transversus muscles), the lacunar ligament, and the reflected inguinal ligament. Medial to the pubic tubercle is the **pubic crest,** which extends to the medial end of the bone. It affords attachment to the inguinal falx, and to the rectus abdominis and pyramidalis muscles. To the point of junction of the pubic crest with the medial border of the bone, as well as to the symphysis pubis, is attached the superior crus of the superficial inguinal ring. The **medial border** of the superior ramus is articular, participating in the formation of the symphysis pubis. It is oval, and marked by eight or nine transverse ridges, separated by grooves which serve for the attachment of a thin layer of cartilage. The cartilage intervenes between the bone and the interpubic fibrocartilaginous lamina or disc. The **lateral border** presents a sharp margin, the **obturator crest,** which forms part of the circumference of the obturator foramen and affords attachment to the obturator membrane.

The **lateral portion** of the superior ramus presents three *surfaces,* superior, inferior, and dorsal. Along the **superior surface** is found the **iliopectineal line,** a continuation of the pecten pubis, or pectineal line, which commences at the pubic tubercle. Ventral to this line, the surface of the bone is triangular in shape, and wider laterally than medially. It is covered by the pectineus muscle. The superior surface is bounded laterally by the **iliopectineal eminence,** which indicates the point of junction of the ilium and pubis, and by a prominent ridge which extends from the acetabular notch to the pubic tubercle. The **inferior surface** forms the superior boundary of the obturator foramen. Laterally, it presents an oblique groove for the passage of the obturator vessels and nerve. Medially, a sharp margin, the **obturator crest,** forms part of the circumference of the obturator foramen and gives attachment to the obturator membrane. The **dorsal surface** is intrapelvic in its orientation. It constitutes part of the ventral boundary of the lesser pelvis, and along this surface arise some fibers of the obturator internus muscle.

The **inferior ramus** of the pubis is thin and flattened (Fig. 4-182). It passes laterally and inferiorly from the medial end of the superior ramus. Becoming narrower as it descends, the inferior ramus of the pubis joins with the ramus of the ischium along the obturator foramen. It has two *surfaces,* external and internal, and two *borders,* medial and lateral. The **external surface** is rough, for the origin of muscles—the gracilis along its medial border, a portion of the obturator externus where the inferior ramus enters into the formation of the obturator foramen, and between these two, the adductors brevis and magnus, the former being the more medial. The **internal surface** is smooth, and gives origin to the obturator internus and, close to the medial margin, the sphincter urethrae muscles. The **medial border** is thick, rough, and everted, especially in females. It presents two ridges, separated by an intervening space. The ridges extend downward and are continuous with similar ridges on the ramus of the ischium. To the more external of the two ridges is attached the superficial perineal fascia (Colles' fascia), and to the internal ridge the inferior fascia of

the urogenital diaphragm. The **lateral border** is thin and sharp, forms part of the circumference of the obturator foramen, and gives attachment to the obturator membrane.

The **acetabulum** (Figs. 4-182; 4-184) is a deep, cup-shaped, hemispherical cavity located on the lateral aspect of the hip bone, near its center. It is oriented laterally and anteroinferiorly and is formed superiorly by the ilium, inferolaterally by the ischium, and anterosuperiorly by the pubis. A little less than two-fifths of the acetabulum is contributed by the ilium, a little more than two-fifths by the ischium, and the remaining one-fifth by the pubis. It is bounded by a prominent uneven rim, which is thick and strong superiorly, and serves for the attachment of the acetabular labrum, which contracts the orifice and deepens the surface for articulation. The acetabulum is deficient inferiorly and presents a deep indentation, the **acetabular notch** (Fig. 4-184), which is continuous with a *circular nonarticular depression,* the **acetabular fossa,** at the bottom of the hemispherical cavity. The acetabular fossa is perforated by numerous apertures and lodges a mass of fat. The acetabular notch is converted into a foramen by the transverse acetabular ligament whose fibers blend inseparably with the acetabular labrum. The nutrient vessels and nerves traverse the foramen to enter the joint. The margins of the acetabular notch serve for the attachment of the ligament of the head of the femur. The rest of the acetabulum is formed by a *curved articular surface,* the **lunate surface,** for articulation with the head of the femur.

The **obturator foramen** (Figs. 4-182 to 4-184) is a large, rounded aperture, situated below the acetabulum and between the ischium and pubis. It is limited *above* by the **obturator groove** of the superior pubic ramus, *below* by the ischial ramus, *medially* by the inferior ramus and body of the pubis, and *laterally* by the body of the ischium. The obturator foramen is large and oval in the male and somewhat smaller and more triangular in the female. Covering the foramen and attached to its uneven margin is the fibrous obturator membrane. The membrane completely covers the obturator foramen except superiorly where a deep groove, called the **obturator groove,** allows a communication between the pelvis and the thigh. This communication is converted into a canal by a specialized part of the obturator membrane, which takes the form of a ligamentous band and attaches to two tubercles. One is the **posterior obturator tubercle,** on the medial border of the ischium, just anterior to the acetabular notch, while the other is the **anterior obturator tubercle,** on the obturator crest of the superior ramus of the pubis. Through the canal the obturator vessels and nerve pass out of the pelvis and into the thigh.

STRUCTURE. The thicker parts of the hip bone consist of cancellous bone, enclosed between two layers of compact bone. The thinner parts, such as those at the bottom of the acetabulum and at the center of the iliac fossa, are usually translucent and composed entirely of compact bone.

OSSIFICATION (Fig. 4-186). The hip bone is ossified from *three primary centers,* one each for the ilium, ischium, and pubis. The centers appear in the following order: in the *lower part of the ilium,* immediately above the greater sciatic notch, about the eighth or ninth week of fetal life; in the *lower portion of the body of the ischium,* about the third month; in the *superior ramus of the pubis,* between the fourth and fifth months. Additionally, five secondary centers appear: at the crest of the ilium, the anterior inferior spine, the ischial tuberosity, the symphysis

By eight centers {Three primary (Ilium, Ischium, and Pubis) / Five secondary

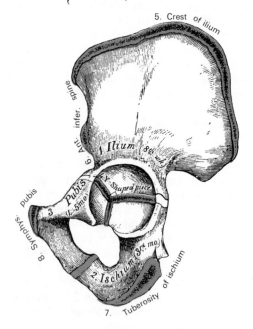

FIG. 4-186. Plan of ossification of the hip bone. The three primary centers unite through a Y-shaped piece about puberty. Epiphyses appear about puberty, and unite about the twenty-fifth year.

pubis, and the Y-shaped strip in the floor of the acetabulum, called the **triradiate cartilage.** At birth, the three primary centers are quite separate, the crest, the bottom of the acetabulum, the ischial tuberosity, the ramus of the ischium, and the inferior ramus of the pubis being still cartilaginous. In a 3½-year-old child the ramus of the ischium and the inferior ramus of the pubis are still connected by cartilage (Fig. 4-187), but by the seventh or eighth year they become almost completely united by bone. About the thirteenth or fourteenth year, the three primary centers have extended their growth into the bottom of the acetabulum, and are there separated from each other by the Y-shaped triradiate cartilage, which now presents traces of ossification, often by two or more centers. One of these, the os acetabuli, appears about the age of twelve, between the ilium and pubis, and fuses with them about the age of eighteen. It forms the pubic part of the acetabulum. The ilium and ischium then become joined, and lastly the pubis and ischium, through the intervention of the triradiate portion. At about the age of puberty, ossification takes place in each of the remaining portions, and they join with the rest of the bone between the twentieth and twenty-fifth years. Separate centers are frequently found for the pubic tubercle and the ischial spine, and for the pubic crest.

ARTICULATIONS. The hip bone articulates with its fellow of the opposite side, and with the sacrum and femur.

BONY PELVIS

The term **pelvis,** which means a basin, is used here to define the skeletal structure interposed between the movable part of the vertebral column, which it supports, and the lower limbs, upon which it rests. Thus, the principal functions of the bony pelvis are to afford attachment of the large musculature of the trunk and lower limbs, to protect the pelvic viscera, and to withstand the pressures and yet allow the functions of urination, defecation, and parturition. The pelvis is stronger and more massively constructed than the wall of the cranial or thoracic cavities, and is composed of four bones: the two hip bones laterally and ventrally and the sacrum and coccyx dorsally (Figs. 4-188 to 4-192).

The pelvis is divided into the **greater** and the **lesser pelvis** by an oblique plane passing through the promontory of the sacrum posteriorly, the arcuate line of the ilium, pecten pubis (iliopectineal line), and the superior margin of the symphysis pubis. The circumference of this plane is termed the **linea terminalis** or **pelvic brim.**

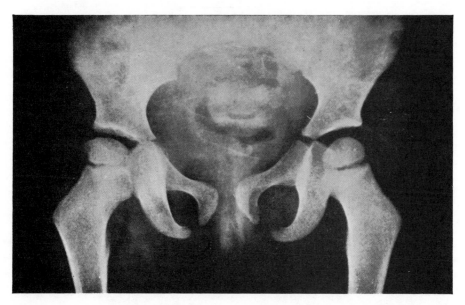

FIG. 4-187.   Pelvis of a child aged 3½ years. The epiphysis for the head of the femur is well formed, but the center for the greater trochanter has not yet appeared. The rami of the pubis and ischium are still connected by cartilage and the triradiate cartilage in the acetabulum is wide.

The **greater** or **false pelvis** is the expanded portion of the cavity situated above and ventral to the pelvic brim. It is bounded on each side by the ala of the ilium. Ventrally it is incomplete, presenting between the anterior borders of the ilia a wide interval, which is occupied in the intact body by the anterior wall of the abdomen. Posteriorly there is a deep notch between the ilium and the base of the sacrum on each side.

The **lesser** or **true pelvis** is that part of the pelvic cavity which is situated below the pelvic brim. The lesser pelvis is continuous with the greater pelvis, but its bony walls are more complete. The cavity of the lesser pelvis is of extreme importance in obstetrics, and measurements of its diameters are of clinical significance. The inlet to the cavity of the lesser pelvis is bounded by the superior circumference, and its outlet bounded below by the inferior circumference.

The **superior circumference** is formed by the brim of the pelvis, the enclosed area being called the **superior aperture** or **inlet**

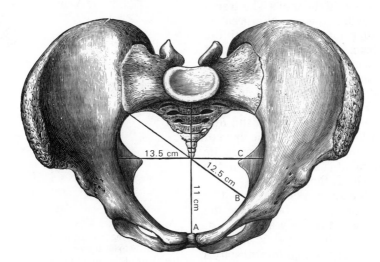

FIG. 4-188.   Diameters of superior aperture of lesser pelvis (female). *A*, Anteroposterior or conjugate diameter. *B*, Oblique diameter. *C*, Transverse diameter.

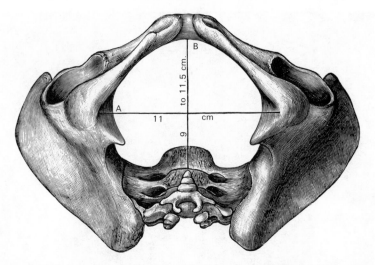

FIG. 4-189. Diameters of inferior aperture of lesser pelvis (female). *A,* Transverse diameters. *B,* Anteroposterior diameter.

(Fig. 4-188). It is formed laterally by the arcuate line and pecten pubis, anteriorly by the crests of the pubes, and posteriorly by the anterior margin of the base of the sacrum and sacrovertebral angle. The superior aperture is rounded or oval and somewhat heart-shaped. It is obtusely pointed anteriorly, diverges on each side, and is encroached upon dorsally by the projection of the promontory of the sacrum. It has three principal diameters: anteroposterior, transverse, and oblique (Fig. 4-188). The **anteroposterior** or **conjugate diameter** extends from the sacrovertebral angle or promontory to the upper border of the symphysis pubis. Its average measurement is about 11 cm in the female and 10 cm in the male. The **transverse diameter** extends across the greatest width of the superior aperture, from the middle of the brim or iliopectineal line on one side to the same point on the opposite. Its average measurement is about 13.5 cm in the female and 12.5 cm in the male. The **oblique diameter** extends from the sacroiliac joint of one side directly across to the diametrically opposite site on the pelvic brim of the other side (iliopectineal eminence). Its average measurement is about 12.5 cm in the female and 12.0 cm in the male.

The **cavity** of the **lesser pelvis** (Fig. 4-172) is bounded *anteriorly and below* by the pubic symphysis and the body and rami of the pubes, *posteriorly* by the pelvic surfaces of the sacrum and coccyx, and *laterally* by a broad, smooth, quadrangular area of bone, on the inner surfaces of the bodies of the

ischium and ilium. The cavity of the lesser pelvis is short and curved, being considerably deeper on its posterior than on its anterior wall. In the intact body, it contains the pelvic colon, rectum, bladder, and certain reproductive organs. The rectum courses along the dorsum of the pelvis in the curve of the sacrum and coccyx, while the bladder lies against the dorsal surface of the pubic

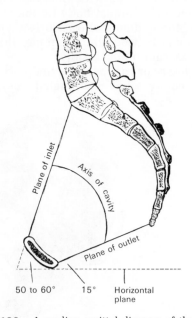

FIG. 4-190. A median sagittal diagram of the female pelvis showing that the plane of the pelvic inlet forms an angle of 50 to 60 degrees with the horizontal plane, while the plane of the outlet forms an angle of 15 degrees. Observe the curvature of the pelvic axis between the planes of inlet and outlet.

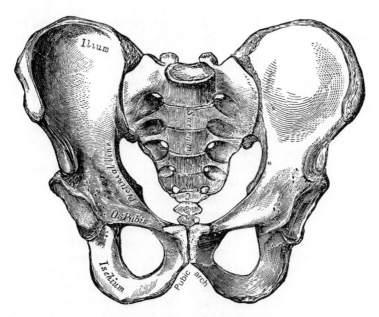

FIG. 4-191.   Male pelvis. Anterior aspect.

symphysis. In the female the uterus and vagina occupy the interval between these viscera.

The **inferior circumference** of the pelvis is more irregular than the superior circumference. The outlet enclosed by its perimeter is named the **inferior aperture** (Fig. 4-189). It is bounded dorsally by the sacrum and coccyx, and laterally by the ischial tuberosities. These bony landmarks are separated by three notches, one of which, the **pubic arch,** lies anterior to the tuberosities and is formed by the convergence of the inferior rami of the pubes. Two dorsal notches, one on each side, extend deeply between the sacrum and the ischium. They are called the **sciatic notches.** In the intact body each sciatic notch is converted into the **greater** and **lesser sciatic foramina** by the sacrotuberous and sacrospinous ligaments. With the ligaments in position, the inferior aperture of the pelvis is nearly diamond-shaped, bounded *anteriorly* by the pubic arcuate ligament and the inferior rami of the pubes, *laterally* by

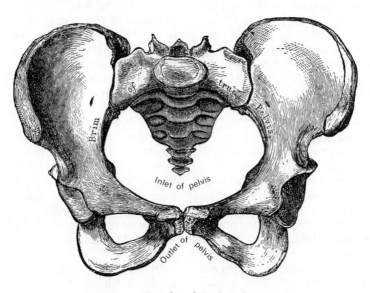

FIG. 4-192.   Female pelvis. Anterior aspect.

the ischial tuberosities, and *dorsally* by the sacrotuberous ligaments and the tip of the coccyx.

The two **diameters** of the **outlet** of the **pelvis** are anteroposterior and transverse. The **anteroposterior diameter** extends from the tip of the coccyx to the inferior edge of the symphysis pubis at the midline anteriorly. Its measurement varies between 9 and 11.5 cm in the female (in comparison to 10.5 to 11 cm in the male) because of the varying length of the coccyx. Additionally, its length may be increased or diminished by the mobility of that bone. The **transverse diameter,** measured between the dorsal parts of the ischial tuberosities, is about 11 cm in the female and 9 cm in the male.

PELVIC AXIS. The **pelvic axis** is an imaginary curved line passing through the lesser pelvis at right angles to the plane of the superior aperture and inferior aperture at their centers, and all the planes between them (Fig. 4-190). A line at right angles to the plane of the superior aperture at its center would, if prolonged, pass through the umbilicus and the middle of the coccyx and is therefore directed downward and backward. The axis of the inferior aperture, if prolonged, would touch the promontory of the sacrum. Thus, the axis of the pelvic cavity between the superior and inferior apertures is curved like the cavity itself, corresponding to the concavity of the sacrum and coccyx. An appreciation of the direction of these axes is especially relevant to an understanding of the course of the fetus in its passage through the pelvis during parturition.

POSITION OF THE PELVIS. In the standing position, the pelvis is so tilted forward that the plane of the superior aperture forms an angle of 50 to 60 degrees with the horizontal plane, and that of the inferior aperture an angle of about 15 degrees (Fig. 4-190). This **inclination of the pelvis** causes the pelvic surfaces of the symphysis pubis to be directed superiorly and posteriorly, while the concavity of the sacrum and coccyx is oriented inferiorly and anteriorly. The position of the pelvis in the erect posture may be indicated by holding it so that the anterior superior iliac spines and the top of the symphysis pubis are in the same plane.

DIFFERENCES BETWEEN THE MALE AND FEMALE PELVIS. The morphological differences encountered between the male and female pelvis reflect differences in function. The reproductive function in the female and the more muscular nature of the male become to some extent manifested in pelvic anatomy. The female pelvis (Fig. 4-192) when compared with that of the male (Fig. 4-191) is more delicate in its bone structure. Because the female pelvis is less massive, its overall size and dimensions are smaller, and its muscular impressions are less marked. The curvature of the ilia is more pronounced in the male in such a manner that its angle of approach to the pubis is more acute and less rounded. Thus, in the female the anterior iliac spines are more widely separated, causing the greater lateral prominence of the hips. The superior aperture of the lesser pelvis in the female is larger than in the male. It is also more nearly circular, and its obliquity is greater. **In the female,** the *true pelvic cavity* is shallower and wider; the *sacrum* is shorter, wider, and less curved; the *obturator foramina* are more triangular in shape and smaller in size; the *inferior aperture* is larger and the *coccyx* more movable; the *sciatic notches* are wider and shallower; the *ischial spines* project less; the *acetabula* are smaller and face more ventrally; the *ischial tuberosities* and the *acetabula* are wider apart, and the former are more everted; the *pubic symphysis* is less deep, and the *pubic arch* is wider and more rounded, and its angle less acute (Figs. 4-191; 4-192).

The **size of the pelvis** varies not only in the two sexes, but also in different members of the same sex. Pelvic size does not appear to be influenced by an individual's height, although women of short stature, as a rule, have broad pelves. Occasionally the pelvis is equally contracted in all its dimensions, so that all its diameters measure 1.25 cm less than average, even in otherwise well-formed women of average height. The principal divergences, however, are found in the shape of the superior aperture, affecting the relationship between the anteroposterior and transverse diameters. Thus the superior aperture may be elliptical in either a transverse or an anteroposterior direction. In the former, the transverse diameter, and in the latter, the anteroposterior diameter greatly exceed the other diameters. In other instances the superior aperture is almost circular.

In the fetus, and for several years after birth, the pelvis is smaller in proportion to the body than it is in the adult, and the projection of the sacrovertebral angle less marked. The characteristic differences between the male and female pelvis are distinctly indicated as early as the fourth month of fetal life, especially in the nature of the pubic arch and subpubic angle.

THE FEMUR

The **femur** or thigh bone is the longest and strongest bone in the skeleton, and is almost perfectly cylindrical in the greater part of its extent. Each femur articulates proximally with the hip bone at the acetabulum, and distally with the tibia at the knee joint. Thus, in the erect posture the two femora are more widely separated above than at the knees, and by inclining gradually in an oblique manner inferomedially, they bring the knee joint near the line of gravity of the body. The degree of this inclination varies somewhat, and is greater in the female than in the male because of the greater breadth of the pelvis. The femora transmit the weight of the trunk to the tibias, and they are so covered with muscles that only their upper and lower

ends are palpable. Similar to other long bones, the femur presents two extremities separated by a body or shaft.

PROXIMAL EXTREMITY. The proximal extremity (Fig. 4-193) consists of a head, a neck, and a greater and a lesser trochanter.

The **head of the femur** is globular, forming rather more than half a sphere. Its surface is smooth and coated with hyaline cartilage, except over an ovoid depression or pit near its center, the **fovea** of the head of the femur, where the ligament of the head of the femur is attached.

The **neck of the femur** is a pyramidal process of bone connecting the femoral head with the body, set at an angle of about 125 degrees in the adult. The angle is wider in infancy, and becomes lessened during growth. By puberty the neck forms a gentle curve from the axis of the body of the bone. Variation in the angular shape of the proximal extremity is observed to be dependent on the development of the pelvis and on stature. Because of the increased width of the pelvis in the female, the neck of the femur forms a less obtuse angle (about 110 to 115

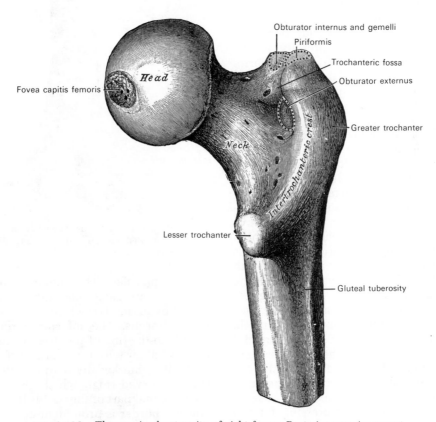

FIG. 4-193.   The proximal extremity of right femur. Posterior superior aspect.

degrees) with the shaft than it does in the male. The angle varies considerably in persons of different height. In the long-legged male the femoral shaft is more vertical and the femoral neck is more in line (about 140 degrees). In addition to projecting superiorly and medially, the femoral neck also projects somewhat anteriorly from the body of the femur. The amount of this projection is quite variable, but averages 12 to 14 degrees.

The neck is flattened anteroposteriorly, constricted in the middle, and broader laterally than medially. The vertical diameter of the *lateral half* is increased by the obliquity of the lower edge, which slopes to join the body at the level of the lesser trochanter. The *medial half* is smaller and more cylindrically shaped. The **anterior surface** of the neck is perforated by numerous vascular foramina. Along the line of junction of the anterior surface with the head is a shallow groove, best marked in elderly subjects, which lodges the orbicular or circular fibers of the capsule of the hip joint. The **posterior surface** is smooth, and it is broader and more concave than the anterior. The posterior part of the capsule of the hip joint is attached to the femoral neck about 1 cm above the intertrochanteric crest. The **superior border** is short and thick. It ends laterally at the greater trochanter, and its surface is perforated by large foramina. The **inferior border** is long and more narrow and it curves downward, ending at the lesser trochanter.

The **greater trochanter** (Figs. 4-193 to 4-197) is a large, irregular, quadrilateral projection surmounting the junction between the neck and the shaft. It has two surfaces and four borders. The **lateral surface** is broad, rough, convex, and marked by a diagonal impression for the insertion of the tendon of the gluteus medius muscle. Above this impression is a triangular surface which sometimes is rough for part of the gluteus medius tendon, and at other times smooth for the interposition of a bursa between the tendon and the bone. Below the diagonal impression is a smooth triangular surface, over which the gluteus maximus muscle glides, a bursa being interposed. The **medial surface** of the greater trochanter is less extensive than the lateral surface. At its base is found a deep depression, the **trochanteric fossa,** where the obturator externus inserts, and anterior to this an impression for the insertion of the obturator internus and the

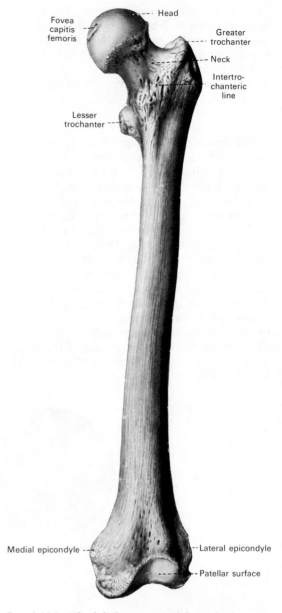

Fig. 4-194. The left femur viewed from the anterior aspect. (From Benninghoff and Goerttler, 1975.)

gemelli muscles. The **superior border** is thick and irregular, and marked near the center by an impression for the insertion of the piriformis. The **inferior border** corresponds to the line of junction of the base of the trochanter with the lateral surface of the shaft. It is marked by a rough, prominent, slightly curved ridge, which gives origin to the proximal part of the vastus lateralis. The **anterior border** is prominent and somewhat irregular, and affords insertion at its lateral

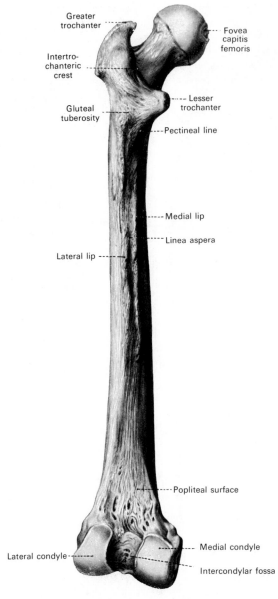

Greater
trochanter

Intertro-
chanteric
crest

Gluteal
tuberosity

Fovea
capitis
femoris

Lesser
trochanter

Pectineal line

Medial lip

Linea aspera

Lateral lip

Popliteal surface

Medial condyle

Lateral condyle

Intercondylar fossa

FIG. 4-195.   The left femur viewed from the posterior aspect. (From Benninghoff and Goerttler, 1975.)

the middle division of the linea aspera. The summit of the trochanter is roughened for the insertion of the tendon of the psoas major. More inferiorly, the iliacus muscle inserts along the junction of the lesser trochanter with the femoral shaft.

The **intertrochanteric line** (Figs. 4-194; 4-196) courses obliquely from the greater trochanter to the lesser on the anterior surface of the proximal extremity. This ridge commences with a tubercle of variable size at the junction of the greater trochanter and the distal part of the neck of the femur. The intertrochanteric line, forming the junction of the neck with the femoral shaft on the anterior surface, winds beyond the lesser trochanter and ends about 5 cm below this eminence in the linea aspera. Its proximal half is rough and affords attachment to the iliofemoral ligament, while its lower half, which is less prominent, gives origin to the proximal part of the vastus medialis.

The **intertrochanteric crest** (Figs. 4-195; 4-197) is a prominent ridge along the posterior surface of the proximal extremity extending in an oblique curve from the summit of the greater trochanter to the lesser. It marks the junction between the neck and the shaft posteriorly. Slightly above its middle is an elevation called the **quadrate tubercle,** which receives the insertion of the quadratus femoris and a few fibers of origin of the adductor magnus.

BODY OR SHAFT.   The **body** or **shaft** is almost cylindrical, but a little broader proximally than in the center and broadest and somewhat flattened distally (Figs. 4-194 to 4-197). It is slightly arched, convex anteriorly and concave posteriorly, where it is strengthened by a prominent longitudinal ridge, the **linea aspera.** It has three *borders*—posterior, lateral, and medial, and four *surfaces*—popliteal, anterior, lateral, and medial.

The **linea aspera** (Fig. 4-195), which forms the prominent **posterior border** of the shaft, projects as an elevated longitudinal ridge or crest along the posterior aspect of the middle third of the bone. This crest consists of a **medial** and a **lateral lip,** and a narrow, rough, intervening band. *The linea aspera is prolonged proximally* by three ridges. The *lateral ridge* is very rough and courses almost vertically upward to the base of the greater trochanter. It is termed the **gluteal tuberosity,** and gives attachment to the gluteus max-

part to the gluteus minimus. The **posterior border** is very prominent and appears as a free, rounded edge which bounds the trochanteric fossa.

The **lesser trochanter** (Figs. 4-193 to 4-197) is a conical eminence which projects medially and posteriorly from the base of the neck. Three well-marked **borders** extend from its apex. The medial is continuous with the lower border of the neck, the lateral with the intertrochanteric crest, the inferior with

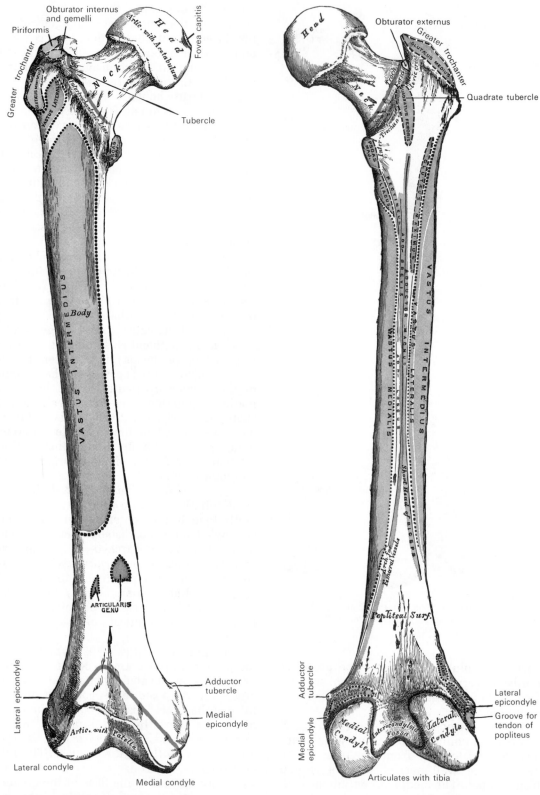

FIG. 4-196. Right femur. Anterior surface. Articular capsule outlined in blue.

FIG. 4-197. Right femur. Posterior surface.

imus. The proximal part of the gluteal tuberosity is often elongated into a rough crest, on which a more or less well-marked, rounded tubercle, the **third trochanter,** is occasionally developed. The *intermediate ridge* or **pectineal line** is continued to the base of the lesser trochanter and gives attachment to the pectineus muscle. The *medial ridge* winds anteriorly to the intertrochanteric line distal to the area where the iliacus muscle is inserted. *The linea aspera is prolonged distally* into two ridges, enclosing between them a triangular area, the **popliteal surface,** which forms the floor of the upper part of the popliteal fossa. Of the two descending ridges, the *lateral ridge* is more prominent, extending to the summit of the lateral condyle. The *medial ridge* is less marked, especially at its proximal part, where it is crossed by the femoral artery. This ridge ends distally at the summit of the medial condyle as the **adductor tubercle,** which receives the insertion of the tendon of the adductor magnus muscle.

The **medial lip of the linea aspera** and its prolongations proximally and distally give origin to the vastus medialis, while from the **lateral lip** and its proximal prolongation arises the vastus lateralis (Fig. 4-197). The adductor magnus is inserted into the linea aspera, and into its lateral prolongation proximally, and its medial prolongation distally. Between the vastus lateralis and the adductor magnus, two muscles are attached, the gluteus maximus above, and the short head of the biceps femoris below. Between the adductor magnus and the vastus medialis four muscles are inserted: the iliacus and pectineus proximally, and the adductor brevis and adductor longus distally. The linea aspera is perforated a little below its center by the **nutrient canal,** which is directed obliquely upward.

The other two borders of the femur are only slightly marked. The **lateral border** extends from the anteroinferior angle of the greater trochanter to the anterior extremity of the lateral condyle, and the **medial border** stretches from the intertrochanteric line, at a point opposite the lesser trochanter, to the anterior extremity of the medial condyle.

The **anterior surface,** which lies between the lateral and medial borders, is smooth, rounded, and convex. The vastus intermedius arises from the upper three-quarters of the anterior surface. The lower one-quarter of the bony surface is separated from the muscle by the intervention of the synovial membrane from the knee joint and by the suprapatellar bursa. Above the bursa arises the articularis genu muscle. The **lateral surface** extends between the lateral border and the linea aspera and is continuous with the corresponding surface of the greater trochanter proximally, and with that of the lateral condyle distally. The vastus intermedius takes origin from its proximal three-fourths. The **medial surface** is smooth like the anterior and lateral surfaces, and stretches between the medial border and the linea aspera. This surface is continuous with the distal border of the neck proximally, and with the medial side of the medial condyle distally. No muscles attach to the medial surface, but it is covered by the vastus medialis.

DISTAL EXTREMITY. The distal extremity of the femur (Figs. 4-194 to 4-198) is greatly expanded in all directions to form two prominences, the **medial** and **lateral condyles,** which are joined together anteriorly but separated posteriorly by the **intercondylar fossa** or **notch** (Fig. 4-198). The distal surface is coated with hyaline cartilage for articulation with the tibia. The articular surface of the lateral condyle is longer and narrower than that of the medial condyle. Both condyles are nearly flat anteroposteriorly, but greatly curved on their posterior surface.

The **medial condyle** diverges more from the anteroposterior plane and projects somewhat farther distally than the **lateral condyle** when the femoral shaft is held perpendicularly. When, however, the femur is in its natural oblique position, the two condyles lie practically in the same horizontal plane. At the junction of the articular surfaces of the two condyles anteriorly, there is a shallow depression, the **patellar surface.** Between the condyles posteriorly is a deep notch, the **intercondylar fossa.** The walls of the fossa are rough and concave, and the fossa is limited proximally by the **intercondylar line** and distally by the margin of the patellar surface. The posterior cruciate ligament of the knee joint is attached to the anterior part of the lateral surface of the medial condyle, which forms the medial wall of the fossa. This ligament also extends to the floor of the fossa. The anterior cruciate ligament is attached to an impression on the proximal and posterior part of the lateral wall of the fossa, formed by the medial surface of the lateral condyle.

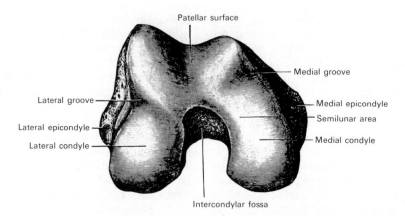

FIG. 4-198. Lower extremity of right femur viewed from below.

The **epicondyles** are two roughened convex prominences which surmount the condyles proximally. The **medial epicondyle** receives the attachment of the tibial collateral ligament of the knee joint. At the upper part of the medial epicondyle is the **adductor tubercle** (Figs. 4-196; 4-197) for the insertion of the adductor magnus, and posterior to this, the medial head of the gastrocnemius takes origin. The **lateral epicondyle,** smaller and less prominent than the medial, gives attachment to the fibular collateral ligament of the knee joint. Directly below the lateral epicondyle is a small depression from which a smooth well-marked groove curves upward and backward to the posterior extremity of the condyle. This groove is separated from the articular surface of the condyle by a prominent lip across which a second, shallower groove runs vertically from the depression. In the intact body these grooves are covered with cartilage. The popliteus muscle arises from the depression, and its tendon lies in the oblique groove when the knee is flexed and in the vertical groove when the knee is extended. Above and posterior to the lateral epicondyle is an area for the origin of the lateral head of the gastrocnemius, and above this arises the plantaris muscle.

The **articular surface** of the distal end of the femur occupies the anterior, distal, and posterior surfaces of the condyles. Its anterior part, the **patellar surface,** articulates with the patella. This surface presents a median groove which extends toward the intercondylar fossa, and two convexities, the lateral of which is broader, more prominent, and extends farther proximally than the medial. The distal and more posterior parts of the articular surface constitute the **tibial surfaces** for articulation with the corresponding condyles of the tibia and menisci. These surfaces are separated from the patellar surface by faint grooves which extend obliquely across the condyles (Fig. 4-198). The **lateral groove** is the better marked and can be seen coursing laterally and anteriorly from the intercondylar fossa, and expanding to form a triangular depression. When the knee joint is fully extended, the triangular depression rests upon the anterior portion of the lateral meniscus. The medial part of the lateral groove comes into contact with the medial margin of the lateral articular surface of the tibia, anterior to the lateral tubercle of the tibial intercondylar eminence. The **medial groove** is less distinct than the lateral. It does not reach as far as the intercondylar fossa, and therefore exists only on the medial part of the condyle. It receives the anterior edge of the medial meniscus when the knee joint is extended. Where the medial groove ceases laterally is seen a **semilunar area** which is a posterior continuation of the patellar surface close to the anterior part of the intercondylar fossa (Fig. 4-198). This semilunar area articulates with the medial vertical facet of the patella in forced flexion of the knee joint. The tibial surfaces of the condyles are convex both from side to side and anteroposteriorly. The anteroposterior length of the tibial surface of the medial condyle is slightly greater than that of the lateral condyle.

ARCHITECTURE. The femur obeys the mechanical laws that govern other elastic bodies under stress (Koch, 1917). Computations have been made of the stress sustained internally in different parts of the femur relative to the load placed on the femoral head. It has been found that the internal structure of the different portions of the femur is in very close agreement with the theoretical relations that should exist between stress and structure for maximum economy and efficiency (Figs. 4-199; 4-200).

The spongy bone at the *proximal end of the femur* is composed of two distinct systems of trabeculae arranged in curved paths. One system, called the **compressive system,** has its origin on the medial side of the shaft, and its trabeculae curve upward in a fan-like radiation to the opposite side of the bone. The other, called the **tensile system,** having origin on the lateral portion of the shaft, arches upward and medially to end in the upper surface of the greater trochanter, neck, and head. The trabeculae of these two systems intersect each other at right angles. The trabeculae of the compressive and tensile systems lie exactly in the paths of the maximum tensile and compressive stresses exerted on the femur. The thickness and spacing of the trabeculae vary with the intensity of the maximum stresses at various points, being thickest and most closely spaced in the regions where the greatest stresses occur. The amount of bony material in the spongy bone varies in proportion to the intensity of the shearing force at the various sections. The arrangement of the trabeculae in the positions of maximum stresses is such that the greatest strength is secured with a minimum of bony material.

In the shaft the shearing stresses are at a minimum. Very little if any material is in the central space, as practically the only material near the neutral plane is in the compact bone. The hollow shaft of the femur is an efficient structure for resisting **bending stresses,** because all of the material in the shaft is at a considerable distance relative to the neutral axis. It provides resistance to bending movement caused by the load on the femoral head as well as by any other loads tending to produce bending in other planes. The structure of the shaft is such that great strength exists in a relatively small amount of material.

In the distal extremity of the femur there are two main systems of trabeculae, a **longitudinal** and a **transverse system.** The longitudinal trabeculae arise from the inner wall of the shaft and continue in straight parallel lines to the epiphyseal line. They then continue in slightly curved lines to meet the articular surface of the femur at the knee joint at right angles. The transverse trabeculae are lighter in structure and are braced at right angles to the longitudinal trabeculae. Thus, with practically no increase in the amount of bony material used, there is a greatly increased stability produced by the expansion of the lower femur from a hollow shaft of compact bone to a structure of much larger cross-section composed almost entirely of spongy bone.

OSSIFICATION. (Figs. 4-201 to 4-203). The femur is ossified from five centers: one each for the

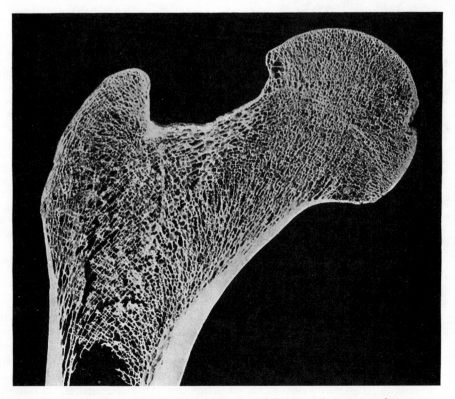

FIG. 4-199. Frontal longitudinal midsection of the upper femur; natural size.

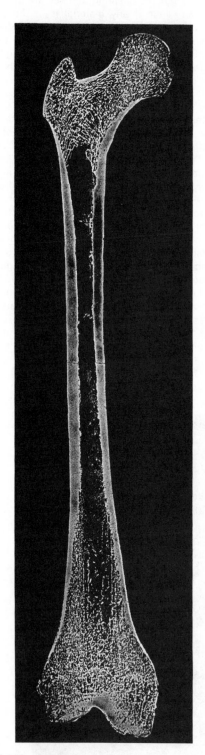

FIG. 4-200. Frontal longitudinal midsection of a right femur; ⁴⁄₉th natural size. (From Koch.)

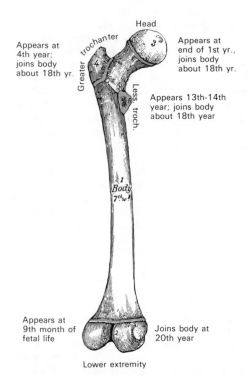

FIG. 4-201. Plan of ossification of the femur. From five centers.

body, the head, the greater trochanter, the lesser trochanter, and the distal extremity. Of all the long bones, except the clavicle, the femur is the first to show traces of ossification. This commences in the *middle of the body,* at about the seventh week of fetal life, and rapidly extends proximally and distally. The centers in the epiphyses appear in the following order: in the *distal end of the bone,* at the ninth month of fetal life (from this center the condyles and epicondyles are formed); in the *head,* at the end of the first year after birth; in the *greater trochanter,* during the fourth year; and in the *lesser trochanter,* between the thirteenth and fourteenth years. The order in which the epiphyses are joined to the body is the reverse of that of their appearance. They are not united until after puberty, and the lesser trochanter is the first to join, then the greater trochanter, then the head, and finally the distal extremity, which is not united until the twentieth year.

THE PATELLA

The **patella,** commonly called the knee cap (Figs. 4-204; 4-205), is a flat, smooth, triangular bone, situated on the front of the knee joint and embedded on the deep aspect of the tendon of the quadriceps femoris. It is the largest of the sesamoid bones, resembling them by (1) being developed in a ten-

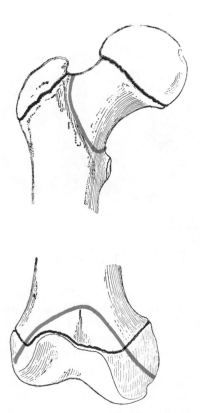

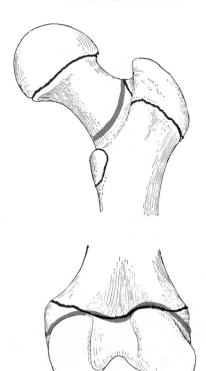

FIG. 4-202. Epiphyseal lines of femur in a young adult. Anterior aspect. The lines of attachment of the articular capsules are in blue.

FIG. 4-203. Epiphyseal lines of femur in a young adult. Posterior aspect. The lines of attachment of the articular capsules are in blue.

don; (2) presenting a knotty, tuberculated appearance as a result of its ossification; and (3) being composed mainly of dense cancellous tissue. Measuring about 5 cm in diameter, it serves to protect the front of the knee joint and increases the leverage of the quadriceps femoris by making it act at a greater angle. It has an anterior and a posterior surface, three borders, and a pointed **apex** that is directed inferiorly. The lower limit of the apex lies about 1 cm above the level of the knee joint in the erect lower extremity.

The **anterior surface** (Fig. 4-204) is convex,

perforated by small apertures for the passage of nutrient vessels, and marked by numerous rough, longitudinal striae. This surface is covered, in the intact body, by an expansion from the tendon of the quadriceps femoris which is continuous distally with the superficial fibers of the ligamentum patellae. It is easily palpable and separated from the integument by a bursa. The **posterior surface** (Fig. 4-205) presents a smooth, oval, *articular area* covered with cartilage, and divided into two facets by a vertical ridge. This ridge lies adjacent to the groove

FIG. 4-204. Right patella. Anterior surface.

FIG. 4-205. Right patella. Posterior (articular) surface.

on the anterior surface of the femur, and the facets of the patella articulate with the medial and lateral aspects of the groove. The lateral patellar facet is broader and deeper than the medial. Distal to the articular surface is a rough, convex, *nonarticular area,* the distal half of which gives attachment to the ligamentum patellae. The proximal half of the nonarticular area is separated from the head of the tibia by adipose tissue.

The **base** or **superior border** is thick and sloped from behind downward and forward. It gives attachment to the portion of the quadriceps femoris muscle that is derived from the rectus femoris and vastus intermedius muscles. The **medial** and **lateral borders** are thinner and converge distally. They give attachment to those portions of the quadriceps femoris that are derived from the vasti lateralis and medialis.

The **apex** is bluntly pointed and gives attachment to the ligamentum patellae, which is a continuation of the quadriceps tendon. The ligamentum patellae, in turn, attaches to the tibia.

STRUCTURE.   The patella consists of nearly uniform dense cancellous bone, covered by a thin compact lamina. The trabeculae immediately beneath the anterior surface are arranged parallel with it. In the remainder of the bone, the trabeculae radiate from the articular surface toward the other parts of the bone.

OSSIFICATION.   The patella ossifies from several centers, which usually make their appearance during the second or third year, but may be delayed until the sixth year. These centers then fuse, and ossification is usually completed by puberty or adolescence (Fig. 4-210).

ARTICULATION.   The posterior surface of the patella actually articulates with the femur, and its tendon, the ligamentum patellae, attaches to the tibia.

THE TIBIA

The **tibia** (Figs. 4-206 to 4-209) or shin bone is situated at the medial side of the leg and, except for the femur, is the longest bone of the skeleton. It lies nearly parallel to the fibula, and it alone transmits the weight of the trunk and thigh to the foot. The tibia is expanded proximally where it enters into the formation of the knee joint. The shaft again enlarges distally, although to a lesser extent, and ends in an inferior projection, the medial malleolus, which extends beyond the rest of the bone. The tibia presents a body or shaft and two extremities.

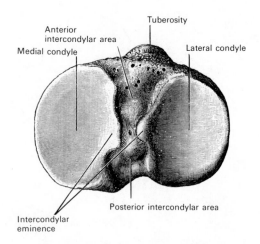

Fɪɢ. 4-206.   Proximal surface of right tibia.

PROXIMAL EXTREMITY.   The **proximal extremity** is expanded to form two eminences, the **medial** and **lateral condyles.** The superior **articular surface** presents two smooth articular facets (Figs. 4-206; 4-211). The medial facet is oval and slightly concave. The lateral facet, nearly circular, is concave from side to side, but slightly convex anteroposteriorly, especially where it extends over the posterior surface for a short distance. The central portions of these facets articulate with the condyles of the femur, while their peripheral portions support the menisci of the knee joint, which here intervene between the two bones (Fig. 4-207). Between the articular facets, but nearer the posterior than the anterior aspect of the bone, is a prominent spine, the **intercondylar eminence.** This eminence is surmounted on each side by the **medial** and **lateral intercondylar**

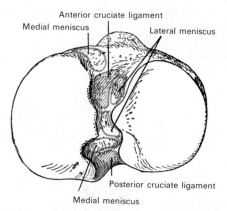

Fɪɢ. 4-207.   An outline of Figure 4-206 showing the attachment of the menisci and cruciate ligaments.

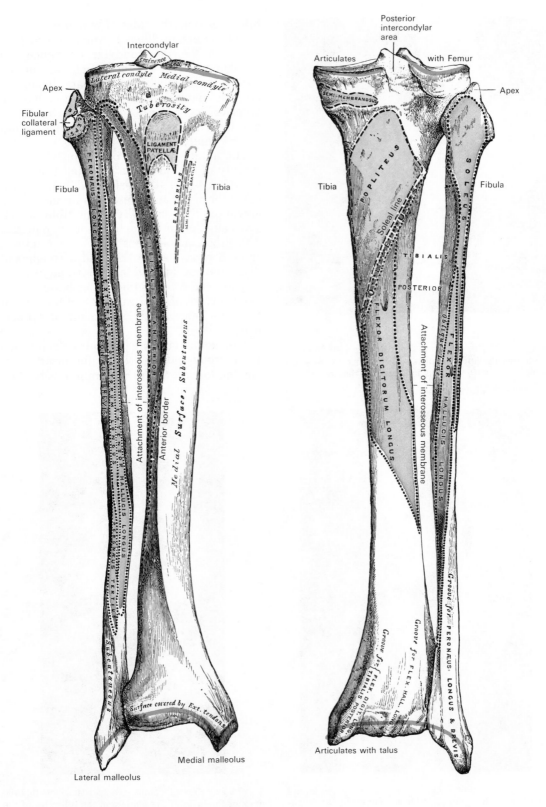

Fig. 4-208. Bones of the right leg. Anterior surface. Articular capsules outlined in blue.

Fig. 4-209. Bones of the right leg. Posterior surface.

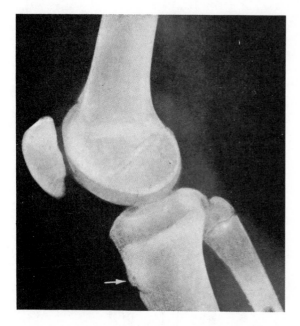

FIG. 4-210.   Knee of a boy aged 16 years. Lateral view. Note that the upper epiphysis of the tibia includes the tibial tuberosity, which is indicated by the arrow.

tubercles onto the sides of which the articular facets are prolonged. Anterior and posterior to the intercondylar eminence are roughened areas for the attachment of the anterior and posterior cruciate ligaments and the menisci. These sites are called the **anterior** and **posterior intercondylar areas,** and they intervene between the medial and lateral condyles (Figs. 4-206; 4-207).

The anterior surface of the proximal extremity of the bone is a flattened triangular area (Fig. 4-208). It is perforated by large vascular foramina and presents a broad oblong elevation, the **tuberosity of the tibia,** which gives attachment to the ligamentum patellae. The tuberosity is divisible into a rounded, smooth proximal portion and a roughened distal portion into which the ligament actually inserts. A bursa intervenes between the deep surface of the ligament and the part of the bone immediately proximal to the tuberosity.

The **medial condyle** presents a deep transverse groove *posteriorly* for the insertion of the tendon of the semimembranosus (Fig.

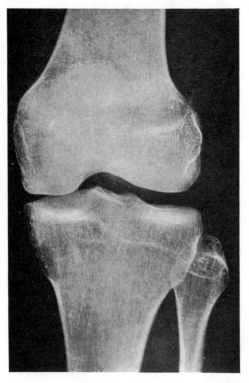

FIG. 4-211.   Adult knee. The gap between the lateral condyles of the femur and tibia is occupied by the articular cartilage of the two bones and the lateral semilunar cartilage.

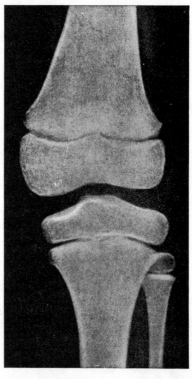

FIG. 4-212.   Knee of a child aged 7½ years. Note that the apex of the head of the fibula and the tubercles of the intercondylar eminence of the tibia are still cartilaginous and therefore cannot be recognized.

4-209). Its *medial surface* is convex and rough, and gives attachment to the tibial collateral ligament. The **lateral condyle** presents a flat, nearly circular, articular facet *posteriorly,* for articulation with the head of the fibula (Fig. 4-209). A rough eminence is situated on a level with the proximal border of the tuberosity at the junction of its anterior and lateral surfaces for the attachment of the iliotibial band. Just distal to this, a part of the extensor digitorum longus takes origin and a slip from the tendon of the biceps femoris is inserted.

BODY OR SHAFT. The **body** or **shaft** of the tibia is triangular in cross-section and considerably expanded proximally. It gradually tapers from the condyles downward to a point about two-thirds along the shaft, where it is thinnest. It then gradually widens again. The shaft presents medial, lateral, and posterior surfaces, which are separated by anterior, medial, and interosseous borders.

The **anterior border** (Fig. 4-208) is the most prominent of the three borders, and is situated subcutaneously throughout its length. This border is referred to as the "skin," and it commences proximally at the tibial tuberosity and ends distally at the anterior margin of the medial malleolus. It is sinuous and more prominent in the proximal two-thirds of its extent, but becomes smoother and rounded distally. It gives attachment to the deep fascia of the leg.

The **medial border** is smooth and rounded both proximally and distally, but sharp and more prominent in the center. It begins below the posterior part of the medial condyle, and ends at the posterior border of the medial malleolus. Its proximal part for about 5 cm gives attachment to the tibial collateral ligament of the knee joint, and below this insert some fibers of the popliteus muscle. From the middle third of this border arise fibers of the soleus and flexor digitorum longus muscles.

The **interosseous** or **lateral border** is thin and prominent, especially its central part, and gives attachment to the interosseous membrane. It commences just below the fibular articular facet of the lateral condyle and bifurcates distally to form the boundaries of a triangular rough surface for the attachment of the interosseous ligament connecting the tibia and fibula. Thus, the interosseous border extends inferiorly to the inferior tibiofibular joint.

The **medial surface** (Fig. 4-208) is smooth, convex, and broad, and is bounded in front by the anterior border and behind by the medial border. Its proximal third, directed anteriorly and medially, is covered by the aponeurosis derived from the tendon of the sartorius, and by the tendons of the gracilis and semitendinosus muscles, all of which are inserted nearly as far anteriorly as the anterior border. The remainder of this surface lies immediately beneath the skin and superficial fascia and bears no muscular attachments.

The **lateral surface** (Fig. 4-208) is somewhat narrower than the medial and lies between the anterior and interosseous borders. Its proximal two-thirds present a shallow concavity for the origin of the tibialis anterior. Its distal third is smooth and convex and curves gradually forward to the anterior aspect of the bone. Although this lower part of the lateral surface has no muscular attachments, it is crossed by the tendons of the tibialis anterior, extensor hallucis longus, and extensor digitorum longus muscles, arranged in this order from medial to lateral. Between the latter two tendons course the anterior tibial vessels and nerve.

The **posterior surface** (Fig. 4-209) at its proximal part presents a prominent ridge, the **soleal line,** which extends obliquely downward from the posterior part of the articular facet for the fibula to the medial border. It marks the distal limit of the insertion of the popliteus, serves for the attachment of the fascia covering this muscle, and gives origin to part of the soleus, flexor digitorum longus, and tibialis posterior muscles. The triangular area proximal to this line receives the insertion of the popliteus. The middle third of the posterior surface presents a **vertical line,** which commences at the soleal line and is well marked proximally and less distinct distally. It separates the origins of the flexor digitorum longus medially and the tibialis posterior laterally (Fig. 4-209). The remaining part of the posterior surface is smooth and covered by the tibialis posterior, flexor digitorum longus, and flexor hallucis longus muscles. Immediately distal to the soleal line is an obliquely directed vascular groove which leads to a rather large **nutrient foramen.**

DISTAL EXTREMITY. The distal extremity is less massive than the proximal, and it presents five surfaces: inferior, anterior, posterior, lateral, and medial. On its medial side there projects distally a strong process, the medial malleolus.

The **inferior surface** is quadrilateral and smooth for articulation with the talus. It is concave anteroposteriorly, broader in front than behind, and medially it is continuous with the articular surface of the medial malleolus.

The **anterior surface** of the distal extremity of the tibia is smooth, rounded proximally, and covered by the tendons of the extensor muscles. Distally the anterior surface extends beyond and slightly overhangs the inferior surface, and its distal margin presents a rough transverse depression for the attachment of the articular capsule of the ankle joint.

The **posterior surface** is traversed by a shallow, longitudinal groove that is directed obliquely downward and medially. It is continuous with a similar groove on the posterior surface of the talus and serves for the passage of the tendon of the flexor hallucis longus.

The **lateral surface** of the distal tibial extremity is triangular and presents the **fibular notch,** which serves for articulation with the distal end of the fibula. The upper circumference of the notch is somewhat roughened for the attachment of the interosseous ligament, which connects the two bones. The lateral surface is bounded by two prominent borders continuous above with the interosseous border. They afford attachment to the anterior and posterior ligaments of the lateral malleolus.

The **medial surface** is prolonged distally to form a short pyramid-shaped process, the **medial malleolus.** The *medial surface* of this process is convex and immediately subcutaneous. Its *lateral* or *articular surface* is smooth, slightly concave, and articulates with the talus. The *anterior border* of the medial malleolus is rough, for the attachment of the anterior fibers of the deltoid ligament of the ankle joint, while its *posterior border* presents a broad groove, the **malleolar sulcus,** which occasionally is double and lodges the tendons of the tibialis posterior and flexor digitorum longus. The *summit* of the medial malleolus is marked by a rough depression for the attachment of the deltoid ligament.

STRUCTURE.   The structure of the tibia is like that of other long bones. The compact wall of the body is

thickest at the junction of the middle and lower thirds of the bone.

OSSIFICATION.   The tibia is ossified from three centers: one for the body and one for each extremity (Figs. 4-213; 4-214). Ossification begins in the center of the shaft at about the seventh week of fetal life, and gradually extends toward the extremities. The center for the upper epiphysis appears before or shortly after birth. It is flat in form (Fig. 4-212) and has a thin tongue-shaped process in front which forms the tuberosity (Fig. 4-214). The center for the lower epiphysis appears in the second year, and the lower epiphysis joins the shaft at about the eighteenth year, while the upper epiphysis joins the shaft at about the twentieth year. Two additional centers occasionally exist, one for the tongue-shaped process of the upper epiphysis, which forms the smooth part of the tibial tuberosity, and one for the medial malleolus.

THE FIBULA

The **fibula** (Figs. 4-208; 4-209) is situated on the lateral aspect of the leg and articulates with the tibia proximally and the tibia and talus distally. In proportion to its length, the fibula is the most slender of all the long bones. Because the fibula does not participate in the formation of the knee joint, it is not a weight-bearing bone. It does, however, afford attachment to numerous muscles which tend to twist its shape, and it does participate importantly in the structure of

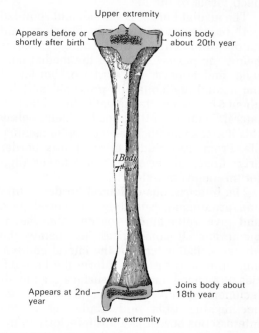

Fig. 4-213.   Plan of ossification of the tibia. From three centers.

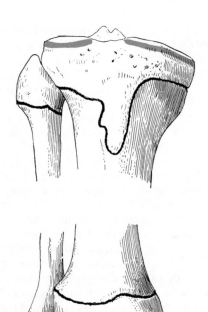

FIG. 4-214. Epiphyseal lines of tibia and fibula in a young adult are shown in black. Lines of attachment of the articular capsules are shown in blue. Anterior aspect.

the ankle joint. Placed parallel to the tibia, the fibula has two extremities and a shaft or body. Its *proximal extremity* is called the **head of the fibula,** and it is placed slightly posterior to the head of the tibia, below the level of the knee joint. Its *distal extremity* inclines a little anteriorly and projects beyond the tibia, forming the lateral part of the ankle joint. It projects laterally as the **lateral malleolus.**

THE HEAD. The head of the fibula (Figs. 4-208; 4-209) is irregular in form and presents a flattened surface, directed upward, forward, and medially, for articulation with a corresponding surface on the lateral condyle of the tibia. On the lateral side is a thick and rough prominence continued posteriorly into a pointed eminence, the **apex of the head** (styloid process). This gives attachment to the tendon of the biceps femoris muscle and to the fibular collateral ligament of the knee joint, the ligament dividing the tendon into two parts. The remaining part of the circumference of the head is rough, for the attachment of muscles and ligaments. *Anteriorly,* the head of the fibula presents a tubercle for the origin of the proximal fibers of the peroneus longus and extensor digi-

torum longus muscles (Fig. 4-208), as well as a surface for the attachment of the anterior ligament of the head. *Posteriorly,* another tubercle gives attachment to the posterior ligament of the head and the origin of the proximal fibers of the soleus muscle (Fig. 4-209).

The slightly narrowed region just below the head of the fibula is sometimes referred to as the **neck of the fibula.** As the common peroneal nerve leaves the popliteal fossa, it courses around the lateral surface of the fibular neck to achieve the anterior aspect of the leg. At this site the nerve lies in a vulnerable position superficially beneath the skin and adjacent to bone.

BODY OR SHAFT. The long, narrow shaft of the fibula is almost entirely surrounded by muscles. It presents *three borders*—anterior, interosseous, and posterior, and *three surfaces*—medial, posterior, and lateral.

The **anterior border** (Fig. 4-215) begins proximally in front of the fibular head. It courses distally to a point just below the middle of the bone, where it curves somewhat laterally and bifurcates to embrace a triangular subcutaneous surface immediately above the lateral malleolus. This border gives attachment to an intermuscular septum, which separates the extensor muscles on the anterior aspect of the leg from the peroneus longus and peroneus brevis on the lateral aspect.

The **interosseous border** (Fig. 4-215) is situated on the medial side of the anterior border and runs nearly parallel with it in the proximal third of the fibula, but diverges from it in the distal two-thirds. It begins just below the head of the fibula and ends at the apex of a rough triangular surface immedi-

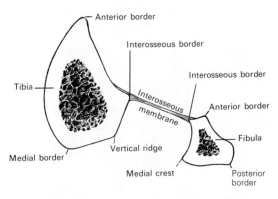

FIG. 4-215. A transverse section through the right tibia and fibula, showing the attachment of the crural interosseous membrane. Viewed from above.

ately proximal to the articular facet of the lateral malleolus (Fig. 4-216). The interosseous border serves for the attachment of the interosseous membrane, which separates the extensor muscles anteriorly from the flexor muscles posteriorly.

The **posterior border** (Fig. 4-215) begins at the apex of the fibula and ends in the posterior aspect of the lateral malleolus. It gives attachment to an aponeurosis which separates the peroneal muscles on the lateral surface from the flexors on the posterior surface.

The **medial surface** is the interval between the anterior and interosseous borders. It is extremely narrow and flat in the proximal third of its extent, but broader and grooved longitudinally in its distal third. It serves for the origin of three muscles: the extensor digitorum longus, extensor hallucis longus, and peroneus tertius (Fig. 4-209).

The **posterior surface** of the fibula is the space between the posterior and interosseous borders. Its proximal three-fourths are separated into posteromedial and posterolateral portions by a longitudinal ridge, the **medial crest** (Fig. 4-215). This crest begins on the medial side of the fibular head and ends by becoming continuous with the interosseous crest at the distal one-fourth of the bone. The medial crest gives attachment to an aponeurosis which separates the tibialis posterior from the soleus and flexor hallucis longus muscles. Thus, the **posterolateral portion** of the posterior surface is included between the posterior border and the medial crest. This portion is continuous distally with the triangular area proximal to the articular surface of the lateral malleolus. Its proximal third is rough, for the origin of the soleus; its distal part presents a triangular surface connected to the tibia by a strong interosseous ligament; the intervening part of the surface is covered by the fibers of origin of the flexor hallucis longus. Near the middle of this surface is the **nutrient foramen.** The **posteromedial portion** is the interval included between the interosseous border and the medial crest. It is grooved for the origin of the tibialis posterior.

The **lateral surface** is the space between the anterior and posterior borders. It is broad, often deeply grooved, and gives origin to the peroneus longus and peroneus brevis.

LATERAL MALLEOLUS.   The lateral malleolus (Figs. 4-216; 4-217) is the expanded distal portion of the fibula, an enlargement not unlike the head. Its point is less narrow and tapered than the apex, however, and its articular surface broader and more oval in shape with a deep longitudinally running groove, the **lateral malleolar fossa,** beside it. The lateral malleolus is more prominent and descends farther distally than does the medial malleolus on the distal extremity of the tibia (Fig. 4-217). The **lateral surface** is convex, subcutaneous, and continuous with the triangular, subcutaneous surface on the lateral side of the fibular body. The **medial surface** (Fig. 4-216) presents a smooth oval facet anteriorly which articulates with a corresponding surface on the lateral side of the talus. Posterior to this **malleolar articular surface** is a rough depression, which gives attachment to the posterior talofibular ligament. The **anterior border** is thick and rough, and marked by a depression for the attachment of the anterior talofibular ligament. The **posterior border** is broad and presents the shallow malleolar sulcus for the passage of the tendons of the peroneus longus and brevis (Fig. 4-209). The inferior tip is rounded and gives attachment to the calcaneofibular ligament.

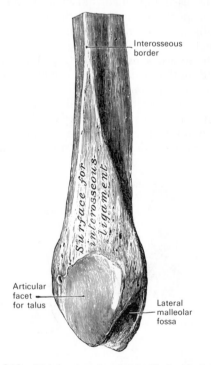

Interosseous border

Surface for interosseous ligament

Articular facet for talus

Lateral malleolar fossa

FIG. 4-216.   Distal extremity of right fibula. Medial aspect.

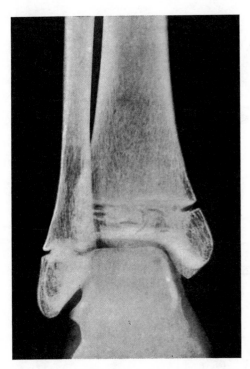

Fig. 4-217.   Ankle of a child aged 10 years. Note that the inferior epiphyseal line of the fibula is opposite the ankle joint.

OSSIFICATION.   The fibula is ossified from three centers: one for the shaft and one for each end (Fig. 4-218). Ossification begins in the shaft about the eighth week of fetal life, and extends toward the extremities. At birth the ends are cartilaginous. Ossification commences in the distal end in the second year, and in the proximal about the third or fourth year. The distal epiphysis, the first to ossify, unites with the body about the twentieth year; the proximal epiphysis, about the twenty-fifth year.

The Foot

Because the bones of the foot must bear the weight of the body as well as transmit thrust in locomotion, it is important that they be structured to lend maximal resilience and elasticity. This is accomplished by their arched arrangement. Each foot forms a half dome with its medial side arched high and the lateral side flattened. Not only transverse, but longitudinal arches as well, contribute to the stability of the foot, and to its ability to withstand shock and distribute weight. The skeleton of the foot (Figs. 4-219; 4-220) consists of three parts: tarsus, metatarsus, and phalanges.

Fig. 4-218.   Plan of ossification of the fibula. From three centers.

THE TARSUS

There are seven **tarsal bones:** the talus, calcaneus, cuboid, navicular, and three cuneiforms.

THE TALUS.   The talus (Figs. 4-221 to 4-224) is the second largest of the tarsal bones and it occupies the proximal part of the tarsus. It supports the tibia, rests upon the calcaneus, and articulates on each side with the malleoli. Distally it articulates with the navicular. The talus consists of a body, a neck, and a head.

The **body** of the talus is cuboidal in shape. Its **superior** or **proximal portion,** the **trochlea,** is convex and covered by a smooth articular surface, which forms a large arc for articulation with the tibia (Figs. 4-221; 4-223). Its lateral surface articulates with the lateral malleolus of the fibula, and its medial with the medial malleolus of the tibia.

Inferiorly, the body of the talus presents two articular areas, the **posterior** and **middle calcaneal surfaces,** separated by a deep groove, the **sulcus tali** (Figs. 4-222; 4-224). In the articulated foot it lies above a similar groove upon the proximal surface of the calcaneus, and forms the **tarsal sinus,** occupied by the interosseous talocalcaneal ligament. The posterior calcaneal articular surface

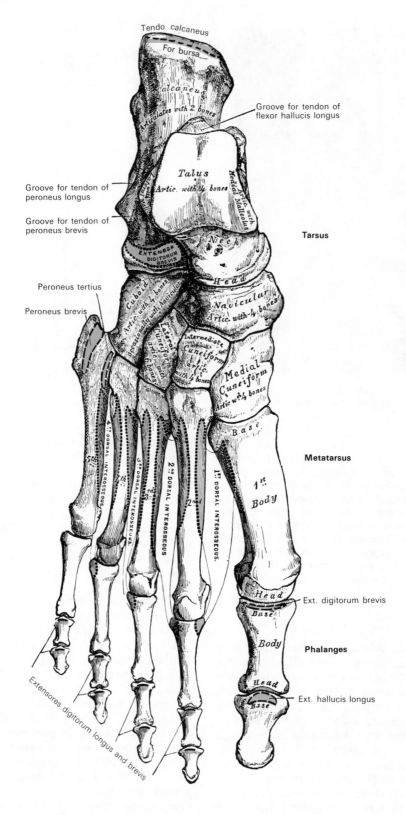

FIG. 4-219. Bones of the right foot. Dorsal surface.

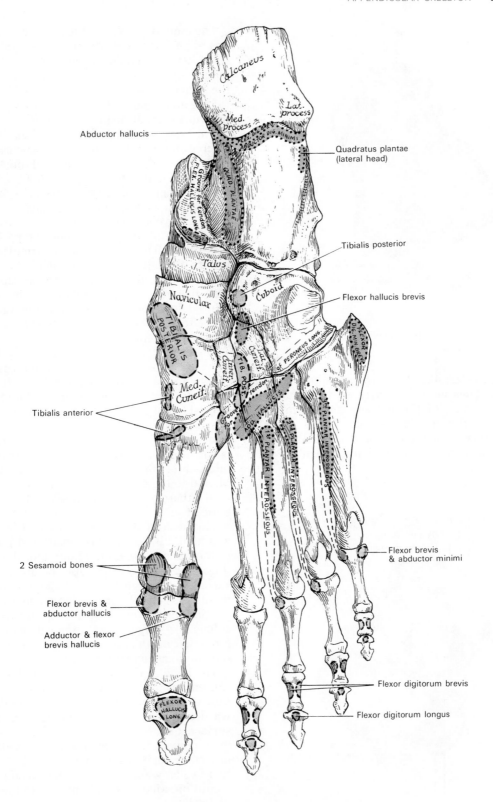

Fig. 4-220. Bones of the right foot. Plantar surface.

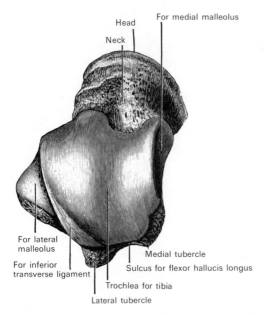

FIG. 4-221.   Left talus. Proximal aspect.

(Figs. 4-222; 4-224) is large and oblong, and articulates with the corresponding facet on the surface of the calcaneus. The middle calcaneal articular surface (Figs. 4-222; 4-224) is small and oval, and articulates with the facet on the sustentaculum tali of the calcaneus.

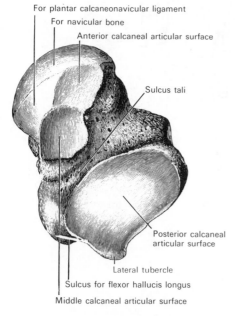

FIG. 4-222.   Left talus. Distal aspect.

The **medial surface** of the body is continuous superiorly with the trochlea and presents a pear-shaped articular facet for the medial malleolus (Figs. 4-221, 4-223). Below the articular surface is a rough depression for the attachment of the deep portion of the deltoid ligament of the ankle joint.

The **lateral surface** of the body, which also is continuous with the trochlea, carries a large triangular facet for articulation with the lateral malleolus (Figs. 4-221; 4-224). Anteriorly, there is a rough depression for the attachment of the anterior talofibular ligament. Between the posterior half of the lateral border of the trochlea and the posterior part of the base of the fibular articular surface is a triangular facet, which comes into contact with the transverse inferior tibiofibular ligament during flexion of the ankle joint. Below the base of this facet is a groove that affords attachment to the posterior talofibular ligament.

Posteriorly, the surface of the body of the talus is narrow and is traversed by a groove for the tendon of the flexor hallucis longus, which lies between two tubercles (Fig. 4-223). Lateral to the groove is a prominent tubercle, the **posterior process,** to which the posterior talofibular ligament is attached. This process is sometimes separated from the rest of the talus, and is then known as the **os trigonum.** Medial to the groove is a second less prominent tubercle to which attaches the medial talocalcaneal ligament.

The **neck of the talus** is directed forward and medially and is the constricted portion of the bone between the body and the head (Figs. 4-223; 4-224). Its superior and medial surfaces are rough for the attachment of ligaments. Its lateral surface is concave and is continuous below with a deep groove, the sulcus tali. Together with the calcaneus this groove forms the tarsal sinus for the interosseous talocalcaneal ligament.

The **head of the talus** is the rounded anterior end of the bone (Figs. 4-223; 4-224). It is directed distally and oriented slightly downward and medially. The **anterior** or **distal surface** of the head is covered by a large, oval, convex facet for articulation with the navicular. Its **inferior surface** (Fig. 4-222) has two facets, which are best seen in the fresh condition. Medially and anterior to the middle calcaneal facet is a semi-oval surface that rests on the plantar calcaneonavicular ligament. Laterally is located the flattened

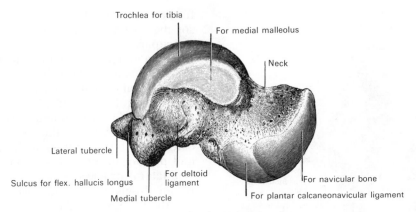

Trochlea for tibia

For medial malleolus

Neck

Lateral tubercle

Sulcus for flex. hallucis longus

For deltoid ligament

Medial tubercle

For navicular bone

For plantar calcaneonavicular ligament

FIG. 4-223.   Left talus. Medial surface.

anterior calcaneal articular surface, which articulates with the facet on the superior surface of the anteromedial part of the calcaneus.

Articulations.   The talus articulates with four bones: tibia, fibula, calcaneus, and navicular.

THE CALCANEUS.   The calcaneus (Figs. 4-225 to 4-228) is the largest and strongest bone in the foot. It is situated in the posterior part of the foot and extends backward beyond the tibia and fibula to form the heel. The calcaneus serves to transmit the weight of the body to the ground and provides a strong lever for the muscles of the calf. It is somewhat cuboidal, with its long axis directed forward and laterally, and presents six surfaces for examination. These are the **superior, distal, lateral, medial, anterior,** and **posterior surfaces.**

The **superior** or **proximal surface** (Fig. 4-225) extends behind onto the portion of the calcaneus that forms the heel. It varies in length in different individuals and is concave from front to back and convex from side to side. It is divisible into two parts, anterior and posterior. The *anterior part* contains three facets which are oriented proximally for articulation with the talus. The **posterior articular surface** is the largest of the three and is oval in shape and faces somewhat anteriorly. Medial to this facet is the **middle articular surface,** which is supported by the sustentaculum tali. The **anterior articular facet** is anterior and lateral to the middle facet and is in close proximity or joined with it. All three facets articulate with those of similar name on the distal surface of the talus. The **calcaneal sulcus** is a deep depression between the posterior and middle articular facets. In the intact foot it combines with a similar groove on the talus, called the sulcus tali, to form the **tarsal sinus** which contains the interrosseous talocalcaneal ligament. Anterior and lateral to the calcaneal sulcus and the three articular facets, the rough surface provides attachment

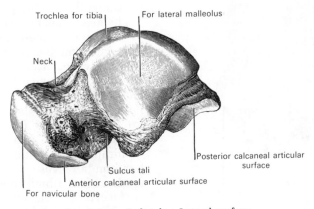

Trochlea for tibia

For lateral malleolus

Neck

Posterior calcaneal articular surface

Sulcus tali

Anterior calcaneal articular surface

For navicular bone

FIG. 4-224.   Left talus. Lateral surface.

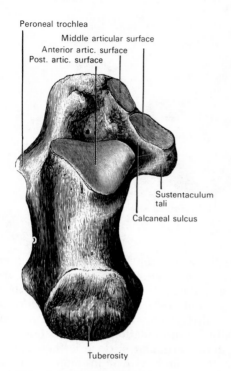

Peroneal trochlea
Middle articular surface
Anterior artic. surface
Post. artic. surface

Sustentaculum tali

Calcaneal sulcus

Tuberosity

FIG. 4-225. Left calcaneus. Superior or proximal surface.

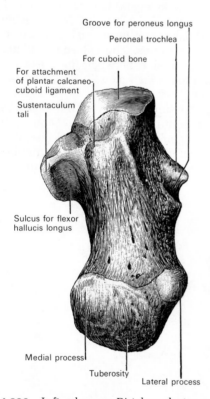

Groove for peroneus longus

Peroneal trochlea

For cuboid bone

For attachment of plantar calcaneo-cuboid ligament

Sustentaculum tali

Sulcus for flexor hallucis longus

Medial process

Tuberosity

Lateral process

FIG. 4-226. Left calcaneus. Distal or plantar surface.

for ligaments and for the origin of the extensor digitorum brevis. The *posterior part* of the superior surface is a rather smooth convex area extending from the articular facets to the calcaneal tuberosity.

The **distal** or **plantar surface** (Fig. 4-226) is wider behind than in front and convex from side to side. It is bounded posteriorly by a transverse elevation, the **calcaneal tuberosity,** which is depressed in the middle and prolonged on each side into a process. The **lateral process**—small, prominent, and rounded—gives origin to part of the abductor digiti minimi. The **medial process** is broader and larger, and gives attachment by its prominent medial margin to the abductor hallucis, the flexor digitorum brevis, and the plantar aponeurosis. The depression between the processes gives origin to the remainder of the abductor digiti minimi. The rough surface anterior to the processes gives attachment to the long plantar ligament, and to the lateral head of the quadratus plantae muscle. The plantar calcaneocuboid ligament is attached to a prominent tubercle nearer the anterior part of this surface, as well as to a transverse groove anterior to the tubercle.

The **lateral surface** (Fig. 4-227) is broad behind and narrow in front, flat, and almost subcutaneous. At its upper and anterior part, the lateral surface gives attachment to the lateral talocalcaneal ligament. Near its center is a tubercle for the attachment of the calcaneofibular ligament. Anterior to the tubercle is a narrow surface marked by two oblique grooves and separated by an elevated ridge, the **peroneal trochlea,** which varies much in size in different bones. The superior groove transmits the tendon of the peroneus brevis, while the inferior groove transmits that of the peroneus longus.

The **medial surface** is deeply concave, and the concavity is made more apparent by the prominent medial process of the calcaneal tuberosity posteriorly and the overhanging horizontal eminence, the **sustentaculum tali,** anterosuperiorly (Fig. 4-228). Within this hollow course the plantar vessels and nerves into the sole of the foot. The medial surface affords origin to the medial head of the quadratus plantae muscle, while the sustentaculum tali gives attachment to a slip of the tendon of the tibialis posterior. The *superior surface* of the sustentaculum tali contains the facet for articulation with the middle

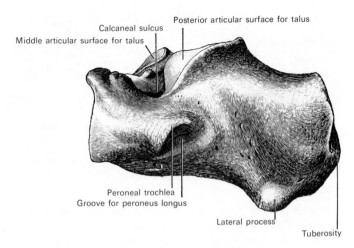

Calcaneal sulcus

Middle articular surface for talus

Posterior articular surface for talus

Peroneal trochlea
Groove for peroneus longus

Lateral process

Tuberosity

FIG. 4-227.   Left calcaneus. Lateral surface.

calcaneal articular surface of the talus. The *inferior surface* is grooved for the tendon of the flexor hallucis longus; its *anterior margin* gives attachment to the plantar calcaneonavicular ligament, and its *medial margin* to a part of the deltoid ligament of the ankle joint.

The **anterior surface** of the calcaneus (Fig. 4-226) is small and somewhat triangular in shape. It consists almost entirely of a facet for articulation with the cuboid bone. The plantar calcaneonavicular ligament is attached to its medial border.

The **posterior surface** protrudes as the prominence of the heel (Figs. 4-226; 4-228). It is divisible into three areas. The *inferior area* is rough and covered by the fatty, fi-brous tissue of the heel. The *middle area* also is rough and forms the calcaneal tuberosity onto which inserts the tendo calcaneus and plantaris. The *superior portion* is smooth, and is covered by a bursa that intervenes between it and the tendo calcaneus.

Articulations.   The calcaneus articulates with two bones: the talus and cuboid.

CUBOID BONE.   The cuboid bone (Figs. 4-229; 4-230) is placed on the lateral aspect of the foot, distal to the calcaneus and proximal to the fourth and fifth metatarsal bones. The **dorsal surface** is directed laterally and is roughened for the attachment of ligaments. The **plantar surface** presents a deep

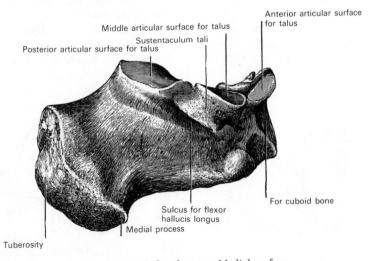

Middle articular surface for talus

Sustentaculum tali

Posterior articular surface for talus

Anterior articular surface for talus

Sulcus for flexor hallucis longus

Medial process

For cuboid bone

Tuberosity

FIG. 4-228.   Left calcaneus. Medial surface.

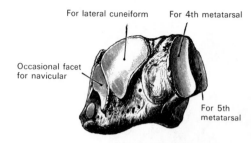

FIG. 4-229.   The left cuboid. Anteromedial view.

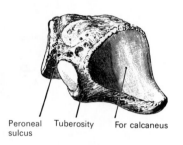

FIG. 4-230.   The left cuboid. Posterolateral view.

oblique groove, the **sulcus for the peroneus longus tendon.** This groove is bounded posteriorly by a prominent ridge, to which the long plantar ligament is attached. This ridge ends laterally in the **tuberosity of the cuboid bone,** the surface of which presents an oval facet over which a sesamoid bone or a cartilage in the tendon of the peroneus longus glides. The surface of bone posterior to the groove is rough, for the attachment of the plantar calcaneocuboid ligament, a few fibers of the flexor hallucis brevis, and a fasciculus from the tendon of the tibialis posterior. The **lateral surface** presents a deep notch formed by the commencement of the peroneal sulcus. The **posterior** or **proximal surface** is smooth and triangular in shape for articulation with the anterior surface of the calcaneus (Fig. 4-230). Its medial plantar angle projects posteriorly as a process that underlies and supports the anterior end of the calcaneus. The **anterior** or **distal surface** (Fig. 4-229) is divided by a vertical ridge into two facets: the medial facet is quadrilateral in shape and articulates with the fourth metatarsal; the lateral is triangular in shape and articulates with the fifth. The **medial surface** is broad and irregularly quadrilateral in shape. It presents at its middle and superior

part a smooth oval facet for articulation with the lateral cuneiform. Posterior to this is occasionally seen a smaller facet for articulation with the navicular. The remainder of this surface is nonarticular, but it is roughened for the attachment of strong interosseous ligaments.

Articulations. The cuboid articulates with four bones: the calcaneus, lateral cuneiform, and fourth and fifth metatarsals. Occasionally it articulates with a fifth bone, the navicular.

NAVICULAR BONE. The navicular bone (Figs. 4-231; 4-232) is situated at the medial side of the tarsus, between the talus proximally and the cuneiform bones distally. The **anterior** or **distal surface** is convex from side to side and subdivided by two ridges into three facets, for articulation with the three cuneiform bones (Fig. 4-231). The **posterior** or **proximal surface** is oval and concave and articulates with the rounded head of the talus. The **dorsal surface** is convex and rough for the attachment of ligaments. The **plantar surface** is irregular, and also rough for the attachment of ligaments. The **medial surface** presents the rounded **navicular tuberosity,** the lower part of which affords the

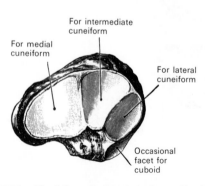

FIG. 4-231.   The left navicular. Anterior or distal view.

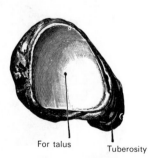

FIG. 4-232.   The left navicular. Posterior or proximal view.

principal attachment to the tendon of the tibialis posterior. The **lateral surface** is rough and irregular for the attachment of ligaments, and occasionally presents a small facet for articulation with the cuboid bone.

Articulations. The navicular articulates with four bones: the talus and the three cuneiforms. Occasionally it articulates with a fifth bone, the cuboid.

MEDIAL CUNEIFORM BONE. The medial cuneiform (wedge-shaped) bone is the largest of the three cuneiforms (Figs. 4-233; 4-234). It is situated on the medial side of the foot, between the navicular proximally, and the base of the first metatarsal distally. The **medial surface** is subcutaneous, broad, and quadrilateral, and at its anterior plantar angle is a smooth oval impression, into which part of the tendon of the tibialis anterior is inserted (Fig. 4-233). The rest of its extent is rough for the attachment of ligaments. The **lateral surface** is concave, presenting along its superior and posterior borders a narrow L-shaped surface which articulates with the intermediate cuneiform. The dorsal and distal part of the lateral surface presents a small facet which articulates with the medial surface of the base of the second metatarsal bone (Fig. 4-234). The rest of this surface is rough for the attachment of ligaments and part of the tendon of the peroneus longus. The **anterior** or **distal surface** is kidney-shaped and articulates with the first metatarsal bone. The **posterior** or **proximal surface** is triangular and concave and articulates with the most medial and largest of the three facets on the anterior surface of the navicular. The **plantar surface** is rough and

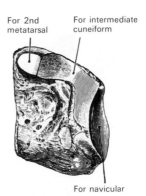

For 2nd metatarsal   For intermediate cuneiform

For navicular

FIG. 4-234.   The left medial cuneiform. Lateral aspect.

forms the base of the wedge. At its posterior part is a tuberosity for the insertion of part of the tendon of the tibialis posterior. More anteriorly and laterally a slip from the tendon of the tibialis anterior also inserts on this surface. The **dorsal surface** is the narrow end of the wedge and is rough for the attachment of ligaments.

Articulations. The first cuneiform articulates with four bones: the navicular, intermediate cuneiform, and first and second metatarsals.

INTERMEDIATE CUNEIFORM BONE. The intermediate cuneiform bone (Figs. 4-235; 4-236) is the smallest of the three cuneiforms and presents a regular wedge shape, the thin end being the plantar. It is situated between the other two cuneiform bones and articulates with the navicular proximally and the second metatarsal distally. Thus, the **anterior** or **distal surface,** triangular in form, articulates with the base of the second metatarsal bone. The **posterior** or **proximal surface,** also triangular in shape, articulates with the intermediate facet on the anterior surface of the navicular. The **medial surface**

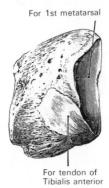

For 1st metatarsal

For tendon of Tibialis anterior

FIG. 4-233.   The left medial cuneiform. Medial aspect.

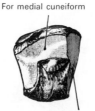

For medial cuneiform

For 2nd metatarsal

FIG. 4-235.   The left intermediate cuneiform. Medial and distal view.

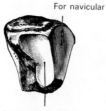

For navicular

For lateral cuneiform

FIG. 4-236.   The left intermediate cuneiform. Lateral and proximal view.

carries an L-shaped articular facet, running along the superior and posterior borders, for articulation with the medial cuneiform (Fig. 4-235); it is rough in the rest of its extent for the attachment of ligaments. The **lateral surface** presents posteriorly a smooth facet for articulation with the lateral cuneiform bone (Fig. 4-236). The **dorsal surface** forms the base of the wedge; it is quadrilateral and rough for the attachment of ligaments. The **plantar surface,** sharp and tuberculated, is also rough for the attachment of ligaments, and for the insertion of a slip from the tendon of the tibialis posterior.

Articulations.   The intermediate cuneiform articulates with four bones: the navicular, medial and lateral cuneiforms, and second metatarsal.

LATERAL   CUNEIFORM   BONE.   The lateral cuneiform bone (Figs. 4-237; 4-238) is also wedge-shaped and interposed between the cuboid and the intermediate cuneiform in the distal row of tarsal bones. The **distal** or **anterior surface** of the lateral cuneiform bone articulates with the base of the third metatarsal. Its **proximal** or **posterior surface** articulates with the lateral facet on the anterior surface of the navicular, and is rough inferiorly for the attachment of ligamentous fibers. The **medial surface** (Fig. 4-237) pre-

sents a distal and a proximal articular facet, separated by a rough depression. The distal facet, sometimes divided, articulates with the lateral aspect of the base of the second metatarsal bone. The proximal facet skirts the proximal margin and articulates with the intermediate cuneiform. The rough depression separating the articular facets gives attachment to an interosseous ligament. The **lateral surface** (Fig. 4-238) also presents two articular facets, separated by a rough nonarticular area. The distal facet, situated at the superior angle of the bone, is small and semi-oval, and articulates with the medial side of the base of the fourth metatarsal bone. The proximal and larger facet is triangular or oval, and articulates with the cuboid. The rough, nonarticular area between these serves for the attachment of an interosseous ligament. The three facets for articulation with the three metatarsal bones are continuous with one another; those for articulation with the intermediate cuneiform and navicular are also continuous, but that for articulation with the cuboid is usually separate. The **dorsal surface** is rectangular in shape and prolonged proximally, forming the base of the wedge. The **plantar surface** is the narrow, rounded margin of the wedge, and serves for the attachment of part of the tendon of the tibialis posterior, part of the flexor hallucis brevis, and for ligaments.

Articulations.   The lateral cuneiform articulates with six bones: the navicular, intermediate cuneiform, cuboid, and second, third, and fourth metatarsals.

THE METATARSUS

The **metatarsus** consists of five bones situated between the tarsal bones and the phalanges (Figs. 4-219; 4-220; 4-245; 4-246). They

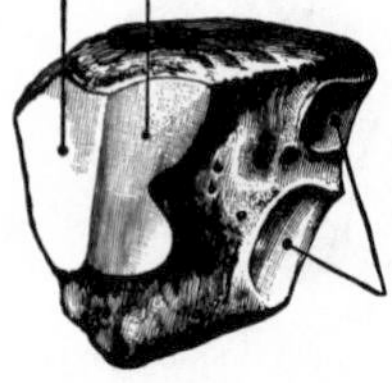

For navicular   For intermediate cuneiform

For 2nd metatarsal

FIG. 4-237.   The left lateral cuneiform. Medial and proximal aspect.

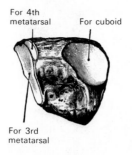

For 4th metatarsal   For cuboid

For 3rd metatarsal

FIG. 4-238.   The left lateral cuneiform. Lateral and distal aspect.

are numbered from one to five from the medial side. Thus, the first metatarsal is that which articulates with the phalanges of the large toe, while the fifth metatarsal joins the phalanges of the small toe. Similar to the metacarpals, each metatarsal is a long bone in miniature and presents for examination a base, body, and head.

COMMON CHARACTERISTICS OF THE METATAR-
SAL BONES. The **body** is long and slender, tapers gradually from the tarsal to the phalangeal extremity, and is curved longitudinally, so as to be convex dorsally and concave on the plantar aspect. The **base** or **proximal extremity** is wedge-shaped, articulating with the tarsal bones and, by its sides, with the contiguous metatarsal bones. The **dorsal** and **plantar surfaces** of the metatarsals are rough for the attachment of ligaments. The **head** or **distal extremity** articulates with the proximal phalanx of each toe and presents a convex articular surface, the plantar portion of which extends farther proximally than the dorsal. The sides of the heads are flattened, and on each is a depression, surmounted by a tubercle, for ligamentous attachment. The plantar surface of the heads is grooved anteroposteriorly for the passage of the flexor tendons, and marked on either side by an articular eminence continuous with the distal articular surface.

CHARACTERISTICS OF THE INDIVIDUAL META-
TARSAL BONES. The individual metatarsal bones possess distinctive anatomical features which allow them to be readily identified.

First Metatarsal Bone. The first metatarsal bone (Fig. 4-239) is remarkable for its great thickness and is the shortest of the metatarsal bones. The **body** or **shaft** is strong, and is markedly prismatic in form. Its lateral surface is triangular and flat and gives origin to the medial head of the first dorsal interosseous muscle. The **base** presents, as a rule, no articular facets on its sides, but occasionally on the lateral aspect there is an oval facet that articulates with the second metatarsal. Its proximal articular surface is large and kidney-shaped for articulation with the medial cuneiform bone. The circumference of this articular surface is grooved, for the tarsometatarsal ligaments. Medially it receives the insertion of part of the tendon of the tibialis anterior, and laterally it presents a rough oval prominence for the insertion of the tendon of the peroneus longus. The **head**

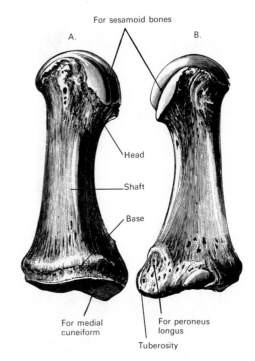

Fig. 4-239. The left first metatarsal bone. *A*, Medial aspect. *B*, Lateral aspect.

is large and on its plantar surface are two grooved facets, on which glide sesamoid bones. The facets are separated by a smooth central elevation.

Second Metatarsal Bone. The second metatarsal bone (Fig. 4-240) is the longest of the metatarsal bones, being prolonged proximally to fit into the recess formed by the three cuneiform bones. Its **base** is broad dorsally, narrow and rough below. It presents four articular surfaces: a posterior triangular facet for articulation with the intermediate cuneiform; one at its medial surface for articulation with the medial cuneiform; and two on its lateral surface separated by a rough nonarticular interval. Each of these lateral articular surfaces is divided into two facets by a vertical ridge. The two anterior facets articulate with the third metatarsal, while the two posterior (sometimes continuous) articulate with the lateral cuneiform. A fifth facet is occasionally present for articulation with the first metatarsal. It is oval in shape, and is situated on the medial side of the body, just distal to the base. The plantar surface of the base of the second metatarsal affords insertion to a slip of the tibialis posterior tendon. From the medial surface of the shaft arises the lateral head of the first dorsal

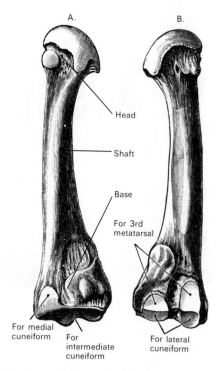

FIG. 4-240. The left second metatarsal bone. *A,* Medial aspect. *B,* Lateral aspect.

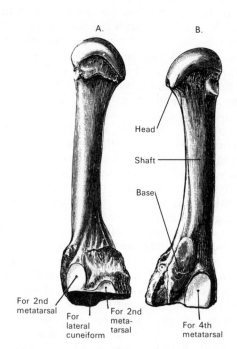

FIG. 4-241. The left third metatarsal bone. *A,* Medial aspect. *B,* Lateral aspect.

interosseous muscle, while the lateral surface gives origin to the medial head of the second dorsal interosseous.

Third Metatarsal Bone. The third metatarsal bone (Fig. 4-241) is shorter than the second, and its **base** articulates proximally with the lateral cuneiform by means of a triangular smooth surface. Medially the base articulates, by two facets, with the second metatarsal, and laterally, by a single facet, with the fourth metatarsal. This last facet is situated at the dorsal angle of the base. The base allows insertion of a small slip from the tibialis posterior. From the shaft arise the lateral head of the second dorsal interosseous muscle and the medial head of the third dorsal interosseous muscle, and from the plantar surface arises the first plantar interosseous muscle.

Fourth Metatarsal Bone. The fourth metatarsal bone (Fig. 4-242) is smaller in size than the preceding. Its **base** presents an oblique quadrilateral surface for articulation with the cuboid. A smooth facet on the medial side is divided by a ridge into an anterior portion for articulation with the third

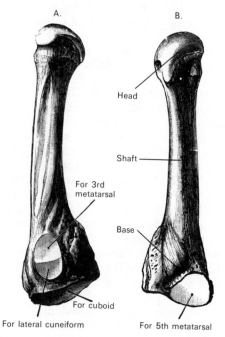

FIG. 4-242. The left fourth metatarsal bone. *A,* Medial aspect. *B,* Lateral aspect.

metatarsal, and a posterior portion for articulation with the lateral cuneiform. On the lateral side a single facet is for articulation with the fifth metatarsal. The shaft of the fourth metatarsal gives origin on its medial aspect to the lateral head of the third dorsal interosseous muscle and on its lateral aspect to the medial head of the fourth dorsal interosseous muscle. Additionally, on its plantar surface arises the second plantar interosseous muscle.

Fifth Metatarsal Bone. The fifth metatarsal bone (Fig. 4-243) is recognized by a rough eminence, the **tuberosity,** on the lateral aspect of its base. The base articulates posteriorly with the cuboid and medially with the fourth metatarsal. On the medial part of its dorsal surface is inserted the tendon of the peroneus tertius and on the dorsal surface of the tuberosity that of the peroneus brevis. A strong band of the plantar aponeurosis connects the projecting part of the tuberosity with the lateral process of the tuberosity of the calcaneus. The plantar surface of the base is grooved for the tendon of the abductor digiti minimi, and gives origin to the flexor digiti minimi brevis. On the medial aspect of the shaft arises the lateral head of the fourth dorsal interosseous muscle, while from its plantar aspect arises the third plantar interosseous muscle.

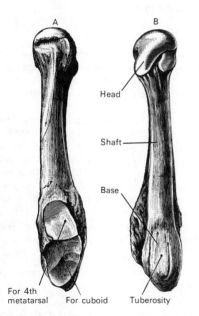

A

B

Head

Shaft

Base

For 4th
metatarsal   For cuboid   Tuberosity

Fɪɢ. 4-243.   The left fifth metatarsal bone. *A,* Medial aspect. *B,* Lateral aspect.

ARTICULATIONS.   The base of each metatarsal bone articulates with one or more of the tarsal bones, and the head of each metatarsal articulates with the first phalanx of the corresponding toe. The first metatarsal articulates with the medial cuneiform, the second with all three cuneiforms, the third with the lateral cuneiform, the fourth with the lateral cuneiform and the cuboid, and the fifth with the cuboid. In addition the bases of the metatarsals articulate with those of adjacent metatarsals.

PHALANGES OF THE FOOT

The **phalanges** of the foot (Figs. 4-219; 4-220; 4-245; 4-246) correspond in number and general arrangement to those of the hand. There are two in the great toe and three in each of the other toes. They differ from the phalanges of the hand because of their smaller size. The shafts are much reduced in length and they are laterally compressed, especially the phalanges of the proximal row.

PROXIMAL ROW (*first row*).   The phalanges of the proximal row are compressed in the middle and expanded at each extremity. The **shafts** are convex dorsally and concave on their plantar surface. The **base** is concave, articulating with the head of the corresponding metatarsal bone. The **head** presents a trochlear surface for articulation with the middle phalanx.

MIDDLE ROW (*second row*).   The phalanges of the second row are remarkably small and short, but somewhat broader than those of the proximal row.

DISTAL ROW.   The **ungual phalanges** resemble those of the fingers, but are smaller and more flattened. Each presents a broad base for articulation with the corresponding bone of the middle row, and an expanded distal extremity, called the **ungual tuberosity,** for the support of the toenail dorsally and the pulp of the tip of the toe distally.

ARTICULATIONS.   In the second, third, fourth, and fifth toes, the phalanges of the proximal row articulate with the metatarsal bones proximally and the middle phalanges distally. The middle phalanges articulate with the proximal and distal phalanges, while the ungual phalanges articulate with the middle. The first toe has but two phalanges, the proximal articulates with the first metatarsal bone and with the distal phalanx. The distal phalanx articulates only with the proximal phalanx.

OSSIFICATION OF THE BONES OF THE FOOT (Fig. 4-244).   The **tarsal bones** are each ossified from a

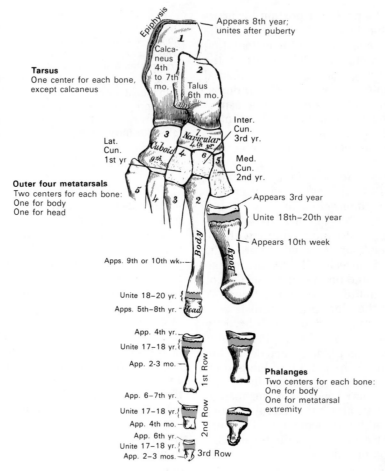

Tarsus
One center for each bone, except calcaneus

Appears 8th year; unites after puberty

Epiphysis

1. Calca-neus 4th to 7th mo.

2 Talus 6th mo.

Inter. Cun. 3rd yr.

Lat. Cun. 1st yr

3 *Cuboid* *4* 9th.

7 *Navicular* 4th yr.

6 *5*

Med. Cun. 2nd yr.

Outer four metatarsals
Two centers for each bone:
One for body
One for head

5 *4* *3* *2*

Body

Appears 3rd year

Unite 18th–20th year

Appears 10th week

Apps. 9th or 10th wk.

Body

Unite 18–20 yr.
Apps. 5th–8th yr. *Head*

App. 4th yr.
Unite 17–18 yr.

App. 2-3 mo.

1st Row

Phalanges
Two centers for each bone:
One for body
One for metatarsal extremity

App. 6–7th yr.
Unite 17–18 yr.
App. 4th mo.

2nd Row

App. 6th yr.
Unite 17–18 yr.
App. 2–3 mos. 3rd Row

FIG. 4-244. Plan of ossification of the foot.

single center, except the calcaneus, which has an epiphysis for its posterior extremity. The centers make their appearance in the following order: *calcaneus,* between the fourth and seventh month of fetal life; *talus,* about the sixth month or earlier; *cuboid,* at the ninth month; *lateral cuneiform,* during the first year; *medial cuneiform,* in the second year; *intermediate cuneiform* and *navicular,* in the third year. The epiphysis for the posterior extremity of the calcaneus appears at the eighth year and unites with the rest of the bone soon after puberty. The posterior process of the talus is sometimes ossified from a separate center and may remain distinct from the main mass of the bone. When this occurs, it is named the *os trigonum.*

The **metatarsal bones** corresponding to the second, third, fourth, and fifth toes are each ossified from two centers, one for the body or shaft, and one for the head. The first metatarsal ossifies from centers for its shaft and base. Ossification commences in the centers of the shafts during the ninth or tenth week and extends toward either extremity. The ossification center of the epiphysis at the base of the first metatarsal appears about the third year, while the centers for the heads of the other bones appear between the fifth and eighth years. The epiphyses unite with the shafts between the eighteenth and twentieth years.

The **phalanges** are each ossified from two centers, a primary center for the body or shaft and an epiphysial center for the base. The primary ossification centers for the shafts of the proximal and distal phalanges appear between the eighth and twelfth fetal week, while those for the shafts of the middle phalanges appear at about the sixteenth intrauterine week. Epiphysial centers in the bases of the phalanges appear between the second and tenth years. The epiphyses join the shafts at about the eighteenth year.

SESAMOID BONES

Sesamoid bones are rounded formations of bone embedded in certain tendons where the latter are subjected to compression as well as to their usual tensile stresses. Generally they measure only a few millimeters in diameter; however, the patella is an example of a large sesamoid bone which is consistently observed in the tendon of the quadriceps femoris muscle. The sesamoid bones are genuine parts of the skeleton and are found in the fetus as well as in phylogenetically distantly related animal forms, often in greater abundance than in the adult human body. The precise function of sesamoid bones is uncertain, and indeed may vary at different sites. Their association with the physical stresses, tensions, and friction produced by tendons is generally acknowledged.

In the *upper limb,* the sesamoid bones occur normally only in relation to the joints of the palmar surface of the hand. The two most constant are in the tendons of the two heads of the flexor pollicis brevis at the metacarpophalangeal joint, the medial or deep head having the larger bone. A corresponding one in the flexor digiti minimi brevis is frequently present at the metacarpophalangeal joint and one (or two) at the same joint of the index finger. Additionally, a sesamoid bone is often found in the insertion tendon of the adductor pollicis muscle. Sesamoids are also found occasionally at the metacarpophalangeal joint of the middle and ring fingers, at the interphalangeal joint of the thumb, and at the distal interphalangeal joint of the index finger. Sesamoid bones apart from joints are seldom found in the tendons of the upper limb, although one is seen at times in the tendon of the biceps brachii opposite the radial tuberosity.

In the *lower limb,* the largest sesamoid bone is the patella, in the tendon of the quadriceps femoris at the knee. Two constant sesamoids occur in the tendons of the flexor hallucis brevis at the metatarsophalangeal joint, the medial being the larger. One is occasionally found at the same joint of the second and fifth toes, or even of the third and fourth toes, and one at the interphalangeal joint of the great toe. Sesamoid bones in the lower extremity are seen in several tendons at sites other than at joints. One is in the tendon of the peroneus longus, where it glides on the cuboid. Another, appearing late in life, is found in the tendon of the tibialis anterior, opposite the smooth facet of the medial cuneiform bone. Other sesamoid bones are found in the tendon of the tibialis posterior, opposite the medial side of the head of the talus; in the lateral head of the gastrocnemius, behind the lateral condyle of the femur; and in the tendon of the psoas major, where it glides over the pubis. Occasionally, sesamoid bones are also found in the tendons that wind around the medial and lateral malleoli, and in the tendon of the gluteus maximus, as it passes over the greater trochanter.

COMPARISON OF THE BONES OF THE HAND AND FOOT

The hand and foot are constructed on somewhat similar principles, each consisting of a proximal part, the carpus or the tarsus, an intermediate portion, the metacarpus or the metatarsus, and a distal portion, the phalanges. The *proximal part* consists of a series of more or less cubical bones which allow a slight amount of gliding on one another and are chiefly concerned in distributing forces transmitted to or from the bones of the arm or leg. The *intermediate part* is made up of slightly movable long bones which assist the carpus or tarsus in distributing forces and also give greater breadth for the reception of such forces. The separation of the individual bones from one another allows the attachments of the interosseous muscles and protects the dorsal-to-palmar and dorsal-to-plantar vascular anastomoses. The *distal portion* is the most movable, and its separate elements enjoy a varied range of movements, the most common of which are flexion and extension.

Although comparable in structure, the hand and foot show modifications in their detailed anatomy which meet the requirements of their very different functions. In essence, the hand serves as the appendage of the upper extremity and is controlled in its function as a grasping, holding, and presenting organ. In contrast, the foot forms a firm basis of support for the body in the erect posture, and is therefore more solid. Its component parts are less movable than are those of the hand.

In the case of the phalanges the differences are readily noticeable. Those of the foot are smaller and their movements are more limited than those of the hand. Very much more marked is the difference between the metacarpal bone of the thumb and the metatarsal bone of the great toe. The *metacarpal bone of the thumb* is constructed to permit great mobility, is directed at an acute angle from that of the index finger, and is capable of a considerable range of movement at its articulation with the carpus. The thumb is rotated, thereby allowing for its ability to oppose each of the other fingers. The *metatarsal bone of the great toe* assists in supporting the weight of the body, is constructed with great solidity, lies parallel with and in the same plane as the other metatarsals, and has a very limited degree of mobility.

The carpus is small in proportion to the size of the rest of the hand. The carpal bones are placed in line with the forearm, and form a transverse arch, the concavity of which constitutes a bed for the flexor tendons entering the hand from the forearm and for the palmar vessels and nerves. In contrast, the tarsus is large and forms a considerable part of the foot. It is placed at right angles to the leg, a position that is almost peculiar to man and facilitates his erect posture. In order to allow the support of the body and to distribute its weight properly, the tarsus, metatarsus, and phalanges of the foot are constructed as a series of both **longitudinal** and **transverse arches,** the nature of which will be considered after the articulations of the foot have been described (Chapter 5).

Histologic Structure of Connective Tissues

The connective tissues are a broadly distributed group of body elements of many varieties derived from embryonic mesoderm. Widely ranging structures, such as fascia, fat, ligaments, and tendons are formed of connective tissue, as are the more specialized tissues of bone, cartilage, and blood. Connective tissue always contains in various proportions three principal components: **cells,** extracellular **fibers,** and interstitial **ground substance.** The extracellular fibers and ground substance form an intercellular matrix which may vary in consistency. It may be liquid, as in tissue fluid, or gelatinous, or it may be solid as in bone and cartilage. The cellular components of connective tissue are surrounded by the intercellular matrix. There are a variety of

connective tissue cell types, both fixed and wandering, but all are derived from mesodermal elements.

CONNECTIVE TISSUE CELLS. The most numerous fixed cells in connective tissue and those principally responsible for the formation of extracellular fibers and ground substance are the **fibroblasts** (Fig. 4-245). Some authors, when discussing the mature body, refer to these same cells as fibrocytes; however, due to the dynamic, ever-changing conditions within the tissues of the adult as well as the fetus, it is customary, regardless of age, to call them fibroblasts. Fibroblasts appear frequently along strands of collagen. They are polymorphic, but frequently appear as long, slender cells with several cytoplasmic processes, which give them a spindle-like appearance. Their nuclei are relatively large, oval-shaped, and basophilic, and their cytoplasm contains the usual organelles such as mitochondria. Although they have little or no visible specializations when found in fibrous connective tissue, they may receive specific names such as osteoblasts and osteocytes, or chondroblasts and chondrocytes in bone and cartilage.

Many other types of cells are found in the connective tissues. One of the more common is the ubiquitous **histiocyte** or **fixed macrophage** (Fig. 4-246). This cell type possesses a remarkable ability to ingest foreign particulate matter and, because of this property, is similar to other phagocytic cells occurring in several organs. The many phagocytic cell types throughout the body comprise the

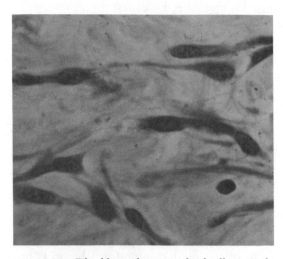

FIG. 4-245. Fibroblasts along strands of collagen in the loose connective tissue of the prepubertal female breast. Hematoxylin and eosin stain. 600 ×.

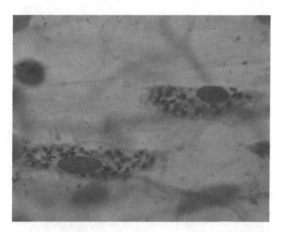

FIG. 4-246. Two macrophages or histiocytes among other elements in this subcutaneous connective tissue spread from a rat. The animal had been injected with trypan blue intraperitoneally in order to demonstrate the phagocytic property of these cells. Note the ingested dye within the cytoplasm. Resorcin-fuchsin and axocarmine. 1000 ×.

great phagocytic system called the **reticulo-endothelial system.** Another cell type found in connective tissue is the **mast cell** (Fig. 4-247). These are widely distributed throughout the body and are abundantly found along the course of small blood vessels. The

cytoplasm of mast cells is filled with granules believed to contain at least three important substances: *heparin,* which acts as an anticoagulant; *histamine,* which is a vasodilator and increases capillary permeability; and *serotonin,* which acts as a vasoconstrictor. Mast cells react to tissue injury by the release of their granules into the surrounding tissue. **Plasma cells,** which are known to contribute to the formation of antibodies, are at times seen in connective tissue, but can more readily be found in lymphatic organs and in subepithelial lymphatic tissue of digestive and respiratory organs. Other cells, such as the **polymorphonuclear leukocytes,** accumulate in an area where there is acute inflammation, and **lymphocytes** where the disease is more chronic. Occasionally, the other types of white blood cells—**monocytes, eosinophils,** and **basophils**—are seen in connective tissue, but red blood corpuscles are usually not found in the connective tissue unless there has been an injury to the blood vessels.

CONNECTIVE TISSUE FIBERS. Three types of fibers are found in adult connective tissue—collagenous, reticular, and elastic. Of these, the **collagenous fibers** (Fig. 4-248) are the most abundant. Fibroblasts produce molecules of tropocollagen which then assemble to form fine fibrils of collagen. The nature and arrangement of collagen lend both great strength and flexibility to the formation of fibrous connective tissue. Collagenous fibers are white and form the bulk of tendons, ligaments, and fascia, where they are pliable

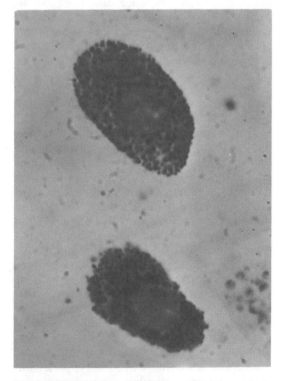

FIG. 4-247. Two mast cells from a subcutaneous connective tissue spread stained with neutral red. Note that the cytoplasm is filled with granules. 1800 ×.

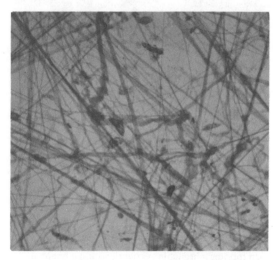

FIG. 4-248. Collagen fibers stained in a teased spread of loose connective tissue. Hematoxylin and eosin. 250 ×.

and enduring, but they resist stretching and are practically not extensible. **Reticular fibers** (Fig. 4-249) are found forming the networks around fat cells and beneath the endothelium of capillaries, as well as forming the fibrous supporting structure of lymphoid and blood-forming organs. They are argyrophilic and are composed of fibrils similar to collagenous fibers, but of a smaller diameter. The fibers appear coated with polysaccharides, which probably give them the property of reacting with silver salts. **Elastic fibers** (Fig. 4-250) contain protein and differ in color from collagen, appearing yellow in the fresh state. Elastic fibers are thin and wavy, branch repeatedly, and in connective tissue form a network among the collagenous fibers and cells. They vary in thickness and chemically consist of *elastin*. They stretch easily but allow the tissue to reassume its original form following stretching. Generally less numerous than collagen in connective tissue, elastic fibers are plentiful in the walls of arteries.

GROUND SUBSTANCE. The **ground substance** or **matrix** surrounding both cells and fibers may be fluid, semifluid, or solid, and although not composed of living material such as protoplasm, it is not inert. In ordinary fibrous connective tissue, the ground substance is a viscous solution. It contains

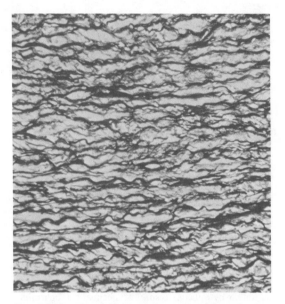

FIG. 4-250. Elastic fibers as observed in the wall of the human aorta. Stained with resorcin fuchsin. 144 ×.

long chained carbohydrates called *acid mucopolysaccharides* (more accurately called glycosaminoglycans) of which *hyaluronic acid* and *chondroitin sulfuric acid* are the most common. Ground substance has a great capacity to bind and hold water. Through it substances are diffused between the capillaries and the cells of tissues and organs, making it a highly important metabolic interface in the body.

CONNECTIVE TISSUE TYPES

The various types of so-called proper connective tissue are classified according to the nature of their cellular and intercellular components. When the proportion of fibers is highly predominant over the cellular and intercellular elements, the tissue is referred to as **dense fibrous connective tissue.** When the fibers are oriented in a parallel manner, the tissue is called dense **regular** connective tissue, in contrast to that containing randomly oriented fiber bundles, which is called dense **irregular** connective tissue. Differing from dense connective tissue, **loose fibrous connective tissue** is a fine network of loosely woven elements, in which the fibers are not densely packed. Also of importance are several other more special types of connective tissue—**reticular connective tissue, adipose tissue, mucous connective tissue,**

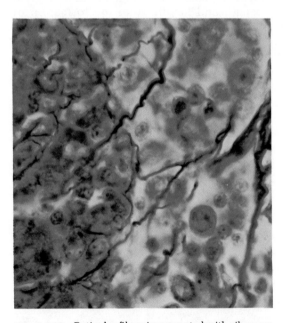

FIG. 4-249. Reticular fibers impregnated with silver are seen forming a branching network within a lymph node. 900 ×.

and **elastic connective tissue. Cartilage** and **bone** will be considered separately.

DENSE FIBROUS CONNECTIVE TISSUE. Dense regular connective tissue is the principal constituent forming the tendons, aponeuroses, and ligaments. The collagenous fibers are arranged in compact parallel bundles and impart a glistening white color that is characteristic of the tissue. Structures formed from dense regular connective tissue are pliable and inelastic, and withstand great tensional stress, particularly in one direction, but can endure little compressional stress. **Dense irregular connective tissue** has its collagenous fibers interwoven rather than arranged into parallel bundles. This tissue forms the fascial membranes, corium or dermis of the skin, periosteum, and capsules of organs. These structures are able to resist strong tensional stress in many directions and, in this respect, differ from the unidirectional resistance capacity of tendons and ligaments. As might be expected, the distinction between regular and irregular connective tissue is at times arbitrary. Some membranes have the collagenous bundles in layers rather than interwoven and thus may have a white, glistening appearance similar to that of aponeuroses, as for example in the dura mater and certain of the named fascias.

Tendons are white, glistening fibrous bands that attach muscles to bones. They vary in length and thickness, are sometimes round, sometimes flattened, of great tensile strength, flexible, and practically inelastic. They consist of collagenous bundles which are firmly united together and whose fibrils have a parallel course (Fig. 4-251). The fibroblasts are arranged in linear fashion between the parallel collagenous bundles. Except where they are attached, the tendons have a sheath or delicate irregular fibrous and elastic connective tissue, and larger tendons have a stroma of thin internal septa as well. They are sparingly supplied with blood vessels, the smaller tendons presenting no trace of them in their interior. They are supplied with sensory nerves whose fibers have specialized terminations called the **neurotendinous endings of Golgi,** which mediate a special stereognostic sensibility.

Aponeuroses are fibrous membranes, of a pearly white color, iridescent, and glistening, and represent very much flattened tendons. They consist of closely packed, parallel, collagenous bundles, and by this characteristic may be differentiated from the fibrous membranes of fascia, which have their collagenous bundles more irregularly interwoven. An aponeurosis usually consists of several layers of collagenous bundles arranged in such a manner that the bundles of adjacent layers are oriented in different directions. Aponeuroses are only sparingly supplied with blood vessels.

Ligaments are structures that strengthen joints and tend to restrain abnormal movements. On the other hand, they are pliant and flexible so as to allow perfect freedom of movement of the joint in its normal directions. Ligaments resist tensions, and their collagenous bundles accordingly are

FIG. 4-251. Longitudinal section of a human tendon showing the parallel orientation of the collagenous bundles. Note also the linear orientation of the fibroblast nuclei. Hematoxylin and eosin. 170 ×.

mainly parallel. They lack the glistening whiteness of tendons, however, because there is a greater admixture of elastic and fine collagenous fibers woven among the parallel bundles. Ligaments are as strong as tendons and have in addition a slight amount of elasticity. They are of different shapes and are formed as cords, bands, or sheets. Ligaments may be well-defined bands of fibers that appear as local thickenings in the capsule of a joint. In addition to capsular ligaments, others may be quite independent of the joint capsule. Their deep surface may form part of the synovial membrane of a joint, but their superficial surface is covered with a fibroelastic tissue that blends with the surrounding connective tissue. Certain ligaments are composed primarily of yellow elastic fibers. Examples are the *ligamenta flava,* which connect the laminae of adjacent vertebrae, and the *ligamentum nuchae* in the dorsal neck of grazing animals. In these cases the elasticity of the ligament is intended to act as a substitute for muscle power.

Fascia is the term used in gross anatomy for all dissectable fibrous connective tissue not specifically organized as tendons, aponeuroses, and ligaments. It varies in thickness and density according to functional demands, and is usually in the form of membranous sheets. In places where two fibrous sheets of fascia are easily separated, a loose areolar connective tissue intervenes. The site is frequently referred to as a fascial cleft. Often a fascial sheet may have a specific name; at other times it is referred to generally as *investing fascia, superficial fascia,* or *deep fascia.* Fascia separates muscles, invests blood vessels and nerves, forms a subcutaneous layer of tissue throughout the body, and contains between its fibers varying amounts of fat.

LOOSE FIBROUS CONNECTIVE TISSUE. Loose fibrous connective tissue is the most pervasive of all the connective tissues. It accompanies blood vessels and nerves throughout the body, extends into most organs, glands, and muscles, and is found beneath mucous and serous membranes (Fig. 4-252). It has a strong binding power, but is very pliable and somewhat stretchable. It contains elastic fibers among its collagenous bundles which are loosely interwoven and easily displaced, in contrast to the compact structure seen in dense connective tissue. Between the fibers

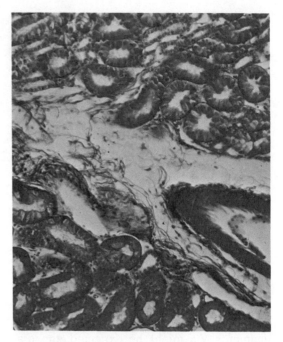

FIG. 4-252. Loose (areolar) connective tissue (stained blue) around blood vessels and between segments of the kidney. Note the extension of wisps of connective tissue between some of the tubules. Trichrome stain. 144 ×.

there may be comparatively wide interstices or areolae which are sponge-like and filled with liquid ground substance or tissue fluid. When this tissue is more closely woven, as in the capsules of organs, it is called **fibroelastic tissue.** When it is loosely woven and weak, as in the fascial clefts, it is called fibroareolar or **areolar tissue.**

RETICULAR CONNECTIVE TISSUE. Reticular connective tissue is composed of delicate reticular fibrils woven into a loose meshwork which surrounds individual or small groups of cells not belonging to the connective tissue. Its most typical form is found in such organs as the spleen, liver, lymph nodes, and bone marrow, where the individual lymphocytes are retained in its meshes. The fibrils are difficult to see in the usually prepared histological slides stained with hematoxylin and eosin because they are masked by the cells. They can be made visible with silver impregnation, however, and in these preparations the delicate fibrils merge with the collagenous bundles of the capsule of the organ (Fig. 4-249). The same delicate meshwork of reticular fibrils surrounds individual muscle fibers, glandular acini, nerve fibers, and capillary blood vessels in other parts of the body, but at these sites is usually given the specific name of reticular tissue. The cells of reticular connective tissue appear as primitive looking mesenchyme cells. Their processes course along the reticular fibers, and certain of these cells exhibit phagocytic properties.

ADIPOSE TISSUE. Adipose tissue is the name given to tissue in which there is an accumulation of connective tissue cells, each containing a large vesicle of fat (Fig. 4-253). In an obese individual these accumulations may occur wherever there is loose fibrous connective tissue. In all but emaciated individuals, however, they are found in certain areas where they act as packing or padding tissue. Examples of the latter are the orbit, perirenal fat pad, and various areas of the subcutaneous tissue. Adipose tissue is not inert and does not simply store fat over long periods of time. Fat cells show a considerable metabolic activity and there is a rapid and continuous turnover of lipid. They are able to synthesize triglycerides from free fatty acids and to store the lipid in the form of fat droplets which then are broken down again under the influence of

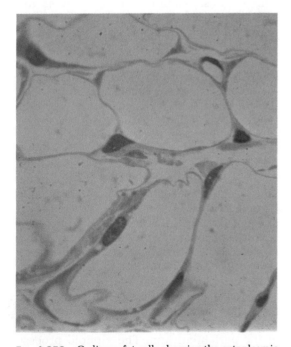

FIG. 4-253. Ordinary fat cells showing the cytoplasmic rim which encloses large intracytoplasmic spaces that in life contained fat droplets. The fat has been extracted during tissue preparation. Hematoxylin and eosin. 900 ×.

hormones. The storage and release of fat is importantly influenced by insulin and epinephrine.

Cartilage

Cartilage is essentially a nonvascular tissue because it has no capillary network of its own, nor does it contain nerves. It is found in various parts of the body—in adult life chiefly in the joints, in the walls of the thorax, and in the walls of various tubular structures that must be kept permanently open, such as the larynx, trachea and bronchi, nose, and ears. In the fetus, at an early period, the greater part of the skeleton is cartilaginous. This temporary cartilage is eventually replaced by bone, in contradistinction to more permanent cartilage, which remains unossified during the whole of life.

Cartilage cells, called **chondrocytes,** are surrounded by a compact matrix. The nature of this matrix and the abundance and relative proportion of collagenous and elastic fibers it contains are used to distinguish three types of cartilage: **hyaline cartilage, yellow** or **elastic fibrocartilage,** and **white fibrocartilage.**

HYALINE CARTILAGE. Hyaline cartilage consists of a gristly mass with firm consistency, which is considerably elastic, translucent, and of a pearly blue color. Except where it coats the articular ends of bones, hyaline cartilage is covered externally by a fibrous membrane, the **perichondrium,** from the vessels of which it imbibes its nutritive substances. Histologically, hyaline cartilage is composed of cells of a rounded or bluntly angular form, lying in groups of two or more in a granular or almost homogeneous matrix (Fig. 4-254). These small clusters of chondrocytes are referred to as **isogenous groups,** because the several cells in each group are derived from a single chondrocyte. Chondrocytes in isogenous groups generally have straight outlines where they are in contact with each other, whereas the rest of their circumference is rounded. These cells contain a clear translucent cytoplasm in which is seen a well-developed endoplasmic reticulum. Their nuclei are rounded or ovoid, containing several nucleoli. The cavities in the matrix, which contain the chondrocytes, are called **cartilage lacunae.** The matrix around the lacunae containing the chondrocytes stains more basophilic because of its higher

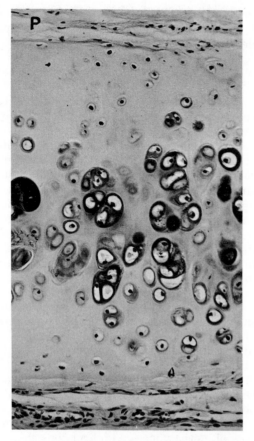

FIG. 4-254. Hyaline cartilage. Note the isogenous groups of chrondrocytes separated by cartilage matrix. Photomicrograph taken of nasal alar cartilage. *P,* Perichondrium. (From Elias and Pauly, 1966.)

different cytoplasmic structures of chondrocytes into the territorial matrix adjacent to the lacunae. Eventually all the radioactive proline metabolized from the cells into the matrix.

Hyaline cartilage at the **articular surfaces** of bone shows no tendency to ossify. The matrix is finely granular and the cells are small and disposed parallel to the surface superficially, while more deeply and nearer the bone they are arranged in vertical rows. The free surface of articular cartilage, where it is exposed to friction, is not covered by perichondrium, although a layer of connective tissue continuous with the synovial membrane can be traced over a small part of its circumference. At this interface the cartilage cells are more or less branched and pass insensibly into the branched connective tissue elements of the synovial membrane. Articular cartilage forms a thin covering layer over the joint surfaces of the bones, and its elasticity enables it to break the force of concussions, while its smoothness affords ease and freedom of movement.

Cartilage at articular surfaces varies as much as 0.5 cm, or more, in thickness according to the shape of the bony surface on which it lies. When the bone is convex the cartilage is thickest at the center, the reverse being the case on concave articular surfaces. Cartilage appears to derive its nutriment partly from the vessels of the neighboring synovial membrane and partly from those of the bone upon which it is encapsulated. The vascularized tufts of osteogenic marrow in the cancellous bone dilate and form arches as they approach the cartilage layer, and then return into the substance of the bone.

The cells of the hyaline cartilage which compose the **cartilages of the ribs** are large, and the usually homogeneous matrix has a tendency to fibrous striation, especially in old age. In the thickest parts of the costal cartilages a few large vascular channels may be detected. At first sight this would appear to be an exception to the statement that cartilage is a nonvascular tissue, but this is not so because the vessels give no branches to the cartilage substance itself, and the channels rather should be looked on as involutions of the perichondrium. The xiphoid process and the cartilages of the nose, larynx, and trachea (except the epiglottis and corniculate cartilages of the larynx, which are composed of elastic fibrocartilage) resemble the costal

concentration of acid mucopolysaccharides. This more basophilic zone is termed the **territorial** or **capsular matrix,** while the remaining, less basophilic matrix is called the **interterritorial matrix.**

The matrix is translucent and appears grossly to be structureless or granular, resembling ground glass. However, ultrastructural and radioautographic studies have revealed that a significant part of the matrix of cartilage consists of fibrils, and these have the characteristics of collagen in articular cartilage. Further, it has been shown that chondrocytes form the precursors of the matrix. Belanger (1954) showed that radioactive sodium sulfate becomes incorporated first in chondrocytes and then in cartilage matrix. Studies by Revel and Hay (1963) have clinched this story. They showed the progressive incorporation of proline from cartilages in their microscopic anatomy. The arytenoid cartilage of the larynx shows a transition from hyaline cartilage at its base to elastic cartilage at the apex.

Structures that are normally composed of hyaline cartilage are prone to calcify in old age. Their matrix becomes permeated by calcium salts without any appearance of true bone. This process often is manifested by the deposition of calcified crusts on the surfaces of the cartilage and occurs frequently in some cartilage such as that of the trachea and larynx and in the costal cartilages. At times it may, however, be succeeded by conversion into true bone.

YELLOW OR ELASTIC FIBROCARTILAGE. Yellow or elastic fibrocartilage is more opaque than hyaline cartilage, and is more flexible and pliant. These characteristics are imparted by the elastic fibers embedded in the cartilage matrix (Fig. 4-255). The lacunae containing the cartilage cells are rounded, similar to those of hyaline cartilage, and have a capsule of clear matrix. The cells are found singly or in isogenous groups. The elastic fibers may be sparsely scattered or densely arranged, and they course in all directions, interlacing into a network. The fibers at the periphery are continuous with the elastic fibers of the perichondrium. Elastic cartilage is found in the external ear, the cartilage of the auditory tube, the epiglottis, and the corniculate and cuneiform cartilages of the larynx.

WHITE FIBROCARTILAGE. White fibrocartilage consists of a combination of dense collagenous fibers and cartilage cells in various proportions. To the former of these constituents it owes its flexibility and toughness, and to the latter its elasticity. The collagenous fibers are arranged in bundles, with cartilage cells between the bundles (Fig. 4-256). The

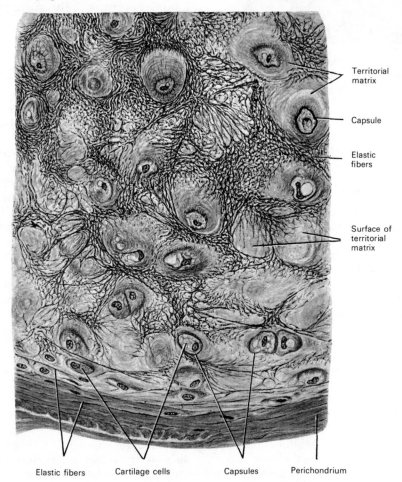

Territorial matrix

Capsule

Elastic fibers

Surface of territorial matrix

Elastic fibers   Cartilage cells   Capsules   Perichondrium

FIG. 4-255. Elastic cartilage of the epiglottis of a 13-year-old boy. Stained with elastin H, Delafield's hematoxylin and eosin. 435 ×. (From Copenhaver, Bunge, and Bunge, 1971.)

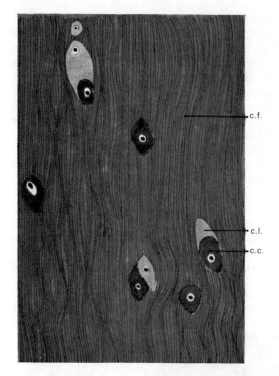

Fig. 4-256. Fibrocartilage from human intervertebral disc. *c.c.*, cartilage cell; *c.f.*, collagenous fibrils embedded in cartilaginous matrix; *c.l.*, cartilage lacuna. 1000 ×. (Redrawn from Sobotta.)

cartilage cells are enclosed in capsules around which is a small concentrically striated area of hyaline matrix. To a certain extent the cells resemble those of tendons, but are distinguished from them by their surrounding matrix and by being less flattened. White fibrocartilage forms a number of structures which may be grouped as **interarticular, connecting, circumferential,** and **stratiform.**

The **interarticular fibrocartilages** (such as the *menisci* of the knee joint) are flattened fibrocartilaginous plates, of a round, oval, triangular, or sickle-like form, interposed between the articular cartilages of certain joints. They are free on both surfaces, usually thinner toward the center than at the circumference, and held in position by the attachment of their margins and extremities to the surrounding ligaments. The synovial surfaces of the joints are prolonged over them. They are found in the temporomandibular, sternoclavicular, acromioclavicular, wrist, and knee joints. Thus, they are found in those joints that are most exposed to violent concussion and subject to frequent movement. Their actual function is little

understood but it is conjectured that they (1) obliterate the intervals between opposed surfaces in their various motions; (2) increase the depths of the articular surfaces and give ease to the gliding movements; (3) moderate the effects of great pressure and deaden the intensity of the shocks to which the parts may be subjected; and (4) act to spread synovial fluid. The presence of interarticular fibrocartilages also serve to increase the variety of movement in a joint. Thus, in the knee joint, although it is a hinge joint in which, as a rule, only one variety of motion is permitted, there are two kinds of motion—angular movement and rotation. Angular movement takes place between the condyles of the femur and the interarticular cartilages, while rotation occurs between the cartilages and the head of the tibia.

The **connecting fibrocartilages** are interposed between the bony surfaces of those joints that have only slight mobility, as between the bodies of the vertebrae. At these sites the fibrocartilages form discs, which are closely adherent to the opposed surfaces. Each *intervertebral disc* is composed of concentric rings of fibrous tissue, with cartilaginous laminae interposed. The fibrous tissue predominates toward the disc circumference, while the cartilaginous material is more evident toward the center.

The **circumferential fibrocartilages** consist of rims of fibrocartilage, which surround the margins of some of the articular cavities, such as the *glenoidal labrum* of the hip, and of the shoulder. The annular lips serve to increase the articular surface of contact between the bones by deepening the articular cavities. They also protect the bony edge of the socket.

The **stratiform fibrocartilages** are those that form a thin coating to osseous grooves through which the tendons of certain muscles glide. Small masses of fibrocartilage are also developed in the tendons of some muscles, where they glide over bones, as in the tendons of the peroneus longus and tibialis posterior muscles.

Bone

Like other connective tissues, bone consists of cells that are surrounded by fibers and ground substance. The calcification of its matrix into a hard and rigid substance that serves to support and protect the softer body structures distinguishes it from other connective tissues. Additionally, the skeleton provides sites of attachment for the muscles that allow the organism motor expression, such as locomotion, as well as other bodily movements. Although bone may appear at first glance to be an inert tissue, it is, in fact, metabolically active and dynamic. Bone is a highly vascularized tissue and has an excellent capacity for self-repair.

MACROSCOPIC STRUCTURE OF BONE

If a bone from an adult limb is sectioned longitudinally and examined grossly, it is found to be a tubular structure, the walls of which are formed of osseous material and the central cavity of which contains yellow or red **bone marrow.** The general contours of the bone resemble the enlarged fully grown model of its miniature prenatal form. Its surface, excepting the cartilage-covered articular ends, is closely invested by a fibrous, tough **periosteum,** which is highly vascularized and freely innervated. Other macroscopic features reveal that the shaft, or **diaphysis,** is capped at both ends by the **epiphyses,** and that these are separated from the diaphysis by an epiphyseal plate, which is cartilaginous.

PHYSICAL FEATURES AND PROPERTIES. Bone is a hard, mineralized tissue which is slightly elastic and resilient and at the same time relatively light in weight for its volume. With the exception of the enamel and dentin in the teeth, bone is the hardest structure in the body. It withstands tension and compression to a remarkable degree, but can readily be fractured by forces applied suddenly or when it becomes demineralized because of metabolic illness or aging.

Fully grown, mature healthy bone is composed of two kinds of osseous tissue. One is dense in texture, like ivory, and is termed **compact bone.** The other consists of slender spicules called *trabeculae,* which are joined into a spongy structure; because of its resemblance to latticework, this structure is called **cancellous** or **spongy bone.** The compact bone is always located on the exterior, surrounding the cancellous in the interior of the bone. The relative quantity of these two kinds of tissue varies in different bones and in different parts of the same bone, according to functional requirements. Examination

of compact bone with a hand lens shows it to be porous, so that the difference in structure between it and the cancellous bone depends principally upon the relative amount of solid matter, and the size and number of spaces in each. The cavities are small in compact bone and the solid matter between them abundant, while in the cancellous bone the spaces are large and the solid matter is in smaller quantity (Fig. 4-257). The flat tabular bones of the skull are composed of two plates of compact bone, called *outer* and *inner tables,* between which is a narrow space of spongy bone called the *diploë.*

PERIOSTEUM. The periosteum (Fig. 4-257) adheres to the surface of the bone in every part except the cartilaginous extremities. When strong tendons or ligaments are attached to a bone, the periosteum becomes blended with them. The periosteum consists of two closely united layers. The outer layer is formed chiefly of collagenous tissue, containing occasionally a few fat cells. The inner layer, which can itself be separated into thinner layers, is formed of fine elastic fibers composing dense membranous networks. During life bone is permeated by blood vessels, which course through the periosteum to achieve hard tissue. If the periosteum is stripped from the surface of living bone, small bleeding points mark the paths of periosteal vessels. In young bones the periosteum is thick and exceedingly vascular and is intimately connected at both ends of the bone with the epiphyseal cartilages, but less closely with the shaft of the bone, from which it is separated by a layer of soft tissue containing a number of bone-forming cells, the **osteoblasts.** These cells are responsible for the process of ossification on the exterior of the young bone. Later in life the periosteum is thinner and less vascular, and the osteoblasts are converted into an epithelioid layer of resting osteoprogenitor cells, which remain in contact with the surface of the bone. If a bone is injured, these cells become active bone-forming elements, assisting in regeneration and repair. Because of the rich distribution of periosteal vessels to the underlying bone, it is understandable why bone denuded of its periosteal layer is more subject to exfoliation and necrosis. In summary, it should be emphasized that although the superficial outer fibrous layer of periosteum acts as a limiting membrane upon which muscles, tendons, and ligaments can

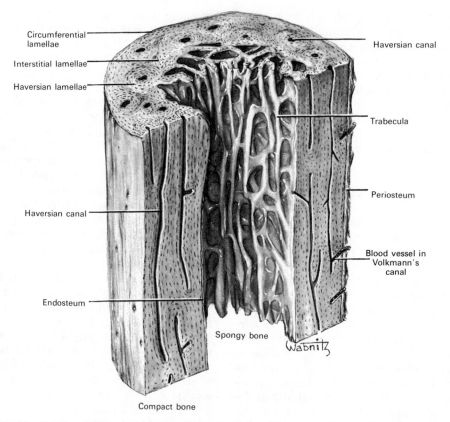

Circumferential
lamellae

Interstitial lamellae

Haversian lamellae

Haversian canal

Endosteum

Haversian canal

Trabecula

Periosteum

Blood vessel in
Volkmann's
canal

Spongy bone

Wabnitz

Compact bone

FIG. 4-257. Portion of finger bone from which marrow has been removed to show spongy bone. 10 ×.

attach, the inner layer is a more dynamic and functionally oriented tissue, interfacing with the underlying bone and assisting in the support of its metabolic processes. Fine nerves and lymphatics, which generally accompany the arteries, may also be demonstrated in the periosteum.

BONE MARROW.   Bone marrow is the soft, fatty tissue found in the medullary cavities of bones as well as in their cancellous extremities. Bone marrow is widely dispersed and two types exist, yellow and red. **Yellow marrow** is found in the large cavities of the long bones. It consists for the most part of fat cells and a few primitive blood cells, but it may be replaced by red marrow in anemia. **Red marrow** is a hemopoietic tissue and is an important site for the production of granular leukocytes (neutrophil, eosinophil, and basophil leukocytes) and red blood cells (erythrocytes). It is found in the flat and short bones, the articular ends of the long bones, the bodies of the vertebrae, the cranial diploë, and the sternum and ribs. Red marrow consists, for the most part, of mye-

loid cells, which are primitive blood cells in immature stages. Macrophages, fat cells, and megakaryocytes (giant cells) are also always present. Both types of marrow consist of a framework of reticular connective tissue fibers through which course fine vascular sinusoids which are derived from the nutrient vessels of the bone. The sinusoids are lined with phagocytic elements, and the myeloid cells are arranged in cord-like rows outside the sinusoids. The newly formed blood cells are able to gain access to the sinusoids and thereby great numbers of recently matured blood cells (both red and white) pass into the blood stream. The mechanism whereby the mature cells enter the blood stream and immature ones are held back is not known. The maturation of the various blood cell types from the more primitive undifferentiated elements is a subject that is still unclear. Most accounts invoke a pluripotential cell type or **stem cell** which then differentiates into the precursor cells of the different myeloid elements. Additional research is also required to explain the various metabolic influences that hormones and other systems in the body have on the process of hemopoiesis.

VESSELS AND NERVES OF BONE. The **arterial blood vessels** that supply bone are numerous. Three sources of blood to bone are usually identified: (1) a limited supply is derived from the **periosteal arteries,** (2) the majority of blood comes from the **nutrient artery,** and (3) in long bones, a significant supply comes from the **epiphyseal arteries.** Because the periosteum contains a dense network of vessels ramifying its structure, at one time it was thought that most of the blood supply to the cortex of a bone came from the *periosteal vessels* and filtered into the Haversian canals from these peripheral sites. Recent studies, however, have shown that the circulation to at least two-thirds of the cortex comes from the other direction by way of the nutrient artery. However, it does appear that at a few sites, periosteal arteries supply more significant amounts of blood to the cortex. Additionally, if the nutrient artery is injured, the cortex is then supplied principally from periosteal vessels by way of anastomosing channels. The *nutrient artery* enters the shaft of a bone through a foramen in the compact layer. It then branches into longitudinally oriented vessels, which supply the bone marrow and spongy tissue as far as the epiphyseal plates. A single nutrient artery may exist as in the tibia, but there may be several as in the femur. The nutrient artery is usually accompanied by one or two veins. The ends of long bones are supplied by **metaphyseal** and **epiphyseal arteries.** These epiphyseal vessels pass into the marrow cavities of the epiphyses, and they, along with the metaphyseal arteries, anastomose in the region of the epiphyseal plate with fine branches from the nutrient artery.

The **venous blood vessels** that drain bone are quite comparable anatomically to the arteries that supply it. In long bones these emerge in three places: (1) one or two **nutrient veins** accompany the artery; (2) numerous large and small veins emerge at the articular extremities as **epiphyseal veins,** and (3) many small veins pass out of the compact substance as **periosteal veins.** It appears that the periosteal veins drain a greater percentage of blood than once was thought, and the nutrient veins drain less. Thus, the drainage pattern is physiologically different from the arterial supply, even though anatomically

comparable vessels are found. In the flat cranial bones the veins are large, very numerous, and run in tortuous canals in the diploic tissue. The sides of these canals are formed by thin lamellae of bone which are perforated, allowing the passage of adjacent veins. This pattern is similar to that seen in other cancellous tissue—finely coated veins being enclosed and supported by osseous material. When a bone is divided, the vessels remain patent, and do not contract in the canal in which they are contained.

Lymphatic vessels can be identified in the periosteal covering of bone accompanying the periosteal vessels, but are much more difficult to demonstrate in bone tissue itself. There have been claims, however, of their existence in the Haversian canals.

Nerve fibers are distributed freely to the periosteum, and accompany blood vessels into the interior of the bone. They appear to be most numerous in the articular extremities of the long bones, in the vertebrae, and in the larger flat bones. Both myelinated and unmyelinated fibers have been demonstrated, and it is probable that they are not only vasomotor but sensory in function.

MICROSCOPIC STRUCTURE OF BONE

A transverse slice of compact bone may be further ground until it is sufficiently thin to be examined under the light microscope. At low magnification, the bone will be seen to be mapped out into a number of circular units, each consisting of a central opening surrounded by a number of concentric rings (Fig. 4-258). These units are termed **Haversian systems,** and the central opening is an **Haversian canal.** The rings are layers of bony tissue arranged concentrically around the central canal, and termed **lamellae.** Moreover, on closer examination it is found that between these lamellae, and therefore also arranged concentrically around the central canal, are a number of small spaces, the **lacunae,** and that these lacunae are connected with each other and with the central Haversian canal by many fine channels, which radiate like the spokes of a wheel and are called **canaliculi** (Fig. 4-259). Within the lacunae are the cell bodies of the bone cells, or **osteocytes,** and the canaliculi contain the cytoplasmic processes of the osteocytes. Between the more regularly oriented circular (in transverse section) Haversian systems are

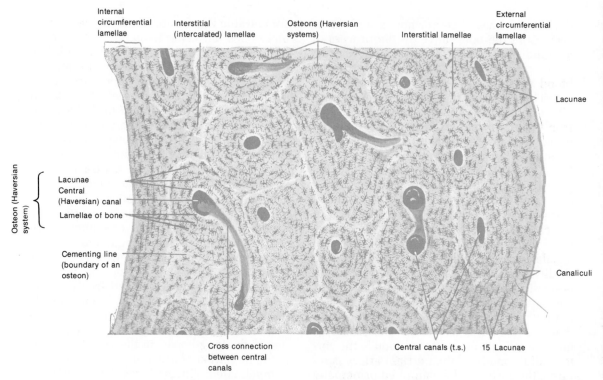

FIG. 4-258. Transverse section of compact dried bone taken from the diaphysis of the tibia. Observe that bone consists of lamellae arranged in concentric layers which form the Haversian systems and between these systems are the interstitial lamellae. Inner and outer circumferential lamellae form the surface layers of bone. Stain: Aniline blue. 80 ×. (From di Fiore, *Atlas of Human Histology,* Lea & Febiger, Philadelphia, 1981.)

FIG. 4-259. Cross section of an Haversian system showing its central canal, the lacunae which contain cell bodies of osteocytes and the intercommunicating canaliculi. Specimen is from human femur. (From Greep and Weiss, *Histology,* 3rd Edition, McGraw-Hill, New York, 1973.)

irregularly shaped bony intervals, which also contain lacunae and canaliculi filled with osteocytes and their processes. These are called the **interstitial systems,** and they also reveal a lamellar structure created by the more or less curved **interstitial lamellae.** Additionally, other lamellae are found on the surface of the bone and are arranged parallel to its circumference. They are termed **circumferential,** or **primary lamellae,** to distinguish them from those surrounding the axes of the Haversian canals, which may then be termed **secondary lamellae** (Fig. 4-258).

HAVERSIAN CANALS. The Haversian canals, which in cross section are observed as openings in the center of each Haversian system, appear as true canals in a longitudinal section (Fig. 4-260). The canals course parallel with the longitudinal axis of the bone for a short distance. The Haversian canals communicate with each other and with the surfaces of the bone by oblique or transversely crossing channels called **Volkmann's canals.** Haversian canals vary considerably in size, some measuring as much as 0.12 mm in diameter, but their average diameter is about 0.05 mm. The canals nearer the medullary cavity are larger than those closer to the surface of the bone. Each canal contains one or two small blood vessels surrounded by a small quantity of delicate connective tissue and some nerve filaments. Those Haversian canals near the surface of the bone open upon it by minute orifices, and those near

the medullary cavity similarly open into this space. Thus, it can be appreciated that the entire bone is permeated by a system of blood vessels that course through the bony canals in the centers of the Haversian systems. The blood vessels branch and course through the Volkmann canals, thereby forming an interconnecting vascular network.

BONY LAMELLAE. One of the most characteristic features of bone is its layered structure. These layers, or **lamellae,** are thin plates of bone tissue arranged around a central hollow cylinder called the Haversian canal. The lamellae are composed of collagenous bundles arranged in a parallel direction within each layer. The bundles of collagen in adjacent lamellae, however, are oriented in different directions. It does not appear that a simple perpendicular arrangement of collagen bundles exists in adjacent layers. Instead it is believed that the collagen fibers are arranged in a helical manner in each layer, and that the pitch of the helix changes in bundles of adjacent layers. Nonetheless, in a cross section of stained decalcified bone, lamellae with bundles cut longitudinally appear striated, and these alternate with lamellae in which the bundles of fibers are cut transversely, giving a more punctate appearance. In addition to the **Haversian system lamellae,** bones contain **interstitial lamellae** occupying the spaces between the Haversian systems, as well as the **outer** and **inner circumferential lamellae,** which form an external encasement of a bone as well as

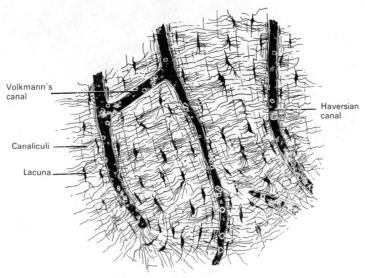

Volkmann's canal

Canaliculi

Lacuna

Haversian canal

FIG. 4-260. Longitudinal section of compact bone tissue. Note that the Haversian canals communicate with each other by means of transverse channels called Volkmann's canals.

the bony layers lying subjacent to the endosteal lining in the central core of a bone. In all these sites the microanatomy of the lamellae is similar. Between the bundles of collagenous fibers and between the lamellae are found a darker staining amorphous extracellular substance containing deposits of calcareous salts. In many places the lamellae appear to be pinned together by **perforating fibers,** which run obliquely through them. Similar fibers, called **Sharpey's fibers,** extend from the surface of a bone into the substance of the periosteum, anchoring the latter to the bone. These perforating fibers may be demonstrated in the dissecting room by detaching the scalp from the calvaria, which then is covered by fine projections (Sharpey's fibers) as well as coarser projections containing minute blood vessels.

LACUNAE AND CANALICULI. The **lacunae** are oblong spaces situated between the lamellae. In a microscopic section of dried bone, viewed by transmitted light, lacunae appear as fusiform opaque spots. Each lacuna is occupied during life by a bone cell or **osteocyte,** processes of which extend into the **canaliculi,** which are exceedingly minute channels that cross the lamellae and connect the lacunae with neighboring lacunae and also with the Haversian canal (Fig. 4-259). In the center of an Haversian system, a number of canaliculi radiate from the Haversian

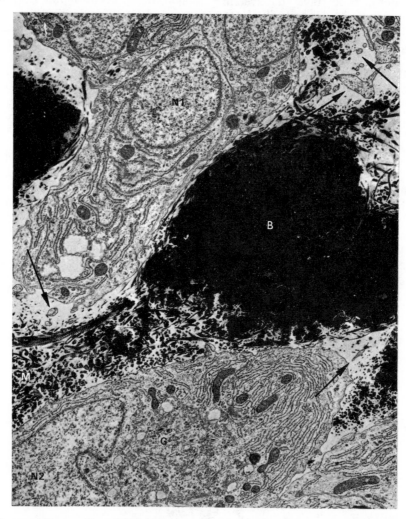

FIG. 4-261. Two osteoblasts, with nuclei identified N1 and N2, are located above and to the left and in the lower part of this electron micrograph. The cells lie adjacent to a mass of fully calcified bone B, which then continues to the left as more recently deposited incompletely calcified preosseous matrix. Observe the profuse amount of endoplasmic reticulum in the cytoplasm of the osteoblasts along with a prominent Golgi complex, G. Arrows point to processes of osteoblasts which extend into the area of bone calcification. 9200 ×. (From Hancox, N.M., *Biology of Bone,* Cambridge University Press, London, 1972.)

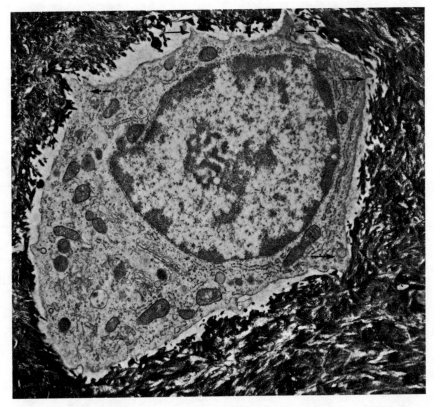

FIG. 4-262.   An osteocyte within a lacuna surrounded by fully developed bone. Note that the endoplasmic reticulum is sparse and the Golgi complex not well developed. Arrows point to irregularities on the cell surface which are sites of extensions of osteocyte processes destined toward the canaliculi. The lacunar wall in this specimen is not fully mineralized. 10,000 ×. (From Cameron, D.A. in *The Biochemistry and Physiology of Bone*, Vol. 1, ed. G.H. Bourne, Academic Press, New York, 1972.)

canal and open into the first set of lacunae between the first and second lamellae. From these lacunae a second set of canaliculi proceed to the next series of lacunae, and so on until the periphery of the Haversian system is reached. The canaliculi radiating from the most peripheral series of lacunae do not as a rule communicate with the lacunae of neighboring Haversian systems, but after passing outward for a short distance, they form loops and return to their own lacunae. Thus, every part of an Haversian system is supplied with nutrient fluids derived from the vessels in the Haversian canal and distributed through the canaliculi and lacunae.

CELLS OF BONE.   The cellular elements in bone are distinguishable as probably four functional variants in the same cellular lineage. Bone cells transform from **osteoprogenitor cells** to **osteoblasts,** which then become **osteocytes,** the principal cells of adult bone. **Osteoclasts,** which are responsible for bone resorption, constitute the fourth cell variant in bone. Although their origin is not certain, it is felt that osteoclasts, which are giant multinucleated cells, arise from the coalescence of many mononuclear osteoprogenitor cells.

Like all other cells in connective tissue, bone cells develop from the mesenchyme. **Osteoprogenitor cells** are relatively undifferentiated cells which line the inner aspect of the periosteum as well as help to form the endosteal lining of the inner bone cavity. These cells also line the Haversian canals. Osteoprogenitor cells undergo mitosis and can be transformed into osteoblasts or they can coalesce to form osteoclasts. They are flat cells and lie directly on the surface of bone. They have oval nuclei and a relatively scanty cytoplasm.

Osteoblasts are bone-forming cells and therefore are always found at the interface between bone and cartilage in developing or growing bone or at the site of repair in an injured bone. During an active phase, these

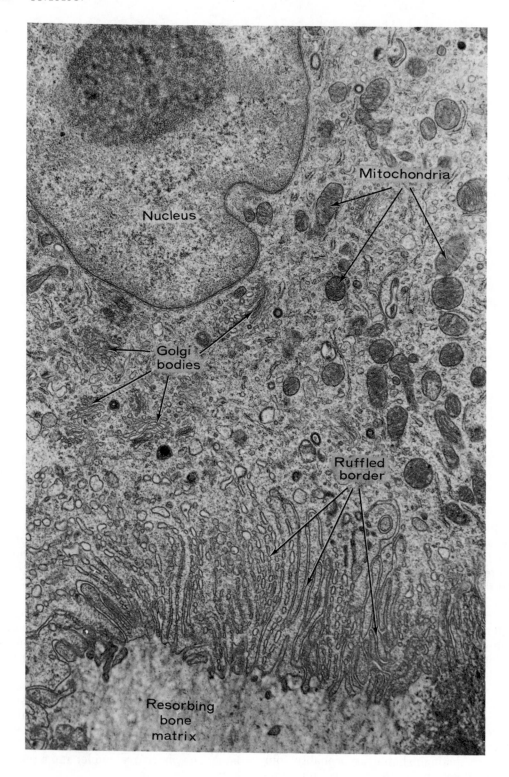

Fig. 4-263. A portion of an osteoclast including part of a nucleus, several Golgi elements and mitochondria. Observe in the lower part of the micrograph the ruffled border of the osteoclast closely applied to an area of bony matrix which is being resorbed. (From Bloom and Fawcett, *A Textbook of Histology*, W. B. Saunders Co., Philadelphia, 1975. Electron micrography by Dr. P. R. Garant.)

cells appear as a layer of low columnar cells and they assume an epithelial-like structure. Osteoblasts are active secretory cells, and therefore it is not surprising that electron microscopic studies have revealed their extensive rough endoplasmic reticulum and prominent Golgi complex (Fig. 4-261). Both cytoplasm and nucleus of the osteoblasts are intensely basophilic. Upon completion of bone growth or bone regeneration, osteoblasts become surrounded by the matrix they secrete and become encased in lacunae, thereby assuming the characteristics of osteocytes.

Osteocytes are the principal cells found in fully developed bone. They occupy the lacunae and their processes pass into the surrounding canaliculi. With the light microscope only the nucleus is easily discernible with the lacuna. They vary in size, being somewhat larger in younger bone than older. With the electron microscope it has been shown that the processes of adjacent osteocytes form contacts with each other, an indication of their functional interrelationship rather than an isolated action. Osteocytes have a more limited amount of rough endoplasmic reticulum than do osteoblasts (Fig. 4-262), and it is believed that they participate in the passage of calcium from blood to bone. They apparently respond to circulating parathyroid hormone, which induces an increased removal of calcium from bone in a process that has been called osteolysis.

Osteoclasts are giant multinucleated cells that appear to be actively involved in the process of bone resorption. They are generally found in shallow spaces called **Howship's lacunae.** Osteoclasts possess finger-like infolded extensions of the cytoplasm adjacent to the bone surface being resorbed, forming what has been termed their fluffy or ruffled border (Fig. 4-263). The portion of the cell away from the ruffled border is more apt to contain the many nuclei of the cell. The cytoplasm near the nuclei also contains multiple Golgi complexes. The manner by which these cells resorb bone is still unclear, but it is believed that they can secrete hydrolytic enzymes that are capable of disintegrating and digesting bone.

CHEMICAL COMPOSITION OF BONE.   Bone consists of an organic and an inorganic part intimately combined.

The *organic part* may be isolated by immersing a bone for a considerable time in dilute mineral acid, after which the bone retains its original shape but becomes completely flexible. This flexibility may be demonstrated, for example, by a long bone such as a rib, which can easily be tied into a knot after it has been demineralized. If then a transverse section is made, the same general arrangement of the Haversian canals, lamellae, lacunae, and canaliculi is seen. The organic part of bone consists of the cells, the collagen fibers, and the amorphous ground substance which contains glycoproteins.

The *inorganic part* may be separately obtained by calcination, a process by which all the organic constituents are completely burnt away. The bone will still retain its original form, but it will be white and brittle, will have lost about one-third of its original weight, and will crumble with the slightest force. The structure of the principal mineral in bone is most closely similar to the substance *hydroxyapatite*. This is a form of calcium phosphate, but bone also contains carbonate and citrate ions as well. Elements such as fluoride, strontium, and isotopes of calcium and phosphorus can be substituted into the chemical makeup of bone mineral. The crystals of mineral are extremely small but may be visualized with the electron microscope. The inorganic part confers to bone its hardness and rigidity, while the organic part gives bone its resilience and tenacity.

References

(References are listed not only to those articles and books cited in the text, but to others as well which are considered to contain valuable resource information for the student who desires it.)

General References

Bloom, W., and D. W. Fawcett. 1975. *A Textbook of Histology.* 10th ed., W. B. Saunders Co., Philadelphia, 1033 pp.

Di Fiore, M. S. 1981. *Atlas of Human Histology.* 5th ed., Lea & Febiger, Philadelphia, 267 pp.

Elias, H. M., and J. E. Pauly. 1966. *Human Microanatomy.* 3rd ed., F. A. Davis, Philadelphia, 380 pp.

Eycleshymer, A. C., and T. Jones. 1925. *Hand Atlas of Clinical Anatomy.* 2nd ed., Lea & Febiger, Philadelphia, 425 pp.

Ferner, H., and J. Staubesand. 1975. *Benninghoff und Goerttler's Lehrbuch der Anatomie des Menschen.* 11th ed., Urban and Schwarzenberg, Munich, 3 vols.

Ham, A. W. 1974. *Histology.* 7th ed., J. B. Lippincott, Philadelphia. 1006 pp.

Healey, J. E., and W. D. Seybold. 1969. *A Synopsis of Clinical Anatomy.* W. B. Saunders Co., Philadelphia, 324 pp.

Hollinshead, W. H. 1968. *Anatomy for Surgeons.* 2nd ed., Harper and Row, Hagerstown, Md. 3 vols.

Langman, J. 1975. *Medical Embryology. Human Development, Normal and Abnormal.* 3rd ed., Williams and Wilkins, Baltimore, 421 pp.

Moore, K. L. 1973. *The Developing Human. Clinically Oriented Embryology.* W. B. Saunders Co., Philadelphia, 374 pp.

Porter, K. R., and M. A. Bonneville. 1973. *Fine Structure of Cells and Tissues.* 4th ed., Lea & Febiger, Philadelphia, 204 pp.

Spalteholz, W. 1896. *Handatlas der Anatomie des Menschen.* S. Hirzel, Leipzig, 3 vols.

Thorek, P. 1962. *Anatomy in Surgery.* 2nd ed., J. B. Lippincott, Philadelphia, 904 pp.

Osteology

Breathnach, A. S. 1960. *Frazer's Anatomy of the Human Skeleton.* 5th ed., J. and A. Churchill, London, 247 pp.

Clark, W. R. Le Gros. 1964. *The Fossil Evidence for Human Evolution.* University of Chicago Press, Chicago, 201 pp.

Evans, F. G. 1957. *Stress and Strain in Bones. Their Relation to Fractures and Osteogenesis.* Charles C Thomas, Springfield, Ill. 245 pp.

Francis, C. C., P. P. Werle, and A. Behm. 1939. The appearance of centers of ossification from birth to 5 years. Amer. J. Phys. Anthrop., 24:273-299.

Hancox, N. M. 1972. *The Biology of Bone.* Cambridge University Press, London, 199 pp.

Hill, A. H. 1939. Fetal age assessment by centers of ossification. Amer. J. Phys. Anthrop., 24:251-272.

Koch, J. C. 1917. The laws of bone architecture. Amer. J. Anat., 21:177-298.

Latimer, H. B., and E. W. Lowrance. 1965. Bilateral asymmetry in weight and in length of human bones. Anat. Rec., 152:217-224.

Lowrance, E. W., and H. B. Latimer. 1957. Weights and linear measurements of 105 human skeletons from Asia. J. Anat., 101:445-460.

Mall, F. P. 1906. On ossification centers in human embryos less than one hundred days old. Amer. J. Anat., 5:433-458.

McLean, F. C., and M. R. Urist. 1968. *Bone: Fundamentals of the Physiology of Skeletal Tissue.* 3rd ed. University of Chicago Press, Chicago, 314 pp.

Merz, A. L., M. Trotter, and R. R. Peterson. 1956. Estimation of skeleton weight in the living. Amer. J. Phys. Anthrop., 14:589-610.

Noback, C. R. 1954. The appearance of ossification centers and the fusion of bones. Amer. J. Phys. Anthrop., 12:63-70.

Noback, C. R., and G. G. Robertson. 1951. Sequences of appearance of ossification centers in the human skeleton during the first five prenatal months. Amer. J. Anat., 89:1-28.

O'Rahilly, R., and E. Gardner. 1972. The initial appearance of ossification in staged human embryos. Amer. J. Anat., 134:291-301.

Trotter, M., and G. C. Gleser. 1958. A re-evaluation of estimation of stature based on measurements of stature taken during life and of long bones after death. Amer. J. Phys. Anthrop., 16:79-124.

Histology, Cytology, Chondrogenesis, and Osteogenesis

Barbour, E. P. 1950. A study of the structure of fresh and fossil human bone by means of the electron microscope. Amer. J. Phys. Anthrop., 8:315-330.

Bassett, A. L. 1962. Current concepts of bone formation. J. Bone Joint Surg., 44-A:1217-1244.

Belanger, L. F. 1954. Autoradiographic visualization of entry and exit S³⁵ in cartilage, bone and dentine of young rats and the effects of hyaluronidase in vitro. Canad. J. Biochem. Physiol., 32:161-169.

Bernard, G. W., and D. C. Pease. 1969. An electron microscope study of initial intramembranous osteogenesis. Amer. J. Anat., 125:271-290.

Bevelander, G., and P. L. Johnson. 1950. A histochemical study of the development of membrane bone. Anat. Rec., 108:1-22.

Bohatirchuk, F. 1963. Microradiology of mammalian bone. J. Canad. Assoc. Radiol., 14:29-38.

Bonucci, E. 1967. Fine structure of early cartilage calcification. J. Ultrastruct. Res., 20:33-50.

Cameron, D. A. 1968. The Golgi apparatus in bone and cartilage. Clin. Orthop., 58:191-211.

Cameron, D. A. 1972. The ultrastructure of bone. In *The Biochemistry and Physiology of Bone.* vol. 1. Edited by G. H. Bourne. Academic Press, New York, pp. 191-236.

Cameron, D. A., and R. A. Robinson. 1958. Electron microscopy of epiphyseal and articular cartilage matrix in the femur of the newborn infant. J. Bone Joint Surg., 40-A:163-170.

Carniero, J., and C. P. Le Blond. 1950. Role of osteoblasts and odontoblasts in secreting the collagen of bone and dentin as shown by radioautography in mice given tritium-labeled glycine. Exp. Cell Res., 18:291-300.

Clarke, I. C. 1971. Articular cartilage: a review and scanning electron microscope study. I. The interterritorial fibrillar architecture. J. Bone Joint Surg., 53-B:732-750.

Clarke, I. C. 1974. Articular cartilage: a review and scanning electron microscope study. II. The territorial fibrillar architecture. J. Anat. (Lond), 118:261-280.

Cohen, J., and W. H. Harris. 1958. The three-dimensional anatomy of Haversian systems. J. Bone Joint Surg., 40-A:419-434.

Decker, J. D. 1966. An electron microscope investigation of osteogenesis in the embryonic chick. Amer. J. Anat., 118:591-614.

Dudley, H. R., and D. Spiro. 1961. The fine structure of bone cells. J. Biophys. Biochem. Cytol., 11:627-649.

Enlow, D. H. 1962. A study of the post-natal growth and remodeling of bone. Amer. J. Anat., 110:79-101.

Enlow, D. H. 1962. Functions of the Haversian system. Amer. J. Anat., 110:269-305.

Epkar, B. N., and H. M. Frost. 1966. Periosteal appositional bone growth from age two to age seventy in man. Anat. Rec., 154:573-578.

Evans, F. G. 1958. Relations between the microscopic structure and tensile strength of human bone. Acta Anat. (Basel), 35:285-301.

Evans, F. G. 1976. Mechanical properties and histology of cortical bone from younger and older man. Anat. Rec., 185:1-12.

Fischman, D. A., and E. D. Hay. 1962. Origin of osteoclasts from mononuclear leukocytes in regenerating newt limbs. Anat. Rec., 143:329-337.

Frasca, P., R. A. Harper, and J. L. Katz. 1976. Isolation of single osteons and osteon lamellae. Acta Anat., 95:122–129.

Frost, H. M. 1964. *Mathematical Elements of Lamellar Bone Remodeling*. Charles C Thomas, Springfield, Ill., 127 pp.

Frost, H. M. 1964. *The Laws of Bone Structure*. Charles C Thomas, Springfield, Ill., 167 pp.

Glimcher, M. J., A. J. Hodge, and F. O. Schmitt. 1957. Macromolecular aggregation states in relation to mineralization: the collagen-hydroxyapatite system as studied in vitro. Proc. Nat. Acad. Sci., 43:860–867.

Gonzales, F., and M. J. Karnovsky. 1961. Electron microscopy of osteoclasts in healing fractures of rat bone. J. Biophys. Biochem. Cytol., 9:299–316.

Hall, B. K. 1975. The origin and fate of osteoclasts. Anat. Rec., 183:1–12.

Jande, S. S. 1971. Fine structural study of osteocytes and their surrounding bone matrix with respect to their age in young chicks. J. Ultrastruct. Res., 37:279–300.

Jande, J. J., and L. F. Belanger. 1973. The life cycle of the osteocyte. Clin. Orthop., 94:281–305.

Kallio, D. M., P. R. Garant, and C. Minkin. 1971. Evidence of coated membranes in the ruffled border of the osteoclast. J. Ultrastruct. Res., 37:169–177.

Kallio, D. M., P. R. Garant, and C. Minkin. 1972. Ultrastructural effects of calcitonin on osteoclasts in tissue culture. J. Ultrastruct. Res., 39:205–216.

Lacroix, P. 1951. *The Organization of Bones*. Trans. by S. Gilder. The Blakiston Co., Philadelphia, 235 pp.

Lufti, A. M. 1970. Mode of growth, fate and functions of cartilage canals. J. Anat. (Lond.), 106:135–145.

Owen, M. 1970. The origin of bone cells. Int. Rev. Cytol., 28:213–238.

Pawlicki, R. 1975. Bone canaliculus endings in the area of the osteocyte lacuna. Acta Anat., 91:292–304.

Pawlicki, R. 1976. Correlation between the structure of the wall of the bone lacuna and the localization of the osteocyte within this lacuna. Acta Anat., 95:421–433.

Revel, J. P., and E. D. Hay. 1963. An autoradiographic and electron microscopic study of collagen synthesis in differentiating cartilage. Z. Zellforsch. mikrosk. Anat., 61:110–144.

Scott, B. L. 1967. Thymidine-³H electron microscopic radioautography of osteogenic cells in the fetal rat. J. Cell Biol., 35:115–126.

Scott, B. L., and D. C. Pease. 1956. Electron microscopy of the epiphyseal apparatus. Anat. Rec., 126:465–495.

Shepard, N., and N. Mitchell. 1977. The localization of articular cartilage proteoglycan by electron microscopy. Anat. Rec., 187:463–475.

Silberberg, R., M. Silberberg, and D. Feir. 1964. Life cycle of articular cartilage cells: an electron microscope study of the hip joint of the mouse. Amer. J. Anat., 114:17–48.

Stillwell, D. L., Jr., and D. J. Gray. 1954. The microscopic structure of periosteum in areas of tendinous contact. Anat. Rec., 120:663–678.

Tappen, N. C. 1977. Three-dimensional studies of resorption spaces and developing osteons. Amer. J. Anat., 149:301–332.

Thorpe, E. J., B. B. Bellony, and R. F. Sellers. 1963. Ultrasonic decalcification of bone. J. Bone Joint Surg., 45-A:1257–1259.

Tonna, E. A., and E. P. Cronkite. 1968. Skeletal cell labeling following continuous infusion with tritiated thymidine. Lab. Invest., 19:510–515.

Tonna, E. A., and E. P. Cronkite. 1962. Use of tritiated thymidine for the study of the origin of the osteoclast. Nature (Lond.), 190:495–496.

Weinstock, A., and C. P. Le Blond. 1971. Elaboration of the matrix glycoprotein of enamel by secretory ameloblasts of the rat incisor as revealed by radioautography after galactose-³H injection. J. Cell Biol., 51:26–51.

Weinstock, A., M. Weinstock, and C. P. Le Blond. 1972. Autoradiographic detection of ³H-fucose incorporation into glycoprotein by odontoblasts and its deposition at the site of calcification front in dentin. Calcif. Tissue Res., 8:181–189.

Weinstock, M., and C. P. Le Blond. 1973. Radioautographic visualization of the deposition of a phosphoprotein at the mineralization front in the dentin of the rat incisor. J. Cell Biol., 56:838–845.

Williams, R. G. 1962. Comparison of living autogenous and homogenous grafts of cancellous bone heterotopically placed in rabbits. Anat. Rec., 143:93–105.

Williams, R. G. 1963. Studies of cartilage and osteoid arising spontaneously and experimental attempts to induce their formation in ear chambers. Anat. Rec., 146:93–108.

Young, R. W. 1962. Cell proliferation and specialization during endochondral osteogenesis in young rats. J. Cell Biol., 14:357–370.

Vertebral Column and Ribs

Arkin, A. M. 1949. The mechanisms of the structured changes in scoliosis. J. Bone Joint Surg., 31-A:519–528.

Bardeen, C. R. 1908. Early development of the cervical vertebrae and the base of the occipital bone in man. Amer. J. Anat., 8:181–186.

Batson, O. V. 1940. The function of the vertebral veins and their role in the spread of metastases. Ann. Surg., 112:138–149.

Bick, E. M., and J. W. Copel. 1950. Longitudinal growth of the human vertebra: contribution to human ostogeny. J. Bone Joint Surg., 32-A:803–814.

Brannon, E. W. 1963. Cervical rib syndrome. J. Bone Joint Surg., 45-A:977–998.

Buchwalter, J. A., R. R. Cooper, and J. A. Maynard. 1976. Elastic fibers in human intervertebral discs. J. Bone Joint Surg., 58-A:73–76.

Coventry, M. B., R. K. Ghormley, and J. W. Kernohan. 1945. The intervertebral disc: its microscopic anatomy and pathology. Part I. Anatomy, development and physiology. J. Bone Joint Surg., 27:105–112.

Crissman, R. S., and F. N. Low. 1974. A study of fine structural changes in the cartilage to bone transition within developing chick vertebra. Amer. J. Anat., 140:451–470.

Davis, P. R. 1959. The medial inclination of the human thoracic articular facets. J. Anat. (Lond.), 93:68–74.

Davis, P. R. 1961. Human lower lumbar vertebrae: some mechanical and osteological considerations. J. Anat. (Lond.), 95:337–344.

Evans, F. G., and H. R. Lissner. 1959. Biomechanical studies on the lumbar spine and pelvis. J. Bone Joint Surg., 41-A:278–290.

Fawcett, E. 1911. Some notes on the epiphyses of the ribs. J. Anat. (Lond.), 45:172–178.

Fawcett, E. 1931. A note on the identification of the lumbar vertebrae of man. J. Anat. (Lond.), 66:384–386.

Fielding, J. W. 1964. Normal and selected abnormal motion of the cervical spine from the second cervical vertebra to the seventh cervical vertebra based on cineroentgenography. J. Bone Joint Surg., 46-A:1779–1781.

Francis, C. C. 1955. Dimensions of the cervical vertebrae. Anat. Rec., *122*:603–610.

Gregersen, G. G., and D. B. Lucas. 1967. An in vivo study of the axial rotation of the human thoraco-lumbar spine. J. Bone Joint Surg., *49-A*:247–262.

Haas, S. L. 1939. Growth in length of the vertebrae. Arch. Surg., *38*:245–249.

Hadley, L. A. 1964. *Anatomico-Roentgenographic Studies of the Spine*. Charles C Thomas, Springfield, Ill., 545 pp.

Haines, R. W. 1946. Movements of the first rib. J. Anat. (Lond.), *80*:94–100.

Hohl, M. 1964. Normal motions in the upper portion of the cervical spine. J. Bone Joint Surg., *46-A*:1777–1779.

Holtzer, H., and S. R. Detwiler. 1953. An experimental analysis of the development of the spinal column. J. Exp. Zool., *123*:335–370.

Kaplan, E. B. 1945. The surgical and anatomic significance of the mammillary tubercle of the last thoracic vertebra. Surgery, *17*:78–92.

Keegan, J. J. 1953. Alterations of the lumbar curve related to posture and seating. J. Bone Joint Surg., *35-A*: 589–603.

Kendrick, G. S., and N. L. Biggs. 1963. Incidence of the ponticulus posticus of the first cervical vertebra between ages of six to seventeen. Anat. Rec., *145*: 449–453.

Lanier, R. R. 1954. Some factors to be considered in the study of lumbosacral fusion. Amer. J. Phys. Anthrop., *12*:363–372.

Lowrance, E. W., and H. B. Latimer. 1967. Weights and variability of components of the human vertebral column. Anat. Rec., *159*:83–88.

Markolf, K. L., and J. M. Morris. 1974. The structural components of the intervertebral disc. A study of their contributions to the ability of the disc to withstand compressive forces. J. Bone Joint Surg., *56-A*: 675–687.

Miles, M., and W. E. Sullivan. 1961. Lateral bending at the lumbar and lumbosacral joints. Anat. Rec., *139*: 387–398.

Mitchell, G. A. G. 1934. The lumbosacral junction. J. Bone Joint Surg., *16*:233–254.

Mitchell, G. A. G. 1936. The significance of lumbosacral transitional vertebrae. Br. J. Surg., *24*:147–158.

Peacock, A. 1952. Observations on the postnatal structure of the intervertebral disc in man. J. Anat. (Lond.), *86*:162–179.

Rowe, G. G., and M. B. Roche. 1953. The etiology of separate neural arch. J. Bone Joint Surg., *35-A*:102–110.

Russell, H. E., and G. T. Aitken. 1963. Congenital absence of the sacrum and lumbar vertebrae with prosthetic management. J. Bone Joint Surg., *45-A*:501–508.

Sensenig, E. C. 1949. The early development of the human vertebral column. Contr. Embryol. Carneg. Instn., *33*:21–41.

Stewart, T. D. 1954. Metamorphosis of the joints of the sternum in relation to age changes in other bones. Amer. J. Phys. Anthrop., *12*:519–536.

Taylor, J. R. 1975. Growth of human intervertebral discs and vertebral bodies. J. Anat. (Lond.), *120*:49–68.

Trotter, M. 1947. Variations of the sacral canal: their significance in the administration of caudal analgesia. Anesth. Analg. (Cleve.), *26*:192–202.

Veleanu, C. 1975. Contributions to the anatomy of the cervical spine. Functional and pathogenetic significance of certain structures of the cervical vertebrae. Acta Anat., *92*:467–480.

Wells, L. H. 1963. Congenital deficiency of the vertebral pedicle. Anat. Rec., *145*:193–196.

Willis, T. A. 1949. Nutrient arteries of the vertebral bodies. J. Bone Joint Surg., *31-A*:538–540.

The Skull

Ashley-Montagu, M. F. 1935. The premaxilla in the primates. Q. Rev. Biol., *10*:32–59; 181–208.

Brodie, A. G. 1941. On the growth of the human head, from the third month to the eighth year. Amer. J. Anat., *68*:209–262.

Chase, S. W. 1942. The early development of the human premaxilla. J. Amer. Dent. Assoc., *29*:1991–2001.

Cobb, W. M. 1943. The cranio-facial union and the maxillary tuber in mammals. Amer. J. Anat., *72*:39–111.

Di Dio, L. J. A. 1962. The presence of the eminentia orbitalis in the os zygomaticum of Hindu skulls. Anat. Rec., *142*:31–39.

Etter, L. E. 1955. *Atlas of Roentgen Anatomy of the Skull*. Charles C Thomas, Springfield, Ill., 215 pp.

Fawcett, E. 1905. Ossification of the lower jaw in man. J.A.M.A., *45*:696–705.

Fawcett, E. 1911. The development of the human maxilla, vomer and paraseptal cartilage. J. Anat. Physiol., *45*:378–405.

Ford, E. H. R. 1956. The growth of the foetal skull. J. Anat. (Lond.), *90*:63–72.

Kölliker, T. 1885. Zur Odontologie der Kieferspalte bei der Hasenscharte. Biol. Centralbl., *5*:371–373.

Graupp, E. 1906. Die Entwickelung des Kopfskelletes. In *Handbuch der Vergleichenden und Experimentellen Entwicklungslehre der Wirbeltiere*. Bd. 3, Teil 2, Gustav Fischer, Jena, pp. 573–874.

Hamparian, A. M. 1973. Blood supply of the human fetal mandible. Amer. J. Anat., *136*:67–75.

Horowitz, S. L., and H. H. Shapiro. 1955. Modification of skull and jaw architecture following removal of the masseter muscle in the rat. Amer. J. Phys. Anthrop., *13*:301–308.

Kendrick, G. S., and H. L. Risinger. 1967. Changes in anteroposterior dimensions of the human male skull during the third and fourth decade of life. Anat. Rec., *159*:77–81.

Latham, R. A. 1970. Maxillary development and growth: the septo-premaxillary ligament. J. Anat. (Lond.), *107*: 471–478.

Latham, R. A. 1971. The development, structure and growth pattern of the human mid-palatal suture. J. Anat. (Lond.), *108*:31–41.

Low, A. 1909. Further observations on the ossification of the human lower jaw. J. Anat. Physiol., *44*:83–95.

Markens, I. S. 1975. Embryonic development of the coronal suture in man and rat. Acta Anat., *93*:257–273.

Mednick, L., and S. L. Washburn. 1956. The role of the sutures in the growth of the braincase of the infant pig. Amer. J. Phys. Anthrop., *14*:175–192.

Miller, W. A. 1965. *The Keys to Orthopedic Anatomy*. Charles C Thomas, Springfield, Ill., 155 pp.

Noback, C. R., and M. L. Moss. 1953. The topology of the human pre-maxillary bone. Amer. J. Phys. Anthrop., *11*:181–187.

Olivier, G. 1975. Biometry of the human occipital bone. J. Anat. (Lond.), *120*:507–518.

Ortiz, M. H., and A. G. Brodie. 1949. On the growth of the human head from birth to the third month of life. Anat. Rec., *103*:311–333.

Oschinsky, L. 1960. Two recently discovered human

mandibles from Cape Dorset sites on Sugluk and Mansel Islands. Anthropologica N.S., 11:1–16.

Parsons, F. G. 1909. The topography and morphology of the human hyoid bone. J. Anat. Physiol., 43:279–291.

Roche, A. F., K. Manuel, and F. S. Seward. 1965. Unusual patterns of growth in the frontal and parietal bones. Anat. Rec., 152:459–464.

Shiller, W. R., and O. B. Wiswell. 1954. Lingual foramina of the mandible. Anat. Rec., 119:387–390.

Sinclair, J. B., and J. Mc Kay. 1945. Median hare lip, cleft palate and glossal agenesis. Anat. Rec., 91:155–160.

Van Der Klaavw, C. J. 1963. *Projections, Deepenings and Undulations of the Skull in Relation to the Attachment of Muscle.* North Holland Publishing Company, Amsterdam, 247 pp.

Walensky, N. A. 1964. A re-evaluation of the mastoid region of contemporary and fossil man. Anat. Rec., 149:67–72.

Warwick, R. 1950. The relation of the direction of the mental foramen to the growth of the human mandible. J. Anat. (Lond.), 84:116–120.

Woo, J. K. 1949. Ossification and growth of the human maxilla, premaxilla and palate bone. Anat. Rec., 105:737–761.

Wood, N. K., L. E. Wragg, and O. H. Stuteville. 1967. The premaxilla: embryological evidence that it does not exist in man. Anat. Rec., 158:485–490.

Wood, N. K., L. E. Wragg, O. H. Stuteville, and R. J. Oglesby. 1969. Osteogenesis of the human upper jaw. Proof of the non-existence of a separate premaxillary center. Arch. Oral Biol., 14:1331–1341.

Young, R. W. 1962. Autoradiographic studies on postnatal growth of the skull in young rats injected with tritiated glycine. Anat. Rec., 143:1–14.

Upper Limb

Bizarro, A. H. 1921. On sesamoid and supernumerary bones of the limbs. J. Anat. (Lond.), 55:256–268.

Carroll, S. E. 1963. A study of the nutrient foramina of the humeral diaphysis. J. Bone Joint Surg., 45-B: 176–181.

Cave, A. J. E. 1961. The nature and morphology of the costoclavicular ligament. J. Anat. (Lond.), 95:170–179.

Christensen, J. B., J. P. Adams, K. O. Cho, and L. Miller. 1968. A study of the interosseous distance between the radius and ulna during rotation of the forearm. Anat. Rec., 160:261–272.

Cleveland, M. 1948. Fracture of the carpal scaphoid. Surg. Gynecol. Obstet., 84:769–771.

Evans, F. G., A. Alfaro, and S. Alfaro. 1950. An unusual anomaly of the superior extremities in a Tarascan Indian girl. Anat. Rec., 106:37–48.

Frantz, C. H., and R. O'Rahilly. 1961. Congenital skeletal limb deficiencies. J. Bone Joint Surg., 43-A:1202–1224.

Gardner, E. 1968. The embryology of the clavicle. Clin. Orthop., 58:9–16.

Gardner, E., and D. J. Gray. 1953. Prenatal development of the human shoulder and acromioclavicular joints. Amer. J. Anat., 92:219–275.

Garn, S. M., and C. G. Rohmann. 1960. Variability in the order of ossification of the bony centers of the hand and wrist. Amer. J. Phys. Anthrop., 18:219–230.

Gray, D. L. 1969. The prenatal development of the human humerus. Amer. J. Anat., 124:431–445.

Gray, D. J., and E. Gardner. 1969. The prenatal development of the human humerus. Amer. J. Anat., 124:431–445.

Gray, D. J., E. Gardner, and R. O'Rahilly. 1957. The prenatal development of the skeleton and joints of the human hand. Amer. J. Anat., 101:169–224.

Greulich, W. W., and S. I. Pyle. 1950. *Radiographic Atlas of Skeletal Development of the Hand and Wrist.* Stanford University Press, Stanford, California. 190 pp.

Handforth, J. R. 1950. Polydactylism of the hand in southern Chinese. Anat. Rec., 106:119–125.

Joseph, J. 1951. The sesamoid bones of the hand and the time of fusion of the epiphyses of the thumb. J. Anat. (Lond.), 85:230–241.

Kimura, K. 1976. Growth of the second metacarpal according to chronological age and skeletal maturation. Anat. Rec., 184:147–158.

Lachman, E. 1953. Pseudo-epiphyses in hand and foot. Amer. J. Roentgenol., 70:149–151.

Laing, P. G. 1956. The arterial supply of the adult humerus. J. Bone Joint Surg., 38-A:1105–1116.

Lewis, O. J., R. J. Hamshere, and T. M. Bucknill. 1970. The anatomy of the wrist joint. J. Anat. (Lond.), 106:539–552.

Mac Conaill, M. A. 1941. The mechanical anatomy of the carpus and its bearing on some surgical problems. J. Anat. (Lond.), 75:166–175.

Martin, B. F. 1958. The annular ligament of the superior radio-ulnar joint. J. Anat. (Lond.), 92:473–482.

Meekle, M. C. 1975. The influence of function on chondrogenesis at the epiphyseal cartilage of a growing long bone. Anat. Rec., 182:387–400.

Moseley, H. F. 1968. The clavicle: its anatomy and function. Clin. Orthop., 58:17–27.

O'Rahilly, R. 1956. Developmental deviations in the carpus and tarsus. Clin. Orthop., 10:9–18.

Singh, I. 1959. Variations in the metacarpal bones. J. Anat. (Lond.), 93:262–267.

Stewart, M. J. 1954. Fractures of the carpal navicular (scaphoid). A report of 436 cases. J. Bone Joint Surg., 36-A:998–1006.

Wise, K. S. 1975. The anatomy of the metacarpophalangeal joint with observations of the aetiology of ulnar drift. J. Bone Joint Surg., 57-B:485–490.

Whitson, R. O. 1954. Relation of the radial nerve to the shaft of the humerus. J. Bone Joint Surg., 36-A:85–88.

Lower Limb

Badi, M. H. 1972. Calcification and ossification of fibrocartilage in the attachment of the patellar ligament in the rat. J. Anat. (Lond.), 112:415–421.

Barnett, C. H. 1962. The normal orientation of the human hallux and the effect of footwear. J. Anat. (Lond.), 96:489–494.

Barnett, C. H. 1953. Locking at the knee joint. J. Anat. (Lond.), 87:91–95.

Cahill, D. R. 1965. The anatomy and function of the contents of the human tarsal sinus and canal. Anat. Rec., 153:1–17.

Cave, E. F., and C. R. Rowe. 1950. The patella: its importance in derangement of the knee. J. Bone Joint Surg., 32-A:542–553.

Claffey, T. J. 1960. Avascular necrosis of the femoral head. An anatomical study. J. Bone Joint Surg., 42-B: 802–809.

Crelin, E. S. 1960. The development of bony pelvic sexual dimorphism in mice. Ann. N. Y. Acad. Sci., 84: 479–512.

Crock, H. V. 1965. A revision of the anatomy of the arteries supplying the upper end of the human femur. J. Anat. (Lond.), *99*:77–88.

Davis, G. G. 1927. Os Vesalianum pedis. Amer. J. Roentgenol., *17*:551–553.

Evans, F. G., and H. R. Lissner. 1955. Studies on pelvic deformities and fractures. Anat. Rec., *121*:141–166.

Frangakis, E. K. 1966. Intracapsular fractures of the neck of the femur. Factors influencing non-union and ischaemic necrosis. J. Bone Joint Surg., *48-B*:17–30.

Gardner, E., and D. J. Gray. 1950. Prenatal development of the human hip joint. Amer. J. Anat., *87*:163–211.

Haxton, H. 1945. The function of the patella and the effects of its excision. Surg. Gynecol. Obstet., *80*:389–395.

Henderson, R. S. 1963. Os intermetatarseum and a possible relationship to hallux valgas. J. Bone Joint Surg., *45-B*:117–121.

Howe, W. W., Jr., T. Lacey, and R. P. Schwartz. 1950. A study of the gross anatomy of the arteries supplying the proximal portion of the femur and the acetabulum. J. Bone Joint Surg., *32-A*:856–866.

Irani, R., and M. Sherman. 1963. The pathological anatomy of club foot. J. Bone Joint Surg., *45-A*:45–52.

Johnson, E. W., Jr., and H. A. Patterson. 1966. Fractures of the os calcis. Arch. Surg., *92*:848–852.

Kaplan, E. B. 1962. Some aspects of the functional anatomy of the human knee joint. Clin. Orthop., *23*:18–29.

Kraus, B. S. 1961. Sequence of appearance of primary centers of ossification in the human foot. Amer. J. Anat., *109*:103–115.

Laing, P. G. 1953. The blood supply of the femoral shaft: anatomical study. J. Bone Joint Surg., *35-B*:462–466.

Last, R. J. 1948. Some anatomical details of the knee joint. J. Bone Joint Surg., *30-B*:683–688.

Laurenson, R. D. 1964. The primary ossification of the human ilium. Anat. Rec., *148*:209–218.

La Velle, C. L. B. 1974. An analysis of the human femur. Amer. J. Anat., *141*:415–426.

Lewis, O. J. 1958. The tubercle of the tibia. J. Anat., *92*:587–592.

McDougall, A. 1955. The os trigonum. J. Bone Joint Surg., *37-B*:257–265.

Morton, D. J. 1935. *The Human Foot—Its Evolution, Physiology, and Functional Disorders*. Columbia University Press, New York. 244 pp.

Nelson, E. M. 1963. A report of a 7-toed foot. Anat. Rec., *147*:1–3.

Ogden, J. A., R. F. Hempton, and W. O. Southwick. 1975.

Development of the tibial tuberosity. Anat. Rec., *182*:431–446.

Pyle, S. I., and N. L. Hoerr. 1955. *Radiographic Atlas of Skeletal Development of the Knee. A Standard of Reference*. Charles C Thomas, Springfield, Ill., 82 pp.

Roberts, W. H. 1962. Femoral torsion in normal human development as related to dysplasia. Anat. Rec., *143*:369–375.

Roche, A. F. 1964. Epiphyseal ossification and shaft elongation in human metatarsal bones. Anat. Rec., *149*:449–452.

Rogers, L. 1928. The styloid epiphysis of the fifth metatarsal bone. J. Bone Joint Surg., *10*:197–199.

Sevitt, S., and R. G. Thompson. 1965. The distribution and anastomoses of arteries supplying the head and neck of the femur. J. Bone Joint Surg., *47-B*:560–573.

Shands, A. R., and I. W. Wentz. 1953. Congenital anomalies, accessory bones, and osteochondritis in the feet of 850 children. Surg. Clin. North Amer., December 1643–1666.

Singh, I. 1959. Squatting facets on the talus and tibia in Indians. J. Anat. (Lond.), *93*:540–550.

Singh, I. 1960. Variations in the metatarsal bones. J. Anat. (Lond.), *94*:345–350.

Smith, J. W. 1956. Observations on the postural mechanisms of the human knee joint. J. Anat. (Lond.), *90*:236–260.

Tobin, W. J. 1955. The internal architecture of the femur and its clinical significance. The upper end. J. Bone Joint Surg., *37-A*:57–72.

Trueta, J. 1957. The normal vascular anatomy of the human femoral head during growth. J. Bone Joint Surg., *39-B*:358–394.

Tucker, F. R. 1949. Arterial supply to the femoral head and its clinical importance. J. Bone Joint Surg., *31-A*:82–93.

Walensky, N. A. 1965. A study of the anterior femoral curvature in man. Anat. Rec., *151*:559–570.

Weinert, C. R., J. H. Mc Master, and R. J. Ferguson. 1973. Dynamic function of the human fibula. Amer. J. Anat., *138*:145–149.

Whitehouse, W. J., and E. D. Dyson. 1974. Scanning electron microscope studies of trabecular bone in the proximal end of the human femur. J. Anat. (Lond.), *118*:417–444.

Wray, J. B., and C. N. Herndon. 1963. Hereditary transmission of congenital coalition of the calcaneus to the navicular. J. Bone Joint Surg., *45-A*:365–372.

5

The Joints

Articulations, or joints, are specialized anatomical structures at which the ends of certain bones are joined or the borders of other bones juxtaposed. These osseous junctions are secured by ligaments, fibrous capsules, and other binding tissues which restrict movement or permit varying degrees of movement. Joints vary widely in their structure, frequently presenting unique morphological features adapted to specific functional requirements.

The study of joints is called **arthrology,** and it is initially of practical importance for the student to realize that movement does not occur in all joints. In *immovable* articulations, the bones are held together by several fibrous layers. These joints are called sutures and they are peculiar to the skull. The adjacent margins of the skull bones are serrated and interlocked, being separated merely by the thin, fibrous sutural ligament. In certain regions at the base of the skull this fibrous membrane is replaced by a layer of cartilage. These are **joints adapted for growth** of the brain. In the developing and maturing infant the brain is enclosed by a membrane destined to ossify. The membranes of the cranial bones are in contact peripherally and ossify from more central-

ized sites. After the completion of brain growth, the sutures close and become unyielding, thereby forming the rigid cranial vault. A comparable, but not identical, situation exists in many limb bones between the ossifying epiphyses and diaphyses. These junctions, temporarily cartilaginous, become ossified after allowing full growth of the maturing limb.

In contrast to joints adapted for growth are **joints adapted for movement.** At certain joints *slight movement* combined with great strength is required. In most of these articulations the osseous surfaces are united by tough and flexible fibrocartilages as in the joints between the vertebral bodies, and in the interpubic articulation. In the *freely movable* joints the bony surfaces are completely separated, and the ends of the bones are expanded or so contoured as to facilitate the connection. Additionally, the osseous surfaces are covered by cartilage and enveloped by capsules of fibrous tissue. The cells lining the interior of the fibrous capsule form the synovial membrane which secretes a lubricating fluid. The joints are strengthened by strong fibrous bands, the ligaments, which extend between the bones.

Development of the Joints

The development of the joints naturally is related to the development of the skeletal elements that they interconnect. Because the skeleton forms from mesoderm, it follows that the tissues forming the joints are also derived from this germ layer. The development of the intervertebral joints occurs simultaneously with the development of the vertebrae. As discussed in Chapter 4, these bones form from a column of mesoderm which becomes divided into sclerotomes along the central axis of the embryo. The intervertebral discs, which become interposed between the vertebral bodies, also form from the sclerotomes. From the sides of the embryo, limb buds develop at the end of the fourth week, and shortly after their appearance, central cores of mesoderm form continuous axes throughout their length. At first the mesoderm from which the different parts of the skeleton will form shows no differentiation between the masses that will correspond to the individual bones. During the fifth week, however, circumscribed condensations of mesenchyme appear within the mesodermal core of the limb bud and present the first indications of the bones of the limbs. Within these condensations of mesenchyme, chondrogenic centers appear, one for each future bone. The mesenchyme intervening between the chondrifying skeletal elements is initially a homogeneous undifferentiated region called the **interzone** (Fig. 5-1,A).

The manner in which the interzone mesenchyme develops further into a joint depends on the type of joint being formed. The mesenchyme may be converted into white fibrous tissue, as in the case of the skull bones, a **fibrous joint** being the result. In contrast, it may form hyaline cartilage or fibrocartilage, in which case a slightly movable **cartilaginous joint** is formed. The interzone of a developing, freely movable, **synovial joint** consists of three layers, *two of which are chondrogenic,* each continuous with the perichondrium and capping the ends of the developing cartilage. Between these two layers is an *intermediate loose layer* of mesenchyme (Fig. 5-1,B). Surrounding the interzone are layers of condensing mesenchyme, which become the **articular fibrous capsule,** and interposed between the interzone and the capsule is the synovial mesenchyme. During the sixth week small cavities appear in the intermediate mesenchyme layer of the interzone and in the synovial mesenchyme (Fig. 5-1,C). As these cavities coalesce, the joint cavity is formed. On the inner surface of the fibrous capsule, the mesenchymal cells differentiate into a mesothelial lining, forming the **synovial membrane** (Fig. 5-1,D).

The tissue surrounding the original mesodermal core forms the perichondrial and periosteal fibrous sheaths for the developing bones. These sheaths are continuous with the articular fibrous capsule which overlies the synovial membrane and encases the joint cavity. All parts of a joint capsule are not of uniform thickness, so that in them may be recognized especially strengthened bands, which become ligaments. Additional ligamentous bands are at times formed by remnants of tendons surrounding the joint.

In several of the movable joints the intermediate layer mesenchyme of the interzone between the ends of the bones does not become completely absorbed, but persists to form fibrous or cartilaginous elements. Some

of these elements, such as the menisci of the knee joint or the temporomandibular joint, become intimately associated in their development with the muscles surrounding the joint. Other cartilaginous structures, such as the articular disc of the sternoclavicular joint, significantly strengthen the articulation, intervening between the ends of the bones in such a manner as to form two joint cavities.

Classification of Joints

It has already been mentioned that joints may be classified according to those *adapted for growth* and those *adapted for movement.* The former are immovable joints, while among the latter are included the slightly movable and freely movable joints. Although all joints can be listed within these two functional categories, little information about the anatomical nature of the joints is conveyed by such a classification. Because of the uniqueness of many articulations, however, no completely adequate classification scheme has been devised that covers the many individual differences among the joints. Perhaps the most frequently utilized classification system, and one officially recognized by the *Nomina Anatomica,* is dependent strictly on the morphological characteristics of the joints. This schema recognizes that principally three types of tissue separate the ends of bones in the different joints: fibrous tissue, cartilage, or synovial membranes. Thus, it distinguishes: (1) **fibrous joints** (many of which are immovable; *synarthroses*); (2) **cartilaginous joints** (slightly movable; *amphiarthroses*); and (3) **synovial joints** (freely movable; *diarthroses*).

FIBROUS JOINTS

The fibrous joints include all those articulations in which the surfaces of the bones are nearly in direct contact, and in which the adjoining bones are fastened together by fibrous connective tissue. There are three varieties of fibrous joints: **syndesmoses, sutures,** and **gomphoses.** In many of these joints there is no appreciable movement, as in those articulations between the bones of the skull which form the cranial vault. How-

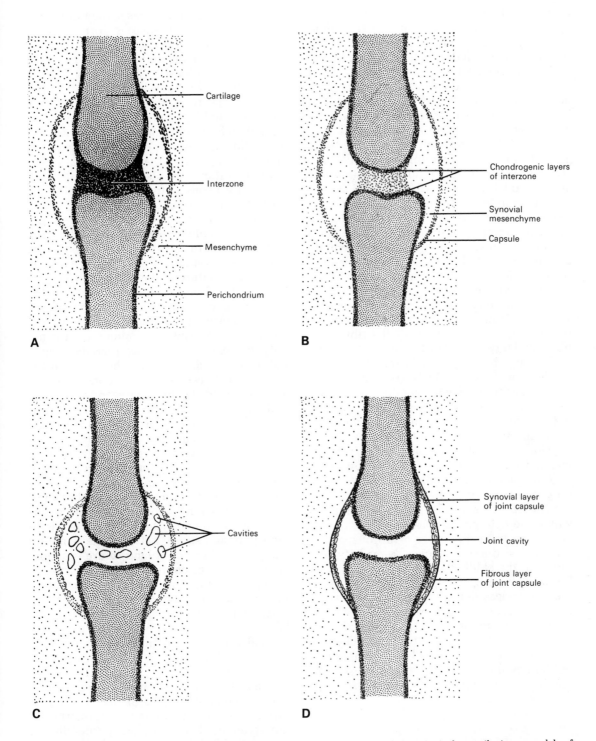

Fig. 5-1. Schematic diagrams showing the developmental stages of a synovial joint. In *A*, the cartilaginous models of the developing bone are separated by a homogeneous region, the interzone. As shown in *B*, the interzone soon becomes transformed into two chondrogenic layers continuous with the perichondrium and separated by an intermediate layer of mesenchyme. Condensation of the mesenchyme laterally forms a primitive joint capsule within which small cavities appear, *C*, and then coalesce, *D*, to form the joint cavity, while the articular capsule differentiates into an inner synovial layer and an outer fibrous layer. (Modified from Langman, J., *Medical Embryology*, Williams and Wilkins, 1963.)

ever, very limited movement is permitted in syndesmoses (such as the inferior tibiofibular joint) and in the gomphoses (which are the joints between the roots of the teeth and their bony sockets).

SYNDESMOSES.   A syndesmosis is usually described as a type of articulation in which two bones are joined by an interosseous ligament formed of fibrous connective tissue. Such a joint generally allows a limited degree of movement. Examples of syndesmoses are the inferior tibiofibular joint and the articulation between the base of the stapes and the margin of the vestibular window in the middle ear. More broadly considered, the syndesmoses could include any number of slightly movable skeletal unions which are interconnected by fibrous ligaments and which do not contain synovial sacs or intervening cartilaginous discs. Thus, it would also be reasonable to include as syndesmoses such ligamentous unions as those between the parallel borders of the tibia and fibula and those between the radius and ulna formed by the interosseous membranes of the leg and forearm. Similarly, the coracoclavicular union, as well as others, could be regarded as syndesmoses. These latter unions, however, were not classified as joints in the *Nomina Anatomica,* but were considered simply as ligamentous attachments. Under this exceedingly narrow definition, the inferior tibiofibular joint is the only syndesmosis in the human body.

SUTURES.   A suture is an articulation in which the contiguous margins of the bones are united by a thin layer of fibrous tissue (Fig. 5-2). Sutures are found only between bones in the skull (cranial and facial). The fibrous sutural ligament interposed between the bony edges derives from the nonossifying portions of the membranes within which the cranial bones develop. It should be noted that after maturation, active bone deposition, or *synostosis,* partially or completely obliterates the sutural line, a process that continues with aging. Study of the histologi-cal structure of sutures has shown that each apposing periosteal edge is formed by a layer of osteogenic cells overlaid by fibrous tissue. Between the two periosteal fibrous coverings is a looser vascularized fibrous tissue (Pritchard, Scott, and Girgis, 1956).

In certain sutures the margins of the bones are connected by a series of indented processes which are interlocked. Three varieties of interlocking sutures can be distinguished: denticulated, serrated, and limbous. A **denticulate suture,** as its name implies, is formed by a consecutive series of tooth-like, projecting processes from each side, which frequently measure 5 mm or more, and which interconnect securely across the joint. The sagittal suture between the two parietal bones is an example of a denticulate suture. In a **serrate suture,** as between the two portions of the frontal bone, the edges of the apposing bones show somewhat more regular and shorter projections, resembling the teeth of a fine saw. A **limbous suture** not only presents interlocking serrated edges, but also a certain degree of bevelling of the articular surfaces, so that the bones overlap, as in the coronal suture between the parietal and frontal bones.

In other sutures the articulation is formed by the rough surfaces of bones without serrated edges, which are placed in apposition to one another. There are two types of these nonindented sutures, called squamous and plane. A **squamous suture** is characterized by an overlapping of the contiguous bones along broad bevelled margins, as in the squamosal suture between the temporal and parietal bones. A **plane suture** is simply an apposition of contiguous rough surfaces, which are neither bevelled nor indented, as in the articulation between the maxillae, or between the horizontal plates of the palatine bones.

Finally, the term **schindylesis** is used for a type of suture in which a thin plate of one bone is received into a cleft or fissure formed by the separation of two laminae in another bone. Examples of this type of suture are the articulations of the rostrum of the sphenoid bone and the perpendicular plate of the ethmoid bone with the vomer. Also, the reception of the vomer in the fissures between the maxillae and between the palatine bones is considered a schindylesis.

GOMPHOSES.   A gomphosis is an articulation that results from the insertion of a coni-

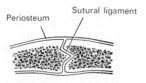

Periosteum   Sutural ligament

FIG. 5-2.   Vertical section through the sagittal suture.

cal process into a socket. This more specialized type of articulation does not occur between bones, strictly speaking, but is seen in the fibrous union of the roots of the teeth with the alveoli of the mandible and maxilla.

CARTILAGINOUS JOINTS

In a cartilaginous joint the bones are united either by a plate of hyaline cartilage or by a fibrocartilaginous disc. There are two varieties: **synchondroses** and **symphyses**.

SYNCHONDROSES. A typical synchondrosis is a temporary form of joint (Fig. 5-3), for the intervening hyaline cartilage is converted into bone before adult life. Such joints are found between the epiphyses and diaphyses of long bones, between the occipital and the sphenoid bones at birth and for some years after, and between the petrous portion of the temporal bone and the jugular process of the occipital bone. Still another example of a synchondrosis is the Y-shaped union of the ilium, ischium, and pubis that forms the cup of the acetabulum.

SYMPHYSES. A symphysis is the union of two contiguous bony surfaces connected by a broad, flattened disc of fibrocartilage, as in the articulations between the bodies of the vertebrae or the two pubic bones (Fig. 5-4). Frequently, the fibrocartilaginous disc is separated from the bone by thin layers of hyaline cartilage. Although limited movement occurs at a symphysis, it appears to depend on the thickness of the fibrocartilaginous disc as well as on its degree of elasticity. The slight movements that occur between the bodies of the vertebrae result in the flexibility of the vertebral column. Mobility of the interpubic joint is also of some importance in parturition. The anatomical structure of intervening discs can vary quite widely in different symphyses. An intervertebral disc contains a soft gelatinous center,

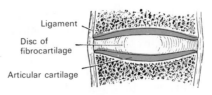

FIG. 5-4. Diagrammatic section of a symphysis.

called the nucleus pulposus, which lends resilience to the joint. In contrast, the interpubic joint sometimes contains a small fluid-filled cavity. Another symphysis in the body is the manubriosternal joint.

SYNOVIAL JOINTS

Synovial joints are highly evolved articulations which permit free movement. Because the human lower limbs are concerned with locomotion and the upper limbs provide a great versatility of movement, it is not surprising that most of the joints in the extremities are of the synovial type. In contrast to fibrous and cartilaginous joints where the ends of the bones are bound in continuity with intervening tissue, the ends of the bones in a synovial joint are in contact, but separate. Because the bones are not bound internally, the integrity of a synovial joint results from its ligaments and capsule, which bind the articulation externally, and to some extent from the surrounding muscles. The contiguous bony surfaces are covered with hyaline cartilage, and the joint cavity is surrounded by a fibrous capsule, the inner surface of which is lined by a synovial layer containing cells that are thought to secrete the viscous lubricating **synovial fluid** (Fig. 5-5).

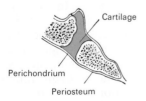

FIG. 5-3. Section through occipitosphenoid synchondrosis of an infant.

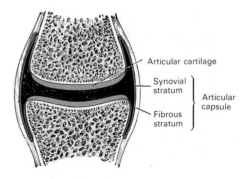

FIG. 5-5. Diagrammatic section of a synovial (diarthrodial) joint. The black area represents the joint cavity.

In certain synovial joints, the joint or **synovial cavity** may be divided, completely or incompletely, by an **articular disc** or **meniscus** (Fig. 5-6). Such discs are fibrocartilaginous, but unlike those seen in cartilaginous joints, intra-articular discs of synovial joints are not connected to the ends of the bones, but are continuous at their periphery with the fibrous capsule. The nonarticular surfaces of such a disc are covered by the synovial membrane, and thus when the joint is completely divided by a disc, as is the temporomandibular joint, two joint cavities are found within the fibrous capsule. Synovial joints involving two bones and containing a single joint cavity are sometimes referred to as **simple joints;** joints that contain an articular disc, thereby forming two joint cavities, are called **composite joints.** The term **compound joint** is used for those articulations in which more than a single pair of articulating surfaces are present.

The varieties of synovial joints have been classified by the types of active motion that they permit (uniaxial, biaxial, and polyaxial), but are differentiated further according to their principal morphological features (hinge, pivot, condyloid). In some joints the movement is *uniaxial,* that is to say, all movements take place around one axis. Among these are the **ginglymus** or hinge joint, in which the axis of movement is, for all practical purposes, transverse to the axes of the bones, and the **trochoid** or pivot joint, in which the axis is longitudinal. In other varieties the movements are *biaxial,* or around two axes at a right angle or any other angle to each other. These include the **condyloid,** the **ellipsoid,** and the **saddle joints.** There is one type of joint in which the movements are *polyaxial,* the **spheroidal** or ball-and-socket joint, at which movements are permitted in an infinite number of axes. Finally, there are the **plane** or gliding joints.

GINGLYMUS OR HINGE JOINT. The articular surfaces of a ginglymus joint are molded to each other in such a manner as to permit motion in only one plane, around the transverse axis. Although these joints are uniaxial, they frequently permit a considerable degree of motion. Flexion at the elbow joint, for example, can reduce the angle between the forearm and arm from about 180° to less than 45° in the fully flexed position. The direction that the distal bone takes in this motion is seldom in exactly the same plane as that of the axis of the proximal bone, because there is frequently a certain deviation from the straight line during flexion. The bones of a hinge joint are usually connected by strong collateral ligaments, which form their chief bond of union. In addition to the *elbow joint,* other examples of ginglymus joints include the *interphalangeal joints* of both the fingers and toes.

TROCHOID OR PIVOT JOINT. Movement in a trochoid or pivot joint also occurs around a single axis, which in these joints is the longitudinal axis. Trochoid joints are formed by a bony pivot-like process surrounded by a ligamentous ring, which may also be partly bony. Movement of the joint occurs as the pivot turns within the ring, or as the ring turns around the pivot, in both instances around the longitudinal axis. There are several pivot joints in the human body. In the *proximal radioulnar articulation,* the ring is formed by the radial notch of the ulna and the annular ligament. The head of the radius forms the pivot, which rotates within the ring. In the *articulation of the dens of the axis with the atlas,* the ring is formed in front by the anterior arch of the atlas, and posteriorly by the transverse ligament of the atlas. In this joint the ring turns around the bony pivot or dens. The distal radioulnar joint is also of the trochoid type.

CONDYLAR JOINT. In a condylar joint movement occurs principally in one plane, and although a slight degree of rotation can also occur at an axis at right angles to that of the principal movement, the rotation is severely limited by the disposition of the ligaments and tendons associated with the joint. In these joints, a pair of convex condyles are received in appropriately shaped concavities. The *tibiofemoral articulation of the*

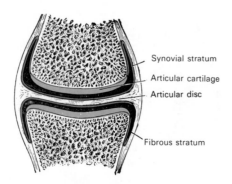

FIG. 5-6. Diagrammatic section of a synovial joint with an articular disc. The two black areas represent the joint cavities separated by the disc.

Synovial stratum

Articular cartilage

Articular disc

Fibrous stratum

knee joint is an example of the condylar joint.

ELLIPSOID JOINT.   Ellipsoid joints allow movements around two principal axes which are at right angles to each other. In these joints an oval convexity is received into an elliptical concavity, allowing a flexion-extension mobility as well as abduction-adduction movements. A combination of these movements results in circumduction, but there is little if any rotation around the third axis because of the ellipsoid shape of the articulating surfaces. Examples of ellipsoid joints include, among others, the *radio-carpal* and *metacarpophalangeal joints.*

SADDLE JOINT.   In saddle (or sellar) joints the articular end of the proximal bone is concave in one axis and convex in a perpendicular axis. These surfaces fit reciprocally into convex and concave surfaces of the distal bone. Movements at such a reciprocally apposed concavoconvex articular surface are the same as in ellipsoid joints, that is, flexion, extension, adduction, abduction, and circumduction are allowed. A modest degree of axial rotation is seen as well, especially in conjunction with the principal movements. The best example of a saddle joint is the *carpometacarpal joint* of the thumb.

SPHEROID OR BALL-AND-SOCKET JOINT.   A spheroid or ball-and-socket joint is one in which the distal bone is capable of motion around an indefinite number of axes, which have one common center. It is formed by the reception of a globular head into a cup-like cavity; however, the so-called sphere does not have the same diameter in all directions, and hence is slightly ellipsoid in shape. The greater the depth of the socket, the more stable the joint, but a deeper socket also results in more limited movement. Examples of this form of articulation are found in the *hip and shoulder joints.*

PLANE OR GLIDING JOINT.   A plane or gliding joint allows only a slight slipping or sliding of one bone over the other. Such a joint is formed by the apposition of plane surfaces, or of one slightly concave and the other slightly convex surface. The amount of motion between the surfaces is limited by the ligaments or osseous processes that surround the articulation. It is the form present in the joints between the *articular processes of certain vertebrae,* the *carpal joints* (except that of the capitate with the scaphoid and lunate), and the *intermetatarsal joints.*

Movements at Joints

The movements admissible in joints may be divided into four types: **gliding** and **angular movements, circumduction,** and **rotation.** Often, however, these movements are combined in varying degrees within joints to produce an infinite variety, and seldom is only a single type of motion found in any particular joint.

GLIDING MOVEMENT.   Gliding movement is the simplest type of motion that can take place in a joint, one surface gliding or moving over another without any significant angular or rotatory movement. The gliding movement is frequently combined with other movements at many joints, but in some, as in most of the intercarpal and intermetatarsal articulations, it is the principal motion permitted. This movement is not confined to plane surfaces, but may exist between any two contiguous bony surfaces, of whatever shape.

ANGULAR MOVEMENT.   Angular movements increase or decrease the angle between two adjoining bones. Flexion-extension and abduction-adduction are the common examples, each pair of movements acting in different axes which are at right angles.

Flexion occurs when the angle between two bones is decreased. A typical example is the bending of the upper extremity at the elbow. In some instances rather arbitrary definitions must be given. For example, bringing the femur forward or ventrally at the hip is flexion, but this motion includes the entire arc beginning in a position with the limb held back and then brought forward, as in kicking an American football, and also includes approximating the thigh against the anterior aspect of the trunk. Flexion at the knee approximates the calf and the posterior thigh, while the term *dorsiflexion* is used for the movement at the ankle joint that attempts to bring the dorsal foot in proximity to the anterior leg. Flexion of the interphalangeal joint of the thumb tends to approximate the palmar surfaces of the two segments of that digit, while flexion of the metacarpophalangeal joint of the thumb brings that digit across the surface of the palm.

Extension occurs when the angle between the bones is increased, as when the forearm is aligned to the arm, and generally is the action opposite flexion. *Hyperextension* is the term used for moving through an arc in the same direction, but beyond the straight position, as in bringing the arm back before launching a bowling ball. *Plantar flexion* of the foot at the ankle joint is the opposite of dorsiflexion, and by this action an attempt is made to approximate the plantar surface of the foot with the calf.

Abduction occurs when a limb is moved away from the midsagittal plane of the body, or when the fingers and toes are moved away from the median longitudinal axis of the hand or foot. An exception is abduction at the carpometacarpal joint of the thumb, where abduction is that action by which the thumb is elevated anterior to the palm.

Adduction occurs when a limb is moved toward or beyond the midsagittal plane or when the fingers or toes are moved toward the median longitudinal axis of the hand or foot. Once again the plane of movement of the carpometacarpal joint of the thumb is different. In the anatomical position, adduction carries the abducted thumb in a posterior direction toward the plane of the palm.

CIRCUMDUCTION. Circumduction is a form of motion that takes place when the head of a long bone revolves within its cup-like cavity. The bone thereby circumscribes a conical space, the base of which is directed distally, while the apex of the cone is in the articular cavity. Circumduction is a sequence of flexion, abduction, extension, and adduction. This type of movement is best seen at the shoulder and hip joints.

ROTATION. Rotation is a form of movement in which a bone moves around a central axis without undergoing any other displacement. A bone may rotate around its own longitudinal axis, as in the rotation of the humerus at the shoulder joint, or alternatively, the central axis of rotation may lie in a separate bone, as in the case of the pivot formed by the dens of the axis around which the atlas turns. The axis of rotation may not even be exactly parallel to the long axis of the bone, as in the movement of the radius on the ulna during pronation and supination of the hand. In this latter instance, the axis of rotation is represented by a line connecting the center of the head of the radius proximally with the center of the head of the ulna distally.

In the anatomical position the upper extremities are oriented along the sides of the trunk with the palms facing anteriorly, or in the **supine** position. If the radius is rotated diagonally across the ulna and the palm is made to face dorsally, the hand is said to be in the **prone** position. **Supination** and **pronation** of the forearm and hand are those rotatory movements of the radius in the forearm that achieve these alternate positions of the hand.

Structures of Synovial Joints

ARTICULAR CARTILAGE.   Firmly adherent to the articular surfaces of a majority of bones is a smooth, glistening and resilient layer of hyaline cartilage. Conforming to the shape of the bony surfaces upon which it rests, **articular cartilage** may vary in thickness from 0.5 mm on smaller bones of older individuals to as much as 6 or 7 mm on long bones during maturation. Normally articular cartilage is not visible roentgenographically and principally accounts for the space between the bones seen in roentgenograms. In a few instances, as in the temporomandibular joint, the articular surfaces are covered by white fibrocartilage. Generally that portion of the articular cartilage adjacent to the bone tends to show calcification. Articular cartilage is neither innervated nor supplied with blood vessels. It probably receives its nourishment both from the synovial fluid and by diffusion from fine vessels that supply the synovial membrane and the adjacent bone, as well as from those that nourish the capsule.

LIGAMENTS.   Ligaments are composed mainly of bundles of **collagenous fibers** oriented in parallel, or closely interwoven to present a white shining, iridescent appearance. They are pliant and flexible to allow perfect freedom of movement, but they are also strong, tough, and inextensible, so as not to yield readily to applied force. Some ligaments consist entirely of **yellow elastic tissue,** as the ligamenta flava which interconnect the laminae of adjacent vertebrae, and the ligamentum nuchae seen in the dorsal neck of grazing animals. In these latter instances the elasticity of the ligament acts as a substitute for muscular power. Ligaments are frequently distinct anatomical structures, such as the collateral ligaments of the elbow and knee joints. At times, however, ligaments are closely applied to, or are simply thickened areas of, the articular capsule and serve as a direct reinforcement to the capsule. Examples of the latter are the glenohumeral ligament of the shoulder joint and the iliofemoral, pubofemoral, and ischiofemoral ligaments of the hip joint.

ARTICULAR CAPSULE.   The articular capsule forms a complete envelope for a freely movable joint. This capsule consists of two layers: an **external fibrous layer** composed of white fibrous connective tissue and an **internal synovial layer,** which is specialized in structure and function and usually is described separately as the synovial membrane.

The **fibrous layer** of the articular capsule is attached to the periosteum along the entire circumference of the articular end of each bone entering the joint, and thus completely surrounds the articulation. Its flexibility permits movement, yet its strength protects the joint from dislocation. Along lines of severe tension, the fibrous capsule is reinforced by ligaments, and it is pierced by nerves and blood vessels which supply it and other intracapsular structures.

The **synovial membrane** covers the inner surface of the fibrous capsule, forming a closed sac called the **synovial cavity.** It is composed of loose connective tissue, cellular in some places, fibrous in others, and it has a free surface of small finger-like projections called the synovial villi. The membrane is thought to elaborate a thick viscous, glairy fluid, similar to the white of an egg and, therefore, termed **synovia** or **synovial fluid.** The synovial membrane covers tendons that pass through certain joints, such as the tendon of the popliteus muscle in the knee and the long head of the biceps in the shoulder. The membrane is not closely applied to the inner surface of the fibrous capsule, but is thrown into folds, fringes, or projections that are composed of connective tissue, fat, and blood vessels. These folds surround the margin of the articular cartilage, filling in clefts and crevices and, in some joints, such as the knee, forming large intra-articular fat pads. The synovial cavity of a normal joint contains only enough synovial fluid to moisten and lubricate the synovial surfaces, but in an injured or inflamed joint, the fluid may accumulate in painful amounts. Part of the lining of the synovial cavity is provided by the surface of articular cartilage that is moistened by synovial fluid but not covered by the synovial membrane.

Similar to the synovial cavities of true joints are the synovial tendon sheaths and synovial bursae, which have an inner lining equivalent to the synovial membrane of a joint and are lubricated by a fluid similar to synovial fluid.

SYNOVIAL TENDON SHEATHS.   Synovial tendon sheaths facilitate the gliding of tendons that pass through fibrous and bony tunnels, such as those under the flexor retinaculum of the wrist. These sheaths are closed sacs, the outer or **parietal layer** of the synovial membrane lining the tunnel, and the inner or **visceral layer** reflected over the surface of the tendon. Between these concentrically oriented layers is a fine film of synovial fluid.

SYNOVIAL BURSAE.   Synovial bursae are located in clefts of connective tissue between muscles, tendons, ligaments, and bones. They are closed sacs of synovial membrane similar to those of a true joint, and may in some cases be continuous, through an opening in the wall, with the lining of a joint cavity. They facilitate the gliding of muscles or tendons over bony or ligamentous prominences, and are named according to their location: subcutaneous, submuscular, or subtendinous. Subcutaneous bursae are found beneath the skin overlying the olecranon at the elbow and the patella at the knee. An example of a submuscular bursa is found deep to the gluteus maximus muscle where it overlies the greater trochanter. Among many other sites, subtendinous bursae are found between the tendon of the obturator internus muscle and the ischium, as well as between the tendon of the semimembranosus and the medial condyle of the tibia.

LIMITATION OF MOVEMENTS AT JOINTS.   Several factors effectively limit the movement capability at a joint. These include: (1) the tension developed by ligaments, (2) the tension developed by muscles, (3) the natural resistance to continued movement offered by the approximation of soft tissue, and (4) the limitation imposed by the contoured ends of the articulating bones in certain joints.

The **tension developed in ligaments** plays an important part in the prevention of excessive movement at many joints. However, because ligaments are fibrous, they can be stretched by persistent stress and may become incompetent, as is seen when the arches of the foot are no longer supported. Ligaments can, however, prevent a sudden unexpected abnormal movement. An example can be seen in the knee joint where its collateral ligaments are relaxed in flexion, and become taut in extension, thereby helping to limit this latter movement.

The **tension developed by muscles** is perhaps the most important factor in maintaining the stability of a joint as well as in limiting its movement. The conditions at one joint of a limb may limit the movement of another joint in that same limb because of the muscles passing over the joints. An example of this is the limitation imposed on flexion at the hip joint by the tension of the hamstring muscles when the leg is extended at the knee. When the leg is flexed at the knee joint, however, full flexion at the hip joint can be achieved. This is possible because flexion at the knee joint relaxes the hamstrings, thereby allowing the thigh to overcome the limitation of movement at the hip joint imposed by the taut hamstrings. This phenomenon has been called the *passive insufficiency* or *ligamentous action of muscles.*

The **approximation of soft tissues** can also be a factor limiting movement at a joint. A simple example of this can be seen in the flexion of the forearm at the elbow joint. As the forearm achieves full flexion, its soft tissues help to limit continued flexion because they press upon the soft tissues on the anterior aspect of the arm.

Finally, in a few joints the **natural contour of the bony articular surfaces** abut in such a manner that continued movement is prevented. Although not encountered as frequently as other limiting mechanisms at joints, it does occur occasionally—for example, at the ankle joint.

Articulations of the Axial Skeleton

ARTICULATION OF THE MANDIBLE

TEMPOROMANDIBULAR JOINT. Each temporomandibular joint is generally considered to be ellipsoid. It is a synovial joint which is divided into upper and lower synovial cavities by a complete fibrous articular disc. The structures entering into its formation on each side are the anterior part of the mandibular fossa and the articular tubercle of the temporal bone above, and the condyle of the mandible below (Figs. 5-7; 5-8; 5-9).

Ligaments. The ligaments of the joint are the following:

Articular capsule
Lateral
Sphenomandibular
Articular disc
Stylomandibular

The **articular capsule** is a thin fibrous envelope, attached above to the circumference of the mandibular fossa and the articular tubercle, and below to the neck of the con-

dyle of the mandible. The fibrous disc attaches to the inner surface of the articular capsule. Above the attachment, the capsule is loose in order to allow the disc to slide easily along the temporal bone, whereas below the disc the capsule is more tautly stretched.

The **lateral ligament** (formerly called temporomandibular ligament) strengthens the anterolateral aspect of the fibrous capsule (Fig. 5-7). It is a short, narrow fibrous band attached superiorly to the lower lateral border of the zygomatic process of the temporal bone. Its fibers pass obliquely downward and backward to the lateral surface and posterior border of the neck of the mandible. It is covered by the parotid gland and by the integument.

The **sphenomandibular ligament** (Fig. 5-8) is a flat, thin band of fibrous tissue attached to the spine of the sphenoid bone, and becoming broader as it descends, it is fixed inferiorly to the lingula of the mandibular foramen. It lies somewhat medial to the capsule of the joint, and its upper lateral surface is in relation to the lateral pterygoid muscle and the auriculotemporal nerve. More inferiorly it is separated from the neck of the condyle by the maxillary vessels. Still lower, near its attachment to the mandible, the inferior alveolar vessels and nerve and a lobule of the parotid gland lie between it and the ramus of the mandible. Its medial surface is related inferiorly to the medial pterygoid muscle, while superiorly it lies lateral to the levator veli palatini muscle and the pharyngeal wall. This ligament develops from the sheath of Meckel's cartilage (of the mandibular arch), and some of its ligamentous fibers even penetrate the petrotympanic fissure to attach to the malleus.

The **articular disc** (Fig. 5-9), formed of fibrous tissue, is a thin, oval plate, placed between the condyle of the mandible and the mandibular fossa. Its superior surface is concavoconvex from front to back, to accommodate itself to the form of the mandibular fossa and the articular tubercle. Its inferior surface, in contact with the condyle, is concave. Its circumference is connected to the articular capsule, and anteriorly it attaches to the tendon of the lateral pterygoid muscle. Both medially and laterally the disc is secured to the condyle by fibrous bands. It is thicker at its periphery than at its center. The fibers composing the disc are arranged concentrically, and they are more

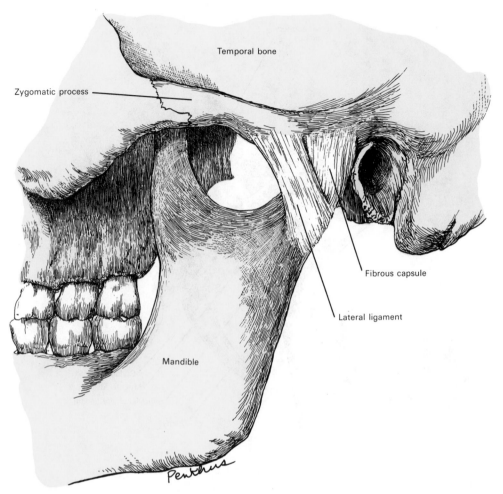

Temporal bone

Zygomatic process

Fibrous capsule

Lateral ligament

Mandible

Penthus

Fig. 5-7.   The left temporomandibular joint. Lateral view.

apparent at the circumference than at the center. The disc completely divides the joint into two cavities, each of which is lined by a synovial membrane. The two synovial membranes are placed one above, and the other below, the articular disc. The upper one, the larger and looser of the two, is continued from the margin of the cartilage covering the mandibular fossa and articular tubercle onto the superior surface of the disc. The separate cavity below the disc passes from the undersurface of the disc to the neck of the condyle, being prolonged a little farther inferiorly behind than in front. The articular disc is sometimes perforated in its center, and the two cavities then communicate with each other.

The **stylomandibular ligament** (Fig. 5-8) is a specialized band of the cervical fascia and extends from near the apex of the styloid process of the temporal bone to the

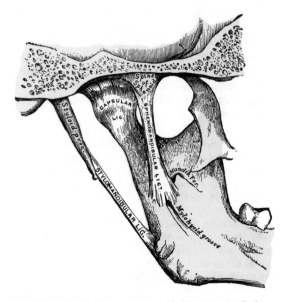

Fig. 5-8.   The left temporomandibular joint. Medial aspect.

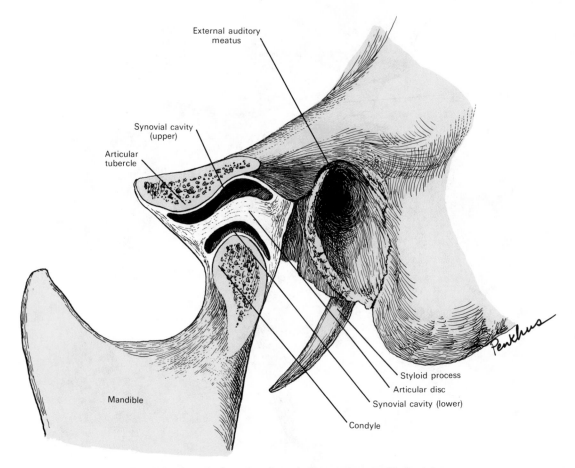

Fig. 5-9. A sagittal section through the temporomandibular joint.

angle and posterior border of the ramus of the mandible, between the masseter and medial pterygoid muscles. This ligament separates the parotid from the submandibular gland, and from its deep surface some fibers of the styloglossus muscle take origin. Although classed among the ligaments of the temporomandibular joint, it can only be considered as accessory to it.

Nerves and Arteries. The **nerves** of the temporomandibular joint are derived from the auriculotemporal and masseteric branches of the mandibular nerve, and the **arteries** come from the superficial temporal and maxillary branches of the external carotid artery.

Movements. Two points should be made initially. It is the mandible that moves, because the maxilla is firmly articulated by sutures to the other bones of the skull, *and* when the mandible is at rest, the maxillary and mandibular teeth are slightly separated. On contact, the teeth assume their **occlusal position.** The mandible may be **depressed** (jaw opening) or **elevated** (jaw closing), as well as **protruded** or **retracted. Lateral displacement** and some **rotation** of the lower jaw also occur.

It should be borne in mind that there are two parts to this articulation—one between the condyle and the articular disc, and the other between the disc and the mandibular fossa. When the jaw is depressed or elevated, by which the mouth is opened or closed, motion takes place in both parts. The disc glides anteriorly on the articular tubercle, and the condyle moves on the disc like a hinge, causing the mandible to rotate around a horizontal axis whose center of suspension traverses the rami of the mandible near their middle. Near this axis at the lingula, which is adjacent to the mandibular foramen, is attached the sphenomandibular ligament and the sling formed by the masseter and medial pterygoid muscles. When the jaw is opened, the angle of the mandible moves posteriorly while the condyle glides forward as the short arm of a lever, and the chin, as the long arm of the lever, describes a wide arc. The motion between the condyle and the articular disc is largely one of accommodation to the change in position. When the jaw is closed, some of the force is applied to the condyle as a fulcrum, especially when biting with the incisors; but when chewing with the molars, the pressure comes more directly between the teeth, and the condyle acts more as a guide than as a fulcrum. In protrusion of the mandible, both discs glide forward in the mandibular fossa. During lateral displacement of the mandible, one disc glides forward while the other remains in place. During grinding or chewing movements, there is first a lateral displacement of the mandible by a forward movement of one condyle, and then the mandible is brought back into place by the action of the closing muscles and the meshing of the teeth. The condyles may be displaced alternately, or the same one may be displaced repeatedly as when chewing with the teeth of one side.

The *mandible is depressed,* and the mouth thereby opened, by the lateral pterygoid muscles, which are assisted by the digastric, mylohyoid, and geniohyoid muscles. Conversely, the *mandible is elevated,* and the mouth closed, by the masseter, medial pterygoid, and temporalis muscles. The *mandible is protruded* by the simultaneous action of the lateral pterygoids of both sides and the synergetic action of the closing muscles. It is *retracted* by the posterior fibers of the temporalis, and *displaced laterally* by the action of the opposite lateral pterygoid muscle.

ARTICULATION OF THE VERTEBRAL COLUMN WITH THE CRANIUM

ATLANTO-OCCIPITAL JOINTS (Figs. 5-10; 5-11). The atlanto-occipital joints, one on each side, interconnect the concave superior articular facets of the lateral masses of the atlas and the convex condyle of the occipital bone. These two synovial joints are ellipsoid and the articular surfaces are reciprocally curved.

Ligaments. The ligaments connecting the bones are:

Two articular capsules
Anterior atlanto-occipital membrane
Posterior atlanto-occipital membrane

The **articular capsules** invest each of the two joints and surround the condyles of the occipital bone, connecting them with the superior articular facets of the atlas. Although they are loosely fit, they are quite distinct. On their medial side, however, they are thin and formed only of loose fibers, whereas laterally they are reinforced by

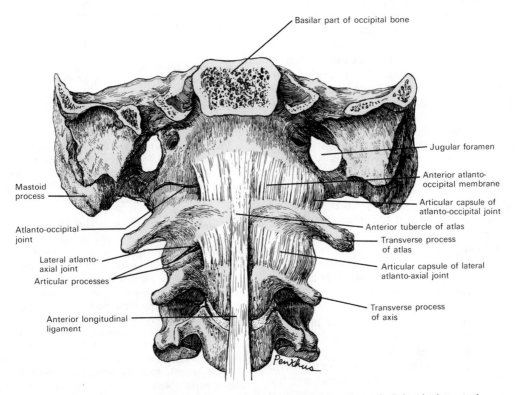

FIG. 5-10. The atlanto-occipital and lateral atlantoaxial joints. Anterior aspect. On the left side the articular capsule and part of the atlanto-occipital membrane covering the atlantoaxial joint have been removed, as has the capsule surrounding the lateral atlantoaxial joint.

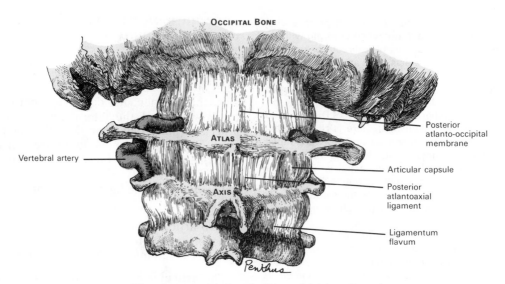

OCCIPITAL BONE

Posterior
atlanto-occipital
membrane

ATLAS

Vertebral artery

Articular capsule

Posterior
atlantoaxial
ligament

AXIS

Ligamentum
flavum

Penthus

FIG. 5-11. The atlanto-occipital and atlantoaxial joints. Posterior aspect.

oblique bands (sometimes called the lateral atlanto-occipital ligaments), which are directed upward and medialward.

The **anterior atlanto-occipital membrane** (Fig. 5-10) is about 2 cm wide and is composed of densely woven fibers, which pass between the anterior margin of the foramen magnum above, and the cranial border of the anterior arch of the atlas below. Laterally it is continuous with the articular capsules, and anteriorly it is strengthened in the midline by a strong, round cord that is continuous with the anterior longitudinal ligament and connects the basilar part of the occipital bone to the tubercle on the anterior arch of the atlas. Deep to the membrane is the ventral aspect of the dura mater covering the upper cervical spinal cord at its medullary junction, while anterior to it are found the alar ligaments and the lateral and anterior rectus capitis muscles.

The **posterior atlanto-occipital membrane** (Fig. 5-11) is a little broader than but not so dense as the anterior membrane. It is a thin fibrous layer, connected above to the posterior margin of the foramen magnum and below to the superior border of the posterior arch of the atlas. On either side this membrane is defective over the groove for the vertebral artery (Fig. 5-11), and forms, with this groove, an opening for the passage of the artery into the central nervous system and for the first cervical spinal nerve (suboccipital nerve) to exit from the underlying

spinal cord. The free border of the membrane, arching over the artery and nerve, is sometimes ossified. Dorsal to the membrane are found the rectus capitis posterior minor and obliquus capitis superior muscles, while ventrally the membrane is intimately adherent to the dura mater of the vertebral canal.

Synovial Membranes. There are two synovial membranes: one lining each of the articular capsules. The atlanto-occipital joints frequently communicate medially with the joint between the dorsal surface of the dens and the transverse ligament of the atlas.

Nerves and Arteries. The **nerve supply** to the atlanto-occipital joints comes from sensory fibers of the first cervical (suboccipital) nerve. Its **arterial supply** comes from branches of the vertebral artery and from meningeal branches of the ascending pharyngeal artery.

Movements. The movements permitted in this joint are (1) flexion and extension, which give rise to the ordinary forward and backward nodding of the head, and (2) slight lateral tilting of the head to one or the other side. **Flexion** is produced mainly by the action of the longus capitis and rectus capitis anterior muscles. **Extension** results from the action of the rectus capitis posterior major and minor muscles, as well as the obliquus capitis superior, semispinalis capitis, splenius capitis, sternocleidomastoideus, and upper fibers of the trapezius. **Lateral flexion** (lateral tilting movement) is produced by the rectus capitis lateralis muscle assisted by the trapezius, splenius capitis, semispinalis capitis, and sternocleidomastoid muscles of the same side, all acting together.

Ligaments Connecting the Axis with the Occipital Bone. Although the occipital bone and the axis are not directly in contact, they are united by ligaments that are observable at dissection after the removal of the posterior arches of the atlas and axis. These occipitoaxial ligaments include:

Tectorial membrane
Two alar
Apical

The **tectorial membrane** (Figs. 5-12; 5-14) is a broad, strong band which covers the dens and its ligaments posteriorly. It courses within the vertebral canal and appears to be a cranial prolongation of the posterior longitudinal ligament of the vertebral column. It is fixed below to the dorsal surface of the body of the axis and, expanding as it ascends, is attached to the basilar portion of the occipital bone ventral to the foramen magnum, where it blends with the cranial dura mater. Its ventral surface is in relation with the transverse ligament of the atlas, and its dorsal surface with the dura mater.

The **alar ligaments** (Figs. 5-12; 5-13) are strong, round cords, which arise one on each side of the cranial part of the dens and, passing obliquely upward and lateralward, are inserted into the rough depressions on the medial sides of the condyles of the occipital bone. The alar ligaments limit rotation of the cranium and therefore are sometimes referred to as **check ligaments.**

The **apical ligament of the dens** (Figs. 5-13; 5-14) is a fibrous cord in the triangular interval between the two alar ligaments. It extends from the tip of the dens to the anterior margin of the foramen magnum, being intimately blended with the deep portion of the anterior atlanto-occipital membrane and superior crus of the transverse ligament of the atlas. It is regarded as a rudimentary intervertebral disc, and in it traces of the notochord may persist.

In addition to the ligaments that unite the atlas and axis to the skull, the **ligamentum nuchae** must be regarded as one of the ligaments connecting the vertebral column with the cranium.

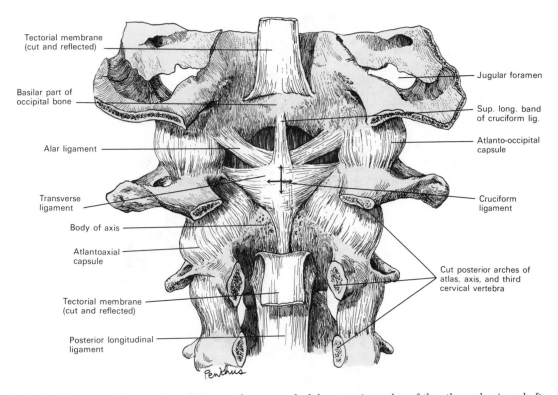

Tectorial membrane (cut and reflected)

Basilar part of occipital bone

Alar ligament

Transverse ligament

Body of axis

Atlantoaxial capsule

Tectorial membrane (cut and reflected)

Posterior longitudinal ligament

Jugular foramen

Sup. long. band of cruciform lig.

Atlanto-occipital capsule

Cruciform ligament

Cut posterior arches of atlas, axis, and third cervical vertebra

Fig. 5-12. Exposure of the cruciform ligament after removal of the posterior arches of the atlas and axis and after cutting and reflecting the tectorial membrane. Observe that the cruciform ligament consists of a longitudinal band and a transverse ligament which lie posterior to the dens and its alar and apical (hidden) ligaments.

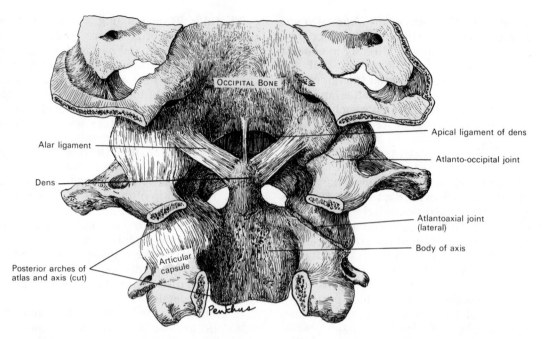

FIG. 5-13. The alar ligaments and the apical ligament of the dens are exposed after the removal of the cruciform ligament. Note that this is a posterior approach to the dens after cutting through the posterior arches of the atlas and axis.

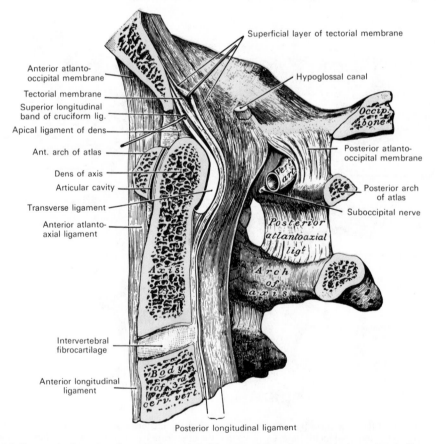

FIG. 5-14. Median sagittal section through the occipital bone and first three cervical vertebrae. (From Spalteholz.)

ARTICULATIONS OF THE ATLAS
WITH THE AXIS

The atlas and the axis articulate by means of three joints. On each side the inferior articular facet of the lateral mass of the atlas articulates with the superior articular facet of the axis to form the **lateral atlantoaxial joints.** In addition, the dens of the axis articulates with the anterior arch and transverse ligament of the atlas to form the **median atlantoaxial joint.**

LATERAL ATLANTOAXIAL JOINT (Figs. 5-12; 5-13).   These two articulations, one on each side, are classified as plane joints. Surrounded by an articular capsule, each flat or slightly convex inferior articular facet of the atlas fits on the smooth upwardly directed superior articular facets of axis in order to allow rotation of the atlas on the axis.

Ligaments. Although the only officially recognized ligaments of these joints *per se* are their capsules, the atlantoaxial extensions of two other ligaments of the spinal column may be considered as helping to secure these two bones, thereby lending additional integrity to the joints. The ligaments to be described therefore are:

Two articular capsules
Anterior atlantoaxial
Posterior atlantoaxial

The **articular capsules** (Figs. 5-12; 5-13) are thin and loose, and connect the margins of the lateral masses of the atlas with those of the posterior articular surfaces of the axis. Each is strengthened at its posterior and medial part by an **accessory ligament,** which is attached below to the body of the axis near the base of the dens, and above to the lateral mass of the atlas near the transverse ligament. The inner aspect of the capsules is lined by a synovial membrane.

The **anterior atlantoaxial ligament** (Figs. 5-10; 5-14) is a strong membrane fixed to the inferior border of the anterior arch of the atlas, and to the ventral surface of the body of the axis. It is strengthened in the midline by a round cord, which connects the tubercle on the anterior arch of the atlas to the body of the axis and is a *continuation cranially of the anterior longitudinal ligament.* Immediately anterior and coursing vertically along each lateral border of the ligament are the longus capitis muscles.

The **posterior atlantoaxial ligament** (Figs. 5-11; 5-14) is a broad, thin membrane attached above to the inferior border of the posterior arch of the atlas, and below to the superior edges of the laminae of the axis. At the atlantoaxial level, *this ligament is comparable to the intervertebral ligamenta flava* found connecting the posterior arches of other lower vertebrae.

MEDIAN ATLANTOAXIAL JOINT.   This pivot joint contains two synovial cavities located between the dens of the axis and a ring formed by the anterior arch of the atlas and the transverse ligament of the atlas (Fig. 5-15). One synovial cavity lies between the dens and the anterior arch of the atlas, while the other is interposed between the dens and the transverse ligament of the atlas.

Ligaments. The **transverse ligament of the atlas** (Figs. 5-12; 5-14; 5-15) is a thick, strong band which arches across the ring of the atlas and retains the dens in contact with the anterior arch. It is broader and thicker in the middle than at the ends, and is firmly attached on each side to a small tubercle on

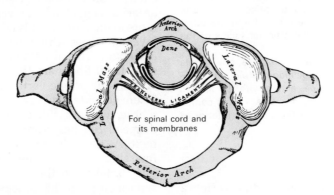

FIG. 5-15.   The median atlantoaxial joint viewed from above the atlas. Note the two synovial cavities, one anterior and the other posterior to the upwardly projecting dens of the axis.

the medial surface of the lateral mass of the atlas. As it crosses the dens, a small fasciculus is prolonged cranially, and another caudally, from the superficial or posterior fibers of the ligament, forming a **longitudinal band.** The superior band is attached to the basilar part of the occipital bone, closely related to the tectorial membrane, while the inferior band is fixed to the posterior surface of the body of the axis. Together with the transverse ligament, these two parts of the longitudinal band form a cross-like structure named the **cruciate ligament of the atlas** (Fig. 5-12). The transverse ligament divides the entire circular opening in the atlas into two unequal parts: the *posterior and larger portion* contains the spinal cord, its membranes and vessels, and the spinal trunks of the accessory nerves; the *anterior and smaller part* contains the dens. The neck of the dens is constricted where it is embraced posteriorly by the transverse ligament, so that this ligament suffices to hold it in position after all the other ligaments have been divided.

Synovial Membranes. There is a synovial membrane for each of the lateral atlantoaxial joints, and two synovial cavities associated with the median atlantoaxial joint. The joint cavity between the dens and the transverse ligament of the atlas may be continuous laterally with those of the atlanto-occipital articulations.

Movements. The atlantoaxial joints allow rotation of the atlas (and with it the skull) upon the axis, the extent of rotation being limited by the alar ligaments.

The opposed articular surfaces of the lateral atlantoaxial joints are not reciprocally curved, because both are convex in their long axes. Thus, when the superior facet glides ventrally on the inferior, it also descends. The fibers of the articular capsule are then relaxed in a vertical direction, thereby permitting movement. As the atlas and the skull pivot around the dens, the lateral masses of the atlas at the lateral atlantoaxial joints glide, one forward and the other backward, upon the superior articular facets of the axis.

The principal **muscles** by which these movements are produced are the sternocleidomastoid and semispinalis capitis *of one side,* acting with the longus capitis, splenius, longissimus capitis, rectus capitis posterior major, and the obliquus capitis inferior *of the other side.*

ARTICULATIONS OF THE VERTEBRAL COLUMN

Interconnecting each succeeding vertebra from the axis to the sacrum are: **a series of cartilaginous joints** that unite each adjacent pair of vertebral bodies by means of longitudinal ligaments and intervertebral discs; and **a series of synovial joints,** reinforced by ligaments that unite the vertebral arches of the vertebrae at their interfacing articular processes.

JOINTS BETWEEN THE VERTEBRAL BODIES. The articulations between the bodies of the vertebrae are separated by fibrocartilaginous intervertebral discs and are classified as symphyses. The individual vertebrae move only slightly on each other, but because this slight degree of movement between the pairs of bones takes place in all the joints of the vertebral column, the total range of movement is considerable.

Ligaments. The ligaments of these articulations are the following:

> Anterior longitudinal
> Posterior longitudinal
> Intervertebral discs

The **anterior longitudinal ligament** (Figs. 5-10; 5-14; 5-16; 5-19) is a broad and strong band of fibers which extends along the anterior surfaces of the bodies of the vertebrae, from the axis to the sacrum. Its extension to the occipital bone may (as it is in this book) be referred to as the anterior atlantoaxial ligament. Being relatively narrower above, the anterior longitudinal ligament broadens somewhat as it descends. It is a bit narrower (but thicker) in its passage over the vertebral bodies than over the intervertebral discs. It consists of dense longitudinal fibers that are intimately adherent to the intervertebral discs and the prominent margins of the vertebrae, but not attached firmly to the middle parts of the bodies. In the latter situation the ligament is thick and serves to fill the concavities on the ventral surfaces, giving the anterior aspect of the vertebral column a more even contour. It is composed of several layers of fibers which vary in length, but are closely interlaced with each other. The most superficial fibers are the longest and extend across four or five vertebrae. A second, subjacent set extends between two or three vertebrae, while a third set, the shortest and deepest, reaches from one vertebra to the next. At the sides of the vertebral bodies, a few short fibers of the ligament pass from one vertebra to the next. These fibers are separated from the concavities of the vertebral bodies by oval apertures for the passage of vessels.

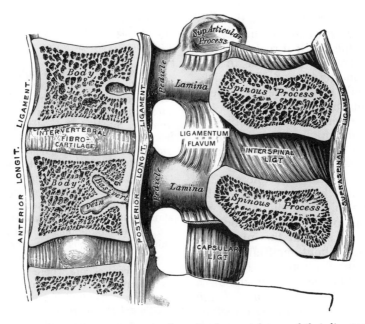

FIG. 5-16. Median sagittal section of two lumbar vertebrae and their ligaments.

The **posterior longitudinal ligament** (Figs. 5-12; 5-14; 5-16; 5-17) extends along the dorsal surfaces of the bodies of the vertebrae from the axis to the sacrum. It lies *within the vertebral canal* (Fig. 5-16), and it is continued above the axis as the tectorial membrane. It is broader above than below, and thicker in the thoracic region than in the cervical and lumbar regions. The ligament broadens somewhat and becomes more intimately adherent as it passes over each of the intervertebral discs and contiguous margins of the vertebrae. The ligament is smooth superiorly, but in the thoracic and lumbar regions it presents a series of dentations with intervening concave margins. It is narrow and thick over the centers of the bodies, from which it is separated by the basivertebral veins. The ligament is composed of longitudinal fibers, which are denser and more compact than those of the anterior ligament. Its superficial layers occupy the interval between three or four vertebrae, while the deeper layers extend between adjacent vertebrae.

The **intervertebral discs** (Figs. 5-16; 5-20) are interposed between the adjacent surfaces of the vertebral bodies from the axis to the sacrum, forming their chief bonds of connection. They vary in shape and thickness in different parts of the vertebral col-

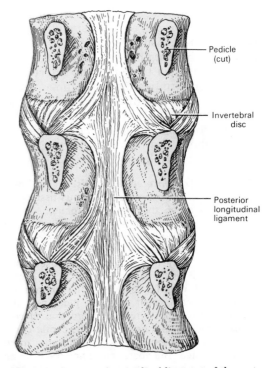

Pedicle (cut)

Invertebral disc

Posterior longitudinal ligament

FIG. 5-17. Posterior longitudinal ligament of the vertebrae in the lumbar region. Interior of spinal column with the dorsal arches cut and viewed from the posterior aspect.

umn. Their shape corresponds to the surfaces of the bodies between which they are placed, except in the cervical region, where they are slightly smaller from side to side than the corresponding bodies. Their thickness varies not only in the different regions of the column, but in different parts of the same disc. In the cervical and lumbar regions the discs are thicker ventrally than dorsally, and thus contribute to the anterior convexities of these parts of the column. The discs are of nearly uniform thickness in the thoracic region, and the anterior concavity of this part of the column is due almost entirely to the shape of the vertebral bodies. From the axis to the sacrum, the intervertebral discs constitute about one-fourth of the length of the vertebral column. Being thicker in the cervical and lumbar regions, the discs account for a greater proportion of the length of the spinal column in these regions than in the thoracic region. Correspondingly, the cervical and lumbar regions show a greater pliancy and freedom of movement. The intervertebral discs are adherent to thin layers of hyaline cartilage which cover the superior and inferior surfaces of the bodies of the vertebrae. Between the lower cervical vertebrae, however, small joints lined by synovial membranes are occasionally present between the superior surfaces of the bodies and the margins of the discs on either side. The rounded border of the intervertebral discs are firmly attached to the anterior and posterior longitudinal ligaments. In the thoracic region the discs are joined laterally, by means of interarticular ligaments, to the heads of those ribs that articulate with the two adjacent vertebrae.

Structure of the Intervertebral Discs. Each disc is composed of outer laminae of fibrous tissue and fibrocartilage called the **anulus fibrosus**, and an inner core of soft, gelatinous and highly elastic substance called the **nucleus pulposus**.

The laminae forming the **anulus fibrosus** are arranged in concentric rings and the outermost consist of ordinary fibrous tissue; those closer to the center are formed of white fibrocartilage. The laminae are not quite vertical in their direction, those near the circumference being curved outward and closely approximated, while those nearest the center curve in the opposite direction, and are somewhat more widely separated. The fibers composing the laminae pass obliquely between the two adjacent vertebrae and are firmly attached to them. Greater stability is achieved in the disc because the fibers of each adjacent lamina pass in opposite directions, crossing one another like the limbs of the letter X. This laminar arrangement characterizes the outer half of each fibrocartilage.

The yellowish **nucleus pulposus** is more abundant in the cervical and lumbar regions than at thoracic levels, and it projects considerably above the surrounding anulus fibrosis if the disc is divided horizontally. Situated somewhat behind the center of the disc, the nucleus pulposus consists of a fine fibrous matrix within which are found some polymorphic cells. Although at birth the nucleus pulposus can be considered as a remnant of the notochord, it undergoes such structural changes with maturation that by adolescence its notochordal cells have been almost entirely replaced by cartilaginous cells from the surrounding anulus fibrosus. With age, the gelatinous substance is replaced by fibrocartilage, and there is more of a blending of the anulus fibrosus with the central core. There is also a decrease in the water content of the nucleus pulposus along with a reduction in its elasticity.

Clinical Correlation. Intervertebral discs cushion mechanical shocks imposed on the vertebral column, thereby distributing the pressure more evenly. Upon receiving a jolting impact, the highly elastic nucleus pulposus becomes flatter and broader and pushes the more resistant fibrous laminae outward in all directions. The nucleus and anulus then return to their original position after the impact is dissipated. Discs are secured between the vertebrae primarily by the anterior and posterior longitudinal ligaments. In older individuals an unusual strain may result in the disc proper, or the nucleus pulposus only, being extruded beyond its intervertebral position and then not returning to its proper site. Such a situation generally produces pain because the protruded disc presses upon spinal roots or nerves. Most commonly this condition is seen in the lower lumbar region with accompanying radiating pain to the gluteal region and lower extremity.

JOINTS BETWEEN THE VERTEBRAL ARCHES. Joints between articular processes of the vertebrae are plane or gliding joints, enveloped by capsules lined by synovial membrane.

Ligaments. Articular capsules (Fig. 5-16) are thin, loose, and attached to margins of the articular processes of adjacent vertebrae. They are longer and looser in the cervical than in the thoracic and lumbar regions.

The laminae, spinous, and transverse processes of the vertebrae are connected by the following ligaments:

Ligamenta flava
Supraspinal
Ligamentum nuchae
Interspinal
Intertransverse

The **ligamenta flava** (Figs. 5-16; 5-18; 5-23) connect the laminae of adjacent vertebrae from the axis to the first segment of the sacrum. Although best seen from the interior of the vertebral canal, when viewed from the outer surface they appear short because they are overlapped by the laminae. Each ligament consists of two lateral portions, which commence on each side at the roots of the articular processes surrounded by the capsule and extend dorsally to the point where the laminae meet to form the spinous process. The dorsal margins of the two portions are in contact except for slight intervals at the midline through which pass vertebral veins. Each ligamentum flavum consists of yellow elastic tissue, the fibers of which, almost perpendicular in direction, are attached to the *ventral surface of the lamina above and to the dorsal surface and superior margin of the lamina below*. In the cervical region the ligaments are thin, but broad and long. They are thicker in the thoracic region and thickest in the lumbar region. Their marked elasticity permits separation of the laminae during flexion of the vertebral column, and also serves to preserve the upright posture.

The **supraspinal ligament** (Figs. 5-16; 5-23) is a strong fibrous cord which connects together the apices of the spinous processes from the seventh cervical vertebra to the sacrum. At its points of attachment to the tips of the spinous processes, fibrocartilage is developed in the ligament. It is thicker and broader in the lumbar than in the thoracic region, and intimately blended in both situations with the neighboring fascia. The most superficial fibers of this ligament extend over three or four vertebrae; those coursing more deeply pass between two or three vertebrae, while the deepest connect the spinous processes of neighboring vertebrae and become continuous with the interspinal ligaments. Above the seventh cervical vertebra it is continued to the external occipital protuberance and median nuchal line as the **ligamentum nuchae.**

The **ligamentum nuchae** is a fibroelastic membrane in the dorsal neck which corresponds to the supraspinal ligament of the lower vertebrae. It extends from the external occipital protuberance and median nuchal line to the spinous process of the seventh cervical vertebra. In its course a fibrous lamina from the ligamentum nuchae attaches initially to the posterior tubercle of the atlas and the spinous processes of the cervical vertebrae, and then forms a septum between the muscles on both sides of the neck. In man, it is merely the rudiment of an important elastic ligament seen in some of the grazing animals which serves to sustain the weight of the head and to assist in flexion of the cervical vertebrae.

The **interspinal ligaments** (Fig. 5-16) are thin and membranous and interconnect adjoining spinous processes. Their attachment extends from the root to the apex of each spinous process and, thereby, they meet the ligamenta flava ventrally and the supraspinal ligament dorsally. They are narrow and elongated in the thoracic region, but broad, thick, and quadrilateral in shape in the lumbar region. These ligaments are only slightly developed in the neck.

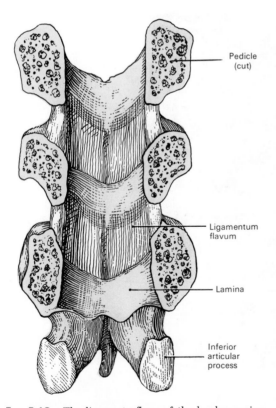

Pedicle (cut)

Ligamentum flavum

Lamina

Inferior articular process

Fig. 5-18. The ligamenta flava of the lumbar region. The interior of the spinal column is seen with the vertebral bodies cut away. This demonstrates the dorsal laminae viewed from the anterior aspect.

The **intertransverse ligaments** (Figs. 5-19; 5-23) are interposed between the transverse processes. In the cervical region they consist of a few irregular, scattered fibers; in the thoracic region they are rounded cords intimately connected with the deep muscles of the back; in the lumbar region they are thin and membranous.

MOVEMENTS OF THE VERTEBRAE. The movements permitted in the vertebral column are: **flexion, extension, lateral flexion, circumduction,** and **rotation.**

In **flexion,** or bending ventrally, the anterior longitudinal ligament is relaxed, and the anterior portions of the intervertebral fibrocartilages are compressed. The posterior longitudinal ligament, the ligamenta flava, and the interspinal and supraspinal ligaments are stretched, as also are the dorsal fibers of the intervertebral discs. The interspaces between the laminae are widened, and the inferior articular processes glide upward upon the superior articular processes of the subjacent vertebrae. Tension of the extensor muscles of the back is the most important factor in limiting the movement. Flexion is the most extensive of all the movements of the vertebral column, and it is freest in the cervical region.

In **extension,** or bending dorsally, an opposite disposition of the parts takes place. Thus, the distance between the anterior borders of the vertebrae is increased while their posterior borders are moved closer together. This movement is limited by the anterior longitudinal ligament and by an approximation of the spinous processes. It is freest in the cervical and lumbar regions.

In **lateral flexion,** the sides of the intervertebral discs are compressed, the extent of motion being

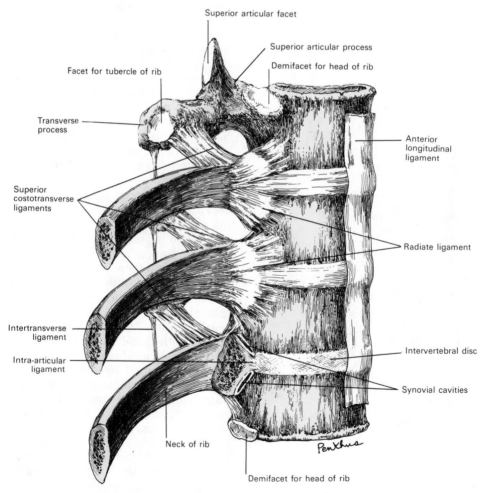

FIG. 5-19. The costovertebral joints viewed from the right lateral aspect. Note the two synovial cavities exposed in the lowest joint from which have been removed much of the radiate ligament and the head of the rib. Observe that each rib articulates with two adjacent vertebral bodies.

limited by the resistance offered by opposing muscles and by the surrounding ligaments. This movement may take place in any part of the column, but is freest in the cervical and lumbar regions, and is always associated with a certain degree of rotation.

Circumduction is limited, and is merely a succession of the preceding movements.

Rotation is produced by the turning of adjacent vertebrae at the joints between the vertebral arches, which then produces a twisting of the intervertebral discs. Although only slight rotation can occur between any two vertebrae, a considerable extent of movement is allowed when it takes place at the same time in the whole length of the column. This movement occurs most freely in the thoracic region, but also occurs to a slight extent at cervical and lumbar levels of the spinal column.

The extent and variety of the movements are influenced by the shape and direction of the articular surfaces. In the **cervical region** the upward inclination of the superior articular surfaces allows free flexion and extension. *Extension* can be carried farther than flexion. At the cranial end of the cervical region, it is checked by the locking of the posterior edges of the superior atlantal facets in the condyloid fossae of the occipital bone. At the caudal end of the cervical region, extension is limited by a mechanism whereby the inferior articular processes of the seventh cervical vertebra slip into grooves behind and below the superior articular processes of the first thoracic. *Flexion* is arrested just beyond the point where the cervical convexity is straightened. The movement is checked by the apposition of the projecting lower lips of the bodies of the vertebrae with the shelving surfaces on the bodies of the subjacent vertebrae. *Lateral flexion* and *rotation* are relatively free in the cervical region, but the two movements are always combined. The inclinations of the superior articular surfaces impart a rotatory movement during lateral flexion, while pure rotation is prevented by the slight medial slope of these surfaces.

In the **thoracic region,** notably in its upper part, all the movements are limited in order to minimize interference with respiration. The almost complete absence of an upward inclination of the superior articular surfaces prohibits any marked *flexion,* whereas *extension* is checked by the contact of the inferior articular margins with the laminae, and the contact of the spinous processes with one another. The mechanism at the caudal end of the cervical region that limits extension also serves to limit flexion of the thoracic region when the neck is extended. *Rotation* is free in the thoracic region. The position of the articular processes allows rotation around a vertical axis that passes through the bodies of the midthoracic vertebrae, but anterior to the vertebral bodies of the upper and lower thoracic vertebrae. The direction of the articular facets would allow free lateral flexion, but this movement is considerably limited by the resistance of the ribs and sternum.

In the **lumbar region** *extension* is free and wider in range than *flexion.* The inferior articular facets are not in close apposition with the superior facets of the subjacent vertebrae, and because of this a considerable degree of lateral flexion is permitted. For the same reason a slight amount of rotation can be achieved, but this is checked so soon by the interlocking of the articular surfaces that it is negligible.

MUSCLE ACTIONS ON THE VERTEBRAE. The principal muscles that produce **flexion** are the sternocleidomastoid, the longus capitis, the longus colli, the scalenes, the abdominal muscles, and the psoas major. **Extension** is produced by the intrinsic muscles of the back (erector spinae), assisted in the neck by the splenius, the semispinales, and the multifidi. **Lateral flexion** is produced by the intrinsic muscles of the back, by the splenius, the scalenes, the quadratus lumborum, and the psoas major when the muscles of one side only are acting. **Rotation** results from the action of the following muscles of one side only, the sternocleidomastoid, the longus capitis, the scalenes, the multifidi, the rotatores, the semispinalis capitis, and the abdominal muscles.

COSTOVERTEBRAL JOINTS

The articulations of the ribs with the vertebral column may be divided into two sets, one connecting the heads of the ribs with the bodies of the vertebrae, and the other uniting the necks and tubercles of the ribs with the transverse processes.

ARTICULATIONS OF THE HEADS OF THE RIBS (Fig. 5-19). These constitute a series of plane or gliding joints, and are formed by the articulation of the heads of the typical ribs with the facets on the contiguous margins of the bodies of the thoracic vertebrae, and with the intervertebral discs between them. Whereas the first, tenth, eleventh, and twelfth ribs each articulate with a single vertebra, the heads of the second to the ninth ribs each articulate by means of two facets, a superior and an inferior, with two adjacent vertebral bodies. The articulation of the head of each rib, however, is considered a single joint because there is but one articular capsule, even though there may be two synovial sacs.

Ligaments. The ligaments at each joint are:

Articular capsule
Radiate
Intra-articular

The **articular capsule** surrounds the joint and is composed of short, strong fibers that connect the head of the rib with the circumference of the articular cavity formed by the intervertebral disc and the adjacent vertebrae. The capsule is most distinct at the superior and inferior parts of the articulation. Some of its upper fibers pass through the intervertebral foramen to the back of the intervertebral disc, while the posterior fibers of the capsule are continuous with the costotransverse ligament.

The **radiate ligament** (Fig. 5-19) connects the anterior part of the head of each rib with the side of the bodies of two vertebrae, and the intervertebral disc between them. It consists of several flat fasciculi which are attached to the anterior part of the head of the rib, just beyond the articular surface. The more superior fasciculi are connected with the body of the vertebra above, while the inferior fibers attach to the body of the vertebra below. The middle fibers are short and less distinct and course horizontally to attach to the intervertebral disc. The radiate ligaments along the spinal column are in relationship anteriorly to the thoracic ganglia of the sympathetic trunk, the pleura, and—on the right side—the azygos vein. Immediately behind the radiate ligaments at each level are the intra-articular ligament and the synovial membranes (Fig. 5-19).

In the case of the first rib, the fibers of the radiate ligament are attached to the body of the last cervical vertebra, as well as to that of the first thoracic. In the articulations of the heads of the tenth, eleventh, and twelfth ribs, each of which articulates with a single vertebra, the fibers of the radiate ligament in each case connect to the vertebra above, as well as to that with which the rib articulates.

The **intra-articular ligament** (Fig. 5-19) is situated in the interior of the joint. It consists of a short, flat band of fibers, attached by one extremity to the crest separating the two articular facets on the head of the rib, and by the other to the intervertebral disc. It divides the joint into two cavities, each of which is lined by a synovial membrane. In the joints of the first, tenth, eleventh, and twelfth ribs, the intra-articular ligament does not exist because there is but one cavity and one synovial membrane at each of these articulations.

COSTOTRANSVERSE JOINTS (Fig. 5-20).   The articular surface on the tubercle of the upper ten ribs forms a plane or gliding joint with the articular facet on the adjacent transverse process of the corresponding vertebra. In the eleventh and twelfth ribs this articulation is wanting.

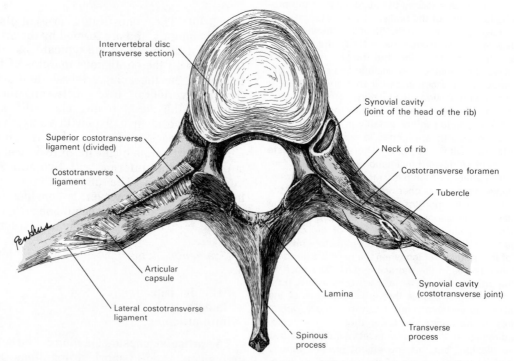

Fig. 5-20.   The costovertebral joints viewed from the superior aspect. Note that on the right side the synovial cavities of the joint of the head of the rib and the costotransverse joint have been exposed.

Ligaments. The ligaments of the joint are:

Articular capsule
Superior costotransverse
Costotransverse
Lateral costotransverse

The **articular capsule** is a thin, fibrous membrane attached to the circumference of the articular surfaces. As in other plane joints, the capsule is lined by a synovial membrane.

The **superior costotransverse ligament** (Figs. 5-19; 5-20; 5-23) is attached to the sharp crest on the superior border of the neck of the rib, and passes obliquely upward and laterally to the lower border of the transverse process immediately above. The *anterior layer* of this ligament blends laterally with the internal intercostal membrane and is crossed by the intercostal vessels and nerves. Its *posterior layer* blends laterally with the external intercostal muscle. Its medial border is thickened and free, and bounds an aperture that transmits the posterior branches of the intercostal vessels and nerves.

The first rib has no superior costotransverse ligament. The neck of the twelfth rib is connected to the base of the transverse process of the first lumbar vertebra by a band of fibers, the **lumbocostal ligament,** which actually lies in series with the superior costotransverse ligaments. It is merely a thickened portion of the lumbocostal aponeurosis or anterior layer of the thoracolumbar fascia.

The **costotransverse ligament** (Figs. 5-20; 5-23) is sometimes called the ligament of the neck of the rib and consists of short but strong fibers connecting the rough surface on the back of the neck of the rib with the anterior surface of the adjacent transverse process. It is observed best with a vertebra cut in transverse section, where it can be seen to fill the **costotransverse foramen.** A rudimentary ligament only may be present in the case of the eleventh and twelfth ribs.

The **lateral costotransverse ligament** (Figs. 5-20; 5-23) is a short, thick, and strong fasciculus, which passes obliquely from the apex of the transverse process of the vertebra to the rough and nonarticular portion of the tubercle of the corresponding rib. For this

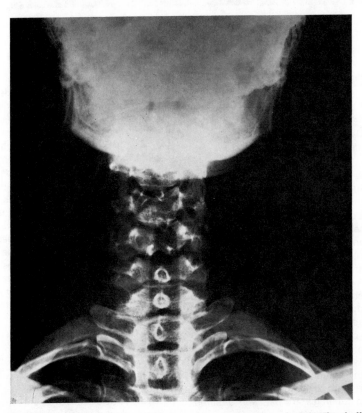

Fɪɢ. 5-21. Anteroposterior view of lower cervical and upper thoracic vertebrae of adult. The skull obscures the upper cervicals. (Courtesy of Doctors Moreland and Arndt.)

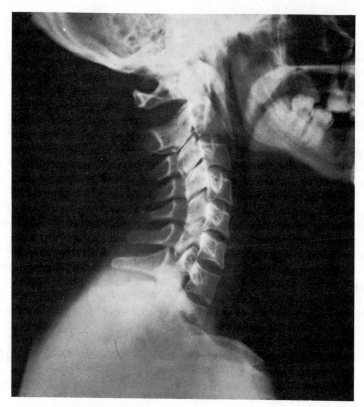

FIG. 5-22. Lateral view of adult neck. Notice spacing and slope between vertebral bodies, articular process, and spinous processes. (Courtesy of Doctors Moreland and Arndt.)

reason it has also been called the ligament of the tubercle of the rib. The ligaments attached to the upper ribs ascend from the transverse processes, and are shorter and more oblique than those attached to lower ribs, which descend slightly.

Movements at the Costotransverse Joints. The heads of the ribs are so closely connected to the bodies of the vertebrae by the radiate and intra-articular ligaments that only slight gliding movements of the articular surfaces on one another can take place. Similarly, the strong ligaments binding the necks and tubercles of the ribs to the transverse processes limit the movements of the costotransverse joints to slight gliding, the nature of which is determined by the shape and direction of the articular surfaces. In the upper six ribs the articular surfaces on the tubercles are *oval in shape and convex from above downward.* They fit into corresponding concavities on the *anterior surfaces* of the transverse process, so that upward and downward movements of the tubercles are associated with rotation of the rib neck on its long axis. On the seventh to the tenth ribs the articular surfaces on the tubercles are *flat, and directed obliquely downward, medialward, and backward.* The surfaces with which they articulate are placed

on the *upper margins* of the transverse processes, and therefore, when the tubercles are drawn up, they are at the same time carried backward and medialward. The joints of the heads of the ribs and the costotransverse joints move simultaneously and in the same directions, the total effect being that the neck of the rib moves as if on a single joint, of which the two articulations form the ends. The necks of the upper six ribs only move slightly upward and downward. Their chief movement is one of rotation around their own long axes. Thus, rotation backward of the neck of the rib results in depression of the anterior end of the rib, whereas rotation forward of the neck causes elevation of the anterior end. The necks of the seventh to tenth ribs may move either upward, backward, and medialward, or downward, forward, and lateralward, thereby altering the shape of the rib cage. Only slight rotation accompanies these movements.

STERNOCOSTAL, INTERCHONDRAL, AND COSTOCHONDRAL JOINTS

STERNOCOSTAL JOINTS (Fig. 5-24). At the anterior aspect of the thoracic wall, the bony portion of each rib becomes continuous with

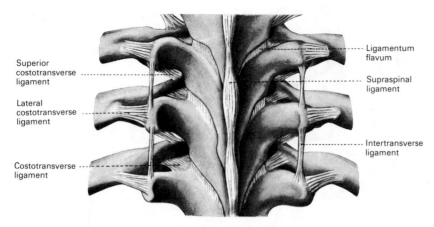

FIG. 5-23. A part of the thoracic spinal column shown from the posterior aspect. Observe the necks of the ribs and the costotransverse joints with their associated ligaments. (From Benninghoff and Goerttler, 1975.)

the lateral end of a strip of cartilage, the **costal cartilage,** resulting in a series of *costochondral unions.* The medial ends of the costal cartilages fit into slight concavities along the lateral borders of the sternum, forming the **sternocostal joints.** With the exception of the first rib, the sternocostal articulations are plane or gliding joints with synovial cavities. The costal cartilage of the first rib is directly united with the sternum and forms a typical synchondrosis.

Ligaments.   The ligaments at the sternocostal joints are:

Articular capsules
Radiate sternocostal
Intra-articular sternocostal
Costoxiphoid

Articular capsules surround the joints between the sternum and the cartilages of the second to seventh ribs. They are thin, intimately blended with the radiate sternocostal ligaments, and strengthened at the upper and lower parts of the articulations by a few fibers that connect the cartilages to the side of the sternum.

The **radiate sternocostal ligaments** (Fig. 5-24) consist of broad and thin membranous bands that radiate from the ventral and dorsal surfaces of the sternal ends of the cartilages of the true ribs to the anterior and posterior surfaces of the sternum. They are composed of fibers that course obliquely superiorly, transversely and obliquely inferiorly, intermingling with fibers of the ligaments above and below them, with those of the opposite side, and ventrally with the ten-

dinous fibers of origin of the pectoralis major muscle. Together these structures form a thick fibrous membrane that envelops the sternum. This is more distinct over the inferior than over the superior part of the bone.

The **intra-articular sternocostal ligaments** are found constantly only between the second costal cartilages and the sternum. The cartilage of the *second rib* is connected with the sternum by means of an intra-articular ligament, attached at its lateral end to the cartilage of the rib, and medially to the fibrocartilage that unites the manubrium and body of the sternum. The second sternocostal joint contains two synovial membranes. Between these is the intra-articular ligament, which then continues medially within the manubriosternal joint. Occasionally the cartilage of the *third rib* is connected with the first and second sternebrae of the body of the sternum by a similar ligament. Intra-articular ligaments may rarely be seen in the lower sternocostal joints in association with the original sternebrae that formed the sternal body. At times the intra-articular ligament obliterates the synovial cavity completely, thereby transforming the joint into a synchondrosis.

The **costoxiphoid ligaments** connect the anterior and posterior surfaces of the seventh costal cartilage, and sometimes those of the sixth, to the dorsal and ventral surfaces of the xiphoid process. They vary in length and breadth in different subjects, the dorsal ligaments being less distinct than the ventral.

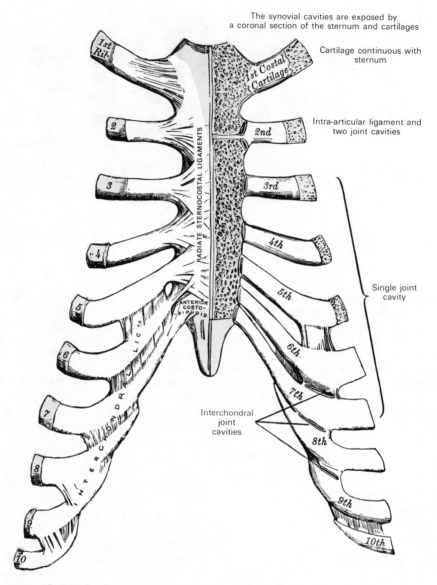

The synovial cavities are exposed by
a coronal section of the sternum and cartilages

Cartilage continuous with
sternum

Intra-articular ligament and
two joint cavities

Single joint
cavity

Interchondral
joint
cavities

FIG. 5-24. Sternocostal and interchondral articulations. Anterior view.

Synovial Membranes. There is no synovial membrane between the first costal cartilage and the sternum because this cartilage is directly continuous with the manubrium. There are generally two in the articulation of the second costal cartilage and usually one in each of the other joints. The synovial cavities of the sixth and seventh sternocostal joints are sometimes absent, and when an intra-articular ligament is present, there are two synovial cavities. After middle life the articular surfaces of the cartilages of most of the ribs become continuous with the sternum, and the joint cavities are consequently obliterated.

Movements. Slight gliding movements are permitted in the sternocostal articulations.

Arteries and Nerves. The **arterial supply** to the sternocostal joints is derived from perforating branches of the internal thoracic arteries. The **nerves** come from the anterior perforating branches of the intercostal nerves.

INTERCHONDRAL ARTICULATIONS (Fig. 5-24). The contiguous borders between the sixth and seventh, the seventh and eighth, the eighth and ninth, and sometimes those of the ninth and tenth costal cartilages articulate with each other by small, smooth, oblong facets. Each joint is enclosed in a thin **articular capsule**, lined by **synovial membrane**

and strengthened laterally and medially by fibrous **interchondral ligaments** which pass from one cartilage to the other. Sometimes the fifth costal cartilages, more rarely the ninth and tenth, articulate by their lower borders with the adjoining cartilages by small oval facets. More frequently, however, this connection consists simply of a few ligamentous fibers.

COSTOCHONDRAL ARTICULATIONS. The lateral end of each costal cartilage is received into a slight depression in the sternal end of the rib. The cartilage and bone are bound together by the periosteum of the bone, which becomes continuous with the perichondrium of the cartilage.

ARTICULATIONS OF THE STERNUM

MANUBRIOSTERNAL ARTICULATION. The inferior surface of the manubrium is united with the superior border of the body of the sternum by a fibrocartilaginous disc, which is shaped to conform to the hyaline cartilage-covered bony surfaces. The lateral margins of this symphysis are contiguous with the second sternocostal joints. At times, the center of the disc softens, becomes absorbed, and forms a cavity, thereby assuming the character of a synovial joint. At their articulation site, the manubrium and sternal body usually present a slight anteriorly projecting ridge, called the **sternal angle.** This is generally palpable beneath the skin and serves as an important surface landmark to the clinician. After maturation, this joint becomes ossified in about 10% of individuals (Trotter, 1934).

XIPHISTERNAL ARTICULATION. The articulation between the xiphoid process and the inferior border of the sternal body is cartilaginous. Bound by longitudinal fibers anteriorly and posteriorly, this joint is also secured laterally by radiating fibers of the sternocostal ligaments. Generally, the seventh costal cartilage articulates with the sternum at the lateral margins of the xiphisternal junction. By the thirtieth year, the xiphisternal joint has usually become ossified.

MECHANISM OF THE THORAX. Each rib possesses its own range of movements, but in combination they allow the respiratory excursions of the thorax. Each rib may be regarded as a lever, the fulcrum of

which is situated immediately lateral to the costotransverse joint so that when the shaft of the rib is elevated, the neck is depressed and *vice versa.* Because there is a disproportion in length of the arms of the lever, a slight movement at the vertebral end of the rib is greatly magnified at the anterior extremity.

The anterior ends of the ribs lie at a more inferior plane than the posterior ends, and thus, when the shaft of the rib is elevated, the anterior extremity is also thrust forward. Because the middle of the shaft of the rib lies at a plane inferior to one passing through the rib's two extremities, the shaft is elevated at the same time as it is thrust forward from the median plane of the thorax (Fig. 5-25). Further, each rib forms the segment of a curve which is greater than that of the rib immediately above, and therefore the elevation of a rib also increases the transverse diameter of the thorax. The modifications of the rib movements at their vertebral ends have already been described (p. 354). Further modifications in their movement result from the attachments of their ventral extremities. It is convenient, therefore, to consider the movements of three different groups of ribs separately according to their anterior attachment: vertebrosternal, vertebrochondral, and vertebral.

VERTEBROSTERNAL RIBS (Figs. 5-25; 5-26). The **first rib** differs from the others of this group because its attachment to the sternum is rigid. Its head, however, possesses no interarticular ligament, and is therefore somewhat more movable. The first pair of ribs moves with the manubrium as a single piece, the ventral portion being elevated by rotatory movements at the necks of the ribs near the verte-

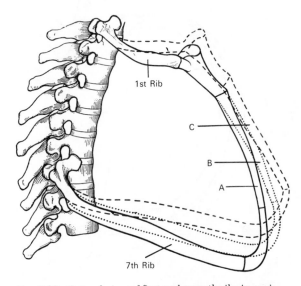

FIG. 5-25. Lateral view of first and seventh ribs in position, showing the movements of the sternum and ribs in *A,* ordinary expiration; *B,* quiet inspiration; *C,* deep inspiration.

bral extremities. In normal quiet respiration, the movement of this arc is practically *nil*. When some movement does occur, the ventral part is raised and carried forward, increasing both diameters of this region of the chest. Movement of the **second rib** is also slight in normal respiration because its anterior extremity is fixed to the manubrium and therefore prevented from moving upward. The sternocostal articulation, however, allows the middle of the body of the second rib to be drawn up, and in this way the transverse thoracic diameter is increased. Elevation of the **third to the sixth ribs** raises and thrusts their anterior extremities forward, the greater part of the movement being effected by rotation of the rib neck. The manubriosternal joint allows the relative positions of the manubrium and the body of the sternum to be adjusted when the anteroposterior thoracic diameter is increased. This movement is soon arrested, however, and the force is then expended in raising the middle part of the shaft of the rib, everting its lower border, and increasing the costochondral angle.

VERTEBROCHONDRAL RIBS (Fig. 5-27). The **seventh rib** is included along with the **eighth, ninth, and tenth ribs** in this group because it conforms more closely to their type. Although the movements of these ribs assist in enlarging the thorax for respiratory purposes, they are also concerned in increasing the upper abdominal space for viscera displaced by the action of the diaphragm. The costal cartilages of the sixth through the tenth ribs articulate with one another, so that each pushes up the costal cartilage above it, the final thrust being directed to pushing

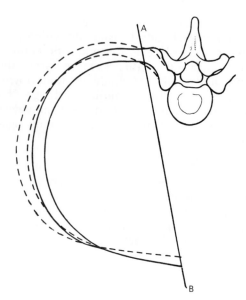

FIG. 5-27. Diagram showing the axis of movement (*AB*) of a vertebrochondral rib. The interrupted lines indicate the position of the rib during inspiration.

forward and upward the lower end of the body of the sternum. The amount of elevation of the anterior extremities is limited because of the very slight rotation of the rib neck. Elevation of the rib shaft is accompanied by an outward and backward movement. The outward movement everts the anterior end of the rib and widens the infrasternal angle, whereas the backward movement pulls back the anterior extremity and counteracts the forward thrust due to its elevation; this latter is most noticeable in the lower ribs, which are the shortest. The total result is a considerable increase in the transverse, and a diminution in the median anteroposterior, diameter of the upper part of the abdomen. At the same time, however, the lateral anteroposterior diameters of the abdomen are increased.

VERTEBRAL RIBS.   Because the **eleventh and twelfth ribs** have free anterior extremities and only costovertebral articulations with no intra-articular ligaments, they are capable of slight movements in all directions. When the other ribs are elevated, these are depressed and fixed by muscles to form stable points of action for the diaphragm.

LUMBOSACRAL, SACROCOCCYGEAL, AND INTERCOCCYGEAL JOINTS

LUMBOSACRAL JOINT.   The **ligaments** that articulate the fifth lumbar vertebra with the sacrum are similar to those that join any other two adjacent vertebrae in the spinal column. These include: (1) the continuation

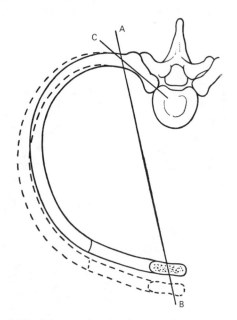

FIG. 5-26. Diagram showing the axes of movement (*AB* and *CD*) of a vertebrosternal rib. The interrupted lines indicate the position of the rib during inspiration.

caudally of the *anterior and posterior longitudinal ligaments*; (2) a thick *intervertebral disc*, which connects the body of the fifth lumbar to that of the first sacral vertebra; (3) *ligamenta flava*, which unite the laminae of the fifth lumbar vertebra with those of the first sacral vertebra; (4) *articular capsules*, which surround the articular processes; and (5) *interspinal and supraspinal ligaments.*

Additionally, on each side, the **iliolumbar ligament** connects the pelvis with the vertebral column.

The **iliolumbar ligament** (Fig. 5-29) is attached to the lateral tip and to the anterior and inferior aspects of the transverse process of the fifth lumbar vertebra. It radiates as it passes laterally and is attached by two main bands to the pelvis. The more inferior band, called the **lumbosacral ligament,** courses to the base of the sacrum and blends with the anterior sacroiliac ligament. The more superior band is attached to the crest of the ilium immediately anterior to the sacroiliac articulation, and is continuous with the thoracolumbar fascia. The psoas major muscle lies anterior to the iliolumbar ligament, while the muscles occupying the vertebral groove are posterior to it. The quadratus lumborum, which partially arises from the ligament, extends superiorly.

SACROCOCCYGEAL JOINT. The sacrococcygeal joint is a slightly movable symphysis interposed between the inferiorly directed, oval apex of the sacrum and the cranially oriented base of the coccyx. Between the bones is a fibrocartilaginous disc, but at a more advanced age, the joint may become ankylosed. The sacrococcygeal joint is homologous with those between the bodies of the vertebrae, and is secured by similar ligaments.

Ligaments. The ligaments at the sacrococcygeal joint are:

Ventral sacrococcygeal
Dorsal sacrococcygeal
Lateral sacrococcygeal
Disc of fibrocartilage
Interarticular

The **ventral sacrococcygeal ligament** consists of a few irregular fibers that descend from the ventral surface of the sacrum to that of the coccyx, blending with the periosteum. This ligament forms a caudal extension of the anterior longitudinal ligament.

The **dorsal sacrococcygeal ligament** is a flat band that arises from the margin of the distal orifice of the sacral canal, and descends to be inserted into the dorsal surface of the coccyx. This ligament completes the distal part of the sacral canal and is divisible into a short **deep portion,** which is a direct continuation of the posterior longitudinal ligament of the spinal column, and a longer **superficial part.** The gluteus maximus muscle overlies the ligament dorsally.

The **lateral sacrococcygeal ligament** is found on each side and, similar to the intertransverse ligaments, connects the transverse process of the first coccygeal vertebra to the lower lateral angle of the sacrum. With these two bones it completes the foramen for the fifth sacral nerve.

A **disc of fibrocartilage** is interposed between the contiguous surfaces of the sacrum and coccyx. It is oval in shape and measures about 2 cm in diameter. This disc differs from those between the bodies of the vertebrae because it is thinner and its central part is firmer in texture. It is somewhat thicker ventrally and dorsally than at the sides. Occasionally the coccyx is freely movable on the sacrum, especially during pregnancy, and in these instances a synovial membrane is present.

The **intercornual ligaments** are thin fibrous bands that unite the cornua of the coccyx with the sacral horns.

INTERCOCCYGEAL JOINTS. The different segments of the coccyx are interconnected by the extension distally of the ventral and dorsal sacrococcygeal ligaments. In younger subjects, thin annular discs of fibrocartilage are interposed between the coccygeal segments. In the adult male, all the segments become ossified together at a comparatively early period, but in the female this does not commonly occur until a later period of life.

MOVEMENTS. The movements that take place between the sacrum and coccyx, and between the different pieces of the latter bone, are quite limited and simply forward and backward in direction. During defecation and parturition, the coccyx is pushed backward.

ARTERIES AND NERVES. The lumbosacral, sacrococcygeal, and intercoccygeal joints are supplied by the iliolumbar, middle sacral, and lateral sacral arteries. The nerve supply is derived from sensory branches of the corresponding segmental spinal nerves.

ARTICULATIONS OF THE PELVIS

The ligaments connecting the bones of the pelvis with each other may be divided into three groups: (1) those connecting the sacrum and ilium; (2) two accessory ligaments passing between the sacrum and ischium; and (3) those between the two pubic bones.

SACROILIAC JOINT (Figs. 5-28 to 5-30; 5-32). The sacroiliac joint is a synovial joint that permits little movement. It connects the auricular surfaces of the sacrum and the ilium, and the interfacing borders of each bone are covered with cartilage. Covering the sacral surface is a layer of hyaline cartilage, which is thicker in depth than the fibrocartilage layer overlying the articular surface of the ilium. The joint surfaces are irregularly shaped, and the interconnecting bones fit snugly, thus restricting movement and buttressing the weight-bearing function of the joint. In advanced life, the cavity of the joint may be reduced in size or even nonexistent because of fibrous adhesions and synostosis.

This seems especially true in males. Surrounding the articular surfaces is a fibrous **articular capsule.**

Ligaments. The bones are held together by the following ligaments:

Ventral sacroiliac
Dorsal sacroiliac
Interosseous

The **ventral sacroiliac ligament** (Fig. 5-29) consists of numerous thin fibrous bands which unite the base and lateral part of the sacrum to the margin of the auricular surface of the ilium and to the preauricular sulcus. It blends with the periosteum on the pelvic surface and reaches the arcuate line.

The **dorsal sacroiliac ligament** (Fig. 5-30) is quite strong and forms an important band of union between the sacrum and ilium. It is situated posteriorly in a deep depression between the two bones and consists of numerous fasciculi which course in various directions. The deeper part is called the **short posterior sacroiliac ligament,** and it is

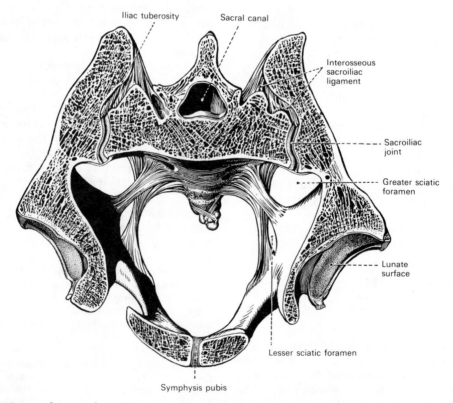

Fig. 5-28. A coronal section through the pelvis showing the sacroiliac joint through the anterior sacral segment. Note the interosseous sacroiliac ligament filling the cleft above and behind the joint cavity. (From Benninghoff and Goerttler, 1975.)

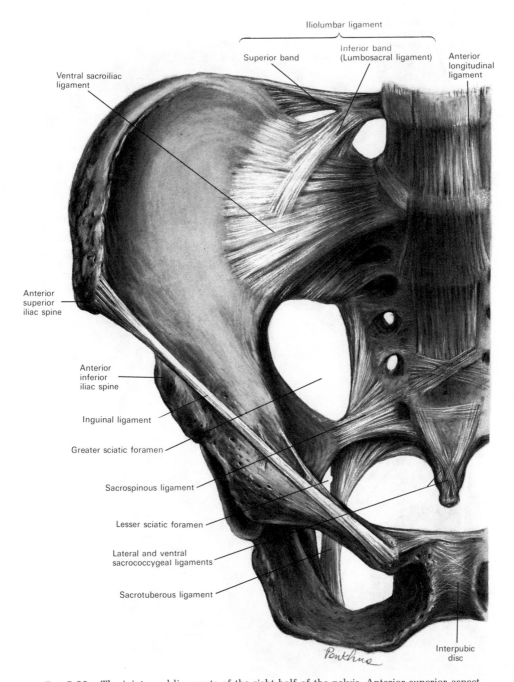

Iliolumbar ligament

Superior band

Inferior band
(Lumbosacral ligament)

Anterior
longitudinal
ligament

Ventral sacroiliac
ligament

Anterior
superior
iliac spine

Anterior
inferior
iliac spine

Inguinal ligament

Greater sciatic foramen

Sacrospinous ligament

Lesser sciatic foramen

Lateral and ventral
sacrococcygeal ligaments

Sacrotuberous ligament

Interpubic
disc

Fig. 5-29. The joints and ligaments of the right half of the pelvis. Anterior superior aspect.

nearly horizontal in direction, passing inferiorly and medially from the first and second
transverse tubercles on the dorsum of the
sacrum to the tuberosity of the ilium. The
more superficial portion is oriented longitudinally and is called the **long posterior sacroiliac ligament.** It partially covers the short
ligament and attaches above to the posterior

superior iliac spine and to the second, third,
and fourth articular tubercles of the sacrum.
Descending a bit obliquely, it becomes continuous with the sacrotuberous ligament.

The **interosseous sacroiliac ligament** (Fig.
5-28) is the strongest of the ligamentous connections between the two bones. It lies deep
to the dorsal sacroiliac ligament, and con-

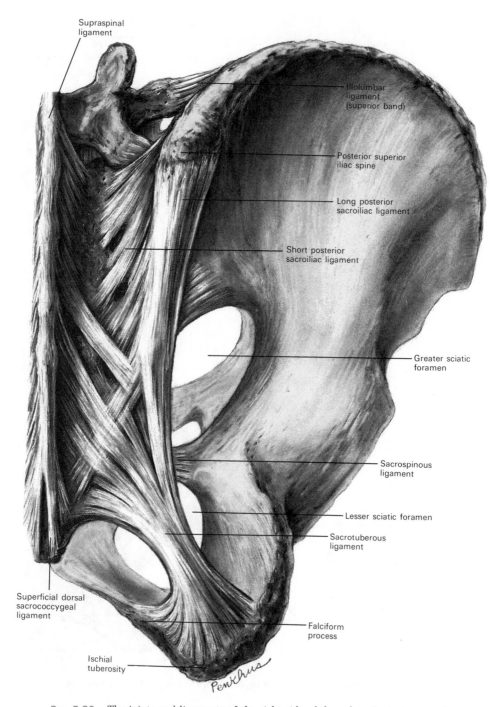

Supraspinal
ligament

Iliolumbar
ligament
(superior band)

Posterior superior
iliac spine

Long posterior
sacroiliac ligament

Short posterior
sacroiliac ligament

Greater sciatic
foramen

Sacrospinous
ligament

Lesser sciatic foramen

Sacrotuberous
ligament

Superficial dorsal
sacrococcygeal
ligament

Falciform
process

Ischial
tuberosity

FIG. 5-30.   The joints and ligaments of the right side of the pelvis. Posterior aspect.

sists of a series of short, strong fibers connecting the tuberosities of the sacrum and
ilium. By filling the narrow cleft above and
behind the articular surfaces, its more superficial fibers blend with those of the short
posterior sacroiliac ligament.

**TWO ACCESSORY LIGAMENTS PASSING BE-
TWEEN THE SACRUM AND THE ISCHIUM.**   The
lower posterolateral wall of the pelvis is not
enclosed by bone. Instead two important
accessory ligaments attach from the sacrum
and lower ilium to the ischium, converting

the large sciatic notch into the **greater** and **lesser sciatic foramina.** The two ligaments are:

Sacrotuberous
Sacrospinous

The **sacrotuberous ligament** (Figs. 5-29; 5-30) is a broad, flat, fan-shaped complex of fibers attaching above to the posterior superior and posterior inferior iliac spines, the fourth and fifth transverse tubercles of the sacrum, and the caudal part of the lateral margin of the sacrum and upper coccyx. Passing obliquely downward and laterally, the ligament is fixed to the medial margin of the ischial tuberosity. Before the converging fibers reach the tuberosity they first form a strong, thick band, and then again fan out as they attach to the tuberosity. Some of its fibers are prolonged forward along the ramus of the ischium as the **falciform process,** to which the obturator fascia is attached. The caudal border of the ligament is directly continuous with the tendon of origin of the biceps femoris muscle, and may be considered to be the proximal end of this tendon, interrupted in its projection by the tuberosity of the ischium. Many of the proximal fibers of the sacrotuberous ligament blend with those of the long posterior sacroiliac ligament. The *posterior surface* of the sacrotuberous ligament gives origin to the gluteus maximus muscle; its *anterior surface* is, in part, united to the sacrospinous ligament and, in part, gives origin to some fibers of the piriformis muscle.

The **sacrospinous ligament** (Figs. 5-29; 5-30) is thin and triangular and is attached by its broad base to the lateral margins of the sacrum and coccyx, where its fibers are intermingled with those of the intrapelvic surface of the sacrotuberous ligament. Its apex extends laterally to the spine of the ischium. The *anterior surface* of the sacrospinous ligament gives attachment to the coccygeus muscle, while *posteriorly,* it is covered by the sacrotuberous ligament and crossed by the internal pudendal vessels and nerve. The sacrospinous ligament transforms the sciatic notch into the **greater** and **lesser sciatic foramina,** its *superior border* forming the lower boundary of the greater sciatic foramen, and its *inferior border,* part of the upper margin of the lesser sciatic foramen.

GREATER AND LESSER SCIATIC FORAMINA (Figs. 5-28 to 5-30). The **greater sciatic foramen** is bounded *in front* and *above* by the posterior border of the hip bone, *behind* by the sacrotuberous ligament, and *below* by the sacrospinous ligament. It is partially occupied by the piriformis muscle, which leaves the pelvis through it. Above this muscle, the superior gluteal vessels and nerve emerge from the pelvis, and below it, the inferior gluteal vessels and nerve, the internal pudendal vessels and nerve, the sciatic and posterior femoral cutaneous nerves, and the nerves to the obturator internus and quadratus femoris all make their exit from the pelvis. The **lesser sciatic foramen** is bounded *in front* by the tuberosity and body of the ischium, *above* by the spine of the ischium and sacrospinous ligament, and *behind* by the sacrotuberous ligament. It transmits the tendon of the obturator internus muscle, its nerve, and the internal pudendal vessels and nerve.

PUBIC SYMPHYSIS (Fig. 5-31). This cartilaginous articulation in the midline between the pubic bones is a slightly movable joint. Interconnecting the two oval, articular symphysial surfaces of the bones is a fibrocartilaginous disc. In the female the joint is somewhat shorter and broader than in the male.

Ligaments. The ligaments of the pubic symphysis are:

Superior pubic
Arcuate pubic
Interpubic disc

The **superior pubic ligament** (Fig. 5-31) connects the two pubic bones superiorly and extends laterally as far as the pubic tubercles. From its yellowish fibers arises the rectus abdominis muscle.

The **arcuate pubic ligament** (Fig. 5-31) is a thick, triangular band of ligamentous fibers connecting the two pubic bones inferiorly, and forming the boundary of the pubic arch. Above, it is blended with the interpubic fibrocartilaginous disc. Laterally, it is attached to the inferior rami of the pubic bones, and its inferior free border is separated from the fascia of the urogenital diaphragm by an opening through which the deep dorsal vein of the penis or clitoris passes into the pelvis.

The **interpubic disc** of fibrocartilage is interposed between the adjacent surfaces of the pubic bones (Fig. 5-31). Both surfaces are covered by a thin layer of hyaline cartilage firmly joined to the bone. A series of nipple-like processes snugly fits into corresponding

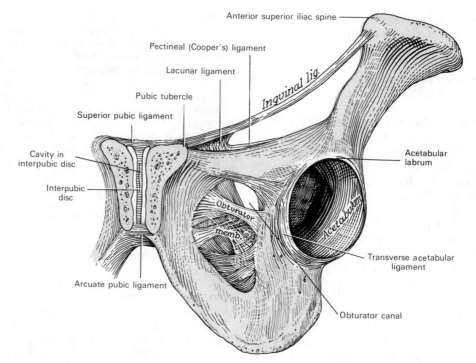

FIG. 5-31. Symphysis pubis exposed by a coronal section.

depressions on the osseous surfaces. The fibrocartilaginous disc intervening between the opposed cartilage-covered surfaces varies in width, but usually is thicker in females than in males. Frequently, a cavity is seen in the upper posterior part of the joint (shown as a fine line in Fig. 5-31). This cavity probably is formed by the softening and absorption of fibrocartilage, since it rarely appears before the tenth year of life and is not lined by synovial membrane. The disc is strengthened anteriorly by decussating fibers which pass obliquely from one bone to the other and interlace with fibers of the aponeuroses of the external oblique muscles and the medial tendons of origin of the rectus abdominis muscles.

MECHANISM OF THE PELVIS. The pelvic girdle supports and protects the contained viscera; in addition, it affords surfaces for the attachment of trunk and lower limb muscles. Because of our upright posture, the most important mechanical function of the human pelvis is to transmit the weight of the trunk and upper limbs to the lower limbs. From the last lumbar vertebra, the weight of the upper body is transmitted through the sacrum to the ilium and then to the head of the femur to the acetabulum.

The pelvic girdle may be divided into a posterior and an anterior arch by a vertical plane passing through the acetabular cavities. The *posterior arch* is the one chiefly concerned in the transmission of weight. Its essential parts are the upper three sacral vertebrae and two strong pillars of bone extending from the sacroiliac joints to the acetabular cavities. The *anterior arch* is formed by the bodies of the pubic bones and their superior rami, which articulate in the middle of the arch at the symphysis pubis.

Thus, for the diffusion of weight, each acetabular cavity is strengthened not only by the superior ramus of the pubis, but also by the upper portion of the body of the ramus. The sacroiliac joints, interposed between the sacrum and the ilia, lessen the concussion to rapid changes in the distribution of weight. Additionally, the symphysis pubis in the center of the anterior arch helps to dissipate sudden weight displacements.

The sacrum forms the summit of the posterior arch. The weight transmitted from the upper part of the body is received by the sacrum at the lumbosacral articulation and, theoretically, is transmitted by two components in different directions. One component of the force is expended in driving the sacrum downward and backward between the iliac bones, while the other thrusts the upper end of the sacrum downward and forward toward the pelvic cavity.

The movements of the sacrum are regulated by its form. Viewed in its entirety, it presents the shape of a wedge with its base oriented upward and forward. The first component of the force is therefore acting against the resistance of the wedge, and its tendency to separate the iliac bones is resisted by the sacroiliac and iliolumbar ligaments and by the ligaments of the pubic symphysis.

If a series of coronal sections of the sacroiliac joints is made, it is possible to divide the articular portion of the sacrum into three segments, anterior, middle, and posterior. In the **first or anterior segment** (Fig. 5-32, *A*), which involves the first sacral vertebra, the articular surfaces are nearly parallel, although the distance between their posterior margins is slightly greater than that between their anterior margins. This segment therefore presents a slight wedge shape with the truncated apex downward and forward. The **second or middle segment** (Fig. 5-32, *B*) is a narrow band across the centers of the articulations. Its posterior width is distinctly greater than its anterior, so that the segment is more definitely wedge-shaped, the truncated apex being again directed downward and forward. In the center of each sacral articular surface is seen a marked concavity into which a corresponding convexity of the iliac articular surface fits, forming an interlocking mechanism. In the **third or posterior segment** (Fig. 5-32, *C*), the anterior width is greater than the posterior, so that the wedge form is the reverse of those of the other segments, and the truncated apex is directed backward and upward. The sacral articular surfaces are only slightly concave.

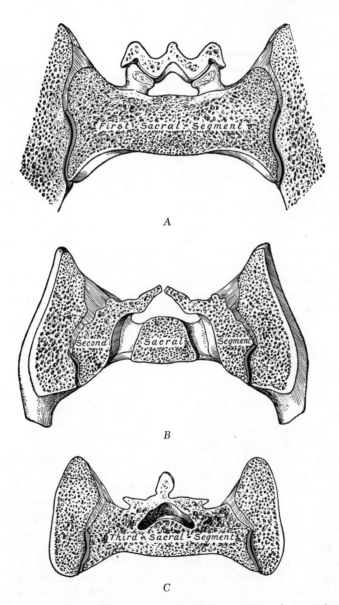

A

B

C

FIG. 5-32. Coronal sections through the sacroiliac joint: *A*, through the anterior or first sacral segment; *B*, through the middle or second sacral segment; *C*, through the posterior or third sacral segment.

Dislocation downward and forward of the sacrum by the second component of the force applied to it is prevented, therefore, by the second or middle segment, which interposes the resistance of its wedge shape and that of the interlocking mechanism on its surfaces. As a result, a rotatory movement is produced by which the anterior or first segment is tilted downward and the posterior or third segment is tilted upward. The axis of this rotation passes through the posterior part of the second or middle segment. Movement of the anterior or first segment is slightly limited by its wedge form, but chiefly by the dorsal and interosseous sacroiliac ligaments. Similarly, movement of the posterior or third segment is checked to a slight extent by its wedge form, but principally by the sacrotuberous and sacrospinous ligaments. In all these movements the effect of the sacroiliac and iliolumbar ligaments and the ligaments of the symphysis pubis in resisting the separation of the iliac bones must be recognized.

During pregnancy the pelvic joints become somewhat more mobile because of the softening of the symphysial cartilage and the relaxation of the ligaments. Thus, the degree of sacral rotation allowable increases slightly, and during childbirth this also may afford a slight increase in the diameter of the pelvis. When the fetus is being expelled, force is applied from within the pelvis to the sacrum. Significant upward dislocation, however, is prevented by the interlocking mechanism of the second or middle segment. As the fetal head passes the anterior segment, the latter is rotated upward, enlarging slightly the dorsoventral diameter of the pelvic inlet. When the head reaches the posterior segment of the sacrum, this also is pressed upward against the resistance of its wedge, the movement being possible only by the laxity of the joints and the stretching of the sacrotuberous and sacrospinous ligaments.

Articulations of the Upper Extremity

The **shoulder girdle** is a term referring to the means by which the upper limb is appended to the trunk. It consists of the scapulae and clavicles, and anteriorly the girdle is completed by the manubrium of the sternum, to which the clavicles are joined at the **sternoclavicular joints.** Posteriorly, the scapulae do not articulate with the axial skeleton, but are attached to ribs and vertebral column by muscles. Laterally, the scapula is attached on each side to the clavicle at the **acromioclavicular joint.** The humerus articulates with the scapula at the **shoulder joint** and is also joined below to the bones of the forearm at the **elbow joint.** The **radioulnar joints** lend stability to the forearm and af-

ford the hand increased mobility. The **wrist joint** interposes the forearm and the hand, and between the carpal bones of the wrist are found the **intercarpal joints.** The skeleton of the hand consists of metacarpal bones which articulate (1) proximally, with the carpal bones at the **carpometacarpal joints,** (2) with each other by means of **intermetacarpal joints,** and (3) distally, with the phalanges of the digits at the **metacarpophalangeal joints.** Finally, the segments of the fingers and thumb articulate at the **interphalangeal joints.**

SHOULDER GIRDLE

The shoulder girdle significantly increases the mobility of the upper limb. Interconnecting its structures are the sternoclavicular and acromioclavicular joints.

STERNOCLAVICULAR JOINT.   The sternoclavicular joint (Fig. 5-33) is a double gliding joint. Entering into its formation are the sternal end of the clavicle, the superior and lateral part of the manubrium sterni at the clavicular notch, and the cartilage of the first rib. The articular surface of the clavicle is much larger than that of the sternum, and it is invested with a layer of fibrocartilage which is thicker than that on the sternum. The bones are not well adapted for articulation and actually are nowhere in contact, being separated completely by an articular disc.

Ligaments.   The ligaments of this joint are:

Articular capsule
Anterior sternoclavicular
Posterior sternoclavicular
Interclavicular
Costoclavicular
Articular disc

The **articular capsule** surrounds the articulation and varies in thickness and strength. In front and behind, it is of considerable thickness and is reinforced by the anterior and posterior sternoclavicular ligaments. Superiorly, the capsule is strengthened somewhat by the interclavicular ligament, but inferiorly, it is thin and partakes more of the character of areolar tissue.

The **anterior sternoclavicular ligament** (Fig. 5-33) is a broad band of fibers covering the anterior surface of the joint. It is at-

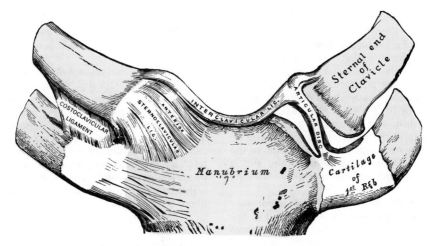

FIG. 5-33. The sternoclavicular joints. Anterior aspect.

tached above to the upper and front part of the sternal end of the clavicle. Passing obliquely downward and medialward, the ligament is attached below to the front and upper part of the manubrium sterni. This ligament is covered by the sternal portion of the sternocleidomastoid muscle and behind it lies the articular capsule, the articular disc, and the two synovial membranes.

The **posterior sternoclavicular ligament** is a band of fibers similar to the anterior ligament, but covering the dorsal surface of the articulation. It is attached to the superior part of the sternal end of the clavicle, and passing obliquely medialward, it is fixed below to the back of the upper part of the manubrium sterni. Adjacent to the ligament anteriorly are the articular disc and synovial membranes, while posteriorly lie the sternohyoid and sternothyroid muscles.

The **interclavicular ligament** (Fig. 5-33) is a flattened band of varying size in different individuals which considerably strengthens the superior part of the capsule. It passes in a curved manner from the upper part of the sternal end of one clavicle to that of the other, and is also attached to the superior margin of the sternum, forming the floor of the jugular notch. Ventral to the ligament are the sternocleidomastoid muscles and the skin, while dorsal to it are the sternohyoids.

The **costoclavicular ligament** (Fig. 5-33) is a short, flat, strong and rhomboid-shaped band located outside the capsule, lateral to the joint. It is attached below to the upper and medial part of the cartilage of the first rib, and ascends obliquely backward and laterally to the costal tuberosity on the inferior surface of the clavicle. Immediately anterior to the ligament is the tendon of origin of the subclavius muscle, while behind it courses the subclavian vein.

The **articular disc** is flat, nearly circular, and interposed between the articulating surfaces of the sternum and clavicle. Above, it is attached to the dorsal superior border of the articular surface of the clavicle and, below, to the cartilage of the first rib near its junction with the sternum. By the rest of its circumference the disc adheres to the interclavicular and anterior and posterior sternoclavicular ligaments. It is thicker near its rounded margin, especially its upper and posterior part, than at its center. The disc divides the joint into two cavities, each of which is furnished with a synovial membrane.

Synovial Membranes. The more lateral of the two synovial cavities is lined by a membrane that is reflected from the sternal end of the clavicle, over the adjacent surface of the articular disc, and around the margin of the facet on the cartilage of the first rib. The synovial membrane lining the more medial cavity is attached to the margin of the articular surface of the sternum and clothes the adjacent surface of the articular disc. The lateral cavity is often a bit larger than the medial and its capsular covering somewhat looser. The two cavities may communicate by way of a perforation in the articular disc.

Arteries and Nerves. The **arteries** supplying twigs to the sternoclavicular joint include: (1) the internal thoracic (from subclavian), (2) the suprascapular (from thyrocervical trunk), and (3) the supreme thoracic or clavicular branch of the thoracoacromial artery (from subclavian). The nerves to the sternoclavicular joint are derived from the more medial of the supraclavicular nerves, and a branch from the nerve to the subclavius muscle.

Movements. The sternoclavicular joint allows limited movement in nearly every direction. When movements occur in this joint, the clavicle carries the scapula with it, and in turn the scapula glides over the surface of the thorax. This joint therefore forms the center from which all movements of the supporting arch of the shoulder originate, and is the only articulation of the shoulder girdle with the trunk.

Elevation and depression of the shoulder take place between the clavicle and the articular disc, the bone rotating upon the ligament on an axis drawn through its own articular facet. When the shoulder is *moved forward* (*protraction*) or *backward* (*retraction*), the clavicle and the articular disc roll to and fro on the articular surface of the sternum, revolving with a sliding movement around an axis drawn nearly vertically through the sternum. In circumduction at the shoulder joint, an action that combines the other movements, the clavicle revolves upon the articular disc, which, with the clavicle, rolls upon the sternum. Elevation of the shoulder is limited principally by the costoclavicular ligament, and depression, by the interclavicular ligament and articular disc.

The **muscles** that *elevate the shoulder* are the cranial fibers of the trapezius, the levator scapulae, and the clavicular head of the sternocleidomastoid, assisted to a certain extent by the rhomboids. *Depression of the shoulder* is principally effected by gravity assisted by the subclavius, pectoralis minor, and caudal fibers of the trapezius. The shoulder is *protracted* (drawn forward) by the serratus anterior and pectoralis minor, and it is *retracted* (drawn backward) by the rhomboids and the middle and caudal fibers of the trapezius.

ACROMIOCLAVICULAR JOINT. The acromioclavicular joint (Figs. 5-34; 5-35) is a small synovial joint of the plane or gliding type between the acromial end of the clavicle and the medial margin of the acromion of the scapula. Both bony surfaces, covered with fibrocartilage, slope downward, and the clavicle has a tendency to override the acromion.

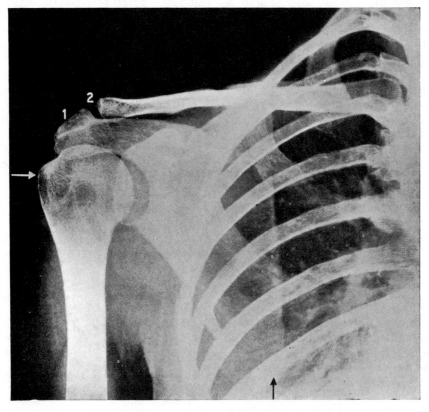

Fig. 5-34.   Roentgenogram of adult shoulder, *1*, acromion; *2*, acromioclavicular joint. The lower arrow indicates the inferior angle of the scapula, the upper arrow the greater tuberosity of the humerus. Note that the shadow of the head of the humerus overlaps the shadow of the acromial angle and a part of the glenoid cavity.

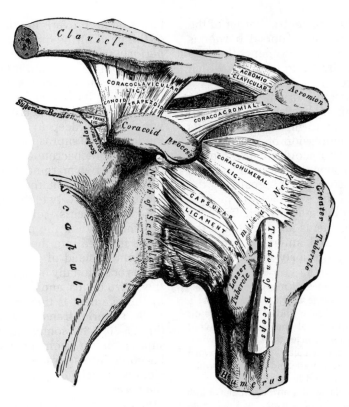

FIG. 5-35.   The left shoulder and acromioclavicular joints viewed from the anterior aspect.

Ligaments.   The joint is surrounded by an articular capsule and its ligaments are:

Articular capsule
Acromioclavicular
Articular disc
Coracoclavicular
    Trapezoid ligament
    Conoid ligament

The **articular capsule** completely surrounds the articular margins and is strengthened above by the acromioclavicular ligament. It is somewhat lax so that the clavicle is not bound tightly to the acromion.

The **acromioclavicular ligament** (Fig. 5-35) is a quadrilateral band of parallel fibers that strengthens the superior part of the articular capsule, extending between the acromial end of the clavicle and the adjoining part of the acromion. It interlaces with the aponeuroses of the trapezius and deltoid and is in contact with the articular disc when this is present. Inferiorly, this ligament is thinner and is adjacent to the tendon of the supraspinatus muscle.

The **articular disc,** which is occasionally present, only partially separates the articular surfaces and occupies the superior part of the articulation, where it attaches to the capsule. More rarely, it completely divides the joint into two cavities. There is usually only one synovial membrane in this articulation, but when a complete articular disc is present, there are two.

The **coracoclavicular ligament** (Fig. 5-35) connects the clavicle with the coracoid process of the scapula. Although placed at some distance medial to this articulation, it forms the most powerful union between the scapula and the clavicle. It consists of two parts which, because of their shapes, are called the **trapezoid** and **conoid ligaments.** These two ligamentous bands are separated by a bursa near their attachment anteriorly and medially along the coracoid process.

The **trapezoid ligament,** placed anterolaterally, is broad, thin, and quadrilateral, and it courses obliquely between the coracoid process and the clavicle. It is attached below to the superior surface of the coracoid process and above to the oblique ridge on the inferior surface of the clavicle. Directed nearly horizontally, the ventral border of the ligament is free, while its dorsal border is joined with the conoid ligament, the two forming an angle by their junction which projects backward.

The **conoid ligament** is oriented in a posterior and medial direction. It is a dense band of fibers, conical in form, attached by its apex to a rough impression on the posteromedial edge of the coracoid process, medial to the trapezoid ligament. Its expanded base is directed to the coracoid tuberosity on the inferior surface of the clavicle, and along a short line proceeding medialward from the tubercle. These ligaments are adjacent to the subclavius and deltoid anteriorly, and the trapezius muscle posteriorly.

Movements. Two types of movement are permitted between the clavicle and the scapula: (1) a gliding motion of the articular end of the clavicle on the acromion; (2) rotation of the scapula forward and backward upon the clavicle, the extent of which is limited by the coracoclavicular ligament. Forward rotation is limited by the trapezoid ligament, while backward rotation is limited by the conoid.

The acromioclavicular and sternoclavicular joints function cooperatively whenever the scapula changes its position on the posterior chest wall. The acromion is held by the clavicle at a fixed distance from the sternum, and thus, the clavicle serves as the radius of a circle through which the acromion must move. Because the diameter of this circle is greater than that of the curvature of the chest wall, the vertebral border of the scapula would be forced to swing away from the chest when the scapula is moved ventrally. The acromioclavicular joint allows the scapula to adjust its position so that it remains in close contact with the chest.

INTRINSIC LIGAMENTS OF THE SCAPULA. The following intrinsic ligaments of the scapula pass between different parts of that bone, and do not interconnect different bones at any joint:

Coracoacromial
Superior transverse scapular
Inferior transverse scapular

The **coracoacromial ligament** (Fig. 5-35) is a strong triangular band extending between the coracoid process and the acromion. Its apex is attached to the summit of the acromion just anterior to the articular surface for the clavicle, and its broad base is fixed to the whole length of the lateral border of the coracoid process. This ligament, together with the coracoid process and the acromion, forms an arch overlying the superior aspect of the head of the humerus. It is related *superiorly* to the clavicle and inner surface of the deltoid muscle, and *inferiorly* to the tendon of the supraspinatus with a bursa being interposed. Its lateral border is continuous with a dense connective tissue lamina, which passes deep to the deltoid upon the tendons of the supraspinatus and infraspinatus. The coracoacromial ligament sometimes consists of two marginal bands and a thinner intervening portion, the two bands being attached respectively to the tip and base of the coracoid process, joining together at the acromion. When the pectoralis minor muscle is inserted into the capsule of the shoulder joint instead of into the coracoid process, as occasionally occurs, it passes between these two bands, and the intervening portion of the ligament is then deficient.

The **superior transverse scapular ligament** (Figs. 5-35; 5-36), sometimes referred to as the *suprascapular ligament,* converts the scapular notch into a foramen. It is a thin, flat fasciculus, attached by one end to the base of the coracoid process, and by the other to the upper border of the scapula on its dorsal surface near the medial end of the scapular notch. The ligament is sometimes ossified, thereby forming a bony foramen. The suprascapular nerve courses through the foramen, and the suprascapular vessels cross over the ligament. Medially, some fibers of the inferior belly of the omohyoid muscle attach to the ligament.

The **inferior transverse scapular ligament,** when present, is a weak membranous band stretching from the lateral border of the scapular spine to the margin of the glenoid cavity. It forms an arch under which the suprascapular vessels and nerve enter the infraspinatous fossa.

SHOULDER JOINT

The **shoulder joint** (Figs. 5-34 to 5-38) is a spheroid or ball-and-socket joint in which an elegant freedom of movement is allowed at some expense to its strength and stability. The bones entering into its formation are the hemispherical head of the humerus and the shallow glenoid cavity of the scapula. Some protection of the joint against displacement is afforded by its ligaments and by the tendons and muscles that surround it. The liga-

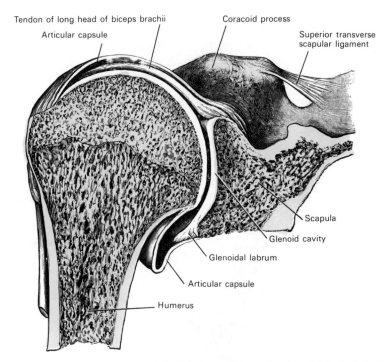

Tendon of long head of biceps brachii

Articular capsule

Coracoid process

Superior transverse
scapular ligament

Scapula

Glenoid cavity

Glenoidal labrum

Articular capsule

Humerus

FIG. 5-36. A frontal section through the head of the left humerus and shoulder joint. Anterior half, viewed from behind.

ments alone, however, do not maintain the joint surfaces in apposition, because when the muscles and tendons are removed, the humerus can be separated to a considerable extent from the glenoid cavity. The *ligamentous protection* supplied by the muscles and tendons effectively limits the degree of movement allowed at the joint. Additional protection superiorly is supplied by the arch formed by the coracoid process, acromion, and coracoacromial ligament. Both articular surfaces are covered with hyaline cartilages, although a small amount of fibrocartilage is seen in the glenoid cavity. The cartilage on the head of the humerus is thicker at the center than at the circumference, the reverse being the case with the articular cartilage of the glenoid cavity. The glenoid cavity is deepened by a fibrocartilaginous rim attached to its bony margin.

Ligaments. The ligaments of the shoulder joint are:

Articular cartilage
Coracohumeral
Glenohumeral
Glenoidal labrum
Transverse humeral

The **articular capsule** (Figs. 5-35 to 5-37) completely encases the joint, being attached *medially* to the circumference of the glenoid cavity beyond the glenoidal labrum, reaching the neck of the scapula, as far as the root of the coracoid process. *Laterally,* the capsule is fixed to the anatomical neck of the humerus, approaching the articular cartilage more closely superiorly than inferiorly, because its attachment below extends along the neck of the bone about 10 to 12 mm. It is thicker above and below than at its middle, and is so remarkably loose and lax that it does not keep the bones in contact, allowing them to be separated by as much as 2.5 cm. The looseness of the articular capsule is an important factor in the extreme freedom of movement peculiar to the shoulder joint. The capsule is reinforced *above* by the supraspinatus; *below* by the long head of the triceps; *behind* by the tendons of the infraspinatus and teres minor; and *in front* by the tendon of the subscapularis. In the passage of the tendons of the subscapularis, supraspinatus, infraspinatus, and teres minor over the fibrous capsule to their insertions on the tubercles of the humerus, they blend together with the capsule to form a musculo-

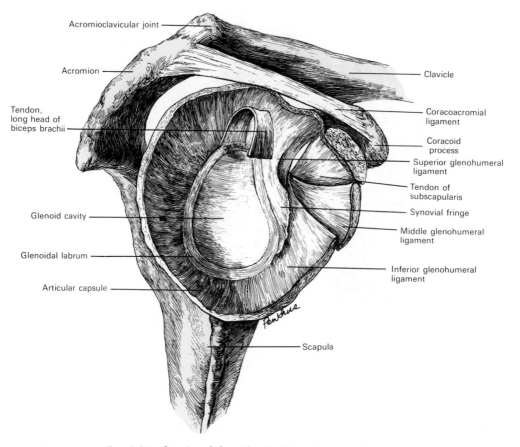

Acromioclavicular joint

Acromion

Tendon,
long head of
biceps brachii

Glenoid cavity

Glenoidal labrum

Articular capsule

Clavicle

Coracoacromial
ligament

Coracoid
process

Superior glenohumeral
ligament

Tendon of
subscapularis

Synovial fringe

Middle glenohumeral
ligament

Inferior glenohumeral
ligament

Scapula

FIG. 5-37. Interior of the right shoulder joint. Lateral aspect.

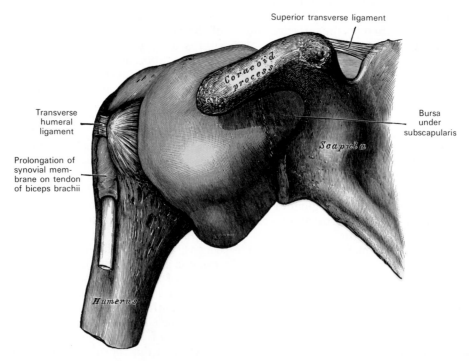

Superior transverse ligament

Transverse
humeral
ligament

Prolongation of
synovial mem-
brane on tendon
of biceps brachii

Coracoid
process

Scapula

Bursa
under
subscapularis

Humerus

FIG. 5-38. Synovial membrane of shoulder joint (distended). Anterior aspect.

tendinous or rotator cuff. The cuff is least complete inferiorly, where the joint is quite weak. There are usually three openings in the capsule. One is anteriorly located below the coracoid process, and establishes a communication between the joint and a bursa beneath the tendon of the subscapularis. The second, which is not constant, lies posteriorly, where an opening sometimes exists between the joint and a bursal sac under the tendon of the infraspinatus. The third is located anterolaterally between the tubercles of the humerus and allows the passage of the long tendon of the biceps brachii muscle.

The **coracohumeral ligament** is a broad band that strengthens the upper part of the capsule (Fig. 5-35). From the lateral border of the root of the coracoid process, it takes a slightly spreading and oblique course laterally to the anterior border of the greater tubercle of the humerus, lateral to the tendon of the long head of the biceps. At its humeral attachment it blends with the tendon of the supraspinatus superficial to it, and with the capsule deep to it (except for its anterior free border, which overlaps the capsule).

The **glenohumeral ligaments** are robust thickenings of the articular capsule over the anterior part of the joint. They are described as three ligaments, but their individuality is sometimes difficult to demonstrate. They may be seen in dissection, however, if the joint is disarticulated (Fig. 5-37), or if the capsule is opened posteriorly and the head of the humerus is removed, allowing the observer to view the inner surface of the anterior capsule from behind. The **superior glenohumeral ligament** is attached to the apex of the glenoid cavity, courses laterally along the medial edge of the biceps tendon (long head), and becomes fixed to a small depression at the top of the lesser tubercle. Separated from the coracohumeral ligament by the biceps tendon, the superior glenohumeral ligament courses between the long head tendon and the tendon of the subscapularis. The **middle glenohumeral ligament** is attached to the medial edge of the glenoid cavity of the scapula and stretches across the joint to the lower part of the lesser tubercle. When there is a communication between the synovial cavity and the subscapular bursa, it is found between the superior and middle ligaments. The **inferior glenohumeral ligament** extends from the inferior edge of the glenoid cavity to the lower part of the anatomical neck of the humerus.

The **transverse humeral ligament** (Fig. 5-38) is a sheet of short transverse fibers passing from the lesser to the greater tubercle of the humerus. It is located on the portion of the bone that lies above the epiphyseal line, and converts the intertubercular groove into a short tunnel for the tendon of the long head of the biceps.

The **glenoidal labrum** is a fibrocartilaginous rim attached around the margin of the glenoid cavity. It is triangular in section, the base being fixed to the circumference of the cavity, while the free edge is thin and sharp. Measuring about 6 mm in depth superiorly and inferiorly, the labrum is somewhat more shallow on the sides of the cavity. It is continuous with the tendon of the long head of the biceps, which gives off two fasciculi to blend with the fibrous tissue of the labrum. It deepens the articular cavity, and protects the edges of the bone (Fig. 5-37).

Synovial Membrane (Fig. 5-38). The synovial membrane is attached to the margin of the glenoid cavity and is reflected over the inner surface of the articular capsule. It covers the lower part and sides of the anatomical neck of the humerus as far as the articular cartilage on the humeral head. The tendon of the long head of the biceps passes through the capsule and is enclosed in a tubular sheath of synovial membrane, which descends to it from the summit of the glenoid cavity and continues around the tendon into the intertubercular groove as far as the surgical neck of the humerus (Fig. 5-38). The tendon thus traverses the shoulder joint, but it is not contained within the synovial cavity.

Bursae Associated with the Shoulder. A number of clinically important bursae are found in the region of the shoulder joint and separate the bony structures from the surrounding muscles and soft tissues. The **subscapular bursa** is found constantly between the tendon of the subscapularis and the underlying joint capsule. It usually communicates with the synovial cavity through an opening between the superior and middle glenohumeral ligaments in the anterior part of the capsule. The **subdeltoid bursa** is a large bursa between the deep surface of the deltoid muscle and the joint capsule, over the upper and lateral aspect of the humerus. It does not communicate with the synovial cavity. The **subacromial bursa** lies between the deep surface of the acromion and tendon of the supraspinatus muscle overlying the joint capsule. It usually extends under the coracoacromial ligament and frequently is continuous with the subdeltoid bursa. A **subcoracoid bursa** may lie between the coracoid process and the capsule, or it may be an extension from the subacromial bursa. The **coracobrachialis bursa** intervenes between the coracobrachialis and subscapularis muscles, but is inconstant. The **infraspinatus** bursa, between the tendon of the infraspinatus muscle and the capsule, may communicate with the synovial cavity. A subtendinous **bursa of the teres major** is found between that muscle and the long head of the triceps near their insertions on the humerus. **Bursae of the latissimus dorsi** are found both superficial and deep to its tendon of insertion. A **bursa of the pectoralis major** muscle is frequently found between its tendon of insertion and the long head of the biceps muscle. The **subcutaneous acromial bursa** is of considerable size and lies between the skin and the superficial surface of the acromion.

Muscles. The shoulder joint is anatomically related to the following muscles: *superior,* the supraspinatus; *inferior,* the long head of the triceps; *anterior,* the subscapularis; *posterior,* the infraspinatus and teres minor. Coursing through the joint and surrounded by the reflected synovial membrane is the long head of the biceps muscle. Finally, the deltoid covers the articulation anteriorly, posteriorly, and laterally.

Blood Supply and Nerves. The **arteries** that supply the shoulder joint from the scapular end include the *suprascapular artery* and the *circumflex scapular* branch of the subscapular artery. From the humeral end of the joint are the *anterior* and *posterior humeral circumflex arteries.*

Nerves to the shoulder joint are derived principally from the *axillary nerve* (from posterior cord of brachial plexus), the *suprascapular nerve* (from upper trunk of brachial plexus through junction of fifth and sixth cervical nerves), and the *lateral pectoral nerve* (from lateral cord of brachial plexus). At times a branch may come directly from the posterior cord of the brachial plexus or from the radial nerve (Gardner, 1948).

Movements. The *shoulder joint* is capable of every variety of movement: flexion, extension, abduction, adduction, circumduction, and rotation. The humerus is **flexed** (drawn forward) by the clavicular part of the pectoralis major, the anterior fibers of the deltoid, the coracobrachialis, and, when the forearm is flexed, the biceps brachii. When the limb is fully extended posterior to the trunk, the sternocostal portion of the pectoralis major assists in flexing the upper extremity until it reaches the side of the body at the anatomical position. The humerus is **extended** (drawn backward) by the latissimus dorsi, the teres major, the posterior fibers of the deltoid, and, when the forearm is extended, the triceps brachii. **Abduction** of the humerus is achieved by the deltoid and supraspinatus, whereas **adduction** is performed by the subscapularis, pectoralis major, latissimus dorsi, and teres major, and by the weight of the limb. **Lateral rotation** of the humerus occurs by action of the infraspinatus, teres minor, and posterior fibers of the deltoid. **Medial rotation** is accomplished by the subscapularis, latissimus dorsi, teres major, pectoralis major, and anterior fibers of the deltoid.

The *scapula* is capable of being moved superiorly and inferiorly, anteriorly and posteriorly, or, by a combination of these movements, circumducted on the posterior wall of the chest. The muscles that **elevate the scapula** are the superior fibers of the trapezius, the levator scapulae, and the rhomboids. Those that **depress the scapula** are the inferior fibers of the trapezius, the pectoralis minor, and, by way of the clavicle, the subclavius. The scapula is **drawn posteriorly** by the rhomboid and the middle and inferior fibers of the trapezius, and it is **drawn anteriorly** by the serratus anterior and pectoralis minor, assisted, when the arm is fixed, by the pectoralis major. This mobility of the scapula is considerable, and greatly assists the movements of the upper limb at the shoulder joint.

Raising of the arm above the head, either anterior to the trunk as in flexion or lateral to the trunk as in abduction, is brought about by a combined action at the shoulder joint and rotation of the scapula on the chest wall. Although it is customary to separate this movement into two parts, i.e., the elevation of the upper limb 90° to the horizontal (due to movement at the shoulder joint) and the continued raising of the arm overhead from 90° to 180° and beyond (due to rotation of the scapula), such a differentiation is artificial. In fact, during practically all of both parts of this movement, the ratio of motion in the two articulations is two parts glenohumeral to one part scapulothoracic. If motion in the glenohumeral articulation is destroyed by ankylosis, only one-third of the whole movement, i.e., 60° of the 180°, will be retained as the compensatory motion of the scapulothoracic articulation.

Special Anatomical Features. The most striking peculiarities of the shoulder joint are: (1) the large size of the head of the humerus when compared with the depth of the glenoid cavity, even though the cavity is deepened by the glenoidal labrum; (2) the looseness of the capsule of the joint; (3) the intimate relationship of the capsule with the muscles attached to the head of the humerus; (4) the peculiar association of the tendon of the long head of the biceps brachii with the joint.

It is in consequence of the *relative sizes of the two articular surfaces and the looseness of the articular capsule* that the joint enjoys such free movement in all directions. The upper limb has considerably increased mobility by the movements of the scapula, which involve the acromioclavicular and sternoclavicular joints as well. These joints are regarded as accessory structures to the shoulder joint. The movements of the scapula are considerable, especially in extreme elevation of the upper limb, a movement best accomplished when the arm is thrown somewhat forward and outward because the margin of the head of the humerus is ovoid and not circular. The greatest diameter of the humeral head is directed downward, medialward, and backward from the intertubercular groove, and the greatest elevation of the arm can be obtained by rolling its articular surface in the direction of this diameter. When the upper limb is elevated 90° to a right angle from the trunk, the great width of the central portion of the humeral head also allows free horizontal movement. In this movement the arch formed by the acromion, the coracoid process, and the coracoacromial ligament may be conceived as a supplemental articular cavity for the head of the humerus.

The *looseness of the capsule* is such that the humeral head falls away from the scapula about 2.5 cm when the muscles are dissected from around the joint. The articular surfaces of the two bones are held in contact, not so much by the capsule as by the surrounding muscles. This arrangement allows easy movement at the joint, especially when the muscles are not under tension. In all ordinary positions of the joint, the capsule is not stretched, but extreme movements are checked by the tension of appropriate portions of the capsule, as well as by certain ligaments and muscles.

The *intimate union of the tendons* of the supraspinatus, infraspinatus, teres minor, and subscapularis, in the formation with the capsule of a musculotendinous or rotary cuff, converts these muscles into elastic and spontaneously acting ligaments of the joint.

The *peculiar relationship of the tendon of the long head of the biceps brachii to the shoulder joint* appears to serve various purposes: (1) by its action at both the shoulder and elbow, the muscle harmonizes the motion at the two joints, and performs as a muscular ligament in all positions; (2) it strengthens the upper part of the articular cavity and prevents the head of the humerus from being pressed against the acromion when the deltoid contracts, and in this manner fixes the head of the humerus as the center of motion in the glenoid cavity; and (3) by the passage along the intertubercular groove, the tendon assists in steadying the head of the humerus in various other movements of the arm. When the upper limb is raised from the side, it assists the supraspinatus and infraspinatus in rotating the head of the humerus in the glenoid cavity. The long biceps tendon also holds the head of the humerus firmly in contact with the glenoid cavity, and prevents its slipping or being displaced by the action of the latissimus dorsi and pectoralis major, as in climbing and many other movements.

Clinical Relevance. **Dislocation at the shoulder joint** occurs more frequently than at any other joint in the body. It usually involves the dislodgement of the head of the humerus inferiorly in the *subglenoid position,* where the capsule is least protected. At the subglenoid site, the triceps muscle usually prevents the humeral head from pushing backward, and thus secondarily, the bone thrusts anteriorly into a subcoracoid or more rarely a subclavicular site in the axilla. At times, the subglenoid dislocation may in fact be secondarily pushed posteriorly to a subacromial or subspinous site.

Rotation of the scapula is of special importance in ankylosis of the shoulder joint because scapular mobility allows some control of the humerus, thereby compensating to some extent for the immobility at the shoulder joint.

ELBOW JOINT

In this description, the elbow joint will be considered as that articulation between the lower end of the humerus and the proximal ends of the ulna and radius. Although it is appreciated that within the capsule of the elbow joint is also located the proximal radioulnar articulation, this latter joint will be considered in the next section with the other radioulnar joints.

The **elbow joint** is a compound ginglymus or hinge joint at which are articulated three bones—the humerus, the radius, and the ulna (Figs. 5-39 to 5-42). The **trochlea** of the humerus fits into the **trochlear notch** of the ulna to form a *humeroulnar articulation,* and the **capitulum** of the humerus articulates with the slightly cupped proximal surface on the **head** of the radius to form a *humeroradial articulation.* The articular surfaces are bound together by a **capsule,** which surrounds the joint and is thickened medially and laterally into the **ulnar collateral** and the **radial collateral** ligaments.

Ligaments. Thus, the ligaments to be considered are:

Articular capsule
Ulnar collateral
Radial collateral

The **articular capsule** completely invests the joint (Figs. 5-40 to 5-42). Its **anterior part** is a thin fibrous layer attached *above* to the front of the medial epicondyle and to the front of the humerus immediately above the coronoid and radial fossae. *Below*, it adheres to the anterior surface of the coronoid process of the ulna and to the annular ligament of the proximal radioulnar joint. *At the sides,* the anterior part of the capsule is continuous with the ulnar and radial collateral ligaments. Its superficial fibers pass obliquely from the medial epicondyle of the humerus to the annular ligament. The middle fibers are directed vertically, and pass from the proximal part of the coronoid fossa of the humerus to insert mainly onto the anterior surface of the coronoid process of the ulna. The deeper fibers course transversely and intersect the others at right an-

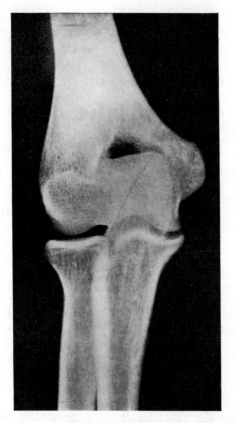

FIG. 5-39. Anteroposterior roentgenogram of an adult elbow joint. The shadow of the olecranon extends upward to the olecranon fossa and is superimposed on the outline of the trochlea. To the left, the head of the radius approximates the capitulum of the humerus, with the humeroradial gap intervening.

gles. The brachialis muscle covers most of the anterior part of the articular capsule, its deep fibers blending with those of the capsule.

The **posterior part** of the articular capsule is thin and membranous, and consists of transverse and oblique fibers (Fig. 5-42). It is attached *above* to the humerus immediately behind the capitulum and close to the medial margin of the trochlea, to the margins of the olecranon fossa, and to the back of the lateral epicondyle, a short distance from the trochlea. Below, it is fixed to the proximal and lateral margins of the olecranon, to the posterior part of the annular ligament, and to the ulna posterior to the radial notch. The transverse fibers of the capsule form a strong band, which bridges the olecranon fossa. Beneath the band is a pouch of synovial membrane and a fat pad, which project into the upper part of the fossa when the joint is

extended. The capsule is related posteriorly to the tendon of insertion of the triceps brachii and to the anconeus muscle.

The **ulnar collateral ligament** is a thick triangular band consisting of two portions, an anterior and posterior, united by a thinner intermediate portion (Fig. 5-40). The **anterior portion** is directed obliquely forward. It is attached *above* by its apex to the front of the medial epicondyle of the humerus and *below* by its broad base to the medial margin of the coronoid process. The **posterior portion,** also triangular in shape, is attached *above* to the lower and posterior part of the medial epicondyle and *below* to the medial margin of the olecranon. Between these two bands a few intermediate fibers descend from the medial epicondyle to blend with a **transverse band** that stretches between the attachments of the anterior and posterior parts on the olecranon and coronoid process of the ulna. Adjacent to this ligament are the triceps brachii and flexor carpi ulnaris muscles *and the ulnar nerve.* Its anterior portion gives origin to part of the flexor digitorum superficialis muscle.

The **radial collateral ligament** (Fig. 5-41) is a short, narrow, triangular-shaped fibrous band that is somewhat less distinct than the ulnar collateral ligament. It is attached *above* by its apex to a depression distal to the lateral epicondyle of the humerus and *below* to the annular ligament. Some of its most posterior fibers pass over this ligament to be inserted into the lateral margin of the ulna. It is intimately blended with the origins of the supinator and extensor carpi radialis brevis muscles.

Synovial Membrane (Figs. 5-43; 5-44). The synovial membrane extends from the margin of the articular surface of the humerus and lines the coronoid, radial, and olecranon fossae on that bone. It is reflected over the inner surface of the articular capsule and forms a pouch between the radial notch of the ulna, the deep surface of the annular ligament, and the circumference of the head of the radius. A crescentic fold of synovial membrane projects between the radius and ulna into the joint cavity from behind, partially dividing the joint into humeroradial and humeroulnar portions. The synovial membrane projects distal to the annular ligament as the **sacciform recess** (Fig. 5-43).

Between the fibrous capsule and the synovial membrane are three fat pads. The largest, over the olecranon fossa, is pressed into the fossa by the triceps brachii during the flexion. A second, over the

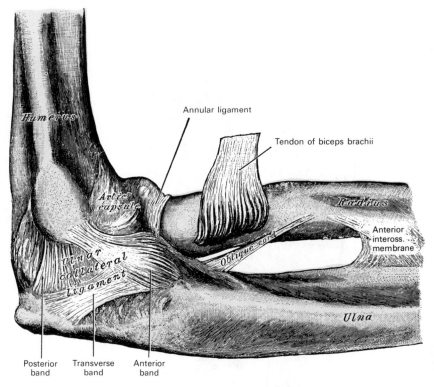

FIG. 5-40. The left elbow joint. Medial aspect.

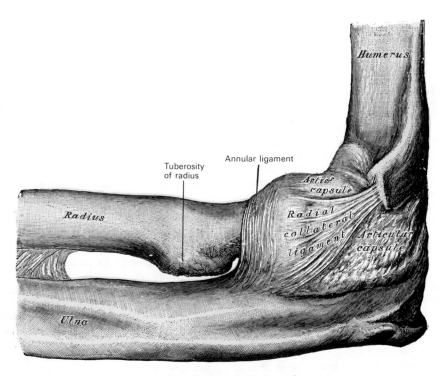

FIG. 5-41. The left elbow joint. Lateral aspect.

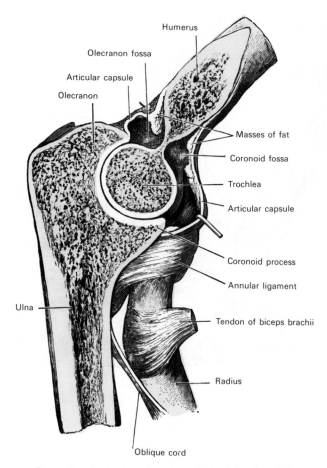

Humerus

Olecranon fossa

Articular capsule

Olecranon

Masses of fat

Coronoid fossa

Trochlea

Articular capsule

Coronoid process

Annular ligament

Ulna

Tendon of biceps brachii

Radius

Oblique cord

FIG. 5-42. Sagittal section through the left elbow joint.

coronoid fossa, and a third, over the radial fossa, are pressed by the brachialis into their respective fossae during extension.

Muscles. In relationship to the elbow joint are the following muscles: *anteriorly,* the brachialis; *posteriorly,* the triceps brachii and anconeus; *laterally,* the supinator, and the common tendon of origin of the extensor muscles; and *medially,* the common tendon of origin of the flexor muscles, including the flexor carpi ulnaris.

Blood Supply and Nerves. The **arteries** supplying the joint are derived from an anastomosis involving the following *descending vessels:* the radial and middle collateral branches of the profunda brachial artery and the superior and inferior ulnar collateral branches from the brachial artery; as well as the following *ascending vessels:* the anterior and posterior ulnar recurrent branches from the ulnar artery, the interosseous recurrent branch from the posterior interosseous artery, and the radial recurrent branch of the radial artery.

The **nerve supply** to the elbow joint is derived from the *musculocutaneous nerve* by way of its

branch to the brachialis muscle, and from the *radial nerve* by way of its branch to the medial head of the triceps muscle and the branch to the anconeus muscle. Additionally, the *ulnar and median nerves* frequently send filaments to the joint.

Movements. The elbow joint comprises three different articular appositions. These are the humeroulnar and humeroradial appositions, and the proximal radioulnar joint. All of these articular surfaces are enveloped by a continuous synovial membrane, and the movements within the entire joint should be studied together. The coordinated movements of flexion or extension of the forearm, with those of pronation or supination of the hand, are ensured by the two being performed at the same joint. This coordination is essential for the accuracy of the various minute movements of the hand.

The portion of the joint between the humerus and ulna is a simple hinge joint, and allows movements of flexion and extension only. Owing to the obliquity of the trochlea of the humerus, this movement does not take place in the anteroposterior plane of the humeral shaft. When the forearm is extended and

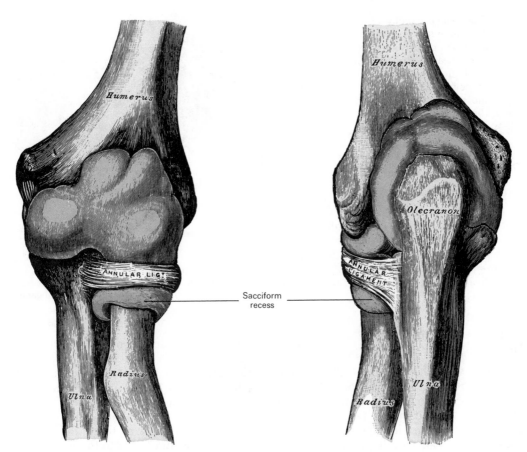

Sacciform recess

FIG. 5-43. Synovial membrane of the left elbow joint. Anterior aspect. The articular cavity was distended and the fibrous capsule removed. Observe the sacciform recess protruding distal to the annular ligament.

FIG. 5-44. Synovial membrane of the left elbow joint. Posterior aspect of the specimen demonstrated in Figure 5-43. Note the limit to which the synovial cavity extends in the olecranon fossa beyond the bony prominence of the olecranon.

supinated, the arm and forearm are not in the same line. Instead, they form an obtuse angle called the *carrying angle,* so that the hand and forearm are directed somewhat laterally to the central axis of the humerus. During flexion at the elbow joint, however, the forearm comes more into line with the central axis of the arm, enabling the hand to be easily carried to the face. The accurate adaptation of the prominences and depressions of the trochlea of the humerus to the semilunar notch of the ulna prevents any lateral movement at the elbow joint. **Flexion** is produced by the action of the biceps brachii and brachialis, assisted by the brachioradialis and the muscles arising from the medial epicondyle of the humerus. **Extension** of the forearm is performed principally by the triceps brachii and anconeus, and to a minor extent by the extensors of the wrist, the extensor digitorum, and the extensor digiti minimi.

The joint between the head of the radius and the capitulum of the humerus is a gliding joint. The bony surfaces are so shaped that they would constitute a spheroid or ball-and-socket joint and allow movement in all directions, were it not for the annular

ligament by which the head of the radius is bound to the radial notch of the ulna. However, this ligament prevents any separation of the two bones laterally and, in fact, gives the head of the radius its security from dislocation, which would otherwise tend to occur because of the shallowness of the cup-like surface on the head of the radius. Were it not for the annular ligament, the tendon of the biceps brachii would be able to pull the head of the radius out of the joint. The radial head is not in complete contact with the capitulum of the humerus in all positions of the joint, so that in complete extension a part of it can be plainly felt projecting at the back of the articulation. In full flexion, movement of the radial head is somewhat hampered by compression of the surrounding soft parts. Thus, the freest rotatory movement of the radius on the humerus (pronation and supination) takes place with the forearm in semiflexion, when the two articular surfaces are in most intimate contact.

The radius, carrying the hand with it, can be rotated in the proximal radioulnar joint when the forearm is in any degree of flexion, or even if it is fully extended. The hand is directly articulated only to the lower surface of the radius, and the ulna is excluded from the wrist joint directly by an articular disc. Additionally, the lower end of the ulna, on its lateral side, is firmly bound to the radius in the ulnar notch. Thus, when the head of the radius rotates, it imparts a circular movement to the hand through a considerable arc. This allows supination and pronation of the hand.

RADIOULNAR JOINTS

The articulations of the radius with the ulna occur at the proximal and distal extremities of the bones by means of synovial joints and along their shafts by means of a fibrous interosseous ligament which firmly unites the two bones in the form of a syndesmosis. Three joints interconnect the bones of the forearm:

Proximal radioulnar joint
Middle radioulnar union
Distal radioulnar joint

PROXIMAL RADIOULNAR JOINT. This articulation is a trochoid or pivot joint between the circumference of the head of the radius and the ring formed by the radial notch of the ulna and the *annular ligament*.

Ligaments. The **annular ligament** (Figs. 5-45; 5-46) is a strong band of fibers which encircles the head of the radius and retains it in contact with the radial notch of the ulna. It forms about four-fifths of the osseofibrous ring, and is attached to the anterior and posterior margins of the notch. For a short distance distal to the radial notch, a few fibers of the annular ligament actually form a com-

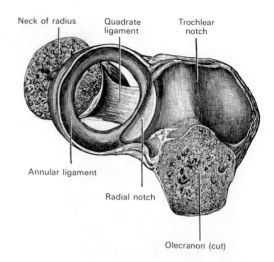

Fig. 5-46. Annular ligament of the proximal radioulnar joint, superior aspect. The head of the radius has been removed and the bone dislodged from the ligament.

plete fibrous ring. Proximally, the annular ligament blends with the anterior and posterior aspects of the articular capsule of the elbow joint, while from its distal border a thin loose membrane is attached to the neck of the radius. A thick band that extends from the inferior border of the annular ligament below the radial notch to the neck of the radius is known as the **quadrate ligament.** The superficial surface of the annular ligament is strengthened by the radial collateral ligament of the elbow (Fig. 5-41), and affords origin to part of the supinator muscle. Its deep surface is smooth and lined by synovial membrane that is continuous with that of the elbow joint.

Movements. The movements allowed in this articulation are limited to rotatory movements of the head of the radius within the ring formed by the annular ligament and the radial notch of the ulna. Rotation of the radius, which moves the thumb from lateral to medial, is called *pronation;* rotation in the opposite direction is *supination.* Supination is performed by the biceps brachii and supinator. Pronation is performed by the pronator teres and pronator quadratus.

MIDDLE RADIOULNAR UNION. The shafts of the radius and ulna are held together in syndesmosis by two interosseous ligaments.

Ligaments. The ligaments of this union are:

Interosseous membrane
Oblique cord

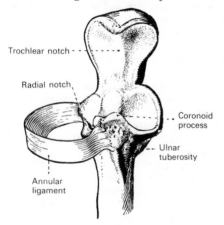

Fig. 5-45. Proximal part of the ulna with the annular ligament of the proximal radioulnar joint. (From Benninghoff and Goerttler, *Lehrbuch der Anatomie des Menschen,* Urban and Schwarzenberg, Munich, 1980).

The **interosseous membrane** (Fig. 5-40) is a broad and thin plane of fibrous tissue oriented downward and medially from the entire length of the interosseous crest of the radius to that of the ulna. The lower part of the membrane is attached to the posterior of the two ridges into which the interosseous crest of the radius divides distally. The membrane is deficient proximally, and originates about 2.5 cm below the tuberosity of the radius. It is broader in the middle than at either end, and presents an oval aperture a little above its distal margin for the passage of the anterior interosseous vessels to the back of the forearm. This membrane serves to increase the extent of surface for the attachment of the deep muscles as well as to connect the bones. Between its upper border and the oblique cord is a gap through which the posterior interosseous vessels pass. Although the direction of the fibrous bundles in the membrane is oblique, extending from the radius proximally to the ulna distally, there are two or three bands occasionally found on the dorsal surface which course obliquely in the opposite direction. *Anteriorly,* the upper three-fourths of the interosseous membrane is related to the flexor pollicis longus on the radial side, and to the flexor digitorum profundus on the ulnar. Lying in the interval between these two muscles are the anterior interosseous vessels and nerve. Over the lower one-fourth of the anterior surface stretches the pronator quadratus. *Posteriorly,* the interosseous membrane is related to the supinator, abductor pollicis longus, extensor pollicis brevis, extensor pollicis longus, and extensor indicis muscles and, near the wrist, with the anterior interosseous artery and posterior interosseous nerve.

The **oblique cord** (Figs. 5-40; 5-42) is a small, flattened band extending inferolaterally from the lateral aspect of the tuberosity of the ulna at the base of the coronoid process to the radius, a little distal to the radial tuberosity. Its fibers run in a direction opposite those of the interosseous membrane. It is sometimes absent.

DISTAL RADIOULNAR JOINT. This pivot joint is formed between the convex head of the ulna and the grooved ulnar notch on the lower end of the radius. The articular surfaces are enclosed by a relatively loose articular capsule that is lined by a synovial membrane. The joint is further secured by an articular disc.

Ligaments. The **articular capsule** (Figs. 5-47; 5-48) is composed of transversely oriented bands of fibers attached to the margins of the ulnar notch and to the head of the ulna. Although the anterior and posterior parts of the capsule are somewhat thickened, it is still weak and loosely surrounds the joint. The capsule is lined by a synovial membrane that extends as a pouch upward a short distance between the radius and ulna to form a **sacciform recess.**

The **articular disc** (Fig. 5-49) is triangular in shape, and is placed transversely beneath the head of the ulna, binding the lower ends of the ulna and radius firmly together. Its periphery is thicker than its center, which is occasionally perforated. It is attached by its *apex* to a depression between the styloid process and the head of the ulna, and by its thin *base* to the prominent edge of the radius, which separates the ulnar notch from the carpal articular surface. Its margins are united to the ligaments of the wrist joint. Its **proximal surface** is smooth and concave, and articulates with the head of the ulna to form a gliding joint. Its **distal surface** forms part of the wrist joint and articulates with the medial part of the lunate bone and, when the hand is adducted, with the triquetral bone.

Synovial Membrane (Fig. 5-49). Both proximal and distal surfaces of the articular disc are covered with a synovial layer. The proximal surface of the disc enters into the radioulnar joint, and the joint cavity that is formed appears L-shaped in longitudinal section. Its synovial membrane extends proximally beyond the articular surfaces of the radius and ulna to form the **sacciform recess.** The distal surface of the disc enters into the formation of the wrist joint, and it is also lined by a synovial membrane.

Vessels and Nerves. The **arteries** to the distal radioulnar joint are derived from the anterior and posterior interosseous arteries. The **nerves** come from the anterior interosseous branch of the median nerve and the posterior interosseous branch of the radial nerve.

Movements. The movements at the distal radioulnar joint consist of rotation of the distal end of the radius around a longitudinal axis that passes through the center of the head of the ulna. When the radius and the attached hand rotate obliquely across the anterior aspect of the ulna, **pronation** of the forearm and hand is the result. The reverse movement results in **supination,** and in the fully supine position, the radius and ulna lie parallel in the forearm. Thus, in pronation and supination the radius rotates around the circumference of a segment of a cone, *the axis of which extends from the center of*

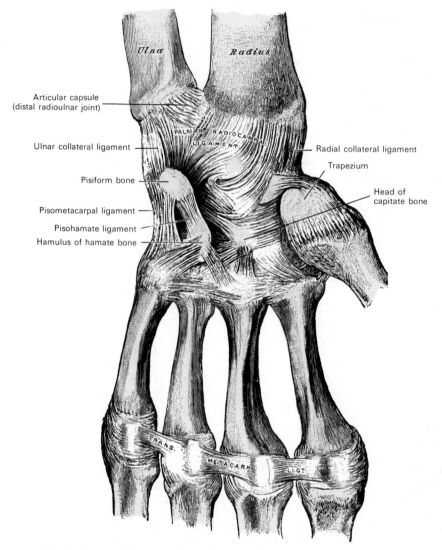

Ulna

Radius

Articular capsule
(distal radioulnar joint)

PALMAR RADIOCARPAL
LIGAMENT

Ulnar collateral ligament

Radial collateral ligament

Pisiform bone

Trapezium

Pisometacarpal ligament

Head of
capitate bone

Pisohamate ligament

Hamulus of hamate bone

TRANS.

METACARP.

LIG.T

FIG. 5-47. The ligaments of the left wrist and metacarpus. Palmar aspect.

*the head of the radius to the middle of the head of
the ulna.* In this movement the head of the ulna is
stationary, but appears to describe a curve in a direction opposite that taken by the head of the radius.

RADIOCARPAL OR WRIST JOINT

The **wrist joint** (Figs. 5-47 to 5-49) is a simple synovial articulation, usually classified
as an ellipsoid joint. The parts forming it are
the distal end of the radius and lower surface of the articular disc above, and the
scaphoid, lunate, and triquetral bones
below. The articular surface of the radius
and the lower surface of the articular disc
form together a transversely elliptical concavity. The superior articular surfaces of the
scaphoid, lunate, and triquetral bones form
a smooth convex surface that is received
into the concavity. The ulna is excluded
from the joint by the intervening articular
disc. The joint is surrounded by an articular
capsule.

Ligaments.   The ligaments of the wrist
include:

Articular capsule
Palmar radiocarpal
Dorsal radiocarpal
Ulnar collateral
Radial collateral

The **articular capsule** is lined by a synovial layer; it is extensive and loosely binds

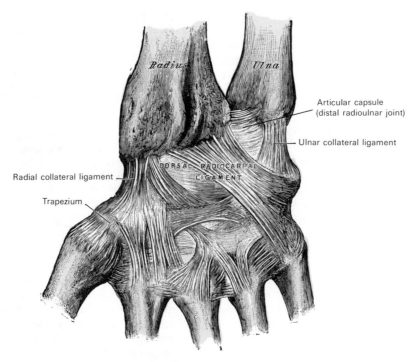

FIG. 5-48. The ligaments of the left wrist. Dorsal aspect.

the joint. The capsule is reinforced by the palmar and dorsal radiocarpal ligaments and the ulnar and radial collateral ligaments, which actually cover the greater part of its external surface.

The **palmar radiocarpal ligament** (Fig. 5-47) is a broad membranous band attached

above to the anterior margin of the lower end of the radius, to its styloid process, and to the palmar aspect of the lower end of the ulna. Its fibers pass downward and medially to be inserted *below* into the palmar surfaces of the scaphoid, lunate, and triquetral bones, some being continued to the capitate. In

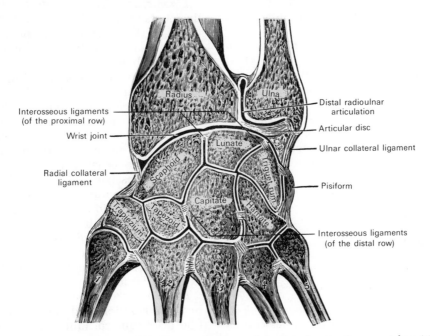

FIG. 5-49. Vertical section through the articulations at the wrist, showing the synovial cavities.

addition to this broad membrane, there is a rounded fasciculus, superficial to the rest, which reaches from the base of the styloid process of the ulna to the lunate and triquetral bones. The ligament is perforated by apertures for the passage of vessels. Its palmar surface is related *anteriorly* to the tendons of the flexor digitorum profundus and flexor pollicis longus, and *posteriorly* this ligament is closely adherent to the palmar surface of the articular disc in the distal radioulnar joint.

The **dorsal radiocarpal ligament** is thin and membranous (Fig. 5-48), and it is weaker than the palmar radiocarpal ligament. It is attached *above* to the posterior border of the lower end of the radius. Its fibers, directed obliquely downward and medially, are fixed *below* to the dorsal surfaces of the scaphoid, lunate, and triquetral bones, and are continuous with those of the dorsal intercarpal ligaments. *Posteriorly*, the ligament is related to the extensor tendons of the fingers, while *anteriorly*, it is blended with the articular disc.

The **ulnar collateral ligament** (Figs. 5-47 to 5-49) is a fan-shaped cord with its apex attached *above* to the end of the styloid process of the ulna. *Below*, it divides into two fasciculi, one of which is attached to the medial aspect of the triquetral bone, the other to the pisiform bone and transverse carpal ligament. The tendon of the flexor carpi ulnaris lies anterior to the ligament, while that of the extensor carpi ulnaris lies posterior.

The **radial collateral ligament** (Figs. 5-47 to 5-49) extends from the tip of the styloid process of the radius to the radial side of the scaphoid, with some of its fibers being prolonged to the trapezium and the flexor retinaculum. It is related to the radial artery, as the vessel curves around the lateral aspect of the wrist, separating the ligament from the tendons of the abductor pollicis longus and extensor pollicis brevis.

Synovial Membrane (Fig. 5-49). The synovial layer lines the inner surfaces of the ligaments described above, extending from the margin of the lower end of the radius and articular disc to the margins of the articular surfaces of the carpal bones below. It is loose and lax, and presents numerous folds, especially dorsally.

Vessels and Nerves. The **arteries** supplying the joint are branches from the anterior interosseous ar-

tery as well as palmar and dorsal carpal branches of the radial and ulnar, the palmar and dorsal metacarpals, and some ascending branches from the deep palmar arch. The **nerves** to the wrist joint are derived from all three principal nerves of the upper extremity. These are the anterior interosseous branch of the median, the posterior interosseous branch of the radial, and the dorsal and deep branches of the ulnar.

Movements. The movements permitted in this joint are flexion, extension, abduction, adduction, and circumduction. They will be described with those of the carpus, with which they are combined. The wrist joint is covered in front by the flexor tendons and dorsally by the extensor tendons.

INTERCARPAL JOINTS

The intercarpal joints are articulations between the carpal bones and are subdivided into three groups: (1) the joints between the bones of the **proximal row,** that is, the joints connecting the scaphoid, lunate, and triquetral bones, and the articulation of the pisiform bone; (2) the joints between the bones of the **distal row;** and (3) the transverse articulation between the two rows of carpal bones called the **midcarpal joint.**

INTERCARPAL JOINTS OF THE PROXIMAL ROW: SCAPHOID, LUNATE, AND TRIQUETRAL. These are gliding joints, and the *scaphoid, lunate,* and *triquetral bones* are interconnected by dorsal, palmar, and interosseous ligaments.

Ligaments. The two **dorsal intercarpal ligaments** are placed transversely behind the bones of the first row. They connect the scaphoid and lunate, and the lunate and triquetral bones.

The two **palmar intercarpal ligaments** interconnect the scaphoid and lunate, and the lunate and triquetral bones on the anterior aspect of the carpus. They are not so strong as the dorsal intercarpal ligaments, and are placed deep to the flexor tendons and the palmar radiocarpal ligament.

The **interosseous ligaments** (Fig. 5-49) are two narrow bundles, one connecting the lunate and scaphoid bones, and the other joining the lunate and triquetral bones. They are on a level with the proximal surfaces of these bones, and their surfaces are smooth, forming part of the convex articular surface of the wrist joint.

ARTICULATION OF THE PISIFORM BONE. The pisiform is articulated to the triquetral bone (the pisotriquetral joint) by a thin **articular**

capsule, the synovial cavity of which remains separate and distinct from those of the other carpal bones.

Ligaments. Additionally, the pisiform bone is firmly connected proximally by the ulnar collateral and the palmar radiocarpal ligaments (Fig. 5-47). Extending distally from the pisiform bone are two strong fibrous bands, the pisohamate and pisometacarpal ligaments.

The **pisohamate ligament** attaches the pisiform bone to the hook of the hamate bone, while the **pisometacarpal ligament** connects the pisiform to the base of the fifth metacarpal bone (Fig. 5-47). These ligaments transmit the pull of the flexor carpi ulnaris tendon from the pisiform to the distal carpal bones and to the metacarpals, and may be considered as prolongations of that tendon.

INTERCARPAL JOINTS OF THE DISTAL ROW. These also are gliding joints, and the four bones of this row are connected by dorsal, palmar, and interosseous ligaments.

Ligaments. The **dorsal intercarpal ligaments** are three in number and they extend transversely from one bone to another on the dorsal surface, connecting the trapezium with the trapezoid, the trapezoid with the capitate, and the capitate with the hamate.

The **palmar intercarpal ligaments,** also three in number, have a similar arrangement on the palmar surface.

The **interosseous ligaments** of the distal row are much thicker than those of the proximal row. There are three of these, one is placed between the capitate and the hamate, the second between the capitate and the trapezoid, and the third between the trapezium and trapezoid (Fig. 5-49). The first is much the strongest, and the third is sometimes absent.

MIDCARPAL JOINT. The transverse articulation between the scaphoid, lunate, and triquetral bones proximally, and the second row of carpal bones distally, is the important **midcarpal joint.** It is a compound joint and consists of three distinct portions. In the center, the head of the capitate and the proximal angle on the superior surface of the hamate articulate in a deep concavity formed by the scaphoid and lunate to constitute a compound saddle joint. On the radial side of the capitate, the trapezium and trapezoid articulate with the scaphoid, and on the ulnar side the more medial aspect of the hamate articulates with the triquetral bone. These

latter joints are usually considered to be of the plane or gliding type.

Ligaments. The midcarpal joint is secured by **palmar and dorsal intercarpal** and **ulnar and radial carpal collateral ligaments.**

The **palmar intercarpal ligaments** consist of short fibers that pass, for the most part, from the palmar surface of the head of the capitate bone in a radiating manner to the surrounding carpal bones of the proximal row (scaphoid, lunate, and triquetral). Collectively, these ligaments are frequently referred to as the **carpal radiate ligament.**

The **dorsal intercarpal ligaments** consist of short, irregular bundles passing between the dorsal surfaces of the bones of the first and second rows.

The **carpal collateral ligaments** are short; one is placed on the radial, the other on the ulnar side of the carpus. The radial carpal collateral ligament is the stronger and more distinct, and connects the scaphoid and trapezium, while the ulnar carpal collateral ligament joins the triquetral bone and hamate. They are continuous with the collateral ligaments of the wrist joint.

In addition to these ligaments, a slender interosseous band sometimes connects the capitate and the scaphoid.

Synovial Membrane. The synovial membrane of the carpus is extensive (Fig. 5-49), and bounds a synovial cavity of irregular shape. The proximal portion of the cavity intervenes between the distal surfaces of the scaphoid, lunate, and triquetral bones and the proximal surfaces of the bones of the second row. It sends *two proximal prolongations*—between the scaphoid and lunate, and the lunate and triquetral bones—and *three distal prolongations,* each of which intervenes between adjacent bones of the second row. The prolongation of the synovial membrane between the trapezium and trapezoid or that between the trapezoid and capitate, because of the absence of interosseous ligaments, is often continuous with the cavity of the carpometacarpal joints. At times this continuity is seen with the joint cavity between the carpal bones and the second, third, fourth, and fifth metacarpal bones; at other times, with that of the second and third metacarpal bones only. In the latter condition the joint between the hamate and the fourth and fifth metacarpal bones has a separate synovial membrane, and is separated from the rest of the carpometacarpal joint cavity by an interosseous ligament (Fig. 5-49). The synovial cavities of the carpometacarpal joints are prolonged for a short distance between the bases of the metacarpal bones. There is a separate synovial membrane between the pisiform and triquetral bones.

Movements. The radiocarpal and midcarpal joints involve four articular surfaces: (1) the distal surfaces of the radius and articular disc; (2) the proximal surfaces of the scaphoid, lunate, and triquetral bones, the pisiform having no essential part in the movement of the hand; (3) the S-shaped inferior surfaces of the scaphoid, lunate, and triquetral, and (4) the reciprocal upper surfaces of the carpal bones of the second row. These four surfaces form two principal joints: (1) a proximal, **the wrist joint** proper, and (2) a distal, **the midcarpal joint** (Fig. 5-50).

1. The **wrist joint** proper is an ellipsoid articulation, permitting all movements but rotation to occur. *Flexion* and *extension* are the most free, and of these a greater degree of extension than of flexion of the hand is permitted, because the articulating surfaces extend farther on the dorsal than on the palmar surfaces of the carpal bones. In these movements the carpal bones swing on a transverse axis drawn between the tips of the styloid processes of the radius and ulna. A certain degree of *adduction* (or ulnar flexion) and *abduction* (or radial flexion) is also permitted. A considerably greater degree of adduction can be achieved because of the more proximal position of the styloid process of the ulna, abduction being quite limited by contact of the styloid process of the radius with the trapezium. Abduction and adduction take place about an anteroposterior axis drawn through the capitate bone. Finally, *circumduction* is permitted by the combined and consecutive movements of adduction, extension, abduction, and flexion. Although rotation does not occur at the wrist or midcarpal joints, the effect of rotation is obtained by pronation and supination of the radius on the ulna.

2. At the **midcarpal joint,** flexion, some extension, abduction, and a minor degree of rotation are permitted. In *flexion* and *extension,* the trapezium and trapezoid on the radial side and the hamate on the ulnar side glide forward and backward on the scaphoid and triquetral bones, respectively. Of these two movements, flexion is more freely enjoyed at the midcarpal joint. *Abduction* is accompanied by a slight *rotation* of the head of the capitate and the superior surface of the hamate in the cup-shaped cavity of the scaphoid and lunate.

Muscles Producing Movements. The movement of **flexion** of the hand is performed by the flexor carpi radialis, the flexor carpi ulnaris, and the palmaris longus when the fingers are extended. When the fingers are flexed, these muscles are aided by the flexors digitorum superficialis and profundus, flexor pollicis longus, and abductor pollicis longus. **Extension** at the wrist is performed by the extensors carpi radiales longus and brevis and the extensor carpi ulnaris when the fingers are flexed. When the fingers are extended, these muscles are aided by the extensor digitorum, extensor digiti minimi, extensor indicis, and extensor pollicis longus. **Adduction** (ulnar flexion) is performed by the flexor carpi ulnaris and the extensor carpi ulnaris, while **abduction** (radial flexion) is produced by the extensor carpi radialis longus and brevis and the flexor carpi radialis.

Vessels and Nerves. The **arteries** supplying the intercarpal joints are derived from a palmar and dorsal carpal anastomosis contributed to by (1) the palmar and dorsal carpal branches of the radial and ulnar arteries; (2) carpal recurrent branches of the deep volar arch; and (3) the dorsal carpal branches of the anterior (and posterior) interosseous arteries.

The **nerve supply** to the intercarpal joints comes from carpal branches of the ulnar, median, and posterior interosseous nerves.

CARPOMETACARPAL JOINTS

The distal surfaces of the distal row of carpal bones interface with the expanded surfaces of the bases of the five metacarpal bones. These five joints may be divided into two sets: (1) carpometacarpal joint of the thumb, and (2) carpometacarpal joints of the four medial fingers.

CARPOMETACARPAL JOINT OF THE THUMB (Figs. 5-47 to 5-50). This highly mobile joint

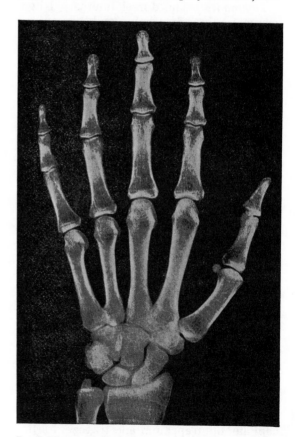

Fɪɢ. 5-50. Radiograph of adult hand showing the radiocarpal and midcarpal joints along with those of the fingers.

between the base of the first metacarpal bone and the trapezium is the best example in the body of a saddle (or sellar) joint. Its great freedom of movement results from the configuration of its articular surfaces. These are reciprocally saddle-shaped, and surrounded by a capsule that is thick but loose, and passes from the circumference of the base of the first metacarpal bone to the rough edge bounding the articular surface of the trapezium.

Ligaments. The capsule is thickest laterally and dorsally, and is reinforced by **lateral** (radial), **palmar** and **dorsal** (anterior and posterior oblique) **carpometacarpal ligaments of the thumb** (Haines, 1944). These ligaments are attached to the trapezium above and pass distally to converge onto the base of the first metacarpal bone.

Movements. When the thumb is in the resting anatomical position, its palmar surface is oriented medially and the dorsal surface directed laterally. At the carpometacarpal joint of the thumb, the movements permitted are *flexion* and *extension* in a plane parallel to the palm of the hand, *abduction* and *adduction* in a plane at right angles to the palm, a small degree of *rotation, circumduction,* and *opposition.* Flexion is generally accompanied by a slight medial rotation of the first metacarpal, while extension is accompanied by its lateral rotation. Medial rotation occurs because the *dorsal* carpometacarpal ligament tenses the ulnar side of the first metacarpal base during flexion, leaving the radial side of the capsule loose and the metacarpal head free to move. Likewise, during extension, lateral rotation of the metacarpal head results from tension on the ulnar side of the metacarpal base brought about by the *palmar* carpometacarpal ligament, once again allowing the more lax radial side of the metacarpal head freedom to move. Flexion and medial rotation, when combined by a partial abduction of the thumb, results in opposition. It is by the movement of opposition that the tip of the thumb is brought into contact with the palmar surfaces of the slightly flexed tips of the other fingers.

Muscles Producing Movements. **Flexion** at the carpometacarpal joint of the thumb is produced by the flexors pollicis brevis and longus, as well as the opponens pollicis. **Extension** of the thumb is effected by the extensors pollicis brevis and longus, as well as the abductor pollicis longus muscle. **Abduction** is produced by the abductors pollicis longus and brevis and to some extent by the opponens pollicis. **Adduction** is carried out by the adductor pollicis. **Opposition** is achieved by the combined action of the flexor pollicis brevis, opponens pollicis, and the abductors which flex, medially rotate, and partially abduct the thumb. In this manner the meta-

carpal bone is rotated to a position in which the palmar surface of the thumb faces the palmar surface of the fingers. After the thumb is in the opposed position, the strong pressure between thumb and fingers is produced by the long flexors. **Circumduction** occurs by the consecutive action of the flexors, abductors, extensors, and adductors.

Vessels and Nerves. The **arteries** to the carpometacarpal joint of the thumb are derived from the radial, the first palmar metacarpal, and the first dorsal metacarpal. The **nerves** come from branches of the median.

CARPOMETACARPAL JOINTS OF THE MEDIAL FOUR FINGERS. The joints between the carpus and the bases of the second, third, fourth, and fifth metacarpal bones are of the plane or gliding type. The bones are united by articular capsules, which are reinforced by dorsal, palmar, and interosseous carpometacarpal ligaments.

Ligaments. The **dorsal carpometacarpal ligaments** are the strongest and most distinct, and connect the carpal and metacarpal bones on their dorsal surfaces. The *second metacarpal* bone receives three fasciculi, one from each of the bones with which it articulates—the trapezium, the trapezoid, and the capitate. The *third metacarpal* receives two, one each from the trapezoid and the capitate; the *fourth* two, one each from the capitate and hamate. The *fifth metacarpal* receives a single fasciculus from the hamate, and this is continuous with a similar ligament on the palmar surface, and forms an incomplete capsule.

The **palmar carpometacarpal ligaments** have an arrangement a bit different from the dorsal. The *second metacarpal* receives one strong band from the trapezium. The *third metacarpal* is attached by three bands to the carpus: a lateral one from the trapezium, situated superficial to the sheath of the tendon of the flexor carpi radialis, an intermediate one from the capitate, and a medial one extending from the hamate. The *fourth* and *fifth metacarpals* receive one palmar ligament each from the hamate. That of the fifth metacarpal is strengthened by fibers extending from the flexor carpi ulnaris tendon.

The **interosseous carpometacarpal ligaments** (Fig. 5-49) consist of short, thick fibers which are limited to two parts of the carpometacarpal articulation. One interosseous ligament connects the distal edge of the trapezium with the lateral margin of the base of the second metacarpal bone. The other connects the contiguous inferior angles of the capitate and hamate with the adjacent surfaces of the third and fourth metacarpal bones.

Synovial Membrane. The synovial cavity is a continuation of that of the intercarpal joints. Occasionally, the joint between the hamate and the fourth and fifth metacarpal bones has a separate synovial cavity, being bound medially by one of the interosseous carpometacarpal ligaments (Fig. 5-49).

Movements. The movements permitted at the carpometacarpal articulations of the fingers are limited to slight gliding of the articular surfaces upon each other, the extent of which varies in the different joints. The metacarpal bone of the little finger has the greatest range of movement, more than that of the ring finger. The metacarpal bones of the index and middle fingers are almost immovable.

INTERMETACARPAL JOINTS

The metacarpal bone of the thumb is not articulated with any other metacarpal bone. In contrast, the bases of the second, third, fourth, and fifth metacarpal bones articulate with one another by small surfaces covered with cartilage, and are connected together by dorsal, palmar, and interosseous intermetacarpal ligaments.

Ligaments. The **dorsal metacarpal ligaments** are short fibrous bands that pass transversely from bone to bone on the dorsal surface of the bases of the metacarpal bones.

The **palmar metacarpal ligaments** are also short bands, somewhat less defined than the dorsal, which interconnect the bases of adjacent metacarpal bones on their palmar surface.

The **interosseous metacarpal ligaments** interconnect the contiguous surfaces of the bases of the metacarpal bones just distal to their intermetacarpal collateral articular facets.

Synovial Membranes. The synovial cavities of the intermetacarpal joints are continuous with those of the carpometacarpal articulations.

METACARPOPHALANGEAL JOINTS

These articulations are ellipsoid in type and are formed by the reception of the rounded heads of the metacarpal bones into shallow cavities on the proximal ends, or bases, of the first phalanges. On their dorsal surfaces the metacarpophalangeal joints are covered by the expansions of the extensor tendons, together with some loose areolar tissue which connects the inner aspect of the tendons of the bones. The joints form the prominent proximal knuckles of the clenched fist. On the palmar aspect of these joints, the rounded articular heads of the metacarpal bones are partly divided to appear similar to condyles. The palmar surfaces are crossed by the flexor tendons. The head of the metacarpal bone of the thumb is less convex, wider, and flatter than that of the other fingers; on each side are facets for sesamoid bones.

Ligaments. Each metacarpophalangeal joint has a loose articular capsule, and palmar and collateral ligaments. Additionally, the deep transverse metacarpal ligament interconnects the palmar surfaces of the metacarpophalangeal joints of the medial four fingers.

The **palmar ligaments** (Fig. 5-51) are thick, dense, fibrocartilaginous plates placed upon the palmar surfaces of the joints in the intervals between the collateral ligaments, to which they are connected. They are loosely united to the metacarpal bones, but are firmly attached to the bases of the first phalanges. Their *palmar surfaces* are intimately blended with the deep transverse metacarpal ligament, and present grooves for the passage of the flexor tendons. The fibrous sheaths of these tendons are connected to the sides of the grooves. Their *deep surfaces* form parts of the articular facets for the heads of the metacarpal bones.

The **collateral ligaments** (Fig. 5-52) are strong, rounded cords placed on the sides of the joints, reinforcing the articular capsule. Each is attached by its proximal extremity to the *dorsal* tubercle and adjacent depression on the side of the head of the metacarpal bone. The fibers of the ligament then run obliquely across the side of the joint to attach distally on the *palmar* aspect of the base of the proximal phalanx. During flexion of the metacarpal phalangeal joint, the collateral ligaments become tense, thereby preventing much abduction or adduction of the flexed finger.

The **deep transverse metacarpal ligament** (Figs. 5-47; 5-51) consists of a series of three narrow fibrous bands that connect the pal-

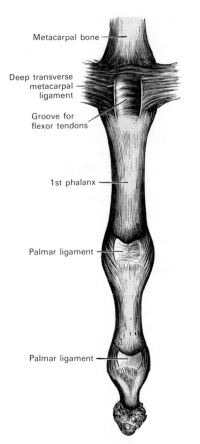

Metacarpal bone

Deep transverse metacarpal ligament

Groove for flexor tendons

1st phalanx

Palmar ligament

Palmar ligament

FIG. 5-51. The metacarpophalangeal and interphalangeal joints of a finger. Palmar aspect.

mar surfaces of the heads of the second, third, fourth, and fifth metacarpal bones. It is blended with the palmar ligaments of the metacarpophalangeal joints, and its palmar surface is concave where the flexor tendons pass. Behind the ligament, the tendons of the interossei pass to their insertions.

Movements. Movements that occur in joints are flexion, extension, adduction, abduction, and circumduction. Abduction and adduction are performed by the interossei, but these movements are exceedingly limited when the fingers are fully flexed.

INTERPHALANGEAL JOINTS

The interphalangeal articulations are purely hinge joints (Figs. 5-51; 5-52), and the arrangement of their ligaments is similar to that seen in the metacarpophalangeal articulations. Each joint is surrounded by a thin fibrous articular capsule, which is reinforced on its palmar surface by thick palmar

ligaments over which glide the flexor tendons. The dorsal aspect of the capsule is largely formed by the expanded hoods of the extensor tendon, which effectively replaces the dorsal ligament. On each side the joint is fixed by obliquely oriented collateral ligaments. There are two interphalangeal joints in each of the four lateral fingers and one in the thumb.

Movements. The only movements permitted in the interphalangeal joints are flexion and extension. These movements are more extensive between the proximal and middle phalanges than between the middle and distal. The degree of flexion is considerable, but extension is limited by ligamentous action of the flexor tendons and by the palmar and collateral ligaments.

Muscles Acting on the Metacarpophalangeal and Interphalangeal Joints. **Flexion** of the *metacarpophalangeal joints* of the four medial fingers is effected by the flexors digitorum superficialis and profundus, the lumbricals, and the interossei, assisted in the case of the little finger by the flexor digiti minimi. **Extension** of the metacarpophalangeal joint is produced by the extensor digitorum in all

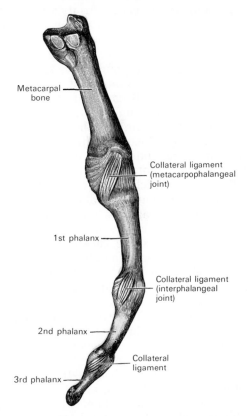

Metacarpal bone

Collateral ligament (metacarpophalangeal joint)

1st phalanx

Collateral ligament (interphalangeal joint)

2nd phalanx

Collateral ligament

3rd phalanx

FIG. 5-52. Metacarpophalangeal and interphalangeal joints of a finger viewed from the ulnar or medial aspect.

four medial fingers, assisted by the extensor indicis in the index finger and the extensor digiti minimi in the little finger.

Flexion of the *interphalangeal joints* of the fingers is accomplished by the flexor digitorum profundus acting on both the proximal and distal joints and by the flexor digitorum superficialis acting on the proximal joints. **Extension** of the interphalangeal joint is now believed to be the result of action by not only the lumbricals and interossei, which are intrinsic hand muscles, but the long extensor digitorum muscle as well.

Flexion of the *metacarpophalangeal joint of the thumb* is effected by the flexors pollicis longus and brevis, and **extension** is produced by the extensor pollicis longus and brevis.

Flexion of the *interphalangeal joint of the thumb* is accomplished by the flexor pollicis longus, and extension is produced by the extensor pollicis longus.

Articulations of the Lower Limb

The articulations of the lower limb include those forming the **pelvic girdle** and those of the free limb. The pelvis, which consists of the two hip bones interconnected at the **pubic symphysis,** is firmly attached to the axial skeleton by way of the **sacroiliac joints.** These structures, along with the connections of the sacrum, below to the coccyx **(sacrococcygeal joint)** and above to the fifth lumbar vertebra **(lumbosacral joint),** form the pelvic girdle. Because of these strong connections to the trunk by way of the vertebral column, the joints of the pelvic girdle were described along with those of the axial skeleton. What follows are descriptions of the remaining joints of the lower extremity.

The femur, which is the longest and largest bone in the body, articulates above with the pelvic girdle at the acetabulum to form the **hip joint.** Inferiorly, the femur joins the tibia and patella at the **knee joint,** which is the largest and most complicated of all the joints. Below the knee, the tibia and fibula are interconnected by **tibiofibular joints** in a manner somewhat similar to the bones in the forearm. At the **ankle** or **talocrural joint,** the tibia and fibula join the talus. In the foot, joints are located between the tarsal bones—**intertarsal joints,** and between the tarsal and metatarsal bones—**tarsometatarsal joints.** The metatarsal bones, in addition to articulating with the tarsus, interconnect with each other at the **intermetatarsal joints** and with the proximal phalanges of the toes—**metatarsophalangeal joints.** As in the hand, each toe consists of phalanges between which are located the **interphalangeal joints.**

HIP JOINT

This articulation is a ball-and-socket, or spheroid, joint formed by the reception of the head of the femur into the cup-shaped cavity of the acetabulum. Unlike the shoulder joints which depend on their strength from surrounding muscles, the hip joints are inherently strong, and their structure is well adapted to supporting the weight of the trunk as well as permitting a significant degree of mobility. The head of the femur forms more than half a sphere, and its articular cartilage, thicker at the center than at the circumference, covers its entire surface, except the shallow fovea of the femoral head, to which the ligament of the head of the femur is attached. The inner surface of the acetabulum is characterized by a horseshoe-shaped region, the **lunate surface,** which also is covered by articular cartilage. Within the lunate surface is a circular depression devoid of cartilage and occupied by fatty tissue covered by synovial membrane. To the rim of the acetabulum is attached a fibrocartilaginous lip, the acetabular labrum, which significantly deepens the socket, and the joint is encased in an articular capsule which is reinforced by ligaments.

Ligaments. The ligamentous structures of the hip joint are:

Articular capsule
Iliofemoral
Pubofemoral
Ischiofemoral
Ligament of the head of the femur
Acetabular labrum
Transverse acetabular

The **articular capsule** (Figs. 5-53; 5-54; 5-56) is strong and dense. It attaches to the pelvis **above.** *Posteriorly,* its pelvic attachment extends to the margin of the acetabulum 5 to 6 mm beyond the acetabular labrum. *Anteriorly,* it is attached to the outer margin of the labrum and, opposite the acetabular notch where the margin of the cavity is deficient, it is connected to the transverse acetabular ligament and to the edge of the obturator foramen. **Below,** the capsule surrounds the neck of the femur. This femoral attach-

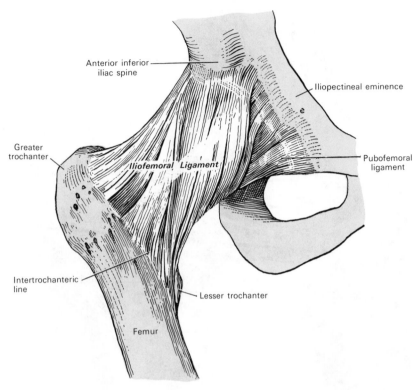

Anterior inferior iliac spine

Iliopectineal eminence

Greater trochanter

Iliofemoral Ligament

Pubofemoral ligament

Intertrochanteric line

Lesser trochanter

Femur

FIG. 5-53. Right hip joint from the front.

ment, *anteriorly,* is to the intertrochanteric line; *superiorly,* to the base of the neck; *posteriorly,* to the neck, about 1.25 cm above the intertrochanteric crest; and *inferiorly,* to the lower part of the neck, close to the lesser trochanter. From its anterior femoral attachment along the intertrochanteric line, some of the capsular fibers are reflected proximally along the neck as longitudinal bands, termed **retinacula.** The capsule is much thicker at the upper and anterior part of the joint, where the greatest amount of resistance is required. Posteriorly and below, it is thin and more loosely attached. Blood vessels that supply the trochanters, neck, and head of the femur penetrate through the capsular attachments on the femur to achieve its upper extremity. The capsule consists of two sets of fibers, circular and longitudinal. The circular fibers, **zona orbicularis,** are deeper and more abundant at the distal and posterior part of the capsule, and form a sling or collar around the neck of the femur (Figs. 5-56; 5-57). Most of the capsular fibers are oriented longitudinally between the pelvis and the femur. They are greatest in number at the proximal and anterior part of the cap-

sule, where they are reinforced by an exceedingly strong ligamentous band, the **iliofemoral ligament.** The articular capsule is also strengthened by the pubofemoral and the **ischiofemoral ligaments.** The external surface of the capsule is rough, covered by numerous muscles, and separated anteriorly from the psoas major and iliacus by a bursa, which at times communicates by means of a circular aperture with the joint cavity.

The **iliofemoral ligament** (Fig. 5-53) is a triangular band of great strength which lies anterior to the joint and is blended intimately with the capsule. Its apex is attached *superiorly* to the lower part of the anterior inferior iliac spine, and to the body of the ilium between this spine and the acetabular rim. *Distally,* it divides into two bands, the more medial one of which passes downward to be fixed to the lower part of the intertrochanteric line. The other band is directed downward and laterally and is attached to the proximal part of the same line. Between the two bands the capsule is thinner. In some instances there is no division, and the ligament stretches to become a flat triangular band, which is attached to the whole length

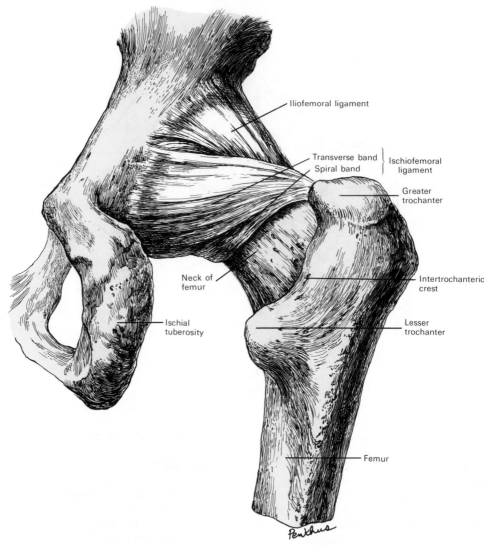

Iliofemoral ligament

Transverse band
Spiral band

Ischiofemoral
ligament

Greater
trochanter

Intertrochanteric
crest

Neck of
femur

Ischial
tuberosity

Lesser
trochanter

Femur

FIG. 5-54. The right hip joint. Posterior aspect.

of the intertrochanteric line. This ligament is frequently called the Y-shaped ligament because it resembles an inverted Y, and its lateral band is sometimes named the **ilio-trochanteric ligament.** The iliofemoral ligament strongly limits extension at the hip joint and resists the strain put upon the anterior part of the fibrous capsule in the usual standing position, when the body weight tends to roll the pelvis backward upon the two femoral heads.

The **pubofemoral ligament** (Fig. 5-53) arises *above* from the body of the pubis near the acetabulum and from the adjacent superior pubic ramus. *Below* it passes anterior to the head of the femur to reach the femoral

neck. It blends with the capsule and with the deep surface of the more medial band of the iliofemoral ligament. The pubofemoral ligament, similar to the iliofemoral, assists in prevention of hyperextension, as well as checking excessive abduction of the thigh.

The **ischiofemoral ligament** (Fig. 5-54) consists of a triangular band of fibers that springs from the body of the ischium below and behind the acetabulum to blend with the circular fibers of the capsule. Its upper fibers are oriented horizontally across the joint, while its lower fibers spiral upward and laterally, attaching to the femoral neck just medial to the greater trochanter. Superiorly its fibers blend with the iliofemoral ligament. As is the case with the other capsular ligaments, the ischiofemoral ligament becomes tense during extension of the femur.

The **ligament of the head of the femur** (Figs. 5-55; 5-56) is a triangular, somewhat flattened band implanted by its apex into the fovea or pit on the head of the femur. Its base is attached by two bands, one on each end of the acetabular notch. Between these attachments at the notch spans the transverse acetabular ligament. The ligament of the head of the femur is ensheathed by the synovial membrane, and varies greatly in strength in different subjects. Occasionally only the synovial fold exists, and in rare cases even this is absent. The ligament is made tense when the thigh is semiflexed and then adducted, and becomes relaxed when the limb is abducted. It is often too weak to

act appreciably as a ligament, but it generally contains a small artery that may assist in supplying the head of the femur in adults.

The **acetabular labrum** (Fig. 5-56) is a fibrocartilaginous rim attached to the margin of the acetabulum, thereby deepening its cavity. It protects the bony rim of the cup and smooths its surface. The labrum bridges the acetabular notch as the **transverse acetabular ligament,** thereby forming a complete circle, which closely surrounds the head of the femur and assists in holding it in place. On cross-section it is triangular in shape, and its base is attached to the margin of the acetabulum, while its opposite edge is free and sharp. Its surfaces are invested by synovial membrane. Externally it is in contact with the capsule, while internally it is inclined inwardly, narrowing the acetabu-

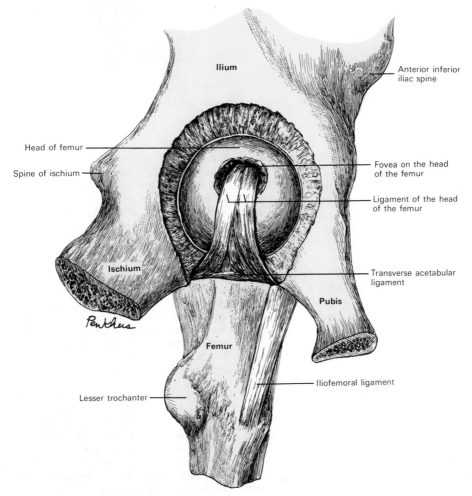

FIG. 5-55.   The left hip joint which has been opened from within the pelvis. The articular surface of the head of the femur has been exposed by the removal of the floor of the acetabulum.

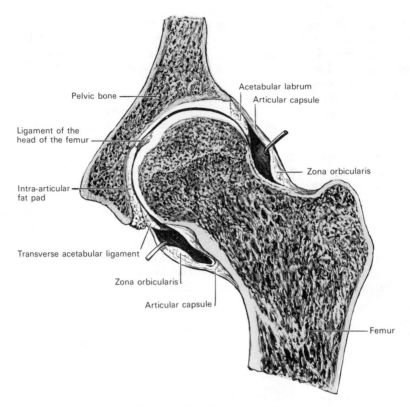

Fig. 5-56. A section through the hip joint.

lum and embracing the cartilaginous surface of the head of the femur.

The **transverse acetabular ligament** (Figs. 5-55; 5-56) is in reality a portion of the acetabular labrum, though differing from it in having no cartilage cells among its fibers. It arises from the margin of the acetabulum on each side of the acetabular notch and consists of strong, flattened fibers which cross the notch and convert it into a foramen through which the nutrient vessels enter the joint. The two bands of the ligament of the head of the femur are attached to the transverse acetabular ligament, one to each end.

Synovial Membrane (Fig. 5-57). The synovial surface is extensive. Commencing at the margin of the cartilaginous surface of the head of the femur, it covers the portion of the neck that is contained within the joint capsule. From the neck it is reflected as the internal surface of the capsule to cover both surfaces of the labrum and the mass of fat in the depression of the acetabular fossa. It ensheathes the ligament of the head of the femur. The joint cavity sometimes communicates with the subtendinous iliac bursa lying deep to the tendon of the psoas major. When this occurs, the communication is found anteriorly between the pubofemoral ligament and the more medially oriented band of the iliofemoral ligament.

Relationships of the Hip Joint (Fig. 5-58). The hip joint is related to the muscles and tendons that surround it as well as to the vessels and nerves that pass from the pelvis into the lower extremity. **Anteriorly**, in direct contact with the joint capsule are bursae, which underlie the iliacus and the tendon of the psoas major. Anteromedially, a few of the more lateral fibers of the pectineus muscle are in contact with the capsule. More superficially are located the femoral vessels and nerves as well as the rectus femoris, sartorius, and tensor fasciae latae muscles. **Superiorly**, the reflected head of the rectus femoris muscle is adjacent to the more medial aspect of the capsule, while the gluteus minimus is closely adherent to its lateral part. **Inferiorly**, the obturator externus, the lateral fibers of the pectineus, and the medial femoral circumflex artery are in close relationship to the joint. **Posteriorly**, and closely related to the articular capsule, are the obturator externus and internus, the gemelli, and the piriformis muscles, while more superficially are found the quadratus femoris and gluteus maximus muscles, as well as the sciatic and posterior femoral cutaneous nerves. Covering all of these is the gluteus maximus.

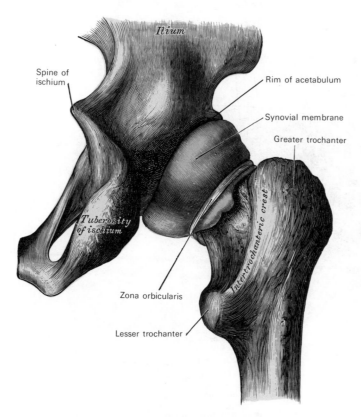

FIG. 5-57. Synovial membrane of capsule of hip joint (distended). Posterior aspect.

Arteries and Nerves. The **arteries** to the hip joint are derived from (1) the medial femoral circumflex branch of the profunda femoral artery; (2) the obturator artery by way of the acetabular branch, which sends a small vessel along the ligament of the head of the femur; (3) the superior and inferior gluteal vessels. At times the lateral femoral circumflex artery contributes directly to the joint as well.

In the adult, vessels are seen to ascend to the femoral head from the neck of the femur in the reflected retinacular fibers of the joint capsule, anastomosing with each other and with twigs derived from the artery of the ligament of the femoral head. This latter vessel is of great importance until puberty, but as age advances and anastomotic channels develop among the other vessels, the artery accompanying the ligament of the head of the femur becomes smaller and less significant to the vascular integrity of the hip joint.

The **nerves** to the hip joint come from the femoral, obturator, and accessory obturator (when present) nerves to the quadratus femoris and superior gluteal.

Movements. The movements at the hip joint are extensive and consist of flexion, extension, adduction, abduction, circumduction, and rotation. Not only does the thigh move on the hip bone (as in walking), but also of importance are those move-ments of the pelvis on the thigh, such as elevating the trunk from the supine position.

Although the hip joint is a freely movable ball-and-socket joint, it is essential to observe that the femur does not rotate around the long axis of the shaft because this would necessitate dislocating the femoral head from its socket. Rotation of the thigh, therefore, takes place around an imaginary axis, that is, a line drawn from the point of attachment of the ligament of the femoral head above, to the medial condyle of the femur at the knee below. Further explanations of this peculiarity rest with the fact that the neck of the femur is a segment of bone of appreciable length and that it joins the shaft at an abrupt angle (the *angle of inclination,* which averages about 125° in adults). All movements of the thigh must be translated into movements of the femoral neck rather than the shaft. For example, **flexion** and **extension** of the thigh as a whole are caused by rotation of the head and neck of the femur around a nearly *transverse axis*. **Rotation** of the femur occurs about a *vertical axis* which extends along a line from the head of the femur through the center of the medial condyle. During rotation of the thigh, the femoral neck swings backward and forward, while the femoral head rotates in the acetabulum around the vertical axis.

Abduction and **adduction** of the thigh move *both*

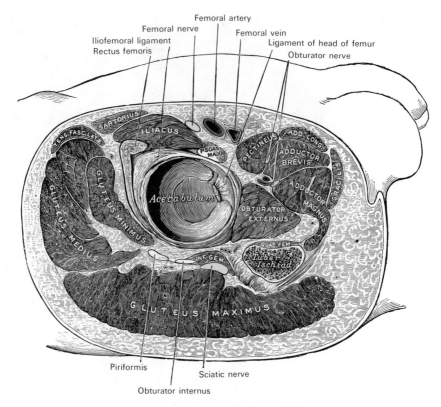

FIG. 5-58.   Structures surrounding right hip joint.

the femoral head and neck around an *anteroposterior axis*. Abduction is somewhat freer than adduction.

The hip joint presents a striking contrast to the shoulder joint in that its mechanical arrangements ensure security and a limitation of movements. In the shoulder, the head of the humerus far exceeds the size of the articular surface of the glenoid cavity, and the ordinary movements of the humeral head are hardly restrained by the loose joint capsule. In the hip joint, on the contrary, the head of the femur is closely fitted to the acetabulum for an area extending over nearly half a sphere. At the margin of the bony cup the femoral head is still more closely embraced by the glenoid labrum, so that it is held in place by that ligament even when the fibers of the capsule have been divided. The iliofemoral ligament is the strongest of all the ligaments in the body, and it becomes stretched by any attempt to extend the femur beyond a straight line with the trunk. This ligament is the chief structure that maintains an erect position without muscular fatigue. A vertical line that descends through the center of gravity of the trunk falls behind the centers of rotation in the hip joints. The pelvis would therefore tend to fall backward, but is prevented from doing so by the tension of the iliofemoral ligaments. Even though the head of the femur and the acetabulum are di-

rectly united through the ligamentum capitis, it is doubtful whether this ligament has much influence on the security of the hip joint.

When the knee is flexed, flexion at the hip joint is arrested when the soft parts of the thigh and anterior abdominal wall are brought into contact. When the knee is extended, flexion at the hip joint is limited by the tension developed through the ligamentous action of the hamstring muscles. Extension at the hip joint is checked by the tension of the iliofemoral ligament. Adduction is limited by the thighs coming into contact, while adduction with flexion is arrested by tension developed in the lateral band of the iliofemoral ligament and the lateral part of the capsule. Although abduction is a relatively free movement, it becomes limited by the medial band of the iliofemoral ligament and the pubofemoral ligament. Lateral rotation is checked by the lateral band of the iliofemoral ligament, whereas medial rotation becomes restricted by the ischiofemoral ligament and the posterior part of the capsule.

Muscles Producing Movements.   **Flexion** of the femur on the pelvis is principally produced by the psoas major, iliacus, and pectineus. Additionally, most of the other muscles that attach to the femur and the anterior aspect of the pelvis assist in flexion of the thigh. These include the rectus femoris, sartorius, tensor fasciae latae, the adductors longus and brevis, the anterior part of the adductor magnus, the gracilis, and the anterior portions of the gluteus minimus and medius. **Extension** is mainly performed by the gluteus maximus, assisted by the hamstring muscles, including the ischial head of the adductor magnus. Accessory extensors include the gluteus medius and piriformis and even the obturator internus.

Adduction is produced by the adductors longus and brevis and the obturator portion of the adductor magnus, as well as the pectineus and gracilis. Accessory adductors include the gluteus maximus, obturator externus and quadratus femoris, and the hamstrings, including the ischial head of the adductor magnus and the iliopsoas. **Abduction** is chiefly performed by the gluteus medius and minimus. In addition, the tensor fasciae latae, piriformis, and sartorius also assist in abduction.

Medial rotation is a rather weak movement, but those muscles whose line of tension passes anterior to the vertical axis of the femoral head and neck act as medial rotators of the thigh, regardless of their insertion. These include the anterior fibers of gluteus medius and minimus, the tensor fasciae latae, the adductors longus, brevis, and magnus, the pectineus, and the iliacus and psoas major. **Lateral rotation** is a powerful movement, and results from the actions of muscles that pass across the hip joint behind the vertical axis around which the femoral head and neck rotate. The external rotators include the posterior fibers of gluteus medius and minimus, the piriformis, obturators externus and internus, gemellus superior, gemellus inferior, quadratus femoris, gluteus maximus, and sartorius.

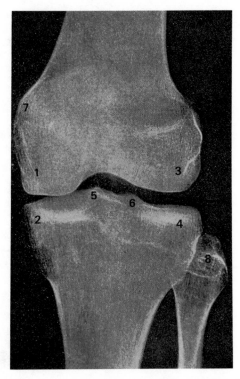

FIG. 5-59. Anteroposterior roentgenogram of the adult knee. 1, Medial condyle of femur. 2, Medial condyle of tibia. 3, Lateral condyle of femur. 4, Lateral condyle of tibia. 5, Medial intercondylar tubercle of tibia. 6, Lateral intercondylar tubercle of tibia. 7, Adductor tubercle of femur. 8, Head of fibula. The gap between the lateral condyles of the femur and tibia is occupied by the articular cartilage of the two bones and by the lateral meniscus.

KNEE JOINT

The knee joint is the largest joint in the body. Although structurally it resembles a hinge joint, its movements are more complicated than those at a simple hinge joint, because rotation and gliding movements also occur at the knee. The knee joint is vulnerable because of the incongruence of the apposing articular surfaces (Figs. 5-59; 5-60), but its great stability is maintained by the arrangement and strength of its ligaments and the muscles and tendons that pass across the joint. The knee joint is a compound joint, and might conveniently be regarded as consisting of three articulations with a common joint cavity. There are two condyloid joints, one between each condyle of the femur and the corresponding condyle of the tibia, and a third between the patella and the femur, classified as a saddle or sellar type of joint (Fig. 5-60). Each of the condyloid joints is partially divided by a fibrocar-

tilaginous meniscus interposed between the corresponding articular condyles. Some phylogenetic evidence supports this concept of knee joint structure. In some lower mammals the knee joint consists of three separate synovial cavities that are either entirely distinct or connected by only small communicating channels and correspond to the three articulations in the human knee cavity. This view is strengthened by the existence of two cruciate ligaments in the middle of the knee joint that can be regarded as collateral ligaments of the medial and lateral condylar joints.

Articular Surfaces. The **tibial condyles** present two cartilage-covered articular surfaces, which are nearly flat but slightly concave, and which are separated by the intercondylar area (Fig. 5-59). Although the facet of the medial condyle lies entirely upon the superior surface of the tibia, the facet of the

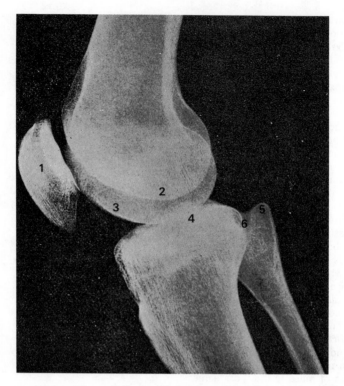

Fig. 5-60.   Lateral roentgenogram of the partially flexed adult knee. *1,* Patella. *2,* Margin of the lateral condyle of the tibia. *3,* Margin of the medial condyle of the tibia. *4,* Intercondylar eminences of the tibia. *5,* Apex of the head of the fibula. *6,* Tibiofibular articulation (proximal). Observe that the medial condyle of the femur descends slightly lower than the lateral condyle and that the superior surface of the tibia and the femoral condyles present incongruous bony contours.

lateral condyle is extended somewhat over the posterior condylar margin where the tendon of the popliteus muscle attaches to the lateral meniscus. The two **femoral condyles** are rounded (Fig. 5-60) and they are curved more posteriorly than anteriorly. They are convex both anteroposteriorly and from side to side, and the condyles are separated posteriorly by the deep intercondylar notch. Anteriorly the condyles fuse to form the patellar surface of the femur. The two femoral condyles are of different shapes. The distal surface of the medial condyle is narrower and longer than that of the lateral condyle.

Attached principally to the periphery of the tibial condyles are the fibrocartilaginous **menisci,** which only slightly deepen the concavities on the superior surface of the tibia. The menisci are importantly involved in the rotation of the femur at the knee. The femur and tibia are not in direct alignment vertically, because the femur slants slightly inward at the knee, whereas the tibia is nearly vertical.

The articular surface of the patella fits rather imperfectly within the concavity of the patellar surface of the femur (Fig. 5-60). A longitudinal ridge divides the articular surface of the patella into a large lateral and a smaller medial part. The latter is further subdivided by another vertical ridge into two smaller areas.

Ligaments.   The ligaments interconnecting the bones at the knee are:

Articular capsule
Patellar
Oblique popliteal
Arcuate popliteal
Tibial collateral
Fibular collateral
Anterior cruciate
Posterior cruciate
Medial meniscus
Lateral meniscus
Transverse

The **articular capsule** (Figs. 5-61; 5-62) is a complex structure which does not form a

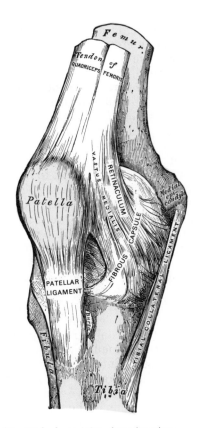

FIG. 5-61. Right knee joint. Anterior view.

tibial condyles. **Posteriorly,** the capsule consists of vertical fibers which arise above from the margins of the femoral condyles and from the borders of the intercondylar fossa of the femur. The capsule descends to attach to the posterior margins of the tibial condyles. The lines of attachment of the posterior part of the articular capsule are situated on the sides of and anterior to the cruciate ligaments; these are, therefore, excluded from the joint cavity. Blending with the capsular fibers, posteriorly and above, are the tendons of origin of the two heads of the gastrocnemius muscle. A bit lower stretches the oblique popliteal ligament, which strengthens the capsule as it expands upward and laterally from the insertion of the tendon of the semimembranosus (Fig. 5-62). **Laterally,** capsular fibers stretch from the border of the lateral femoral condyle above the origin of the popliteus muscle to the lateral aspect of the tibial condyle and the head of the fibula. The fibular collateral ligament remains distinct from the capsule. **Medially,** the articular capsule attaches above to the margin of the medial femoral condyle and below to the corresponding margin of the medial tibial

complete fibrous sac as is generally found in other joints. Instead, throughout most of its extent the fibrous capsule consists of muscle tendons or tendinous expansions between which a few true capsular fibers are found interconnecting the articulating bones. Its inner surface is lined by a synovial membrane, but often this synovial lining is separated from the fibrous-ligamentous layer by various other structures within the joint such as fat pads and the menisci. **Anteriorly,** the fibrous layer of the capsule is completely lacking above the patella, and deep to the tendon of the quadriceps femoris is directly found the synovial lining. At the level of the patella and inferior to the patella, the anteromedial and anterolateral aspects of the fibrous capsule blend with expansions of the tendons of the vastus medialis and vastus lateralis muscles, as well as the fascia lata and its iliotibial band (Fig. 5-61). These fill the intervals between the patellar ligament and collateral ligaments and constitute the **medial** and **lateral patellar retinacula,** which descend to attach to the anterior rim of the

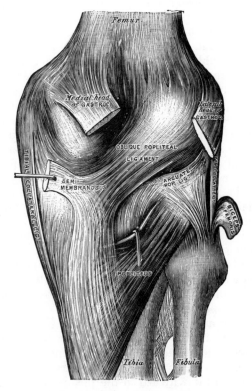

FIG. 5-62. The right knee joint. Posterior aspect.

condyle. Fibrous expansions from the sartorius and semimembranosus pass proximally to the tibial collateral ligament, strengthening the capsule. On the deep side of the capsule medially is a thickening that extends from the medial epicondyle of the femur to the medial meniscus (Last, 1948). Considered today as a deep portion of the tibial collateral ligament, it holds the medial convexity of the meniscus firmly to the femur.

The **patellar ligament** (Fig. 5-61) is the central portion of the common tendon of the quadriceps femoris, which is continued from the patella to the tuberosity of the tibia. It is an extremely strong, flat band, about 8 cm in length, attached *above* to the apex and adjoining margins of the patella and the rough depression on its posterior surface, and *below* to the tuberosity of the tibia. Its superficial fibers are continuous over the front of the patella with those of the tendon of the quadriceps femoris. The medial and lateral portions of the tendon of the quadriceps pass down on each side of the patella, to be inserted into the proximal extremity of the tibia along the sides of its tuberosity. These portions merge into the capsule, to form the **medial** and **lateral patellar retinacula.** The posterior surface of the patellar ligament is separated from the synovial membrane of the joint by a large infrapatellar pad of fat, and from the tibia by a bursa.

The **oblique popliteal ligament** (Fig. 5-62) is a broad, flat fibrous band of great strength which blends into and reinforces the posterior part of the joint capsule. Its obliquely coursing fibers are attached laterally and proximally to the margin of the intercondylar fossa and posterior surface of the lateral condyle of the femur, and distally to the posterior margin of the medial condyle of the tibia. Many of its more superficial fibers are derived from the tendon of the semimembranosus muscle, passing upward and laterally from the insertion site of that muscle. The oblique popliteal ligament is perforated by branches of the middle genicular vessels as well as articular nerve filaments. It forms part of the floor of the popliteal fossa, and the popliteal artery rests upon it.

The **arcuate popliteal ligament** (Fig. 5-62) is a triangular band of capsular fibers attached above to the posterior aspect of the lateral condyle of the femur. It arches medially and downward in the posterior part of the knee joint capsule, over the tendon of the popliteus muscle, to attach to the posterior border of the intercondylar area of the tibia and to the posterior surface of the apex of the fibular head.

The **tibial collateral ligament** (Figs. 5-61 to 5-63; 5-66) is a broad, flat band situated on the medial aspect of the knee joint, nearer the posterior part of the joint than the anterior. It is attached *above* to the medial epicondyle of the femur immediately below the adductor tubercle, and *below,* to the medial condyle and medial surface of the shaft of the tibia. The fibers of the posterior part of the ligament are short and are inserted into the medial condyle of the tibia proximal to the groove for the semimembranosus. The anterior part of the ligament is a flattened band, about 10 cm long, which inclines forward as it descends to be inserted into the medial surface of the shaft of the tibia about 2.5 cm distal to the level of the condyle. It is crossed, at its distal part, by the tendons of the sartorius, gracilis, and semitendinosus, a bursa being interposed. Inferiorly, its deep surface covers the inferior medial genicular vessels and nerve and the anterior portion of the tendon of the semimembranosus, to which it is connected by a few fibers.

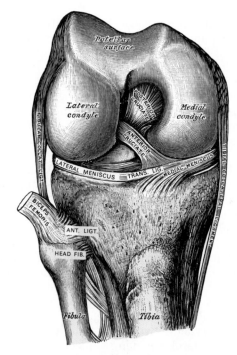

FIG. 5-63.   The right knee joint. Flexed and dissected anteriorly.

The anterior margin of the ligament is free, but its posterior margin is firmly attached to the medial meniscus as it descends past the mid-region of the joint.

The **fibular collateral ligament** (Figs. 5-62; 5-63; 5-66) is a strong, rounded fibrous cord, attached *above* to the posterior part of the lateral epicondyle of the femur, immediately above the groove for the tendon of the popliteus muscle, and *below* to the lateral side of the head of the fibula anterior to its apex. The greater part of its lateral surface is covered by the tendon of the biceps femoris, although the tendon divides at its insertion into two parts which are separated by the ligament. Deep to the ligament are the tendon of the popliteus muscle (Fig. 5-64) and the inferior lateral genicular vessels and nerve. The ligament lies free from the capsule and has no attachment to the lateral meniscus.

The **cruciate ligaments** (Figs. 5-63 to 5-68) are two intra-articular ligaments of considerable strength, situated in the middle of the joint, nearer its posterior than its anterior surface. They are called **cruciate** because they cross each other somewhat like the lines of the letter X, and have received the names **anterior** and **posterior,** from the position of their attachments to the tibia.

The **anterior cruciate ligament** (Figs. 5-63 to 5-65) is attached to a depression in the anterior intercondylar area on the surface of the tibial plateau. This tibial attachment lies in front of the anterior intercondylar tubercle and is blended with the anterior extremity of the lateral meniscus. It passes upward, backward, and laterally, to be fixed into the posterior part of the medial surface of the lateral condyle of the femur (Figs. 5-64; 5-65).

The anterior cruciate ligament limits hyperextension of the knee as well as prevents the backward sliding of the femur on the tibial plateau. Its limitation of extension of the lateral condyle, to which it is attached, then causes medial rotation of the femur as the unencumbered medial condyle continues to extend, until it achieves its full extension. The collateral ligaments of the knee along with the oblique popliteal ligament then become taut, thereby limiting further rotation as the knee joint becomes locked. This "screw home" action, to which the anterior cruciate ligament importantly contributes, allows the knee to attain rigidity.

In contrast, when the foot is solidly on the ground, thereby fixing the leg, the anterior cruciate ligament assists in limiting medial rotation.

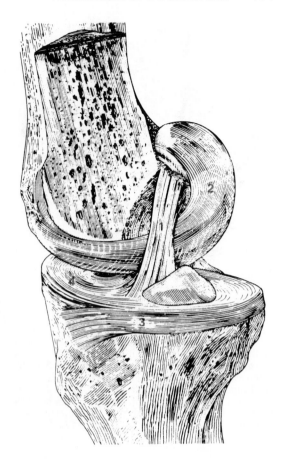

FIG. 5-64.   The anterior cruciate ligament during **extension** in the right knee joint. Medial view. The lower medial half of the femur, including the medial femoral condyle, has been removed to expose fully the ligament. *1,* Anterior cruciate ligament. *2,* Lateral femoral condyle. *3,* Medial meniscus. *4,* Lateral meniscus. (From Girgis, Marshall and Al-Monajem, 1975.)

The **posterior cruciate ligament** (Figs. 5-66 to 5-68) is stronger, but shorter and less oblique in its direction than the anterior. It is attached to a depression on the posterior intercondylar area of the tibia, and extends slightly onto the contiguous posterior surface of the tibia. It also is attached to the posterior extremity of the lateral meniscus.

Passing upward, forward, and medially, the posterior cruciate ligament is fixed to the lateral surface of the medial femoral condyle (Figs. 5-67; 5-68).

The posterior cruciate ligament prevents hyperflexion of the knee, and its attachments prevent the femur from sliding forward on the superior tibial surface when the knee is flexed. This function of the ligament is especially important in walking down a

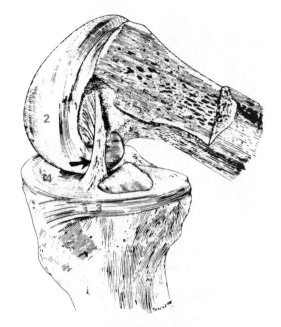

Fig. 5-65. The anterior cruciate ligament during **flexion** in the right knee joint. Medial view. As in Figure 5-64, the medial femoral condyle has been removed. Note the tightness of the anteromedial band of the anterior cruciate ligament (*arrow*) during flexion and the looseness of the major part of the ligament. *1,* Anterior cruciate ligament. *2,* Lateral femoral condyle. *3,* Medial meniscus. *4,* Lateral meniscus. (From Girgis, Marshall and Al-Monajem, 1975.)

steep incline or down steps, when the body weight is transferred to the tibia at the same time as the knee joint flexes.

The **menisci** (Fig. 5-69) are two crescentic lamellae that serve to deepen the surfaces of the articular plateau on the head of the tibia for reception of the condyles of the femur. The peripheral outer border of each meniscus is thick and convex, while the opposite inner border tapers to a thin free edge. The upper surfaces of the menisci are concave, and in contact with the condyles of the femur, and their lower surfaces are flat, and rest upon the head of the tibia. Both surfaces are smooth, and adjacent to synovial membrane. Each meniscus covers approximately the peripheral two-thirds of the corresponding articular surface of the tibia. The peripheral rim of each meniscus is connected to the margin of the tibial head by portions of the fibrous capsule that are called the **coronary ligaments.**

The **medial meniscus** (Figs. 5-63; 5-66; 5-69; 5-70) is nearly semicircular in form, and broader posteriorly than in front. Its *anterior end,* thin and pointed, is attached to the anterior intercondylar area of the tibia, in front of the anterior cruciate ligament (Fig. 5-63),

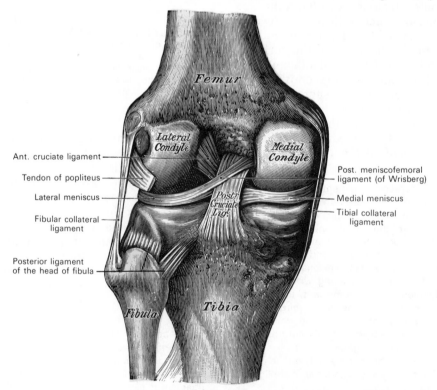

Fig. 5-66. Left knee joint, posterior aspect, showing interior ligaments.

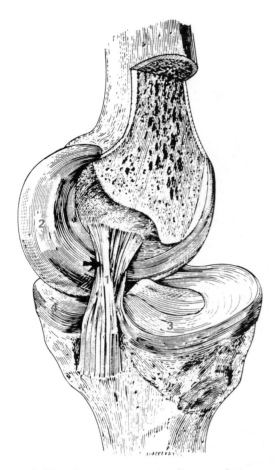

rior end is attached in front of the intercondylar eminence of the tibia, lateral to and behind the anterior cruciate ligament with which it blends. Additionally, the anterior convex margin of the lateral meniscus attaches to the transverse ligament of the knee (Fig. 5-69). The *posterior end* is attached behind the intercondylar eminence of the tibia and in front of the posterior end of the medial meniscus. The anterior attachment of the lateral meniscus is twisted on itself and rests on a sloping shelf of bone on the lateral process of the intercondylar eminence. Close to its posterior attachment it sends off a strong fasciculus, the *posterior meniscofemoral ligament (of Wrisberg),* which passes upward and medially to be inserted into the medial condyle of the femur, immediately behind the attachment of the posterior cruciate ligament (Figs. 5-66; 5-69). Occasionally, another small fasciculus, the *anterior meniscofemoral ligament (of Humphrey),* also attaches to the posterior part of the lateral meniscus, and passes in front of the posterior cruciate ligament to be inserted into the medial condyle of the femur. One or the other of the meniscofemoral ligaments are found in about 70% of

FIG. 5-67. The posterior cruciate ligament during **extension** of the right knee joint. Posterolateral view. The lower lateral part of the femur, including the lateral condyle, has been removed to reveal the attachments of the ligament. Only the posterior band of the ligament (*arrow*) becomes taut during extension. *1,* Posterior cruciate ligament. *2,* Medial femoral condyle. *3,* Lateral meniscus. *4,* Medial meniscus. (From Girgis, Marshall and Al-Monajem, 1975.)

and a band of fibers behind the tibial attachment is continuous with the transverse ligament of the knee (Figs. 5-66; 5-69). Its *posterior end* is fixed to the posterior intercondylar area of the tibia, between the attachments of the lateral meniscus and the posterior cruciate ligament. The circumference of the medial meniscus is attached to the joint capsule and to the deep surface of the tibial collateral ligament.

The **lateral meniscus** (Figs. 5-63; 5-66; 5-69) is nearly circular in form. It covers a larger portion of the tibial articular surface, but is less firmly attached than the medial. The meniscus is grooved laterally for the tendon of the popliteus muscle, which separates it from the fibular collateral ligament. Its *ante-*

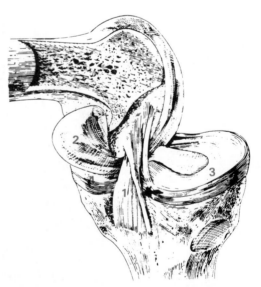

FIG. 5-68. The posterior cruciate ligament during **flexion** of the right knee joint. Posterolateral view. As in Figure 5-66, the lateral femoral condyle has been removed. During flexion the small posterior band of the ligament (*arrow*) is loose while the remaining major portion of the ligament is taut. *1,* Posterior cruciate ligament. *2,* Medial condyle. *3,* Lateral meniscus. *4,* Medial meniscus. (From Girgis, Marshall and Al-Monajem, 1975.)

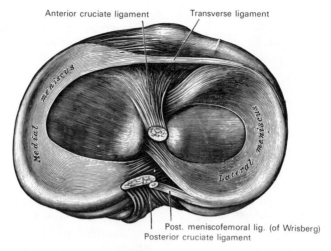

Anterior cruciate ligament   Transverse ligament

Medial meniscus

Lateral meniscus

Post. meniscofemoral lig. (of Wrisberg)
Posterior cruciate ligament

FIG. 5-69. Head of right tibia seen from above, showing the menisci and the attachments onto the tibia of the cruciate ligaments.

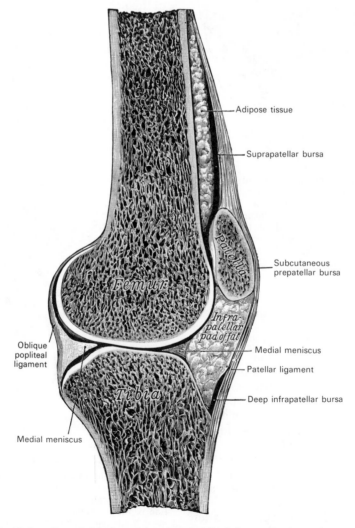

Adipose tissue

Suprapatellar bursa

Patella

Subcutaneous prepatellar bursa

Femur

Infra-patellar pad of fat

Medial meniscus

Patellar ligament

Deep infrapatellar bursa

Oblique popliteal ligament

Tibia

Medial meniscus

FIG. 5-70. Sagittal section of right knee joint, showing the infrapatellar fat pad and certain bursae.

knee joints (Girgis, Marshall, and Al-Mona-jem, 1975).

The **transverse ligament** of the knee (Fig. 5-69) connects the anterior convex margin of the lateral meniscus to the anterior end of the medial meniscus. Its thickness varies considerably in different subjects, and it is sometimes absent.

Synovial Membrane. The synovial membrane of the knee joint is the most extensive and compli-cated joint in the body. Originating at the proximal border of the patella, it forms a large sac beneath the quadriceps femoris (Fig. 5-71) on the distal part of the front of the femur. Although frequently referred to as the **suprapatellar bursa,** it is not a true fric-tional bursa, but only an extension of the synovial cavity above the patella. This large pouch of synovial membrane is supported during movements of the knee by a small muscle, the articularis genu, which inserts into it. On both sides of the patella, the syno-vial membrane extends beneath the aponeuroses of the vasti muscles, but more especially beneath that of the vastus medialis. Distal to the patella it is sepa-

rated from the patellar ligament by a considerable quantity of fat, known as the **infrapatellar fat pad.** From both the medial and lateral borders of the artic-ular surface of the patella, the synovial membrane projects as two fringe-like folds to the interior of the joint. These are called the **alar folds,** which con-verge and continue as a single band, the **infrapatel-lar synovial fold,** to attach to the anterior aspect of the intercondylar fossa of the femur. At the upper posterior aspect of the joint, the synovial membrane forms two pouches that lie deep to the origins of the gastrocnemius muscles (Fig. 5-72).

On both sides of the joint, the synovial membrane passes downward from the femur, lining the capsule to its point of attachment to the menisci. It may then be traced over the proximal surfaces of the menisci to their free borders, which are not covered by syno-vial membrane, and then along their distal surfaces to the tibia (Figs. 5-71; 5-72). At the posterior part of the lateral meniscus, the synovial membrane forms a small sac, the **subpopliteal recess,** between the groove on the surface of the meniscus and the tendon of the popliteus muscle. The membrane is reflected anteriorly across the cruciate ligaments, which are therefore excluded from the synovial cavity.

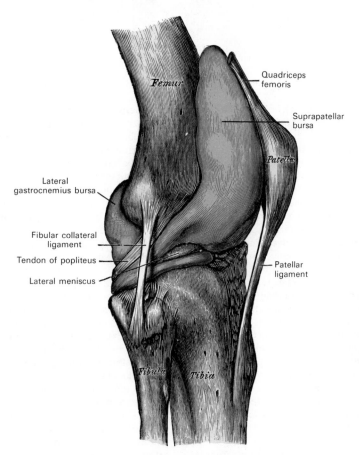

Femur

Quadriceps femoris

Suprapatellar bursa

Patella

Lateral gastrocnemius bursa

Fibular collateral ligament

Tendon of popliteus

Lateral meniscus

Patellar ligament

Fibula   Tibia

FIG. 5-71. Synovial membrane of the articular capsule of right knee joint (distended and colored blue). Lateral aspect.

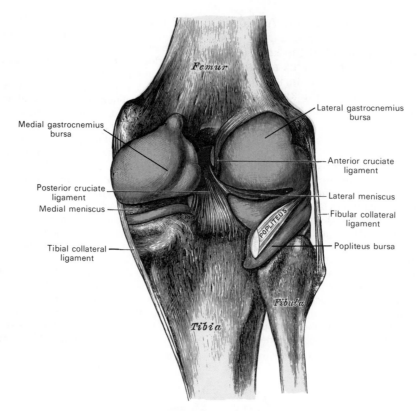

FIG. 5-72. Synovial membrane of the articular capsule of right knee joint (distended and colored blue). Posterior aspect.

Bursae (Figs. 5-70 to 5-72). The bursae related to the knee joint are the following.

Anteriorly there are *four* bursae: (1) the large **subcutaneous prepatellar bursa,** interposed between the patella and the skin; (2) the smaller **deep infrapatellar bursa** between the upper part of the tibia and the patellar ligament; (3) the **subcutaneous infrapatellar bursa** between the lower part of the tuberosity of the tibia and the skin; (4) the **suprapatellar bursa** between the anterior surface of the lower part of the femur and the deep surface of the quadriceps femoris, which usually communicates with the knee joint.

Laterally there are *four* bursae: (1) the **lateral gastrocnemius bursa,** which sometimes communicates with the joint, between the lateral head of the gastrocnemius muscle and the capsule; (2) the **inferior biceps femoris bursa** between the fibular collateral ligament and the tendon of the biceps femoris muscle; (3) one between the fibular collateral ligament and the tendon of the popliteus muscle (this is sometimes only an expansion from the next bursa); (4) the **popliteus bursa** between the tendon of the popliteus and the lateral condyle of the femur, which is usually an extension from the synovial membrane of the joint.

Medially, there are *five* bursae: (1) the **medial gastrocnemius bursa** between the medial head of the gastrocnemius and the capsule; this sends a prolongation between the tendon of the medial head of the gastrocnemius and the tendon of the semimembranosus and often communicates with the joint; (2) the **anserine bursa,** superficial to the tibial collateral ligament, and between it and the tendons of the sartorius, gracilis, and semitendinosus; (3) one deep to the tibial collateral ligament, and between it and the tendon of the semimembranosus (this is sometimes only an expansion from the next bursa); (4) one between the tendon of the semimembranosus and the medial condyle of the tibia; (5) occasionally there is a bursa between the tendons of the semimembranosus and semitendinosus muscles.

Structures Around the Joint. **Anteriorly,** the knee joint is covered by the tendon of the quadriceps femoris. Expanding anteromedially and anterolaterally are the tendons of the vastus medialis and vastus lateralis, forming the medial and lateral patellar retinacula. **Laterally** in relationship to the joint are the tendons of the biceps femoris and popliteus muscles and the common peroneal nerve. **Medially** are found the sartorius, gracilis, semitendinosus, and semimembranosus muscles. **Posteriorly** and within the popliteal fossa are located the popliteal artery and vein, and the tibial nerve along with lymph nodes and fat. Also found posteriorly are the popliteus and plantaris muscles and the medial and lateral heads of the gastrocnemius muscle.

Arteries and Nerves. The **arteries** supplying the knee joint form an anastomosis that includes the descending genicular branch of the femoral artery, the genicular branches of the popliteal, the recurrent branches of the anterior tibial, and the descending branch from the lateral femoral circumflex branch of the deep (profunda) femoral artery.

The **nerves** that supply the knee joint are sensory articular branches from the obturator, femoral, common peroneal, and tibial nerves.

Movements. The movements that take place at the knee joint are flexion and extension, and medial and lateral rotation.

The movements of flexion and extension at the knee differ from those in a typical hinge joint, such as the elbow, because (1) the femoral condyles are curved in such a way that the axis around which flexion and extension occur is not fixed, but shifts somewhat forward during extension and backward during flexion of the leg on the thigh; and (2) *with the foot firmly fixed on the ground,* the commencement of flexion (as in the assumption of a sitting position from a standing position) is accompanied by some lateral rotation of the **femur,** and at the end of extension (as in arising from a sitting position), a considerable medial rotation of the **femur** occurs. *With the foot off the ground,* extension (as in swinging of the leg in the air while sitting in a chair) is accompanied by lateral rotation of the **tibia,** and flexion with medial rotation of the **tibia.**

The movement from full flexion to full extension with the foot firmly stabilized on the ground (as in arising from a sitting position) may be described by the following phases:

1. In the fully flexed position, the posterior parts of the femoral condyles rest on the corresponding surfaces of the tibial articular plateau, and their superimposed menisci.

2. During extension of the thigh at the knee joint from the fully flexed position, *the tibia and the menisci glide forward,* so that the axis, which at the initiation of movement is represented by a line through the medial and lateral condyles of the femur, gradually shifts anteriorly.

3. The lateral condyle of the femur is brought almost to rest by the tightening of the anterior cruciate ligament; however, it moves slightly anterior and medialward, pushing before it the anterior part of the lateral meniscus. Because the articular surface of the medial condyle is prolonged farther anteriorly than that of the lateral condyle, as the anterior movement of the condyles is checked by the anterior cruciate ligament, continued muscular action causes the medial condyle to drag with it the medial meniscus and to travel posteriorly and medially, thus producing a medial rotation of the femur on the tibia.

4. When the position of full extension is reached, medial rotation of the femur causes tautness in the oblique popliteal ligament as well as in the two collateral ligaments, preventing further medial rotation. The knee joint, now fully extended and the femur medially rotated, is rigid and locked in the so-called ''screw home'' position.

5. In standing erect, the weight of the body falls in front of a line carried across the centers of the knee joints, and therefore tends to produce overextension of the articulations. This, however, is prevented by the tension of the anterior cruciate, oblique popliteal, and collateral ligaments.

The movement of flexion from the fully extended position results from the converse of the above phases. Therefore, it is preceded by a lateral rotation of the femur which unlocks the extended joint. This initial lateral rotation of the femur at the commencement of flexion is produced by the popliteus muscle (Last, 1950).

In addition to flexion and extension and the passive rotation essential for their initiation or completion (conjunct rotation), a certain degree of active voluntary rotation at the knee joint (adjunct rotation) can be achieved *when the knee is partially flexed,* because the cruciate and collateral ligaments are not fully tightened at that time. Voluntary rotation cannot be performed when the knee is in full extension.

Movements of the Patella. The articular surface of the patella is indistinctly divided into seven facets—proximal, middle, and distal horizontal pairs, and a medial perpendicular facet (Fig. 5-73). When the knee is forcibly flexed, the medial perpendicular facet is in contact with the crescenteric semilunar area on the lateral part of the medial condyle of the femur. As the leg is carried from the flexed to the extended position, first the proximal pair, then the middle pair, and last the distal pair of horizontal facets is successively brought into contact with the patellar surface of the femur. In the fully extended position, when the quadriceps femoris is relaxed, the patella lies loosely covering the anterior aspect of the distal surface of the femur, extending slightly distal to the tibial rim.

Role of Ligaments During Movement. During **flexion,** the patellar ligament is stretched, and in extreme flexion much of the posterior cruciate and the oblique popliteal ligaments are tense, but the two collateral ligaments and the major part of the anterior cruciate ligament are relaxed. Flexion is checked by the contact of the soft structures of the calf with the thigh. In full **extension** of the knee joint, the oblique popliteal and collateral ligaments, the anterior cruciate ligament, and most of the posterior cruciate ligament, are rendered tense. In the act of extending the leg, the patellar ligament is tightened by the quadriceps femoris, but in full extension with the heel supported, the patellar ligament is relaxed and the patella can be easily moved.

When the leg is partially flexed, all ligaments with the exception of the anterior part of the tibial collateral ligament are somewhat more relaxed than in full extension. In this situation, up to 50° of voluntary

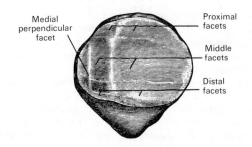

Medial perpendicular facet — Proximal facets — Middle facets — Distal facets

FIG. 5-73. Posterior surface of the right patella, showing its different areas of contact with the femur in varying positions of the knee joint.

rotation (adjunct rotation) of the leg at the knee can occur. Thus, **medial rotation** of the tibia tends to twist the cruciate ligaments, making them taut, and thereby checking the movement. **Lateral rotation** is checked by the collateral ligaments.

The main function of the **cruciate ligaments** is to act as a direct bond between the tibia and femur and to prevent the tibial plateau from being carried too far backward or forward during movements of the leg at the knee joint. The anterior cruciate ligament prevents forward gliding of the tibia on the femur, while the posterior cruciate ligament prevents backward gliding of the tibia. The cruciate ligaments also assist the collateral ligaments in resisting any bending of the joint to the sides. The **menisci** adapt the surface of the tibial plateau to the shape of the femoral condyles, but their role during movement at the knee joint is limited to rotation and not to flexion or extension. They also appear to spread synovial fluid, thereby assisting in lubrication of the knee joint. The **patella** affords important protection to the front of the knee joint, and distributes the pressure during kneeling which would otherwise fall upon the prominent ridges of the femoral condyles. It also affords leverage to the quadriceps femoris muscle, and strength to its tendinous attachment.

Role of Muscles During Movement. Muscles are an indispensable factor in maintaining the stability of the knee joint. Their role in movements at the joint will simply be summarized at this point. **Extension** of the leg on the thigh is performed principally by the quadriceps femoris. **Flexion** of the leg at the knee joint results from action of the biceps femoris, semitendinosus, and semimembranosus, assisted by the gracilis, sartorius, gastrocnemius, popliteus, and plantaris. **Lateral rotation** of the free non-weight-bearing leg is effected by the biceps femoris, and its **medial rotation** by the popliteus, semitendinosus, and, to a slight extent, the semimembranosus, the sartorius, and the gracilis.

The *popliteus muscle* comes into action at the commencement of the movement of flexion of the knee. By its contraction the leg is rotated medialward, or, if the tibia is fixed, the thigh is rotated lateralward, and the knee joint is unlocked.

Clinical Interest. The articular structures of the knee joint are not contoured in a manner that would lend stability to the joint. This poor adaptation of the

bony surfaces of the joint is helped considerably by the strength of the ligaments and overlying muscles which do, in fact, allow the joint considerable strength and weight-bearing durability. However, their inextensible nature and their tautness frequently result in tearing injuries of the ligaments and the menisci when excessive strains are accidentally sustained. The most common athletic injury of the knee involves a tear of the tibial collateral ligament and/or the medial meniscus, to which this ligament is securely attached. This occurs when the foot of the victim is firmly planted on the ground and the knee is semiflexed. An impact, especially from behind, could cause a shift in the weight of the body to medially rotate the femur severely. The leg, which is forced into abduction, shears the tibial collateral ligament or tears the medial meniscus, or both.

TIBIOFIBULAR JOINTS

The tibia and fibula are connected at their proximal extremities by means of a synovial joint, and at their distal extremities by a fibrous joint. Additionally an interosseous membrane joins the shafts of the two bones.

SUPERIOR TIBIOFIBULAR JOINT. This articulation is a plane or gliding joint between the lateral condyle of the tibia and the head of the fibula. The contiguous surfaces of the bones present flat, oval facets covered with cartilage and connected together by an articular capsule, which is lined by a synovial membrane and strengthened by the anterior and posterior ligaments of the head of the fibula.

Ligaments. The **articular capsule** surrounds the joint and is attached around the margins of the articular facets on the tibia and fibula. It is much thicker anteriorly than posteriorly. The capsule is lined by a synovial membrane, the cavity of which is closed off completely from the knee joint anteriorly. In some instances the synovial cavity may communicate with that of the knee joint posterolaterally by means of the popliteal bursa deep to the tendon of the popliteus muscle.

The **anterior ligament of the head of the fibula** (Fig. 5-63) consists of two or three broad, flat bands which pass obliquely upward from the front of the head of the fibula to the anterior part of the lateral condyle of the tibia.

The **posterior ligament of the head of the fibula** (Fig. 5-66) is a single thick, broad band which passes obliquely upward from the back of the head of the fibula to the posterior part of the lateral condyle of the tibia. It is covered by the tendon of the popliteus muscle.

CRURAL INTEROSSEOUS MEMBRANE. An interosseous membrane extends between the interosseous crests of the tibia and fibula, thereby separating the muscles on the anterior from those on the posterior aspect of the leg. The membrane is a thin but strong lamina composed of oblique fibers, which for the most part course downward and laterally, but a few fibers pass in the opposite direction. Proximally, its margin does not quite reach the superior tibiofibular joint, but presents a free concave border, above which is a large, oval aperture for the passage of the anterior tibial artery and vein which supply and drain the front of the leg. Distally, there is another oval opening in the membrane for passage of the perforating anterior peroneal vessels. The membrane is continuous inferiorly with the interosseous ligament of the inferior tibiofibular joint and presents numerous perforations for the passage of small vessels. It is related anteriorly to the tibialis anterior, extensor digitorum longus, extensor hallucis longus, and peroneus tertius muscles, and the anterior tibial vessels and deep peroneal nerve. Posteriorly, it lies adjacent to the tibialis posterior and flexor hallucis longus muscles.

INFERIOR TIBIOFIBULAR JOINT (Fig. 5-74). This articulation is formed by a rough, triangular, convex surface on the medial side of the distal end of the fibula, and the rough concave fibular notch on the lateral side of the distal tibia. For the most part this is a fibrous joint of the syndesmosis type with its surfaces interconnected by an interosseous ligament. The bony surfaces of the distal one-half centimeter of the joint, however, are smooth, covered with cartilage, and lined by a synovial membrane that is continuous with that of the ankle joint.

Ligaments. The ligaments of this joint are:

Anterior tibiofibular
Posterior tibiofibular
Inferior transverse
Interosseous

The **anterior tibiofibular ligament** (Figs. 5-74; 5-76) is a flat band of fibers, broader below than above, which extends obliquely downward and laterally between the adjacent margins of the tibia and fibula on the anterior aspect of the joint. Anterior to the

ligament is found the peroneus tertius muscle, the aponeurosis of the leg, and the integument. Posteriorly, it lies in contact with the interosseous ligament and the cartilage covering the talus.

The **posterior tibiofibular ligament** (Figs. 5-74; 5-77) is somewhat smaller than the anterior ligament, but disposed in a similar manner across the posterior surface of the joint. Inferiorly, it is continuous with the more deeply (i.e., anteriorly) placed inferior transverse ligament.

The **inferior transverse ligament** (Fig. 5-77) lies in front of and below the posterior tibiofibular ligament. It is a strong, thick band of yellowish fibers that passes transversely across the posterior aspect of the joint, from the medial surface of the lateral malleolus of the fibula to the posterior border of the articular surface of the tibia, almost as far as its malleolar process. This ligament, containing yellow elastic fibers, projects below the margin of the bones, and forms part of the articulating surface for the talus.

The **interosseous ligament** (Fig. 5-78) consists of numerous short, strong, fibrous bands which pass between the contiguous rough surfaces of the tibia and fibula, and constitute the chief bond of union between the bones. It is continuous proximally with the crural interosseous membrane.

TALOCRURAL OR ANKLE JOINT

The ankle joint (Figs. 4-74 to 4-78) is principally uniaxial and classified as a ginglymus, or hinge joint. The structures entering into its formation above are the lower end of the tibia and its malleolus, the malleolus of the fibula, and the inferior transverse ligament, which together form a deep cavity for the reception of the proximal convex surface of the talus and its medial and lateral facets.

Ligaments. The bones are connected by the following ligaments:

Articular capsule
Deltoid
Anterior talofibular
Posterior talofibular
Calcaneofibular

The **articular capsule** surrounds the joint and is attached *above* to the borders of the articular surfaces of the tibia and malleoli, and *below* to the talus around its articular

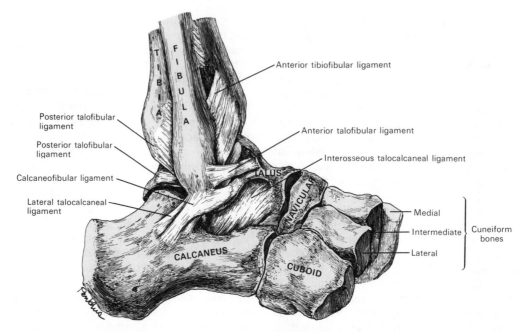

FIG. 5-74.   The inferior tibiofibular joint and the lateral aspect of the ankle joint of the right lower limb.

surface. The **anterior part** of the capsule is a broad, thin, membranous layer which stretches from the anterior margin of the distal end of the tibia to the rough dorsal surface of the neck of the talus, somewhat in front of its superior articular surface. Lying in front of the anterior part of the capsule are the extensor tendons of the toes, the tendons of the tibialis anterior and peroneus tertius, and the anterior tibial vessels and deep peroneal nerve. The **posterior part** of the capsule is thin, and consists principally of transverse fibers. It is attached above to the margin of the articular surface of the tibia, blending with the transverse ligament, and extends below to the talus. It is somewhat thickened laterally, and attaches near the lateral malleolar fossa of the fibula. The inner aspect of the capsule is lined by a synovial membrane that invests the deep surfaces of the ligaments, and sends a small process proximally between the distal ends of the tibia and fibula.

The **deltoid (or medial) ligament** (Figs. 5-75; 5-77) is a strong, flat, triangular band, attached above to the apex and anterior and posterior borders of the medial malleolus. It consists of two sets of fibers, superficial and deep. Of the *superficial fibers* the most anterior (**tibionavicular part**) pass forward to be inserted into the tuberosity of the navicular

bone, and immediately behind this they blend with the medial margin of the plantar calcaneonavicular ligament. The middle fibers (**tibiocalcaneal part**) descend almost perpendicularly to be inserted into the whole length of the sustentaculum tali of the calcaneus. The posterior fibers (**posterior tibiotalar part**) pass backward and laterally to be attached to the inner side of the talus, and to the prominent tubercle on its posterior surface, medial to the groove for the tendon of the flexor hallucis longus. The *deep fibers* (**anterior tibiotalar part**) are attached above to the tip of the medial malleolus, and below to the medial surface of the talus. Crossing the surface of the deltoid ligament are the tendons of the tibialis posterior and the flexor digitorum longus muscles.

The *lateral aspect of the ankle joint* is reinforced by three quite distinct ligamentous bands, the anterior and posterior talofibular ligaments and the calcaneofibular ligament. Some descriptions consider these as three fascicles of a single **lateral ligament** of the ankle joint.

The **anterior talofibular ligament** (Figs. 5-74; 5-76), the shortest of the three, passes forward and medially from the anterior margin of the fibular malleolus to the talus, where it attaches in front of its lateral articular facet and laterally on its neck.

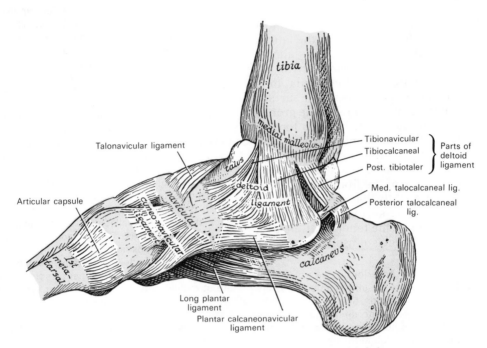

FIG. 5-75. The ligaments of the right ankle and tarsus. Medial aspect.

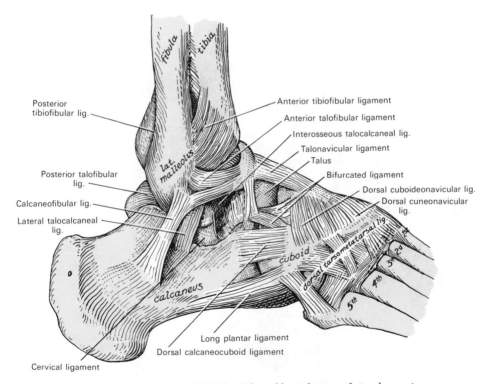

FIG. 5-76. The ligaments of the right ankle and tarsus. Lateral aspect.

The **posterior talofibular ligament** (Figs. 5-74; 5-76; 5-77) is the strongest and most deeply seated of the ligaments on the lateral aspect of the ankle joint. It courses almost horizontally from the posterior part of the lateral malleolar fossa of the fibula to a prominent tubercle on the posterior surface of the talus immediately lateral to the groove for the tendon of the flexor hallucis longus.

The **calcaneofibular ligament** (Figs. 5-74; 5-76) is the longest of the three lateral ligaments and extends as a narrow, rounded cord from the apex of the fibular malleolus, downward and slightly backward, to a tubercle on the lateral surface of the calcaneus. It is covered by the tendons of the peroneus longus and brevis muscles.

Relations. Located **anterior** to the ankle joint from medial to lateral are the tibialis anterior, extensor hallucis longus, anterior tibial vessels, deep peroneal nerve, extensor digitorum longus, and peroneus tertius. Over the **posterior** aspect of the joint from medial to lateral are found the tibialis posterior, flexor digitorum longus, posterior tibial vessels, tibial nerve, flexus hallucis longus, and, in the groove behind the fibular malleolus, the tendons of the peroneus longus and brevis.

Arteries and Nerves. The **arteries** supplying the joint are derived from the malleolar branches of the anterior tibial and the peroneal. The **nerves** come from the deep peroneal and tibial.

Movements. When the body is in the erect position, the foot is at a right angle to the leg. The movements of the ankle joint are those of dorsiflexion and plantar flexion.

Dorsiflexion results from an approximation of the foot to the front of the leg, whereas during **plantar flexion** the heel is raised and the toes pointed downward. The range of plantar flexion is about 55°; that of dorsiflexion is about 35°; however, considerable variation occurs among different individuals. These movements take place around a *transverse axis,* which is slightly oblique and extends between the apex of the lateral malleolus to a point just distal to the medial malleolus. The malleoli tightly embrace the talus in all positions of the joint, so that any slight degree of side-to-side movement can occur only if the ligaments of the inferior tibiofibular joint are stretched and the fibula is bent. The superior articular surface of the talus is broader in front than behind. In dorsiflexion, therefore, greater space is required between the two malleoli. This is obtained by a slight lateral rotation of the lower end of the fibula, which is accomplished by a stretching of the ligaments of the inferior tibiofibular joint, thereby allowing a slight gliding at the tibiofibular articulation.

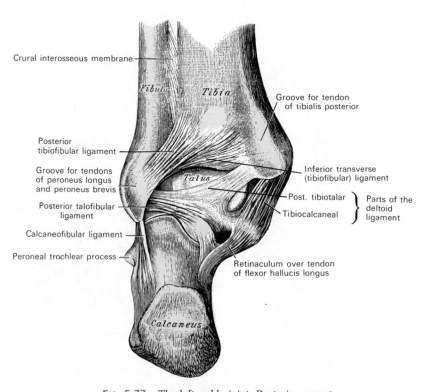

FIG. 5-77. The left ankle joint. Posterior aspect.

The deltoid ligament is of great strength, so much so that it usually resists a force powerful enough to fracture the process of bone to which it is attached. Its middle portion, together with the calcaneofibular ligament, binds the bones of the leg firmly to the foot, resisting displacement in every direction. Its anterior and posterior fibers assist in limiting plantar flexion and dorsiflexion. The posterior talofibular ligament assists the calcaneofibular in resisting the displacement of the foot backward, and deepens the cavity for the reception of the talus. The anterior talofibular is a security against the displacement of the foot forward, and helps limit plantar flexion of the joint.

Muscles Producing Movements. **Plantar flexion** of the foot is produced principally by the gastrocnemius, soleus, and plantaris, and to a lesser extent by the tibialis posterior, flexor digitorum longus, and flexor hallucis longus muscles. **Dorsiflexion** results from the action of the tibialis anterior, which is assisted by the peroneus tertius, extensor digitorum longus, and extensor hallucis longus.

Clinical Interest. A "sprained ankle" may result when the weight of the body is forcibly exerted upon the foot as it is inverted or everted. Most sprains are of the *inversion type,* which results in the tearing of ligaments at the lateral aspect of the ankle joint. In these instances the calcaneofibular as well as the anterior talofibular ligaments most often rupture, but a portion of the posterior talofibular ligament may also tear. *Eversion sprains* may rupture a portion of the deltoid ligament, but frequently the medial malleolus will fracture under these circumstances because of the great strength of the ligament.

INTERTARSAL JOINTS

The intertarsal joints will be described in the following order:

Subtalar (talocalcaneal) joint
Talocalcaneonavicular joint
Calcaneocuboid joint
Transverse tarsal joint
Ligaments connecting the calcaneus and
    navicular
Cuneonavicular joint
Cuboideonavicular joint
Intercuneiform joints
Cuneocuboid joint

SUBTALAR (TALOCALCANEAL) JOINT. Two articulations exist between the calcaneus and talus, an anterior and posterior. Because the anterior joint forms part of the talocalcaneonavicular joint, it will be described with that articulation.

The more posterior joint, the subtalar, is formed between the concave posterior calcaneal facet on the inferior surface of the talus, and the convex posterior facet on the superior surface of the calcaneus. It is a gliding joint, and the two bones are connected by an articular capsule and by anterior, posterior, lateral, medial, and interosseous talocalcaneal ligaments (Figs. 5-78; 5-79).

Ligaments. The **articular capsule** envelops the joint, and consists for the most part of short fibers that are split into distinct slips. Between these there is only a weak fibrous investment. A synovial membrane lines the capsule of the joint, and the joint cavity is distinct from all others in the foot because it has no intercommunicating channels.

The **anterior talocalcaneal ligament** extends from the anterior and lateral surface of the neck of the talus to the superior surface of the calcaneus (Fig. 5-79). It forms the posterior boundary of the talocalcaneonavicular joint and is sometimes described as the **anterior interosseous ligament.**

The **posterior talocalcaneal ligament** (Fig. 5-75) connects the lateral tubercle of the talus with the proximal and medial part of the calcaneus. It is a short band, and its fibers radiate downward to the calcaneus from their narrow attachment to the talus.

The **lateral talocalcaneal ligament** (Figs. 5-74; 5-76; 5-79) is a short, strong fasciculus, passing from the lateral surface of the talus immediately beneath its fibular facet to the lateral surface of the calcaneus. Its fibers course in the same direction as those of the calcaneofibular ligament, but it is placed on a deeper plane than that ligament.

The medial talocalcaneal ligament (Figs. 5-75; 5-80) connects the medial tubercle on the back of the talus with the sustentaculum tali. Its fibers blend with those of the plantar calcaneonavicular ligament.

The **cervical ligament** (Fig. 5-76) occupies the sinus tarsi, a small concavity on the anterolateral aspect of the foot between the talus and the calcaneus. The ligament is a strong band of fibers attached above to the inferolateral aspect of the neck of the talus and below to the upper surface of the calcaneus. The cervical ligament is important because it becomes taut during inversion, helping to prevent the foot from an excessive degree of that movement (Jones, 1944).

The **interosseous talocalcaneal ligament**

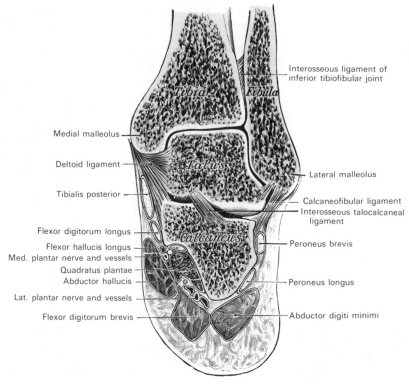

Interosseous ligament of
inferior tibiofibular joint

Tibia   Fibula

Medial malleolus

Deltoid ligament

Tibialis posterior

Talus

Lateral malleolus

Calcaneofibular ligament
Interosseous talocalcaneal
ligament

Flexor digitorum longus
Flexor hallucis longus
Med. plantar nerve and vessels
Quadratus plantae
Abductor hallucis
Lat. plantar nerve and vessels
Flexor digitorum brevis

Calcaneus

Peroneus brevis

Peroneus longus

Abductor digiti minimi

FIG. 5-78.   Coronal section through right talocrural and talocalcaneal joints showing the interosseous talocalcaneal ligament and the relationship of structures coursing beneath the malleoli.

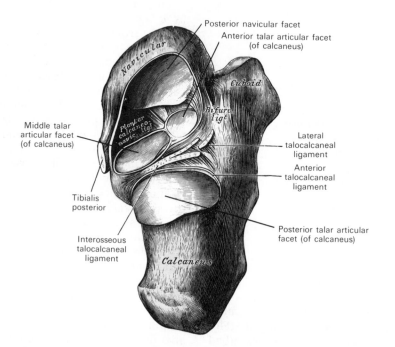

Posterior navicular facet
Anterior talar articular facet
(of calcaneus)

Navicular

Cuboid

Bifurc.
lig't

Middle talar
articular facet
(of calcaneus)

Plantar
calcaneo-
navic. lig't

Lateral
talocalcaneal
ligament

Anterior
talocalcaneal
ligament

Tibialis
posterior

Interosseous
talocalcaneal
ligament

Posterior talar articular
facet (of calcaneus)

Calcaneus

FIG. 5-79.   Talocalcaneal and talocalcaneonavicular articulations exposed from above by removing the talus.

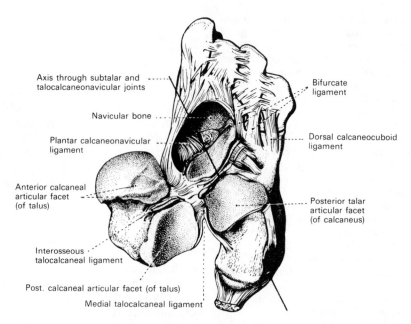

Axis through subtalar and
talocalcaneonavicular joints

Navicular bone

Plantar calcaneonavicular
ligament

Anterior calcaneal
articular facet
(of talus)

Interosseous
talocalcaneal ligament

Post. calcaneal articular facet (of talus)

Medial talocalcaneal ligament

Bifurcate
ligament

Dorsal calcaneocuboid
ligament

Posterior talar
articular facet
(of calcaneus)

FIG. 5-80. The subtalar and talocalcaneonavicular joints of the right foot viewed from above. The talus has been removed from the joint cavity and rotated medially to expose its inferior articular surfaces. Inversion and eversion of the foot occurs around the indicated axis. (From Benninghoff and Goerttler, *Lehrbuch der Anatomie des Menschen,* Urban and Schwarzenberg, Munich, 1980).

(Figs. 5-74; 5-78; 5-79; 5-82) forms the chief bond of union between the talus and calcaneus. It is, in fact, a portion of the united capsules of the talocalcaneonavicular and the talocalcaneal joints, and consists of two partially united layers of fibers, one belonging to the former and the other to the latter joint. It is attached to the groove between the articular facets on the distal surface of the talus, and to a corresponding depression on the proximal surface of the calcaneus. It is thick and strong, being at least 2.5 cm in breadth from side to side, and serves to bind the calcaneus and talus firmly together.

TALOCALCANEONAVICULAR JOINT. This articulation is a compound, multiaxial, synovial joint of the spheroid type. The rounded head of the talus is received into a concavity or socket (Fig. 5-79) formed by *two bones,* the posterior surface of the navicular, the anterior articular surface of the calcaneus, and *two ligaments,* the plantar calcaneonavicular (spring) ligament medially, and the calcaneonavicular part of the bifurcated ligament laterally. These bony and ligamentous structures are all surrounded by an articular capsule. Dorsally, the joint is reinforced by the dorsal talonavicular ligament. At this point only the articular capsule and (dorsal) talonavicular ligament will be described.

Ligaments. The **articular capsule** is imperfectly developed on the plantar aspect of the joint because of the ligamentous and bony socket. However, posteriorly it is considerably thickened and forms the anterior component of the strong interosseous talocalcaneal ligament which fills the sinus tarsi, formed by the opposing grooves on the calcaneus and talus. A synovial membrane lines the inner aspect of the capsule of this joint.

The **(dorsal) talonavicular ligament** (Figs. 5-75; 5-76) is a broad, thin band that connects the neck of the talus to the dorsal surface of the navicular bone. Medially, it blends with the ligament of the deltoid, and it is covered by the extensor tendons.

The plantar calcaneonavicular ligament and the calcaneonavicular part of the bifurcated ligament form the plantar and lateral ligaments of this joint.

Movements. The subtalar (talocalcaneal) and talocalcaneonavicular joints permit a considerable range of gliding and rotation. The movements at these joints (not to be confused with those at the ankle joint) result in inversion or eversion of the plantar surface of the foot.

Inversion of the foot is a movement in which the medial border of the foot is raised and the lateral border lowered, so that the sole of the foot is turned medially. In **eversion** the lateral border is raised and the medial border is lowered, and as a result the sole of the foot is turned laterally. The greatest range of inversion and eversion occurs at the subtalar joint, and usually inversion is combined with some degree of plantar flexion and adduction, whereas eversion is frequently associated with dorsiflexion and abduction. These movements can be performed either with the foot off the ground or with the weight of the body transmitted to the foot, as in walking along a surface that is sloped sideways or in turning quickly, which requires a leaning of the legs to the side while the sole of the foot is flat on the ground. The axis around which inversion and eversion occur is an oblique line approximately between the centers of the subtalar and talocalcaneonavicular joints. The axis passes from the dorsomedial side of the foot, plantarward, backward, and laterally to emerge on the lateral aspect of the heel (Fig. 5-80).

Muscles Producing Movements. *Inversion* is produced by the tibialis anterior and tibialis posterior muscles, assisted by the extensor and flexor hallucis longus muscles. *Eversion* results from the action of the peroneus longus and brevis muscles.

CALCANEOCUBOID JOINT. This joint resembles a saddle joint, but does not have the usual freedom of movement of such a joint. Its articular surfaces are reciprocally concavoconvex, with the distal (anterior) surface of the calcaneus forming the concave saddle that receives the convex proximal (posterior) border of the cuboid (Fig. 5-82). Five ligaments connect the calcaneus with the cuboid: the articular capsule, the dorsal calcaneocuboid, the calcaneocuboid portion of the bifurcated ligament, the long plantar, and the plantar calcaneocuboid.

Ligaments. The **articular capsule** of the calcaneocuboid joint completely surrounds the joint, but it is an imperfectly developed investment, containing certain strengthened bands which form the other ligaments of the joint.

The **dorsal calcaneocuboid ligament** is a thin but broad fasciculus which passes between the contiguous surfaces of the calcaneus and cuboid on the dorsal surface of the joint (Figs. 5-76; 5-80).

The **bifurcated ligament** (Figs. 5-76; 5-79; 5-80) is a strong band, attached behind to the upper anterior surface of the calcaneous. The ligament courses anteriorly, and divides into two parts: calcaneocuboid and calcaneonavicular. The **calcaneocuboid part** is fixed to the dorsomedial aspect of the cuboid and forms one of the principal bonds between the first and second rows of the tarsal bones. The **calcaneonavicular part** is attached to the dorsolateral surface of the navicular.

The **long plantar ligament** (Figs. 5-75; 5-76; 5-81) is the longest of all the ligaments of the tarsus. It is attached *posteriorly* to the plantar surface of the calcaneus in front of the calcaneal tuberosity. *Anteriorly*, its deep fibers stretch to the tuberosity on the plantar surface of the cuboid bone, while its more superficial fibers are continued distally to the bases of the second, third, fourth, and fifth metatarsal bones. These anteriorly extending fibers of the long plantar ligament convert the groove on the plantar surface of the cuboid into a canal for the tendon of the peroneus longus muscle. By extending nearly the entire length of the plantar surface of the lateral longitudinal arch, the strong, long plantar ligament reinforces that arch, and helps prevent its flattening.

The **plantar calcaneocuboid ligament** (Fig. 5-81) is sometimes called the *short plantar ligament*. It lies nearer the bones, or deep to the long plantar ligament, from which it is separated by a little areolar tissue. It is a short but wide band of great strength which extends from the plantar aspect of the anterior tubercle of the calcaneus and the depression in front of it to the plantar surface of the cuboid bone, behind the peroneal groove.

Movements. The movements permitted at the calcaneocuboid joint are principally of a gliding and rotational nature, and they accommodate the movements in the subtalar and talocalcaneonavicular joints during inversion and eversion of the foot.

TRANSVERSE TARSAL JOINT. The transverse tarsal joint is a collective term used for the combined calcaneocuboid and talonavicular part of the talocalcaneonavicular joints, because together their joint cavities extend in an irregular transverse plane across the foot. Although the two joint cavities do not communicate, a slight degree of dorsiflexion and plantar flexion of the anterior part of the foot is permitted along this transverse tarsal plane. At the same time, the sole of the foot is slightly inverted or everted. The bones anterior to the transverse tarsal joint are the cuboid and navicular, while those behind it are the calcaneus and talus.

LIGAMENTS CONNECTING THE CALCANEUS AND NAVICULAR. Although the calcaneus and navicular do not articulate directly, they are connected by two ligaments: the calcaneonavicular part of the bifurcated, and the plantar calcaneonavicular.

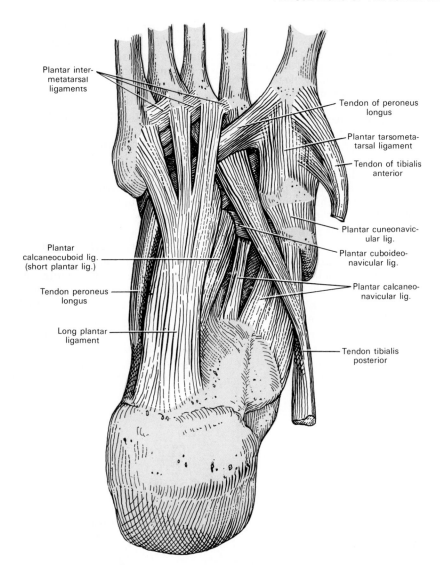

Plantar inter-
metatarsal
ligaments

Tendon of peroneus
longus

Plantar tarsometa-
tarsal ligament

Tendon of tibialis
anterior

Plantar cuneonavic-
ular lig.

Plantar cuboideo-
navicular lig.

Plantar
calcaneocuboid lig.
(short plantar lig.)

Plantar calcaneo-
navicular lig.

Tendon peroneus
longus

Long plantar
ligament

Tendon tibialis
posterior

FIG. 5-81. Ligaments of the sole of the foot, with the tendons of the peroneus longus, tibialis posterior, and tibialis anterior muscles.

The **calcaneonavicular part of the bifur-cated ligament,** already described, attaches the anterior part of the calcaneus to the adjacent lateral surface of the navicular bone.

The **plantar calcaneonavicular ligament** (Figs. 5-75; 5-81), commonly called the *spring ligament,* is a broad and thick band of fibers that connects the anterior margin of the sustentaculum tali of the calcaneus to the plantar surface of the navicular. This ligament not only serves to connect the calcaneus and navicular, but supports the head of the talus, forming part of the articular socket into which it is received. The *dorsal surface* of the ligament presents a fibrocartilaginous facet, lined by synovial membrane, upon

which a portion of the head of the talus rests. Its *plantar surface* is supported by the tendon of the tibialis posterior medially, and the tendons of the flexor hallucis longus and flexor digitorum longus laterally. These tendons augment the supporting function of the ligament. Its *medial border* is blended with the anterior part of the deltoid ligament of the ankle joint.

Clinical Interest. The plantar calcaneonavicular ligament, by supporting the head of the talus, is principally concerned in maintaining the medial longitudinal arch of the foot. When it yields, the head of the talus is pressed medially and distally by the weight of the body, and the foot becomes pronated, everted, and turned lateralward, exhibiting the condition known as *flat-foot*. This ligament contains considerable numbers of elastic fibers, in order to give elasticity to the arch and spring to the foot. Hence, it is sometimes called the **spring ligament**. It is supported, on its plantar surface, by the tendon of the tibialis posterior, which spreads out at its insertion into a number of fasciculi, to be attached to most of the tarsal and metatarsal bones.

CUNEONAVICULAR JOINT. The convex distal (anterior) articular surface of the navicular bone is joined to the concave proximal (posterior) surfaces of the three cuneiform bones (Fig. 5-83). The navicular surface is characterized by three facets into which the cuneiform bones fit.

Ligaments. The joint is surrounded by an articular capsule within which a continuous synovial cavity extends across the joint, as well as between the adjacent cuneiform bones (Fig. 5-82). The joint cavity also extends distally between the lateral cuneiform and the cuboid. In addition to the capsule, both dorsal and plantar cuneonavicular ligaments interconnect the bones.

The **dorsal cuneonavicular ligaments** (Figs. 5-75; 5-76) are three small bundles, one attached to each of the cuneiform bones. The bundle connecting the navicular with the medial cuneiform is continuous around the medial side of the articulation with the plantar ligament which unites these two bones.

The **plantar cuneonavicular ligaments** are arranged similarly to the dorsal ligaments. They are stronger bands than the dorsal, which are reinforced by slips from the tendon of the tibialis posterior (Fig. 5-81).

CUBOIDEONAVICULAR JOINT. The rounded lateral surface of the navicular bone is joined to the posterior part of the medial surface of the cuboid. At times the joint contains a synovial cavity which is continuous with that of the cuneonavicular joint, but more often it is a fibrous joint of the syndesmosis type (Fig. 5-82). The two bones are interconnected by dorsal, plantar, and interosseous ligaments.

Ligaments. The **dorsal cuboideonavicular ligament** (Fig. 5-76) extends obliquely for-

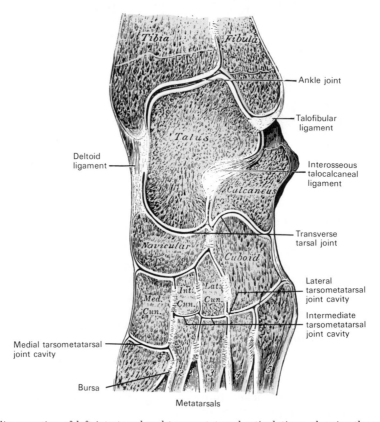

FIG. 5-82. Oblique section of left intertarsal and tarsometatarsal articulations, showing the synovial cavities.

ward and laterally from the navicular to the cuboid bone on their dorsal surfaces.

The **plantar cuboideonavicular ligament** (Fig. 5-81) passes nearly transversely between the cuboid and navicular bones on their plantar surfaces.

The **cuboideonavicular interosseous ligament** (Fig. 5-82) consists of strong transverse fibers interconnecting the rough nonarticular portions of the adjacent lateral surface of the cuboid and medial surface of the navicular bones.

INTERCUNEIFORM JOINTS. The three cuneiform bones are united by two synovial intercuneiform joints (Fig. 5-82). Each of the two joints interconnects adjacent surfaces of the bones, the intermediate cuneiform being placed between the medial and lateral cuneiform bones. Their synovial cavities are continuous with that of the cuneonavicular joint. The joints are connected by dorsal, plantar, and interosseous ligaments.

Ligaments. The **dorsal intercuneiform ligaments,** two in number, are weak transverse bands. One connects the medial with the intermediate cuneiform and the other, the intermediate with the lateral cuneiform.

The **plantar intercuneiform ligaments** are arranged on the plantar surface similarly to those on the dorsal surface. They are stronger than the dorsal ligaments and are reinforced by slips from the tendinous insertion of the tibialis posterior muscle.

The **intercuneiform interosseous ligaments,** also two in number, consist of strong transverse fibers that pass between the rough nonarticular portions of the adjacent surfaces of the bones (Fig. 5-82). They are strong, deeply placed, fibrous ligaments which extend between the bones nearly the entire length of their apposed borders.

CUNEOCUBOID JOINT. This joint connects the smooth rounded lateral border of the lateral cuneiform bone with the adjacent medial border of the cuboid. Its small synovial cavity usually opens posteriorly into the cuneonavicular joint (Fig. 5-82). Similar to the intercuneiform joints, the cuneocuboid joint is interconnected by dorsal, plantar, and interosseous ligaments.

The **dorsal cuneocuboid ligament** is a relatively weak transverse band extending from the dorsal surface of the lateral cuneiform to the cuboid.

The **plantar cuneocuboid ligament** is considerably stronger than the dorsal and ex-

tends transversely between the lateral cuneiform and the cuboid bones on their plantar surfaces. A slip from the tendon of the tibialis posterior muscle may strengthen the ligament.

The **interosseous cuneocuboid ligament** is the strongest bond between the two bones. Its transverse fibers are deeply placed between the apposed borders of the bones in a manner similar to those of the intercuneiform ligaments (Fig. 5-82).

Movements. The cuneonavicular, cuboideonavicular, intercuneiform, and cuneocuboid joints permit slight gliding and rotational movements. These assist the more proximal intertarsal joints during inversion and eversion of the foot, and assist the bare foot especially to achieve a firmer grasp when necessary. The synovial cavities of all these joints are usually intercommunicating, and their ligaments help in the maintenance of the transverse arch of the foot.

TARSOMETATARSAL JOINTS

These are gliding joints. The bones entering into their formation are the medial, intermediate, and lateral cuneiforms, and the cuboid, which articulate with the bases of the five metatarsal bones. The **first metatarsal** bone articulates with the anterior surface of the medial cuneiform. The **second metatarsal** is wedged between the medial and lateral cuneiforms. It articulates medially with the lateral side of the medial cuneiform, by its base with the intermediate cuneiform, and laterally with the medial side of the lateral cuneiform. The **third metatarsal** articulates with the lateral cuneiform bone. The **fourth metatarsal** articulates with the cuboid and lateral cuneiform bone. The **fifth metatarsal** articulates only with the cuboid (Fig. 5-83).

Although five metatarsal bones articulate with four tarsal bones (the cuboid and the three cuneiforms), only three joint cavities are formed (Fig. 5-82). Medially, the joint between the first metatarsal and the medial cuneiform has its own distinct synovial sac. This **medial tarsometatarsal joint cavity** usually lies more distal to the others and is separated from them by an interosseous ligament which connects the medial cuneiform to the second metatarsal (Fig. 5-82). The **intermediate tarsometatarsal joint cavity** is found between the intermediate and lateral cuneiforms and the second and third metatarsal

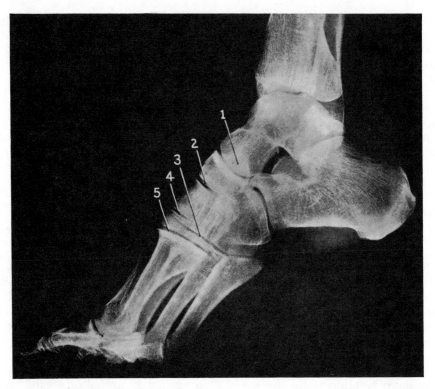

FIG. 5-83.   Adult foot. 1, Tuberosity of navicular bone, partly obscured by the shadow of the head of the talus. 2, Cuneonavicular joint. 3, Joint between metatarsal III and the lateral cuneiform bone. 4, Joint between metatarsal II and the intermediate cuneiform bone. 5, Joint between metatarsal I and the medial cuneiform bone.

bones. This cavity is continuous proximally with those of the intercuneiform joints and often with the cavity of the cuneonavicular joint. Additionally, two prolongations of the cavity extend distally (forward), one between the adjacent sides of the second and third metatarsal bones and the other between the sides of the third and fourth metatarsals. The **lateral tarsometatarsal joint cavity** has a distinct synovial sac between the cuboid and the bases of the fourth and fifth metatarsal bones. This cavity is also prolonged forward between the fourth and fifth metatarsals, but is separated from the intermediate joint cavity by an interosseous ligament which connects the lateral cuneiform and the base of the fourth metatarsal bone.

The bones are connected by dorsal and plantar tarsometatarsal ligaments and by interosseous cuneometatarsal ligaments.

Ligaments. The **dorsal tarsometatarsal ligaments** are strong, flat bands which pass between the adjacent surfaces of the tarsal and metatarsal bones on their dorsal aspect (Fig. 5-76). The first metatarsal is joined to the medial cuneiform by a thin articular

capsule (Fig. 5-75). To the dorsal surface of the second metatarsal are attached three bands, one from each cuneiform bone. The third metatarsal has one dorsal ligament, and this attaches to the lateral cuneiform. The fourth receives one from the lateral cuneiform and one from the cuboid; and the fifth, one from the cuboid.

The **plantar tarsometatarsal ligaments** consist of longitudinal and oblique bands disposed with less regularity than the dorsal ligaments. Those for the first and second metatarsals are the strongest (Fig. 5-81). The second and third metatarsals are joined by oblique bands to the medial cuneiform, while the fourth and fifth metatarsals are connected by a few fibers to the cuboid. The articulation of the medial cuneiform and the first metatarsal (tarsometatarsal joint of the large toe) is reinforced by the insertion of the peroneus longus tendon laterally and that of the tibialis anterior medially. Additionally, strands from the tendon of the tibialis posterior muscle strengthen the plantar surfaces of the tarsometatarsal joints of the second, third, and fourth toes, while the long plantar ligament, as well, helps support the third and fourth tarsometatarsal joints by passing across their plantar aspect.

The **interosseous cuneometatarsal ligaments** are three in number. The first is constant and is the strongest, passing from the lateral surface of the medial cuneiform to the adjacent angle of the second metatarsal (Fig. 5-82). The second, which is the smallest and is less constant, passes from the lateral cuneiform to the lateral aspect of the second metatarsal. The third connects the lateral angle of the lateral cuneiform with the adjacent side of the base of the fourth metatarsal.

Movements. The movements permitted at the tarsometatarsal joints are limited to slight gliding of the bones upon each other. These joints, along with the others in the anterior part of the foot, move as a unit at the transverse tarsal joint, allowing the forepart of the foot a degree of rotatory movement which might be called supination and pronation in contrast to the inversion and eversion seen at the subtalar and talocalcaneonavicular joints.

Nerve Supply. The intertarsal and tarsometatarsal joints are supplied by the medial and lateral plantar nerves on their plantar aspect and by the deep peroneal nerve on their dorsal aspect.

INTERMETATARSAL JOINTS

JOINTS BETWEEN THE BASES OF THE METATARSAL BONES. Even though the base of the first metatarsal bone lies adjacent to the proximal part of the shaft of the second metatarsal, there are no ligaments between the bones, but a bursa is frequently found at this site. In contrast, the bases of the second and third, third and fourth, and fourth and fifth metatarsals are interconnected by dorsal, plantar, and interosseous intermetatarsal ligaments.

Ligaments. The **dorsal intermetatarsal ligaments** pass transversely between the dorsal surfaces of the bases of the adjacent four lateral metatarsal bones.

The **plantar intermetatarsal ligaments** (Fig. 5-81) are strong bands that connect the bases of the bones on the plantar surface in a pattern similar to the dorsal intermetatarsal ligaments.

The **interosseous intermetatarsal ligaments** are usually three in number and consist of strong transverse fibers that connect the rough nonarticular portions of the adjacent surfaces of the bases of the lateral four metatarsal bones.

Synovial Membranes (Fig. 5-82). The synovial membranes between the bases of the second and third and the third and fourth metatarsal bones are continuous with that of the intermediate tarsometatarsal joint cavity. The synovial cavity between the fourth and fifth metatarsals is a prolongation of the lateral tarsometatarsal joint cavity (Fig. 5-82).

Movements. The movement permitted between the tarsal ends of the metatarsal bones is limited to a slight gliding of the articular surfaces upon one another. This allows for an elevation or a slight widening of the transverse arch, adding some flexibility to the forepart of the foot.

UNION OF THE HEADS OF THE METATARSAL BONES. The heads of all the metatarsal bones are interconnected on their plantar aspect by the deep transmetatarsal ligaments.

Ligaments. The **deep transverse metatarsal ligaments** are four short, narrow bands which course across and interconnect the heads of all the metatarsal bones on their plantar surfaces. They blend anteriorly with the plantar (glenoid) ligaments of the metatarsophalangeal joints. Along their concave plantar surfaces in the middle of each toe course the tendons of the long flexors. Between the toes, the plantar surfaces of the ligaments are in contact with the lumbrical muscles and the digital vessels and nerves. The dorsal surfaces of the ligaments are in relation to the interosseous muscles whose tendons course along both sides of the metatarsophalangeal joints of the four lateral toes. The deep transverse metatarsal ligaments differ from the deep transverse metacarpal ligaments in that the first metatarsal is connected to the second, whereas in the hand, the first metacarpal is not attached by means of a deep metacarpal ligament.

SYNOVIAL CAVITIES OF THE TARSAL AND TARSOMETATARSAL JOINTS (Fig. 5-82). Six synovial cavities are found in the articulations of the tarsus and metatarsus: *one* for the subtalar joint; a *second* for the talocalcaneonavicular joint; a *third* for the calcaneocuboid joint; a *fourth* for the cuneonavicular, intercuneiform, and cuneocuboid joints, and the articulations of the intermediate and lateral cuneiforms with the bases of the second and third metatarsal bones, and the adjacent surfaces of the bases of the second, third, and fourth metatarsal bones; a *fifth* for the joint between the medial cuneiform and the metatarsal bone of the great toe; and a *sixth* for the articulation of the cuboid with the fourth and fifth metatarsal bones. A small synovial cavity is sometimes found between the contiguous surfaces of the navicular and cuboid bones.

METATARSOPHALANGEAL JOINTS

The metatarsophalangeal articulations may be considered ellipsoid in type and are formed by the reception of the oval rounded heads of the metatarsal bones in shallow cavities on the bases of the proximal phalanges. A loose fibrous capsule lined with a synovial membrane surrounds the metatarsophalangeal joint on each toe. The capsule is reinforced by a plantar ligament and two collateral ligaments. The dorsal surface of the capsule is thin, but it is strengthened by expansions from the extensor tendons.

Ligaments. The **plantar ligaments,** sometimes called *glenoid ligaments,* are thick, dense, fibrous structures. They are placed on the plantar surfaces of the joints in the intervals between the collateral ligaments, to which they are connected. They are loosely united to the metatarsal bones, but firmly attached to the bases of the first phalanges. Their plantar surfaces are intimately blended with the deep transverse metatarsal ligaments, and in addition, they are grooved for the passage of the flexor tendons, whose sheaths are connected to the sides of the grooves. The deep surfaces of the plantar ligaments form part of the articular facets for the heads of the metatarsal bones, and are lined by synovial membrane. Two sesamoid bones lie in the plantar ligament of the large toe.

The **collateral ligaments** are strong fan-shaped bands placed on both sides of each metatarsophalangeal joint. Each ligament is attached by one end to the posterior tubercle on the side of the head of the metatarsal bone and, by the other, to the contiguous extremity of the phalanx.

Movements. The movements permitted in the metatarsophalangeal articulations are *flexion, extension, abduction,* and *adduction.* Flexion is freer than extension and hence has a greater range. Flexion is limited by the extensor tendons and their expansions, whereas extension is limited by the flexor tendons and the plantar ligaments. Abduction and adduction are referenced to a line passing through the second toe.

INTERPHALANGEAL JOINTS

The interphalangeal joints of the toes are ginglymus or hinge joints. The distal surface of the proximal bone at each joint resembles a pulley, and the proximal surface of each distal bone has a reciprocally curved double concavity that fits in the pulley. The phalanges of the toes are smaller than those in the fingers, and frequently the joints between the middle and distal phalanges in the more lateral toes are ankylosed and nonfunctional.

Ligaments. Each joint has an articular capsule, a plantar and two collateral ligaments. The arrangement of these ligaments is similar to that in the metatarsophalangeal articulations. The extensor tendons and their dorsal expansions replace the dorsal ligaments.

Movements. The only movements permitted in the joints of the digits are flexion and extension. These movements have a greater range between the proximal and middle phalanges than between the middle and distal. The degree of flexion is quite considerable, but extension is limited by the flexor tendons and the plantar ligaments.

Arches of the Foot

The bones of the foot, their interconnecting ligaments, and the musculature are arranged in a manner consistent with the important functions of support, weight distribution, and locomotion of the body. An imprint of a wet foot on a floor shows that the heel, the lateral margin, the region beneath the heads of the metatarsal bones (the ball of the foot), and the pads of the distal phalanges touch the floor, whereas the medial margin of the foot is arched and does not touch the floor. The bony substructure of this **medial longitudinal arch,** when visualized on the articulated skeleton of the foot, is seen to consist of the calcaneus, talus, navicular, the three cuneiform bones, and the associated three medial metatarsal bones. The concavity of the medial arch is supported on two pillars, a posterior or calcaneal pillar, and an anterior or metatarsal pillar which consists of the heads of the first, second, and third metatarsal bones. The flatter **lateral longitudinal arch,** parallel to the medial, is composed also of the calcaneus posteriorly, the cuboid bone, and the fourth and fifth metatarsal bones anteriorly. Its posterior pillar is the calcaneal tuberosity and its anterior pillar consists of the heads of the lateral two metatarsal bones (Figs. 5-84; 5-85).

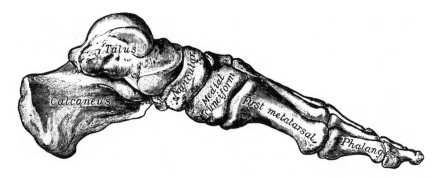

FIG. 5-84.   Bones on the medial aspect of the foot. Observe the concave medial longitudinal arch.

The stability of the medial longitudinal arch is dependent, to a large extent, on the functional integrity of the plantar aponeurosis and its subjacent long plantar and plantar calcaneonavicular (or spring) ligaments. However, the muscles are also of importance in maintaining the medial arch during locomotion. These especially include the tendon of the flexor hallucis longus, along with the flexor digitorum longus, the tibialis posterior, and the intrinsic foot muscles. In the static standing position, however, electromyographic studies have shown that these muscles are relatively inactive.

The lateral longitudinal arch transmits weight to the ground along the lateral aspect of the foot, but the height of its arch is lower than the medial. Its slighter concavity is supported by the lateral part of the long plantar ligament along with the plantar calcaneocuboid (short plantar) ligament. Additionally, the peroneus longus muscle and the plantar muscles on the lateral aspect of the sole of the foot help maintain the integrity of the

lateral arch. It has been stated that some special difference in function is served by the two longitudinal arches. The greater curvature, resiliency, and elasticity of the medial arch is believed to be of more importance in sustaining marked and immediate forces, such as those received during jumping. The more slightly curved lateral arch is less elastic as well, and is constructed to transmit thrust and weight to the ground, in contrast to the medial arch which absorbs such forces. The lateral arch appears to serve a more significant role in walking and running.

In addition to being arched longitudinally, the foot is also arched transversely. Actually, each foot forms half a transverse dome which is completed when the two feet are placed side by side. The more proximal **transverse** or **metatarsal arch** is formed by the navicular, the three cuneiforms, the cuboid, and the bases of the five metatarsal bones. The maintenance of the transverse arch is more importantly due to ligaments

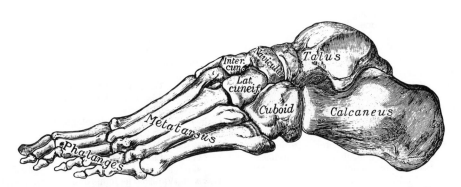

FIG. 5-85.   Bones on the lateral aspect of the foot. Observe that the concavity of the lateral longitudinal arch is less pronounced than that of the medial arch shown in Figure 5-84.

and tendons which bind the cuneiforms and the bases of the metatarsal bones together, than to the construction of the bony parts. The tendon of the peroneus longus muscle is of particular significance in the integrity of the transverse arch, because it stretches across the arch and tends to approximate the bones. More distally the heads of the metatarsal bones form a shallow but complete transverse arch in each foot. These latter arches are maintained by the deep transverse metatarsal ligaments.

Clinical Interest. The most common abnormality of the foot is a flattening of its plantar surface. In **flat foot**, the weight becomes distributed over the entire plantar surface, because the contour of the medial longitudinal arch (and consequently also the transverse arch) is lost. This depression of the medial arch is brought about by stretching the ligamentous connections at the talonavicular and cuneonavicular joints. As these ligaments become incompetent, the head of the talus pushes the soft plantar tissues to the ground, thereby flattening the medial longitudinal arch.

References

(References are listed not only to those articles and books cited in the text, but to others as well which are considered to contain valuable resource information for the student who desires it.)

Development, Histology and Physiology of Joints

Andersen, H. 1962. Histochemical studies of the development of the hip joint. Acta Anat., 48:258–292.

Andersen, H., and E. Bro-Rassmussen. 1961. Histochemical studies on the histogenesis of the joints in human fetuses with special reference to the development of the joint cavities in the hand and foot. Amer. J. Anat., 108:111–122.

Bailey, J. P., J. B. Cubberley, C. A. L. Stephens, Jr., and A. B. Stanfield. 1963. Vascular networks from explants of human synovialis in vitro. Anat. Rec., 147:525–531.

Barland, P., A. B. Novikoff, and D. Hamerman. 1962. Electron microscopy of the human synovial membrane. J. Cell Biol., 14:207–220.

Barnett, C. H., D. V. Davies, and M. A. MacConaill. 1961. *Synovial Joints, Their Structure and Mechanics.* Charles C Thomas, Springfield, Ill., 304 pp.

Bauer, W., M. W. Ropes, and H. Waine. 1940. The physiology of articular structures. Physiol. Rev., 20:272–312.

Bywaters, E. G. L. 1937. The metabolism of joint tissues. J. Path. Bact., 44:247–268.

Coggeshall, H. C., C. F. Warren, and W. Bauer. 1940. The cytology of normal human synovial fluid. Anat. Rec., 77:129–144.

Corbin, K. B., and J. C. Hinsey. 1939. Influence of the nervous system on bone and joints. Anat. Rec., 75:307–317.

Crelin, E. S., and W. E. Koch. 1965. Development of mouse pubic joint *in vivo* following initial differentiation *in vitro*. Anat. Rec., 153:161–165.

Crelin, E. S., and W. O. Southwick. 1964. Changes induced by sustained pressure in the knee joint articular cartilage of adult rabbits. Anat. Rec., 149:113–133.

Davies, D. V. 1942. The staining reaction of normal synovial membrane with special reference to the origin of synovial mucin. J. Anat. (Lond.), 77:160–169.

Davies, D. V. 1945. The cell content of synovial fluid. J. Anat. (Lond.), 79:66–73.

Davies, D. V. 1946. The lymphatics of the synovial membrane. J. Anat. (Lond.), 80:21–23.

Dintefass, L. 1963. Lubrication in synovial joints: A theoretical analysis. J. Bone Joint Surg., 45-A:1241–1256.

Drachman, D. B. 1969. Normal development and congenital malformations of joints. Bull. Rheum. Dis., 19:536–540.

Drachman, D. B., and L. Sokoloff. 1966. The role of movement in embryonic joint development. Dev. Biol., 14:401–420.

Elliott, H. C. 1936. Studies on articular cartilages. I. Growth mechanisms. Amer. J. Anat., 58:127–145.

Gardner, E. 1963. Physiology of joints. J. Bone Joint Surg., 45-A:1061–1066.

Gardner, E., and D. J. Gray. 1950. Prenatal development of the human hip joint. Amer. J. Anat., 87:163–211.

Gardner, E., and D. J. Gray. 1953. Prenatal development of the human shoulder and acromioclavicular joints. Amer. J. Anat., 92:219–276.

Gardner, E., and R. O'Rahilly. 1968. The early development of the knee joint in staged embryos. J. Anat. (Lond.), 102:289–299.

Ghadially, F. N., and S. Roy. 1969. *The Ultrastructure of Synovial Joints in Health and Disease.* Butterworth, London, 186 pp.

Gray, D. J., and E. Gardner. 1950. Prenatal development of the human knee and superior tibiofibular joints. Amer. J. Anat., 86:235–287.

Gray, D. J., and E. Gardner. 1951. Prenatal development of the human elbow joint. Amer. J. Anat., 88:429–469.

Gray, D. J., E. Gardner, and R. O'Rahilly. 1957. The prenatal development of the skeleton and joints of the human hand. Amer. J. Anat., 101:169–223.

Hackett, G. S. 1965. *Joint Ligament Relation Treated by Fibro-Osseous Proliferation.* Charles C Thomas, Springfield, Ill., 97 pp.

Haines, R. W. 1953. The early development of the femorotibial and tibiofibular joints. J. Anat. (Lond.), 87:192–206.

Haines, R. W. 1947. The development of joints. J. Anat. (Lond.), 81:33–55.

Lever, J. D., and E. H. R. Ford. 1958. Histological, histochemical and electron microscopic observations on synovial membranes. Anat. Rec., 132:525–540.

Lewis, R. W. 1955. *The Joints of the Extremities, a Radiographic Study.* Charles C Thomas, Springfield, Ill., 108 pp.

Moffet, B. C., Jr. 1957. The prenatal development of the human temporomandibular joint. Contr. Embryol. Carneg. Inst., 36:19–28.

Morton, D. J. 1952. Human Locomotion and Body Form. Williams and Wilkins, Baltimore, 285 pp.

O'Rahilly, R. 1951. The early prenatal development of the human knee joint. J. Anat. (Lond.), 85:166–170.

O'Rahilly, R. 1957. The development of joints. Ir. J. Med. Sci., 382:456–461.

Peacock, A. 1951. Observations on the prenatal development of the intervertebral disc in man. J. Anat. (Lond.), 85:260–274.

Pritchard, J. J., J. H. Scott, and F. G. Girgis. 1956. The structure and development of cranial and facial sutures. J. Anat. (Lond.), 90:73–87.

Sensenig, E. C. 1949. The early development of the vertebral column. Contr. Embryol. Carneg. Inst., 33: 21–42.

Sensenig, E. C. 1957. The development of the occipital and cervical segments and the associated structures in human embryos. Contr. Embryol. Carneg. Inst., 36:141–152.

Smith, J. W. 1962. The relationship of epiphysial plates to stress in some bones of the lower limb. J. Anat. (Lond.), 96:58–78.

Symons, N. B. B. 1952. The development of the human mandibular joint. J. Anat. (Lond.), 86:326–332.

Whillis, J. 1940. The development of synovial joints. J. Anat. (Lond.), 74:277–283.

Joints of the Axial Skeleton and Skull

Angle, J. L. 1948. Factors in temporomandibular joint form. Amer. J. Anat., 83:223–246.

Ashley, G. T. 1954. The morphological and pathological significance of synostosis at the manubriosternal joint. Thorax, 9:159–166.

Briscoe, C. 1925. The interchondral joints of the human thorax. J. Anat. (Lond.), 59:432–437.

Brooke, R. 1923. The sacroiliac joint. J. Anat. (Lond.), 58:299–305.

Cave, A. J. E. 1934. On the occipito-atlanto-axial articulations. J. Anat. (Lond.), 68:416–423.

Corner, E. M. 1906. The physiology of the lateral atlanto-axial joints. J. Anat. (Lond.), 41:149–154.

Coventry, M. B., R. K. Ghormley, and J. W. Kernohan. 1945. The intervertebral disc: Its microscopic anatomy and pathology. J. Bone Joint Surg., 27:105–112; 233–247; 460–474.

Davis, P. R. 1959. The medial inclination of the human thoracic intervertebral articular facets. J. Anat. (Lond.), 93:68–74.

Davis, P. R. 1961. Human lower lumbar vertebrae: Some mechanical and osteological considerations. J. Anat. (Lond.), 95:337–344.

Davis, P. R., J. D. G. Troup, and J. H. Burnard. 1965. Movements of the thoracic and lumbar spine when lifting: a chronocyclophotographic study. J. Anat. (Lond.), 99:13–26.

Doherty, J. L., and J. A. Doherty. 1937. Dislocation of the mandible. Amer. J. Surg., 38:480–484.

Donisch, E. W., and W. Trapp. 1971. The cartilage end-plates of the human vertebral column. Anat. Rec., 169:705–715.

Francis, C. C. 1955. Variations in the articular facets of the cervical vertebrae. Anat. Rec., 122:589–609.

Ganguly, D. N., and K. K. Roy. 1964. A study on the craniovertebral joint in man. Anat. Anz., 114:433–452.

Gray, D. J. and E. D. Gardner. 1943. The human sternochondral joints. Anat. Rec., 87:235–253.

Greene, E. 1937. Temporomandibular joint: Dental aspect. Ann. Otol. Rhinol. Laryngol., 46:150–157.

Haines, R. W. 1946. Movements of the first rib. J. Anat. (Lond.), 80:94–100.

Hohl, M. 1964. Normal motions of the upper portion of the cervical spine. J. Bone Joint Surg., 46-A:1777–1779.

Leonhart, G. P. 1914. A case of stylo-hyoid ossification. Anat. Rec., 8:325–332.

Lewin, T. 1968. Anatomical variations in lumbo-sacral synovial joints with particular references to subluxation. Acta Anat., 71:229–248.

Miles, M., and W. E. Sullivan. 1961. Lateral bending at the lumbar and lumbosacral joints. Anat. Rec., 139:387–398.

Mitchell, G. A. G. 1934. The lumbosacral junction. J. Bone Joint Surg., 16:233–254.

Moffet, B. C., Jr., L. C. Johnson, J. B. McCabe, and H. C. Askew. 1964. Articular remodeling in the adult human temporomandibular joint. Amer. J. Anat., 115:119–142.

Nachemson, A. 1960. Lumbar intradiscal pressure. Experimental studies on postmortem material. Acta Orthop. Scand. [Suppl. 43], 104 pp.

Nachemson, A. 1965. The effect of forward leaning on lumbar intradiscal pressure. Acta Orthop. Scand., 35:314–328.

Odgers, P. N. B. 1932. The lumbar and lumbosacral diarthrodial joints. J. Anat. (Lond.), 67:301–317.

Peacock, A. 1952. Observations on the postnatal structure of the intervertebral disc in man. J. Anat. (Lond.), 86:162–179.

Powell, T. V., and A. G. Brodie. 1963. Closure of the sphenoccipital synchondrosis. Anat. Rec., 147:15–23.

Rees, L. A. 1954. The structure and function of the mandibular joint. Br. Dent. J., 96:125–133.

Robinson, M. 1946. The temporomandibular joint: theory of reflex controlled nonlever action of the mandible. J. Amer. Dent. Assoc., 33:1260–1271.

Rogers, L. C., and E. E. Payne. 1961. The dura mater at the craniovertebral-junction. J. Anat. (Lond.), 95:586–588.

Sarnat, B. G. 1951. The Temporomandibular Joint. 2nd ed., Charles C Thomas, Springfield, Ill., 260 pp.

Shapiro, H. H., and R. C. Truex. 1943. The temporomandibular joint and the auditory junction. J. Amer. Dent. Assoc., 30:1147–1168.

Singh, S. 1965. Variations of the superior articular facets of atlas vertebrae. J. Anat. (Lond.), 99:565–571.

Spurling, R. G. 1956. Lesions of the Cervical Intervertebral Disc. Charles C Thomas, Springfield, Ill., 134 pp.

Stein, M. R. 1939. The "mandibular sling." Dental Survey, 15:883–887.

Stewart, T. D. 1938. Accessory sacro-iliac articulations in the higher primates and their significance. Amer. J. Phys. Anthropol., 24:43–59.

Stilwell, D. L., Jr. 1956. The nerve supply of the vertebral column and its associated structures in the monkey. Anat. Rec., 125:139–170.

Sylvin, B. 1951. On the biology of the nucleus pulposus. Acta Orthop. Scand., 20:275–279.

Trotter, M. 1934. Synostosis between the manubrium and body of the sternum in Whites and Negroes. Amer. J. Phys. Anthropol., 18:439–442.

Trotter, M. 1937. Accessory sacroiliac articulations. Amer. J. Phys. Anthropol., 22:247–261.

Trotter, M. 1940. A common anatomical variation in the sacro-iliac region. J. Bone Joint Surg., 22:293–299.

Wakeley, C. P. G. 1939. The surgery of the temporomandibular joint. Surgery, 5:697–706.

Wakeley, C. P. G. 1948. The mandibular joint. Ann. R. Coll. Surg., 2:111–120.

Weisl, H. 1954. The ligaments of the sacroiliac joint examined with particular reference to their function. Acta Anat., 20:201–213.

Weisl, H. 1954. The articular surfaces of the sacroiliac joint and their relation to the movements of the sacrum. Acta Anat., 22:1–14.

Weisl, H. 1955. Movements of the sacroiliac joint. Acta Anat., 23:80–91.

Wiles, P. 1935. Movements of the lumbar vertebrae during flexion and extension. Proc. Roy. Soc. Med., 28:647–651.

Clavicular and Shoulder Joints

Abbott, L. C., and J. B. deC. M. Saunders. 1939. Acute traumatic dislocation of the long head of the biceps brachii. Surgery, 6:817–840.

Bailey, R. W. 1967. Acute and recurrent dislocation of the shoulder. J. Bone Joint Surg., 49-A:767–773.

Basmajian, J. V., and F. J. Bazant. 1959. Factors preventing downward dislocation of the adducted shoulder joint. An electromyographic and morphological study. J. Bone Joint Surg., 41-A:1182–1186.

Bearn, J. G. 1967. Direct observation on the function of the capsule of the sternoclavicular joint in clavicular support. J. Anat. (Lond.), 101:159–170.

Cave, A. J. E. 1961. The nature and morphology of the costoclavicular ligament. J. Anat. (Lond.), 95:170–179.

De Palma, A. F. 1957. Degenerative Changes in the Sternoclavicular and Acromioclavicular Joints in Various Decades. Charles C Thomas, Springfield, Ill., 178 pp.

Flecker, H. 1929. Roentgenographic study of movements of abduction at the normal shoulder joint. Med. J. Aust., 2:123–124.

Gardner, E. 1948. The innervation of the shoulder joint. Anat. Rec., 102:1–18.

Inman, V. T., J. B. deC. M. Saunders, and L. G. Abbott. 1944. Observations on the function of the shoulder joint. J. Bone Joint Surg., 26:1–30.

Johnston, T. B. 1937. The movements of the shoulder joint. Br. J. Surg., 25:252–260.

Kaplan, E. B. 1943. The coraco-humeral ligament of the human shoulder. Bull. Hosp. Joint Dis., 4:62–65.

Leslie, J. T., Jr., and T. J. Ryan. 1962. The anterior axillary incision to approach the shoulder joint. J. Bone Joint Surg., 44-A:1193–1196.

Lewis, O. J. 1959. The coraco-clavicular joint. J. Anat. (Lond.), 93:296–303.

Lockhart, R. D. 1930. Movements of the normal shoulder joint. J. Anat. (Lond.), 64:288–302.

Lockhart, R. D. 1933. A further note on movements of the shoulder joint. J. Anat. (Lond.), 68:135–137.

Martin, C. P. 1940. The movements of the shoulder-joint, with special reference to rupture of the supraspinatus tendon. Amer. J. Anat., 66:213–234.

Moseley, H. F. 1952. Ruptures of the Rotator Cuff. Charles C Thomas, Springfield, Ill., 90 pp.

Sheving, L. E., and J. E. Pauly. 1959. An electromyographic study of some muscles acting on the upper extremity of man. Anat. Rec., 135:239–245.

Elbow, Wrist, and Hand Joints

Basmajian, J. V., and A. Latif, 1957. Integrated actions and functions of the chief flexors of the elbow. A detailed electromyographic analysis. J. Bone Joint Surg., 39-A:1106–1118.

Bradley, K. C., and S. Sunderland. 1953. The range of movement at the wrist joint. Anat. Rec., 116:139–145.

Eliason, E. L., and R. B. Brown. 1937. Posterior dislocation at the elbow with rupture of the radial and ulnar arteries. Ann. Surg., 106:1111–1115.

Gad, P. 1967. The anatomy of the volar part of the capsules of the finger joints. J. Bone Joint Surg., 49-B:362–367.

Gardner, E. 1948. The innervation of the elbow joint. Anat. Rec., 102:161–174.

Gray, D. J., and E. Gardner. 1965. The innervation of the joints of the wrist and hand. Anat. Rec., 151:261–266.

Haines, R. W. 1944. The mechanisms of rotation at the first carpometacarpal joint. J. Anat. (Lond.), 78:44–46.

Haines, R. W. 1951. The extensor apparatus of the finger. J. Anat. (Lond.), 85:251–259.

Hakstian, R. W., and R. Tubiana. 1967. Ulnar deviation of the fingers. The role of joint structure and function. J. Bone Joint Surg., 49-A:299–316.

Halls, A. A., and A. Travill. 1964. Transmission of pressures across the elbow joint. Anat. Rec., 150:243–247.

Haxton, H. A. 1945. A comparison of the action of extension of the knee and elbow joints in man. Anat. Rec., 93:279–286.

Hewitt, D. 1928. The range of active motion at the wrist of women. J. Bone Joint Surg., 10:775–787.

Horwitz, T. 1940. An anatomic and roentgenologic study of the wrist joint. Surgery, 7:773–783.

Joseph, J. 1951. Further studies of the metacarpophalangeal and interphalangeal joints of the thumb. J. Anat. (Lond.), 85:221–229.

Kaplan, E. B. 1936–37. Extension deformities of the proximal interphalangeal joints of the fingers. J. Bone Joint Surg., 18:781–983. Correction, 19:1144.

Kilburn, P., J. G. Sweeney, and F. F. Silk. 1962. Three cases of compound posterior dislocation of the elbow with rupture of the brachial artery. J. Bone Joint Surg., 44-B:119–121.

Kropp, B. N. 1945. A note on the piso-triquetral joint. Anat. Rec., 92:91–92.

Kuczynski, K. 1974. Carpometacarpal joint of the human thumb. J. Anat. (Lond.), 118:119–126.

Landsmeer, J. M. F. 1955. Anatomical and functional investigations on the articulation of the human fingers. Acta Anat., Suppl. 24, 25:1–69.

Lewis, O. J. 1970. The development of the human wrist joint during the fetal period. Anat. Rec., 166:499–515.

Lewis, O. J., R. M. Hamshere, and T. M. Bucknill. 1970. The anatomy of the wrist joint. J. Anat. (Lond.), 106:539–552.

MacConaill, M. A. 1941. The mechanical anatomy of the carpus and its bearings on some surgical problems. J. Anat. (Lond.), 75:166–175.

Martin, B. F. 1958. The annular ligament of the superior radioulnar joint. J. Anat. (Lond.), 92:473–482.

Martin, B. F. 1958. The oblique cord of the forearm. J. Anat. (Lond.), 92:609–615.

Napier, J. R. 1955. The form and function of the carpometacarpal joint of the thumb. J. Anat. (Lond.), 89:362–369.

Ray, R. D., R. J. Johnson, and R. M. Jameson. 1951. Rotation of the forearm. J. Bone Joint Surg., 33-A:993–996.

Robbins, H. 1963. Anatomical study of the median nerve in the carpal tunnel and etiologies of the carpal-tunnel syndrome. J. Bone Joint Surg., 45-A:953–966.

Roston, J. B., and R. W. Haines. 1947. Cracking (audible) in the metacarpophalangeal joint. J. Anat. (Lond.), 81:165–173.

Smith, R. D., and G. R. Holcomb. 1958. Articular surface interrelationships in finger joints. Acta Anat., 32:217–229.

Stecher, R. M. 1958. Ankylosis of the finger joints in rheumatoid arthritis. Ann. Rheum. Dis., 17:365–375.

Stopford, J. S. B. 1921. The nerve supply of the interphalangeal and metacarpophalangeal joints. J. Anat. (Lond.), 56:1–11.

Travill, A. A. 1964. Transmission of pressure across the elbow joint. Anat. Rec., 150:243–247.

Waugh, R. L., and A. G. Yancey. 1948. Carpometacarpal dislocations. J. Bone Joint Surg., 30-A:397–404.

Wright, R. D. 1935. A detailed study of movement of the wrist joint. J. Anat. (Lond.), 70:137–142.

Hip Joint

Boscoe, A. R. 1932. The range of active abduction and lateral rotation at the hip joint of men. J. Bone Joint Surg., 14:325–331.

Campbell, W. C. 1940. Surgery of the hip joint from the physiologic aspect. Surgery, 7:167–186.

Chandler, S. B., and P. H. Krenscher. 1932. A study of the blood supply of the ligamentum teres and its relation to the circulation of the head of the femur. J. Bone Joint Surg., 14:834–846.

Claffey, T. J. 1960. Avascular necrosis of the femoral head. An anatomical study. J. Bone Joint Surg., 42-B:802–809.

Crock, H. V. 1965. A revision of the anatomy of the arteries supplying the upper end of the human femur. J. Anat. (Lond.), 99:77–88.

Gardner, E. 1948. The innervation of the hip joint. Anat. Rec., 101:353–371.

Ghormley, J. W. 1944. Hip motions. Amer. J. Surg., 66:24–30.

Hart, V. L. 1952. Congenital Dysplasia of the Hip Joint and Sequelae in the Newborn and Early Postnatal Life. Charles C Thomas, Springfield, Ill., 187 pp.

Howe, W. W., Jr., T. Lacey, and R. P. Schwartz. 1950. A study of the gross anatomy of the arteries supplying the proximal portion of the femur and the acetabulum. J. Bone Joint Surg., 32-A:856–866.

Inman, V. T. 1947. Functional aspects of the abductor muscles of the hip. J. Bone Joint Surg., 29:607–619.

Joseph, J., and J. P. Williams. 1957. Electromyography of certain hip muscles. J. Anat. (Lond.), 91:286–294.

Milch, H. 1961. The measurement of pelvi-femoral motion. Anat. Rec., 140:135–145.

Moore, J. B., and J. O. Vaughn. 1928. The range of active motion at the hip joint of men. J. Bone Joint Surg., 10:248–257.

Naffziger, H. C., and N. C. Norcross. 1942. The surgical approach to lesions of the upper sciatic nerve and the posterior aspect of the hip joint. Surgery, 12:929–932.

Roberts, W. H. 1963. The locking mechanism of the hip joint. Anat. Rec., 147:321–324.

Stanisavljevic, S. 1964. Diagnosis and Treatment of Congenital Hip Pathology in the Newborn. Williams and Wilkins, Baltimore, 94 pp.

St. Clair Strange, F. G. 1965. The Hip. Heinemann, London, 284 pp.

Sutro, C. J. 1936. The pubic bones and their symphysis. Arch. Surg., 32:823–841.

Trueta, J. 1957. The normal vascular anatomy of the human femoral head during growth. J. Bone Joint Surg., 39-B:358–394.

Trueta, J., and M. H. M. Harrison. 1953. The normal vascular anatomy of the femoral head in adult man. J. Bone Joint Surg., 35-B:442–461.

Tucker, F. R. 1949. Arterial supply to the femoral head and its clinical importance. J. Bone Joint Surg., 31-B:82–93.

Weathersby, H. T. 1959. The origin of the artery of the ligamentum teres femoris. J. Bone Joint Surg., 41-A:261–263.

Wertheimer, L. G. 1952. The sensory nerves of the hip joint. J. Bone Joint Surg., 34-A:477–487.

Young, J. 1940. Relaxation of the pelvic joints in pregnancy. J. Obstet. Gynaec. Brit. Emp., 47:493–524.

Knee Joint

Barnett, C. H. 1953. Locking at the knee joint. J. Anat. (Lond.), 87:91–95.

Brantigan, O. C., and A. F. Voshell. 1941. The mechanics of the ligaments and menisci of the knee joint. J. Bone Joint Surg., 23:44–66.

Brantigan, O. C., and A. F. Voshell. 1943. The tibial collateral ligament: Its function, its bursae, and its relation to the medial meniscus. J. Bone Joint Surg., 25:121–131.

Freeman, M. A. R., and B. Wyke. 1967. The innervation of the knee joint. J. Anat. (Lond.), 101:505–532.

Fullerton, A. 1916. The surgical anatomy of the synovial membrane of the knee joint. Br. J. Surg. 4:191–200.

Gardner, E. 1948. The innervation of the knee joint. Anat. Rec., 101:109–130.

Gardner, E., and D. J. Gray. 1968. The innervation of the joints of the foot. Anat. Rec., 161:141–148.

Gardner, E., and R. O'Rahilly, 1968. The early development of the knee joint in staged human embryos. J. Anat. (Lond.), 102:289–299.

Girgis, F. G., J. L. Marshall, and A. R. S. Al-Monajam. 1975. The cruciate ligaments of the knee joint. Anatomical, functional and experimental analysis. Clin. Orthop., 106:216–231.

Heller, L., and J. Langman. 1964. The meniscofemoral ligaments of the human knee. J. Bone Joint Surg., 46-B:307–313.

Kaplan, E. B. 1962. Some aspects of the functional anatomy of the human knee joint. Clin. Orthop., 23:18–29.

King, D. 1936. The function of semilunar cartilages. J. Bone Joint Surg., 18:1069–1076.

Last, R. J. 1948. Some anatomical details of the knee joint. J. Bone Joint Surg., 30-B:683–688.

Last, R. J. 1950. The popliteus muscle and the lateral meniscus: With a note on the attachment of the medial meniscus. J. Bone Joint Surg., 32-B:93–99.

MacConaill, M. A. 1931. The function of intraarticular fibrocartilages with special reference to the knee and inferior radioulnar joints. J. Anat. (Lond.), 66:210–227.

Ross, R. F. 1932. A quantitative study of rotation of the knee joint in man. Anat. Rec., 52:209–223.

Scapinelli, R. 1968. Studies on the vasculature of the human knee joint. Acta Anat., 70:305–331.

Shaw, N. E., and B. F. Martin. 1962. Histological and histochemical studies on mammalian knee joint tissues. J. Anat. (Lond.), 96:359–373.

Smith, J. W. 1956. Observations on the postural mechanism of the human knee joint. J. Anat. (Lond.), 90:236–260.

Ankle and Foot Joints

Barnett, C. H., and J. R. Napier. 1952. The axis of rotation at the ankle joint in man. J. Anat. (Lond.), 86:1–9.

Basmajian, J. V., and G. Stecko. 1963. The role of muscles in arch support of the foot. An electromyographic study. J. Bone Joint Surg., 45-A:1184–1190.

Bunning, P. S. C., and C. H. Barnett. 1965. A comparison of the adult and foetal talocalcaneal articulations. J. Anat. (Lond.), 99:71–76.

Cahill, D. R. 1965. The anatomy and function of the contents of the human tarsal sinus and canal. Anat. Rec., 153:1–17.

Hardy, R. H. 1951. Observations on the structure and properties of the plantar calcaneonavicular ligament in man. J. Anat. (Lond.), 85:135–139.

Hicks, J. H. 1953. The mechanics of the foot. I. The joints. J. Anat. (Lond.), 87:345–357.

Hicks, J. H. 1954. The mechanics of the foot. II. The plantar aponeurosis and the arch. J. Anat. (Lond.), 88:25–30.

Hicks, J. H. 1955. The foot as a support. Acta Anat., 25:34–45.

Inkster, R. G. 1938. Inversion and eversion of the foot and the transverse tarsal joint. J. Anat. (Lond.), 72:612–613.

Jones, F. W. 1949. *Structure and Function as Seen in the Foot.* 2nd ed., Baillière, Tindall and Cox, London, 333 pp.

Jones, R. L. 1941. The human foot. An experimental study of its mechanics, and the role of its muscles and ligaments in the support of the arch. Amer. J. Anat., 68:1–40.

MacConaill, M. A. 1945. The postural mechanisms of the human foot. Proc. Roy. Ir. Acad. L., B, 14:265–278.

Manter, J. T. 1941. Movements of the subtalar and transverse tarsal joints. Anat. Rec., 80:397–410.

Manter, J. T. 1946. Distribution of compression forces in the joints of the human foot. Anat. Rec., 96:313–321.

Morton, D. J. 1935. *The Human Foot.* Columbia Press, New York, 244 pp.

Shepard, E. 1951. Tarsal movements. J. Bone Joint Surg., 33-B:258–263.

Smith, J. W. 1957. The forces acting at the ankle joint during standing. J. Anat. (Lond.), 91:545–564.

Smith, J. W. 1958. The ligamentous structures in the canalis and sinus tarsi. J. Anat. (Lond.), 92:616–620.

6

Muscles

and

Fasciae

Contractility, a fundamental property characteristic of protoplasm, is especially well developed in cells of muscular tissue. Aggregations of these cells in various parts of the body form the **muscles.** These structures are capable of contracting, and muscles are responsible not only for voluntary motion, such as that of the limbs, but also for the contractions of the heart as well as the less obvious movements of the viscera. This chapter deals with the somatic muscles. Cardiac muscle and the visceral muscles of internal organs will be discussed in subsequent chapters.

The ability to move provides organisms with a means of reacting to the environment. The contractions of somatic muscles are made mechanically effective by means of the tendons, aponeuroses, and fasciae, which secure the ends of the muscles and control the direction of their pull. Many muscles are attached to bones, but others, such as the superficial muscles of the face, have no connections to the skeleton. Muscles of the limbs and trunk underlie the skin and

fascial layers, giving substance and contour to these regions. Muscle tissue accounts for approximately 40% of the body weight, and the individual muscles vary greatly in size. The gastrocnemius forms the bulk of the calf of the leg; the sartorius is nearly 61 cm in length; and the stapedius, a tiny muscle of the middle ear, weighs 0.1 gm and is 2 to 3 mm in length.

Somatic muscles have also been called **voluntary, skeletal,** or **striated** muscles, although these terms are not always appropriate. Examples of somatic muscles that are not under complete voluntary control are found in the pharynx and upper esophagus; some somatic muscles, such as certain facial muscles, are not attached to the skeleton, and both visceral and cardiac muscles show striations. In general, however, most somatic muscles are under voluntary control, are attached to skeletal structures, and show prominent longitudinal and cross striations.

Development of Muscles

Most muscles throughout the body, with the exception of a few that are of ectodermal origin, arise from mesoderm. The intrinsic muscles of the trunk are derived from the myotomes, while the muscles of the head and limbs differentiate directly from the mesoderm.

MUSCLES DERIVED FROM MYOTOMES. The intrinsic muscles of the trunk, which are derived directly from the myotomes, are conveniently treated in two groups, the deep muscles of the back and the thoracoabdominal muscles.

The **deep muscles of the back** extend from the sacral to the occipital region and vary much in length and size. They act chiefly on the vertebral column. The shorter muscles, such as the interspinales, intertransversarii, the deeper layers of the multifidus, the rotatores, levatores costarum, obliquus capitis inferior, and obliquus capitis superior, which *extend between adjoining vertebrae,* and the rectus capitis posterior minor, which extends between the adjoining atlas and occipital bone, all retain the primitive segmental character of the myotomes. Other muscles, such as the splenius capitis, splenius cervicis, sacrospinalis, erector spinae, superficial and intermediate layers of the multifidus, iliocostalis, longissimus, spinales, semispinales, and rectus capitis posterior major, which *extend over several vertebrae,* are formed by the fusion of successive myotomes and then split into longitudinal columns. A thick fascial plane, the thoracolumbar fascia, develops between these true myotomic muscles and the more superficial ones which migrate over the back, such as the trapezius, the rhomboids, and the latissimus dorsi.

The **anterior vertebral muscles**—the longus colli, longus capitis, rectus capitis anterior, and rectus capitis lateralis—are derived from the ventral part of the cervical myotomes, as are probably also the scalenes.

The **thoracoabdominal muscles** arise through the ventral extension of the thoracic myotomes into the body wall. This process takes place coincidentally with the ventral extension of the ribs. In the thoracic region the primitive myotomic segments persist as the intercostal muscles, but over the abdomen these ventral myotomic processes fuse into a sheet that splits in various ways to form the rectus abdominis, the external and internal oblique, and the transversus. Such muscles as the pectoralis major and minor and the serratus anterior do not belong to the aforementioned group.

VENTROLATERAL MUSCLES OF THE NECK. The intrinsic muscles of the *tongue,* the *infrahyoid muscles,* and the *diaphragm* are derived from a more or less continuous premuscle mass, formed by the **occipital myotomes,** which extends on each side from the tongue into the lateral region of the upper half of the neck. Into this developing mesoderm extend the hypoglossal nerves and branches of the upper cervical nerves. The two halves, which form the infrahyoid muscles and the diaphragm, are at first widely separated from each other by the heart. As the heart descends into the thorax, the diaphragmatic portion of each lateral muscle mass is carried with its phrenic nerve down into the thorax, while the laterally placed infrahyoid muscles move toward the midventral line of the neck.

MUSCLES OF THE SHOULDER GIRDLE AND UPPER EXTREMITY. The *trapezius* and *sternocleidomastoid* arise from a common premuscle mass in the occipital region just caudal to the last branchial arch. As the mass increases in size, it spreads caudally to the shoulder girdle to which it later becomes attached, and dorsally to the spinous processes, gaining attachment on these at a still later period. The *levator scapulae, serratus anterior,* and the *rhomboids* arise from premuscle tissue in the lower cervical region and undergo extensive migration. The *latissimus dorsi* and *teres major* are associated in their origin from the premuscle sheath of the arm, as are also the two *pectoral muscles,* when the arm bud lies in the lower cervical region.

The remaining *intrinsic muscles of the upper extremity* develop *in situ* from mesoderm of the arm bud and probably do not derive cells or buds from the myotomes. The nerves to these muscles enter the arm bud when it still lies in the cervical region and, as the arm shifts caudally over the thorax, these cervical nerves unite to form the brachial plexus.

MUSCLES OF THE LOWER EXTREMITY. The *muscles of the lower extremity,* like those of the upper, develop *in situ* from the mesoderm of the leg bud, the myotomes apparently taking no part in their formation.

MUSCLES OF THE HEAD. The *extraocular muscles* in the orbit arise from the mesoderm over the dorsal and caudal sides of the optic stalk. This mesenchyme gives rise to what may be called three *preotic myotomes,* each of which is supplied by its own cranial nerve (oculomotor, trochlear, and abducens). The *muscles of mastication* arise from the mesoderm of the **mandibular arch.** The mandibular division of the trigeminal nerve enters this premuscle mass before the temporal, masseter, and pterygoid muscles split to develop as separate structures. The *facial muscles of expression* arise from the mesoderm of the **hyoid arch.** The facial nerve enters this mass before the individual muscles begin to differentiate, and as the muscle mass spreads out over the face, head, and neck, it splits more or less incompletely into the various muscles.

MUSCLE-NERVE MIGRATION. Initial differentiation of the muscular system proceeds independently of the nervous system and only later does it appear that muscles require the functional stimuli of the nerves for their continued existence and growth. Owing to their early attachments to muscle rudiments, the nerves in a general way may indicate the position of origin of many of the muscles, and likewise, in many instances, the nerves reveal the paths along which the developing muscles have migrated during development. The muscle mass that forms the diaphragm, for example, has its origin in the region of the fourth and fifth cervical segments. The phrenic nerve centers this developing mesoderm in the cervical region and is drawn out as the diaphragm migrates through the thorax. Likewise, the trapezius and sternocleidomastoid arise in the lateral occipital region as a common muscle mass, into which the accessory nerve grows at a very early period. As the muscle mass migrates and expands caudally, the nerve is carried with it. The pectoralis major and minor arise in the cervical region, receive their nerves while in this position and, as the muscle mass migrates and expands caudally over the thorax, the nerves are carried along. The latissimus dorsi and serratus anterior are excellent examples of migrating muscles whose nerve supply indicates their origin in the cervical region. The rectus abdominis and the other abdominal muscles also migrate or shift from a lateral to a ventrolateral or abdominal position, carrying with them the nerves.

The facial nerve, which at an early stage enters the common facial muscle mass of the second branchial or hyoid arch, migrates with the differentiating muscle as it spreads over the head, face, and neck. As the muscle mass splits into the various muscles of facial expression, the nerve is correspondingly split. Similarly, the mandibular division of the trigeminal nerve enters the muscle mass of the mandibular arch, and as this mass splits and migrates to form the muscles of mastication, the nerve splits into its various branches. The muscles supplied by the oculomotor nerve arise from a single mass in the orbital region, and the lingual muscles arise from a common mass supplied by the hypoglossal nerve. These examples show that the nerve supply may serve as a key to understanding the common origin of certain muscle groups.

Fasciae

A number of noncontractile connective tissue elements are required to organize the contractile muscle fibers into effective mechanical structures. Large numbers of individual muscle fibers are bound together into fascicles by collagenous connective tissue septa, collectively called the **perimysium.**

Surrounding all the fascicles of a muscle is a thicker connective tissue layer, the **epimysium.** The ends of the muscle are attached to bones by tendons and aponeuroses, and individual muscles and groups of muscles are held in place and separated by connective tissue sheets called **fasciae.** The term fascia is used for the dissectible, fibrous connective tissues of the body other than specifically organized structures such as tendons, aponeuroses, and ligaments. In a more restricted sense, the term fascia also is used to indicate local connective tissue membranes that enclose a part of the body or invest organs. It should be appreciated that the fasciae are part of the general connective tissues which form a functional as well as morphological system in which the connective tissue varies in thickness, in density, in accumulation of fat, and in relative amounts of collagenous fibers, elastic fibers, and tissue fluids.

Three general types of fasciae are usually described: the **superficial fascia,** the **deep fascia,** and the **subserous fascia.** The superficial fascia is a loose, fat-filled layer which intervenes between the skin and the deep fascia. It is variable in thickness. The deep fascia is the principal somatic fascia which invests and penetrates between the various structures that form the body wall and the limbs. It is the most extensive of the three types, and occupies much of the dissector's time in the separation of anatomical structures. The subserous fascia lies within the body cavities, forming the fibrous layer of the serous membranes (pleura, pericardium, and peritoneum), which covers and supports the viscera and attaches the parietal layer of the serous membranes to the deep fascia of the internal surface of the body wall.

SUPERFICIAL FASCIA. The superficial fascia is continuous over the entire body between the skin and the deep fascial investment of the specialized structures of the body, such as muscles. It is composed of two layers. The *outer* one, sometimes called *panniculus adiposus,* normally contains an accumulation of fat that may be several centimeters thick or, in emaciated individuals, may be almost entirely lacking. Although the amount of fat and its distribution in the superficial fascia varies between individuals, generally it is somewhat more abundant in females. In certain regions of the body, such as the external nose, the auricle of the ear, the eyelids, the labia minora, the penis, and the scrotum, the superficial fascia contains virtually no fat. The *inner* layer is a thin membrane that ordinarily has no fat, but contains a generous amount of elastic tissue. The two layers are quite adherent in most regions, but are separable by careful dissection, particularly in the lower anterior abdominal wall. Between the two layers frequently course the superficial arteries, veins, nerves, and lymphatics. The mammary glands, most of the facial muscles, the platysma muscle, as well as the dartos muscle of the scrotum, are all found in the superficial fascia. The superficial fascia acts as an insulator for the body and helps to prevent the loss of body heat.

In many parts of the body the superficial fascia glides freely over the deep fascia, producing the characteristic movability of the skin in a region such as the back of the hand. At these sites the superficial and deep fasciae can be dissected apart easily by a probing finger or blunt instrument, as they are separated by a *fascial cleft.* At certain other points on the body surface, especially over bony prominences, the two fasciae are closely adherent. They retain their individuality even at these sites, however, and are not continuous.

DEEP FASCIA.   Deep to the superficial fascia is a dense stratum of fibrous tissue which immediately covers the muscles and is called the deep fascia. It consists of a rather intricate series of sheets and bands which hold the individual muscles and other structures in their proper relative positions, separating them for independent function as well as joining them together into an integrated whole. The intrinsic connective tissue of the capsules or stroma of these structures is not included in the definition of deep fascia. In the case of a muscle, the epimysium may be fused with the overlying deep fascia and even lose its identity, as in the triceps, or the epimysium may be separated from the deep fascia by a cleft and retain its individuality, as in the biceps brachii.

The sheets of deep fascia are organized into a continuous system and frequently attach to the periosteum of bones or the perichondrium of cartilage. The ligaments also may assist in establishing this continuity. On occasion a sheet of deep fascia may split in order to surround or *invest* muscles, only to unite again into a single sheet beyond the muscles. Splitting and fusing are important because in this manner any sheet of fascia can be traced to other sheets and can be shown to make eventual attachment to the skeleton.

The mechanical function of fascia is particularly well developed in the deep fascia and is responsible for its many regional variations and specializations. A fascial sheet may be thickened, either for strength or padding. It may be fused with another sheet or split into several layers. It may be separated from another fascial sheet by a plane of cleavage, or several sheets may combine to form compartments for groups of muscles.

The thickening of a fascial sheet for greater strength, especially if it receives the direct pull of a muscle, occurs through the addition of parallel bundles of collagenous fibers which impart to it the white, glistening appearance of an aponeurosis. Deep fascia of this type may separate muscles of different functional groups by attaching along the longitudinal axis of a bone between the muscles. These fascial sheets are called **intermuscular septa,** and examples are shown in Figure 6-1, where the flexors in the arm (biceps brachii and brachialis) are separated both medially and laterally from the extensor (triceps) in this manner. At another site, a strengthened fascial sheet may cover a muscle and be used by it as a surface for attachment, for example, the **investing layer** of the forearm. An extreme instance is that of the fascia lata of the thigh, where the iliotibial band is in fact the principal tendon of insertion for the gluteus maximus and the tensor fasciae latae muscles.

A band of deep *fascia may act as a ligament.* This is the case with the greatly strengthened portion of the clavipectoral fascia called the costocoracoid membrane. By its attachment to the coracoid process and ribs, this membrane serves as a ligament for the articulations of the clavicle. Other specializations of the deep fascia are the annular ligaments and retinacula at the wrist and ankle which provide tunnels for the long tendons of the hand and foot.

At some sites a laminated fascial sheet may be thickened without a corresponding increase in strength. The fascia splits into two leaves which are separated by a pad of connective tissue containing fat and an occasional blood vessel or lymph node. An example of this is the lamination of the outer cervical fascia above the sternum which is called the suprasternal space.

A **fascial compartment** is a region of the

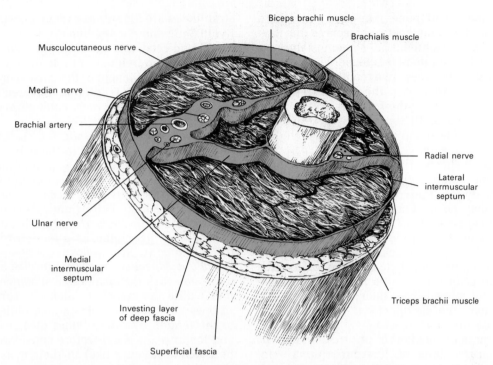

Musculocutaneous nerve

Median nerve

Brachial artery

Ulnar nerve

Medial
intermuscular
septum

Investing layer
of deep fascia

Superficial fascia

Biceps brachii muscle

Brachialis muscle

Radial nerve

Lateral
intermuscular
septum

Triceps brachii muscle

Fıg. 6-1. Cross section through the right arm a little below the middle of the humeral shaft. Note the deep fascia of the arm and observe that the investing layer of brachial fascia surrounds the arm, while the medial and lateral intermuscular septa separate the flexor (biceps brachii and brachialis muscles) from the extensor (triceps brachii muscle) compartment.

body that is walled off by deep fascia. Characteristically, it contains a muscle or a group of muscles, but in some instances other structures are included. A typical example, described above, is the flexor and extensor compartments of the arm. The arm, enclosed by brachial fascia, is divided into two compartments by the medial and lateral intermuscular septa. In many descriptions of fascia, such compartments are ambiguously called spaces or potential spaces. For example, the mediastinum is sometimes called a "space" containing the heart, great vessels, esophagus, and other structures, whereas in reality it is a compartment enclosed by mediastinal fascia.

The **fascial cleft,** an important specialization, is a cleavage that separates two contiguous fascial surfaces. At times it is a stratum rich in fluid but poor in traversing fibers, allowing the two fascial surfaces to move more or less freely and making separation easy at dissection. The degree of separation may vary from an almost complete detachment to a comparatively strong adhesion, apparently depending on the need for motion between the parts. The cleft between the superficial fascia and the deep fascia has

already been mentioned. The cleft between a muscle and an overlying, restraining fascial layer, like the biceps brachii and brachial fascia, actually separates the epimysium of the muscle and the true fascial sheet. The cleft between two adjacent muscles is likely to separate the epimysium of each muscle, but the latter may, in some places, be thickened into a true fascial sheet.

A **bursa** is a specialized structure that allows an efficient freedom of motion between contiguous connective tissue surfaces. It is a relatively small, circumscribed pocket of complete separation, the lining of which provides two opposed, lubricated surfaces similar to the synovial membranes of a joint. When a tendon must glide freely immediately adjacent to bone, a bursa is characteristically interposed between the tendon and the periosteum. The **synovial tendon sheaths** of the hand and foot are highly specialized bursae.

Just as there are adaptations of deep fascia for the separation of structures, there are also fascial *adaptations for structural attachments.* Different deep fascial sheets may fuse with each other, as is the case with the outer investing and middle cervical fasciae near the hyoid bone. They may attach to bones, as the clavipectoral fascia to the clavicle. The attachment in some instances is secure, but in others it is loose. Thus, the relationship between two contiguous fascial layers may vary from the complete separation at a bursa, or the functional separation of a fascial cleft, through progressive degrees of adhesion and attachment, including complete fusion.

Fascial sheets are named most commonly and most appropriately for the regions of the body that they occupy or the structures that they cover. For example, the *brachial fascia* encloses the arm and the *deltoid fascia* covers the deltoid muscle. Some fasciae have descriptive names, such as the *fascia lata* from its broad extent on the thigh or the *fascia cribrosa* from its many openings. A few are named from their attachments, for example, the *clavipectoral fascia* and the *coracoclavicular fascia.*

A more detailed description of the various fascial structures of different regions is included with the discussions of the appropriate muscle groups and organs.

Form and Attachments
of Muscles

There is a considerable variation in the manner by which muscle fibers are arranged to form whole muscles. The different architectural types of muscle are adapted so that the arrangement and shape of the fasciculi best meet the direction and force required of the contracted muscle as it functions. Although there are many muscle forms and shapes, certain general categories can be identified.

STRUCTURAL TYPES OF MUSCLES. The arrangement of the fasciculi and the manner in which they approach their tendons are used as a means of classifying muscles. The following types will be described: **parallel, fusiform, oblique, triangular,** and **overlapping** (crossed and spiral).

Parallel Muscle. In this simplest arrangement the muscle fibers are parallel with the longitudinal or transverse axis of the muscle and terminate at the ends in flat tendons (Fig. 6-2). These muscles may be short and flat like the thyrohyoid, somewhat longer and straplike as are the sternothyroid and sternohyoid, long and narrow like the sartorius, or broad and relatively thin like the rhomboid major. The individual muscle fibers may course the entire length of the muscle or may interdigitate into **tendinous intersections,** such as is seen in the rectus abdominis and semispinalis capitis muscles.

Fusiform Muscle. In some muscles the fleshy portion is rounded in the center and tapering at the ends to form a spindle shaped belly which terminates in a tendon at one or both ends. These fusiform muscles (Fig. 6-2) may have a single belly (lumbricals or biceps brachii: one belly from two heads) or two bellies interrupted by a tendon (digastric and omohyoid muscles).

Oblique Muscle. In many muscles a tendon courses for considerable distances either within the muscle or along its surface. In these muscles the fibers attach obliquely to the tendon like the plumes of a feather, and the muscles are therefore considered to be **pennate** (feather-like) in their construction. If the muscle fibers converge obliquely on one side of the tendon, the muscle is called **unipennate,** but should be called semipennate (flexor pollicis longus in the hand and extensor digitorum longus and peroneus tertius in the leg). If the fleshy muscle fibers arise from a broader area and attach obliquely from several sides to a centrally placed tendon, the muscle is considered to be a **bipennate** type (peroneus longus, flexor hallucis longus, dorsal interossei). In **multipennate** muscles, there are several tendons in the muscle mass with fibers attaching obliquely from many directions, such as is seen in the deltoid muscle (Fig. 6-2).

Triangular Muscle. A muscle in which the fiber bundles attach to a broad tendon and converge to a more narrow apex may give, more or less, the surface appearance of a triangle. Examples of triangular type muscles include the pectoralis minor and anconeus muscles.

Overlapping Muscle: Crossed or Spiral. In certain muscles an overlapping course of the muscle bundles allows large numbers of fibers arising from a broad origin to converge in a laminar or twisting fashion onto the tendon(s) of insertion. Some of the bundles course in a curved direction toward the insertion, while others are straight, as is the

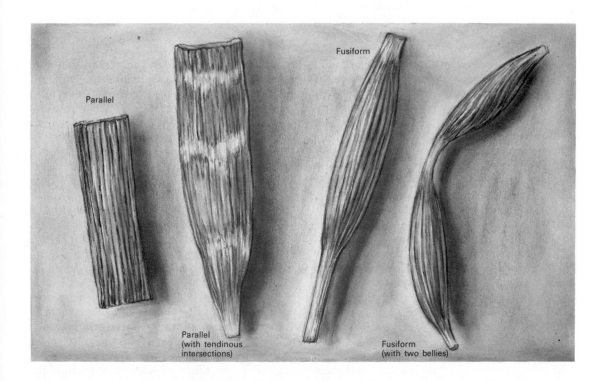

Fig. 6-2.   Some of the morphological types of muscles.

pectoralis major muscle. A variation of this type of muscle form allows fibers arising from several sites and planes to cross, thereby becoming oriented toward the site of insertion from different directions (examples include the masseter and adductor pollicis muscles). In other muscles the fiber bundles may be arranged in a course forming, more or less, a spiral type of structure such as is seen in the supinator muscle of the forearm (Fig. 6-2).

Finally, it should be appreciated that frequently a muscle may show characteristics of several of these various muscle forms.

These arrangements of the fasciculi are correlated with the power that muscles can generate. Those with comparatively few fasciculi, extending the length of the muscle, have a greater range of motion but not much power. Penniform muscles, with a large number of fasciculi distributed along their tendons, have greater power, but a smaller range of motion, that is, their pull will be shorter but more powerful.

NAMES OF MUSCLES. Muscles have been named according to (1) their **location** (brachialis, pectoralis, supraspinatus); (2) their **direction** (rectus, obliquus, and transversus abdominis); (3) their **action** (flexors, extensors); (4) their **shape** (deltoid, trapezius, rhomboid); (5) the **number of divisions** (biceps, triceps, quadriceps); (6) their **points of attachment** (sternocleidomastoid, omohyoid).

ORIGIN AND INSERTION OF MUSCLES. The attachments of the two ends of a muscle are called the **origin** and the **insertion**, and it is customary to describe the muscle as *arising* from the origin and *inserting* at the insertion. The origin usually is the more fixed and proximal end, while the insertion is the more movable and distal end. For example, the pectoralis major arises or has its origin from the sternum, ribs, and clavicle, and its insertion is into the humerus. If the individual were climbing a tree, however, the origin and insertion would be reversed, since the upper extremity would be more fixed and the body more movable. The designations of origin and insertion in the following descriptions are, therefore, somewhat arbitrary and a matter of convention among anatomists. **Illustrations of areas of attachment** of muscles are found with the descriptions of the bones and the origins marked in red, the insertions in blue.

Actions of Muscles

At one time muscle action was determined simply by the effect observed through manipulation of the muscle being studied in the cadaver or the deficits in muscle function seen in patients with peripheral nerve injuries. The introduction of electrical stimulation of muscles added much to our knowledge of their action, and the current use of electromyographic recordings of motor unit discharge in the muscles of human subjects is now refining our understanding and more precisely defining muscle actions.

MOTOR UNIT. Physiologically, a muscle contracts because a motor nerve conducts impulses to myoneural junctions where a chemical transmitter, **acetylcholine,** is released from the presynaptic nerve terminals. This transmitter apparently binds with protein molecules called **acetylcholine receptors,** found on the membrane surface of the muscle. The binding then presumably causes biophysical changes in the membrane, increasing its permeability and allowing a rapid influx of sodium ions into the muscle fiber. This initiates a postsynaptic excitation of the membranes of the somatic skeletal muscle cells.

A single motor nerve fiber innervates more than one skeletal muscle fiber. The nerve fiber and all the muscle fibers that it supplies are a functional unit, which is called the **motor unit.** In general, muscles that are small and react quickly, such as certain laryngeal and extraocular muscles, have ten or fewer muscle fibers innervated by each nerve fiber. In contrast, large muscles, such as the gastrocnemius or the deep muscles of the back, which do not require fine central nervous system control, may have as many as 500 or 1000 muscle fibers in each motor unit. Under normal resting circumstances, certain motor units in a muscle may be discharging while others are quiescent. Soon the activity of these units subsides and that of others commences, and it is this type of motor unit activity pattern that serves as the background against which muscular contraction for the performance of a purposeful act is set.

ISOTONIC AND ISOMETRIC CONTRACTIONS. When muscles contract, their fibers act upon certain movable structures in order to effect the desired action. Movements result from the summation of increasing numbers of motor units in certain groups of muscles and the simultaneous relaxation of motor units in others. Movement that results from muscular contraction causes the lengths of muscles to change. Under these circumstances, the tension generated within the muscle mass remains constant and the contraction is referred to as **isotonic.** If, for some reason, movement does not occur as a result of muscular contraction, and the lengths of the muscles remain constant while the tension generated within the muscles is elevated, the contraction is said to be **isometric.** An example of isometric muscular contraction is encountered in the activities of the muscles of the calf during standing. The center of gravity would tend to allow the body to fall forward; however, this is prevented by the sustained activity of motor units in the calf muscles, causing them to contract without altering their lengths, thus, isometrically. Isotonic contractions may result in a shortening of the muscle **(concentric contraction),** or at times in its lengthening **(eccentric contraction).**

GROUP ACTIONS OF MUSCLES. With rare exceptions, an individual muscle does not act alone in the performance of movement. It is, however, of practical importance to understand the individual actions of muscles. For example, the surgeon who is trying to diagnose and treat a displaced part of a limb due to a fracture must depend on his knowledge of individual muscle functions to guide his decisions.

Most bodily movements require the combined action of a number of muscles. It is customary to use the term **prime mover** for those muscles that act directly to bring about the desired movement. Practically every muscle acting upon a joint is matched with another muscle that has an opposite action. Each muscle of such a pair is the **antagonist** of the other; for example, the biceps brachii, a flexor, and the triceps, an extensor, are antagonists at the elbow. An antagonist does not significantly counteract the action of a prime mover and, in fact, only shows a brief and transient discharge pattern at the initiation of movement, relaxing then until the intended action decelerates. In some instances, the antagonist actually controls the smoothness of performance. For example, if the abducted arm is allowed to lower passively to the side of the trunk, gravity serves as a substitute for the prime mover. The an-

tagonist, which is the deltoid in this instance, simultaneously lengthens, controlling the movement and retarding the force of gravity. At times prime movers and antagonists may contract together, serving to stabilize a joint or hold a part of the body in an appropriate position. Under these conditions, the muscles are said to behave as **fixators.**

Muscles that contract at the same time as prime movers are called **synergists.** The term may refer not only to muscles that potentiate the action of the prime mover, but, more meaningfully, to fixators or stabilizing muscles that contract at the same time as a prime mover, thereby preventing some unwanted movement that would be counterproductive. If antagonists contract during a given movement, even they are acting as synergists. For example, in closing the fist, the *prime movers* are the flexors digitorum superficialis and profundus, the flexor pollicis longus, and the small muscles of the thumb. The *fixation muscles* are the triceps, the biceps, the brachialis, and the muscles about the shoulder that hold the arm in position. The *synergetic muscles* are the extensors carpi radialis and ulnaris, which prevent flexion of the wrist. In some instances, when an act is performed with extreme force, muscles that are not required for a moderate performance come to the assistance of the prime movers. An example of this might be the flexors of the fingers, which may flex the wrist in an emergency situation.

Individual muscles cannot always be treated as single mechanical units with regard to their actions. Different parts of the same muscle may have different or even antagonistic actions. An example of this is the anterior part of the deltoid, which flexes the arm while the posterior portion extends it. Two adjacent muscles, such as the infraspinatus and the teres minor, on the other hand, may have the same action.

Muscles and Fasciae of the Head

The muscles of the head may be arranged in groups, of which the following two will be described in this chapter:

Facial muscles
Muscles of mastication

In addition to these two groups, other muscles found in the head are described regionally in future chapters. These include the: (1) extraocular muscles, (2) muscles of the auditory ossicles, (3) muscles of the tongue, and (4) muscles of the pharynx.

FACIAL MUSCLES

The **facial muscles** of expression are cutaneous muscles, lying within the layers of superficial fascia. They are derived from the mesoderm of the second pharyngeal arch and migrate to their adult locations. Generally, they arise either from fascia or from the bones of the face, and insert into the skin. The individual muscles seldom remain separate and distinct throughout their length, having a tendency to merge, especially at their sites of attachment. They may be grouped into the: (1) muscle of the scalp, (2) extrinsic muscles of the ear, (3) muscles of the eyelid, (4) muscles of the nose, (5) muscles of the mouth. An additional muscle, the platysma, really belongs to the facial group but will be described with the muscles of the neck.

Fasciae and Muscle of the Scalp

Superficial fascia
Epicranius
　Occipitofrontalis
　Temporoparietalis
　Galea aponeurotica

SUPERFICIAL FASCIA. The superficial fascia of the head invests the facial muscles and carries the superficial blood vessels and nerves. It varies considerably in thickness and texture in different regions, but everywhere has an abundant blood and nerve supply. Above the superior nuchal and temporal lines, and anterior to the masseter muscle, there is no deep fascia other than the periosteum of the bones. Beneath the skin of the scalp, the superficial fascia is thick and tough, and over the cranial vault, the muscular layer is replaced by the broad epicranial aponeurosis or **galea aponeurotica** (Fig. 6-3). A fascial cleft like those commonly found deep to the superficial fascia in the rest of the body is prominent in this region and separates the galea from the **pericranium** or **cranial periosteum.** This cleft accounts for the movability of the scalp and makes possible the sudden accumulation of large amounts of blood that form hematomas following blows on the head. Over the forehead, the superficial layers of the fascia are much thinner than over the superior aspect of the cranial vault, and the skin is closely attached to the underlying frontal belly of the occipitofrontalis muscle. Over the eyelids, the fascia is devoid of fat and is composed of a loose areolar tissue which is easily distended and infiltrated with tissue fluid in edema, or blood in ecchymosis or hemorrhage. On the cheeks and lips, the superficial fascia contains a considerable amount of fat and is tougher and more fibrous, especially in men. In contrast, the superficial fascia is virtually absent over the cartilages of the nose and external ear, the skin being closely bound to the underlying **perichondrium.** The superficial fascia of the face is directly continuous over the mandible with that of the anterolateral neck, and that of the scalp merges posteriorly with a similar fibrous layer of the posterior neck.

The skin of the scalp is thicker than that in any other part of the body. The hair follicles are closely set together, have numerous sebaceous glands, and extend deep into the superficial fascia. The subcutaneous fat is broken into granular lobules and is mattressed into a firm layer by the many fibrous bands that secure the skin to the deeper layers of superficial fascia.

THE EPICRANIUS. The epicranius consists of the occipitofrontalis, its aponeurosis, called the galea aponeurotica, and the two thin temporoparietalis muscles. This muscular and tendinous layer covers the top and sides of the skull from the occipital bone to the eyebrow, and extends laterally as far as the squamae of the temporal bones.

The **occipitofrontalis muscle** (Fig. 6-4) consists of four thin, broad muscular bellies, two occipital and two frontal, connected by an extensive intermediate aponeurosis, the **galea aponeurotica.** The **occipital belly,** quadrilateral in form, *arises* by short tendinous fibers from the lateral two-thirds of the superior nuchal line of the occipital bone, and from the mastoid part of the temporal bone. The muscular fasciculi ascend in a parallel course toward the vertex and *end* in the galea aponeurotica. Between the muscles of the two sides of the scalp is a consider-

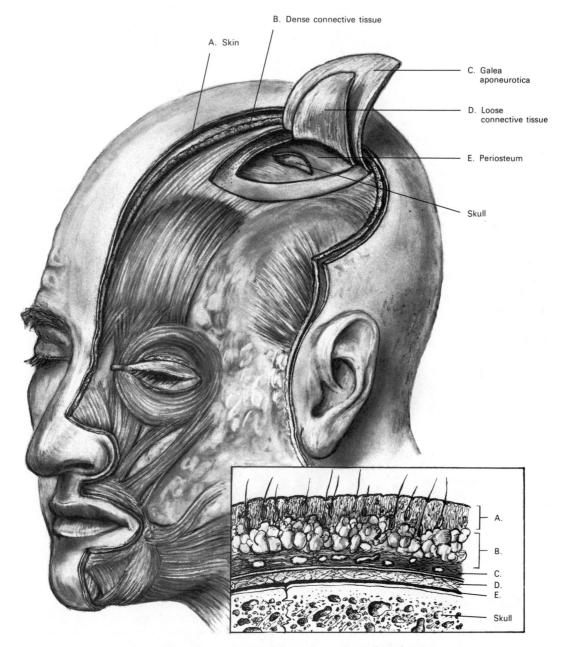

FIG. 6-3.   The five layers of the scalp overlying the calvaria.

able, though variable, interval occupied by a posterior prolongation of the galea. The **frontal belly** consists of a thin quadrilateral muscular layer on each side. It is broader than the occipital belly and its fasciculi are longer, finer, and paler in color, and it has no bony attachments. Its *medial fibers* are continuous with those of the procerus; its *intermediate fibers* blend with the corrugator supercilii and orbicularis oculi; and its *lat-*

eral fibers are blended with the latter muscle also, over the zygomatic process of the frontal bone. The fibers are directed upward, and join the galea aponeurotica anterior to the coronal suture. The medial margins of the muscles of the two sides are joined together for some distance above the root of the nose.

The **temporoparietalis muscle** (Fig. 6-4) is a broad, thin, variable sheet of muscle on each side of the scalp. It *arises* from the tem-

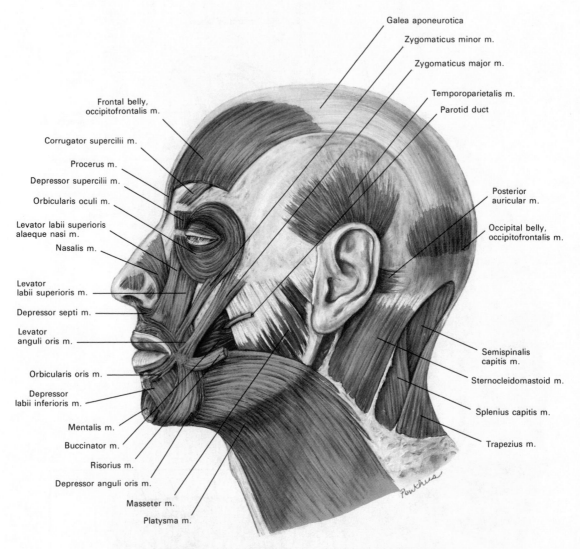

Galea aponeurotica
Zygomaticus minor m.
Zygomaticus major m.
Temporoparietalis m.
Parotid duct

Frontal belly,
occipitofrontalis m.

Corrugator supercilii m.

Procerus m.

Depressor supercilii m.

Orbicularis oculi m.

Levator labii superioris
alaeque nasi m.

Nasalis m.

Levator
labii superioris m.

Depressor septi m.

Levator
anguli oris m.

Orbicularis oris m.

Depressor
labii inferioris m.

Mentalis m.

Buccinator m.

Risorius m.

Depressor anguli oris m.

Masseter m.

Platysma m.

Posterior
auricular m.

Occipital belly,
occipitofrontalis m.

Semispinalis
capitis m.

Sternocleidomastoid m.

Splenius capitis m.

Trapezius m.

Penthus

FIG. 6-4. Superficial muscles of the face, scalp, and neck. Left lateral view.

poral fascia just above and anterior to the superior margin of the auricula of the external ear. It spreads out like a fan over the temporal fascia, and *inserts* into the skin and temporal fascia somewhat higher on the side of the head, nearly reaching the lateral border of the galea aponeurotica.

The **galea aponeurotica** (Figs. 6-4; 6-6), also called the **epicranial aponeurosis,** covers the upper part of the cranium between the frontal and occipital bellies of the occipitofrontalis. In addition to its attachment to these muscle bellies, its fibers extend behind, in the interval between the two occipital bellies, to the external occipital protuberance and the highest nuchal line of the

occipital bone. In front, it forms a short, narrow prolongation between the two frontal bellies. On each side it loses its aponeurotic character, and continues over the temporal fascia as a layer of laminated areolar tissue. It is closely attached to the integument by the firm, dense, adipose layer of superficial fascia and is separated more deeply from the pericranium by a fascial cleft. This allows the aponeurosis, carrying with it the integument, to move the scalp quite freely over the underlying bony calvaria.

Action. The occipital and frontal bellies of the **occipitofrontalis** acting together draw back the scalp, raising the eyebrows and wrinkling the forehead as in an expression of surprise. The frontal bellies acting alone raise the eyebrow, on one or both sides. The **temporoparietalis** tightens the scalp and draws back the skin of the temples; it combines with the occipitofrontalis to wrinkle the forehead and widen the eyes in an expression of fright or horror. The temporoparietalis raises the auricula.

Nerves. The frontal belly of the occipitofrontalis and the temporoparietalis are supplied by the temporal branches, and the occipital belly by the posterior auricular branch of the **facial nerve.**

Variations. Both frontal and occipital bellies may vary considerably in size and in extent, and either may be absent. The muscles of the two sides may fuse in the midline, and the occipital belly may fuse with the auricularis posterior muscle.

A thin muscular slip, the **transversus nuchae** or **occipitalis minor,** is present in about 25% of the bodies. It *arises* from the external occipital protuberance or from the superior nuchal line, either superficial or deep to the trapezius. Frequently, it is *inserted* with the auricularis posterior muscle, but may join the posterior edge of the sternocleidomastoid.

Extrinsic Muscles of the Ear

Auricularis anterior
Auricularis superior
Auricularis posterior

AURICULARIS ANTERIOR. The auricularis anterior muscle is thin, pale, delicate, and indistinct. It is the smallest of the three and *arises* from the anterior portion of the fascia in the temporal area. It is fan-shaped and its fibers converge to be *inserted* into a spine on the front of the helix.

AURICULARIS SUPERIOR. The auricularis superior muscle is thin and fan-shaped. Its fibers *arise* from the fascia of the temporal area and converge to be *inserted* by a thin, flattened tendon into the upper part of the cranial surface of the auricula. It is the largest of the auricular muscles.

AURICULARIS POSTERIOR. The auricularis posterior muscle (Fig. 6-4) consists of two or three fleshy fasciculi which *arise* from the mastoid portion of the temporal bone by short aponeurotic fibers. They are *inserted* onto the medial (cranial) surface of the auricula at the convexity of the concha.

Actions. The **auricularis anterior** draws the auricula forward and upward; the **auricularis superior** draws it upward; and the **auricularis posterior** draws it backward. In man these muscles seem to act more in conjunction with the occipitofrontalis to move the scalp than to move the auricula, but some

individuals can use them to execute voluntary movements of the auricula.

Nerves. The auriculares anterior and superior are supplied by the temporal branches, and the auricularis posterior by the posterior auricular branch of the **facial nerve.**

Muscles of the Eyelids

Levator palpebrae superioris
Orbicularis oculi
Corrugator supercilii

The **levator palpebrae superioris** is described with the anatomy of the eye.

ORBICULARIS OCULI. The orbicularis oculi (Figs. 6-4 to 6-6) *arises* from the nasal part of the frontal bone, from the frontal process of the maxilla in front of the lacrimal groove, and from the anterior surface and borders of a short fibrous band, the **medial palpebral ligament.** From this origin, the fibers are directed lateralward, forming a broad and thin layer which occupies the eyelids or palpebrae, surrounds the circumference of the orbit, and spreads over the temple and down the cheek. The orbicularis oculi consists of **palpebral, orbital,** and **lacrimal parts.**

The **palpebral part** of the muscle is thin and pale (Fig. 6-6). It *arises* from the bifurcation of the medial palpebral ligament and spreads in concentric curves over the eyelids anterior to the orbital septum. Laterally, the muscle fibers again converge to *insert* into the lateral palpebral raphe. A few superficial muscle fibers along the free margin of the lids behind the eyelashes are called the **ciliary bundle** (of Riolan).[1]

The **orbital part** is thicker and of a reddish color. Its fibers form a complete ellipse without interruption at the lateral palpebral commissure, and spread onto the forehead, temple, and cheek. The upper fibers of the orbital part blend with the frontal belly of the occipitofrontalis muscle and the corrugator supercilii. Some fibers of the orbital part insert into the skin of the eyebrow, and draw the eyebrow downward. The latter fibers constitute the **depressor supercilii.**

The **lacrimal part** is small and thin, about 6 mm in breadth and 12 mm in length, and situated behind the medial palpebral ligament and lacrimal sac (Fig. 6-5). It *arises* from the posterior crest and adjacent part of

---

[1]Jean Riolan (1577–1657): A French physician, anatomist, and botanist (Paris).

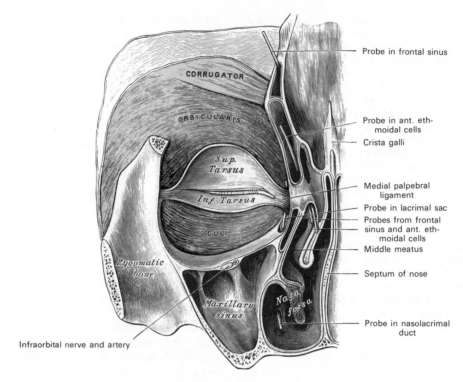

Probe in frontal sinus

CORRUGATOR

ORBICULARIS

Probe in ant. eth-
moidal cells

Crista galli

*Sup.
Tarsus*

Medial palpebral
ligament

Inf. Tarsus

Probe in lacrimal sac

Probes from frontal
sinus and ant. eth-
moidal cells

OCULI

Middle meatus

*Zygomatic
bone*

Septum of nose

*Maxillary
sinus*

*Nasal
fossa*

Probe in nasolacrimal
duct

Infraorbital nerve and artery

FIG. 6-5. Left orbicularis oculi, seen from behind.

the orbital surface of the lacrimal bone, and passing behind the lacrimal sac, divides into an upper and a lower slip, which are *inserted* into the superior and inferior tarsi medial to the puncta lacrimalia.

The **medial palpebral ligament** (Fig. 6-6), about 4 mm in length and 2 mm in breadth, is attached to the frontal process of the maxilla in front of the lacrimal groove. Crossing the lacrimal sac, it divides into upper and lower parts, each attached to the medial end of the corresponding tarsus. As the ligament crosses the lacrimal sac, it gives off from its posterior surface a strong aponeurotic lamina which expands over the sac and is attached to the posterior lacrimal crest.

The **lateral palpebral raphe** is a much weaker structure than the medial palpebral ligament. It is formed by an interlacing of the orbicularis oculi muscle fibers at the lateral angle of the eye. The raphe is attached to the margin of the frontosphenoidal process of the zygomatic bone, and passes medialward to the lateral commissure of the eyelids. Here it divides into two slips, which are attached to the margins of the respective tarsi. Deep to the raphe is the **lateral palpebral ligament.**

CORRUGATOR SUPERCILII. The corrugator supercilii (Figs. 6-4 to 6-6) is a small, narrow, pyramidal muscle, placed at the medial end of the eyebrow, beneath the frontal belly of the orbitofrontalis and orbicularis oculi. It *arises* from the medial end of the superciliary arch, and its fibers pass upward and lateralward, between the palpebral and orbital portions of the orbicularis oculi, and are *inserted* into the deep surface of the skin, above the middle of the orbital arch.

Actions. The **orbicularis oculi** is the sphincter muscle of the eyelids. The palpebral portion closes the lids gently, as in blinking or in sleep. The orbital portion is also used to close the eye, especially when a more forceful contraction is necessary, such as in winking one eye. When the entire muscle is brought into action, the skin of the forehead, temple, and cheek is drawn toward the medial angle of the orbit, and the eyelids are firmly closed. The skin thus drawn upon is thrown into folds, especially radiating from the lateral angle of the eyelids; these folds become permanent in old age, and form the so-called ''crow's feet.'' The levator palpebrae superioris is the direct antagonist of the orbicularis oculi, because it raises the upper eyelid and exposes the front of the bulb of the eye. Each time the eyelids are closed through the action of the orbicularis, the medial palpebral ligament is tightened, the wall of the lacrimal sac is thus drawn lateralward and forward, and the tears are directed into the lacrimal canals. The lacrimal part of the orbicularis oculi draws the eyelids and the ends of the lacrimal canals medialward and compresses them against the surface of the globe of the eye, thus placing them in the most favorable position for receiving the tears. The **corrugator supercilii** draws the eyebrow downward and medialward, producing the vertical wrinkles of the forehead. It is the "frowning" muscle, and may be regarded as the principal muscle in the expression of suffering.

Nerves. These muscles are innervated by the temporal and zygomatic branches of the **facial nerve.**

Muscles of the Nose

Procerus
Nasalis
Depressor septi

THE PROCERUS. The procerus muscle (Figs. 6-4; 6-6) is a small pyramidal slip *arising* on each side by tendinous fibers from the fascia covering the lower part of the nasal bone and upper part of the lateral nasal cartilage. Coursing superiorly over the bridge of the nose, it is *inserted* into the skin over the lower part of the forehead between the two eyebrows, its fibers decussating with those of the frontal belly of the occipitofrontalis muscle.

THE NASALIS. The nasalis muscle consists of two parts, transverse and alar. The **transverse part** (*compressor naris*) *arises* from the maxilla, above and lateral to the incisive fossa (Fig. 6-6). Its fibers proceed upward and medialward, expanding into a thin aponeurosis which is continuous on the bridge of the nose with that of the muscle of the opposite side, and with the aponeurosis of the procerus. The **alar part** (*dilatator naris*) *arises* from the maxilla above the lateral incisor tooth medial to the transverse part and is *inserted* into the lower part of the cartilaginous ala of the nose.

DEPRESSOR SEPTI. The depressor septi (Fig. 6-4) *arises* from the incisive fossa of the maxilla. Its fibers ascend to be *inserted* into the mobile part of the nasal septum. It lies between the mucous membrane and muscular structure of the lip.

Actions. The **procerus** draws down the medial angle of the eyebrows and produces transverse wrinkles over the bridge of the nose. The **transverse part of the nasalis** depresses the cartilaginous part of the nose and draws the ala toward the septum.

The **alar part of the nasalis** enlarges the aperture of the nares. Its action in ordinary breathing is to resist the tendency of the nostrils to close from atmospheric pressure, but in difficult breathing, as well as in some emotions, such as anger, it contracts strongly. The **depressor septi** draws the ala of the nose downward, and thereby constricts the aperture of the nares.

Nerves. These muscles are innervated by the buccal branches of the **facial nerve.**

Muscles of the Mouth

Levator labii superioris
Levator labii superioris alaeque nasi
Levator anguli oris
Zygomaticus minor
Zygomaticus major
Risorius
Depressor labii inferioris
Depressor anguli oris
Mentalis
Transversus menti
Orbicularis oris
Buccinator
    Pterygomandibular raphe
    Buccal fat pad

LEVATOR LABII SUPERIORIS. This thin quadrangular muscle (Figs. 6-4; 6-6) has a rather broad *origin* from the lower margin of the orbit immediately above the infraorbital foramen; some of its fibers are attached to the maxilla, others to the zygomatic bone. Its fibers converge to be *inserted* into the muscular substance of the upper lip between the levator anguli oris and the levator labii superioris alaeque nasi.

LEVATOR LABII SUPERIORIS ALAEQUE NASI. This narrow muscle (Figs. 6-4; 6-6) *arises* from the upper part of the frontal process of the maxilla, passes obliquely downward and lateralward, and divides into two slips. One of these is *inserted* into the greater alar cartilage and skin of the nose, while the other is prolonged into the upper lip, blending with the levator labii superioris.

LEVATOR ANGULI ORIS. This flat quadrilateral muscle (Figs. 6-4; 6-6) *arises* from the canine fossa, immediately below the infraorbital foramen. Its fibers are *inserted* into the angle of the mouth, intermingling with those of the zygomaticus major, depressor anguli oris, and orbicularis oris.

ZYGOMATICUS MINOR. This small bundle of muscle fibers *arises* from the malar surface of the zygomatic bone immediately behind the zygomaticomaxillary suture,

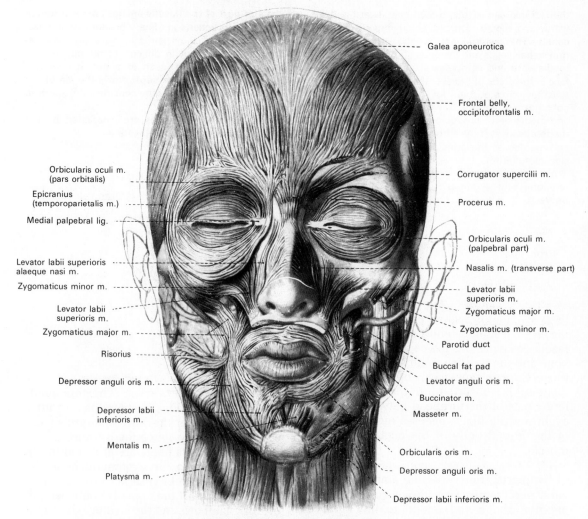

Galea aponeurotica

Frontal belly,
occipitofrontalis m.

Orbicularis oculi m.
(pars orbitalis)

Epicranius
(temporoparietalis m.)

Medial palpebral lig.

Corrugator supercilii m.

Procerus m.

Levator labii superioris
alaeque nasi m.

Zygomaticus minor m.

Levator labii
superioris m.

Zygomaticus major m.

Risorius

Depressor anguli oris m.

Depressor labii
inferioris m.

Mentalis m.

Platysma m.

Orbicularis oculi m.
(palpebral part)

Nasalis m. (transverse part)

Levator labii
superioris m.

Zygomaticus major m.

Zygomaticus minor m.

Parotid duct

Buccal fat pad

Levator anguli oris m.

Buccinator m.

Masseter m.

Orbicularis oris m.

Depressor anguli oris m.

Depressor labii inferioris m.

FIG. 6-6. The facial muscles, anterior view. On the right (reader's left) is shown the more superficial layer, while on the left are the deeper muscles. (From Benninghoff and Goerttler, *Lehrbuch der Anatomie des Menschen*, Urban & Schwarzenberg, 11th Ed., Vol. 1, 1975.)

passes downward and medialward as a narrow slip, and is *inserted* into the upper lip between the levator labii superioris and the zygomaticus major (Figs. 6-4; 6-6).

ZYGOMATICUS MAJOR.   Somewhat larger than the preceding muscle, the zygomaticus major *arises* from the zygomatic bone, in front of the zygomaticotemporal suture. It descends obliquely with a medial inclination to be *inserted* into the angle of the mouth, where it blends with the fibers of the levator and depressor anguli oris and the orbicularis oris muscle (Figs. 6-4; 6-6).

Actions.   The **levator labii superioris** raises the upper lip, carrying it at the same time a little forward. The **levator labii superioris alaeque nasi**

raises the upper lip and also dilates the nostril. These two muscles, together with the **zygomaticus minor,** form the nasolabial furrow, from the side of the nose to the upper lip, which is deepened in expressions of sadness. When these three muscles act in conjunction with the **levator anguli oris,** the furrow is deepened into an expression of contempt or disdain. The **zygomaticus major** draws the angle of the mouth upward and backward in laughing.

Nerves.   All five of the preceding muscles are supplied by buccal branches of the **facial nerve.**

THE RISORIUS.   The risorius muscle (Figs. 6-4; 6-6) usually *arises* in the parotid fascia overlying the masseter muscle, but may arise more superiorly or posteriorly. It passes horizontally forward, superficial to the platysma, to be *inserted* into the skin at the angle of the mouth. The risorius varies greatly in size.

DEPRESSOR LABII INFERIORIS.   This depressor of the lower lip is a small quadrilateral muscle that *arises* from the oblique line of the mandible, between the symphysis menti and the mental foramen (Figs. 6-4; 6-6). It passes upward and medialward, to be *inserted* into the skin of the lower lip, its fibers blending with the orbicularis oris, and with those of the depressor labii superioris of the opposite side. At its origin it is continuous with the fibers of the platysma, and considerable fat is intermingled with the fibers of this muscle.

DEPRESSOR ANGULI ORIS.   This muscle *arises* from the oblique line of the mandible, lateral to and below the depressor labii inferioris. Its fibers ascend in a curved direction and are *inserted* as a narrow fasciculus into the angle of the mouth (Figs. 6-4; 6-6). At its origin, it is continuous with the platysma, and at its insertion, with the orbicularis oris and risorius. Some of its fibers are directly continuous with those of the levator anguli oris.

THE MENTALIS.   The mentalis is a small conical fasciculus, situated at the side of the frenulum of the lower lip. It *arises* from the incisive fossa of the mandible, and descends to be *inserted* into the integument of the chin (Figs. 6-4; 6-6).

TRANSVERSE MENTI.   This muscle, found in more than half the bodies, is small and crosses the midline just under the chin. It is frequently continuous with the depressor anguli oris.

Actions.   The **risorius** retracts the angle of the mouth. The **depressor labii inferioris** draws the lower lip directly downward and a little lateralward, producing the expression of irony. The **depressor anguli oris** depresses the angle of the mouth, being the antagonist of the levator anguli oris and zygomaticus major; acting with the levator, it draws the angle of the mouth medialward. The **mentalis** raises and protrudes the lower lip, and at the same time wrinkles the skin of the chin, as in pouting or expressing doubt or disdain.

Nerves.   These muscles of the lower lip are supplied by the mandibular and buccal branches of the **facial nerve.**

ORBICULARIS ORIS.   The orbicularis oris (Figs. 6-4; 6-6) is not a simple sphincter muscle like the orbicularis oculi. It consists of numerous strata of muscular fibers surrounding the orifice of the mouth and coursing in different directions. Some fibers of the orbicularis oris are derived from other facial muscles which insert into the lips, while other fibers are intrinsic to the muscle itself. Of the former, a considerable number are derived from the buccinator and form the deeper stratum of the orbicularis oris. Some of the buccinator fibers—namely, those near the middle of the muscle—decussate at the angle of the mouth, those arising from the maxilla pass to the lower lip, and those from the mandible to the upper lip (Fig. 6-7). The uppermost and lowermost fibers, derived from the buccinator, pass across the lips from side to side without decussation. Superficial to this stratum is a second layer, formed on each side by fibers from the levator and depressor anguli oris which cross each other at the angle of the mouth (Fig. 6-7). Those from the levator pass to the lower lip, and those from the depressor to the upper lip. Coursing along the lips, these fibers are inserted into the skin near the median line. Other muscle fibers of the orbicularis oris are derived from the levator labii superioris, the zygomaticus major, and the depressor labii inferioris. These intermingle with the transverse fibers described above, and have principally an oblique direction. The proper fibers of the lips are oblique, and pass from the undersurface of the skin to the mucous membrane through the thickness of the lip. Finally, there are fibers by which the muscle is connected with the *maxillae* and the *septum of the nose* above, and with the *mandible* below. In the upper lip these consist of lateral and medial bands on each side of the midline. The lateral band, called the **incisivus labii superioris** (Fig. 6-7), *arises*

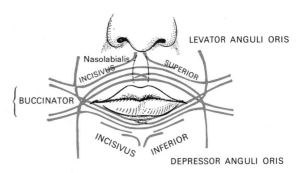

FIG. 6-7.   Diagram showing arrangement of fibers of orbicularis oris.

from the alveolar border of the maxilla, opposite the lateral incisor tooth, and arching lateralward, is continuous with the other muscles at the angle of the mouth. The more medial band, the **nasolabialis muscle,** connects the upper lip to the back of the septum of the nose. The interval between the two nasolabial bands corresponds to a depression, the **philtrum,** seen on the upper lip beneath the septum of the nose. Similar fibers for the lower lip constitute the **incisivus labii inferioris** on each side of the midline. This *arises* from the mandible, lateral to the mentalis, and intermingles with the other muscles at the angle of the mouth.

Actions. The **orbicularis oris,** in its ordinary action, effects the direct closure of the lips. By its deep and oblique fibers, it closely applies the lips to the teeth and the alveolar arches. The superficial part, consisting principally of the decussating fibers, brings the lips together and also protrudes them forward. Additionally, the orbicularis oris is an important muscle for speech because it alters the shape of the mouth as well as assists feeding and drinking.

Nerves. The orbicularis oris is supplied by the buccal branches of the **facial nerve.**

THE BUCCINATOR.   The buccinator is the principal muscle of the cheek and forms the lateral wall of the oral cavity (Fig. 6-8). It lies deeper than the other facial muscles, is

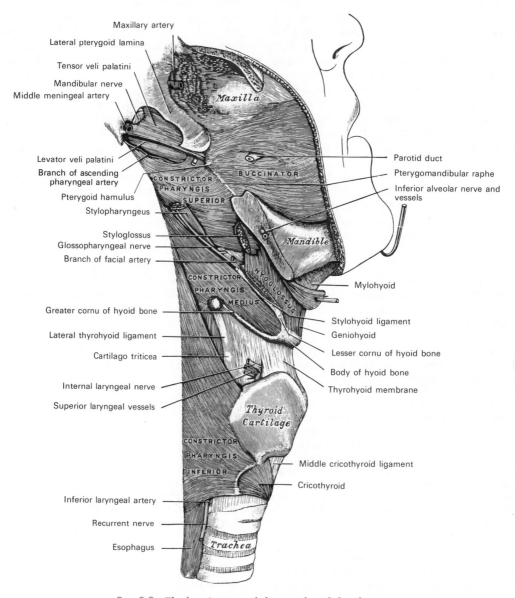

FIG. 6-8. The buccinator and the muscles of the pharynx.

quadrilateral in form, and occupies the interval between the maxilla and mandible, lateral to the teeth. It *arises* from the outer surfaces of the alveolar processes of the maxilla above and the mandible below, opposite the three molar teeth. Posteriorly, between these two bony attachments, the buccinator also *arises* from the pterygomandibular raphe, a tendinous inscription which gives origin to the buccinator anteriorly and to the superior pharyngeal constrictor posteriorly. The fibers of the upper and lower portions follow a slightly converging course forward, and *insert* by blending with the deeper stratum of muscle fibers at the corresponding angle of the lips. The fibers of the central portion converge toward the angle of the mouth and decussate, those from above becoming continuous with the orbicularis oris of the lower lip, those from below with that of the upper lip.

The superficial surface of the buccinator is covered by the **buccopharyngeal fascia** and the **buccal fat pad.** Its deep surface is adjacent to the buccal glands and the mucous membrane of the mouth. The buccinator is pierced by the **duct of the parotid gland** opposite the upper second molar tooth.

Action. The **buccinator** compresses the cheek and is, therefore, an important accessory muscle of mastication, holding the food under the immediate pressure of the teeth. When the cheeks are distended with air, the buccinators compress them and force the air out between the lips as in blowing a trumpet (Latin: buccinator = a trumpet player). Electromyographic studies show that the orbicularis oris acts with the buccinator to compress the cheek.

Nerve. The motor fibers to the buccinator come from the lower buccal branches of the **facial nerve.**

PTERYGOMANDIBULAR RAPHE. The pterygomandibular raphe (Fig. 6-8) is a tendinous inscription extending superiorly from the hamulus of the medial pterygoid plate to the posterior end of the mylohyoid line of the mandible below. It gives origin to the *superior pharyngeal constrictor muscle,* which extends posteriorly from the raphe to form part of the pharyngeal wall, and the *buccinator muscle,* which courses anteriorly to form the lateral wall of the oral cavity. The *medial surface* of the raphe is covered by the mucous membrane of the mouth, while its *lateral surface* is separated from the ramus of the mandible by a quantity of adipose tissue, the **buccal fat pad.** If it were not

for the raphe, the superior pharyngeal constrictor, the buccinator, and the orbicularis oris would form a continuous sphincter-like band of muscle.

The **buccopharyngeal fascia** covers the lateral surface of the buccinator, the raphe, and the superior pharyngeal constrictor muscles. This provides a fascial cleft between the muscles and the more superficial structures overlying them, thereby allowing the muscles to move freely.

BUCCAL FAT PAD. This circumscribed, encapsulated mass of fat lies superficial to the buccinator at the anterior border of the masseter (Fig. 6-6). A well-defined fascial cleft separates it from the superficial fascia and facial muscles. From this main mass of adipose tissue, narrow prolongations extend deeply between the masseter and temporalis muscles and upward under the deep temporal fascia. Some of the tissue continues still more deeply into the infratemporal fossa, separating the lateral pterygoid and temporalis muscles from the maxilla, filling spaces between the various soft structures and the bony fossae. The mass superficial to the buccinator is particularly prominent in infants, and is called the *suctorial pad* because of its supposed, but dubious, assistance in the act of sucking.

MUSCLES OF MASTICATION

The muscles of mastication directly responsible for the movements of the mandible during chewing and speech include:

Temporalis
Masseter
Medial pterygoid
   Mandibular sling
Lateral pterygoid

Overlying these muscles are the following fascial sheets:

Temporal fascia
Parotid-masseteric fascia
Pterygoid fascia

TEMPORAL FASCIA. The temporal fascia is a strong, fibrous sheet, aponeurotic in appearance, which covers the temporalis and is used by it for the attachment of some of its fibers. It is the most cranial extension of the deep fascia, because farther cranially the deep fascia is represented only by the pericranium. It is covered by the skin and superficial fascia, which includes the galea aponeurotica, the auriculares anterior and superior, and, more anteriorly, the orbicularis oculi. Just anterior to the ear, the temporal fascia is crossed by the superficial temporal vessels and auriculotemporal nerve. *Superiorly,* this fascia is a thin, single sheet, attached to the entire extent of the superior temporal line. More *inferiorly,* near its attachment to the zygomatic arch, the temporal fascia splits into two layers. The inner, or **deep layer,** ends by attaching to the medial surface of the zygomatic arch, while the outer or **superficial layer,** after attaching to the lateral border of the arch, is continued below as the masseteric fascia. Between the leaves is a small quantity of fat, the zygomatico-orbital branch of the superficial temporal artery, and a filament from the zygomaticotemporal branch of the maxillary nerve.

PAROTID-MASSETERIC FASCIA. The parotid-masseteric fascia covers the lateral surface of the masseter muscle and splits to enclose the parotid gland. It is attached to the zygomatic arch cranially and is continuous with the investing cervical fascia over the sternocleidomastoid posteriorly. The sheet that covers the lateral surface of the parotid gland is fused with the dense and tough overlying superficial fascia, is intimately mingled with the capsule of the gland, and sends numerous irregular septa into the glandular substance so that, unlike the submandibular gland, the parotid cannot be easily shelled from its connective tissue bed. The fascial layer on the deep surface of the gland continues behind the ramus of the mandible and there fuses with the fascia of the posterior belly of the digastric muscle into a strong band, the **stylomandibular ligament.** The fascia covering the masseter muscle, the **masseteric fascia,** terminates anteriorly by encircling the ramus of the mandible and becomes continuous with the fascia of the medial pterygoid muscle deep to the bone. The masseteric fascia is attached to the border of the mandible inferiorly and posteriorly, completing a compartment that encloses all of the muscle except its upper, deep portion, where there is a communication with the tissue spaces about the insertion of the temporalis.

PTERYGOID FASCIA. The pterygoid fascia invests the medial and lateral pterygoid muscles. It is *continuous below,* at the angle of the mandible, with the masseteric and investing cervical fasciae, both of which become attached to the bone. In this region also, it is continuous with a thickened band, the stylomandibular ligament. The pterygoid fascia extends *upward and forward* along the deep surface of the medial pterygoid muscle to be attached with the origin of the muscle to the pterygoid process of the sphenoid bone. Inferiorly, the pterygoid fascia is attached to the mandible at both borders of the inferior half of the medial pterygoid muscle, but as this muscle angles away from the mandible toward its origin, the fascia wraps around the muscle, forming a sheet on its lateral or superficial surface. This superficial sheet, continuing upward, splits to invest the lateral pterygoid muscle, becoming attached to the skull with this muscle's origin.

The fascia between the two pterygoid muscles is attached to the skull along a line extending from the lateral pterygoid plate to the spine of the sphenoid bone. The part attached to the spine is thickened into a strong band which is attached below to the lingula of the mandible, forming the **sphenomandibular ligament.** Another band, the **pterygospinous ligament,** extends from the spine, between the two pterygoid muscles to the posterior margin of the lateral pterygoid plate. Occasionally this band is ossified, creating, between its upper border and the skull, a **pterygospinous foramen** which transmits the branches of the mandibular division of the trigeminal nerve to the muscles of mastication. Between the sphenomandibular ligament and the neck of the mandible, there is an interval that allows passage of the maxillary vessels into the infratemporal fossa.

The fascia on the lateral surface of the lateral pterygoid muscle is related to the pterygoid plexus of veins. Deep to the pterygoid and deep temporal fasciae, the layer of soft adipose tissue, which is an extension of the buccal fat pad, separates these fasciae from the buccopharyngeal fascia overlying the buccinator and superior pharyngeal constrictor muscles.

The masticator compartment contains the four muscles of mastication along with the ramus and posterior part of the body of the mandible. The compartment is enclosed laterally by the masseteric fascia and the superficial layer of the temporal fascia, and deeply by the pterygoid and deep temporal fasciae.

TEMPORALIS MUSCLE.   The temporalis is a broad, radiating muscle, situated at the side of the head (Fig. 6-9). It *arises* from the whole of the temporal fossa and from the deep surface of the temporal fascia. Its fibers converge as they descend in a tendon that passes deep to the zygomatic arch, through a space between the arch and the lateral aspect of the skull. The tendon *inserts* into the medial surface, apex, and anterior border of the coronoid process, and the anterior border of the ramus of the mandible nearly as far as the last molar tooth.

Action.   The **temporalis muscle** elevates the mandible and thereby closes the jaws, while its posterior fibers retract the mandible.

Nerves.   The temporalis muscle is supplied by anterior and posterior deep temporal nerves from the anterior trunk of the mandibular division of the **trigeminal nerve.**

MASSETER MUSCLE.   This thick, quadrilateral muscle frequently is described as consisting of two parts, but actually consists of three—superficial, intermediate, and deep, which are superimposed and fuse anteriorly (Fig. 6-10). The **superficial part,** which is the largest, *arises* by a thick, tendinous aponeurosis from the zygomatic process of the maxilla and from the anterior two-thirds of the inferior border of the zygomatic arch. Its fibers pass inferiorly and posteriorly, to be *inserted* into the angle and lower half of the lateral surface of the ramus of the mandible. The **intermediate part** *arises* from the inner surface of the anterior two-thirds of the zygomatic arch and *inserts* into the ramus of the mandible. The **deep part** *arises* from the posterior third of the inner surface of the zygomatic arch in continuity with the intermediate part. Its fibers pass anteriorly and inferiorly, to be *inserted* into the superior half of the ramus of the mandible, and the lateral surface of its coronoid process. The deep part of the muscle is partly concealed anteriorly by the more superficial parts, while posteriorly it is covered by the parotid gland. The fibers of the three parts blend and are continuous at their insertion.

Action.   The **masseter muscle** closes the jaws by elevating the mandible.

Nerve.   The masseter muscle is supplied by the masseteric branch of the mandibular division of the **trigeminal nerve.**

MEDIAL PTERYGOID MUSCLE.   The medial pterygoid (Figs. 6-11; 6-12) is a thick, quadrilateral muscle that occupies a position on the inner aspect of the ramus of the mandible, opposite that of the masseter on the

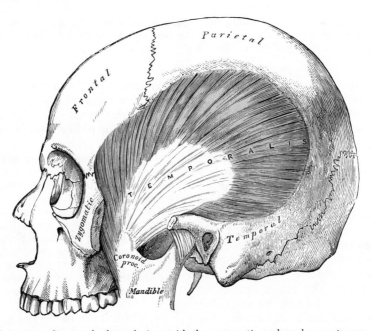

FIG. 6-9.   The temporalis muscle, lateral view with the zygomatic arch and masseter muscle removed.

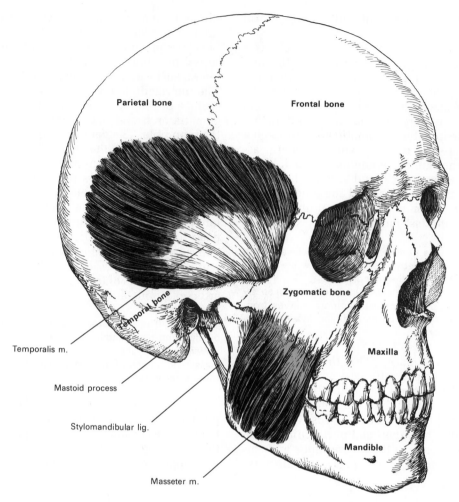

FIG. 6-10. A lateral view of the skull showing the origin and insertion of the masseter muscle, as well as the temporalis muscle and the stylomandibular ligament. (After Benninghoff and Goerttler, *Lehrbuch der Anatomie des Menschen,* Urban & Schwarzenberg, 11th Ed., Vol. 1, 1975.)

outer aspect. It *arises* from the medial surface of the lateral pterygoid plate and the grooved surface of the pyramidal process of the palatine bone. A second, more lateral, slip of the muscle *arises* from the lateral surfaces of the pyramidal process of the palatine and tuberosity of the maxilla. The second slip lies superficial to the lateral pterygoid, while the main mass of the muscle lies deeper. Its fibers pass inferiorly, laterally, and posteriorly, and are *inserted* by a strong tendinous lamina into the lower and back part of the medial surface of the ramus and angle of the mandible, as high as the mandibular foramen. The upper portion of the muscle is separated from the mandible by the sphenomandibular ligament, the maxillary vessels, the inferior alveolar ves-

sels and nerve, and the lingual nerve. The medial surface of the muscle is closely related to the tensor veli palatini muscle and to the superior pharyngeal constrictor.

Action. The **medial pterygoid** assists in closing the jaws by elevating the mandible. The two medial pterygoids protrude the mandible, while acting with the lateral pterygoids. When one of the medial pterygoids acts with the lateral pterygoid of the same side, the mandible is protruded forward and to the opposite side. An alternation of these movements is important in the act of chewing food.

Nerve. The medial pterygoid branch from the mandibular division of the **trigeminal nerve** supplies the medial pterygoid muscle.

MANDIBULAR SLING.   The masseter and the medial pterygoid are so placed that they suspend the angle of the mandible in a muscular sling (Fig. 6-12). They form a functional articulation between the mandible and the maxilla, with the temporomandibular joint acting as a guide, in a manner similar to the articulation between the scapula and the thorax, where the clavicle acts as a guide. When the mouth is opened and closed, the mandible moves about a center of rotation established by the attachment of the sling and the sphenomandibular ligament.

LATERAL PTERYGOID MUSCLE. The lateral pterygoid (Figs. 6-11; 6-12) is a short, thick muscle of two heads. It is somewhat conical in form, and extends almost horizontally between the infratemporal fossa and the condyle of the mandible. The upper or **superior head** *arises* from the infratemporal crest and inferior part of the lateral surface of the great wing of the sphenoid bone, while the **inferior head** *arises* from the lateral surface of the lateral pterygoid plate. Its fibers pass horizontally backward and lateralward, to be *inserted* into a depression in the anterior part of the neck of the condyle of the mandible, and into the anterior margin of the articular disc of the temporomandibular joint.

Action. The **lateral pterygoid** muscle opens the mouth by drawing the condyle of the mandible and the articular disc of the temporomandibular joint forward. When the lateral pterygoids act in conjunction with the elevators of the mandible, the jaws are protruded so that the lower incisor teeth project in front of the upper teeth. With the medial pterygoid of the same side, the lateral pterygoid protrudes the mandible, and the jaw (and chin) is rotated to the opposite side, producing a grinding movement between the upper and lower teeth.

Nerve. The lateral pterygoid nerve from the mandibular division of the **trigeminal nerve** supplies the lateral pterygoid muscle.

Group Actions. The **temporalis, masseter, and medial pterygoid** close the jaws. Biting with the incisor teeth is performed by the masseter and medial pterygoid primarily, and to some extent by the anterior portion of the temporalis. Biting or chewing with the molars calls all three into maximal action.

Opening of the jaws is performed primarily by the **lateral pterygoid** pulling forward on the condyle and rotating the mandible about the center of rotation near the angle. It is assisted, at the beginning of the action, by the mylohyoid, digastric, and geniohyoid. When the mouth is opened against great resistance, in addition to the above, the infrahyoid muscles act to fix the hyoid, and other suprahyoid muscles probably come into action. The lateral pterygoid protrudes the jaw when accompanied by appropriate synergistic action of the closing muscles. The medial pterygoid assists in protrusion of the jaw only as a synergist, along with the other closing muscles, since they prevent rotation that opens the jaws widely. If the lateral pterygoid of one side acts, the corresponding side of the mandible is drawn forward while the opposite condyle remains comparatively fixed, allowing side-to-side movements to occur, such as in the trituration of food. The mandible is retracted by the posterior fibers of the temporalis.

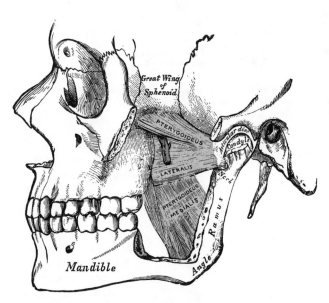

FIG. 6-11. The lateral and medial pterygoid muscles, left side, lateral view. The zygomatic arch and a portion of the ramus of the mandible have been removed, showing the position of the maxillary artery.

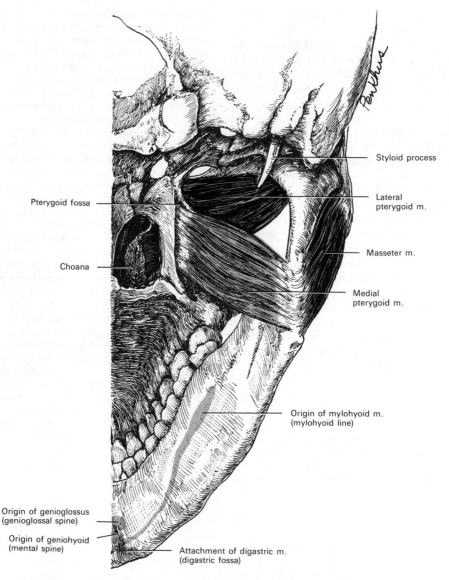

FIG. 6-12. The pterygoid and masseter muscles viewed from below, showing how the medial pterygoid and masseter form a sling suspending the angle of the mandible. (After Benninghoff and Goerttler, *Lehrbuch der Anatomie des Menschen,* Urban & Schwarzenberg, 11th Ed., Vol. 1, 1975.)

Anterolateral Muscles and the Fasciae of the Neck

The **anterolateral muscles** of the neck may be arranged into the following groups:

Superficial cervical
Lateral cervical
Suprahyoid
Infrahyoid
  Triangles of the neck

Anterior vertebral
Lateral vertebral

The **fascial layers** of the neck will be described according to the following schema:

Superficial cervical fascia
Deep cervical fascia
  Superficial investing fascia
    (anterior layer)
  Pretracheal fascia (middle layer)
  Prevertebral fascia (posterior layer)
  Carotid sheath

SUPERFICIAL CERVICAL FASCIA

The **superficial fascia** (Fig. 6-13) in the anterior and lateral regions of the neck is considerably thinner and less dense than the superficial fascia over the face, with which it is continuous above the border of the mandible and over the parotid gland. It has embedded in its deeper layers the fibers of the **platysma muscle,** and it is separated from the deep fascia by a distinct fascial cleft, which facilitates the action of this muscle and increases the movability of the skin in this region. It is continuous over the clavicle with the superficial fascia of the pectoral and deltoid regions. Posteriorly, it is continuous with the superficial fascia of the back of the neck which, in contrast, is thick, tough, fibrous, and tightly adherent to the deep fascia.

DEEP CERVICAL FASCIA

The **deep cervical fascia** (Fig. 6-13) forms certain compartments and fascial clefts that are important surgically because the neck contains structures that descend and ascend between the head and thorax, as well as structures that extend into the upper extremity.

Because descriptions of the deep cervical fascia vary throughout the literature, an understanding of the subject, at first glance, might be considered difficult. Many of these discussions are unnecessarily complex. There is, however, general agreement as to the existence of three principal layers of cervical fascia, the **investing fascia** (anterior layer), the **pretracheal fascia** (middle layer), and the **prevertebral fascia** (posterior layer) (Fig. 6-13). To these is usually added the **carotid sheath,** which contains the common carotid artery, the internal jugular vein, and the vagus nerve. Fascial layers characteristically split as they surround muscles and other organs and then fuse once again and attach firmly to adjacent bone. The layers of cervical fascia follow this pattern as they envelop the structures of the neck in their paths.

INVESTING FASCIA (Anterior Layer of Deep Cervical Fascia). The investing layer

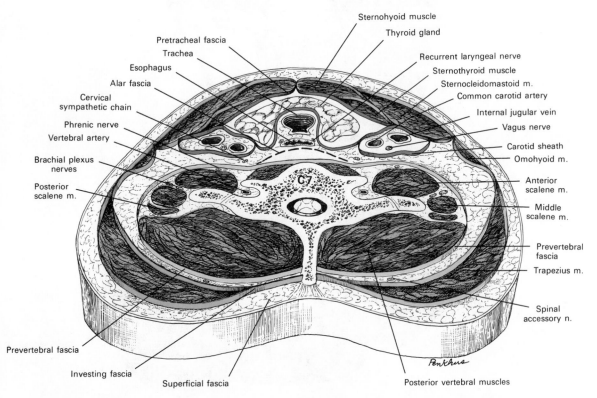

FIG. 6-13.   Cross section of the neck at the level of the seventh cervical vertebra. Observe the superficial fascia and the layers of deep cervical fascia.

attaches posteriorly on each side to the ligamentum nuchae and spinous processes of
the cervical vertebrae, and coursing anteriorly, it completely encircles the neck (Figs.
6-13; 6-14). It splits to enclose two prominent
superficial muscles on each side of its path,
the trapezius and the sternocleidomastoid.
Between these two muscles it covers the posterior triangle of the neck as a single sheet,
except that it envelops the inferior belly of
the omohyoid muscle as that structure
crosses the posterior triangle. Extending
around the neck anterior to the sternocleidomastoid muscles, the investing layer passes

in front of the strap muscles of the anterior
triangle. It then extends to the midline
where the corresponding halves of the two
sides of this fascial layer are continuous.

Superiorly, the investing fascia fuses with
the periosteum over the external occipital
protuberance and the superior nuchal line of
the occipital bone and the mastoid process
of the temporal bone. More anteriorly, it
attaches along the lower border of the body
of the mandible to its angle. Between the
mandibular angle and the mastoid process
the fascia extends superiorly and splits to
enclose the parotid gland as the **parotid fas-**

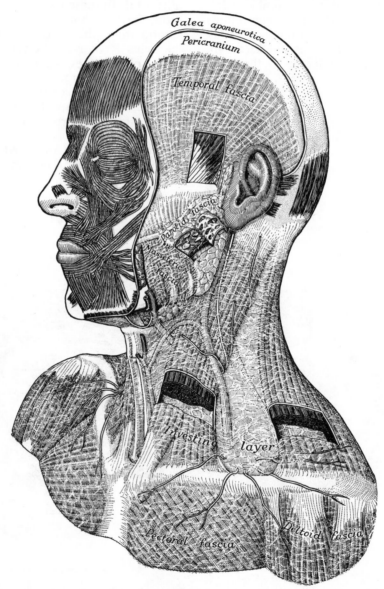

Fig. 6-14.

cia (Fig. 6-14). The layer that passes deep to the parotid is thickened between the styloid process and the angle of the mandible and is called the **stylomandibular ligament.** This sheet separates the tail of the parotid gland from the submandibular salivary gland. Anteriorly, the fascia is attached to the hyoid bone, but in the suprahyoid region the investing layer continues upward to encapsulate the submandibular gland and to attach along the length of both halves of the body of the mandible. The fascia extends inferiorly to attach to the acromion and spine of the scapula, to the clavicle, and to the manubrium sterni. Just above the manubrium and between the two sternocleidomastoids and the heads of the clavicles, the fascia remains split into two layers which attach to the anterior and posterior borders of the jugular notch. The shallow interval between the two laminae is called the **suprasternal space** (of Burns). This space contains the lower portions of the anterior jugular veins and their transverse connecting branch, the **jugular venous arch,** the sternal heads of the sternocleidomastoid muscles, and sometimes one or more lymph nodes. The anterior jugular veins, in order to reach the external jugular veins, traverse extensions of the laminated interval which are prolonged laterally, deep to the heads of the sternocleidomastoid muscles. The external and anterior jugular veins, through most of their course in the neck, appear to lie between the superficial fascia and the investing layer of deep fascia, but actually are embedded in the superficial surface of the investing sheet.

PRETRACHEAL FASCIA (Middle Layer of the Deep Cervical Fascia).   The pretracheal fascia (Fig. 6-13) is a thin, somewhat delicate layer of connective tissue that passes across the front of the neck, anterior to the trachea and thyroid gland and deep to the strap muscles. According to some accounts, it is the epimysium of the strap muscles, but this is an inadequate description. It splits to enclose the thyroid gland and contributes to its capsule. Extending above, it attaches to and is limited superiorly by the hyoid bone. Inferiorly, it extends behind the manubrium and is continuous with the fibrous pericardium along the great vessels. Laterally, the pretracheal fascia is continuous with the anterior part of the carotid sheath, and the thyroid vessels penetrate it to supply and drain the thyroid gland.

PREVERTEBRAL FASCIA (Posterior Layer of the Deep Cervical Fascia).   The prevertebral fascia (Fig. 6-13), similar to the investing layer, is a closely fitting fascial sheet stretching across the neck *behind* the esophagus and trachea and *anterior* to the vertebral column and the anterior vertebral muscles (longus colli, longus capitis, and others). As it extends laterally, the prevertebral fascia on each side is loosely attached to the posterior part of the carotid sheath, and it then courses beneath the sternocleidomastoid muscles. It then continues laterally around the neck to form, with the immediately overlying investing fascia, the tough fascial floor of the posterior triangle, and as such it closely invests the lateral vertebral muscles (the scalene muscles and the levator scapulae). That part of the prevertebral fascia overlying the scalene muscles is frequently and appropriately called the **scalene fascia.** Deep to the trapezius, the prevertebral fascia thins and extends posteriorly to invest the posterior vertebral muscles (splenius capitis, splenius cervicis, semispinalis capitis, and others), and finally attaches to the spines of cervical vertebrae in a manner similar to that of the investing fascial layer.

Superiorly, the prevertebral fascia reaches to the base of the skull, and inferiorly the fascia extends along the surface of the longus colli anterior to the vertebral column and into the superior mediastinum of the thorax, where it becomes continuous with the anterior longitudinal ligament. An anterior lamina of the fascia detaches in the superior mediastinum and fuses with the posterior surface of the esophagus at about the level of the sternal angle. In the posterior triangle, the roots of the brachial plexus emerge between the anterior and middle scalene muscles. As these nerves and the subclavian artery course into the axilla, they carry with them beneath the clavicle a prolongation of the prevertebral fascial layer (the "scalene fascia"), which forms the **axillary neurovascular sheath.** This prolongation is carried outward into the axilla as a sleeve around the plexus and the subclavian artery.

On the outer surface of the pharynx, anterior to the prevertebral fascia, is a filmy layer of loose areolar tissue (the epimysium of the pharyngeal constrictor muscles). Some authors have identified this as the "buccopharyngeal fascia." Last (1978), however, points out that this is not a properly developed fascial membrane and, if it were, it would serve to impede or prevent the peristaltic movements of the pharynx; thus, it clearly does not warrant the designation "fascia." Anterior to the vertebral bodies, some authors have described an additional fascial layer, the alar fascia, between the prevertebral fascia and the pretracheal fascia.

A number of important neural structures are anatomically related to the prevertebral fascia. In the posterior triangle the **spinal accessory nerve** courses along the superficial surface of the prevertebral fascia. The roots of the **cervical plexus** and **brachial plexus** arise deep to the fascia from the vertebral column, and the cervical cutaneous nerves pierce the fascia along the posterior border of the sternocleidomastoid muscle in their course to the anterior triangle. The trunks of the brachial plexus become surrounded by this same fascia as they course beneath the clavicle into the axilla. The **phrenic nerve** descends in the neck toward the thorax along the front of the anterior scalene muscle *behind the prevertebral fascia.* The **cervical sympathetic trunk** ascends in the neck along the anterior surface of the longus colli muscle, and is embedded in the prevertebral fascia.

CAROTID SHEATH. The carotid sheath (Fig. 6-13) is a condensation of dense connective tissue which forms a tubular investment for the common carotid artery (more superiorly, the internal carotid artery), the vagus nerve, and the internal jugular vein. The sheath is much thinner over the vein than around the other structures. The sheath is, in fact, divided into three separate compartments, one for each of its contents. It is dense around the common carotid artery and loose around the internal jugular vein. Anterolaterally, it is fused with the investing fascia, anteromedially with the pretracheal fascia, and posteriorly it attaches loosely to the prevertebral fascia. According to some authors, the carotid sheath extends superiorly as far as the base of the skull, where the internal carotid artery enters the lower end of the carotid canal. Other descriptions place its upper limit at the bifurcation of the common carotid artery. Inferiorly, the sheath reaches into the thorax as far as the brachiocephalic artery on the right and the aortic arch on the left, from which the common carotid arteries

are derived, and it is often described as the upward continuation of the fibrous pericardium, indeed, a form of tunica externa for the common carotid artery.

SUPERFICIAL CERVICAL MUSCLE

THE PLATYSMA. The platysma (Figs. 6-4; 6-6) is a broad sheet of muscle *arising* in the superficial fascia covering the superior parts of the pectoralis major and deltoid muscles. Its fibers ascend over the clavicle, and proceed obliquely upward and medialward along the side of the neck. The most anterior fibers interlace across the midline, inferior and posterior to the symphysis menti, with those of the muscle of the opposite side. The posterior fibers cross the border of the mandible superficial to the facial vessels, and some are *inserted* into the mandible below the oblique line, while others *insert* into the skin and subcutaneous tissue of the lower part of the face. Many of these latter fibers blend with the muscles about the angle and lower part of the mouth. Under cover of the platysma, the external jugular vein descends from the angle of the mandible to the clavicle. The platysma may be composed of delicate, pale, scattered fasciculi, or may form a broad layer of robust dark fibers. It may be deficient or reach well below the clavicle, and it may extend into the face for a short distance or may continue as high as the zygoma. An additional band, the **occipitalis minor,** may join the platysma from the superficial fascia over the occipital bone and mastoid process to the fascia over the insertion of the sternocleidomastoid muscle.

Action. The **platysma muscle** draws the lower lip and corner of the mouth laterally and inferiorly, partially opening the mouth, as in an expression of surprise or horror. When all the fibers act maximally, the skin over the clavicle is wrinkled and drawn toward the mandible, increasing the diameter of the neck, such as is seen during the intensive respiration of a sprinting runner. Electromyography has shown that it is not used in laughing or in motions of the jaw or the head.

Nerve. The platysma is supplied by the cervical branch of the **facial nerve.**

LATERAL CERVICAL MUSCLES

Trapezius
Sternocleidomastoid

The **trapezius** muscle is described with the muscles connecting the upper limb to the vertebral column.

THE STERNOCLEIDOMASTOID. The sternocleidomastoid muscle (Fig. 6-15) passes obliquely across the side of the neck, dividing the cervical region on each side into two triangular areas, the anterior and posterior triangles of the neck. The **anterior triangle** is bounded by the midline *anteriorly,* the medial border of the sternocleidomastoid *laterally,* and the body of the mandible *superiorly.* The **posterior triangle** is bounded by the clavicle *inferiorly,* and by the interfacing posterior border of the sternocleidomastoid *anteriorly,* and the anterior border of the trapezius *posteriorly.*

The sternocleidomastoid is thick and narrow at its central part, but is broad and thin at each end. It *arises* below from the sternum and clavicle by two heads. The **medial** or **sternal head** is rounded and tendinous ventrally, but fleshy dorsally, and it *arises* from the upper part of the ventral surface of the manubrium sterni, being directed upward and dorsally. The **lateral** or **clavicular head,** composed of fleshy and aponeurotic fibers, *arises* from the superior border and anterior surface of the medial third of the clavicle. It is directed almost vertically upward. The two heads are separated at their origins by a triangular interval, but gradually blend below the middle of the neck, so that the clavicular head passes behind the sternal head. These form a thick, rounded muscle which is *inserted* by a strong tendon into the lateral surface of the mastoid process, from its apex to its superior border, and by a thin aponeurosis into the lateral half of the superior nuchal line of the occipital bone.

Action. The **sternocleidomastoid muscle** of one side bends the head laterally, approximating the ear toward the shoulder of the same side. At the same time it rotates the head, directing the face upward and to the opposite side. The anterior fibers of both muscles acting together flex the head, bringing it ventrally, whereas the posterior fibers extend the head since they insert behind the atlanto-occipital joint (Last, 1954). If the head is fixed, the two muscles assist in elevating the thorax during forced inspiration.

Nerves. The sternocleidomastoid is supplied by the spinal part of the **accessory nerve** and branches from the anterior rami of the **second and third cervical nerves.**

Variations. The sternocleidomastoid varies in the extent of its origin from the clavicle: in some cases the clavicular head may be as narrow as the sternal; in others it may be as much as 7.5 cm in breadth. When the clavicular origin is broad, it is occasionally subdivided into several slips, separated by narrow intervals. More rarely, the adjoining margins of the sternocleidomastoideus and trapezius have been found in contact.

The **supraclavicularis** is a relatively rare anomalous muscle which extends from the upper edge of the manubrium laterally to about the middle of the clavicle along its upper surface.

SUPRAHYOID MUSCLES

Digastric
Stylohyoid
    Stylohyoid ligament
Mylohyoid
Geniohyoid

DIGASTRIC MUSCLE. The digastric muscle (Fig. 6-15) consists of two fleshy bellies united by an intermediate rounded tendon. It lies below the body of the mandible, and extends, in a curved form, from the mastoid process to the symphysis menti. The **posterior belly,** longer than the anterior, *arises* from the mastoid notch of the temporal bone and passes anteriorly and inferiorly. The **anterior belly** *arises* from a depression on the inner side of the inferior border of the mandible, close to the symphysis, and passes posteriorly and inferiorly. The two bellies end in an intermediate tendon which perforates the stylohyoid muscle, and which is connected to the side of the body and greater horn of the hyoid bone by a fibrous loop, which is sometimes lined by a synovial sheath. A broad aponeurotic layer, called the **suprahyoid aponeurosis,** extends from the tendon of the digastric and attaches to the body and greater horn of the hyoid bone.

Action. The **digastric muscle** elevates the hyoid bone, and the two digastric muscles acting together assist the lateral pterygoid muscles in opening the mouth by depressing the mandible.

Nerves. The anterior belly of the digastric is supplied by the mylohyoid nerve from the inferior alveolar branch of the mandibular division of the **trigeminal nerve.** The posterior belly is supplied by a branch of the **facial nerve.**

The two bellies are innervated by different nerves because they are derived from the mesenchyme of different branchial arches; the anterior belly forms

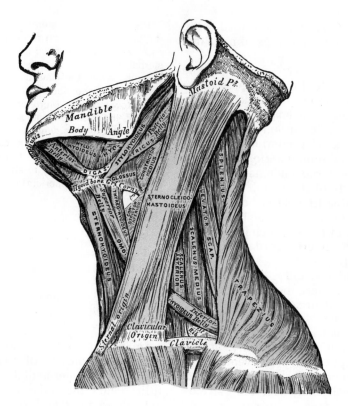

FIG. 6-15.   Muscles of the neck. Lateral view.

from the first or **mandibular arch**, and the posterior belly from the second or **hyoid arch**.

Variations.   The posterior belly may arise partly or entirely from the styloid process, or be connected by a slip to the middle or inferior constrictor. The anterior belly may be doubled, or extra slips from this belly may pass to the jaw or mylohyoid, or decussate with a similar slip on the opposite side. The anterior belly may be absent and the posterior belly inserted into the middle of the jaw or hyoid bone. The intermediate tendon between the two bellies may pass in front of or, more rarely, behind the stylohoid.

At times an anomalous muscle, the **mentohyoid**, passes from the body of the hyoid bone to the chin.

STYLOHOID MUSCLE.   The stylohyoid (Fig. 6-15) is a slender muscle lying anterior and superior to the posterior belly of the digastric. It *arises* by a small tendon from the posterior and lateral surface of the styloid process near its base. Passing anteriorly and inferiorly, the stylohyoid is *inserted* into the body of the hyoid bone at its junction with the greater horn and just above the omohyoid. It is perforated, near its insertion, by the tendon of the digastric.

Action.   The **stylohyoid** elevates and retracts the hyoid bone, thereby elongating the floor of the mouth.

Nerve.   The stylohyoid is supplied by a branch of the **facial nerve**.

Variations.   Occasionally the stylohyoid may be absent or doubled, or it may lie beneath the carotid artery, or be inserted into the omohyoid, thyrohyoid, or mylohyoid.

STYLOHYOID LIGAMENT.   The stylohyoid ligament is a fibrous cord that attaches to the tip of the styloid process of the temporal bone *above*, and to the lesser horn of the hyoid bone *below*. In its course it gives origin to some fibers of the middle pharyngeal constrictor and its lower portion lies deep to the hyoglossus muscle. It represents an unossified part of the skeleton of the second pharyngeal arch, frequently containing a little cartilage in its center. The ligament may be partially ossified, and in many animals it forms a distinct bone, the **epihyal**.

MYLOHYOID MUSCLE.   The mylohyoid is a flat, triangular muscle located immediately superior to the anterior belly of the digastric, and forms, with its fellow of the opposite side, the muscular floor of the oral cavity. It *arises* from the entire length of the mylohyoid line of the mandible, extending from the symphysis menti to the last molar tooth. The **posterior fibers** pass medially and slightly downward, to be *inserted* into the body of the hyoid bone. The **middle** and **anterior fibers** are *inserted* into a median fibrous raphe extending from the symphysis menti to the hyoid bone, where they join at an angle with the fibers of the opposite muscle. This median raphe is sometimes wanting, and then the fibers of the two muscles are continuous.

Action. The **mylohyoid muscles** raise the floor of the mouth during swallowing. They elevate the hyoid bone, thereby pushing the tongue upward as in swallowing or protrusion of the tongue (Whillis, 1946). Additionally, the mylohyoids may depress the mandible and, therefore, assist in opening the mouth. Thus, they are active in mastication, deglutition, sucking and blowing (Lehr et al., 1971).

Nerve. The mylohyoid is supplied by the mylohyoid branch of the inferior alveolar nerve. The latter nerve is derived from the mandibular division of the **trigeminal nerve.**

Variations. At times the mylohyoid muscle may be fused with the anterior belly of the digastric, and accessory slips of the mylohyoid to other hyoid muscles are frequent.

GENIOHYOID MUSCLE.   The geniohyoid is a narrow muscle, situated above the medial border of the mylohyoid. (*Note:* In dissection from the cervical approach, the geniohyoid lies deep to the mylohyoid.) It *arises* from the inferior mental spine on the inner surface of the symphysis menti, and runs posteriorly and slightly inferiorly, to be *inserted* into the anterior surface of the body of the hyoid bone. It lies in contact with its fellow of the opposite side.

Action. The **geniohyoid muscles** slightly elevate and draw the hyoid bone forward, shortening the floor of the oral cavity for the reception of food during feeding. If the hyoid bone remains fixed, the geniohyoids help to retract and depress the mandible.

Nerve. The geniohyoid is supplied by fibers of the **first cervical nerve,** that are carried along the hypoglossal nerve.

Variations. At times the muscle fibers of the two geniohyoids actually fuse. Muscular slips from the geniohyoid to the greater horn of the hyoid bone and the genioglossus also occur.

Group Actions. The **suprahyoid muscles** perform two very important actions. During deglutition they raise the hyoid bone. When the hyoid bone is fixed by its depressors, the infrahyoid muscles, the suprahyoid muscles assist in opening the mouth by depressing the mandible.

INFRAHYOID MUSCLES

Sternohyoid
Sternothyroid
Thyrohyoid
Omohyoid

STERNOHYOID MUSCLE.   The sternohyoid (Figs. 6-15; 6-16) is a thin, narrow strap muscle which *arises* from the posterior surface of the medial end of the clavicle, the posterior sternoclavicular ligament, and the upper and back part of the manubrium sterni. Passing superiorly and medially, it is *inserted* by short, tendinous fibers into the lower border of the body of the hyoid bone. The two sternohyoids are separated inferiorly by a considerable interval, but their medial borders come into contact about midway along their course, thereafter remaining contiguous above until their insertions. The sternohyoid sometimes presents a transverse tendinous inscription immediately above its origin.

Action. The **sternohyoid** depresses the hyoid bone after a bolus of food has been swallowed. It also acts importantly as an antagonist of the elevators of the hyoid bone and larynx.

Nerve. The sternohyoid is supplied by a branch of the ansa cervicalis, containing fibers from the **first, second, and third cervical nerves.**

Variations. The muscle may be absent or duplicated on one or both sides, or an additional muscular slip, the **cleidohyoid muscle,** may be present extending from the clavicle to the hyoid.

STERNOTHYROID MUSCLE.   The sternothyroid (Fig. 6-16) is shorter and wider than the sternohyoid and is situated immediately deep to it in the neck. It *arises* from the posterior surface of the manubrium sterni, below the origin of the sternohyoid, and from the edge of the cartilage of the first rib, and sometimes that of the second rib. It ascends to be *inserted* into the oblique line on the lamina of the thyroid cartilage. The two sternothyroids are in close contact in the

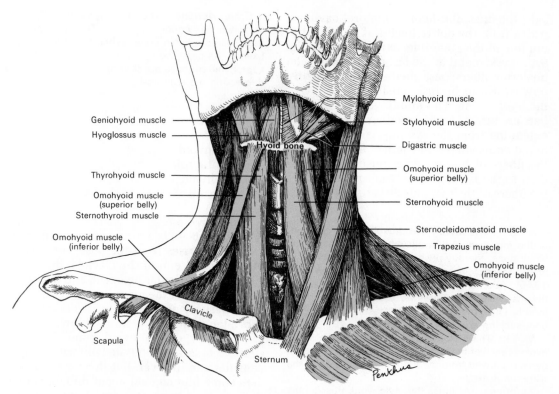

FIG. 6-16.   Muscles of the neck. Anterior view. On the right side certain of the more superficial muscles have been removed.

lower part of the neck, but diverge somewhat as they ascend. Occasionally the sternothyroid is crossed by a transverse or oblique tendinous inscription.

Action. The **sternothyroid** draws the larynx downward after it has been elevated during deglutition.

Nerve. The sternothyroid is supplied by a branch of the ansa cervicalis, containing fibers from the **first three cervical nerves.**

Variations. The sternothyroid may occasionally be absent or duplicated. It also may send accessory slips to the thyrohyoid, inferior pharyngeal constrictor, or the carotid sheath.

THYROHYOID MUSCLE. The thyrohyoid (Figs. 6-15; 6-16) is a small, quadrilateral muscle which appears to be an upward continuation of the sternothyroid. It *arises* from the oblique line on the lamina of the thyroid cartilage, and is *inserted* into the inferior border of the greater horn of the hyoid bone.

Action. The **thyrohyoid** depresses the hyoid bone or, if the latter is fixed, draws the thyroid cartilage superiorly.

Nerve. The thyrohyoid is supplied by fibers from the **first cervical nerve,** which course with the hypoglossal nerve and branch from it before the superior root of the ansa cervicalis descends from the hypoglossal.

OMOHYOID MUSCLE. The omohyoid (Figs. 6-15; 6-16) consists of two fleshy bellies united by a central tendon. It *arises* from the upper border of the scapula, and occasionally from the superior transverse ligament which crosses the scapular notch. Its extent of attachment to the scapula varies from a few millimeters to 2.5 cm. From this origin, the **inferior belly** forms a flat, narrow fasciculus, which inclines forward and slightly upward across the lower part of the neck, being attached to the clavicle by a fibrous expansion. It then passes deep to the sternocleidomastoid, ending in its intermediate tendon. Forming an obtuse angle with the inferior belly, the **superior belly** passes almost vertically upward from the tendon, close to the lateral border of the sternohyoid, to be inserted into the lower border of the body of the hyoid bone, lateral to the insertion of the sternohyoid. The central tendon of this muscle varies somewhat in length and form. It is held in position at the level of the cricoid cartilage by a process of the superficial investing layer of the deep cervical fascia, which sheathes it, and is prolonged caudally to be attached to the clavicle and first rib. It is by this fascial attachment that the angular form of the muscle is maintained.

Action. The **omohyoid** helps to depress the hyoid bone after it has been elevated.

Nerves. Both bellies of the omohyoid are supplied by branches from the ansa cervicalis, containing fibers from the **first, second, and third cervical nerves.**

Group Actions. The **infrahyoid muscles** depress the larynx and hyoid after these structures have been elevated with the pharynx in the act of deglutition. The omohyoids not only depress the hyoid bone, but carry it dorsalward and to one or the other side. These latter muscles are concerned especially in prolonged inspiratory efforts because they render the lower part of the cervical fascia tense, thereby lessening the inward suction of the soft parts which would otherwise compress the great vessels and the apices of the lungs.

Triangles of the Neck

The cervical region is divided by the sternocleidomastoid muscle into an anterior and a posterior triangle.

The **anterior triangle of the neck** is bounded by the anterior midline, the anterior border of the sternocleidomastoid, and the lower border of the mandible. This triangle is further subdivided into three triangles.

1. The **submandibular triangle,** found above the hyoid bone, is bounded *superiorly* by the inferior border of the mandible, *inferiorly* by the posterior belly of the digastric, and *anteriorly* by the anterior belly of the digastric.

2. The **carotid triangle,** located both above and below the hyoid bone, is bounded *superiorly* by the posterior belly of the digastric, *posteriorly* by the anterior border of the sternocleidomastoid, and *inferiorly* by the superior belly of the omohyoid.

3. The **muscular triangle,** located below the hyoid bone, is bounded *superiorly* by the superior belly of the omohyoid, *inferiorly* by the anterior border of the sternocleidomastoid muscle, and *medially* by the anterior midline of the neck.

The **posterior triangle of the neck** is bounded below by the middle third of the clavicle and by the interfacing borders of the trapezius and sternocleidomastoid muscles. It is further subdivided by the inferior belly of the omohyoid into two triangles.

1. The **occipital triangle,** found above, is bounded *inferiorly* by the inferior belly of the omohyoid, *anteriorly* by the posterior border of the sternocleidomastoid muscle, and *posteriorly* by the anterior border of the trapezius.

2. The **omoclavicular** (subclavian or supraclavicular) **triangle,** located below, is bounded *superiorly* by the inferior belly of the omohyoid, *anteriorly* by the sternocleidomastoid muscle, and *inferiorly* by the clavicle.

Finally, the area between the right and left anterior bellies of the digastric muscles and the body of the hyoid bone forms the **submental triangle,** which is unpaired.

ANTERIOR VERTEBRAL MUSCLES

Longus colli
Longus capitis
Rectus capitis anterior
Rectus capitis lateralis

LONGUS COLLI MUSCLE. The longus colli is situated along the anterior surface of the vertebral column, between the atlas and the third thoracic vertebra (Fig. 6-17). It is broad in the middle, narrow and more pointed at each end, and consists of three portions: a superior oblique, an inferior oblique, and a vertical portion. The **superior oblique portion** *arises* from the anterior tubercles of the transverse processes of the third, fourth, and fifth cervical vertebrae. It ascends obliquely with a medial inclination and is *inserted* by a narrow tendon into the tubercle on the anterior arch of the atlas. The **inferior oblique portion,** the smallest part of the muscle, *arises* from the anterior surface of the bodies of the first two or three thoracic vertebrae and, ascending obliquely in a lateral direction, is *inserted* into the anterior tubercles of the transverse processes of the fifth and sixth cervical vertebrae. The **vertical portion,** interposed between the other two parts, *arises* from the anterolateral surface of the bodies of the first three thoracic and last three cervical vertebrae, and is *inserted* into the anterior surface of the bodies of the second, third, and fourth cervical vertebrae.

Action. The **longus colli** is a rather weak flexor of the neck. It also slightly rotates and laterally bends the cervical portion of the vertebral column.

Nerves. The longus colli is supplied by branches from ventral rami of the **second to the sixth cervical nerves.**

LONGUS CAPITIS MUSCLE.   The longus capitis (Fig. 6-17) is narrow below but becomes broader and thicker above. It *arises* by four tendinous slips from the anterior tubercles of the transverse processes of the third, fourth, fifth, and sixth cervical vertebrae, and ascends, converging medially toward the opposite longus capitis, to be *inserted* into the inferior surface of the basilar part of the occipital bone, lateral to the pharyngeal tubercle.

Action. The **longus capitis** flexes the head and the upper cervical spine.

Nerves. The longus capitis is supplied by branches from the **first, second, and third cervical nerves.**

RECTUS CAPITIS ANTERIOR MUSCLE.   This short, flat muscle is situated immediately deep to the superior part of the longus capitis (Fig. 6-17). It *arises* from the anterior surface of the lateral mass of the atlas, and from the root of its transverse process. It passes upward and slightly medially, to be *inserted* into the inferior surface of the basilar part of the occipital bone immediately anterior to the foramen magnum. It can be considered an upward continuation of the anterior intertransverse muscles seen between lower cervical vertebrae.

Action. The **rectus capitis anterior** flexes the head and helps to stabilize the atlanto-occipital joint.

Nerve. This muscle is supplied by fibers from the anterior rami of the **first** (and possibly the second) **cervical nerve.**

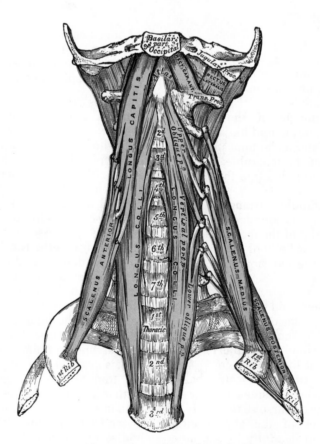

FIG. 6-17.   The anterior and lateral vertebral muscles.

RECTUS CAPITIS LATERALIS MUSCLE. The rectus capitis lateralis is also a short, flat strap muscle (Fig. 6-17). It *arises* from the superior surface of the transverse process of the atlas, and is *inserted* into the inferior surface of the jugular process of the occipital bone. This muscle is comparable to the posterior cervical intertransverse muscle.

Action. The **rectus capitis lateralis** bends the head laterally to the same side and assists in stabilizing the atlanto-occipital joint.

Nerve. The rectus capitis lateralis, similar to the anterior, is supplied by a branch of the loop between the **first and second cervical nerves.**

Group Actions. The **longus capitis** and **rectus capitis anterior** are the direct antagonists of the muscles at the back of the neck and restore the head to its natural position after it has been drawn backward. These muscles also flex the head, and due to their obliquity, rotate it, turning the face to one or the other side. The **rectus capitis lateralis**, acting on one side, bends the head laterally. The **longus colli** flexes and slightly rotates the cervical portion of the vertebral column.

LATERAL VERTEBRAL MUSCLES

Scalenus anterior
Scalenus medius
Scalenus posterior

SCALENUS ANTERIOR MUSCLE.
anterior (Fig. 6-17) lies at the side of the neck, deep to the sternocleidomastoid muscle. It *arises* by four slender tendons, one each from the anterior tubercles of the transverse processes of the third, fourth, fifth, and sixth cervical vertebrae. Descending almost vertically, it is *inserted* by a narrow, flat tendon onto the scalene tubercle on the inner border of the first rib, and into the ridge on the upper surface of the rib anterior to the groove for the subclavian artery.

Action. If the neck is fixed, the **scalenus anterior** elevates the first rib as an accessory muscle of respiration. If the first rib is fixed, this muscle bends the neck forward and laterally, and rotates it to the opposite side.

Nerves. The scalenus anterior is innervated by branches from the anterior rami of the **fifth and sixth (and sometimes the fourth) cervical nerves.**

SCALENUS MEDIUS MUSCLE.
medius (Fig. 6-17) is the largest and longest of the three scalene muscles. It *arises* from the posterior tubercles of the transverse processes of the last six cervical vertebrae, and from the atlas as well in over 50% of the cases. Descending along the side of the vertebral column, it is *inserted* by a broad attachment into the cranial surface of the first rib, between the tubercle and the groove for the subclavian artery.

Action. The **scalenus medius**, similar to the scalenus anterior, raises the first rib from above, or bends and slightly rotates the neck from below.

Nerves. The scalenus medius is supplied by branches from anterior rami of the **third through the eighth cervical nerves.**

SCALENUS POSTERIOR MUSCLE.
smallest and most deeply seated of the three scalene muscles (Fig. 6-17). It *arises* by separate tendons from the posterior tubercles of the transverse processes of the fourth, fifth, and sixth cervical vertebrae, and is *inserted* by a thin tendon into the outer surface of the second rib, deep to the attachment of the serratus anterior. It is occasionally blended with the scalenus medius.

Action. The **scalenus posterior** raises the second rib or bends and slightly rotates the neck.

Nerves. The scalenus posterior is supplied by branches of ventral primary divisions of the **last three cervical nerves.**

Group Actions. When the scalenes are fixed superiorly, they elevate the first and second ribs and are, therefore, inspiratory muscles. Acting from below, they bend the vertebral column to the same side. If the muscles of both sides act, the vertebral column is flexed slightly.

Variations. The scalene muscles vary considerably in their attachments and in the arrangement of their fibers. A slip from the scalenus anterior may pass behind the subclavian artery. The scalenus posterior may be absent or extend to the third rib.

The **scalenus pleuralis muscle** or **scalenus minimus** extends from the transverse process of the seventh cervical vertebra to the fascia supporting the dome of the pleura and inner border of the first rib.

Fasciae and Muscles of the Trunk

The muscles of the trunk may be arranged in six groups:

Deep muscles of the back
Suboccipital muscles

Muscles of the thorax
Muscles of the abdomen
Muscles of the pelvis
Muscles of the perineum

DEEP MUSCLES OF THE BACK

The deep or intrinsic muscles of the back consist of a complex, serially arranged group of muscles that extends from the pelvis to the skull. They may be divided into four subgroups: the **splenius,** the **erector spinae,** the **transversospinalis,** and the **interspinal-intertransverse.**

The **splenius group** (Fig. 6-20) is situated on the posterior aspect of the neck and upper thorax, and extends and rotates the head and neck. The two muscles are the:

Splenius capitis muscle
Splenius cervicis muscle

The **erector spinae** (sacrospinalis) is a great mass of muscle that stretches from the sacrum to the skull (Fig. 6-19). This mass splits in the upper lumbar region into lateral, intermediate, and medial columns of muscles that lie parallel. These columns are further subdivided into three muscles according to the vertebral region to which each attaches superiorly. Muscles comprising the erector spinae primarily extend and laterally bend the vertebral column. They are the:

Iliocostalis muscle (lateral column)
    Iliocostalis lumborum
    Iliocostalis thoracis
    Iliocostalis cervicis
Longissimus muscle (intermediate column)
    Longissimus thoracis
    Longissimus cervicis
    Longissimus capitis
Spinalis muscle (medial column)
    Spinalis thoracis
    Spinalis cervicis
    Spinalis capitis

The **transversospinalis group** (Figs. 6-19; 6-20) is so named because generally the fibers of these muscles extend from the transverse processes of vertebrae upward and medially to the spinous processes of higher vertebrae. In this group are the:

Semispinalis muscle
    Semispinalis thoracis

Semispinalis cervicis
    Semispinalis capitis
Multifidus muscles
Rotatores muscles
    Rotatores thoracis
    Rotatores cervicis
    Rotatores lumborum

The **interspinal-intertransverse group** (Fig. 6-20) includes those muscles that pass from the spinous process of one vertebra to the spine of the next vertebra, or the transverse process of one vertebra to the transverse process of the next vertebra. In this group are the:

Interspinalis muscles
    Interspinalis cervicis
    Interspinalis thoracis
    Interspinalis lumborum
Intertransversarii muscles
    Intertransversarii anterior cervicis
    Intertransversarii posterior cervicis
        Medial and Lateral
    Intertransversarii thoracis
    Intertransversarii lumborum
        Medial and Lateral

Overlying these muscles is an extensive layer of deep fascia. Although the deep fascia covering cervical, thoracic, and lumbar structures can be considered as a continuous layer, two portions of it will be described separately. These are the:

Nuchal fascia
Thoracolumbar fascia

NUCHAL FASCIA. The nuchal fascia is the cervical portion of the more extensive deep fascia of the back, and it is continuous on each side of the neck with the prevertebral layer of cervical fascia covering the scalene muscles (the **scalene fascia**). It covers the splenius capitis and splenius cervicis muscles and, near the skull, the upper portion of the semispinalis capitis. With these muscles, it is attached to the skull just inferior to the superior nuchal line. Medially, it fuses with the ligamentum nuchae and the spinous processes of the seventh cervical and first six thoracic vertebrae. In the cranial part of the neck it is more or less adherent to the superficial investing layer of cervical fascia on the deep surface of the trapezius. More caudally, a distinct fascial cleft separates it from the fascia of the serratus posterior superior and rhomboid muscles.

The deep muscles of the neck are enclosed by fascial septa which form compartments for each muscle. A fascial cleft separates the splenius capitis and splenius cervicis from the semispinalis capitis. A considerable layer of adipose tissue and a fascial cleft intervene between the semispinalis capitis and semispinalis cervicis. In this adipose layer are found the deep cervical blood vessels. The fascia covering the semispinalis cervicis continues superiorly from the atlas to form the thick adherent covering of the suboccipital muscles.

THORACOLUMBAR FASCIA. The thoracolumbar or lumbodorsal fascia (Figs. 6-18; 6-29; 6-42), in general terms, is the deep fascial complex that forms the *sheath of the erector spinae muscle*. In the **thoracic region** this fascial layer is thin, gray, and transparent, attaching medially to spines of the vertebrae and the supraspinal ligaments, and laterally, at the margin of the erector spinae, to the angles of the ribs and to the fascia covering the intercostal muscles. Becoming thinner as it ascends in the thorax, the fascia continues superiorly, deep to the rhomboids, and gradually fades out above the first rib.

In the **lumbar region,** the thoracolumbar fascia is considerably thickened and encases the erector spinae by its **posterior** and **anterior layers.** *Medially,* the white glistening posterior layer, which covers the dorsal surface of the erector spinae, attaches to the spinous processes of the lumbar and sacral vertebrae. The anterior layer, which stretches between the ventral surface of the erector spinae and the quadratus lumborum, attaches to the transverse processes of the lumbar vertebrae. Extending *laterally* from these bony attachments, the two laminae blend with the aponeuroses of the latissimus dorsi, the internal oblique, and the transversus abdominis muscles. The aponeurosis of the latissimus dorsi blends with the posterior layer, while the aponeuroses of the internal oblique and transversus abdominis

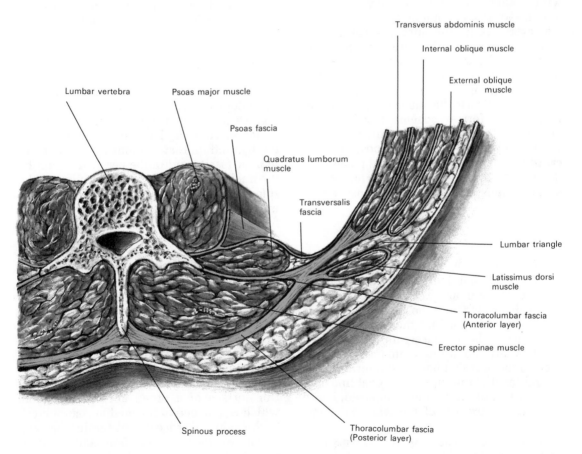

FIG. 6-18. Cross section through the posterior abdominal wall, showing the posterior and anterior layers of the thoracolumbar fascia along with the investing transversalis fascia. (After Sobotta.)

blend with the anterior layer. *Inferiorly*, the layers attach not only to the medial crest of the sacrum, but also to the lateral sacral crests, the iliolumbar ligaments, and the iliac crests. Superiorly, the posterior layer extends over the thorax, as described above, while the anterior layer extends to the twelfth rib and attaches to it and to the transverse processes of the first two lumbar vertebrae as the **lumbocostal ligament.**

A third fascial layer usually discussed in relation to the thoracolumbar fascia covers the anterior surface of the quadratus lumborum muscle. It is the thinnest of the layers in the lumbar region and is attached medially to the anterior surfaces of the transverse processes of the lumbar vertebrae posterior to the lateral part of the psoas major muscle. Some authors describe this layer as a continuation of the **fascia transversalis** extending over the inner aspect of the posterior abdominal muscles, while others describe this layer as joining laterally with the aponeurosis of the transversus abdominis muscle in conjunction with the fascial layer covering the anterior surface of the erector spinae. In the latter schema the lumbodorsal fascia is described as consisting of *posterior* (dorsal to erector spinae), *middle* (ventral to the erector spinae), and *anterior* (ventral to the quadratus lumborum) layers. The former schema (used in this book) calls the two layers in contact with the erector spinae the anterior and posterior layers of the lumbodorsal fascia and recognizes the fascia ventral to the quadratus lumborum as a continuation of the fascia transversalis.

Splenius Muscle Group

SPLENIUS CAPITIS.   The splenius capitis muscle (Fig. 6-20) is a flat muscle *arising* from the caudal half of the ligamentum nuchae, the spinous process of the seventh cervical vertebra, and the spinous processes of the first three or four thoracic vertebrae. The fibers of the muscle are directed upward and laterally and *inserted* into the rough surface on the occipital bone just inferior to the lateral third of the superior nuchal line and, under cover of the sternocleidomastoid, into the mastoid process of the temporal bone.

Action.   The **splenius capitis** draws the head dorsally and laterally and so rotates the head, turning the face to the same side. When both splenius capitis muscles act together, they extend the head.

Nerves.   The splenius capitis is supplied by lateral branches of the dorsal rami of the **middle cervical nerves.**

SPLENIUS CERVICIS.   The splenius cervicis (Fig. 6-20) is a narrow band of muscle that *arises* from the spinous processes of the third to the sixth thoracic vertebrae. Ascending in the posterior neck, it is *inserted* by tendinous fasciculi into the posterior tubercles of the transverse processes of the upper two or three cervical vertebrae.

Action.   The **splenius cervicis** draws the neck dorsally and laterally and also rotates the cervical vertebrae, thereby assisting to turn the head to the same side. Acting together, the two splenius cervicis muscles extend and arch the neck.

Nerves.   The splenius cervicis is innervated by lateral branches of the dorsal primary divisions of the **lower cervical nerves.**

Variations.   Accessory slips of the splenius muscles are frequently found, especially one that attaches to the first cervical vertebra, the atlas.

Erector Spinae (Sacrospinalis)

The **erector spinae** muscle (Fig. 6-19) and its prolongations in the thoracic and cervical regions lie in the groove on the side of the vertebral column. They are covered in the lumbar and thoracic regions by the thoracolumbar fascia, and in the cervical region by the nuchal fascia, over which are found the splenius and serratus posterior superior muscles above and the serratus posterior inferior muscle below. This large muscular and tendinous mass varies in size and structure at different parts of the vertebral column. In the sacral region it is narrow and pointed, and at its origin below chiefly tendinous. In the lumbar region it is larger, and forms a thick fleshy mass which, in its ascending course, is subdivided into three vertical columns. These gradually diminish in mass as parts of the muscle insert successively into the vertebrae and ribs.

The erector spinae *arises* from the anterior surface of a broad and thick tendon, which is attached to the median sacral crest, to the spinous processes of the lumbar and lower two thoracic vertebrae and their supraspinal ligaments, to the inner aspect of

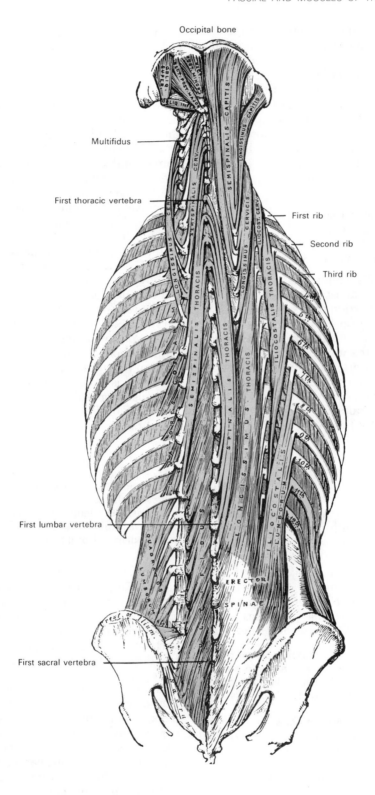

Occipital bone

Multifidus

First thoracic vertebra

First rib

Second rib

Third rib

First lumbar vertebra

First sacral vertebra

FIG. 6-19.   Deep muscles of the back. The erector spinae.

the dorsal part of the iliac crests, and to the lateral crests of the sacrum, where it blends with the sacrotuberous and posterior sacro-iliac ligaments. Some of its fibers are continuous with the fibers of origin of the gluteus maximus. The muscular fibers form a large fleshy mass which splits, in the upper lumbar region, into three columns: a *lateral,* the **iliocostalis;** an *intermediate,* the **longissimus;** and a *medial,* the **spinalis.** Each of these columns consists of three parts (Fig. 6-19).

Iliocostalis (lateral column)
  Iliocostalis lumborum
  Iliocostalis thoracis
  Iliocostalis cervicis

The iliocostalis column is a substantial mass of muscle that ascends as the most laterally oriented portion of the erector spinae.

ILIOCOSTALIS LUMBORUM.   The iliocostalis lumborum (Fig. 6-19) splits from the erector spinae mass at the upper lumbar level and ascends to be *inserted* by six or seven flattened tendons into the inferior borders of the angles of the last six or seven ribs.

ILIOCOSTALIS THORACIS.   The iliocostalis thoracis (Fig. 6-19) *arises* by flattened tendons from the upper borders of the angles of the lower six ribs medial to the tendons of insertion of the iliocostalis lumborum. These become muscular, and are *inserted* into the upper borders of the angles of the first six ribs and into the dorsum of the transverse process of the seventh cervical vertebra.

ILIOCOSTALIS CERVICIS.   The iliocostalis cervicis (Figs. 6-19; 6-21) *arises* from the angles of the third, fourth, fifth, and sixth ribs, medial to the insertions of the iliocostalis thoracis, and is *inserted* into the posterior tubercles of the transverse processes of the fourth, fifth, and sixth cervical vertebrae.

Action. The **iliocostalis muscles** are all extensors of the vertebral column and are also able to bend the column to one side (lateral flexion) and assist in its rotation. Additionally, the iliocostalis lumborum can depress the ribs.

Nerves. The iliocostalis muscles are supplied segmentally by branches of the dorsal primary divisions of the **spinal nerves.**

Longissimus (intermediate column)
  Longissimus thoracis
  Longissimus cervicis
  Longissimus capitis

The longissimus is placed intermediately between the iliocostalis and the spinalis. It is the largest and the longest of the three columns comprising the erector spinae.

LONGISSIMUS THORACIS.   The longissimus thoracis (Fig. 6-19) separates from the iliocostalis lumborum and spinalis thoracis in the upper lumbar region. Below this level it is still blended with the rest of the erector spinae, and some of its fibers are attached to the whole length of the posterior surfaces of the transverse processes and the accessory processes of the lumbar vertebrae, and to the anterior layer of the thoracolumbar fascia. In the thoracic region it is *inserted* by rounded tendons onto the tips of the transverse processes of all the thoracic vertebrae, and by fleshy processes to the lower nine or ten ribs between their tubercles and angles.

LONGISSIMUS CERVICIS.   The longissimus cervicis (Figs. 6-19; 6-21), situated medial to the longissimus thoracis, *arises* by long thin tendons from the tips of the transverse processes of the upper four or five thoracic vertebrae, and is *inserted* by similar tendons into the posterior tubercles of the transverse processes of the second to the sixth cervical vertebrae.

Action. The longissimus thoracis and longissimus cervicis extend the vertebral column and bend it to one side. The longissimus thoracis is also able to depress the ribs.

Nerves. The longissimus thoracis and longissimus cervicis are supplied by branches of the dorsal primary divisions of the spinal nerves.

LONGISSIMUS CAPITIS.   The longissimus capitis (Figs. 6-19; 6-21) lies between the longissimus cervicis and the semispinalis capitis. It *arises* by tendons from the transverse processes of the upper four or five thoracic vertebrae, and the articular processes of the last three or four cervical vertebrae. It is *inserted* into the posterior margin of the mastoid process, deep to the splenius capitis and sternocleidomastoid. In the upper part of the neck, where the longissimus capitis and cervicis diverge toward their insertions, they are crossed by the splenius cervicis. The insertion of the splenius cervicis separates the insertion of the longissimus capitis from that of the longissimus cervicis. The longissimus capitis generally has a transverse tendinous inscription near its insertion.

Action. The **longissimus capitis** extends the head. The muscle of one side acting alone bends the head to the same side and rotates the face toward that side.

Nerves. The longissimus capitis is supplied by branches of the dorsal primary divisions of the **middle and lower cervical nerves.**

Spinalis (medial column)
   Spinalis thoracis
   Spinalis cervicis
   Spinalis capitis

The spinalis column of muscles is the least massive of the three divisions of the erector spinae, and it is situated most medially and adjacent to the vertebral column.

SPINALIS THORACIS. The spinalis thoracis (Fig. 6-19) is the medial continuation of the erector spinae and is scarcely separable as a distinct muscle. It is situated at the medial side of the longissimus thoracis, with which it is intimately blended. It *arises* by three or four tendons from the spinous processes of the first two lumbar and the last two thoracic vertebrae. These unite to form a long, slender muscle which is *inserted* by separate tendons into the spinous processes of the upper thoracic vertebrae, the number varying from four to eight. It is intimately united with the semispinalis thoracis situated deep to it.

SPINALIS CERVICIS. The spinalis cervicis (Fig. 6-21) is an inconstant muscle that *arises* from the caudal part of the ligamentum nuchae, the spinous process of the seventh cervical, and sometimes from the spines of the first and second thoracic vertebrae, and is *inserted* into the spinous process of the axis, and occasionally into the spines of the two vertebrae caudal to it.

SPINALIS CAPITIS. The spinalis capitis is usually inseparably connected with the medial part of the semispinalis capitis. It *arises* and *inserts* with the semispinalis capitis, but it may have some separate attachments below to the upper thoracic vertebral spines.

Actions. The **spinalis muscles** extend the vertebral column.

Nerves. The spinalis muscles are innervated by branches of the dorsal primary divisions of the **spinal nerves.**

Transversospinalis Group

The muscles in this group are more deeply placed than the erector spinae and they fill the concave region between the spinous processes and the transverse processes of the vertebrae. They consist of muscular fascicles that course obliquely upward and medially from the transverse processes to higher spinous processes. The semispinalis, multifidus, and rotatores muscles are included in this group.

Semispinalis muscles
   Semispinalis thoracis
   Semispinalis cervicis
   Semispinalis capitis

The semispinalis muscles are found only in the thoracic and cervical regions extending to the head. They underlie the spinalis and longissimus columns of the erector spinae.

SEMISPINALIS THORACIS. The semispinalis thoracis (Fig. 6-19) consists of thin, narrow, fleshy fasciculi, interposed between tendons of considerable length. It *arises* by a series of tendons from the transverse processes of the sixth to the tenth thoracic vertebrae, and is *inserted* by tendons into the spinous processes of the first four thoracic and last two cervical vertebrae.

SEMISPINALIS CERVICIS. This muscle (Figs. 6-19; 6-21), thicker than the preceding, *arises* by a series of tendinous and fleshy fibers from the transverse processes of the first five or six thoracic vertebrae. Ascending in the neck and becoming thicker and more muscular, it is *inserted* into the cervical spinous processes from the axis to the fifth cervical vertebra inclusive. The fasciculus connected with the axis is the largest, and is chiefly muscular in structure. This muscle is almost entirely covered by the semispinalis capitis.

Action. The **semispinalis thoracis and semispinalis cervicis** extend the vertebral column and rotate it toward the opposite side.

Nerves. The semispinalis thoracis and semispinalis cervicis are supplied by branches of the dorsal primary divisions of **thoracic and cervical spinal nerves.**

SEMISPINALIS CAPITIS. The semispinalis capitis (Figs. 6-19; 6-20) is a large muscle situated on the posterior aspect of the neck, deep to the splenius, and medial to the lon-

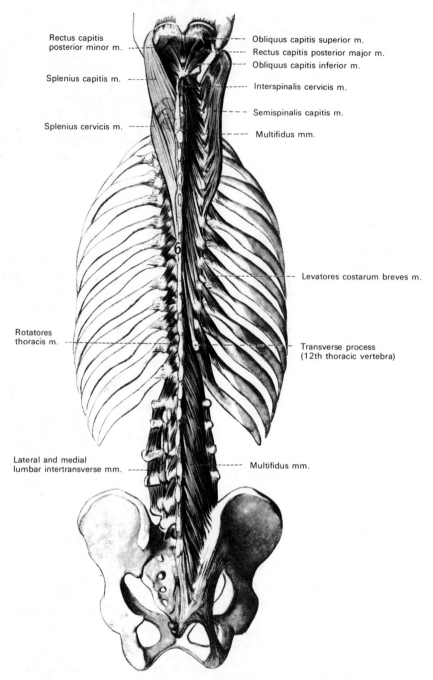

Rectus capitis posterior minor m.

Splenius capitis m.

Splenius cervicis m.

Obliquus capitis superior m.
Rectus capitis posterior major m.
Obliquus capitis inferior m.

Interspinalis cervicis m.

Semispinalis capitis m.

Multifidus mm.

Levatores costarum breves m.

Rotatores thoracis m.

Transverse process (12th thoracic vertebra)

Lateral and medial lumbar intertransverse mm.

Multifidus mm.

FIG. 6-20. The deep back muscles, showing some of the transversospinalis and interspinalis-intertransverse muscles. (From Benninghoff and Goerttler, *Lehrbuch der Anatomie des Menschen*, Urban & Schwarzenberg, 11th Ed., Vol. 1, 1975.)

gissimus cervicis and longissimus capitis. It *arises* by a series of tendons from the tips of the transverse processes of the first six or seven thoracic and the seventh cervical vertebrae, and from the articular processes of the fourth, fifth, and sixth cervical vertebrae.

The tendons unite to form a broad muscle, which passes almost vertically upward to be *inserted* between the superior and inferior nuchal lines of the occipital bone. The medial part, usually more or less distinct from the remainder of the muscle, is frequently indistinguishable from the **spinalis capitis** muscle. This part is also called the **biventer cervicis** because it is traversed by an imperfect tendinous intersection.

Action. The **semispinalis capitis** extends the head and rotates it so that the face is turned toward the opposite side.

Nerves. The semispinalis capitis receives innervation from branches of the dorsal primary divisions of the **cervical nerves.**

MULTIFIDUS MUSCLE. The multifidus muscle (Figs. 6-20; 6-21) consists of a number of fleshy and tendinous fasciculi that fill the groove on both sides of the spinous processes of the vertebrae from the *sacrum to the axis.* The fascicles lie deep to the erector spinae in the lumbar region and deep to the semispinalis thoracis and semispinalis cervicis more superiorly. In the *sacral region,* these fasciculi *arise* from the back of the sacrum, as low as the fourth sacral foramen, from the aponeurosis of origin of the erector spinae, from the medial surface of the posterior superior iliac spine, and from the posterior sacroiliac ligaments; in the *lumbar region* (Fig. 6-20), from all the mamillary processes; in the *thoracic region,* from all the transverse processes; and in the *cervical region* (Fig. 6-21), from the articular processes of the lower four cervical vertebrae. Each fasciculus ascends obliquely, crossing over two to four vertebrae in its course toward the midline, and is *inserted* onto the spinous process of a higher vertebra. The fascicles vary in length and depth of position, but insertions are made onto the spines of all the vertebrae from the lowest lumbar to the axis. The longest and most superficial pass from one vertebra to the fifth or fourth above; those somewhat deeper are shorter and cross three vertebrae; the deepest and shortest cross two.

Action. Acting alone, the **multifidus muscle** bends (or laterally flexes) the vertebral column and rotates it to the opposite side. When the multifidus of each side acts together, the vertebral column is extended.

Nerves. The multifidus is innervated segmentally by branches of the dorsal primary divisions of the **spinal nerves.**

Rotatores muscles
    Rotatores thoracis
    Rotatores cervicis
    Rotatores lumborum

The rotatores are the deepest muscles of the transversospinalis group. Some cross one vertebra in their oblique course (*rotatores longi*), while others insert in the next succeeding vertebra and course in an almost horizontal direction (*rotatores brevi*). Although they are found along the entire length of the vertebral column, they are only developed to any degree in the thoracic region.

ROTATORES THORACIS. These narrow rectangular muscles (Fig. 6-20) form the deepest layer in the groove between the spinous and transverse processes of the thoracic vertebrae. They arise from the transverse process of one vertebra and insert at the base of the spine of the thoracic vertebra immediately above, or the one above that. They lie deep to the multifidus and cannot be distinguished readily from its deepest fibers.

ROTATORES CERVICIS AND LUMBORUM. In the cervical and lumbar regions the rotatores are less well defined and more variable. In the cervical region these muscular fascicles course from the articular processes below to the base of the spines above, while in the lumbar region they attach to the mamillary process below and to the inferior margin of the lamina above.

Action. The **rotatores** extend the vertebral column and rotate it toward the opposite side.

Nerves. The rotatores are supplied by branches of the dorsal primary divisions of the **spinal nerves.**

Interspinal-Intertransverse Group

In this group are muscles that pass segmentally from one vertebra to the next. The muscles considered in this group are the interspinales and the intertransversarii. Certain of the rotators also pass between adjacent vertebrae (rotatores brevi), but they are better classified with the transversospinal muscles because of their attachments and rotator action.

Interspinales muscles
    Interspinales cervicis
    Interspinales thoracis
    Interspinales lumborum

The interspinales are short, paired muscular fasciculi, placed between the spinous processes of the contiguous vertebrae, one on each side of the interspinal ligament.

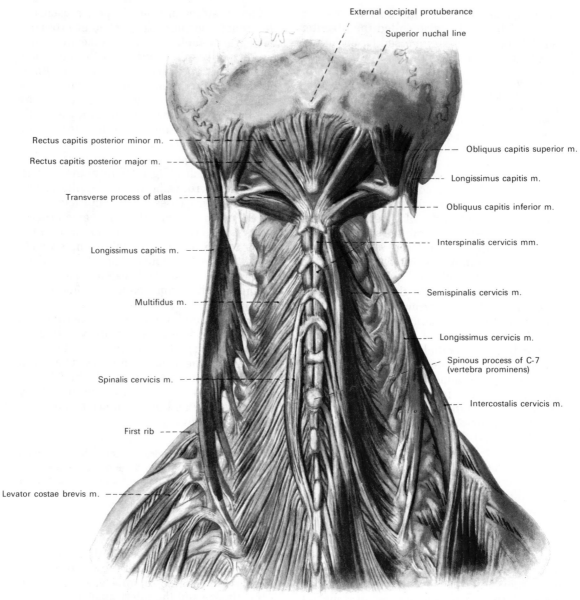

External occipital protuberance

Superior nuchal line

Rectus capitis posterior minor m.

Rectus capitis posterior major m.

Transverse process of atlas

Longissimus capitis m.

Multifidus m.

Spinalis cervicis m.

First rib

Levator costae brevis m.

Obliquus capitis superior m.

Longissimus capitis m.

Obliquus capitis inferior m.

Interspinalis cervicis mm.

Semispinalis cervicis m.

Longissimus cervicis m.

Spinous process of C-7 (vertebra prominens)

Intercostalis cervicis m.

FIG. 6-21.   The deep muscles of the posterior cervical region. (From Benninghoff and Goerttler, *Lehrbuch der Anatomie des Menschen,* Urban & Schwarzenberg, 11th Ed., Vol. 1, 1975.)

INTERSPINALES CERVICIS.   In the cervical region (Figs. 6-20; 6-21), the interspinales are most distinct and consist of six pairs, the first being situated between the axis and third vertebra, and the last between the seventh cervical and the first thoracic. They are small narrow bundles, attached to the apices of the adjacent spinous processes.

INTERSPINALES THORACIS.   In the thoracic region the interspinales are found between the first and second vertebrae, and sometimes between the second and third, and be-

tween the eleventh and twelfth. They are generally undeveloped between the other thoracic spines.

INTERSPINALES LUMBORUM.   In the lumbar region, there are four pairs in the intervals between the five lumbar vertebrae. There is also occasionally one pair between the last thoracic and first lumbar, and another between the fifth lumbar and the sacrum.

Action.   The **interspinales** extend the vertebral column.

Nerves. The interspinales are supplied by branches of the dorsal primary divisions of the **spinal nerves.**

The **extensor coccygis** is a slender muscular fasciculus that is not always present. It extends over the caudal part of the posterior surface of the sacrum and coccyx, and *arises* by tendinous fibers from the last segment of the sacrum, or first piece of the coccyx. It passes caudally to be *inserted* into the lower part of the coccyx, and is a rudiment of the extensor muscle of the caudal vertebrae in lower animals.

Intertransversarii muscles
　Intertransversarii anterior cervicis
　Intertransversarii posterior cervicis
　　Medial and Lateral
　Intertransversarii thoracis
　Intertransversarii medial lumborum
　Intertransversarii lateral lumborum

The intertransversarii are small, rounded, tendinous muscles placed between the transverse processes of contiguous vertebrae.

INTERTRANSVERSARII CERVICIS. The intertransversarii are best developed in the cervical region where at each segment there are paired anterior and posterior muscles. The **anterior cervical intertransverse muscles** interconnect the anterior tubercles of contiguous transverse processes of the cervical vertebrae, while those interconnecting the posterior tubercles are the **posterior cervical intertransverse muscles.** The ventral primary division of the segmental cervical nerve lies in a groove between the anterior and posterior muscles at each level. The posterior intertransverse muscles each consist of a medial and lateral part. The *medial part* belongs to the intrinsic muscles of the back and lies between the transverse processes, close to the vertebral body, and therefore is supplied by the posterior primary ramus of the spinal nerve. The *lateral part* attaches the transverse processes more laterally and, along with the anterior intertransverse muscles, is homologous to the intercostal muscles. Both of these latter elements are therefore supplied by the anterior primary ramus of the spinal nerve. There are seven pairs of cervical intertransverse muscles, the most superior being between the atlas and axis, and the lowest pair between the seventh cervical and first thoracic vertebrae.

INTERTRANSVERSARII THORACIS. In the thoracic region, the intertransverse muscles consist of single slips on each side. They are present between the transverse processes of the last three thoracic vertebrae, and between the transverse processes of the last thoracic and the first lumbar. They are comparable to the medial part of the posterior cervical intertransverse muscles and are therefore supplied by the posterior primary rami of the thoracic nerves.

INTERTRANSVERSARII LUMBORUM. In the lumbar region (Fig. 6-20) the intertransverse muscles are once again arranged in pairs on each side of the vertebral column. One muscle of each pair, the **medial intertransverse muscle,** passes from the accessory process of one vertebra to the mammillary process of the vertebra below. These are intrinsic deep back muscles supplied by the dorsal primary division of the lumbar nerves. The other muscle of each pair is the **lateral intertransverse muscle,** and it occupies the entire interspace between the transverse processes of contiguous vertebrae. The lateral intertransverse muscles are themselves separable into posterior and anterior parts, which are homologues of the intercostal muscles, and both parts are innervated by branches of the anterior primary rami of the lumbar nerves.

Action. The **intertransversarii** bend the vertebral column laterally when acting on one side and extend the vertebral column when acting on both sides.
Nerves. Supplied by **ventral primary rami of spinal nerves** are the:
　(a) anterior cervical intertransverse muscles;
　(b) lateral part of the posterior cervical intertransverse muscles;
　(c) lateral (lumbar) intertransverse muscles.
Supplied by **dorsal primary rami of spinal nerves** are the:
　(a) medial part of the posterior cervical intertransverse muscles;
　(b) thoracic intertransverse muscles;
　(c) medial (lumbar) intertransverse muscles.

SUBOCCIPITAL MUSCLES

The suboccipital muscles, four on each side, lie deep to the semispinalis capitis in the uppermost part of the dorsal cervical region.

Rectus capitis posterior major
Rectus capitis posterior minor
Obliquus capitis inferior
Obliquus capitis superior

RECTUS CAPITIS POSTERIOR MAJOR. This muscle (Figs. 6-21; 6-22) *arises* by a pointed tendon from the spinous process of the axis. Becoming broader as it ascends, it is *inserted* into the lateral part of the inferior nuchal line of the occipital bone and the surface of the bone immediately inferior to the line. As the muscles of the two sides pass upward and laterally, the triangular space between them is occupied by the rectus capitis posterior minor muscles.

Action. The **rectus capitis posterior major** extends the head and rotates it to the same side.

Nerve. The rectus capitis posterior major is supplied by a branch of the dorsal ramus of the **suboccipital nerve** (first cervical nerve).

RECTUS CAPITIS POSTERIOR MINOR. This muscle (Figs. 6-21; 6-22) *arises* by a narrow pointed tendon from the tubercle on the posterior arch of the atlas. Widening as it ascends, it is *inserted* into the medial part of the inferior nuchal line of the occipital bone and the bony surface between the line and the foramen magnum.

Action. The **rectus capitis posterior minor** extends the head.

Nerve. This muscle also is supplied by a branch of the dorsal primary division of the **suboccipital nerve.**

OBLIQUUS CAPITIS INFERIOR. This muscle is the larger of the two oblique muscles and

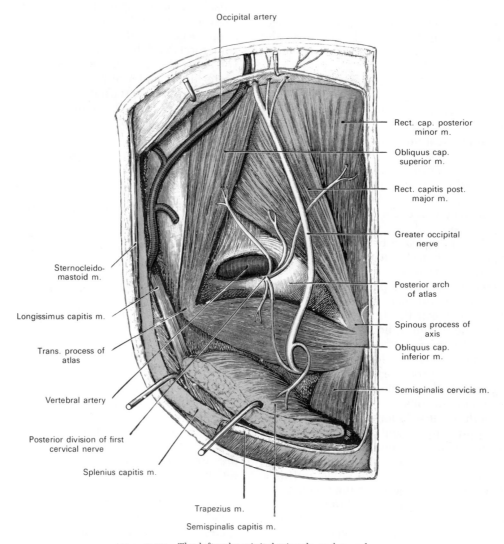

Occipital artery

Rect. cap. posterior minor m.

Obliquus cap. superior m.

Rect. capitis post. major m.

Greater occipital nerve

Posterior arch of atlas

Spinous process of axis

Obliquus cap. inferior m.

Semispinalis cervicis m.

Sternocleido-mastoid m.

Longissimus capitis m.

Trans. process of atlas

Vertebral artery

Posterior division of first cervical nerve

Splenius capitis m.

Trapezius m.

Semispinalis capitis m.

FIG. 6-22. The left suboccipital triangle and muscles.

it *arises* from the apex of the spinous process of the axis (Figs. 6-21; 6-22). Passing laterally and slightly upward, it is *inserted* into the inferior and dorsal part of the transverse process of the atlas.

Action. The **obliquus capitis inferior** rotates the atlas and thereby turns the face toward the same side.

Nerve. This muscle is supplied by a branch of the dorsal primary division of the **suboccipital nerve**.

OBLIQUUS CAPITIS SUPERIOR. This muscle (Figs. 6-21; 6-22) *arises* by narrow tendinous fibers from the superior surface of the transverse process of the atlas, and thus joins with the insertion of the obliquus capitis inferior. It passes upward and slightly medially to be *inserted* into the occipital bone, between the superior and inferior nuchal lines, lateral to the semispinalis capitis.

Action. The **obliquus capitis superior** extends the head and bends it laterally.

Nerve. As are the other muscles in this group, the obliquus capitis superior is supplied by a branch of the dorsal primary division of the **suboccipital nerve**.

SUBOCCIPITAL TRIANGLE. Between the obliqui and the rectus capitis posterior major is the **suboccipital triangle**. It is bounded *superiorly* and *medially* by the rectus capitis posterior major; *superiorly* and *laterally* by the obliquus capitis superior; *inferiorly* and *laterally* by the obliquus capitis inferior. The triangle is covered by dense, fibrous and fatty tissue, and it is situated deep to the semispinalis capitis. The floor of the triangle is formed by the posterior atlanto-occipital membrane and the posterior arch of the atlas. In the deep groove on the surface of the posterior arch of the atlas are the *vertebral artery* and the *first cervical* or *suboccipital nerve* (Fig. 6-22).

MUSCLES OF THE THORAX

The muscles considered in this section are all attached to the ribs and concerned with rib movements, especially in relation to respiration. Although many other muscles also attach to the thorax, these are described in sections dealing with the upper extremity, abdomen, back, and neck, with which they are more specifically identified.

Intercostales externi
Intercostales interni
Intercostales intimi
Subcostales

Transversus thoracis
Levatores costarum
Serratus posterior superior
Serratus posterior inferior
Diaphragm
    Diaphragmatic openings

Fasciae

The superficial fascia and the outer layers of deep fascia of the thorax are described with the pectoral region and back. The thoracic cage proper, composed of ribs and intercostal muscles, is covered internally and externally by thin layers of deep fascia. The outer layer essentially forms a covering over the external intercostal muscles, while the internal layer, consisting of loose areolar tissue, is called the **endothoracic fascia.**

ENDOTHORACIC FASCIA. The endothoracic fascia is a deep investing fascial layer that lines the internal aspect of the thoracic cage. It covers the inner surface of the intercostal muscles and the intervening ribs, along with the subcostal and transverse thoracic muscles and the diaphragm. It lies between the parietal pleura and the thoracic cage, which in the absence of adhesions can therefore be easily separated. Dorsally, the endothoracic fascia is continuous over the bodies of the vertebrae and the intervertebral discs. In this mediastinal region, certain structures such as the azygous and hemiazygous veins, the thoracic duct, the sympathetic chain, and the splanchnic nerves are partially or completely surrounded by the fascia. Superiorly, it extends over the apices of the lungs where, somewhat thickened, it forms the **suprapleural membrane,** which is also called Sibson's fascia.[1] Inferiorly, it thins considerably over the superior surface of the diaphragm, but is continuous with the internal investing fascia of the abdominal cavity, the **fascia transversalis,** dorsal to the diaphragm at the lumbocostal arches and through the aortic hiatus.

Muscles

The **intercostales muscles** (Figs. 6-23; 6-24) are thin layers of muscular and tendinous fibers occupying each of the intercostal

[1]Francis Sibson (1814–1876): An English physician and anatomist (London).

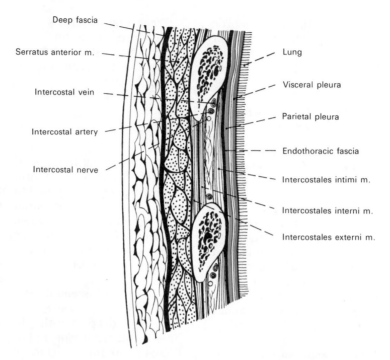

Deep fascia

Serratus anterior m.

Intercostal vein

Intercostal artery

Intercostal nerve

Lung

Visceral pleura

Parietal pleura

Endothoracic fascia

Intercostales intimi m.

Intercostales interni m.

Intercostales externi m.

FIG. 6-23. Longitudinal section of the thoracic wall. Note the position of the intercostal vein, artery and nerve in the sulcus of the rib, and that these structures course around the trunk between the innermost (intimi) and internal intercostal muscle layers. (From Benninghoff and Goerttler, *Lehrbuch der Anatomie des Menschen,* Urban & Schwarzenberg, 11th Ed., Vol. 1, 1975.)

spaces. They are named **external, internal,** and **innermost** (intimi) from their surface relations, the external being the most superficial and the innermost being the deepest, while the internal intercostal muscles lie between the other two.

INTERCOSTALES EXTERNI. There are eleven external intercostal muscles on each side. They extend in the intercostal spaces from the tubercles of the ribs dorsally to the cartilages of the ribs ventrally (Fig. 6-23), where each ends in a thin, fibrous sheet, the **external intercostal membrane,** which continues forward to the sternum. Each external intercostal muscle *arises* from the lower border of a rib and is *inserted* into the upper border of the rib below. In the lowest two spaces the muscles extend to the ends of the cartilages, and in the uppermost two or three spaces they do not quite reach the ends of the ribs. They are thicker than the intercostales interni, and their fibers are directed obliquely inferolaterally on the dorsum of the thorax, and inferomedially and somewhat ventrally on the anterior aspect of the thorax.

Action. The actions of the **external intercostal muscles** are not clearly understood. There is reasonable evidence to support the view that at least some of these muscles are active during inspiration and that they elevate the ribs. Apparently, however, they are also active during the early phase of expiration.

Nerves. The external intercostal muscles are innervated by the **intercostal nerves,** numbered sequentially according to the interspace, e.g., the fourth intercostal nerve supplies the muscles occupying the fourth intercostal space between the fourth and fifth ribs.

Variations. At times muscular slips join the external intercostal muscles to the external oblique or serratus anterior muscle. A **supracostalis muscle,** extending from the anterior end of the first rib down to the second, third, or fourth rib, occasionally is seen.

INTERCOSTALES INTERNI. There are also eleven internal intercostal muscles on each side. They commence anteriorly at the sternum, in the interspaces between the cartilages of the true ribs, and at the anterior extremities of the cartilages of the false ribs. They extend posteriorly as far as the angles of the ribs, where they are continued to the vertebral column by thin aponeuroses, the **internal intercostal membranes.** Each muscle *arises* from the ridge on the inner surface of a rib, as well as from the corresponding costal cartilage, and is *inserted* into the upper border of the rib below (Figs. 6-23; 6-24). Their fibers are also directed obliquely, but pass in a direction perpendicular to those of the external intercostal muscles.

Action. The **internal intercostal muscles** are also involved in respiratory movements. The portion of the upper four or five muscles that interconnects the costal cartilages is thought to elevate the ribs (as occurs with the external intercostals) during inspiration. The more lateral and dorsal portions of the muscles, where the fibers are oriented more obliquely in direction (inferiorly and posteriorly), are active during expiration. Some doubt still exists about these generally described actions.

Nerves. The **intercostal nerves** innervate the internal as well as the external intercostal muscles.

INTERCOSTALES INTIMI. These innermost intercostal muscles are frequently considered to be the deepest parts of the internal intercostal muscles. They lie deep to the intercostal vessels and nerves, being interposed between these structures and the parietal pleura (Fig. 6-23). They extend between the costal groove of the rib above to the inner lip of the upper margin of the rib below. They are not well developed and may, in fact, even be absent in the upper four or five interspaces. Their fibers generally course in the same direction as those of the internal intercostal muscles.

Action. It may be that the action of the **innermost intercostals** is similar to that of the internal intercostals, but conclusive data in this regard do not exist.

Nerves. Like the other intercostal muscles, the innermost muscles are supplied by the **intercostal nerves.**

SUBCOSTALES MUSCLES. The subcostal muscles consist of oblique muscular and aponeurotic fasciculi which are usually well developed only in the lower part of the thorax. Each *arises* from the inner surface of one rib, dorsally near its angle, and is *inserted* onto the inner surface of the second or third rib below. Their fibers run in the same direction as those of the innermost intercostals and similarly separate the intercostal vessels and nerves from the pleura.

Action. The **subcostales** probably draw adjacent ribs together. With the last rib fixed by the quadratus lumborum, they may lower the ribs, decreasing the volume of the thoracic cavity.

Nerves. Like the intercostal muscles, the subcostales are supplied by the **intercostal nerves.**

TRANSVERSUS THORACIS. The transversus thoracis muscle is a thin plane of muscular and tendinous fibers situated on the inner surface of the anterior wall of the chest (Fig. 6-24). It *arises* on each side from the caudal third of the inner surface of the body of the sternum, from the inner surface of the xiphoid process, and from the inner surface of the costal cartilages of the third through the sixth ribs at their sternocostal junctions. Its fibers diverge as they pass superiorly and laterally, to be *inserted* by slips into the caudal borders and inner surfaces of the costal cartilages of the second, third, fourth, fifth, and sixth ribs. The lowest fibers of this muscle are horizontal and are continuous with those of the transversus abdominis; the intermediate fibers are oblique, while the highest fibers are almost vertical. This muscle varies in its attachments, not only in different subjects, but on opposite sides of the same subject.

Action. The **transversus thoracis** draws the ribs downward.

Nerves. This muscle is also supplied by branches of the **intercostal nerves.**

LEVATORES COSTORUM. There are twelve levatores costorum muscles on each side of the posterior surface of the thoracic cage (Figs. 6-19 to 6-21). They are tendinous and fleshy bundles that *arise* from the ends of the transverse processes of the seventh cervical and upper eleven thoracic vertebrae. The muscles pass obliquely inferolaterally like the fibers of the external intercostals in this dorsal region, and each is *inserted* into the outer surface of the rib immediately below the vertebra from which it takes origin, between the tubercle and the angle **(levatores costarum breves).** Each of the four most inferior muscles divides into two fasciculi, one of which is inserted as described above; the other descends to the second rib below its origin **(levatores costarum longi).**

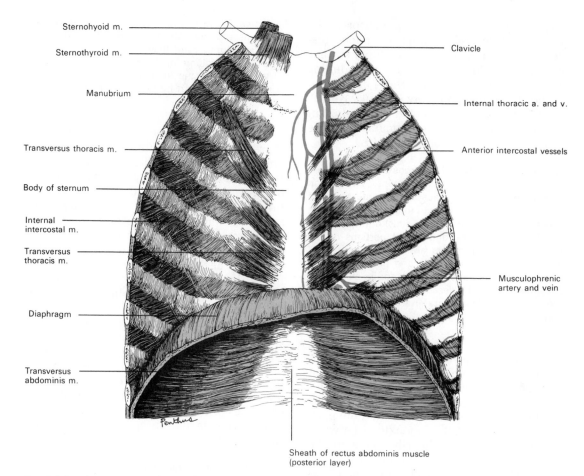

Sternohyoid m.

Sternothyroid m.

Manubrium

Transversus thoracis m.

Body of sternum

Internal intercostal m.

Transversus thoracis m.

Diaphragm

Transversus abdominis m.

Clavicle

Internal thoracic a. and v.

Anterior intercostal vessels

Musculophrenic artery and vein

Sheath of rectus abdominis muscle (posterior layer)

FIG. 6-24.   The posterior surface of the anterior thoracic wall viewed from within. Note the transverse thoracic muscle on each side, as well as the descending internal thoracic vessels from which branch the anterior intercostal arteries and veins.

Action.   The **levatores costorum** elevate the ribs and bend the vertebral column laterally, rotating it slightly toward the opposite side.

Nerves.   The levatores costorum are supplied by branches from the **intercostal nerves.**

SERRATUS POSTERIOR SUPERIOR.   This thin, quadrilateral muscle is situated on the dorsal and upper part of the thorax. It *arises* by a thin and broad aponeurosis from the lower part of the ligamentum nuchae, from the spinous processes of the seventh cervical and first two or three thoracic vertebrae, and from the supraspinal ligament. Inclining inferiorly and laterally, it becomes muscular and is *inserted* by four, fleshy digitations into the cranial borders of the second, third, fourth, and fifth ribs, a little beyond their angles.

Action.   The **serratus posterior superior** elevates the upper ribs.

Nerves.   This muscle is supplied by branches of the ventral primary divisions of the **first four thoracic nerves.**

Variations.   There may be an increase or decrease in size and number of slips or an entire absence of the muscle.

SERRATUS POSTERIOR INFERIOR.   This thin, irregularly quadrilateral muscle is situated at the junction of the thoracic and lumbar regions. It is broader than the preceding muscle and is separated from it by about four ribs. The serratus posterior inferior *arises* by a thin aponeurosis from the spinous processes of the last two thoracic and first two or three lumbar vertebrae, and from the supraspinal ligament. Passing obliquely upward and laterally, it becomes fleshy and divides into four flat digitations which are *inserted* into the inferior borders of the last four ribs, a little lateral to their angles. The thin aponeurosis of origin is intimately blended with the thoracolumbar fascia and the aponeurosis of the latissimus dorsi.

Action. The **serratus posterior inferior** draws the ribs to which it is attached outward and downward, counteracting the inward pull of the diaphragm.

Nerves. This muscle is supplied by branches of the ventral primary divisions of the **ninth to twelfth thoracic nerves**.

THE DIAPHRAGM. The diaphragm is a dome-shaped musculofibrous septum that separates the thoracic from the abdominal cavity (Figs. 6-24; 6-25). Its convex upper surface forms the floor of the thorax, and its concave inferior surface forms the roof of the abdomen. Its peripheral part consists of muscular fibers that take origin from the cir-

cumference of the thoracic outlet and converge to be *inserted* into a central tendon.

The muscular fibers may be grouped according to their origins into three parts—sternal, costal, and lumbar. The **sternal part** *arises* by two fleshy slips from the dorsum of the xiphoid process; the **costal part** *arises* from the inner surfaces of the cartilages and adjacent portions of the last six ribs on each side, interdigitating with the transversus abdominis; and the **lumbar part** *arises* from two aponeurotic arches on each side, named the **medial** and **lateral arcuate ligaments**, and from the lumbar vertebrae by two pillars or **crura**. The two crura are interconnected at the midline by a tendinous arch called the **median arcuate ligament** which lies just in front of the aorta.

The **medial arcuate ligament** (*medial lumbocostal arch*) is a tendinous arch in the fascia covering the upper part of the psoas major. *Medially,* it is continuous with the lateral tendinous margin of the correspond-

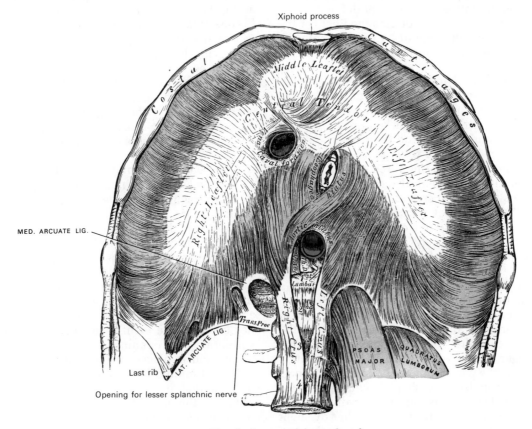

FIG. 6-25. The diaphragm. Abdominal surface.

ing crus, and is attached to the side of the body of the first or second lumbar vertebra. Laterally, it is fixed to the front of the transverse process of the first, and sometimes also to that of the second, lumbar vertebra.

The **lateral arcuate ligament** (*lateral lumbocostal arch*) is a thickened tendinous band across the upper part of the quadratus lumborum. It is attached *medially* to the ventral surface of the transverse process of the first lumbar vertebra, and *laterally* to the tip and inferior margin of the twelfth rib.

The **crura of the diaphragm** are tendinous in structure, and blend with the anterior longitudinal ligament of the vertebral column. The **right crus,** somewhat wider and longer than the left, *arises* from the anterior surfaces of the bodies and intervertebral fibrocartilages of the first three lumbar vertebrae, while the **left crus** *arises* from the corresponding parts of the first two lumbar vertebrae only. The medial tendinous margins of the crura pass anteriorly and medially, meet in the midline, and form the median arcuate ligament, which crosses the ventral aspect of the aorta and is often poorly defined.

From this series of origins the fibers of the diaphragm converge to be *inserted* into the central tendon. The fibers arising from the xiphoid process are short and occasionally aponeurotic. Those from the medial and lateral arcuate ligaments, and more especially those from the ribs and their cartilages, are longer and describe marked curves as they ascend and converge to their insertion. The fibers of the crura diverge as they ascend, the most lateral being directed upward and laterally to the central tendon. The medial fibers of the right crus ascend on the *left side* of the esophageal hiatus, and occasionally a fasciculus of the left crus crosses the aorta and runs obliquely through these fibers of the right crus toward the vena caval foramen.

The **central tendon** of the diaphragm is a thin but strong aponeurosis situated near the center of the muscle, but somewhat closer to the anterior than to the dorsal part of the thorax, so that the dorsal muscle fibers are longer. The tendon is situated immediately below the pericardium, with which it is partially blended. It is shaped somewhat like a clover leaf, consisting of three divisions or leaflets separated from one another by slight indentations. The right leaflet is the largest; the middle is directed toward the xiphoid process and is the next in size; and the left is the smallest. In structure the tendon is composed of several planes of fibers, which intersect at various angles and unite into straight or curved bundles, an arrangement that affords considerable strength to the muscle.

Diaphragmatic Openings. The diaphragm is pierced by a series of apertures that permit the passage of structures between the thorax and the abdomen. Three large openings, the **aortic,** the **esophageal,** and the **vena caval,** are found along with a series of smaller ones (Fig. 6-25).

The **aortic hiatus** is the most inferior and posterior of the large apertures. It lies just to the left of the midline at the level of the twelfth thoracic vertebra. Strictly speaking, it is not an aperture in the diaphragm but an osseoaponeurotic opening between it and the vertebral column, and therefore behind the diaphragm. Occasionally some tendinous fibers are prolonged across the bodies of the vertebrae from the medial parts of the lower ends of the crura and pass behind the aorta, thus converting the hiatus into a true fibrous ring. The hiatus is bounded anteriorly by the crura and posteriorly by the body of the first lumbar vertebra, and through it pass the aorta and the thoracic duct. Usually the aortic hiatus also transmits a vein connecting the right ascending vein to the azygos vein, while the azygos vein passes through the right crus. At times, however, the azygos vein passes through the hiatus itself to the right of the aorta.

The **esophageal hiatus** is elliptical in shape and is situated in the muscular part of the diaphragm at the level of the tenth thoracic vertebra. It is placed superior, anterior, and a little to the left of the aortic hiatus, and transmits the esophagus, the vagus nerves, and esophageal branches of the left gastric artery and vein.

The **vena caval foramen** is the most superior of the large apertures and is situated at about the level of the intervertebral disc that lies between the eighth and ninth thoracic vertebrae. The opening is quadrilateral in shape, and is placed at the junction of the right and middle leaflets of the central tendon, so that its margins are tendinous. It transmits the inferior vena cava, the wall of which is adherent to the margins of the foramen and some branches of the right phrenic nerve.

Of the **lesser apertures,** two in the right crus transmit the greater and lesser right splanchnic nerves, while three in the left crus give passage to the greater and lesser left splanchnic nerves and the hemiazygos vein. The ganglionated sympathetic chains usually enter the abdominal cavity behind the diaphragm and deep to the medial arcuate ligaments.

On each side two small intervals exist at which the muscular fibers of the diaphragm are deficient and are replaced by areolar tissue. One, between the sternal and costal parts, transmits the superior epigastric branch of the internal thoracic artery and some lymphatics from the abdominal wall and the convex surface of the liver. The other, between fibers springing from the medial and lateral arcuate ligaments, is less constant. When this interval does exist, the upper posterior part of the kidney is separated from the pleura by areolar tissue only.

Action. Acting from their attachments on the ribs and lumbar vertebrae, the muscle fibers of the **diaphragm** draw the central tendon downward and forward during inspiration. This action has two effects: (1) it tends to increase the volume and decrease the pressure within the thoracic cavity, and (2) it tends to decrease the volume and increase the pressure within the abdominal cavity. During inspiration, lowering of the diaphragm decreases the pressure within the thorax, and air is forced into the lungs through the open glottis of the larynx by the higher pressure of the atmosphere. At the same time, the descending diaphragm presses against the abdominal viscera, forcing them downward against the passive resistance of the abdominal and pelvic muscles, and causing the anterior abdominal wall to protrude slightly.

The pressure within the abdominal cavity is greatly increased when the abdominal muscles and the diaphragm contract actively at the same time. This increase in pressure tends to make the abdominal or pelvic viscera discharge their contents as in micturition, defecation, emesis, and parturition.

Nerve. The diaphragm is innervated by the **phrenic nerve** from the cervical plexus, containing mainly fibers from the fourth, but also some from the third and fifth cervical nerves.

Variations. The sternal portion of the diaphragm is sometimes absent, and more rarely defects occur in the lateral part of the central tendon or adjoining muscle fibers.

Movements of Respiration. The respiratory cycle of inspiration and expiration is the result of increases and decreases in the capacity of the thoracic cavity. **Inspiration,** which is the correlate of an increase in the volume of the cavity, results from muscular action and is brought about in two ways: (1) by the descent of the diaphragm due to contraction of its muscle, which increases the vertical dimension of the thorax and (2) by the action of certain muscles on the ribs, sternum, and vertebral column which increases the transverse and anteroposterior dimensions of the thorax. **Expiration,** associated with a decrease in the volume of the thoracic cavity, is primarily a passive process due to the elastic recoil of the thoracic wall and the tissues of the lungs and bronchi. Thoracic volume may also decrease, however, as the result of muscular action, in which case (1) the abdominal muscles force the diaphragm upward by increasing the abdominal pressure, and (2) the thoracic wall is actively contracted by the action of certain muscles on the ribs and vertebral column.

Quiet inspiration results from diaphragmatic contraction which increases the vertical diameter of the thoracic cavity. The first and second ribs remain fixed by the inertia and resistance of the cervical structures, but the remaining ribs, except the last two, are brought upward toward them by the contraction of the external intercostal muscles. The upward motion of the ribs, due to their position and to the obliquity of their axis of rotation, enlarges the anteroposterior and transverse diameters of the thorax. Thus, quiet inspiration consists of both diaphragmatic and costal components. Diaphragmatic breathing is also called abdominal because the visible result of contraction of the diaphragm is protrusion of the abdominal wall. To contrast with this, costal breathing has also been called thoracic breathing.

In the newborn child, the ribs are virtually horizontal and, therefore, are already in the position of full inspiration. Breathing in the baby is entirely dependent upon diaphragmatic action and abdominal movements may easily be perceived. As the child develops, the ribs become more oblique, and costal movements with respiration result in the development of thoracic breathing. In the adult, both abdominal and thoracic breathing are observed, but one or the other may predominate. Breathing patterns can be altered with training and exercise.

Electromyographic studies indicate that both external and internal intercostal muscles are slightly active during quiet inspiration, but have no rhythmic increase and decrease, acting more to keep the ribs in a constant position than to expand the thorax. The scalene muscles, however, do show rhythmic activity (Jones, Beargie and Pauly, 1953).

Quiet expiration follows the relaxation of the inspiratory muscles, and the normal resting position of the thorax is that seen at the end of a quiet expiration. This position is restored without muscular effort by the recoil of structures that were displaced by the inspiratory act. The displacement of the anterior abdominal wall is overcome by the tonus of the abdominal muscles. The ribs are restored from their displacement by the elasticity of the ligaments and cartilages that hold them in place. The extensive network of elastic fibers that permeates lung tissue retracts the lungs. The bronchial tree, which has been elongated by the descent of the diaphragm, helps to return the lungs to their resting position by its elastic recoil.

Deep inspiration requires the intensification of all the actions seen in quiet inspiration. Additionally, the first two ribs are raised by the scalene and sternocleidomastoid muscles, and the remaining ribs are raised more forcibly by the actions of the levatores costarum and the serratus posterior superior. The ribs are further elevated by a straightening of the vertebral column through contraction of the erector spinae muscle. After the abdominal viscera have been forced downward by the diaphragm to a considerable extent, the abdominal muscles offer resistance to further displacement. The viscera then may act as a point of fixation for the diaphragm, so that any further contraction raises the ribs.

Forced inspiration is observed in patients with great air hunger, and all the muscles of the body seem to coordinate to assist in breathing. The levator scapulae, the trapezius, and the rhomboids elevate and fix the scapula, which is then used as an origin by the pectoralis minor to draw the ribs upward. The patient frequently further fixes the shoulder girdle by grasping the back of a chair or end of the bed, so that the pectoralis major and the serratus anterior further elevate the ribs for a greater inspiratory effort.

Forced expiration, such as that necessary for sneezing and coughing, unlike quiet expiration, requires the action of muscles. The last two ribs are pulled downward and fixed by the quadratus lumborum, and the other ribs are also drawn downward by the internal intercostals and the serratus posterior inferior muscles. The muscles of the abdominal wall, by pressing on the abdominal viscera, force the diaphragm upward, and by flexing the vertebral column, assist in lowering the ribs.

Position of the Diaphragm. The height of the diaphragm is constantly changing during respiration, and it also varies with the degree of distension of the stomach and the intestines and with the size of the liver. After a forced expiration, the right cupola of the diaphragm is on a level anteriorly with the fourth costal cartilage, laterally with the fifth, sixth, and seventh ribs, and posteriorly with the eighth rib. The left cupola is a little lower than the right. The absolute range of movement of the diaphragm between deep inspiration and deep expiration averages in both males and females 30 mm on the right side and 28 mm on the left. In quiet respiration, the average movement is 12.5 mm on the right side and 12 mm on the left.

Radiography shows that the height of the diaphragm in the thorax varies considerably with the position of the body. It stands highest when the body is horizontal and the patient on his back, and in this position the diaphragm performs the largest respiratory excursions with normal breathing. When the body is erect, the dome of the diaphragm descends and its respiratory movements become smaller. The dome falls still lower when the sitting posture is assumed, and in this position, diaphragmatic excursions are the smallest. These facts explain why patients suffering from severe dyspnea are most comfortable and least short of breath when they sit up. When the body is horizontal and the patient on his side, the two halves of the diaphragm do not behave alike. The uppermost half sinks to a level lower than that when the patient sits, and moves little with respiration; the lower half rises higher in the thorax than it does when the patient is supine, and its respiratory excursions are greatly increased. In unilateral disease of the pleura or lungs, analogous interference with the position or movement of the diaphragm can generally be observed radiographically.

The position of the diaphragm in the thorax depends on three main factors: (1) the elastic retraction of the lung tissue, which tends to pull it upward; (2) the pressure exerted on its inferior surface by the viscera, which naturally tends to be a negative pressure or downward suction, when the patient sits or stands, and positive or upward pressure, when he reclines; (3) the intra-abdominal tension due to the abdominal muscles.

The following figures represent the average thoracic changes that occur during deepest possible inspiration. The manubrium sterni moves 30 mm in an upward and 14 mm in a forward direction; the width of the subcostal angle is increased by 26 mm at a level of 3 cm below the articulation between the body of the sternum and the xiphoid process; the umbilicus is retracted and drawn upward for a distance of 13 mm.

Diaphragmatic Hernias. At times abnormal openings in the diaphragm allow abdominal viscera to herniate into the thorax. The most common **congenital diaphragmatic hernia** results from an incomplete separation of the pleural and peritoneal cavities in the posterolateral region of the diaphragm. This occurs more frequently on the left than the right (Butler and Claireaux, 1962; Gray and Skandalakis, 1972), and allows the small intestine or other organs to enter the thorax through the incompletely closed pleuroperitoneal canal (of Bochdalek).[1]

Most **acquired diaphragmatic hernias** occur at the esophageal hiatus (hiatal hernia). In this condition the upper end of the stomach herniates into the thorax, so that it "slides" for some distance up and down through the esophageal hiatus. In other hiatal hernias, the cardiac end of the stomach may herniate into the thorax and carry with it the lower esophagus, thereby producing a U-shaped redundancy of the lower esophagus.

[1]Vincent A. Bochdalek (1801–1883): A Czechoslovakian anatomist (Prague).

FASCIAE AND MUSCLES OF THE ABDOMEN

The muscles of the abdomen may be divided into two groups: the **anterolateral muscles** and the **posterior muscles.**

Anterolateral Abdominal Muscles

Obliquus externus abdominis
Obliquus internus abdominis
Cremaster
Transversus abdominis
Rectus abdominis
Pyramidalis

SUPERFICIAL ABDOMINAL FASCIA

The **superficial fascia** (Fig. 6-28) of the anterior abdominal wall is soft and movable, and likely to contain fat. It is continuous, *superiorly,* with the superficial fascia of the thorax; *inferiorly,* with that of the thigh and external genitalia; and *laterally,* it gradually becomes tougher and more resistant as it continues as the fascia of the back. Below the umbilicus the superficial fascia is unusually easy to separate into superficial and deep layers. Between the two layers are found the superficial vessels and nerves and the superficial lymph nodes. This uncommon divisibility and certain other peculiar features require that the two layers be described separately.

SUPERFICIAL LAYER (FASCIA OF CAMPER[1]).

The superficial layer is thick, areolar in texture, and contains a variable amount of fat. It is a genuine panniculus adiposus, and may be several centimeters thick in obese individuals, in which case it also is likely to be irregularly divisible into laminae. It is continuous over the inguinal ligament with a similar and corresponding layer of the thigh. In the male, it continues down on the penis and scrotum, loses its fat and fuses with the deep layer. It assists in the formation of a special pale reddish fascia of these genital organs called the **dartos,** which also contains some nonstriated muscle fibers. In the female, it retains some adipose tissue as it is continued into the labia majora. In both sexes, it is prolonged dorsally in the crease between the external genitalia and the thigh

and becomes continuous with the outermost layer of superficial fascia in the perineum and medial surface of the thigh.

DEEP LAYER (FASCIA OF SCARPA[2]).

The deep layer of superficial fascia is a membranous sheet that usually contains little or no adipose tissue. It is composed, in considerable part, of yellow elastic fibers, and probably corresponds to the tunica abdominalis, an elastic layer that contributes to the support of the inguinal mammae in some lower mammals. It forms a continuous sheet across the midline, but attaches to the linea alba as it crosses. *Superiorly and laterally,* it loses its identity as a special layer of the superficial fascia of the upper abdomen and back. *Inferiorly,* it passes over the inguinal ligament and is *securely attached,* either to the ligament itself or to the fascia lata of the thigh just beyond it. Inferior to the ligament, a corresponding layer, called the **fascia cribrosa,** covers and fills an opening, the **fossa ovalis,** for the passage of the great saphenous vein, where it enters the femoral vein. At the medial end of the inguinal ligament, Scarpa's fascia passes over the superficial inguinal ring without being attached and continues into the penis and scrotum. It continues along the crease between the scrotum (or labium majus) and thigh into the perineum where it is called the **fascia of Colles.**[3] As the superficial fascia courses deep to the skin of the scrotum and penis, its two layers are fused to form the dartos. Because the dartos also contains a layer of scattered smooth muscle cells, the scrotal skin is thrown into folds or rugae. In the midline over the symphysis pubis, Scarpa's fascia is thickened in the male by the addition of numerous closely set, strong bands which extend down to the dorsum and sides of the penis, forming the **fundiform ligament of the penis.**

The fascial cleft that separates the deep layer of the superficial fascia (Scarpa's) from the true deep fascia over the caudal portion of the *aponeurosis* of the obliquus externus muscle is quite definite and of considerable clinical interest because of its continuity with a similar cleft in the perineum. Thus,

---

[1]Petrus Camper (1722–1789): A Dutch physician and anatomist (Gröningen).

[2]Antonio Scarpa (1747–1832): A Venetian anatomist and ophthalmologist (Pavia).

[3]Abraham Colles (1773–1843): An Irish surgeon and anatomist (Dublin).

the possibility exists that extravasated urine from a ruptured urethra in the perineum can find its way into the scrotum, deep to the dartos, and then onto the anterior abdominal wall deep to Scarpa's fascia.

This fascial cleft is limited superiorly near the umbilicus because the superficial fascia (Scarpa's) becomes adherent to the deep fascia, and laterally because Scarpa's fascia attaches closely to the *muscular portion* of the external oblique. The cleft is limited, inferiorly and laterally, by the firm attachment of Scarpa's fascia either to the inguinal ligament or to the fascia lata just below it. Over the medial portion of the inguinal ligament and the superficial inguinal ring, however, Scarpa's fascia and the deep fascia are not attached, so that the cleft follows along the narrow crease between the scrotum (or labium) and thigh, and is continuous with the cleft between the superficial fascia of the perineum (Colles') and the deep fascia, called the inferior fascia of the urogenital diaphragm or the perineal membrane. From this groove, it is continuous medially with the cleft beneath the movable dartos of the scrotum and penis, but the cleft is abruptly limited laterally by an attachment to the deep fascia over the pubic ramus where the adductor muscles of the medial thigh originate. In obese individuals there may be accumulations of fat in the cleft deep to Scarpa's fascia. The superficial epigastric and superficial circumflex iliac blood vessels lie between Camper's and Scarpa's fasciae, but are attached to the superficial surface of Scarpa's layer.

DEEP ABDOMINAL FASCIA

The **outer investing layer** of deep fascia is easily identified in the lateral portion of the anterior abdominal wall where it covers the fleshy fibers of the external oblique muscle. It is continuous with the deep fascia covering the latissimus dorsi and pectoralis major. More medially, over the aponeurosis of the external oblique, the deep fascia is so firmly adherent that it may escape recognition. Its presence is easily demonstrated in dissection, however, by scraping it back until the glistening fibers of the aponeurosis beneath are revealed. *Superiorly,* it covers the upper end of the rectus sheath and is continuous with the pectoral fascia. *Inferiorly,* it is firmly attached to the inguinal ligament and joins the deep fascia emerging from beneath that ligament to become the fascia lata of the thigh. It covers the superficial inguinal ring as a distinct and separate layer, and there, reinforced by the fascia of the inner surface of the external oblique aponeurosis, gives rise to a tubular prolongation, the external spermatic fascia, which is the coat of the spermatic cord and testis just deep to the dartos. Near the midline, the deep fascia is attached to the pubic bone and is then continuous with deep fascia investing the penis. At the lower end of the linea alba, the deep fascia is thickened into a strong, fibrous triangle, the **suspensory ligament of the penis,** which attaches the dorsum of the penis to the symphysis pubis and arcuate pubic ligament. At the medial end of the inguinal ligament, it is attached to the pubic ramus and arcuate pubic ligament and is then continuous posteriorly, over the ischiocavernosus muscle, with the deep fascia of the perineum. Laterally, beyond its attachment to the pubic ramus, it is continuous with the fascia covering the adductor muscles of the medial side of the thigh.

MUSCLES

OBLIQUUS EXTERNUS ABDOMINIS. The external oblique muscle, situated on the lateral and anterior parts of the abdomen, is the largest and the most superficial of the three flat muscles in this region (Fig. 6-26). It is broad, thin, and irregularly quadrilateral, and its muscular portion occupies the lateral wall of the abdomen, while its aponeurosis stretches across the anterior abdominal wall. It *arises* by eight fleshy digitations from the external surfaces and inferior borders of the lower eight ribs. These digitations are arranged sequentially and form an oblique line that runs downward and backward; the cranial ones are attached close to the cartilages of the corresponding ribs, the intermediate ones to the ribs at some distance from their cartilages, and the most caudal to the apex of the cartilage of the last rib. The five superior digitations increase in size from above downward and are received between corresponding muscular slips of the serratus anterior. The three lower ones diminish in size from above downward and receive between them corresponding muscular digitations from the latissimus dorsi. From these attachments, the fleshy fibers proceed in a diverging manner to their insertions. The fasciculi from the last two ribs are nearly vertical and are *inserted* into the anterior half of the outer lip of the iliac crest; the upper and middle fasciculi, directed downward and forward, terminate in a broad aponeurosis along a line extending from the prominence of the ninth costal cartilage to the anterior superior iliac spine.

Action. The **obliquus externus abdominis** compresses the abdominal contents and assists in micturition, defecation, emesis, parturition, and forced expiration. Both sides acting together flex the vertebral column by drawing the pubis toward the xiphoid process. One side alone bends the vertebral column laterally and rotates it, bringing the shoulder of the same side forward.

Nerves. This muscle is supplied by branches of the **seventh to twelfth intercostal nerves**.

Aponeurosis of Obliquus Externus Abdominis. The aponeurosis of the external oblique muscle is a strong membrane formed by tendinous bundles in continuity with the muscle fibers of the external oblique. Initially, the aponeurotic fibers course downward and medially, covering the entire ventral surface of the abdomen and lying superficial to the rectus abdominis muscle, thereby helping to form its sheath. The fibers of the aponeurosis on each side are arranged in two layers, superficial and deep, which pass obliquely perpendicular to each other. Additionally, as the aponeurotic fibers cross the midline, they interlace with those of the opposite side in such a manner that the superficial fibers of one side extend as the direct continuation of the fibers of the deep layer of the opposite external oblique aponeurosis (Rizk, 1976). As the fibers pass across the midline, their sites of intersection form the **linea alba,** which extends from the xiphoid process to the symphysis pubis. The uppermost part of the aponeurosis serves as the origin for the inferior fibers of the pectoralis major. The lowermost portion ends in a tendinous border called the **inguinal ligament.** Although it is aponeurotic rather than ligamentous in function, it extends from the anterior superior spine of the ilium at its superolateral end to the pubic tubercle inferomedially (Figs. 6-26 to 6-31). Near its medial end, the free border is curled under

like a sling to support the spermatic cord. Its attachment to the pubic bone is fanned out along the pectineal line beyond the pubic tubercle to leave a crescentic, free border named the **lacunar ligament** (Fig. 6-27). In a small percentage of cases, other fibers may double back superiorly toward the linea alba deep to the main aponeurosis in a small triangular sheet called the **reflected inguinal ligament.** The tendinous bundles of the aponeurosis just above the inguinal ligament separate from each other at the pubic tubercle, leaving a narrow triangular opening called the **superficial inguinal ring,** through which passes the spermatic cord in a male or the round ligament in a female. The bundles bordering the ring are called the **superior and inferior crura.** Lateral to the opening the aponeurosis contains, in addition to the bundles running in the usual direction, some scattered, transverse, reinforcing strands that sweep superiorly and toward the midline in curved lines. These are called **intercrural fibers.** They are part of the aponeurosis and should not be confused with the intercrural fascia. In the foregoing description the aponeurosis has been treated as if it belonged solely to the external oblique muscle. This is advantageous for the presentation of the subcutaneous inguinal ring and associated structures, but it gives an incomplete picture because the aponeurosis serves also as the insertion of the internal oblique and transversus muscles, and it contributes, as well, to the formation of the sheath for the rectus abdominis muscle.

Inguinal Ligament. This ligament, formed by the lower border of the aponeurosis of the external oblique muscle, extends from the anterior superior iliac spine to the pubic tubercle. Its lateral half is firmly attached to a band in the iliac fascia where the latter is fused with the fascia innominata and transversalis fascia and continues into the thigh as the fascia lata. The medial portion of the ligament is a free border, and the femoral sheath and vessels pass deep to it on their way into the thigh. The medial fibers of the ligament curl under the adjacent aponeurosis to reach their insertion into the pectineal fascia and pubic bone lateral to the tubercle, thereby forming a narrow sling for support of the spermatic cord. The curved free border at this attachment is named the **lacunar ligament** (Fig. 6-27). At their attachment

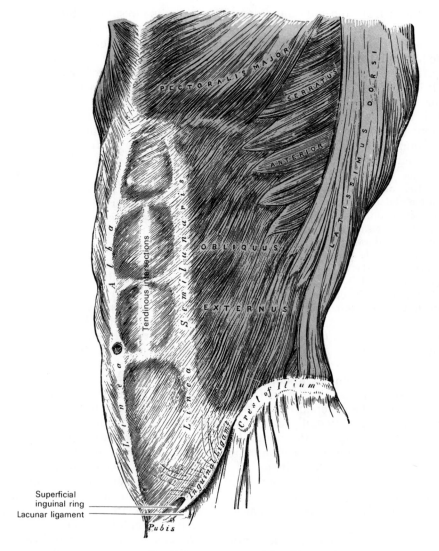

FIG. 6-26. The obliquus externus abdominis muscle.

some fibers of the lacunar ligament continue lateralward along the pectineal line, producing a strong tendinous ridge called the **pectineal** (Cooper's[1]) **ligament** (Fig. 5-31).

Lacunar Ligament. The lacunar ligament (Figs. 5-31; 6-27) is that part of the medial end of the inguinal ligament which is rolled under the spermatic cord and is attached along the pectineal fascia just lateral to the pubic tubercle. When it is viewed through the superficial inguinal ring, after the spermatic cord has been removed, the lacunar ligament appears to be a triangular fibrous

membrane. Its crescentic free margin forms the base of the triangular membrane, and it is concave laterally. Its apex is directed medially at the pubic tubercle (Fig. 6-27), and the ligament measures slightly less than two centimeters from apex to base. It lies almost horizontally when the body is erect, and the spermatic cord rests on its superior surface. Against its concave lateral border lies the medial wall of the femoral canal. The strong fibrous pectineal ligament extends from the base of the lacunar ligament along the pecten pubis.

Reflected (Inguinal) Ligament. The reflected ligament (Figs. 6-27; 6-30) is a triangular, tendinous sheet, 2 or 3 cm wide, extend-

[1]Sir Astley Paston Cooper (1768–1841): An English anatomist and surgeon (London).

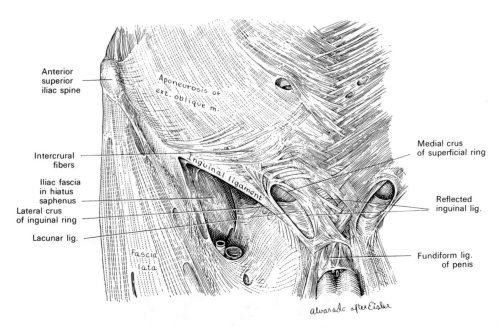

Anterior
superior
iliac spine

Aponeurosis of
ext. oblique m.

Intercrural
fibers

Iliac fascia
in hiatus
saphenus

Lateral crus
of inguinal ring

Lacunar lig.

Inguinal ligament

fascia
lata

Medial crus
of superficial ring

Reflected
inguinal lig.

Fundiform lig.
of penis

alvarado after Eisler

FIG. 6-27.   The aponeurosis of the obliquus externus abdominis and the ligaments of the inguinal region. (Redrawn from Eisler.)

ing from the medial part of the inguinal ring to the linea alba. The fibers arise from the attachment of the inguinal ligament to the pectineal line and course upward and medially, deep to the main aponeurosis of the external oblique, and interlace with the fibers of that aponeurosis in the linea alba. The reflected ligament may be independent, but more often it is fused either with the aponeurosis of the external oblique or with the falx inguinalis, which lies deep to it. Frequently it is entirely lacking.

Superficial Inguinal Ring.   The superficial inguinal ring (Figs. 6-26 to 6-28) is the opening in the aponeurosis of the external oblique just above and lateral to the pubic crest. This aperture is narrow and triangular, pointing laterally and superiorly in the direction of the fibers of the aponeurosis. The base of the triangular opening is at the crest of the pubis, while the sides form the margins of the aperture in the aponeurosis and are called the crura of the ring. The **lateral (inferior) crus** is the stronger and is formed by the portion of the inguinal ligament that is attached to the pubic tubercle; it is curved and turned under into a narrow sling upon which the spermatic cord rests. The **medial (superior) crus** is thin and flat, and represents merely the part of the aponeurosis next to the opening; its descending fibers are attached to the front of the symphysis pubis. Reinforcing the superficial inguinal ring laterally are arching bands of **intercrural fibers,** which are part of the aponeurosis and course at right angles to the aponeurotic fibers forming the medial and lateral crura (Figs. 6-27; 6-28). The triangular hiatus in the aponeurosis is converted by fascia into an oval ring, 2.5 cm long and 1.25 cm wide. The fascia on the superficial surface of the aponeurosis fuses with the fascia on its deep surface, filling in the angular, lateral portion of the opening. A tubular prolongation of this fascia is continued down into the scrotum, enclosing the spermatic cord and testis in a sheath called the **external spermatic fascia.** The superficial inguinal ring gives passage to the ilioinguinal nerve in both sexes, as well as to the spermatic cord in males or the round ligament of the uterus in females. The ring is larger in men than in women.

Pectineal Ligament.   The pectineal ligament (of Cooper) is a narrow band of strong aponeurotic fibers that continues laterally from the lacunar ligament along the pectineal line of the pubis (pecten pubis). It is firmly attached to the bone along this line and its medial end is continuous with the lacunar ligament. It diminishes in size gradually toward its lateral extremity at the iliopectineal eminence. It is aponeurotic rather

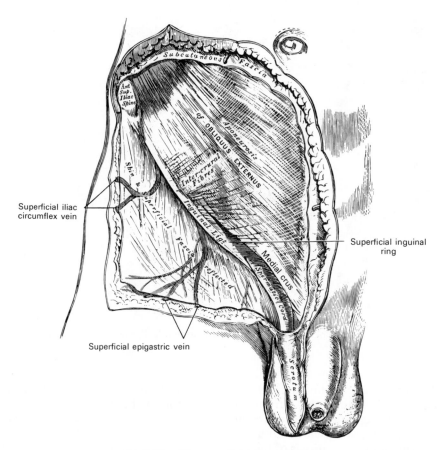

Superficial iliac
circumflex vein

Superficial inguinal
ring

Superficial epigastric vein

FIG. 6-28.   The superficial inguinal ring.

than fascial in origin, and to it are attached parts of the iliopectineal, pectineal, and transversalis fasciae. It serves well as a site for the attachment of sutures in the repair of femoral hernias and, at times, even of inguinal hernias.

OBLIQUUS INTERNUS ABDOMINIS. The internal oblique muscle (Figs. 6-29; 6-30), situated in the lateral and ventral part of the abdominal wall, is of an irregularly quadrilateral form, and is smaller and thinner than the external oblique under which it lies. It *arises* by fleshy fibers from the lateral half of the inguinal ligament and the nearby iliac fascia, from the anterior two-thirds of the middle lip of the iliac crest, and from the lower portion of the posterior layer of the thoracolumbar fascia near the crest. The posterior fasciculi pass upward and laterally to be *inserted* into the inferior borders of the cartilages of the last three or four ribs by fleshy digitations which appear to be continuations of the internal intercostal muscles. The remainder of the fasciculi, which arises from the iliac crest, diverges as it spreads

over the side of the abdomen and, in the region of the linea semilunaris, terminates in an aponeurosis that fuses with the aponeuroses of the externus and the transversus at a variable distance from the midline. The fibers of the aponeurosis assist in the formation of the rectus sheath, and in the upper two-thirds of the anterior abdominal wall, some of the aponeurosis passes ventrally and some dorsally to the rectus abdominis muscle. Over the lower one-third of the abdomen, the entire aponeurosis of the internal oblique passes anterior to the rectus muscle, since the posterior part of the rectus sheath is wanting at this level. The fasciculi arising from the inguinal ligament are less compact and of paler color. Descending, they form an arch over the spermatic cord in the male or the round ligament in the female. These muscle fibers terminate in a tendinous sheet, which they share with the others from the transversus muscle. This sheet is not fused with the aponeurosis of the externus, but independently is inserted into the pubis and medial part of the pectineal line, behind

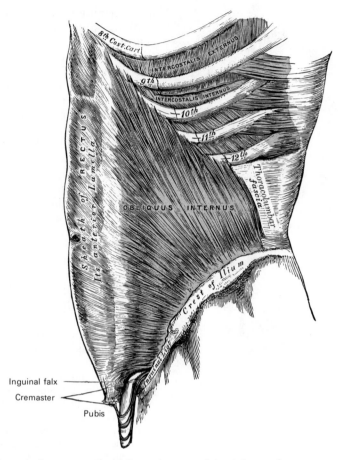

FIG. 6-29. The obliquus internus abdominis muscle.

the lacunar ligament, where it becomes the **conjoined tendon** (*inguinal falx*).

Action. The **obliquus abdominis internus** compresses the abdominal contents, assisting in micturition, defecation, emesis, parturition, and forced expiration. Both sides acting together flex the vertebral column, drawing the costal cartilages downward toward the pubis. One side acting alone bends the vertebral column laterally, rotating it and bringing the shoulder of the opposite side anteriorly.

Nerves. The internal oblique is supplied by branches of the **eighth to twelfth intercostal nerves** and by the iliohypogastric and ilioinguinal branches of the **first lumbar nerve.**

Variations. Occasionally, tendinous inscriptions to the internal oblique muscle occur from the tips of the tenth or eleventh cartilages or even from the ninth. An additional slip to the ninth cartilage is sometimes found, and separation between iliac and inguinal parts of the internal oblique may occur.

CREMASTER MUSCLE. The cremaster (Figs. 6-29; 6-30) is a thin muscular layer whose fasciculi are separate and spread out over the spermatic cord in a series of loops. It *arises* from the middle of the inguinal ligament as a continuation of the internal oblique muscle, with some attachments also from the transversus, and is *inserted* by a small pointed tendon into the tubercle and crest of the pubis and into the front of the sheath of the rectus abdominis. The fasciculi form a compact layer as they lie within the inguinal canal. After passing through the superficial inguinal ring, the muscle fibers form a series of loops, the longest of which extends down as far as the testis, and these are attached to the tunica vaginalis. The intervals between the loops are occupied by fascia, called the **cremasteric fascia,** which is a fused continuation of the fasciae over the deep and superficial surfaces of the internal oblique. The muscular loops and the fascia together form a single layer, which constitutes the middle fascial tunic of the spermatic cord.

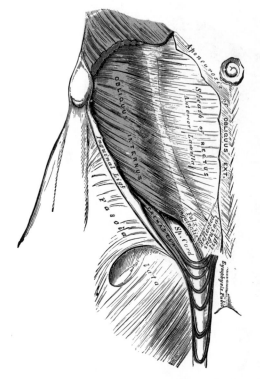

FIG. 6-30.   The cremaster, inguinal falx, and reflected inguinal ligament.

Action.   The **cremaster muscle** draws the testis up toward the superficial inguinal ring.

Nerve.   The cremaster muscle is supplied by the genital branch of the **genitofemoral nerve,** which is derived from the **first and second lumbar nerves.**

TRANSVERSUS ABDOMINIS.   The transversus abdominis muscle (Figs. 6-31; 6-32) derives its name from the direction of its fibers. It is the most internal of the flat muscles of the abdomen, being placed immediately deep to the internal oblique. It *arises* by fleshy fibers from the lateral third of the inguinal ligament, from the anterior three-fourths of the inner lip of the iliac crest, from the thoracolumbar fascia between the iliac crest and the twelfth rib, and from the inner surface of the cartilages of the last six ribs. The origin from the ribs is by means of digitations, which are separated from similar digitations of the diaphragm by a narrow fibrous raphe. Viewed from the interior of the abdomen, the two muscles appear to be components of a single stratum of muscle (Fig. 6-24). The fasciculi of the transversus, except the most inferior, pass horizontally across the abdominal wall

and terminate in an aponeurosis that fuses with the aponeurosis of the internal oblique. These join the aponeurosis of the opposite side to form the *insertion* of the muscle into the linea alba. The aponeurosis of the transversus muscle assists in the formation of the sheath of the rectus abdominis as follows: the upper three-quarters of the aponeurosis of the transversus pass entirely posterior to the rectus muscle, and contribute to the formation of its posterior sheath. The lower one-quarter, beyond a curved fibrous border called the **arcuate line,** approximately half way between the umbilicus and the pubis, passes entirely anterior to the rectus abdominis muscle. The fasciculi of the lowest portion of the muscle pass downward and medially and terminate as the **conjoined tendon** along with the lowest fasciculi of the internal oblique. Inferiorly, the muscle presents a free border that forms an arch extending from the lateral part of the inguinal ligament to the pubis, a short distance above the deep inguinal ring and spermatic cord.

Action.   The **transversus abdominis** constricts the abdomen, compressing the contents and assisting in micturition, defecation, emesis, parturition, and forced expiration.

Nerves.   The transversus abdominis muscle is supplied by branches of the **seventh to twelfth intercostal nerves** and by the iliohypogastric and ilioinguinal branches of the **first lumbar nerve.**

Variations.   The transversus abdominis may be more or less fused with the internal oblique or may be absent. The spermatic cord may pierce its lower border. Slender muscle slips may be found from the iliopectineal line to the transversalis fascia, the aponeurosis of the transversus abdominis, or the outer end of the arcuate line.

Conjoined Tendon (Inguinal Falx).   The conjoined tendon (Figs. 6-29 to 6-32) is the inferior terminal portion of the common aponeurosis of the internal oblique and transversus abdominis muscles. It is inserted into the crest and pecten of the pubis immediately deep to the superficial inguinal ring, giving strength from behind to a potentially weak point in the anterior abdominal wall. There is a wide variation in its width, strength, composition, and degree of union with neighboring aponeurotic and fascial structures. It may be narrow with a high arch, scarcely reaching the lateral part of the superficial inguinal ring, or it may be a broad, strong band, arching close to the in-

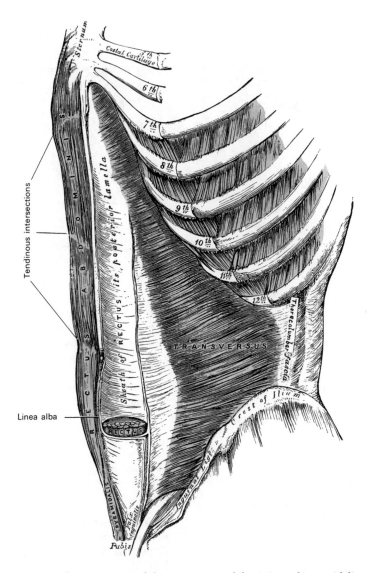

FIG. 6-31. The transversus abdominis, rectus abdominis, and pyramidalis.

guinal ligament, and greatly reinforcing the abdominal wall in the region of the ring. In many cases it could be called the conjoined muscle instead of the conjoined tendon because the muscular fasciculi continue almost to the pubis. Frequently it is intimately fused with the reflected inguinal ligament, which lies between it and the aponeurosis of the external oblique. It may be reinforced on its deep surface by a lateral fascial expansion of the lowermost part of the rectus tendon called **Henle's ligament.**[1]

[1]Friedrich Gustav Jacob Henle (1809–1885): A German anatomist and pathologist (Göttingen).

RECTUS ABDOMINIS. The rectus abdominis (Figs. 6-31; 6-32) is a long flat muscle that extends along the whole length of the ventral aspect of the abdomen. It is broader but thinner superiorly, and is separated from its fellow of the opposite side by the linea alba. It *arises* inferiorly by two tendons. The lateral tendon of origin is the larger and it is attached to the crest of the pubis, while the medial tendon interlaces with its fellow of the opposite side and is connected with the ligaments covering the front of the symphysis pubis. The muscle is *inserted* by three fascicles of unequal size into the cartilages of the fifth, sixth, and seventh ribs. The su-

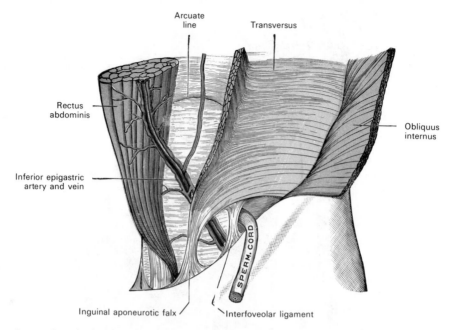

FIG. 6-32.   The interfoveolar ligament, ventral aspect. The spermatic cord passes through the deep inguinal ring lateral to the ligament. (Modified from Braune.)

perior fascicle, attached principally to the cartilage of the fifth rib, usually has some fibers of insertion into the ventral extremity of the rib itself. Some fibers are occasionally attached to the costoxiphoid ligaments, and to the side of the xiphoid process.

The rectus abdominis muscle is crossed by fibrous bands, three in number, which are named the **tendinous intersections.** One intersection is usually situated opposite the umbilicus, one at the free extremity of the xiphoid process, and the third about midway between the xiphoid process and the umbilicus. These inscriptions pass transversely or obliquely across the muscle in a zigzag course. They rarely extend completely through its substance and may pass only halfway across it, and they are intimately adherent to the sheath of the muscle anteriorly. Sometimes one or two additional inscriptions, generally incomplete, are present below the umbilicus.

Action. The **rectus abdominis** flexes the vertebral column, particularly the lumbar portion, drawing the sternum toward the pubis. It also tenses the anterior abdominal wall and assists in compressing the abdominal contents.

Nerves. The rectus abdominis is innervated by branches of the **seventh to twelfth intercostal nerves.** The seventh supplies the portion above the first tendinous intersection, the eighth innervates the

portion between the first and second intersections, and the ninth the portion between the lower two intersections.

Variations. The rectus may insert as high as the fourth or third rib or may be short and fail to reach the fifth. Fibers may spring from the lower part of the linea alba. Both the aponeurotic composition and the position of the arcuate line may vary considerably.

Sheath of the Rectus Abdominis. The rectus abdominis is enclosed in a sheath (Figs. 6-29; 6-31; 6-33) that holds it in position but does not restrict its motion during contraction because it is separated from the muscle by a fascial cleft. The sheath is formed by the aponeuroses of the external oblique, internal oblique, and transversus muscles, which preserve their identity in some regions but fuse or interlace in others. At the lateral border of the rectus, the aponeuroses form two membranes, the anterior and posterior lamellae of the sheath.

The *aponeurosis of the externus* keeps its position anterior to the rectus muscle throughout its entire length, but fuses with the aponeurosis of the internus at a variable distance from the midline (Fig. 6-33, *A* to *C*). The line of fusion between the aponeuroses of the external and internal obliques is close to the linea alba at the pubis and below the umbilicus, and gradually angles laterally as well as superiorly, but remains medial to the lateral border of the rectus. Thus, a surgical incision over the rectus, below the umbilicus, will go through the aponeuroses of the externus and internus as separate layers before it reaches the rectus muscle itself.

The *aponeurosis of the internus,* above the umbilicus, splits into two layers, one of which passes anterior to the rectus and fuses with the externus as just described (Fig. 6-33, *B*). The other layer passes posterior to the rectus and fuses with the aponeurosis of the transversus to form the posterior layer of the sheath. Above the costal margin, to which the internus muscle is attached, the costal cartilages and xiphoid process of the sternum take the place of the dorsal layer of the aponeurotic sheath. The posterior layer of the internus aponeurosis, above the costal margin, is attached to the cartilages, while the anterior layer ends abruptly in a fibrous band. Thus, the rectus muscle above the costal margin is covered anteriorly only by the aponeurosis of the external oblique.

The *aponeurosis of the transversus,* above the umbilicus, passes entirely dorsal to the rectus and fuses with the aponeurosis of the internus as described above (Fig. 6-33, *B*).

Below the umbilicus the disposition of the aponeuroses is somewhat more complicated and variable. At a variable distance between the pubis and umbilicus, usually about half way, the contribution of the aponeuroses to the posterior layer of the rectus sheath ceases abruptly in a curved, nearly semicircular line, called the **arcuate line** (of Douglas[1]). Below this line the aponeurotic fibers of all three muscles pass anterior to the rectus muscle (Fig. 6-33, *C*). Between the umbilicus and the arcuate line, the aponeuroses of the internus and transversus fuse and interdigitate, and fiber bundles from both may pass either anterior or posterior to the rectus. Usually a short distance above the arcuate line, the internus terminates its contribution to the posterior lamella and only the aponeurosis of the transversus remains posterior to the muscle. Thus, in the majority of instances, the arcuate line itself is formed by aponeurotic fibers of the transversus alone.

Below the arcuate line the posterior layer of the rectus sheath consists only of trans-

---

[1] James Douglas (1675–1742): An Irish physician and anatomist, who worked in London.

versalis fascia, since the aponeuroses of all three flat abdominal muscles pass anterior to the rectus abdominis. This posterior portion of the sheath is occasionally reinforced by scattered tendinous bundles from the transversus and by thickened laminae of the subserous fascia. In addition to the rectus muscle, the sheath contains the pyramidalis muscle, the superior and inferior epigastric arteries and veins, and branches of the intercostal nerves.

PYRAMIDALIS MUSCLE. The pyramidalis is a small triangular muscle, placed at the lower part of the abdomen, in front of the rectus, and contained in the sheath of that muscle (Fig. 6-31). It *arises* by tendinous fibers from the ventral surface of the pubis and the anterior pubic ligament. The fleshy portion of the muscle extends superiorly, diminishing in size as it ascends, and ends by a pointed extremity which is *inserted* into the linea alba, midway between the umbilicus and pubis.

Action. The **pyramidalis** muscle tenses the linea alba.

Nerve. The pyramidalis is supplied by a branch of the **twelfth thoracic nerve.**

Variations. The pyramidalis may vary from 1.5 to 12 cm in length, but it averages between 6 and 7 cm. It is occasionally doubled on one or both sides, and the muscles of the two sides may be unequal in size. The pyramidalis is absent in 10% of bodies, in which case the inferior end of the rectus is proportionately larger.

Linea Alba. The linea alba (Fig. 6-26) is the name given to the portion of the anterior abdominal aponeurosis or rectus sheath in the midline. Extending from the xiphoid process to the symphysis pubis, it is formed by an interlacing and fusion of fibers of the aponeuroses of the external oblique, internal oblique, and transversus muscles of both sides. It is broader above where the recti are separated from each other by a considerable interval. A surgical incision above the umbilicus in the midline will go through the linea as a single aponeurotic layer. It is narrower below, where the recti are more closely placed. A surgical incision in the midline below the umbilicus usually penetrates the anterior and posterior lamellae of the rectus sheath as individual layers. The inferior end of the linea alba has a double attachment. The superficial one passes in front of the medial heads of the recti to the

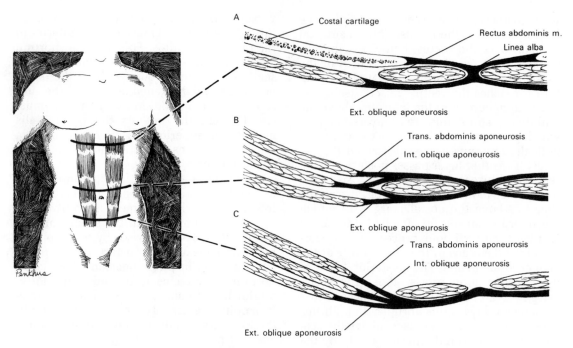

Fig. 6-33. Diagrams of cross sections of the anterior abdominal wall, showing the formation of the sheath of the rectus abdominis muscle above the costal margin, *A*, just above the umbilicus, *B*, and below the arcuate line, *C*.

symphysis pubis. The deeper attachment spreads out into a triangular sheet dorsal to the recti, attaches to the posterior lip of the crest of the pubis, and is named the **adminiculum lineae albae.** The **umbilicus,** which is an aperture for the passage of the umbilical vessels, the urachus, and during the first trimester, the vitelline duct in the fetus, is closed in the adult and presents a hard fibrous ring or plate of scar tissue within the linea alba.

Arcuate Line. The arcuate line (semicircular line of Douglas) is a curved, upwardly convex, tendinous band in the posterior lamella of the rectus sheath below the umbilicus (Fig. 6-32). It marks the inferior limit of the aponeurotic portion of the posterior lamella, the latter being composed of only transversalis fascia inferior to the line. Arching from the lateral border of the sheath, the arcuate line may be found 4 to 13 cm, average 8 cm, from the pubis. From each side, the arcuate line terminates at the linea alba, although occasionally as far down as the pubic crest. Its tendinous bundles are usually derived from the aponeurosis of the transversus abdominis, but may be from the internal oblique or from an interlacing of fibers from both the transversus and the internus. A secondary line may be found

superior to the primary one, especially in those instances where the primary line is formed by the transversus. In these cases the secondary line is formed by the aponeurotic fibers of the internal oblique.

Linea Semilunaris. The linea semilunaris (Fig. 6-26), a slightly curved line on the anterior abdominal wall, courses approximately parallel to the midline and lies about half way between the linea alba and the side of the body. It marks the lateral border of the rectus abdominis muscle, and can be seen as a shallow groove in the living subject when that muscle is tensed. The abdominal wall is thin along this line, particularly its inferior part, where it is composed only of the aponeuroses of the oblique and transversus muscles and their fasciae. At its superior end the linea semilunaris is thicker where the muscular fasciculi of the transversus extend deep to the rectus for a variable distance.

Group Actions. If pelvis and thorax are fixed, the abdominal muscles, assisted by the descent of the diaphragm, compress the abdominal viscera by constricting the cavity of the abdomen. By these means assistance is given in expelling the feces from the rectum, the urine from the bladder, the fetus from the uterus, and the contents of the stomach in vomiting.

If the pelvis and vertebral column are fixed, these

muscles compress the inferior part of the thorax, significantly assisting expiration. If the pelvis alone is fixed, the thorax is bent directly anteriorly when the muscles of both sides act. When the muscles of only one side contract, the trunk is bent toward that side and rotated toward the opposite side.

If the thorax is fixed, the muscles acting together draw the pelvis upward as in climbing. Acting singly, they draw the pelvis upward and bend the vertebral column to one side or the other. The recti, acting from below, depress the thorax and consequently flex the vertebral column. When acting from above, the recti flex the pelvis upon the vertebral column. The abdominal muscles are the **flexors of the vertebral column**, an important action that is frequently overlooked.

TRANSVERSALIS FASCIA.   The internal investing layer of deep fascia that lines the entire wall of the abdomen is called the **transversalis fascia** (Fig. 6-18). It may be defined even to include the pelvic portion of this internal layer, although more frequently, within the pelvis, this layer is called the *endopelvic fascia*. Formerly, the name was applied only to the deep fascia covering the internal surface of the transversus abdominis muscle, but because this muscle occupies a large proportion of the surface of the abdominal cavity, the name has gradually become adopted for the entire internal sheet. Various parts of it are still referred to by more specific designations such as the *iliac, psoas,* or *obturator fascia,* used for the portions covering these muscles. This internal fascia is of surgical interest, and in different areas it covers muscles, aponeuroses, bones, and ligaments. The transversalis fascia completely encases the abdominal cavity, but it may be thin and adherent in one place or thickened and independent in another. It gives rise to specialized structures such as tubular investments, and from it are derived certain components of important extra-abdominal fasciae. It is a gray, felt-like membrane, sometimes transparent, sometimes thickened into strong bands, but seldom aponeurotic in appearance and, except in obese individuals, contains no fat. Between this membrane and the peritoneum there is a layer of subserous fascia that may contain fat.

A definitive part of the transversalis fascia, that covering the internal surface of the muscular portion of the transversus muscle, is readily identifiable at dissection. Over the aponeurosis of this muscle, however, it is

thin and so closely adherent that only with difficulty can it be dissected free. Extending *superiorly* from the muscular portion of the transversus, the transversalis fascia continues onto the undersurface of the diaphragm, covering its entire abdominal dimension. It is thin and adherent over the diaphragmatic muscular portion as well as over the central tendon. *Anteriorly,* the fascia over the transversus aponeurosis of one side is continuous across the midline with that of the other side. *Posteriorly,* as it leaves the muscular fasciculi of the transversus, it continues for a short distance over the aponeurosis of origin of this muscle and then covers the abdominal surface of the quadratus lumborum and the psoas muscles. From the psoas it extends over the crura of the diaphragm and over the bodies and discs of the lumbar vertebrae, and is then continuous with the fascia over the psoas of the other side. As the dorsal origin of the diaphragm crosses the quadratus lumborum and psoas, the fascia is thickened into the strong, fibrous medial and lateral arcuate ligaments. *Inferiorly,* from the muscular portions of the transversus and quadratus lumborum, the fascia is attached to bone along the crest of the ilium and continues into the greater pelvis on the surface of the iliacus muscle. From the iliacus and psoas muscles, the transversalis fascia is continuous with the endopelvic fascia (described in the section dealing with the muscles of that region).

The fascia on the internal surface of the anterior abdominal wall *below the umbilicus* requires a more detailed description. In a medial direction from the muscular fasciculi of the transversus, the transversalis fascia, below the umbilicus, continues on the surface of the aponeurosis toward the midline. The portion *below the arcuate line,* however, splits at the lateral border of the rectus into two sheets. The thin anterior sheet continues on the aponeurosis of the transversus, and passes with it anterior to the rectus. A thick posterior sheet forms the posterior lamina of the rectus sheath, serving to separate the lower portion of the rectus muscle from abdominal viscera. Lateral to the lower part of the rectus, the transversalis fascia continues downward on the transversus aponeurosis to its lowest limit. At this site it covers the inner surface of the conjoined tendon (inguinal falx), and passing over the free border of the transverse aponeurosis, it forms an arch extending from the crest of the pubic bone to the lateral part of the inguinal ligament. The fascial membranes on both the deep and superficial surfaces of the transversus fuse into a single sheet at this free border, providing reinforcement of the transversalis fascia as it bridges the interval between the lower arched border of the transversus muscle and the inguinal ligament (Fig. 6-34).

At a point just above the middle of the inguinal ligament, a tubular prolongation of this reinforced fascia is carried outward on the ductus deferens and testicular vessels as they leave the abdominal cavity. This tubular investment is the inner coat of the spermatic cord and testis and is known as the **internal spermatic fascia** (Fig. 6-34). The ductus deferens (or round ligament) and vessels leave the abdominal cavity at a point which is called the **deep** (abdominal or internal) **inguinal ring,** and they follow an oblique course through the abdominal wall in a tunnel called the **inguinal canal.** The lateral part of this canal, that is, before it has begun to penetrate the internal oblique muscle, has transversalis fascia forming its posterior wall. The fascia is loosely attached to the inguinal ligament and has two thickenings near the ring, one extending cranialward

called the **interfoveolar ligament,** one extending caudalward called the **deep crural arch.**

Interfoveolar Ligament (of Hesselbach[1]). The interfoveolar ligament (Fig. 6-32) is a thickening in the transversalis fascia which forms a crescentic medial boundary for the deep inguinal ring. Connected above to the transversus muscle and below to the inguinal ligament, it may be poorly defined or it may be a strong band whose fibers fan out medially as it follows the upward course of the deep inferior epigastric artery. It forms a slight ridge, usually palpable from the interior of the abdominal cavity, which extends upward from the middle of the inguinal ligament, dividing the shallow fossa above the ligament into **medial** and **lateral inguinal fovea.** The deep inguinal ring, through which the ductus deferens or round ligament of the uterus leaves the abdominal cavity, is in the lateral fovea, and it is here that the sac of an *indirect inguinal hernia* penetrates the abdominal wall. In the medial fovea, a triangular area is marked out by the inguinal ligament, the lateral boundary of the rectus abdominis, and the inferior deep

[1]Franz Kaspar Hesselbach (1759–1816): A German surgeon (Würzburg).

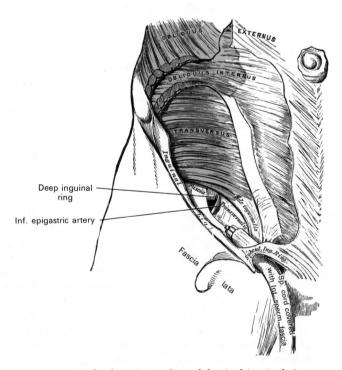

FIG. 6-34.   The deep, internal, or abdominal inguinal ring.

epigastric artery. This is the **inguinal triangle** (of Hesselbach) and a **direct inguinal hernia** penetrates through it.

Deep Crural Arch. The deep crural arch (sometimes called the iliopubic tract) is a downward extension of the transversalis fascia from the region of the internal inguinal ring. It arches across the external iliac vessels as they pass beneath the inguinal ligament and marks the transition from transversalis fascia to femoral sheath (Clark and Hashimoto, 1946). Laterally, it is attached to the iliac fascia where the latter gives origin to the inferior fibers of the transversus muscle. Medially, it follows the downward curve of the lacunar ligament. It may be a strong band or it may be poorly defined and appear to be merely the proximal part of the femoral sheath.

Deep Inguinal Ring. The deep inguinal ring (abdominal or internal inguinal ring) is that interruption in the transversalis fascia where the spermatic cord (round ligament in the female) penetrates the anterior abdominal wall (Fig. 6-33), carrying with it a sleeve-like investment of the transversalis fascia called the internal spermatic fascia. It is situated 1.25 cm above the inguinal ligament, midway between the anterior superior iliac spine and the symphysis pubis. It is oval in form, the long axis of which is vertical. It varies in size, being larger in males than in females, and fits rather closely about the penetrating structures, unless the ring has been distended by the sac of an indirect inguinal hernia. It is bounded, *superiorly,* by the arched inferior margin of the transversus abdominis; *medially,* by the interfoveolar ligament accompanying the inferior deep epigastric vessels; and *inferiorly,* by the deep crural arch.

Inguinal Canal. The inguinal canal is the tunnel in the lower anterior abdominal wall through which the spermatic cord (round ligament in the female), as well as the ilioinguinal nerve, pass. Its internal or lateral end is at the deep inguinal ring, and its external or medial end is at the superficial inguinal ring. The canal is about 4 cm long and takes an oblique course parallel with and a little above the inguinal ligament. It is bounded *superficially* by the skin, subcutaneous fascia, aponeurosis of the external oblique and, over its lateral third, by the internal oblique. The canal is bounded *deeply* from its medial to lateral ends by the reflected inguinal ligament, the inguinal falx,

the transversalis fascia, the subserous fascia, and the peritoneum. *Proximally* are found the arched fibers of the internal oblique and transversus, and *distally* are located the inguinal ligament and, at its medial end, the lacunar ligament. Through the canal pass the spermatic cord (or round ligament), the ilioinguinal nerve, the testicular vessels, and the cremaster muscle.

SUBSEROUS FASCIA. Intervening between the transversalis fascia and the peritoneum is fibroelastic connective tissue that supports the peritoneum. This subserous fascia, also referred to as subperitoneal fascia or extraperitoneal connective tissue, resembles the subcutaneous superficial fascia in that it supports a free surface epithelial layer. It commonly contains adipose tissue in varying thicknesses, and it is frequently separated from the transversalis fascia by a fascial cleft. Not only does the subserous fascia have a parietal portion supporting the peritoneum of the abdominal wall, but it also has a visceral portion which continues over the viscera at the peritoneal reflections and accompanies the blood vessels into the mesenteries. The subserous fascia has localized thickenings and accumulations of fat that are associated with the particular requirements of the different regions; these are described in sections dealing with the viscera. The subserous fascia in the pelvis is uninterruptedly continuous with that of the abdomen, and will be described with the pelvic viscera.

Posterior Muscles of the Abdomen

Psoas major
Psoas minor
Iliacus
Quadratus lumborum

The psoas major, the psoas minor, and the iliacus, with the fasciae covering them, will be described with the muscles of the lower limb.

FASCIA COVERING THE QUADRATUS LUMBORUM. The transversalis (internal investing) fascia covers the lateral portion of the quadratus lumborum on its ventral surface (Fig. 6-18). Since the medial portion of the muscle is overlapped by the psoas muscles, the fascia continues medially between the muscles and is attached to the bases of the transverse processes of the lumbar vertebrae. Its superior portion is thickened into a strong band, called the lateral arcuate ligament (Fig. 6-25), which is attached to the transverse process of the first lumbar vertebra and the apex and inferior border of the last rib. Inferiorly, the fascia is attached to the crest of the ilium, and is then continuous with the iliac fascia. At the lateral border of the quadratus lumborum, the fascia fuses with the combined fascia and aponeurosis of origin of the transversus muscle. The latter, extending medially from this point of fusion, covers the dorsal surface of the quadratus. The more medial portion of this aponeurotic sheet lies between the quadratus lumborum and the erector spinae and is the anterior layer of the thoracodorsal fascia (Fig. 6-18).

QUADRATUS LUMBORUM. The quadratus lumborum (Figs. 6-18; 6-19; 6-25) is irregularly quadrilateral in shape, but it is broader inferiorly. It *arises* below by aponeurotic fibers from the iliolumbar ligament and for about 5 cm along the adjacent portion of the iliac crest. It is *inserted* above into the medial one-half of the inferior border of the last rib, and by four small tendons into the apices of the transverse processes of the first four lumbar vertebrae. Occasionally a second portion of this muscle is found anterior to the preceding. When present, it arises from the superior borders of the transverse processes of the last three or four lumbar vertebrae, and is inserted into the inferior margin of the last rib. Anterior to the quadratus lumborum are the colon, the kidney, the psoas major and minor, and the diaphragm. Between the muscle and the transversalis fascia are found the twelfth thoracic, ilioinguinal, and iliohypogastric nerves.

Action. When the pelvis is fixed, the **quadratus lumborum** laterally flexes the lumbar vertebral column toward the same side. It also fixes the twelfth rib during deep inspiration, thereby preventing its elevation by the diaphragm. If both muscles act together, they help extend the lumbar vertebral column.

Nerves. The quadratus lumborum is innervated by the **twelfth thoracic and the upper three or four lumbar nerves.**

MUSCLES AND FASCIAE OF THE PELVIS

The muscles within the pelvis may be divided into two groups. The first group consists of the true pelvic muscles.

Levator ani
Coccygeus

The second group consists of the muscles of the lower limb. These originate within the pelvis and thus form part of the pelvic wall.

Obturator internus
Piriformis

The muscles of the second group will be described later with the muscles of the lower limb, but their fasciae will be considered in this section because they form an important part of the pelvic fascia.

PELVIC DIAPHRAGM. The pelvic diaphragm (Figs. 6-36; 6-39) is composed of the levator ani and coccygeus muscles together with the fasciae covering their superior (internal) and inferior (external) surfaces. It stretches across the pelvic cavity like a hammock. It forms the lowest portion of the body wall, closing the abdominopelvic cavity below, restraining the abdominal contents, and giving support to the pelvic viscera. It is pierced by the anal canal, the urethra, and the vagina, and is reinforced in the perineum by the special muscles and fasciae associated with these structures. The pelvic diaphragm and the structures of the perineum are intimately associated both structurally and functionally, and an accurate knowledge of one cannot be obtained without study of the other.

LEVATOR ANI MUSCLE. The levator ani is a broad, thin muscle attached to the internal surface of the pelvis (Fig. 6-35). The muscles of the two sides are separated from each other anteriorly, but are inserted into a raphe posteriorly. They function as a single sheet across the midline, forming the principal part of the pelvic diaphragm. It *arises,* anteriorly, from the inner surface of the superior ramus of the pubis lateral to the symphysis, and posteriorly, from the inner surface of the spine of the ischium. Between these two points the muscle arises from the **tendinous arch of the levator ani.** This arch is a thickened band of the obturator fascia attached posteriorly to the spine of the ischium and anteriorly to the pubic bone at the ventral margin of the obturator membrane. From their origin the muscle fascicles sweep downward and medially in the floor of the pelvis to be *inserted* into the side of the last two segments of the coccyx, the anococcygeal raphe, the external anal sphincter, and the central tendinous point of the perineum. The **anococcygeal raphe** is a narrow fibrous band extending from the coccyx to the posterior margin of the anus where the muscles of the two sides join at the midline. The levator ani consists of two principal parts, usually separated by a triangular gap and generally more distinct in lower mammals than in man, the pubococcygeus and the iliococcygeus.

The **pubococcygeus** *arises* from the dorsal surface of the pubis along an oblique line extending from the lower part of the symphysis to the obturator canal. The muscular fasciculi pass posteriorly, more or less parallel with the midline, and terminate by joining the fibers from the other side. The lateral margin of the pubococcygeus may be separated from the iliococcygeus by a narrow interval, or it may be overlapped by the latter muscle. Its medial margin is separated from the muscle of the other side by an interval known as the genital hiatus, which allows the passage of the urethra, vagina, and rectum. The most anterior fasciculi, which are also the most medial, pass behind the prostate in males and insert into the central tendinous point just in front of the anus and are called the **levator prostatae.** In females a similar muscular fascicle passes along the sides of the vagina to insert behind the vagina into the central tendinous point of the perineum. This constitutes an important sphincter of the vagina, the **pubovaginalis.** The majority of the fasciculi of the pubococcygeus pass horizontally backward beside the anal canal. The more superficial ones join the anococcygeal raphe, while the deeper ones join the muscle of the other side to form a U-shaped loop or sling around the rectum called the **puborectalis.**

The **iliococcygeus** arises from the tendinous arch of the levator ani muscle and the spine of the ischium, and *inserts* into the last two segments of the coccyx and the anococcygeal raphe. Its fasciculi pass medially as well as inferiorly, which helps to distinguish them from the fascicles of the pubococcygeus.

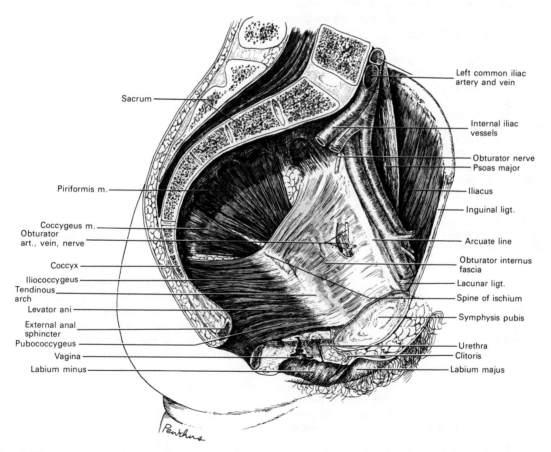

FIG. 6-35.   Interior of the female pelvis in a sagittal section, showing the muscles of the left lateral wall, the obturator fascia, and the tendinous arch of the levator ani muscle.

Action. The **levator ani** supports and slightly raises the pelvic floor, resisting increased intra-abdominal pressure, as during forced expiration. The pubococcygeus draws the lower end of the rectum toward the pubis and constricts it as well as the vagina.

Nerves. The levator ani is supplied by branches of the pudendal plexus, containing fibers from the **fourth** and sometimes also the **third or fifth sacral nerves.**

Variations. The degree of distinctness of the pubococcygeus and iliococcygeus varies, and the latter may be replaced by fibrous tissue.

THE COCCYGEUS. The coccygeus (Fig. 6-35) is located posterior to the levator ani but contiguous with it and on the same plane. It is a triangular sheet of muscular and tendinous fibers, *arising* by its apex from the spine of the ischium and sacrospinous ligament, and *inserted* by its base into the margin of the coccyx and into the side of the last piece of the sacrum. It assists the levator ani and piriformis in closing the posterior part of the pelvic outlet. By position it could be considered the "ischiococcygeus" in continuity with the pubo- and iliococcygeus portions of the levator ani.

Action. The **coccygeus** draws the coccyx forward after it has been pushed backward during parturition or defecation. It also supports the pelvic floor against intra-abdominal pressure.

Nerves. The coccygeus is supplied by branches of the pudendal plexus that contain fibers from the **fourth and fifth sacral nerves.**

Variations. The **iliosacralis** is an occasional band of muscle, ventral to the coccygeus, attached to the iliopectineal line and the lateral border of the sacrum. The **sacrococcygeus ventralis** is an occasional muscular or tendinous slip from the anterior surface of the lower sacral vertebrae to the coccyx. The **sacrococcygeus dorsalis** is an inconstant muscular slip that extends from the dorsal aspect of the sacrum to the coccyx.

PELVIC FASCIA. This internal investing layer of fascia covers the levator ani and coccygeus and the intrapelvic portions of the obturator internus and piriformis muscles. It is the fascia lining the pelvis that is comparable to the transversalis fascia of the abdomen and is directly continuous with the latter over the brim of the lesser pelvis, where it is attached to or fused with the periosteum of the symphysis and superior ramus of the pubis, the ilium along the arcuate line, and the promontory of the sacrum. From

these attachments it sweeps downward and across the midline, attaching to the anococcygeal raphe dorsally and blending with the fasciae of the anal canal and urogenital structures ventrally. Although it is a continuous sheet, for convenience of description it is divided into (1) piriformis fascia, (2) obturator fascia, and (3) superior fascia of the pelvic diaphragm (supra-anal fascia).

The **piriformis fascia** is very thin and is attached to the anterior surface of the sacrum and the sides of the greater sciatic foramen. It is prolonged outward through that foramen, joins the fascia on the external surface of the muscle at its distal border, and, becoming extrapelvic, forms part of the deep gluteal fascia. At its sacral attachment around the margins of the anterior sacral foramina, it ensheathes the nerves emerging from these foramina. Hence the sacral nerves are frequently described as lying behind this fascia. The internal iliac vessels and their branches, on the other hand, lie in the extraperitoneal tissue in front of the fascia, and their branches to the gluteal region emerge in special sheaths of this tissue, proximal and distal to the piriformis muscle.

The **obturator fascia** covers the obturator internus muscle and is partly intrapelvic and partly extrapelvic (Figs. 6-35; 6-36; 6-39). The intrapelvic portion has incorporated in it the aponeurosis of origin of the levator ani. It is as if the levator had at one time been attached to the pelvic brim above the obturator (a condition found in lower primates), but had slipped downward for a variable distance, usually about half the length of the muscle. At this site the origin of the levator is visible as a thickened band called the **tendinous arch of the levator ani muscle** (Fig. 6-36). Posteriorly, the arch always ends by attaching to the spine of the ischium. Anteriorly, although more variable, the arch usually attaches to the anterior margin of the obturator membrane or the pubic bone medial to it. Since the levator closes the aperture of the pelvis, the arch marks the boundary between the intra- and extrapelvic portions of the obturator fascia. The intrapelvic portion usually is not aponeurotic in appearance and may be quite thin except at the tendinous arch. *Anteriorly,* the obturator fascia is attached to the superior border of the obturator membrane or the pubic bone just in front of the obturator canal. It con-

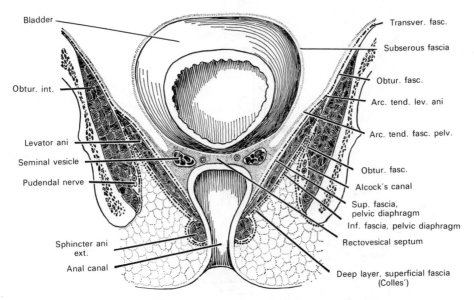

Bladder

Obtur. int.

Levator ani

Seminal vesicle

Pudendal nerve

Sphincter ani ext.

Anal canal

Transver. fasc.

Subserous fascia

Obtur. fasc.

Arc. tend. lev. ani

Arc. tend. fasc. pelv.

Obtur. fasc.

Alcock's canal

Sup. fascia, pelvic diaphragm

Inf. fascia, pelvic diaphragm

Rectovesical septum

Deep layer, superficial fascia (Colles')

Fig. 6-36. Fasciae of pelvis and anal region of perineum. Diagram of frontal section.

tributes a tubular investment for the obturator nerve and vessels as they leave the pelvis through that canal. Its attachment to the bone gradually angles cranialward from the obturator membrane until it reaches the arcuate line near the sacroiliac articulation. *Posteriorly,* it is attached to the margin of the greater sciatic notch down to the spine of the ischium. At the tendinous arch of the levator ani, the obturator fascia splits into three sheets, the outer one continues as the extrapelvic (perineal) obturator fascia, the other two cover the superior (intrapelvic) and inferior (perineal) surfaces of the pelvic diaphragm. The extrapelvic, perineal portion of the obturator fascia follows the muscle downward into the ischiorectal fossa and will be described with the perineum.

The **superior fascia of the pelvic diaphragm** (supra-anal fascia) covers the superior (intrapelvic or internal) surface of the levator ani and coccygeus muscles (Figs. 6-36; 6-39). *Anteriorly,* above the pubococcygeus part of the levator, it is attached to the pubic bone. *Laterally,* above the iliococcygeus, it is continuous with the intrapelvic obturator fascia at the tendinous arch of the levator ani muscle. *Posteriorly,* it covers the coccygeus and becomes continuous with the fascia of the piriformis. Behind the rectum, the fascia of each side is continuous across the midline over the anococcygeal raphe. In front of the rectum, in the region of

the genital hiatus where the medial borders of the pubococcygei of the two sides do not meet, the superior fascia of the pelvic diaphragm joins the inferior fascia of the pelvic diaphragm to assist in the formation of the deep layer of the urogenital diaphragm.

SUBSEROUS FASCIA.   The subserous fascia (Figs. 6-36; 6-37; 6-39), continuous with the subserous fascia of the abdomen, not only covers the parietal wall of the pelvis and the viscera, but also acts as a padding tissue for the viscera in the lower part of the pelvis. It forms important fascial ligaments and folds that extend from the pelvic wall to assist in the support of the pelvic viscera. A thickened band of subserous fascia, the **tendinous arch of the pelvic fascia** on the inner surface of the superior fascia of the pelvic diaphragm (Figs. 6-36; 6-38), marks the site of attachment of the lateral ligament of the bladder. Other supporting subserous fascial folds of this type are described in the chapter dealing with the pelvic viscera.

MUSCLES AND FASCIAE OF THE PERINEUM

The perineum is a diamond-shaped region that lies immediately beneath the outlet of the pelvis. Bony structures form its boundaries as follows: *anterior,* the pubic arch and the arcuate ligament of the pubis; *posterior,* the tip of the coccyx; and *on each side,* the

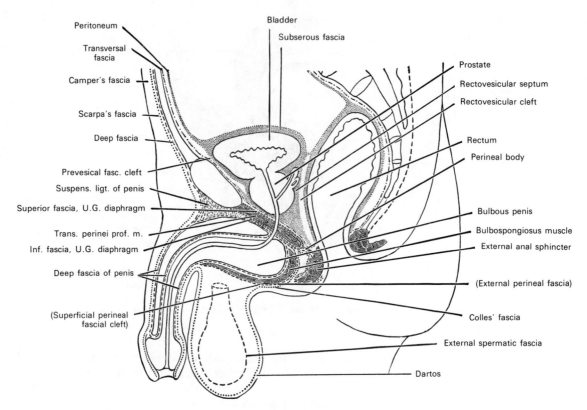

Peritoneum

Transversal fascia

Camper's fascia

Scarpa's fascia

Deep fascia

Prevesical fasc. cleft

Suspens. ligt. of penis

Superior fascia, U.G. diaphragm

Trans. perinei prof. m.

Inf. fascia, U.G. diaphragm

Deep fascia of penis

(Superficial perineal fascial cleft)

Bladder

Subserous fascia

Prostate

Rectovesicular septum

Rectovesicular cleft

Rectum

Perineal body

Bulbous penis

Bulbospongiosus muscle

External anal sphincter

(External perineal fascia)

Colles' fascia

External spermatic fascia

Dartos

FIG. 6-37. Diagram of fasciae of pelvis and perineum in median sagittal section.

inferior rami of the pubis and ischium, and the sacrotuberous ligament. The perineum is limited on the surface of the body by the external genitalia ventrally, by the buttocks dorsally, and laterally by the medial side of the thigh. A line drawn transversely between the ischial tuberosities divides the diamond-shaped region into two portions or triangular parts. The posterior contains the termination of the anal canal and is known as the **anal region or triangle,** while the anterior, which contains the external urogenital organs, is termed the **urogenital region or triangle.**

Perineal Fasciae

SUPERFICIAL FASCIA. The superficial fascia of the perineum is divisible into two layers, superficial and deep, which are similar to and continuous with the corresponding layers of the anterior abdominal wall (Figs. 6-37; 6-38). The deep layer is called the fascia of Colles instead of Scarpa. The superficial layer, corresponding to Camper's fascia,

usually is not specifically named, although it has been called Cruveilhier's fascia.[1]

Superficial Layer (of Superficial Fascia). The most superficial layer of fascia in the anterior part of the perineum, that is, over the urogenital region, contains a considerable amount of adipose tissue and is likely to be irregularly laminated. In the groove between the scrotum or labium majus and the thigh, it is directly continuous with the corresponding layer on the anterior abdominal wall. More medially, it is combined with the deep layer into the dartos tunic of the scrotum in the male, while in the female, it forms the greater part of the labium majus. Laterally, it is continuous with the superficial layer of the thigh, and posteriorly, it becomes the most superficial layer of fascia in the anal region.

The superficial layer in the anal region is greatly thickened into a mass of fat, which occupies the ischiorectal fossa. It is continu-

[1]Jean Cruveilhier (1791–1874): A French anatomist and pathologist (Paris).

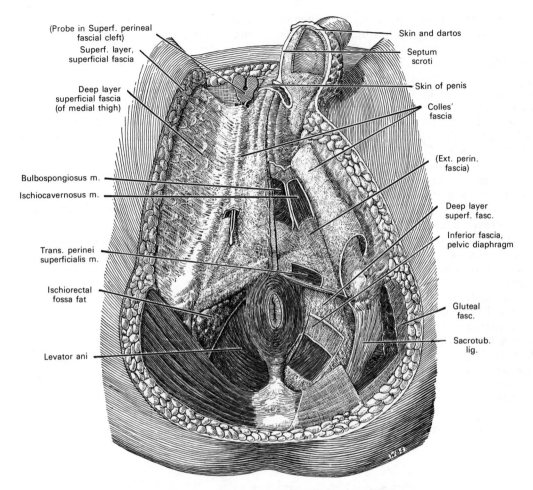

(Probe in Superf. perineal fascial cleft)

Superf. layer, superficial fascia

Deep layer superficial fascia (of medial thigh)

Bulbospongiosus m.

Ischiocavernosus m.

Trans. perinei superficialis m.

Ischiorectal fossa fat

Levator ani

Skin and dartos

Septum scroti

Skin of penis

Colles' fascia

(Ext. perin. fascia)

Deep layer superf. fasc.

Inferior fascia, pelvic diaphragm

Gluteal fasc.

Sacrotub. lig.

FIG. 6-38.   Fasciae of perineum. Dissection of superficial layers.

ous posteriorly with the superficial fascia of the gluteal region and posterior thigh. Covering the ischial tuberosities, the fibrous tissue is increased in amount, forming a tough pad.

The **ischiorectal fossa** (Fig. 6-38) is somewhat prismatic and wedge-shaped, with its *base* at the surface of the perineum and its *apex* deeply placed in the perineum at the line of meeting of the levator ani and obturator internus muscles. The fossa is bounded *medially* by the inferior fascia of the pelvic diaphragm covering the levator ani and external anal sphincter; *laterally,* by the obturator fascia over the extrapelvic portion of the obturator internus, and by the tuberosity of the ischium; *anteriorly,* by the posterior borders of the transversus perinei superficialis and profundus; and *posteriorly,* by the fascia over the gluteus maximus and the sac-

rotuberous ligament. The superficial boundary is the skin, and the fossa is occupied by a mass of adipose tissue belonging to the most superficial layer of fascia. The superficial and deep layers of the superficial fascia are not separable in the fossa and both are securely attached to the deep fascia covering the entire inner surface of the fossa. The dorsal portion of the transversus perinei profundus is separated from the levator ani for a short distance, so that the ischiorectal fossa has an **anterior recess** which extends beneath the posterior border of the transversus muscles. The adipose tissue within the fossa is traversed by fibrous bands and incomplete septa. The ischiorectal fossa is crossed transversely from lateral to medial by the inferior rectal vessels and nerves. In the posterior part of the fossa are the perineal and perforating cutaneous branches of the pudendal plexus, while from the anterior part of the fossa emerge the posterior scrotal (or labial) vessels and nerves. The internal pudendal vessels and pudendal nerve achieve the fossa by means of the **pudendal canal** (of Alcock[1]), which passes deep to the obturator fascia in a special reduplication of that fascia on the lateral wall of the fascia.

Deep Layer of Superficial Fascia (Colles' Fascia). In the urogenital region, but not in the anal region, the deep layer of superficial fascia is a distinctive structure (Figs. 6-36 to 6-39). It is a strong membrane but does not have the white glistening appearance of an aponeurosis. Instead, it has a slightly yellow color due to its content of elastic fibers, and it is smooth in texture, its fibrous nature not being detectable with the naked eye. This characteristic texture frequently is of assistance in differentiating it from the deep fascia in the region. *Anteriorly,* it is directly continuous with the deep layer of superficial abdominal fascia (Scarpa's fascia) in the groove between the scrotum or labium majus and thigh. More *medially,* it joins the superficial layer in the formation of the

[1]Benjamin Alcock (1801–?): An Irish anatomist (Cork, then America).

dartos tunic of the scrotum (Fig. 6-37). In the midline, the deep layer of superficial fascia is attached to the superficial layer along the raphe and continues ventrally into the **scrotal septum,** which divides that sac into two cavities. *Laterally,* it is firmly adherent to the medial surface of the thigh along the ischiopubic ramus at the origin of the adductor muscles. In the anterior part of this area, it is continuous with the fascia cribrosa which covers the saphenous opening. *Posteriorly,* it dips inward toward the ischiorectal fossa around the posterior border of the transversus perinei superficialis and becomes firmly attached to the deep fascia along the posterior border of the transversus perinei profundus. It is attached also, with all the other layers, to the central tendinous point of the perineum.

A fascial cleft, *not recognized in most other accounts of the perineal fascia,* has been described between Colles' fascia and the external perineal fascia (*a deep fascial layer also not usually recognized*), overlying the bulbocavernosus, ischiocavernosus, and the transverse perinei superficialis muscles. This **superficial perineal cleft** is thought by some to be continuous with the cleft deep to Scarpa's fascia on the anterior abdominal wall, but it is closed off laterally and posteriorly by the attachments described above.

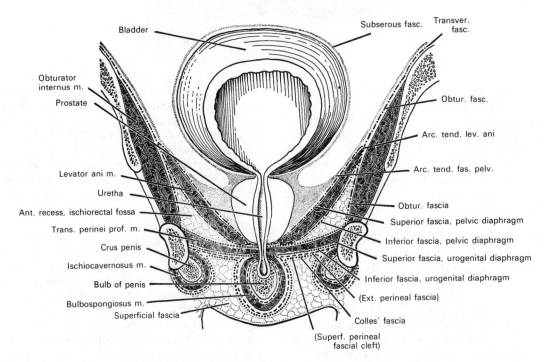

FIG. 6-39. Fasciae of pelvis and urogenital region of perineum. Diagram of frontal section.

In the anal region, the deep layer of the superficial fascia is adherent both to the superficial layer and to the deep fascia. Although not easily dissected in this region, its attachment to the posterior border of the transversus perinei profundus, mentioned previously, continues deeply into the ischiorectal fossa in close apposition to the inferior fascia of the pelvic diaphragm covering the inferior surface of the levator ani muscle. At the origin of this muscle, the deep layer of superficial fascia folds back sharply over the extrapelvic portion of the obturator fascia. Posteriorly, it leaves the ischiorectal fossa along the border of the gluteal region and posterior thigh.

DEEP FASCIAE.   The deep fasciae of the perineum consist of the obturator fascia and the inferior fascia of the pelvic diaphragm in the anal region, and in the urogenital region, of the superior and inferior fascial layers of the urogenital diaphragm, along with the external perineal fascial layer (frequently not cited in other descriptions). The phenomena of splitting, fusion, and cleavage are prominent features of fascia in this region. The clinical interest and importance of these fasciae have made them the subject of numerous treatises in which the various parts are frequently segregated and given special names. In the following description these parts will be considered individually, but an attempt will be made to correlate the diverse terminology into a consistent description.

The **obturator fascia** (perineal part) covers the extrapelvic portion of the obturator internus muscle and forms the lateral wall of the ischiorectal fossa. Superiorly, it meets the inferior fascia of the pelvic diaphragm at a sharp angle in the deepest part of the fossa. Its inferior portion, extending from the lesser sciatic foramen to the ischial tuberosity, is thickened and splits to enclose the pudendal nerve and internal pudendal vessels in a fibrous tunnel called the **pudendal canal** (of Alcock). Anteriorly, it is attached to the ischiopubic ramus, and the fibrous sheath of the pudendal canal merges with the inferior fascia of the urogenital diaphragm (and perhaps with the external perineal fascia) where branches of the nerve and vessels enter the superficial and deep perineal spaces or compartments.

The **inferior fascia of the pelvic diaphragm,** also known as the infra-anal fascia, covers the inferior or perineal (external, or superficial) surface of the levator ani and coccygeus muscles (Figs. 6-36 to 6-39). It adheres to the muscle throughout. Superiorly, the portion of this fascia subjacent to the iliococcygeus is continuous with the extrapelvic obturator fascia at the tendinous arch of the levator ani. The portion underlying the pubococcygeus is attached to the ischiopubic ramus and the pubic bone at the origin of the muscle. Between the medial borders of the two pubococcygei, the infra-anal fascia joins the supra-anal fascia to form a thick sheet in the genital hiatus. Behind the rectum, the inferior fascia of the pelvic diaphragm is continuous from side to side over the anococcygeal raphe where it is firmly attached. More laterally, it bridges the slight interval between the levator ani and coccygeus, and then binds the latter muscle to the inferior edge of the sacrospinous ligament. From this ligament, it passes outward from the ischiorectal fossa along the overhanging distal border of the gluteus maximus, becoming continuous with the gluteal fascia. Near the anus, it invests the external anal sphincter, and just anterior to the anus, it is firmly attached to the other perineal layers at the central tendinous point.

If the infra-anal fascia is traced anteriorly from the ischiorectal fossa, it will be seen to split into two sheets at a transverse line connecting the anterior extremities of the ischial tuberosities (three sheets if the external perineal fascia is considered). The deepest of these continues on the surface of the levator ani, and lying between this muscle and the transversus perinei profundus, it is called the **superior layer of the urogenital diaphragm.** Another of the fascial sheets covers the superficial surface of the transversus profundus and is called the **inferior layer of the urogenital diaphragm.** If the external perineal fascia is considered a true fascial plane, it would be the most superficial of these deep fascial layers in the urogenital region which split from the inferior fascia of the pelvic diaphragm in the mid-perineum.

UROGENITAL DIAPHRAGM.   The transversus perinei profundus muscle is covered superiorly (internally) and inferiorly (externally) by fascial membranes that are called, respectively, the superior and inferior layers of the urogenital diaphragm (Fig. 6-39). The transversus perinei profundus muscle and the two fascial layers taken together constitute the **urogenital diaphragm.** The urogenital diaphragm is synonymous with the **deep perineal compartment** (pouch or space) and its contents, which will be described. The urogenital diaphragm, as it has just been defined, is assisted in its role as a supporting structure by the portion of the levator ani over which it lies and by the superficial perineal muscles and their fasciae.

The **genital hiatus** is the interval in the midline between the medial borders of the two pubococcygeus portions of the levator ani muscle. It is through this hiatus that the urethra passes in both sexes and the vagina in the female. The fibrous tissue composing the region of the genital hiatus is derived from the superior and inferior fasciae of the pelvic diaphragm and the superior layer of the urogenital diaphragm.

The **superior fascia of the urogenital diaphragm** (also called deep or internal fascia) is a flat, triangular membrane stretching across the interval between the ischiopubic rami (Figs. 6-37; 6-39). It lies between the transversus perinei profundus and the pubococcygeus portions of the levator ani and represents the fused fascial membranes of both these muscles. It represents also, in the genital hiatus between the medial borders of the two pubococcygei, a fusion with still a third membrane, the superior fascia of the pelvic diaphragm. It is securely attached to the symphysis pubis anteriorly and joins the other perineal layers at the central tendinous point posteriorly. Laterally, it is attached to the medial borders of the ischiopubic rami, and there it is continuous with the obturator fascia. The middle portion of the fascia, which occupies the genital hiatus, is thickened to fill in the gap between the two pubococcygei and binds their medial borders together. At this site the fascia is pierced by the urethra and vagina, blending with their walls as they pass through. The prostate gland rests on the pelvic surface of the fascia in the male, and the connective tissue of the inferior portion of the prostatic capsule also intimately blends with the fascia. The attachment of the fascia to the pubic bone and its blending with the glandular capsule make the anterior part of the fascia a true **ligament of the prostate.** The tissue attachments may form three strands, a **middle puboprostatic ligament** to the symphysis and two **lateral puboprostatic ligaments,** one to each pubic bone, at the sites where the anterior ends of the tendinous arches of the

levator ani muscles are attached. At the posterior border of the deep transversus, the fascia is fused with the superficial layer of the urogenital diaphragm, closing the deep perineal compartment.

The **inferior fascia of the urogenital diaphragm** (also called superficial or external fascia) is a flat, triangular membranous sheet which, like the superior, bridges the angular interval between the ischiopubic rami (Figs. 6-37; 6-39). It is attached laterally to the medial borders of the rami from the arcuate pubic ligament to the ischial tuberosities. Along these same borders, the superior layer of the diaphragm is attached more deeply and the crura of the penis more superficially (Fig. 6-39). The middle portion of the fascia is pierced by the urethra and vagina and blends with their walls. It is perforated also by the arteries to the bulb, the ducts of the bulbourethral glands, the deep arteries of the penis, and the dorsal arteries and nerves of the penis. The part posterior to the urethra blends with the fascia of the bulb and, because of its attachment to the rami laterally, is sometimes called the **ligament of the bulb.** At these lateral attachments, the fascia also blends with that covering the crura of the penis. The part anterior to the urethra arches across the subpubic angle and is sometimes called the **transverse ligament of the pelvis.** The latter is separated from the symphysis pubis and arcuate pubic ligament by an opening for the passage of the dorsal vein of the penis. At the posterior border of the deep transversus muscle, the superficial and deep layers of the urogenital diaphragm fuse into a single layer and blend with the inferior fascia of the pelvic diaphragm from within the ischiorectal fossa.

The *external perineal fascia* (Figs. 6-36 to 6-39) is a fascial sheet, the existence of which is controversial, although it is described by some anatomists. Its defenders claim it to be an external investing layer of deep fascia in the urogenital region (in contrast to the inferior fascia of the urogenital diaphragm, which is thought by them to be the internal investing layer). It is considered to be the most superficial layer of the deep fascia in the urogenital region of the perineum, and to cover approximately the same triangular area as the inferior fascia of the urogenital diaphragm. Not being flat, the fascia is described as accommodating itself to the contours of the superficial perineal muscles. Those denying the existence of this fascial plane consider the aforementioned connective tissue to be simply the epimysium of the superficial perineal muscles.

A detailed account of the external perineal fascia will be omitted from this edition because it is not a generally accepted structure and because it is not recognized in the *Nomina Anatomica*. The importance of considering it at all relates to the fact that those authors claiming it to be a fascial sheet have further described it, rather than Colles' fascia, as the superficial limiting boundary of the superficial perineal compartment. Additionally, its proponents claim the existence of a superficial perineal fascial cleft between the external perineal fascia and Colles' fascia.

PERINEAL BODY.   The perineal body, often called the central tendinous point of the perineum, is a mass of fibromuscular tissue in the midline between the anus and the bulb of the penis in the male and between the anus and vagina in the female. It is approximately 2 cm in width and depth in the male and about twice this dimension in the female. It is composed predominately of fibrous tissue since it represents a fusion of the superior and inferior fasciae of the urogenital diaphragm and Colles' fascia (and perhaps the external perineal fascia). It contains a few muscular fibers also, principally from the pubococcygeal portions of the two levator ani muscles and the external anal sphincter. Besides these muscles, the perineal body also has attached to it the transversus perinei profundus and superficialis muscles of both sides as well as the bulbospongiosus muscle. It is directly continuous anteriorly with the fibrous tissue that fills in the genital hiatus between the two pubococcygei. It continues deep into the pelvis with the rectoprostatic and rectovesical septum in the male and the rectovaginal septum in the female. The perineum is a time-honored surgical approach to the bladder and prostate, and it is the site of the perineal tears that frequently result from childbirth.

DEEP PERINEAL COMPARTMENT (space or pouch).   The deep perineal compartment is that region between the superior and inferior fascial layers of the urogenital diaphragm. The compartment is principally occupied by the transversus perinei profundus, but contains also the sphincter urethrae and the membranous portion of the urethra,

the bulbourethral glands (vestibular glands in the female) and their ducts, the internal pudendal vessels, the deep dorsal vein of the penis, and the dorsal nerve of the penis. The internal pudendal artery enters the deep perineal compartment posteriorly and its branches to the penile bulb, to the urethra, and to the deep and the dorsal arteries of the penis pierce the inferior fascia of the urogenital diaphragm to reach their destination.

SUPERFICIAL PERINEAL COMPARTMENT (space or pouch).   The superficial perineal compartment is bounded by the inferior layer of the urogenital diaphragm and the deep layer of superficial fascia (Colles' fascia). (The proponents of the existence of the external perineal fascia state that it forms the superficial boundary of the pouch, instead of Colles' fascia.) The superficial perineal pouch contains the bulbospongiosus, ischiocavernosus, and transversus perinei superficialis muscles, and it is traversed by the perineal vessels and nerve.

CLINICAL CONSIDERATIONS.   Extreme trauma may result in rupture of the urethra or even the bladder. Depending on the site of rupture (i.e., whether it occurs above or below the urogenital diaphragm), urine accumulates in either the pelvis or the perineum. Thus, because it is located above the urogenital diaphragm, if the bladder or prostatic urethra is ruptured, as is frequently the case with a fracture of the pelvis, extravasated urine collects in the extraperitoneal tissues of the pelvis and is prevented from spreading into the perineum by the pelvic diaphragm.

Rupture of the spongy or bulbous part of the urethra, which is below the urogenital diaphragm, is a common result of straddle injuries that occur when the pubic arch forcibly strikes an unyielding object such as a fencetop or a saddle. In these instances, urine is extravasated into the perineum, since the urogenital diaphragm limits its movement into the pelvis. If the deep fascia of the penis (Buck's fascia[1]) remains intact, urine is confined by this fascial plane to the penis only. More often, however, the deep fascia of the penis is also punctured, and urine escapes into the superficial perineal compartment. When this occurs, the urine is limited in its extravasation by the attachments of Colles' fascia and the fasciae with which it is continuous, the dartos layer over the scrotum, the superficial fascia of the penis, and the deep layer of superficial fascia over the anterior abdominal wall (Scarpa's fascia). Urine is usually prevented from spreading posteriorly into the anal region of the perineum or laterally into the thigh

[1]Gurdon Buck (1807–1877): An American surgeon (New York).

by the attachments of the inferior fascia of the urogenital diaphragm, Colles' and Scarpa's fasciae. Instead, urine extravasates into the scrotum, penis, and anterior abdominal wall, and in poorly attended cases, it may reach as high as the anterior thoracic wall.

Perineal Muscles

The muscles of the perineum may be divided into two groups:

> Muscles of the urogenital region
> In the male
> In the female
> Muscles of the anal region

MUSCLES OF THE UROGENITAL REGION IN THE MALE

Superficial Group
  Transversus perinei superficialis
  Bulbospongiosus
  Ischiocavernosus
Deep Group
  Transversus perinei profundus
  Sphincter urethrae

TRANSVERSUS PERINEI SUPERFICIALIS. This muscle is a narrow slip on each side that passes more or less transversely across the perineal area in front of the anus (Figs. 6-38; 6-40). It *arises* by tendinous fibers from the inner and anterior part of the tuberosity of the ischium, and, coursing medially, is *inserted* centrally into the perineal body, joining at this site with the muscle of the opposite side, the external anal sphincter posteriorly, and the bulbospongiosus anteriorly. In some instances, the fibers of the deeper layer of the external anal sphincter decussate in front of the anus and are continued into this muscle. Occasionally, the superficial transverse perineal muscle gives off fibers that join with the bulbospongiosus of the same side.

Actions. The simultaneous contraction of the two muscles serves to fix the centrally located perineal body.
Variations. The transversus perinei superficialis may be absent or doubled, or it may insert into the bulbospongiosus or external anal sphincter.

THE BULBOSPONGIOSUS. This muscle (formerly called accelerator urinae or bulbocavernosus) is placed in the midline of the perineum, ventral to the anus (Fig. 6-40). It consists of two symmetrical parts, united along the median line by a tendinous raphe. It *arises* from the perineal body and its ventral extension into the median raphe. Its fibers diverge like the barbs of a feather. The most *posterior fibers* form a thin layer, which is lost on the inferior fascia of the urogenital diaphragm. The *middle fibers* encircle the bulk and adjacent parts of the corpus spongiosum penis, and join with the fibers of the opposite side, on the upper part of the corpus cavernosum penis, in a strong aponeurosis. The *anterior fibers* spread out over the side of the corpus cavernosum penis and are inserted partly into that body, anterior to the ischiocavernosus, occasionally extending to the pubis, and partly end in a tendinous expansion that covers the dorsal vessels of the penis. The latter fibers can be seen by dividing the muscle longitudinally and reflecting it from the surface of the corpus spongiosum penis.

Actions. The **bulbospongiosus** serves to empty the urethra, after the bladder has expelled its contents. During the greater part of the act of micturition its fibers are relaxed, and it only comes into action at the end of the process. The middle fibers assist in the erection of the corpus spongiosum penis by compressing the erectile tissue of the bulb. The anterior fibers also contribute to the erection of the penis by compressing the deep dorsal vein of the penis as these fibers insert into the penile fascia, which encases the dorsal vessels of the penis.

THE ISCHIOCAVERNOSUS. This muscle (at one time called the erector penis) is paired, each covering one of the crura of the penis (Fig. 6-40). It is an elongated muscle, broader in the middle than at either end, and situated at the lateral boundary of the perineum. It *arises* by tendinous and fleshy fibers from the inner surface of the tuberosity of the ischium dorsal to the crus penis, and from the rami of the pubis and ischium on each side of the crus. From these points, fleshy fibers course ventrally along the crus and end in an aponeurosis, which is *inserted* into the sides and undersurface of the crura as they become the body of the penis.

Action. The **ischiocavernosus** compresses the crus penis, retarding the return of the blood through the veins, and thus serves to maintain penile erection.

TRANSVERSUS PERINEI PROFUNDUS. This muscle *arises* from the inferior rami of the

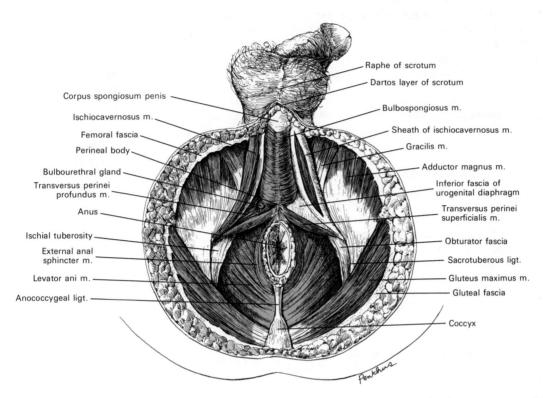

Corpus spongiosum penis

Ischiocavernosus m.

Femoral fascia

Perineal body

Bulbourethral gland

Transversus perinei
profundus m.

Anus

Ischial tuberosity

External anal
sphincter m.

Levator ani m.

Anococcygeal ligt.

Raphe of scrotum

Dartos layer of scrotum

Bulbospongiosus m.

Sheath of ischiocavernosus m.

Gracilis m.

Adductor magnus m.

Inferior fascia of
urogenital diaphragm

Transversus perinei
superficialis m.

Obturator fascia

Sacrotuberous ligt.

Gluteus maximus m.

Gluteal fascia

Coccyx

Fig. 6-40.   Muscles of the male perineum. The inferior fascia of the urogenital diaphragm has been removed on the right to reveal the transversus perinei profundus muscle and the bulbourethral gland. (After Sobotta.)

ischium and runs to the median line, where it interlaces in a tendinous raphe with its fellow of the opposite side (Figs. 6-37; 6-39; 6-40). Lying in the same plane as the sphincter urethrae, these two muscles are interposed between the superior and inferior fascial layers of the urogenital diaphragm and form much of the bulk of that structure. Formerly these two muscles were described together as the *constrictor urethrae.*

SPHINCTER URETHRAE.   In the male, this muscle surrounds the whole length of the membranous portion of the urethra and is enclosed in the fasciae of the urogenital diaphragm. Its *external* fibers *arise* laterally along about 2 cm of the ischiopubic ramus and from the neighboring fasciae. They arch across the front of the urethra and bulbourethral glands. Passing around the urethra, the superficial fibers unite posterior to the urethra with muscle fibers from the opposite side by means of a tendinous raphe. Its *internal* fibers form a continuous circular investment for the membranous urethra.

Actions.   The **sphincter urethrae** muscles of both sides act together as a sphincter, compressing the membranous portion of the urethra. During micturition, like the bulbospongiosus, the sphincter urethrae muscles are relaxed, and only come into action at the end of the process to eject the last drops of urine. They also function during ejaculation.

Nerve Supply.   All the muscles of the male urogenital region are supplied by the second, third, and fourth sacral nerves by way of the **perineal branch of the pudendal nerve.**

MUSCLES OF THE
UROGENITAL REGION IN THE FEMALE

Transversus perinei superficialis
Bulbospongiosus
Ischiocavernosus
Transversus perinei profundus
Sphincter urethrae

TRANSVERSUS PERINEI SUPERFICIALIS.   In the female this muscle is a narrow slip of fibers that *arises* by a small tendon from the inner and ventral part of the tuberosity of the ischium and is *inserted* into the perineal body (Fig. 6-41). At this centrally located site of the perineum, as in the male, it joins the muscle of the opposite side, the external anal sphincter, and the bulbospongiosus.

Action. The simultaneous contraction of the two **transverse perinei superficialis** muscles serves to fix the perineal body.

THE BULBOSPONGIOSUS. This muscle has also been called the sphincter vaginae, and it surrounds the orifice of the vagina. It covers the lateral parts of the vestibular bulb and is attached posteriorly to the perineal body, where it blends with the external anal sphincter (Fig. 6-41). Its fibers pass anteriorly on each side of the vagina to be inserted into the corpgra cavernosa clitoridis. A small muscular fasciculus also crosses over the body of the clitoris and compresses its deep dorsal vein.

Actions. The **bulbospongiosus** diminishes the orifice of the vagina. The anterior fibers contribute to the erection of the clitoris, being inserted into and continuous with the fascia of the clitoris and thereby compressing the deep dorsal vein during the contraction of this muscle.

THE ISCHIOCAVERNOSUS. In the female, this muscle (formerly called the erector clitoridis) is smaller than the corresponding muscle in the male. It covers the unattached surface of the crus clitoridis and is elongated, broader at the middle than at either end, and situated along the lateral boundary of the perineum (Fig. 6-41). It *arises* by tendinous and fleshy fibers from the inner surface of the tuberosity of the ischium, from the surface of the crus clitoridis, and from the adjacent portion of the ramus of the ischium. From these points fleshy fibers extend anteriorly and end in an aponeurosis, which is *inserted* into the sides and undersurface of the crus clitoridis.

Actions. The **ischiocavernosus** compresses the crus clitoridis, retarding the return of blood through the veins, and thus serves to maintain erection of the clitoris.

UROGENITAL DIAPHRAGM IN THE FEMALE. The urogenital diaphragm in the female, as in the male, is formed of two layers of fascia between which are interposed the deep transverse perineal muscle and the urethral sphincter. The fascial layers are not so strong in the female as in the male. Being attached to the pubic arch, the apices of the

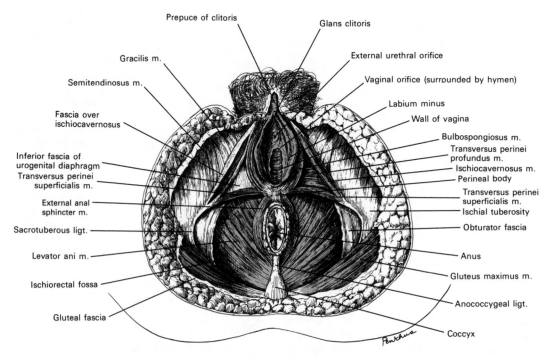

FIG. 6-41. Muscles of the female perineum. The fat pad of the ischiorectal fossa has been removed to reveal the levator ani muscle.

fascial layers are connected to the arcuate pubic ligament. Divided in the midline by the aperture of the vagina, the fasciae blend with the external coat of the vagina. Anterior to this the urogenital diaphragm is perforated by the urethra. The posterior borders of the fasciae, as in the male, are continuous with the deep layer of the superficial fascia around the transversus perinei superficialis.

Between the two fascial layers are found the deep dorsal vein of the clitoris, a portion of the urethra, the deep transverse perineal muscle and the sphincter urethrae muscle, the greater vestibular glands and their ducts, the internal pudendal vessels, the dorsal nerves of the clitoris, the arteries and nerves of the vestibular bulb, and a plexus of veins.

TRANSVERSUS PERINEI PROFUNDUS. In the female, the deep transverse perineal muscle courses from the inferior ramus of the ischium to the side of the vagina, meeting fibers of the muscle from the opposite side.

Action. The transversus perinei profundus helps to fix the perineal body.

SPHINCTER URETHRAE. Like the corresponding muscle in the male, the sphincter urethrae in the female consists of external and internal fibers. The *external* fibers *arise* on each side from the margin of the inferior ramus of the pubis and the transverse perineal ligament. They are directed across the pubic arch in front of the urethra, and pass around it to blend with the muscular fibers of the opposite side, between the urethra and vagina. The *internal* fibers encircle the lower end of the urethra.

Action. The sphincter urethrae of the two sides act together as a constrictor of the urethra.
Nerve Supply. All the muscles of the female urogenital region are supplied by the second, third, and fourth sacral nerves by way of the perineal branch of the pudendal nerve.

MUSCLES OF THE ANAL REGION

Corrugator cutis ani
Sphincter ani externus
Sphincter ani internus

CORRUGATOR CUTIS ANI. Around the anus is a thin stratum of involuntary muscle and yellow elastic fibers that radiates from the orifice. *Medially* the fibers fade into the submucous tissue, while *laterally* they blend with the true skin.

Action. By its contraction the corrugator cutis ani muscle puckers the skin, raising it into ridges around the anal orifice.

EXTERNAL ANAL SPHINCTER. The external anal sphincter is a flat plane of muscular fibers, elliptical in shape and intimately adherent to the integument surrounding the margin of the anus. It measures about 8 to 10 cm from its anterior to its posterior extremity, and is about 2.5 cm broad opposite the anus. It surrounds the last 2 cm of the anal canal and consists of three parts, subcutaneous, superficial, and deep. The subcutaneous lamina is a band of fibers that lies beneath the skin surrounding the anal orifice. Some fibers are attached anteriorly to the perineal body and posteriorly to the anococcygeal ligament. The superficial lamina lies deep to the subcutaneous lamina and constitutes the main portion of the muscle. It also arises from the narrow tendinous anococcygeal ligament, which stretches from the tip of the coccyx to the posterior margin of the anus. The muscle fibers then encircle the anus and meet in front to be inserted into the perineal body, joining with fibers from the transversus perinei superficialis, the levator ani, and the bulbospongiosus. The deep layer forms a complete sphincter to the anal canal. Its fibers surround the canal, closely applied to the internal anal sphincter, and its deep fibers are interlaced with fibers of the puborectalis muscle. In front, the deep part of the external anal sphincter blends with the other muscles at the perineal body. In many individuals, the fibers decussate in front of the anus and are continuous with those of both transversus perinei superficialis muscles. Posteriorly, they are not attached to the coccyx, but are continuous with those of the opposite side behind the anal canal. The upper edge of the muscle is ill defined because some fibers join with those of the levator ani muscle.

Actions. The action of the external anal sphincter is peculiar. Like some muscles, this muscle is always in a state of tonic contraction, and having no antagonistic muscle, it serves to keep the anal canal and orifice closed. Under voluntary control, the muscle can be made to contract more firmly, thereby occluding the anal aperture in expiratory efforts unconnected with defecation. Considering its bony attachment by way of the anococcygeal ligament at the coccyx, it helps to fix the perineal body so that the bulbospongiosus may act from this fixed point. During defecation the muscle relaxes, thereby allowing the contents of the anal canal and rectum to be eliminated.

Nerve Supply. The external anal sphincter is supplied by a branch from the fourth sacral and twigs from the inferior rectal branch of the **pudenal nerve.**

INTERNAL ANAL SPHINCTER. The internal sphincter of the anus is a muscular ring that surrounds about 2.5 cm of the anal canal. Its inferior border is in contact with, but quite separate from, the external sphincter. It is about 5 mm thick, and is formed by an aggregation of the involuntary circular fibers of the intestine. Its distal border is about 6 mm from the orifice of the anus.

Actions. The action of the **internal anal sphincter** is entirely involuntary. It helps the external sphincter to occlude the anal aperture and aids in the expulsion of the feces.

Nerve Supply. The internal anal sphincter is supplied by fibers of the **autonomic nervous system** as is the musculature surrounding many of the abdominal and pelvic organs.

Muscles and Fasciae of the Upper Limb

The structures of the upper limb will be described under the following headings:

Muscles connecting the upper limb to the vertebral column
Muscles connecting the upper limb to the thoracic wall
Muscles of the shoulder
Muscles of the arm
Muscles of the forearm
Fasciae, retinacula, synovial sheaths, and compartments of the wrist and hand
Muscles of the hand

MUSCLES CONNECTING THE UPPER LIMB TO THE VERTEBRAL COLUMN

Trapezius
Latissimus dorsi
Rhomboid major
Rhomboid minor
Levator scapulae

SUPERFICIAL FASCIA. The superficial fascia of the back is a thick fibrous and fatty layer that extends from the scalp to the gluteal region as a comparatively uniform sheet. At the sides of the neck and trunk, it gradually changes into a thinner or softer fascia characteristic of the more anterior regions. The dermal fibrous layer is thick and is bound down to the deeper layers by numerous heavy bands and septa, which divide the fat into small granular lobules and mattress the entire layer into a tough, resilient pad. The fascial cleft, usually present between the superficial and deep fascia, is lacking in the dorsal regions, making the superficial fascia firmer and less movable than that in the anterior parts of the trunk and neck.

DEEP FASCIA. The **investing layer of deep fascia** is attached in the midline to the ligamentum nuchae, the supraspinal ligament, and the spinous processes of all vertebrae caudal to the sixth cervical. It splits to enclose the trapezius and the fleshy portion of the latissimus dorsi, but in the neck it covers the posterior triangle as a single layer and there becomes continuous with the investing layer of fascia of the anterior neck. It is attached, over the shoulder, to the acromion and spine of the scapula, and is then continuous with the deltoid fascia. Laterally from the latissimus, it is continuous with the axillary fascia above and with the deep fascia covering the external oblique muscle of the abdominal wall. Over the fleshy portions of the trapezius and latissimus, it is gray and felt-like, but strong and adherent both to the superficial fascia and to the muscles. In a triangular area bounded by the trapezius, deltoid, and latissimus, it is white and glistening, forming an aponeurotic fascia over the infraspinatus. In the lumbar region, the deep fascia is greatly strengthened by having incorporated in it the aponeurosis of origin of the latissimus dorsi. This portion constitutes the posterior layer of the thoracolumbar fascia.

The investing fascia lining the deep surface of the trapezius is thickened by an accumulation of adipose tissue similar to that under the pectoralis major. Within it course the transverse cervical artery, its superficial (ascending) branch, and the accessory nerve. The fascia lining the deep surface of the fleshy portion of the latissimus is similar to that of the trapezius, but over the aponeurosis of the latissimus it loses its identity as a separate layer and is fused with the thoracolumbar fascia.

The rhomboid muscles and the levator

scapulae are enclosed in their own proper fascial sheaths, which are attached to the vertebrae and to the vertebral border of the scapula. The fascia along the superficial surface of these muscles, after attaching to the border of the scapula, continues laterally as the supra- and infraspinatus fasciae. The fascia deep to these muscles continues laterally on the deep surface of the serratus anterior, and a distinct fascial cleft separates both superficial and deep surfaces of the rhomboids from contiguous structures. The fascia surrounding the levator scapulae is more adherent to surrounding structures and is continuous with the scalene fascia over the anterolateral aspect of the neck.

THE TRAPEZIUS.   The trapezius is the flat and triangular muscle that covers the posterior part of the neck, shoulders, and thorax (Fig. 6-42). It *arises* from the external occipital protuberance and the medial third of the superior nuchal line of the occipital bone, from the ligamentum nuchae, the spinous process of the seventh cervical, and the spinous processes of all the thoracic vertebrae, and from the corresponding supraspinal ligaments. From this origin, the superior fibers proceed downward and laterally, the inferior upward and laterally, and the middle horizontally. The superior fibers are *inserted* into the posterior border of the lateral third of the clavicle; the middle fibers into the medial margin of the acromion and into the superior lip of the posterior border of the spine of the scapula; the inferior fibers converge near the scapula and end in an aponeurosis that glides over the smooth triangular surface on the medial end of the scapular spine and are inserted into a tubercle at the apex of this surface. At its occipital origin, the trapezius is connected to the bone by a thin fibrous lamina, firmly adherent to the skin. The middle part is connected to the spinous processes by a long triangular aponeurosis that reaches from the sixth cervical to the third thoracic vertebra, and joins with that of the opposite muscle to form a diamond-shaped tendinous area, observable as a hollow area in the middle of the back of the living subject. The rest of the muscle arises by numerous short tendinous fibers. The two trapezius muscles together resemble a trapezium, or diamond-shaped quadrangle: the two lateral angles correspond to the shoulders; the third to the occipital protuberance superiorly; and the fourth to the spinous process of the twelfth thoracic vertebra inferiorly.

Action.   The **trapezius muscle**, especially the upper and lower fibers, functions most importantly in *rotation of the scapula,* so that the glenoid fossa is made to face upward as is necessary in full abduction of the upper extremity. This movement is performed by the upper fibers pulling superiorly on the acromion and lateral aspect of the scapular spine and the lower fibers pulling inferiorly on the medial aspect of the scapular spine. Rotation is assisted by the serratus anterior muscle.

The upper part acting alone *elevates the shoulder,* thereby bracing the shoulder girdle when a weight is being carried either on the shoulder or by the hand (Mortensen and Wiedenbauer, 1952; Bearn, 1961). The lower part *draws the scapula downward.* With both muscles acting together, the scapulae can be adducted and the head drawn directly backward.

Nerve.   The trapezius muscle receives its motor nerve supply from the spinal part of the **spinal accessory nerve.** It also receives sensory branches (proprioceptive) from the ventral primary divisions of the **third and fourth cervical nerves.**

Variations.   Variations in the structure of the trapezius muscle can occur. Its attachments to the thoracic vertebrae are often reduced and the inferior ones absent. The occipital attachment may be small or wanting, and the clavicular attachment may be reduced, but is more often enlarged, covering most of the posterior triangle. The cervical and thoracic portions may be separate. The muscles of the two sides are seldom symmetrical, and aberrant bundles are not uncommon. Complete absence of the trapezius has been described.

LATISSIMUS DORSI.   The latissimus dorsi is a large triangular muscle that covers the lumbar and lower half of the posterior thoracic region (Fig. 6-42). Its *origin* is principally in a broad aponeurosis, which joins the posterior layer of the thoracolumbar fascia (Fig. 6-18). By this means it is attached to the spinous processes of the lower six thoracic, the lumbar, and the sacral vertebrae; to the supraspinal ligament; and to the posterior part of the crest of the ilium. It also *arises* by muscular fasciculi from the external lip of the crest of the ilium lateral to the margin of the erector spinae, and from the caudal three or four ribs by fleshy digitations that are interposed between similar processes of the external abdominal oblique muscle (Fig. 6-26). From this extensive origin, the fasciculi converge toward the upper humerus. Those of the upper part are almost horizontal and, as they pass over the inferior angle of the scapula, are joined by additional fasciculi

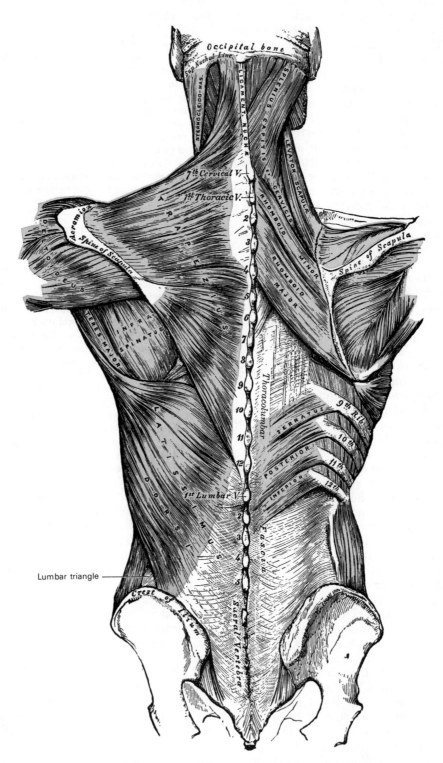

Lumbar triangle

FIG. 6-42.   Muscles connecting the upper limb to the vertebral column.

arising from this bone. The muscle curves around the lower border of the teres major, twisting upon itself so that the superior fibers become initially posterior and then inferior, and the vertical fibers at first are anterior and then are superior. The muscle ends in a quadrilateral tendon about 7 cm long, which passes anterior to the tendon of the teres major and is *inserted* into the bottom of the intertubercular groove of the humerus. Its insertion extends farther proximally on the humerus than that of the tendon of the pectoralis major. The inferior border of the latissimus tendon is united with that of the teres major, their tendons being separated near their insertions by a bursa underlying the tendon of the latissimus dorsi (Fig. 6-47). Another bursa is sometimes interposed between the muscle and the inferior angle of the scapula. At its insertion, the latissimus tendon gives off an expansion to the deep fascia of the arm.

Action. The **latissimus dorsi** extends, adducts, and rotates the humerus medially. It is a powerful swimming muscle, allowing the upper limbs to downstroke against the resistance of the water. With its insertion into the upper extremity fixed, this muscle is one of the most important climbing muscles because it elevates the trunk toward the arms when they are stretched above the head.

The latissimus also aids in the forced expiratory movements of sneezing and coughing, as well as in deep inspiration. Additionally, through its attachment onto the humerus, it can rotate the scapula downward, such movement being required when the shoulders must bear the weight of the body in the use of crutches.

Nerve. The latissimus dorsi is supplied by the **thoracodorsal nerve** (C6,7,8) from the posterior cord of the brachial plexus.

Variations. The number of thoracic vertebrae to which the latissimus dorsi is attached varies from four to eight. The number of costal attachments varies, and the muscle fibers may or may not reach the crest of the ilium.

A muscular slip, the **axillary arch**, varying from 7 to 10 cm in length and from 5 to 15 mm in breadth, occasionally springs from the upper border of the latissimus dorsi about the middle of the posterior fold of the axilla. When present, it crosses the axilla in front of the axillary vessels and nerves, and joins the deep surface of the tendon of the pectoralis major, the coracobrachialis, or the fascia over the biceps. Because the axillary arch crosses the axillary artery just above the site usually selected for application of a ligature, it may mislead the surgeon during that procedure. It is present in about 7% of subjects and may be easily recognized by the transverse direction of its fibers.

A fibrous slip usually passes from the inferior border of the tendon of the latissimus dorsi, near its insertion, to the long head of the triceps brachii. This slip is occasionally muscular and is the representative of the *dorsoepitrochlearis brachii* of apes.

Clinical Considerations. The **lumbar triangle** (of Petit[1]) is a small triangular interval that separates the lateral margin of the lower portion of the latissimus dorsi from the external abdominal oblique muscle just above the ilium. The base of the triangle is the iliac crest, and its floor is formed by the internal abdominal oblique muscle. It is occasionally the site of an abdominal hernia.

The **triangle of auscultation** is a landmark of clinical importance associated with the upper portion of the latissimus. This triangle is bounded superiorly by the trapezius, inferiorly by the latissimus dorsi, and laterally by the vertebral border of the scapula. The floor is partly formed by the rhomboid major. If the scapula is drawn anteriorly by folding the arms across the chest, and the trunk bent forward, parts of the sixth and seventh ribs and their interspace become subcutaneous and accessible for auscultation.

RHOMBOID MAJOR. The rhomboid major (Fig. 6-42) *arises* by tendinous fibers from the spinous processes of the second, third, fourth, and fifth thoracic vertebrae and the supraspinal ligament. Its fibers course inferolaterally to be *inserted* by a narrow tendinous arch to the medial border of the scapula between the triangular surface at the root of the scapular spine above and the inferior angle of the scapula below. Usually the muscle fibers terminate in a tendinous arch, which then is attached to the scapula by a thin membrane. Occasionally, however, the muscle fibers are inserted directly into the scapula.

RHOMBOID MINOR. The rhomboid minor (Fig. 6-42) *arises* from the inferior part of the ligamentum nuchae and from the spinous processes of the seventh cervical and first thoracic vertebrae. Its fibers course inferolaterally just superior to the rhomboid major to be *inserted* into the base of the triangular smooth surface at the root of the spine of the scapula. The rhomboid minor is usually separated from the rhomboid major by a slight interval, but the adjacent margins of the two muscles are occasionally united.

Action. The **rhomboid major and minor** muscles adduct the scapula by pulling it medially toward the vertebral column. They also support the scapula by elevating it, especially when weights are placed

[1]Jean Louis Petit (1674–1750): French anatomist and surgeon (Paris).

on the shoulder or carried by the upper extremity. The lower part of the rhomboid major rotates the scapula to depress the lateral angle, assisting in adduction of the upper extremity.

Nerve. The rhomboid muscles are supplied by the **dorsal scapular nerve** from the brachial plexus, containing fibers from the fifth cervical nerve.

Variations. The vertebral and scapular attachments of the two rhomboid muscles vary in extent. A small slip, which is called the **rhomboid occipitalis**, is occasionally found extending from the scapula to the occipital bone.

LEVATOR SCAPULAE. This narrow elongated muscle is situated at the dorsal and lateral part of the neck (Fig. 6-42). It *arises* by tendinous slips from the transverse processes of the atlas and axis and from the posterior tubercles of the transverse processes of the third and fourth cervical vertebrae. It descends deep to the sternocleidomastoid, along the floor of the posterior triangle of the neck, and then under cover of the trapezius to be *inserted* into the vertebral border of the scapula, between the superior angle and the triangular smooth surface at the root of the spine.

Action. The **levator scapulae** elevates the superior angle of the scapula, tends to draw the scapula medially, and rotates it so as to lower the lateral angle. With the scapula fixed, the levator scapulae bends the neck laterally and rotates it slightly toward the same side.

Nerves. The levator scapulae is supplied directly by branches of the **third and fourth cervical nerves** from the cervical plexus. Frequently the lower portion of the muscle receives a branch of the **dorsal scapular nerve**, containing fibers from the fifth cervical nerve.

Variations. The number of vertebral attachments of the levator scapulae varies. A slip may extend to the occipital or mastoid bones; to the trapezius, scalene, or serratus anterior; or to the first or second rib. The muscle may be subdivided into several distinct parts from origin to insertion. A slip from the transverse processes of one or two upper cervical vertebrae to the outer end of the clavicle corresponds to the **levator claviculae** muscle of lower animals. More or less union with the serratus anterior is not uncommon.

MUSCLES CONNECTING
THE UPPER LIMB
TO THE THORACIC WALL

Pectoralis major
Pectoralis minor

Subclavius
Serratus anterior

SUPERFICIAL FASCIA. The superficial fascia of the pectoral region is continuous with that of the neck superiorly, the abdomen inferiorly, and the axilla and upper extremity laterally. The fasciculi of the platysma muscle extend down from the neck for a variable distance between its superficial and deep layers. In the female, the adipose tissue is increased and molded into a rounded mass, which gives the bulk and form to the mammary gland. The parenchyma of the gland is embedded in this fat, and thus the entire gland is located within the superficial fascia. The connective tissue stroma, distributed between the lobes of the gland, is thickened into fibrous bands called the **suspensory ligaments** (of Cooper), which secure the skin to the deep layer of superficial fascia. A fascial cleft separates the superficial fascia from the deep fascia, allowing the mammary gland in its entirety to be easily removed from the underlying pectoralis major muscle. Through these fascial structures, carcinoma of the breast may make its presence known either by pulling on Cooper's ligaments and dimpling the skin like an orange peel, or by interfering with the normal movability of the gland due to adhesions between the superficial and deep fasciae.

PECTORAL FASCIA. The pectoral fascia (Fig. 6-43) is a membranous sheet of deep fascia that consists of external investing fascia superficial to the pectoralis major and a deeper layer enclosing its inner surface. Its bony attachments are consistent throughout with those of the pectoralis major muscle, i.e., to the clavicle and sternum and to the humerus. The external layer is continuous *medially*, across the midline, with the pectoral fascia of the other side; *superiorly* and *laterally*, with the brachial fascia that covers the coracobrachialis and biceps muscles; and *inferiorly*, with the axillary, thoracic, and abdominal investing fasciae. At the more inferior sternocostal and abdominal origins of the pectoralis major muscle, the fascia is aponeurotic and is blended with the sheath of the rectus abdominis. The layer covering the deep surface of the muscle is thickened by a considerable pad of fat in which the thoracoacromial blood vessels and pectoral nerves are embedded. A definite fascial cleft separates this layer from the underlying

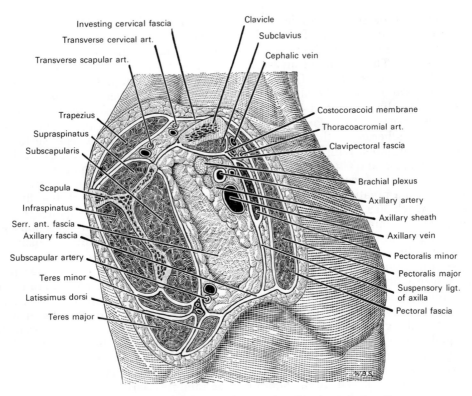

Investing cervical fascia
Transverse cervical art.
Transverse scapular art.
Clavicle
Subclavius
Cephalic vein
Trapezius
Supraspinatus
Subscapularis
Scapula
Infraspinatus
Serr. ant. fascia
Axillary fascia
Subscapular artery
Teres minor
Latissimus dorsi
Teres major
Costocoracoid membrane
Thoracoacromial art.
Clavipectoral fascia
Brachial plexus
Axillary artery
Axillary sheath
Axillary vein
Pectoralis minor
Pectoralis major
Suspensory ligt. of axilla
Pectoral fascia

FIG. 6-43. Fasciae of axillary and pectoral regions in sagittal section.

clavipectoral fascia (Fig. 6-43). The external and the deeper layers fuse at the borders of the muscle, forming a closed compartment. Beyond the superior border, the fascia separates again to enclose the deltoid muscle. Beyond the inferolateral border, however, a single sheet is formed which immediately becomes the **axillary fascia.**

CLAVIPECTORAL FASCIA. The clavipectoral fascia is a stratum of deep fascia that lies between the deep layer of pectoral fascia and the thoracic wall (Fig. 6-43). It invests the pectoralis minor and subclavius muscles and extends from the clavicle to the axillary fascia. It attaches to the clavicle by means of two membranes which lie superficial and deep to the subclavius and are separated from each other by the insertion of the muscle. The two sheets form a compartment for the subclavius by fusing at the muscle's inferior border into a single sheet, which stretches across the interval between the subclavius and the pectoralis minor. Extending laterally to the coracoid process and medially to the first rib, this fascia becomes strengthened by the addition of fibrous bundles to form the **costocoracoid ligament.**

This strong band helps to stabilize and brace the clavicle, thereby serving to strengthen the clavicular articulations both medially and laterally. From its attachment to the coracoid process, the fascia passes downward as a thin sheet, investing the pectoralis minor on both its surfaces and attaching to the ribs with the origin of this muscle. Between the ribs the clavipectoral fascia is continuous with the external intercostal fascia. At the superior border of the pectoralis minor, the two layers investing the muscle fuse into the single sheet that bridges the triangular interval between the pectoralis minor and the subclavius. This single sheet is thin below the ligament and has one or more openings through which pass the thoracoacromial artery and vein, the cephalic vein, and the lateral pectoral nerve. At the inferior or lateral border of the pectoralis minor, the two layers that enclose the muscle combine again into a single sheet, which then passes into the axilla and fuses with the deep surface of the axillary fascia a short distance from the lateral border of the pectoralis major. The axillary sheath, which encloses the axillary vessels and the nerves of the brachial plexus, passes deep to the clavi-pectoral fascia and is separated from it by some of the fat in the deep axilla.

AXILLARY FASCIA.   The portion of investing fascia called the axillary fascia (Fig. 6-43) crosses the interval between the lateral borders of the pectoralis major and latissimus dorsi and dips inward to form the hollow of the armpit. It is adherent to the superficial fascia and is continuous with the pectoral, latissimus, serratus, and brachial fasciae. Fused with its deep surface in the hollow of the armpit is the termination of the clavipectoral fascia, which has continued laterally from the pectoralis minor and is named the **suspensory ligament of the axilla.** It is believed that the hollow of the armpit, seen when the arm is abducted, is produced mainly by the traction of this suspensory ligament on the axillary floor.

THE AXILLA.   The axilla comprises more than the externally visible armpit. It is a region, shaped like a pyramid, between the medial side of the arm and the lateral surface of the chest wall. Its *base* is directed inferolaterally and is formed by the skin of the armpit, while its *apex* is bounded by the approximation of three bones—the first rib, the clavicle, and the coracoid process. The axilla is filled with adipose tissue through which course the important vessels and nerves that supply the upper extremity. Additionally, many lymphatics draining the upper extremity and the anterior chest wall (including the mammary gland) are located in the axilla. The *walls of the axilla* are formed by the fascial coverings over the following muscles: *anteriorly,* the pectoralis major and minor; *posteriorly,* the latissimus dorsi, teres major, and subscapularis; *medially,* the serratus anterior; and *laterally,* the coracobrachialis and biceps (Fig. 6-45). At the apex of the axilla, the adipose padding is continuous with a mass of similar tissue in the posterior triangle of the neck, and the fascia covering the first two ribs and first intercostal space is continuous with the scalene fascia of the cervical region. The latter fascia descends from the neck, deep to the clavicle, with the large vessels and nerves of the axilla to form a tubular investment for these structures called the **axillary sheath.** This sheath, partially adherent to the clavipectoral fascia on the deep surface of the subclavius and pectoralis minor as it passes under them, traverses the lateral wall of the axilla along the coracobrachialis to become fused with the anterior surface of the medial intermuscular septum of the arm.

PECTORALIS MAJOR.   The pectoralis major is a thick, fan-shaped muscle situated on the anterior and superior aspect of the chest (Fig. 6-44). It *arises* from the anterior surface of the sternal half of the clavicle; from half the breadth of the anterior surface of the sternum, as far down as the attachment of the cartilage of the sixth or seventh rib; from the cartilages of all the true ribs, with the exception, frequently, of the first and/or seventh; and from the aponeurosis of the external abdominal oblique muscle. The muscle fibers, from this extensive origin, converge toward the anterior wall of the axilla. The fibers of the **clavicular part** pass obliquely downward and laterally and are usually separated from the rest by a slight interval. The fibers of the **sternocostal part** course horizontally from the mid-sternum and its attached ribs, or ascend from the lower sternum and its associated ribs and costal cartilages. The fibers composing the **abdominal part** ascend almost vertically toward the axilla. All three parts of the muscle end in a flat tendon, about 5 cm broad, which is *inserted* into the lower end of the lateral lip of the intertubercular sulcus of the humerus. This tendon terminates in three laminae, which are oriented longitudinally and insert parallel to each other on the bone, the posterior two blending together below in the form of a U. The anterior lamina is the thickest and receives the clavicular fibers of the muscle. These descend directly and insert in the same order as that by which they arise, that is, the most lateral of the clavicular fibers are inserted at the most superior part of the anterior lamina and the most medial clavicular fibers insert most inferiorly in the lamina. The fibers of the sternocostal part fold upon themselves to form the middle and posterior laminae of insertion, which are continuous with each other inferiorly. The upper sternal fibers from the manubrium form the middle lamina, whereas the sternocostal fibers arising below the sternal angle and the fibers of the abdominal part form the posterior lamina of insertion. Because the lowermost fibers of the pectoralis major insert highest on the posterior lamina and those arising progressively higher on the sternum insert progressively lower on the posterior lamina, the sternocostal and

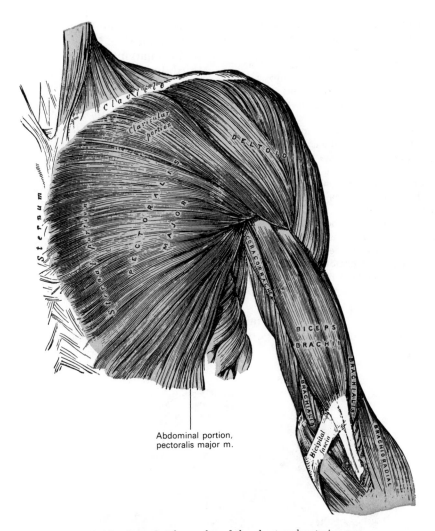

Abdominal portion,
pectoralis major m.

Fɪɢ. 6-44. Superficial muscles of the chest and anterior arm.

abdominal parts of the muscle give the twisted structure of the muscle at the anterior axillary fold. The posterior lamina extends more superiorly on the humerus than either of the other two. From this posterior lamina an expansion is given off which covers the intertubercular groove and blends with the capsule of the shoulder joint. From the deepest fibers of the posterior lamina at its insertion, a second expansion is given off which lines the intertubercular groove, while from the inferior border of the tendon a third expansion passes distally to the fascia of the arm.

Action. The primary action of the **pectoralis major** is adduction of the arm; however, it is also a medial rotator of the humerus, especially if resistance must be overcome (Scheving and Pauly, 1959).

The clavicular portion (along with the coracobrachialis and anterior fibers of the deltoid) helps to flex the humerus and adduct it medially across the midline. In contrast, the sternocostal portion (with the latissimus dorsi and teres major) can extend, against resistance, the flexed humerus to the side of the trunk.

Nerves. The pectoralis major muscle is supplied by the **medial and lateral pectoral nerves** from the brachial plexus, containing fibers from the fifth, sixth, seventh, eighth cervical, and first thoracic nerves.

Variations. The more frequent variations in the structure of the pectoralis major are a greater or lesser extent of attachment to the ribs and sternum, varying size of the abdominal part or its absence, a greater or lesser extent of separation of the sternocostal and clavicular parts, fusion of clavicular part with deltoid, or decussation of the two muscles in front of the sternum. Deficiency or absence of the sternocostal part is not uncommon, but absence of the clavicular part is less frequent; rarely, the whole muscle is wanting.

The **costocoracoid** is a muscular band occasionally found arising from the ribs or aponeurosis of the external oblique between the pectoralis major and latissimus dorsi and inserting onto the coracoid process.

The **chondroepitrochlearis** is a muscular slip occasionally found arising from the costal cartilages or from the aponeurosis of the external oblique below the pectoralis major or from the pectoralis major itself. The insertion is variable onto the inner side of the arm to the brachial fascia, intermuscular septum, or medial epicondyle.

The **sternalis** is a small superficial muscle at the sternal end of the pectoralis major at right angles to its fibers and parallel with the margin of the sternum. It is supplied by the pectoral nerves, and when it is present, it is probably a misplaced part of the pectoralis.

PECTORALIS MINOR.   The pectoralis minor is a thin, triangular muscle, situated on the upper part of the thorax, deep to the pectoralis major (Fig. 6-45). It *arises* from the superior margins and outer surfaces of the third, fourth, and fifth ribs, near their cartilages, and from the aponeuroses covering the intercostal muscles. The fibers pass upward and laterally and converge to form a flat tendon, which is *inserted* into the medial border and superior surface of the coracoid process of the scapula.

Action. The **pectoralis minor** draws the scapula forward and downward, a movement called protraction. It also rotates the scapula so as to lower the lateral angle during adduction of the arm. If the scapula is fixed by the levator scapulae, the pectoralis minor raises the third, fourth, and fifth ribs during forced inspiration.

Nerve. The pectoralis minor muscle is supplied by the **medial pectoral nerve** from the brachial plexus, containing fibers from the eighth cervical and first thoracic nerves.

Variations. The origin of the pectoralis minor

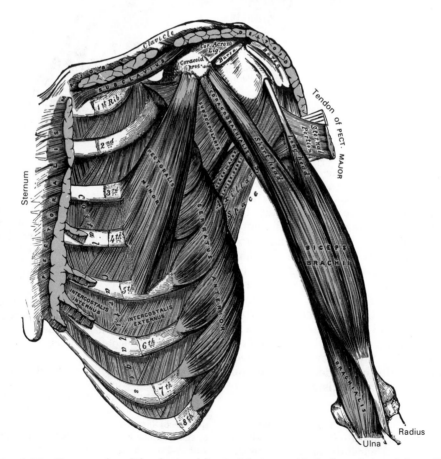

FIG. 6-45.   Deep muscles of the chest and front of the arm, with the boundaries of the axilla.

may extend from the second through the fifth ribs. The tendon of insertion may extend over the coracoid process to the greater tubercle of the humerus. The muscle may be split into several parts, but its complete absence is rare.

The **pectoralis minimus** is a slip from the cartilage of the first rib to the coracoid process and is found rarely.

THE SUBCLAVIUS. This small cylindrical muscle is placed between the clavicle and the first rib (Fig. 6-45). It *arises* by a short, thick tendon from the first rib and its cartilage at their junction. The fleshy fibers proceed obliquely upward and laterally to be *inserted* into the groove on the inferior surface of the middle third of the clavicle between the costoclavicular and conoid ligaments.

Action. The **subclavius** depresses and pulls forward the lateral end of the clavicle, thereby helping to stabilize that bone during movements at the shoulder joint.

Nerve. The subclavius is supplied by a special nerve from the lateral trunk of the brachial plexus which contains fibers from the **fifth and sixth cervical nerves.**

Variations. Insertion of the subclavius may occur onto the coracoid process of the scapula instead of the clavicle or onto both bones.

At times a slip of muscle called the sternoclavicularis extends from the manubrium to the clavicle between the pectoralis major and coracoclavicular fascia.

SERRATUS ANTERIOR. The serratus anterior is a thin muscular sheet that is situated between the ribs and the scapula, spreading over the lateral part of the chest (Fig. 6-45). It *arises* by fleshy digitations from the outer surfaces and superior borders of the first eight or nine ribs, and from the aponeuroses covering the intervening intercostal muscles. Each digitation (except the first) arises from the corresponding rib. The first springs from the first and second ribs and from the fascia covering the first intercostal space. The lower four slips interdigitate at their origins with the upper five slips of the external abdominal oblique muscle. From this extensive attachment the fibers pass backward, closely applied to the chest wall, to be *inserted* onto the ventral surface of the medial (vertebral) border of the scapula. The first digitation is inserted into a triangular area on the anterior surface of the superior angle. The next two digitations spread out to form

a thin, triangular sheet, the base of which is directed posteriorly and is inserted into nearly the whole length of the anterior surface of the vertebral border. The lower five or six digitations converge like a fan, the apex of which is inserted by muscular and tendinous fibers into a triangular impression on the anterior surface of the inferior angle.

Action. The **serratus anterior** draws the medial border of the scapula anteriorly close to the thoracic wall, thereby preventing the bone from protruding backward (winging) in reaching and pushing movements. Its upper fibers help suspend the scapula, while the lower fibers assist in upward rotation of the scapula by laterally rotating the inferior angle. This latter action raises the point of the shoulder, thereby helping in abduction of the arm. During the first 90° of abduction of the humerus, the serratus anterior helps to brace the scapula, allowing the deltoid to abduct the arm. When the arm is abducted to 90°, further abduction to 180° (raising the arm straight overhead) can only be accomplished by lateral rotation of the inferior angle of the scapula (i.e., upward rotation of the glenoid fossa), performed by the serratus anterior below and by both the upper and lower fibers of the trapezius.

Nerve. The serratus anterior is supplied by the **long thoracic nerve** from the brachial plexus, containing fibers from the fifth, sixth, and seventh cervical nerves.

Variations. Varying attachments of the serratus anterior include attachment to the tenth rib, absence of attachment to the first rib or to one or more of the lower ribs, attachment with the levator scapulae, external intercostals, or external abdominal oblique muscles. The muscle may be divided into three parts, and the middle part may be defective or absent.

MUSCLES OF THE SHOULDER

Deltoid
Subscapularis
Supraspinatus
Infraspinatus
Teres minor
Teres major

These six muscles of the shoulder all extend from the scapula to the humerus and act on the shoulder joint. The deltoid, which also attaches to the clavicle, overlies the other five and itself is covered by its deep fascial layer, the deltoid fascia.

DELTOID FASCIA. This deltoid portion of the investing fascia covers the deltoid muscle and is attached superiorly to the clavicle, acromion, and spine of the scapula. Inferiorly, it is continuous with the brachial fascia, and anteriorly, it bridges the narrow triangular region between the adjacent borders of the deltoid and the pectoralis major. In this interval, called the **deltopectoral triangle,** the fascia is thick but is pierced by the cephalic vein and deltoid branch of the thoracoacromial artery. The deltoid fascia is stronger posteriorly and is continuous with the infraspinatus fascia. Along both borders of the deltoid, the investing layer fuses with the fascia lining the deep surface of the muscle to form a closed compartment. The deep layer may contain adipose tissue and is separated by a distinct fascial cleft from the underlying humerus, subdeltoid bursa, shoulder joint, and associated tendons and ligament.

THE DELTOID. The deltoid is a large, thick, triangular muscle that covers the shoulder joint anteriorly, posteriorly, and laterally (Fig. 6-44). It *arises* from the anterior and superior surfaces of the lateral third of the clavicle, from the lateral margin and superior surface of the acromion, and from the lower lip of the posterior border of the spine of the scapula, as far dorsally as the triangular surface at its medial end. From this extensive origin the fibers converge toward their insertion, the middle passing vertically, the anterior obliquely backward and laterally, and the posterior obliquely forward and laterally. They unite in a thick tendon, which is *inserted* into the deltoid tuberosity on the lateral aspect of the body of the humerus, midway along the shaft. At its insertion the muscle gives off an expansion to the deep fascia of the arm. The deltoid is remarkably coarse in texture, and the arrangement of its fibers is somewhat peculiar. The central portion of the muscle, arising from the acromion, consists of oblique fibers that are multipennate in form. These spring from the sides of interspersed tendons, generally four in number, which are attached above to the acromion and pass distally parallel to one another in the substance of the muscle. The fibers are attached to these four tendons above, and then attached to three similarly interspersed tendons which ascend from the deltoid tuberosity below. The portions of the muscle arising from the clavicle and spine of the scapula are not arranged in this manner, but attach directly into the margins of the tendon of insertion.

Action. The principal function of the **deltoid muscle** is abduction of the arm, an action that is initiated with the supraspinatus. Abduction is effected primarily by the multipennate acromial portion of the muscle, with the anterior and posterior fibers bracing the limb to steady the abducted arm. The clavicular portion aids the pectoralis major in flexing the arm, and the spinous portion aids the latissimus dorsi in extending the arm.

Nerve. The deltoid is supplied by the **axillary nerve** from the brachial plexus, containing fibers from the fifth and sixth cervical nerves.

Variations. Large variations in the structure of the deltoid muscle are uncommon, but more or less splitting is common. Continuation into the trapezius, fusion with the pectoralis major, additional slips from the vertebral border of the scapula, infraspinous fascia, and axillary border of scapula are not uncommon variations. Its insertion varies in extent and rarely is prolonged to the origin of the brachioradialis.

SUBSCAPULAR FASCIA. The subscapular fascia is a thin membrane attached to the entire circumference of the subscapular fossa, affording attachment by its deep surface to some of the fibers of the subscapularis muscle.

THE SUBSCAPULARIS. The subscapularis (Fig. 6-45) is a large triangular muscle that fills the subscapular fossa of the scapula; it *arises* from the medial two-thirds of this fossa and from the lower two-thirds of the groove on the lateral axillary border of the bone. Some fibers arise from tendinous laminae that intersect the muscle and are attached to ridges on the bone; others arise from an aponeurosis that separates the muscle from the teres major and the long head of the triceps brachii. The fibers pass laterally and, gradually converging, end in a tendon that is *inserted* into the lesser tubercle of the humerus and the anterior part of the capsule of the shoulder joint. The tendon of the muscle is separated from the neck of the scapula by a large bursa, which usually communicates with the cavity of the shoulder joint through an aperture in the capsule (Fig. 6-47).

Action. The **subscapularis**, acting with the other scapular muscles of the musculotendinous rotator cuff (supraspinatus, infraspinatus, teres minor), gives stability to the shoulder joint by maintaining the humeral head in the glenoid fossa. It especially helps to prevent anterior displacement of the humerus. When the arm is by the side of the body, the subscapularis acts as a medial rotator of the humerus.

Nerves. The subscapularis is supplied by the **upper and lower subscapular nerves** from the brachial plexus, containing fibers from the fifth and sixth cervical nerves.

SUPRASPINATUS FASCIA. The supraspinatus fascia completes the osseofibrous sheath in which the supraspinatus muscle is contained. It affords attachment, on its deep surface, to some of the fibers of the muscle. The fascia is thick medially, but thinner laterally under the coracoacromial ligament.

THE SUPRASPINATUS. This muscle occupies the whole of the supraspinatus fossa, *arising* from its medial two-thirds and from the strong supraspinatus fascia (Fig. 6-46). The muscular fibers converge to form a tendon that crosses the upper part of the shoulder joint, deep to the acromion, and is *inserted* into the highest of the three impressions on the greater tubercle of the humerus. The tendon is closely adherent to the tendon of the infraspinatus and to the capsule of the shoulder joint.

Action. The **supraspinatus**, with the other muscles forming the rotator cuff, especially strengthens the shoulder joint by drawing the humerus toward the glenoid fossa. Along with the deltoid, the supraspinatus assists and, perhaps, initiates abduction of the arm, and it is also a lateral rotator of the humerus.

Nerve. The supraspinatus is supplied by branches of the **suprascapular nerve** from the brachial plexus that contain fibers from the fifth cervical nerve.

INFRASPINATUS FASCIA. This fascia is a dense fibrous membrane that covers the infraspinatus muscle and is fixed to the circumference of the infraspinatus fossa. It affords attachment, by its deep surface, to some fibers of that muscle and is intimately attached to the deltoid fascia along the overlapping border of the deltoid muscle.

THE INFRASPINATUS. This thick triangular muscle occupies most of the infraspinatus fossa (Fig. 6-46). It *arises* by fleshy fibers from the medial two-thirds of the fossa and by tendinous fibers from the ridges on its

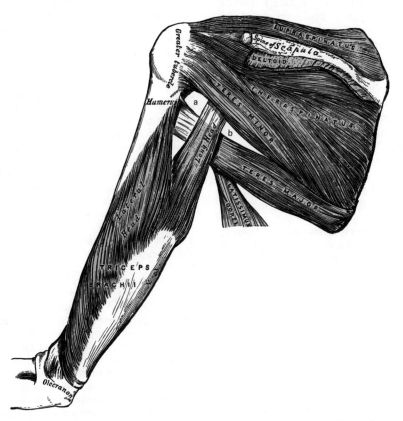

FIG. 6-46. Muscles on the posterior aspect of the scapula, and the triceps brachii, showing the quadrangular, *a*, and triangular, *b*, spaces between the two teres muscles.

surfaces It also arises from the infraspinatus fascia, which covers the muscle and separates it from the teres major and minor. The fibers converge to form a tendon that glides over the lateral border of the spine of the scapula and, passing across the posterior part of the capsule of the shoulder joint, is *inserted* into the middle impression on the greater tubercle of the humerus, where it is sometimes fused with its neighbors. The tendon of this muscle is sometimes separated from the capsule of the shoulder joint by a bursa, which may communicate with the joint cavity.

Action. The **infraspinatus,** as one of the muscles forming the rotator cuff, helps to stabilize the shoulder joint by bracing the head of the humerus in the glenoid fossa. It is also a lateral rotator of the humerus.

Nerve. The infraspinatus is supplied by the **suprascapular nerve** from the brachial plexus, which contains fibers from the fifth and sixth cervical nerves.

TERES MINOR. The teres minor is a cylindrical, elongated muscle that *arises* from the upper two-thirds of the dorsal surface of the lateral or axillary border of the scapula and from two aponeurotic laminae, one of which separates it from the infraspinatus, the other from the teres major (Fig. 6-46). Its fibers course obliquely upward and laterally. The upper fibers end in a tendon that is *inserted* into the most inferior of the three impressions on the greater tubercle of the humerus, and the lower fibers are inserted directly into the humerus immediately distal to this impression. The tendon of this muscle passes across, and is united with, the posterior part of the capsule of the shoulder joint.

Action. The **teres minor** rotates the arm laterally and weakly adducts it. As do the other muscles forming the rotator cuff, it draws the humerus toward the glenoid fossa, strengthening the shoulder joint.

Nerve. The teres minor is supplied by a branch of the **axillary nerve,** which comes from the posterior cord of the brachial plexus and contains fibers from the fifth cervical root.

Variations. The teres minor is sometimes inseparable from the infraspinatus.

TERES MAJOR. The teres major is a thick but somewhat flattened muscle that *arises* from the oval area on the dorsal surface of the inferior angle of the scapula and from the fibrous septa interposed between the muscle and the teres minor and infraspinatus. The fibers are directed superiorly and laterally and end in a flat tendon, about 5 cm long, which is *inserted* into the medial lip of the intertubercular sulcus of the humerus. The tendon, at its insertion, lies beneath that of the latissimus dorsi, from which it is separated by a bursa (Fig. 6-47), except where the two tendons unite along their lower borders for a short distance.

Action. The **teres major** adducts as well as assists in extending the arm when it is in a flexed position. It is also a medial rotator. In function it is opposite to those muscles inserting into the rotator cuff.

Nerve. The teres major is supplied by a branch of the **lower subscapular nerve** from the brachial plexus, containing fibers from the fifth and sixth cervical nerves.

Group Action of Muscles About the Shoulder. *Flexion of the arm* is brought about by the anterior part of the deltoid, coracobrachialis, and short head of the biceps, acting at the shoulder joint, and by the trapezius and serratus anterior rotating the inferior angle of the scapula laterally on the posterior thoracic wall to raise the point of the shoulder. The clavicular part of the pectoralis major also assists in flexion until the arm is raised above the shoulder.

Extension of the arm is brought about by the latissimus dorsi, acting on both the shoulder joint and the scapula, and it is strongly assisted by the sternocostal part of the pectoralis major except in hyperextension. The teres major, posterior deltoid, and long head of the triceps extend the arm at the shoulder joint. The *scapula is protracted* (rotated forward and downward) by the pectoralis minor and *retracted* (drawn backward) by the rhomboids and trapezius.

Abduction of the arm is carried out by the deltoid and supraspinatus, acting on the shoulder joint, and by the trapezius and serratus anterior rotating the scapula to raise the shoulder.

Adduction of the arm is performed principally by the pectoralis major and latissimus dorsi, and by the coracobrachialis and teres major acting on the shoulder joint, by the pectoralis minor protracting the scapula, and by the lower portion of the trapezius, which draws the scapula downward.

Medial rotation of the humerus is brought about primarily by the subscapularis and teres major when it is performed as a voluntary act, but the pectoralis major and latissimus dorsi have strong medial rotating power incidental to their contraction.

Lateral rotation of the humerus is performed by the infraspinatus, supraspinatus, and teres minor.

Independent movements of the scapula are *elevation* by the levator scapulae and upper part of the trapezius; *depression* by the pectoralis minor, lower trapezius, and lower serratus anterior; *abduction of*

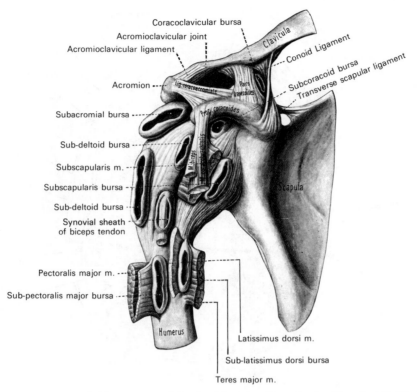

FIG. 6-47. Muscular insertions and bursae around the shoulder joint. (From Rauber-Kopsch, 19th Ed., Vol. 1, Georg Thieme Verlag, Stuttgart.)

the scapula (drawing it forward) by the serratus anterior, as in pushing; *adduction of the scapula* (drawing it backward) by the rhomboids and trapezius.

MUSCLES OF THE ARM

The arm is defined as that part of the upper limb extending between the shoulder joint and the elbow joint. The arm contains four muscles:

Coracobrachialis
Biceps brachii
Brachialis
Triceps brachii

BRACHIAL FASCIA.   The portion of the investing fascia that covers the arm is a strong membrane, but it is not, for the most part, distinctly aponeurotic. It is called the **brachial fascia** and is continuous proximally with the deltoid, pectoral, and axillary fasciae. Attaching distally to the epicondyles of the humerus and the olecranon, the brachial fascia is then continuous with the antebrachial fascia. Beginning at the attachment to the

epicondyles and prolonged proximally into the arm are two intermuscular septa, medial and lateral, which divide the arm into flexor and extensor compartments (Fig. 6-48). The **lateral brachial intermuscular septum** is attached along the lateral supracondylar ridge and is fused with the internal surface of the investing brachial fascia. The septum extends distally to the lateral epicondyle and proximally to the insertion of the deltoid, where it continues into the deltoid fascia. Its posterior surface is used by the triceps for the origin of some of its fibers, while the anterior surface gives origin to some fibers of the brachialis, brachioradialis, and extensor carpi radialis longus. Its distal portion is pierced by the radial nerve and the radial collateral branch of the deep brachial artery. The **medial brachial intermuscular septum** is attached to the medial supracondylar ridge below, and it extends from the medial epicondyle distally to the insertions of the teres major and latissimus dorsi proximally. Some fibers of the triceps originate from its dorsal surface, and some fibers of the brachialis from its ventral surface. It is

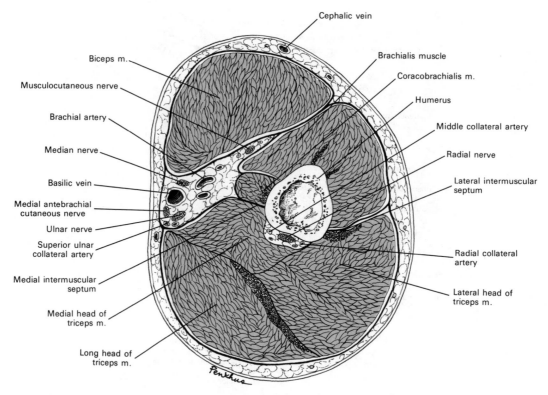

FIG. 6-48. Cross section through the middle of the right arm. Viewed from the superior aspect.

pierced near the epicondyle by the ulnar nerve, the superior ulnar collateral artery, and the posterior branch of the inferior ulnar collateral artery. The medial septum appears much thicker than the lateral because the axillary sheath, containing the main vessels and nerves of the arm, blends with its ventral surface, and the nerves and vessels continue this close association down to the elbow. The two intermuscular septa and the investing fascia of the posterior aspect of the arm form the posterior or **extensor compartment,** which contains the triceps, radial nerve, and deep brachial artery (Fig. 6-48). The anterior or **flexor compartment** contains the biceps, the brachialis, part of the coracobrachialis, the brachial vessels, and the median and ulnar nerves.

The relationship of the investing fascia to the muscles is different on the dorsal and ventral aspects of the arm. The fascia over the triceps is adherent to the muscle and is used in part for its origin. In contrast, the fascia over the biceps is separated from the muscle by a distinct fascial cleft which continues around its deep surface, thereby also separating the biceps from the brachialis.

Just below the middle of the arm, the anterior investing fascia is pierced medially by the basilic vein and at several sites by filaments of the brachial cutaneous nerves.

THE CORACOBRACHIALIS. The coracobrachialis is the smallest of the three muscles in this region (Figs. 6-45; 6-49). Situated at the upper and medial part of the arm, it *arises* from the apex of the coracoid process, in common with the short head of the biceps brachii, and from the intermuscular septum between the two muscles. It is *inserted* by means of a flat tendon into an impression at the middle of the medial surface and border of the body of the humerus between the origins of the triceps brachii and brachialis. It is perforated by the musculocutaneous nerve.

Action. The **coracobrachialis** muscle flexes and adducts the arm.

Nerve. The coracobrachialis is supplied by a branch of the **musculocutaneous nerve** that contains fibers from the sixth and seventh cervical nerves.

Variations. The most frequent variation in the attachment of the coracobrachialis is for the most superficial portion to reach as far as the medial epicondyle. More rarely a short head is found inserting on the lesser tubercle.

BICEPS BRACHII.   The biceps brachii is a long fusiform muscle, placed on the anterior aspect of the arm, that consists of two heads, a fact implied by its name (Figs. 6-45; 6-48; 6-49). The **short head** *arises* by a thick flattened tendon from the apex of the coracoid process, in common with the coracobrachialis. The **long head** *arises* within the capsule of the shoulder joint from the supraglenoid tuberosity at the upper margin of the glenoid cavity, and is continuous with the glenoidal labrum. This tendon, enclosed in a special sheath of the synovial membrane of the shoulder joint, arches over the head of the humerus. It emerges from the capsule through an opening between the two tubercles, deep to the transverse humeral ligament, and descends in the intertubercular sulcus. It is retained in the sulcus by the transverse humeral ligament and by a fibrous prolongation from the tendon of the pectoralis major. The tendon of each head is succeeded by an elongated muscular belly, and the two bellies, although closely applied to each other, can readily be separated down to about 7 cm above the elbow joint. Here they end in a flattened tendon, which is *inserted* into the rough posterior surface of the

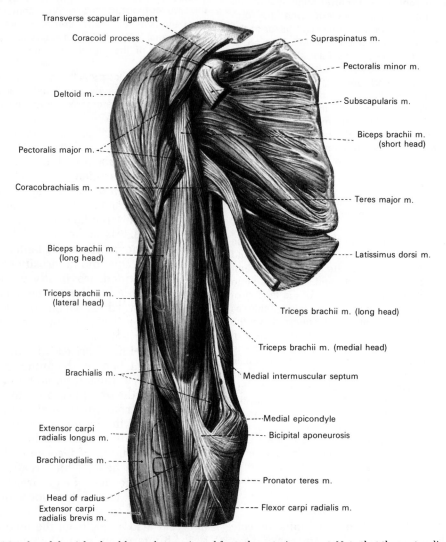

FIG. 6-49.   Muscles of the right shoulder and arm, viewed from the anterior aspect. Note that the pectoralis major and minor muscles have been cut near their insertions to reveal the underlying structures. (From Benninghoff and Goerttler, *Lehrbuch der Anatomie des Menschen,* Urban & Schwarzenberg, 11th Ed., Vol. 1, 1975.)

tuberosity of the radius, a bursa being interposed between the tendon and the anterior part of the tuberosity. As the tendon of the muscle approaches the radius it is twisted, so that its anterior surface becomes lateral at its insertion to the tuberosity of the radius. Opposite the bend of the elbow the tendon gives off a broad aponeurosis from its medial side, the **bicipital aponeurosis** (lacertus fibrosus), which passes obliquely downward and medially across the brachial artery and is continuous with the deep fascia covering the origins of the flexor muscles of the forearm (Fig. 6-49).

Action. When the forearm is partially flexed, the **biceps brachii** is a powerful supinator of the forearm. This action is less powerful when the forearm is extended. It is also an important flexor of the supinated forearm, and upon continued action, it can flex, although not so powerfully, the arm at the shoulder joint. The biceps does not appear to act as a flexor of the pronated forearm.

One can conceive of the action of the biceps in the use of a screwdriver, placed in the right hand and turning a screw in a clockwise manner. It has also cleverly been described as the muscle that puts in the corkscrew and pulls out the cork. The tendon of the long head also helps to prevent the elevation of the head of the humerus during abduction by the deltoid.

Nerves. The biceps brachii is supplied by branches of the **musculocutaneous nerve** that contain fibers from the fifth and sixth cervical nerves.

Variations. A third head to the biceps brachii frequently is noted (10%), arising at the upper and medial part of the brachialis. Its fibers are continuous with those of the brachialis and are inserted into the bicipital fascia and medial side of the tendon of that muscle. In most cases, this additional slip lies deep to the brachial artery in its course down the arm. In some instances the third head consists of two slips, which pass downward, one superficial and the other deep to the artery, concealing the vessel in the lower half of the arm. More rarely, a fourth head is seen arising from the lateral aspect of the humerus, from the intertubercular groove, or from the greater tubercle. Other heads are occasionally found. Slips sometimes pass from the inner border of the muscle over the brachial artery to the medial intermuscular septum or the medial epicondyle, or more rarely to the pronator teres or brachialis. The long head may be absent or may arise from the intertubercular groove.

THE BRACHIALIS. The brachialis muscle covers the lower half of the humerus and the anterior aspect of the elbow joint (Figs. 6-45; 6-48; 6-49). It *arises* from the distal half of the anterior surface of the humerus, commencing at the insertion of the deltoid, which it embraces by two angular processes. Its origin extends distally to within 2.5 cm of the margin of the articular surface of the elbow joint. It also arises from the intermuscular septa, but more extensively from the medial than the lateral. Distally, it is separated from the lateral intermuscular septum by the brachioradialis and extensor carpi radialis longus. Its fibers converge to a thick tendon, which is *inserted* into the tuberosity of the ulna and the rough depression on the anterior surface of the coronoid process.

Action. The **brachialis** is a powerful flexor of the forearm.

Nerve. The brachialis is supplied by a branch of the **musculocutaneous nerve**, containing fibers from the fifth and sixth cervical nerves. Usually a small branch of the radial nerve supplies some of the lateral part of the muscle; this may be a sensory nerve.

Variations. Occasionally the brachialis may be doubled or, more rarely, additional muscular slips to the supinator, pronator teres, biceps, bicipital fascia, or radius are found.

TRICEPS BRACHII. The triceps is situated on the dorsal aspect of the arm within the extensor compartment and extends along the entire length of the dorsal surface of the humerus (Figs. 6-46; 6-48). It is a large muscle that arises by three heads: long, lateral, and medial, hence its name.

The **long head** *arises* by a flattened tendon from the infraglenoid tuberosity of the scapula. It is blended proximally with the capsule of the shoulder joint. The muscle fibers pass downward between the two other heads of the muscle, joining them in the tendon of insertion.

The **lateral head** *arises* from the posterior surface of the body of the humerus, between the insertion of the teres minor and the proximal part of the oblique groove for the radial nerve, and from the lateral border of the humerus and the lateral intermuscular septum. From this origin, the fibers converge toward the tendon of insertion.

The **medial head,** which really might be considered the deep head, *arises* from the posterior surface of the body of the humerus, distal to the groove for the radial nerve. It also arises from the medial border of the humerus and from the whole length of the medial intermuscular septum. The medial head is narrow and pointed proximally, and it extends from the insertion of the teres major to within 2.5 cm of the trochlea. Some of the fibers are directed downward to the olecranon, while others converge to the tendon of insertion.

The tendon of the triceps begins about the middle of the muscle. It consists of two aponeurotic laminae, one of which is immediately subcutaneous and covers the superficial surface of the distal half of the muscle, and the other more deeply seated in the substance of the muscle. After receiving the attachment of the muscular fibers, the two tendinous layers join a short distance above the elbow, and are *inserted,* for the most part, into the posterior portion of the proximal surface of the olecranon. A band of fibers is continued downward, on the lateral side, over the anconeus muscle to blend with the deep fascia of the forearm.

Triangular and Quadrangular Spaces. The long head of the triceps descends between the teres minor and teres major, dividing the triangular space between these two muscles and the humerus into two smaller spaces, one triangular and the other quadrangular (Fig. 6-46). The **triangular space** transmits the scapular circumflex vessels. It is bounded *superiorly* by the teres minor; *inferiorly,* by the teres major; and *laterally,* by the long head of the triceps. The **quadrangular space** transmits the posterior humeral circumflex vessels and the axillary nerve. *Superiorly,* it is bounded by the teres minor and capsule of the shoulder joint; *inferiorly,* by the teres major; *medially,* by the long head of the triceps; and *laterally,* by the humerus.

Action. All three heads of the **triceps brachii** extend the forearm at the elbow joint. The long and lateral heads are especially active against resistance. The long head, arising from the scapula, is also able to extend the humerus at the shoulder joint.

Nerves. The triceps brachii is supplied by branches of the **radial nerve** that contain fibers from the seventh and eighth cervical nerves.

Variations. At times this muscle receives a fourth head from the inner part of the humerus. A muscular slip, the **dorsoepitrochlearis brachii** at times extends from the latissimus dorsi to the long head of the triceps.

Articularis Cubiti. Also called the subanconeus, this structure consists only of a few fibers that spring from the deep surface of the lower part of the triceps and are inserted into the posterior aspect of the articular capsule of the elbow joint.

MUSCLES OF THE FOREARM

ANTEBRACHIAL FASCIAE. The portion of the investing fascia that covers the proximal forearm is a strong aponeurotic sheet, closely adherent to the underlying muscles (Fig. 6-50). Proximally, it is continuous with the brachial fascia and is attached to the epicondyles of the humerus and the olecranon. Distally, it is attached to the lower portions of the radius and ulna and is continued into the fascia of the hand. It is attached to the dorsal border of the ulna through most of its length, closing off the flexor and extensor compartments of the forearm. In the proximal two-thirds of the forearm, the underlying muscles utilize the deep surface of the investing fascia for attachment of some muscle fibers. The fascia also extends deep between the muscles toward the radius and ulna, forming strong intermuscular septa which also allow attachment of muscle fibers. The anterior or volar aspect of the antebrachial fascia is visibly thickened by collagenous bundles derived from the tendon of the biceps, which fans out medially as an aponeurosis of the biceps brachii muscle. On the posterior aspect of the forearm, the fascia is even thicker, receiving similar collagenous bundles from the triceps tendon. Near the distal ends of the radius and ulna, the fascia is abruptly thickened by the addition of prominent, annular, collagenous bundles, which form the **flexor and extensor retinacula.** In the distal third of the forearm, the muscles and tendons are separated from the overlying fascia by fascial clefts, whereas more proximally, the fascia adheres to the underlying muscle tissue. The antebrachial fascia is pierced in several places by vessels and nerves. The largest aperture, anterior to the elbow, is penetrated by the branch of the median cubital vein that communicates with the deep veins in the antecubital fossa. The radius, ulna, and interosseous membrane form a septum dividing the forearm into an anterior flexor and a posterior extensor compartment.

In the **extensor compartment,** an intermuscular fascial cleft separates the superficial from the deep muscles, especially near the wrist, but it is closed medially and laterally and does not communicate with other clefts. In the **flexor compartment,** the intermuscular fascial cleft is more extensive and may communicate with the fascial clefts of the palm. This cleft is especially prominent between the anterior surface of the pronator quadratus and the overlying muscles and tendons, and at this site it is called the **anterior interosseous cleft.** It continues proximally between the flexors digitorum profundus and superficialis and may follow

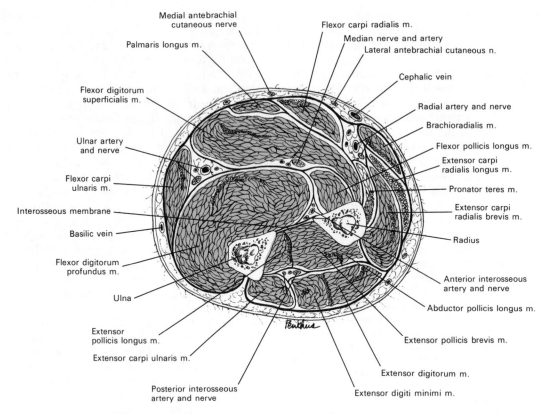

Medial antebrachial cutaneous nerve

Palmaris longus m.

Flexor digitorum superficialis m.

Ulnar artery and nerve

Flexor carpi ulnaris m.

Interosseous membrane

Basilic vein

Flexor digitorum profundus m.

Ulna

Extensor pollicis longus m.

Extensor carpi ulnaris m.

Posterior interosseous artery and nerve

Flexor carpi radialis m.

Median nerve and artery

Lateral antebrachial cutaneous n.

Cephalic vein

Radial artery and nerve

Brachioradialis m.

Flexor pollicis longus m.

Extensor carpi radialis longus m.

Pronator teres m.

Extensor carpi radialis brevis m.

Radius

Anterior interosseous artery and nerve

Abductor pollicis longus m.

Extensor pollicis brevis m.

Extensor digitorum m.

Extensor digiti minimi m.

FIG. 6-50. Transverse section through the middle of the right forearm, viewed from above.

along the ulnar vessels and nerve to the antecubital fossa. At the wrist, the anterior interosseous cleft continues distally deep to the flexor tendon sheaths to communicate with the middle palmar cleft of the hand. The antebrachial or forearm muscles may be divided into an **anterior group,** which occupies the flexor compartment, and a **posterior group** in the extensor compartment.

Anterior Antebrachial Muscles

These muscles are divided for convenience of description into two groups, superficial and deep.

SUPERFICIAL FLEXOR GROUP

Pronator teres
Flexor carpi radialis
Palmaris longus
Flexor carpi ulnaris
Flexor digitorum superficialis

The superficial group of anterior antebrachial muscles forms the rounded contour

on the medial aspect of the upper forearm (Fig. 6-51). They take origin from the medial epicondyle of the humerus by a common tendon, and from the deep fascia of the forearm near the elbow, as well as from the septa, which pass from this fascia between the individual muscles.

PRONATOR TERES. The pronator teres has two heads of origin, humeral and ulnar (Fig. 6-51). The **humeral head,** larger and more superficial, *arises* immediately above the medial epicondyle, and also from the tendon common to the origin of the flexor muscles, from the intermuscular septum between it and the flexor carpi radialis, and from the antebrachial fascia. The **ulnar head** is a thin fasciculus that *arises* from the medial side of the coronoid process of the ulna and joins the humeral head at an acute angle. *The median nerve enters the forearm between the two heads of the muscle and is separated from the ulnar artery by the ulnar head. The muscle passes obliquely across the forearm and ends in a flat tendon that is inserted* on a rough impression about mid-

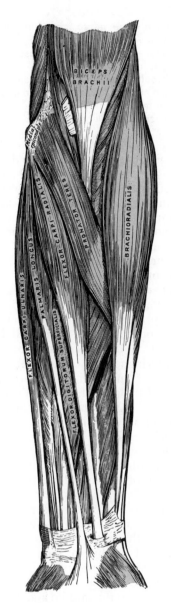

FIG. 6-51.   Anterior aspect of left forearm. Superficial muscles.

way along the lateral surface of the body of the radius. The lateral border of the pronator teres forms the medial boundary of a triangular hollow, the **cubital fossa,** situated on the anterior aspect of the upper extremity in front of the elbow joint. Among other structures, the fossa contains the median cubital vein, the brachial artery, the median nerve, and the tendon of the biceps muscle.

Action.   The **pronator teres** rotates the radius on the ulna, thereby pronating the forearm and the

hand. It also assists in flexing the forearm at the elbow joint.

Nerve.   The pronator teres is supplied by a branch of the **median nerve** that contains fibers from the sixth and seventh cervical nerves.

Variations.   At times the ulnar head may be absent. Additional slips may be derived from the medial intermuscular septum, the biceps, and occasionally from the brachialis muscle.

FLEXOR CARPI RADIALIS.   The flexor carpi radialis lies on the medial or ulnar side of the pronator teres (Figs. 6-50; 6-51). It *arises* from the medial epicondyle by the common flexor tendon, from the antebrachial fascia, and from the intermuscular septa between it and the pronator teres laterally, the palmaris longus medially, and the flexor digitorum superficialis which lies beneath. The muscle is slender and aponeurotic in structure at its origin; it increases in size and ends in a tendon a little more than halfway down the forearm. The tendon passes through a canal in the lateral part of the flexor retinaculum and courses along a groove on the trapezium, which becomes converted into a canal by fibrous tissue and is lined by a synovial sheath. The tendon is *inserted* into the base of the second metacarpal bone and sends a slip to the base of the third metacarpal bone. In the distal part of the forearm, the radial artery lies between the tendon of the flexor carpi radialis and the brachioradialis.

Action.   The **flexor carpi radialis** flexes the hand at the wrist joint and helps to abduct (radially flex) it.

Nerve.   The flexor carpi radialis is supplied by a branch of the **median nerve**, containing fibers from the sixth and seventh cervical nerves.

Variations.   Occasionally slips from the tendon of the biceps, the bicipital fascia, the coronoid, and the radius have been found extending to the flexor carpi radialis. Its insertion often varies and may be extended to the trapezium or the fourth metacarpal. The muscle may be absent.

PALMARIS LONGUS.   The palmaris longus is a slender, fusiform muscle lying on the medial aspect of the flexor carpi radialis (Figs. 6-50; 6-51). It *arises* from the medial epicondyle of the humerus by the common flexor tendon, from the intermuscular septa between it and the adjacent muscles, and from the antebrachial fascia. It ends in a slender, flattened tendon, which passes superficial to the flexor retinaculum and is *inserted* into the anterior aspect of the distal part of the flexor retinaculum and into the palmar aponeurosis; it frequently sends a tendinous slip to the short muscles of the thumb.

Action. The **palmaris longus** flexes the hand at the wrist.

Nerve. The palmaris longus is innervated by a branch of the **median nerve**, containing fibers from the sixth and seventh cervical nerves.

Variations. The palmaris longus is one of the most variable muscles in the body and is often (10%) absent. It may be tendinous proximally and muscular distally, or it may be muscular in the center with a tendon above and below. It may present two muscular bundles with a central tendon and might even consist solely of a tendinous band. The muscle may be doubled. Slips of origin from the coronoid process or from the radius have been seen. Partial or complete insertion into the antebrachial fascia, into the tendon of the flexor carpi ulnaris and pisiform bone, into the scaphoid, or into the hypothenar muscles has been observed.

FLEXOR CARPI ULNARIS. This muscle lies along the ulnar side of the forearm (Figs. 6-50; 6-51). It is formed from two heads, humeral and ulnar, connected by a tendinous arch, beneath which the ulnar nerve descends in the forearm and the posterior ulnar recurrent artery ascends to participate in the anastomosis around the elbow joint. The **humeral head** *arises* from the medial epicondyle of the humerus by the common flexor tendon. The **ulnar head** *arises* from the medial margin of the olecranon and from the upper two-thirds of the posterior border of the ulna by an aponeurosis common to it and the extensor carpi ulnaris and flexor digitorum profundus. It also arises from the intermuscular septum between it and the flexor digitorum superficialis. The fibers of the flexor carpi ulnaris end in a tendon that occupies the anterior part of the lower half of the muscle and, prolonged to the hamate and fifth metacarpal bones by the pisohamate and pisometacarpal ligaments, is *inserted* into the pisiform bone. It is also attached by a few fibers to the flexor retinaculum. Along the distal two-thirds of the forearm, the ulnar vessels and nerve lie on the lateral aspect of the tendon of this muscle.

Action. The **flexor carpi ulnaris** flexes and adducts (ulnar flexes) the hand.

Nerve. The flexor carpi ulnaris is supplied by a branch of the **ulnar nerve,** containing fibers from the eighth cervical and first thoracic nerves.

Variations. Occasionally the muscle receives a slip of origin from the coronoid.

The **epitrochleoanconeus** is a small muscle that is often seen coursing over the ulnar nerve from the back of the medial condyle of the humerus to the medial aspect of the olecranon.

FLEXOR DIGITORUM SUPERFICIALIS. The flexor digitorum superficialis lies deep to the other muscles in the superficial flexor group (Figs. 6-50; 6-51). It is the largest of the muscles of the superficial group and is formed from two heads, humeroulnar and radial. The **humeroulnar head** *arises* from the medial epicondyle of the humerus by the common flexor tendon, from the ulnar collateral ligament of the elbow joint, from the intermuscular septa between it and the preceding muscles, and from the medial aspect of the coronoid process, proximal to the ulnar origin of the pronator teres. The **radial head** is broad but thin. It *arises* from an oblique line on the anterior surface of the radius, and this line extends from the radial tuberosity to the insertion of the pronator teres. From its origin the flexor digitorum superficialis quickly separates into two planes of muscular fibers, superficial and deep. The *superficial plane* divides into two parts, which end in tendons for the middle and ring fingers. The *deep plane* gives off a muscular slip to join the portion of the superficial plane that is associated with the tendon of the ring finger; it then divides into two parts, which end in tendons for the index and little fingers. As the four tendons thus formed pass beneath the flexor retinaculum into the palm of the hand, they are arranged in pairs, the superficial pair going to the middle and ring fingers, the deep pair to the index and little fingers. The tendons diverge in the palm, passing dorsal to the superficial palmar arch and digital branches of the median and ulnar nerves. Opposite the bases of the first phalanges, each tendon divides into two slips to allow the passage to each finger of the corresponding tendon of the flexor digitorum profundus. The two slips then reunite and form a grooved channel for the reception of the accompanying tendon of the flexor digitorum profundus. Finally, each superficialis tendon divides again to be *inserted* on the sides of the second phalanx of the corresponding fingers.

Action. The **flexor digitorum superficialis** flexes the second phalanx of each finger and, by continued action, flexes the first phalanx at the metacarpophalangeal joint and the hand at the wrist joint.

Nerves. The flexor digitorum superficialis is supplied by branches of the **median nerve,** containing fibers from the seventh and eighth cervical and first thoracic nerves.

Variations. The radial head, or little finger portion, of the flexor digitorum superficialis may be absent. Accessory slips may occur from the tuberosity of the ulna to the index and middle finger portions. Another slip may be derived from the humeroulnar head to the flexor profundus, and variations occur in the origin of muscular slips to the little finger.

DEEP FLEXOR GROUP

Flexor digitorum profundus
Flexor pollicis longus
Pronator quadratus

FLEXOR DIGITORUM PROFUNDUS. The flexor digitorum profundus is situated on the ulnar side of the forearm, immediately beneath the superficial flexors (Figs. 6-50; 6-52). It *arises* from the upper three-fourths of the anterior and medial surfaces of the body of the ulna, embracing the insertion of the brachialis above and extending downward to within a short distance of the pronator quadratus. It also *arises* from a depression on the medial aspect of the coronoid process; from the upper three-fourths of the dorsal border of the ulna by way of an aponeurosis in common with the flexor and extensor carpi ulnaris; and from the ulnar half of the interosseous membrane. The muscle ends in four tendons, which enter the hand beneath the transverse carpal ligament, deep to the tendons of the flexor digitorum superficialis. The portion of the muscle for the index finger is usually distinct throughout, but the tendons for the middle, ring, and little fingers are connected by areolar tissue and tendinous slips, as far as the palm of the hand. On the palmar aspect of the first phalanges, the tendons pass through the openings in the tendons of the flexor digitorum superficialis to finally become *inserted* into the bases of the last phalanges. Four small muscles, the lumbricals, are connected with the tendons of the flexor digitorum profundus in the palm. These will be described with the muscles of the hand.

FIBROUS SHEATHS OF THE FLEXOR TENDONS. Distal to the metacarpophalangeal joints, the

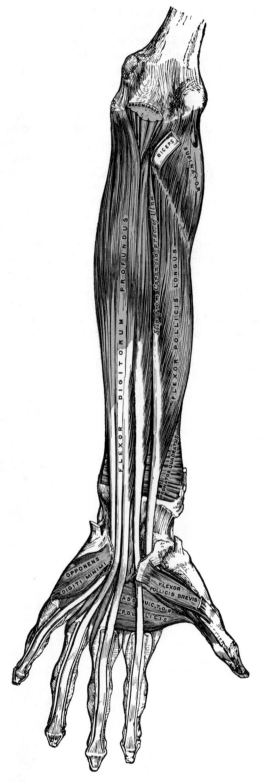

Fig. 6-52.   Anterior aspect of the left forearm. Deep muscles.

tendons of the flexors digitorum superficialis and profundus lie in strong ligamentous tunnels, the digital **fibrous tendon sheaths.** Each tunnel is lined by a **synovial tendon sheath,** which is a lubricated synovial layer reflected on the contained tendons. Within each digital sheath, the tendons of the superficialis and profundus are connected to each other and to the phalanges by tendinous bands called **vincula tendinum** (Fig. 6-53). There are two types of vincula. (1) The **vincula brevia,** two in number in each finger, are fan-shaped expansions near the termination of the tendons. One connects the superficialis tendon to the palmar surface of the proximal interphalangeal joint and the head of the first phalanx, and the other connects the profundus tendon to the palmar surface of the distal interphalangeal joint and the head of the second phalanx. (2) The **vincula longa** are slender, independent bands that are found in two positions. One pair in each finger connects the deep surface of the pro-

fundus tendon to the subjacent superficialis tendon just distal to the site where the profundus tendon traverses the split in the superficialis tendon. Another pair, or a single vinculum longum, connects the superficialis tendon to the proximal end of the first phalanx.

Action. The **flexor digitorum profundus** flexes the terminal phalanx of each finger and, by continued action, flexes the other phalanges and to some extent the hand. This muscle is the largest and most powerful of the forearm muscles.

Nerves. The flexor digitorum profundus is supplied by the anterior interosseous branch of the **median nerve** and a branch of the **ulnar nerve,** containing fibers from the eighth cervical and first thoracic nerves.

Variations. The index finger portion of the flexor digitorum profundus may arise partly from the proximal portion of the radius. Accessory slips may arise from the humeroulnar head of the flexor superficialis, the medial epicondyle, the coronoid process of the ulna, or the flexor pollicis longus.

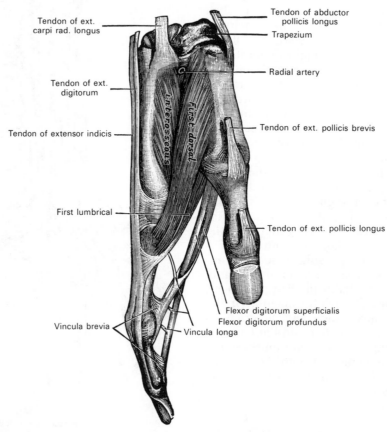

FIG. 6-53. Tendons of the index finger of the right hand, viewed from the lateral (radial) aspect. Note the vincula longa and brevia.

FLEXOR POLLICIS LONGUS. The flexor pollicis longus is situated on the radial side of the forearm, lateral to but on the same plane as the flexor digitorum profundus (Figs. 6-50; 6-52). It *arises* from the grooved volar surface of the body of the radius, which extends from immediately below the tuberosity and oblique line to within a short distance of the pronator quadratus. It also *arises* from the adjacent part of the interosseous membrane and, generally, by a fleshy slip from the coronoid process of the ulna or from the medial epicondyle of the humerus. The fibers end in a flattened tendon, which passes deep to the flexor retinaculum, being then lodged between the lateral part of the flexor pollicis brevis and the oblique part of the adductor pollicis. Entering an osseoaponeurotic canal, similar to those for the flexor tendons of the fingers, the muscle is *inserted* into the base of the distal phalanx of the thumb. The anterior interosseous (branch of the median) nerve and vessels pass downward in the forearm on the anterior aspect of the interosseous membrane between the flexor pollicis longus and flexor digitorum profundus.

Action. The **flexor pollicis longus** flexes the second phalanx of the thumb and, by continued action, may flex the first phalanx at the metacarpophalangeal joint.

Nerve. The flexor pollicis longus is supplied by the anterior interosseous branch of the **median nerve,** containing fibers from the eighth cervical and first thoracic nerves.

Variations. The flexor pollicis longus may be connected by additional muscular slips to the flexor superficialis or profundus, or to the pronator teres. An additional tendon to the index finger is sometimes found.

PRONATOR QUADRATUS. This flat, small, quadrilateral muscle extends across the anterior aspect of the distal parts of the radius and ulna (Fig. 6-52). It *arises* from the medial aspect of the anterior surface along the distal one-fourth of the ulna and from an oblique ridge marking this part of the bone. It also *arises* from a strong aponeurosis that covers the medial third of the muscle. The fibers pass laterally and slightly downward, to be *inserted* into the distal fourth of the lateral border and anterior surface of the shaft of the radius. The deeper fibers of the muscle are inserted into a triangular area above the ulnar notch of the radius.

Action. The **pronator quadratus** is the principal pronator of the hand. It is assisted by the pronator teres when additional power is required to pronate the hand against resistance (Basmajian and Travill, 1961).

Nerve. The pronator quadratus is supplied by the anterior interosseous branch of the **median nerve,** containing fibers from the eighth cervical and first thoracic nerves.

Variations. The muscle may split into two or three layers, or it may have increased attachments proximally or distally.

Posterior Antebrachial Muscles

These muscles are divided for convenience of description into two groups, superficial and deep.

SUPERFICIAL EXTENSOR GROUP

Brachioradialis
Extensor carpi radialis longus
Extensor carpi radialis brevis
Extensor digitorum
Extensor digiti minimi
Extensor carpi ulnaris
Anconeus

THE BRACHIORADIALIS. The brachioradialis is the most superficial muscle on the radial side of the forearm (Figs. 6-50; 6-54). It *arises* from the upper two-thirds of the lateral supracondylar ridge of the humerus, between the brachialis and triceps muscles, and from the lateral intermuscular septum. Interposed between the brachioradialis and the brachialis are the radial nerve and the anastomosis between the radial collateral branch of the profunda artery and the radial recurrent artery. The muscle fibers end above the middle of the forearm in a flat tendon, which is *inserted* into the lateral side of the base of the styloid process of the radius. The tendon is crossed near its insertion by the tendons of the abductor pollicis longus and extensor pollicis brevis. On its ulnar side is the radial artery.

Action. Embryologically, the **brachioradialis** develops with the extensor muscles and is, in fact, even supplied by the radial nerve. However, it acts as a *flexor of the forearm* at the elbow joint. This action is especially evident when the forearm is semipronated and flexion is carried out against resistance. The brachioradialis usually does not act when the forearm is supinated or the flexion is not intense.

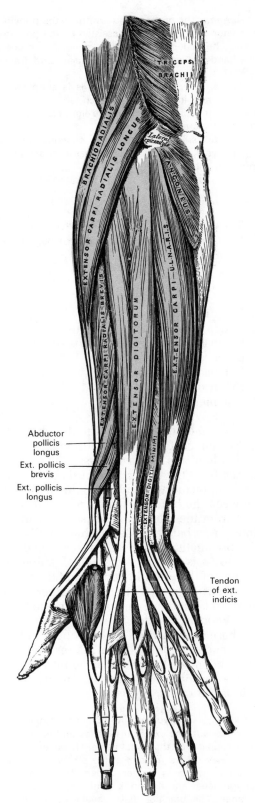

FIG. 6-54. Posterior aspect of the forearm and hand. Superficial muscles.

Nerve. The brachioradialis is supplied by a branch of the **radial nerve,** containing fibers from the fifth and sixth cervical nerves.

Variations. The brachioradialis may fuse with the brachialis. Its tendon of insertion may be divided into two or three slips, and the muscle may partially or completely insert into the middle of the radius. Additional muscular slips may extend to the tendon of the biceps, the tuberosity or oblique line of the radius, the extensor carpi radialis longus, or the abductor pollicis longus. The muscle may be absent. In rare instances it is doubled.

EXTENSOR CARPI RADIALIS LONGUS. This muscle is placed partly deep to the brachioradialis (Figs. 6-50; 6-54). It *arises* in line with the brachioradialis from the distal third of the lateral supracondylar ridge of the humerus, from the lateral intermuscular septum, and by a few fibers from the common tendon of origin of the extensor muscles of the forearm. The muscle fibers end at the upper third of the forearm in a flat tendon, which runs along the lateral border of the radius, beneath the abductor pollicis longus and extensor pollicis brevis. It then passes deep to the extensor retinaculum, where it lies in a groove with the extensor carpi radialis brevis on the back of the radius, immediately behind the styloid process. It is *inserted* into the dorsal surface of the base of the second metacarpal bone, on its radial side (Fig. 6-53).

Action. The **extensor carpi radialis longus** is capable of extending and abducting the hand at the wrist joint. This muscle, however, usually acts as a synergist during flexion of the fingers. Its effectiveness as an extensor of the hand becomes appreciated when a quick, intense extensor action is required. It is thought that this muscle, like the brachioradialis, is capable of flexing the forearm, although weakly.

Nerve. The extensor carpi radialis longus is supplied by a branch of the **radial nerve** that contains fibers from the sixth and seventh cervical nerves.

EXTENSOR CARPI RADIALIS BREVIS. This muscle is shorter, thicker, and partially covered by the extensor carpi radialis longus (Figs. 6-50; 6-54). It *arises* from the lateral epicondyle of the humerus, by a tendon common to it and other extensor muscles, from the radial collateral ligament of the elbow joint, from a strong aponeurosis which covers its surface, and from the intermuscular septa between it and the adjacent muscles. The fibers end about the middle of the forearm in a flat tendon, which is closely connected with that of the extensor carpi radialis longus and accompanies it to the wrist. Passing beneath the abductor pollicis longus and extensor pollicis brevis, and then beneath the extensor retinaculum, it is *inserted* into the dorsal surface of the base of the third metacarpal bone on its radial side. Under the extensor retinaculum the tendon lies on the dorsum of the radius in a shallow groove, to the ulnar side of that which lodges the tendon of the extensor carpi radialis longus and separated from it by a faint ridge. The tendons of both the extensor carpi radialis longus and brevis pass through the same compartment of the dorsal carpal ligament in a single synovial sheath.

Action. The **extensor carpi radialis brevis** extends the hand and may also abduct (radially flex) the hand at the wrist joint.

Nerve. The extensor carpi radialis brevis is supplied by a branch of the **radial nerve,** containing fibers from the sixth and seventh cervical nerves.

Variations. The extensor carpi radialis longus and brevis may split into two or three tendons of insertion to the second and third or even the fourth metacarpal. The two muscles may form a single belly with two tendons. Cross slips between the two muscles may occur.

The **extensor carpi radialis intermedius** rarely arises as a distinct muscle from the humerus, but is not uncommon as an accessory slip from one or both muscles to the second or third or both metacarpals.

The **extensor carpi radialis accessorius** is occasionally found arising from the humerus with or below the extensor carpi radialis longus and inserted into the first metacarpal, or attached to the abductor pollicis brevis, the first dorsal interosseous, or elsewhere.

EXTENSOR DIGITORUM. The extensor digitorum is the common extensor of the fingers. Lying centrally in the posterior forearm (Figs. 6-50; 6-54), it *arises* from the lateral epicondyle of the humerus by the common extensor tendon, from the intermuscular septa between it and the adjacent muscles, and from the antebrachial fascia. It divides distally into four tendons, which pass, together with that of the extensor indicis, through a separate compartment deep to the extensor retinaculum, within a synovial sheath. The tendons then diverge on the back of the hand, each going to a different finger where it is *inserted* into the middle and distal phalanges. The tendon to the

index finger is accompanied by the extensor indicis, which lies on its ulnar side. As the tendons to the middle, ring, and little fingers descend along the dorsum of the hand, they are fastened together by three bands called the **intertendinous connections.** Occasionally the tendon of the index finger is connected to that of the middle finger by a thin transverse band.

Opposite the metacarpophalangeal joint, each tendon is bound by fibrous tissue to the collateral ligaments and thereby serves as the dorsal ligament of this joint. Beyond the metacarpophalangeal joint, each tendon spreads out into a broad aponeurosis that covers the dorsal surface of the first phalanx. These are called the **extensor hoods** and are reinforced by the tendons of the interossei and lumbrical muscles. Opposite the proximal interphalangeal joint, this aponeurosis divides into three slips, an intermediate and two collateral. The intermediate slip is inserted into the base of the middle phalanx. The two collateral slips continue distally along the sides of the middle phalanx, unite by their contiguous margins, and are inserted into the dorsal surface of the distal phalanx. As the tendons cross the interphalangeal joints, they furnish them with dorsal ligaments.

Action. The primary action of the **extensor digitorum** is extension of the fingers as in opening the clenched fist. In addition, it abducts the index, ring, and little finger in a spreading action away from the middle finger. Upon continued action this muscle can also extend the hand at the wrist.

Nerve. The extensor digitorum is supplied by a branch of the deep **radial nerve,** containing fibers from the sixth, seventh, and eighth cervical nerves.

Variations. An increase or decrease in the number of tendons is common, and an additional slip to the thumb is sometimes present.

EXTENSOR DIGITI MINIMI. The extensor digiti minimi is a slender muscle placed on the medial side of the extensor digitorum, with which it is generally connected (Figs. 6-50; 6-54). It *arises* from the common extensor tendon by a thin tendinous slip and from the intermuscular septa between it and the adjacent muscles. The muscle descends in the forearm between the extensor digitorum and the extensor carpi ulnaris and ends in a tendon that courses through a compartment of the extensor retinaculum dorsal to the distal radioulnar joint. The tendon divides into two as it crosses the hand, and finally joins the extensor hood and the tendon of the extensor digitorum on the dorsum of the first phalanx of the little finger.

Action.   The **extensor digiti minimi** extends the little finger and, upon continued action, can aid in extending the hand at the wrist.

Nerve.   The extensor digiti minimi is innervated by a branch of the **deep radial nerve,** containing fibers from the sixth, seventh, and eighth cervical nerves.

Variations.   An additional fibrous slip may be found from the lateral epicondyle. The tendon of insertion may not divide, or it may send a slip to the ring finger. The muscle may be fused with the extensor digitorum, but its complete absence is rare.

EXTENSOR CARPI ULNARIS.   The extensor carpi ulnaris lies on the ulnar side of the forearm (Figs. 6-50; 6-54). It *arises* from the common extensor tendon on the lateral epicondyle of the humerus, from the dorsal border of the ulna by way of an aponeurosis in common with the flexor carpi ulnaris and the flexor digitorum profundus, and from the deep fascia of the forearm. It ends in a tendon that lies in a groove between the head and styloid process of the ulna, passing through a separate compartment of the extensor retinaculum. The tendon is *inserted* onto the prominent tubercle at the ulnar side of the base of the fifth metacarpal bone.

Action.   The **extensor carpi ulnaris** extends and adducts the hand. Similar to the extensors carpi radialis longus and brevis, it also acts as a synergist to the flexors of the fingers, keeping the wrist extended in order to give additional strength in maintaining a clenched fist or in grasping objects with the hand.

Nerve.   The extensor carpi ulnaris is supplied by a branch of the **deep radial nerve,** containing fibers from the sixth, seventh, and eighth cervical nerves.

Variations.   In many cases a slip is continued from the insertion of the tendon anteriorly over the opponens digiti minimi to the fascia covering that muscle, the fifth metacarpal bone, the capsule of the fifth metacarpophalangeal articulation, or the proximal phalanx of the little finger. This slip may be replaced by a muscular fasciculus arising from or near the pisiform. It may also insert partially onto the fourth metacarpal bone. The muscle may at times be reduced to a tendinous band, or it may be doubled.

THE ANCONEUS.   This small triangular muscle, which lies on the dorsum of the elbow joint, appears to be a continuation of the triceps brachii (Fig. 6-54). It *arises* by a separate tendon from the dorsal part of the lateral epicondyle of the humerus. Its fibers diverge and are *inserted* into the lateral aspect of the olecranon and into the upper one-fourth of the dorsal surface of the body of the ulna.

Action.   The **anconeus** assists in extending the forearm.

Nerves.   The anconeus is supplied by a branch of the **radial nerve** that contains fibers from the seventh and eighth cervical nerves.

DEEP EXTENSOR GROUP

Supinator
Abductor pollicis longus
Extensor pollicis brevis
Extensor pollicis longus
Extensor indicis

THE SUPINATOR.   The supinator, a broad muscle, is curved around the upper one-third of the radius. It consists of two planes

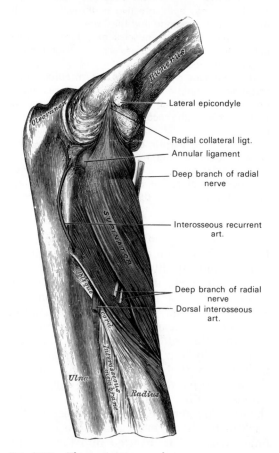

Lateral epicondyle

Radial collateral ligt.
Annular ligament

Deep branch of radial nerve

Interosseous recurrent art.

Deep branch of radial nerve
Dorsal interosseous art.

FIG. 6-55.   The supinator muscle.

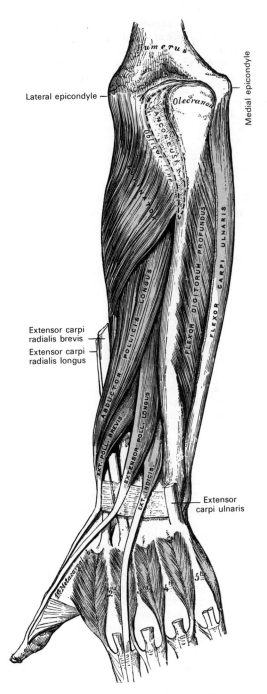

Extensor carpi
radialis brevis
Extensor carpi
radialis longus

Extensor
carpi ulnaris

FIG. 6-56. Posterior aspect of the forearm. Deep muscles.

of fibers, between which the deep branch of the radial nerve lies (Figs. 6-55; 6-56). The two planes *arise* in common—the superficial one by tendinous and the deeper by muscular fibers—from the lateral epicondyle of the humerus, from the radial collateral ligament

of the elbow joint and the annular ligament of the proximal radioulnar joint, from the supinator crest on the ulna, which runs obliquely downward from the dorsal end of the radial notch, from a triangular depression distal to the notch, and from a tendinous aponeurosis that covers the surface of the muscle. The superficial fibers surround the upper part of the radius and are *inserted* into the lateral edge of the radial tuberosity and the oblique line of the radius, as far down as the insertion of the pronator teres. The proximal fibers of the deeper plane encircle the neck of the radius to the radial tuberosity and are *inserted* onto the back part of its medial surface. The greater part of the deep portion of the muscle is *inserted* onto the dorsal and lateral aspects of the upper third of the radial shaft.

Action. As its name implies, the **supinator** rotates the radius to supinate the forearm and hand. It is not as powerful as the biceps in this action but it is effective whether the forearm is flexed or extended, whereas the biceps principally supinates the flexed forearm.

Nerve. The supinator is supplied by the **deep branch of the radial nerve**, containing fibers from the sixth cervical nerve.

ABDUCTOR POLLICIS LONGUS.   The abductor pollicis longus lies immediately below the supinator (Fig. 6-56) and is sometimes united with it. It *arises* from the lateral part of the posterior surface of the body of the ulna distal to the insertion of the anconeus, from the interosseous membrane, and from the middle third of the posterior surface of the body of the radius. Passing obliquely downward and laterally, it ends in a tendon that courses through a groove on the lateral side of the lower end of the radius, accompanied by the tendon of the extensor pollicis brevis. It is *inserted* into the radial side of the base of the first metacarpal bone (Fig. 6-53). It usually gives off two slips near its insertion: one to the trapezium and the other to blend with the origin of the abductor pollicis brevis.

Action. The primary action of the **abductor pollicis longus** is a combined abduction and extension of the thumb at the carpometacarpal joint. It is also an important abductor (radial flexor) of the wrist, and its insertion allows it also to help in *flexing* the wrist.

Nerve. The abductor pollicis longus is supplied

by the posterior interosseous branch of the **deep radial nerve**, containing fibers from the sixth and seventh cervical nerves.

EXTENSOR POLLICIS BREVIS. The extensor pollicis brevis lies medial to and contiguous with the abductor pollicis longus (Fig. 6-56). It *arises* from the posterior surface of the body of the radius below the abductor pollicis longus, and from the interosseous membrane. Its direction is similar to that of the abductor pollicis longus, and its tendon passes through the same groove on the lateral side of the distal end of the radius. It is *inserted* into the base of the first phalanx of the thumb (Fig. 6-53).

Action. The **extensor pollicis brevis** extends the proximal phalanx at the metacarpophalangeal joint of the thumb. By continued action it can extend the first metacarpal bone at the carpometacarpal joint. It can also radially flex (abduct) the hand at the wrist.

Nerve. The extensor pollicis brevis is supplied by the posterior interosseous branch of the **deep radial nerve**, containing fibers from the sixth and seventh cervical nerves.

Variations. At times this muscle is absent, or its tendon is fused with that of the extensor pollicis longus.

EXTENSOR POLLICIS LONGUS. The extensor pollicis longus is much larger than the extensor pollicis brevis, the origin of which it partly covers (Figs. 6-50; 6-56). It *arises* from the middle third of the dorsolateral surface of the ulna distal to the origin of the abductor pollicis longus, and from the interosseous membrane. It ends in a tendon that passes through a separate compartment in the extensor retinaculum, lying in a narrow, oblique groove on the dorsum of the lower end of the radius. It then crosses obliquely the tendons of the extensor carpi radialis longus and brevis, and is separated from the extensor pollicis brevis by a triangular interval, commonly referred to as the *"anatomical snuff-box,"* in which the radial artery is found. Its tendon is finally *inserted* into the base of the last phalanx of the thumb. The radial artery is crossed by the tendons of the abductor pollicis longus and of the extensors pollicis longus and brevis before it penetrates to the palmar side of the hand between the two heads of the dorsal interosseous muscle.

Action. The **extensor pollicis longus** extends the distal phalanx of the thumb and, by continued action, can extend the proximal phalanx and the first metacarpal bone. It also may assist in extension of the hand at the carpometacarpal joint.

Nerve. The extensor pollicis longus is supplied by the posterior interosseous branch of the **deep radial nerve**, containing fibers from the sixth, seventh, and eighth cervical nerves.

EXTENSOR INDICIS. This narrow, elongated muscle is placed medial to, and parallel with, the extensor pollicis longus. It *arises* from the posterior surface of the body of the ulna below the origin of the extensor pollicis longus, and from the interosseous membrane. Its tendon passes under the extensor retinaculum in the same compartment that contains the tendons of the extensor digitorum. Opposite the head of the second metacarpal bone, the tendon of the extensor indicis joins the ulnar side of the tendon of the extensor digitorum coursing to the index finger, thereby *inserting* into the extensor hood of the second digit.

Action. The **extensor indicis** allows extension of the index finger, independent of the other fingers. It assists in adduction of the index finger and can also help in extending the hand at the wrist joint.

Nerve. The extensor indicis is supplied by the posterior interosseous branch of the **deep radial nerve**, containing fibers from the sixth, seventh, and eighth cervical nerves.

Variations. A slip from the tendon of the extensor indicis may pass to the middle finger, and the muscle may, at times, be doubled.

Group Actions of Forearm Muscles. *Flexion of the wrist joint* is brought about by the flexor carpi ulnaris, flexor carpi radialis, palmaris longus, and abductor pollicis longus. If the fingers are kept extended and prevented from flexing, the flexors digitorum superficialis and profundus can assist in voluntary flexion at the wrist joint.

Extension at the wrist joint is performed by the extensors carpi radialis longus and brevis and by the extensor carpi ulnaris. If the fingers are kept flexed as in a clenched fist, the extensor digitorum and extensor pollicis longus can assist in extension at the wrist joint.

Abduction (radial flexion) *at the wrist joint* principally results from the action of the abductor pollicis longus and the extensor pollicis brevis; however, the flexor carpi radialis, the extensors carpi radialis longus and brevis, and the extensor pollicis longus all assist in this action.

Adduction (ulnar flexion) *at the wrist joint* is performed by the combined action of the flexor carpi ulnaris and extensor carpi ulnaris.

Pronation of the hand is brought about by the pronator quadratus and pronator teres muscles.

Supination of the hand is performed by the supinator muscle and more strongly by the biceps brachii muscle.

FASCIAE, RETINACULA, TENDON SHEATHS AND COMPARTMENTS OF THE WRIST AND HAND

The tendons of the muscles in the forearm, which attach at the wrist or continue into the hand, along with the descending vessels and nerves, are retained close to bone by strong fascial structures. Additionally, the tendons are surrounded by lubricating synovial sheaths contained within compartments, that serve to further maximize efficiency of function. In the hand, fascial spaces of clinical significance separate the muscle groups and other structures.

Superficial Fascia

The superficial fascia on the anterior aspect of the forearm changes its character abruptly at the distal crease of the wrist from a delicate movable tissue into the tough cushion that covers the palm and palmar surface of the digits. The superficial fascia of the hand contains a considerable amount of fat, but cannot be separated readily into superficial and deep layers. The adipose tissue is permeated by strong fibrous bands and septa, which divide it into small lobules and bind it securely to the deep fascia as well as to the skin. The dermis is compact; it not only protects the underlying structures, but also offers resistance to the progress of infectious processes seeking to achieve the surface. The vertical direction of the fibrous bands tends to guide the spread of infection into deeper layers. The superficial fascia is adherent to the deep fascia over the entire palm, but the union is especially strong at the skin creases of the wrist, the major creases of the palm, and the creases of the digits. At the medial and lateral borders of the hand and digits, the fascia changes its character rather abruptly as it becomes continuous with the superficial fascia that covers the dorsum of the hand.

The superficial fascia over the dorsum of the hand and fingers is delicate and movable, similar to that of the forearm with which

it is continuous. Its two layers can be identified: the superficial layer is thin but may contain a small amount of fat; the deep layer of superficial fascia is a definite fibrous sheet and supports the superficial veins and cutaneous nerves. It is separated from the deep fascia by a distinct fascial cleft, the **dorsal subcutaneous cleft,** which imparts the characteristic movability to the skin of the back of the hand.

Deep Fascia

The antebrachial fascia at the wrist is thickened into an annular band or cuff, which holds the tendons of the forearm muscles close against the wrist. This distal portion of the investing antebrachial fascia is strengthened by strong transverse collagenous bundles. It is attached medially and laterally to the styloid processes of the ulna and radius, and it is frequently called the *superficial part of the flexor retinaculum*. At one time it was called the palmar (or volar) carpal ligament, but distally it actually merges with the flexor retinaculum. The ulnar nerve and vessels emerge from beneath this layer of fascia to enter the hand superficial to the flexor retinaculum.

FLEXOR RETINACULUM. The flexor retinaculum (Fig. 6-57) is a thick fibrous band that arches over the deep groove on the palmar surface of the carpal bones, forming the **carpal tunnel** through which the long flexor tendons and the median nerve pass. At one time this retinaculum was called the *transverse carpal ligament*. It is attached, *medially*, to the pisiform and the hamulus of the hamate, and *laterally*, to the tuberosity of the scaphoid and the medial part of the palmar surface and ridge of the trapezium. Its proximal border is partly merged with the distal portion of the investing antebrachial fascia, but the latter belongs to a different and more superficial stratum and is separated from the flexor retinaculum by the ulnar vessels and nerve. The flexor retinaculum is attached to the palmar aponeurosis, which lies superficial to it, and contributes oblique crossed fibers to the deep surface of the aponeurosis. It has two attachments to the trapezium, one on each side of the groove in which the tendon of the flexor carpi radialis lies. The flexor carpi ulnaris, at its insertion, contributes tendinous fibers to the flexor retinacu-

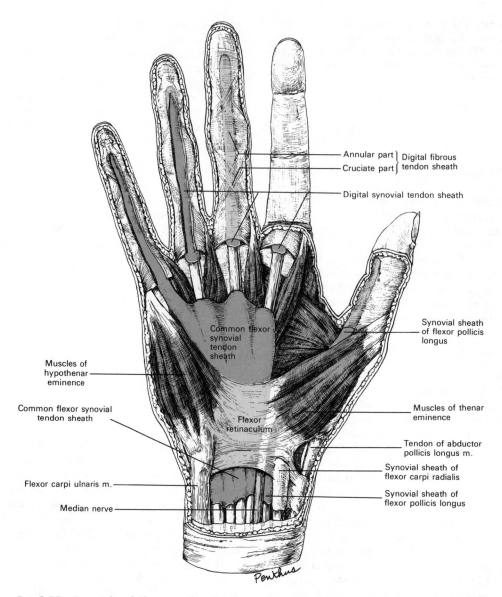

FIG. 6-57. Synovial and fibrous tendon sheaths on the palmar aspect of the right wrist and hand.

lum; to a large extent, the thenar and hypothenar muscles arise from it.

EXTENSOR RETINACULUM. The extensor retinaculum is a strong fibrous band under which the extensor tendons lie (Fig. 6-58). It is the distal portion of the investing antebrachial fascia, which is thickened at the dorsum of the wrist by the addition of transverse collagenous bundles. These take a somewhat oblique course, extending distalward as they cross from the radial to the ulnar side. The retinaculum is attached *medially* to the styloid process of the ulna and to the triquetral and pisiform bones, and *lat-*erally to the lateral margin of the radius. Between these medial and lateral borders, it is attached to the ridges on the dorsal surface of the radius.

Tendon Sheaths

As the forearm tendons descend through the wrist and into the hand and then along the digits, they are encased by synovial sheaths. These tubular structures, which resemble the bursae found at other joints, are formed by inner visceral and outer parietal membranous layers with a potential space intervening that contains lubricating fluid. The two layers are interconnected by delicate bands called mesotendons, which carry the fine blood vessels to the tendons. Sheaths are found surrounding both flexor and extensor tendons at the wrist.

SYNOVIAL SHEATHS OF FLEXOR TENDONS. As the flexor tendons pass deep to the flexor retinaculum, they are enclosed in two synovial sheaths (Fig. 6-57). The larger one contains all the tendons of the flexor digitorum superficialis and flexor digitorum profundus, and is called the **common flexor tendon sheath** (formerly called the ulnar bursa). The smaller one, found on the radial side, is called the **tendon sheath of the flexor pollicis longus** (radial bursa), and contains the tendon of the long flexor of the thumb. The two sheaths extend into the forearm for about 2.5 cm proximal to the flexor retinaculum. The tendon sheath for the flexor pollicis longus extends distally to the terminal phalanx of the thumb, where the muscle inserts. The common flexor tendon sheath is continuous distally beyond the middle of the palm as the digital sheath only for the little finger. It is greatly reduced in diameter at the midpalmar region by the formation of terminal diverticula about the tendons coursing to the second, third, and fourth digits. The tendons to these three fingers are without synovial sheaths for a short distance in the middle of the palm, but they have independent **digital sheaths,** beginning proximally over the heads of their metacarpal bones and continuing distally to the terminal phalanges, where the tendons of the flexor profundus insert.

DIGITAL TENDON SHEATHS. The tendons of the flexors digitorum superficialis and profundus are held in position along the digits by strong **fibrous sheaths.** These tunnels or canals are formed by the palmar surfaces of the phalanges and by strong collagenous bands that arch over the tendons and are attached to the margins of the phalanges on both sides. Opposite the middle of the proximal and second phalanges, the fibrous sheaths are reinforced by strong transversely coursing fibers, which form the **annular part** of the fibrous sheath. Opposite the joints, the fibrous sheaths are much thinner, and consist of obliquely coursing fibers. These form the **cruciate part** of the fibrous sheath (Fig. 6-57). At their proximal ends, the digital sheaths merge with the deeper parts

of the palmar aponeurosis. Within each of the five fibrous digital sheaths, there is a **synovial tendon sheath.** The fibrous sheath for the thumb is continuous with the tendon sheath of the flexor pollicis longus (radial bursa), and that for the little finger with the common flexor tendon sheath (ulnar bursa). Those for the other three fingers are closed proximally at the metacarpophalangeal joints (Fig. 6-57).

SYNOVIAL SHEATHS OF EXTENSOR TENDONS. Between the extensor retinaculum and the carpal bones, six tunnels are formed for the passage of tendons, each tunnel having a separate synovial sheath (Fig. 6-58). One is found in each of the following positions: (1) on the lateral (radial) side of the styloid process of the radius, for the tendons of the abductor pollicis longus and extensor pollicis brevis; (2) dorsal to the styloid process, for the tendons of the extensors carpi radialis longus and brevis; (3) about the middle of the dorsal surface of the radius, for the tendon of the extensor pollicis longus; (4) more medially, for the tendons of the extensor digitorum and extensor indicis; (5) opposite the interval between the radius and ulna, for the extensor digiti minimi; (6) between the head and styloid process of the ulna, for the tendon of the extensor carpi ulnaris. The sheaths lining these tunnels all begin proximal to the extensor retinaculum. Those for the tendons of the abductor pollicis longus, extensor pollicis brevis, extensors carpi radialis longus and brevis, and extensor carpi ulnaris stop immediately proximal to the bases of the metacarpal bones, while the sheaths for the extensor digitorum, extensor indicis, and extensor digiti minimi are prolonged an additional centimeter or more along the dorsum of the metacarpal bones.

Deep Fascia of the Hand

DEEP FASCIA OF THE PALM. The investing layer of deep fascia in the palm is continuous with the antebrachial fascia on the volar aspect of the wrist. At the borders of the hand, the deep palmar fascia is continuous with the fascia of the dorsum, attaching to the fifth metacarpal bone medially, and the first and second metacarpal bones laterally, as it passes over them. The thenar fascia, which covers the muscular eminence at the radial side of the hand, and the hypothenar fascia, which covers the eminence of the

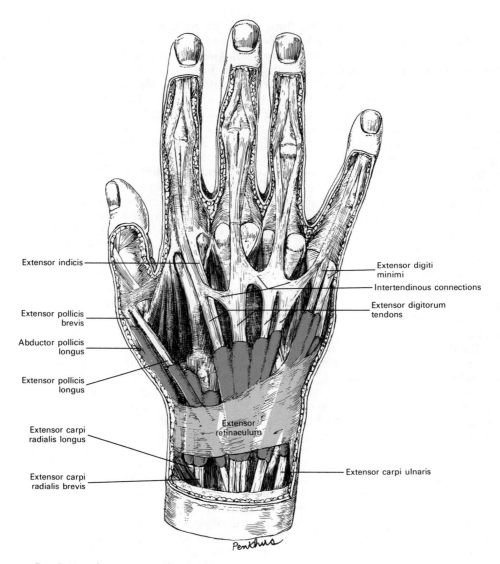

Extensor indicis

Extensor digiti minimi

Intertendinous connections

Extensor digitorum tendons

Extensor pollicis brevis

Abductor pollicis longus

Extensor pollicis longus

Extensor carpi radialis longus

Extensor retinaculum

Extensor carpi radialis brevis

Extensor carpi ulnaris

Penthus

FIG. 6-58.   The synovial sheaths and tendons on the dorsal aspect of the wrist and hand.

ulnar side, are similar in texture to the ante-brachial fascia. The central part of the palm, however, is greatly strengthened into what is called the palmar aponeurosis.

The **palmar aponeurosis** (Fig. 6-59) is made up of two components: (1) a thick su-perficial stratum of longitudinal bundles, which are the direct continuation of the ten-don of the palmaris longus, and (2) a thinner deep stratum of transverse fibers, which is continuous with the flexor retinaculum. The two strata are intimately fused and partly interwoven. The deeper portion is securely attached to the flexor retinaculum, which may contribute obliquely running fibers to the aponeurosis. The longitudinal bundles of

the superficial stratum form a uniform layer in the proximal part of the palm, but distally they fan out into divergent bands, which extend toward the bases of the digits and cover the long flexor tendons. The four bands to the fingers are heavier and more constant than the one to the thumb. Each of these bands has a double termination, super-ficially attaching to the skin and, more deeply, ending on the fibrous flexor tendon sheath. The most superficial fibers attach to the skin at the curved distal crease of the palm, while other superficial fibers termi-nate at the transverse crease at the base of the digits. The deeper portion of each band contributes to the fibrous tendon sheath in

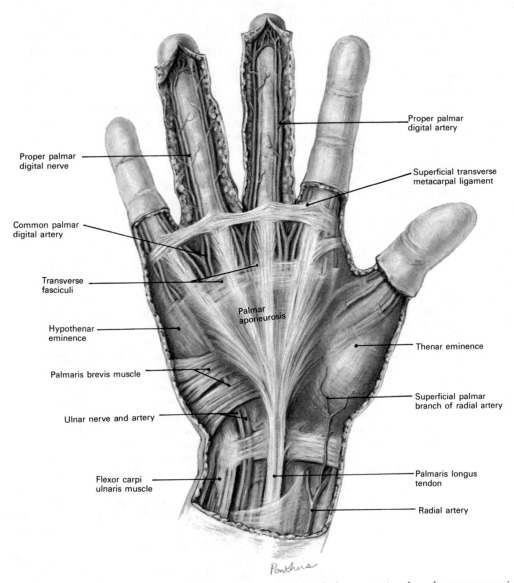

Proper palmar digital nerve

Common palmar digital artery

Transverse fasciculi

Hypothenar eminence

Palmaris brevis muscle

Ulnar nerve and artery

Flexor carpi ulnaris muscle

Proper palmar digital artery

Superficial transverse metacarpal ligament

Palmar aponeurosis

Thenar eminence

Superficial palmar branch of radial artery

Palmaris longus tendon

Radial artery

FIG. 6-59. Superficial dissection of the palmar surface of the hand, demonstrating the palmar aponeurosis.

two ways: some of the fibers continue distally into the digit, assisting in the formation of the digital sheath, but the greater number of fibers form two arching ligamentous bands, which penetrate deeply toward the metacarpal bone on each side of the tendon. These bands attach to the bones, thus completing the formation of tunnels at the heads of the metacarpal bones. In the central part of the palm, as the longitudinal bands diverge, the intervals between them are occupied by the deeper transverse fibers. These **transverse fasciculi** extend as far distally as the heads of the metacarpal bones (Fig. 6-59).

From this site to the webs between the digits, the intervals are not covered by aponeurosis. In this distal, uncovered portion, therefore, the digital vessels and nerves, as well as the tendons of the lumbrical muscles, are more readily accessible to the surgeon. The intervals are closed distally by other transverse fibers, which occupy and support the webs between the digits. These more distal transverse fibers collectively constitute the **superficial transverse metacarpal ligament** (also called *superficial transverse ligament of the fingers, interdigital ligament, natatory ligament*). They are usually distinct from the transverse fasciculi, which form the deep stratum of the palmar aponeurosis, and they attach to the digital sheaths at the bases of the proximal phalanges, merging into fibrous septa at the sides of the fingers. The digital vessels and nerves enter the fingers deep to the superficial transverse metacarpal ligament (Fig. 6-59).

As mentioned above, the band of longitudinal fibers that extends from the palmar aponeurosis toward the thumb is not as robust as the other four. Some of these longitudinal fibers attach to the longitudinal crease of the palm medial to the thumb, many fuse with the fascia of the thenar eminence, and a few assist in the formation of the sheath for the flexor pollicis longus tendon. When the palmaris longus is absent (about 13%), the attachment of the palmar aponeurosis to the flexor retinaculum is strengthened in order to compensate for the loss of continuity with the palmaris longus tendon. The palmaris brevis is a small but constant muscle that lies superficial to the hypothenar fascia. It has its origin at the ulnar border of the palmar aponeurosis and inserts into the skin at the ulnar border of the palm. The palmar aponeurosis is fused at its radial border with the fascia covering the thenar eminence. From this line of union, a fascial layer is continued deep into the palm and is attached to the first metacarpal bone, forming the **thenar septum.** Similarly, the aponeurosis is fused with the hypothenar fascia, and from this union a septum is continued deep to the fifth metacarpal to form the **hypothenar septum.** These two septa divide the palm into three compartments: a **thenar,** a **hypothenar,** and a **central compartment** (Fig. 6-60).

DEEP FASCIA OF THE DORSUM OF THE HAND. The investing layer of deep fascia on the dorsum of the hand is directly continuous with the antebrachial fascia, which is thickened at the wrist by the addition of annular collagenous bundles into the extensor retinaculum. On the ulnar side of the hand, the deep fascia of the dorsum is continuous with the hypothenar fascia beyond its attachment to the dorsum of the fifth metacarpal bone, and at the radial border of the hand, it is continuous with the thenar fascia dorsal to the first metacarpal bone on which the fascia attaches. The deep fascia on the dorsum forms the superficial boundary of a flat compartment, which contains the tendons of the extensors of the digits.

The deep fascia on the dorsum of the thumb, after being attached to the dorsum of the second metacarpal bone, continues laterally over the first dorsal interosseous muscle, in the web between the thumb and the index finger, and is attached to the ulnar border of the first metacarpal bone. After forming a compartment for the extensor tendons of the thumb, it attaches to the radial border of the first metacarpal, where it becomes continuous with the thenar fascia.

Compartments of the Hand

Due to the attachments of the palmar fascia, the palm of the hand may be considered to be divided into four compartments: the **thenar, hypothenar, central,** and **interosseous-adductor compartments.** On the dorsum of the hand is found the **dorsal tendon compartment.** These compartments are enclosed by fascial layers and contain muscles, bones, and other structures. Between the compartments (and fascial layers) are the **fascial spaces** or **clefts.** An appreciation of these compartments and spaces is still of significant clinical importance in an understanding of the spread of infections and abscesses of the hand.

THENAR COMPARTMENT. The thenar compartment accounts for the thenar eminence of the palm and contains the short muscles of the thumb with the exception of the two heads of the adductor pollicis (Fig. 6-60). The *inferomedial* boundary is formed by the investing layer of deep fascia and its inward continuation, the thenar septum, which lies between the adductor and the flexor brevis. The compartment is closed by the attachment of this fascia *proximally* to the carpal bones and the flexor retinaculum; *distally,* to the first phalanx at the insertion of the enclosed muscles; *dorsally,* along the subcutaneous border of the first metacarpal bone; and *ventrally,* along the adductor. In addition to the abductor brevis, the flexor brevis, and the opponens, other structures occupying this compartment include the first metacarpal bone, the superficial palmar branch of the radial artery, and a portion of the tendon of the flexor pollicis longus enclosed in its tendon sheath. Within the compartment, the muscles are enclosed in their individual fascial sheaths and, for the most part, are separated from each other by fascial clefts, but these clefts do not communicate with each other nor with the major fascial spaces of the palm.

HYPOTHENAR COMPARTMENT. This compartment on the ulnar side of the hand forms the hypothenar eminence and contains the short muscles of the little finger (Fig. 6-60). It is enclosed by the hypothenar investing fascia and the hypothenar septum, which lies between the flexor digiti minimi and the third palmar interosseous muscles. The fascia forms a closed compartment by attaching along the fifth metacarpal bone on both the dorsal and palmar aspects of the muscles that it surrounds. Within the compartment are found the hypothenar muscles and branches of the ulnar nerve and artery.

CENTRAL COMPARTMENT. The central compartment is bounded *medially* and *laterally* by the thenar and hypothenar fascial septa; *superficially,* by the palmar aponeurosis; and *deeply,* by a fascial lamina, which covers the deep surface of the long flexor tendon mass and overlies the interossei and metacarpal bones (Fig. 6-60). The compartment contains the tendons of the flexor digitorum superficialis and flexor digitorum profundus, the lumbrical muscles, the superficial palmar arch, the palmar branch of the median nerve, and the superficial branch of the ulnar nerve. The compartment is narrow proximally, but widens distally as the tendons diverge toward their fingers. Its fascial attachments close the compartment proximally, but the tendon sheaths within it extend back into the forearm, and its tissues merge distally with those of the webs and digits.

INTEROSSEOUS-ADDUCTOR COMPARTMENT. Dorsal to the central compartment and separating the compartments of the palm and the dorsum of the hand is a region that contains principally the interossei and the adductor pollicis muscles; accordingly, this region is called the interosseous-adductor compartment. The compartment is enclosed by two fascial layers that are continuous with each other around its medial and lateral borders. The lamina on the dorsal surface, called the **dorsal interosseous fascia,** covers and is adherent to the dorsal surfaces of the second to fifth metacarpal bones and the intervening dorsal interossei (Fig. 6-61). The palmar

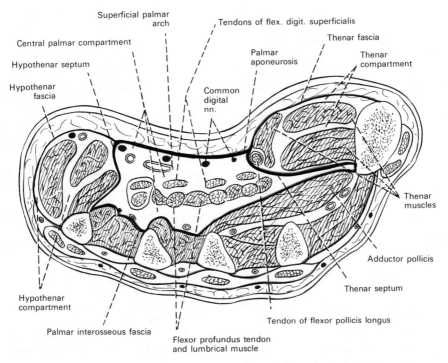

FIG. 6-60. Cross section of the hand, outlining (in red) the boundaries of the thenar, central palmar, and hypothenar compartments. Deep to the thenar and central palmar compartments of the palm, observe the large adductor-interosseous compartment. (From Hollinshead, *Anatomy for Surgeons*, 2nd Ed., Harper & Row, 1969.)

surface is covered by the **palmar interosseous fascia** on the ulnar half and by the fascia
anterior to the adductor on the radial half.
The compartment is closed by the attachment of these membranes at the origins of
the muscles proximally and at their insertions distally. It contains the second, third,
fourth, and fifth metacarpal bones, all the
interossei, the transverse and oblique heads
of the adductor pollicis, the deep palmar
arch, and the deep branch of the ulnar nerve.
The adductor and the interossei have been
placed in the same rather than in separate
compartments because the entire muscle
mass lies deep to the palmar compartments
and is separated from them by the major fascial spaces of the palm. A fascial cleft may
separate the two heads of the adductor from
each other and from the first dorsal interosseous, and these clefts may communicate
with the thenar cleft or with the tissue
spaces of the web between the thumb and
index finger.

DORSAL TENDON COMPARTMENT. On the
dorsum of the hand, the deep fascia splits
into two layers, enclosing the dorsal tendon
compartment which contains the extensor
tendons. The more dorsal of these layers,
called the **supratendinous layer,** is in continuity proximally with the extensor retinaculum. The deeper layer, which is called the
infratendinous layer, lies between the extensor tendons and the dorsal interosseous fascia. The dorsal tendon compartment is
closed at the sides of the hand by the fusion
of the two membranes into a single sheet
where both are attached to the dorsum of the
second and fifth metacarpal bones. It is
closed distally by a fusion of the two layers
at the webs between the fingers and their attachment to the joint capsules and tendinous
expansions of the digits.

*Fascial Spaces (Clefts)
in the Hand*

Fascial spaces or clefts are planes of cleavage *between* fascial layers and should not be
confused with fascial compartments, which
are enclosures formed by fascial layers and
contain muscles, bones, and other structures.
Fascial spaces of less clinical importance
may occur within compartments, but those
of great clinical importance lie between the
compartments.

The major fascial space of the palm lies

deep to the fascia covering the deep surface
of the long flexor tendon mass and anterior
to the fascia covering the adductor pollicis
muscle medially, and the interosseous muscles laterally, that is, it lies between the central palmar compartment and the interos-
seus-adductor compartment. Delicate septa
attach to the metacarpal bones distally and
extend toward the wrist, more or less subdividing the space. An **intermediate palmar
septum,** which is attached to the middle
metacarpal bone, is more constant and better developed than the rest and is commonly
described as subdividing the larger space
into two parts, the **middle palmar space** and
the **thenar space** (Kanavel, 1939).

MIDDLE PALMAR SPACE. The middle palmar
space or cleft is triangular in shape and separates the deep surface of the long flexor tendons from the interossei in the central part
of the palm (Fig. 6-61). It lies between the
palmar interosseous fascia and the fascia
covering the deep surface of the long flexor
tendon mass. The cleft may be closed at the
wrist by adhesion between the fascial layers
at the flexor retinaculum, or it may be continued *proximally,* deep to the flexor tendons, and communicate with the deep fascial space of the forearm that lies between
the tendons of the flexor muscles and the
pronator quadratus (Parona's space[1]). It is
closed, *medially,* by the attachment of the
hypothenar septum to the fifth metacarpal
bone, and *laterally,* from the thenar cleft by
the transparent, fibrous intermediate palmar
septum which is attached along the middle
metacarpal bone. This septum, instead of
attaching to the flexor tendon mass directly
over the middle metacarpal, may take an
oblique course toward the region of the second metacarpal, causing the middle palmar
space to extend into the radial portion of the
palm, thereby overlapping to some extent
the thenar space. The membrane between
the middle palmar and the thenar cleft may
be incomplete proximally, allowing the two
spaces to communicate. Closely associated
with the middle palmar space and acting as
diverticula, are the clefts, called the **lumbrical canals,** that surround the second, third,
and fourth lumbrical muscles.

THENAR SPACE. The thenar space overlies
the palmar surface of the adductor pollicis

---

[1]Francesco Parona (19th century): An Italian surgeon
(Novara).

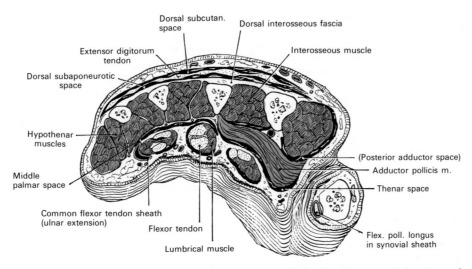

Dorsal subcutan. space

Dorsal interosseous fascia

Extensor digitorum tendon

Interosseous muscle

Dorsal subaponeurotic space

Hypothenar muscles

Middle palmar space

(Posterior adductor space)

Adductor pollicis m.

Thenar space

Common flexor tendon sheath (ulnar extension)

Flexor tendon

Lumbrical muscle

Flex. poll. longus in synovial sheath

FIG. 6-61. Transverse section across the hand, showing (in black) the fascial spaces. (After Kanavel, 1939.)

(Fig. 6-61). It is bounded, *medially,* by the intermediate palmar septum, which is attached to the middle metacarpal bone and separates it from the middle palmar space. The thenar space is bounded *laterally* by the thenar septum and the first metacarpal bone; *proximally,* by the flexor retinaculum; and *distally,* by the extent of the transverse head of the adductor pollicis muscle. The thenar space is commonly continuous with the cleft between the two heads of the adductor pollicis, and between the adductor and the first dorsal interosseous. It may communicate with the middle palmar space proximally, and it usually has a diverticulum, the lumbrical canal, extending along the first lumbrical muscle.

LUMBRICAL CANALS. This is a term used to designate the fascial clefts that separate the lumbrical muscles from the denser connective tissue that surrounds them. These canals are distal extensions or diverticula of the fascial spaces of the palm. The canal associated with the first lumbrical muscle (index finger) projects from the thenar space, while the lumbrical canals of the second, third, and fourth muscles extend distally from the middle palmar space.

DORSAL SUBCUTANEOUS SPACE. On the dorsum of the hand, the dorsal subcutaneous space separates the superficial fascia from the deep fascia (Fig. 6-61). It extends *distally* into the fingers and *proximally* into the forearm. It is closed at the radial and ulnar borders of the hand by the attachment of the superficial fascia to the deep fascia of the

palm. Although of less clinical importance than the palmar spaces, it allows pus to spread quite easily over the entire dorsum, if an infection is present.

DORSAL SUBAPONEUROTIC SPACE. Deep to the investing fascia, and between it and the dorsal interosseous fascia, lies the dorsal subaponeurotic space (Fig. 6-61). It is closed at the sides of the hand by the fusion of the two fascial layers near their attachment to the second and fifth metacarpal bones. It is closed distally by the fusion of the two layers and their union with the joint capsules and extensor expansions of the digits. Proximally, it is limited by the attachment of the tendon sheaths to the bones and ligaments of the wrist. This space does not communicate with the dorsal subcutaneous space, nor with the palmar spaces, and it does not pass from the hand into the forearm.

Clinical Considerations. The relationship of the fascial spaces to each other and to the tendon sheaths may be discussed by reviewing the probable spread of an infection, independent of the participation of blood vessels or lymphatics.

A *subcutaneous abscess on the dorsum* of the hand or in the webs of the fingers is generally directed toward the surface locally because of the softness of the tissues. A *subcutaneous abscess in the palm* might reach the surface or spread to the webs, but would not be likely to penetrate the palmar aponeurosis. An *abscess of the index finger,* after penetrating to the deeper tissues, might progress proximally until it arrived in the lumbrical canal of that finger and, through this path, reach the thenar space. *Abscesses of the middle, ring, or little fingers* might reach the palmar space by a similar path. Infections that have reached either of the deep palmar spaces may spread to the other space by penetrating the intermediate palmar septum, or they may ascend to the deep fascial space (of Parona) above the wrist.

Infections of the flexor tendons of the little finger or thumb for a period of time are confined to the common flexor tendon sheath or to the tendon sheath of the flexor pollicis longus. The infection then ascends and ruptures into the deep fascial space of the forearm. An infection in the digital tendon sheath of the index finger can be expected to rupture into the thenar space; if it is in the middle or fourth finger, it will rupture into the middle palmar space.

An *infection in the subaponeurotic space* can be expected to spread throughout the space and then eventually rupture into the webs of the fingers or at the sides of the hand.

MUSCLES OF THE HAND

The muscles of the hand are subdivided into three groups: (1) those of the thumb, which occupy the radial side and produce the **thenar eminence;** (2) those of the little finger, which occupy the ulnar side and give rise to the **hypothenar eminence;** (3) those in the middle of the palm, and the interosseous muscles between the metacarpal bones.

Thenar Muscles

The thenar muscles are the intrinsic muscles of the thumb. They include:

Abductor pollicis brevis
Opponens pollicis
Flexor pollicis brevis
Adductor pollicis
    Oblique head
    Transverse head

ABDUCTOR POLLICIS BREVIS. The abductor pollicis brevis is a thin, flat muscle, placed most superficially on the radial side of the thenar eminence (Figs. 6-62; 6-64). It *arises* from the flexor retinaculum, the tuberosity of the scaphoid, and the ridge of the trapezium, frequently by two distinct slips. Coursing laterally and distally, it is *inserted* by a thin, flat tendon into the radial side of the base of the proximal phalanx of the thumb and the capsule of the metacarpophalangeal articulation.

The abductor pollicis brevis is sometimes divided into outer and inner parts, and accessory slips to the muscle may come from the abductor pollicis longus tendon or the palmaris longus. More rarely, slips to the muscle may come from the extensor carpi radialis longus, the styloid process of the radius, the opponens pollicis, or the fascia over the thenar muscles.

Action. The **abductor pollicis brevis** abducts the thumb at the carpometacarpal and metacarpophalangeal joint by drawing it away in a plane at right angles to that of the palm of the hand.

Nerve. The abductor pollicis brevis is supplied by a branch of the **median nerve,** containing fibers from the eighth cervical and first thoracic nerves.

OPPONENS POLLICIS. The opponens pollicis is a small triangular muscle placed beneath the abductor pollicis brevis (Figs. 6-62 to 6-64). It *arises* from the ridge on the trapezium and from the flexor retinaculum, passes distally and laterally, and is *inserted* into the whole length of the metacarpal bone of the thumb on its radial side.

Action. The **opponens pollicis** abducts, flexes, and medially rotates the first metacarpal bone, bringing the thumb across the palm from lateral to medial. This action, called **opposition,** occurs primarily at the carpometacarpal joint and allows the pad of the thumb to come into contact with the pads of the other fingers.

Nerve. The opponens pollicis is innervated by a branch of the **median nerve,** containing fibers from the eighth cervical and first thoracic nerves.

FLEXOR POLLICIS BREVIS. The flexor pollicis brevis lies medial and distal to the abductor pollicis brevis (Figs. 6-62; 6-64) and consists of superficial and deep parts. The **superficial part** is more lateral and larger than the deep part and *arises* from the distal border of the flexor retinaculum and the distal part of the tubercle on the trapezium. It passes along the radial side of the tendon of the flexor pollicis longus and, becoming tendinous, is *inserted* into the radial side of the base of the proximal phalanx of the thumb. In its tendon of insertion is found the radial sesamoid bone of this joint. The **deep part** is considerably smaller and located medial to the superficial part. It *arises* from the trapezoid and capitate bones, and its fibers take an oblique course to join the fibers of the superficial part, *inserting* by a common tendon onto the radial side of the proximal phalanx of the thumb.

There has been some disagreement with regard to the identity of the deep head of this

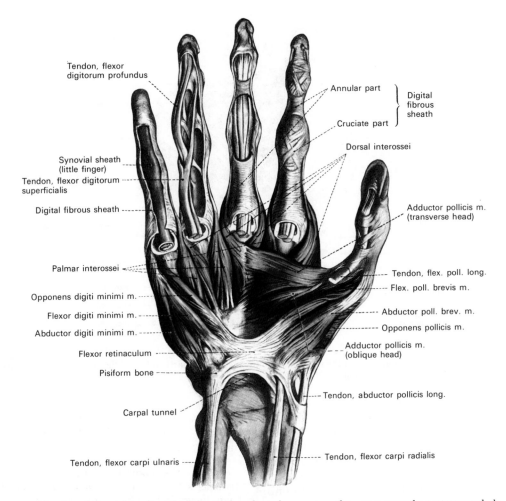

Tendon, flexor
digitorum profundus

Annular part

Cruciate part

Digital
fibrous
sheath

Dorsal interossei

Synovial sheath
(little finger)

Tendon, flexor digitorum
superficialis

Digital fibrous sheath

Adductor pollicis m.
(transverse head)

Palmar interossei

Tendon, flex. poll. long.

Flex. poll. brevis m.

Opponens digiti minimi m.

Flexor digiti minimi m.

Abductor digiti minimi m.

Abductor poll. brev. m.

Opponens pollicis m.

Adductor pollicis m.
(oblique head)

Flexor retinaculum

Pisiform bone

Tendon, abductor pollicis long.

Carpal tunnel

Tendon, flexor carpi radialis

Tendon, flexor carpi ulnaris

Fig. 6-62. Muscles of the right palm. The long tendons have been cut as they pass over the metacarpophalangeal joints. (From Benninghoff and Goerttler, *Lehrbuch der Anatomie des Menschen*, Urban and Schwarzenberg, Munich, 1980).

muscle, but the study of Day and Napier (1961) showed that nearly 80% of the hands that they examined followed the foregoing description. The deep head may be absent.

Action. The **flexor pollicis brevis** flexes the proximal phalanx at the metacarpophalangeal joint and then, indirectly, medially rotates the metacarpal bone of the thumb at the carpometacarpal joint.

Nerve. The superficial part of the flexor pollicis brevis is supplied by a branch of the **median nerve,** containing fibers from the eighth cervical and first thoracic nerves. The deep part usually receives innervation from the deep branch of the **ulnar nerve,** which also contains fibers from the eighth cervical and first thoracic nerves.

ADDUCTOR POLLICIS. The adductor pollicis consists of oblique and transverse heads. The **oblique head** (Figs. 6-62; 6-63) *arises* by

several slips from the capitate bone, the bases of the second and third metacarpals, the intercarpal ligaments, and the sheath of the tendon of the flexor carpi radialis. It may also receive a slip from the flexor retinaculum. From its origin most of the fibers pass obliquely downward and converge in a tendon that contains a sesamoid bone and is *inserted* into the ulnar side of the base of the proximal phalanx of the thumb. Some fibers usually pass more obliquely beneath the tendon of the flexor pollicis longus to join the lateral portion of the flexor pollicis brevis and the abductor pollicis brevis.

The **transverse head** of the adductor pollicis is the most deeply situated of the thenar muscles (Figs. 6-62 to 6-64). It is triangular in shape and *arises* by a broad base from the distal two-thirds of the palmar sur-

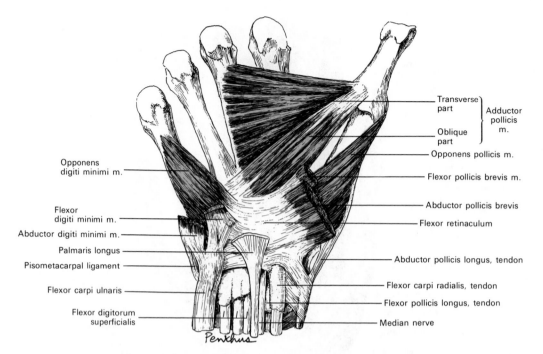

Transverse part ⎫
Oblique part ⎬ Adductor pollicis m.

Opponens pollicis m.

Flexor pollicis brevis m.

Abductor pollicis brevis

Flexor retinaculum

Abductor pollicis longus, tendon

Flexor carpi radialis, tendon

Flexor pollicis longus, tendon

Median nerve

Opponens digiti minimi m.

Flexor digiti minimi m.

Abductor digiti minimi m.

Palmaris longus

Pisometacarpal ligament

Flexor carpi ulnaris

Flexor digitorum superficialis

Fig. 6-63. A dissection of the palm of the right hand, showing the two parts of the adductor pollicis muscle, as well as the opponens pollicis and opponens digiti minimi muscles.

face of the third metacarpal bone. Its converging fibers form a distal border that is nearly transverse. Coursing toward the thumb, they join the fibers of the oblique head to be *inserted* onto the ulnar side of the base of the proximal phalanx of the thumb.

Action. The **adductor pollicis** brings the abducted proximal phalanx toward the palm of the hand, thereby adducting the thumb.

Nerve. The adductor pollicis is supplied by the deep palmar branch of the **ulnar nerve**, containing fibers from the eighth cervical and first thoracic nerves.

Hypothenar Muscles

The hypothenar muscles are the intrinsic muscles of the little finger. They are the:

Palmaris brevis
Abductor digiti minimi
Flexor digiti minimi brevis
Opponens digiti minimi

PALMARIS BREVIS. The palmaris brevis is a thin, quadrilateral superficial muscle, placed beneath the integument of the ulnar side of the hand (Fig. 6-59). It *arises* by tendinous fasciculi from the flexor retinaculum and

palmar aponeurosis, and it *inserts* into the skin on the ulnar border of the palm of the hand.

Action. The **palmaris brevis** draws the skin at the ulnar side of the palm toward the middle of the palm, increasing the height of the hypothenar eminence, as in clenching the fist. It also holds the hypothenar subcutaneous pad in place, as in catching a ball.

Nerve. The palmaris brevis is supplied by the superficial branch of the **ulnar nerve,** containing fibers from the eighth cervical and first thoracic nerves.

ABDUCTOR DIGITI MINIMI. This abductor of the little finger is situated on the ulnar border of the palm of the hand (Figs. 6-62 and 6-64). It *arises* from the pisiform bone and from the tendon of the flexor carpi ulnaris; it ends in a flat tendon, which divides into two slips: one is *inserted* into the ulnar side of the base of the first phalanx of the little finger; the other is *inserted* into the ulnar border of the dorsal digital aponeurosis of the extensor digiti minimi.

Accessory slips may join the muscle from the tendon of the flexor carpi ulnaris, the flexor retinaculum, the fascia of the distal

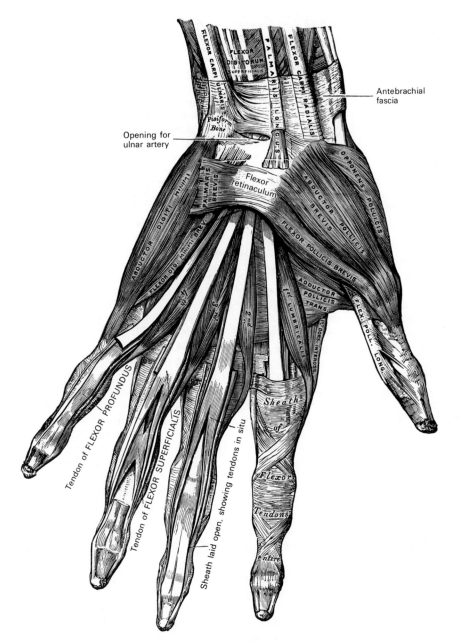

Antebrachial
fascia

Opening for
ulnar artery

FIG. 6-64.   The muscles of the left hand. Palmar surface.

forearm, or the tendon of the palmaris longus. A part of the muscle may insert into the fifth metacarpal bone.

Action.   The **abductor digiti minimi** abducts the little finger away from the ring finger, and it also flexes the proximal phalanx at the metacarpophalangeal joint.

Nerve.   The abductor digiti minimi is innervated by the deep branch of the **ulnar nerve**, containing fibers from the eighth cervical and first thoracic nerves.

FLEXOR DIGITI MINIMI BREVIS.   This short flexor of the little finger lies on the same plane as the preceding muscle, on its radial side (Figs. 6-62; 6-64). It *arises* from the convex surface of the hamulus of the hamate bone and the palmar surface of the flexor retinaculum, and it is *inserted* into the ulnar side of the base of the proximal phalanx of the little finger. It is separated from the abductor, at its origin, by the deep branches of the ulnar artery and nerve. This muscle is sometimes wanting, in which case the abductor is unusually large.

Action. The **flexor digiti minimi brevis** flexes the little finger at the metacarpophalangeal joint.

Nerve. The flexor digiti minimi brevis is supplied by the deep branch of the **ulnar nerve**, containing fibers from the eighth cervical and first thoracic nerves.

OPPONENS DIGITI MINIMI. The opponens digiti minimi is triangular in shape and placed immediately beneath the abductor and flexor (Figs. 6-62; 6-63). It *arises* from the convexity of the hamulus of the hamate bone and contiguous portion of the flexor retinaculum. It is *inserted* onto the whole length of the metacarpal bone of the little finger, along its ulnar margin.

Action. The **opponens digiti minimi** abducts, flexes, and laterally rotates the fifth metacarpal, thereby bringing the little finger in opposition to the thumb.

Nerve. The opponens digiti minimi is supplied by the deep branch of the **ulnar nerve**, containing fibers from the eighth cervical and first thoracic nerves.

Intermediate Muscles

Lumbricals
Interossei

THE LUMBRICALS. The lumbrical muscles (Fig. 6-64) are four small fleshy fasciculi that are associated with the tendons of the flexor digitorum profundus. The first and second *arise* from the radial sides and palmar surfaces of the tendons of the index and middle fingers, respectively; the third lumbrical *arises* from the contiguous sides of the tendons of the middle and ring fingers; and the fourth lumbrical *arises* from the contiguous sides of the tendons of the ring and little fingers. Each passes to the radial side of the corresponding finger, and opposite the metacarpophalangeal joint is *inserted* into the tendinous expansion of the extensor digitorum, which is called the extensor hood and covers the dorsal aspect of the finger.

Action. Each **lumbrical** muscle flexes the proximal phalanx at the corresponding metacarpophalangeal joint and extends the two distal phalanges at the interphalangeal joints.

Nerves. The first and second lumbrical muscles are supplied by twigs from the third and fourth digital branches of the **median nerve**, containing fibers from the eighth cervical and first thoracic nerves. The third and fourth lumbricals receive twigs from the deep palmar branch of the **ulnar nerve**, which also contains fibers from the eighth cervical and first thoracic segments. The third lumbrical may receive filaments from both nerves, or even all its supply from the median nerve.

Variations. The lumbricals may vary in number from two to five or six, and there is considerable variation in their attachments.

THE INTEROSSEI

The interossei are so named because they occupy the intervals between the metacarpal bones. They are divided into two sets, dorsal and palmar.

DORSAL INTEROSSEI. Each of the *four* dorsal interossei (Figs. 6-62; 6-65) is a bipennate muscle that *arises* by two heads from the adjacent sides of the metacarpal bones, but more extensively from the metacarpal bone of the finger into which the muscle attaches distally. Each interosseous muscle is *inserted* into the base of the proximal phalanx and into the corresponding aponeurosis that forms the dorsal hood of the tendon of the extensor digitorum. Between the double origin of each of these muscles is a narrow triangular interval. Through the first of these intervals the radial artery passes, and through each of the other three a perforating branch from the deep palmar arch is transmitted.

The **first dorsal interosseous** (also called the **abductor indicis**) is larger than the others. It is flat and triangular in form, and its two heads of origin are separated by a fibrous arch for the passage of the radial artery from the dorsum to the palm of the hand. The **lateral head** *arises* from the proximal half of the ulnar border of the first metacarpal bone, while the **medial head** *arises* from almost the entire length of the radial border of the second metacarpal bone. Its tendon is *inserted* almost entirely onto the radial side of the proximal phalanx of the index finger, although some fibers may reach the dorsal extensor aponeurosis.

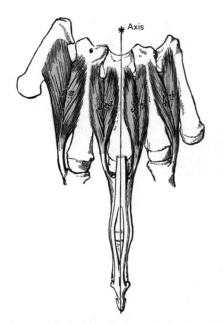

FIG. 6-65. The dorsal interossei of left hand. Line down middle finger represents an imaginary axis. Abduction and adduction are determined according to this reference.

The **second** and **third dorsal interosseous** muscles both *insert* onto the middle finger, the former to its radial side and the latter to its ulnar side. The tendon of the second muscle inserts about equally onto the proximal phalanx and the dorsal extensor aponeurosis, while most of the tendon of the third muscle attaches to the dorsal extensor aponeurosis.

The tendon of the **fourth dorsal interosseous** muscle *inserts* onto the ulnar side of the ring finger. Its fibers are attached about equally to the proximal phalanx and the dorsal extensor aponeurosis.

Action. The **dorsal interossei** abduct the fingers from an imaginary line drawn longitudinally through the axis of the middle finger (Fig. 6-65). They also flex the fingers at the metacarpophalangeal joints and extend the two distal phalanges at the interphalangeal joints.

Nerves. All the dorsal interossei are supplied by the deep palmar branch of the **ulnar nerve**, containing fibers from the eighth cervical and first thoracic nerves.

PALMAR INTEROSSEI. The palmar interossei are smaller than the dorsal interossei,

and they are placed upon the palmar surfaces of the metacarpal bones rather than between them. Three palmar interossei are clearly distinct, but some authors describe four muscles. Each *arises* from the entire length of the metacarpal bone of one finger and is *inserted* into the side of the base of the proximal phalanx and aponeurotic expansion of the extensor digitorum tendon to the same finger.

The **first** *arises* from the ulnar side of the second metacarpal bone and is *inserted* into the same side of the proximal phalanx of the index finger. The **second** *arises* from the radial side of the fourth metacarpal bone and is *inserted* into the same side of the ring finger. The **third** *arises* from the radial side of the fifth metacarpal bone and is *inserted* into the same side of the little finger.

Some authors add another palmar interosseous muscle to their description, attaching the term *first palmar interosseous* to a few fibers that pass from the base of the first metacarpal to the base of the proximal phalanx. In their descriptions, the other muscles in the foregoing discussion become the second, third, and fourth muscles, respectively. Since the thumb has a large adductor muscle of its own, these fibers have been considered as part of that muscle in the present account.

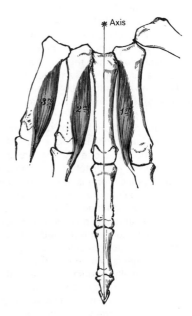

FIG. 6-66. The palmar interossei of left hand.

Action. The **palmar interossei** adduct the fingers toward an imaginary longitudinal line through the axis of the middle finger (Fig. 6-66). They also flex the proximal phalanx at the metacarpophalangeal joint and extend the two distal phalanges at the interphalangeal joints.

Nerves. The palmar interossei are supplied by twigs from the deep palmar branch of the **ulnar nerve,** containing fibers from the eighth cervical and first thoracic nerves.

Group Actions. Flexion of the fingers in grasping an object is performed by the flexors digitorum superficialis and profundus. The wrist extensors contract synergistically to prevent flexion of the wrist while the fingers are being flexed. The action of both flexors on the two terminal phalanges can be performed independent of flexing the proximal phalanx by calling into play the synergistic action of the extensor digitorum on the proximal phalanx. Flexion of the proximal phalanx at the same time as extension of the two distal joints is performed by the dorsal and palmar interossei and the lumbrical muscles.

Extension of the fingers is performed by the extensor digitorum, extensor indicis, and extensor digiti minimi. The wrist flexors contract synergistically to prevent extension at the wrist while the fingers are being extended. The long extensors have a relatively weak action on the two terminal joints and are assisted in this action by the lumbricals and the interossei.

Abduction of the fingers is performed by the dorsal interossei and the abductor digiti minimi, considering the longitudinal axis of the middle finger as the center of the hand. Full abduction can be carried out only if the fingers are extended.

Adduction of the fingers is performed by the palmar interossei, and this action can be carried out with the fingers either flexed or extended.

The thumb is so placed that its plane of flexion and extension is at right angles to that of the fingers. Flexion and extension of the thumb, therefore, are in the same plane as abduction and adduction of the fingers; abduction and adduction of the thumb are in the same plane as flexion and extension of the fingers.

Flexion of the distal phalanx of the thumb is performed by the flexor pollicis longus, and flexion of the proximal phalanx alone by the flexor pollicis brevis. *Extension* of the distal phalanx is performed by the extensor pollicis longus, the proximal phalanx by the extensor pollicis brevis. *Abduction of the thumb* is by the abductors pollicis longus and brevis; adduction by the adductor pollicis. It is seldom that extension or abduction is performed independently of each other, most normal activity being a mixture of the two movements. When the thumb is used in grasping, it is initially abducted by the abductors and rotated by the opponens so that its palmar surface faces the palm of the hand. The actual grasping then is performed largely by the flexor pollicis longus.

Fasciae and Muscles of the Lower Limb

The fasciae and muscles of the lower limb will be subdivided into groups corresponding to the different regions of the limb.

Muscles of the iliac region
Muscles of the thigh
Muscles of the leg
Muscles of the foot

MUSCLES AND FASCIAE OF THE ILIAC REGION

Psoas major
Psoas minor
Iliacus

The fascia covering the intra-abdominal surface of the iliacus and psoas muscles is part of the endoabdominal or internal investing layer of deep fascia; it forms a continuation over these muscles of the transversalis fascia that lines the abdominal cavity. Following the iliac muscles under the inguinal ligament and into the thigh, it becomes continuous with the fascia lata, which is a portion of the external investing fascia.

ILIAC FASCIA. The posterior abdominal portion of the iliac fascia is attached superiorly to the entire length of the inner lip of the crest of the ilium, along with the muscle. Above the iliac crest it is continuous with the definitive transversalis fascia, and near the vertebral column, it is continuous with the fascia covering the abdominal surface of the quadratus lumborum muscle. It is continuous medially with the psoas fascia, and after attaching to the arcuate line of the ilium, it continues down into the lesser pelvis as the obturator internus fascia. At the inguinal ligament it fuses with the transversalis fascia of the lateral portion of the anterior abdominal wall, and follows the surface of the iliacus under the ligament and into the thigh. It is securely attached to the ligament as it passes beneath it, and in the thigh, it becomes continuous with the part of the fascia lata overlying the sartorius muscle laterally and with the iliopectineal fascia medially.

PSOAS FASCIA. The superior extremity of the psoas fascia is intimately blended with the medial arcuate ligament of the diaphragm, which stretches from the bodies to the transverse processes of the first or second lumbar vertebrae. It is attached medially, by a series of arched processes, to the intervertebral discs and prominent margins of the vertebrae and to the upper part of the sacrum. The intervals left between these arched processes and the constricted bodies of the vertebrae transmit the lumbar arteries and veins and the filaments of the sympathetic trunk. It is continuous laterally with the fasciae covering the quadratus lumborum and iliacus muscles. At the inguinal ligament and in the thigh, the psoas and iliac fasciae help to form part of the iliopectineal fascia.

ILIOPECTINEAL FASCIA.   The iliopectineal fascia is formed by a blending of: (1) the fascia covering the femoral portions of the iliacus and psoas; (2) the fascia over the proximal portion of the pectineus; and (3) a thickened band, called the **iliopectineal arch,** which dips between the psoas major and the femoral vessels as they pass under the inguinal ligament. Thus, the iliopectineal fascia is continuous, under the inguinal ligament, with the iliac, the psoas, and the transversalis fasciae. The fascial layer covering the iliacus, psoas major, and pectineus muscles forms a fascial sheet that stretches across the floor of the femoral (Scarpa's) triangle. At the junction of the iliopsoas and pectineal portions of the iliopectineal fascia, it attaches firmly to the iliopectineal eminence of the ilium and to the pubocapsular ligament of the hip joint. From the attachment to the iliopectineal eminence, the fascia passes outward as the thickened band mentioned above, the iliopectineal arch, which separates the psoas major muscle and the femoral vessels and courses toward the inguinal ligament to which it becomes attached. This band divides the interval deep to the inguinal ligament, that is, between the ligament and the pelvic bone, into two parts known as the lacuna musculorum and the lacuna vasorum. The **lacuna musculorum** contains the iliacus and psoas major muscles and the femoral nerve. The **lacuna vasorum** contains the femoral artery and vein and the femoral canal.

The **iliopsoas** muscle is frequently regarded as a single structure because it is a blending of two muscles, the psoas major and the iliacus. Its two parts, however, will be described separately.

PSOAS MAJOR.   The psoas major is a long fusiform muscle lateral to the lumbar region of the vertebral column and brim of the lesser pelvis (Fig. 6-67). It *arises* (1) from the anterior surfaces and lower borders of the transverse processes of all the lumbar vertebrae; (2) by five slips from the sides of the bodies and the corresponding intervertebral discs of the last thoracic and all the lumbar vertebrae (each slip is attached to the adjacent upper and lower margins of two vertebrae, and to the intervertebral disc); (3) from a series of tendinous arches that extends across the constricted parts of the bodies of the lumbar vertebrae between the previously described slips. The lumbar arteries and veins and filaments from the sympathetic trunk pass beneath these tendinous arches. The muscle proceeds downward across the brim of the lesser pelvis and, diminishing gradually in size, passes beneath the inguinal ligament, anterior to the capsule of the hip joint. It ends in a tendon that also receives nearly all of the fibers of the iliacus and is *inserted* onto the lesser trochanter of the femur. The large **subtendinous iliac bursa,** which may communicate with the cavity of the hip joint, separates the tendon from the pubis and the capsule of the joint (Fig. 6-70).

Action.   The **psoas major** (in conjunction with the iliacus) is a powerful flexor of the thigh at the hip joint. If both psoas major muscles are fixed from below, they act as important flexors of the trunk on the hip, as in sitting up from the supine position. The psoas major may assist in maintaining posture by preventing hyperextension at the hip joint. Although mechanically it appears to be a medial rotator of the thigh, electromyographic studies contradict this interpretation and, in fact, indicate that it may be active in lateral rotation.

Nerves.   The psoas major is innervated by branches of the lumbar plexus, containing fibers from the **second, third, and fourth lumbar nerves.**

PSOAS MINOR.   The psoas minor is a long, slender muscle that is placed anterior to the psoas major on the posterior abdominal wall (Fig. 6-67). It *arises* from the sides of the bodies of the twelfth thoracic and first lumbar vertebrae and from the intervertebral disc between them. It ends in a long, flat tendon which is *inserted* into the pectineal line, the iliopectineal eminence, and, laterally, the iliac fascia. This muscle is absent bilaterally in over half the subjects studied.

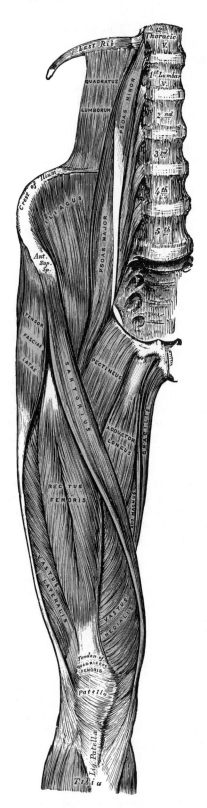

Fig. 6-67.   Muscles of the iliac and anterior femoral regions.

Action.   The **psoas minor** is a weak flexor of the trunk and lumbar spinal column.

Nerve.   The psoas minor is supplied by a branch of the **first lumbar nerve.**

THE ILIACUS.   The iliacus is a flat, triangular muscle that fills the iliac fossa (Fig. 6-67). It *arises* from the superior two-thirds of this fossa and from the inner lip of the iliac crest; dorsally from the anterior sacroiliac and the iliolumbar ligaments and base of the sacrum; and anteriorly from as far as the anterior superior and anterior inferior iliac spines, and the notch between them. Most of the fibers converge inferiorly to be *inserted* into the lateral aspect of the tendon of the psoas major, but some fibers extend to the body of the femur for about 2.5 cm distal and anterior to the lesser trochanter.

Action.   The **iliacus** (along with the psoas major) is a powerful flexor of the thigh at the hip joint. It joins the psoas in a number of other actions as the iliopsoas.

Nerves.   The iliacus is supplied by branches of the **femoral nerve,** containing fibers from the second and third lumbar nerves.

Variations.   The **iliacus minor** or **iliocapsularis,** a small detached part of the iliacus, is frequently present. It arises from the anterior inferior spine of the ilium and is inserted into the lower part of the intertrochanteric line of the femur or into the iliofemoral ligament.

FASCIAE AND MUSCLES OF THE THIGH

The fascial layers of the thigh and buttocks overlie four functional groups of muscles which will be described as follows:

Anterior femoral muscles
Medial femoral muscles
Muscles of the gluteal region
Posterior femoral muscles

The fascial layers include the superficial fascia and the fascia lata; the latter represents the deep fascia, which continues over the gluteal region as the gluteal fascia.

SUPERFICIAL FASCIA.   The superficial fascia forms a prominent layer over the entire thigh. It usually contains a considerable amount of fat, but it varies in thickness in different regions. It is continuous with the superficial fascia of the abdomen, that of the leg, and, above the gluteal region, with the superficial fascia of the back. It may be separated into a superficial fatty layer and a deeper membranous layer between which are found the superficial vessels and nerves, the superficial inguinal lymph nodes, and the great saphenous vein. In well-nourished individuals, the fatty tissue of the superficial layer is usually divided by fibrous membranes into two or three subsidiary layers, which are associated with the emergence of the superficial nerves. The deep or fibrous layer is adherent to the fascia lata a little below the inguinal ligament and along the upper medial portion of the thigh. It is attached to the margin of the **saphenous hiatus** (fossa ovalis) and fills the opening itself with an irregular layer of spongy tissue, which is called the **cribriform fascia** because it is pierced by numerous openings for the passage of the great saphenous vein and other blood and lymphatic vessels.

A large subcutaneous bursa is found in the superficial fascia over the patella.

FASCIA LATA.   The external investing or deep fascia of the thigh is named the fascia lata from its broad extent. Its thick, lateral portion is commonly taken to be typical of its texture, but it is thin in some areas where it has not been reinforced by fibrous contributions from the tendons. *Superiorly,* it is on the same plane as the external oblique aponeurosis and thoracolumbar fasciae after being attached to the pelvis and inguinal ligament; *distally,* it is continuous with the fascia of the leg.

The **medial portion** of the fascia lata overlies the adductor group of muscles. It is thin and gray, and it is not aponeurotic. It is attached to the ischial tuberosity and ischiopubic ramus, and beyond this it is on the same plane as the deep fascia of the perineum. At the knee it is thick and aponeurotic, having been strengthened by fibers from the tendon of the sartorius muscle.

The **anterior portion** of the fascia lata is attached to the pubic tubercle, inguinal ligament, and anterior superior iliac spine. Just below the lateral half of the inguinal ligament, it is a single sheet formed by the fusion of three abdominal fasciae. Of these, the most superficial is the deep fascia overlying the aponeurosis of the external oblique (fascia innominata), which passes superficial to the inguinal ligament; the middle one is the transversalis fascia on the inner aspect of the anterior abdominal wall, which passes under the inguinal ligament; the internal one

also passes under the inguinal ligament and is the continuation of the iliac fascia. Below the medial half of the inguinal ligament, the middle and internal layers mentioned above fail to join the fascia lata at the ligament and continue into the thigh as the anterior and posterior portions of the femoral sheath. The fascia lata in this ventromedial region over the femoral vessels is thickened and laminated and has an opening through it called the saphenous hiatus for the passage of the great saphenous vein. The outer lamina is attached to the pubic tubercle along with the inguinal ligament. It has a free falciform margin which crosses the proximal end of the great saphenous vein and spirals distally around the vein to join the deep lamina medial to the vein. The two laminae are separated by a pad of fatty tissue. The anterior portion of the fascia lata is thicker than the medial, but is truly aponeurotic only near the knee, where it is reinforced by fibers from the tendons of the quadriceps femoris muscle.

The **lateral portion** of the fascia lata is a thick strong aponeurosis containing the tendinous fibers of insertion of the gluteus maximus and the tensor fasciae latae muscles. It is attached proximally to the crest of the ilium and the dorsum of the sacrum. Between the iliac crest and the superior border of the gluteus maximus, it is thickened by vertical tendinous bundles and is known as the **gluteal aponeurosis** (Fig. 6-72), which is used by the gluteus medius for part of its origin. At the border of the gluteus maximus, the gluteal aponeurosis splits to enclose the muscle. Thus, the thin external layer of this **gluteal fascia** is closely bound to the superficial fascia and the muscle, and it sends septa down between large bundles of the muscle. In the region over the greater trochanter, the muscular fasciculi end in a broad tendon, which becomes fused with the fascia lata and is called the **iliotibial tract.** Below the anterior part of the iliac crest, the fascia splits to enclose the tensor fasciae latae, which is inserted into the iliotibial tract below the gluteus maximus. The iliotibial tract is separated from the underlying vastus lateralis by a distinct fascial cleft. It is inserted into the tibia and is blended with fibrous expansions from the vastus lateralis and biceps femoris muscles.

The **posterior portion** of the fascia lata is formed proximally by the union of the two layers of fascia enclosing the gluteus maximus at its inferior border. It descends to cover the hamstring muscles and the popliteal fossa.

Two strong intermuscular septa (Fig. 6-69) connect the deep surface of the fascia lata with the linea aspera of the femur. The **lateral intermuscular septum** is the stronger. It separates the vastus lateralis from the biceps femoris and is used by both muscles for the origin of some of their fibers. It extends from the insertion of the gluteus maximus to the lateral condyle. The **medial intermuscular septum** lies between the vastus medialis and the adductors and pectineus. Its superficial portion near the fascia lata splits to enclose the sartorius muscle, and in its descent the medial septum contributes to the formation of the adductor canal, which transmits the femoral vessels.

The **saphenous hiatus** is an oval aperture in the fascia lata in the proximal part of the thigh, a little below the medial end of the inguinal ligament (Fig. 6-68). The great saphenous vein passes through the hiatus just before it joins the femoral vein. The fascia lata in this part of the thigh is laminated into two leaves separated by fat. The superficial layer is attached to the inguinal ligament and pubic tubercle. It ends abruptly in a free border, the **falciform margin of the fossa,** which forms a spiral rim beginning at the pubic tubercle and extending around the superior, lateral, and then inferior aspects of the saphenous vein. Medial to the vein, the superficial leaf merges with the deep leaf. The upper and lateral part of the falciform margin is called the **superior cornu,** while the medial and lower part is called the **inferior cornu.** The deep leaf is formed by the pectineal, iliopectineal, and iliac fasciae. The hiatus is filled and covered by a thickened pad derived from the deep layer of superficial fascia, called the **cribriform fascia.**

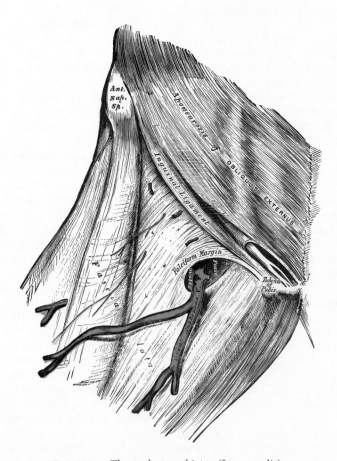

Fig. 6-68. The saphenous hiatus (fossa ovalis).

Anterior Femoral Muscles

Sartorius
Quadriceps femoris
   Rectus femoris
   Vastus lateralis
   Vastus medialis
   Vastus intermedius
Articularis genus

THE SARTORIUS. The sartorius is the longest muscle in the body (Figs. 6-67; 6-69). It is narrow and ribbon-like, and it *arises* by tendinous fibers from the anterior superior iliac spine and the upper half of the notch below it. The sartorius passes obliquely across the upper anterior part of the thigh from lateral to medial, and it then descends vertically as far as the medial side of the knee, passing posterior to the medial condyle of the femur. It ends in a tendon that curves obliquely anteriorly and expands into a broad aponeurosis, which is *inserted* into the proximal part of the medial surface of the body of the tibia, anterior to the gracilis and semitendinosus, and nearly as far forward as the anterior crest. A bursa separates the insertion of the sartorius from that of the underlying gracilis. The proximal part of the aponeurosis is curved backward over the upper edge of the tendon of the gracilis so as to be inserted behind it. A tendinous expansion from the upper margin of its insertion blends with the capsule of the knee joint; another, from its

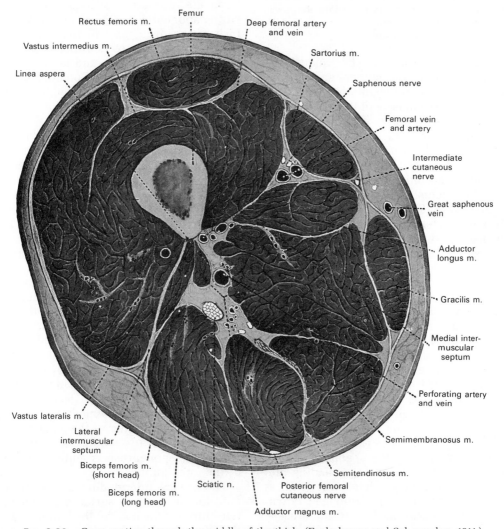

Femur
Rectus femoris m.
Deep femoral artery and vein
Vastus intermedius m.
Sartorius m.
Linea aspera
Saphenous nerve
Femoral vein and artery
Intermediate cutaneous nerve
Great saphenous vein
Adductor longus m.
Gracilis m.
Medial inter-muscular septum
Perforating artery and vein
Vastus lateralis m.
Lateral intermuscular septum
Semimembranosus m.
Biceps femoris m. (short head)
Semitendinosus m.
Sciatic n.
Posterior femoral cutaneous nerve
Biceps femoris m. (long head)
Adductor magnus m.

FIG. 6-69. Cross section through the middle of the thigh. (Eycleshymer and Schoemaker, 1911.)

lower border, blends with the fascia on the medial side of the leg.

Action. The **sartorius** flexes, abducts, and laterally rotates the thigh at the hip joint. It also flexes the leg at the knee joint and, after flexing, rotates it slightly medially. Its complicated actions are thought to draw the lower extremity into a sitting position, with the leg crossed in such a way that the heel of one limb is placed on the knee of the opposite limb, the so-called sitting tailor's position.

Nerves. The sartorius is supplied by branches of the **femoral nerve,** usually two in number, containing fibers from the second and third lumbar nerves. These generally arise from the femoral nerve with its anterior cutaneous branches.

Variations. At times slips of origin to the sartorius are derived from the outer end of the inguinal ligament, the notch of the ilium, the iliopectineal line, or the pubis. The muscle may be split into two parts, with one part being inserted into the fascia lata, the femur, the ligament of the patella, or the tendon of the semitendinosus.

QUADRICEPS FEMORIS. The quadriceps femoris consists of four component muscles or heads of origin on the anterior thigh. It is the great extensor muscle of the leg, forming a large fleshy mass that covers the front and sides of the femur. It is subdivided into separate portions, which have received distinctive names. One, occupying the middle of the thigh and connected above with the ilium, is called the **rectus femoris** because of its straight course (Fig. 6-67). The other three lie in immediate connection with the body of the femur, which they cover from the trochanters to the condyles. The portion on the lateral side of the femur is termed the **vastus lateralis;** that covering the medial side, the **vastus medialis;** and that between, the **vastus intermedius.**

The **rectus femoris** is situated in the middle of the anterior thigh (Fig. 6-67). It is fusiform in shape, and its superficial fibers are arranged in a bipennate manner, whereas its deep fibers course straight down to the deep aponeurosis. It *arises* by two tendons: one, the anterior or straight, attaches to the anterior inferior iliac spine; the other, the posterior or reflected, attaches to a groove above the posterior brim of the acetabulum. The two unite at an acute angle, spreading into an aponeurosis from which the muscle fibers arise; this aponeurosis is prolonged downward on the anterior surface of the muscle. The muscle ends in a broad and thick aponeurosis. This occupies the lower two-thirds of its posterior surface, and it gradually becomes narrowed into a flattened tendon, which is *inserted* into the base of the patella.

The **vastus lateralis** is the largest component of the quadriceps femoris (Fig. 6-67). It *arises* by a broad aponeurosis from the proximal part of the intertrochanteric line, the anterior and inferior borders of the greater trochanter, the lateral lip of the gluteal tuberosity, and the proximal half of the lateral lip of the linea aspera. This aponeurosis covers the upper three-fourths of the muscle, and from its deep surface many additional fibers take origin. A few other fibers arise from the tendon of the gluteus maximus and from the lateral intermuscular septum between the vastus lateralis and short head of the biceps femoris. The fibers form a large fleshy mass, which is attached to a strong aponeurosis that lies on the deep surface of the distal part of the muscle. This aponeurosis becomes contracted and thickened into a flat tendon and is *inserted* into the lateral border of the patella, blending with the quadriceps femoris tendon and giving a tendinous expansion to the capsule of the knee joint.

The vastus medialis and vastus intermedius appear to be inseparably united, but when the rectus femoris has been reflected, a narrow cleft will be observed extending proximally from the medial border of the patella between the two muscles. The separation may be continued as far as the lower part of the intertrochanteric line, where, however, the two muscles are frequently fused.

The **vastus medialis** (Fig. 6-67) *arises* from the lower half of the intertrochanteric line, the medial lip of the linea aspera, the upper part of the medial supracondylar line, the tendons of the adductor longus and adductor magnus, and the medial intermuscular septum. Its fibers are directed downward and forward and are chiefly attached to an aponeurosis that lies on the deep surface of the muscle. The vastus medialis is *inserted* into the medial border of the patella and the quadriceps femoris tendon, and an expansion of the aponeurosis is sent to the capsule of the knee joint.

The **vastus intermedius** *arises* from the anterior and lateral surfaces of the upper two-thirds of the body of the femur and from the lower part of the lateral intermuscular septum. Its fibers end in an aponeurosis on the anterior surface of the muscle, which forms the deep part of the quadriceps femoris tendon.

The **tendons** of the different portions of the quadriceps unite at the distal part of the thigh to form a single strong tendon that is inserted into the base of the patella, although a few fibers pass over it to blend with the ligamentum patellae. More properly, the patella may be regarded as a sesamoid bone, developed in the tendon of the quadriceps; the ligamentum patellae, which is continued from the apex of the patella to the tuberosity of the tibia, may be regarded as the proper tendon of insertion of the muscle, and the medial and lateral patellar retinacula as expansions from its borders. The **suprapatellar bursa,** which usually communicates with the cavity of the knee joint, is situated between the femur and the portion of the quadriceps tendon above the patella. The **deep infrapatellar bursa** is interposed between the tendon and the upper part of the anterior tibia. A third bursa, the **prepatellar bursa,** is placed superficial to the patella itself.

Action. The entire **quadriceps** muscle extends the leg at the knee joint. The **rectus femoris** also flexes the thigh at the hip joint. Additionally, the rectus femoris assists the iliopsoas muscle in flexing the trunk on the thigh.

Nerves. The quadriceps femoris is supplied by branches of the **femoral nerve,** containing fibers from the second, third, and fourth lumbar nerves.

ARTICULARIS GENUS.   The articularis genus is a small muscle, usually distinct from the vastus intermedius but occasionally blended with it. It *arises* from the anterior surface of the lower part of the body of the femur and is *inserted* into the upper part of the synovial membrane of the knee joint. It sometimes consists of several separate muscular bundles.

Action. The **articularis genus** draws the synovial membrane of the knee joint upward as the leg is extended, thereby preventing folds of the membrane from being compressed within the joint.

Nerve. The articularis genus is supplied by fibers of the **femoral nerve** through its branch to the vastus intermedius muscle.

Medial Femoral Muscles

Gracilis
Pectineus

Adductor longus
Adductor brevis
Adductor magnus

THE GRACILIS.   The gracilis, the most superficial muscle on the medial aspect of the thigh (Figs. 6-67; 6-69), is thin and flat, broad proximally, narrow and tapering distally. It *arises* by a thin aponeurosis from the lower part of the body of the pubis near the symphysis and from the adjoining pubic ramus. The fibers run vertically downward and end in a rounded tendon, which passes behind the medial condyle of the femur and then curves around the medial condyle of the tibia. At this site, it becomes flattened and is *inserted* into the upper part of the medial surface of the body of the tibia, below the tibial condyle. A few of the fibers of the distal part of the tendon are prolonged into the deep fascia of the leg. At its insertion the tendon is situated immediately above that of the semitendinosus, and its upper edge is overlapped by the tendon of the sartorius, with which it is in part blended. The tendons of insertion of the gracilis, sartorius, and semitendinosus blend to form a tendinous expansion called the **pes anserinus.** The **anserine bursa** lies between these tendons and the tibia.

Action. The **gracilis** adducts the thigh. It also flexes the leg at the knee and, after it is flexed, assists in its medial rotation.

Nerve. The gracilis is supplied by a branch of the anterior division of the **obturator nerve,** containing fibers from the second and third lumbar nerves.

THE PECTINEUS.   The pectineus is a flat quadrangular muscle, situated at the anterior part of the upper and medial aspect of the thigh (Fig. 6-67). It *arises* from the pectin pubis (pectineal line) and, to a slight extent, from the surface of bone anterior to it, between the iliopectineal eminence and the tubercle of the pubis. Some fibers also arise from the fascia covering the anterior surface of the muscle itself. The fibers pass downward, backward, and lateralward, to be *inserted* along a rough line leading from the lesser trochanter to the linea aspera.

Action. The **pectineus** flexes and adducts the thigh, and rotates it medially.

Nerve. The pectineus is usually supplied by a branch of the **femoral nerve,** containing fibers from the second, third, and fourth lumbar nerves. When an **accessory obturator nerve** is present, one of its branches is distributed to the pectineus. The muscle may receive a branch from the **obturator nerve.**

Variations. The pectineus may consist of two incompletely separated strata. The lateral stratum is supplied by a branch of the femoral nerve or the accessory obturator, if present, and the medial stratum, when present, is supplied by the obturator nerve. The muscle may attach to the capsule of the hip joint.

ADDUCTOR LONGUS. This muscle is the most anterior of the three adductors, and it is triangular in shape and lies in the same plane as the pectineus (Fig. 6-71). It *arises* by a flat, narrow tendon from the anterior pubis, at the angle of junction of the crest with the symphysis, and soon expands into a broad fleshy belly. This passes downward, backward, and laterally to be *inserted* by an aponeurosis on the middle third of the femur along the linea aspera. Its insertion attaches between the origin of the vastus medialis and the insertion of the adductor magnus, muscles with which its fibers are usually blended.

Action. The **adductor longus** adducts, flexes, and tends to rotate the thigh medially.

Nerve. The adductor longus receives innervation from a branch of the anterior division of the **obturator nerve,** containing fibers from the second, third, and fourth lumbar nerves.

Variations. The adductor longus may be doubled, it may extend to the knee, or it may be more or less united with the pectineus muscle.

ADDUCTOR BREVIS. The adductor brevis is situated immediately deep to the pectineus and adductor longus (Fig. 6-71). It is somewhat triangular in form and *arises* by a narrow attachment from the outer surface of the inferior ramus of the pubis, between the gracilis and obturator externus. Its fibers, passing backward, laterally, and downward, are *inserted* by an aponeurosis into the femur along a line leading from the lesser trochanter to the linea aspera and into the upper part of the linea aspera, immediately behind the pectineus and proximal part of the adductor longus.

Action. Similar to the adductor longus, the **adductor brevis** adducts, flexes, and tends to rotate the thigh medially at the hip joint.

Nerve. The adductor brevis is innervated by a branch of the **obturator nerve,** usually from its anterior division, containing fibers from the second, third, and fourth lumbar nerves.

Variations. The adductor brevis may be divided into two or three parts, or it may be united with the adductor magnus.

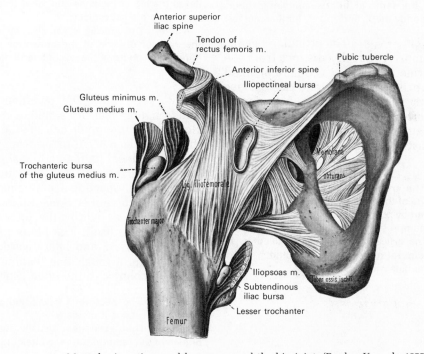

FIG. 6-70. Muscular insertions and bursae around the hip joint. (Rauber-Kopsch, 1955.)

aspect of the tuberosity of the ischium. The fibers that arise from the ramus of the pubis are short and horizontal, and they are *inserted* into the rough line leading from the greater trochanter to the linea aspera, medial to the gluteus maximus. The **adductor minimis** is the name given to this part of the adductor magnus when it forms a distinct muscle. The fibers arising from the ramus of the ischium are directed downward and laterally, to be *inserted* by means of a broad aponeurosis into the linea aspera and the upper part of the medial supracondylar line. The medial portion of the muscle, composed principally of the fibers arising from the tuberosity of the ischium, forms a thick fleshy mass consisting of coarse bundles. These descend almost vertically and form, in the distal third of the thigh, a rounded tendon that is *inserted* into the adductor tubercle on the medial condyle of the femur. It is also connected by a fibrous expansion to the medial supracondylar line leading proximally from the tubercle to the linea aspera. Along the insertion of the muscle, there is a series of openings in the aponeurosis formed by tendinous arches attached to the bone. The upper four openings are small and give passage to the perforating branches of the profunda femoris artery. The lowest opening, called the **hiatus tendineus,** is large and transmits the femoral vessels to the popliteal fossa.

Action. The **adductor magnus** is a powerful adductor of the thigh. Its upper portion also weakly flexes and medially rotates the thigh, while the lower portion extends and laterally rotates the thigh.

Nerves. The adductor magnus is supplied by branches of the posterior division of the **obturator nerve,** containing fibers from the second, third, and fourth lumbar nerves, and by a branch from the **sciatic nerve.** The ischiocondylar portion, which is attached to the adductor tubercle, is derived from the flexor or hamstring muscles of lower forms and is the portion supplied by the sciatic nerve.

Variations. The adductor magnus may be fused with the quadratus femoris or with either the adductor longus or brevis.

Muscles of the Gluteal Region

The muscles of the gluteal region are the principal extensors and lateral rotators of the thigh at the hip joint. They include the:

Gluteus maximus
Gluteus medius
Gluteus minimus

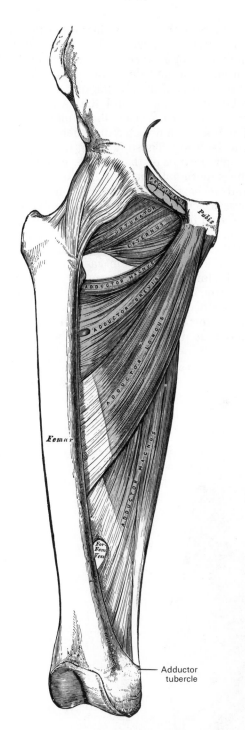

Adductor tubercle

FIG. 6-71. Deep muscles of the medial femoral region.

ADDUCTOR MAGNUS.   The adductor magnus is a large triangular muscle, situated on the medial side of the thigh (Fig. 6-71). It *arises* from a small part of the inferior ramus of the pubis, from the inferior ramus of the ischium, and from the inferior and lateral

Tensor fasciae latae
Piriformis
Obturator internus
Gemellus superior
Gemellus inferior
Quadratus femoris
Obturator externus

GLUTEUS MAXIMUS.   The gluteus maximus is the largest and most superficial muscle in the gluteal region. It is a broad and thick fleshy mass, quadrilateral in shape, which forms the prominence of the buttocks (Fig. 6-72). Its large size is one of the most characteristic features of the human muscular system, connected as it is with the ability and power of maintaining the trunk in the erect posture. The muscle is remarkably coarse in structure, being composed of fasciculi lying parallel and collected into large bundles separated by fibrous septa. It *arises* from the posterior gluteal line of the ilium and the rough portion of bone, including the crest, immediately superior and dorsal to it; from the posterior surface of the lower part of the sacrum and the side of the coccyx; from the aponeurosis of the erector spinae, the sacrotuberous ligament, and the fascia (gluteal aponeurosis) covering the gluteus medius. The fibers are directed obliquely downward and laterally. Those forming the upper and larger portion of the muscle, together with the superficial fibers of the lower portion, end in a thick tendinous lamina, which passes across the greater trochanter and is *inserted* into the iliotibial band of the fascia lata. The deeper fibers of the lower portion of the muscle are inserted into the gluteal tuberosity between the vastus lateralis and adductor magnus.

Three bursae are usually found in relation to the deep surface of this muscle. One of these, the **trochanteric bursa of the gluteus maximus,** is large and generally multilocu-

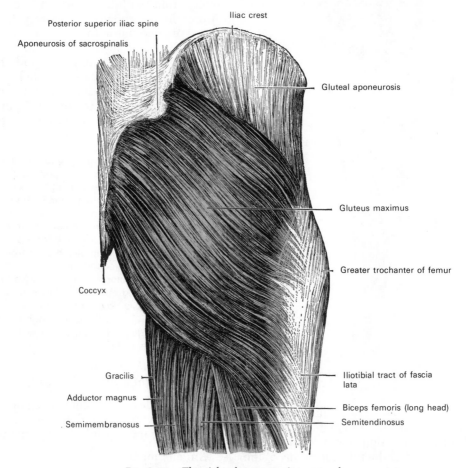

Posterior superior iliac spine
Aponeurosis of sacrospinalis
Iliac crest
Gluteal aponeurosis
Gluteus maximus
Greater trochanter of femur
Coccyx
Gracilis
Adductor magnus
Semimembranosus
Iliotibial tract of fascia lata
Biceps femoris (long head)
Semitendinosus

FIG. 6-72.   The right gluteus maximus muscle.

lar, and separates the muscle from the greater trochanter. A second, the **ischial bursa of the gluteus maximus,** is often absent, but when present, it separates the muscle from the ischial tuberosity. A third bursa is found between the tendon of the muscle and that of the vastus lateralis.

Action. The **gluteus maximus** powerfully extends and laterally rotates the thigh at the hip joint, when its fixed site is the pelvis. Its upper fibers assist in abduction of the thigh, while the lower fibers adduct the thigh. It is relatively inactive when the limbs are motionless in standing or in easy normal walking, but it is very active in running or jumping. Acting by way of its attachment to the iliotibial tract, the gluteus maximus braces the fully extended knee. When the fixed site is at its insertion, the gluteus maximus extends the trunk on the lower extremity, as is necessary to rise from a sitting position or to climb stairs.

Nerve. The gluteus maximus is supplied by the **inferior gluteal nerve,** containing fibers from the fifth lumbar and first and second sacral nerves.

GLUTEUS MEDIUS. The gluteus medius is a broad, thick, radiating muscle, situated on the outer surface of the pelvis (Fig. 6-73). Its posteromedial third is covered by the gluteus maximus, its anterolateral two-thirds by the gluteal aponeurosis, which separates it from the superficial fascia and integument. It *arises* from the outer surface of the ilium between the iliac crest and posterior gluteal line dorsally, and the anterior gluteal line ventrally; it also *arises* from the gluteal aponeurosis covering its outer surface. The fibers converge in a fan-shaped manner to a strong, flattened tendon, which is *inserted* into the oblique ridge on the lateral surface of the greater trochanter. The **trochanteric bursa of the gluteus medius** separates the tendon of the muscle from the surface of the trochanter, over which it glides (Fig. 6-70).

Action. The **gluteus medius** is a strong abductor and medial (anterior fibers) rotator of the thigh. It is an important muscle (as is the gluteus minimus) in walking because it helps to maintain balance and an upright posture. As a foot is elevated from the ground with each step, all the body's weight is placed on the opposite limb, which should result in a marked sagging of the pelvis on the unsupported side. The action of the gluteus medius and minimus of the supported side prevents this tilting. If the gluteus medius and minimus are paralyzed, the trunk sways from side to side with each step in an attempt to prevent the downward tilting of the pelvis on the unsupported side.

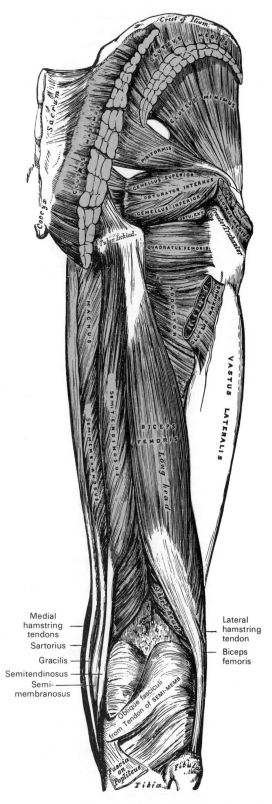

Medial hamstring tendons
Sartorius
Gracilis
Semitendinosus
Semi-membranosus

Lateral hamstring tendon
Biceps femoris

FIG. 6-73. Muscles of the gluteal and posterior femoral regions.

Nerve. The gluteus medius is supplied by branches of the **superior gluteal nerve,** containing fibers from the fourth and fifth lumbar and first sacral nerves.

Variation. At times the posterior border of the gluteus medius is blended with the piriformis muscle.

GLUTEUS MINIMUS. The gluteus minimus, the smallest of the three glutei, is placed immediately deep to the gluteus medius (Fig. 6-71). It is fan-shaped, *arising* from the outer surface of the ilium, between the anterior and inferior gluteal lines, and behind the margin of the greater sciatic notch. The fibers converge to the deep surface of an aponeurosis, which then ends in a tendon that is inserted into an impression on the anterior border of the greater trochanter and gives an expansion to the capsule of the hip joint. The **trochanter bursa of the gluteus minimus** is interposed between the tendon and the greater trochanter. *Separating the gluteus medius and gluteus minimus are the deep branches of the superior gluteal vessels and the superior gluteal nerve.* Beneath the deep surface of the gluteus minimus is found the reflected tendon of the rectus femoris and the capsule of the hip joint.

Action. The **gluteus minimus** abducts and rotates the thigh medially. It also assists the gluteus medius during walking, as already described.

Nerve. The gluteus minimus is supplied by a branch of the **superior gluteal nerve,** containing fibers from the fourth and fifth lumbar and first sacral nerves.

Variations. The gluteus minimus may be divided into anterior and posterior parts, or it may send muscular slips to the piriformis, the superior gemellus, or the vastus lateralis.

TENSOR FASCIAE LATAE. The tensor fasciae latae (Fig. 6-67) *arises* from the anterior part of the outer lip of the iliac crest; from the outer surface of the anterior superior iliac spine, and part of the notch below it, between the gluteus medius and sartorius; and from the deep surface of the fascia lata. It is *inserted* between the two layers of the iliotibial tract of the fascia lata about at the junction of the middle and upper thirds of the thigh.

Action. The **tensor fasciae latae** flexes and abducts (and possibly rotates) the thigh. By contraction, it also tenses the iliotibial tract, thereby helping

to support the femur on the tibia during erect posture.

Nerve. The tensor fasciae latae is supplied by a branch of the **superior gluteal nerve** to the gluteus minimus, containing fibers from the fourth and fifth lumbar and first sacral nerves.

THE PIRIFORMIS. The piriformis is a flat muscle, pyramidal in shape, lying almost parallel with the posterior margin of the gluteus medius (Fig. 6-73). It is situated partly against the posterior wall on the interior of the pelvis (Fig. 6-74), and partly in the gluteal region behind the hip joint. The piriformis *arises* from the anterior sacrum by three fleshy digitations attached to the portions of bone between the first, second, third, and fourth anterior sacral foramina, and to the grooves leading from the foramina. A few fibers also *arise* from the margin of the greater sciatic foramen and from the anterior surface of the sacrotuberous ligament. The muscle passes out of the pelvis through the greater sciatic foramen, the upper part of which it fills. Coursing laterally, the piriformis is *inserted* by a rounded tendon into the superior border of the greater trochanter posterior to, but often partly blended with, the common tendon of the obturator internus and gemelli (Fig. 6-73).

Action. The **piriformis** laterally rotates the extended thigh. When the thigh is flexed, it abducts the femur. The piriformis assists in stabilizing the hip joint and helps to hold the femoral head in the acetabulum.

Nerve. The piriformis is supplied by either the **first or second sacral nerve,** but more frequently by branches from both nerves.

Variations. The piriformis is frequently pierced by the common peroneal nerve and, thus, becomes divided more or less into two parts. It may be united with the gluteus medius, or send fibers to the gluteus minimus, or receive fibers from the superior gemellus. The piriformis may have only one or two sacral attachments or be inserted into the capsule of the hip joint.

OBTURATOR INTERNUS. The obturator internus is situated both within the true or lesser pelvis (Fig. 6-74) and posterior to the hip joint (Fig. 6-73). It *arises* from the inner surface of the anterolateral wall of the pelvis where it surrounds the greater part of the obturator foramen. It is attached to the inferior rami of the pubis and ischium. Laterally it attaches to the inner pelvic surface, inferior to and behind the pelvic brim. It reaches

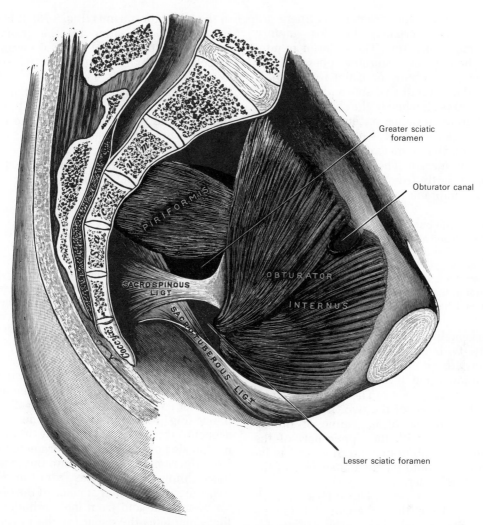

Greater sciatic foramen

Obturator canal

PIRIFORMIS

OBTURATOR

INTERNUS

SACROSPINOUS LIGT

SACRO-TUBEROUS LIGT

Coccyx

Lesser sciatic foramen

FIG. 6-74.   The left obturator internus. Pelvic aspect.

from the upper part of the greater sciatic foramen above and behind, to the obturator foramen below and in front. It also *arises* from the pelvic surface of the obturator membrane except in the posterior part, from the tendinous arch that completes the canal for the passage of the obturator vessels and nerve, and to a slight extent from the obturator fascia, which covers the muscle. The fibers converge rapidly toward the lesser sciatic foramen and end in four or five tendinous bands on the deep surface of the muscle. These bands make a right angle bend around the grooved surface between the spine and tuberosity of the ischium. This bony surface is covered by smooth cartilage and serves as a pulley, being separated from the tendon by a bursa and presenting one or

more ridges corresponding with the furrows between the tendinous bands. These bands leave the pelvis through the lesser sciatic foramen and unite into a single flattened tendon. This passes horizontally across the capsule of the hip joint and, after receiving the attachments of the gemelli, is *inserted* into the anterior part of the medial surface of the great trochanter, proximal to the trochanteric fossa. The **subtendinous bursa of the obturator internus,** narrow and elongated in form, is usually found between the tendon and the capsule of the hip joint. It occasionally communicates with the bursa between the tendon and the ischium (**ischiadic bursa of the obturator internus**).

The **obturator membrane** (Fig. 5-31) is a thin fibrous sheet that almost completely closes the obturator foramen. Its fibers are arranged in interlacing bundles, mainly transverse in direction. The superior bundle is attached to the obturator tubercles and completes the obturator canal for the passage of the obturator vessels and nerve. The membrane is attached to the sharp margin of the obturator foramen except at its inferior lateral angle, where it is fixed to the pelvic surface of the ramus of the ischium, i.e., within the margin. The two obturator muscles arise partly from the opposite surfaces of this membrane.

Action. The **obturator internus** rotates the extended thigh and abducts the thigh when it is flexed.

Nerves. The obturator internus is supplied by a special nerve from the sacral plexus, containing fibers from the lumbosacral trunk. The **nerve to the obturator internus** contains fibers from the fifth lumbar and the first and second sacral segments.

The **gemelli** are two small muscles that lie parallel with the tendon of the obturator internus after it emerges from the lesser sciatic foramen (Fig. 6-73).

SUPERIOR GEMELLUS. The superior gemellus is the smaller of the two. It *arises* from the outer surface of the spine of the ischium, blends with the upper part of the tendon of the obturator internus, and is *inserted* with it onto the medial surface of the greater trochanter.

INFERIOR GEMELLUS. The inferior gemellus *arises* from the tuberosity of the ischium, immediately below the groove for the obturator internus tendon. It blends with the lower part of the tendon of the obturator internus and is *inserted* with it into the medial surface of the greater trochanter.

Action. The **gemelli** rotate the extended thigh laterally and abduct the flexed thigh.

Nerves. The superior gemellus is supplied by a branch of the **nerve to the obturator internus**.

The inferior gemellus receives its innervation from a branch of the **nerve to the quadratus femoris**.

Variations. The gemelli may vary in size. The superior is more frequently absent, and the inferior is more frequently bound intimately to the obturator internus. The superior may be fused with the piriformis or gluteus minimus, the inferior with the quadratus femoris.

QUADRATUS FEMORIS. This small, flat, quadrilateral muscle is found between the gemellus inferior and the proximal margin of the adductor magnus (Fig. 6-73). It is separated from the latter by the transverse branches of the medial femoral circumflex vessels. It *arises* from the upper part of the external border of the tuberosity of the ischium and, passing laterally, is *inserted* into a small tubercle (quadrate tubercle) found on the posterior aspect of the femur. The insertion continues distally for a short distance along the intertrochanteric crest. A bursa is often found between this muscle and the lesser trochanter.

Action. The **quadratus femoris** rotates the thigh laterally.

Nerve. The quadratus femoris is supplied by a special nerve, the **nerve to the quadratus femoris**, which contains fibers from the lumbosacral trunk (fourth and fifth lumbar) and first sacral nerves.

Variations. Absence of the quadratus femoris has been reported in 1 or 2%, but this may be more apparent than real because in these instances it is usually fused either with the gemellus inferior or the adductor magnus. It may be doubled at its insertion.

OBTURATOR EXTERNUS. The obturator externus is a flat, triangular muscle that covers the outer surface of the anterior wall of the pelvis (Fig. 6-75). It *arises* from the anterior aspect of the pelvis immediately around the medial side of the obturator foramen, that is, from the rami of the pubis and the ramus of the ischium. It also *arises* from the medial two-thirds of the outer surface of the obturator membrane and from the tendinous arch that completes the canal for the passage of the obturator vessels and nerves. A few fibers near the pubic arch *arise* from the inner surface of the pelvis between the margin of the foramen and the attachment of the obturator membrane. The fibers converge as they pass laterally and end in a tendon that courses behind the neck of the femur and the distal part of the capsule of the hip joint, to be *inserted* into the trochanteric fossa of the femur. The obturator vessels lie between the muscle and the obturator membrane. The anterior branch of the obturator nerve reaches the thigh by passing anterior to, or in front of, the muscle, and the posterior branch by piercing it.

Action. The **obturator externus** rotates the thigh laterally.

Nerve. The obturator externus is supplied by a branch of the **obturator nerve**, containing fibers from the third and fourth lumbar nerves. This branch

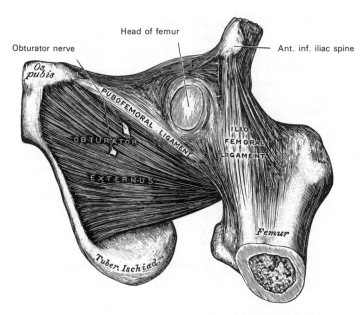

Obturator nerve

Head of femur

Ant. inf. iliac spine

Os pubis

PUBOFEMORAL LIGAMENT

OBTURATOR

EXTERNUS

ILIO FEMORAL LIGAMENT

Femur

Tuber. Ischiad.

FIG. 6-75.   The obturator externus.

comes either from the obturator nerve before it divides or from the posterior branch of the obturator after it divides.

Group Muscle Actions at the Hip Joint. *Extension of the thigh* is performed by the gluteus maximus and adductor magnus; the former is most powerful in a position of lateral rotation, the latter in medial rotation. The gluteus medius acts synergetically during extension to neutralize the adduction of the magnus and the lateral rotation of the gluteus maximus. The hamstring muscles can also extend the thigh, but are used for this action only if it accompanies flexion of the leg.

Flexion of the thigh is performed by the iliopsoas, tensor fasciae latae, pectineus, and sartorius. The action is initiated by the tensor, pectineus, and sartorius, but stronger and final action is by the iliopsoas, with assistance from the adductor muscles. The rectus femoris also flexes the hip, but only if the muscle is acting as an extensor at the knee.

Abduction of the thigh is principally performed by the gluteus medius and minimus. Some assistance in this action comes from the tensor fasciae latae and sartorius, and even from the obturator internus, piriformis, and lower fibers of the gluteus maximus.

Adduction of the thigh can result from the action of the three adductors, the pectineus, the obturator externus, the gracilis, and the hamstring muscles, as well as from the quadratus femoris and the upper fibers of the gluteus maximus.

Lateral rotation of the thigh is performed by the obturator externus and internus, gemelli, piriformis, quadratus femoris, gluteus maximus, posterior fibers of the gluteus medius, and sartorius. Electromyographic studies indicate that the iliopsoas muscle is also active during lateral rotation.

Medial rotation of the thigh is performed by the tensor fasciae latae, pectineus, gluteus minimus, and the lower fibers of the adductor magnus.

Posterior Femoral Muscles

The posterior femoral muscles cross the hip and knee joints and act primarily as flexors of the leg at the knee joint and extensors of the thigh at the hip. They are frequently referred to as the hamstring muscles and include the:

Biceps femoris
Semitendinosus
Semimembranosus

BICEPS FEMORIS.   The biceps femoris is situated on the posterior and lateral aspect of the thigh (Fig. 6-73). Its two heads of origin are called long and short. The **long head** *arises* from an inferomedial impression on the posterior part of the ischial tuberosity by a tendon in common with the semitendinosus muscle, and from the lower part of the sacrotuberous ligament. The **short head** does not cross the hip joint, but instead *arises* from the lateral lip of the linea aspera of the femur between the adductor magnus and vastus lateralis. It extends upward almost as high as the insertion of the gluteus maximus, and is prolonged down beyond the linea aspera to a point about 5 cm above the lateral condyle. It also *arises* from the lateral intermuscular septum. The fibers of the long head form a fusiform belly, which descends obliquely laterally across the sciatic nerve to end in an aponeurosis covering the posterior surface of the muscle. This aponeurosis receives on its deep surface the fibers of the short head. This aponeurosis becomes gradually contracted into a tendon, which is *inserted* into the lateral aspect of the head of the fibula, and a small tendinous slip, which attaches to the lateral condyle of the tibia. At its insertion the tendon divides into two portions, and these embrace the fibular collateral ligament of the knee joint. From the posterior border of the tendon a thin expansion extends to the fascia of the leg. The tendon of the biceps femoris forms the **lateral hamstring,** and the common peroneal nerve descends along its medial border.

Action. The **biceps femoris** flexes the leg and, after it is flexed, rotates the tibia laterally on the femur. The long head also extends the thigh and tends to rotate it laterally.

Nerves. The long head of the biceps femoris is supplied by branches, usually two, from the **tibial portion of the sciatic nerve,** containing fibers from the first three sacral nerves. The short head of the biceps femoris receives innervation from the peroneal portion of the sciatic nerve, containing fibers from the fifth lumbar and first two sacral nerves.

Variations. The short head of the biceps femoris may be absent. Additional heads may arise from the ischial tuberosity, the linea aspera, or the medial supracondylar ridge of the femur. A muscular slip may pass to the gastrocnemius.

THE SEMITENDINOSUS. The semitendinosus, remarkable because of the great length of its tendon of insertion, is situated at the posterior and medial aspect of the thigh (Fig. 6-73). It *arises* from the inferomedial impression of the ischial tuberosity by a tendon common to it and the long head of the biceps femoris. It also *arises* from an aponeurosis that connects the adjacent surfaces of the two muscles for an extent of about 7.5 cm from their origin. The muscle is fusiform and ends a little distal to the middle of the thigh in a long round tendon, which lies along the surface of the semimembranosus muscle, medial to the popliteal fossa. It then curves around the medial condyle of the tibia and passes over the tibial collateral ligament of the knee joint, from which it is separated by the anserine bursa. The tendon then is *inserted* into the proximal part of the medial surface of the body of the tibia, nearly as far anteriorly as its anterior crest. At its insertion it gives off from its distal border a prolongation to the deep fascia of the leg. Lying posterior to the tendon of the sartorius and distal to that of the gracilis, the semitendinosus tendon unites with tendons of these other muscles to form a flattened aponeurosis known as the *pes anserinus.* A tendinous intersection is usually observed about the middle of the muscle.

Action. The **semitendinosus** flexes the leg and, after it is flexed, rotates the tibia medially. It also is able to extend the femur at the hip joint.

Nerves. The semitendinosus is supplied by the **tibial portion of the sciatic nerve** (usually by two branches), containing fibers from the fifth lumbar and first two sacral nerves.

THE SEMIMEMBRANOSUS. This muscle, so named because of its membranous tendon of origin, is situated at the back and medial side of the thigh (Fig. 6-73). It *arises* by a thick tendon from the upper and lateral impression on the ischial tuberosity adjacent to the origin of the biceps femoris and semitendinosus. The tendon of origin expands into an aponeurosis that covers the proximal part of the anterior surface of the muscle. From this aponeurosis, muscular fibers arise and converge to another aponeurosis. The latter aponeurosis covers the distal part of the posterior surface of the muscle, which then becomes the tendon of insertion. The semimembranosus is *inserted* mainly into the horizontal groove on the posteromedial aspect of the medial condyle of the tibia. The tendon of insertion gives off a number of fibrous expansions; one, of considerable size, passes upward and laterally to be *inserted* into the posterior aspect of the lateral condyle of the femur, forming part of the oblique popliteal ligament of the knee joint. Another is continued distally to the fascia that covers the popliteus muscle, while a few fibers join the tibial collateral ligament of the joint and the fascia of the leg. The muscle overlaps the proximal part of the popliteal vessels. The tendons of the semitendinosus and semimembranosus form the **medial hamstrings.**

Action. The **semimembranosus** flexes the leg and, after it is flexed, tends to rotate it medially. It also extends the thigh at the hip joint.

Nerves. The semimembranosus is supplied by

several branches from the **tibial portion of the sciatic nerve,** containing fibers from the fifth lumbar and first two sacral nerves.

Variations. The semimembranosus may be reduced, absent, or doubled. It may arise primarily from the sacrotuberous ligament and give a slip to the femur or adductor magnus muscle.

Group Actions at the Knee. Extension of the leg is performed by the quadriceps femoris, i.e., the rectus femoris, vastus lateralis, vastus medialis, and vastus intermedius.

Flexion of the leg is performed by the hamstring muscles, i.e., the biceps femoris, semitendinosus, and semimembranosus, and by the popliteus. The sartorius and gracilis act in full flexion and against resistance. The gastrocnemius may assist in flexion of the leg, but is used more to protect the knee joint from hyperextension.

Lateral rotation of the leg, when the knee is partially flexed, is performed by the biceps femoris muscle.

Medial rotation of the leg is performed by the gracilis and popliteus. The semimembranosus and semitendinosus also can act as medial rotators of the partially flexed leg.

MUSCLES AND FASCIAE OF THE LEG

The portion of the lower limb between the knee and ankle joints is known as the **leg.** Attached to its bones, the **tibia** and **fibula,** are three groups of muscles—anterior, posterior, and lateral.

The superficial fascia covering the leg is both fatty and fibrous, and coursing through it are the cutaneous nerves and superficial vessels. Continuous with the superficial fascia of the thigh above and the foot below, this fascia can be dissected rather easily from the underlying deep fascia.

DEEP FASCIA OF THE LEG. Also known as the **crural fascia,** the deep fascia of the leg forms a complete investment for the muscles and becomes fused with the periosteum over the subcutaneous surfaces of the bones. It is continuous *superiorly* with the fascia lata, and is attached around the knee to the patella, the patellar ligament, the condyles of the femur, the tuberosity and condyles of the tibia, and the head of the fibula. *Posteriorly,* it forms the **popliteal fascia,** covering the popliteal fossa, where it is strengthened by transverse fibers and perforated by the small saphenous vein. *Medially,* at the knee, the crural fascia receives expansions from the tendons of the sartorius, gracilis, semitendinosus, and semimembranosus, while *later-*

ally, at the knee, it receives an expansion from the tendon of the biceps femoris muscle. The deep fascia blends *anteriorly* with the periosteum covering the subcutaneous surface of the tibia; *inferolaterally,* with the periosteum covering the head and malleolus of the fibula. *Inferiorly,* the crural fascia is continuous with the superior extensor retinaculum and flexor retinaculum. It is thick and dense in the upper anterior part of the leg, and gives attachment by its deep surface to the tibialis anterior and extensor digitorum longus muscles. In the mid-posterior aspect of the calf, it is thinner where it covers the gastrocnemius and soleus. It gives off from its deep surface, on the lateral aspect of the leg, two strong intermuscular septa, the **anterior** and **posterior crural intermuscular septa,** which enclose the peroneus longus and brevis muscles and separate them from the muscles of the anterior and posterior crural regions. Other septa in the anterior and posterior crural regions separate the individual muscles. A broad transverse intermuscular septum, called the **deep transverse fascia of the leg,** intervenes between the superficial and deep posterior crural muscles.

The muscles of the leg will be described in three groups:

Anterior crural
Posterior crural
Lateral crural

Anterior Crural Muscles

The anterior crural muscles occupy the **anterior compartment** of the leg, which lies between the tibia and the anterior intermuscular septum. The four muscles of this compartment are:

Tibialis anterior
Extensor hallucis longus
Extensor digitorum longus
Peroneus tertius

TIBIALIS ANTERIOR. The tibialis anterior muscle, situated on the lateral aspect of the tibia, is thick and fleshy proximally, but tendinous distally (Fig. 6-76). It *arises* from the lateral condyle and upper half or two-thirds of the lateral surface of the body of the tibia; from the adjoining part of the interosseous membrane; from the deep surface of the crural fascia; and from an intermuscular sep-

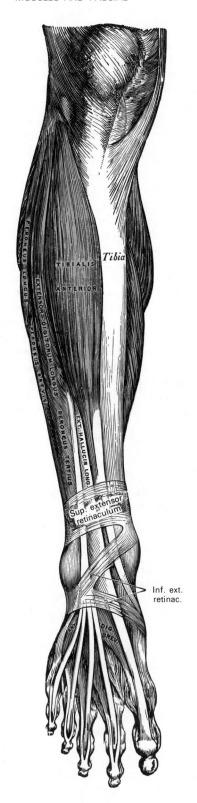

Tibia

PERONEUS LONGUS

TIBIALIS

ANTERIOR

EXTENSOR DIGITORUM LONGUS

PERONEUS BREVIS

PERONEUS TERTIUS

EXT. HALLUCIS LONG.

Sup. extensor retinaculum

Inf. ext. retinac.

DIG. BREV.

FIG. 6-76. Muscles of the anterior leg.

tum between it and the extensor digitorum longus. The fibers course vertically downward to end in a tendon that becomes apparent on the anterior surface of the muscle along the lower third of the leg. After passing through the most medial compartments of the superior and inferior extensor retinacula, it is *inserted* into the medial and plantar surfaces of the medial cuneiform bone, and the base of the first metatarsal bone. This muscle overlaps the anterior tibial vessels and deep peroneal nerve in the proximal part of the leg.

Action. The **tibialis anterior** dorsally flexes the foot at the talocrural joint. It also inverts and adducts (supinates) the foot at the subtalar and midtarsal joints. These actions are especially important as the foot leaves the ground during walking and running.

Nerve. The tibialis anterior is supplied by a branch of the **deep peroneal nerve,** containing fibers from the fourth and fifth lumbar and first sacral nerves.

Variations. In rare instances, a deep portion of the tibialis anterior muscle is inserted into the talus, or a tendinous slip may pass to the head of the first metatarsal bone or the base of the proximal phalanx of the great toe. The **tibiofascialis anterior** is a small muscle sometimes seen extending from the distal part of the tibia to the superior or inferior extensor retinacula or deep fascia.

EXTENSOR HALLUCIS LONGUS. The extensor hallucis longus is a thin muscle, situated between the tibialis anterior and the extensor digitorum longus. Its muscular belly is largely covered by these two muscles, and its tendon emerges to a superficial position in the lower third of the leg (Fig. 6-76). The extensor hallucis longus *arises* from the medial surface of the fibula along the middle two-fourths of its extent, medial to the origin of the extensor digitorum longus, and from the interosseous membrane. The anterior tibial vessels and deep peroneal nerve lie between it and the tibialis anterior. The fibers descend and end in a tendon that occupies the anterior border of the muscle. It then passes deep to the superior extensor retinaculum, through a distinct compartment in the inferior extensor retinaculum, crosses from the lateral to the medial side of the anterior tibial vessels near the talocrural joints, and is *inserted* into the base of the distal phalanx of the great toe. Opposite the metatarsophalangeal articulation, the tendon gives off a thin prolongation on both sides to cover the dorsal surface of the joint. An expansion from the medial side of the tendon is usually inserted into the base of the proximal phalanx.

Action. The **extensor hallucis longus** extends the proximal phalanx of the great toe. It also dorsally flexes and supinates the foot.

Nerve. The extensor hallucis longus is supplied by a branch of the **deep peroneal nerve**, containing fibers from the fourth and fifth lumbar and first sacral nerves.

Variations. The origin of the extensor hallucis longus is occasionally united with that of the extensor digitorum longus. The **extensor ossis metatarsi hallucis** is a small muscle that is sometimes found as a slip from the extensor hallucis longus, or from the tibialis anterior, or from the extensor digitorum longus, or as a distinct muscle traversing the same compartment of the superior extensor retinaculum as the extensor hallucis longus.

EXTENSOR DIGITORUM LONGUS. The extensor digitorum longus is a penniform muscle situated in the lateral part of the anterior leg (Figs. 6-76; 6-79). It *arises* from the lateral condyle of the tibia, from the upper three-fourths of the anterior surface of the body of the fibula, from the proximal part of the interosseous membrane, from the deep surface of the fascia, and from the intermuscular septa between it and the tibialis anterior on the medial aspect and the peroneal muscles on the lateral aspect. Between the extensor digitorum longus and the tibialis anterior in the upper leg are the proximal portions of the anterior tibial vessels and deep peroneal nerve. The tendon passes beneath the superior and inferior extensor retinacula in company with the peroneus tertius. It then divides into four slips, which course forward on the dorsum of the foot and are *inserted* into the middle and distal phalanges of the four lesser toes. The tendons to the second, third, and fourth toes are each joined laterally, opposite the metatarsophalangeal articulation, by a tendon of the extensor digitorum brevis. Each of the tendons receives a fibrous expansion from the interossei and lumbricals before spreading outward into a broad aponeurosis called the digital extensor hood, which covers the dorsal surface of the proximal phalanx. This aponeurosis, at the articulation of the proximal with the middle phalanx, divides into three slips—an intermediate, which is inserted into the base

of the middle phalanx, and two collateral slips, which, after uniting on the dorsal surface of the middle phalanx, are continued onward, to be inserted into the base of the distal phalanx (Fig. 6-76).

Action. The **extensor digitorum longus** extends the proximal phalanges of the four lateral toes. It also dorsally flexes and everts (tends to pronate) the foot.

Nerve. The extensor digitorum longus is supplied by branches of the **deep peroneal nerve**, containing fibers from the fourth and fifth lumbar and first sacral nerves.

Variations. The extensor digitorum longus muscle varies considerably in its mode of origin and in the arrangement of its various tendons. The tendons to the second and fifth toes may be doubled, or extra slips may be given off from one or more tendons to their corresponding metatarsal bones, to the short extensor, or to one of the interosseous muscles. A slip to the great toe from the innermost tendon has also been reported.

PERONEUS TERTIUS. The peroneus tertius (Figs. 6-76; 6-79) is part of the extensor digitorum longus and might be described as its fifth tendon. The muscle fibers attaching to this tendon *arise* from the distal third or more of the anterior surface of the fibula, from the distal part of the interosseous membrane, and from an intermuscular septum between it and the peroneus brevis. The tendon, after passing deep to the superior and inferior extensor retinacula in the same canal as the extensor digitorum longus, is *inserted* into the dorsal surface of the base of the metatarsal bone of the little toe (Fig. 6-80).

Action. The **peroneus tertius** dorsally flexes and assists in everting the foot.

Nerve. The peroneus tertius is supplied by a branch of the **deep peroneal nerve**, containing fibers from the fifth lumbar and first sacral nerves.

Posterior Crural Muscles

The muscles of the posterior leg are subdivided into two groups—superficial and deep. Those of the superficial group constitute a powerful muscular mass, forming the calf of the leg. Their large size is one of the most characteristic features of the muscular apparatus in man, and this size bears a direct relation to his erect posture and his mode of locomotion.

SUPERFICIAL GROUP

Gastrocnemius
Soleus
Plantaris

The two heads of the gastrocnemius and the soleus muscle are collectively called the **triceps surae.** Its common tendon of insertion is the **tendo calcaneus,** more popularly known as the Achilles' tendon.

THE GASTROCNEMIUS.   The gastrocnemius is the most superficial muscle of the calf (Fig. 6-77). Forming the prominent contour of the posterior leg, it arises by two heads and these are connected to the condyles of the femur by strong, flat tendons. The **medial head,** the larger of the two, *arises* from a depression at the upper and posterior part of the medial condyle and from the adjacent popliteal surface of the femur above the medial condyle. The **lateral head** *arises* from an impression on the side of the lateral condyle and from the posterior surface of the femur immediately above the lateral part of the condyle. Both heads also *arise* from the subjacent part of the capsule of the knee joint. Each tendon spreads out into an aponeurosis that covers the posterior surface of the respective head. From the anterior surfaces of these tendinous expansions, muscular fibers arise, those of the medial head being thicker and extending more distally than those of the lateral. The fibers unite at an angle in the midline of the muscle, forming a tendinous raphe that expands into a broad aponeurosis on the anterior surface of the muscle. Into this raphe the remaining fibers are inserted. The aponeurosis, gradually contracting, unites with the tendon of the soleus and forms with it the **tendo calcaneus.**

Action.   The **gastrocnemius** plantar flexes the foot (points the toe), flexes the leg, and tends to supinate the foot.

Nerves.   The **gastrocnemius** is supplied by branches of the **tibial nerve,** containing fibers from the first and second sacral nerves.

Variations.   Absence of the lateral head or of the entire muscle has been observed. An extra slip from the popliteal surface of the femur is a more frequent variation.

THE SOLEUS.   The soleus muscle is broad and flat and lies immediately deep to the gastrocnemius (Figs. 6-77; 6-79). It *arises* by tendinous fibers from the posterior surface

of the head of the fibula, from the proximal third of the posterior surface of the body of the fibula, from the soleal line, and from the middle third of the medial border of the tibia. Some fibers also *arise* from a tendinous arch placed between the tibial and fibular origins of the muscle, under which the popliteal vessels and tibial nerve run.

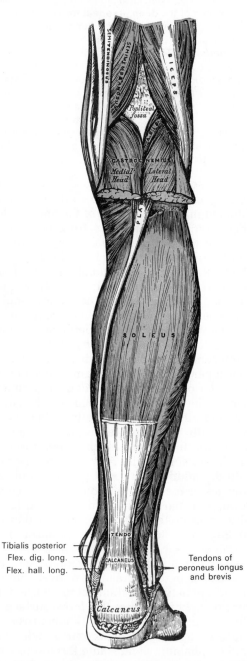

Tibialis posterior
Flex. dig. long.
Flex. hall. long.

Tendons of peroneus longus and brevis

FIG. 6-77.   Muscles of the posterior compartment of the leg. Superficial layer.

The fibers end in an aponeurosis, which covers the posterior surface of the muscle. This aponeurosis gradually becomes thicker and narrower and joins the *insertion* of the gastrocnemius to form with it the **tendo calcaneus.**

Action. The **soleus** muscle plantar flexes the foot, but it is an important postural muscle as well, appearing to be constantly active during ordinary quiescent standing. Because the center of gravity of the body passes anterior to the ankle joints, this muscle helps to prevent the body from falling forward.

Nerve. The soleus is supplied by a branch of the **tibialis nerve,** containing fibers from the first and second sacral nerves.

Variations. At times an accessory head is seen at its lower and inner part, usually ending in the tendo calcaneus, the calcaneus, or the flexor retinaculum.

TENDO CALCANEUS.   The calcaneal tendon (also called the tendon of Achilles) is the common tendon of the gastrocnemius and soleus and is the thickest and strongest in the human body (Figs. 6-77; 6-79). It is about 15 cm long and begins near the middle of the leg, but its anterior surface receives fleshy fibers almost to its distal end. It gradually contracts as it descends, so that it is rounded and narrowest at a point about 4 cm above its insertion. Below this it expands once again, spreading somewhat to insert into the middle part of the posterior surface of the calcaneus. The **calcaneal subtendinous bursa** is interposed between the tendon and the bone. It is covered by the fascia and the integument, and it is separated from the deep muscles and vessels by a considerable interval that is filled with areolar and adipose tissue. Along its lateral aspect, but superficial to it, is the small saphenous vein.

THE PLANTARIS.   The plantaris is a small muscle placed between the gastrocnemius and soleus (Figs. 6-77; 6-79). It arises from the distal part of the lateral prolongation of the linea aspera and from the oblique popliteal ligament of the knee joint. It forms a small fusiform belly, from 7 to 10 cm long, ending in a long slender tendon that crosses obliquely between the gastrocnemius and soleus and runs along the medial border of the tendo calcaneus, to be *inserted* with it into the posterior part of the calcaneus. This muscle is sometimes doubled, and at other times wanting. Occasionally, its tendon

fuses with the flexor retinaculum or with the fascia of the leg. The plantaris is a rudimentary muscle and is frequently compared to the palmaris longus in the forearm.

Action. The **plantaris** is a weak flexor of the leg at the knee joint and plantar flexor of the foot at the ankle joint.

Nerve. The plantaris is supplied by a branch of the **tibial nerve,** containing fibers from the fourth and fifth lumbar and first sacral nerves.

DEEP GROUP

Popliteus
Flexor hallucis longus
Flexor digitorum longus
Tibialis posterior

DEEP TRANSVERSE FASCIA.   The deep transverse fascia is an intermuscular fibrous septum that stretches across the posterior leg between the superficial and deep posterior crural muscles. Extending transversely, it is connected at the sides to the margins of the tibia and fibula. Above, where it covers the popliteus muscle, it is thick and dense, and receives an expansion from the tendon of the semimembranosus. It is thinner in the middle of the leg, but distally, where it covers the tendons passing behind the malleoli, it is thickened and continuous with the **flexor retinaculum.**

THE POPLITEUS.   The popliteus is a thin, flat, triangular muscle that forms the distal part of the floor of the popliteal fossa. The lateral portion of the muscle *arises* by a strong tendon about 2.5 cm long, from a depression at the anterior part of the groove on the lateral condyle of the femur. Its more medial fibers *arise* from the arcuate popliteal ligament and the capsule of the knee joint. The muscle descends across the upper leg to *insert* into the medial two-thirds of the triangular surface proximal to the soleal line on the posterior surface of the tibia, and into the tendinous expansion covering the surface of the muscle.

Action. The **popliteus** flexes and medially rotates the leg. More importantly, however, when the tibia is fixed, as occurs when the limb is supporting weight, the popliteus rotates the femur laterally on the tibia to "unlock" the knee joint.

Nerve. The popliteus is supplied by a branch of the **tibial nerve,** containing fibers from the fourth and fifth lumbar and first sacral nerves.

Variations. An additional slip of the popliteus muscle may arise from the sesamoid bone in the lateral head of the gastrocnemius. More rarely, a **popliteus minor** muscle arises from the femur on the inner side of the plantaris and inserts into the posterior ligament of the knee joint. In 14% of limbs, a muscle arises on the inner side of the head of the fibula and inserts into the upper end of the oblique line of the tibia, lying beneath the popliteus muscle.

FLEXOR HALLUCIS LONGUS. The flexor hallucis longus is situated on the fibular side of the leg (Fig. 6-78). It *arises* from the inferior two-thirds of the posterior surface of the body of the fibula (except the most distal 2.5 cm), from the distal part of the interosseous membrane, from an intermuscular septum between the flexor hallucis longus and the peronei, laterally, and from the fascia covering the tibialis posterior, medially. The fibers pass obliquely downward, ending in a tendon that occupies nearly the whole length of the posterior surface of the muscle. This tendon lies in a groove that crosses the posterior surface of the lower end of the tibia, the posterior surface of the talus, and the inferior surface of the sustentaculum tali of the calcaneus. In the sole of the foot it runs forward between the two heads of the flexor hallucis brevis and is *inserted* into the base of the terminal phalanx of the large toe. The grooves on the talus and calcaneus, which contain the tendon of the muscle, are converted by tendinous fibers into distinct canals, lined by a synovial sheath. As the tendon courses distally in the sole of the foot, it passes beneath the tendon of the flexor digitorum longus to which it is connected by a strong tendinous slip.

Action. The **flexor hallucis longus** flexes the distal phalanx of the great toe. It plantar flexes and supinates the foot, and it importantly contributes to the propulsion of the foot at its "take off" during walking and running.

Nerve. The flexor hallucis longus is supplied by a branch of the **tibial nerve,** containing fibers from the fifth lumbar and first and second sacral nerves.

Variations. The flexor hallucis longus usually sends a slip to the flexor digitorum longus, and frequently an additional slip courses from the flexor digitorum longus to the flexor hallucis. On rare occasions, an aberrant muscle, the **peroneocalcaneus internus,** takes origin below or lateral to the flexor hallucis longus from the posterior aspect of the fibula, passes under the sustentaculum tali with the flexor hallucis longus, and inserts on the calcaneus.

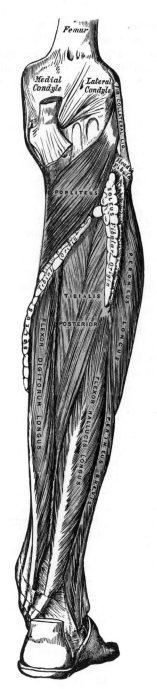

Fɪɢ. 6-78. Muscles of the posterior compartment of the leg. Deep layer.

FLEXOR DIGITORUM LONGUS. The flexor digitorum longus, situated on the tibial side of the leg, is thin and pointed at its origin, but it gradually increases in size as it descends (Fig. 6-78). It *arises* from the posterior surface of the body of the tibia, medial to the origin of the tibialis posterior. Its attachment extends from immediately below the soleal line to within 7 or 8 cm of the lower tibial extremity. It also *arises* from the fascia covering the tibialis posterior. The fibers end in a tendon that extends nearly the entire length of the posterior surface of the muscle. The tendon passes behind the medial malleolus in a groove common to it and the tendon of the tibialis posterior, but separated from the latter by a fibrous septum. Each tendon is contained in a special compartment lined by a separate synovial sheath. The tendon of the flexor digitorum longus passes obliquely downward and laterally, superficial to the deltoid ligament of the ankle joint, into the sole of the foot (Fig. 6-81), where it crosses superficial to the tendon of the flexor hallucis longus, from which it receives a strong tendinous slip. It then expands, is joined by the quadratus plantae muscle, and finally divides into four tendons, which are *inserted* into the bases of the distal phalanges of the second, third, fourth, and fifth toes, each tendon passing through an opening in the corresponding tendon of the flexor digitorum brevis opposite the base of the proximal phalanx.

Action. The **flexor digitorum longus** flexes the terminal phalanges of the four small toes, thereby allowing them to grip the ground more securely during walking. It also plantar flexes and supinates the foot at the ankle joint.

Nerve. The flexor digitorum longus is supplied by a branch of the **tibial nerve**, containing fibers from the fifth lumbar and first sacral nerves.

Variations. Not infrequently, a muscular slip called the **flexor accessorius longus digitorum** arises from the fibula, tibia, or deep fascia and ends in a tendon that passes deep to the flexor retinaculum and joins the tendon of the flexor digitorum longus or the quadratus plantae tendon.

TIBIALIS POSTERIOR. The tibialis posterior lies between the flexor digitorum longus and the flexor hallucis longus in the upper half of the leg, and it is the most deeply situated muscle in this flexor group (Fig. 6-78). The proximal end of the muscle consists of two pointed processes, separated by an angular interval through which the anterior tibial vessels pass to the anterior leg. It *arises* from the whole of the posterior surface of the interosseous membrane, except its lowest part; from the lateral portion of the posterior surface of the body of the tibia, between the

commencement of the soleal line proximally and the junction of the middle and lower thirds of the tibial shaft distally; and from the proximal two-thirds of the medial surface of the fibula. Some fibers also *arise* from the deep transverse fascia, and from the intermuscular septa separating it from the adjacent muscles. In the lower one-fourth of the leg, its tendon passes anterior, that is, deep, to that of the flexor digitorum longus and lies with it in a groove behind the medial malleolus, but enclosed in a separate sheath. It next passes deep to the flexor retinaculum and superficial to the deltoid ligament. As it enters the foot, it turns forward superficial to the plantar calcaneonavicular ligament, and at this site the tendon contains a sesamoid fibrocartilage. It is *inserted* into the tuberosity of the navicular bone and gives off fibrous expansions, one of which passes backward to the sustentaculum tali of the calcaneus, while others course distally and laterally to the three cuneiforms, the cuboid, and the bases of the second, third, and fourth metatarsal bones.

Action. When the foot is not bearing weight, the **tibialis posterior** is a plantar flexor of the foot. Because it also attaches to the lateral tarsal bones, it inverts and adducts (tends to supinate) the foot as well. When the foot is bearing weight, as in standing or walking, the tibialis posterior, along with other leg muscles, helps to distribute weight on the foot, thereby assisting in maintaining balance.

Nerve. The tibialis posterior is supplied by the **tibial nerve**, containing fibers from the fifth lumbar and first sacral nerves.

Lateral Crural Muscles

The lateral crural muscles, two in number, occupy the lateral compartment of the leg, and they arise from the external surface of the fibula. They are separated from the anterior and posterior crural muscles by the anterior and posterior intermuscular septa. Innervated by the superficial peroneal nerve, these muscles act as everters of the nonweight-bearing foot. The two lateral crural muscles are:

Peroneus longus
Peroneus brevis

PERONEUS LONGUS. The peroneus longus is situated proximally on the lateral aspect of the leg and is the more superficial of the two muscles. It *arises* from the head and upper two-thirds of the lateral surface of the body of the fibula, from the deep surface of the fascia, and from the intermuscular septa between it and the muscles on the anterior and posterior leg. Occasionally, a few fibers also *arise* from the lateral condyle of the tibia. Between its attachments to the head and to the body of the fibula, there is a gap through which the common peroneal nerve passes to the anterior leg. The peroneus longus ends in a long tendon, which runs behind the lateral malleolus, in a groove common to it and the tendon of the peroneus brevis, behind which it lies. The groove is converted into a canal by the superior peroneal retinaculum, the tendons being contained in a common synovial sheath. The tendon of the peroneus longus then extends obliquely forward across the lateral aspect of the calcaneus, below the peroneal trochlea and the tendon of the peroneus brevis, and under cover of the inferior peroneal retinaculum. It crosses the lateral aspect of the cuboid, and then runs on the plantar surface of that bone in a groove that is converted into a canal by the long plantar ligament. The tendon then crosses the sole of the foot obliquely, and is *inserted* into the lateral aspect of the base of the first metatarsal bone and the lateral aspect of the medial cuneiform. Occasionally, it sends a slip to the base of the second metatarsal bone. The tendon of the peroneus longus changes direction at two points: first, behind the lateral malleolus; second, on the cuboid bone. At both sites the tendon is thickened, and along the cuboid bone a sesamoid fibrocartilage, which sometimes ossifies, is usually developed in its substance.

Action. The **peroneus longus** everts and, at the same time, abducts the foot in an action referred to as pronation. It may also act as a weak plantar flexor of the foot. During normal standing the muscle does not appear to be active, but is importantly involved as the foot takes off from the ground during walking. Additionally, the oblique course of the tendon across the plantar aspect of the foot helps to maintain the integrity of the lateral longitudinal and transverse arches of the foot.

Nerve. The peroneus longus is supplied by a branch of the **superficial peroneal nerve** that contains fibers from the fourth and fifth lumbar and first sacral nerves.

PERONEUS BREVIS.   The peroneus brevis lies under cover of the peroneus longus and is a shorter and smaller muscle. It *arises* from the distal two-thirds of the lateral surface of the body of the fibula, its upper fibers being anterior to the lower fibers of the peroneus longus. It also *arises* from the intermuscular septa separating it from the adjacent muscles on the anterior and posterior leg. The fibers pass vertically downward, ending in a tendon that courses behind the lateral malleolus, along with but anterior to that of the peroneus longus. The two tendons are enclosed in the same compartment and are lubricated by a common synovial sheath. The peroneus brevis then turns forward on the lateral aspect of the calcaneus, above the peroneal trochlea and the tendon of the peroneus longus, and is *inserted* into the tuberosity at the base of the fifth metatarsal bone, on its lateral aspect.

On the lateral surface of the calcaneus, the tendons of the peroneus longus and peroneus brevis occupy separate osseoaponeurotic canals formed by the calcaneus and the inferior peroneal retinaculum. Each tendon is enveloped by an individual distal prolongation of the common synovial sheath.

Action. The **peroneus brevis** everts and abducts (pronates) the foot. It also acts as a plantar flexor of the foot at the ankle joint. During walking, it compensates for inversion, thereby helping to balance the foot in order to support properly the weight of the body.

Nerve. The peroneus brevis is supplied by a branch of the **superficial peroneal nerve** that contains fibers from the fourth and fifth lumbar and first sacral nerves.

Variations. Fusion of the two peroneus muscles is rare. A slip from the peroneus longus to the base of the third, fourth, or fifth metatarsal bone, or to the adductor hallucis, is occasionally seen. Several additional muscle variants have been described:

1. **Peroneus accessorius** arises from the fibula between the longus and brevis, and joins the tendon of the longus in the sole of the foot.

2. **Peroneus digiti minimi** is rare, but when present arises from the lower fourth of the fibula under the brevis and inserts into the extensor aponeurosis of the little toe. It is more common as a slip of the tendon of the peroneus brevis.

3. **Peroneus quartus** is present in about 13% of specimens and arises from the back of the fibula between the brevis and the flexor hallucis to insert into the peroneal trochlea of the calcaneum **(peroneocalcaneus externum)** or, less frequently into the tuberosity of the cuboid **(peroneocuboideus).**

Group Actions at the Ankle. *Plantar flexion of the foot* is brought about by the gastrocnemius, soleus, plantaris, peroneus longus and brevis, and tibialis posterior. The peronei and tibialis posterior act alone if there is no resistance to be overcome, whereas the flexor digitorum longus and flexor hallucis longus come into action to counter resistance.

Dorsal flexion of the foot is performed by the tibialis anterior, extensor digitorum longus, extensor hallucis longus, and peroneus tertius. The peroneus brevis, along with the extensor digitorum longus and peroneus tertius, acts to neutralize the inversion of the tibialis anterior and extensor hallucis longus.

Supination of the foot (combined adduction and inversion) is performed principally by the tibialis anterior and tibialis posterior. The tibialis posterior adducts more powerfully, whereas the tibialis anterior inverts more powerfully. The gastrocnemius also tends to supinate the foot.

Pronation of the foot (combined abduction and eversion) is performed by the peronei. The peroneus brevis abducts more strongly and acts slightly before the longus, whereas the peroneus longus everts more strongly. The peroneus tertius also tends to pronate as does the extensor digitorum longus.

Fascia Around the Ankle

Embedded in or fused with the deep fascia of the lower leg are fibrous bands that form retinacula and bind down the tendons as they cross the ankle joint in their passage into the foot. These fibrous bands are more distinct in some bodies than in others. Generally, however, there are (1) two retinacula binding the extensor muscles, the **superior and inferior extensor retinacula;** (2) one retinaculum located medially over the flexor tendons, the **flexor retinaculum;** and (3) two retinacula located laterally over the tendons of the peroneal muscles, the **superior and inferior peroneal retinacula.**

SUPERIOR EXTENSOR RETINACULUM. The superior extensor retinaculum (Figs. 6-76; 6-79; 6-80), also known as the *transverse crural ligament,* binds down the tendons of the tibialis anterior, extensor hallucis longus, extensor digitorum longus, and peroneus tertius just above the anterior aspect of the ankle joint. Coursing with the tendons as they enter the foot deep to the retinaculum are the anterior tibial vessels and deep peroneal nerve. This retinaculum is attached laterally to the distal end of the fibula, and medially to the tibia, and at this site only the tendon of the tibialis anterior has a synovial sheath.

INFERIOR EXTENSOR RETINACULUM. The inferior extensor retinaculum (Figs. 6-76; 6-79 to 6-81), which is also called the *cruciate crural ligament,* is a Y-shaped band placed anterior to the ankle joint. The stem of the Y is attached laterally to the upper surface of the calcaneus, adjacent to the calcaneal sulcus, which contains the interosseous talocalcaneal ligament. It is directed medially as a double layer, which surrounds the tendons of the peroneus tertius and the extensor digitorum longus, one layer passing superficial to the tendons, the other deep to them. At the medial border of the extensor digitorum longus, the two layers join to form a compartment in which the tendons are enclosed. Continuing medially, the retinaculum diverges to form the two limbs of the Y. The *upper band* is directed proximally and medially to be attached to the medial malleolus. It passes superficial to the tendon of the extensor hallucis longus, the anterior tibial vessels, and the deep peroneal nerve, but encloses the tibialis anterior by a splitting of its fibers. The *lower band* extends distally and medially, to be attached to the border of the plantar aponeurosis. It passes superficial to the tendons of the extensor hallucis longus and tibialis anterior and also the dorsalis pedis artery and deep peroneal nerve.

FLEXOR RETINACULUM. The flexor retinaculum (Fig. 6-81) has also been called the *laciniate ligament.* It is a strong fibrous band that extends from the medial malleolus of the tibia proximally, to the margin of the calcaneus distally, where it converts a series of bony grooves into canals for the passage of the tendons of the flexor muscles, the posterior tibial vessels, and the tibial nerve into the sole of the foot. Its upper border is continuous with the deep fascia of the leg, and its lower border blends with the plantar aponeurosis and the fibers of the origin of the abductor hallucis muscle. From medial to lateral, the four canals formed by the flexor retinaculum transmit: (1) the tendon of the tibialis posterior, (2) the tendon of the flexor digitorum longus, (3) the posterior tibial vessels and tibial nerve, which course through a broad space beneath the ligament, and (4) in a canal formed partly by the talus, the tendon of the flexor hallucis longus.

SUPERIOR PERONEAL RETINACULUM. The superior peroneal retinaculum (Fig. 6-79) binds down the tendons of the peroneus

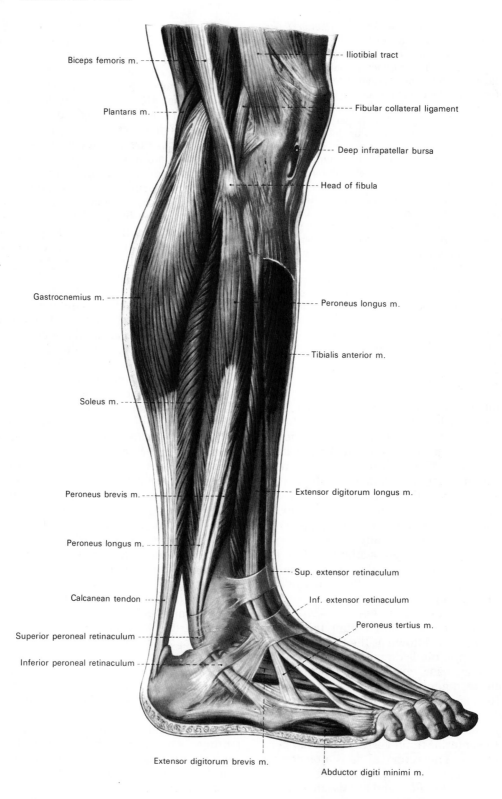

Biceps femoris m.

Plantaris m.

Gastrocnemius m.

Soleus m.

Peroneus brevis m.

Peroneus longus m.

Calcanean tendon

Superior peroneal retinaculum

Inferior peroneal retinaculum

Iliotibial tract

Fibular collateral ligament

Deep infrapatellar bursa

Head of fibula

Peroneus longus m.

Tibialis anterior m.

Extensor digitorum longus m.

Sup. extensor retinaculum

Inf. extensor retinaculum

Peroneus tertius m.

Extensor digitorum brevis m.

Abductor digiti minimi m.

FIG. 6-79. Muscles of the right leg, viewed from the lateral aspect. (From Benninghoff and Goerttler, *Lehrbuch der Anatomie des Menschen*, Urban & Schwarzenberg, 11th Ed., Vol. 1, 1975.)

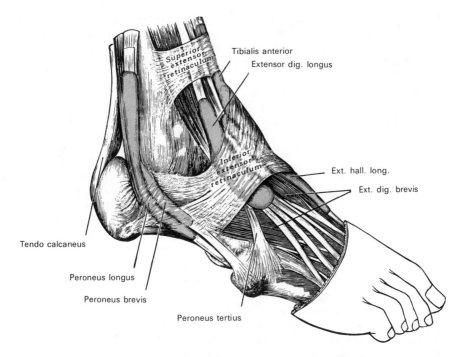

FIG. 6-80. The synovial sheaths of the tendons around the ankle. Lateral aspect.

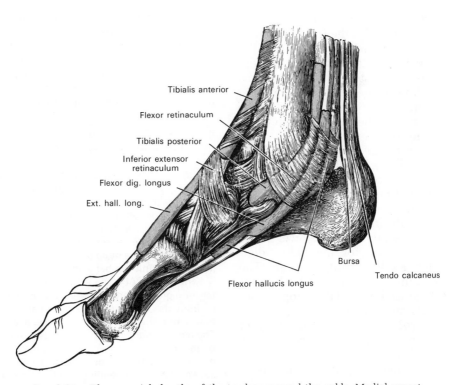

FIG. 6-81. The synovial sheaths of the tendons around the ankle. Medial aspect.

longus and peroneus brevis as they course around the lateral aspect of the ankle. The fibers are attached above to the lateral malleolus and below to the lateral surface of the calcaneus.

INFERIOR PERONEAL RETINACULUM. The inferior peroneal retinaculum (Fig. 6-79) is continuous with the fibers of the inferior extensor retinaculum above and is attached behind to the lateral surface of the calcaneus. Some of the fibers are attached to the peroneal trochlea and form a septum between the tendons of the peroneus longus and peroneus brevis.

Synovial Sheaths of the Tendons Around the Ankle

All of the tendons that cross the ankle joint are enclosed for part of their length in synovial sheaths (Figs. 6-80; 6-81). These have an almost uniform length of about 8 cm each. On the anterior aspect of the ankle joint, the sheath for the tibialis anterior extends from the proximal margin of the superior extensor retinaculum to the interval between the diverging limbs of the inferior retinaculum. The common sheath for the extensor digitorum longus and peroneus tertius and the sheath for the extensor hallucis longus reach proximally to just above the level of the malleoli, the former being the higher. The sheath of the extensor hallucis longus is prolonged to the base of the first metatarsal bone, while that of the extensor digitorum longus reaches only to the level of the base of the fifth metatarsal.

On the *medial aspect of the ankle joint* (Fig. 6-81), the sheath for the tibialis posterior extends most proximally—to about 4 cm above the medial malleolus—while distally it ends just above the tuberosity of the navicular. The sheath for the flexor hallucis longus commences above at the level of the malleolus, while that for the flexor digitorum longus starts slightly higher. The former is continued distally to the base of the first metatarsal, but the latter stops opposite the medial cuneiform bone.

On the *lateral aspect of the ankle joint* (Fig. 6-80), a sheath that is single for the greater part of its extent encloses the tendons of the peroneus longus and brevis. It extends proximally 4 cm above the medial malleolus before it splits into separate sheaths for the two tendons lateral to the calcaneus. Each sheath then continues distally for about another 4 cm.

MUSCLES AND FASCIAE OF THE FOOT

The skin and superficial fascia on the dorsum of the foot are thin and loosely adherent. The underlying **deep fascia,** also a thin layer, is continuous above with the superior and inferior extensor retinacula. On both sides of the foot, it blends with the plantar aponeurosis, and anteriorly it invests the tendons on the dorsum of the foot.

Dorsal Muscle of the Foot

EXTENSOR DIGITORUM BREVIS. The extensor digitorum brevis (Fig. 6-80) is a broad, thin muscle that *arises* from the superior (proximal) surface of the calcaneus, anterior and lateral to the calcaneal sulcus and distal to the groove for the peroneus brevis. Some of its fibers also arise from the lateral talocalcaneal ligament and from the common limb of the inferior extensor retinaculum. It passes obliquely across the dorsum of the foot and ends in four tendons. The most medial, which is the largest, crosses the dorsalis pedis artery and is *inserted* into the dorsal surface of the base of the proximal phalanx of the great toe. It is frequently described as a separate muscle, the **extensor hallucis brevis.** The other three are *inserted* into the lateral surfaces of the tendons of the extensor digitorum longus muscle of the second, third, and fourth toes.

Action. The **extensor digitorum brevis** extends the proximal phalanx of the great toe, and by its attachments to the tendons of the extensor digitorum longus, it extends the phalanges of the second, third, and fourth toes.

Nerve. The extensor digitorum brevis is supplied by a branch of the **deep peroneal nerve,** containing fibers from fifth lumbar and first sacral nerves.

Variations. Accessory slips of origin to the extensor digitorum brevis from the talus and navicular, or from the lateral cuneiform and third metatarsal bones, or from the cuboid bone have been observed. The tendons vary in number and position. They may be reduced to two, or one of them may be doubled, or an additional slip may pass to the little toe. A supernumerary slip ending on one of the metatarsophalangeal articulations or joining a dorsal interosseous muscle is not uncommon.

Plantar Aponeurosis

The plantar aponeurosis is of great strength and consists of pearly white glistening fibers disposed, for the most part, longitudinally. It is divided into central, lateral, and medial portions.

The **central portion** is the thickest. It is narrow behind and attached to the medial process of the tuberosity of the calcaneus, proximal to the origin of the flexor digitorum brevis (Fig. 6-82). It becomes broader and thinner distally and divides near the heads of the metatarsal bones into five processes, one for each of the toes. Each of these processes separates opposite the metatarsophalangeal joint into two strata, superficial and deep. The superficial stratum is attached to the skin of the transverse sulcus, which separates the toes from the sole. The deeper stratum divides into two slips and these embrace the sides of the flexor tendons of the toes, blending with the sheaths of the tendons and with the transverse metatarsal ligament, and forming a series of arches through which the tendons of the short and long flexors pass to the toes. At the intervals between the five processes, the digital vessels and nerves and the tendons of the lumbrical muscles become superficial. At the point of division of the aponeurosis, numerous transverse fasciculi add to the strength of the aponeurosis by binding the processes together, and connecting them with the integument. The central portion of the plantar aponeurosis is continuous with the lateral and medial portions. At these lines of junction, two strong vertical intermuscular septa, broader distally than proximally, are sent deeper into the foot. These separate the intermediate from the lateral and medial groups of plantar muscles. From these vertical septa are derived thinner transverse septa, which separate the various layers of muscles in this region. The deep surface of this aponeurosis gives origin proximally to some fibers of the flexor digitorum brevis. The lateral and medial portions of the plantar aponeurosis, which cover the sides of the sole of the foot, are thinner than the central portion.

The **lateral portion** lies superficial to the abductor digiti minimi muscle. It is thin distally and thick proximally, where it forms a strong band between the lateral process of the tuberosity of the calcaneus and the base of the fifth metatarsal bone. It is continuous medially with the central portion of the plantar aponeurosis, and laterally with the fascia on the dorsum of the foot.

The **medial portion** is thin and lies superficial to the abductor hallucis muscle. It is attached behind to the flexor retinaculum and is continuous medially, around the side of the foot, with the fascia on the foot dorsum and laterally with the central portion of the plantar aponeurosis.

Plantar Muscles of the Foot

The muscles in the plantar region of the foot can be divided into three groups—medial, lateral, and intermediate, like those

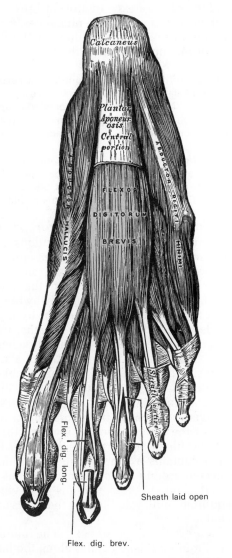

FIG. 6-82. The first layer of plantar muscles. Right foot.

in the hand. The medial plantar muscles comprise the intrinsic muscles of the great toe and correspond to those of the thumb. The lateral plantar muscles consist of the intrinsic muscles of the small toe and correspond to those of the little finger. The intermediate plantar muscles, including the lumbricals, the interossei, the flexor digitorum brevis, and the quadratus plantae, are connected with the tendons intervening between the two former groups. In order to facilitate the description of these muscles, it is more convenient to divide them into four layers, in the sequence encountered during dissection.

FIRST LAYER

This is the most superficial of the plantar layers and consists of three muscles, all of which arise from the calcaneus. They lie deep to the plantar aponeurosis and are surrounded by their own fascial covering.

Abductor hallucis
Flexor digitorum brevis
Abductor digiti minimi

ABDUCTOR HALLUCIS. The abductor hallucis (Fig. 6-82) lies along the medial border of the foot, covering the entrance of the plantar vessels and nerves into the sole. It *arises* from the medial process of the tuberosity of the calcaneus, from the flexor retinaculum, from the plantar aponeurosis, and from the intermuscular septum between it and the flexor digitorum brevis. The fibers end in a tendon that is *inserted,* together with the medial tendon of the flexor hallucis brevis, into the medial side of the base of the proximal phalanx of the great toe.

Action. Although the **abductor hallucis** can abduct the great toe at the metatarsophalangeal joint, it probably acts more consistently as a flexor at that joint.

Nerve. The abductor hallucis is supplied by a branch of the **medial plantar nerve,** containing fibers from the fifth lumbar and first sacral nerves.

FLEXOR DIGITORUM BREVIS. The flexor digitorum brevis lies in the middle of the sole of the foot, immediately deep to the central part of the plantar aponeurosis, to which many of its fibers are firmly united (Fig. 6-82). Its deep surface is separated from the lateral plantar vessels and nerves by a thin

layer of fascia. It *arises* by a narrow tendon from the medial process of the tuberosity of the calcaneus, as well as from the central part of the plantar aponeurosis and from the intermuscular septa between it and the adjacent muscles. It passes distally and divides into four tendons, one for each of the four lesser toes. Opposite the bases of the proximal phalanges, each tendon divides into two slips through which passes the corresponding tendon of the flexor digitorum longus. The two portions of the tendon then reunite and form a grooved channel for the accompanying long flexor tendon. Finally, the tendon of the flexor digitorum brevis divides a second time and is *inserted* onto both sides of the middle phalanx. The mode of division of the tendons of the flexor digitorum brevis, and of their insertion into the phalanges, is comparable to that of the tendons of the flexor digitorum superficialis in the hand.

Action. The **flexor digitorum brevis** flexes the second phalanges of the four lateral toes.
Nerve. The flexor digitorum brevis is supplied by a branch of the **medial plantar nerve,** containing fibers from the fifth lumbar and first sacral nerves.
Variations. The tendon to the little toe is frequently absent (38%), or it may be replaced by a small fusiform muscle arising from the long flexor tendon (33%), or from the quadratus plantae.

FIBROUS SHEATHS OF THE FLEXOR TENDONS. The terminal portions of the tendons of the long and short flexor muscles are contained in osseous-aponeurotic canals similar in their arrangement to those in the fingers. These canals are limited dorsally by the phalanges and on the plantar aspect by fibrous bands called **digital fibrous sheaths,** which arch across the tendons and are attached on both sides to the margins of the phalanges. Opposite the bodies of the proximal and middle phalanges, the **annular part** of the fibrous bands is strong, and the fibers are transverse. Opposite the joints, the **cruciate part** of the fibrous sheaths is much thinner and the fibers are directed obliquely. Each canal contains a synovial sheath, which is reflected around the contained tendons and is attached to the tendons by **vincula tendinum** similar to those in the digits of the hand.

ABDUCTOR DIGITI MINIMI. The abductor digiti minimi (Fig. 6-82) lies along the lateral border of the foot, its medial margin being related to the lateral plantar vessels and nerves. It *arises,* by a broad origin, from the lateral process and distal part of the medial process of the tuberosity of the calcaneus, as well as from the intervening plantar surface of the bone. It also *arises* from the plantar aponeurosis and from the intermuscular septum between it and the flexor digitorum brevis. Its tendon, after gliding over a smooth facet on the plantar surface of the base of the fifth metatarsal bone, is *inserted,* with the flexor digiti minimi brevis, into the lateral aspect of the base of the proximal phalanx of the little toe.

Action. The **abductor digiti minimi** abducts and helps to flex the small toe at the metatarsophalangeal joint.

Nerve. The abductor digiti minimi is supplied by a branch of the **lateral plantar nerve**, containing fibers from the second and third sacral nerves.

Variations. Usually some fibers of this muscle insert onto the tuberosity at the base of the fifth metatarsal. At times these fibers actually constitute a separate muscle, the **abductor ossis metatarsi quinti**, with fibers arising on the lateral process of the tuberosity of the calcaneus and inserting onto the tuberosity of the fifth metatarsal bone in common with or beneath the outer margin of the plantar fascia.

SECOND LAYER

The intrinsic muscles of the second layer are all attached to the tendons of the flexor digitorum longus as that muscle courses through the sole of the foot on its way to the digits. The muscles of this layer are:

Quadratus plantae
Lumbricales

QUADRATUS PLANTAE. The quadratus plantae (Fig. 6-83), also known as the flexor accessorius, alters the posteromedial pull of the tendon of the flexor digitorum longus into an almost straight posteriorly oriented pull. Between it and the muscles of the first layer are found the lateral plantar vessels and nerve. It *arises* by two heads, which are separated from each other by the long plantar ligament. The larger, muscular **medial head** *arises* from the medial concave surface of the calcaneus, below the groove that lodges the tendon of the flexor hallucis longus. The **lateral head,** which is flat and tendinous, *arises* from the lateral border of the calcaneus, distal to the lateral process of

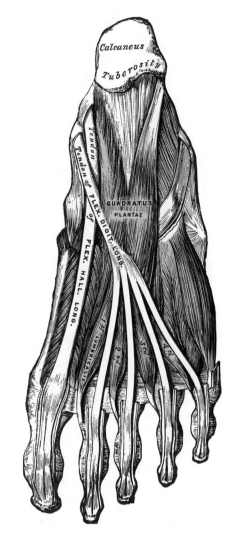

FIG. 6-83. The second layer of plantar muscles. Right foot.

its tuberosity, and from the long plantar ligament. The two portions join at an acute angle and end in a flattened muscular band, which is *inserted* deep to and along the lateral margin of the tendon of the flexor digitorum longus near the middle of the plantar region. The deeper part of the quadratus plantae usually sends slips to those tendons of the flexor digitorum longus that pass to the second, third, and fourth toes.

Action. The **quadratus plantae** helps the flexor digitorum longus to flex the terminal phalanges of the four small toes.

Nerve. The quadratus plantae is supplied by a branch of the **lateral plantar nerve**, containing fibers from the second and third sacral nerves.

Variations. The lateral head or sometimes the

entire muscle is absent. The muscle varies with respect to the number of digital tendons receiving
muscular slips. Occasionally, tendons to only two
toes receive muscle fibers; sometimes those to the
fourth are absent, and in many cases, fibers are sent
to the fifth.

THE LUMBRICALS.   The lumbricals are four
small muscles, accessory to the tendons of
the flexor digitorum longus, numbered from
the medial side of the foot (Fig. 6-83). They
arise from these tendons, as far back as their
angles of division. Each lumbrical springs
from two tendons except the first, which
arises along the medial surface of the first
tendon. The muscles end in tendons, which
pass distally along the medial sides of the
four lesser toes and are *inserted* into the expansions of the tendons of the extensor digitorum longus on the dorsal surfaces of the
proximal phalanges.

Action.   The **lumbricals** in the foot, similar to
those in the hand, flex the proximal phalanges at
the metatarsophalangeal joints and extend the
two distal phalanges at the interphalangeal joints
of the four lateral toes.
Nerves.   The first lumbrical is supplied by a
branch of the **medial plantar nerve**, containing
fibers from the fifth lumbar and first sacral
nerves. The other three lumbricals receive branches
of the **lateral plantar nerve**, containing fibers from
the second and third sacral nerves.
Variations.   Absence of one or more lumbrical
muscles has been noted, and doubling of the third or
fourth has been reported. At times they may be inserted partly or wholly onto the bone of the first phalanges instead of the dorsal expansions.

THIRD LAYER

Two short flexors, one for the large toe and
one for the little toe, and the adductor of the
large toe comprise the muscles of the third
laqer. They are:

Flexor hallucis brevis
Adductor hallucis
Flexor digiti minimi brevis

FLEXOR   HALLUCIS   BREVIS.   Throughout
most of its extent, the flexor hallucis brevis
lies adjacent to the plantar surface of the
first metatarsal bone (Fig. 6-84). It *arises* by a
pointed tendinous process from the medial
part of the plantar surface of the cuboid
bone, from the contiguous portion of the lateral cuneiform, and from the prolongation of
the tendon of the tibialis posterior, which is

FIG. 6-84.   The third layer of plantar muscles. Right
foot.

attached to that bone. It divides distally into
medial and lateral parts, which are *inserted*
into the medial and lateral aspects of the
base of the proximal phalanx of the large
toe, a sesamoid bone being present in each
tendon at its insertion. The **medial part** is
blended with the abductor hallucis of the
first layer just before its insertion, while the
lateral part inserts near the tendon of the
adductor hallucis. The tendon of the flexor
hallucis longus lies in a groove between the
two parts.

Action.   The **flexor hallucis brevis** flexes the
proximal phalanx of the large toe at the metatarsophalangeal joint.
Nerve.   The flexor hallucis brevis is supplied by a

branch of the **medial plantar nerve,** containing fibers from the fifth lumbar and first sacral nerves.

Variations. The origin of the flexor hallucis brevis may vary. It often receives fibers from the calcaneus or long plantar ligament, and the attachment to the cuboid may be absent. An additional slip may insert onto the proximal phalanx of the second toe.

ADDUCTOR HALLUCIS. The adductor hallucis consists of oblique and transverse heads (Fig. 6-84). The **oblique head** is a large, thick, fleshy mass, which occupies an otherwise hollow space superficial to the first, second, third, and fourth metatarsal bones. It *arises* from the bases of the second, third, and fourth metatarsal bones, and from the sheath of the tendon of the peroneus longus. Crossing obliquely to the medial side of the foot, it is *inserted,* together with the lateral portion of the flexor hallucis brevis, into the lateral aspect of the base of the proximal phalanx of the large toe. The **transverse head** is a narrow, flat fasciculus that arises from the plantar metatarsophalangeal ligaments of the third, fourth, and fifth toes (sometimes only from the third and fourth) and from the transverse metatarsal ligaments of these same digits. Coursing medially, its fibers are *inserted* into the lateral aspect of the base of the proximal phalanx of the large toe, its fibers blending with the tendon of insertion of the oblique head.

Action. The **adductor hallucis** adducts the large toe, drawing it toward the longitudinal axis of the foot. It also flexes the proximal phalanx of the large toe at the metatarsophalangeal joint and helps to maintain the transverse metatarsal arch.

Nerve. The adductor hallucis is supplied by a branch of the **lateral plantar nerve,** containing fibers from the second and third sacral nerves.

Variations. A small slip of the muscle may insert onto the base of the proximal phalanx of the second toe. At times a portion of the muscle may insert onto the first metatarsal bone, constituting an **opponens hallucis** muscle.

FLEXOR DIGITI MINIMI BREVIS. The short flexor of the little toe lies superficial to the fifth metatarsal bone and resembles one of the interossei (Fig. 6-84). It *arises* from the base of the fifth metatarsal bone and from the sheath of the peroneus longus. Its tendon is *inserted* into the lateral aspect of the base of the proximal phalanx of the fifth toe. Occasionally a few of the deeper fibers are in-

serted into the lateral part of the distal half of the fifth metatarsal bone, being described by some authors as a distinct muscle, the **opponens digiti minimi.**

Action. The **flexor digiti minimi brevis** flexes the proximal phalanx of the small toe at the metatarsophalangeal joint.

Nerve. The flexor digiti minimi brevis is supplied by a branch of the **lateral plantar nerve,** containing fibers from the second and third sacral nerves.

FOURTH LAYER

The fourth layer of muscles on the plantar aspect of the foot consists of the **interossei.** These are similar to the interossei in the hand, except that their action is considered relative to the midline of the *second digit,* through which the longitudinal axis of the foot courses, in contrast to the midline of the *third digit,* as is the case in the hand. There are seven interossei in the foot and they are arranged in two groups:

Dorsal interossei
Plantar interossei

DORSAL INTEROSSEI. The dorsal interossei (Fig. 6-85), situated between the metatarsal bones, are four bipennate muscles, each *arising* by two heads from the adjacent sides of the metatarsal bones between which it is

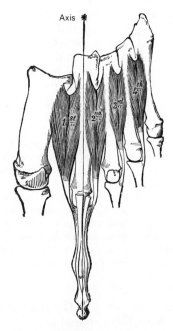

FIG. 6-85. The dorsal interossei. Left foot, dorsal view.

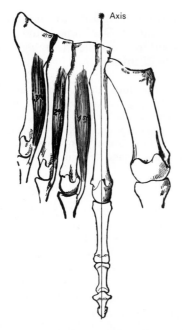

FIG. 6-86.   The plantar interossei. Left foot, plantar view.

placed. Their tendons are *inserted* into the bases of the proximal phalanges and into the dorsal digital aponeuroses of the tendons of the extensor digitorum longus. The first is *inserted* into the medial side of the second toe, while the other three are *inserted* into the lateral aspects of the second, third, and fourth toes. In the angular interval left between the heads of each of the three lateral dorsal interosseous muscles, one of the perforating arteries passes to the dorsum of the foot. Through the space between the two heads of the first muscle, the deep plantar branch of the dorsalis pedis artery enters the sole of the foot.

Action.   The **dorsal interossei** abduct the toes from the longitudinal axis, which extends through the second toe. They also flex the proximal phalanx and extend the distal phalanges.

Nerves.   The dorsal interossei are supplied by twigs from the **lateral plantar nerve**, containing fibers from the second and third sacral nerves. The fourth dorsal interosseous receives nerve fibers from the superficial branch, while the other three receive their innervation from the deep branch of the lateral plantar nerve.

PLANTAR INTEROSSEI.   The plantar interossei, which are three in number, lie along the plantar surface of the metatarsal bones,

rather than between them. Each is connected with but one metatarsal bone. They *arise* from the bases and medial sides of the bodies of the third, fourth, and fifth metatarsal bones; they are *inserted* into the medial sides of the bases of the proximal phalanges of the same toes and into the dorsal digital aponeuroses of the tendons of the extensor digitorum longus.

Action.   The **plantar interossei** adduct the third, fourth, and fifth toes toward the axis of the second toe. They also flex the proximal and extend the distal phalanges.

Nerve.   The plantar interossei are supplied by branches of the **lateral plantar nerve** that contain fibers from the second and third sacral nerves. The third (most lateral) interosseous muscle is supplied by the superficial branch of the lateral plantar nerve, while the other two are supplied by the deep branch.

Group Actions of Muscles at the Metatarsophalangeal Joints of the Toes.   *Flexion* of the toes at these joints is performed by the flexor digitorum longus, the quadratus plantae, the flexor digitorum brevis, the flexor hallucis longus and brevis, the flexor digiti minimi brevis, as well as the lumbricals and interossei. *Extension* at the metatarsophalangeal joint is performed by the extensors digitorum longus and brevis and the extensor hallucis longus. *Abduction* of the toes results from the action of the dorsal interossei, the abductor hallucis, and the abductor digiti minimi. *Adduction* of the toes is performed by the plantar interossei and the adductor hallucis.

Group Actions of Muscles at the Interphalangeal Joints of the Toes.   *Flexion* of the toes at the interphalangeal joints occurs through the action of the flexor digitorum longus with the assistance of the quadratus plantae. The flexor digitorum brevis flexes the interphalangeal joints between the middle and proximal phalanges, while the flexor hallucis longus flexes the interphalangeal joint of the large toe. *Extension* of the interphalangeal joints is performed by the extensor digitorum longus, the lumbricals, and the interossei, as well as the extensor digitorum brevis and extensor hallucis longus.

Structure of Skeletal Muscle

The contractile property of muscle tissue allows for movement of limbs and other body parts. Although the fundamental units of muscle tissue are called **muscle fibers,** these elements are in fact elongated cells containing nuclei and cytoplasm. There are two general types of muscle, smooth and striated. **Smooth muscle** is generally found in the walls of viscera and is under the neural control of motor fibers in the visceral nervous system (the autonomic nervous system). **Striated muscle,** however, is further subdivided into **cardiac striated muscle,** which is found in the walls of the heart, and **skeletal striated muscle.** The latter is connected to the bony skeleton and forms the bulk of the flesh in the body; it is under voluntary neural control by way of the somatic motor nerve fibers in the cranial and spinal nerves. All the muscles described in this chapter are formed of skeletal striated muscle. Because the chapter on the heart will describe cardiac striated muscle, and chapters on the viscera will elaborate the structure of smooth muscle, this discussion will be limited to the general organization, cellular features, and ultrastructural characteristics of skeletal striated muscle.

GENERAL ORGANIZATION OF SKELETAL MUSCLE. Skeletal muscle is designed principally for mobility, serving to enable an organism to react appropriately to environmental conditions. Yet, skeletal muscles benefit the body in more ways than only the production of movement. They also produce body heat, and to function effectively they must limit movement, as well as counteract the force of gravity in order to maintain posture. Most skeletal muscles are connected to bones, cartilages, or ligaments, usually through the intervention of tendons or aponeuroses. These tendinous attachments to bone may be long (insertion of semitendinosus or origin of long head of biceps brachii), or the muscle fibers may attach quite directly to the bone (origins of lateral and medial heads of triceps brachii). Muscles may also attach to the skin (facial muscles), to a mucous membrane (certain tongue muscles), or to a fibrous plate (extraocular muscles), or they may form circular bands (sphincter muscles).

The long, cylindrical, striated skeletal muscle fibers (Fig. 6-87) are arranged, more or less, in a parallel manner to form bundles or **fascicles.** Between the individual muscle fibers ramifies a delicate network of connective tissue called the **endomysium** (Fig. 6-87). Each bundle or fascicle is surrounded by a more distinct, but still thin, layer of collagenous and elastic connective tissue, the **perimysium,** which binds the fascicles into larger groups and forms the fibrous septa throughout a muscle (Fig. 6-87). Investing the entire muscle is the connective tissue **epimysium.** This latter covering may be well developed, or it may be quite delicate, as it is when a muscle glides freely beneath a strong fascial sheet. At times the epimysium may lose its identity and even become fused with the surrounding fascial layer, especially if the adjacent fascia is used by some of the muscle fibers for their attachment. Such an elaborate meshwork of stromal connective

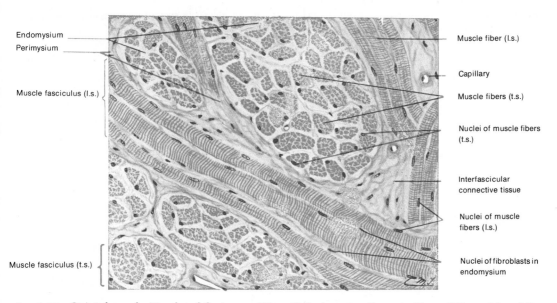

Endomysium
Perimysium
Muscle fasciculus (l.s.)
Muscle fasciculus (t.s.)

Muscle fiber (l.s.)
Capillary
Muscle fibers (t.s.)
Nuclei of muscle fibers (t.s.)
Interfascicular connective tissue
Nuclei of muscle fibers (l.s.)
Nuclei of fibroblasts in endomysium

FIG. 6-87. Striated muscle. Muscles of the tongue. 320 ×. Stain: hematoxylin-eosin. (From di Fiore, *Atlas of Human Histology,* 5th Ed., Lea & Febiger, 1981.)

tissue elements allows for a more independent action of the individual fascicles and muscle fiber groups.

In principle, the ends of muscle fibers attach in a similar manner whether the fasciculi end in a tendon or are connected directly to bones or cartilage. Each muscle fiber has a distinct termination, which is rounded, conical, or truncated, and is covered by the muscle fiber membrane. The collagen fibrillae of the endomysium become thickened and closely adherent to the end of the fiber as they pass over it (Fig. 6-88). The collagen fibers then course beyond the muscle fiber into the tendon, where they become the actual substance of the tendon (Goss, 1944). In muscles that appear to attach directly to bones, cartilages, or other structures, the fibers of the endomysium become continuous with the periosteum or other fibrous layers in the same manner as with a tendon. When the muscle fibers terminate within a fasciculus, as in long muscles, the ends of the fibers may be either blunt or long and tapering, and they are secured in place by the merging of their terminal endomysium with the endomysium of neighboring fibers. These findings have been confirmed by electron microscopy (Mackey et al., 1969).

Skeletal muscle is well supplied with both vessels and nerves. Arterial and venous vessels course through the perimysium, eventually dividing into fine capillaries which ramify among the individual muscle fibers, anastomosing freely and forming a rich bed for the exchange of nutrients and metabolites. Accompanying the vessels in the perimysium are branches of larger nerve trunks, which together with the vessels frequently form a neurovascular network in the connective tissue septa between the fascicles of a muscle. Finer branching of the nerve trunks eventually results in both the motor and sensory innervation of the individual skeletal muscle fibers. A single alpha motor nerve fiber may supply anywhere from three to over two thousand skeletal muscle fibers. In muscles serving to produce fine movements under discriminating voluntary control, such as the extraocular muscles in the orbit, each nerve fiber innervates a small number of muscle fibers. In contrast, muscles producing gross movements have many hundreds of muscle fibers supplied by a single nerve fiber. A single motor nerve fiber **(alpha efferent)** and all the muscle fibers that it innervates constitute a functional subdivision of muscle and collectively are called the **motor unit.** Thus, when a nerve impulse travels along a motor nerve fiber that it supplies, all of the muscle fibers in the motor unit contract together. Muscle is also supplied with sensory fibers, most of which course from receptor sites in neuromuscular spindles and Golgi[1] tendon organs.

Neuromuscular spindles are composed of three to fewer than twenty striated muscle fibers surrounded by a thick connective tissue capsule. These intracapsular muscle fi-

[1]Camillo Golgi (1843–1926): An Italian anatomist (Pavia and Siena).

Reticular fibrils

Tendon fibrils

Recticular fibrils

Muscle nucleus

Blood corpuscles

Tendon

Fɪɢ. 6-88. Attachment of a muscle fiber from the flexor digitorum superficialis of a monkey. Reticular connective tissue fibrils are blackened with Masson's silver stain. 900 ✕. (From Goss, 1944.)

bers are called **intrafusal** to differentiate them from the extrafusal muscle fibers in the remainder of the muscle (Fig. 6-89). Surrounded by afferent nerve fibers, the intrafusal fibers of muscle spindles receive motor innervation by a special group of nerve fibers called the **gamma efferents.** Muscle spindles serve as receptors in many somatic muscles and are capable of responding to a stretching of the entire muscle, or a part of it, thereby initiating a sensory discharge in their afferent fibers. This feedback information to the central nervous system is importantly involved in muscle control and the maintenance of muscle tone.

CELLULAR FEATURES OF SKELETAL MUSCLE. The principal parenchymal unit in skeletal muscle is the striated skeletal muscle cell. These may be large, exceedingly long cells, like those in the sartorius muscle, extending many centimeters in length and measuring from 10 to 100 microns in width, or they may be only a few millimeters in length and extremely slender. If teased apart, skeletal muscle cells can be within the limit of visibility of the naked eye.

Each striated muscle cell is multinucleated, the numbers of nuclei being proportional to the volume of the cell. The nuclei are flat, elongated discs with their long axis oriented in the direction of the muscle fiber. Although they are numerous and irregularly placed along the muscle fiber, they are located close to the inner surface of the outer cell membrane or **sarcolemma** (Fig. 6-90). In special preparations, the mitochondria and the Golgi apparatus can be demonstrated adjacent to the poles of each skeletal muscle cell nucleus. In addition to the nuclei of the skeletal muscle cell, striated muscle tissue contains nuclei of fibroblasts, as delicate wisps of connective tissue forming the endomysium are interposed between the muscle cells.

Within the sarcoplasm (cytoplasm) of skeletal muscle cells are found many longitudinally oriented parallel units called **myofibrils.** These are visible with the light microscope and measure less than one micron to about two or three microns in diameter. When muscle fibers are cut in cross section, the myofibrils appear as clusters of fine dots that occupy nearly all of the intracellular volume (Fig. 6-90). Between the myofibrils, however, are small amounts of sarcoplasm that slightly separate them from each other. In longitudinal sections, the most remarkable characteristic of skeletal muscle cells is their **cross striations** (Fig. 6-91). The cross striated appearance is due to the fact that the myofibrils consist of alternating dark **(A band)** and light **(I band)** staining segments. When observed with the polarizing microscope, the bands that appear dark with the

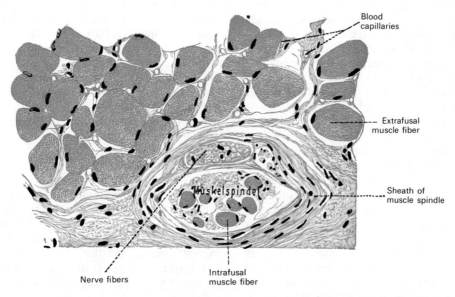

FIG. 6-89. Cross section of part of a lumbrical muscle from a human hand, showing extrafusal and intrafusal muscle fibers, endomysium, blood capillaries, and a muscle spindle. Approximately 400 ×. (From Rauber-Kopsch, *Lehrbuch u. Atlas d. Anatomie d. Menschen,* 19th Ed., Vol. I, Georg Thieme Verlag, 1955.)

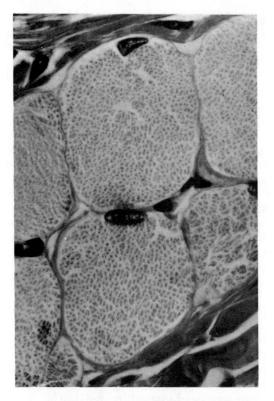

FIG. 6-90. Photomicrograph of striated muscle fibers cut in cross section. Note that within each muscle fiber are many myofibrils and that the nuclei are placed near the surface of the fiber membrane. (From Ham, *Histology*, 7th Ed., J. B. Lippincott, 1974.)

ordinary light microscope are doubly refractile (birefringent) or **anisotropic** (hence called A bands). Contrastingly, the light staining bands of customary histologic preparations appear **isotropic** with polarizing light (hence called I bands).

The cross striations of adjacent myofibrils within a skeletal muscle cell observed in longitudinal section are not exactly aligned, and the intervening sarcoplasm does not show striations. They are, however, evenly registered and thereby give the muscle fiber striations the appearance of extending across the entire fiber. Well-fixed light microscopic preparations of longitudinally oriented skeletal muscle cells reveal a dark transverse line that bisects each lightly stained I band. This line, called the **Z line** (zwischenscheibe), may be conceived of as a septum across the myofibril and is designated as the limits of a **sarcomere.** The sarcomere is the repeating structural unit of a myofibril as well as its functional unit of

contraction. Thus, between two Z lines, the sarcomere includes one A band, plus one half of an I band on each of its ends (Fig. 6-92).

There appears to be morphological correlation to the speed of contraction of muscle fibers. Two main types of fibers have been identified: small-diameter **red fibers,** which are rich in mitochondria and myoglobin and are surrounded by more blood capillaries, and large-diameter **white fibers,** which contain little myoglobin and few mitochondria (Gauthier, 1970). Frequently, muscles consist of a mixture of the fiber types, but in certain muscles one fiber type predominates over the other. The duration of contraction of red muscle fibers may last ten times that of white muscle fibers. The soleus muscle, which must function for extended periods of time as an antigravity muscle, principally consists of red muscle fibers. It contracts slowly, but the duration of its contraction is quite long (100 msec). In contrast, extraocular muscles, which rapidly fixate the eyeballs, consist of white fibers that allow muscle contractions to be rapid but of short duration (10 msec).

ULTRASTRUCTURAL FEATURES OF SKELETAL MUSCLE. Although the A bands, I bands, and Z lines can quite readily be observed with ordinary microscopy, the electron microscope has significantly contributed to an understanding of skeletal muscle. First of all, the dark staining A band is crossed by a lighter band called the **H zone** (Hensen's zone[1]). In turn, the lighter H zone is bisected by a slightly darker M line (Fig. 6-92). The individual myofibrils within a skeletal muscle cell are themselves composed of longitudinally oriented **myofilaments.** Two types of myofilaments have been identified according to their size and chemical composition. The thicker ones, between 100 Å to 150 Å in diameter, are confined to the A band and are **myosin myofilaments,** about 1.5 microns long. The thinner myofilaments are between 50 Å and 80 Å in diameter and attach at their middle to the Z line, from which they extend for about one micron in either longitudinal direction. Chemically, they are the **actin myofilaments,** and they form the I bands. Additionally, these thin actin filaments extend from the I bands into the adjacent A

[1]Victor Hensen (1835–1924): A German anatomist and physiologist (Kiel).

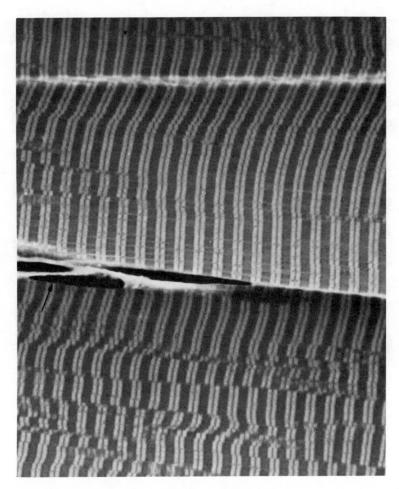

FIG. 6-91. Photomicrograph of longitudinally sectioned skeletal muscle fibers. Arrow points to a nucleus just beneath the sarcolemma of one of the muscle fibers. Note the cross striations. The wide, dark striations are the A bands; the light striations are the I bands. Each I band is bisected by a narrow dark Z line. 2000 ×. (Modified from Copenhaver, Kelly, and Wood, *Bailey's Textbook of Histology*, 17th Ed., Williams and Wilkins, 1978.)

bands, interdigitating with the thicker myosin filaments in a regular manner (Fig. 6-93). Each thick myosin filament in the A band is surrounded by six thin actin filaments, spaced evenly and hexagonally around the myosin filament (Fig. 6-94).

Ultrastructural studies have shown that the thin actin filaments of *relaxed muscle fibers* extend into the A bands from each side, but they do not meet at the center. The central part of the A band in relaxed muscle, into which the actin filaments do not invade, is the lighter H zone. *When muscle fibers contract,* the thin actin filaments slide further into the A band, meeting in the center or even sliding past each other, thereby reducing the size of the H zone or eliminating it altogether (Fig. 6-95). Because the former H zone contains both types of filaments in a contracted fiber, it becomes as dense as the rest of the A band. The Z lines are simultaneously pulled closer to the A bands, and therefore the I bands also become reduced in size. Although the length of the sarcomere is reduced, contraction involves not a reduction in the size of either the actin or myosin protein components, but an attraction of these two elements for each other, so that their surface areas can achieve the maximum contact for the formation of the **actomyosin complex.** This process requires the energy supplied by the breakdown of ATP (adenosintriophosphate), found in the mitochondria, to ADP (adenosindiphosphate).

Other important ultrastructural features of skeletal muscle cells are beginning to elu-

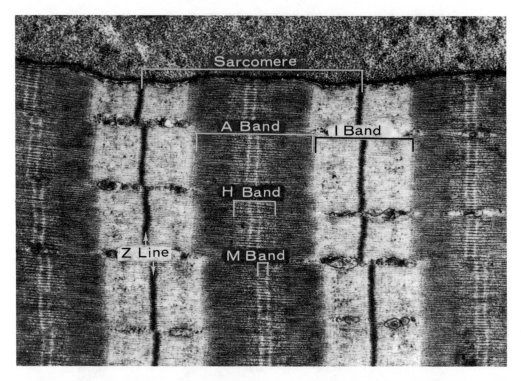

FIG. 6-92.   Electron micrograph of five longitudinally oriented myofibrils in relaxed skeletal muscle. Observe that the limit of a sarcomere is defined as a region of a myofibril between two Z lines. This includes an A band capped at each end by one-half of an I band. (From Bloom and Fawcett, *A Textbook of Histology,* 10th Ed., W. B. Saunders Co., 1975.)

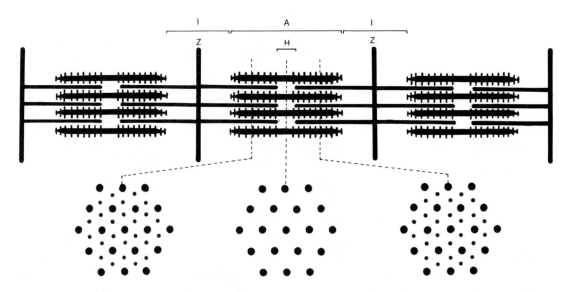

FIG. 6-93.   Diagrammatic representation of the structure of striated muscle showing the overlapping arrays of the thinner, actin- and thicker myosin-containing filaments. Lower diagrams show cross-sectional patterns of the filaments through the A and H bands. (From H. E. Huxley, 1969. Science, *164:* 1356. Copyright 1969 by the American Association for the Advancement of Science.)

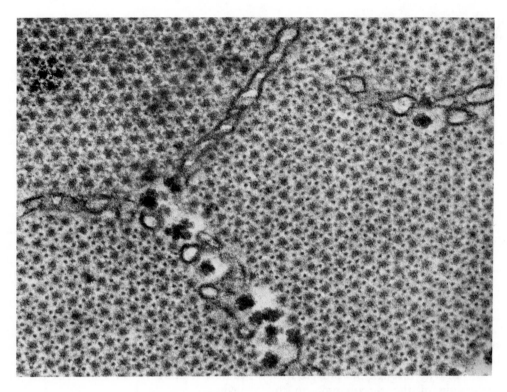

FIG. 6-94. Electron micrograph of frog sartorius myofibrils showing A bands cut in cross section. Note that the thick myosin filaments within the myofibril are each surrounded by six thin actin filaments. 150,000 ×. (Modified from H. E. Huxley, 1968.)

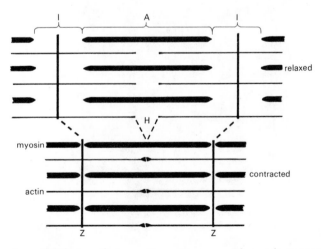

FIG. 6-95. This diagram shows the changes that occur in a sarcomere with muscle contraction. During contraction (low part), the thin actin filaments slide toward the center of the A band. This diminishes the width of the I bands and the H band seen in relaxed muscle (upper part). (From Windle, *Textbook of Histology,* 4th Ed., McGraw-Hill, 1969. Used with permission of McGraw-Hill Book Company.)

cidate how the contraction of myofibrils is initiated by arriving nerve impulses to the surface membrane or sarcolemma of the muscle cell. Myofibrils are encased longitudinally by a parallel array of interconnected membranous tubules and cisternae that composes the **sarcoplasmic reticulum** (Figs. 6-96; 6-97). Extending along both the A band and the I band, the sarcoplasmic reticulum fuses to form a cisterna at the H zone. Near both ends of the sarcomere at the Z lines, the tubules expand and coalesce to form terminal cisternae. Each terminal cisterna interfaces another, located on the opposite side of the Z line. At this site (in amphibian muscle) or at the junction of the A band and I bands (in mammalian muscle), the terminal cisternae of the sarcoplasmic reticulum become associated with a transverse tubular system that courses along the Z line called the **T system.** The T tubule and the two terminal cisternae are collectively referred to as the **triad** (Figs. 6-96; 6-97). It has been shown that the T tubule, which passes all the way through the muscle fiber from one side of the sarcolemma to the

other, opens extracellularly. It is in contact with the exterior of the cell and contains extracellular fluid (Fig. 6-98).

With the T tubules in continuity with the extracellular space and the sarcoplasmic reticulum intimately associated with the intracellular contractile components of the muscle fiber, a morphological system exists that can communicate surface events occurring on the sarcolemma to the contractile substructure in the interior of the myofibril. It has been proposed that the contractile wave of depolarization initiated at the myoneural junction spreads not only along the skeletal muscle surface, but also through the T tubule system. In some manner, this excitation is transmitted across the common membranous wall shared by the T tubule and the terminal cisternae of the sarcoplasmic reticulum, which contains calcium ions. The release of calcium within the sarcoplasmic reticulum probably initiates contraction, after which calcium becomes bound once again, perhaps by a relaxation factor, leading to relaxation of the myofibrils.

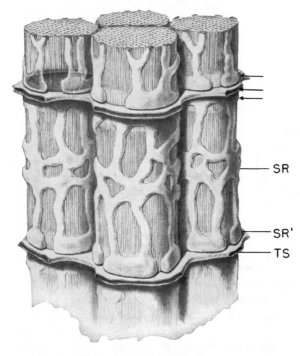

SR

SR'

TS

FIG. 6-96. This drawing depicts the relationship between the myofibrils (four are shown) of a muscle fiber, the associated sarcoplasmic reticulum (SR), and the T system (TS). The sarcoplasmic reticulum expands into terminal cisterns (SR') at the I bands, and forms at the Z lines a triad (see three arrows) with the T tubule. (From Porter and Bonneville, *Fine Structure of Cells and Tissues,* 4th Ed., Lea & Febiger, 1973.)

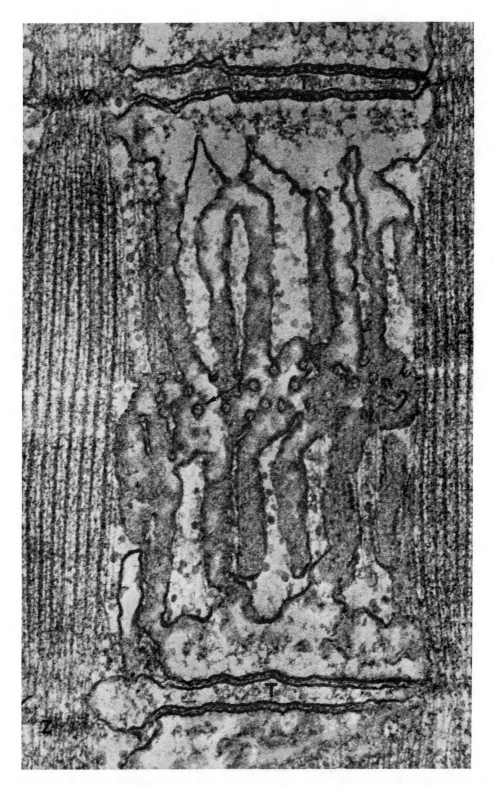

Fig. 6-97. Electron micrograph of the sarcoplasmic reticulum and T tubules of a goldfish skeletal muscle myofibril (78,300 ×). T: T tubules; Z: Z lines. Note the fenestrations (f) or pores in the sarcoplasmic reticulum at the H zone of the sarcomere. (From Uehara, Campbell, and Burnstock, *Muscle and Its Innervation*, Edward Arnold Publishers, Ltd., 1976.)

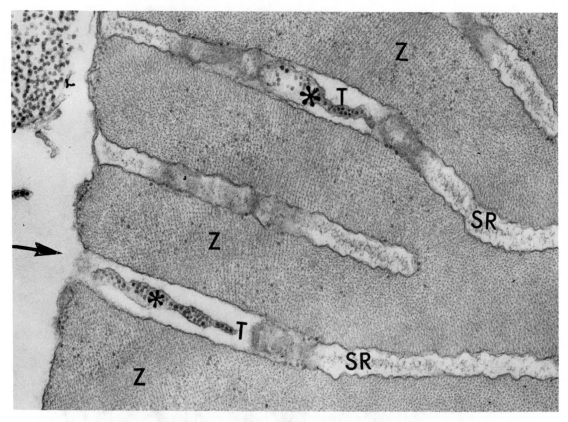

FIG. 6-98. Transverse section of a skeletal muscle fiber cutting across the Z line (Z) and showing the T tubules (T). The arrow indicates that the T tubule is in direct continuity with the external environment. Where the section cuts at the level of the I band, the terminal sacs of the sarcoplasmic reticulum (SR) can be seen. 38,500 ×. (From Franzini-Armstrong and Porter, J. Cell Biol., 1964.)

References

(References are listed not only to those articles and books cited in the text, but to others as well which are considered to contain valuable resource information for the student who desires it.)

General References

Bourne, G. H. (ed.). 1973. *The Structure and Function of Muscle.* 2nd ed. Academic Press, New York, 4 vols.

Close, R. I. 1972. Dynamic properties of mammalian skeletal muscles. Physiol. Rev., *52*:129–197.

Eycleshymer, A. C., and D. M. Shoemaker. 1911. *A Cross-Section Anatomy.* Appleton–Century–Crofts, New York, 373 pp.

Ferner, H., and J. Staubesand. 1975. *Benninghoff und Goerttler's Lehrbuch der Anatomie des Menschen.* 11th ed. Urban and Schwarzenburg, Munich, 3 vols.

Healey, J. E., and W. D. Seybold. 1969. *A Synopsis of Clinical Anatomy.* W. B. Saunders, Philadelphia, 324 pp.

Hollinshead, W. H. 1968. *Anatomy for Surgeons.* 2nd ed. Hoeber–Harper, New York, 3 vols.

Huxley, A. F. 1974. Muscular contraction. J. Physiol. (Lond.), *243*:1–43.

Last, R. J. 1978. *Anatomy. Regional and Applied.* 6th ed. Churchill Livingstone, New York, 598 pp.

MacConaill, M. A., and J. V. Basmajian. 1969. *Muscles and Movements.* Williams and Wilkins, Baltimore, 325 pp.

Rasch, P. J., and R. K. Burke. 1963. *Kinesiology and Applied Anatomy.* 2nd ed. Lea & Febiger, Philadelphia, 503 pp.

Rauber-Kopsch, Fr. 1955. *Lehrbuch und Atlas der Anatomie des Menschen.* 19th ed. Georg Thieme Verlag, Stuttgart, 2 vols.

Muscle Development

Arey, L. B. 1938. The history of the first somite in human embryos. Carneg. Instn., Contr. Embryol., *27*:235–270.

Beckham, C., R. Dimond, and T. K. Greenlee, Jr. 1977. The role of movement in the development of a digital flexor tendon. Amer. J. Anat., *150*:443–460.

Bintliff, S., and B. E. Walker. 1960. Radioautographic study of skeletal muscle regeneration. Amer. J. Anat., *106*:233–245.

Butler, N., and A. E. Claireaux. 1963. Congenital diaphragmatic hernia as a cause of perinatal mortality. Lancet, 1:659–663.

Chaplin, D. M., and T. K. Greenlee. 1975. The development of human digital tendons. J. Anat. (Lond.), 120:253–274.

Gamble, J. F., and G. Allsopp. 1978. Electron microscopic observations of human fetal striated muscle. J. Anat. (Lond.), 126:567–590.

Gasser, R. F. 1967. The development of the facial muscles in man. Amer. J. Anat., 120:357–376.

Godman, J. C. 1957. On the regeneration and redifferentiation of mammalian striated muscles. J. Morph., 100:27–81.

Goldspink, G. 1970. The proliferation of myofibrils during muscle fiber growth. J. Cell Sci., 6:593–603.

Gray, S. W., and J. E. Skandalakis. 1972. Embryology for Surgeons. W. B. Saunders Co., Philadelphia, 918 pp.

Hitchcock, S. E. 1970. The appearance of a functional contractile apparatus in developing muscle. Devel. Biol., 23:399–423.

Hitchcock, S. E. 1971. Detection of actin filaments in homogenates of developing muscle using heavy meromyosin. Devel. Biol., 25:492–501.

McKenzie, J. 1962. The development of the sternomastoid and trapezius muscles. Carneg. Instn., Contr. Embryol., 37:123–129.

Maier, A., and E. Eldred. 1974. Postnatal growth of extra- and intrafusal fibers in the soleus and medial gastrocnemius of the cat. Amer. J. Anat., 141:161–178.

Ontell, M., and R. F. Dunn. 1978. Neonatal muscle growth: a quantitative study. Amer. J. Anat., 152:539–556.

Speidel, C. C. 1938. Growth, injury and repair of striated muscle. Amer. J. Anat., 62:179–235.

Spyropoulos, M. N. 1977. The morphogenetic relationship of the temporal muscle to the coronoid process in human embryos. Amer. J. Anat., 150:395–410.

Straus, W. L., and M. E. Rawles. 1953. An experimental study of the origin of the trunk musculature and ribs in the chick. Amer. J. Anat., 92:471–509.

Swinyard, C. A. (Ed.). 1969. Limb Development and Deformity. Charles C Thomas, Springfield, Ill., 672 pp.

Tomanek, R. J., and A. Colling-Salton. 1977. Cytological differentiation of human fetal skeletal muscle. Amer. J. Anat., 149:227–246.

Walker, S. M., and M. B. Edge. 1971. The sarcoplasmic reticulum and development of Z lines in skeletal muscle fibers of fetal and postnatal rats. Anat. Rec., 169:661–678.

Wells, L. J. 1948. Observations on the development of the diaphragm in the human embryo. Anat. Rec., 100:778.

Wells, L. J. 1954. Development of the human diaphragm and pleural sacs. Carneg. Instn., Contr. Embryol., 35:107–134.

Histology and Electron Microscopy

Bloom, W., and D. W. Fawcett. 1975. A Textbook of Histology. 10th ed. W. B. Saunders Co., Philadelphia, 1033 pp.

Chiakulas, J. J., and J. E. Pauly. 1965. A study of postnatal growth of skeletal muscle in the rat. Anat. Rec., 152:55–61.

Comer, R. D. 1956. An experimental study of the "laws" of muscle and tendon growth. Anat. Rec., 125:665–682.

Constantin, L. L., C. Franzini-Armstrong, and R. J. Podolsky. 1965. Localization of calcium accumulating structures in striated muscle. Science, 147:158–160.

Copenhaver, W. M., D. E. Kelly, and R. L. Wood. 1978. Bailey's Textbook of Histology. 17th ed. Williams and Wilkins, Baltimore, 800 pp.

Di Fiore, M. S. 1974. Atlas of Human Histology. 4th ed. Lea & Febiger, Philadelphia, 252 pp.

Eaton, B. L., and F. A. Pepe. 1974. Myosin filaments showing a 430 Å axial repeat periodicity. J. Molec. Biol., 82:421–423.

Ebashi, S. 1972. Calcium ions and muscle contraction. Nature, 240:217–218.

Eisenberg, B. R., A. M. Kuda, and J. B. Peter. 1974. Stereological analysis of mammalian skeletal muscle. I. Soleus muscle of the adult guinea pig. J. Cell Biol., 60:732–754.

Farrell, P. R., and M. R. Fedde. 1969. Uniformity of structural characteristics throughout the length of skeletal muscle fibers. Anat. Rec., 164:219–229.

Fawcett, D. W. 1960. The sarcoplasmic reticulum of skeletal and cardiac muscle. Circulation, 24:336–348.

Franzini-Armstrong, C. 1970. Studies of the triad. I. Structure of the junction in frog twitch fibers. J. Cell Biol., 47:488–499.

Franzini-Armstrong, C. 1971. Studies of the triad. II. Penetration of tracers into the junctional gap. J. Cell Biol., 49:196–203.

Franzini-Armstrong, C., and K. R. Porter. 1964. Sarcolemmal invaginations constituting the T system in fish muscle fibers. J. Cell Biol., 22:675–696.

Gauthier, G. F. 1970. The ultrastructure of three fiber types in mammalian skeletal muscle fibers. In The Physiology and Biochemistry of Muscle as a Food. Edited by E. J. Briskey, R. G. Cassens, and B. B. Marsh. The University of Wisconsin Press, Madison.

Gay, A. J., Jr., and T. E. Hunt. 1954. Reuniting of skeletal muscle fibers after transection. Anat. Rec., 120:853–872.

Goss, C. M. 1944. The attachment of skeletal muscle fibers. Amer. J. Anat., 74:259–290.

Ham, A. W. 1974. Histology. 7th ed. J. B. Lippincott, Philadelphia, 1006 pp.

Hanson, J., and J. Lowy. 1963. The structure of F-actin and of actin filaments isolated from muscle. J. Molec. Biol., 6:46–60.

Hanson, J., and J, Lowy. 1964. The structure of actin filaments and the origin of the axial periodicity in the I-substance of vertebrate striated muscle. Proc. Roy. Soc. Lond., B160:449–458.

Hay, E. D. 1963. The fine structure of differentiating muscle in salamander tail. Zeitschr. f. Zellforsch., 59:6–34.

Hess, A. 1970. Vertebrate slow muscle fibers. Physiol. Rev., 50:40–62.

Huxley, A. F., and R. Niedergerke. 1954. Structural changes in muscle during contraction. Nature, 173:971–993.

Huxley, A. F., and R. E. Taylor. 1958. Local activation of striated muscle. J. Physiol. (Lond.), 144:426–441.

Huxley, H. E. 1957. The double array of filaments in cross-striated muscle. J. Biophys. Biochem. Cytol., 3:431–648.

Huxley, H. E. 1963. Electron microscopic studies on the structure of natural and synthetic protein filaments from striated muscle. J. Molec. Biol., 7:281–308.

Huxley, H. E. 1964. Evidence for continuity between the central elements of the triads and extracellular space in frog sartorius muscle. Nature, 202:1067–1071.

Huxley, H. E. 1966. The fine structure of striated muscle and its functional significance. Harvey Lect., 60:85–118.

Huxley, H. E. 1968. Structural difference between resting and rigor muscle; evidence from intensity changes in the low angle equatorial x-ray diagram. J. Molec. Biol., 37:507–520.

Huxley, H. E. 1969. The mechanism of muscular contraction. Science, 164:1356–1366.

Huxley, H. E. 1971. The structural basis of muscular contraction. Proc. Roy. Soc. Lond., B178:131–149.

Huxley, H. E. 1973. Structural changes in the actin- and myosin-containing filaments during contraction. In The Mechanism of Muscle Contraction. Cold Spring Harb. Symp. Quant. Biol., 37:361–376.

Huxley, H. E., and J. Hanson. 1954. Changes in the cross striations of muscle during contraction and stretch and their interpretation. Nature, 173:973–976.

Kelly, D. E. 1969. The fine structure of skeletal muscle triad junctions. J. Ultrastruct. Res., 29:37–49.

Kelly, D. E. 1969. Myofibrillogenesis and Z band differentiation. Anat. Rec., 163:403–425.

Kelly, D. E., and M. A. Cahill. 1972. Filamentous and matrix components of skeletal muscle Z disks. Anat. Rec., 172:623–642.

MacKay, B., T. J. Harrop, and A. R. Muir. 1969. The fine structure of the muscle tendon junction in the rat. Acta Anat., 73:588–604.

Mair, W. G. P., and F. M. S. Tomé. 1972. The ultrastructure of the adult and developing human myotendinous junction. Acta Neuropath., 21:239–252.

Morimoto, K., and W. F. Harrington. 1974. Evidence for structural changes in vertebrate thick filaments induced by calcium. J. Molec. Biol., 88:693–710.

Peachy, L. D. 1965. The sarcoplasmic reticulum and transverse tubules of the frog's sartorius. J. Cell Biol., 25:209–230.

Pepe, F. A. 1966. Some aspects of the structural organization of the myofibril as revealed by antibody staining. J. Cell Biol., 28:505–525.

Pogogeff, I. A., and M. R. Murray. 1946. Form and behavior of adult mammalian skeletal muscle in vitro. Anat. Rec., 95:321–335.

Porter, K. R. 1956. The sarcoplasmic reticulum in muscle cells of amblystoma larvae. J. Biophys. Biochem. Cytol., Suppl., 2:163–170.

Porter, K. R., and M. A. Bouneville. 1973. Fine Structure of Cells and Tissues. 4th ed. Lea & Febiger, Philadelphia, 204 pp.

Porter, K. R., and G. E. Palade. 1957. Studies on the endoplasmic reticulum. III. Its form and distribution in striated muscle cells. J. Biophys. Biochem. Cytol., 3:269–299.

Reger, J. F., and A. S. Craig. 1968. Studies on the fine structures of muscle fibers and associated satellite cells in hypertrophic human deltoid muscle. Anat. Rec., 162:483–499.

Revel, J. P. 1962. The sarcoplasmic reticulum of the bat cricothyroid muscle. J. Cell Biol., 12:571–588.

Shaffino, S., V. Hanzlikova, and S. Pierobon. 1970. Relations between structure and function in rat skeletal muscle fibers. J. Cell Biol., 47:107–119.

Smith, R. D. 1950. Studies on rigor mortis. I. Observations on the microscopic and submicroscopic structure. Anat. Rec., 108:185–206.

Strickholm, A. 1966. Local sarcomere contraction in fast muscle fibers. Nature, 212:835–836.

Szepsenwol, J. 1946. A comparison of growth, differentiation, activity and action currents of heart and skeletal muscle in tissue culture. Anat. Rec., 95:125–146.

Uehara, Y., G. R. Campbell, and G. Burnstock. 1976. Muscle and Its Innervation. Edward Arnold Publishers, Ltd., London, 526 pp.

Walker, S. M., and M. B. Edge. 1971. The sarcoplasmic reticulum and development of Z lines in skeletal muscle fibers of fetal and postnatal rat. Anat. Rec., 169:661–677.

Walker, S. M., and G. R. Schrodt. 1974. I segment lengths and thin filament periods in skeletal muscle fibers of the rhesus monkey and the human. Anat. Rec., 178:63–81.

Windle, W. F. 1969. Textbook of Histology. 4th ed. McGraw-Hill, New York, 551 pp.

Winegrad, S. 1965. The location of muscle calcium with respect to the myofibrils. J. Gen. Physiol., 48:997–1002.

Muscles; Tendons; Muscle Actions

Anson, B. J., L. E. Beaton, and C. B. Mc Vay. 1938. The pyramidalis muscle. Anat. Rec., 72:405–411.

Ashley, G. T. 1952. The manner of insertion of the pectoralis major muscle in man. Anat. Rec., 113:301–307.

Baba, M. A. 1954. The accessory tendon of the abductor pollicis longus muscle. Anat. Rec., 119:541–548.

Basmajian, J., and G. Stecko. 1963. The role of muscles in arch support of the foot. J. Bone Jt. Surg., 45-A:1184–1190.

Basmajian, J., and A. Travill. 1961. Electromyography of the pronator muscles in the forearm. Anat. Rec., 139:45–49.

Basmajian, J. V. 1974. Muscles Alive. 3rd Ed., Williams and Wilkins, Baltimore, 525 pp.

Basmajian, J. V., and J. W. Bentzon. 1954. An electromyographic study of certain muscles of the leg and foot in the standing position. Surg. Gynec. Obstet., 98:662.

Basmajian, J. V., and C. R. Dutta. 1961. Electromyography of the pharyngeal constrictors and levator palati in man. Anat. Rec., 139:561–563.

Basmajian, J. V., T. P. Harden, and E. M. Regenos. 1972. Integrated actions of the four heads of quadriceps femoris: an electromyographic study. Anat. Rec., 172:15–19.

Bearn, J. G. 1961. An electromyographic study of the trapezius, deltoid, pectoralis major, biceps and triceps muscles, during static loading of the upper limb. Anat. Rec., 140:103–107.

Bearn, J. G. 1963. The history of the ideas on the functions of the biceps brachii muscle as a supinator. Medical History, 7:32–42.

Beaton, L. E., and B. J. Anson. 1937. The relation of the sciatic nerve and of its subdivisions to the piriformis muscle. Anat. Rec., 70:1–6.

Boivin, G., G. E. Wadsworth, J. M. Landsmeer, and C. Long, II. 1969. Electromyographic kinesiology of the hand: Muscles driving the index finger. Arch. Phys. Med. Habil., 50:17–26.

Bojsen-Moller, F. 1976. Osteoligamentous guidance of the movements of the human thumb. Amer. J. Anat., 147:71–80.

Boyd, W., H. Blincoe, and J. C. Hayner. 1965. Sequence of

action of the diaphragm and quadratus lumborum during quiet breathing. Anat. Rec., 151:579–581.

Broome, H. L., and J. V. Basmajian. 1971. The function of the teres major muscle: An electromyographic study. Anat. Rec., 170:309–310.

Campbell, E. J. M. 1955. The role of scalene and sternomastoid muscles in breathing in normal subjects. J. Anat. (Lond.), 89:378–386.

Carr, N. D., J. D. O'Callaghan, and R. Vaughn. 1977. An unusual flexor of the fifth finger. Acta Anat., 98:376–379.

Clark, J. H., and E. I. Hashimoto. 1946. Utilization of Henle's ligament, iliopubic tract, aponeurosis of transversus abdominis and Cooper's ligament in inguinal herniorrhaphy: A report of 162 consecutive cases. Surg. Gynec. Obstet., 82:480–484.

Crawford, G. N. C. 1971. The effect of temporary limitation of movement on the longitudinal growth of voluntary muscle. J. Anat. (Lond.), 111:143–150.

Cummins, E. J., B. J. Anson, B. W. Carr, and R. R. Wright. 1946. The structure of the calcaneal tendon (of Achilles) in relation to orthopedic surgery. Surg. Gynec. Obstet., 83:107–116.

Dahlgard, D. L., and G. E. Kauth. 1965. An anomalous arrangement of the flexor musculature of the forearm and hand. Anat. Rec., 152:251–255.

Davis, P. R., and J. D. G. Troup. 1966. Human thoracic diameters at rest and during activity. J. Anat. (Lond.), 100:397–410.

Day, M. H., and J. R. Napier. 1961. The two heads of flexor pollicis brevis. J. Anat. (Lond.), 95:123–130.

De Sousa, O. M., J. Lacaz De Moraes, and L. De Moraes Vieira. 1961. Electromyographic study of the brachioradialis muscle. Anat. Rec., 139:125–131.

Donisch, E. W., and J. V. Basmajian. 1972. Electromyography of deep back muscles in man. Amer. J. Anat., 133:25–36.

Froimson, A., and K. S. Alfred. 1961. Sesamoid bone of the subscapularis tendon. J. Bone Jt. Surg., 43-A:881–884.

Furlani, J. 1976. Electromyographic study of m. biceps brachii in movements of the glenohumeral joint. Acta Anat., 96:270–284.

George, R. 1953. Co-incidence of palmaris longus and plantaris muscles. Anat. Rec., 116:521–524.

Grant, P. G. 1973. Lateral pterygoid: Two muscles? Amer. J. Anat., 138:1–10.

Gray, D. J. 1945. Some anomalous hamstring muscles. Anat. Rec., 91:33–38.

Gray, E. G., and J. V. Basmajian. 1968. Electromyography and cinematography of leg and foot ("normal and flat") during walking. Anat. Rec., 161:1–15.

Greig, H. W., B. J. Anson, and J. M. Budinger. 1952. Variations in the form and attachments of the biceps brachii muscle. Quart. Bull., Northwestern Univ. Med. School, 26:241–244.

Grodinsky, M. 1930. A study of the tendon sheaths of the foot and their relation to infection. Surg. Gynec. Obstet., 51:460–468.

Hrycyshyn, A. W., and J. V. Basmajian. 1972. Electromyography of the oral stage of swallowing in man. Amer. J. Anat., 133:333–340.

James, N. T. 1971. The distribution of muscle fiber types in fasciculi and their analysis. J. Anat. (Lond.), 110:335–342.

Jensen, R. H., and W. K. Metcalf. 1975. A systemic approach to the quantitative description of musculoskeletal geometry. J. Anat. (Lond.), 119:209–222.

Johnson, C. E., J. V. Basmajian, and W. Dasher. 1972.

Electromyography of sartorius muscle. Anat. Rec., 173:127–130.

Jones, D. S., R. J. Beargie, and J. E. Pauly. 1953. An electromyographic study of some muscles of costal respiration in man. Anat. Rec., 117:17–24.

Kanavel, A. B. 1939. Infections of the Hand. 7th ed. Lea & Febiger, Philadelphia, 503 pp.

Kaplan, E. B. 1945. Surgical anatomy of the flexor tendons of the wrist. J. Bone Jt. Surg., 27:368–372.

Landsmeer, J. M. F. 1949. The anatomy of the dorsal aponeurosis of the human finger and its functional significance. Anat. Rec., 104:31–44.

Last, R. J. 1954. The muscles of the mandible. Proc. Roy. Soc. Med., 47:571–578.

Latham, R. A., and T. G. Deaton. 1976. The structural basis of the philtrum and the contour of the vermilion border: A study of the musculature of the upper lip. J. Anat. (Lond.), 121:151–160.

Lehr, R. P., P. L. Blanton, and N. L. Biggs. 1971. An electromyographic study of the mylohyoid muscle. Anat. Rec., 169:651–659.

Lewis, O. J. 1962. The comparative morphology of M. flexor accessorius and the associated long flexor tendons. J. Anat. (Lond.), 96:321–333.

Lewis, O. J. 1964. The tibialis posterior tendon in the primate foot. J. Anat. (Lond.), 98:209–218.

Lewis, O. J. 1965. The evolution of Mm. interossei in the primate hand. Anat. Rec., 153:275–287.

Lieb, F. J., and J. Perry. 1968. Quadriceps function. An anatomical and mechanical study using amputated limbs. J. Bone Jt. Surg., 50-A:1534–1548.

Lord, F. P. 1937. Movements of the jaw and how they are effected. Internat. J. Orthodontia, 23:557–571.

Lovejoy, J. R., Jr., and T. P. Harden. 1971. Popliteus muscle in man. Anat. Rec., 169:727–730.

Mc Namara, J. A., Jr. 1973. The independent functions of the two heads of the lateral pterygoid muscle. Amer. J. Anat., 138:197–206.

Manter, J. T. 1945. Variations of the interosseous muscles of the human foot. Anat. Rec., 93:117–124.

Marmor, L., C. D. Bechtol, and C. B. Hall. 1961. Pectoralis major muscle. Function of sternal portion and mechanism of rupture of normal muscle. J. Bone Jt. Surg., 43-A:81–87.

Martin, B. F. 1964. Observations of the muscles and tendons of the medial aspect of the sole of the foot. J. Anat. (Lond.), 98:437–453.

Martin, B. F. 1968. The origins of the hamstring muscles. J. Anat. (Lond.), 102:345–352.

Martin, C. P. 1940. The movements of the shoulder joint with special reference to rupture of the supraspinatus tendon. Amer. J. Anat., 66:213–234.

Matheson, A. B., D. C. Sinclair, and W. G. Skene. 1970. The range and power of ulnar and radial deviation of the fingers. J. Anat. (Lond.), 107:439–458.

Mehta, H. J., and W. U. Gardner. 1961. A study of lumbrical muscles in the human hand. Amer. J. Anat., 109:227–238.

Morris, J. M., G. Benner, and D. B. Lucas. 1962. An electromyographic study of the intrinsic muscles of the back in man. J. Anat. (Lond.), 96:509–520.

Mortonson, O. A., and M. Wiederbauer. 1952. An electromyographic study of the trapezius muscle. Anat. Rec., 112:366.

Moyers, R. E. 1950. An electromyographic analysis of certain muscles involved in temporomandibular movement. Amer. J. Orthodont., 36:481.

Pauly, J. E. 1966. An electromyographic analysis of certain movements and exercises. I. Some deep muscles of the back. Anat. Rec., *155*:223–234.

Pauly, J. E., J. L. Rushing, and L. E. Scheving. 1967. An electromyographic study of some muscles crossing the elbow joint. Anat. Rec., *159*:47–53.

Pfuhl, W. 1937. Die gefiederten Muskeln, ihre Form und ihre Wirkungsweise. Zeitsch. f. Anat. u. Entwicke-lungsgesch., *106*:749–769.

Reimann, A. F., E. H. Daseler, B. J. Anson, and L. E. Beaton. 1944. The palmaris longus muscle and ten-don. A study of 1600 extremities. Anat. Rec., *89*:495–505.

Rizk, N. 1976. A new description of the anterior abdomi-nal wall. Anat. Rec., *184*:515.

Roberts, D. 1978. A mechanism for passive mandibular depression. Acta Anat., *101*:160–169.

Scheving, L. E., and J. E. Pauly. 1959. An electromyo-graphic study of some muscles acting on the upper extremity of man. Anat. Rec., *135*:239–245.

Seib, G. A. 1938. The m. pectoralis minor in American whites and American negroes. Amer. J. Phys. Anthrop., *23*:389–419.

Stanier, D. I. 1977. The function of muscles around a simple joint. J. Anat. (Lond.), *123*:827–830.

Steendijk, R. 1948. On the rotating function of the iliopsoas muscle. Acta Neerl. Morph. Norm. et Pa-thol., *6*:175.

Stein, A. H., Jr. 1951. Variations of the tendons of inser-tion of the abductor pollicis longus and the extensor pollicis brevis. Anat. Rec., *110*:49–55.

Straus, W. L., Jr. 1942. The homologies of the forearm flexors: Urodeles, lizards, mammals. Amer. J. Anat., *70*:281–316.

Sullivan, W. E., O. A. Mortensen, M. Miles, and L. S. Greene. 1950. Electromyographic studies of M. bi-ceps brachii during normal voluntary movement at the elbow. Anat. Rec., *107*:243–252.

Sunderland, S. 1945. The actions of the extensor digi-torum communis, interosseous and lumbrical mus-cles. Amer. J. Anat., *77*:189–217.

Takebe, K., M. Vitti, and J. V. Basmajian. 1974. Electro-myography of pectineus muscle. Anat. Rec., *180*:281–283.

Takebe, K., M. Vitti, and J. V. Basmajian. 1974. The func-tions of semispinalis capitis and splenius capitis muscles: An electromyographic study. Anat. Rec., *179*:477–480.

Travill, A. 1962. Electromyographic study of the extensor apparatus of the forearm. Anat. Rec., *144*:373–376.

Travill, A., and J. V. Basmajian. 1961. Electromyography of the supinators of the forearm. Anat. Rec., *139*:557–560.

Vitti, M., and J. V. Basmajian. 1977. Integrated actions of masticatory muscles: simultaneous EMG from eight intramuscular electrodes. Anat. Rec., *187*:173–190.

Vitti, M., M. Fujiwara, J. V. Basmajian, and M. Iida. 1973. The integrated roles of longus colli and sternocleido-mastoid muscles: An electromyographic study. Anat. Rec., *177*:471–484.

Waters, R. L., and J. M. Morris. 1971. Electrical activity of muscles of the trunk during normal walking. J. Anat. (Lond.), *111*:191–200.

Whillis, J. 1946. Movements of the tongue in swallowing. J. Anat. (Lond.), *80*:115–116.

Wright, R. R., W. Greig, and B. J. Anson. 1946. Accessory tendinous (peroneal) origin of the first dorsal inter-osseous muscle. A study of 125 specimens of lower extremity. Quart. Bull., Northwestern Med. School, *20*:339–341.

Blood Vessels and Nerves of Muscles

Barrnett, R. J. 1966. Ultrastructural histochemistry of normal neuromuscular junctions. Ann. N. Y. Acad. Sci., *135*:27–34.

Bowden, R. E. M., and J. R. Napier. 1961. The assessment of hand function after peripheral nerve injuries. J. Bone Jt. Surg., *43-B*:481–492.

Brockis, J. G. 1953. The blood supply of the flexor and extensor tendons of the fingers in man. J. Bone Jt. Surg., *35-B*:131–138.

Bruns, R. R., and G. E. Palade. 1968. Studies on blood capillaries. I. General organization of blood capil-laries in muscle. J. Cell Biol., *37*:244–276.

Duchen, L. W. 1971. An electron microscopic comparison of motor end-plates of slow and fast skeletal muscle fibers of the mouse. J. Neurol. Sci., *14*:37–46.

Edwards, D. A. W. 1946. The blood supply and lymphatic drainage of tendons. J. Anat. (Lond.), *80*:147–152.

Grant, R. T., and H. P. Wright. 1970. Anatomical basis for non-nutritive circulation in skeletal muscle exem-plified by blood vessels of rat biceps femoris tendon. J. Anat. (Lond.), *106*:125–134.

Harness, D., and E. Sekeles. 1971. The double anasto-motic innervation of thenar muscles. J. Anat. (Lond.), *109*:461–466.

Harness, D., E. Sekeles, and J. Chaco. 1974. The double motor innervation of the opponens pollicis muscle: An electromyographic study. J. Anat. (Lond.), *117*:329–332.

Harrison, V. F., and O. A. Mortensen. 1962. Identification and voluntary control of single motor unit activity in the tibialis anterior muscle. Anat. Rec., *144*:109–116.

Hollinshead, W. H., and J. E. Markee. 1946. The multiple innervation of limb muscles in man. J. Bone Jt. Surg., *28*:721–731.

Ip, M. C., and K. S. F. Chang. 1968. A study on the radial supply of the human brachialis muscle. Anat. Rec., *162*:363–371.

McKinney, Jr., R. V., B. Singh, and P. D. Brewer. 1977. Fenestrations in regenerating skeletal muscle capil-laries. Amer. J. Anat., *150*:213–218.

Markee, J. E., W. B. Stanton, and R. N. Wrenn. 1952. The intramuscular distribution of the nerves to the mus-cles of the inferior extremity. Anat. Rec., *112*:457.

Marlow, C. D., R. K. Winkelmann, and J. A. Gibilisco. 1965. General sensory innervation of the human tongue. Anat. Rec., *152*:503–511.

Morrison, A. B. 1954. The levatores costarum and their nerve supply. J. Anat. (Lond.), *88*:19–24.

Rakhawy, M. T., S. H. Shehata, and Z. H. Badawy. 1976. The points of nerve entry and the intramuscular nerve branchings in the human muscles of mastica-tion. Acta Anat., *94*:609–616.

Robbins, H. 1963. Anatomical study of the median nerve in the carpal tunnel and etiologies of the carpal-tunnel syndrome. J. Bone Jt. Surg., *45-A*:953–965.

Stillwell, D. L. 1957. The innervation of tendons and apo-neuroses. Amer. J. Anat., *100*:289–317.

Sunderland, S. 1945. The innervation of the flexor digi-torum profundus and lumbrical muscles. Anat. Rec., *93*:317–321.

Sunderland, S. 1946. The innervation of the first dorsal interosseous muscle of the hand. Anat. Rec., *95*:7–10.

Telford, I. R. 1941. Loss of nerve endings in degenerated skeletal muscles of young vitamin E deficient rats.

Anat. Rec., *81*:171–182.

Thornton, M. W., and M. R. Schweisthal. 1969. The phrenic nerve: Its terminal divisions and supply to the crura of the diaphragm. Anat. Rec., *164*:283–289.

Zweifach, B. W. 1973. Microcirculation. Ann. Rev. Physiol., *35*:117–150.

Fascia

Anson, B. J., and C. B. Mc Vay. 1938. The fossa ovalis, and related blood vessels. Anat. Rec., *72*:399–404.

Anson, B. J., E. H. Morgan, and C. B. Mc Vay. 1960. Surgical anatomy of the inguinal region based upon a study of 500 body halves. Surg. Gynec. Obstet., *111*:707–725.

Bellocq, P.,and P. Meyer. 1957. Contribution à l'étude de l'aponévrose dorsale du pied (Fascia dorsalis pedis, P.N.A.). Acta Anat., *30*:67–80.

Chandler, S. B. 1950. Studies on the inguinal region. III. The inguinal canal. Anat. Rec., *107*:93–102.

Chouke, K. S. 1935. The constitution of the sheath of the rectus abdominis muscle. Anat. Rec., *61*:341–349.

Coller, F. A., and L. Yglesias. 1937. The relation of the spread of infection to fascial planes in the neck and thorax. Surgery, *1*:323–333.

Congdon, E. D., J. N. Edson, and S. Yanitelli. 1946. Gross structure of the subcutaneous layer of the anterior and lateral trunk in the male. Amer. J. Anat., *79*:399–429.

Congdon, E. D., and H. S. Fish. 1953. The chief insertion of the bicipital aponeurosis is on the ulna. A study of collagenous bundle patterns of antebrachial fascia and bicipital aponeurosis. Anat. Rec., *116*:395–408.

Cooper, G. W. 1952. Fascial variants of the trigonum lumbale (Petití). Anat. Rec., *114*:1–8.

Davies, J. W. 1934. The pelvic outlet---its practical application. Surg. Gynec. Obstet., *58*:70–78.

Dorling, G. C. 1944. Fascial sling of the scapula and clavicle for dropped shoulder and winged scapula. Brit. J. Surg., *32*:311–315.

Doyle, J. F. 1971. The superficial inguinal arch. A reassessment of what has been called the inguinal ligament. J. Anat. (Lond.), *108*:297–304.

Gaughran, G. R. L. 1964. Suprapleural membrane and suprapleural bands. Anat. Rec., *148*:553–559.

Grayson, J. 1941. The cutaneous ligaments of the digits. J. Anat. (Lond.), *75*:164–165.

Grodinsky, M. 1929. A study of the fascial spaces of the foot and their bearing on infections. Surg. Gynec. Obstet., *49*:737–751.

Grodinsky, M., and E. A. Holyoke. 1938. The fasciae and fascial spaces of the head, neck, and adjacent regions. Amer. J. Anat., *63*:367–408.

Grodinsky, M., and E. A. Holyoke. 1941. The fasciae and fascial spaces of the palm. Anat. Rec., *79*:435–451.

Henke, Jak. Wilhelm. 1872. Untersuchung der Ausbreitung des Bindegewebes mittelst Kunstlicher. Infiltration. Beiträge zur Anatomie des Menschen mit Beziehung auf Bewegung. Hefte I. C. F. Winter'schen Verlags durch handlung, Leipzig.

Jones, F. W. 1942. *The Principles of Anatomy As Seen in the Hand.* 2nd ed. Williams and Wilkins, Baltimore, 417 pp.

Kamel, R., and F. B. Sakla. 1961. Anatomical compartments of the sole of the human foot. Anat. Rec., *140*:57–60.

Kaplan, E. B. 1953. *Functional and Surgical Anatomy of the Hand.* 2nd ed. J. B. Lippincott Co., Philadelphia, 337 p.

Kaplan, E. B. 1958. The iliotibial tract. Clinical and morphological significance. J. Bone Jt. Surg., *40-A*:817–832.

Larsen, R. D., and J. L. Posch. 1958. Dupuytren's contracture. J. Bone Jt. Surg., *40-A*:773–792.

Lee, F. C. 1941. Description of a fascia situated between the serratus anterior muscle and the thorax. Anat. Rec., *81*:35–41.

Lee, F. C. 1944. Note on a fascia underneath the pectoralis major muscle. Anat. Rec., *90*:45–49.

Mc Vay, C. B., and B. J. Anson. 1940. Aponeurotic and fascial continuities in the abdomen, pelvis and thigh. Anat. Rec., *76*:213–231.

Mc Vay, C. B., and B. J. Anson. 1940. Composition of the rectus sheath. Anat. Rec., *77*:213–225.

Roberts, W. H., J. Habenicht, and G. Krishingner. 1964. The pelvic and perineal fasciae and their neural and vascular relationships. Anat. Rec., *149*:707–720.

Roberts, W. H., and W. H. Taylor. 1970. The presacral component of the visceral pelvic fascia and its relation to the pelvic splanchnic innervation of the bladder. Anat. Rec., *166*:207–212.

Tobin, C. E. 1944. The renal fascia and its relation to the transversalis fascia. Anat. Rec., *89*:295–311.

Tobin, C. E., and J. A. Benjamin. 1944. Anatomical study and clinical consideration of the fasciae limiting urinary extravasation from the penile urethra. Surg. Gynec. Obstet., *79*:195–204.

Tobin, C. E., and J. A. Benjamin. 1945. Anatomical and surgical restudy of Denonvilliers' fascia. Surg. Gynec. Obstet., *80*:373–388.

Tobin, C. E., and J. A. Benjamin. 1949. Anatomical and clinical re-evaluation of Camper's, Scarpa's and Colles' fasciae. Surg. Gynec. Obstet., *88*:545–559.

Tobin, C. E., J. A. Benjamin, and J. C. Wells. 1946. Continuity of the fasciae lining the abdomen, pelvis, and spermatic cord. Surg. Gynec. Obstet., *83*:575–596.

Washburn, S. L. 1957. Ischial callosities as sleeping adaptations. Amer. J. Phys. Anthrop., *15*:269–276.

Wesson, M. B. 1953. What are Buck's and Colles' fasciae? J. Urol., *70*:503–511.

Uhlenhuth, E. 1953. *Problems in the Anatomy of the Pelvis.* J. B. Lippincott Co., Philadelphia, 206 pp.

7

The

Heart

The **heart** is the central organ of the blood circulatory system. By its rhythmic contraction, this dynamic muscular structure pumps the blood through a system of **arteries** to all parts of the body. Blood flows in a continuous circulation, delivering oxygen and nutrients to all organs, including the heart itself. This is accomplished through an enormous ramification of the arteries into minute vessels called **arterioles,** which in turn form a network of **capillaries** that are microscopic in size and have very thin walls. Allowing an exchange of substances with the tissues, the capillaries collect as **venules,** which then drain into larger **veins** that progressively increase in diameter and eventually return blood to the heart.

The human heart is a four-chambered organ consisting of two **ventricles** and two **atria.** The powerful pumping portions of the heart are the right and left ventricles, which are separated by the muscular **interventricular septum.** The ventricles have thick walls and are made to function efficiently by being quickly and forcibly filled with blood by contraction of the right and left atria. As blood is returned to the heart through the **superior** and **inferior venae cavae,** it enters the **right atrium** and is then forced into the **right ventricle.** From here it is pumped through the **pulmonary arteries** and into the capillaries of the lungs, where it is refreshed as carbon dioxide is removed and oxygen absorbed. Blood is returned by the **pulmonary veins** to the **left atrium,** which forces it into the **left ventricle.** The left ventricle propels oxygenated blood through the aorta and systemic arteries to the capillaries, and then back to the heart again through the venous system. The right side of the heart, the pulmonary arteries, the capillary system in the lungs, and the pulmonary veins constitute the **lesser** or **pulmonary circulation.** The left side of the heart, the aorta, the ramifying arteries to all the organs (except the lungs), and the returning veins (except the pulmonary veins) constitute the **systemic circulation** (Fig. 7-1).

Although the systemic circulation to and from most organs characteristically involves only one set of capillaries within the various tissues, an exception is found in the vessels of the abdominal organs. The blood supplied to the spleen, pancreas, stomach, and intestines by the systemic arteries collects into the large **portal vein,** which enters the liver and ramifies within it. As the blood passes through the capillary-like **hepatic sinusoids,** it exchanges nutrient materials with the liver cells, and is then collected into the hepatic veins (Fig. 7-1). These drain into the large inferior vena cava, which is transporting systemic venous blood, just before that vessel opens into the right atrium. This diversion of blood by the way of the portal vein from certain abdominal organs, through the liver and back into the inferior vena cava, constitutes the **portal circulation.**

The field of study of the vascular system is called **angiology.** It includes the study of blood vessels as well as vessels constituting the lymphatic system. Although the lymphatic system drains into the veins, making the two systems interdependent, there are sufficient differences morphologically and functionally to make their separate treatment advisable. For convenience of description, the field of angiology will be treated under four headings: the heart, the arteries, the veins, and the lymphatic system. These constitute the next four chapters.

Development of the Heart

FORMATION OF THE HEART TUBE. The first primordium of the **vascular system** can be recognized soon after the initial appearance

606

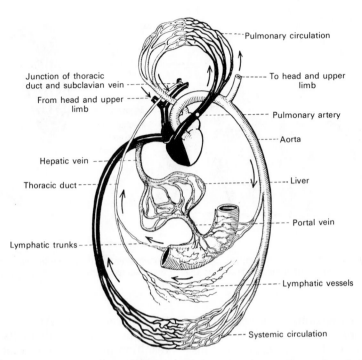

Pulmonary circulation

To head and upper limb

Junction of thoracic duct and subclavian vein

From head and upper limb

Pulmonary artery

Aorta

Hepatic vein

Thoracic duct

Liver

Portal vein

Lymphatic trunks

Lymphatic vessels

Systemic circulation

FIG. 7-1.   Schematic diagram of the blood vascular and lymphatic systems. (From Benninghoff and Goerttler, *Lehrbuch der Anatomie des Menschen*, 10th Ed., Vol 2, Urban & Schwarzenberg, 1975.)

of the **coelom** or body cavity within the embryo. This occurs on about the nineteenth or twentieth prenatal day, at a time when the mesoderm has been formed by the primitive streak, but before it has been organized into the primitive segments or somites. This first coelom is a U-shaped cavity encircling the cephalic end of the neural folds (Fig. 7-2). It is the primordium of the **pericardial cavity.** Its outer layer, the **somatic mesoderm,** will become the parietal pericardium; its inner layer, the **splanchnic mesoderm,** will develop into the myocardium and epicardium. Scattered cells of

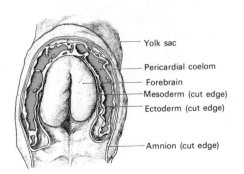

Yolk sac

Pericardial coelom

Forebrain

Mesoderm (cut edge)

Ectoderm (cut edge)

Amnion (cut edge)

FIG. 7-2.   Human embryo with first somite forming, 1.5 mm in length. Dorsal view with amnion removed and ectoderm and mesoderm cut away to show pericardial cavity. (Redrawn from Davis, 1927.)

mesodermal origin, called **mesenchyme,** migrate between the compact laminae of the splanchnic mesoderm and the entoderm. This mesenchyme proliferates rapidly and forms strands and sheets, which are the primordium of the **endocardium.**

The earliest stages of development of the human heart are imperfectly known because of the scarcity of available human embryos of this age. It is helpful, however, that the early development of the heart in lower mammalian embryos, such as the rat, is sufficiently similar to that seen in man to permit the use of timed specimens in these animals (Figs. 7-3 to 7-6) to elucidate the period between the human first somite stage (Fig. 7-2) and that of the human eight-somite stage (Fig. 7-7). It is convenient to use the number of somites as an index of the early stages of development because they can be counted accurately and they retain a relatively stable growth relationship to the rest of the embryo.

By the time the embryo has acquired its **first three somites** or primitive body segments and before the neural folds have closed to form the neural tube, the **endocardial mesenchyme** has spread in a thin sheet across the midline in the cranial portion of the embryo (Fig. 7-3). It also extends downward on each side of the still shallow foregut invagination, and a lumen has begun to form

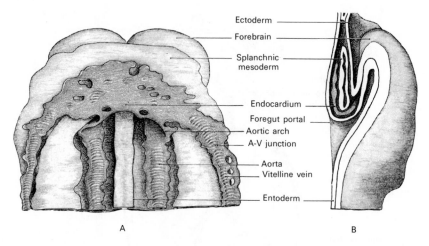

FIG. 7-3. Rat embryo with 3 somites. *A,* Ventral view with entoderm removed. Lumen of heart and vessels is incomplete. *B,* Median sagittal section of same embryo. Drawings of wax reconstruction made at 400 × magnification. (From Goss, 1952.)

in these lateral extensions. The splanchnic mesoderm becomes thickened where it is in contact with the endocardium and is thus identifiable as the primordium of the **myocardium.** The myocardial layer is also deeply grooved along the lateral endocardial tubes. Thus, the two lateral endocardial tubes with their partial cloak of myocardium constitute the **primitive lateral hearts,** where the first contractions occur. The site of the future atrioventricular junction is marked by a constriction of the lumen of the lateral tube slightly caudal to the level of the foregut portal.

The median sheet of mesenchyme is quickly differentiated into endocardium by having the lumen extend into it from the lateral tubes. At the same time, the sheet appears to pull in its outlying parts, resulting in a sac rather than a tube, which, with its adjacent myocardial mesoderm, constitutes the **primitive ventricle.** As the endocardial sac expands, it sinks deep into a pocket in the mesoderm. The pericardial cavity enlarges to accommodate this growth and in a **four-somite embryo** the entire cardiac complex bulges ventrally at the foregut portal (Fig. 7-4).

Until this time the myocardium has been entirely dorsal to the endocardium. The rapid expansion of the pericardial cavity in a ventral direction allows the myocardium to protrude ventrally over the endocardial sac, until, in a **five-somite**

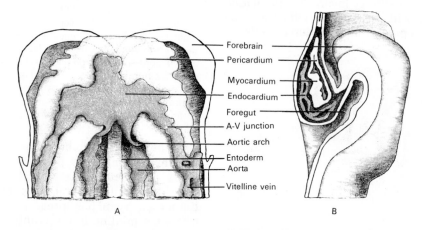

FIG. 7-4. Rat embryo with 4 somites. *A,* Ventral view with entoderm removed. Compare endocardium with that in Figure 7-3. *B,* Median sagittal section of same embryo. (From Goss, 1952.)

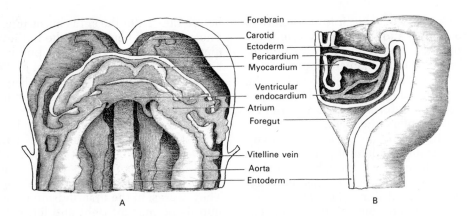

Forebrain
Carotid
Ectoderm
Pericardium
Myocardium
Ventricular
endocardium
Atrium
Foregut
Vitelline vein
Aorta
Entoderm

A   B

FIG. 7-5. Rat embryo with 5 somites. *A*, Ventral view with entoderm removed. Ventral part of pericardium cut away to show myocardium. Dotted line indicates extent of endocardium inside ventricle. *B*, Median sagittal section of same embryo. Compare with Figures 7-3 and 7-4. (From Goss, 1952.)

embryo, the heart lies in a plane at right angles to the axis of the embryo with the arterial end more dorsally placed and the venous end more ventrally placed (Fig. 7-5).

By a continued rapid growth, the ventricular myocardium extends over the entire ventral surface of the endocardium (Fig. 7-6). In a **six-somite embryo,** it encloses the endocardium except for a narrow interval along the midline dorsally. This change allows the heart to assume a more tubular shape with cranial and caudal extremities. The opening of the aortic sac is always dorsal rather than cranial to the heart and the whole tube maintains a curvature with ventral convexity. The result is the marked ventral protrusion of the cardiac complex which is characteris-

tic of early mammalian embryos. When the embryo has acquired **seven somites,** the myocardium completely surrounds the endocardium. The middle portion of the tubular ventricle becomes free and only the ends are attached: the *arterial end* by its continuity with the aortic sac and first aortic arches, and the *venous end* by the atria and vitelline veins (Figs. 7-7; 7-9,*A* and *B*).

DEVELOPMENT OF THE CARDIAC LOOP.   The length of the tubular heart increases much more rapidly than the longitudinal extent of the pericardium. The heart tube therefore continues to bend, this time in a loop with convexity to the right, called the **bulboventricular loop** (Figs. 7-7; 7-9,*C*). The resulting groove on the left side is called the **bulboventricular sulcus** and establishes a

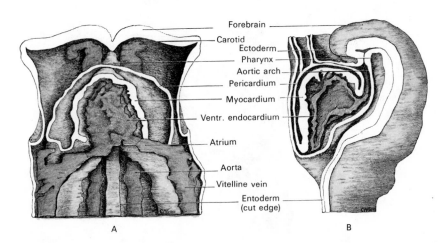

Forebrain
Carotid
Ectoderm
Pharynx
Aortic arch
Pericardium
Myocardium
Ventr. endocardium
Atrium
Aorta
Vitelline vein
Entoderm
(cut edge)

A   B

FIG. 7-6. Rat embryo with 6 somites. *A*, Ventral view with entoderm removed. Ventral part of pericardium and myocardium also removed to show endocardium. *B*, Median sagittal section of same embryo. (From Goss, 1952.)

subdivision of the tube into the **bulbus cordis** (or conus) and the **primitive ventricle** (Figs. 7-8; 7-9,D). The ventricular portion increases rapidly in all diameters, protruding ventrally and causing another bend at the atrioventricular junction. The opening of the atrium narrows to form the **atrioventricular canal,** which enters the ventricle dorsally and from the left. The pericardial cavity expands over the two lateral portions of the atrium, which are then brought closer together, and a constriction is formed between the veins and the atrium (Fig. 7-8).

The proximal portions of the veins become part of a common chamber and are called the right and left **horns of the sinus venosus.** The umbilical and common cardinal veins (from the embryo proper) have joined the vitelline veins on each side just

before they enter the sinus horns (Fig. 7-9,D). The continued bending of the bulboventricular loop eventually folds the bulbus cordis against the atrium on the dorsal and cranial aspect of the heart (Fig. 7-10). The bulbus cordis, which later becomes the conus arteriosus and truncus arteriosus, has remained a relatively narrow tube, and as it presses against the median region of the atrium, the lateral portions bulge out on each side. The resulting two expansions of the atrium are the **primitive right** and **left atria** (Fig. 7-10). They represent the first division of the heart into its permanent right and left sides.

The tissue between the pericardial cavity and the foregut portal becomes thickened into a mass called the **septum transversum** (Figs. 7-8; 7-10). As the liver cords grow into the septum there is a reorganization of the

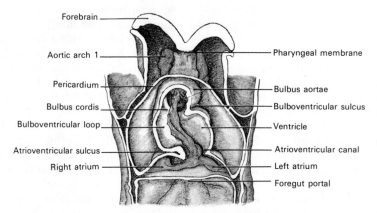

FIG. 7-7. Human embryo with 8 somites, 2 mm in length. Ventral view of plaster model of heart and pericardial region. Pericardial wall and myocardium removed to expose endocardium. (Redrawn from Davis, 1927.)

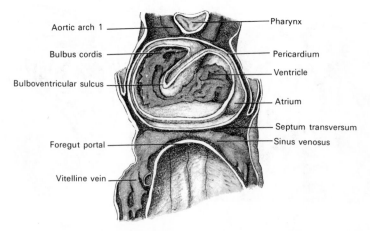

FIG. 7-8. Human embryo with 11 to 12 somites, 3.09 mm in length. Ventral view of plaster model of heart and pericardial region. Pericardial wall and myocardium removed to expose endocardium. (Redrawn from Davis, 1927.)

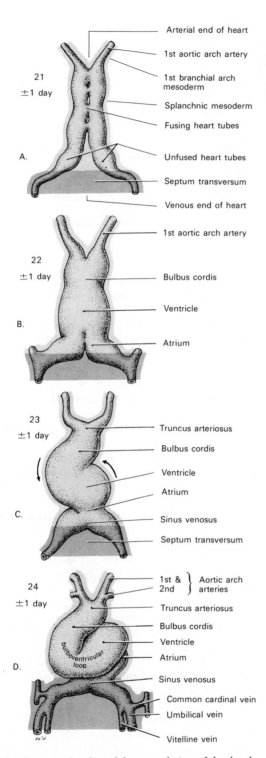

21
±1 day

Arterial end of heart

1st aortic arch artery

1st branchial arch mesoderm

Splanchnic mesoderm

Fusing heart tubes

A.

Unfused heart tubes

Septum transversum

Venous end of heart

22
±1 day

1st aortic arch artery

Bulbus cordis

Ventricle

B.

Atrium

23
±1 day

Truncus arteriosus

Bulbus cordis

Ventricle

Atrium

C.

Sinus venosus

Septum transversum

24
±1 day

1st & 2nd } Aortic arch arteries

Truncus arteriosus

Bulbus cordis

Ventricle

bulboventricular loop

Atrium

D.

Sinus venosus

Common cardinal vein

Umbilical vein

Vitelline vein

FIG. 7-9. Four sketches of the ventral view of the developing heart during the fourth week, showing fusion of the heart tubes and bending of the single heart tube. (From Moore, K. L., *The Developing Human*, 2nd Ed., W. B. Saunders Co., 1971.)

veins. The right vein gradually takes over the drainage of blood from the vitelline, umbilical, and posterior cardinal circulations of both sides and is thus the **precursor of the inferior vena cava.** The resultant enlargement of the right horn of the sinus venosus shifts the sinoatrial opening to the right. The right anterior cardinal vein also becomes dominant and shifts the opening somewhat cranialward.

PARTITIONING OF THE ATRIA. In the narrow part of the atrium between the two expansions to the right and left of the conus arteriosus, a thin crescent-shaped fold of tissue grows from the dorsocranial wall of the atrium toward the atrioventricular opening. This is known as **septum I** or the **septum primum** (Figs. 7-11,*A* and *B*; 7-12). As the septum primum progresses toward the ventricles, two mesenchyme covered endocardial swellings appear in the wall of the atrioventricular canal. These **endocardial cushions** (Fig. 7-11,*C* and *D*: AV canal cushions) soon join across the opening, making a division into a right and left atrioventricular opening. The narrowing open space between the advancing septum primum and the endocardial cushions is the **ostium I** or **ostium primum** (Fig. 7-11,*B* and *C*). Before the septum primum completely closes the opening between the two atria, which would shut off all flow of blood between them, the substance of the septum primum, near its attachment to the cranial part of the atrial wall, becomes perforated and forms **ostium II** or the **ostium secundum** (Fig. 7-11,*C* and *D*).

In the right atrium, a new partition now grows down from the cranial and ventral part of the atrial wall on the right side of septum primum, thus covering the new opening in the first septum. This second crescentic membrane is called the **septum II** or **septum secundum** (Figs. 7-11,*D* to *F*; 7-13 to 7-15). Its growth ceases before it is complete, leaving an opening called the **foramen ovale** (Fig. 7-11,*E* and *F*). Its crescentic border persists as the **limbus fossae ovalis,** which can still be seen in the adult right atrium. After septum primum has joined with the endocardial cushions, its remaining part becomes thinner and persists as a flap over the foramen ovale below the crescentic edge of septum secundum. At this stage it acts as the **valve of the foramen ovale** (Fig. 7-11,*F*). After birth, when the foramen ovale is no

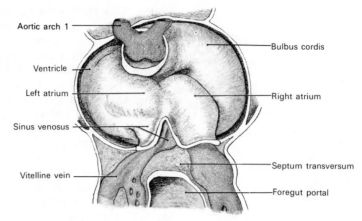

Aortic arch 1

Ventricle

Left atrium

Sinus venosus

Vitelline vein

Bulbus cordis

Right atrium

Septum transversum

Foregut portal

FIG. 7-10. Human embryo with 20 somites, 3.01 mm in length. Dorsal view of plaster model of heart. (Redrawn from Davis, 1927.)

longer functional, this flap becomes adherent to the limbus and seals the opening. During this development the **pulmonary veins** evaginate from the dorsal wall of the left atrium, but they are of insignificant size because the lungs are still not functioning as aerating organs. The left heart, therefore, receives the bulk of its blood through the foramen ovale rather than from these veins.

CHANGES IN THE SINUS VENOSUS.   While the interatrial septa are forming, the slit-like opening of the **sinus venosus** into the right atrium is guarded by two valves, the **right** and **left venous valves** (Figs. 7-12 to 7-15). Above the opening, the two venous valves unite as a single fold, the **septum spurium,** which is a prominent feature at this stage, but is of no significance for the eventual partitioning of the heart (Figs. 7-11; 7-13 to 7-15).

As the right atrium increases in size it absorbs the sinus venosus into its walls (Fig. 7-16,C and D). The upper part of the sinus then becomes the opening for the **superior vena cava.** The inferior part becomes divided by the further growth of the right venous valve into two openings, the **inferior vena cava** and the **coronary sinus** (Fig. 7-16). The left venous valve disappears except for the lower part, which is absorbed into the septum secundum. The upper part of the right venous valve and the septum spurium almost disappear, but are retained in the adult heart as the **crista terminalis** (Fig. 7-11, E and F). The lower part of the right valve persists and is divided into two parts by the growth of a transverse fold, the **sinus septum** (Fig. 7-15). These become the **valve of** the **inferior vena cava** and the **valve of the coronary sinus.** Of the horns of the sinus venosus, the right becomes much more prominent because it furnishes the inferior and superior venae cavae. The left horn remains small, receiving only the left common cardinal vein. The left anterior cardinal dwindles into the **oblique vein** (Fig. 7-16,F) and the **vestigial fold of Marshall.**[1] The left common cardinal also loses its connection with the body wall and becomes the **coronary sinus,** draining blood only from the substance of the heart (Fig. 7-16).

DIVISION OF THE VENTRICLES AND PARTITIONING OF THE TRUNCUS ARTERIOSUS.   The primitive ventricle becomes divided by the growth of the **interventricular septum** (Fig. 7-11,A to D). This muscular ridge develops from the most prominent part of the ventricular wall, and its position on the surface of the heart is indicated by a furrow. The dorsal part of the septum grows more rapidly than the ventral and fuses with the dorsal part of the fused endocardial cushions. The opening between the two ventricles remains for some time, but at about the end of the seventh week, it ultimately is closed by the fusion of the **membranous part** of the interventricular septum, derived from the right side of the united endocardial cushions, and the **bulbar ridges,** which appear in the bulbus cordis prior to the partitioning of the truncus arteriosus (Figs. 7-11,D to F; 7-17).

[1]John Marshall (1818–1891): An English anatomist and surgeon (London).

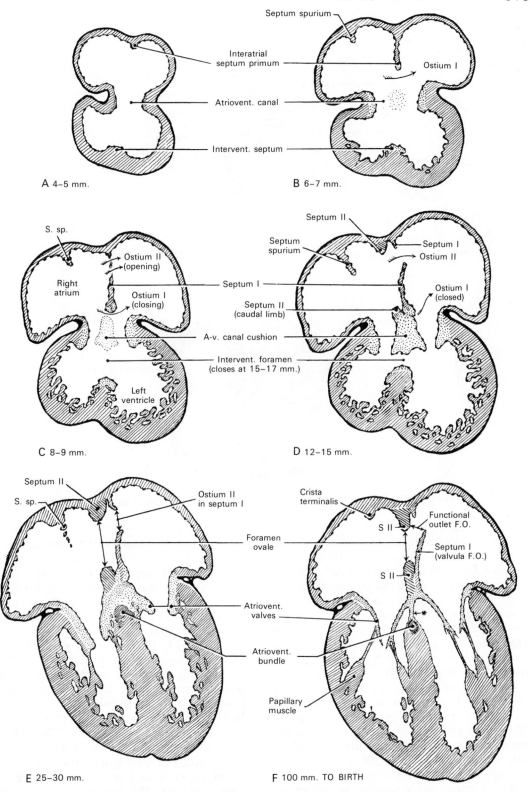

FIG. 7-11. Frontal sections of human embryonic heart giving a sequential picture of its progressive partitioning at six different stages of development. *Stippled areas* indicate endocardial cushion tissue, and the *diagonal hatching* is the myocardium. In *F*, the asterisk identifies the membranous part of the interventricular septum. (From Patten, B. M., *Human Embryology,* 3rd Ed. McGraw-Hill, 1968.)

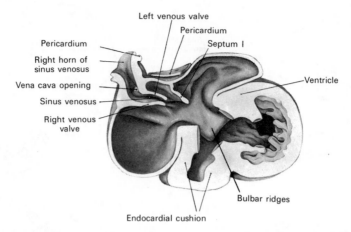

FIG. 7-12.   Model of heart of 6.5-mm human embryo, interior of lower half seen from above. (From Tandler, 1912.)

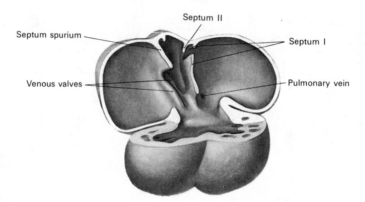

FIG. 7-13.   Model of heart of 9-mm human embryo, ventral view of interior of atria. (From Tandler, 1912.)

Closely related to the separation of the ventricles and the closure of the interventricular foramen is the partitioning of the truncus arteriosus into the pulmonary artery and aorta. The conus arteriosus (bulbus cordis) was at first separated from the ventricle by a deep fold, the bulboventricular sulcus. This fold gradually recedes, until the conus and ventricle open freely into each other. This has the effect of allowing the **truncus aortae** to open directly into the ventricle, and it comes to lie in line with the path of the growing interventricular septum. The portion of the ventricular wall that had been the conus remains to the right of the interventricular septum and becomes a part of the adult right ventricle.

The **truncus arteriosus** or **primitive ascending aorta** has at this time developed both its fourth and its sixth, or pulmonary, pairs of arches. The partitioning of the

truncus into the **aorta** and **pulmonary trunk** begins by the growth of two thickenings, called **truncal ridges,** on the interior of the truncus arteriosus at its more cephalic end in the region of the fourth pair of arches. The growth of these ridges toward the bulbus cordis (or conus arteriosus), where they have been called **bulbar ridges,** is such that they follow a spiral course (Fig. 7-18). As it extends away from the heart, the right ridge passes ventrally and then to the left, while the ridge on the left side of the truncus below passes dorsally and to the right as it extends farther from the heart. Their growth inward toward the lumen of the truncus finally results in their fusion, thereby forming a spirally oriented **aorticopulmonary septum** (Fig. 7-18). The truncus arteriosus then becomes divided into the pulmonary trunk and the aorta, but because the aorticopulmonary septum takes the form of a spiral,

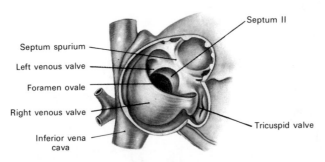

FIG. 7-14. Interior of the right atrium in the heart of a 25-mm human embryo. (From Licata, 1954.)

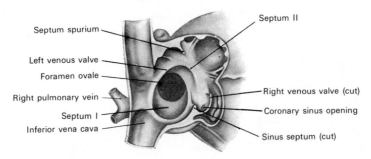

FIG. 7-15. Interior of the right atrium in the same heart as Figure 7-14 with the right venous valve removed. (From Licata, 1954.)

the two vessels become twisted around each other. Adjacent to the heart, the pulmonary trunk comes to lie anterior and to the right of the aorta, while more distally the pulmonary trunk lies to the left and behind the aorta (Fig. 7-18). The right ventricle becomes connected with the pulmonary arches that go to the lungs, while the left ventricle becomes connected with the fourth arch which forms the aorta.

While the aorticopulmonary septum is partitioning the truncus arteriosus, three small local swellings of endocardial tissue develop in both the aorta and pulmonary trunk at the level of transition between the truncus arteriosus and its ventricular extension, the conus arteriosus. These special swellings become the **semilunar valves.** Two of the valve cusps in each vessel are formed from tissue of the ridges that partition the truncus, and they retain this association of position throughout the subsequent rotation and twisting of the vessels (Fig. 7-19,*A* and *B*). The third cusp in each vessel is located opposite the points of fusion of the aorticopulmonary septum and *develops independently* from dorsal and ventral intercalated swellings (Fig. 7-19,*C* to *E*).

BEGINNING OF CONTRACTION. The first contractions of the heart have not been observed in human embryos, but it can be assumed that they follow the same sequence as in other mammals. In a rat embryo with three somites (Fig. 7-3), two lateral heart tubes are recognizable. On the ventricular side of a slight constriction that marks the future atrioventricular junction, a small group of cells in the splanchnic mesoderm begins contracting with a regular rhythm. In a rat embryo, the left lateral heart initiates the contraction at a rate of 20 to 30 per minute. A few hours later, the right heart commences with a slower rate and a regular rhythm independent of the left side. As the contraction involves the median portion of the ventricle, the rate increases gradually, the left side acting as pacemaker. By the time the rate has reached 90 per minute, a small part of the atrium adjacent to the atrioventricular junction has begun contraction and acts as the pacemaker, but with a distinct interval between the atrial and ventricular contraction (Goss, 1940, 1942). It is reasonable to conclude from these studies on subhuman mammals that *the developing human heart begins contracting at the end of the*

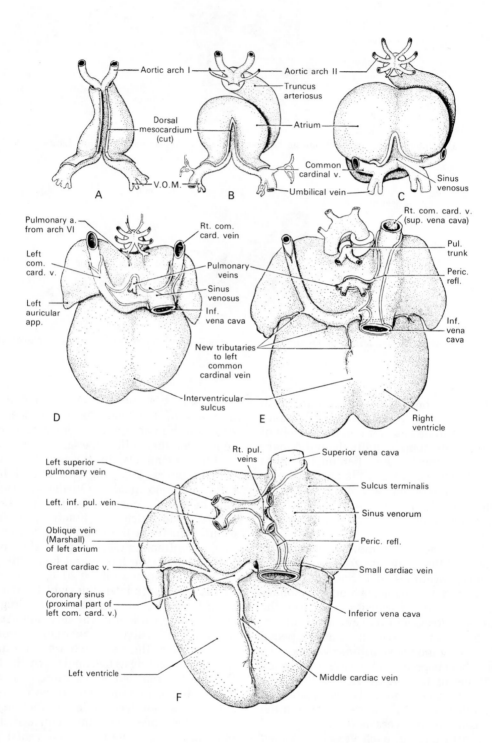

FIG. 7-16. Six stages in the development of the heart seen from the dorsal aspect and showing the changing relations of the sinus venosus and the great veins entering the heart. *A*, 2½ weeks, 8 to 10 somites; *B*, 3 weeks, 12 to 14 somites; *C*, 3½ weeks, 17 to 19 somites; *D*, 5 weeks, 6 to 8 mm; *E*, 8 weeks, about 25 mm; *F*, 11 weeks, about 60 mm. (From Patten, B. M., *Human Embryology*, McGraw-Hill, 1968.)

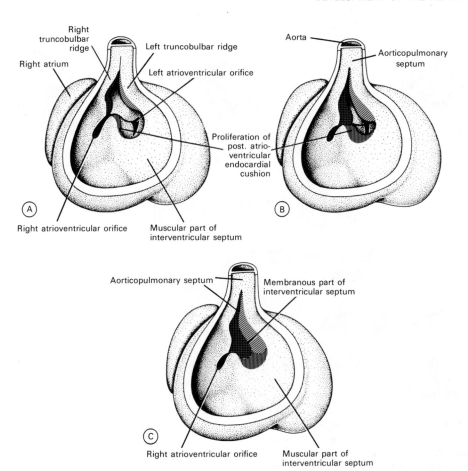

FIG. 7-17. Schematic drawings showing the way in which the truncobulbar ridges, combined with the proliferation of the posterior atrioventricular cushion, close the interventricular foramen and form the membranous portion of the interventricular septum. *A*, 6 weeks, 12 mm; *B*, beginning of 7th week, 14.5 mm; *C*, end of 7th week, 20 mm. (From Langman, J., *Medical Embryology,* 3rd Ed. Williams and Wilkins, 1975.)

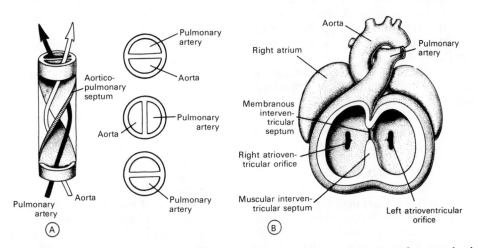

FIG. 7-18. *A*, Diagram to show the spiral shape of the aorticopulmonary septum. *B*, Position of aorta and pulmonary artery at 25-mm stage (8th week). Note that the aorta and pulmonary artery twist around each other. (From Langman, J., *Medical Embryology,* 3rd Ed. Williams and Wilkins, 1975.)

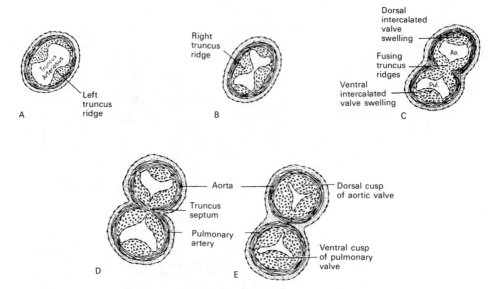

FIG. 7-19. Schematic diagrams of the partitioning of the truncus arteriosus and the origin of the aortic and pulmonary semilunar valves. (From Kramer, 1942.)

third prenatal week or the beginning of the fourth week, i.e., on the twenty-first or twenty-second day.

BEGINNING OF CIRCULATION. The primitive heart, as it enlarges and differentiates, continues to contract after the three-somite stage, but in embryonic rats it is ineffective for circulation until the embryo has eight or nine somites (Fig. 7-7). By that time the atrium is well established, the ventricle is an effective muscular pump, the cells in the blood islands of the yolk sac have acquired hemoglobin, and a continuous lumen has been established through the aortae, the umbilical and omphalomesenteric arteries, the yolk sac, and the placental capillaries, and back through the corresponding veins to the heart (Goss, 1942).

The development of the arteries and veins is described at the beginning of the corresponding chapters.

Peculiarities in the Vascular System of the Fetus

FETAL HEART. The chief peculiarities of the fetal heart are the direct communication between the atria and the foramen ovale, and the large size of the valve of the inferior vena cava. Among other peculiarities the following may be noted: (1) In early fetal life the heart lies immediately below the mandibular arch and is relatively large in size. As development proceeds, it is gradually drawn within the thorax, but at first it lies in the midline; toward the end of pregnancy it gradually becomes oriented obliquely toward the left. (2) For a time the atrial portion exceeds the ventricular in size, and the walls of the ventricles are of equal thickness; toward the end of fetal life the ventricular portion becomes the larger and the wall of the left ventricle exceeds that of the right in thickness. (3) The size of the fetal heart is large in relation to the size of the rest of the body, the proportion at the second fetal month being 1 to 50, and at birth, 1 to 120, while in the adult the average is about 1 to 160.

The **foramen ovale,** situated at the lower part of the atrial septum, forms a free communication between the atria until the end of fetal life. A septum **(septum secundum)** grows down from the upper wall of the atrium to the right of **septum primum,** in which the foramen ovale is situated. Shortly after birth, the two septa fuse, and the foramen ovale is obliterated.

The **valve of the inferior vena cava** serves to direct the blood from that vessel through the foramen ovale into the left atrium.

FETAL ARTERIAL SYSTEM. The principal peculiarities of the arterial system of the fetus are the communication between the pulmonary artery and the aorta by means of the ductus arteriosus, and the continuation of the internal iliac arteries as the umbilical arteries to the placenta.

The **ductus arteriosus** is a short tube, about 1.25 cm in length at birth, and 4.4 mm in diameter. In the early condition it forms the continuation of the pulmonary artery and opens into the aorta just beyond the origin of the left subclavian artery (Fig. 7-20). Most of the blood that leaves the right ventricle is conducted by way of the ductus arteriosus into the aorta. When the branches of the pulmonary artery become larger than the ductus arteriosus, the latter is chiefly connected to the left pulmonary artery.

The **internal iliac arteries** course along the sides of the bladder and then turn upward on the inner aspect of the anterior abdominal wall to reach the umbilicus. At the umbilicus they pass out of the abdomen and are continued as the **umbilical arteries** in the umbilical cord to the placenta (Fig. 7-20). They convey blood from the fetal arterial system to the placenta.

FETAL VENOUS SYSTEM. The peculiarities in the venous system of the fetus are the communications established between the placenta and the liver and portal vein, through the **umbilical vein,** and between the umbilical vein and the inferior vena cava through the **ductus venosus** Fig. 7-20).

FETAL CIRCULATION

In early fetuses blood is returned from the placenta to the umbilical orifice by a single umbilical vein, which then divides into right and left umbilical veins. The *right umbilical vein* becomes atrophic early and then completely disappears, leaving the *left umbilical vein,* which remains functional until birth (Fig. 7-20). This vein enters the abdomen at the umbilicus and passes upward along the free margin of the falciform ligament of the liver to the caudal surface of that organ, where it gives off several branches, a large one to the left lobe and others to the quadrate lobe and caudate lobe. At the **porta hepatis** it divides into two branches, of which the larger is joined by the portal vein and enters the right lobe of the liver and the smaller is continued cranially as the **ductus venosus** to join the inferior vena cava (Fig. 7-20). The blood that traverses the umbilical vein, therefore, passes to the inferior vena

cava in three different ways: (1) a considerable quantity circulates through the liver with the portal venous blood before entering the inferior vena cava by the hepatic veins; (2) some blood enters the liver directly, and is carried to the inferior vena cava by the hepatic veins; and (3) the remainder passes directly into the inferior vena cava through the ductus venosus.

In the inferior vena cava, the blood carried by the ductus venosus and hepatic veins becomes mixed with that returning from the lower extremities and abdominal wall. Entering the right atrium, and guided by the valve of the inferior vena cava, it passes through the foramen ovale into the left atrium, where it mixes with a small quantity of blood returned from the lungs by the pulmonary veins. From the left atrium it passes into the left ventricle and then into the aorta, through which it is distributed almost entirely to the heart, head, and upper extremities; a small quantity is probably also carried into the descending aorta. From the head and upper limbs the blood is returned by the superior vena cava to the right atrium, where it mixes with a small portion of the blood from the inferior vena cava. From the right atrium it passes into the right ventricle, and thence into the pulmonary artery. Because the lungs of the fetus are inactive, only a *small quantity* of the blood in the pulmonary artery is distributed to them by the right and left pulmonary arteries. The blood that does go to the lungs is returned by the pulmonary veins to the left atrium. By far the *larger quantity* of blood in the pulmonary artery passes through the ductus arteriosus into the aorta, where it mixes with a small quantity of the blood transmitted by the left ventricle into the aorta. By means of this vessel, it is in part distributed to the lower limbs and the viscera of the abdomen and pelvis, but the greater amount is conveyed by the umbilical arteries to the placenta (Fig. 7-20).

From the preceding account of the circulation of the blood in the fetus, which has been based on solid research evidence, the following facts can be documented: (1) The placenta serves the purposes of nutrition and excretion, receiving impure blood that is low in oxygen from the fetus and returning it oxygenated and charged with additional nutritive material. (2) About half of the blood of the umbilical vein traverses the

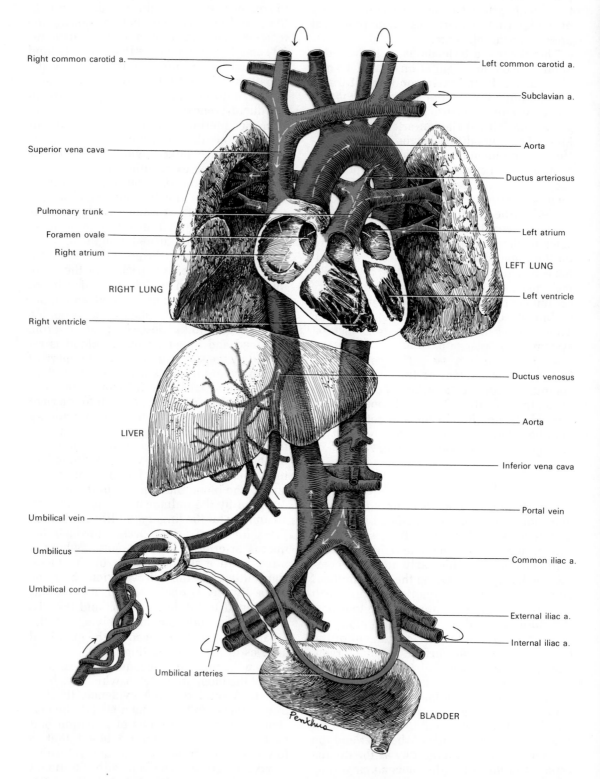

Right common carotid a.

Left common carotid a.

Subclavian a.

Superior vena cava

Aorta

Ductus arteriosus

Pulmonary trunk

Left atrium

Foramen ovale

Right atrium

RIGHT LUNG

LEFT LUNG

Left ventricle

Right ventricle

Ductus venosus

LIVER

Aorta

Inferior vena cava

Umbilical vein

Portal vein

Umbilicus

Common iliac a.

Umbilical cord

External iliac a.

Internal iliac a.

Umbilical arteries

BLADDER

Penthus

FIG. 7-20. A simplified schema of the fetal circulation. Colors indicate degree of oxygen saturation of the blood. Brightest red is highest oxygen saturation, while deepest blue is the lowest oxygen saturation.

liver before entering the inferior vena cava; this accounts for the large size of the liver, especially at an early period of fetal life. The remainder of the blood in the umbilical vein bypasses the liver and enters the inferior vena cava through the ductus venosus. (3) The right atrium receives blood from two sources, the inferior vena cava and the superior vena cava. At an early period of fetal life, it is quite probable that these two streams remain nearly distinct. Most of the blood flowing in the inferior vena cava is guided by the valve of this vessel directly through the foramen ovale and into the left atrium, but a small amount does seem to mix with blood returning to the right atrium from the head and upper limbs by way of the superior vena cava. It is from these latter sources that blood enters the right ventricle and exits from the pulmonary artery. Because the pressure in the right atrium exceeds that in the left atrium (which receives only a small amount of blood from the lungs by means of the pulmonary veins), the valve of the foramen ovale (septum primum) opens to the left, allowing blood from the inferior vena cava to flow from the right atrium to the left atrium. (4) The highly oxygenated (about 80% O_2 saturated) blood carried from the placenta to the fetus by the umbilical vein, mixed with the blood from the portal vein and inferior vena cava, passes almost directly to the arch of the aorta, and is distributed by the branches of that vessel to the head for the developing brain and to the upper limbs. (5) The blood contained in the descending aorta, derived chiefly from that which has already circulated through the head and upper limbs, together with a small quantity from the left ventricle, is distributed to the abdomen and lower limbs.

CHANGES IN THE VASCULAR SYSTEM AT BIRTH

When breathing commences in the newborn child, an increased amount of blood from the pulmonary artery passes through the lungs, and the placental circulation is eliminated by ligation of the umbilical cord. The **foramen ovale** gradually decreases in size during the first month, but a small opening frequently persists until the latter part of the first year. The valve of the foramen ovale adheres to the margin of the foramen for the greater part of its circumference, and thereby the septum primum begins to fuse with septum secundum. A slit-like opening, however, may persist between the two atria. Unless this opening is large, it is of no functional significance.

The **ductus arteriosus** begins to contract immediately after respiration is established, and its lumen gradually becomes obliterated. It ultimately becomes an impervious cord, the **ligamentum arteriosum,** which connects the left pulmonary artery to the arch of the aorta (Figs. 7-21; 7-26).

The portions of the **internal iliac arteries** that extend from the sides of the bladder to the umbilicus become obliterated between the second and fifth days after birth, and persist as fibrous cords covered with peritoneum, the **medial umbilical ligaments.** These form ridges on the inner surface of the anterior abdominal wall.

The **umbilical vein** and **ductus venosus** no longer receive blood after the umbilical cord has been severed. The umbilical vein becomes the **ligamentum teres,** while the ductus venosus becomes the **ligamentum venosus** of the liver. The hepatic half of the ductus venosus may remain open, receive tributaries from the liver, and thus function as a hepatic vein in the adult.

The Pericardium

Except for its continuity with the roots of the great vessels, the heart is unattached to other organs in the thoracic cavity. It is maintained in its proper position by these vessels and by an enclosing fibroserous sac, the **pericardium,** which allows the living heart an appropriate freedom of movement required for its contraction cycles (Figs. 7-21; 7-24). The pericardium consists of two quite different components, the membranous **serous pericardium** and the thicker **fibrous pericardium.**

SEROUS PERICARDIUM. The serous pericardium is composed of a single layer of mesothelial cells. It lines the inner surface of the fibrous pericardium and is continuous at the roots of the great vessels with the epicardium that forms the outermost layer of the heart wall itself. Its characteristics are such that it provides the interfacing pericar-

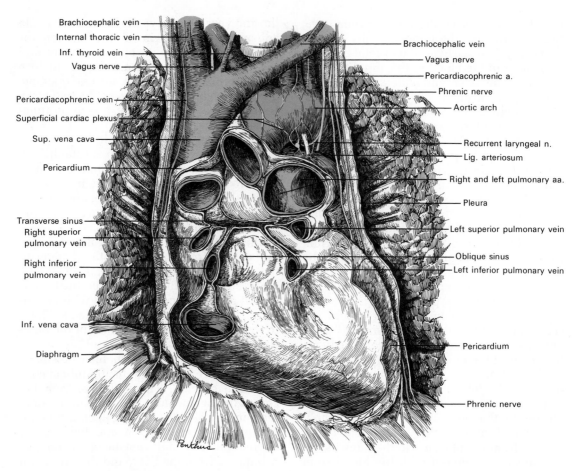

Brachiocephalic vein
Internal thoracic vein
Inf. thyroid vein
Vagus nerve

Pericardiacophrenic vein
Superficial cardiac plexus
Sup. vena cava
Pericardium

Transverse sinus
Right superior pulmonary vein
Right inferior pulmonary vein

Inf. vena cava
Diaphragm

Brachiocephalic vein
Vagus nerve
Pericardiacophrenic a.
Phrenic nerve
Aortic arch
Recurrent laryngeal n.
Lig. arteriosum
Right and left pulmonary aa.
Pleura
Left superior pulmonary vein
Oblique sinus
Left inferior pulmonary vein
Pericardium
Phrenic nerve

Penthus

FIG. 7-21.   Anterior view of the dorsal aspect of the pericardial sac with the heart removed. Observe especially the oblique and transverse pericardial sinuses and the eight cut blood vessels that enter or leave the heart. Pulmonary veins and aorta colored red to indicate oxygenated blood; superior and inferior vena cavae and pulmonary artery colored blue to indicate blood with low oxygen levels. (After Sobotta.)

dial sac and heart with smooth and glistening surfaces that are completely free and movable. The serous layer of the heart itself may be considered the **visceral pericardium.** It covers the atria and ventricles and extends beyond them along the great vessels for 2 or 3 cm (Fig. 7-21). The serous layer lining the fibrous sac may be considered the **parietal pericardium.** Along the roots of the great vessels this continuous serous membrane folds upon itself, and the point at which the visceral layer becomes the parietal layer is called the **reflection of the pericardium.**

PERICARDIAL CAVITY.   Under normal conditions the two serous membranes are closely apposed, being separated only by enough serous or watery fluid to moisten their surfaces. This allows the heart to move easily during its contraction in systole and its re-

laxation in diastole. Because the two surfaces are not fused, there is a potential space between them called the **pericardial cavity,** not unlike the pleural cavity. After injury or due to disease, fluid may exude into the cavity, causing a wide separation between the epicardium and the fibroserous pericardium.

The extension of the epicardium on the great vessels is in the form of two tubular prolongations. One encloses the aorta and pulmonary trunk and is called the **arterial mesocardium.** The other encloses the superior and inferior venae cavae and the four pulmonary veins and is called the **venous mesocardium.** The reflection of the venous mesocardium forms a U-shaped cul-de-sac in the dorsal wall of the pericardial cavity called the **oblique pericardial sinus** (Fig. 7-21). Between the arterial mesocardium and the venous mesocardium is a serous-lined passage named the **transverse pericardial sinus** (Fig. 7-21).

FIBROUS PERICARDIUM. The fibrous pericardium forms a cone-shaped sac, the neck of which is closed by its attachment to the great vessels just beyond the reflection of the serous pericardium (Fig. 7-21). It is a tough, dense membrane, much thicker than the parietal pleura. Its outer surface is adherent in varying degrees to all the structures surrounding it. It is attached *anteriorly* to the manubrium of the sternum by a fibrous condensation, the **superior pericardiosternal ligament,** and to the xiphoid process by the **inferior pericardiosternal ligament.** The fibrous tissue intervening *posteriorly* between the pericardial sac and the vertebral column is the pericardiovertebral ligament. The sac is securely attached *inferiorly* to the central tendon and muscular part of the left side of the dome of the diaphragm. A thickening of this fibrous attachment in the region of the inferior vena cava has been called the **pericardiophrenic ligament.**

LIGAMENT OF THE LEFT VENA CAVA. This small fibrous strand is covered by a triangular fold of serous pericardium, and it stretches from the left pulmonary artery to the atrial wall or the subjacent pulmonary vein. It is formed by the remnant of the left common cardinal vein (or left duct of Cuvier) which becomes obliterated during fetal life. If well developed, it may remain as a fibrous band stretching from the highest left intercostal vein to a small vein draining into the coronary sinus, the **oblique vein of the left atrium** (vein of Marshall).

RELATIONS OF THE PERICARDIAL SURFACES. *Laterally,* the surfaces of the pericardial sac are apposed to the mediastinal parietal pleura. The pericardium and the mediastinal pleura are adherent but not fused, and the phrenic nerve with its accompanying blood vessels is held between them as it descends in the thorax. Because of the rounded contour of the heart, the pleural cavity partly encircles the pericardial sac, extending *anteriorly* between it and the chest wall except for a small triangular area on the left side (Figs. 3-14; 3-17). This area corresponds to the caudal portion of the body of the sternum and the medial ends of the left fourth and fifth costal cartilages. As mentioned above, in percussion of the chest for physical diagnosis, the triangle is called the area

of superficial cardiac dullness because no lung is present to give it resonance. This site is also important clinically as the area through which a needle can be introduced into the pericardial cavity for removal of excess fluid, without traversing the pleural cavity or lungs. More superiorly, the anterior surface may be in contact with the thymus in children. *Posteriorly,* the pericardial sac is in relation to the bronchi, esophagus, and descending thoracic aorta. *Inferiorly,* the pericardium is attached to the dome of the diaphragm.

General Features of the Heart

The **heart** is a hollow muscular organ shaped like a blunt cone and comparable in size to a human fist. It rests on the diaphragm between the lower part of the two lungs (Figs. 7-22; 7-23). It is enclosed in a fibroserous sac, the pericardium, and occupies a topographical compartment of the thorax called the **middle mediastinum** (Fig. 7-24). Its position relative to the chest wall is described in Chapter 3 and shown diagrammatically in Figure 7-22, and the position and extent of the cardiac shadow in a roentgenogram are shown in Figure 7-23. The heart is partially covered anteriorly by the mediastinal borders of the lungs, although a triangular region, which can be outlined by percussion, is not covered with lung tissue and is called the **area of superficial cardiac dullness.** The heart lies deep to the sternum and adjoining parts of the third to sixth costal cartilages. Its **apex** is directed downward, anteriorly, and to the left. Projecting farther into the left half of the thorax than the right, two-thirds of the heart lies to the left and one-third lies to the right of the median plane (Fig. 7-24).

SIZE. The average adult heart measures about 12 cm in length, 8 to 9 cm transversely at the broadest part, and 6 cm anteroposteriorly in thickness. Its weight, in the male, varies from 280 to 340 grams; in the female, from 230 to 280 grams. The heart frequently continues to increase in weight and size until an advanced period of life. This increase is more marked in men than in women, and in many instances is pathological.

HEART WALL. The wall of the heart is composed of three layers: the outermost is the

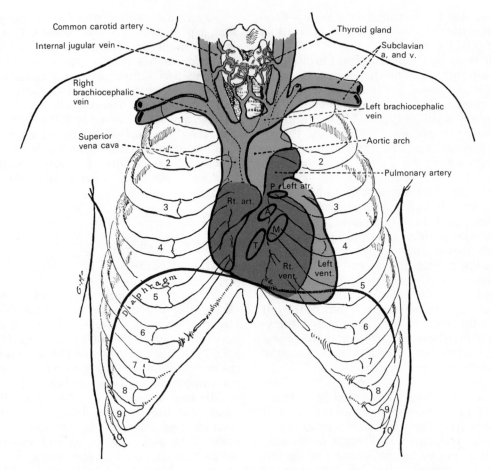

Fig. 7-22.   The heart and cardiac valves projected on the anterior chest wall, showing their relation to the ribs, sternum, and diaphragm. P = pulmonary valve; A = aortic valve; T = tricuspid valve; M = mitral valve. (From Eycleshymer and Jones.)

epicardium, deep to this is the **myocardium,** and the innermost is the **endocardium.**

The surface layer, or **epicardium,** is a serous membrane that is continuous with the inner lining of the fibrous pericardium and reflects over the heart wall at the roots of the great vessels, and can, therefore, be considered the **visceral pericardium.** It consists of a single sheet of squamous mesothelial cells resting on a lamina propria of delicate connective tissue. Between the epicardium and the myocardium is a layer of heavier fibroelastic connective tissue, which is interspersed with fatty tissue that fills in the crevices and sulci to give the heart a smooth, rounded contour. The larger blood vessels and the nerves are also found in this connective tissue. The dark reddish color of the myocardium is visible through the epicar-

dium except where fat has accumulated. The amount of fat varies greatly, but is seldom absent except in emaciated individuals. It may almost completely obscure the myocardium in the very obese.

The **myocardium** is composed of layers and bundles of cardiac muscle with a minimum of other tissue except for the blood vessels. Its detailed structure is described in a later section.

The **endocardium** is the inner lining of the heart. Its surface layer is composed of squamous endothelial cells, and it is continuous with the endothelial lining of the contiguous blood vessels. The connective tissue between this endothelial lining and the myocardium is quite thin and transparent over the muscular walls of the ventricles, but is thickened in the atria and at the attachments

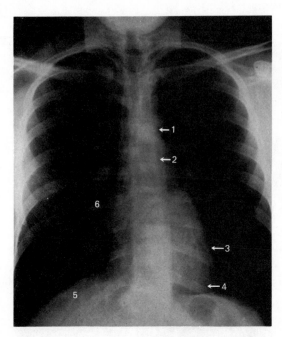

FIG. 7-23. Anterior view of the thorax of an adult showing the contours of the heart and great vessels. *1*, Arch of aorta. *2*, Left atrium. *3*, Left ventricle. *4*, Apex. *5*, Diaphragm. *6*, Vascular shadow in lung. (Courtesy of Dr. James Collins, Department of Radiology, UCLA School of Medicine, Los Angeles.)

of the valves. It contains small blood vessels, parts of the specialized conduction system, and a few bundles of smooth muscle.

External Features of the Heart

The heart consists of four chambers: two large ventricles with thick muscular walls making up the bulk of the organ, and two smaller atria with thin muscular walls. The atria are directed toward the **base** of the heart, while the ventricles are oriented toward the **apex.** The septum that separates the ventricles also extends between the atria, subdividing the whole heart into what frequently are called **left** and **right halves or sides** of the heart. As they lie in the body, however, the right side is mostly ventral or anterior and the left side largely dorsal or posterior.

SURFACE SULCI. The atria are separated from the ventricles on the surface of the heart by the **coronary sulcus** (Figs. 7-25; 7-26). This groove encircles the heart and is occu-

pied by the coronary vessels that supply the heart. The sulcus is deficient anteriorly because the root of the pulmonary trunk arises from the right ventricle at this site. The **interatrial groove,** which separates the two atria, is scarcely marked on the posterior surface, while anteriorly it is hidden by the pulmonary trunk and the aorta. The line of separation between the two ventricles is marked by the **anterior interventricular sulcus** on the sternocostal surface and by the **posterior interventricular sulcus** on the diaphragmatic surface. Within these two sulci course the anterior and posterior interventricular branches of the coronary arteries and their accompanying veins (Figs. 7-25; 7-26). The two grooves become continuous just to the right of the apex at a notch termed the **apical incisure** of the heart.

THE APEX. The **apex** of the heart points forward, to the left, and downward. Although its position changes continually during life, the tip remains close to the following point under usual circumstances: deep to the left fifth intercostal space, 8 or 9 cm from the midsternal line. This is about 4 cm below and 2 cm medial to the left nipple in the male. Because the site of the nipple can vary somewhat in females, this latter means of approximating the position of the apex may be less useful. The apex is overlapped by an extension of the pleura and lungs as well as by the structures of the anterior thoracic wall.

THE BASE. Although the base of the heart is somewhat more difficult to visualize than the apex, it can be conceived as the base of a blunt cone oriented in a direction opposite the apex. It faces backward, to the right, and upward. It is nearly quadrilateral in shape and is formed mainly by the left atrium, part of the right atrium, and the proximal parts of the great vessels (Fig. 7-27). The base of the heart extends *superiorly* as far as the bifurcation of the pulmonary trunk, while *inferiorly* it is bounded by the posterior part of the coronary sulcus containing the coronary sinus. To the *right* it is limited by the sulcus terminalis of the right atrium and to the *left* by the oblique vein of the left atrium. The descending thoracic aorta, the esophagus, and the thoracic duct intervene between the base of the heart and the bodies of the fifth to eighth thoracic vertebrae. At the base of the heart, the four pulmonary veins, two on

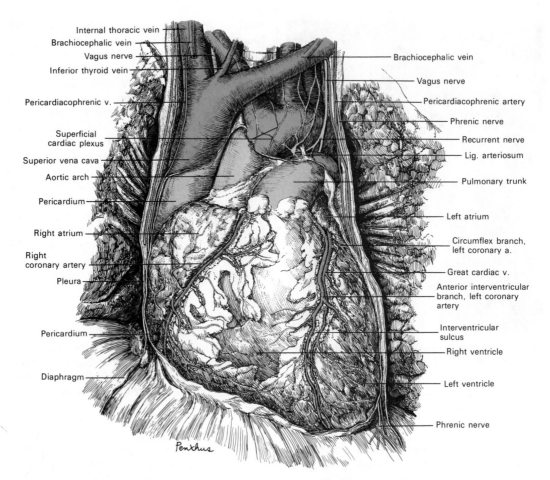

Internal thoracic vein
Brachiocephalic vein
Vagus nerve
Inferior thyroid vein
Pericardiacophrenic v.
Superficial cardiac plexus
Superior vena cava
Aortic arch
Pericardium
Right atrium
Right coronary artery
Pleura
Pericardium
Diaphragm

Brachiocephalic vein
Vagus nerve
Pericardiacophrenic artery
Phrenic nerve
Recurrent nerve
Lig. arteriosum
Pulmonary trunk
Left atrium
Circumflex branch, left coronary a.
Great cardiac v.
Anterior interventricular branch, left coronary artery
Interventricular sulcus
Right ventricle
Left ventricle
Phrenic nerve

Penthus

Fɪɢ. 7-24.   The anterior or sternocostal surface of the heart and great vessels after removal of the anterior part of the pericardium.

each side, open into the left atrium, while the superior vena cava opens into the upper part and the inferior vena cava into the lower part of the right atrium.

SURFACES AND MARGINS.   The heart presents several surfaces and margins (or borders) which upon clinical examination can be useful in describing surface relationships and possible pathological distortion of its normal shape and contours.

The **sternocostal surface** is formed by the right atrium, and especially its auricular appendage, and the right ventricle, to the left of which is a small part of the left ventricle (Fig. 7-25). The right upper aspect of this surface is crossed obliquely by the coronary sulcus, while the anterior interventricular sulcus passes from above downward and forms a line of separation between the right and left ventricles.

The **diaphragmatic surface** is formed prin-cipally by the left ventricle, but also to some extent by the right ventricle (Fig. 7-26). It rests chiefly upon the central tendon of the diaphragm, and it is crossed obliquely by the posterior interventricular sulcus. Superiorly, it is separated from the base of the heart by the coronary sulcus.

Right Margin.   The right margin of the heart is longer than the left, and it describes an arc from the superior vena cava to the apex. Its upper part defines the margin of the right atrium, while the lower part of the right margin outlines the border of the right ven-tricle. Because the right ventricle lies nearly horizontally on the diaphragm, so does the lower part of the right margin. Coursing along the line of attachment of the dia-phragm to the anterior chest wall, the *ven-tricular part* of the right margin tends to be thin and sharp. It extends from the sternal end of the sixth right costal cartilage to the

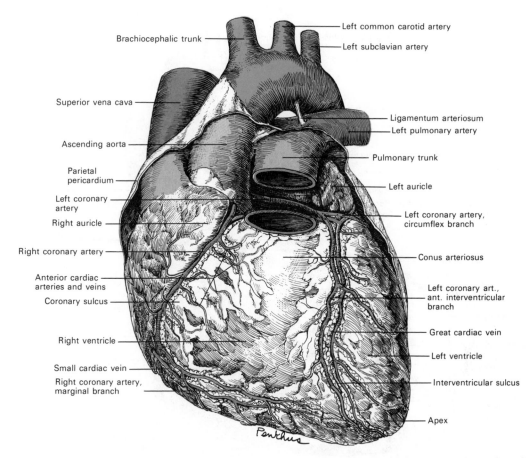

FIG. 7-25.   The sternocostal surface of the heart viewed from the anterior aspect. (Modified after Sobotta.)

cardiac apex and is sometimes called the **acute margin,** or even the **inferior border,** of the heart. The *atrial part* of the right margin is almost vertical and is situated behind the third, fourth, and fifth costal cartilages, 1.25 cm to the right of the margin of the sternum.

Left Margin.   The left margin is shorter and more rounded, and is formed mainly by the left ventricle and to a small extent by the left atrium. It extends obliquely downward to the apex from a point 2.5 cm from the sternal margin in the second left interspace, describing a curve that is convex laterally to the left.

Chambers of the Heart

The interior of the human heart consists of four chambers, two **atria** superiorly, and two **ventricles** inferiorly. A continuous septum, passing from the apex to the base, separates the atria **(interatrial septum)** and the ventricles **(interventricular septum).** On each side of the heart the passage between the atrium and the ventricle is guarded by valves **(right** and **left atrioventricular valves).** Other valves **(the aortic valve** and the **pulmonary valve)** are interposed between the ventricles and the principal arteries leaving the heart. The interior of each of the four chambers presents specialized structures that contribute to its function.

RIGHT ATRIUM

The right atrium (Fig. 7-28) is a somewhat cuboid chamber that has a capacity of about 57 ml. It appears larger than the left, and its wall, which is about 2 mm in thickness, is thinner than that of the left atrium. It consists of two parts: a principal cavity, situated posteriorly, called the **sinus venarum,** and a more anterior, smaller, conical pouch called the **right auricle** (Fig. 7-28).

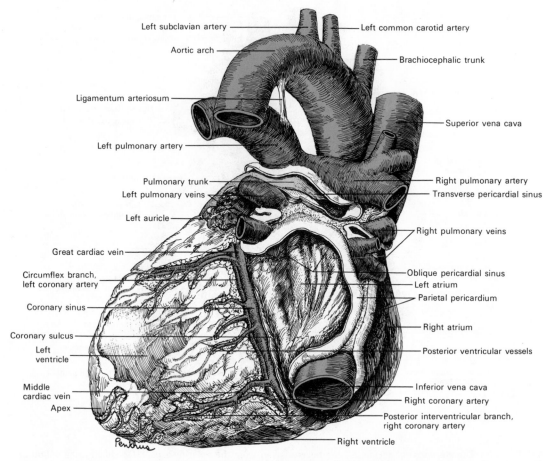

Left subclavian artery

Aortic arch

Ligamentum arteriosum

Left pulmonary artery

Pulmonary trunk

Left pulmonary veins

Left auricle

Great cardiac vein

Circumflex branch,
left coronary artery

Coronary sinus

Coronary sulcus

Left
ventricle

Middle
cardiac vein

Apex

Left common carotid artery

Brachiocephalic trunk

Superior vena cava

Right pulmonary artery

Transverse pericardial sinus

Right pulmonary veins

Oblique pericardial sinus

Left atrium

Parietal pericardium

Right atrium

Posterior ventricular vessels

Inferior vena cava

Right coronary artery

Posterior interventricular branch,
right coronary artery

Right ventricle

FIG. 7-26.   The diaphragmatic surface and the base of the heart viewed from the dorsal aspect. The visceral pericardium still overlies the heart, but observe the cut edges of its reflection as the parietal pericardium over the large vessels. (After Sobotta.)

SINUS VENARUM.   The sinus venarum is the part of the right atrium located between the superior and inferior venae cavae and the atrioventricular opening (Fig. 7-28). Its wall merges with these two large veins, and its interior surface is quite smooth except for certain rudimentary structures to be described.

RIGHT AURICLE.   The right auricle is a small, conical, muscular pouch which is an extension of the right atrium (Fig. 7-28). It is directed upward and to the left between the superior vena cava and the right ventricle, overlapping the right side of the root of the aorta. The junction of the right auricle with the sinus venarum is marked on the exterior of the atrium by a groove, the **sulcus terminalis** (Fig. 7-27), which corresponds to a ridge on the atrial inner surface, the **crista terminalis.** The internal surface of the auricula has its muscular bundles raised into

distinctive parallel ridges resembling the teeth of a comb and named, therefore, the **musculi pectinati.**

In addition to the crista terminalis and the musculi pectinati, the interior of the right atrium presents for examination a number of openings, valves, and other specialized structures.

OPENING OF THE SUPERIOR VENA CAVA.   The superior vena cava opens into the upper posterior part of the sinus venarum (Fig. 7-28). Its orifice is oriented downward and forward so that blood entering the atrium through it is directed toward the atrioventricular opening. The superior vena cava returns blood from the cranial half of the body, and there is no valve at its opening in the right atrium.

OPENING OF THE INFERIOR VENA CAVA.   The inferior vena cava opens into the most caudal part of the sinus venarum near the inter-

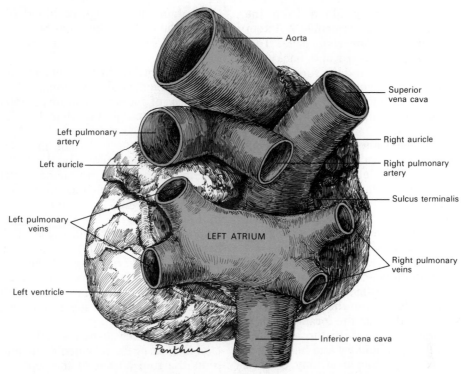

Aorta

Superior
vena cava

Left pulmonary
artery

Right auricle

Left auricle

Right pulmonary
artery

Sulcus terminalis

Left pulmonary
veins

LEFT ATRIUM

Right pulmonary
veins

Left ventricle

Inferior vena cava

Penthus

FIG. 7-27.   The base of the heart viewed from the dorsal and superior aspect. Pulmonary artery shown in blue because of low blood oxygenation level; pulmonary veins shown in red because of high blood oxygenation level.

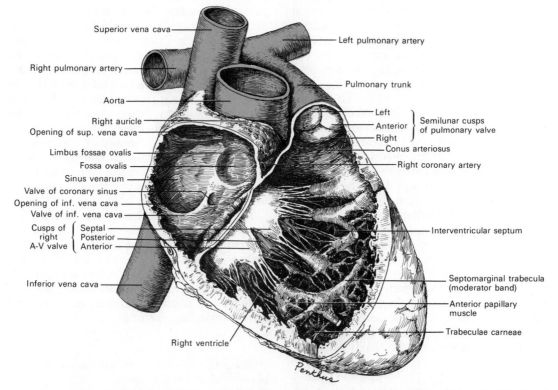

Superior vena cava

Left pulmonary artery

Right pulmonary artery

Pulmonary trunk

Aorta

Left ⎫
Anterior ⎬ Semilunar cusps
Right ⎭ of pulmonary valve

Right auricle

Opening of sup. vena cava

Conus arteriosus

Limbus fossae ovalis

Right coronary artery

Fossa ovalis

Sinus venarum

Valve of coronary sinus

Opening of inf. vena cava

Valve of inf. vena cava

Interventricular septum

Cusps of
right
A-V valve { Septal
Posterior
Anterior

Inferior vena cava

Septomarginal trabecula
(moderator band)

Anterior papillary
muscle

Trabeculae carneae

Right ventricle

Penthus

FIG. 7-28.   Interior of the right atrium and right ventricle after the removal of the sternocostal wall of the heart. Pulmonary trunk and pulmonary arteries shown in blue to indicate the low level of oxygenation of the blood in these vessels.

atrial septum. Returning blood from the lower half of the body, its orifice is larger than that of the superior vena cava. This opening is directed upward and backward toward the fossa ovalis, and the rudimentary **valve of the inferior vena cava** extends a short distance superiorly from the opening on the interatrial septum.

VALVE OF THE INFERIOR VENA CAVA. This valve (sometimes called the Eustachian valve[1]) is a single crescentic fold attached along the anterior and left margin of the orifice of the inferior vena cava. Its concave free margin ends in two cornua, of which the left is continuous with the anterior margin of the limbus fossae ovalis and the right spreads out on the atrial wall. The valve consists of a fold of the membranous lining of the atrium containing a few muscular fibers. In the fetus the valve is more prominent and tends to direct blood *from the inferior cava to the left atrium* through the foramen ovale, which is open prenatally. In the adult the valve is usually rudimentary and has little if any functional significance. It may be thin and fenestrated, very small, or even entirely absent.

OPENING OF THE CORONARY SINUS. The coronary sinus opens into the right atrium between the orifice of the inferior vena cava and the atrioventricular opening. It returns most of the blood from the substance of the heart itself and the opening is protected by a thin valve.

VALVE OF THE CORONARY SINUS. This valve is a single semicircular fold of the lining membrane of the right atrium, and it is attached to the right and inferior lips of the orifice of the coronary sinus. It may have a cribriform structure and contain many perforations or it may be doubled. The valve is seldom effective for more than partial closure of the orifice during contraction of the atrium.

FORAMINA VENARUM MINIMARUM. These are openings of small veins, the venae cordae minimi (or thebesian veins), which drain blood from the heart muscle directly into the atrial cavity. A few larger orifices can usually be seen in the septal wall, including those of the anterior cardiac veins. These openings were used by early scientists to explain the passage of blood from one side of the heart to the other before William Harvey[2] discovered the circulation of blood in 1610 A.D.

Another orifice in the right atrium is the right atrioventricular opening, a large oval aperture of communication between the atrium and the right ventricle. It is described with the right ventricle.

INTERATRIAL SEPTUM. The interatrial septum forms the dorsal wall of the right atrium. It contains rudimentary structures which were of significance in the fetus. The **fossa ovalis** (Fig. 7-28) is an oval depression in the lower part of the septal wall and corresponds to the foramen ovale of the fetal heart. It lies within a triangular area bounded by the openings of the two venae cavae and the coronary sinus. Its floor is formed by the remnant of the septum primum that initially separated the atria in the fetus.

The **limbus fossae ovalis** is the prominent oval margin of the foramen ovale and represents the free border of the septum secundum. The latter fetal structure forms just to the right of the septum primum in the fetus, and from it is derived most of the muscular part of the interatrial septum in the adult—that is, all except the floor of the fossa ovalis, which persists from the septum primum. The limbus is distinct cranially and at the sides, but it is deficient caudally. Frequently the cranial part of the two fetal septi do not fuse, leaving a slit-like opening in the adult interatrial septum through which a probe may be passed into the left atrium. This is spoken of as *probe patency* of the foramen and is found in 20 to 25% of all hearts. Larger openings are not uncommon, but these interatrial defects are seldom of functional significance.

The **intervenous tubercle** (of Lower[3]) is a small raised area in the septal wall of the right atrium between the fossa ovalis and the orifice of the superior vena cava. It is more distinct in the hearts of certain quadrupeds than in man, and in fetal life it may serve to direct blood from the superior cava into the atrioventricular opening.

[1]Bartolommeo Eustachio (1520–1574): An Italian anatomist (Rome).

[2]William Harvey (1578–1657): An English physician, physiologist, and embryologist.
[3]Richard Lower (1631–1691): An English physician and physiologist.

RIGHT VENTRICLE

The right ventricle occupies a large triangular part of the anterior or sternocostal surface of the heart and extends from the right atrium almost to the apex (Fig. 7-28). Its right boundary on the surface is the coronary sulcus, while its left is the anterior longitudinal sulcus. Superiorly, the part of the right ventricle called the **conus arteriosus** joins the pulmonary trunk. Inferiorly its wall forms the acute margin of the heart and extends for some distance around the diaphragmatic surface. The wall of the right ventricle is thickest near the base of the heart and gradually becomes thinner toward the apex. The right ventricular wall is about one-third the thickness of that forming the left ventricle,

but their capacities are the same, about 85 ml (Latimer, 1953).

The interior of the right ventricle presents a number of openings and specialized parts (Figs. 7-29; 7-30) that require further description.

RIGHT ATRIOVENTRICULAR ORIFICE. This opening or **ostium** is a large oval aperture through which the right atrium communicates with the right ventricle (Figs. 7-28; 7-29). It measures about 4 cm in diameter and is large enough to admit the tips of four fingers. It is surrounded by a strong fibrous ring that is covered by the endocardial lining of the heart, and it is guarded by the right atrioventricular or tricuspid valve.

RIGHT ATRIOVENTRICULAR VALVE. The right atrioventricular or **tricuspid valve** encircles

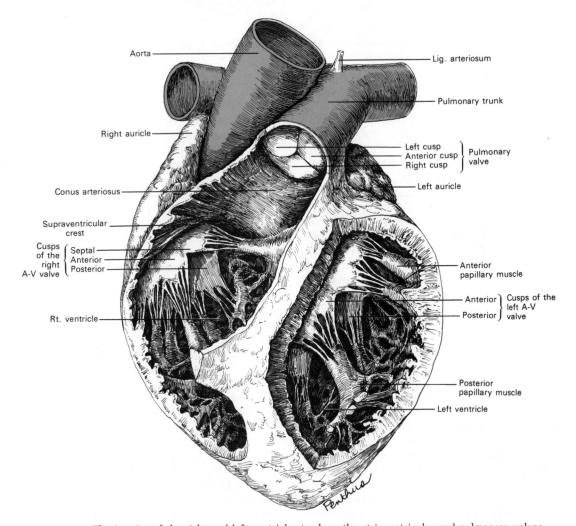

FIG. 7-29. The interior of the right and left ventricles to show the atrioventricular and pulmonary valves.

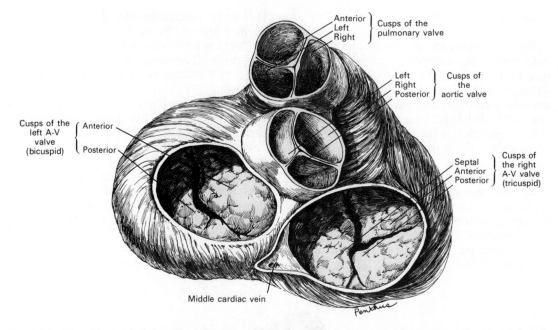

Fig. 7-30.   The atrioventricular, pulmonary, and aortic valves of the heart viewed from above after removal of the atria as well as the pulmonary trunk and aorta.

the right atrioventricular orifice and is notched into three roughly triangular leaflets or cusps. These are attached to the right ventricular wall, and when the valve opens, the cusps project into the ventricle. The three cusps are named **anterior, posterior,** and **septal,** and they are of unequal size (Figs. 7-28 to 7-30).

The **anterior** (or infundibular) **cusp** is the largest (Figs. 7-28; 7-29) and is interposed between the atrioventricular orifice and the anterior wall of the right ventricle, to which it is attached in the region of the conus arteriosus (or infundibulum). The **posterior** (or marginal) **cusp,** which is the smallest, extends from the atrioventricular orifice to the ventricular wall adjoining the sternocostal and diaphragmatic surfaces, sometimes called the acute margin of the heart (Fig. 7-29). The **septal** (or medial) **cusp** extends between the interventricular orifice and the right side of the interventricular septal wall. Small intercalated leaflets may occur between the three larger ones (Fig. 7-29).

Each cusp is composed of strong fibrous tissue that is thick in the central part, but thin and translucent near the margin. The bases of the cusps are attached to the fibrous ring surrounding the atrioventricular orifice, while their apices project into the ventricu-

lar cavity. The atrial surface of the leaflets is smooth, but the ventricular surface is more irregular and the free border of each cusp presents a ragged edge for the attachment of delicate tendinous cords called the **chordae tendineae.** In addition to their attachment to the cusps, the chordae tendineae are secured to papillary muscles in the ventricle. This makes the valve as a whole competent enough to withstand the back pressure of the blood during ventricular systole, thereby preventing the regurgitation of blood into the atrium.

CHORDAE TENDINEAE.   These strong fibrous cords are delicate but inelastic. They are attached to the ventricular end of the valve cusps at their apices and margins, and then are anchored to the muscular ventricular wall. They number about twenty and are of different length and thickness. The majority are attached to projections of the **trabeculae carneae** called **papillary muscles** (Figs. 7-28; 7-29).

TRABECULAE CARNEAE.   The trabeculae carneae are irregular bundles and bands of muscle that project from the entire inner surface of the ventricle except at the conus arteriosus, where the surface is smooth (Fig. 7-28). They are of three types: some are attached along their entire length to form ridges on the inner ventricular wall; others are fixed at their two extremities to the heart wall and extend across the lumen for short distances; and still others form the more specialized structures called **papillary muscles,** which project from the heart wall at one end and attach to the chordae tendineae at the other (Fig. 7-28). Although the inner surface of the conus arteriosus is quite smooth, its limit is marked by a ridge of muscular tissue, the **supraventricular crest,** which extends across the dorsal ventricular wall toward the pulmonary trunk from the attachment of the ventral cusp at the atrioventricular ring (Fig. 7-29).

The **septomarginal trabecula** is a stout bundle of heart muscle in the central or apical part of the right ventricle. It crosses the ventricular lumen from the base of the anterior papillary muscle to the interventricular septum (Fig. 7-28). It is of variable size and constancy in the human heart, but is more prominent in some larger mammals where it is supposed to resist overdistention of the ventricle. For this reason it is also called the **moderator band.** When present, it usually contains the right limb of the atrioventricular conduction bundle (Truex and Copenhaver, 1947).

PAPILLARY MUSCLES.   The papillary muscles are conical projections of ventricular muscle, the apices of which afford attachment to the chordae tendineae. Although their size and number are somewhat variable, the two principal ones are named the anterior and posterior papillary muscles.

The **anterior papillary muscle** is the larger of the two prominent papillary muscles, protruding partly from the anterior and partly from the septal wall of the right ventricle (Fig. 7-28). Its chordae tendineae are attached to the anterior and posterior cusps of the tricuspid valve, and frequently a portion of this muscle constitutes the moderator band.

The **posterior papillary muscle** usually consists of two or three parts which arise from the posterior ventricular wall and to which are attached the chordae tendineae of the posterior and septal cusps. Another small papillary muscle arises near the septal end of the supraventricular crest. It is called the *papillary muscle of the conus arteriosus,* and it is attached to the anterior and septal cusps. Other small *septal papillary muscles* spring directly from the interventricular sep-

tum and attach to individual chordae tendineae.

ORIFICE OF THE PULMONARY TRUNK.   At the summit of the conus arteriosus is a circular opening leading to the pulmonary trunk (Figs. 7-28; 7-29). It is close to the interventricular septum, above and to the left of the atrioventricular aperture, and is guarded by the pulmonary valve.

PULMONARY VALVE.   This valve consists of three **semilunar cusps** (or valvula) formed by duplications of the endocardial lining which are reinforced by fibrous tissue (Figs. 7-28 to 7-30). They are attached by their curved margins to the wall of the pulmonary trunk at its site of junction with the conus arteriosus. The free borders of the cusps are directed upward toward the lumen of the pulmonary trunk. Behind each cusp, that is, between the cusp and the wall of the pulmonary trunk, is a pocket called the **pulmonary sinus** (sinus of Valsalva[1]). The point at which the attachments of two adjacent cusps come together is called a **commissure.** Each cusp is marked by a small thickened fibrocartilaginous **nodule** (corpus of Arantius[2]) at the center of the free margin. From the nodule tendinous fibers radiate through the cusps to the commissure of attachment. These strengthen both the free and attached margins of the cusp. Two thin crescentic arcs, free of fibers, are placed one on each side of the nodule immediately adjoining the free margin. These are called the **lunules.** The line of contact between the cusps, when the valve is closed, is not at the free margin, but by the nodule and lunulae.

The three semilunar cusps forming the pulmonary valve are named the **right** (right anterior), **left** (posterior), and **anterior** (left anterior).

Two methods are used to name the semilunar cusps of the pulmonary valve. One is based on their position in the adult body. The other, and currently recommended terminology in the *Nomina Anatomica,* is based on the position of the cusps in the fetus prior to rotation of the pulmonary artery and the aorta. In this book, the latter terminology has been presented as preferred and is printed in **bold type** above, while the former terminology is shown in parentheses.

---

[1]Antonio M. Valsalva (1666–1723): An Italian anatomist.
[2]Giulio C. Arantius (1530–1589): An Italian physician and anatomist.

LEFT ATRIUM

The left atrium is rather smaller than the right and the wall is thicker, measuring about 3 mm. It forms a large part of the base and dorsal part of the heart (Figs. 7-26; 7-27). Its separation from the right atrium is not identifiable on the posterior heart surface except when the organ is distended, but the aorta and pulmonary trunk lie between the two atria on the anterior surface. The left atrium consists of two parts, a **principal cavity** and the **left auricle.**

PRINCIPAL CAVITY.   The principal cavity of the left atrium contains the openings of the four pulmonary veins, two on each side of the chamber, and they are not guarded by valves. Frequently the two left veins have a common opening. Additionally, the principal cavity opens into the left ventricle by means of the left atrioventricular orifice, which is a little smaller than the right and is guarded by the **left atrioventricular** or **mitral valve.** The surface of the principal cavity is smooth. The interatrial septum contains a depression that is bounded below by a crescentic ridge. This is the edge of the **valve of the foramen ovale,** the remnant of the septum primum which became fused over the opening of the foramen ovale at birth.

LEFT AURICLE.   The left auricle is somewhat constricted at its junction with the principal cavity. It is longer, narrower, and more curved than the right auricle, and its margins are more deeply indented. It curves anteriorly around the base of the pulmonary trunk, only its tip being visible on the sternocostal aspect of the heart (Fig. 7-25). At this site it also overlies the commencement of the left coronary artery. The inner surface of the auricle is marked by muscular ridges, similar to those in the right auricle, called musculi pectinati.

LEFT VENTRICLE

The left ventricle occupies a small part of the sternocostal and about half of the diaphragmatic surface of the heart (Figs. 7-25; 7-26). It forms part of the left margin of the heart and its tip forms the apex. The left ventricle is longer and more conical in shape than the right, and its walls are about three times as thick. In a cross section its cavity is oval or circular in outline. The interior presents two openings: the atrioventricular orifice guarded by the mitral valve and the aortic aperture guarded by the aortic valve.

LEFT ATRIOVENTRICULAR ORIFICE.   The left atrioventricular orifice is located below and somewhat posterior to the aortic opening. It is smaller than the corresponding orifice in the right heart and its breadth allows only two fingers to pass. It is surrounded by a dense fibrous ring that supports the **left atrioventricular valve,** which is also called the **mitral** or **bicuspid valve.**

LEFT ATRIOVENTRICULAR VALVE.   The left atrioventricular or mitral valve is attached to the fibrous ring surrounding the circumference of the left atrioventricular orifice. It is bicuspid, and its two cusps, which are of unequal size, extend down into the left ventricle in a manner similar to those of the valve on the opposite side (Fig. 7-29). The cusps are named anterior and posterior according to their position in the ventricle.

The larger **anterior cusp** is placed ventrally and to the right, adjacent to the aortic opening. The smaller or posterior cusp is located behind and to the left of the aortic aperture. Two smaller cusps are sometimes found at the angles of junction between the main cusps. The chordae tendineae are thicker and stronger, but less numerous than those in the right ventricle (Fig. 7-29).

TRABECULAE CARNEAE.   There are three types of trabeculae carneae, similar to those described for the right ventricle, but they are more numerous and more densely packed, especially at the apex and on the dorsal wall.

PAPILLARY MUSCLES.   There are two principal papillary muscles in the left ventricle. One is attached to the anterior or sternocostal wall and is therefore called the **anterior papillary muscle.** The other, attached to the posterior or diaphragmatic wall, is the **posterior papillary muscle.** Both are of large size and end in rounded extremities to which the chordae tendineae are attached. Each papillary muscle has chordae tendineae which go to both cusps of the valve.

AORTIC ORIFICE.   The aortic opening is a circular aperture located anterior and to the right of the atrioventricular orifice, from which it is separated by the anterior cusp of the mitral valve. The aortic orifice is guarded by the **aortic valve,** which consists of three semilunar cusps. It is slightly over 2.5 cm (one inch) in diameter, and the portion of the ventricle immediately adjacent to the orifice is termed the **aortic vestibule,** which consists largely of fibrous tissue rather than cardiac muscle.

AORTIC VALVE. The aortic valve is composed of three semilunar cusps, similar to those of the pulmonary valve but larger, thicker, and stronger (Fig. 7-31). The cusps surround the aortic orifice and they contain **lunules** that are more distinct and nodules that are thicker and more prominent than those of the pulmonary valve. Between the cusps and the aortic wall there are dilated pockets called the **aortic sinuses** (of Valsalva), which are larger than those of the pulmonary trunk, and from the aortic wall bounding two of these sinuses the coronary arteries take origin.

The cusps of the aortic valve are named **posterior** (right posterior), **right** (anterior), and **left** (left posterior), based on the embryological development of the aortic septum by division of the truncus arteriosus into the aorta and pulmonary trunk. (In parentheses are the names of the cusps according to their position in the adult heart, but this is no longer recommended by the *Nomina Anatomica.*) The preferred terminology (in **bold face** above) is consistent with that used for the pulmonary valve and has the added correlation that behind the left and right cusps, the coronary arteries arise from the aorta. The posterior cusp, therefore, would be the non-coronary cusp.

INTERVENTRICULAR SEPTUM. The left ventricle is separated from the right ventricle by the interventricular septum. This septum is directed obliquely backward and to the right, and it also has a curvature with convexity to the right, thus completing the oval of the thick left ventricle and encroaching on the cavity of the right ventricle. Its margins correspond to the anterior and posterior interventricular sulci. The greater part of it is thick and muscular and is called the **muscular part** of the interventricular septum. The upper portion of the interventricular septum, which separates the aortic vestibule from the lower part of the right atrium and the upper part of the right ventricle, is thinner and fibrous and is called the **membranous part** of the interventricular septum. This is the last part of the septum to close in embryological development (Fig. 7-17) and is the usual site of the defect in a condition known as *patent interventricular septum.*

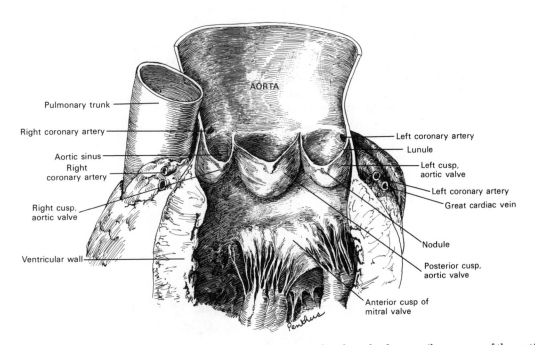

FIG. 7-31. The aortic entrance from the left ventricle has been opened to show the three semilunar cusps of the aortic valve. Within the ventricle a part of the left atrioventricular (mitral) valve is also exposed.

Structure of the Cardiac Wall

The wall of the heart consists principally of **cardiac muscle** and a **fibrous skeleton,** the latter serving, in part, for the attachment of some cardiac muscle fibers and the heart valves. The heart wall is covered by mesothelial linings, both externally by the **epicardium** and internally by the **endocardium.**

FIBROUS CARDIAC SKELETON. The fibrous skeleton of the heart consists of fibrous or fibrocartilaginous tissue which surrounds the atrioventricular, aortic, and pulmonary apertures with fibrous rings. This more resistant tissue merges with the membranous part of the interventricular septum and serves for the attachment of valves and heart muscle fibers. In the heart of certain large animals, such as the ox, this fibrous tissue may contain cartilage and even bone. Between the right margin of the **left atrioventricular ring,** and the **aortic ring** and **right atrioventricular ring,** the dense tissue forms a triangular mass, the **right fibrous trigone** (Fig. 7-32). This mass also forms the basal thickening of the membranous septum. A fibrous band from the right trigone and right atrioventricular ring to the posterior side of the conus arteriosus and anterior aspect of the aortic ring is called the **tendon of the conus.** A similar but smaller mass of fibrous tissue, the **left fibrous trigone,** lies between the aortic and left atrioventricular rings. The atrioventricular rings and the trigones separate the muscular walls of the atria from those of the ventricles. The bundles of muscular fibers of both chambers, however, use the fibrous tissue for their attachment.

HEART MUSCULATURE. The muscular structure of the heart consists of bands of fibers that present an exceedingly intricate interlacement. Distinct bundles of muscle fibers form the walls of the atria and the ventricles, but those bundles forming the atrial chambers are functionally separated from those forming the ventricles by the fibrous rings. An additional special bundle of muscle fibers, the atrioventricular bundle (of His[1]), which forms the conducting system of the heart, does constitute a functional connection between the atrial and ventricular chambers and will be discussed separately.

The **musculature of the atria** can be dissected into the two layers, superficial and deep. The more *superficial bundles* extend across both atria, encircling them in single or double loops. The *deeper bundles* are more generally confined to each atrium, although some fibers pass through the interatrial septum to encircle the other atrium as well. The atrial bundles are attached to the fibrous trigones and the atrioventricular rings, and they extend for short distances to encircle the roots of the venae cavae on the right side and the pulmonary veins on the left.

The **musculature of the ventricles** consists of bundles that probably all arise from the fibrotendinous structures at the base of the heart, converge in spiral courses toward the apex for varying distances, and then turn spirally upward to be inserted on the opposite side of these same fibrotendinous structures. The **superficial bundles** spiral to their vortex at or near the apex of the left ventricle before they turn

[1] Wilhelm His, Jr. (1863–1934): A German anatomist (Leipzig and Berlin).

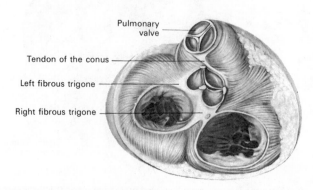

FIG. 7-32. The fibrous skeleton of the heart. The atria have been removed and the heart is viewed toward the ventricles. (Redrawn from Tandler, 1912.)

upward, while the **deeper bundles** turn upward at varying distances without reaching the apex.

The *superficial bulbospiral bundle* arises from the conus arteriosus, left side of the aortic septum, aortic ring, and left atrioventricular ring, and passes toward the apex of the heart. At their origin the fibers form a broad thin sheet that becomes thick and narrow at the apex where the bundle twists on itself. It then continues upward and posteriorly in a spiral manner on the inner surface of the left ventricle, spreading out into a thin sheet that is inserted on the opposite side of the fibrotendinous structures from which it arose. These fibers nearly make a double circle that is open at the top, somewhat like a figure 8, around the heart. As the fibers pass toward the apex they lie superficial to the deep bulbospiral bundle, and as they pass upward from the apex they partly blend and partly pass on the inner side of the deep bundle in directions nearly at right angles to the descending superficial fibers.

The *superficial sinuspiral bundle* arises as a thin layer from the posterior aspect of the right and left atrioventricular rings. The fibers pass more horizontally around the heart to the apex than do those of the superficial bulbospiral bundle. They encircle the right ventricle and converge as they approach the apex, passing upward into the papillary muscles on the inner wall of the ventricles. They become attached to the fibrous rings either by way of the chordae tendineae or directly by the fibers themselves.

The *deep bulbospiral bundle* arises immediately beneath the superficial bulbospiral bundle from the left side of the aortic and left atrioventricular rings. The fibers pass downward to the right and enter the septum through the posterior longitudinal sulcus. They then encircle the left ventricle without reaching the apex. Turning upon themselves on the apical side of the ring, they blend with the fibers of the superficial bundle as they pass spirally upward to be inserted on the right side of the fibrous aortic and left atrioventricular rings. These fibers also seem to form an open figure 8, with both loops of about the same size.

The *deep sinuspiral bundle* is more especially concerned with the right ventricle, although its fibers communicate freely with the papillary muscles of both ventricles. Its fibers arise from the posterior part of the left atrioventricular ring and pass diagonally into the deeper layer of the wall of the right ventricle, where they turn upward to the conus arteriosus and membranous part of the interventricular septum.

The *interventricular bundles* are represented in part by the longitudinal bundle of the right ventricle, which passes through the interventricular septum and must be cut in order to unroll the heart, and in part by the interpapillary bands. The *circular bands of the conus* simply extend from one side of the tendon of the conus arteriosus around the root of the pulmonary trunk and conus to the opposite side of the tendon.

Conduction System of the Heart

The conduction system of the heart consists of the sinuatrial node, the atrioventricular node, the atrioventricular bundle, and the terminal conducting fibers or Purkinje[1] fibers. Although these structures differ from each other in certain details, they are all composed of specially differentiated cardiac muscle fibers that have the capacity to conduct electrical impulses and thus are more highly developed than fibers in the rest of the heart. The musculature of both the ventricles and the atria has an intrinsic capability to contract spontaneously, independent of any nervous influence. The conduction system, however, initiates and superimposes a rhythmicity to this contraction, which has a rapid rate and is transmitted to all parts of the heart. The rate of the cardiac rhythm is regulated by the autonomic nerves which are under the control of the central nervous system.

SINUATRIAL NODE.   The sinuatrial node (S–A node) is a small mass of specialized heart muscle situated in the crista terminalis at the junction of the superior vena cava and right atrium. It receives its name from the fact that this region develops from the margin of the sinus venosus in the embryo. The sinuatrial node consists of very narrow fusiform muscle fibers, which are assembled in a crescent-shaped region less than a centimeter in length and slightly more than a millimeter in breadth. The node is difficult to identify from surrounding muscle tissue at dissection, but may be located by tracing the course of the small sinunodal artery, which usually branches from the right coronary artery and supplies the region. The contraction of the heart is initiated by the node, and it is therefore called the **pacemaker of the heart.** No specialized bundle for the conduction of the impulses from this node to the ventricle has, as yet, been identified morphologically. The fibers of the node merge with the surrounding atrial muscle fibers, and these alone appear to be responsible for transmitting the contractile wave across the atria and the impulses to the atrioventricular node.

[1]Johannes E. Purkinje (1787–1869): A German and then Czechoslovakian anatomist (Breslau and Prague).

ATRIOVENTRICULAR NODE. The atrioventricular node (A–V node) is slightly smaller than the sinuatrial node and lies near the orifice of the coronary sinus in the septal wall of the right atrium (Fig. 7-33). Its muscle fibers, not being encapsulated by connective tissue, can seldom be recognized except that they are continuous with those of the atrioventricular bundle below and the atrial musculature above. The node is supplied by autonomic nerve fibers and usually receives its blood supply from the posterior interventricular artery.

ATRIOVENTRICULAR BUNDLE. The atrioventricular bundle (A–V bundle) begins at the atrioventricular node and follows along the membranous part of the interventricular septum toward the left atrioventricular opening, a distance of about one centimeter. Toward the middle of the septum, the A–V bundle splits into right and left crura, which straddle the summit of the muscular part of the interventricular septum. The **right crus** of the A–V bundle continues under the endocardium toward the apex, spreading to all parts of the right ventricle (Fig. 7-33). It

breaks up into small bundles of fibers called the **terminal conducting Purkinje fibers,** which become continuous with the cardiac muscle fibers of the right ventricle. A large fascicle of terminal fibers enters the **septomarginal** (or moderator) **band,** if this is present, to reach the opposite wall of the ventricle and the anterior papillary muscle. The **left crus** of the A–V bundle penetrates the fibrous septum and comes to lie just under the endocardium of the left ventricle (Fig. 7-34). At first it is a flattened band which soon fans out on the septal wall more quickly than the right crus and breaks up into bundles of the terminal conducting Purkinje fibers, which are distributed throughout the left ventricle.

The two crura are surrounded by more or less distinct connective tissue sheaths and may be visible, therefore, on gross examination of the heart. They are more easily seen in a fresh specimen than in a preserved one, and the left branch frequently shows more clearly than the right. The bundle and branches may be demonstrated by injecting India ink into the connective tissue sheath of

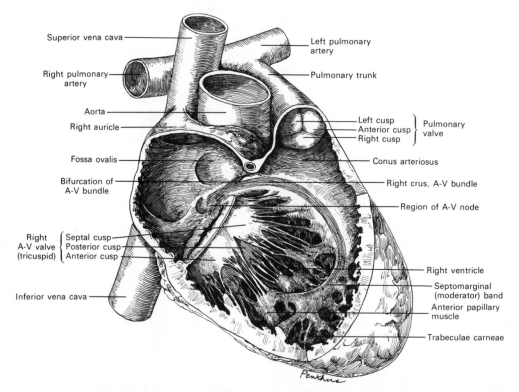

FIG. 7-33.   The conduction system of the heart. The interior of the right side of the heart has been opened to show (in yellow) the right crus of the atrioventricular bundle (of His).

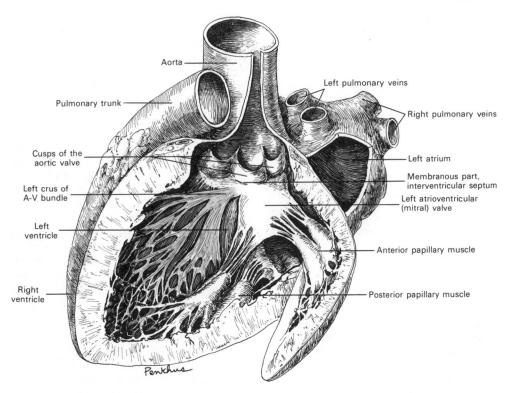

FIG. 7-34. The left crus of the atrioventricular bundle exposed after opening the left ventricle. The aorta has been opened to reveal the cusps of the aortic valve and the membranous part of the interventricular septum.

a fresh specimen, a procedure that is particularly effective with a sheep or ox heart.

The terminal conducting Purkinje fibers are different histologically from the nodal fibers, merging with the nodal tissue at one end of the A–V bundle and with the regular heart muscle at the other. The individual fibers can only be identified with the aid of a microscope. Coursing initially in small bundles under the endocardium, they then penetrate and ramify as individual fibers throughout the ventricular musculature.

CARDIAC CYCLE

The complete cardiac cycle commences at the end of one heart contraction and extends to the end of the next contraction. It therefore includes first a period of cardiac relaxation called **diastole,** which is then followed by the contraction phase called **systole.** Because each cardiac muscle fiber possesses the ability to contract rhythmically, it remains the important function of the conducting system of the heart to initiate the contraction wave, which has its initial site of generation at the sinuatrial node. After spreading across the atria, the contraction wave allows the muscle fibers of both atria to contract as a unit. Then, after the electrical impulse reaches the

atrioventricular node, the conduction system of the heart transmits it through both ventricles, causing the cardiac muscle fibers forming the walls of these chambers to contract efficiently and in unison. It is the changes in pressure within the chambers, caused by the contraction and relaxation of the cardiac muscle, that open and close the atrioventricular valves as well as the pulmonary and aortic valves. Closure of the valves accounts for the audible heart sounds that can be heard over the anterior chest wall.

The graphic representation of the electrical action currents in the heart that can be recorded from the body surface is the commonly used electrocardiogram (Fig. 7-35). Each cardiac cycle, when visualized on the electrocardiogram, is seen to consist of P, QRS, and T waves. The **P wave** is generated by the spread of depolarization from the sinuatrial node across the two atria. Atrial contraction soon occurs, causing blood to flow through the open atrioventricular valves. Approximately 0.16 seconds after the onset of the P wave, the electrocardiogram displays the **QRS wave complex.** This signals that the depolarization wave is spreading across the two ventricles, causing them to contract and thereby to elevate intraventricular pressure. The **first heart sound** is caused by the simultaneous closure of the two atrioventricular valves as ventricular pressure exceeds atrial pressure. Closure of the atrioventricular valves and continued ventricular contraction force the

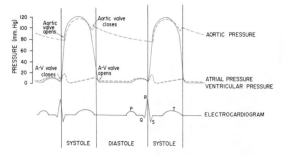

Fig. 7-35. Events in the cardiac cycle shown in a diagrammatic manner. The upper three traces show aortic, atrial, and left ventricular pressure in millimeters of mercury. The bottom trace is the corresponding electrocardiogram showing the P, QRS, and T waves. This diagram carries the cardiac cycle through two systoles, interrupted by a diastole. (Modified from Guyton, 1974.)

valves of the pulmonary artery and aorta to open and blood to flow into these vessels. Slightly before the end of ventricular contraction, the electrocardiogram displays the **T wave,** which represents the initiation of the phase of ventricular relaxation. As a result of blood leaving the ventricles and with the completion of ventricular contraction, the pulmonary and aortic valves close, causing the shorter but sharper **second heart sound.** With continuing diminished ventricular pressure, the atrioventricular valves open and the cycle is repeated.

Histology and Ultrastructure of Cardiac Muscle

The myocardium is composed of muscle fibers that show many similarities to the fibers of striated muscle (Fig. 7-36). Cardiac muscle fibers consist of myofibrils that are separated by sarcoplasm, and they also show **A, I, Z, H, and M transverse striations.** For many years it was thought that cardiac muscle was not composed of individual muscle cells, but that the cardiac fibers were joined in such a way as to form a syncytium. Ultrastructural studies have shown this not to be true. It is now recognized that the **intercalated discs** that cross cardiac muscle fibers in a transverse manner represent sites where individual muscle fibers are joined in an end-to-end arrangement of interdigitating membranes and show specialized surface membrane complexes (Fig. 7-38).

Cardiac muscle fibers are oriented in parallel columns similar to skeletal muscle; however, unlike skeletal muscle, cardiac fibers tend to bifurcate and anastomose (Fig. 7-36). Another difference between skeletal

and cardiac muscle is the fact that the elongated nuclei of heart muscle fibers lie deep within the substance of the muscle cells rather than directly beneath the sarcolemma of the fiber as is seen in skeletal muscle (Fig. 7-37). Between adjacent cardiac fibers is found loose connective tissue containing nerve fibers and an abundance of capillaries and lymphatics. The myofibrillar substructure of cardiac fibers consists of both myosin and actin myofilaments, accounting for the similarity of the transverse striations to those observed in skeletal muscle.

Cardiac muscle fibers show a great abundance of large **mitochondria,** presumably because of the high metabolic and energy requirements of the tissue (Fig. 7-38). The sarcoplasm is also rich in glycogen and lipids, helping to serve the energy resources of the muscle fibers. Invaginations of the sarcolemma of heart muscle fibers give rise to a **T system** which, similar to skeletal muscle, allows the internal structure of cardiac fibers to be more closely related to the extracellular space. In heart muscle, however, the transverse tubules are found at the Z line rather than at the A–I junctions and they are somewhat wider in diameter than those of skeletal muscle. The **sarcoplasmic reticulum** is neither as elaborate nor as abundant as that in skeletal muscle (Fig. 7-39). Although the sarcoplasmic reticulum network extends over the surface of the myocardial sarcomere, terminal cisterns similar to those of skeletal muscle are not seen. In their place are small cisternal expansions that are more irregularly placed and come into individual contact with the T tubular membrane to form a dyad coupling, in contrast to the triad configuration seen in skeletal muscle.

Intercalated discs have long been recognized as a characteristic microscopic feature of cardiac muscle fibers. They are easily identified by the light microscope in ordinary hematoxylin and eosin sections, and their visibility may even be enhanced by special stains. Electron micrographs have revealed that the darkly stained intercalated discs observed with the light microscope are junctional intercellular complexes between adjacent cardiac muscle fibers (Fig. 7-40). At the intercalated disc, apposed surfaces of the cardiac fibers are covered by their own cell membranes, which are joined together by specialized complexes that maintain intercellular cohesion. The **transverse parts** of intercalated discs are arranged in a stepwise fashion, crossing the fibers at right angles, while the **lateral parts** of the discs are longitudinally oriented and seen to course parallel to the myofibrils.

CONDUCTION SYSTEM. The **sinuatrial node** consists of slender fusiform cells largely filled with sarcoplasm but containing a few striated fibrillae. The cells are smaller than the surrounding atrial muscular fibers and they are intermeshed in connective tissue and arranged, more or less, in a parallel manner and in relation to the nodal artery. At the periphery of the node the fusiform nodal fibers merge with the atrial musculature. Closely associated with the sinuatrial node are a number of parasympathetic ganglia associated with the vagus nerve, but there are also postganglionic sympathetic fibers nearby. The **atrioventricular node** is a small mass of slender specialized cells found in the lower part of the interatrial septum. Its fibers are branched and irregularly arranged, and they merge with the atrial musculature on the one hand and continue into the atrioventricular bundle on the other. In the atrioventricular bundle and its first two branches, the specialized fibers remain slender, but in the more peripheral branches of the A–V bundle the fibers assume the histo-

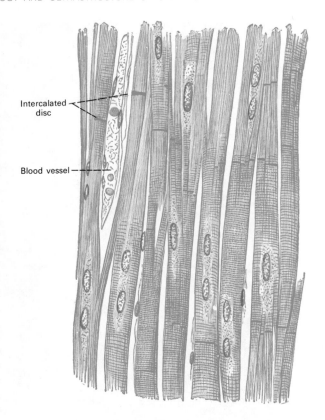

FIG. 7-36. Longitudinal section of human cardiac muscle fibers showing their branching manner. 400 × magnification. (From Rauber-Kopsch, 1955.)

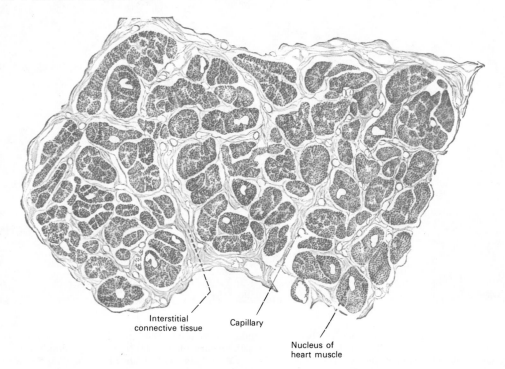

FIG. 7-37. Cross section of human cardiac muscle showing the nuclei located deeply within the interior of the fibers rather than immediately beneath the sarcolemma as is seen in skeletal muscle. (From Rauber-Kopsch, 1955.)

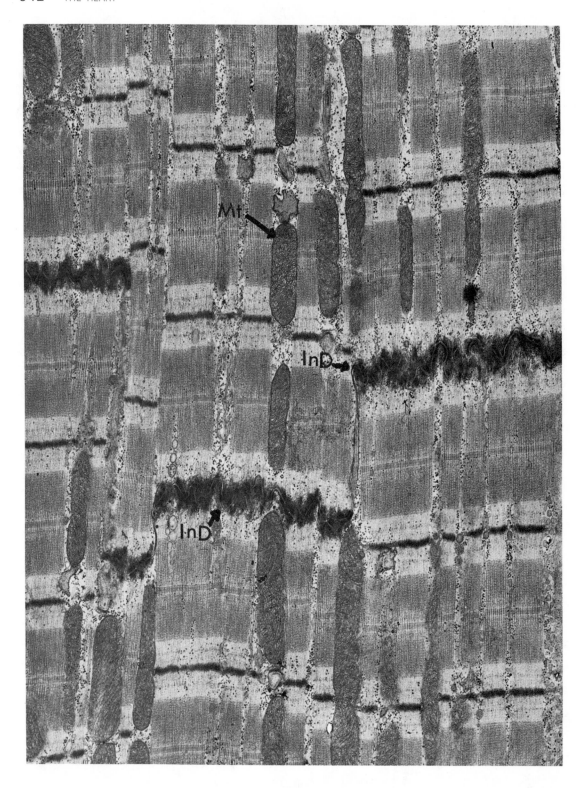

FIG. 7-38.   Electron micrograph of cardiac fibers in longitudinal section. The cells are joined end-to-end by a typical steplike intercalated disc (InD). Mitochondria (Mt) divide the contractile substance, and lipid droplets are found between the ends of the mitochondria. (From Fawcett, D. W., *The Cell, An Atlas of Fine Structure*, W. B. Saunders Co., 1966.)

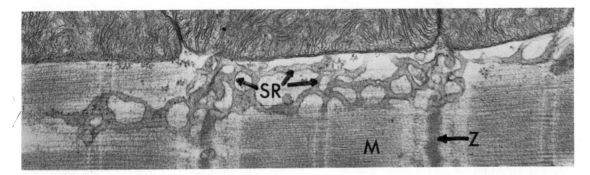

Fig. 7-39. The sarcoplasmic reticulum (SR) consisting of a simple network of anastomosing tubules. The section passes tangential to the inner surface of a mass of myofilaments (M), and the sarcoplasmic reticulum continues across the Z line (Z arrow) without terminal cisternae (From Fawcett and McNutt, J. Cell Biol., 42:1–45, 1969.)

logical characteristics of the Purkinje fibers. These latter fibers are larger in diameter than ordinary cardiac muscle. They contain relatively few peripherally placed myofibrillae, have abundant sarcoplasm, and the centrally placed nuclei are larger and more vesicular than those of the usual cardiac muscle. As the ramifications of the bundles spread into the ventricular wall, the individual Purkinje fibers become continuous with fibers of the main cardiac musculature.

VESSELS AND NERVES. The **arteries** supplying the heart are the right and left coronary arteries, and the **veins** are mostly tributaries of the coronary sinus. The **lymphatics** consist of deep and superficial plexuses, which form right and left trunks and end in the tracheobronchial nodes. Detailed descriptions of the blood vessels and lymphatics of the heart can be

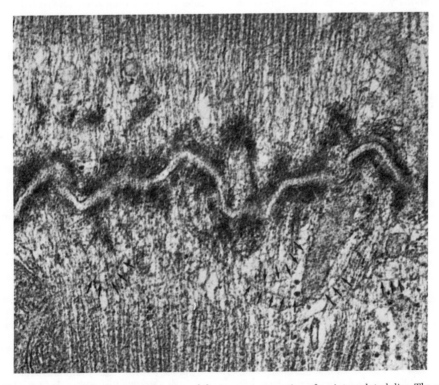

Fig. 7-40. Electron micrograph of restricted region of the transverse portion of an intercalated disc. The cell surfaces at the disc are approximately 200 to 300 Å apart. Actin filaments are seen to enter accumulations of dense material in the sarcoplasm subjacent to the cell membrane. Other actin filaments near the disc are seen coursing transverse to the prevailing direction of the myofilaments (at arrows). 70,000 × (From Fawcett and McNutt, J. Cell Biol., 42:1–45, 1969.)

found in the chapters dealing with these systems.

The **nerves** that supply the heart are cardiac branches of the vagus nerve and sympathetic trunks. These nerves join to form the cardiac plexuses and their ramifications, the coronary plexuses, which accompany the coronary arteries. The sympathetic fibers are postganglionics from the cervical and upper thoracic ganglia. The vagus fibers are preganglionics, whose ganglion cells form clusters in the connective tissue of the epicardium of the atria and the interatrial septum. The autonomic innervation of the heart is described in more detail in the chapter on the nervous system.

CONGENITAL MALFORMATIONS OF THE HEART

Developmental abnormalities may result from the suppression or overgrowth in varying degrees of the individual parts of the heart. The more prevalent abnormalities, however, are the result of partial failure in the partitioning of the heart. **Interatrial defects** that allow the passage of a probe through the foramen ovale are quite common (20%), but are of no functional significance. Larger defects of the septum are less common and may involve the atrial or ventricular septa or both. Abnormality in the growth of the septum and the endocardial cushions responsible for separating the truncus arteriosus into the aorta and pulmonary trunk results in what are called transposition complexes. **Transposition** refers to a shifting of the arterial trunks from their proper origins. The abnormalities are designated according to the degree of displacement as (*a*) overriding, (*b*) partial transposition, and (*c*) complete transposition. In *overriding,* one of the arterial trunks straddles the ventricular septum in which there is a defect. In *partial transposition,* both arterial trunks arise from the same ventricle. In *complete transposition,* the aorta arises from the right ventricle and the pulmonary trunk from the left ventricle. When abnormalities occur, they are often seen in groups or complexes. One of the better known complexes, called the tetralogy of Fallot,[1] has a combination of the following four abnormalities: (1) overriding of the aorta, (2) a defect in the ventricular septum, (3) pulmonary stenosis, and (4) hypertrophied right ventricle. The **Eisenmenger complex**[2] is similar to the tetralogy, having (1) an overriding aorta, (2) septal defect, (3) hypoplasia of the aorta, and (4) right ventricular hypertrophy. These and many other congenital malformations of the heart interfere with the pumping of the proper amount of blood into the pulmonary circulation. In consequence, the blood sent into the systemic arteries is poorly oxygenated, the predominance of venous blood causes cyanosis, and the individual is popularly known as a "blue baby." Detailed descriptions of cardiac anomalies have been published by Gould (1953), Lev (1953), and Gray and Skandalakis (1972).

[1] Etienne Louis Arthur Fallot (1850–1911): A French physician (Marseilles).

[2] Victor Eisenmenger (1864–1932): A German physician.

References

(References are listed not only to those articles and books cited in the text, but to others as well which are considered to contain valuable resource information for the student who desires it.)

Embryology of the Heart, Changes at Birth and Anomalies

Anderson, R. C. 1954. Causative factors underlying congenital heart malformations. Pediatrics, 14:143–152.

Barnard, C. N., and V. Shire. 1968. *Surgery of the Common Congenital Cardiac Malformations.* Staples Press, London, 179 pp.

Bedford, D. E. 1960. The anatomical types of atrial septal defect. Amer. J. Cardiol., 6:568–574.

Blalock, A., and H. B. Taussig. 1945. The surgical treatment of malformations of the heart in which there is pulmonary stenosis or pulmonary atresia. J.A.M.A., 128:189–202.

Campbell, M. 1970. Natural history of atrial septal defect. Brit. Heart J., 32:820–826.

Christie, G. A. 1963. The development of the limbus fossae ovalis in the human heart; a new septum. J. Anat. (Lond.), 97:45–54.

Cobb, W. M. 1944. Apical pericardial adhesion resembling the reptilian gubernaculum cordis. Anat. Rec., 89:87–91.

Cooper, M. H., and R. O'Rahilly. 1971. The human heart at seven postovulatory weeks. Acta Anat., 79:280–299.

Copenhaver, W. M. 1945. Heteroplastic transplantation of the sinus venosus between two species of Amblystoma. J. Exp. Zool., 100:203–216.

Davis, C. L. 1927. Development of the human heart from its first appearance to the stage found in embryos of twenty paired somites. Carneg. Instn., Contr. Embryol., 19:245–284.

Duckworth, J. W. A. 1967. Embryology of congenital heart disease. In *Heart Disease in Infancy and Childhood.* 2nd ed., Edited by J. D. Kieth, R. D. Rowe, and P. Vlad. Macmillan Co., New York, 1967, pp. 136–158.

Everett, N. B., and R. J. Johnson. 1951. A physiological and anatomical study of the closure of the ductus arteriosus in the dog. Anat. Rec., 110:103–111.

Fales, D. E. 1946. A study of double hearts produced experimentally in embryos of Amblystoma punctatum. J. Exp. Zool., 101:281–298.

Fox, M. H., and C. M. Goss. 1958. Experimentally produced malformations of the heart and great vessels in rat fetuses. Transposition complexes and aortic arch abnormalities. Amer. J. Anat., 102:65–92.

Goss, C. M. 1935. Double hearts produced experimentally in rat embryos. J. Exp. Zool., 72:33–49.

Goss, C. M. 1940. First contractions of the heart without cytological differentiation. Anat. Rec., 76:19–27.

Goss, C. M. 1942. The physiology of the embryonic mammalian heart before circulation. Amer. J. Physiol., *137*:146–172.

Goss, C. M. 1952. Development of the median coordinated ventricle from the lateral hearts in rat embryos with three to six somites. Anat. Rec., *112*:761–796.

Gould, S. E. 1953. *Pathology of the Heart.* Charles C Thomas, Springfield, Ill., 1023 p.

Gray, S. W., and J. E. Skandalakis. 1972. *Embryology for Surgeons.* W. B. Saunders Co., Philadelphia, 918 pp.

de Haan, R. L. 1965. Morphogenesis of the vertebrate heart. In *Organogenesis.* Edited by R. L. de Haan and H. Ursprung. Holt, Rhinehart and Winston, New York, pp. 377–420.

Hay, D. A. and F. N. Low. 1972. The fusion of dorsal and ventral endocardial cushions in the embryonic chick heart: a study in fine structure. Amer. J. Anat., *133*:1–23.

Hoffman, J. I. E., and A. M. Randolf. 1965. The natural history of ventricular septal defects in infancy. Amer. J. Cardiol., *16*:634–653.

Keen, E. N. 1955. The postnatal development of the human cardiac ventricles. J. Anat., *89*:484–502.

Kerr, A., Jr. and C. M. Goss. 1956. Retention of embryonic relationship of aortic and pulmonary valve cusps and a suggested nomenclature. Anat. Rec., *125*:777–782.

Kramer, T. C. 1942. The partitioning of the truncus and conus and the formation of the membranous portion of the interventricular septum in the human heart. Amer. J. Anat., *71*:343–370.

Langman, J. 1975. *Medical Embryology.* 3rd ed. Williams and Wilkins, Baltimore, 421 pp.

Lev, M. 1953. *Autopsy Diagnosis of Congenitally Malformed Hearts.* Charles C Thomas, Springfield, Ill., 194 pp.

Licata, R. H. 1954. The human embryonic heart in the ninth week. Amer. J. Anat., *94*:73–126.

Maron, B. J., and G. M. Hutchins. 1972. Truncus arteriosus malformation in a human embryo. Amer. J. Anat., *134*:167–173.

Mayer, F. E., A. S. Nadas, and P. A Ongley. 1957. Ebstein's anomaly: Presentation of ten cases. Circulation, *16*:1057–1069.

Mitchell, S. C., H. W. Berendes, and W. M. Clark, Jr. 1967. The normal closure of the ventricular septum. Amer. Heart J., *73*:334–338.

Mitchell, S. C., S. B. Korones, and H. W. Berendes. 1971. Congenital heart disease in 56,109 births; incidence and natural history. Circulation, *43*:323–332.

Moore, K. L. 1973. *The Developing Human.* W. B. Saunders Co., Philadelphia, 347 pp.

Moss, A. J., and F. H. Adams. 1977. *Heart Disease in Infants, Children and Adolescents.* Williams and Wilkins, Baltimore, 2nd ed., 757 pp.

O'Rahilly, R. 1971. The timing and sequence of events in human cardiogenesis. Acta Anat., *79*:70–75.

Orts-Llorca, F. 1970. Curvature of the heart: its first appearance and determination. Acta Anat., *77*:454–468.

Patten, B. M. 1931. The closure of the foramen ovale. Amer. J. Anat., *48*:19–44.

Patten, B. M. 1968. *Human Embryology.* McGraw-Hill Co., New York, 3rd ed., 651 pp.

Reemtsma, K., and W. M. Copenhaver. 1958. Anatomic studies of the cardiac conduction system in congenital malformations of the heart. Circulation, *17*:271–276.

Shaner, R. F. 1949. Malformation of the atrioventricular endocardial cushions of the embryo pig and its relation to defects of the conus and truncus arteriosus. Amer. J. Anat., *84*:431–455.

Shaner, R. F. 1963. Abnormal pulmonary and aortic semilunar valves in embryos. Anat. Rec., *147*:5–13.

Smith, R. B. 1970. The development of the intrinsic innervation of the human heart between 10 and 70 mm stages. J. Anat. (Lond.), *107*:271–279.

Stalsberg, H., and R. L. de Haan. 1969. The precardiac areas and the formation of the tubular heart in the chick embryo. Devel. Biol., *19*:128–159.

Tandler, J. 1912. The development of the heart. In *Manual of Human Embryology.* Edited by F. Keibel and F. P. Mall. J. B. Lippincott, Philadelphia. *2*:534–570.

Taussig, H. B. 1963. Cardiac abnormalities. In *Birth Defects.* Edited by M. Fishbein. J. B. Lippincott, Philadelphia, pp. 268–276.

Troyer, J. R. 1961. A multiple anomaly of the human heart and great veins. Anat. Rec., *139*:509–513.

Vernall, D. G. 1962. The human embryonic heart in the seventh week. Amer. J. Anat., *111*:17–24.

de Vries, P. A., and J. B. de C. M. Saunders. 1962. Development of the ventricles and spiral outflow tract in the human heart. Carneg. Instn., Contr. Embryol., *37*:87–114.

Yoshihara, H., and H. Maisel. 1973. An acardiac human fetus. Anat. Rec., *177*:209–218.

Histology and Electron Microscopy

Bacon, R. L. 1948. Changes with age in the reticular fibers of the myocardium of the mouse. Amer. J. Anat., *82*:469–496.

Copenhaver, W. M., and R. C. Truex. 1952. Histology of the atrial portion of the cardiac conduction system in man and other mammals. Anat. Rec., *114*:601–625.

Ellison, J. P. and R. G. Hibbs. 1973. The atrioventricular valves of the guinea pig. I. A light microscopic study. Amer. J. Anat., *138*:331–345.

Fawcett, D. W. and N. S. Mc Nutt. 1969. The ultrastructure of the cat myocardium. I. Ventricular papillary muscle. J. Cell Biol., *42*:1–45.

Forssmann, W. G., and L. Girardier. 1970. A study of the T system in rat heart. J. Cell Biol., *44*:1–19.

Goss, C. M. 1931. "Slow-motion" cinematographs of the contraction of single cardiac muscle cells. Proc. Soc. Exp. Biol. (N.Y.), *29*:292–293.

Goss, C. M. 1933. Further observations on the differentiation of cardiac muscle in tissue cultures. Arch. f. exp. Zellforsch., *14*:175–201.

Hibbs, R. G., and J. P. Ellison. 1973. The atrioventricular valves of the guinea pig. II. An ultrastructural study. Amer. J. Anat., *138*:347–369.

Jamieson, J. D., and G. E. Palade. 1964. Specific granules in atrial muscle cells. J. Cell Biol., *23*:151–172.

Leak, L. V., and J. F. Burke. 1964. The ultrastructure of human embryonic myocardium. Anat. Rec., *149*:623–650.

Manasek, F. J. 1970. Histogenesis of the embryonic myocardium. Amer. J. Cardiol., *25*:149–168.

Mc Nutt, N. S., and D. W. Fawcett. 1969. The ultrastructure of the cat myocardium. II. Atrial muscle. J. Cell Biol., *42*:46–67.

Muir, A. R. 1957. An electron microscope study of the embryology of the intercalated disc in the heart of the rabbit. J. Biophys. Biochem. Cytol., *3*:193–202.

Muir, A. R. 1965. Further observations of the cellular structure of the cardiac muscle. J. Anat. (Lond.), 99:27–46.

Nelson, D. A., and E. S. Benson. 1963. On the structural continuities of the transverse tubular system of rabbit and human myocardial cells. J. Cell Biol., 16:297–313.

Peine, C. J., and F. N. Low. 1975. Scanning electron microscopy of cardiac endothelium of the dog. Amer. J. Anat., 142:137–157.

Rostgaard, J., and O. Behnke. 1965. Fine structural localization of adenine nucleoside phosphatase activity in the sarcoplasmic reticulum and T system of the rat myocardium. J. Ultrastruct. Res., 12:579–591.

Sachs, H. G., J. A. Colgan, and M. L. Lazarus. 1977. Ultrastructure of the aging myocardium: a morphometric approach. Amer. J. Anat., 150:63–71.

Sinclair, J. G. 1957. Synchronous mitosis on a cardiac infarct. Tex. Rep. Biol. Med., 15:347–352.

Sjostrand, F. S., E. Andersson-Cedergren, and M. Dewey. 1958. The ultrastructure of the intercalated discs of frog, mouse and guinea pig cardiac muscle. J. Ultrastruct. Res., 1:271–287.

Sommer, J. R., and E. A. Johnson. 1968. Cardiac muscle. A comparative study of Purkinje fibers and ventricular fibers. J. Cell Biol., 36:497–526.

Song, S. H. 1977. Endocardial surface structures of the feline heart observed with a scanning electron microscope. Acta Anat., 99:67–75.

Truex, R. C. 1972. Myocardial cell diameters in primate hearts. Amer. J. Anat., 135:269–280.

Truex, R. C., and W. M. Copenhaver. 1947. Histology of the moderator band in man and other mammals with special reference to the conduction system. Amer. J. Anat., 80:173–202.

Van Winkle, W. B. 1977. The fenestrated collar of mammalian cardiac sarcoplasmic reticulum: a freeze-fracture study. Amer. J. Anat., 149:277–282.

Conduction System of the Heart

Anderson, R. H., and R. A. Latham. 1971. The cellular architecture of the human atrioventricular node, with a note on its morphology in the presence of a left superior vena cava. J. Anat. (Lond.), 109:443–456.

Baerg, R. D., and D. L. Bassett. 1963. Permanent gross demonstration of the conduction tissue in the dog heart with palladium iodide. Anat. Rec., 146:313–317.

Davies, F., and E. T. B. Francis. 1946. The conducting system of the vertebrate heart. Biol. Rev., 21:173–188.

Davies, F., and E. T. B. Francis. 1952. The conduction of the impulse for cardiac contraction. J. Anat. (Lond.), 86:130–143.

Davies, F., E. T. B. Francis, and T. S. King. 1952. Neurological studies of the cardiac ventricles of mammals. J. Anat. (Lond.), 86:139–143.

Erickson, E. E., and M. Lev. 1952. Aging changes in the human atrioventricular node, bundle, and bundle branches. J. Geront., 7:1–12.

Field, E. J. 1951. The development of the conducting system in the heart of the sheep. Brit. Heart J., 13:129–147.

James, T. N. 1961. Anatomy of the human sinus node. Anat. Rec., 141:109–139.

James, T. N. 1961. Morphology of the human atrioventricular node, with remarks pertinent to its electrophysiology. Amer. Heart J., 62:756–771.

Kugler, J. H., and J. B. Parkin. 1956. Continuity of Purkinje fibres with cardiac muscle. Anat. Rec., 126:335–341.

Muir, A. R. 1954. The development of the ventricular part of the conducting tissue in the heart of the sheep. J. Anat. (Lond.), 88:381–391.

Muir, A. R. 1957. Observations on the fine structure of the Purkinje fibers in the ventricles of the sheep's heart. J. Anat. (Lond.), 91:251–258.

Rhodin, J. A., P. del Missier, and L. C. Reid. 1961. The structure of the specialized impulse-conducting system of the steer heart. Circulation, 24:348–367.

Rybicka, K. 1977. Sarcoplasmic reticulum in the conducting fibers of the dog heart. Anat. Rec., 189:237–262.

Stotler, W. A., and R. A. Mc Mahon. 1947. The innervation and structure of the conductive system of the human heart. J. Comp. Neurol., 87:57–84.

Thaemert, J. C. 1970. Atrioventricular node innervation in ultrastructural three dimensions. Amer. J. Anat., 128:239–263.

Thaemert, J. C. 1973. Fine structure of the atrioventricular node as viewed in serial sections. Amer. J. Anat., 136:43–65.

Titus, J. L., G. W. Daugherty, and J. Edwards. 1963. Anatomy of the normal human atrioventricular conduction system. Amer. J. Anat., 113:407–415.

Truex, R. C., and M. Q. Smythe. 1967. Reconstruction of the human atrioventricular node. Anat. Rec., 158:11–20.

Truex, R. C., M. Q. Smythe, and M. J. Taylor. 1967. Reconstruction of the human sinoatrial node. Anat. Rec., 159:371–378.

Walls, E. W. 1945. Dissection of the atrioventricular node and bundle in the human heart. J. Anat. (Lond.), 79:45–47.

Walls, E. W. 1947. The development of the specialized conducting tissue of the human heart. J. Anat. (Lond.), 81:93–110.

Wenink, A. C. G. 1976. Development of the human cardiac conducting system. J. Anat. (Lond.), 121:617–632.

White, P. D. 1957. The evolution of our knowledge about the heart and its diseases since 1628. Circulation, 15:915–923.

Widran, J., and M. Lev. 1951. The dissection of the atrioventricular node, bundle and bundle branches in the human heart. Circulation, 4:863–867.

Gross Anatomy, Blood and Nerve Supply

Armour, J. A. and W. C. Randall. 1975. Functional anatomy of canine cardiac nerves. Acta Anat., 91:510–528.

Blalock, A. 1947. The technique of creation of artificial ductus arteriosus in the treatment of pulmonic stenosis. J. Thorac. Surg., 16:244–257.

Brooks, C. Mc C. and H. Lu. 1972. The Sinoatrial Pacemaker of the Heart. Charles C Thomas, Springfield, Ill., 179 pp.

Ellison, J. P. 1974. The adrenergic cardiac nerves of the cat. Amer. J. Anat., 139:209–225.

Ellison, J. P., and T. H. Williams. 1969. Sympathetic nerve pathways to the human heart, and their variations. Amer. J. Anat., *124*:149–162.

Gardner, E., and R. O'Rahilly. 1976. The nerve supply and conducting system of the human heart at the end of the embryonic period proper. J. Anat. (Lond.), *121*:571–588.

Gray, H., and E. Mahan. 1943. Prediction of heart weight in man. Amer. J. Phys. Anthrop., *1*:271–287.

Gross, L. 1921. *The Blood Supply to the Heart*. P. B. Hoeber, New York, 171 pp.

Gross, L., and M. A. Kugel. 1933. The arterial blood vascular distribution to the left and right ventricles of the human heart. Amer. Heart J., *9*:165–177.

Gross, R. E. 1947. Complete division for the patent ductus arteriosus. J. Thorac. Surg., *16*:314–327.

Guyton, A. C. 1976. Heart muscle; the heart as a pump. Chapter 13 in *Textbook of Medical Physiology*. 5th ed. W. B. Saunders, Philadelphia, pp. 160–175.

Hoar, R. M., and J. L. Hall. 1970. The early pattern of cardiac innervation in the fetal guinea pig. Amer. J. Anat., *128*:499–507.

James, T. N. 1960. The arteries of the free ventricular walls in man. Anat. Rec., *136*:371–384.

James, T. N., and G. E. Burch. 1958. The atrial coronary arteries in man. Circulation, *17*:90–98.

James, T. N., and G. E. Burch. 1958. Blood supply of the human interventricular septum. Circulation, *17*:391–396.

James, T. N., and G. E. Burch. 1958. Topography of the human coronary arteries in relation to cardiac surgery. J. Thorac. Surg., *36*:656–664.

Jönsson, L., G. Johansson, N. Lannek, and P. Lindberg. 1974. Intramural blood supply of porcine heart. A postmortem angiographic study. Anat. Rec., *178*:647–656.

Kirk, G. R., D. M. Smith, D. P. Hutcheson, and R. Kirby. 1975. Postnatal growth of the dog heart. J. Anat. (Lond.), *119*:461–470.

Langer, G. A., and A. J. Brady. 1974. *The Mammalian Myocardium*. John Wiley, New York, 310 pp.

Latimer, H. B. 1953. The weight and thickness of the ventricular walls in the human heart. Anat. Rec., *117*:713–724.

Leak, L. V., A. Schannahan, A. Scully, and W. M. Daggett. 1978. Lymphatic vessels of the mammalian heart. Anat. Rec., *191*:183–202.

Mikhail, Y. 1970. Intrinsic nerve supply of the ventricles of the heart. Acta Anat., *76*:289–298.

Mikhail, Y., and I. Kamel. 1972. Nerve endings in the atrial muscle of the heart. Acta Anat., *82*:138–144.

Mitchell, G. A. G. 1956. *Cardiovascular Innervation*. Livingstone, Edinburgh, 356 pp.

Papez, J. W. 1920. Heart musculature of the atria. Amer. J. Anat., *27*:255–286.

Randall, W. C., J. A. Armour, D. C. Randall, and A. A. Smith. 1971. Functional anatomy of the cardiac nerves in the baboon. Anat. Rec., *170*:183–198.

Rauber-Kopsch, Fr. 1955. *Lehrbuch und Atlas der Anatomie des Menschen*. 19th ed. George Thieme Verlag, Stuttgart, 2 vols.

Smith, R. B. 1971. Innervation of the parietal pericardium in the human infant, various mammals and birds. J. Anat. (Lond.), *108*:109–114.

Smith, R. B. 1971. Intrinsic innervation of the atrioventricular and semilunar valves in various mammals. J. Anat. (Lond.), *108*:115–122.

Smith, R. B. 1971. The occurrence and location of intrinsic cardiac ganglia and nerve plexuses in the human neonate. Anat. Rec., *169*:33–40.

Tandler, J. 1913. Anatomie des Herzens. In *von Bardeleben's Handbuch der Anatomie des Menschen*. Gustav Fischer, Jena, 292 pp.

Thomas, C. E. 1957. The muscular architecture of the ventricles of hog and dog hearts. Amer. J. Anat., *101*:17–58.

Tomanek, R. J. 1970. Effects of age and exercise on the extent of the myocardial capillary bed. Anat. Rec., *167*:55–62.

Truex, R. C., and A. W. Angulo. 1952. Comparative study of the arterial and venous systems of the ventricular myocardium with special reference to the coronary sinus. Anat. Rec., *113*:467–491.

Walmsley, R. 1958. The orientation of the heart and the appearance of its chambers in the adult cadaver. Brit. Heart J., *20*:441–458.

Woolard, H. H. 1926. The innervation of the heart. J. Anat. (Lond.), *60*:345–373.

The Arteries

After reaching the ventricles of the heart, blood is then propelled to all parts of the body through the arteries. Two large arteries leave the heart: the **pulmonary trunk** and the **aorta.** The pulmonary trunk divides into a **right** and **left pulmonary artery,** each of which courses transversely in the thorax to its corresponding lung. These vessels subdivide profusely within the lung tissue to form a capillary bed, which exchanges carbon dioxide in the blood for oxygen in the alveoli. The pulmonary arteries, along with the pulmonary veins which return oxygenated blood to the left atrium, constitute the **pulmonary circulation.** The aorta and its many branches also ramify into capillary beds within all the body organs. These arteries and the veins returning blood to the heart form the **systemic circulation.** The continuous ramification of the systemic arteries results in the distribution of capillaries to all the tissues except the nails, epidermis, cornea, mucous membranes, and most cartilage. These latter avascular structures receive their nourishment through the processes of diffusion.

Arteries may show various patterns in their mode of branching. A short trunk may subdivide into several branches at the same point, such as occurs to the celiac and thyrocervical trunks. Several branches may be given off in succession and the main trunk continue, as in the arteries of the limbs. The division may be dichotomous as at the bifurcation of the abdominal aorta into the common iliac arteries. Although the branches of an artery are smaller than their trunk, the combined cross-sectional area of the two resulting arteries is generally greater than that of the parent trunk before the division. The combined cross-sectional area of all the arterial branches, therefore, greatly exceeds that of the aorta. It has been estimated that the total cross-sectional area of all the capillaries in the body is nearly 1,000 times that of the aorta, from which they derive their arterial blood.

Throughout the body generally the larger arterial branches pursue a fairly straight course, but at certain sites they are tortuous. The facial artery and the arteries supplying the lips, for example, are extremely tortuous, being accommodated to the movements of the parts. The uterine arteries also are coiled and tortuous, allowing them to accommodate the increase in size that the uterus undergoes during pregnancy. The larger arteries usually are deeply placed and occupy other more protected locations such as along the flexor surface of the limbs where they are less exposed to injury.

ANASTOMOSIS OF ARTERIES. Arterial branches do not always terminate in capillaries. In many parts of the body they open into branches of other arteries of similar size, forming what are called **anastomoses.** Anastomosis may take place between larger arteries forming vascular arches such as those in the palm of the hand or the arcades of the intestines. At the base of the brain, the two vertebral arteries join to form the basilar artery, while branches from the two internal carotid arteries and the basilar artery freely join to form the vital cerebral arterial circle of Willis.[1] More frequently, however, anastomoses form between smaller arteries of one millimeter or less in diameter. In the limbs, anastomoses are quite numerous around the joints, the branches of an artery above the joint uniting with branches of the vessels below. Definite patterns of anastomosis between neighboring arteries occur throughout the body, but in any single body all the different anastomoses that are de-

[1]Thomas Willis (1621–1675): An English physician (London).

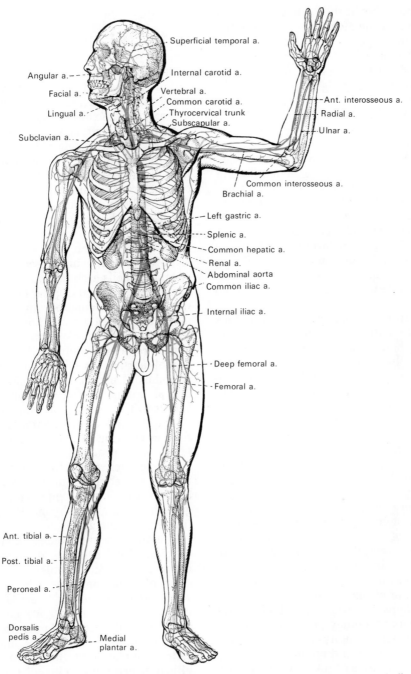

FIG. 8-1. A schematic representation of the arterial system in the adult male. (From Benninghoff and Goerttler, *Lehrbuch der Anatomie des Menschen*, 10th Ed., Vol. 2, Urban & Schwarzenberg, 1975.)

scribed will not be equally well developed nor will they be easily demonstrable at dissection.

COLLATERAL CIRCULATION.   When the supply of blood to the distal part of an artery that has been occluded is effected through anastomosis of its branches, the resulting vascular paths constitute a **collateral circulation.** It is frequently necessary for a surgeon to tie off or ligate an artery in order to prevent excessive bleeding. The anastomoses with branches of adjacent arteries may be more numerous in one part of the artery than in another, and the surgeon will endeavor to place his ligature where advantage can be taken of the most numerous or effective anastomoses. Collateral circulation may occur through anastomoses with branches of the proximal part of the same artery or with branches of a large neighboring artery. In some regions, the collateral circulation is free, as it would be if either the radial or ulnar artery of the forearm were ligated. At other sites the collateral circulation might be carried by small arteries only. If the vessels in a part of the body have sustained trauma, and the region can be placed at rest, a small flow of blood through small anastomosing channels will keep the tissues alive until the anastomoses expand. The arteries retain their power of growth throughout life, and over time a small anastomosing artery may enlarge into a main trunk, completely restoring the circulation to the initially injured region.

Arteries in certain parts of the body lack anastomoses and are therefore called **end arteries.** When vessels such as these are blocked by thrombosis or embolism, the tissues of the region are left without blood supply and become necrotic, resulting in a condition called an **infarct.** A prime example of an end artery is the central artery of the retina, occlusion of which results in permanent blindness. The collateral circulation of vessels in other organs may exist only at the capillary level, and if a sudden blockage of a main artery occurs, degeneration of tissue results before revascularization is effective. Notwithstanding the free anastomosing of vessels on the surface of the brain, ineffective collateral circulation exists in the substance of the brain. Blockage of a vessel within the brain, therefore, causes the clinical features of **stroke.** Similar microcirculatory conditions are found in other organs such as the lungs, kidneys, and spleen and at the transitional metaphyseal interface between the diaphysis and epiphysis of a long bone. The capillaries of the heart do anastomose, but this condition becomes more pronounced as individuals age. These anastomoses may not be capable of maintaining life if a major occlusion occurs in the heart of a young person, whereas a comparable occlusion in a much older person may be survived because of the years that the heart has had to develop anastomoses and collateral circulation.

Development of the Arteries

Blood vessels commence development within the mesoderm of the yolk sac and body stalk by the formation of **blood islands** as early as the third week (about 16 days). Shortly thereafter, spaces appear within the islands, and the angioblasts become arranged in cords and form the endothelial linings of the earliest vessels. These vascular channels then continue to enlarge by endothelial outgrowth and fuse with other nearby developing channels.

The primordium of the aorta appears toward the end of the third week at about the same time as the heart (see Fig. 7-3). Two strands of cells arch dorsally from the endocardial mesenchyme, pass on each side of the foregut invagination, and turn caudally along the neural groove. The strands beside the foregut are the primordia of the first aortic arches and their continuations are the **dorsal aortae.** The latter acquire isolated stretches of lumen at the same time as the lateral hearts, i.e., in embryos with three somites. The arches become patent some time later, at seven or eight somites, just before circulation begins. The umbilical arteries also probably arise independently from mesenchyme. After these main channels have become connected and the circulation is established, however, no further growth appears to be by local differentiation from the mesenchyme.

The circulation becomes functional when the first blood cells are washed out of the blood islands of the yolk sac at about the stage of nine somites, early in the fourth week. At this time the heart is contracting vigorously, capillaries have connected the aortae with the blood islands which, in turn, drain into the vitelline veins, and the blood cells have elaborated hemoglobin. The umbilical arteries and veins also have become connected, and oxygenation of the blood takes place in the primitive placenta. From this time forward, the circulatory system develops by budding from existing trunks, the formation of new capillary networks, and the selection of parts of the network as arteries and veins. As the bulk of an embryonic part increases in size, the capillaries are lengthened. Soon the hemodynamic forces result in the selection of certain channels as arterioles and arteries and others as venules and veins. Later the pattern of the arteries and veins changes continually according to the growth requirements of the embryo. Not only do some vessels grow larger, but well-established vessels may be superseded by others more favorably placed and, in consequence, either regress or disappear. As the endocardial tubes fuse, beyond the pericardial cavity the ventral parts of the two aortae join to form the **aortic sac.**

AORTIC ARCHES.   The first **aortic arch** functions until the neural tube is closed and the pharynx begins to differentiate into pouches (Figs. 8-2; 8-3). As the second pharyngeal pouch forms, a sprout from the aortic sac joins the dorsal aorta passing between the first and second pouches. The first aortic arch diminishes as this **second arch** develops (Figs. 8-4; 8-5). The **third, fourth,** and **sixth arches** are formed in a similar manner and disappear or become modified into various adult structures. The **fifth arch** is never more than a questionable rudiment in the human embryo.

By the time the embryo is 4 mm in length (toward the latter part of the fourth week), the first arch has about disappeared; the second arch has formed, reached its maximum development, and diminished in size; and the third arch is well developed (Figs. 8-4; 8-5). Sprouts may be present for the fourth and pulmonary (sixth) arches. In a 5-mm embryo (beginning of the fifth week), the third and fourth arches have reached a maximum and the dorsal and ventral sprouts of the pulmonary or sixth arches are nearly joined (Figs. 8-6; 8-7). The pulmonary arches are usually complete in a 6-mm embryo (middle of the fifth week). The right one soon begins to regress and by the end of the fifth week has disappeared in 12- to 13-mm embryos. The third arches also have become incomplete by this time (Figs. 8-8; 8-9; 8-10).

Although the aortic arches do not persist as such, remnants of them remain as important parts of the arterial system (Fig. 8-11). The **first arch,** even at the earliest stages, has an extension into the region of the forebrain, the primordium of the **internal carotid artery** (Figs. 8-5; 8-7). *The first arch disappears and*

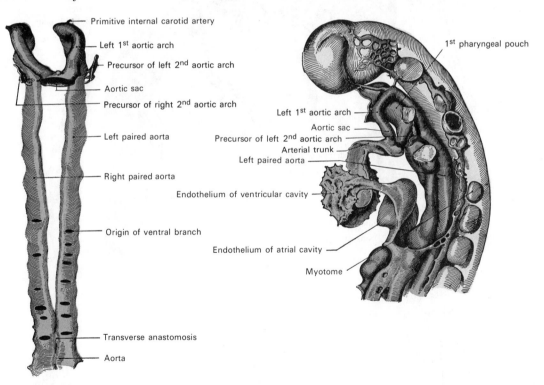

Primitive internal carotid artery

Left 1st aortic arch

Precursor of left 2nd aortic arch

Aortic sac

Precursor of right 2nd aortic arch

Left paired aorta

Right paired aorta

Origin of ventral branch

Transverse anastomosis

Aorta

1st pharyngeal pouch

Left 1st aortic arch

Aortic sac

Precursor of left 2nd aortic arch

Arterial trunk

Left paired aorta

Endothelium of ventricular cavity

Endothelium of atrial cavity

Myotome

FIGS. 8-2 and 8-3.   Ventral and lateral views of the cranial portion of the arterial system at about the end of the third week in a human embryo of 3 mm. The first aortic arch is at its maximum development and the dorsal and ventral outgrowths, which are to aid in the formation of the second arch, are just appearing. (From Congdon, 1922.)

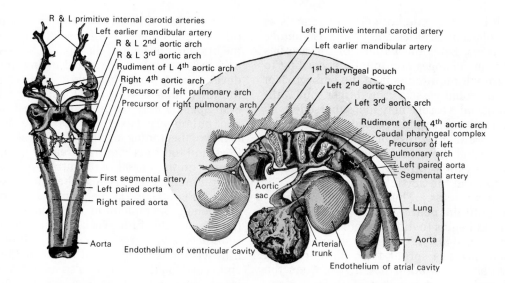

R & L primitive internal carotid arteries
Left earlier mandibular artery
R & L 2nd aortic arch
R & L 3rd aortic arch
Rudiment of L 4th aortic arch
Right 4th aortic arch
Precursor of left pulmonary arch
Precursor of right pulmonary arch

First segmental artery
Left paired aorta
Right paired aorta

Aorta

Endothelium of ventricular cavity

Left primitive internal carotid artery
Left earlier mandibular artery
1st pharyngeal pouch
Left 2nd aortic arch
Left 3rd aortic arch
Rudiment of left 4th aortic arch
Caudal pharyngeal complex
Precursor of left pulmonary arch
Left paired aorta
Segmental artery

Aortic sac

Lung

Arterial trunk

Aorta

Endothelium of atrial cavity

Figs. 8-4 and 8-5. Ventral and lateral views of an embryo, 4 mm in length, near the end of the fourth week. The first arch has receded, the second is much reduced, and the third well developed. Dorsal and ventral outgrowths for the fourth and probably the pulmonary arch (sixth) are present. (From Congdon, 1922.)

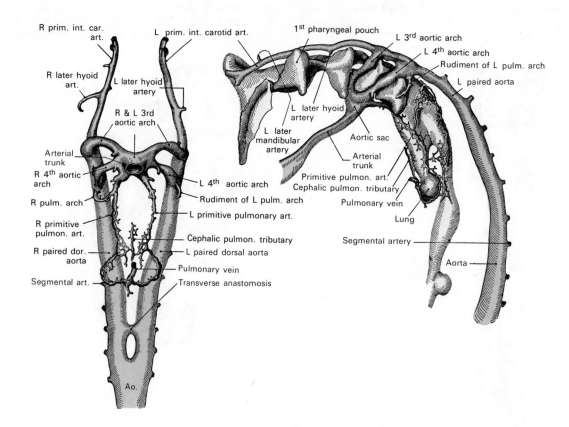

R prim. int. car. art.
R later hyoid art.
R & L 3rd aortic arch
Arterial trunk
R 4th aortic arch
R pulm. arch
R primitive pulmon. art.
R paired dor. aorta
Segmental art.

L prim. int. carotid art.
L later hyoid artery
L 4th aortic arch
Rudiment of L pulm. arch
L primitive pulmonary art.
Cephalic pulmon. tributary
L paired dorsal aorta
Pulmonary vein
Transverse anastomosis

Ao.

1st pharyngeal pouch
L 3rd aortic arch
L 4th aortic arch
Rudiment of L pulm. arch
L paired aorta
L later hyoid artery
L later mandibular artery
Aortic sac
Arterial trunk
Primitive pulmon. art.
Cephalic pulmon. tributary
Pulmonary vein
Lung
Segmental artery
Aorta

Figs. 8-6 and 8-7. Ventral and lateral views of a 5-mm embryo during the early part of the fifth week. The third and fourth arches are in a condition of maximum development. The dorsal and ventral sprouts for the pulmonary arches have nearly met, and the primitive pulmonary arches are already of considerable length. (From Congdon, 1922.)

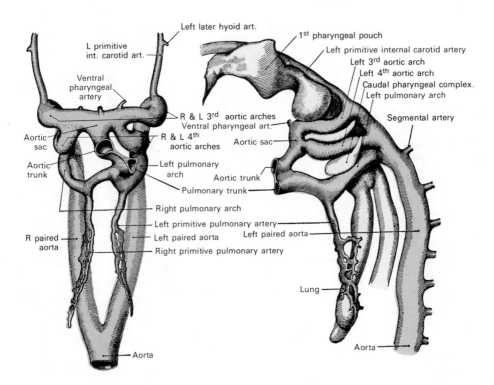

FIGS. 8-8 and 8-9. Ventral and lateral views of an 11-mm embryo estimated to be between 33 and 35 days old (about 5 weeks). The pulmonary arches are complete and the right is already regressing. The third arch is bent cranially at its dorsal end. (From Congdon, 1922.)

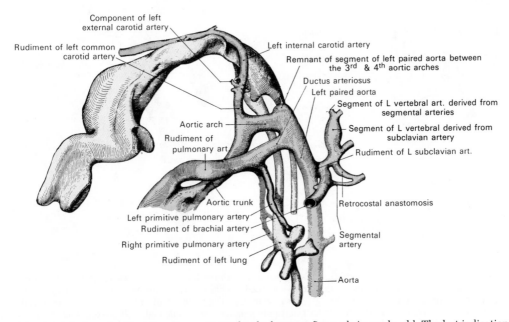

FIG. 8-10. Lateral view of a 14-mm embryo estimated to be between five and six weeks old. The last indications of the aortic arch system are just disappearing. (From Congdon, 1922.)

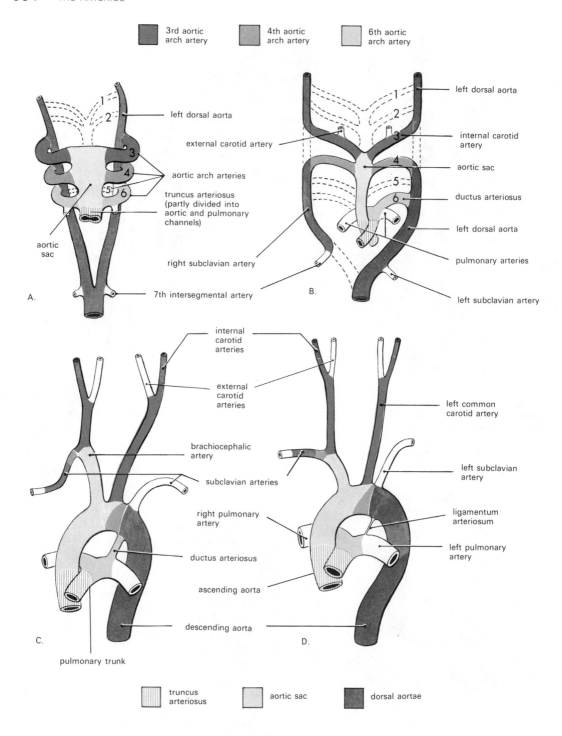

FIG. 8-11. Schematic drawings illustrating the changing nature of the truncus arteriosus, aortic sac, aortic arch arteries, and the dorsal aorta. The vessels that are not shaded or colored are not derived from these structures. *A.* The first two pairs of aortic arch arteries have largely disappeared. *B.* The carotid and subclavian vessels are forming. The parts of the dorsal aortae and aortic arches that normally disappear are indicated with broken lines. *C.* The arterial arrangement after the second prenatal month. *D.* Sketch of the arterial vessels of a 6-month-old infant. (From Moore, K. L., *The Developing Human,* W. B. Saunders Co., 1973.)

the primitive carotid becomes a cranial continuation of the dorsal aorta. The **second arch** (or hyoid arch artery) also disappears early, but its dorsal end gives rise to the **stapedial artery,** which atrophies in man but persists in some mammals. Passing through the ring of the stapes, the stapedial artery divides into supraorbital, infraorbital, and mandibular branches, which follow the three divisions of the trigeminal nerve. The infraorbital and mandibular arise from a common stem, the terminal part of which anastomoses with the external carotid. On the obliteration of the **stapedial artery,** this anastomosis enlarges and forms the **maxillary artery.** The common stem of the infraorbital and mandibular branches passes between the two roots of the auriculotemporal nerve and becomes the **middle meningeal artery.** The original supraorbital branch of the stapedial artery remains as the small orbital branches of the middle meningeal. The **third** (or carotid) **arch** forms the **common carotid artery** and the first part of the **internal carotid artery.** The proximal segment of this arch is connected with the aortic sac and persists as the common carotid. The remainder of the internal carotid forms from the cranial extension of the dorsal aorta (Fig. 8-11, *A* and *B*). The manner of formation of the external carotid artery is not well understood, but portions of it may be contributed by each of the first three arches. The arch itself persists, keeping its connection with the cranial extension of the dorsal aorta, but losing its connection caudally toward the fourth arch (Fig. 8-11, *B*). The **left fourth arch** persists and provides the basis for the adult **arch of the aorta,** which extends between the aortic sac and the ductus arteriosus. The **right fourth arch** persists as the proximal part of the **right subclavian artery.** Both **fifth arch** vessels are at best rudimentary and disappear early. The **sixth arches** are frequently referred to as the **pulmonary arches.** The proximal or ventral part of the **right sixth arch** persists as the **right pulmonary artery,** while the distal or dorsal part of it degenerates and disappears. The proximal part of the **left sixth arch** becomes absorbed into the **pulmonary trunk,** forming the region of its bifurcation. The distal segment of the left sixth arch persists as the **ductus arteriosus** until after birth, at which time it contracts and gradually be-

comes fibrosed into the **ligamentum arteriosum** (Fig. 8-11, *C* and *D*).

DORSAL AORTAE.   The two dorsal aortae remain separate for a short time, but toward the end of the third week, at about the 3-mm stage, they fuse to form a single trunk caudal to the eighth or ninth somites. The segment of the aorta between the third and fourth arches disappears on both sides. The **left dorsal aorta** becomes the **descending aorta,** continuing caudally from the left fourth arch. The **right dorsal aorta** retains its continuity with the left fourth arch, and the part that extends as far as the seventh intersegmental artery becomes the right subclavian artery (Fig. 8-11, *B* to *D*). The part of the right dorsal aorta between the seventh intersegmental artery and its original site of junction with the left dorsal aorta disappears (Fig. 8-11, *B*). A slight constriction called the **aortic isthmus** is frequently seen in that part of the aorta between the origin of the left subclavian and the attachment of the ductus arteriosus. A more serious congenital narrowing of the aorta at this site results in the malformation called *coarctation of the aorta.*

Changes in the location of the heart during development produce certain changes in the aortic arches. The heart originally lies ventral to the most cranial part of the pharynx. It later recedes into the thorax, drawing the aortic arches with it. On the right side the fourth arch recedes to the root of the neck; on the left it is withdrawn into the thorax. The recurrent laryngeal branches of the vagus nerve originally pass caudal to the sixth or pulmonary arches. When the heart descends into the thorax, the nerves are pulled down by these arches. On the right side, however, the sixth arch disappears, allowing the nerve to slip up to the next arch (i.e., the fourth, because the fifth arch also disappears), and it thus loops around the adult right subclavian artery. On the left side the sixth arch becomes the ductus arteriosus, which in the adult is represented by the ligamentum arteriosum. This explains why the left recurrent laryngeal nerve loops around the ligamentum arteriosum at the part of the aorta to which the ligamentum is attached.

SUBCLAVIAN AND VERTEBRAL ARTERIES.   Segmental arteries arise from the primitive dorsal aortae and anastomose between successive segments (Figs. 8-7; 8-9). The **seventh segmental artery** is of special interest because it forms the lower end of the **vertebral artery** and, when the forelimb bud appears, sends a branch to it which becomes the **subclavian artery.** From the paired seventh segmental arteries the entire left subclavian and the greater part of the right subclavian are formed. The **second pair of segmental arteries** accompanies the hypoglossal nerves to the brain and is named the **hypoglossal arteries.** Each sends forward a branch which forms the cerebral part of the **vertebral artery** and anastomoses with the posterior branch of the internal carotid. The two vertebrals unite on the ventral surface of the hindbrain to form the **basilar artery.** Later the hypoglossal artery atrophies and the ver-

tebral is connected with the first segmental artery. The cervical part of the vertebral is developed from a longitudinal anastomosis between the first seven segmental arteries, so that the seventh of these ultimately becomes the source of the artery. As a result of the growth of the upper limb, the subclavian artery increases greatly in size, and the vertebral then appears to spring from it.

Several segmental arteries contribute branches to the upper limb bud and form in it a free capillary anastomosis. Of the segmental branches, only one, that derived from the **seventh segmental artery,** persists to form the **subclavian artery.** The subclavian artery is prolonged into the upper limb during its development and becomes its arterial stem. Although this **axis artery** (Fig. 8-12)

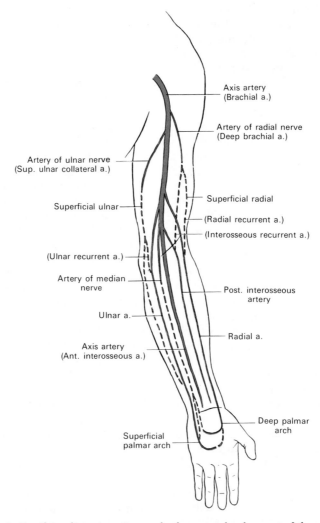

FIG. 8-12.   Diagram illustrating the embryonic pattern and subsequent development of the arteries to the upper limb. Names in parentheses refer to the adult structures.

is a continuous vessel, its parts are named topographically subclavian, axillary, and brachial as far as the forearm. The direct continuation of the axis artery in the forearm is the anterior interosseous artery. A branch that accompanies the median nerve, called the **median artery,** soon increases in size and forms the main vessel of the forearm, while the anterior interosseous diminishes. Later the radial and ulnar arteries are developed as branches of the brachial part of the stem and, coincidentally with their enlargement, the median artery recedes. Occasionally the median artery persists as a vessel of some considerable size and accompanies the median nerve into the palm of the hand.

DESCENDING AORTA. The **segmental arteries** caudal to the seventh develop around the body wall, retaining their segmental character and becoming the **intercostal** and **lumbar arteries.** In the early embryo, paired ventral branches of the aorta grow out on the yolk sac as the **vitelline arteries.** At its caudal end, the ventral branches accompanying the allantois become the **umbilical arteries.** As the gut develops, a number of ventral visceral branches grow into it. At first these are paired, but with the formation of a mesentery the pairs fuse into single stems. These ventral branches are irregularly spaced, and by a process of shifting along a rich anastomosis, the three main trunks, the **celiac,** the **superior,** and the **inferior mesenteric,** are finally selected. Lateral branches of the aorta grow into the mesonephros. The long course of the **testicular** and **ovarian arteries** is the result of the caudal migration of the gonads after their arteries had become established. Later similar lateral branches grow into the suprarenal glands and the kidneys. In this manner the **suprarenal** and **renal arteries** develop.

The primary arterial trunk or **axis artery** of the embryonic lower limb *arises from the dorsal root of the umbilical artery and courses along the posterior surface of the thigh, knee and leg* (Fig. 8-13). The **femoral artery** springs from the external iliac and forms a new channel along the anterior aspect of the thigh to its communication with the axis artery above the knee. As this channel increases in size, that part of the axis artery (called the **ischiadic artery**) proximal to the communication disappears, except its upper end, which persists as the inferior glu-

teal artery. Two other segments of the axial artery persist: one forms the proximal part of the popliteal artery, and the other forms a part of the peroneal artery (Fig. 8-13).

Pulmonary Arterial Vessels

The **pulmonary arterial vessels** convey deoxygenated blood, i.e., venous blood, from the right ventricle of the heart to the lungs. Consistent with the relatively thin wall of the right ventricle in comparison to that of the left ventricle, the pulmonary arterial vessels also have walls only one-third the thickness of vessels of comparable size in the systemic arterial circulation. This is explained by the differences in blood pressure within the pulmonary and systemic arterial vessels. In a young adult, the systolic pressure in the systemic arteries is about 120 mm Hg, whereas in the pulmonary arterial vessels the systolic pressure averages between 20 and 25 mm Hg.

The pulmonary arterial vessels decrease in diameter as they divide. The large pulmonary trunk, which originates in the right ventricle, divides into two pulmonary arteries, one directed to each lung. These vessels divide and then subdivide in association with the bronchi, eventually to send arteries to all of the bronchopulmonary segments that form the different lobes of each lung.

PULMONARY TRUNK

The pulmonary trunk (see Figs. 7-24; 7-25; 8-14) arises from the conus arteriosus of the right ventricle at the pulmonary orifice. Being short (5 cm in length) and 3 cm wide in diameter, and ascending obliquely backward, it passes at first in front of and then to the left of the ascending aorta. In the rounded concavity of the aortic arch at about the level of the fibrocartilage between the fifth and sixth thoracic vertebrae, the pulmonary trunk divides into right and left pulmonary arteries which are of nearly equal size.

RELATIONS. The entire pulmonary trunk is contained within the pericardium. It is enclosed with the ascending aorta in a single tube of the visceral layer of the serous pericardium, which is continued up-

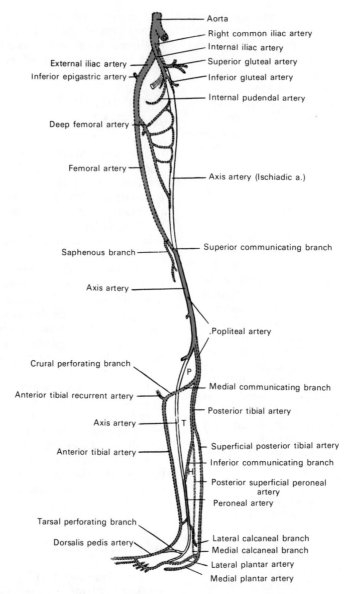

Aorta
Right common iliac artery
Internal iliac artery
Superior gluteal artery
Inferior gluteal artery
Internal pudendal artery
External iliac artery
Inferior epigastric artery
Deep femoral artery
Femoral artery
Axis artery (Ischiadic a.)
Saphenous branch
Superior communicating branch
Axis artery
Popliteal artery
Crural perforating branch
Medial communicating branch
Anterior tibial recurrent artery
Posterior tibial artery
Axis artery
Anterior tibial artery
Superficial posterior tibial artery
Inferior communicating branch
Posterior superficial peroneal artery
Peroneal artery
Tarsal perforating branch
Dorsalis pedis artery
Lateral calcaneal branch
Medial calcaneal branch
Lateral plantar artery
Medial plantar artery

FIG. 8-13. A diagram illustrating the general development of the arteries of the lower limb. The letter *P* indicates the position of the popliteus; *T*, that of the tibialis posterior; *H*, that of the flexor hallucis longus. (From Senior, 1919.)

ward upon them from the base of the heart. The fibrous layer of the pericardium is gradually lost upon the external coats of the two pulmonary arteries. *Anteriorly,* the pulmonary trunk is separated from the sternal end of the second left intercostal space by the pleura and left lung, in addition to the pericardium. Initially, the pulmonary artery lies ventral to the ascending aorta, and a bit more superiorly, it lies anterior to the left atrium, at a plane behind the ascending aorta. On *each side* of its origin is the auricle of the corresponding atrium and a coronary artery. In the first part of its course, the left coronary artery passes behind the pulmonary trunk.

The superficial part of the cardiac plexus lies above its bifurcation, between it and the arch of the aorta.

RIGHT PULMONARY ARTERY. The right pulmonary artery (Fig. 8-14) is longer and slightly larger than the left. Originating at the bifurcation of the pulmonary trunk, it curves around the ascending aorta and proceeds horizontally to the right. In its course toward the lung, the right pulmonary artery lies behind the aorta and superior vena cava and in front of the right bronchus. At the

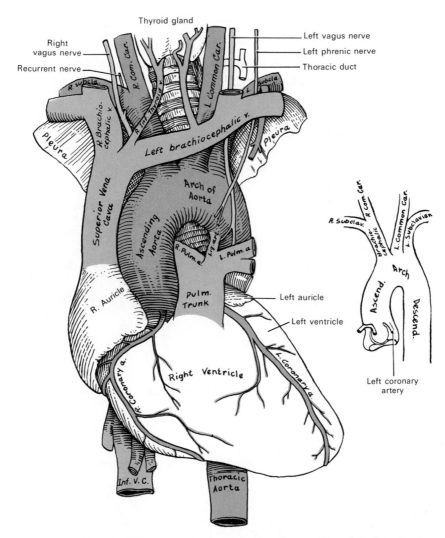

Thyroid gland

Right vagus nerve

Recurrent nerve

Left vagus nerve

Left phrenic nerve

Thoracic duct

R. subcla.

R. Com. Car.

R. Inf. thyroid v.

L. Common Car.

L. subcla.

R. Brachiocephalic v.

Pleura

Left brachiocephalic v.

Pleura

Superior Vena Cava

Ascending Aorta

Arch of Aorta

Lig. art.

R. Pulm. a.

L. Pulm. a.

R. Auricle

Pulm. Trunk

Left auricle

Left ventricle

R. Coronary a.

Right Ventricle

L. coronary a.

Inf. V.C.

Thoracic Aorta

R. Subclav.

Brachiocephalic

R. Com. Car.

L. Common Car.

L. Subclavian

Ascend.

Arch

Descend.

Left coronary artery

FIG. 8-14. The arch of the aorta, and its branches. (Insert: Plan of the branches.)

root of the lung it divides into one large branch, the **anterior trunk,** which is the principal but not the only artery to the superior lobe of the right lung, and a somewhat larger **interlobar branch** (Fig. 8-15). The principal vessel to the superior lobe of the right lung crosses in front of the superior lobe bronchus, which in consequence is frequently called the **eparterial bronchus.** The interlobar trunk, after sending one or more other branches to the superior lobe, furnishes the middle and inferior lobe arteries.

The **anterior trunk** divides into *apical, posterior,* and *anterior segmental arteries,* which supply the respective bronchopulmonary segments of the upper lobe. Both the

anterior and posterior segmental arteries split into ascending and descending branches (Fig. 8-15).

The **interlobar trunk** lies near the bottom of the horizontal fissure that separates the superior and middle lobes. It represents the continuation of the pulmonary artery as far as the origin of the middle lobe artery, and it supplies one or two important ascending branches to the posterior segment of the superior lobe. These latter vessels arise from the interlobar trunk more deeply in the horizontal fissure and beyond the site where the anterior trunk branches to ascend to the upper lobe.

The **middle lobe artery** is shown (Fig. 8-15)

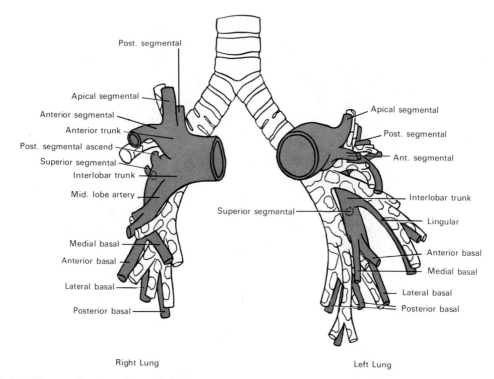

Post. segmental

Apical segmental

Anterior segmental

Anterior trunk

Post. segmental ascend

Superior segmental

Interlobar trunk

Mid. lobe artery

Medial basal

Anterior basal

Lateral basal

Posterior basal

Apical segmental

Post. segmental

Ant. segmental

Interlobar trunk

Lingular

Anterior basal

Medial basal

Lateral basal

Posterior basal

Superior segmental

Right Lung

Left Lung

FIG. 8-15.   Diagram showing relation of the branches of the pulmonary artery to the bronchi. (Redrawn after Boyden, *Segmental Anatomy of the Lungs,* 1955.)

as a single artery dividing into a *medial* and a *lateral segmental artery,* but it is quite common for the two segmental arteries to arise separately. The lateral segmental artery divides into a posterior and an anterior branch, which supply the lateral bronchopulmonary segment of the middle lobe. The medial segmental artery, which supplies the medial bronchopulmonary segment of the middle lobe, divides into a superior and an inferior branch.

The interlobar artery continues into the inferior lobe of the right lung. Its initial branch to the inferior lobe is the *superior segmental artery,* which supplies the superior bronchopulmonary segment of that lobe. This vessel divides into a lateral branch and a superomedial branch.

The interlobar artery terminates as the **basal segmental artery,** which supplies the four basal segments of the lower lobe. It branches into the *medial basal, lateral basal, anterior basal,* and *posterior basal segmental arteries,* each of which courses to its respective bronchopulmonary segment. The individual segmental arteries of these four basal segments are quite variable in

their origin and in the number and size of accessory branches.

LEFT PULMONARY ARTERY.   The left pulmonary artery is shorter and somewhat smaller than the right. It courses horizontally to the left from the bifurcation of the pulmonary trunk, crossing in front of the descending aorta to lie anterior to the left primary bronchus at the hilum of the left lung (Fig. 8-15). In the fetus, the left pulmonary artery is larger and more important than the right because from it branches the ductus arteriosus. In the adult, the ductus has regressed to become a short fibrous cord, the **ligamentum arteriosum,** connecting the left pulmonary artery with the arch of the aorta just distal to the reflection of the pericardium (Figs. 8-14; 8-16). To the left of the ligamentum arteriosum is found the left recurrent nerve, and to the right of the ligamentum is located the superficial part of the cardiac autonomic plexus. Inferiorly the left pulmonary artery is joined to the superior left pulmonary vein by the **ligament of the left vena cava.**

Although the left pulmonary artery has many similarities with the right, it differs in both general and specific details of branch-

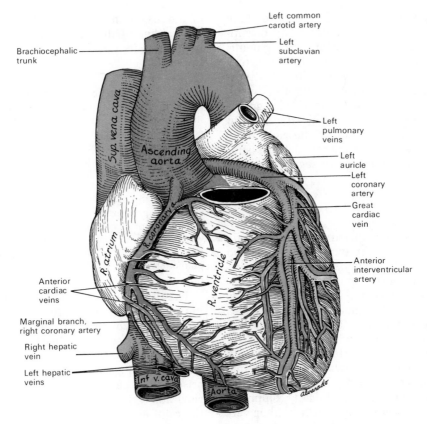

Left common carotid artery

Left subclavian artery

Brachiocephalic trunk

Left pulmonary veins

Left auricle

Left coronary artery

Great cardiac vein

Anterior interventricular artery

Sup. vena cava

Ascending aorta

R. atrium

R. coronary

R. ventricle

Anterior cardiac veins

Marginal branch, right coronary artery

Right hepatic vein

Left hepatic veins

Inf. v. cava   *Aorta*   *alvarado*

FIG. 8-16. The ascending aorta and the coronary blood vessels on the sternocostal aspect of the heart. The pulmonary trunk and pulmonary valve have been removed. (Redrawn from Rauber-Kopsch.)

ing. It tends to have more separate branches than the right. Within the oblique fissure, the left pulmonary artery gives off a series of vessels (from four to eight in number) to the upper lobe and does not have a common vessel comparable to the anterior trunk of the right pulmonary artery. Thus, there is no eparterial bronchus at the right hilum.

The apical and posterior bronchopulmonary segments of the superior lobe, which some authors consider to be a single apicoposterior segment, are supplied by separate apical and posterior arteries. The *apical segmental artery* arises from the anterior surface of the left pulmonary artery as the latter arches over the bronchus, and from there the apical segmental artery runs along the bronchus. Rather than a single vessel, there may be two or more branches, one of which may combine with branches to the posterior or anterior segments. The *posterior segmental artery* arises distal to the apical vessel and continues along the bronchus to the posterior segment. The *anterior segmental artery*

may be single or multiple, arising from the superior part of the arched portion of the main artery. Frequently, this segmental artery may have both ascending and descending branches.

The remainder of the left pulmonary artery is named the **interlobar portion,** and it, in fact, supplies the two lingular bronchopulmonary segments of the superior lobe and all five of the segments of the inferior lobe.

The **lingular artery** arises from the anterior or medial part of the interlobar artery, usually as a single branch. It quickly divides into *superior and inferior lingular segmental arteries*, which course to the two lingular segments of the superior lobe. At times these vessels may arise directly from the interlobar portion as separate arteries.

The *superior segmental artery* of the left inferior lobe arises unexpectedly from the interlobar artery between the origin of the apical and posterior arteries (or apicoposterior artery) of the superior lobe and that of the lingular artery. It divides into three branches, which may combine with or supply branches to other inferior lobe segments.

The remainder of the interlobar artery continues as the **basal artery.** From this vessel branch the *medial basal, lateral basal, anterior basal,* and *posterior basal segmental arteries* in a pattern somewhat similar to those of the inferior lobe of the right lung. Each of these basal segmental arteries supplies its bronchopulmonary segment.

The Aorta

The **aorta** is the main trunk of the systemic arteries which convey oxygenated blood to the tissues of the body. At its commencement from the aortic opening of the left ventricle, it is about 3 cm in diameter. It ascends toward the neck for about 5 cm, then arches posteriorly and to the left over the root of the left lung. It descends within the thorax on the left side of the vertebral column and, gradually inclining toward the midline, passes through the aortic hiatus of the diaphragm into the abdominal cavity. Opposite the caudal border of the fourth lumbar vertebra and considerably diminished in size (about 1.75 cm in diameter), it bifurcates into the two common iliac arteries. The parts of the aorta are the **ascending aorta,** the **arch of the aorta,** and the **descending aorta.** The descending aorta is further divided into **thoracic** and **abdominal** portions of the aorta.

ASCENDING AORTA

The **ascending aorta** (Fig. 8-15) is about 5 cm in length. It is contained within fibrous pericardium and covered by the visceral pericardium, which encloses it in a common sheath with the pulmonary trunk. It commences at the aortic opening found at the base of the left ventricle on a level with the lower border of the third costal cartilage posterior to the left half of the sternum. The ascending aorta initially curves obliquely to the right, being oriented along the axis of the heart. It ascends as high as the upper border of the second right costal cartilage, lying about 6 cm deep to the inner surface of the sternum. At its origin, opposite the cusps of the aortic valve, are three small dilatations called the **aortic sinuses.** At the continuation of the ascending aorta into the aortic arch, the caliber of the vessel is increased by a bulging of its right wall, called the **bulb of the aorta.**

RELATIONS. The commencement of the ascending aorta is surrounded by the pulmonary artery and the right auricle, except anteriorly at the origin of the coronary artery (Fig. 8-14). More superiorly, its *anterior aspect* is separated from the sternum by the pericardium, the right pleura, the anterior border of the right lung, some loose areolar tissue, and the remains of the thymus. *Posteriorly* it rests upon the left atrium and right pulmonary artery, behind which is the right principal bronchus. On the *right side,* the ascending aorta is related to the superior vena cava, which lies partly behind the aorta, and the right atrium. On the *left side* are the left atrium and the pulmonary trunk.

The two arteries derived from the ascending aorta are the vital **right** and **left coronary arteries,** which supply the musculature and other structures of the heart.

RIGHT CORONARY ARTERY. The right coronary artery (Figs. 7-25; 7-26; 8-14; 8-16 to 8-18) arises from the aorta behind the right (anterior) aortic sinus (of Valsalva). It passes to the right and toward the apex, between the conus arteriosus and right auricle, into the coronary sulcus. It follows the sulcus first to the right around the right margin of the heart and then to the left on the diaphragmatic surface. It ends in two or three branches beyond the interventricular sulcus. In its course it supplies small branches to the right atrium, to the right ventricle by way of its **marginal branch,** and to both ventricles on their diaphragmatic surface by means of the **posterior interventricular branch.**

The **atrial branches** of the right coronary artery are directed superiorly to the musculature of the right atrium. One of these, the **sinoatrial nodal artery,** passes between the right atrium and the superior vena cava to supply the sinoatrial node.

The **marginal branch** is quite large and arises from the main artery at the right margin of the heart, and courses along the surface of the right ventricle, along the acute margin, to the apex. It terminates beyond the apex on the posterior surface of the right ventricle. Thus, the marginal branch supplies both the anterior and posterior surfaces of the right ventricle.

The **posterior interventricular branch** of the right coronary artery courses down the

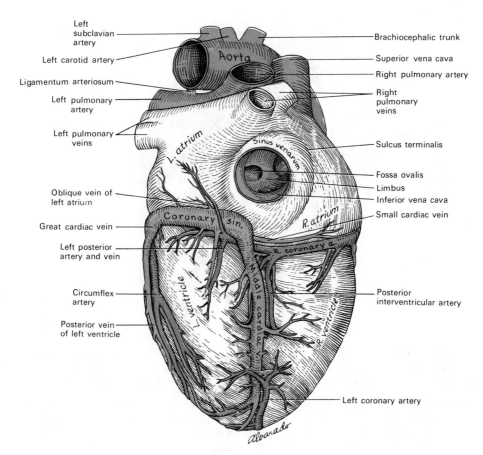

Left subclavian artery
Left carotid artery
Ligamentum arteriosum
Left pulmonary artery
Left pulmonary veins
Oblique vein of left atrium
Great cardiac vein
Left posterior artery and vein
Circumflex artery
Posterior vein of left ventricle

Aorta
L. atrium
Coronary sin.
L. ventricle

Brachiocephalic trunk
Superior vena cava
Right pulmonary artery
Right pulmonary veins
Sinus venarum
Sulcus terminalis
Fossa ovalis
Limbus
Inferior vena cava
Small cardiac vein
R. atrium
R. coronary a.
Middle cardiac v.
R. ventricle
Posterior interventricular artery
Left coronary artery

Alvarado

FIG. 8-17.   The coronary blood vessels of the diaphragmatic aspect of the heart. (Redrawn from Rauber-Kopsch.)

posterior interventricular sulcus two-thirds of the way to the apex, supplying branches to both ventricles as it descends. It anastomoses with the anterior interventricular branch of the left coronary artery near the apex.

LEFT CORONARY ARTERY.   The left coronary artery (see Figs. 7-25; 7-26; 8-14; 8-16 to 8-18) arises from the aorta behind the left (left posterior) aortic sinus (of Valsalva). After a short course under cover of the left auricle, it bifurcates into the **anterior interventricular branch** and the **circumflex branch.** In those instances in which the right coronary artery does not supply the sinoatrial node, a small **sinoatrial nodal branch** may be given off by the left coronary artery near its origin.

The **anterior interventricular branch** of the left coronary artery passes to the left between the pulmonary trunk and the left auricle to the anterior interventricular sulcus, which it follows to the apex, supplying branches to both ventricles. It anastomoses

with the posterior interventricular branch of the right coronary artery.

The **circumflex branch** of the left coronary artery follows the left part of the coronary sulcus, running first to the left and then, around the left margin, it courses backward to the diaphragmatic surface of the heart and to the right, reaching nearly as far as the posterior interventricular sulcus. It supplies branches to the left atrium and ventricle. It anastomoses with the right coronary artery.

VARIATIONS.   The left coronary artery or its descending branch supplying the left ventricle may arise from the left pulmonary artery. This anomaly can be diagnosed clinically and usually leads to death in infancy. Rarely both coronaries arise from the pulmonary artery.

Single coronary arteries have been described. The left coronary may be a branch of the right or both arteries may arise in the same aortic sinus. There may be three coronary arteries, the accessory vessel giving off some of the branches usually derived from the normal coronary arteries.

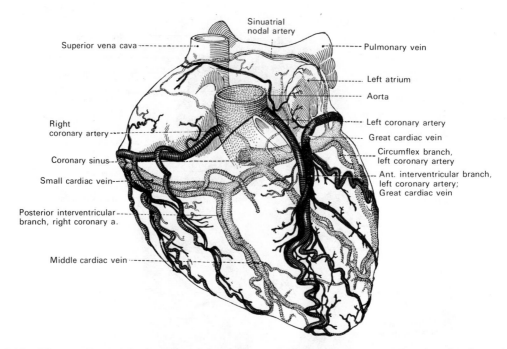

FIG. 8-18.   The coronary arteries (in red) and the cardiac veins (in black). The more intensely colored vessels are shown on the sternocostal surface of the heart, while those on the diaphragmatic surface are shown in lighter shades. (From Benninghoff and Goerttler, *Lehrbuch der Anatomie des Menschen*, Urban and Schwarzenberg, Munich, 1980).

Three **patterns of distribution** of the coronary arteries have been described. In half the hearts, the right coronary predominates; in a third, the two vessels are equally balanced; and in the rest, the left coronary predominates. The *sinuatrial node* is supplied by a branch of the right coronary in 70%, by the left in 25%, by both in 7%. The *atrioventricular node* is supplied by the right in 92%. The *right atrioventricular bundle branch* generally is supplied by the anterior interventricular branch of the left coronary; the *left bundle branch* comes from septal branches of the left coronary artery and small vessels from the right coronary artery (Gregg, 1950).

ARCH OF THE AORTA

The **arch of the aorta** (Figs. 8-14; 8-16) begins at the level of the upper border of the right second sternocostal articulation and runs at first upward, backward, and to the left, anterior to the trachea. It is then directed posteriorly on the left side of the trachea and finally curves downward to the left side of the body of the fourth thoracic vertebra, at the caudal border of which it becomes the descending aorta. It thus forms two curvatures: one with its convexity oriented superiorly, the other with its convexity directed anteriorly and to the left. The upper border of the aortic arch reaches superiorly to about the middle of the manubrium, and the entire arch, which measures about 4.5 cm in length, is contained within the lower half of the superior mediastinum.

RELATIONS.   *Anteriorly,* the arch of the aorta is covered by the pleura, the anterior margins of the lungs, and by the remains of the thymus. As the vessel courses posteriorly, its *left side* is in contact with the left lung and left mediastinal pleura. Additionally, to the left of the aortic arch are four nerves coursing downward in the thorax. In order, i.e., from front to back, these are the left phrenic, the inferior cervical cardiac branch of the left vagus, the superior cervical cardiac branch of the left sympathetic, and the trunk of the left vagus nerve. As the left vagus nerve crosses the arch, it gives off its recurrent branch, which hooks around the vessel and then passes superiorly on its right side. The left superior intercostal vein runs obliquely on the left side of the arch, between the phrenic and vagus nerves. On the *right side* of the aortic arch are the deep cardiac plexus, the right recurrent nerve, the esophagus, and the thoracic duct. The trachea lies posterior to and to the right of the vessel. *Superiorly* are the brachiocephalic, left common carotid, and left subclavian arteries, all of which arise from the convexity of the arch and are crossed close to their origins by the left brachiocephalic vein. *Inferiorly* are the bifurcation of the pulmonary trunk, the left principal bronchus, the ligamentum arteriosum, the superficial cardiac plexus, and the left recurrent nerve. As stated above, the ligamentum arteriosum connects the pulmonary artery near its commencement to the aortic arch.

Between the origin of the left subclavian artery and the attachment of the ductus arteriosus, the lumen of the aorta in the fetus is considerably narrowed, forming what is termed the **aortic isthmus,** while immediately beyond the ductus arteriosus the vessel presents a fusiform dilation which has been called the **aortic spindle.** The point of junction of the two parts is marked in the concavity of the arch by an indentation or angle. This situation persists to some extent in the adult, in whom the average diameter of the spindle exceeds that of the isthmus by about 3 mm.

VARIATIONS IN THE COURSE OF THE ARCH OF THE AORTA. The height to which the aorta ascends in the thorax is usually about 2.5 cm below the cranial border of the sternum. This may vary, however, and the summit of the aorta may not ascend as high as this site, or it may rise as high as the upper border of the sternum. Sometimes the aorta arches over the root of the right lung (right aortic arch) instead of the left, passing along the right side of the vertebral column, a condition that is normal in birds. In such cases, the thoracic and abdominal viscera are usually transposed. Less frequently the aorta, after arching over the root of the right lung, passes dorsal to the esophagus to reach its usual position on the left side of the vertebral column. This variation is not accompanied by transposition of the viscera. As in some quadrupeds, the aorta occasionally divides into an ascending and a descending trunk. In these instances the ascending trunk is directed vertically upward and subdivides into three branches, which supply the head and upper limbs. Sometimes the aorta subdivides near its origin into two branches, which soon reunite. When this variation occurs, the esophagus and trachea are found to pass through the interval between the two branches. This is the normal condition of the vessel in the reptilian class of vertebrates.

BRANCHES FROM THE ARCH OF THE AORTA. Arising from the aortic arch are vessels that principally supply the neck, head, upper extremity, and, to some extent, the upper thoracic wall. In 83 to 94% of cadavers, there are three branches derived from the arch of the aorta: the **brachiocephalic trunk,** the **left common carotid artery,** and the **left subclavian artery** (Fig. 8-14). The latter two vessels will be discussed with the vasculature of the head, neck, and upper extremity. The brachiocephalic trunk will be described following a discussion of the possible variations in the branching pattern of these great vessels.

VARIATIONS IN THE BRANCHING PATTERN OF THE ARCH OF THE AORTA. The branches may arise from the commencement of the arch or upper part of the ascending aorta, instead of from the highest part of the arch. The distance between the branches at their origins may be increased or diminished, the most frequent change in this respect being the approximation of the left common carotid artery toward the brachiocephalic trunk.

The number of the primary branches may be reduced to one, or more frequently two (7%), the left common carotid artery arising from the brachiocephalic trunk. More rarely, the common carotid and subclavian arteries of the left side arise from a left brachiocephalic artery. The number of branches from the aortic arch may be increased to four when the right common carotid and subclavian arteries arise directly from the aorta, the brachiocephalic trunk being absent. In most instances the right subclavian has been found to be the first branch and to arise from the left end of the arch, but in a few specimens it is the second or third branch from the arch. Another variation in which there are four primary branches from the aortic arch is that in which the left vertebral artery arises from the arch between the left common carotid and subclavian arteries. Finally the number of trunks from the arch may be increased to five or six, in which cases the external and internal carotid arteries arise separately from the arch, the common carotid being absent on one or both sides. In a few instances both vertebral arteries have been found to arise from the aortic arch.

When the aorta arches to the right side, the three branches have a reverse arrangement, with a brachiocephalic trunk coursing to the left and the right common carotid and subclavian arteries arising separately. In other instances, when the aorta takes its usual course to the left, the two carotids may arise as a common trunk, with the subclavian arteries originating separately from the arch and the right subclavian generally arising from the left end of the arch.

In some instances, other arteries branch from the arch of the aorta. Of these the most common are one or both of the **bronchial arteries** and the **thyroid ima artery.** Although rarely seen, it has been reported that the internal thoracic and inferior thyroid arteries may arise from the aortic arch.

BRACHIOCEPHALIC TRUNK

The brachiocephalic trunk (formerly called the innominate artery) is the largest of the three branches arising from the arch of the aorta, and it measures 4 to 5 cm in length (Fig. 8-14). It arises near the commencement of the arch of the aorta, on a plane anterior to the origin of the left carotid. At its origin the brachiocephalic trunk is positioned behind the manubrium sterni at about the level of the upper border of the right second costal cartilage. The vessel courses superiorly, somewhat posteriorly, and obliquely to the right to the level of the upper border of the right sternoclavicular articulation, where it divides into the **right common carotid** and **right subclavian arteries.**

RELATIONS. *Anteriorly,* the **brachiocephalic trunk** is separated from the manubrium sterni by the sternohyoid and sternothyroid muscles, the remains of the thymus, the left brachiocephalic and right inferior thyroid veins which cross its root, and sometimes the cardiac branches of the right vagus. *Posteriorly* lies the trachea, which the brachiocephalic trunk crosses obliquely. On the *right side* are the right brachiocephalic vein, the superior vena cava, the right phrenic nerve, and the pleura. On the *left side* are found the remains of the thymus, the origin of the left common carotid artery, the inferior thyroid veins, and more superiorly, the trachea.

The brachiocephalic trunk usually does not give off branches, but occasionally (in about 10% of cases) a small vessel, the **thyroid ima artery,** arises from it. At times a **thymic artery** or **bronchial artery** may arise from the brachiocephalic trunk.

THYROID IMA ARTERY. The thyroid ima artery, when present, ascends anterior to the trachea to the inferior part of the thyroid gland, which it supplies. It varies greatly in size and appears to compensate for a deficiency or absence of one of the other thyroid arteries. It occasionally arises from the aorta on the right common carotid, subclavian, or internal thoracic arteries.

VARIATIONS. The brachiocephalic trunk sometimes divides above the level of the sternoclavicular joint, less frequently below it. It may be absent, in which case the right subclavian and the right common carotid then arise directly from the aorta. When the aortic arch is on the right side, the brachiocephalic trunk is directed toward the left.

Arteries of the Head and Neck

The principal arteries supplying the structures of the head and neck are the two **common carotids.** These vessels, one on each side, ascend in the neck, and at the level of the upper border of the thyroid cartilage (of the larynx) each divides into two branches. These branches are the **external carotid artery,** which supplies the exterior of the head, the face, and the greater part of the neck, and the **internal carotid artery,** which supplies to

a great extent the structures within the cranial and orbital cavities. The **vertebral arteries,** which branch from the subclavian arteries, also ascend in the neck and assist in supplying the brain. These will be described with the subclavian artery.

COMMON CAROTID ARTERY

The **common carotid arteries** differ in length and in their mode of origin. The **right common carotid artery** begins at the bifurcation of the brachiocephalic trunk behind the right sternoclavicular joint and is confined to the neck. The **left common carotid artery** branches from the highest part of the arch of the aorta immediately to the left and slightly behind the brachiocephalic trunk. It therefore consists of a thoracic and a cervical portion.

The **thoracic portion of the left common carotid artery** is 2.5 to 3.5 cm in length and ascends from the arch of the aorta through the superior mediastinum to the level of the left sternoclavicular joint, where it becomes the cervical portion.

RELATIONS. *Anteriorly,* the thoracic portion of the left common carotid artery is separated from the manubrium sterni by the sternohyoid and sternothyroid muscles, the anterior portions of the left pleura and lung, the left brachiocephalic vein, and the remains of the thymus. *Posteriorly,* the vessel is related to the trachea, esophagus, left recurrent laryngeal nerve, and thoracic duct. To its *right side* are the brachiocephalic trunk, the trachea, the inferior thyroid veins, and the remains of the thymus. To its *left side* are the left vagus and phrenic nerves, left pleura, and lung. The left subclavian artery is posterior and slightly lateral to the left common carotid artery in the thorax.

The **cervical portions** of the two common carotid arteries resemble each other so closely that one description will apply to both. Each vessel passes obliquely from behind the sternoclavicular articulation, to the level of the upper border of the thyroid cartilage, where it divides into the external and internal carotid arteries (Figs. 8-19; 8-20; 8-24).

In the lower part of the neck the two common carotid arteries are separated by a narrow interval which contains the trachea. More superiorly in the neck, however, the thyroid gland, the larynx, and the pharynx

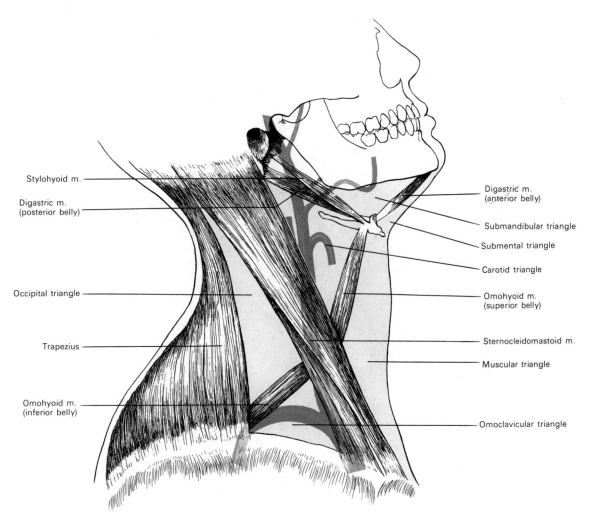

Stylohyoid m.

Digastric m.
(posterior belly)

Occipital triangle

Trapezius

Omohyoid m.
(inferior belly)

Digastric m.
(anterior belly)

Submandibular triangle

Submental triangle

Carotid triangle

Omohyoid m.
(superior belly)

Sternocleidomastoid m.

Muscular triangle

Omoclavicular triangle

FIG. 8-19.   The triangles of the neck. The sternocleidomastoid muscle separates the anterior from the posterior triangle. Observe the course of the carotid arteries in the neck.

project anteriorly between the two vessels. The common carotid artery is contained in the **carotid sheath,** which is derived from the deep cervical fascia. The sheath also encloses the internal jugular vein and vagus nerve, the vein lying lateral to the artery, and the nerve between the artery and vein on a plane posterior to both vessels. When the sheath is opened, each of the structures is seen to have a separate fibrous investment.

RELATIONS.   In the *lower cervical region,* the common carotid artery lies quite deep, being covered by the integument, superficial fascia and platysma muscle, deep cervical fascia, and the sternocleidomastoid, sternohyoid, sternothyroid, and omohyoid muscles. *Above the superior belly of the omohyoid muscle,* the common carotid artery is more superficial, being covered only by the integument, the su-

perficial fascia and platysma, deep cervical fascia, and the medial margin of the sternocleidomastoid muscle (Fig. 8-19). When this muscle is drawn backward, the artery is seen in a triangular space, the **carotid triangle,** bounded posteriorly by the sternocleidomastoid, superiorly by the stylohyoid and posterior belly of the digastric, and inferiorly by the superior belly of the omohyoid muscle (Fig. 8-19). This part of the artery is crossed obliquely, from its medial to its lateral side, by the sternocleidomastoid branch of the superior thyroid artery; it is also crossed by the superior and middle thyroid veins which flow into the internal jugular. Usually superficial to the carotid sheath, but occasionally contained within it are the superior and inferior roots of the **ansa cervicalis,** which descend in the neck and form an interconnecting nerve loop from which the strap muscles of the neck are innervated (Fig. 8-20). The superior thyroid vein crosses the common carotid artery near its termination, and the middle thy-

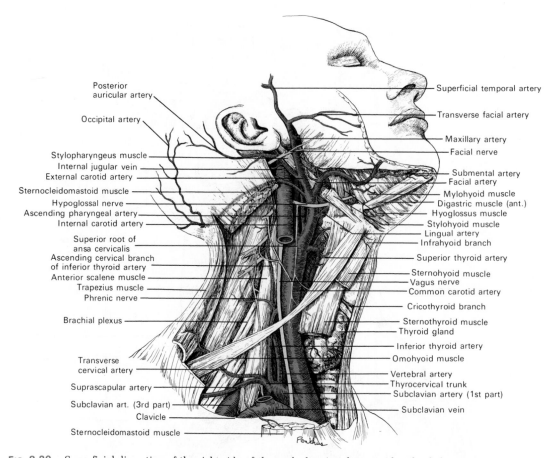

Posterior
auricular artery
Occipital artery
Stylopharyngeus muscle
Internal jugular vein
External carotid artery
Sternocleidomastoid muscle
Hypoglossal nerve
Ascending pharyngeal artery
Internal carotid artery
Superior root of
ansa cervicalis
Ascending cervical branch
of inferior thyroid artery
Anterior scalene muscle
Trapezius muscle
Phrenic nerve
Brachial plexus
Transverse
cervical artery
Suprascapular artery
Subclavian art. (3rd part)
Clavicle
Sternocleidomastoid muscle

Superficial temporal artery
Transverse facial artery
Maxillary artery
Facial nerve
Submental artery
Facial artery
Mylohyoid muscle
Digastric muscle (ant.)
Hyoglossus muscle
Stylohyoid muscle
Lingual artery
Infrahyoid branch
Superior thyroid artery
Sternohyoid muscle
Vagus nerve
Common carotid artery
Cricothyroid branch
Sternothyroid muscle
Thyroid gland
Inferior thyroid artery
Omohyoid muscle
Vertebral artery
Thyrocervical trunk
Subclavian artery (1st part)
Subclavian vein

FIG. 8-20. Superficial dissection of the right side of the neck showing the carotid and subclavian arteries and their branches. Note that the submandibular and parotid glands have been removed, as well as much of the sterno-cleidomastoid muscle and internal jugular veins in order to reveal the underlying arteries and nerves.

roid vein crosses the artery a little below the level of the cricoid cartilage. The anterior jugular vein crosses the artery just above the clavicle, but is separated from it by the sternohyoid and sternothyroid muscles. *Posteriorly,* the artery is separated from the transverse processes of the fourth, fifth, and sixth cervical vertebrae by the longus colli and longus capitis muscles, the sympathetic trunk being interposed between the common carotid and the muscles. The inferior thyroid artery crosses behind the lower part of the common carotid artery. *Medially,* the common carotid artery courses in relation to the esophagus, trachea, thyroid gland (which overlaps it), larynx, and pharynx, the inferior thyroid artery and recurrent laryngeal nerve being interposed. *Lateral* to the artery are the internal jugular vein and vagus nerve.

In the lower neck the right recurrent laryngeal nerve crosses obliquely dorsal to the artery. The right internal jugular vein diverges from the artery in the lower neck, but the left vein approaches and often overlaps it.

CAROTID BODY. The carotid body lies deep to the bifurcation of the common carotid artery or somewhat between the two branches. It is a small, flattened, oval structure, 2 to 5 mm in diameter, with a characteristic structure composed of epithelioid cells, which are in close relation to capillary sinusoids, and an abundance of nerve fibers. Surrounding the carotid body is a delicate fibrous capsule. It is part of the visceral afferent system of the body, containing chemoreceptor endings that respond to low levels of oxygen in the blood or high levels of carbon dioxide and lowered pH of the blood. It is supplied by nerve fibers from both the glossopharyngeal and vagus nerves.

CAROTID SINUS. The carotid sinus is a slight dilatation in the wall of the terminal portion of the common carotid artery and of the internal carotid artery at its origin. The dilatation extends for about 1 cm in length and may be limited to the wall of the internal carotid artery only. The carotid sinus is an important organ for the regulation of systemic blood pressure. Special receptorlike nerve endings capable of responding to alterations in blood pressure are found in the arterial wall. The nerve fibers, which constitute the carotid branch of the glossopharyngeal nerve, transmit these impulses to the medulla oblongata of the brain stem. Cardiovascular reflex changes result from this homeostatic mechanism designed to return blood pressure to its more normal level by altering the rate of the heart beat.

VARIATIONS. The **right common carotid artery** arises above the level of the superior border of the sternoclavicular joint in about 12% of cadavers. This vessel may arise as a separate branch from the arch of the aorta, or in conjunction with the left common carotid. The **left common carotid artery** varies more frequently in its origin than the right. In the majority of abnormal specimens, it arises with the brachiocephalic trunk; if the latter vessel is absent, the two common carotids may arise as a single trunk. The site of division of the common carotid artery into its external and internal branches may vary. More frequently this occurs higher than normal (at about the level of the hyoid bone). More rarely division occurs more caudally, opposite the middle of the larynx or at the lower border of the cricoid cartilage. Very rarely, the common carotid ascends in the neck without dividing, and thus either the external or the internal carotid is absent. In a few instances the common carotid has been found to be absent, the external and internal carotids arising directly from the arch of the aorta.

The common carotid artery usually does not give off any branches before its bifurcation, but occasionally it does give origin to the superior thyroid or its laryngeal branch, the ascending pharyngeal, the inferior thyroid, or, more rarely, the vertebral artery.

COLLATERAL CIRCULATION. After ligature of one common carotid, collateral circulation can become perfectly established, owing to the free communication that exists between the carotid arteries of the two sides, both within the cranium and in the face and neck. Also there is enlargement of the branches of the subclavian artery on the side on which the common carotid artery has been tied. The chief communications outside the skull take place between the superior and inferior thyroid arteries, and between the deep cervical branch of the costocervical trunk and the descending branch of the occipital artery. Within the cranium, the vertebral artery supplies blood to the branches of the ligated internal carotid artery.

External Carotid Artery

The **external carotid artery** (Fig. 8-20) begins opposite the upper border of the thyroid cartilage, and passing upward, it curves somewhat anteriorly and then inclines posteriorly to the space behind the neck of the mandible, where, in the substance of the parotid gland, it divides into the superficial temporal and maxillary arteries. It rapidly diminishes in size in its course up the neck, owing to the number and large size of its branches. In the child, it is somewhat smaller than the internal carotid, but in the adult, the two vessels are of nearly equal size. At its origin, the external carotid is more superficial and nearer the midline than the internal carotid, and is contained within the carotid triangle.

RELATIONS. Within the carotid triangle, the **external carotid artery** lies deep to the skin, superficial fascia, platysma, deep fascia, and anterior margin of the sternocleidomastoid muscle. It is crossed by the hypoglossal nerve and its vena comitans and by the lingual, facial, and superior thyroid veins. Just above the carotid triangle it lies deep to the posterior belly of the digastric and stylohyoid muscles. More superiorly, it penetrates into the substance of the parotid gland, where it lies deep to the facial nerve and the junction of the superficial temporal and maxillary veins. Medial (or deep) to the artery are found initially the hyoid bone, the wall of the pharynx, and the superior laryngeal nerve. More superiorly it is separated from the internal carotid artery by the styloglossus and stylopharyngeus muscles, the glossopharyngeal nerve, the pharyngeal branch of the vagus nerve, and part of the parotid gland.

Branches of the External Carotid Artery

Eight branches arise from the external carotid artery, usually in the following order:

Superior thyroid
Ascending pharyngeal
Lingual
Facial
Occipital
Posterior auricular
Superficial temporal
Maxillary

SUPERIOR THYROID ARTERY

The **superior thyroid artery** (Figs. 8-20; 8-21; 8-24) arises from the anterior aspect of the external carotid artery just caudal to the level of the greater horn of the hyoid bone and ends in the thyroid gland. In 16% of cadavers the superior thyroid artery arises from the common carotid artery. From its origin deep to the anterior border of the sternocleidomastoid muscle, it initially courses upward and forward for a short distance in the carotid triangle, where it is covered by the skin, platysma, and fascia. The superior thyroid artery then arches downward deep to the omohyoid, sternohyoid, and sternothyroid muscles. To its medial side are the inferior pharyngeal constrictor muscle and the external branch of the superior laryngeal nerve.

Branches of the Superior Thyroid Artery

The superior thyroid artery distributes small vessels to adjacent muscles, and usually two main branches to the thyroid gland. One of these, the **anterior branch,** is larger and supplies principally the ventral surface of the gland. At the isthmus of the thyroid gland, the anterior branch anastomoses with the corresponding artery of the opposite side. Also coursing to the thyroid gland is the **posterior branch,** which descends on the dorsal surface of the gland and anastomoses with the inferior thyroid artery.

In addition to the glandular branches and small muscular twigs, the branches of the superior thyroid are:

Infrahyoid
Sternocleidomastoid
Superior laryngeal
Cricothyroid

INFRAHYOID BRANCH. The infrahyoid branch of the superior thyroid artery is small and runs along the lower border of the hyoid bone (Fig. 8-20). It lies on the thyrohyoid membrane, deep to the thyrohyoid muscle, and anastomoses with the suprahyoid branch of the lingual artery and with the infrahyoid branch of the opposite side.

STERNOCLEIDOMASTOID BRANCH. The sternocleidomastoid branch of the superior thyroid artery courses downward and laterally across the carotid sheath and supplies the middle part of the sternocleidomastoid muscle and neighboring muscles and integument (Fig. 8-24). It frequently arises as a separate branch from the external carotid artery.

SUPERIOR LARYNGEAL ARTERY. The superior laryngeal artery, larger than either of the preceding, passes medially and accompanies the internal laryngeal branch of the superior laryngeal nerve, deep to the thyrohyoid muscle (Fig. 8-21). It pierces the thyrohyoid membrane and supplies the muscles, mucous membrane, and glands of the larynx, anastomosing the inferior laryngeal branch of the inferior thyroid artery with the superior laryngeal artery from the opposite side. It sometimes (13%) arises separately from the external carotid.

CRICOTHYROID BRANCH. The cricothyroid branch of the superior laryngeal artery is small and runs transversely across the cricothyroid membrane, anastomosing with the artery of the opposite side (Figs. 8-20; 8-24).

ASCENDING PHARYNGEAL ARTERY

The ascending pharyngeal artery (Figs. 8-20; 8-23; 8-24) is the smallest branch of the external carotid artery. Being a long, slender vessel, it is deeply placed in the neck, dorsal to the other branches of the external carotid and to the stylopharyngeus. It arises from the posterior part of the external carotid artery, near the origin of that vessel, and ascends vertically to the inferior surface of the base of the skull, between the internal carotid and the side of the pharynx, anterior to the longus capitis muscle. In 14% of cadavers it arises from the occipital artery, and occasionally it may arise from the internal carotid artery.

Branches of the Ascending Pharyngeal Artery

The branches of this vessel are small and somewhat variable. In its ascent, it helps to supply the pharynx, the soft palate, the rectus capitis and longus capitis muscles, the superior cervical ganglion, the medial wall of the tympanic cavity, and the dura mater covering the brain. It anastomoses with the ascending palatine branch of the facial artery. Its branches are:

Pharyngeal
Palatine

Prevertebral
Inferior tympanic
Meningeal

PHARYNGEAL BRANCHES. Three or four pharyngeal branches arise from the ascending pharyngeal artery. These supply the superior and middle constrictor muscles, as well as the mucous membrane lining the pharynx. One branch supplies the stylopharyngeus muscles. These branches anastomose with vessels from the superior thyroid artery.

PALATINE BRANCH. The palatine branch (which is considered by some authors as one of the pharyngeal branches) may take the place of the ascending palatine branch of the facial artery, when that vessel is small. It passes beyond the upper border of the superior constrictor muscle and sends ramifications to the soft palate and tonsil, supplying also a branch to the auditory tube.

PREVERTEBRAL BRANCHES. These are numerous small vessels that supply the longus capitis and longus colli muscles, the sympathetic trunk, the hypoglossal and vagus nerves, and a number of the deep cervical lymph nodes. These small arteries anastomose with the ascending cervical branch of the inferior thyroid artery.

INFERIOR TYMPANIC ARTERY. The inferior tympanic artery is a small vessel that passes, with the tympanic branch of the glossopharyngeal nerve, through a minute foramen, the tympanic canaliculus, in the petrous portion of the temporal bone. Entering the middle ear, it supplies the medial wall of the tympanic cavity and anastomoses with the other tympanic arteries.

MENINGEAL BRANCHES. Several small meningeal vessels branch from the ascending pharyngeal artery to supply the dura mater. One, the **posterior meningeal artery,** enters the cranium through the jugular foramen; a second passes through the foramen lacerum; and occasionally a third passes through the canal for the hypoglossal nerve.

LINGUAL ARTERY

The lingual artery (Figs. 8-20; 8-24) arises from the external carotid between the superior thyroid and facial arteries, opposite the tip of the greater horn of the hyoid bone. Supplying structures in the floor of the oral cavity, it is the principal artery to the tongue. Initially it courses obliquely upward and medially above the greater horn of the hyoid bone. It then curves downward and anteriorly to form a loop, which is crossed by the hypoglossal nerve. Coursing deep to the posterior belly of the digastric and stylohyoid muscles, the lingual artery then runs horizontally under cover of (i.e., medial to) the hyoglossus muscle to ascend perpendicularly to the tongue. The artery then turns anteriorly as the **deep lingual artery** and courses along the inferior surface of the tongue as far as the tip.

Frequently, the lingual and facial arteries arise as a common vessel, the **linguofacial trunk,** from the external carotid artery. The trunk courses a short distance superomedially before dividing into the lingual and facial arteries.

RELATIONS. The *first (or oblique) part* of the **lingual artery** lies superficially and is contained within the carotid triangle. It rests on the middle pharyngeal constrictor muscle and is covered by the skin, platysma, and fascia of the neck. Its *second (or curved) part* also rests on the middle pharyngeal constrictor, being covered at first by the tendon of the digastric muscle and by the stylohyoid muscle, and more anteriorly by the hyoglossus. Its *third (or horizontal) part* lies between the hyoglossus and genioglossus. The *fourth (or terminal) part* of the vessel, called the **deep lingual artery** (*ranine artery*), runs along the inferior surface of the tongue to its tip. At this site it is quite superficial, being covered only by the mucous membrane. Deep to the artery are the inferior longitudinal muscle of the tongue and, on its medial side, the genioglossus muscle. The **hypoglossal nerve** crosses the first part of the lingual artery, but is separated from the second part by the hyoglossus muscle.

Branches of the Lingual Artery

Four named branches of the lingual artery are usually identified. They are:

Suprahyoid
Dorsal lingual
Sublingual
Deep lingual

SUPRAHYOID BRANCH. The suprahyoid branch of the lingual artery is quite small and courses along the superior border of the hyoid bone, supplying some of the muscles attached to it (Fig. 8-21). The vessel anastomoses with the infrahyoid branch of the superior thyroid artery and with the suprahyoid branch of the opposite side.

DORSAL LINGUAL BRANCHES.   There are usually two or three small dorsal lingual branches of the lingual artery, and they arise medial to and under cover of the hyoglossus muscle (Fig. 8-24). They ascend to the posterior part of the dorsum of the tongue and supply the mucous membrane of this region, the palatoglossal arch, the tonsil, soft palate, and epiglottis. There is some anastomosis with tonsillar arteries and with vessels of the opposite side.

SUBLINGUAL ARTERY.   The sublingual artery arises at the anterior margin of the hyoglossus muscle and courses anteriorly between the genioglossus and mylohyoid muscles to achieve the sublingual gland. It supplies the gland and gives branches to the

mylohyoid and neighboring muscles and to the mucous membrane of the mouth and gums. Small vessels supply the sides of the tongue, and the artery to the frenulum of the tongue is usually derived from the sublingual artery. One branch, running medial to the alveolar process of the mandible in the substance of the gum, anastomoses with a similar artery from the other side; another pierces the mylohyoid muscle and anastomoses with the submental branch of the facial artery.

DEEP LINGUAL ARTERY.   The deep lingual artery (also known as the *ranine artery*) is the terminal portion of the lingual artery; it pursues a tortuous course, running along the inferior lingual surface, between the inferior longitudinal muscle of the tongue and the mucous membrane (Fig. 8-24). Lying on the lateral side of the genioglossus muscle, it is

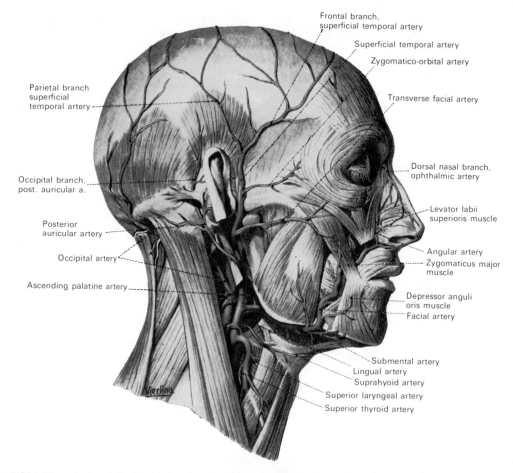

FIG. 8-21.   The arteries of the face and scalp viewed from the right side. Both the submandibular and parotid glands have been removed to reveal the course of the external carotid artery behind the ramus of the mandible. (From Benninghoff and Goerttler, *Lehrbuch der Anatomie des Menschen,* 1975.)

accompanied by the lingual vein and the terminal part of the lingual nerve. It anastomoses with the artery of the opposite side at the tip of the tongue.

FACIAL ARTERY

The facial artery (Figs. 8-20; 8-21; 8-23) arises in the carotid triangle a little superior to the lingual artery and, sheltered by the ramus of the mandible, passes obliquely deep to the posterior belly of the digastric and the stylohyoid muscles. It then arches anteriorly to enter a groove on the posterior surface of the submandibular gland where it joins the facial vein. Becoming quite superficial, the facial artery winds around the inferior border of the mandible at the anterior edge of the masseter muscle to enter the face. It crosses the cheek lateral to the angle of the mouth and, as the **angular artery,** ascends along the side of the nose, to end at the medial palpebral commissure of the eye. At this site it anastomoses with the dorsal nasal branch of the ophthalmic artery. The facial artery, both in the neck and on the face, is remarkably tortuous. In the neck, this may assist the vessel to accommodate itself to movements of the pharynx in deglutition, and in the face, to movements of the mandible, lips, and cheeks.

RELATIONS.   *In the neck,* the origin of the facial artery is quite superficial, being covered by the integument, platysma, and fascia. It then passes deep to the posterior belly of the digastric and the stylohyoid muscle, part of the submandibular gland, and frequently the hypoglossal nerve. It lies on the medial and superior pharyngeal constrictor muscles, the latter separating the vessel at the summit of its arch from the tonsil. *On the face,* where the facial artery passes over the body of the mandible, it is also superficial, lying immediately beneath the platysma. In its course over the face, it is covered by the integument, the fat of the cheek, and near the angle of mouth, by the platysma, risorius, and zygomaticus major muscles. It rests on the buccinator and levator anguli oris, and passes either superficial or deep to the levator labii superioris. The anterior facial vein lies lateral to the artery and takes a more direct course across the face. Branches of the facial nerve cross the artery.

Branches of the Facial Artery

The branches of the facial artery may be divided into two sets: those given off in the neck, **cervical,** and those on the face, **facial.**

Cervical Branches
  Ascending palatine
  Tonsillar
  Glandular
  Submental
Facial Branches
  Inferior labial
  Superior labial
  Lateral nasal
  Angular

Additionally, *muscular branches* are given off both in the neck and in the face.

ASCENDING PALATINE ARTERY.   The ascending palatine artery (Figs. 8-21; 8-23) arises close to the origin of the facial artery and passes upward between the styloglossus and stylopharyngeus muscles to the side of the pharynx, along which it continues between the superior pharyngeal constrictor and the medial pterygoid muscles toward the base of the skull. It divides near the levator veli palatini muscle into *two branches.* One branch follows the course of this muscle, and, winding over the upper border of the superior pharyngeal constrictor, supplies the soft palate and the palatine glands, anastomosing with its fellow of the opposite side and with the greater palatine branch of the maxillary artery. The other branch pierces the superior pharyngeal constrictor muscle and supplies the palatine tonsil and auditory tube, anastomosing with the tonsillar branch of the facial artery and the ascending pharyngeal branch of the external carotid artery.

TONSILLAR BRANCH.   The tonsillar branch of the facial artery (Fig. 8-24) ascends between the medial pterygoid and styloglossus muscles. On the side of the pharynx, at the level of the palatine tonsil, it perforates the superior pharyngeal constrictor to ramify in the substance of the palatine tonsil and root of the tongue. Penetrating the lower pole of the tonsillar bed, this vessel is usually the primary arterial supply to the tonsil.

GLANDULAR BRANCHES.   As the facial artery courses near the submandibular gland, three or four large branches are distributed to the submandibular gland, some being prolonged to the neighboring muscles, lymph nodes, and integument. Generally, a small branch supplies the submandibular duct (of Wharton[1]).

[1]Thomas Wharton (1610-1673): An English anatomist (London).

SUBMENTAL ARTERY. The submental artery, being the largest of the cervical branches, comes off the facial artery just as that vessel emerges from beneath the submandibular gland (Figs. 8-20; 8-21). It runs anteriorly upon the mylohyoid muscle just below the body of the mandible and deep to the anterior belly of the digastric. It supplies the surrounding muscles and anastomoses with the sublingual branch of the lingual artery and with the mylohyoid branch of the inferior alveolar artery. At the symphysis menti, the submental artery turns upward over the border of the mandible and divides into a superficial and a deep branch. The *superficial branch* passes between the skin and the levator labii inferioris muscle, and anastomoses with the inferior labial artery. The *deep branch* runs between this same muscle and the bone, supplies the lip, and anastomoses with the inferior labial and mental arteries.

INFERIOR LABIAL ARTERY. The inferior labial artery arises near the angle of the mouth (Figs. 8-22; 8-23). It passes anteriorly beneath the depressor anguli oris and, penetrating the orbicularis oris, runs a tortuous course along the edge of the lower lip, between this muscle and the mucous membrane. It supplies the labial glands, the mucous membrane, and the muscles of the lower lip. The inferior labial artery anastomoses with the same vessel of the opposite side and with the mental branch of the inferior alveolar artery.

SUPERIOR LABIAL ARTERY. The superior labial artery is larger and more tortuous than the inferior (Figs. 8-22; 8-23). It follows a comparable course along the edge of the upper lip, passing deep to the zygomaticus major muscle to lie between the mucous membrane and the orbicularis oris. It sup-

plies the upper lip and anastomoses with the artery of the opposite side. Additionally, the superior labial artery sends at least two branches that ascend to the nose. These are the **septal branch,** which ramifies on the nasal septum as far as the point of the nose, and the **alar branch,** which supplies the ala of the nose.

LATERAL NASAL BRANCH. The lateral nasal branch of the facial artery is derived from the larger trunk as that vessel ascends alongside the nose. It supplies the ala and dorsum of the nose, anastomosing with its fellow from the other side, with the septal and alar branches of the superior labial artery, with the dorsal nasal branch of the ophthalmic artery, and with the infraorbital branch of the maxillary artery.

ANGULAR ARTERY. The angular artery (Figs. 8-21; 8-23) is actually the terminal part of the facial. It ascends in the groove lateral to the nose to the medial angle of the orbit, embedded in the fibers of the levator labii superioris alaeque nasi and is accompanied by the angular vein. The branches of the angular artery anastomose with the infraorbital artery, and after supplying the lacrimal sac and orbicularis oculi muscle, it ends by anastomosing with the dorsal nasal branch of the ophthalmic artery.

MUSCULAR BRANCHES. Unnamed separate muscular arteries frequently branch directly from the main trunk of the facial artery. From the cervical part of the vessel a branch generally ascends to the medial pterygoid muscle in the face and another supplies the stylohyoid muscle. Muscular branches arising from the artery on the face supply the masseter and buccinator muscles as well as the muscles of expression.

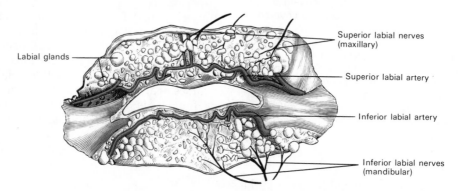

Labial glands

Superior labial nerves (maxillary)

Superior labial artery

Inferior labial artery

Inferior labial nerves (mandibular)

FIG. 8-22.   The labial arteries, the glands of the lips, and the nerves of the right side seen from the inner surface of the oral cavity after removal of the mucous membrane. (From Poirier and Charpy.)

ANASTOMOSES OF THE FACIAL ARTERY. The anastomoses of the facial artery are quite numerous, not only across the midline with branches of the same vessel of the opposite side, but *in the neck,* with the sublingual branch of the lingual, with the ascending pharyngeal, and by its ascending palatine and tonsillar branches with the greater palatine branch of the maxillary. *On the face,* branches of the facial artery anastomose with the mental branch of the inferior alveolar artery as it emerges from the mental foramen, with the transverse facial branch of the superficial temporal artery, with the infraorbital branch of the maxillary artery, and with the dorsal nasal branch of the ophthalmic artery.

VARIATIONS. The facial and lingual arteries frequently arise as a common vessel, the **linguofacial trunk,** which quickly divides into separate facial and lingual branches. The facial artery may vary in size and in the extent to which it supplies the face. It occasionally ends by forming the submental artery at the level of the chin, and not infrequently extends only as high as the angle of the mouth or nose. The deficiency is then compensated for by an enlargement of the neighboring arteries.

OCCIPITAL ARTERY

The occipital artery arises from the posterior part of the external carotid artery, opposite the origin of the facial artery near the lower margin of the posterior belly of the digastric muscle (Figs. 8-20; 8-21; 8-23). It is generally a vessel of considerable size and ends in the posterior part of the scalp.

COURSE AND RELATIONS. At its origin, the **occipital artery** is covered by the posterior belly of the digastric and the stylohyoid muscles. It is crossed by the hypoglossal nerve as that nerve winds around the vessel in its course toward the tongue. The vessel crosses the internal carotid artery, the internal jugular vein, and the vagus and accessory nerves. The artery next ascends to the interval between the transverse process of the atlas and the mastoid process of the temporal bone, passing backward and upward along the occipital groove of the latter bone. In its course, the occipital artery is covered by the sternocleidomastoid, splenius capitis, and longissimus capitis muscles, and the posterior belly of the digastric muscle, and rests sequentially on the rectus capitis lateralis, the obliquus capitis superior, and the semispinalis capitis muscles. Posteriorly at the base of the skull, the artery changes its course and runs vertically upward, piercing the fascia connecting the cranial attachments of the trapezius and sternocleidomastoid muscles. The occipital artery then ascends in a tortuous course within the superficial fascia of the scalp, where it divides into numerous branches. These branches reach as high as the vertex of the skull, anastomosing with the posterior auricular and superficial temporal arteries. The terminal portion of the occipital artery is accompanied by the greater occipital nerve.

Branches of the Occipital Artery

Arising from the occipital artery are the following branches:

Muscular
Sternocleidomastoid
Auricular
Mastoid
Meningeal
Descending
Terminal (Occipital)

MUSCULAR BRANCHES. Unnamed muscular branches of the occipital artery supply the digastric, stylohyoid, splenius, and longissimus capitis muscles.

STERNOCLEIDOMASTOID ARTERY. The sternocleidomastoid artery generally arises from the occipital artery close to its origin, but sometimes springs directly from the external carotid (Fig. 8-23). It passes downward and backward over the hypoglossal nerve and enters the substance of the muscle, in company with the accessory nerve. This artery anastomoses with the sternocleidomastoid branch of the superior thyroid artery. Frequently two sternocleidomastoid vessels arise from the occipital artery.

AURICULAR BRANCH. The auricular branch of the occipital artery ascends over the mastoid process of the temporal bone to supply the back of the cartilaginous concha of the external ear. It sometimes replaces the posterior auricular artery.

MASTOID BRANCH. The mastoid branch arises either from the occipital artery or from its auricular branch. It is a small vessel that enters the skull through the mastoid foramen and supplies the diploë, the dura mater, and the mastoid air cells. It anastomoses with the middle meningeal artery.

MENINGEAL BRANCHES. One or more long, slender meningeal branches arise from the occipital artery and ascend, with the internal jugular vein, to enter the cranial cavity through the jugular foramen and condyloid canal. They supply the dura mater lining the posterior fossa of the skull.

DESCENDING BRANCH. The descending branch is the largest of the vessels arising from the occipital artery (Fig. 8-24). It courses downward in the posterior neck and divides into a superficial and a deep branch. The **superficial branch** lies deep to the splenius, giving off branches which pierce that muscle to supply the trapezius and anastomose with the superficial branch of the transverse cervical. The **deep branch** descends between the semispinales capitis and cervicis and anastomoses with the vertebral artery and with the deep cervical artery, a branch of the costocervical trunk. The anastomoses among these vessels assist in establishing a collateral circulation after ligation of the common carotid or subclavian artery.

TERMINAL (OR OCCIPITAL) BRANCHES. The terminal (or occipital) branches of the occipital artery are distributed to the scalp on the back of the head. They are very tortuous and lie between the skin and the occipital belly of the occipitofrontalis muscle. Anastomosing with similar vessels from the opposite side and with the posterior auricular and superficial temporal arteries, these terminal branches of the occipital artery supply the occipital belly of the occipitofrontalis muscle, the skin, and the pericranium. One of the terminal branches may give off a meningeal twig, which passes through the parietal foramen to help supply the dura mater.

POSTERIOR AURICULAR ARTERY

The posterior auricular artery is small and arises from the external carotid just above the posterior belly of the digastric and stylohyoid muscles opposite the apex of the styloid process (Figs. 8-20; 8-21; 8-23). The vessel ascends on the styloid process of the temporal bone under cover of the parotid gland to the groove between the cartilage of the external ear and the mastoid process, where it divides into its auricular and occipital branches.

Branches of the Posterior Auricular Artery

In addition to several small branches that the posterior auricular artery supplies to the digastric, stylohyoid, and sternocleidomas-

toid muscles and to the parotid gland, this vessel gives off three branches:

Stylomastoid
Auricular
Occipital

STYLOMASTOID ARTERY. The stylomastoid artery enters the stylomastoid foramen and supplies the mucous membrane of the tympanic cavity, the mastoid antrum and mastoid cells, and the semicircular canals (Fig. 8-23). In young subjects, a branch from this vessel, the **posterior tympanic artery** (along with the anterior tympanic artery from the maxillary), forms a vascular ring that surrounds the inner surface of the tympanic membrane and sends delicate vessels to ramify on that membrane. The stylomastoid artery anastomoses with the petrosal branch of the middle meningeal artery by a twig, which enters the hiatus of the facial canal.

AURICULAR BRANCH. The auricular branch of the posterior auricular artery ascends behind the ear, deep to the auricularis posterior, and is distributed to the back of the auricle, upon which it ramifies minutely. Some branches curve around the margin of the cartilage, while others perforate it to supply the anterior surface. The auricular branch of the posterior auricular artery anastomoses with the parietal and anterior auricular branches of the superficial temporal artery.

OCCIPITAL BRANCH. The occipital branch passes posteriorly over the sternocleidomastoid muscle to the scalp above and behind the ear. It supplies the occipital belly of the occipitofrontalis muscle and the other layers of the scalp in this region, and it anastomoses with the occipital artery (Figs. 8-21; 8-23).

SUPERFICIAL TEMPORAL ARTERY

The superficial temporal artery is the smaller of the two terminal branches of the external carotid and appears to be the continuation of that vessel (Figs. 8-20; 8-21; 8-23). It begins in the substance of the parotid gland posterior to the neck of the mandible and crosses over the posterior root of the zygomatic process of the temporal bone. About 5 cm above this process, the superficial temporal artery divides into its frontal and parietal branches.

RELATIONS. As it crosses the zygomatic process, the superficial temporal artery is covered by the auricularis anterior muscle and by a dense fascia. Within the substance of the parotid gland, the artery is crossed by the temporal and zygomatic branches of the facial nerve and one or two veins. In its course along the side of the head, it is accompanied by the auriculotemporal nerve, which lies immediately posterior to it. Just above the zygomatic process and in front of the auricle, the superficial temporal artery is quite superficial, being covered only by skin and fascia, and can easily be felt pulsating. This artery is often used for determining the pulse, particularly by anesthesiologists.

Branches of the Superficial Temporal Artery

Besides some twigs to the parotid gland, the temporomandibular joint, and the masseter muscle, the superficial temporal artery gives rise to the following branches:

Transverse facial
Middle temporal
Zygomatico-orbital
Anterior auricular
Frontal
Parietal

TRANSVERSE FACIAL ARTERY. The transverse facial artery is the largest branch of the superficial temporal artery and is given off even before the superficial temporal artery emerges from the substance of the parotid gland (Figs. 8-20; 8-21). Initially the transverse facial artery is deeply placed in the parotid gland, and coursing anteriorly, the vessel then passes transversely across the side of the face, between the parotid duct and the inferior border of the zygomatic arch. It divides into numerous branches, which supply the parotid gland and duct, the masseter muscle, and the skin. Branches of the transverse facial artery anastomose with the facial, masseteric, buccal, and infraorbital arteries. This vessel rests on the masseter muscle and is accompanied by one or two branches of the facial nerve.

MIDDLE TEMPORAL ARTERY. The middle temporal artery arises immediately above the zygomatic arch and, perforating the temporal fascia, lies against the squama of the temporal bone, where it sends branches to the temporalis muscle (Fig. 8-23). This vessel anastomoses with the deep temporal branches of the maxillary artery.

ZYGOMATICO-ORBITAL ARTERY. The zygomatico-orbital artery courses between the two layers of temporal fascia along the superior border of the zygomatic arch to the lateral angle of the orbit (Fig. 8-21). This vessel supplies the orbicularis oculi muscle and anastomoses with the lacrimal and palpebral branches of the ophthalmic artery. It may arise from the middle temporal artery instead of branching directly from the superficial temporal artery.

ANTERIOR AURICULAR BRANCHES. Three or four anterior auricular branches are distributed to the anterior portion of the auricle, the tragus, the lobule, and part of the external acoustic meatus. They anastomose with the posterior auricular artery.

FRONTAL BRANCH. The frontal branch of the superficial temporal artery takes a tortuous course anteriorly and upward toward the forehead, supplying the muscles, integument, and pericranium in this region (Figs. 8-21; 8-23). It anastomoses with the supraorbital and supratrochlear branches of the ophthalmic artery.

PARIETAL BRANCH. The parietal branch is larger than the frontal, curves upward and backward on the side of the head, and lies between the skin and the temporal fascia (Figs. 8-21; 8-23). It anastomoses anteriorly with the frontal branch of the superficial temporal artery, posteriorly with the posterior auricular and occipital branches of the external carotid, and across the vertex of the skull with the parietal branch of the other side.

MAXILLARY ARTERY

The maxillary artery is the larger of the two terminal branches of the external carotid (Figs. 8-20; 8-23). It arises behind the neck of the mandible and is at first embedded in the substance of the parotid gland. It passes anteriorly between the ramus of the mandible and the sphenomandibular ligament, and then courses to the pterygopalatine fossa, either superficial or deep to the lateral pterygoid muscle. It supplies the deep structures of the face and may be divided into **mandibular, pterygoid,** and **pterygopalatine portions.**

The **first** or **mandibular portion** lies between the neck of the mandible and the sphenomandibular ligament, courses hori-

zontally forward, nearly parallel to but a little below the auriculotemporal nerve. It is
embedded in the parotid gland and crosses
the inferior alveolar nerve while running
along the inferior border of the lateral pterygoid muscle.

The **second** or **pterygoid portion** arches
forward and upward under cover of the
ramus of the mandible and the insertion of
the temporalis muscle. It lies either superficial or deep to the lower head of the lateral
pterygoid muscle. When it lies superficial to
the latter muscle, the vessel is located between the lateral pterygoid and temporalis
muscles. When the vessel lies more deeply, it
is located between the lateral pterygoid
muscle and branches of the mandibular
nerve, lying close to the lateral pterygoid
plate. In this latter situation, the artery then
penetrates the muscle to gain entrance to the
pterygopalatine fossa. This portion supplies
the muscles of mastication and the buccinator.

The **third** or **pterygopalatine portion** is a
short segment of the maxillary artery, lying
in the pterygopalatine fossa. It gives off several important branches and makes its way
toward the sphenopalatine foramen, passing
beside the pterygopalatine ganglion and terminating as the sphenopalatine artery. The
course of the third part of the maxillary artery is the same in both superficially and
deeply running vessels. It sends branches to
the orbit and anterior face, the paranasal
sinuses, nasal cavity and nasopharynx, and
the upper teeth and palate.

The origin of branches of the first two
parts of the maxillary artery differs somewhat, depending on whether the vessel
courses superficial or deep to the lateral
pterygoid muscle.

**Branches of the
First (or Mandibular)
Portion of the
Maxillary Artery**

The descriptions of the following
branches pertain to those instances in which
the maxillary artery takes a more superficial
course in the deep face:

Deep auricular
Anterior tympanic
Inferior alveolar
Middle meningeal
Accessory meningeal

DEEP AURICULAR ARTERY. The deep auricular is a small vessel that ascends in the substance of the parotid gland, deep to the temporomandibular articulation (Fig. 8-23). It
pierces the cartilaginous or bony wall of the
external acoustic meatus and supplies its
cuticular lining and the outer surface of the
tympanic membrane. It gives a branch to the
temporomandibular joint as it courses behind the articular capsule.

ANTERIOR TYMPANIC ARTERY. The anterior
tympanic artery is small and often arises in
common with the deep auricular artery
(Fig. 8-23). It passes behind the temporomandibular articulation and enters the tympanic
cavity through the petrotympanic fissure.
Within the middle ear it ramifies upon the
inner surface of the tympanic membrane,
forming a vascular circle around the membrane with the stylomastoid branch of the
posterior auricular artery. The anterior tympanic artery also anastomoses with the artery of the pterygoid canal and with the
caroticotympanic branch from the internal
carotid artery.

INFERIOR ALVEOLAR ARTERY. The inferior
alveolar artery arises from the maxillary artery as that vessel courses between the neck
of the mandible and the sphenomandibular
ligament. It then descends with the inferior
alveolar nerve to the mandibular foramen
on the medial surface of the ramus of the
mandible (Fig. 8-23). The artery then enters
the mandibular canal with the nerve and
courses through the canal to the first premolar tooth, where it divides into the mental
and incisor branches. Along its course, the
inferior alveolar artery also gives off mylohyoid and dental branches and a small lingual branch.

The **mylohyoid artery** arises from the inferior alveolar artery just before the latter vessel enters the mandibular foramen. It runs in
the mylohyoid groove on the medial surface
of the mandibular ramus, along with the
mylohyoid nerve, to the inferior surface of
the mylohyoid muscle, which it supplies.

The **incisor branch,** which is one of the
terminal branches of the inferior alveolar
artery, continues in the mandibular canal to
the midline anteriorly, where it gives
branches to the incisor teeth and anastomoses with a similar vessel from the opposite
side.

The **mental branch,** which is the other terminal branch of the inferior alveolar artery,

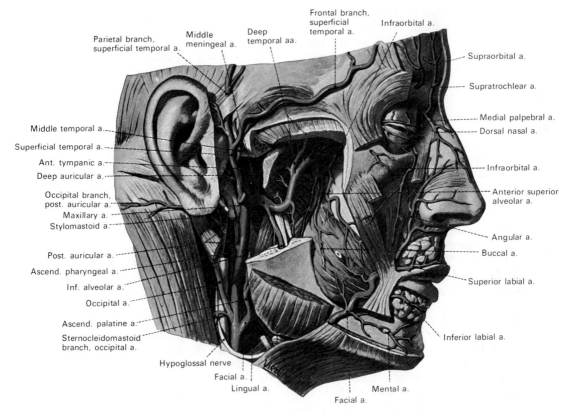

FIG. 8-23. The right maxillary artery exposed in the infratemporal and pterygopalatine regions by the removal of the parotid gland, ramus of mandible, zygomatic arch, and muscles of mastication. Note the facial, occipital, ascending pharyngeal, posterior auricular, and superficial temporal branches of the right external carotid artery. (From Benninghoff and Goerttler, *Lehrbuch der Anatomie des Menschen*, 1975.)

emerges with the mental nerve from the mental foramen to supply the chin (Fig. 8-23). It anastomoses with the submental and inferior labial branches of the facial artery.

The **dental branches** of the inferior alveolar artery correspond in number to the roots of the lower teeth. They enter the minute apertures at the extremities of the roots and supply the pulp of the teeth.

The **lingual branch,** which is small, arises from the inferior alveolar artery near its origin and descends with the lingual nerve to supply the mucous membrane of the mouth.

MIDDLE MENINGEAL ARTERY. The middle meningeal artery is the largest of the arteries that supply the dura mater (Fig. 8-23). It ascends between the sphenomandibular ligament and the lateral pterygoid muscle and between the two roots of the auriculotemporal nerve to the *foramen spinosum of the sphenoid bone,* through which it enters the cranial cavity accompanied by two small veins and some sympathetic nerve fibers.

Finally, in a groove on the great wing of the sphenoid bone that extends to the squamous part of the temporal bone, the middle meningeal artery divides into frontal (or anterior) and parietal (or posterior) branches.

The **frontal (or anterior)** branch is the larger of the two principal terminal branches. It crosses the great wing of the sphenoid to reach the groove, or canal, in the sphenoidal angle of the parietal bone. It then divides into branches that spread out between the dura mater and internal surface of the cranium, some passing upward as far as the vertex, and others backward to the occipital region.

The **parietal (or posterior)** branch curves backward on the squamous part of the temporal bone, and reaching the parietal bone some distance anterior to its mastoid angle, it divides into branches that supply the posterior part of the dura mater and cranium. The branches of the middle meningeal artery are distributed partly to the dura mater, but chiefly to the bones. They anastomose with the arteries of the opposite side and with the anterior meningeal (from the anterior ethmoid) and posterior meningeal (from the vertebral) arteries.

The middle meningeal artery, as it enters the cranium, gives off several other branches. A number of small **ganglionic vessels** supply the trigeminal ganglion, the roots of the trigeminal nerve, and the dura mater nearby. A **petrosal branch,** which enters the hiatus for the greater petrosal nerve, supplies the facial nerve and anastomoses with the stylomastoid branch of the posterior auricular artery. A **superior tympanic artery** runs in the canal for the tensor tympani and supplies this muscle and the lining membrane of the canal. An **anastomotic branch with the lacrimal artery** passes through the superior orbital fissure or through separate canals in the great wing of the sphenoid bone to achieve the orbit. It anastomoses with the recurrent meningeal branch of the lacrimal artery. Occasionally, the lacrimal artery may arise from the middle meningeal artery. **Temporal branches** pass through small foramina in the great wing of the sphenoid and anastomose in the temporal fossa with the deep temporal arteries.

ACCESSORY MENINGEAL ARTERY.   The accessory meningeal artery may arise directly from the maxillary artery or, by means of a common vessel, with the middle meningeal artery. It enters the cranial cavity by passing upward along the course of the mandibular division of the trigeminal nerve *through the foramen ovale.* The accessory meningeal artery supplies muscles in its upward course in the deep face and, within the cranial cavity, supplies the trigeminal ganglion and dura mater.

Branches of the Second (or Pterygoid) Portion of the Maxillary Artery

The descriptions of the following branches pertain to those instances in which the maxillary artery takes a more superficial course in the deep face. The branches of this second portion supply muscles.

Deep temporal
Pterygoid
Masseteric
Buccal

DEEP TEMPORAL BRANCHES.   There are generally two deep temporal branches, an **anterior** and a **posterior** (Fig. 8-23). These ascend between the temporalis muscle and the pericranium, supplying that muscle and anastomosing with the middle temporal artery. The anterior deep temporal artery communicates with the lacrimal artery by means of small branches which perforate the zygomatic bone and great wing of the sphenoid. Certainly, the two temporal branches anastomose with each other.

PTERYGOID BRANCHES.   The pterygoid branches vary in number and origin, and they supply the two pterygoid muscles.

MASSETERIC ARTERY.   The masseteric artery arises from the maxillary between the neck of the mandible and the sphenomandibular ligament. It is a small vessel and passes laterally through the mandibular notch with the masseteric nerve to the deep surface of the masseter muscle. It supplies that muscle and anastomoses with the masseteric branches of the facial artery and with the transverse facial artery.

BUCCAL ARTERY.   The buccal artery is a small vessel that courses, along with the buccal nerve, forward and downward between the medial pterygoid muscle and the insertion of the temporalis muscle to the external surface of the buccinator, which it supplies (Fig. 8-23). It anastomoses with branches of the facial, transverse facial, and infraorbital arteries.

VARIATIONS.   The foregoing descriptions of the first and second parts of the maxillary artery apply to those specimens in which the maxillary artery courses superficial to the lateral pterygoid muscle. In about one-third of all cadavers, however, the vessel *passes deep to the lateral pterygoid,* which results in some variation in the origin of branches from the first two parts. In the latter situation, the deep auricular and anterior tympanic branches are unchanged. The masseteric and posterior deep temporal arteries come from a common trunk with the inferior alveolar artery from the first portion. The middle and accessory meningeal arteries arise from the second portion close to the foramen spinosum. The pterygoid, buccal, and anterior deep temporal branches also arise from the second portion, either separately or from a common trunk, deep to the lateral pterygoid muscle.

Branches of the Third (or Pterygopalatine) Portion of the Maxillary Artery

The branches of the third part of the maxillary artery arise within the pterygopalatine fossa, and accompanied by branches of the maxillary division of the trigeminal nerve, they pass through foramina or bony canals. These branches are:

Posterior superior alveolar
Infraorbital
Descending palatine
Artery of the pterygoid canal
Pharyngeal
Sphenopalatine

POSTERIOR SUPERIOR ALVEOLAR ARTERY. This vessel arises from the maxillary artery, frequently in conjunction with the infraorbital artery, just as the trunk of the maxillary is passing into the pterygopalatine fossa. Descending upon the tuberosity of the maxilla, the posterior superior alveolar artery divides into numerous branches, some of which supply the lining of the maxillary sinus while others enter the alveolar canals to suppy the molar and premolar teeth. Still other branches continue anteriorly on the alveolar process to supply the gums. The vascular branches are accompanied in their course by branches of the posterior superior alveolar nerve, which are derived from the maxillary division of the trigeminal.

INFRAORBITAL ARTERY. The infraorbital artery arises as a branch from the maxillary artery as the latter vessel enters the pterygopalatine fossa, and frequently in conjunction with the posterior superior alveolar artery (Fig. 8-23). It runs along the infraorbital groove and canal with the infraorbital nerve and emerges on the face through the infraorbital foramen. In its course, the infraorbital artery gives off orbital, anterior superior alveolar, and facial branches.

The **orbital branches** arise in the infraorbital canal and they assist in the supply of the lacrimal gland and of the inferior rectus and inferior oblique muscles.

The **anterior superior alveolar branches** of the infraorbital artery, with branches of the middle and anterior superior alveolar nerves, descend through the grooves in the anterior wall of the maxilla (Fig. 8-23). Together the vessels and nerves enter the anterior alveolar canals to supply *dental* branches to the upper incisor and canine teeth. Some branches also supply the mucous membrane of the maxillary sinus.

The **facial branches** of the infraorbital artery arise after the vessel emerges from the infraorbital foramen. On the anterior face some branches course upward to the medial angle of the orbit, supplying the lacrimal sac and anastomosing with the angular branch of the facial artery. Branches also descend toward the nose and anastomose with the dorsal nasal branch of the ophthalmic artery. Other branches descend between the levator labii superioris and the levator labii superioris alaeque nasi muscles and anastomose with the transverse facial and buccal arteries.

DESCENDING PALATINE ARTERY. The descending palatine artery arises from the maxillary artery in the pterygopalatine fossa and descends in the greater palatine canal with the greater palatine nerve, which is derived from the maxillary nerve and pterygopalatine ganglion. After emerging from the greater palatine foramen at the palate, the descending palatine artery becomes the *greater palatine artery*, having already given off several small *lesser palatine branches.*

The **greater palatine artery** courses anteriorly from the greater palatine foramen to the incisive canal in a groove called the greater palatine sulcus near the alveolar border on the oral surface of the hard palate. Branches of the vessel are distributed to the gums, palatine glands, and mucous membrane of the roof of the mouth. A terminal branch anastomoses with the nasopalatine branch of the sphenopalatine artery in the incisive canal.

The **lesser palatine arteries** arise from the descending palatine artery and course through the lesser palatine canals to emerge at the lesser palatine foramina. They then course medially, laterally, and posteriorly to supply the oral surface of the soft palate and the palatine tonsil. These vessels anastomose with the ascending palatine branch of the facial artery.

ARTERY OF THE PTERYGOID CANAL. This artery arises from the maxillary artery in the pterygopalatine fossa and enters the pterygoid canal (Vidian[1] canal) with the corre-

---

[1]Vidus Vidius (Latinized name of Guido Guidi, 1500-1569): An Italian physician and anatomist (Pisa).

sponding nerve. It is a long, slender branch that is distributed to the upper part of the pharynx and anastomoses with the ascending pharyngeal and sphenopalatine arteries. The artery of the pterygoid canal also supplies the auditory tube and sphenoidal sinus and sends into the tympanic cavity a small branch that anastomoses with the other tympanic arteries.

PHARYNGEAL BRANCH.   The pharyngeal branch of the maxillary artery is quite small. It runs posteriorly in the palatovaginal canal with a nerve, the pharyngeal branch of the pterygopalatine ganglion, and is distributed to the auditory tube, the upper part of the pharynx, and the sphenoidal sinus.

SPHENOPALATINE ARTERY.   The sphenopalatine artery is actually the terminal or continuation branch of the maxillary artery. It enters the nasal cavity through the sphenopalatine foramen at the posterior end of the superior meatus, close to the pterygopalatine ganglion, and is accompanied by lateral nasal, septal, and nasopalatine nerve branches from the maxillary nerve and pterygopalatine ganglion. Shortly after traversing the foramen, the sphenopalatine artery divides into posterior lateral nasal and posterior septal arteries.

The **posterior lateral nasal branches** run anteriorly, ramifying over the nasal conchae and meatuses that form the lateral wall of the nasal cavity. They also assist in the supply of the mucosa that lines the frontal, maxillary, ethmoidal, and sphenoidal sinuses.

The **posterior septal branches** arise from the sphenopalatine artery and, arching over the roof of the nasal cavity along the inferior surface of the sphenoid bone, anastomose with branches of the anterior and posterior ethmoidal arteries in the mucosa of the nasal septum. Other vessels ramify anteriorly and inferiorly on the septum. One branch, the **nasopalatine artery,** courses in a groove on the vomer with the nasopalatine nerve to the incisive canal, where it anastomoses with the greater palatine artery.

Internal Carotid Artery

The **internal carotid artery** ascends in the neck, enters the base of the skull, opens into the cranial cavity, and supplies much of the cerebral hemisphere, the pituitary gland, the eyeball and other orbital structures, the anterior aspect of the forehead, and the exter-

nal nose. It consists of cervical, petrous, cavernous, and cerebral portions. The **cervical portion** of the internal carotid artery begins at the bifurcation of the common carotid opposite the upper border of the thyroid cartilage, and runs perpendicularly upward, anterior to the transverse processes of the first three cervical vertebrae, to reach the inferior surface of the petrous portion of the temporal bone (Fig. 8-24). The vessel then enters the carotid canal, becoming the **petrous portion.** It makes an abrupt turn forward and medially to the foramen lacerum. At the end of the carotid canal, the internal carotid artery curves upward, above the cartilage occupying the foramen lacerum, to enter the middle cranial fossa between the lingula and petrosal process of the sphenoid bone. The **cavernous portion** of the internal carotid artery is the first part of its course within the cranial cavity, where it is suspended between the layers of the dura mater that form the cavernous sinus (Fig. 8-25). Ascending toward the posterior clinoid process, the artery curves forward on the lateral aspect of the body of the sphenoid bone in the carotid sulcus. At the medial aspect of the anterior clinoid process, the vessel then perforates the part of the dura mater that forms the roof of the cavernous sinus. Thus, the cavernous portion of the internal carotid artery makes a double curvature like the letter S before piercing the dura to achieve the substance of the brain (Fig. 8-31). The **cerebral portion** of the artery then passes between the optic and oculomotor nerves and approaches the anterior perforated substance at the medial end of the lateral cerebral sulcus, where it gives off its intracerebral branches.

CERVICAL PORTION.   The cervical portion of the internal carotid artery is comparatively superficial at its origin at the bifurcation of the common carotid artery (Fig. 8-24). Here the vessel is contained in the **carotid triangle,** and lies posterior and lateral to the external carotid artery, overlapped by the sternocleidomastoid muscle and covered by the deep fascia, platysma, and integument. It then passes deep to the parotid gland, being crossed by the hypoglossal nerve, the posterior belly of the digastric muscle, the stylohyoid muscle, and the occipital and posterior auricular branches of the external carotid artery. More superiorly, the internal carotid artery is separated from the external carotid by the styloglossus and stylopharyngeus muscles, the tip of the styloid process and the stylohyoid ligament, the glossopha-

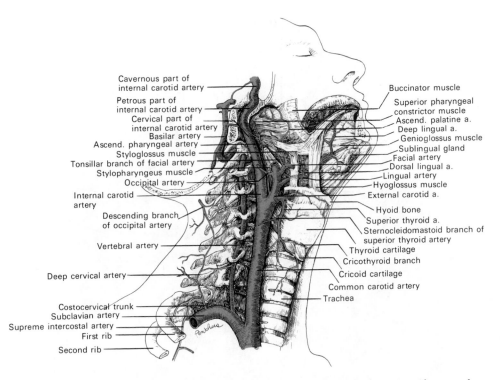

Cavernous part of
internal carotid artery
Petrous part of
internal carotid artery
Cervical part of
internal carotid artery
Basilar artery
Ascend. pharyngeal artery
Styloglossus muscle
Tonsillar branch of facial artery
Stylopharyngeus muscle
Occipital artery
Internal carotid
artery
Descending branch
of occipital artery
Vertebral artery
Deep cervical artery
Costocervical trunk
Subclavian artery
Supreme intercostal artery
First rib
Second rib

Buccinator muscle
Superior pharyngeal
constrictor muscle
Ascend. palatine a.
Deep lingual a.
Genioglossus muscle
Sublingual gland
Facial artery
Dorsal lingual a.
Lingual artery
Hyoglossus muscle
External carotid a.
Hyoid bone
Superior thyroid a.
Sternocleidomastoid branch of
superior thyroid artery
Thyroid cartilage
Cricothyroid branch
Cricoid cartilage
Common carotid artery
Trachea

FIG. 8-24.   This dissection of the deep neck shows the internal carotid and vertebral arteries. Also note the courses and branches of the superior thyroid and lingual arteries.

ryngeal nerve, and the pharyngeal branch of the vagus. *Posteriorly,* it is in relation with the longus capitis muscle, the superior cervical ganglion of the sympathetic trunk with its internal carotid branch to the carotid plexus that surrounds the artery, and the superior laryngeal branch of the vagus nerve. *Laterally,* the internal carotid artery is related to the internal jugular vein and vagus nerve, the nerve lying on a plane posterior to the artery. *Medially,* the vessel is related to the wall of the pharynx, the superior laryngeal nerve, and the ascending pharyngeal artery. At the base of the skull, the glossopharyngeal, vagus, accessory, and hypoglossal nerves lie between the artery and the internal jugular vein.

PETROUS PORTION.   When the internal carotid artery enters the carotid canal in the petrous portion of the temporal bone, it first ascends a short distance, then curves forward and medially, and again ascends as it leaves the canal to enter the cavity of the skull between the lingula and petrosal process of the sphenoid (Fig. 8-24). The artery lies at first anterior to the cochlea and tympanic cavity. It is separated from the tympanic cavity by a thin, bony lamella, which is cribriform in the young subject and often partly absorbed in old age. More anteriorly, it is separated from the trigeminal ganglion by a thin plate of bone that forms the floor of the fossa for the ganglion and the roof of the horizontal portion of the canal. Frequently this bony plate is more or less deficient; in these instances the ganglion is separated from the artery by fibrous membrane. The artery is

surrounded by a plexus of small veins that helps to drain the tympanic cavity and flows into the cavernous sinus and internal jugular vein, and by filaments of the carotid plexus of nerves, derived from the internal carotid branch of the superior cervical ganglion of the sympathetic trunk.

CAVERNOUS PORTION.   In this part of its course, the internal carotid artery is situated between the layers of the dura mater forming the cavernous sinus, but it is covered by the lining membrane of the sinus (Figs. 8-25; 8-31). Initially, it ascends toward the posterior clinoid process, then passes forward by the side of the body of the sphenoid bone, and again curves upward on the medial aspect of the anterior clinoid process, perforating the dura mater that forms the roof of the sinus. This portion of the artery is surrounded by a plexus of sympathetic nerves, and directly on its lateral aspect courses the abducent nerve.

CEREBRAL PORTION.   Having perforated the dura mater on the medial side of the anterior clinoid process, the internal carotid artery passes between the optic and oculomotor nerves toward the anterior perforated substance at the medial extremity of the lateral cerebral sulcus, where it divides into its cerebral branches (Figs. 8-28 to 8-32).

VARIATIONS.   The length of the internal carotid varies according to the length of the neck and according to the point of bifurcation of the common carotid. It arises sometimes from the arch of the aorta. The course of the artery may be tortuous in-

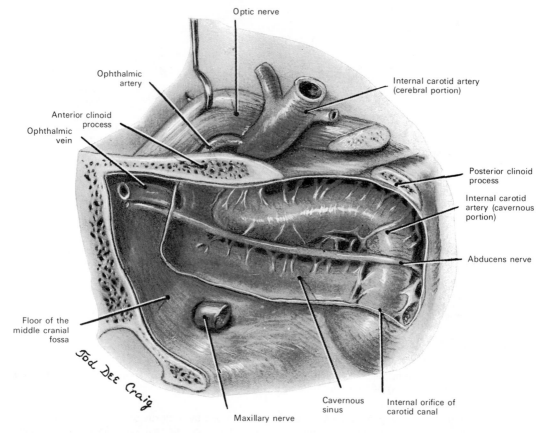

FIG. 8-25. Diagram of the cavernous portion of the internal carotid artery. (From Taveras and Wood, *Diagnostic Neuroradiology*, 1964.)

stead of straight. A few instances have been reported in which this vessel was completely absent. In one such case, the common carotid artery ascended in the neck and gave off the usual branches of the external carotid, but the cranial portion of the internal carotid was replaced by two branches of the maxillary artery, which entered the skull through the foramen rotundum and foramen ovale and joined to form a single vessel.

Branches of the Internal Carotid Artery

The cervical portion of the internal carotid gives off no branches. Those from the other portions are:

Petrous Portion
  Caroticotympanic
  Pterygoid
Cavernous Portion
  Cavernous

Hypophyseal
Ganglionic
Anterior meningeal
Cerebral Portion
  Ophthalmic
  Anterior cerebral
  Middle cerebral
  Posterior communicating
  Anterior choroidal

CAROTICOTYMPANIC ARTERY

The caroticotympanic branch of the internal carotid artery is a small vessel that enters the tympanic cavity through a minute foramen on the posterior wall of the carotid canal. On the surface of the promontory it anastomoses with the anterior tympanic branch of the maxillary and with the stylomastoid branch of the posterior auricular artery.

PTERYGOID BRANCH

The pterygoid branch of the internal carotid artery is a small inconstant vessel that passes into the pterygoid canal. It anastomoses with the pterygoid branch of the maxillary artery; this latter branch enters the opposite end of the canal from the pterygopalatine fossa.

CAVERNOUS BRANCHES

The cavernous branches of the internal carotid artery are small vessels that supply the walls of the cavernous and inferior petrosal sinuses. Some of these arteries anastomose with branches of the middle meningeal artery.

HYPOPHYSIAL ARTERIES

The superior and inferior hypophysial arteries branch from each internal carotid artery to supply the hypophysis (pituitary gland). Although these are small vessels, they are important because they supply a vital endocrine organ.

GANGLIONIC BRANCHES

Several small ganglionic branches come off the internal carotid artery to supply the trigeminal ganglion.

ANTERIOR MENINGEAL BRANCH

The anterior meningeal branch of the internal carotid artery is a small vessel that passes over the lesser wing of the sphenoid bone to supply the dura mater of the anterior cranial fossa. It anastomoses with the meningeal branch from the posterior ethmoidal artery.

OPHTHALMIC ARTERY

The ophthalmic artery arises from the internal carotid, just as that vessel emerges from the dura mater that forms the cavernous sinus, on the medial aspect of the anterior clinoid process (Fig. 8-26). It enters the orbital cavity through the optic canal, inferior and lateral to the optic nerve. Deep to the superior rectus muscle, the ophthalmic artery passes obliquely over the optic nerve from lateral to medial to reach the medial wall of the orbit. The vessel then continues forward along the lower border of the superior oblique muscle to the medial end of the upper palpebral region where it divides into two terminal branches, the **supratrochlear** and **dorsal nasal arteries.** As the ophthalmic artery crosses the optic nerve, it is accompanied by the nasociliary nerve and is separated from the frontal branch of the ophthalmic nerve by the superior rectus and levator palpebrae superioris muscles. The ophthalmic artery crosses deep to the optic nerve, rather than above, in about 15% of cases.

Branches of the Ophthalmic Artery

The branches of the ophthalmic artery may be divided into an **orbital** group, distributed to orbital structures including the extraocular muscles and regions surrounding the orbit, and an **ocular** group, which supplies the eyeball (Fig. 8-26).

Orbital Group
    Lacrimal
    Supraorbital
    Posterior ethmoidal
    Anterior ethmoidal
    Medial palpebral
    Supratrochlear
    Dorsal nasal
    Muscular
Ocular Group
    Central artery of the retina
    Short posterior ciliary
    Long posterior ciliary
    Anterior ciliary

LACRIMAL ARTERY. The lacrimal artery arises from the ophthalmic artery soon after the latter vessel has entered the orbit. Sometimes it arises before the ophthalmic artery enters the orbit. The lacrimal artery is usually the first and often the largest branch derived from the ophthalmic (Fig. 8-26). It accompanies the lacrimal nerve along the superior border of the lateral rectus muscle and supplies the lacrimal gland. Its terminal branches, emerging from the gland, are distributed to the eyelids and conjunctiva. Of those supplying the eyelids, the two **lateral palpebral arteries** are of considerable size. They run medially in the upper and lower lids, respectively, and anastomose with the medial palpebral arteries to form an arterial arch. The lacrimal artery gives off one or two **zygomatic branches.** One of these passes through the zygomaticotemporal foramen to reach the temporal fossa and anastomoses with the deep temporal branch of the maxillary artery. Another zygomatic branch appears on the cheek through the zygomatico-facial foramen and anastomoses with the transverse facial and zygomatico-orbital branches of the superficial temporal artery (Fig. 8-27). A **recurrent meningeal branch** of the lacrimal artery passes backward through the lateral part of the superior orbital fissure to the dura mater where it anastomoses with a branch of the middle meningeal artery. The lacrimal artery is sometimes derived from one of the anterior branches of the middle meningeal artery.

SUPRAORBITAL ARTERY. The supraorbital artery branches from the ophthalmic, the latter vessel crossing over the optic nerve (Fig. 8-26). Initially, it ascends in the orbit along the medial borders of the superior rectus and levator palpebrae superioris muscles to meet the supraorbital nerve. Accompanying the nerve between the periosteum and the levator palpebrae superioris, the artery courses forward in the orbit to the supraorbital foramen. Passing through the foramen, the supraorbital artery divides into a super-ficial and a deep branch, which supply the integument, the muscles, and the pericranium of the forehead, anastomosing with the supra-trochlear branch of the ophthalmic, the frontal branch of the superficial temporal (Fig. 8-27), the angular branch of the facial artery, and the artery of the opposite side. Within the orbit, the supraorbital artery supplies the superior rectus and the levator palpebrae superioris muscles and sends a branch across the pulley of the superior oblique muscle to help supply the region at the upper medial angle of the orbit. As the supraorbital artery courses through the supra-orbital notch, a small branch passes through a foramen to supply the diploë of the frontal bone and the lining of the frontal sinus.

POSTERIOR ETHMOIDAL ARTERY. The posterior ethmoidal artery branches from the ophthalmic artery after the latter vessel has crossed to the medial side of the orbit (Fig. 8-26). It passes through the posterior ethmoidal canal and supplies the posterior ethmoidal air cells. Entering the cranium through a slitlike transverse aperture between the sphenoid bone and cribriform plate, it gives

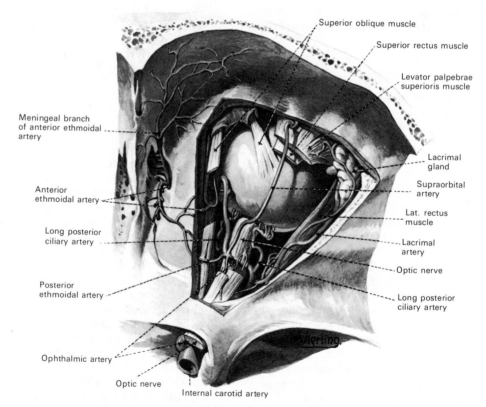

Superior oblique muscle

Superior rectus muscle

Levator palpebrae superioris muscle

Meningeal branch of anterior ethmoidal artery

Lacrimal gland

Supraorbital artery

Anterior ethmoidal artery

Lat. rectus muscle

Long posterior ciliary artery

Lacrimal artery

Optic nerve

Posterior ethmoidal artery

Long posterior ciliary artery

Ophthalmic artery

Optic nerve

Internal carotid artery

FIG. 8-26   The ophthalmic artery and some of its branches. (From Benninghoff and Goerttler, *Lehrbuch der Anatomie des Menschen,* 1975.)

off a meningeal branch to the dura mater. Also, nasal branches descend into the nasal cavity through apertures in the cribriform plate. These anastomose with small vessels that come from the sphenopalatine branch of the maxillary artery.

ANTERIOR ETHMOIDAL ARTERY. The anterior ethmoidal branch of the ophthalmic artery accompanies the nasociliary nerve through the anterior ethmoidal canal. It supplies the anterior and middle ethmoidal cells and the frontal sinus and, entering the cranium, gives off a **meningeal branch** to the dura mater. The **nasal branches** descend into the nasal cavity through a slit lateral to the crista galli and, running along the groove on the inner surface of the nasal bone, supply branches to the lateral wall and septum of the nose. A terminal branch appears on the dorsum of the nose between the nasal bone and the lateral (upper) nasal cartilage.

MEDIAL PALPEBRAL ARTERIES. The medial palpebral arteries, two in number, superior and inferior, arise from the ophthalmic artery deep to the pulley of the superior oblique muscle. They descend from the orbit, one above the medial palpebral ligament and one below the ligament. These ves-

sels then encircle the eyelids near their free margins to form a superior and an inferior arch between the orbicularis oculi muscle and the tarsi. The **superior palpebral arch** anastomoses, at the lateral angle of the orbit, with the zygomatico-orbital branch of the superficial temporal artery, and with the superior of the two lateral palpebral branches from the lacrimal artery (Fig. 8-27). The **inferior palpebral arch** anastomoses, at the lateral angle of the orbit, with the inferior of the two lateral palpebral branches from the lacrimal artery and with the superficial temporal artery, and, at the medial part of the lid, with a twig from the angular branch of the facial artery. From this last anastomosis, a branch passes to the nasolacrimal duct, ramifying in its mucous membrane as far as the inferior meatus of the nasal cavity.

SUPRATROCHLEAR ARTERY. The supratrochlear artery is one of the terminal branches of the ophthalmic artery. It leaves the orbit at its medial angle with the supratrochlear nerve and, ascending on the forehead, helps to supply the integument, muscles, and pericranium in the region (Fig. 8-27). It anastomoses with the supraorbital

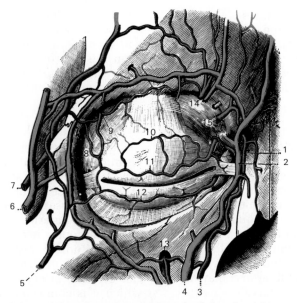

FIG. 8-27. Blood vessels of the eyelids, anterior view. *1*, Branch of dorsal nasal artery that anastomoses with the angular artery. *2*, Angular vein. *3*, Angular branch of facial artery. *4*, Facial vein. *5*, Zygomatico-orbital branch of superficial temporal artery. *6*, Middle temporal vein receiving orbital vein. *7*, Frontal branch of superficial temporal artery. *8*, Anastomosis of superior lateral palpebral branch of lacrimal artery with the anterior branch of superficial temporal artery. *9*, Lacrimal artery giving rise to the superior and inferior lateral palpebral arteries. *10*, Superior palpebral arch. *11*, Superior palpebral artery. *12*, Inferior palpebral artery. *13*, Infraorbital artery. *14*, Supraorbital artery and vein. *15*, Supratrochlear artery.

artery and with the artery of the opposite side.

DORSAL NASAL ARTERY. The dorsal nasal artery is the other terminal branch of the ophthalmic, and it emerges from the orbit above the medial palpebral ligament (Fig. 8-27). After giving a twig to the superior part of the lacrimal sac, it divides into two branches. One of these crosses the root of the nose and anastomoses with the angular branch of the facial artery; the other runs along the dorsum of the nose, supplies its outer surface, and anastomoses with the lateral nasal branch of the facial artery and with the artery of the opposite side.

MUSCULAR BRANCHES. Two groups of muscular branches are derived from the main trunk of the ophthalmic artery. The **superior muscular branches** supply the levator palpebrae superioris, superior rectus, and superior oblique muscles. The **inferior muscular group** passes forward between the optic nerve and the inferior rectus muscle and is distributed to the lateral, medial, and inferior rectus muscles as well as to the inferior oblique muscle. The anterior ciliary arteries are derived from the inferior muscular vessels, and additional muscular branches are given off from the lacrimal and supraorbital arteries.

CENTRAL ARTERY OF THE RETINA. This vessel is the first and one of the smallest arteries that branch from the ophthalmic. It also is the most important because it supplies the retina and because, beyond the optic disc, it is an end artery and does not anastomose with any other vessel. If its function is compromised, permanent retinal damage results. The central artery of the retina branches from the ophthalmic artery within the orbit, but close to the optic foramen. It pierces the dural sheath and courses for a short distance in the sheath with the optic nerve. At about 1.25 cm behind the eyeball it pierces the substance of the optic nerve obliquely and runs forward in the center of the nerve to the retina. Its mode of distribution in the retina will be described with the anatomy of the eye. The central retinal artery may be a branch of the lacrimal artery.

The **ciliary arteries** are divisible into three groups: the short and long posterior, and the anterior.

SHORT POSTERIOR CILIARY ARTERIES. There are usually more than six but fewer than ten short posterior ciliary arteries. They arise from the ophthalmic artery or its branches and, surrounding the optic nerve, run to the posterior part of the eyeball. A number of the vessels may branch at this point before piercing the sclera around the entrance of the nerve in order to supply the choroid layer of the eyeball and the ciliary processes.

LONG POSTERIOR CILIARY ARTERIES. Usually two (but sometimes three) long posterior ciliary arteries arise from the ophthalmic artery on either side of the optic nerve (Fig. 8-26). Coursing forward with the optic nerve and the short posterior ciliary arteries, they pierce the posterior part of the sclera a short distance from the optic nerve and run along either side of the eyeball, between the sclera and choroid, to the ciliary muscle. Reaching the junction of the ciliary body and iris, each vessel divides into an upper and a lower branch. These branches form a vascular circle, the **circulus arteriosus major,** around the circumference of the iris. This circle sends numerous converging branches in the substance of the iris to its pupillary margin, where a second arterial circle, the **circulus arteriosus minor,** is formed.

ANTERIOR CILIARY ARTERIES. The anterior ciliary arteries are derived from muscular branches of the ophthalmic artery. These vessels course to the front of the eyeball along the tendons of the recti muscles, form a vascular zone beneath the conjunctiva, and then pierce the sclera a short distance from the sclerocorneal junction. They end by anastomosing with the circulus arteriosus major around the attached margin of the iris.

ANTERIOR CEREBRAL ARTERY

The anterior cerebral artery is one of the terminal branches of the internal carotid artery. It commences at the medial extremity of the lateral cerebral sulcus (Figs. 8-28; 8-29; 8-31). It passes anteriorly and medially across the anterior perforated substance, above the optic nerve, to the beginning of the longitudinal cerebral fissure. Here it comes into close relationship with the anterior cerebral artery of the opposite side, to which it is connected by a short trunk, the **anterior communicating artery.** From this point the two anterior cerebral arteries run side by side in the longitudinal cerebral fissure, curve around the genu of the corpus callosum and, turning backward, continue along the upper surface of the corpus callo-

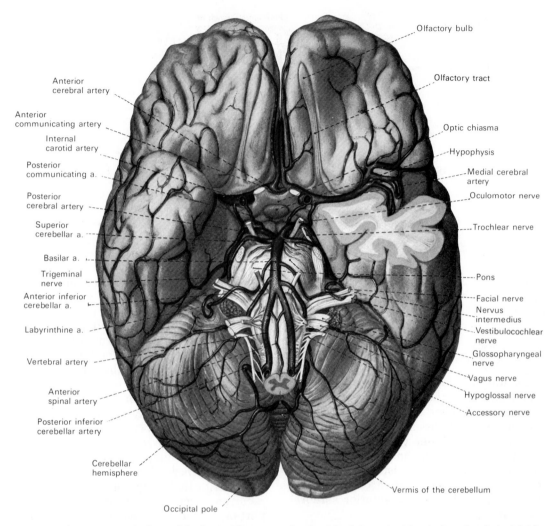

Olfactory bulb

Olfactory tract

Optic chiasma

Hypophysis

Medial cerebral artery

Oculomotor nerve

Trochlear nerve

Pons

Facial nerve

Nervus intermedius

Vestibulocochlear nerve

Glossopharyngeal nerve

Vagus nerve

Hypoglossal nerve

Accessory nerve

Vermis of the cerebellum

Anterior cerebral artery

Anterior communicating artery

Internal carotid artery

Posterior communicating a.

Posterior cerebral artery

Superior cerebellar a.

Basilar a.

Trigeminal nerve

Anterior inferior cerebellar a.

Labyrinthine a.

Vertebral artery

Anterior spinal artery

Posterior inferior cerebellar artery

Cerebellar hemisphere

Occipital pole

FIG. 8-28.   The arteries at the base of the brain. The temporal pole of the left cerebral hemisphere (reader's right) has been removed to reveal the course taken by the middle cerebral artery to the lateral sulcus of the cerebral cortex. (From Benninghoff and Goerttler, *Lehrbuch der Anatomie des Menschen,* 1975.)

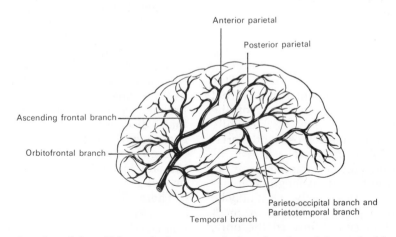

Anterior parietal

Posterior parietal

Ascending frontal branch

Orbitofrontal branch

Parieto-occipital branch and Parietotemporal branch

Temporal branch

FIG. 8-29.   Branches of the *middle cerebral artery* to the lateral surface of the cerebral hemisphere.

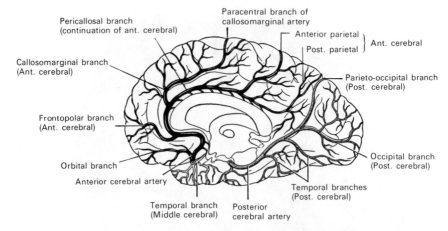

Pericallosal branch
(continuation of ant. cerebral)

Paracentral branch of
callosomarginal artery

Anterior parietal ⎫
Post. parietal ⎭ Ant. cerebral

Callosomarginal branch
(Ant. cerebral)

Parieto-occipital branch
(Post. cerebral)

Frontopolar branch
(Ant. cerebral)

Occipital branch
(Post. cerebral)

Orbital branch

Anterior cerebral artery

Temporal branches
(Post. cerebral)

Temporal branch
(Middle cerebral)

Posterior
cerebral artery

FIG. 8-30.   Medial surface of cerebral hemisphere, showing areas supplied by the *three cerebral arteries*.

sum to its posterior part. Here they end by anastomosing with the posterior cerebral arteries.

Branches of the Anterior Cerebral Artery

In addition to the **anterior communicating artery,** each anterior cerebral artery gives off **central branches** and, within the longitudinal cerebral fissure, several **cortical branches.** These latter vessels can best be seen coursing over the anterior medial surface of the cerebral hemisphere when the brain is sagittally sectioned through the midline. The branches of the anterior cerebral artery are:

Anterior Communicating
    Anteromedial branches
Central Branches
    Recurrent (Medial striate)
Cortical Branches
    Orbital
    Frontal
    Parietal

ANTERIOR COMMUNICATING ARTERY.   This artery connects the two anterior cerebral arteries across the commencement of the longitudinal fissure (Figs. 8-28; 8-32). Usually forming the anterior anastomosis of the **arterial circle of Willis,** the vessel varies in its configuration and may be completely absent, doubled, or even tripled. Its length averages about 4 mm, but varies greatly and it may be slender or wide. It gives off some anteromedial branches that course toward the optic chiasma.

CENTRAL BRANCHES.   The first, or proximal, part of the anterior cerebral artery gives off several small vessels, which pierce the anterior perforated substance and the lamina terminalis, to supply the rostrum of the corpus callosum and the septum pellucidum. Additionally, a larger vessel, the **recurrent artery** (of Heubner[1]), also called the **medial striate artery,** arises just beyond the anterior communicating artery. Phylogenetically, it is one of the oldest of the cerebral arteries. It takes a rather long recurrent course laterally and backward over the anterior perforated substance and pierces the brain. Ascending in the anterior limb of the internal capsule, the recurrent artery supplies the lower anterior portion of the basal ganglia, including the lower part of the head of the caudate nucleus, the lower part of the frontal pole of the putamen, the frontal pole of the globus pallidus, and the anterior limb of the internal capsule up to the dorsal limit of the globus pallidus.

CORTICAL BRANCHES.   Cortical branches of the anterior cerebral artery arise from that vessel as it courses backward over the corpus callosum (Fig. 8-29). They ramify over the anterior medial surface of the cerebral hemisphere and are named according to their distribution. Two or three **orbital branches** are distributed to the orbital surface of the frontal lobe, where they supply the olfactory lobe, gyrus rectus, and medial orbital gyrus.

There are at least two large **frontal**

[1]Johann O. Heubner (1843–1926): A German pediatric neurologist (Berlin).

branches, the frontopolar and calloso-marginal arteries. The *frontopolar artery* usually arises from the anterior cerebral just below the genu of the corpus callosum, and it supplies the medial aspect of the superior frontal gyrus, extending branches over the edge of the hemisphere to the lateral aspect of the superior and middle frontal gyri. The *callosomarginal artery* usually branches just anterior to the genu of the corpus callosum (Fig. 8-30). It supplies the corpus callosum, the cingulate gyrus, part of the medial surface of the superior frontal gyrus, and the precentral and postcentral gyri.

The anterior cerebral artery continues posteriorly in the cingulate sulcus as far as the parieto-occipital sulcus. Beyond the origin of the callosomarginal artery, the anterior cerebral is often called the *pericallosal artery* (Fig. 8-30). From this are derived the anterior and posterior **parietal branches,** which supply the medial surface of the precuneate gyrus and adjacent areas (Fig. 8-30).

MIDDLE CEREBRAL ARTERY

The middle cerebral artery is the largest branch of the internal carotid and one of its terminal branches (Figs. 8-28; 8-30; 8-31). Initially coursing laterally to enter the lateral sulcus (or Sylvian[1] fissure), it then runs backward and upward on the surface of the insula, where it divides into a number of branches that are distributed to the lateral surface of the cerebral hemisphere.

Branches of the Middle Cerebral Artery

Similar to the anterior cerebral, the middle cerebral artery gives off both central and cortical branches. The central branches are distributed to deep structures, while the cortical branches ramify more superficially in the sulci of the dorsal and lateral aspects of the cerebral cortex. The branches of the middle cerebral artery are subdivided as follows:

Central Branches
    Striate (Lateral striate)
Cortical Branches
    Orbital

    Frontal
    Parietal
    Temporal

CENTRAL BRANCHES. The central branches of the middle cerebral artery arise soon after that vessel enters the lateral sulcus. Usually more than 10 but less than 15 delicate **striate** (lenticulostriate) **arteries** pierce the floor of the lateral sulcus to achieve subcortical structures. These vessels supply the corpus striatum, including the upper part of the head and entire body of the caudate nucleus, and much of the lentiform nucleus. Of the latter structure, the striate arteries supply the internal capsule above the level of the globus pallidus and they even extend into the thalamus. One of the larger striate arteries, which courses between the lentiform nucleus and external capsule and passes through the internal capsule to terminate in the caudate nucleus, is especially prone to hemorrhage. It has been called "the artery of cerebral hemorrhage" by Charcot,[2] and is the vessel frequently involved in cerebral stroke.

CORTICAL BRANCHES. Cortical branches of the anterior cerebral artery arise as the vessel courses posterosuperiorly in the lateral gyrus (Figs. 8-30; 8-31). The **orbital branches,** which include the *orbitofrontal artery,* supply the inferior frontal gyrus (Broca's[3] convolution) and the lateral part of the orbital surface of the frontal lobe. The ascending **frontal branches** supply the precentral gyrus, the middle frontal gyrus, and part of the inferior frontal gyrus. The anterior and posterior **parietal branches** ascend to the postcentral gyrus, the lower part of the superior parietal lobule, and the inferior parietal lobule. *Parietotemporal* and *parieto-occipital branches* supply the supramarginal and angular gyri, extending to the parieto-occipital sulcus (Fig. 8-29). The anterior, middle, and posterior **temporal branches** of the middle cerebral artery are distributed to the lateral surface of the temporal lobe, and they supply the superior, middle, and inferior temporal gyri from the temporal pole anteriorly up to the occipital gyri posteriorly (Fig. 8-29).

---

[1]Franciscis Sylvius (Latin form of François Dubois or de la Boë, 1614–1672): A Dutch physician and anatomist (Leyden).

[2]Jean M. Charcot (1825–1893): A French neurologist (Paris).
[3]Pierre P. Broca (1824–1880): A French surgeon and neurologist (Paris).

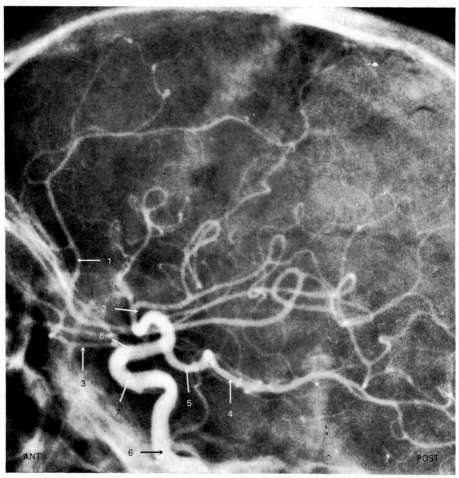

FIG. 8-31. Angiogram of normal left internal carotid artery showing its intracerebral branches, lateral view. *1*, Anterior cerebral artery. *2*, Middle cerebral artery. *3*, Ophthalmic artery. *4*, Posterior cerebral artery. *5*, Posterior communicating artery. *6*, *7*, and *8*, Petrous, cavernous and cerebral portions of internal carotid artery. (From Taveras and Wood, *Diagnostic Neuroradiology, 1964*.)

POSTERIOR COMMUNICATING ARTERY

The posterior communicating artery arises from the posterior aspect of the internal carotid artery just before that vessel terminates into the anterior and middle cerebral arteries (Figs. 8-28; 8-32). It occasionally arises from the middle cerebral artery. Coursing backward over the optic tract and the oculomotor nerve, the posterior communicating artery anastomoses with the posterior cerebral artery, a branch of the basilar. It varies in size, being sometimes small and occasionally so large that the posterior cerebral may be considered as arising from the internal carotid rather than from the basilar. It is frequently larger on one side than on the other and the vessel may even be doubled. **Central branches** of the posterior communicating artery immediately enter the base of the cer-

ebrum by piercing the posterior perforated substance, lateral and posterior to the infundibulum. These supply the anterior one-third of the posterior limb of the internal capsule, the medial thalamus, and the tissue adjacent to the third ventricle.

ANTERIOR CHOROIDAL ARTERY

The anterior choroidal artery arises from the posterior aspect of the internal carotid artery, near the origin of the posterior communicating artery. It is a relatively slender vessel that passes backward along the course of the optic tract and around the cerebral peduncles as far as the lateral geniculate body of the thalamus. It then courses beneath the uncinate gyrus, enters the choroid fissure to achieve the inferior horn of the lat-

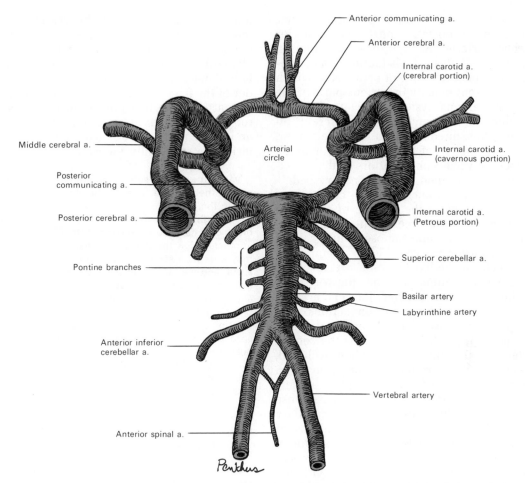

FIG. 8-32.   Diagram of the principal vessels that contribute to and branch from the arterial circulation at the base of the brain.

eral ventricle, and ramifies in the choroid plexus. Along its course, branches are supplied to the optic tract, the cerebral peduncles, the lateral geniculate body, the optic radiation, the tail of the caudate nucleus, the posterior limb of the internal capsule, the globus pallidus, the hippocampus and fibria, and the choroid plexus.

ARTERIAL CIRCLE (OF WILLIS)

Because the mode of distribution of the vessels of the brain has an important bearing on a considerable number of the central nervous system lesions, it is important to consider a little more in detail the manner in which the vessels are distributed.

All the arterial blood to the brain is derived from the internal carotid and vertebral arteries, which at the base of the brain form a remarkable anastomosis known as the **arterial circle of Willis** (Figs. 8-28; 8-32). It is formed *anteriorly* by the anterior cerebral arteries, branches of the internal carotid that are interconnected by the anterior communicating artery, and *posteriorly* by the two posterior cerebral arteries, which are branches of the basilar artery and are connected on either side with the internal carotid artery by the posterior communicating artery (Figs. 8-36; 8-37). The arterial circle underlies the hypothalamus, and the structures it encircles include the lamina terminalis, the optic chiasma, the infundibulum, the tuber cinereum, the mammillary bodies, and the posterior perforated substance. The three trunks that supply the cerebral hemisphere on each side all stem from the arterial circle. From its *anterior part* issue the two anterior cerebral arteries, from its *anterolat-*

eral parts stem the middle cerebral arteries, and from its *posterior part* arise posterior cerebral arteries. Each of these principal arteries contributes vessels that form a continuous complex network of capillaries, which is of different density in various parts of the central nervous system in relation to the varying requirements of the different regions. Thus, in comparison to white matter, the gray matter of the brain has a much higher oxygen consumption and, as expected, has a much denser capillary bed.

The cortical branches of the cerebral arteries that supply the surface of the brain anastomose quite freely, but significantly large arterial anastomoses of central branches are rare. There is no unequivocal evidence for the existence of arteriovenous anastomoses either in the pia or deep within the brain substance. True **end arteries** are arteries that have no connections with neighboring vessels. These are quite rare, but perhaps the best example is the central artery of the retina, which, when occluded, results in central retinal degeneration and a permanent loss of sight. Deeply coursing central vessels in the brain only anastomose with surrounding vessels at five capillary levels. Because of the high vulnerability of brain tissue to a lack of oxygen, a sudden occlusion of a main arterial trunk will result in an ischemic region consistent with the zone supplied by the occluded vessel. The inefficiency of the capillary anastomoses among centrally coursing arteries in the brain and their lack of practical benefit make them **functional end arteries.**

Subclavian Artery

The **subclavian artery** supplies major vessels and branches to the brain, neck, anterior thoracic wall, and shoulder. The main trunk continues through the axilla as the **axillary artery** and achieves the arm, where it is called the **brachial artery.** The subclavian artery arises differently on the two sides. *On the right side* it stems from the brachiocephalic trunk, deep to the right sternoclavicular articulation (Figs. 8-20; 8-24; 8-33; 8-34). *On the left side* it arises from the arch of the aorta. The first part of the two vessels differs in length, course, direction, and relationships to neighboring structures. To facilitate their description, each subclavian artery is divided into three parts. The *first portion* ex-

tends from the origin of the vessel to the medial border of the scalenus anterior muscle; the *second portion* lies deep to this muscle; and the *third portion* extends from the lateral margin of the muscle to the outer border of the first rib where it becomes the axillary artery. The first portion of each vessel requires separate description; the second and third parts of the two subclavian arteries are practically the same.

VARIATIONS. The subclavian arteries vary in their origin, their course, and the height to which they rise in the neck.

The **right subclavian artery** may arise from the brachiocephalic trunk, in some instances above and in others below the sternoclavicular joint. The artery may arise as a separate trunk from the arch of the aorta, and may be either the first, second, third, or even the last branch derived from the arch. Generally, however, it is the first or last. When it is the first branch, the right subclavian occupies the ordinary position of the brachiocephalic trunk. When it is the second or third branch, the artery achieves its usual position by passing behind the right common carotid artery. When the right subclavian arises as the last branch, it stems from the left extremity of the arch and passes obliquely toward the right side, usually dorsal to the trachea, esophagus, and right common carotid (sometimes between the esophagus and trachea) to the upper border of the first rib, from which it follows its usual course. In rare instances, the right subclavian artery arises from the thoracic aorta, as far down as the fourth thoracic vertebra. It may occasionally perforate the scalenus anterior, and more rarely, it passes anterior to that muscle. Sometimes the subclavian vein passes with the artery behind the scalenus anterior. The artery may ascend as high as 4 cm above the clavicle, or may only ascend to the upper border of that bone; usually the right subclavian ascends higher than the left.

The **left subclavian artery** is occasionally joined at its origin with the left common carotid. It is more deeply placed than the right subclavian in the first part of its course and usually does not reach quite as high a level in the neck.

FIRST PART
OF THE RIGHT
SUBCLAVIAN ARTERY

The first part of the right subclavian artery arises from the brachiocephalic trunk, deep to the cranial part of the right sternoclavicular articulation, and passes upward and laterally to the medial margin of the scalenus anterior (Figs. 8-20; 8-24). It arches a little above the clavicle, but the height to which it rises varies in different individuals.

RELATIONS. The first part of the right subclavian artery is covered *anteriorly* by the skin, superficial fascia, platysma, deep fascia, the clavicular origin of the sternocleidomastoid, as well as the sternohyoid, the sternothyroid, and another layer of the deep fascia. It is crossed by the internal jugular and vertebral veins, by the right vagus nerve and the sympathetic and vagal cardiac branches, and by the subclavian loop of the sympathetic trunk, called the **ansa subclavia,** which forms a ring around the vessel. The anterior jugular vein is directed laterally, anterior to the artery, but is separated from it by the sternohyoid and sternothyroid muscles. *Posteriorly* and *inferiorly,* the artery is in contact with the pleura, which separates it from the apex of the lung. Posteriorly is found the sympathetic trunk, the longus colli muscle, and the first thoracic vertebra. The right recurrent laryngeal nerve winds around the proximal part of the vessel.

FIRST PART OF THE LEFT SUBCLAVIAN ARTERY

The first part of the left subclavian artery arises from the arch of the aorta behind the left common carotid artery at the level of the fourth thoracic vertebra (Figs. 8-33; 8-34). It ascends in the superior mediastinum to the root of the neck and then arches laterally to the medial border of the scalenus anterior muscle.

RELATIONS. The first part of the left subclavian artery is related *anteriorly* to the vagus, cardiac, and phrenic nerves which course parallel to it in the thorax. In addition, the left common carotid artery, left internal jugular and vertebral veins, and the commencement of the left brachiocephalic vein lie in front of the thoracic part of the subclavian artery; all these are covered by the sternothyroid, sternohyoid, and sternocleidomastoid muscles. *Posteriorly,* it is related to the esophagus, thoracic duct, left recurrent laryngeal nerve, inferior cervical ganglion of the sympathetic trunk, and longus colli muscle. More superiorly, however, the esophagus and thoracic duct lie to its right side, the latter ultimately arching over the vessel to join the angle of union between the subclavian and internal jugular veins. *Medial* to the left subclavian artery are the trachea, esophagus, thoracic duct, and left recurrent laryngeal nerve, while *lateral* to it are found the left mediastinal pleura and lung.

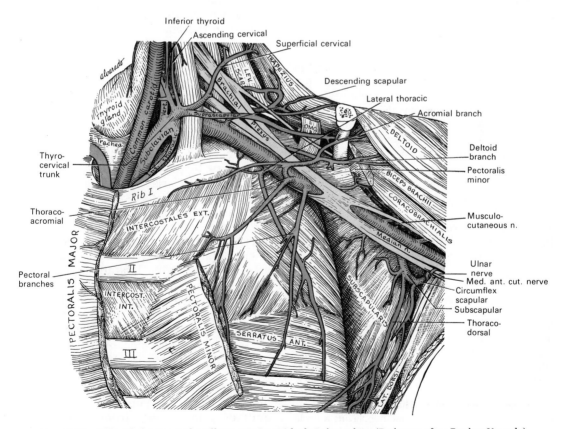

FIG. 8-33. The subclavian and axillary arteries with their branches. (Redrawn after Rauber-Kopsch.)

SECOND PART
OF THE
SUBCLAVIAN ARTERY

The second part of the subclavian artery lies behind the scalenus anterior muscle (Figs. 8-20; 8-33). It is very short, measuring only about 2 cm in length, and forms the highest part of the arch achieved by the vessel in the root of the neck.

RELATIONS. The second part of the subclavian artery is covered *anteriorly* by the skin, superficial fascia, platysma, deep cervical fascia, sternocleidomastoid, and scalenus anterior. On the right side of the neck the phrenic nerve is separated from the second part of the artery by the scalenus anterior, while on the left side the nerve crosses the first part of the artery close to the medial edge of the muscle. *Posterior* to the vessel are the suprapleural membrane, the pleura, and the scalenus medius muscle; *superiorly,* are found the upper two trunks of the brachial plexus; and *inferiorly,* the pleura and lung. The subclavian vein lies anterior and inferior to the artery, separated from it by the scalenus anterior muscle.

THIRD PART
OF THE
SUBCLAVIAN ARTERY

The third part of the subclavian artery continues its course inferolaterally from the lateral margin of the scalenus anterior to the outer border of the first rib, where it becomes the axillary artery (Figs. 8-20; 8-33). This, the most superficial portion of the vessel, crosses the supraclavicular or subclavian triangle (Fig. 8-19). The posterior border of the sternocleidomastoid muscle corresponds closely to the lateral border of the scalenus anterior. Thus, the third part of the subclavian artery is the most accessible for surgical procedures, since it can be achieved immediately lateral to the lower part of the posterior border of the sternocleidomastoid muscle.

RELATIONS. The third part of the subclavian artery is covered *anteriorly* by the skin, the superficial fascia, the platysma, the supraclavicular nerves, and the deep cervical fascia. The external jugular vein crosses its medial part and receives the suprascapular, transverse cervical, and anterior jugular veins, which frequently form a plexus in front of the artery. The nerve to the subclavius muscle descends between the artery and veins. The terminal part of the artery lies behind the clavicle and subclavius muscle and is crossed by the suprascapular vessels. The

subclavian vein is anterior and slightly inferior to the artery. *Posteriorly,* the third part of the subclavian artery lies on the inferior trunk of the brachial plexus, which intervenes between the vessel and the scalenus medius muscle. The upper two trunks of the brachial plexus and the inferior belly of the omohyoid muscle are superior and lateral to the artery. *Inferiorly,* it rests on the upper surface of the first rib.

COLLATERAL CIRCULATION. After ligature of the third part of the subclavian artery, collateral circulation is established mainly by anastomoses between the suprascapular, the descending ramus of the transverse cervical artery, and the subscapular artery. An anastomosis also exists between branches of the internal thoracic, lateral thoracic, and subscapular arteries.

Branches of the
Subclavian Artery

The branches of the subclavian artery and the vessels derived from these branches help to supply the brain, the neck, the thoracic wall, and the shoulder region. They include:

Vertebral
   Basilar
Thyrocervical trunk
Internal thoracic
Costocervical trunk
Descending scapular

On the left side, all the branches except the descending scapular artery arise from the first part of the subclavian artery. *On the right side,* the costocervical trunk more frequently stems from the second part of the subclavian artery. On both sides, the vertebral artery, thyrocervical trunk, and internal thoracic artery arise close together at the medial border of the scalenus anterior. In most instances, there is a free interval of 1.25 to 2.5 cm between the commencement of the artery and the origin of the nearest branch. There is quite a variation in the origin of the vessels, which course laterally at the base of the posterior triangle of the neck, but usually one branch stems directly from the third part of the subclavian artery, which passes laterally behind the brachial plexus.

VERTEBRAL ARTERY

The vertebral artery is the first and usually the largest branch of the subclavian, arising quite deep in the neck from the upper and posterior surface of the vessel (Figs. 8-20;

8-28; 8-32 to 8-36). It angles backward to the transverse process of the *sixth cervical vertebra* and runs superiorly through its foramen and the foramina in the transverse processes of the other upper five cervical vertebrae to the inferior surface of the skull. After penetrating the foramen in the atlas it bends abruptly medialward around the lateral surface of the superior articular process to lie in a groove on the cranial surface of the posterior arch of the atlas. It passes under an arch in the posterior atlanto-occipital membrane (Figs. 5-11; 5-14), makes an abrupt bend to enter the cranial cavity through the

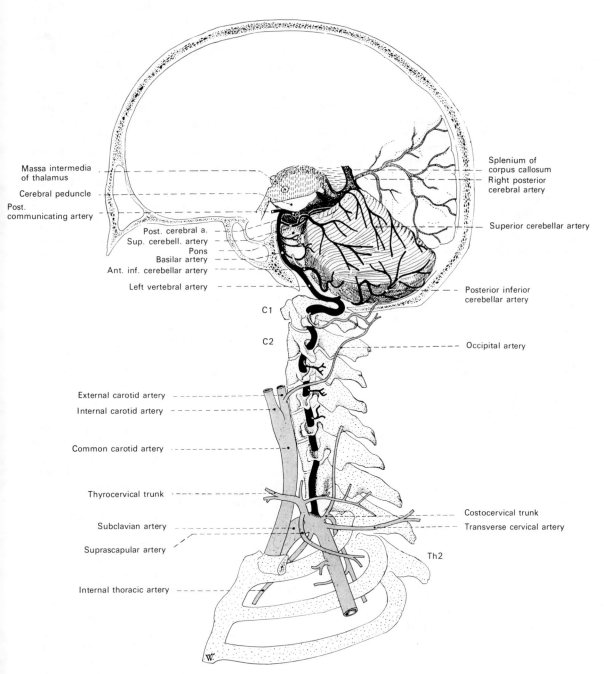

FIG. 8-34. Cervical and intracranial course of the left vertebral artery and its branches. (From Krayenbühl and Yasargil, *Cerebral Angiography,* 1968.)

foramen magnum, and, after a short course, *joins the other vertebral to form the* **basilar artery.**

RELATIONS. The vertebral artery may be divided into four parts. The **first part** courses upward and posteriorly in the neck between the longus colli and the scalenus anterior muscles. Anteriorly it is related to the internal jugular and vertebral veins, and it is crossed by the inferior thyroid artery. The left vertebral artery is crossed also by the thoracic duct. Posterior to the first part of the vertebral artery are the transverse process of the seventh cervical vertebra, the sympathetic trunk, and the cervicothoracic (stellate) ganglion. The **second part** of the vertebral artery ascends in the neck through the foramina in the transverse processes of the first six cervical vertebrae; it is surrounded by branches from the cervicothoracic sympathetic ganglion and by a plexus of veins that unite to form the vertebral vein at the caudal part of the neck. It is situated anterior to the ventral rami of the upper cervical nerves and pursues an almost vertical course as far as the foramen in the transverse process of the axis. After coursing through this foramen, it runs lateralward to the foramen in the transverse process of the atlas. The **third part** of the vertebral artery issues from the latter foramen on the medial aspect of the rectus capitis lateralis muscle and curves behind the lateral mass of the atlas, the anterior ramus of the first cervical nerve being on its medial side. It then lies in the groove on the superior surface of the posterior arch of the atlas and enters the vertebral canal by passing under the arched margin of the posterior atlanto-occipital membrane. This part of the artery is covered by the semispinalis capitis muscle and is contained in the **suboccipital triangle**, a region bounded by the rectus capitis posterior major, the obliquus superior, and the obliquus inferior (Fig. 6-22). The first cervical or suboccipital nerve lies between the artery and the posterior arch of the atlas. The **fourth part** of the vertebral artery pierces the dura mater and inclines medially to the anterior aspect of the medulla oblongata. It is placed between the hypoglossal nerve and the anterior root of the first cervical nerve and beneath the first digitation of the ligamentum denticulatum. At the inferior border of the pons, it unites with the vessel of the opposite side to form the **basilar artery.**

Branches of the Vertebral Artery

The branches of the vertebral artery may be divided into two sets: those given off in the neck and those within the cranium.

Cervical Branches
  Spinal
  Muscular

Cranial Branches
  Meningeal
  Posterior spinal
  Anterior spinal
  Posterior inferior cerebellar
  Medullary

SPINAL BRANCHES. Spinal branches of the vertebral artery enter the vertebral canal through the intervertebral foramina, and each divides into two branches. Of these, one passes along the roots of the nerves to supply the spinal cord and its membranes, anastomosing with the other arteries of the spinal cord. The other divides into an ascending and a descending branch, which unite with similar branches from arteries of adjacent segments above and below, so that two lateral anastomotic chains are formed on the dorsal surfaces of the bodies of the vertebrae, near the attachment of the pedicles. From these anastomotic chains some branches are supplied to the periosteum and the bodies of the vertebrae, while others form communications with similar branches from the opposite side. Arising from these communications are small twigs that join similar branches, both above and below, to form a central anastomotic chain on the dorsal surface of the bodies of the vertebrae.

MUSCULAR BRANCHES. Muscular branches are given off the vertebral artery to the deep muscles of the neck as that vessel curves around the lateral mass of the atlas (Fig. 8-35). They anastomose with the occipital artery and with the ascending and deep cervical arteries.

MENINGEAL BRANCH. One or two small meningeal branches arise from the vertebral artery at the level of the foramen magnum. They ramify between the bone and dura mater in the cerebellar fossa, and they supply the falx cerebelli.

POSTERIOR SPINAL ARTERY. The posterior spinal artery may originate from the intracranial portion of the vertebral artery at the side of the medulla oblongata, or from the posterior inferior cerebellar artery. Passing dorsally, it descends on the medulla beyond the spinomedullary junction, lying ventral to the dorsal roots of the spinal nerves. As the vessel descends on the posterolateral aspect of the spinal cord, it becomes reinforced by a succession of spinal branches and is thereby continued to the end of the spinal cord and the cauda equina. Branches from the posterior spinal arteries form a free anastomosis around the dorsal roots of the spinal nerves, communicating by means of tortuous transverse branches with vessels of the opposite side. Close to its origin each posterior spinal artery gives off an ascending branch that ramifies in the lateral wall of the fourth ventricle.

ANTERIOR SPINAL ARTERY.   The anterior spinal artery is a vessel formed by the junction of two branches, one from each vertebral artery (Figs. 8-28; 8-32; 8-35; 8-36). The branches arise near the point at which the two vertebral arteries join to form the basi-

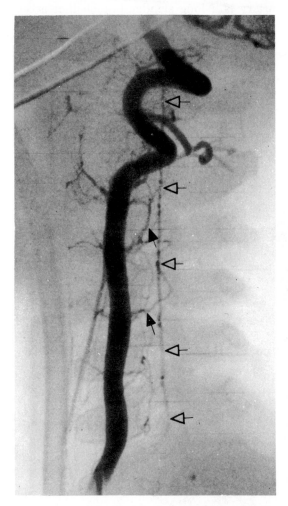

FIG. 8-35. Angiogram of the vertebral artery showing the anterior spinal artery descending in the spinal cord (open arrows) and anastomotic radicular branches (solid arrows) which enter the cord with the cervical roots of the spinal nerves. (From Newton and Mani, *Radiology of the Skull and Brain*, 1974.)

lar artery. Each branch descends ventral to the medulla and generally unites within about 2 cm from its origin at about the level of the foramen magnum. The single trunk thus formed descends in the anterior median fissure of the spinal cord, and it is reinforced by a succession of small branches that enter the vertebral canal through the intervertebral foramina. These branches are derived from the vertebral and the ascending cervical branch of the inferior thyroid artery in the neck, the intercostal arteries in the thorax, and the lumbar, iliolumbar, and lateral sacral arteries of the posterior abdominal and pelvic wall. These vessels anastomose by means of ascending and descending branches to form a single anterior median artery, which extends as far as the lower part of the spinal cord and is continued as a slender twig on the filum terminale. Coursing within the pia mater, the anterior spinal artery helps to supply that membrane and the substance of the spinal cord, and it also sends off branches at its caudal part to be distributed to the cauda equina.

POSTERIOR INFERIOR CEREBELLAR ARTERY. The posterior inferior cerebellar artery is the largest branch of the vertebral artery (Figs. 8-28; 8-32; 8-33; 8-36). Although this artery varies greatly in its course and distribution, it most often arises from the vertebral approximately 1.5 cm before the vertebrals unite to form the basilar, but occasionally it is absent. It courses at first laterally and then posteriorly around the lower end of the olive of the medulla oblongata. Passing between the rootlets of the glossopharyngeal and vagus nerves, over the inferior peduncle to the inferior surface of the cerebellum, the artery divides into two branches. The **medial branch** is continued posteriorly to the notch between the two hemispheres of the cerebellum, while the **lateral branch** supplies the inferior surface of the cerebellum as far as its lateral border. Here it anastomoses with the anterior inferior cerebellar and the superior cerebellar branches of the basilar artery. Branches from the posterior inferior cerebellar artery also supply penetrating branches to the medulla as well as to the choroid plexus of the fourth ventricle.

MEDULLARY BRANCHES.   Several small medullary vessels branching from the vertebral artery are distributed to the medulla oblongata.

BASILAR ARTERY

The basilar artery is formed by the junction of the two vertebral arteries (Figs. 8-28; 8-32; 8-34; 8-36). As a single vascular trunk, it courses superiorly in a shallow median groove on the anterior surface of the pons. It extends from the pontomedullary sulcus, which demarcates the rostral medulla and caudal pons, to the level of the emerging oculomotor nerves from the caudal midbrain. The basilar artery is often tortuous and may be slightly curved in its rostral course along the brain stem. Its length varies from 2.5 to 4.0 cm, and after passing between the two oculomotor nerves, the artery terminates by dividing into the two posterior cerebral arteries.

Branches of the Basilar Artery

In its course the basilar artery gives off the following branches on both sides:

Pontine
Labyrinthine
Anterior inferior cerebellar
Superior cerebellar
Posterior cerebral

PONTINE BRANCHES. The pontine branches of the basilar artery consist of a number of small vessels that supply the pons and midbrain (Fig. 8-32). Median vessels arise at right angles from the posterior aspect of the basilar and enter the pons along its anterior median groove. Transverse branches arise from the posterolateral and lateral aspects of the basilar, and in their circumferential course around the brain stem, they send penetrating vessels into the pons.

LABYRINTHINE ARTERY. Also known as the internal auditory artery, the labyrinthine artery is a long slender branch that arises near the middle of the basilar artery (Figs. 8-28; 8-32). It accompanies the facial and vestibulocochlear nerves through the internal acoustic meatus and is distributed to the internal ear. It often arises from the anterior inferior cerebellar artery.

ANTERIOR INFERIOR CEREBELLAR ARTERY. The anterior inferior cerebellar artery generally arises from the lower third of the basilar artery (Figs. 8-28; 8-34; 8-36). Its course is

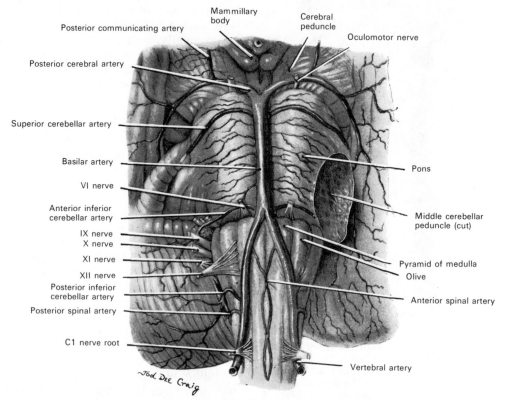

Fig. 8-36. The ventral surface of the medulla and pons showing the vertebral and basilar arteries and their branches. (From Taveras and Wood, *Diagnostic Neuroradiology,* 1964.)

somewhat variable, but initially it usually courses laterally and posteriorly and lies in contact with either the dorsal (33%) or ventral (67%) aspect of the abducens nerve. Continuing its downward course, the vessel frequently lies ventral to the roots of the facial and vestibulocochlear nerves. Often it forms a loop in relation to these nerves at the level of the internal acoustic meatus. The main trunk of the anterior inferior cerebellar artery sends small branches into the lateral aspect of the pons. At the cerebellopontine angle, the artery divides into **lateral** and **medial branches,** which are distributed to the anterolateral and anteromedial parts of the inferior surface of the cerebellum. The anterior inferior cerebellar artery anastomoses with the posterior inferior cerebellar branch of the vertebral artery.

SUPERIOR CEREBELLAR ARTERY. The superior cerebellar artery arises a few millimeters proximal to the termination of the basilar artery (Figs. 8-28; 8-34; 8-36). It passes lateralward immediately caudal to the oculomotor nerve, which separates it from the posterior cerebral artery. It then winds around the cerebral peduncle, close to the trochlear nerve. Arriving at the superior surface of the cerebellum, which it supplies, it divides into branches that ramify in the pia mater and anastomose with those of the inferior cerebellar arteries. Several branches go to the midbrain, the pineal body, the anterior medullary velum, and the tela choroidea of the third ventricle.

POSTERIOR CEREBRAL ARTERY. Embryologically, the posterior cerebral artery develops as a branch of the internal carotid artery and, in some instances, remains a branch of the internal carotid in the adult. In most instances, however, the posterior cerebral arteries are the terminal branches of the basilar artery, and they maintain their attachment to the internal carotid on each side by way of the posterior communicating artery, forming the posterior part of the arterial circle of Willis. The posterior cerebral artery is larger than the superior cerebellar artery, and it is separated from that vessel near its origin by the oculomotor nerve (Figs. 8-28; 8-32; 8-36). Passing laterally, parallel to the superior cerebellar artery and receiving the posterior communicating artery from the internal carotid, it winds around the cerebral peduncle and reaches the tentorial surface of the occipital lobe of the cerebrum,

where it breaks up into branches for the supply of the temporal and occipital lobes. The posterior cerebral artery gives off central, posterior choroidal, and cortical branches.

CENTRAL BRANCHES. The central branches of the posterior cerebral artery can be divided into mesencephalic and thalamic branches. The **mesencephalic branches** arise from the posterior surface of each posterior cerebral artery and enter the posterior perforated substance to supply the interpeduncular region of the midbrain as well as the cerebral peduncles themselves. Such important structures as the oculomotor nuclei, mesencephalic reticular formation, substantia nigra, and the corticospinal paths at the midbrain level are supplied by these vessels. Additionally, *circumflex mesencephalic branches* pass around the midbrain to supply perforating arteries to the peduncles and to the substantia nigra, as well as longer branches to dorsal tegmental structures. Anterior and posterior **thalamic branches** arise from the posterior cerebral artery to perforate and supply both hypothalamic and thalamic centers (Fig. 8-37). The *anterior thalamic branches* supply the more caudal parts of the hypothalamus, including the mammillary bodies, as well as the optic tract. Certain of these vessels ascend through the hypothalamus to achieve the thalamus and thereby supply its anterior and ventromedial nuclei. Some *posterior thalamic branches* ascend through the posterior perforated substance to supply the more posterior thalamic nuclei, while posterior thalamogeniculate vessels course more superiorly to supply the geniculate bodies (Newton and Potts, 1974).

POSTERIOR CHOROIDAL BRANCHES. Two or more posterior choroidal branches arise from the posterior cerebral artery. The *(posterior) medial choroidal artery* branches near the origin of the posterior cerebral and courses dorsally and medially around the midbrain, supplying the corpora quadrigemina and the pineal gland (Figs. 8-37; 8-38). It then courses forward to supply the roof of the third ventricle, ramifies within the choroid plexus of that ventricle, and sends penetrating branches to the dorsal medial nucleus of the thalamus and frequently to the pulvinar. The *(posterior) lateral choroidal artery* may arise singly or as multiple vessels from the posterior cerebral, but it may also arise from the parieto-occipital branch of the posterior cerebral artery

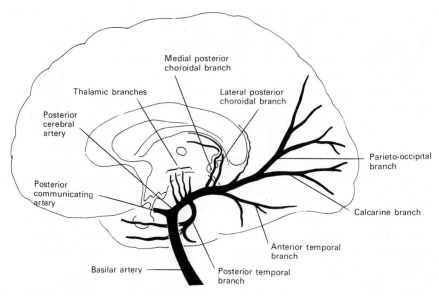

Fɪɢ. 8-37. The branches of the posterior cerebral artery as viewed on the medial aspect of the right cerebral hemisphere. (From Krayenbühl and Yasargil, *Cerebral Angiography*, 1968.)

(Figs. 8-37; 8-38). One branch of the lateral choroidal artery courses laterally and enters the choroid fissure to supply the choroid plexus of the temporal horn. Another branch courses back behind the pulvinar to supply the choroid plexus of the lateral ventricle. Branches perforate into the thalamus to supply the dorsomedial nucleus, the pulvinar, and the lateral geniculate bodies. Branches of the posterior medial and lateral choroidal arteries anastomose with each other (Newton and Potts, 1974).

CORTICAL BRANCHES. Generally there are

four cortical branches of the posterior cerebral artery that supply the medial surface of the temporal and occipital lobes. They are the anterior temporal, posterior temporal, parieto-occipital, and occipital (calcarine) branches (Figs. 8-30; 8-31; 8-37). The *anterior temporal branch* extends laterally and anteriorly beneath the temporal lobe to supply the uncus and parahippocampal gyrus, where it anastomoses with the anterior temporal branch of the middle cerebral artery. The *posterior temporal branch* courses back along the parahippocampal gyrus to supply

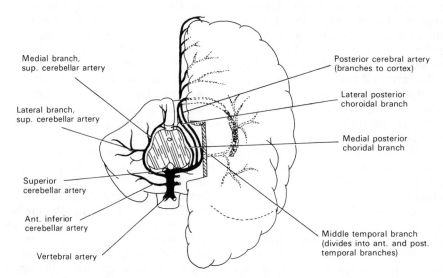

Fɪɢ. 8-38.   The course and branches of the posterior cerebral artery after it arises from the basilar artery. (From Wilson, *The Anatomic Foundation of Neuroradiology of the Brain*, 1972.)

the medial and lateral occipitotemporal gyri. The *parieto-occipital branch* usually divides into a number of smaller vessels that course superiorly and posteriorly on the medial surface of the cerebral hemisphere to supply the precuneate and cuneate gyri. The *occipital (calcarine) branch* supplies the primary visual cortex by following the calcarine sulcus as far as the occipital pole. It supplies the lingual and cuneate gyri and frequently the occipital and parieto-occipital branches overlap somewhat in their distribution to the visual cortical receiving area.

THYROCERVICAL TRUNK

The thyrocervical trunk is a short thick vessel that arises from the first portion of the subclavian artery, close to the medial border of the scalenus anterior muscle (Figs. 8-20; 8-33; 8-34; 8-39; 8-40). Almost immediately it divides into the following three arteries:

Inferior thyroid
Suprascapular
Transverse cervical

Inferior Thyroid Artery

The inferior thyroid artery courses cranially anterior to the vertebral artery and longus colli muscle. It then loops medially dorsal to the carotid sheath and its contents, as well as to the sympathetic trunk; the middle cervical ganglion often rests on the vessel (Figs. 8-20; 8-33; 8-39; 8-40). Reaching the lower border of the thyroid gland, the inferior thyroid artery divides into two branches, which supply the inferior parts of the gland and anastomose with the superior thyroid artery and with the inferior thyroid artery of the opposite side. The recurrent laryngeal nerve courses superiorly in the neck, generally dorsal but occasionally ventral to the artery.

Branches of the Inferior Thyroid Artery

In addition to the terminal branches that supply the thyroid gland, the inferior thyroid artery gives off the following branches:

Inferior laryngeal
Tracheal
Esophageal
Ascending cervical
Muscular

INFERIOR LARYNGEAL ARTERY. The inferior laryngeal artery ascends upon the trachea to the dorsal part of the larynx under cover of the inferior pharyngeal constrictor muscle and in company with the recurrent laryngeal nerve. It supplies the muscles and mucous membrane of this region of the larynx, anastomosing with the branch from the opposite side and with the superior laryngeal branch of the superior thyroid artery.

TRACHEAL BRANCHES. Tracheal branches of the inferior thyroid artery are distributed to the trachea and anastomose inferiorly with the bronchial arteries and superiorly with tracheal branches from the superior thyroid artery.

ESOPHAGEAL BRANCHES. The esophageal branches of the inferior thyroid artery supply the esophagus and anastomose with esophageal branches that arise directly from the aorta.

ASCENDING CERVICAL ARTERY. The ascending cervical artery arises from the inferior thyroid as that vessel passes dorsal to the carotid sheath (Figs. 8-20; 8-33; 8-39; 8-40). It may arise directly from the thyrocervical trunk. It ascends in the neck on the anterior tubercles of the transverse processes of the cervical vertebrae in the interval between the scalenus anterior and longus capitis muscles. Lying parallel and medial to the phrenic nerve, the ascending cervical artery supplies some deep muscles of the neck with twigs that anastomose with branches of the vertebral artery. In addition, it sends one or two **spinal branches** into the vertebral canal through the intervertebral foramina to be distributed to the spinal cord and its membranes, and to the bodies of the vertebrae in the same way as spinal branches are distributed from the vertebral artery. It anastomoses with the ascending pharyngeal and occipital arteries, with branches of the external carotid artery, and with the vertebral artery.

MUSCULAR BRANCHES. Muscular branches of the inferior thyroid artery supply the infrahyoid, the longus colli, the scalenus anterior, and the inferior pharyngeal constrictor muscles.

Suprascapular Artery

The suprascapular artery passes initially downward and laterally across the scalenus anterior and the phrenic nerve, being cov-

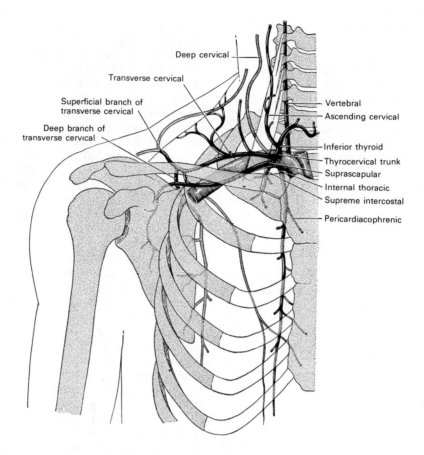

Deep cervical

Transverse cervical

Superficial branch of
transverse cervical

Deep branch of
transverse cervical

Vertebral

Ascending cervical

Inferior thyroid

Thyrocervical trunk

Suprascapular

Internal thoracic

Supreme intercostal

Pericardiacophrenic

FIG. 8-39. The subclavian artery and its branches showing the so-called *Type I* pattern in which the *transverse cervical artery* branches from the thyrocervical trunk, crosses the lower neck, and divides into a *superficial branch,* which ascends in the neck, and a *deep branch,* which descends to the medial scapular region.

ered by the sternocleidomastoid muscle (Figs. 8-20; 8-33; 8-34; 8-39; 8-40). It then crosses in front of the subclavian artery and the brachial plexus, and running deep to and parallel with the clavicle and subclavius muscle and deep to the inferior belly of the omohyoid muscle, it reaches the superior border of the scapula. It passes over the superior transverse scapular ligament by which it is separated from the suprascapular nerve to enter the supraspinous fossa. Within the fossa it lies directly on the bone and sends branches to the overlying supraspinatus muscle, which it supplies. It then winds laterally around the neck of the scapula, descending through the great scapular notch under cover of the inferior transverse scapular ligament to reach the infraspinous fossa. Here it supplies the infraspinatus muscle and anastomoses with the scapular circumflex branch of the subscapular artery and the deep branch of the transverse cervi-

cal (descending scapular) artery. Besides distributing branches to the sternocleidomastoid, subclavius, and neighboring muscles, the suprascapular artery gives off a **suprasternal branch,** which crosses over the sternal end of the clavicle to the skin of the chest, and an **acromial branch,** which pierces the trapezius to supply the skin over the acromion, anastomosing with the acromial branch of the thoracoacromial artery. As the artery passes over the superior transverse scapular ligament, it sends a branch into the subscapular fossa, where it ramifies beneath the subscapularis muscle and anastomoses with the subscapular artery and deep branch of the transverse cervical (descending scapular) artery. It also sends **articular branches** to the acromioclavicular and shoulder joint and **nutrient branches** to the clavicle and scapula. In about 25% of cadavers studied, the suprascapular artery arises from the third part of the subclavian artery.

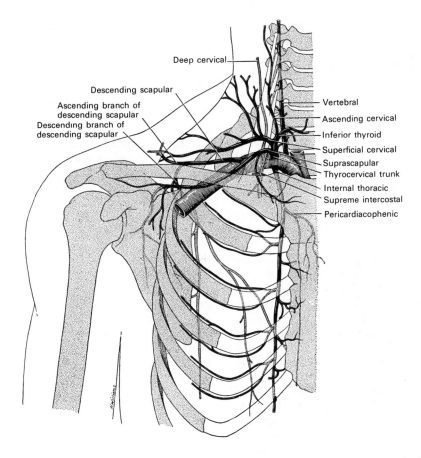

Deep cervical
Descending scapular
Ascending branch of
descending scapular
Descending branch of
descending scapular

Vertebral
Ascending cervical
Inferior thyroid
Superficial cervical
Suprascapular
Thyrocervical trunk
Internal thoracic
Supreme intercostal
Pericardiacophenic

Fig. 8-40. The subclavian artery and its branches showing the so-called *Type II* pattern in which the superficial cervical artery is derived from the thyrocervical trunk and the descending scapular artery arises from the third part of the subclavian. These vessels are distributed to the same regions as the superficial and deep branches of the transverse cervical artery of the *Type I* pattern.

Transverse Cervical Artery

The transverse cervical artery, in more than half the specimens studied, *arises as a single vessel* and eventually divides into two main branches, a superficial and a deep branch (Fig. 8-39). In these instances, the transverse cervical artery stems from the thyrocervical trunk, but it may occasionally arise as a single vessel directly from the third part of the subclavian artery rather than from the thyrocervical trunk. When the transverse cervical artery does arise from the thyrocervical trunk, it passes laterally across the posterior cervical triangle in a position somewhat above and behind the suprascapular artery. Medially, it crosses superficial to the scalenus anterior and phrenic nerve and lies deep to the sternocleidomastoid muscle. Laterally, it crosses the trunks of the brachial plexus and is covered only by the platysma, the investing layer of deep fascia, and the inferior belly of the omohyoid. As the artery approaches the anterior margin of the trapezius, it divides into a **superficial branch** and a **deep branch.**

In another large percentage of cadavers studied, *there is no transverse cervical artery* as such, branching either from the thyrocervical trunk or from the third part of the subclavian artery (Fig. 8-40). In these instances, the regions supplied by the superficial and deep branches of the transverse cervical artery become supplied by other arteries. Thus, the *superficial branch of the transverse cervical artery* becomes replaced by the **superficial cervical artery,** and the *deep branch of the transverse cervical artery* becomes replaced by the **descending scapular artery** (also known as the dorsal scapular artery, a term no longer found in the *Nomina Anatomica*).

Because these two different vascular patterns are about equally represented in the specimens studied, the descriptions of the superficial and deep branches of the transverse cervical artery will be followed by descriptions of the superficial cervical artery and the descending scapular artery.

SUPERFICIAL BRANCH OF TRANSVERSE CERVICAL ARTERY. The superficial branch of the transverse cervical artery is the principal blood supply to the trapezius muscle (Fig. 8-39). Lying along the deep surface of this muscle, the artery divides into an ascending and a descending branch. The *ascending branch* courses superiorly along the anterior border of the trapezius, distributing branches to it and to neighboring muscles and anastomosing with the superficial part of the descending branch of the occipital artery. The *descending branch* accompanies the accessory nerve along the deep surface of the trapezius, supplying the muscle.

DEEP BRANCH OF TRANSVERSE CERVICAL ARTERY. The deep branch of the transverse cervical artery supplies vessels to the levator scapulae and neighboring deep cervical muscles (Fig. 8-39). Passing deep to the levator scapulae muscle, it reaches the superior angle of the scapula and passes down along the vertebral border to the inferior angle, lying deep to the rhomboid muscles and accompanying the dorsal scapular nerve. It supplies *muscular branches* to the rhomboid, serratus posterior superior, subscapularis, supraspinatus, and infraspinatus muscles and to other neighboring muscles. These branches anastomose with the suprascapular, subscapular, and circumflex scapular arteries.

SUPERFICIAL CERVICAL ARTERY. The superficial cervical artery is seen in cadavers that do not have a transverse cervical artery, and it corresponds to the superficial branch of the transverse cervical (Fig. 8-40). It arises from the thyrocervical trunk, as would the transverse cervical artery, but no vessel comparable to the deep branch of the transverse cervical artery stems from the superficial cervical artery. It passes laterally across the posterior triangle of the neck, deep to the sternocleidomastoid muscle and superficial to the scalenus anterior muscle and phrenic nerve. It then courses cranially across the trunks of the brachial plexus to the anterior border of the trapezius muscle, where its *ascending branch* supplies the upper part of the trapezius and neighboring muscles, anastomosing with the superficial descending branch of the occipital artery, and its *descending branch* lies against the deep surface of the trapezius, supplying it with branches and accompanying the accessory nerve.

DESCENDING SCAPULAR ARTERY. The descending scapular artery (sometimes still carelessly called the dorsal scapular artery) is observed in specimens in which there is no transverse cervical artery. Its distribution corresponds to that of the deep branch of the transverse cervical artery (Fig. 8-40). When present, the descending scapular artery has an independent origin from the third part of the subclavian artery instead of from the thyrocervical trunk. Occasionally it may arise from the second part of the subclavian artery. It passes upward for a short distance, then loops around the brachial plexus, frequently passing between the anterior and posterior divisions of the upper trunk. The vessel then courses downward toward the scapula and, at the border of the levator scapulae, divides into an ascending and a descending branch. The *ascending branch* supplies the levator scapulae and neighboring deep cervical muscles. The *descending branch,* usually double, passes deep to the levator scapulae to the superior angle of the scapula, and then descends anterior to the rhomboid muscles along the medial border of the bone as far as its inferior angle. When there are two descending branches, the more medial is accompanied by the dorsal scapular nerve, while the larger, more lateral vessel lies on the costal surface of the serratus anterior muscle. The descending branch(es) supplies the rhomboid, latissimus dorsi, and trapezius muscles and anastomoses with the suprascapular and subscapular arteries, and with the posterior branches of some of the intercostal arteries.

INTERNAL THORACIC ARTERY

The internal thoracic artery arises from the first portion of the subclavian artery about 2 cm above the clavicle, opposite the thyrocervical trunk and close to the medial border of the scalenus anterior muscle (Fig. 8-41). It descends behind the cartilages of the upper six ribs at a distance of about 1.25 cm from the margin of the sternum. At the level of the sixth intercostal space, the internal thoracic artery terminates by dividing into the **musculophrenic** and **superior epigastric** arteries.

RELATIONS. As it descends toward the sternum, the **internal thoracic artery** lies deep to the sternal end of the clavicle, the subclavian and internal jugular veins, and the first costal cartilage. As it enters the thorax, it passes close to the lateral aspect of the brachiocephalic vein, and the phrenic nerve crosses it obliquely from its lateral to its medial aspect. Below the first costal cartilage, it descends almost vertically to its point of bifurcation. It is covered anteriorly by the cartilages of the first six ribs, by the intervening internal intercostal muscles, and by external intercostal membranes. The vessel is crossed by the terminal portions of the first six intercostal nerves. Between the artery and the pleura is a deep fascial plane as far as the third costal cartilage, and below this level the transversus thoracis muscle separates the vessel from the pleura. It is accompanied by a chain of lymph nodes and by a pair of veins that unite to form a single vessel, which ascends medial to the artery and ends in the corresponding brachiocephalic vein.

Branches of the Internal Thoracic Artery

Among other structures, the branches of the internal thoracic artery help to supply the pericardium, pleura, thymus, and bronchi, as well as the transverse thoracis, intercostal, and pectoralis major muscles, the rectus abdominis and other anterior abdominal muscles, the diaphragm, and even the mammary gland in the female. Its branches are:

Pericardiacophrenic
Mediastinal
Thymic
Bronchial
Sternal
Anterior intercostal
Perforating
Lateral costal branch
Musculophrenic
Superior epigastric

PERICARDIACOPHRENIC ARTERY. The pericardiacophrenic artery is a long, slender, and somewhat tortuous vessel that arises from the internal thoracic artery just after the latter vessel has entered the thorax (Figs. 7-24; 8-39; 8-40). Accompanying the phrenic nerve to the diaphragm, which it helps to supply, the pericardiacophrenic artery courses between the pleura and pericar-

dium, and it sends branches to both these structures. It anastomoses with the musculophrenic and the superior and inferior phrenic arteries (Fig. 7-24).

MEDIASTINAL BRANCHES. The mediastinal branches of the internal thoracic artery are small vessels, distributed to the areolar tissue and lymph nodes in the anterior mediastinum, and they also supply the upper part of the sternocostal surface of the pericardium.

THYMIC BRANCHES. In the adult, the thymic branches are small and supply the remains of the thymus. In the infant and child, however, they are more significant vessels.

BRONCHIAL BRANCHES. The bronchial branches from the internal thoracic artery, when present, are distributed to the bronchi and lower part of the trachea.

STERNAL BRANCHES. Sternal branches of the internal thoracic artery are distributed to the transversus thoracis muscle and to the posterior surface of the sternum.

The mediastinal, pericardial, bronchial, and sternal branches, together with some twigs from the pericardiacophrenic, anastomose with branches from the posterior, intercostal, and bronchial arteries from the thoracic aorta to form a **subpleural mediastinal plexus.**

ANTERIOR INTERCOSTAL ARTERIES. The anterior intercostal branches of the internal thoracic artery supply the first five or six intercostal spaces, the remaining more caudal spaces being supplied by the musculophrenic branch. In each space, two small arteries pass laterally from the internal thoracic, one each at the upper and lower margins of the intercostal space (Fig. 8-41). They supply the intercostal muscles and finally anastomose with the external intercostal arteries from the aorta. At first the anterior intercostal arteries lie between the pleura and the internal intercostal muscles, and then more laterally between the innermost and internal intercostal muscles. Some branches perforate through the external intercostal muscles to supply the pectoral muscles and the mammary gland.

PERFORATING BRANCHES. The perforating branches from the internal thoracic artery pierce the anterior thoracic wall in the first five or six intercostal spaces close to the sternum (Fig. 8-41). These vessels course with the anterior cutaneous branches of the intercostal nerves and penetrate the intercostal

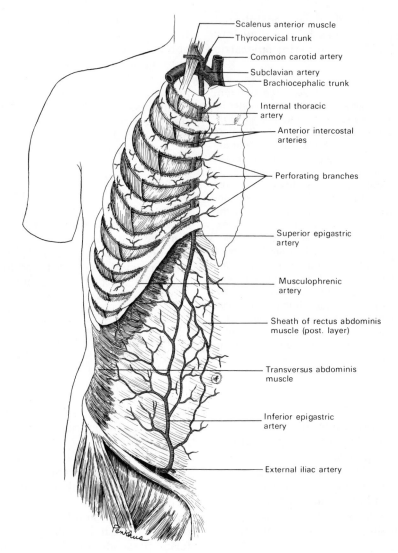

Scalenus anterior muscle
Thyrocervical trunk
Common carotid artery
Subclavian artery
Brachiocephalic trunk
Internal thoracic artery
Anterior intercostal arteries
Perforating branches
Superior epigastric artery
Musculophrenic artery
Sheath of rectus abdominis muscle (post. layer)
Transversus abdominis muscle
Inferior epigastric artery
External iliac artery

FIG. 8-41.   The left internal thoracic artery and its branches. Observe the anastomosis between the superior and inferior epigastric arteries.

muscles and pectoral fascia to achieve the deep surface of the pectoralis major muscle. They curve laterally as they become more superficial and supply the pectoralis major muscle and the skin. The arteries of the second, third, and fourth spaces give **mammary branches** to the breast in the female and become greatly enlarged during lactation.

LATERAL COSTAL BRANCH.   Although frequently absent or very small, the lateral costal branch arises at times from the internal thoracic artery near the first rib. When present (about 25% of cadavers), it descends lateral to the costal cartilages and behind the ribs and anastomoses with the upper anterior intercostal arteries.

MUSCULOPHRENIC ARTERY.   The musculophrenic artery is one of the two terminal branches of the internal thoracic artery, and it is directed obliquely downward and laterally, behind the cartilages of the eighth, ninth, and tenth ribs (Fig. 8-41). It perforates the diaphragm near the eighth or ninth costal cartilage, and ends, considerably reduced in size, opposite the last intercostal space. It gives off two anterior intercostal branches to the seventh, eighth, and ninth intercostal spaces. These diminish in size as the intercostal spaces decrease in length and are distributed in a manner precisely similar to the anterior intercostals that branch directly from the internal thoracic. Some other branches of the musculophrenic artery course to the lower part of the pericardium, some course dorsally to the diaphragm, and still others course down to the abdominal muscles.

SUPERIOR EPIGASTRIC ARTERY. The superior epigastric artery is a terminal branch of the internal thoracic artery, and it continues in the original direction of the parent vessel (Fig. 8-41). It descends through the interval between the costal and sternal attachments of the diaphragm and enters the sheath of the rectus abdominis muscle. At first the superior epigastric muscle lies behind the rectus muscle, but then perforates and supplies the muscle and thereby anastomoses with the inferior epigastric artery from the external iliac. Branches eventually perforate the anterior layer of the rectus sheath and supply other anterior abdominal wall muscles and the overlying skin. A small branch passes in front of the xiphoid process and anastomoses with the artery of the opposite side. The superior epigastric artery also gives some twigs to the diaphragm, while from the artery on the right side, small branches extend into the falciform ligament of the liver and anastomose with the hepatic artery.

COSTOCERVICAL TRUNK

On the left side the costocervical trunk arises from the first part of the subclavian artery, medial to the scalenus anterior muscle. On the right side, however, it arises from the dorsal aspect of the second part of the subclavian artery, deep to the serratus anterior muscle. It courses upward and then arches backward above the cupula of the cervical pleura (Figs. 8-24; 8-34). Passing downward in front of the neck of the first rib, it divides into the supreme (highest) intercostal and deep cervical arteries.

SUPREME INTERCOSTAL ARTERY. The supreme or highest intercostal artery descends under the pleura anterior to the necks of the first and second ribs and anastomoses with the most superior of the intercostal arteries derived from the aorta, the one that courses in the third intercostal space (Figs. 8-24; 8-39, 8-40). As the supreme intercostal artery crosses the neck of the first rib, it lies medial to the ventral ramus of the first thoracic nerve and lateral to the cervicothoracic ganglion of the sympathetic trunk. The **first pos-**terior intercostal artery branches from the supreme intercostal artery within the first intercostal space, and its distribution is the same as that of the lower posterior intercostal arteries that arise from the aorta. The supreme intercostal artery then descends, usually to become the **second posterior intercostal artery,** being joined by a branch from the highest of the aortic posterior intercostal arteries (the third). This pattern is more common on the right side, but at times the second posterior intercostal artery is derived from the aorta.

DEEP CERVICAL ARTERY. The deep cervical artery arises in most cases from the costocervical trunk and is analogous to the dorsal branch of an aortic posterior intercostal artery (Figs. 8-24; 8-39; 8-40). Occasionally it springs as a separate branch from the subclavian artery. Passing dorsally above the eighth cervical nerve and between the transverse process of the seventh cervical vertebra and the neck of the first rib, it ascends along the dorsum of the neck, between the semispinales capitis and cervicis muscles, as high as the axis. It supplies these and adjacent muscles and anastomoses with the deep division of the descending branch of the occipital artery and with branches of the vertebral. It gives off a spinal twig, which enters the vertebral canal through the intervertebral foramen between the seventh cervical and first thoracic vertebrae.

DESCENDING SCAPULAR ARTERY

In about half the cadavers studied, a descending scapular artery branches from the third part of the subclavian artery (Fig. 8-40). When this vessel is found, its distribution is usually similar to that of the deep branch of the transverse cervical artery. The vascular pattern in this region and the course of the descending scapular artery are described with the transverse cervical artery.

Arteries of the Upper Limb

The axis artery, which supplies the upper limb, continues from its origin to the elbow as a single trunk, but different portions of it are named according to the regions through which they pass. The part of the axis artery that extends from its origin to the outer border of the first rib is termed the **subclavian artery.** Beyond this point and as far as the distal border of the axilla, formed by the lower margin of the teres major, it is named the **axillary artery.** From the axilla to a point just beyond the elbow joint in the proximal forearm, the vessel is called the **brachial artery;** the brachial artery ends by dividing into two branches, the **radial** and **ulnar arteries.**

THE AXILLA

The **axilla** is a cone-shaped or pyramidal region located between the upper lateral part of the chest wall and the medial aspect of the arm. Its anatomy has great relevance because through the axilla course the *neurovascular structures* that supply the upper extremity and because its anterior surface is relatively unprotected and *vulnerable to injury.* Furthermore, the axilla receives many of the *lymphatic channels* that drain the anterior thoracic wall, making it a region of strategic importance to disease processes that occur in the *mammary gland* of females.

BOUNDARIES OF THE AXILLA. The **apex** of the axilla is directed superiorly toward the root of the neck and corresponds to the interval between the outer border of the first rib, the superior border of the scapula, and the posterior surface of the clavicle. Through the apex pass the axillary vessels and accompanying nerves, which descend behind the clavicle from the neck. The **base** of the axilla is directed downward and laterally and is formed by the skin and fascia of the armpit. It is narrow at the arm, but broader at the chest. The thick **axillary fascia,** which helps to form the base, extends between the lower border of the pectoralis major muscle anteriorly and the lower border of the latissimus dorsi muscle posteriorly. The **anterior wall** of the axilla is formed by the pectoralis major and minor muscles, the former covering the whole of this wall and the latter only its central part. The space between the upper border of the pectoralis minor and the clavicle is occupied by the **clavipectoral fascia.** The **posterior wall,** which extends somewhat more caudally than the anterior wall, is formed by the subscapularis muscle above and the teres major and latissimus dorsi muscles below. The **medial wall** is formed by the first four ribs with their corresponding intercostal muscles and part of the serratus anterior. The narrowed **lateral wall** of the axilla, where the anterior and posterior walls converge, is bounded by the intertubercular sulcus (bicipital groove) of the humerus, the coracobrachialis, and the biceps brachii.

CONTENTS OF THE AXILLA. Within the axilla are found the axillary vessels, the infraclavicular part of the brachial plexus and its branches, the lateral cutaneous branches of the intercostal nerves, and a large number of lymph nodes, together with a quantity of fat and loose areolar tissue. The axillary artery and vein, with the brachial plexus of nerves, extend obliquely through the axilla from its apex to its base and are placed much nearer the anterior wall than the posterior wall. The axillary vein lies superficial to and on the thoracic aspect of the artery and partially conceals it. In the anterior part of the axilla and in contact with the pectoral muscles are the thoracic branches of the axillary artery. Coursing downward along the lateral aspect of the chest wall, and frequently along the lateral margin of the pectoralis minor artery, extends the lateral thoracic artery. The subscapular vessels and nerves and the thoracodorsal nerve course along the posterior wall of the axilla, in contact with the subscapularis muscle. The scapular circumflex artery, which is a branch of the subscapular artery, winds around the lateral border of the subscapularis muscle to the dorsal aspect of the shoulder. Near the neck of the humerus, the posterior humeral circumflex vessels and the axillary nerve also curve dorsally to the shoulder. The upper medial aspect of the axilla has no large vessels, but does contain lymphatics and a few small branches from the supreme thoracic artery. There are in this region, however, the important long thoracic nerve, descending on the surface of the serratus anterior to which it is distributed, and the intercostobrachial nerve, which perforates the upper anterior aspect of the medial wall and passes across the axilla to the medial aspect of the arm.

The position and arrangement of the lymph nodes are described in the chapter on lymphatics.

Axillary Artery

The **axillary artery** is the continuation of the subclavian, and it commences at the outer border of the first rib and ends at the distal border of the tendon of the teres major muscle, where it is called the **brachial artery** (Figs. 8-42; 8-43). Its direction varies with the position of the limb: thus, the vessel is nearly straight when the arm is directed at right angles to the trunk, but concave superiorly when the arm is elevated, and convex superiorly when the arm lies by the side. At its origin the artery is quite deeply situated, but near its termination becomes superficial, being covered only by the skin and fascia. Because the pectoralis minor crosses the axillary artery anteriorly, for descriptive purposes it is convenient to divide the vessel into three portions: the first part lies proximal, the second part deep, and the third part distal to the pectoralis minor. Its branches are subject to considerable variation.

RELATIONS OF THE FIRST PART OF THE AXILLARY ARTERY. The first part of the axillary artery is covered *anteriorly* by the skin, the superficial fascia, the clavicular portion of the pectoralis major, and the clavipectoral fascia. The artery is crossed by the lateral pectoral nerve and the thoracoacromial and cephalic veins. *Posterior* to the first part of the axillary artery are the first intercostal space, the corresponding external intercostal muscle, the first and second digitations of the serratus anterior, the long thoracic and medial pectoral nerves, and the medial cord of the brachial plexus. *Laterally,* the vessel is in relation to the lateral and posterior cords of the brachial plexus, from which it is separated by a little areolar tissue. *Medially,* the axillary vein closely overlaps the artery. The first part of the axillary artery is enclosed, together with the axillary vein and brachial plexus, in the fibrous **axillary sheath,** which is continuous above with the prevertebral layer of the deep cervical fascia.

RELATIONS OF THE SECOND PART OF THE AXILLARY ARTERY. The second part of the axillary artery is covered *anteriorly* by the skin, the superficial and deep fasciae, the pectoralis major muscle, and, immediately anterior to the artery, the pectoralis minor muscle. *Dorsal* to the second part of the artery are the posterior cord of the brachial plexus and some areolar tissue, both of which intervene between the vessel and the subscapularis muscle. *Medially,* the axillary vein is separated from the artery by the medial cord of the brachial plexus and the medial pectoral nerve. *Laterally* is found the lateral cord of the brachial plexus. The cords of the brachial plexus thus surround the artery on three sides and separate it from direct contact with the vein and adjacent muscles.

RELATIONS OF THE THIRD PART OF THE AXILLARY ARTERY. The third part of the axillary artery extends from the lateral border of the pectoralis minor to the distal border of the tendon of the teres major. Its proximal portion is covered *anteriorly* by the pectoralis major muscle, its distal portion by only the integument and fascia. *Posteriorly* are found the subscapularis muscle and the tendons of the latissimus dorsi and teres major. On its *lateral aspect* is the coracobrachialis muscle and on its *medial* or thoracic *aspect* the axillary vein. The *nerves of the brachial plexus* bear the following relations to the third part of the artery: on the *lateral aspect* are the lateral root and trunk of the median nerve, and the musculocutaneous nerve for a short distance. On the *medial aspect* (between the axillary vein and artery) are found the ulnar nerve and (to the medial aspect of the vein) the medial brachial cutaneous nerve. *Anteriorly* are the medial root of the median nerve and the medial antebrachial cutaneous nerve, and posteriorly the radial and axillary nerves, the latter extending only as far as the distal border of the subscapularis muscle.

COLLATERAL CIRCULATION AFTER LIGATURE OF THE AXILLARY ARTERY. If the axillary artery is tied proximal to the origin of the thoracoacromial artery, the collateral circulation achieved is the same as that which results from ligature of the third part of the subclavian. If the ligature is made more distally between the thoracoacromial and the subscapular, the latter vessel, by its free anastomosis with the suprascapular and transverse cervical branches of the subclavian, becomes the chief agent in carrying on the circulation. If the lateral thoracic artery is distal to the ligature, it materially contributes to the circulation by its anastomoses with the intercostal and internal thoracic arteries. A ligature made distal to the origin of the subscapular artery is probably also distal to the origins of the two humeral circumflex arteries, in which case the chief vessels providing blood to more distal structures in the arm are the subscapular and the two humeral circumflex arteries anastomosing with branches of the deep brachial artery.

Branches of the
Axillary Artery

Although quite variable in their pattern, usually one branch is derived from the first part of the axillary artery, two from the second, and three from the third. They are:

First Part:
   Supreme thoracic
Second Part:
   Thoracoacromial
   Lateral thoracic
Third Part:
   Subscapular
   Posterior humeral circumflex
   Anterior humeral circumflex

SUPREME THORACIC ARTERY

The supreme thoracic artery is a small vessel which usually arises from the axillary artery just below the clavicle, but it may branch from the thoracoacromial artery or may be absent. It passes medially, generally behind the axillary vein, to the side of the chest in the first intercostal space (Figs. 8-42; 8-43). Coursing between the pectoralis minor and major, the supreme thoracic artery supplies branches to these muscles, and to the wall of the thorax, anastomosing with the internal thoracic and upper one or two intercostal arteries.

THORACOACROMIAL ARTERY

The thoracoacromial artery is a short trunk that arises anteriorly from the axillary artery, its origin being generally overlapped by the upper (or medial) edge of the pectoralis minor (Figs. 8-35; 8-42; 8-43). At the border of this muscle, the vessel pierces the clavipectoral fascia and divides into four branches: pectoral, acromial, clavicular, and deltoid.

PECTORAL BRANCHES. The pectoral branches of the thoracoacromial artery descend between the two pectoral muscles and are distributed to them and to the breast, anastomosing with the intercostal branches of the internal thoracic and with the lateral thoracic artery (Fig. 8-33).

ACROMIAL BRANCH. The acromial branch courses laterally, superficial to the coracoid process and deep to the deltoid muscle to which it gives branches (Figs. 8-33; 8-44). It then pierces that muscle and ends on the acromion in an arterial network, called the **acromial rete,** formed by branches from the suprascapular, the deltoid branch of the thoracoacromial artery, and the posterior humeral circumflex artery.

CLAVICULAR BRANCH. The clavicular branch of the thoracoacromial artery courses upward and medially beneath the clavicle to the sternoclavicular joint. It supplies this articulation and the subclavius muscle and anastomoses with the suprascapular artery.

DELTOID BRANCH. The deltoid branch frequently arises with the acromial branch and crosses laterally, superficial to the pectoralis minor (Fig. 8-33). It courses in the deltopectoral groove with the cephalic vein, passing between the pectoralis major and deltoid, and supplying branches to both muscles.

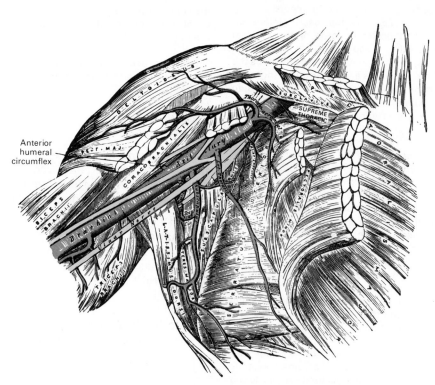

Anterior humeral circumflex

FIG. 8-42.   The axillary artery and its branches. This vascular pattern shows all six branches of the artery derived from the main stem.

LATERAL THORACIC ARTERY

The lateral thoracic artery (Figs. 8-42; 8-43) follows the lateral or lower border of the pectoralis minor to the side of the chest. It supplies the serratus anterior and both pectoral muscles and sends branches across the axilla to the axillary lymph nodes and subscapularis muscle. The lateral thoracic artery anastomoses with the internal thoracic, subscapular, and intercostal arteries, and with the pectoral branches of the thoracoacromial. The lateral thoracic artery is often (60%) a branch either of the thoracoacromial artery or of the subscapular artery. It comes directly from the axillary only in about 30% of cases.

LATERAL MAMMARY BRANCHES. In the female, the lateral mammary branches (sometimes called the external mammary branches) can be of considerable size. They turn around the free lateral margin of the pectoralis major and supply the breast.

SUBSCAPULAR ARTERY

The subscapular artery is usually the largest branch of the axillary artery and it arises at the distal border of the subscapularis muscle. Lying on the anterior surface of the subscapularis muscle, it descends under cover of the latissimus dorsi. After a short course of about 4 cm, the subscapular artery divides into the circumflex scapular and thoracodorsal arteries (Figs. 8-30; 8-42; 8-43).

CIRCUMFLEX SCAPULAR ARTERY. The circumflex scapular artery is generally larger than the thoracodorsal artery. It curves around the lateral border of the scapula, traversing the **triangular space** bordered below by the teres major, above by the subscapularis, and laterally by the long head of the triceps. The vessel then enters the infraspinous fossa between the teres minor and the scapula, remaining close to the bone (Fig. 8-44). It supplies branches to the infraspinatus and anastomoses with the suprascapular

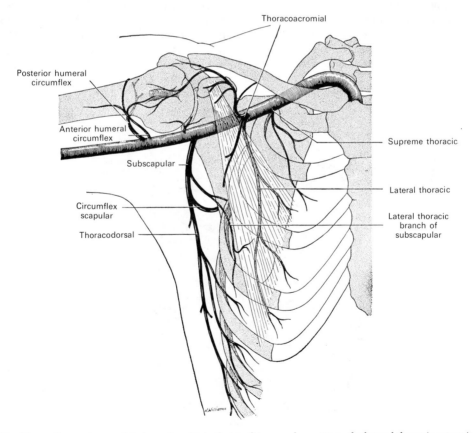

FIG. 8-43. The axillary artery and its branches. Note that in this vascular pattern, the lateral thoracic artery is derived from the thoracoacromial artery rather than directly from the axillary, and a lateral thoracic branch also comes from the subscapular artery. This pattern is seen in about 33% of cases.

artery and the deep branch of the transverse cervical artery (which in some cadavers is the descending branch of the descending scapular artery). In the triangular space, the circumflex scapular artery gives a branch to the subscapularis muscle. Another large branch continues along the lateral border of the scapula between the teres major and minor, supplying these muscles as well as the long head of the triceps and the deltoid. Finally, it anastomoses with the deep branch of the transverse cervical artery (or descending branch of the descending scapular) at the inferior angle of the scapula.

THORACODORSAL ARTERY. The thoracodorsal artery (Figs. 8-33; 8-43) is the continuation inferiorly of the subscapular artery through the axilla along the anterior border of the latissimus dorsi muscle. The vessel courses with the thoracodorsal nerve; it gives branches to the subscapularis and is the principal supply to the latissimus. It anastomoses with the circumflex scapular artery and the deep branch of the transverse cervical artery (or descending branch of the descending scapular). One or two sizable branches cross the axilla to supply the serratus anterior and intercostal muscles, anastomosing with the intercostal, lateral thoracic, and thoracoacromial arteries. When the lateral thoracic artery is small or lacking, a branch of the thoracodorsal may take its place.

POSTERIOR HUMERAL CIRCUMFLEX ARTERY

The posterior humeral circumflex artery arises from the axillary artery at the distal border of the subscapularis muscle (Figs. 8-42 to 8-44). It curves dorsally with the axillary nerve through the **quadrangular space** bounded by the subscapularis and teres minor muscles above, the teres major below, the long head of the triceps brachii medially, and the surgical neck of the humerus laterally (Fig. 8-44). The artery winds around the neck of the humerus and is distributed to the deltoid muscle and the shoulder joint, anastomosing with the anterior humeral circumflex and deep brachial arteries.

ANTERIOR HUMERAL CIRCUMFLEX ARTERY

The anterior humeral circumflex artery is considerably smaller than the posterior humeral circumflex, and it arises from the lateral aspect of the axillary artery near the lower border of the subscapularis muscle and opposite the posterior humeral circumflex artery (Figs. 8-42 to 8-44). It may arise in

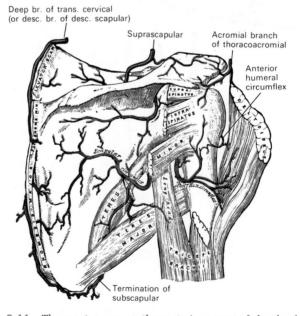

Deep br. of trans. cervical
(or desc. br. of desc. scapular)

Suprascapular

Acromial branch
of thoracoacromial

Anterior
humeral
circumflex

Termination of
subscapular

FIG. 8-44.   The anastomoses on the posterior aspect of the shoulder.

common with the posterior humeral circumflex or be represented by three or four very small branches. The vessel runs horizontally beneath the coracobrachialis muscle and short head of the biceps brachii, anterior to the neck of the humerus. As it reaches the intertubercular sulcus, it gives off a branch that ascends in the sulcus to supply the head of the humerus and the shoulder joint. The main stem of the artery then continues laterally, deep to the long head of the biceps bra-

chii and the deltoid, to anastomose with the posterior humeral circumflex artery.

Brachial Artery

The **brachial artery** is the continuation of the axillary artery and originates at the lower margin of the tendon of the teres major muscle. It passes down the arm and ends about 1 cm distal to the bend of the elbow, where it divides into the **radial** and **ulnar arteries** (Figs. 8-45; 8-46). At first the

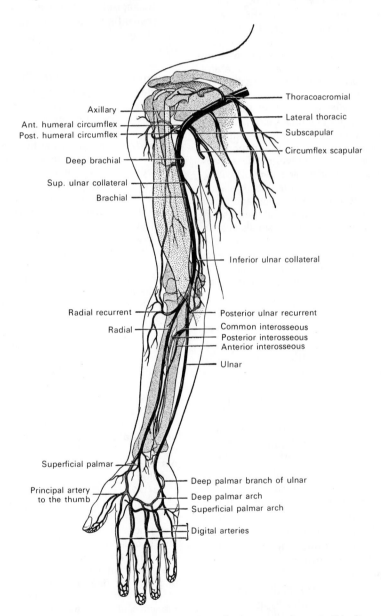

FIG. 8-45. The major arteries of the upper extremity. The branching pattern shown in this diagram is the one most commonly observed.

brachial artery lies medial to the humerus, but as it descends, it gradually curves to the front of the arm, and at the antecubital fossa anterior to the elbow joint, it lies midway between its two epicondyles.

RELATIONS.   The **brachial artery** is superficial throughout its entire course, being covered *anteriorly* by the skin and the superficial and deep fasciae. The bicipital aponeurosis, which extends from the biceps brachii muscle, lies anterior to the artery in front of the elbow, separating it from the median cubital vein. The median nerve crosses the brachial artery from lateral to medial, opposite the insertion of the coracobrachialis muscle (Fig. 8-46). *Posteriorly,* the brachial artery is separated from the long head of the triceps brachii by the radial nerve and the deep brachial artery. It then lies successively on the medial head of the triceps brachii, on the insertion of the coracobrachialis, and finally on the brachialis muscle. *Laterally,* the brachial artery is in relation more proximally in the arm with the median nerve and the coracobrachialis muscle and more distally with the biceps brachii, the two muscles overlapping the artery to a considerable extent. *Medially,* the proximal half of the brachial artery is in relation with the medial antebrachial cutaneous and ulnar nerves, while medial to the distal half of the artery is found the median nerve. The basilic vein also lies to its medial side, but is separated from it in the distal part of the arm by the deep fascia. The artery is accompanied by two venae comitantes, which lie in close contact to it and are interconnected at intervals by short transverse branches.

CUBITAL FOSSA.   On the anterior aspect of the bend at the elbow, the brachial artery sinks deeply into a triangular interval, the **cubital fossa.** The *base* of the triangle is directed proximally and is represented by a line connecting the two epicondyles of the humerus. The *sides* of the triangle are formed by the medial edge of the brachioradialis muscle and the lateral margin of the pronator teres. The *apex* of the triangle is directed inferiorly at the point where the two muscles meet. The *floor* of the cubital fossa is formed by the brachialis and supinator muscles. This fossa contains the tendon of the biceps brachii muscle, the brachial artery with its accompanying veins, the stems of the radial and ulnar arteries, and a part of the median nerve.

The brachial artery occupies the middle of the cubital fossa and divides opposite the neck of the radius into the radial and ulnar arteries. It is covered *anteriorly* by the skin, superficial fascia, and median cubital vein,

the last being separated from the artery by the bicipital aponeurosis. *Posteriorly,* it is the brachialis muscle that separates the artery from the elbow joint. The median nerve lies close to the *medial aspect* of the artery in the upper part of the fossa, but is separated from it more distally by the ulnar head of the pronator teres. The tendon of the biceps brachii lies to the *lateral aspect* of the artery.

The radial nerve normally lies just outside the fossa, between the supinator and the brachioradialis muscles, but may be exposed by entering through the fossa and deflecting the brachioradialis laterally.

VARIATIONS OF THE BRACHIAL ARTERY.   The brachial artery, accompanied by the median nerve, may leave the medial border of the biceps brachii and descend toward the medial epicondyle of the humerus. In such cases it usually passes behind a *supracondylar process* of the humerus, from which a fibrous arch is thrown over the vessel. It then runs beneath or through the substance of the pronator teres to the bend of the elbow. This variation bears considerable analogy with the normal condition of the artery in some of the carnivores.

A frequent variation is the presence of a **superficial brachial artery,** which may continue into the forearm to form a **superficial antebrachial artery** or may rejoin the brachial distally. Frequently the brachial artery divides at a higher level than usual, and the three vessels concerned in this high division are the radial, ulnar, and common interosseous arteries. Most frequently, the radial artery branches high in the arm, while the other limb of the bifurcation consists of the ulnar and common interosseous arteries. In some instances, the ulnar arises above the ordinary level, and the radial and common interosseous form the other limb of the division. Occasionally the common interosseous arises at a higher level (see Fig. 8-12).

Sometimes long, slender vessels called **vasa aberrantia** connect the brachial or axillary artery with one of the arteries of the forearm, or with branches from them. Such vessels usually join the radial artery.

COLLATERAL CIRCULATION.   After the application of a ligature to the brachial artery in the *upper third of the arm,* blood can circulate through branches from the anterior and posterior humeral circumflex and subscapular arteries, which anastomose with ascending branches from the deep brachial artery. If the main stem of the brachial artery is tied *below* the origin of the deep brachial artery and the superior ulnar collateral, circulation is maintained by branches of these two vessels, which anastomose with the inferior ulnar collateral, the radial and ulnar recurrent arteries, and the posterior interosseous artery.

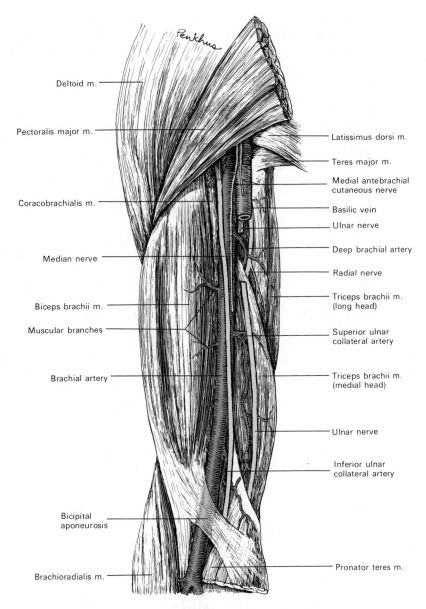

Deltoid m.

Pectoralis major m.

Coracobrachialis m.

Median nerve

Biceps brachii m.

Muscular branches

Brachial artery

Bicipital
aponeurosis

Brachioradialis m.

Latissimus dorsi m.

Teres major m.

Medial antebrachial
cutaneous nerve

Basilic vein

Ulnar nerve

Deep brachial artery

Radial nerve

Triceps brachii m.
(long head)

Superior ulnar
collateral artery

Triceps brachii m.
(medial head)

Ulnar nerve

Inferior ulnar
collateral artery

Pronator teres m.

FIG. 8-46. The right brachial artery and some of its branches.

CLINICAL NOTE.   To control excessive bleeding, the brachial artery should be compressed laterally against the humerus in the upper third of the arm, laterally and backward in the middle third, and directly backward in the lower third as the vessel courses anterior to the humerus in this region.

Branches of the Brachial Artery

As it descends in the arm, the brachial artery and its branches supply the humerus, muscles, and other tissues of the region, as well as participate in the anastomosis around the elbow joint. The branches of the brachial artery are:

Deep brachial
Principal nutrient artery of humerus
Superior ulnar collateral
Inferior ulnar collateral
Muscular

DEEP BRACHIAL ARTERY

The deep brachial artery (profound brachii) is the largest branch of the brachial artery. It arises from the medial and posterior part of the main stem, just distal to the lower border of the teres major muscle (Figs. 8-45; 8-46). Initially, the deep brachial artery passes backward into the arm between the long and lateral heads of the triceps brachii. It then accompanies the radial nerve in the spiral groove between the lateral and medial heads of the triceps, on the posterior aspect of the humerus. The vessel terminates by dividing into the radial collateral and middle collateral arteries. The deep brachial artery has four named branches in addition to muscular vessels:

Deltoid (ascending)
Radial collateral
Middle collateral
Nutrient

DELTOID (ASCENDING) BRANCH. The deltoid branch of the deep brachial artery ascends between the long and lateral heads of the triceps brachii muscle. It anastomoses with a descending branch of the posterior humeral circumflex artery and helps to supply the brachialis and deltoid muscles.

RADIAL COLLATERAL ARTERY. The radial collateral artery, which is frequently described as the terminal portion of the deep brachial

artery, continues into the forearm with the radial nerve. It lies deep to the lateral head of the triceps until it reaches the lateral supracondylar ridge of the humerus. At this site, it pierces the lateral intermuscular septum, descends between the brachioradialis and the brachialis muscles to the palmar aspect of the lateral epicondyle, and ends by anastomosing with the radial recurrent artery. Just before the radial collateral artery pierces the intermuscular septum, it gives off a branch that descends to the posterior aspect of the lateral epicondyle and there joins the anastomosis around the elbow.

MIDDLE COLLATERAL ARTERY. The middle collateral artery is usually larger than the radial collateral branch (Fig. 8-47). It enters the substance of the long and medial heads of the triceps and descends along the posterior aspect of the humerus to the elbow, where it anastomoses with the interosseous recurrent and joins into the anastomosis around the elbow.

NUTRIENT BRANCH. A nutrient branch, which usually is accessory to the principal nutrient artery from the brachial, enters a nutrient canal of the humerus posterior to the deltoid tuberosity. It may be absent.

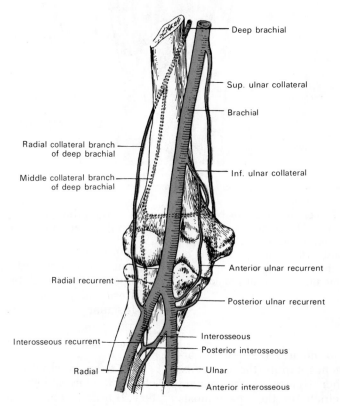

FIG. 8-47. Diagram of the anastomosis around the elbow joint.

PRINCIPAL NUTRIENT ARTERY OF HUMERUS

The main or principal nutrient artery of the humerus arises from the brachial about at the middle of the arm and enters a nutrient canal near the insertion of the coracobrachialis muscle.

SUPERIOR ULNAR COLLATERAL ARTERY

The superior ulnar collateral artery is a long and slender vessel that arises from the brachial a little below the middle of the arm (Figs. 8-45 to 8-47). It frequently springs from the proximal part of the deep brachial artery. It pierces the medial intermuscular septum and descends, with the ulnar nerve, on the surface of the medial head of the triceps brachii to the space between the medial epicondyle and olecranon. The vessel ends deep to the flexor carpi ulnaris by anastomosing with the posterior ulnar recurrent and inferior ulnar collateral arteries. It sometimes sends a branch anterior to the medial epicondyle, to anastomose with the anterior ulnar recurrent artery.

INFERIOR ULNAR COLLATERAL ARTERY

The inferior ulnar collateral artery arises about 5 cm above the elbow (Figs. 8-45 to 8-47). Initially, it passes medially upon the brachialis, and piercing the medial intermuscular septum, it then winds around the dorsum of the humerus between the triceps brachii and the bone. By its junction with the middle collateral branch of the deep brachial artery, it forms an arch just above the olecranon fossa. As the vessel lies on the brachialis, it gives off branches that ascend to anastomose with the superior ulnar collateral, and others that descend in front of the medial epicondyle to anastomose with the anterior ulnar recurrent. Behind the medial epicondyle a branch anastomoses with the superior ulnar collateral and posterior ulnar recurrent arteries.

MUSCULAR BRANCHES

The brachial artery gives off three or four muscular branches which are distributed to the coracobrachialis, biceps brachii, and brachialis muscles.

ANASTOMOSIS AROUND THE ELBOW JOINT. The vessels contributing to this anastomosis may be conveniently divided into those anterior to and those posterior to the medial and lateral epicondyles of the humerus (Fig. 8-47). The branches anastomosing *anterior to the medial epicondyle* are the anterior branch of the inferior ulnar collateral, the anterior ulnar recurrent, and the anterior branch of the superior ulnar collateral. Those *posterior to the medial epicondyle* are the inferior ulnar collateral, the posterior ulnar recurrent, and the posterior branch of the superior ulnar collateral. The branches anastomosing *anterior to the lateral epicondyle* are the radial recurrent and the radial collateral branch of the deep brachial artery. Those situated *posterior to the lateral epicondyle* are the inferior ulnar collateral, the interosseous recurrent, and the middle collateral branch of the deep brachial artery. There is also a transverse arch of anastomosis, which crosses the posterior aspect of the arm just proximal to the olecranon, formed by the interosseous recurrent joining with the inferior ulnar collateral and posterior ulnar recurrent (Fig. 8-47).

Ulnar Artery

The **ulnar artery** is the larger of the two terminal branches of the brachial and, like the radial artery, it begins about 1 cm distal to the bend of the elbow (Figs. 8-48; 8-50 to 8-52). Passing obliquely downward and medially, it reaches the ulnar side of the forearm at a point about midway between the elbow and the wrist. It then runs along the ulnar border to the wrist, crosses the flexor retinaculum on the radial side of the ulnar nerve and pisiform bone, and, immediately beyond this bone, becomes the **superficial palmar arch.**

RELATIONS OF THE ULNAR ARTERY. In the **proximal half of the forearm**, the ulnar artery is deeply seated, being covered by the pronator teres, flexor carpi radialis, palmaris longus, and flexor digitorum superficialis muscles. It lies upon the brachialis and flexor digitorum profundus muscles (Fig. 8-51). For about 2.5 cm from its commencement at the elbow, the ulnar artery courses lateral to the median nerve, and then the median nerve crosses anterior to the artery but is separated from it by the ulnar half of the pronator teres muscle. In the **distal half of the forearm,** the ulnar artery lies upon the flexor digitorum profundus, being covered by the skin and the superficial and deep fasciae, and being placed between the flexor carpi ulnaris and flexor digitorum superficialis (Figs. 8-48; 8-51). The artery is accompanied by two venae comitantes, and it is overlapped in its middle third by the flexor carpi ulnaris. The ulnar nerve lies on the medial side of the distal two-thirds of the artery, and the palmar cutaneous branch of the ulnar nerve descends on the distal part of the vessel to the palm of the hand.

At the wrist the ulnar artery lies on the flexor retinaculum and is covered by skin, the fasciae, and the palmaris brevis (Figs. 8-48; 8-50 to 8-52). On its medial side is the pisiform bone and, slightly dorsal to the artery, the ulnar nerve.

VARIATIONS. The ulnar artery varies in its origin in somewhat less than 10% of cases. It may arise about 5 to 7 cm below the elbow, but more frequently higher. Variations in the position of this vessel are more common than in the radial. When its origin from the brachial is normal, the course of the vessel is rarely changed. When it arises higher up, its course is almost invariably superficial to the flexor muscles in the forearm, lying commonly beneath the deep fascia, more rarely between the fascia and skin. In a few instances, its position has been reported to be subcutaneous in the upper part of the forearm and beneath the deep fascia in the lower part.

Branches of the Ulnar Artery

The branches of the ulnar artery may be arranged into three groups, like those of the radial artery.

In the forearm:
    Anterior ulnar recurrent
    Posterior ulnar recurrent
    Common interosseous
        Anterior interosseous
        Posterior interosseous
    Muscular
At the wrist:
    Palmar carpal
    Dorsal carpal
In the hand:
    Deep palmar
    Superficial palmar arch
        Common palmar digital

ANTERIOR ULNAR RECURRENT

The anterior ulnar recurrent artery arises immediately below the elbow joint from the medial aspect of the ulnar artery (Fig. 8-51). It courses upward between the brachialis muscle and the pronator teres, supplying these muscles and, in front of the medial epicondyle, anastomoses with the inferior ulnar collateral artery.

POSTERIOR ULNAR RECURRENT

The posterior ulnar recurrent artery is much larger and arises somewhat more distally than the anterior ulnar recurrent, but

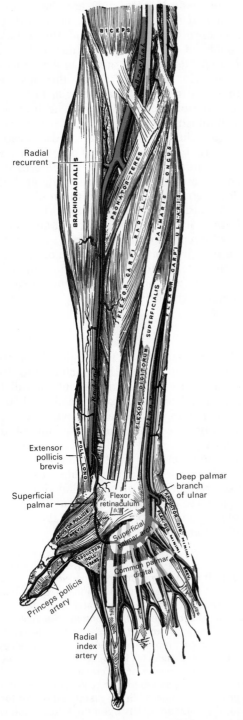

FIG. 8-48. The ulnar and radial arteries and the superficial palmar arch.

also from the medial side of the ulnar artery (Figs. 8-49; 8-51). It passes medially and backward between the flexor digitorum superficialis and flexor digitorum profundus, and then it ascends behind the medial epicondyle of the humerus. Between this process and the olecranon it lies beneath the flexor carpi ulnaris and ascends, together with the ulnar nerve, between the two heads of that muscle. The posterior ulnar recurrent artery, which supplies the neighboring muscles and the elbow joint, anastomoses with the superior and inferior ulnar collateral and the interosseous recurrent arteries.

COMMON INTEROSSEOUS ARTERY

The common interosseous artery is a short, thick vessel, only one centimeter long, which arises from the lateral and posterior aspect of the ulnar artery, immediately distal to the tuberosity of the radius (Fig. 8-51). The vessel passes dorsal to the upper border of the interosseous membrane, where it divides into two branches, the anterior and posterior interosseous arteries.

ANTERIOR INTEROSSEOUS ARTERY. The anterior interosseous artery passes down the forearm on the anterior surface of the interosseous membrane (Figs. 8-49; 8-51). It is accompanied by the palmar interosseous branch of the median nerve and is overlapped by the contiguous margins of the flexor digitorum profundus and flexor pollicis longus, giving off **muscular branches** and **nutrient arteries** to the radius and ulna. At the proximal border of the pronator quadratus it gives off a small branch that descends dorsal to the pronator to reach the *palmar carpal network* (Fig. 8-51), and then the anterior interosseous artery pierces the interosseous membrane to enter the posterior compartment of the forearm, where it anastomoses with the dorsal interosseous artery (Fig. 8-49). On the posterior aspect of the interosseous membrane in the lower forearm, the anterior interosseous artery descends in company with the terminal portion of the posterior interosseous nerve (from the deep branch of the radial nerve) to the dorsum of the wrist to join the *dorsal carpal network.*

The **median artery** is a long, slender branch that arises from the proximal part of the anterior interosseous artery. It passes forward to the median nerve, which it accompanies and supplies in its course down the forearm. This artery is sometimes much enlarged, and when so, it courses with the median nerve into the palm of the hand and enters into the formation of the superficial palmar arch.

POSTERIOR INTEROSSEOUS ARTERY. The posterior interosseous artery, smaller than the anterior, arises from the common interosseous artery and passes dorsally either over or between the oblique cord and the proximal border of the interosseous membrane (Figs. 8-49; 8-51). It appears between the contiguous borders of the supinator and the abductor pollicis longus and descends in the posterior compartment of the forearm between the superficial and deep layers of muscles, sending branches to both groups. Coursing along the surface of the abductor pollicis longus and the extensor pollicis brevis, it is accompanied by the posterior interosseous nerve, which is derived from the deep branch of the radial nerve. In the distal forearm, it anastomoses with terminal branches of the anterior interosseous artery and with the *dorsal carpal network.*

The **interosseous recurrent artery** branches from the posterior interosseous artery near its origin, or at times it may arise directly from the common interosseous artery (Fig. 8-49). It ascends on or through the fibers of the supinator to the interval between the lateral epicondyle and olecranon, but deep to the anconeus muscle. Here it anastomoses with the middle collateral branch of the deep brachial artery, the posterior ulnar recurrent artery, and the inferior ulnar collateral artery.

MUSCULAR BRANCHES

Many muscular branches arise from the ulnar artery in the forearm (Fig. 8-51). These supply the superficial and deep flexors of the fingers and muscles on the ulnar aspect of the forearm, including the flexors carpi radialis and ulnaris, as well as the pronator teres.

PALMAR CARPAL BRANCH

The palmar carpal branch is a small vessel that arises from the ulnar artery opposite the carpal bones. It crosses the palmar aspect of the wrist joint, deep to the tendons of the flexor digitorum profundus (Fig. 8-51), and anastomoses with the corresponding palmar carpal branch of the radial artery.

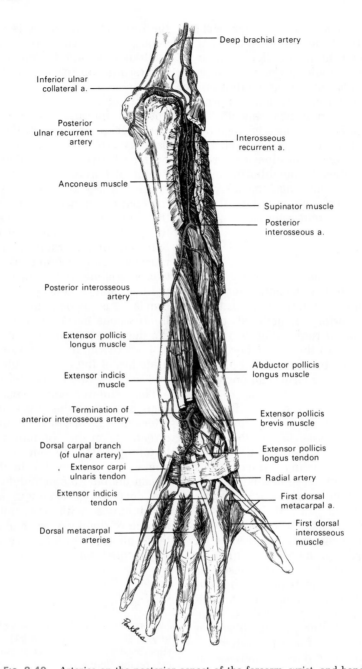

FIG. 8-49.   Arteries on the posterior aspect of the forearm, wrist, and hand.

DORSAL CARPAL BRANCH

The dorsal carpal branch of the ulnar artery arises immediately above the pisiform bone and winds dorsally beneath the tendon of the flexor carpi ulnaris (Fig. 8-49). It then passes across the dorsal surface of the wrist beneath the extensor tendons, to anastomose with the corresponding dorsal carpal branch of the radial artery. Just beyond its origin, it gives off a small branch that courses along the ulnar aspect of the fifth metacarpal bone, supplying the ulnar aspect of the dorsal surface of the little finger.

DEEP PALMAR BRANCH

The deep palmar branch arises from the ulnar artery at the level of the flexor retinaculum (Figs. 8-48; 8-50 to 8-52). It courses medially between the abductor digiti minimi and flexor digiti minimi brevis and through

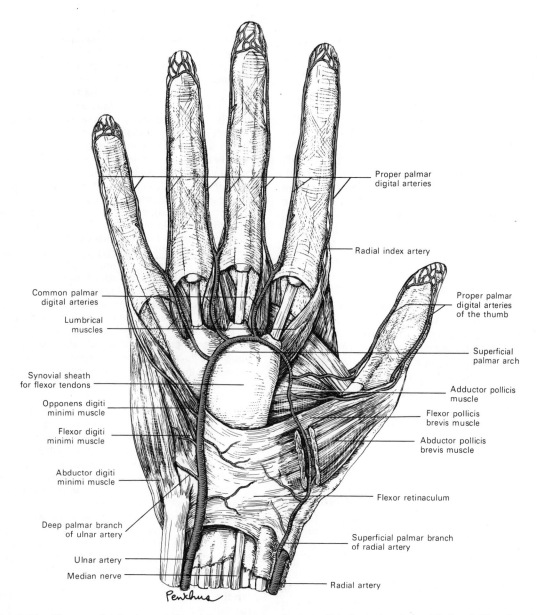

FIG. 8-50.   The superficial palmar arch and the common and proper digital arteries of the right hand. Note that the ulnar artery is the principal contributor to the superficial palmar arch.

the origin of the opponens digiti minimi, all of which are muscles of the hypothenar eminence. The vessel then curves laterally into the palm, accompanied by the deep branch of the radial nerve, to anastomose with the radial artery and complete the **deep palmar arch.**

SUPERFICIAL PALMAR ARCH

The superficial palmar arch is formed primarily by the deep palmar branch of the ulnar artery and is usually completed by a branch that comes from the radial artery (Figs. 8-48; 8-50). Most frequently this is the superficial palmar branch of the radial, but it may be the radial index artery or the princeps pollicis artery. More rarely, the median artery, which is a branch of the anterior interosseous artery, descends to help form the arch. The superficial palmar arch curves across the palm with its convexity oriented distally. It is covered by the skin, the palmaris brevis, and the palmar aponeurosis, and it lies upon the flexor retinaculum, the flexor digiti minimi brevis and opponens digiti minimi, the tendons of the flexor digitorum superficialis, the lumbrical muscles, and branches of the median and ulnar nerves. The superficial palmar arch gives rise to three common palmar digital arteries.

COMMON PALMAR DIGITAL ARTERIES. Three common palmar digital arteries arise from the convexity of the superficial palmar arch and proceed distally on the second, third, and fourth lumbrical muscles (Figs. 8-48; 8-50 to 8-52). Each is joined by the corresponding palmar metacarpal artery from the deep palmar arch, and then divides into a pair of **proper palmar digital arteries** (Figs. 8-50 to 8-52). These latter vessels course along the contiguous sides of the index, middle, ring, and little fingers dorsal to the corresponding digital nerves. They anastomose freely in the subcutaneous tissue of the finger tips and by smaller branches near the interphalangeal joints. Each gives off two dorsal branches, which anastomose with the dorsal digital arteries, and supply the soft parts on the dorsum of the middle and distal phalanges, including the matrix of the fingernail. The proper palmar digital artery for the ulnar aspect of the little finger springs from the ulnar artery under cover of the palmaris brevis.

Radial Artery

Due to its direction, the **radial artery** appears to be the direct continuation of the brachial trunk, but it is smaller in caliber than the ulnar artery, which is the other terminal branch of the brachial artery (Figs. 8-45; 8-48; 8-51). It originates at the bifurcation of the brachial artery opposite the neck of the radius, about 1 cm below the bend of the elbow, and passes along the radial aspect of the forearm to the wrist. It then winds around the lateral aspect of the carpus to the dorsum of the wrist, deep to the tendons of the abductor pollicis longus and the extensors pollicis longus and brevis, reaching the proximal end of a space between the metacarpal bones of the thumb and index finger. Finally it passes between the two heads of the first dorsal interosseous muscle into the palm of the hand, joining the deep palmar branch of the ulnar artery to form the deep palmar arch. Hence, the radial artery has three parts, one part in the forearm, another at the wrist, and a third part in the hand.

RELATIONS OF THE RADIAL ARTERY. The radial artery **in the forearm** extends from the neck of the radius to the anterior aspect of its styloid process, being placed to the medial aspect of the radial shaft proximally and anterior to the bone distally (Figs. 8-48; 8-51). The upper part of the artery is overlapped by the fleshy belly of the brachioradialis muscle, while the rest of the artery is superficial, being covered only by skin and the superficial and deep fasciae. In its course through the forearm, the radial artery lies first upon the biceps tendon, and then successively on the supinator muscle, the pronator teres muscle, the radial origin of the flexor digitorum superficialis muscle, the flexor pollicis longus muscle, the pronator quadratus muscle, and the distal end of the radius. In the proximal third of its course, it lies between the brachioradialis and the pronator teres muscles; in the distal third of its course it lies between the brachioradialis and flexor carpi radialis muscles. The superficial branch of the radial nerve is close to the lateral aspect of the artery in the middle third of its course, and some filaments of the lateral antebrachial cutaneous nerve run along the lower part of the artery as it winds around the wrist. The vessel is accompanied by a pair of venae comitantes throughout its whole course. Distally the radial artery lies on the lower anterior aspect of the radius and, being very superficial, is an *ideal site for taking the pulse.*

At the wrist, the radial artery reaches the dorsum of the carpus by passing between the radial collateral ligament of the wrist and the tendons of the abductor pollicis longus and extensor pollicis brevis muscles (Fig. 8-49). It then descends on the scaphoid and trapezium, and before disappearing between the heads of the first dorsal interosseous muscle, it is crossed by the tendon of the extensor pollicis longus. In the interval between the two extensors of the thumb, sometimes called the "anatomical snuff box," the radial artery is crossed by the dorsal digital nerves as they pass to the thumb and index finger; these nerves are derived from the superficial branch of the radial nerve.

In the hand, the radial artery passes through the proximal end of the first interosseous space, between the heads of the first dorsal interosseous muscle (Figs. 8-50; 8-52). It then crosses the palm transversely between the oblique and transverse heads of the adductor pollicis muscle. The vessel then pierces the transverse head and reaches the base of the metacarpal bone of the little finger, where it anastomoses with the deep palmar branch from the ulnar artery, completing the **deep palmar arch** (Figs. 8-51; 8-52).

VARIATIONS. The origin of the radial artery is higher than usual in about 12% of cases. In the forearm it deviates less frequently from its normal position than does the ulnar artery. It has been observed lying on the deep fascia of the forearm, instead of beneath it. It has also been seen on the surface of the brachioradialis instead of deep to its medial bor-

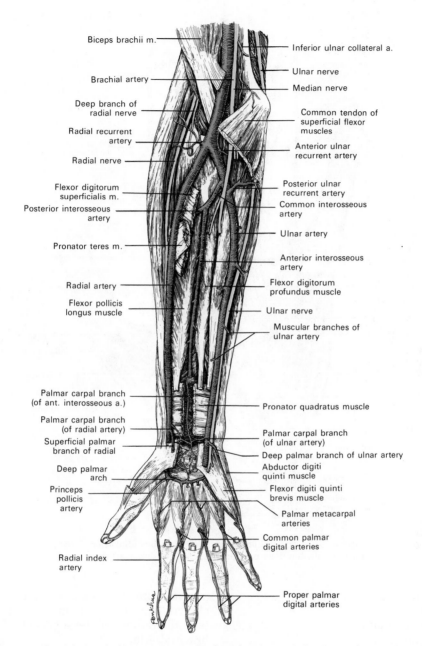

Biceps brachii m.

Brachial artery

Deep branch of
radial nerve

Radial recurrent
artery

Radial nerve

Flexor digitorum
superficialis m.

Posterior interosseous
artery

Pronator teres m.

Radial artery

Flexor pollicis
longus muscle

Palmar carpal branch
(of ant. interosseous a.)

Palmar carpal branch
(of radial artery)

Superficial palmar
branch of radial

Deep palmar
arch

Princeps
pollicis
artery

Radial index
artery

Inferior ulnar collateral a.

Ulnar nerve

Median nerve

Common tendon of
superficial flexor
muscles

Anterior ulnar
recurrent artery

Posterior ulnar
recurrent artery

Common interosseous
artery

Ulnar artery

Anterior interosseous
artery

Flexor digitorum
profundus muscle

Ulnar nerve

Muscular branches of
ulnar artery

Pronator quadratus muscle

Palmar carpal branch
(of ulnar artery)

Deep palmar branch of ulnar artery

Abductor digiti
quinti muscle

Flexor digiti quinti
brevis muscle

Palmar metacarpal
arteries

Common palmar
digital arteries

Proper palmar
digital arteries

FIG. 8-51.   The radial and ulnar arteries in the deep forearm and the deep palmar arch in the hand.

der. In its course around the wrist, the radial artery has been seen lying on the extensor tendons of the thumb instead of beneath them. The presence of a large **median artery** may replace the radial in the formation of the palmar arches.

Branches of the Radial Artery

The branches of the radial artery may be divided into three groups, which correspond to the three regions in which the vessel is situated.

In the forearm:
  Radial recurrent
  Muscular
  Palmar carpal
  Superficial palmar
At the wrist:
  Dorsal carpal

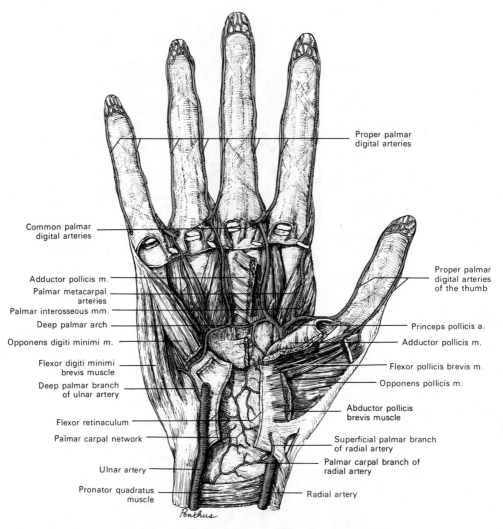

Proper palmar
digital arteries

Common palmar
digital arteries

Adductor pollicis m.
Palmar metacarpal
arteries
Palmar interosseous mm.
Deep palmar arch
Opponens digiti minimi m.
Flexor digiti minimi
brevis muscle
Deep palmar branch
of ulnar artery
Flexor retinaculum
Palmar carpal network
Ulnar artery
Pronator quadratus
muscle

Proper palmar
digital arteries
of the thumb

Princeps pollicis a.
Adductor pollicis m.
Flexor pollicis brevis m.
Opponens pollicis m.
Abductor pollicis
brevis muscle
Superficial palmar branch
of radial artery
Palmar carpal branch of
radial artery
Radial artery

Penthus

FIG. 8-52.   The deep palmar arch and princeps pollicis artery. Note that the radial artery is the principal contributor to this arch, but that it anastomoses with the deep palmar branch of the ulnar artery.

First dorsal metacarpal
In the hand:
  Princeps pollicis
  Radial index
  Deep palmar arch
    Palmar metacarpal
    Perforating
    Recurrent

RADIAL RECURRENT ARTERY

The radial recurrent artery arises immediately below the elbow (Figs. 8-48; 8-51). It ascends between the branches of the radial nerve, lying on the supinator muscle and then between the brachioradialis and brach-

ialis muscles. The vessel supplies these muscles and the elbow joint and anastomoses with the radial collateral branch of the deep brachial artery.

MUSCULAR BRANCHES

The radial artery sends muscular branches to the brachioradialis and pronator teres muscles, as well as to other muscles on the radial aspect of the forearm.

PALMAR CARPAL BRANCH

The palmar carpal branch of the radial artery (Figs. 8-51; 8-52) is a small vessel that arises near the distal border of the pronator quadratus and, running across the palmar aspect of the wrist, anastomoses with the palmar carpal branch of the ulnar artery. This anastomosis is joined by a branch from the anterior interosseous artery proximally and by recurrent branches from the deep palmar arch distally, thus forming a **palmar carpal network,** which supplies the articulations of the wrist and the carpal bones (Fig. 8-52).

SUPERFICIAL PALMAR BRANCH

The superficial palmar branch arises from the radial artery just above the wrist, where the radial artery is about to wind laterally around the carpus (Figs. 8-48; 8-50 to 8-52). It passes through, but occasionally over, the muscles of the thenar eminence, which it supplies. The vessel then anastomoses with the terminal portion of the ulnar artery, completing the **superficial palmar arch** (Figs. 8-48; 8-50). The superficial palmar branch of the radial artery varies considerably in size. Frequently it is very small and ends in the muscles of the thumb, but sometimes it is as large as the continuation of the radial.

DORSAL CARPAL BRANCH

The dorsal carpal branch of the radial artery is a small vessel that arises deep to the extensor tendons of the thumb. Crossing the carpus transversely toward the medial border of the hand, it anastomoses with the dorsal carpal branch of the ulnar artery and with terminal branches of the anterior and posterior interosseous arteries to form a **dorsal carpal network.** From this network are given off three slender **dorsal metacarpal arteries,** which run distally on the second, third, and fourth dorsal interosseous muscles (Fig. 8-49). The metacarpal arteries bifurcate into **dorsal digital branches,** which then supply the adjacent sides of the index, middle, ring, and little fingers. The dorsal digital vessels anastomose with the proper palmar digital branches of the superficial palmar arch. Near their origins the dorsal metacarpal arteries anastomose with the deep palmar arch by the **proximal perforating arteries,** and near their points of bifurcation, they also anastomose with the common

palmar digital vessels of the superficial palmar arch by the **distal perforating arteries.**

FIRST DORSAL METACARPAL ARTERY

The first dorsal metacarpal artery arises just before the radial artery passes between the two heads of the first dorsal interosseous muscle (Fig. 8-49). It divides almost immediately into two branches, which supply the adjacent sides of the thumb and index finger. The radial aspect of the thumb receives a branch directly from the radial artery.

PRINCEPS POLLICIS ARTERY

The princeps pollicis artery arises from the radial artery as it turns medially toward the deep part of the palm (Figs. 8-48; 8-51; 8-52). It descends between the first dorsal interosseous muscle and the oblique head of the adductor pollicis, along the ulnar aspect of the metacarpal bone of the thumb to the base of the first phalanx. Here the princeps pollicis artery lies beneath the tendon of the flexor pollicis longus and divides into two branches. These make their appearance between the medial and lateral insertions of the oblique head of the adductor pollicis muscle and run along the sides of the thumb (Figs. 8-50; 8-52). On the palmar surface of the distal phalanx, they form an arch from which branches are distributed to the skin and subcutaneous tissue of the thumb.

RADIAL INDEX ARTERY

The radial index artery (arteria radialis indicis) arises from the radial artery close to the princeps pollicis artery, between the first dorsal interosseous muscle and the transverse head of the adductor pollicis muscle (Figs. 8-48; 8-50; 8-51). It courses along the radial aspect of the index finger to its extremity, where it anastomoses with the proper digital artery supplying the ulnar aspect of the finger. At the distal border of the transverse head of the adductor pollicis, this vessel anastomoses with the princeps pollicis and gives a communicating branch to the superficial palmar arch. The princeps pollicis and radial index arteries may spring from a common trunk, which is then called the **first palmar metacarpal artery.**

DEEP PALMAR ARCH

The deep palmar arch is the terminal part of the radial artery and it anastomoses with the deep palmar branch of the ulnar artery (Figs. 8-51; 8-52). It lies upon the carpal extremities of the metacarpal bones and on the interosseous muscles, being covered by the oblique head of the adductor pollicis muscle, the flexor tendons of the fingers, and the lumbrical muscles. In nearly two-thirds of cases, the deep palmar arch lies deep to the ulnar nerve; in one-third of cases, it lies superficial to the nerve (to the palmar side). Occasionally it is doubled and encircles the ulnar nerve. Arising from the deep palmar arch are the **palmar metacarpal arteries, perforating branches,** and **recurrent branches.**

PALMAR METACARPAL ARTERIES.   There are three or four palmar metacarpal arteries that arise from the convexity of the deep palmar arch (Figs. 8-51; 8-52). They course distally upon the interosseous muscles in the second, third, and fourth intermetacarpal spaces and anastomose, at the clefts of the fingers, with the common digital branches of the superficial palmar arch.

PERFORATING BRANCHES.   Three perforating branches pass dorsally from the deep palmar arch, through the second, third, and fourth interosseous spaces, and between the heads of the corresponding dorsal interosseous muscles to anastomose with the dorsal metacarpal arteries.

RECURRENT BRANCHES.   The recurrent branches arise from the concavity of the deep palmar arch. They ascend on the palmar aspect of the wrist, supply the intercarpal articulations, and end in the palmar carpal network.

Arteries of the Thorax and Abdomen

From the aortic arch, the descending aorta courses along the posterior wall of the trunk through both the thoracic and abdominal cavities. Descending in the posterior mediastinum, the vessel is called the **thoracic aorta;** as it penetrates the diaphragm, it becomes the posterior abdominal aorta.

THORACIC AORTA

The **thoracic aorta,** contained in the posterior mediastinum, descends along the posterior wall of the thorax (Fig. 8-53). It begins at the lower border of the fourth thoracic vertebra, where it is continuous with the aortic arch, and ends anterior to the lower border of the twelfth thoracic vertebra, at the aortic hiatus in the diaphragm. At its origin, the thoracic aorta is situated on the left of the vertebral column, but as it descends, it approaches the median line. At its termination, the vessel lies directly anterior to the column. In its descent, the thoracic aorta describes a curve that is concave ventrally, and because its thoracic branches are small, there is no significant decrease in its diameter (Fig. 8-53).

RELATIONS.   The **thoracic aorta** is related *anteriorly* to the root of the left lung, the pericardium, the esophagus, and the diaphragm, and *posteriorly* to the vertebral column and the hemiazygos veins. *On the right side* of the thoracic aorta are the azygos vein and thoracic duct, and *on the left side,* the left pleura and lung. The esophagus, with its accompanying plexus of nerves, lies to the right side of the upper thoracic aorta. More inferiorly in the thorax it is placed anterior to the aorta, but as it nears the diaphragm, it is situated to the left side of the aorta.

Branches of the Thoracic Aorta

The thoracic aorta gives off visceral branches to organ structures in the thorax as well as parietal branches to the walls of the thoracic cavity. They are:

Visceral Branches
  Pericardial
  Bronchial
  Esophageal
  Mediastinal
Parietal Branches
  Posterior intercostal
  Subcostal
  Superior phrenic

PERICARDIAL BRANCHES

The pericardial branches of the thoracic aorta consist of two or three small vessels that are distributed to the posterior surface

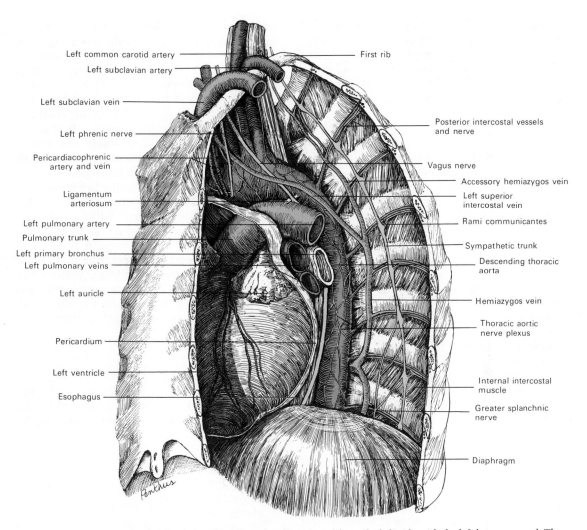

Left common carotid artery

Left subclavian artery

Left subclavian vein

Left phrenic nerve

Pericardiacophrenic artery and vein

Ligamentum arteriosum

Left pulmonary artery

Pulmonary trunk

Left primary bronchus

Left pulmonary veins

Left auricle

Pericardium

Left ventricle

Esophagus

First rib

Posterior intercostal vessels and nerve

Vagus nerve

Accessory hemiazygos vein

Left superior intercostal vein

Rami communicantes

Sympathetic trunk

Descending thoracic aorta

Hemiazygos vein

Thoracic aortic nerve plexus

Internal intercostal muscle

Greater splanchnic nerve

Diaphragm

FIG. 8-53.   The aortic arch and the descending thoracic aorta viewed from the left side with the left lung removed. The pericardium has been opened to expose the left side of the heart.

of the pericardium. They also supply some of the posterior mediastinal lymph nodes and anastomose with the pericardiacophrenic branches of the internal thoracic artery.

BRONCHIAL ARTERIES

The bronchial arteries vary in number, size, and origin. There is as a rule only *one* **right bronchial artery,** which arises from the third posterior intercostal artery (the first intercostal artery that comes from the aorta) or from the upper left bronchial artery. There are usually *two* **left bronchial arteries** and these arise from the thoracic aorta. The upper left bronchial arises oppo-

site the fifth thoracic vertebra, while the lower left bronchial artery stems from the aorta just below the level of the left bronchus. Each vessel courses along the posterior surface of its bronchus, dividing and subdividing along the bronchial tubes, supplying them, the areolar tissue of the lungs, and the bronchial lymph nodes. They also send a few twigs to the esophagus, to bronchial lymph nodes, and to the pericardium.

ESOPHAGEAL ARTERIES

There are four or five esophageal arteries, which arise from the ventral aspect of the aorta and pass obliquely downward to the esophagus, on which they form a chain of anastomoses. Additionally, these vessels anastomose with the esophageal branches of the inferior thyroid arteries and with ascending branches from the left inferior phrenic and left gastric arteries.

MEDIASTINAL BRANCHES

Numerous small mediastinal branches arise from the thoracic aorta to supply lymph nodes, vessels, nerves, and loose areolar tissue in the posterior mediastinum, as well as some of the adjacent mediastinal pleura.

POSTERIOR INTERCOSTAL ARTERIES

There usually are nine pairs of posterior intercostal arteries which branch from the thoracic aorta. They arise from the dorsal aspect of the aorta and are distributed to the lower nine intercostal spaces. The first two intercostal spaces are supplied by the supreme intercostal artery, which is a branch of the costocervical trunk from the subclavian artery. The **right posterior intercostal arteries** are longer than the left because the aorta is positioned to the left side of the vertebral column. They pass across the bodies of the vertebrae dorsal to the esophagus, thoracic duct, and azygos vein and are covered by the right lung and pleura. The **left posterior intercostal arteries** initially course dorsally, curving along the sides of the vertebral bodies and being covered by the left lung and pleura. The two uppermost vessels are crossed by the left superior intercostal vein, the lower vessels by the accessory hemiazygos and hemiazygos veins (Fig. 8-53). The further course of the posterior intercostal arteries is practically the same on both sides. Opposite the heads of the ribs, the sympathetic trunk passes anterior to them, and the splanchnic nerves descend anterior to the lower posterior intercostal arteries.

Each posterior intercostal artery crosses its corresponding intercostal space obliquely toward the angle of the next rib above and is continued ventrally in the costal groove. (This portion of the artery at one time was called the *ventral ramus*.) It is placed at first between the pleura and the internal posterior intercostal membrane. It then courses anteriorly around the intercostal space, lying between the internal inter-

costal muscle and the innermost intercostal muscle. Anteriorly, it anastomoses with the anterior intercostal branch of the internal thoracic or of the musculophrenic artery. Each artery is accompanied by a vein and a nerve, the former placed above and the latter below the artery (Fig. 8-53). This relationship exists except in the posterior part of the upper intercostal spaces, where the artery, which must ascend to the space, initially lies below the nerve and then between the nerve and vein. The third posterior intercostal artery (first aortic) anastomoses with the supreme intercostal artery, which branches from the costocervical trunk, and may form the chief supply of the second intercostal space. The lowest two posterior intercostal arteries are continued anteriorly from the intercostal spaces into the anterior abdominal wall and anastomose with the subcostal, superior epigastric, and lumbar arteries.

Branches of the Posterior Intercostal Artery

In its course from the aorta around the thoracic wall, each posterior intercostal artery gives off several branches. These are the following:

    Dorsal
    Collateral intercostal
    Muscular
    Lateral cutaneous
    Mammary

DORSAL BRANCH. The dorsal branch of each posterior intercostal artery is a large vessel that courses backward through a space that is bounded above and below by the necks of the ribs, medially by the body of a vertebra, and laterally by a superior costotransverse ligament. It gives off a **spinal branch** which enters the vertebral canal through the intervertebral foramen and is distributed to the vertebrae, the spinal cord, and its membranes. The dorsal branch then courses over the transverse process of the vertebra, accompanied by the dorsal ramus of the segmentally corresponding thoracic spinal nerve. It supplies branches to the muscles of the back and cutaneous vessels which course with corresponding cutaneous branches of the dorsal ramus of the spinal nerve.

COLLATERAL INTERCOSTAL BRANCH. The collateral intercostal branch comes off from the posterior intercostal artery near the angle of the rib and descends to the upper border of the rib below, along which it courses to anastomose with the intercostal branch of the internal thoracic artery.

MUSCULAR BRANCHES. Muscular branches arise from the posterior intercostal artery to supply the intercostal and pectoral muscles and the serratus anterior. They anastomose with the supreme thoracic and lateral thoracic branches of the axillary artery.

LATERAL CUTANEOUS BRANCHES. Lateral cutaneous branches arise from the posterior intercostal artery at about the midaxillary line. They penetrate laterally through the intercostal and serratus anterior muscles in company with the lateral cutaneous branches of the thoracic nerves. Achieving the superficial fascia, the vessels divide into *anterior and posterior branches* that course in a segmental manner to supply the skin and superficial fascia overlying an intercostal space. In the third, fourth, and fifth intercostal regions the anterior branches reach the lateral part of the mammary gland in the female and are sometimes called the *lateral or external mammary branches.*

MAMMARY BRANCHES. After giving off the lateral cutaneous branches, the posterior intercostal arteries continue around within the intercostal space, eventually to penetrate the skin and superficial fascia more anteriorly. In the third, fourth, and fifth spaces, (medial) mammary branches arise to reach the medial part of the base of the breast. They increase considerably in size during the period of lactation.

SUBCOSTAL ARTERIES

The subcostal arteries are so named because they lie below the last ribs. As such, they cannot be considered true intercostal arteries, but they are the lowest pair of arteries derived from the thoracic aorta, and they are in series with the other pairs of posterior intercostal arteries. Each artery passes along the lower border of the twelfth rib dorsal to the kidney and ventral to the quadratus lumborum muscle; it is accompanied by the ventral ramus of the twelfth thoracic or subcostal nerve. It then pierces the posterior aponeurosis of the transversus abdominis, and passing between this muscle and the internal oblique, it anastomoses with the superior epigastric, the lowest true intercos-

tal, and the lumbar arteries. Each subcostal artery gives off a dorsal branch, which has a distribution similar to those of posterior intercostal arteries.

SUPERIOR PHRENIC ARTERIES

The superior phrenic arteries are small vessels that arise from the lower part of the thoracic aorta. They are distributed to the dorsal part of the upper surface of the diaphragm and anastomose with the musculophrenic and pericardiacophrenic arteries.

A small **aberrant artery** is sometimes found arising from the right side of the thoracic aorta near the origin of the right bronchial. It passes upward and to the right behind the trachea and the esophagus and may anastomose with the right supreme intercostal artery. It represents the remains of the **right dorsal aorta,** and in a small proportion of cases, it is enlarged to form the first part of the right subclavian artery.

ABDOMINAL AORTA

The abdominal aorta begins in the midline at the aortic hiatus of the diaphragm, ventral to the lower border of the body of the last thoracic vertebra. Descending anterior to the vertebral column, it ends on the body of the fourth lumbar vertebra, commonly a little to the left of the midline, by dividing into the two common iliac arteries. It diminishes rapidly in size as it descends in the abdomen because of the many large vessels that branch from it. As the aorta lies upon the bodies of the vertebrae, it displays a curve that is convex anteriorly. The summit of the convexity corresponds to the third lumbar vertebra. The aortic bifurcation may be roughly judged on the surface of the abdomen to occur slightly to the left of the midline at a site about 2.54 cm below the umbilicus. This is approximately at the level of a line drawn between the most superior extent of the iliac crests.

RELATIONS. The upper part of the **abdominal aorta** is covered *anteriorly* by the lesser omentum and stomach, dorsal to which are the branches of the celiac artery and the celiac plexus. Below these are the splenic vein, the pancreas, the left renal vein, the inferior part of the duodenum, the mesentery, and the aortic plexus. *Posteriorly,* the abdominal aorta is separated from the lumbar vertebrae and intervertebral discs by the anterior longitudinal ligament and the second, third, and fourth left lumbar veins. On its *right side,* the abdominal aorta is related above to the azygos vein, cisterna chyli, thoracic duct, and the right crus of the diaphragm, the latter separating it from the proximal part of the inferior vena cava and the right celiac ganglion. More caudally in the abdomen, the right side of the aorta is in contact with the inferior vena cava. On the *left side,* the upper part of the abdominal aorta is related to the left crus of the diaphragm and the left celiac ganglion; somewhat lower are the fourth or ascending part of the duodenum, the duodenojejunal junction, and some coils of the small intestine.

COLLATERAL CIRCULATION. A degree of collateral circulation is provided by several anastomoses interconnecting vessels that arise above the aorta with others that arise below, or anastomoses between higher and lower branching vessels from the aorta itself. Thus, an anastomosis is found on the anterior wall of the trunk between the internal thoracic artery from the subclavian and the inferior epigastric artery, which branches from the external iliac. There is a free communication between the superior and inferior mesenterics, if an obstruction or ligature is placed between these vessels. When the point of ligature is below the origin of the inferior mesenteric (as is more common), collateral circulation is achieved between the inferior mesenteric artery and the internal pudendal artery of the pelvis. Other anastomoses exist between the lumbar arteries and the branches of the internal iliac. To these examples of collateral circulation can be added many others that protect regions of the body and organs that suffer diminished circulation.

Branches of the Abdominal Aorta

The branches of the abdominal aorta may be divided into three sets: visceral, parietal, and terminal (Fig. 8-54).

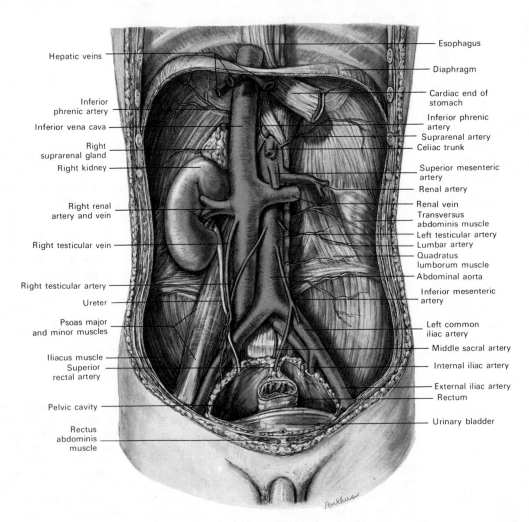

Hepatic veins
Inferior phrenic artery
Inferior vena cava
Right suprarenal gland
Right kidney
Right renal artery and vein
Right testicular vein
Right testicular artery
Ureter
Psoas major and minor muscles
Iliacus muscle
Superior rectal artery
Pelvic cavity
Rectus abdominis muscle

Esophagus
Diaphragm
Cardiac end of stomach
Inferior phrenic artery
Suprarenal artery
Celiac trunk
Superior mesenteric artery
Renal artery
Renal vein
Transversus abdominis muscle
Left testicular artery
Lumbar artery
Quadratus lumborum muscle
Abdominal aorta
Inferior mesenteric artery
Left common iliac artery
Middle sacral artery
Internal iliac artery
External iliac artery
Rectum
Urinary bladder

FIG. 8-54. The abdominal aorta and its principal branches in the male.

Visceral Branches
  Celiac trunk
  Superior mesenteric
  Inferior mesenteric
  Middle suprarenal
  Renal
  Testicular
  Ovarian
Parietal Branches
  Inferior phrenic
  Lumbar
  Middle sacral
Terminal Branches
  Common iliac

The **visceral branches** are distributed to the organs in the abdomen, and of these the celiac trunk and the superior and inferior mesenteric arteries are unpaired, whereas the middle suprarenals, renals, testicular, and ovarian are paired. The **parietal branches** supply the diaphragm from below as well as the posterior abdominal wall, and of these the inferior phrenics and lumbars are paired, while the middle sacral is unpaired. The **terminal branches** descend to supply the organs of the pelvis, the pelvic wall, and the lower extremities; of course, the common iliacs are paired (Fig. 8-54).

CELIAC TRUNK

The celiac trunk is a short thick vessel that ranges from 7 to 20 mm in diameter and arises from the aorta just below the aortic hiatus of the diaphragm (Figs. 8-54; 8-55). It is covered by the peritoneum of the dorsal wall of the lesser sac (omental bursa). *On its right side* are the right celiac ganglion, the right crus of the diaphragm, and the caudate lobe of the liver. *On its left side* are the left celiac ganglion, the left crus of the diaphragm, and the cardiac end of the stomach. Below it is related to the upper border of the pancreas and the splenic vein. A short distance below its origin from the aorta (i.e., from 1 mm to 2.2 cm, but averaging about 1.25 cm), the aorta gives origin to the superior mesenteric artery.

Branches of the
Celiac Trunk

After a short course of about 1 to 2 cm, the celiac trunk usually divides into three large branches (Figs. 8-55; 8-57). These are the:

Left gastric

Hepatic
Splenic

VARIABILITY IN THE BRANCHING OF THE CELIAC TRUNK. The most constant feature of this artery is the variability of its branching and of the routes by which blood reaches the organs that it principally supplies. According to Michels (1955), whose descriptions and statistics are followed in this account, the stomach, liver, pancreas, duodenum, and spleen are the organs supplied, and they can conveniently be used to designate the important variations.

A *gastrohepatosplenic trunk* has the classic branches: a left gastric, a common hepatic, and a splenic artery. This is the usual pattern (89%), and it gives origin to all its expected sub-branches in 64.5%, and in the remainder there are supernumerary or accessory branches.

An *hepatosplenic trunk* (3.5%) has hepatic and splenic arteries, but the left gastric arises independently from the aorta or from the hepatic or splenic arteries.

A *gastrosplenic trunk* (5.5%) has the left gastric and splenic arteries, but the hepatic arises from the aorta or from the superior mesenteric artery.

An *hepatogastric trunk* (1.5%) has the left gastric and hepatic arteries, but the splenic arises from the aorta or from the superior mesenteric artery.

In rare instances the celiac trunk and superior mesenteric arteries are combined.

Left Gastric Artery

The left gastric artery most commonly is the first branch of the celiac trunk, arising near its middle part (Figs. 8-55; 8-57). Being 4 to 5 mm in diameter, it is the smallest of the three branches off the celiac trunk, but it is much larger than the right gastric artery, which usually derives from the common hepatic artery. Frequently (25%), the left gastric arises from the termination of the celiac trunk together with the hepatic and splenic arteries. The vessel is covered by that part of the parietal peritoneum of the dorsal abdominal wall which forms the dorsal layer bounding the omental bursa. Initially, the left gastric artery courses anteriorly, upward, and to the left in a gentle curve. It raises a crescentic ridge of peritoneum called the gastrophrenic fold and reaches the anterior wall of the omental bursa near the cardiac end of the stomach, accompanied by the coronary vein. Here it reverses its direction and, lying between the two leaves of the lesser omentum, gives off a **cardioesophageal branch** before bifurcating into a **ventral** and a **dorsal branch.**

The **cardioesophageal branch** arises before the terminal bifurcation of the left gas-

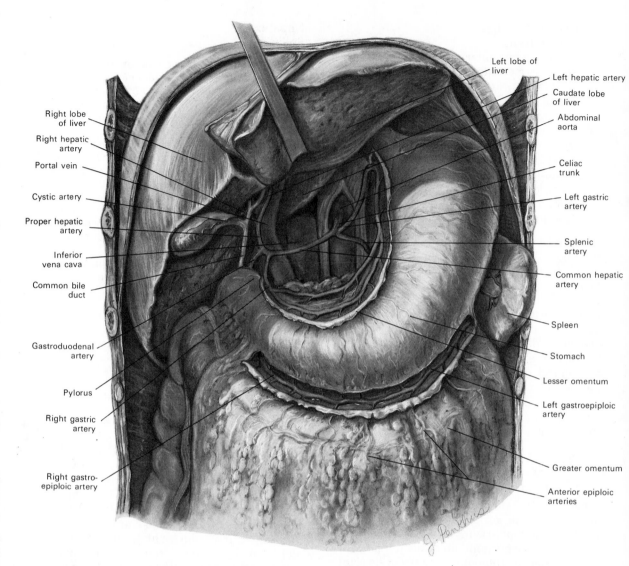

Right lobe of liver

Right hepatic artery

Portal vein

Cystic artery

Proper hepatic artery

Inferior vena cava

Common bile duct

Gastroduodenal artery

Pylorus

Right gastric artery

Right gastro-epiploic artery

Left lobe of liver

Left hepatic artery

Caudate lobe of liver

Abdominal aorta

Celiac trunk

Left gastric artery

Splenic artery

Common hepatic artery

Spleen

Stomach

Lesser omentum

Left gastroepiploic artery

Greater omentum

Anterior epiploic arteries

FIG. 8-55.   The celiac trunk and its branches. The liver has been elevated to reveal the porta hepatis and the gallbladder. Note that the lesser omentum and the anterior layers of the greater omentum have been cut. The arterial supply along the lesser curvature is represented by a double arcade.

tric artery. It is distributed to the cardia of the stomach and the lower esophagus by one to three smaller vessels.

The **ventral branch** of the left gastric is distributed to the anterior surface of the stomach by dividing into two or three smaller arteries.

The **dorsal branch** of the left gastric artery courses along the *lesser curvature of the stomach*, giving branches to the posterior surface of the stomach and usually anastomosing with the right gastric artery (Figs. 8-55; 8-57).

VARIATIONS INVOLVING THE LEFT GASTRIC ARTERY.   The left gastric artery was found to be a branch of the aorta in 2.5% of cadavers studied, of a gastrosplenic trunk in 4.5%, and of an hepatogastric trunk in 1.5%. An **accessory left gastric artery** arose from the splenic in 6%, from the left hepatic in 3%, and from the celiac trunk in 2% of 200 bodies studied by Michels (1953).

An **aberrant left hepatic artery** branches from the left gastric artery just proximal to its bifurcation in 23% of cadavers studied. In considering ligation of the left gastric artery, it is important to realize that these aberrant left hepatic arteries completely replace the regular left hepatic artery in supplying the left lobe of the liver in about half (11.5%) the afore-mentioned cases, and are accessory to the left lobe in the rest. It may be as large as 3 to 5 mm in diameter, and, lying between the two layers of the lesser omentum, it courses upward about 6 cm toward the esophagus. It then crosses the caudate lobe and enters the liver substance at the fissure for the ligamentum venosum. Because of this frequent origin of the left hepatic, the left gastric was named the gastrohepatic artery by Haller as early as 1756.[1]

The inferior phrenic arteries also may arise from the left gastric.

Hepatic Artery

The hepatic artery, arising from the celiac trunk, is slightly smaller (7 to 8 mm) than the splenic artery in the adult, but is the largest branch of the celiac trunk in the fetus (Figs. 8-55 to 8-57). Its first part is horizontal, running from left to right along the upper border of the head of the pancreas to the pylorus or first part of the duodenum, where it usually gives off the **right gastric artery.** Behind the duodenum, the hepatic artery gives rise to the **gastroduodenal artery,** and then the main stem of the vessel turns superiorly to course toward the liver. It is covered by the peritoneum of the dorsal wall of the omental bursa, which is raised thereby into the right hepatopancreatic fold. It then passes between the layers of the hepatoduodenal ligament (lesser omentum) to the porta hepatis where it lies ventral to the epiploic foramen, having the common bile duct to its right and the portal vein dorsal to it. Frequently a distinction is made between two parts of the hepatic artery. That portion coursing from the celiac trunk to where the gastroduodenal artery branches is called the **common hepatic artery,** and the part coursing in the porta hepatis from which stem the lobar hepatic branches is called the **proper hepatic artery** (Figs. 8-55; 8-56). At a variable distance from the liver, the proper hepatic artery gives rise to three terminal branches, the **right, left,** and **middle hepatic arteries.**

VARIATIONS. Of 200 bodies, 41.5% had one or more aberrant hepatic arteries, 50.5% had none, several bodies had two, and a few three aberrants. The common hepatic arose from the aorta in three specimens of the 200, from the superior mesenteric in five, and from the left gastric in one (Michels, 1953).

[1]Albrecht von Haller (1708–1777): A Swiss physiologist and anatomist.

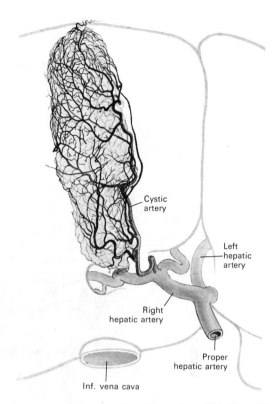

Cystic artery

Left hepatic artery

Right hepatic artery

Proper hepatic artery

Inf. vena cava

FIG. 8-56. The arterial supply to the gallbladder. Observe that the cystic artery usually branches from the right hepatic artery. (From Benninghoff and Goerttler, *Lehrbuch der Anatomie des Menschen*, 1975.)

BRANCHES OF THE HEPATIC ARTERY. The hepatic artery gives rise to the **right gastric, the gastroduodenal, the right hepatic, the left hepatic,** and the **middle hepatic arteries.**

RIGHT GASTRIC ARTERY. The right gastric artery is much smaller and less constant than the left gastric (Figs. 8-55; 8-57). It arises with about equal frequency from the common hepatic artery (40%) or the left hepatic (40.5%). It may arise, however, from the gastroduodenal (8%) or even the right (5.5%) or middle (5%) hepatic arteries. The vessel passes between the layers of the lesser omentum to the pylorus, which it supplies with branches. It runs to the left along the lesser curvature of the stomach, supplying an extensive area with anterior and posterior branches, and anastomoses with the dorsal branch of the left gastric artery.

GASTRODUODENAL ARTERY. The gastroduodenal artery arises from the common hepatic trunk, usually about halfway between its celiac origin and its division into the hepatic branches (Figs. 8-55; 8-57). It may arise from

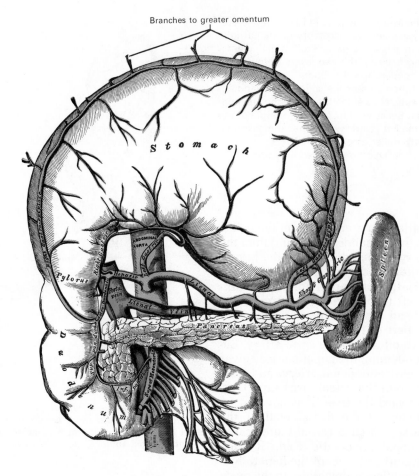

FIG. 8-57. The celiac artery and its branches; the stomach has been raised and the peritoneum removed.

the left hepatic artery (11%), the right hepatic (7%), the middle hepatic (1%), or more rarely from the celiac or superior mesenteric artery. It is important to note that at times the gastroduodenal artery may arise from a hepatic artery that is the sole blood supply to one of the lobes of the liver. Its ligation, therefore, should always be considered in relation to its origin.

The gastroduodenal artery is a short but thick vessel, and it usually lies to the left of the common bile duct. It then crosses the duct either dorsally or ventrally and *descends behind the first part of the duodenum and in front of the neck of the pancreas* to the lower border of the junction between the pylorus and the first part of the duodenum. After giving off branches to the pyloric end of the stomach, the pancreas, and the **supraduodenal** and **retroduodenal branches,** the gastroduodenal artery ends by dividing into the **right gastroepiploic** and **superior pancreaticoduodenal arteries.**

The **supraduodenal artery,** when present, frequently arises from the retroduodenal or gastroduodenal (but may arise from the common or proper hepatic, the right or left hepatic, the superior pancreaticoduodenal, or the right gastric). It is distributed to the upper, anterior, and posterior surfaces of the first 3 cm of the superior (first) part of the duodenum.

The **retroduodenal artery** also arises from the gastroduodenal above the duodenum, but may arise from one of the hepatic arteries (10%). It descends behind the duodenum along the left side of the common bile duct, crosses anterior to the supraduodenal portion of the duct, and forms the U-shaped **dorsal pancreaticoduodenal arcade.** This arcade supplies branches to all four parts of the duodenum, to the head of the pancreas, and to the common bile duct. The retroduodenal artery anastomoses with a dorsal branch of the inferior pancreaticoduodenal from the superior mesenteric artery.

The **right gastroepiploic artery** is the larger of the two terminal branches of the gastroduodenal and is much larger than the left gastroepiploic artery (Figs. 8-55; 8-57). It passes from right to left along the greater curvature of the stomach at a somewhat variable distance from the border of the organ. It lies between the two layers of the gastrocolic ligament or the ventral two layers of the greater omentum when these are not adherent to the colon. It gives off a large ascending **pyloric branch** near its origin, and at its termination usually anastomoses with the left gastroepiploic branch of the splenic artery. It supplies a number of ascending **gastric branches** to the stomach and descending branches to the greater omentum. These latter long anterior **epiploic branches** extend around the free border of the omentum and, in the dorsal layer of the omentum, may anastomose with other gastric or with pancreatic arteries.

The **superior pancreaticoduodenal artery** arises from the gastroduodenal artery as the latter passes dorsal to the first part of the duodenum (Figs. 8-57; 8-58). It forms a loop on the anterior surface of the pancreas and runs along the groove between the pancreas and the descending (second) portion of the duodenum. The vessel is embedded within the substance of the pancreas and, dorsal to the head of the pancreas, anastomoses with the inferior pancreaticoduodenal artery, a branch of the superior mesenteric artery. The loop supplies branches to the ventral surface of the distal three parts (descending, transverse, and ascending) of the duodenum. The **ventral pancreaticoduodenal arcade,** formed by the anastomosis of the superior and inferior pancreaticoduodenal arteries, supplies numerous branches to the pancreas and duodenum (Fig. 8-58).

RIGHT HEPATIC ARTERY. Near the porta hepatis and beyond the origin of the gastroduodenal artery, the right hepatic artery most frequently branches from the proper hepatic artery (Figs. 8-55 to 8-57). Aberrant right hepatic arteries have been found, however, in about one-quarter of cadavers studied. In over three-fifths of the bodies showing an anomalous origin for the right hepatic, the vessel arises from the superior mesenteric artery. The remainder may arise directly from the aorta, the celiac trunk, the gastroduodenal, or even the left hepatic.

At its origin from the proper hepatic artery, the right hepatic artery usually lies anterior to the portal vein and crosses dorsal to the hepatic duct to enter the **cystic triangle** (of Calot[1]), which is bounded by the common hepatic duct on the left, the cystic duct on the right, and the hilum of the liver above. The right hepatic then gives off the **cystic artery** and finally divides into two

[1]Jean F. Calot (1861–1944): A French surgeon.

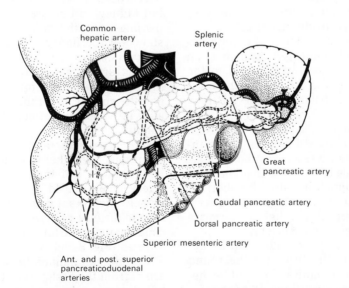

FIG. 8-58. The arterial supply to the pancreas. Note that the head of the pancreas receives the anterior and posterior superior pancreaticoduodenal arteries from the gastroduodenal, whereas the body and tail of the pancreas receive the dorsal, great, and caudal pancreatic branches from the splenic artery. (From Benninghoff and Goerttler, *Lehrbuch der Anatomie des Menschen,* 1975.)

main branches before entering the right lobe of the liver.

The **cystic artery** usually arises from the right hepatic artery to the right of the hepatic duct and within the cystic triangle (for boundaries, see p. 737). Normally it courses a short distance along the cystic duct to reach the neck of the gallbladder (Figs. 8-55; 8-56). At this site the artery divides into a **superficial branch,** which supplies the free peritoneal surface, and a **deep branch,** which supplies the posterior surface of the gallbladder, embedded in the liver. The two branches anastomose and supply twigs to the adjacent liver substance.

There was only a single cystic artery in 75% of the bodies studied, whereas in 25% there was an **accessory cystic artery.** Although 82% of the cystic arteries arose within the cystic triangle, 18% arose to the left of the hepatic duct and not in the triangle. These latter vessels had to cross (either in front of or behind) the hepatic duct to reach the gallbladder. In most instances in which there were two cystic arteries, they served as superficial and deep vessels, and both generally arose within the cystic triangle. However, the superficial vessel was more variable than the deep and, at times, was observed originating from the right hepatic to the left of the duct or from another artery. In only 5% of cases studied did the cystic artery arise from some vessel other than the right hepatic artery. In these, it arose from the proper hepatic artery or, very rarely, from the gastroduodenal artery.

LEFT HEPATIC ARTERY. The left hepatic artery usually divides into two branches, an upper and a lower, supplying twigs to the capsule of the liver and the caudate lobe before entering the substance of the left lobe (Figs. 8-55; 8-56).

Aberrant left hepatic arteries are found in about 25% of bodies studied. In these instances, the aberrant vessel was the only artery to the left lobe of the liver in about 60%, whereas in the remaining 40% the aberrant vessel was an accessory left hepatic artery. Most of the aberrant left hepatic arteries arose from the left gastric artery.

MIDDLE HEPATIC ARTERY. The middle hepatic artery enters the fossa for the round ligament and accompanies the middle hepatic duct to the quadrate lobe. It is the principal supply for this lobe and sends twigs to the round ligament and sometimes to the left lobe. The middle hepatic artery arises from the right hepatic in 45% of cases, from the left hepatic in 45%, and from other sources in 10%.

Splenic (or Lienal) Artery

The splenic or lienal artery is the largest branch of the celiac, and it passes horizontally to the left along the pancreas to reach the spleen. It varies from 8 to 32 cm in length and is usually tortuous (Figs. 8-55; 8-57; 8-58). The *first part,* before it reaches the pancreas, forms a short arc, which swings down and to the right, then across the aorta to the upper border of the pancreas. The *second part* lies along a groove in the upper part of the dorsal surface of the body of the pancreas. It is the most tortuous and undulating part and has one or more loops or coils, but seldom courses far from the pancreas. The *third part* leaves the upper border of the pancreas to cross obliquely the anterior surface of the pancreas. At this point the vessel divides into its superior and inferior arteries in most subjects. The *fourth part* courses between the tail of the pancreas and the hilum of the spleen, and usually the superior and inferior vessels divide into five or more terminal branches at the splenic hilum.

The splenic artery, lying dorsal to the stomach, is covered by the peritoneum of the dorsal wall of the omental bursa. Its fourth part crosses anterior to the upper third of the left kidney and enters the splenic hilum by passing with the splenic vein in the substance of the lienorenal ligament.

BRANCHES OF THE SPLENIC ARTERY. The splenic artery generally supplies blood to three organs—the pancreas, the stomach, and the spleen. Its branches are the **pancreatic, left gastroepiploic, short gastric,** and **splenic.**

PANCREATIC BRANCHES. In addition to numerous twiglike branches to the neck, body, and tail of the pancreas, the splenic artery gives rise to three larger pancreatic vessels: the **dorsal pancreatic,** the **great pancreatic** (arteria pancreatica magna), and the **caudal pancreatic** (arteria caudae pancreatis) **arteries** (Fig. 8-58).

The **dorsal pancreatic artery** frequently arises from the first part of the splenic (39%), but it commonly has other origins, such as from the celiac (22%), one of the hepatic arteries (19%), the superior mesenteric (14%), or the gastroduodenal (2%). It gives a number of twigs to the neck and body of the pancreas. It has two small **right branches,** one curving ventrally to supply the head of the pancreas and anastomosing with branches of the gastroduodenal; the other supplies the uncinate process with a plexus of vessels that anastomoses with the inferior pancreaticoduodenal artery. A branch often courses inferiorly to anastomose with superior mesenteric branches below the pancreas. The principal **left branch** is of considerable size and becomes the **transverse pancreatic artery.** This latter vessel takes a course to the left along the dorsocaudal surface of the pancreas for about two-thirds of its length and enters the substance of the gland to anastomose with other pancreatic branches from the splenic artery. It runs parallel to the main pancreatic duct for some distance, supplying it with branches. Its other long and short branches descend in the dorsal layer of the greater omentum as the **posterior epiploic arteries,** which anastomose with the anterior epiploics from the gastroepiploic arcade.

The **great pancreatic artery** (arteria pancreatica magna) arises from the second part of the splenic artery (Fig. 8-58). It is the largest pancreatic branch (2 to 4 mm diameter) and enters the left part of the body of the gland obliquely. It branches into right and left vessels. These vessels help the transverse splenic artery to supply the main pancreatic duct, and they anastomose with other pancreatic arteries.

The **caudal pancreatic artery** (arteria caudae pancreatis) arises from the third part of the splenic artery or one of its terminal branches (Fig. 8-58). It supplies vessels to the tail of the pancreas, anastomoses with the great and dorsal pancreatic branches, and supplies vessels to an accessory spleen when one is present.

LEFT GASTROEPIPLOIC ARTERY. The left gastroepiploic artery is the largest branch from the splenic artery and arises from its third part or from its inferior terminal artery (Figs. 8-55; 8-57). It reaches the stomach through the pancreaticolienal and gastrolienal ligaments and courses along the greater curvature of the stomach from left to right within the anterior part of the greater omentum. Its branches are distributed to both surfaces of the stomach and to the greater omentum and anastomose with both

gastric and epiploic branches of the right gastroepiploic artery.

SHORT GASTRIC ARTERIES. The short gastric arteries consist of five to seven small branches, which arise near the end of the splenic artery and from its terminal divisions (Fig. 8-57). They pass from left to right, between the layers of the gastrolienal ligament, and are distributed along the greater curvature of the fundus and cardiac end of the stomach. They anastomose with branches of the left gastric, left gastroepiploic, and inferior phrenic arteries.

SPLENIC BRANCHES. The splenic artery divides about 3.5 cm from the spleen into a **superior** and an **inferior terminal branch.** These terminal branches may divide into several branches before entering the spleen, and there may even be intermediate branching. Division of the inferior terminal artery may be more complex than that of the superior, and it may give origin to the left gastroepiploic and inferior polar arteries. Arteries to the poles of the spleen are of frequent occurrence. The **superior polar artery** usually has its origin from the main splenic artery, or it may come directly from the celiac axis. **Inferior polar arteries** are more frequent than superior and arise most frequently from the left gastroepiploic, but may come from the splenic or inferior terminal branches. Cross anastomoses between the larger branches at the hilum are common.

SUPERIOR MESENTERIC ARTERY

The superior mesenteric artery is a large vessel that supplies the whole length of the small intestine except the superior part of the duodenum (Figs. 8-54; 8-57; 8-59; 8-62). It also supplies the cecum and the ascending part of the colon and about one-half of the transverse part of the colon (Fig. 8-59). It arises from the anterior surface of the aorta, about 1.25 cm below the celiac artery, at the level of the first lumbar vertebra, and is crossed at its origin by the splenic vein and the neck of the pancreas. The superior mesenteric artery then passes downward and forward, anterior to the uncinate process of the head of the pancreas and horizontal part of the duodenum, descending between the layers of the mesentery to the right iliac fossa. Here it is considerably diminished in size and anastomoses with one of its own branches, the ileocolic artery. In its descent

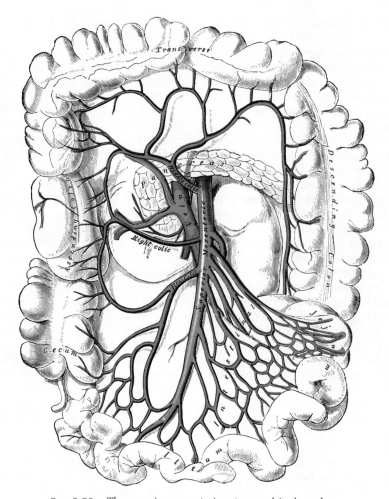

FIG. 8-59. The superior mesenteric artery and its branches.

it crosses ventral to the inferior vena cava, the right ureter, and the psoas major muscle, and forms an arch, the convexity of which is directed forward and downward to the left side. It is accompanied by the superior mesenteric vein, which lies to its right side (Fig. 8-59), and it is surrounded by the superior mesenteric plexus of nerves. Occasionally, the superior mesenteric artery arises from the aorta by a common trunk with the celiac trunk.

Branches of the Superior Mesenteric Artery

The branches of the superior mesenteric artery supply the head of the pancreas, the duodenum, the jejunum and ileum, the cecum and appendix, the ascending and transverse colon, and about half of the descending colon. They are the:

Inferior pancreaticoduodenal
Jejunal and Ileal
Ileocolic
Right colic
Middle colic

Inferior Pancreaticoduodenal Artery

The inferior pancreaticoduodenal artery arises from the superior mesenteric or from its first jejunal branch, opposite the upper border of the horizontal part of the duodenum (Figs. 8-57; 8-58). It courses to the right and frequently divides into anterior and posterior superior pancreaticoduodenal branches. The **anterior branch** ascends in front of the head of the pancreas to anastomose with the anterior branch of the superior pancreaticoduodenal. The **posterior branch** ascends on the dorsal surface of the head of the pancreas to anastomose with the posterior branch of the superior pancreatico-duodenal. From these vessels branches are distributed to the head of the pancreas and to the descending and horizontal parts of the duodenum.

Jejunal and Ileal Arteries

The jejunal and ileal arteries total about 12 to 15 in number (Figs. 8-59; 8-62). These intestinal vessels arise from the convex side of the superior mesenteric artery and are distributed to the jejunum and all of the ileum except its terminal portion. They run nearly parallel to one another between the layers of the mesentery, each vessel dividing into two branches. These unite with adjacent branches, forming a series of arches, the convexities of which are directed toward the intestine (Fig. 8-60). Branches from this first set of arches unite with similar branches from above and below, thus forming a second series of arches. In the ileal region the intestinal arteries may form a third, a fourth, or even a fifth series of arches, which dimin-

ish in size as they approach the intestine. In the short, upper jejunal part of the mesentery only one set of arches exists (Fig. 8-60), but more distally in the ileum, as the depth of the mesentery increases, multiple layers of arches or arcades are developed. From the terminal arches arise numerous small straight vessels, called *vasa recti,* which encircle the intestine and ramify between its coats. From the intestinal arteries small branches are given off to the lymph nodes and other structures between the layers of the mesentery.

Ileocolic Artery

The ileocolic artery is the most inferior of the vessels that arise from the right side of the superior mesenteric artery (Figs. 8-59; 8-61). It passes downward and to the right toward the right iliac fossa, where it divides into a superior and an inferior branch. The **superior branch** of the ileocolic artery passes to the right behind the peritoneum and ascends to about the middle of the ascending colon, where it anastomoses with

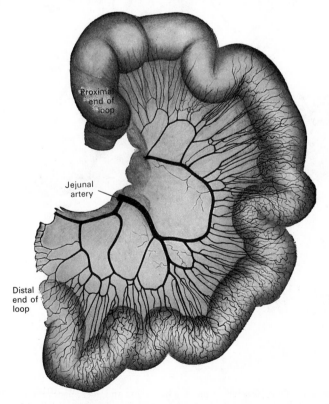

FIG. 8-60. A small loop of jejunum showing the distribution of a jejunal artery coursing in the mesentery. Note that only one or two layers of arterial arches are formed by branches of the jejunal artery before the long vasa recti branch toward the intestine. (From an injected preparation of Hamilton Drummond.)

the descending branch of the right colic artery (Figs. 8-59; 8-61). The **inferior branch** of the ileocolic artery courses toward the superior border of the ileocolic junction and divides into colic, cecal, appendicular, and ileal branches (Fig. 8-61).

The **colic branch** passes upward along the ascending colon to supply several centimeters of its most proximal part. The **anterior** and **posterior cecal** are distributed to the front and back of the cecum. The **appendicular artery** (Fig. 8-61) descends dorsal to the termination of the ileum and enters the mesentery of the vermiform appendix, called the mesoappendix. Although the orientation of the appendix in the lower right quadrant is quite variable, the vessel courses along the free margin of the mesoappendix, ending in several branches that supply the organ. One of these is a recurrent branch, which supplies the posterior surface of the junction of the appendix with the cecum. The **ileal branch** courses to the left and upward along the distal part of the ileum, supplying it and anastomosing with the termination of the superior mesenteric.

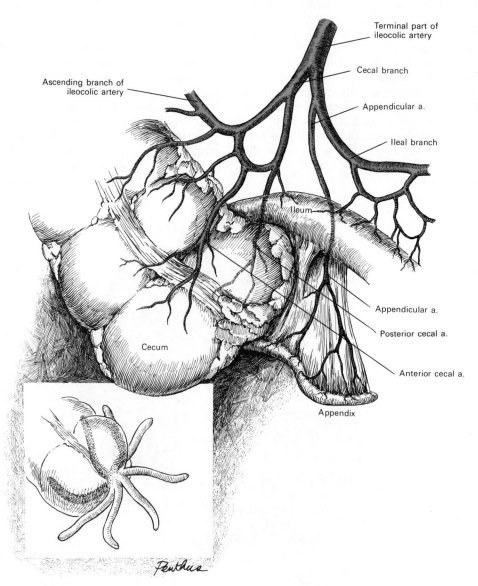

Terminal part of
ileocolic artery

Cecal branch

Appendicular a.

Ileal branch

Ascending branch of
ileocolic artery

Ileum

Appendicular a.

Posterior cecal a.

Anterior cecal a.

Cecum

Appendix

FIG. 8-61. The branches of the iliocolic artery supplying the cecum and the appendix. Lower left insert shows that the appendix may extend in virtually any direction from the cecum. In nearly 65% of cases studied, it projects superiorly behind the cecum and in about 31% inferomedially, as drawn in the larger figure above.

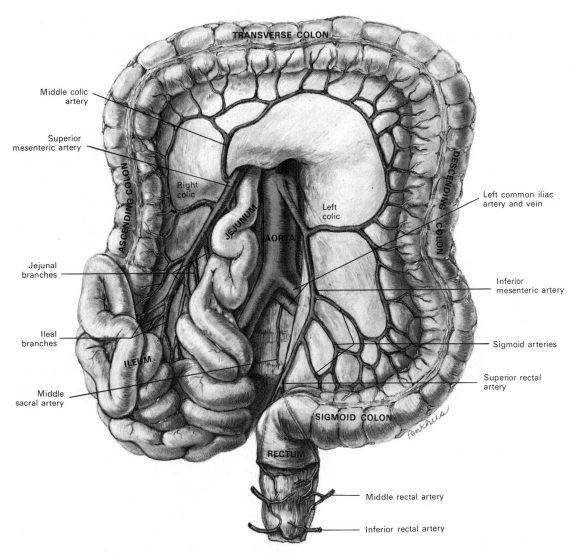

FIG. 8-62.   The arterial supply to the large colon.

Right Colic Artery

The right colic artery arises directly from the right side of the superior mesenteric artery, or from a stem common to it and the ileocolic (Figs. 8-59; 8-62). It passes to the right, behind the peritoneum and anterior to the right testicular or ovarian vessels, the right ureter, and the psoas major muscle, toward the middle of the ascending colon. Sometimes the vessel lies at a higher level, and crosses the descending part of the duodenum and the caudal end of the right kidney. At the ascending colon, the right colic artery divides into a descending branch, which anastomoses with the ileocolic, and an ascending branch, which anastomoses with the middle colic. These branches form arches, from the convexity of which vessels are distributed to the ascending colon, as far as the right colic flexure.

Middle Colic Artery

The middle colic artery arises from the superior mesenteric just below the pancreas (Figs. 8-59; 8-62) and, passing anteriorly between the layers of the transverse mesocolon, divides into right and left branches. The **right branch** anastomoses with the right colic, and the **left branch** with the left colic, which arises from the inferior mesenteric.

The arches thus formed are placed about 4 cm from the transverse colon, to which they distribute branches.

INFERIOR MESENTERIC ARTERY

The inferior mesenteric artery supplies nearly the left half of the transverse colon, the whole of the descending colon, the sigmoid colon, and the greater part of the rectum (Figs. 8-54; 8-62; 8-63). It is smaller than the superior mesenteric artery and arises from the aorta, about 3 or 4 cm above its bifurcation into the common iliac arteries. The origin of the inferior mesenteric artery is close to the lower border of the horizontal part of the duodenum, at the level of the middle of the third lumbar vertebra. The vessel descends behind the peritoneum, lying initially in front of, and then on the left side of, the aorta. It crosses the left common iliac artery, medial to the left ureter, and is continued into the lesser pelvis as the **superior rectal artery**, descending between the two layers of the sigmoid mesocolon and ending on the upper part of the rectum.

Branches of the Inferior Mesenteric Artery

The branches of the inferior mesenteric artery are initially directed to the left or inferiorly. They are the:

Left colic
Sigmoid
Superior rectal

Left Colic Artery

The left colic artery courses to the left, deep to the peritoneum and anterior to the psoas major muscle (Fig. 8-62), and after a short but variable course, it divides into an ascending and a descending branch. The main stem of the artery or its branches cross the left ureter and left testicular or ovarian vessels. The **ascending branch** crosses in front of the left kidney and ends, between the two layers of the transverse mesocolon, by anastomosing with the middle colic artery. The **descending branch** courses down toward the sigmoid flexure and anastomoses with the highest sigmoid artery. From

the arches formed by these anastomoses, branches are distributed to the descending colon and the left part of the transverse colon.

Sigmoid Arteries

Two or three sigmoid arteries branch from the continuing stem of the superior mesenteric artery and run obliquely downward and to the left behind the peritoneum and ventral to the psoas major muscle, the ureter, and the testicular or ovarian vessels (Figs. 8-62; 8-63). Their branches supply the caudal part of the descending colon and the sigmoid colon, anastomosing above with the left colic artery and below with the superior rectal artery.

Superior Rectal Artery

The superior rectal artery (sometimes still called the superior hemorrhoidal artery) is the continuation of the inferior mesenteric artery. It descends into the pelvis between the layers of the sigmoid mesentery and crosses the left common iliac vessels (Figs. 8-62; 8-63). Opposite the third sacral vertebra, it divides into two branches, which descend one on each side of the rectum and, about 10 or 12 cm above the anus, separate into several small branches. These pierce the muscular coat of the bowel and course downward as straight vessels that are placed at regular intervals in the wall of the gut, between its muscular and mucous coats, to the level of the internal anal sphincter. From this site they form a series of loops around the lower end of the rectum and anastomose with the middle rectal branches of the internal iliac and with the inferior rectal branches of the internal pudendal (Figs. 8-62; 8-63).

CLINICAL SIGNIFICANCE. A rich anastomosis of arterial branches that supply the large intestine is formed from the ileocolic, right colic, middle colic, left colic, and sigmoid arteries. These vessels join in the formation of a series of interconnecting arcades, the pattern of which gives rise to a continuous vessel, called the **marginal artery of Drummond**, which courses along the inner perimeter of the large colon, extending from the ileocolic junction to the rectum. From the marginal artery branch straight vessels (arteriae rectae), which are directed to the

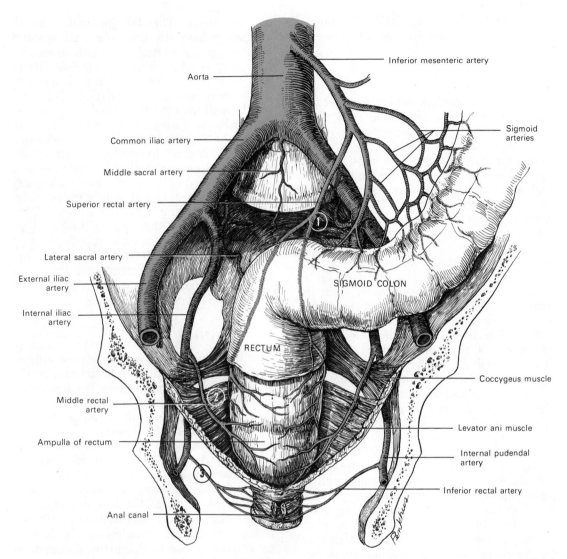

Aorta

Inferior mesenteric artery

Common iliac artery

Sigmoid arteries

Middle sacral artery

Superior rectal artery

Lateral sacral artery

External iliac artery

Internal iliac artery

SIGMOID COLON

RECTUM

Coccygeus muscle

Middle rectal artery

Levator ani muscle

Ampulla of rectum

Internal pudendal artery

Anal canal

Inferior rectal artery

FIG. 8-63.   The inferior mesenteric artery and the blood supply to the sigmoid colon, rectum, and anal canal. Note that the rectum receives its blood supply from three different sources. These are the superior rectal artery (1) from the inferior mesenteric artery; the middle rectal artery (2) from the internal iliac artery; and the inferior rectal artery (3) from the internal pudendal artery.

organ. The clinical benefit of this vascular pattern is appreciated by the surgeon, who frequently is required to ligate one or more branches of the mesenteric vessels in the resection of bowel and yet maintain the vascular integrity of the remainder of the large colon.

A **critical point** of anastomosis for the maintenance of the vascular supply to the distal sigmoid colon, however, is found between the fields supplied by the sigmoid vessels and the superior rectal artery. The pattern of blood supply to this region must be studied carefully because high ligation of the superior rectal artery may, in fact, result in the loss of vascularization to a portion of the distal sigmoid colon.

MIDDLE SUPRARENAL ARTERIES

Although there are three pairs of suprarenal arteries—superior, middle, and inferior, only the middle set of vessels arises directly from the aorta. The superior arises from the inferior phrenic, and the inferior suprarenal arises from the renal arteries. The middle suprarenal arteries are two small vessels that arise, one from each side of the aorta, opposite the superior mesenteric artery. They pass laterally and slightly upward, over the crura of the diaphragm, to the suprarenal glands, where they anastomose with suprarenal branches from the inferior phrenic and renal arteries. In the fetus, these arteries are large.

RENAL ARTERIES

The renal arteries are two large trunks that arise, one on each side of the aorta, immediately below the superior mesenteric artery at the level of the disc between the first and second lumbar vertebrae (Fig. 8-54). Each is directed across the crus of the diaphragm of the corresponding side, thus forming nearly a right angle with the aorta. The **right renal artery** is longer than the left because of the position of the aorta. It passes behind the inferior vena cava, the right renal vein, the head of the pancreas, and the descending part of the duodenum. The **left renal artery** arises somewhat more superiorly than the right. It courses dorsal to the left renal vein, the body of the pancreas, and the splenic vein, and is itself crossed by the inferior mesenteric vein. Before reaching the hilum of the kidney, each artery divides into four or five branches, most of which lie between the renal vein and the ureter. At the renal hilum, the renal vein is most anterior and the pelvis of the kidney is most posterior. The renal artery lies between the two; however, a few of its branches usually enter the kidney behind the renal pelvic origin of the ureter.

Each renal artery gives off some small **inferior suprarenal branches** to the suprarenal gland, the ureter, and the surrounding cellular tissue and muscles.

One or two **accessory renal arteries** are frequently found (23%), especially on the left side. These usually arise from the aorta and come off somewhat more frequently above the main renal artery than below. Instead of entering the kidney at the hilum, they usually pierce the upper or lower part of the organ.

TESTICULAR ARTERIES

The testicular arteries are two long slender vessels that arise from the anterior aspect of the aorta a little below the renal arteries (see Fig. 8-54). Each passes obliquely downward and laterally behind the peritoneum, resting on the psoas major muscle. The **right testicular artery** lies anterior to the inferior vena cava and behind the horizontal portion of the duodenum, the middle colic and ileocolic arteries, and the terminal part of the ileum. The **left testicular artery** courses behind the left colic and sigmoid arteries and the sigmoid colon. Each testicular artery crosses obliquely over the ureter and distal part of the external iliac artery to reach the deep inguinal ring (Fig. 8-70). Passing through the ring, the vessel accompanies the other constituents of the spermatic cord along the inguinal canal to the scrotum. Along its course, the testicular artery becomes tortuous and divides into several branches. Two or three of these accompany the ductus deferens and supply the epididymis, anastomosing with the artery of the ductus deferens. Others pierce the back part of the tunica albuginea and supply the substance of the testis. The testicular artery supplies one or two small branches to the ureter and, in the inguinal canal, gives one or two twigs to the cremaster muscle.

OVARIAN ARTERIES

In the female the ovarian arteries are the homologues of the testicular arteries in the male. They are shorter than the testicular, and because they supply the ovaries, they do not pass out of the abdominal cavity. The origin and course of the first part of each artery are the same as those described for the testicular. Arriving at the superior inlet to the lesser pelvis, the ovarian artery passes medially between the two layers of the suspensory ligament of the ovary and of the broad ligament of the uterus, to be distributed to the ovary. Small branches supply the ureter and the uterine tube, and one passes on to the side of the uterus and anastomoses with the uterine artery. Other branches are distributed to the round ligament of the uterus, through the inguinal canal, and reach the skin of the inguinal region and the labium majus.

In the early stages of development, the testes or ovaries lie by the side of the vertebral column, below the kidneys, and the testicular or ovarian arteries are short. When the gonads descend into the scrotum or lesser pelvis, their arteries gradually lengthen.

INFERIOR PHRENIC ARTERIES

The inferior phrenic arteries are two small vessels that supply the diaphragm, but their origin is subject to some variation (see Fig.

8-54). They may arise separately from the anterior aspect of the aorta, immediately above the celiac artery. In an almost equal number of instances, however, these vessels have been found arising from the celiac trunk. In nearly one-third of cadavers studied, the two inferior phrenic arteries arise by means of a common vessel instead of separately, and this may come from either the aorta or the celiac trunk. Less frequently these vessels may stem from the renal or accessory renal arteries, the left gastric artery, the hepatic arteries, and even the testicular arteries.

From their origin, the inferior phrenic arteries diverge from each other, cross the crura of the diaphragm, and course laterally in an oblique direction upon the abdominal surface of the diaphragm. The **left phrenic artery** passes behind the esophagus and then courses anteriorly on the left side of the esophageal hiatus. The **right phrenic artery** passes behind the inferior vena cava and then along the right side of the diaphragmatic foramen, which transmits that vein. Near the dorsal part of the central tendon of the diaphragm, each vessel divides into a medial and a lateral branch. The **medial branch** curves ventrally and anastomoses with the same vessel of the opposite side, and with the musculophrenic and pericardiacophrenic arteries. The **lateral branch** passes toward the side of the thorax and anastomoses with the lower posterior intercostal arteries and with the musculophrenic artery. The lateral branch of the right phrenic gives off a few vessels to supply the wall of the inferior vena cava, and the left inferior phrenic gives off some branches to the esophagus.

Each inferior phrenic artery gives origin to several **superior suprarenal branches** to the suprarenal gland of its own side. The spleen and the liver also receive a few twigs from the left and right vessels respectively.

LUMBAR ARTERIES

The lumbar arteries are in series with the posterior intercostal arteries. There are usually four on each side, and they arise from the dorsum of the aorta, opposite the bodies of the upper four lumbar vertebrae. A small fifth pair arising from the middle sacral artery is occasionally present, but more frequently lumbar branches of the iliolumbar

arteries take their place. The lumbar arteries run laterally and posteriorly on the bodies of the lumbar vertebrae, behind the sympathetic trunk, to the intervals between adjacent transverse processes of the vertebrae and are then continued into the abdominal wall. The right lumbar arteries pass behind the inferior vena cava, and the upper two on each side course dorsal to the corresponding crus of the diaphragm. The arteries of both sides pass beneath the tendinous arches, extending across the bodies of the lumbar vertebrae, which give origin to slips of the psoas major muscle. The vessels are then continued behind this muscle and the lumbar plexus, and cross the quadratus lumborum, the upper three arteries running behind and the last usually in front of that muscle. At the lateral border of the quadratus lumborum, they pierce the posterior aponeurosis of the transversus abdominis and are carried anteriorly between this muscle and the internal oblique. They anastomose with the lower posterior intercostal, the subcostal, the iliolumbar, the deep iliac circumflex, and the inferior epigastric arteries and with each other.

BRANCHES OF THE LUMBAR ARTERIES. In the interval between the adjacent transverse processes, each lumbar artery gives off a **dorsal ramus,** which is continued backward between the transverse processes and is distributed to the muscles and skin of the back. The dorsal ramus also furnishes a **spinal branch,** which enters the vertebral canal and is distributed similarly to spinal branches of the dorsal rami of the posterior intercostal arteries. **Muscular branches** are supplied from each lumbar artery and from its posterior ramus to the neighboring muscles. The lumbar arteries also send **vertebral branches** to the bodies of the lumbar vertebrae and small **renal branches** to the capsules of the kidney.

MIDDLE SACRAL ARTERY

The middle sacral artery is a small vessel that arises from the dorsal aspect of the aorta a short distance above the bifurcation (Figs. 8-54; 8-62; 8-63). At times it arises from one of the two fifth lumbar arteries. It descends in the midline anterior to the fourth and fifth lumbar vertebrae, the sacrum, and the coccyx. At the tip of the coccyx, the middle sacral artery reaches the coccygeal body, which is a nodular structure consisting of many arteriovenous anastomoses intermeshed in a dense connective tissue. On the last lumbar vertebra it anastomoses with the lumbar branch of the iliolumbar artery; anterior to the sacrum, it anastomoses with the lateral sacral arteries and sends small vessels into the anterior sacral foramina. It is crossed by the left common iliac vein and is accompanied by a pair of venae comitantes. These unite to form a single vessel, which opens into the left common iliac vein.

Arteries of the Pelvis, Perineum, and Gluteal Region

The abdominal aorta terminates at its bifurcation into the two **common iliac arteries** (Figs. 8-54; 8-64; 8-65; 8-75). On each side these vessels divide and subdivide, eventually to supply the organs and walls of the pelvis, the perineum, and the gluteal region.

COMMON ILIAC ARTERIES

Just to the left side of the body of the fourth lumbar vertebra, the abdominal aorta divides into the two common iliac arteries. These vessels diverge laterally as they descend from the end of the aorta, and each divides, opposite the intervertebral fibrocartilage between the last lumbar vertebra and the sacrum, into an **external** and an **internal iliac artery.** The former vessel supplies a good deal of the lower limb, while the latter vessel supplies the viscera and walls of the pelvis.

In addition to the terminal internal and external iliac branches, each common iliac artery gives off small branches to the peritoneum, the psoas major muscle, the ureter, and the surrounding areolar tissue, and occasionally gives origin to the iliolumbar or accessory renal arteries.

Right Common Iliac Artery

The right common iliac artery is usually somewhat longer than the left and passes more obliquely across the body of the last lumbar vertebra (Figs. 8-54; 8-65). It measures about 5 cm in length, and *anterior* to it

are the peritoneum, the loops of small intestine, branches of sympathetic nerves, and, at its point of division, the ureter. *Posteriorly,* it is separated from the bodies of the fourth and fifth lumbar vertebrae and the intervening disc by the terminations of the two common iliac veins and the commencement of the inferior vena cava. More deeply between the fifth vertebra and the psoas major muscle are the obturator nerve, the lumbosacral sympathetic trunk, and the iliolumbar branch of the internal iliac artery. *Laterally,* it is in relation above with the inferior vena cava and below with the right common iliac vein and the psoas major muscle. *Medial* to its upper part is the left common iliac vein.

Left Common Iliac Artery

The left common iliac artery, being a little shorter and thicker than the right, measures about 4 cm in length (see Fig. 8-54). It is related *anteriorly* to the peritoneum, loops of the small intestine, branches of sympathetic nerves, and the superior rectal artery. It is crossed at its point of bifurcation by the left ureter. *Posteriorly* are found the bodies of the fourth and fifth lumbar vertebrae and their intervening disc, and the disc between the fifth lumbar vertebra and the sacrum. Crossing behind the vessel between the fifth lumbar vertebra and the psoas major muscle are the lumbosacral sympathetic trunk, the obturator nerve, and the iliolumbar artery. *Medial* to the left common iliac artery are the left common iliac vein, the hypogastric plexus, and the middle sacral artery, while *lateral* to it is the psoas major muscle.

VARIATIONS RELATED TO THE COMMON ILIAC ARTERIES. The **point of origin** of the common iliac arteries varies according to the bifurcation of the aorta. In three-quarters of a large number of cases, the aorta bifurcated either anterior to the fourth lumbar vertebra or upon the disc between it and the fifth. The bifurcation is located, in one case out of nine, below, and in one out of eleven, above this point. In about 80% of cases studied, the aorta bifurcated within 1.25 cm above or below the level of the crest of the ilium, but more frequently below than above.

The **point of division** of the common iliac arteries into internal and external iliac branches varies greatly. In two-thirds of the cases studied, this division occurred between the last lumbar vertebra and the upper border of the sacrum, being above that point in one of eight cases and below it in one of six

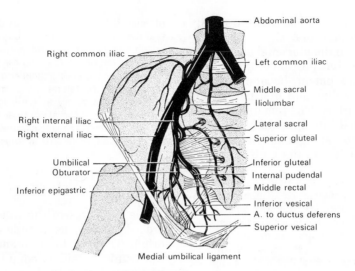

Abdominal aorta

Right common iliac

Left common iliac

Middle sacral

Iliolumbar

Right internal iliac

Right external iliac

Lateral sacral

Superior gluteal

Umbilical

Obturator

Inferior gluteal

Internal pudendal

Inferior epigastric

Middle rectal

Inferior vesical

A. to ductus deferens

Superior vesical

Medial umbilical ligament

FIG. 8-64.   The iliac arteries and their branches.

cases. The left common iliac artery divides lower than the right more frequently.

The **relative lengths** of the two common iliac arteries also vary. The right common iliac was the longer in 37% of cases, the left in 31%; they were equal in 32%. The length of the arteries varied from 3.5 to 7.5 cm in 72% of the cases studied. Of the remaining 28%, 14% were shorter than 3.5 cm and 14% were longer than 7.5 cm. The shortest common iliac artery measured only 1.25 cm, the longest 11 cm. In rare instances, the right common iliac has been absent, and the external and internal iliac arteries arose directly from the aorta.

COLLATERAL CIRCULATION.   The principal vessels contributing to collateral circulation after the application of a ligature to the common iliac are the anastomoses of (1) the middle rectal branch of the internal iliac with the superior rectal from the inferior mesenteric, (2) the uterine, ovarian, and vesical arteries of the opposite side with the same vessels on the ligated side, (3) the lateral sacral artery from the internal iliac and the middle sacral artery from the aorta, (4) the inferior epigastric artery from the external iliac with the internal thoracic artery, the lower posterior intercostal arteries, and the lumbar arteries, (5) the deep iliac circumflex from the external iliac artery and the lumbar arteries from the aorta, (6) the iliolumbar artery from the internal iliac with the last lumbar artery, and (7) the obturator artery, by means of its pubic branch, with the same vessel from the opposite side and with the inferior epigastric arteries, which anastomose with the superior epigastric arteries.

Internal Iliac Artery

The internal iliac artery, also known as the **hypogastric artery,** begins at the bifurcation of the common iliac, anterior to the sacroiliac joint at the level of the lumbosacral disc (Figs. 8-64; 8-65; 8-75). It follows a short, curved course of about 4 cm, descending toward the greater sciatic foramen. Although it is an exceedingly variable artery in terms of the pattern of its branches, in slightly more than half the bodies it divides into an **anterior trunk** and a **posterior trunk.** It supplies the walls of the pelvis, the pelvic viscera, the buttock, the genital organs, and part of the medial thigh.

The internal iliac artery is covered anteriorly by peritoneum and in females is related to the ovary and uterine tube. It is also crossed by the ureter. Posterior to the vessel courses the internal iliac vein, the sacroiliac joint, and the lumbar contribution to the sacral plexus, called the lumbosacral trunk. Lateral to the internal iliac artery is the internal iliac vein, which separates the artery from the psoas major and the obturator nerve. Medial to the right artery are the terminal loops of the ileum, and medial to the left artery is the flexure of the sigmoid colon as it becomes the rectum.

In the fetus, the internal iliac artery is twice as large as the external iliac; it is the direct continuation of the common iliac. It ascends along the side of the bladder and runs superiorly on the inner aspect of the anterior wall of the abdomen to reach the umbilicus, converging toward the same vessel from the opposite side. Having passed through the umbilical opening, the two arteries, which are now called **umbilical arteries,** enter the umbilical cord and become coiled around the umbilical vein, to ramify ultimately in the placenta.

After birth, when the placental circulation ceases, only the pelvic portion of the umbilical artery remains patent, becoming in the adult the internal iliac and first part of the superior vesical artery. The remainder of the vessel is converted into a solid fibrous cord, the **medial umbilical ligament,** which extends from the pelvis to the umbilicus.

COLLATERAL CIRCULATION. After ligature of the internal iliac artery, collateral circulation is accomplished by anastomoses (1) of the ovarian artery from the aorta and the uterine artery; (2) between the vesical arteries of the opposite side and the same vessels on the ligated side; (3) between the middle rectal branches of the internal iliac and the superior rectal artery from the inferior mesenteric; (4) between the obturator artery and the inferior epigastric and medial femoral circumflex arteries, and by means of the pubic branch of the obturator artery and the same vessel from the opposite side; (5) between the circumflex and perforating branches of the deep femoral artery and the inferior gluteal; (6) between the superior gluteal artery and the posterior branches of the lateral sacral arteries; (7) between the iliolumbar artery and the last lumbar; (8) between the lateral sacral artery and the middle sacral; and (9) between the iliac circumflex and the iliolumbar and superior gluteal arteries.

Branches of the Internal Iliac Artery

The branches of the internal iliac artery are subject to considerable variation in their origin. The division into **anterior** and **posterior trunks** is seldom complete, and any two branches may arise from a common trunk. However, the distribution to tissues and organs of the individual branches is more con-

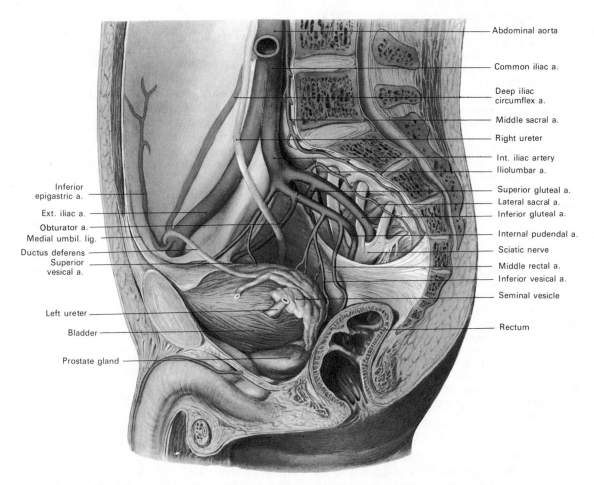

FIG. 8-65.   The right internal and external iliac arteries and their branches viewed from the left side. (From Benninghoff and Goerttler, 1975.)

stant, and the different vessels may be grouped as follows: visceral, anterior parietal, and posterior parietal.

Visceral
    Umbilical
    Inferior vesical
    Middle rectal
    Uterine
    Vaginal
Anterior Parietal
    Obturator
    Internal pudendal
Posterior Parietal
    Iliolumbar
    Lateral sacral
    Superior gluteal
    Inferior gluteal

VARIATIONS IN THE BRANCHING PATTERN OF THE INTERNAL ILIAC ARTERY. In a series of 130 consecutive dissections, Ashley and Anson (1941) recorded the pattern of variation in the branching of the internal iliac artery. In the most common pattern (58%), the inferior gluteal and internal pudendal arteries were derived from a common stem, and the umbilical and superior gluteal arteries arose above them by separate stems. In 17%, the inferior and superior gluteal arteries arose from one stem and the umbilical and pudendal arteries arose from another stem. In 10%, the internal pudendal, inferior gluteal, and umbilical arteries arose by a common stem, and the superior gluteal branched separately. In 8%, the umbilical and internal pudendal arteries arose separately above a common stem for the two gluteal arteries. In these instances the internal pudendal artery appeared to be the continuation of the internal iliac, a condition described by some authors as a more primitive type found in lower animals. Five other patterns were described, but their appearance was less frequent than the above. The obturator was probably the most variable branch, and a few of its patterns are described with that artery.

UMBILICAL ARTERY

The umbilical arteries are generally the first of the visceral branches that arise from the internal iliac arteries (Fig. 8-64). During intrauterine life, these vessels form the principal channels from the fetal internal iliac arteries to the umbilical cord. By adulthood, most of the vessel on each side becomes a fibrous cord called the **medial umbilical ligament;** however, it does retain a patent lumen for a short distance beyond the internal iliac artery. Just proximal to the site where the umbilical artery becomes obliter-

ated, the umbilical artery on each side gives rise to several **superior vesical branches** which supply the bladder. It also may give rise to the **artery of the ductus deferens** and, at times, to the **middle vesical artery** as well.

SUPERIOR VESICAL ARTERIES. From one to four small superior vesical arteries arise from the umbilical artery to supply the upper part of the bladder (Figs. 8-64; 8-65). In nearly 75% of cases, there are two or three superior vesical arteries on each side.

ARTERY OF THE DUCTUS DEFERENS. This slender vessel may arise directly from the umbilical artery, or from one of the superior vesical arteries, or from the middle vesical artery (Fig. 8-64). It accompanies the ductus in its course to the testis, where it anastomoses with the testicular artery.

MIDDLE VESICAL ARTERY. This vessel may arise directly from the umbilical artery or as a branch of the superior vesicle. It is distributed to the fundus of the bladder and the seminal vesicles and anastomoses with the other vesical arteries. Frequently, one or more **ureteric branches** arise from the middle vesical artery or the artery to the ductus deferens to supply the lower end of the ureter. These anastomose with other vessels that supply the more superior parts of the ureter.

INFERIOR VESICAL ARTERY

The inferior vesical artery sometimes arises from the anterior trunk of the internal iliac artery in common with the middle rectal. It may, however, arise from a trunk in common with the internal pudendal and superior gluteal arteries, or as a branch from the internal pudendal (Fig. 8-65). It is usually a single branch and is distributed to the fundus of the bladder, the prostate, and the seminal vesicles. The branches to the prostate communicate with the corresponding vessels of the opposite side.

MIDDLE RECTAL ARTERY

The middle rectal artery may arise from the anterior trunk of the internal iliac with the inferior vesical artery, but more frequently it arises with or from the internal pudendal artery or from the inferior gluteal artery (Figs. 8-64; 8-65). Only rarely is it completely absent. It is distributed to the rectum, anastomosing with the inferior vesical and with the superior and inferior rectal arteries. It gives branches to the seminal vesicles and the prostate.

UTERINE ARTERY

The uterine artery generally arises from the medial surface of the anterior trunk of the internal iliac (Fig. 8-66). It continues medially on the levator ani toward the cervix of the uterus. *At about 2 cm from the cervix it crosses above and in front of the ureter,* to which it supplies a small branch. Reaching the side of the uterus, it ascends in a tortuous manner between the two layers of the broad ligament to the junction of the uterine tube and uterus. It then runs laterally toward the hilum of the ovary and ends by anastomosing with the ovarian artery. It supplies branches to the uterine cervix and others to the vagina. These latter vessels anastomose with branches of the vaginal arteries and form with them the median longitudinal vessels called the **azygos arteries of the vagina.** There are usually two azygos arteries, one of which courses down the ventral and the other down the dorsal surface of the vagina (Fig. 8-66). The uterine artery supplies numerous branches to the body of the uterus, and from its terminal portion twigs are distributed to the uterine tube and the round ligament of the uterus before anastomosing with the ovarian artery (Fig. 8-66).

VAGINAL ARTERIES

The vaginal arteries are represented by one, two, or even three vessels which may arise directly from the anterior trunk of the internal iliac artery (Fig. 8-66). More frequently one or more of the vaginal arteries arises from the uterine artery or even from the inferior vesical artery. Approaching the vagina at its junction with the uterine cervix, the vaginal arteries then descend upon the vagina, supplying its mucous membrane. They send branches to the bulb of the vestibule, the fundus of the bladder, and the contiguous part of the rectum, and assist in forming the azygos arteries of the vagina (Fig. 8-66).

OBTURATOR ARTERY

The obturator artery arises most frequently from the ventral or medial surface of the internal iliac artery or from one of its branches (Figs. 8-64; 8-65; 8-70; 8-74). It

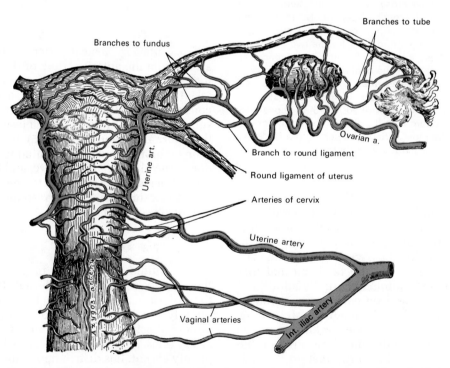

FIG. 8-66. The arteries of the female internal organs of reproduction as seen from behind.

courses ventrally on the lateral wall of the pelvis to the obturator canal, through which it courses as it leaves the pelvic cavity to reach the thigh, where it divides into anterior and posterior branches. Lateral to the obturator artery in the pelvis is the obturator fascia covering the obturator internus muscle. The vessel is crossed medially by the ureter and ductus deferens, and it is covered by the peritoneum. The obturator nerve accompanies the artery and courses above the vessel, while the obturator vein courses below.

Branches of the Obturator Artery

Within the pelvis, the obturator artery gives rise to **iliac, vesical,** and **pubic branches.** *Outside the pelvis,* the vessel divides at the margin of the obturator foramen into an **anterior** and a **posterior branch,** which encircle the foramen under cover of the obturator externus muscle.

ILIAC BRANCHES. The iliac branches of the obturator artery ascend in the iliac fossa and supply the iliacus muscle and send nutrient vessels to the ilium; they also anastomose with branches of the iliolumbar artery.

VESICAL BRANCH. The vesical branch of the obturator courses medially and backward to help supply the bladder.

PUBIC BRANCH. The pubic branch arises from the obturator artery just before it leaves the pelvic cavity. This branch ascends upon the inside of the pubis and communicates with the same vessel from the opposite side and with the inferior epigastric artery.

ANTERIOR BRANCH. The anterior branch of the obturator artery courses ventrally on the outer surface of the obturator membrane and then curves downward along the anterior margin of the foramen. It distributes branches to the obturator externus, pectineus, the adductor muscles, and the gracilis, anastomosing with the posterior branch and with the medial femoral circumflex artery.

POSTERIOR BRANCH. The posterior branch of the obturator artery follows the posterior margin of the foramen and divides into two branches. One branch runs anteriorly on the inferior ramus of the ischium, where it anastomoses with the anterior branch of the obturator. The other branch gives twigs to the muscles attached to the ischial tuberosity and anastomoses with the inferior gluteal. It

also supplies the **acetabular** or articular branch, which enters the hip joint through the acetabular notch, ramifies in the fat at the bottom of the acetabulum, and sends a twig deep to the transverse acetabular ligament accompanying the ligament of the head of the femur.

VARIATIONS OF THE OBTURATOR ARTERY. The obturator artery was found to arise from the internal iliac or one of its branches in nearly 70% of 320 bodies studied (Pick, Anson, and Ashley, 1942). It arose from the main vessel in 23%, from the anterior trunk in 20%, from the posterior trunk in 3%, from the superior gluteal in 11%, from the inferior gluteal in 9%, and from the internal pudendal or external iliac in 2 or 3%. The origin of the obturator artery from the inferior epigastric artery occurred in 27% of cases. In these instances, the artery lies in contact with the external iliac vein. Its course toward the obturator canal is relevant clinically during operations for the repair of a femoral hernia. If its course is lateral to the lacunar ligament, repair is relatively safe. If, however, the obturator artery *curves along the free margin of the lacunar ligament,* it can encircle the neck of a hernial sac and be injured during the hernial repair (Fig. 8-67).

INTERNAL PUDENDAL ARTERY

The internal pudendal artery (Figs. 8-64; 8-65; 8-68) supplies the perineum and external organs of generation, and although the course of the artery is the same in the two sexes, the vessel is smaller in the female than in the male, and the distribution of its branches somewhat different.

Internal Pudendal Artery in the Male

The internal pudendal artery in the male courses downward and laterally toward the lower border of the greater sciatic foramen and emerges from the pelvis between the

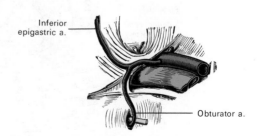

FIG. 8-67. An obturator artery is shown arising from the inferior epigastric and following the free edge of the lacunar ligament to the obturator foramen.

piriform and coccygeus muscles. It then crosses the ischial spine and courses through the medial part of the lesser sciatic foramen to enter a tunnel deep to the fascia of the obturator internus muscle called the pudendal canal (of Alcock[1]). At this point the obturator internus forms the lateral wall of the ischiorectal fossa, and the canal courses about 4 cm above the lower margin of the ischial tuberosity. The artery gradually approaches the margin of the inferior ramus of the ischium, and then passes ventrally between the two layers of the fascia of the urogenital diaphragm. Continuing anteriorly along the medial margin of the inferior ramus of the pubis and about 1.25 cm dorsal

[1]Benjamin Alcock (1801-unknown): An Irish anatomist who moved to America in 1855.

to the arcuate pubic ligament, it divides into the **dorsal** and **deep arteries of the penis,** but it may pierce the superficial fascia of the urogenital diaphragm before doing so (Fig. 8-68).

RELATIONS OF THE MALE INTERNAL PUDENDAL ARTERY. *Within the pelvis,* the internal pudendal artery lies anterior to the piriformis muscle, the sacral plexus of nerves, and the inferior gluteal artery. *Crossing the ischial spine,* it is covered by the gluteus maximus and overlapped by the sacrotuberous ligament. At this latter side, the pudendal nerve lies medial to the artery, and the nerve to the obturator internus muscle lies lateral to the artery. *In the perineum,* the internal pudendal artery lies on the lateral wall of the ischiorectal fossa in the pudendal canal, formed by the splitting of the obturator fascia. It is accompanied by a pair of companion veins and the pudendal nerve.

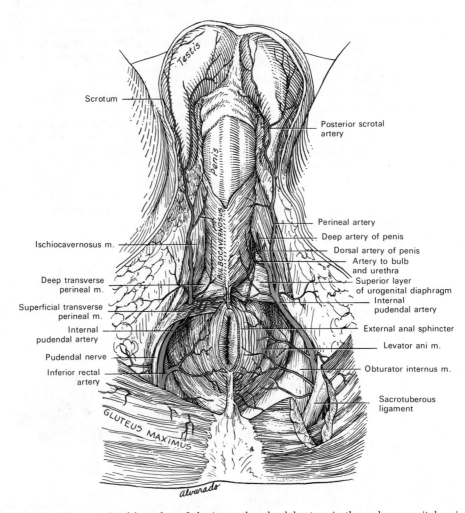

FIG. 8-68.   The superficial branches of the internal pudendal artery in the male urogenital region.

BRANCHES OF THE MALE INTERNAL PUDENDAL ARTERY. Although a few muscular branches arise from the internal pudendal artery within the pelvis, most branches supply structures in both the anal and the urogenital regions of the perineum. The branches are:

Muscular
Inferior rectal
Perineal
Artery of the bulb of the penis
Urethral
Deep artery of the penis
Dorsal artery of the penis

MUSCULAR BRANCHES. Muscular branches from the internal pudendal artery consist of two sets: one given off in the pelvis and the other as the vessel crosses the ischial spine in the gluteal region. Within the pelvis several small unnamed vessels supply the levator ani, the obturator internus, the piriformis, and the coccygeus. The branches given off in the gluteal region are distributed to the adjacent parts of the gluteus maximus and to other lateral rotator muscles; they anastomose with branches of the inferior gluteal artery.

INFERIOR RECTAL ARTERY. Also known as the *inferior hemorrhoidal artery,* this vessel arises from the internal pudendal as it passes above the ischial tuberosity. Piercing the fascial sheath forming the pudendal canal, the artery divides into two or three branches which cross the ischiorectal fossa and are distributed to the muscles and integument of the anal region (Fig. 8-68). Small branches also ascend around the lower edge of the gluteus maximus to the skin of the buttock. Branches of the inferior rectal artery anastomose with the corresponding vessels of the opposite side, with the superior and middle rectal, and with the perineal artery.

PERINEAL ARTERY. The perineal artery arises from the internal pudendal anterior to the inferior rectal branches. It pierces the urogenital diaphragm and crosses either over or under the superficial transverse perineal muscle (Fig. 8-68). Continuing forward in the superficial perineal compartment of the urogenital region, the vessel courses anteriorly, parallel to the pubic arch, in the interspace between the bulbospongiosus and ischiocavernosus muscles. It supplies both these muscles and finally divides into several **posterior scrotal branches,** which are distributed to the skin and dartos tunic of the scrotum. As it pierces the urogenital diaphragm, the perineal artery gives off a small **transverse branch,** which courses medially along the cutaneous surface of the superficial transverse perineal muscle and anastomoses with the corresponding vessel of the opposite side and with the inferior rectal arteries. The transverse branch supplies the muscle and other structures between the anus and the bulb of the penis.

ARTERY OF THE BULB OF THE PENIS. The artery to the bulb is a short vessel of large caliber which arises from the internal pudendal between the two layers of fascia of the urogenital diaphragm (Fig. 8-68). It passes medially, pierces the inferior fascia of the urogenital diaphragm, and gives off branches that ramify in the bulb of the penis and in the posterior part of the corpus spongiosum penis. It gives off a small branch to the **bulbourethral gland** (of Cowper[1]).

URETHRAL ARTERY. The urethral branch of the internal pudendal artery arises a short distance anterior to the artery of the bulb of the penis. It runs medially, pierces the inferior fascia of the urogenital diaphragm, and enters the corpus spongiosum penis in which it is continued forward to the glans penis. It supplies the urethra and its surrounding tissue.

DEEP ARTERY OF THE PENIS. The deep artery of the penis is one of the terminal branches of the internal pudendal artery (Fig. 8-68). It arises from that vessel while it is situated between the two fascial layers of the urogenital diaphragm. It pierces the inferior fascial layer of the urogenital diaphragm and enters the crus penis obliquely. The deep artery continues distally in the center of the corpus cavernosum penis, its branches being distributed to the erectile tissue.

DORSAL ARTERY OF THE PENIS. The dorsal artery of the penis is the other terminal branch of the internal pudendal artery (Fig. 8-68). It arises in the deep perineal compartment, just lateral and deep to the bulbospongiosus muscle. Piercing the inferior fascia of the urogenital diaphragm between the crus penis and the pubic arch, it then passes between the two layers of the suspensory ligament to reach the dorsum of the penis,

[1]William Cowper (1666-1709): An English surgeon and anatomist (London).

where it runs distally to the glans, sending branches to the glans and prepuce. Along the penis, it lies between the dorsal nerve and deep dorsal vein. It supplies the skin and fibrous sheath of the corpora cavernosa penis and anastomoses with the deep artery.

VARIATIONS OF THE MALE INTERNAL PUDENDAL ARTERY.   Sometimes the internal pudendal artery is smaller than usual or fails to give off one or two of its usual branches. In such instances, the deficiency generally is supplied by branches derived from an additional vessel, the **accessory pudendal artery,** which usually arises from the internal pudendal artery before its exit from the greater sciatic foramen. The accessory vessel passes forward along the lower part of the bladder and across the side of the prostate to the root of the penis, where it perforates the urogenital diaphragm, and gives off branches usually derived from the internal pudendal artery. The variation most frequently encountered is that in which the internal pudendal ends as the artery of the urethral bulb, while the dorsal and deep arteries of the penis arise from the accessory pudendal. The internal pudendal artery may also end as the perineal artery, the artery of the urethral bulb being derived, with the other two branches, from the accessory vessel. Occasionally the accessory pudendal artery is derived from one of the other branches of the internal iliac artery, most frequently the inferior vesical or the obturator.

Internal Pudendal Artery in the Female

In the female, the internal pudendal artery is considerably smaller than in the male. Its origin and course, however, are similar and there are comparable branches for the homologous structures. **Posterior labial branches** of the perineal artery supply the labia majora; the **artery of the bulb** supplies the erectile tissue of the vestibular bulb and the vagina; the **deep artery of the clitoris** supplies the corpus cavernosum clitoridis; and the **dorsal artery of the clitoris** supplies the dorsum of that organ, ending in the glans and prepuce of the clitoris.

ILIOLUMBAR ARTERY

The iliolumbar artery is a branch of the internal iliac artery, which usually comes off the posterior trunk (Figs. 8-64; 8-65). It courses upward and laterally beneath the common iliac artery, initially between the lumbosacral trunk and the obturator nerve. It then ascends between the vertebral col-

umn and the psoas major muscle. Just above the pelvic brim and behind the psoas major muscle, the iliolumbar artery divides into a **lumbar** and an **iliac branch.**

LUMBAR BRANCH.   The lumbar branch of the iliolumbar artery supplies the psoas major and quadratus lumborum, and anastomoses with the fourth, which is the last, lumbar artery. It also sends a small **spinal branch** through the intervertebral foramen between the last lumbar vertebra and the sacrum, into the vertebral canal, to supply the cauda equina of the spinal cord.

ILIAC BRANCH.   The iliac branch of the iliolumbar artery descends to supply the iliacus muscle. Some branches, running between the muscle and the bone, anastomose with the iliac branches of the obturator artery. One of these enters an oblique canal to supply the ilium, while others run along the crest of the ilium, distributing branches to the gluteal and abdominal muscles, and anastomosing in their course with the superior gluteal, the circumflex iliac, and the lateral femoral circumflex arteries.

LATERAL SACRAL ARTERIES

The lateral sacral arteries arise from the posterior trunk of the internal iliac; there are usually two, a **superior** and an **inferior** (Figs. 8-64; 8-65).

SUPERIOR LATERAL SACRAL ARTERY.   The superior lateral sacral artery passes medially and, after anastomosing with branches from the middle sacral, enters the first or second pelvic (anterior) sacral foramen. It supplies branches to the contents of the sacral canal and anastomoses with other spinal vessels. It then leaves the spinal canal by passing through the corresponding posterior sacral foramen, and is distributed to the skin and muscles on the dorsum of the sacrum, anastomosing with the superior gluteal artery.

INFERIOR LATERAL SACRAL ARTERY.   The inferior lateral sacral artery courses obliquely across the piriformis muscle and the anterior rami of the sacral nerves to the medial side of the pelvic (anterior) sacral foramina. It descends on the sacrum and anastomoses over the coccyx with the middle sacral and the opposite lateral sacral artery. In its course, it gives off branches that enter the pelvic sacral foramina. After supplying the contents of the sacral canal, these vessels emerge through the posterior sacral foramina and are distributed to the muscles and skin on the dorsal surface of the sacrum. They anastomose with both the superior and inferior gluteal arteries.

SUPERIOR GLUTEAL ARTERY

The superior gluteal artery is the largest branch of the internal iliac artery and is a direct continuation of its posterior trunk (Figs. 8-64; 8-65; 8-74; 8-76). It is a short artery that courses backward between the lumbosacral trunk and the first sacral nerve. Passing out of the pelvis through the greater sciatic foramen, above the superior border of the piriformis muscle, the superior gluteal artery immediately divides into a **superficial** and a **deep branch** (Fig. 8-69). Within the pelvis it gives off muscular branches to the iliacus, piriformis, and obturator internus, and before leaving that cavity, it gives off a nutrient artery which enters the ilium.

SUPERFICIAL BRANCH. The superficial branch of the superior gluteal artery enters the deep surface of the gluteus maximus and divides into numerous branches (Fig. 8-69). Some of these supply the muscle and anastomose with the inferior gluteal artery; others perforate its tendinous origin and supply the skin covering the dorsal surface of the sacrum, anastomosing with the posterior branches of the lateral sacral arteries.

DEEP BRANCH. The deep branch of the superior gluteal artery lies deep to the gluteus medius and almost immediately subdivides into a superior and an inferior division (Fig. 8-69). The **superior division** continues the original course of the vessel and passes along the superior border of the gluteus minimus to the anterior superior spine of the ilium. It anastomoses with the deep iliac circumflex artery and the ascending branch of the lateral femoral circumflex artery. The **inferior division** crosses the gluteus minimus obliquely to the greater trochanter, distributing branches to the gluteal muscles and anastomosing with the lateral femoral circumflex artery. Some branches pierce the

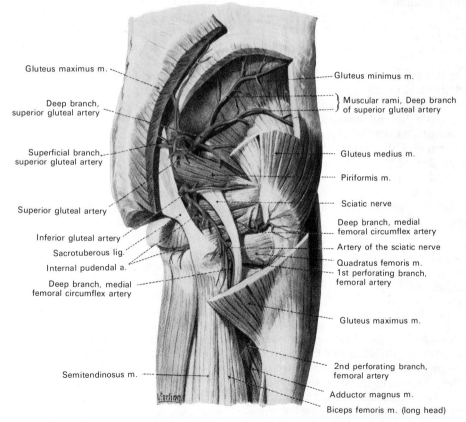

FIG. 8-69. The superior and inferior gluteal arteries and their branches. (From Benninghoff and Goerttler, *Lehrbuch der Anatomie des Menschen*, 1975.)

gluteus minimus to supply the hip joint. Throughout most of its course, the deep branch of the superior gluteal artery ramifies between the gluteus medius and gluteus minimus muscles, helping to define the plane of separation between their fibers.

INFERIOR GLUTEAL ARTERY

The inferior gluteal artery is distributed chiefly to the buttocks and back of the thigh. It passes posteriorly between the first and second sacral nerves, or between the second and third sacral nerves, and then descends between the piriformis and coccygeus muscles through the lower part of the greater sciatic foramen to the gluteal region (Figs. 8-64; 8-65; 8-74; 8-76). It then descends in the interval between the greater trochanter of the femur and tuberosity of the ischium, accompanied by the sciatic and posterior femoral cutaneous nerves and covered by the gluteus maximus. The inferior gluteal artery continues down the posterior thigh, supplying the skin and anastomosing with vessels from the perforating branches of the femoral artery.

Branches of the
Inferior Gluteal Artery

Within the pelvis, the inferior gluteal artery distributes branches to the piriformis, coccygeus, and levator ani. Some branches help to supply the fat around the rectum, and occasionally take the place of the middle rectal artery. Other branches help to supply the fundus of the bladder, the seminal vesicles, and the prostate.

Outside the pelvis (Fig. 8-69), the inferior gluteal artery gives off the following branches:

Muscular
Coccygeal
Artery of the sciatic nerve
Anastomotic
Articular
Cutaneous

MUSCULAR BRANCHES. Muscular branches from the inferior gluteal artery in the gluteal region supply (1) the gluteus maximus, where they anastomose with the superior gluteal artery in the substance of the muscle; (2) the smaller, lateral rotator muscles, anastomosing with the internal pudendal artery;

and (3) the upper part of the hamstring muscles attached to the tuberosity of the ischium. These latter vessels anastomose with the posterior branch of the obturator and the medial femoral circumflex arteries.

COCCYGEAL BRANCHES. The coccygeal branches course medially, pierce the sacrotuberous ligament, and supply the gluteus maximus, the skin, and other structures on the back of the coccyx (Fig. 8-76).

ARTERY OF THE SCIATIC NERVE. This long, slender branch of the inferior gluteal artery accompanies the sciatic nerve for a short distance, then penetrates it and runs in its substance to the distal part of the thigh (Fig. 8-69).

ANASTOMOTIC BRANCH. The anastomotic branch of the inferior gluteal artery descends obliquely and laterally in the gluteal region across the small external rotators of the thigh to assist in forming the so-called **cruciate anastomosis.** This anastomosis around the neck and greater trochanter of the femur involves the first perforating and medial and lateral femoral circumflex branches of the femoral artery in addition to the inferior gluteal artery.

ARTICULAR BRANCH. An articular branch, generally arising from the anastomotic, is distributed to the capsule of the hip joint.

CUTANEOUS BRANCHES. The cutaneous branches of the inferior gluteal artery are distributed to the skin of the buttock and posterior thigh.

External Iliac Artery

The external iliac artery on both the right and left sides is larger than the internal iliac. Each vessel passes obliquely downward and laterally along the medial border of the psoas major muscle. The external iliac artery extends from the bifurcation of the common iliac artery to a point beneath the inguinal ligament, midway between the anterior superior spine of the ilium and the symphysis pubis (Figs. 8-64; 8-65; 8-70; 8-74). Entering the thigh below the inguinal ligament, the vessel then becomes the **femoral artery.**

RELATIONS. *Anteriorly and medially,* the external iliac artery is related to the parietal peritoneum, the subperitoneal areolar tissue, the termination of the ileum, and, frequently, the vermiform appendix on the right side and the sigmoid colon on the left side. The origin of the artery is occasionally crossed by the ureter and, in females, by the ovarian artery and vein. The testicular vessels lie for some distance on the external iliac artery above the inguinal ligament, but near its termination. The vessel is crossed at this site by the genital branch of the genitofemoral nerve and the deep iliac circumflex vein. Curving across its medial aspect in the male is the ductus deferens, or in the female, the round ligament of the uterus. *Posteriorly,* the external iliac artery is related to the medial border of the psoas major muscle, from which it is separated by the iliac fascia. Proximally, the lateral portion of the external iliac vein lies behind the artery, but more distally, the vein courses entirely medial to the artery. *Laterally,* the external iliac artery rests on the iliac fascia overlying the psoas major muscle. Numerous lymphatic vessels and lymph nodes lie on the anterior and medial aspects of the vessel.

COLLATERAL CIRCULATION. The principal anastomoses providing collateral circulation after the application of a ligature to the external iliac are: (1) the iliolumbar with the iliac circumflex, (2) the superior gluteal with the lateral femoral circumflex, (3) the obturator with the medial femoral circumflex, (4) the inferior gluteal with the first perforating and circumflex branches of the profunda artery, and (5) the internal pudendal with the external pudendal. When the obturator artery arises from the inferior epigastric, it receives blood by branches from the internal iliac or the lateral sacral or the internal pudendal artery, and the inferior epigastric receives blood by anastomoses with the internal thoracic and lower intercostal arteries, and from the internal iliac by anastomoses of its branches with the obturator.

Branches of the External Iliac Artery

In addition to several small branches given to the psoas major muscle and the neighboring lymph nodes, the external iliac gives off two branches of considerable size. These are the:

Inferior epigastric
Deep iliac circumflex

INFERIOR EPIGASTRIC ARTERY

The inferior epigastric artery arises from the external iliac immediately above the inguinal ligament (Figs. 8-70; 8-74). It curves anteriorly in the subperitoneal tissue and then ascends obliquely along the medial margin of the deep inguinal ring. Continuing its course upward, the vessel pierces the transversalis fascia, and passing anterior to the arcuate line, it ascends between the rectus abdominis muscle and the posterior layer of its sheath. The inferior epigastric artery finally divides into numerous branches, and these anastomose, above the umbilicus, with the superior epigastric branch of the internal thoracic and with the lower posterior intercostal arteries (see Fig. 8-31). As the inferior epigastric artery passes obliquely upward from its origin, it lies along the lower and medial margins of the

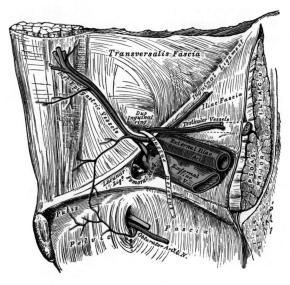

FIG. 8-70. The inferior epigastric artery seen from the inner aspect of the lower abdominal wall. Observe also the relationships of the femoral ring and the deep inguinal ring.

deep inguinal ring and behind the commencement of the spermatic cord. The ductus deferens in the male or the round ligament of the uterus in the female winds around the lateral and dorsal aspects of the artery. In its ascent on the inner aspect of the anterior abdominal wall, it is covered by a reflection of peritoneum called the **lateral umbilical fold.** The following branches arise from the inferior epigastric artery:

Cremasteric
Artery of the round ligament
Pubic
Muscular

CREMASTERIC ARTERY. The cremasteric artery is found in males and it accompanies the spermatic cord, supplying the cremaster muscle and other coverings of the cord. It anastomoses with the testicular, the external pudendal, and the perineal arteries.

ARTERY OF THE ROUND LIGAMENT. In females, the artery that corresponds to the cremasteric is a small vessel that accompanies the round ligament of the uterus.

PUBIC BRANCH. The pubic branch of the inferior epigastric artery courses along the inguinal ligament. It then descends along the medial margin of the femoral ring to the internal surface of the pubis and there anastomoses with the pubic branch of the obturator artery. In almost one-third of the cadavers, the pubic branch is much larger and actually becomes the obturator artery.

MUSCULAR BRANCHES. Some of the muscular branches of the inferior epigastric artery are distributed to the abdominal muscles and peritoneum, anastomosing with the iliac circumflex and lumbar arteries. Other branches perforate the tendon of the external oblique muscle to supply the skin and to anastomose with branches of the superficial epigastric artery.

VARIATIONS. The inferior epigastric artery may arise from any part of the external iliac between the inguinal ligament and a point 6 cm above it. The vessel may also arise from the femoral artery below the inguinal ligament. Frequently it branches from the external iliac by a common trunk with the obturator. Sometimes the inferior epigastric arises from the obturator, the latter vessel being furnished by the internal iliac, or it may be formed by two branches, one derived from the external iliac, the other from the internal iliac.

DEEP ILIAC CIRCUMFLEX ARTERY

The deep iliac circumflex artery arises from the lateral aspect of the external iliac almost opposite the inferior epigastric artery (Fig. 8-74). It ascends laterally to the anterior superior iliac spine, deep to the inguinal ligament. In its course it lies between the transversalis fascia and the peritoneum, or it is contained in a fibrous sheath formed by the union of the transversalis fascia and iliac fascia. In the lateral wall of the pelvis, the deep iliac circumflex artery anastomoses with the ascending branch of the lateral femoral circumflex artery from below. It then pierces the transversalis fascia and passes along the inner lip of the crest of the ilium to about its middle, where it perforates the transversus abdominis, and runs dorsally between that muscle and the internal oblique to anastomose with the iliolumbar and superior gluteal arteries. Opposite the anterior superior iliac spine, it gives off a large **ascending branch,** which courses between the internal oblique and transversus muscles, supplying them and anastomosing with the lumbar and inferior epigastric arteries.

Arteries of the Lower Limb

The artery that supplies the greater part of the lower limb is the direct continuation of the external iliac (Fig. 8-74). It courses as a single trunk from the inguinal ligament to the lower border of the popliteus muscle behind the knee joint, where it divides into two branches, the **anterior** and **posterior tibial.** The upper part of the main trunk courses through the anterior and medial aspect of the thigh and is called the **femoral artery.** Below the adductor canal, the artery passes through a hiatus in the tendon of the adductor magnus muscle in order to reach the popliteal fossa, where the vessel is called the **popliteal artery.**

Femoral Artery

The femoral artery, being the continuation of the external iliac artery, begins immediately below the inguinal ligament, midway

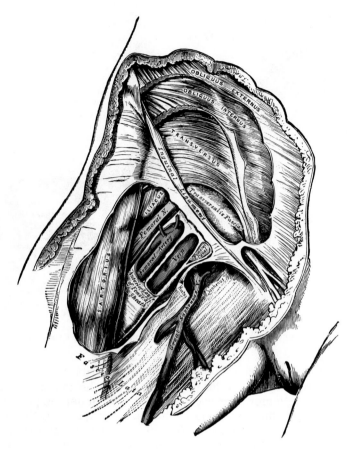

Fig. 8-71. Femoral sheath opened to show its three compartments.

between the anterior superior iliac spine and the symphysis pubis (Figs. 8-71 to 8-75). It passes distally along the anteromedial aspect of the thigh, ending at the junction of the middle and lower thirds of the thigh, where it passes through an opening in the tendon of the adductor magnus muscle to become the **popliteal artery** (Fig. 8-76). In the upper part of the thigh, the femoral artery lies anterior to the hip joint, while in the lower part of its course, it lies medial to the shaft of the femur. Between these two parts, it crosses the angle between the head and shaft, remaining at some distance from the bone. The first 4 cm of the vessel are enclosed, together with the femoral vein, in the fibrous **femoral sheath.** In the upper third of the thigh, the femoral artery is contained in the **femoral triangle** (of Scarpa[1]) and in the

middle third of the thigh, in the **adductor canal** (of Hunter[2]).

FEMORAL SHEATH. The femoral sheath is formed by a prolongation downward under the inguinal ligament of the transversalis fascia anterior to the femoral vessels and the iliac fascia behind them (Figs. 8-71; 8-72). The sheath assumes the form of a short funnel, the wide end of which is directed superiorly, while the inferior, narrow end fuses with the adventitia of the vessels, about 4 cm below the inguinal ligament. It is strengthened anteriorly by a band termed the **deep crural arch.** The *lateral wall* of the sheath is perforated by the femoral branch of the genitofemoral nerve. The *medial wall* is pierced by the great saphenous vein and by some lymphatic vessels. The sheath is divided into three compartments by two verti-

[1]Antonio Scarpa (1747–1842): A Venetian anatomist and ophthalmologist (Pavia).

[2]John Hunter (1728–1793): A Scottish surgeon, anatomist and physiologist (Edinburgh and London).

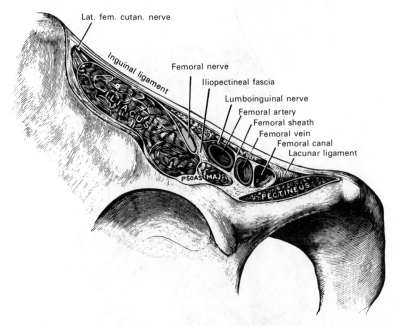

Lat. fem. cutan. nerve

Inguinal ligament

Femoral nerve

Iliopectineal fascia

Lumboinguinal nerve

Femoral artery

Femoral sheath

Femoral vein

Femoral canal

Lacunar ligament

ILIACUS

PSOAS MAJ.

PECTINEUS

FIG. 8-72. The structures passing deep to the inguinal ligament, as seen inferiorly.

cal partitions which stretch between its anterior and posterior walls. The *lateral compartment* contains the femoral artery, and the *intermediate compartment* contains the femoral vein. The *medial compartment* is the smallest and is called the **femoral canal** (Figs. 8-71; 8-72). It contains some lymphatic vessels and a lymph node (of Cloquet[1] or Rosenmüller[2]) embedded in a small amount of areolar tissue. The femoral canal is conical and measures about 1.25 cm in length. Its base is directed superiorly and, although oval in form, is named the **femoral ring** (Fig. 8-70). The long diameter of the ring is directed transversely and it measures about 1.25 cm. The femoral ring (Fig. 8-70) is bounded *anteriorly* by the inguinal ligament, *posteriorly* by the pectineus muscle and its overlying pectineal fascia, *medially* by the crescentic base of the lacunar ligament, and *laterally* by the fibrous septum on the medial aspect of the femoral vein. The spermatic cord in the male and the round ligament of the uterus in the female lie immediately superior to the anterior margin of the ring, and the inferior epigastric vessels are close to its superior and lateral angle (Fig. 8-70). The femoral ring is closed by a somewhat condensed portion of the subperitoneal fatty tissue, called the **femoral septum,** the abdominal surface of which supports a small lymph node and is covered by the parietal layer of the peritoneum. The peritoneum presents a slight depression named the **femoral fossa.** The femoral septum is pierced by numerous lymphatic vessels passing from the deep inguinal to the external iliac lymph nodes. The femoral nerve is not enclosed by the femoral sheath (Fig. 8-71).

FEMORAL TRIANGLE.   The femoral triangle corresponds to the depression seen immediately below the fold of the groin (Fig. 8-73). Its **apex** is directed inferiorly, and the **sides** are formed *laterally* by the sartorius, *medially* by the adductor longus, and *superiorly* by the inguinal ligament. The **floor** of the space is formed from its lateral to medial aspect by the iliacus, psoas major, and pectineus muscles, and in some cases by a small part of the adductor brevis. It is divided into two nearly equal parts by the femoral vessels, which extend from near the middle of its base to its apex. The femoral artery gives off both its more superficial and deep branches, while the femoral vein receives its femoral and great saphenous tributaries. On the lateral aspect of the femoral artery is the

[1]Jules G. Cloquet (1790–1883): A French anatomist and surgeon (Paris).
[2]Johann C. Rosenmüller (1771–1820): A German anatomist (Leipzig).

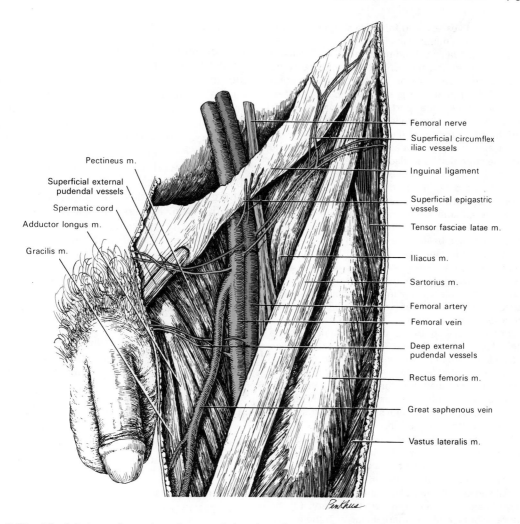

Pectineus m.

Superficial external
pudendal vessels

Spermatic cord

Adductor longus m.

Gracilis m.

Femoral nerve

Superficial circumflex
iliac vessels

Inguinal ligament

Superficial epigastric
vessels

Tensor fasciae latae m.

Iliacus m.

Sartorius m.

Femoral artery

Femoral vein

Deep external
pudendal vessels

Rectus femoris m.

Great saphenous vein

Vastus lateralis m.

FIG. 8-73. The left femoral vessels and some of their branches, and the femoral nerve in the femoral triangle.

femoral nerve and its branches. In addition to the vessels and nerves, the femoral triangle contains some fat and lymphatics.

ADDUCTOR CANAL.   The adductor canal is a fascial tunnel in the middle third of the thigh, extending from the apex of the femoral triangle to the opening in the tendon of the adductor magnus muscle, through which the femoral vessels course to the popliteal fossa (Figs. 8-75; 8-76). The canal is bounded anteriorly and laterally by the vastus medialis muscle, and posteriorly by the adductors longus and magnus. It is covered anteriorly by a strong fascia that lies deep to the sartorius and extends from the vastus medialis across the femoral vessels to the adductors longus and magnus. The canal contains the femoral artery and vein, and the saphenous nerve.

RELATIONS OF THE FEMORAL ARTERY.   In the **femoral triangle,** the artery is situated quite superficially. *Anterior* to it are the skin, the superficial fascia, the superficial inguinal lymph nodes, the superficial iliac circumflex vein, the superficial layer of the fascia lata, and the anterior part of the femoral sheath. The femoral branch of the genitofemoral nerve courses for a short distance within the lateral compartment of the femoral sheath and lies at first lateral and then anterior to the artery. Near the apex of the femoral triangle, the medial cutaneous branches of the femoral nerve cross the artery from its lateral to its medial aspect.

Posterior to the femoral artery are the posterior part of the femoral sheath, the pectineal fascia, the medial part of the tendon of the psoas major muscle, the pectineus muscle, and the adductor longus muscle. The artery is separated from the capsule of the hip joint by the tendon of the psoas major muscle, from the pectineus muscle by the femoral vein and deep femoral vessels, and from the adductor longus muscle by the femoral vein. The nerve to the pectineus passes medially behind the artery. On the *lateral aspect* of the artery, but separated from it by some fibers of the psoas major muscle, is the femoral nerve. The femoral vein is on the *medial aspect* of the artery in the upper part of the femoral triangle, but is posterior to it in the lower part of the triangle.

In the **adductor canal** (Fig. 8-75), the femoral artery is more deeply situated, being covered by the skin, the superficial and deep fasciae, the sartorius, and the fibrous roof of the femoral canal. The saphenous nerve crosses the artery from its lateral to its medial aspect within the adductor canal. *Posterior* to the artery are the adductors longus and magnus, and *anterolateral* to it is the vastus medialis. The femoral vein lies behind the artery in the upper part of the adductor canal, but lateral to it in the distal part of the canal.

VARIATIONS. In rare cases the femoral artery is absent, and its place is taken by an enlarged inferior gluteal artery, which has accompanied the sciatic nerve to the popliteal fossa. When this occurs, the inferior gluteal represents the persistence of the original axis artery (Fig. 8-13), and the external iliac is small, terminating as the deep femoral artery. Occasionally, the femoral vein courses along the medial aspect of the artery throughout the entire extent of the femoral triangle, or it may be split so that a large vein is placed on both sides of the artery for a greater or lesser distance through the thigh.

COLLATERAL CIRCULATION. After ligature of the femoral artery, collateral circulation (Fig. 8-74) results from anastomoses of (1) the superior and inferior gluteal branches of the internal iliac with the medial and lateral femoral circumflex and first perforating branches of the deep femoral artery; (2) the obturator branch of the internal iliac with the medial femoral circumflex branch of the deep femoral; (3) the internal pudendal from the internal iliac with the superficial and deep external pudendal arteries off the femoral; (4) the deep iliac circumflex from the external iliac with the lateral femoral circumflex of the deep femoral and the superficial iliac circumflex of the femoral; and (5) the inferior gluteal of the internal iliac with the perforating branches of the deep femoral artery (Fig. 8-76).

Branches of the Femoral Artery

The branches of the femoral artery are:

Superficial epigastric
Superficial circumflex iliac
Superficial external pudendal
Deep external pudendal
Muscular
Profunda femoris
Descending genicular artery

SUPERFICIAL EPIGASTRIC ARTERY

The superficial epigastric artery arises from the femoral artery about 1 cm below the inguinal ligament (Figs. 8-73; 8-75). It passes through the femoral sheath and the cribriform fascia, turns upward in front of the inguinal ligament, and ascends between the two layers of the superficial fascia, anterior to the external oblique muscle of the abdominal wall, almost to the umbilicus. It distributes branches to the superficial subinguinal lymph nodes, the superficial fascia, and the skin. It anastomoses with branches of the inferior epigastric artery and with the superficial epigastric of the other side.

SUPERFICIAL CIRCUMFLEX ILIAC ARTERY

The superficial circumflex iliac artery is the smallest of the superficial branches of the femoral artery. It arises close to the superficial epigastric, pierces the fascia lata, runs laterally, parallel to the inguinal ligament, to the crest of the ilium (Figs. 8-73; 8-75). It divides into branches that supply the skin of the groin and the superficial subinguinal lymph nodes, and it anastomoses with the deep circumflex iliac, superior gluteal, and lateral femoral circumflex arteries.

SUPERFICIAL EXTERNAL PUDENDAL ARTERY

The superficial external pudendal artery arises from the medial aspect of the femoral artery, close to the preceding vessels (Figs. 8-73; 8-75). After piercing the femoral sheath and the cribriform fascia, it courses medially across the spermatic cord (or round ligament in the female) and is distributed to the skin on the lower part of the abdomen, and to the penis and scrotum in the male, or the labium majus in the female. It anastomoses with branches of the internal pudendal artery.

DEEP EXTERNAL PUDENDAL ARTERY

The deep external pudendal artery, more deeply situated than the preceding vessel, passes medially across the pectineus and adductor longus muscles (Figs. 8-73; 8-75). It is covered by the fascia lata, which it pierces at the medial aspect of the thigh, and is dis-

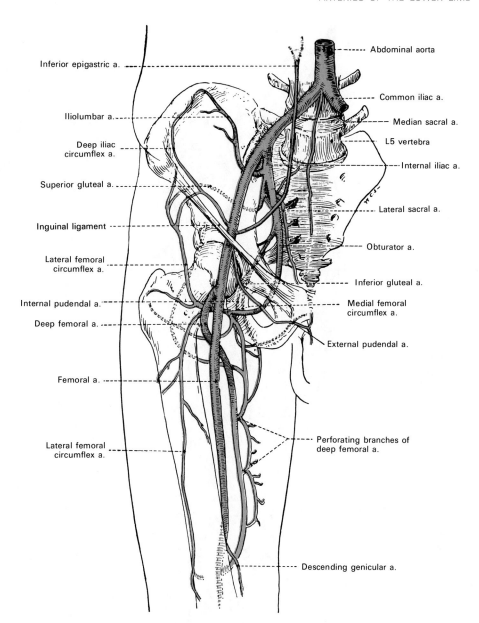

Inferior epigastric a.

Iliolumbar a.

Deep iliac circumflex a.

Superior gluteal a.

Inguinal ligament

Lateral femoral circumflex a.

Internal pudendal a.

Deep femoral a.

Femoral a.

Lateral femoral circumflex a.

Abdominal aorta

Common iliac a.

Median sacral a.

L5 vertebra

Internal iliac a.

Lateral sacral a.

Obturator a.

Inferior gluteal a.

Medial femoral circumflex a.

External pudendal a.

Perforating branches of deep femoral a.

Descending genicular a.

FIG. 8-74. Collateral circulation about the hip and the upper part of the right thigh. (From Eycleshymer and Jones.)

tributed to the skin of the scrotum and perineum in the male or to the labium majus in the female. Its branches anastomose with the posterior scrotal (or labial) branches of the perineal artery.

MUSCULAR BRANCHES

Muscular branches of the femoral artery supply the sartorius, vastus medialis, and adductor muscles.

DEEP FEMORAL ARTERY

The deep (or profunda) femoral artery is a large vessel that arises from the lateral and posterior part of the femoral artery, 2 to 5 cm below the inguinal ligament (Figs. 8-74; 8-75). Initially, it lies lateral to the femoral artery, but then courses behind it and the femoral vein to the medial aspect of the femur. Passing down to the lower third of the thigh, posterior to the adductor longus muscle, the deep femoral artery ends as a small branch.

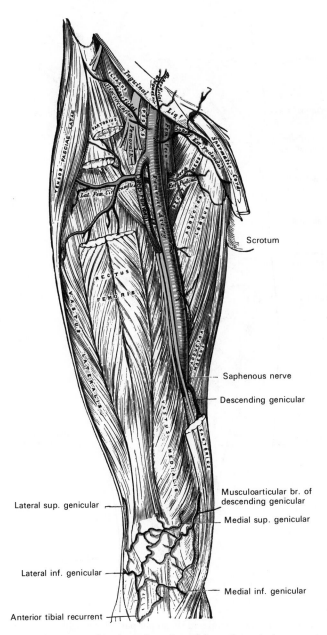

Scrotum

Saphenous nerve

Descending genicular

Musculoarticular br. of
descending genicular

Lateral sup. genicular

Medial sup. genicular

Lateral inf. genicular

Medial inf. genicular

Anterior tibial recurrent

FIG. 8-75.   The femoral artery and its branches viewed from the anterior aspect of the thigh.

This branch pierces the adductor magnus and is distributed to the hamstring muscles on the posterior thigh. The terminal part of the deep femoral artery is sometimes named the **fourth perforating artery,** and it anastomoses with upper branches of the popliteal artery.

RELATIONS OF THE DEEP FEMORAL ARTERY. *Posterior* to the vessel, from above downward, are the iliacus, pectineus, adductor brevis, and adductor

magnus. *Anteriorly,* the deep femoral artery is separated from the femoral artery by the femoral and deep femoral veins above and by the adductor longus muscle below. *Laterally,* the origin of the vastus medialis intervenes between the upper part of the deep femoral artery and the femur.

VARIATIONS. The deep femoral artery sometimes arises from the medial side and, more rarely, from the back of the femoral artery. A more important variation, however, from a surgical point of view, is the level in the thigh at which the vessel arises. In three-fourths of a large number of specimens it

arose from 2.25 to 5 cm below the inguinal ligament. In a few specimens it arose higher, more
rarely opposite the inguinal ligament, and in one
subject, above the ligament from the external iliac
artery. Occasionally the distance between the origin
of the deep femoral artery and the inguinal ligament
exceeds 5 cm.

Branches of the
Deep Femoral Artery

The deep femoral artery supplies much of
the musculature in the anterior and medial
compartments of the thigh. Some of its vessels penetrate through the musculature to
the posterior compartment, thereby contributing to the supply of the hamstrings. Its
branches are:

Medial femoral circumflex
Lateral femoral circumflex
Perforating
Muscular

Medial Femoral
Circumflex Artery

The medial femoral circumflex artery
arises from the medial aspect of the deep
femoral and winds around the medial aspect
of the femur (Fig. 8-74). In about 25% of cadavers it arises directly from the femoral artery. It passes first between the pectineus
and psoas major muscles, and then between
the obturator externus and the adductor
brevis. Continuing into the gluteal region
between the quadratus femoris and adductor magnus muscles, it gives off ascending
and transverse branches. The **ascending
branch** goes to the adductor muscles, the
gracilis, and the obturator externus, and it
anastomoses with the obturator artery. The
transverse branch descends beneath the
adductor brevis to supply it and the adductor magnus.

The continuation of the medial femoral
circumflex artery passes posteriorly and divides into superficial, deep, and acetabular
branches. The **superficial branch** appears
between the quadratus femoris and the
proximal border of the adductor magnus. It
anastomoses with the inferior gluteal, lateral
femoral circumflex, and first perforating arteries, thereby contributing to the cruciate
anastomosis. The **deep branch** runs obliquely upward upon the tendon of the obturator externus, and in front of the quadratus

femoris toward the trochanteris fossa, where
it anastomoses with twigs from the gluteal
arteries. The **acetabular branch** arises opposite the acetabular notch and enters the hip
joint beneath the transverse acetabular ligament in company with the acetabular
branch of the obturator artery. It supplies
the fat in the acetabular fossa and continues
along the round ligament to the head of the
femur.

Lateral Femoral
Circumflex Artery

The lateral femoral circumflex artery
arises from the lateral aspect of the deep
femoral artery (Figs. 8-74; 8-75). In nearly
20% of cadavers studied, it arises independently from the femoral artery. It passes laterally between the divisions of the femoral
nerve and behind the sartorius and rectus
femoris muscles, and divides into ascending,
transverse, and descending branches.

The **ascending branch** passes superiorly,
deep to the tensor fasciae latae, to the lateral
aspect of the hip, and anastomoses with the
terminal branches of the superior gluteal
and deep iliac circumflex arteries.

The **descending branch** of the lateral femoral circumflex artery courses downward
behind the rectus femoris and upon the
vastus lateralis, which it supplies. One long
branch descends in the vastus lateralis muscle as far as the knee and anastomoses with
the superior lateral genicular branch of the
popliteal artery. It is accompanied by the
branch of the femoral nerve to the vastus lateralis.

The **transverse branch** is the smallest
branch, but it is often absent. It passes laterally over the vastus intermedius, pierces the
vastus lateralis, and winds around the
femur, just below the greater trochanter. It
anastomoses on the back of the thigh with
the medial femoral circumflex, inferior gluteal, and first perforating arteries, thereby
participating in the cruciate anastomosis.

Perforating Arteries

Usually three perforating arteries arise
from the deep femoral artery, and the termination of the deep femoral is frequently considered the fourth (Figs. 8-74; 8-76). These
vessels are so named because they perforate
the adductor magnus muscle to reach the

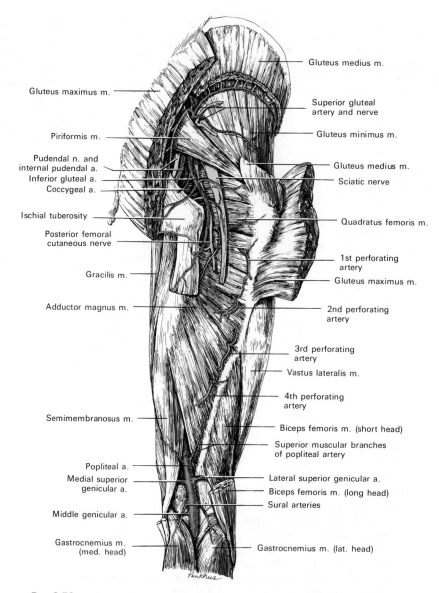

Gluteus medius m.

Gluteus maximus m.

Superior gluteal
artery and nerve

Piriformis m.

Gluteus minimus m.

Pudendal n. and
internal pudendal a.

Gluteus medius m.

Inferior gluteal a.

Sciatic nerve

Coccygeal a.

Ischial tuberosity

Quadratus femoris m.

Posterior femoral
cutaneous nerve

1st perforating
artery

Gracilis m.

Gluteus maximus m.

Adductor magnus m.

2nd perforating
artery

3rd perforating
artery

Vastus lateralis m.

4th perforating
artery

Semimembranosus m.

Biceps femoris m. (short head)

Superior muscular branches
of popliteal artery

Popliteal a.

Lateral superior genicular a.

Medial superior
genicular a.

Biceps femoris m. (long head)

Sural arteries

Middle genicular a.

Gastrocnemius m.
(med. head)

Gastrocnemius m. (lat. head)

FIG. 8-76.   The arteries of the gluteal, posterior femoral, and popliteal regions.

back of the thigh. They pass posteriorly close to the linea aspera of the femur under cover of the small tendinous arches in the muscle. The first is given off above the adductor brevis, the second anterior to that muscle, and the third immediately below it.

The **first perforating artery** passes posteriorly between the pectineus and adductor brevis muscles, and sometimes it perforates the latter (Figs. 8-74; 8-76). It then pierces the adductor magnus close to the linea aspera, giving branches to the adductors brevis and magnus, the biceps femoris, and the gluteus maximus. It anastomoses with the infer-

ior gluteal, medial and lateral femoral circumflex, and second perforating arteries (Fig. 8-74).

The **second perforating artery,** larger than the first, pierces the tendons of the adductors brevis and magnus and divides into ascending and descending branches, which supply the posterior femoral muscles and anastomose with the first and third perforating (Fig. 8-76). The second perforating artery frequently arises in common with the first. The **nutrient artery of the femur** usually branches from the second perforating artery, but when two nutrient arteries exist, they usually are derived from the first and third perforating vessels.

The **third perforating artery** pierces the adductor magnus and divides into branches that supply the posterior femoral muscles (Fig. 8-76). These anastomose above with the higher perforating arteries and below with the terminal branches of the deep femoral artery and the muscular branches of the popliteal. The termination of the deep femoral artery, as was already described, is sometimes termed the **fourth perforating artery.**

Muscular Branches

Numerous muscular branches arise from the deep femoral artery. Some of these end in the adductor muscles; others pierce the adductor magnus, give branches to the hamstrings, and anastomose with the medial femoral circumflex artery and with the superior muscular branches of the popliteal artery.

DESCENDING GENICULAR ARTERY

The descending genicular artery arises from the femoral just before the latter vessel passes through the opening in the tendon of the adductor magnus muscle (Figs. 8-74; 8-75). It immediately divides into a saphenous and an articular branch.

SAPHENOUS BRANCH. The saphenous branch of the descending genicular artery pierces the aponeurotic covering of the adductor canal and accompanies the saphenous nerve to the medial aspect of the knee. It passes between the sartorius and gracilis muscles and, piercing the fascia lata, is distributed to the skin on the upper and medial part of the leg. It anastomoses with the medial inferior genicular artery.

ARTICULAR BRANCH. The articular branch of the descending genicular artery descends in the substance of the vastus medialis, anterior to the tendon of the adductor magnus, to the medial aspect of the knee (Fig. 8-75). Here it anastomoses with the medial superior genicular artery and anterior recurrent tibial artery. Another anastomosing branch crosses the lower thigh above the patellar surface of the femur, forming an arterial arch with the lateral superior genicular artery, from which branches supply the knee joint.

Popliteal Artery

The popliteal artery is the continuation of the femoral artery and it courses through the popliteal fossa (Figs. 8-76; 8-77; 8-81). The vessel commences at an opening in the aponeurosis of the adductor magnus muscle at the lower end of the adductor canal. Extending from the junction of the middle and distal thirds of the posterior thigh, the popliteal artery courses downward to the intercondylar fossa of the femur. It then continues distally to the lower border of the popliteus muscle, where it divides into **anterior** and **posterior tibial arteries.**

POPLITEAL FOSSA. The popliteal fossa is a diamond-shaped region posterior to the knee joint. *Laterally* it is bounded by the biceps femoris above, and by the plantaris and the lateral head of the gastrocnemius below. *Medially* it is limited by the semitendinosus and semimembranosus above, and by the medial head of the gastrocnemius below. The floor of the fossa is oriented anteriorly and is formed by the popliteal surface of the femur, the oblique popliteal ligament of the knee joint, the proximal end of the tibia, and the fascia covering the popliteus muscle. The fossa is covered posteriorly by the popliteal fascia, which is pierced by the small saphenous vein.

Contents of the Popliteal Fossa. The popliteal fossa contains the popliteal vessels, the tibial and the common peroneal nerves, the termination of the small saphenous vein, the distal part of the posterior femoral cutaneous nerve, the articular branch from the obturator nerve, a few small lymph nodes, and a considerable amount of fat. The **tibial nerve** descends through the middle of the fossa, lying most superficially beneath the deep fascia and crossing behind the popliteal vessels from lateral to medial. The **common peroneal nerve** descends laterally along the upper part of the fossa, close to the tendon of the biceps femoris. The popliteal vessels course along the floor of the fossa. The **popliteal artery** is closest to the bone, and the popliteal vein, coursing superficial to the artery, is interposed between it and the nerves. The two vessels are united by dense areolar tissue. The **popliteal vein** is a thick-walled vessel which at first is lateral to the artery and then crosses it posteriorly to gain its medial aspect below. Sometimes the vein is doubled and interconnected by short transverse branches. In these instances, the artery courses between the veins. The **articular branch of the obturator nerve** descends upon the artery to the knee joint. The **popliteal lymph nodes,** six or seven in number, are embedded in the fat. One node lies beneath the popliteal fascia near the termination of the external saphenous vein, another between the popliteal artery and the back of the knee joint, while others are aligned along the popliteal vessels. Arising from the popliteal artery and passing at right angles are its **genicular branches.**

RELATIONS OF THE POPLITEAL ARTERY. *Anterior* to the artery from above downward are the popliteal surface of the femur (which is separated from the vessel by some fat), the back of the knee joint, and the fascia covering the popliteus muscle. *Posteriorly,* the vessel is overlapped by the semimembranosus above, and it is covered by the gastrocnemius and plantaris muscles below. In the middle part of its course, the artery is separated from the skin and fascia by fat, and it is crossed from lateral to medial by the tibial nerve and the popliteal vein, the vein being interposed between the nerve and the artery and closely adherent to the latter. On its *lateral aspect* above are the biceps femoris, the tibial nerve, the popliteal vein, and the lateral condyle of the femur; below are the plantaris and the lateral head of the gastrocnemius. On its *medial aspect* above are the semimembranosus and the medial condyle of the femur; below are the tibial nerve, the popliteal vein, and the medial head of the gastrocnemius.

VARIATIONS. Occasionally the popliteal artery divides into its terminal branches opposite the knee joint. The anterior tibial under these circumstances usually passes anterior to the popliteus. Sometimes the popliteal artery divides into anterior tibial and peroneal branches, the posterior tibial being wanting or very small. Occasionally it divides into three branches, the anterior and posterior tibial and the peroneal arteries.

Branches of the Popliteal Artery

In addition to its terminal branches, the popliteal artery gives rise to muscular, cutaneous, and genicular vessels, the latter forming an anastomosis around the knee. The branches are:

Superior muscular
Sural
Cutaneous
Medial superior genicular
Lateral superior genicular
Middle genicular
Medial inferior genicular
Lateral inferior genicular

SUPERIOR MUSCULAR BRANCHES

Two or three superior muscular branches arise from the proximal part of the popliteal artery and are distributed to the lower parts of the adductor magnus and hamstring muscles (Fig. 8-76). These anastomose with the terminal vessels from the deep femoral artery.

SURAL ARTERIES

The medial and lateral sural arteries are two large branches which are distributed to the gastrocnemius, soleus, and plantaris muscles of the calf (Figs. 8-77; 8-81). They arise from the popliteal artery opposite the knee joint.

CUTANEOUS BRANCHES

Cutaneous branches arise either from the popliteal artery or from some of its branches. They descend between the two heads of the gastrocnemius, and piercing the deep fascia, they are distributed to the skin of the back of the leg. One branch usually accompanies the small saphenous vein.

SUPERIOR GENICULAR ARTERIES

There are usually two superior genicular arteries. These arise one on each side of the popliteal artery and are therefore called medial and lateral superior genicular arteries. They wind around the femur, immediately proximal to its condyles, to the front of the knee joint.

MEDIAL SUPERIOR GENICULAR ARTERY. The medial superior genicular courses anterior to the semimembranosus and semitendinosus above the medial head of the gastrocnemius (Figs. 8-76 to 8-78). Passing deep to the tendon of the adductor magnus, it divides into two branches. One of these supplies the vastus medialis, anastomosing with the descending genicular and medial inferior genicular arteries; the other ramifies close to the surface of the femur, supplying it and the knee joint, and anastomosing with the lateral superior genicular artery. The medial supe-

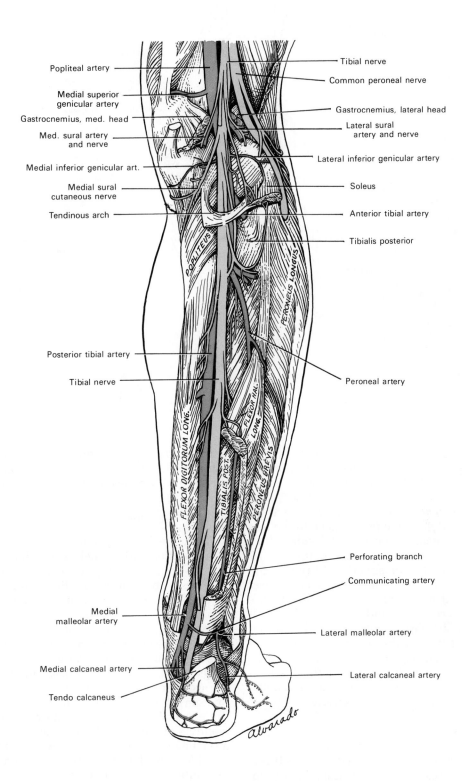

FIG. 8-77.   The popliteal, posterior tibial, and peroneal arteries. Right leg.

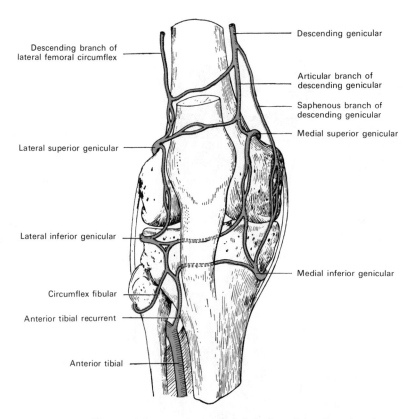

Descending branch of
lateral femoral circumflex

Descending genicular

Articular branch of
descending genicular

Saphenous branch of
descending genicular

Medial superior genicular

Lateral superior genicular

Lateral inferior genicular

Medial inferior genicular

Circumflex fibular

Anterior tibial recurrent

Anterior tibial

FIG. 8-78. The arterial anastomosis around the knee joint. Anterior view.

rior genicular artery is often small, with a resulting compensatory increase in the size of the descending genicular artery.

LATERAL SUPERIOR GENICULAR ARTERY. The lateral superior genicular courses above the lateral condyle of the femur, deep to the tendon of the biceps femoris, and divides into a superficial and a deep branch (Figs. 8-76; 8-78; 8-80; 8-81). The **superficial branch** supplies the vastus lateralis and anastomoses with the descending branch of the lateral femoral circumflex and the lateral inferior genicular arteries. The **deep branch** supplies the lower part of the femur and knee joint and forms an anastomotic arch across the anterior aspect of the joint with the descending genicular, the medial superior genicular, and the inferior genicular arteries.

MIDDLE GENICULAR ARTERY

The middle genicular artery is a small vessel that arises from the popliteal artery opposite the back of the knee joint (Fig. 8-76). It

pierces the oblique popliteal ligament and supplies the ligaments and synovial membrane on the interior of the knee joint.

INFERIOR GENICULAR ARTERIES

Two inferior genicular arteries arise from the popliteal artery below the knee joint. Under cover of the gastrocnemius muscle, one vessel branches from the medial aspect of the main vessel and the other from the lateral.

MEDIAL INFERIOR GENICULAR ARTERY. The medial inferior genicular artery is usually the larger of the two lower genicular vessels (Figs. 8-77 to 8-79; 8-81). It descends medially in an oblique direction along the proximal margin of the popliteus, to which it gives branches. It then passes below the medial condyle of the tibia, deep to the tibial collateral ligament of the knee. At the anterior border of this ligament, the vessel ascends to the anterior and medial aspects of the joint, to supply the upper end of the tibia and the knee joint itself. It anastomoses with the lateral inferior and medial superior genicular arteries.

LATERAL INFERIOR GENICULAR ARTERY. The lateral inferior genicular courses laterally above the head of the fibula to the front of the knee joint, passing deep to the lateral head of the gastrocnemius, the fibular collateral ligament of the knee joint, and the tendon of the biceps femoris (Figs. 8-77 to 8-80). It ends by dividing into branches that anastomose with the medial inferior and lateral superior genicular arteries, the anterior and posterior tibial recurrent, and the circumflex fibular branch of the posterior tibial artery.

ANASTOMOSIS AROUND THE KNEE JOINT. An intricate network of vessels forms a superficial and deep plexus of arteries around the patella and on the contiguous ends of the femur and tibia (Figs. 8-78; 8-80). The **superficial plexus** is situated in the superficial fascia around the patella and forms three well-defined arches. One is found above the upper border of the patella, in the loose connective tissue over the quadriceps femoris muscle. The other two, distal to the patella, are situated in the fat deep to the ligamentum patellae. The **deep plexus** forms a close network of vessels which lies on the lower end of the femur and upper end of the tibia around their articular surfaces, sending numerous branches into the interior of the knee joint. The arteries that form this plexus are the two medial and the two lateral genicular branches of the popliteal, the descending genicular, the descending branch of the lateral femoral circumflex, the anterior and posterior tibial recurrent, and the circumflex fibular.

Anterior Tibial Artery

The anterior tibial artery commences at the lower border of the popliteus muscle at the bifurcation of the popliteal artery (Figs. 8-77; 8-81). It passes anteriorly between the two heads of the tibialis posterior muscle and then through the aperture above the upper border of the interosseous membrane, to achieve the deep part of the front of the leg. Here the anterior tibial artery lies close to the medial aspect of the neck of the fibula. Descending on the anterior surface of the interosseous membrane, the vessel gradually approaches the tibia (Figs. 8-79; 8-80). In the distal part of the leg it lies on the tibia, and then anterior to the ankle joint, it becomes more superficial and is called the **dorsalis pedis artery.**

RELATIONS OF THE ANTERIOR TIBIAL ARTERY. In the upper two-thirds of its course, the anterior tibial artery rests upon the interosseous membrane. The lower one-third of the artery descends upon the front of the tibia and the anterior ligament of the ankle joint. In the upper third of the leg, it lies between the tibialis anterior and the extensor digitorum longus, and in the middle third, between the tibialis anterior and extensor hallucis longus. At the ankle the artery is crossed from lateral to medial by the tendon of the extensor hallucis longus, lying between this latter tendon and the first tendon of the extensor digitorum longus. The vessel is covered in the upper two-thirds of the leg by muscles that lie on both sides and by the deep fascia. In the distal third of the leg it is covered by the skin and fascia, and by the extensor retinaculum.

The anterior tibial artery is accompanied by a pair of venae comitantes which course along the sides of the artery. The deep peroneal nerve, descending obliquely around the lateral aspect of the neck of the fibula, enters the anterior compartment of the leg to lie on the lateral aspect of the anterior tibial artery. At about the middle of the leg, the nerve lies anterior to the artery, but in the lower leg it again is usually on the lateral aspect.

VARIATIONS. The anterior tibial artery may be small or entirely absent, its place being taken by perforating branches from the posterior tibial, or by the perforating branch of the peroneal artery. Occasionally, the artery deviates toward the fibular side of the leg, regaining its usual position anterior to the ankle. More rarely, the vessel becomes superficial as high as the middle of the leg, being covered only by skin and fascia beyond that point.

Branches of the Anterior Tibial Artery

The anterior tibial artery supplies vessels to the anastomosis around the knee joint, to the muscles in the anterior compartment of the leg, and to the anastomosis around the ankle joint, becoming at that point the dorsalis pedis artery. Its branches are:

Posterior tibial recurrent
Anterior tibial recurrent
Muscular
Anterior medial malleolar
Anterior lateral malleolar

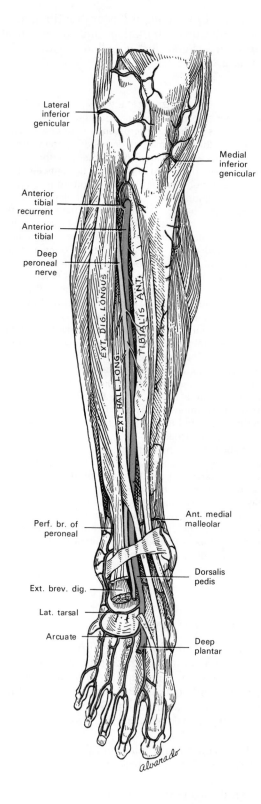

Lateral
inferior
genicular

Medial
inferior
genicular

Anterior
tibial
recurrent

Anterior
tibial

Deep
peroneal
nerve

EXT. DIG. LONGUS.

EXT. HALL. LONG.

TIBIALIS ANT.

Perf. br. of
peroneal

Ant. medial
malleolar

Ext. brev. dig.

Dorsalis
pedis

Lat. tarsal

Arcuate

Deep
plantar

alvarado

FIG. 8-79. Anterior tibial and dorsalis pedis arteries.
Right leg.

POSTERIOR TIBIAL RECURRENT ARTERY

The posterior tibial artery is an inconstant branch, but when present, it arises from the anterior tibial artery *before* that vessel achieves the anterior aspect of the leg. It ascends anterior to the popliteus muscle, which it supplies, and it anastomoses with the inferior genicular branches of the popliteal artery, sending a twig to the tibiofibular joint.

ANTERIOR TIBIAL RECURRENT ARTERY

The anterior tibial recurrent artery arises from the anterior tibial immediately *after* that vessel has reached the anterior aspect of the leg (Figs. 8-78 to 8-80). It ascends in the tibialis anterior muscle, and ramifies on the front and sides of the knee joint, assisting in the formation of the arterial plexus. It anastomoses with the inferior genicular branches of the popliteal artery and with the circumflex fibular artery.

MUSCULAR BRANCHES

Numerous muscular branches arise from the anterior tibial artery. They are distributed to the muscles that lie on both sides of the vessel. Some branches pierce the deep fascia to supply the skin, while others pass deeply through the interosseous membrane to anastomose with branches of the posterior tibial and peroneal arteries.

ANTERIOR MEDIAL MALLEOLAR ARTERY

The anterior medial malleolar artery arises about 5 cm above the ankle joint, but it may arise lower down, even from the dorsalis pedis artery (Fig. 8-79). It passes behind the tendons of the extensor hallucis longus and tibialis anterior, to the medial aspect of the ankle. Its branches ramify over the medial malleolus and contribute to a network of vessels, anastomosing with branches from the posterior tibial and medial plantar arteries.

ANTERIOR LATERAL MALLEOLAR ARTERY

The anterior lateral malleolar artery is generally larger than the anterior medial vessel (Fig. 8-80). It usually arises from the anterior tibial artery distal to the lateral malleolus and may even arise from the dorsalis pedis artery. It passes deep to the tendons of the extensor digitorum longus and peroneus tertius and supplies the lateral aspect of the ankle. It anastomoses with the perforating branch of the peroneal artery and with ascending twigs from the lateral tarsal artery.

ANASTOMOSIS AROUND THE ANKLE JOINT. The arteries around the ankle joint anastomose freely with one another to form networks below the corresponding malleoli. The **medial malleolar network** is formed by the anterior medial malleolar branch of the anterior tibial, the medial tarsal branches of the dorsalis pedis, the posterior medial malleolar and medial calcaneal branches of the posterior tibial, and branches from the medial plantar artery. The **lateral malleolar network** is formed by the anterior lateral malleolar branch of the anterior tibial, the lateral tarsal branch of the dorsalis pedis, the perforating and the lateral calcaneal branches of the peroneal, and twigs from the lateral plantar artery (Fig. 8-80).

Dorsalis Pedis Artery

The dorsalis pedis artery is the continuation of the anterior tibial artery beyond the ankle joint (Figs. 8-79; 8-80). It passes distally along the tibial side of the dorsum of the foot to the proximal part of the first intermetatarsal space. At this site the vessel divides into two terminal branches. These are the **first dorsal metatarsal artery,** which courses between the first and second toes, and the **deep plantar artery,** which penetrates between the two heads of the first dorsal interosseous muscle to enter the sole of the foot, helping to complete the plantar arch.

RELATIONS OF THE DORSALIS PEDIS ARTERY. The dorsalis pedis artery rests upon the anterior aspect of the articular capsule of the ankle joint. It continues distally in front of the talus, navicular, and intermediate cuneiform bones, and the ligaments interconnecting them. The artery is covered by the skin, fascia, and the inferior extensor retinaculum, and it is crossed near its termination by the first tendon of the extensor digitorum brevis. On its tibial or *medial aspect* is the tendon of the extensor hallucis longus. On its fibular or *lateral aspect* are the first tendon of the extensor digitorum longus and the termination of the deep peroneal nerve. The dorsalis pedis artery is usually accompanied by two veins.

VARIATIONS. The dorsalis pedis artery may be larger than usual to compensate for a deficient lateral plantar artery. Its terminal branches to the toes may be absent, the toes then being supplied by the medial plantar, or its place may be taken by a large perforating branch of the peroneal artery (3%). In its descending course, the dorsalis pedis artery may lie lateral to the line between the middle of the ankle and the first interosseous space. In 12% of bodies, the dorsalis pedis is so small that it is essentially absent. The arcuate branch of the dorsalis pedis artery is a vessel of significant size in only 54% of feet studied (Huber, 1941).

Branches of the Dorsalis Pedis Artery

The dorsalis pedis artery supplies the dorsum of the foot with the following branches:

Lateral tarsal
Medial tarsal
Arcuate
First dorsal metatarsal
Deep plantar

LATERAL TARSAL ARTERY

The lateral tarsal artery arises most frequently from the dorsalis pedis as that vessel crosses the head of the talus (Figs. 8-79; 8-80). It passes laterally in an arched course, lying upon the tarsal bones and covered by the extensor digitorum brevis muscle. The lateral tarsal artery supplies this muscle and the articulations of the tarsus. It anastomoses with branches of the arcuate, anterior lateral malleolar, and lateral plantar arteries, and with the perforating branch of the peroneal artery.

MEDIAL TARSAL ARTERIES

Two or three small medial tarsal arteries ramify on the medial border of the foot and join the medial malleolar network (Fig. 8-80).

ARCUATE ARTERY

The arcuate artery arises from the dorsalis pedis artery on the dorsum of the foot, between the medial cuneiform and the base of the second metatarsal bone (Fig. 8-79). It passes laterally over the bases of the second, third, and fourth metatarsal bones, deep to the tendons of the extensor digitorum brevis, its direction being influenced by its point of origin. It anastomoses with the lateral tarsal and lateral plantar arteries. The arcuate ar-

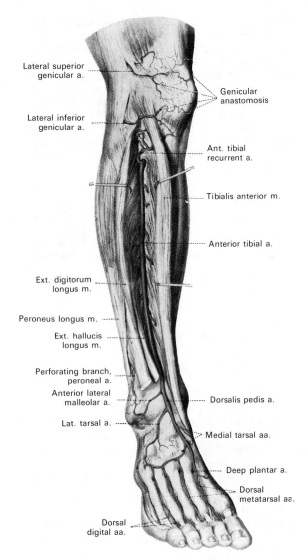

Lateral superior
genicular a.

Genicular
anastomosis

Lateral inferior
genicular a.

Ant. tibial
recurrent a.

Tibialis anterior m.

Anterior tibial a.

Ext. digitorum
longus m.

Peroneus longus m.

Ext. hallucis
longus m.

Perforating branch,
peroneal a.

Anterior lateral
malleolar a.

Dorsalis pedis a.

Lat. tarsal a.

Medial tarsal aa.

Deep plantar a.

Dorsal
metatarsal aa.

Dorsal
digital aa.

FIG. 8-80.   The right anterior tibial and dorsalis pedis arteries and their branches. (From Benninghoff and Goerttler, *Lehrbuch der Anatomie des Menschen*, 1975.)

tery gives off the **second, third,** and **fourth dorsal metatarsal arteries** (Fig. 8-80). These vessels course distally upon the corresponding dorsal interosseous muscles, and in the clefts between the toes, each divides into two **dorsal digital arteries** for the adjoining toes (Fig. 8-80). At the proximal parts of the interosseous spaces, the dorsal metatarsal arteries receive perforating branches from the plantar arch, and at the distal parts of the spaces they are joined by other perforating branches from the plantar metatarsal arteries. The fourth dorsal metatarsal artery gives off a branch that supplies the lateral aspect of the fifth toe.

FIRST DORSAL METATARSAL ARTERY

The first dorsal metatarsal artery courses distally on the first dorsal interosseous muscle and, at the cleft between the first and second toes, divides into two branches. One of these passes deep to the tendon of the extensor hallucis longus and is distributed to the medial border of the great toe; the other bifurcates to supply the adjoining sides of the great and second toes.

DEEP PLANTAR ARTERY

The deep plantar artery penetrates into the sole of the foot between the two heads of the first dorsal interosseous muscle and unites with the termination of the lateral plantar artery, thereby completing the **plantar arch** (Figs. 8-79; 8-80). On the plantar aspect of the foot, it sends a branch along the medial side of the great toe, continuing distally along the first interosseous space as the **first plantar metatarsal artery** (Figs. 8-82; 8-83). This latter vessel bifurcates and thereby supplies the adjacent sides of the first and second toes.

Posterior Tibial Artery

The posterior tibial artery begins at the lower border of the popliteus muscle, opposite the interval between the tibia and fibula (Figs. 8-77; 8-81). It is larger than the anterior tibial artery, and it descends obliquely downward and medially in the posterior compartment of the leg between the superficial and deep muscle layers. As it approaches the tibial side of the leg, lying behind the tibia in the lower part of its course, the vessel is situated midway between the medial malleolus and the medial process of the calcaneal tuberosity. Under cover of the origin of the abductor hallucis it divides into the **medial** and **lateral plantar arteries** (Figs. 8-82; 8-83).

RELATIONS OF THE POSTERIOR TIBIAL ARTERY. The posterior tibial artery lies successively upon the tibialis posterior, the flexor digitorum longus, the tibia, and the posterior part of the ankle joint. It is covered in the upper part of the leg by the deep transverse fascia that separates it from the gastrocnemius and soleus, and at its entrance into the foot, it is covered by the abductor hallucis muscle. In the distal third of the leg, the artery is more superficial, being covered only by skin and fascia, and it courses parallel but anterior to the medial border of the tendo calcaneus. The posterior tibial artery is accompanied by two veins and by the tibial nerve. The latter structure at first lies medial to the artery, but soon crosses the vessel posteriorly, and throughout most of its course lies lateral to the artery.

Deep to the flexor retinaculum behind the medial malleolus, the tendons, blood vessels, and nerve are arranged from medial to lateral in the following order: the tendons of the tibialis posterior and flexor digitorum longus, the posterior tibial artery with a vein on each side, the tibial nerve, and, 1.25 cm nearer the heel, the tendon of the flexor hallucis longus.

VARIATION. The posterior tibial artery quite frequently is smaller than usual or even absent. In these cases its field is supplied by a large peroneal artery, which either joins the small posterior tibial artery or continues alone to the sole of the foot.

Branches of the Posterior Tibial Artery

From the posterior tibial artery arise the following branches:

Circumflex fibular
Peroneal
Nutrient (tibial)
Muscular
Medial malleolar
Communicating
Medial calcaneal
Medial plantar
Lateral plantar

CIRCUMFLEX FIBULAR ARTERY

The circumflex fibular artery usually arises from the upper end of the posterior tibial artery, but may arise from the popliteal artery or even from the anterior tibial artery. It pierces the fibular head of the soleus muscle, winds around the neck of the fibula, and supplies the soleus along with the peroneal muscles arising from the upper fibula. It participates in the anastomosis around the lateral aspect of the knee joint (Fig. 8-78).

PERONEAL ARTERY

The peroneal artery is deeply located on the fibular side of the back of the leg. It arises from the posterior tibial artery about 2.5 cm distal to the lower border of the popliteus muscle (Figs. 8-77; 8-81). It passes

obliquely toward the fibula and then descends along the medial aspect of that bone, contained in a fibrous canal between the tibialis posterior and the flexor hallucis longus, or in the substance of the latter muscle. The peroneal artery then runs behind the tibiofibular syndesmosis, ending in lateral calcaneal branches which ramify over the lateral and posterior surfaces of the calcaneus. In the proximal part of the leg, the peroneal artery is covered by the soleus muscle and the deep transverse fascia. More distally, it lies deep to the flexor hallucis longus muscle.

VARIATIONS. The peroneal artery may arise from the posterior tibial artery higher than normal, or even from the popliteal artery. In contrast, it may arise as far as 7 or 8 cm below the popliteus muscle. If the vessel is larger than normal, it either reinforces the posterior tibial by joining it, or when it is very large, altogether takes the place of the posterior tibial in the distal part of the leg and foot. In the rarer instances, when the peroneal artery is smaller than normal, another branch from the posterior tibial helps to supply its muscular field, and a branch from the anterior tibial compensates for the diminished perforating artery. In 3% of specimens, the perforating branch of the peroneal artery is considerably enlarged and it takes the place of the dorsalis pedis artery.

Branches of the Peroneal Artery

The peroneal artery supplies vessels to structures in the lateral aspect of the leg. The branches are:

Muscular
Nutrient (fibular)
Perforating
Communicating
Lateral malleolar
Lateral calcaneal

MUSCULAR BRANCHES. Muscular branches of the peroneal artery go to the soleus, tibialis posterior, flexor hallucis longus, and peroneal muscles.

NUTRIENT ARTERY. The nutrient artery from the peroneal is directed downward and enters the nutrient canal of the fibula.

PERFORATING BRANCH. The perforating branch of the peroneal artery pierces the interosseous membrane about 5 cm above the lateral malleolus to reach the anterior part of

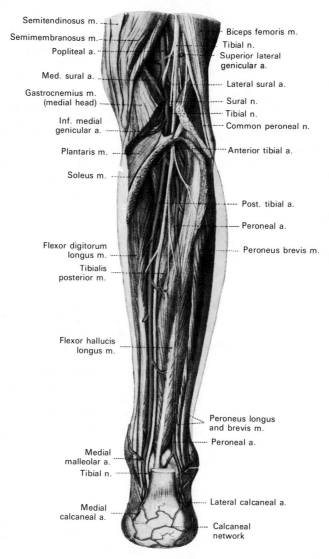

Semitendinosus m.

Semimembranosus m.

Popliteal a.

Med. sural a.

Gastrocnemius m.
(medial head)

Inf. medial
genicular a.

Plantaris m.

Soleus m.

Flexor digitorum
longus m.

Tibialis
posterior m.

Flexor hallucis
longus m.

Medial
malleolar a.

Tibial n.

Medial
calcaneal a.

Biceps femoris m.

Tibial n.

Superior lateral
genicular a.

Lateral sural a.

Sural n.

Tibial n.

Common peroneal n.

Anterior tibial a.

Post. tibial a.

Peroneal a.

Peroneus brevis m.

Peroneus longus
and brevis m.

Peroneal a.

Lateral calcaneal a.

Calcaneal
network

FIG. 8-81.   The right posterior tibial artery and its branches. (From Benninghoff and Goerttler, *Lehrbuch der Anatomie des Menschen,* 1975.)

the leg, where it anastomoses with the anterior lateral malleolar artery (Figs. 8-77; 8-79; 8-80). It then descends anterior to the tibiofibular syndesmosis, gives branches to the tarsus, and anastomoses with the lateral tarsal artery.

COMMUNICATING BRANCH.   A communicating branch arises from the peroneal artery about 5 cm from its distal end (Fig. 8-77). It courses transversely across the leg just posterior to the interosseous membrane to join the communicating branch of the posterior tibial.

LATERAL MALLEOLAR BRANCHES.   Several small lateral malleolar branches arise from the peroneal artery; they wind around the lateral malleolus and join the lateral malleolar network (Fig. 8-77).

LATERAL CALCANEAL BRANCHES.   Lateral calcaneal branches are the terminal vessels of the peroneal artery (Figs. 8-77; 8-81). They pass to the lateral aspect of the heel and communicate with the anterior lateral malleolar branch of the anterior tibial and with the lateral malleolar branches of the peroneal artery. On the back of the heel, they anastomose with the medial calcaneal branches of the posterior tibial.

NUTRIENT ARTERY

The nutrient artery of the tibia arises from the posterior tibial artery near its origin. It pierces the tibialis posterior muscle to which it gives several muscular branches, and then it descends obliquely to enter the nutrient canal of the tibia, located just below the soleal line. Within the bone, the nutrient artery divides into an **ascending branch,** which supplies the superior end of the tibia, and a **descending branch,** which supplies the body and inferior end of the bone. It is the largest nutrient artery to a bone in the body and is accompanied to the bone by a branch of the nerve to the popliteus muscle.

MUSCULAR BRANCHES

Muscular branches of the posterior tibial artery are distributed to the soleus and to the deep muscles along the back of the leg.

MEDIAL MALLEOLAR ARTERY

The medial malleolar artery is a small branch of the posterior tibial artery which winds around the tibial malleolus and joins in the medial malleolar network (Figs. 8-77; 8-81).

COMMUNICATING BRANCH

A communicating branch arises from the posterior tibial artery about 5 cm above the medial malleolus (Fig. 8-77). It courses transversely across the posterior aspect of the tibia and the interosseous membrane. Deep to the flexor hallucis longus muscle, it joins the communicating branch of the peroneal artery.

MEDIAL CALCANEAL ARTERIES

The medial calcaneal arteries are several large vessels that arise from the posterior tibial artery just before it divides into the medial and lateral plantar arteries (Figs. 8-77; 8-80). Frequently, however, they branch from the lateral plantar rather than the posterior tibial. They pierce the flexor retinaculum and are distributed to the fat and skin behind the tendo calcaneus and about the heel, also helping to supply the muscles on the tibial side of the sole of the foot. The medial calcaneal arteries anasto-

mose with the medial malleolar branches of the posterior tibial, and on the back of the heel, with the lateral calcaneal branches of the peroneal artery (Fig. 8-81).

MEDIAL PLANTAR ARTERY

The medial plantar artery is the smaller of the two terminal branches of the posterior tibial artery (Figs. 8-82; 8-83). It passes distally along the medial aspect of the foot and is at first situated deep to the abductor hallucis, and then between it and the flexor digitorum brevis, both of which it supplies. At the base of the first metatarsal bone, and much diminished in size, it passes along the medial border of the first toe, anastomosing with a branch of the first plantar metatarsal artery. Small superficial digital branches accompany the digital branches of the medial plantar nerve, and these join the plantar metatarsal arteries of the first three spaces (Fig. 8-82).

LATERAL PLANTAR ARTERY

The lateral plantar artery is much larger than the medial plantar. Initially, it passes obliquely lateralward to the base of the fifth metatarsal bone, and then it curves medially to the interval between the bases of the first and second metatarsal bones (Figs. 8-82; 8-83). There it unites with the deep plantar branch of the dorsalis pedis artery, which completes the formation of the **plantar arch** (Fig. 8-83). As the lateral plantar artery passes laterally, it is first placed between the calcaneus and abductor hallucis muscle, and then between the flexor digitorum brevis and quadratus plantae muscles. As it runs distally along the base of the little toe, it lies more superficially between the flexor digitorum brevis and abductor digiti minimi, and is covered only by the plantar aponeurosis, superficial fascia, and skin.

PLANTAR ARCH. The remaining portion of the lateral plantar artery is the more deeply situated plantar arch (Fig. 8-83). Extending from the base of the fifth metatarsal bone, the plantar arch penetrates through the layers of the sole of the foot to course medially, beneath the adductor hallucis. It crosses the plantar surface of the bases of the fourth, third, and second metatarsal bones and the corresponding interosseous muscles. At the

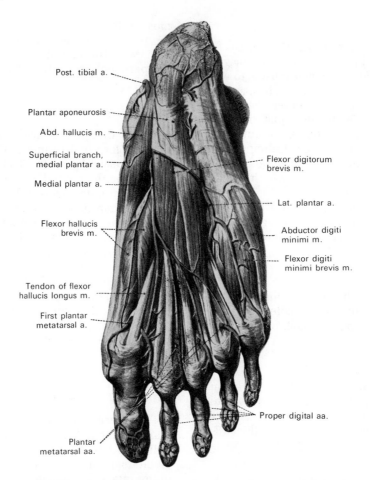

Post. tibial a.

Plantar aponeurosis

Abd. hallucis m.

Superficial branch,
medial plantar a.

Medial plantar a.

Flexor hallucis
brevis m.

Tendon of flexor
hallucis longus m.

First plantar
metatarsal a.

Plantar
metatarsal aa.

Flexor digitorum
brevis m.

Lat. plantar a.

Abductor digiti
minimi m.

Flexor digiti
minimi brevis m.

Proper digital aa.

FIG. 8-82. The more superficial branches of the plantar arteries as seen after removal of the plantar aponeurosis. Right foot. (From Benninghoff and Goerttler, *Lehrbuch der Anatomie des Menschen,* 1975.)

proximal part of the first interosseous space, the plantar arch is completed by its junction with the deep plantar branch of the dorsalis pedis artery. Besides distributing numerous branches to the muscles, integument, and fasciae in the sole, the plantar arch gives rise to three perforating and four plantar metatarsal arteries.

The three **perforating branches** of the plantar arch penetrate through the proximal parts of the second, third, and fourth interosseous spaces, between the heads of the dorsal interosseous muscles, to anastomose with the dorsal metatarsal arteries.

The four **plantar metatarsal arteries** course distally between the metatarsal bones and lie in contact with the interosseous mus-

cle (Figs. 8-82; 8-83). Each vessel divides into two **plantar digital arteries,** which supply the adjacent sides of the toes. Near their points of division each plantar metatarsal artery gives off an **anterior perforating branch** to join the corresponding dorsal metatarsal artery.

The **first plantar metatarsal artery** springs from the junction between the plantar arch and the deep plantar branch of the dorsalis pedis artery (Figs. 8-82; 8-83). It sends a digital branch to the medial aspect of the large toe. The digital branch for the lateral aspect of the little toe arises from the lateral plantar artery near the base of the fifth metatarsal bone, just as the latter vessel curves medially to become the plantar arch.

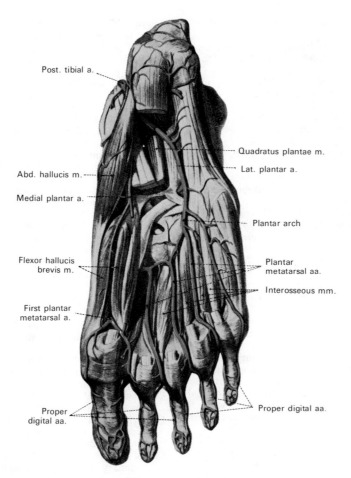

FIG. 8-83. The deep arteries on the plantar aspect of the right foot. (From Benninghoff and Goerttler, *Lehrbuch der Anatomie des Menschen,* 1975.)

References

(References are listed not only to those articles and books cited in the text, but to others as well which are considered to contain valuable resource information for the student who desires it.)

General

Adams, W. E. 1942. The blood supply of nerves. I. Historical review. J. Anat. (Lond.), 76:323–341.

Benninghoff, A., and K. Goerttler. 1975. *Lehrbuch der Anatomie des Menschen.* 11th ed. Edited by H. Ferner and J. Staubesand., Urban and Schwarzenberg, Munich, 3 vols.

Blomfield, L. B. 1945. Intramuscular vascular patterns in man. Proc. Roy. Soc. Med., 38:617–618

Clara, M. 1956. *Die arterio-venosen Anastomosen, Anatomie, Biologie, Pathologie.* 2nd ed. Springer-Verlag, Vienna, 315 pp.

Coates, A. E. 1931. Observations on the distribution of the arterial branches of the peripheral nerves. J. Anat. (Lond.), 66:499–507.

Edwards, D. A. W. 1946. The blood supply and lymphatic drainage of tendons. J. Anat. (Lond.), 80:147–152.

Learmonth, J. 1950. Collateral circulation, natural and artificial. Surg. Gynec. Obstet., 90:385–392.

McDonald, D. A. 1960. *Blood Flow in Arteries.* Williams & Wilkins, Baltimore, 328 pp.

Moore, K. L. 1973. *The Developing Human.* W. B. Saunders Co., Philadelphia, 374 pp.

Orbinson, J. L., and D. E. Smith. 1963. *The Peripheral Blood Vessels.* Williams & Wilkins, Baltimore, 357 pp.

Quiring, D. P. 1949. *Collateral Circulation (Anatomical Aspects).* Lea & Febiger, Philadelphia, 142 pp.

Embryology

Allan, F. D. 1961. An histological study of the nerves associated with the ductus arteriosus. Anat. Rec., 139:531–537.

Auër, J. 1948. The development of the human pulmonary

vein and its major variations. Anat. Rec., 101:581–594.

Boyden, E. A. 1970. The developing bronchial arteries in a fetus of the twelfth week. Amer. J. Anat., 129:357–368.

Goldsmith, J. B., and H. W. Butler. 1937. The development of the cardiac-coronary circulatory system. Amer. J. Anat., 60:185–202.

Hislop, A., and L. Reid. 1972. Intrapulmonary arterial development during fetal life: branching pattern and structure. J. Anat. (Lond.), 113:35–48.

Khodadad, G. 1977. Persistent hypoglossal artery in the fetus. Acta Anat., 99:477–481.

Maron, B. J., and G. M. Hutchins. 1972. Truncus arteriosus malformation in a human embryo. Amer. J. Anat., 134:167–174.

Moffat, D. B. 1961. The development of the posterior cerebral artery. J. Anat. (Lond.), 95:485–494.

Morris, E. D., and D. B. Moffat. 1956. Abnormal origin of the basilar artery from the cervical part of the internal carotid and its embryological significance. Anat. Rec., 125:701–712.

Noback, G. J., and I. Rehman. 1941. The ductus arteriosus in the human fetus and newborn infant. Anat. Rec., 81:505–527.

Padget, D. H. 1948. Development of the cranial arteries in the human embryo. Carneg. Inst., Contr. Embryol., 32:205–262.

Padget, D. H. 1954. Designation of the embryonic intersegmental arteries in reference to the vertebral artery and subclavian stem. Anat. Rec., 119:349–356.

Scammon, R. E., and E. H. Norris. 1918. On the time of the post-natal obliteration of the fetal blood-passages (foramen ovale, ductus arteriosus, ductus venosus). Anat. Rec., 15:165–180.

Sciacca, A., and M. Condorelli. 1960. Involution of the Ductus Arteriosus. S. Karger, Basel, 52 pp.

Senior, H. D. 1919. The development of the arteries of the human lower extremity. Amer. J. Anat., 25:55–95.

Woollard, H. H. 1922. The development of the principal arterial stems in the forelimb of the pig. Carneg. Inst., Contr. Embryol., 14:139–154.

Aorta and Aortic Arches

Barry, A. 1951. The aortic arch derivatives in the human adult. Anat. Rec., 111:221–238.

Blalock, A. 1946. Operative closure of patent ductus arteriosus. Surg. Gynec. Obstet., 82:113–114.

Congdon, E. D. 1922. Transformation of the aortic arch system during the development of the human embryo. Carneg. Inst., Contr. Embryol., 14:47–110.

De Garis, C. F. 1941. The aortic arch in primates. Amer. J. Phys. Anthrop., 28:41–74.

Fox, M. H., and C. M. Goss. 1958. Experimentally produced malformations of the heart and great vessels in rat fetuses. Transposition complexes and aortic arch abnormalities. Amer. J. Anat., 102:65–92.

Gross, R. E., and P. F. Ware. 1946. The surgical significance of aortic arch anomalies. Surg. Gynec. Obstet., 83:435–448.

Hanlon, C. R., and A. Blalock. 1948. Complete transposition of the aorta and the pulmonary artery: Experimental observations on venous shunts as corrective features. Ann. Surg., 127:385–397.

Lewis, C. W. D., and J. N. M. Parry. 1948. Double aortic arch. Anat. Rec., 101:613–615.

Mc Donald, J. J., and B. J. Anson. 1940. Variations in the origin of arteries derived from the aortic arch. Amer. J. Phys. Anthrop., 27:91–108.

Sealy, W. C. 1951. A report of two cases of the anomalous origin of the right subclavian artery from the descending aorta. J. Thoracic Surg., 21:319–324.

Shaner, R. F. 1956. The persisting right sixth aortic arch of mammals, with a note on fetal coarctation. Anat. Rec., 125:171–184.

Sinclair, J. G., and N. D. Schofield. 1944. Anomalies of the cardiopulmonary circuit compensated without a ductus arteriosus. Anat. Rec., 90:209–216.

Sprong, D. H., Jr., and N. L. Cutler. 1930. A case of human right aorta. Anat. Rec., 45:365–375.

Stibbe, E. P. 1929. True congenital diverticulum of the trachea in a subject showing also right aortic arch. J. Anat. (Lond.), 64:62–66.

Woodburne, R. T. 1951. A case of right aortic arch and associated venous anomalies. Anat. Rec., 111:617–628.

Wright, N. L. 1969. Dissection study and measuration of the human aortic arch. J. Anat. (Lond.), 104:377–385.

Coronary Arteries

Abramson, D. I., and H. J. Eisenberg. 1934. The coronary blood supply in the rhesus monkey. J. Anat. (Lond.), 69:520–525.

Ahmed, S. H., M. T. El-Rakhawy, A. Abdalla, and R. G. Harrison. 1973. A new conception of coronary artery preponderance. Acta Anat., 83:87–94.

Chander, S., and I. Jit. 1957. Single coronary artery. J. Anat. Soc. India, 6:116–118.

Chase, R. E., and C. F. De Garis. 1939. Arteriae coronariae (cordis) in the higher primates. Amer. J. Phys. Anthrop., 24:427–448.

Chinn, J., and M. A. Chinn. 1961. Report of an accessory coronary artery arising from the pulmonary artery. Anat. Rec., 139:23–28.

Colborn, G. L. 1966. The gross morphology of the coronary arteries of the common squirrel monkey. Anat. Rec., 155:353–368.

Esperança-Pina, J. A. 1974. Injection-corrosion-fluorescence in the study of human coronary arterial anastomoses. Acta Anat., 90:481–488.

Gregg, D. E. 1950. Coronary Circulation in Health and Disease. Lea & Febiger, Philadelphia, 227 pp.

Gross, L. 1921. The Blood Supply of the Heart in Its Anatomical and Clinical Aspects. Paul B. Hoeber, Inc., New York, 171 pp.

Gross, L., and M. A. Kugel. 1933. The arterial blood vascular distribution to the left and right ventricles of the human heart. Amer. Heart J., 9:165–177.

Halpern, M. H., and M. M. May. 1958. Phylogenetic study of the extracardiac arteries to the heart. Amer. J. Anat., 102:469–480.

Hill, W. C. O. 1945. Atrial arteries in a human heart. J. Anat. (Lond.), 79:41–43.

Hutchinson, M. C. E. 1978. A study of the atrial arteries in man. J. Anat. (Lond.), 125:39–54.

James, T. N. 1960. The arteries of the free ventricular walls in man. Anat. Rec., 136:371–384.

James, T. N. 1961. Anatomy of the Coronary Arteries. Paul B. Hoeber, Inc., New York, 211 pp.

James, T. N., and G. E. Burch. 1958. The atrial coronary arteries in man. Circulation, 17:90–98.

Jordan, R. A., T. J. Dry and J. E. Edwards. 1950. Cardiac

clinics: CXXXVI. Anomalous origin of the right coronary artery from the pulmonary trunk. Proc. Staff Meet., Mayo Clinic, 25:673–678.

Ryback, R., and N. J. Mizeres. 1965. The sinus node artery in man. Anat. Rec., 153:23–30.

Tomanek, R. J. 1970. Effects of age and exercise on the extent of the myocardial capillary bed. Anat. Rec., 167:55–62.

Velican, C., and D. Velican. 1977. Studies on human coronary arteries. I. Branch pads or cushions. Acta Anat., 99:377–385.

Velican, D., and C. Velican. 1978. Human coronary arteries. II. Branching anatomical pattern and arterial wall microarchitecture. Acta Anat., 100:258–267.

White, N. K., and J. E. Edwards. 1948. Anomalies of the coronary arteries: Report of four cases. Arch. Path., 45:766–771.

Arteries of the Head and Neck

Abbie, A. A. 1932. The blood supply of the lateral geniculate body with a note on the morphology of the choroidal arteries. J. Anat. (Lond.), 67:491–521.

Abbie, A. A. 1933. The clinical significance of the anterior choroidal artery. Brain, 56:233–246.

Abbie, A. A. 1933. The morphology of the forebrain arteries, with especial reference to the evolution of the basal ganglia. J. Anat. (Lond.), 68:433–470.

Adams, W. E. 1955. The carotid sinus complex, "parathyroid" III and thymo-parathyroid bodies, with special reference to the Australian opossum, Trichosurus vulpecula. Amer. J. Anat., 97:1–58.

Adams, W. E. 1957. On the possible homologies of the occipital artery in mammals, with some remarks on the phylogeny and certain anomalies of the subclavian and carotid arteries. Acta Anat., 29:90–113.

Ahmed, D. S., and R. H. Ahmed. 1967. The recurrent branch of the anterior cerebral artery. Anat. Rec., 157:699–700.

Allan, F. D. 1952. An accessory or superficial inferior thyroid artery in a full term infant. Anat. Rec., 112:539–542.

Atkinson, W. J. 1949. The anterior cerebellar artery. Its variations, pontine distribution and significance in the surgery of cerebellopontine angle tumors. J. Neurol. Psychiat., 12:137–151.

Baldwin, B. A. 1964. The anatomy of the arterial supply to the cranial regions of the sheep and ox. Amer. J. Anat., 115:101–117.

Baumel, J. J., and D. Y. Beard. 1961. The accessory meningeal artery of man. J. Anat. (Lond.), 95:386–402.

Bell, R. H., L. L. Swigart, and B. J. Anson. 1950. The relation of the vertebral artery to the cervical vertebrae. Quart. Bull. Northwest. Univ. Med. Sch., 24:184–185.

Blair, C. B., Jr., K. Nandy, and G. H. Bourne. 1962. Vascular anomalies of the face and neck. Anat. Rec., 144:251–257.

Boemer, L. C. 1937. The great vessels in deep infections of the neck. Arch. Otolaryngol., 25:465–472.

Borodulya, A. V., and E. K. Pletchkova. 1973. Distribution of cholinergic and adrenergic nerves in the internal carotid artery: A histochemical study. Acta Anat., 86:410–425.

Boyd, G. I. 1934. Abnormality of subclavian artery associated with presence of the scalenus minimus. J. Anat. (Lond.), 68:280–281.

Boyd, J. D. 1933. Absence of the right common carotid artery. J. Anat. (Lond.), 68:551–557.

Cairney, J. 1925. Tortuosity of the cervical segment of the internal carotid artery. J. Anat. (Lond.), 59:87–96.

Carpenter, M. B., C. R. Noback, and M. L. Moss. 1954. The anterior choroidal artery. Its origins, course, distribution and variations. Arch. Neurol. Psychiat., 71:714–722.

Chakravorty, B. G. 1971. Arterial supply of the cervical spinal cord (with special reference to the radicular arteries). Anat. Rec., 170:311–330.

Chandler, S. B., and C. F. Derezinski. 1935. The variations of the middle meningeal artery within the middle cranial fossa. Anat. Rec., 62:309–319.

Chanmugam, P. K. 1936. Note on an unusual ophthalmic artery associated with other abnormalities. J. Anat. (Lond.), 70:580–582.

Curtis, G. M. 1930. The blood supply of the human parathyroids. Surg. Gynec. Obstet., 51:805–809.

Dahl, E. 1973. The innervation of the cerebral arteries. J. Anat. (Lond.), 115:53–64.

Dandy, W. E., and E. Goetsch. 1911. The blood supply of the pituitary gland. Amer. J. Anat., 11:137–150.

De La Torre, E., and M. G. Netsky. 1960. Study of persistent primitive maxillary artery in human fetus: Some homologies of cranial arteries in man and dog. Amer. J. Anat., 106:185–195.

Evans, T. H. 1956. Carotid canal anomaly: Other instances of absent internal carotid artery. Med. Times, 84:1069–1072.

Furlani, J. 1973. The anterior choroidal artery and its blood supply to the internal capsule. Acta Anat., 85:108–112.

Gillian, L. A. 1961. The collateral circulation of the human orbit. Arch. Ophthal., 65:684–694.

Gillian, L. A. 1968. The arterial and venous blood supplies to the forebrain (including the internal capsule) of primates. Neurology, 18:653–670.

Hayreh, S. S. 1962. The ophthalmic artery. III. Branches. Brit. J. Ophthal., 46:212–247.

Hayreh, S. S., and R. Dass. 1962. The ophthalmic artery. I. Origin and intra-cranial and intra-canicular course. Brit. J. Ophthal., 46:65–98.

Hayreh, S. S., and R. Dass. 1962. The ophthalmic artery. II. Intraorbital course. Brit. J. Ophthal., 46:165–185.

Herman, L. H., A. Z. Ostrowski, and E. S. Gurdjian. 1963. Perforating branches of the middle cerebral artery; an anatomical study. Arch. Neurol., 8:32–34.

Jackson, B. B. 1967. The external carotid as a brain collateral. Amer. J. Surg., 113:375–378.

Jones, F. W. 1939. The anterior superior alveolar nerve and vessels. J. Anat. (Lond.), 73:583–591.

Kassell, N. F., and T. W. Langfitt. 1965. Variations in the circle of Willis in Macaca mulatta. Anat. Rec., 152:257–264.

Kirchner, J., E. Yanagisawa, and E. S. Crelin. 1961. Surgical anatomy of the ethmoidal arteries. Arch. Otolaryng., 74:382–386

Krayenbühl, H. A., and M. G. Yasargil. 1968. Cerebral Angiography. Butterworth and Co., London, 401 pp.

Lasker, G. W., D. L. Opdyke, and H. Miller. 1951. The position of the internal maxillary artery and its questionable relation to the cephalic index. Anat. Rec., 109:119–126.

Lazorthes, G., and G. Salamon. 1971. The arteries of the thalamus: an anatomical and radiological study. J. Neurosurg., 34:23–26.

Lee, I. N. 1955. Anomalous relationship of the inferior thyroid artery. Anat. Rec., 122:499–506.

Maisel, H. 1958. Some anomalies of the origin of the left vertebral artery. S. Afr. Med. J., 32:1141–1142.

McCullough, A. W. 1962. Some anomalies of the cerebral arterial circle (of Willis) and related vessels. Anat. Rec., 142:537–543.

Newton, T. H. 1968. The anterior and posterior meningeal branches of the vertebral artery. Radiology, 91:271–279.

Newton, T. H., and R. L. Mani. 1974. The vertebral artery. Chapter 67 in Radiology of the Skull and Brain. Edited by T. H. Newton and D. G. Potts. C. V. Mosby Co., St. Louis, Vol. 2, Book 2, pp.1659–1709.

Newton, T. H., and D. G. Potts. 1974. In Radiology of the Skull and Brain: Angiography. C. V. Mosby Co., St. Louis, Vol. 2.

Page, R. B., and R. M. Bergland. 1977. The neurohypophyseal capillary bed. I. Anatomy and arterial supply. Amer. J. Anat 148:345–357.

Read, W. T., and M. Trotter. 1941. The origins of transverse cervical and transverse scapular arteries in American whites and Negroes. Amer. J. Phys. Anthrop., 28:293–247.

Reed, A. F. 1943. The relations of the inferior laryngeal nerve to the inferior thyroid artery. Anat. Rec., 85:17–24.

Ring, B. A. 1962. Middle cerebral artery: Anatomical and radiographic study. Acta Radiol., 57:289–300.

Rogers, L. 1947. The function of the circulus arteriosus of Willis. Brain, 70:171–178.

Salamon, G., J. Gonzales, J. Faure, and G. Giudicelli. 1972. Topographic investigation of the cortical branches of the middle cerebral artery. Acta Radiol., 13:226–232.

Scarlett, H. 1935. Occlusion of the central artery of the retina. Ann. Surg., 101:318–323.

Scharrer, E. 1940. Arteries and veins in the mammalian brain. Anat. Rec., 78:173–196.

Seydel, H. G. 1964. The diameters of the cerebral arteries of the human fetus. Anat. Rec., 150:79–88.

Shanklin, W. M., and N. A. Azzam. 1963. A study of valves in the arteries of the rodent brain. Anat. Rec., 147:407–413.

Smith, S. D., and R. S. Benton. 1978. A rare origin of the superior thyroid artery. Acta Anat., 101:91–93.

Stephens, R. B., and D. L. Stilwell. 1969. Arteries and Veins of the Human Brain. Charles C Thomas, Springfield, Ill., 181 pp.

Stewart, J. D. 1932. Circulation of the human thyroid. Arch. Surg., 25:1157–1165.

Sunderland, S. 1945. The arterial relations of the internal auditory meatus. Brain, 68:23–27.

Taveras, J. M., and E. H. Wood. 1964. Diagnostic Neuroradiology. Williams and Wilkins, Baltimore, 960 pp.

Watt, J. C., and A. N. Mc Killop. 1935. Relation of arteries to roots of nerves in the posterior cranial fossa in man. Arch. Surg., 30:336–345.

Watts, J. W. 1933. A comparative study of the anterior cerebral artery and the circle of Willis in primates. J. Anat. (Lond.), 68:534–550.

Westberg, G. 1963. The recurrent artery of Heubner and the arteries of the central ganglia. Acta Radiol. (Diagn.), 1:949–954.

Westberg, G. 1966. Arteries of the basal ganglia. Acta Radiol. (Diagn.), 5:581–596.

Williams, D. J. 1936. The origin of the posterior cerebral artery. Brain, 59:175–180.

Wilson, Mc Clure. 1972. The Anatomic Foundation of Neuroradiology of the Brain. 2nd ed. Little, Brown and Co., Boston, 390 pp.

Windle, W. F., F. R. Zeiss, and M. S. Adamski. 1927. Note on a case of anomalous right vertebral and subclavian arteries. J. Anat. (Lond.), 62:512–514.

Wislocki, G. B. 1938. The vascular supply of the hypophysis cerebri of the rhesus monkey and man. Res. Publ. Assn. Nerv. Ment. Dis., 17:48–68.

Arteries of the Thorax

Alexander, W. F. 1946. The course and incidence of the lateral costal branch of the internal mammary artery. Anat. Rec., 94:446.

Anson, B. J. 1936. The anomalous right subclavian artery: Its practical significance; with a report of three cases. Surg. Gynec. Obstet., 62:708–711.

Anson, B. J., R. R. Wright, and J. A. Wolfer. 1939. Blood supply of the mammary gland. Surg. Gynec. Obstet., 69:468–473.

Appleton, A. B. 1945. The arteries and veins of the lungs. I. Right upper lobe. J. Anat. (Lond.), 79:97–120.

Boyden, E. A. 1955. Segmental Anatomy of the Lungs: A Study of the Patterns of the Segmental Bronchi and Related Pulmonary Vessels. McGraw-Hill, New York, 276 pp.

Boyden, E. A. 1970. The time lag in the development of bronchial arteries. Anat. Rec., 166:611–614.

Boyden, E. A., and J. G. Scannell. 1948. An analysis of variations in the bronchovascular pattern of the right upper lobe of fifty lungs. Amer. J. Anat., 82:27–73.

Cauldwell, E. W., R. G. Siekert, R. A. Lininger, and B. J. Anson. 1948. The bronchial arteries: An anatomic study of 150 human cadavers. Surg. Gynec. Obstet., 86:395–412.

Cole, F. H., F. H. Alley, and R. S. Jones. 1951. Aberrant systemic arteries to the lower lung. Surg. Gynec. Obstet., 93:589–596.

Douglass, R. 1948. Anomalous pulmonary vessels. J. Thoracic Surg., 17:712–716.

Ferry, R. M., Jr., and E. A. Boyden. 1951. Variations in the bronchovascular patterns of the right lower lobe of fifty lungs. J. Thoracic Surg., 22:188–201.

Jackson, C. L., and J. F. Huber. 1943. Correlated applied anatomy of the bronchial tree and lungs with a system of nomenclature. Dis. Chest, 9:319–326.

Kropp, B. N. 1951. The lateral costal branch of the internal mammary artery. J. Thoracic Surg., 21:421–425.

Maliniac, J. W. 1943. Arterial blood supply of the breast: Revised anatomic data relating to reconstructive surgery. Arch. Surg., 47:329–343.

Miller, W. S. 1907. The vascular supply of the pleura pulmonalis. Amer. J. Anat., 7:389–407.

O'Rahilly, R., H. Debson, and T. S. King. 1950. Subclavian origin of bronchial arteries. Anat. Rec., 108:227–238.

Pitel, M., and E. A. Boyden. 1953. Variations in the bronchovascular patterns of the left lower lobe of fifty lungs. J. Thoracic Surg., 26:633–653.

Potter, S. E., and E. A. Holyoke. 1950. Observations on the intrinsic blood supply of the esophagus. Arch. Surg., 61:944–948.

Shapiro, A. L., and G. L. Robillard. 1950. The esophageal arteries. Their configurational anatomy and variations in relation to surgery. Ann. Surg., 131:171–185.

Swigart, L. L., R. G. Siekert, W. C. Hambley, and B. J. Anson. 1950. The esophageal arteries: An anatomic study of 150 specimens. Surg. Gynec. Obstet., 90:234–243.

Tobin, C. E. 1952. The bronchial arteries and their connections with other vessels in the human lung. Surg. Gynec. Obstet., 95:741–750.

Tobin, C. E. 1960. Some observations concerning the pulmonic vasa vasorum. Surg. Gynec. Obstet., 111:297–303.

Arteries of the Upper Limb

Blunt, M. J. 1959. The vascular anatomy of the median nerve in the forearm and hand. J. Anat. (Lond.), 93:15–22.

Brockis, J. G. 1953. The blood supply of the flexor and extensor tendons of the fingers in man. J. Bone Jt. Surg., 35B:131–138.

Coleman, S. S., and B. J. Anson. 1961. Arterial patterns in the hand based upon a study of 650 specimens. Surg. Gynec. Obstet., 113:409–424.

Daseler, E. H., and B. J. Anson. 1959. Surgical anatomy of the subclavian artery and its branches. Surg. Gynec. Obstet., 108:149–174.

De Garis, C. F., and W. B. Swartley. 1928. The axillary artery. Amer. J. Anat., 41:353–397.

Edwards, E. A. 1960. Organization of the small arteries of the hand and digits. Amer. J. Surg., 99:837–846.

Huelke, D. F. 1959. Variation in the origins of the branches of the axillary artery. Anat. Rec., 135:33–41.

Huelke, D. F. 1962. The dorsal scapular artery—A proposed term for the artery to the rhomboid muscles. Anat. Rec., 142:57–61.

Marcarian, H. Q., and R. D. Smith. 1968. A quantitative study of the vasa nervorum in the ulnar nerve of cats. Anat. Rec., 161:105–110.

Markee, J. E., J. Wray, J. Nork, and F. McFalls. 1961. A quantitative study of the vascular beds of the hand. J. Bone Jt. Surg., 43A:1187–1196.

McCormack, L. J., E. W. Cauldwell, and B. J. Anson. 1953. Brachial and antebrachial arterial patterns. A study of 750 extremities. Surg. Gynec. Obstet., 96:43–54.

Miller, R. A. 1939. Observations upon the arrangement of the axillary artery and brachial plexus. Amer. J. Anat., 64:143–163.

Misra, B. D. 1955. The arteria mediana. J. Anat. Soc. India, 4:48.

O'Rahilly, R., H. Debson, and T. S. King. 1950. Subclavian origin of bronchial arteries. Anat. Rec., 108:227–238.

Pick, J. 1958. The innervation of the arteries in the upper limb of man. Anat. Rec., 130:103–124.

Rohlich, K. 1934. Uber die Arteria transversa colli des Menschen. Anat. Anz., 79:37.

Slager, R. F., and K. P. Klassen. 1958. Anomalous right subclavian artery arising distal to a coarctation of the aorta. Ann. Surg., 147:93–97.

Sunderland, S. 1945. Blood supply of the nerves of the upper limb in man. Arch. Neurol. Psychiat., 53:91–115.

Taleisnik, J., and P. J. Kelly. 1966. The extraosseous and intraosseous blood supply of the scaphoid bone. J. Bone Jt. Surg., 48A:1125–1137.

Weathersby, H. T. 1955. The artery of the index finger. Anat. Rec., 122:57–64.

Weathersby, H. T. 1956. Unusual variation of the ulnar artery. Anat. Rec., 124:245–248.

Arteries of the Abdomen

Anson, B. J., E. W. Cauldwell, J. W. Pick, and L. E. Beaton. 1947. The blood supply of the kidney, suprarenal gland and associated structures. Surg. Gynec. Obstet., 84:313–320.

Anson, B. J., G. A. Richardson, and W. L. Minear. 1936. Variations in the number and arrangement of the renal vessels: A study of the blood supply of four hundred kidneys. J. Urol., 36:211–219.

Barlow, T. E., F. H. Bentley, and D. N. Walder. 1951. Arteries, veins and arteriovenous anastomoses in the human stomach. Surg. Gynec. Obstet., 93:657–671.

Basmajian, J. V. 1954. The marginal anastomoses of the arteries to the large intestine. Surg. Gynec. Obstet., 99:614–616.

Baylin, G. J. 1939. Collateral circulation following an obstruction of the abdominal aorta. Anat. Rec., 75:405–408.

Benton, R. S., and W. B. Cotter. 1963. A hitherto undocumented variation of the inferior mesenteric artery in man. Anat. Rec., 145:171–173.

Browne, E. Z. 1940. Variations in origin and course of the hepatic artery and its branches: Importance from surgical viewpoint. Surgery, 8:424–445.

Cadete-Leite, A. 1973. The arteries of the pancreas of the dog. An injection-corrosion and microangiographic study. Amer. J. Anat., 137:151–157.

Cauldwell, E. W., and B. J. Anson. 1943. The visceral branches of the abdominal aorta: Topographical relationships. Amer. J. Anat., 73:27–57.

Chambers, G., E. Eldred, and C. Eggett. 1972. Anatomical observations on the arterial supply to the lumbosacral spinal cord of the cat. Anat. Rec., 174:421–433.

Clark, K. 1959. The blood vessels of the adrenal glands. J. Roy. Coll. Surg. Edin., 4:257–262.

Clausen, H. J. 1955. An unusual variation in origin of the hepatic and splenic arteries. Anat. Rec., 123:335–340.

Cokkins, A. J. 1930. Observations on the mesenteric circulation. J. Anat. (Lond.), 64:200–205.

Crelin, E. S., Jr. 1948. An unusual anomalous blood vessel connecting the renal and internal spermatic arteries. Anat. Rec., 102:205–212.

Daseler, E. H., B. J. Anson, W. C. Hambley, and A. F. Reimann. 1947. The cystic artery and constituents of the hepatic pedicle: A study of 500 specimens. Surg. Gynec. Obstet., 85:47–63.

Drummond, H. 1914. The arterial supply of the rectum and pelvic colon. Brit. J. Surg., 1:677–685.

Edwards, L. F. 1941. The retroduodenal artery. Anat. Rec., 81:351–355.

El-Eishi, H. I., S. F. Ayoub, and M. Abd-El-Khalek. 1973. The arterial supply of the human stomach. Acta Anat., 86:565–580.

Elias, H., and D. Petty. 1952. Gross anatomy of the blood vessels and ducts within the human liver. Amer. J. Anat., 90:59–111.

Falconer, C. W. A., and E. Griffiths. 1950. The anatomy of the blood vessels in the region of the pancreas. Brit. J. Surg., 37:334–344.

Fine, H., and E. N. Keen. 1966. The arteries of the human kidney. J. Anat. (Lond.), 100:881–894.

Fried, L. C., and J. Doppman. 1974. The arterial supply to the lumbosacral spinal cord in the monkey: A comparison with man. Anat. Rec., 178:41–48.

Fuller, P. M., and D. F. Huelke. 1973. Kidney vascular supply in the rat, cat, and dog. Acta Anat., *84*:516–522.

George, R. 1935. Topography of the unpaired visceral branches of the abdominal aorta. J. Anat. (Lond.), *69*:196–205.

Goligher, J. C. 1949. The blood supply to the sigmoid colon and rectum: With reference to the technique of rectal resection with restoration of continuity. Brit. J. Surg., *37*:157–162.

Goligher, J. C. 1954. The adequacy of the marginal blood-supply to the left colon after high ligation of the inferior mesenteric artery during excision of the rectum. Brit. J. Surg. *41*:351–353.

Graves, F. T. 1956. The aberrant renal artery. J. Anat. (Lond.), *90*:553–558.

Graves, F. T. 1971. *The Arterial Anatomy of the Kidney: The Basis of Surgical Technique.* Williams and Wilkins, Baltimore, 101 pp.

Greenberg, M. W. 1950. Blood supply of the rectosigmoid and rectum. Ann. Surg., *131*:100–108.

Harrison, R. G. 1951. A comparative study of the vascularization of the adrenal gland in the rabbit, rat and cat. J. Anat. (Lond.), *85*:12–23.

Healey, J., P. Schroy, and R. Sorensen. 1953. The intrahepatic distribution of the hepatic artery in man. J. Internat. Coll. Surg., *20*:133–148.

Laufman, H., R. E. Berggren, T. Finley, and B. J. Anson. 1960. Anatomical studies of the lumbar arteries: With reference to the safety of translumbar aortography. Ann. Surg., *152*:621–634.

Michaels, N. A. 1951. The hepatic, cystic and retroduodenal arteries and their relations to the biliary ducts: With samples of the entire celiacal blood supply. Ann. Surg., *133*:503–524.

Michaels, N. A. 1953. Variational anatomy of the hepatic, cystic, and retroduodenal arteries. A.M.A. Arch. Surg., *66*:20–32.

Michels, N. A. 1953. Collateral arterial pathways to the liver after ligation of the hepatic artery and removal of the celiac axis. Cancer, *6*:708–724.

Michels, N. A. *Blood Supply and Anatomy of the Upper Abdominal Organs, with a Descriptive Atlas.* J. B. Lippincott Co., Philadelphia, 581 pp.

Michels, N. A. 1962. The anatomic variations of the arterial pancreaticoduodenal arcades: Their import in regional resection involving the gall-bladder, bile ducts, liver, pancreas, and parts of the small and large intestines. J. Int. Coll. Surg., *37*:13–40.

Michels, N. A., P. Siddharth, P. L. Kornblith, and W. W. Parke. 1965. The variant blood supply to the descending colon, rectosigmoid and rectum based on 400 dissections. Its importance in regional resections: a review of medical literature. Dis. Colon Rectum, *8*:251–278.

Milloy, F. J., B. J. Anson, and D. K. McAffee, 1960. The rectus abdominis muscle and the epigastric arteries. Surg. Gynec. Obstet., *110*:293–302.

Mitra, S. K. 1966. The terminal distribution of the hepatic artery with special reference to the arterio-portal anastomosis. J. Anat. (Lond.), *100*:651–663.

Noer, R. J. 1943. The blood vessels of the jejunum and ileum: A comparative study of man and certain laboratory animals. Amer. J. Anat., *73*:293–334.

Pick, J. W., and B. J. Anson. 1940. The inferior phrenic artery: Origin and suprarenal branches. Anat. Rec., *78*:413–427.

Pick, J. W., and B. J. Anson. 1940. The renal vascular pedicle: An anatomical study of 430 body-halves. J. Urol., *44*:411–434.

Pierson, J. M. 1943. The arterial blood supply of the pancreas. Surg. Gynec. Obstet., *77*:426–432.

Ross, J. A. 1950. Vascular patterns of small and large intestine compared. Brit. J. Surg., *39*:330–333.

Shah, M. A., and M. Shah. 1946. The arterial supply of the vermiform appendix. Anat. Rec., *95*:457–460.

Shah, P. M., H. A. Scarton, and M. J. Tsapogas. 1978. Geometric anatomy of the aortic-common iliac bifurcation. J. Anat. (Lond.), *126*:451–458.

Sunderland, S. 1942. Blood supply of the distal colon. Austral. New Zealand J. Surg., *11*:253–263.

Vandamme, J. P. J., and G. Van Der Schuren. 1976. Re-evaluation of the colic irrigation from the superior mesenteric artery. Acta Anat., *95*:578–588.

Wharton, G. K. 1932. The blood supply of the pancreas, with special reference to that of the islands of Langerhans. Anat. Rec., *53*:55–81.

Wilmer, H. A. 1941. The blood supply of the first part of the duodenum: With description of the gastroduodenal plexus. Surgery, *9*:679–687.

Woodburne, R. T., and L. L. Olsen. 1951. The arteries of the pancreas. Anat. Rec., *111*:255–270.

Yule, E. 1926. The arterial supply to the duodenum. J. Anat. (Lond.), *61*:344–345.

Arteries of the Pelvis

Ashley, F. L., and B. J. Anson. 1941. The hypogastric artery. Amer. J. Phys. Anthrop., *28*:381–395.

Braithwaite, J. L. 1952. The arterial supply of the male urinary bladder. Brit. J. Urol., *24*:64–71.

Braithwaite, J. L. 1953. Variations in origin of the parietal branches of the internal iliac artery. J. Anat. (Lond.), *86*:423–430.

Daniel, O., and R. Schackman. 1952. The blood supply of the human ureter in relation to ureterocolic anastomosis. Brit. J. Urol., *24*:334–343.

Dees, J. E. 1940. Anomalous relationship between ureter and external iliac artery. J. Urol., *44*:207–215.

Flocks, R. H. 1937. The arterial distribution within the prostate gland: Its role in transurethral prostatic resection. J. Urol., *37*:524–548.

Gordon, K. C. D. 1967. A comparative anatomical study of the distribution of the cystic artery in man and other species. J. Anat. (Lond.), *101*:351–359.

Harrison, R. G. 1949. The distribution of the vasal and cremasteric arteries of the testis and their functional importance. J. Anat. (Lond.), *83*:267–282.

Harrison, R. G., and A. E. Barclay. 1948. Distribution of testicular artery (internal spermatic artery) to human testis. Brit. J. Urol., *20*:57–66.

Lindenbaum, E., J. M. Brandes, and J. Itskovitz. 1978. Ipsi- and contralateral anastomosis of the uterine arteries. Acta Anat., *102*:157–161.

Mansfield, A. O., and J. M. Howard. 1964. Absence of both common iliac arteries. A case report. Anat. Rec., *150*:363–381.

Shehata, R. 1976. The arterial supply of the urinary bladder. Acta Anat., *96*:128–134.

Arteries of the Lower Limb

Assar, Y. H. 1938. The saphenous artery. J. Anat. (Lond)., *73*:194.

Blair, C. B. Jr., and K. Nandy. 1965. Persistence of the

axis artery of the lower limb. Anat. Rec., 152:161–172.

Chandler, S. B., and P. H. Kreuscher. 1932. A study of the blood supply of the ligamentum teres and its relation to the circulation of the head of the femur. J. Bone Jt. Surg., 14:834–846.

Chavatzas, D. 1974. Revision of the incidence of congenital absence of dorsalis pedis artery by an ultrasonic technique. Anat. Rec., 178:289–290.

Crock, H. V. 1965. A revision of the anatomy of the arteries supplying the upper end of the human femur. J. Anat. (Lond.), 99:77–88.

Howe, W. W., Jr., T. Lacey, and R. P. Schwartz. 1950. A study of the gross anatomy of the arteries supplying the proximal portion of the femur and the acetabulum. J. Bone Jt. Surg., 32A:856–866.

Huber, J. F. 1941. The arterial network supplying the dorsum of the foot. Anat. Rec., 80:373–391.

Keen, J. A. 1961. A study of the arterial variations in the limbs, with special reference to symmetry of vascular patterns. Amer. J. Anat., 108:245–261.

Laing, P. G. 1953. The blood supply of the femoral shaft. J. Bone Jt. Surg., 35B:462–466.

Mustalish, A. C., and J. Pick. 1964. On the innervation of the blood vessels in the human foot. Anat. Rec., 149:587–590.

Nelson, G. E., Jr., P. J. Kelly, L. F. A. Peterson, and J. M. Janes. 1960. Blood supply of the human tibia. J. Bone Jt. Surg., 42A:625–636.

Pick, J. W., B. J. Anson, and F. L. Ashley. 1942. The origin of the obturator artery. Amer. J. Anat., 70:317–343.

Pierson, H. H. 1925. Seven arterial anomalies of the human leg and foot. Anat. Rec., 30:139–145.

Rogers, W. M., and H. Gladstone. 1950. Vascular foramina and arterial supply of the distal end of the femur. J. Bone Jt. Surg., 32A:867–874.

Rook, F. W. 1953. Arteriography of the hip joint for predicting end results in intracapsular and intertrochanteric fractures of the femur. Amer. J. Surg., 86:404–409.

Sanders, R. J. 1963. Relationships of the common femoral artery. Anat. Rec., 145:169–170.

Scapinelli, R. 1967. Blood supply of the human patella: Its relation to ischaemic necrosis after fracture. J. Bone Jt. Surg., 49B:563–570.

Senior, H. D. 1919. An interpretation of the recorded anomalies of the human leg and foot. J. Anat. (Lond.), 53:130–171.

Senior, H. D. 1929. Abnormal branching of the human popliteal artery. Amer. J. Anat., 44:111–120.

Sevitt, S., and R. G. Thompson. 1965. The distribution and anastomoses of arteries supplying the head and neck of the femur. J. Bone Jt. Surg., 47B:560–573.

Sunderland, S. 1945. Blood supply of the sciatic nerve and its popliteal divisions in man. Arch. Neurol. Psychiat., 54:283–289.

Trueta, T. 1957. The normal vascular anatomy of the human femoral head during growth. J. Bone Jt. Surg., 39B:358–394.

Trueta, J., and M. H. M. Harrison. 1953. The normal vascular anatomy of the femoral head in adult man. J. Bone Jt. Surg., 35B:442–461.

Vann, H. M. 1943. A note on the formation of the plantar arterial arch of the human foot. Anat. Rec., 85:269–275.

Weathersby, H. T. 1959. The origin of the artery of the ligamentum teres femoris. J. Bone Jt. Surg., 41A:261–263.

Williams, G. D., C. H. Martin, and L. R. McIntire. 1934. Origin of the deep and circumflex femoral group of arteries. Anat. Rec., 60:189–196.

Woollard, H. H., and G. Weddell. 1934. Arterial vascular patterns. J. Anat. (Lond.), 69:25–37.

Woollard, H. H., and G. Weddell. 1934. The composition and distribution of vascular nerves in the extremities. J. Anat. (Lond.), 69:165–176.

The Veins

The veins are the most variable part of the cardiovascular system. Variations in venous patterns are more frequent in vessels that drain the more superficial parts of the body than in the deeper coursing veins which often accompany the arteries. At times veins exist in pairs, one lying on each side of an artery. These are called **venae comitantes** and are often seen in the medium-sized vessels of the arm, forearm, and leg (brachial, radial, ulnar, and tibial). The larger arteries (axillary, femoral, and popliteal) usually have only one accompanying vein. In certain organs and body regions, however, the deep veins do not accompany arteries; this is true of the venous sinuses in the skull and vertebral canal, the hepatic veins, and the large veins draining blood from the long bones.

Development of the Veins

The development of veins actually commences when minute plexuses of vessels receive blood from the earliest developing capillaries. These plexuses give rise to tributaries that finally unite to form trunks. Passing toward the heart, the trunks increase in size as they receive other tributaries or as they join other veins.

The further development of the venous system in the human embryo will be considered by dividing the veins into two groups: the **parietal veins,** which drain the body cavity, extremities, and head and form the large inferior and superior venae cavae, and the **visceral veins,** which drain the yolk sac and placenta and join to form the sinus venosus.

PARIETAL VEINS

The parietal veins, which drain most of the developing embryonic structures, are initially represented by four large vessels: the two **anterior cardinal** (or precardinal) **veins** and the two **posterior cardinal** (or postcardinal) **veins** (Figs. 9-1, *A*; 9-3). The two anterior cardinal veins, one lying on each side of the embryo, drain the more cranial venous tributaries, returning blood from the head, and soon become the primitive jugular veins. The two posterior cardinal veins, also one on each side, drain the caudal venous tributaries of the embryo, returning

The venous system conveys blood from the capillaries to the heart. There are two distinct sets of veins, the **pulmonary** and the **systemic.** These serve two different vascular circuits: the **pulmonary veins** return blood to the heart from the lungs, and the **systemic veins** return blood to the heart from the rest of the body.

The pulmonary veins return oxygenated blood from the capillaries of the lungs to the left atrium of the heart. The blood then continues through the left ventricle and aorta to supply all the body organs.

The systemic veins belong to the larger systemic circulation and convey deoxygenated blood from all the organs except the lungs to the right atrium of the heart. Blood from all the systemic veins except those of the heart eventually flows through the superior and inferior venae cavae to the right atrium. The **coronary veins** flow into the coronary sinus, which then also opens into the right atrium.

The **portal vein** is interposed between certain abdominal viscera and the liver. Its special feature is that it both begins and ends in capillaries. The portal vein receives venous blood from the capillaries of the digestive tube, spleen, and pancreas and delivers it to the venous capillary-like sinusoids of the liver. The blood then makes its way into the general systemic venous system through the tributaries of the hepatic veins, which flow into the inferior vena cava.

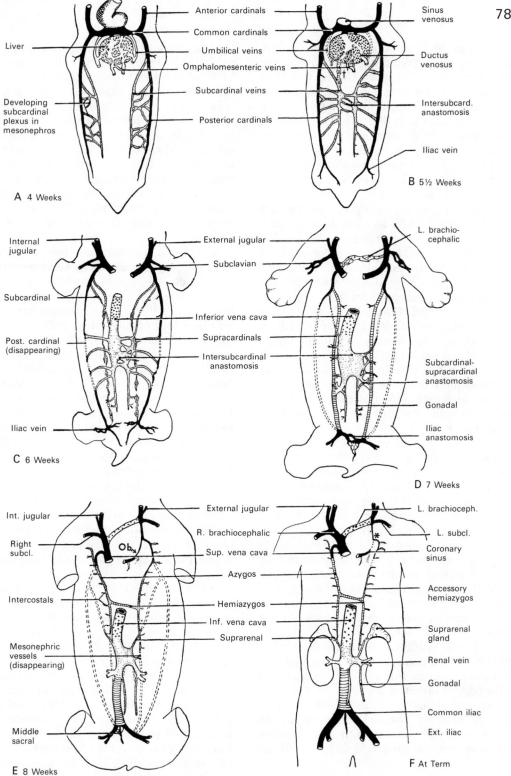

FIG. 9-1. Schematic diagrams showing certain stages in the development of the inferior vena cava between the fourth and eighth prenatal weeks and at term, based on the work of McClure and Butler (1925). Anterior and posterior cardinal veins are shown in solid black, supracardinal veins are hatched horizontally. † (in B) = vessels from the upper part of the right subcardinal vein and liver that commence forming the hepatic segment of the inferior vena cava; Ob. (in E) = oblique vein of the left atrium; * (in F) = left superior intercostal vein. (From Patten, B. M., *Human Embryology*, 3rd ed., McGraw-Hill, New York, 1968.)

blood from the lower body wall, the pelvis, the mesonephroi (Wolffian[1] bodies), and the lower limb buds. The anterior and posterior cardinal veins on each side join to form a symmetrically arranged longitudinally oriented cardinal system (Fig. 9-1, *A*). Two short transverse veins, the **common cardinal veins** (ducts of Cuvier[2]), interconnect the anterior and posterior cardinal veins of each side to the corresponding horns of the sinus venosus (Figs. 9-1, *B*; 9-4, *A*).

To this relatively simple pattern are added, during the fifth week of development, the **subcardinal veins.** These veins develop parallel and ventral to the posterior cardinal veins in the ventromedial part of the mesonephric ridges and gradually take over the drainage of the mesonephroi from the posterior cardinal veins (Fig. 9-1, *B*). During the seventh week are added the **supracardinal veins,** which form dorsomedial to the posterior cardinal veins (Fig. 9-1, *D*).

This primitive double (right and left) plan of veins then is converted to the single caval system, in which blood is returned to the heart through the inferior and superior venae cavae.

INFERIOR VENA CAVA. The principal changes that occur to form the vena caval systems in the embryo culminate in the alteration of the bilaterally symmetrical venous plan, initiated by the formation of the anterior and posterior cardinal veins along with the subcardinal veins, into large unpaired vessels found on the right side of the body (the superior and inferior venae cavae). The peripheral channels in the head and limbs that flow into these centrally unpaired veins still retain much of their symmetry.

The inferior vena cava, which drains blood from the lower trunk, pelvis, and lower limbs, gradually develops by replacing the various cardinal veins or by utilizing portions of them. Initially, the subcardinal veins begin to replace partially the posterior cardinal veins as the drainage channels of the mesonephroi. These channels soon form many cross connections with the posterior cardinal veins on each side (Fig. 9-1, *A*). As

the mesonephroi grow, they approximate medially, and consequently the subcardinal veins of the two sides become nearly adjacent. By five and one-half weeks, the two subcardinal veins establish an interconnection called the **intersubcardinal anastomosis** (Fig. 9-1, *B*), which then enlarges to form the **subcardinal sinus.** Soon blood begins to flow medially into this sinus from the many small vessels that had interconnected the posterior cardinal and the subcardinal veins. This results in a disappearance of the more caudal parts of the posterior cardinal veins (Fig. 9-1, *C*).

As the subcardinal sinus forms, the veins of the superior pole of the right mesonephros, which lie close to the developing liver, anastomose with some of the vessels flowing through the liver. This anastomosis occurs within a fold of mesentery called the **caval fold** (or mesentery), into which have grown capillaries from the liver above and some capillaries from the right subcardinal veins below (Fig. 9-1, *B* and *C*). Soon this anastomosis enlarges within the substance of the liver and becomes a principal venous channel. It achieves the surface of the liver as the **hepatic part of the inferior vena cava.** The right omphalomesenteric vein enlarges to form the **suprahepatic portion of the inferior vena cava,** which is the segment between the liver and the right atrium. The portion of the inferior vena cava arising above the kidneys, which can be called the **prerenal part,** consists of the renal segment (subcardinal sinus), the hepatic segment, and the suprahepatic segment.

Below the level of the kidneys, the two **supracardinal veins** participate in the formation of the **postrenal part** of the inferior vena cava (Fig. 9-1, *C* and *D*). These veins arise during the seventh week, and they are seen initially as a periganglionic plexus of veins associated with the developing sympathetic ganglionic chain. The supracardinal veins develop as a pair of longitudinally oriented vessels that drain the posterior body wall dorsal to the posterior cardinal veins. Near the level of the middle part of the developing mesonephros, the supracardinal veins merge with the subcardinal sinus.

Above the subcardinal sinus, together with the persisting proximal part of the right postcardinal vein, the supracardinal vein forms the **azygos system of veins** (Fig. 9-1, *E*

---

[1]Kasper F. Wolff (1733–1794): A German embryologist and physiologist, who worked in St. Petersburg, Russia.
[2]Georges L. Cuvier (1769–1832): A French zoologist and natural historian (Paris).

and *F*). The proximal part of the left supracardinal vein, together with some of the upper part of the left postcardinal vein, forms the **hemiazygos, accessory hemiazygos,** and the **left superior intercostal vein.**

Below the subcardinal sinus, blood from the lower limbs is collected by the right and left iliac veins, which in the earlier stages of development open into the corresponding right and left posterior cardinal veins. Later a transverse **iliac anastomosis,** which at this stage connects the right and left iliac vein, eventually develops into the **left common iliac vein** (Fig. 9-1, *C* to *F*). Blood returning from the lower extremities increasingly is shunted from the lower end of the right postcardinal vein to the right supracardinal vein. As the postcardinal vein gradually disappears, the lower part of the right supracardinal vein enlarges and forms the **postrenal part of the inferior vena cava** (Fig. 9-1, *E* and *F*).

Above the subcardinal sinus, the left posterior cardinal-subcardinal vein disappears, except the part immediately above the renal vein, which is retained as the **left suprarenal vein.** The fate of the left posterior cardinal-subcardinal system also accounts for the fact that the left gonadal vein (testicular or ovarian) opens into the left renal vein, while the vein on the right side drains into the postrenal segment of the inferior vena cava (Fig. 9-1, *E* and *F*).

SUPERIOR VENA CAVA. The superior vena cava develops at the junction of the two brachiocephalic veins, which in turn receive blood from the head by way of the internal and external jugular veins and from the upper limb by means of the subclavian vein. The primitive **internal jugular veins,** which drain the cranial end of the embryo on each side, initially are simply the anterior cardinal veins, whereas the **external jugular veins** develop in the region of the mandible by the junction of a number of smaller, more superficial vessels (Fig. 9-1, *B* to *D*).

At four weeks the lower portion of the anterior cardinal vein joins the posterior cardinal vein to form the common cardinal vein (Figs. 9-1, *A*; 9-2, *A*). Because the heart as well as the common cardinal veins lie far to the cranial end of the developing embryo at the time that the upper limb buds appear, the initial veins that drain the upper limb buds flow into the posterior cardinal veins (Fig. 9-1, *B*). As development continues, the heart descends relative to the upper limb buds, and soon the limb bud veins (axillary) begin to drain into the more caudal part of the anterior cardinal vein (internal jugular). When this occurs (sixth to seventh weeks), the primitive pattern of two brachiocephalic veins (right and left superior venae cavae) is formed (Figs. 9-1, *C*; 9-2, *B*). About this same time, an anastomosis forms between the two anterior cardinal veins, which results in the formation of a transverse connecting vein in front of the trachea as a confluence of thymic and thyroid veins. This anastomosis becomes the **left brachiocephalic vein,** and it allows blood returning from the cranial left side of the embryo to be shunted to the right side of the heart, where it must enter the right atrium (Figs. 9-1, *D*; 9-2, *B*). The junction of the two brachiocephalic veins eventually forms the superior

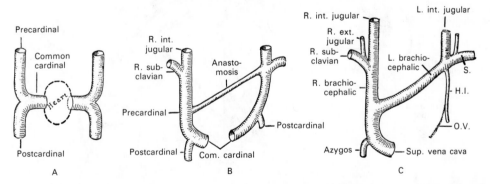

Fig. 9-2. Diagrams showing the transformation of the cardinal veins into the superior vena cava in human embryos at six weeks (*A*), eight weeks (*B*), and in the adult (*C*). H.I. = left superior intercostal vein (highest intercostal); O.V. = oblique vein of the left atrium; S = left subclavian vein. (From Arey, L. B., *Developmental Anatomy,* 7th ed., W. B. Saunders Co., Philadelphia, 1974.)

vena cava (Fig. 9-1, *D* to *F*). The remainder of the left anterior cardinal vein, below the left brachiocephalic, becomes much smaller, but it contributes to the formation of the **left superior intercostal vein,** in conjunction with the small remaining cranial end of the left posterior cardinal vein. The right anterior cardinal vein between the azygos vein and the union with the left brachiocephalic vein becomes the *upper part* of the adult **superior vena cava.** The *lower part* of the superior cava, that is, the part between the azygos vein and the heart, is formed by the original right common cardinal vein (duct of Cuvier). The left anterior cardinal vein, caudal to the transverse left brachiocephalic vein and left superior intercostal vein, and the left common cardinal vein regress. The upper part becomes the **vestigial fold of Marshall**[1] (Figs. 9-1, *E* and *F*; 9-2), and the proximal part of the left common cardinal vein becomes the **oblique vein of the left atrium** (of Marshall). The oblique vein of the left atrium drains across the dorsal surface of the left atrium to open into the **coronary**

[1]John Marshall (1818–1891): An English anatomist and surgeon (London).

sinus, which represents the persistent left horn of the sinus venosus. Both the right and left superior venae cavae are present in some animals, and occasionally are found as anomalies in adult human beings.

VISCERAL VEINS

The visceral veins are the two **vitelline** or **omphalomesenteric veins** which bring blood to the embryo from the yolk sac, and the two **umbilical veins** which return oxygenated fetal blood to the embryo from the placenta. These four veins open into the **sinus venosus.**

VITELLINE VEINS.   The paired vitelline veins follow the stalk of the yolk sac, and thereby enter the body of the embryo. They course cranially, at first ventral to and then on either side of the developing foregut, each on its way to its respective horn of the sinus venosus. The vitelline veins unite on the ventral aspect of the intestinal canal, and more cranially, they are connected to each other by two anastomotic branches, one on the dorsal and the other on the ventral aspect of the duodenal portion of the intestine,

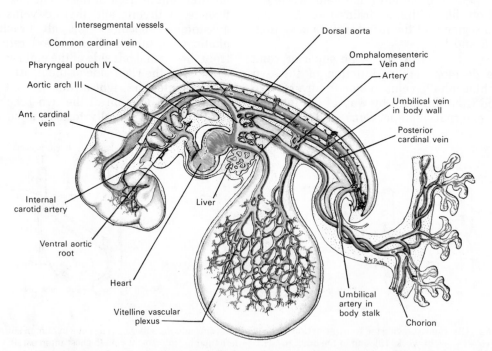

FIG. 9-3.   Semischematic diagram showing the basic vascular plan of the human embryo at the end of first month. For the sake of simplicity, the paired vessels are shown only on the side toward observer. (From Patten, B. M., *Human Embryology,* 3rd ed., McGraw-Hill, New York, 1968.)

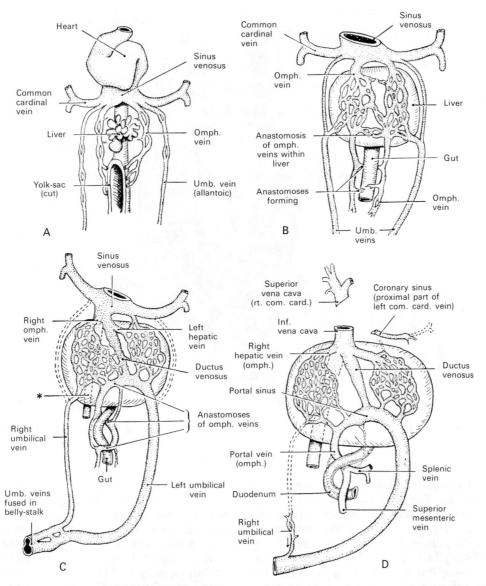

FIG. 9-4. Diagrams showing steps in the development of the hepatic-portal circulation from the omphalomesenteric veins, and the changes by which blood returning in the umbilical veins from the placenta is rerouted through the liver. *A,* fourth week; *B,* fifth week; *C,* sixth week; *D,* seventh week and older. The asterisk in *C* indicates the hepatic part of the inferior vena cava. (From Patten, B. M., *Human Embryology,* 3rd ed., McGraw-Hill, New York, 1968.)

which is thus encircled by two venous rings (Fig. 9-4, *B*). Into the middle or dorsal anastomosis opens the superior mesenteric vein, which arises in the dorsal mesentery of the intestine. Within the septum transversum, the portions of the veins above the upper ring are interrupted by the developing liver and are broken up by it into a plexus of small capillary-like vessels termed **sinusoids.** The vessels conveying blood to this plexus become the tributaries of the **portal vein,** while the vessels draining the plexus from the liver into the sinus venosus form the future **hepatic veins** (Figs. 9-4, *C* and *D*; 9-5). Ultimately the left vitelline (hepatic) veins no longer communicate directly with the sinus venosus, but open into the developing right hepatic veins. The persistent part of the upper venous ring, above the opening of the superior mesenteric vein, forms the trunk of the portal vein.

UMBILICAL VEINS. The two umbilical veins

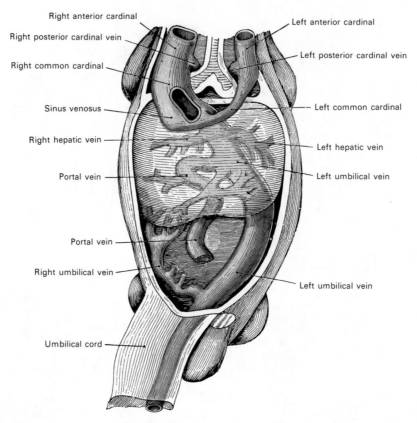

Right anterior cardinal

Right posterior cardinal vein

Right common cardinal

Sinus venosus

Right hepatic vein

Portal vein

Portal vein

Right umbilical vein

Umbilical cord

Left anterior cardinal

Left posterior cardinal vein

Left common cardinal

Left hepatic vein

Left umbilical vein

Left umbilical vein

Fɪɢ. 9-5. Human embryo with heart and anterior body wall removed to show the sinus venosus and its tributaries. (After His.)

return oxygenated blood to the embryo from the placenta. They form a single trunk in the body stalk, but remain separate within the embryo and pass forward to the sinus venosus in the side walls of the embryonic body (Fig. 9-4, B). Like the vitelline veins, their direct connection with the sinus venosus becomes interrupted by the developing liver, and thus at this stage (6 mm, fifth week), all the blood from the yolk sac and placenta passes through the substance of the liver before reaching the heart. During the sixth week, the two umbilical veins commence to fuse in the umbilical cord, and the left umbilical channel becomes the principal course for the returning blood (Figs. 9-4, C; 9-5). In the embryo the right umbilical and right vitelline veins shrivel and disappear, while the left umbilical vein enlarges and opens into the upper venous ring of the vitelline veins. When the yolk sac atrophies, the left vitelline vein also undergoes atrophy and disappears (Fig. 9-5). Finally, a direct branch is established between this ring and the right

hepatic vein. This branch enlarges rapidly into a wide channel named the **ductus venosus** (Fig. 9-4, C and D). Through the ductus venosus most of the blood is returned from the placenta and carried directly to the heart without passing through the liver. The left umbilical vein and the ductus venosus undergo atrophy and obliteration after birth, and form respectively the **ligamentum teres** and **ligamentum venosum** of the liver.

The Pulmonary Veins

The **pulmonary veins,** two from each lung, return oxygenated blood to the left atrium of the heart (Fig. 9-6). The capillary networks in the walls of the air sacs join together to form one vessel for each lobule. These vessels unite successively to form a single trunk for each lobe, three for the right and two for the left lung. The vein from the middle lobe of the right lung usually unites with that from the superior lobe so that ultimately two

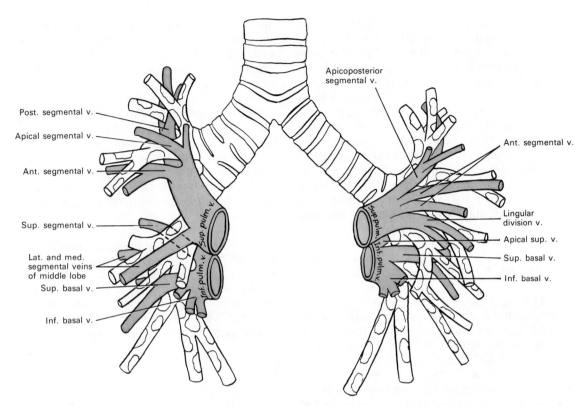

Post. segmental v.

Apical segmental v.

Ant. segmental v.

Sup. segmental v.

Lat. and med. segmental veins of middle lobe

Sup. basal v.

Inf. basal v.

Apicoposterior segmental v.

Ant. segmental v.

Lingular division v.

Apical sup. v.

Sup. basal v.

Inf. basal v.

FIG. 9-6. Diagrammatic representation of the branches of the superior and inferior pulmonary veins on each side of the body. (Redrawn after *Nomina Anatomica,* 1968, Excerpta Medica Foundation.)

trunks are formed from each lung. Near the heart, the four pulmonary veins perforate the fibrous layer of the pericardium and open separately into the superior part of the left atrium (Fig. 9-7). The three veins from the right lung may remain separate, and at times the two left veins join and enter the heart by a common opening.

At the roots of the lungs, the superior pulmonary veins lie anterior and a little inferior to the pulmonary arteries. The inferior veins are situated in the most caudal part of the hilum on a plane posterior to the superior veins. The pulmonary veins are about 1.5 cm in length, and on the right side they pass behind the right atrium and superior vena cava to reach the left atrium. The left pulmonary veins course anterior to the descending thoracic aorta on their way to the heart.

Within the lung the ramifications of the veins do not always accompany the bronchi and arteries in the central regions of the bronchopulmonary segments. More superficially near the pleura, they are found between bronchopulmonary segments, or more deeply, they are found intrasegmentally. The

branchings of both the larger and the smaller tributaries are quite variable. A list of the bronchopulmonary segments is found in Chapter 15.

RIGHT SUPERIOR PULMONARY VEIN

The right superior pulmonary vein has three tributaries for the superior lobe and one for the middle lobe (Fig. 9-6).

APICAL SEGMENTAL VEIN. The apical segmental vein lies near the mediastinal pleura. It has an *intrasegmental* tributary draining the apex and an *infrasegmental* tributary separating the apical from the anterior segments.

POSTERIOR SEGMENTAL VEIN. The posterior segmental vein is the largest vein of the superior lobe and has both intralobar and infralobar parts. The *intralobar part* has an apical, a posterior, and an intersegmental tributary, which is a prominent vein of the posterior surface. The *infralobar part* lies near the pleura between the superior and middle lobes.

ANTERIOR SEGMENTAL VEIN. The anterior segmental vein has two tributaries, *intrasegmental* and *infrasegmental*. These accompany the two branches of the anterior segmental bronchus of the right superior lobe.

RIGHT MIDDLE LOBE VEIN. The right middle lobe vein is usually a single tributary of the right superior pulmonary vein, but its two segmental veins may remain separate. The *lateral segmental vein* splits into a tributary for the lateral segment and a tributary separating the segments. The *medial segmental vein* lies near the mediastinal surface and has two tributaries, one subpleural and one intrasegmental.

RIGHT INFERIOR PULMONARY VEIN

The right inferior pulmonary vein occupies the most inferior part of the hilum of the right lung, and has two main tributaries, a superior (or apical) segmental vein and a common basal vein (Fig. 9-6).

SUPERIOR (OR APICAL) SEGMENTAL VEIN. The superior segmental vein begins near the medial side of the middle lobe bronchus and is formed by two or three tributaries with intrasegmental and infrasegmental tributaries.

COMMON BASAL VEIN. The common basal vein is a large short trunk that drains the four basal segments of the lower lobe of the right lung. It has two principal tributaries, a superior basal and an inferior basal. The **superior basal** tributary lies deep to the surface of the inferior lobe and drains both its anterior basal and lateral basal bronchopulmonary segments by *intrasegmental* and *infrasegmental* tributaries. The **inferior basal** tributary drains the posterior basal and lateral basal bronchopulmonary segments. The medial basal bronchopulmonary segment is drained by several tributaries that open into both the superior basal and inferior basal veins.

LEFT SUPERIOR PULMONARY VEIN

The left superior pulmonary vein drains blood from the two bronchopulmonary segments of the superior lobe, as well as from the two segments of the lingula of the superior lobe. It is formed by a confluence of the apicoposterior segmental vein, the anterior segmental vein, and the lingular division vein. The latter vein is formed by the junction of the superior and inferior lingular veins (Fig. 9-6).

APICOPOSTERIOR SEGMENTAL VEIN. The apicoposterior segmental vein is formed by the junction of an *intrasegmental* tributary and an *infrasegmental* tributary.

ANTERIOR SEGMENTAL VEIN. The anterior segmental vein drains the anterior segment of the left superior lobe, and is formed by *intrasegmental* and *infrasegmental* tributaries.

LINGULAR DIVISION VEIN. The lingular division vein drains the superior lingular and inferior lingular segments. The superior lingular vein has posterior and anterior intersegmental tributaries. The tributaries of the inferior lingular vein are superior and inferior. The latter drains into the inferior pulmonary vein in some instances.

LEFT INFERIOR PULMONARY VEIN

The left inferior pulmonary vein receives a smaller apical superior segmental vein and a larger common basal segmental vein (Fig. 9-6).

APICAL SUPERIOR SEGMENTAL VEIN. The apical superior segmental vein lies on the medial side of the superior lobe bronchus and is itself formed by a central *intrasegmental* and two *intersegmental* tributaries, the latter separating the segment from its neighbors.

COMMON BASAL VEIN. The common basal vein receives the **superior basal vein,** which largely drains the anterior basal segment by *intrasegmental* and *infrasegmental tributaries,* and the **inferior basal vein,** which drains the posterior basal and lateral basal segments by *intrasegmental* and *infrasegmental* tributaries. The medial basal bronchopulmonary segment is drained by two venous channels, one intrasegmental and the other intersegmental; the latter separates the medial basal segment from the posterior basal segment. One or both of these veins drains into the common basal vein.

The *anterior basal vein,* which drains into the superior basal division of the common basal vein, has an intrasegmental and an infrasegmental tributary; the latter usually lies deep to the pulmonary ligament and separates the medial basal from the lateral basal bronchopulmonary segments. The *lateral basal vein,* which drains into the inferior basal division of the common basal vein, is primarily an intersegmental vessel in the depth of the lobe between the lateral basal and posterior basal bronchi, but it also has an intrasegmental tributary. The *posterior basal vein,* which also drains into the inferior basal division of the common basal vein, is intrasegmental; it lies in the hollow between the branches of the posterior basal segmental bronchus and drains these regions by two tributaries.

The Systemic Veins

The **veins** commence as minute plexuses that receive blood from the capillaries. The tributaries arising from these plexuses unite into trunks, and these, in their passage toward the heart, constantly increase in size as they receive tributaries or join other veins. The veins are larger in caliber and more numerous than the arteries, giving the venous system a larger capacity than that of the arterial system. The capacity of the pulmonary veins, however, only slightly exceeds that of the pulmonary arteries. The veins, when full, are cylindrical like the arteries, but their walls are thin and they collapse when empty. In the cadaver, they may be interrupted at intervals by nodular enlargements that indicate the existence of **valves** in their interior.

Veins anastomose freely, especially in certain regions of the body. These communications exist between large trunks as well as between smaller branches. Between the venous sinuses of the cranium, and between the veins of the neck, where obstruction would bring imminent danger to the cerebral venous system, large and frequent anastomoses are found. The same free communication exists between the veins throughout the entire extent of the vertebral canal, and between the veins composing the various venous plexuses in the abdomen and pelvis. Three types of venous channels can be identified according to their location in the body: superficial veins, deep veins, and venous sinuses.

SUPERFICIAL VEINS. The superficial veins are found within the substance of the superficial fascia immediately beneath the skin. They drain the capillary networks near the surface of the body and return this blood to deeper veins by perforating the deep fascia. Some of the larger superficial veins include the facial vein, which drains the superficial structures of the face, the cephalic and basilic veins of the upper extremity, the thoracoepigastric vein of the anterior trunk, and the great and small saphenous veins of the lower extremity.

DEEP VEINS. The deep veins generally accompany the deeper coursing arteries and usually are enclosed in the same sheaths as the vessels. The larger arteries, such as the axillary, subclavian, popliteal, and femoral, usually have only one accompanying vein. In the case of the smaller arteries, such as the radial, ulnar, brachial, tibial, and peroneal, the veins exist generally in pairs, one lying on each side of the vessel, and are called **venae comites** or **venae comitantes.** In certain organs of the body, however, the deep veins do not accompany the arteries. Examples of these are the veins in the skull and vertebral canal, the hepatic veins in the liver, and the larger veins that return blood from the bones.

VENOUS SINUSES. Throughout the body are veins to which the term *sinus* has been given. At times a sinus is a vessel that does not have the same layers in its walls as other veins, but this definition cannot be applied consistently. Within the cranial cavity of the skull are found venous sinuses that consist of canals formed by a separation of the two layers of the dura mater. Their outer coat consists of fibrous tissue, their inner of an endothelial layer continuous with the lining of the veins. The coronary sinus is the vessel into which drain the veins of the heart. Small venous capillary channels in the liver are called venous sinusoids. The sinus venosus is a venous channel found in the embryonic heart.

The systemic veins may be arranged in four groups: (1) the cardiac veins or veins of the heart; (2) the superior vena cava and its tributaries, which include the veins of the head, neck, upper limbs, and thorax; (3) the inferior vena cava and its tributaries, which include the veins of the lower limbs, some veins of the abdomen, and the veins of the pelvis; and (4) the portal system of veins.

Veins of the Heart

The veins of the heart, or the **cardiac veins,** with the exception of certain small vessels that open directly into the chambers, are tributaries of the coronary sinus.

CORONARY SINUS

The coronary sinus is a wide venous channel, about 2.25 cm in length, situated in the posterior part of the coronary sulcus of the heart, and usually covered by muscular fibers from the left atrium (Fig. 9-7). It ends in the right atrium, between the opening of the inferior vena cava and the atrioventricular aperture; its orifice is guarded by a single semilunar valve, the **valve of the coronary sinus** (*valve of Thebesius*[1]). This valve does not appear to have any functional significance in the adult heart.

Most of the venous blood returning from heart tissue courses in veins that open into the coronary sinus. These tributaries include the:

Great cardiac vein
Small cardiac vein
Middle cardiac vein
Posterior vein of the left ventricle
Oblique vein of the left atrium

Each of these veins, except the last, is provided with a valve at its orifice.

GREAT CARDIAC VEIN. The great cardiac vein begins at the apex of the heart and ascends along the anterior interventricular sulcus to the base of the ventricles (Figs. 8-16; 8-18). It then curves to the left in the coronary sulcus, and reaching the back of the heart, opens into the left extremity of the coronary sinus (Fig. 9-7). It receives tributaries from the left atrium and from both ventricles. One of these, the **left marginal vein,** is of considerable size and ascends along the left margin of the heart (Fig. 9-7).

SMALL CARDIAC VEIN. The small cardiac vein courses in the coronary sulcus between the right atrium and ventricle. It flows into the right extremity of the coronary sinus,

[1] Adam C. Thebesius (1686–1732): A German physician and anatomist.

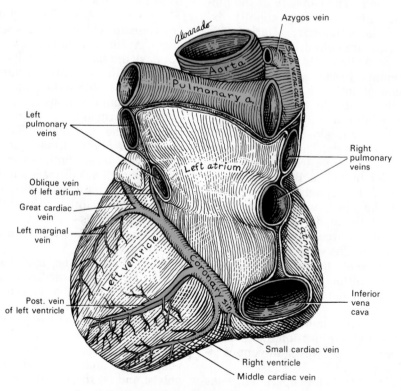

FIG. 9-7. Base and diaphragmatic surface of heart.

near the site where the coronary sinus terminates in the right atrium (Fig. 9-7). The small cardiac vein receives blood from the back of the right atrium and ventricle. The **right marginal vein,** which ascends from near the apex and along the right margin of the heart, joins the small cardiac vein in the coronary sulcus, or opens directly into the right atrium.

MIDDLE CARDIAC VEIN. The middle cardiac vein commences at the apex of the heart, ascends in the posterior interventricular sulcus, receives tributaries from both ventricles, and ends in the coronary sinus near its right extremity (Fig. 9-7).

POSTERIOR VEIN OF THE LEFT VENTRICLE. Ascending along the diaphragmatic surface of the left ventricle, the posterior vein of the left ventricle accompanies the circumflex branch of the left coronary artery. Usually it opens into the coronary sinus to the left of the middle cardiac vein, but it may course farther to the left and open into the great cardiac vein (Fig. 9-7).

OBLIQUE VEIN OF THE LEFT ATRIUM. The oblique vein of the left atrium is a small vessel that descends obliquely on the back of the left atrium and ends in the coronary sinus near its left extremity (Fig. 9-7). It is continuous superiorly with the **ligament of the left vena cava,** and the two structures form the remnant of the left common cardinal vein.

Several cardiac veins do not end in the coronary sinus, but are directed into the chambers of the heart. These are the:

Anterior cardiac veins
Smallest cardiac veins

ANTERIOR CARDIAC VEINS

The anterior cardiac veins consist of three or four small vessels that collect blood from the ventral aspect of the right ventricle and open into the right atrium. The right marginal vein frequently opens into the right atrium, and is therefore sometimes regarded as belonging to this group.

SMALLEST CARDIAC VEINS

These vessels (also called the venae cordis minimi) consist of a number of minute veins that arise in the muscular wall of the heart. Most of them open into the atria, but a few end in the ventricles.

Superior Vena Cava and its Tributaries

Draining into the superior vena cava are venous tributaries that collect blood from the head, neck, upper extremity, the thorax (including the thoracic wall), and even from much of the abdominal wall.

SUPERIOR VENA CAVA

The superior vena cava drains venous blood from the superior half of the body (Fig. 9-8). Its caliber of about 2 cm in diameter is second in size in the human body only to the inferior vena cava, which is slightly larger. The superior vena cava measures about 7 cm in length and is formed by the junction of the two brachiocephalic veins at the level of the first intercostal space, close behind the sternum on the right side. From this junction the superior vena cava transports blood inferiorly, lying deep to the second rib cartilage and intercostal space at the right border of the sternum, and empties into the upper part of the right atrium at the level of the third right costal cartilage. As it descends, the superior vena cava curves gently with its convexity to the right. The part of the vessel nearer the heart, amounting to about half its length, lies within the pericardial sac and it is covered by the serous pericardium.

RELATIONS OF THE SUPERIOR VENA CAVA. *Anterior* to the superior vena cava are the anterior margins of the right lung and pleura, with the pericardium intervening below. These separate it from the first and second intercostal spaces and from the second and third right costal cartilages. *Posterior* to it are the root of the right lung and the right vagus nerve. *On the right side* of the superior vena cava are the phrenic nerve and right pleura, and *on its left side,* is the commencement of the brachiocephalic artery and the ascending aorta, the latter overlapping the large vein. Just before piercing the pericardium, the superior vena cava receives the azygos vein and several small veins from the pericardium and other structures in the mediastinum. The superior vena cava has no valves.

VARIATIONS IN THE SUPERIOR VENA CAVA. At times the brachiocephalic veins open separately into the right atrium. In these cases the right brachioce-

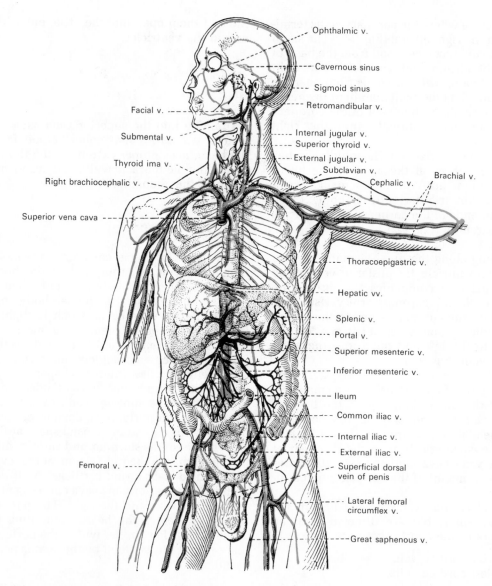

Ophthalmic v.

Cavernous sinus

Sigmoid sinus

Retromandibular v.

Facial v.

Submental v.

Internal jugular v.
Superior thyroid v.

Thyroid ima v.

External jugular v.
Subclavian v.

Right brachiocephalic v.

Cephalic v.

Brachial v.

Superior vena cava

Thoracoepigastric v.

Hepatic vv.

Splenic v.

Portal v.

Superior mesenteric v.

Inferior mesenteric v.

Ileum

Common iliac v.

Internal iliac v.

External iliac v.

Superficial dorsal
vein of penis

Femoral v.

Lateral femoral
circumflex v.

Great saphenous v.

FIG. 9-8.   The large veins of the body. Note that the portal system of veins in the abdomen is represented in black. (From Benninghoff and Goerttler, *Lehrbuch der Anatomie des Menschen*, Urban and Schwarzenberg, 1975.)

phalic vein follows the normal course of the superior vena cava, while the left vein, called the **left superior vena cava,** passes ventral to the root of the left lung, and turning to the dorsum of the heart, ends in the right atrium. This occasional condition in the adult is due to the persistence of the early fetal condition (see Fig. 9-1, *C* and *D*), but is normal in adult birds and some mammals. In some instances of persistent left superior vena cava, a small branch may interconnect the two large veins.

TRIBUTARIES OF THE SUPERIOR VENA CAVA. The two brachiocephalic veins join to form

the superior vena cava (Fig. 9-8). Additionally, the superior vena cava receives the azygos vein and several small veins from the mediastinum and pericardium. Further discussion of the tributaries of the superior vena cava relates to its regional drainage. Thus, the following sections are described sequentially:

Veins of the head and neck
Veins of the upper limb
Veins of the thorax

VEINS OF THE HEAD AND NECK

The veins of the head and neck may be divided into the following groups:

Veins of the face
   Superficial veins of the face
   Deep veins of the face
Veins of the cranium
   Veins of the brain
   Sinuses of the dura mater
   Diploic veins
   Emissary veins
Veins of the neck
   External jugular vein
   Internal jugular vein
   Vertebral vein

Superficial Veins of the Face

Facial
Superficial temporal
Posterior auricular
Occipital
Retromandibular

These veins are tributaries of the internal and external jugular veins.

FACIAL VEIN

The facial vein drains blood from the superficial structures of the face (Figs. 9-9 and 9-10). Its first part, named the **angular vein**,

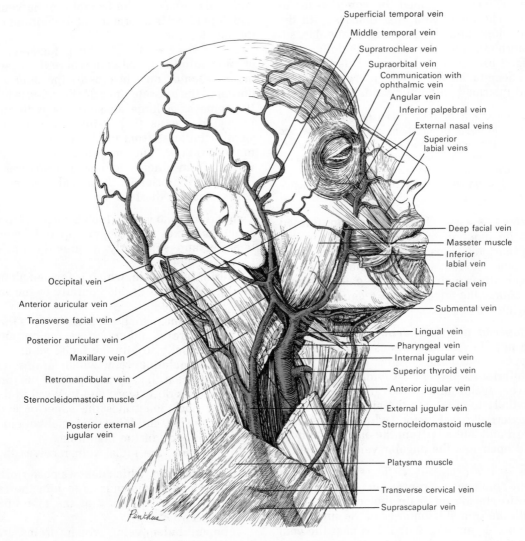

Fig. 9-9. The superficial veins of the head and neck.

begins at the medial angle of the eye by a union of the supratrochlear and supraorbital veins. It accompanies the facial artery along the lateral border of the nose, but has a less tortuous course. Descending on the face, deep to the zygomaticus major and minor muscles, the facial vein follows the border of the masseter muscle to the inferior border of the body of the mandible. It then curves around this bone to enter the neck. In the neck, the facial vein communicates with the anterior jugular vein and with the external jugular before emptying into the internal jugular at about the level of the hyoid bone. Formerly, the communication with the external jugular was named the *posterior facial* and the segment emptying into the internal jugular was named the *common facial vein.* The facial vein anastomoses with the cavernous sinus by way of the angular, supraorbital, and superior ophthalmic veins (Fig. 9-10), and also by way of its tributary, the **deep facial vein,** to the pterygoid plexus and inferior ophthalmic vein.

CLINICAL CORRELATION. The facial vein has no valves. In cases of infections about the nose and mouth, the blood in the vein may become thrombosed. An infected thrombus can progress against the flow of blood, finally reaching the cavernous sinus through its anastomoses and causing meningitis.

ANGULAR VEIN. The angular vein, formed by the junction of the supratrochlear and supraorbital veins at the root of the nose, becomes the facial vein near the level of the zygomaticus major muscle (Fig. 9-9). During its course the angular vein receives the following tributaries:

superior palpebral vein, which arises in the upper eyelid and opens into the lateral aspect of the angular vein;

inferior palpebral vein, which arises in the lower eyelid and passes downward and medially to open into the angular vein; and

external nasal veins, which course upward and laterally from the alae of the nose and open into the angular vein (Fig. 9-9).

SUPRATROCHLEAR VEIN. The supratrochlear vein commences from a venous plexus on the forehead and scalp that communicates with the frontal tributaries of the superficial temporal vein (Fig. 9-9). It lies near the vein of the opposite side in its course to the root of the nose. At this site, the two supra-

trochlear veins communicate by means of a transverse nasal arch before each joins the supraorbital vein of that side. Occasionally the two supratrochlear veins join on the upper forehead to form a prominent single vein that bifurcates at the root of the nose and joins the two angular veins.

SUPRAORBITAL VEIN. The supraorbital vein also begins by anastomosing with a frontal tributary of the superficial temporal vein (Fig. 9-9). It lies superficial to the frontal belly of the occipitofrontalis muscle, which it drains, and joins the supratrochlear vein at the medial angle of the orbit. As it passes over the supraorbital notch in the frontal bone, the supraorbital vein forms an anastomosis with the superior ophthalmic vein, and it also receives the frontal diploic vein, which perforates a foramen at the bottom of the notch.

DEEP FACIAL VEIN. The deep facial vein empties into the facial vein where the zygomaticus major muscle crosses the anterior border of the masseter muscle (Figs. 9-9; 9-10). It begins as a large anastomosis with the pterygoid plexus (Fig. 9-10), and receives small tributaries from the buccinator, zygomaticus major, and masseter muscles, as well as from other neighboring structures.

Other tributaries of the facial vein as it courses across the face are:

superior labial vein, which is formed from a venous plexus on the upper lip and drains laterally and superiorly to join the facial vein (Fig. 9-9);

inferior labial vein, which is formed along the lower lip and drains laterally to the facial vein (Fig. 9-9);

parotid veins, which are small vessels that drain the parotid and masseteric regions and course medially to join the facial vein;

external palatine vein, which drains the soft palate near the superior pole of the tonsil. It courses downward and forward in the tonsillar bed, penetrates the superior constrictor muscle, and joins the facial vein just below the body of the mandible.

In the neck, the facial vein receives the:

submental vein, which courses posteriorly along the mylohyoid muscle to join the facial vein below the body of the mandible (Fig. 9-9); and

submandibular vein, which drains the submandibular gland and courses laterally to join the facial vein.

SUPERFICIAL TEMPORAL VEIN

The superficial temporal vein begins from a broad plexus of veins on the vertex and side of the head, where it anastomoses with the supratrochlear, supraorbital, posterior auricular, and occipital veins (Fig. 9-9). Over the crown of the skull, it anastomoses with the superficial temporal vein of the other side. Anterior and posterior tributaries unite in front of the ear to form the trunk of the superficial temporal vein, which then crosses the posterior root of the zygomatic arch and enters the substance of the parotid gland. Here it unites with the maxillary vein to form the **retromandibular vein.**

TRIBUTARIES OF THE SUPERFICIAL TEMPORAL VEIN. The superficial temporal vein receives the following tributaries:

middle temporal vein, which receives smaller vessels in the lateral orbital region, courses deep to the temporal fascia, and joins the superficial temporal vein just above the zygomatic arch (Fig. 9-9);

transverse facial vein, which courses laterally across the face superficial to the masseter muscle and joins the superficial temporal vein at about the level of the ear lobe (Fig. 9-9); and

anterior auricular vein, which returns blood from the external ear and joins the superficial temporal vein from behind (Fig. 9-9).

POSTERIOR AURICULAR VEIN

The posterior auricular vein begins on the side of the head from a venous plexus that includes anastomoses with the occipital and superficial temporal veins (Fig. 9-9). It descends behind the external ear and joins the retromandibular vein within or below the parotid gland to form the external jugular vein. The posterior auricular vein receives tributaries from the back of the ear and the **stylomastoid vein.**

OCCIPITAL VEIN

The occipital vein begins in a plexus on the posterior part of the scalp, where it anastomoses with posterior auricular and superficial temporal venous tributaries (Fig. 9-9). From this plexus the occipital vein becomes a single vessel that courses with the occipital artery and greater occipital nerve, and with

them pierces the cranial attachment of the trapezius muscle. It enters the suboccipital triangle, and forming a plexus, joins the deep cervical and vertebral veins. The **parietal emissary vein** connects it with the superior sagittal sinus, the **mastoid emissary** vein connects it with the transverse sinus, and through the **occipital diploic vein** it is connected with the confluens sinuum. Occasionally the occipital vein follows the occipital artery as far as the carotid sheath and empties into the internal jugular, or it may join the posterior auricular vein and drain into the external jugular vein.

RETROMANDIBULAR VEIN

The retromandibular vein is formed from the junction of the superficial temporal and maxillary veins (Figs. 9-9; 9-10). It courses through the substance of the parotid gland between the ramus of the mandible and the upper part of the anterior border of the sternocleidomastoid muscle. The retromandibular vein lies superficial to the external carotid artery, but deep to the facial nerve. The retromandibular vein has a large communicating branch with the facial vein that was formerly called the *posterior facial vein,* and the common trunk from these two veins was called the *common facial vein* (Fig. 9-10). It receives small tributaries from the parotid gland and the masseter muscle, and joins the posterior auricular vein to form the beginning of the **external jugular vein.**

Deep Veins of the Face

> Maxillary
> Pterygoid plexus

These veins are tributaries of the internal and external jugular veins.

MAXILLARY VEIN

The maxillary vein is a short trunk that accompanies the first part of the maxillary artery; it passes backward between the neck of the mandible and the sphenomandibular ligament. It is formed by a confluence of the veins in the pterygoid plexus and ends by joining the superficial temporal vein to form the retromandibular vein (Fig. 9-10).

PTERYGOID PLEXUS

The pterygoid plexus is an extensive network of veins situated between the temporalis and lateral pterygoid muscles and between the two pterygoid muscles, and it extends outward between surrounding structures in the infratemporal fossa (Fig. 9-10). It receives tributaries that correspond to branches of the maxillary artery. These include the:

inferior alveolar vein, which drains the lower jaw and lower teeth;

middle meningeal vein, which drains the dura mater located within the cranial cavity;

deep temporal veins, through which the pterygoid plexus anastomoses with the temporal plexus;

masseteric vein, which drains the masseter muscle;

buccal vein, which drains the buccinator muscle;

posterior superior alveolar veins, which drain the upper posterior teeth and gums;

pharyngeal veins, which drain the superior pharyngeal constrictor and the upper pharynx;

descending palatine vein, which drains the palate;

infraorbital vein, which assists in the drainage of the inferior structures of the orbit and the infraorbital region of the face;

vein of the pterygoid canal, which links the pterygopalatine fossa with the cranial cavity; and

sphenopalatine vein, which helps to drain the nasal cavity and nasal septum.

The sphenopalatine vein, accompanying the artery, drains the veins of the nasal sep-

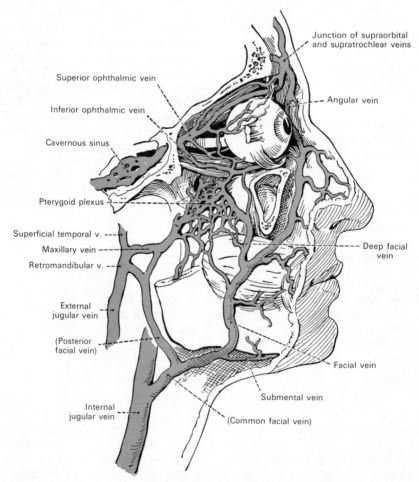

FIG. 9-10. Principal veins of face and orbit. Two labels in parentheses utilize old terminology for which new terms were not included in the *Nomina Anatomica*. (From Eycleshymer and Jones.)

tum and the extensive plexuses of the erectile tissue in the mucous membrane of the nasal conchae. Its tributaries anastomose with the inferior ophthalmic and ethmoidal veins and may communicate through the cribriform plate and foramen cecum of the ethmoid bone with the superior sagittal sinus.

The pterygoid plexus communicates with the cavernous sinus through the foramen of Vesalius, the rete foraminae ovalis, emissary veins of the foramen lacerum, and inferior ophthalmic vein. It communicates with the facial vein through the deep facial and angular veins.

Veins of the Brain

The veins of the brain are tributaries of the venous sinuses that flow into the internal jugular veins. They possess no valves, and their walls, due to the absence of muscular tissue, are extremely thin. As they leave the brain, they pierce the arachnoid membrane and the inner or meningeal layer of the dura mater, and open into the cranial venous sinuses. The veins of the brain do not accompany the arteries. They may be subdivided into the:

Cerebral veins
Cerebellar veins
Veins of the brainstem

CEREBRAL VEINS

The cerebral veins include both **external veins,** which drain the outer surfaces of the brain and empty into the venous sinuses, and the **internal veins,** which drain the inner parts of the cerebral hemispheres and empty into the great cerebral vein (of Galen[1]). The external veins are the **superior, superficial middle,** and **inferior cerebral veins.**

[1]Claudius Galen (120–210 A.D.): A Greek physician, physiologist, and anatomist who practiced in Rome.

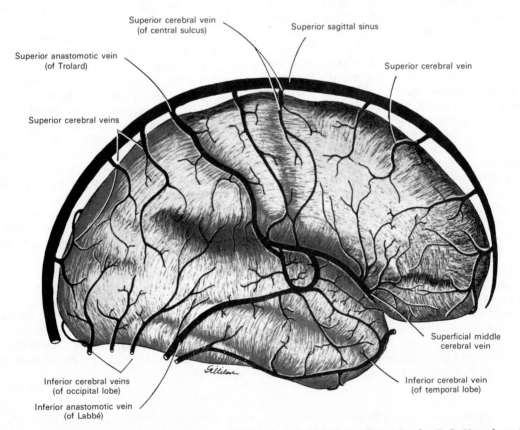

FIG. 9-11. The superficial cerebral veins as seen from the right side of the brain. (From Crosby, E. C., Humphrey, T. and Lauer, E. W., *Correlative Anatomy of the Nervous System,* Macmillan Publishing Co., Inc., New York, 1962.)

Superior Cerebral Veins

The superior cerebral veins, 10 to 16 in number, drain the superolateral and medial surfaces of the hemispheres (Fig. 9-11). Most are lodged in the sulci of the cerebral cortex, but some run across the gyri. All of the superior cerebral veins open into the superior sagittal sinus after a course of one or more centimeters in the dural wall. The anterior veins course nearly at right angles to the sinus, while the posterior and larger veins are directed more obliquely forward and open into the sinus in a direction opposed to the current of the blood contained within it (this may be of little functional significance).

Superficial Middle Cerebral Vein

The superficial middle cerebral vein begins on the lateral surface of the hemisphere and usually ends in the cavernous sinus (Fig. 9-11). It drains most of the lateral surface of the cerebral cortex, and courses in the lateral (Sylvian) sulcus. Although there may be multiple veins of smaller diameter within the sulcus, the superficial middle cerebral vein anastomoses both superiorly toward the midline and inferolaterally over the temporal lobe. It is connected above with the superior sagittal sinus by way of the **superior anastomotic vein** (of Trolard[1]), which then becomes one of the superior cerebral veins that open into the sinus. Inferolaterally, the superficial middle cerebral vein is joined to the transverse sinus by the **inferior anastomotic vein** (of Labbé[2]). Thus, the superficial middle cerebral vein is directly connected with the cavernous, superior sagittal, and tranverse venous sinuses of the skull.

Inferior Cerebral Veins

The inferior cerebral veins are small and drain the inferior surfaces of the hemispheres (Fig. 9-11). Those on the orbital surface of the frontal lobe join the superior cerebral veins, and through these open into the superior sagittal sinus. Others, which drain the temporal lobe, anastomose with the middle cerebral and basal veins, and join the cavernous, sphenoparietal, superior petrosal, and transverse sinuses.

Located more deeply in the brain and draining the internal structures of the cerebral hemispheres are the **great cerebral vein** and its principal tributaries, the **internal cerebral veins,** and the **basal vein.**

Great Cerebral Vein

The great cerebral vein is formed by the union of the two internal cerebral veins (Figs. 9-12; 9-13). It is a short (about 2 cm) median trunk into which flow the veins that drain the deep, medially oriented structures of the basal ganglia, thalamus, hypothalamus, and midbrain. It curves posteriorly and upward around the splenium of the corpus callosum, and ends by joining the inferior sagittal sinus at the anterior extremity of the straight sinus.

INTERNAL CEREBRAL VEINS. The two internal cerebral veins drain the deep parts of the hemisphere (Figs. 9-12; 9-13). Each is formed near the interventricular foramen by the union of the **thalamostriate** and **choroid veins.** The internal cerebral veins run backward, parallel with one another, between the layers of the tela choroidea of the third ventricle, and beneath the splenium of the corpus callosum, where they unite to form the great cerebral vein. Just prior to their union, each internal cerebral vein receives the corresponding basal vein of that side.

THALAMOSTRIATE VEIN. The thalamostriate vein commences in the groove between the corpus striatum and thalamus, and receives numerous veins from both of these regions (Figs. 9-12; 9-13). On each side the thalamostriate vein unites behind the anterior columns of the fornix with the choroid vein, to form the corresponding internal cerebral vein.

CHOROID VEIN. The choroid vein courses along the entire length of the choroid plexus, and receives veins from the hippocampus, the fornix, and the corpus callosum. Ascending to the interventricular foramen (of Monro[3]), the choroid vein is joined by the thalamostriate vein to form the internal cerebral vein.

---

[1]Paulin Trolard (1842–1910): A French anatomist.
[2]Leon Labbé (1832–1916): A French surgeon (Paris).

[3]Alexander Monro II (1733–1817): A Scottish anatomist (Edinburgh).

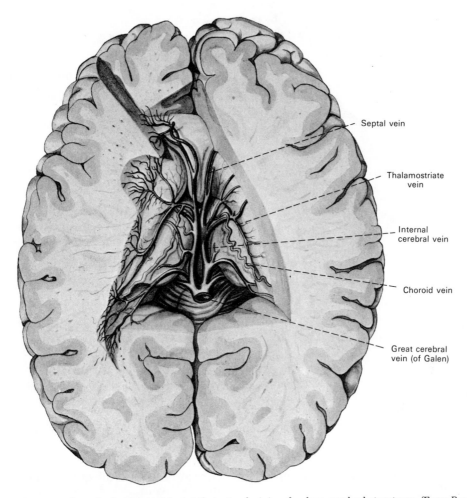

Septal vein

Thalamostriate
vein

Internal
cerebral vein

Choroid vein

Great cerebral
vein (of Galen)

FIG. 9-12. The internal cerebral veins and their tributaries draining the deep cerebral structures. (From Benninghoff and Goerttler, *Lehrbuch der Anatomie des Menschen*, Urban and Schwarzenberg, 1975.)

BASAL VEIN

The basal vein is formed at the anterior perforated substance by the union of a small **anterior cerebral vein** which accompanies the anterior cerebral artery, the **deep middle cerebral vein** which receives tributaries from the insula and neighboring gyri and runs in the lower part of the lateral cerebral sulcus, and the **inferior striate veins** which leave the corpus striatum through the anterior perforated substance (Fig. 9-13). The basal vein passes posteriorly around the cerebral peduncle, and ends either in the internal cerebral vein near its junction with the great cerebral vein, or in the great cerebral vein itself. It receives tributaries from the interpeduncular fossa, the inferior horn of the lateral ventricle, the parahippocampal gyrus, and the midbrain.

CEREBELLAR VEINS

The cerebellar veins are located on the surface of the cerebellum and consist of two sets, superior and inferior. Certain of the **superior cerebellar veins** pass forward and medially, across the superior vermis, and end in the straight sinus or in the great cerebral vein, while others course laterally to the transverse and superior petrosal sinuses. The **inferior cerebellar veins** are larger than the superior cerebellar; some course anteriorly and laterally and end in the superior petrosal or transverse sinuses, while others

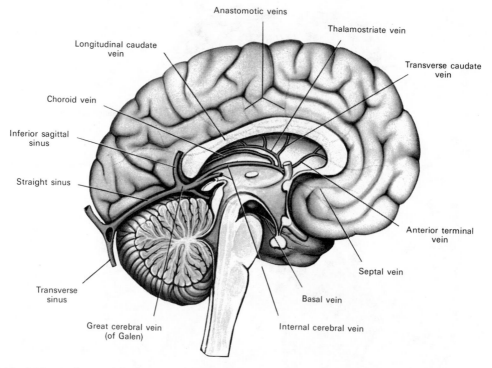

Anastomotic veins

Thalamostriate vein

Longitudinal caudate
vein

Transverse caudate
vein

Choroid vein

Inferior sagittal
sinus

Straight sinus

Anterior terminal
vein

Septal vein

Transverse
sinus

Basal vein

Great cerebral vein
(of Galen)

Internal cerebral vein

Fɪɢ. 9-13. Midsagittal view of the deeply located veins of the cerebrum. Note the relationship of the internal cerebral veins, the great cerebral vein, and the straight sinus. (Adapted from Truex, R. C. and Carpenter, M. B., *Human Neuroanatomy*, 6th ed., Williams and Wilkins, Baltimore, 1969.)

course directly backward and open into the occipital sinus.

VEINS OF THE BRAINSTEM

The veins of the midbrain, pons, and medulla form a superficial plexus which lies deep to the arteries. From the plexus anterior and posterior median longitudinal channels form along the *medulla oblongata*. The anterior vein drains superiorly into the pontine plexus, inferiorly into the vertebral veins and the corresponding veins of the spinal cord, and laterally into radicular veins which leave the cranial cavity with the lower cranial nerves. The posterior vein drains into the inferior petrosal and basilar sinuses, as well as along the cranial nerve rootlets. Over the *pons*, the superficial venous channels form more laterally situated longitudinal veins. These drain laterally into the inferior petrosal sinuses or through the foramina with the cranial nerves. The venous drainage of the *midbrain* is directed

superiorly toward the great cerebral vein or the basal vein.

Sinuses of the Dura Mater

The sinuses of the dura mater are venous channels which drain blood from the brain and the bones that form the cranial cavity and carry it into the internal jugular vein. Situated between the two layers of the dura mater, they are devoid of valves but are lined by endothelium continuous with that of the veins into which they drain. They may be divided into a posterosuperior and an anteroinferior group.

The **posterosuperior group** of intracranial venous sinuses consists of the following:

Superior sagittal sinus
Inferior sagittal sinus
Straight sinus
Transverse sinuses
Sigmoid sinuses
Petrosquamous sinuses
Occipital sinus
Confluence of sinuses

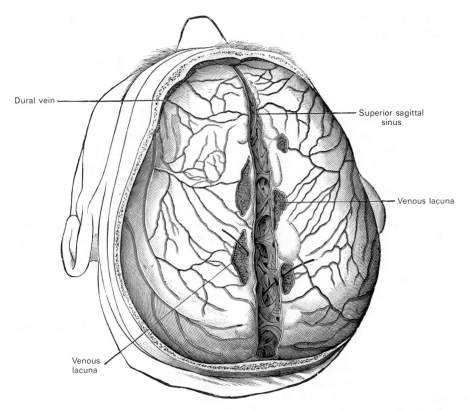

Dural vein

Superior sagittal sinus

Venous lacuna

Venous lacuna

FIG. 9-14. Superior sagittal sinus laid open after removal of the skull cap. The fibrous cords (of Willis) are clearly seen. The venous lacunae are also well shown; from two of them probes have been passed into the superior sagittal sinus. (From Poirier and Charpy.)

SUPERIOR SAGITTAL SINUS

The superior sagittal sinus occupies the attached convex margin of the falx cerebri (Figs. 9-14; 9-15; 9-19). Commencing at the foramen cecum, through which it may receive a vein from the nasal cavity, the sinus is directed posteriorly, grooving the inner surface of the frontal bone, the adjacent margins of the two parietal bones, and the inner aspect of the squamous part of the occipital bone along the superior part of the cruciate eminence. Near the internal occipital protuberance, it deviates to one side, usually the right, and is continued as the corresponding transverse sinus. The superior sagittal sinus is triangular in cross section, and gradually enlarges as it passes posteriorly. Its inner surface presents the openings of the superior cerebral veins. Their orifices are concealed by fibrous folds and numerous fibrous bands (chords of Willis[1]) that extend

transversely across the inferior angle of the sinus (Fig. 9-14). Small openings in the wall communicate with irregularly shaped spaces, called **lateral venous lacunae,** in the dura mater near the sinus. There are usually three lacunae on each side, a small frontal, a large parietal, and an occipital which is intermediate in size. Most of the cerebral veins from the outer surface of the hemisphere open into these lacunae, and numerous **arachnoid granulations** (Pacchionian[2] bodies) project into them. These latter structures, evaginations of arachnoid mesothelium into the sinus, communicate with the subarachnoid space, and provide a pathway by which cerebrospinal fluid can flow directly into the venous blood. The superior sagittal sinus receives the superior cerebral veins and veins from the diploë, and near the posterior extremity of the sagittal suture, also the anastomosing emissary veins from the pericranium, which pass through the

[1]Thomas Willis (1621–1675): An English physician (London).

[2]Antonio Pacchioni (1665–1726): An Italian anatomist (Rome).

parietal foramina, and veins from the dura mater. Numerous anastomoses exist between this sinus and the veins of the nose and scalp.

INFERIOR SAGITTAL SINUS

The inferior sagittal sinus is contained in the posterior half or two-thirds of the free margin of the falx cerebri. This is found in the deepest region of the longitudinal fissure that separates the two hemispheres of the brain, and just above the white commissure, called the corpus callosum, which interconnects the two hemispheres across the midline. The inferior sagittal sinus is cylindrical, increases in size as it passes posteriorly, and ends in the straight sinus (Figs. 9-15; 9-19). It receives several veins from the falx cerebri, and occasionally a few from the medial surface of the hemispheres.

STRAIGHT SINUS

The straight sinus is situated at the line of junction of the falx cerebri and the tentorium cerebelli (Figs. 9-13; 9-15; 9-19). It is triangular in cross section, and it enlarges as it proceeds posteriorly from the end of the inferior sagittal sinus. It opens into the transverse sinus of the side opposite that in which the superior sagittal sinus terminates. Besides receiving blood from the inferior sagittal sinus, the straight sinus receives the great cerebral vein and the superior cerebellar veins. A few transverse fibrous bands cross its interior.

TRANSVERSE SINUSES

The transverse sinuses are large and begin at the internal occipital protuberance (Figs. 9-15; 9-16; 9-19). One, generally the right, is the direct continuation of the superior sagittal sinus, the other of the straight sinus. Each transverse sinus passes horizontally lateralward and then forward, describing a slight curve, to the base of the petrous portion of the temporal bone. In this part of its course, it lies in the attached margin of the tentorium cerebelli. As it leaves the tentorium, it becomes the sigmoid sinus. The transverse sinuses are frequently of unequal size, the one formed by the superior sagittal sinus being the larger. At the base of the petrous portion of the temporal bone, where the transverse sinuses become the sigmoid sinuses, they receive the superior petrosal sinuses. During their course, the transverse sinuses receive some of the inferior cerebral veins and the inferior anastomotic veins from the cerebral cortex, and some of the inferior cerebellar veins, as well as the posterior temporal and occipital diploic veins.

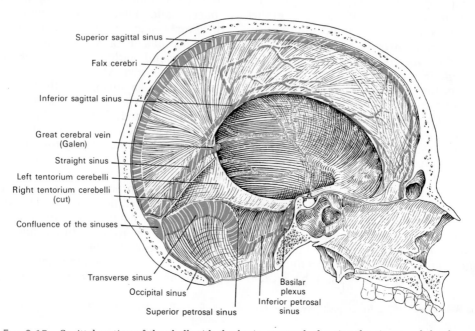

FIG. 9-15.　Sagittal section of the skull with the brain removed, showing the sinuses of the dura.

SIGMOID SINUSES

The sigmoid sinuses are directly continuous with the transverse sinuses and commence where the transverse sinuses emerge from the tentorium cerebelli. Each sigmoid sinus curves inferiorly and medially in an S-shaped manner, in order to reach the posterior part of the jugular foramen, where it ends in the internal jugular vein (Fig. 9-16). In its course it rests initially on the mastoid part of the temporal bone, and then just before its termination, on the jugular process of the occipital bone. As it passes through the jugular foramen, the sigmoid sinus is separated by the vagus and spinal accessory nerves from the inferior petrosal sinus, which emerges through the anterior part of the jugular foramen.

PETROSQUAMOUS SINUSES

The petrosquamous sinuses are found only occasionally. When present, each is located in a groove that courses along the junction of the squamous and petrous parts of the temporal bone. On each side this sinus opens posteriorly into the transverse sinus, close to where the latter becomes the sigmoid sinus. Anteriorly the petrosquamous sinus communicates with the retromandibular vein behind the ramus of the mandible.

OCCIPITAL SINUS

The occipital sinus is one of the smallest of the cranial sinuses (Figs. 9-15; 9-19). It is situated in or near the midline, along the attached margin of the falx cerebelli, and is generally single but occasionally paired. It commences around the margin of the foramen magnum by several small venous channels, one of which joins the terminal part of the sigmoid sinus. Inferiorly it communicates with the internal vertebral venous plexuses, and superiorly it opens into one of the transverse sinuses or, more frequently, into the confluence of sinuses.

CONFLUENCE OF SINUSES

The confluence of sinuses (also known as the torcular Herophili[1]) is the dilated junction of three tributary sinuses, the superior

[1]Herophilus (about 350 B.C.): A Greek physician and anatomist, who worked in Alexandria, Egypt.

sagittal, the straight, and the occipital, with the two large transverse sinuses (Fig. 9-15). It is situated on the inner surface of the cranium, usually just to the right of the internal occipital protuberance. The direction of currents within the confluence is such that the right transverse sinus usually receives the greater part of its blood from the superior sagittal sinus, and the left transverse sinus receives blood from the straight sinus.

The **anteroinferior group** of intracranial venous sinuses includes the following:

Cavernous sinuses
  Ophthalmic veins
Sphenoparietal sinuses
Intercavernous sinus
Superior petrosal sinuses
Inferior petrosal sinuses
Basilar plexus

CAVERNOUS SINUSES

The cavernous sinuses are situated to the right and left of the body of the sphenoid bone, and communicate with each other both in front of and behind the hypophysis (Fig. 9-16). They are about 1 cm wide and are traversed by numerous interlacing fibrous filaments which present them with irregular venous spaces internally, and a trabeculated, spongy character that slows the flow of venous blood. They are larger behind than in front, and each sinus extends about 2 cm from the superior orbital fissure anteriorly to the apex of the petrous portion of the temporal bone posteriorly. Each cavernous sinus drains posteriorly into the superior petrosal and inferior petrosal sinuses.

Along the medial wall of each cavernous sinus courses the internal carotid artery, surrounded by the carotid plexus of sympathetic nerve fibers (Fig. 9-17), and just inferior and lateral to the artery is found the abducens nerve. *On the lateral wall* of the sinus are the oculomotor and trochlear nerves, along with the ophthalmic and maxillary divisions of the trigeminal nerve (Fig. 9-17). These structures are separated from the blood flowing through the sinus by the endothelial lining of the sinus wall.

The cavernous sinus receives the superior and inferior ophthalmic veins, the superficial middle cerebral vein, several of the inferior cerebral veins, and the small sphenoparietal sinus. It communicates with the *transverse sinus* by means of the superior

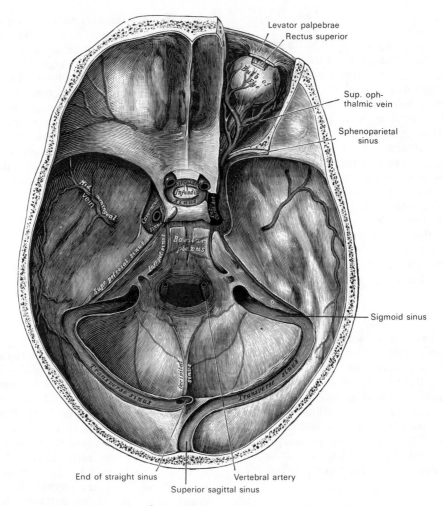

Levator palpebrae
Rectus superior
Sup. oph-
thalmic vein
Sphenoparietal
sinus
Sigmoid sinus
End of straight sinus
Vertebral artery
Superior sagittal sinus

FIG. 9-16.   The venous sinuses at the base of the skull.

petrosal sinus; with the *internal jugular vein* by way of the inferior petrosal sinus; with the *pterygoid plexus* by veins that pass through the sphenoidal emissary foramen (of Vesalius[1]), the foramen ovale, and foramen lacerum; and with the *angular and facial veins* through the superior ophthalmic vein. The two cavernous sinuses anastomose with each other through the *anterior and posterior intercavernous sinuses.*

Superior Ophthalmic Vein

The superior ophthalmic vein begins at the medial angle of the upper eyelid by a confluence of tributaries of the supratrochlear, supraorbital, and angular veins (Figs.

[1]Andreas Vesalius (1514–1564): A great Flemish anatomist who did his finest work in Padua, Italy.

9-16; 9-19). It passes posteriorly through the superior part of the orbit, and receives tributaries that correspond to the branches of the ophthalmic artery. Forming a short single trunk, it passes between the two heads of the

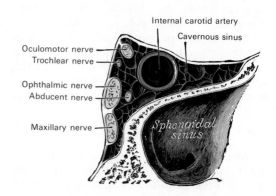

Internal carotid artery
Cavernous sinus
Oculomotor nerve
Trochlear nerve
Ophthalmic nerve
Abducent nerve
Maxillary nerve
Sphenoidal sinus

FIG. 9-17.   Oblique section through the cavernous sinus.

lateral rectus muscle and then through the medial part of the superior orbital fissure below the abducens nerve, and opens into the cavernous sinus.

Inferior Ophthalmic Vein

The inferior ophthalmic vein begins in a venous network on the anterior part of the floor and medial wall of the orbit (Fig. 9-19). It receives veins from the inferior rectus and inferior oblique muscles, as well as from the lacrimal sac and eyelids. Coursing posteriorly in the inferior part of the orbit, it divides into two branches. One of these passes through the inferior orbital fissure and joins the pterygoid venous plexus. The other enters the cranial cavity through the superior orbital fissure and ends in the cavernous sinus, either separately, or more frequently in common with the superior ophthalmic vein.

SPHENOPARIETAL SINUSES

Each of the two sphenoparietal sinuses, one on each side, begins from a small meningeal vein near the lateral tip of the lesser wing of the sphenoid bone. It then courses medially in a groove on the inferior surface of the lesser wing of the sphenoid bone and drains into the anterior part of the cavernous sinus. In addition to receiving a few small meningeal venous tributaries, it usually receives the anterior temporal diploic vein.

INTERCAVERNOUS SINUSES

The anterior and posterior intercavernous sinuses interconnect the cavernous sinuses across the midline. The anterior intercavernous sinus passes in front of the hypophysis, while the posterior passes behind the gland (Fig. 9-16). Together with the cavernous sinuses, they form a venous circle, sometimes called the **circular sinus,** around the hypophysis. The anterior intercavernous sinus is usually the larger of the two, but either one may be absent.

SUPERIOR PETROSAL SINUSES

The superior petrosal sinuses are small and narrow, and on each side they connect the cavernous with the transverse sinus (Figs. 9-15; 9-16; 9-19). Each superior petrosal

sinus courses laterally and backward from the posterior end of the cavernous sinus. Passing over the trigeminal nerve, it lies in the attached margin of the tentorium cerebelli, along a groove in the superior border of the petrous portion of the temporal bone. It joins the transverse sinus where the latter curves downward and becomes the sigmoid sinus on the inner surface of the mastoid part of the temporal. It receives certain of the cerebellar and inferior cerebral veins, and veins from the tympanic cavity.

INFERIOR PETROSAL SINUSES

Each of the two inferior petrosal sinuses is situated in the corresponding inferior petrosal sulcus that is formed by the junction of the petrous part of the temporal bone and the basilar part of the occipital bone (Figs. 9-15; 9-16; 9-19). Commencing at the posteroinferior part of the cavernous sinus, the inferior petrosal sinus passes to the anterior part of the jugular foramen, and it ends in the superior bulb of the *internal jugular vein*. The inferior petrosal sinus receives the **labyrinthine veins,** and also veins from the medulla oblongata, pons, and inferior surface of the cerebellum.

The relationship of the vessels and nerves passing through the jugular foramen is as follows: the inferior petrosal sinus lies anteromedially, with the meningeal branch of the ascending pharyngeal artery, and is directed obliquely downward and backward; the sigmoid sinus is situated at the lateral and posterior part of the foramen, with a meningeal branch of the occipital artery; between the two sinuses are the glossopharyngeal, vagus, and accessory nerves. These three sets of structures are separated by two processes of fibrous tissue. The junction of the inferior petrosal sinus and the internal jugular vein takes place on the lateral aspect of the nerves.

BASILAR PLEXUS

The basilar plexus consists of several interlacing venous channels located between the layers of the dura mater over the basilar part of the occipital bone (Fig. 9-16). It serves to connect the two inferior petrosal sinuses, and it communicates with the anterior vertebral venous plexus.

Diploic Veins

The diploë of the cranial bones contain bone marrow and are drained by venous channels called the diploic veins. These veins are large and at irregular intervals exhibit pouch-like dilatations. Their walls are thin, consisting only of endothelium resting on a layer of elastic tissue. While the cranial bones are separated from one another, these veins are confined to particular bones, but when the sutures are obliterated, the veins unite and generally increase in size. Thus, they are enlarged in the elderly, but virtually nonexistent in the newborn, only gradually developing during early childhood. The diploic veins communicate with the meningeal veins and the sinuses of the dura mater, with the veins of the pericranium, and with each other. Usually the following diploic veins are found in the skull:

Frontal
Anterior temporal
Posterior temporal
Occipital

FRONTAL DIPLOIC VEIN

The frontal diploic vein is found in the anterior part of the frontal bone (Fig. 9-18). Through the inner table of the skull, it communicates with the superior sagittal sinus. The veins of the two sides usually join anteriorly to form a single vessel, which perforates the outer table in the roof of the supraorbital notch and anastomoses with the supraorbital vein.

ANTERIOR TEMPORAL DIPLOIC VEIN

There are usually two anterior temporal diploic veins, one found in front of the coronal suture in the frontal bone, and the other behind the suture in the parietal bone (Fig. 9-18). Internally they communicate with the sphenoparietal sinus, while externally they anastomose with one of the deep temporal veins through an aperture in the great wing of the sphenoid bone.

POSTERIOR TEMPORAL DIPLOIC VEIN

The posterior temporal diploic vein courses downward in the parietal bone toward the mastoid region of the temporal bone (Fig. 9-18). It either perforates the inner table through an aperture at the mastoid angle of the parietal bone, or it courses through the mastoid foramen and terminates in the transverse sinus.

OCCIPITAL DIPLOIC VEIN

The occipital diploic vein, the largest of the diploic veins, is confined to the occipital bone (Fig. 9-18). It opens either externally

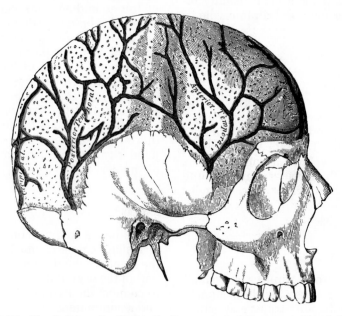

Fig. 9-18. The diploic veins displayed after removal of the outer table of the skull.

into the occipital vein, or internally into the transverse sinus or the confluence of sinuses.

Emissary Veins

The **emissary veins** pass through various foramina and openings in the cranial wall and establish anastomoses between the si-

nuses of the dura inside the skull and the veins on the exterior of the skull (Fig. 9-19). Some of the following emissary veins are always present, others only occasionally.

MASTOID EMISSARY VEIN.   The mastoid emissary vein is usually present. It courses through the mastoid foramen and may tra-

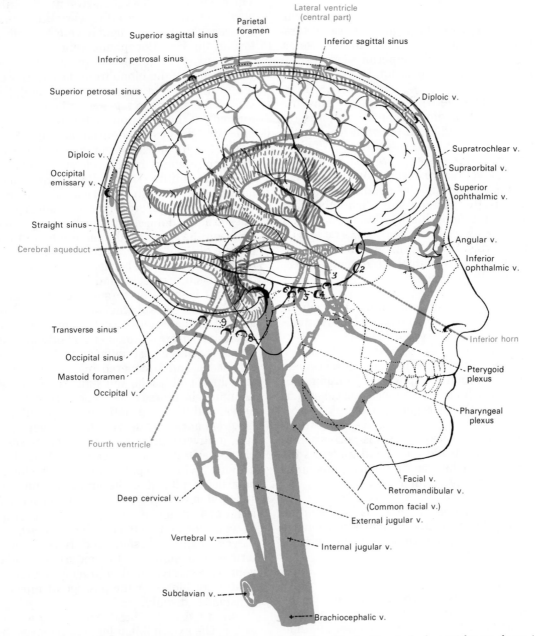

FIG. 9-19.   The veins of the head and neck projected onto the skull and brain. The cerebral ventricles are shown in red. The extracranial portions of the veins are shown in solid blue, while the intracranial portions of the veins are in hatched blue. Numbers identify the foramina through which veins traverse the cranium. 1, Superior orbital fissure. 2, Inferior orbital fissure. 3, Foramen ovale. 4, Foramen spinosum. 5, Foramen lacerum. 6, Carotid canal. 7, Jugular foramen. 8, Hypoglossal canal. 9, Condyloid canal. Black inverted crescents indicate openings through which emissary veins pass. (From Eycleshymer and Jones.)

verse a diploic vein for a short distance, connecting the sigmoid sinus with the posterior auricular or with the occipital vein.

OCCIPITAL EMISSARY VEIN. The occipital emissary vein courses through a small foramen at the occipital protuberance and links the confluence of sinuses with the occipital vein.

PARIETAL EMISSARY VEIN. The parietal emissary vein passes through the parietal foramen, anastomoses with the diploic veins, and interconnects the superior sagittal sinus with the veins of the scalp.

VENOUS PLEXUS OF THE HYPOGLOSSAL CANAL. This network of small veins traverses the hypoglossal canal and joins the sigmoid sinus with the vertebral vein or the internal jugular vein.

CONDYLAR EMISSARY VEIN. The condylar emissary vein is inconstant, but when present, it passes through the condylar canal and connects the sigmoid sinus with the deep veins in the upper neck.

VENOUS PLEXUS OF THE FORAMEN OVALE. This plexus of small veins interconnects the cavernous sinus with the pterygoid plexus through the foramen ovale.

EMISSARY VEINS OF THE FORAMEN LACERUM. Two or three small veins course through the foramen lacerum and connect the cavernous sinus with the pterygoid plexus.

EMISSARY VEIN OF THE SPHENOID FORAMEN. In about half of the skulls, an emissary vein courses through the sphenoidal foramen (of Vesalius), which also connects the cavernous sinus with the pterygoid plexus.

INTERNAL CAROTID VENOUS PLEXUS. An internal carotid plexus of veins accompanies the internal carotid artery and traverses the carotid canal between the cavernous sinus and the internal jugular vein.

EMISSARY VEIN OF THE FORAMEN CECUM. In some specimens a vein is transmitted through the foramen cecum, which connects the superior sagittal sinus with the veins of the nasal cavity.

Other anastomoses between the dural sinus and veins outside the cranial cavity that are not classed as emissary veins are the connection between the cavernous sinus and the facial vein through the superior ophthalmic and angular veins, and the communication between the inferior petrosal sinus and the vertebral veins by way of the basilar plexus.

External Jugular Vein

The **external jugular vein** is one of three large veins that course in the neck and return blood from structures in the cranial cavity, face, and neck. The other two are the **vertebral vein** and the **internal jugular vein**. The external jugular vein flows into the subclavian vein, as does the vertebral vein, whereas the internal jugular vein is a tributary of the brachiocephalic vein.

The external jugular vein receives the greater part of the blood from the exterior of the cranium, including the scalp and the superficial and deep regions of the face. It is formed by the junction of the retromandibular and the posterior auricular veins (Figs. 9-20; 9-22). It commences in the substance of the parotid gland, on a level with the angle of the mandible. Descending perpendicularly in the neck, the external jugular vein obliquely crosses the superficial surface of the sternocleidomastoid muscle, along a line between the angle of the mandible and the middle of the clavicle. At the clavicle, it lies near the posterior border of the sternocleidomastoid muscle, and within the subclavian triangle it perforates the deep fascia and ends in the subclavian vein, lateral or ventral to the anterior scalene muscle. It is separated from the sternocleidomastoid muscle by the investing layer of deep cervical fascia, and is covered by the platysma, the superficial fascia, and skin. The external jugular vein crosses the transverse cervical nerve, and its upper half courses parallel to the great auricular nerve. The vein varies in size, in inverse proportion to the other veins of the neck; it is occasionally doubled. It is provided with two pairs of valves, the lower pair being placed at its entrance into the subclavian vein; the upper pair in most cases is about 4 cm above the clavicle. The portion of vein located between the two sets of valves is often dilated, and is sometimes termed the **sinus of the external jugular vein.** These valves do not prevent the regurgitation of blood, or the passage of injected substances distally.

TRIBUTARIES OF THE EXTERNAL JUGULAR VEIN. The external jugular vein receives the posterior external jugular vein, and near its termination, the anterior jugular, transverse scapular, and suprascapular veins. It also is joined by a large communicating branch from the internal jugular vein, and occasionally it receives the occipital vein.

POSTERIOR EXTERNAL JUGULAR VEIN

The posterior external jugular vein begins in the occipital region and returns blood from the skin and superficial muscles in the upper and posterior part of the neck. It lies between the splenius and trapezius muscles (Fig. 9-23), runs down the posterior part of the neck, and opens into the middle third of the external jugular vein.

Anterior Jugular Vein

The anterior jugular vein begins near the hyoid bone, and is formed by the confluence of several superficial veins from the submandibular region (Fig. 9-20). It descends between the midline of the neck and the anterior border of the sternocleidomastoid muscle. In the lower neck, the anterior jugular vein passes deep to the sternocleidomas-toid muscle and opens into the terminal part of the external jugular, or in some instances, into the subclavian vein (Fig. 9-20). It varies considerably in size, usually in inverse proportion to the size of the external jugular vein. Most frequently there are two anterior jugular veins, a right and left, but sometimes only one. It usually receives tributaries from some laryngeal veins and occasionally from a small thyroid vein, and it communicates with the internal jugular vein. Just above the sternum the two anterior jugular veins are united by a transverse trunk and form the **jugular venous arch,** which receives tributaries from the inferior thyroid veins. The anterior jugular vein contains no valves.

TRANSVERSE CERVICAL VEIN

The transverse cervical vein forms from tributaries that drain from beneath the trapezius (Figs. 9-20; 9-22). It is directed anteromedially, across the posterior triangle of the neck, and opens into the external jugular

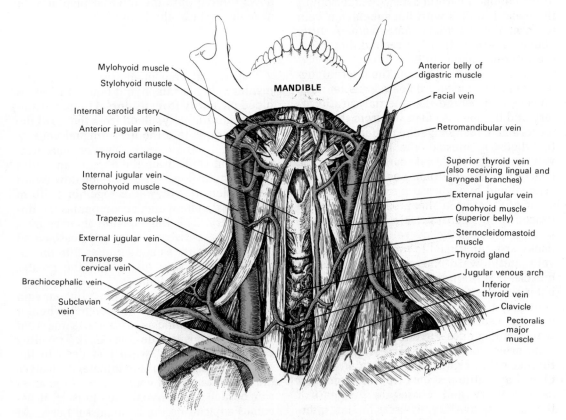

FIG. 9-20. The superficial veins of the neck. The right sternocleidomastoid muscle has been removed to reveal the course of the internal jugular vein. (After Sobotta.)

vein near its termination, but it may drain into the subclavian vein. It usually accompanies the transverse cervical artery.

SUPRASCAPULAR VEIN

The suprascapular (or transverse scapular) vein accompanies the artery of the same name and usually opens into the external jugular near its termination. It may empty into the subclavian vein.

Internal Jugular Vein

The internal jugular vein collects blood from the brain, the superficial parts of the face, and the neck (Figs. 9-20; 9-22). It is directly continuous with the sigmoid sinus in the posterior compartment of the jugular foramen at the base of the skull. At its origin is a dilatation called the **superior jugular bulb.** The internal jugular vein courses downward in the neck, lying at first lateral to the internal carotid artery, and then lateral to the common carotid artery. At the root of the neck, it unites with the subclavian vein to form the brachiocephalic vein. A little above its termination is a second dilatation, the **inferior jugular bulb.**

Superiorly near its origin, the internal jugular vein lies on the rectus capitis lateralis muscle, and behind the internal carotid artery and the nerves emerging from the jugular foramen. *Just above the posterior belly of the digastric muscle,* the internal jugular vein and the internal carotid artery lie nearly in the same plane as the glossopharyngeal and hypoglossal nerves, which pass anteriorly between them. The vagus nerve descends between the vein and the artery, and the accessory nerve runs obliquely backward, superficial to the internal jugular vein and toward the sternocleidomastoid muscle. It is frequently stated that the internal jugular vein is surrounded throughout its course by the carotid sheath. The vein, however, is usually superficial to the sheath, which in fact does surround the internal carotid artery and vagus nerve. *At the root of the neck,* the right internal jugular vein is placed at a little distance from the common carotid artery, and crosses the first part of the subclavian artery, whereas the left internal jugular vein usually overlaps the common carotid artery. The left vein is generally smaller than the right, and each contains a

pair of valves that are placed about 2.5 cm above the termination of the vessel.

TRIBUTARIES OF THE INTERNAL JUGULAR VEIN. During its course, the internal jugular vein receives the inferior petrosal sinus and the facial, lingual, pharyngeal, superior and middle thyroid, and sometimes occipital veins. The thoracic duct on the left side and the right lymphatic duct on the right side open into the angle of union of the internal jugular and subclavian veins.

Because the inferior petrosal sinus and the facial and occipital veins were described earlier in the chapter, they will not be elaborated on further here. However, note that the **inferior petrosal sinus** leaves the skull through the anterior part of the jugular foramen, and joins the superior bulb of the internal jugular vein; the **facial vein** joins the internal jugular vein just below the angle of the mandible; and the **occipital vein** usually joins the posterior auricular vein and flows into the external jugular vein, but at times it follows the course of the occipital artery and flows directly into the internal jugular vein near the level of the hyoid bone.

LINGUAL VEIN

The lingual vein opens into the internal jugular vein near the hyoid bone. It receives blood primarily from its **dorsal lingual** tributaries. Additionally, blood from the tip of the tongue drains into the **deep lingual** vein.

DORSAL LINGUAL VEINS. Two or more dorsal lingual veins drain the dorsum and sides of the tongue (Fig. 9-21). Coursing downward and posteriorly, they join together to form the lingual vein, which accompanies the lingual artery. The lingual vein then courses posteriorly between the genioglossus and hyoglossus muscles and empties into the internal jugular vein just above the greater horn of the hyoid bone.

DEEP LINGUAL VEIN. The deep lingual vein commences at the tip of the tongue, where it forms an anastomosis with the same vein from the opposite side (Fig. 9-21). Traveling downward and backward just deep to the mucous membrane of the tongue, it receives the small **sublingual vein,** and near the anterior border of the hyoglossus muscle, it becomes adjacent to the hypoglossal nerve. At this point it is called the **accompanying vein of the hypoglossal nerve,** also known as the *ranine vein.* Passing backward with the hy-

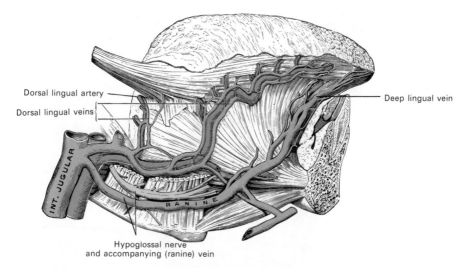

FIG. 9-21.   Veins of the tongue. The hypoglossal nerve has been displaced downward. (From Testut.)

poglossal nerve along the lateral (superficial) border of the hyoglossus muscle, this vein usually opens into the facial vein, but it may join the lingual vein or even open directly into the internal jugular vein.

PHARYNGEAL VEINS

The pharyngeal veins begin in the pharyngeal plexus on the outer surface of the pharynx, and after receiving some posterior meningeal veins and the vein coursing with the artery of the pterygoid canal, they end in the internal jugular vein. They occasionally open into the facial, lingual, or superior thyroid vein.

SUPERIOR THYROID VEIN

The superior thyroid vein begins in the substance and on the surface of the thyroid gland by tributaries that correspond to the branches of the superior thyroid artery, and ends in the middle part of the internal jugular vein, just below the hyoid bone (Figs. 9-20; 9-22). It receives the superior laryngeal and cricothyroid veins. At times the superior thyroid vein opens into the facial vein.

MIDDLE THYROID VEIN

The middle thyroid vein collects blood from the lower lateral part of the thyroid gland, and is joined by some veins from the larynx and trachea. It crosses laterally in front of the common carotid artery and deep to the strap muscles, and opens into the internal jugular vein just above the level of the lower border of the thyroid gland.

Vertebral Vein

The vertebral vein **does not** emerge from the cranial cavity through the foramen magnum with the vertebral artery. Instead, it is formed in the suboccipital triangle from numerous small tributaries that spring from the internal vertebral venous plexuses and leave the vertebral canal above the posterior arch of the atlas (Fig. 9-23). These vessels unite with small veins from the deep muscles in the upper part of the dorsal neck, and form a vessel that enters the foramen in the transverse process of the atlas. Forming a dense plexus around the vertebral artery, they descend in a canal formed by the transverse foramina of the cervical vertebrae. This plexus ends in a single trunk, the vertebral vein, which emerges usually from the transverse foramen of the sixth (but occasionally from the seventh) cervical vertebra. In the root of the neck, the vertebral artery opens into the dorsal part of the brachiocephalic vein near its origin, its opening being guarded by a pair of valves. On the right side, the vertebral vein crosses the first part of the subclavian artery.

TRIBUTARIES OF THE VERTEBRAL VEIN. The vertebral vein communicates with the sigmoid sinus by a vein that passes through the condylar canal, when that canal exists. It also receives tributaries from the occipital

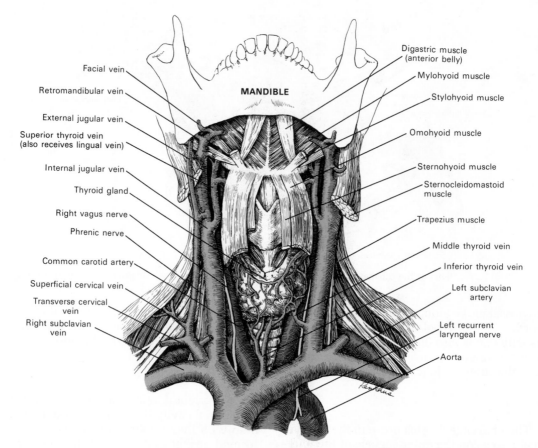

Facial vein

Retromandibular vein

External jugular vein

Superior thyroid vein
(also receives lingual vein)

Internal jugular vein

Thyroid gland

Right vagus nerve

Phrenic nerve

Common carotid artery

Superficial cervical vein

Transverse cervical
vein

Right subclavian
vein

MANDIBLE

Digastric muscle
(anterior belly)

Mylohyoid muscle

Stylohyoid muscle

Omohyoid muscle

Sternohyoid muscle

Sternocleidomastoid
muscle

Trapezius muscle

Middle thyroid vein

Inferior thyroid vein

Left subclavian
artery

Left recurrent
laryngeal nerve

Aorta

FIG. 9-22. The internal jugular system of veins in the deeper neck. Note the superior middle and inferior thyroid veins. (After Sobotta.)

vein, the prevertebral muscles, the internal and external vertebral venous plexuses, and the anterior vertebral and the deep cervical veins. Near its termination it is sometimes joined by the first intercostal vein.

ANTERIOR VERTEBRAL VEIN

The anterior vertebral vein (ascending cervical vein) commences in a plexus around the transverse processes of the cranial cervical vertebrae (Fig. 9-23). It descends in the neck with the ascending cervical artery, between the anterior scalene and longus capitis muscles, and opens into the terminal part of the vertebral vein.

DEEP CERVICAL VEIN

The deep cervical vein accompanies its artery between the semispinalis capitis and cervicis (Fig. 9-23). It begins in the suboccipital region as communicating branches from the occipital vein and as small veins

from the deep muscles at the back of the neck. It receives tributaries from the plexuses around the spines of the cervical vertebrae, and terminates in the inferior part of the vertebral vein.

ACCESSORY VERTEBRAL VEIN

The accessory vertebral vein, when present, arises as a small vein from the venous plexus surrounding the vertebral artery, in the canal formed by the transverse foramina of the cervical vertebrae. It emerges from the transverse foramen of the seventh cervical vertebra, courses anteriorly in the lower neck, and opens into the terminal part of the vertebral vein.

VEINS OF THE UPPER LIMB

The veins of the upper limb can be divided into two sets, **superficial** and **deep.** Distinguished by their topographic position,

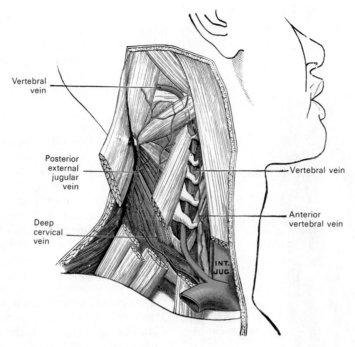

FIG. 9-23.   The vertebral vein and other veins in the neck, lateral view. (From Poirier and Charpy.)

they anastomose with each other at frequent sites, and thus form several parallel channels of drainage from any single region. The superficial veins are distributed immediately beneath the skin, within the superficial fascia. The deep veins accompany the arteries, are the venae comitantes, and usually have the same name as their arteries. Both sets of veins are provided with valves, which are more numerous in the deep veins than in the superficial.

Superficial Veins of the Upper Limb

Two large veins, the cephalic and the basilic, receive blood from the smaller superficial veins of the upper limb, and convey it to the axillary vein and then into the subclavian and brachiocephalic veins (Fig. 9-24). The cephalic vein courses along the radial side of the forearm and the lateral side of the arm. The basilic vein occupies the ulnar aspect of the forearm and medial side of the arm. The **superficial veins of the upper limb** are:

Cephalic
Basilic
Median antebrachial
Venous network of the dorsal hand
Superficial venous palmar arch

CEPHALIC VEIN

The cephalic vein begins on the radial aspect of the dorsal venous network of the hand (Fig. 9-24). As it ascends, it winds around the radial border of the forearm, thereby receiving tributaries from both palmar and dorsal surfaces of the hand. Just below the antecubital fossa, it communicates with the basilic vein by means of the **median cubital vein,** which also receives a branch from the deep forearm veins. The main stem of the cephalic vein then continues proximally, along the lateral side of the antecubital fossa in the groove between the brachioradialis and the biceps brachii muscles, where it crosses superficial to the lateral antebrachial cutaneous branch of the musculocutaneous nerve. Its ascent in the arm continues in the groove along the lateral border of the biceps brachii. In the proximal third of the arm, it passes between the pectoralis major and deltoid muscles, where it is accompanied by the deltoid branch of the thoracoacromial artery. From the interval between these two muscles and the clavicle, called the **deltopectoral triangle,** the cephalic vein passes behind the clavicular head of the pectoralis major muscle. It then pierces the clavipectoral fascia, and crossing the axillary artery, it enters the axillary vein

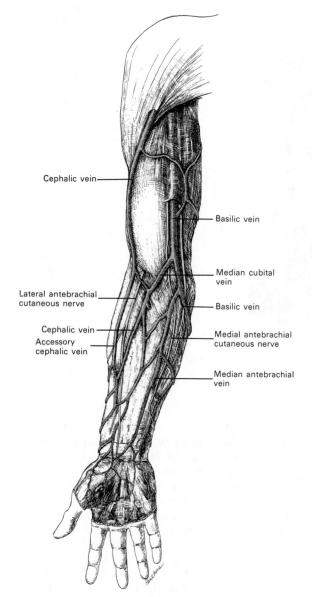

Cephalic vein

Basilic vein

Median cubital vein

Lateral antebrachial cutaneous nerve

Basilic vein

Cephalic vein

Medial antebrachial cutaneous nerve

Accessory cephalic vein

Median antebrachial vein

FIG. 9-24.   The superficial veins of the upper limb. Note the radial or lateral course of the cephalic vein and the ulnar or medial ascent of the basilic vein.

just below the clavicle (Fig. 9-27). Sometimes it communicates with the external jugular vein by a tributary that ascends in front of the clavicle.

ACCESSORY CEPHALIC VEIN.   The accessory cephalic vein arises either from a small tributary plexus on the dorsum of the forearm or from the ulnar side of the dorsal venous network (Fig. 9-24). It remains on the radial side of the cephalic vein and joins it just below the elbow. The accessory cephalic

vein may spring from the cephalic vein, proximal to the wrist, and join it once again higher in the forearm. A large oblique anastomosis frequently connects the basilic and cephalic veins on the back of the forearm.

BASILIC VEIN

The basilic vein begins on the ulnar aspect of the dorsal venous network of the hand. It ascends on the posterior surface of the ulnar side of the forearm, and is inclined toward the anterior surface just below the elbow, where it is joined by the median cubital vein (Fig. 9-24). It then courses obliquely upward in the groove between the biceps brachii and pronator teres and crosses the brachial artery, from which it is separated by the bicipital aponeurosis. Filaments of the medial antebrachial cutaneous nerve pass both in front of and behind this portion of the vein. Continuing upward in the arm, the basilic vein then travels along the medial border of the biceps brachii. It perforates the deep fascia a little below the middle of the arm, and ascending on the medial side of the brachial artery to the lower border of the teres major, it joins the brachial vein to form the **axillary vein** (Fig. 9-27).

MEDIAN ANTEBRACHIAL VEIN

The median antebrachial vein drains the venous plexus on the palmar surface of the hand (Fig. 9-24). It ascends slightly toward the ulnar side of the anterior forearm, and ends either in the basilic vein or in the median cubital vein. At times it divides into two vessels, one of which joins the basilic vein and the other the cephalic vein below the antecubital fossa. The median antebrachial vein also anastomoses with the deep veins of the forearm.

There is great variation in the pattern of the superficial veins of the forearm. Frequently there is a reciprocal relationship in the size of the cephalic and basilic veins. Either one may predominate or even be lacking. The median antebrachial vein may be absent as a definite vessel. The **median cubital vein** may split into a distinct Y, with one arm of the Y draining into the cephalic and the other into the basilic vein. In this case, one branch is called the **median cephalic vein,** the other the **median basilic vein.**

CLINICAL CONSIDERATIONS. The veins of the proximal forearm are usually the ones employed in venipuncture. One of the veins of the median cubital complex regularly has a large vessel anastomosing with the deep veins (Fig. 9-27). This anastomosis holds the superficial vein in place and prevents it from slipping away from the point of the needle.

VENOUS NETWORK OF THE DORSAL HAND

The venous network of the dorsal hand forms initially from **dorsal digital veins** which pass along the sides of the fingers. These are joined to one another by oblique communicating branches, and the digital veins from the adjacent sides of the fingers unite to form three **dorsal metacarpal veins** which end in the venous network on the back of the hand (Fig. 9-25). The *radial part of the network* is joined by the dorsal digital vein from the radial side of the index finger and by the dorsal digital veins of the thumb; it is prolonged proximally as the **cephalic vein.** The *ulnar part of the network* receives the dorsal digital vein of the ulnar side of the little finger, and is continued proximally as the **basilic vein.** An anastomosing vein frequently makes an additional connection with either the cephalic or basilic vein in the middle of the forearm.

SUPERFICIAL VENOUS PALMAR ARCH

The superficial venous palmar arch is more delicate than the venous network of the dorsal hand and is formed by **palmar**

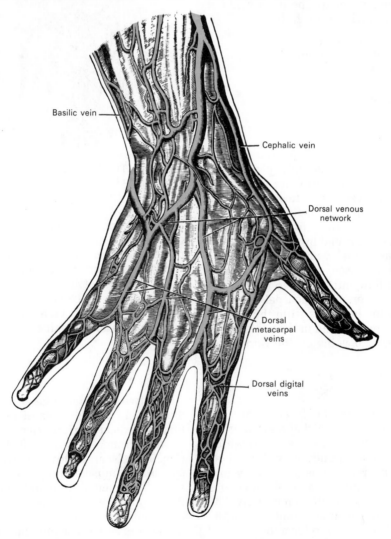

Basilic vein

Cephalic vein

Dorsal venous network

Dorsal metacarpal veins

Dorsal digital veins

FIG. 9-25.   The veins on the dorsum of the hand.

digital veins that drain into venous networks situated over the thenar and hypothenar eminences. These, in turn, flow proximally over the palmar surface of the wrist and help to form the **median antebrachial vein.** The palmar digital veins also are interconnected in the folds between the fingers to the venous network of the dorsal hand by the **intercapitular veins.**

Deep Veins of the Upper Limb

The **deep veins** follow the course of the arteries, accompanying them as their **venae comitantes.** Accompanying veins are generally arranged in pairs, situated on both sides of the corresponding artery, and several short, transverse anastomotic channels generally interconnect them. The deep veins of the upper limb include the following:

Deep veins of the hand
Deep veins of the forearm
Brachial veins
Axillary veins
Subclavian vein

DEEP VEINS OF THE HAND

The superficial and deep palmar arterial arches are each accompanied by a pair of venae comitantes that form the **superficial and deep palmar venous arches** and receive the veins corresponding to the branches of the arterial arches. Thus, the **common palmar digital veins**, formed by the union of the **proper palmar digital veins,** open into the superficial venous arch, and the **palmar metacarpal veins** flow into the deep palmar venous arch. The **dorsal metacarpal veins** receive perforating branches from the palmar metacarpal veins, and end in the radial veins and the superficial veins on the dorsum of the wrist.

DEEP VEINS OF THE FOREARM

The deep veins of the forearm are the venae comitantes of the radial and ulnar arteries and constitute the vessels into which the palmar venous arches flow. The deep venous palmar arch drains principally into the radial veins, while the superficial venous palmar arch drains into the ulnar veins. The radial and ulnar venae comitantes all unite at the bend of the elbow to form the **brachial veins** (Fig. 9-26). The radial veins are smaller

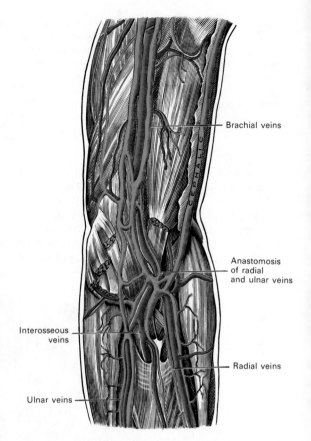

FIG. 9-26. The deep veins of the upper forearm and lower arm in the left upper limb.

than the ulnar, and also receive the dorsal metacarpal veins. The ulnar veins also receive small tributaries from the deep palmar venous arch and communicate with the superficial veins at the wrist. Near the elbow the ulnar veins receive the anterior and posterior interosseous veins and send a large communicating branch to the median cubital vein.

BRACHIAL VEINS

The two brachial veins are placed on the medial and lateral sides of the brachial artery, and receive tributaries that correspond to the branches given off by the artery (Fig. 9-26). Near the lower margin of the subscapularis muscle, they join the axillary vein. The medial brachial vein frequently flows into the basilic vein. During the remainder of its ascent into the axillary vein, the basilic vein takes the place of the medial brachial accompanying vein. These deep veins have numerous anastomoses, not only with each other, but also with the superficial veins.

AXILLARY VEIN

The axillary vein begins at the junction of the basilic and brachial veins, which is located usually near the lower border of the teres major muscle, and it ends at the outer border of the first rib, where it becomes the subclavian vein (Fig. 9-27). In addition to the tributaries that correspond to the branches of the axillary artery, it receives the cephalic vein near its termination. It lies on the medial side of the artery, which it partly overlaps. Between the two vessels are the medial cord of the brachial plexus, the median, the ulnar, and the medial pectoral nerves. Adjacent to the axillary vein are the medial brachial and antebrachial cutaneous nerves, and closely related to the axillary neurovascular bundle are the lateral axillary lymph nodes. The axillary vein is provided with a pair of valves opposite the distal border of the subscapularis muscle, and valves also are found at the ends of the cephalic and subscapular veins.

SUBCLAVIAN VEIN

The subclavian vein is the continuation of the axillary vein, and it extends from the lateral border of the first rib to the sternal end of the clavicle, where it unites with the internal jugular vein and forms the **brachiocephalic vein.** It is in relation *anteriorly* with the clavicle and subclavius muscle, and *posteriorly* and *superiorly* with the subclavian artery, from which it is separated medially by the anterior scalene muscle, and on the right side, by the phrenic nerve. *Inferiorly* it rests in a depression on the first rib and on the pleura. It usually is provided with a pair of valves that are situated about 2.5 cm from its termination. The subclavian vein occasionally ascends in the neck to the level of the third part of the subclavian artery, and at times passes with this vessel behind the anterior scalene muscle.

TRIBUTARIES OF THE SUBCLAVIAN VEIN. The subclavian vein usually receives only the external jugular vein, but sometimes it also receives the anterior jugular vein, and occasionally also a small tributary from the cephalic vein, which ascends ventral to the clavicle. At its angle of junction with the in-

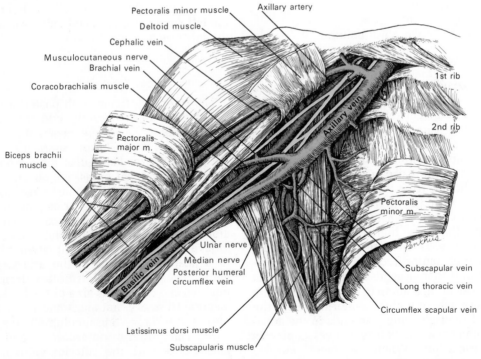

FIG. 9-27.   The veins of the right axilla, showing their relationship to the axillary artery and some of the nerves and muscles of the region.

ternal jugular, the *left subclavian vein* receives the **thoracic duct,** and the *right subclavian vein* receives the **right lymphatic duct.**

VEINS OF THE THORAX

The special vein of the thorax that drains the blood from the thoracic wall and intercostal spaces is the **azygos vein** (Fig. 9-28). It is the first tributary of the superior vena cava, joining the latter as it passes ventral to the root of the lung. The other principal tributaries of the superior vena cava are the **brachiocephalic veins,** two large collecting trunks that receive blood from the head, neck, and upper limbs (Fig. 9-28).

Because the anatomy of the superior vena cava, pulmonary veins, and cardiac veins has already been discussed, only the following **veins of the thorax,** which contribute blood to the superior vena cava, will be described at this point:

Azygos
Brachiocephalic
  Right brachiocephalic
  Left brachiocephalic
    Internal thoracic
    Inferior thyroid
    Superior intercostal
Veins of the vertebral column

Azygos Vein

The azygos vein in the adult is the vessel that persists from the fetal right posterior cardinal and right supracardinal veins (Figs. 9-1; 9-2). It usually commences opposite the first or second lumbar vertebra by a union of the ascending lumbar vein and the right subcostal vein, but it also may be joined at times by a tributary from the right renal vein, or a tributary from the inferior vena cava (Fig. 9-28). The azygos vein may enter the thorax through the aortic hiatus of the diaphragm, or it may pass into the thorax separately, through or behind the right crus of the diaphragm. It then ascends along the right side of the vertebral column to the fourth thoracic vertebra, where it arches anteriorly over the root of the right lung, and ends in the superior vena cava, just before that vessel pierces the pericardium. When the azygos vein traverses the aortic hiatus, it lies on the right side of the aorta, with the thoracic duct. In

the thorax it lies on the right posterior intercostal arteries, to the right of the aorta and thoracic duct, and is partly covered by pleura.

TRIBUTARIES OF THE AZYGOS VEIN. The azygos vein receives the following tributaries:

Right superior intercostal vein
Right fifth to eleventh posterior intercostal veins
Right subcostal vein
Hemiazygos vein
Accessory hemiazygos vein
Esophageal, mediastinal, and pericardial veins

A few imperfect valves are found in the azygos vein, but the valves of its tributaries are complete.

RIGHT SUPERIOR INTERCOSTAL VEIN

Generally the right first intercostal vein drains into the right brachiocephalic vein or into one of its tributaries, such as the vertebral vein or the right subclavian vein. The second, third, and fourth right intercostal veins (and sometimes also the first) join to form a single trunk, called the **right superior intercostal vein,** which usually empties into the azygos vein as the latter arches over the hilum of the right lung. As will be described, the **left superior intercostal vein** drains into the left brachiocephalic vein.

RIGHT FIFTH TO ELEVENTH POSTERIOR INTERCOSTAL VEINS

The right fifth to eleventh posterior intercostal veins drain directly into the azygos vein at their respective segmental levels (Fig. 9-28). The posterior intercostal veins are so named to distinguish them from the anterior intercostal veins, which are tributaries of the internal thoracic veins. Accompanying the intercostal arteries (one in each intercostal space), the intercostal veins lie along the upper border of the space, in a groove along the lower margin of the rib and in a position superior to the artery. Each vein receives a **dorsal tributary** from the muscles and cutaneous area of the back, and a **spinal tributary** that anastomoses with the plexuses of the vertebral column and spinal cord. The posterior intercostal veins also anastomose with the anterior veins of the internal thoracic, and with the lateral thoracic or thoracoepigastric veins.

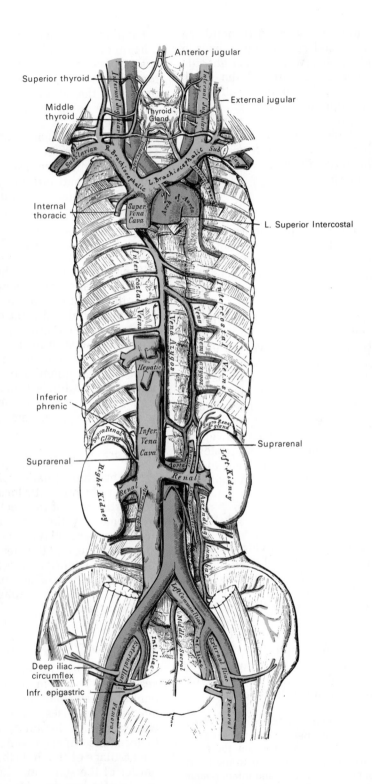

FIG. 9-28. The venae cavae and azygos veins, with their tributaries. This pattern shows an expanded left superior intercostal vein, no accessory hemiazygos vein, and the left seventh posterior intercostal vein (not labeled) draining directly into the azygos vein.

RIGHT SUBCOSTAL VEIN

The right subcostal vein corresponds to the twelfth intercostal vein. It commences in the anterior abdominal wall on the right side and courses posteriorly around the trunk along the lower border of the twelfth rib. Dorsally, as it approaches the right lateral side of the body of the first lumbar vertebra, the right subcostal vein usually joins the right ascending lumbar vein to form the azygos vein. The **left subcostal vein** usually joins the hemiazygos vein.

HEMIAZYGOS VEIN

The hemiazygos vein is even more variable in its origin than the azygos vein, but most frequently it begins by the junction of the left subcostal vein and the ascending lumbar vein (Fig. 9-28). Either of these two vessels alone, however, may form the origin of the hemiazygos vein. It then enters the thorax through the left crus of the diaphragm with the left greater splanchnic nerve, but it may go through the aortic hiatus. Ascending on the left side of the vertebral column as high as the ninth thoracic vertebra, the hemiazygos vein passes horizontally across the vertebral column, dorsal to the aorta, esophagus, and thoracic duct, and ends in the azygos vein. It receives the caudal four or five left posterior intercostal veins, the subcostal vein of the left side, and some esophageal and mediastinal veins.

ACCESSORY HEMIAZYGOS VEIN

The accessory hemiazygos vein descends on the left side of the vertebral column. Typically, it receives the fourth to the eighth left posterior intercostal veins, and either crosses the body of the eighth thoracic vertebra to join the azygos vein or ends in the hemiazygos vein. It varies inversely in size with the left superior intercostal vein; when it is small, or altogether wanting, the left superior intercostal vein may extend as low as the fifth or sixth intercostal space.

CLINICAL NOTE. During obstruction of the inferior vena cava, the azygos and hemiazygos veins, as well as the vertebral veins, form important routes by which venous blood can be returned to the heart. These veins connect the superior and inferior venae cavae and communicate below with the common iliac veins via the ascending lumbar veins and with many of the tributaries of the inferior vena cava.

ESOPHAGEAL, MEDIASTINAL AND PERICARDIAL VEINS

Esophageal veins, draining the thoracic portion of the esophagus, open into the azygos and hemiazygos veins. These veins anastomose with upper esophageal vessels that drain into the left brachiocephalic vein, and with lower esophageal veins that also communicate with branches from the left gastric vein. Many small posterior **mediastinal veins** drain into the azygos and hemiazygos veins. In addition to draining into the internal thoracic and inferior phrenic veins, the **pericardial veins** communicate with the azygos and hemiazygos veins.

BRONCHIAL VEINS

There are usually two bronchial veins on each side, and these vessels return blood from the structures at the roots of the lungs. The **right bronchial veins** open into the azygos vein near its termination, while the **left bronchial veins** drain into the left superior intercostal or the accessory hemiazygos vein. A large quantity of the blood that is carried to the lungs through the bronchial arteries is returned to the left side of the heart through the pulmonary veins.

Brachiocephalic Veins

The brachiocephalic veins are two large trunks, right and left, located one on each side of the root of the neck (Fig. 9-28). They are formed by the junction of the internal jugular and subclavian veins of the corresponding side, and they terminate by uniting to form the superior vena cava. Because the superior vena cava is situated toward the right, the left brachiocephalic vein is longer than the right.

RIGHT BRACHIOCEPHALIC VEIN

The right brachiocephalic vein is a short vessel, about 2.5 cm in length, that begins behind the sternal end of the clavicle (Figs. 9-8; 9-28). It passes almost vertically downward and joins the left brachiocephalic vein to form the superior vena cava just below the cartilage of the first rib, close to the right border of the sternum. The right brachiocephalic vein lies anterior and to the right of the brachiocephalic artery; on its right side are the phrenic nerve and the pleura, which are interposed between it and the apex of the lung. This vein, at its commencement, receives the right vertebral vein, and more caudally, the right internal thoracic and right inferior thyroid veins, and occasionally the vein from the first intercostal space.

LEFT BRACHIOCEPHALIC VEIN

The left brachiocephalic vein measures about 6 cm in length. From deep to the sternal end of the left clavicle, it courses obliquely downward and to the right, behind the upper half of the manubrium sterni (Fig. 9-28). At the sternal end of the first right costal cartilage, it unites with the right brachiocephalic vein to form the superior vena cava. It is separated from the manubrium sterni by the sternohyoid and sternothyroid muscles, the thymus or its remains, and some loose areolar tissue. Dorsal to the left brachiocephalic vein are the large brachiocephalic, left common carotid, and left subclavian arteries which arise from the aortic arch. Also behind the vein are the trachea and the left vagus and phrenic nerves. At times the brachiocephalic vein courses more superiorly and lies in the root of the neck at the level of the jugular notch.

TRIBUTARIES OF THE BRACHIOCEPHALIC VEINS. Both brachiocephalic veins receive as tributaries the vertebral, the internal thoracic, and the inferior thyroid veins. The left brachiocephalic vein also receives the left superior intercostal veins, and occasionally some thymic and pericardiac veins. The right brachiocephalic vein may receive the vein from the first intercostal space. Because the vertebral vein was described with the deep veins of the neck, the following vessels will be considered now also:

Internal thoracic vein
Inferior thyroid veins
Superior intercostal veins

Internal Thoracic Veins

On each side the internal thoracic vein is the accompanying vein of the internal thoracic artery, and it receives tributaries corresponding to the branches of the artery (Fig. 9-28). Uniting to form a single trunk just below the manubriosternal joint, the internal thoracic vein ascends on the medial side of its artery and opens into the corresponding brachiocephalic vein. The **pericardiaco-phrenic vein,** which accompanys the pericardiacophrenic artery, usually flows into the internal thoracic vein.

Inferior Thyroid Veins

Usually there are at least two, but frequently three or four, inferior thyroid veins. These arise in the venous plexus of the thyroid gland and communicate with the middle and superior thyroid veins (Fig. 9-28). At the level of the jugular notch, they form a plexus anterior to the trachea and deep to the sternothyroid muscles. From this plexus a left inferior thyroid vein descends and joins the left brachiocephalic vein, and a right vein passes obliquely downward and to the right across the brachiocephalic artery, and opens into the right brachiocephalic vein, close to its junction with the superior vena cava. Sometimes both inferior thyroid veins open by a common trunk, either into the left brachiocephalic vein or into the superior vena cava. The inferior thyroid veins drain the esophageal, tracheal, and inferior laryngeal veins, and are provided with valves at their terminations.

Superior Intercostal Veins

The right and left superior intercostal veins drain blood from the upper two or three intercostal spaces. The **right superior intercostal vein** passes downward and opens into the azygos vein. The **left superior intercostal vein** runs across the arch of the aorta and the origins of the left subclavian and left common carotid arteries, and opens into the left brachiocephalic vein (Fig. 9-28). It usually receives the left bronchial vein, and sometimes the left pericardiacophrenic vein. It has a prominent anastomosis with the accessory hemiazygos vein, and when the latter is missing, it drains all the upper left spaces.

Veins of the Vertebral Column

The veins that drain blood from the vertebral column, some adjacent musculature, and the meninges of the spinal cord form intricate plexuses that extend along the entire length of the column. These plexuses may be divided into two groups, external and internal, according to whether they are located outside or within the vertebral canal.

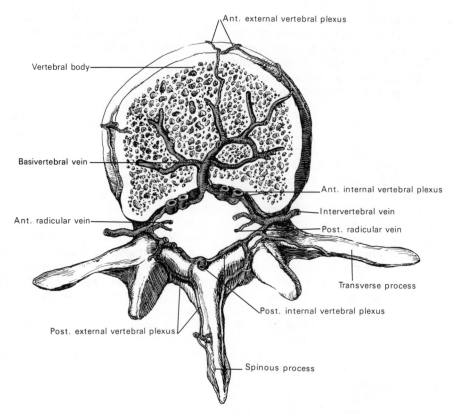

Ant. external vertebral plexus

Vertebral body

Basivertebral vein

Ant. internal vertebral plexus

Intervertebral vein

Ant. radicular vein

Post. radicular vein

Transverse process

Post. internal vertebral plexus

Post. external vertebral plexus

Spinous process

FIG. 9-29. Transverse section through the body of a thoracic vertebra, showing the vertebral venous plexuses. (After Netter.)

The plexuses of the two groups anastomose freely with each other and terminate in the intervertebral veins. The plexuses and veins of the vertebral system are:

External vertebral venous plexuses
Internal vertebral venous plexuses
Basivertebral veins
Intervertebral veins
Veins of the spinal cord

EXTERNAL VERTEBRAL VENOUS PLEXUSES

The external vertebral venous plexuses are best developed in the cervical region and consist of anterior and posterior plexuses which anastomose freely with each other (Figs. 9-29; 9-30). The **anterior external plexuses** lie ventral to the bodies of the vertebrae, communicate with the basivertebral and intervertebral veins, and receive tributaries from the bodies of the vertebrae. The **posterior external plexuses** are placed partly on the dorsal surface of the vertebral arches and their processes, and partly between the deep dorsal muscles. In the cervical region the external vertebral plexuses

anastomose with the vertebral, occipital, and deep cervical veins. More inferiorly along the vertebral column, the external plexuses anastomose with the posterior intercostal veins of the thorax and with the lumbar veins of the posterior abdominal wall.

INTERNAL VERTEBRAL VENOUS PLEXUSES

The internal vertebral venous plexuses lie within the vertebral canal between the dura mater and the vertebrae, and receive tributaries from the bones and the spinal cord (Figs. 9-29; 9-30). They form a closer network than the external plexuses, and course mainly in a vertical direction, forming four longitudinal veins, two anterior and two posterior. The **anterior internal plexuses** consist of two large veins that lie along the posterior surface of the vertebral bodies and intervertebral discs on each side of the posterior longitudinal ligament. Under cover of this ligament, they are connected by transverse branches into which the basivertebral veins open. The two **posterior internal plexuses** are placed on each side of the midline

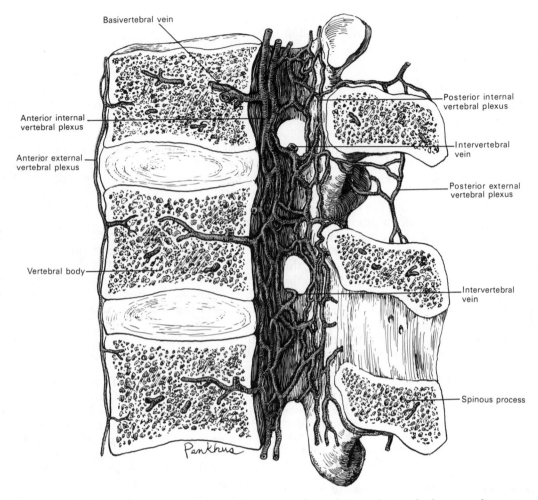

FIG. 9-30.   Median sagittal section of three thoracic vertebrae, showing the vertebral venous plexuses.

ventral to the vertebral arches and ligamenta flava, and they anastomose by veins that pass through these ligaments with the posterior external plexuses. The anterior and posterior plexuses communicate freely with each other within the vertebral canal by a series of **venous rings,** one opposite each vertebra. Around the foramen magnum, the internal plexuses form an intricate network that opens into the vertebral veins and is connected above with the occipital and sigmoid sinuses, the basilar plexus, the condylar emissary vein, and the venous plexus in the hypoglossal canal.

BASIVERTEBRAL VEINS

The basivertebral veins emerge from the foramina on the dorsal surface of the vertebral bodies (Figs. 9-29; 9-30). They are thin-walled vessels contained in large, tortuous

channels in the cancellous tissue of the bodies of the vertebrae, similar in every respect to those found in the diploë of the cranial bones. They communicate through small openings anteriorly and laterally on the bodies of the vertebrae with the anterior external vertebral plexuses. They also converge posteriorly in the vertebral canal as single or doubled vessels that join the transverse branches uniting the anterior internal vertebral plexuses. The openings of the basivertebral veins are valved, and these vessels become greatly enlarged in advanced age.

INTERVERTEBRAL VEINS

The intervertebral veins accompany the spinal nerves through the intervertebral foramina (Figs. 9-29; 9-30). They receive blood from the veins of the spinal cord, drain the internal and external vertebral plexuses, and flow into the vertebral, posterior inter-costal, lumbar, and lateral sacral veins; their orifices are provided with valves.

VEINS OF THE SPINAL CORD

The veins of the spinal cord are situated in the pia mater, within which they form a tor-tuous venous plexus. They emerge chiefly from the median fissures of the cord and are largest in the lumbar region. In this plexus there are two **median longitudinal spinal veins,** one in front of the anterior median fis-sure and the other behind the posterior me-dian sulcus of the cord. Additionally, there are two **anterolateral** and two **posterolateral longitudinal spinal veins** that course along the outer surface of the cord, deep to the dorsal and ventral roots. They drain into the intervertebral veins. Near the base of the skull they unite and form two or three small trunks. These communicate with the verte-bral veins and then end in the inferior cere-bellar veins or in the inferior petrosal si-nuses.

CLINICAL CONSIDERATIONS. Batson (1940) dis-covered that the venous plexuses of the vertebral column constitute a system that parallels the caval system, and that if he injected a radiopaque sub-stance into the dorsal vein of the penis in a cadaver, the substance found its way readily into the veins of the entire vertebral column, the skull, and the inte-rior of the cranium. The material drained from the dorsal vein of the penis into the prostatic plexus and then followed communications with the veins of the sacrum, ilium, lumbar vertebrae, upper femur, and the venae vasorum of the large femoral blood ves-sels, without traversing the main caval tributaries. Similarly, material injected into a small breast vein found its way into the veins of the clavicle, the inter-costal veins, the head of the humerus, cervical verte-brae, and dural sinuses, without following the caval paths. Radiopaque material injected into the dorsal vein of the penis of an anesthetized monkey drained into the caval system when the animal was undis-turbed, but if its abdomen was put under pressure, simulating the increased intra-abdominal pressure of coughing or straining, the material drained into the veins of the vertebrae.

There are certain instances in which metastases from pelvic organ tumors appear earlier in the pelvic bones, bodies of vertebrae, and central nervous sys-tem regions than in the lungs. According to Batson, this occurs during periods of increased abdomino-pelvic pressure (such as coughing, urination, and defecation), when considerable blood returns through the vertebral plexuses rather than through the inferior caval system.

Inferior Vena Cava
and its Tributaries

Draining into the inferior vena cava are most of the veins of the abdomen and pelvis and all of the veins of the lower extremities. Certain veins that receive blood from the lower anterior abdominal wall (inferior epi-gastric vein, superficial epigastric vein, ex-ternal pudendal vein, and superficial cir-cumflex iliac vein) drain inferiorly into the femoral or great saphenous vein, and thereby achieve the inferior vena cava. However, the superficial and deep veins, which drain the upper anterior abdominal and thoracic wall (superior epigastric vein, lateral thoracic vein), as well as the ascend-ing lumbar veins that drain the posterior abdominal wall, flow into tributaries that drain into the superior vena cava.

INFERIOR VENA CAVA

The inferior vena cava returns blood from structures below the diaphragm to the right atrium of the heart (Fig. 9-31). It is formed by the junction of the two common iliac veins, anterior and slightly to the right of the body of the fifth lumbar vertebra. It ascends in front of the vertebral column, on the right side of the aorta, and when it reaches the liver, it lies in a groove on the dorsal surface of that organ. The inferior vena cava then perforates the diaphragm between the me-dian and right portions of its central tendon. Subsequently, it inclines ventrally and me-dially for about 2.5 cm, and after piercing the fibrous pericardium, it becomes covered by a reflection of serous pericardium and opens into the right atrium near the level of the ninth thoracic vertebra. Anterior to its atrial orifice is a semilunar valve, the **valve of the inferior vena cava,** which is rudimen-tary in the adult but is large and exercises an important function in the fetus.

RELATIONS OF THE INFERIOR VENA CAVA. The **ab-dominal portion** of the inferior vena cava is overlaid *anteriorly* at its commencement by the right com-mon iliac artery. It is crossed by the root of the mes-entery and by the right ovarian or testicular artery. It is covered anteriorly by peritoneum in its ascent along the posterior abdominal wall as far as the hori-zontal (third) part of the duodenum. Losing its peri-toneum, the inferior vena cava courses behind the

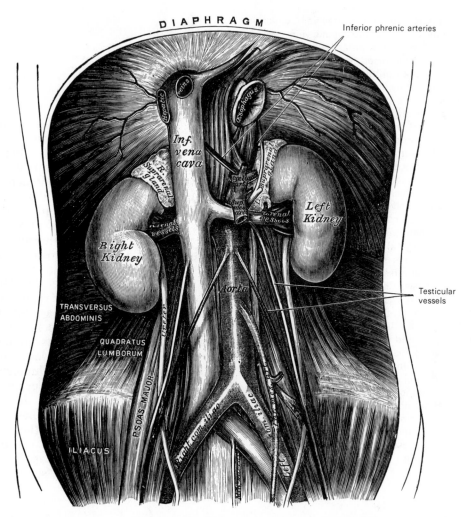

FIG. 9-31. The inferior vena cava and some of its tributaries.

horizontal part of the duodenum, as well as behind the head of the pancreas, the superior (first) part of the duodenum, the common bile duct, and the portal vein. Behind the hilum of the liver, it is again briefly covered by peritoneum until it achieves the posterior surface of the liver, which then partly overlaps and occasionally completely surrounds it. *Posterior* to the inferior vena cava are the bodies of the upper four lumbar vertebrae, the right psoas major muscle, the right crus of the diaphragm, the right inferior phrenic, suprarenal, renal, and lumbar arteries, right sympathetic trunk and right celiac ganglion, and the medial part of the right suprarenal gland. *On the right side,* the inferior vena cava is related to the right kidney and ureter (Fig. 9-31), while *on the left side* it is related to the aorta, right crus of the diaphragm, and the caudate lobe of the liver.

The **thoracic portion** is only about 2.5 cm in length and is situated partly inside and partly outside the pericardial sac. The *extrapericardial part* of the inferior vena cava is separated from the right pleura and lung by the phrenic nerve and by a fibrous band, the **right phrenicopericardiac ligament.** This ligament, often feebly marked, is attached to the margin of the caval opening in the diaphragm and to the pericardium at the root of the right lung. The *intrapericardiac part* of the inferior vena cava is short, and is covered by the serous layer of the pericardium.

VARIATIONS IN THE INFERIOR VENA CAVA. Because the inferior vena cava is formed by a rather complex series of developmental changes, it is not surprising that it has many structural anomalies. Cases have been reported in which there is an **absence of the hepatic or prerenal part of the inferior vena cava.** This results from the failure of the tributaries of the right subcardinal vein to join the veins of the liver between the fifth and sixth prenatal week. In these instances, the postrenal part of the inferior vena cava joins the azygos or hemiazygos vein, which is then large. Thus, except for the hepatic veins, which pass directly to the right atrium, the superior vena cava receives all of the blood from the body before transmitting it to the right atrium.

More frequently, however, anomalies of the inferior vena cava involve the postrenal portion. These include **double inferior vena cava,** which results from persistence of the left subcardinal vein; **persistent left inferior vena cava,** with the absence of the right vessel; **preureteral inferior vena cava,** in which the right ureter passes behind and around the inferior vena cava to reach the bladder; and **persistent renal collar,** which is seen in cases of double venae cavae and involves two left renal veins interconnecting the venae cavae, one passing in front of and the other behind the aorta to form a ring (Gray and Skandalakis, 1972).

COLLATERAL CIRCULATION.   The inferior vena cava below the renal arteries is sometimes ligated because of thrombosis. Channels of anastomosis for collateral circulation under these circumstances include the following: (1) the vertebral veins, (2) anastomoses between the lumbar veins of the inferior vena caval system and the ascending lumbar of the azygos system, (3) anastomoses between the superior, middle, and inferior rectal veins, (4) the thoracoepigastric vein, which connects the superficial inferior epigastric with the lateral thoracic vein, and (5) several anastomoses between vessels in the portal system of veins with vessels in the superior vena caval system, including left gastric-esophageal anastomosis and anastomosis between hepatic and diaphragmatic vessels along the bare area of the liver.

VEINS OF THE ABDOMEN

The veins of the abdomen flow directly into the inferior vena cava, or achieve the inferior vena cava indirectly by draining into the portal system of veins that courses through the liver and out the hepatic veins to the inferior vena cava. Thus, the abdominal veins will be described under two headings: the parietal and posterior abdominal visceral veins, and the portal system of veins.

Parietal and Posterior Abdominal Visceral Veins

The inferior vena cava receives blood from the two common iliac veins and from the following vessels:

Lumbar
Testicular
Ovarian
Renal
Suprarenal
Inferior phrenic
Hepatic

LUMBAR VEINS

The lumbar veins, four on each side, collect blood by dorsal tributaries from the muscles and integument of the lumbar region of the back, and by abdominal tributaries from the walls of the abdomen, where they communicate with the epigastric veins. Adjacent to the vertebral column, they receive veins from the vertebral plexuses, and then pass anteriorly around the sides of the bodies of the vertebrae, posterior to the psoas major muscle, and end in the dorsal part of the inferior vena cava. The left lumbar veins are longer than the right, and pass behind the aorta to reach the inferior vena cava. The lumbar veins are connected by a longitudinal vein, called the **ascending lumbar vein** (Fig. 9-28), which passes anterior to the transverse processes of the lumbar vertebrae. It is most frequently the origin of the azygos vein on the right side and of the hemiazygos vein on the left. These vessels provide anastomoses between the common iliac and iliolumbar veins of the inferior vena caval system, and between the azygos and hemiazygos veins of the superior vena caval system.

TESTICULAR VEINS

The testicular veins emerge from the back of the testis, receive tributaries from the epididymis, and unite to form a convoluted mass of vessels called the **pampiniform plexus,** which constitutes the greater part of the spermatic cord (Fig. 9-32). The numerous vessels composing this plexus ascend along the cord, anterior to the ductus deferens. Just below the superficial inguinal ring, the vessels unite to form three or four veins that pass along the inguinal canal. Entering the abdomen through the deep inguinal ring, the veins coalesce to form two vessels that ascend on the psoas major. Under cover of the peritoneum and lying on either side of the testicular artery, the two veins unite to form a single testicular vein on each side. The **right testicular vein** opens into the inferior vena cava at an acute angle, while the **left testicular vein** flows into the left renal vein

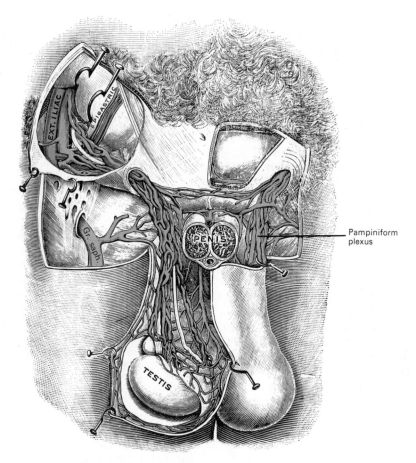

Fig. 9-32.   Testicular veins. (From Testut.)

at a right angle (Figs. 9-28; 9-31). The testicular veins are provided with valves. The left testicular vein passes behind the lower portion of the descending colon, and thus is exposed to pressure from the contents of that part of the bowel. The right testicular vein passes behind the distal part of the ileum and the horizontal or third part of the duodenum on its path to the inferior vena cava.

OVARIAN VEINS

The ovarian veins correspond to the testicular veins in the male. Each forms a plexus in the broad ligament near the ovary and uterine tube, which communicates freely with the uterovaginal plexus of veins. From the ovarian plexus in the broad ligament on each side emerge at least two vessels which unite to form a single vein. These terminate in a manner similar to that of the testicular veins, the **right ovarian vein** opening into the

inferior vena cava, and the **left ovarian vein** flowing into the left renal vein. Valves are usually found in these veins, especially near their junctures with the inferior vena cava and left renal vein. Like the uterine veins, the ovarian veins become much enlarged during pregnancy.

RENAL VEINS

The renal veins are large and pass anterior to the renal arteries (Figs. 9-28; 9-31). The **left renal vein** is longer (8.5 cm) than the right (3.5 cm), and it passes in front of the aorta, just below the origin of the superior mesenteric artery. It receives the left testicular (or ovarian) and left inferior phrenic veins, and usually the left suprarenal vein. The left renal vein, due to the more superior position of the left kidney, opens into the inferior vena cava at a slightly higher level than the right renal vein. The short, thick **right renal vein** courses medially to the inferior vena cava behind the descending (second) part of the duodenum. At times it sends a communicating tributary that helps to form the origin of the azygos vein.

SUPRARENAL VEINS

Usually a single suprarenal vein emerges from each suprarenal gland, even though multiple arteries supply the glands (Fig. 9-28). The short **right suprarenal vein** passes transversely and opens directly into the inferior vena cava on its posterior aspect, while the longer **left suprarenal vein** courses inferomedially, behind the body of the pancreas, and opens into the left renal vein from above.

INFERIOR PHRENIC VEINS

The inferior phrenic veins follow the course of the inferior phrenic arteries along the inferior surface of the diaphragm. The **right** vein ends in the inferior vena cava, while the **left** vein is represented by two branches, one of which opens into the left renal or suprarenal vein, and the other of which passes anterior to the esophageal hiatus in the diaphragm and empties into the inferior vena cava.

HEPATIC VEINS

The hepatic veins commence in the substance of the liver by the coalescence of many **hepatic sinusoids** in the center of the hepatic lobules to form the central or **intralobular veins.** These latter vessels, as they leave the hepatic lobules, join one another to form the **sublobular veins,** which are the tributaries to the hepatic veins. The hepatic veins become arranged in two groups and emerge from the posterior surface of the liver to flow into the inferior vena cava. The **upper group** usually consists of three large veins: left, middle, and right. These converge toward the posterior surface of the liver and open into the inferior vena cava, which courses along its groove on the dorsal part of the liver (Figs. 9-28; 9-31). The veins of the **lower group** vary in number, are small, and come from the right and caudate lobes. The tributaries that form these hepatic veins run singly, are in direct contact with the hepatic tissue, and do not contain valves.

Portal System of Veins

The **portal system** (Fig. 9-33) includes all of the veins that drain blood from the abdominal part of the digestive tube (except the lower part of the rectum) and from the spleen, pancreas, and gallbladder. From these, visceral blood is conveyed to the liver by the **portal vein.** In the liver this vein ramifies like an artery and ends in capillary-like vessels termed **sinusoids,** from which the blood is conveyed to the inferior vena cava by the **hepatic veins.** Thus, the blood in the portal system passes through two sets of minute vessels that allow an exchange of substances: (a) the capillaries of the digestive tube, spleen, pancreas, and gallbladder, and (b) the sinusoids of the liver. In the adult, the portal vein and its tributaries do not have valves, but in the fetus and for a short time after birth, valves can be demonstrated in the tributaries of the portal vein. As a rule these valves soon atrophy and disappear, but in some subjects their remnants can still be observed.

PORTAL VEIN

The portal vein, about 8 cm in length, is formed at the level of the second lumbar vertebra, anterior to the inferior vena cava and behind the neck of the pancreas, by the junction of the superior mesenteric and splenic veins (Fig. 9-33). It passes behind the superior or first part of the duodenum and then ascends in the right border of the lesser omentum to the right portion of the porta hepatis. At this point it divides into a right and a left branch, each accompanying the corresponding branches of the hepatic artery into the substance of the liver. Within the lesser omentum, the portal vein lies posterior to the common bile duct and the hepatic artery, positioned in a plane between the two structures, with the duct oriented more to the right and the artery to the left. The portal vein is surrounded by the hepatic plexus of nerves, and is accompanied by numerous lymphatic vessels and some lymph nodes. The **right branch** of the portal vein courses almost transversely to the right at the porta hepatis, and generally receives the cystic vein from the gallbladder before entering the right lobe of the liver. The **left branch** is longer but of smaller caliber than the right. It ascends more vertically in the porta hepatis, gives branches to the caudate and quadrate lobes, and then enters the left lobe of the liver. In its ascent, the left branch of the portal vein is joined anteriorly by a fibrous cord, the **ligamentum teres,** which represents the remains of the **obliterated umbilical vein.** The left branch is also united with the left hepatic vein, which then flows into the inferior vena cava by a second fibrous cord, the **ligamentum venosum,** which is the remnant of the fetal **ductus venosus.** The fetal umbilical vein is the principal channel by which oxygenated blood is returned to the fetus from the placenta. Blood is shunted to the inferior vena cava by means of the ductus venosus so that placental blood can reach the heart for distribution to the rest of the body.

TRIBUTARIES OF THE PORTAL VEIN. The portal vein forms at the junction of the splenic and superior mesenteric veins, but it also directly receives vessels that drain the lesser curvature of the stomach and the gallbladder. Additionally, communicating veins from the umbilical region of the anterior abdominal wall drain into the portal vein. Its tributaries include the following:

Splenic (lienal)
Superior mesenteric
Left gastric
Right gastric
Prepyloric
Cystic
Paraumbilical

Splenic Vein

The splenic (or lienal) vein commences as five or six tributaries that return blood from the spleen (Fig. 9-33). These unite to form a single vessel located medial to the hilum of the spleen; initially the vessel lies, with the splenic artery and the tail of the pancreas, between the leaves of the phrenicosplenic (lienorenal) ligament. The splenic vein then passes across the posterior abdominal wall, anterior to the upper third of the left kidney and behind the body of the pancreas, which it grooves and from which it receives many small veins. In its course it lies caudal to the lienal artery, and ends posterior to the neck of the pancreas by uniting at a right angle with the superior mesenteric vein to form the portal vein. The splenic vein is large, but is not tortuous like the artery.

TRIBUTARIES OF THE SPLENIC VEIN. The splenic vein receives the short gastric veins, the left gastroepiploic vein, the pancreatic veins, and the inferior mesenteric veins.

Short Gastric Veins. The short gastric veins, four or five in number, drain the fundus and left part of the greater curvature of the stomach (Fig. 9-33). These pass between the two layers of the gastrosplenic ligament and end in the splenic vein or in one of its large tributaries.

Left Gastroepiploic Vein. The left gastroepiploic vein receives tributaries from both the ventral and dorsal surfaces of the stomach and from the greater omentum (Fig. 9-33). It courses from right to left along the greater curvature of the stomach, with the left gastroepiploic artery, and usually flows into the splenic vein near its junction with the superior mesenteric vein.

Pancreatic Veins. The pancreatic veins consist of several small vessels that drain the body and tail of the pancreas and open into the trunk of the splenic vein.

Inferior Mesenteric Vein. The inferior mesenteric vein returns blood from the rectum and from the sigmoid and descending parts of the colon (Fig. 9-33). It begins in the rectum as the **superior rectal vein,** with its origin in the rectal plexus, and through this plexus it communicates with the middle and inferior rectal veins. The superior rectal vein then receives the **sigmoid veins** and leaves the lesser pelvis, accompanied by the superior rectal artery. After ascending across the common iliac vessels, the superior rectal vein receives the large **left colic vein,** which drains the left colic flexure and descending colon, and then continues its ascent as the inferior mesenteric vein. This vein lies to the left of its artery, and ascends under cover of the peritoneum, ventral to the left psoas major muscle. It then passes behind the body of the pancreas and opens into the splenic vein. Sometimes (10% of cases) the inferior mesenteric vein ends in the angle of union of the splenic and superior mesenteric veins, or it drains into the superior mesenteric vein.

Superior Mesenteric Vein

The superior mesenteric vein returns blood from the small intestine, the cecum, and the ascending and transverse portions of

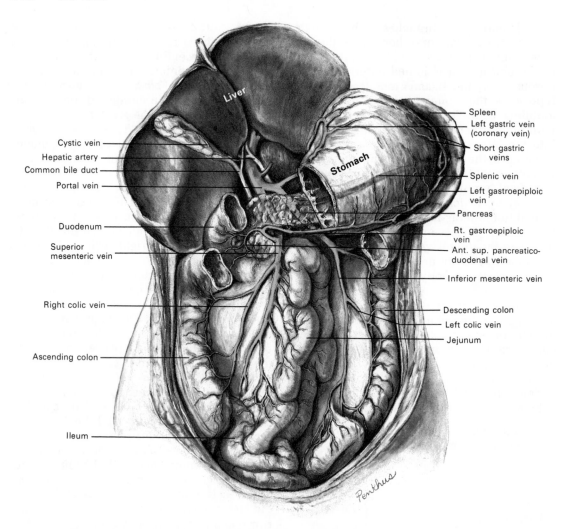

Cystic vein
Hepatic artery
Common bile duct
Portal vein

Liver

Duodenum

Superior
mesenteric vein

Right colic vein

Ascending colon

Ileum

Stomach

Spleen
Left gastric vein
(coronary vein)
Short gastric
veins
Splenic vein
Left gastroepiploic
vein
Pancreas
Rt. gastroepiploic
vein
Ant. sup. pancreatico-
duodenal vein
Inferior mesenteric vein
Descending colon
Left colic vein
Jejunum

FIG. 9-33.   The portal vein and its tributaries. Note that the pyloric end of the stomach and the transverse colon have been removed to reveal the underlying mesenteric veins.

the colon (Fig. 9-33). It begins in the right iliac fossa by the union of the veins that drain the terminal part of the ileum, the cecum, and the vermiform appendix, and it ascends between the two layers of the mesentery on the right side of the superior mesenteric artery. In its upward course, the superior mesenteric vein passes in front of the right ureter, the inferior vena cava, the horizontal part of the duodenum, and the uncinate process of the pancreas. Behind the neck of the pancreas it unites with the splenic vein to form the portal vein.

TRIBUTARIES OF THE SUPERIOR MESENTERIC VEIN.   In addition to the tributaries that correspond to the branches of the superior mesenteric artery (the **jejunal, ileal, ileocolic, right colic,** and **middle colic veins**), the supe-

rior mesenteric vein is joined by the **right gastroepiploic** and the **pancreaticoduodenal veins.**

Right Gastroepiploic Vein.   The right gastroepiploic vein receives tributaries from the greater omentum, parts of both ventral and dorsal surfaces of the stomach, and the anterior superior pancreaticoduodenal vein. It anastomoses with the left gastroepiploic vein and courses from left to right between the anterior two layers of the greater omentum, along the greater curvature of the stomach (Fig. 9-33).

Pancreaticoduodenal Veins.   The superior and inferior pancreaticoduodenal veins correspond to their accompanying arteries and form venous arcades anterior and posterior to the head of the pancreas. The **anterior**

superior pancreaticoduodenal vein drains into the right gastroepiploic vein, while the **posterior superior** pancreaticoduodenal vein flows directly into the portal vein (Fig. 9-33). The **anterior** and **posterior inferior** usually drain singly or as a common trunk into the superior mesenteric vein or its first jejunal branch.

Left Gastric Vein

The left gastric vein frequently is called the *coronary vein* of the stomach, and it courses upward and from right to left along the lesser curvature (Fig. 9-33). It lies between the two layers of the lesser omentum and receives venous tributaries from both the anterior and posterior surfaces of the stomach. As the left gastric vein reaches the esophageal end of the stomach, it receives lower esophageal veins and then turns to the right and courses inferiorly, behind the peritoneum of the lesser omental sac (lesser omentum), and terminates in the portal vein.

Right Gastric Vein

The right gastric vein, at times called the *pyloric vein,* is a small vessel coursing between the two layers of the lesser omentum, along the lesser curvature of the stomach with the right gastric artery. It anastomoses with the left gastric vein midway along the lesser curvature, and it flows toward the right to empty most often into the portal vein. It may, however, terminate in the superior mesenteric vein or even in the right gastroepiploic vein. It is frequently joined by the **prepyloric vein,** which drains the stomach in the region of the pyloric canal.

Cystic Veins

The cystic veins consist of a number of small vessels that vary considerably in their drainage pattern of the gallbladder. As many as six or eight individual vessels flow directly along the attached (to the liver) anterior surface of the organ into the substance of the liver. The vessels draining the exposed serosal or posterior surface join at the neck of the gallbladder to form either single or double cystic veins (Fig. 9-33). These flow along the cystic duct and upward along the hepatic ducts and drain into the liver. Inferiorly, they anastomose with veins that drain the common bile duct, and only rarely do the cystic veins drain directly into the portal vein or its main branches.

Paraumbilical Veins

The paraumbilical veins extend along the ligamentum teres of the liver and the median umbilical ligament. They are small vessels that establish an anastomosis between the veins of the anterior abdominal wall and the portal and internal iliac veins. The best marked of these small veins is one that commences at the umbilicus and courses dorsally and upward, between the layers of the falciform ligament along the surface or within the ligamentum teres, and ends in the left branch of the portal vein.

COLLATERAL CIRCULATION BETWEEN THE PORTAL AND CAVAL VENOUS SYSTEMS. Clinical conditions, such as cirrhosis of the liver, valvular heart disease, or the physical pressure of tumors, may obstruct or diminish the portal circulation through the liver. In these situations several collateral routes become available for the return of portal venous blood: (a) Tributaries of the left gastric veins of the portal system anastomose with tributaries of the lower esophageal veins that drain through the azygos and hemiazygos veins into the superior vena cava. (b) The superior rectal branches of the inferior mesenteric vein (portal system) anastomose with the middle rectal veins and the inferior rectal branches of the internal pudendal veins, which are tributaries of the internal iliac veins (inferior vena caval system). (c) In the umbilical region, small venous tributaries of the portal vein course along the ligamentum teres and the falciform ligament. These paraumbilical vessels anastomose with the superior epigastric and internal thoracic veins (superior vena cava) and with the inferior epigastric veins (inferior vena cava). Varicosities of these paraumbilical veins, radiating from the umbilicus, constitute a clinical condition called *caput medusae.* (d) Veins along the surface of the bare area of the liver (portal system) anastomose with small tributaries of the inferior phrenic veins that drain the inferior surface of the diaphragm (inferior vena cava). (e) Tributaries of intestinal veins that course in the roots of the mesenteries (portal system) anastomose with tributaries of retroperitoneal veins that drain the posterior abdominal wall.

VEINS OF THE PELVIS AND PERINEUM

Venous blood from the organs and much of the pelvic wall and from the muscles and external genitalia in the perineum drains into veins that join on each side to form the

internal iliac vein. This vessel joins its corresponding **external iliac vein** to form the **common iliac vein.** At the junction of the right and left common iliac veins, the inferior vena cava is formed (Figs. 9-28; 9-34).

Common Iliac Veins

The common iliac veins, formed by the union of the internal and external iliac veins, commence anterior to the sacroiliac joint. Passing obliquely upward to the right side of the aorta and the body of the fifth lumbar vertebra, the two common iliac veins end by uniting with each other at an acute angle to form the inferior vena cava. The **right common iliac vein** is shorter than the left, nearly vertical in direction, and ascends dorsal and then lateral to its corresponding artery. The **left common iliac vein** is longer and more oblique in its course, and is situated at first on the medial side of its corresponding artery and later dorsal to the right common iliac artery. Each common iliac vein receives the **iliolumbar vein,** and sometimes the **lateral sacral veins.** Additionally, the left common iliac vein receives the **median sacral vein.** No valves are found in these veins.

VARIATIONS OF THE COMMON ILIAC VEINS. The left common iliac vein, instead of joining the right in its usual position, occasionally ascends on the left side of the aorta as high as the kidney, where, after receiving the left renal vein, it crosses over the aorta and joins the right vein to form the inferior vena cava. In these bodies, the two common iliacs are connected by a small communicating branch at the site where usually they are united. This variation represents a persisting left posterior cardinal or supracardinal vein (Fig. 9-1, *C* and *D*).

Iliolumbar Vein

The iliolumbar vein, coursing on each side with its corresponding artery, commences superiorly in the lower part of the posterior abdominal wall, where it anastomoses with the lower lumbar veins, and laterally in the iliac fossa just below the iliac crest (Fig. 9-34). Its *lumbar* tributary passes downward into the pelvis deep to the psoas major muscle, while its *iliac* tributary drains medially over the iliacus muscle. The two tributaries join to form the iliolumbar vein, which then courses inferiorly and medially and empties into the lateral aspect of the common iliac vein.

Median Sacral Vein

The median sacral vein is formed by the junction of smaller vessels on each side that drain medially within the hollow on the anterior aspect of the sacrum. The vein ascends with its accompanying artery nearly in the midline and ends in the left common iliac vein, sometimes in the angle of junction of the two iliac veins.

INTERNAL ILIAC VEIN

The internal iliac vein, formerly called the *hypogastric vein,* begins near the upper part of the greater sciatic foramen. It passes upward and backward in the pelvis, slightly medial to the corresponding artery. At the brim of the pelvis and anterior to the sacroiliac joint, the internal iliac vein joins the external iliac to form the common iliac vein (Fig. 9-34).

TRIBUTARIES OF THE INTERNAL ILIAC VEIN. With the exception of the iliolumbar vein which usually joins the common iliac vein, the tributaries of the internal iliac vein correspond to the branches of the internal iliac artery. It receives the following:

Superior gluteal
Inferior gluteal
Internal pudendal
Obturator
Lateral sacral
Middle rectal
Dorsal veins of penis and clitoris
Vesical
Uterine
Vaginal

Superior Gluteal Veins

The superior gluteal veins are venae comitantes of the superior gluteal artery. They receive tributaries corresponding with the branches of the artery from the gluteal structures, and enter the pelvis above the piriformis muscle through the greater sciatic foramen. Frequently the venae comitantes unite as a single vessel on each side before ending in the internal iliac vein.

Inferior Gluteal Veins

The inferior gluteal veins are venae comitantes of the inferior gluteal artery. They begin on the proximal part of the posterior thigh, where they anastomose with the me-

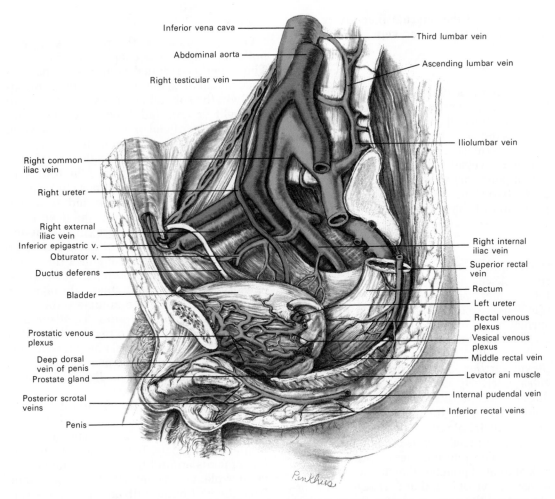

Inferior vena cava

Abdominal aorta

Right testicular vein

Third lumbar vein

Ascending lumbar vein

Iliolumbar vein

Right common iliac vein

Right ureter

Right external iliac vein

Inferior epigastric v.

Obturator v.

Ductus deferens

Bladder

Prostatic venous plexus

Deep dorsal vein of penis

Prostate gland

Posterior scrotal veins

Penis

Right internal iliac vein

Superior rectal vein

Rectum

Left ureter

Rectal venous plexus

Vesical venous plexus

Middle rectal vein

Levator ani muscle

Internal pudendal vein

Inferior rectal veins

FIG. 9-34. The veins in the right half of the male pelvis viewed from the left side. (After Spalteholz.)

dial femoral circumflex and first perforating veins. Entering the pelvis through the lower part of the greater sciatic foramen, below the piriformis muscle, they join to form a single stem that opens on each side into the distal part of the internal iliac vein.

Internal Pudendal Veins

The internal pudendal veins are the venae comitantes of the internal pudendal artery. Venous blood draining the corpora cavernosa penis along the deep dorsal veins usually flows directly posteriorly and enters the prostatic plexus of veins. From this plexus arise the vessels that become the internal pudendal veins, accompanying the internal pudendal artery from this point and then uniting on each side to form a single vessel, which ends in the internal iliac vein. At times the deep dorsal veins of the penis tend

to drain laterally, following more closely the deep arterial supply to the penis. In these instances tributaries of the deep dorsal veins themselves form the origin of the internal pudendal veins. The internal pudendal veins receive tributaries from the crura and bulb of the penis, as well as from the scrotal (or labial) and inferior rectal veins. Frequently the portion of the internal pudendal vein that extends from the penis to the opening of the pudendal canal in the ischiorectal fossa is called the *perineal vein.*

Obturator Vein

The obturator vein begins in the adductor region of the thigh and enters the pelvis through the upper part of the obturator foramen (Fig. 9-34). It courses dorsally and superiorly on the lateral wall of the pelvis below the obturator artery, then passes between the ureter and the internal iliac artery, and ends in the internal iliac vein.

Lateral Sacral Veins

The lateral sacral veins accompany the lateral sacral arteries on the anterior surface of the sacrum and open into the internal iliac vein. As the lateral sacral veins ascend along the posterolateral pelvic wall, they are joined by vessels that emerge through the pelvic (anterior) sacral foramina. The veins of the two sides anastomose and join in the formation of the *sacral venous plexus.*

Middle Rectal Vein

The middle rectal vein originates in the rectal plexus and receives tributaries from the bladder, prostate, and seminal vesicle (Fig. 9-34). It courses laterally on the pelvic surface of the levator ani muscle and ends in the internal iliac vein.

RECTAL VENOUS PLEXUS. The rectal venous plexus surrounds the rectum and communicates anteriorly with the vesical plexus in the male or the uterovaginal plexus in the female (Fig. 9-34). It consists of two parts, an **internal rectal plexus** in the submucosa, and an **external rectal plexus** outside the muscular coat. The internal plexus presents a series of dilated pouches that are connected by transverse vessels and are arranged in a circle around the tubular rectum, immediately above the anal orifice. The lower part of the external plexus is drained by the **inferior rectal veins** into the internal pudendal vein; the middle part drains into the **middle rectal vein** which joins the internal iliac vein; and the upper part is drained by the **superior rectal vein** which forms the commencement of the inferior mesenteric vein, a tributary of the portal vein. A free communication between the portal and systemic venous systems is established through the rectal plexus.

CLINICAL CORRELATION. The veins that form the internal rectal plexus are contained in very loose connective tissue; thus they get less support from surrounding structures than do most other veins, and are less capable of resisting increased blood pressure. Frequently they become dilated and varicosed, forming internal hemorrhoids or piles. Varicosities of the external rectal veins also cause painful external hemorrhoids. Enlargement of these veins can occur when the portal vein is obstructed or in childbearing women.

Dorsal Veins of the Penis and Clitoris

On the dorsum of the penis are located both superficial and deep veins (Fig. 9-35). The **superficial dorsal vein** drains the prepuce and skin of the penis. It courses proximally in the subcutaneous tissue, inclines to the right or left, and opens into the corresponding superficial external pudendal vein, a tributary of the great saphenous vein. The **deep dorsal vein** lies beneath the deep fascia of the penis, but superficial to the tunica albuginea of the corpora cavernosa. It receives blood from the glans penis and from the corpora cavernosa by way of the *deep veins of the penis.* It courses proximally in the midline between the two dorsal arteries. Near the root of the penis it enters the pelvis by passing between the two parts of the suspensory ligament and then through an aperture just below the arcuate pubic ligament of the symphysis pubis. The deep dorsal vein then divides into two branches that enter the **prostatic plexus,** one on each side. The deep dorsal vein also anastomoses with the internal pudendal vein, distal to the symphysis pubis.

The **dorsal vein of the clitoris** drains the erectile tissue of that organ and courses backward into the pelvis in a manner similar to that in the male. It terminates in the venous plexus around the bladder, the **vesical venous plexus.** In contrast, the venous blood from the vestibular bulb and the other structures in the urogenital region of the female, along with the tissue within the ischiorectal fossa and surrounding the anus, drains backward and laterally into the internal pudendal veins. Thus, venous blood from most of the female perineum enters the pelvis along routes comparable to those in the male (Fig. 9-36).

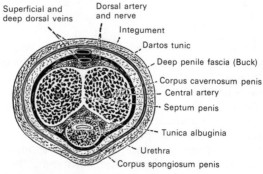

Superficial and deep dorsal veins
Dorsal artery and nerve
Integument
Dartos tunic
Deep penile fascia (Buck)
Corpus cavernosum penis
Central artery
Septum penis
Tunica albuginia
Urethra
Corpus spongiosum penis

FIG. 9-35. The penis in transverse section, showing the blood vessels.

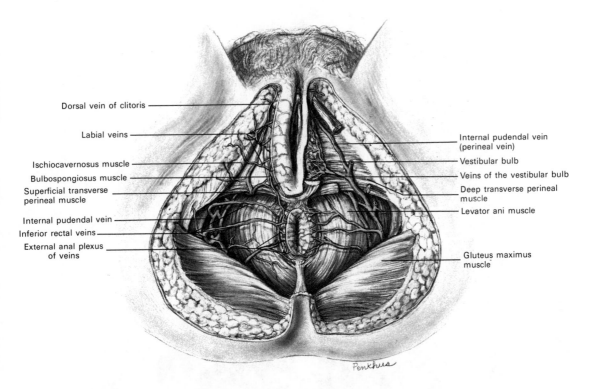

Dorsal vein of clitoris

Labial veins

Ischiocavernosus muscle

Bulbospongiosus muscle

Superficial transverse perineal muscle

Internal pudendal vein

Inferior rectal veins

External anal plexus of veins

Internal pudendal vein (perineal vein)

Vestibular bulb

Veins of the vestibular bulb

Deep transverse perineal muscle

Levator ani muscle

Gluteus maximus muscle

FIG. 9-36.   The veins of the anal and urogenital regions of the female perineum.

PROSTATIC VENOUS PLEXUS.   The prostatic venous plexus lies behind the arcuate pubic ligament and the symphysis pubis, and in front of the bladder and prostate. Its chief tributary is the deep dorsal vein of the penis, but it also receives tributaries from the bladder and prostate. It anastomoses with the vesical plexus, the internal pudendal vein, and the vertebral veins, and drains into the vesical and internal iliac veins. The prostatic plexus lies partly in the fascial sheath of the prostate and partly between the sheath and the prostatic capsule.

Vesical Veins

The vesical veins envelop the caudal part of the bladder and the base of the prostate. They form the **vesical venous plexus** and anastomose with the prostatic plexus in the male or with the uterine and vaginal plexuses in the female. The vesical venous plexus drains by means of several vesical veins that frequently unite into one or two vessels on each side before entering the internal iliac vein.

Uterine Veins

A complex network of uterine veins courses along the sides of the uterus between the two layers of the broad ligament (Fig. 9-37). The veins form the **uterine venous plexuses** and anastomose with the ovarian and vaginal plexuses. From the inferior part of the uterine plexuses, at the level of the external orifice of the cervix, arises a pair of uterine veins on each side. These course laterally and posteriorly and open into the corresponding internal iliac vein.

Vaginal Veins

The venous drainage of the vagina occurs by means of **vaginal plexuses** along the sides of the vagina (Fig. 9-37). These are in continuity with the uterine plexuses and they also communicate with the vesical and rectal plexuses. The vaginal plexuses are drained by at least one, and at times two vaginal veins on each side that flow into the internal iliac veins either directly or through connections with the internal pudendal veins.

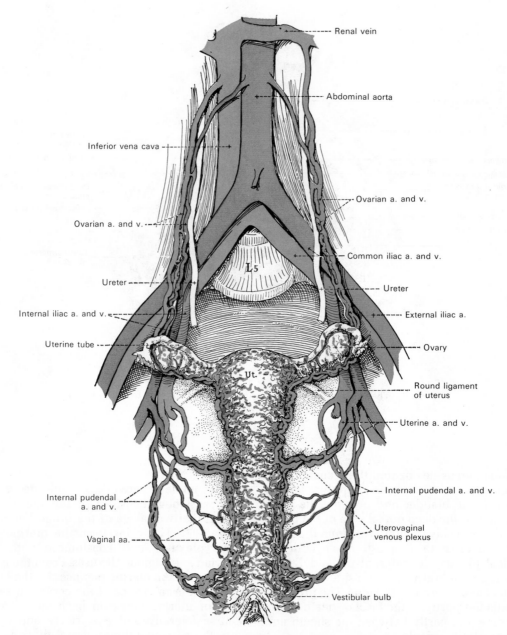

Fig. 9-37. The blood vessels of the female pelvis, showing the chief source of blood supply to the uterus and vagina. (From Eycleshymer and Jones.)

EXTERNAL ILIAC VEIN

The external iliac vein is the upward continuation of the femoral vein and therefore receives blood from the lower limb (Figs. 9-28; 9-34). It also drains the lower part of the anterior abdominal wall. The femoral vein becomes the external iliac vein at the level of the inguinal ligament. Passing superiorly along the brim of the lesser pelvis, the external iliac vein ends opposite the sacroiliac joint by uniting with the internal iliac vein to form the common iliac vein. *On the right side*, it lies at first medial to the external iliac artery, but as it passes upward it gradually inclines behind the artery. *On the left side*, it lies altogether on the medial side of the artery. It frequently contains one, sometimes two, valves.

TRIBUTARIES OF THE EXTERNAL ILIAC VEIN. As it ascends, the external iliac vein receives the following veins:

Inferior epigastric
Deep circumflex iliac
Pubic

Inferior Epigastric Vein

The inferior epigastric vein is formed by the union of the venae comitantes of the inferior epigastric artery. At their source superiorly, the venae comitantes anastomose with the superior epigastric vein. The inferior epigastric vein joins the external iliac vein about 1.25 cm above the inguinal ligament.

Deep Circumflex Iliac Vein

The deep circumflex iliac vein is formed by the union of the venae comitantes of the deep circumflex iliac artery, and joins the external iliac vein about 2 cm above the inguinal ligament.

Pubic Vein

The pubic vein interconnects the obturator vein in the obturator foramen and the external iliac vein. It ascends on the inner aspect of the pubis and accompanies the pubic branch of the obturator artery. In about 33% of cases, it is enlarged, functionally taking the place of the obturator vein. In these instances it becomes a dangerous vessel when operations are made in the region of the femoral ring.

VEINS OF THE LOWER LIMB

The veins of the lower limb are subdivided into superficial and deep sets in a manner similar to those of the upper limb. The **superficial veins** course in the superficial fascia just beneath the skin, while the **deep veins** accompany the arteries. Both sets of veins are provided with valves, which are more numerous in the deep veins than in the superficial. A greater number of valves is also found in the veins of the lower limb than in veins of the upper limb.

Superficial Veins of the Lower Limb

The superficial veins of the lower limb are the **great** and **small saphenous veins,** their tributaries, and the **superficial veins of the foot.**

GREAT SAPHENOUS VEIN

The great saphenous vein is the longest vein in the body, beginning in the superficial vein that courses along the dorsomedial margin of the foot. It terminates in the femoral vein about 3 cm below the inguinal ligament (Fig. 9-38). It ascends anterior to the tibial malleolus and along the medial side of the leg together with the saphenous nerve. Passing upward behind the medial condyles of the tibia and femur, it then ascends along the medial side of the anterior thigh to a point about 3 cm below the inguinal ligament, where it dips through the **saphenous hiatus** (fossa ovalis), penetrates the lower end of the femoral sheath, and opens into the femoral vein.

CLINICAL CORRELATION. The great saphenous vein contains between 10 and 20 valves. These are more numerous in the leg than in the thigh. Frequently this vein and its branches become dilated and then varicosed. This results when valves that normally prevent the flow of venous blood from the long deep veins to those more superficial in the lower limb become incompetent. Dangerous thrombi may develop, especially from the portion of the great saphenous vein which extends from the midregion of the calf to the upper third of the thigh.

TRIBUTARIES OF THE GREAT SAPHENOUS VEIN. *At the ankle,* the great saphenous vein receives tributaries from the sole of the foot by way of the medial marginal vein. *In the leg,* it anastomoses freely with the small saphenous vein, communicates with the anterior and posterior tibial veins from deeper in the leg, and receives many cutaneous veins. *In the thigh,* it anastomoses with the femoral vein and receives numerous tributaries. Those from the medial and posterior parts of the thigh frequently unite to form a large **accessory saphenous vein** that joins the main vein at a variable level. *Near the saphenous hiatus* it is joined by the **superficial external**

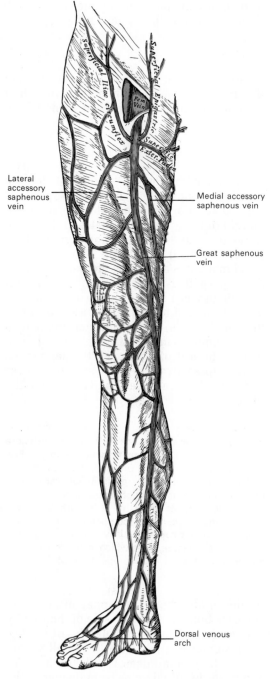

Lateral accessory saphenous vein

Medial accessory saphenous vein

Great saphenous vein

Dorsal venous arch

FIG. 9-38. The great saphenous vein and its tributaries.

pudendal, **superficial circumflex iliac,** and **superficial epigastric veins** (Fig. 9-38). Additionally, through the superficial epigastric vein it forms an anastomosis with veins of the upper limb by means of the **thoraco-epigastric vein.** Of these tributaries the following will be described further:

Accessory saphenous
Superficial external pudendal
Superficial circumflex iliac
Superficial epigastric
   Thoracoepigastric

Accessory Saphenous Vein

The accessory saphenous vein is the name frequently given to the superficial vein that arises along the posteromedial border of the thigh. It ascends around the medial aspect of the thigh, receiving tributaries primarily from the posterior surface, and joins the saphenous vein on its medial aspect. At times this vessel is called the **medial accessory saphenous vein,** to differentiate it from the **lateral accessory saphenous vein,** which arises anteriorly just above the knee (Fig. 9-38). Receiving tributaries from the lateral and anterior surfaces of the thigh, the lateral accessory saphenous vein ascends on the front of the thigh and opens into the saphenous vein on its lateral aspect (Daseler et al., 1946).

Superficial External Pudendal Vein

The superficial external pudendal vein forms from tributaries that laterally drain the superficial fascia from the upper medial thigh, the lower medial inguinal region, and the scrotum (or labia majora) (Fig. 9-38). It usually (90% of cases) opens into the great saphenous vein, but at times (10%) flows into the medial accessory saphenous vein.

Superficial Circumflex Iliac Vein

The superficial circumflex iliac vein forms from tributaries that drain the superficial fascia over the uppermost lateral surface of the anterior thigh, extending even from the anterior abdominal wall above the lateral part of the inguinal ligament (Fig. 9-38). The vein descends medially in an oblique course and opens directly into the great saphenous vein (60% of cases) or into the lateral accessory saphenous vein (40%).

Superficial Epigastric Vein

The superficial epigastric vein arises from tributaries that drain the superficial fascia over a rather wide region of the lower anterior abdominal wall (Fig. 9-38). These tributaries descend directly from the pubic region or obliquely from the more lateral inguinal region, and become a single vessel that opens into the lateral (80% of cases) or medial (20%) aspect of the great saphenous vein or into the corresponding accessory saphenous vein. The superficial epigastric vein anastomoses by way of the **thoracoepigastric vein** with vessels draining superiorly over the thoracic wall.

THORACOEPIGASTRIC VEIN.   The thoracoepigastric vein is a longitudinally oriented vessel whose tributaries form an anastomo-

sis over the anterior abdominal and thoracic walls (Fig. 9-39). It connects venous channels that drain into the axillary vein, usually by means of the lateral thoracic vein, with others that drain into the superficial epigastric and superficial circumflex iliac veins, which flow into the great saphenous-femoral venous complex. Normally the vein is a relatively small vessel that simply contributes to the drainage of the superficial fascia. However, the thoracoepigastric vein provides an important collateral channel in clinical conditions that obstruct either the inferior vena

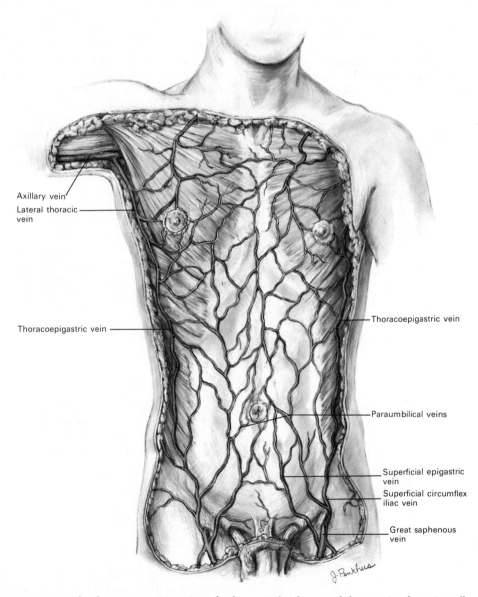

Axillary vein
Lateral thoracic vein

Thoracoepigastric vein

Thoracoepigastric vein

Paraumbilical veins

Superficial epigastric vein
Superficial circumflex iliac vein

Great saphenous vein

FIG. 9-39.   The thoracoepigastric vein and other superficial veins of the anterior thoracic wall.

cava or the portal vein. Under these circumstances this vein can become greatly enlarged and tortuous.

SMALL SAPHENOUS VEIN

The small saphenous vein begins behind the lateral malleolus as a continuation of the marginal vein that courses along the dorsolateral aspect of the foot. At first it ascends along the lateral border of the calcaneal tendon and then crosses this tendon to reach the middle of the back of the leg. Running directly upward, it perforates the deep fascia in the distal part of the popliteal fossa, and ends in the popliteal vein between the two heads of the gastrocnemius muscle (Fig. 9-40). It anastomoses with the deep veins on the dorsum of the foot, and receives numerous large tributaries from the superficial fascia on the back of the leg. Before the small

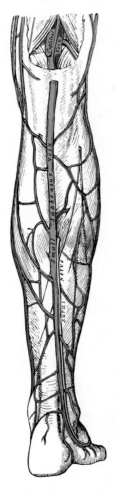

FIG. 9-40.   The small saphenous vein.

saphenous vein pierces the deep fascia, it joins a vessel that courses superiorly and anteriorly to anastomose with the great saphenous vein. In its ascent, the small saphenous vein possesses from 9 to 12 valves, one of which is always found near its termination in the popliteal vein. In the distal third of the leg the small saphenous vein is accompanied by the sural nerve, and in the proximal two-thirds by the medial sural cutaneous nerve.

SUPERFICIAL VEINS OF THE FOOT

On the *dorsum of the foot,* the **dorsal digital veins** communicate with the plantar digital veins in the clefts between the toes. The adjacent dorsal digital veins join to form the **dorsal metatarsal veins.** These latter vessels unite across the distal ends of the metatarsal bones in a **dorsal venous arch** (Fig. 9-38). Proximal to this arch is an irregular venous network that receives tributaries from the deep veins. This network is joined at the sides of the foot by **medial** and **lateral marginal veins,** which are formed mainly by the union of branches from the superficial parts of the sole of the foot.

On the *sole of the foot,* the superficial veins form a **plantar cutaneous venous arch** that extends across the roots of the toes and opens at the sides of the foot into the medial and lateral marginal veins. Proximal to this arch is a **plantar cutaneous venous network** that is especially dense in the fat beneath the heel. This network communicates with the cutaneous venous arch and with the deep veins, but drains chiefly into the medial and lateral marginal veins.

Deep Veins of the Lower Limb

The deep veins and their tributaries in the lower limb accompany the arteries and their branches, and they possess numerous valves. Sequentially they include:

Deep veins of the foot
Deep veins of the leg
Popliteal vein
Deep veins of the thigh

DEEP VEINS OF THE FOOT

The **plantar digital veins** arise from plexuses on the plantar surfaces of the digits, and after sending communicating vessels to join the dorsal digital veins, they unite to form four **plantar metatarsal veins.** These course proximally in the metatarsal spaces and anastomose by means of perforating veins with the veins on the dorsum of the foot. The plantar metatarsal veins unite to form the **deep plantar venous arch,** which accompanies the plantar arterial arch across the plantar aspect of the foot between the oblique part of the adductor hallucis muscle and the interossei muscles. From the deep plantar venous arch, the **medial** and **lateral plantar veins** run proximally near their corresponding arteries. They communicate by anastomosing vessels with the great and small saphenous veins on the two sides of the foot. The medial and lateral plantar veins then unite to form the **posterior tibial veins** behind the medial malleolus.

DEEP VEINS OF THE LEG

The deep veins of the leg ascend with their accompanying arteries in both the anterior and posterior compartments. The principal deep veins include:

Posterior tibial
  Peroneal
Anterior tibial

Posterior Tibial Veins

The posterior tibial veins accompany the posterior tibial artery. Paired as venae comitantes in the lower leg, the vessels at times join to form a single vein, which with its artery ascends in the fascial plane between the superficial and deep muscles of the posterior compartment. The posterior tibial veins drain blood from these posterior compartment muscles and also receive communicating vessels from the more superficial veins. Two-thirds of the way up the leg, the posterior tibial veins receive the **peroneal veins** (Fig. 9-41).

PERONEAL VEINS.   The peroneal veins ascend obliquely in the same plane as the posterior tibial veins from the lateral aspect of the leg. They accompany the peroneal artery and open into the posterior tibial veins distal to the popliteal fossa, just below the lower border of the popliteus muscle (Fig. 9-41). The peroneal veins drain the lateral compartment muscles and also muscles of the posterior compartment.

FIG. 9-41.   The popliteal vein and the junction of the peroneal and posterior tibial veins.

Anterior Tibial Veins

The anterior tibial veins accompany the anterior tibial artery and deep peroneal nerve, and they are the upward continuation of the venae comitantes of the dorsalis pedis artery. Ascending above the ankle joint in the lower leg between the tibialis anterior and extensor hallucis longus muscles, the anterior tibial veins course deeply on the interosseous membrane. In the upper leg the neurovascular bundle lies between the tibialis anterior and the extensor digitorum longus. The veins leave the anterior leg by passing between the tibia and fibula through the upper part of the interosseous membrane. They unite with the posterior tibial veins just below the popliteal fossa in the posterior compartment to form the **popliteal vein.**

POPLITEAL VEIN

The popliteal vein is formed by the junction of the anterior and posterior tibial veins at the distal border of the popliteus muscle.

It ascends through the popliteal fossa to the aperture in the adductor magnus muscle, called the adductor hiatus, where it becomes the femoral vein in the adductor canal. In the lower part of its course within the popliteal fossa, it is placed medial to the popliteal artery (Fig. 9-41); between the heads of the gastrocnemius muscle, it lies superficial to that vessel; above the knee joint, it remains superficial to the popliteal artery but courses along its lateral side. The popliteal vein receives tributaries corresponding to the branches of the popliteal artery, as well as the small saphenous vein, and in its course it usually contains four valves.

DEEP VEINS OF THE THIGH

The deep veins of the thigh accompany the deep arteries. The principal deep vein is the **femoral vein,** and into this flow the **deep femoral vein,** the **saphenous vein,** and the **medial** and **lateral femoral circumflex veins.** Additionally, the **perforating veins** which drain the posterior structures of the thigh are large tributaries of the deep femoral vein.

Femoral Vein

The femoral vein, accompanying the femoral artery, ascends from the superior end of the popliteal fossa through the adductor canal, lateral to and somewhat behind the artery. Proximal to the adductor canal in the lower part of the femoral triangle, the femoral vein lies behind the artery and as it ascends, it gradually crosses to the medial side of the artery. At the level of the inguinal ligament it occupies the medial compartment of the femoral sheath, and the femoral vein becomes the external iliac vein above the ligament. The femoral vein receives numerous muscular tributaries, and about 4 cm or more below the inguinal ligament, it is joined on its posterior aspect by the deep femoral vein and somewhat higher on its anterior aspect by the saphenous vein. Also the medial and lateral femoral circumflex veins may open either into the femoral vein or into the deep femoral vein. There are usually three or four valves in the femoral vein; one is generally found (in 90% of cases) just below the entrance of the deep femoral vein (Basmajian, 1952).

DEEP FEMORAL VEIN. The deep femoral vein is formed by the junction of three or four **perforating veins** that correspond to the perforating branches of the deep femoral artery. Through the perforating veins, anastomoses are established by the femoral vein with the inferior gluteal vein above and with the popliteal vein below. The perforating veins return blood from muscles in all three compartments of the thigh, as well as from the femur.

MEDIAL AND LATERAL FEMORAL CIRCUMFLEX VEINS. The medial and lateral femoral circumflex veins do not terminate in a pattern consistent with the origin of the corresponding arteries. In 86% of cases studied, the two femoral circumflex veins terminate in the femoral vein, whereas in 2% both veins terminate in the deep femoral vein. In the remaining 12% either the medial or the lateral femoral circumflex vein joins the femoral vein, while the other joins the deep femoral vein (Baird and Cope, 1933).

Histology of the Blood Vessels

ARTERIES

The arterial system consists of an elaborate system of tubular structures that conducts blood away from the heart. All arteries are composed of three coats: an inner coat called the **tunica intima,** an intermediate coat, the **tunica media,** and an external coat, the **tunica adventitia** (Fig. 9-42). The composition and relative thickness of these coats differ in arteries of different size; however, in all of these the functional cell type of the tunica intima is the endothelial cell, that of the tunica media is the smooth muscle cell, while the principal cell type of the tunica adventitia is connective tissue.

TUNICA INTIMA. The tunica intima, or innermost coat, consists of a thin lining of **endothelial cells** that is longitudinally oriented and continuous with the endothelium of capillaries on the one hand and with the endocardium of the heart on the other. The endothelium is a single layer of simple squamous or plate-like cells, polygonal, oval, or fusiform in shape, with rounded oval or flattened nuclei. Although the outlines of the cells can be brought out by treatment with silver nitrate, studies in recent years utilizing the electron microscope have shown that endothelial cells are closely apposed, being separated by only about 200 Å, and contain little or no intercellular cement at their junctions. The endothelial cells rest on a **basal lamina,** and in arteries larger than arteriolar capillaries, delicate strands of connective tissue form a **subendothelial layer** that intervenes between the endothelium and the **internal elastic membrane.** The latter is easily identified in muscular arteries. It consists of a network of elastic fibers arranged more or less longitudinally, leaving elongated apertures or perforations that give it a fenestrated appearance. Fine cytoplasmic processes on the basal surface of the endothelial cells project through the fenestrations of the internal elastic membrane to achieve contact with the overlying smooth muscle cells of the tunica media. The internal elastic membrane forms the chief thickness of the tunica intima, and it is thrown into folds when arteries are fixed histologically. In microscopic

cross sections the internal elastic membrane appears as a wavy line, glassy and almost unstained when using routine histological techniques such as hematoxylin and eosin; however, it stains heavily with special substances such as orcein, which characteristically reacts with elastic tissue.

TUNICA MEDIA. The tunica media, or middle coat, is principally composed of thin, cylindrical smooth muscle cells and elastic tissue, and it accounts for the bulk of the wall of most arteries. The smooth muscle cells are disposed in circular layers around the vessel, and the thickness of the tunica media, as well as its composition, varies with the size of the vessel. The smooth muscle cells are held together by an abundant amount of intercellular glycoprotein, which contains collagen fibrils and reticular and elastic fibers. The intercellular material is probably formed by the smooth muscle cells themselves, because virtually no other cells

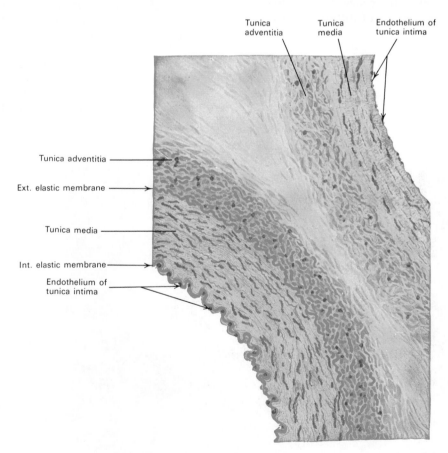

FIG. 9-42. Cross section of parts of the walls of adjacent medium-sized artery and vein. Artery, lower left; vein, upper right.

are found in the tunica media. The external margin of the tunica media frequently is marked by a prominent layer of elastic tissue called the **external elastic membrane.**

TUNICA ADVENTITIA.   The tunica adventitia, or outer coat, consists of areolar connective tissue that contains fibroblasts within a fine meshwork of elastic fibers and bundles of collagen. As described previously, the elastic tissue is more abundant adjacent to the tunica media, where it is called the external elastic membrane. The tunica adventitia varies in thickness in different types of arteries, but usually it is not as thick as the tunica media.

The tunica adventitia contains the **vasa vasorum,** which are the arteries and veins that supply the vessel walls. It also contains fine lymphatic vessels, as well as the nerve fibers that supply the arteries. The latter appear to terminate near or within the external elastic membrane.

As arteries extend distally, there is a gradual transition in their size. Certain features characterize arteries of different sizes.

Large Elastic Arteries

Large elastic arteries such as the aorta and its major branches, as well as the pulmonary trunk and its two principal branches, conduct blood from the heart to the smaller distributing arteries (Fig. 9-43). They have a thick **tunica intima,** and although the *endothelium* is still only a single layer in thickness, the *subendothelial layer* contains considerable connective tissue that is composed of bundles of collagen and some smooth muscle cells along with occasional macrophages. Surrounding the subendothelial connective tissue is a *fenestrated elastic tissue* layer, which in smaller vessels is considered to be the internal elastic membrane. In large arteries, however, it is difficult to determine the external boundary between the tunica intima and the internal border of the tunica media, because the **tunica media** contains many concentric layers of fenestrated elastic tissue (Fig. 9-43). There may be 50 or more elastic laminae in the tunica media, and interspersed between these elastic fibers are smooth muscle cells, collagenous fibers, and extracellular ground substance (Fig. 9-44).

The **tunica adventitia** of large arteries is thin relative to the tunica media and the line of transition may not be easily apparent. Characteristically, the connective tissue of the external coat is more fibrous in nature and contains more collagen and less elastic tissue. The tunica adventitia of large vessels also contains numerous small vessels

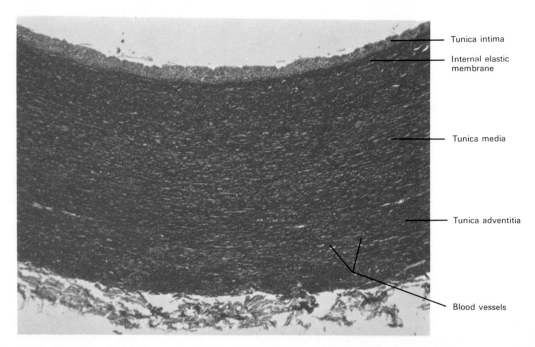

Tunica intima

Internal elastic membrane

Tunica media

Tunica adventitia

Blood vessels

FIG. 9-43.   Photomicrograph of a section of the wall of the aorta, stained for elastic fibers. Note the relative widths of the three layers. 24 ×.

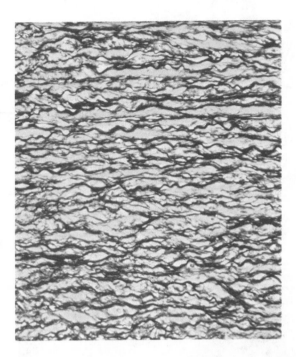

FIG. 9-44. Photomicrograph of the tunica media of the human aorta showing many lamellae of elastic fibers, which characterize the microscopic anatomy of that coat in the walls of large elastic arteries. 144 ×.

and nerves, the *vasa vasorum* and *nervi vasorum,* and its external layers gradually merge with the surrounding connective tissue.

Medium-Sized Muscular Arteries

The medium-sized arteries include most of the named arteries that distribute blood to the different organs and parts of the body. The **tunica intima** is formed by a single layer of endothelium resting on a basal lamina. Deep to this is a thin subendothelial layer that lies adjacent to a well-defined fenestrated internal elastic lamina. The muscular coat, or **tunica media,** is especially well developed and the physiology of the smooth muscle cells is under the influence of neurohumors liberated by the vasomotor nerves. This mechanism controls the flow of blood by altering the diameter of the lumen of these arteries. In the somewhat smaller arteries the tunica media is almost entirely muscular, but in larger ones more and more elastic tissue is interposed with the smooth muscle cells (Fig. 9-45). The thickness of the **tunica adventitia** is variable. In arteries located in well-protected sites such as those in the abdominal or cranial cavities, it is relatively thin, but it is much thicker in exposed regions such as the limbs. The bundles of

nerve fibers are numerous in the tunica adventitia of many muscular arteries, and on some arteries supplying the abdominal organs they are so profuse that they form extensive plexuses that coat the vessels almost completely.

Small Arteries and Arterioles

As the named muscular arteries are distributed peripherally, they branch into smaller muscular arteries; these divide into **muscular arterioles,** which then branch into **terminal arterioles** less than 50 μm in diameter (Figs. 9-46; 9-47). The **tunica intima** of arterioles consists of a layer of flattened, elongated endothelial cells resting on a basal lamina. In cross sections of larger arterioles, a corrugated internal elastic membrane surrounds the endothelium; however in arterioles of small diameter, less elastic tissue is seen and this membrane is incomplete. In the smallest terminal arterioles, the elastic tissue disappears completely (Fig. 9-47).

The **tunica media** of arterioles consists of smooth muscle cells, and its thickness varies directly with the diameter of the vessels. In the larger arterioles, perhaps three or more layers of smooth muscle cells surround the endothelium in a circumferential pattern, whereas in the terminal arterioles the tunica

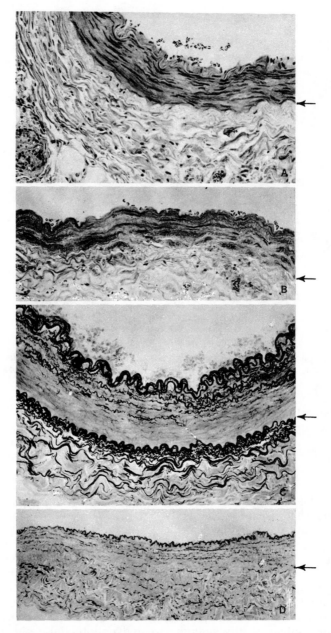

FIG. 9-45.   Photomicrographs of intercostal artery (A, C) and vein (B, D). A and B are stained with hematoxylin and eosin, while C and D are stained for elastic fibers. Arrows mark the boundary between the tunica media (above) and tunica adventitia (below) in each photomicrograph. 130 ×. (From Windle, W. F., *Textbook of Histology,* 4th ed., McGraw-Hill, New York, 1969.)

media consists of only a single layer of smooth muscle. The **tunica adventitia** of larger arterioles consists of loose connective tissue with some collagenous and elastic fibers and fibroblasts, but this outer coat becomes less well defined and greatly reduced in the smallest arterioles (Fig. 9-47).

Capillaries

From the smallest arteriolar vessels, fine networks of endothelium-lined tubes permeate most tissues and organs (Fig. 9-48). These minute tubes are the capillaries and their networks form the functional and morpho-

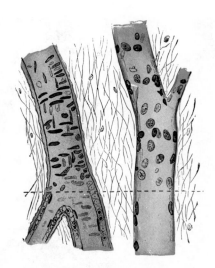

FIG. 9-46. Small artery and vein, pia mater of sheep. 250 ×. Surface view above the interrupted line; longitudinal section below. Artery in red; vein in blue.

logic connections between the arterial and venous systems. Frequently and closely associated with the outer surface of endothelial cells are macrophages or mesenchymal perivascular cells called **pericytes.** Although there is some variation in their size, the average diameter of capillaries is about 8 µm, roughly equivalent to that of a red blood corpuscle (Fig. 9-48). Functionally, capillaries carry nutrient substances to tissues and organs and take away the metabolic products of cells. These processes require the transfer of substances across the capillary wall, utilizing the permeable nature of the endothelial lining. The anatomical basis of capillary permeability and the sites through which substances pass across the endothelial lining has been a subject of intensive study both physiologically and with the electron microscope. On morphological grounds, two types of capillaries have been described, the **continuous** and the **fenestrated.**

CONTINUOUS CAPILLARIES.   Continuous capillaries are usually seen in muscle, skin, and connective tissue and in the brain and spinal cord. Along both the luminal and outer surfaces of the endothelial cells in this type of capillary are found numerous pinocytotic vesicular invaginations, called **caveolae,** which measure between 500 and 700 Å (Fig. 9-49). It is theorized that these vesicles may be sites for the entrapment of large molecules, and that the vesicles then pinch off and traverse the endothelial cell cytoplasm to the opposite surface, where they fuse with the membrane again and their contents are discharged outside the endothelial cell. Continuous capillaries have walls of endothelial cells whose basal laminae are continuous and whose intercellular junctions are of the occludens type. These junctional complexes have intercellular spaces that measure only about 150 to 200 Å; these narrow spaces are believed to be wide enough to allow water and simple solutions to pass through them, but not larger molecules such as proteins (Fig. 9-50).

FENESTRATED CAPILLARIES.   Fenestrated capillaries are found in the glomeruli of the kidney, the intestinal villi, the endocrine glands, and other tissues and organs where rapid transfer of fluids occurs. In addition to containing the pinocytic vesicular caveolae, the endothelial cells that form these capillaries have cytoplasmic zones, frequently adjacent to the nuclei of the cells, that characteristically are exceedingly narrow. At these sites the endothelial cells are so thin that a series of fenestrations or pores about 1,000 Å in diameter perforates the cells. The perforations are actually covered by a fine diaphragm, which is thinner than the cell membrane and effectively allows water and solutions to pass readily through the cell.

VEINS

The veins return blood to the heart from tissues and organs throughout the body. Their walls are thinner but their lumina are larger than those of their accompanying arteries, and although the structure of veins varies, usually they have the same three layers as do arteries: **tunica intima, tunica media,** and **tunica adventitia.** Venous blood is under less pressure than arterial blood and it flows more slowly; therefore, at various sites veins contain thin membranous **valves** to prevent the blood from flowing backward toward the capillary bed and away from the heart. Like arteries, veins are supplied with nutrient vessels, the **vasa vasorum.** They also are innervated, but they receive fewer nerve fibers than arteries. Generally, veins are distinguished by size as either small-caliber venules, medium-sized veins, or large venous trunks.

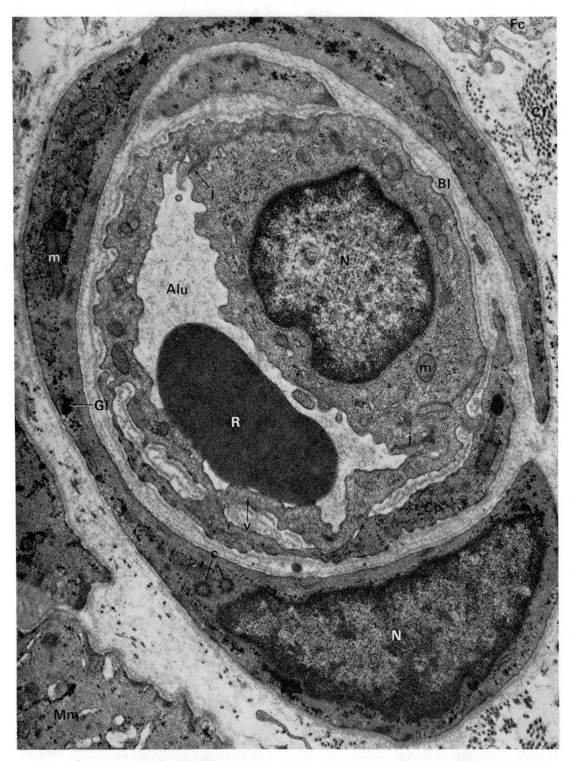

FIG. 9-47.   Electron micrograph of an arteriole from the human pyloric antrum. The lumen of the arteriole (Alu) is lined by two endothelial cells interconnected at junctions (j). A smooth muscle cell is shown outside the basal lamina (Bl) of the endothelium. Surrounding this are the basal lamina and collagenous connective tissue layer (Cf). N, nucleus; v, vesicles; Cp, cytoplasmic processes; c, centrioles; m, mitochondria; Fc, fibrocyte; Gl, glycogen; Mm, smooth muscle of stomach wall; R, red blood cell. (From Ebe, T. and Kobayashi, S., *Fine Structure of Human Cells and Tissue,* John Wiley and Sons, New York, 1972.)

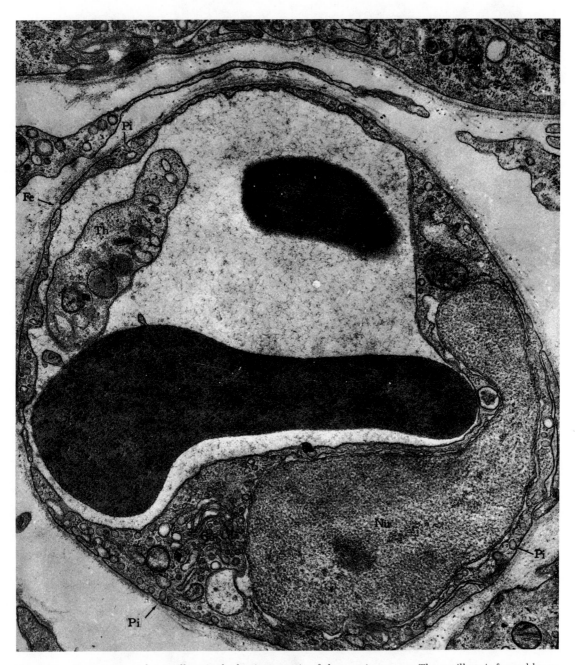

FIG. 9-48. Cross section of a capillary in the lamina propria of the gastric mucosa. The capillary is formed by an endothelial cell surrounded by a basement membrane (Bm). Note the numerous pinocytotic vesicles (Pi) along the membrane of the endothelial cell and the fenestrations (Fe), each of which is bridged by a thin diaphragm. Nu, nucleus; Go, Golgi complex; Mi, mitochondria; Er, erythrocyte; Th, thrombocyte. 30,000 ×. (From Rhodin, J. A. G., *An Atlas of Ultrastructure,* W. B. Saunders Co., Philadelphia, 1963.)

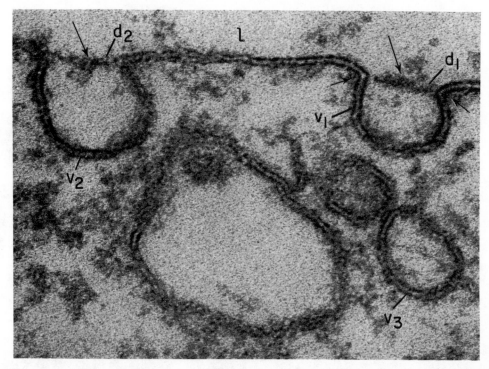

FIG. 9-49. Electron micrograph of a portion of endothelial membrane in a capillary, showing surface pinocytotic vesicles (V_1, V_2) bridged by diaphragms consisting of a single layer (d_1) or probably more than one layer (d_2). A closed vesicle within the endothelial cytoplasm is shown as V_3. (From Bruns and Palade, *J. Cell Biol.*, 37: 244–276, 1968.)

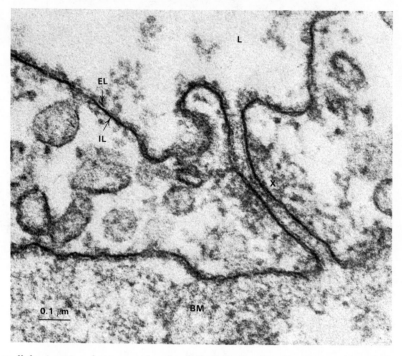

FIG. 9-50. Intercellular junctions between pairs of endothelial cells, showing regions of constriction (X), the basement membrane (BM), the external (EL) and internal (IL) leaflets of the unit membrane. The lumen of the capillary is indicated by L. 160,000 ×. (From Luft, J. H., Capillary Permeability in *The Inflammatory Process*, 2nd ed., Zweifach, B. W., Grant, L., and McCluskey, R. T. (eds.), Vol. 2, Academic Press, New York, 1973.)

Venules

The venules are the smallest veins, and they are formed by the union of several capillaries. They are larger than capillaries, measuring as much as 20 or 25 μm, and they are importantly involved in the transfer of substances between the body fluids and the vascular system. Although their walls resemble those of capillaries, there is a thin layer of collagenous fibrils and fibroblasts oriented longitudinally outside the endothelium. It is difficult, however, to distinguish the three separate layers in the walls of the smallest venules. In the walls of slightly larger venules (i.e., those with diameters of 50 to 60 μm) are found a few smooth muscle cells interposed between the endothelial cells and the connective tissue strands that form the tunica adventitia. In many of these larger venules the smooth muscle cells do not form a complete layer around the endothelium. In veins that measure 200 to 400 μm in diameter, the three coats become recognizable and even thin elastic fibers are seen interwoven with the collagenous fibers of the tunica adventitia.

Veins of Medium Size

The walls of medium-sized veins may vary somewhat in their structure, but usually the three coats can be identified (Figs. 9-42; 9-51). The endothelial cells of the **tunica intima** in these veins contain nuclei that are more oval and less flattened than those in arteries. These cells are supported by delicate connective tissue, external to which is a network of elastic fibers that takes the place of the fenestrated membrane of the arteries. The **tunica media** is usually thinner than that of accompanying arteries, and it is composed of circularly arranged smooth muscle cells, intermingled with collagenous and elastic fibers (Fig. 9-45). At times there is a blending of the connective tissue elements of the innermost and middle coats, and a true definition of the border between them is not readily seen. The **tunica adventitia** consists of loose connective tissue with longitudinal elastic fibers, and it is usually the thickest coat (Figs. 9-42; 9-45).

Large Veins

In large veins the **tunica adventitia** may be several times thicker than the tunica media and it constitutes a major part of the vessel wall. The tunica adventitia consists of loose connective tissue that contains elastic fibers and longitudinally oriented collagenous fibers. Within the adventitia are also found layers of longitudinally coursing smooth muscle fibers. These are most distinct in the inferior vena cava (especially at its termination in the heart), in the trunks of hepatic veins, in all the large trunks of the portal vein, and in the external iliac, renal, and azygos veins. In the inferior vena cava, renal, and portal veins, the muscle fibers extend through the whole thickness of the outer coat, but in the other veins mentioned, a layer of connective and elastic tissue is found external to the muscular fibers. The large veins that open into the heart are covered for a short distance by a layer of cardiac muscle from the heart. *Muscular tissue is lacking* in the following: (a) veins of the maternal part of the placenta, (b) venous sinuses of the dura mater and veins of the pia mater of the brain and spinal cord, (c) veins of the retina, (d) veins of the cancellous tissue of bones, and (e) venous spaces of the corpora cavernosa. Veins at these sites consist of an internal endothelial lining supported on one or more layers of loose connective tissue.

VALVES OF THE VEINS. Many veins are provided with valves to prevent the reflux of blood (Fig. 9-52). Each valve is formed by a reduplication of the tunica intima which is strengthened by connective tissue and elastic fibers. The valve is covered on both surfaces with endothelium, the arrangement of which differs on the two surfaces. On the surface of the valve next to the wall of the vein, the endothelial cells are arranged transversely, while on the other surface, over which the current of blood flows, the cells are arranged longitudinally in the direction of the current. Most commonly two such valves are found placed opposite each other. This is especially so in the smaller veins or in the larger trunks at the points where they are joined by smaller tributaries; occasionally there are three valves and sometimes only one. The valves are semilunar, and are attached by their convex edges to the wall of the vein. The concave margins are free, directed in the course of the venous current, and lie in close apposition with the wall of the vein so long as the current of blood flows in its natural course toward the heart. If, however, any regurgitation occurs, the valves become distended, their opposed

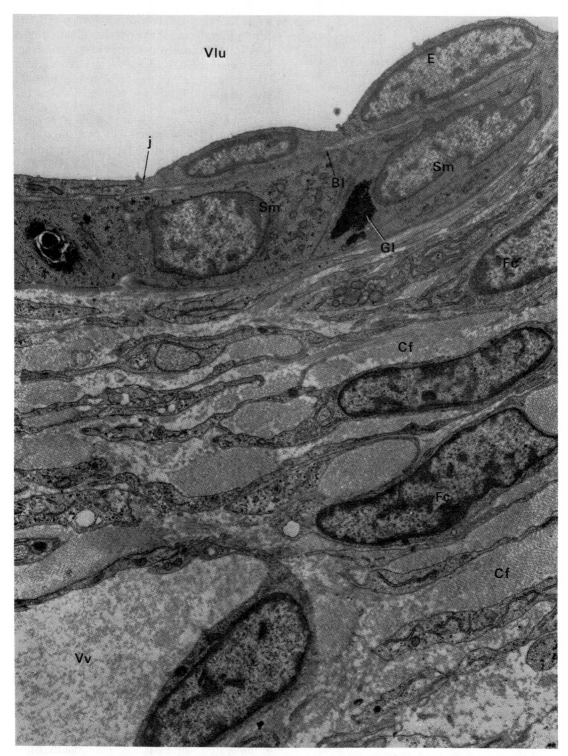

FIG. 9-51. Electron micrograph of a posterior abdominal vein in a seven-month-old fetus. This vein shows quite clearly the three layers forming the wall of the vessel. The endothelial cells (E) of the tunica intima rest upon a basal lamina (Bl). Surrounding these is a single layer of smooth muscle cells (Sm) that forms the tunica media. The tunica adventitia is a thicker layer of collagenous connective tissue fibers (Cf), among which are fibrocytes (Fc). Vlu, lumen; Vv, vas vasorum; j, junction between endothelial cells. 15,000 ×. (From Ebe, Y. and Kobayashi, S., *Fine Structure of Human Cells and Tissues,* John Wiley and Sons, Inc., New York, 1972.)

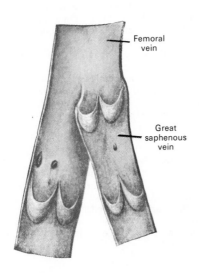

Femoral
vein

Great
saphenous
vein

FIG. 9-52.   The proximal portions of the femoral and great saphenous veins laid open to show the valves. About two-thirds natural size.

edges are brought into contact, and the backward flow is interrupted. The wall of the vein on the cardiac side of the point of attachment of each valve is expanded into a pouch or sinus that gives the vessel a knotted appearance when injected or distended with blood. Valves are numerous in the veins of the limbs, especially the lower limbs, because these vessels must conduct blood against the force of gravity. They are absent in the very small veins and also in the venae cavae, hepatic, renal, uterine, and ovarian veins. A few valves are found in the testicular veins, and one also at their points of junction with the renal vein or inferior vena cava. The cerebral and spinal veins, the veins of cancellous bone, the pulmonary veins, and the umbilical vein and its branches are also destitute of valves. A few valves are occasionally found in the azygos and intercostal veins. Rudimentary valves are found in the tributaries of the portal venous system.

References

(References are listed not only to those articles and books cited in the text, but to others as well which are considered to contain valuable resource information for the student who desires it.)

Development of Veins

Arey, L. B. 1974. *Developmental Anatomy.* Rev. 7th ed. W. B. Saunders, Philadelphia, 695 pp.

Aüer, J. 1948. The development of the human pulmonary vein and its major variations. Anat. Rec., *101*:581–594.

Bremer, J. L. 1914. The earliest blood vessels in man. Amer. J. Anat., *16*:447–476.

Butler, E. G. 1927. The relative role played by the great embryonic veins in the development of the mammalian vena cava posterior. Amer. J. Anat., *39*:267–353.

Butler, H. 1957. The development of certain human dural venous sinuses. J. Anat., (Lond.), *91*:510–526.

Dickson, A. D. 1957. The development of the ductus venosus in man and the goat. J. Anat. (Lond.), *91*:358–368.

Fraser, R. S., J. Dvorkin, R. E. Rossall, and R. Eidem. 1961. Left superior vena cava. A review of associated congenital heart lesions, catheterization data and roentgenologic findings. Amer. J. Med., *31*:711–716.

Friedman, S. M. 1945. Report of two unusual venous abnormalities (left postrenal inferior vena cava; postaortic left innominate vein). Anat. Rec., *92*:71–76.

Gladstone, R. J. 1910. A case in which the right ureter passed behind the inferior vena cava. J. Anat. (Lond.), *45*:225–231.

Gladstone, R. J. 1929. Development of the inferior vena cava in the light of recent research, with especial reference to certain abnormalities, and current descriptions of the ascending lumbar and azygos veins. J. Anat. (Lond.), *64*:70–93.

Gray, S. W., and J. E. Skandalakis. 1972. *Embryology for Surgeons.* W. B. Saunders Co., Philadelphia, 918 pp.

Karrer, H. E. 1960. Electron microscope study of developing chick embryo aorta. J. Ultrastruct. Res., *4*:420–454.

Kessler, R. E., and D. S. Zimmon. 1966. Umbilical vein angiography. Radiology, *87*:841–844.

Mc Clure, C. F. W., and C. G. Butler. 1925. The development of the vena cava inferior in man. Amer. J. Anat., *35*:331–383.

Neil, C. A. 1956. Development of the pulmonary veins. Pediatrics, *18*:880–887.

Padget, D. H. 1956. The cranial venous system in man in reference to development, adult configuration, and relation to the arteries. Amer. J. Anat., *98*:307–356.

Padget, D. H. 1957. The development of cranial venous system in man, from the viewpoint of comparative anatomy. Carneg. Inst., Contr. Embryol., *36*:79–140.

Patten, B. M. 1966. *Human Embryology.* 3rd ed., McGraw-Hill Book Co., New York, 651 pp.

Reagan, F. P. 1927. The supposed homology of vena azygos and vena cava inferior considered in the light of new facts concerning their development. Anat. Rec., *35*:129–148.

Reis, R. H., and G. Esenther. 1959. Variations in the pattern of renal vessels and their relation to the type of posterior vena cava in man. Amer. J. Anat., *104*:295–318.

Rosenquist, G. C. 1971. The common cardinal veins in the chick embryo: their origin and development as studied by radioautographic mapping. Anat. Rec., *169*:501–508.

Shaner, R. F. 1961. The development of the bronchial veins, with special reference to anomalies of the pulmonary veins. Anat. Rec., *140*:159–165.

Stewart, W. B. 1923. The ductus venosus in the fetus and in the adult. Anat. Rec., *25*:225–235.

Streeter, G. L. 1915. The development of the venous sinuses of the dura mater in the human embryo. Amer. J. Anat., *18*:145–178.

Williams, A. F. 1953. The formation of the popliteal vein. Surg. Gynecol. Obstet., *97*:769–772.

Veins of the Deep Head, Face, Neck and Upper Extremity

Armstrong, L. D., and A. Horowitz. 1971. The brain venous system of the dog. Amer. J. Anat., *132*:479–490.

Balo, J. 1950. The dural venous sinuses. Anat. Rec., *106*:319–326.

Batson, O. V. 1942. Veins of the pharynx. Arch. Otolaryngol., *36*:212–219.

Bedford, T. H. B. 1934. The great vein of Galen and the syndrome of increased intracranial pressure. Brain, *57*:1–24.

Boemer, L. C. 1937. The great vessels in deep infection of the neck. Arch. Otolaryngol., *25*:465–472.

Bol'shakov, O. P. 1964. Macroscopic and microscopic structural features of cavernous sinus. Fed. Proc., *23*:308–311.

Browder, J., A. Browder, and H. A. Kaplan. 1973. Anatomical relationships of the cerebral and dural venous systems in the parasagittal area. Anat. Rec., *176*:329–332.

Brown, S. 1941. The external jugular vein. Amer. J. Phys. Anthrop., *28*:213–226.

Browning, H. C. 1953. The confluence of dural venous sinuses. Amer. J. Anat., *93*:307–330.

Campbell, E. H. 1933. The cavernous sinus: anatomical and clinical considerations. Ann. Otol. Rhinol. Laryngol., *42*:51–63.

Charles, C. M. 1932. On the arrangement of the superficial veins of the cubital fossa. Anat. Rec., *54*:9–14.

Clark, W. E. Le Gros. 1920. On the Pacchionian bodies. J. Anat. (Lond.), *55*:40–48.

Coates, A. E. 1934. A note on the superior petrosal sinus and its relation to the sensory root of the trigeminal nerve. J. Anat. (Lond.), *68*:428.

Hayner, J. C. 1949. Variations in the torcular Herophili and transverse sinuses. Anat. Rec., *103*:542.

Holdorff, B., and G. B. Bradac. 1974. The intrapanencymatous ponto-mesencephalic veins. A radio-anatomical study. Acta Anat., *89*:333–344.

Jefferson, G., and D. Stewart. 1928. On the veins of the diploë. Br. J. Surg., *16*:70–88.

Kaplan, T., and A. Katz. 1937. Thrombosis of the axillary vein: case report with comments on etiology, pathology and diagnosis. Amer. J. Surg., *37*:326–333.

Krayenbühl, H. A., and M. G. Yasargil. 1968. *Cerebral Angiography.* Butterworths and Co. Ltd., London, 401 pp.

Kumar, A. J., G. M. Hochwald, and I. Kricheff. 1976. An angiographic study of the carotid arterial and jugular venous systems in the cat. Amer. J. Anat., *145*:357–369.

Neuhof, H. 1938. Excision of the axillary vein in the radical operation for carcinoma of the breast. Ann. Surg., *108*:15–20.

Newton, T. H., and D. G. Potts. 1974. Veins. In *Radiology of the Skull and Brain,* Vol. 2: Angiography. Book 3, Part XIII. C. V. Mosby, St. Louis, pp. 1851–2254.

O'Connell, J. E. A. 1934. Some observations on cerebral veins. Brain, *57*:484–503.

Pikkieff, E. 1937. On subcutaneous veins of the neck. J. Anat. (Lond.), *72*:119–127.

Sargent, P. 1910. Some points in the anatomy of the intracranial blood sinuses. J. Anat. (Lond.), *45*:69–72.

Schlesinger, B. 1939. The venous drainage of the brain, with special reference to the galenic system. Brain, *62*:274–291.

Schwadron, L., and B. C. Moffett. 1950. Relationships of cranial nerves to Meckel's cave and the cavernous sinus. Anat. Rec., *106*:131–137.

Stopford, J. S. B. 1929. The functional significance of the arrangement of the cerebral and cerebellar veins. J. Anat. (Lond.), *64*:257–261.

Woodhall, B., and A. E. Seeds. 1936. Cranial venous sinuses: correlation between skull markings and roentgenograms of the occipital bone. Arch. Surg., *33*:867–875.

Pulmonary Veins, Superior Vena Cava, and Other Veins of the Thorax

Appleton, A. B. 1945. The arteries and veins of the lungs: I. Right upper lobe. J. Anat. (Lond.), *79*:97–120.

Atwell, W. J., and P. Zoltowski. 1938. A case of left superior vena cava without a corresponding vessel on the right side. Anat. Rec., *70*:525–532.

Brody, H. 1942. Drainage of the pulmonary veins into the right side of the heart. Arch. Pathol., *33*:221–240.

Butler, H. 1951. The veins of the esophagus. Thorax, *6*:276–296.

Carlson, H. A. 1934. Obstruction of the superior vena cava: an experimental study. Arch. Surg., *29*:669–677.

Chouké, K. S. 1939. A case of bilateral superior vena cava in an adult. Anat. Rec., *74*:151–157.

Coakley, J. B., and T. S. King. 1959. Cardiac muscle relation of the coronary sinus, the oblique vein of the left atrium and the left precaval vein in mammals. J. Anat. (Lond.), *93*:30–35.

Conn, L. C., J. Calder, J. W. Mac Gregor, and R. F. Shaner. 1942. Report of a case in which all the pulmonary veins from both lungs drain into the superior vena cava. Anat. Rec., *83*:335–340.

Esperança-Pina, J. A. 1975. Morphological study of the human anterior cardiac veins, venae cardis anteriores. Acta Anat., *92*:145–159.

Esperança-Pina, J. A., and A. Dos Santos Ferreira. 1974. Microangiographic aspects of the thebisian veins. Acta Anat., *88*:156–160.

Esperança-Pina, J. A., M. Correia, and J. G. O'Neill. 1975. Morphological study on the thebisian veins of the right cavities of the heart in the dog. Acta Anat., *92*:310–320.

Fieldstein, L. E., and J. Pick. 1942. Drainage of the coronary sinus into the left auricle: report of a rare congenital cardiac anomaly. Amer. J. Clin. Pathol., *12*:66–69.

Kegaries, D. L. 1934. The venous plexus of the oesophagus: its clinical significance. Surg. Gynecol. Obstet., *58*:46–51.

Lindsay, H. C. 1925. An abnormal vena hemiazygos. J. Anat. (Lond.), *59*:438.

Massopust, L. C., and W. D. Gardner. 1950. Infrared photographic studies of the superficial thoracic veins in the female. Surg. Gynecol. Obstet., *91*:717–727.

Mc Cotter, R. E. 1916. Three cases of the persistence of the left superior vena cava. Anat. Rec., *10*:371–383.

Morton, W. R. M. 1948. Pre-aortic drainage of the hemiazygos veins: Report of two cases. Anat. Rec., *101*:187–191.

Nandy, K., and C. B. Blair, Jr. 1965. Double superior venae cavae with completely paired azygos veins. Anat. Rec., *151*:1–9.

Papez, J. W. 1938. Two cases of persistent left superior vena cava in man. Anat. Rec., *70*:191–198.

Prows, M. S. 1943. Two cases of bilateral superior venae cavae, one draining a closed coronary sinus. Anat. Rec., *87*:99–106.

Reeves, J. T. 1967. Microradiography of intrapulmonary bronchial veins of the dog. Anat. Rec., *159*:255–262.

Sanders, J. M. 1946. Bilateral superior vena cavae. Anat. Rec., *94*:657–662.

Smith, W. C. 1916. A case of a left superior vena cava without a corresponding vessel on the right side. Anat. Rec., *11*:191–198.

Thompson, I. M. 1929. Venae cavae superiores dextra et sinistra of equal size in an adult. J. Anat. (Lond.), *63*:496–497.

Troyer, J. R. 1961. A multiple anomaly of the human heart and great veins. Anat. Rec., *139*:509–513.

Truex, R. C., and A. W. Angulo. 1952. Comparative study of the arterial and venous systems of the ventricular myocardium with special reference to the coronary sinus. Anat. Rec., *113*:467–491.

Vajda, J., M. Tomcsik, and W. J. Van Doorenmaalen. 1973. Connections between the venous system of the heart and the epicardiac lymphatic network. Acta Anat., *83*:262–274.

Inferior Vena Cava and Veins of the Abdomen, Pelvis, Perineum, and Lower Extremity

Amsler, F. R., Jr., and M. C. Wilber. 1967. Intraosseous vertebral venography as a diagnostic aid in evaluating intervertebral-disc disease of the lumbar spine. J. Bone Joint Surg., *49-A*:703–712.

Anson, B. J., E. W. Cauldwell, J. W. Pick, and L. E. Beaton. 1948. The anatomy of the pararenal system of veins, with comments on the renal arteries. J. Urol., *60*:714–737.

Baird, R. D., and J. S. Cope. 1933. On the termination of the circumflex veins of the thigh and their relations to the origins of the circumflex arteries. Anat. Rec., *57*:325–337.

Barlow, T. E., F. H. Bentley, and D. N. Walder. 1951. Arteries, veins and arteriovenous anastomoses in the human stomach. Surg. Gynecol. Obstet., *93*:657–671.

Batson, O. V. 1940. The function of the vertebral veins and their rôle in the spread of metastases. Ann. Surg., *112*:138–149.

Batson, O. V. 1942. The rôle of the vertebral veins in metastatic processes. Ann. Intern. Med., *16*:38–45.

Batson, O. V. 1957. The vertebral vein system. Amer. J. Roentgenol., *78*:195–212.

Becker, F. F. 1962. A singular left sided inferior vena cava. Anat. Rec., *143*:117–120.

Blasingame, F. J. L., and C. H. Burge. 1937. A case of left postrenal inferior vena cava without transposition of viscera. Anat. Rec., *69*:465–470.

Boruchow, I. B., and J. Johnson. 1972. Obstructions of the vena cava. Review. Surg. Gynecol. Obstet., *134*:115–121.

Boyer, C. C. 1953. Anomalous inferior vena cava formed from supracardinal complexes. Anat. Rec., *117*:829–839.

Brantigan, O. C. 1947. Anomalies of the pulmonary veins; their surgical significance. Surg. Gynecol. Obstet., *84*:653–658.

Brunschwig, A., and T. S. Walsh. 1949. Resection of the great veins on the lateral pelvic wall. Surg. Gynecol. Obstet., *88*:498–500.

Charles, C. M., T. L. Finley, R. D. Baird, and J. S. Cope. 1930. On the termination of the circumflex veins of the thigh. Anat. Rec., *46*:125–132.

Conn, L. C., J. Calder, J. W. Mac Gregor, and R. F. Shaner. 1942. Report of a case in which all pulmonary veins from both lungs drain into the superior vena cava. Anat. Rec., *83*:335–340.

Crock, H. V. 1962. The arterial supply and venous drainage of the bones of the human knee joint. Anat. Rec., *144*:199–217.

Daniel, P. M., and M. M. L. Prichard. 1951. Variations in the circulation of the portal venous blood within the liver. J. Physiol. (Lond.), *114*:521–537.

Daseler, E. H., B. J. Anson, A. F. Reimann, and L. E. Beaton. 1946. The saphenous venous tributaries and related structures in relation to the technique of high ligation: based chiefly upon a study of 550 dissections. Surg. Gynecol. Obstet., *82*:53–63.

Di Dio, L. J. A. 1961. The termination of the vena gastrica sinistra in 220 cadavers. Anat. Rec., *141*:141–144.

Douglass, B. E., A. H. Baggestoss, and W. H. Hollinshead. 1950. The anatomy of the portal vein and its tributaries. Surg. Gynecol. Obstet., *91*:562–576.

Edwards, E. A., and J. D. Robuck, Jr. 1947. Applied anatomy of the femoral vein and its tributaries. Surg. Gynecol. Obstet., *85*:547–557.

Ferraz De Carvalho, C. A. 1971. Considerations on the muscularis mucosae and its relations with veins at the esophago-gastric transition zone. Acta Anat., *78*:559–573.

Gerber, A. B., M. Lev, and S. L. Goldberg. 1951. The surgical anatomy of the splenic vein. Amer. J. Surg., *82*:339–343.

Gibson, J. B. 1959. The hepatic veins in man and their sphincter mechanisms. J. Anat. (Lond.), *93*:368–379.

Glasser, S. T. 1943. An anatomic study of venous variations at the fossa ovalis. The significance of recurrences following ligation. Arch. Surg., *46*:289–295.

Gonzalo-Sanz, L. M., and R. Insansti. 1976. The structure of the suprarenal venous system and its possible functional significance. Acta Anat., *95*:309–318.

Grady, E. D., and E. M. Colvin. 1953. Treatment of venous insufficiency of the lower extremities with a note on the use of ascending phlebography. Amer. Surg., *19*:936–945.

Heath, T., and B. House. 1970. Origin and distribution of portal blood in the cat and rabbit. Amer. J. Anat., *127*:71–80.

Heller, R. E. 1942. The circulation in normal and varicose veins. Surg. Gynecol. Obstet., *74*:1118–1127.

Hollinshead, W. H., and J. A. Mc Farlane. 1953. The collateral venous drainage from the kidney following occlusion of the renal vein in the dog. Surg. Gynecol. Obstet., *97*:213–219.

Huseby, R. A., and E. A. Boyden. 1941. Absence of the hepatic portion of the inferior vena cava with bilateral retention of the supracardinal system. Anat. Rec., *81*:537–544.

Keck, S. W., and P. J. Kelly. 1965. The effect of venous stasis on intraosseous pressure and longitudinal bone growth in the dog. J. Bone Joint Surg., *47-A*:539–544.

Keen, J. A. 1941. The collateral venous circulation in a case of thrombosis of the inferior vena cava, and its embryological interpretation. Br. J. Surg., *29*:105–114.

Kosinki, C. 1926. Observations on the superficial venous system of the lower extremity. J. Anat. (Lond.), *60*:131–142.

Kreider, P. G. 1933. The anatomy of the veins of the gall bladder. Their relation to an impacted stone. Surg. Gynecol. Obstet., *57*:475–482.

Kuster, G., E. F. Lofgren, and W. H. Hollinshead. 1968. Anatomy of the veins of the foot. Surg. Gynecol. Obstet., *127*:817–823.

Mavor, G. E., and J. M. D. Galloway. 1967. Collaterals of the deep venous circulation of the lower limb. Surg. Gynecol. Obstet., *125*, 561–571.

Mullarky, R. E. 1965. *The Anatomy of Varicose Veins.* Charles C Thomas, Springfield, Ill., 89 pp.

O'Keeffe, A. F., R. Warren, and G. A. Donaldson. 1951. Venous circulation in lower extremities following femoral vein interruption. Surgery, *29*:267–270.

Pollack, A. A., and E. H. Wood. 1949. Venous pressure in the saphenous vein at the ankle in man during exercise and changes in posture. J. Appl. Physiol., *1*:649–662.

Schnug, E. 1943. Ligation of the superior mesenteric vein. Surgery, *14*:610–616.

Severn, C. B. 1972. A morphological study of the development of the human liver. II. Establishment of liver parenchyma, extrahepatic ducts and associated venous channels. Amer. J. Anat., *133*:85–107.

Shah, A. C., and Srivastava. 1966. Fascial canal for the small saphenous vein. J. Anat. (Lond.), *100*:411–413.

Simonds, J. P. 1936. Chronic occlusion of the portal vein. Arch. Surg., *33*:397–424.

Suh, T. H., and L. Alexander. 1939. Vascular system of the human spinal cord. Arch. Neurol. Psychiat., *41*:659–677.

Van Cleave, C. D. 1931. A multiple anomaly of the great veins and interatrial septum in a human heart. Anat. Rec., *50*:45–51.

Wagner, J. B., Jr., and P. A. Herbut. 1949. Etiology of primary varicose veins: histologic study of one hundred saphenofemoral junctions. Amer. J. Surg., *78*:876–880.

Wermuth, E. G. 1939. Anastomoses between the rectal and uterine veins forming a connexion between the somatic and portal venous system in the rectouterine pouch. J. Anat. (Lond.), *74*:116–126.

Yelin, G. 1940. Retroaortic renal vein. J. Urol., *44*:406–410.

Histology and Ultrastructure of Arteries and Veins

Abraham, A. 1969. *Microscopic Innervation of the Heart and Blood Vessels in Vertebrates Including Man.* Pergamon Press, Oxford, 433 pp.

Bennett, H. S., Luft, J. H., and J. C. Hampton. 1957. Morphological classification of vertebrate blood capillaries. Amer. J. Physiol., *196*:381–390.

Bierring, F., and T. Kobayashi. 1963. Electron microscopy of the normal rabbit aorta. Acta Path. Microbiol. Scand., *57*:154–168.

Brightman, M. W., and T. S. Reese. 1969. Junctions between intimately apposed cell membranes in the vertebrate brain. J. Cell Biol., *40*:648–677.

Bruns, R. R., and G. E. Palade. 1968. Studies on blood capillaries. I. General organization of blood capillaries in muscle. J. Cell Biol., *37*:244–276.

Bruns, R. R., and G. E. Palade. 1968. Studies on blood capillaries. II. Transport of ferritin molecules across the wall of muscular capillaries. J. Cell Biol., *37*:277–299.

Buck, R. C. 1958. The fine structure of endothelium of large arteries. J. Biophys. Biochem. Cytol., *4*:187–190.

Chambers, R., and B. W. Zweifach. 1944. Topography and function of the mesenteric capillary circulation. Amer. J. Anat., *75*:173–205.

Chambers, R., and B. W. Zweifach. 1947. Intercellular cement and capillary permeability. Physiol. Rec., *27*:436–463.

Clark, E. R., and E. L. Clark. 1943. Caliber changes in minute blood-vessels observed in the living mammal. Amer. J. Anat., *73*:215–250.

Ebe, T., and S. Kobayashi. 1972. *Fine Structure of Human Cells and Tissues.* John Wiley & Sons, New York, 266 pp.

Epling, G. E. 1966. Electron microscopic observations of pericytes of small blood vessels in the lungs and hearts of normal cattle and swine. Anat. Rec., *155*:513–530.

Farquhar, M. 1961. Fine structure and function in capillaries of the anterior pituitary gland. Angiology, *12*:270–292.

Fawcett, D. W. 1959. The fine structure of capillaries, arterioles and small arteries. In *The Microcirculation.* Edited by S. R. M. Reynolds and B. W. Zweifach. Univ. Illinois Press, Urbana, pp. 1–27.

Fawcett, D. W. 1963. Comparative observations on the fine structure of blood capillaries. In *The Peripheral Blood Vessels.* Edited by J. L. Orbison and D. E. Smith. Internat. Acad. Pathol. Monograph No. 4. Williams & Wilkins, Baltimore, pp. 17–44.

Fernando, N. V. P., and H. Z. Movat. 1964. The capillaries. Exp. Mol. Pathol., *3*:87–97.

Fernando, N. V. P., and H. Z. Movat. 1964. The smallest arterial vessels: terminal arterioles and metarterioles. Exp. Mol. Pathol., *3*:1–9.

Florey, H. 1961. Exchange of substances between the blood and tissues. Nature, *192*:908–912.

Florey, H. 1966. The endothelial cell. Br. Med. J., *2*:487–490.

Haust, M. D., R. H. More, S. A. Bencosme, and J. V. Balis. 1965. Elastogenesis in human aorta: an electron microscope study. Exp. Mol. Pathol., *4*:508–524.

Hibbs, R. G., G. E. Burch, and J. H. Phillips. 1958. The fine structure of the small blood vessels of normal human dermis and subcutis. Amer. Heart J., *56*:662–670.

Hüttner, I., M. Boutet, and R. H. More. 1973. Gap junctions in arterial endothelium. J. Cell Biol., *57*:247–252.

Iwayama, T. 1971. Nexuses between areas of the surface membrane of the same arterial smooth muscle cell. J. Cell Biol., *49*:521–525.

Kaech, M. K. 1960. Electron microscopy of the normal rat aorta. J. Biophys. Biochem. Cytol., *7*:533–538.

Karnovsky, M. J. 1967. The ultrastructural basis of capillary permeability studied with peroxide as a tracer. J. Cell Biol., 35:213–236.

Karnovsky, M. J., and R. S. Cotran. 1966. The intercellular passage of exogenous peroxidase across endothelium and mesothelium. Anat. Rec., 154:365.

Luft, J. H. 1963. The fine structure of the vascular wall. In *Evolution of the Arteriosclerotic Plaque*. Edited by R. J. Jones., Univ. Chicago Press, Chicago, pp. 3–28.

Luft, J. H. 1973. Capillary permeability. In *The Inflammatory Process*. Vol. 2. Edited by B. W. Zweifach, L. Grant, and R. T. Mc Cluskey. 2nd ed. Academic Press, New York, pp. 47–93.

Majno, G. 1965. Ultrastructure of the vascular membrane. In *Handbook of Physiology*. Sect. 2, Circulation, Vol. 3. Edited by W. F. Hamilton and P. Dow. American Physiological Society, Washington, D.C., pp. 2293–2376.

Movat, H. Z., and N. V. P. Fernando. 1963. Small arteries with an internal elastic lamina. Exp. Mol. Pathol., 2:549–563.

Movat, H. Z., and N. V. P. Fernando. 1964. The venules and their perivascular cells. Exp. Mol. Pathol., 3:98–114.

Palade, G. E. 1960. Transport in quanta across the endothelium of blood capillaries. Anat. Rec., 136:254.

Palade, G. E. 1961. Blood capillaries of the heart and other organs. Circulation, 24:368–384.

Palade, G. E. 1968. Small pore and large pore systems in capillary permeability. In *Hemorheology*. Edited by A. L. Copley. Pergamon Press, New York, pp. 703–720.

Parker, F. 1958. An electron microscope study of coronary arteries. Amer. J. Anat., 103:247–273.

Paule, W. J. 1965. Electron microscopy of the newborn rat aorta. J. Ultrastruct. Res., 8:219–235.

Pease, D. C., and S. Molinari. 1960. Electron microscopy of muscular arteries: pial vessels of the cat and monkey. J. Ultrastruct. Res., 3:447–468.

Pease, D. C., and W. J. Paule. 1960. Electron microscopy of elastic arteries: the thoracic aorta of the rat. J. Ultrastruct. Res., 3:469–483.

Phelps, P. C., and J. H. Luft. 1969. Electron microscopical study of relaxation and constriction in frog arterioles. Amer. J. Anat., 125:399–428.

Rhodin, J. A. G. 1962. The diaphragm of capillary endothelial fenestrations. J. Ultrastruct. Res., 6:171–185.

Rhodin, J. A. G. 1962. Fine structure of the vascular wall in mammals, with special reference to the smooth muscle component. Physiol. Res., 42 (Suppl. 5):48–81.

Rhodin, J. A. G. 1963. *An Atlas of Ultrastructure*. W. B. Saunders Co., Philadelphia, 222 pp.

Rhodin, J. A. G. 1967. The ultrastructure of mammalian arterioles and precapillary sphincters. J. Ultrastruct. Res., 18:181–223.

Rhodin, J. A. G. 1968. Ultrastructure of mammalian venous capillaries, venules and small collecting veins. J. Ultrastruct. Res., 25:452–500.

Schwartz, S. M., and E. P. Benditt. 1972. Studies on

aortic intima. Amer. J. Pathol., 66:241–264.

Shea, S. M., and M. S. Karnovsky. 1966. Brownian movement: a theoretical explanation for the movement of vesicles across endothelium. Nature, 212:353–355.

Simionescu, M., N. Simionescu, and G. E. Palade. 1974. Morphometric data on the endothelium of blood capillaries. J. Cell Biol., 60:128–152.

Smith, V., J. W. Ryan, D. D. Michie, and D. Smith. 1971. Endothelial projections, as revealed by scanning electron microscopy. Science, 173:925–927.

Takada, M. 1970. Fenestrated venules of the large salivary glands. Anat. Rec., 166:605–610.

Takada, M., and I. Gore. 1968. Capillary endothelial fenestrations in the lamina propria of the guinea pig tongue. Anat. Rec., 161:465–469.

Takada, M., and S. Hattori. 1972. Presence of fenestrated capillaries in the skin. Anat. Rec., 173:213–220.

Windle, W. F. 1976. *Textbook of Histology*. 5th ed. McGraw-Hill Book Co., New York, 561 pp.

Zweifach, B. W. 1959. The microcirculation of the blood. Sci. Amer., 200:54–60.

Zweifach, B. W., L. Grant, and R. T. Mc Cluskey (eds.) 1973. *The Inflammatory Process*. 2nd ed. Academic Press, New York, 3 vols.

Valves in Veins

Basmajian, J. V. 1952. The distribution of valves in the femoral, external iliac, and common iliac veins and their relationship to varicose veins. Surg. Gynecol. Obstet., 95:537–542.

Edwards, E. A. 1936. The orientation of venous valves in relation to body surfaces. Anat. Rec., 64:369–385.

Edwards, E. A., and J. E. Edwards. 1943. The venous valves in thromboangiitis obliterans. Arch. Pathol., 35:242–252.

Kampmeier, O. F., and C. Birch. 1927. The origin and development of the venous valves, with particular reference to the saphenous district. Amer. J. Anat., 38:451–499.

Kelly, R. E. 1930. Is it true that the valves in a vein necessarily become incompetent when the vein dilates? Br. J. Surg., 18:53.

Lofgren, E. P., T. T. Meyers, K. A. Lofgren, and G. Kuster. 1968. The venous valves of the foot and ankle. Surg. Gynecol. Obstet., 127:289–290.

Luke, J. C. 1951. The deep vein valves: a venographic study in normal and postphlebetic states. Surgery, 29:381–386.

Powell, T., and R. B. Lynn. 1951. The valves of the external iliac, femoral, and upper third of the popliteal veins. Surg. Gynecol. Obstet., 92:453–455.

Reuther, T. F. 1940. The valves and anastomoses of the hemorrhoidal and related veins. Amer. J. Surg., 49:326–334.

Van Cleave, C. D., and R. L. Holman. 1954. A preliminary study of the number and distribution of valves in normal and varicose veins. Amer. Surg., 20:533–537.

10

The Lymphatic System

The **lymphatic system** is a vascular network of thin-walled capillaries and larger lymphatic vessels that drains lymph from the tissue spaces of most organs and returns it to the venous system for recirculation. Peripherally, the lymphatic vessels do not communicate with the blood vessels, but the lymph eventually empties into the bloodstream at the junction of the jugular and subclavian veins at both sides of the neck. The endothelium of the veins at these points of junction is continuous with that of the lymphatic vessels.

Lymphatic fluid is an ultrafiltrate of blood that contains plasma proteins, and during its course it circulates through lymphatic tissue and lymph nodes that are lined with phagocytes capable of removing foreign matter. Additionally, many lymphocytes enter the lymphatic fluid at these nodes, as do the immunoglobulins or antibodies that are so important for immunologic protection. Lymph draining from different organs can vary significantly in its constituents. From the intestines it may have a white milky appearance and contain a high level of digested fatty substances, whereas from the liver it may be rich in proteins.

The pervasive nature of the lymphatic capillary system, so vital for the system to perform its beneficial functions, may also prove to be dangerous because its channels serve as a means by which bacteria are able to spread quickly throughout the body, with the result of generalized infection. Likewise, the lymphatic system furnishes pathways for the spread of malignant diseases, allowing routes for the establishment of metastatic tumors in other tissues and organs from a primary cancer site. The enormous clinical importance of the lymphatic system is widely recognized today, and yet the study of the system in the dissecting room is unsatisfactory because the thinness of its vascular walls and the small size of its channels make them indistinguishable from the surrounding connective tissues. Thus, most of the information concerning the anatomy of the system has been obtained by laborious special investigations using the injection of colored substances into the minute vessels. Injection into the larger vessels is ineffective because of the presence of numerous valves.

The lymphatic system consists of: (1) an extensive capillary network that collects lymph peripherally from various organs and tissues; (2) an elaborate system of collecting vessels that carries the lymph from the lymphatic capillaries to the bloodstream through an opening into the great veins at the root of the neck; (3) a number of firm, rounded bodies called lymph nodes, placed like filters in the paths of the collecting vessels; (4) certain lymphatic organs that resemble the lymph nodes, such as the tonsils and the aggregated lymphatic tissue found in the walls of the alimentary tract; (5) the spleen; and (6) the thymus. Another element that might be added is the lymphoid tissue, recognizable only with the aid of a microscope and consisting of reticular or areolar connective tissue that contains an accumulation of lymphocytes but lacks the organization of lymphatic nodules.

The spleen serves as an important lymphoid organ, performing phagocytic functions as well as being an organ in which circulating antigens activate appropriate lymphocytes to become functioning immunologic cells. The thymus, important early in life as a lymphoid organ, produces the so-called T lymphocytes, which participate in cell-mediated immunologic responses. Providing for the proper development of a competent immunologic system, the organ is of greater significance for the maturing child than for the adult.

The lymphatic capillaries and collecting vessels are lined throughout by a continuous layer of endothelial cells, thus forming a closed system. The lymphatic vessels of the small intestine receive the special designation of **lacteals** because they transport the milk-white fluid called **chyle,** but they do not differ morphologically from other lymphatic vessels.

Lymphatic vessels are not found in avascular structures such as the nails, hair, cornea, or epidermis, nor are they found in the central nervous system, meninges, eyeball, internal ear, or spleen.

Development of Lymphatic System

The precise manner by which the development of the lymphatic system commences is still incompletely understood; however, two views have been forwarded. According to one view, the earliest lymphatic vessels arise from the endothelium of veins as sprouts that coalesce to form **primitive lymph sacs.** Proponents of this view feel that these initial lymphatic plexuses, derived from the veins, then spread to surrounding tissues and organs to form a lymphatic network (Sabin, 1902, 1904, 1912, 1916; Yoffey and Courtice, 1970). Another view maintains that the initial lymphatic sacs arise in mesenchyme independent of the veins, and only secondarily establish venous connections (Huntington and McClure, 1910; Huntington, 1908). Regardless of the specific manner by which the earliest lymphatic sacs are formed, the two **jugular lymph sacs,** one on each side, form during the sixth week near the junction of the subclavian and anterior cardinal veins (Fig. 10-1). From these sacs lymphatic capillary plexuses spread to the neck, head, arms, and thorax. The more direct channels of the plexuses enlarge and form the lymphatic vessels. The larger vessels acquire smooth muscular coats with nerve connections and exhibit contractility. Each jugular sac establishes at least one con-

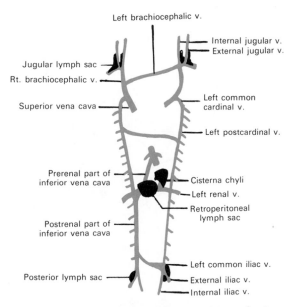

Left brachiocephalic v.

Internal jugular v.
External jugular v.

Jugular lymph sac
Rt. brachiocephalic v.

Superior vena cava

Left common cardinal v.

Left postcardinal v.

Prerenal part of inferior vena cava

Cisterna chyli
Left renal v.
Retroperitoneal lymph sac

Postrenal part of inferior vena cava

Left common iliac v.

Posterior lymph sac

External iliac v.
Internal iliac v.

FIG. 10-1. Schema showing relative positions of primary lymph sacs based on the description given by Sabin (1916).

nection with its jugular vein, and from the left sac the upper part of the thoracic duct develops.

At a slightly later stage (eighth week), another lymphatic sac, the retroperitoneal lymph sac, forms near the primitive inferior vena cava and mesonephric veins. From it lymphatics spread to the abdominal viscera and diaphragm, and the sac establishes connections with the cisterna chyli by the end of the ninth week. Also during the eighth week, another unpaired lymphatic sac, the **cisterna chyli,** forms near the Wolffian bodies. Giving rise to the adult cisterna chyli and the lower part of the thoracic duct, it joins that part of the duct that develops from the left jugular sac (Fig. 10-2).

By the end of the eighth week, the paired **posterior lymph sacs** appear near the junctions of the primitive iliac veins and the posterior cardinal veins. From these sacs, lymphatics spread to the abdominal wall, pelvic region, and legs. The posterior lymphatic sacs join each other by the end of the ninth week and also link superiorly with the cisterna chyli (Fig. 10-2). The adult main lymphatic duct forms by means of a diagonal junction of the inferior part of the right lymphatic trunk (connected caudally with the dilated cisterna chyli) and the upper part of the left lymphatic trunk (continuous superiorly with the left jugular sac).

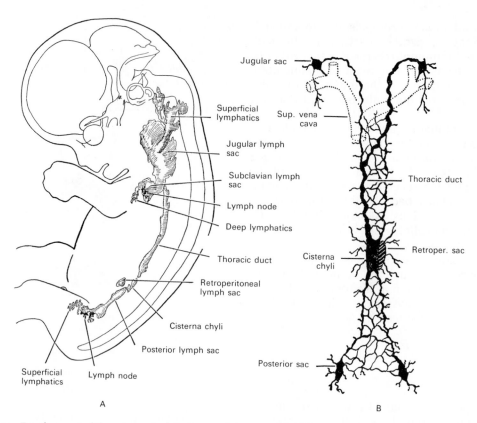

Jugular sac

Superficial lymphatics

Sup. vena cava

Jugular lymph sac

Subclavian lymph sac

Lymph node

Deep lymphatics

Thoracic duct

Thoracic duct

Retroper. sac

Cisterna chyli

Retroperitoneal lymph sac

Cisterna chyli

Posterior lymph sac

Posterior sac

Superficial lymphatics   Lymph node

A

B

FIG. 10-2. Development of the primary lymphatic vessels in man. *A*, Profile reconstruction of the lymphatic system in a nine-week-old human embryo (Sabin, 1916). *B*, Diagram of the definitive thoracic duct emerging from a lymphatic plexus. (From Arey, L. B., *Developmental Anatomy*, 7th ed., W. B. Saunders Co., Philadelphia, 1965.)

DEVELOPMENT OF LYMPH NODES. The primary lymph nodes begin to develop during the third month in capillary lymphatic plexuses that are formed from the large lymph sacs. Secondary lymph nodes develop later, and even after birth, in peripherally located capillary lymphatic plexuses. Lymphocytes are already present in the bloodstream and tissues long before lymph nodes begin to appear. The earliest stages of lymph node development are obscure, but it seems probable that lymphocytes lodge in special regions of the capillary plexuses and multiply to form larger and larger masses that become the cortical lymphoid nodules and medullary cords. Parallel with this multiplication of lymphocytes, the surrounding mesenchyme forms a connective tissue capsule from which trabeculae carrying blood vessels grow into the lymphoid tissue. Collagenous fibers form in the larger trabeculae, and these continue as reticular fibrils into the smaller trabeculae. The mesenchymal cells

of the larger trabeculae become fibroblasts, and those accompanying the reticular fibrils are known as reticular cells. Just what happens to the lymphatic endothelium of the capillary plexus within the developing node is obscure. Some authors believe that it forms the cortical and medullary sinuses. Others believe that the endothelial lining becomes incomplete and that the reticular cells are washed by the lymph. With this idea goes the theory that some of the reticular cells become spherical to form the first lymphocytes of the developing node and that they retain this potency throughout life.

Distribution of Lymphatic Vessels

The vessels that form the lymphatic system consist of extensive **capillary plexuses** that drain into an intricate system of **collect-**

ing lymphatic vessels. These, in turn, flow into larger **lymphatic trunks** that end in the **cisterna chyli,** the **thoracic duct,** or the **right lymphatic duct** (Fig. 10-3).

LYMPHATIC CAPILLARY PLEXUSES.   The capillary plexuses that collect lymph from the intercellular fluid constitute the beginning of the lymphatic system, because it is from these plexuses that the lymphatic collecting vessels arise. These collecting vessels conduct the lymph centrally through one or more lymph nodes to the lymphatic trunks, cisterna chyli, thoracic duct, or right lymphatic duct. The number, size, and richness of the capillary plexuses differ in different regions and organs. Where abundant, they usually are arranged in two or more anastomosing layers. Most lymphatic capillaries do not contain valves.

Lymphatic capillaries are especially abundant in the **dermis of the skin.** They form a continuous network beneath the epidermis throughout the entire body, with the exception of the cornea. The dermis has a superficial plexus, without valves, that is connected by many anastomoses to a somewhat wider, coarser deep plexus with a few valves. The superficial plexus sends blind ends into the dermal papillae. The plexuses are especially rich over the palmar surface of the hands

and fingers, the plantar surface of the feet and toes, the conjunctiva, the scrotum, the vulva, and around orifices where the skin becomes continuous with mucous membranes (Fig. 10-4).

Lymphatic capillary plexuses are abundant in the **mucous membranes of the respiratory and digestive systems.** They form a continuous network from the nares and lips to the anus. In most places there is a subepithelial plexus in the mucosa that anastomoses freely with a coarser plexus in the submucosa. Blind ends extend between the tubular glands of the stomach and into the villi of the intestine. The latter are the **lacteals.** They have a smooth muscle coat and are contractile. Those portions of the alimentary canal that are covered by peritoneum have a subserous capillary plexus beneath the mesothelium that anastomoses with the submucosal set.

The **lungs** have a rich subserous plexus that connects with the deep plexuses within the lungs. The latter accompany the bronchi and bronchioles but their capillaries do not extend to the alveoli.

The **salivary glands, pancreas,** and **liver** possess deep lymphatic plexuses. They are perilobular and do not extend between the epithelial cells. The **gallbladder,** and the **cys-**

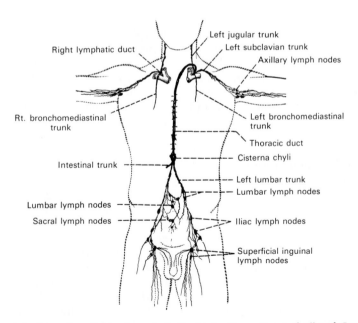

FIG. 10-3.   Schema of the larger lymphatic channels in the body. (From Benninghoff and Goerttler, *Lehrbuch der Anatomie des Menschen,* 10th ed., Urban and Schwarzenberg, 1975.)

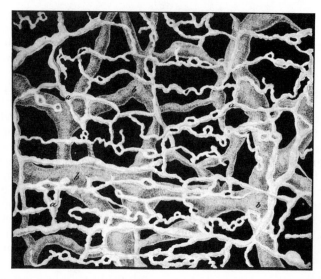

FIG. 10-4. Lymphatic capillaries of the dermis from the medial border of the sole of the human foot. Note the finer capillaries of the superficial plexus (a) and the larger vessels in the deeper plexus (b). 30 ×. (From Teichmann.)

tic, **hepatic,** and **common bile ducts** have rich plexuses in the mucosa. The deep lymphatics of the liver deliver a copious supply of lymph. The liver and gallbladder have rich subserous plexuses.

The **kidney** has a rich network in the capsule and a deep plexus between the tubules of the parenchyma. The renal pelvis and the ureters have rich networks in the mucosa and muscular layer that are continuous with similar plexuses of the **urinary bladder.** The male **urethra** has a dense plexus in the mucosa. The lymphatic capillaries are especially abundant around the navicular fossa.

The **testis, epididymis, ductus deferens, seminal vesicles,** and **prostate** have superficial capillary plexuses. Both the testis and prostate have deep interstitial plexuses.

The **ovary** has a rich capillary plexus in the parenchyma. Capillaries are absent in the tunica albuginea. The **uterine tubes** and **uterus** have mucous and muscular plexuses, as well as serous and subserous ones. The **vagina** has a rich fine-meshed plexus in the mucosa and a coarser one in the muscular layer.

Lymphatic capillary plexuses are abundant beneath the **mesothelial lining** of the pleural, peritoneal, and pericardial cavities and the joint capsules. The dense connective tissues of **tendons, ligaments,** and **periosteum** are richly supplied with plexuses. Very fine capillary plexuses around the fibers of

skeletal muscle have been described. Their occurrence in **bone** and **bone marrow** is not settled.

The **heart** has a rich subepicardial plexus as well as a subendocardial plexus. The myocardium has a uniform capillary plexus that anastomoses with both the subendocardial and subepicardial ones. There are no collecting trunks in the myocardium.

COLLECTING LYMPHATIC VESSELS AND LYMPHATIC TRUNKS.   Lymph drains from the lymphatic capillary plexuses into collecting lymphatic channels that resemble venules and enter lymph nodes as afferent vessels. From these initial nodes, several collecting lymphatic vessels join to form slightly larger collecting channels, which then accompany larger blood vessels and enter a series of lymph nodes in succession and anastomose frequently. These larger collecting lymphatic vessels often extend over long distances without a change of caliber. They are exceedingly delicate, and their coats are so transparent that lymph can readily be seen through them. The walls of collecting vessels are interrupted at intervals by constrictions, which correspond to the location of valves on their interior, and give the vessels a knotted or beaded appearance.

At certain sites in the body, groups of collecting vessels course in a parallel manner to form **lymphatic trunks.** These are especially seen along routes leading to the cisterna chyli and toward the termination of the right lymphatic and thoracic ducts. Leading into the cisterna chyli are the **intestinal trunk** and the **right** and **left lumbar trunks,** while the right lymphatic duct and the thoracic duct both receive the **bronchomediastinal, jugular,** and **subclavian trunks** on their respective sides, near the junctions of the two ducts with the venous system.

Lymphatic vessels are arranged in **superficial** and **deep sets.** The **superficial lymphatic vessels** are placed immediately beneath the integument, accompanying the superficial veins. They join the deep lymphatic vessels at certain sites by perforating the deep fascia. The **deep lymphatic vessels,** fewer in number but larger than the superficial, accompany the deep blood vessels.

In the interior of the body, lymphatic vessels lie in the submucous areolar tissue throughout the whole length of the digestive, respiratory, and genitourinary tracts, and in the subserous tissue of the thoracic and abdominal walls. Plexiform networks of minute lymphatic vessels are found interspersed among the parenchymal elements and blood vessels of the several tissues. The lymphatic vessels that compose the network, as well as the meshes between them, are much larger than those of the capillary plexus. From these networks small vessels emerge and pass either to a neighboring node or join some larger lymphatic trunk.

The lymphatic vessels in any region of the body or in any organ exceed the veins in number, but they are much smaller in diameter. Their anastomoses, especially those of the large trunks, are also more frequent. These anastomoses are effected by vessels equal in diameter to those that they connect, and the continuing trunks retain the same diameter.

Microscopic Structure of Lymphatic Vessels and Lymph Nodes

The lymphatic vessels are channels through which lymphatic fluid from tissue spaces passes in a one-way direction into the venous system. Along the course of the lymphatic vessels are located encapsulated accumulations of lymphatic tissue, the lymph nodes, within which the lymphocytes are formed.

STRUCTURE OF LYMPHATIC VESSELS

LYMPHATIC CAPILLARIES. Lymphatic capillaries are thin, endothelial-lined channels whose walls are even more delicate than those of capillaries of the blood vascular system (Fig. 10-5). Frequently the contiguous edges of adjacent endothelial cells overlap, and fine *anchoring filaments* of collagen appear to attach the outer surface of the endothelium to the surrounding tissue. Lymphatic capillaries do not possess a well-developed basal lamina, and not only do they allow the passage of extracellular fluids through their walls, but they also absorb back into the circulation considerable amounts of proteins that have escaped into the interstitial spaces.

LARGER LYMPHATIC VESSELS. Lymphatic vessels whose diameters measure 1 to 2 mm are considered to be collecting vessels, and their walls are somewhat thicker and better defined than those of lymphatic capillaries. Nonetheless, they are readily distinguishable from blood vessels because their walls are considerably thinner than those of arteries and veins of comparable diameter. The walls of lymphatic vessels that measure more than 2 mm in diameter are composed of three layers, similar to those of blood vessels, but the boundaries of these tunics are not easily distinguishable.

The *internal coat* is thin, transparent, and slightly elastic, consisting of a layer of elongated endothelial cells with wavy margins by which the contiguous cells are dovetailed. The endothelial cells are supported on an elastic membrane. The *middle coat* is composed of smooth muscle cells that are disposed in a circular or oblique manner, and fine elastic fibers that are oriented in a transverse direction. The *external coat* is fairly well developed and consists of connective tissue intermixed with smooth muscle fibers that course longitudinally or obliquely. This adventitia forms a protective covering over the other coats, and connects the vessel with the neighboring structures.

The **thoracic duct** is the largest of the lymphatic vessels, measuring about 0.5 cm in diameter. Its wall is somewhat more complex than that of other lymphatic vessels, having a *tunica intima* that consists of endothelial cells, deep to which are collagenous and elastic fibers. The subendothelial elastic fibers form an *internal elastic membrane*

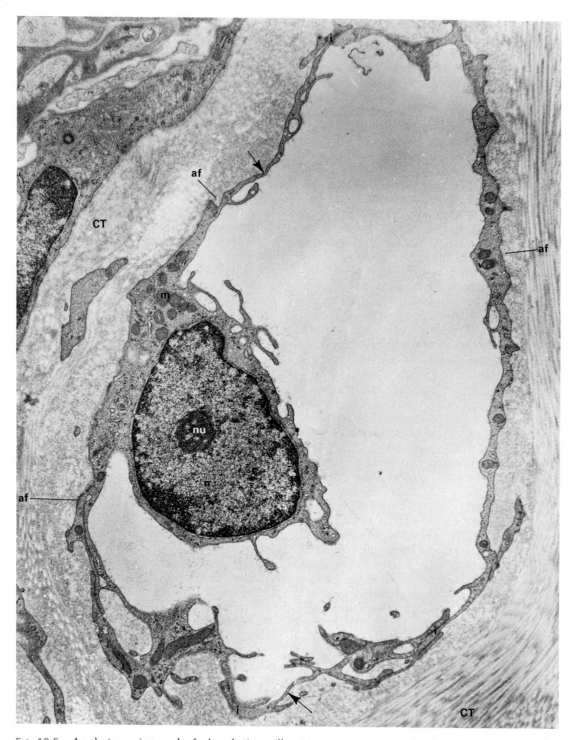

Fig. 10-5.   An electron micrograph of a lymphatic capillary in cross-section. Note the close association of the surrounding connective tissue (CT) with the capillary wall which is maintained by anchoring filaments (af). The extreme narrowing (arrows) achieved by the endothelium is illustrated at several regions of the capillary wall. Observe that the nucleus (n) protrudes into the capillary lumen. nu: nucleolus; i: several intercellular junctions; m: mitochondria. 11,000 ×. (From Leak, 1970).

similar to that found in the arteries. In the *tunica media* there is, in addition to muscular and elastic fibers, a layer of connective tissue with longitudinally arranged fibers. The *tunica adventitia* consists of both connective tissue and bundles of smooth muscle fibers, and it tends to blend with surrounding connective tissue.

The larger lymphatic vessels are supplied by **nutrient vessels,** which are distributed to their outer and middle coats, as well as by unmyelinated nerve fibers, which form fine plexuses in their wall.

The **valves** of lymphatic vessels are formed by thin layers of fibrous connective tissue, covered on both surfaces by endothelium, and having the same arrangement as the valves in veins. The valves consist of semilunar cusps that are attached by their convex edges to the wall of the vessel, the concave edges being free and directed along the course of the contained current. Usually two such cusps of equal size are found opposite each other, but occasionally exceptions occur, especially at or near the anatomoses of lymphatic vessels. In such cases one cusp may be small and the other proportionately larger. At times, a valve may consist of three cusps. Valves are placed at much shorter intervals in lymphatic vessels than in the veins. They are most numerous near lymph nodes, and are found more frequently in lymphatic vessels of the neck and upper extremity than in those of the lower extremity. The wall of the lymphatic vessel immediately above the point of attachment of each cusp of a valve is expanded into a pouch or sinus that gives to these vessels, when distended, a characteristic knotted or beaded appearance. Valves are wanting in the capillaries that compose the plexiform network, from which the larger lymphatic vessels originate. **Lymph** is propelled by contractions of the walls in the larger lymphatic vessels, or by the motion of surrounding structures in smaller vessels. The valves prevent the backward flow of lymph.

STRUCTURE OF LYMPH NODES

Lymph nodes are small oval or bean-shaped bodies that are situated along the course of lymphatic vessels so that lymph will filter through them on its way to the blood. Each node generally has a slight depression on one side, the **hilus,** through which the blood vessels enter and leave (Figs. 10-6; 10-7). The **efferent lymphatic vessels** also emerge from the hilus, while the **afferent lymphatic vessels** enter lymph nodes at various places on the periphery. When a fresh lymph node is transected, it displays a lighter, outer **cortex,** surrounding an inner, darker **medulla.** The boundary between these is indefinite, but at the hilus the cortex is deficient and the medulla extends to the nodal surface (Fig. 10-6). The afferent lymph vessels carry fluid to the nodal periphery, where it achieves the cortex, and after traversing the node, lymph leaves the medulla by way of the efferent lymphatic vessels at the hilus (Fig. 10-7).

Lymph nodes measure 1 to about 30 mm in diameter, and each consists of enormous numbers of **lymphocytes, plasma cells,** and **macrophages,** densely packed into masses that are partially subdivided into a series of cortical nodules and medullary cords by anastomosing connective tissue trabeculae, bordered by lymph sinuses. The **trabeculae** extend into the node from the collagenous connective tissue **capsule** and the hilus. An intricate network of **reticular fibers** extends from the trabeculae to form a supporting meshwork for the lymphoid elements (Fig. 10-8). A continuous narrow channel, probably lined by endothelial cells and macrophages, is found just beneath the capsule. From this **subcapsular** or **marginal sinus** arise other radially-oriented sinuses that border the trabeculae; these course through the cortex and medulla and are appropriately called **cortical sinuses** and **medullary sinuses** (Figs. 10-6 to 10-8). The afferent lymphatic vessels open into the subcapsular sinus, and the efferent vessels arise from the medullary sinuses. As lymph slowly passes through this system of channels from afferent to efferent vessels, its flow is retarded by the large numbers of reticular fibers that project across the sinuses in all directions. The elegantly intricate architecture of the lymph node allows the lymphatic fluid an abundant exposure to the lymphoid elements that surround the sinuses, thereby creating an effective means by which foreign elements in the tissue fluids may be detected, attacked, and phagocytosed. Numerous flattened stellate endothelial cells and macrophages cling to the reticular fibers.

The supporting stromal structures of a

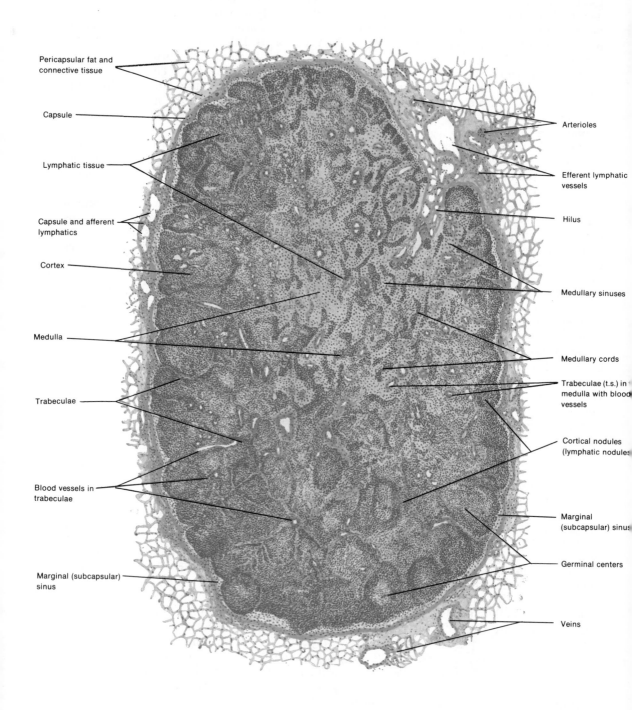

Pericapsular fat and
connective tissue

Capsule

Lymphatic tissue

Capsule and afferent
lymphatics

Cortex

Medulla

Trabeculae

Blood vessels in
trabeculae

Marginal (subcapsular)
sinus

Arterioles

Efferent lymphatic
vessels

Hilus

Medullary sinuses

Medullary cords

Trabeculae (t.s.) in
medulla with blood
vessels

Cortical nodules
(lymphatic nodules)

Marginal
(subcapsular) sinus

Germinal centers

Veins

FIG. 10-6.   A cross-section view through a lymph node. Stained with hematoxylin and eosin. 32 × . (From di Fiore, M., *Atlas of Human Histology*, 5th ed., Lea & Febiger, Philadelphia, 1981.)

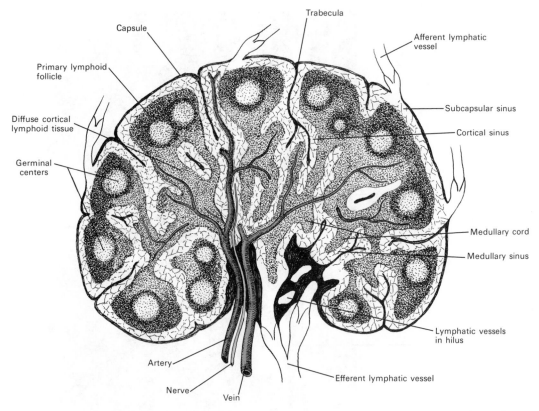

FIG. 10-7. Diagram of a lymph node indicating how the afferent lymphatic vessels open into the subcapsular sinus. Lymph then flows through the cortical sinuses and the medullary sinuses before leaving by way of the efferent lymphatic vessels. (After Bloom and Fawcett, *A Textbook of Histology,* 9th ed., 1968.)

lymph node include the collagenous connective tissue capsule, which also contains networks of elastic fibers and a few smooth muscle cells, its trabecular extensions, and the finer reticular fibers (Figs. 10-8; 10-9). The capsule is thicker at the hilum, where blood vessels enter and leave the node along with their accompanying vasomotor nerve fibers.

The parenchymal tissue of a lymph node is organized within the cortex into dense spherical masses of lymphoid cells, called **primary lymphoid follicles** or **primary nodules** (Figs. 10-6 to 10-8). Within these follicles are frequently found central areas where mitotic nuclei indicate multiplication of the lymphocytes. These areas are termed **germinal centers** or **secondary nodules**. The cells that compose them have more abundant protoplasm than the peripheral cells, and consist of large and medium-sized lymphocytes. Other, more diffusely disseminated lymphoid tissue is found between the primary follicles in the cortex of the node (Figs. 10-8; 10-9). Within the medulla, the lymph-

oid tissue is even more diffuse and helps to form the **medullary cords** that are oriented toward the hilus. These cords are aggregations of lymphocytes, plasma cells, and macrophages that usually surround blood vessels and reticular fibers.

HEMAL NODES.   Although a few red blood cells are normally found in lymph nodes, hemal nodes contain exceptionally large numbers of erythrocytes. They are found in certain animals, notably ruminants, but probably only occur as pathologic structures in man. Hemal nodes are quite small and resemble lymph nodes in structure, but are likely to be without afferent and efferent lymphatic vessels and are engorged with blood.

LYMPH.   Although lymph should be defined as the fluid within the walls of a closed system consisting of the lymphatic vessels, it should be understood that the composition of the lymph found in a regional set of lymphatics is nearly identical with that of the interstitial or tissue fluid of the region being

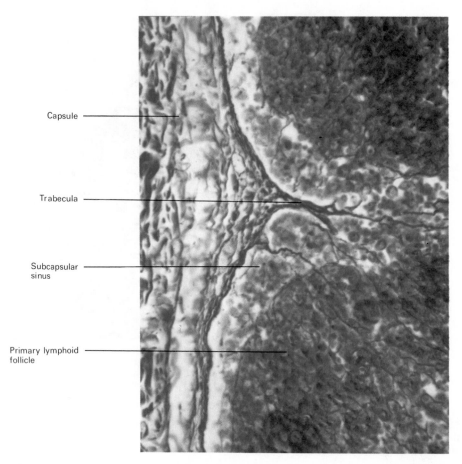

Capsule

Trabecula

Subcapsular
sinus

Primary lymphoid
follicle

FIG. 10-8. This is a photomicrograph of a human lymph node showing the capsule, the subcapsular sinus and portions of two primary lymphoid follicles which are separated by a darkly impregnated connective tissue trabecula. 360 ×. (Counterstained silver preparation.)

drained. Thus, lymphatic fluids derived from different parts of the body have different chemical compositions. Interstitial fluid has an *average* protein concentration of 2 gm/100 ml, as does lymph; however, lymph drained from the liver and intestines may have protein levels two to three times higher, and after a fatty meal, lymph from the intestinal lacteals may have a fat concentration high enough to make the fluid milky white.

Frequently, lymph has been called an ultrafiltrate of blood plasma, but the important issue is that lymphatic capillaries are capable of reabsorbing proteins, a process that does not occur to any appreciable degree at the venule end of the blood capillary system. Although venules reabsorb water, the proteins that leak into the interstitial spaces remain behind in the tissues. Unless reabsorption of proteins by the lymphatic capil-

laries occurs, death can follow in one or two days.

Lymph that drains from the limbs is generally a colorless or slightly yellow watery fluid with a specific gravity slightly over 1. When examined under the microscope, leukocytes of the lymphocyte class are found floating in the transparent fluid. They always increase in number after the passage of the lymph through lymphoid tissue, as in lymph nodes.

Large Lymphatic Vessels

THORACIC DUCT

The **thoracic duct** conveys a greater part of the lymph in the body back into the blood vascular system (Figs. 10-10; 10-22). It is the

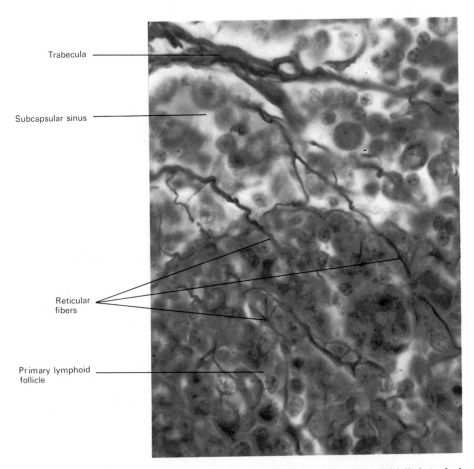

Trabecula

Subcapsular sinus

Reticular
fibers

Primary lymphoid
follicle

FIG. 10-9. High magnification of a portion of the subcapsular sinus and primary lymphoid follicle in the lymph node shown in Figure 10-8. Note the meshwork of reticular fibers and dense masses of lymphoid tissue which characterize the follicle and the looser lymphoid elements in the subcapsular sinus. 900 ×. (Counterstained silver preparation.)

common trunk of all the lymphatic vessels *except* those on the right side of the head, neck, and thorax, the right upper limb, the right lung, right side of the heart, and the diaphragmatic surface of the liver. In the adult, the thoracic duct varies in length from 38 to 45 cm and extends from the second lumbar vertebra to the root of the neck. It *begins* in the abdomen by a dilatation, the **cisterna chyli,** which is situated on the anterior surface of the body of the first and second lumbar vertebrae, to the right side of and behind the aorta, and adjacent to the right crus of the diaphragm. The duct then enters the thorax through the aortic hiatus of the diaphragm, and ascends through the posterior mediastinum between the aorta and azygos vein. *Dorsal* to it in this region are the vertebral column, the right intercostal arteries, and the hemiazygos veins, which cross the midline to open into the azygos

vein. *Ventral* to it are the diaphragm, the esophagus, and the pericardium, the latter being separated from the duct by a recess in the right pleural cavity.

Opposite the fifth thoracic vertebra, the thoracic duct inclines toward the left side and enters the superior mediastinum. Its ascent in the thorax continues to the thoracic inlet, dorsal to the aortic arch and the thoracic part of the left subclavian artery, and between the left side of the esophagus and the left pleura.·

Passing into the root of the neck on the left side, the thoracic duct forms an arch that rises about 3 or 4 cm above the clavicle, crossing anterior to the subclavian artery, the vertebral artery and vein, and the thyrocervical trunk or its branches. It also passes ventral to the phrenic nerve and the medial border of the scalenus anterior muscle, and dorsal to the left common carotid artery,

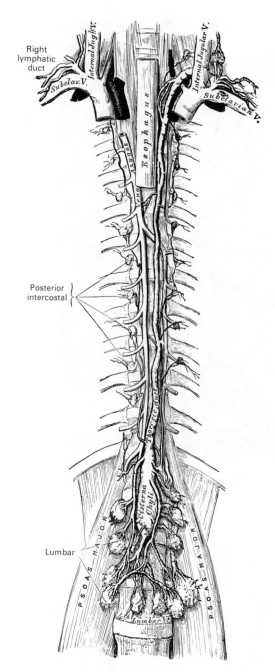

FIG. 10-10. The thoracic and right lymphatic ducts.

termination. It is generally flexuous and constricted at intervals and presents a varicose appearance. Not infrequently, it divides in the middle of its course into two vessels of unequal size that soon reunite, or into several branches that form a plexiform interlacement. It occasionally divides in the upper part of its course into two branches, right and left, the left ending in the usual manner, while the right opens into the right subclavian vein with the right lymphatic duct.

The thoracic duct has several valves. At its junction with the venous system, it has a bicuspid valve, and the free borders of the cusps of this valve are directed toward the vein to prevent the reflux of venous blood into the duct.

CISTERNA CHYLI. The cisterna chyli is a dilated sac anterior to the bodies of the two uppermost lumbar vertebrae; it forms the origin of the thoracic duct (Figs. 10-10; 10-11). It lies lateral to the right crus of the diaphragm, and it receives the right and left lumbar lymphatic trunks, along with the intestinal lymphatic trunks. The **lumbar trunks** are formed by the union of the efferent vessels from the lateral aortic lumbar lymph nodes. They receive lymph from the lower limbs, the walls and viscera of the pelvis, the kidneys, suprarenal glands, ovaries or testes, and the deep lymphatics of the greater part of the abdominal wall. The **intestinal trunks** consist of large lymphatic vessels that receive lymph from the stomach and intestine, the pancreas and spleen, and the visceral surface of the liver.

Tributaries of the Thoracic Duct

Opening into the thoracic duct near the cisterna chyli, on each side, is a descending trunk from the posterior intercostal lymph nodes that drains the lower six or seven intercostal spaces. On each side of the thorax, the duct is joined by a trunk that drains the upper lumbar lymph nodes and pierces the crus of the diaphragm. It also receives efferent vessels from the posterior mediastinal lymph nodes and from the posterior intercostal lymph nodes of the upper six spaces on the left side. In the neck, it is frequently joined by the **left jugular trunk,** which drains the left side of the head and neck, and the **left subclavian trunk** from the left upper extremity. Sometimes it also receives the **left**

vagus nerve, and internal jugular vein. It *terminates* by opening into the angle of junction of the left subclavian vein and the left internal jugular vein.

The thoracic duct is about 3 to 5 mm in diameter at its commencement, but its caliber diminishes considerably in the middle of the thorax, and again dilates just before its

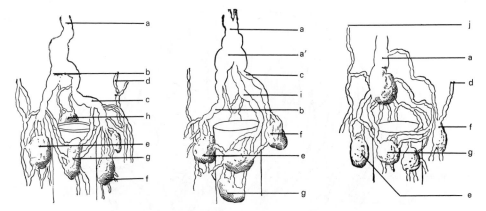

FIG. 10-11.   Modes of origin of thoracic duct. *a*, Thoracic duct. *a′*, Cisterna chyli. *b*, *c*, Efferent trunks from lateral aortic node. *d*, An efferent vessel which pierces the left crus of the diaphragm. *e*, *f*, Lateral aortic nodes. *h*, Retroaortic nodes. *i*, Intestinal trunk. *j*, Descending branch from intercostal lymphatics. (Poirier and Charpy.)

bronchomediastinal trunk; however, this vessel usually opens independently into the junction of the left subclavian and internal jugular veins.

RIGHT LYMPHATIC DUCT

The right lymphatic duct measures about 1.25 cm in length. It courses along the medial border of the right scalenus anterior muscle at the root of the neck, and ends in the right subclavian vein, at its angle of junction with the right internal jugular vein. Its orifice is guarded by a bicuspid valve, the two semilunar cusps of which prevent the passage of venous blood into the duct.

Tributaries of the Right Lymphatic Duct

The right lymphatic duct receives lymph from the right side of the head and neck through the **right jugular trunk,** from the

right upper extremity through the **right subclavian trunk,** and from the right side of the thorax, right lung, right side of the heart, and part of the convex surface of the liver through the **right bronchomediastinal trunk.** Although a single right lymphatic duct, formed by the junction of these three trunks, occurs at times, more frequently these trunks open separately near the angle of junction of the right internal jugular and right subclavian veins (Fig. 10-12).

Lymphatic Vessels and Nodes of the Head and Neck

For the most part, lymph that drains from the superficial structures in the head initially flows into superficial vessels and nodes in the upper neck (Fig. 10-13). It is then directed into the deep cervical nodes, which also re-

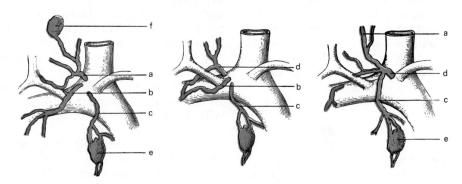

FIG. 10-12.   Terminal collecting trunks of right side. *a*, Jugular trunk. *b*, Subclavian trunk. *c*, Bronchomediastinal trunk. *d*, Right lymphatic trunk. *e*, Node of internal mammary chain. *f*, Node of deep cervical chain. (Poirier and Charpy.)

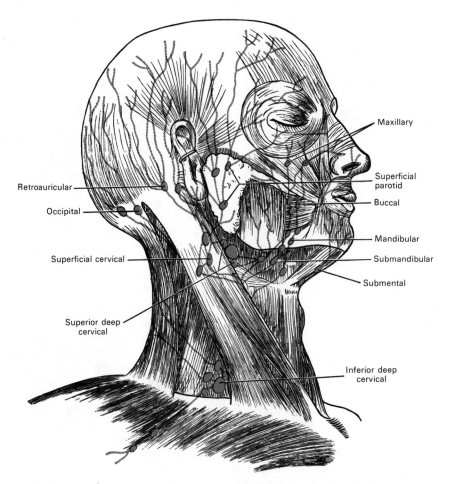

FIG. 10-13. Superficial lymph nodes and lymphatic vessels of head and neck.

ceive lymphatic channels from deeper structures in the face and head, as well as from the deeper viscera of the neck. In the deep neck the lymphatic vessels join to form the jugular trunk, which on the right side joins the right lymphatic duct or opens directly into the right subclavian vein, and on the left side joins the thoracic duct. The following description will consider first the lymphatic vessels and lymph nodes of the head, and then the lymphatic vessels and nodes of the neck.

LYMPHATIC VESSELS OF THE HEAD

The lymphatic vessels of the head include those that drain the scalp, external ear, face, walls of the nasal cavity, and mouth, as well as those of the walls of the oral cavity, palatine tonsils, and tongue.

LYMPHATIC VESSELS OF THE SCALP. The lymphatic vessels that drain the scalp include those from the temporoparietal region and those that drain the occipital region (Fig. 10-13). The **temporoparietal vessels** end in superficial parotid and retromandibular nodes, while the **occipital vessels** terminate partly in the occipital nodes and partly in a trunk that runs downward along the posterior border of the sternocleidomastoid muscle and ends in the inferior deep cervical nodes. Lymphatic vessels that drain the frontal region of the scalp above the nose will be considered with the facial channels.

LYMPHATIC VESSELS OF THE EXTERNAL EAR. The lymphatic vessels draining the auricula of the external ear and the walls of the external acoustic meatus are also divisible into three groups: anterior, posterior, and inferior. The **anterior auricular vessels** drain the lateral surface of the pinna and the anterior wall of the meatus and flow to the anterior auricular nodes, situated immediately in front of the tragus. The **posterior auricular vessels** drain the margin of the auricula and the upper part of its cranial surface, and the internal surface and posterior wall of the meatus, and flow to the posterior auricular and superior deep cervical nodes. The **inferior auricular vessels** from the floor of the external acoustic meatus and the lobule of the auricula drain downward into superficial and superior deep cervical nodes.

LYMPHATIC VESSELS OF THE FACE. The lymphatic vessels of the face are more numerous than those of the scalp. The **palpebral vessels** from the eyelids and the **conjunctival vessels** drain partly downward to the submandibular nodes, but principally laterally and posteriorly to the parotid nodes (Figs. 10-13; 10-14). Lymphatic vessels from the posterior part of the cheek also pass to the parotid nodes, while those from the anterior portion of the cheek, the side of the nose, the upper lip, and the lateral portions of the lower lip end in the submandibular nodes. The **deep temporal and infratemporal vessels** pass to the deep facial and superior deep cervical nodes. The deeper vessels of the cheek and lips, like the more superficial lymphatics, terminate in the submandibular nodes. Both **superficial** and **deep labial ves-**sels from the central part of the lower lip course to the submental nodes.

LYMPHATIC VESSELS OF THE NASAL CAVITY. From the anterior parts of the walls of the nasal cavity, lymphatic vessels communicate with channels from the integument of the nose, and therefore, end in the submandibular nodes. Lymphatic vessels from the posterior two-thirds of the nasal cavity and the paranasal air sinuses pass partly to the retropharyngeal and partly to the superior deep cervical nodes.

LYMPHATIC VESSELS OF THE MOUTH. Lymphatic vessels of the *gums* pass to the submandibular nodes. Channels that drain the *hard palate* are continuous in front with those of the upper gum, but pass backward to pierce the superior pharyngeal constrictor muscle and end in the superior deep cervical and subparotid nodes. Lymphatic vessels of the *soft palate* pass backward and laterally, and end partly in the retropharyngeal and subparotid nodes, and partly in the superior deep cervical nodes. The vessels of the anterior part of the *floor of the mouth* pass either directly to the inferior nodes of the superior deep cervical group, or indirectly through the submental nodes. Lymphatics from the remainder of the floor of the mouth pass to the submandibular and superior deep cervical nodes.

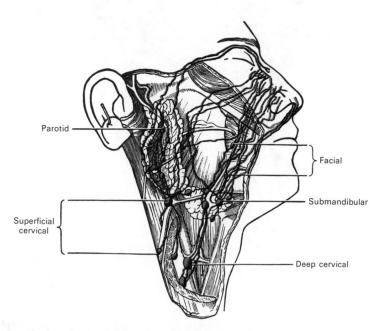

Parotid

Facial

Submandibular

Superficial cervical

Deep cervical

FIG. 10-14. The lymphatics and lymph nodes of the lateral aspect of the face. (After Küttner.)

Lymphatic Vessels from Palatine Tonsil. Usually three to five lymphatic vessels arise from the palatine tonsil. These pierce the buccopharyngeal fascia and the superior pharyngeal constrictor muscle, and pass between the stylohyoid muscle and the internal jugular vein to the most superior deep cervical node (Fig. 10-15). They end in a node, called the **jugulodigastric node,** which lies at the side of the posterior belly of the digastric muscle and adjacent to the internal jugular vein. Occasionally one or two additional vessels course to small nodes on the lateral side of the internal jugular vein, under cover of the sternocleidomastoid muscle.

Lymphatic Vessels from the Tongue. Lymphatic vessels that arise in the tongue drain chiefly into the deep cervical nodes that lie between the posterior belly of the digastric muscle and the superior belly of the omohyoid muscle (Fig. 10-15). One lymph node, situated at the bifurcation of the common carotid artery and posterior to the internal jugular vein and the omohyoid muscle, is intimately associated with these vessels; it is called the **jugulo-omohyoid node** and has frequently been referred to as the principal node of the tongue. The lymphatic vessels of the tongue can be divided into four groups: **apical** (vessels from the tip of the tongue that course to the suprahyoid nodes and the jugulo-omohyoid node); **lateral** or **marginal** (vessels from the margin of the tongue, some of which pierce the mylohyoid muscle and end in the submandibular nodes, while others pass downward on the hyoglossus muscle to the superior deep cervical nodes) (Fig. 10-16); **basal** (channels from the region of the vallate papillae that travel to the superior deep cervical nodes); and **median** or **central** (a few vessels from the more anterior central region) that perforate the mylohyoid muscle to reach the submental and submandibular nodes, although the majority turn around the posterior bor-

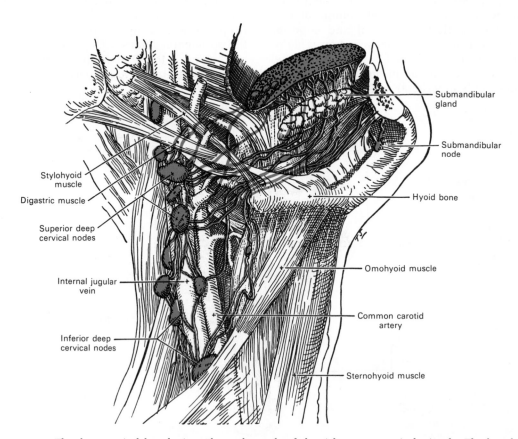

Submandibular gland

Submandibular node

Stylohyoid muscle

Digastric muscle

Superior deep cervical nodes

Hyoid bone

Internal jugular vein

Omohyoid muscle

Common carotid artery

Inferior deep cervical nodes

Sternohyoid muscle

FIG. 10-15.   The deep cervical lymphatic nodes and vessels of the right upper cervical triangle. The lymphatic drainage of the tongue is shown. (Eycleshymer and Jones.)

der of the muscle and enter both the superior and inferior deep cervical lymph nodes.

It is important to realize that lymphatics from the apical, median, and basal parts of the tongue course to deep cervical nodes on *both sides* of the neck, whereas lymphatic vessels from the lateral margins of the tongue initially flow to nodes on the *same side* (Fig. 10-16).

LYMPH NODES OF THE HEAD

The lymph nodes of the head are arranged in the following groups:

Occipital
Retroauricular
Superficial parotid
Deep parotid
Facial
Deep facial
Lingual
Retropharyngeal

OCCIPITAL LYMPH NODES.   The occipital lymph nodes, one to three in number, are located on the back of the head, close to the margin of the trapezius muscle, and rest on the insertion of the semispinalis capitis (Figs. 10-13; 10-18). Their afferent vessels drain the occipital region of the scalp, while their efferent vessels pass to the superior deep cervical nodes.

RETROAURICULAR LYMPH NODES.   The retroauricular lymph nodes, usually two in number, are situated on the mastoid insertion of the sternocleidomastoid muscle, deep to the posterior auricular muscle (Figs. 10-13; 10-18). Their afferent vessels drain the posterior part of the temporoparietal region, the upper part of the cranial surface of the auricula or pinna, and the posterior part of the external acoustic meatus; efferent vessels pass to the superior deep cervical nodes.

SUPERFICIAL PAROTID LYMPH NODES.   The superficial parotid or preauricular nodes usually number three or more, and they lie immediately anterior to the tragus of the external ear (Figs. 10-13; 10-14). Their afferent vessels drain the lateral surface of the auricula, external acoustic meatus, root of the nose, eyelids, and frontotemporal region of the scalp, while their efferent vessels pass to the superior deep cervical nodes.

DEEP PAROTID LYMPH NODES.   The deep parotid lymph nodes are arranged in two groups. One group is embedded in the substance of the parotid gland, and a second group, the subparotid nodes, is located deep to the gland and lies on the lateral wall of the pharynx. These deep parotid nodes receive lymphatic afferents from the parotid gland, the posterior parts of the palate, the floor of the nasal cavity, and the walls of the tympanic cavity of the middle ear; the efferent vessels of these nodes pass to the superior deep cervical nodes. Other afferent vessels to the subparotid nodes drain the nasal part of the pharynx and the posterior parts of the nasal cavities, while the efferent vessels also pass to the superior deep cervical nodes.

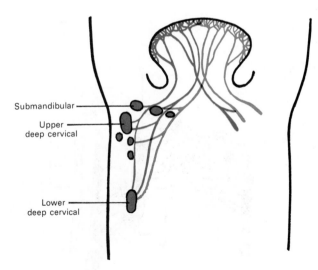

FIG. 10-16.   A diagram to show the course of the marginal and central lymphatic vessels of the tongue to the lymph nodes on both sides of the neck. (Jamieson and Dobson.)

FACIAL LYMPH NODES.   The facial lymph nodes consist of maxillary, buccal, and mandibular nodes (Fig. 10-13). The **maxillary nodes** are scattered over the infraorbital region, from the groove between the nose and cheek to the zygomatic arch. The **buccal nodes,** one or more, are placed over the buccinator muscle, opposite the angle of the mouth. The **mandibular nodes** lie on the outer surface of the mandible, in front of the masseter muscle and in contact with the facial artery and vein. Their afferent vessels drain the eyelids, the conjunctiva, and the skin and mucous membrane of the nose and cheek, and their efferent vessels pass to the submandibular nodes (Figs. 10-13; 10-14).

DEEP FACIAL LYMPH NODES.   The deep facial or internal maxillary nodes are placed deep to the ramus of the mandible, on the outer surface of the lateral pterygoid muscle, in relation to the maxillary artery. Their afferent vessels drain the temporal and infratemporal fossae and the nasal part of the pharynx, while their efferents pass to the superior deep cervical nodes.

LINGUAL LYMPH NODES.   Two or three small lingual nodes are located along the surface of the hyoglossus muscle and under the genioglossus muscle. They serve as substations in the course of the lymphatic vessels of the tongue to the deep cervical nodes.

RETROPHARYNGEAL LYMPH NODES.   The retropharyngeal lymph nodes, one to three in number, lie in the buccopharyngeal fascia,

behind the upper part of the pharynx and anterior to the arch of the atlas, from which they are separated by the longus capitis muscle (Fig. 10-17). Their afferent vessels drain the nasal cavity, nasal part of the pharynx, and auditory tubes, while their efferent vessels pass to the superior deep cervical nodes.

LYMPHATIC VESSELS OF THE NECK

In addition to draining the skin and muscles of the neck, the lymphatic vessels of this region serve the pharynx and upper esophagus, the larynx and upper trachea, and the thyroid gland. Lymphatics from the **skin and strap muscles** of the anterolateral neck drain into the deep cervical nodes.

The lymphatic vessels from the **pharynx** generally drain backward, laterally, and downward, while those from the nasopharynx and pharyngeal tonsil drain to the lateral retropharyngeal nodes (Fig. 10-17). From the oropharynx and palatine tonsil, lymphatics drain to the superior deep cervical chain of nodes, and especially to the jugulodigastric node (Figs. 10-15; 10-18). From the laryngeal pharynx and upper esophagus, lymphatic vessels course in the piriform sinus and paratracheal region to the inferior deep cervical nodes.

Numerous lymphatic vessels drain the **larynx.** They are divided into two sets, a superior and an inferior, by the vocal fold, which

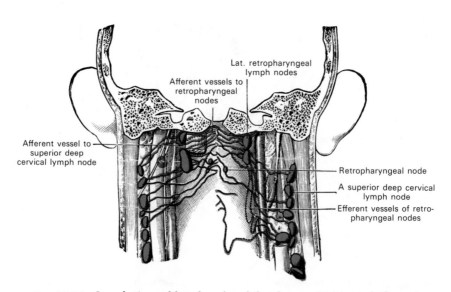

FIG. 10-17.   Lymphatics and lymph nodes of the pharynx. (Poirier and Charpy.)

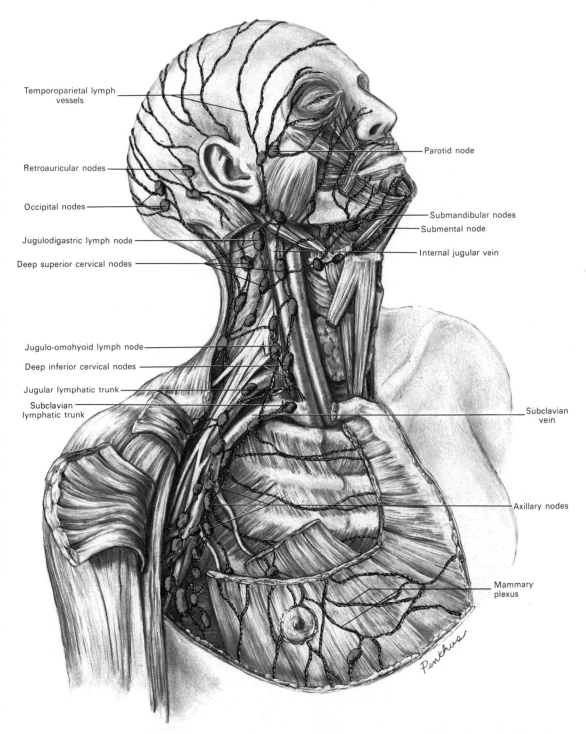

Temporoparietal lymph vessels

Retroauricular nodes

Occipital nodes

Jugulodigastric lymph node

Deep superior cervical nodes

Jugulo-omohyoid lymph node

Deep inferior cervical nodes

Jugular lymphatic trunk

Subclavian lymphatic trunk

Parotid node

Submandibular nodes

Submental node

Internal jugular vein

Subclavian vein

Axillary nodes

Mammary plexus

Penkhus

FIG. 10-18. The deep lymphatic vessels and nodes on the right side of the head, neck, and axilla.

itself does not contain numerous vessels. The vessels of the *superior set* pierce the thyrohyoid membrane, course with the superior laryngeal artery, and join the superior deep cervical nodes. The *inferior set* of la-

ryngeal lymphatic vessels courses along branches of the inferior thyroid artery. Some pierce the conus elasticus and join the pretracheal and prelaryngeal nodes, while others course between the cricoid and first tracheal ring and descend to the inferior deep cervical nodes.

The lymphatic vessels of the **thyroid gland** also consist of superior and inferior sets; these follow the branches of the superior and inferior thyroid artery. The *superior set* drains the upper margin of the isthmus and the upper parts of the lateral lobes of the gland, and enters the superior deep cervical nodes. The *inferior set* drains most of the isthmus and the lower parts of the lateral lobes. Both sets descend to the deep inferior cervical nodes by way of both the pretracheal and paratracheal nodes, the latter accompanying the recurrent laryngeal nerves. Lymphatic vessels from the cervical portion of the trachea also drain into these paratracheal nodes. Some lymphatic vessels of the thyroid gland may open directly into the large veins in the root of the neck, or into the thoracic duct.

LYMPH NODES OF THE NECK

The lymph nodes of the neck include the following groups:

Submandibular
Submental
Superficial cervical
Anterior cervical
Deep cervical

SUBMANDIBULAR LYMPH NODES.   The submandibular chain of lymph nodes, three to six in number, are located beneath the body of the mandible in the submandibular triangle (Figs. 10-13 to 10-15; 10-18). They usually rest on the superficial surface of the submandibular salivary gland, but may extend as far forward as the border of the anterior belly of the digastric muscle. One node, which lies on the facial artery as it turns over the mandible, is the largest and most constant of the series. Additional small lymph nodes are sometimes found on the deep surface of the submandibular salivary gland. Afferent lymphatic vessels to the submandibular nodes drain the medial palpebral commissure of the eye, the cheek, the side of the nose, the upper lip, the lateral part of the lower lip, the gums and teeth, the anterior part of the margin of the tongue (except the tip), the frontal, maxillary, and ethmoidal sinuses, as well as the efferent vessels from the facial and submental nodes. Thus,

not only superficial integumentary structures, but also mucous membranes, drain into the submandibular nodes, and in turn, their efferent vessels pass to both superficial cervical and superior deep cervical nodes.

SUBMENTAL LYMPH NODES.   Three or four small submental lymph nodes are found in the submental triangle between the anterior bellies of the digastric muscles (Figs. 10-13; 10-18). Their afferents drain a centrally placed wedge of tissue that includes the incisor region of the lower gum and lips, tip of the tongue, central part of the floor of the mouth, and skin of the chin. Efferent vessels from the submental nodes pass partly to the submandibular nodes, and partly to the deep cervical chain, especially the juguloomohyoid node.

SUPERFICIAL CERVICAL LYMPH NODES.   The superficial cervical lymph nodes lie in close relationship with the external jugular vein as it emerges from the parotid gland, and therefore are located superficial to the sternocleidomastoid muscle (Figs. 10-13; 10-14). Their afferent vessels drain the lower parts of the external ear and parotid region, while their efferent channels pass around the anterior margin of the sternocleidomastoid muscle to join the superior deep cervical nodes.

ANTERIOR CERVICAL LYMPH NODES.   The anterior cervical lymph nodes form an irregular and inconstant group located below the hyoid bone and in front of the larynx, trachea, and thyroid gland. They may be divided into a superficial and a deep set. The **superficial set** is placed on the anterior jugular vein and drains the skin of the anterior region of the neck below the hyoid bone. The **deep set** is further subdivided into *prelaryngeal nodes,* which lie on the middle cricothyroid ligament, and *pretracheal nodes,* which are located ventral to the trachea. This deeper set drains the lower part of the larynx, the thyroid gland, and the cervical part of the trachea. Its efferent vessels pass to the most inferior of the superior deep cervical nodes.

DEEP CERVICAL LYMPH NODES.   The deep cervical lymph nodes are numerous and large, forming a chain of 20 to 30 nodes along the carotid sheath and around the internal jugular vein. Lying by the side of the pharynx, esophagus, and trachea, the deep cervical nodes extend from the base of the skull to the root of the neck. They are usually described in two groups, the superior and inferior deep cervical nodes.

Superior Deep Cervical Lymph Nodes. The superior deep cervical nodes are located above the level at which the omohyoid muscle crosses the common carotid artery (Figs. 10-13 to 10-15; 10-17; 10-18). They lie deep to the sternocleidomastoid muscle, in close relationship to the accessory nerve and the internal jugular vein; some of the nodes lie anterior, and others posterior, to the vessel. The superior deep cervical nodes drain the occipital portion of the scalp, the auricula, the back of the neck, a considerable part of the tongue, and the larynx, thyroid gland, trachea, nasal part of the pharynx, nasal cavities, palate, and esophagus. They also receive the efferent vessels from all the other nodes of the head and neck, except the inferior deep cervical nodes. The most superior, and one of the most important of these deep cervical nodes, lies just below the site where the posterior belly of the digastric muscle crosses the internal jugular vein and internal carotid artery. It is called the **jugulodigastric lymph node** and is located in a rather superficial position that allows it to be palpated with ease in the living patient (Fig. 10-18). It is sometimes called the *subdigastric* or *tonsillar node,* and it becomes swollen when the tonsil or pharynx is inflamed, or when there is a cancer on the posterior third of the tongue or the palatine tonsil.

Inferior Deep Cervical Lymph Nodes. The inferior deep cervical nodes extend beyond the posterior margin of the sternocleidomastoid muscle and into the supraclavicular triangle, where they are in close relationship to the brachial plexus and subclavian vein (Figs. 10-13; 10-15; 10-18). One large and relatively constant node in this group is the **jugulo-omohyoid lymph node** (Fig. 10-18). It lies on or immediately behind the internal jugular vein, just superior to the tendon of the omohyoid muscle. The inferior deep cervical nodes drain the back of the scalp and neck, the superficial pectoral region, part of the arm, and occasionally part of the superior surface of the liver. In addition, they receive vessels from the tongue and the superior deep cervical nodes. These latter nodes pass partly to the inferior deep cervical nodes, and partly to a trunk that unites with the efferent vessel of the inferior deep cervical nodes to form the **jugular trunk.** On the right side, this trunk ends in the junction of the internal jugular and subclavian veins, while on the left side it joins the thoracic

duct. A few minute **paratracheal nodes** are situated alongside the recurrent laryngeal nerves on the lateral aspects of the trachea and esophagus.

Lymphatic Vessels and Nodes of the Upper Limb

The lymphatic drainage of the upper limb is directed along the routes of the major blood vascular channels. The superficial lymphatics generally follow the veins, while the deep lymphatics course with the arteries. For the most part, lymph from the upper limb drains to axillary nodes because the arm, forearm, and hand possess very few nodes and those few are exceedingly small.

LYMPHATIC VESSELS OF THE UPPER LIMB

The lymphatic vessels of the upper limb consist of two sets, superficial and deep.

SUPERFICIAL LYMPHATIC VESSELS. The superficial lymphatic vessels of the upper extremity commence in the complex lymphatic plexus that pervades the skin throughout the entire limb. The meshes of the plexus are much denser in the palm of the hand and on the flexor aspect of the digits than elsewhere. The **digital plexuses** are drained by vessels that course along the sides of each finger and incline backward to reach the dorsum of the hand (Fig. 10-19). From the dense **palmar plexus,** vessels pass in different directions, proximally toward the wrist, distally to join the digital vessels, medially to join the vessels on the ulnar border of the hand, and laterally to those on the thumb. Several vessels from the central part of the palmar plexus unite to form a trunk that passes around the metacarpal bone of the index finger to join the vessels on the back of that digit and on the back of the thumb. Continuing proximally from around the wrist, the lymphatic vessels are collected, more or less as parallel channels, into **radial, median,** and **ulnar** groups, which accompany, respectively, the cephalic, median, and basilic veins in the forearm (Fig. 10-20). A few of the ulnar lymphatics end in the small **supratrochlear nodes** that are located superficially just above the medial aspect of the cubital fossa, but the majority

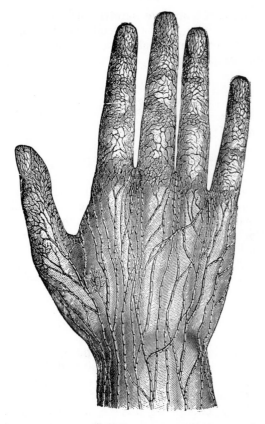

Fig. 10-19. Lymphatic vessels of the dorsal surface of the hand. (Sappey.)

pass directly to the lateral group of **axillary nodes.** Some of the radial vessels are collected into a trunk that ascends with the cephalic vein to the **deltopectoral nodes** (Figs. 10-20; 10-21). The efferents from this group may penetrate the clavipectoral fascia, with the cephalic vein, to enter the apical group of axillary nodes in the infraclavicular region, or they may ascend as far as the lower deep cervical nodes above the clavicle.

DEEP LYMPHATIC VESSELS. The deep lymphatic vessels accompany the deep blood vessels. In the forearm, they consist of four sets, which correspond to the radial, ulnar, anterior, and posterior interosseous arteries. They communicate at intervals with the superficial lymphatics, and some channels end in nodes that are occasionally found with these arteries. Along their ascent proximally in the arm, these deeper lymphatic vessels follow the course of the brachial artery. A few may even terminate in small nodes along these vessels, but most of them pass to the lateral group of axillary nodes.

LYMPH NODES OF THE UPPER LIMB

For the most part the lymph nodes of the upper limb are grouped in the axilla. However, there exists a set of somewhat isolated, outlying, and more superficial nodes, which are small and receive afferent lymphatic vessels from some of the ascending channels of the forearm and arm. Therefore, the lymph nodes of the upper extremity will be described as consisting of superficial and deep sets.

SUPERFICIAL LYMPH NODES. The superficial lymph nodes in the upper extremity are few and small, and they are located either deep to the epidermis or in the underlying superficial fascia, closely associated with large superficial veins. They include both supratrochlear and deltopectoral nodes.

Supratrochlear Nodes. One or as many as five small supratrochlear nodes are present in the superficial fascia, above the medial epicondyle of the humerus, medial to the basilic vein (Fig. 10-20). Their afferents drain the middle, ring, and little fingers, the medial portion of the hand, and the superficial area over the ulnar side of the forearm, but these vessels are in free communication with the other lymphatic vessels of the forearm. Their efferent vessels also accompany the basilic vein and some join the deeper vessels.

Deltopectoral Nodes. One or two deltopectoral nodes are found beside the cephalic vein and within the deltopectoral triangle, between the pectoralis major and deltoid muscles, immediately below the clavicle.

DEEP LYMPH NODES. The deep lymph nodes are chiefly grouped in the axilla, although a few tiny nodes may be found in the forearm, along the course of the radial, ulnar, and interosseous arteries, and in the arm along the medial side of the brachial artery.

The Axillary Nodes. The axillary lymph nodes (Fig. 10-21) are large, vary from 20 to 30 in number, and are arranged in the following groups:

Lateral
Pectoral (anterior)
Subscapular (posterior)
Central
Apical (subclavicular)

Lateral Group. The lateral group consists of four to six axillary nodes that are located

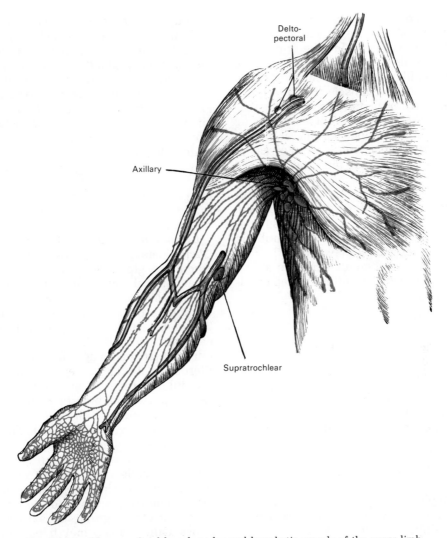

Delto-
pectoral

Axillary

Supratrochlear

FIG. 10-20. The superficial lymph nodes and lymphatic vessels of the upper limb.

medial and posterior to the axillary vein (Fig. 10-21). The afferents of these nodes drain the whole upper limb with the exception of that part whose vessels accompany the cephalic vein. The efferent vessels pass partly to the central and apical groups of axillary nodes, and partly to the inferior deep cervical nodes.

Pectoral Group. The pectoral or *anterior group* of four or five nodes lies along the lateral border of the pectoralis minor muscle, in relationship to the lateral thoracic artery (Fig. 10-21). Their afferent vessels drain the skin and muscles of the anterior and lateral thoracic walls, and the central and lateral parts of the mammary gland; their efferent vessels pass partly to the central and partly to the apical groups of axillary nodes.

Subscapular Group. The subscapular or *posterior group* consists of six or seven axillary nodes placed along the lower margin of the posterior wall of the axilla in the course of the subscapular artery (Fig. 10-21). The afferents of this group drain the skin and muscles of the dorsal part of the neck and the posterior thoracic wall. Their efferents pass to the central group of axillary nodes.

Central Group. A central group of three or four large axillary lymph nodes is embedded in the adipose tissue near the base of the axilla (Fig. 10-21). This group of nodes does not drain any specific part of the upper limb directly, but receives vessels from the lateral pectoral and subscapular groups of nodes. Its efferent vessels pass upward and medially to the apical nodes.

Apical Group. The apical or **subclavicular group** consists of 6 to 12 axillary lymph nodes situated partly behind the upper part of the pectoralis minor and partly in the apex of the axilla, medial and superior to the upper border of that muscle (Fig. 10-21). Its only direct territorial afferents are those that accompany the cephalic vein, and one or more that drain the upper peripheral part of the mammary gland; however, it receives the efferent vessels from all the other axillary nodes. The efferent vessels of the subclavicular group unite to form the **subclavian trunk,** which opens either directly into the junction of the internal jugular and subclavian veins or into the jugular lymphatic trunk; on the left side it may end in the thoracic duct. A few efferents from the apical nodes usually pass to the inferior deep cervical nodes.

Lymphatic Vessels and Nodes of the Thorax

In considering the lymphatic drainage of the thorax, initially the *lymphatic vessels* and then the *lymph nodes* will be described.

LYMPHATIC VESSELS OF THE THORAX

The lymphatic vessels of the thorax drain both the *thoracic wall* and the *viscera* located within the thoracic cavity.

Lymphatic Vessels of the Thoracic Wall

The lymphatic vessels of the thoracic wall include the more *superficial* lymphatic channels that drain the skin and superficial fascia (which on the anterior thoracic wall includes the mammary gland) and the *deeper* lymphatic vessels that drain the musculature of the thorax, the intercostal spaces, and the diaphragm.

SUPERFICIAL LYMPHATIC VESSELS OF THE THORACIC WALL. The superficial lymphatic channels of the thoracic wall ramify beneath the skin and converge on the axillary nodes. These vessels drain the back region superficial to the trapezius and latissimus dorsi muscles, and they course anteriorly and unite to form about 10 or 12 trunks that terminate in the **subscapular group** of axillary

lymph nodes (Fig. 10-21). The vessels that drain the pectoral region, including channels from the skin covering the peripheral part of the mammary gland, course dorsally, but those over the serratus anterior muscle course superiorly, to the **pectoral group** of axillary nodes. Others near the lateral margin of the sternum pass inward, between the rib cartilages, and end in the **parasternal nodes;** the vessels of the two sides anastomose across the anterior surface of the sternum. A few vessels from the upper part of the pectoral region ascend over the clavicle to the supraclavicular group of cervical nodes.

Lymphatic Vessels of the Mammary Gland. The lymphatic vessels that drain the mammary gland are numerous and their great clinical importance stems from their role in the spread of malignant cells from carcinoma of the breast. Within the structure of the gland, the lymphatic vessels originate as channels that course along the interlobular spaces and the walls of the lactiferous ducts. Those vessels that drain the glandular tissue and overlying skin of the *central part* of the breast pass to an intricate plexus situated beneath the areola called the **subareolar plexus** (Fig. 10-21). This plexus also receives lymphatic vessels from the areola, as well as from the nipple.

Efferent collecting vessels from both the superior and inferior aspects of the subareolar plexus course laterally and then superiorly to the **pectoral group** of axillary lymph nodes. Other vessels may bypass the pectoral nodes and course laterally and dorsally to the **subscapular group** of axillary nodes, while a few channels from the superior part of the breast may course to the **apical group** of nodes (Fig. 10-21).

Because most of the lymph draining the breast courses initially to the various groups of axillary nodes, this route is considered the primary direction of drainage. However, some lymphatic vessels from the *medial part* of the gland are directed medially, pierce the thoracic wall, and end in **parasternal lymph nodes** that course retrosternally with the internal thoracic artery and vein. Some channels communicate directly with the **intercostal nodes,** and other efferent vessels from the *lower part* of the breast may communicate inferiorly with diaphragmatic and abdominal lymphatic channels, as well as with some from the opposite breast.

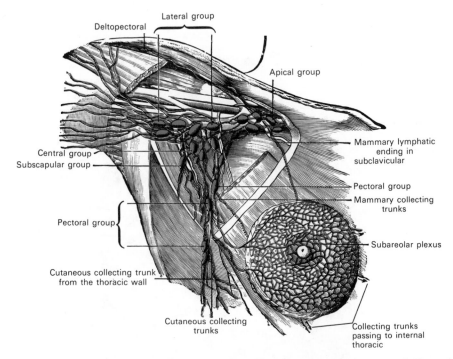

FIG. 10-21.   Lymphatics of the mammary gland and the axillary nodes. (Poirier and Charpy.)

DEEP LYMPHATIC VESSELS OF THE THORACIC WALL.   The deep lymphatic vessels of the thoracic wall collect lymph from the deeper tissues of the thoracic cage, including channels from the following three sources: thoracic muscles, intercostal spaces, and the diaphragm.

Lymphatic Vessels from Thoracic Muscles.   Most of the lymphatic vessels that are derived from muscles attached to the ribs end in the axillary nodes, but some vessels from the medial part of the pectoralis major muscle pass to the parasternal lymph nodes.

Lymphatic Vessels from Intercostal Spaces.   Lymphatic vessels within the intercostal spaces drain the intercostal muscles and the parietal pleura. Those vessels that drain the external intercostal muscles course dorsally, and after receiving the vessels that accompany the posterior branches of the intercostal arteries, end in the **intercostal nodes** (Fig. 10-22). Those lymphatic vessels that drain the internal intercostal muscles and parietal pleura consist of a single trunk in each space. Coursing ventrally around the intercostal spaces in the subpleural tissue, the upper six trunks open separately into parasternal nodes or into the vessels that unite them. Those of the lower spaces unite

to form a single trunk that terminates in the most caudal of the parasternal nodes.

Lymphatic Vessels of the Diaphragm.   The lymphatic vessels of the diaphragm form two plexuses, one on its thoracic and another on its abdominal surface. These plexuses anastomose freely with each other, and are most numerous on the parts covered, respectively, by the pleura and peritoneum. On the **thoracic surface** of the diaphragm, the lymphatic vessels communicate with those of the costal and mediastinal parts of the pleura. Their efferents consist of three groups: *anterior,* passing to **anterior diaphragmatic nodes** that lie near the junction of the seventh rib and its cartilage; *middle,* coursing to nodes on the esophagus and to those around the termination of the inferior vena cava; and *posterior,* which terminate in nodes that surround the aorta at the point where this vessel leaves the thoracic cavity. On the **abdominal surface** of the diaphragm, a plexus of fine lymphatic vessels anastomoses with those of the liver, and at the periphery of the diaphragm, with those of the subperitoneal tissue. The efferents from the right half of this plexus terminate partly in a group of nodes on the trunk of the corresponding inferior phrenic artery, while oth-

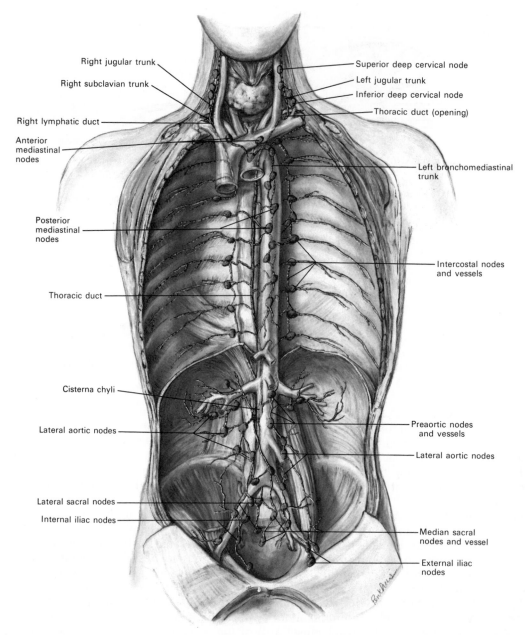

FIG. 10-22.   Diagrammatic view of the deep nodes of the thorax and abdomen. Observe the cisterna chyli in the abdomen and the ascending courses of the thoracic duct and the right lymphatic duct.

ers end in the right **lateral aortic nodes**. Those from the left half of the plexus pass to **preaortic** and left lateral aortic nodes, and to the nodes on the terminal portion of the esophagus.

Visceral Lymphatic Vessels of the Thorax

Within the thoracic cavity, visceral plexuses of lymphatic vessels drain the heart, lungs, pleura, thymus, and esophagus.

LYMPHATIC VESSELS OF THE HEART. The lymphatic vessels of the heart are arranged in three plexuses, one located deep to the endocardium, one in the myocardium, and one beneath the epicardium (deep to the visceral pericardium). Drainage occurs from deeper lymphatic vessels toward the more superficial surfaces of the heart; thus, the **subendocardial** vessels drain into the **myocardial** lymphatics, which, in turn, drain into the **subepicardial** plexus. Efferent vessels from the subepicardial plexus form right and left collecting lymphatic trunks. The **left trunks,** two to three in number, ascend in the anterior interventricular sulcus, and receive lymphatic vessels from both ventricles. When they reach the coronary sulcus, they are joined by a large lymphatic trunk that ascends in the posterior interventricular sulcus on the diaphragmatic surface of the heart. The junction of these two lymphatic channels, which primarily drain the left side of the heart, forms a single vessel that ascends between the pulmonary artery and the left atrium and ends in one of the right superior tracheobronchial lymph nodes. The **right trunk** receives its afferents from the right atrium, and from the right border and diaphragmatic surface of the right ventricle. It courses upward in the coronary sulcus, adjacent to the right coronary artery, and then continues its ascent in front of the ascending aorta and terminates in one of the upper left **anterior mediastinal nodes**.

LYMPHATIC VESSELS OF THE LUNGS.   The lymphatic vessels of the lungs originate in superficial and deep plexuses. The **superficial plexus** is located on the surface of the lung just beneath the pulmonary pleura, while the **deep plexus** accompanies the branches of the pulmonary vessels and the ramifications of the bronchi. In the case of the larger bronchi, the deep plexus consists of a *submucous network*, located in the walls of the bronchi beneath the mucous membrane, and a *peribronchial network* found outside the walls of the bronchi. In the smaller bronchi there is only a single plexus, which extends as far as the bronchioles but fails to reach the alveoli, in the walls of which there are no lymphatic vessels. Efferents from the superficial plexus course around the borders of the lungs and along the margins of the fissures separating the lobes. The lymphatics converge to end in **bronchopulmonary nodes** situated at the hilum. Efferents of the deep lymphatics are conducted to the hilum along the pulmonary vessels and bronchi, and end in both **bronchopulmonary** and **tracheobronchial nodes** (Fig. 10-23). Little or no anastomosis occurs between the superficial and deep lymphatics of the lungs, except in the region of the hilum.

LYMPHATIC VESSELS OF THE PLEURA.   The lymphatic vessels of the pleura consist of channels located in both the visceral and parietal layers. The **visceral set** of lymphatics drains into the superficial efferent vessels of the lungs, deep to the visceral pleura. The **parietal set** of lymphatics has three modes of termination: lymphatics of the *costal pleura* join the lymphatics of the internal intercostal muscles and so reach the **parasternal nodes;** those of the *diaphragmatic pleura* are drained by the efferent vessels on the thoracic surface of the diaphragm (see "Lymphatic Vessels of the Diaphragm"); those of the *mediastinal pleura* terminate in the **posterior mediastinal lymph nodes.**

LYMPHATIC VESSELS OF THE THYMUS.   The lymphatic vessels of the thymus end in the anterior mediastinal, tracheobronchial, and parasternal nodes.

LYMPHATIC VESSELS OF THE ESOPHAGUS.   The efferent lymphatic vessels from the *cervical part* of the esophagus drain into **paratracheal** and **deep cervical nodes;** those from the *thoracic part* end in **posterior mediastinal nodes;** and those from the *abdominal part* drain into the **left gastric lymph nodes**.

LYMPH NODES OF THE THORAX

The lymph nodes of the thorax are divided into nodes that drain the thoracic wall and those that drain the thoracic viscera.

Lymph Nodes of the Thoracic Wall

The lymph nodes of the thoracic wall drain the soft and hard tissues of the thoracic cage, and are at times referred to as the **parietal nodes** of the thorax. They include the following:

Parasternal nodes
Intercostal nodes
Diaphragmatic nodes

PARASTERNAL NODES.   The parasternal lymph nodes are located at the anterior ends of the intercostal spaces, and form a longitudinal chain of four to six nodes along the course of the internal thoracic artery. These important nodes derive afferent vessels from the mammary gland, from the deeper structures of the anterior abdominal wall above the level of the umbilicus, from the diaphragmatic surface of the liver through a small group of nodes that lies behind the xiphoid process, and from the deeper parts of the anterior thoracic wall. Their efferent vessels usually unite to form a single trunk on each side of the sternum. This trunk may open directly into the junction of the subclavian and internal jugular veins, or, on the right side, it may join the main lymphatic duct on the subclavian duct, and on the left side, the thoracic duct. At times it may join lymphatic vessels from the tracheobronchial and anterior mediastinal nodes to form the **bronchomediastinal trunk,** which then enters the venous system either directly or through the large lymphatic ducts at the root of the neck.

INTERCOSTAL NODES. The intercostal lymph nodes are located in the dorsal part of the thorax, within the intercostal spaces near the heads of the ribs (Fig. 10-22). There are one or more in each interspace and they lie in relationship to the intercostal vessels. The intercostal lymph nodes receive the deep lymphatics from the posterolateral aspect of the chest, as well as from certain channels from the breast; some of these vessels are interrupted by small lateral intercostal nodes. The efferent vessels of the nodes in the lower four or five spaces unite to form a trunk that descends and opens either into the cisterna chyli or into the commencement of the thoracic duct. The efferent channels of the nodes in the upper spaces of the left side end in the thoracic duct, and those of the corresponding right spaces, in the right lymphatic duct.

DIAPHRAGMATIC NODES. The diaphragmatic nodes lie on the thoracic aspect of the diaphragm and consist of three sets: anterior, lateral, and posterior.

Anterior Diaphragmatic Nodes. The anterior diaphragmatic set consists of two or three small nodes, located dorsal to the base of the xiphoid process, that receive afferents from the convex surface of the liver, and one or two nodes, located on each side near the junction of the seventh rib with its cartilage, that receive lymphatic vessels from the anterior part of the diaphragm. The efferent vessels of the anterior set pass to the **parasternal nodes**.

Lateral Diaphragmatic Nodes. The lateral diaphragmatic set consists of two or three small nodes on each side, close to the site where the phrenic nerves enter the diaphragm. On the right side, some of the nodes of this group lie within the fibrous sac of the pericardium, anterior to the termination of the inferior vena cava; on the left side, they

extend to the esophageal hiatus. The afferent vessels of the lateral diaphragmatic nodes are derived from the middle region of the diaphragm; those on the right side also receive afferents from the convex surface of the liver. Their efferents pass to the **parasternal nodes** by way of the anterior diaphragmatic nodes, to the upper **anterior mediastinal nodes** by a vessel that courses superiorly with the phrenic nerve, and to the **posterior mediastinal nodes**.

Posterior Diaphragmatic Nodes. The posterior diaphragmatic set consists of a few nodes situated on the back of the crura of the diaphragm, and communicates on the one hand with the lateral aortic set of lumbar nodes, and on the other with the posterior mediastinal nodes.

Lymph Nodes of the Thoracic Viscera

Lymphatic vessels that drain the organs of the thorax pass initially through groups of **visceral nodes**. Usually three groups are described:

Anterior mediastinal
Posterior mediastinal
Tracheobronchial

ANTERIOR MEDIASTINAL NODES. The anterior mediastinal nodes are located in the anterior part of the superior mediastinum, in front of the brachiocephalic veins, the aortic arch, and the large arterial trunks that arise from the aorta (Fig. 10-22). They receive afferent lymphatic vessels from the thymus, heart, and pericardium, and from the lower end of the thyroid gland. Additionally, the anterior mediastinal nodes receive lymph from the mediastinal pleura and anterior part of the hilus of the lung, from the parasternal nodes, and from the diaphragm and liver by way of the lateral diaphragmatic nodes. Their efferent vessels unite with those of the tracheobronchial nodes to form the **right** and **left bronchomediastinal trunks,** which ascend in the thorax to open into the main lymphatic ducts or into the venous system directly in the root of the neck.

POSTERIOR MEDIASTINAL NODES. The posterior mediastinal nodes lie behind the pericardium in relationship to the esophagus and descending thoracic aorta. There are eight to ten nodes in this group, and their afferent vessels are derived from the esophagus, the posterior part of the pericardium, the diaphragm, and the convex surface of the liver. Their efferent channels mostly end in the **thoracic duct,** but some join the **tracheobronchial nodes**.

TRACHEOBRONCHIAL NODES. The tracheobronchial nodes are assembled in five groups:

Tracheal
Superior tracheobronchial
Inferior tracheobronchial
Bronchopulmonary
Pulmonary

Tracheal Nodes. The tracheal nodes extend along both sides of the thoracic part of the trachea, and are therefore sometimes called the paratracheal nodes (Fig. 10-23).

Superior Tracheobronchial Nodes. The superior tracheobronchial nodes are found on each side, superior and lateral to the angle at which the trachea bifurcates into the two primary bronchi. The chains continue laterally along the superior border of the two principal bronchi (Fig. 10-28).

Inferior Tracheobronchial Nodes. The inferior tracheobronchial nodes lie in the angle below the bifurcation of the trachea, and their chains continue inferolaterally along the lower border of the principal bronchi (Fig. 10-23).

Bronchopulmonary Nodes. The bronchopulmonary nodes are found at the hilus of each lung, at the site of division of the principal bronchi and pulmonary vessels into the lobar bronchi and vessels (Fig. 10-23).

Pulmonary Nodes. The pulmonary lymph nodes are found in the lung substance along the course of the segmental bronchi.

The *tracheal nodes* in the thorax receive afferent vessels that drain the trachea and

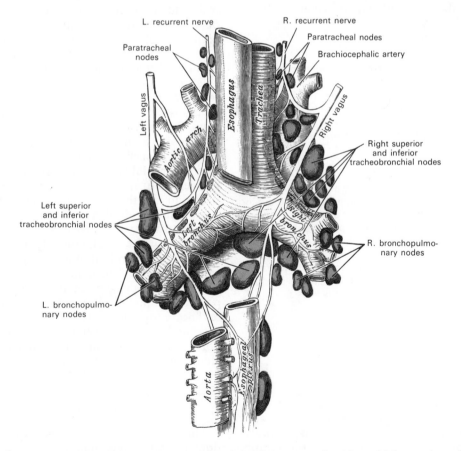

FIG. 10-23. Posterior view of the deep visceral nodes of the thorax. Note the right and left superior and inferior tracheobronchial nodes. The superior nodes of these groups lie above the tracheobronchial angle on each side, while the inferior tracheobronchial nodes are located below the angle of division of the trachea into the two primary bronchi. (After Hallé.)

the upper thoracic portion of the esophagus. The remaining groups of tracheobronchial nodes form continuous chains, and it is difficult to define the limits of each group. Certainly, however, the superficial and deep lymphatics of the lungs and the bronchial trees drain into the *pulmonary* and *bronchopulmonary nodes,* which, in turn, drain into the *superior* and *inferior tracheobronchial nodes.* The latter also receive some afferent vessels from other posterior mediastinal organs.

The efferent vessels from the tracheobronchial nodes ascend with those of the tracheal nodes, which unite with efferents of the parasternal and anterior mediastinal nodes to form the **right** and **left bronchomediastinal trunks**. The right bronchomediastinal trunk may join the right lymphatic duct, and the left may join the thoracic duct. More frequently, however, they open independently of these ducts into the junction of the internal jugular and subclavian veins of their own side.

CLINICAL NOTE. There are continually being swept into the bronchi and alveoli of all city dwellers large quantities of inhaled dust and black carbonaceous pigment. Particles that are not picked up by dust cells, caught in the mucus, and discharged are taken up by phagocytes and transported in the lymphatic vessels to the lymph nodes. This results in an enlargement and a blackened coloration of the lymph nodes in the tracheobronchial groups. The work environment can markedly enhance this condition. The lymph nodes and lungs of coal miners invariably contain large amounts of entrapped black particles (anthracosis). Other types of dust particles may also be inhaled chronically as a result of occupational activities and entrapped in these lymph nodes. Thus, the lungs of iron workers show deposits of iron particles (siderosis), and those of stone masons show stone dust or sand (silicosis).

Lymphatic Vessels and Nodes of the Abdomen and Pelvis

In describing the lymphatic vessels and nodes of the abdomen and pelvis, initially the lymphatic vessels and then the lymph nodes will be considered.

LYMPHATIC VESSELS OF THE ABDOMEN AND PELVIS

The lymphatic vessels of the abdomen and pelvis include the *parietal* channels that drain the abdominal and pelvic walls, and the *visceral* vessels of the abdominal and pelvic organs.

Lymphatic Vessels of the Abdominal and Pelvic Wall

Both superficial and deep lymphatic vessels drain the walls of the abdomen and pelvis. Together they constitute the parietal vessels of these regions of the trunk.

SUPERFICIAL LYMPHATIC VESSELS OF THE ABDOMINAL AND PELVIC WALLS. The lymphatic vessels in the skin and superficial fascia of the anterior abdominal wall *above the umbilicus* drain upward, into the superficial collecting channels of the anterior thoracic wall and therefore into the **axillary nodes**. *Below the umbilicus,* the superficial lymphatic vessels follow the course of the superficial blood vessels and converge to the **superficial inguinal nodes**. Lymphatic channels from the superficial tissue overlying the rectus abdominis muscle follow the course of the superficial epigastric vessels; those from the lateral regions of the abdominal wall pass along the crest of the ilium with the superficial iliac circumflex vessels. The superficial lymphatic vessels of the gluteal region turn horizontally around the buttock, and join the upper and lower groups of superficial inguinal lymph nodes.

DEEP LYMPHATIC VESSELS OF THE ABDOMINAL AND PELVIC WALLS. The deep lymphatic vessels of the abdominal and pelvic walls also accompany the principal blood vessels. Those of the parietes of the pelvis accompany the branches of the internal iliac artery and drain into the **internal iliac** and **common iliac nodes,** and eventually into the *lateral aortic* group of **lumbar lymph nodes** on each side (Figs. 10-24; 10-34 to 10-36).

Lymphatic Vessels of the Abdominal and Pelvic Viscera

The visceral lymphatic vessels of the abdomen and pelvis include the lymphatic channels of the:

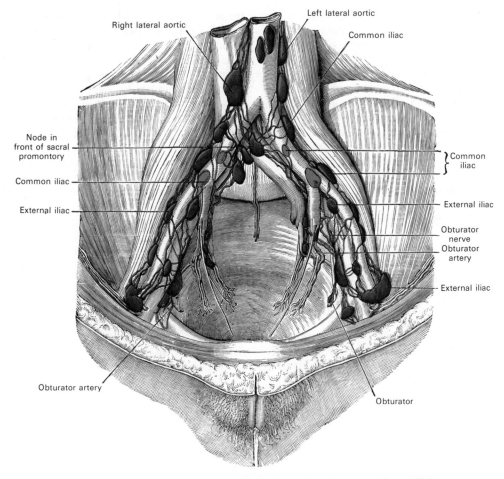

FIG. 10-24. The parietal lymph nodes of the pelvis. (Cunéo and Marcille.)

Subdiaphragmatic portion of the digestive tube and its associated glands
Spleen and suprarenal glands
Urinary tract
Reproductive organs
Perineum and external genitalia

LYMPHATIC VESSELS FROM THE SUBDIAPHRAGMATIC PORTION OF THE DIGESTIVE TUBE AND ITS ASSOCIATED GLANDS

The lymphatic vessels of the digestive tube below the level of the diaphragm are situated partly in the mucosa and partly in the submucosal, muscular, and serosal coats. Because the vessels in the mucosa drain into the deeper layers, the lymphatics will be considered as a single system in any part of the gastrointestinal tube. The following regional visceral lymphatics will be described:

Stomach
Duodenum
Greater omentum
Jejunum and ileum
Appendix and cecum
Colon
Rectum and anal canal
Liver
Gallbladder
Pancreas

LYMPHATIC VESSELS OF THE STOMACH. The lymphatic vessels of the stomach are continuous at the cardiac orifice with those of the esophagus, and at the pylorus with those of the duodenum (Figs. 10-25; 10-26). They mainly follow the blood vessels and are arranged in four sets. Lymphatic vessels of the *first set* accompany the branches of the left gastric artery, receive tributaries from a large area on both anterior and posterior surfaces of the stomach, and terminate in the **left gastric lymph nodes**. Lymphatic channels of the *second set* drain the fundus and body of the stomach to the left of a line drawn vertically from the esophagus. They accompany, more or less closely, the short gastric and left gastroepiploic arteries and end in the **pancreaticosplenic lymph nodes**. The vessels of the *third set* drain the right portion of the greater curvature of the stomach as far as the pylorus, and end in the **right gastroepiploic nodes**, the efferents of which pass to the **pyloric nodes**. Lymphatic vessels of the *fourth set* drain the pyloric portion and pass to the **hepatic, pyloric,** and **left gastric lymph nodes** (Figs. 10-25; 10-26).

LYMPHATIC VESSELS OF THE DUODENUM. The lymphatic vessels of the duodenum consist of *anterior* and *posterior sets* that open into a series of small **pyloric nodes,** located on the anterior and posterior aspects of the groove between the head of the pancreas and the duodenum (Figs. 10-25; 10-26). The efferents of these nodes course in two directions, superiorly to the **hepatic nodes** and inferiorly to the **preaortic group of lumbar nodes,** around the origin of the superior mesenteric artery.

LYMPHATIC VESSELS OF THE GREATER OMENTUM. The lymphatic vessels of the greater omentum course longitudinally with the omental blood vessels that branch from the right and left gastroepiploic arteries and veins. The lymphatic channels ascend to empty into lymphatic vessels and nodes along the greater curvature of the stomach. On the right side these drain into the **right gastroepiploic** and **pyloric nodes,** but on the left side they drain into the **pancreaticosplenic nodes.**

LYMPHATIC VESSELS OF THE JEJUNUM AND ILEUM. The lymphatic vessels of the jejunal and ileal regions of the small intestine are termed **lacteals** because of the milk-white lymph they contain during intestinal digestion after a fatty meal. They course between the layers of the mesentery and enter the **mesenteric nodes,** the efferents of which end in the **superior mesenteric set** of **preaortic nodes.**

LYMPHATIC VESSELS OF THE APPENDIX AND CECUM. The ileocolic, appendicular, and cecal regions of the gastrointestinal tract

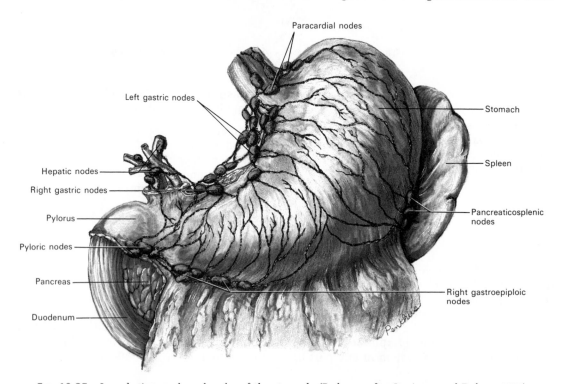

FIG. 10-25.   Lymphatic vessels and nodes of the stomach. (Redrawn after Jamieson and Dobson, 1907.)

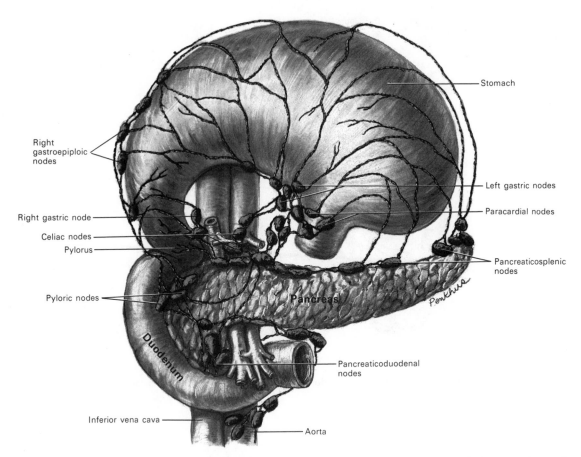

Fig. 10-26. The lymphatic vessels of the stomach, duodenum, and pancreas demonstrated with the stomach turned upward. (After Jamieson and Dobson, 1907b.)

contain many lymphatic vessels because there is much lymphoid tissue in the walls of these parts of the gastrointestinal tract. From the body and tail of the vermiform appendix, 8 to 15 vessels ascend between the layers of the mesentery of the appendix, several being interrupted by the small (appendicular) nodes that lie between the peritoneal layers. These vessels unite to form three or four channels that end partly in the lower and partly in the upper nodes of the **ileocolic chain** (Figs. 10-27; 10-28). The vessels from the root of the appendix and from the cecum form anterior and posterior groups. The anterior vessels pass in front of the cecum, and end in the **anterior ileocolic nodes** and in the upper and lower nodes of the ileocolic chain (Fig. 10-27). The posterior vessels ascend over the dorsal aspect of the cecum, and terminate in the **posterior ileocolic nodes** and in the lower nodes of the ileocolic chain (Fig. 10-28).

LYMPHATIC VESSELS OF THE COLON.   The lymphatic vessels of the ascending and transverse colon drain through nodes located along the course of the marginal artery (paracolic nodes), as well as through the **right colic** and **middle colic nodes,** on their way to the **superior mesenteric set of preaortic nodes** (Fig. 10-29). Those of the descending and sigmoid parts of the colon are interrupted by the small nodes on the branches of the left colic and sigmoid arteries, and ultimately end in the **inferior mesenteric set of preaortic nodes**.

LYMPHATIC VESSELS OF THE RECTUM AND ANAL CANAL. The lymphatic vessels of the rectum have important clinical significance because of the high incidence of carcinoma in this lower region of the large intestine. As in other parts of the intestinal tract, the rectum contains both submucosal and muscular networks of lymphatic vessels. The *mucocutaneous* or *pectinate line* forms a boundary

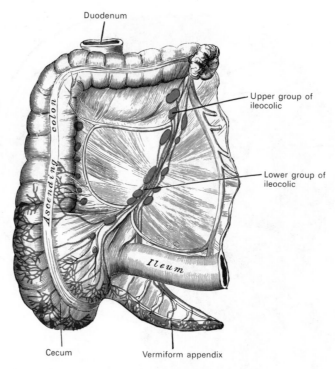

FIG. 10-27.   The lymphatics of cecum and vermiform appendix from the front. (Jamieson and Dobson, 1907a.)

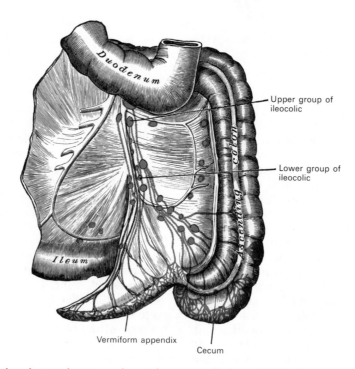

FIG. 10-28.   The lymphatics of cecum and vermiform appendix from behind. (Jamieson and Dobson, 1907a.)

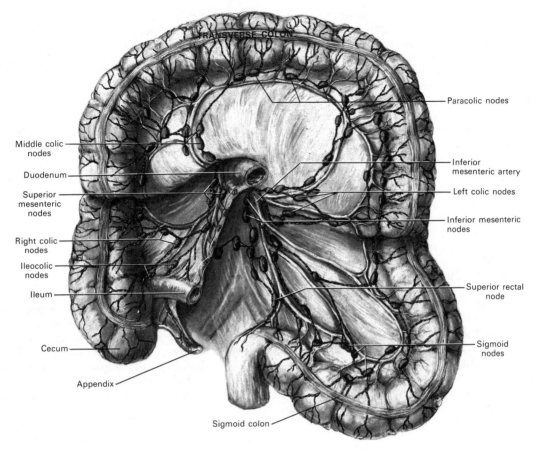

FIG. 10-29. Lymphatic vessels and nodes of the large intestine. (After Jamieson and Dobson, 1909.)

between those lymphatics above, which drain into pelvic and lower abdominal nodes, and those lymphatics below, which drain outside the pelvis and with the perineal lymphatics to the superficial inguinal nodes. The lymphatic vessels of the rectum can be divided into superior, middle, and inferior channels.

The *superior lymphatics* of the rectum join those of the sigmoid colon and drain initially through **pararectal nodes,** which lie parallel to the rectum, and then superiorly along the course of the superior rectal artery and vein, to end in the **inferior mesenteric group of preaortic nodes**. The *middle group* of lymphatics drains approximately 10 cm of rectum above the mucocutaneous line. These channels run *downward,* along the course of the inferior rectal blood vessels, and terminate in nodes that accompany their branches; *laterally,* along the course of the middle rectal artery and vein, to terminate in

the **internal iliac nodes;** and *superiorly,* to join the uppermost channels of the rectum. The *inferior group* of rectal lymphatics drains the anal canal distal to the mucocutaneous line. These vessels descend to the skin and superficial fascia of the perineum, and course to the **superficial inguinal nodes**. From these nodes lymphatic vessels enter the pelvis by way of the external and common iliac nodes.

LYMPHATIC VESSELS OF THE LIVER. The lymphatic vessels of the liver are divisible into two sets, superficial and deep.

Superficial Lymphatic Vessels. The superficial lymphatic vessels of the liver course beneath the peritoneum and over the entire surface of the organ; they may be grouped into those located on the convex **diaphragmatic surface** and those on the **visceral surface**.

Diaphragmatic Surface. The *dorsal part* of the diaphragmatic surface of the liver is drained by lymphatic vessels on its right, middle, and left aspects. One or two vessels along the *right side* course backward on the abdominal surface of the diaphragm, and after crossing the right crus, they end in the **celiac group of preaortic nodes.** Five or six channels along the middle part of the diaphragmatic surface pass through the vena caval foramen in the diaphragm, and end in one or two lymph nodes that are situated around the terminal part of the inferior vena cava. A few lymphatic vessels from the *left side* pass dorsally, toward the esophageal hiatus, and terminate in the **paracardial group of left gastric nodes.**

From the *ventral part* of the diaphragmatic surface of the right and left lobes adjacent to the falciform ligament, the lymphatic vessels converge to form two trunks. One of these accompanies the inferior vena cava through the diaphragm and ends in the nodes around the terminal part of this vessel, while the other courses downward and anteriorly, and turning around the inferior sharp margin of the liver, accompanies the upper part of the ligamentum teres to end in the **hepatic nodes.** A few additional vessels from this anterior surface also turn around the inferior sharp margin to reach the hepatic nodes.

Visceral Surface. The lymphatic vessels from the visceral surface of the liver mostly converge toward the porta hepatis. They accompany the deep lymphatics that emerge from the liver at this site, and end in the **hepatic nodes.** One or two channels from the posterior parts of the right and caudate lobes accompany the inferior vena cava through the diaphragm and end in nodes near the termination of this vein.

Deep Lymphatic Vessels. The deep lymphatic vessels from the internal substance of the liver join to form ascending and descending trunks. The ascending trunks accompany the hepatic veins and pass through the diaphragm, to end in lymph nodes around the terminal part of the vein. The descending trunks emerge from the porta hepatis and end in the **hepatic nodes.**

LYMPHATIC VESSELS OF THE GALLBLADDER. The lymphatic vessels that drain the gallbladder pass to the **hepatic nodes** in the porta hepatis. There is evidence that interconnections exist between lymphatic channels of the gallbladder and the liver, which possibly helps to explain the spread of infec-tions and other diseases between these two organs. Lymphatic vessels of the **common bile duct** course to the **hepatic nodes,** alongside the duct, and to the upper **pancreatico-duodenal nodes.**

LYMPHATIC VESSELS OF THE PANCREAS. The lymphatic vessels of the pancreas follow the course of its blood vessels (Fig. 10-26). Most of them enter the **pancreatico-splenic nodes,** but some end in nodes located along the pancreaticoduodenal vessels. Other pancreatic lymphatic channels open into the **superior mesenteric set of preaortic nodes.**

LYMPHATIC VESSELS FROM THE
SPLEEN AND SUPRARENAL GLANDS

LYMPHATIC VESSELS OF THE SPLEEN. The lymphatic vessels of the spleen are principally derived from the capsule and from the connective tissue that forms the trabeculae. These pass to **splenic nodes,** located between the layers of the gastrosplenic ligament at the hilus, and to **pancreaticosplenic nodes,** located along the superior border of the tail of the pancreas (Fig. 10-25).

LYMPHATIC VESSELS OF THE SUPRARENAL GLANDS. The lymphatic vessels of the suprarenal glands usually accompany the suprarenal veins and end in the **lateral aortic set of lumbar nodes.** Occasionally some vessels pierce the crura of the diaphragm and end in the **posterior mediastinal nodes.**

LYMPHATIC VESSELS OF THE URINARY TRACT

LYMPHATIC VESSELS OF THE KIDNEY. The lymphatic vessels of the kidney form three plexuses: one in the substance of the renal cortex, a second beneath the fibrous capsule of the kidney, and a third in the perinephric fat.

The subcapsular and perinephric plexuses communicate freely with each other. The lymphatic vessels from the plexus in the substance of the kidney converge to form four or five trunks that emerge at the hilus. Here they are joined by vessels from the subcapsular plexus, and following the course of the renal vein, they end in the **lateral aortic set of lumbar nodes.** The perinephric plexus drains directly into the more superior of the **lateral aortic nodes.**

LYMPHATIC VESSELS OF THE URETER. The lymphatic vessels of the ureter commence in plexuses within the muscular and adventitial layers of the organ. The channels from the *proximal portion* end partly in the efferent vessels of the kidney, and in part directly in the **lateral aortic set of lumbar nodes**. Those from the *middle portion* (above the pelvic brim) drain into **common iliac nodes,** whereas the vessels from the *intrapelvic portion* of the tube (below the pelvic brim) drain into **external and internal iliac nodes**, and more distally may join efferent vessels from the bladder.

LYMPHATIC VESSELS OF THE BLADDER. The lymphatic vessels of the bladder commence in muscular and extramuscular plexuses, and it is claimed that the mucous membrane is devoid of lymphatic channels. The efferent vessels emerge from the bladder in three groups: those from the anterior aspect, the

posterior aspect, and the region of the trigone.

The vessels from the *anterior part* of the inferolateral surface and the *anterior part* of the superior surface pass laterally along the course of the obliterated umbilical artery and end in the **external iliac nodes** (Fig. 10-30). Located along their course are minute nodes that are arranged in front of the bladder in an **anterior vesical group,** and a **lateral vesical group,** adjacent to the lateral umbilical ligament.

Most of the vessels from the *posterior part* of the inferolateral surface pass laterally and backward to **common iliac and internal iliac nodes,** but a few vessels course backward, even beyond the bifurcation of the aorta, to one or two nodes in front of the sacral promontory (Fig. 10-30). The channels from the

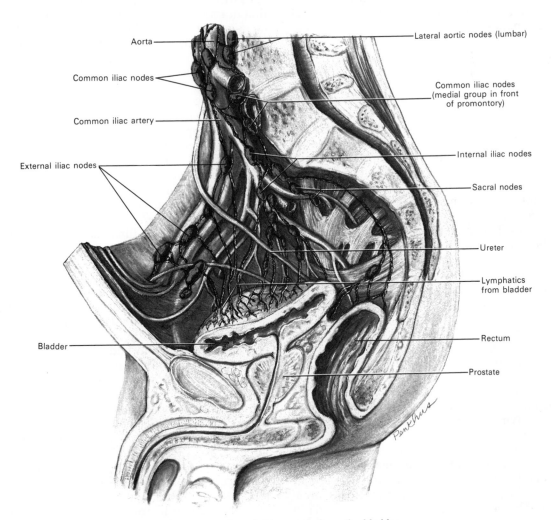

FIG. 10-30. Lymphatic vessels from the bladder.

posterior part of the superior surface also course laterally to the **common iliac and internal iliac nodes**.

Lymphatic vessels from the *trigone of the bladder* course upward and laterally to both the **external and internal iliac nodes**.

LYMPHATIC VESSELS OF THE URETHRA. The lymphatic vessels of nearly the *entire female urethra* and of the *prostatic* and *membranous parts of the male urethra* pass principally to the **internal iliac lymph nodes**. Lymphatics from the distal extremity of the female urethra drain with those from the skin of the female external genitalia to the superficial inguinal nodes.

Lymphatic vessels from the *spongy portion of the male urethra* accompany those of the glans penis and terminate in the **deep inguinal nodes,** adjacent to the femoral artery. Other vessels course upward along the inguinal canal and drain into the **external iliac nodes**.

LYMPHATIC VESSELS OF THE MALE REPRODUCTIVE ORGANS

LYMPHATIC VESSELS OF THE TESTIS. The lymphatic vessels of the testis consist of two sets, superficial and deep. The **superficial vessels** commence on the surface of the tunica vaginalis, while the **deep vessels** are derived from the epididymis and body of the testis. They form four to eight collecting trunks that ascend with the testicular blood vessels in the spermatic cord. At the abdominal inguinal ring, they continue upward on the anterior surface of the psoas major muscle to the point where the testicular artery and vein cross the ureter, and they end in the **lateral and preaortic sets of lumbar nodes**.

LYMPHATIC VESSELS OF THE DUCTUS DEFERENS AND SEMINAL VESICLES. Only superficial lymphatic vessels that drain the *ductus deferens* have been described, and these terminate in the **external iliac nodes**.

Both superficial and deep lymphatic vessels drain the *seminal vesicles,* and these drain into the **external iliac and internal iliac lymph nodes** (Fig. 10-31).

LYMPHATIC VESSELS OF THE PROSTATE. The lymphatic vessels of the prostate gland terminate principally in the **internal iliac and sacral lymph nodes** (Fig. 10-31). One lymph vessel from the posterior surface of the prostate, however, courses laterally and ends in

the **external iliac nodes,** while another from the anterior surface of the gland joins the vessels that drain the membranous part of the urethra and terminates in the **internal iliac nodes**.

LYMPHATIC VESSELS OF THE FEMALE REPRODUCTIVE ORGANS

LYMPHATIC VESSELS OF THE OVARY. A rich plexus of lymphatic vessels drains each ovary, and when they leave the hilum, they follow the course of the ovarian artery and vein to terminate in the **lateral and preaortic sets of the lumbar nodes**.

LYMPHATIC VESSELS OF THE UTERINE TUBI The lymphatic vessels of the upper and middle parts of the uterine tube follow those from the ovary and end in the **lateral and preaortic sets of lumbar nodes**. Channels that drain the lowest part of the uterine tube course laterally between the leaves of the broad ligament to the **internal iliac nodes,** and with the round ligament, through the inguinal canal to the **superficial inguinal nodes**.

LYMPHATIC VESSELS OF THE UTERUS. The lymphatic vessels of the uterus consist of two sets, **superficial** and **deep,** the former placed beneath the peritoneum, the latter in the wall of the uterus.

Efferent vessels from the **cervix of the uterus** course in three directions: *laterally* with the uterine artery to the **external iliac nodes;** *posterolaterally,* passing behind the ureter, to the **internal iliac nodes;** and *posteriorly,* to the **common iliac and lateral sacral nodes** (Figs. 10-32; 10-33).

The majority of the lymphatic vessels of the **body and fundus of the uterus** pass laterally in the broad ligaments with the uterine artery to **internal iliac nodes**. Others are continued superiorly with the ovarian vessels and end in the **lateral and preaortic** sets of **lumbar nodes**. A few channels course to the **external iliac nodes,** and one or two vessels traverse the inguinal canal and end in the **superficial inguinal nodes**. In the nongravid uterus, the lymphatic vessels are very small, but during gestation they are greatly enlarged.

LYMPHATIC VESSELS OF THE VAGINA. Three groups of lymphatic vessels drain the vagina (Fig. 10-33). The *upper* (cervicovaginal) channels course laterally with branches of

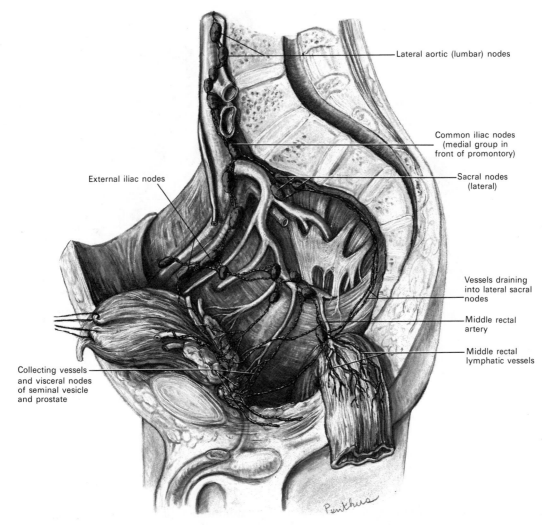

Lateral aortic (lumbar) nodes

Common iliac nodes
(medial group in
front of promontory)

Sacral nodes
(lateral)

External iliac nodes

Vessels draining
into lateral sacral
nodes

Middle rectal
artery

Middle rectal
lymphatic vessels

Collecting vessels
and visceral nodes
of seminal vesicle
and prostate

Fig. 10-31. Lymphatic vessels of the prostate and middle rectal region.

the uterine artery to the **external and internal iliac nodes**. Lymphatic vessels from the *middle* region of the vagina follow the branches of the vaginal artery to the **internal iliac nodes,** along the course of which are interspersed other minute nodes. Some lymphatic vessels from the *lower* part of the vagina join those of the middle portion and course toward nodes in the lateral wall of the pelvis, while others near the vaginal orifice join those of the vulva and pass to the **superficial inguinal nodes**. The lymphatics of the vagina interconnect with those of the cervix of the uterus, the rectum, and the external genitalia, but not with those of the bladder.

LYMPHATIC VESSELS OF THE PERINEUM AND EXTERNAL GENITALIA

LYMPHATIC VESSELS OF THE UROGENITAL REGION. The lymphatic vessels of the skin and superficial fascia in the *urogenital* region of the perineum, along with the superficial lymphatic vessels from the skin of the scrotum and penis and those of the labia majora and labia minora, course laterally with the external pudendal artery and vein and terminate in the **upper and lower superficial inguinal nodes**.

The lymphatic vessels of the glans penis and clitoris course with the deep dorsal vein and drain laterally to the deep inguinal

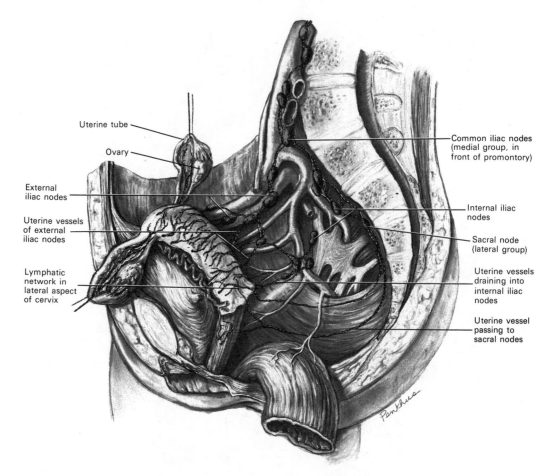

Uterine tube

Ovary

External
iliac nodes

Uterine vessels
of external
iliac nodes

Lymphatic
network in
lateral aspect
of cervix

Common iliac nodes
(medial group, in
front of promontory)

Internal iliac
nodes

Sacral node
(lateral group)

Uterine vessels
draining into
internal iliac
nodes

Uterine vessel
passing to
sacral nodes

FIG. 10-32. Lymphatic vessels draining the uterus.

nodes (Fig. 10-34). These vessels may extend superiorly, even through the femoral canal, to the **external iliac nodes**. Other channels course upward through the inguinal canal also to achieve the **external iliac nodes**.

LYMPHATIC VESSELS OF THE ANAL REGION. The superficial lymphatic vessels in the *anal region* of the perineum also course to the **superficial inguinal lymph nodes**. Some channels, those deeper in the ischiorectal fossa, follow the course of the inferior rectal and internal pudendal blood vessels and thereby enter the pelvis and end in the **internal iliac nodes**.

LYMPH NODES OF THE ABDOMEN AND PELVIS

The lymph nodes of the abdomen and pelvis may be divided, according to their location, into those nodes that lie behind the peritoneum, adjacent to the **walls** of the abdominopelvic cavity in close association with the large blood vessels, and those lymph nodes that are more intimately related to the abdominopelvic **viscera** and to the vessels more directly supplying the organs.

Lymph Nodes of the Abdominal and Pelvic Walls

The lymph nodes located adjacent to the walls of the abdominal and pelvic cavities are frequently referred to as parietal nodes. These nodes include the following groups:

External iliac
Common iliac
Epigastric
Circumflex iliac
Internal iliac

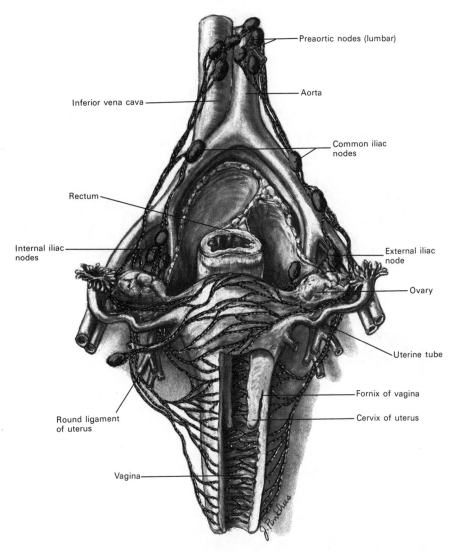

FIG. 10-33.   Lymphatic vessels of the vagina and uterus.

Sacral
Lumbar
   Lateral aortic
   Preaortic
   Retroaortic

EXTERNAL ILIAC NODES.   The external iliac nodes, eight to ten in number, lie along the external iliac vessels (Figs. 10-24; 10-35; 10-36). They are arranged in three groups, one on the lateral aspect, another on the medial aspect, and a third, which is sometimes absent, on the anterior aspect of the vessels. The external iliac nodes receive afferent vessels from:

(a) *the lower extremity,* by way of the superficial and deep inguinal nodes, as well as from some lymphatics from the adductor region of the thigh;

(b) *the lower anterior abdominal wall,* by way of the epigastric and circumflex iliac nodes;

(c) *the perineum,* by way of lymphatic channels from the glans penis or clitoris that reach the external iliac nodes through the inguinal canal, and by means of vessels that course laterally from the genitalia and then superiorly through the femoral canal;

(d) *the pelvis,* through direct channels from the fundus of the bladder, the prostate,

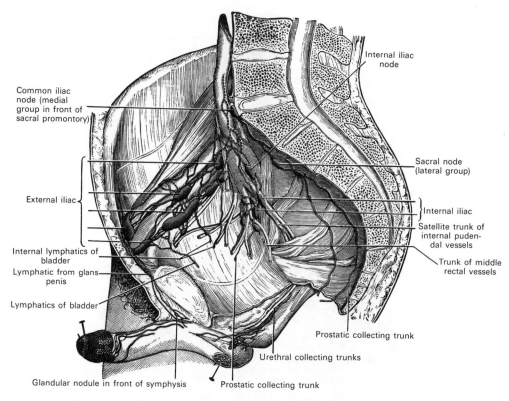

Common iliac
node (medial
group in front of
sacral promontory)

External iliac

Internal lymphatics of
bladder

Lymphatic from glans
penis

Lymphatics of bladder

Glandular nodule in front of symphysis

Internal iliac
node

Sacral node
(lateral group)

Internal iliac

Satellite trunk of
internal puden-
dal vessels

Trunk of middle
rectal vessels

Prostatic collecting trunk

Urethral collecting trunks

Prostatic collecting trunk

FIG. 10-34.   Lymphatic vessels draining the glans penis and the spongy and membranous portions of the urethra.

and membranous urethra in the male, or from the upper vagina and uterine cervix in the female.

Efferent vessels leave the external iliac nodes and ascend to common iliac and lumbar nodes.

COMMON ILIAC NODES.   The six to eight common iliac nodes are located medial, lateral, and posterior to the common iliac vessels (Figs. 10-24; 10-35; 10-36), and they extend from the external iliac nodes to the bifurcation of the aorta. Several of the **medial nodes** are located in front of the fifth lumbar vertebra and the promontory of the sacrum; these extend downward from the bifurcation of the aorta. The **lateral and posterior nodes** directly extend the chain of parietal pelvic nodes from the external and internal iliac sets to the lateral lumbar nodes.

EPIGASTRIC NODES.   Three or four epigastric nodes follow the course of the lower part of the inferior epigastric vessels, and their efferent channels drain into the external iliac nodes.

CIRCUMFLEX ILIAC NODES.   Two to four cir-

cumflex iliac nodes, when present, are situated along the course of the deep circumflex iliac vessels. Their efferent channels also drain into the external iliac nodes.

INTERNAL ILIAC NODES.   The internal iliac nodes surround the internal iliac vessels, and they receive lymphatics that correspond to the distribution of the branches of the internal iliac artery (Figs. 10-30; 10-32 to 10-36). Lying deep to the pelvic fascia, they receive lymphatics from all the pelvic viscera, from the deeper parts of the perineum, including the membranous and spongy portions of the urethra, and from the gluteal region and dorsum of the thigh. An **obturator node** is sometimes seen in the upper part of the obturator foramen. The efferent vessels of the internal iliac group end in the common iliac nodes.

SACRAL NODES.   Five or six sacral nodes lie in the cavity of the sacrum, in relationship to the median and lateral sacral arteries (Figs. 10-30 to 10-32; 10-34). They receive lymphatics from the rectum and posterior wall of the pelvis, and pass to the internal iliac and lumbar nodes.

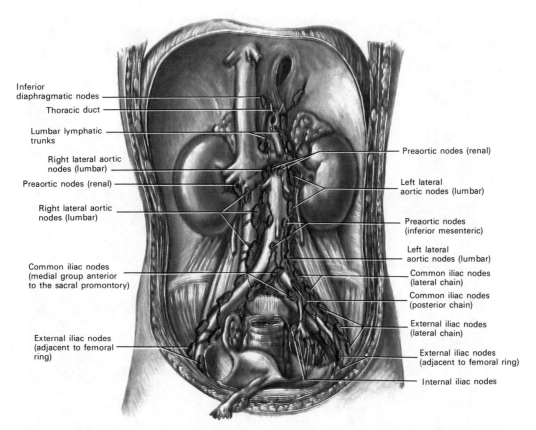

Inferior
diaphragmatic nodes

Thoracic duct

Lumbar lymphatic
trunks

Right lateral aortic
nodes (lumbar)

Preaortic nodes (renal)

Right lateral aortic
nodes (lumbar)

Common iliac nodes
(medial group anterior
to the sacral promontory)

External iliac nodes
(adjacent to femoral
ring)

Preaortic nodes (renal)

Left lateral
aortic nodes (lumbar)

Preaortic nodes
(inferior mesenteric)

Left lateral
aortic nodes (lumbar)

Common iliac nodes
(lateral chain)

Common iliac nodes
(posterior chain)

External iliac nodes
(lateral chain)

External iliac nodes
(adjacent to femoral ring)

Internal iliac nodes

FIG. 10-35.   Lymphatic nodes of the pelvis and posterior abdominal wall. (From Rouvier, H., *Anatomie Humaine,* 7th ed., Masson, Paris, 1954.)

LUMBAR NODES.   The lumbar nodes are numerous and consist of right and left lateral aortic, preaortic, and retroaortic groups.

Right and Left Lateral Aortic Nodes.   The *right lateral aortic nodes* are situated partly anterior to the inferior vena cava, near the termination of the renal vein, and partly posterior to it on the origin of the psoas major muscle, and on the right crus of the diaphragm. The *left lateral aortic nodes* form a chain on the left side of the abdominal aorta, in front of the origin of the psoas major and left crus of the diaphragm (Figs. 10-22; 10-24; 10-30; 10-31; 10-35; 10-36).

On both sides of the midline, the lateral aortic nodes receive lymphatic vessels from the common iliac nodes; from the testis in the male or from the ovary, uterine tube, and body of the uterus in the female; from the kidney and suprarenal glands; and from lymphatics draining the lateral abdominal muscles and accompanying the lumbar veins. Most of the efferent vessels of the lateral aortic nodes converge to form the **right**

and left lumbar trunks, which join the cisterna chyli, but some enter the preaortic and retroaortic nodes, and others pierce the crura of the diaphragm to join the lower end of the thoracic duct.

Preaortic Nodes.   The preaortic nodes are divided into **celiac, superior mesenteric,** and **inferior mesenteric** groups, arranged around the origins of the corresponding arteries (Figs. 10-22; 10-33; 10-35; 10-36). They receive a few vessels from the lateral aortic nodes, but their principal afferent vessels are derived from the viscera supplied by the three arteries with which they are associated. Some of their efferent vessels pass to the retroaortic nodes, but the majority unite to form the **intestinal trunk,** which enters the cisterna chyli.

Retroaortic Nodes.   The retroaortic nodes are placed below the cisterna chyli on the bodies of the third and fourth lumbar vertebrae. They receive lymphatic trunks from the lateral and preaortic nodes, and their efferent vessels end in the cisterna chyli.

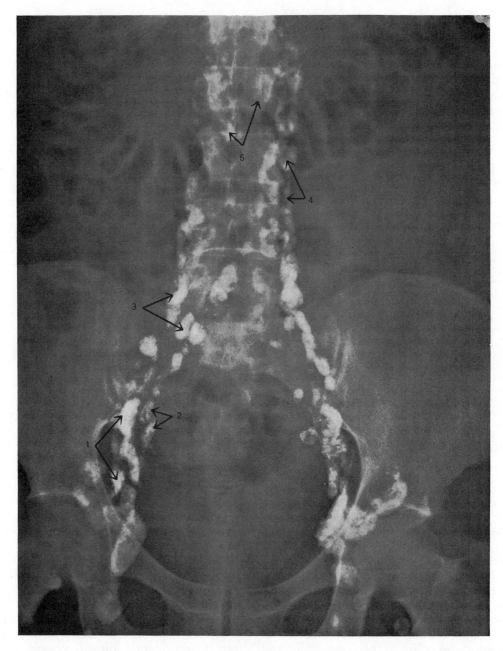

Fɪɢ. 10-36. Lymphadenogram showing the parietal lymph nodes of the pelvis and of the posterior abdominal wall. 1, External iliac nodes; 2, internal iliac nodes; 3, common iliac nodes; 4, lateral aortic (lumbar) nodes; 5, preaortic (lumbar) nodes. (From Gooneratne, B. W. M., *Lymphography—Clinical and Experimental,* Butterworth and Co. Ltd., London, 1974.)

Lymph Nodes of the Abdominal Organs

The lymph nodes of the abdominal organs are frequently referred to as the **visceral nodes** of the abdomen. They are closely associated with the blood vessels, either being clustered in groups at the origin of the ves- sels, or forming chains along the courses of their branches. The three principal groups of nodes into which lymph from most of the abdominal visceral nodes eventually drains before entering the cisterna chyli or the tho- racic duct are the **celiac, superior mesen- teric,** and **inferior mesenteric** sets of *preaortic nodes.* These surround the roots of the large and similarly named unpaired arteries, allowing convenient classification of the outlying chains of nodes along branches or subbranches of these vessels (Fig. 10-22).

CELIAC NODES

The celiac nodes include those nodes that drain the stomach, greater omentum, liver, gallbladder, and spleen, and most but not all of the lymph from the pancreas and duodenum. The celiac nodes are divided into the following sets:

Left gastric
Right gastric
Hepatic
Pyloric
Pancreaticosplenic

LEFT GASTRIC NODES. The left gastric nodes follow the course of the left gastric artery and consist of the following three groups: upper, lower, and paracardial (Figs. 10-25; 10-26). The **upper left gastric nodes** lie at the root of the left gastric artery, where it branches from the celiac trunk. The **lower left gastric nodes** accompany the descending branches of the artery along the cardiac half of the lesser curvature of the stomach, between the two layers of the lesser omentum. The **paracardial** group of celiac nodes is located around the cardiac end of the stomach in a radial manner, comparable to a chain of beads. The left gastric nodes receive afferent vessels from the stomach, and their efferent channels pass to the **celiac group of preaortic nodes**.

RIGHT GASTRIC NODES. The right gastric nodes lie between the leaves of the lesser omentum, along the course of the right gastric artery. They drain the lower end of the lesser curvature and their efferent vessels pass to the **hepatic nodes** (Figs. 10-25; 10-26).

HEPATIC NODES. The hepatic nodes are located between the two layers of the lesser omentum and course along the common hepatic artery and its right and left hepatic branches as far as the porta hepatis (Fig. 10-25). One node of this group is placed near the neck of the gallbladder and is known as the **cystic node.** A few additional nodes descend along the common bile duct as far as the head of the pancreas. The nodes of the hepatic chain receive afferent vessels from the stomach, duodenum, liver, gallbladder, the bile ducts, and pancreas. Their efferent channels join the **celiac group of preaortic nodes**.

PYLORIC NODES. The pyloric nodes lie adjacent to the head of the pancreas and along the medial border of the superior end of the descending part of the duodenum (Figs. 10-25; 10-26). They are located in close relationship to the pylorus of the stomach, and at the site of bifurcation of the gastroduodenal artery. They receive afferent vessels from the pylorus of the stomach, the superior part of the duodenum, as well as from the **right gastroepiploic nodes,** which course between the layers of the greater omentum, along the greater curvature of the stomach with the right gastroepiploic artery (Fig. 10-25). Their efferent channels drain to the **hepatic nodes** and the **celiac group of preaortic nodes,** as well as to the **superior mesenteric group of preaortic nodes**.

PANCREATICOSPLENIC NODES. The pancreaticosplenic nodes accompany the splenic artery and are located adjacent to the posterior surface and upper border of the pancreas (Figs. 10-25; 10-26). One or two nodes of this group are found in the gastrosplenic ligament. The nodes at the hilus of the spleen, in addition to receiving channels from the splenic hilum, receive lymphatic vessels from the **left gastroepiploic nodes**. These latter nodes course between the layers of the greater omentum, along the greater curvature of the stomach. Thus, the pancreaticosplenic nodes receive afferent vessels from the stomach, spleen, and pancreas, and their efferent vessels join the **celiac group of preaortic nodes**.

SUPERIOR MESENTERIC NODES

The superior mesenteric nodes, located around the root of the superior mesenteric artery, are numerous and large. They drain the visceral field supplied by branches of the superior mesenteric artery, and this includes part of the head of the pancreas, a portion of the duodenum, the entire jejunum, ileum, appendix, cecum, ascending colon, and most of the transverse colon. These nodes can be divided into the following groups:

Mesenteric
Ileocolic
Right colic
Middle colic

MESENTERIC NODES. The mesenteric nodes lie between the layers of the mesentery. They vary in number from 100 to 200, and may be grouped into three sets: one lying close to the wall of the small intestine, among the terminal twigs of the superior mesenteric artery; a second in relation to the loops and primary branches of the vessels; and a third along the proximal portion of the inferior mesenteric arterial trunk. They drain the jejunum and ileum along the intestinal branches of the superior mesenteric artery. Also, a part of the head of the pancreas and a portion of the duodenum are drained by nodes that course along the inferior pancreaticoduodenal branch of the superior mesenteric artery. Efferent vessels from the more proximal mesenteric nodes drain into the superior mesenteric group of preaortic nodes.

ILEOCOLIC NODES. Ten to twenty ileocolic lymph nodes form a chain around the ileocolic artery (Figs. 10-27; 10-28), but show a tendency to divide into two groups, one near the duodenum and another on the lower part of the trunk of the artery. Where the ileocolic artery divides into its terminal branches, the chain is broken up into several groups: **ileal,** in relation to the ileal branch of the artery; **anterior ileocolic,** usually three nodes in the ileocecal fold near the wall of the cecum; **posterior ileocolic,** mostly placed in the angle between the ileum and the colon, but partly lying dorsal to the cecum at its junction with the ascending colon; and a single **appendicular** node, between the layers of the mesenteriole of the vermiform appendix. These nodes receive lymph from the terminal portion of the ileum, the appendix and cecum, and from several inches of the proximal portion of the ascending colon. Their efferent channels drain into the **inferior mesenteric group of pancreatic nodes**.

RIGHT COLIC NODES. The right colic group of superior mesenteric nodes follows the course of the right colic artery and its branches as far as the medial border of the ascending colon. These nodes receive afferent vessels from **paracolic nodes** that are spaced along the medial border of the ascending colon. Their efferent vessels pass to the **superior mesenteric group of preaortic nodes**.

MIDDLE COLIC NODES. The middle colic nodes are numerous and accompany the middle colic artery and its branches to the transverse colon, between the layers of the transverse mesocolon (Fig. 10-29). They are best developed in the area of the right and left colic flexures. The middle colic nodes receive afferent vessels from the paracolic nodes that lie along the mesenteric border of the transverse colon. Their efferent vessels course to the **superior mesenteric group of preaortic nodes**.

INFERIOR MESENTERIC NODES

The inferior mesenteric nodes are clustered around the stem of the inferior mesenteric artery and its branches. They drain the descending colon, sigmoid colon, and the upper part of the rectum, a field consistent with the blood supply of the inferior mesenteric artery. They are divided into the following groups:

Left colic
Sigmoid
Superior rectal

LEFT COLIC NODES. The left colic nodes lie along the course of the left colic branch of the inferior mesenteric artery, as far as the wall of the descending colon (Fig. 10-29). Along the medial border of the descending colon are found the paracolic nodes, which directly drain the large intestine. Efferent vessels from these nodes lead to the left colic nodes, which, in turn, send their efferent branches to the **inferior mesenteric group of preaortic nodes**.

SIGMOID NODES. The sigmoid lymph nodes are located between the layers of the sigmoid mesocolon, along the sigmoid branches of the inferior mesenteric artery (Fig. 10-29). They drain the sigmoid colon through the paracolic nodes that are located along the mesenteric border of the viscus. Efferent vessels from the sigmoid nodes terminate in the **inferior mesenteric group of preaortic nodes**.

SUPERIOR RECTAL NODES. The superior rectal group of lymph nodes courses along the superior rectal branch of the inferior mesenteric artery (Fig. 10-29). Lymphatic vessels from the rectum course to **pararectal nodes,** which are oriented longitudinally along the rectum and in contact with its muscular coat. From these nodes, efferent vessels pass to the superior rectal nodes, which then drain to the **inferior mesenteric group of preaortic nodes**.

Lymphatic Vessels and Nodes of the Lower Limb

Most of the lymphatic vessels of the lower limb ascend either superficial or deep to groups of inguinal lymph nodes that lie in the upper anterior thigh, just below the inguinal ligament. Intermediate nodes are also found in the anterior leg and popliteal fossa, along the course of major blood vessels. In this section are described initially the lymphatic vessels of the lower limb and then the groups of lymph nodes.

LYMPHATIC VESSELS OF THE LOWER LIMB

The lymphatic vessels of the lower limb consist of two sets, superficial and deep, and their distribution corresponds closely to that of the veins.

SUPERFICIAL LYMPHATIC VESSELS. The superficial lymphatic vessels of the lower limb lie in the superficial fascia, and are most numerous on the sole of the foot. They are divisible into two groups: a medial, which follows the course of the great saphenous vein, and a lateral, which accompanies the small saphenous vein.

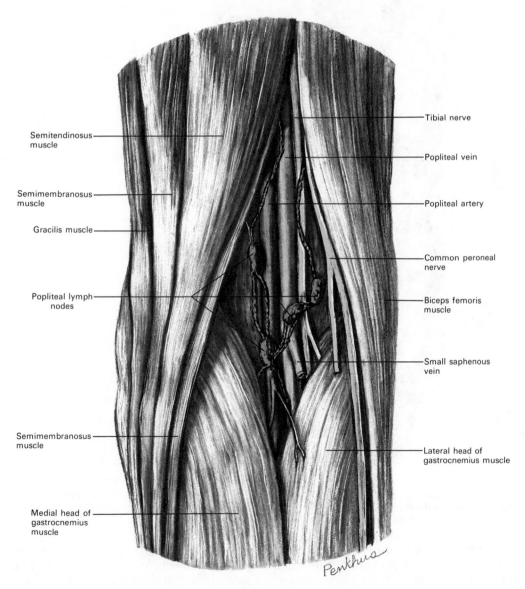

Semitendinosus muscle

Semimembranosus muscle

Gracilis muscle

Popliteal lymph nodes

Semimembranosus muscle

Medial head of gastrocnemius muscle

Tibial nerve

Popliteal vein

Popliteal artery

Common peroneal nerve

Biceps femoris muscle

Small saphenous vein

Lateral head of gastrocnemius muscle

FIG. 10-37.   Lymph nodes of the popliteal fossa.

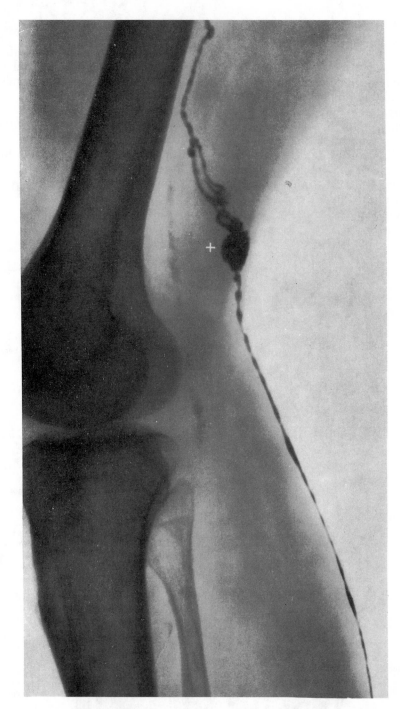

FIG. 10-38.   Lateral view of the right popliteal region. A contrast medium was injected into a superficial lymphatic vessel behind the lateral malleolus. This outlined one of the lymphatic vessels ascending to a popliteal lymph node (marked by a cross) in the upper superficial region of the popliteal fossa. (From Battezzati and Ippolito, *The Lymphatic System,* John Wiley and Sons, New York, 1972.)

The **medial group** of superficial vessels is larger and more numerous than the lateral group, and commences on the tibial side and dorsum of the foot (Fig. 10-39). The vessels ascend both anterior and posterior to the me-dial malleolus, course up the leg with the great saphenous vein, and pass with it behind the medial condyle of the femur to the groin. These vessels end in the **lower set of superficial inguinal lymph nodes** (Figs. 10-39; 10-40).

The **lateral group** of superficial vessels commences on the fibular side of the foot. Some of these vessels ascend on the anterior aspect of the leg, cross the tibia just below the knee, join the lymphatics on the medial side of the thigh, and end in the lower set of superficial inguinal nodes. Others pass behind the lateral malleolus, and accompanying the small saphenous vein, enter the **popliteal nodes** (Figs. 10-37; 10-38).

The superficial lymphatic vessels of the dorsal thigh curve around both the lateral and medial aspects of the limb to open into the *upper and lower* sets of the **superficial inguinal nodes**. From the lateral gluteal region, superficial lymphatic vessels course around the lateral hip region to reach the *upper* superficial inguinal nodes, while vessels from the anus and external genitalia (except the glans penis or clitoris) pass medially to the *lower* set of superficial inguinal nodes.

DEEP LYMPHATIC VESSELS. The deep lymphatic vessels are few in number and accompany the deep blood vessels. In the leg, they consist of three sets, the anterior tibial, posterior tibial, and peroneal, that accompany the corresponding blood vessels. Each of the sets consists of two or three lymphatic channels, and these enter the **popliteal lymph nodes**.

The deep lymphatic vessels of the gluteal and ischial regions follow the course of the corresponding blood vessels. Those accompanying the superior gluteal vessels end in a node that lies on the intrapelvic portion of the superior gluteal artery, near the upper border of the greater sciatic foramen. Those following the inferior gluteal vessels traverse one or two small nodes that lie below the piriformis muscle and end in the **internal iliac nodes**.

LYMPH NODES OF THE LOWER LIMB

The lymph nodes of the lower limb consist of the following:

Anterior tibial node
Popliteal nodes
Superficial inguinal nodes
    Upper group
    Lower group
Deep inguinal nodes

ANTERIOR TIBIAL NODE. The anterior tibial node is small and inconstantly found. When present, it lies on the interosseous mem-

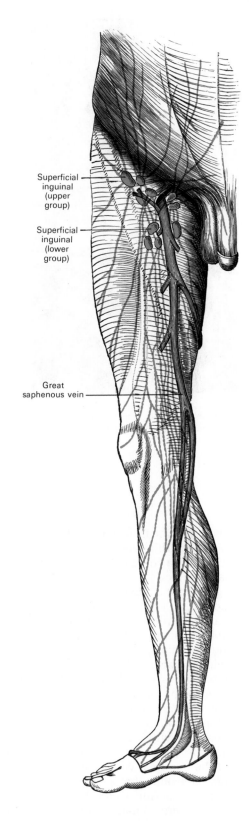

Superficial inguinal (upper group)

Superficial inguinal (lower group)

Great saphenous vein

FIG. 10-39.   The superficial inguinal lymph nodes and lymphatic vessels of the lower limb.

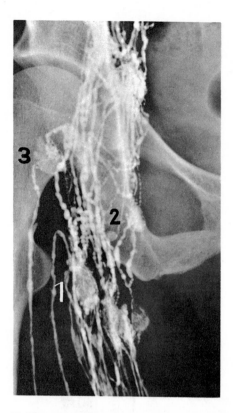

FIG. 10-40. Normal lymphogram of the right inguinal region. 1, Superficial upper inguinal nodes; 2, deep inguinal nodes; 3, superficial lower inguinal nodes. (From de Roo, T., *Atlas of Lymphography,* J. B. Lippincott, Philadelphia, 1975.)

brane, between the tibia and fibula, adjacent to the anterior tibial vessels. It constitutes a substation in the course of the anterior tibial lymphatic trunks.

POPLITEAL NODES. Six or seven small popliteal nodes are embedded in the fat of the popliteal fossa (Figs. 10-37; 10-38). One lies immediately beneath the popliteal fascia, near the terminal part of the small saphenous vein, and drains the region from which this vein derives its tributaries. Another is placed between the popliteal artery and the posterior surface of the knee joint; it receives the lymphatic vessels from the knee joint, together with those that accompany the genicular arteries. The others lie at the sides of the popliteal vessels, and receive as afferents the trunks that accompany the anterior and posterior tibial blood vessels. Most of the efferent channels of the popliteal nodes course upward along the femoral ves-

sels to the **deep inguinal nodes,** but a few may accompany the great saphenous vein and end in the **superficial inguinal nodes**.

INGUINAL NODES

The inguinal lymph nodes are located below the inguinal ligament in the upper anterior and medial thigh (Figs. 10-39; 10-40). These are the principal terminal nodes of the lower limb and they receive most of the efferent lymphatic channels from the foot, leg, and thigh. There are 12 to 20 inguinal nodes and they are divided into **superficial and deep inguinal sets**. Of these, the superficial inguinal nodes are more numerous and consist of *upper and lower groups.*

SUPERFICIAL INGUINAL NODES. The superficial inguinal lymph nodes lie in the fat between the skin and deep fascia, distal to the inguinal ligament. These nodes are divided into an **upper group,** forming a chain parallel to the inguinal ligament and extending from the anterior superior iliac spine to the femoral vessels, and a **lower group,** found in the superficial fascia surrounding the terminal part of the great saphenous vein (Figs. 10-39; 10-40).

The **upper group** of superficial inguinal nodes, consisting of five or six nodes, receives afferent lymphatic channels from the superficial structures of the gluteal region and from the superficial fascia of the anterior abdominal wall, below the umbilicus. Additionally, they receive superficial afferent channels from both the external genitalia and from the perianal region of the perineum.

The **lower group** of four or five superficial inguinal nodes (formerly called superficial subinguinal nodes) receives most of the superficial afferent lymphatic channels of the lower limb, as well as a few superficial vessels from both the urogenital and anal regions of the perineum.

Some efferent vessels from the superficial inguinal nodes drain to the **deep inguinal nodes** through the cribriform fascia, but most efferent channels pass into the pelvis by way of the femoral canal to the **external iliac nodes**.

DEEP INGUINAL NODES. The deep inguinal lymph nodes vary from one to three in number, and are located deep to the fascia lata, on the medial side of the femoral vein (Fig.

10-40). When three are present, the lowest node is situated just below the junction of the great saphenous and femoral veins, the intermediate node lies within the femoral canal, and the highest node (also known as the node of **Cloquet**[1] or **Rosenmüller**[2]) is located in the lateral part of the femoral ring. Of these, the intermediate node is the most inconstant of the three, but the highest is also frequently absent. The deep inguinal nodes receive afferent channels from the deep lymphatic vessels that accompany the femoral artery and vein, other lymphatic vessels from the glans penis or glans clitoris, and some efferent channels from the lower group of superficial lymph nodes. Efferent channels from the deep inguinal nodes pass through the femoral canal to the **external iliac nodes**.

The Spleen

The spleen, a freely movable organ that is largely invested by peritoneum, is situated principally in the left hypochondriac region of the abdomen. Its posterior extremity extends into the epigastric region, and it is usually located between the fundus of the stomach and the diaphragm. It is soft and friable, highly vascular, and dark purplish. Although during fetal life and shortly after birth the spleen gives rise to new red blood corpuscles, this is not thought to occur normally in the adult human organ. Hemopoiesis does occur, however, in the spleens of many other adult vertebrates. The spleen, like certain other lymphoid organs, transforms lymphocytes into cells that produce antibodies or participate in cell-mediated immunologic reactions. Additionally, the spleen contains many macrophages that destroy old red blood cells, the hemoglobin of which is transformed into bilirubin and reused by the body.

The size and weight of the spleen are liable to extreme variations at different periods of life, in different individuals, and in the same individual under different conditions. In the adult, it is usually about 12 cm

in length, 7 cm in breadth, and 3 or 4 cm in thickness. The spleen increases in weight from 17 gm or less during the first year after birth to 170 gm at 20 years, and then slowly decreases to 122 gm at 76 to 80 years of age. Male spleens weigh more than female, and the variation in weight of adult spleens is from 100 to 250 gm, and in extreme instances, 50 to 400 gm. The size of the spleen increases during and after digestion, and varies according to the state of nutrition of the body, being large in well fed, and small in starved animals. In a patient with malarial fever it becomes much enlarged, weighing occasionally as much as 9 kg.[1]

DEVELOPMENT OF THE SPLEEN

The spleen appears initially as several clusters of mesenchymal cells between the layers of the dorsal mesogastrium, near the tail of the pancreas during the fifth week (6-mm stage). These cellular aggregates become vascularized and soon fuse to form a lobulated structure. Evidence of this lobulation is still indicated in the adult spleen by notches usually present along the superior margin. With the rotation of the stomach, which results in the greater curvature being directed toward the left, the spleen is also carried to the left, and lies behind the stomach and in contact with the left kidney. The part of the dorsal mesogastrium that intervened between the spleen and the kidney becomes the **lienorenal ligament,** while that part between the spleen and the greater curvature of the stomach forms the **gastrosplenic ligament**.

ANOMALIES IN DEVELOPMENT. Simple **agenesis** of the spleen, splenic agenesis associated with cardiac defects, the occurrence of several spleens, each of good size (polysplenia), the existence of a normal spleen with **accessory small spleens,** and the ectopic location of splenic tissue in the scrotum **(splenogonadal fusion)** that descended with the testis are abnormalities in development that have been described (Gray and Skandalakis, 1972).

---

[1] Jules Germain Cloquet (1790–1883): A French anatomist and surgeon (Paris).
[2] Johann Christian Rosenmüller (1771–1820): A German anatomist and surgeon (Leipzig).

[1] These data are taken from the monograph entitled *The Structure and Use of the Spleen* by Henry Gray, which was published in 1854 and won the triennial "Astley Cooper Prize" of £300 in 1853.

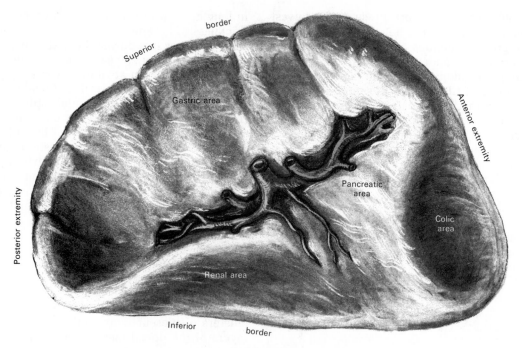

FIG. 10-41.   The visceral surface of the spleen.

RELATIONSHIPS OF THE ADULT SPLEEN

The spleen has *two surfaces,* diaphragmatic and visceral; *two margins,* superior and inferior; and *two extremities,* anterior and posterior (Figs. 10-41; 10-42).

DIAPHRAGMATIC SURFACE.   The diaphragmatic surface of the spleen is smooth, convex, and directed upward, backward, and to the left, except toward its posterior extremity, where it is directed slightly medially. This surface is adjacent to the abdominal aspect of the diaphragm, which separates the spleen from the ninth, tenth, and eleventh ribs of the left side, and the intervening caudal border of the left lung and pleura (Fig. 10-42,C).

VISCERAL SURFACE.   The visceral surface of the spleen is oriented toward the abdominal cavity, and it has gastric, renal, and colic areas (Fig. 10-41).

Gastric Area.   The gastric area of the visceral surface of the spleen is directed anteriorly, upward, and medially (Fig. 10-41). It is broad and concave, and is in contact with the dorsal wall of the stomach, and below this, with the tail of the pancreas. It has near

its medial border a long fissure, or more frequently a series of depressions, termed the **hilus.** This is pierced by several irregular apertures for the entrance and exit of vessels and nerves.

Renal Area.   The renal area of the visceral surface of the spleen is somewhat flattened and separated from the gastric area by an elevated ridge along the hilus. It is directed medially and downward and it is considerably narrower than the gastric area (Fig. 10-41). It is in anatomic relationship to the upper part of the anterior surface of the left kidney and occasionally with the upper aspect of the left suprarenal gland.

Colic Area.   The colic area of the visceral surface of the spleen is flat and triangular. It rests on the left flexure of the large colon and the phrenicocolic ligament, and it is frequently in contact with the tail of the pancreas (Fig. 10-41).

BORDERS OF THE SPLEEN.   The **superior border** is free, sharp, thin, and often notched, and it separates the diaphragmatic surface from the gastric area of the visceral surface. The **inferior border** is more rounded and more blunt than the superior border

(Fig. 10-41). It separates the renal area of the visceral surface from the diaphragmatic surface, and it corresponds to the lower border of the eleventh rib, lying between the diaphragm and left kidney.

EXTREMITIES OF THE SPLEEN. The **posterior extremity** is rounded and directed toward the vertebral column. The **anterior extremity** is broader and more expanded than the posterior (Fig. 10-41), extending between the superior and inferior borders and separating the colic area of the visceral surface from the diaphragmatic surface.

PERITONEAL ATTACHMENTS. The capsule of the spleen is completely covered, except at the hilus, by a closely adherent layer of visceral peritoneum. Developing between the layers of the dorsal mesogastrium, the spleen divides this mesentery into two portions. One extends between the stomach and the spleen and is called the **gastrosplenic ligament**. The other is interposed between the spleen and the dorsal body wall and is called the **phrenicolienal ligament** (Fig. 10-42,A).

The lower part of the phrenicolienal ligament, called the **lienorenal ligament,** extends to the left kidney, and between its layers are found the tail of the pancreas and the splenic vessels and nerves that course through the hilus of the spleen (Fig. 10-42,B). Between the leaves of the gastrosplenic ligament are found the short gastric and left gastroepiploic vessels.

Because these ligaments form the left boundary of the omental bursa (lesser sac), their outer surfaces are covered by peritoneum that surrounds the greater sac and their inner surfaces are covered by peritoneum that lines the omental bursa. The spleen itself is covered by greater sac peritoneum only, unless the gastrolienal and phrenicolienal ligaments fail to meet at the hilum, in which condition a small area within the hilum may be covered by peritoneum that lines the omental bursa.

As pointed out previously, near the spleen are frequently found small nodules of splenic tissue. This is especially so in the gastrosplenic ligament and greater omentum, and these nodules may be either isolated or connected to the spleen by thin bands of splenic tissue. These are called accessory spleens, and their size varies from that of a pea to that of a plum.

MICROSCOPIC STRUCTURE OF THE SPLEEN

The spleen is an encapsulated lymphatic organ, but it is unlike the lymph nodes because through the sinuses of the spleen venous blood filters, whereas through the sinuses of the lymph nodes the lymphatic fluid filters.

CAPSULE AND TRABECULAE

The spleen is invested by two coats: an external serous layer and an internal fibroelastic capsule. From the latter, connective tissue trabeculae penetrate the substance of the spleen, separating it into many small compartments (Fig. 10-43).

EXTERNAL SEROUS COAT. The external serous coat is not an intrinsic part of the spleen because it is derived from the peritoneum. It is a thin, smooth layer of mesothelial cells, and it is intimately adherent to the underlying fibroelastic capsule. It invests the entire organ, except at the hilum and along the lines of reflection of the phrenicolienal and gastrosplenic ligaments.

SPLENIC CAPSULE. The spleen is enclosed in a rather heavy fibroelastic capsule that is composed of dense fibrous connective tissue, is rich in elastic fibers, and contains scattered fibroblasts and some smooth muscle cells. The capsule closely invests the organ, and at the hilum is prolonged inward upon the vessels in the form of sheaths. From these sheaths, as well as from the inner surface of the fibroelastic capsule, numerous small fibrous bands, called trabeculae, are given off in all directions. Because of the high content of elastic fibers in the capsule, the vascular sheaths, and the trabeculae, the spleen possesses a considerable ability to undergo variations in size.

TRABECULAE. The trabeculae are flattened strands of connective tissue that divide the spleen into many small intercommunicating compartments, each several millimeters in diameter (Fig. 10-43). Containing an even higher number of elastic fibers than the capsule, the trabeculae are scattered throughout the spleen, thereby forming a complex meshwork. The trabeculae arise from the branching of larger connective tissue sheaths that penetrate the organ at the hilus, while other trabecular strands branch from

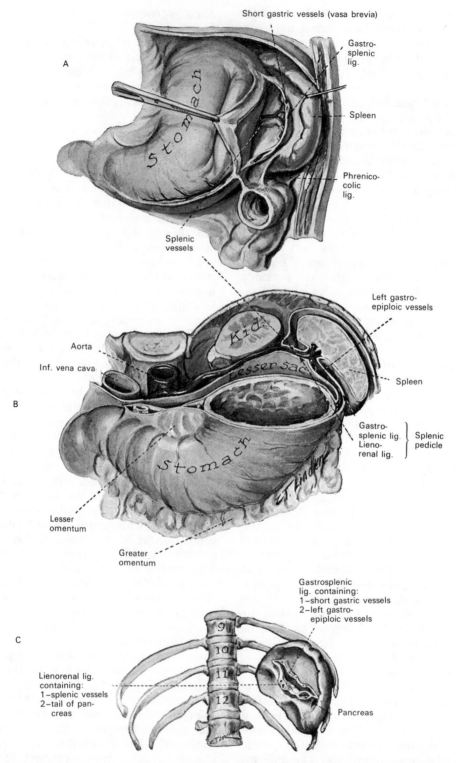

Short gastric vessels (vasa brevia)

A

Gastro-
splenic
lig.

Spleen

Phrenico-
colic
lig.

Splenic
vessels

Stomach

Left gastro-
epiploic vessels

Aorta

Inf. vena cava

Kid.

Lesser sac

Spleen

B

Gastro-
splenic lig.
Lieno-
renal lig. } Splenic
pedicle

stomach

Lesser
omentum

Greater
omentum

Gastrosplenic
lig. containing:
1—short gastric vessels
2—left gastro-
   epiploic vessels

C

9
10
11
12

Lienorenal lig.
containing:
1—splenic vessels
2—tail of pan-
   creas

Pancreas

Fig. 10-42.   The splenic ligaments. *A*, The gastrosplenic ligament (part of the greater omentum) has been opened to reveal the splenic vessels and the short gastric vessels. *B*, Cross section through the spleen and fundus of the stomach; observe the gastrosplenic ligament oriented anteriorly and the lienorenal ligament oriented posteriorly. *C*, The spleen in relationship to the ninth, tenth, and eleventh ribs. Note the cut edges of the gastrosplenic and lienorenal ligaments. (From Thorek, P., *Anatomy in Surgery*, 2nd ed., J. B. Lippincott, Philadelphia, 1962.)

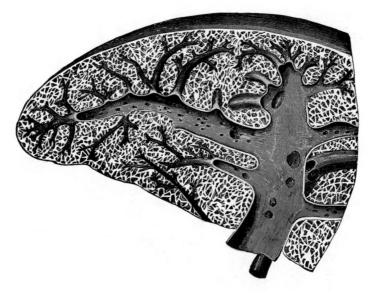

Fig. 10-43.   Transverse section of the spleen, showing the trabecular tissue and the splenic vein and its tributaries. (From Henry Gray, 1854).

the internal capsular surface. Within the compartments outlined by the trabeculae is found the **splenic pulp**.

SPLENIC PULP

In a cross section of freshly cut spleen, small whitish to gray regions, measuring 0.25 to 1 mm in diameter, are seen scattered throughout a mass of soft, dark reddish-brown tissue. These lighter regions are collectively called the **white pulp** and individually are known as *lymphatic nodules* (or Malpighian[1] bodies), while the surrounding larger mass of darker tissue is called the **red pulp** (Fig. 10-44).

WHITE PULP.   The white pulp of the spleen consists of many small masses of periarterial lymphocytes and other related cells. Arteries that enter the hilum of the spleen course along the trabecular meshwork and branch into smaller vessels. Supporting these smaller vessels are sheaths of reticular fibers, in the meshes of which are infiltrations of lymphocytes. As the arteries branch, they leave the trabeculae and eventually penetrate the red pulp. As the periarterial lymphatic sheaths surrounding the vessels leave the trabeculae, they become enlarged at certain sites to form the lymphatic nodules of

[1] Marcello Malpighi (1628–1694): A great Italian microscope anatomist and embryologist (Bologna, Pisa, and Messina).

the white pulp (Fig. 10-44). The microscopic structure of these nodules, which have compact masses of lymphocytes enmeshed within a network of reticular fibers, is not significantly different from that of the germinal centers of lymph nodes.

RED PULP.   The red pulp forms the main substance of the spleen (Fig. 10-44). It is composed of rich plexuses of tortuous venous sinuses. Between the sinuses are splenic cords that contain diffuse lymphatic cells, which include many macrophages, all the white blood cell types, many erythrocytes, and plasma cells, all enmeshed in a fine network formed by the processes of stellate reticular cells. Attached by perpendicular processes to the outer endothelial walls of the venous sinuses, the reticular cell bodies lie suspended within the splenic cords, and come into direct contact with the free cellular elements. The macrophages are often large and contain ingested and partly digested red blood cells and pigment.

Between the lymphoid tissue of the white pulp and the sinuses and splenic cords of the red pulp is a **marginal zone** that shows characteristics transitional between the two types of splenic tissue.

BLOOD VESSELS OF THE SPLEEN

ARTERIES.   Derived as one of the branches of the celiac trunk, the **splenic artery** is remarkable for its size, being quite large in

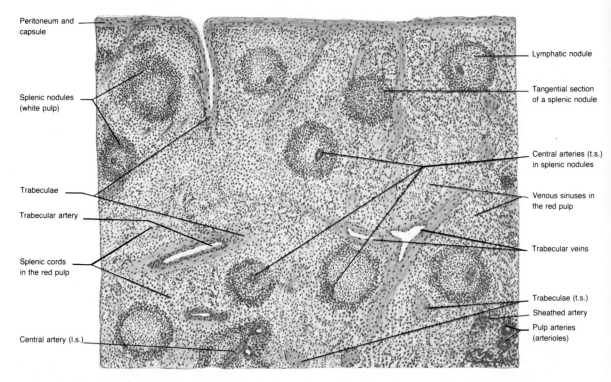

Peritoneum and capsule

Splenic nodules (white pulp)

Trabeculae

Trabecular artery

Splenic cords in the red pulp

Central artery (l.s.)

Lymphatic nodule

Tangential section of a splenic nodule

Central arteries (t.s.) in splenic nodules

Venous sinuses in the red pulp

Trabecular veins

Trabeculae (t.s.)

Sheathed artery

Pulp arteries (arterioles)

Fig. 10-44. A section through a portion of the human spleen, stained with hematoxylin and eosin. 50 × . (From di Fiore, M., *Atlas of Human Histology*, 5th ed., Lea & Febiger, Philadelphia, 1981.)

proportion to the size of the organ, and also for its tortuous course. It divides into six or more branches, which enter the hilum and ramify throughout its substance. Each of these branches supplies chiefly that region of the spleen in which it ramifies, having no anastomosis with the majority of other branches. They course in the transverse axis of the organ and diminish in size during transit.

These initial branches of the splenic artery further subdivide into smaller muscular **trabecular arteries,** which course along the trabecular system (Fig. 10-45). Their adventitial coat blends with the connective tissue of the trabeculae, and they frequently are described as having trabecular sheaths, which similarly invest the veins emerging at the hilum and the incoming nerves. The trabecular arteries ultimately leave the trabecular sheaths and their tunica adventitia is replaced by packs of lymphocytes, enmeshed in reticular fibers, which form the white pulp. These vessels are then called the **central** (or follicular) **arteries,** and they contain one or two layers of smooth muscle. Central arteries are not accompanied by veins, and despite their name they are usually dis-

placed to an eccentric position in the splenic lymphatic nodules (Figs. 10-44; 10-45).

As the central arteries branch, they become reduced in diameter, and their sheaths of lymphoid tissue become considerably thinner. When central arteries become reduced to about 50 μ in diameter, each vessel divides into two to six straight branches called **penicillar arteries**. These latter vessels, still surrounded by a single or double layer of lymphocytes, enter the red pulp, course for less than 1 mm, and branch into two or three arterial capillaries. At this point, the arterial capillaries are lined by endothelial cells and have no muscle or adventitial coats, but they do acquire a characteristic thick sheath that consists of concentrically arranged macrophages and reticular fibers. The diameter of these **sheathed capillaries** is about 6 to 8 μ. Although found in the human spleen, these capillaries are more prominent in other animals, such as the dog.

After a short course, sheathed capillaries branch into normal capillaries. Blood from these fine arterial capillaries reaches the venous sinuses, but exactly how this occurs is still conjectural. Some authors claim that the arterial capillaries are connected with

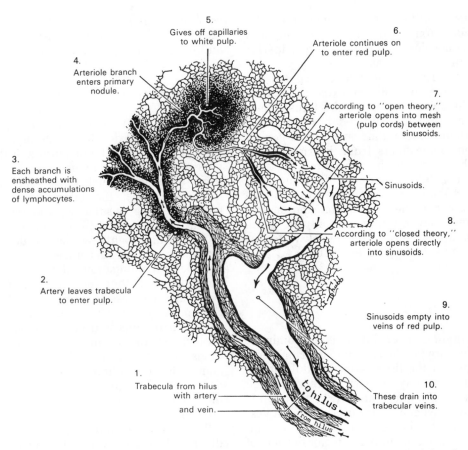

5.
Gives off capillaries
to white pulp.

6.
Arteriole continues on
to enter red pulp.

4.
Arteriole branch
enters primary
nodule.

7.
According to "open theory,"
arteriole opens into mesh
(pulp cords) between
sinusoids.

3.
Each branch is
ensheathed with
dense accumulations
of lymphocytes.

Sinusoids.

8.
According to "closed theory,"
arteriole opens directly
into sinusoids.

2.
Artery leaves trabecula
to enter pulp.

9.
Sinusoids empty into
veins of red pulp.

1.
Trabecula from hilus
with artery
and vein.

to hilus

from hilus

10.
These drain into
trabecular veins.

Fig. 10-45.   This diagram illustrates the course of blood within the spleen from arteries at the hilus, through the splenic pulp, the sinusoids, and back through the veins. The labels should be read sequentially from 1 to 10. (From Ham, A. W., *Histology*, 7th ed., J. B. Lippincott, Philadelphia, 1974.)

the venous sinuses in a *closed* system, and that the endothelial lining between them is complete. Other authors maintain that the endothelial connections are incomplete and in this *open* type of circulation, the arterial blood is thought to enter the splenic lymphatic cords and then to diffuse through the endothelium and into the venous sinuses (Fig. 10-45). The flow through the splenic pulp is controlled by rhythmic contraction and relaxation of individual and groups of arterioles.

VENOUS SINUSES.   The venous sinuses of the red pulp are tortuous channels measuring 12 to 40 μ in diameter. Their walls are formed by long, narrow endothelial cells that are oriented longitudinally along the sinusoid wall. Their nuclei bulge into the lumen; however, their cytoplasmic processes do not form true intercellular junctions. More recent evidence indicates that they do not behave as fixed macrophages, as was

once thought. The endothelial cells are supported by circularly arranged reticular fibers that are continuous with the collagen fibers of the trabeculae. The reticular fibers form rings around the sinusoids.

VEINS.   The venous sinuses open into the slender, thin-walled **veins of the splenic pulp**. These join to form the **trabecular veins,** which in turn unite to form the six or more branches of the **splenic vein** emerging from the hilus (Fig. 10-45).

LYMPHATIC VESSELS AND NERVES

LYMPHATIC VESSELS.   Although few, if any, lymphatic vessels have been found in the lymphoid tissue of the splenic pulp in the human spleen, they have been described in some other mammals. Lymphatic vessels do exist in the capsule of the human spleen and along the larger trabeculae. They emerge at the hilus with the veins and open into splenic nodes found at the hilus and into pancreaticosplenic nodes along the superior border of the pancreas.

NERVES. Nerve fibers to the spleen are derived from the celiac plexus and course along the splenic artery to the hilus. These are primarily unmyelinated postganglionic sympathetic fibers that are vasomotor in function. Some fibers from the right vagus nerve, however, also have been found coursing to the spleen. The nerve fibers enter the spleen at the hilus and course along the arteries as far as their trabecular and central branches. They terminate not only along the smooth musculature of the vessel walls, but within the splenic pulp as well.

The Thymus

The thymus in an infant is a prominent organ, occupying the anterior superior mediastinum, but the thymus in an adult of advanced years may be scarcely recognizable because of atrophic changes (Fig. 10-46). It is situated between the sternum and great vessels, resting on the pericardium and overlapping somewhat the upper anterior borders of both lungs. At one time, the thymus was thought simply to function as a producer of lymphocytes. Only since the 1960s has the enormous importance of the thymus in the maturation of a competent immunologic system come to light. Although much still needs to be learned, the thymus is now recognized as perhaps the key organ that contributes to the ability of the body to produce antibodies and to reject foreign tissues and cells.

DEVELOPMENT OF THE THYMUS

During the latter part of the fifth prenatal week, when the developing embryo is only about 10 mm long, the primordia of the thymus appear on the anterior aspect of the third pharyngeal pouch, one on each side. At this same time, the inferior parathyroid glands develop as small solid masses on the dorsal aspect of the third pouch. Between the sixth and eighth weeks the developing thymus is initially recognized as two hollow endodermal elongations that descend into the thorax and then fuse in front of the developing great vessels. At this time, the inferior parathyroid glands usually separate from the thymus and come to rest on the posterior surface of the thyroid gland. Sometimes, however, the inferior parathyroid glands do not separate and their continued development occurs within the substance of the thymus, where they come to lie in the anterior mediastinum.

The fusion medially of the two thymic diverticula involves only the mesenchymal connective tissue into which the thymic tissue has grown, and there is no fusion of the parenchyma of the two developing lobes. The pharyngeal opening of each diverticulum, the **thymopharyngeal duct,** is soon obliterated, but this stalk may persist for a period of time as a cellular cord. Some in-

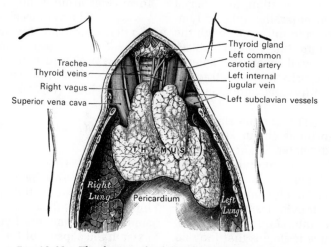

FIG. 10-46.   The thymus of a full-term fetus; exposed *in situ*.

vestigators have claimed that the anterior aspect of the fourth pharyngeal pouch also develops into a small amount of thymic tissue, but this has not been definitely established as a routine occurrence.

As the lobules continue to increase in size, their peripheral zones form the **cortex** of the thymus, in which many small lymphocytes are concentrated, and their central zones form the **medulla,** in which the cellular reticulum predominates. The thymus attains a weight of 12 to 14 gm before birth, but it does not reach its greatest relative size until the age of two years. It continues to grow until puberty, at which time it reaches its greatest absolute size, weighing about 35 gm. After adolescence it begins its process of involution.

HISTOGENESIS OF THE THYMUS. The cellular differentiation of the thymus commences when endodermal epithelial cells form solid cords that grow caudally into the surrounding mesenchyme tissue. Soon the tubular masses of epithelial cells are penetrated by blood vessels, and the cells change their shape into stellate cells. They still remain attached to each other by cellular processes, but what results is a cytoreticular meshwork arrangement. Some of the epithelial cells, instead of contributing to the cytoreticular meshwork, become arranged in concentric, spherical clusters called **thymic corpuscles** (or Hassall's[1] bodies), the centers of which appear to show degenerative changes (Fig. 10-49).

At about the eighth prenatal week, lymphocytes appear in the developing epithelial cytoreticular meshwork, filling the interstitial spaces between the epithelial cells. At one time some investigators believed that these lymphocytes were derived from the epithelial cells themselves, while others thought that they came from the mesenchymal cells. The weight of evidence now indicates, however, that the thymus (like the developing spleen, liver, and bone marrow) receives stem cells from the yolk sac in the embryo, whereas after birth the stem cells come from the bone marrow. These reach the thymus by way of the blood stream and they begin to differentiate into small lymphocytes, which in the thymus are called **thymocytes.** During embryogenesis, the pro-

liferation of thymocytes occurs at a high rate and large numbers of these leave the thymus to reside at many sites elsewhere in the body.

RELATIONSHIPS OF THE THYMUS

The thymus consists of two lateral lobes that are held in close contact by connective tissue, which also encloses the whole organ in a distinct capsule. It is situated partly in the thorax and partly in the neck, extending from the fourth costal cartilage to the lower border of the thyroid gland (Fig. 10-46). *In the neck,* it lies anterior and lateral to the trachea, and deep to the origins of the sternohyoid and sternothyroid muscles. *In the thorax,* it occupies the anterior portion of the superior mediastinum; superficial to it is the sternum, and deep to it are the great vessels and the upper part of the fibrous pericardium. The two lobes generally differ in size and shape, the right frequently overlapping the left. The thymus has a pinkish gray color, is soft and lobulated, and at birth measures approximately 5 cm in length, 4 cm in width, and 6 mm in thickness. Of greater importance, however, is its weight at different periods of maturation from birth through adulthood and senility (described previously), at which times the organ alters significantly in size.

MICROSCOPIC STRUCTURE OF THE THYMUS

The two lobes of the thymus are composed of numerous lobules, varying from 0.5 to 2 mm in diameter, which are incompletely separated from each other by delicate connective tissue. The parenchymal cells of one lobule are joined to those of adjacent ones by thin cellular interconnections. The lobules can be seen on stained sections with the unaided eye. Using low power magnification, two zones of tissue can be seen within each lobule, an outer **cortex** and an inner **medulla** (Fig. 10-47).

The *cortical part* consists primarily of a network of stellate **reticular cells** and their processes, which differentiated from endodermal epithelial cells, and numerous closely packed aggregations of lymphocytes called **thymocytes,** which occupy the spaces within the cytoreticulum (Fig. 10-48). The

---

[1]Arthur H. Hassall (1817–1894): An English physician and botanist (London).

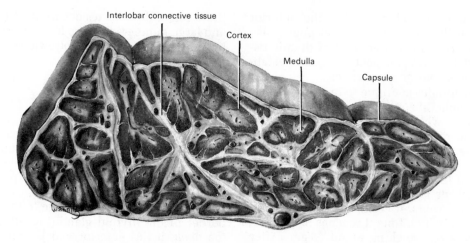

Interlobar connective tissue

Cortex

Medulla

Capsule

FIG. 10-47.   Section of thymus from full-term fetus, lightly stained with hematoxylin. 5 ×.

medullary portion of a thymic lobule contains reticular cells of varying shapes. Whereas some maintain the stellate characteristics of reticular cells found in the cortex, others become elongated, somewhat flattened, and circularly arranged in lamellae to form the **thymic corpuscles,** which measure 0.03 to 0.10 mm in diameter (Fig. 10-49). The central core of these corpuscles consists of granular cells that appear to be degenerating. The medulla contains many fewer lymphocytes than the cortex.

The *stroma* of the thymus consists of a thin investing capsule and interlobular connective tissue septa, in which are found the blood vessels and nerves supplying the organ.

VESSELS AND NERVES.   The **thymic arteries** are derived from mediastinal branches of the internal thoracic artery, and from branches of the inferior thyroid arteries. The **thymic veins** drain into the left brachiocephalic, the internal thoracic, and the inferior thyroid veins. From the interlobular septa, smaller arteries enter the depths of the lobular parenchyma along the cortical-medullary border. From these vessels capillaries penetrate and ramify within the cortex, while other small vessels course into the medulla. The returning capillaries and venules merge in a comparable manner and return to the interlobular septa in a similar course.

The **lymphatic vessels** that drain the thymus terminate in parasternal, anterior mediastinal, and tracheobronchial lymph nodes.

The **nerves** that supply the thymus are very small and are derived from both sym-

pathetic and vagal branches. Also distributed to the capsule of the thymus, but not penetrating the substance of the gland, are small branches from the phrenic nerves and from the superior root of the ansa cervicalis.

INVOLUTION OF THE THYMUS.   The thymus commences gradual involutionary changes in early childhood, and this process is accelerated after puberty. This is a normal change and it is called **age involution,** but it may be superseded by a rapid **accidental involution** due to excessive stress, starvation, ionizing radiation, or acute disease. The small lymphocytes of the cortex disappear and the reticular tissue becomes compressed. The disappearing thymic tissue is likely to be replaced by adipose tissue, but the connective tissue capsule persists, retaining its original shape. With age the weight of the thymus is usually reduced significantly, and in an older individual what appears to be a yellowish-colored thymus, when sectioned, will reveal only small islands of thymic tissue surrounded by fat.

FUNCTIONS OF THE THYMUS.   Only recently has the importance of the thymus been appreciated. One of the most exciting fields of research in the latter half of the twentieth century has been the work that has elucidated the cellular and humoral mechanisms by which the body achieves and maintains its immunologic defense. At one time it was thought that the thymus quite simply produced lymphocytes. Although this is one of its most important functions, it was not then realized how essential these specific lymphocytes, and probably certain humoral se-

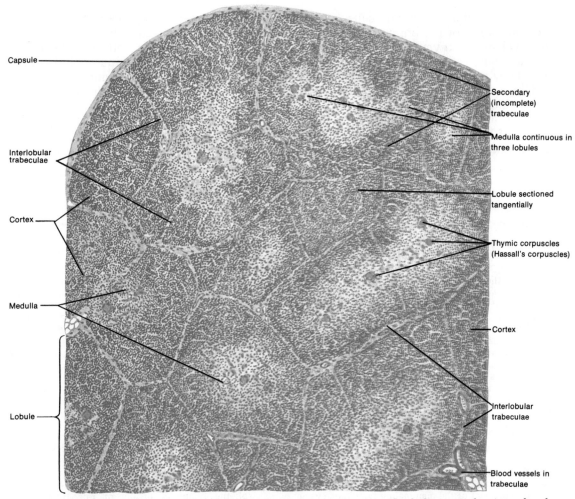

Capsule

Secondary (incomplete) trabeculae

Medulla continuous in three lobules

Interlobular trabeculae

Lobule sectioned tangentially

Cortex

Thymic corpuscles (Hassall's corpuscles)

Medulla

Cortex

Lobule

Interlobular trabeculae

Blood vessels in trabeculae

FIG. 10-48. Low power view of the thymus showing its lobular structure. Note the darker cortical regions of each lobule surrounding the lighter medullary regions. Hematoxylin and eosin, 40 ×. (From di Fiore, M., *Atlas of Human Histology,* 5th ed., Lea & Febiger, Philadelphia, 1981.)

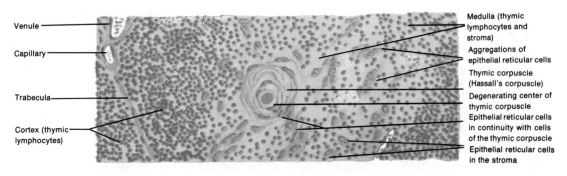

Venule

Capillary

Trabecula

Cortex (thymic lymphocytes)

Medulla (thymic lymphocytes and stroma)

Aggregations of epithelial reticular cells

Thymic corpuscle (Hassall's corpuscle)

Degenerating center of thymic corpuscle

Epithelial reticular cells in continuity with cells of the thymic corpuscle

Epithelial reticular cells in the stroma

FIG. 10-49. A medium-power view of the medullary region of a thymic lobule showing a thymic corpuscle (Hassall's body). Hematoxylin and eosin, 250 ×. (From di Fiore, M., *Atlas of Human Histology,* 5th ed. Lea & Febiger, Philadelphia, 1981.)

cretions liberated by thymic cells, were to the maturation of the immunologic system. Removal of the thymus at birth results in animals that are immunologically incompetent. They cannot reject transplanted grafts, and being defenseless to the normal antigenic challenges of the environment, they soon die. Further, these animals show a deficiency of lymphocytes in the spleen, lymph nodes, and blood.

The epithelial (reticular) cells of the thymus probably have an endocrine function. It is thought that these cells secrete a hormone(s) that induces a capacity in lymphocytes, both of the thymus itself and elsewhere, and in plasma cells to form antibodies in response to antigens.

References

(References are listed not only to those articles and books cited in the text, but to others as well which are considered to contain valuable resource information for the student who desires it.)

Embryology of the Lymphatic System

Ackerman, G. A. 1967. Developmental relationship between the appearance of lymphocytes and lymphopoietic activity in the thymus and lymph nodes of the fetal cat. Anat. Rec., 158:387–392.

Bailey, R. P., and L. Weiss. 1975. Ontogeny of human fetal lymph nodes. Amer. J. Anat., 142:15–28.

Cash, J. R. 1921. On the development of the lymphatics in the stomach of the embryo pig. Carneg. Inst., Contr. Embryol., 13:1–15.

Clark, E. R., and E. L. Clark. 1920. On the origin and early development of the lymphatic system in the chick. Carneg. Inst., Contr. Embryol., 45:449–482.

Clark, E. R., and E. L. Clark. 1932. Observations on the new growth of lymphatic vessels as seen in transparent chambers introduced into the rabbit's ear. Amer. J. Anat., 51:49–87.

Gray, S. W., and J. E. Skandalakis. 1972. Embryology for Surgeons. W. B. Saunders Co., Philadelphia, 918 pp.

Huntington, G. S. 1908. The genetic interpretation of the development of the mammalian lymphatic system. Amer. J. Anat., 2:19–45.

Huntington, G. S. 1911. The Anatomy and Development of the Systemic Lymphatic Vessels in the Domestic Cat. American Anatomical Memoir No. 1, Wistar Institute of Anatomy and Biology, Philadelphia, 175 pp.

Huntington, G. S., and C. F. W. McClure. 1910. The anatomy and development of the jugular lymph sac in the domestic cat (Felis domestica). Amer. J. Anat., 10:177–311.

Kampmeier, O. F. 1928. The genetic history of the valves in the lymphatic system of man. Amer. J. Anat., 40:413–457.

Kampmeier, O. F. 1928. On the lymph flow of the human heart, with reference to the development of channels and first appearance, distribution, and physiology of their valves. Amer. Heart J., 4:210–222.

Lewis, F. T. 1906. The development of the lymphatic system in rabbits. Amer. J. Anat., 5:95–111.

Lewis, F. T. 1909. On the cervical veins and lymphatics in four human embryos. With an interpretation of anomalies of the subclavian and jugular veins in the adult. Amer. J. Anat., 9:33–42.

McClure, C. F. W. 1915. The development of the lymphatic system in the light of the more recent investigations in the field of vasculogenesis. Anat. Rec., 9:563–579.

Sabin, F. R. 1902. On the origin of the lymphatic system from the veins, and the development of the lymph hearts and thoracic duct in the pig. Amer. J. Anat., 1:367–389.

Sabin, F. R. 1904. On the development of the superficial lymphatics in the skin of the pig. Amer. J. Anat., 3:183–195.

Sabin, F. R. 1916. The origin and development of the lymphatic system. Johns Hopkins Hosp. Rep., 17:347–440.

Lymph Nodes and Lymphocytes

Bailey, R. P., and L. Weiss. 1975. Light and electron microscopic studies of postcapillary venules in developing human fetal lymph nodes. Amer. J. Anat., 143:43–58.

Baker, B. L., D. J. Ingle, and C. R. Li. 1951. The histology of the lymphoid organs of rats treated with adrenocorticotropin. Amer. J. Anat., 88:313–349.

Borum, K., and M. H. Claesson. 1971. Histology of the induction phase of the primary immune response in lymph nodes of germfree mice. Acta Pathol. Microbiol. Scand. (A), 79:561–568.

Brahim, F., and D. G. Osmond. 1973. The migration of lymphocytes from bone marrow to popliteal lymph nodes demonstrated by selective bone marrow labeling with ^{3}H-thymidine in vivo. Anat. Rec., 175:737–746.

Clark, S. L., Jr. 1962. The reticulum of lymph nodes in mice studied with the electron microscope. Amer. J. Anat., 110:217–258.

Cohen, S., P. Vassalli, B. Benacerraf, and R. T. McCluskey. 1966. The distribution of antigenic and nonantigenic compounds within draining lymph nodes. Lab. Invest., 15:1143–1155.

Conway, E. A. 1937. Cyclic changes in lymphatic nodules. Anat. Rec., 69:487–508.

Ehrich, W. E. 1929. Studies on lymphoid tissue, I: The anatomy of secondary nodules and some remarks on the lymphatic and lymphoid tissues. Amer. J. Anat., 43:347–384.

Farr, A. G., and P. P. H. De Bruyn. 1975. Macrophage-lymphocyte clusters in lymph nodes: a possible substrate for cellular interactions in the immune response. Amer. J. Anat., *144*:209–232.

Farr, A. G., and P. P. H. De Bruyn. 1975. The mode of lymphocyte migration through postcapillary venule endothelium in lymph node. Amer. J. Anat., *143*:59–92.

Ford, W. L., and J. L. Gowans. 1969. The traffic of lymphocytes. Semin. Hematol., *6*:67–83.

Fujita, T., M. Miyoshi, and T. Murakami. 1972. Scanning electron microscope observation of the dog mesenteric lymph node. Z. Zellforsch. mikrosk. Anat., *133*:147–162.

Furuta, W. J. 1948. The histologic structure of the lymph node capsule at the hilum. Anat. Rec., *102*:213–223.

Gowans, J. L. 1966. Lifespan, recirculation and transformation of lymphocytes. Internat. Rev. Exp. Pathol., *5*:1–24.

Gowans, J. L., and E. J. Knight. 1964. The route of recirculation of lymphocytes in the rat. Proc. Roy. Soc. Lond. B, *159*:257–282.

Han, S. S. 1961. The ultrastructure of the mesenteric lymph node of the rat. Amer. J. Anat., *109*:183–225.

Handley, R. S., and A. C. Thackray. 1947. Invasion of the internal mammary lymph glands in carcinoma of the breast. Brit. J. Cancer, *1*:15–20.

Handley, R. S., and A. C. Thackray. 1949. The internal mammary lymph chain in carcinoma of the breast. Lancet, *2*:276–278.

Harris, T. N., K. Hummeler, and S. Harris. 1966. Electron microscopic observations on the antibody-producing lymph node cells. J. Exp. Med., *123*:161–172.

Kelly, R. H., B. M. Balfour, J. A. Armstrong, and S. Griffiths. 1978. Functional anatomy of lymph nodes. II. Peripheral lymph-borne monocular cells. Anat. Rec., *190*:5–22.

Marchesi, V. T., and J. L. Gowans. 1964. The migration of lymphocytes through the endothelium of venules and lymph nodes. Proc. Roy. Soc. Lond. B, *159*:283–290.

Meyer, A. W. 1917. Studies on hemal nodes. VII. The development and function of hemal nodes. Amer. J. Anat., *21*:375–406.

Pinto, S. 1946. Technica de la injection de linfaticos en el animal vivo. Medicine, Madrid, *14*:453.

Pressman, J. J., and M. B. Simon. 1961. Experimental evidence of direct communications between lymph nodes and veins. Surg. Gynecol. Obstet., *113*:537–541.

Pressman, J. J., M. B. Simon, K. Hand, and J. Miller. 1962. Passage of fluids, cells and bacteria via direct communications between lymph nodes and veins. Surg. Gynecol. Obstet., *115*:207–214.

Reinhardt, W. O. 1946. Growth of lymph nodes, thymus and spleen, and output of thoracic duct lymphocytes in the normal rat. Anat. Rec., *94*:197–211.

Roitt, I. M., M. F. Greaves, G. Torrigiani, J. Brostoff, and J. H. L. Playfair. 1969. The cellular basis of immunological responses. Lancet, *2*:367–371.

Slonecker, C. E. 1969. H³-leucine incorporation in antigenically stimulated rat popliteal lymph node cells. Anat. Rec., *165*:363–378.

Sorenson, G. D. 1960. An electron microscopic study of popliteal lymph nodes from rabbits. Amer. J. Anat., *107*:73–96.

Stibbe, E. P. 1918. The internal mammary lymphatic glands. J. Anat. (Lond.), *52*:257–264.

Turner, D. R. 1969. The vascular tree of the haemal node in the rat. J. Anat. (Lond.), *104*:481–494.

White, A., and T. F. Dougherty. 1946. The role of lymphocytes in normal and immune globulin production and the mode of release of globulin from lymphocytes. Ann. N.Y. Acad. Sci., *46*:859–882.

Yoffey, J. M. 1964. The lymphocyte. Ann. Rev. Med., *15*:125–148.

Lymphatic System and Vessels

Allen, L. 1943. The lymphatics of the parietal tunica vaginalis propria of man. Anat. Rec., *85*:427–433.

Allen, L. 1967. Lymphatics and lymphoid tissue. Ann. Rev. Physiol., *29*:197–224.

Bartlett, R. W., G. Crile, Jr., and E. A. Graham. 1935. A lymphatic connection between the gall bladder and liver. Surg. Gynecol. Obstet., *61*:363–365.

Batlezzati, M., and D. Ippolito. 1972. *The Lymphatic System.* John Wiley & Sons, New York, 496 pp.

Blair, J. B., E. A. Holyoke, and R. R. Best. 1950. A note on the lymphatics of the middle and lower rectum and anus. Anat. Rec., *108*:635–644.

Boggon, R. P., and A. J. Palfrey. 1973. The microscopic anatomy of human lymphatic trunks. J. Anat. (Lond.), *114*:389–406.

Burch, G. E. 1939. Superficial lymphatics of human eyelids observed by injection *in vivo.* Anat. Rec., *73*:443–446.

Busch, F., and E. S. Sayegh. 1963. Roentgenographic visualization of human testicular lymphatics: A preliminary report. J. Urol., *89*:106–110.

Butler, H., and K. Balankura. 1952. Preaortic thoracic duct and azygos veins. Anat. Rec., *113*:409–419.

Cain, J. C., J. H. Grindlay, J. L. Bollman, E. V. Flock, and F. C. Mann. 1947. Lymph from liver and thoracic duct: An experimental study. Surg. Gynecol. Obstet., *85*:559–562.

Coller, F. A., E. B. Kay, and R. S. McIntyre. 1941. Regional lymphatic metastases of carcinoma of the stomach. Arch. Surg., *43*:748–761.

Cuneo, B., and L. Marcille. 1901. Lymphatiques de l'ombilic. Bull. Mém. Soc. Anat. (Paris), *76*:580–583.

Cuneo, B., and L. Marcille. 1901. Note sur les lymphatiques du clitoris. Bull. Mém. Soc. Anat. (Paris), *76*:624–625.

Cuneo, B., and L. Marcille. 1901. Topographie des ganglions ilio-pelviens. Bull. Mém. Soc. Anat. (Paris), *76*:653–663.

Davis, H. K. 1915. A statistical study of the thoracic duct in man. Amer. J. Anat., *17*:211–244.

De Roo, T. 1975. *Atlas of Lymphography.* J. B. Lippincott, Philadelphia, 190 pp.

Dixon, F. W., and N. L. Hoerr. 1944. Lymphatic drainage of paranasal sinuses. Laryngoscope, *54*:165–175.

Edwards, D. A. W. 1946. The blood supply and lymphatic drainage of tendons. J. Anat. (Lond.), *80*:147–152.

Eliška, O., and M. Elišková. 1976. Lymph drainage of sinu-atrial node in man and dog. Acta Anat., *96*:418–428.

Engeset, A. 1959. The route of peripheral lymph to the blood stream. An x-ray study of the barrier theory. J. Anat. (Lond.), *93*:96–100.

Gilchrist, R. K., and V. C. David. 1938. Lymphatic spread of carcinoma of the rectum. Ann. Surg., *108*:621–642.

Glover, R., and J. M. Waugh. 1946. The retrograde lymphatic spread of carcinoma of the "rectosigmoid region": Its influence on surgical problems. Surg. Gynecol. Obstet., *82*:434–448.

Gooneratne, B. W. M. 1972. Lymphatic system in cats outlined by lymphography. Acta Anat., *81*:36–41.

Gooneratne, B. W. M. 1972. The lymphatic system in rhesus monkeys (*Macaca mulatta*) outlined by lower limb lymphography. Acta Anat., *81*:602–608.

Gooneratne, B. W. M. 1974. *Lymphography—Clinical and Experimental*. Butterworths, London, 194 pp.

Gray, J. H. 1937. The lymphatics of the stomach. J. Anat. (Lond.), *71*:492–496.

Gray, J. H. 1939. Studies of the regeneration of lymphatic vessels. J. Anat. (Lond.), *74*:309–335.

Grinnell, R. S. 1942. The lymphatic and venous spread of carcinoma of the rectum. Ann. Surg., *116*:200–216.

Grinnell, R. S. 1950. Lymphatic metastases of carcinoma of the colon and rectum. Ann. Surg., *131*:494–506.

Gusev, A. M. 1964. Lymph vessels of human conjunctiva. Fed. Proc., *23*:T1099–T1102.

Halsell, J., R. Smith, C. Bentlage, O. Park, and J. Humphreys. 1965. Lymphatic drainage of the breast demonstrated by vital dye staining and radiography. Ann. Surg., *162*:221–226.

Holmes, M. J., P. J. O'Morchoe, and C. C. O'Morchoe. 1977. Morphology of the intrarenal lymphatic system. Capsular and hilar communications. Amer. J. Anat., *149*:333–351.

Jamieson, J. K., and J. F. Dobson. 1907a. The lymphatic system of the cecum and appendix. Lancet, *1*:1137–1143.

Jamieson, J. K., and J. F. Dobson. 1907b. The lymphatic system of the stomach. Lancet, *1*:1061–1066.

Jamieson, J. K., and J. F. Dobson. 1909. The lymphatics of the colon. Proc. Roy. Soc. Med., *2* (Part 1):149–174.

Jamieson, J. K., and J. F. Dobson. 1920. The lymphatics of the tongue: with particular reference to the removal of lymphatic glands in cancer of the tongue. Brit. J. Surg., *8*:80–87.

Kampmeier, O. F. 1969. *Evolution and Comparative Morphology of the Lymphatic System*. Charles C Thomas, Springfield, Ill., 620 pp.

Kinmonth, J. B. 1952. Lymphangiography in man. A method of outlining lymphatic trunks at operation. Clin. Sci., *11*:13–20.

Kinmonth, J. B. 1964. Some general aspects of the investigation and surgery of the lymphatic system. J. Cardiovasc. Surg., *5*:680–682.

Leak, L. V. 1970. Electron microscopic observations of lymphatic capillaries and the structural components of the connective tissue-lymph interface. Microvasc. Res., *2*:361–391.

Leak, L. V. 1972. The transport of exogenous peroxidase across the blood-tissue-lymph interface. J. Ultrastruct. Res., *39*:24–42.

Leak, L. V., and J. F. Burke. 1968. Ultrastructural studies on the lymphatic anchoring filaments. J. Cell Biol., *36*:129–149.

Leak, L. V., A. Schannahan, H. Scully, and W. M. Daggett. 1978. Lymphatic vessels of the mammalian heart. Anat. Rec., *191*:183–202.

Lee, F. C. 1923. On the lymph vessels of the liver. Carneg. Inst., Contr. Embryol., *15*:63–72.

Miller, W. S. 1896. The lymphatics of the lung. Anat. Anz., *12*:110–114.

Nesselrod, J. P. 1936. An anatomic restudy of the pelvic lymphatics. Ann. Surg., *104*:905–916.

Papp, M., P. Rohlich, I. Rusznyák, and I. Törö. 1964. Central chyliferous vessel of intestinal villus. Fed. Proc., *23*:T155–T158.

Parker, A. E. 1936. The lymph collectors from the urinary bladder and their connections with the main posterior lymph channels of the abdomen. Anat. Rec., *65*:443–460.

Parker, A. E. 1936. The lymph vessels from the posterior urethra, their regional lymph nodes and relationships to the main posterior abdominal lymph channels. J. Urol., *36*:538–557.

Parker, A. E. 1940. Lymph collectors from the ureters, their regional nodes and relations to posterior abdominal lymph channels. J. Urol., *43*:811–830.

Patek, P. R. 1939. The morphology of the lymphatics of the mammalian heart. Amer. J. Anat., *64*:203–249.

Pflug, J. J., and J. S. Calnin. 1971. The normal anatomy of the lymphatic system in the human leg. Brit. J. Surg., *58*:925–930.

Pierce, E. C. 1944. Renal lymphatics. Anat. Rec., *90*:315–335.

Powell, T. O. 1944. Studies in the lymphatics of the female urinary bladder. Surg. Gynecol. Obstet., *78*:605–609.

Reiffenstuhl, G. 1964. *The Lymphatics of the Female Genital Organs*. J. B. Lippincott, Philadelphia, 165 pp.

Roenberger, A., and H. Abrams. 1971. Radiology of the thoracic duct. Amer. J. Roentgenol., *111*:807–820.

Rouviere, H. 1954. *Anatomie Humaine, Descriptive et Topographique*. 7th ed. vol. 2. Masson, Paris.

Rusznyák, I., M. Földi, and G. Szabo. 1960. *Lymphatics and Lymph Circulation, Physiology and Pathology*. Pergamon Press, New York, 853 pp.

Schatzki, P. F. 1978. Electron microscopy of lymphatics of the porta hepatis. Acta Anat., *102*:54–59.

Shore, L. R. 1929. The lymphatic drainage of the human heart. J. Anat. (Lond.), *63*:291–313.

Shrewsbury, M. M., Jr., and W. O. Reinhardt. 1955. Relationships of adrenals, gonads, and thyroid to thymus and lymph nodes, and to blood and thoracic duct leukocytes. Blood, *10*:633–645.

Simer, P. H. 1935. Omental lymphatics in man. Anat. Rec., *63*:253–262.

Spirin, B. A. 1964. Internal lymphatic system of penis. Fed. Proc., *23*:T159–T167.

Teichmann, L. 1861. *Das Saugadersystem von anatomischen Standpunkte*. W. P. Engelmann, Leipzig.

Tobin, C. E. 1957. Human pulmonic lymphatics. Anat. Rec., *127*:611–624.

Turner-Warwick, R. T. 1959. The lymphatics of the breast. Brit. J. Surg., *46*:574–582.

Vadja, J., and M. Tomcsik. 1971. The structure of the valves of the lymphatic vessels. Acta Anat., *78*:521–531.

Vadja, J., M. Tomcsik, and W. J. Van Doorenmaalen. 1973. Connections between the venous system of the heart and the epicardiac lymphatic network. Acta Anat., *83*:262–274.

Van Pernis, P. A. 1949. Variations of the thoracic duct. Surgery, *26*:806–809.

Wallace, S., et al. 1961. Lymphangiograms: their diagnostic and therapeutic potential. Radiology, *76*:179–199.

Weinberg, J., and E. M. Greaney. 1950. Identification of regional lymph nodes by means of a vital staining dye during surgery of gastric cancer. Surg. Gynecol. Obstet., *90*:561–567.

Wislocki, G. B., and E. W. Dempsey. 1939. Remarks on the lymphatics of the reproductive tract of the female rhesus monkey (Macaca mulatta). Anat. Rec., *75*:341–362.

Yoffey, J. M., and F. C. Courtice. 1956. *Lymphatics,*

Lymph and Lymphoid Tissue. Harvard University Press, Cambridge, 510 pp.

Yoffey, J. M., and C. M. Drinker. 1939. Some observations on the lymphatics of the nasal mucous membrane in the cat and monkey. J. Anat. (Lond.), 74:45–52.

Spleen

Barcroft, J., and J. G. Stephens. 1936. Observations on the size of the spleen. J. Physiol. (Lond.), 64:1–22.

Bennett-Jones, M. J., and C. A. St. Hill. 1952. Accessory spleen in the scrotum. Brit. J. Surg., 40:259–262.

Braithwaite, J. L., and D. J. Adams. 1957. The venous drainage of the rat spleen. J. Anat. (Lond.), 91:352–357.

Burke, J. S., and G. T. Simon. 1970. Electron microscopy of the spleen. I. Anatomy and microcirculation. Amer. J. Pathol., 58:127–155.

Burke, J. S., and G. T. Simon. 1970. Electron microscopy of the spleen. II. Phagocytosis of colloidal carbon. Amer. J. Pathol., 58:157–181.

Chen, L. T., and L. Weiss. 1972. Electron microscopy of the red pulp of human spleen. Amer. J. Anat., 134:425–458.

Curtis, G. M., and D. Movitz. 1946. The surgical significance of the accessory spleen. Ann. Surg., 123:276–298.

Edwards, V. D., and G. T. Simon. 1970. Ultrastructural aspects of red cell destruction in the normal rat spleen. J. Ultrastruct. Res., 33:187–201.

Galindo, B., and J. A. Freeman. 1963. Fine structure of splenic pulp. Anat. Rec., 147:25–29.

Galindo, B., and T. Imaeda. 1962. Electron microscope study of the white pulp of the mouse spleen. Anat. Rec., 143:399–416.

Gray, H. 1854. *On the Structure and Use of the Spleen.* J. W. Parker & Son, London, 380 pp.

Halpert, B., and F. Györkey. 1959. Lesions observed in the accessory spleens of 311 patients. Amer. J. Clin. Pathol., 32:165–168.

Holyoke, E. A. 1936. The role of the primitive mesothelium in the development of the mammalian spleen. Anat. Rec., 65:333–349.

Jacobsen, G. 1971. Morphological-histochemical comparison of dog and cat splenic ellipsoid sheaths. Anat. Rec., 169:105–113.

Knisely, M. H. 1936. Spleen studies. I. Microscopic observations of the circulatory system of living unstimulated mammalian spleens. Anat. Rec., 65:23–50.

Knisely, M. H. 1936. Spleen studies. II. Microscopic observations of the circulatory system of living traumatized spleens, and of dying spleens. Anat. Rec., 65:131–148.

Krumbhaar, E. B., and S. W. Lippincott. 1939. Postmortem weight of "normal" human spleen at different ages. Amer. J. Med. Sci., 197:344–358.

Lewis, O. J. 1957. The blood vessels of the adult mammalian spleen. J. Anat. (Lond.), 91:245–250.

Mac Kenzie, D. W., Jr., A. O. Whipple, and M. P. Wintersteiner. 1941. Studies on microscopic anatomy and physiology of living trans-illuminated mammalian spleens. Amer. J. Anat., 68:397–456.

Mall, F. P. 1900. The architecture and blood vessels of the dog's spleen. Z. Morphol., 2:1–42.

Mall, F. P. 1903. On the circulation through the pulp of the dog's spleen. Amer. J. Anat., 2:315–332.

Michels, N. A. 1942. The variational anatomy of the spleen and splenic artery. Amer. J. Anat., 70:21–72.

Miyoshi, M., and T. Fujita. 1971. Stereo-fine structure of the splenic red pulp. A combined scanning and transmission electron microscope study on dog and rat spleen. Arch. Histol. Jpn., 33:225–246.

Miyoshi, M., T. Fujita, and J. Tokunaga. 1970. The red pulp of the rabbit spleen studied under the scanning electron microscope. Arch. Histol. Jpn., 32:289–306.

Murphy, J. W., and W. A. Mitchell. 1957. Congenital absence of the spleen. Pediatrics, 20:253–256.

Peck, H. M., and N. L. Hoerr. 1951. The intermediary circulation in the red pulp of the mouse spleen. Anat. Rec., 109:447–477.

Sneath, W. A. 1912. An apparent third testicle consisting of scrotal spleen. J. Anat. (Lond.), 47:340–342.

Snook, T. 1950. A comparative study of the vascular arrangements in mammalian spleens. Amer. J. Anat., 87:31–77.

Thiel, G. A., and H. Downey. 1921. The development of the mammalian spleen, with special reference to its hematopoietic activity. Amer. J. Anat., 28:279–339.

Thomas, C. E. 1967. An electron and light microscopic study of sinus structure in perfused rabbit and dog spleens. Amer. J. Anat., 120:527–552.

Weiss, L. 1957. A study of the structure of splenic sinuses in man and the albino rat with the light microscope and the electron microscope. J. Biophys. Biochem. Cytol., 3:599–610.

Weiss, L. 1963. The structure of the intermediate vascular pathways in the spleen of rabbits. Amer. J. Anat., 113:51–91.

Weiss, L. 1973. The development of the primary vascular reticulum in the spleen of human fetuses (38- to 57-mm crown-rump length). Amer. J. Anat., 136:315–338.

Williams, R. G. 1961. Studies of the vasculature in living autografts of spleen. Anat. Rec., 140:109–121.

Thymus

Abe, K., and T. Ito. 1970. Fine structure of small lymphocytes in the thymus of the mouse: qualitative and quantitative analysis by electron microscopy. Z. Zellforsch. mikrosk. Anat., 110:321–335.

Ackerman, G. A., and J. R. Hostetler. 1970. Morphological studies of the embryonic rabbit thymus: The *in situ* epithelial versus the extrathymic derivation of the initial population of lymphocytes in the embryonic thymus. Anat. Rec., 166:27–46.

Ackerman, G. A., and R. A. Knuoff. 1964. Lymphocyte formation in the thymus of the embryonic chick. Anat. Rec., 149:191–215.

Auerbach, R. 1961. Experimental analysis of the origin of cell types in the development of the mouse thymus. Devel. Biol., 3:336–354.

Burnet, F. M. 1962. The immunological significance of the thymus: an extension of the clonal selection theory of immunity. Australas. Ann. Med., 11:79–91.

Castlemann, B. 1966. The pathology of the thymus gland in myasthenia gravis. Ann. N.Y. Acad. Sci., 135:496–505.

Clark, S. L., Jr. 1963. The thymus in mice of strain 129/J, studied with the electron microscope. Amer. J. Anat., 112:1–33.

Colley, D. G., A. Malakian, and B. H. Waksman. 1970. Cellular differentiation in the thymus. II. Thymus-specific antigens in rat thymus and peripheral lymphoid cells. J. Immunol., 104:585–592.

Colley, D. G., A. Y. Shih Wu, and B. H. Waksman. 1970. Cellular differentiation in the thymus. III. Surface properties of rat thymus and lymph node cells separated on density gradients. J. Exp. Med., *132*:1107–1121.

Csaba, G. Y., I. Törö, and E. Kapa. 1960. Provoked tissue reaction of the thymus in tissue culture. Acta Morphol. Acad. Sci. Hung., *9*:197–202.

Dung, H. C. 1973. Electron microscopic study of involuting thymus of "lethargic" mutant mice. Anat. Rec., *177*:585–606.

Goldschneider, I., and D. D. McGregor. 1968. Migration of lymphocytes and thymocytes in the rat. I. The route of migration from blood to spleen and lymph nodes. J. Exp. Med., *127*:155–168.

Good, R. A., et al. 1962. The role of the thymus in the development of immunologic capacity in rabbits and mice. J. Exp. Med., *116*:773–795.

Good, R. A., and A. E. Gabrielsen (eds.). 1964. *The Thymus in Immunobiology.* P. B. Hoeber, New York, 778 pp.

Grègoire, C. 1943. Regeneration of the involuted thymus after adrenalectomy. J. Morphol., *72*:239–261.

Hoshino, T., M. Takeda, K. Abe, and T. Ito. 1969. Early development of thymic lymphocytes in mice, studied by light and electron microscopy. Anat. Rec., *164*:47–66.

Izard, J. 1966. Ultrastructure of thymic reticulum in guinea pig. Cytological aspects of the problem of the thymic secretion. Anat. Rec., *155*:117–132.

Karetzky, M., and L. E. Rudolf. 1964. Current status of the thymus gland. Surg. Gynecol. Obstet., *119*:129–138.

Keynes, G. 1954. The physiology of the thymus gland. Brit. Med. J., *2*:659–663.

Mandel, T. 1968. The development and structure of Hassall's corpuscles in the guinea pig. Z. Zellforsch. mikrosk. Anat., *89*:180–192.

Mandel, T. 1968. Ultrastructure of epithelial cells in the cortex of guinea pig thymus. Z. Zellforsch. mikrosk. Anat., *92*:159–168.

Miller, J. F. A. P. 1962. Effect of neonatal thymectomy on the immunological responsiveness of the mouse. Proc. Roy. Soc. Biol. (Lond.), *156*:415–428.

Miller, J. F. A. P. 1963. Role of the thymus in immunity. Brit. Med. J., *5355*:459–464.

Miller, J. F. A. P. 1964. The thymus and the development of immunologic responsiveness. Science, *144*:1544–1551.

Miller, J. F. A. P., and D. Osoba. 1967. Current concepts of the immunological function of the thymus. Physiol. Rev., *47*:437–520.

Mitchell, G. F., and J. F. A. P. Miller. 1968. Immunological activity of thymus and thoracic-duct lymphocytes. Proc. Nat. Acad. Sci. (USA), *59*:296–303.

Moore, M. A. S., and J. J. T. Owen. 1967. Experimental studies on the development of the thymus. J. Exp. Med., *126*:715–726.

Murray, R. G., A. Murray, and A. Pizzo. 1965. The fine structure of the thymocytes of young rats. Anat. Rec., *151*:17–40.

Norris, E. H. 1938. The morphogenesis and histogenesis of the thymus gland in man. Carneg. Inst., Contr. Embryol., *27*:191–207.

Osoba, D. 1972. Thymic function, immunologic deficiency and autoimmunity. Med. Clin. North Amer., *56*:319–335.

Sainte-Marie, G. 1974. Tridimensional reconstruction of the rat thymus. Anat. Rec., *179*:517–526.

Sainte-Marie, G., and F. S. Peng. 1971. Emigration of thymocytes from the thymus. A review and study of the problem. Rev. Canad. Biol., *30*:51–78.

Sanel, F. T., 1967. Ultrastructure of differentiating cells during thymus histogenesis. A light and electron microscopic study of epithelial and lymphoid cell differentiation during thymus histogenesis in C57 black mice. Z. Zellforsch. mikrosk. Anat., *83*:8–29.

Shier, K. J. 1963. The morphology of the epithelial thymus: Observations on the lymphocyte-depleted and fetal thymus. Lab. Invest., *12*:316–326.

Small, M., and N. Trainin. 1971. Contribution of a thymic humoral factor to the development of an immunologically competent population from cells of mouse bone marrow. J. Exp. Med., *134*:786–800.

Smith, C., and H. T. Parkhurst. 1949. Studies on the thymus of the mammal. II. A comparison of the staining properties of Hassall's corpuscles and of thick skin of the guinea pig. Anat. Rec., *103*:649–673.

Törö, I. 1961. The cytology of the thymus gland. Folia Biol. (Krakow), *7*:145–149.

Weissman, I. L. 1967. Thymus cell migration. J. Exp. Med., *126*:291–304.

Weller, G. L., Jr. 1933. Development of the thyroid, parathyroid, and thymus glands in man. Carneg. Inst., Contr. Embryol., *24*:93–142.

Wilson, A., A. R. Obrist, and H. Wilson. 1953. Some effects of extracts of thymus glands removed from patients with myasthenia gravis. Lancet, *2*:368–371.

11

Developmental and Gross Anatomy of the Central Nervous System

The human nervous system, having evolved from that of lower animals, accounts more nearly for the behaviors characteristic of people than does any other organ system in the body. Its role in the control and management of most functions of other organs and tissues allows for appropriate reactions to alterations in the environment and, thus, for survival. In this latter respect it functions comparably to the nervous systems of other mammals, but in its highly evolved integrative, communicative, reflective, creative, and acutely perceptive functions, the human nervous system lends to Man those abilities that have truly elevated our species above all others. In addition to these more abstract functions of the human brain, which are common to all healthy members of our species, our individualities and the distinctive characteristics that give to each of us an identity and separate us, one from another, are also based in the structure and functions of our own personal nervous system.

In this chapter emphasis is placed on the developing nervous system and the gross anatomy of the spinal cord and brain. The detailed internal structure of the nervous system and the broad discipline of neurobiology and the other neurosciences will not be extensively recounted, for there are many excellent specialized textbooks in which these fascinating fields are handled more appropriately.

Development of the Nervous System

The architectural and functional development of the nervous system and the underlying principles involved collectively form one of the most exciting chapters in all of biology. The descriptive and experimental neuroembryologist has had the task of explaining how and why the nervous system develops its infinite complexity. Similarities among the early developmental stages of many of the higher vertebrates have allowed us to gain much information about early human neurogenesis.

Formation of Presumptive Neural Derivatives

Each of us develops from the single-celled, fertilized ovum. Through successive mitotic divisions, the ovum becomes a fluid-filled ball of cells known as the **blastula,** within which exists an **inner cell mass** that is destined to become the embryo proper. This mass of cells is surrounded by an outer, flattened, cellular shell known as the **trophoblast,** which rapidly proliferates and invades the maternal uterine tissue. The use of vital staining techniques has made possible the discovery of regions of the living blastula that become the three primary germ layers: the **ectoderm, mesoderm,** and **endoderm.** Along the mid-dorsal line of the embryo at this early stage, there appears a **primitive streak,** an area of ectoderm where a faster

933

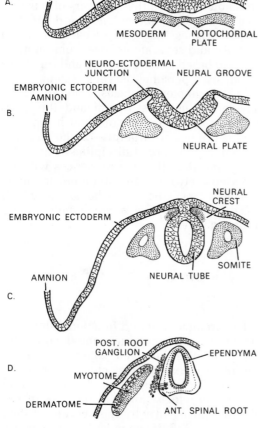

FIG. 11-1. Schematic sections through the neural groove and neural tube in successively older embryos. (Redrawn after W. J. Hamilton and H. W. Mossman, *Human Embryology*, 4th ed., Williams and Wilkins, Baltimore, 1972.)

proliferation of cells occurs. Because the entire nervous system is ectodermal in origin, it is at this time (15 to 17 days after fertilization) that the presumptive neural regions of the human embryo commence differential transformation.

In front of the primitive streak appears the **primitive knot** (Hensen's[1] node), from which extends the developing **notochord.** The flattened layer of ectoderm on the dorsal aspect of the embryo now becomes stratified, and this unequal growth of the **neural plate** leads to an infolding of the ectoderm and thus the formation of the **neural groove** (Fig. 11-1,A

[1] Victor Hensen (1835–1924): A German embryologist and physiologist (Kiel).

and B). Coursing along the longitudinal axis of the embryo, the neural groove then becomes surrounded by the **neural folds.** As the process of invagination continues, the neural folds fuse to form the **neural tube,** the rostral part of which is broader and eventually develops into the brain, while the narrow caudal part forms the spinal cord (Fig. 11-1,C and D). The central lumen of the neural tube is retained as the ventricular system rostrally and as the central canal of the spinal cord caudally.

NEURAL INDUCTION. At these early developmental stages, the organization of the portion of the embryo that is destined to become the nervous system, i.e. essentially the neural plate and folds, appears to be under the inductive influence of underlying mesodermal elements. In their classic experiments, Spemann and Mangold (1924) first emphasized the necessity for the maintenance of physical contact between the neural ectoderm and the subjacent mesodermal tissue for normal neural development. When a particular part of an embryo is capable of liberating a chemical substance that causes the surrounding tissue to develop in a certain manner, the process is called **induction.** The biochemical substance is called the **evocator,** and the property exhibited by the surrounding tissue in response to the evocator is called **competence.**

Because mesoderm at the dorsal lip of the blastopore (and probably tissue in other regions of the primitive streak) is capable of directly inducing the formation of central neural tissue in the embryo, it is called a **primary organizer.** Although the chemical nature of the evocator is still unknown, several types of substances have been shown to have inductive effects.

The great significance of these early periods to the normal development of the complete mature central nervous system should be stressed. The shifting of only a few cells in the wrong direction, or the failure in growth or closure of a small part of the neural tube in an embryo 2 to 3 mm long, can result in embryonic death or serious malformation. For example, if the neural groove fails to close, or if having closed, it fails to separate completely from the overlying ectoderm, the development of the associated vertebrae is also abnormal, and a condition known as spina bifida results.

Histogenesis of the Neural Tube

The neural tube eventually separates from the layer of ectoderm that is destined to become skin and continues to develop subcutaneously. Rapid division and growth in the cephalic end of the neural tube occurs and soon the accumulations of nuclear masses in the walls of the neural tube become recognizable as forebrain, midbrain, and hindbrain.

In recent decades the histogenesis of the cellular elements that comprise the central nervous system has become a subject of enhanced scientific interest. For many years the pseudostratified, columnar, neuroepithelial structure of the very early neural tube, even before its closure, was described as consisting of several cell types. His (1889, 1904) proposed that from these cell types three layers develop: ependymal, mantle, and marginal. In His' schema, the innermost **ependymal layer** contains *germinal cells* that divide to form *neuroblasts*. These migrate through a wider layer of *spongioblasts* (precursors to the neuroglia) to form the **mantle layer** or gray matter region. The outermost **marginal layer** does not contain many cell bodies, but instead consists of many nerve processes (fibers) that eventually form the white matter.

FOUR-ZONE SCHEMA RECOMMENDED.   A more recent and welcome clarification of the terminology, and a variation on the classic schema of His regarding histogenesis of the neural tube, is based on modern electron microscopic and autoradiographic findings (Boulder Committee, 1970). This proposal states that four geographic zones form in the developing vertebrate central nervous system: ventricular, subventricular, intermediate, and marginal (Fig. 11-2).

Within the **ventricular zone,** which borders the central lumen, a single pseudostratified columnar cell type called the ventricular cell exists (Fig. 11-2,*A*). This cellular element is the source of all neurons and macroglia (astrocytes and oligodendroglia) in the central nervous system. The germinal cells and the spongioblasts, which His considered different cell types, are in fact mitotic and intermitotic forms of the ventricular cell. The Fugitas (1963a, b; 1964) and others have shown that as cells in this innermost ventricular zone (Fugita's matrix layer) pass through intermitotic and mitotic phases, their nuclei "elevate" away from and then return to the innermost region of the neural tube. During their intermitotic period (DNA synthetic phase), the ventricular cells have an elongated columnar form, and the nuclei are located away from the lumen in

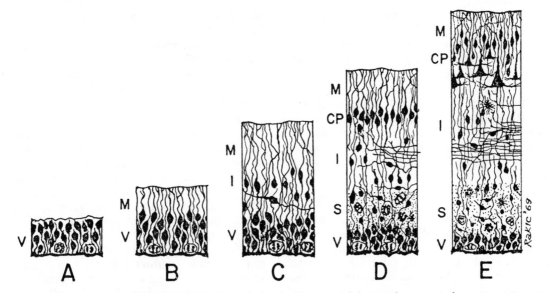

Fig. 11-2. Schematic drawing of five stages (*A-E*) in the development of the vertebrate central nervous system, as described in the text. Abbreviations: CP, cortical plate; I, intermediate zone; M, marginal zone; S, subventricular zone; V, ventricular zone. (From Boulder Committee, Anat. Rec., *116*:257–262, 1970.)

the outer limit of the ventricular zone. When these columnar cells are ready to divide, thereby replicating their DNA, the cells become spherical, the nuclei move inward toward the surface of the ventricular lumen, and at this deepest portion in the wall of the neural tube, the cells divide.

Shortly after the formation of the ventricular zone, an outermost **marginal zone,** composed principally of the cytoplasmic processes of the ventricular cells, is recognizable (Fig. 11-2,B). As the ventricular cells continue to divide, some newly formed cells migrate outward from the ventricular zone to create an **intermediate zone,** interposed between the marginal and ventricular zones (Fig. 11-2,C). Certain cells that form this intermediate zone are immature nerve cells that have ceased to divide. As they mature, these cells develop axons and dendrites that course either within the intermediate zone or enter the more outlying marginal zone.

In certain parts of the developing neuraxis, such as the anterior region of the spinal cord, the intermediate zone remains relatively uniform. In the developing cerebrum and cerebellum, however, cell bodies within this zone migrate to form other layers, such as the cortical plate illustrated in Figure 11-2,D and E.

This more recent description of the process also recognizes a **subventricular zone** that eventually forms between the ventricular and intermediate zones (Fig. 11-2,D and E). Migrating to the subventricular zone are round proliferating cells that do not elevate and descend during their intermitotic and mitotic phases. These cells give rise to the adult subependymal neuronal elements, as well as to the adult astrocytes and oligodendroglia, but not to the ependymal cells. From this subventricular zone, the neuroglial elements are thought to migrate to all other zones of the embryonic neuraxis.

Development of the Spinal Cord

As a result of the proliferation and migration of ventricular cells into the intermediate zone, the lateral walls of the neural tube increase in thickness. In contrast, the roof plate and floor plate become quite thin, and the central lumen is rather narrow and slitlike (Fig. 11-3,B). The ventral part of each lateral wall further enlarges, and the middle portion of the lumen becomes altered on

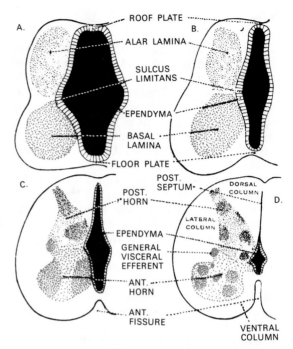

FIG. 11-3. Diagrams illustrating four successive stages in the development of the alar (in black) and basal (in red) laminae of the spinal cord region. (From W. J. Hamilton and H. W. Mossman, *Human Embryology*, 4th ed., Williams and Wilkins, Baltimore, 1972.)

each side by the formation of a longitudinal groove, the **sulcus limitans** (Fig. 11-3). The sulcus indicates the area on the wall of the neural tube at which the dorsal or alar lamina is separated from the ventral or basal lamina. These laminae represent an early functional differentiation: the **alar lamina** or **alar plate** will become the sensory portion of the gray substance, and the **basal lamina** or **basal plate** will become the motor portion (Figs. 11-3; 11-4).

The basal lamina thickens, and in a cross section the intermediate zone (mantle layer) appears as an expanded region between the marginal and ventricular zones. This thickening is the rudiment of the **ventral gray column** or **ventral horn.** Its immature nerve cells send their axons out through the marginal layer to form the ventral roots of the spinal nerves (Fig. 11-4). The alar lamina also thickens gradually, becoming the **dorsal gray column** or **dorsal horn.** Later, axons from many of these cells grow rostrally and assist in the formation of the **dorsal and lateral white columns** or **funiculi** (Fig. 11-4). A few fibers cross to the opposite side and form the ventral white commissure.

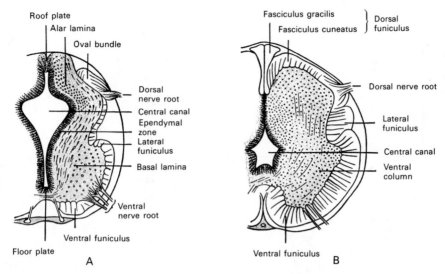

FIG. 11-4. Transverse sections through the spinal cords of human embryos. *A*, Age about four and a half weeks; *B*, age about three months. (W. His.)

At the end of about the fourth week, nerve fibers begin to appear in the marginal layer. The first to develop are the short intersegmental fibers, which come from differentiating nerve cells in the intermediate zone. Fibers from the cells in the spinal ganglia also grow into the marginal zone and by the sixth week have formed a well-defined **oval bundle** in the peripheral part of the alar lamina (Fig. 11-4). This bundle gradually increases in size and spreads toward the midline from each side; it forms the rudiments of the dorsal funiculi, which are separated by the **dorsal median sulcus.** Long intersegmental fibers appear about the third month, and corticospinal fibers appear about the fifth month. Initially, all the nerve fibers are devoid of myelin, and the various bundles of fibers become myelinated at different times. Thus, the dorsal and ventral roots acquire myelin about the fifth month, and the corticospinal fibers, after the ninth month. The expanding growth of the **ventral horn** and **ventral funiculus** of the two sides causes these parts to bulge laterally beyond the floor plate, leaving a deep groove, the **ventral median fissure,** between the two ventral funiculi.

SEGMENTATION. Segmentation is the repetition of structural patterns at successive levels and is one of the earliest vertebrate characteristics of the human embryo. The first somites that are formed are the most ce- phalic ones; they appear behind the auditory placode during the third week of development. Thereafter, the number of somites increases rapidly, so that by the fifth week there are already 38 pairs. The most constant of these developing myotomes are the eight cervical, twelve thoracic, five lumbar, and five sacral somites. These become innervated by the developing spinal nerves and create the segmental organization of the body, providing the basis for segmental reflex activities.

Corresponding to the development of the somites, the thickening of the wall of the neural tube is not uniform. This enlargement is somewhat greater at the sites at which the spinal nerves attach to the cord than in between, producing a beaded appearance and establishing the **neuromeres** or **segments.** The relative thickness of the spinal segments varies according to the size of the nerves and the bulk of the peripheral field that the nerve supplies. There exists no valid evidence to indicate that segmentation is a characteristic inherent in the early neuraxis. In contrast, experimental evidence has shown that segmentation in the nervous system is determined to a large extent by segmentation of the mesodermal somites. On the basis of the segmentation of the axial mesoderm, it is generally accepted that spinal nerve and ganglion segmentation is acquired secondarily. In later development, the long projection

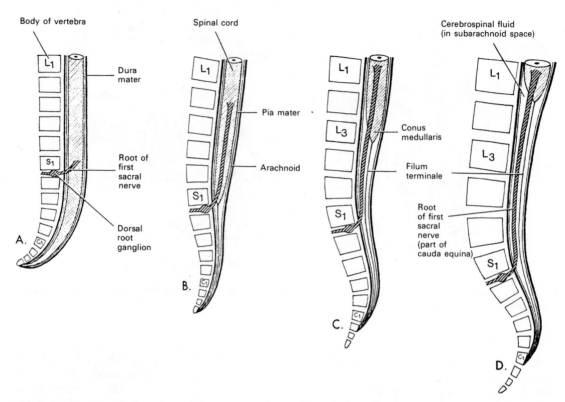

FIG. 11-5.   Diagrams showing four successive stages in the differential growth of the spinal cord in relation to the vertebral column and meninges. Note the elevation in the position of the caudal end of the cord and the increasing length and obliquity of the first sacral nerve, used here for reference. *A*, Eighth prenatal week; *B*, twenty-fourth prenatal week; *C*, newborn infant; *D*, adult. (From K. L. Moore, *The Developing Human*, W. B. Saunders, Philadelphia, 1973.)

tracts grow into the marginal zone, thereby obliterating the segmental appearance of the spinal cord.

GROWTH OF THE SPINAL CORD WITHIN THE VERTEBRAL CANAL.   In embryos during the *somite stages* and until the third prenatal month, the spinal cord forms a long, slender tube curved to the shape of the embryo and extending its entire length as far as the coccygeal region (Fig. 11-5). The segmentally arranged spinal nerve roots emerge and enter the spinal cord at levels directly in line with their sites of origin. By the *sixth month* the caudal end of the cord reaches only as far as the upper part of the sacrum, while *at birth* it is on a level with the third lumbar vertebra. These positional changes continue during postnatal maturation, so that in the adult, the end of the spinal cord lies at the level of the intervertebral disc between the first and second lumbar vertebrae (Fig. 11-5).

This uneven growth and the marked cephalic flexure of the brain keep the rostral end of the maturing nervous system within the cranial cavity. The adult spinal cord gives the appearance of having been pulled upward in the spinal canal, and the spinal nerves below the cervical level become more and more longitudinally displaced within the vertebral canal. Because these nerves must still emerge at their somatic structures through their respective intervertebral foramina, the lower end of the vertebral canal becomes filled with the long coursing roots of the lower spinal segments. These roots are located intradurally, and this region of the cord is called the **cauda equina.** The finely tapered end of the cord itself, still attached to the most distal part of the vertebral canal in the coccyx, is called the **filum terminale** (Fig. 11-5).

Development of the Spinal Nerves

The continued maturation of young nerve cells in the ventrolateral portion of the intermediate zone of the spinal cord results in the formation of the **ventral roots,** whereas the

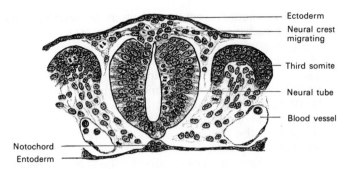

FIG. 11-6. Neural crest formation. Transverse section through third somite of nine somite rat embryo.

immature neurons in the dorsal root ganglia grow into the cord to form the **dorsal roots.**

VENTRAL ROOTS. As groups of nerve cells that form the gray matter in the basal plate continue to differentiate during the fifth week, they sprout delicate processes that emerge from the ventral horn and form the rudiments of the ventral roots. Many of these fibers, motor in function, course toward masses of differentiating striated muscle cells and eventually make contact.

DORSAL ROOTS. The dorsal roots, which develop differently than the ventral roots, form differentiating nerve cells in the dorsal root ganglia. During the sixth and seventh prenatal weeks, these young neurons send processes both centrally into the posterior gray matter of the spinal cord and peripherally to structures that are developing within their respective myotomes. The tissue

that forms the spinal ganglia is a ridge of neuroectodermal cells called the **neural crest** or **ganglion ridge.**

The Neural Crest. The neural crest forms from bands of ectodermal cells that originate early during the development of the nervous system and are located on the sides of the neural folds (Figs. 11-6; 11-7). As the folds begin to fuse, the neural crest cells from both sides form an aggregation of cells along the midline, dorsal to the neural tube and extending from the caudal part of the cephalic region all along the developing neuraxis (Fig. 11-7). After a short period of time, the neural crest cells once again become displaced bilaterally, undergo metameric segmentation, and begin to differentiate. The proliferating neural crest cells are gathered into paired spherical groups that correspond to each segment and will become the primi-

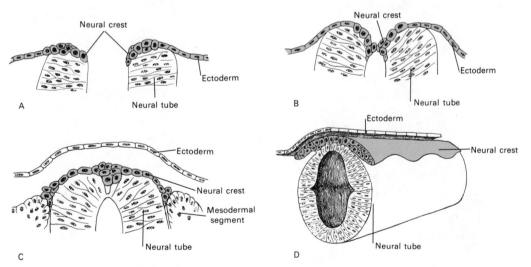

FIG. 11-7. Development of the human neural crest. *A,* 2.5 mm; *B,* 5.0 mm; *C,* 7.5 mm; *D,* longitudinal diagram of the early spinal cord showing the neural crest as a plate of tissue overlying the neural tube. The neural crest then divides into right and left portions, each of which will further subdivide segmentally. (After L. B. Arey, *Developmental Anatomy,* revised 7th ed., W. B. Saunders, Philadelphia, 1974.)

tive spinal ganglia. Some of the neural crest cells from the ventral part of each primitive ganglion separate and migrate peripherally to form the sympathetic ganglia and other neural crest derivatives.

In addition to the spinal ganglia and the ganglia of the sympathetic system, the neural crest gives origin to the parasympathetic ganglia in the head, along with the sensory ganglia associated with the trigeminal, facial, glossopharyngeal, and vagus nerves. Also derived from the neural crest are the neurilemma sheath cells of the peripheral nerves. Other neural crest cells become dispersed with the general mesenchyme and are responsible for the melanocytes in the skin, the cells of the adrenal medulla, the chromaffin cell system, and probably certain cartilages in the head.

Neurons of the Spinal Ganglia. As just stated, proliferation of the neural crest results in the formation of cellular aggregations, lateral to the neural tube, that are destined to become the sensory ganglia (Fig. 11-8,*A*). The developing neurons within these segmentally arranged ganglia give rise to the sensory fibers of the spinal nerves. During differentiation these nerve cells are initially **bipolar** due to the development of a primary process at either end. As their maturation continues, a majority of the cells in mammals become **unipolar** as the two primary processes fuse into a single stem, which emerges from the neuronal soma (Fig. 11-8,*B*). Beyond the point of fusion, the axon divides like a T into two primary branches, one of which is directed centrally, and the other, peripherally. The centrally oriented branch grows into the neural tube as a sensory root fiber. As they penetrate the dorsolateral wall of the neural tube, the sensory fibers bifurcate into ascending and descending branches within the marginal zone of the spinal cord. The fibers then enter the developing intermediate zone (gray matter) and come into synaptic connection with other maturing nerve cells there. The peripherally directed branches of the spinal ganglion cells become the afferent fibers in the cerebrospinal nerves.

Within the ganglia, some of the neural crest cells form capsules that surround the ganglion cell bodies. These supporting elements, which are disc-like and flattened, are called **satellite cells.** In contrast to what occurs in other ganglia, note that neurons in the sensory ganglia of the vestibulocochlear nerve remain bipolar throughout life.

NERVE FIBER SHEATHS. The axis cylinders in the developing peripheral nerves soon become surrounded by protective sheaths that are also ectodermal in origin. In the path

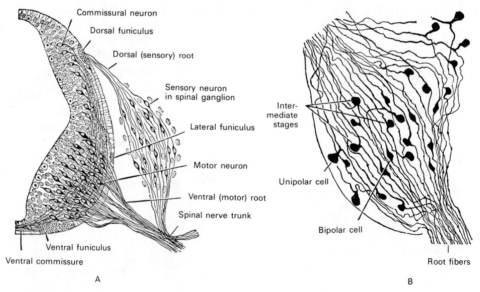

FIG. 11-8. Development of a human spinal nerve. *A,* Transverse section of the spinal cord and the dorsal and ventral roots in a 7-mm embryo. *B,* Longitudinal section of a spinal ganglion in an embryo of ten weeks, showing the transformation of bipolar ganglion cells into unipolar cells. These two figures come from the studies of S. Ramon y Cajal. (From L. B. Arey, *Developmental Anatomy,* revised 7th ed., W. B. Saunders, Philadelphia, 1974.)

of the outgrowing axons can be seen numerous spindle-shaped ectodermal cells that have migrated from the neural tube and the neural crest along the course of the ventral and dorsal roots. These are the **Schwann**[1] **cells;** they form not only a prominent morphologic feature in a developing peripheral nerve, but they also have an important role in the differentiation of the nerve fibers. They form the nucleated sheath, or **neurilemma,** of the peripheral nerve, and because they form myelin, they are comparable to the oligodendroglia of the central nervous system. The layers of **myelin** that form around the developing nerve fibers have been shown by electron microscopy not to be an extracellular substance, as was once believed, but alternating layers of lipids and protein which are plasma membranes of the Schwann cells wrapped around the axis cylinders in a spiral manner.

Although there is no question that the Schwann cells associated with the dorsal roots have a neural crest origin, it has been experimentally inferred that certain sheath cells accompanying the ventral root fibers may originate in the neural tube.

Development of the Brain

The cephalic portion of the neural tube enlarges and begins to change its shape even before it is completely closed. The unequal growth in size and thickness of the different parts, together with the formation of certain flexures, establishes at an early stage three recognizable regions within the primitive brain: the **rhombencephalon** (or hindbrain), the **mesencephalon** (or midbrain), and the **prosencephalon** (or forebrain). During the fourth prenatal week in embryos of 3 to 4 mm, the first of the flexures to appear is the **midbrain** (or cephalic) **flexure** in the region of the mesencephalon (Fig. 11-9,A). By means of this flexure the forebrain makes a U-shaped bend ventrally over the rostral end of the notochord and foregut, causing the midbrain for a period of time to protrude dorsally as the most prominent part of the brain. The second bend, the **cervical flexure,** appears at the junction of the hindbrain and spinal cord at nearly the same time as does

the midbrain flexure. The cervical flexure continues to develop until the hindbrain and spinal cord form a right angle with each other (Fig. 11-9,A). It then gradually diminishes as the body posture changes and the head becomes erect, and it disappears some time after the sixth week. The third bend is named the **pontine flexure** because it occurs in the region of the future pons (Fig. 11-9,B). In contrast to the midbrain and cervical flexures, the convexity of the pontine flexure is directed anteriorly.

The sulci limitans, which divide the lateral walls of the spinal cord into alar and basal laminae, clearly extend rostrally through the hindbrain and midbrain. They also divide these brain stem regions of the neuraxis into alar (or dorsal) and basal (or ventral) laminae.

The early developing brain, being tubular in nature, contains a cavity that is continuous throughout the entire neuraxis, from the forebrain above to the lowest regions of the spinal cord. As development progresses and changes occur in the external shape of different regions of the central nervous system, the shape of the inner cavity also undergoes regional changes. In the forebrain the cavity becomes the **lateral ventricles** of the cerebral hemispheres and the **third ventricle** of the diencephalon (Fig. 11-9,C). The lumen narrows throughout the midbrain, eventually becoming the **cerebral aqueduct,** and opens once again as the **fourth ventricle** in the hindbrain. At the rostral end of the hindbrain, a constriction of the cavity results from a transverse invagination called the **dorsal rhombencephalic sulcus.** This region of the neuraxis is sometimes known as the **rhombencephalic isthmus,** and its cavity is continuous caudally with the fourth ventricle, which, in turn, becomes the **central canal** of the spinal cord (Fig. 11-9,C).

THE RHOMBENCEPHALON

When the midbrain flexure appears, the rhombencephalon, or hindbrain, is as long as the forebrain and midbrain combined (Figs. 11-11; 11-12). For descriptive purposes, the rhombencephalon is divided into a more caudal portion called the **myelencephalon,** which forms the **medulla oblongata,** and a more cranial part, the **metencephalon,** which forms the **cerebellum** and **pons.**

THE MYELENCEPHALON. The caudal part of

[1]Theodore Schwann (1810–1882): A German histologist and physiologist who was a professor at Liège in Belgium.

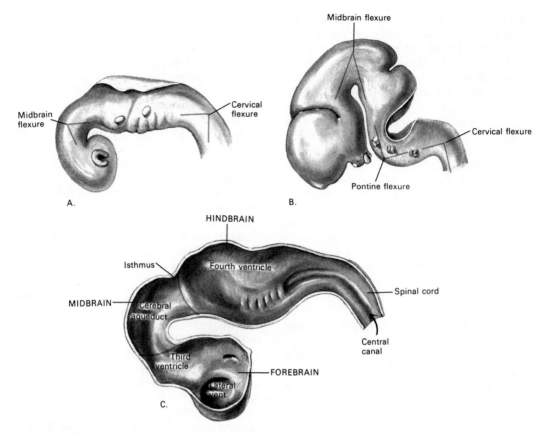

FIG. 11-9. *A,* Flexures of the human brain in 6-mm embryo (early part of fifth week). *B,* Flexures of the human brain in 14-mm embryo (early part of sixth week). *C,* Longitudinal hemisection of human brain in an 11-mm embryo (end of fifth week), showing the internal cavities. (Redrawn after L. B. Arey, *Developmental Anatomy,* revised 7th ed., W. B. Saunders, Philadelphia, 1974.)

the hindbrain is the myelencephalon, and it forms the medulla oblongata. At first the medulla is similar to the spinal cord, consisting of a roof plate, floor plate, and lateral plates of alar and basal laminae. The roof plate becomes greatly stretched and thinned at an early stage (Fig. 11-10,*B* and *C*), but the floor plate holds the lateral plates together so that they diverge and flatten simultaneously. The sulcus limitans persists, but now it separates a laterally placed alar lamina from a more medially placed basal lamina. The functional pattern remains similar to the spinal cord with a sensory alar plate, a motor basal plate, and autonomic nuclei along the sulcus limitans (Fig. 11-10,*C, D,* and *E*).

Sensory fibers from the neural crest ganglia of the glossopharyngeal and vagus nerves form a bundle opposite the sulcus limitans. Called the **tractus solitarius** (Fig. 11-10,*B*), it corresponds to the oval bundle of the primitive spinal cord (Fig. 11-4,*A*). At first it is in contact with the outer surface of

the alar lamina, but it later is buried by the overgrowth of neighboring parts (Fig. 11-10,*D*). At about five weeks, the part of the alar lamina next to the floor plate bends over laterally, forming the **rhombic lip** (Figs. 11-10,*E;* 11-13). The rhombic lip folds downward, over the main part of the alar lamina, fuses with it, and buries the tractus solitarius and spinal root of the trigeminal nerve. Later, the nodulus and flocculus of the cerebellum develop from the rhombic lip.

Although the basal plate corresponds in function to the ventral horn of the spinal cord because both give rise to motor fibers, in the basal plate the cells become arranged in groups or nuclei instead of continuous columns. In addition, neuroblasts migrate from the alar plate and rhombic lip into the basal lamina and become aggregated in the **olivary nuclei** (Fig. 11-10,*D*). Many of the fibers from these cells cross the midline of the floor plate and thus constitute the rudiment of the **raphe of the medulla.** The accumula-

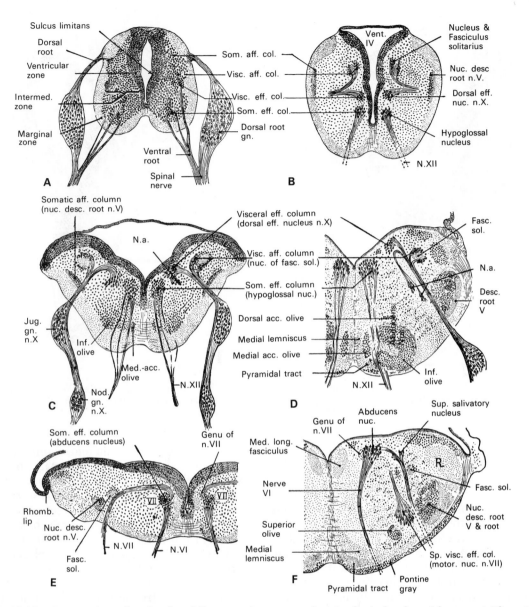

Fig. 11-10. Cross sections of cord and medulla comparing manner of origin of spinal and cranial nerves. *A*, Thoracic cord of 14.8-mm embryo to show origin of typical spinal nerve. *B*, Lower part of medulla for comparison with cord structure. *C*, Medulla of 15-mm embryo at level of upper rootlets of nerves X and XII. *D*, Medulla of 73-mm embryo at level of upper rootlets of nerves X and XII. Compare with *C*. *E*, Medulla of 15-mm embryo at level of pons, showing the roots of nerves VI and VII. *F*, Medulla of 73-mm embryo at level of pons, showing the roots of nerves VI and VII. Compare with *E*.

Abbreviations: Acc., accessory; Aff., afferent; Col., column; Desc., descending; Eff., efferent; Fasc. Sol., fasciculus solitarius; Gn., ganglion; Inf., inferior; Jug., jugular; N.A., nucleus ambiguus (note that in the younger stage (*C*), it is not separated from the general afferent column as happens later (*D*) and since it supplies skeletal muscle of branchiomeric origin rather than smooth muscle it is classified as a *special* visceral efferent nucleus); Nuc., nucleus; R., restiform body; Rhomb., rhombic; Sp., special; Visc., visceral. (Courtesy of B. M. Patten, *Human Embryology*, 3rd ed., McGraw-Hill, New York, 1968.)

tion of these cells and fibers in the ventral part of the basal plate pushes the motor nuclei deep into the interior, and in the adult they are found close to the interior lumen. The change is further accentuated by the

development of the pyramids at about the fourth month and by the fiber connections of the cerebellum. On the floor of the fourth ventricle, the rhomboid fossa—a series of six temporary transverse **rhombic grooves**—ap-

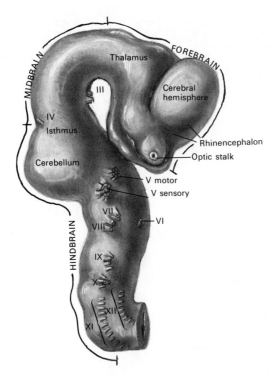

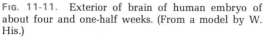

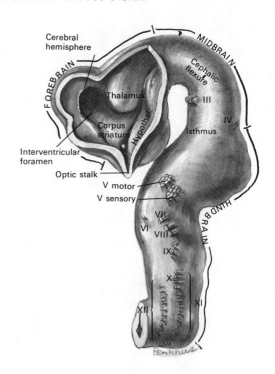

FIG. 11-11. Exterior of brain of human embryo of about four and one-half weeks. (From a model by W. His.)

FIG. 11-12. Brain of a human embryo of about four and one-half weeks, showing the interior of the forebrain. (From a model by W. His.)

pears. The most cephalic, or first and second, grooves lie over the trigeminal nucleus; the third groove lies over the facial nucleus; the fourth, over the abducent nucleus; the fifth, over the glossopharyngeal nucleus; and the sixth, over the nucleus of the vagus nerve.

THE METENCEPHALON. The metencephalon is the rostral portion of the hindbrain, and it extends from the pontine flexure to the rhombencephalic isthmus. Its cavity is the rostral part of the fourth ventricle. Two regions of the neuraxis, the **pons** ventrally and the **cerebellum** dorsally, are derived from the metencephalon (Fig. 11-14).

The Pons. The pons consists of a dorsal, more primitive, and phylogenetically older part called the **pontine tegmentum,** which forms a thick floor for the fourth ventricle. Its structure in many ways resembles that of the medulla oblongata and it contains the upward continuation of many structures found in the medulla. In contrast, the **basilar** or **ventral portion of the pons** is a more recent and distinctly mammalian acquisition, since through it course fiber systems that interconnect the cerebral cortex and the cerebellar regions.

The *roof plate* of the embryonic pons forms a thin sheet of white matter in front of the cerebellar cortex called the **medullary velum.** The *basal plates,* located ventral to the sulci limitans, as in the myelencephalon and spinal cord, form motor neurons (Fig. 11-15). In the pons these contribute general somatic efferent fibers for the **abducens nerve,** which supplies the lateral rectus muscle in the orbit, and special visceral efferent neurons for the **trigeminal** and **facial nerves,** which supply the muscles derived from the first and second branchial arches. Additionally, preganglionic parasympathetic general visceral motor neurons are formed in the **superior salivatory nucleus.** Their axons also emerge with the facial nerve to synapse peripherally in the pterygopalatine and submandibular ganglia. The basal plate also contributes neuronal elements to the **reticular formation.**

The *alar plates,* located dorsal to the sulci limitans, form the relay nuclei for sensory components in the **trigeminal, facial,** and **vestibulocochlear nerves** (Fig. 11-15). The **sensory nucleus of the trigeminal nerve,** found in the pontine tegmentum and also

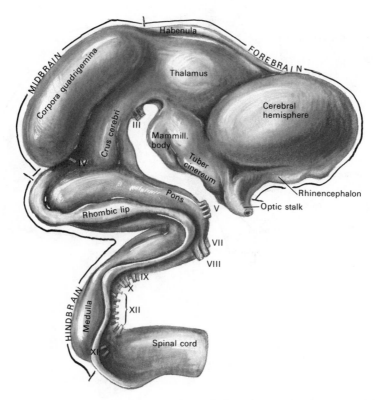

FIG. 11-13.   Exterior of the brain of a human embryo of about five weeks. (From a model by W. His.)

extending caudally into the medulla oblong-ata, receives general sensory afferent trigem-inal fibers. Special and general visceral af-ferent fibers course in the pons with the facial nerve to enter the **tractus** and **nucleus solitarius,** and the **vestibular** and **cochlear nuclei** receive special sensory afferent fibers from the vestibulocochlear nerve. The *floor plate* of the developing pons forms a midline raphe, as it does in the myelencephalon.

The ventrally located basilar portion of the pons contains the large **pontine nuclei,** which are formed by cells that migrate from the alar plates of both the myelencephalon and metencephalon. Fibers from the cere-bral cortex enter these nuclei, whereas fibers from the pontine nuclei cross the midline and course to the cerebellum by way of the **middle cerebellar peduncles.**

The Cerebellum.   Principally due to the pontine flexure, the dorsolateral parts of the alar plates in the cephalic portion of the hindbrain bend laterally and posteriorly to form the **rhombic lips** (Figs. 11-13; 11-16). The rostral part of each rhombic lip be-comes thickened and, together with the alar plates in the rostral rhombencephalon,

forms the rudiments of the cerebellum (Fig. 11-14).   Initially, the two rhombic lips partially project into the fourth ventricle and partially onto the dorsal surface of the metencephalon above the attachment of the roof plate. As the pontine flexure deepens, the rhombic lips become compressed, which causes them to fuse, thereby forming the **transverse cerebellar plate.** After the third month, this plate becomes differentiated into an unpaired midline portion, which be-comes the **vermis,** and symmetrically shaped lateral portions, which develop into the **cerebellar hemispheres.** At this stage the developing cerebellum assumes a "dumb-bell" appearance, and further growth is prin-cipally the result of enlargement of its extra-ventricular part. Because of this so-called eversion, the cerebellum becomes essen-tially an extraventricular structure, overly-ing the fourth ventricle.

The outer surface of the cerebellum is ini-tially smooth. Soon after the third month, however, the first of the cerebellar fissures appears, separating the **nodulus** and the **floc-culus** from the rest of the vermis and the lat-eral cerebellar lobes. This is called the **pos-**

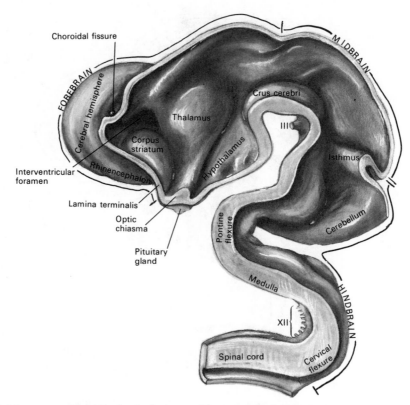

FIG. 11-14.   Interior of the brain of a human embryo of about five weeks. (From a model by W. His.)

terolateral fissure, behind which is the phylogenetically older **flocculonodular node,** importantly related to the vestibular system, and in front of which is the more rapidly growing cerebellar hemispheres. During the fourth and fifth months this rapid growth results in the formation of a number of fis-

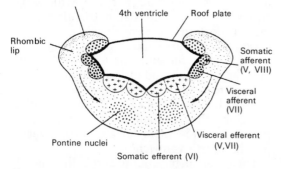

FIG. 11-15.   Diagrammatic representation of the basal and alar plates of the metencephalon. Note the motor nuclei of the basal plates, the sensory nuclei of the alar plates, and the pontine nuclei located more ventrally. (From J. Langman, *Medical Embryology*, Williams and Wilkins, Baltimore, 1963.)

sures that separate newly formed lobules in the cerebellar hemispheres. These include the **primary fissure,** which courses transversely and deeply to separate the paleocerebellar **anterior lobe** from the neocerebellar **posterior lobe,** as well as several other transverse fissures, including (from back to front) the **secondary fissure,** the **prepyramidal fissure,** the **horizontal fissure,** and the **posterior superior fissure.**

THE MESENCEPHALON

The mesencephalon, or midbrain, develops around a primitive neural cavity that eventually is reduced in diameter to form the **cerebral aqueduct** (of Sylvius[1]). The narrow canal joins the lumen of the diencephalon (the third ventricle) rostrally with that of the rhombencephalon (the fourth ventricle) caudally. Three major regions of the embryonic neural tube become differentiated in

[1]Franciscus Sylvius (Latinized form of François Dubois, or de la Boë (1614–1672): A Dutch physician and anatomist (Leyden).

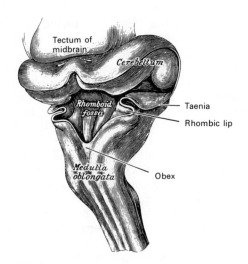

FIG. 11-16. Hindbrain of a human embryo of three months—viewed from behind and partly from the left side. (From a model by W. His.)

this part of the brain stem. These are the **tectum,** which develops dorsal to the cerebral aqueduct, and the midbrain **tegmentum** and **cerebral peduncles,** which form ventral and lateral to the aqueduct.

The midbrain tectum develops from the two alar laminae that invade the roof plate of the original neural tube. As in other regions of the neuraxis, cells from the alar plates develop into structures related to sensory functions, and in the midbrain tectum, they form the **corpora quadrigemina.** These four bodies are the paired inferior and superior colliculi, which serve as important relay centers for the integration of reflexes initiated by both auditory and visual impulses (Fig. 11-17).

The basal plate forms the tegmentum of the midbrain and from it originate the motor

nuclei of the **oculomotor** and **trochlear nerves** (Fig. 11-17). These general somatic efferent fibers supply the extraocular musculature that is formed within the preotic myotomes. Also formed by the basal plate are the general visceral efferent neurons of the **Edinger[2]-Westphal[3] nucleus,** preganglionic parasympathetic nerve cells that emerge with the oculomotor nerve to synapse in the ciliary ganglion. From this ganglion, postganglionic fibers supply the sphincter of the pupil.

The ventrolateral aspect of the basal plates receives large numbers of descending corticofugal fibers destined for the pons, medulla oblongata, and spinal cord. These form the **crura cerebri** or cerebral peduncles, which are separated from the tegmental region on both sides by nuclear masses called the **substantia nigra** (Fig. 11-17). The cells that form the **mesencephalic reticular formation,** as well as the **red nucleus,** are thought by some investigators to migrate from the alar plates into the tegmental regions. Others have claimed that these midbrain elements develop *in situ* from basal plate cells. There is also conjecture relating to the origin of the cells that form the substantia nigra.

THE PROSENCEPHALON

At early stages, transverse sections of the prosencephalon, or **forebrain,** display the same parts as those that are seen at similar

[2]Ludwig Edinger (1855–1918): A German anatomist and neurologist (Frankfurt-am-Main).
[3]Karl F. O. Westphal (1833–1890): A German neurologist and psychiatrist (Berlin).

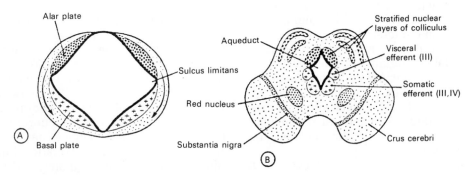

FIG. 11-17. Diagrams showing the differentiation of midbrain structures from alar and basal plates. The arrows in *A* indicate the migration of alar plate cells, which form the red nucleus and substantia nigra. (From J. Langman, *Medical Embryology,* Williams and Wilkins, Baltimore, 1963.)

stages of development in the spinal cord and medulla: namely, thick lateral walls connected by thin roof and floor plates. Although many investigators believe that the sulcus limitans does not extend forward beyond the midbrain and that the prosencephalon is formed primarily from the alar laminae, others believe that the hypothalamus is derived from the basal laminae and that the rest of the forebrain forms from the alar laminae. The latter point to the **hypothalamic sulcus,** which continues forward as far as the optic recess and might be considered analogous to the sulcus limitans. Irrespective of this matter, the telencephalon gives origin to all of the central nervous system cranial to the midbrain.

At a very early period, even before the closure of the cranial part of the neural tube, the **optic vesicles** appear as diverticula on each side of the forebrain. Initially they communicate with the forebrain vesicle by relatively wide openings. Later the proximal part of the vesicle is narrowed into the **optic stalk** (Figs. 11-11 to 11-14), and the peripheral part is expanded into the **optic cup.** The cavity of the vesicle eventually disappears, and the stalk is invaded by nerve fibers and becomes the **optic nerve** and tract. The optic cup becomes the **retina.**

After closure of the neural tube, the lateral walls of the forebrain grow anteriorly and ventrally much more rapidly than the median portion, resulting in the formation of a large pouch on each side, the **cerebral hemispheres** (Figs. 11-11; 11-13). The cavities within the pouches are the rudiments of the **lateral ventricles** and their intercommunications become the future **interventricular foramina of Monro**[4] (Figs. 11-12; 11-14). The median portion of the wall of the forebrain vesicle, formed by the anterior part of the roof plate, consists of a thin lamina that stretches from the interventricular foramina to the recess at the base of the optic stalk called the **lamina terminalis.** The rostral part of the forebrain, including the hemispheres, is the **telencephalon** or **endbrain,** and the caudal part is called the **diencephalon** or **intermediate brain.** The cavity of the median part of the forebrain vesicle becomes the **third ventricle.**

THE DIENCEPHALON. The diencephalon

gives rise to the **thalamus, metathalamus, epithalamus,** and **hypothalamus** (Figs. 11-12; 11-14; 11-18). A groove in the lateral wall of this part of the forebrain vesicle, the **hypothalamic sulcus,** separates the epithalamus, metathalamus, and thalamus, located more dorsally, from the hypothalamus, located ventrally. The *caudal* extent of the diencephalon consists of a plane drawn between the posterior commissure dorsally and the mammillary bodies ventrally, while *rostrally* it extends to the cerebral hemispheres and forms the telencephalon. At the base of the brain the diencephalon extends rostrally to a site a few millimeters anterior to the optic chiasma.

The **thalamus** arises as a thickening of the cephalic two-thirds of the alar plate. For some time it is visible as a prominence on the external surface of the brain (Figs. 11-11; 11-13), but it later becomes buried by the overgrowth of the hemispheres (Fig. 11-18). The thalami of the two sides protrude medially into the ventricular cavity and eventually join across the midline in a small commissure-like junction of gray substance called the **interthalamic adhesion.** This was formerly termed the **massa intermedia.**

The **metathalamus** consists of the **medial** and **lateral geniculate bodies,** which appear as slight prominences on the outer surface of the alar lamina. The medial geniculate bodies are relay stations for the transmission of auditory impulses, while the lateral geniculate bodies function as relay centers in the visual pathways.

The **epithalamus** includes the **pineal body,** the **posterior commissure,** and the **habenular trigone.** The pineal body arises as an evagination of the roof plate between the thalamus and the colliculi (Fig. 11-18). The habenular trigone, consisting of the habenular nuclei and the habenular commissure, develops in the roof plate just rostral to the pineal gland, and the **posterior commissure** develops just caudal to it. The roof plate of the diencephalon rostral to the pineal gland remains thin and epithelial in character and becomes invaginated by the **choroid plexus** of the third ventricle. The choroid plexus consists of modified ependymal cells that form a close functional relationship with blood vessels in the pia mater, and as elsewhere in the ventricular system, it secretes cerebrospinal fluid.

The **hypothalamus** is formed from the

---

[4]Alexander Monro Secundus (1733–1817): A Scottish anatomist and surgeon (Edinburgh).

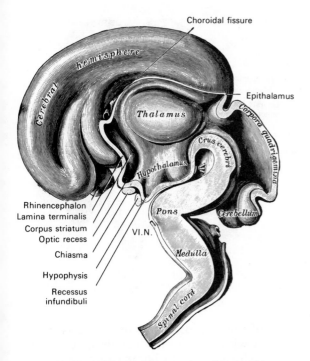

Choroidal fissure

Epithalamus

Thalamus

Corpora quadrigemina

Crus cerebri

Hypothalamus

Rhinencephalon
Lamina terminalis
Corpus striatum
Optic recess

Chiasma

Hypophysis

Recessus
infundibuli

Pons

Cerebellum

VI.N.

Medulla

Spinal cord

FIG. 11-18.   Median sagittal section of brain of human embryo of three months. (From a model by W. His.)

portion of the lateral walls of the diencephalon located below the hypothalamic sulcus (Figs. 11-12; 11-14; 11-18). One group of hypothalamic nuclei that develops becomes importantly related to many visceral and endocrine functions, as well as to the innate behaviors of feeding, drinking, and sleeping. Caudally the **mammillary bodies** are formed; initially these arise as a single ventrally protruding structure that later during the third prenatal month becomes divided by a midline furrow (Figs. 11-13; 11-14).

Anterior to the mammillary bodies the tuber cinereum is formed, and in front of this a diverticulum develops in the hypothalamic floor and gives rise to the **infundibulum** and **posterior lobe of the pituitary gland** (Fig. 11-18). In contrast, the anterior lobe of the pituitary gland develops from an ectodermal diverticulum called **Rathke's**[1] **pouch.** This latter outgrowth appears in the roof of the stomodeum during the middle of the third week. It extends upward toward the floor of the hypothalamus, and as it makes contact with the posterior lobe, it loses its

[1]Martin H. Rathke (1793–1860): A German anatomist and zoologist (Königsburg).

attachment on the pharyngeal roof during the sixth week.

The optic part of the hypothalamus initially is represented by two diverticula, the **optic vesicles.** Peripherally the optic cup forms the retina, whereas the optic stalk is retained as the pathway through which the ingrowing optic nerve fibers from the ganglionic layer of the retina form the **optic chiasma** and **optic tract.** The optic chiasma, located rostrally in the hypothalamic floor, forms the boundary between the diencephalon and telencephalon at the base of the brain.

THE TELENCEPHALON.   The telencephalon is the most rostral portion of the developing neuraxis, and early during its formation it consists of two lateral diverticula and an interconnecting median part, the **lamina terminalis,** which is continuous with the diencephalon. The cavities of the diverticula represent the **lateral ventricles** and their walls become thickened to form the substance of the **cerebral hemispheres.** The cavity of the median portion is the most rostral extent of the third ventricle. Initially, the lateral ventricles widely communicate with the third ventricle, but eventually, as the cerebral hemispheres grow, these communications become the slit-like interventricular foramina.

The cerebral hemispheres expand rapidly after the sixth week, and by the fourth month almost completely cover the diencephalon, mesencephalon, and cerebellum (Fig. 11-18). The medial surfaces of the two hemispheres come into contact, which accounts for their flattened appearance and results in the formation of the **longitudinal cerebral fissure,** which intervenes between the hemispheres. This enormous growth of the cerebral hemispheres is characteristic of mammalian brains, especially the human brain. It is convenient and functionally relevant to describe the further development of the following principal parts of each cerebral hemisphere: the **rhinencephalon** and **archeopallium,** the **corpus striatum,** and the **neopallium.**

Rhinencephalon and Archeopallium.   Continued use of the term **rhinencephalon** (olfactory brain) can be justified only when the structures incorporated within its definition are clearly stated. Confusion arises because authors have frequently used the term loosely to include many structures that are not directly related to olfaction, but are associated instead with emotional and visceral functions (such as mating behavior, feeding, aggressivity, and fear)—functions that, especially in lower vertebrates, depend importantly on olfactory cues. In this book, the term rhinencephalon is used strictly to include the **olfactory bulb** and those forebrain regions that receive nerve fibers directly from it. Thus, in addition to the olfactory bulb, the rhinencephalon includes the **olfactory tracts,** the **medial, intermediate,** and **lateral olfactory striae,** the **olfactory trigone** (formed where the striae diverge), the **anterior olfactory nucleus,** the **olfactory tubercle** (anterior perforated substance), the **corticomedial group of nuclei in the amygdaloid complex,** and the **prepiriform** and **periamygdaloid cortex** of the piriform lobe. In essence, this represents the olfactory bulb, its tracts, and the **paleopallium,** the so-called olfactory cortex. Because the **hippocampal formation** in mammals does not appear to receive major projections of olfactory fibers directly, its structures are included in the term **archeopallium.**

The rhinencephalon and archeopallium are phylogenetically the oldest portions of the telencephalon, and in fishes, amphibia, and reptiles they constitute virtually the entire cerebral hemisphere. The rhinencephalic primordia first appear on the medial aspect of the anteroventral surface of each developing cerebral hemisphere during the latter part of the fifth week (Figs. 11-13; 11-14). Externally, a slight longitudinal ridge or diverticulum appears which corresponds to a furrow that forms on the inner aspect of the hemisphere, close to the lamina terminalis. The rhinencephalon is separated from the lateral surface of the hemisphere by the **rhinal sulcus** and is continuous caudally with the portion of the hemisphere that will form the anteromedial end of the temporal lobe. As development continues, the longitudinal ridge becomes divided by a groove into an anterior and a posterior part. The rostral part of the ridge is the primitive **olfactory lobe,** while the caudal part is the **piriform lobe** (Fig. 11-19). *From the rostral part of the longitudinal ridge,* a hollow shaft grows forward, the cavity of which initially retains a communication with the lateral ventricle. During the third month, the stalk loses its cavity and becomes solid, forming the primitive olfactory bulb and tract. The proximal

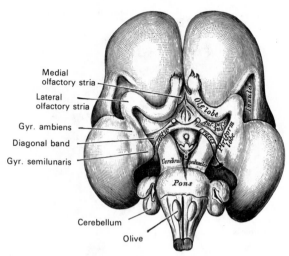

Medial olfactory stria
Lateral olfactory stria
Gyr. ambiens
Diagonal band
Gyr. semilunaris
Cerebellum
Olive

FIG. 11-19.   Inferior surface of brain of embryo at beginning of fourth month. (From Kollmann.)

part of the primitive olfactory lobe is connected with the parolfactory or subcallosal area, adjacent to the lamina terminalis on the medial surface of the hemisphere, by the **medial olfactory gyrus** (Fig. 11-19).

From the caudal part of the primitive olfactory ridge develop the anterior perforated substance and piriform lobe (Fig. 11-19). At the beginning of the fourth month, the piriform lobe forms a curved elevation that is continuous behind with the medial surface of the future temporal lobe. Interconnected rostrally to the olfactory lobe by the **lateral olfactory gyrus** (or prepiriform region), the surface of the piriform lobe is marked at this stage by the gyrus ambiens and the gyrus semilunaris (Fig. 11-19). These develop into the corticomedial part of the amygdala and the **lumen insulae.** The adult piriform lobe, bounded laterally by the rhinal sulcus, is divided into the prepiriform, periamygdaloid, and entorhinal regions. The prepiriform and periamygdaloid regions receive direct olfactory bulb fibers by way of the lateral olfactory tract, while the entorhinal region, which forms a large part of the parahippocampal gyrus, receives projections from the prepiriform cortex, but none directly from the olfactory bulb.

The adult **archeopallium** consists of the **hippocampus** (cornu ammonis), the **dentate gyrus,** and the **subicular part of the parahippocampal gyrus.** Together these structures are also frequently called the **hippocampal formation,** and before the appearance of the corpus callosum, they develop on the medial wall of the cerebral hemisphere along a line parallel with, but above and in front of, the choroid fissure. Becoming folded into the cavity of the lateral ventricle, so that the prominence bulging into the ventricle forms the hippocampus, the invagination is accompanied by the development of a corresponding fissure, the **hippocampal sulcus,** on the medial surface of the cerebral hemisphere. Soon the marginal zone of the invagination differentiates to form the dentate gyrus, while inferiorly and laterally the hippocampus merges with the cortex of the parahippocampal gyrus by means of the subiculum. As the infolding proceeds, the hippocampal fissure is elongated; the subiculum becomes located on one side of the hippocampal sulcus and the dentate gyrus on the other side. Eventually, the hippocampus (cornu ammonis) curves around the hippocampal fissure, and the dentate gyrus lies within the concave aspect of the curve.

The outer gray substance of the primordium of the archeopallial cortex ends within the bulging prominence between the hippocampal fissure and the choroid fissure, leaving the white matter, called the **alveus,** exposed along the edge of the hippocampus adjacent to the epithelial cells of the ependyma. These fibers join to form the fimbria of the fornix. With the downward and forward growth of the temporal lobe, the hippocampal fissure and parts associated with it extend from the interventricular foramen to the end of the inferior horn of the lateral ventricle.

Corpus Striatum.   The corpus striatum appears as a thickening in the floor of the telencephalon between the optic stalk and the interventricular foramen, adjacent to the thalamus (Figs. 11-12; 11-14; 11-20; 11-21). By the second month it has increased in size and bulges from the floor of the lateral ventricle into the ventricular lumen, extending as far as the occipital pole of the primitive hemisphere.

As the expansion of the cerebral hemispheres gradually covers the diencephalon, a part of the corpus striatum is carried into the lateral ventricular wall and downward into the roof of the inferior horn of the lateral ventricle as the tail of the caudate nucleus. During the fourth and fifth months, the corpus striatum is invaded and incompletely subdivided by large numbers of fibers of the internal capsule into two masses: an inner or dorsomedial part adjacent to the ventricle, the **caudate nucleus,** and an outer part, ventrolateral to the fibers of the internal capsule, the **lentiform nucleus** (Fig. 11-21). This latter nucleus develops into the larger **putamen** laterally and the smaller **globus pallidus** medially.

The Neopallium.   The neopallium, or non-olfactory part of the cerebral cortex, forms the greater part of the hemisphere. Early in development its cavity is the primitive **lateral ventricle.** This is enclosed by a thin wall of neural tissue from which the cortex of the hemisphere is derived, and it expands in all directions but especially dorsally and caudally. By the third month the hemispheres cover the diencephalon; by the sixth month they overlap the midbrain; and by the eighth month they reach the hindbrain. The median lamina uniting the two hemispheres does not share in their expansion and remains as the thin roof of the third ventricle. Thus, between the hemispheres there develops a deep cleft, the forerunner of the **longitudinal cerebral fissure.** The cleft becomes occupied by a septum of mesodermal tissue that constitutes the primitive **falx cerebri.** As the cerebral hemispheres expand, each lateral ventricle is gradually drawn outward onto three prolongations that represent the future anterior, inferior, and posterior horns.

The part of the wall along the medial border of the primitive lateral ventricle, which is immediately continuous with the roof plate of the diencephalon, lies over the primitive interventricular foramen and extends caudally. This border remains thin and has an ependymal or epithelial character. It is invaginated into the medial wall of the lateral ventricle to form the **choroid fissure** (Figs. 11-14; 11-22). Mesodermal tissue from the outer surface of the hemisphere, the rudiment of the pia mater, spreads into the fissure between the two layers of the ependyma to form the rudiment of the **tela choroidea.** The blood vessels accompanying the mesoderm form the **choroid plexus,** which almost completely fills the cavity of the ventricle for several months (Figs. 11-20; 11-21). The tela choroidea, lying over the roof of the diencephalon, also invaginates the thin epithelial covering to form the choroid plexus of the third ventricle.

As described previously, the medial wall

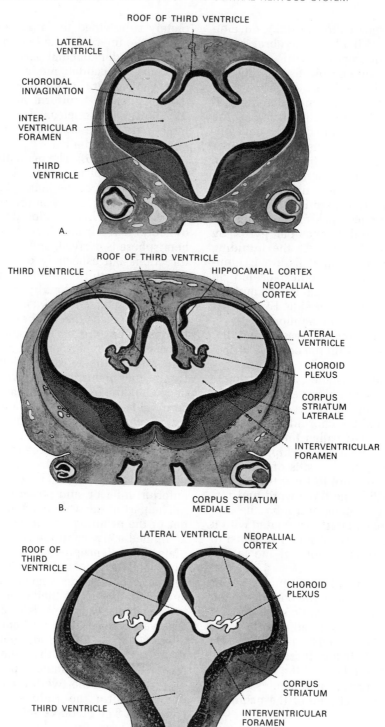

FIG. 11-20. Frontal sections through the forebrain of (A) 15-mm, (B) 17-mm, and (C) 19-mm human embryos, showing various stages in the development of the lateral ventricles and the choroid plexuses. Embryonic ages: early, mid, and latter part of the sixth week. (From W. J. Hamilton and H. W. Mossman, *Human Embryology*, 4th ed., Williams and Wilkins, Baltimore, 1972.)

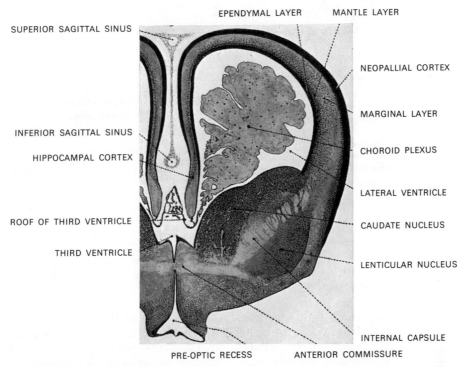

FIG. 11-21. Frontal section through the left cerebral hemisphere of a 73-mm human fetus (fourth month). Observe the developing ventricular system and the corpus striatum. (From W. J. Hamilton, and H. W. Mossman, *Human Embryology*, 4th ed., Williams and Wilkins, Baltimore, 1972.)

of the cerebral hemisphere becomes folded into the cavity of the lateral ventricle and develops as the rudiment of the hippocampal formation. Growth of the remaining cerebral cortex is so great that the structure is thrown into folds or convolutions (gyri) between which are formed the sulci or fissures. Additionally, groups of nerve fibers pass from one hemisphere to the other, forming the commissures.

The Commissures. The development of the **posterior commissure** was mentioned in the section describing the embryology of the diencephalon. The other commissures interconnecting the two hemispheres are the large **corpus callosum,** the **anterior commissure,** and the small **commissure of the fornix,** which all arise from the lamina terminalis (Fig. 11-22). At about the fourth month a small thickening appears in this lamina, immediately in front of the interventricular foramen. The *lower part* of this thickening soon becomes constricted, and fibers grow into it to form the anterior commissure. These fibers interconnect the two olfactory bulbs and certain other paleopallial structures of the two sides, such as the piriform and prepiriform regions and the amygdaloid

nuclei. The *upper part* continues to grow caudally with the hemispheres and is invaded by two sets of fibers. Transverse fibers, extending between the two cerebral hemispheres, pass into its dorsal part and become differentiated as the corpus callosum. Longitudinal fibers from the hippocampus pass into the ventral part of the lamina terminalis, and arching over the thalamus to the mammillary bodies, develop into the fornix. The anterior portion between the corpus callosum and fornix is not invaded by commissural fibers and becomes the **septum pellucidum.**

Sulci and Gyri. The outer surface of the cerebral hemisphere is at first smooth. Later it exhibits a number of convolutions, or gyri, that are separated from each other by sulci, which appear during the sixth or seventh month of fetal life. The term *complete sulcus* is applied to grooves that involve the entire thickness of the cerebral wall and thus produce corresponding eminences in the ventricular cavity. The other sulci affect only the superficial part of the wall and therefore leave no impressions in the ventricle.

The complete sulci are the **choroidal** and **hippocampal,** which have already been de-

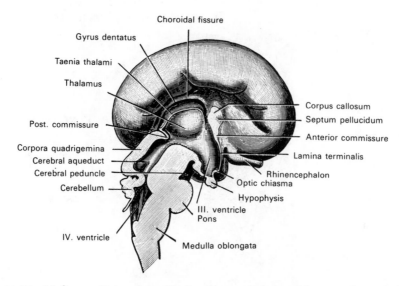

Choroidal fissure
Gyrus dentatus
Taenia thalami
Thalamus
Post. commissure
Corpora quadrigemina
Cerebral aqueduct
Cerebral peduncle
Cerebellum
IV. ventricle
III. ventricle
Pons
Medulla oblongata
Corpus callosum
Septum pellucidum
Anterior commissure
Lamina terminalis
Rhinencephalon
Optic chiasma
Hypophysis

FIG. 11-22. Median sagittal section of brain of human embryo of four months. (Marchand.)

scribed; the **calcarine,** which produces the swelling known as the **calcar avis** in the posterior horn of the ventricle; and the **collateral sulcus,** with its corresponding eminence in the ventricular cavity. The **central sulcus** (the fissure of Rolando[1]) develops in two parts; the **intraparietal sulcus,** in four parts; and the **cingulate sulcus,** in two or three parts. The **lateral cerebral** (or Sylvian) **sulcus** differs from all the other sulci in its mode of development. It appears toward the end of the third month as a depression, the **Sylvian fossa,** on the lateral surface of the hemisphere (Fig. 11-23). The floor of this fossa becomes the **insula,** and its adherence to the subjacent corpus striatum prevents this part of the cortex from expanding at the same rate as the portions that surround it. The neighboring parts of the hemisphere, therefore, gradually grow over and cover the insula, constituting the **temporal, parietal, frontal,** and **orbital opercula** of the adult brain. The frontal and orbital opercula are the last to form, but by the end of the first year after birth, the insula is completely submerged. The sulci separating the opposed margins of the opercula constitute the adult lateral cerebral sulcus.

Development of the Autonomic Nervous System

The autonomic nervous system consists of **sympathetic** and **parasympathetic** divisions. It is a two-motor neuron system, of which

[1]Luigi Rolando (1773–1831): An Italian anatomist (Turin).

the cell body of the first, or **preganglionic,** neuron lies within the substance of the brain or spinal cord, while the cell body of the second, or **postganglionic,** neuron is located more peripherally in one of the autonomic ganglia outside the central nervous system.

SYMPATHETIC NERVOUS SYSTEM. The cell bodies of the preganglionic neurons in the sympathetic division of the autonomic nervous system are located in the lateral gray matter of the spinal cord in a column extending from the first thoracic segment through the second lumbar segment. Occupying a position between the afferent groups of the dorsal horn and the somatic efferent groups of the ventral horn, the preganglionic

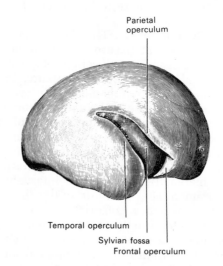

Parietal operculum
Temporal operculum
Sylvian fossa
Frontal operculum

FIG. 11-23. Outer surface of cerebral hemisphere of human embryo of about five months.

sympathetic neurons develop axons that grow out through the ventral roots with the somatic motor fibers. Eventually these preganglionic elements leave the spinal nerves by way of the white rami communicantes to synapse with postganglionic sympathetic neurons within the ganglia of the **sympathetic trunk** or in the other outlying collateral ganglia found in the abdomen.

Sympathetic Ganglia. Most investigators today believe that the ganglion cells that form the sympathetic ganglia are derived from the **neural crest.** These cells commence migrating along the dorsal roots of the spinal nerves during the fifth prenatal week, and they eventually form segmentally paired neuronal groups in the embryonic thoracic region dorsolateral to the aorta. Soon the fibers grow between the masses of cells to form the sympathetic chains. Other ganglion cells grow from the cervical and lumbar regions of the neural crest to form the cervical and lumbosacral ganglia. Certain sympathetic ganglion cells migrate beyond the sympathetic trunk to form ganglia adjacent to the branches of the abdominal aorta. These outlying **collateral ganglia** develop shortly after those of the sympathetic trunk and include the **celiac, superior mesenteric,** and **inferior mesenteric ganglia.** The nerve fibers of the postganglionic sympathetic neurons are either unmyelinated or surrounded by very thin myelin sheaths.

PARASYMPATHETIC NERVOUS SYSTEM. The parasympathetic nervous system consists of **preganglionic** nerve cells that arise in the visceral efferent nuclei of the brain, and their axons grow out of the brain by way of the oculomotor, facial, glossopharyngeal, and vagus nerves. Additionally, other preganglionic parasympathetic neurons develop in the lateral gray column of the second, third, and fourth sacral segments and leave the spinal cord in the ventral roots of the sacral nerves. The oculomotor, facial, and glossopharyngeal parasympathetic neurons course to ganglia in the orbit, deep face, and lower jaw, from which **postganglionic** parasympathetic fibers course to the viscera in the head. The vagus nerve descends to supply preganglionic fibers to terminal ganglia within or near the viscera of the neck, thorax, and abdomen. Sacral parasympathetic fibers grow as far as terminal ganglia in the lower part of the gastrointestinal tract and in the pelvic organs.

Parasympathetic Ganglia. The precise origin of the ganglion cells that form the **ciliary, pterygopalatine, submandibular,** and **otic** ganglia of the head is still not completely understood. Some believe that the ganglion cells migrate from the trigeminal ganglion along the developing divisions of the trigeminal nerve. Others feel that all the cells are derived from the central nervous system and migrate along the specific cranial nerves that carry the preganglionic fibers. Still other investigators feel that the parasympathetic ganglia in the head receive cells not only from the sensory ganglion of the trigeminal nerve, but also from the geniculate ganglion of the facial nerve and the inferior ganglion of the glossopharyngeal nerve.

The origin of the postganglionic neurons that synapse with the preganglionic vagal fibers is also uncertain. It has been proposed that they derive from several sources, including the ganglia of the vagus nerve, and from the central nervous system by migrating along the vagal trunks. The ganglion cells associated with the sacral parasympathetic fibers are thought to arise from the neural crest, which forms adjacent to the sacral segments of the neural tube.

Development of the Cranial Nerves

The development of the olfactory, optic, and vestibulocochlear nerves is described in more detail in Chapter 13. The pattern of development of the other cranial nerves is quite similar to that of the spinal nerves. The cranial *sensory or afferent nerves* are derived from cells in the ganglion rudiments of the neural crest. Evidence from research in lower animal forms also indicates that ectodermal thickenings that overlie the ganglia make some contribution to the formation of the sensory ganglia of the **trigeminal, facial, glossopharyngeal,** and **vagus** nerves. These have been called **epibranchial placodes.** The central processes of these ganglion cells grow into the brain and form the sensory elements in the roots of the nerves, while the peripheral processes extend distally to their sensory fields of distribution. In considering the development of the medulla oblongata, it appears that the **tractus solitarius** (Fig. 11-10) is derived from the fibers that grow inward from the ganglion rudiments of the glossopharyngeal and vagus nerves, and that it is the homologue of the **oval bundle** in the cord, which has its origin in the dorsal nerve roots. During the development of the bulbar

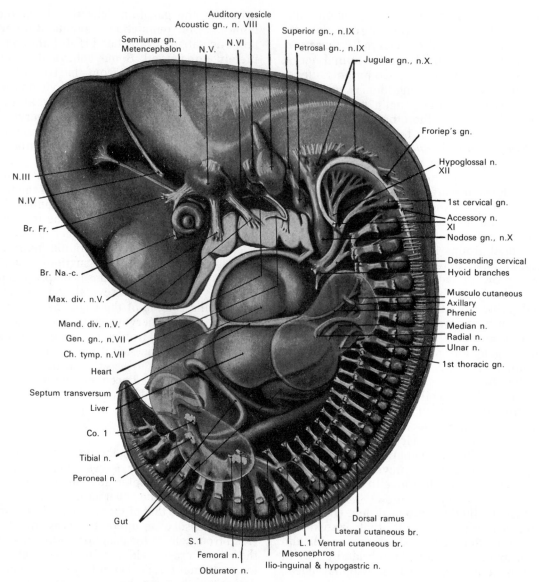

FIG. 11-24. Reconstruction of the nervous system in a 10-mm human embryo during the fifth prenatal week (from the studies of G. L. Streeter). Note the developing cranial and spinal nerves. (Taken from B. M. Patten, *Human Embryology*, 3rd ed., McGraw-Hill, New York, 1968.)

part of the **accessory nerve** and the **hypoglossal nerve,** strands of ganglion cells extend along the rootlets of these cranial nerves. These are derived from an embryonic sensory ganglion, known as **Froriep's[2] ganglion,** which develops between the jugular ganglion of the vagus and the first cervical ganglion (Fig. 11-24). Most of Froriep's ganglion disappears in the adult, although occasionally ganglion cells can be observed

[2]August von Froriep (1849–1917): A German anatomist.

along the course of the accessory and hypoglossal nerves.

The *motor or efferent* components of the cranial nerves arise as outgrowths of the neuroblasts that are situated in the basal laminae of the midbrain and hindbrain. In contrast to the spinal motor nerve rootlets that develop from the basal lamina in single sets at each segmental level, the motor rootlets of the cranial nerves form two sets as they emerge from the brainstem. One group of rootlets develops from the medial and the

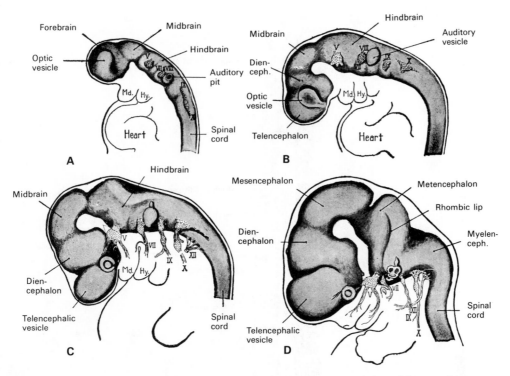

FIG. 11-25. Four stages in the development of the brain and cranial nerves (adapted from various sources, but principally from the studies of G. L. Streeter). *A,* 20-somite stage (about three and one-half weeks); *B,* 4-mm stage (about four weeks); *C,* 8-mm stage (about four and one-half weeks); *D,* 17-mm stage (about six and one-half weeks). The cranial nerves are indicated by the appropriate Roman numerals. Abbreviations: Hy., hyoid arch; Md., mandibular arch. (Taken from B. M. Patten, *Human Embryology,* 2nd ed., McGraw-Hill, New York, 1953.)

other from the lateral parts of the basal lamina. From the medial part of the basal lamina develop the *somatic efferent fibers* of the oculomotor, trochlear, abducent, and hypoglossal nerves that supply the striated musculature formed from the somites (Fig. 11-24). The lateral part of the basal lamina forms the accessory nerve and the *branchial motor fibers* of the trigeminal, facial, glossopharyngeal and vagus nerves that supply the voluntary musculature formed from the branchial arches (Figs. 11-24; 11-25).

Gross Anatomy of the Central Nervous System

The central nervous system, divided for convenience of description into the **spinal cord** (or medulla spinalis) and **brain** (or encephalon), is a single, continuous organ system from which also are derived the 12 pairs of cranial nerves and the 31 pairs of spinal nerves (Fig. 11-26). The cranial and spinal nerves, along with their ganglia and those of the autonomic nervous system, constitute the **peripheral nervous system.** The central nervous system functions to integrate information that it receives from the environment by way of **receptors** and the **sensory nerves.** Its reactions are transmitted to the **effectors** through the **motor nerves.**

The spinal cord and brain are composed of groups of neuron cell bodies, which together with their dendritic processes form the **gray substance.** Elongated processes, frequently myelinated and extending from the nerve cells, collect in bundles to form the **white substance** and the **tracts.** Enmeshed among the neurons and the processes are large numbers of neuroglial cells and capillaries that maintain the appropriate chemical environment and provide the nutrients necessary for proper neuronal metabolism.

Because the internal neuroanatomy of the central nervous system is a subject more appropriately handled in specialized textbooks and taught as a discipline separate from gross anatomy, most of the remainder

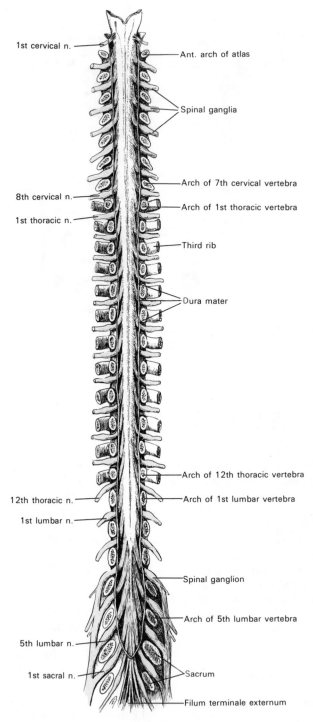

1st cervical n.

Ant. arch of atlas

Spinal ganglia

Arch of 7th cervical vertebra

8th cervical n.

Arch of 1st thoracic vertebra

1st thoracic n.

Third rib

Dura mater

Arch of 12th thoracic vertebra

12th thoracic n.

Arch of 1st lumbar vertebra

1st lumbar n.

Spinal ganglion

Arch of 5th lumbar vertebra

5th lumbar n.

1st sacral n.

Sacrum

Filum terminale externum

FIG. 11-26.   The dorsal aspect of the spinal cord exposed within the vertebral canal. The spinous processes and the laminae of the vertebrae have been removed and the dura mater has been opened longitudinally. (Redrawn after Sobotta.)

of this chapter deals with the external and gross anatomy of the spinal cord and brain. Reference is made to the internal anatomy of the central nervous system on those occa-

sions when it will lead to a better understanding or description of the external form.

The principal subdivisions of the central nervous system include the **spinal cord,** rhombencephalon (hindbrain), **mesencepha-lon** (midbrain), and **prosencephalon** (fore-brain). The rhombencephalon consists of the **myelencephalon** (medulla oblongata) and the **metencephalon** (pons and cerebellum), while the prosencephalon includes the **dien-cephalon** (thalamus and hypothalamus) and **telencephalon** (cerebral hemispheres).

SPINAL CORD

The **spinal cord** (*medulla spinalis*) is the elongated, nearly cylindrical part of the central nervous system that lies within most of the vertebral canal (Fig. 11-26). The diameter of the spinal cord in the midthoracic region is about 1 cm, but it is significantly greater at the lower cervical and midlumbar levels. Its average length in the male is about 45 cm, and in the female, from 42 to 43 cm, while its weight amounts to about 30 gm. The spinal cord extends from the upper border of the atlas, where it is continuous through the foramen magnum with the medulla oblongata of the brain stem, to the lower border of the first or upper border of the second lumbar vertebra, where it tapers to a point, forming the **conus medullaris** (Figs. 11-27; 11-30).

The position of the spinal cord varies somewhat with the movements of the vertebral column, being drawn upward slightly when the column is flexed. The degree to which the spinal cord fills the vertebral column also varies at different periods of life. Until the third prenatal month the spinal cord is as long as the vertebral column, but from this stage onward the vertebral column elongates more rapidly than the spinal cord. By the end of the fifth prenatal month, the spinal cord terminates at the base of the sacrum, and at birth it ends at about the third lumbar vertebra. It gradually recedes during childhood until it reaches the adult position (Fig. 11-5). Surrounding the spinal cord are three protective membranes called **menin-ges**—the *dura mater,* the *arachnoid,* and the *pia mater* (Figs. 11-29; 11-31). Between the latter two is found the *cerebrospinal fluid.* These membranes are described more completely elsewhere in this chapter.

The **filum terminale** is a delicate filament that continues down the vertebral canal from the apex of the conus medullaris to the first segment of the coccyx (Figs. 11-26; 11-27; 11-30). It is approximately 20 cm in length. Its upper 15 cm, contained within the

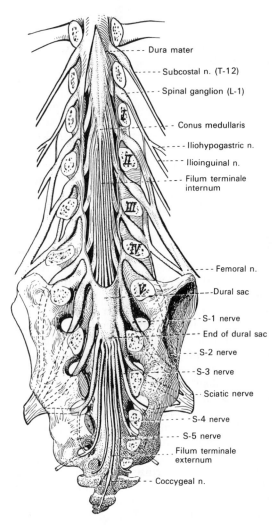

FIG. 11-27. The cauda equina of the spinal cord showing the conus medullaris, the filum terminale internum, and the filum terminale externum. (From Benninghoff and Goerttler, *Lehrbuch der Anatomie des Menschen,* Vol. 3, 9th ed., Urban and Schwarzenberg, Munich, 1975.)

tubular sheath of the dura mater and surrounded by the nerves of the cauda equina, is called the **filum terminale internum.** The lower part, the **filum terminale externum,** is closely invested by the dura mater to which it adheres. Extending beyond the apex of the sheath, it finally attaches to the dorsal part of the first coccygeal segment (Fig. 11-27). The filum consists mainly of fibrous tissue continuous with the pia mater. A few nerve fibers adherent to the outer surface probably represent rudimentary second and third coccygeal nerves. The central canal of the spinal cord continues down into the filum for 5 or 6 cm.

Relationships and Form of the Spinal Cord

VERTEBRAL COLUMN AND CAUDA EQUINA. It was just stated that the neural substance of the adult spinal cord does not occupy the entire length of the vertebral canal, as was the case during the third month of prenatal development, because the vertebral column grows at a faster pace than the spinal cord. Because the cord is continuous with the developing brain, and thereby firmly anchored rostrally, this differential growth pattern causes the spinal cord to be drawn upward in the vertebral canal. There results a great intravertebral lengthening of the lumbar and sacral nerve roots because they still emerge from the vertebral column through their segmentally appropriate intervertebral foramina below (Fig. 11-28). These caudally placed nerve roots are directed in a progressively more oblique manner, so that they pass almost vertically downward within the dura mater for some distance before reaching their foramina of exit. The resulting collection of rootlets beyond the termination of the neural substance of the cord is called the **cauda equina** (Figs. 11-26 to 11-28).

EPIDURAL SPACE. The dimensions of the vertebral canal within the spinal column are considerably greater than the diameter of the enclosed spinal cord and its meningeal coverings. This permits the required movements of the cord during flexion and extension of the vertebral column without a resultant spinal contusion or compression. Between the bony inner wall of the vertebral canal and the outer surface of the dura mater is found the **epidural space,** and this is loosely packed with strands of **epidural fat.** Closely adhering to the wall of the vertebral canal are the internal vertebral venous plexuses. Additionally, fine dorsal and ventral epidural branches of the spinal arteries ramify within the epidural fat to supply the vertebrae and the external surface of the spinal dura mater.

ENLARGEMENTS OF THE CORD. The spinal cord is slightly flattened dorsoventrally and its diameter is increased by enlargements in two regions: cervical and lumbar (Fig. 11-30). The **cervical enlargement** extends from about the fifth cervical to the second thoracic vertebra, its maximum circumference (about 38 mm) being at the level of the sixth pair of cervical nerves. It corresponds to the origin of the large nerves that supply the upper limbs. The **lumbar enlargement** be-

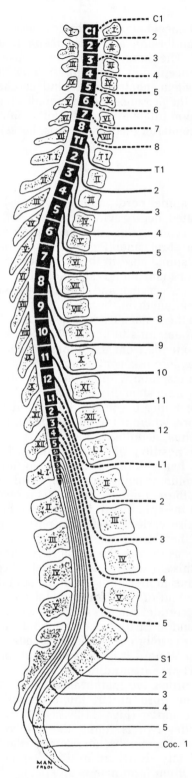

FIG. 11-28. Diagram showing the relationship of the spinal cord segments and the level of emergence of the spinal nerves to the bodies and spinous processes of the vertebrae. (From W. Haymaker and B. Woodhall, *Peripheral Nerve Injuries,* 2nd ed., W. B. Saunders, Philadelphia, 1953.)

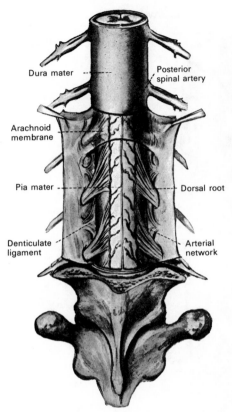

FIG. 11-29. The dura mater has been opened to reveal the dorsal aspect of the spinal cord. Note that the thin arachnoid membrane has been left intact in the upper part of the exposure. (From P. Thorek, *Anatomy in Surgery,* 2nd ed., J. B. Lippincott, Philadelphia, 1962.)

gins at the level of the ninth thoracic vertebra and reaches its maximum circumference (about 33 mm) opposite the last thoracic vertebra, after which it tapers rapidly into the conus medullaris. These vertebral levels correspond to the sites of attachment of nerve roots that extend from the first lumbar to the third sacral segment. From these segments arise the nerves that supply the lower limb.

FISSURES AND SULCI. After removal of the meningeal coverings, especially the pia mater, the surface of the spinal cord is marked by several longitudinal furrows. Two of these, the **anterior median fissure** and the **posterior median sulcus,** nearly divide the spinal cord longitudinally into symmetrical right and left halves. The two sides of the cord remain interconnected, however, by commissural bands of gray and white matter. Within the gray commissural bands is located the central canal. The dorsal surface of the cord also shows a **posterolateral sulcus,** as well as a **posterior intermediate sulcus** (Figs. 11-30 to 11-32).

Anterior Median Fissure. The anterior median fissure traverses the entire extent of the anterior surface of the spinal cord along its midline (Figs. 11-30 to 11-32). It penetrates an average depth of about 3 mm, being somewhat deeper more caudally. It contains adjacent folds of pia mater and the floor of the fissure is formed by a band of transversely crossing white fibers, the **ventral**

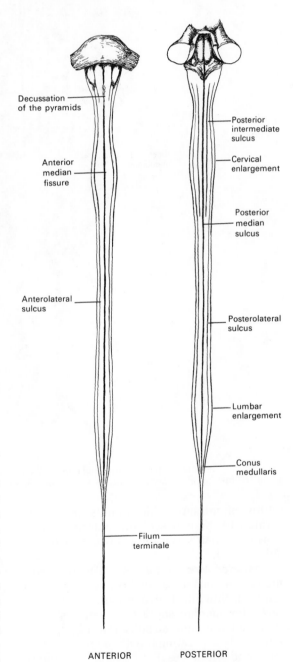

FIG. 11-30. Anterior and posterior diagrams of the spinal cord.

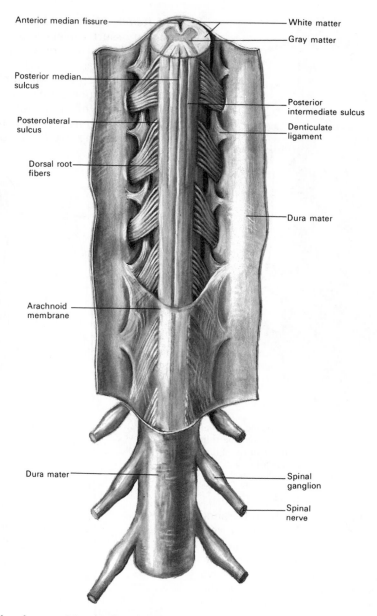

Fig. 11-31. The dorsal aspect of the spinal cord. In the lower region the dura mater is intact, while in the upper region all of the meningeal coverings have been removed. (After Sobotta.)

white commissure. This commissure is perforated by branches of the anterior spinal vessels, which serve the central part of the spinal cord.

Posterior Median Sulcus. The posterior median sulcus is a shallow groove from which a thin sheet of neuroglial tissue, the **posterior median septum,** penetrates more than half way into the substance of the cord, effectively separating the dorsal portion into right and left halves (Fig. 11-32). This septum varies in depth from 4 to 6 mm, but diminishes considerably in the lower part of the spinal cord.

Posterolateral Sulcus. The posterolateral sulcus is a longitudinal furrow located on each side of the posterior median sulcus along the line of attachment of the dorsal roots of the spinal nerves (Figs. 11-30; 11-32). Between the posterolateral sulcus and the posterior median sulcus on each side is found the **posterior funiculus.** Together the posterior funiculi are frequently called the posterior white columns of the spinal cord.

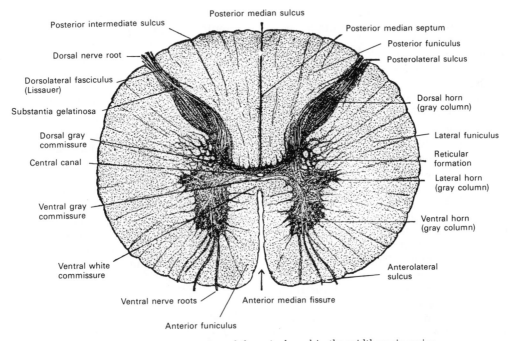

FIG. 11-32. Transverse section of the spinal cord in the midthoracic region.

Posterior Intermediate Sulcus. The posterior intermediate sulcus is a groove found in the upper thoracic and cervical regions of the spinal cord and interposed between the posterior median sulcus and the posterolateral sulcus along the posterior funiculus (Figs. 11-30; 11-32). It marks the line of separation between two ascending large tracts of white nerve fibers, the more medially located **fasciculus gracilis,** and the **fasciculus cuneatus** found more laterally.

Another less prominent groove, the *anterolateral sulcus,* is sometimes described as being in relationship to the emerging ventral roots (Figs. 11-30; 11-32). The relationship is frequently indistinct, however, because the ventral rootlets arise in bundles and not in a linear series and they emerge over an area of some slight width. The white matter of the spinal cord between the posterolateral sulcus and the anterior median fissure is divided into a **lateral funiculus** and an **anterior funiculus** by the ventral roots (Fig. 11-32). In the upper part of the cervical spinal cord, in addition to the dorsal and ventral roots, a series of nerve roots passes outward through the lateral funiculus. These unite to form the **spinal portion of the accessory nerve** (XI), which ascends through the foramen magnum to enter the cranial cavity. There it contributes the motor fibers of the eleventh cranial nerve that supply the sternocleido-mastoid and trapezius muscles.

SEGMENTS OF THE SPINAL CORD. Thirty-one pairs of spinal nerves originate from the spinal cord, each with a **ventral** (or anterior) **root** and a **dorsal** (or posterior) **root.** The pairs are grouped as follows: 8 *cervical,* 12 *thoracic,* 5 *lumbar,* 5 *sacral,* and 1 *coccygeal* (Fig. 11-28). The cord itself also is divided for convenience of description into cervical, thoracic, lumbar, and sacral regions to correspond with the nerve roots. It is customary also to speak of the part of the cord giving rise to each pair of nerves as a **spinal segment** or **neuromere,** although the distinguishing marks, except for the roots, are not retained after embryonic life. Arbitrarily defined, each spinal segment is considered to extend from the uppermost filament of one root to the uppermost filament of the next root. The segments vary in their extent along the cord; in the cervical region they average 13 mm and in the midthoracic region, 26 mm, whereas their length diminishes rapidly in the lumbar and sacral regions from 15 to 4 mm.

SPINAL NERVE ROOTS. Each spinal nerve has two roots: a **dorsal** (or posterior) **root,** also called the sensory root because its fibers transmit impulses to the cord, and a **ventral** (or anterior) **root,** also called the motor root because its fibers carry impulses outward from the cord to muscles and other structures.

Dorsal (Posterior) Roots. The dorsal roots are attached in linear series along the posterolateral sulcus of the cord, and each segmental root consists of six or eight **rootlets** (fila radicularia) (Fig. 11-31). They contain the central processes of spinal ganglion cells whose peripheral processes extend outward to the sensory endings in the skin, muscles, tendons, and viscera. The rootlets are divided into two groups as they enter the cord: a *medial bundle* containing mainly large myelinated fibers and a *lateral bundle* containing finely myelinated and unmyelinated fibers. Arising from the spinal ganglion cells, the roots are accompanied by delicate radicular branches of the spinal arteries and veins. Within the vertebral canal, the dorsal roots penetrate the dura mater and the underlying arachnoid membrane and then course for varying distances, depending on their segmental level, within the subarachnoid space before entering the spinal cord at the root-cord junction.

Ventral (Anterior) Roots. The ventral roots emerge from the spinal cord in two or three irregular rows of rootlets spread over a strip about 2 or 3 mm wide. Like the dorsal roots, the ventral roots are usually accompanied by branches of the spinal vessels. They course within the subarachnoid space prior to becoming surrounded by the tubular prolongations of the dura mater and arachnoid, which encase the spinal nerves and the dorsal root ganglia at the intervertebral foramina. The ventral roots join the dorsal roots to form the mixed spinal nerve at each segmental level. The ventral roots contain the axons of the cells in the anterior and lateral cell columns of the central gray matter of the cord.

Arterial Blood Supply of the Spinal Cord

The basic pattern of the arterial blood supply to the spinal cord consists of three longitudinally coursing vessels that arise just rostral to the cervical region of the cord and descend on its surface as far as the conus medullaris, and numerous feeder arteries and radicular vessels that gain entrance to the vertebral canal segmentally by coursing through the intervertebral foramina with the spinal nerves (Figs. 11-33 to 11-36). Anastomoses between the longitudinally oriented and the segmentally derived vessels occur on the surface of the spinal cord and result in the formation of a rich vascular plexus from which medullary vessels penetrate the substance of the cord to reach both the white and gray matter. Vessels that penetrate the spinal cord substance are thought, for the most part, to be end arteries and, therefore, they do not anastomose further. Although there is significant variation with respect to the segmental levels at which the feeder vessels enter the vertebral canal, there is a general consistency in the overall vascular pattern, namely, that longitudinally oriented vessels descending from above are supplemented by branches from the medially coursing spinal arteries at varying segmental levels.

LONGITUDINAL ARTERIES SUPPLYING THE SPINAL CORD

There are usually three longitudinally oriented spinal arteries. These are a single **anterior spinal artery,** which descends in the anterior median fissure, and two **posterior spinal arteries,** which descend along or near the posterolateral sulci of the spinal cord (Figs. 11-33 to 11-36).

ANTERIOR SPINAL ARTERY. The anterior spinal artery is sometimes called the *anterior median longitudinal artery of the spinal cord.* It usually forms rostrally from the junction of two anterior spinal branches, one derived from each of the two vertebral arteries (Figs. 11-33; 11-34). The union of these two vessels generally occurs intracranially at or near the midline on the anterior aspect of the medulla oblongata at about the level of the olivary nuclei. From this site, the anterior spinal artery descends without interruption to the tip of the conus medullaris, lying just ventral to the anterior median fissure. Slightly dorsal and to one side of the artery courses the anterior median longitudinal venous trunk of the spinal cord (Suh and Alexander, 1939). Frequently the anterior spinal artery is duplicated in the cervical cord region because of an incomplete fusion of its two fetal precursors (Aminoff, 1976). Occasionally, one of the paired vertebral branches is much smaller or even absent, and at other times two or more very slender spinal vessels may branch from one of the vertebral arteries to unite with a large

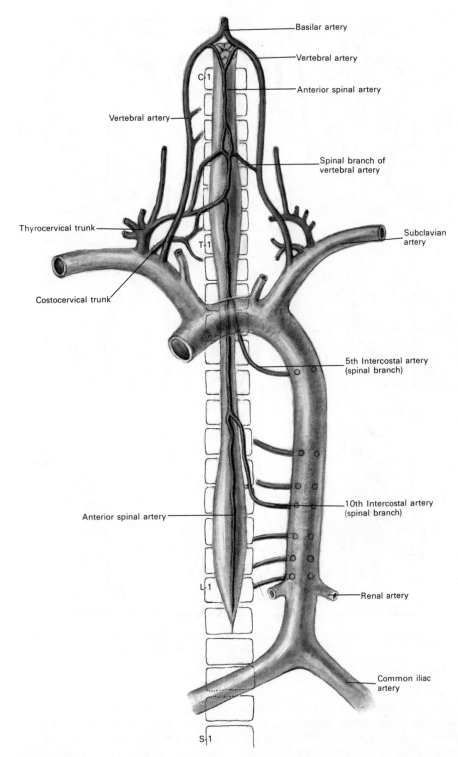

FIG. 11-33. Diagrammatic representation of the anterior spinal artery. Note that the anterior feeder vessels are derived from the intercostal and vertebral arteries, as well as from spinal branches from vessels coming off the thyrocervical and costocervical trunks. Rostrally, the anterior spinal artery is formed from anterior spinal branches derived from each vertebral artery. (After Sobotta.)

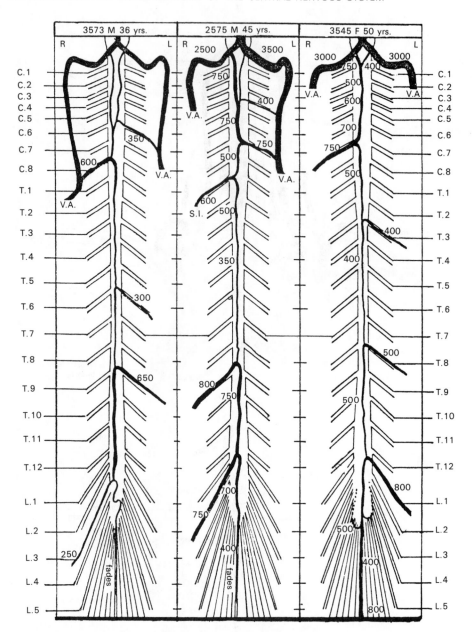

FIG. 11-34. The vascular pattern of the anterior spinal artery and the anterior medullary feeder arteries in three adult human specimens. The numbers within the figure indicate the approximate diameters of the different vessels in micrometers. V.A., vertebral artery; S.I., superior intercostal artery. (From G. F. Dommisse, *The Arteries and Veins of the Human Spinal Cord from Birth,* Churchill-Livingstone, Edinburgh, 1975.)

branch from the other side (Tveten, 1976). Along its course the anterior spinal artery varies considerably in its diameter, but its caliber is consistently larger in the cervical and lower thoracic regions. Along the mid-portion of the thoracic spinal cord (i.e., between T3 and T9), the vessel is quite slender (0.35 to 0.50 mm), and this region of the cord has been considered a *vulnerable zone* with

respect to its circulation (Dommisse, 1975). At the conus medullaris and along the filum terminale, the anterior spinal artery is reduced in size to a tiny vessel that frequently communicates by anastomotic branches with the posterior longitudinal arteries.

The anterior spinal artery is reinforced at a number of segmental levels by **feeder arterial branches** from the segmental arteries

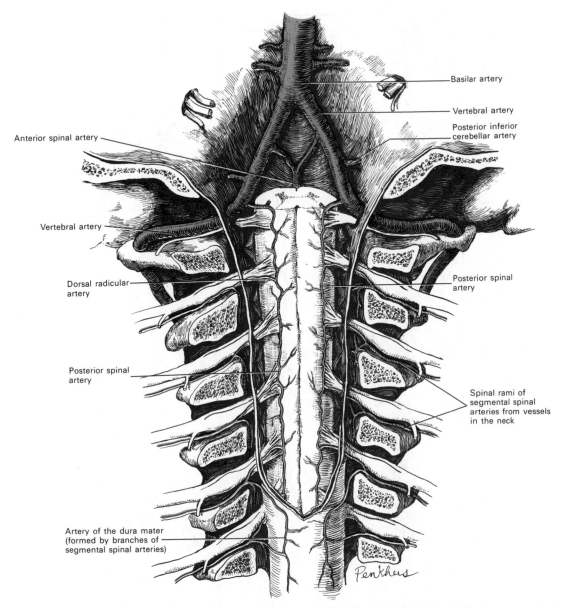

Basilar artery

Vertebral artery

Posterior inferior cerebellar artery

Anterior spinal artery

Vertebral artery

Dorsal radicular artery

Posterior spinal artery

Posterior spinal artery

Spinal rami of segmental spinal arteries from vessels in the neck

Artery of the dura mater (formed by branches of segmental spinal arteries)

FIG. 11-35.   The arterial blood supply of the spinal cord is derived from anterior and posterior spinal arteries, which usually branch from the vertebral artery, as well as from radicular and medullary feeder branches coming from segmental vessels. (Redrawn after Netter.)

that enter the vertebral canal, accompanying the spinal nerves and their roots (Fig. 11-34). The largest of these feeder vessels invariably increases the diameter of the anterior spinal artery beyond its point of junction.

POSTERIOR SPINAL ARTERIES.   The two posterior spinal arteries are sometimes called the *paired posterolateral longitudinal arterial trunks of the spinal cord.* On each side the vessel usually originates directly from the intracranial part of the vertebral

artery, somewhat lower than the origins of the branches that unite to form the anterior spinal artery (Figs. 11-35; 11-36). At times the posterior spinal arteries do not arise directly from the vertebral arteries, but from their posterior inferior cerebellar branches instead. Each vessel descends on the posterior surface of the spinal cord along the posterolateral sulcus in close relationship to the line of entrance of the dorsal roots. They are smaller in diameter and more irregular than

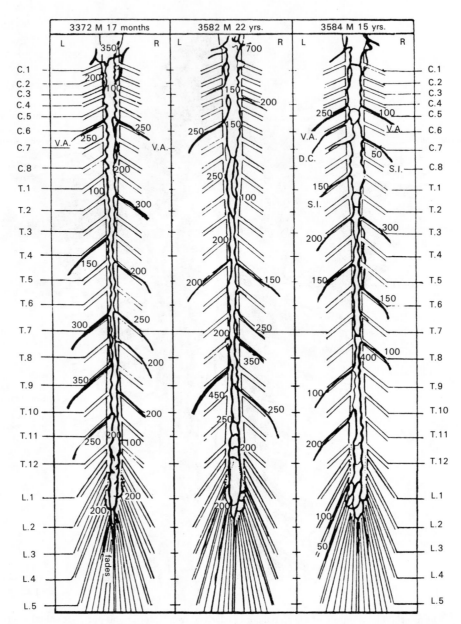

Fig. 11-36. The vascular pattern of the posterior spinal arteries and the posterior medullary feeder arteries in three human specimens (15 months, 22 years, 15 years). The numbers within the figures indicate the approximate diameters of the different vessels in micrometers. V.A., vertebral artery; D.C., deep cervical artery; S.I., superior intercostal artery. (From G. F. Dommisse, *The Arteries and Veins of the Human Spinal Cord from Birth,* Churchill-Livingstone, Edinburgh, 1975.)

the anterior spinal artery and communicate by many anastomosing channels with incoming segmental arteries and across the midline with each other. To some extent the pattern thus formed resembles an embryonic plexus, but usually the longitudinal continuity of the vessels can be discerned throughout the length of the cord (see Fig.

11-37). Along the conus medullaris, the posterior spinal arteries communicate freely with branches from the anterior spinal artery to form what is frequently called the **cruciate anastomosis** of the conus medullaris. Throughout its course, each posterior spinal artery gives off branches that penetrate the cord to supply the white matter of

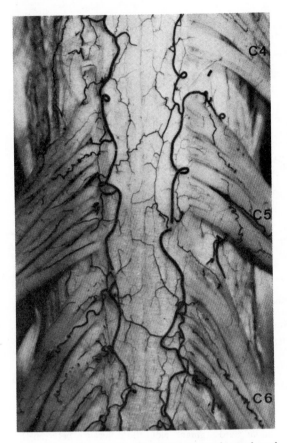

Fig. 11-37. A posterior view of the cervical spinal cord showing an arterial injection of the posterior spinal arteries, certain radicular arteries, several larger vessels along the roots which are posterior medullary feeder arteries, and the anastomotic network of the dorsal cord. (Maillot and Koritke, 1970.)

the posterior columns, the dorsal gray matter, and the superficial dorsal aspect of the lateral columns.

SEGMENTAL ARTERIES SUPPLYING THE SPINAL CORD

The spinal cord receives segmental arteries bilaterally at every intervertebral level and these vessels enter the spinal canal by way of the intervertebral foramina. In the upper three to five *cervical* segments, they are medially coursing branches of the vertebral artery, whereas in the lower cervical region, they may be derived from the ascending cervical branch of the inferior thyroid artery or from the deep cervical branch of the costocervical trunk (Fig. 11-35). In the *thoracic* region, the segmental vessels sup-

plying the spinal cord come from the posterior intercostal arteries, which stem directly from the posterior aspect of the thoracic aorta. In the *lumbar* region, they branch from the lumbar arteries, while the segmental arteries supplying the *sacral* region usually stem from the lateral sacral arteries or, less commonly, from the middle sacral artery (Gillilan, 1958).

RADICULAR AND MEDULLARY FEEDER BRANCHES OF THE SEGMENTAL ARTERIES. As they enter the intervertebral foramina, the segmental spinal arteries at *every level* give rise to the true **dorsal and ventral radicular branches,** which course with and supply the dorsal and ventral roots. Additionally, at *certain* of the vertebral levels, **medullary feeder branches** arise from the segmental spinal arteries to join the anterior and posterolateral longitudinally coursing spinal arteries (Figs. 11-34 to 11-38). These reinforce the circulation descending from above and are most important in maintaining an appropriate level of blood supply to the cord. The medullary feeder arteries vary both in number and in the levels at which they arise in different individuals.

Anterior Medullary Feeder Arteries. There is an *average* total of eight anterior medullary feeder arteries (both sides included in the total) that enter the vertebral canal segmentally to join the anterior (longitudinal) spinal artery (Fig. 11-34); however, the number varies from 2 to 17 in different individuals (Dommisse, 1975). Their diameter varies from 0.2 to 0.8 mm. One especially large (1.0 to 1.3 mm) and important vessel, the **great anterior medullary artery** (or artery of Adamkiewicz[1]), which reinforces the circulation to the lumbar enlargement, usually enters the vertebral canal at the lower thoracic or upper lumbar level. This vessel, however, has been observed to enter as high as the seventh thoracic and as low as the fourth lumbar level. It is a unilateral artery and in 77% of the specimens studied it entered the spinal cord from the left side (Dommisse, 1975).

Three anterior medullary feeder arteries (range one to six) join the ventral spinal artery in the *cervical* region. The largest of these is most frequently at the fifth or sixth

[1]Albert Adamkiewicz (1850–1921): A Polish experimental pathologist (Krakow).

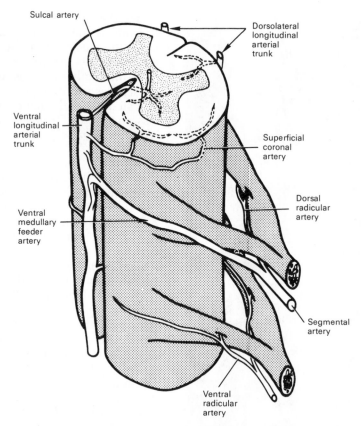

FIG. 11-38. Diagram of the arterial vascular pattern of a spinal cord segment. Note the difference between the medullary feeder branches and the radicular branches of the segmental spinal arteries. (From W. F. Windle, ed., *The Spinal Cord and Its Reaction to Injury*, Marcel Dekker, New York, 1980.)

cervical level. The *thoracic* region receives an average of three feeder arteries (range one to five), and the lumbar spinal cord, an average of two feeder vessels.

Posterior Medullary Feeder Arteries. The posterior medullary feeder arteries are smaller in diameter but more numerous and more evenly distributed than the anterior (Fig. 11-36). In the series of dissections reported by Dommisse (1975), the average total of posterior feeder arteries observed (both sides included) was 12, with a range from 6 to 25. Gillilan (1958) reported a range of 10 to 20 vessels, while Tveten (1976) found a range of 14 to 25. Thus, in the cervical region, an average number of three posterior medullary feeder arteries can be expected, with the highest incidence of vessels found at vertebral levels C6 to C8. An average of five posterior feeder vessels has been found between spinal segments T2 to T9, and four vessels between T10 and S5. The diameter of the posterior medullary feeder arteries ranges from 0.05 to 0.35 mm.

INTRAMEDULLARY ARTERIES OF THE SPINAL CORD

Most of the gray and white substance of the spinal cord receives arterial blood by way of branches from the anterior and posterior spinal arteries. A small amount of the parenchyma of the cord, however, is supplied by the five radicular branches of the segmental spinal arteries that course intimately with the dorsal and ventral roots. Some of these latter vessels penetrate the surface of the cord and accompany the roots through the root-cord junction (Figs. 11-38; 11-39).

INTRAMEDULLARY BRANCHES FROM THE ANTERIOR SPINAL ARTERY.
The anterior spinal artery gives rise to many short and straight branches that frequently stem at right angles to the main trunk and course posteriorly through the anterior median fissure to reach the anterior commissural region of the spinal cord. These are called **sulcal** (or **central**) **arteries,** and most arise singly (Fig. 11-38). Reaching the cord at the depth of the fissure,

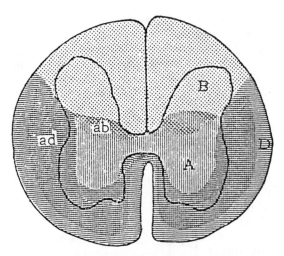

Fig. 11-39. Diagram of a cross section of the spinal cord showing the regions supplied by the various intramedullary arteries. *A*, Region of distribution of the sulcal (central) artery; *D*, region supplied by the penetrating arteries from the lateral and ventral pial plexus; *ad*, the region that may be supplied by either the sulcal artery or the lateral and ventral penetrating arteries; *B*, the region supplied by the penetrating arteries from the posterior plexus; *ab*, the region that may be supplied by either the sulcal or the posterior penetrating arteries. (L. A. Gillilan, 1958.)

each sulcal artery turns to the left or right but does not bifurcate to supply both sides. Successive sulcal arteries frequently turn alternately, but occasionally consecutive vessels turn to the same side. The number of sulcal arteries supplying each segment varies from five to nine, being more numerous in those segments that form the spinal cord enlargements. Counts ranging from 184 to 228 sulcal arteries (average 210) for all 31 segments of the cord have been reported (Tvetan, 1976).

In addition to the sulcal branches, the anterior spinal artery gives rise to a number of delicate **coronal arteries,** which course laterally around the outer surface of the cord (Fig. 11-38). They help to form a superficial pial plexus and they anastomose with each other and with superficial branches from the posterior spinal arteries. From the coronal arteries and the pial plexus, penetrating branches enter the white matter of the ventral and lateral columns.

The sulcal arteries and the penetrating branches from the coronal arteries together supply the anterior two-thirds of the spinal cord (Fig. 11-39). The anterior gray column, the base of the posterior gray column, and the deeper portions of the white matter in the anterior and lateral columns are supplied by sulcal vessels, while the more superficial regions of white substance are supplied by branches from the coronal vessels and the surface pial plexus.

INTRAMEDULLARY BRANCHES FROM THE POSTERIOR SPINAL ARTERIES. The two posterior spinal arteries, coursing in the posterolateral sulci, give off penetrating branches that supply the posterior one-third of the spinal cord (Fig. 11-39). At each level, some of these vessels accompany the fibers of the dorsal roots into the posterior gray substance of the dorsal horn. Entering the spinal cord, they branch at an acute angle and quickly subdivide into precapillary and capillary vessels (Gillilan, 1958). Other branches from the posterior spinal artery are directed medially on the surface of the cord, and from these vessels stem small penetrating arteries that supply the white matter of the posterior column on each side. Some of these vessels enter the cord directly on its posterior surface, while others enter the posterior median sulcus to penetrate the posterior columns medially.

Although there may be some anatomic continuity between a few capillaries derived from the anterior spinal artery and those derived from the posterior spinal artery, the intramedullary branches from these vessels do not form significant functional anastomoses. During life, their "functional distribution is as if they were end arteries" (Gillilan, 1958).

RADICULAR ARTERIES. Many authors present a confusing description of the radicular arteries because they do not differentiate between the anterior and posterior medullary feeder branches and the true dorsal and ventral radicular branches, all of which stem from the segmentally derived spinal vessels after the latter have entered the intervertebral foramina. The true radicular arteries, which nourish the nerve roots, are found at every level and on both sides of the cord (Fig. 11-38). At times some of these vessels achieve the root-cord junction and penetrate the spinal cord to assist in the supply of the most superficial parts of the gray matter in both the dorsal and ventral horns.

Venous Drainage of the Spinal Cord

The veins that drain the spinal cord parenchyma are described as two principal groups: the **intramedullary veins,** which form from the intrinsic capillaries of the spinal cord, and the intradurally located **surface veins of the spinal cord,** which receive the intramedullary veins and drain extradurally into veins in the neck, thorax, abdomen, pelvis, and spinal column, as well as the communicating venous sinuses of the cranial cavity. In general, the veins of the spinal cord follow a pattern consistent with that of the spinal arteries, but a number of differences exist. The veins of the spinal cord should not be confused with the veins of the vertebral column. These latter include the internal and external vertebral plexuses (of Batson[1]), which are extradural but communicate with the veins that drain the neural tissue of the spinal cord on the one hand and the systemic veins of the thorax, abdomen, and pelvis, as well as the dural sinuses of the skull on the other.

INTRAMEDULLARY VEINS OF THE SPINAL CORD

Venous blood from the spinal cord drains from within the neural tissue to the more superficial channels located on the surface of the cord by means of an anterior median group of **sulcal (or central) veins,** which opens into the longitudinally coursing anterior median vein, and by a **radial group of veins,** which arises from the periphery of the gray matter or from the white matter and courses radially toward the superficial coronal plexus of veins in the pia mater (Fig. 11-40) (Gillilan, 1970).

SULCAL (OR CENTRAL) VEINS. Each sulcal (or central) vein is formed by the confluence of several venules just deep to the anterior median sulcus (fissure) within the substance of the anterior white commissure (Figs. 11-40; 11-41). Their location corresponds to the sulcal arteries, but contrary to the unilaterally directed arteries, the sulcal veins receive blood from both halves of the cord and even anastomose above and below with venules that form adjacent sulcal veins. The sulcal veins are less numerous and smaller than the sulcal arteries, and they open into the anterior median spinal vein, which is

[1]Oscar V. Batson (1894–1979): An American anatomist at the University of Pennsylvania (Philadelphia).

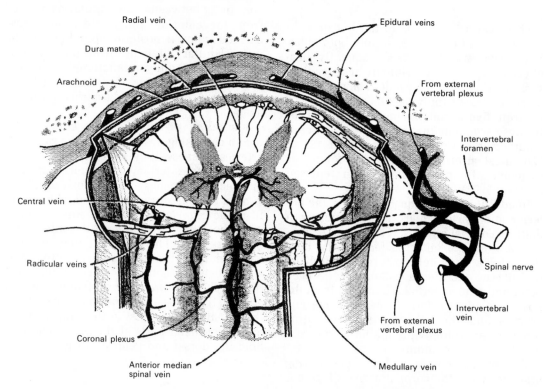

FIG. 11-40. This figure shows the sulcal (central) and radial veins within the spinal cord and their relationship to the pial and epidural venous plexuses. Convergence of the intramedullary, radicular, and epidural veins at the intervertebral foramen is shown on the right. (L. A. Gillilan, 1970).

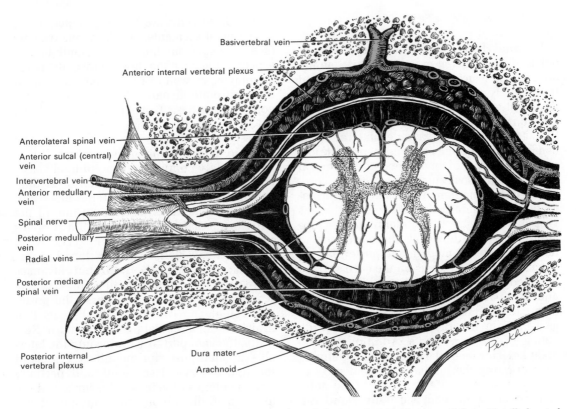

Basivertebral vein

Anterior internal vertebral plexus

Anterolateral spinal vein

Anterior sulcal (central) vein

Intervertebral vein

Anterior medullary vein

Spinal nerve

Posterior medullary vein

Radial veins

Posterior median spinal vein

Posterior internal vertebral plexus

Dura mater

Arachnoid

FIG. 11-41. Cross section of the spinal cord and vertebral canal showing the spinal veins and the epidurally located internal vertebral plexuses of veins. (After Netter.)

located in the anterior median fissure, slightly dorsal to and to one side of the anterior spinal artery. Within the fissure there are also frequent venous anastomoses. The

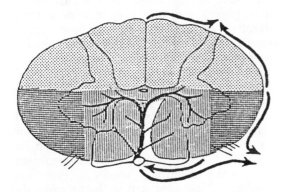

FIG. 11-42. Diagrammatic representation of a cross section of the spinal cord showing its venous drainage pattern. The posterior half of the cord (stippled) is drained by way of radial veins into the coronal plexus on the cord surface and then into the posterior medullary veins. The anterolateral quadrant on each side (cross hatching) is drained by radial veins into the anterior medullary veins, while the anteromedial quadrant drains into the sulcal (central) veins, which are tributaries of the anterior median vein. (L. A. Gillilan, 1970.)

sulcal veins drain the anterior gray and white commissures, the medial cell columns of the ventral horns, and the white matter of the anterior funiculi (Fig. 11-42) (Gillilan, 1970).

RADIAL GROUP OF INTRAMEDULLARY VEINS. Arising from capillaries near the peripheral aspect of the gray matter or from the white matter are many radially oriented intramedullary veins that are directed outward toward the surface of the spinal cord, where they join the coronal venous plexus of the pia mater (Figs. 11-40; 11-41). These radial veins are more numerous in the white matter of the posterior and lateral funiculi, but they are also found in the anterior funiculi. At certain cervical and thoracic levels, the radial veins are more prominent and drain laterally from the gray matter of the lateral horn, as well as posteriorly from the dorsal nucleus of Clarke.[1] Other radial veins course toward the dorsal surface at or near the midline within the posterior median sulcus.

[1]Jacob A. L. Clarke (1817–1880): An English anatomist (London).

SURFACE VEINS OF THE SPINAL CORD

The superficial venous pattern of the spinal cord is somewhat more complex than the arterial pattern, and of the longitudinal trunks, only the **anterior median spinal vein** is consistently observed as a continuous vessel, coursing the entire length of the cord (Fig. 11-43). Frequently, **anterolateral veins** can be seen coursing longitudinally just posterior to the line of emergence of the ventral roots. On the posterior surface of the cord, the **posterior median and posterior intermediate veins** generally do not form continuous vessels (Fig. 11-43). When identifiable as longitudinal trunks, however, these are located near the posterior median and posterolateral sulci. The venous plexus on the cord surface is called the coronal plexus, while **anterior and posterior medullary veins,** which are comparable to the medullary feeder arteries, as well as the true **radicular veins,** follow the course of the nerve roots. Additionally, the **great anterior medullary vein** can usually be seen draining the lumbar enlargement.

ANTERIOR MEDIAN SPINAL VEIN.   The anterior median spinal vein is usually found as a single longitudinally oriented trunk, located slightly dorsal to and to one side of the anterior spinal artery in the region of the anterior median fissure (Fig. 11-43). At times it may be duplicated, especially in the cervical and thoracic regions. Like its corresponding artery, the anterior median spinal vein does not have a uniform size, and it varies from 0.2 mm to as much as 0.9 mm in diameter throughout its course. Rostrally, this vein is continuous with the venous plexus located on the anterior aspect of the medulla oblongata, while caudally it tapers to form a delicate vessel along the conus medullaris. In addition to receiving the sulcal veins, the anterior spinal vein receives small pial vessels from the anterior and lateral aspects of the cord surface that help to form the coronal plexus. The anterior spinal vein is drained laterally on either side at certain segmental levels by a group of medullary veins that course toward the intervertebral foramina with the ventral roots in a manner similar to that of the anterior medullary feeder arteries.

ANTEROLATERAL SPINAL VEINS.   When present, the anterolateral spinal veins are longitudinally oriented vessels, located in line with but slightly posterior to the ventral roots. These veins are best developed along the cervical segments, and they receive surface vessels from the anterior and lateral aspect of the cord that help to form the coronal plexus, as well as radial veins that drain the ventrolateral parenchyma of the spinal cord (Lazorthes, Gouazé, and Djindjian, 1973).

POSTERIOR MEDIAN SPINAL VEIN.   The posterior median spinal vein is usually an incompletely continuous longitudinal vessel located within or near the posterior median sulcus (Fig. 11-43). Its diameter is more uniform and sometimes larger than that of the anterior median spinal vein, and it receives surface collateral vessels from the coronal venous plexus. It is best developed in the cervical region of the cord but becomes more delicate and thinner along the thoracic segments. Throughout its course the posterior spinal vein receives radial veins that drain the dorsomedial part of the cord. Rostrally, the vein partially drains into the laterally directed posterior medullary vessels of the upper cervical segments and partially into the plexus of veins on the surface of the lower medulla oblongata. Caudally it extends to the conus medullaris.

POSTERIOR INTERMEDIATE SPINAL VEINS.   Longitudinally oriented posterior intermediate spinal veins are sometimes observed overlying the posterolateral sulci in a line along the sites of entrance of the dorsal roots. These vessels frequently course for only several segments, join the surface venous plexus, and then may become longitudinally oriented once again. They usually receive radial veins that drain the dorsolateral part of the spinal cord parenchyma.

ANTERIOR AND POSTERIOR MEDULLARY VEINS.   The superficial surface veins of the spinal cord are drained laterally at certain segmental levels by anterior and posterior medullary veins that course with some of the dorsal and ventral roots but do not drain the roots (Fig. 11-40). These veins are comparable to the medullary feeder arteries and should not be confused with the radicular veins, even though some authors do not differentiate between the medullary veins, which drain the cord, and the radicular veins, which drain the roots. The anterior and posterior medullary veins are largest and most constant at the cervical and lumbar enlargements.

There are 8 to 14 **anterior medullary veins,**

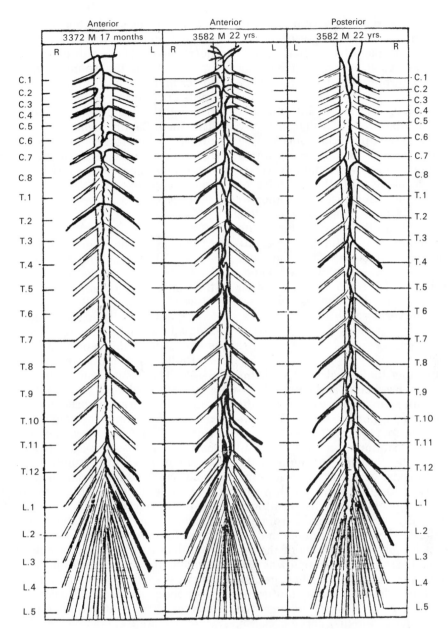

Anterior	Anterior	Posterior
3372 M 17 months	3582 M 22 yrs.	3582 M 22 yrs.

FIG. 11-43. The anterior (left and middle) and posterior (right) longitudinal surface veins of the human spinal cord. Note that anterior and posterior medullary veins, comparable to the medullary feeder arteries, are found at various segmental levels coursing as tributaries along certain roots from the longitudinally oriented veins. (From G. F. Dommisse, *The Arteries and Veins of the Human Spinal Cord from Birth*, Churchill-Livingstone, Edinburgh, 1975.)

which stem from either the anterior median spinal vein or from the coronal plexus of veins on the anterior and anterolateral cord surface. These vessels are visible without magnification and frequently arise at segmental levels that do not have medullary feeder arteries. The **posterior medullary veins** are generally more numerous than the anterior medullary veins, especially in the cervical region (Gillilan, 1970).

The **great anterior medullary vein** usually arises on the anterior cord surface in the lower thoracic or upper lumbar segments. This vessel descends in the cauda equina to its corresponding intervertebral foramen, along one of the anterior roots. A comparable **great posterior medullary vein** is sometimes observed arising on the posterior surface of the conus medullaris. It descends along one of the posterior roots of the cauda equina to its appropriate intervertebral foramen.

RADICULAR VEINS. Delicate anterior and posterior radicular veins, which actually drain the anterior and posterior roots, can be found on both sides of virtually every spinal segment. They course toward the intervertebral foramina, along with the fascicles of nerve fibers forming the roots. The radicular veins are usually microscopic in size; they generally arise from the gray matter of the dorsal and ventral horns and emerge at the spinal cord-root junction. The radicular veins sometimes terminate in the venous plexus on the cord surface, but more often they terminate in the venous plexuses within the intervertebral foramina.

MEDULLA OBLONGATA

Interconnecting the spinal cord with the cerebellum, diencephalon, and cerebral cortex is the **brain stem,** which consists of the medulla oblongata, pons, and mesencephalon (Figs. 11-44 to 11-46; 11-53; 11-54). The direct upward continuation of the spinal cord through the foramen magnum is the **medulla oblongata,** which is itself continuous rostrally with the pons. Within the human skull, the brain stem is positioned nearly vertically, and it extends from the upper border of the atlas to a plane that interconnects the jugular tubercles of the occipital bone. These limits are somewhat arbitrary, however, because there is a gradual transition in the internal anatomy of the spinal cord to that of the medulla oblongata at the upper border of the first cervical nerve. The rearrangement of gray matter and fiber tracts more rostrally within the medulla, or **bulb** as it is commonly known, is such that at the level of the decussation of the pyramids, or slightly higher at the level of the olive, a transverse section of the brain stem is totally different from one through the spinal cord.

External Anatomy of the Medulla Oblongata

The shape of the medulla oblongata is comparable to that of a truncated cone, with its smaller end directed inferiorly toward the spinal cord and its broader extremity directed superiorly toward the pons. Its approximate measurements are 3 cm longitudinally, 2 cm transversely, and 1.25 cm anteroposteriorly. The central canal continues upward from the spinal cord through its lower half, but in its upper half the medulla splits open at the dorsal median sulcus, expanding the canal into the **fourth ventricle.** The medulla oblongata may therefore be divided into a lower *closed part* containing the central canal, and an upper *open part* corresponding with the lower portion of the fourth ventricle.

FISSURES AND SULCI

Similar to the spinal cord, the external surface of the medulla oblongata is marked by several longitudinally oriented sulci and the anterior median fissure (Figs. 11-44; 11-54).

ANTERIOR MEDIAN FISSURE. The anterior median fissure continues rostrally from the spinal cord in the ventral midline along the entire length of the medulla oblongata (Fig. 11-45). The fissure reaches the lower border of the pons, where it terminates in a small pit called the **foramen caecum.** In the lower medulla just above the spinal cord, the anterior median fissure becomes indistinct because it is crossed by a series of obliquely oriented nerve fiber bundles that form the **decussation of the pyramids** (Figs. 11-45; 11-47). These fibers constitute the **lateral corticospinal tracts,** which descend from the medulla oblongata to the spinal cord and cross to the opposite side. Somewhat above this decussation, delicate **anterior external arcuate fibers** emerge from the anterior median fissure and curve laterally and upward over the surface of the medulla to join the inferior cerebellar peduncle (Fig. 11-47).

POSTERIOR MEDIAN SULCUS. The posterior median sulcus of the medulla oblongata is a narrow groove in the dorsal midline that represents the continuation rostrally of the posterior median sulcus of the spinal cord (Figs. 11-44; 11-54). It exists only in the more caudal, closed portion of the medulla, becoming more shallow as it ascends. Finally, the posterior median sulcus ends at about the middle of the medulla, where the central canal expands into the cavity of the fourth ventricle.

ANTERIOR LATERAL AND POSTERIOR LATERAL SULCI. In the medulla oblongata, these sulci are the rostral continuation of the anterolateral and posterolateral sulci of the spinal

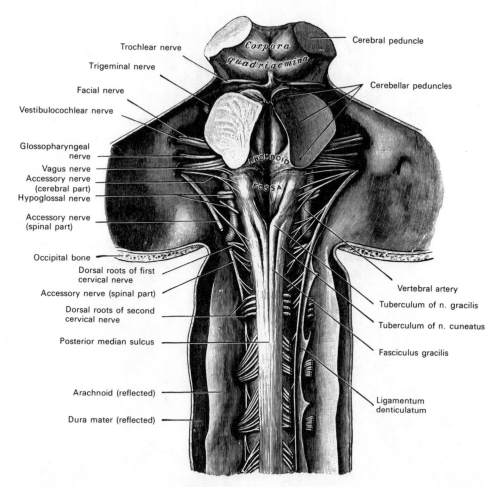

Trochlear nerve

Trigeminal nerve

Facial nerve

Vestibulocochlear nerve

Glossopharyngeal
nerve

Vagus nerve

Accessory nerve
(cerebral part)

Hypoglossal nerve

Accessory nerve
(spinal part)

Occipital bone

Dorsal roots of first
cervical nerve

Accessory nerve (spinal part)

Dorsal roots of second
cervical nerve

Posterior median sulcus

Arachnoid (reflected)

Dura mater (reflected)

Corpora
quadrigemina

RHOMBOID

FOSSA

Cerebral peduncle

Cerebellar peduncles

Vertebral artery

Tuberculum of n. gracilis

Tuberculum of n. cuneatus

Fasciculus gracilis

Ligamentum
denticulatum

FIG. 11-44. The dorsal aspect of the upper spinal cord and brain stem, showing the lower eight cranial nerves and the upper cervical nerves.

cord, where they mark the lines of emergence of the ventral rootlets and the entrance of the dorsal rootlets of the spinal nerves (Fig. 11-54). In the medulla, the rootlets of the hypoglossal nerve are located in linear series with the ventral roots of the spinal cord, and they emerge from the brain stem along the anterior lateral sulcus (Fig. 11-45). Similarly, the attachments of the accessory vagus and glossopharyngeal nerves to the medulla oblongata are positioned in the depths of the posterior lateral sulcus in a line corresponding with the entrance of the dorsal roots in the spinal cord.

Because the anterior median fissure and posterior median sulcus, which are located in the midline, and the anterior lateral and posterior lateral sulci on each side are longitudinally oriented, the external surface of the medulla oblongata is divided into right and left anterior, lateral, and posterior regions. These roughly correspond to the anterior, lateral, and posterior funiculi of the spinal cord. Although these three regions appear to be directly continuous with the corresponding funiculi of the spinal cord, they do not necessarily contain the same nerve fibers, because some of the tracts from the spinal cord end in the medulla while others alter their course as they pass through it.

ANTERIOR REGION OF THE MEDULLA

The anterior region of the medulla on each side is named the **pyramid;** it lies between the anterior median fissure and the anterior lateral sulcus (Fig. 11-45). Its upper end is slightly constricted and, at the pontomedullary junction between the pyramid and the pons, the fibers of the abducens nerve emerge from the brain stem. Slightly below the pons, the pyramid becomes enlarged and

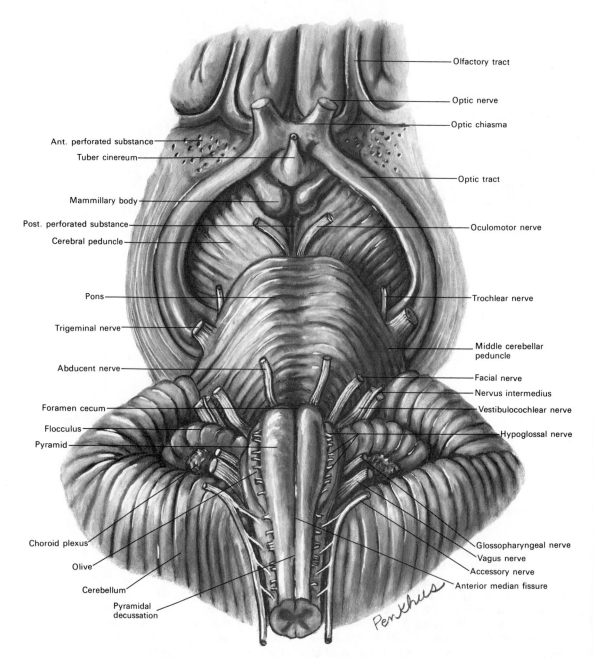

FIG. 11-45.   The anterior aspect of the brain stem.

prominent, whereas in the lower medulla it tapers into the anterior funiculus of the spinal cord, with which it falsely appears to be directly continuous (Fig. 11-45).

The two pyramids contain the descending motor fibers that course from the cerebral cortex to the spinal cord; these are the important **corticospinal** (or pyramidal) **fibers** (Figs. 11-47 to 11-50). When the corticospinal fibers are traced downward, it is found that

nearly 90% leave the pyramids in large, obliquely interdigitating bundles and cross to the other side in the anterior median fissure, where they form the **decussation of the pyramids.** This decussation almost obscures the anterior median fissure. The *crossed* pyramidal fibers descend in the posterior part of the lateral funiculus of the spinal cord as the **lateral corticospinal tract.** Some of the fibers in the lateral part of each pyramid that do not cross in the pyramidal decussation descend in the lateral funiculus of the spinal cord as *uncrossed* fibers in the lateral corticospinal tract, while the remaining descend in the anterior funiculus of the spinal cord as the uncrossed **anterior corticospinal tract.** The pyramidal fibers that supply the cervical enlargement of the spinal cord and, thus, are functionally related to the musculature of the upper extremity, cross in the more rostral part of the pyramidal decussation. Those that descend to lumbosacral levels of the spinal cord, thereby supplying spinal cord levels that innervate musculature in the lower limbs, cross more caudally in the decussation.

LATERAL REGION OF THE MEDULLA

The lateral region of the medulla oblongata is bounded in front, or *anteriorly,* by the anterior lateral sulcus, within which emerge the roots of the **hypoglossal nerve,** and behind, or *posteriorly,* by the posterior lateral sulcus, which contains the roots of the **accessory, vagus,** and **glossopharyngeal nerves** (Fig. 11-45). Its upper part consists of a prominent oval mass that is named the **olive,** while its lower part is the same width as the lateral funiculus of the spinal cord and appears on the surface to be a direct continuation of it (Fig. 11-45). Actually, only a portion of the lateral funiculus of the spinal cord is continued upward into the lateral region of the medulla, because the lateral corticospinal tract is formed from the pyramidal fibers of the opposite side and most of the fibers in the **posterior spinocerebellar tract** course into the inferior cerebellar peduncle in the posterior region of the medulla oblongata. The **anterior spinocerebellar tract** continues superiorly on the lateral surface of the medulla oblongata in the same relative position that it occupies in the spinal cord (Fig. 11-47), until it passes under cover of the external arcuate fibers just dorsal to the olive and ventral to the roots of the vagus and glossopharyngeal nerves. This tract then continues upward through the pons and winds around the dorsolateral edge of the lateral lemniscus; it enters the cerebellum by traveling along the dorsal surface of the superior cerebellar peduncle (Figs. 11-48; 11-49). The remainder of the lateral funiculus of the spinal cord consists of the **lateral fasciculus proprius,** which in the medulla dips deep to

the olive and disappears from the surface, except for a small strand that remains superficial to the olive. In a depression at the upper end of this strand is found the vestibulocochlear nerve.

THE OLIVE.   The olive is a prominent oval mass, measuring about 1.25 cm long, that is situated lateral to the pyramid, from which it is separated by the anterior lateral sulcus of the medulla oblongata and the rootlets of the hypoglossal nerve (Figs. 11-45; 11-48; 11-49). Posteriorly, its boundary is formed by the posterior lateral sulcus and the attachments of the glossopharyngeal, vagus, and accessory nerves. In a depression between the upper end of the olive and the pons are the roots of the vestibulocochlear and facial nerves. The anterior external arcuate fibers are conspicuous on the surface of the medulla as they wind across the lower part of the pyramid and olive to enter the inferior cerebellar peduncle. In cross section, the neural mass that forms the olive consists of a large, irregularly convoluted accumulation of gray matter, which includes the principal **inferior olivary nucleus** and both the **medial** and **dorsal accessory olivary nuclei** (Figs. 11-48; 11-49). On the surface of the olive are delicate bundles of nerve fibers, the anterior external arcuate fibers, which leave the anterior median fissure, wind across the lower part of the pyramid and olive, and enter the inferior cerebellar peduncle.

POSTERIOR REGION OF THE MEDULLA

The posterior region of the medulla lies behind the posterior lateral sulcus and the roots of the accessory, vagus, and glossopharyngeal nerves; like the lateral region, it is divisible into caudal and rostral parts.

The *caudal part* of the posterior region of the medulla is bounded behind by the posterior median sulcus and is the direct continuation of the posterior funiculus of the spinal cord; it consists of the **fasciculus gracilis** and the **fasciculus cuneatus** (Fig. 11-47). The fasciculus gracilis is located parallel and adjacent to the posterior median sulcus, and it is separated from the fasciculus cuneatus by the posterior intermediate sulcus and septum, which extend into the medulla from the upper cervical cord. The fasciculi gracilis and cuneatus are at first oriented vertically, but at the caudal extent of the **rhomboid fossa** they diverge from the midline in a V-

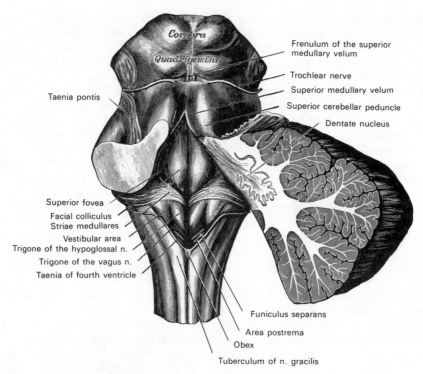

Frenulum of the superior medullary velum

Trochlear nerve

Superior medullary velum

Superior cerebellar peduncle

Dentate nucleus

Taenia pontis

Superior fovea
Facial colliculus
Striae medullares
Vestibular area
Trigone of the hypoglossal n.
Trigone of the vagus n.
Taenia of fourth ventricle

Funiculus separans

Area postrema

Obex

Tuberculum of n. gracilis

Fig. 11-46.   The dorsal aspect of the brain stem, showing the floor of the fourth ventricle or rhomboid fossa after the structures of the roof have been cut or pulled aside.

shaped manner, and each presents an elongated enlargement. The swelling on the fasciculus gracilis is called the **tuberculum of the nucleus gracilis;** it is produced by the gray matter of the subjacent **nucleus gracilis** (Figs. 11-46; 11-54). The enlargement on the fasciculus cuneatus is called the **tuberculum of the nucleus cuneatus,** which is likewise formed by a mass of underlying gray matter, the **nucleus cuneatus** (Fig. 11-54). The nerve fibers of these two ascending spinal fasciculi terminate on the cells of these two nuclei in a pattern that suggests a somatotopic projection. A third elevation, called the **tuberculum cinereum,** is produced by the rostral continuation of the **substantia gelatinosa,** and is present in the lower part of the posterior region of the medulla. It is narrow below and gradually expands above to end 1.25 cm below the pontomedullary junction, and it is formed by an accumulation of gray matter called the **nucleus of the spinal tract of the trigeminal nerve.** Lying on the lateral aspect of the fasciculus cuneatus, it is separated from the surface of the medulla oblongata by a band of nerve fibers that forms the spinal tract of the trigeminal nerve.

The *rostral part* of the posterior region of the medulla oblongata is occupied on each side by the **inferior cerebellar peduncle** (sometimes called the *restiform body*), a thick, curved bundle situated between the lower part of the fourth ventricle and the roots of the glossopharyngeal and vagus nerves (Figs. 11-49; 11-50; 11-53). As they ascend, the inferior cerebellar peduncles diverge from each other and help to form the lateral boundaries of the fourth ventricle. More rostrally they become directed posteromedially, each passing to its corresponding cerebellar hemisphere. Near their entrance into the cerebellum, the inferior cerebellar peduncles are crossed by several strands of fibers, called the **striae medullares of the fourth ventricle,** which course to the median sulcus of the rhomboid fossa (Figs. 11-46; 11-54).

A demonstration of the changes that occur in the morphology of the neuraxis through the transition from the cervical spinal cord to the medulla oblongata is most commonly presented in cross sections of the brain stem at various bulbar levels. Three such cross sections of the medulla are presented in Figures 11-47 to 11-49. These show the lower medulla at the level of the decussation of the pyramids (Fig. 11-47), the midbulbar region across the middle of the olive (Fig. 11-48), and the upper medulla near the pontomedullary junction at the level of the entrance of the vestibulocochlear nerve (Fig. 11-49).

THE PONS

The pons is continuous above with the midbrain, being bounded anteriorly and laterally from the cerebral peduncles of the midbrain by a pronounced sulcus, sometimes called the *superior pontine sulcus*. Inferiorly, the pons is continuous with the medulla oblongata at the pontomedullary junction, where anteriorly and laterally another transverse furrow, frequently called the *inferior pontine sulcus,* marks the brain stem and contains the attachments of the abducens, facial, and vestibulocochlear cranial nerves (Figs. 11-44; 11-45; 11-54).

DORSAL SURFACE OF THE PONS

The **dorsal** or **posterior surface** of the pons has a triangular shape and forms the upper part of the floor of the rhomboid fossa. The overlying **cerebellum,** which covers much of the posterior surface of the brain stem, is separated from the dorsal pons by the fourth ventricle, the thin pontine roof of which is formed by a delicate and thin sheet of white matter called the **superior medullary velum** (Figs. 11-46; 11-51 to 11-54). The narrow central canal of the midbrain, called the **cerebral aqueduct,** opens inferiorly into the cavity of the fourth ventricle at the upper border of the dorsal pons.

VENTRAL SURFACE OF THE PONS

The **ventral** or **anterior surface** of the pons lies against the basilar part of the occipital bone as far caudally as its jugular tubercles; rostrally, the pons extends along the clivus, beyond the spheno-occipital synchondrosis almost to the dorsum sellae on the body of the sphenoid bone. The ventral pons is cushioned from the bony skull by a sizable space called the **cisterna pontis,** which contains subarachnoid fluid (Fig. 11-52). Anteriorly, the pons forms a prominent bulge in the

brain stem that is markedly convex from side to side, but is less so from above downward (Fig. 11-45). It consists of transversely coursing pontocerebellar fibers that arch across the midline and gather on each side into a compact mass that forms the **middle cerebellar peduncle.** Superficially, these fibers give the ventral pons a transversely striated appearance.

Coursing longitudinally along the midline of the anterior pons is a shallow groove called the **basilar sulcus** (Fig. 11-45). The **basilar artery,** which lies within the cisterna pontis, frequently occupies this sulcus, but at times this vessel may curve somewhat and course to one or the other side of the groove. Several **pontine branches** (on each side) arise from the basilar artery as it ascends ventral to the pons. These course laterally, adhering closely to the surface of the ventral pons and then penetrating the surface to supply the pons with arterial blood. At times, the **labyrinthine artery** also arises from the basilar artery, ventral to the lower pons, although frequently these vessels branch from the anterior inferior cerebellar artery. The labyrinthine arteries also pass laterally, and near the pontomedullary junction, they join the facial and vestibulocochlear nerves to enter the internal acoustic meatus, eventually to supply the inner ear. The basilar sulcus is bounded on either side by an eminence caused by the descent of the corticospinal fibers through the substance of the pons.

The only cranial nerve that is attached to the pons is the fifth or **trigeminal nerve** (Fig. 11-51), although the **abducens, facial,** and **vestibulocochlear nerves** attach at the pontomedullary junction below, and the **trochlear nerve** emerges from the brain stem opposite the inferior colliculus just above the pons in the lower midbrain. The trigeminal nerve attaches on the upper lateral aspect of the ventral pons on each side just beyond the eminences formed by the corticospinal tracts. Each trigeminal nerve consists of a smaller, more medially and superiorly located **motor root,** and a much larger, more laterally and inferiorly placed **sensory root.** Together the two roots of the trigeminal nerve form the largest of all the cranial nerves. Vertical lines drawn immediately beyond the trigeminal nerves may be taken as the boundaries between the ventral surface of the pons and the middle cerebellar peduncle, and frequently the point of attach-

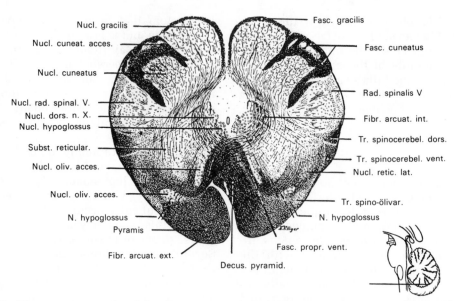

Fig. 11-47. Transverse section of medulla oblongata at the decussation of the pyramids. Nuclear groups represented diagrammatically in color. (From Villiger-Addison, courtesy of J. B. Lippincott Co.)

ment of the trigeminal nerve to the pons is described as the site at which the middle cerebellar peduncle arises from the pons.

STRUCTURE OF THE PONS

Transverse sections through the pons show that it is composed of two parts that differ in appearance and structure (Figs. 11-50; 11-51). These are the **basilar** or **ventral part of the pons** and the dorsally located **pontine tegmentum.** The *basilar part* consists largely of fibers arranged in transverse and longitudinal bundles, together with small groups of nerve cells that constitute the **nuclei pontis** scattered between the transverse fibers. The descending bundles of longitudinally coursing corticospinal fibers

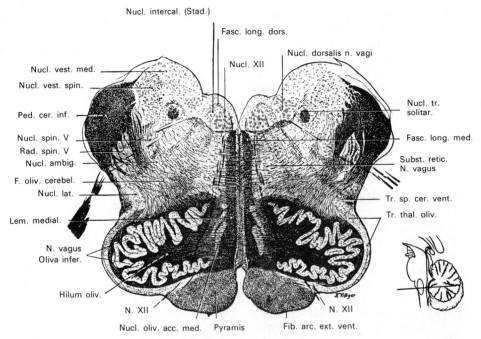

Fig. 11-48. Transverse section of medulla oblongata at the middle of the olive. Nuclear groups represented diagrammatically in color. (From Villiger-Addison, courtesy of J. B. Lippincott Co.)

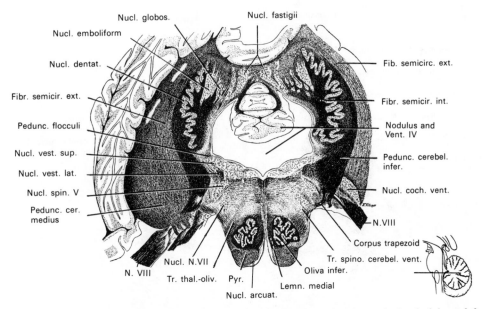

FIG. 11-49. Transverse section through medulla oblongata and cerebellar peduncles at the level of the eighth cranial nerve. Nuclear groups represented diagrammatically in color. (From Villiger-Addison, courtesy of J. B. Lippincott Co.)

are more interspersed by the transverse fascicles of pontine fibers in the upper pons, whereas they aggregate conspicuously in the lower pons and move gradually nearer the ventral surface to form the anteriorly located pyramids of the rostral medulla. Contrastingly, the *pontine tegmentum* is a direct continuation of the medulla oblongata, and its structure reflects this fact. It consists principally of the upward extension of the retic-

ular formation and gray substance of the medulla and of important nuclear groups related to the cranial nerves and cerebellum.

Fourth Ventricle

The fourth ventricle is the flattened, diamond-shaped cavity of the hindbrain and it contains the cerebrospinal fluid. It is situated ventral to the cerebellum and dorsal to

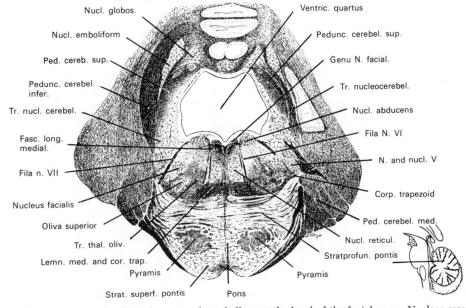

FIG. 11-50. Transverse section through pons and cerebellum at the level of the facial nerve. Nuclear groups represented diagrammatically in color. (From Villiger-Addison, courtesy of J. B. Lippincott Co.)

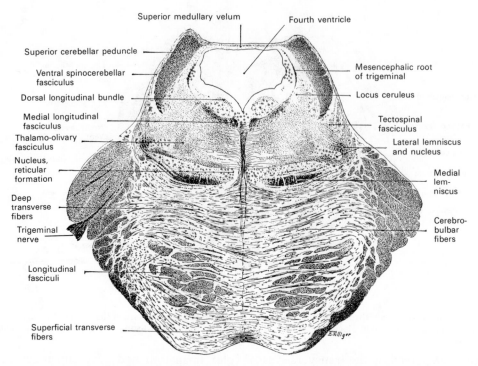

FIG. 11-51.   Transverse section through rostral part of pons. Nuclear groups represented diagrammatically in color. (From Villiger-Addison, courtesy of J. B. Lippincott Co.)

the pons and upper half of the medulla oblongata, and it is lined by a continuous sheet of epithelial cells called the **ependyma** (Fig. 11-52). *Superiorly,* the fourth ventricle extends the upper border of the pons, where it forms an angle of less than 90° (Fig. 11-54). Here its cavity is continuous with the cerebral aqueduct of the mesencephalon, through which communication is made with the cavity of the third ventricle. *Inferiorly,* the limit of the fourth ventricle is level with the lower end of the olive, where it forms an angle of nearly 90°. Its cavity opens below into the central canal of the medulla oblongata, which more inferiorly is continuous with that of the spinal cord (Fig. 11-54). Midway along the longitudinal axis of the fourth ventricle, on a level with the **striae medullares,** the ventricular cavity is prolonged outward in the form of two narrow **lateral recesses.** These are situated on each side between the inferior cerebellar peduncle and the flocculus of the cerebellum, and they extend laterally as far as the attachments of the glossopharyngeal and vagus nerves.

LATERAL BOUNDARIES OF THE FOURTH VENTRICLE.   The *lower part* of each lateral boundary of the fourth ventricle is formed by the tubercula of the cuneate and gracile nuclei, the fasciculus cuneatus, and the inferior cerebellar peduncle. The *upper part* of each lateral boundary consists of the middle and superior cerebellar peduncles (Fig. 11-54).

ROOF OF THE FOURTH VENTRICLE.   The *superior portion* of the roof of the fourth ventricle on each side is formed by the superior cerebellar peduncle and the superior medullary velum.

The *superior cerebellar peduncles,* as they emerge from the central white substance of the cerebellum, pass cranially and anteriorly and form, initially, the lateral boundaries of the upper portion of the fourth ventricle (Figs. 11-53; 11-54). As they approach the inferior colliculi, the two superior cerebellar peduncles converge and form part of the superior roof of the cavity of the fourth ventricle by means of their overlapping medial portions.

The **superior medullary velum** is a thin, transparent lamina of white substance that stretches between the two superior cerebellar peduncles (Figs. 11-53; 11-54). On the dorsal surface of its lower half are prolonged the folia of the lingula of the cerebellum, where the superior medullary velum is con-

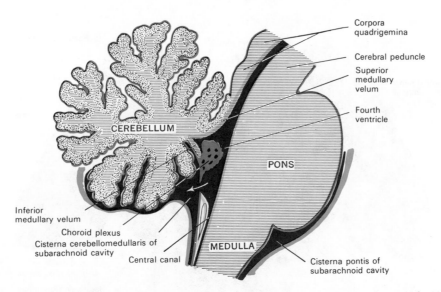

Corpora quadrigemina

Cerebral peduncle

Superior medullary velum

Fourth ventricle

CEREBELLUM

PONS

Inferior medullary velum

Choroid plexus

Cisterna cerebellomedullaris of subarachnoid cavity

Central canal

MEDULLA

Cisterna pontis of subarachnoid cavity

FIG. 11-52. Diagram showing a midsagittal view of the brain stem, cerebellum, and fourth ventricle. Note that the white arrow is located in the median aperture of the fourth ventricle (foramen of Magendie). Red, pia mater and vessels in the choroid plexus; yellow, ependymal lining; black, cerebrospinal fluid; blue, dura mater.

tinuous with the cerebellar white substance. A slightly elevated ridge, the **frenulum of the superior medullary velum,** descends upon the upper part of the superior medullary velum from between the inferior colliculi, and the trochlear nerve emerges adjacent to the sides of this frenulum.

The *inferior portion* of the roof of the fourth ventricle is formed primarily by the inferior medullary velum, tela choroidea of the fourth ventricle and choroid plexus, the taenia of the fourth ventricle, and the obex (Figs. 11-52; 11-53).

The **inferior medullary velum** is a thin sheet of white matter located on each side of the cerebellar nodule and consists of a layer of white matter lined on its inner (ventricular) surface by ependymal cells and on its external surface by the pia mater. It extends from the undersurface of the cerebellum, near the line of origin of the superior medullary velum, where its white matter is continuous with the peduncle of the flocculus, to its attachment on the medulla oblongata along the taenia of the fourth ventricle to the obex (Fig. 11-53). Each lateral angle of the diamond-shaped cavity of the fourth ventricle is prolonged into the lateral recess, which lies over the striae medullares and under the choroid plexus.

Openings in the Roof of the Fourth Ventricle. In the inferior part of the roof of the fourth ventricle are three openings or foram-

ina: the median aperture and two lateral apertures (Figs. 11-52; 11-53). Through these openings, spinal fluid produced by the choroid plexuses in the ventricular cavity can escape into the subarachnoid space.

The **median aperture** (or foramen of Magendie[1]) is a midline slit of variable size situated immediately above the obex at the inferior angle of the fourth ventricle (Fig. 11-53). It is approximately 2 cm long and usually a small extension of choroid plexus penetrates the aperture. This opening empties into the cerebellomedullary cistern. The **lateral apertures** (or foramina of Luschka[2]) are found at the extremities of the lateral recesses, just caudal to the point at which the middle cerebellar peduncle on each side courses into the cerebellum (Fig. 11-53). Located between the flocculus of the cerebellum and the vagus nerve on each side, the lateral apertures of the fourth ventricle are frequently occupied by small projections of choroid plexus that protrude into the subarachnoid space.

CHOROID PLEXUSES. The choroid plexuses of the fourth ventricle are formed by two highly vascular, elongated tufts of capillary size that are extensions of the tela choroidea and are covered by a secretory epithelium.

[1]François Magendie (1783–1855): A French experimental physiologist and pathologist (Paris).
[2]Hubert Luschka (1820–1875): A German anatomist (Tübingen).

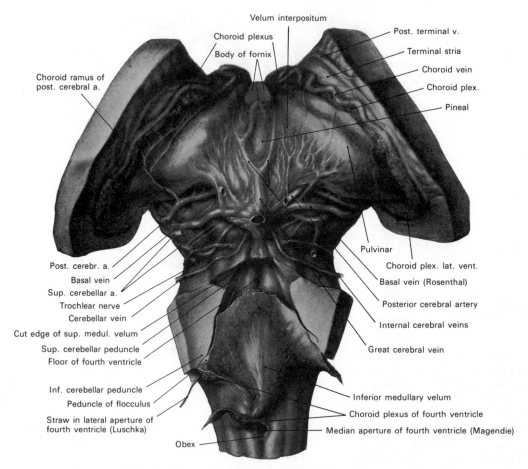

FIG. 11-53. The dorsal surface of the inferior medullary velum and the foramina of the fourth ventricle as seen after removal of the cerebellum. (From F. A. Mettler, *Neuroanatomy*, C. V. Mosby, St. Louis, 1942.)

These become invaginated into the cavity of the fourth ventricle, protruding inward from the roof, and each consists of a vertical and a horizontal portion. The two vertical portions lie close to the midline on each side, whereas the horizontal portions pass laterally at right angles to the vertical portions, coursing toward the lateral recesses and even projecting beyond their apices through the lateral apertures. Thus, the right and left horizontal halves are joined in the midline and the entire structure presents the form of the letter T, the vertical limb of which, however, is doubled.

FLOOR OF THE FOURTH VENTRICLE OR RHOMBOID FOSSA. The anterior wall of the fourth ventricle constitutes the floor of that cavity; it has a rhomboid shape and, therefore, is called the **rhomboid fossa** (Figs. 11-44; 11-46; 11-54). Formed by the dorsal surface of the pons and the more rostral open half of the

medulla oblongata, it is covered by a thin layer of gray substance that is directly continuous with that which surrounds the central canal of the medulla oblongata below and the cerebral aqueduct above. Superficial to this is a thin layer of neuroglia, the surface of which is lined by ependymal cells.

The rhomboid fossa consists of three parts: superior, intermediate, and inferior. The **superior part** has a triangular shape, and it is limited laterally by the superior cerebellar peduncles (Fig. 11-54). The apex of this triangular region is directed cranially and is continuous with the tissue forming the wall of the cerebral aqueduct. The base is represented by an imaginary line across the brain stem at the level of the upper end of two small depressions called the **superior foveae.** The **intermediate part** extends caudally from this level to another imaginary line at the level of the horizontal portions of the

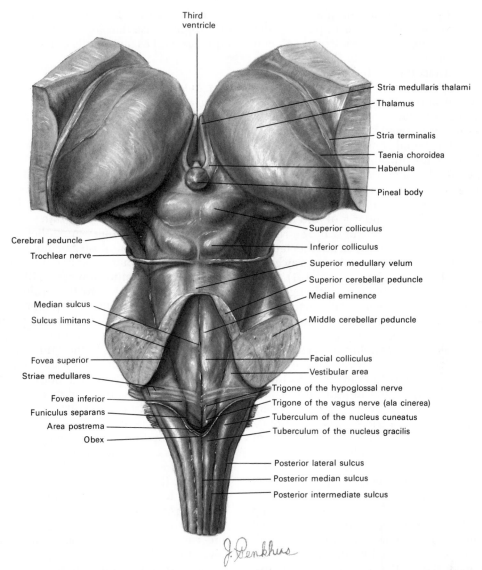

Third ventricle

Stria medullaris thalami

Thalamus

Stria terminalis

Taenia choroidea

Habenula

Pineal body

Superior colliculus

Cerebral peduncle

Trochlear nerve

Inferior colliculus

Superior medullary velum

Superior cerebellar peduncle

Medial eminence

Median sulcus

Sulcus limitans

Middle cerebellar peduncle

Fovea superior

Striae medullares

Facial colliculus

Vestibular area

Trigone of the hypoglossal nerve

Fovea inferior

Funiculus separans

Area postrema

Obex

Trigone of the vagus nerve (ala cinerea)

Tuberculum of the nucleus cuneatus

Tuberculum of the nucleus gracilis

Posterior lateral sulcus

Posterior median sulcus

Posterior intermediate sulcus

FIG. 11-54. The dorsal aspect of the brain stem following the removal of the cerebellum to expose the floor of the fourth ventricle. (Redrawn and slightly modified from F. A. Mettler, *Neuroanatomy*, C. V. Mosby, St. Louis, 1942.)

taenia of the fourth ventricle (Fig. 11-46). This part narrows somewhat above, where it is limited laterally on each side by the middle cerebellar peduncles, but it widens caudally where it is prolonged into the lateral recess of the fourth ventricle. The **inferior part** is triangular and its apex, which is directed caudally, is continuous with the wall of the central canal of the closed part of the medulla oblongata (Figs. 11-46; 11-54). The apex of this inferior part, which bears some resemblance to the point of a quill pen, is called the **calamus scriptorius.**

The symmetrical right and left halves of

the rhomboid fossa are divided in the midline by a **median sulcus** (Fig. 11-54). This longitudinal furrow reaches from the upper apex to the lower apex of the fossa, but it is deeper caudally than above. On each side of the median sulcus is an elevation, the **medial eminence,** which is itself bounded laterally by another sulcus called the **sulcus limitans** (Fig. 11-54). In the more rostral part of the rhomboid fossa, each medial eminence has a width equal to that of the entire corresponding half of the fossa, but opposite the superior fovea it forms an elongated swelling called the **facial colliculus.** This latter structure overlies the nucleus of the abducens nerve and is formed, at least in part, by the ascending portion of the root of the facial nerve. In the inferior part of the floor of the fourth ventricle, the medial eminence forms a triangular area called the **trigone of the hypoglossal nerve.** Careful examination reveals that this trigone consists of medial and lateral zones separated by a series of oblique furrows. The medial zone corresponds to the rostral part of the **nucleus of the hypoglossal nerve,** while the lateral zone corresponds to the **nucleus intercalatus** (Figs. 11-48; 11-54).

As just described, the **sulcus limitans** forms the lateral boundary of the medial eminence. In the superior part of the rhomboid fossa, it corresponds with the lateral limit of the fossa and has a bluish-gray zone named the **locus ceruleus** (Fig. 11-51). This region owes its color to an underlying patch of deeply pigmented nerve cells called the **nucleus pigmentosus pontis** or **substantia ferruginea;** many of these cells are now known to contain catecholamines, especially norepinephrine. At the level of the facial colliculus, the sulcus limitans widens into a flattened depression, the **superior fovea,** and in the inferior part of the fossa there appears in the floor of the ventricle a distinct dimple called the **inferior fovea** (Fig. 11-54). Lateral to these foveae is a rounded elevation, the **vestibular area,** which extends into the lateral recess as the **cochlear tubercle** and indicates the positions of the underlying vestibular and dorsal cochlear nuclei. Winding around the inferior cerebellar peduncle and crossing the vestibular area and medial eminence are strands of white nerve fibers, the **striae medullares** (Fig. 11-54). These course to the midline and penetrate the brain stem through the median sulcus.

Below the inferior fovea and between the trigone of the hypoglossal nerve and the lower part of the vestibular area is another triangular dark field called the **trigone of the vagus nerve** (previously called the *ala cinerea*). This indicates the position of the subjacent **dorsal** (efferent) **nucleus of the vagus nerve.** The lower end of the trigone of the vagus nerve is crossed by a translucent ridge, the **funiculus separans,** and between this and the tuberculum of the nucleus gracilis is a small tongue-shaped region called the **area postrema** (Fig. 11-54). The funiculus separans is formed by a strip of thickened ependyma, while the area postrema consists of a highly vascular zone of neuroglial cell types, as well as nerve cells of moderate size.

THE CEREBELLUM

The cerebellum is located dorsal to the pons and medulla oblongata and occupies the space between the brain stem and the occipital lobes of the cerebral cortex (Fig. 11-55). The **cerebellar hemispheres** rest on the floor of the posterior cranial fossa, and the median cerebellar region, called the **vermis,** is separated from the brain stem by the fourth ventricle. Intervening between the cerebellum and the posterior portion of the cerebral hemispheres is a dense sheet of dura mater, the **tentorium cerebelli.** The cerebellum is connected to the brain stem on each side by three peduncles: superior, middle, and inferior. The **superior cerebellar peduncle** (brachium conjunctivum) connects the cerebellum to the midbrain, while the **middle peduncle** (brachium pontis) and **inferior peduncle** (restiform body) attach the cerebellum to the pons and medulla oblongata, respectively (Figs. 11-62; 11-63).

Although the cerebellum probably does not initiate voluntary movement, it appears to serve as a suprasegmental coordinator of muscular activity, especially those motor functions that require sequential, repetitive, or complex movements. Additionally, it helps to regulate muscular tone and to maintain a proper balance for standing, walking, and running (Gilman, Bloedel, and Lectenberg, 1981). Once developed and fully matured, cerebellar connections appear to be capable of influencing the functions of virtually all parts of the central nervous system, and yet the cerebellum is not essential for life, since certain individuals "who are born without a cerebellum do not betray themselves in daily life by any defects" (Brodal, 1981). Which brain regions, through the processes of plasticity, assume the normal cerebellar functions under these circumstances is unknown. In contrast, individuals whose central nervous system has matured with a normally functioning cerebellum frequently suffer severe motor symptoms as a result of cerebellar disease.

The cerebellum is somewhat oval, is constricted in its median portion, and nearly

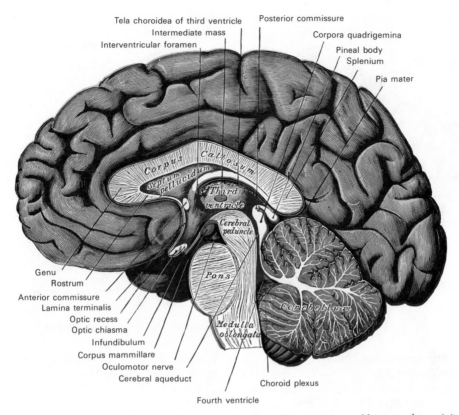

Tela choroidea of third ventricle
Intermediate mass
Interventricular foramen
Posterior commissure
Corpora quadrigemina
Pineal body
Splenium
Pia mater

Genu
Rostrum
Anterior commissure
Lamina terminalis
Optic recess
Optic chiasma
Infundibulum
Corpus mammillare
Oculomotor nerve
Cerebral aqueduct
Choroid plexus
Fourth ventricle

FIG. 11-55. Median sagittal section of brain. Observe that the cerebellum is interposed between the occipital lobes of the cerebral cortex and the brain stem.

flattened from above downward; its greatest diameter is from side to side. The surface of the cerebellum is not convoluted as is that of the cerebrum, but it is traversed by numerous curved furrows or fissures. The **cerebellar fissures** vary in depth in the different regions and result in the formation of the **cerebellar folia;** these present to the cerebellar surface a leaf-like, laminated appearance. The average weight of the cerebellum in the male is 150 gm. In the adult the proportion between the cerebellum and cerebrum is about 1 to 8, while in the infant it is about 1 to 20.

Gross Form and Topography of the Cerebellum

Although terms implying a functional or even phylogenetic subdivision of the cerebellum have been used, anatomically the cerebellum consists of a narrow median portion, called the **vermis,** which is located between two laterally and posteriorly protrud-

ing **cerebellar hemispheres** (Figs. 11-56 to 11-60). On the inferior surface at the rostral end of each cerebellar hemisphere is found a small, but somewhat separate, lateral projection called the **flocculus.** This attaches medially to the **nodulus** portion of the vermis by means of the **floccular peduncle** (Figs. 11-58 to 11-60,A).

FISSURES OF THE CEREBELLUM

The surface of the cerebellum is marked by a series of transversely directed fissures that are somewhat curved and extend into the cerebellar substance for a considerable distance. These divide the cerebellum into a series of lobes and lobules.

HORIZONTAL CEREBELLAR FISSURE. The horizontal fissure is the largest, deepest, and perhaps the most distinctive of the cerebellar fissures (Figs. 11-57 to 11-59). It commences in front of the pons near the middle cerebellar peduncles and passes horizontally and laterally on each side. It then

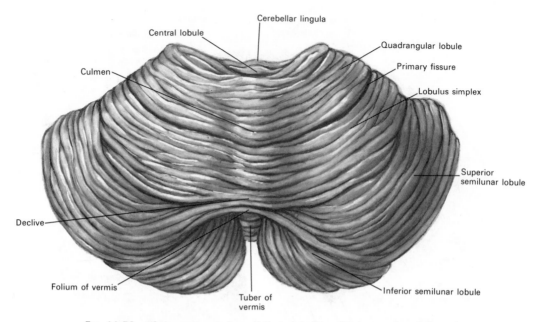

FIG. 11-56.   The superior surface of the cerebellum. (Redrawn after Sobotta.)

sweeps entirely around the free margin of the cerebellar hemispheres to the posterior cerebellar notch located in the midline behind. The horizontal fissure is used as the landmark that divides the cerebellum into its **superior** and **inferior surfaces** (Figs. 11-57; 11-60, A, B). Several smaller, but nonetheless deep, fissures divide the cerebellum into lobes or lobules, and these are further subdivided by more shallow furrows into individual folia. The most clearly visible fis-

sure on the superior surface is the primary fissure (Figs. 11-56; 11-60; 11-61).

PRIMARY FISSURE.   The primary fissure separates the anterior and posterior lobes of the cerebellum (Fig. 11-60). Those parts of the cerebellum rostral to this fissure constitute the anterior lobe and those parts behind the primary fissure, as far as the flocculonodular lobe, constitute the posterior lobe. When viewed from the superior surface, the primary fissure is V-shaped, with its apex

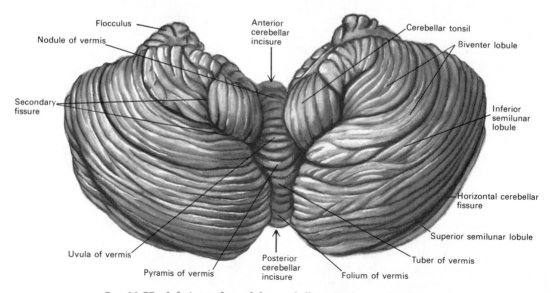

FIG. 11-57.   Inferior surface of the cerebellum. (Redrawn after Sobotta.)

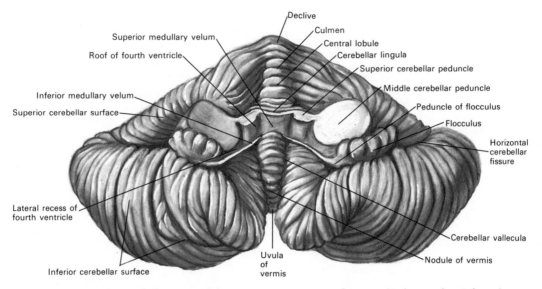

FIG. 11-58. The cerebellum viewed from its anterior or ventral aspect. (Redrawn after Sobotta.)

directed backward toward the median line and the limbs of the V directed anterolaterally as far as the horizontal fissure (Fig. 11-56).

POSTEROLATERAL FISSURE. The posterolateral fissure is located rostrally on the inferior surface of the cerebellum (Figs. 11-60; 11-61). It lies behind the flocculus and the floccular peduncle on each side and extends medially behind the nodulus, thereby separating the flocculonodular lobe from the remainder of the cerebellar hemispheres. It is the earliest of the cerebellar fissures to appear during embryonic development.

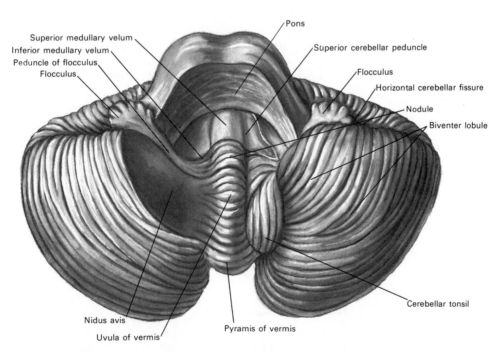

FIG. 11-59. The anteroinferior aspect of the cerebellum and pons. The right (reader's left) tonsil and a part of the biventer lobule have been removed and the pons has been sectioned transversely, exposing the flocculonodular lobe and the bed of the cerebellar tonsil. (Redrawn after Sobotta.)

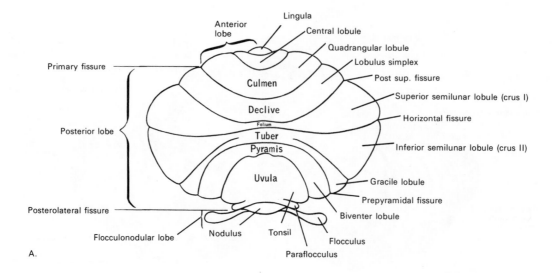

A.

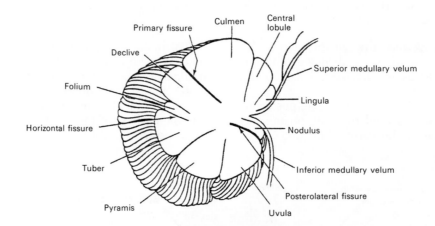

B.

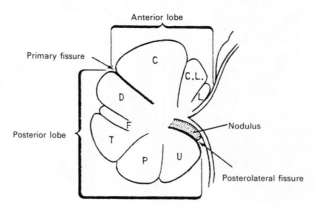

C.

FIG. 11-60.   Diagrammatic representations of the lobes of the cerebellum (A and C) and their constituent lobules (A) and divisions of the vermis (A, B, and C). Note also the principal fissures. (From E. L. House, B. Pansky, and A. Siegel, *A Systematic Approach to Neuroscience,* McGraw-Hill, New York, 1979.)

OTHER CEREBELLAR FISSURES.   Other frequently named cerebellar fissures visible from the superior surface include the **postlingual fissure,** located behind the lingula of the vermis; the **postcentral fissure,** lying between the alae of the central lobule and the quadrangular lobules; and the **posterior superior fissure** (also called the postclival or postlunate fissure), seen between the lobulus simplex and the superior semilunar lobule (Fig. 11-61). Visible on the inferior surface of the cerebellum are the **prepyramidal fissure,** located between the inferior semilunar lobule and the biventer lobule, as well as the **secondary fissure** (also called the retrotonsillar fissure or postpyramidal fissure), which is found between the uvula and pyramid medially and between the biventer lobule and the tonsil more laterally (Fig. 11-61).

VERMIS OF THE CEREBELLUM

The **vermis** of the cerebellum lies between the cerebellar hemispheres and is a continuous but unpaired narrow median region visible on both the superior and inferior cerebellar surfaces (Figs. 11-56 to 11-59; 11-60, B; 11-61). Its name is derived from the Latin word *vermis* (meaning a worm) because the surface shape of this median cerebellar region resembles a worm bent nearly completely back onto itself to form a slender circular structure. On the superior surface, the **superior vermis** forms a midline elevation that throughout its length nearly blends with the cerebellar hemispheres (Figs. 11-56; 11-61). In contrast, the **inferior vermis** is sunken and lies at the bottom of a longitudinally oriented hollow space called the **cerebellar vallecula,** located between the hemispheres (Figs. 11-57 to 11-59).

SUPERIOR VERMIS.   The superior vermis is subdivided by short, deep, transverse fissures that form, from front to back, parts of the vermis called the **lingula, central lobule, culmen, declive,** and **folium vermis** (Figs. 11-56; 11-60; 11-61). Of these, only the lingula is not joined laterally to a portion of the cerebellar hemispheres; instead, the white matter of its four or five folia is continuous with the superior medullary velum (Fig. 11-61). Consisting of a small tongue-shaped process, the *lingula* lies in front of and is concealed by the central lobule. It is separated from the central lobule by the postlingual fissure. The *central lobule* is continuous laterally with the alae of the central lobule, while the *culmen,* lying just in front of the primary fissure, is joined laterally on each side to the quadrangular lobule. The portion of the vermis directly behind the primary fissure is the *declive,* and this is laterally contiguous with the lobulus simplex on each side. The most caudal part of the vermis visible from the superior surface is the *folium vermis,* which is joined to the superior semilunar lobules laterally. The folium vermis is a short, narrow, and somewhat concealed band that seems to consist of a single folium, but in reality is marked on its upper and lower surfaces by secondary fissures. The folium vermis and the superior semilunar lobules on each side lie immediately above the horizontal fissure.

INFERIOR VERMIS.   The inferior vermis is similarly divided into smaller segments by transversely coursing fissures. Progressing rostrally from the most caudal part, these are called the **tuber vermis, pyramis** (vermis), **uvula** (vermis), and **nodulus** (Figs. 11-57 to 11-61). The *tuber vermis* is small and continuous laterally with the inferior semilunar lobules. These are located just below the horizontal fissure. Rostral to the tuber vermis is the *pyramis* (vermis), which is joined on each side to the biventer lobules. In front of the pyramis is the *uvula* (vermis), bounded laterally by the nearly spherical cerebellar tonsils. The uvula forms a considerable portion of the inferior vermis. It is separated on each side from the tonsil by the **sulcus valleculae,** at the bottom of which it is connected to the tonsil by a ridge of gray matter that is indented on its surface by shallow furrows and hence called the **furrowed band.** Finally, the most rostral part of the inferior vermis is called the *nodulus.* This is attached on each side to the flocculus by the floccular peduncles. Together the nodulus, the two flocculi and their peduncles form the flocculonodular lobe, which is separated from the cerebellar tonsils and the uvula (vermis) by the posterolateral fissure.

CEREBELLAR HEMISPHERES

In addition to the centrally located vermis, it is customary to consider the cerebellum as composed of two large laterally projecting masses called the **cerebellar hemispheres.** As described previously, these are continuous with the vermis on both the superior and

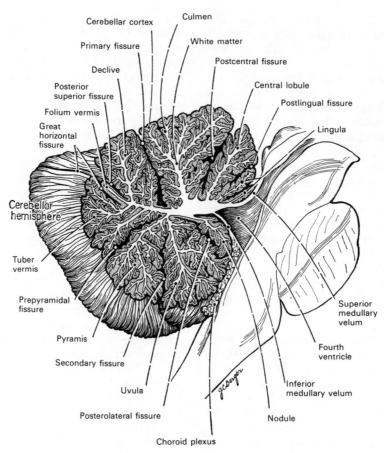

FIG. 11-61. Sagittal section through the vermis of the cerebellum. (From E. C. Crosby, T. Humphrey, and E. W. Lauer, *Correlative Anatomy of the Nervous System*, Macmillan Publishing Co., New York, 1962.)

inferior cerebellar surfaces (Figs. 11-56 to 11-59). Each hemisphere is composed of eight lobules, arranged in pairs so that the total shape of the two hemispheres is nearly symmetrical. Four pairs of lobules lie above the horizontal fissure and, therefore, are visible on the superior cerebellar surface, while four pairs of lobules lie below the horizontal fissure and can only be clearly seen if the cerebellum is viewed from below or if it is detached from its brain stem connections and turned to expose the inferior cerebellar surface.

The eight cerebellar lobules that form each cerebellar hemisphere are named, from front to back on the superior surface, the **ala of the central lobule,** the **quadrangular lobule,** the **lobulus simplex,** and the **superior semilunar lobule,** and from back to front on the inferior surface, the **inferior semilunar lobule,** the **biventer lobule,** the **cerebellar**

tonsil, and the **flocculus** (Figs. 11-56 to 11-60,*A*).

The most rostral part of the cerebellar hemispheres on the anterior surface forms the *alae of the central lobule.* These thin, wing-like prolongations of the central lobule of the vermis extend laterally along the upper and anterior part of each hemisphere. Behind the alae of the central lobule and separated from them by the postcentral fissure on each side are the *quadrangular lobules* (Fig. 11-56). The surface of these lobules is roughly four-sided; medially they are continuous with the culmen, and caudally they are separated from the lobulus simplex by the primary fissure.

Behind the primary fissure are the relatively large *lobuli simplex.* These lobules are continuous medially with the declive, and their superior surfaces are also nearly four-sided, although their caudal borders are curved. They are bounded behind by the posterior superior fissure (also called the postclival fissure or postlunate fissure). The most caudal part of the cerebellar hemispheres as seen from above is formed by the large half moon-shaped *superior semilunar lobules.* Expanding laterally from the small medially located folium of the vermis, they form the entire posterior one-third of the superior surface of the cerebellum. The superior semilunar lobules are bounded below by the horizontal fissure (Figs. 11-56; 11-58; 11-60,*A*).

The posterior two-thirds of the inferior surface of each cerebellar hemisphere consists of the *inferior semilunar lobule* (Fig. 11-57). Expanding laterally from the small, medially placed tuber vermis, these large lobules lie below the horizontal fissure and behind the prepyramidal fissure. In front of the inferior semilunar lobules are the *biventer lobules.* On their surface these are somewhat triangular, with their apices projected backward, and medially they are joined to the pyramis (vermis). The *cerebellar tonsils* are rounded masses situated medially and slightly rostral to the biventer lobules, and each lies in a deep fossa on the posterior cerebellar surface called the **nidus avis** (bird's nest) (Fig. 11-59). The cerebellar tonsils are continuous medially with the uvula. Finally, each cerebellar hemisphere has appended to it the *flocculus.* Each flocculus is a prominent, irregular lobule situated in front of the biventer lobule and behind the middle cerebellar peduncle. It is subdivided by furrows into a few small folia and it is attached medially to the nodulus. The flocculus is also connected to the inferior medullary velum by its central white core (Figs. 11-58; 11-59).

LOBES OF THE CEREBELLUM

A commonly used and sometimes convenient way to describe cerebellar development and adult cerebellar anatomy and physiology is to divide the cerebellum into the flocculonodular lobe and the corpus cerebelli. As described previously, the **flocculonodular lobe** includes the flocculi, their peduncles, and the centrally located nodulus. It is bounded behind by the posterolateral fissure, which both phylogenetically and ontogenetically is the first fissure to appear in the cerebellum. The remainder of the cerebellum is considered the *corpus cerebelli,* which consists of all of the vermis except the nodulus and all of the cerebellar hemispheres except the flocculi. *The primary fissure is the important landmark in dividing the corpus cerebelli into its anterior and posterior lobes* (Figs. 11-56; 11-60,*C*).

The **anterior lobe** is the part of the cerebellum that extends between the superior medullary velum and the primary fissure. It includes the lingula, central lobule and culmen of the vermis, and the alae of the central lobules, as well as the quadrangular lobules (Figs. 11-56; 11-60, *C*). Phylogenetically, it is the oldest part of the corpus cerebelli, but it does not appear as early as the flocculonodular lobe.

The remainder of the corpus cerebelli is considered the **posterior lobe.** It includes the declive, folium vermis, tuber vermis, pyramis (vermis), and uvula (vermis) portions of

Parts of the Human Cerebellum

Cerebellar Lobes	Cerebellar Lobule	Vermis
Anterior Lobe	—	Lingula
	Ala of the central lobule	Central lobule
	Quadrangular lobule	Culmen
Posterior Lobe	Lobulus simplex	Declive
	Superior semilunar lobule	Folium vermis
	Inferior semilunar lobule	Tuber vermis
	Biventer lobule	Pyramis (vermis)
	Cerebellar tonsil	Uvula (vermis)
Flocculonodular Lobe	Flocculus	Nodulus

the centrally located vermis, and the lobuli simplex, superior and inferior semilunar lobules, the biventer lobules, and the cerebellar tonsils (Figs. 11-57 to 11-60). The posterior lobe is much larger than the anterior lobe and most of it has a more recent phylogenetic history than does the anterior lobe.

The flocculonodular lobe, which essentially subserves the vestibular system in its function, is often called the **archeocerebellum** (Figs. 11-58; 11-59). The anterior lobe along with the pyramis (vermis) and the uvula (vermis) of the posterior lobe are generally considered to constitute the **paleocerebellum,** which receives essentially afferent projections from the spinal cord by way of the spinocerebellar septum. The remainder of the cerebellum—i.e., the posterior lobe minus the pyramis and uvula—is a relatively late phylogenetic acquisition, often called the **neocerebellum,** and it receives projections from the cerebral cortex by way of the relay nuclei in the pons. Although not strictly consistent with all experimental data, these three regions could be considered vestibulocerebellar (archeocerebellum), spinocerebellar (paleocerebellum), and pontocerebellar (neocerebellum) with respect to the origin of their afferent connections.

CEREBELLAR ATTACHMENTS TO THE BRAIN STEM

The cerebellum is physically connected to the brain stem by way of the superior medullary velum, the inferior medullary velum, and the inferior, middle, and superior cerebellar peduncles.

SUPERIOR MEDULLARY VELUM. The superior medullary velum is a single, thin, transparent lamina of white matter that stretches between the superior cerebellar peduncles (Figs. 11-58; 11-59; 11-61). It is narrow above, where it passes beneath the inferior colliculi of the midbrain and broader below, where it is continuous with the white matter of the superior vermis. Together with the superior cerebellar peduncles, it forms the upper part of the roof or posterior wall of the fourth ventricle. A slightly elevated ridge, the **frenulum of the superior medullary velum,** descends upon its upper part from between the inferior colliculi, and on either side of the frenulum, the trochlear nerves emerge from the brain stem.

INFERIOR MEDULLARY VELA. The inferior medullary vela are two thin layers of white matter, one on each side, which are attached to the front and lower surface of the nodule and are then extended laterally along the peduncles of the flocculi (Figs. 11-58; 11-59; 11-61). Each inferior medullary velum has a nearly semilunar shape and its neural elements extend only as far as its crescenteric lower border; beyond this, as it covers the inferior part of the fourth ventricle, the inferior medullary velum is formed essentially of a layer of ependymal epithelium supported by a backing of pial tissue, the **tela choroidea.** This attaches inferiorly to the surface of the medulla along the taenia of the fourth ventricle. It is somewhat deficient in its lower part where the median aperture of the fourth ventricle allows that cavity to communicate directly with the subarachnoid space on the surface of the brain.

INFERIOR CEREBELLAR PEDUNCLES. The inferior cerebellar peduncles, or *restiform bodies,* are a pair of prominent, rounded masses of nerve fibers that interconnect the rostral part of the posterior region of the medulla oblongata and the cerebellum (Figs. 11-62; 11-63). Each is situated between the floor of the fourth ventricle and the emerging roots of the glossopharyngeal and vagus nerves. Initially, the inferior cerebellar peduncles pass upward and then diverge laterally to form a part of the lateral walls of the fourth ventricle. After they ascend to the cerebellum and reach the anterior cerebellar notch, they bend abruptly backward and each passes to its corresponding cerebellar hemisphere. As it enters the cerebellum, the inferior cerebellar peduncle lies between the superior cerebellar peduncle medially and the middle cerebellar peduncle laterally. The inferior cerebellar peduncles carry both afferent and efferent fibers with respect to the cerebellum, and these interconnect the cerebellum principally to structures in the spinal cord and medulla oblongata.

MIDDLE CEREBELLAR PEDUNCLES. The middle cerebellar peduncles, or *brachia pontis,* are the largest of the three pairs of cerebellar peduncles. They arch upward to the cerebellum from the brain stem, where they are located in the dorsolateral, midpontine region just rostral to the penetration sites of the facial and vestibulocochlear nerves (Figs. 11-58; 11-62; 11-63). Consisting principally of afferent fiber tracts from the pontine nuclei which cross from the opposite side in the pons, the middle cerebral peduncles enter

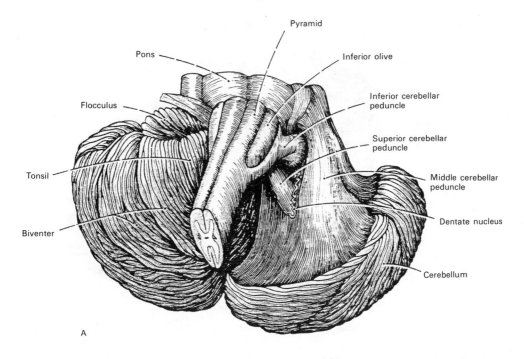

Pyramid

Pons

Inferior olive

Inferior cerebellar peduncle

Flocculus

Superior cerebellar peduncle

Tonsil

Middle cerebellar peduncle

Dentate nucleus

Biventer

Cerebellum

A

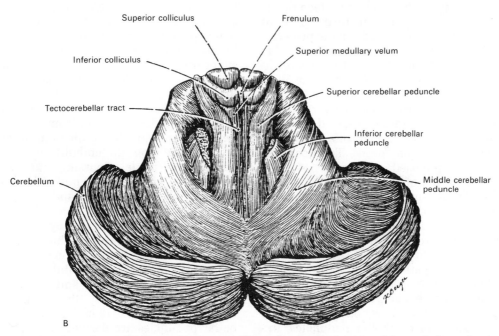

Superior colliculus

Frenulum

Inferior colliculus

Superior medullary velum

Superior cerebellar peduncle

Tectocerebellar tract

Inferior cerebellar peduncle

Cerebellum

Middle cerebellar peduncle

B

FIG. 11-62. Drawings of partially dissected brain stems illustrating the relationships of the cerebellar peduncles. (A) From the anteroinferior aspect; (B) from the posterosuperior aspect. (From E. C. Crosby, T. Humphrey, and E. W. Lauer, *Correlative Anatomy of the Nervous System,* The Macmillan Publishing Co., New York, 1962.)

the rostral part of the inferior surface of the cerebellum lateral to the attachments of the superior and inferior peduncles.

The fibers that form the middle cerebellar peduncle are arranged in three fascicles: superior, inferior, and deep. The **superior fasciculus** is formed by the more rostral transverse fibers of the pons, and it is directed backward and laterally and is more superficial to the other two fasciculi. Its cer-

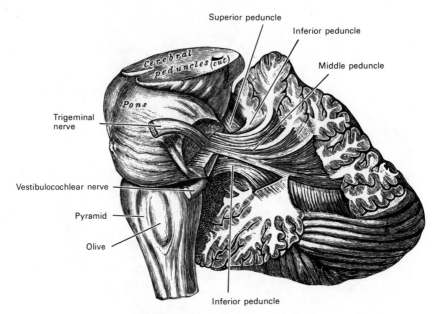

FIG. 11-63.   Dissection showing the projection fibers of the cerebellum. (After E. B. Jamieson.)

ebellar afferent fibers are distributed mainly to the lobules on the inferior surface of the cerebellar hemispheres and to the parts of the superior surface that adjoin the posterior and lateral margins. The **inferior fasciculus** is formed by the most caudal transverse fibers of the pons. It passes from the pons under cover of the superior fasciculus, and continues downward and backward, parallel to the superior fasciculus, and becomes distributed to the cerebellar folia of the inferior surface near the vermis. The **deep fasciculus** is composed principally of deep transverse fibers of the pons. Initially, it is covered by the superior and inferior fasciculi, but it then crosses obliquely to appear on the medial side of the superior fasciculus, from which it receives a bundle. Its fibers then spread out to pass to the cerebellar folia in the anterior part of the superior surface. The fibers of this fasciculus cover those of the inferior cerebellar peduncle.

SUPERIOR CEREBELLAR PEDUNCLES.   Each of the two superior cerebellar peduncles, frequently called a *brachium conjunctivum,* emerges from the cerebellum medial to the middle peduncle and medial and superior to the inferior peduncle (Figs. 11-58; 11-59; 11-62; 11-63). It joins the cerebellum and the midbrain and contains the most important efferent fiber bundles from the cerebellum. The two superior peduncles are joined to each other across the midline by the superior medullary velum, and from the cerebel-

lum they can be followed superiorly as far as the inferior colliculi, under which they disappear. Along their ascent the superior cerebellar peduncles first form the dorsolateral boundaries of the fourth ventricle, but more superiorly they converge on the dorsal aspect of the ventricle and thus assist in forming its roof.

Most of the fibers that form the superior cerebellar peduncles arise from cells in the dentate nuclei of the cerebellum, although others come from the emboliform, globose, and fastigial nuclei. Coursing into the pontine tegmentum and lower midbrain, most of the fibers course medially and cross to the opposite side ventral to the cerebral aqueduct. They terminate in many brain stem and thalamic structures, including the red nucleus, the reticular formation of the midbrain, the pons, and the medulla oblongata, the oculomotor nucleus, and several thalamic nuclei, especially the ventrolateral nucleus. Afferent fibers in the superior cerebellar peduncles are derived from the ventral spinocerebellar tract, as well as from neurons in the colliculi of the midbrain.

THE MIDBRAIN

The midbrain or mesencephalon is that part of the neuraxis located between the pons inferiorly and the thalamus and hypothalamus superiorly (Fig. 11-64). It is a short,

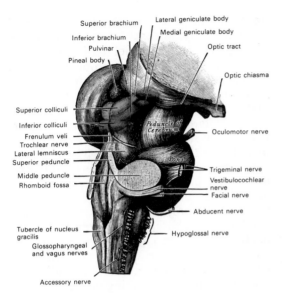

Superior brachium
Lateral geniculate body
Inferior brachium
Medial geniculate body
Pulvinar
Optic tract
Pineal body
Optic chiasma
Superior colliculi
Peduncle of Cerebrum
Inferior colliculi
Oculomotor nerve
Frenulum veli
Trochlear nerve
Lateral lemniscus
Superior peduncle
Pons
Middle peduncle
Trigeminal nerve
Rhomboid fossa
Vestibulocochlear nerve
Facial nerve
Abducent nerve
Tubercle of nucleus gracilis
Hypoglossal nerve
Glossopharyngeal and vagus nerves
Accessory nerve

FIG. 11-64. A dorsolateral view of brain stem.

constricted segment of the brain stem, measuring only about 2 cm. *Inferiorly,* the boundary of the midbrain is represented by a transverse plane passing just caudal to the inferior colliculi posteriorly and just rostral to the ventral pons anteriorly, where the superior pontine sulcus marks the upper limit of the transverse pontine fibers. *Superiorly,* the midbrain is limited by another transverse plane that passes through the posterior commissure dorsally and just behind the mammillary bodies of the posterior hypothalamus ventrally.

By definition, each half of the midbrain is called a **cerebral peduncle,** and this is further divided into a ventral part, called the **crus cerebri,** and a dorsal part, called the **midbrain tegmentum.** In a cross section of the midbrain, it can be seen that the crus cerebri and the tegmentum are separated on both sides by a layer of pigmented nerve cells, the **substantia nigra.** A slender canal, the **cerebral aqueduct,** passes longitudinally through the midbrain and interconnects the third ventricle of the diencephalon above with the fourth ventricle below. Surrounding the cerebral aqueduct is a layer of gray matter called the periaqueductal central gray substance. The portion of the midbrain dorsal to the cerebral aqueduct is called the **tectum** and it consists principally of four rounded swellings or eminences—a pair of **superior colliculi** and a pair of **inferior colliculi** (Fig. 11-64). Also called the *corpora quadrigemina,* the superior and inferior

colliculi on each side of the dorsal midbrain surface are easily identifiable and made more prominent because they are separated by a shallow sulcus.

Basal Part of the Midbrain

The basal part of the midbrain consists of the crus cerebri and the substantia nigra on each side (Figs. 11-65; 11-67; 11-68).

CRUS CEREBRI. Coursing along the ventrolateral aspect of the midbrain on both sides are longitudinally oriented fissures that mark the boundary between the tegmentum dorsally and the crus cerebri ventrolaterally. Each crus cerebri consists of large bundles of nerve fibers that originate in the cerebral cortex and descend as the corticobulbar and corticospinal tracts through the internal capsule and terminate either in the brain stem below the mesencephalon or in the spinal cord. The crura cerebri also contain corticopontine fibers that arise from the frontal, parietal, temporal, and occipital lobes and end in the pontine nuclei. After emerging from the cerebrum, the two crura cerebri form the most ventral part of the midbrain, and on each side they appear as broad, compact bands that are semilunar in transverse section (Fig. 11-65).

Between the crura cerebri is located the wedge-shaped **interpeduncular fossa,** the dorsal floor of which is formed of gray matter and called the **posterior perforated substance** because branches of the posterior cerebral artery pierce the midbrain at this site. During their descent through the mesencephalon, the two crura cerebri converge, and as they enter the rostral region of the ventral pons, they meet on each side of the midline, being separated only by the emerging oculomotor nerves. At this site, the medial surface of each crus is marked by a longitudinal furrow, the **medial sulcus of the crus cerebri,** along which the **oculomotor nerves** pass (Fig. 11-68).

The ventral surface of each crus is crossed from medial to lateral by the superior cerebellar and posterior cerebral branches of the basilar artery. The lateral surfaces of the crura are in relationship to the parahippocampal gyri of the cerebral hemispheres, and the **trochlear nerves,** which leave the brain stem dorsally, wind around the sides of the crura from dorsal to ventral. Also, the lateral surface of each crus is marked by a longitudinal furrow called the **lateral sulcus**

of the mesencephalon. The fibers of the lateral lemniscus come to the surface in this sulcus and then turn dorsally to pass to the inferior colliculus and its brachium. As the crura emerge from the cerebral hemispheres to enter the midbrain, they course deep (dorsal) to the optic tracts, which are themselves directed laterally and somewhat caudally to their connections in the lateral thalamus (Fig. 11-65).

When viewing the crescenteric shape of the crura cerebri in transverse section, it is frequently stated that the middle three-fifths of each crus is occupied by corticospinal and corticonuclear (corticobulbar) fibers arranged in a somatotopic pattern. These are the great motor pathways. Beyond the midbrain, the **corticospinal fibers** descend through the pons to form the pyramids of the medulla oblongata; most (70 to 90%) then decussate to form the crossed lateral corticospinal tracts, while the remainder descend uncrossed into the cord to form the ventral corticospinal tract. The **corticonuclear fibers** descend and terminate in or near the motor nuclei of the cranial nerves. Most of these decussate but some supply certain motor nuclei ipsilaterally. Within this middle region of the crus, descending corticospinal fibers to the lower extremity are located more laterally, while corticonuclear fibers serving the motor nuclei that control the musculature of the face, oropharynx, and larynx are located more medially. Interposed between these are the corticospinal fibers that serve the upper extremity.

Other efferent fibers from the cerebral cortex occupy the medial one-fifth and lateral one-fifth of the crura cerebri. These are the **corticopontine tracts,** which descend to synapse in the pontine nuclei, from which pontocerebellar fibers then cross transversely in the pons and project to the contralateral cerebellar hemisphere. From the frontal lobes, ipsilateral frontopontine fibers occupy the medial one-fifth of each crus cerebri, while temporopontine, parietopontine, and occipitopontine fibers originating in the temporal, parietal, and occipital lobes of the cerebral cortex are found in the lateral one-fifth of each crus.

SUBSTANTIA NIGRA. Located just dorsal to the crus cerebri on each side of the basal part of the midbrain is a layer of gray substance, the substantia nigra, which in transverse section also presents a semilunar shape with its concavity directed dorsally toward the tegmentum (Figs. 11-65; 11-67;

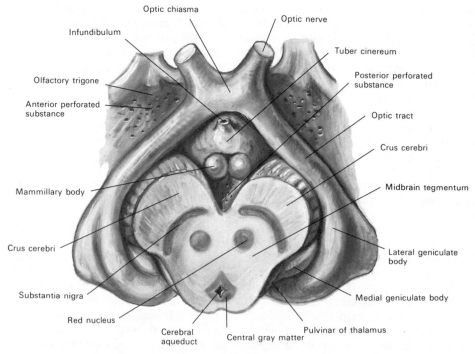

Optic chiasma

Infundibulum

Optic nerve

Tuber cinereum

Olfactory trigone

Posterior perforated substance

Anterior perforated substance

Optic tract

Crus cerebri

Mammillary body

Midbrain tegmentum

Crus cerebri

Lateral geniculate body

Substantia nigra

Medial geniculate body

Red nucleus

Cerebral aqueduct

Central gray matter

Pulvinar of thalamus

FIG. 11-65.   Caudal view of the transversely sectioned midbrain, the ventral surface of the hypothalamus, and the optic tract. (After Sobotta.)

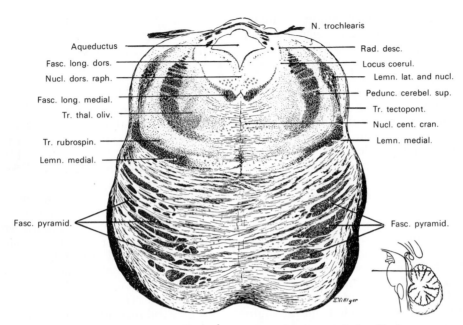

N. trochlearis

Aqueductus

Fasc. long. dors.

Nucl. dors. raph.

Fasc. long. medial.

Tr. thal. oliv.

Tr. rubrospin.

Lemn. medial.

Fasc. pyramid.

Rad. desc.

Locus coerul.

Lemn. lat. and nucl.

Pedunc. cerebel. sup.

Tr. tectopont.

Nucl. cent. cran.

Lemn. medial.

Fasc. pyramid.

E.Villiger

FIG. 11-66. Transverse section through transition between pons and mesencephalon. Nuclear groups represented diagrammatically in color. (From Villiger-Addison, courtesy of J. B. Lippincott Co.)

11-68). This nuclear mass extends the entire length of the midbrain, and it contains numerous deeply pigmented multipolar nerve cells that are distributed in two zones, a more dorsally located black *compact zone* containing melanin pigment, and a more ventrally situated reddish-brown reticular zone that contains no melanin but is rich in iron pigments.

The reticular zone is located just dorsal to the crus cerebri and prolongations of its cells penetrate downward between the bundles of fibers in the crura. Although there is some pigment in the substantia nigra at birth, it increases greatly in amount between the fifth and eighth years of life and seems to accumulate further, but much more slowly, in subsequent years. The exact function of this pigment is unknown, although in recent years strong experimental evidence, utilizing the techniques of histochemical fluorescence, has shown that cells in the substantia nigra produce the neural transmitter dopamine. The substantia nigra receives descending corticonigral fibers from the cerebral cortex, strionigral fibers from the caudate nucleus and putamen, pallidonigral fibers from the globus pallidus, and thalamonigral fibers from the ventral thalamus; it also receives ascending serotonergic fibers from the raphe nuclei of the pons. Efferent fibers from the substantia nigra course to the cau-

date nucleus, putamen, globus pallidus, ventrolateral and anterior ventral nuclei of the thalamus, and the reticular formation in the midbrain tegmentum. It is believed that dopamine synthesized in the substantia nigra is transported intra-axonally to the neostriatum. Disruption of this neurochemical system may be responsible for the condition called paralysis agitans or Parkinson's[1] disease.

Tegmentum of the Midbrain

The midbrain **tegmentum** is the region of the mesencephalon that in transverse sections extends dorsally from the substantia nigra as far as the level of the cerebral aqueduct (Figs. 11-67; 11-68). Posterior to the level of the cerebral aqueduct in the most dorsal part of the midbrain is the midbrain **tectum,** which consists principally of the superior and inferior colliculi.

The tegmentum of the mesencephalon is continuous below with the tegmentum of the pons, and it contains continuations of many of the same nuclear groups and fiber systems as well as other different structures (Fig. 11-66). Its substance also extends superiorly into the subthalamic region of the diencephalon without a sharp line of demarcation.

[1]James Parkinson (1755–1824): An English physician (London).

The tegmentum consists of bundles of longitudinally and transversely oriented fiber tracts, as well as masses of diffuse gray matter. It contains aggregations of nerve cells that are frequently assembled in cell groups and collectively are referred to as the tegmental nuclei. In the more dorsal tegmental region, ventral to the cerebral aqueduct and its surrounding gray substance, are found the nuclei of the trochlear and oculomotor nerves, while the mesencephalic nucleus of the trigeminal nerve is situated more dorsolaterally (Figs. 11-67; 11-68).

MESENCEPHALIC NUCLEUS OF THE TRIGEMINAL NERVE. This nucleus is formed by the cell bodies of primary sensory neurons and is located in the lateral part of the central gray substance that surrounds the aqueduct. Its unipolar, rounded neurons are arranged in small clusters that extend along the lateral angle of the fourth ventricle from the main sensory nucleus of the trigeminal nerve in the pons to the upper midbrain. The axons of its cells gather at the lateral border of the nucleus to form the tract of the mesencephalic nucleus, from which collaterals form monosynaptic connections with neurons in the trigeminal motor nucleus. These sensory nerve cells are thought to serve a proprioceptive function for the muscles of mastication, the teeth, and possibly even certain facial and extraocular muscles.

TROCHLEAR NUCLEUS. The trochlear nucleus is located at the level of the inferior colliculus; it consists of a compact group of cells in the ventral part of the central gray substance, close to the midline and just dorsal to the medial longitudinal fasciculus (Fig. 11-67). The trochlear fibers arch dorsally and laterally around the central gray substance, decussate in the superior medullary velum, and emerge at the dorsal surface of the brain stem just behind the inferior colliculus (Fig. 11-66).

OCULOMOTOR NUCLEUS. The oculomotor nucleus is a complex of nuclear groups found just ventral to the cerebral aqueduct and its surrounding central gray substance in a wedge-shaped region bounded by the medial longitudinal fasciculi of the two sides, immediately rostral to the trochlear nucleus (Figs. 11-67; 11-68). It extends as far as the rostral limit of the superior colliculus, and its cells are somewhat separated by fibers of the medial longitudinal fasciculus. The fibers that form the oculomotor nerve course ventrally through the midbrain tegmentum, medial to the red nucleus, with some fibers traversing that nucleus. They emerge ventrally from the midbrain into the interpeduncular fossa along the medial sulcus of the crus cerebri.

RED NUCLEUS. The red nucleus is a prominent structure of the midbrain tegmentum;

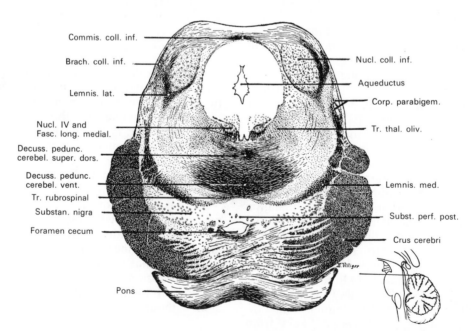

FIG. 11-67. Transverse section through mesencephalon at inferior colliculus. Nuclear groups represented diagrammatically in color. (From Villiger-Addison, courtesy of J. B. Lippincott Co.)

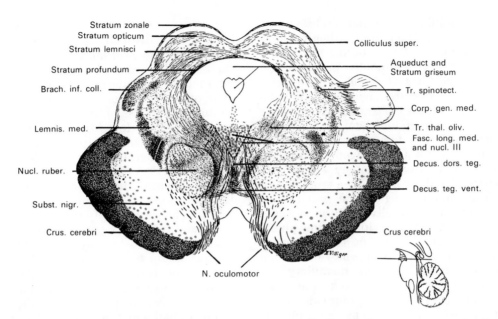

Stratum zonale
Stratum opticum
Stratum lemnisci
Stratum profundum
Brach. inf. coll.
Lemnis. med.
Nucl. ruber.
Subst. nigr.
Crus. cerebri
N. oculomotor

Colliculus super.
Aqueduct and Stratum griseum
Tr. spinotect.
Corp. gen. med.
Tr. thal. oliv.
Fasc. long. med. and nucl. III
Decus. dors. teg.
Decus. teg. vent.
Crus cerebri

FIG. 11-68. Transverse section through mesencephalon at superior colliculus. Nuclear groups represented diagrammatically in color. (From Villiger-Addison, courtesy of J. B. Lippincott Co.)

it is located dorsomedial to the substantia nigra and just lateral to the descending fibers of the oculomotor nerve, some of which actually traverse the nucleus (Fig. 11-68). It has a reddish-pink color in freshly sectioned brains and its cells form a large egg-shaped mass extending from the lower limit of the superior colliculus caudally into the subthalamic region of the diencephalon rostrally. The human red nucleus is composed of a **parvocellular** (or neorubral) **portion** consisting principally of small- and medium-sized cells, and this is capped caudally by a large-celled component, the **magnocellular** (or paleorubral) **part,** which is more prominent in lower mammals but, nonetheless, clearly is present in man. Afferent projections to the red nucleus are derived mainly from the contralateral cerebellar nuclei by way of the superior cerebellar peduncle and from the ipsilateral precentral gyrus of the cerebral cortex. Efferent fibers from the red nucleus project to the spinal cord (rubrospinal), to the cerebellar nuclei (rubrocerebellar), and to certain brain stem structures (rubrobulbar), such as the facial nucleus, the motor nucleus of the trigeminal nerve, and the inferior olivary nucleus. The red nucleus also has connections with other brain centers, such as the thalamus, globus pallidus, and reticular formation, the details of which go beyond the purposes of this text.

OTHER MIDBRAIN NUCLEI. Other frequently described nuclear groups found in the midbrain tegmentum include the *dorsal tegmental nucleus,* located dorsal to the nucleus of the trochlear nerve; the *ventral tegmental nucleus,* found ventral to the medial longitudinal fasciculus; the *interpeduncular nucleus,* located just posterior to the interpeduncular fossa; the *interstitial nucleus (of Cajal[1]),* found in the rostral midbrain just lateral to the medial longitudinal fasciculus; the *nuclei of the posterior commissure,* located in the rostral midbrain in relation to that commissure; the *nucleus of Darkschewitsch,[2]* lateral and ventral to the cerebral aqueduct; and the *Edinger-Westphal nucleus,* located dorsal to the oculomotor nucleus.

Tectum of the Midbrain

The mesencephalic **tectum** is the most dorsal part of the midbrain and is externally marked by four rounded eminences, the superior and inferior colliculi (or corpora quadrigemina), which project posteriorly from the **tectal** (or quadrigeminal) **plate,** a lamina of neural tissue that forms the dorsal

[1]Santiago Ramon y Cajal (1852–1934): A brilliant Spanish histologist and neuroanatomist (Madrid).
[2]Liverij O. Darkschewitsch (1858–1925): A Russian neurologist.

roof of the midbrain over the cerebral aqueduct (Fig. 11-64). The paired colliculi are situated rostral to the superior medullary velum and superior cerebellar peduncle and caudal to the third ventricle and posterior commissure. They are covered by the splenium of the corpus callosum and are partly overlapped on both dorsolateral sides by the pulvinar region of the posterior thalamus. The four colliculi are arranged in pairs, superior and inferior, and they are clearly demarcated and separated by a cruciate sulcus. The midline longitudinal part of this sulcus expands rostrally to form a slight depression in which is located the **pineal body** (Fig. 11-64). From the caudal end of the longitudinal sulcus, a white band—the **frenulum of the superior medullary velum**—is prolonged downward to the superior medullary velum. On either side of this band the trochlear nerve emerges from the brain stem and then passes ventrally on the lateral aspect of the superior cerebellar peduncle to the base of the brain (Fig. 11-54), coursing parallel to but between the superior cerebellar and posterior cerebral branches of the basilar artery.

The **superior colliculi** are somewhat oval and larger and darker than the inferior colliculi (Figs. 11-54; 11-64). They serve primarily as reflex centers that alter the position of the eyes and head in response to sensory stimuli. The **inferior colliculi** have a hemispherical shape and are somewhat more prominent than the superior colliculi; they possibly function as auditory reflex centers as well as nuclei for the transmission of auditory signals to thalamic centers.

Extending ventrolaterally from the superior colliculus and passing between the pulvinar and medial geniculate body to reach the lateral geniculate body and the optic tract is the **brachium of the superior colliculus** (Fig. 11-64). It contains fibers from the optic tract, the lateral geniculate body, and the cerebral cortex that are destined for the superior colliculus. The **brachium of the inferior colliculus** is a thick band of tissue that passes rostrally and laterally from the inferior colliculus to the medial geniculate body (Fig. 11-64). It contains fibers of the auditory pathway from the lateral lemniscus and inferior colliculus to the medial geniculate body and possibly some fibers to and from the auditory cortex interconnecting with the inferior colliculus.

INFERIOR COLLICULUS. The inferior colliculi are two distinctive hemispherical neural masses that form the greater part of the caudal midbrain tectum (Figs. 11-54; 11-64). Internally, each inferior colliculus consists of an outer capsule of myelinated fibers and an inner oval mass of nerve cells called the **nucleus of the inferior colliculus.** More deeply located still is a thin layer of myelinated fibers interposed between the nucleus and the cerebral aqueduct.

The **nucleus of the inferior colliculus** (Fig. 11-67) is formed of medium and large-sized neurons that are densely packed and interconnected with the nucleus of the opposite inferior colliculus by way of a commissure that lies closer to the surface of the tectum than to the cerebral aqueduct. The **outer capsule** of myelinated fibers is formed principally of incoming projections of the lateral lemniscal pathway from the cochlear nuclei, the superior olive, and the trapezoid body. These fibers terminate on the cells of the nucleus of the inferior colliculus. They form the principal afferent pathway to the inferior colliculus and some of the lateral lemniscal fibers cross in the commissure to the inferior colliculus of the opposite side. Other incoming fibers to the inferior colliculus are derived from the ipsilateral medial geniculate body and from the auditory cortex. Connections with the superior colliculus also exist. Efferent fibers leave the inferior colliculus through its brachium to project to the ipsilateral medial geniculate body and even to the cerebral cortex. Other fibers from the inferior colliculus project back to relay nuclei in the auditory pathway, to the superior colliculus, as well as to the opposite inferior colliculus. The inferior colliculus functions as an important auditory center and may participate in the integration of auditory reflexes.

SUPERIOR COLLICULUS. The superior colliculi are two rounded elevations that form most of the rostral half of the mesencephalic tectum (Figs. 11-54; 11-64). The human superior colliculus is a complex laminated structure that reflects the anatomy of its inframammalian homologue, the optic lobe. Although it is difficult to generalize functional studies done in subhuman mammals to the functions served by the human superior colliculus, the available information indicates that the optic tectum is a center for the integration of visual, acoustic, and other envi-

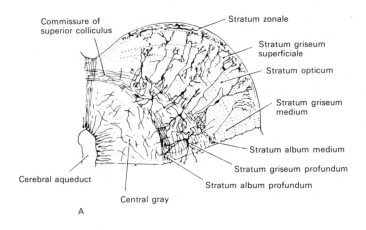

Commissure of superior colliculus

Stratum zonale

Stratum griseum superficiale

Stratum opticum

Stratum griseum medium

Stratum album medium

Stratum griseum profundum

Stratum album profundum

Cerebral aqueduct

Central gray

A

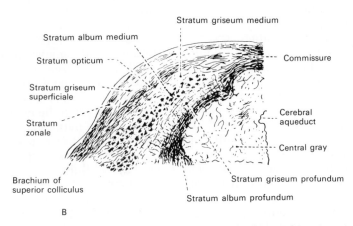

Stratum griseum medium

Stratum album medium

Stratum opticum

Stratum griseum superficiale

Stratum zonale

Brachium of superior colliculus

Commissure

Cerebral aqueduct

Central gray

Stratum griseum profundum

Stratum album profundum

B

FIG. 11-69.   The laminae of the superior colliculus based on reconstruction from Golgi preparations (*A*) of an 8-month fetus to show the cellular organization, and of Weigert-stained adult brain (*B*) to show the myelinated fiber structure. (From M. B. Carpenter, *Human Neuroanatomy*, 7th ed., Williams and Wilkins, Baltimore, 1976.)

ronmental stimuli so that appropriate reactive movements of the eyes, head, and body can be made. Internally, the structure of the superior colliculus consists of alternating layers of gray substance and white matter. These laminae have been variously named, but the terms most frequently used today are the stratum zonale, stratum griseum superficiale (or stratum cinereum), the stratum opticum, and the stratum lemnisci (Fig. 11-69). Because the stratum lemnisci is frequently subdivided into four layers itself— the intermediate gray and white layers and the deep gray and white layers—it is not uncommon to describe the superior colliculus as consisting of seven strata.

The **stratum zonale** (I) is composed of fine, transversely coursing, myelinated and unmyelinated nerve fibers. These fibers form the external corticotectal pathway, which is derived principally from the occipital lobe and enters the superior colliculus by way of its brachium. The **stratum griseum superficiale** (II), or superficial gray layer (frequently called stratum cinereum), is much thicker than the stratum zonale and it consists of many small multipolar nerve cells. The neurons are arranged in such a way that they receive terminals from both the overlying stratum zonale and the underlying zonum opticum, and their axons project inward to deeper layers. The **stratum opticum** (III) is a fiber-rich layer that is thinner in primates, including man, than in infraprimate mammals. The fibers are derived mostly from ganglion cells of the retina, enter the superior colliculus by way of its brachium, and terminate at least in the two adjacent gray layers—the superficial and intermediate gray layers.

The **stratum lemnisci** is subdivided into intermediate gray, intermediate white, deep gray, and deep white layers. The **intermediate gray** (IV) and **intermediate white** (V) layers (stratum griseum medium and stratum album medium) are made up of masses of neurons and bundles of nerve fibers and they constitute important sites for the reception of corticotectal fibers from the occipital and preoccipital cortical regions. These fibers achieve the superior colliculus by way of the internal corticotectal pathway; they are importantly involved in the coordination of eye movements that allow the eye to follow a moving object across the field of vision. These layers also receive input from the spinal cord (spinotectal and spinothalamic paths), substantia nigra, and ventral thalamus. The **deep gray** (VI) layer and the **deep white** (VII) layer (stratum griseum profundum and stratum album profundum) are the strata closest to the periaqueductal gray substance. The deep gray layer consists of medium- to large-sized neurons that receive input from more superficial layers, while the deep white layer is composed of fibers that form the efferent paths from the superior colliculus (Fig. 11-69).

The superior colliculus receives afferent input from the retina, the cerebral cortex, the spinal cord, the ventral thalamus, the inferior colliculus, and the substantia nigra, while its efferent paths are distributed to the visual motor nuclei of the brain stem, the spinal cord, the pontine tegmentum, other sites in the midbrain, and the thalamus.

Pretectal Region

The pretectal region of the brain stem is the zone of transition between the midbrain and the diencephalon. It is located just rostral to the superior colliculus at the level of the posterior commissure, and its **pretectal nucleus** receives fibers from the optic tract on each side. The pretectal nucleus also receives input from the occipital and preoccipital regions of the cerebral cortex and possibly even from the frontal cortex. Efferent fibers leave the pretectal nucleus to terminate in the Edinger-Westphal nucleus.

The pretectal region is the important central relay station mediating the pupillary light reflex. This reflex is initiated by a bright light stimulus to the retina, and it is manifested through the parasympathetic elements in the oculomotor nerve that synapse in the ciliary ganglion. This parasympathetic effect causes a contraction of the sphincter of the iris, thereby decreasing the diameter of the pupil.

Blood Supply to the Cerebellum and Brain Stem

The cerebellum and brain stem, including the midbrain, receive their blood supply by way of the vertebrobasilar arterial system. Each of the two **vertebral arteries** ascends in the deep neck from its origin on the posterior aspect of the subclavian artery, perforates the posterior occipitoatlantal membrane, and enters the posterior cranial fossa through the foramen magnum. On the lateral aspects of the first cervical segment of the spinal cord, the vertebral arteries pierce the dura mater at the level at which the first cervical nerves arise. The vessels then incline ventrally and medially and ascend nearly parallel to the longitudinally-oriented ventrolateral sulci of the medulla oblongata that separate the pyramids from the inferior olives. Each of the arteries passes ventral to the rootlets of the hypoglossal nerves; near or at the midline, ventral to the pontomedullary junction, the two vessels unite to form the **basilar artery** (Fig. 11-70).

One of the two vertebral arteries, more frequently the left, is larger than the other. The **posterior inferior cerebellar arteries** usually arise from the lateral aspects of the vertebral arteries before the two vertebrals join. Each vertebral artery also gives rise to a vessel that joins a similar branch from the other side to form the **anterior spinal artery;** likewise, the **posterior spinal arteries** arise from the vertebral arteries. The rami of the anterior spinal artery branch ventral to, while the posterior spinal arteries arise somewhat caudal to, the origins of the posterior inferior cerebellar arteries.

BASILAR ARTERY

The basilar artery ascends along the basilar sulcus on the ventral aspect of the brain stem from the pontomedullary junction to the rostral border of the pons, where the vessel divides into the **right** and **left posterior cerebral arteries** (Fig. 11-70). The paired **anterior inferior cerebellar arteries** branch from the basilar artery near the caudal border of the pons, and the two **superior cerebellar arteries** arise more rostrally near the

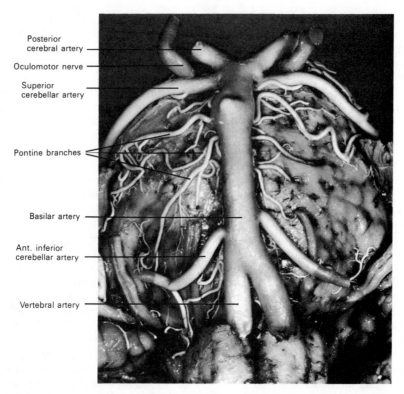

Posterior
cerebral artery

Oculomotor nerve

Superior
cerebellar artery

Pontine branches

Basilar artery

Ant. inferior
cerebellar artery

Vertebral artery

FIG. 11-70. Photograph of the human basilar artery and its branches seen on the ventral aspect of the pons. The junction of the vertebral arteries in this specimen is located slightly rostral to the pontomedullary junction. (From M. Takahashi, in *Radiology of the Skull and Brain: Angiography,* edited by T. H. Newton and D. G. Potts, Vol. 2, Book 2, C. V. Mosby, St. Louis, 1974.)

pontomesencephalic junction, just behind the point at which the basilar artery divides into the posterior cerebral arteries. Often the course of the basilar artery is tortuous or curved and it may show an S-shaped ascent on the ventral pons, in which case it may slightly deviate from the midline. Its length varies from about 2.5 cm to nearly 4.0 cm and its average diameter is about 4.1 mm (Busch, 1966).

Between the origins of the anterior inferior and superior cerebellar arteries, the basilar artery gives off a large number of transversely coursing **pontine branches** that supply the pons and lowest parts of the midbrain (Fig. 11-70). Penetrating branches derived from the vertebral arteries, the upper part of the anterior and posterior spinal arteries, and small vessels from the posterior inferior cerebellar arteries supply the medulla oblongata. Most of the arteries supplying the brain stem from these parent vessels are unnamed, but they can be categorized as **median** (or paramedian) **branches** (see Figs.

11-74 to 11-79), which almost immediately penetrate the ventral median zone of the brain stem as they emerge from the larger arteries, and as the longer coursing **circumferential branches,** which pass laterally around the brain stem and from which short penetrating rami supply the paramedian, lateral, and dorsal zones of the brain stem (Gillilan, 1972). Usually four to six pairs of circumferential arteries branch from the lateral aspects of the basilar artery to supply the pons (Kaplan and Ford, 1966). The basilar artery also gives rise to the paired **labyrinthine arteries,** which course laterally and enter the internal acoustic meatus on each side to supply the vestibulocochlear and facial nerves, but do not contribute any blood supply to the brain stem.

BLOOD SUPPLY TO THE CEREBELLUM

The cerebellum receives its blood supply from three paired arterial branches that arise from the vertebrobasilar system: the posterior inferior cerebellar arteries, the anterior inferior cerebellar arteries, and the superior cerebellar arteries. Although there is some variation in the origin and course of these vessels, it is fair to say that most blood to the cerebellum comes by way of the vertebral and basilar arteries. The posterior communicating arteries, which link the vertebrobasilar system with the carotid system anteriorly, may be significant vessels, but they are usually small and their average diameter is less than 1 mm (Wollschlaeger and Wollschlaeger, 1974).

POSTERIOR INFERIOR CEREBELLAR ARTERY. The posterior inferior cerebellar artery usually arises from the vertebral artery and supplies much of the medulla oblongata, as well as the lower part of the floor of the fourth ventricle and its choroid plexus, and much of the inferior aspect of the cerebellum, including the inferior vermis, inferior semilunar lobule, biventer lobule, and tonsil (Figs. 11-71; 11-76). It also supplies the dentate, interpositus, and fastigial nuclei of the cerebellum, in addition to sending branches to the inferior cerebellar peduncle.

The origin and course of the posterior inferior cerebellar artery vary. Commonly, it branches from the vertebral artery on each side, about 1.5 cm below the junction of the two vertebrals, and usually it arises rostral (about 1 cm) to the foramen magnum (Fig. 11-71). The vessel may arise, however, lower on the vertebral artery or closer to the vertebral junction, or even from the basilar artery itself. It may be absent (15 to 20%), in which case the territory of the posterior inferior artery is usually supplied by the anterior inferior cerebellar artery (Margolis and Newton, 1974).

From its origin, the posterior inferior cerebellar artery courses laterally and superiorly between the rootlets of the hypoglossal and vagus nerves and across the inferior cerebellar peduncle, along the groove between the lateral aspect of the medulla and the cerebellum (Fig. 11-71). As it approaches the lateral angle of the fourth ventricle, it makes a loop around the lateral ventricular aperture (foramen of Luschke), adjacent to the cerebellar tonsil. It then becomes directed once again between the brain stem and the cerebellar surface, sending branches to both the choroid plexus of the fourth ventricle and the inferior surface of the cerebellum.

In addition to the many small branches of the posterior inferior cerebellar artery that supply the posterior and lateral aspects of the medulla, its terminal branches include the lateral and medial tonsillohemispheric branches, which course along the inferior surface of the cerebellar hemispheres, and the vermis branches, which course medially along the inferior vermis (Fig. 11-71, B). The **lateral and medial tonsillohemispheric branches** of the posterior inferior cerebellar artery supply the tonsil, the biventer lobule, and much of the inferior semilunar lobule. The **vermis branches** course along the cerebellar vallecula between the vermis and the cerebellar hemisphere and supply much of the inferior vermis from the tuber vermis to the uvula (Fig. 11-71, C).

ANTERIOR INFERIOR CEREBELLAR ARTERY. The anterior inferior cerebellar artery arises from the basilar artery but its precise site of branching from the main trunk varies. Most commonly (about 75%), it arises from the lower third of the basilar artery about 1 cm above the union of the two vertebral arteries (Fig. 11-72). The manner by which the paired vessels arise and their comparative size also vary. Frequently, the anterior inferior cerebellar artery and the labyrinthine artery arise as a common stem. When the vessels of the two sides are uneven in size, there is a compensatory enlargement or diminution in the size of the corresponding posterior inferior cerebellar artery (Atkinson, 1949).

From its site of origin, the anterior inferior cerebellar artery courses laterally and somewhat caudally on the ventral surface of the pons, just above the sulcus marking the pontomedullary junction. It lies more frequently ventral than dorsal to the abducens nerve (Fig. 11-72), and it passes to the pontocerebellar angle ventral to the facial and vestibulocochlear nerves also. In its course toward the cerebellum, it gives off fine branches to the pons, root branches to the nearby cranial nerves, a branch to the choroid plexus, and an anastomotic branch to the posterior inferior cerebellar artery. Beyond the cerebellopontine angle, the anterior inferior cerebellar artery sends **lateral and medial branches** to supply the flocculus, the biventer lobule, and the inferior semilunar lobule, as well as the middle and inferior cerebellar peduncles (Fig. 11-72). If the posterior inferior cerebellar artery is small, the region supplied by the anterior inferior cerebellar artery is reciprocally larger.

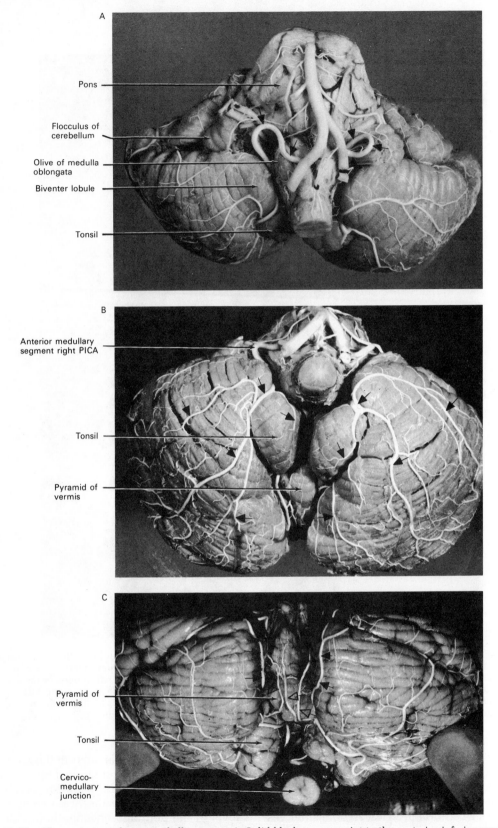

FIG. 11-71. The posterior inferior cerebellar artery. *A,* Solid black arrows point to the posterior inferior cerebellar arteries, which arise symmetrically from the vertebrals on the two sides. *B,* Shows the distribution of the tonsillohemispheric branches (black arrows) of the posterior inferior cerebellar arteries on the inferior surface of the cerebellum. *C,* The vermis branches of the posterior inferior cerebellar arteries are indicated by the open arrows and the medial tonsillohemispheric branches are identified by the solid black arrows. (From M. T. Margolis and T. H. Newton, in *Radiology of the Skull and Brain: Angiography,* edited by T. H. Newton and D. G. Potts, Vol. 2, Book 2, C. V. Mosby, St. Louis, 1974.)

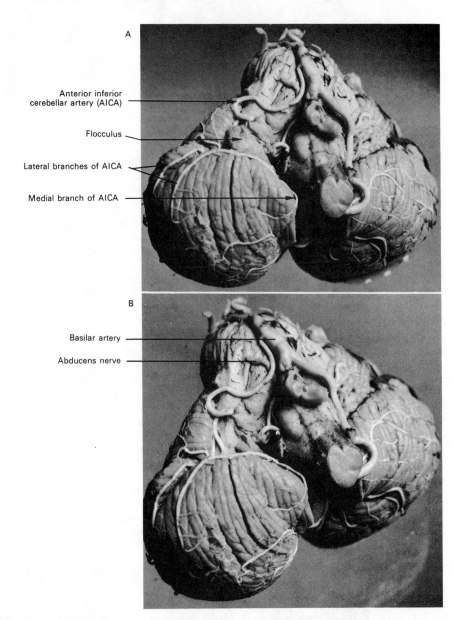

Anterior inferior
cerebellar artery (AICA)

Flocculus

Lateral branches of AICA

Medial branch of AICA

Basilar artery

Abducens nerve

FIG. 11-72. Pictured here is the anterior inferior cerebellar artery (AICA) and its relationship to the ventral aspect of the brain stem and inferior surface of the cerebellum. Note that the two vertebral arteries join to form the basilar artery and that the right (reader's left) AICA arises from the basilar artery. A, The AICA is partially hidden by the flocculus, but its lateral and medial branches can be seen. B, The flocculus has been removed. Note also that the AICA courses ventral to the abducens nerve in this specimen. (From M. Takahashi, in *Radiology of the Skull and Brain: Angiography*, edited by T. H. Newton and D. G. Potts, Vol. 2, Book 2, C. V. Mosby, St. Louis, 1974.)

SUPERIOR CEREBELLAR ARTERY. The superior cerebellar arteries (Fig. 11-73) arise from the basilar artery several millimeters caudal to the division of the basilar trunk into the posterior cerebral arteries. At times the vessels may be duplicated on one or even both sides. The vessel's course on each side initially parallels that of the posterior cerebral artery, from which it is separated by the oculomotor nerve. Directed at first laterally from their origins, the superior cerebellar arteries then encircle the brain stem at the level of the pontomesencephalic junction to achieve the dorsal aspect of the neuraxis (Fig. 11-73). Along this course around the upper pons, perforating arteries are supplied to the pons and lower midbrain, including the crus cerebri, and anastomoses are effected rostrally with branches of the posterior cerebral arteries. In order to supply the cerebellar lobules and vermis of the superior part of the cerebellum, each superior cerebellar artery divides into marginal and hemispheric (sometimes called cortical) branches, which are directed laterally over the superior cerebellar surface, and the superior vermis branches, which are more medially oriented.

The **marginal branch** frequently arises from the superior cerebellar artery on the lateral aspect of the pons (Fig. 11-73). It continues to course posteriorly, alongside the main trunk of the artery, and then it turns acutely laterally to reach the upper lateral margin of the cerebellum. Just above the horizontal fissure of the cerebellum, the marginal branch courses backward to supply the lateral portions of the quadrangular lobule, lobulus simplex, and superior semilunar lobule. Two or three **hemispheric**

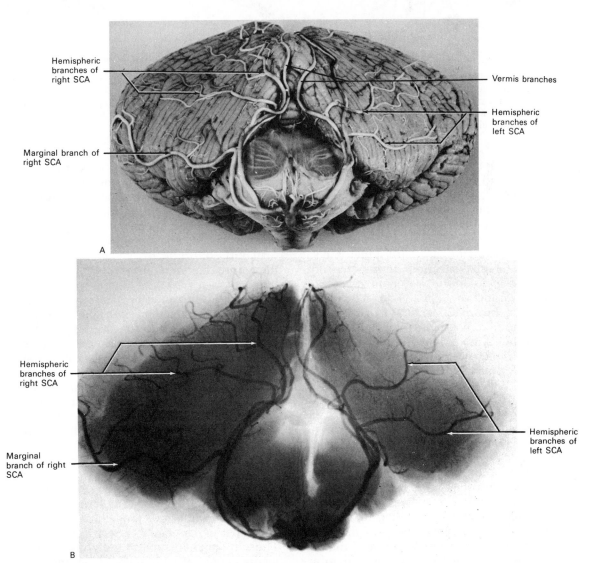

FIG. 11-73. The superior cerebellar artery (SCA). The superior surface of the cerebellum and the transversely severed upper pons are viewed from above. *A*, Observe the two superior cerebellar arteries branching from the basilar artery, encircling the brain stem, and dividing into marginal, hemispheric, and vermis branches on the superior surface of the cerebellum. *B*, Roentgenogram of the specimen in *A*. (From H. B. Hoffman, M. T. Margolis, and T. H. Newton, in *Radiology of the Skull and Brain: Angiography*, edited by T. H. Newton and D. G. Potts, Vol. 2, Book 2, C. V. Mosby, St. Louis, 1974.)

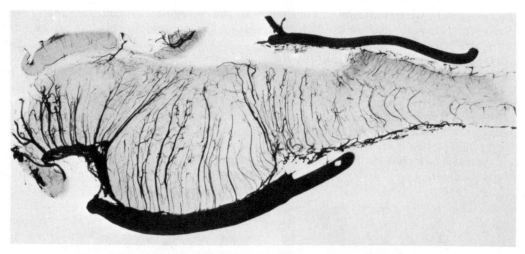

FIG. 11-74. Angiogram of the midbrain, pons, and medulla seen in a longitudinal section of the human brain stem along the midline. Note the different angles formed by the median arteries as they supply the different levels. In the lower midbrain and upper pontine regions they course more caudally; in the midpontine region they converge toward the fourth ventricle at nearly right angles, while in the caudal pons and upper medulla the vessels are oriented in a somewhat rostral direction. (From O. Hassler, Neurology, *17*: 368–375, 1967.)

branches of the superior cerebellar artery arise beyond the origin of the marginal branch (Fig. 11-73). After reaching the superior cerebellar surface, they course backward over the cerebellar cortex to supply the medial two-thirds of all the lobules superior to the horizontal fissure. Finally, each superior cerebellar artery terminates dorsally and medially in two or three **superior vermis branches** (Fig. 11-73, *A*). These course backward along the upper surface of the superior vermis, close to the midline, and send penetrating branches into the superior vermis from the lingula rostrally to the folium vermis caudally.

BLOOD SUPPLY TO THE BRAIN STEM

The blood supply to the medulla, pons, and midbrain is derived from **median arteries** that pierce the ventral brain stem near its midline and from other vessels that penetrate the brain stem more laterally and dorsally (Fig. 11-74). In this manner, a pattern of four zones has been described that can be generalized for cross sections of the brain stem at any level. These zones are a **median zone,** which is supplied by the median arteries that enter the brain stem ventrally near the midline, and the **paramedian, lateral,** and **dorsal zones,** which are supplied by vessels that pierce the brain stem ventrolaterally, laterally, and dorsolaterally re-

spectively (Fig. 11-75). These vessels may be circumferential branches coursing around the brain stem from their origin ventrally, or they may be short penetrating arteries derived from larger named vessels coursing through the region.

MEDULLA OBLONGATA. The median zone of the entire medulla oblongata receives its blood supply by way of median vessels that branch from the anterior spinal artery (Fig. 11-76). The paramedian and lateral regions of the lower medulla also receive penetrating vessels from the vertebral artery. Likewise, the paramedian zone of the upper half of the medulla receives its branches from the vertebral artery. The dorsal region of the medulla receives its penetrating arteries by way of the posterior spinal arteries and by way of the posterior inferior cerebellar arteries (Fig. 11-77).

PONS. The pons is supplied by median, short circumferential, and long circumferential branches of the basilar artery, as well as by dorsolaterally penetrating branches from the anterior inferior cerebellar artery and the superior cerebellar artery (Fig. 11-78). Thus, the median zone of the pons receives median penetrating branches from the basilar artery that enter the basilar sulcus of the ventral brain stem near the midline. The paramedian zone receives the short circumferential branches of the basilar artery, which initially course to the ventrolateral

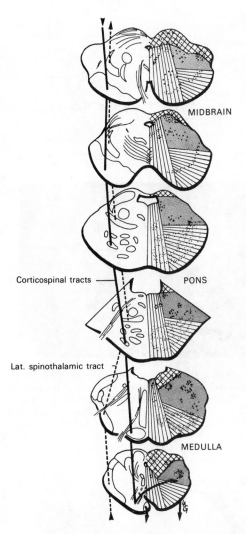

MIDBRAIN

Corticospinal tracts

PONS

Lat. spinothalamic tract

MEDULLA

FIG. 11-75. Diagrammatic cross sections through the brain stem illustrating representative levels of the medulla (lower two), pons (middle two), and midbrain (uppermost two) on the left, and the four intrinsic arterial zones on the right. Cross sections to be compared with those in Figures 11–77 to 11–79. Vertical lines, *median zone;* horizontal lines, *paramedian zone;* stippled region, *lateral zone;* cross-hatched region, *dorsal zone.* (From L. A. Gillilan, in *Handbook of Clinical Neurology,* edited by J. P. Vinken and G. W. Bruyn, American Elsevier, New York, 1972.)

aspect of the pons before piercing the neural tissue. The lateral zone in the lower pons is supplied by lateral penetrating branches of the anterior inferior cerebellar artery; in the midpontine region, the lateral zone receives long circumferential branches of the basilar artery, and in the upper pontine region, it receives penetrating branches from the superior cerebellar artery. No large dorsal

zone comparable to that seen in the lower medulla and midbrain exists in the pons because of the position of the fourth ventricle, but its dorsolateral region is also reached by long circumferential vessels from the basilar artery or by penetrating branches from the superior cerebellar artery.

MIDBRAIN. The mesencephalon receives its arterial supply through branches derived mostly from the superior cerebellar artery and the posterior cerebral artery, but some vessels from the posterior communicating arteries and from the posterior choroidal branch of the posterior cerebral artery also supply it.

The median zone of the midbrain receives its median penetrating branches from three sources: the basilar artery at its point of bifurcation, the proximal parts of the two posterior cerebral arteries, and the posterior communicating arteries (Fig. 11-79). The paramedian zone of the mesencephalon receives short circumferential arteries and penetrating branches from the proximal parts of the posterior cerebral and superior cerebellar arteries. The lateral mesencephalic region also receives long circumferential branches from the posterior cerebral arteries and penetrating branches from the superior cerebellar artery as that vessel courses around the midbrain (Fig. 11-80). The dorsal or tectal zone of the midbrain receives long circumferential branches from the posterior cerebral artery, penetrating branches from the superior cerebellar artery, and other penetrating branches from the posterior choroidal branch of the posterior cerebral artery.

VENOUS DRAINAGE OF THE BRAIN STEM AND CEREBELLUM

The veins that drain the substance of the brain are thin-walled and contain no valves. Venous blood courses toward the surface where the venous channels are frequently more superficial to the arterial vessels. These usually drain into larger named channels, which eventually flow into the sinuses of the dura mater. For the most part, venous blood from the posterior cranial fossa leaves the cranial cavity by way of the internal jugular vein or the basilar plexus, which flows caudally into the internal vertebral plexus of veins within the spinal column.

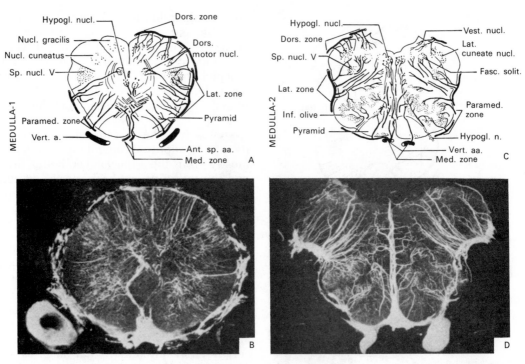

FIG. 11-76. The intrinsic arterial blood supply to the medulla oblongata. A, Diagrammatic representation of the median and other penetrating vessels supplying the median, paramedian, lateral, and dorsal zones of the medulla at the level of the nuclei gracilis and cuneatus. B, Angiogram of the medulla about 2.0 cm below the pontomedullary junction. Note that the median artery bends around the decussation of the pyramids. C, Schema of the intrinsic arteries at the level of the inferior olive in the more rostral medulla. D, Angiogram taken in the upper part of the medulla 3 to 7 mm below the pontomedullary junction. (A and C from L. A. Gillilan, 1972; B and D from O. Hassler, 1967.)

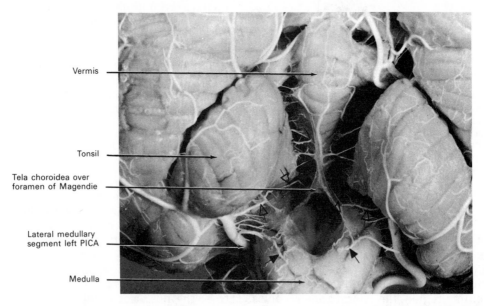

FIG. 11-77. The dorsal aspect of the medulla oblongata at the level of the obex, viewed from above with the cerebellum elevated and its hemispheres spread apart. Observe the penetrating branches of the posterior inferior cerebellar arteries supplying the dorsolateral medulla (solid arrows) and other branches (open arrows) supplying the tela choroidea. (From M. T. Margolis and T. H. Newton in Radiology of the Skull and Brain: Angiography, edited by T. H. Newton and D. G. Potts, Vol. 2, Book 2, C. V. Mosby, St. Louis, 1974.)

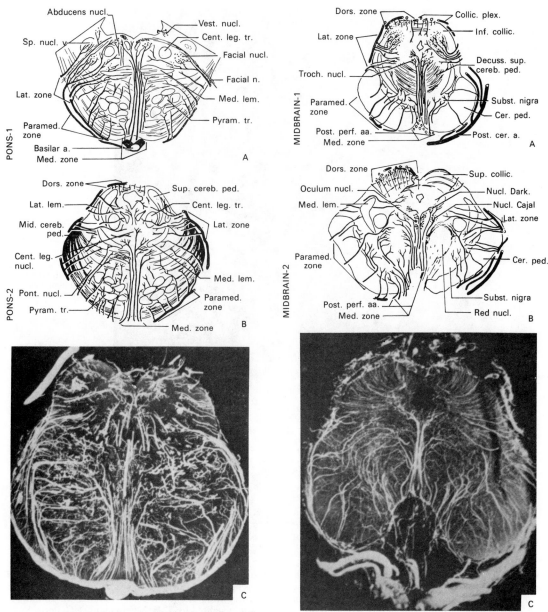

FIG. 11-78. The intrinsic arterial blood supply to the pons. *A* and *B*, Diagrammatic cross sections of the pons at the level of the abducens nucleus (*A*) and more rostrally (*B*), showing its median and other penetrating arteries, which supply the median, paramedian, lateral, and dorsal zones. *C*, Angiogram of the pons seen in cross section about 1 cm below the pontomesencephalic border. (*A* and *B* from L. A. Gillilan, 1972; *C* from O. Hassler, 1967.)

FIG. 11-79. The intrinsic arterial blood supply to the midbrain. *A* and *B*, Diagrammatic cross sections of the midbrain at the level of the inferior colliculus (*A*) and superior colliculus (*B*), showing the blood supply to the median, paramedian, lateral, and dorsal zones. *C*, Cross section of an angiogram of the midbrain, about 4 mm above the pontomesencephalic border. (*A* and *B* from L. A. Gillilan, 1972; *C* from O. Hassler, 1967.)

BRAIN STEM. Most of the blood from the substance of the midbrain drains laterally and anteriorly into the basal vein; however, venous drainage from the dorsal midbrain region is into the great cerebral vein. Venous drainage from the pons is into superficial plexuses, which then drain laterally and ventrally into the inferior petrosal sinuses and the basilar plexus of veins in the floor of the cranial cavity. Venous blood from the

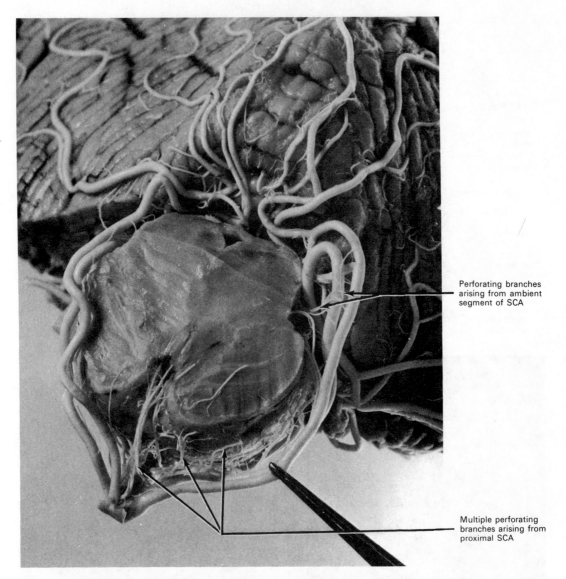

Perforating branches
arising from ambient
segment of SCA

Multiple perforating
branches arising from
proximal SCA

Fɪɢ. 11-80. In this specimen (viewed from above), the brain stem has been severed near the pontine-mesencephalic junction. Observe the multiple perforating (penetrating) branches of the superior cerebellar arteries, which supply the median, paramedian, and lateral zones of the upper pons and lower midbrain. (From H. B. Hoffman, M. T. Margolis, and T. H. Newton in *Radiology of the Skull and Brain: Angiography*, edited by T. H. Newton and D. G. Potts, Vol. 2, Book 2, C. V. Mosby, St. Louis, 1974.)

medulla oblongata drains into the anterior and posterior spinal veins inferiorly, as well as into the occipital and basilar sinuses.

CEREBELLUM. The veins from the substance of the cerebellum drain toward the cerebellar surfaces to form superior and inferior median veins that drain the vermal regions and the medial portions of the cerebellar hemispheres. The superior median vein drains into the great cerebral vein, while the inferior median vein flows into the straight sinus. The veins on the more lateral aspects of the superior and inferior cerebellar surfaces drain into the transverse, superior petrosal, and occipital sinuses.

THE DIENCEPHALON

The **diencephalon** is the region of the brain that is interposed between the mesencephalon, or uppermost part of the brain stem, and the **telencephalon,** which consists principally of the greatly expanded cerebral hemispheres (Figs. 11-81; 11-82). Together, the diencephalon and telencephalon form the **prosencephalon,** which develops from the forebrain vesicle around a cavity that is to become the **third ventricle** and its intercommunicating lateral ventricles. The upper surface of the diencephalon is concealed beneath the corpus callosum and is covered by a fold of pia mater called the tela choroidea of the third ventricle; inferiorly the diencephalon reaches the base of the brain. The diencephalon extends from the pretectal region caudally, just behind the mammillary bodies and at the level of the posterior commissure, to a plane rostrally as far as the interventricular foramen and optic chiasma. On each side the diencephalon is separated from other prosencephalic structures situated laterally and rostrally by the internal capsule (Fig. 11-84). Also located dorsolaterally are the tail of the caudate nucleus and the stria terminalis. The right and left halves of the diencephalon are separated by the narrow, vertically oriented, slit-like cavity of the third ventricle, with the exception of the **interthalamic adhesion,** which is located just caudal to the interventricular foramen, where the two median surfaces are frequently, but not always, joined (Fig. 11-84).

Coursing longitudinally along the lateral wall of the third ventricle is the **hypothalamic sulcus.** This sulcus, which can best be seen in a median sagittal view of the forebrain, extends forward from the lateral wall of the cerebral aqueduct at the level of the posterior commissure and to the rostral end of the diencephalon as far as the interventricular foramen (Fig. 11-82). Serving as an anatomic landmark and perhaps even as the rostral extension of the sulcus limitans separating the alar and basal plates, the hypothalamic sulcus divides the diencephalon on each side into dorsal and ventral parts. The *dorsal part of the diencephalon* consists of the **dorsal thalamus** (frequently referred to simply as the **thalamus**), the **metathalamus,** consisting of medial and lateral geniculate bodies, and the **epithalamus;** the *ventral part of the diencephalon* includes the **subthalamus** and the **hypothalamus.**

Dorsal Thalamus

On each side the dorsal thalamus consists of a large ovoid mass about 4 cm long that forms most of the lateral wall of the third ventricle and extends caudally for some distance beyond that cavity. Its external form can be described as consisting of anterior and posterior extremities and of superior, medial, inferior, and lateral surfaces (Fig. 11-83). The medial and superior thalamic surfaces are exposed to the ventricles, whereas the inferior and lateral surfaces lie adjacent to other neural structures.

The *anterior or rostral extremity* is narrow, lies close to the midline, and on each side forms the posterior boundary of the interventricular foramen.

The *posterior extremity* is expanded and, on each side, directed backward and laterally to extend over the superior colliculus and its brachium. The **pineal body** is situated between the two rounded, medially prominent, posterior portions called the **pulvinars.** The inferolateral aspect of the posterior extremity presents an oval swelling, the **lateral geniculate body,** while beneath the pulvinar and separated from it by the brachium of the superior colliculus is the **medial geniculate body.** The geniculate bodies constitute the metathalamus.

The *superior or dorsal surface* (Fig. 11-81) is slightly convex and freely adjoins the ventricular cavities. It is covered by a layer of white matter called the **stratum zonale** and is separated laterally from the caudate nucleus by a band of white fibers, the **stria terminalis,** and by the **thalamostriate vein.** The superior surface is divided into medial and lateral parts by an oblique shallow furrow that corresponds in position to the lateral margin of the overlying fornix. The broader lateral part of the superior surface forms the floor of the lateral ventricle and is covered by the ependymal lining of that cavity. The narrower medial part is covered by the tela choroidea of the third ventricle, which also separates it from the overlying fornix. Anteriorly, the superior surface is separated from the medial surface by a raised ridge formed by the reflection of the ependyma of the third ventricle, called the **taenia thalami,** which becomes slightly more emphasized by the underlying **stria medullaris thalami.** Posteriorly, the taenia thalami and stria medullaris converge toward the midline near the stalk of the **pineal body** (Fig. 11-81).

The *inferior surface* of the thalamus rests on and is continuous with the upward prolongation of the tegmentum of the midbrain, the subthalamus; more rostrally, it rests on the hypothalamus.

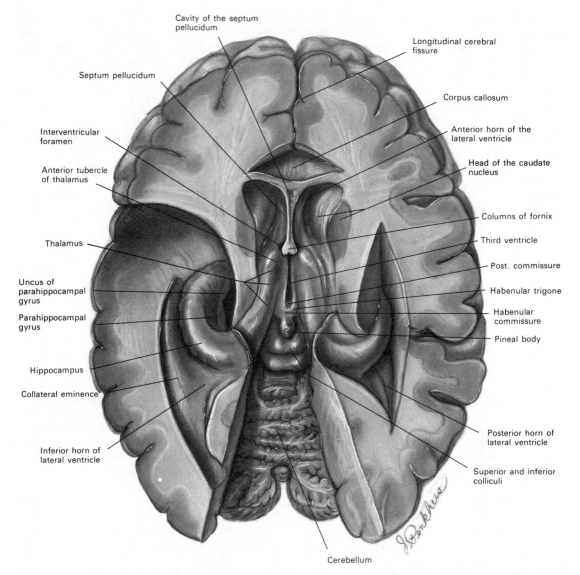

FIG. 11-81. A horizontal section through both cerebral hemispheres, exposing the dorsal thalamus and the third ventricle as viewed from above. Most of the corpus callosum, along with the columns of the fornix and the tela choroidea of the third ventricle, has been removed. The left temporal lobe has been dissected as deep as the inferior horn of the lateral ventricle, thereby exposing more completely the hippocampal formation. (After Sobotta.)

The protruding *medial surface* of the thalamus on each side constitutes the upper part of the lateral walls of the third ventricle. As mentioned previously, the two medial surfaces are frequently joined by a flattened gray band, the interthalamic adhesion (formerly called the massa intermedia). This surface is also marked by the longitudinally oriented hypothalamic sulcus.

The *lateral surface* of the thalamus is in contact with a thick band of white fibers that forms the occipital part or posterior limb of the internal capsule; the thalamus is thereby separated from the lentiform nucleus of the corpus striatum.

GENERAL STRUCTURE AND NUCLEAR GROUPS. The thalamus consists chiefly of gray substance, but its superior surface is covered by a layer of white substance, named the **stratum zonale.** Laterally, next to the reticular nucleus and the internal capsule, lies a similar thin layer of white substance termed the **external medullary lamina.** Internally, the thalamic gray substance

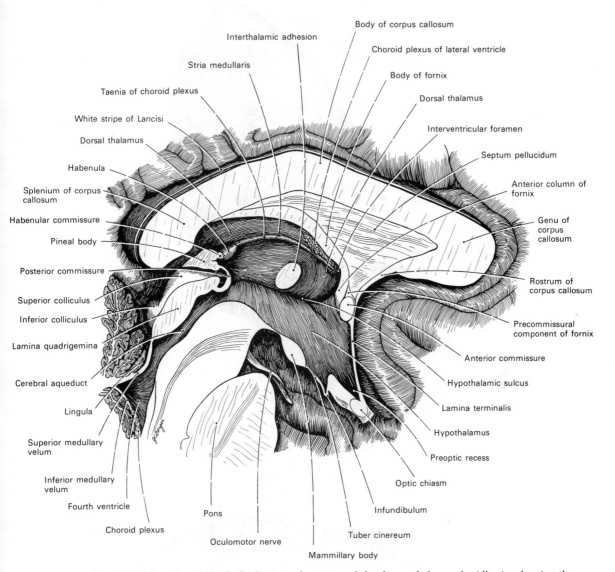

FIG. 11-82. A midsagittal section through the brain in the region of the diencephalon and midbrain, showing the thalamic and hypothalamic structures that bound the third ventricle. Note the cerebral aqueduct interconnecting the fourth ventricle with the third ventricle. (From E. C. Crosby, T. Humphrey and E. W. Lauer, *Correlative Anatomy of the Nervous System.* The Macmillan Co., New York, 1962.)

is subdivided into three parts—anterior, medial, and lateral—by a vertical septum of white matter, the **internal medullary lamina.** As this septum passes forward, it bifurcates and then partially surrounds the anterior thalamic region. Although each of the three principal thalamic parts contains large neuronal masses, they are further subdivided into smaller nuclear groups to facilitate their description and understanding. The plans by which the thalamic nuclei have been subdivided vary somewhat according to different authors, but generally the nuclear groups can be listed as follows: anterior, midline, medial, intralaminar, lateral, ventral, and the thalamic reticular nucleus (Fig. 11-83).

NUCLEI OF THE DORSAL THALAMUS

ANTERIOR THALAMIC NUCLEI. There are three nuclei usually described in the anterior group: **anterior dorsal, anterior ventral,** and **anterior medial;** they form a distinct knob-like swelling on the anterior dorsal surface of the thalamus called the anterior thalamic tubercle (Fig. 11-81). The anterior

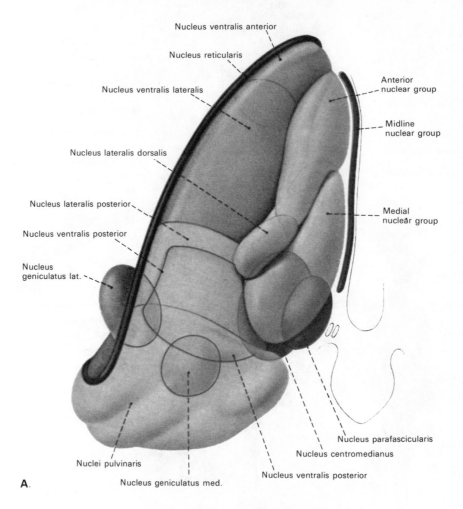

Nucleus ventralis anterior

Nucleus reticularis

Nucleus ventralis lateralis

Anterior nuclear group

Midline nuclear group

Nucleus lateralis dorsalis

Medial nuclear group

Nucleus lateralis posterior

Nucleus ventralis posterior

Nucleus geniculatus lat.

Nucleus parafascicularis

Nucleus centromedianus

Nuclei pulvinaris

Nucleus geniculatus med.

Nucleus ventralis posterior

A.

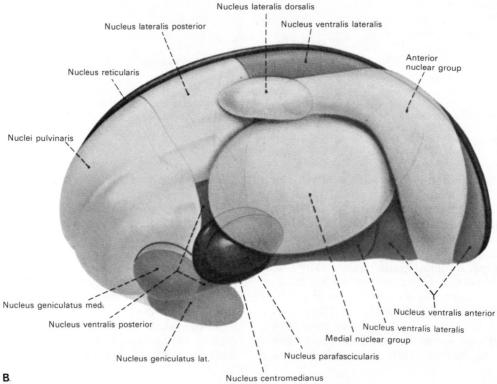

Nucleus lateralis dorsalis

Nucleus lateralis posterior

Nucleus ventralis lateralis

Nucleus reticularis

Anterior nuclear group

Nuclei pulvinaris

Nucleus geniculatus med.

Nucleus ventralis posterior

Nucleus geniculatus lat.

Nucleus ventralis anterior

Nucleus ventralis lateralis

Medial nuclear group

Nucleus parafascicularis

Nucleus centromedianus

B.

ventral nucleus is the largest of the nuclei in the anterior group, and the group as a whole is separated from the remainder of the thalamus by the diverging limbs of the **internal medullary lamina.** The distinctness of the three nuclei in the anterior group is more marked in lower animals, but in the subhuman primate and in the human thalamus, the demarcation between the anterior ventral and anterior medial nuclei is less clear. The anterior dorsal nucleus forms a long, narrow group of cells on the dorsomedial surface of the anterior ventral nucleus.

The anterior thalamic nuclei receive fibers from the mammillary bodies by way of the mammillothalamic tracts (Fig. 11-84) and probably send some thalamomammillary fibers back to the mammillary bodies in the same bundle. They also receive large numbers of fibers directly from the fornix and communicate intrathalamically with nuclei in both the medial and lateral thalamic groups. Thalamocortical efferent fibers from the anterior nuclei project to areas 23, 24, and 32 of the cingulate gyrus by way of the anterior limb of the internal capsule. Thus, the anterior nuclei form a relay station for impulses from the mammillary bodies to the cingulate gyrus; corticothalamic fibers also project back to the anterior nuclei from the cingulate gyrus. Other fibers from the cingulum course to the entorhinal cortex, which in turn projects to the hippocampus. Impulses coursing along this latter path could subsequently influence cells in the anterior nucleus by way of the fornix-mammillary body-mammillothalamic path.

MIDLINE NUCLEI. The nuclei of the midline (Fig. 11-83) consist of clusters of cells that form the periventricular gray substance bordering the third ventricle; they also contribute to the structure of the interthalamic adhesion. They are considerably less well developed in the thalami of subhuman primates and man than in those of lower vertebrates, where, in fact, they form a large part of the thalamus. In the human thalamus, small nuclear masses representative of the more prominent nuclei in submammalian species can frequently be recognized. These include the **parataenial nucleus,** found ventral and medial to the stria medullaris; the **paraventricular nuclei** (anterior and posterior), juxtaposed to the third ventricle on the dorsomedial wall of the thalamus; and remnants of the **nucleus reuniens, nucleus centralis medialis,** and the **nucleus rhomboidalis,** located near or within the interthalamic adhesion.

The midline nuclei have some interconnections with the preoptic and hypothalamic zones ventral to the thalamus and with other nuclei of the thalamus, such as the dorsomedial nucleus and the nuclei of the intralaminar group. In lower forms, where the midline nuclei are better developed, there appears to be evidence of interconnections with the corpus striatum and probably with other regions of the neuraxis as well. Although much more needs to be learned about the anatomy and physiology of these phylogenetically old structures, there is conjecture that they may in some manner be related to other structures in the brain that assist in the regulation of visceral functions.

MEDIAL NUCLEI. The medial thalamic nuclear group (Fig. 11-83) consists of those cellular masses that are bounded laterally by the internal medullary lamina and medially by small collections of neurons that lie adjacent to the third ventricle. The nucleus medialis dorsalis (frequently called the dorsomedial nucleus) is the largest of the nuclei in this group; it is a prominent mass of cells that extends rostrocaudally for a distance of nearly 2 cm. In some descriptions, the centromedian, parafascicularis, paracentralis, and centralis lateralis nuclei are included in the medial group; however, they are classified in the intralaminar nuclear group in this account. Thus, from this nuclear group, only the nucleus medialis dorsalis is detailed further.

The **nucleus medialis dorsalis** (Fig. 11-85) consists of a magnocellular portion, containing relatively large polygonal cells located rostromedially, and a parvocellular part, containing clusters of small cells found caudolaterally. This nucleus is more prominent in primates than in lower forms. It be-

FIG. 11-83. Diagrammatic representations of the human thalamus, showing the principal nuclei from a dorsal view (A) and a medial view (B). The anterior, midline and medial nuclei are shown as groups: Dark blue in A and dark gold in B; Nuclei that project to association regions of the cerebral cortex: Light blue in A and yellow in B; Specific relay nuclei: Orange in A and purple in B. (From Benninghoff and Goerttler, *Lehrbuch der Anatomie des Menschen*, Urban and Schwarzenberg, Munich, 1980.)

gins rostrally near the level of the anterior medial nucleus and extends caudally as far as the habenular complex. It is known to interconnect with the nucleus centromedianus, the amygdala, the parolfactory area, the anterior perforated substance, and the corpus striatum. Its interconnections with the hypothalamus, preoptic region, and especially the frontal cortex, however, are most significant. The medial portion of the nucleus medialis dorsalis both receives and sends fibers to the preoptic region and to the hypothalamus by way of the diencephalic periventricular system, and it also has connections with cortical areas 11 and 12 on the orbital surface of the frontal lobe. Profuse projections from the parvocellular portion of the nucleus medialis dorsalis are directed by way of the inferior thalamic radiation to the entire prefrontal cortex on the lateral surface of the cerebral hemisphere (areas 10, 11, 44, 45, 46, and 47), as well as to the cortex of the temporal operculum (areas 40 and 42). Likewise, there are massive corticothalamic projections to the nucleus medialis dorsalis from the prefrontal cortex.

INTRALAMINAR NUCLEI. The intralaminar nuclei are so designated because they are cellular groups that anatomically are adjacent to or intermingled with the fibers of the internal medullary lamina and thereby separate the medial from the lateral thalamic masses, and because they are considered not to have specific projections to the cerebral cortex. In this classification, the intralaminar nuclear group includes the centromedian, parafascicular, paracentral, central lateral, and limitans nuclei. Certain other descriptions include several of these nuclei in the medial thalamic nuclear group.

The **nucleus centromedianus** is situated in the middle third of the thalamus and is surrounded by fibers of the internal medullary lamina, except medially where it merges without any definite boundary with the cells of the parafascicular nucleus (Figs. 11-82; 11-85). This nucleus is more prominent in subhuman primates and man than in lower mammals. Some authors divide the nucleus into different parts on the basis of its cellular components, but the significance of this is less than clear. The centromedian nucleus is not believed to send direct fiber connections to the cerebral cortex, even though physiologic studies show a profound functional relationship between this nucleus and the cer-

ebral cortex; it does, however, project to the caudate nucleus and the putamen. Afferent fibers to the nucleus appear to come from many sources, including ascending fibers from the reticular formation of the brain stem, perhaps from the spinothalamic paths and the trigeminal system, and from the cerebellum, as well as descending fibers from the cerebral cortex and globus pallidus.

The **nucleus parafascicularis** is a prominent collection of neurons located medial to the posterior part of the nucleus centromedianus and ventral to the caudal portion of the nucleus medialis dorsalis (Fig. 11-83). The fasciculus retroflexus or habenulopeduncular tract crosses the medial part of this nucleus, but the two structures are not thought to be functionally related. The **paracentral nucleus** consists of scattered cells located along the internal medullary lamina, anterior to the nucleus centromedianus. The **nucleus centralis lateralis** is found within the internal medullary lamina behind the paracentral nucleus and lateral to the nucleus medialis dorsalis. The **nucleus limitans** is a band of neurons found between the suprageniculate nucleus and the nucleus medialis dorsalis. The anatomic connections of these smaller intralaminar nuclei also have not, as yet, been determined with satisfaction.

LATERAL NUCLEI. The lateral group of thalamic nuclei in this description includes the dorsolateral region of the thalamus, which extends rostrocaudally from a site somewhat behind the anterior limit of the thalamus to its extreme posterior pole. This thalamic region contains the nucleus lateralis dorsalis, the nucleus lateralis posterior, and the greatly expanded pulvinar.

The **nucleus lateralis dorsalis** and the **nucleus lateralis posterior** lie in the dorsolateral thalamus, lateral to the internal medullary lamina, with the nucleus lateralis dorsalis anterior to the nucleus lateralis posterior (Fig. 11-83). There is good evidence to indicate that the nucleus lateralis dorsalis projects to the inferior parietal and posterior cingulate regions of the cerebral cortex, while the nucleus lateralis posterior projects largely to the parietal cortex, behind the postcentral gyrus. Both nuclei also receive corticothalamic fibers as well as interconnecting fibers from other adjacent thalamic nuclei.

The **pulvinar** is a large nuclear mass that

forms nearly the entire caudal end of the thalamus (Fig. 11-83). It overlies the geniculate bodies as well as the dorsolateral aspect of the mesencephalic tectum. A phylogenetically recent structure, it is best developed in the subhuman primate and in the human brain, although it can also be seen in the thalami of carnivores and other mammals. The pulvinar is usually divided into lateral, medial, and inferior nuclei on the basis of topography and cytoarchitecture. The *lateral nucleus* comprises the lateral part of the pulvinar and it lies dorsal to the lateral geniculate body. The *medial nucleus* is larger and nearly twice the volume of the lateral nucleus, while the *inferior nucleus* is also relatively large but lies more ventrally between the medial and lateral geniculate bodies. The pulvinar probably receives most of its afferent fibers from other thalamic nuclei such as the medial and lateral geniculate bodies, the nucleus lateralis posterior, and possibly even the nearby intralaminar nuclei. The pulvinar projects widely to the cerebral cortex of the parietal, occipital, and temporal lobes.

VENTRAL NUCLEI. The ventral thalamic nuclear mass is most frequently divided into three nuclei: the nucleus ventralis anterior, the nucleus ventralis lateralis, and the nucleus ventralis posterior. The large and important nucleus ventralis posterior is subdivided further into the nucleus ventralis posteromedialis (VPM) and the nucleus ventralis posterolateralis (VPL).

The **nucleus ventralis anterior** is the most rostral part of the ventral nuclear group; it is located lateral to the anterior nuclear group (Fig. 11-83). It is bounded medially by the internal medullary lamina (which separates it from the anterior nuclei) and ventrolaterally by the reticular nucleus. Rostrocaudally, it extends from nearly the anterior end of the thalamus to only the midregion level of the anterior nuclei, where it is then bounded behind by the nucleus ventralis posterior. The nucleus ventralis anterior receives afferent fibers from the globus pallidus by way of the thalamic fasciculus. Additionally, this nucleus has connections with the intralaminar and midline thalamic nuclei and it receives other afferent fibers from the reticular formation and from the substantia nigra of the mesencephalon. Although there are conflicting views with respect to the thalamocortical projections from the nucleus

ventralis anterior, many authors have offered evidence to suggest projections to areas 6 and 4, the frontal lobes, and the insular lobe. In turn, corticothalamic fibers to the nucleus have also been described.

The **nucleus ventralis lateralis** is located behind the nucleus ventralis anterior and gradually replaces it without a sharp line of demarcation. Laterally it is bounded by the nucleus reticularis and the external medullary lamina; medially are found the internal medullary lamina and the intralaminar nuclei (Fig. 11-83). Behind the nucleus ventralis lateralis are located the parts of the nucleus ventralis posterior, i.e., the large nucleus ventralis posterolateralis and the smaller nucleus ventralis posteromedialis. Some authors have described various subdivisions of the nucleus ventralis lateralis based on cytoarchitecture, but these will not be detailed here. The nucleus ventralis lateralis has interconnections with midline and intralaminar nuclei in the thalamus. Afferent fibers to the nucleus ventralis lateralis are derived from the contralateral dentate nucleus of the cerebellum and course through the superior cerebellar peduncle. Other ascending fibers are derived from the red nucleus, while descending fibers reach the nucleus ventralis lateralis from the globus pallidus as well as from cortical areas 4 and 6. Efferent thalamocortical fibers project to motor and premotor areas 4 and 6 in a somatotopic manner.

The **nucleus ventralis posterior** is the largest of the nuclear masses in the ventral part of the thalamus. It commences somewhat rostral and lateral to the nucleus centromedianus and continues back to the pulvinar as the posterior part of the ventral thalamus (Fig. 11-83). Its lateral boundary is the external medullary lamina and the nucleus reticularis. The nucleus ventralis posterior serves as an important sensory relay nucleus, and as such is subdivided into the smaller ventral posteromedial nucleus and the large ventral posterolateral nucleus.

The *nucleus ventralis posteromedialis* (VPM) lies adjacent to the lateral and ventral aspects of the rounded nucleus centromedianus and therefore was formerly called the semilunar or arcuate nucleus (Figs. 11-83; 11-85). It is located medial to the nucleus ventralis posterolateralis and ventral to the nucleus lateralis posterior. On the basis of cytoarchitectural and physiologic data, a

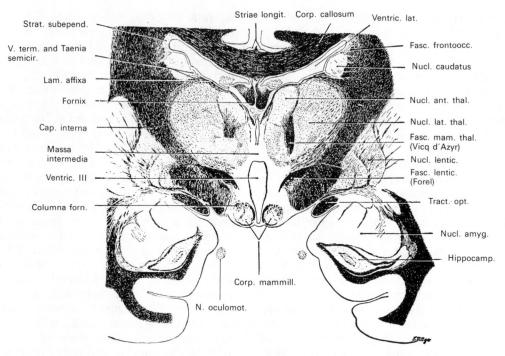

FIG. 11-84. Frontal section through the thalamus and corpus striatum at the level of the mammillary body and mammillothalamic tract. Weigert myelin stain. (From Villiger-Addison, courtesy of J. B. Lippincott Co.)

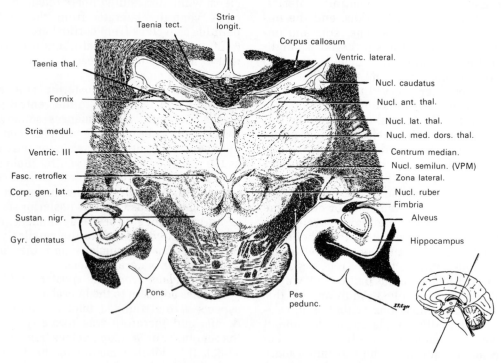

FIG. 11-85. Frontal section through the thalamus and rostral midbrain at the level of the red nucleus, as indicated by the diagram at the lower right. Weigert myelin stain. (From Villiger-Addison, courtesy of J. B. Lippincott Co.)

small-celled or parvocellular region is now considered to be the thalamic relay nucleus for secondary neurons of the taste pathway; these neurons probably originate in the nucleus solitarius of the lower brain stem or from a site in the pons called the pontine taste region. The remainder of VPM contains larger neurons and serves as the thalamic relay nucleus for both crossed and uncrossed secondary sensory trigeminal system fibers, which transmit somatosensory information from the face and head; this information is then received in the nucleus somatotopically. Efferent thalamocortical fibers from VPM course in the posterior limb of the internal capsule and project somatotopically to the head and face sensory receiving areas of the postcentral gyrus in the cerebral cortex.

The *nucleus ventralis posterolateralis* (VPL) forms the large lateral part of the nucleus ventralis posterior. It is situated rostral to the pulvinar and immediately lateral to the nucleus ventralis posteromedialis; it consists of large, sparsely distributed, easily stained neurons. It is to the VPL thalamic nucleus that the fibers of the medial lemniscus and the spinothalamic tracts project, but it should be stated that the terminations of the fibers from these tracts account for only a small percentage of the synapses in VPL. Other afferent fibers in VPL are derived from the mesencephalic reticular formation, the midline and reticular nuclei of the thalamus, the medial telencephalic fasciculus (medial forebrain bundle), as well as many corticofugal fibers from the somatosensory cortex. Thus, according to Brodal (1981), it would be an oversimplication to consider the VPL as purely a relay nucleus for impulses from the dorsal column nuclei and spinothalamic tracts. The pattern by which fibers transmitting impulses of the various modalities terminate on VPL neurons is a question that has been complicated over the past 25 years by the use of various anesthetics and by an attempt to compare findings from different species. Efferent fibers from the nucleus ventralis posterolateralis project to the sensory receiving areas of the postcentral gyrus, as do those from the nucleus ventralis posteromedialis.

RETICULAR NUCLEUS OF THE THALAMUS. The reticular nucleus of the thalamus is a thin, curved band of large nerve cells, situated between the external medullary lamina and the internal capsule (Fig. 11-83). It forms a cellular shell that envelopes much of the anterior two-thirds of the thalamus on its dorsolateral, lateral, and ventrolateral aspects. Although the reticular nucleus is frequently described as part of the dorsal thalamus, it is a derivative of the ventral thalamus (subthalamus) and its substance is continuous ventrally with the zona incerta. Many afferent fibers reach the nerve cells of the reticular nucleus from the cerebral cortex in an apparently topographic manner, but the efferent projection from the reticular nucleus is not to the cortex. Instead its neurons send fibers caudally to the reticular formation of the mesencephalon and to other thalamic nuclei. The precise manner by which this nucleus alters cortical activity is obscure, yet its electrical stimulation causes marked changes in the pattern of the electroencephalogram.

The Metathalamus

The metathalamus is the region of the diencephalon composed principally of the medial and lateral geniculate bodies, which serve as important relay nuclei in the auditory and visual pathways. Some authors classify the geniculate bodies as structures in the posterior part of the dorsal thalamus, along with the pulvinar; however, because they lie at the interface between the midbrain and the diencephalon in a region that also contains a certain amount of pretectal tissue, the term metathalamus will be retained in this description.

MEDIAL GENICULATE BODY. The medial geniculate body is a small oval tubercle that lies, on each side, under the pulvinar of the dorsal thalamus and on the lateral aspect of the superior colliculus (Figs. 11-64; 11-65; 11-83). It is, however, connected to the inferior colliculus by a prominent bundle of fibers called the **brachium of the inferior colliculus.** Internally, the nerve cells that form the **medial geniculate nucleus** serve to relay impulses along the central auditory pathway from fibers ascending in the lateral lemniscus and the brachium of the inferior colliculus to the auditory receiving areas of the temporal lobe. The efferent medial geniculocortical projection reaches the cerebral cortex by way of the auditory radiation in the posterior portion of the internal capsule.

Careful cytoarchitectural studies of the medial geniculate complex have demonstrated the probability that morphologic and functional subdivisions exist within the medial geniculate nucleus. Analyses of Golgi-impregnated material (in the cat) led Morest (1964, 1965) to subdivide the medial geniculate nucleus into medial, dorsal, and ventral parts, all of which are nonlaminated except for a portion of the ventral part. Cells in the laminated region of the ventral part serve as the principal auditory relay station, receiving input from the central nucleus of the inferior colliculus and projecting di-

rectly to the primary auditory cortex in a tonotopic manner. The nonlaminated regions of the medial geniculate nucleus receive input from multiple structures, including the cerebellum, and impulses of other sensory modalities in addition to the auditory sense. These regions also appear to project more diffusely to the temporal cortex, but much detail still needs to be learned about the role of the medial geniculate complex in hearing.

LATERAL GENICULATE BODY.   The lateral geniculate body is an oval elevation on the lateral aspect of the posterior end of the thalamus (Figs. 11-64; 11-65; 11-83; 11-85; 11-86). It is physically connected to the superior

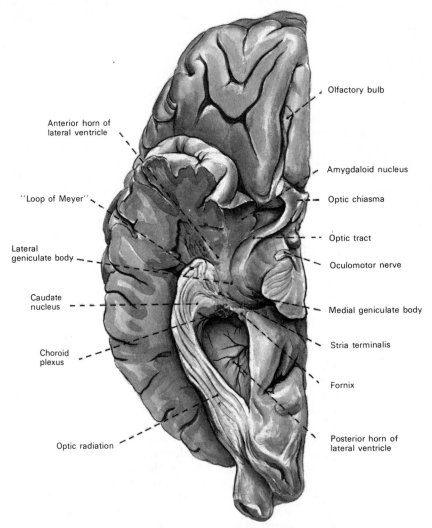

Olfactory bulb

Anterior horn of lateral ventricle

Amygdaloid nucleus

Optic chiasma

"Loop of Meyer"

Optic tract

Lateral geniculate body

Oculomotor nerve

Caudate nucleus

Medial geniculate body

Stria terminalis

Choroid plexus

Fornix

Posterior horn of lateral ventricle

Optic radiation

FIG. 11-86. A dissection of the right cerebral hemisphere, viewed from below, exposing the lateral geniculate body and the spiraled, fan-shaped optic radiation coursing in the posterior limb of the internal capsule to the visual cortex in the occipital lobe. (Courtesy of Dr. Charles F. Bridgman, who drew it from a dissection made by the late Professor A. T. Rasmussen.)

colliculus by the **brachium of the superior colliculus,** which is formed by primary optic nerve fibers that do not terminate in the lateral geniculate body but, instead, project to the pretectal region and superior colliculus. These optic fibers function importantly through the tectospinal tracts and the nuclei innervating the extraocular muscles in the visual motor reflexes. Most of the fibers in the human optic tract, however, do project to the lateral geniculate body, where they synapse with neurons in the lateral geniculate nucleus.

The **lateral geniculate nucleus** actually consists of a small ventral nucleus and a somewhat larger dorsal nucleus. In the human brain, the *ventral nucleus* is a small cluster of cells that is often called the **pregeniculate nucleus.** Its neurons are oriented toward the zona incerta of the ventral thalamus, and in lower mammals, the ventral nucleus projects not only to the zona incerta but to the pretectal region and superior colliculus and to other structures. Although the pregeniculate nucleus has been shown to receive retinal fibers in several species, its cells do not project to the cerebral cortex.

The *dorsal* or *principal lateral geniculate nucleus* is a laminated structure that in the human brain consists of six easily recognized curved layers of cells oriented in a dome-shaped mound, with its blunted crest directed dorsolaterally and its shallow hilus directed ventromedially (Fig. 11-87). The lay-

ers of neurons are separated by bands of optic nerve fibers and the cellular laminae are numbered from 1 to 6, beginning at the hilus and continuing dorsally toward the crest of the geniculate body. The first two layers consist of large cells and sometimes are referred to as the magnocellular layers, while layers 3 to 6 are composed of smaller neurons. Nerve fibers that cross the optic chiasma (i.e. from the contralateral retina) terminate in layers 1, 4, and 6, while the optic nerve fibers that do not cross (and are derived from the ipsilateral retina) synapse in layers 2, 3, and 5. An enormous amount of research has been reported on the retinotopic organization of the lateral geniculate nucleus, and one morphologic feature, the *synaptic glomerulus,* emphasizes the fact that lateral geniculate neurons contain dendritic synaptic pools with terminals from interneurons and corticofugal fibers, in addition to those from primary afferent ganglion cells of the retina. The lateral geniculate nucleus projects to the striate area (Brodmann's area 17) on the occipital cortex by way of the optic radiation (Fig. 11-86). The lateral geniculocortical fibers pass to the cerebral cortex in the retrolenticular, or most posterior, portion of the internal capsule.

The Epithalamus

The epithalamus is situated on the dorsomedial aspect of the caudal diencephalon and includes the **habenular trigone** and the underlying habenular nuclei and commissure, the **pineal body,** and the **posterior commissure.**

HABENULAR TRIGONE. When viewed from above, the habenular trigone is a small, depressed triangular area located in front of the superior colliculus on each side of the epithalamus. Beneath its surface are found the **medial and lateral habenular nuclei,** which form a spherical mass of cells on each side. The habenular nuclei are functionally related to the olfactory pathways, and between the nuclei of the right and left sides, rostral to the stalk of the pineal body, courses the **habenular commissure.** As might be expected, the habenulae are relatively smaller in man than in macrosmotic animals. Many afferent fibers reach the habenular nuclei by way of the **stria medullaris thalami** and some of these cross in the habenular commissure to the opposite side.

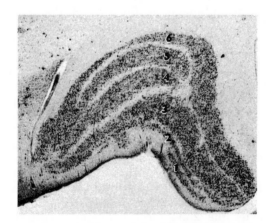

FIG. 11-87. Photomicrograph of the lateral geniculate nucleus showing the six cellular laminae, numbered from the hilus to the dorsal surface of the nucleus. Cresyl violet, 10 ×. (From E. C. Crosby, T. Humphrey, and E. W. Lauer, *Correlative Anatomy of the Nervous System,* Macmillan Publishing Co., New York, 1962.)

Afferent connections are derived from the amygdala by way of the stria terminalis, and from the hippocampus, the septal nuclei and preoptic region, and the anterior thalamic nuclei. The principal efferent pathway from the habenular nuclei is the **habenulopeduncular tract** (also called the fasciculus retroflexus). This pathway arises from both medial and lateral nuclei, courses through the caudal part of the nucleus medialis dorsalis of the thalamus, through the red nucleus, and finally terminates in the interpeduncular nucleus. Other fibers terminate in the mesencephalic reticular formation.

PINEAL BODY. The pineal body (or epiphysis) is a small, conical, gland-like structure that lies in a depression at the midline between the two superior colliculi. It is placed beneath the splenium of the corpus callosum, from which it is separated by the tela choroidea of the third ventricle. It measures about 6 to 8 mm in length and about 4 mm in width; it is attached by a stalk that is composed of a superior lamina contiguous with the habenular commissure and an inferior lamina adjacent to the posterior commissure (Fig. 11-88).

Surrounded by pia mater from which it receives its blood supply, the pineal body consists of cords of epithelioid cells called **pinealocytes,** which contain the usual organelles as well as large numbers of microtubules. Additionally, the organ contains **interstitial cells** that possess large numbers of fine filaments and resemble astrocytic neuroglial cells. Although the precise function of the pineal body is still not clear, it does appear to be capable of suppressing the secretion of gonadotrophin from the pituitary gland. Precocious puberty has been observed in children in whom the pineal body has lesions or has been destroyed. The pineal body contains melatonin and large quantities of serotonin, and it has been suggested that the organ plays an active role in the regulation of certain diurnal rhythms.

POSTERIOR COMMISSURE. The posterior commissure is a rounded bundle of white nerve fibers that crosses the midline at a site just rostral to the superior colliculi and immediately dorsal to the cerebral aqueduct, where, in fact, the aqueduct becomes continuous with the third ventricle (Figs. 11-81; 11-82). The fibers are derived from several sources, including clusters of neurons within and around the posterior commissure called the interstitial and dorsal nuclei of the poste-

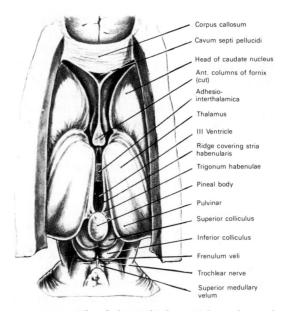

FIG. 11-88. The thalami, third ventricle, and pineal body exposed from above. The trunk and splenium of the corpus callosum, most of the septum pellucidum, the body of the fornix, the tela choroidea with its contained plexuses, and the epithelial roof of the third ventricle have all been removed.

rior commissure and from the nucleus of Darkschewitsch, the interstitial nucleus of Cajal, the pretectal nuclei, and the superior colliculi. The precise functions served by the various fibers that cross in the posterior commissure are not well understood.

The Subthalamus

The subthalamus (or ventral thalamus) lies ventral to the hypothalamic sulcus and is a transitional zone between the dorsal thalamus and the tegmentum of the midbrain. It is continuous structurally with the mesencephalon and both the red nucleus and the substantia nigra are prolonged upward into the most caudal subthalamic region. The subthalamus lies medial to the internal capsule, which separates it from the globus pallidus, and it is located caudal and lateral to the hypothalamus. In addition to the red nucleus and substantia nigra, the subthalamus contains the subthalamic nucleus, the zona incerta, the nuclei of the tegmental fields of Forel,[1] and the entopeduncular nucleus (nucleus of the ansa lenticularis). The fiber bundles coursing with

[1] August H. Forel (1848–1931): A Swiss neurologist (Zurich).

the subthalamus include the ansa lenticularis, the fibers of the tegmental field (field H of Forel), the fasciculus thalamicus (field H_1 of Forel), the fasciculus lenticularis (field H_2 of Forel), and the fasciculus subthalamicus, as well as others.

SUBTHALAMIC NUCLEUS. The subthalamic nucleus is an oval, biconvex mass of gray substance seen only in mammals and located in the more caudal part of the subthalamus. It is positioned along the medial aspect of the internal capsule, which separates the nucleus from the globus pallidus. In the caudal diencephalon, it lies over the dorsal surface of the crus cerebri and dorsolateral to the substantia nigra. The subthalamic nucleus receives fibers from the lateral segment of the globus pallidus and sends fibers through the internal capsule, by way of the **subthalamic fasciculus,** to the medial segment of the globus pallidus. Additionally, other connections of the subthalamic nucleus with the putamen, the caudate nucleus, the mesencephalic tegmentum, and the substantia nigra have been described but not ascertained. Lesions of this nucleus result in contralateral hemiballismic movements, which seem to be alleviated if secondary lesions are placed in the medial globus pallidus.

ZONA INSERTA. The zona inserta is a thin layer of diffuse gray substance that is located immediately ventral to the dorsal thalamus and extends nearly the entire rostrocaudal length of the diencephalon. It is continuous laterally with the reticular thalamic nucleus and caudally with the mesencephalic reticular formation. It lies just above the fasciculus lenticularis and below the thalamic fasciculus (field H_1 of Forel). The zona inserta probably receives nerve fibers from the globus pallidus and the precentral region of the cerebral cortex, and some evidence exists to indicate that efferent fibers are distributed to the mesencephalic tegmentum. The precise function of this region and the manner in which it might influence the great motor pathways, however, are still unknown.

FIELDS OF FOREL. Rostral to the red nucleus is a dense mass of longitudinally running fibers known as the **tegmental field** (*field H*) **of Forel.** The medial fibers are composed of dentato-, rubro-, and reticulothalamic connections. They continue rostrally as the **thalamic fasciculus** (field H_1 of Forel), which enters the medial part of the external medullary lamina and ends in the nucleus ventralis lateralis of the thalamus. The lateral fibers of the tegmental field come from field H_2 of Forel, which is located rostrally and laterally. It is composed of fibers from the globus pallidus in the **ansa lenticularis** and **fasciculus lenticularis** that curve around the medial edge of the internal capsule on the way to the red nucleus and reticular substance.

The Hypothalamus

The hypothalamus is located ventral to the thalamus and it forms the floor and the lower lateral walls of the third ventricle (Fig. 11-82). Strictly speaking, the hypothalamus is considered to extend from the mammillary bodies caudally to the optic chiasma rostrally (Fig. 11-89). The preoptic region, which lies dorsal to and in front of the optic chiasma, is a telencephalic derivative; however, it is usually included in descriptions of the hypothalamus because the functions of the two regions are so closely interrelated. The hypothalamus, consisting of slightly more than 4 gm of tissue, occupies only a small part of the entire brain, and yet it functions diversely and importantly in the regulation of many autonomic and behavioral reactions. Through its connections with the hypophysis, the hypothalamus exerts an enormous influence over the endocrine system and thereby over the general metabolic functions of many organs. The hypothalamus assists in the regulation of body temperature, in feeding, drinking, and mating behavior, in functions relating to biologic rhythms such as the reproductive cycles, and probably even in the onset of sleep. Its role in the maturation of behavior associated with the genders is actively being investigated. Additionally, this part of the diencephalon participates importantly in the behavioral translation of emotional feelings such as fear and rage, and within the hypothalamus are found high concentrations of many neurally active substances such as noradrenaline, serotonin, dopamine, and acetylcholinesterase. As might be expected, the nuclear patterns and the many connections of the hypothalamus with other regions of the central nervous system are complex. For descriptive purposes, the hypothalamus can be divided *longitudinally* into lateral and medial zones that extend throughout its length, or *transversely* into preoptic, supraoptic, tuberal (or infundibu-

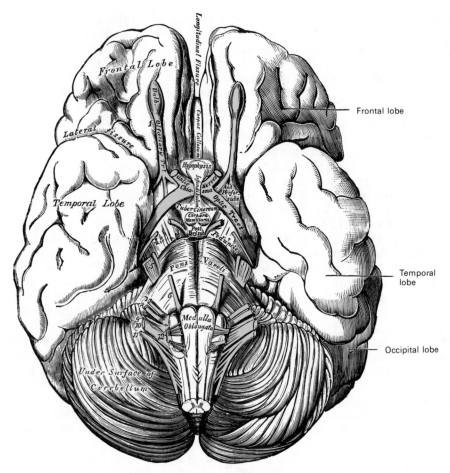

Fig. 11-89. The inferior surface of the brain, showing the hypophysis and infundibulum and the tuber cinereum and mammillary bodies of the hypothalamus. The cranial nerves are shown in yellow.

lar), and posterior (or mammillary body) regions.

Although the precise boundaries of these hypothalamic regions are quite arbitrarily defined, certain longitudinally oriented nerve fiber bundles, such as the fornix and the mammillothalamic tract, have been conveniently used to separate the lateral and medial hypothalamic zones. In this account, however, each of the four transverse regions listed above will be described, and these descriptions include both medial and more lateral structures.

PREOPTIC REGION

The preoptic region occupies a part of the basal forebrain in front of the optic chiasma (Fig. 11-90). Although it is derived from the telencephalon rostral to the hypothalamus, the preoptic region is usually described with

the hypothalamus because it functions importantly in the regulation of certain autonomic mechanisms, frequently in concert with the hypothalamus. Even though the optic chiasma serves as an arbitrary boundary between the preoptic region and the anterior hypothalamus, the gray matter of the **medial preoptic zone** extends forward from the anterior hypothalamus, with no sharp line of demarcation. The **preoptic periventricular nucleus** is located adjacent to the wall of the preoptic recess of the third ventricle and the **medial preoptic nucleus** is found ventral to the anterior commissure.

In the **lateral preoptic zone** are located the **lateral preoptic nucleus** and fascicles of fibers belonging to the **medial forebrain bundle.** Both preoptic zones are interconnected with olfactory centers, the hypothalamus, and the anterior and medial thalamic nuclei. The preoptic region is importantly related to temperature regulation, probably by initiating mechanisms that increase the loss of body heat, such as sweating, vasodilation, and in certain animals, panting. Acute lesions in this region frequently result in a rapid rise in body temperature.

SUPRAOPTIC REGION

The supraoptic region of the hypothalamus lies dorsal to the optic chiasma, ventral to the fornix, and caudal to the preoptic region. It extends behind the anterior commissure and contains the paraventricular nucleus, suprachiasmatic nucleus, and the anterior hypothalamic nucleus medially, and the supraoptic nucleus and the rostral part of the lateral hypothalamic nucleus more laterally (Fig. 11-90).

The **paraventricular nucleus** is a pyramidally shaped group of nerve cells located immediately ventral to the fornix. It is adjacent to the **periventricular nucleus,** which forms a thin vertical column of cells located just deep to the ependyma of the third ventricle and extends nearly to the base of the brain. The **suprachiasmatic nucleus** is a small collection of neurons located directly dorsal to the optic chiasma, and the more diffusely cellular **anterior hypothalamic nucleus** lies adjacent to the paraventricular nucleus in the more medial part of the anterior hypothalamus and blends rostrally with the medial preoptic zone.

Overlying the optic tract near its divergence posterolaterally from the optic chiasma is the well defined **supraoptic nucleus.** It consists of deeply staining, moderately large cells with a rich capillary bed, and the nucleus is easily differentiated from the less compact gray substance of the **lateral hypothalamic nucleus,** which merges forward imperceptibly with the lateral preoptic zone.

The cells of the paraventricular and supraoptic nuclei send fibers to the posterior lobe of the hypophysis by way of the infundibular stalk to form the paraventriculo-hypophysial and supraopticohypophysial tracts (Fig. 11-91). The cells in these nuclei synthesize neurosecretory material that passes down the axons to the infundibulum and the posterior lobe. It is believed that these neurosecretions are precursors of the important hormones vasopressin (the antidiuretic hormone) and oxytocin (the uterine-contracting and milk-releasing hormone),

which are secreted by the posterior lobe. The suprachiasmatic nucleus receives a projection of fibers directly from the optic nerve and retina; these fibers are thought to be an important link relating the light-dark diurnal cycles to the secretion of hormones and the physiologic circadian rhythms. The more diffuse anterior hypothalamic regions, both medially and laterally, participate with the preoptic zones in temperature regulation, as well as with other hypothalamic regions in the secretion of releasing factors and inhibiting factors that influence the production of anterior lobe hormones from the hypophysis.

TUBERAL (OR INFUNDIBULAR) REGION

The tuberal or infundibular region occupies the intermediate portion of the hypothalamus between the more anteriorly located supraoptic region and the caudally placed mammillary region. It is marked on its ventral surface by the tuber cinereum, the median eminence, and the infundibulum, the latter interconnecting the brain and the hypophysis (Fig. 11-89). The **tuber cinereum** is an elevated mound of gray substance situated between the optic chiasma anteriorly and the mammillary bodies posteriorly. Laterally, the tuber cinereum is bounded by the optic tracts and the cerebral peduncles, and in the midline just behind the optic chiasma, a hollow conical process, the **infundibulum,** projects ventrally and rostrally to end in the posterior lobe of the **hypophysis cerebri,** or pituitary gland. At the floor of the third ventricle is located a slightly raised region, the **median eminence,** which encircles the site at which the infundibulum becomes the stalk of the pituitary gland.

Surrounding the lateral walls of the third ventricle in the tuberal region are collections of small neurons that are oriented vertically; these collectively constitute the **periventricular nucleus.** More ventrally, in the region of the infundibulum, these cells become more compact, and in transverse or coronal sections they may be identified as an arch-shaped group of neurons, the **arcuate (or infundibular) nucleus,** which forms the floor of the third ventricle (Fig. 11-90). Occupying most of the medial part of the tuberal region and located between the lateral wall of the third ventricle and the fornix are the ventromedial and dorsomedial nuclei of the hypothalamus. The **ventromedial nucleus** con-

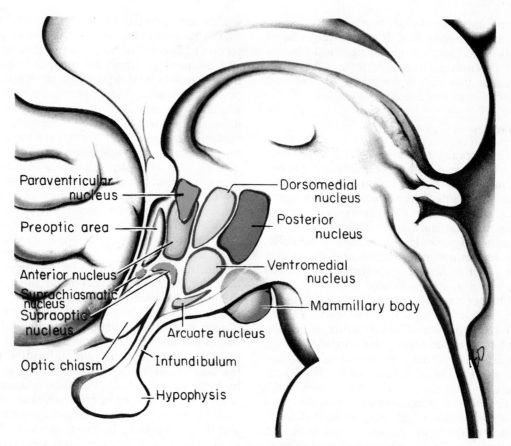

FIG. 11-90. Schematic diagram of the more medially located hypothalamic nuclei as seen from a midsagittal view of the diencephalon. (From M. B. Carpenter, *Human Neuroanatomy*, 7th ed., Williams and Wilkins, Baltimore, 1976.)

sists of an oval mass of densely packed cells that is more conspicuous in lower mammals than in humans, where it is less well defined. Overlying the ventromedial nucleus is the somewhat less distinct **dorsomedial nucleus** (Fig. 11-90), and dorsal to this latter nucleus is an even more diffuse band of cells and fibers that is frequently referred to as the dorsal hypothalamic nucleus (region or area).

Located lateral to the fornix in the tuberal region of the hypothalamus is the **lateral hypothalamic region.** This is continuous both rostrally and caudally with the lateral parts of the hypothalamus at the suprachiasmatic and mammillary levels, which similarly consist of a loose mixture of both fibers and cells. Also found in the lateral hypothalamic region between the level of the infundibulum and the more caudal plane of the mammillary nuclei are several small columns of cells called the **tuberomammillary nucleus** and the **lateral tuberal nuclei.** The latter form small enlargements that are visi-

ble on the ventral surface of the tuber cinereum. These nuclei are believed to send fibers to the infundibular stalk.

The intake of food and fluids is markedly altered by lesions involving various sites in the hypothalamus. Additionally, stimulation of certain hypothalamic regions as well as other CNS sites produces striking changes in overt behavior. Bilateral lesions in the ventromedial nuclei result in an excessive appetite for food, and the resulting *hyperphagia* soon leads to obesity. Animals with such lesions also display varying degrees of savage behavior and inappropriate rage reactions to relatively nonthreatening stimuli. Lesions placed laterally at approximately the same hypothalamic levels tend to result in a loss of appetite, and the resulting *aphagia* leads to emaciation. Destruction of the infundibulum or the more rostrally located supraoptic nucleus, which gives rise to the hypothalamohypophysial tract concerned with the production of vasopressin, results in an excessive intake of fluids and a greatly increased urinary output, a condition known as *diabetes insipidus*. Electrical stimulation of anterior hypothalamic sites induces drinking behavior, and increases in blood pressure and violent attack behavior can be observed by stimulation of the region around the fornix.

The role of the supraoptic and paraventricular nuclei in the production of hormones by the posterior lobe of the pituitary gland has already been mentioned; however, there is considerable evidence that the hypothalamus also plays an important role in the secretion of anterior lobe hypophysial hormones. Although there are no direct neural connections between the hypothalamus and the anterior pituitary, it is believed that these influences result from humoral agents, and that neurosecretions are transported along a fine network of vascular sinusoids that drain from the hypothalamus to the anterior lobe (Fig. 11-91). Overlapping zones in the hypothalamic region between the preoptic level and the level of the mammillary

bodies are thought to be importantly involved in the production of releasing factors (probably peptides) that influence the secretion of adrenocorticotrophic, thyrotrophic, and somatotrophic hormones, the follicle-stimulating hormone, the luteinizing hormone, as well as the prolactin-inhibiting factor.

POSTERIOR (OR MAMMILLARY) REGION

Behind the tuberal region is found the posterior part of the hypothalamus, which includes the posterior hypothalamic nucleus and the mammillary complex. The **posterior hypothalamic nucleus** (or area) is a moderately extensive zone that receives fibers from the medial forebrain bundle, the fornix, and the periventricular system, and it merges on its dorsal aspect with the midline nuclei of the thalamus. Caudally, it extends over the mammillary body to blend with the tegmentum of the midbrain.

Projecting prominently on each side of the ventral surface of the posterior hypothala-

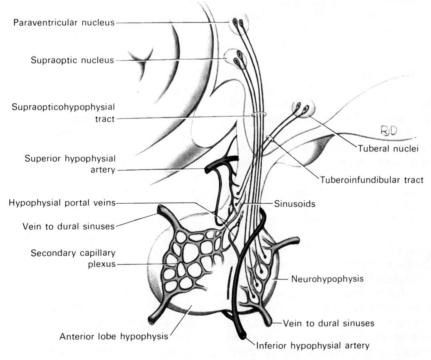

FIG. 11-91. Diagrammatic representation of the hypothalamo-hypophysial neural and vascular connections. Note that the paraventricular and supraopticohypophysial neural pathways carry neurosecretions directly from the hypothalamus to the posterior lobe of the pituitary gland. In contrast, the tuberoinfundibular tract carries neurosecretions to the infundibulum and these are then transported from the infundibulum to the anterior lobe of the pituitary gland by the sinusoids of the hypophysial portal system of veins. (From M. B. Carpenter, *Human Neuroanatomy*, 7th ed., Williams and Wilkins, Baltimore, 1976.)

mus near the midline are the **mammillary bodies** (Figs. 11-82; 11-89; 11-90). In transverse section, this mammillary region consists of the mammillary complex and scattered cells in the premammillary and supramammillary areas (Fig. 11-84). The **medial mammillary nucleus** is a large, ovoid mass of cells producing most of the protuberance of the mammillary body. The **lateral mammillary nucleus** is smaller and fits into the angle between the medial nucleus and the ventrolateral surface of the brain. The **nucleus intercalatus** is a separate group of smaller neurons that is located dorsolateral to the medial nucleus. Both medial and lateral mammillary nuclei receive fibers from the fornix of each side, the fibers decussating just above the mammillary bodies. They also receive fibers from the medial forebrain bundle, from the thalamus, and from the inferior mammillary peduncle. This latter tract arises in the mesencephalic tegmentum and terminates in the lateral mammillary nucleus, but it is rather small in the human brain. Efferent fibers from the mammillary nuclei form the mammillothalamic and mammillotegmental tracts. Initially, the two tracts course together dorsally as the **fasciculus mammillaris princeps,** but a short distance above the mammillary body they diverge. Some of the fibers form the **mammillothalamic tract** (bundle of Vicq d'Azyr[1]), which courses to the anterior nuclei of the thalamus; there they synapse with neurons that project to the cingulate gyrus (Fig. 11-84). Other efferent fibers in the fasciculus mammillaris princeps form the **mammillotegmental tract,** which curves caudally to course to neurons in the mesencephalic tegmentum. Some of the fibers that form this latter tract are actually descending collateral branches of the mammillothalamic fibers.

Stimulation of the posterior hypothalamus usually elicits a sympathetic response, in contrast to stimulation of anterior and preoptic zones, from which parasympathetic responses are achieved. There is some evidence to indicate that neural mechanisms exist in the posterior hypothalamus, some of which increase the conservation of body heat and others of which increase the generation of heat. Heat conservation is accomplished through mechanisms that cause surface capillaries to constrict and sweating to diminish, while heat production is increased by the phenomenon of shivering and by piloerection ("goose-bumps" in man), but the exact role of the posterior hypothalamus in these temperature regulating functions has not yet been well established. Because mammillary nuclei receive large numbers of fibers from the subiculum, by way of the fornix, and from the septum and send fibers to those thalamic nuclei that project to the cingulum, it has been assumed that they participate in pathways intimately related to emotional expression. Specific lesions localized to the mammillary nuclei only, however, have not clarified their functional role in the so-called emotional circuit.

Third Ventricle

The third ventricle is a narrow, vertical median cleft between the thalami of the two hemispheres (Figs. 11-81; 11-88). It is lined by ependyma and filled with spinal fluid. Rostrally, it communicates with the two lateral ventricles by means of an opening on each side called the **interventricular foramen.** Caudally, the third ventricle communicates with the fourth ventricle through the **cerebral aqueduct.** The third ventricle is limited by a roof, a floor, anterior and posterior boundaries, and a pair of lateral walls.

ROOF. The roof of the third ventricle is formed by a thin layer of ependyma that stretches between the upper edges of the lateral walls of the cavity and is continuous with the ependymal lining of the ventricle (Fig. 11-92). The outer surface of this ependymal roof is covered by a fold of pia mater called the **tela choroidea of the third ventricle.** From the inner ventricular surface of the ependymal roof can be seen two delicate vascular folds that are invaginated into the ventricle, one projecting downward on each side of the midline. These vascular processes carry with them the invaginated ependymal layer and form the **choroid plexuses of the third ventricle.**

FLOOR. The floor of the third ventricle is composed of structures that constitute the hypothalamus, and its anterior end slopes ventrally over the optic chiasma to form the **supraoptic recess.** Behind the optic chiasma the ventricular floor is prolonged downward into the infundibulum as the funnel-shaped

[1]Felix Vicq d'Azyr (1748–1794): A French anatomist (Paris).

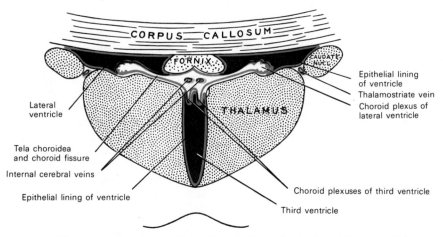

FIG. 11-92. Coronal section of the third and lateral ventricles (diagrammatic).

infundibular recess. Posteriorly, the floor is formed by the posterior perforated substance and by the tegmenta of the cerebral peduncles.

ANTERIOR BOUNDARY. The anterior boundary of the third ventricle is formed *inferiorly* by the **lamina terminalis,** a thin layer of gray substance that stretches from the upper surface of the optic chiasma to the rostrum of the corpus callosum. *Superiorly,* the anterior boundary is formed by the columns of the fornix and by the anterior commissure. Located at the junction of the anterior boundary with the ventricular floor is the small angular diverticulum called the supraoptic recess. At the junction of the anterior boundary and the roof of the third ventricle, situated between the anterior tubercles of the thalami behind and the columns of the fornix in front, are found the **interventricular foramina.** These narrow openings, one on each side, allow the third ventricle to communicate with the two lateral ventricles.

POSTERIOR BOUNDARY. The posterior boundary of the third ventricle is formed by the pineal body, the posterior commissure, and the cerebral aqueduct. The **pineal recess** is a small prolongation of the third ventricle within the stalk of the pineal gland, while the **suprapineal recess,** located in front of and above the pineal body, is a small diverticulum of the epithelium that forms the ventricular roof.

LATERAL WALLS. Each lateral wall of the third ventricle consists of an *upper portion,* formed by the medial surface of the anterior two-thirds of the thalamus, and a *lower portion,* formed by the medial surface of the hypothalamus and subthalamus and the upward continuation of the gray substance of the ventricular floor (Fig. 11-92). These two parts of the lateral ventricular wall are separated by a furrow called the **hypothalamic sulcus,** which extends on each side from the interventricular foramen to the cerebral aqueduct (Fig. 11-82). The lateral wall is limited above by the taenia thalami. The columns of the fornix curve ventrally in front of the interventricular foramina and then course in the lateral walls of the third ventricle, where initially they form distinct prominences, but subsequently they course more deeply in the hypothalamus and are not visible on the lateral ventricular surface. Above the hypothalamic sulci, the lateral walls of the third ventricle are interconnected across the ventricular cavity by medial extensions of gray matter that form the **interthalamic adhesion.**

THE TELENCEPHALON

The telencephalon or "end brain" consists of the **pars optica hypothalami,** into which the anterior part of the third ventricle projects, and the two **cerebral hemispheres,** within which are contained the lateral ventricles. The pars optica hypothalami is the portion of the basal forebrain immediately rostral to the optic chiasma and it includes principally the preoptic region. This has been described with the hypothalamus because of its similarities in function. The cerebral hemispheres are large, rounded expansions of the forebrain that occupy most of the cranial cavity. They are incompletely divided by a deep sagittal groove, the **longitudinal cerebral fissure,** but remain interconnected by the great central white commissure, the **corpus callosum,** located immediately under this fissure. The internal structures of the hemispheres merge with those of the diencephalon and further continuity with the brain stem is established through the cerebral peduncles.

Cerebral Hemispheres

The two cerebral hemispheres constitute the largest part of the brain, and, when viewed together from above, they assume the form of an ovoid mass that is broader behind than in front and whose greatest transverse diameter is across the parietal lobes. Perhaps the most distinguishing feature of the external cerebral surface is the presentation of its substance in a series of convolutions called **gyri,** between which are located grooves of varying depth called **sulci** or **fissures.** This pattern is not exhibited in all mammals, however, for some animals, such as the rat and rabbit, have a smooth or *lissencephalic* cerebral surface. It is assumed that the differential growth rates of some regions of the brain over those of other regions result in the formation of the folded, or *gyrencephalic,* cerebral surface that is characteristic of the human brain and the brains of the higher mammals. Each cerebral hemisphere consists of an extensive covering of outer gray substance called the **cerebral cortex,** the underlying fascicles of **cerebral white substance** (sometimes called the centrum semiovale), several internally situated masses of central gray and white substance that are collectively called the **basal ganglia,** and the **lateral ventricle.**

The cerebral hemispheres present for inspection **three surfaces**—superolateral, medial, and inferior; **three poles**—frontal, temporal, and occipital; and **five lobes** or subdivisions—frontal, parietal, temporal, occipital, and insular. Additionally, a group of phylogenetically older structures is subsumed in the term **rhinencephalon** (or limbic lobe), and the subcortical **basal ganglia** include the corpus striatum, the claustrum, and the amygdaloid body. The fascicles of **white matter of the cerebral hemispheres** may be classified into three types: nerve fibers of projection, which interconnect the cerebral cortex with subcortical centers; commissural fibers, which interconnect cortical regions across the two hemispheres; and association fibers, which interconnect different cerebral cortical regions of the same side. All of these morphologic structures encase the cavities of the **lateral ventricles.**

LONGITUDINAL CEREBRAL FISSURE. The longitudinal cerebral fissure is a deep cleft located in the median sagittal plane between the two cerebral hemispheres, and it contains a sickle-shaped process of dura mater called the **falx cerebri.** The rostral and caudal parts of this fissure extend from the superior to the inferior surfaces of the hemispheres and thereby completely separate them. Its middle portion, however, separates the hemispheres for only about one-half of their vertical extent, reaching down only to a large horizontally oriented sheet of white matter, the corpus callosum, which interconnects the two hemispheres across the midline.

SUPEROLATERAL SURFACE OF THE CEREBRUM

The superolateral surface of each cerebral hemisphere is convex and its shape conforms to the rounded contour of the cranial vault in which it is contained (Fig. 11-93). Superomedially, this surface is sharply bounded from the medial surface of the hemisphere by the **superior border** of the cerebrum, while inferolaterally the **lateral border** forms a more rounded boundary between the superolateral and inferior cerebral surfaces. When the cerebrum is viewed from above, its blunted anterior end forms the **frontal pole;** this is in contrast to the more acutely shaped posterior end, which is called the **occipital pole.** Laterally the anterior limit of the **temporal lobe** forms the **temporal pole** of the cerebrum.

The gyri and sulci on the superolateral surface underlie the flattened parts of the frontal, parietal, occipital, and temporal bones and their disposition initially appears to present an irregular pattern. Although certain differences can be identified between any two brains, close inspection reveals a basic arrangement of the gyri and sulci, within which limited variations exist. Certain sulci form the boundaries between large subdivisions, or lobes, of the cerebrum. Two such sulci are the lateral sulcus and the central sulcus.

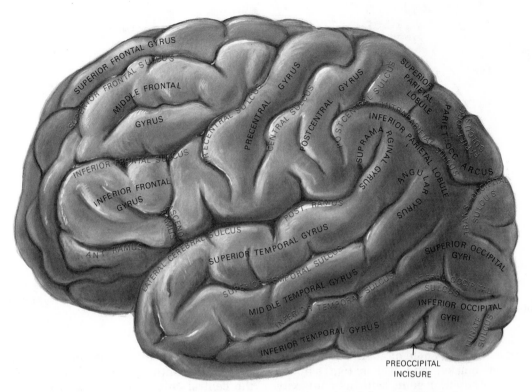

Fig. 11-93. A lateral view of the superolateral surface of the left cerebral hemisphere.

LATERAL SULCUS. The lateral sulcus, some-times called the Sylvian fissure, is a well marked cleft situated on the lateral and infe-rior surfaces of the hemisphere, and it sepa-rates the larger superior masses of the fron-tal and parietal lobes from the temporal lobe (Figs. 11-93; 11-94). It is a complete fissure, isolating much of the temporal lobe, and the floor of its posterior portion is formed by the limen insulae and the insula. The lateral sul-cus consists of a short **stem,** which begins on the inferior surface of the brain in a depres-sion at the lateral angle of the anterior perfo-rated substance. From this point, it extends between the anterior part of the temporal lobe and the orbital surface of the frontal lobe to achieve the lateral border of the hemisphere. This part of its course corre-sponds in direction to the posterior border of the smaller wing of the sphenoid bone. After it reaches the inferolateral aspect of the cer-ebrum, the lateral sulcus divides into three rami: anterior horizontal, anterior ascend-ing, and posterior (Figs. 11-93; 11-94). The **an-terior horizontal ramus** passes forward for about 2.5 cm into the inferior frontal gyrus, while the **anterior ascending ramus** extends

upward into the same convolution for about an equal distance. The **posterior ramus** is the longest of the rami (about 7 cm) and it repre-sents the posterior continuation of the lat-eral sulcus. It is directed horizontally back-ward and upward to terminate in the inferior parietal lobule. Within the lateral sulcus course the middle cerebral artery and vein.

CENTRAL SULCUS. The central sulcus (also called Rolandic fissure) is an important land-mark on the superolateral surface of the cer-ebrum because it forms the boundary be-tween the frontal and parietal lobes and it also separates the primary somatomotor re-gion of the cerebral cortex, located just in front of the sulcus, from the primary soma-tosensory region just behind it (Figs. 11-93; 11-94; 11-96). The sulcus begins on the me-dial surface of the hemisphere, a little be-hind the midpoint between the frontal and occipital poles (Fig. 11-95). It follows an oblique course downward and forward, along the lateral surface of the cerebrum, and it ends a little above the posterior ramus of the lateral sulcus and about 2.5 cm behind the ascending ramus of that sulcus. Its

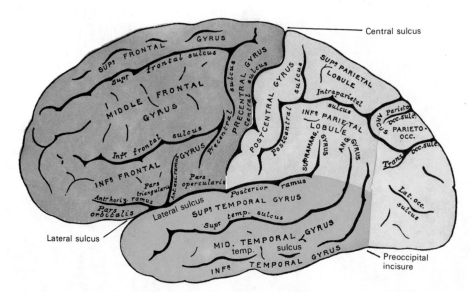

FIG. 11-94. Lateral view of the superolateral surface of the left cerebral hemisphere. Pink, frontal lobe; blue, parietal lobe; yellow, occipital lobe; green, temporal lobe.

course forms an angle of about 70° with the superior border, and it has two sinuous curves, a **superior genu,** with its convexity directed posteriorly, and an **inferior genu,** with its convexity directed anteriorly; additional curves are frequently seen in different brains. The length of the central sulcus is about 9 cm if measured in a straight line, or 10 cm if measured along the curves. Its depth varies along its course from 1.0 to 2.0 cm.

This variable depth occurs because during the sixth month of prenatal development, the central sulcus appears as two sulci—a short upper and a longer lower, which are interrupted by a transversely oriented interlobar gyrus. After further development, the two sulci usually fuse and the transversely oriented gyrus is buried in the floor of the central sulcus, thus altering the sulcal depth.

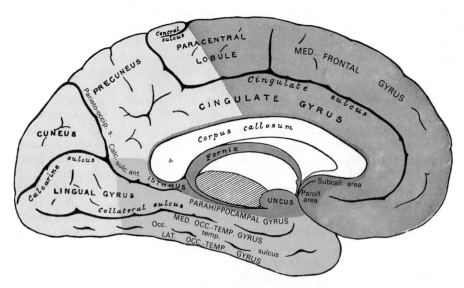

FIG. 11-95. Medial surface of the left cerebral hemisphere with the brain stem and cerebellum removed. Pink, frontal lobe; blue, parietal lobe; yellow, occipital lobe; green, temporal lobe.

Sulci and Gyri on the Superolateral Surface of the Frontal Lobe

The frontal lobe is the most anterior and largest of the lobes in the human cerebrum and its convex superolateral surface extends from the frontal pole to the central sulcus, the latter separating it from the parietal lobe. Usually the superolateral surface of the frontal lobe is traversed by three sulci, which divide it into four gyri. These are the *precentral, superior frontal,* and *inferior frontal sulci* and the *precentral, superior frontal, middle frontal,* and *inferior frontal gyri.*

PRECENTRAL SULCUS. The precentral sulcus courses parallel to and in front of the central sulcus, the two sulci being separated by the precentral gyrus (Figs. 11-93; 11-94; 11-96). During development and maturation, the precentral sulcus is formed from three grooves, the upper two of which usually fuse so that in the fully matured brain, this sulcus generally is seen to be divided into a superior and an inferior segment; the two segments, however, may become continuous.

SUPERIOR FRONTAL SULCUS. The superior frontal sulcus commences posteriorly at about the middle of the upper segment of the precentral sulcus and it courses forward in the superior part of the frontal lobe (Figs. 11-93; 11-94; 11-96). It is directed perpendicular to the precentral gyrus and nearly parallel to the superior border of the cerebrum. The superior frontal sulcus separates the superior frontal gyrus from the middle frontal gyrus and may exist as a continuous sulcus or be divided into two or more parts (over 50% of cases); it extends almost to the frontal pole. Small sulci extend from the superior frontal sulcus and project into the convolutions above and below.

INFERIOR FRONTAL SULCUS. The inferior frontal sulcus also courses horizontally in the frontal lobe, somewhat below and roughly parallel to the superior frontal sulcus (Figs. 11-93; 11-94; 11-96). Typically, it begins posteriorly as a furrow projecting forward from the inferior segment of the precentral sulcus. It may course as an uninterrupted sulcus forward to the lower or orbital part of the anterior frontal pole, but more frequently it is disconnected and forms two or more furrows. The inferior frontal sulcus separates the middle frontal gyrus above from the inferior frontal gyrus below.

PRECENTRAL GYRUS. The precentral gyrus is bounded in front by the precentral sulcus and behind by the central sulcus. It is positioned parallel to these sulci and extends obliquely across the superolateral surface of the cerebral cortex from the superior border of the hemisphere to the posterior ramus of the lateral fissure (Figs. 11-93; 11-94; 11-96). On the medial surface of the cerebral hemisphere, the precentral gyrus is continuous with the anterior part of the paracentral lobule (Fig. 11-95). The precentral gyrus and the anterior surface of the central sulcus function as the primary motor region of the cerebral cortex, in which the lower extremity is represented on the superior aspect and medial surface of the gyrus, followed by the upper extremity and hand on the superolateral surface, and the face, jaw, and tongue more inferolaterally.

SUPERIOR FRONTAL GYRUS. The superior frontal gyrus is situated above the superior frontal sulcus and it extends horizontally forward from the precentral sulcus to the frontal pole (Figs. 11-93; 11-94; 11-96). It is a relatively wide, somewhat uneven convolution and it is continued onto the medial surface of the hemisphere. The portion on the lateral surface is frequently incompletely subdivided into an upper and a lower part by a horizontal sulcus called the **paramedial sulcus.**

MIDDLE FRONTAL GYRUS. The middle frontal gyrus is a wide convolution that extends anteroinferiorly from the precentral gyrus and is bounded above by the superior frontal sulcus and below by the inferior frontal sulcus (Figs. 11-93; 11-94; 11-96). A horizontally oriented sulcus frequently divides the middle frontal gyrus into upper and lower parts. Inferolaterally, the middle frontal gyrus extends as far as the orbital surface of the frontal lobe.

INFERIOR FRONTAL GYRUS. The inferior frontal gyrus lies below the inferior frontal sulcus and extends forward from the lower part of the precentral gyrus (Figs. 11-93; 11-94; 11-96). Inferolaterally, it is continuous with the lateral and posterior orbital gyri on the inferior surface of the frontal lobe. Helping to form the superior wall of the lateral sulcus below the precentral gyrus, the inferior frontal gyrus becomes subdivided into three regions by the anterior horizontal and ascending rami of the lateral sulcus: the **orbital part** below the anterior horizontal ramus, the **triangular part** between the anterior horizontal and the ascending rami, and the **opercular part** above and behind the ascending ramus (Fig. 11-94). The opercular part and a portion of the triangular part of the inferior frontal gyrus are frequently referred to as Broca's[1] **speech area** because they are importantly related to the motor aspects of speech behavior.

Sulci and Gyri on the Superolateral Surface of the Parietal Lobe

The parietal lobe is separated anteriorly from the frontal lobe by the central sulcus, but its boundaries below and behind are somewhat less definite. Posteriorly, its boundary with the occipital lobe commences at the superior border of the cerebrum by the parieto-occipital sulcus, and then it continues as an arbitrary line that starts at the lower lateral tip of this sulcus and is carried across the lateral surface of the hemisphere to a shallow indentation on the ventrolateral border of the brain called the **preoccipital incisure** (Fig. 11-94).

Inferiorly, the parietal lobe is bounded by the posterior ramus of the lateral sulcus and by another arbitrary line drawn from the site at which the posterior ramus turns to ascend into the parietal lobe to the point at which this line intersects the posterior boundary line. Two sulci, the *postcentral sulcus* and the *intraparietal sulcus,* subdivide the superolateral surface of the parietal lobe into the *postcentral gyrus,* the *superior parietal lobule,* and the *inferior parietal lobule.*

POSTCENTRAL SULCUS. The postcentral sulcus is oriented parallel to the central sulcus and lies posterior to it, and the two sulci are separated by the postcentral gyrus (Figs. 11-93; 11-94). The sulcus may be a continuous fissure, but at times it is interrupted and has two parts, a superior and an inferior sulcus. The postcentral sulcus extends downward along the superolateral surface of the parietal lobe from a site close to the longitudinal cerebral fissure above. It ends inferiorly just above the posterior ramus of the lateral sulcus at a site slightly anterior to the upward turn of that ramus. Anterior to the postcentral sulcus is located the postcentral gyrus, while posterior to it are the superior

and inferior parietal lobules, which are themselves separated by the intraparietal sulcus.

INTRAPARIETAL SULCUS. The intraparietal sulcus is an anteroposteriorly oriented fissure that commences at about the middle of the postcentral sulcus and extends horizontally backward toward the occipital pole (Figs. 11-93; 11-94; 11-96). It divides the posterior part of the parietal lobe into an upper portion called the superior parietal lobule and a lower part called the inferior parietal lobule. The intraparietal sulcus extends posteriorly into the occipital lobe to join the transverse occipital sulcus at nearly a right angle.

POSTCENTRAL GYRUS. The postcentral gyrus extends along the superolateral surface of the parietal lobe from the longitudinal cerebral fissure above to the posterior ramus of the lateral sulcus below (Figs. 11-93; 11-94; 11-96). Its oblique anteroinferior course on the hemisphere is limited in front by the central sulcus and behind by the postcentral sulcus. It lies nearly parallel to the precentral gyrus, with which it is physically connected around the lower end of the central sulcus. The postcentral gyrus serves as the primary somesthetic cortex, in which general sensory projections are represented in a characteristic somatotopic pattern. The anal, genital, foot, and leg regions are localized on the medial surface of the postcentral gyrus, followed by the thigh, trunk, neckhead, upper limb, hand, face, tongue, pharynx, and intra-abdominal regions, in that order, on the superolateral convexity of the gyrus. The territories along the gyrus that are devoted to the hand and face are disproportionately large in comparison to those for other parts of the body.

SUPERIOR PARIETAL LOBULE. The superior parietal lobule is located between the superomedial border of the hemisphere and the intraparietal sulcus (Figs. 11-93; 11-94; 11-96). Anteriorly, it is bounded by the superior part of the postcentral sulcus, around the upper end of which it is continuous with the postcentral gyrus. Posteriorly, the superior parietal lobule is continuous with the **parieto-occipital arcus** or gyrus, a U-shaped convolution that curves around the lateral limit of the **parieto-occipital sulcus.**

INFERIOR PARIETAL LOBULE. The inferior parietal lobule lies below the intraparietal sulcus and behind the lower part of the post-

---

[1]Pierre P. Broca (1824–1880): A French surgeon, neurologist, and anthropologist (Paris).

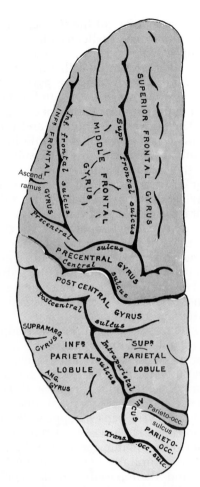

FIG. 11-96. The superolateral surface of the left cerebral hemisphere, viewed from above. Pink, frontal lobe; blue, parietal lobe; yellow, occipital lobe.

central sulcus (Figs. 11-93; 11-94; 11-96). It extends posteriorly as far as the arbitrary line that defines the posterior limit of the parietal lobe. The inferior parietal lobule is divided into the supramarginal gyrus, the angular gyrus, and an additional posterior convolution. The **supramarginal gyrus** arches over the upturned end of the lateral sulcus and it is continuous in front with the lower end of the postcentral gyrus and posteroinferiorly with the superior temporal gyrus. The **angular gyrus** caps the posterior end of the superior temporal sulcus, behind and below which it is continuous with the middle temporal gyrus. An additional convolution of the inferior parietal lobule often lies behind the angular gyrus and it is continuous below with the inferior temporal gyrus.

Sulci and Gyri on the Superolateral Surface of the Occipital Lobe

The occipital lobe is small in comparison to the other cerebral lobes, and it has a pyramidal shape, with the apex of the pyramid forming the occipital pole of the hemisphere (Fig. 11-94). Its superolateral surface is limited in front by the lateral part of the parieto-occipital sulcus and by an arbitrary line from that sulcus to the preoccipital incisure. Although there is considerable variability with respect to the sulcal and gyral pattern on this convex surface of the occipital lobe, the most frequently observed sulci include the **transverse occipital sulcus,** the **lateral occipital sulcus,** and the **lunate sulcus** (Fig. 11-93). The convolutions on the superolateral surface are also quite variable, but usually the lateral sulcus divides much of the occipital lobe into the **superior** and **inferior lateral occipital gyri** (Fig. 11-93). The convolution capping the parieto-occipital sulcus, which helps to form the border between the parietal and occipital lobes, is the **parieto-occipital arcus.**

TRANSVERSE OCCIPITAL SULCUS. The transverse occipital sulcus is one of the most constant of the sulci on the superolateral surface of the occipital lobe. It commences at the superior border of the cerebrum, and when viewed from the side, it descends on the surface of the cerebrum a variable distance (about 2.0 cm), just behind the posterior part of the parieto-occipital sulcus (Figs. 11-93; 11-94; 11-96). When viewed from behind, the course of the sulcus appears to be more transverse because of the slope of the occipital lobe. About midway along its course, it is usually joined, at nearly a right angle, by the posterior end of the intraparietal sulcus, which as mentioned previously extends posteriorly into the occipital lobe.

LATERAL OCCIPITAL SULCUS. The lateral occipital sulcus, when present, is always a short sulcus that extends horizontally (anteroposteriorly) on the lateral surface of the occipital lobe, dividing it into the superior and inferior occipital gyri (Figs. 11-93; 11-94). It is usually located in front of the lunate sulcus, and its course anteriorly is perpendicular to that of the lunate sulcus.

LUNATE SULCUS. The lunate sulcus is a short, semilunar-shaped sulcus oriented vertically on the lateral surface of the occipital lobe, a short distance in front of the occipital pole (Fig. 11-93). This crescenteric sulcus approximately limits the striate area anteriorly on the lateral surface; it is present in about 70% of human occipital lobes. At times, the lunate sulcus is joined posteriorly by the anterior end of the calcarine sulcus.

PARIETO-OCCIPITAL ARCUS. The parieto-occipital arcus is a U-shaped gyrus that caps the lateral end of the parieto-occipital sulcus (Figs. 11-93; 11-94; 11-96). This small, arched convolution is usually bounded below by the occipital part of the intraparietal sulcus and behind by the superior part of the transverse occipital sulcus. The anterior limb of the gyrus is arbitrarily considered to be in the parietal lobe, and its posterior limb, in the occipital lobe because the parieto-occipital sulcus, around which the gyrus arches, forms the boundary between these two lobes.

SUPERIOR AND INFERIOR LATERAL OCCIPITAL GYRI. The horizontally oriented lateral occipital sulcus divides the lateral aspect of the occipital lobe into the superior and inferior occipital gyri (Fig. 11-93). These gyri lie behind the middle temporal sulcus, extend to the posterior border of the cerebrum, and vary in number and in shape. The superior lateral occipital gyri lie above the lateral occipital sulcus, while the inferior gyri lie below the sulcus.

Sulci and Gyri on the Lateral Surface of the Temporal Lobe

The temporal lobe is located below the lateral sulcus and in front of the occipital lobe; it is limited posteriorly by an arbitrary line drawn from the preoccipital incisure to the site at which the parieto-occipital sulcus crosses the superomedial margin of the cerebrum, about 5 cm in front of the occipital pole (Fig. 11-94). Its location on the cerebrum becomes apparent as early as the second fetal month, when a slight depression appears on the smooth surface of the developing cortex; the depression later deepens to form the **lateral cerebral fossa** and the **lateral sulcus.** The margins of this fossa continue to grow, forming the overlapping folds of the **opercula,** and the sulcus itself elongates in both directions to separate the temporal lobe from the frontal and parietal lobes anteriorly and superiorly. No specific cleft ever forms to separate the temporal lobe from the occipital lobe.

The lateral aspect of the temporal lobe is divided into three parallel gyri by two sulci. The names of the gyri present no confusion because they are simply called the **superior, middle,** and **inferior temporal gyri.** The sulcus between the superior and middle temporal gyri is the **superior temporal sulcus,** and the sulcus between the middle and inferior temporal gyri is now called the **inferior temporal sulcus.** Formerly, this latter sulcus was called the middle temporal sulcus and the term inferior temporal sulcus was reserved for the fissure located below the inferior temporal gyrus. Current terminology recognizes only two long temporal sulci on the lateral surface of the temporal lobe (the superior and inferior temporal sulci), and the sulcus below the inferior temporal gyrus (or medial to the gyrus, if it is located on the inferior surface of the lobe) is now called the **occipitotemporal sulcus.**

SUPERIOR TEMPORAL SULCUS. The superior temporal sulcus courses below the superior temporal gyrus and follows an oblique course upward and backward across the temporal lobe. It extends from a site near the temporal pole into the parietal lobe, where the angular gyrus caps its uppermost posterior end (Figs. 11-93; 11-94). It separates the superior from the inferior temporal gyri and its course is initially parallel to the lateral sulcus and then parallel to the posterior ramus of the lateral sulcus. The superior temporal sulcus is most frequently a continuous fissure, but at times it forms a discontinuous line that is separated into two or three sections.

INFERIOR TEMPORAL SULCUS. The inferior temporal sulcus separates the middle temporal gyrus from the inferior temporal gyrus (Figs. 11-93; 11-94). It follows the same general direction as the superior temporal sulcus, but it is rarely a continuous fissure. Most frequently it consists of two to six fragmented segments (the average is four), which usually appear as a disconnected line that commences at a site near the inferolateral border of the temporal lobe and may extend into the inferior parietal lobule. Often, however, it terminates in the posterior part of the temporal lobe before reaching the parietal lobe.

SUPERIOR TEMPORAL GYRUS. The superior temporal gyrus is bounded by the posterior ramus of the lateral sulcus and by the superior temporal sulcus (Figs. 11-93; 11-94). It extends from the temporal pole to the anterior part of the inferior parietal lobule, where it is continuous with parts of the supramarginal and angular gyri. The anterosuperior bank of this gyrus forms the **temporal operculum** and helps to overlap the **insula,** which lies in the floor of the lateral sulcus. This opercular surface of the superior temporal gyrus is marked by three or four **transverse temporal sulci** which define several short, obliquely oriented convolutions called the **transverse temporal gyri** (of Heschl[1]). These serve as the primary auditory receiving areas of the cerebral cortex.

MIDDLE TEMPORAL GYRUS.   The middle temporal gyrus is located between the superior and inferior temporal sulci, and it courses in a somewhat convoluted pattern backward and upward, roughly parallel to the superior temporal gyrus (Figs. 11-93; 11-94). It commences at the temporal pole and it ends in the inferior parietal lobule, where it is continuous with part of the angular gyrus and with the posterior convolution of that lobule.

INFERIOR TEMPORAL GYRUS.   The inferior temporal gyrus courses backward from the temporal pole below the inferior temporal sulcus, and it forms the inferolateral border of the temporal lobe (Figs. 11-93; 11-94). Posteriorly, it becomes continuous with the inferior lateral occipital gyri of the occipital

[1]Richard L. Heschl (1824–1881): An Austrian pathologist and anatomist (Vienna).

lobe. The inferior temporal gyrus also extends onto the inferior or ventral surface of the cerebral hemisphere and is limited there by another anteroposteriorly coursing fissure called the **occipitotemporal sulcus.** The portion of the inferior temporal gyrus that extends beyond the inferolateral border of the temporal lobe and is located on the inferior temporal surface lateral to the occipitotemporal sulcus is designated the **lateral occipitotemporal gyrus.**

Insula or Central Lobe

The insula lies hidden in the depths of the lateral sulcus on the lateral aspect of the cerebrum, and it can be seen only if the lips of the sulcus are bent backward or are cut away (Fig. 11-97). The cortical gyri that overlap this central or insular lobe of the cerebral cortex are derived from the overgrowth of the frontal, parietal, and temporal lobes. These gyri are separated by the anterior (horizontal), anterior ascending, and posterior rami of the lateral sulcus, and they are called the frontal, frontoparietal, and temporal opercula.

OPERCULA OF THE INSULA.   The **frontal operculum** actually includes two parts, an *orbital part,* which lies below the anterior (horizontal) ramus of the lateral sulcus and consists of the **pars orbicularis** of the inferior frontal gyrus, and a *frontal part,* which lies between the anterior (horizontal) and anterior ascending rami and consists of the

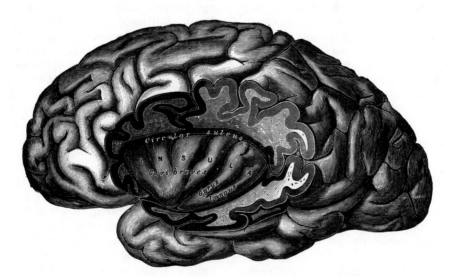

FIG. 11-97.   The insula of the left side, exposed by removing the opercula.

pars triangularis of the inferior frontal gyrus (Fig. 11-94). The **frontoparietal operculum** is bounded by the anterior ascending ramus of the lateral sulcus and the upturned end of the posterior ramus, and it consists of the **pars opercularis** of the inferior frontal gyrus, plus the lower parts of the precentral and postcentral gyri, and the anterior part of the supramarginal gyrus (Fig. 11-94). The **temporal operculum** is the largest of the opercula of the insula and, located below the posterior ramus of the lateral sulcus, it consists of the superior temporal gyrus and the transverse temporal gyri.

SULCI AND GYRI OF THE INSULA. The insula is encircled and separated from the opercula by a deep furrow called the **circular sulcus of the insula** (Fig. 11-97). This is sometimes called the limiting sulcus of Reil.[1] When the opercula are removed, the insula is seen as a pyramidal eminence, the apex of which is called the **limen insulae.** This apex is found on the basal surface of the cerebrum and it points toward the anterior perforated substance. From it the sulci and gyri of the insula are directed superiorly in a radial manner. The deepest sulcus, called the **central sulcus of the insula,** divides the insula into a large anterior and a smaller posterior part. The *anterior part* is subdivided by several shallow **sulci breves** (short sulci) into three to five **gyri breves insulae,** while the more elongated *posterior part* is formed by a single, long gyrus, the **gyrus longus insulae,** which is usually divided by a slight furrow that bifurcates its upper end (Fig. 11-97). The cortical gray substance of the insula is continuous with the gray substance in the cortex of the surrounding opercula, and the deep surface of the insula overlies the putamen of the neostriatum. Stimulation of the insular cortex in the human brain results in visceral sensations and autonomic responses.

MEDIAL SURFACE OF THE CEREBRUM

The two cerebral hemispheres are apposed at the median sagittal plane and their medial surfaces are flattened, oriented vertically, and interconnected in their central regions by a large commissure of white matter called the corpus callosum (Figs. 11-95; 11-98). Immediately surrounding much of the corpus callosum on the medial cerebral surface are the sulcus of the corpus callosum, the cingulate gyrus, and the cingulate sulcus. Beyond these are the outer sulci and gyri that mark the medial aspect of the frontal, parietal, and occipital lobes. Below are located the convolutions and sulci on the inferomedial surface of the temporal lobe, but these are described with the inferior surface of the cerebrum.

CORPUS CALLOSUM. The sectioned corpus callosum presents the appearance of a broad-arched band; it forms one of the most distinguishing features on the medial surface of the cerebrum (Fig. 11-98). If the two cerebral hemispheres are observed when they are still interconnected, the corpus callosum can be located in the floor of the longitudinal cerebral fissure. It consists of fascicles of myelinated nerve fibers whose main function appears to be the transmission of information between the neocortical portions of the two hemispheres during the learning process. Below the anterior and middle parts of the corpus callosum, however, the cerebral hemispheres are also joined by three other tracts—the hippocampal commissure, the anterior commissure, and the crossed fibers of the optic chiasma.

When visualized on the medial surface of one cerebral hemisphere in a split brain, sectioned along the median sagittal plane, the corpus callosum is seen to consist of three divisions: (1) a curved anterior end called the **genu,** which gradually tapers into a thinner portion called the **rostrum,** the latter being continued downward and backward in front of the anterior commissure to join the **lamina terminalis;** (2) a thick, rounded, posterior end termed the **splenium,** situated dorsal to the pineal body and the corpora quadrigemina, from which it is separated by the tela choroidea of the third ventricle; and (3) a middle portion or **trunk** of the corpus callosum, which arches dorsally between these anterior and posterior extremities (Figs. 11-98; 11-99).

Central to the curvature of the anterior half of the corpus callosum can be seen the laminae of the **septum pellucidum.** These are two thin plates of gray and white matter that extend on each side between the corpus callosum and another curved, flattened fascicle of white matter, the **fornix** (Figs. 11-98; 11-99).

---

[1]Johann C. Reil (1759–1813): A German physician, neurologist, and histologist.

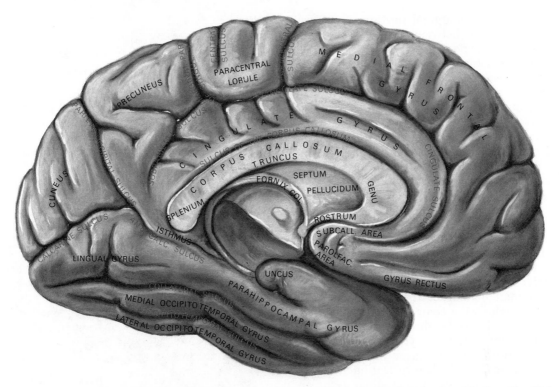

FIG. 11-98. The medial surface of the left cerebral hemisphere in a brain that has been sectioned in the midsagittal plane. The brain stem and cerebellum have been removed by an oblique section through the thalamus to expose the inferomedial surface of the temporal lobe and the inferior surface of the occipital lobe. (After Sobotta.)

SULCUS OF THE CORPUS CALLOSUM. The sulcus of the corpus callosum (callosal sulcus) is a slit-like fissure that separates this commissure of white matter from the overlying cerebral cortical substance (Fig. 11-98). It commences below or ventral to the rostrum of the corpus callosum and then encircles the convexity of the callosal trunk; finally, it curves around the splenium, where it then becomes continuous with the hippocampal sulcus. Immediately surrounding the corpus callosum and separated from it by the callosal sulcus is the cingulate gyrus.

CINGULATE GYRUS. The cingulate gyrus is an arched convolution that lies in its entirety adjacent to the corpus callosum and is separated from it by the sulcus of the corpus callosum (Fig. 11-98). It begins below the rostrum, curves around in front of the genu, extends along the dorsal surface of the trunk, and then turns downward behind the splenium of the corpus callosum, where it is connected by the narrow **isthmus of the cingulate gyrus** to the parahippocampal gyrus. The cingulate gyrus is separated from the **medial frontal gyrus** by the cingulate sul-

cus and from the **precuneus** by the more variable **subparietal** (or suprasplenial) **sulcus.** The cingulate gyrus has many reciprocal interconnections with the anterior thalamic nuclei, and frequently it is considered to be a part of the "limbic system."

CINGULATE SULCUS. Visible on the medial surface of the cerebral hemisphere is the cingulate sulcus, a fissure that begins below the anterior end of the corpus callosum, from which it is separated by the cingulate gyrus (Fig. 11-98). This sulcus courses forward and then upward and backward, paralleling the rostrum, genu, and the trunk of the corpus callosum, as well as the callosal sulcus. Posteriorly, the cingulate sulcus ascends a short distance behind the upper end of the central sulcus to the superomedial border of the hemisphere, where it separates the precuneus on the medial surface of the parietal lobe from the paracentral lobule. This upturned posterior end of the cingulate sulcus is sometimes called the **marginal portion of the cingulate sulcus.** Another branch, the **paracentral sulcus,** usually curves dorsally from the cingulate sulcus, at some distance rostral to the marginal portion, and marks the boundary between the medial frontal gyrus and the paracentral lobule (Fig. 11-98).

The variable **subparietal sulcus** is located more posteriorly, but follows the same direction as the cingulate sulcus beyond the site at which the marginal portion curves dorsally. This sulcus sometimes is directly continuous with the cingulate sulcus, but frequently it is not, and it incompletely separates the more posterior part of the cingulate gyrus from the precuneus (Fig. 11-98).

Medial Surface of the Frontal Lobe

On its medial surface, the frontal lobe stretches from the frontal pole anteriorly to a vertically oriented arbitrary line posteriorly. This line extends downward from the superomedial border of the cerebrum, at the site at which the central sulcus curves onto the medial surface, through the **paracentral lobule** and cingulate gyrus to the sulcus of the corpus callosum, near the midpoint of the callosal trunk. The rostral halves of the paracentral lobule and cingulate gyrus and the medial frontal gyrus form most of the medial surface of the frontal lobe (Figs. 11-95; 11-98).

The **medial frontal gyrus** is clearly separated inferiorly from the portion of the cingulate gyrus located within the frontal lobe by the cingulate sulcus, while the posterior end of the medial frontal gyrus and the anterior half of the paracentral lobule are separated by an ascending branch of the cingulate sulcus that is called the paracentral sulcus (Fig. 11-98). The convolution found between the paracentral and central sulci, i.e., much of the anterior half of the paracentral lobule, actually is the medial extension of the precentral gyrus from the lateral cerebral surface, and it functions as the somatomotor region for much of the lower extremity.

Anteriorly the medial frontal and cingulate gyri curve nearly 180° backward around the genu of the corpus callosum, thereby becoming oriented posteriorly under the rostrum of the corpus callosum. The medial frontal gyrus then becomes continuous with the **subcallosal** (or parolfactory) **area**, being incompletely separated from it by the rather shallow **anterior parolfactory sulcus** (Fig. 11-99). Behind the subcallosal area, between the **lamina terminalis** and the **posterior**

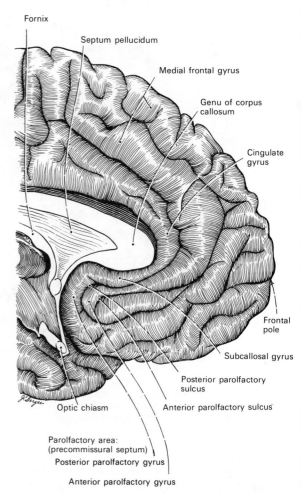

Fornix

Septum pellucidum

Medial frontal gyrus

Genu of corpus callosum

Cingulate gyrus

Frontal pole

Subcallosal gyrus

Posterior parolfactory sulcus

Optic chiasm

Anterior parolfactory sulcus

Parolfactory area: (precommissural septum)

Posterior parolfactory gyrus

Anterior parolfactory gyrus

Fig. 11-99. The medial surface of the anterior part of the left cerebral hemisphere, showing the subcallosal and parolfactory regions. (From E. C. Crosby, T. Humphrey, and E. W. Lauer, *Correlative Anatomy of the Nervous System,* Macmillan Publishing Co., New York, 1962.)

parolfactory sulcus, is the **paraterminal gyrus,** which forms the anteroinferior continuation of the **indusium griseum,** a thin lamina of gray matter that overlies the superior surface of the corpus callosum.

Medial Surface of the Parietal Lobe

The anterior boundary of the parietal lobe on the medial surface of the cerebrum is formed by the central sulcus, from which then is continued an imaginary line extending from the termination of this sulcus ventrally to the sulcus of the corpus callosum (Fig. 11-95). The posterior boundary is formed by the prominent parieto-occipital sulcus. Between these boundaries are located the posterior portions of both the paracentral lobule and the cingulate gyrus, and the entire **precuneus** (Figs. 11-95; 11-98). The marginal branch of the cingulate sulcus separates the paracentral lobule from the precuneus, and the **parieto-occipital sulcus** courses between the precuneus and the anterior limit of the medial surface of the occipital lobe. The cingulate sulcus and its caudal continuation, the subparietal sulcus, limit both the paracentral lobule and the precuneus ventrally from the cingulate gyrus (Fig. 11-98).

The posterior part of the paracentral lobule consists of the extension of the postcentral gyrus from the lateral surface onto the medial cerebral surface, and it functions as the primary somatosensory receiving region for much of the lower limb. The precuneus, located on the dorsomedial part of the hemisphere, represents the medial extension from the superolateral surface of the superior parietal lobule.

Medial Surface of the Occipital Lobe

The medial surface of the occipital lobe is marked by the parieto-occipital and calcarine sulci, which form the boundaries of a wedge-shaped lobule called the **cuneus.** Extending backward into the occipital lobe from the inferior surface of the temporal lobe is the posterior end of the **collateral sulcus.** Between the calcarine sulcus and the collateral sulcus is located the **lingual gyrus,** which is the caudal continuation of the parahippocampal gyrus (Figs. 11-95; 11-98).

PARIETO-OCCIPITAL SULCUS. The parieto-occipital sulcus commences on the convex superolateral aspect of the cerebral hemisphere, about 5 cm in front of the occipital pole. After coursing superiorly for slightly more than 1 cm, it continues around the superomedial border of the cerebrum and, as a deep cleft, it forms the rostral boundary of the occipital lobe on the medial surface (Fig. 11-95). Its course is directed downward and slightly forward, where it then joins the calcarine sulcus at an angle of about 45°. Beyond this junction the sulcus, which continues forward, is called the anterior part of the calcarine sulcus. The parieto-occipital sulcus separates the precuneus from the cuneus, and in the human brain, it almost always joins the calcarine sulcus, thereby forming a Y-shaped configuration (Fig. 11-98). If the junction between these two sulci is opened widely, a submerged transitional gyrus, the **anterior cuneolingual gyrus,** which was buried by the overgrowth of the surrounding gyri, can be exposed.

CALCARINE SULCUS. The calcarine sulcus begins slightly above the occipital pole, usually on the medial surface but at times on the lateral surface of the occipital lobe. Along the medial aspect of the cerebrum it courses forward between the cuneus and the lingual gyrus, and it is frequently characterized by a dorsal curve, in front of which it is joined by the descending parieto-occipital sulcus (Figs. 11-95; 11-98). It continues anteriorly below the isthmus of the cingulate gyrus and splenium of the corpus callosum, where, for a short distance, it forms the upper boundary of the parahippocampal gyrus before terminating.

The portion of the calcarine sulcus in front of the junction with the parieto-occipital sulcus is frequently called the **anterior part of the calcarine sulcus** (Fig. 11-98), and it gives rise to a prominence called the **calcar avis,** which is visible on the medial wall of the posterior horn of the lateral ventricle. The primary visual receiving area, the **striate cortex,** is located on the upper and lower banks, as well as in the depth of the **posterior part** of the calcarine sulcus (i.e., the portion of the sulcus behind its junction with the parieto-occipital sulcus). Another submerged gyrus, the **posterior cuneolingual gyrus,** can be seen crossing the floor of the sulcus near the occipital pole. This can be exposed by separating the lips of the posterior part of the calcarine sulcus.

INFERIOR SURFACE OF THE CEREBRUM

The inferior or ventral surface of the cerebral hemisphere consists anteriorly of the orbital surface of the frontal lobe and, more posteriorly, of the inferior surface of the temporal and occipital lobes. The stem of the lateral sulcus commences on the inferior surface of the cerebrum, near the anterior perforated substance, and forms a natural division between the orbital surface of the frontal lobe and the inferior surface of the temporal lobe. The inferior surface of the occipital lobe and that of the more posterior part of the temporal lobe rest on the tentorium of the cerebellum, while the anterior part of the inferior temporal surface lies on the floor of the middle cranial fossa.

Inferior or Orbital Surface of the Frontal Lobe

The inferior or orbital surface of the frontal lobe is concave and rests on the orbital plate of the frontal bone laterally, on the cribriform plate of the ethmoid bone medially, and on the lesser wing of the sphenoid bone posteriorly. Together these bony structures form the **anterior cranial fossa** on the inner aspect of the neurocranium. Most medially, the inferior surface of the frontal lobe presents a straight and deep anteroposteriorly oriented sulcus called the **olfactory sulcus** (Fig. 11-100), within which lies the olfactory bulb and the olfactory tract. Medial to this sulcus is found the **gyrus rectus,** a narrow, straight gyrus that occupies the region along the ventromedial margin of the frontal lobe. The gyrus rectus is continuous anteriorly with the medial frontal gyrus and posteriorly it meets the parolfactory area (Figs. 11-98; 11-100).

Lateral to the olfactory sulcus, the orbital surface contains the somewhat variable **orbital sulci.** These often assume an H-shape, the lateral and medial sides of which course parallel to the gyrus rectus. Thus, the **lateral, medial,** and **transverse orbital sulci** divide this region into the lateral, anterior, posterior, and medial orbital gyri (Fig. 11-100).

The **lateral orbital gyrus** is continuous around the inferolateral margin of the cerebral cortex with the inferior frontal gyrus.

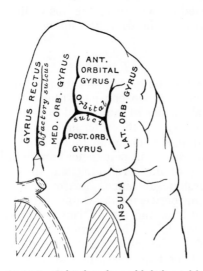

FIG. 11-100. Orbital surface of left frontal lobe.

The **anterior orbital gyrus** is located in front of the transverse orbital sulcus and represents the orbital surface continuation of the middle frontal gyrus around the frontal pole of the cerebrum. The **posterior orbital gyrus** is located behind the transverse orbital sulcus and it extends backward nearly to the pyramidal apex of the insula, called the limen insula. The **medial orbital gyrus** lies medial to the medial orbital sulcus and becomes continuous with the superior frontal gyrus around the frontal pole of the cerebrum (Fig. 11-100).

Inferomedial Surface of the Temporal Lobe

The inferomedial surface of the temporal lobe has a concave shape and is continuous posteriorly with the tentorial surface of the occipital lobe. It is traversed by the **collateral** and **occipitotemporal sulci,** which course posteroanteriorly, as well as by the **rhinal sulcus.** These sulci divide the region into **medial** and **lateral occipitotemporal gyri** and the **parahippocampal gyrus,** the anteromedial extremity of which curves back to form a hook called the **uncus** (Figs. 11-95; 11-98).

COLLATERAL SULCUS. The collateral sulcus is a relatively long fissure that commences posteriorly near the occipital pole and courses rostrally along the tentorial surface of the occipital lobe and the inferomedial surface of the temporal lobe, where it may be continuous with the rhinal sulcus but more frequently is not (Figs. 11-95; 11-98). Caudally, it separates the lingual gyrus from the medial occipitotemporal gyrus (formerly called the fusiform gyrus), and more anteriorly it courses between the parahippocampal gyrus, which is the rostral continuation of the lingual gyrus, and the medial occipitotemporal gyrus. The **rhinal sulcus,** which is directed anteriorly in the same line as the collateral sulcus, separates the terminal part of the parahippocampal gyrus and its anteromedial continuation, the uncus, which is archeopallial in origin, from the remainder of the temporal lobe more laterally located.

OCCIPITOTEMPORAL SULCUS. The occipitotemporal sulcus courses parallel but lateral to the collateral sulcus. Its origin, like that of the collateral sulcus, is in the occipital lobe but it commences more anterior to the occipital pole than does the lateral sulcus (Figs. 11-95; 11-98). The occipitotemporal sulcus is usually a discontinuous fissure of three or more segments, and throughout its extent, it separates the medial occipitotemporal gyrus from the lateral occipitotemporal gyrus. In some older descriptions, this sulcus is called the inferior temporal sulcus because the lateral occipitotemporal gyrus is continuous around the inferolateral margin of the temporal lobe with the inferior temporal gyrus.

PARAHIPPOCAMPAL GYRUS. The parahippocampal gyrus (formerly called the hippocampal gyrus) is located on the upper aspect of the inferomedial surface of the temporal lobe, where it arches forward toward the temporal pole from the isthmus of the cingulate gyrus, with which it is continuous (Figs. 11-95; 11-98). Posteriorly, the parahippocampal gyrus is continued into the occipital lobe as the lingual gyrus, while anteromedially its extension forms the uncus, a rounded convolution that has a hooked shape because it is reflected posteriorly and medially upon itself. Separating the parahippocampal gyrus from the medial occipitotemporal gyrus are the rhinal and collateral sulci. Above and posteriorly, the parahippocampal gyrus is bounded by the anterior part of the calcarine sulcus, while internally it is bounded by the hippocampal sulcus, which separates the parahippocampal gyrus from the dentate gyrus.

MEDIAL OCCIPITOTEMPORAL GYRUS. The medial occipitotemporal gyrus (formerly called the fusiform gyrus) is a long, posteroanteriorly oriented convolution that lies on the tentorial or basal surface of the cerebrum and separates the parahippocampal and lingual gyri from the lateral occipitotemporal gyrus (Figs. 11-95; 11-98). It is bounded by the collateral and rhinal sulci on its medial side and by the occipitotemporal sulcus laterally, and it extends from nearly the occipital pole to the temporal pole, where it merges with the anterior ends of the three temporal gyri from the lateral surface of the cerebrum.

LATERAL OCCIPITOTEMPORAL GYRUS. The lateral occipitotemporal gyrus lies parallel to the medial occipitotemporal gyrus, and although anteriorly it reaches as far as the temporal pole, it does not extend posteriorly into the occipital lobe nearly as far as does the medial occipitotemporal gyrus (Figs. 11-95; 11-98). It is separated medially from the medial occipitotemporal gyrus by the occipitotemporal sulcus. Laterally it forms

the inferolateral margin of the cerebral hemisphere and it is continuous around this border with the inferior temporal gyrus.

Inferior Surface of the Occipital Lobe

The inferior surface of the occipital lobe overlies the arched crescenteric fold of dura mater, called the tentorium cerebelli. Because the tentorium covers the cerebellum, this surface is sometimes referred to as the tentorial surface of the occipital lobe. By accommodating the superior surface of the cerebellum, its shape is somewhat concave and it extends backward all the way to the occipital pole from the temporal lobe, beyond an arbitrary line drawn between the occipital incisure and the isthmus of the cingulate gyrus. Marking the inferior surface of the occipital lobe are the posterior extensions of the collateral and occipitotemporal sulci, which were described above. The gyri are also continuous with those of the temporal lobe across the arbitrary temporo-occipital boundary; they include the posterior part of the parahippocampal gyrus, called the **lingual gyrus,** and the posterior segments of the medial and lateral occipitotemporal gyri (Fig. 11-98).

THE RHINENCEPHALON

The term rhinencephalon is derived from two Greek words, *rhis* (rhin), meaning nose, and *enkephalos,* meaning brain. It refers to the olfactory brain and, in agreement with Brodal (1981), it is used strictly here to incorporate only the olfactory bulb and those regions that receive nerve fibers from the olfactory bulb. This more classical use of the term avoids some of the confusion in the literature that has arisen during the last several decades and focuses on the structures that are principally related to the afferent pathways for the sense of smell in the microsmotic infrahuman primate and human brain. In certain lower vertebrates, the rhinencephalon is large, and in these macrosmotic mammals much of the pallium or cortex is occupied by olfactory structures. In the subhuman primate and human brain, however, the rhinencephalon is relegated to more or less hidden regions on the medial and inferior surfaces of the hemispheres due to the enormous development of the neocortex.

In essence, the rhinencephalon consists of structures derived from the primitive olfactory and piriform lobes (Fig. 11-19) and includes the associated prepiriform and periamgydaloid cortical regions, or **paleocortex.** In this description the following structures will be considered as comprising the rhinencephalon: the olfactory bulb and tracts, including the medial, intermediate, and lateral olfactory striae; the anterior olfactory nucleus; the olfactory trigone; the anterior perforated substance (or olfactory tubercle) and diagonal band; the prepiriform, periamygdaloid, and entorhinal cortical regions of the piriform lobe; as well as the corticomedial nuclei of the amygdaloid complex.

Olfactory Bulb

The olfactory bulb is a flat, ovoid mass of gray substance that rests on the lateral border of the cribriform plate of the ethmoid bone. Its anterior and inferior surfaces receive the olfactory nerve fibers, called **fila olfactoria.** These pass upward through the cribriform plate from the specialized olfactory epithelium, which contains the neurosensory olfactory receptor cells located in the superior part of the nasal cavity. The smooth superior surface of the olfactory bulb underlies the rostral part of the olfactory sulcus, a fissure that courses between the gyrus rectus and the medial orbital gyrus on the orbital surface of the frontal lobe (Fig. 11-89). Because the olfactory bulb develops in the embryo around an ependymal cell-lined cavity that extends anteriorly and ventrally from the lateral ventricle, its central core in the adult still contains some ependymal cells that mark the site of the primitive olfactory ventricle.

The olfactory bulb is the primary relay station for olfactory impulses generated in the olfactory receptor cell. Although the human olfactory bulb does not have so distinct a laminar structure as that seen in certain other mammals, its organization into layers is still recognizable (Fig. 11-101). Commencing on its superficial aspect, the olfactory bulb consists of the following six laminae.

OLFACTORY NERVE FIBER LAYER. The olfactory nerve fiber layer is composed of axons derived from olfactory receptor bipolar neurons, whose cell bodies are located in a rather evenly distributed pattern in the olfactory epithelium of the nasal cavity. The

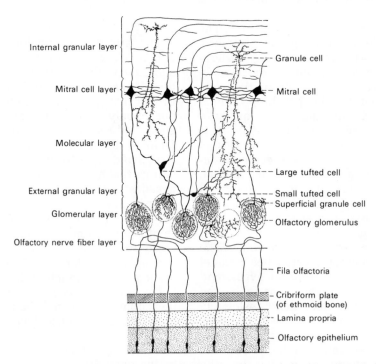

Internal granular layer
Mitral cell layer
Molecular layer
External granular layer
Glomerular layer
Olfactory nerve fiber layer

Granule cell
Mitral cell
Large tufted cell
Small tufted cell
Superficial granule cell
Olfactory glomerulus
Fila olfactoria
Cribriform plate (of ethmoid bone)
Lamina propria
Olfactory epithelium

FIG. 11-101.   The neuronal structure (right) and layers (left) of the olfactory bulb. (From Benninghoff and Goerttler, *Lehrbuch der Anatomie des Menschen*, 8th ed., Vol. 3, Urban and Schwarzenberg, Munich, 1967.)

fibers are collected into about 20 small nerve bundles and most of these enter the anterior and inferior aspect of the olfactory bulb to form a shallow superficial lamina.

GLOMERULAR LAYER. Immediately deep to the olfactory nerve fiber layer is located the glomerular layer. This consists of a band of synaptic spheres called **olfactory glomeruli,** within which the incoming olfactory receptor fibers synapse, primarily with descending dendrites from **mitral cells.** The olfactory glomeruli, which measure as much as 0.1 mm in diameter, are complex synaptic pools that contain hundreds of synaptic processes from tufted cells and multipolar neurons, called periglomerular cells, in addition to the incoming olfactory receptor fibers and dendrites from the mitral cells.

EXTERNAL GRANULAR LAYER. Deep to the glomerular layer is located a lamina of rounded cells that forms the external granular layer. These cells are small and multipolar and are considered periglomerular cells because their processes participate in the formation of the glomeruli.

MOLECULAR LAYER. Deep to the external granular layer and separating it from the mitral cell layer is a rather wide lamina of nerve cell processes called the molecular layer. Many of these processes are dendrites from mitral cells and they course superficially toward the synaptic glomeruli, but others come from granule cells and tufted cells that also participate in the glomerular synaptic pools. A few scattered nerve cell bodies are also seen in the molecular layer.

MITRAL CELL LAYER. The mitral cell layer is a thin lamina consisting of a single row of sparsely placed mitral cell bodies that have a pyramidal shape. These are the largest neurons in the olfactory bulb and they send their dendrites toward the surface to the olfactory glomeruli and their axons inward to the core of the olfactory bulb to form many of the fibers in the olfactory tract.

INTERNAL GRANULAR LAYER. Deep to the mitral cell layer is the internal granular layer, which consists of rounded or stellate neuronal cell bodies whose dendritic processes extend superficially into the molecular layer, and some may even reach the olfactory glomeruli. The cells of the internal granular layer also synapse with collaterals from the mitral cells. Deep to the internal granular layer, near the central core of the olfactory bulb, are small collections of ependymal cells that remain as vestiges of the lining of the primitive olfactory ventricle. Also at this level are the axons of the mitral and tufted cells that form the fibers of the olfactory tract (Fig. 11-101).

Anterior Olfactory Nucleus

In its caudal part, the olfactory bulb loses its laminated structure and the cells of the internal granular layer are replaced by scattered groups of medium-sized, pyramidal-shaped neurons that extend diffusely along the olfactory tract and collectively are called the **anterior olfactory nucleus** (Fig. 11-102). The neuronal clusters of the anterior olfactory nucleus in the human brain also extend behind the olfactory tracts along the olfactory striae to the olfactory trigone, the anterior perforated substance, the parolfactory area, and laterally to the prepiriform cortex. Some fibers in the olfactory tract synapse with the cells in the anterior olfactory nucleus, and then the axons of the postsynaptic cells enter the olfactory tract to join those olfactory tract fibers that do not synapse.

Olfactory Tract, Trigone, and Striae

OLFACTORY TRACT. The olfactory tract is about 20 mm long and courses backward from the olfactory bulb as a narrow white band, triangular in section, that lies below the olfactory sulcus along the inferior surface of the frontal lobe (Fig. 11-103). It consists of axons of the mitral and tufted cells of the ipsilateral olfactory bulb, other fibers from cells in the anterior olfactory nucleus, recurrent collaterals from cells in the ipsilateral olfactory bulb, and efferent axons from the opposite olfactory bulb. These latter fibers cross in the anterior commissure to synapse on neurons of the anterior olfactory nucleus, as well as on mitral cells in the contralateral olfactory bulb. Just rostral to the anterior perforated substance, the olfactory tract joins the posteroinferior surface of the frontal lobe at a triangular expansion, the olfactory trigone. The olfactory tract then diverges into three strands: medial, intermediate, and lateral olfactory striae.

OLFACTORY TRIGONE. The olfactory trigone is a small, smooth, triangular wedge of cortical gray matter, located on both sides of the brain just above the optic nerve; it consists of the diverging fibers of the olfactory tract

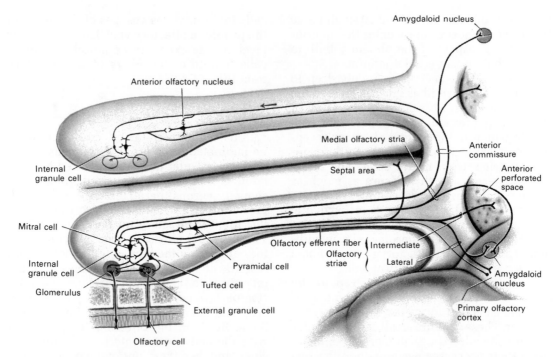

FIG. 11-102. The olfactory pathways. Note the synaptic connections in the anterior olfactory nucleus and diverging olfactory tract fibers that form the olfactory striae. (From C. R. Noback, *The Human Nervous System*, McGraw-Hill, New York, 1967.)

(Fig. 11-103). With its apex directed forward and its base oriented posteriorly toward the anterior perforated substance, the olfactory trigone is bounded medially and laterally by the roots of the olfactory tracts, which continue to divide to form the medial and lateral olfactory striae.

OLFACTORY STRIAE. As the olfactory tract diverges caudally, its fibers form the **medial** and **lateral olfactory striae** (or tracts). These initially course along the medial and lateral sides of the olfactory trigone and enclose two of its three sides. A small third fascicle of fibers, called the **intermediate olfactory stria,** is formed between the medial and lateral striae (Fig. 11-103).

Medial Olfactory Stria. The medial olfactory stria passes medially along the rostral border of the anterior perforated substance and is augmented by a thin layer of gray matter to form the **medial olfactory gyrus.** Near the ventromedial border of the frontal lobe, it is accompanied in its course by the diagonal band (of Broca). These superficially located fibers ascend on the medial aspect of the hemisphere, immediately in front of the paraterminal body, which separates them from the upper end of the laminae terminalis. The fibers of the diagonal band enter

the paraterminal (or subcallosal) gyrus and the medial olfactory stria becomes continuous with the parolfactory area. Beyond the olfactory trigone, the medial olfactory stria supplies fibers to the medial part of the anterior perforated substance, the parolfactory area (of Broca), and the paraterminal gyrus. Other fibers course across the anterior commissure to terminate in the opposite olfactory bulb.

Lateral Olfactory Stria. The lateral olfactory stria becomes the superficial layer of the olfactory cortex and forms the **lateral olfactory gyrus.** It passes immediately laterally and caudally, around the lateral border of the anterior perforated substance and toward the anteroinferior corner of the insula (the limen insula), where it bends sharply backward and medially toward the region of the uncus. The fibers of the lateral olfactory stria terminate in the lateral part of the anterior perforated substance, the prepiriform and periamygdaloid areas, the corticomedial nuclei of the amygdaloid complex, and even in portions of the entorhinal cortex.

Intermediate Olfactory Stria. The intermediate olfactory stria arises from the olfactory tract between the medial and lateral ol-

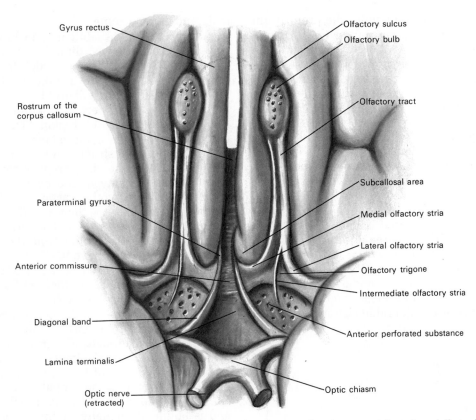

Gyrus rectus

Rostrum of the
corpus callosum

Paraterminal gyrus

Anterior commissure

Diagonal band

Lamina terminalis

Optic nerve
(retracted)

Olfactory sulcus

Olfactory bulb

Olfactory tract

Subcallosal area

Medial olfactory stria

Lateral olfactory stria

Olfactory trigone

Intermediate olfactory stria

Anterior perforated substance

Optic chiasm

FIG. 11-103. The basal aspect of the forebrain with the optic chiasm and optic nerves reflected caudally to show the olfactory structures. (Redrawn after M. B. Carpenter, *Human Neuroanatomy,* 7th ed., Williams and Wilkins, Baltimore, 1976.)

factory striae. It passes directly posteriorly and its fibers terminate within the ipsilateral anterior perforated substance.

Anterior Perforated Substance

The anterior perforated substance is a flat, smooth, nearly triangular area of gray matter located on the basal surface of the cerebral hemisphere, just behind the olfactory trigone and olfactory striae (Fig. 11-103); in macrosmotic mammals it is marked by a small, oval elevation, the **olfactory tubercle.** It is named for the numerous minute orifices on its basal surface that are formed by many small penetrating branches of the middle cerebral artery, which supplies the region, and many small perforations become evident after removal of the pia mater. The fibers of the intermediate olfactory stria enter the anterior perforated substance directly from the olfactory tract. Posteromedially, the anterior perforated substance is bounded by the diagonal band of Broca and the optic nerve, while posterolaterally it is limited by the lat-

eral olfactory stria, as that bundle turns medially toward the prepiriform cortex.

The anterior perforated substance consists of three well defined layers in its middle region, but its rostral and caudal parts are poorly differentiated and show only an outer non-neurocellular plexiform zone that overlies a thin layer of gray matter. In the middle region, however, deep to the **external plexiform zone,** is found an intermediate **pyramidal cell layer,** and below this lies a **polymorphic layer** of cells. Additionally, significant collections of pyramidal and granule cells, called the *islands of Calleja,*[1] are scattered within the anterior perforated substance; their function is unknown.

Piriform Lobe

The piriform lobe is so named because it is pear-shaped and it retains that appearance in many macrosmotic lower mammals. From

[1]Camillo Calleja y Sanchez (?–1913): A Spanish anatomist (Madrid).

rostral to caudal it includes the prepiriform area and the periamygdaloid (or piriform) area (Figs. 11-104; 11-105), which occupy the upper anterior aspect of the uncus and the entorhinal area. The latter constitutes much of the parahippocampal gyrus in the human brain (area 28). Whereas the prepiriform and periamygdaloid cortical areas are characterized by three layers, the entorhinal cortex, which is much larger in the human brain than in the brains of lower mammals, is a somewhat specialized five-layered region; it probably represents a transitional form of cortical structure, somewhat simpler than the typical six-layered isocortex.

The **prepiriform area,** frequently called the lateral olfactory gyrus, is the anterior part of the piriform lobe. It consists of a thin layer of gray substance underlying the inner border of the superficial lateral olfactory stria. It is of considerable size in lower ani-

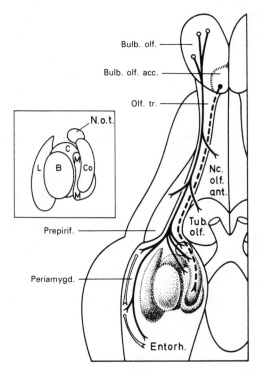

Fig. 11-104. A diagrammatic representation of the sites of termination of some of the fibers in the olfactory tract. Insert diagram shows subdivisions of the amygdaloid complex: N.o.t., nucleus of the olfactory tract. Amygdaloid nuclei: M, medial; Co, cortical; C, central; B, basal; L, lateral. (From A. Brodal, *Neurological Anatomy,* 3rd ed., Oxford University Press, New York, 1981, after H. J. Lammers, in *Neurobiology of the Amygdala,* Plenum Press, New York, 1972.)

mals and is also quite evident in the fetal human brain. In the adult human brain its relative size becomes diminished and it forms only a slight elevated ridge that initially curves laterally toward the insula and then stretches caudomedially, blending imperceptibly with the periamygdaloid area in the rostral amygdaloid region. The **periamygdaloid area** is a three-layered paleocortical region that structurally is similar to, but extends behind, the prepiriform area and overlies the amygdaloid complex (Fig. 11-105). Unlike all other primary afferent cortical receiving areas, the prepiriform and periamygdaloid areas do not receive their afferent fibers as projections from the thalamus, and they do not have large numbers of stellate (or granule) cells, which are typical in the primary visual, auditory, and somesthetic cortical receiving areas.

The **entorhinal area** is the cortical region of the most posterior part of the piriform lobe in lower mammals, and in the human brain it receives its maximal development, forming a large part of the exposed surface of the rostral parahippocampal gyrus (area 28). It is limited laterally by the rhinal sulcus and is bounded medially by the hippocampal sulcus (Fig. 11-105). The entorhinal area is a five-layered cortical region that until recently was thought to not receive direct olfactory bulb fibers. It now appears that in several animal species the entorhinal area not only receives projections from the prepiriform and periamygdaloid areas, but from the olfactory bulb as well. The entorhinal region is continuous medially by way of the subicular zones with the beginning of Ammon's horn.

For many years it has been known that the prepiriform and periamygdaloid areas receive direct projections from the olfactory bulb by way of the lateral olfactory stria, and they have been considered, therefore, the primary olfactory cortex. In contrast, the entorhinal area was thought to receive only olfactory impulses through synaptic connections in the prepiriform and periamygdaloid areas, and was therefore considered a secondary olfactory cortical region. Recent research utilizing new anatomic methods on several animal species has offered evidence that certain parts of the entorhinal area receive direct olfactory bulb projections, as well as input from the known primary olfactory cortical regions.

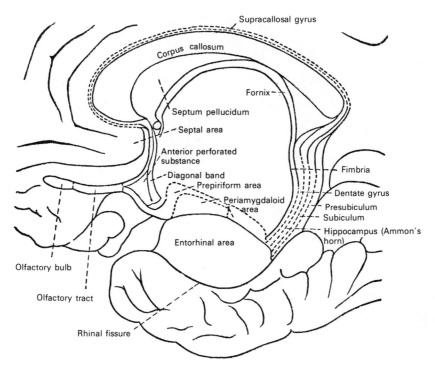

FIG. 11-105. A schematic drawing showing the rhinencephalic structures of the right hemisphere as seen from a medial view. Note the prepiriform, periamygdaloid, and entorhinal parts of the piriform lobe, bounded laterally by the rhinal fissure. (From O. Larsell, *Anatomy of the Nervous System,* 2nd ed., Appleton-Century Crofts, New York, 1951.)

Amygdaloid Body

The amygdaloid body consists of a group of nuclei, located in the dorsomedial part of the rostral temporal lobe, that underlies or forms a bulge called the uncus (Fig. 11-106). The uncus is partially covered by the periamygdaloid cortex. The amygdala overlies the rostral extremity of the temporal horn of the lateral ventricle (Fig. 11-107), and it receives its name from its almond-like shape. Its nuclear masses are divisible into two major groups: the **corticomedial nuclei** and the **basolateral nuclei.** The central amygdaloid nucleus, the boundaries of which are somewhat obscure, is usually included in the corticomedial group.

The **corticomedial nuclear group** in the human amygdaloid body is positioned in the dorsomedial part of the complex because of the nature of the rotation of the temporal lobe. It includes the *anterior amygdaloid area,* the *nucleus of the lateral olfactory tract,* and the *central, medial,* and *cortical amygdaloid nuclei* (Fig. 11-104). The **basolateral nuclear group** consists of the *lateral amygdaloid nucleus,* the *basal amyg-*

daloid nucleus, and dorsomedial to the basal nucleus and indistinctly separated from the overlying cortex, the *accessory basal amygdaloid nucleus* (Fig. 11-104). As one ascends the phylogenetic scale, the portion of the amygdaloid body consisting of the basolateral nuclear group increases in size in comparison with the portion that represents the corticomedial nuclear group. This is clearly the case when the amygdala in the human brain is compared with the same structure in infraprimate mammals such as the rat and rabbit, and these differences seem to be reflected both in the nature of the neural connections and in the functions served by the two nuclear groups.

AFFERENT FIBER CONNECTIONS OF THE AMYG-DALOID NUCLEI. The nucleus of the lateral olfactory tract and the cortical nucleus in the corticomedial group receive direct projections from the olfactory bulb by way of the lateral olfactory stria, and thus they may be considered "rhinencephalic" structures. The basolateral nuclei also receive olfactory impulses, but these are not direct olfactory bulb projections because they come, instead, through synaptic connections in the primary

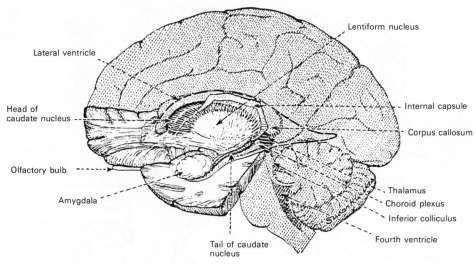

FIG. 11-106. A schematic drawing of a brain in which the left hemisphere has been dissected to expose the amygdaloid body and the caudate and lentiform nuclei. The left lateral ventricle is shown with a solid line where the wall is intact and with an interrupted line where it has been dissected away. Note the position of the amygdala and compare with Figure 11-107. (From S. W. Ranson and S. L. Clark, *The Anatomy of the Nervous System,* 10th ed., W. B. Saunders, Philadelphia, 1959).

olfactory cortex of the piriform lobe. Additionally, the amygdaloid nuclei receive afferent fibers from several neocortical regions. The inferior temporal cortical region has been reported to project to the central and lateral amygdaloid nuclei, while orbitofrontal cortical regions have been reported to project to both the medial and lateral parts of the amygdala in certain species.

The amygdala also receives afferent fibers by means of pathways that formerly were considered only efferent amygdaloid tracts. Thus, the **stria terminalis** contains afferent fibers coursing to both corticomedial and

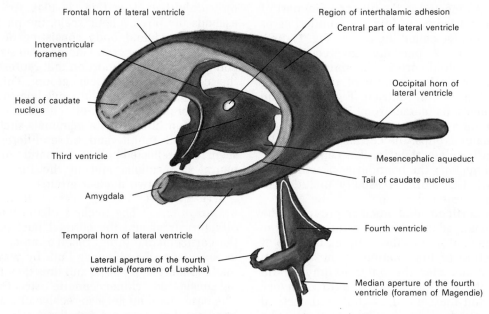

FIG. 11-107. A diagrammatic representation of the ventricular system in the brain to illustrate the relationship of the amygdaloid body and the caudate nucleus to the lateral ventricle. (Redrawn after E. C. Crosby, T. Humphrey, and E. W. Lauer, *Correlative Anatomy of the Nervous System,* Macmillan Publishing Co., New York, 1962.)

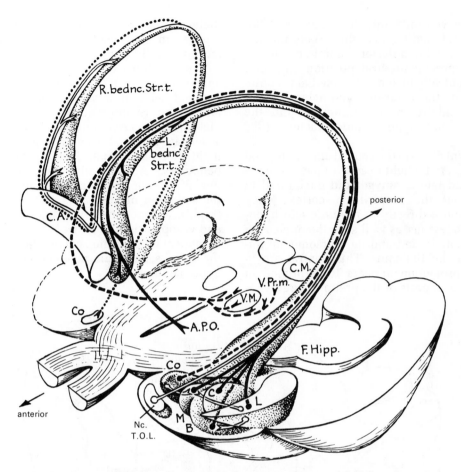

FIG. 11-108. The dorsal amygdalofugal pathway. Fibers that arise in the amygdala follow the curved postcommissural course of the stria terminalis (solid line) to terminate in the bed nucleus of the stria terminalis on the same side (L. bednc. Str.T.), as well as the preoptic area (A.P.O.). Other fibers (dashed line) course in front of the anterior commissure, curve caudally, and terminate in the ventromedial nucleus of the hypothalamus (V.M.) and the ventral premammillary region (V.Pr.M.). A few fibers (dotted line) cross in the anterior commissure (C.A.) and project to the opposite amygdala. (From H. J. Lammers in *The Neurobiology of the Amygdala*, edited by B. E. Eleftheriou, Plenum Press, New York, 1972.)

basolateral nuclear groups from the preoptic and anterior hypothalamic regions and probably also from certain thalamic nuclei. Even the so-called ventral amygdalofugal pathway is known now to contain amygdalopetal fibers from preoptic and rostral hypothalamic zones. Furthermore, the amygdaloid nuclei receive input from the nucleus medialis dorsalis of the thalamus by way of the inferior thalamic peduncle and also from the raphe nuclei and the reticular formation in the brain stem.

EFFERENT FIBER CONNECTIONS OF THE AMYGDALOID NUCLEI. Efferent fibers emerge from the amygdaloid nuclei by way of a **dorsal amygdalofugal tract,** which contributes to the stria terminalis and for the most part is thought to arise in the corticomedial nuclei (Fig. 11-108), *and* by a **ventral amygdalofugal tract,** which principally consists of efferent fibers from the basolateral nuclei (Fig. 11-109).

The stria terminalis arises from the posterior aspect of the amygdaloid nuclei and follows a curved course, along with the thalamostriate vein, in a groove that is visible on the wall of the lateral ventricle between the thalamus and the tail of the caudate nucleus. As the stria terminalis approaches the posterior aspect of the anterior commissure, many of its fibers enter groups of neurons that are collectively called the **bed nucleus of the stria terminalis,** while other postcommissural fibers continue ventrally into the anterior hypothalamus. Other fibers in the stria terminalis course over the anterior commissure to assume a precommissural position and then curve ventrally to end in the medial preoptic region, in the anterior hypothalamus, and more caudally in the region around the ventromedial hypothalamic nucleus (Fig. 11-108).

Efferent fibers in the ventral amygdalofugal pathway are thought to arise in part from the basal and lateral amygdaloid nuclei and in part from the prepiriform cortex. They course more directly to preoptic and hypothalamic regions, as well as to the nucleus of the diagonal band and the dorsomedial nucleus of the thalamus. This ventral amygdaloid projection appears to terminate in more lateral parts of the preoptic and hypothalamic regions, in contrast to the projections from the stria terminalis, which end more medially (Fig. 11-109).

The functions of the amygdaloid nuclei appear to be many and varied, and a wide range of behavioral, visceral, and endocrine alterations are observed after electrical stimulation or selective ablation. These include arrest, arousal, and orienting responses, and flight, defense, attack, feeding, and sexual behaviors. Additionally, the amygdaloid nuclei are reported to exert an influence on the secretion of hormones such as gonadotrophin, adrenocorticotrophin, thyrotrophin, and vasopressin. In many of these functions, it appears that the amygdala functions through its anatomic connections with the hypothalamus.

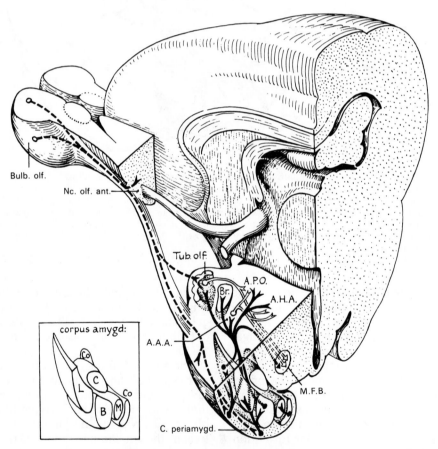

Fig. 11-109. The ventral amygdalofugal pathway. Fibers that arise in the basal and lateral amygdaloid nuclei, as well as others that arise in the piriform and periamygdaloid cortex (solid lines) project ventromedially to the preoptic (A.P.O.) and anterior hypothalamic (A.H.A.) regions. Other fibers project to the olfactory tubercle (Tub. olf.), the diagonal band of Broca (Br.), and the anterior amygdaloid area (A.A.A.). From these latter sites nerve fibers descend in the medial forebrain bundle (M.F.B.). Also note (dashed lines) the projections from the olfactory bulb to the primary olfactory cortex. (From H. J. Lammers in *The Neurobiology of the Amygdala*, edited by B. E. Eleftheriou, Plenum Press, New York, 1972.)

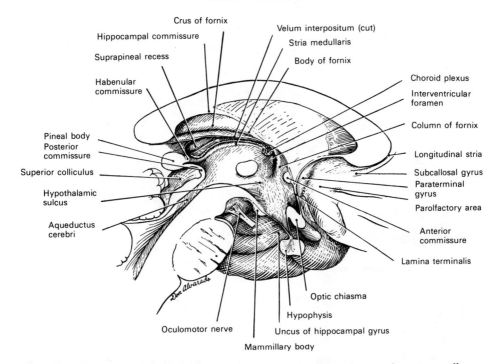

FIG. 11-110. Median sagittal section of the brain stem and diencephalon, showing the corpus callosum, septum pellucidum (not labelled), lamina terminalis, and subcallosal and parolfactory regions.

SEPTAL REGION

The septal region includes the **septum pellucidum,** which occupies the nearly triangular and vertically-oriented interval between the fornix and the arch of the corpus callosum, and the so-called **septal area** or *septum verum* (true septum), which is located rostral to the anterior commissure and lamina terminalis. The true septum corresponds to the medially located cortical regions called the paraterminal gyrus and the subcallosal area (Figs. 11-110; 11-111).

SEPTUM PELLUCIDUM. The septum pellucidum lies above the anterior commissure and therefore is sometimes referred to as the supracommissural septum. On each side, it consists of a thin, vertical partition; together these partitions separate the anterior horns of the two lateral ventricles. Each of these two laminae extends backward from the rostrum and genu of the corpus callosum and is prolonged under the body of the corpus callosum to the splenium, where it terminates. Inferiorly, each vertical lamina is attached to the fornix, and the medial surfaces of the two laminae may be fused in the midline or may enclose a narrow median cavity called

the **cavum septum pellucidi,** which is sometimes referred to as the fifth ventricle but is not known to communicate with the ventricular system. The laminae are thin sheets of neural tissue composed of a layer of "gray substance," which is located medially, adjacent to the cavum, and a layer of white matter, oriented laterally toward the ventricle, over which is spread an ependymal lining.

SEPTAL AREA. In a median sagittal view of the brain can be seen a region, located immediately rostral to the anterior commissure and lamina terminalis, that constitutes the paraterminal gyrus or paraterminal body (Fig. 11-110). In fact, it is the rostral continuation of a thin layer of gray substance, called the **indusium griseum** (or hippocampal rudiment), that covers the superior surface of the corpus callosum. In front of the paraterminal gyrus are the subcallosal and parolfactory areas (Figs. 11-110; 11-111). The **septal area** includes the paraterminal gyrus and the adjacent parolfactory area as well as the subcallosal area, and it consists of nuclear groups that are usually divided into the **medial** and **lateral septal nuclei.** The medial septal nucleus is coextensive with the **nucleus** and **tract of the diagonal band,** while

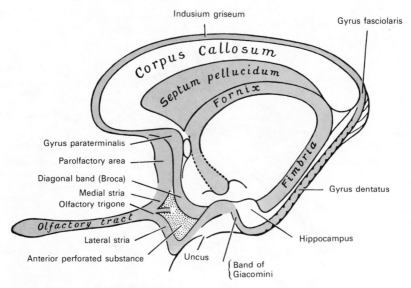

FIG. 11-111. Schema of the rhinencephalon and hippocampal formation. Note the septum pellucidum, paraterminal gyrus, and parolfactory area.

the lateral septal nucleus occupies the region anterodorsal to the anterior commissure and is continuous with the gray lamina of the septum pellucidum. Two other nuclear groups in this region include the **bed nucleus of the stria terminalis** and the **nucleus accumbens.**

Afferent fibers to the septal nuclei are derived from several forebrain and brain stem sites. Projections that appear to be topographically organized reach the lateral septal nucleus from the subicular part of the hippocampal formation by way of the precommissural fornix. Other afferent fibers reach the septal nuclei from the amygdala, after they pass through the stria terminalis and the diagonal band. Still other fibers from the locus ceruleus, raphe nuclei, mesencephalic reticular formation, and lateral hypothalamus course to the septum via ascending catecholaminergic fibers in the medial forebrain bundle. There also may be some input to the septum from the cingulate gyrus.

Efferent projections from the septal nuclei course back to the hippocampus through the fornix. These appear to be cholinergic fibers and they arise mainly from the medial septal nucleus. A second important septal projection travels along the medial forebrain bundle to the lateral hypothalamus, the mammillary bodies, and the tegmentum of the midbrain. Other efferent septal fibers enter

the stria medullaris and pass to the habenular and interpeduncular nuclei, and still others course along the stria terminalis to the central and medial amygdaloid nuclei. As a result of these interconnections, it can readily be appreciated that the septal area is a forebrain structure that strategically interacts with many important brain stem, diencephalic, and telencephalic structures, and it is not surprising that septal lesions can influence a variety of behaviors. The so-called septal syndrome, which is observed in some animals following septal lesions, is characterized by a transient increased aggressivity and a heightened reactivity and irritability to environmental stimuli; however, many more subtle effects are also observed in these animals when they are studied in conditioning paradigms.

ANTERIOR (ROSTRAL) COMMISSURE

The anterior commissure is a bundle of myelinated nerve fibers that crosses the midline just rostral to the columns of the fornix (Figs. 11-102; 11-108; 11-110). Located within the upper portion of the lamina terminalis, it overlies the preoptic area and thereby helps to form the anterior wall of the third ventricle. In sagittal section it is oval, with the long diameter of the oval oriented vertically. Its constituent fiber bundles are twisted like the

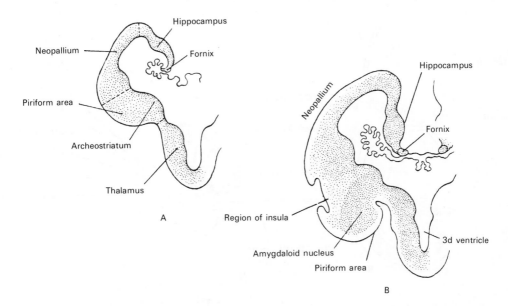

FIG. 11-112. Coronal sections of the human forebrain at two stages of development. In *A*, the neopallium occupies a restricted region in the lateral surface of the hemisphere between the hippocampus and the piriform area. In *B*, the neopallium has expanded, thereby displacing the hippocampus onto the medial surface of the cerebrum, close to the ependymal roof of the diencephalon. (From J. Davies, *Human Developmental Anatomy*, Ronald Press, New York, 1963.)

strands of a rope, and these curve backward and laterally, forming a deep groove on the inferior aspect of the rostral part of the lentiform nucleus. Across the anterior commissure course fibers that interconnect the olfactory bulbs and anterior olfactory nuclei, the anterior perforated substance, the diagonal band and the olfactory tubercles, the amygdaloid nuclei, the prepiriform cortex, the entorhinal area, the parahippocampal gyri and neocortical portions of the cortex in the temporal lobe, and some fibers from the frontal lobe.

HIPPOCAMPAL FORMATION

The hippocampal formation develops along the fringe of the pallium on the lower medial aspect of the cerebral hemisphere; it commences its differentiation even before the corpus callosum is formed. During the second prenatal month, the lower part of the medial pallial wall becomes thin near the site at which the pallium attaches to the diencephalon and is reduced to a single layer of ependymal cells, into which a layer of vascular mesenchyme grows (Fig. 11-112, *A*). Together, the ependymal cell layer and its associated developing blood vessels invagi-

nate into the cavity of the cerebral vesicle on each side and form the **choroid plexuses** of the lateral ventricles; the line of invagination is called the **choroid fissure.**

Meanwhile, the medial pallial wall along the upper curved border of the choroid fissure begins to thicken and develop into the hippocampal formation (Fig. 11-112, *B*). Further growth in its ventromedial part forces an infolding to occur, thereby creating the hippocampal sulcus around which the pallium arches over itself to form the hippocampus. As the hippocampal sulcus deepens, the invaginated pallium becomes the cornu ammonis (Ammon's horn), and the adjacent innermost lip arches to form the dentate gyrus while the outer lip forms the subicular zones and the parahippocampal gyrus (Fig. 11-113). With expansion of the temporal lobe, the enlarging hippocampal formation migrates farther downward and laterally into the lateral ventricle. The collateral sulcus, the rostral part of which is called the rhinal sulcus, invaginates the inferomedial surface of the hemisphere to form the border between the parahippocampal gyrus or entorhinal area and the medial occipitotemporal gyrus. Thus, in the fully developed brain, the arched contour of

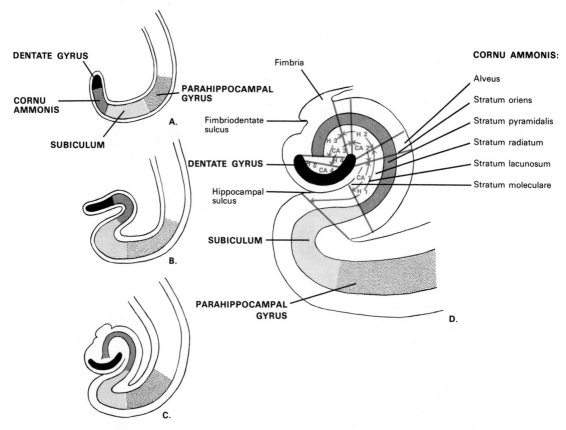

FIG. 11-113. Diagrammatic coronal sections of the developing hippocampal formation. In *A*, *B*, and *C* note the migration of the pallial regions, which eventually form the dentate gyrus (1), cornu ammonis (2), subiculum (3), and parahippocampal gyrus (4). The fully formed hippocampal formation is shown in *D*. (From B. Pansky and D. J. Allen, *Review of Neuroscience*, Macmillan Co., New York, 1980.)

the hippocampal formation consists of phylogenetically old cortex (archeopallium), which blends through the parahippocampal gyrus (paleopallium) with the greatly expanded neopallium. Commencing at the choroidal fissure, the three principal zones of the hippocampal formation include the **dentate gyrus,** the **cornu ammonis,** and the **subiculum.**

Efferent fibers arise from the pyramidal cells of the hippocampus to form the **fornix,** a bundle that courses forward above the choroid fissure. Anteriorly, the fibers of the fornix enter the lamina terminalis and then turn inferiorly and grow into the region of the mammillary body of the hypothalamus. At the end of about the fourth month, fibers from neurons in the neocortical areas begin to interconnect the two cerebral hemispheres by passing across the lamina terminalis, thereby commencing the formation of the **corpus callosum** (Fig. 11-114). Prolific

growth of this great commissure, however, soon results in its extension caudally beyond the lamina terminalis. This causes a separation of the upper part of the hippocampal formation, the thin rudiment of which, the **indusium griseum,** is seen in the adult brain above the corpus callosum. Additionally, this process also causes a splitting of the fornix fibers into a dorsal component, the **medial** and **lateral longitudinal striae,** and a ventral component, which in the adult becomes the two columns of the fornix proper (Fig. 11-115).

Indusium Griseum

The indusium griseum, sometimes called the supracallosal gyrus, is a thin sheet of gray matter that covers the superior surface of the corpus callosum. Laterally, it is continuous on both sides with the cortex of the cingulate gyrus. Posteriorly, over the sple-

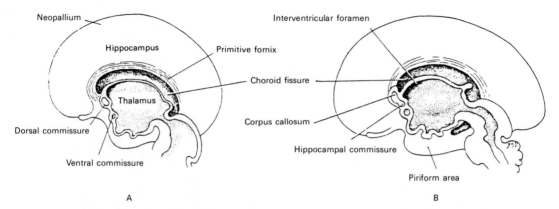

FIG. 11-114. Sagittal sections of the human fetal brain, showing stages in the formation of the corpus callosum. In A there is a massing of commissural fibers from the neopallium, dorsal to the dorsal commissure in the lamina terminalis. The callosal fibers then extend caudally (B), thereby separating the fornix fibers that lie in their path into a dorsal portion and a ventral portion. (From J. Davies, *Human Developmental Anatomy*, Ronald Press, New York, 1963.)

nium of the corpus callosum, the indusium griseum consists of right and left parts, each becoming the slender **gyrus fasciolaris,** a delicate layer of gray substance that courses ventrolaterally and forward to merge with the posterior end of the dentate gyrus (Fig. 11-111). Elevated above the surface of the indusium griseum, as well as embedded

within it, are the **medial** and **lateral longitudinal striae** (of Lancisii[1]). These are two longitudinal bundles of fibers on each side, the medial stria lying close to the median plane and the lateral stria hidden under the cingulate gyri on the floor of the callosal sul-

[1]Giovanni N. Lancisi (1654–1720): An Italian physician.

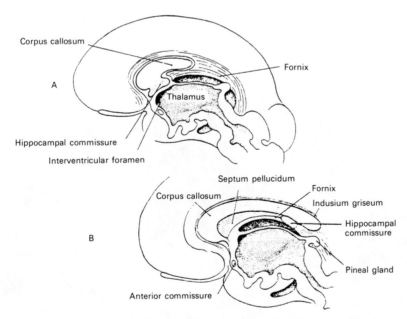

FIG. 11-115. Two later stages in the development of the corpus callosum. From the caudal extension of the callosal fibers (A), there results a splitting of the fornix fibers into a dorsal component (the indusium griseum and the medial and lateral longitudinal striae) and a ventral component (the columns of the fornix). Some fibers in the latter pass across the midline in the dorsal commissure, which now becomes the hippocampal commissure and is displaced caudally out of the lamina terminalis by the backward extension of the corpus callosum (B). Part of the lamina terminalis becomes stretched and incorporated into the septum pellucidum. (From J. Davies, *Human Developmental Anatomy*, Ronald Press, New York, 1963.)

cus. Anteriorly, the medial and lateral longitudinal striae on each side pass over the genu of the corpus callosum and emerge as a single band that becomes continuous with the paraterminal gyrus, which is contiguous with the diagonal band (of Broca). Posteriorly, the striae join the gyrus fasciolaris.

The Hippocampus

The hippocampus is a curved elevation, about 5 cm long, that extends forward from its position under the posterior part of the corpus callosum to the uncus, thereby occuping the entire length of the floor of the temporal (inferior) horn of the lateral ventricle (Fig. 11-116). The picturesque name of the hippocampus implies that its shape in the forebrain is similar to that of a seahorse seen in profile. Its anterior or lower extremity appears somewhat dilated, and the lateral margin of this expanded portion is crossed by three or four shallow digitations that cause it to resemble an animal's paw; therefore it is called the **pes hippocampi.** The hippocampus is formed by a specialized type of cortex, and it consists of the dentate gyrus and the cornu ammonis. It is continuous ventrally with the subiculum of the parahippocampal gyrus (Fig. 11-117).

As described above, the hippocampus is an archeocortical gyrus that is infolded upon itself, so that its ependymal cell-lined, convex, ventricular surface is its deepest layer; this layer consists of a fasciculus of myelinated fibers called the **alveus.** The axons in the alveus converge dorsomedially to form the fimbria of the fornix (fimbria hippocampi), a flat band of white fibers that lies above the dentate gyrus and immediately below the choroid fissure. In addition to serving as an efferent path to the fornix for pyramidal and other hippocampal neurons, the alveus is also a pathway for afferent fibers to the hippocampus from the medial part of the entorhinal region and the parasubiculum.

THE SUBICULUM. The cortical regions that are located ventral and medial to the dentate gyrus and form an extension of the parahippocampal gyrus to the cornu ammonis are collectively called the **subiculum.** More precisely, this region is described as consisting of four zones: the parasubiculum, presubiculum, subiculum, and prosubiculum. The parasubiculum is contiguous with the entorhinal cortex (area 28) laterally and the other zones follow sequentially the curvature of the parahippocampal gyrus toward the cornu ammonis (Fig. 11-117).

The archeopallial nature of the subiculum

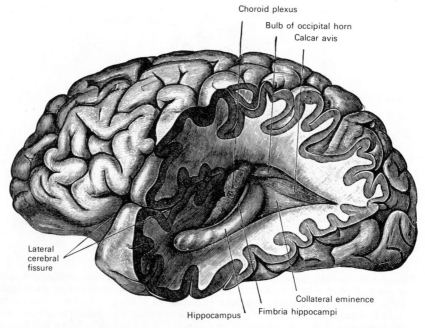

Choroid plexus

Bulb of occipital horn
Calcar avis

Lateral cerebral fissure

Hippocampus Fimbria hippocampi

Collateral eminence

FIG. 11-116. Lateral view of a temporal lobe dissection in the left cerebral hemisphere. The lateral ventricle is opened, exposing its occipital (posterior) and temporal (inferior) horns, as well as the hippocampus and fimbria.

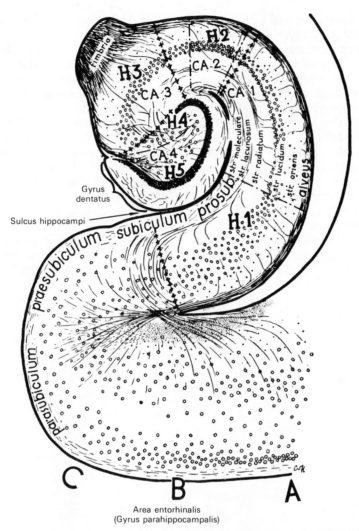

FIG. 11-117. A schematic drawing of a transverse section through the parahippocampal gyrus, subiculum, cornu ammonis, and dentate gyrus, showing the subdivisions of these regions. Note the laminae of the cornu ammonis (the stratum pyramidalis is called the stratum lucidum in this figure), the hippocampal subdivisions according to Rose (H1 to H5), and the cornu ammonis fields (CA 1 to CA 4) according to Lorente de Nó. (From H. J. Lammers and H. Gastaut in *Physiologie de l' Hippocampe*, Editions du Centre National de la Recherche Scientifique, No. 107, Paris, 1962.)

is manifested by its three-layered structure, a feature that gradually develops from the adjacent five-layered entorhinal area through the parasubicular and presubicular zones. This fundamental trilaminar cortical structure, which is characteristic of the cornu ammonis and dentate gyrus as well, includes an outer **molecular,** an intermediate **pyramidal,** and a deep **polymorphic layer.** Many afferent fibers to the subiculum, cornu ammonis, and dentate gyrus arise in the entorhinal cortex. Fibers that originate in the medial part of the entorhinal cortex and the parasubiculum enter the hippocampus, close to its ven-

tricular surface, by way of the alveus (the **alvear path**). Fibers from the lateral portion of the entorhinal cortex course through the subicular zones along a **perforant path** that crosses the alvear path and becomes distributed widely within the cornu ammonis and dentate gyrus. These pathways constitute the so-called direct temporoammonic tracts.

CORNU AMMONIS.   Continuous with the prosubicular zone at its open end and interlocking with the concavity of the dentate gyrus at the other, the cornu ammonis is characterized in transverse sections of the hippocampal formation by its reversed C or comma-shaped orientation and by its tightly packed pyramidal cell layer (Fig. 11-117). Although the cornu ammonis displays the basic trilaminar archeocortical structure, commencing at the ventricular surface the following strata or sublayers are readily recognizable: (1) the *ependymal cells* of the lateral ventricle; (2) the afferent and efferent fiber bundle called the *alveus;* (3) the *stratum oriens,* containing mainly basal dendrites of pyramidal cells, but also containing nonmyelinated axon collaterals that leave to join the alveus, incoming fibers that pass in the opposite direction, irregularly-shaped polymorphic neurons, and basket cells; (4) the *stratum pyramidalis,* consisting of both large and small pyramidal cells that are oriented so that the axon, coming from the base

of each pyramidal cell or from a basal dendrite, courses through the stratum oriens, into the alveus, and out the fimbria of the fornix. Long axon collateral branches (the so-called Schaeffer collaterals) turn back through the pyramidal layer to the stratum moleculare and terminate near the base of the apical dendrites of other nearby pyramidal cells (Fig. 11-119). (5) The *stratum radiatum* consists principally of the unbranched apical dendrites of the pyramidal cells; (6) the *stratum lacunosum* has a rich plexus of pyramidal cell dendrites, various incoming axons, axon collaterals, and the cell bodies of interneurons; and (7) the *stratum moleculare* consists of fine terminal branches of apical dendrites from pyramidal cells and some incoming fibers (Fig. 11-117).

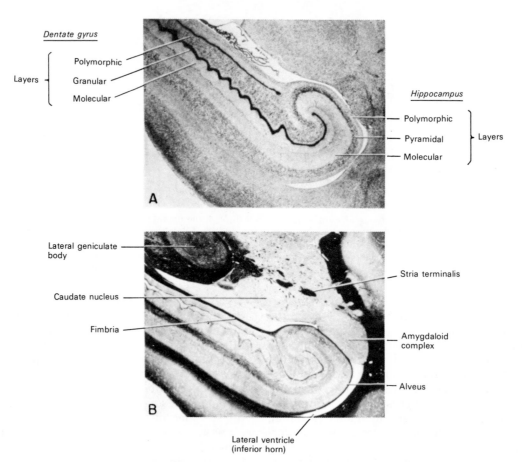

FIG. 11-118. Sagittal sections through the hippocampal formation in the monkey (*Macaca mulatta*). In *A,* the cellular layers of the hippocampus (cornu ammonis) and dentate gyrus are identified, while in *B,* the relationship of the alveus and fimbria to the lateral ventricle and the nearby amygdaloid body and caudate nucleus is demonstrated. (From M. B. Carpenter, *Human Neuroanatomy,* 7th ed., Williams and Wilkins, Baltimore, 1976.)

Thus, the basic trilaminar arrangement (Fig. 11-118) of the cornu ammonis might be said to be composed of the following sublayers:

Polymorphic layer:
  Ependymal cells
  Alveus
  Stratum oriens
Pyramidal layer:
  Stratum pyramidalis
  Stratum radiatum
Molecular layer:
  Stratum lacunosum
  Stratum moleculare

Detailed analysis of cytoarchitectural differences between the various regions of the hippocampus has led certain investigators to divide it into a series of fields. Rose (1926) designated these as **hippocampal fields H_1 through H_5,** and Lorente de Nó (1934) divided Ammon's horn into **cornu ammonis fields CA 1 through CA 4** and pointed out their special variations in cell morphology, fiber distribution, and connections (Fig. 11-117). These two analyses are not identical, because Rose's H fields extend from the subicular transitional region through the cornu ammonis and include the dentate gyrus, while Lorente de Nó's CA fields subdivide the cornu ammonis only. Field CA 1 contains the smallest pyramidal cells and is continuous with the subiculum, whereas the largest pyramids are located in field CA 4, which lies within the curve of the dentate gyrus.

DENTATE GYRUS. The dentate gyrus is a narrow band of archeocortex that also consists of three basic cytoarchitectonic layers: a superficial plexiform or molecular layer, a granular layer, and a deep polymorphic layer of cells (Fig. 11-118). The general configuration of the dentate gyrus resembles closely a wide-angled U, and its open concavity, which is oriented toward the fimbria, is nearly filled by the CA 4 field of the cornu ammonis. The **molecular layer** contains many interwoven neuronal processes, as well as a few polymorphic neurons, including an occasional Golgi type II cell or granule cell. The **granular layer** is composed of closely positioned spherical or ovoid granule cells and some small pyramidal cells. The dendrites of these cells are oriented toward the molecular layer, while their axons, called **mossy fibers,** course through the polymorphic layer to enter the cornu ammonis, where they synapse with the dendrites of the pyramidal cells (Fig. 11-119). The **polymorphic layer** contains some pyramidal cells and basket cells distributed within an interlacing fiber network. The dentate gyrus appears to be interposed between certain incoming afferent projections to the hippocampus and the pyramidal cell layers of the CA fields of the cornu ammonis, because afferent impulses reaching the granule cells of the dentate gyrus are clearly relayed to the hippocampal pyramids.

HIPPOCAMPAL CONNECTIONS AND FUNCTIONS. The hippocampus receives its principal *afferent connections* from the entorhinal cortex and subiculum along the alvear and perforant pathways of the temporoammonic tracts. These fibers terminate on the outer portions of the dendrites of the

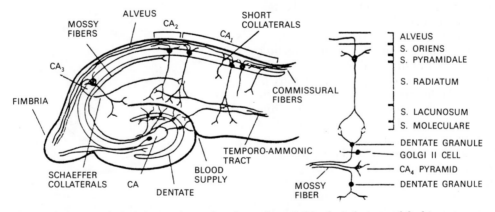

FIG. 11-119. The diagram on the left is a schema that shows the main histologic features of the hippocampus seen in transverse section. Indicated on the right are the strata and cellular structures through the cornu ammonis (CA 2) and dentate gyrus. (From J. D. Green, Physiol. Rev., *44,* 1964.)

pyramidal and granule cells. Additionally, fibers from the cingulate gyrus project by way of the cingulum through synaptic connections in the subicular region, but the existence of direct projections to the hippocampus from the cingulate gyrus has been questioned. Incoming fibers from the septum reach the hippocampus along the fornix from the medial septal nucleus. These are not present in great numbers, but they seem to terminate diffusely throughout the hippocampal fields and are cholinergic in nature. Afferent fibers to the hippocampus also arise from the supramammillary region in the posterior hypothalamus, the nucleus reuniens, and the anterior nuclei of the thalamus, and from the locus ceruleus and raphe nuclei in the brain stem. Some of these latter fiber groups are adrenergic, dopaminergic, and serotonergic projections. Finally, many transhippocampal projections course through the commissure of the fornix and interconnect the two hippocampal formations.

The principal *efferent pathway* from the hippocampus is by way of the **fornix system** (Fig. 11-120). In addition to its afferent fibers, this thick nerve bundle contains efferent projections that achieve sites outside the hippocampal formation and commissural projections that course to the opposite hippocampus. At one time it was believed that most of the efferent fibers in the fornix represent axons of pyramidal cells. It now appears that, in addition to the pyramidal cell axons, many fibers in the fornix are derived from the subiculum and prosubiculum (Swanson and Cowan, 1977; Meibach and Siegel, 1977). On each of the two sides of the brain, fibers from these sources initially pass toward the ventricular surface to form the **alveus.** They then converge on the medial aspect of the hippocampus as the **fimbria,** a flat band of white fibers that lies above the dentate gyrus and below the lower part of the choroidal fissure. Anteriorly, the fimbria extends into the hook of the uncus, while posteriorly it ascends below the splenium of the corpus callosum, arches upward over the thalamus, and inclines toward the midline as the **crus of the fornix** (Fig. 11-120).

The two crura are connected by a lamina of transverse fibers that passes between the hippocampal formation of the two hemispheres to form the **commissure of the fornix** or hippocampal commissure. This commissure presents the appearance of a thin, triangular lamina called the **psalterium** or **lyra.** At the highest part of their arch, the two crura fuse to become the **body of the fornix** (Fig. 11-120). The body of the fornix lies above the tela choroidea and the ependymal roof of the third ventricle and is attached to the undersurface of the corpus callosum and, more anteriorly, to the inferior borders of the laminae of the septum pellucidum. Laterally, the body of the fornix medially overlies the upper surface of the thalamus and the choroid fissure. Anteriorly, above the interventricular foramen, the body of the fornix divides again into the **columns** (or anterior pillars) **of the fornix,** which bend ventrally toward the anterior commissure and form the anterior boundary of the interventricular foramen.

About half of the fibers in the fornix descend behind the anterior commissure as the **postcommissural fornix,** while the remaining fibers descend rostral to the commissure as the **precommissural fornix.** Below the anterior commissure, a number of the fornix fibers reassemble and the bundle curves posteriorly, coursing through the gray matter in the lateral wall of the third ventricle to terminate in the medial nucleus of the mammillary body. Evidence now indicates that many of the fibers in the postcommissural fornix are derived from the subiculum and terminate in the mammillary bodies, whereas most of the fornix fibers that originate in the hippocampus proper course into the precommissural fornix and terminate in the septum.

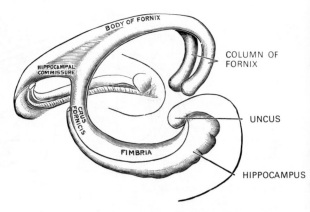

FIG. 11-120. Diagrammatic representation of the fornix system, viewed from above and from the right side.

Along its course toward the septum and mammillary bodies, the fornix system also distributes fibers to other regions. A small group of fibers from the fimbria passes dorsal to the splenium of the corpus callosum as the medial and lateral longitudinal striae of the indusium griseum. These fibers are sometimes referred to as constituting the dorsal fornix and most of them either terminate on scattered cells in the septum pellucidum or they penetrate the corpus callosum to rejoin the main fornix bundle. The fornix also distributes fibers to the anterior thalamus. These also appear to be derived from the subiculum rather than from the hippocampal fields, to pass through the postcommissural fornix, and to project to the anterior thalamic nuclei. Projections to the lateral preoptic and anterior hypothalamic regions from the fornix course by way of the precommissural route and also represent subicular connections rather than axons from the hippocampus itself. In addition to contributing many fibers that emerge from the hippocampus along the fornix system, the subicular region also projects axons to the adjacent entorhinal cortex and to the cingulate gyrus.

The hippocampal formation at one time was considered predominantly a structure that functioned in the olfactory pathway, and this may still be true in certain macrosmotic animals. In the microsmotic human brain and in anosmotic animals such as the whale, however, the hippocampus is well developed, does not receive direct projections from the olfactory bulb, and probably functions minimally as an olfactory structure. Many alterations in physiologic and behavioral phenomena have been reported as resulting from techniques commonly utilized to stimulate or to produce lesions in and around the hippocampus. It has not been possible, however, to identify any hippocampal function that is separate and distinct from functions that are derived from hippocampal interconnections with the temporal lobe, septal, diencephalic, and brain stem structures. Because of the profound elaboration of forebrain mechanisms due to the encephalization process, observations made on lower mammals and even on subhuman primates must be carefully assessed when the functions of the human hippocampus are discussed. The role of the hippocampus in altering aggressive behavior, sexual behavior, endocrine functions, attention, alertness, learning and memory have all been documented in various species. There may be some evidence that the human hippocampus functions importantly in the retention of short-term memory, but additional definitive data are required and the mechanisms involved need further exploration. The functions of the hippocampus have frequently been incorporated with those of the so-called "limbic system," a rather imprecise term popularly used to interconnect several forebrain and hypothalamic structures that are phylogenetically older, concerned with visceral functions, and appear to be related to the elaboration of emotional phenomena.

BASAL NUCLEI

The basal nuclei, or **basal ganglia,** are large masses of gray substance deeply located in the basal regions of the cerebral hemispheres (Fig. 11-121). In part they are separated medially from the diencephalon by the internal capsule, but in front of the thalamus they form the wall of the lateral ventricle, with the internal capsule coursing through their substance. As the subcortical nuclei of the telencephalon, the basal ganglia are separated laterally and superiorly from the gray substance of the cerebral cortex by the white matter underlying the cortical laminae.

The basal ganglia include the **globus pallidus, caudate nucleus, putamen,** and **amygdaloid body** (Fig. 11-122). The latter structure has been described with the rhinencephalon and will not be considered further. The globus pallidus is frequently referred to as the **paleostriatum,** or more simply, the **pallidum,** while the caudate nucleus and putamen together constitute the **neostriatum,** commonly called the **striatum.** More inclusive is the term **corpus striatum,** which refers collectively to both the paleostriatum (globus pallidus) and the neostriatum (putamen and caudate nucleus). The corpus striatum is so named because of its striped appearance, which results from the mixture of diverging white fibers and gray substance, characteristic of its more rostral part. Reference is often made also to the **lentiform** (or lenticular) **nucleus,** a term that includes the larger, more laterally located putamen and the medially placed globus pallidus. The following diagram clarifies the use of these terms:

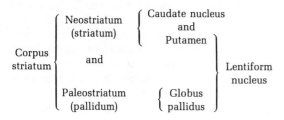

The following diagram shows:

Corpus striatum
{
  Neostriatum (striatum)
  {
    Caudate nucleus and Putamen
  } Lentiform nucleus
  and
  Paleostriatum (pallidum)
  {
    Globus pallidus
  }
}

Caudate Nucleus

The caudate nucleus is an elongated mass of gray substance that arches over itself and, throughout its entire extent, is related closely to the wall of the lateral ventricle. Its prominent and enlarged rostral extremity, called the **head** of the caudate nucleus, is pear-shaped and occupies most of the lateral wall of the frontal (anterior) horn of the lateral ventricle. The remainder of the caudate nucleus is at first directed backward on the lateral side of the thalamus as the **body** of the caudate nucleus and is separated medially from the thalamus by the stria terminalis and thalamostriate vein. The nucleus gradually becomes drawn into a long and more slender **tail,** which initially arches downward into the temporal lobe in the roof of the temporal (inferior) horn of the lateral

ventricle, but then curves forward toward the amygdaloid body (Fig. 11-122). Around the rostral border of the internal capsule, the head of the caudate nucleus is fused with the putamen of the lentiform nucleus, while its tail terminates in close relationship to the central nucleus of the amygdala (Fig. 11-122). In most coronal sections that pass through the corpus striatum in the cerebrum, the caudate nucleus is cut twice because of this arched form.

Lentiform Nucleus

The lentiform nucleus lies lateral to both the thalamus and the head of the caudate nucleus, and when it is sectioned horizontally, it has somewhat the shape of a biconvex lens. In totality, it is a wedge-shaped structure, shorter in rostro-caudal length than the caudate nucleus, and it does not extend as far forward (Fig. 11-122). The lentiform nucleus is located in the white substance of the cerebrum, deep to the insula; it is separated laterally from that cortical region by a lamina of white matter called the **external capsule** and by a thin layer of gray substance, the **claustrum.** Its anterior end is continuous with the lower part of the head of the caudate nucleus and with the anterior perforated substance. The lentiform nucleus is divided by a layer of white matter called the **lateral medullary lamina** into two parts: the

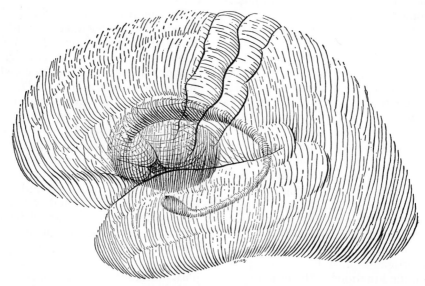

FIG. 11-121. Phantom of the corpus striatum within the cerebral hemisphere. (Courtesy of W. J. Krieg's *Functional Neuroanatomy*, 2nd ed., Blakiston Div., McGraw-Hill, New York, 1953.)

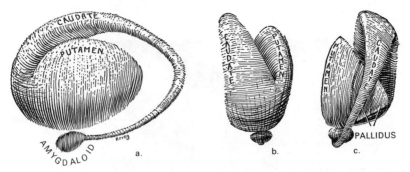

Fɪɢ. 11-122. The corpus striatum of the left side with the internal capsule omitted; (*a*) lateral aspect; (*b*) rostral aspect; (*c*) caudal aspect. (From W. J. Krieg, *Functional Neuroanatomy,* 2nd ed., Blakiston Div., McGraw-Hill, New York, 1953.)

putamen, which is larger and more laterally located, and the **globus pallidus.**

THE PUTAMEN. The putamen is the largest of the basal ganglia, and its most rostral part is located lateral to the head of the caudate nucleus, being separated from it by the anterior limb of the internal capsule. A frontal section through this most rostral level (Fig. 11-123) shows a small part of the globus pallidus interposed ventromedially between the putamen and the internal capsule. Slightly more caudally (Fig. 11-124), the putamen assumes more the shape of a thick,

laterally located shell-like covering to the globus pallidus ventrally and the internal capsule dorsally. At this point the body of the caudate nucleus is already displaced a bit more dorsally in its position adjacent to the lateral ventricle. A horizontal section of the brain reveals the entire rostro-caudal extent of the putamen and its relationship to the external capsule, claustrum, and insula, located laterally (Fig. 11-125).

GLOBUS PALLIDUS. The globus pallidus forms the more medial part of the lentiform nucleus (Fig. 11-122, *C*) and it is separated

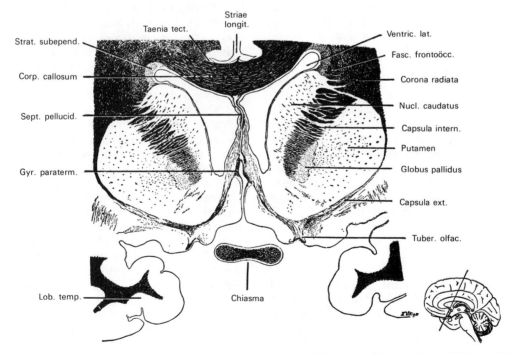

Fɪɢ. 11-123. Frontal section through rostral part of corpus striatum. Weigert myelin stain. (From Villiger-Addison, courtesy of J. B. Lippincott Co.)

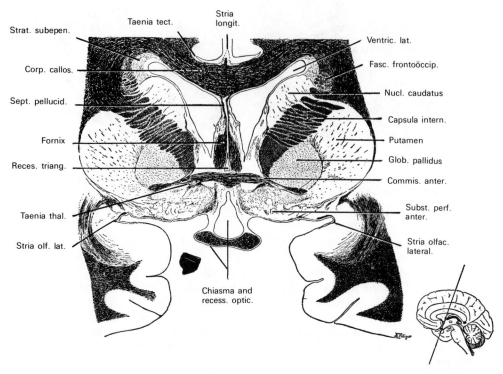

Strat. subepen.
Corp. callos.
Sept. pellucid.
Fornix
Reces. triang.
Taenia thal.
Stria olf. lat.

Taenia tect.

Stria
longit.

Ventric. lat.
Fasc. frontoöccip.
Nucl. caudatus
Capsula intern.
Putamen
Glob. pallidus
Commis. anter.
Subst. perf.
anter.
Stria olfac.
lateral.

Chiasma and
recess. optic.

FIG. 11-124. Frontal section through basal part of cerebral hemisphere at the anterior commissure. Weigert myelin stain. (From Villiger-Addison, courtesy of J. B. Lippincott Co.)

from the putamen by a thin layer of white matter called the **lateral medullary lamina.** Phylogenetically, it is the oldest part of the corpus striatum, and it receives its name from the fact that the many myelinated fibers that traverse its structure give it a lighter color than that of the putamen or caudate nucleus when observed in fresh preparations. The globus pallidus is subdivided into medial and lateral parts by another thin layer of white matter called the internal or **medial medullary lamina.** The posterior limb of the internal capsule lies on the dorsomedial border of the globus pallidus, thereby separating the lentiform nucleus from the thalamus (Fig. 11-125).

Fiber Connections of the Corpus Striatum

As a general principle, the putamen and caudate nucleus can be said to receive many afferent connections, and cells in these nuclei project to the globus pallidus, while the globus pallidus gives rise to many efferent fibers that course to lower centers in the neuraxis. Additional afferent and efferent

pathways interconnect the striatum and pallidum with other regions of the brain (see schematic, Fig. 11-126).

AFFERENT FIBERS TO THE NEOSTRIATUM. For a long time there was some uncertainty about whether or not cortical regions project to the striatum; however, the recent use of more sophisticated neuroanatomic methods has shown that the caudate nucleus and putamen receive afferent fibers from nearly all regions of the cerebral cortex (Fig. 11-126). These **corticostriate projections** appear to be organized in a topographic manner, although there is considerable overlapping. The sensorimotor areas and the more rostral regions of the cerebral cortex in the frontal lobe project more heavily to the neostriatum than does the occipital lobe. A great majority of the fibers from these more rostral cortical zones interconnect with the more anterior portion of the head of the caudate nucleus and with the rostral parts of the putamen, while projections from the parietal and occipital lobes flow progressively into the more caudal regions of the striatum. Most corticostriate fibers project to the ipsilateral striatum, but some fibers project across the

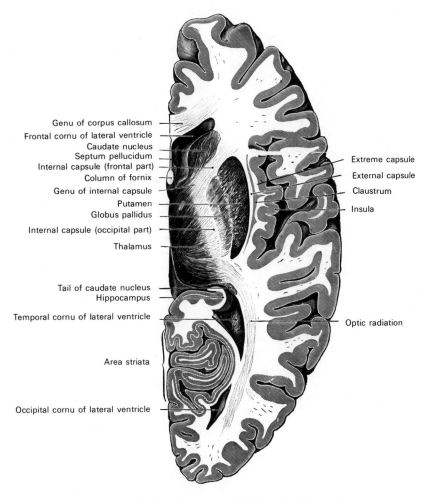

Genu of corpus callosum
Frontal cornu of lateral ventricle
Caudate nucleus
Septum pellucidum
Internal capsule (frontal part)
Column of fornix
Genu of internal capsule
Putamen
Globus pallidus
Internal capsule (occipital part)
Thalamus

Tail of caudate nucleus
Hippocampus
Temporal cornu of lateral ventricle

Area striata

Occipital cornu of lateral ventricle

Extreme capsule
External capsule
Claustrum
Insula

Optic radiation

FIG. 11-125.   Horizontal section of right cerebral hemisphere.

corpus callosum to the contralateral caudate nucleus and putamen.

A second source of afferent fibers to the striatum is the substantia nigra. This **nigro-striatal projection** was definitely established with the use of fluorescence techniques that demonstrate the pathways of neurons containing monoamines such as dopamine (Fig. 11-126). Most nigrostriatal fibers arise from nerve cells in the pars compacta of the substantia nigra, but some also come from the pars reticulata. Those from the more caudal parts of the substantia nigra project to the putamen, while those from the more rostral parts course to the head of the caudate nucleus. The nigrostriatal pathway is considered the principal source of dopamine for the corpus striatum, and its destruction is thought to be importantly related to the etiology of Parkinson's disease.

A third afferent pathway to the striatum originates in the intralaminar nuclei of the thalamus. Other thalamic nuclei, such as the midline nuclei and the nucleus medialis dorsalis, also contribute fibers to this **thalamo-striatal projection.** This projection appears to be topographically organized, with the centromedian and parafascicular nuclei sending fibers to more caudal parts of the putamen and the other intralaminar nuclei projecting to the caudate nucleus more anteriorly in the striatum (Fig. 11-126). Finally, other afferent fibers to the striatum arise in the **raphe nuclei** in the mesencephalon of the brain stem. This latter pathway is probably involved in the maintenance of the high level of serotonin found in the putamen.

EFFERENT FIBERS FROM THE NEOSTRIATUM. Efferent fibers from the striatum course to the globus pallidus and the substantia nigra

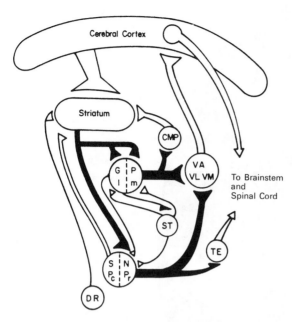

FIG. 11-126. Schematic diagram of the neuronal circuitry of the basal ganglia and related structures. Open pathways and arrows represent probable excitatory connections and the solid black pathways and terminals represent probable inhibitory connections. CMP, centromedian-parafascicular complex; DR, dorsal raphe nucleus; GP globus pallidus (1, lateral; m, medial); P^c, pars compacta, P^r, pars reticulata; SN, substantia nigra; ST, subthalamus; TE, midbrain tectum; VA, nucleus ventralis anterior of thalamus; VL, nucleus ventralis lateralis of thalamus; VM, nucleus ventralis medialis of thalamus. (From S. T. Kitai, 1981.)

(Fig. 11-126). Many fine striatofugal fibers are derived from neurons in the putamen and terminate in both the lateral and medial segments of the globus pallidus. These appear to project topographically and to collect as small bundles that converge radially on the globus pallidus. Some fibers terminate in the globus pallidus, while others course through and emerge from the medial segment of the pallidum and, having given off collaterals to pallidal cells, project to the substantia nigra. There appears to be a mediolateral organization of the striatopallidal pathway with fibers from the lateral part of the putamen projecting to the lateral segment of the globus pallidus, while fibers from the medial part of the putamen and caudate nucleus project to the medial segment of the globus pallidus.

Striatofugal fibers to the substantia nigra are also topographically organized. Some terminate in the pars compacta, but the projection to the pars reticulata of the substan-

tia nigra is more profuse. Fibers from the ventral and medial parts of the caudate nucleus project to the more rostral and medial portions of the substantia nigra and projections from the caudal parts of the putamen end more caudally and laterally in the substantia nigra.

AFFERENT FIBERS TO THE GLOBUS PALLIDUS. As mentioned above, the principal incoming fibers to the globus pallidus are derived from the caudate nucleus and putamen. These connections are topographically organized, with the caudate nucleus projecting through the internal capsule to dorsal and rostral parts of the globus pallidus, and the putamen sending fibers directly medially to the ventral and more caudal parts of the pallidum. Striatopallidal projections are believed to use gamma aminobutyric acid (GABA) as their neurotransmitter because the globus pallidus is rich in this substance, as well as in the enzyme that synthesizes it, glutamic acid decarboxylase (GAD).

Another source of afferent input to the globus pallidus is the subthalamic nucleus (Carpenter and Strominger, 1967). Confirmation of this long suspected pathway comes from a convincing study by Nauta and Cole (1978), based on the axoplasmic transport of tritiated amino acids. These latter authors showed that large numbers of subthalamic fibers course to both segments of the globus pallidus in a manner that suggests that the projection is topographically organized.

EFFERENT FIBERS FROM THE GLOBUS PALLIDUS. Pallidofugal fibers to various sites in the diencephalon and midbrain form the principal efferent pathways from the globus pallidus (Fig. 11-126). Diencephalic projections from the globus pallidus include direct pathways across the ventral thalamus to the subthalamic nucleus and pathways to the thalamus by way of the ansa lenticularis and lenticular fasciculus (Fig. 11-127). The lateral pallidal segment gives origin to the pallidosubthalamic pathway, while the outer portion of the medial pallidal segment gives rise to the **ansa lenticularis,** which projects to the nucleus ventralis anterior of the thalamus, and the inner portion of the medial pallidal segment gives rise to the **lenticular fasciculus,** which projects to the nucleus ventralis lateralis of the thalamus (Carpenter, 1981). Other diencephalic efferent pathways from the globus pallidus course to the lateral habenular nucleus.

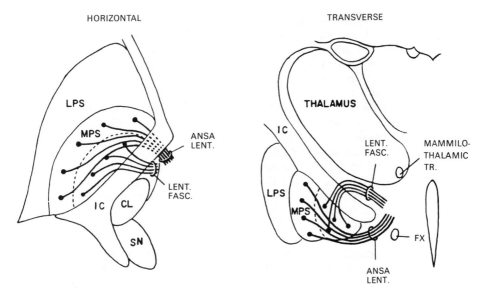

FIG. 11-127. Schematic diagrams showing the origin of fibers forming the ansa lenticularis (ansa lent.) and lenticular fasciculus (lent. fasc.) in the monkey. The ansa lenticularis arises from the outer portion of the medial pallidal segment, lateral to the medial medullary lamina (dashed line), and courses rostrally, ventrally, and medially. The lenticular fasciculus arises from the medial pallidal segment, medial to the medial medullary lamina, and courses dorsally and medially through the internal capsule. CL, subthalamic nucleus (corpus Luys); FX, fornix; IC, internal capsule; LPS, lateral pallidal segment; MPS, medial pallidal segment; SN, substantia nigra. (From Kuo and Carpenter, 1973.)

Pallidofugal fibers to the mesencephalon include projections from the globus pallidus to the pars reticulata of the substantia nigra and to the large-celled red nucleus. No pallidofugal fibers have been traced to more caudal levels of the brain stem or to the spinal cord.

FUNCTIONAL COMMENT. The motor functions of the basal nuclei were initially recognized by studies of patients suffering from a large number of movement disorders such as tremors, involuntary movements, and akinesias, and then correlating these clinical observations with the existence of lesions in these forebrain structures. Because the pathways involved are outside the classic pyramidal or great efferent motor pathway, the basal ganglia were considered the principal parts of the so-called "extrapyramidal motor system," and their effects on motor behavior were thought to result primarily from descending influences on lower motor centers. Today, however, it is known that the basal ganglia more closely interact with the cerebral cortex and thalamus than with motor centers in the brain stem and spinal cord, and that the influences exerted by the corpus striatum on motor function occur much higher in the neuraxis than was previously suspected. Much also has been learned in recent years about the transmitters and neurochemical mechanisms that underlie the functions of the basal ganglia, but a true understanding of how the basal ganglia participate in integrated normal motor behavior awaits elucidation.

The Claustrum

The claustrum is a thin layer of gray matter interposed between the insula and the putamen (Fig. 11-125). It is separated medially from the putamen by a lamina of white matter called the **external capsule,** and its lateral surface, being somewhat irregular and following the gyri and sulci of the insula, is separated from the outer gray cortex of the insula by a layer of white fibers called the **extreme capsule.** The claustrum has been variously considered a part of the insula or a part of the corpus striatum, and although its functional significance is unknown, there is good evidence to suggest that widespread regions of the cerebral cortex project onto the claustrum in a manner similar to projections from the cerebral cortex to the striatum.

CEREBRAL CORTEX

The cerebral cortex is a mantle of gray substance that covers the cerebral hemispheres. Study of its substance has generated great interest among neuroscientists because many of the more abstract neural functions appear to be related to the activity of its cellular layers. Its complexity, however, still defies a true understanding of the mechanisms by which information received by the brain is analyzed, stored, retrieved, and utilized to effect an appropriate response. Sophisticated microscopic and ultrastructural anatomic techniques have led to detailed descriptions of the cellular geometry of cortical elements, estimates of the volume of cortical substance, and even computations of the numbers of neurons that the cerebral cortex contains. The fact that cortical neurons are arranged in certain laminar patterns and various regions of cortex display differences in these patterns has been recognized for well over a century. It is customary to consider cortical regions that are phylogenetically older as comprising the **archeocortex,** or **paleocortex,** and others that are phylogenetically more recent as the **neocortex.** Because the cortices of the archeopallium and paleopallium have a rather wide variation in histologic structure in different regions, they are often referred to as **allocortex.** The neocortex, in contrast, shows a more similar histologic pattern in different regions and is called **isocortex.**

The archeocortex is characterized by a basic three-layered structure (polymorphic, pyramidal, and molecular), and it includes the cornu ammonis, dentate gyrus, and the subicular extension of the parahippocampal gyrus (Figs. 11-117; 11-118). During development it commences on the dorsomedial aspect; it then moves ventrally and becomes folded to form the hippocampus (Figs. 11-112; 11-113). The primordium of the paleocortex arises more ventrolaterally in the developing cortex and moves medially with the elaboration of the neocortex to form a transitional cortical region that includes the three-layered prepiriform and periamygdaloid (or piriform) areas, as well as the five-layered entorhinal area (area 28) which constitutes much of the parahippocampal gyrus. With continued development, the more primitive cortical structures migrate to ventromedially located sites, so that most of the dorsal and lateral walls of the pallium located between the archeocortex and paleocortex become neocortex (or neopallium) (Fig. 11-112), and this is characterized by a typical six-layered structure.

Structure of the Neocortex

In man, the largest part of the cerebral cortex consists of neocortex. Although certain differences are found in various cortical re-

gions, a six-layered schema can usually be recognized in sections that have been cut vertically through the gray substance and have been stained to show the morphology of the nerve cell bodies (such as with the Nissl[1] staining technique) (Figs. 11-128; 11-129). From the surface of the brain inward, the layers are as follows:

I. **MOLECULAR LAYER.** This outermost layer, also called the *plexiform layer* or *lamina zonalis,* is found immediately beneath the pia mater. It is a narrow zone of white substance composed largely of tangentially coursing fibers derived from nerve cells of deeper layers and directed parallel to the surface of the cortex (Fig. 11-128). Within the molecular layer are the terminal fibers of the apical dendrites of the pyramidal cells, as well as other fibers of cells from the subjacent layers. Scattered among the nerve fibers are a small number of nerve cells, which include cells with short axons as well as the **horizontal cells** (of Cajal), the axes of which are oriented parallel to the cortical surface among the processes of the apical dendrites.

II. **EXTERNAL GRANULAR LAYER.** The external granular layer contains large numbers of rather densely packed granule cells and some small pyramidal cells, the apical dendrites of which course superficially into the molecular layer and the axons of which descend into deeper cortical layers (Figs. 11-128; 11-129). Also coursing through the external granular layer are the apical dendrites from pyramidal cells of deeper layers, which are bound for the outermost or molecular layer.

III. **EXTERNAL PYRAMIDAL LAYER.** The external pyramidal layer consists principally of medium-sized and large pyramidal cells. These are arranged so that the smaller pyramidal cells are in the more superficial part, closer to the external granular layer, while the larger pyramidal neurons are located more deeply in this layer (Figs. 11-128; 11-129). The apical dendrites of the pyramidal cells also course superficially toward the molecular layer, and among the pyramidal cells in the external pyramidal layer are some granule cells and the through-going apical dendrites of even more deeply located pyramidal cells. Also located in this layer, as well as in others, are some small multipolar cells with short branching dendrites called

---

[1]Franz Nissl (1860–1919): A German neurologist and neuroanatomist (Heidelberg).

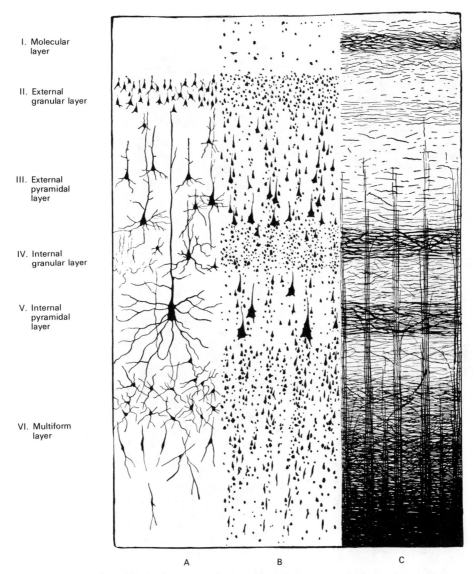

I. Molecular layer

II. External granular layer

III. External pyramidal layer

IV. Internal granular layer

V. Internal pyramidal layer

VI. Multiform layer

A   B   C

FIG. 11-128. A diagram showing the layers of cells and fibers in the gray substance of the cortex of the human cerebral hemisphere, according to the histologic methods of Golgi, Nissl, and Weigert. A, Stained by the method of Golgi; B, by that of Nissl; C, by that of Weigert. (After Brodmann.)

the **cells of Martinotti,**[1] whose axons course toward the molecular layer with the apical dendrites of the pyramidal cells.

IV. INTERNAL GRANULAR LAYER. The internal granular layer is a densely packed lamina containing many granule or stellate cells that have short axons (Fig. 11-128). Among these cells is found a scattering of small pyramidal cells and some vertically oriented neuronal processes traversing the internal granular layer on their way to more superficial layers. Additionally, the internal granular layer contains a well-defined collection of horizontally arranged nerve fibers that is called the **external band of Baillarger**[2] (Fig. 11-128).

V. INTERNAL PYRAMIDAL LAYER. Sometimes called the ganglionic layer, the internal pyramidal layer is composed principally of large and medium-sized pyramidal cells, the apical dendrites of which are directed toward the molecular layer (Fig. 11-128). This layer also contains many of the fine basal

[1]Giovanni Martinotti (1857–1928): An Italian physician.

[2]Jules G. F. Baillarger (1809–1890): A French psychiatrist (Paris).

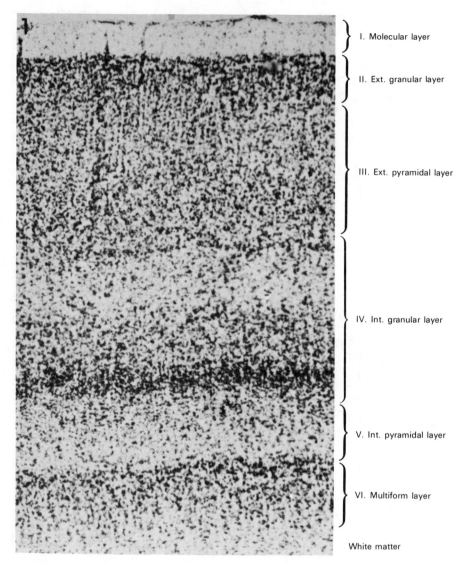

I. Molecular layer

II. Ext. granular layer

III. Ext. pyramidal layer

IV. Int. granular layer

V. Int. pyramidal layer

VI. Multiform layer

White matter

FIG. 11-129. Nissl stained section showing the laminae of the striate cortex (area 17, perimacular region) of a monkey, *Macaca mulatta*. (From J. S. Lund, 1973.)

dendrites of the pyramidal cells, processes of neurons from other layers, a few granule cells, and some cells of Martinotti. The largest of the pyramidal cells in the precentral motor cortex are called the **giant cells of Betz**[1] (Fig. 11-130). The internal pyramidal layer also contains a well-defined plexus of horizontally oriented nerve fibers referred to as the **internal band of Baillarger.** The axons of the large pyramidal cells enter the white matter as the corticofugal projection fibers.

VI. MULTIFORM LAYER. The multiform layer contains many neurons of different

[1]Vladimir A. Betz (1834–1894): A Russian anatomist (Kiev).

shapes, including some spindle-shaped cells, cells of Martinotti, and other neurons that have a fusiform shape (Fig. 11-128). At times this layer is subdivided into an outer or superficial lamina and an inner, deeper lamina which gradually merges with the subcortical white matter. Many axons of the cells in the multiform layer enter the subjacent white matter.

Cytoarchitectonics and Myeloarchitectonics

Various regions of the cerebral cortex show differences in morphologic structure, and on the basis of these variations, detailed

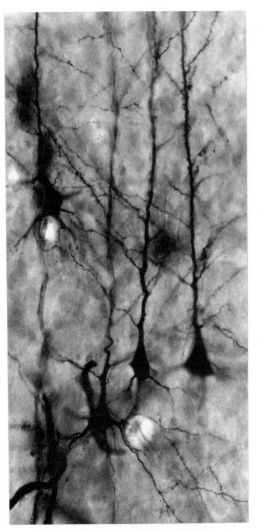

Campbell (1905) identified six cortical layers according to the morphology of the principal neuronal cell type found within each layer, and he recognized about 20 different cortical fields. Brodmann (1909) also differentiated six cortical layers that generally are similar to those of Campbell, and it is Brodmann's schema that is most commonly utilized today and is adopted here. Additionally, Brodmann mapped 47 different cortical fields as the result of his cytoarchitectonic studies: these are shown in Figure 11-131, A and B. Vogt and Vogt (1919) and von Economo and Koskinas (1925) elaborated on the Brodmann maps and identified far greater numbers of cortical fields. The Vogts described over 200 cortical areas on the basis of differences in the patterns made by the myelinated fibers, and von Economo and Koskinas listed over 100 areas. Later von Economo (1927) grouped the neocortical fields into five fundamental types based principally on the cytoarchitectural patterns of the granule and pyramidal cells. These were called **agranular, frontal, parietal, polar,** and **atypical granular** types (Fig. 11-132). The frontal, parietal, and polar types all have six distinguishable laminae that may vary in thickness, but are generally similar in cytoarchitecture and, therefore, are called **homotypical neocortex.** The cytoarchitecture of the agranular and atypical granular cortical areas varies markedly from the pattern seen in the homotypical neocortex and sometimes the term heterotypical neocortex is used for them.

AGRANULAR TYPE. The agranular cortex (type 1) is characterized by its great thickness, by the lack of distinct layers of granule cells (layers II and IV), and by the large size and number of pyramidal cells (Fig. 11-132). The granular cells in layers II and IV have been replaced predominately by pyramidal cells. In addition to other sites, the agranular type of neocortex is found in the posterior part of the frontal lobe, just rostral to the central sulcus, and is generally considered the cortical type forming the motor area of the precentral gyrus (Fig. 11-133).

FRONTAL TYPE. The frontal type (type 2) of neocortex shows the typical six-layered pattern; it is relatively thick and has medium- to large-sized pyramidal cells in layers III and V (Fig. 11-132). Distinct granular layers (II and IV) are seen, but these are rather narrow. Although this architectonic pattern appears typically in regions of the frontal

FIG. 11-130. Photomicrograph showing a group of giant pyramidal cells whose neuron cell bodies are located in the internal pyramidal layer (layer V). Note that these nerve cells extend their apical dendrites upward through the more superficial layers of the cortex and many reach the molecular layer (Golgi method.)

studies have identified numerous cortical areas with respect to differences in the cellular structure of the various neuronal layers (cytoarchitectonics) or variations in the arrangement of the myelinated fibers in different regions (myeloarchitectonics). Utilizing light microscopic techniques for the study of stained sections of the cerebral cortex, investigators such as Campbell (1905), Brodmann (1909), Vogt and Vogt (1919), von Economo (1929), Rose (1935), and Bailey and von Bonin (1951) distinguished different numbers of cortical areas based on the cellular and fiber patterns of the cortical laminae.

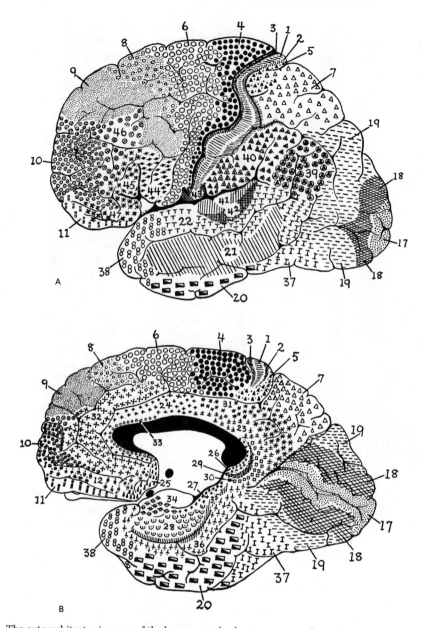

FIG. 11-131. The cytoarchitectonic areas of the human cerebral cortex, as seen from the superolateral (*A*) and medial (*B*) surfaces of the cerebral hemisphere as described by Brodmann. The different areas are defined by various symbols and enumerated in the order that Brodmann studied them. (From K. Brodmann, *Vergleichende Localisationslehre der Grosshirnrinde,* J. A. Barth, Leipzig, 1909.)

lobe rostral to the motor areas, it also appears in other cortical regions (Fig. 11-133).

PARIETAL TYPE. The parietal type (type 3) of neocortex has wider and well developed granular layers (Fig. 11-132). In contrast, the pyramidal cell layers are thin and contain pyramidal cells that are smaller and more slender than those in the frontal type of cortex. This cortical type is found in both the parietal and temporal lobes (Fig. 11-133).

POLAR TYPE. The polar type (type 4) is found at both the frontal and occipital poles of the cerebrum (Fig. 11-133). Although it shows a characteristic six-layered pattern, this cortex is thinner overall, the pyramidal cell layers being especially reduced in width and the granular cell layers rather well developed (Fig. 11-132). The polar cell layers of the frontal lobe have more pyramidal cells than those in the occipital lobe.

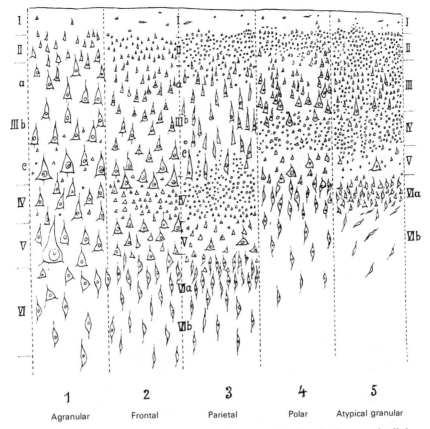

FIG. 11-132. The five types of cortical structure, showing differences in laminar thickness and cellular constituents of the various layers. See text for further details. (From C. von Economo, *Zellaufbau der Grosshirnrinde des Menschen*, J. Springer, Berlin, 1927.)

ATYPICAL GRANULAR CORTEX. Also known as **koniocortex,** the atypical granular cortex (type 5) contains large numbers of densely packed granule cells in layers II, III, and IV (Fig. 11-132). Layer III (external pyramidal layer), which in the typical six-layered neocortex contains large numbers of pyramidal cells, contains only a small number of these cells in granulous cortex. Overall, the koniocortex is the thinnest of the five neocortical types, and in addition to being found at other sites, it can be seen bounding the calcarine fissure and occupying the most rostral part of the postcentral gyrus (Fig. 11-133).

Principal Functional Cortical Areas: Sensory

The localization of function in different cortical regions has been a field of great interest in neurophysiology for many decades. Although neuroscientists have been unable to apply a functional significance to all of the many anatomically recognized parcels of cerebral cortex that have been described by cytoarchitectonic and myeloarchitectonic methods, it can be said that primary sensory receiving areas characteristically have well-developed granular layers that contain great numbers of densely packed small cells, while the motor areas have many large pyramidal cells. Cortical regions do not function in isolation, and the diagrammatic maps frequently used to illustrate cortical localization must be accepted with a degree of caution. To what extent the cytoarchitecture of different cortical regions and the morphologic geometry of cortical neurons are related to cortical function is unclear, but it would be surprising if the variations in laminar structure from region to region have little to do with the mode of operation of the cerebral cortex. Appearing in recent years are reports of detailed studies of cortical neurons and their precise somatosensory or visual inputs and analyses of unitary discharge patterns of pyramidal neurons associated with specific movements. Such efforts

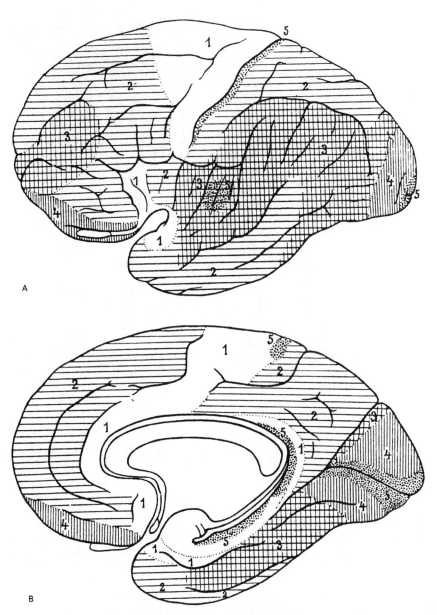

FIG. 11-133. Distribution of the five types of cortex seen in Figure 11-132, as shown on the superolateral (A) and medial (B) surfaces of the cerebral hemisphere. The types of cortex are: agranular (1), frontal (2), parietal (3), polar (4), and atypical granular (5). (From C. von Economo, *Zellaufbau der Grosshirnrinde des Menschen*, J. Springer, Berlin, 1927.)

are attempts to unravel the complexities of cortical neuronal function in relation to the laminar organization observable anatomically.

Various regions of the cerebral cortex receive impulses arising from receptors located in particular sense organs throughout the body or from nerve fibers supplying the skin and internal organs. These include visual, auditory, olfactory, and gustatory impulses of special sensation, as well as impulses of somatic sensibility or somesthesis. The latter term incorporates various sensations associated with touch and pressure (mechanoreception), warmth and cold, position sense and movement of joints, and pain. The sensory receiving areas in the cerebral cortex have been categorized as **primary receiving areas,** to which project the principal thalamocortical pathways from the thalamic relay nuclei (Fig. 11-134), and **secondary receiving areas,** which are located near the primary areas but are distinctly separated from the cortical zones receiving the principal thalamic projection (Fig. 11-135). The secondary areas are usually smaller than the primary areas and do not appear to be organized in the same manner. Primary somesthetic, visual, auditory, and olfactory areas have been localized, as have secondary somesthetic, visual, and auditory areas. Taste and vestibular projections to the cerebral cortex have not been localized as clearly, but are believed to exist.

PRIMARY SOMESTHETIC RECEIVING AREA. The primary somatic sensory area (SI or SmI) is located in the postcentral gyrus and consists of areas 3, 1 and 2, which Brodmann identified as having different cytoarchitectural patterns (Fig. 11-136). In the human brain, area 3 of the primary somesthetic area extends to the floor of the central sulcus, where it becomes continuous with the primary motor cortex (Brodmann's area 4; MsI). Sometimes this transitional zone is called area 3a, while the main part of area 3 is called area 3b. Caudal to area 3 and oriented longitudinally along the crest of the postcentral gyrus is area 1. Behind this is located another longitudinally oriented narrow strip,

area 2, which extends along the posterior wall of the postcentral gyrus to the postcentral sulcus. This latter sulcus separates the primary somesthetic area anteriorly from Brodmann's areas 5 and 7 in the superior parietal lobule posteriorly (Fig. 11-131). Although the primary somesthetic area is a six-layered cortex, it is thinner than the adjacent motor cortex and contains well-developed granular layers. Area 3 differs from areas 1 and 2 because layer III is composed almost completely of granule cells and layer V is poorly developed—characteristics typically seen in cortical sensory receiving areas. Thus, areas 2 and 1 are thicker than area 3, have somewhat less densely arranged granule cells, and contain more pyramidal cells in layers III and V.

It has been well established that the primary somesthetic area receives direct and precise projections from the nucleus ventralis posterolateralis (VPL) and the nucleus ventralis posteromedialis (VPM), which relay afferent impulses to the cortex from the ascending spinothalamic and trigeminal tracts and from the medial lemniscus. The representation of the body in the postcentral gyrus is arranged somatotopically, so that the contralateral lower limb is localized uppermost in the gyrus and extends onto the

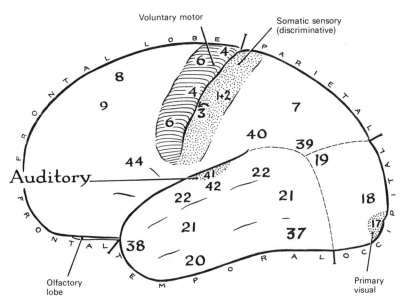

FIG. 11-134.   The lobes of the left hemisphere, showing a few of Brodmann's cortical areas by numbers. Indicated by stippling are the primary cortical receiving areas for somatosensory, visual, and auditory impulses, while the primary motor area is lined. Note the areas are only shown from this superolateral view of the cerebrum and this diagram should be compared with Figure 11-135. (From W. Penfield and L. Roberts, *Speech and Brain Mechanisms,* Princeton University Press, Princeton, 1959.)

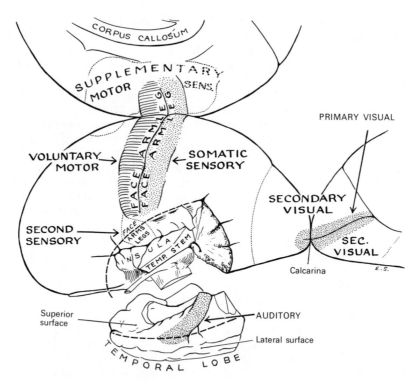

FIG. 11-135. The primary sensory receiving areas (stippled) and motor areas (lined) of the left hemisphere. Secondary somatosensory and visual areas and the supplementary motor area are also indicated on the medial surface of the cerebrum. The temporal lobe has been cut and turned down to reveal the primary auditory area. (From W. Penfield and L. Roberts, *Speech and Brain Mechanisms,* Princeton University Press, Princeton, 1959.)

medial surface of the cerebrum, where the representation of the genitalia and the bladder is also located. These are followed laterally on the convexity of the gyrus by the areas of localization for the contralateral trunk, upper limb, face, oral cavity, pharynx, and intra-abdominal organs, in that order (Fig. 11-137). The cortical representation of the different regions of the body is not uniform; for example, the foot, hand, and face areas are disproportionately large. The cortical volume allotted to a specific body part is related to the density of the peripheral innervation and not to the size of the region.

The general plan of representation of different body parts has now been extended by detailed studies utilizing microelectrode recordings of the activity of single cortical neurons. It has been shown that neurons in area 2 of the postcentral gyrus respond principally to input from receptors that are activated by the rotation of joints or by the mechanical stimulation of deep tissues, such as the periosteum or fascia, and nerve cells in areas 3b and 1 respond, respectively, to slowly adapting cutaneous receptors. Neu-

rons in area 3a, located adjacent to the motor cortex, respond to stretch afferents. Thus, an anteroposterior representation of modalities exists in the postcentral gyrus (Mountcastle and Powell, 1959a and b; Powell and Mountcastle, 1959). Additionally, there is a complete representation of the body mediolaterally in each of the cytoarchitectonic subdivisions, 3b, 1, and 2, and probably even in 3a of the postcentral gyrus (Fig. 11-138) (Merzenich et al., 1978; Kaas et al., 1979).

SECONDARY SOMESTHETIC RECEIVING AREA.
A secondary somatosensory area (SII) is located in the cortex of the parietal lobe, behind the lower part of the postcentral sulcus and along the superior bank of the lateral sulcus (Figs. 11-134; 11-146). It is smaller than the primary sensory area and is also somatotopically organized; however, the localization is less discrete than that in SI and there is considerable overlap of the dermatomes. Another distinguishing feature of the secondary somatic sensory area is that both ipsilateral and contralateral body parts are represented in a single somatotopic pattern, in contrast to the strictly contralateral repre-

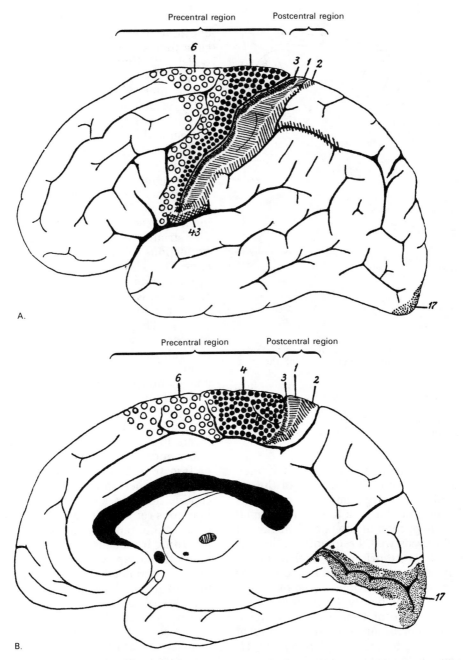

Precentral region   Postcentral region

A.

Precentral region   Postcentral region

B.

FIG. 11-136.   Postcentral cortical areas, 3, 1, and 2 and precentral areas 4 and 6 are shown on the superolateral (A) and medial (B) surfaces of the human cerebrum, described by Brodmann on the basis of different cytoarchitectonic features. (From K. Brodmann, *Vergleichende Localisationslehre der Grosshirnrinde,* J. A. Barth, Leipzig, 1909.)

sentation in SI. The existence of SII was initially discovered in the cat by Adrian (1941), but its presence has subsequently been verified in infrahuman primates and in man. There is also evidence that reciprocal neural connections exist between the primary and secondary somesthetic areas.

PRIMARY VISUAL RECEIVING AREA.   The primary visual receiving area (V-I) in man and other primates corresponds to area 17 of Brodmann, which is located along the superior and inferior banks and in the depths of the calcarine fissure, and it extends into the cuneus and lingual gyrus (Figs. 11-131; 11-

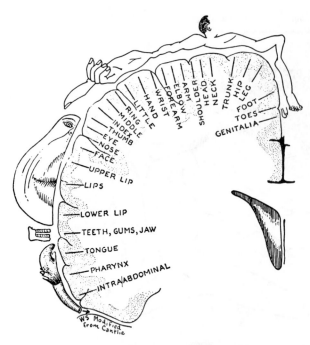

FIG. 11-137. Somatotopic pattern of representation in the primary somesthetic receiving area of the postcentral gyrus in the human cerebral cortex. Note the differences in relative size of the cortical regions from which sensations localized to different body parts have been achieved by electrical stimulation. (After W. Penfield and T. Rasmussen, *The Cerebral Cortex of Man,* Macmillan Publishing Co., New York, 1950.)

134 to 11-136). In man, this area is located principally on the medial surface of the occipital lobe, but frequently it continues around the occipital pole to occupy a small but variable portion of the lateral surface of the cerebrum as well. Frequently called the **striate cortex,** the primary visual receiving area may be clearly identified because of the presence of a conspicuous layer of myelinated fibers, limited only to this area, which is called the line of Gennari.[1] This myelinated stria corresponds to the middle of three sublayers in layer IV (Figs. 11-129; 11-139). The striate cortex is quite thin in comparison to other cortical regions, and it is especially rich in granule cells. It contains well developed outer and inner granular layers (II and IV), but the pyramidal cells, except the large **solitary cells of Meynert,**[2] are small. These latter elements, mostly found in

[1]Francisco Gennari (1750–1797?): An Italian anatomist and physician (Parma).
[2]Theodor H. Meynert (1833–1892): An Austrian neurologist (Vienna).

layer V, have a stellate shape and may be considered somewhat modified pyramidal cells that are oriented in a line; they have widely ramifying dendrites and axons that project to the superior colliculus.

The internal granular layer (IV) of the primary visual area (area 17) receives fibers from the lateral geniculate body by way of the geniculocalcarine tract, and the striate area on each side receives input from the contralateral visual half of the visual field, i.e., the temporal half of the ipsilateral retina and the nasal half of the contralateral retina (Fig. 11-140). The macular region of the retina is represented in the more caudal part of the striate region. The geniculocalcarine projections to the striate cortex retain their laterality. The input from the contralateral retina terminates in lateral geniculate layers 1, 4, and 6, while the ipsilateral retinal fibers terminate in lateral geniculate layers 2, 3, and 5. The geniculocalcarine projection courses to layer IV of the striate cortex and is organized in such a manner that neurons from the different geniculate layers terminate in alternating portions of layer IV, each of which measures as much as 0.5 mm wide. Additionally, the striate cortex is organized vertically in columns of cells through all the layers in a manner that also reflects ocular dominance in a right eye-left eye alternating pattern (Fig. 11-141). In a series of elegant studies, Hubel and Wiesel mapped the receptive fields of striate cortical neurons to luminous retinal stimulation. Unlike the circular retinal receptive fields encountered by recording from cells in the lateral geniculate body, the receptive fields for cortical neurons are arranged linearly in the retina, so that cortical cells respond with more or less complexity to slits of light or to edges between light and dark.

VISUAL AREAS II AND III. Surrounding the primary visual receiving area are two cortical zones from which evoked potentials can be recorded when the retina is stimulated. These prestriate cortical areas occupy the remainder of the occipital lobes and extend rostrally into the parietal and temporal lobes; they correspond roughly to Brodmann's areas 18 and 19, but are not exactly coextensive with them (Fig. 11-131). Sometimes considered as the visual association cortex, these areas have been shown to be interconnected at the cortical level with the primary visual receiving cortex of area 17. A micro-

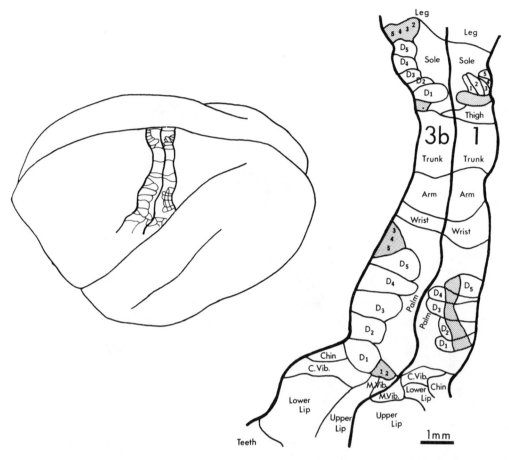

FIG. 11-138. Microelectrode studies of areas 3b and 1 in the contralateral postcentral gyrus have shown that a complete representation of the body parts exists in each of these two cytoarchitectonic fields. Similar functionally distinct fields have also been reported to characterize the organization of area 2 and are now suspected to exist in area 3a, but these latter two cytoarchitectonic areas are not shown in this figure. This particular map displays the primary somesthetic area in the owl monkey, *Aotus trivirgatus,* but similar results have been reported for many species of monkeys. (From Merzenich, Kaas, Sur, and Lin, 1978.)

electrode analysis of single cell responses has revealed that the retinotopic pattern in area 18 (V-II) is a mirror image of that in area 17. Within area 19, still a third visual receiving area (V-III) exists, and its retinotopic organization also appears to be a mirror image of area 17. Most of the neurons in visual areas II and III respond as "complex" or "hypercomplex" cells according to Hubel and Wiesel, and few if any appear to belong in the "simple" category that has been described in the striate cortex. It is believed that visual areas II and III function by elaborating on the visual input received in the striate cortex and they serve to transfer this information both to other cortical areas and back to brain stem sites. Additionally, they may also play a role in analyzing and coordi-

nating visual information for appropriate environmental and spatial adjustments.

AUDITORY RECEIVING AREAS (AI AND AII). The auditory receiving areas are located along the anterior and posterior transverse temporal gyri and correspond to Brodmann's areas 41 and 42 (Fig. 11-131). In man and monkeys, these gyri form part of the temporal operculum and extend from the upper bank of the superior temporal gyrus to the floor of the lateral sulcus, where they lie hidden. To see these areas in the human brain, it is necessary to separate the temporal and parietal opercula, as one does to expose the insula (Fig. 11-135). Much of what is known about auditory receiving areas in the human brain, therefore, has been gained from studies on cats, because these cortical

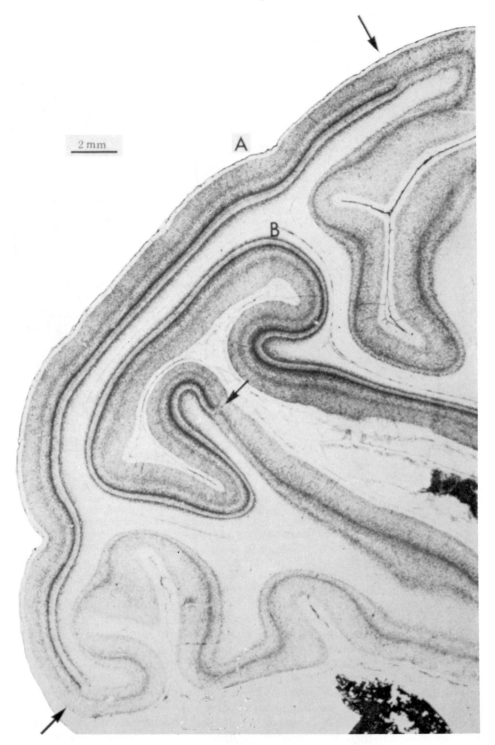

Fɪɢ. 11-139. A low power photomicrograph of a parasagittal section of monkey visual cortex stained by the Nissl method for neuronal cell bodies. Note that the visual cortex is a continuous sheet of neurons about 2 mm thick, and in this section it consists of an outer gyrus (A) and a buried fold (B). The arrows indicate the borders between areas 17 and 18. (From D. Hubel and T. Wiesel, 1977.)

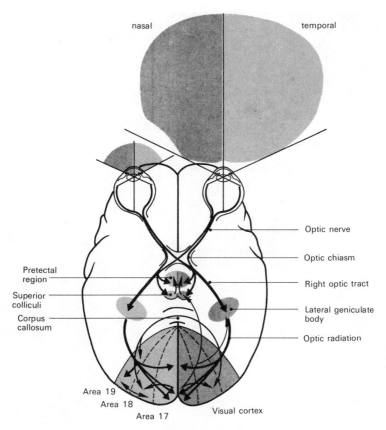

nasal

temporal

Optic nerve

Optic chiasm

Pretectal
region

Right optic tract

Superior
colliculi

Lateral geniculate
body

Corpus
callosum

Optic radiation

Area 19

Area 18

Area 17

Visual cortex

FIG. 11-140. A simplified diagram of the visual pathway in the human brain. Note that some of the efferent connections between the visual cortex and subcortical structures are shown on the right and that transcallosal fibers interconnect the right and left visual areas. (From R. F. Schmidt, *Fundamentals of Sensory Physiology,* Springer, New York, 1978.)

areas in the cat are located on the exposed lateral surface of the cortex and are easily accessible.

The primary auditory cortex that constitutes area 41 may be described as a highly cellular, six-layered koniocortex with rather thick and densely packed granular layers, similar to the cortex seen in other primary receiving cortical areas. This region (in the cat) is nearly surrounded by a belt of at least three auditory cortical fields (Fig. 11-142), which also respond to auditory stimuli and constitute the secondary auditory area (AII). These latter fields in the cat may be comparable to area 42 in the human brain, located adjacent to area 41. Unlike area 41, however, the six-layered cortex in area 42 is of the parietal type (type 3).

The auditory cortex receives its principal afferent input from the medial geniculate nucleus. These geniculotemporal fibers course by way of the sublenticular part of

the internal capsule and project to both the primary and secondary auditory areas. In the anesthetized monkey or chimpanzee, there is a precise tonotopic localization within the primary auditory receiving area. Studied most extensively in the cat, high tonal frequencies are represented anteriorly in AI, and progressively lower frequencies are represented as one proceeds posteriorly in the primary auditory receiving area. In AII, the tonotopic representation appears to be reversed from that seen in AI, with the higher frequencies being represented more posteriorly and the lower frequencies anteriorly. Some questions have been raised about the precision of this tonotopic representation, however, because microelectrode studies in unanesthetized animals have revealed a more complex response of many cortical neurons in the auditory region than the precise tonotopic organization observed in anesthetized animals. More detailed infor-

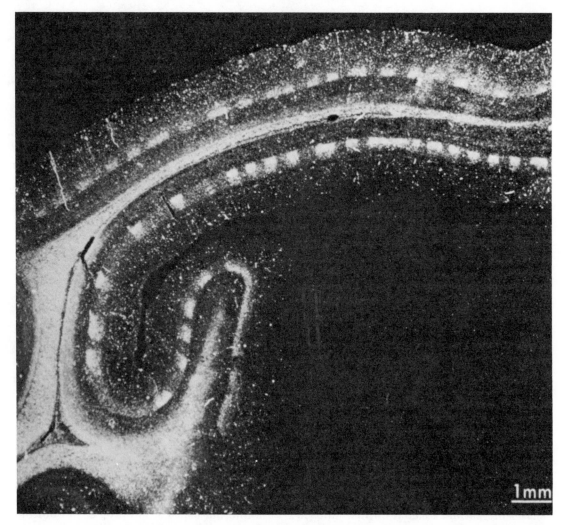

FIG. 11-141. A dark field autoradiograph taken of area 17 in a macaque monkey in which the ipsilateral eyeball had been injected with tritiated proline-fucose two weeks earlier. The labelled regions stand out in a periodic pattern of white patches in cortical layer IVc as linearly alternating columns, which actually are curving slabs of cortex cut in cross section to the brain surface in this specimen. This demonstrates that the radioactive material injected into the ipsilateral eye had been carried from the retina to alternating specific columns of cells in the ipsilateral striate cortex, between which are gaps of similar width that were not labelled and presumably represent ocular columns for the noninjected contralateral retina. (From D. Hubel and T. Wiesel, 1977.)

mation is necessary before definitive comments can be made about the nature of the organization of the primary and secondary human auditory receiving areas. Interest is currently focused on the columnar organization of the auditory cortex and the manner by which binaural input is processed at a cortical level.

TASTE OR GUSTATORY RECEIVING AREA. It is well established that the peripheral pathways mediating special sensory impulses of taste course along special visceral afferent nerve fibers in the facial, glossopharyngeal,

and vagus nerves; however, the central pathways traveled by these impulses are somewhat less clearly understood. Afferent impulses reach the nucleus of the tractus solitarius in the medulla oblongata and are then relayed to higher brain stem, diencephalic, and telencephalic levels. Some evidence appears to indicate that the nucleus of the tractus solitarius projects to a region in the pons located in the parabrachial nuclei within the superior cerebellar peduncle; this has been called the pontine taste area (Norgren and Pfaffmann, 1975). From the

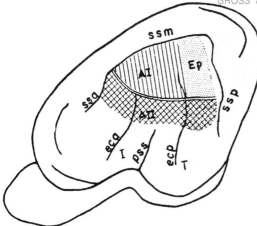

Fig. 11-142. The primary (AI, vertical lines) and secondary (AII, cross hatch) cortical auditory receiving areas in the cat. Caudal to the posterior ectosylvian sulcus (ecp) is the posterior ectosylvian area (Ep, dotted), which is structurally distinct from AI and from which only small evoked potentials can be elicited from cochlear nerve stimulation. Below this is the caudal portion of the secondary auditory area (cross hatched and dotted). I, insular region; T, temporal region; ssa, ssm, and ssp, anterior, middle, and posterior parts of the suprasylvian sulcus; eca, anterior ectosylvian sulcus; pss, pseudosylvian sulcus. (From J. E. Rose and C. N. Woolsey, in *Biological and Biochemical Bases of Behavior,* edited by H. F. Harlow and C. N. Woolsey, University of Wisconsin Press, Madison, 1958.)

pons, nerve fibers project to the lateral hypothalamus, the central nucleus of the amygdala, and the medial part of the nucleus ventralis posteromedialis (VPM) in the thalamus. The hypothalamic and amygdaloid projections may be related to feeding behavior, because lesions in these regions markedly alter food intake, whereas the pathway to the thalamus projects to the cerebral cortex.

The cortical areas responsible for the perception of taste are not precisely known; however, a close relationship appears to exist between the cortical receiving areas for general somatic sensations from the tongue and those to which the taste pathways project from the thalamus. Recent information points to the opercular part of the postcentral gyrus, with extensions into the lateral fissure, as one of the cortical taste receiving areas. This region is either within or near the primary somesthetic tongue area. Ablations of the frontal and parietal opercular regions in primates have been reported to result in a loss of taste, while stimulation of the opercular cortex of the parietal lobe in conscious human patients has resulted in the perception of taste sensations.

VESTIBULAR CORTICAL REPRESENTATION. There is reason to believe that impulses traveling along pathways from the vestibular nuclei in the brain stem reach the cerebral cortex because human beings have the ability to perceive consciously their orientation in space. Although studies have revealed cortical cells capable of responding to vestibular input, there does not appear to be a cortical area in which the cells respond only to vestibular impulses. Experiments in cats have implicated a region anterior, but close, to the auditory receiving area. This area includes parts of the anterior ectosylvian and anterior suprasylvian gyri adjacent to the anterior part of the suprasylvian sulcus in this species. More recent experiments in monkeys, however, point to a vestibular receiving area near or within the somatosensory face area in the postcentral gyrus. Cortical neurons in this latter region show a convergence of somatosensory impulses originating from muscles and joints and vestibular impulses. These units also appear to interact with neurons in the motor cortex. The precise pathway(s) by which vestibular information reaches cortical levels is not known, but it is presumed that thalamic or other nuclei lower in the neuraxis are interposed between the primary vestibular nuclei and the cerebral cortex.

Principal Functional Cortical Areas: Motor

In addition to its functions related to sensation, the cerebral cortex is importantly involved in the control of movement. Axons arise in deep layers of the cerebral cortex and project to the basal ganglia, thalamus, brain stem, and spinal cord, and many are associated with motor function. Since the initial experiments of Fritsch and Hitzig in 1870, it has been known that electrical stimulation of certain cortical sites in animals and in man can elicit the contraction of individual muscles or muscle groups or inhibit ongoing motor activity. Motor localization studies have identified primary, secondary, or supplementary cortical motor areas, as well as motor areas associated with eye movements and speech. It must be understood that sensory and motor functions of cortical regions are so intimately interrelated that often motor responses can be elicited from stimulation of the so-called sensory areas, and sensory impulses are

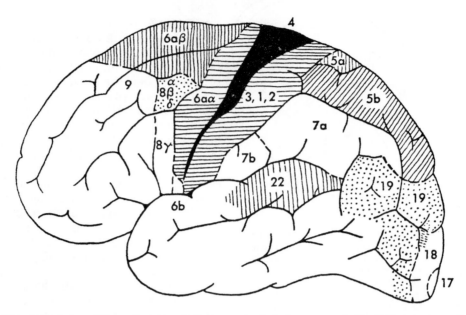

FIG. 11-143. Lateral view of the cortical areas in the human brain that were reported by O. Foerster to result in motor responses when electrically stimulated. Stimulation of the primary motor cortex (area 4, indicated in solid black) and other parts of the precentral gyrus and the postcentral gyrus (areas 6a α, 3, 1, and 2; horizontally lined areas) resulted in discrete movements that could be elicited at relatively low thresholds. Stronger electrical stimulation of other regions in the frontal, temporal, and parietal lobes (areas 6a β, 5a and 5b, and 22) resulted in more complex mass movements and turning toward the opposite side; Foerster considered these to be extrapyramidal in their mechanism. Note that Foerster superimposed his results on the cytoarchitectural numbering system utilized by the Vogts, and this differed somewhat from that of Brodmann. (From W. Penfield and T. Rasmussen, 1950, after O. Foerster, 1935).

recordable within the motor regions. Notwithstanding the safety of assuming that the nervous system functions in its entirety in the execution of behavior, there still is ample evidence in support of the concept that a high degree of localization of motor control exists at a cortical level.

PRIMARY SOMATOMOTOR AREA. The primary somatomotor area is located in the posterior part of the frontal lobe along the anterior bank of the central sulcus, and it includes most of the adjoining precentral gyrus (Figs. 11-131; 11-134 to 11-136; 11-143). It corresponds to Brodmann's area 4, and its structure is characterized by large numbers of pyramidal cells in layers III and V, by the presence of giant pyramidal cells (of Betz) in layer V, and by its poorly developed external and internal granular layers (II and IV). The primary motor cortex is usually quite thick (3.5 to 4.0 mm) and it is categorized as agranular, belonging to the type 1 grouping. It extends onto the medial surface of the cerebrum, where it occupies the anterior part of the paracentral lobule (Fig. 11-136,B), and on the lateral surface of the cerebrum, it reaches nearly to the lateral fissure but

tapers more narrowly before it ends (Fig. 11-143). It has been estimated that there are between 30,000 and 35,000 Betz cells in the primary motor area, and although their axons join the corticospinal or pyramidal tract, they form only a small percentage of the estimated one million fibers located within that great motor pathway. The giant pyramidal cells may reach sizes of 60 μm wide and 120 μm high, but many are considerably smaller, especially in the inferolateral or opercular part of the precentral gyrus.

Contraction of specific muscle groups that results in discrete movements of contralateral body parts may be induced by electrical stimulation on the surface of the primary somatomotor area. Much of the information available on the somatotopic pattern of organization in the human motor cortex has been achieved from direct electrical stimulation of the brain during neurosurgical procedures. These studies and others on subhuman primates and other mammals have established that excitation of discrete points of foci will elicit specific motor responses and, by this manner, it has been possible to construct somatotopic maps of the cortical

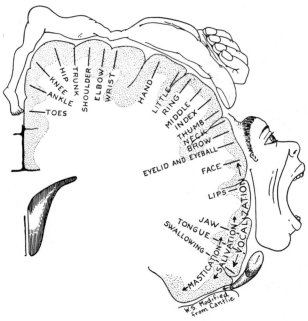

regions, and the fibers in the corticospinal tract are derived from neurons in the primary somatosensory areas and probably also from cortical areas in occipital and temporal lobes, as well as from other sites in the parietal and frontal lobes. The movements that are elicited by electrical stimulation of the primary motor area are the result of straightforward muscular contractions and never take the form of complex or skilled purposeful actions.

Not much information exists on the functional organization of the motor area; however, it is believed that the cellular elements are organized in a columnar manner as they are in other cortical regions. Recordings from individual units during the performance of a motor task have revealed a complex pattern of interaction among cortical cells. What appears to be represented in the motor cortex is related more to the joint at which the movement occurs and the nature of that movement than to the individual muscles being contracted. Certainly many factors, such as the afferent inputs to the motor region from other cortical fields and other levels of the neuraxis, and the previous experience and learned behavior of the individual, all contribute to the functional organization of the motor cortex.

PREMOTOR AREA. The premotor area is a region of the cerebral cortex that extends on both the lateral and medial surfaces of the frontal lobe immediately rostral to the primary motor area. Cytoarchitecturally, it is an agranular, six-layered cortex quite similar to that seen in the motor area, except that it lacks the giant pyramidal cells characteristic of area 4. This entire cortical region, lying behind the frontal association cortex and frontal eye fields and in front of the precentral motor area, was anatomically delineated as area 6 by Brodmann (Figs. 11-131; 11-136); physiologically, however, this region does not appear to function as a single entity. The portion of area 6 that is on the medial surface of the hemisphere has been designated as part of the supplementary motor cortex (Figs. 11-134; 11-145), and that on the lateral cortical surface has been divided into several subareas on both functional and anatomic grounds.

Electrical stimulation of the premotor area gives rise to complex movements, frequently consisting of deviation of the eyes, head, and trunk to the contralateral side, and these are

FIG. 11-144. Motor homunculus indicating parts of the body from which motor responses were obtained when area 4 in the precentral gyrus of human patients was electrically stimulated during neurosurgical procedures. (After W. Penfield and T. Rasmussen, *The Cerebral Cortex of Man,* Macmillan Publishing Co., New York, 1950.)

surface on which are represented sequentially the various regions of the body that respond to electrical stimulation (Fig. 11-144). Although some differences have been reported, the lower extremity is represented on the medial surface of the cerebrum in the paracentral lobule and it extends onto the superior part of the precentral gyrus. From this point and continuing laterally along the precentral gyrus, the trunk, arm, forearm, hand, neck, face, lips, jaw, tongue, larynx, and pharynx are represented in sequence. As is true in the primary somatosensory receiving area, there is a disproportionately greater representation in the motor cortex for the hand and face than would be expected on the basis of the size of these body parts (Fig. 11-146). It appears that the more intricate and delicate the capability of motor performance, the greater is the volume of cortical substance required for motor control. The motor field for movements of the thumb is especially large.

It must be emphasized that movement can also be achieved by stimulation of cortical areas other than those in the so-called motor probably mediated through extrapyramidal paths. Lesions of this region in monkeys and man result in forced grasping. Although there is no general agreement about the portions of area 6 that are responsible for this clinical sign, the best evidence indicates that forced grasping results from lesions of the supplementary motor area located on the medial surface of the hemisphere rather than that part on the lateral cerebral surface. Lesions in the premotor area have also been reported to result in perseveration of a given motor task in monkeys and the inability of human beings to perform correct movements in response to a verbal command.

SUPPLEMENTARY MOTOR AREA. The supplementary motor area is located in the medial frontal gyrus on the medial surface of the cerebrum just above the cingulate gyrus and just rostral to the primary motor region that controls movements of the lower extremity (Figs. 11-134; 11-145). It has no specific cytoarchitectural borders and does not contain giant pyramidal cells. It is a physiologically defined region that corresponds to the part of Brodmann's area 6 located on the medial surface of the cerebrum. Electrical stimulation of this region elicits movements of the contralateral extremities in both mon-

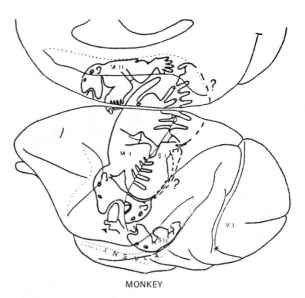

MONKEY

FIG. 11-146. Diagram of monkey cerebral cortex showing the general organization plan of the precentral motor cortex (MI), the supplementary motor cortex on the medial surface of the cerebrum (MII), the postcentral primary somesthetic area (SI), and the secondary somesthetic area (SII). (From C. N. Woolsey, in *Biological and Biochemical Bases of Behavior*, edited by H. F. Harlow and C. N. Woolsey, University of Wisconsin Press, Madison, 1958.)

keys and man, but the threshold for these responses is usually higher than that in the primary motor cortex.

In the monkey, the supplementary motor area seems to be organized in a topographic pattern nearly as precise as that seen in the primary motor cortex (Fig. 11-146), whereas in the human brain electrical stimulation appears to elicit complex motor synergies of the eyes, upper extremities and trunk, along with vocalization, and no precise somatotopic pattern has been reported. Lesions in the supplementary motor area of monkeys results in hypertonia, spasticity, and forced grasping without paralysis, while human patients with lesions confined to this region show more subtle motor symptoms, including a reduction in speech, but these are usually transient and recovery of deficits occurs.

FRONTAL AND OCCIPITAL EYE FIELDS. It is well established that areas in both the frontal and occipital lobes are functionally related to movements of the eyes. Rostral to the premotor area is a cortical region that produces conjugate eye movements when electrically stimulated. In the human brain, the **frontal eye field** corresponds approximately to Brodmann's area 8, located in front

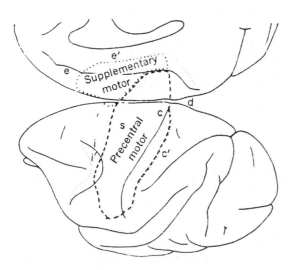

FIG. 11-145. Diagrammatic representation of the medial (above) and superolateral surfaces of the left cerebrum in the monkey, showing the extent of the primary somatomotor area (precentral motor) and the supplementary motor area. Note the localization pattern within these areas shown in Figure 11-146. (From Woolsey et al., 1952.)

of the precentral sulcus and occupying the caudal part of the middle frontal gyrus and a portion of the adjacent caudal part of the inferior frontal gyrus (Fig. 11-131). Because the frontal eye field is located immediately anterior to the precentral motor areas that control movements of other facial and oral musculature, some investigators have considered this field a rostral extension of the motor area. It is unclear, however, how this cortical region receives its sensory input, and its functional relationships to brain stem centers that innervate the extraocular musculature are also not well understood. Electrical stimulation of the frontal eye field results in a conjugate deviation of the eyes to the contralateral side of the body. A lesion in this cortical region results in the loss of this motor function and the eyes become deviated toward the side of the lesion because the eye field in the opposite cortex is functioning unopposed.

Electrical stimulation of cortical regions located in the occipital lobe also results in conjugate deviation of the eyes to the opposite side of the body. One **occipital eye field** is more extensive than that found in the frontal lobe and appears to occupy most of striate area 17, and others are located in parts of areas 18 and 19. Stimulation of the striate region above the calcarine fissure appears to evoke lateral and downward movements of the eyes, while stimulation below the fissure results in lateral and upward movements. Efferent projections from the occipital cortex to the superior colliculus and pretectal regions probably serve as the initial part of this corticofugal pathway to the extraocular muscles. The occipital eye fields are thought to be importantly involved in ocular movements required to fix upon and follow a moving object.

MOTOR SPEECH AREAS. Although it is believed that certain animals have an ability to communicate by utilizing vocalization and other types of motor performance, communication by the use of language and the motor functions of speech are uniquely human. In most individuals, the cortical areas important for the perception of language and the motor control of speech are confined principally to one of the two hemispheres. Right-handed people have a left cortical dominance for motor skills and most, likewise, have a left hemispheric dominance for the organization and control of speech. Left-handed or ambidextrous people have a right cortical dominance for most motor skills, but many still have a left hemispheric dominance with respect to speech. Thus, recent studies have indicated that the left hemisphere serves especially for functions related to language and speech, while the right hemisphere serves more importantly for visuospatial and conceptual functions and for other nonverbal skills. As yet, however, there is little definitive morphologic evidence of differences between the two cortical sides. Our understanding of the localization of speech function in the cerebral cortex has principally resulted from studies in people who sustained brain lesions and from reports of electrical stimulation of the brain in conscious patients during neurosurgical procedures.

The **motor speech area of Broca** is located near the operculum of the frontal lobe within the inferior frontal gyrus, just in front of the precentral sulcus (Figs. 11-147; 11-148). It includes the region that coincides approximately with Brodmann's area 44, as well as part of area 45 (Fig. 11-131). Broca's area lies rostral to the primary motor region that controls the laryngeal and oropharyngeal musculature essential for vocalization. Electrical stimulation of this region in the dominant hemisphere may result in the arrest of speech, in aphasic mistakes, or simply in the utterance of vowel sounds. Additionally, alterations in speech behavior have been achieved by stimulation of at least two other cortical regions. One of these is the supplementary motor area on the medial surface of the frontal lobe in front of the primary motor region for the lower extremity, and the other is located in the dominant hemisphere on the lateral surface of the parietal and temporal lobes, the so-called **secondary motor speech area of Wernicke**[1] (Figs. 11-147; 11-148). This latter region occupies a rather large area below the lateral sulcus within the posterior part of the superior temporal gyrus and extends into the supramarginal and angular gyri. It lies adjacent to the auditory receiving area and it is thought to function in speech recognition, because lesions at this site result in receptive aphasia or failure to understand a previously known language.

[1]Karl Wernicke (1848–1905): A German neurologist (Berlin).

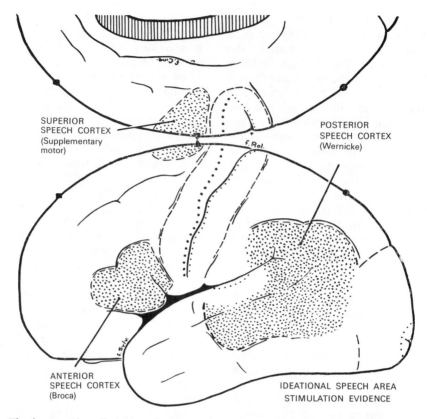

FIG. 11-147. The dominant (usually left) cerebral hemisphere, showing areas devoted to the elaboration of speech as determined by evidence from electrical stimulation of the human brain during neurosurgical procedures. (From W. Penfield and L. Roberts, *Speech and Brain Mechanisms,* Princeton University Press Princeton, 1959.)

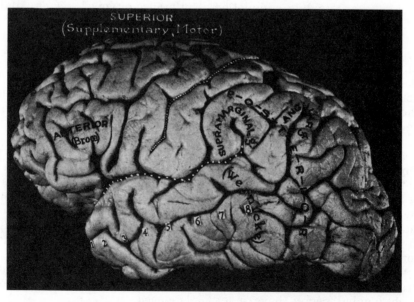

FIG. 11-148. Photograph of the dominant left hemisphere of the human brain on the surface of which are indicated the areas devoted to speech. Local injury or electrical stimulation of these areas produces aphasia. White dots outline the central and lateral sulci and the numbers in the temporal lobe mark the centimeters back from the tip of that lobe. (From W. Penfield and L. Roberts, *Speech and Brain Mechanisms,* Princeton University Press, Princeton, 1959.)

WHITE SUBSTANCE OF THE CEREBRAL HEMISPHERE

The white substance of the cerebral hemisphere underlies the outer laminated mantle of cells that forms the cortical gray substance. Consisting of a great mass of myelinated fibers, this cerebral white substance intervenes between the cortical gray matter and the gray matter of the more deeply located basal ganglia. If a horizontal section of the brain is made deep to the level of the cortical sulci and gyri, but superficial to the corpus callosum, the subcortical white substance on each side may be seen to form a semioval mass that is called the **centrum semiovale** (Fig. 11-149). The fibers that compose the centrum semiovale are the same as those that form the **corona radiata** (see below) when cut transversely, and they consist of **projection fibers,** which connect the gray matter of the cerebral cortex with the caudal parts of the brain and spinal cord, and association or **arcuate cerebral fibers,** which interconnect different parts of the cerebral cortex of the same side. A third group of fibers, called **commissural fibers,** interconnects regions of the two cerebral hemispheres across the median plane. The

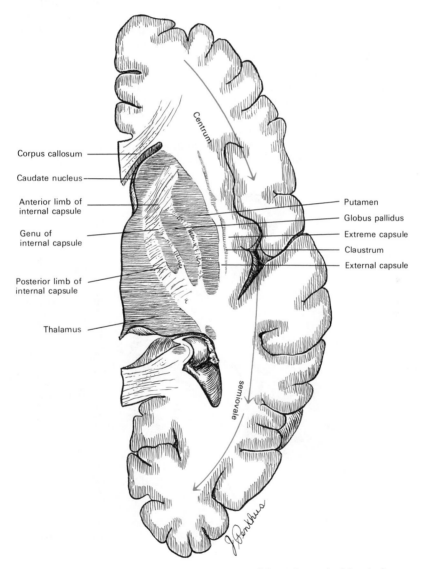

FIG. 11-149. A diagrammatic representation of a frontal section of the right cerebral hemisphere seen from above. Note the semioval mass of white matter that forms the centrum semiovale (red arrows) and the limbs and genu of the internal capsule.

commissural fibers comprising the corpus callosum intermesh with the projection and arcuate cerebral fibers at the level of the centrum semiovale, but they can best be seen passing across the midline at the depth of the longitudinal cerebral fissure. Commissural fibers forming the commissure of the fornix (hippocampal commissure), the epithalamic commissure (posterior commissure), and the rostral commissure (anterior commissure) are found more deeply in the cerebrum.

Projection Fibers

The projection fibers of the cerebral hemisphere include both afferent fibers, which originate in subcortical centers and terminate in the cerebral cortex **(corticopetal fibers),** and efferent fibers, which arise in the cerebral cortex and terminate in lower centers of the neuraxis **(corticofugal fibers).** Certain groups of projection fibers have already been described in previous sections of this chapter, such as the afferent and efferent fibers that constitute the fimbria and fornix

system of the hippocampal formation. Mention has also been made of the cortical interconnections with the brain stem, thalamus, and basal ganglia. Above or superior (rostral) to the thalamus and basal ganglia, these latter interconnections and most of the other fibers conveying information to and from the gray matter of the cerebral cortex become arranged in a radiating mass of white matter called the **corona radiata** (Fig. 11-150). The arrangement of the corona radiata can best be visualized when a careful dorsoventral dissection of the cerebrum is made in the same orientation as that followed by the course of the corticopetal and corticofugal fibers. The corona radiata is continuous caudally with the more compactly formed internal capsule.

INTERNAL CAPSULE.   The internal capsule is a thick mass of white matter that is bounded laterally by the lentiform nucleus and medially by the head of the caudate nucleus, the dorsal thalamus, and the tail of the caudate nucleus. Caudally (or inferiorly), many of the fibers that comprise the internal capsule form the cerebral peduncles, while

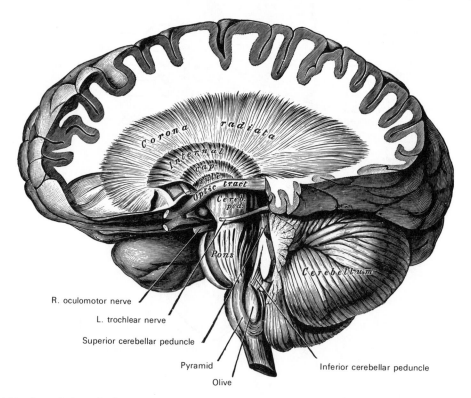

FIG. 11-150.   Lateral view of a dissection of the left cerebral hemisphere, showing the fanlike arrangement assumed by the corticopetal and corticofugal fibers throughout the hemisphere as they form the internal capsule and the corona radiata.

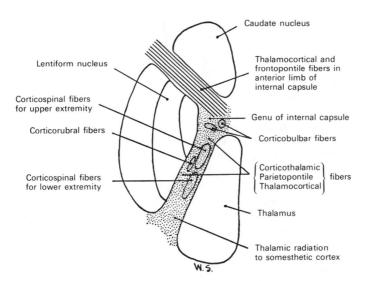

Caudate nucleus

Thalamocortical and
frontopontile fibers in
anterior limb of
internal capsule

Lentiform nucleus

Corticospinal fibers
for upper extremity

Corticorubral fibers

Genu of internal capsule

Corticobulbar fibers

Corticospinal fibers
for lower extremity

⎧ Corticothalamic ⎫
⎨ Parietopontile ⎬ fibers
⎩ Thalamocortical ⎭

Thalamus

Thalamic radiation
to somesthetic cortex

W.S.

FIG. 11-151. Diagram showing the locations of the various functional groups of fibers in the internal capsule as seen in a horizontal section. (N. B. Everett, *Functional Neuroanatomy,* Lea & Febiger, Philadelphia, 1965.)

rostrally (or superiorly) the internal capsule assumes a more fan-like arrangement and thereby forms the corona radiata (Fig. 11-150). Between the caudate nucleus and the dorsal thalamus, the internal capsule is bent approximately to a right angle around the medial margin of the lentiform nucleus (Fig. 11-151). This angular part of the internal capsule is called the **genu of the internal capsule.** When the internal capsule is studied in horizontal section, it can be seen that the portion rostral to the genu, the **anterior limb** (*crus anterius*), lies between the caudate nucleus medially and the lentiform nucleus laterally. The portion caudal to the genu, the **posterior limb** (*crus posterius*), is interposed between the thalamus medially and the lentiform nucleus laterally. The largest part of the posterior limb is called the **thalamolenticular part** (*pars thalamolenticularis*). The portion of the posterior limb that curves around the caudal end of the lentiform nucleus is called the **retrolenticular part** (*pars retrolenticularis*), and the portion that lies ventral to the lentiform nucleus is the **sublenticular part** (*pars sublenticularis*).

The **anterior limb of the internal capsule** contains *thalamocortical fibers* that comprise the **anterior thalamic radiation.** These fibers are derived from the anterior and medial thalamic nuclei and course to the frontal lobe of the cerebral cortex (Fig. 11-151). *Corticothalamic* fibers coursing from frontal lobe areas to the thalamus and

frontopontine fibers interconnecting the frontal lobe to the pontine nuclei are also located in the anterior limb of the internal capsule.

The **genu of the internal capsule** contains, in addition to corticothalamic and thalamocortical fibers, several fascicles of **corticobulbar fibers** to the motor nuclei of the cranial nerves (Fig. 11-151). One group of fibers is derived from the frontal eye fields of cortical areas 8 and 9 and courses to the oculomotor nucleus bilaterally, the trochlear nucleus ipsilaterally, and the abducens nucleus contralaterally. Another corticobulbar fascicle consists of fibers from the face, jaw, and oral cavity regions in the primary somatomotor cortex of area 4. These latter fibers terminate in the motor nucleus of the trigeminal nerve bilaterally, the facial nucleus bilaterally, the nucleus ambiguus bilaterally, and the hypoglossal nucleus contralaterally.

The **thalamolenticular part of the posterior limb of the internal capsule** is the part that is oriented posterolateral to the genu (Fig. 11-151). It contains the **central thalamic radiation,** which consists of thalamocortical somesthetic fibers to both the precentral and postcentral gyri, and corticofugal fibers from the postcentral gyrus (areas 3, 1, and 2) to the thalamus. Also passing through the thalamolenticular part of the internal capsule are the corticospinal and corticorubral fibers. The *corticospinal fibers* are derived from the upper and lower extremity and from trunk regions of the somatomotor cortex of area 4, and those fibers coursing to the motor nuclei of the muscles of the arm are nearer the genu than are those coursing to the motor nuclei of the muscles of the leg. The *corticorubral fibers* arise in area 6 and course to the ipsilateral red nucleus.

The **retrolenticular part of the posterior limb** contains some fibers of the **posterior thalamic radiation,** including certain geniculocalcarine fibers and others that interconnect the occipital cortex and the pulvinar region of the thalamus. It also contains many corticofugal fibers from areas 18 and 19 that project to visual motor centers in the rostral brain stem. Other corticofugal fibers in the retrolenticular part are derived from association areas located in the parietal and occipital lobes and course to the pontine nuclei.

The **sublenticular part of the posterior limb** contains corticofugal fibers from the parietal and temporal lobes that course to the pontine nuclei, and corticotectal fibers that project to the superior colliculus in the mesencephalon. It also contains the **auditory radiation** from the medial geniculate body, which courses to the auditory receiving area in the transverse temporal gyrus; the **optic radiation** (geniculocalcarine tract) from the lateral geniculate body, which courses to the visual receiving area in the striate cortex of the occipital lobe; and corticofugal fibers from the occipital cortex, which project to the superior colliculus.

A few of the fibers that compose the optic radiation pass in the retrolenticular part of the internal capsule, but most of the geniculocalcarine fibers pass through the sublenticular part. There are some differences in the pathways taken by fibers that course to area 17 from the lateral geniculate body. Geniculocalcarine fibers destined for the superior bank of the calcarine fissure leave the dorsal part of the lateral geniculate body and course almost directly posteriorly to the striate cortex. Fibers from the more ventral parts of the geniculate body, destined for the inferior bank of the calcarine fissure, initially loop forward and downward in the temporal lobe, forming the so-called loop of Meyer[1] or loop of Meyer and Archambault[2]

[1] Adolf Meyer (1866–1950): An American psychiatrist (New York and Baltimore).
[2] LaSalle Archambault (1879–1940): An American neurologist.

(Figs. 11-86; 11-152) before turning back to join the other fibers in the optic radiation.

Arcuate Cerebral Fibers (Association Fibers)

The arcuate cerebral fibers, or association fibers, interconnect the different cortical regions of the same hemisphere. By definition, they are all ipsilateral because the fibers that interconnect cortical areas of the two hemispheres cross the median plane to form the corpus callosum and are referred to as commissural fibers. Arcuate cerebral fibers are of two types: **short arcuate fibers** that interconnect adjacent cortical gyri, and **long arcuate fibers** that interconnect cortical gyri separated by greater distances (Fig. 11-153).

SHORT ARCUATE FIBERS. The short arcuate fibers are of two kinds: intracortical and subcortical. The **intracortical short arcuate fibers** commence and terminate within the gray substance. They form bands of both myelinated and unmyelinated fibers within certain cortical layers that interconnect adjacent gyri; they do not enter the subcortical white matter (Fig. 11-128,C). The **subcortical short arcuate fibers** are groups of fibers that leave the gray substance in one gyrus, enter the subcortical white matter, form a U-shaped loop, and curve upward to join the gray substance of an adjacent gyrus (Fig. 11-153).

LONG ARCUATE FIBERS. The long arcuate fibers form bundles of varying sizes and lengths and are deeply situated in the subcortical white matter. The longest fibers, which join more widely separated gyri, are generally located in the deepest portions of these fasciculi. Several distinct bundles can be dissected readily and have been given the following descriptive names: the uncinate fasciculus, the cingulum, the superior longitudinal fasciculus, the inferior longitudinal fasciculus, the superior occipitofrontal fasciculus, the inferior occipitofrontal fasciculus, and the perpendicular occipital fasciculus.

The **uncinate fasciculus** has a dorsal and a ventral component. Its ventral part interconnects the gyri on the orbital aspect of the frontal lobe with the cortex of the parahippocampal gyrus and other gyri on the medial surface of the temporal lobe (Figs. 11-152; 11-153). Its dorsal component unites other gyri on the superolateral aspect of the frontal lobe with the cortex of the temporal gyri

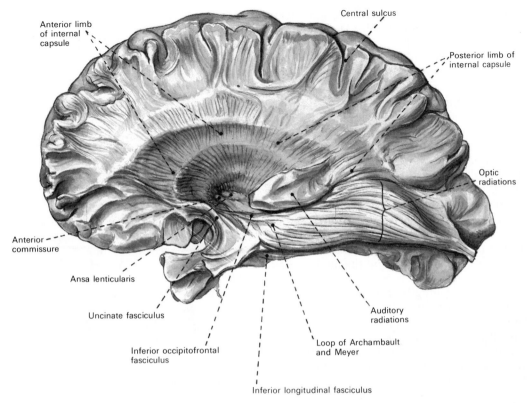

FIG. 11-152. Lateral view of a dissection of the left cerebral hemisphere that exposes the anterior and posterior limbs of the internal capsule, as well as the optic radiations. (Courtesy of Dr. Charles F. Bridgman, who drew it from a dissection made by the late Professor A. T. Rasmussen.)

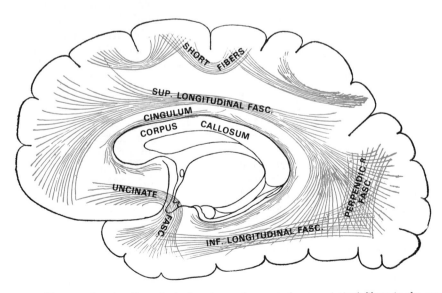

FIG. 11-153. Diagram showing the principal systems of arcuate (or association) fibers in the cerebrum.

or the more lateral aspect of the temporal lobe nearer the temporal pole. The uncinate fasciculus is a hook-shaped bundle that curves around the stem of the lateral sulcus deep to the point at which that fissure divides into its rami.

The **cingulum** is a prominent fasciculus that courses along the medial aspect of the cerebral hemisphere and forms much of the white substance within the cingulate gyrus (Figs. 11-153; 11-154). It courses from the subcallosal and parolfactory gyri rostrally, follows the curve of the corpus callosum, and joins the parahippocampal gyrus caudally (Figs. 11-153 to 11-155). Although some fibers course along its entire length, others join the cingulum or leave it to enter other gyri located more dorsally on the medial surface of the cerebrum.

The **superior longitudinal fasciculus** is a large bundle of arcuate fibers that is located deep in the white matter of the cerebrum, along the dorsolateral border of the lentiform nucleus. It courses along virtually the entire anteroposterior extent of the cerebrum. Commencing in the frontal lobe, the superior longitudinal fasciculus extends backward above the insula and through the parietal lobe to the occipital lobe, where it arches downward and forward to disperse within the temporal lobe. The arching nature of this bundle led to its former name of **arcuate fasciculus.**

The **inferior longitudinal fasciculus** interconnects the cortex of areas 18 and 19 in the occipital lobe with the superior, middle, and inferior temporal gyri, the medial and lateral occipitotemporal gyri, and the parahippocampal gyrus of the temporal lobe. It courses near the lateral walls of the temporal (inferior) and occipital (posterior) horns of the lateral ventricle and is separated from them by the optic radiation (Figs. 11-152 to 11-155).

The **superior occipitofrontal fasciculus** (or fronto-occipital fasciculus) passes between the lateral aspect of the caudate nucleus and the site at which the fibers of the corpus callosum interdigitate with those of the internal capsule to form the corona radiata (Fig. 11-154). This bundle of arcuate fibers connects gyri in the occipital and temporal regions with others in the frontal lobe and insula.

The **inferior occipitofrontal fasciculus** lies lateral to the claustrum in the inferior part of the extreme capsule, just dorsal to the uncinate fasciculus (Figs. 11-152; 11-154). It in-

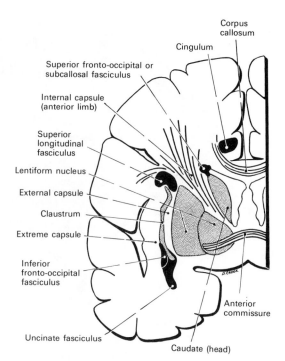

FIG. 11-154. A diagram of a coronal section through the brain at the level of the anterior commissure, showing (in solid black) the approximate positions of several of the major long arcuate fiber bundles. (From E. C. Crosby, T. Humphrey, and E. W. Lauer, *Correlative Anatomy of the Nervous System,* Macmillan Publishing Co., New York, 1962.)

terconnects the gyri on the inferolateral aspect of the occipital and temporal lobes with gyri on the inferolateral portion of the frontal lobe.

The **perpendicular occipital fasciculus** runs vertically through the rostral part of the occipital lobe and connects the inferior parietal lobule with the lateral occipitotemporal gyrus of the temporal lobe (Fig. 11-153).

Commissural Fibers

The commissural fibers in the cerebral white substance include those in the corpus callosum, in the commissure of the fornix, in the anterior (rostral) commissure, and in the posterior (epithalamic) commissure.

LATERAL VENTRICLES

The lateral ventricles are irregularly shaped cavities located within the lower and medial parts of each cerebral hemisphere, one on either side of the midline (Figs. 11-156; 11-157). Developing prenatally as lateral

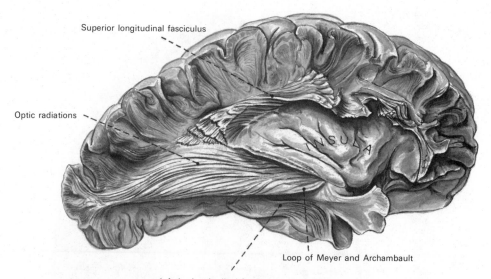

FIG. 11-155. Lateral view of a dissection of the right cerebral hemisphere, exposing the superior and inferior longitudinal fasciculi and the right optic radiations. (Courtesy of Dr. Charles F. Bridgman, who drew it from a dissection made by the late Professor A. T. Rasmussen.)

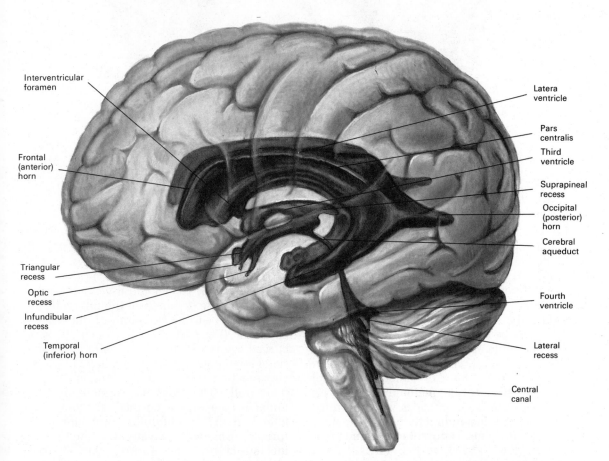

FIG. 11-156. The ventricles of the brain, projected from within the cerebrum and visualized from the left lateral aspect.

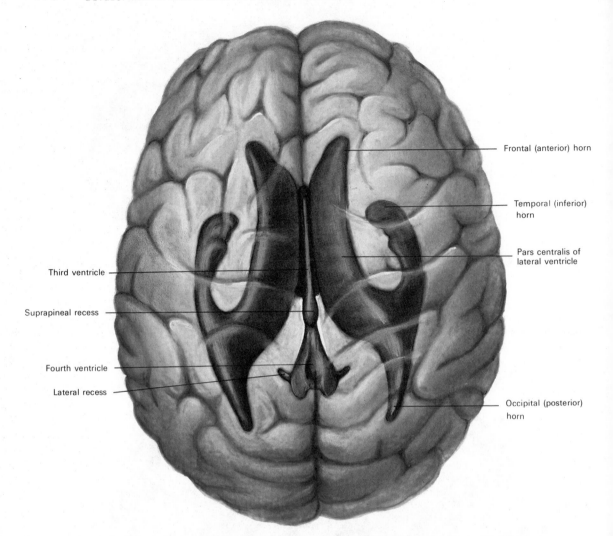

Frontal (anterior) horn

Temporal (inferior) horn

Pars centralis of lateral ventricle

Third ventricle

Suprapineal recess

Fourth ventricle

Lateral recess

Occipital (posterior) horn

FIG. 11-157. The ventricles of the brain, projected from within the cerebrum and visualized from above.

dilatations from the primary cerebral vesicle in the prosencephalon, they are lined by a ciliated epithelial covering called the **ependyma** and they contain cerebrospinal fluid.

In different individuals the lateral ventricles may vary considerably in size and, within the same person, the ventricular walls may be nearly in apposition in one region and quite apart in another (Fig. 11-161). The two lateral ventricles are separated from each other by a thin median vertical partition, the **septum pellucidum** (Fig. 11-162), but each communicates with the third ventricle and thus indirectly with each other through the **interventricular foramen** (of Monro) (Figs. 11-156; 11-158; 11-160). Each lateral ventricle consists of a **central part** with three prolongations called the **cornua**

or horns. These are the **frontal** (or anterior) **horn,** the **occipital** (or posterior) **horn,** and the **temporal** (or inferior) **horn** (Figs. 11-156 to 11-159).

CENTRAL PART.   The central part of the lateral ventricle extends from the interventricular foramen to the splenium of the corpus callosum (Figs. 11-157 to 11-159). It is a curved cavity that lies principally in the parietal lobe; it has a triangular shape in transverse section, with a roof, a floor, and a medial wall. The *roof* is formed by the undersurface of the corpus callosum. The *floor* is really an oblique inferolateral wall formed from medial to lateral by the following structures: the fornix, the choroid plexus, the lateral part of the dorsal surface of the thalamus, the thalamostriate vein, the

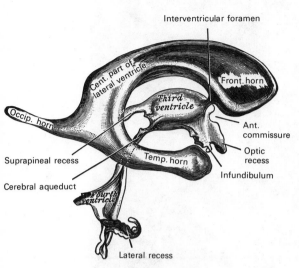

FIG. 11-158.   Drawing of a cast of the ventricular cavities, viewed from the side. (Retzius.)

stria terminalis, and the caudate nucleus. The medial wall is formed by the septum pellucidum, which separates it from the opposite lateral ventricle.

FRONTAL (ANTERIOR) HORN. From the central part of the lateral ventricle, the frontal or anterior horn passes forward and laterally with a slight inclination downward beyond the interventricular foramen into the frontal lobe (Figs. 11-156 to 11-159). As it curves around the anterior end of the caudate nu-

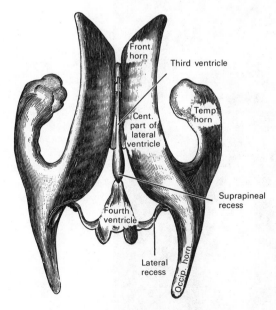

FIG. 11-159.   Drawing of a cast of the ventricular cavities, viewed from above. (Retzius.)

cleus, its *rostral boundary* is formed by the genu and rostrum of the corpus callosum. The frontal horn is bounded *medially* by the rostral part of the septum pellucidum, and its *lateral wall, inferolateral angle* and much of its *floor* are formed by the head of the caudate nucleus (Fig. 11-162). The remaining medial portion of its floor consists of the curved rostrum of the corpus callosum, and its *roof* is formed by the rostral part of the trunk of the corpus callosum. In frontal section, the cavity of the frontal horn is triangular.

OCCIPITAL (POSTERIOR) HORN. The occipital horn of the lateral ventricle is a somewhat narrowed posterior extension of the central part (Figs. 11-156 to 11-159). Its direction is initially backward and laterally and, as it continues backward, it then curves medially before coming to a point in the occipital lobe (Fig. 11-162). Its *roof* is formed by the tapetum of the corpus callosum, which consists of fibers passing to the cortex of the temporal and occipital lobes. On the *medial wall* of the occipital horn is located a longitudinal eminence, the **calcar avis,** which is formed by the deep infolding of the calcarine sulcus (Figs. 11-163; 11-164). The forceps occipitalis of the corpus callosum sweeps around to enter the occipital lobe and forms another projection located above the calcar avis called the **bulb of the occipital horn** (Figs. 11-163; 11-164).

TEMPORAL (INFERIOR) HORN. The temporal or inferior horn is the largest of the three cornua. It traverses the temporal lobe of the brain, forming a curve around the posterior end of the thalamus during its course (Figs. 11-157 to 11-159). It passes initially posteriorly, laterally, and downward, and then it curves forward to extend within 2.5 cm of the temporal pole. Its direction is fairly well indicated on the surface of the brain by the course of the superior temporal sulcus. Its *roof* is formed chiefly by the inferior surface of the tapetum of the corpus callosum, but the tail of the caudate nucleus and the stria terminalis also extend forward in the roof of the temporal horn to its rostral extremity. From lateral to medial along the *medial wall* and *floor* of the temporal horn are found the following structures: (1) the longitudinally oriented **collateral eminence** (Figs. 11-163; 11-164), which is an elongated elevation caused by the collateral sulcus that is continuous posteriorly with the flattened **collateral**

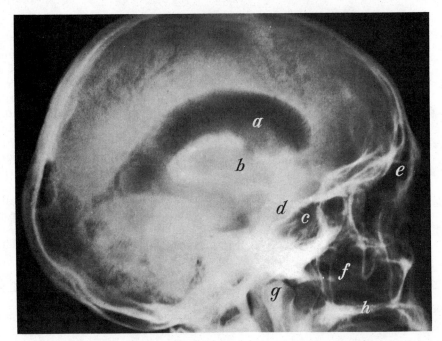

FIG. 11-160. Adult head. *a*, Lateral ventricle injected with *air*; *b*, third ventricle, interventricular foramen midway between *a* and *b*; *c*, sphenoidal sinus; *d*, sella turcica; *e*, frontal sinus; *f*, maxillary sinus; *g*, condyle of mandible; *h*, hard palate. (Department of Radiology, University of Pennsylvania.)

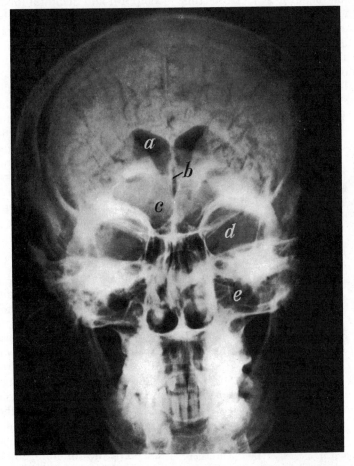

FIG. 11-161. Brain ventricles injected with air. Note extension of air into sulci. *a*, Lateral ventricle; *b*, third ventricle; *c*, frontal sinus; *d*, orbit; *e*, maxillary sinus. (Department of Radiology, University of Pennsylvania.)

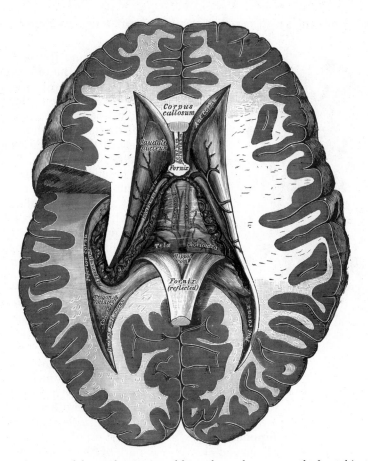

Fig. 11-162. Horizontal section of the cerebrum, viewed from above, that exposes the frontal (anterior) and occipital (posterior) horns of the right lateral ventricle and the occipital and temporal (inferior) horns of the left lateral ventricle. Note also the tela choroidea of the third ventricle and the choroid plexus of the left lateral ventricle.

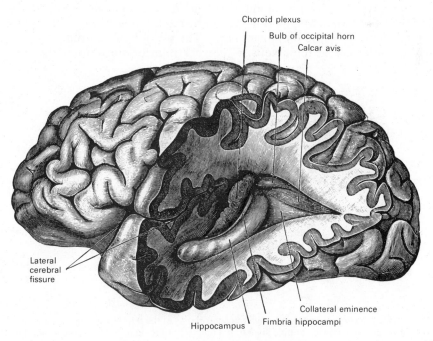

Fig. 11-163. The occipital (posterior) and temporal (inferior) horns of the left lateral ventricle, exposed from the left side.

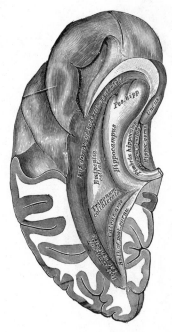

FIG. 11-164. The occipital (posterior) and temporal (inferior) horns of the left lateral ventricle, opened and viewed from above.

Partly covering the upper surface of the hippocampus is the inferior or temporal extension of the **choroid plexus** (Figs. 11-162; 11-163). If the choroid plexus is removed, a cleft-like opening is left along the medial wall of the temporal horn. This is the inferior part of the **choroid fissure.**

CHOROID PLEXUS OF THE LATERAL VENTRICLE. The choroid plexus of the lateral ventricle, similar to that found in the fourth ventricle, is a highly vascular, fringe-like process of pia mater covered by an epithelial layer of ependymal cells (Fig. 11-165). Projecting into the cavity of the lateral ventricle, it extends posteriorly from the interventricular foramen through the central part of the lateral ventricle, where it lies upon the upper surface of the thalamus (Fig. 11-162). At the interventricular foramen, it is continuous with the choroid plexus of the opposite ventricle (Fig. 11-165), and from the central part of the lateral ventricle it courses downward within the ventricular cavity into the temporal (inferior) horn (Figs. 11-162; 11-163). It does not extend into the frontal (anterior) or occipital (posterior) horns.

The portion of choroid plexus in the central part of the lateral ventricle forms the vascular fringed margin of a triangular process of pia mater called the **tela choroidea of the third ventricle.** The choroid plexus projects into the lateral wall of the lateral ventricle from under the lateral edge of the fornix through a cleft called the **choroid fissure.** The portion of choroid plexus in the temporal (inferior) horn lies in the concavity of the

trigone (Fig. 11-164); (2) the **hippocampus,** which is visible as a prominent, curved elevation that is narrow posteriorly but expands anteriorly, where its surface is marked by radial grooves called **hippocampal digitations** (Fig. 11-164); (3) the **fimbria** of the hippocampus, which is located along the medial border of the hippocampus (Fig. 11-164) and is continuous with the hippocampal layer of white matter called the **alveus.**

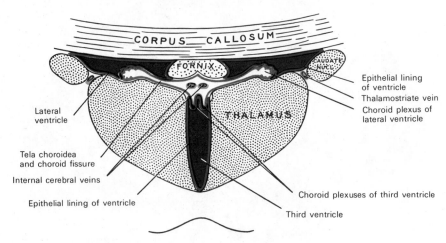

FIG. 11-165. Diagram of a coronal section through the central part of the lateral ventricle and the third ventricle. The pia mater and the tela choroidea (fused double layer of pia mater) are shown in red and the epithelial ependymal lining is shown in green.

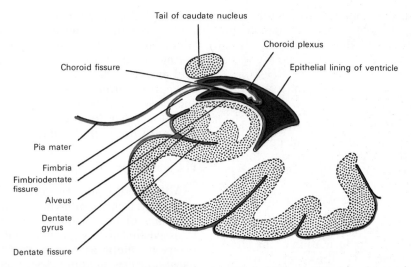

Tail of caudate nucleus

Choroid plexus

Choroid fissure

Epithelial lining of ventricle

Pia mater

Fimbria

Fimbriodentate fissure

Alveus

Dentate gyrus

Dentate fissure

Fɪɢ. 11-166.   Diagram of a coronal section through the temporal (inferior) horn of the lateral ventricle. The pia mater is shown in red and the epithelial ependymal lining is shown in green.

hippocampus and overlaps the fimbria hippocampi (Fig. 11-163). From the lateral edge of the fimbria the epithelium of the choroid plexus is reflected onto the roof of the temporal horn (Fig. 11-166).

The choroid plexus consists of minute villous processes or tufts of blood vessels brought into the tela choroidea by the pia mater. The tufts are covered everywhere by a layer of epithelial cells derived from the ependyma. The arteries of the choroid plexus in the lateral ventricle are (a) the **anterior choroidal artery,** a branch from the cavernous portion of the internal carotid artery that enters the choroid plexus at the end of the temporal (inferior) horn, and (b) several **posterior choroidal arteries,** which are branches of the posterior cerebral arteries and pass forward under the splenium of the corpus callosum. The **choroidal veins** draining the choroid plexus unite to form a tortuous vein that flows rostrally to the interventricular foramen. Here they join the **thalamostriate vein** to form the **internal cerebral vein.**

The **choroid fissure** is the cleft-like space that is produced when the choroid plexus is pulled away and the continuity between the ependyma of the choroid plexus and the epithelium lining the ventricle (i.e., the line of invagination of the choroid plexus) is severed. Like the choroid plexus, the choroid fissure extends from the interventricular foramen to the tip of the temporal (inferior) horn (Fig. 11-166). The upper part of the fissure nearest the interventricular foramen is situated between the lateral edge of the fornix and the upper surface of the thalamus (Fig. 11-165). More posteriorly, near the commencement of the temporal (inferior) horn, the choroid fissure lies between the fimbria hippocampi and the caudal end of the thalamus. In the remainder of the temporal (inferior) horn, the fissure is located between the fimbria in the ventricular floor and the stria terminalis in the roof (Fig. 11-166).

The **tela choroidea of the third ventricle** is a fused double fold of pia mater, triangular in shape, that lies beneath the fornix (Fig. 11-162). The lateral portions of its lower surface rest upon the thalami, while its medial portion is in contact with the ependymal roof of the third ventricle (Fig. 11-165). The apex of this triangular fold is situated at the interventricular foramen, while its base is located posteriorly at the level of the splenium of the corpus callosum and occupies the interval between the corpus callosum above and the colliculi of the mesencephalic tectum and pineal body below. Its lateral margins are modified to form the highly vascular choroid plexuses of the lateral ventricles (Figs. 11-162; 11-165). After removal of the tela choroidea, an interval sometimes called the **transverse fissure of the brain,** is opened between the splenium above and the mesencephalic tectum below.

From the inferior surface of the tela choroidea, a pair of vascular fringed processes, the **choroid plexuses of the third ventricle,** project downward, one on either side of the midline (Fig. 11-165). These invaginate the ependymal roof of the third ventricle to achieve the cavity of that ventricle.

Blood Supply of the Prosencephalon

The prosencephalon receives its arterial blood supply from the internal carotid arteries as they enter the cranial cavity and from the posterior cerebral arteries, formed by the bifurcation of the basilar artery on the ventral aspect of the brain stem at the pontine-mesencephalic junction (Fig. 11-167). Thus, the neuraxis rostral to the midbrain is served by branches of vessels that participate in the formation of the **circulus arteriosus cerebri** or the cerebral arterial circle of Willis.[1] The vessels that supply the

[1]Thomas Willis (1621–1675): An English physician, neurologist, and anatomist (London).

diencephalon and the telencephalon are derived from the **posterior cerebral** branches of the *basilar artery,* the **posterior communicating, anterior choroidal, middle cerebral,** and **anterior cerebral** branches of the *internal carotid artery,* as well as a few penetrating vessels that branch directly from the internal carotid to supply the infundibular region of the hypothalamus.

Posterior Cerebral Artery

Although the posterior cerebral arteries in the adult brain commence as the terminal branches of the basilar artery, it should be stressed that as a supratentorial vessel, this artery develops as a branch of the internal carotid artery. In about 20% of patients studied by cerebral angiograms, the posterior cerebral artery originates as a branch of the internal carotid artery, thereby reflecting its

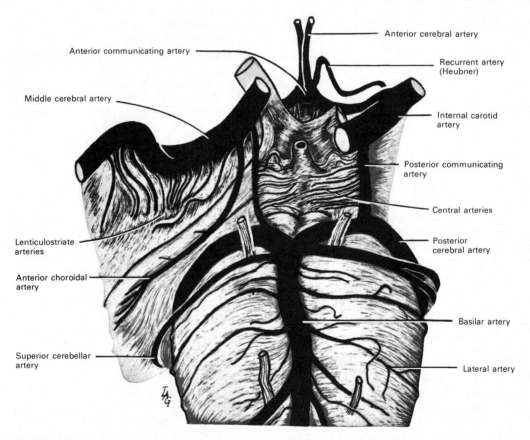

FIG. 11-167.   The arterial cerebral circle (of Willis), formed by the posterior, middle, and anterior cerebral arteries and the posterior and anterior communicating arteries. Note the centrally penetrating vessels from the posterior communicating artery, the anterior choroidal artery, the anterolateral central arteries (lenticulostriate), and the long central artery (recurrent artery of Heubner). (From L. A. Gillilan in *Correlative Anatomy of the Nervous System,* edited by E. C. Crosby, T. Humphrey and E. C. Lauer, Macmillan Publishing Co., New York, 1962.)

developmental origin (Fig. 11-168). It makes its appearance rather late in development, coincident with the period of posterior expansion of the forebrain. As prenatal brain growth continues, the posterior communicating arteries regress in size, and the stem and terminal branches of the posterior cerebral arteries gradually shift caudally and become dependent on the vertebrobasilar system.

In the adult the posterior cerebral artery supplies the upper pons and midbrain, the posterior hypothalamus, posterior thalamus, much of the inferomedial aspect of the temporal lobe, and most of the occipital lobe. Perhaps more than any other vessel, it supplies those regions of the brain that are responsible for visual motor behavior and visual perception. It has been pointed out (Hoyt, Newton and Margolis, 1974) that the posterior cerebral artery serves the comprehensive functions of seeing and looking by supplying: (1) the mesencephalic and pontine regions concerned with pupillary reflexes, convergence, vertical and horizontal eye movements, vestibulo-ocular reflexes, and various corneal, palpebral, and facial reflexes; (2) the lateral geniculate body and pulvinar, which transmit and modulate retinal impulses; (3) the splenium of the corpus callosum, which is the portion of the large interhemispheric commissure that transfers visual information across the cerebral hemispheres; (4) the inferior part of the temporal lobe that is concerned with the storage and retrieval of visual data (memory); (5) cortical area 17, the visual or striate cortex, which receives and processes visual input; and (6) parts of cortical areas 18 and 19, the peristriate cortex, which elaborate on the visual input and, in turn, transmit information to other cortical areas and additionally act on

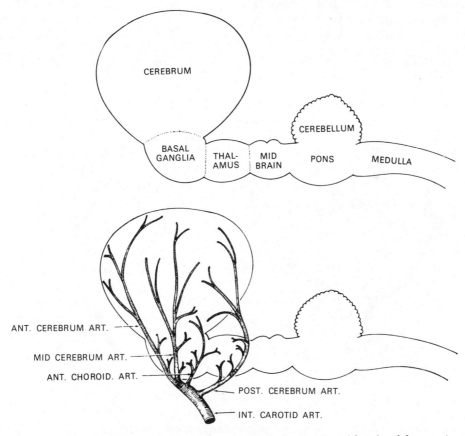

Fig. 11-168. Diagrammatic representations of the basic neural tube brain regions (above) and the superimposed four large branches of the internal carotid artery (below). Note that during early development the posterior cerebral artery depends on the internal carotid circulation, but later it shifts to the vertebrobasilar circulation (latter not shown.) Observe that the ventromedial and ventrolateral perforating vessels from the large branches of the internal carotid artery supply the basal cell mass, while dorsally directed vessels supply the cerebral cortex, (From H. A. Kaplan, in *Handbook of Clinical Neurology*, Vol. 11, edited by J. P. Vinkin and G. W. Bruyn, American Elsevier, New York, 1972.)

the mesencephalic oculomotor systems. The initial or circular segment of the posterior cerebral artery consists of a *precommunical part* and a *postcommunical part,* while the cortical segment is called the *terminal part.*

PRECOMMUNICAL PART. The precommunical part of the circular segment of the posterior cerebral artery is usually less than 1.0 cm long; it extends from its origin at the basilar bifurcation laterally to the point of junction of the posterior communicating artery (Fig. 11-167). From this part the **central posteromedial arteries** are derived. These include: (a) three to six *interpeduncular branches,* which penetrate the floor of the interpeduncular fossa through the posterior perforated substance to supply the upper midbrain (substantia nigra, red nucleus, oculomotor nucleus, and oculomotor and trochlear nerves) and the caudal hypothalamus; and (b) the *quadrigeminal branches,* which course around the brain stem to supply the region of the superior colliculus and extend rostrally to help supply the medial geniculate nucleus.

Posterior thalamoperforating and **medial posterior choroidal arteries** usually arise from the precommunical part of the posterior cerebral artery, but may arise more laterally from the postcommunical part. The posterior thalamoperforating arteries penetrate the posterior perforated substance to supply the posterior part of the thalamus. After branching from the posterior cerebral artery, the medial posterior choroidal artery parallels the course of the posterior cerebral artery around the brain stem and sends branches to the quadrigeminal plate and the pineal gland. It then courses rostrally along the roof of the third ventricle, where it helps to supply the choroid plexuses of the lateral and third ventricles and the tela choroidea.

POSTCOMMUNICAL PART. The postcommunical part of the circular segment of the posterior cerebral artery commences at the junction of the posterior communicating artery and courses posteriorly around the brain stem to reach the tentorial surface of the cerebrum, where it becomes the terminal part after dividing into cortical branches (Figs. 11-167; 11-169). As mentioned previously, at times thalamoperforating branches and the medial posterior choroidal artery arise from the postcommunical part.

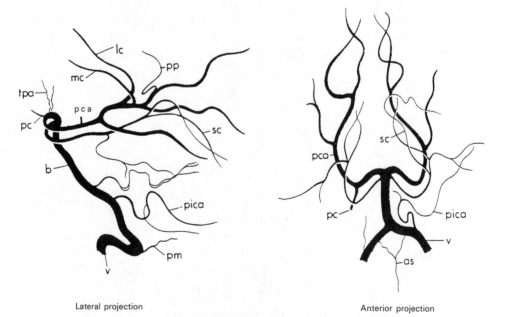

Lateral projection   Anterior projection

FIG. 11-169. Vertebral arteriogram showing both posterior and anterior projections of the posterior cerebral artery (pca), as well as other vessels derived from the vertebral artery that supply the brain stem, cerebellum, and spinal cord. As, Anterior spinal artery; b, basilar artery; lc, lateral posterior choroidal artery; mc, medial posterior choroidal artery; pc, posterior communicating artery; pca, posterior cerebral artery; pica, posterior inferior cerebellar artery; pm, posterior meningeal artery; pp, posterior pericallosal artery; sc, superior cerebellar artery; tpa, thalamoperforating arteries; v, vertebral artery. (From M. M. Schechter, in *Neurological Surgery,* Vol. 1, edited by J. R. Youmans, W. B. Saunders, Philadelphia, 1982.)

Peduncular and **lateral posterior choroidal arteries** nearly always arise from the postcommunical part of the posterior cerebral artery (Fig. 11-169). Just beyond the junction of the posterior communicating artery near the lateral aspect of the base of the cerebral peduncles, the peduncular arteries branch from the posterior cerebral artery. These course dorsally and somewhat laterally to help supply the parahippocampal gyrus and the inferomedial cortex of the temporal lobe. Other branches arise that supply the lateral geniculate body, the pulvinar, the dentate gyrus, and the hippocampus (Krayenbühl and Yaşargil, 1968). One or two lateral posterior choroidal arteries arise from the ventrolateral aspect of the posterior cerebral trunk, become directed around the brain stem, enter the choroid fissure, and with branches of the anterior choroidal artery, help supply the choroid plexus in the inferior horn of the lateral ventricle. The lateral posterior choroidal arteries also send prominent branches to the thalamus.

TERMINAL PART. The terminal part or cortical segment of the posterior cerebral artery commences on the dorsolateral aspect of the brain stem, dorsal to the edge of the tentorium and caudal to the lateral geniculate bodies. At this point, the posterior cerebral artery usually divides into several cortical branches. In some descriptions, these vessels are called the anterior temporal, posterior temporal, parieto-occipital and occipital (or calcarine) arteries (see Chapter 8). The terminology adopted by the *Nomina Anatomica* (1977) subdivides the terminal part of the posterior cerebral artery into lateral and medial occipital arteries. According to this classification, the smaller **lateral occipital artery** divides into **anterior, intermediate,** and **posterior temporal branches,** which supply the inferomedial and inferolateral aspects of the temporal and occipital lobes, including the uncus, parahippocampal gyrus, and the more rostral parts of the medial and lateral occipitotemporal gyri, as well as the inferior temporal gyrus (Fig. 11-170). The larger **medial occipital artery** divides into a **dorsal callosal branch** and **parietal, parieto-occipital, calcarine,** and **occipitotemporal branches** (Fig. 11-173). The dorsal callosal and parieto-occipital branches supply the isthmus of the cingulate gyrus, the splenium of the corpus callosum, and the cortex of the cuneus and lingual gyrus. The

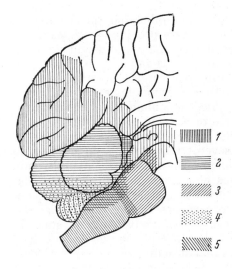

(1) Posterior cerebral artery

(2) Superior cerebellar artery

(3) Basilar artery and superior cerebellar artery

(4) Posterior inferior cerebellar artery

(5) Vertebral artery (posterior inferior cerebellar artery, anterior spinal artery)

FIG. 11-170. The medial surface of the posterior part of the cerebral hemisphere, indicating the regions supplied by different vessels from the vertebrobasilar arterial system. Note that the posterior cerebral artery, which includes its posterior choroidal branches (vertically hatched zone 1), supplies the posterior part of the diencephalon and telencephalon. (From H. Krayenbühl and M. G. Yaşargil, in *Handbook of Clinical Neurology*, Vol. 11, edited by J. P. Vinken and G. W. Bruyn, American Elsevier, New York, 1972.)

parietal, parieto-occipital, and calcarine branches supply the medial surface of the occipital lobes from the lingual gyrus through the cuneus to the parieto-occipital sulcus, including the occipital pole (Fig. 11-173).

Posterior Communicating Artery

The posterior communicating artery is approximately 1.5 cm long. It interconnects the posterior cerebral and internal carotid arteries and thereby participates in the formation of the circulus arteriosus cerebri (Fig. 11-167). It is generally smaller in diameter than the posterior cerebral artery, but it varies in size in different individuals and, rarely, it may be absent or doubled. More often than not, the two posterior communicating arteries in the same individual are about equal in caliber. The posterior communicating artery provides an anastomosis between the internal carotid and vertebro-basilar circulations and also an arterial link between the two cerebral hemispheres.

The posterior communicating artery usually supplies an **oculomotor nerve ramus** in its course dorsal and medial to that nerve. Near its internal carotid junction it also sends several **rami to the optic chiasma**, while more caudally small **hypothalamic branches** course to the neural tissue that forms the wall of the third ventricle, and **thalamic branches** supply the medial thalamic region. Frequently, another ramus, the **branch to the tail of the caudate nucleus**, supplies that structure as well as the posterior limb of the internal capsule.

Anterior Choroidal Artery

The anterior choroidal artery is a small (0.6 to 1.0 mm in diameter) but important branch of the internal carotid artery; it arises usually 2 to 4 mm distal to the posterior communicating-internal carotid junction (Fig. 11-167). Absence of this vessel is rare. From its origin lateral to the optic chiasma, the anterior choroidal artery courses posteriorly and assumes a position along the medial border of the optic tract, which it follows as far as the rostral margin of the lateral geniculate body. The artery then passes laterally toward the medial surface of the temporal lobe, and coursing around the uncus, enters the tip of the temporal (inferior) horn of the lateral ventricle through the choroidal fissure.

The number and size of the branches of the anterior choroidal artery vary considerably, and thus, its region of distribution also varies somewhat. Branches to the following structures and regions, however, are frequently observed and have been described (Goldberg, 1974): (1) perforating branches to the caudal two-thirds of the *optic tract*, which continue through the *anterior perforated substance* to terminate in the lateral and medial segments of the *globus pallidus* and the *genu of the internal capsule*; (2) laterally directed branches to the *piriform cortex* and *uncus* of the temporal lobe with perforating branches to the underlying *amygdaloid body*, the anterior part of the *hippocampal formation*, and the *tail of the caudate nucleus*; (3) medially directed branches to the *cerebral peduncle*, some of which course more deeply to the *substantia*

nigra and *red nucleus* in the rostral midbrain and subthalamus and to nuclei in the ventral part of the *thalamus*; (4) distal branches that arise at the level of the *lateral geniculate body*, helping to supply it as well as the *posterior limb* and the *retrolenticular regions of the internal capsule*; these branches anastomose with rami from the lateral posterior choroidal branch of the posterior cerebral artery; and (5) branches to the *choroid plexus in the temporal (inferior) horn of the lateral ventricle*, which also anastomose with vessels from the lateral posterior choroidal artery.

Middle Cerebral Artery

The middle cerebral artery commences with the division of the internal carotid artery into its anterior cerebral and middle

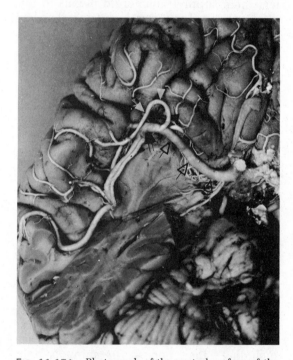

Fɪɢ. 11-171. Photograph of the ventral surface of the right cerebral hemisphere in which the tip of the temporal lobe has been resected. Note the sphenoid or horizontal part of the middle cerebral artery (open arrows) coursing laterally from its origin at the internal carotid artery to the site at which it enters the lateral sulcus (double solid black arrow). The lateral frontobasal (orbitofrontal) branch (solid black arrow), coursing to the inferior surface of the frontal lobe, and the anterior temporal branch (solid white arrows) looping laterally and backward to the temporal lobe arise from the middle cerebral just proximal to its entrance into the lateral sulcus. (From B. A. Ring, in *Radiology of the Skull and Brain: Angiography*, edited by T. H. Newton and D. G. Potts, Vol. 2, Book 2, C. V. Mosby Co., St. Louis, 1974.)

cerebral branches (Figs. 11-167; 11-171). Although the middle cerebral artery is considered the largest branch of the internal carotid artery, for all practical purposes it appears as the lateral continuation of the internal carotid along the base of the brain above the lesser wing of the sphenoid bone. This initial horizontal segment is called the **sphenoid part** of the middle cerebral artery and it averages 1.5 cm in length. Its course laterally crosses the inferior surface of the anterior perforated substance and this places it almost immediately between the temporal lobe and the apex of the insula (Fig. 11-171). As it enters the lateral cerebral sulcus at the limen insulae, the **insular part** of the middle cerebral artery turns dorsally and caudally and quickly divides into two (90% of cases) or three (10% of cases) branches. Continuing upward and backward within the depth of the lateral cerebral sulcus, the cortical or **terminal part** of the middle cerebral artery gives rise to many branches that emerge from the lateral cerebral sulcus and course anteriorly, superiorly,

and posteriorly to supply the lateral surface of most of the frontal lobe, nearly all of the parietal lobe, and most of the temporal lobe (Fig. 11-172).

SPHENOID PART. The sphenoid part of the middle cerebral artery gives rise only to **anterolateral central arteries** and generally does not supply any part of the cerebral cortex. Occasionally, however, it does give rise to the **medial frontobasal artery** (medial orbitofrontal artery), but usually this vessel arises from the anterior cerebral artery (Fig. 11-173). The anterolateral central arteries are also known as the *anterolateral thalamostriate arteries* or the *lenticulostriate arteries* and consist of 10 to 15 delicate vessels that are divided into medial and lateral groups (Fig. 11-167).

The **medial striate arteries** arise from the proximal part of the middle cerebral artery, perforate the medial part of the anterior perforated substance, and supply both segments of the globus pallidus, the caudate nucleus, the internal capsule, and perhaps, even the lateral aspect of the thalamus. The **lateral**

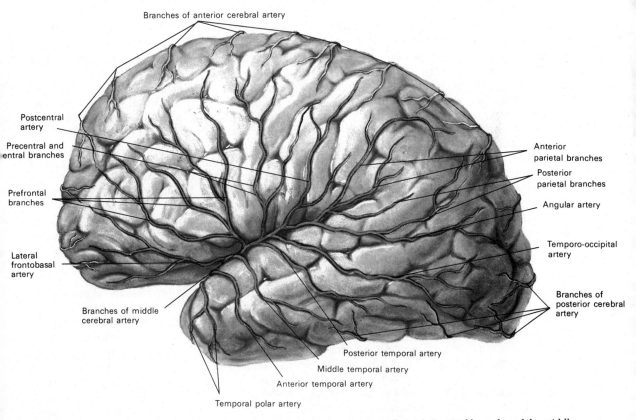

Branches of anterior cerebral artery

Postcentral artery

Precentral and central branches

Prefrontal branches

Lateral frontobasal artery

Branches of middle cerebral artery

Anterior parietal branches

Posterior parietal branches

Angular artery

Temporo-occipital artery

Branches of posterior cerebral artery

Posterior temporal artery

Middle temporal artery

Anterior temporal artery

Temporal polar artery

Fig. 11-172.   Arteries on the superolateral aspect of the cerebral cortex (lateral view). Cortical branches of the middle cerebral artery are shown in red; anterior cerebral artery, in yellow; posterior cerebral artery, in green.

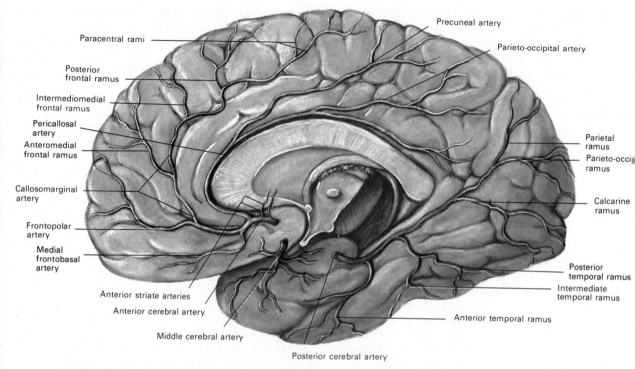

Precuneal artery

Parieto-occipital artery

Paracentral rami

Posterior
frontal ramus

Intermediomedial
frontal ramus

Pericallosal
artery

Anteromedial
frontal ramus

Parietal
ramus

Parieto-occi
ramus

Callosomarginal
artery

Calcarine
ramus

Frontopolar
artery

Medial
frontobasal
artery

Posterior
temporal ramus

Intermediate
temporal ramus

Anterior striate arteries

Anterior cerebral artery

Anterior temporal ramus

Middle cerebral artery

Posterior cerebral artery

Fig. 11-173. Arteries on the medial surface of the cerebral hemisphere. Branches of the middle cerebral artery are shown in red; anterior cerebral artery, in yellow; posterior cerebral artery, in green.

striate arteries arise from the middle cerebral artery just distal to the origins of the medial striate vessels. They penetrate the lateral part of the anterior perforated substance to supply the putamen, the internal capsule, and the caudate nucleus.

INSULAR PART. From the insular part of the middle cerebral artery arise insular, lateral frontobasal (lateral orbitofrontal), and anterior, intermediate, and posterior temporal arteries. Penetrating branches of the **insular arteries** (also called operculofrontal arteries) supply the limen insulae, the gyri brevi, and the gyrus longus, which form the insular lobe. These also penetrate more deeply to supply the extreme and external capsules and the claustrum. As the sphenoid part of the middle cerebral artery approaches the limen insulae, it already has bifurcated or trifurcated. These two or three branches give rise to a group of five to eight vessels that ascends in the lateral cerebral sulcus and then courses laterally over the surface of the insula as insular arteries. Because these vessels branch symmetrically in a fan-like manner and are visible as such in an angiogram, the term *candelabrum* is frequently used to identify them as a group.

After supplying the insula, these arteries follow the curvatures of the opercula to emerge from the lateral cerebral sulcus and continue their ascent over the lateral surface of the frontal and parietal lobes.

The **lateral frontobasal artery** (lateral orbitofrontal artery) arises from the middle cerebral artery just beyond the origins of the lateral striate arteries (Figs. 11-171; 11-172; 11-174). It almost immediately emerges from the lateral cerebral sulcus and courses anteriorly to supply the inferolateral and lateral orbital surfaces of the frontal lobe.

The anterior, intermediate, and posterior temporal arteries also emerge from the lateral cerebral sulcus and then course posteriorly and inferiorly to supply the lateral surface of the temporal lobe (Figs. 11-172; 11-174). Of these latter vessels, the **anterior temporal artery** arises from the middle cerebral artery most proximally and at about the same level as the lateral frontobasal artery. Near its origin, it emerges immediately from the lateral cerebral sulcus, loops over the posterior bank of that sulcus, courses inferiorly over the temporal lobe, and becomes distributed to the anterior third of the superior, middle, and inferior temporal gyri.

Most of the time a branch of the anterior temporal artery called the *temporal polar artery* passes forward to supply the tip of the temporal lobe (Figs. 11-172; 11-174).

The **intermediate temporal artery** arises from the middle cerebral artery at about the level of the pars opercularis of the inferior frontal gyrus. It is frequently quite small, and it quickly leaves the lateral cerebral sulcus and curves backward over the margin of the superior temporal gyrus to supply the superior and middle temporal gyri behind the cortical field served by the anterior temporal artery (Fig. 11-174).

The **posterior temporal artery** arises from the posterior trunk of the middle cerebral artery if a bifurcated or trifurcated middle cerebral pattern exists (Michotey, Moscow and Salamon, 1974). It emerges from the lateral cerebral sulcus at a point just below the postcentral gyrus and courses in an oblique posteroinferior direction over the temporal lobe. The posterior temporal artery supplies the middle and posterior parts of the superior temporal gyrus and the more posterior portions of the middle and inferior temporal gyri (Fig. 11-174).

TERMINAL PART. The terminal part of the middle cerebral artery gives rise to a group of ascending vessels that emerge from the lateral cerebral sulcus and course over its anterosuperior bank to supply the superior and lateral surfaces of the frontal and parietal lobes. These vessels include the prefrontal, precentral, central, anterior and posterior parietal, and angular arteries (Fig. 11-172).

The **prefrontal artery** leaves the lateral cerebral sulcus adjacent to the pars orbitalis of the inferior frontal gyrus and courses anterosuperiorly over the rostral part of the inferior, middle, and superior frontal gyri (Figs. 11-172). It supplies the anterolateral aspect of the frontal lobe in front of the precentral sulcus (Fig. 11-175), partly overlapping the fields of the lateral frontobasal artery below and that of the precentral artery behind. The **precentral artery** arises from the anterior trunk if the middle cerebral artery is bifurcated. It crosses the insula and then curves backward to emerge from the lateral cerebral sulcus at the level of the pars orbitalis of the inferior frontal gyrus, in close proximity to the prefrontal artery (Fig. 11-172). Its main trunk usually ascends on the lateral surface of the cerebral cortex within

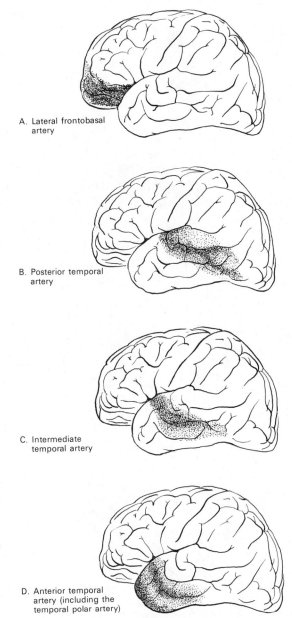

A. Lateral frontobasal artery

B. Posterior temporal artery

C. Intermediate temporal artery

D. Anterior temporal artery (including the temporal polar artery)

FIG. 11-174. Cortical fields of distribution of the lateral frontobasal (*A*), posterior temporal (*B*), intermediate temporal (*C*), an anterior temporal with temporal polar (*D*) branches of the middle cerebral artery. (Modified from Michotey, Moscow, and Salamon, 1974.)

the precentral sulcus and branches course forward to supply the posterior parts of the superior, middle, and inferior frontal gyri and backward to supply the precentral gyrus (Fig. 11-175). It partially overlaps the field of the prefrontal artery anteriorly and that of the central artery posteriorly.

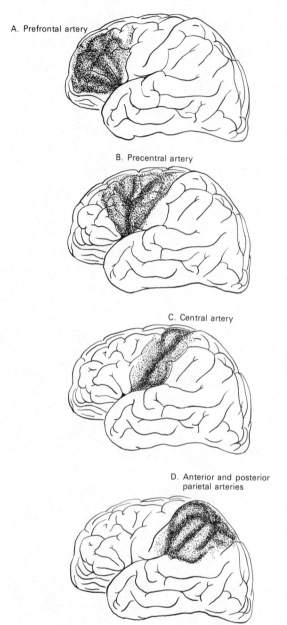

A. Prefrontal artery

B. Precentral artery

C. Central artery

D. Anterior and posterior parietal arteries

FIG. 11-175. Cortical fields of distribution of the prefrontal (A), precentral (B), central (C), and anterior and posterior parietal (D) branches of the middle cerebral artery. (Modified from Michotey, Moscow, and Salamon, 1974.)

The **central artery** arises from the anterior trunk of a bifurcated middle cerebral artery or the middle trunk of a trifurcated vessel. If the middle cerebral artery remains a single vessel, the central artery may arise from a common branch with the anterior parietal artery (Michotey, Moscow and Salamon, 1974). The central artery emerges from the

lateral cerebral sulcus just below the central sulcus (Fig. 11-172). It ascends on the lateral surface of the cerebral cortex within the central sulcus, and it sends branches to both the anterior and posterior banks of the central sulcus to supply the upper part of the precentral gyrus and the lower two-thirds of the postcentral gyrus (Fig. 11-175). Posteriorly it overlaps part of the field supplied by the postcentral and anterior parietal arteries, and anteriorly it supplies the region of the precentral gyrus not reached by the precentral artery. The **postcentral artery** courses along the postcentral sulcus.

The **anterior and posterior parietal arteries** arise near the summit of the lateral cerebral sulcus and ascend over the inferior and superior parietal lobules on the lateral surface of the parietal lobe, branching as far backward as the parieto-occipital sulcus (Fig. 11-172). The anterior parietal artery partially supplies the postcentral gyrus and the anterior parts of the inferior and superior parietal lobules. The posterior parietal artery supplies the supramarginal gyrus and the posterior parts of the two parietal lobules. Its field does not generally extend into the occipital lobe (Fig. 11-175).

The angular and temporo-occipital arteries emerge from the lateral cerebral sulcus opposite the supramarginal gyrus (Fig. 11-172). The **angular artery** is often the largest of the branches of the middle cerebral artery, and it courses posteriorly to supply the supramarginal and angular gyri and the lateral surface of the occipital lobe below the parieto-occipital sulcus and above the occipital pole. It also helps to supply the posterior part of the superior temporal gyrus (Fig. 11-176). The **temporo-occipital artery** courses posteroinferiorly from the lateral

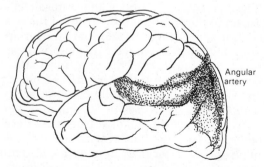

Angular artery

FIG. 11-176. The cortical field of distribution of the angular branch of the middle cerebral artery. (Modified from Michotey, Moscow, and Salamon, 1974.)

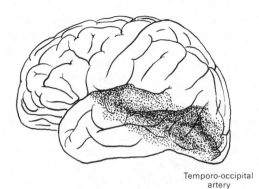

Temporo-occipital
artery

FIG. 11-177. The cortical field of distribution of the temporo-occipital branch of the middle cerebral artery. (Modified from Michotey, Moscow, and Salamon, 1974.)

cerebral sulcus, and it may arise as an individual vessel from the middle cerebral artery or from a common stem with the angular artery (Fig. 11-172). It supplies the posterior portions of the temporal gyri and the inferolateral surface of the occipital lobe, but it does not extend as far as the occipital pole (Fig. 11-177). The angular and temporo-occipital arteries supply the lateral cerebral surfaces of areas 18 and 19, but not the striate cortex of area 17. The latter area is served by the posterior cerebral artery.

Anterior Cerebral Artery

When the internal carotid arteries enter the cranial cavity on each side at the base of the brain, they are located on the medial aspect of the lateral cerebral sulcus. Each internal carotid artery terminates by bifurcating into the anterior and middle cerebral arteries, of which the smaller is the anterior cerebral artery (Fig. 11-167). From its origin, the anterior cerebral artery courses medially and slightly anteriorly across the base of the brain, dorsal to the optic nerve and just in front of the optic chiasma (Fig. 11-178). Upon reaching the longitudinal cerebral fissure that separates the cerebral hemispheres, the two anterior cerebral arteries are joined across the midline by the **anterior communicating artery** (Fig. 11-178). Beyond the anterior communicating artery, the two anterior cerebral arteries ascend side by side in the longitudinal cerebral fissure, just in front of the lamina terminalis. Each vessel then passes along the medial surface of its respective hemisphere by curving upward and backward around the genu of the corpus cal-

losum (Figs. 11-173; 11-179). The principal trunk of the anterior cerebral artery continues backward along the upper surface of the corpus callosum, extending nearly to the splenium. During its course, several vessels branch from the principal stem to ramify along the medial surface of the cerebrum and thereby help to supply the frontal and parietal lobes (Fig. 11-173). The portion of the anterior cerebral artery that extends from the internal carotid bifurcation to the anterior communicating artery (about 1.5 cm in length) is called the **precommunical part** (proximal segment or horizontal part), while beyond the anterior communicating artery, the main trunk of the anterior cerebral artery is referred to as the **postcommunical part** (or pericallosal part).

PRECOMMUNICAL PART. The precommunical part of the anterior cerebral artery supplies a number of central arteries that penetrate the base of the brain along the horizontal course of the vessel from the in-

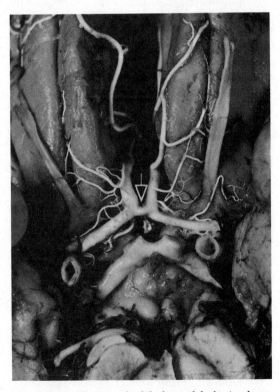

FIG. 11-178. Photograph of the base of the brain, showing the precommunical part of the anterior cerebral artery and the anterior communicating artery (arrow). (From J. P. Lin and I. I. Kricheff, in *Radiology of the Skull and Brain: Angiography*, Vol. 2, Book 2, edited by T. H. Newton and D. G. Potts, C. V. Mosby, St. Louis, 1974.)

ternal carotid bifurcation to the anterior communicating artery. These include the anteromedial central arteries (anteromedial thalamostriate arteries) that come directly from the anterior cerebral and the central anteromedial rami of the anterior communicating artery.

The **anteromedial central arteries** are usually divided into short central arteries, of which there are both medial and lateral perforators, and the long central artery (also called the recurrent artery of Heubner[1]). The **short medial and lateral central arteries** arise from both the dorsal and ventral aspects of the precommunical part. Those arising ventrally supply the superior surface of the optic nerve and optic chiasma. From the dorsal aspect of the precommunical part of the anterior cerebral artery arise 8 to 12 more laterally located short perforating central arteries and anywhere from 4 to 10 more medially located perforators. All of these vessels penetrate the ventral aspect of the basal forebrain, with some entering the anterior perforated substance and others, the olfactory trigone and olfactory striae to achieve deeper forebrain structures. They supply the preoptic region and the anterior hypothalamus caudally to the infundibulum, the genu of the corpus callosum, the septal region and anterior commissure, the fornix and the anteroinferior aspect of the striatum. The **long central artery** (recurrent artery of Heubner) originates medially near the anterior communicating artery and it then courses laterally for about 2 or 3 cm before ascending and entering the anterior perforated substance, just medial to the short central arteries (Fig. 11-167). It may send a branch anteriorly to supply part of the medial aspect of the orbital surface of the frontal lobe and olfactory bulb. Its principal region of supply is the anterior limb of the internal capsule, the anterior and medial aspects of the head of the caudate nucleus, and the anterior parts of the putamen and globus pallidus.

ANTERIOR COMMUNICATING ARTERY. The anterior communicating artery is a short vessel (0.1 to 3.0 mm long) that interconnects the two anterior cerebral arteries across the longitudinal cerebral fissure and forms the rostral part of the cerebral arterial circle of

Willis (Figs. 11-167; 11-178). This artery has a variable pattern, and it may be doubled, Y-shaped, reticulated, very short and thick, or quite slender and long. Several small perforating **central anteromedial rami** arise from the anterior communicating artery and supply the optic chiasma, the preoptic region, and the hypothalamus in the region of the infundibulum. In about 10% of specimens, a vessel called the *anterior middle cerebral artery* (median artery of the corpus callosum) arises from the anterior communicating artery and ascends over the corpus callosum in the midline to supply—in addition to the rostrum, genu, and anterior part of the trunk of the corpus callosum—the septal nuclei, the septum pellucidum, and the fornix. In certain cases when the anterior cerebral artery is absent or hypoplastic, the anterior middle cerebral artery has large branches that supply cortical areas normally served by the anterior cerebral artery.

POSTCOMMUNICAL PART. The postcommunical part of the anterior cerebral artery (also called the pericallosal segment or artery) is the portion of the anterior cerebral artery distal to the anterior communicating artery. Ascending around the genu of the corpus callosum, it continues posteriorly above the corpus callosum to supply the medial surface of the cerebral hemisphere (Figs. 11-173; 11-179). Its branches include the medial frontobasal artery, the frontopolar artery, the callosomarginal artery, and the paracentral, precuneal, and parietooccipital artery.

The **medial frontobasal artery** (also called the medial orbitofrontal artery) arises several millimeters distal to the anterior communicating artery as the first important branch of the postcommunical part of the anterior cerebral artery. It courses anteriorly along either the inferior part of the medial surface of the frontal lobe or the medial aspect of the orbital surface and it supplies the gyrus rectus and the region of the olfactory tract as far as the olfactory bulb (Figs. 11-173; 11-179). The **frontopolar artery** arises from the main stem of the anterior cerebral artery at about the level of the genu of the corpus callosum. At times a single branch arises from the anterior cerebral artery, which then divides into the medial frontobasal artery and the frontopolar artery. From its origin the frontopolar artery (sometimes doubled) courses anteriorly in a gradual curve along

[1]Johann Otto L. Heubner (1843–1926): A German pediatrician (Leipzig, Berlin, and Dresden).

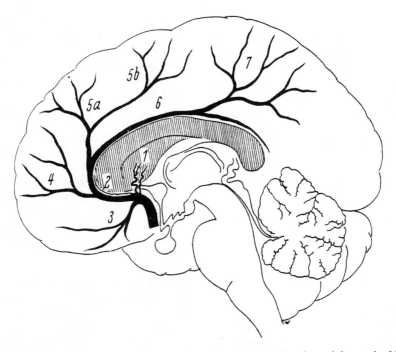

FIG. 11-179. The anterior cerebral artery and its branches along the medial surface of the cerebral hemisphere. The numbered branches are (1) short medial and lateral central arteries and the long central artery (recurrent of Heubner); (2) anterior middle central artery (actually from the anterior communicating artery); (3) medial frontobasal artery; (4) frontopolar artery: (5a and 5b) callosomarginal artery and its anteromedial, intermediomedial, and posteromedial frontal rami and the cingular ramus; (6) pericallosal artery (continuation of the anterior cerebral trunk; also called artery of the corpus callosum); (7) paracentral, precuneal, and parieto-occipital terminal branches. (From H. Krayenbühl and M. G. Yaşargil, in *Handbook of Clinical Neurology,* Vol. 11, edited by J. P. Vinken and G. W. Bruyn, American Elsevier, New York, 1972.)

the medial surface of the frontal lobe toward the frontal pole (Figs. 11-173; 11-179). It supplies the anterior part of the medial frontal gyrus and it extends over the superior border of the cerebrum to supply the anterior part of the superior frontal gyrus on its superolateral surface.

Several centimeters beyond the origin of the frontopolar artery, the main stem of the anterior cerebral artery gives off a large branch called the **callosomarginal artery** (Figs. 11-173; 11-179). Arising near the summit of the genu of the corpus callosum, the callosomarginal artery courses along the medial surface of the cerebrum over the cingulate gyrus. Its path along the cingulate sulcus is nearly parallel to the main trunk of the anterior cerebral artery, which lies in the sulcus of the corpus callosum. From the callosomarginal artery arise **anteromedial, intermediomedial,** and **posteromedial frontal rami,** as well as a **cingular ramus;** these supply the medial surface of the posterior part of the frontal lobe and the rostral part of

the parietal lobe, as well as part of the cingulate gyrus (Figs. 11-173; 11-179).

Continuing caudally, the main stem of the anterior cerebral artery branches into terminal vessels called the **paracentral, precuneal,** and **parieto-occipital arteries** (Figs. 11-173; 11-179). These arteries supply the more caudal part of the medial surface of the parietal lobe, including the paracentral lobule and the precuneus, and their vascular fields extend caudally to the parieto-occipital sulcus.

Meninges of the Brain and Spinal Cord

The brain and spinal cord are enclosed by three membranes: (1) an outer tough, protective membrane, the **dura mater,** (2) a subjacent delicate, spider web-like structure called the **arachnoid mater,** and (3) an inner, finely structured fibrous membrane that carries blood vessels to the brain and spinal cord called the **pia mater.** The dura mater is also frequently called the *pachymenix,* while the arachnoid and pia mater are collectively called the *leptomeninges.*

DURA MATER

The dura mater is a thick, dense, and tough, inelastic membrane composed of fibrous connective tissue with collagenous bundles arranged in interlacing layers. The portion of the dura mater that encloses the brain, the **dura mater encephali,** differs in several essential particulars from that which surrounds the spinal cord, the **dura mater spinalis,** and therefore it is necessary to describe them separately. It must be emphasized, however, that these two regional parts of the dura mater form one complete membrane and are continuous with each other through the foramen magnum.

Cranial Dura Mater (Dura Mater Encephali)

The cranial dura mater serves as both an internal periosteum for the bones that encase the cranial cavity and a membrane for the protection of the brain. It may be considered to consist of two layers: an outer or **endosteal layer** and an inner or **meningeal layer.** These two layers are tightly fused except in certain places where they are separated to provide space for the venous sinuses, within which venous blood drains from the cranial cavity. The outer surface of the dura mater is rough and fibrillated, and it adheres closely to the inner surface of the cranial bones. It sends many fine fibrous and vascular projections into the bony substance; these are severed when the dura is stripped from the bones. Because the adhesions are most marked along the sutures and at the base of the skull, the dural attachments are most secure at these sites. The internal surface of the dura is smooth, unattached, and covered with a layer of squamous mesothelial cells that line the subdural space.

The outer or endosteal layer is continuous through the suture lines and the foramina at the base of the skull with the pericranium that lines the external surface of the cranial bones. The inner meningeal layer (a) separates from the endosteal layer to form the venous sinuses (described in Chapter 9); (b) forms a sheath that surrounds the cranial nerves through their sites of exit at the foramina in the base of the skull; (c) sends inwardly three partitions, called the falx cerebri, the tentorium cerebelli, and the falx cerebelli, which separate large divisions of the brain; (d) forms a roof for the sella turcica, called the diaphragma sellae, which overlies the hypophysis; and (e) forms the cavum trigeminale, which lodges the trigeminal ganglion.

FALX CEREBRI. The falx cerebri, named because of its sickle-like shape, is a strong, arched membrane that extends vertically downward in the longitudinal fissure between the two cerebral hemispheres (Fig. 11-180). It is narrow anteriorly, where it is attached to the crista galli of the ethmoid bone, and broad posteriorly, where it is continuous with the tentorium cerebelli. Its upper margin is convex and attached superiorly to the inner surface of the cranial vault in the midline as far back as the internal occipital protuberance. At this attachment it is separated from the outer layer of dura, leaving space for the superior sagittal sinus (Fig. 11-180). The free inferior margin is concave, and overlying the corpus callosum, it contains the inferior sagittal sinus. Posteriorly, the inferior border of the falx cerebri attaches to the tentorium cerebelli. Along this attachment is the straight sinus (Figs. 11-180; 11-181).

TENTORIUM CEREBELLI. The tentorium cerebelli is a transversely oriented semilunar fold of dura mater that covers the superior surface of the cerebellum and underlies the occipital lobes of the cerebrum (Figs. 11-180; 11-181). Because the upper surface of the cerebellum is convex, the center of the tentorium is much higher than its sides. Its anterior border is free and sharply concave and it bounds a large oval opening called the **tentorial incisure,** which surrounds the midbrain and through which the posterior cerebral arteries pass in their supratentorial course to the cerebral hemispheres.

The outer, broad, and convex border of the tentorium cerebelli attaches posterolaterally to the edges of the sulcus for the transverse sinus on the inner surface of the occipital bone; thus, it encloses the transverse sinuses (Fig. 11-181). Anterolaterally, it is attached along the superior margin of the petrous part of the temporal bone and encloses the superior petrosal sinus on both sides. Anteromedially, near the apex of the petrous part of the temporal bone, the free and attached borders of the tentorium cerebelli meet and cross each other. The free border passes above and is fixed to the anterior

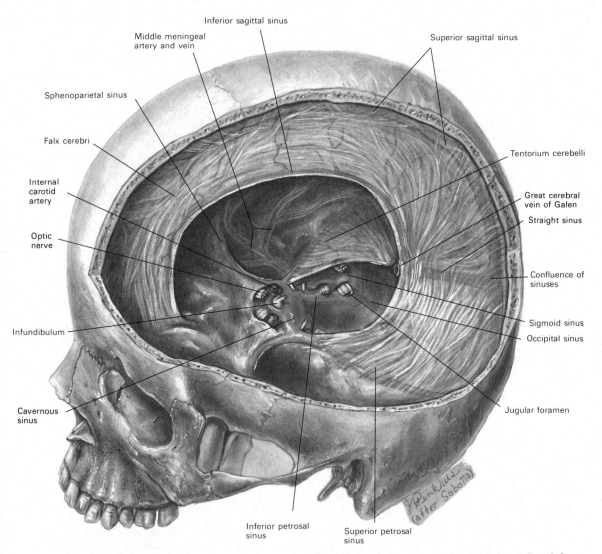

Inferior sagittal sinus

Middle meningeal artery and vein

Superior sagittal sinus

Sphenoparietal sinus

Falx cerebri

Tentorium cerebelli

Internal carotid artery

Great cerebral vein of Galen

Straight sinus

Optic nerve

Confluence of sinuses

Sigmoid sinus

Occipital sinus

Infundibulum

Cavernous sinus

Jugular foramen

Inferior petrosal sinus

Superior petrosal sinus

FIG. 11-180.   The cerebral dura mater and the dural sinuses, seen after removal of the left side of the skull and the brain. Note the cerebral falx and the tentorium cerebelli. (After Sobotta.)

clinoid process of the sphenoid bone, while the attached border passes below to become anchored to the posterior clinoid process of the same bone. The posterior portion of the inferior border of the falx cerebri attaches to the upper surface of the tentorium cerebelli along the midline, thereby enclosing the straight sinus.

FALX CEREBELLI.   The falx cerebelli is a small sickle-shaped, somewhat triangular fold of dura mater located between the hemispheres of the cerebellum. Its base is attached in the median plane above to the inferior and posterior parts of the tentorium. It has a free anterior border that projects forward into the posterior cerebellar notch be-

tween the cerebellar hemispheres. Its posterior border, which contains the occipital sinus, is attached along the vertically oriented internal occipital crest on the inner surface of the occipital bone, thereby incompletely dividing the posterior cranial fossa into two halves.

DIAPHRAGMA SELLAE.   The diaphragma sellae is a small, circular, horizontal fold of dura mater that attaches to the clinoid processes and forms a roof for the sella turcica, which contains the hypophysis (Fig. 11-181). A circular opening in the center transmits the infundibulum of the hypothalamus, which is surrounded by the intercavernous sinus (Fig. 11-180). Partially covering the

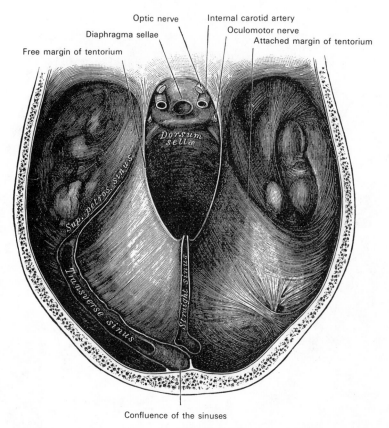

Optic nerve
Internal carotid artery
Diaphragma sellae
Oculomotor nerve
Attached margin of tentorium
Free margin of tentorium

Dorsum sellæ

Sup. petros. sinus

Transverse sinus

Straight sinus

Confluence of the sinuses

FIG. 11-181. The tentorium cerebelli, viewed from above with the brain removed. Observe that its anterior free margin bounds an oval opening called the tentorial incisure. The transverse sinus (which continues as the sigmoid sinus) lies in the posterolateral attached margin of the tentorium.

dorsal surface of the diaphragma sellae is the optic chiasma.

CAVUM TRIGEMINALE. \ The cavum trigeminale (trigeminal cave of Meckel[1]) is a cleft-like recess located between the endosteal and meningeal dural layers in the middle cranial fossa near the apex of the petrous portion of the temporal bone, just below the superior petrosal sinus. It lodges the posterior part of the trigeminal ganglion and the roots of the trigeminal nerve, and it is fused anteriorly with the dura that forms the lateral wall of the cavernous sinus.

STRUCTURE OF THE DURA MATER. The dura mater is formed chiefly of dense fibrous connective tissue, mixed with some elastic fibers, and as its name suggests, it has a tough consistency. Its connective tissue fibers are arranged in flattened laminae that are incompletely separated into two layers, endosteal

[1]Johann Friedrich Meckel (1714–1774): German anatomist, gynecologist, and botanist (Berlin).

and meningeal. The **endosteal layer** is the internal periosteum for the cranial bones, and it contains the blood vessels for their supply. At the margin of the foramen magnum, the endosteal layer of the cranial dura mater is continuous with the periosteum lining the vertebral canal. The **meningeal layer** overlies the arachnoid, and it is lined on its inner surface with a flattened, nucleated layer of cells that appears to be mesothelial and similar to that found lining serous membranes.

ARTERIES AND VEINS OF THE DURA MATER. The **arteries** that course within the dura mater are numerous and derived from several sources; they principally supply the cranial bones. Those in the *anterior fossa* are the anterior meningeal branches of the anterior and posterior ethmoidal arteries and of the internal carotid artery, as well as a branch from the middle meningeal artery. The vessels that supply the dura mater in the *middle fossa* are the middle and accessory meningeal branches of the maxillary artery; the meningeal branch of the ascending pharyngeal artery, which enters the skull through the foramen lacerum; branches from the internal carotid artery; and the recurrent meningeal branch of the lacrimal artery. Arteries that supply the dura mater in the *posterior fossa* include two meningeal branches from the occipital artery, one that enters the skull through the jugular foramen and another that enters through the mastoid foramen; the posterior meningeal branches from the vertebral artery; other meningeal branches from the ascending pharyngeal artery that enter the skull through the jugular foramen and hypoglossal canal; and a branch from the middle meningeal artery.

The **veins** returning blood from the cranial dura mater anastomose with the diploic veins and end in the various sinuses. Many of the meningeal veins do not open directly into the sinuses, but indirectly through a series of ampullae termed **venous lacunae.** These are found on either side of the superior sagittal sinus, especially near its middle portion, and are often invaginated by the arachnoid granulations. Venous lacunae are also found near the transverse and straight sinuses. The veins in the dura mater communicate with the underlying cerebral veins, as well as with the diploic and emissary veins.

NERVES OF THE DURA MATER. The cranial dura mater is richly supplied by sensory nerve fibers and sympathetic fibers; most of the latter course to the dural vessels. The sensory fibers are derived from the trigeminal ganglion, from each of the three divisions of the trigeminal nerve, as well as from the upper two or three cervical segments. The latter appear both to course with the vagus and hypoglossal nerves and to ascend directly through the foramen magnum to supply the dura mater of the posterior fossa. Sympathetic fibers to the dural vessels course along the arteries from the vertebral and carotid sympathetic plexuses.

Spinal Dura Mater (Dura Mater Spinalis)

The spinal dura mater forms a loose sheath around the spinal cord and represents only the inner or meningeal layer of the cranial dura mater (Fig. 11-182). The outer or periosteal layer of the cranial dura mater becomes the periosteum of the vertebrae below the foramen magnum and forms the lining of the vertebral canal. A considerable interval, the **epidural cavity,** intervenes between the spinal dura and the periosteal lining of the vertebral canal. It contains a quantity of loose areolar tissue and a plexus of veins that corresponds in position with the cranial dural sinuses. The spinal dura is attached to the circumference of the foramen magnum, to the second and third cervical vertebrae, and by fibrous slips to the posterior longitudinal ligament, especially near the caudal end of the vertebral canal.

The substance of the spinal cord diminishes significantly caudal to the second lumbar vertebra and the cord ends at this level in a conical extremity called the **conus medullaris.** From the apex of the conus medullaris, a delicate fibrous filament called the **filum terminale** descends for about 20 cm as far as the first coccygeal vertebral segment (Figs. 11-26; 11-27). Between the second lumbar vertebra and the level of the second sacral vertebra, the conus medullaris and filum terminale are surrounded by the longitudinally oriented roots of the spinal nerves that course intradurally. This portion of the cord is called the **cauda equina** and it is surrounded by spinal dura mater. Below the second sacral vertebral level, the dura mater closely invests the final 5 cm of the filum terminale (filum terminale externum) and attaches caudally to the dorsum of the first coccygeal vertebra by blending with the periosteum (Fig. 11-27).

The intradural volumetric capacity is considerably greater than that required by the spinal cord and leptomeninges. Much of the additional space is occupied by cerebrospinal fluid located deep to the arachnoid membrane in the so-called subarachnoid space (Fig. 11-182). An especially spacious portion of the subarachnoid space exists deep to the dura mater and arachnoid membrane between the apex of the conus medullaris (third lumbar vertebra) and the second sacral vertebral level along the course of the upper 15 cm of the filum terminale (filum terminale internum). This is a convenient site for the sampling of cerebrospinal fluid (lumbar puncture) or the injection of anesthetics (spinal anesthesia).

Between the dura mater and the arachnoid membrane is found a potential space, called the **subdural space,** that contains a small amount of serous fluid (Fig. 11-182). The subdural space terminates at the lower border of the second sacral vertebra, where the filum terminale externum becomes invested closely by the dura mater. The dura mater forms tubular prolongations along both the roots of the spinal nerves and the proximal parts of the spinal nerves themselves as they pass through the intervertebral foramina on each side. These **dural sheaths** are short in the upper part of the vertebral column, but gradually they become longer in the lower part because of the greater interval that exists between the root-cord junction and the intervertebral foramina.

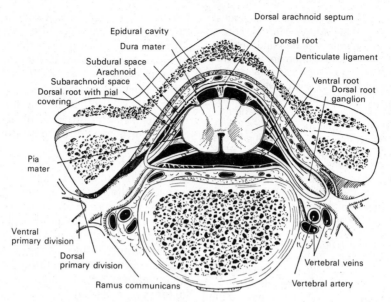

FIG. 11-182. Cross section of spinal cord in the spinal canal, showing its meningeal coverings and the manner of exit of the spinal nerves. (After Rauber in N. B. Everett, *Functional Neuroanatomy,* 6th ed., Lea & Febiger, Philadelphia, 1971.)

THE ARACHNOID

The arachnoid (Figs. 11-182; 11-183; 11-185; 11-186) is a delicate, avascular membrane lying between the dura mater and pia mater. It is separated from the dura mater by the **subdural space** and from the pia mater by the **subarachnoid space,** which contains the cerebrospinal fluid. The arachnoid, together with the underlying pia mater, forms the **leptomeninges.** Although the pia and arachnoid are clearly different and separable membranes, fine trabecular strands extend inward to the pia mater from the arachnoid across the subarachnoid space (Fig. 11-185). This meshwork, along with the delicate texture of arachnoid tissue, gives a spider web-like appearance to the arachnoid mater, hence its name.

Cranial Arachnoid

The cranial arachnoid is closely applied to the inner surface of the cranial dura but is separated from it by a thin film of fluid in the subdural space. Its surface adjacent to the dura is covered with a layer of mesothelium. It does not dip into the sulci or fissures except to follow the falx and tentorium. The inner surface of the arachnoid membrane attaches to the underlying pia mater by delicate fibrous threads, the **arachnoid trabeculae,** which traverse the subarachnoid space intervening between these two membranes (Fig. 11-185).

Spinal Arachnoid

The spinal arachnoid is a thin, delicate tubular sheath that loosely invests the spinal cord (Fig. 11-186). It is continuous above with the cranial arachnoid and it widens below to invest the cauda equina and its emerging nerves. Like the cranial arachnoid, the spinal arachnoid is separated from the spinal dura mater by the subdural space. As the cranial arachnoid surrounds the roots of the cranial nerves along their intracranial course between the foramina and the brain, so also the spinal arachnoid encloses the spinal nerves and forms loose sheaths that surround these nerves along their course within the vertebral canal between their sites of attachment to the spinal cord and their exits through the intervertebral foramina.

SUBARACHNOID SPACE

The subarachnoid space is the cavity between the arachnoid and the pia mater (Figs. 11-182; 11-183). It contains cerebrospinal fluid located within intercommunicating channels that are separated by delicate connective tissue trabeculae extending from the arachnoid to the pia. The inner surface of the arachnoid is lined by mesothelial-like cells that form a continuous layer over the trabeculae and likewise cover the outer surface of the pia mater. Because the subarachnoid space can be defined as a cavity that separates two parts of a continuous membrane, the leptomeninx, it may be considered an intraleptomeningeal space.

The subarachnoid space is narrow over the surface of the cerebral hemispheres and especially so over the summits of the cortical gyri, where the arachnoid and pia are in close contact. The arachnoid, however, bridges the sulci, whereas the pia follows the cortical surface closely and dips into the sulci. This results in a wider subarachnoid space along these intergyral sites. Over certain regions of the brain, the arachnoid is somewhat thicker and is separated from the pia mater by wide intervals that communicate freely with each other. These are called the **subarachnoid cisternae** and they contain little subarachnoid trabecular tissue (Fig. 11-183).

Subarachnoid Cisternae

The subarachnoid cisternae are cavities of considerable size located within the subarachnoid space. They are found in regions where the arachnoid adheres closely to a loosely fitting meningeal dura mater that bridges the contours of the brain, while the pia mater remains intimately attached to the surface of the brain and follows its contours. Most of the cisternae are found adjacent to the ventral and dorsal surfaces of the brain stem and the ventral surface of the hypothalamus (Figs. 11-183; 11-184). Because they represent only enlargements within the common subarachnoid cavity, they contain cerebrospinal fluid. They have a clinical relevance because blood and exudates may accumulate at these sites, and certain cisternae may be punctured to obtain samples of their contents for analysis.

CEREBELLOMEDULLARY CISTERN. The cerebellomedullary cistern, often called the *cisterna magna,* is the largest and most accessible of the subarachnoid cisternae in the skull (Figs. 11-183; 11-184). It is found deep to the arachnoid that bridges the interval between the caudal part of the cerebellum and the dorsal surface of the medulla oblongata; the cerebellomedullary cistern is directly continuous below with the subarachnoid space of the spinal cord.

The ventricular system of the brain communicates with the subarachnoid space by way of three foramina. These open between the fourth ventricle and the cerebellomedullary cistern and include the **median aperture of the fourth ventricle** (foramen of Magendie), located in the midline at the inferior part of the roof of the fourth ventricle, and two **lateral apertures of the fourth ventricle** (foramina of Luschka), found at the extremities of the lateral recesses of the fourth ventricle between the flocculus of the cerebellum and the roots of the glossopharyngeal nerve (Fig. 11-183). Cerebellomedullary cisternal puncture is done by carefully introducing a needle through the posterior atlanto-occipital membrane and entering the cistern at the foramen magnum.

PONTINE CISTERN. The pontine cistern is a large space on the ventral aspect of the pons; within it courses the basilar artery longitudinally (Figs. 11-183; 11-184). It is continuous caudally with the subarachnoid space adjacent to the ventral surface of the medulla oblongata and spinal cord, and rostrally with the interpeduncular space.

INTERPEDUNCULAR AND CHIASMATIC CISTERNAE. The interpeduncular and chiasmatic cisternae are sometimes together called the *cisterna basalis* (Fig. 11-183). They are located rostral to the pontine cistern and adjacent to the ventral surfaces of the mesencephalon and hypothalamus. The optic chiasma separates the smaller, more rostrally located chiasmatic cistern from the wider and more extensive interpeduncular cistern. The interpeduncular cistern contains the circulus arteriosus and the posterior and middle cerebral arteries, while the chiasmatic cistern has coursing through it the anterior cerebral artery.

CISTERNAE OF THE LATERAL CEREBRAL FOSSAE. The cisternae of the lateral cerebral fossae, one on each side, are located at the base of the cerebrum, just medial to and in front of the temporal lobe. They are formed as the arachnoid membrane bridges the lateral sulci, and each contains the middle cerebral artery as that vessel courses toward and enters the lateral sulcus.

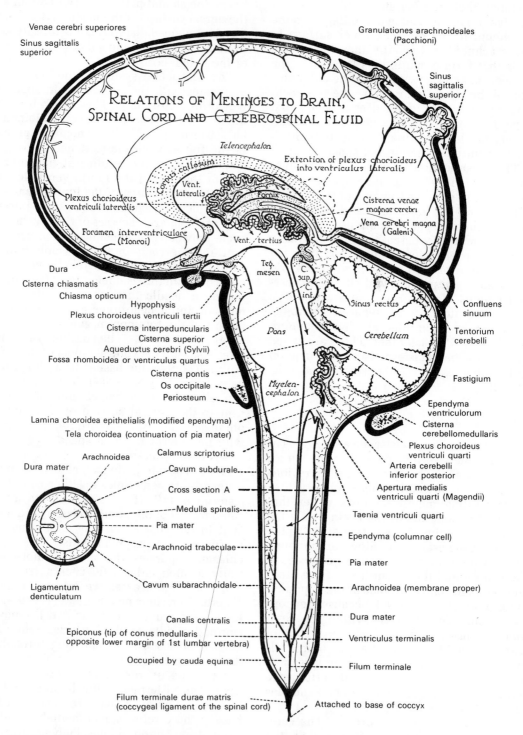

FIG. 11-183. A median sagittal section of the brain and spinal cord that shows the subarachnoid space and the cisternae and ventricles and the relationships of the meninges to the neural tissue. The arrows indicate the direction of cerebrospinal fluid flow. (From A. T. Rasmussen, *The Principal Nervous Pathways*, 4th ed., Macmillan Publishing Co., New York, 1952.)

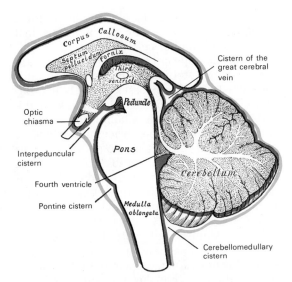

FIG. 11-184.   A diagram showing the locations of the principal subarachnoid cisterns in the skull. Blue, arachnoid membrane; red, pia mater.

CISTERN OF THE GREAT CEREBRAL VEIN.   The cistern of the great cerebral vein occupies the interval between the splenium of the corpus callosum, the pineal body, the superior surface of the cerebellum, and the colliculi of the mesencephalon (Figs. 11-183; 11-184). It reaches rostrally between the layers of the tela choroidea of the third ventricle, and through it courses the great cerebral vein (of Galen[1]).

As mentioned previously, the ventricular system of the brain communicates with the subarachnoid space by way of the median and lateral apertures of the fourth ventricle. The subarachnoid space surrounding the cerebral hemispheres and the intracranial cisternae are directly continuous with the subarachnoid space surrounding the spinal cord.

Spinal Part of the Subarachnoid Space

The spinal part of the subarachnoid space is a relatively wide interval that is most spacious in the caudal part of the vertebral canal, where the arachnoid surrounds the lower roots of the spinal cord that form the cauda equina (Fig. 11-183). A longitudinally oriented **subarachnoid septum** incompletely divides the spinal subarachnoid space. This

septum connects the arachnoid to the pia mater along the posterior median sulcus of the cord; it is less complete in the upper cervical region but more definitive in the lower cervical and thoracic regions. On each side, the **denticulate ligaments** also tend to partition the spinal part of the subarachnoid space (see discussion on pia mater). The subarachnoid space ends at the level of the second sacral vertebra.

Arachnoid Granulations

The arachnoid granulations (sometimes called the arachnoid villi or Pacchionian[2] bodies) are ovoid tufts of arachnoid tissue, visible to the naked eye, that protrude outward from the arachnoid membrane into the dura mater that forms the superior sagittal sinus, the transverse sinus, and at times certain other sinuses (Fig. 11-183). Each arachnoid granulation consists of a finger-like projection of arachnoid mesothelium that evaginates into the lumen of the sinus, so that the arachnoid mesothelium comes to lie adjacent to the vascular endothelium. The inner cavity of the granulation or villus is continuous with the subarachnoid space, and the cerebrospinal fluid within that space becomes separated from the venous blood only by the mesothelium of the arachnoid, which fuses with the endothelial lining of the sinus. Return of cerebrospinal fluid to the blood plasma is thought to be effected by osmosis across the mesothelial lining of the arachnoid villi and the vascular endothelium.

Although fully developed arachnoid granulations are not seen in the newborn infant and are rarely found before the third year, many minute elevations that will become arachnoid granulations can be seen quite early. They become obvious after the seventh year and they increase in number and size as age advances. During their growth, the arachnoid granulations push against the thinned dura and eventually cause absorption of bone; this results in the formation of ovoid pits or depressions on the inner table of the calvaria. The arachnoid granulations can readily be seen to occupy these depressions as the skull cap is removed during dissection of the adult cranial cavity.

---

[1]Claudius Galen (130–201 A.D.): A Greek physician born in Asia Minor who practiced medicine in Rome.

[2]Antonio Pacchioni (1665–1726): An Italian anatomist (Rome and Tivoli).

Cerebrospinal Fluid

The cerebrospinal fluid is a clear and colorless, watery liquid that fills the ventricles of the brain and occupies the subarachnoid space. It has a specific gravity of 1.006 to 1.009, and it normally contains 1.8 leukocytes/ml, some proteins, glucose, sodium, chloride, and lipids, as well as many other substances in small quantities. Although it may serve a nutritive role during development, cerebrospinal fluid probably functions in a protective capacity for the adult central nervous system. Because the brain and spinal cord are totally suspended in cerebrospinal fluid, they are buffered from the rigid bony skeleton of the skull and spinal column and are less subject to mechanical and concussive injury. Cerebrospinal fluid is constantly being formed by the choroid plexuses within the ventricles and it communicates with the subarachnoid space surrounding the brain and spinal cord. At any one time there may be as much as 140 ml of cerebrospinal fluid within and around the central nervous system, although it has been calculated that about 500 ml are formed daily, the excess being reabsorbed through the arachnoid villi into the venous system.

Cerebrospinal fluid circulates from its sites of formation in the lateral ventricles, through the interventricular foramina into the third ventricle, and then through the cerebral aqueduct into the fourth ventricle (Fig. 11-183). It passes into the cerebromedullary cistern (cisterna magna) and the pontine cistern by way of the medial and lateral apertures in the roof of the fourth ventricle and, thus, into the subarachnoid space. Within the subarachnoid space it continues to circulate around the surfaces of the cerebrum, cerebellum, and spinal cord and eventually it passes back into the venous circulation at the arachnoid granulations. Davson (1967) has estimated that of the 140 ml of cerebrospinal fluid present in the central nervous system at any one time, 23 ml are located within the ventricles, 30 ml are in the subarachnoid space of the spinal cord, and the rest is in the cerebral cisternae and the intracranial subarachnoid space.

As cerebrospinal fluid circulates in the subarachnoid space, it is interposed between the larger blood vessels and their penetrating branches on the one hand and the connective tissue of the pia mater on the other.

As the penetrating vessels divide to form capillaries, however, the pia mater is lost and the endothelial cells forming the capillary wall lie adjacent to the neuropil and are almost completely surrounded by the endfeet of astrocytic processes (Fig. 11-185). Because certain substances circulating in the blood stream, such as alcohol and glucose, quite freely enter the extracellular space or cerebrospinal fluid and come quickly to equilibrium, and other substances, such as proteins and certain vital dyes do not, a concept has developed that there exists a functional barrier between the blood and the brain that selectively allows the passage of some substances but restricts others. This physiologic mechanism has been called the **blood-brain** or **hematoencephalic barrier.** It is now thought that the tight junctions that interconnect the endothelial cells lining the capillaries in the central nervous system serve to limit the passage of certain substances across the capillary wall. This mechanism, in addition to others, may help to explain the blood-brain barrier phenomenon.

Obstruction of the flow of cerebrospinal fluid can cause an increased accumulation of intracranial cerebrospinal fluid, frequently characterized by an enlargement of the ventricles and an increased intracranial pressure. This condition is called **hydrocephalus,** and when it occurs before the closure of the fontanels of the skull of a child, the increase in intracranial pressure results in an enlargement of the head and, if severe enough, in the destruction of neural tissue. Because the skull is unable to expand in adults, obstruction of the passage of cerebrospinal fluid frequently results in ventricular dilatation. If flow is restricted through the cerebral aqueduct or through the medial and lateral apertures in the roof of the fourth ventricle, there results a noncommunicative obstruction because the passage of fluid between the ventricular system and the subarachnoid space is prevented. A communicating obstruction may occur somewhere in the subarachnoid space, such as in the cerebellomedullary or pontine cisternae.

Lumbar puncture is performed to sample spinal fluid that is circulating in the subarachnoid space surrounding the cauda equina of the spinal cord. A needle inserted in the midline between the spines of the third and fourth lumbar vertebrae will enter the sub-

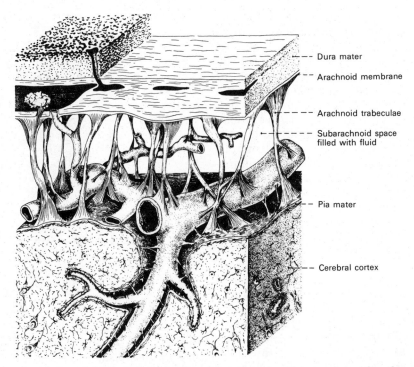

Dura mater
Arachnoid membrane
Arachnoid trabeculae
Subarachnoid space filled with fluid
Pia mater
Cerebral cortex

FIG. 11-185. Diagrammatic representation of a portion of the subarachnoid space surrounding the cerebral cortex. Note the arachnoid trabeculae between the arachnoid membrane and the pia mater, and the fact that the subarachnoid space still exists around the vessels as they penetrate the neural substance. As these vessels branch into the capillary network of the brain, however, they lose their arachnoid covering and the connective tissue of the pia mater, while the neuroglia attach directly to the basement membrane of the endothelial cells. (From Benninghoff and Goerttler, *Lehrbuch der Anatomie des Menschen,* Vol. 3, 9th ed., Urban and Schwarzenberg, Munich, 1975.)

arachnoid space, avoiding the spinal cord and insinuating itself among the nerves of the cauda equina (Fig. 11-27). **Cisternal puncture** is performed by inserting a needle between the atlas and the occipital bone and entering the cisterna cerebellomedullaris (cisterna magna). **Spinal anesthesia** is administered by introducing the anesthetic into the subarachnoid space by lumbar puncture.

PIA MATER

The innermost meningeal layer, the pia mater, is a delicate connective tissue and vascular membrane that closely adheres to the brain and spinal cord (Figs. 11-29; 11-185; 11-186). It carries a rich network of blood vessels that supply the neural tissue and it consists of loose interlacing bundles of collagenous and elastic fibers covered by a layer of flattened mesothelial-like squamous cells, similar to those seen in the arachnoid. The basement membrane on the innermost surface of the endothelial cells forming the

capillaries allows attachment of the astrocytic neuroglial processes, and together they form a **pia-glial membrane.** The astrocytes and their end-feet attachments make up a nearly complete sheath around the capillaries and may play an important role in the transfer of metabolic substances between the blood and the brain. On its external surface, the pia mater is attached to the overlying arachnoid by fine trabecular extensions of the arachnoid that cross the subarachnoid space (Fig. 11-185). The pia mater also forms sheaths for the cranial and spinal nerves and, together with the dura mater, blends with the perineurium as these nerves leave the cranial cavity and the vertebral canal. The pia mater and the arachnoid are often referred to collectively as the **leptomeninges.**

Cranial Pia Mater

The cranial pia mater invests the entire surface of the brain, dipping into the fissures and sulci of the cerebral and cerebellar hemispheres (Fig. 11-185). It extends into the transverse cerebral fissure, where it forms the *tela choroidea of the third ventricle* and combines with the ependyma to form the *choroid plexuses of the third and lateral ventricles.* It also passes over the roof of the fourth ventricle and forms the *tela choroidea and choroid plexus of the fourth ventricle.*

Over the surface of the cerebrum, the pia mater contains the blood vessels that supply the brain, and from these vessels large numbers of branches arise that penetrate the brain substance (Fig. 11-185). For a certain distance the connective tissue of the pia mater continues to surround these penetrating vessels, as it does the larger vessels on the brain surface. *Between the connective tissue of the pia mater and the penetrating vessels* is found a true **perivascular space** that is directly continuous with the subarachnoid space. As the penetrating vessels branch to form the capillary networks, however, the *connective tissue of the pia mater is lost.* The basement membrane of the endothelial cells of the capillaries lies immediately adjacent to the neural tissue and is almost surrounded by the perivascular endfeet of astrocytes. Helping to complete this pia-glial perivascular sheath are a few processes from oligodendrocytes and perhaps even a few neuronal processes.

Spinal Pia Mater

The spinal pia mater (Figs. 11-29; 11-186) is thicker, firmer, and less vascular than the cranial pia mater. Like the cranial pia, it consists of an outer layer formed by collagenous and elastic connective tissue and an inner layer formed by the basement membrane of the endothelial cells of the capillaries onto which attach the perivascular neuroglial processes (the pia-glial membrane). The spinal pia mater is intimately adherent to the surface of the spinal cord, and on the anterior spinal surface it dips into the ventral median fissure and thereby lines the two medial surfaces of that fissure. Longitudinally along the surface of the cord at the ventral median fissure, the pia mater is somewhat thickened and forms a band of connective tissue, called the **linea splendens,** that courses along the length of the fissure. Other concentrations of fibers along each side form the **denticulate ligaments** (see following).

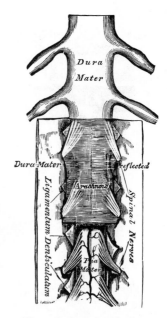

Fig. 11-186. The spinal cord and its meninges. Upper portion shows the dura mater intact; intermediate portion shows the dura reflected and the arachnoid intact; lower portion shows the dura reflected and the arachnoid removed to reveal the pia mater.

From the caudal tip of the conus medullaris, the pia mater continues as a long, slender filament called the **filum terminale.** This descends within the vertebral canal and is centrally located among the spinal roots that form the cauda equina (Fig. 11-27). The filum terminale blends with the dura mater at the lower border of the second sacral vertebra, and then it extends downward as far as the base of the coccyx, where it fuses with the periosteum. Because it is thought to assist in maintaining the spinal cord in place during movements of the trunk, the filum terminale has sometimes been called the central ligament of the spinal cord.

DENTICULATE LIGAMENTS. The denticulate ligaments are two narrow, longitudinally oriented fibrous bands of pia mater, one situated on each of the two sides of the spinal cord. Each ligament spreads laterally toward the dura mater from the entire length of the spinal cord, midway between the dorsal-ventral sites of attachment of the posterior and anterior roots (Figs. 11-29; 11-186). Thus, the medial border of each ligament is continuous with the pia mater at the side of the spinal cord, while the scalloped lateral border presents a series of triangular, tooth-like processes, the points of which are fixed at intervals to the dura mater. There are 21 of these pointed processes and each extends laterally through the arachnoid to become attached to the inner surface of the dura mater. The uppermost is attached to the dura opposite the margin of the foramen magnum, between the vertebral artery and the hypoglossal nerve. Each successive process extends laterally between the spinal nerves intersegmentally, so that the most caudal is found between the sites of exit of the last thoracic and first lumbar nerves. These ligaments, along with the nerve roots, assist in suspending the spinal cord within the subarachnoid cavity.

References

(References are listed not only to those articles and books cited in the text, but to others as well which are considered to contain valuable resource information for the student who desires it.)

General

Benninghoff, A., and K. Goerttler. 1975. *Lehrbuch der Anatomie des Menschen.* Vol. 3. 9th ed. Edited by H. Ferner. Urban and Schwarzenberg, Munich, 469 pp.

Brodal, A. 1981. *Neurological Anatomy in Relation to Clinical Medicine.* 3rd ed. Oxford University Press, New York, 1053 pp.

Carpenter, M. B. 1976. *Human Neuroanatomy.* 7th ed. Williams and Wilkins, Baltimore, 741 pp.

Crosby, E. C., T. Humphrey, and E. W. Lauer (eds.). 1962. *Correlative Anatomy of the Nervous System.* Macmillan Publishing Co., New York, 731 pp.

Haymaker, W., and B. Woodhall. 1953. *Peripheral Nerve Injuries.* 2nd ed. W. B. Saunders, Philadelphia, 333 pp.

House, E. L., B. Pansky, and A. Siegel. 1979. *A Systemic Approach to Neuroscience.* 3rd ed. McGraw-Hill Book Co., New York, 576 pp.

Larsell, O. 1951. *Anatomy of the Nervous System.* 2nd ed. Appleton, Century, Crofts, Inc., New York, 520 pp.

Mettler, F. A. 1948. *Neuroanatomy.* 2nd ed. C. V. Mosby, St. Louis, 536 pp.

Mountcastle, V. B. 1980. *Medical Physiology.* 2 Vols. C. V. Mosby, St. Louis, 1999 pp.

Noback, C. R. 1981. *The Human Nervous System.* McGraw-Hill Book Co., New York, 591 pp.

Pansky, B., and D. J. Allen. 1980. *Review of Neuroscience.* Macmillan Publishing Co., New York, 622 pp.

Ranson, S. W., and S. L. Clark. 1959. *The Anatomy of the Nervous System.* 10th ed. W. B. Saunders, Philadelphia, 622 pp.

Thorek, P. 1962. *Anatomy in Surgery.* 2nd ed. J. B. Lippincott, Philadelphia, 904 pp.

Development of the Nervous System

Adinolfi, A. M. 1977. The postnatal development of the caudate nucleus: A Golgi and electron microscopic study of kittens. Brain Res., *133*:251–266.

Anker, R. L. 1977. The prenatal development of some of the visual pathways in the cat. J. Comp. Neurol., *173*:185–204.

Arey, L. B. 1974. *Developmental Anatomy.* Rev. 7th ed. W. B. Saunders, Philadelphia, 695 pp.

Banker, G. A. 1980. Trophic interactions between astroglial cells and hippocampal neurons in culture. Science, *209*:809–810.

Blakemore, W. F. 1969. The ultrastructure of the subependymal plate in the rat. J. Anat. (Lond.), *104*:423–433.

Bodick, N., and C. Levinthal. 1980. Growing optic nerve fibers follow neighbors during embryogenesis. Proc. Nat. Acad. Sci. (USA) *77*:4374–4378.

Boulder Committee. 1970. Embryonic vertebrate central nervous system: Revised terminology. Anat. Rec., *166*:257–261.

Boydston, W. R., and G. S. Sohal. 1980. Intact peripheral target essential for branching of developing nerve fibers. Exp. Neurol., *70*:173–178.

Brand, S., and P. Rakic. 1979. Genesis of the primate neostriatum: [³H] thymidine autoradiographic analysis of the time of neuron origin in the rhesus monkey. Neuroscience, *4*:767–778.

Brookes, N. 1978. Actions of glutamate on dissociated mammalian spinal neurons *in vitro.* Devel. Neurosci., *1*:203–215.

Brooksbank, B. W. L., M. Martinez, D. J. Atkinson, and R. Balazs. 1978. Biochemical development of the human brain. I. Some parameters of the cholinergic system. Devel. Neurosci., *1*:267–284.

Brown, J. W. 1956. The development of the nucleus of the spinal tract of V in human fetuses of 14 to 21 weeks of menstrual age. J. Comp. Neurol., *106*:393–423.

Brown, J. W. 1974. Prenatal development of the human chief sensory trigeminal nucleus. J. Comp. Neurol., *156*:307–336.

Bueker, E. D. 1943. Intracentral and peripheral factors in the differentiation of motor neurons in transplanted lumbo-sacral spinal cords of chick embryos. J. Exp. Zool., *93*:99–129.

Bueker, E. D. 1948. Implantation of tumors in the hind limb field of the embryonic chick and the developmental response of the lumbosacral nervous system. Anat. Rec., *102*:369–389.

Burr, H. S. 1932. An electro-dynamic theory of development suggested by studies of proliferation rates in the brain of Amblystoma. J. Comp. Neurol., *56*:347–371.

Cohen, A. M., and E. D. Hay. 1971. Secretion of collagen by embryonic neuroepithelium at the time of spinal cord-somite interaction. Devel. Biol., *26*:578–605.

Garber, B. B., and A. A. Moscona. 1972. Reconstruction of brain tissue from cell suspensions. I. Aggregation patterns of cells dissociated from different regions of the developing brain. Devel. Biol., *27*:217–234.

Garber, B. B., and A. A. Moscona. 1972. Reconstruction of brain tissue from cell suspensions. II. Specific enhancement of aggregation of embryonic cerebral cells by supernatant from homologous cell cultures. Devel. Biol., 27:235–243.

Hamilton, W. J., W. Boyd, and H. W. Mossman. 1972. *Human Embryology.* 4th ed. Williams and Wilkins, Baltimore, 646 pp.

Harrison, R. G. 1924. Neuroblast versus sheath cell in the development of peripheral nerves. J. Comp. Neurol. 37:123–205.

His, W. 1889. Die Neuroblasten und deren Entstehung im embryonalen Mark. Arch. Anat. Physiol., Anat. Abt., 1889:249–300.

His, W. 1904. *Die Entwicklung des menschlichen Gehirns wahrend der ersten Monate.* S. Hirzel, Leipzig, 176 pp.

Hogg, I. D. 1944. The development of the nucleus dorsalis (Clarke's column). J. Comp. Neurol., 81:69–95.

Hughes, A. 1976. The development of the dorsal funiculus in the human spinal cord. J. Anat. (Lond.) 122:169–175.

Humphrey, T. 1950. Intramedullary sensory ganglion cells in the roof plate area of the embryonic human spinal cord. J. Comp. Neurol., 92:333–399.

Humphrey, T. 1952. The spinal tract of the trigeminal nerve in human embryos between $7\frac{1}{2}$ and $8\frac{1}{2}$ weeks of menstrual age and its relation to early fetal behavior. J. Comp. Neurol., 97:143–209.

Conel, J. L. 1942. The origin of the neural crest. J. Comp. Neurol., 76:191–215.

Davies, J. 1963. *Human Developmental Anatomy.* Ronald Press, New York, 298 pp.

Donahue, S. 1964. A relationship between fine structure and function of blood vessels in the central nervous system of rabbit fetuses. Amer. J. Anat., 115:17–26.

Dräger, U. C., and J. F. Olsen. 1980. Origins of crossed and uncrossed retinal projections in pigmented and albino mice. J. Comp. Neurol., 191:383–412.

Duncan, D. 1957. Electron microscope study of the embryonic neural tube and notochord. Tex. Rep. Biol. Med., 15:367–377.

Eakin, R. M. 1949. The nature of the organizer. Science, 109:195–197.

Fugita, H. and S. Fugita. 1963. Electron microscopic studies on neuroblast differentiation in the central nervous system of domestic fowl. Z. Zellforsch. mikrosk. Anat., 60:463–478.

Fugita, S. 1963. The matrix cell and cytogenesis in the developing central nervous system. J. Comp. Neurol., 120:37–42.

Fujita, S. 1964. Analysis of neuron differentiation in the central nervous system by tritiated thymidine autoradiography. J. Comp. Neurol., 122:311–328.

Fujita, S. 1973. Genesis of glioblasts in the human spinal cord as revealed by Feulgen cytophotometry. J. Comp. Neurol., 151:25–34.

Garber, B. B., P. R. Huttenlocher, and L. H. Larramendi. 1980. Self-assembly of cortical plate cells in vitro within embryonic mouse cerebral aggregates. Golgi and electron microscopic analysis. Brain Res., 201:255–278.

Humphrey, T. 1963. The development of the anterior olfactory nucleus of human fetuses. In *Progress in Brain Research.* Vol. 3. Edited by W. Bargmann and J. P. Schadé, pp. 170–190.

Jacobson, M. 1978. *Developmental Neurobiology.* 2nd ed. Plenum Press, New York, 562 pp.

Kimmel, D. L. 1941. Development of the afferent components of the facial, glossopharyngeal and vagus nerves in the rabbit embryo. J. Comp. Neurol., 74:447–471.

Langman, J. 1981. *Medical Embryology.* 4th ed. Williams and Wilkins, Baltimore, 384 pp.

Langman, J., and C. C. Haden. 1970. Formation and migration of neuroblasts in the spinal cord of the chick embryo. J. Comp. Neurol., 138:419–432.

Lauder, J. M., and H. Krebs. 1978. Serotonin as a differentiation signal in early neurogenesis. Devel. Neurosci., 1:15–30.

Lund, J. S., R. G. Boothe, and R. D. Lund. 1977. Development of neurons in the visual cortex (area 17) of the monkey (*Macaca nemestrina*): A Golgi study from fetal day 127 to postnatal maturity. J. Comp. Neurol., 176:149–188.

Macchi, G. 1951. The ontogenetic development of the olfactory telencephalon in man. J. Comp. Neurol., 95:245–305.

McConnell, J. A., and J. W. Sechrist. 1980. Identification of early neurons in the brainstem and spinal cord. I. An autoradiographic study in the chick. J. Comp. Neurol., 192:769–783.

Mitzdorf, U., and G. Neumann. 1980. Effects of monocular deprivation in the lateral geniculate nucleus of the cat: an analysis of evoked potentials. J. Physiol. (Lond.), 304:221–230.

Mitzdorf, U., and W. Singer. 1980. Monocular activation of visual cortex in normal and monocularly deprived cats: an analysis of evoked potentials. J. Physiol. (Lond.), 304:203–220.

Moore, K. L. 1982. *The Developing Human.* 3rd ed. W. B. Saunders, Philadelphia, 479 pp.

O'Brien, R. A. D., A. J. Östberg, and G. Vrbová. 1980. The effect of acetylcholine on the function and structure of the developing mammalian neuromuscular junction. Neuroscience, 5:1367–1379.

Okado, N. 1980. Development of the human cervical spinal cord with reference to synapse formation in the motor nucleus. J. Comp. Neurol., 191:495–513.

O'Rahilly, R., and E. Gardner. 1979. The initial development of the human brain. Acta Anat. (Basel), 104:123–133.

Patten, B. M. 1968. *Human Embryology.* 3rd ed. McGraw-Hill Book Co., New York, 651 pp.

Pearson, A. A. 1939. The hypoglossal nerve in human embryos. J. Comp. Neurol., 71:21–39.

Purves, D., and J. W. Lichtman. 1980. Elimination of synapses in the developing nervous system. Science, 210:153–157.

Rakic, P. 1972. Mode of cell migration to the superficial layers of fetal monkey neocortex. J. Comp. Neurol., 145:61–84.

Rakic, P., and R. L. Sidman. 1968. Subcommissural organ and adjacent ependyma: Autoradiographic study of their origin in the mouse brain. Amer. J. Anat., 122:317–336.

Rakic, P., and R. L. Sidman. 1969. Telencephalic origin of pulvinar neurons in the fetal human brain. Z. Anat. Entwicklungsgesch., 129:53–82.

Robertson, T. W., T. L. Hickey, and R. W. Guillery. 1980. Development of the dorsal lateral geniculate nucleus in normal and visually deprived Siamese cats. J. Comp. Neurol., 191:573–579.

Sauer, F. C. 1935. Mitosis in the neural tube. J. Comp. Neurol., 62:377–405.

Sauer, F. C. 1935. The cellular structure of the neural tube. J. Comp. Neurol., 63:13–23.

Schoenwolf, G. C., and R. O. Kelley. 1980. Characterization of intercellular junctions in the caudal portion of the developing neural tube of the chick embryo. Amer. J. Anat., *158*:29–41.

Sidman, R. L., I. L. Miale, and N. Feder. 1959. Cell proliferation and migration in the primitive ependymal zone; an autoradiographic study of histogenesis in the nervous system. Exp. Neurol., *1*:322–333.

Smart, I. 1961. The subependymal layer of the mouse brain and its cell production as shown by radioautography after thymidine-H³ injection. J. Comp. Neurol., *116*:325–347.

Smith, D. E., and A. J. Castro. 1979. Retrograde changes in Clarke's column following neonatal hemicerebellectomy in the rat. Amer. J. Anat., *156*:533–542.

Spemann, H., and H. Mangold. 1924. Uber Induktion von embryonalanlagen durch Implantation artfremder Organisatoren. Arch. f. Entwickl., *100*:599–638.

Stafstrom, C. E., D. Johnston, J. M. Wehner, and J. R. Sheppard. 1980. Spontaneous neural activity in fetal brain reaggregate cultures. Neuroscience, *5*:1681–1689.

Stensaas, L. J., and S. S. Stensaas. 1968. An electron microscope study of cells in the matrix and intermediate laminae of the cerebral hemisphere of the 45 mm rabbit embryo. Z. Zellforsch. mikrosk. Anat., *91*:341–365.

Strong, L. H. 1961. The first appearance of vessels within the spinal cord of the mammal: Their developing patterns as far as partial formation of the dorsal septum. Acta Anat. (Basel), *44*:80–108.

Taber Pierce, E. 1966. Histogenesis of the nuclei griseum pontis, corporis pontobulbaris and reticularis tegmenti pontis (Bechterew) in the mouse. An autoradiographic study. J. Comp. Neurol., *126*:219–240.

Takashima, S., F. Chan, L. E. Becker, and D. L. Armstrong. 1980. Morphology of the developing visual cortex of the human infant. A quantitative and qualitative Golgi study. J. Neuropathol. Exp. Neurol., *39*:487–501.

Tanaka, D., Jr. 1980. Development of spiny and aspiny neurons in the caudate nucleus of the dog during the first postnatal month. J. Comp. Neurol., *192*:247–263.

Vassilopoulos, D. 1977. Fetal development of the human cervical spine and cord. Acta Anat. (Basel), *98*:116–120.

Vaughn, J. E. 1969. An electron microscopic analysis of gliogenesis in rat optic nerves. Z. Zellforsch. mikrosk. Anat., *94*:293–324.

Yntema, C. L., and W. S. Hammond. 1954. The origin of intrinsic ganglia of trunk viscera from vagal neural crest in the chick embryo. J. Comp. Neurol., *101*:515–541.

Spinal Cord

Albe-Fessard, D., J. Boivie, G. Grant, and A. Levante. 1975. Labelling of cells in the medulla oblongata and the spinal cord of the monkey after injections of horseradish peroxidase in the thalamus. Neurosci. Lett., *1*:75–80.

Andersen, P., J. C. Eccles, and T. A. Sears. 1964. Cortically evoked depolarization of primary afferent fibers in the spinal cord. J. Neurophysiol., *27*:63–77.

Austin, G. M. (ed.). 1972. *The Spinal Cord: Basic Aspects and Surgical Considerations*. 2nd ed. Charles C Thomas, Springfield, Ill., 762 pp.

Barron, K. D., M. P. Dentinger, L. R. Nelson, and M. E. Scheibly. 1976. Incorporation of tritiated leucine by axotomized rubral neurons. Brain Res., *116*:251–266.

Barson, A. J. 1970. The vertebral level of termination of the spinal cord during normal and abnormal development. J. Anat. (Lond.), *106*:489–497.

Barson, A. J., and J. Sands. 1977. Regional and segmental characteristics of the human adult spinal cord. J. Anat. (Lond.), *123*:797–803.

Beal, J. A., and M. H. Cooper. 1978. The neurons in the gelatinosal complex (laminae II and III) of the monkey (*Macaca mulatta*): A Golgi study. J. Comp. Neurol., *179*:89–122.

Beal, J. A., and C. A. Fox. 1976. Afferent fibers in the substantia gelatinosa of the adult monkey (*Macaca mulatta*): A Golgi study. J. Comp. Neurol., *168*:113–144.

Bodian, D. 1966. Synaptic types on spinal motoneurons: An electron microscopic study. Bull. Johns Hopkins Hosp., *119*:16–45.

Bodian, D. 1975. Origin of specific synaptic types in the motoneuron neuropil of the monkey. J. Comp. Neurol., *159*:225–243.

Boivie, J. 1979. An anatomical reinvestigation of the termination of the spinothalamic tract in the monkey. J. Comp. Neurol. *186*:343–369.

Bryan, R. N., D. L. Trevino, and W. D. Willis, 1972. Evidence for a common location of alpha and gamma motoneurons. Brain Res., *38*:193–196.

Burke, R., A. Lundberg, and F. Weight. 1971. Spinal border cell origin of the ventral spinocerebellar tract. Exp. Brain Res., *12*:283–294.

Burton, H., and A. D. Loewy. 1976. Descending projections from the marginal cell layer and other regions of the monkey spinal cord. Brain Res., *116*:485–491.

Chambers, W. W., C. N. Liu, G. P. McCouch, and E. D'Aquili. 1966. Descending tracts and spinal shock in the cat. Brain, *89*:377–390.

Clemente, C. D. 1964. Regeneration in the vertebrate central nervous system. Internat. Rev. Neurobiol., *6*:257–301.

Coggeshall, R. E., J. D. Coulter, and W. D. Willis, Jr. 1974. Unmyelinated axons in the ventral roots of the cat lumbosacral enlargement. J. Comp. Neurol., *153*:39–58.

Coulter, J. D., and E. G. Jones. 1977. Differential distribution of corticospinal projections from individual cytoarchitectonic fields in the monkey. Brain Res., *129*:335–340.

Crutcher, K. A., and W. G. Bingham, Jr. 1978. Descending monoaminergic pathways in the primate spinal cord. Amer. J. Anat., *153*:159–164.

Culberson, J. L., D. E. Haines, D. L. Kimmel, and P. B. Brown. 1979. Contralateral projection of primary afferent fibers to mammalian spinal cord. Exp. Neurol., *64*:83–97.

Duggan, A. W. 1974. The differential sensitivity to L-glutamate and L-aspartate of spinal interneurones and Renshaw cells. Exp. Brain Res., *19*:522–528.

Earle, K. M. 1952. The tract of Lissauer and its possible relation to the pain pathway. J. Comp. Neurol., *96*:93–111.

Eccles, J. C. 1964. Presynaptic inhibition in the spinal cord. In *Progress in Brain Research*. Vol. 12. Edited by J. C. Eccles and J. P. Schadé. pp. 65–91.

Eccles, J. C., F. Magni, and W. D. Willis. 1962. Depolarization of central terminals of group I afferent fibers from muscle. J. Physiol. (Lond.), *160*:62–93.

Elliott, H. C. 1942. Studies on the motor cells of the spinal cord. I. Distribution in the normal human cord. Amer. J. Anat., 70:95–117.

Elliott, H. C. 1943. Studies on the motor cells of the spinal cord. II. Distribution in the normal human fetal cord. Amer. J. Anat., 72:29–38.

Gamble, H. J. 1971. Electron microscope observations upon the conus medullaris and filum terminale of human fetuses. J. Anat. (Lond.), 110:173–179.

Goldberger, M. E., and M. Murray. 1974. Restitution of function and collateral sprouting in the cat spinal cord: the deafferented animal. J. Comp. Neurol., 158:37–54.

Granit, R., and B. Holmgren. 1955. Two pathways from brain stem to gamma ventral horn cells. Acta Physiol. Scand., 35:93–108.

Gray, E. G. 1963. Electron microscopy of pre-synaptic organelles of the spinal cord. J. Anat. (Lond.), 97:101–106.

Ha, H. 1971. Cervicothalamic tract in the rhesus monkey. Exp. Neurol., 33:205–212.

Hagbarth, K.-E. 1952. Excitatory and inhibitory skin areas for flexor and extensor motoneurones. Acta Physiol. Scand. (Suppl. 94), 26:1–58.

Hayes, N. L., and A. Rustioni. 1980. Spinothalamic and spinomedullary neurons in macaques: a single and double retrograde tracer study. Neuroscience, 5:861–874.

Hongo, T., N. Ishizuka, H. Mannen, and S. Sasaki. 1978. Axonal trajectory of single group Ia and Ib fibers in the cat spinal cord. Neurosci. Lett., 8:321–328.

Hunt, S. P., J. S. Kelly, and P. C. Emson. 1980. The electron microscopic localization of methionine-enkephalin within the superficial layers (I and II) of the spinal cord. Neuroscience, 5:1871–1890.

Iles, J. F. 1976. Central terminations of muscle afferents on motoneurones in the cat spinal cord. J. Physiol. (Lond.), 262:91–117.

Kawamura, Y. and P. J. Dyck. 1977. Lumbar motoneurons of man: III. The number and diameter distribution of large- and intermediate-diameter cytons by nuclear columns. J. Neuropathol. Exp. Neurol., 36:956–963.

Kitai, S. T., and F. Morin. 1962. Microelectrode study of dorsal spinocerebellar tract. Amer. J. Physiol., 203:799–802.

Kumazawa, T., and E. R. Perl. 1978. Excitation of marginal and substantia gelatinosa neurons in the primate spinal cord: Indications of their place in dorsal horn functional organization. J. Comp. Neurol., 177:417–434.

Kumazawa, T., E. R. Perl, P. R. Burgess, and D. Whitehorn. 1975. Ascending projections from marginal zone (lamina I) neurons of the spinal dorsal horn. J. Comp. Neurol., 162:1–12.

LaMotte, C. 1977. Distribution of the tract of Lissauer and the dorsal root fibers in the primate spinal cord. J. Comp. Neurol., 172:529–561.

LaMotte, C., C. B. Pert, and S. H. Snyder. 1976. Opiate receptor binding in primate spinal cord: Distribution and changes after dorsal root section. Brain Res., 112:407–412.

Lassek, A. M. 1954. The Pyramidal Tract: Its Status in Medicine. Charles C Thomas, Springfield, Ill. 166 pp.

Light, A. R., and E. R. Perl. 1979. Reexamination of the dorsal root projection to the spinal dorsal horn including observations on the differential termination of coarse and fine fibers. J. Comp. Neurol., 186:117–131.

Light, A. R., and E. R. Perl. 1979. Spinal termination of functionally identified primary afferent neurons with slowly conducting myelinated fibers. J. Comp. Neurol., 186:133–150.

Liu, C.-N., and W. W. Chambers. 1958. Intraspinal sprouting of dorsal root axons; development of new collaterals and preterminals following partial denervation of the spinal cord in the cat. Arch. Neurol. Psychiat., 79:46–61.

Liu, C.-N., and W. W. Chambers. 1964. An experimental study of the cortico-spinal system in the monkey (Macaca mulatta). J. Comp. Neurol., 123:257–284.

Liu, C.-N, W. W. Chambers, and G. P. McCouch. 1966. Reflexes in the spinal monkey (Macaca mulatta). Brain, 89:349–358.

Lloyd, D. P. C. 1951. Electrical signs of impulse conduction in spinal motoneurons. J. Gen. Physiol., 35:255–288.

Lundberg, A., and P. Voorhoeve. 1962. Effects from the pyramidal tract on spinal reflex arcs. Acta Physiol. Scand., 56:201–219.

Lundberg, A., and F. Weight. 1971. Functional organization of connexions to the ventral spinocerebellar tract. Exp. Brain Res., 12:295–316.

McCouch, G. P., C.-N. Liu, and W. W. Chambers. 1966. Descending tracts and spinal shock in the monkey (Macaca mulatta). Brain, 89:359–376.

Mehler, W. R. 1962. The anatomy of the so-called "pain tract" in man: an analysis of the course and distribution of the ascending fibers of the fasciculus anterolateralis. In: Basic Research in Paraplegia, Edited by J. D. French and R. W. Porter. Charles C Thomas, Springfield, Ill., pp. 26–55.

Murray, M., and M. E. Goldberger. 1974. Restitution of function and collateral sprouting in the cat spinal cord: The partially hemisected animal. J. Comp. Neurol., 158:19–36.

Nijensohn, D. E., and F. W. Kerr. 1975. The ascending projections of the dorsolateral funiculus of the spinal cord in the primate. J. Comp. Neurol., 161:459–470.

Nyberg-Hansen, R., and A. Brodal. 1963. Sites of termination of corticospinal fibers in the cat. An experimental study with silver impregnation methods. J. Comp. Neurol., 120:369–391.

Nyberg-Hansen, R., and A. Brodal. 1964. Sites and mode of termination of rubrospinal fibres in the cat. An experimental study with silver impregnation methods. J. Anat. (Lond.), 98:235–253.

Petras, J. M. 1968. The substantia gelatinosa of Rolando. Experientia, 24:1045–1047.

Pompeiano, O. 1972. Vestibulospinal relations: Vestibular influences on gamma motoneurons and primary afferents. In Progress in Brain Research. Vol. 37. Edited by A. Brodal and O. Pompeiano. pp. 197–232.

Price, D. D., and D. J. Mayer. 1974. Physiological laminar organization of the dorsal horn of M. mulatta. Brain Res., 79:321–325.

Ralston, H. J., III. 1979. The fine structure of laminae I, II and III of the macaque spinal cord. J. Comp. Neurol., 184:619–642.

Ralston, H. J., III, and D. D. Ralston. 1979. The distribution of dorsal root axons in laminae I, II and III of the macaque spinal cord: A quantitative electron microscope study. J. Comp. Neurol., 184:643–684.

Ralston, H. J., and D. D. Ralston. 1979. Identification of dorsal root synaptic terminals on monkey ventral horn cells by electron microscopic autoradiography. J. Neurocytol., 8:151–166.

Ranson, S. W. 1914. The tract of Lissauer and the substantia gelatinosa Rolandi. Amer. J. Anat., 16:97–126.

Renshaw, B. 1940. Activity in the simplest spinal reflex pathways. J. Neurophysiol., 3:373–387.

Rexed, B. A. 1952. The cytoarchitectonic organization of the spinal cord in the cat. J. Comp. Neurol., 96:415–496.

Rexed, B. A. 1954. Cytoarchitectonic atlas of the spinal cord of the cat. J. Comp. Neurol., 100:297–379.

Rustioni, A. 1977. Spinal neurons project to the dorsal column nuclei of rhesus monkeys. Science, 196:656–658.

Scheibel, M. E., and A. B. Scheibel. 1966. Spinal motoneurons, interneurons and Renshaw cells: A Golgi study. Arch. Ital. Biol., 104:328–353.

Scheibel, M. E., and A. B. Scheibel. 1968. Terminal axonal patterns in cat spinal cord. II. The dorsal horn. Brain Res., 9:32–58.

Scheibel, M. E., and A. B. Scheibel. 1970. Organization of spinal motoneuron dendrites in bundles. Exp. Neurol., 28:106–112.

Sindou, M., C. Quoex, and C. Baleydier. 1974. Fiber organization of the posterior spinal cord-rootlet junction in man. J. Comp. Neurol., 153:15–26.

Sprague, J. M., and H. Ha. 1964. The terminal fields of dorsal root fibers in the lumbosacral spinal cord of the cat, and the dendritic organization of the motor nuclei. In Progress in Brain Research. Vol. 11. Edited by J. C. Eccles and J. P. Schadé. pp. 120–154.

Taub, E., M. Harger, H. Cannon Grier, and W. Hodos. 1980. Some anatomical observations following chronic dorsal rhizotomy in monkeys. Neuroscience, 5:389–401.

Ten Bruggencate, G., and A. Lundberg. 1974. Facilitatory interaction in transmission to motoneurones from vestibulospinal fibres and contralateral primary afferents. Exp. Brain Res., 19:248–270.

Truex, R. C., M. J. Taylor, M. Q. Smythe, and P. L. Gildenberg. 1970. The lateral cervical nucleus of cat, dog and man. J. Comp. Neurol., 139:93–104.

Vierck, C. J., Jr., D. M. Hamilton, and J. I. Thornby. 1971. Pain reactivity of monkeys after lesions to the dorsal and lateral columns of the spinal cord. Exp. Brain Res., 13:140–158.

Warwick, R., and G. A. G. Mitchell. 1956. The phrenic nucleus of the macaque. J. Comp. Neurol., 105:553–585.

Willis, W. D., Jr. 1979. Studies of the spinothalamic tract. Tex. Rep. Biol. Med., 38:1–45.

Willis, W. D., R. B. Leonard, and D. R. Kenshalo, Jr. 1978. Spinothalamic tract neurons in the substantia gelatinosa. Science, 202:986–988.

Willis, W. D., M. A. Weir, R. D. Skinner, and R. N. Bryan. 1973. Differential distribution of spinal cord field potentials. Exp. Brain Res., 17:169–176.

Willis, W. D., and J. C. Willis. 1966. Properties of interneurons in the ventral spinal cord. Arch. Ital. Biol., 104:354–386.

Windle, W. F. (ed.). 1980. The Spinal Cord and Its Relation to Injury. Marcel Dekker, Inc., New York, 384 pp.

Wong, W. C. and C. K. Tan. 1980. The fine structure of the intermediolateral nucleus of the spinal cord of the monkey (Macaca fascicularis). J. Anat. (Lond.), 130:263–277.

Blood Supply of the Spinal Cord

Adamkiewicz, A. 1881. Ueber die mikroskopischen Gefässe des menschlichen Rückenmarkes. Trans. Internat. Med. Cong., 7th Session, London, 1:155–157.

Adamkiewicz, A. 1881. Die Blutgefässe des menschlichen Rückenmarkes. I. Die Gefässe der Rückenmarkssubstanz. Sitzungsber. Akad. Wiss. Wien Math.-Naturwiss. Kl., 84:469–502.

Adamkiewicz, A. 1882. Die Blutgefässe des menschlichen Rückenmarkes. II. Die Gefässe der Rückenmarkoberfläche. Sitzungsber. Akad. Wiss. Wien Math.-Naturwiss. Kl., 85:101–130.

Aminoff, M. J. 1976. Spinal Angiomas. Blackwell Scientific Publications, Oxford, 179 pp.

Aminoff, M. J., R. O. Barnard, and V. Logue. 1974. The pathophysiology of spinal vascular malformations. J. Neurol. Sci., 23:255–263.

Batson, O. V. 1940. The function of the vertebral veins and their rôle in the spread of metastases. Ann. Surg., 112:138–149.

Batson, O. V. 1942. The vertebral vein system as a mechanism for the spread of metastases. Amer. J. Roentgenol. Rad. Therapy, 48:715–718.

Bolton, B. 1939. The blood supply of the human spinal cord. J. Neurol. Psychiatry, 2:137–148.

Chakravorty, B. G. 1971. Arterial supply of the cervical spinal cord (with special reference to the radicular arteries). Anat. Rec., 170:311–329.

Craigie, E. H. 1931. The vascularity of parts of the spinal cord, brain stem, and cerebellum of the wild Norway rat (Rattus norvegicus) in comparison with that in the domesticated Albino. J. Comp. Neurol., 53:309–318.

Crock, H. V., and H. Yoshizawa. 1977. The Blood Supply of the Vertebral Column and Spinal Cord in Man. Springer-Verlag, New York, 130 pp.

DiChiro, G., and L. Wener. 1973. Angiography of the spinal cord. A review of contemporary techniques and applications. J. Neurosurg., 39:1–29.

Dommisse, G. F. 1974. The blood supply of the spinal cord. A critical vascular zone in spinal surgery. J. Bone Joint Surg., 56B:225–235.

Dommisse, G. F. 1975. The Arteries and Veins of the Human Spinal Cord from Birth. Churchill-Livingstone, Edinburgh, 104 pp.

Gillilan, L. A. 1958. The arterial blood supply of the human spinal cord. J. Comp. Neurol., 110:75–103.

Gillilan, L. A. 1962. Blood vessels, meninges, cerebrospinal fluid. In Correlative Anatomy of the Nervous System. Edited by E. C. Crosby, T. Humphrey, and E. W. Lauer. Macmillan Publishing Co., New York, pp. 550–579.

Gillilan, L. A. 1970. Veins of the spinal cord. Anatomic details; suggested clinical applications. Neurology, 20:860–868.

Hassler, O. 1966. Blood supply to human spinal cord; a microangiographic study. Arch. Neurol., 15:302–307.

Herren, R. Y., and L. Alexander. 1939. Sulcal and intrinsic blood vessels of human spinal cord. Arch. Neurol. Psychiatry, 41:678–687.

Knox-Macaulay, H., M. T. Morrell, D. M. Potts, and T. D. Preston. 1960. The arterial supply to the spinal cord of the guinea pig. Acta Anat., 40:249–255.

Lazorthes, G., A. Gouazé, G. Bastide, J. J. Santini, O. Zadeh, and Ph. Burdin. 1966. La vascularisation artérielle de la moelle cervicale. Étude des suppléances. Revue Neurol., 115:1055–1068.

Lazorthes, G., A. Gouazé, G. Bastide, J.-H. Soutoul, O. Zadeh, and J.-J. Santini. 1966. La vascularisation artérielle du renflement lombaire; étude des variations et des suppléances. Revue Neurol., 114:109–122.

Lazorthes, G., A. Gouazé, and R. Djindjian. 1973. *Vascularisation et Circulation de la Moelle Épinière.* Masson et Cie, Paris, 186 pp.

Lazorthes, G., A. Gouazé, J. O. Zadeh, J. J. Santini, Y. Lazorthes, and P. Burdin. 1971. Arterial vascularization of the spinal cord. Recent studies of the anastomotic substitution pathways. J. Neurosurg., 35:253–262.

Lazorthes, G., J. Poulhes, G. Bastide, J. Roulleau, and A. R. Chancholle. 1958. La vascularisation artérielle de la moelle. Recherches anatomiques et applications à la pathologie medullaire et à la pathologie aortique. Neurochirurgie, 4:3–19.

Maillot, C., and J. C. Koritke. 1970. Les origines du tronc arteriel spinal posterieur chez l'homme. C. R. Assoc. Anat. (Nancy), 149:837–847.

Sahs, A. L. 1942. Vascular supply of the monkey's spinal cord. J. Comp. Neurol., 76:403–415.

Suh, T. H., and L. Alexander. 1939. Vascular system of the human spinal cord. Arch. Neurol. Psychiatry, 41:659–677.

Torr, J. B. D. 1957. The arterial supply of the foetal spinal cord. J. Anat. (Lond.), 91:576.

Torr, J. B. D. 1957. The embryological development of the anterior spinal artery in man. J. Anat. (Lond.), 91:587.

Torr, J. B. D. 1957. The dependence of blood supply of the spinal cord on certain aortic segments. J. Anat. (Lond.), 91:612.

Turnbull, I. M., A. Brieg, and O. Hassler. 1966. Blood supply of cervical spinal cord in man: A microangiographic cadaver study. J. Neurosurg., 24:951–965.

Tveten, L. 1976. Spinal cord vascularity. III. The spinal cord arteries in man. Acta Radiol. Diagn. (Stockh.), 17:257–273.

Woollam, D. H. M., and J. W. Millen. 1955. The arterial supply of the spinal cord and its significance. J. Neurol. Neurosurg. Psychiatry, 18:97–102.

Brain Stem

Anderson, F. D., and C. M. Berry. 1956. An oscillographic study of the central pathways of the vagus nerve in the cat. J. Comp. Neurol., 106:163–181.

Armstrong, D. M. 1974. Functional significance of connections of the inferior olive. Physiol. Rev., 54:358–417.

Assaf, S. Y., and J. J. Miller. 1978. The role of a raphe serotonin system in the control of septal unit activity and hippocampal desynchronization. Neuroscience, 3:539–550.

Baker, R. and W. Precht. 1972. Electrophysiological properties of trochlear motoneurons as revealed by IVth nerve stimulation. Exp. Brain Res., 14:127–157.

Baldissera, F., A. Lundberg, and M. Udo. 1972. Stimulation of pre- and postsynaptic elements in the red nucleus. Exp. Brain Res., 15:151–167.

Beckstead, R. M., J. R. Morse, and R. Norgren. 1980. The nucleus of the solitary tract in the monkey: Projections to the thalamus and brain stem nuclei. J. Comp. Neurol., 190:259–282.

Berkley, K. J. 1980. Spatial relationships between the terminations of somatic sensory and motor pathways in the rostral brainstem of cats and monkeys. I. Ascending somatic sensory inputs to lateral diencephalon. J. Comp. Neurol., 193:283–317.

Bijlani, V., and N. H. Keswani. 1970. The salivatory nuclei in the brainstem of the monkey (Macaca mulatta). J. Comp. Neurol., 139:375–384.

Bowsher, D., and J. Westman. 1970. The gigantocellular reticular region and its spinal afferents: A light and electron microscope study in the cat. J. Anat. (Lond.), 106:23–36.

Brodal, P. 1979. The pontocerebellar projection in the rhesus monkey: An experimental study with retrograde axonal transport of horseradish peroxidase. Neuroscience, 4:193–208.

Brown, J. O. 1943. The nuclear pattern of the non-tectal portions of the midbrain and isthmus in the dog and cat. J. Comp. Neurol., 78:365–405.

Buchanan, A. R. 1937. The course of the secondary vestibular fibers in the cat. J. Comp. Neurol., 67:183–204.

Burton, H., and A. D. Loewy. 1977. Projections to the spinal cord from medullary somatosensory relay nuclei. J. Comp. Neurol., 173:773–792.

Cammermeyer, J. 1973. Migration of mast cells through the area postrema. J. Hirnforsch., 14:519–526.

Car, A., A. Jean, and C. Roman. 1975. A pontine primary relay for ascending projections of the superior laryngeal nerve. Exp. Brain Res., 22:197–210.

Carpenter, M. B., and R. J. Pierson. 1973. Pretectal region and the pupillary light reflex. An anatomical analysis in the monkey. J. Comp. Neurol., 149:271–300.

Carpenter, M. B., and J. Pines. 1957. The rubrobulbar tract: Anatomical relationships, course, and terminations in the rhesus monkey. Anat. Rec., 128:171–185.

Carpenter, M. B., B. M. Stein, and P. Peter. 1972. Primary vestibulocerebellar fibers in the monkey: Distribution of fibers arising from distinctive cell groups of the vestibular ganglia. Amer. J. Anat., 135:221–249.

Chan-Palay, V. 1978. The paratrigeminal nucleus. I. Neurons and synaptic organization. J. Neurocytol., 7:405–418.

Chan-Palay, V. 1978. The paratrigeminal nucleus. II. Identification and inter-relations of catecholamine axons, indoleamine axons, and substance P immunoreactive cells in the neuropil. J. Neurocytol., 7:419–442.

Cooper, S. J., and E. T. Rolls. 1974. Relation of activation of neurones in the pons and medulla to brain-stimulation reward. Exp. Brain Res., 20:207–222.

FitzPatrick, K. A. 1975. Cellular architecture and topographic organization of the inferior colliculus of the squirrel monkey. J. Comp. Neurol., 164:185–207.

Flumerfelt, B. A., S. Otabe, and J. Courville. 1973. Distinct projections to the red nucleus from the dentate and interposed nuclei in the monkey. Brain Res., 50:408–414.

Graybiel, A. M. 1977. Direct and indirect preoculomotor pathways of the brainstem: An autoradiographic study of the pontine reticular formation in the cat. J. Comp. Neurol., 175:37–78.

Graybiel, A. M. 1979. Periodic-compartmental distribution of acetylcholinesterase in the superior colliculus of the human brain. Neuroscience, 4:643–650.

Hirosawa, K. 1968. Electron microscopic studies on pigment granules in the substantia nigra and locus coeruleus of the Japanese monkey (*Macaca fuscata yakui*). Z. Zellforsch. mikrosk. Anat., *88*:187–203.

Ingram, W. R., and S. W. Ranson. 1935. The nucleus of Darkschewitsch and nucleus interstitialis in the brain of man. J. Nerv. Ment. Dis., *81*:125–137.

Jean, A., A. Car, and C. Roman. 1975. Comparison of activity in pontine versus medullary neurones during swallowing. Exp. Brain Res., *22*:211–220.

Jensen, D. W. 1979. Reflex control of acute postural asymmetry and compensatory symmetry after a unilateral vestibular lesion. Neuroscience, *4*:1059–1074.

Jensen, D. W. 1979. Vestibular compensation: Tonic spinal influence upon spontaneous descending vestibular nuclear activity. Neuroscience, *4*:1075–1084.

Kalil, K. 1979. Projections of the cerebellar and dorsal column nuclei upon the inferior olive in the rhesus monkey: An autoradiographic study. J. Comp. Neurol., *188*:43–62.

King, J. S., R. C. Schwyn, and C. A. Fox. 1971. The red nucleus in the monkey (Macaca mulatta): A Golgi and an electron microscopic study. J. Comp. Neurol., *142*:75–107.

Kitai, S. T., J. D. Koscis, and T. Kiyohara. 1976. Electrophysiological properties of nucleus reticularis tegmenti pontis cells: Antidromic and synaptic activation. Exp. Brain Res., *24*:295–309.

Kitai, S. T., T. Tanaka, N. Tsukahara, and H. Yu. 1972. The facial nucleus of cat: Antidromic and synaptic activation and peripheral nerve representation. Exp. Brain Res., *16*:161–183.

Klinke, R., and N. Galley. 1974. Efferent innervation of vestibular and auditory receptors. Physiol. Rev., *54*:316–357.

MacLean, P. D., R. H. Denniston, and S. Dua. 1963. Further studies on cerebral representation of penile erection: Caudal thalamus, midbrain, and pons. J. Neurophysiol., *26*:273–293.

Magoun, H. W. 1950. Caudal and cephalic influences of the brainstém reticular formation. Physiol. Rev., *30*:459–474.

McCall, R. B., and G. K. Aghajanian. 1979. Denervation supersensitivity to serotonin in the facial nucleus. Neuroscience, *4*:1501–1510.

Mitchell, R. A. 1980. Site of termination of primary afferents from the carotid body chemoreceptors. Fed. Proc., *39*:2657–2661.

Mizuno, N. 1970. Projection fibers from the main sensory trigeminal nucleus and the supratrigeminal region. J. Comp. Neurol., *139*:457–471.

Nathan, M. A. 1972. Pathways in medulla oblongata of monkeys mediating splanchnic nerve activity. Electrophysiological and anatomical evidence. Brain Res., *45*:115–126.

Norgren, R., and C. Pfaffmann. 1975. The pontine taste area in the rat. Brain Res., *91*:99–117.

Rasmussen, A. T., and W. T. Peyton. 1948. The course and termination of the medial lemniscus in man. J. Comp. Neurol., *88*:411–424.

Rasmussen, G. L. 1946. The olivary peduncle and other fiber projections of the superior olivary complex. J. Comp. Neurol., *84*:141–219.

Rhoton, A. L., Jr. 1968. Afferent connections of the facial nerve. J. Comp. Neurol., *133*:89–100.

Russell, G. V. 1957. The brainstem reticular formation. Tex. Rep. Biol. Med., *15*:332–337.

Sadjadpour, K., and A. Brodal. 1968. The vestibular nuclei in man. A morphological study in the light of experimental findings in the cat. J. Hirnforsch., *10*:299–323.

Schwyn, R. C., and C. A. Fox. 1974. The primate substantia nigra: A Golgi and electron microscopic study. J. Hirnforsch., *15*:95–126.

Siegel, J. M. 1979. Behavioral functions of the reticular formation. Brain Res. Rev., *1*:69–105.

Smith, R. L. 1975. Axonal projections and connections of the principal sensory trigeminal nucleus in the monkey. J. Comp. Neurol., *163*:347–376.

Stern, K. 1938. Note on the nucleus ruber magnocellularis and its efferent pathway in man. Brain, *61*:284–289.

Stotler, W. A. 1953. An experimental study of the cells and connections of the superior olivary complex of the cat. J. Comp. Neurol., *98*:401–431.

Strominger, N. L. 1973. The origins, course and distribution of the dorsal and intermediate acoustic striae in the rhesus monkey. J. Comp. Neurol., *147*:209–233.

Szentágothai, J. 1948. The representation of facial and scalp muscles in the facial nucleus. J. Comp. Neurol., *88*:207–220.

Tarlov, E. 1969. The rostral projections of the primate vestibular nuclei: An experimental study in macaque, baboon and chimpanzee. J. Comp. Neurol., *135*:27–56.

Weisberg, J. A. and A. Rustioni. 1977. Cortical cells projecting to the dorsal column nuclei of rhesus monkeys. Exp. Brain Res., *28*:521–528.

Westlund, K. N., and J. D. Coulter. 1980. Descending projections of the locus coeruleus and subcoeruleus/medial parabrachial nuclei in monkey: Axonal transport studies and dopamine-β-hydroxylase immunocytochemistry. Brain Res. Rev., *2*:235–264.

Whitsel, B. L., L. M. Petrucelli, G. Sapiro, and H. Ha. 1970. Fiber sorting in the fasciculus gracilis of squirrel monkeys. Exp. Neurol., *29*:227–242.

Wurtz, R. H., and M. E. Goldberg. 1971. Superior colliculus cell responses related to eye movements in awake monkeys. Science, *171*:82–84.

Cerebellum

Asanuma, C., W. T. Thach, and E. G. Jones. 1980. Nucleus interpositus projection to spinal interneurons in monkeys. Brain Res., *191*:245–248.

Batton, R. R., III, A. Jayaraman, D. Ruggiero, and M. B. Carpenter. 1977. Fastigial efferent projections in the monkey: An autoradiographic study. J. Comp. Neurol., *174*:281–306.

Brodal, A. 1972. Cerebrocerebellar pathways. Anatomical data and some functional implications. Acta Neurol. Scand. (Suppl.), *51*:153–195.

Brodal, A., and F. Walberg. 1977. The olivocerebellar projection in the cat studied with the method of retrograde axonal transport of horseradish peroxidase. IV. The projection to the anterior lobe. J. Comp. Neurol., *172*:85–108.

Brodal, A., and F. Walberg. 1977. The olivocerebellar projection in the cat studied with the method of retrograde axonal transport of horseradish peroxidase. VI. The projection onto longitudinal zones of the paramedian lobule. J. Comp. Neurol., *176*:281–294.

Carpenter, M. B., G. M. Brittin, and J. Pines. 1958. Isolated lesions of the fastigial nuclei in the cat. J. Comp. Neurol., *109*:65–89.

Carpenter, M. B., B. M. Stein, and P. Peter. 1972. Primary

vestibulocerebellar fibers in the monkey: Distribution of fibers arising from distinctive cell groups of the vestibular ganglia. Amer. J. Anat., 135:221–249.

Chambers, W. W., and J. M. Sprague. 1955. Functional localization in the cerebellum. I. Organization in longitudinal cortico-nuclear zones and their contribution to the control of posture, both extrapyramidal and pyramidal. J. Comp. Neurol., 103:105–129.

Crosby, E. C., J. A. Taren, and R. Davis. 1970. The anterior lobe and the lingula of the cerebellum in monkeys and man. Bibl. Psychiatr. Neurol., 143:22–39.

Dillon, L. S., and D. L. Atkins. 1970. Two unique features of the anterior cerebellum in the higher primates, including man. Anat. Rec., 168:415–431.

Eccles, J. C. 1977. An instruction-selection theory of learning in the cerebellar cortex. Brain Res., 127:327–352.

Frankfurter, A., J. T. Weber, and J. K. Harting. 1977. Brain stem projections to lobule VII of the posterior vermis in the squirrel monkey: As demonstrated by the retrograde axonal transport of tritiated horseradish peroxidase. Brain Res., 124:135–139.

Gilman, S., J. R. Bloedel, and R. Lechtenberg. 1981. *Disorders of the Cerebellum.* F. A. Davis, Philadelphia, 415 pp.

Graybiel, A. M. 1974. Visuo-cerebellar and cerebellovisual connections involving the ventral lateral geniculate nucleus. Exp. Brain Res., 20:303–306.

Haines, D. E. 1975. Cerebellar cortical efferents of the posterior lobe vermis in a prosimian primate (*Galago*) and the tree shrew (*Tupaia*). J. Comp. Neurol., 163:21–39.

Haines, D. E. 1975. Cerebellar corticovestibular fibers of the posterior lobe in a prosimian primate, the lesser bushbaby (*Galago senegalensis*). J. Comp. Neurol., 160:363–397.

Haines, D. E. 1977. Cerebellar corticonuclear and corticovestibular fibers of the flocculonodular lobe in a prosimian primate (*Galago senegalensis*). J. Comp. Neurol., 174:607–630.

Hoddevik, G. H., and A. Brodal. 1977. The olivocerebellar projection studied with the method of retrograde axonal transport of horseradish peroxidase. V. The projections to the flocculonodular lobe and the paraflocculus in the rabbit. J. Comp. Neurol., 176:269–280.

Ito, M. 1978. Cerebellar control mechanisms of movements investigated in connection with vestibular functions. Neuroscience, 3:117–118.

Jansen, J. 1972. Features of cerebellar morphology and organization. Acta Neurol. Scand. (Suppl.), 51:197–217.

Lapresle, J., and M. B. Hamida. 1970. The dentato-olivary pathway. Somatotopic relationship between the dentate nucleus and the contralateral inferior olive. Arch. Neurol., 22:135–143.

Larsell, O. 1947. The development of the cerebellum in man in relation to its comparative anatomy. J. Comp. Neurol., 87:85–129.

Loeser, J. D., R. J. Lemire, and E. C. Alvord, Jr. 1972. The development of the folia in the human cerebellar vermis. Anat. Rec., 173:109–113.

Rakic, P., and R. L. Sidman. 1970. Histogenesis of cortical layers in human cerebellum, particularly the lamina dissecans. J. Comp. Neurol., 139:473–500.

Rivera-Dominguez, M., F. A. Mettler, and C. R. Noback. 1974. Origin of cerebellar climbing fibers in the rhesus monkey. J. Comp. Neurol., 155:331–342.

Seeger, J. F., J. F. Hemmer, and R. G. Quisling. 1975. The great horizontal fissure of the cerebellum: Angiographic appearance. Radiology, 117:321–327.

Somogyi, P., and J. Hámori. 1976. A quantitative electron microscopic study of the Purkinje cell axon initial segment. Neuroscience, 1:361–365.

Tomasch, J. 1968. The overall information carrying capacity of the major afferent and efferent cerebellar cell and fiber systems. Confin. Neurol., 30:359–367.

Tomasch, J. 1969. The numerical capacity of the human cortico-pontocerebellar system. Brain Res., 13:476–484.

Thalamus

Adrianov, O. 1977. The problem of organization of thalamo-cortical connections. J. Hirnforsch., 18:191–221.

Carpenter, M. B. 1967. Ventral tier thalamic nuclei. Mod. Trends Neurol., 5:29–40.

Carpenter, M. B., and N. L. Strominger. 1967. Efferent fibers of the subthalamic nucleus in the monkey. A comparison of the efferent projections of the subthalamic nucleus, substantia nigra and globus pallidus. Amer. J. Anat., 121:41–71.

Dekeban, A. 1953. Human thalamus: An anatomical, developmental and pathological study. I. Division of the human adult thalamus into nuclei by use of the cyto-myelo-architectonic method. J. Comp. Neurol., 99:639–683.

DeVito, J. L., and D. M. Simmons. 1976. Some connections of the posterior thalamus in monkey. Exp. Neurol., 51:347–362.

Gardner, J. H. 1953. Innervation of pineal gland in hooded rat. J. Comp. Neurol., 99:319–329.

Glendenning, K. K., J. A. Hall, I. T. Diamond, and W. C. Hall. 1975. The pulvinar nucleus of *Galago senegalensis*. J. Comp. Neurol., 161:419–458.

Headon, M. B., and T. P. S. Powell. 1973. Cellular changes in the lateral geniculate nucleus of infant monkeys after suture of the eyelids. J. Anat. (Lond.), 116:135–145.

Hickey, T. L., and R. W. Guillery. 1979. Variability of laminar patterns in the human lateral geniculate nucleus. J. Comp. Neurol., 183:221–246.

Hunt, W. E., and J. L. O'Leary. 1952. Form of thalamic response evoked by peripheral nerve stimulation. J. Comp. Neurol., 97:491–514.

Iwahori, N. 1978. A Golgi study on the subthalamic nucleus of the cat. J. Comp. Neurol., 182:383–398.

Jones, E. G., and T. P. S. Powell. 1971. An analysis of the posterior group of thalamic nuclei on the basis of its afferent connections. J. Comp. Neurol., 143:185–216.

Lin, C. S., and J. H. Kaas. 1979. The inferior pulvinar complex in owl monkeys: Architectonic subdivisions and patterns of input from the superior colliculus and subdivisions of visual cortex. J. Comp. Neurol., 187:655–678.

Mathers, L. H. 1972. Ultrastructure of the pulvinar of the squirrel monkey. J. Comp. Neurol., 146:15–42.

Minderhoud, J. M. 1971. An anatomical study of the efferent connections of the thalamic reticular nucleus. Exp. Brain Res., 12:435–446.

Morest, D. K. 1964. The neuronal architecture of the medial geniculate body of the cat. J. Anat. (Lond.), 98:611–630.

Morest, D. K. 1965. The laminar structure of the medial geniculate body of the cat. J. Anat. (Lond.), *99*:143–160.

Nauta, H. J. W., and M. Cole. 1978. Efferent projections of the subthalamic nucleus: An autoradiographic study in monkey and cat. J. Comp. Neurol., *180*:1–16.

Niimi, K., and E. Kuwahara. 1973. The dorsal thalamus of the cat and comparison with monkey and man. J. Hirnforsch., *14*:303–325.

Ogren, M. P., and A. E. Hendrickson. 1979. The morphology and distribution of striate cortex terminals in the inferior and lateral subdivisions of the *Macaca* monkey pulvinar. J. Comp. Neurol., *188*:179–200.

Ogren, M. P., and A. E. Hendrickson. 1979. The structural organization of the inferior and lateral subdivisions of the *Macaca* monkey pulvinar. J. Comp. Neurol., *188*:147–178.

Olivier, A., A. Parent, and L. J. Poirier. 1970. Identification of the thalamic nuclei on the basis of their cholinesterase content in the monkey. J. Anat. (Lond.), *106*:37–50.

Partlow, G. D., M. Colonnier, and J. Szabo. 1977. Thalamic projections of the superior colliculus in the rhesus monkey, *Macaca mulatta.* A light and electron microscopic study. J. Comp. Neurol., *171*:285–318.

Phyllis, J. W. 1971. The pharmacology of thalamic and geniculate neurons. Internat. Rev. Neurobiol., *14*:1–48.

Polley, E. H., and R. W. Guillery. 1980. An anomalous uncrossed retinal input to laminar A of the cat's dorsal lateral geniculate nucleus. Neuroscience, *5*:1603–1608.

Rockel, A. J., C. J. Heath, and E. G. Jones. 1972. Afferent connections to the diencephalon in the marsupial phalanger and the question of sensory convergence in the "posterior group" of the thalamus. J. Comp. Neurol., *145*:105–130.

Saporta, S., and L. Kruger. 1977. The organization of thalamocortical relay neurons in the rat ventrobasal complex studied by the retrograde transport of horseradish peroxidase. J. Comp. Neurol., *174*:187–208.

Scheibel, M. E., and A. B. Scheibel. 1972. Specialized organizational patterns within nucleus reticularis thalami of the cat. Exp. Neurol., *34*:316–322.

Tigges, J., and W. K. O'Steen. 1974. Termination of retinofugal fibers in squirrel monkey: A re-investigation using autoradiographic methods. Brain Res., *79*:489–495.

Toncray, J. E., and W. J. S. Krieg. 1946. The nuclei of the human thalamus: A comparative approach. J. Comp. Neurol., *85*:421–459.

Van Buren, J. M., and R. C. Borke. 1974. Nucleus dorsalis superficialis (lateralis dorsalis) of the thalamus and the limbic system in man. J. Neurol. Neurosurg. Psychiatry, *37*:765–789.

Welker, W. I. 1973. Principles of organization of the ventrobasal complex in mammals. Brain Behav. Evol., *7*:253–336.

White, E. L. 1979. Thalamocortical synaptic relations: A review with emphasis on the projections of specific thalamic nuclei to the primary sensory areas of the neocortex. Brain Res. Rev., *1*:275–311.

Yelnik, J., and G. Percheron. 1979. Subthalamic neurons in primates: a quantitative and comparative analysis. Neuroscience, *4*:1717–1743.

Hypothalamus

Anand, B. K., and J. R. Brobeck. 1951. Hypothalamic control of food intake in rats and cats. Yale J. Biol. Med., *24*:123–140.

Andersson, B., and W. Wyrwicka. 1957. The elicitation of a drinking motor conditioned reaction by electrical stimulation of the hypothalamic drinking area in the goat. Acta Physiol. Scand., *41*:194–198.

Antunes, J. L., P. W. Carmel, and E. A. Zimmerman. 1977. Projections from the paraventricular nucleus to the zona externa of the median eminence of the rhesus monkey: An immunohistochemical study. Brain Res., *137*:1–10.

Antunes, J. L., and E. A. Zimmerman. 1978. The hypothalamic magnocellular system of the rhesus monkey: An immunocytochemical study. J. Comp. Neurol., *181*:539–566.

Bailey, P., and F. Bremer. 1921. Experimental diabetes insipidus. Arch. Intern. Med., *28*:773–803.

Bard, P. 1940. The hypothalamus and sexual behavior. Proc. Assoc. Res. Nerv. Ment. Dis., *20*:551–579.

Barraclough, C. A., and R. A. Gorski. 1961. Evidence that the hypothalamus is responsible for androgen-induced sterility in the female rat. Endocrinology, *68*:68–79.

Beals, J. K. 1976. Synaptogenesis in the ventromedial hypothalamus of the prenatal mouse. J. Comp. Neurol., *167*:361–384.

Bellinger, L. L., L. L. Bernardis, and S. Brooks. 1979. The effect of dorsomedial hypothalamic nuclei lesions on body weight regulation. Neuroscience, *4*:659–665.

Bernardis, L. L., and L. A. Frohman. 1971. Effects of hypothalamic lesions at different loci on development of hyperinsulinemia and obesity in the weanling rat. J. Comp. Neurol., *141*:107–116.

Brawer, J. R. 1971. The role of the arcuate nucleus in the brain-pituitary-gonad axis. J. Comp. Neurol., *143*:411–446.

Brobeck, J. R. 1946. Mechanism of the development of obesity in animals with hypothalamic lesions. Physiol. Rev., *26*:541–559.

Burford, G. D., R. E. J. Dyball, R. L. Moss, and B. T. Pickering. 1974. Synthesis of both neurohypophysial hormones in both the paraventricular and supraoptic nuclei of the rat. J. Anat. (Lond.), *117*:261–269.

Cheung, Y., and J. R. Sladek, Jr. 1975. Catecholamine distribution in feline hypothalamus. J. Comp. Neurol., *164*:339–360.

Conrad, L. C. A., and D. W. Pfaff. 1976. Efferents from medial basal forebrain and hypothalamus in the rat. I. An autoradiographic study of the medial preoptic area. J. Comp. Neurol., *169*:185–220.

Conrad, L. C. A., and D. W. Pfaff. 1976. Efferents from medial basal forebrain and hypothalamus in the rat. II. An autoradiographic study of the anterior hypothalamus. J. Comp. Neurol., *169*:221–262.

Crosby, E. C., and R. T. Woodburne. 1951. The mammalian midbrain and isthmus regions. Part II. The fiber connections. C. The hypothalamotegmental pathways. J. Comp. Neurol., *94*:1–32.

Cruce, J. A. F. 1977. An autoradiographic study of the descending connections of the mammillary nuclei of the rat. J. Comp. Neurol., *176*:631–644.

Daniel, P. M. 1976. Anatomy of the hypothalamus and pituitary gland. J. Clin. Pathol. (Suppl. 7), *30*:1–7.

Fox, C. A. 1943. The stria terminalis, longitudinal association bundle and precommissural fornix fibers in the cat. J. Comp. Neurol., 79:277–295.

Fry, W. J. 1970. Quantitative delineation of the efferent anatomy of the medial mammillary nucleus of the cat. J. Comp. Neurol., 139:321–336.

Green, J. D., and G. W. Harris. 1947. The neurovascular link between the neurohypophysis and the adenohypophysis. J. Endocrinol., 5:136–146.

Haymaker, W., E. Anderson, and W. J. H. Nauta. 1969. The Hypothalamus. Charles C Thomas, Springfield, Ill., 805 pp.

Hoffman, G. E., D. L. Felten, and J. R. Sladek, Jr. 1976. Monoamine distribution in primate brain. III. Catecholamine-containing varicosities in the hypothalamus of Macaca mulatta. Amer. J. Anat., 147:501–513.

Ifft, J. D. 1972. An autoradiographic study of the time of final division of neurons in rat hypothalamic nuclei. J. Comp. Neurol., 144:193–204.

Ingram, W. R. 1940. Nuclear organization and chief connections of the primate hypothalamus. Proc. Assoc. Res. Nerv. Ment. Dis., 20:195–244.

Jenkins, J. S. 1972. The hypothalamus. Brit. Med. J., 2:99–102.

Joseph, S. A., and K. M. Knigge. 1978. The endocrine hypothalamus: Recent anatomical studies. Res. Publ. Assoc. Res. Nerv. Ment. Dis., 56:15–47.

Kuhlenbeck, H. 1954. The Human Diencephalon—A Summary of Development, Structure, Function and Pathology. S. Karger, Basel, 230 pp.

LuQui, I. J., and C. A. Fox. 1976. The supraoptic nucleus and the supraopticohypophysial tract in the monkey (Macaca mulatta). J. Comp. Neurol., 168:7–40.

Magoun, H. W., and M. Ranson. 1942. The supraoptic decussations in the cat and monkey. J. Comp. Neurol., 76:435–459.

McBride, R. L., and J. Sutin. 1977. Amygdaloid and pontine projections to the ventromedial nucleus of the hypothalamus. J. Comp. Neurol., 174:377–396.

McCann, S. M. 1977. Luteinizing-hormone-releasing factor. N. Engl. J. Med., 296:797–802.

Nauta, W. 1960. Limbic system and hypothalamus: Anatomical aspects. Physiol. Rev. (Suppl. 4), 40:102–104.

Norgren, R. 1976. Taste pathways to hypothalamus and amygdala. J. Comp. Neurol., 166:17–30.

Palay, S. L. 1953. Neurosecretory phenomena in the hypothalamo-hypophysial system of man and monkey. Amer. J. Anat., 93:107–141.

Raisman, G. 1966. Neural connections of the hypothalamus. Brit. Med. Bull., 22:197–201.

Raisman, G. 1973. An ultrastructural study of the effects of hypophysectomy on the supraoptic nucleus of the rat. J. Comp. Neurol., 147:181–208.

Ranson, S. W., and H. W. Magoun. 1939. The hypothalamus. Ergebn. Physiol., 41:56–163.

Renaud, L. P. 1978. Neurophysiological organization of the endocrine hypothalamus. Res. Publ. Assoc. Res. Nerv. Ment. Dis., 56:269–301.

Saper, C. B., L. W. Swanson, and W. M. Cowan. 1978. The efferent connections of the anterior hypothalamic area of the rat, cat and monkey. J. Comp. Neurol., 182:575–599.

Sawyer, C. H. 1957. Triggering of the pituitary by the central nervous system. In Physiological Triggers and Discontinuous Rate Processes. Edited by T. H. Bullock. American Physiological Society, Washington, D. C., pp. 164–174.

Sawyer, C. H. 1966. Neural mechanisms in the steroid feedback regulation of sexual behavior and pituitary-gonad function. In Brain and Behavior, Vol. 3. The Brain and Gonadal Function. Edited by R. Gorski and R. Whalen. University of California Press, Los Angeles, pp. 221–255.

Schally, A. V., A. Arimura, and A. J. Kastin. 1973. Hypothalamic regulatory hormones. Science, 179:341–350.

Smialowski, A. 1975. Medial hypothalamic nuclei in the macaque (Macaca mulatta) brain. Acta Anat., 91:261–271.

Swanson, L. W. 1976. An autoradiographic study of the efferent connections of the preoptic region in the rat. J. Comp. Neurol., 167:227–256.

Swanson, L. W., and W. M. Cowan. 1975. The efferent connections of the suprachiasmatic nucleus of the hypothalamus. J. Comp. Neurol., 160:1–12.

Troiano, R., and A. Siegel. 1975. The ascending and descending connections of the hypothalamus in the cat. Exp. Neurol., 49:161–173.

Rhinencephalon

Aggleton, J. P., M. J. Burton, and R. E. Passingham. 1980. Cortical and subcortical afferents to the amygdala of the rhesus monkey (Macaca mulatta). Brain Res., 190:347–368.

Allen, W. F. 1948. Fiber degeneration in Ammon's horn resulting from extirpations of the piriform and other cortical areas and from transection of the horn at various levels. J. Comp. Neurol., 88:425–438.

Allison, A. C. 1953. The structure of the olfactory bulb and its relationship to the olfactory pathways in the rabbit and the rat. J. Comp. Neurol., 98:309–353.

Andersen, P., B. H. Bland, and J. D. Dudar. 1973. Organization of the hippocampal output. Exp. Brain Res., 17:152–168.

Broadwell, R. D. 1975. Olfactory relationships of the telencephalon and diencephalon in the rabbit. I. An autoradiographic study of the efferent connections of the main and accessory olfactory bulbs. J. Comp. Neurol., 163:329–345.

DeFrance, J. F., S. T. Kitai, and T. Shimono. 1973. Electrophysiological analysis of the hippocampal-septal projections. I. Response and topographical characteristics. Exp. Brain Res., 17:447–462.

DeFrance, J. F., S. T. Kitai, and T. Shimono. 1973. Electrophysiological analysis of the hippocampal-septal projections. II. Functional characteristics. Exp. Brain Res., 17:463–476.

DeVito, J. L. 1980. Subcortical projections to the hippocampal formation in squirrel monkey (Saimiri sciureus). Brain Res. Bull., 5:285–290.

Field, P. M. 1972. A quantitative ultrastructural analysis of the distribution of amygdaloid fibres in the preoptic area and the ventromedial hypothalamic nucleus. Exp. Brain Res., 14:527–538.

Green, J. D. 1964. The hippocampus. Physiol. Rev., 44:561–608.

Green, J. D., C. D. Clemente, and J. deGroot. 1957. Rhinencephalic lesions and behavior in cats. J. Comp. Neurol., 108:505–545.

Harrison, J. M., and M. Lyon. 1957. The role of the septal nuclei and components of the fornix in the behavior of the rat. J. Comp. Neurol., 108:121–137.

Heimer, L. 1968. Synaptic distribution of centripetal and centrifugal nerve fibres in the olfactory system of the rat. An experimental anatomical study. J. Anat. (Lond.), 103:413–432.

Hjorth-Simonsen, A. 1972. Projection of the lateral part

of the entorhinal area to the hippocampus and fascia dentata. J. Comp. Neurol., 146:219–232.

Hjorth-Simonsen, A. 1977. Distribution of commissural afferents to the hippocampus of the rabbit. J. Comp. Neurol., 176:495–514.

Jiminez-Castellanos, J. 1949. The amygdaloid complex in monkey studied by reconstructional methods. J. Comp. Neurol., 91:507–526.

Jones, E. G., H. Burton, C. B. Saper, and L. W. Swanson. 1976. Midbrain, diencephalic and cortical relationships of the basal nucleus of Meynert and associated structures in primates. J. Comp. Neurol., 167:385–419.

Kaada, B. R. 1972. Stimulation and regional ablation of the amygdaloid complex with reference to functional representations. In *The Neurobiology of the Amygdala*. Edited by B. E. Eleftheriou. Plenum Press, New York, pp. 205–281.

Krettek, J. E., and J. L. Price. 1977. Projections from the amygdaloid complex to the cerebral cortex and thalamus in the rat and cat. J. Comp. Neurol., 172:687–722.

Krettek, J. E., and J. L. Price. 1977. Projections from the amygdaloid complex and adjacent olfactory structures to the entorhinal cortex and to the subiculum in the rat and cat. J. Comp. Neurol., 172:723–752.

Lammers, H. J. 1972. The neural connections of the amygdaloid complex in mammals. In *The Neurobiology of the Amygdala*. Edited by B. E. Eleftheriou. Plenum Press, New York, pp. 123–144.

Lammers, H. J., and H. Gastaut. 1962. Relations cytoarchitectoniques et enzymo-architectoniques dans l'hippocampe. In *Physiologie de l'Hippocampe*. Editions du Centre National de la Recherche Scientifique, No. 107, Paris, pp. 13–21.

Lorente de Nó, R. 1934. Studies on the structure of the cerebral cortex. II. Continuation of the study of the ammonic system. J. Psychol. Neurol. (Leipzig), 46:113–177.

MacLean, P. D. 1949. Psychosomatic disease and the "visceral brain." Recent developments bearing on the Papez theory of emotion. Psychosom. Med., 11:338–353.

Mehler, W. R. 1980. Subcortical afferent connections of the amygdala in the monkey. J. Comp. Neurol., 190:733–762.

Meibach, R. C., and A. Siegel. 1977. Efferent connections of the hippocampal formation in the rat. Brain Res., 124:197–224.

Nauta, W. J. H. 1956. An experimental study of the fornix system in the rat. J. Comp. Neurol., 104:247–271.

Nauta, W. J. H. 1961. Fibre degeneration following lesions of the amygdaloid complex in the monkey. J. Anat. (Lond.), 95:515–531.

Nauta, W. J. H. 1962. Neural associations of the amygdaloid complex in the monkey. Brain, 85:505–520.

Nicoll, R. A. 1972. Olfactory nerves and their excitatory action in the olfactory bulb. Exp. Brain Res., 14:185–197.

Olds, J., and P. Milner. 1954. Positive reinforcement produced by electrical stimulation of septal areas and other regions of the rat brain. J. Comp. Physiol. Psychol., 47:419–427.

Papez, J. W. 1937. A proposed mechanism of emotion. Arch. Neurol. Psychiat., 38:725–743.

Powell, E. W., and R. B. Leman. 1976. Connections of the nucleus accumbens. Brain Res., 105:389–403.

Price, J. L. 1973. An autoradiographic study of complementary laminar patterns of termination of afferent fibers to the olfactory cortex. J. Comp. Neurol., 150:87–108.

Price, J. L., and T. P. S. Powell. 1970. The afferent connexions of the nucleus of the horizontal limb of the diagonal band. J. Anat. (Lond.), 107:239–256.

Price, J. L., and T. P. S. Powell. 1971. Certain observations on the olfactory pathway. J. Anat. (Lond.), 110:105–126.

Rae, A. S. L. 1969. Histology of the zone of contact between amygdala and hippocampus. Confin. Neurol., 31:330–333.

Raisman, G. 1972. An experimental study of the projection of the amygdala to the accessory olfactory bulb and its relationship to the concept of a dual olfactory system. Exp. Brain Res., 14:395–408.

Scalia, F., and S. S. Winans. 1975. The differential projections of the olfactory bulb and accessory olfactory bulb in mammals. J. Comp. Neurol., 161:31–56.

Siegel, A., H. Edinger, and S. Ohgami. 1974. The topographical organization of the hippocampal projection to the septal area: A comparative neuroanatomical analysis in the gerbil, rat, rabbit, and cat. J. Comp. Neurol., 157:359–378.

Skeen, L. C., and W. C. Hall. 1977. Efferent projections of the main and the accessory olfactory bulb in the tree shrew (*Tupaia glis*). J. Comp. Neurol., 172:1–36.

Stephan, H., and O. J. Andy. 1977. Quantitative comparison of the amygdala in insectivores and primates. Acta. Anat., 98:130–153.

Steward, O. 1976. Topographic organization of the projections from the entorhinal area to the hippocampal formation of the rat. J. Comp. Neurol., 167:285–314.

Swanson, L. W. 1978. The anatomical organization of the septo-hippocampal projections. In *Functions of the Septo-Hippocampal System*. Ciba Foundation Symposium, No. 58, Elsevier, New York, pp. 25–48.

Swanson, L. W., and W. M. Cowan. 1977. An autoradiographic study of the organization of the efferent connections of the hippocampal formation in the rat. J. Comp. Neurol., 172:49–84.

Van Hoesen, G. W., D. L. Rosene, and M. M. Mesulam. 1979. Subicular input from temporal cortex in the rhesus monkey. Science, 205:608–610.

White, L. E., Jr. 1965. Olfactory bulb projections in the rat. Anat. Rec., 152:465–479.

Willey, T. J. 1973. The ultrastructure of the cat olfactory bulb. J. Comp. Neurol., 152:211–232.

Basal Ganglia

Bunney, B. S., and G. K. Aghajanian. 1976. Dopaminergic influence in the basal ganglia: Evidence for striato-nigral feedback regulation. Res. Publ. Assoc. Res. Nerv. Ment. Dis., 55:249–267.

Carey, J. H. 1957. Certain anatomical and functional interrelations between the tegmentum of the midbrain and the basal ganglia. J. Comp. Neurol., 108:57–90.

Carpenter, M. B. 1976. Anatomy of the basal ganglia and related nuclei: A review. Adv. Neurol., 14:7–48.

Carpenter, M. B. 1981. Anatomy of the corpus striatum and brain stem integrating systems. In *Handbook of Physiology*, Section 1, *The Nervous System*, Vol. 2, *Motor Control*, Part 2. Edited by V. B. Brooks. American Physiological Society, Williams and Wilkins, Baltimore, pp. 947–995.

Carpenter, M. B., K. Nakano, and R. Kim. 1976. Nigro-thalamic projections in the monkey demonstrated by autoradiographic technics. J. Comp. Neurol., 165:401–416.

Dray, A,. 1979. The striatum and substantia nigra: A commentary on their relationships. Neuroscience, 4:1407–1439.

Fox, C. A., A. N. Andrade, D. E. Hillman, and R. C. Schwyn. 1971/72. The spiny neurons in the primate striatum: A Golgi and electron microscopic study. J. Hirnforsch., 13:181–201.

Fox, C.A., A. N. Andrade, I. J. LuQui, and J. A. Rafols. 1974. The primate globus pallidus: A Golgi and electron microscopic study. J. Hirnforsch., 15:75–93.

Fox, C. A., A. N. Andrade, R. C. Schwyn, and J. A. Rafols. 1971/72. The aspiring neurons and the glia in the primate striatum: A Golgi and electron microscopic study. J. Hirnforsch., 13:341–362.

Fox, C. A., and J. A. Rafols. 1976. The striatal efferents in the globus pallidus and in the substantia nigra. Res. Publ. Assoc. Res. Nerv. Ment. Dis., 55:37–55.

Hassler, R. 1974. Fiber connections within the extrapyramidal system. Confin. Neurol., 36:237–255.

Hassler, R. 1979. Basal ganglia. Historical perspective. Appl. Neurophysiol., 42:5–8.

Kitai, S. T. 1981. Electrophysiology of the corpus striatum and brain stem integrating systems. In Handbook of Physiology, Section 1, The Nervous System, Vol. 2, Motor Control, Part 2. Edited by V. B. Brooks. American Physiological Society, Williams and Wilkins, Baltimore, pp. 997–1015.

Kuo, J. S., and M. B. Carpenter. 1973. Organization of pallidothalamic projections in the rhesus monkey. J. Comp. Neurol., 151:201–236.

Lauer, E. W. 1945. The nuclear pattern and fiber connections of certain basal telencephalic centers in the Macaque. J. Comp. Neurol., 82:215–254.

Mehler, W. R., and W. J. Nauta. 1974. Connections of the basal ganglia and of the cerebellum. Confin. Neurol., 36:205–222.

Nauta, H. J. W. 1979. A proposed conceptual reorganization of the basal ganglia and telencephalon. Neuroscience, 4:1875–1881.

Parent, A. 1979. Identification of the pallidal and peripallidal cells projecting to the habenula in monkey. Neurosci. Lett., 15:159–164.

Parent, A., and A. Oliver. 1970. Comparative histochemical study of the corpus striatum. J. Hirnforsch., 12:73–81.

Rioch, D. McK., and C. Brenner. 1938. Experiments on the corpus striatum and rhinencephalon. J. Comp. Neurol., 68:491–507.

Szabo, J. 1970. Projections from the body of the caudate nucleus in the rhesus monkey. Exp. Neurol., 27:1–15.

Szabo, J. 1972. The course and distribution of efferents from the tail of the caudate nucleus in the monkey. Exp. Neurol., 37:562–572.

Szabo, J. 1980. Organization of the ascending striatal afferents in monkeys. J. Comp. Neurol., 189:307–321.

Webster, K. E. 1975. Structure and function of the basal ganglia. A non-clinical view. Proc. Roy. Soc. Med. (Lond.), 68:203–210.

Cerebral Cortex

Adrian, E. D. 1941. Afferent discharges to the cerebral cortex from peripheral sense organs. J. Physiol. (Lond.), 100:159–191.

Asanuma, H., and I. Rosén. 1972. Topographical organization of cortical efferent zones projecting to distal forelimb muscles in the monkey. Exp. Brain Res., 14:243–256.

Astruc, J. 1971. Corticofugal connections of area 8 (frontal eye field) in Macaca mulatta. Brain Res., 33:241–256.

Bailey, P., and G. von Bonin. 1951. The Isocortex of Man. University of Illinois Press, Urbana, 301 pp.

Barnard, J. W., and C. N. Woolsey. 1956. A study of localization in the corticospinal tracts of monkey and rat. J. Comp. Neurol., 105:25–50.

Beach, F. A., A. Zitrin, and J. Jaynes. 1955. Neural mediation of mating in male cats. II. Contributions of the frontal cortex. J. Exp. Zool., 130:381–401.

Benevento, L. A., and M. Rezak. 1975. Extrageniculate projections to layers VI and I of striate cortex (area 17) in the rhesus monkey (Macaca mulatta). Brain Res., 96:51–55.

Bindman, L., and O. Lippold. 1981. The Neurophysiology of the Cerebral Cortex. Arnold, London, 495 pp.

Bowden, D. M., P. S. Goldman, H. E. Rosvold, and R. L. Greenstreet. 1971. Free behavior of rhesus monkeys following lesions of the dorsolateral and orbital prefrontal cortex in infancy. Exp. Brain Res., 12:265–274.

Brodmann, K. 1909. Vergleichende Lokalisationslehre der Grosshirnrinde in ihren Prinzipien dargestellt auf Grund des Zellenbaues. J. A. Barth, Leipzig, 324 pp.

Brody, H. 1955. Organization of the cerebral cortex. III. A study of aging in the human cerebral cortex. J. Comp. Neurol., 102:511–556.

Campbell, A. W. 1905. Histological Studies on the Localisation of Cerebral Function. Cambridge University Press, Cambridge, 360 pp.

Campos-Ortega, J. A., and W. R. Hayhow. 1972. On the organisation of the visual cortical projection to the pulvinar in Macaca mulatta. Brain Behav. Evol., 6:394–423.

Carmon, A., J. Mor, and J. Goldberg. 1976. Evoked cerebral responses to noxious thermal stimuli in humans. Exp. Brain Res., 25:103–107.

Coxe, W. S., and W. M. Landau. 1970. Patterns of Marchi degeneration in the monkey pyramidal tract following small discrete cortical lesions. Neurology, 20:89–100.

Cragg, B. G. 1976. Ultrastructural features of human cerebral cortex. J. Anat. (Lond.), 121:331–362.

Emson, P. C., and O. Lindvall. 1979. Distribution of putative neurotransmitters in the neocortex. Neuroscience, 4:1–30.

Evarts, E. V. 1978. Mediation of quick motor responses by motor cortex pyramidal tract neurons in the monkey. Neuroscience, 3:95–98.

Feldman, M. L., and C. Dowd. 1975. Loss of dendritic spines in aging cerebral cortex. Anat. Embryol., 148:279–301.

Felix, D., and M. Wiesendanger. 1971. Pyramidal and non-pyramidal motor cortical effects on distal forelimb muscles of monkeys. Exp. Brain Res., 12:81–91.

Foerster, O. 1936. The motor cortex in man in the light of Hughlings Jackson's doctrines. Brain, 59:135–159.

Friedman, D. P., E. G. Jones, and H. Burton. 1980. Representation pattern in the second somatic sensory area of the monkey cerebral cortex. J. Comp. Neurol., 192:21–41.

Fritsch, G., and E. Hitzig. 1870. Ueber die elektrische

Erregbarkeit des Grosshirns. Arch. Anat. Physiol. Wiss. Med., *37*:300–332.

Frontera, J. G. 1956. Some results obtained by electrical stimulation of the cortex of the island of Reil in the brain of the monkey (*Macaca mulatta*). J. Comp. Neurol., *105*:365–394.

Funkenstein, H. H., and P. Winter. 1973. Responses to acoustic stimuli of units in the auditory cortex of awake squirrel monkeys. Exp. Brain Res., *18*:464–488.

Galaburda, A. M., F. Sanides, and N. Geschwind. 1978. Human brain. Cytoarchitectonic left-right asymmetries in the temporal speech region. Arch. Neurol., *35*:812–817.

Gatter, K. C., and T. P. S. Powell. 1977. The projection of the locus coeruleus upon the neocortex in the macaque monkey. Neuroscience, *2*:441–445.

Geschwind, N. 1979. Specializations of the human brain. Sci. Amer., *241*:180–199.

Gosavi, V. S., and P. N. Dubey. 1972. Projection of striate cortex to the dorsal lateral geniculate body in the rat. J. Anat. (Lond.), *113*:75–82.

Hanaway, J., and R. R. Young. 1977. Localization of the pyramidal tract in the internal capsule of man. J. Neurol. Sci., *34*:63–70.

Harman, P. J., and C. M. Berry. 1956. Neuroanatomical distribution of action potentials evoked by photic stimuli in cat fore- and midbrain. J. Comp. Neurol., *105*:395–416.

Harnarine-Singh, D., G. Geddes, and J. B. Hyde. 1972. Sizes and numbers of arteries and veins in normal human neopallium. J. Anat. (Lond.), *111*:171–179.

Hubel, D. H., and T. N. Wiesel. 1959. Receptive fields of single neurones in the cat's striate cortex. J. Physiol. (Lond.), *148*:574–591.

Hubel, D. H., and T. N. Wiesel. 1962. Receptive fields, binocular interaction and functional architecture in the cat's visual cortex. J. Physiol. (Lond.), *160*:106–154.

Hubel, D. H., and T. N. Wiesel. 1963. Shape and arrangement of columns in cat's striate cortex. J. Physiol. (Lond.), *165*:559–568.

Hubel, D. H., and T. N. Wiesel. 1968. Receptive fields and functional architecture of monkey striate cortex. J. Physiol. (Lond.), *195*:215–243.

Hubel, D. H., and T. N. Wiesel. 1972. Laminar and columnar distribution of geniculo-cortical fibers in the macaque monkey. J. Comp. Neurol., *146*:421–450.

Hubel, D. H., and T. N. Wiesel. 1977. Functional architecture of macaque monkey visual cortex. Proc. Roy. Soc. Biol. (Lond.), *198*:1–59.

Hubel, D. H., T. N. Wiesel, and M. P. Stryker. 1978. Anatomical demonstration of orientation columns in macaque monkey. J. Comp. Neurol., *177*:361–380.

Huttenlocher, P. R. 1979. Synaptic density in human frontal cortex—Developmental changes and effects of aging. Brain Res., *163*:195–205.

Jürgens, U., and D. Ploog. 1970. Cerebral representation of vocalization in the squirrel monkey. Exp. Brain Res., *10*:532–554.

Kaas, J. H., R. J. Nelson, M. Sur, and M. M. Merzenich. 1979. Multiple representations of the body within the primary somatosensory cortex of primates. Science, *204*:521–523.

Kohno, K., V. Chan-Palay, and S. L. Palay. 1975. Cytoplasmic inclusions of neurons in the monkey visual cortex (area 19). Anat. Embryol., *147*:117–125.

Krieg, W. J. S. 1954. *Connections of the Frontal Cortex of the Monkey.* Charles C Thomas, Springfield, Ill., 299 pp.

Künzle, H. 1978. Cortico-cortical efferents of primary motor and somatosensory regions of the cerebral cortex in *Macaca fascicularis.* Neuroscience, *3*:25–39.

Künzle, H., and K. Akert. 1977. Efferent connections of cortical area 8 (frontal eye field) in *Macaca fascicularis.* A reinvestigation using the autoradiographic technique. J. Comp. Neurol., *173*:147–164.

Lammers, H. J., and A. H. Lohman. 1974. Structure and fiber connections of the hypothalamus in mammals. In *Progress in Brain Research.* Vol. 41. Edited by D. F. Swaab and J. P. Schadé, pp. 61–78.

Lund, J. S. 1973. Organization of neurons in the visual cortex area 17 of the monkey (*Macaca mulatta*). J. Comp. Neurol., *147*:455–496.

Lund, J. S., et al. 1975. The origin of efferent pathways from the primary visual cortex, area 17, of the macaque monkey as shown by retrograde transport of horseradish peroxidase. J. Comp. Neurol, *164*:287–303.

McCulloch, W. S. 1949. Cortico-cortical connections. In *The Precentral Motor Cortex,* 2nd ed. Edited by P. Bucy. University of Illinois Press, Urbana, pp. 211–242.

McGuinness, E., D. Sivertsen, and J. M. Allman. 1980. Organization of the face representation in macaque motor cortex. J. Comp. Neurol., *193*:591–608.

Merzenich, M. M., J. H. Kaas, M. Sur, and C. S. Lin. 1978. Double representation of the body surface within cytoarchitectonic Areas 3b and 1 in "S 1" in the owl monkey (*Aotus trivirgatus*). J. Comp Neurol., *181*:41–74.

Montero, V. M. 1980. Patterns of connections from the striate cortex to cortical visual areas in superior temporal sulcus of macaque and middle temporal gyrus of owl monkey. J. Comp. Neurol., *189*:45–59.

Mountcastle, V. B., and T. P. S. Powell. 1959a. Central nervous system mechanisms subserving position sense and kinesthesis. Johns Hopkins Med. J., *105*:173–200.

Mountcastle, V. B., and T. P. S. Powell. 1959b. Neural mechanisms subserving cutaneous sensibility, with special reference to the role of afferent inhibition in sensory perception and discrimination. Johns Hopkins Med. J., *105*:201–232.

Newman, J. D., and D. F. Lindsley. 1976. Single unit analysis of auditory processing in squirrel monkey frontal cortex. Exp. Brain Res., *25*:169–181.

Peele, T. L. 1942. Cytoarchitecture of individual parietal areas in the monkey (*Macaca mulatta*) and the distribution of the efferent fibers. J. Comp. Neurol., *77*:693–737.

Penfield, W. and T. Rasmussen. 1950. *The Cerebral Cortex of Man; A Clinical Study of Localization of Function.* Macmillan Publishing Co., New York, 248 pp.

Penfield, W., and L. Roberts. 1959. *Speech and Brain Mechanisms.* Princeton University Press, Princeton, 286 pp.

Phillis, J. W., and S. Ochs. 1971. Excitation and depression of cortical neurones during spreading depression. Exp. Brain Res., *12*:132–149.

Porter, R. 1973. Functions of the mammalian cerebral cortex in movement. Prog. Neurobiol., *1*:3–51.

Powell, T. P. S., and V. B. Mountcastle. 1959. Some aspects of the functional organization of the cortex of the postcentral gyrus of the monkey: A correlation of findings obtained in a single unit analysis with cytoarchitecture. Johns Hopkins Med. J., 105:133–162.

Ramon y Cajal, S. 1955. *Studies on the Cerebral Cortex (Limbic Structure)*. Translated from Spanish by Lisbeth M. Kraft. Year Book Publishers, Chicago, 179 pp.

Rasmussen, A. T. 1943. The extent of recurrent geniculo-calcarine fibers (loop of Archambault and Meyer) as demonstrated by gross brain dissection. Anat. Rec., 85:277–284.

Rees, S. 1975. A quantitative electron microscopic study of atypical structures in normal human cerebral cortex. Anat. Embryol., 148:303–331.

Rose, J. E., and C. N. Woolsey. 1958. Cortical connections and functional organization of the thalamic auditory system of the cat. In *Biological and Biochemical Bases of Behavior*. Edited by H. F. Harlow and C. N. Woolsey. University of Wisconsin Press, Madison, pp. 127–150.

Rose, M. 1926. Der allocortex bei tier und mensch. J. Psychol. Neurol. (Leipzig), 34:1–111.

Rose, M. 1935. Cytoarchitectonik und Myeloarchitektonik der Grosshirnrinde. In *Handbuch der Neurologie*. Vol. 1. Edited by O. Bumke and O. Foerster. J. Springer-Verlag, Berlin, pp. 588–778.

Rubens, A. B., and M. W. Mahowald, and J. T. Hutton. 1976. Asymmetry of the lateral (Sylvian) fissures in man. Neurology, 26:620–624.

Scheibel, M. E., U. Tomiyasu, and A. B. Scheibel. 1977. The aging human Betz cell. Exp. Neurol., 56:598–609.

Schmidt, R. F. (ed.). 1978. *Fundamentals of Sensory Physiology*. Springer-Verlag, New York, 286 pp.

Streitfeld, B. D. 1980. The fiber connections of the temporal lobe with emphasis on the rhesus monkey. Internat. J. Neurosci., 11:51–71.

Tigges, J., S. Nakagawa, and M. Tigges. 1979. Efferents of area 4 in a South American monkey (*Saimiri*). I. Terminations in the spinal cord. Brain Res., 171:1–10.

Tomasch, J. 1954. Size, distribution, and number of fibres in the human corpus callosum. Anat. Rec., 119:119–135.

Van Essen, D. C. 1979. Visual areas of the mammalian cerebral cortex. Ann. Rev. Neurosci., 2:227–263.

Vogt, C., and O. Vogt. 1919. Allgemeine Ergebnisse unserer Hirnforschung. Vierte Mitteilung: Die physiologische Bedeutung der architektonischen Rindewreizungen. J. Psychol. Neurol. (Leipzig), 25:279–462.

von Economo, C. 1927. *Zellaufbau der Grosshirnrinde des Menschen*. J. Springer, Berlin, 145 pp.

von Economo, C. 1929. *The Cytoarchitectonics of the Human Cerebral Cortex*. Translated by S. Parker. Oxford University Press, London, 186 pp.

Whitsel, B. L., D. A. Dreyer, and J. R. Roppolo. 1971. Determinants of body representation in postcentral gyrus of macaques. J. Neurophysiol., 34:1018–1034.

Witelson, S. F. 1977. Anatomic asymmetry in the temporal lobes: Its documentation, phylogenesis and relationship to functional asymmetry. Ann. N. Y. Acad. Sci., 299:328–354.

Woolsey, C. N., et al. 1952. Patterns of localization in the precentral and "supplementary" motor areas and their relation to the concept of a premotor area. Res. Publ. Assoc. Res. Nerv. Ment. Dis., 30:238–264.

Woolsey, T. A. 1978. Some anatomical bases of cortical somatotopic organization. Brain Behav. Evol., 15:325–371.

Woolsey, T. A., C. Welker, and R. H. Schwartz. 1975. Comparative anatomical studies of the Sm1 face cortex with special reference to the occurrence of "barrels" in layer IV. J. Comp. Neurol., 164:79–94.

Blood Supply of the Brain

Abbie, A. A. 1934. The morphology of the fore-brain arteries, with especial reference to the evolution of the basal ganglia. J. Anat. (Lond.), 68:433–470.

Alexander, L. 1942. The vascular supply of the strio-pallidium. Res. Publ. Assoc. Res. Nerv. Ment. Dis., 21:77–132.

Atkinson, W. J. 1949. The anterior inferior cerebellar artery. Its variations, pontine distribution, and significance in the surgery of cerebello-pontine tumours. J. Neurol. Neurosurg. Psychiatry, 12:137–151.

Billenstien, D. C. 1953. The vascularity of the motor cortex of the dog. Anat. Rec., 117:129–144.

Brightman, M. W., et al. 1973. Osmotic opening of tight junctions in cerebral endothelium. J. Comp. Neurol., 152:317–326.

Busch, W. 1966. Beitrag zur Morphologie und Pathologie der Arteria basalis (Untersuchungsergebnisse bei 1000 Gehirnen). Arch. Psychiatr. Nervenkr., 208:326–344.

Craigie, E. H. 1938. The comparative anatomy and embryology of the capillary bed of the central nervous system. Res. Publ. Assoc. Res. Nerv. Ment. Dis., 18:3–29.

Dahl, E. 1973. The innervation of the cerebral arteries. J. Anat. (Lond.), 115:53–63.

Daniel, P. M. 1966. The blood supply of the hypothalamus and pituitary gland. Brit. Med. Bull., 22:202–208.

DeLand, F. H. 1976. *Cerebral Radionuclide Angiography*. W. B. Saunders, Philadelphia, 309 pp.

Epstein, B. S. 1966. *Pneumoencephalography and Cerebral Angiography*. Year Book Medical Publishers, Chicago, 349 pp.

Gillilan, L. A. 1972. Anatomy and embryology of the arterial system of the brain stem and cerebellum. In *Handbook of Clinical Neurology*, Vol. 11. Edited by P. J. Vinken and G. W. Bruyn. American Elsevier Press, New York, pp. 24–44.

Goldberg, H. I. 1974. The anterior choroidal artery. In *Radiology of the Skull and Brain*, Vol. 2, *Angiography*, Book 2, *Arteries*. Edited by T. H. Newton and D. G. Potts. C. V. Mosby, St. Louis, pp. 1628–1658.

Harnarine-Singh, D., G. Geddes, and J. B Hyde. 1971. Sizes and numbers of arteries and veins in normal human neopallium. J. Anat. (Lond.), 111:171–179.

Hassler, O. 1967. Arterial pattern of the human brainstem. Neurology, 17:368–375.

Heubner, A. 1872. Zur Topographie der Ernährungsgebiete der einzelnen Hirnarterien. Zentralbl. Med. Wiss., 52:817–821.

Hoffman, H. B., M. T. Margolis, and T. H. Newton. 1974. The superior cerebellar artery. Section I. Normal gross and radiographic anatomy. In *Radiology of the Skull and Brain*, Vol. 2, *Angiography*, Book 2, *Arteries*. Edited by T. H. Newton and D. G. Potts. C. V. Mosby, St. Louis, pp. 1809–1830.

Hoyt, W. F., T. H. Newton, and M. T. Margolis. 1974. The posterior cerebral artery. Section I. Embryology and developmental anomalies. In *Radiology of the Skull and Brain*, Vol. 2, *Angiography*, Book 2, *Arteries*. Edited by T. H. Newton and D. G. Potts. C. V. Mosby, St. Louis, pp. 1540–1550.

Jones, E. G. 1970. On the mode of entry of blood vessels into the cerebral cortex. J. Anat. (Lond.), *106*:507–520.

Kaplan, H. A. 1972. Anatomy and embryology of the arterial system of the forebrain. In *Handbook of Clinical Neurology*, Vol. 11. Edited by P. J. Vinken and G. W. Bruyn. American Elsevier Publishing Co., New York, pp. 1–23.

Kaplan, H. A., and D. H. Ford. 1966. *The Brain Vascular System*. American Elsevier Publishing Co., New York, 230 pp.

Krayenbühl, H. A., and M. G. Yaşargil. 1968. *Cerebral Angiography*. Butterworths, London, 401 pp.

Krayenbühl, H., and M. G. Yaşargil. 1972. Radiological anatomy and topography of the cerebral arteries. In *Handbook of Clinical Neurology*, Vol. 11. Edited by P. J. Vinken and G. W. Bruyn. American Elsevier Publishing Co., New York, pp. 65–101.

Lazorthes, G. 1961. *Vascularisation et Circulation Cérébrales*. Masson et Cie., Paris, 323 pp.

Lin, J. P., and I. I. Kricheff. 1974. The anterior cerebral artery complex. Section I. Normal anterior cerebral artery complex. In *Radiology of the Skull and Brain*, Vol. 2, *Angiography*, Book 2, *Arteries*. Edited by T. H. Newton and D. G. Potts. C. V. Mosby, St. Louis, pp. 1391–1410.

Margolis, M. T., and T. H. Newton. 1974. The posterior inferior cerebellar artery. In *Radiology of the Skull and Brain*, Vol. 2, *Angiography*, Book 2, *Arteries*. Edited by T. H. Newton and D. G. Potts. C. V. Mosby, St. Louis, pp. 1710–1774.

Margolis, M. T., T. H. Newton, and W. F. Hoyt. 1974. The posterior cerebral artery. Section II. Gross and roentgenographic anatomy. In *Radiology of the Skull and Brain*, Vol. 2, *Angiography*, Book 2, *Arteries*. Edited by T. H. Newton and D. G. Potts. C. V. Mosby, St. Louis, pp. 1551–1579.

Michotey, P., N. P. Moscow, and G. Salamon. 1974. The middle cerebral artery. Section II. Anatomy of the cortical branches of the middle cerebral artery. In *Radiology of the Skull and Brain*, Vol. 2, *Angiography*, Book 2, *Arteries*. Edited by T. H. Newton and D. G. Potts. C. V. Mosby, St. Louis, pp. 1471–1526.

Moscow, N. P., P. Michotey, and G. Salamon. 1974. The anterior cerebral artery complex. Section II. Anatomy of the cortical branches of the anterior cerebral artery. In *Radiology of the Skull and Brain*, Vol. 2, *Angiography*, Book 2, *Arteries*. Edited by T. H. Newton and D. G. Potts. C. V. Mosby, St. Louis, pp. 1411–1420.

Naidich, T. P., I. I Kricheff, A. E. George, and J. P. Lin. 1976. The normal anterior inferior cerebellar artery. Radiology, *119*:355–373.

Newton, T. H., W. F. Hoyt, and M. T. Margolis. 1974. The posterior cerebral artery. Section III. Pathology. In *Radiology of the Skull and Brain*, Vol. 2, *Angiography*, Book 2, *Arteries*. Edited by T. H. Newton and D. G. Potts. C. V. Mosby, St. Louis, pp. 1580–1627.

Newton, T. H., and M. T. Margolis. 1974. The superior cerebellar artery. Section II. Pathology involving the superior cerebellar artery. In *Radiology of the Skull and Brain*, Vol. 2, *Angiography*, Book 2, *Arteries*. Edited by T. H. Newton and D. G. Potts. C. V. Mosby, St. Louis, pp. 1831–1848.

Nilges, R. G. 1944. The arteries of the mammalian cornu ammonis. J. Comp. Neurol., *80*:177–190.

Nomura, T. 1970. *Atlas of Cerebral Angiography*. Igaku Shoin Ltd., Tokyo, 322 pp.

Otomo, E. 1965. The anterior choroidal artery. Arch. Neurol., *13*:656–658.

Padget, D. M. 1948. The development of the cranial arteries in the human embryo. Carneg. Inst., Contr. Embryol., *32*:205–261.

Ring, B. A. 1974. The middle cerebral artery. Section I. Normal middle cerebral artery. In *Radiology of the Skull and Brain*, Vol. 2, *Angiography*, Book 2, *Arteries*. Edited by T. H. Newton and D. G. Potts. C. V. Mosby, St. Louis, pp. 1442–1470.

Salamon, G. (ed.). 1975. *Advances in Cerebral Angiography: Anatomy, Stereotaxy, Embolization, Computerized Axial Tomography*. Springer-Verlag, Berlin, 378 pp.

Schechter, M. M. 1982. Cerebral angiography. In *Neurological Surgery*, Vol. 1. Edited by J. R. Youmans. W. B. Saunders, Philadelphia, pp. 231–350.

Takahashi, M. 1974. The anterior inferior cerebellary artery. In *Radiology of the Skull and Brain*, Vol. 2, *Angiography*, Book 2, *Arteries*. Edited by T. H. Newton and D. G. Potts. C. V. Mosby, St. Louis, pp. 1796–1808.

Takahashi, M. 1974. The basilar artery. In *Radiology of the Skull and Brain*, Vol. 2, *Angiography*, Book 2, *Arteries*. Edited by T. H. Newton and D. G. Potts. C. V. Mosby, St. Louis, pp. 1775–1795.

Taveras, J. M., and E. H. Wood. 1976. *Diagnostic Neuroradiology*. 2nd ed. Williams and Wilkins, Baltimore, 2 vols.

Wackenheim, A., and J. P. Braun. 1978. *The Veins of the Posterior Fossa: Normal and Pathologic Findings*. Springer-Verlag, Berlin, 157 pp.

Westberg, G. 1963. The recurrent artery of Heubner and the arteries of the central ganglia. Acta Radiol. Diagn. (Stockh.), *1*:949–954.

Westergaard, E., and M. W. Brightman. 1973. Transport of proteins across normal cerebral arterioles. J. Comp. Neurol., *152*:17–44.

Wollschlaeger, G., and P. B. Wollschlaeger. 1974. The circle of Willis. In *Radiology of the Skull and Brain*, Vol. 2, *Angiography*, Book 2, *Arteries*. Edited by T. H. Newton and D. G. Potts. C. V. Mosby, St. Louis, pp. 1171–1201.

Ventricles, Meninges, Cerebrospinal Fluid

Attwood, H. D., and P. A. Wooley. 1974. Arachnoid villi in spinal cord of Dasyurid marsupials. J. Comp. Neurol., *155*:343–354.

Bering, E. A., Jr. 1962. Circulation of the cerebrospinal fluid. J. Neurosurg., *19*:405–413.

Bleier, R. 1975. Surface fine structure of supraependymal elements and ependyma of hypothalamic third ventricle of mouse. J. Comp. Neurol., *161*:555–568.

Bleier, R. 1977. Ultrastructure of supraependymal cells and ependyma of hypothalamic third ventricle of mouse. J. Comp. Neurol., *174*:359–376.

Brawer, J. R. 1972. The fine structure of the ependymal tanycytes at the level of the arcuate nucleus. J. Comp. Neurol., *145*:25–42.

Brightman, M. W. 1965. The distribution within the brain of ferritin injected into cerebrospinal fluid compartments. Amer. J. Anat., *117*:193–219.

Brightman, M. W., and T. S. Reese. 1969. Junctions between intimately apposed cell membranes in the vertebrate brain. J. Cell Biol., *40*:648–677.

Cammermeyer, J. 1971. Median and caudal apertures in

the roof of the fourth ventricle in rodents and primates. J. Comp. Neurol., *141*:499–512.

Cloyd, M. W., and F. N. Low. 1974. Scanning electron microscopy of the subarachnoid space in the dog. I. Spinal cord levels. J. Comp. Neurol., *153*:325–368.

Cutler, R. W. P., L. Page, J. Galicich, and G. V. Watters. 1968. Formation and absorption of cerebrospinal fluid in man. Brain. *91*:707–720.

Davson, H. 1967. *Physiology of the Cerebrospinal Fluid.* J. and A. Churchill, Ltd., London, 445 pp.

DiChiro, G. 1966. Observations on the circulation of the cerebrospinal fluid. Acta Radiol. Diagn. (Stockh.), *5*:988–1002.

DiChiro, G. 1971. *An Atlas of Detailed Normal Pneumo-encephalographic Anatomy,* 2nd ed. Charles C Thomas, Springfield, Ill., 343 pp.

Feindel, W., W. Penfield, and F. McNaughton. 1960. The tentorial nerves and localization of intracranial pain in man. Neurology, *10*:555–563.

Flyger, G., and U. Hjelmquist. 1957. Normal variations in the caliber of the human cerebral aqueduct. Anat. Rec., *127*:151–162.

Fraher, J. P., and R. D. McDougall. 1975. Macrophages related to leptomeninges and ventral nerve roots. J. Anat. (Lond.), *120*:537–549.

Harvey, S. C., and H. S. Burr. 1926. The development of the meninges. Arch. Neurol. Psychiatry, *15*:545–567.

Kasantikul, V., M. G. Netsky, and A. E. James, Jr. 1979. Relation of age and cerebral ventricle size to central canal in man. Morphological analysis. J. Neurosurg., *51*:85–93.

Kido, D. K., D. G. Gomez, A. M. Pavese, Jr., and D. G. Potts. 1976. Human spinal arachnoid villi and granulations. Neuroradiology, *11*:221–228.

Kimmel, D. L. 1961. Innervation of spinal dura mater and dura mater of the posterior cranial fossa. Neurology, *11*:800–809.

Kimmel, D. L. 1961. The nerves of the cranial dura mater and their significance in dural headache and referred pain. Chicago Med. Sch. Quart., *22*:16–26.

Levinger, I. M., and J. Kedem. 1974. A method for the evaluation of the surface area of cerebral ventricles in animals. J. Anat. (Lond.), *117*:481–485.

Malloy, J. J., and F. N. Low. 1974. Scanning electron microscopy of the subarachnoid space in the dog. II. Spinal nerve exits. J. Comp. Neurol., *157*:87–107.

Millen, J. W., and D. H. M. Woollam. 1962. *The Anatomy of the Cerebrospinal Fluid.* Oxford University Press, New York, 151 pp.

Nabeshima, S., T. S. Reese, D. M. D. Landis, and M. W. Brightman. 1975. Junctions in the meninges and marginal glia. J. Comp. Neurol., *164*:127–170.

Nakai, J. 1963. *Morphology of Neuroglia.* Igaku Shoin Ltd., Tokyo, 198 pp.

Newton, T. H. 1968. Cisterns of posterior fossa. Clin. Neurosurg., *15*:190–246.

Oldendorf, W. 1975. Permeability of the blood-brain barrier. In *The Nervous System,* Vol. 1, *The Basic Neurosciences.* Edited by D. Tower. Raven Press, New York, pp. 279–289.

Purdy, J. L., and S. C. Bondy. 1976. Blood-brain barrier: Selective changes during maturation. Neuroscience, *1*:125–129.

Rapoport, S. I. 1976. *Blood-Brain Barrier in Physiology and Medicine.* Raven Press, New York, 316 pp.

Sarnat, H. B., J. F. Campa, and J. M. Lloyd. 1975. Inverse prominence of ependyma and capillaries in the spinal cord of vertebrates: A comparative histochemical study. Amer. J. Anat., *143*:439–450.

Schultz, R. L., E. A. Maynard, and D. C. Pease. 1957. Electron microscopy of neurons and neuroglia of cerebral cortex and corpus callosum. Amer. J. Anat., *100*:369–407.

Strong, L. H. 1956. Early development of the ependyma and vascular pattern of the fourth ventricular choroid plexus in the rabbit. Amer. J. Anat., *99*:249–290.

Sturrock, R. R. 1979. A morphological study of the development of the mouse choroid plexus. J. Anat. (Lond.), *129*:777–793.

Sweet, W. H., and H. B. Locksley. 1953. Formation, flow, and reabsorption of cerebrospinal fluid in man. Proc. Soc. Exp. Biol. Med., *84*:397–402.

Tourtellotte, W. W., and R. J. Shorr. 1982. Cerebrospinal fluid. In *Neurological Surgery,* Vol. 1. Edited by J. R. Youmans. W. B. Saunders, Philadelphia, pp. 423–486.

Westergaard, E. 1972. The fine structure of nerve fibers and endings in the lateral cerebral ventricles of the rat. J. Comp. Neurol., *144*:345–354.

Woollam, D. H. M., and J. W. Millen. 1955. The perivascular spaces of the mammalian central nervous system and their relation to the perineuronal and subarachnoid spaces. J. Anat. (Lond.), *89*:193–200.

Wright, P. M., G. J. Nogueira, and E. Levin. 1971. Role of the pia mater in the transfer of substances in and out of the cerebrospinal fluid. Exp. Brain Res., *13*:294–305.

12

The Peripheral Nervous System

Nervous impulses are conveyed to and from the nervous system by the nerve trunks that form the **peripheral nervous system.** It is composed of nerve fibers, ganglia, and end organs. Afferent or sensory fibers carry toward the central nervous system the impulses arising from stimulation of sensory end organs. Efferent or motor fibers carry impulses from the central nervous system to the muscles and other responsive organs. The somatic fibers, both afferent and efferent, are associated with the general body, typified by the skeletal muscles and skin. The visceral fibers are also both afferent and efferent, and are associated with the internal organs, vessels, and mucous membranes.

The peripheral nervous system consists of (a) the cranial nerves, (b) the spinal nerves, and (c) the sympathetic nervous system. These three morphologic subdivisions are not functionally independent, but combine and communicate with each other to supply both somatic and visceral structures with afferent and efferent fibers. For example, the efferent impulses for control of the viscera are carried partly by the sympathetic system and partly by portions of certain cranial and sacral nerves. The latter are grouped under the name parasympathetic or craniosacral system, and both sympathetic and parasympathetic systems together make up the visceral motor or autonomic nervous system, which is described in later pages.

Cranial Nerves

The **cranial nerves** (*nervi craniales; cerebral nerves*) are attached to the base of the brain (Fig. 11-89) and make their passage from the cranial cavity through various openings or foramina in the skull (Fig. 12-1). There are 12 pairs, and beginning with the most anterior, they are designated by Roman numerals and named as follows:

    I. Olfactory
   II. Optic
  III. Oculomotor
  IV. Trochlear
   V. Trigeminal
  VI. Abducent
 VII. Facial
VIII. Vestibulocochlear
  IX. Glossopharyngeal
   X. Vagus
  XI. Accessory
 XII. Hypoglossal

In addition to the commonly described cranial nerves, there is a small nerve named the **nervus terminalis,** which some comparative anatomists classify **morphologically** as the first cranial nerve. The nervus terminalis (*terminal nerve*) originates from the cerebral hemisphere in the region of the olfactory trigone, courses anteriorly along the medial surface of the olfactory tract and bulb to the lateral surface of the crista galli, and passes through the anterior part of the cribriform plate of the ethmoid bone. It is a compact bundle beside the olfactory tract, a close plexus beside the bulb, and a loose plexus on the crista galli, where it is embedded in the dura mater some distance above the cribriform bone. Within the cranium, fila-

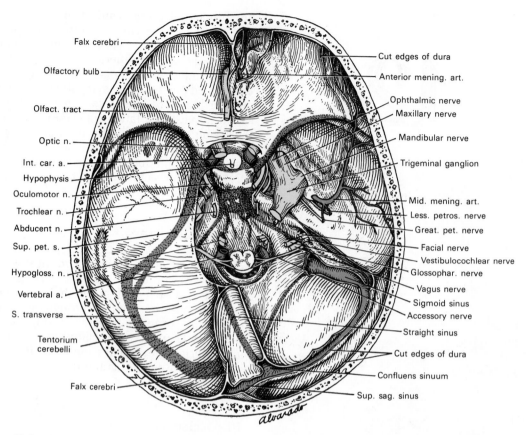

Fig. 12-1.   Interior of base of skull, showing dura mater, dural sinuses and exit of cranial nerves. (Redrawn from Töndury.)

ments join the olfactory bundles and *vomeronasal nerves*, and apparently pass to the mucous membrane of the septum along with them. The majority of filaments form a single strand that passes through the cribriform plate anterior to the vomeronasal nerves and is distributed to the membrane near the anterior superior border of the nasal septum. In the nasal cavity, the nerve communicates with the medial nasal branch of the anterior ethmoidal branch of the ophthalmic division of the trigeminal nerve.

In the embryo, a loose mass of fibers and cells in the portion of the nerve medial to the rostral end of the olfactory bulb is called the **ganglion terminale.** The nerve fibers are unmyelinated and ganglion cells are scattered in groups along the peripheral course of the nerve. The cells in the ganglion and along the nerve have been described by different authors as unipolar, bipolar, and multipolar (Pearson, 1941).

The central connections of the nervus terminalis end in the septal nuclei, the olfactory lobe, the posterior precommissural region, and the anterior portion of the supraoptic region of the brain. Those in the first three areas are sensory; those in the supraoptic region may be preganglionic autonomic fibers associated with the serous glands (of Bowman[1]) or with blood vessels in the olfactory membrane (Larsell, 1950).

I. OLFACTORY NERVE

The **olfactory nerve** (*nervus olfactorius; first nerve*) (Fig. 12-2), or nerve of smell, in its properly restricted sense is represented on both sides of the nasal cavity by bundles of nerve fibers which are the central processes

[1]Sir William Bowman (1816–1892): An English physician (London).

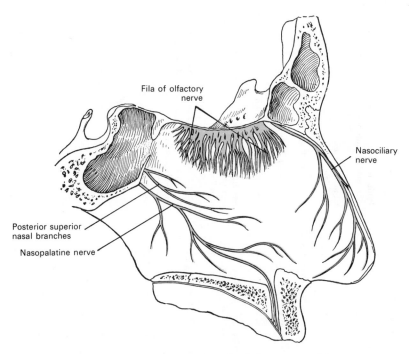

Fila of olfactory
nerve

Nasociliary
nerve

Posterior superior
nasal branches

Nasopalatine nerve

FIG. 12-2. The nerves of the right side of the septum of the nose.

of the neuroepithelial cells in the olfactory mucous membrane of the nose. These cells are described in more detail as part of the organ of smell in Chapter 13. The nerve fibers make a plexiform network in the mucous membrane of the superior nasal concha and the part of the septum opposite, and then are gathered into approximately 20 bundles that pass through the foramina of the cribriform plate of the ethmoid bone in two groups. A lateral group comes from the concha; a medial group, from the septum. The fibers are unmyelinated, and the bundles have connective tissue sheaths derived from the tissues of the dura, arachnoid, and pia mater. After passing through the foramina, the fibers end in the glomeruli of the olfactory bulb, an oval mass approximately 3 mm × 15 mm, which rests against the intracranial surface of the cribriform plate of the ethmoid (Fig. 12-1).

Although the olfactory bulb and its connection with the brain, the olfactory tract, have a gross appearance like that of a nerve, they are more accurately classified as parts of the brain. They are components of the rhinencephalon, the portion of the brain associated with the sense of smell, which is described in Chapter 11.

The **vomeronasal nerve,** which is an important part of the olfactory system in macrosmatic animals, is present in the human fetus but disappears before birth. The nerve fibers originate in the olfactory epithelial cells of the vomeronasal organ (organ of Jacobson[1]), a rudiment of which may persist in man; they pass upward in the submucous tissue of the nasal septum and through the cribriform plate of the ethmoid bone to end in the accessory olfactory bulb.

II. OPTIC NERVE

The **optic nerve** (*n. opticus; second nerve*) (Fig. 12-3), or nerve of sight, consists mainly of the axons or central processes of the cells in the ganglionic layer of the retina. Within the bulb of the eye, these axons lie in the stratum opticum or layer of nerve fibers of the retina (see next chapter). They converge toward the optic papilla, or disc, which is 3 mm medial to the posterior pole of the bulb, where they are gathered into small bundles that pierce the choroid and scleral

[1]Ludwig Levin Jacobson (1783–1843): A Danish anatomist and physician (Copenhagen).

Outline of the Cranial Nerves

Nerves	Components	Function	Central Connection	Cell Bodies	Peripheral Distribution
I. Olfactory	Afferent Special visceral	Smell	Olfactory bulb and tract	Olfactory epithelial cells	Olfactory nerves
II. Optic	Afferent Special somatic	Vision	Optic nerve and tract	Ganglion cells of retina	Rods and cones of retina
III. Oculomotor	Efferent Somatic	Ocular movement	Nucleus III	Nucleus III	Branches to levator palpebrae; rectus superior, medius, inferior; obliquus inferior
	Efferent General visceral	Contraction of pupil and accommodation	Nucleus of Edinger-Westphal	Nucleus of Edinger-Westphal	Ciliary ganglion; ciliaris and sphincter pupillae
	Afferent Proprioceptive	Muscular sensibility	Nucleus mesencephalicus V	Nucleus mesencephalicus V	Sensory endings in ocular muscles
IV. Trochlear	Efferent Somatic	Ocular movement	Nucleus IV	Nucleus IV	Branches to obliquus superior
	Afferent Proprioceptive	Muscular sensibility	Nucleus mesencephalicus V	Nucleus mesencephalicus V	Sensory endings in obliquus superior
V. Trigeminal	Afferent General somatic	General sensibility	Trigeminal sensory nucleus	Trigeminal ganglion (Gasserian)	Sensory branches of ophthalmic, maxillary and mandibular nerves to skin and mucous membranes of face and head
	Efferent Special visceral	Mastication	Motor V nucleus	Motor V nucleus	Branches to temporalis, masseter, pterygoid, mylohyoid, digastric, tensores tympani and palatini muscles
	Afferent Proprioceptive	Muscular sensibility	Nucleus mesencephalicus V	Nucleus mesencephalicus V	Sensory endings in muscles of mastication
VI. Abducent	Efferent Somatic	Ocular movement	Nucleus VI	Nucleus VI	Branches to rectus lateralis
	Afferent Proprioceptive	Muscular sensibility	Nucleus mesencephalicus V	Nucleus mesencephalicus V	Sensory endings in rectus lateralis
VII. Facial	Efferent Special visceral	Facial expression	Motor VII nucleus	Motor VII nucleus	Branches to facial muscles, stapedius, stylohyoid, digastric muscles
	Efferent General visceral	Glandular secretion	Nucleus salivatorius	Nucleus salivatorius	Greater petrosal nerve, pterygopalatine ganglion, with branches of maxillary V to glands of nasal mucosa. Chorda tympani, lingual nerve, submandibular ganglion, submandibular and sublingual glands
	Afferent Special visceral	Taste	Nucleus tractus solitarius	Geniculate ganglion	Chorda tympani, lingual nerve, taste buds, anterior tongue
	Afferent General visceral	Visceral sensibility	Nucleus tractus solitarius	Geniculate ganglion	Great petrosal, chorda tympani and branches

Nerve		Function	Central connection	Central connection	Distribution
VIII. Vestibulo-cochlear	Afferent General somatic	Cutaneous sensibility	Nucleus spinal tract of V	Geniculate ganglion	With auricular branch of vagus to external ear and mastoid region
	Afferent Special somatic	Hearing	Cochlear nuclei	Spiral ganglion	Organ of Corti in cochlea
	Afferent Proprioceptive	Sense of equilibrium	Vestibular nuclei	Vestibular ganglion	Semicircular canals, saccule, and utricle
IX. Glosso-pharyngeal	Afferent Special visceral	Taste	Nucleus tractus solitarius	Inferior ganglion IX	Lingual branches, taste buds, posterior tongue
	Afferent General visceral	Visceral sensibility	Nucleus tractus solitarius	Inferior ganglion IX	Tympanic nerve to middle ear, branches to pharynx and tongue, carotid sinus nerve
	Efferent General visceral	Glandular secretion	Nucleus salivatorius	Nucleus salivatorius	Tympanic, lesser petrosal nerves, otic ganglion, with auriculotemporal V to parotid gland
	Efferent Special visceral	Swallowing	Nucleus ambiguus	Nucleus ambiguus	Branch to stylopharyngeus muscle
X. Vagus	Efferent General visceral	Involuntary muscle and gland control	Dorsal motor nucleus X	Dorsal motor nucleus X	Cardiac nerves and plexus; ganglia on heart. Pulmonary plexus; ganglia, respiratory tract. Esophageal, gastric, celiac plexuses; myenteric and submucous plexuses, muscles and glands of digestive tract down to transverse colon
	Efferent Special visceral	Swallowing and phonation	Nucleus ambiguus	Nucleus ambiguus	Pharyngeal branches, superior and inferior laryngeal nerves
	Afferent General visceral	Visceral sensibility	Nucleus tractus solitarius	Inferior ganglion X	Fibers in all cervical, thoracic, and abdominal branches; carotid and aortic bodies
	Afferent Special visceral	Taste	Nucleus tractus solitarius	Inferior ganglion X	Branches to region of epiglottis and taste buds
	Afferent General somatic	Cutaneous sensibility	Nucleus spinal tract V	Superior ganglion X	Auricular branch to external ear and meatus
XI. Accessory	Efferent Special visceral	Swallowing and phonation	Nucleus ambiguus	Nucleus ambiguus	Bulbar portion, communication with vagus, in vagus branches to muscles of pharynx and larynx
	Efferent Special somatic	Movements of shoulder and head	Lateral column of upper cervical spinal cord	Lateral column of upper cervical spinal cord	Spinal portion, branches to sternocleidomastoid and trapezius muscles
XII. Hypoglossal	Efferent General somatic	Movements of tongue	Nucleus XII	Nucleus XII	Branches to extrinsic and intrinsic muscles of tongue

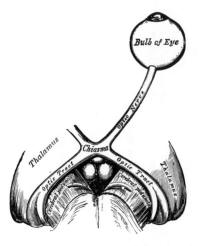

FIG. 12-3.   The left optic nerve and the optic tracts.

coats through the many small foramina of the lamina cribrosa sclerae to form the optic nerve. The optic nerve, as it courses posteriorly toward the brain, traverses the central region of the orbit, passes through the optic canal (foramen), and then, approaching the nerve of the other side, joins it to form the optic chiasma. From the chiasma, the fibers continue in the optic tracts, which diverge from each other and reach the base of the brain near the cerebral peduncle. The central connections of the fibers are described in Chapter 11.

Parts of the Optic Nerve

The optic nerve has four portions, contained in: (a) the bulb, (b) the orbit, (c) the optic canal, and (d) the cranial cavity.

INTRAOCULAR PORTION.   The intraocular portion is very short, about 1 mm, and the nerve fibers within it are unmyelinated until they pass through the lamina cribrosa, at which point they become myelinated and supported by neuroglia.

ORBITAL PORTION.   The orbital portion, 3 to 4 mm in diameter, is 20 to 30 mm long; its course is slightly sinuous, which allows greater length for unrestricted movement of the eyeball. It is invested by sheaths derived from the dura, arachnoid, and pia mater, all three of which fuse and become continuous with the sclera at the lamina cribrosa; the dura extends as far back as the cranial cavity, the arachnoid somewhat farther, and the pia all the way to the chiasma. The pia

closely ensheathes the nerve and sends numerous septa into its substance, carrying the blood supply. Between the sheaths are subarachnoid and subdural spaces similar to and continuous with those of the cranial cavity, the subarachnoid ends in a cul-de-sac at the lamina cribrosa. As it traverses the orbit, the optic nerve is surrounded by the posterior part of the fascia bulbi (Tenon's capsule[1]), the orbital fat, and, in its anterior two-thirds, by the ciliary nerves and arteries. Toward the posterior part of the orbit, it is crossed obliquely by the nasociliary nerve, the ophthalmic artery, the superior ophthalmic vein, and the superior division of the oculomotor nerve. Farther toward the roof of the orbit are the rectus superior and levator palpebrae superioris muscles, and the trochlear and frontal nerves. Inferior to it are the inferior division of the oculomotor nerve and the rectus inferior muscle; medial to it is the rectus medialis muscle; and lateral to it are the abducent nerve and rectus lateralis muscle. In the posterior part of the orbit are the ciliary ganglion and the ophthalmic artery (Fig. 12-6). As it passes into the optic canal, accompanied by the ophthalmic artery, it is surrounded by the common anular tendon (of Zinn[2]), which serves as the origin of the ocular muscles. A short distance behind the bulb, the optic nerve is pierced by the central artery of the retina and its accompanying vein. These vessels enter the bulb through an opening in the lamina cribrosa and supply the retina.

OPTIC CANAL PORTION.   Within the optic canal, the ophthalmic artery lies inferior to the nerve, just after it has branched from the internal carotid artery. Medially, separated by a thin plate of bone, is the sphenoidal air sinus. If the pneumatization of the bone is extensive, the nerve may be almost completely surrounded by the sphenoidal sinus or the posterior ethmoidal cells. Superior to the nerve, the three sheaths—dura, arachnoid, and pia—are fused to each other, to the nerve, and to the periosteum of the bone, thereby fixing the nerve and preventing it from being forced back and forth in the foramen. The subarachnoid and subdural spaces are present only below the nerve.

[1] Jacques René Tenon (1724–1816): A French pathologist and surgeon (Paris).
[2] Johann Gottfried Zinn (1727–1759): A German professor of medicine and botanist (Göttingen).

INTRACRANIAL PORTION. The intracranial portion of the nerve (Fig. 12-1) rests on the anterior part of the cavernous sinus and on the diaphragma sellae, which overlies the hypophysis. The part of the brain above the nerve is the anterior perforated substance. The internal carotid artery approaches the nerve laterally and then is directly inferior to it at the origin of the ophthalmic artery. The anterior cerebral artery crosses superior to it.

Optic Chiasma

The **optic chiasma** (Fig. 12-3), as the name indicates, resembles the Greek letter "X" from the convergence of the optic nerves in front and the divergence of the optic tracts behind. It rests upon the tuberculum sellae of the sphenoid bone and on the diaphragma sellae of the dura. It is continuous superiorly with the lamina terminalis, and posteriorly, with the tuber cinereum and the infundibulum of the hypophysis. Lateral to the optic chiasma, on each side, are the anterior perforated substance and the internal carotid artery.

Optic Tract

The **optic tract** leaves the chiasma and diverges from its fellow of the opposite side until it reaches the cerebral peduncle, where it winds obliquely across the undersurface. It is adjacent to the tuber cinereum and peduncle, thus having contact with the third ventricle. As it terminates, a shallow groove divides it into a medial and lateral root. The lateral root contains fibers of visual function and ends in the lateral geniculate body. The medial root carries fibers that decussate (commissure of Gudden[1] and probably other nonvisual commissures, those of Meynert[2] and Darkschewitsch[3]).

The fibers lying medially in the optic nerve cross in the chiasma and continue in the optic tract of the other side. The lateral fibers are uncrossed and continue to the brain in the optic tract of the same side. The

[1]Bernhard Aloys von Gudden (1824–1886): A Swiss professor of psychiatry (Zurich).
[2]Theodor Herman Meynert (1833–1892): An Austrian professor of neurology and psychiatry (Vienna).
[3]Liverij Osipovich Darkschewitsch (1858–1925): A Russian neurologist (St. Petersburg).

fibers from the two sides become intermingled in the tract.

The optic nerve corresponds to a tract of fibers within the brain rather than to the other cranial nerves because of its embryologic development and its structure. It develops from a diverticulum of the lateral aspect of the forebrain; its fibers probably are third in the chain of neurons from the receptors to the brain. The fibers are supported by neuroglia instead of neurilemmal sheaths, and the nerve has three sheaths prolonged from the corresponding meninges of the brain (see Chapter 13).

III. OCULOMOTOR NERVE

The **oculomotor nerve** (Figs. 12-6; 12-7) supplies the levator palpebrae superioris muscle, the extrinsic muscles of the eye except the obliquus superior and rectus lateralis, and the intrinsic muscles with the exception of the dilator pupillae. Its superficial origin is from the midbrain at the oculomotor sulcus on the medial side of the cerebral peduncle. Its deep origin and central connections within the brain are described in Chapter 11. The oculomotor nerve is traditionally considered a motor nerve, containing special somatic efferent fibers for the ocular muscles and parasympathetic fibers for the ciliary ganglion, but experimental evidence indicates that it contains proprioceptive fibers from cells in the brain stem near the motor cells (Corbin and Oliver, 1942).

As it emerges from the brain into the posterior cranial fossa, the nerve is invested with pia mater, and is bathed in the cerebrospinal fluid of the cisterna interpeduncularis (basalis). It passes between the superior cerebellar and posterior cerebral arteries near the termination of the basilar artery. It is covered with arachnoid lateral to the posterior clinoid process, and between the anterior and posterior clinoid processes it pierces the dura (Fig. 12-1) by passing between the free and attached borders of the tentorium cerebelli. It runs anteriorly, embedded in the lateral wall of the cavernous sinus (Fig. 12-13), superior to the other orbital nerves. It enters the superior orbital fissure between the two heads of the rectus lateralis (Fig. 12-4) in two divisions, a superior and an inferior. Within the fissure, the trochlear, lacrimal, and fron-

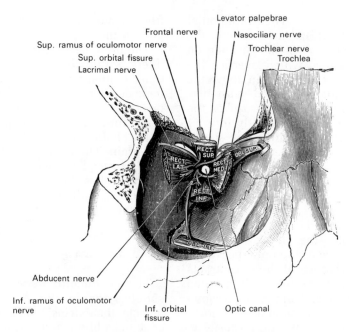

Fig. 12-4.   Dissection showing origins of right ocular muscles, and nerves entering through the superior orbital fissure.

tal nerves lie superior to it, the abducent nerve lies inferior and lateral to it, and the nasociliary nerve passes between its two divisions.

Two **communications** join the oculomotor nerve as it runs along the wall of the cavernous sinus: (1) one from the **cavernous plexus** of the sympathetic system, carrying postganglionic fibers from the superior cervical ganglion; and (2) one from the **ophthalmic division** of the trigeminal nerve.

The **superior division** (Fig. 12-6), smaller than the inferior, passes medialward across the optic nerve and sends branches to the **rectus superior** and the **levator palpebrae superioris muscles.** The branch to the levator contains sympathetic postganglionic fibers from the cavernous plexus, which continue anteriorly to reach the nonstriated muscle attached to the superior tarsus.

The **inferior division** has four branches: to the **rectus medialis,** passing inferior to the optic nerve; to the **rectus inferior;** and to the **obliquus inferior.** The latter runs forward between the inferior and lateral recti muscles and gives off a short, rather thick branch, the **root of the ciliary ganglion.**

CILIARY GANGLION.   The ciliary ganglion (*ophthalmic or lenticular ganglion*) (Figs. 12-5; 12-6) is a small parasympathetic gan-

glion (1 or 2 mm in diameter) whose preganglionic fibers come from the oculomotor nerve and whose postganglionic fibers carry motor impulses to the ciliaris and sphincter pupillae muscles. It is situated about 1 cm from the posterior boundary of the orbit, close to the lateral surface of the optic nerve and between it and the rectus lateralis muscle and the ophthalmic artery.

The **parasympathetic motor root** (*radix oculomotorii; short root*) of the ganglion is a rather short, thick nerve, which may be double. It comes from the branch of the inferior division of the oculomotor nerve that supplies the obliquus inferior and is connected with the posterior inferior angle of the ganglion. It contains preganglionic parasympathetic fibers from the Edinger-Westphal group of cells of the third nerve nucleus, which form synapses in the ganglion with cells whose postganglionic fibers pass through the short ciliary nerves into the bulb.

Two communications were formerly called roots of the ganglion: (1) a communication with the **nasociliary nerve** (*long root; sensory root*) joins the posterior superior angle of the ganglion; it contains sensory fibers that traverse the short ciliary nerves on their way from the cornea, iris, and ciliary body,

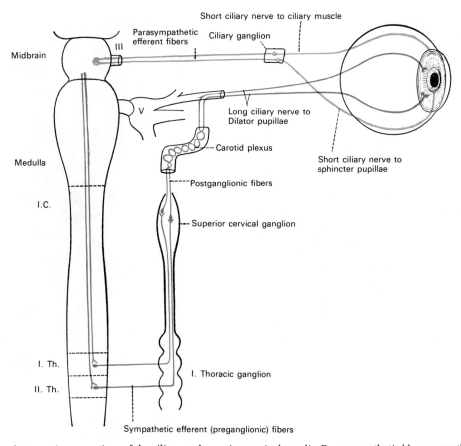

FIG. 12-5. Autonomic connections of the ciliary and superior cervical ganglia. Parasympathetic blue; sympathetic red.

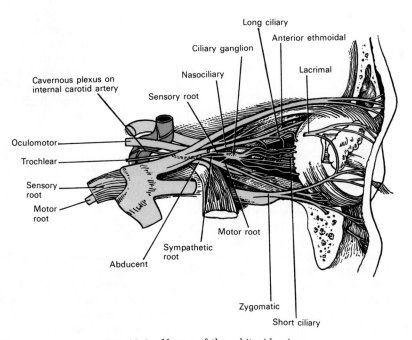

FIG. 12-6. Nerves of the orbit, side view.

without forming synapses in the ganglion; (2) a communication with the **sympathetic** system (*sympathetic root*) that is a slender filament from the cavernous plexus which frequently blends with the communication with the nasociliary nerve. It contains postganglionic fibers from the superior cervical ganglion that pass through the ciliary ganglion without forming synapses and reach the dilator pupillae muscle and the blood vessels of the bulb through the short ciliary nerves.

The branches of the ganglion are the **short ciliary nerves,** which are six to ten delicate filaments that leave the anterior part of the ganglion in two bundles, superior and inferior. They run anteriorly with the ciliary arteries in a wavy course, one set above and the other below the optic nerve, and they are accompanied by the long ciliary nerves from the nasociliary. They pierce the sclera at the posterior part of the bulb, pass anteriorly in delicate grooves on the inner surface of the sclera, and are distributed to the ciliary body, iris, and cornea. The parasympathetic postganglionic fibers supply the ciliaris and sphincter pupillae muscles. The short ciliary nerves also carry sympathetic postganglionic fibers to the dilator pupillae muscle, and sensory fibers from the cornea, iris, and ciliary body.

IV. TROCHLEAR NERVE

The **trochlear nerve** (Fig. 12-7) is the smallest of the cranial nerves and supplies the obliquus superior oculi muscle. Its superficial origin is from the surface of the anterior medullary velum at the side of the frenulum veli, immediately posterior to the inferior colliculus. Its deep origin and central connections are described in Chapter 11.

The nerve is directed laterally across the superior cerebellar peduncle, and it winds around the cerebral peduncle near the pons. It runs anteriorly along the free border of the tentorium cerebelli for 1 or 2 cm (Fig. 12-1), pierces the dura just posterior to the posterior clinoid process, and without changing the direction of its course, takes its place in the lateral wall of the cavernous sinus (Fig. 12-13) between the oculomotor nerve and the ophthalmic division of the trigeminal nerve, to which it becomes firmly attached

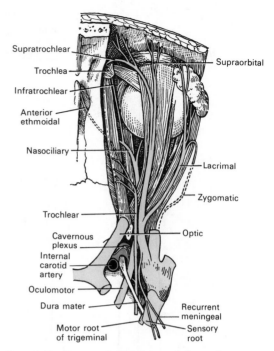

FIG. 12-7. Nerves of the orbit, seen from above.

by connective tissue. It crosses the oculomotor nerve in a slightly upward course, enters the orbit through the superior orbital fissure, and becomes superior to the other nerves. In the orbit it passes medialward, above the origin of the levator palpebrae superioris (Fig. 12-4), and finally enters the orbital surface of the obliquus superior.

In the lateral wall of the cavernous sinus the trochlear nerve forms communications with the cavernous plexus of the sympathetic and with the ophthalmic division of the trigeminal nerve.

VARIATIONS. Occasionally sensory branches appear to come from the trochlear nerve, but they are probably aberrant fibers that have come from the ophthalmic nerve.

V. TRIGEMINAL NERVE

The **trigeminal nerve** is the largest of the cranial nerves. It is the great cutaneous sensory nerve of the face, the sensory nerve to the mucous membranes and other internal structures of the head, and the motor nerve of the muscles of mastication.

Roots

The trigeminal nerve traditionally has two roots, a larger sensory root and a smaller motor root (Fig. 12-6). Investigations have also revealed a third or intermediate root.

The **sensory root** is composed of a large number of fine bundles in close association with each other that enter the pons through the lateral part of its ventral surface. The nerve fibers in the sensory root are the central processes of the ganglion cells in the trigeminal (semilunar) ganglion.

The **motor root** emerges in several bundles from the surface of the pons approximately 2 to 5 mm medial and anterior (rostral) to the sensory root. In addition to the motor fibers, the motor root includes proprioceptive sensory fibers from the mesencephalic central nucleus of the nerve (Corbin and Harrison, 1940).

The **intermediate root,** composed of one or more bundles, has an origin from the surface of the pons quite distinct from that of the other two roots (Jannetta and Rand, 1966). The bundles arise between the motor and sensory roots and are separated from them by 1 to 5 mm (Vidic and Stephanos, 1969). The internal fiber connections, function, and surgical importance of this root are not yet explained.

The three roots pass anteriorly in the posterior cranial fossa, under the shadow of the tentorium where the latter is attached to the petrous portion of the temporal bone, to the trigeminal ganglion. The deep origin and central connections of these roots are described in Chapter 11.

Trigeminal Ganglion

The **trigeminal ganglion** (*ganglion semilunare; Gasserian ganglion*[1]) (Fig. 12-8) lies in a pocket of dura mater (*cavum Meckelii*[2]) that occupies the trigeminal impression near the apex of the petrous portion of the temporal bone (Burr and Robinson, 1925). The ganglionic mass is flat and has a semilunar shape; it is approximately 1 cm × 2 cm (Fig. 12-8). The central processes of the cells leave

the concavity and the peripheral fibers leave the convexity of the crescent.

The ganglion is lateral to the posterior part of the cavernous sinus and the internal carotid artery at the foramen lacerum, and the central fibers pass inferior to the superior petrosal sinus. The motor root, being medial to the sensory, passes beneath the ganglion (that is, between it and the petrous bone), and leaves the skull through the foramen ovale with the mandibular nerve. The greater petrosal nerve also passes between the ganglion and the bone. The ganglion receives filaments from the carotid plexus of the sympathetic system, and it gives off minute branches to the tentorium cerebelli and to the dura mater of the middle cranial fossa.

The peripheral fibers from the ganglion are collected into three large divisions: (1) the ophthalmic, (2) the maxillary, and (3) the mandibular nerves. The ophthalmic and maxillary nerves remain sensory, but the mandibular nerve becomes mixed, being joined by the motor root just outside the skull.

Four small ganglia are associated with these nerves: the *ciliary ganglion* with the ophthalmic, the *pterygopalatine* with the maxillary, and the *otic* and *submandibular* with the mandibular. These ganglia are not a part of the trigeminal complex functionally, however, and the trigeminal fibers that communicate with them pass through without synapses. The ganglia are parasympathetic and are therefore described with the nerves that supply their motor roots with preganglionic fibers (viz., the ciliary ganglion with the oculomotor nerve, the pterygopalatine and submandibular ganglia with the facial nerve, and the otic ganglion with the glossopharyngeal nerve.

Ophthalmic Nerve

The **ophthalmic nerve** (Figs. 12-6; 12-7), or first division of the trigeminal, leaves the anterior superior part of the trigeminal ganglion and enters the orbit through the superior orbital fissure. It is a flat band about 2.5 cm long and lies in the lateral wall of the cavernous sinus, inferior to the oculomotor and trochlear nerves (Fig. 12-13). It is a sensory nerve, supplying the bulb of the eye, conjunctiva, lacrimal gland, part of the mucous membrane of the nose and paranasal

---

[1]Johann Laurenz Gasser (?–1765): An Austrian anatomist (Vienna).
[2]Johann Friedrich Meckel (the Elder, 1714–1774): A German anatomist and obstetrician (Berlin).

sinuses, and the skin of the forehead, eyelids, and nose.

The ophthalmic nerve is joined by communicating sympathetic filaments from the **cavernous plexus** and communicates with the **oculomotor, trochlear,** and **abducent nerves.** It gives off a recurrent filament to the dura mater, and just before it passes through the superior orbital fissure it divides into three other branches: frontal, lacrimal, and nasociliary.

BRANCHES AND COMMUNICATIONS OF OPHTHALMIC NERVE

Tentorial Branch

The tentorial branch (Fig. 12-7, *recurrent filament*) arises near the ganglion, passes across and adheres to the trochlear nerve, and runs between the layers of the tentorium, to which it is distributed.

Lacrimal Nerve

The **lacrimal nerve** (Fig. 12-7) is the smallest of the three branches of the ophthalmic nerve. It passes anteriorly in a separate tube of dura mater, and enters the orbit through the narrowest part of the superior orbital fissure. In the orbit it runs along the superior border of the rectus lateralis muscle close to the periorbita, and enters the lacrimal gland with the lacrimal artery, giving off filaments to supply the gland and adjacent conjunctiva. Finally, it pierces the orbital septum and ends in the skin of the upper eyelid, joining with filaments of the facial nerve.

In the orbit, through a communication with the **zygomatic branch of the maxillary nerve** (Figs. 12-6; 12-8), it receives postganglionic parasympathetic fibers, which are the secretomotor fibers for the lacrimal gland. These fibers pass from their cells of origin in the pterygopalatine ganglion through the pterygopalatine nerves to the maxillary nerve, then along the zygomatic and zygomaticotemporal nerves, finally traversing the communication just mentioned and being distributed with the branches of the lacrimal nerve to the gland. The preganglionic fibers reach the ganglion from the facial nerve by way of the greater petrosal nerve.

VARIATIONS. The lacrimal nerve is occasionally absent, and its place is then taken by the zygomaticotemporal branch of the maxillary nerve. Sometimes the latter is absent and a communication of the lacrimal is substituted for it.

Frontal Nerve

The **frontal nerve** (Fig. 12-7) is the largest branch of the ophthalmic nerve, and may be regarded, both from its size and direction, as the continuation of the nerve. It enters the orbit through the superior orbital fissure, continues rostrally between the levator palpebrae superioris and the periorbita, and at a variable distance approximately halfway to the supraorbital margin, it divides into a large supraorbital and a small supratrochlear branch.

SUPRATROCHLEAR NERVE. The supratrochlear nerve (Fig. 12-7) bends medially to pass superior to the pulley of the obliquus superior muscle and gives off a filament that communicates with the infratrochlear branch of the nasociliary nerve. It pierces the orbital fascia, sends filaments to the conjunctiva and skin of the medial part of the upper lid, passes deep to the corrugator and frontalis muscles, and divides into branches that pierce the muscles to supply the skin of the lower and mesial part of the forehead (Fig. 12-10).

SUPRAORBITAL NERVE. The supraorbital nerve (Fig. 12-7), the continuation of the frontal nerve, leaves the orbit through the supraorbital notch or foramen. It gives filaments to the upper lid and continues upon the forehead, dividing into medial and lateral branches beneath the frontalis muscle (Fig. 12-10). The **medial branch,** sometimes called the **frontal branch,** is smaller; it pierces the muscle and supplies the scalp as far as the parietal bone. The larger **lateral branch** pierces the galea aponeurotica and supplies the scalp nearly as far back as the lambdoidal suture (Fig. 12-18). The supraorbital nerve may divide before leaving the orbit, in which case the lateral branch occupies the supraorbital notch or foramen and the medial or frontal branch may have a notch of its own.

FRONTAL SINUS BRANCH. In the supraorbital notch, a small filament pierces the bone to supply the mucous membrane of the frontal sinus.

Nasociliary Nerve

The **nasociliary nerve** (Fig. 12-7) is intermediate in size between the frontal and lacrimal nerves and is more deeply placed in the orbit. It enters the orbit between the two heads of the rectus lateralis muscle, and between the superior and inferior divisions of the oculomotor nerve. It passes across the optic nerve and runs obliquely inferior to the rectus superior and obliquus superior muscles to the medial wall of the orbital cavity. Here it passes through the anterior ethmoidal foramen as the **anterior ethmoidal nerve** and enters the cranial cavity just superior to the cribriform plate of the ethmoid bone. It runs along a shallow groove on the lateral margin of the plate, and penetrating the bone through a slit at the side of the crista galli, it enters the nasal cavity. It supplies branches to the mucous membrane of the nasal cavity (Figs. 12-2; 12-12). Finally, it emerges between the inferior border of the nasal bone and the lateral nasal cartilage as the external nasal branch.

BRANCHES OF THE NASOCILIARY NERVE. The **communication** with the **ciliary ganglion** (*long or sensory root of the ganglion*) (Fig. 12-6) usually arises from the nasociliary nerve between the two heads of the rectus lateralis muscle. It runs anteriorly on the lateral side of the optic nerve and enters the posterior superior angle of the ciliary ganglion. It contains sensory fibers that pass through the ganglion without synapses and continue on into the bulb by way of the short ciliary nerves. It is sometimes joined by a filament from the cavernous plexus of sympathetic fibers or from the superior ramus of the oculomotor nerve. The ciliary ganglion is described with the oculomotor nerve.

Two or three **long ciliary nerves** are given off from the nasociliary nerve as it crosses the optic nerve. They accompany the short ciliary nerves from the ciliary ganglion, pierce the posterior part of the sclera, and running anteriorly between it and the choroid, are distributed to the iris and cornea. In addition to afferent fibers, the long ciliary nerves probably contain sympathetic fibers from the superior cervical ganglion to the dilator pupillae muscle, which passes through the communication between the cavernous plexus and the ophthalmic nerve.

The **infratrochlear nerve** (Fig. 12-7) is given off from the nasociliary nerve just before it enters the anterior ethmoidal foramen. It runs anteriorly along the superior border of the rectus medialis muscle and is joined near the pulley of the obliquus superior by a filament from the supratrochlear nerve. It then passes to the medial angle of the eye and supplies the skin of the eyelids and side of the nose, the conjunctiva, the lacrimal sac, and the caruncula lacrimalis (Fig. 12-10).

The **ethmoidal branches** supply the mucous membrane of the sinuses. The **posterior ethmoidal nerve** leaves the orbit through the posterior ethmoidal foramen and supplies the posterior ethmoidal and the sphenoidal sinuses. The **anterior ethmoidal branches** are filaments that are given off as the nerve passes through the anterior ethmoidal foramen; they supply the anterior ethmoidal and frontal sinuses.

The **internal nasal branches** supply the mucous membrane of the anterior part of the septum and lateral wall of the nasal cavity (Figs. 12-2; 12-12).

The **external nasal branch** emerges between the nasal bone and the lateral nasal cartilage, passes deep to the nasalis muscle, and supplies the skin of the ala and the apex of the nose (Fig. 12-8).

Maxillary Nerve

The **maxillary nerve** (*superior maxillary nerve*) (Fig. 12-8), or second division of the trigeminal, arises from the middle of the trigeminal ganglion. It is intermediate between the other two divisions in size and position, and like the ophthalmic nerve, it is entirely sensory. It supplies the skin of the middle portion of the face, lower eyelid, side of the nose, and the upper lip (Fig. 12-10); the mucous membrane of the nasopharynx, maxillary sinus, soft palate, tonsil and roof of the mouth; and the upper gums and teeth. It passes horizontally rostralward, at first in the inferior part of the lateral wall of the cavernous sinus and then beneath the dura to the foramen rotundum, through which it leaves the cranial cavity. From the foramen rotundum, it crosses the pterygopalatine fossa, inclines lateralward in a groove on the posterior surface of the maxilla, and enters the orbit through the inferior orbital fissure.

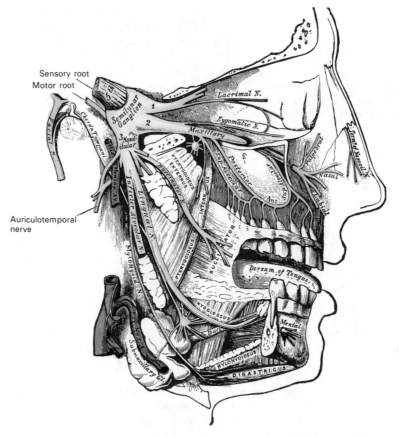

FIG. 12-8.   Distribution of the maxillary and mandibular nerves, and the submandibular ganglion.

In the posterior part of the orbit it becomes the **infraorbital nerve,** lies in the infraorbital groove, and, continuing rostralward, dips into the infraorbital canal. It emerges into the face through the infraorbital foramen, where it is deep to the levator labii superioris, and divides into branches for the skin of the face, nose, lower eyelid, and upper lip.

The branches of the maxillary nerve may be divided into four groups, those given off: (1) in the cranium, (2) in the pterygopalatine fossa, (3) in the infraorbital canal, and (4) in the face.

BRANCHES IN THE CRANIUM

The **middle meningeal nerve** (*meningeal or dural branch*) is given off from the maxillary nerve directly after its origin from the trigeminal (semilunar) ganglion; it accompanies the middle meningeal artery and supplies the dura mater.

BRANCHES IN THE PTERYGOPALATINE FOSSA

Zygomatic Nerve

The **zygomatic nerve** (*temporomalar nerve; orbital nerve*) (Fig. 12-6) arises in the pterygopalatine fossa, enters the orbit by the inferior orbital fissure, and divides into two branches, zygomaticotemporal and zygomaticofacial.

ZYGOMATICOTEMPORAL BRANCH.   The zygomaticotemporal branch runs along the lateral wall of the orbit in a groove in the zygomatic bone, and passing through a small foramen or through the sphenozygomatic suture, enters the temporal fossa. It runs upward between the bone and substance of the temporalis muscle, pierces the temporal fascia about 2.5 cm above the zygomatic arch, is distributed to the skin of the side of the forehead, and communicates with the facial nerve and with the auriculotemporal branch of the mandibular nerve. As it pierces the temporal fascia, it gives off a slender twig, which runs between the two layers of the fascia to the lateral side of the orbit (Fig. 12-18).

Before it leaves the orbit, it sends a communication to the **lacrimal nerve** through which the postganglionic parasympathetic fibers from the pterygopalatine ganglion reach the lacrimal gland (Fig. 12-8).

ZYGOMATICOFACIAL BRANCH. The zygomaticofacial branch (*malar branch*) passes along the inferior lateral angle of the orbit, through the zygomatic bone by way of the zygomaticoörbital and zygomaticofacial foramina, emerges upon the face, and perforating the orbicularis oculi muscle, supplies the skin on the prominence of the cheek. It joins with the facial nerve and with the inferior palpebral branches of the infraorbital nerve (Fig. 12-18).

VARIATIONS. The two branches vary in size, a deficiency in one being made up by the other or by the lacrimal or infraorbital nerves.

Pterygopalatine Nerves

The **pterygopalatine nerves** (*sphenopalatine nerves*) (Figs. 12-8; 12-9) are two short trunks that unite at the pterygopalatine ganglion and then are redistributed into branches. Formerly these trunks were called the sensory roots of the ganglion, and their peripheral branches were listed as the branches of distribution of the ganglion. Since most fibers in the trunks are trigeminal somatic afferents that merely pass beside or through the ganglion without synapses, the branches are listed here as belonging to the maxillary nerve rather than the ganglion. The ganglion is described with the facial nerve.

The pterygopalatine nerves serve also as important functional communications between the ganglion and the maxillary nerve. Postganglionic parasympathetic secretomotor fibers from the ganglion pass through them and back along the main maxillary nerve to the zygomatic nerve, through which they are routed to the lacrimal nerve and lacrimal gland. Other fibers from the ganglion accompany the branches of distribution of the maxillary nerve to the glands of the nasal cavity and palate.

The **branches of distribution** from the pterygopalatine nerves are divisible into four groups: (1) orbital, (2) palatine, (3) posterior superior nasal, and (4) pharyngeal.

ORBITAL BRANCHES. The orbital branches (*ascending branches*) are two or three delicate filaments that enter the orbit by the inferior orbital fissure and supply the periosteum. Filaments pass through foramina in the frontoethmoidal suture to supply the mucous membrane of the posterior ethmoidal and sphenoidal sinuses.

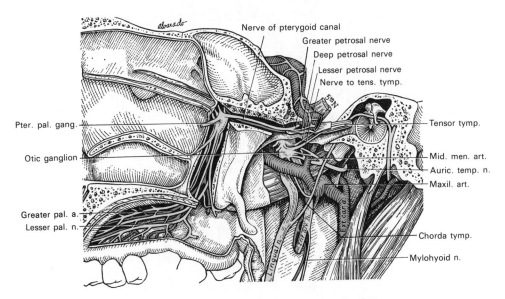

Fɪɢ. 12-9.   The pterygopalatine ganglion and nasal and palatine nerves.

GREATER PALATINE NERVE. The greater palatine nerve (*anterior palatine nerve; descending branch*) (Figs. 12-9; 12-12) passes through the pterygopalatine canal, emerges upon the hard palate through the greater palatine foramen, and divides into several branches, the longest of which passes anteriorly in a groove in the hard palate nearly as far as the incisor teeth. It supplies the gums and mucous membrane of the hard palate and adjacent parts of the soft palate, and communicates with the terminal filaments of the nasopalatine nerve.

Posterior Inferior Nasal Branches. These branches leave the nerve while it is in the canal, enter the nasal cavity through openings in the palatine bone, and ramify over the inferior nasal concha and middle and inferior meatuses.

Lesser Palatine Nerves. These nerves (Fig. 12-9) emerge through the lesser palatine foramina and distribute branches to the soft palate, uvula, and tonsil. They join with the tonsillar branches of the glossopharyngeal nerve to form a plexus around the tonsil (*circulus tonsillaris*). Many of the somatic afferent fibers contained in the lesser palatine nerves belong to the facial nerve, have their cells in the geniculate ganglion, and traverse

the greater petrosal nerve and the nerve of the pterygoid canal (*Vidian*[1] *nerve*).

POSTERIOR SUPERIOR NASAL BRANCHES. The posterior superior nasal branches enter the posterior part of the nasal cavity by the sphenopalatine foramen and supply the mucous membrane covering the superior and middle conchae, the lining of the posterior ethmoid sinuses, and the posterior part of the septum. One branch, longer and larger than the others, and named the **nasopalatine nerve,** passes across the roof of the nasal cavity inferior to the ostium of the sphenoidal sinus to reach the septum. It runs obliquely forward and downward, lying between the mucous membrane and periosteum of the septum, to the incisive canal (Fig. 12-2). It passes through the canal and communicates with the corresponding nerve of the opposite side and with the greater palatine nerve.

PHARYNGEAL BRANCH. The pharyngeal branch (*pterygopalatine nerve*) (Fig. 12-9) leaves the posterior part of the pterygopalatine ganglion. It passes through the pharyn-

---

[1]Vidus Vidius (Guido Guidi, 1500–1569): An Italian physician and professor of medicine (Pisa).

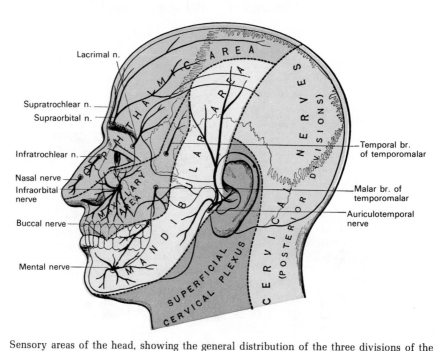

FIG. 12-10. Sensory areas of the head, showing the general distribution of the three divisions of the fifth nerve. (Modified from Testut.)

geal canal with the pharyngeal branch of the maxillary artery, and is distributed to the mucous membrane of the nasal part of the pharynx posterior to the auditory tube.

Posterior Superior Alveolar Branches

The **posterior superior alveolar branches** (*posterior superior dental branches*) (Fig. 12-8) arise from the trunk of the nerve just before it enters the infraorbital groove; there are generally two, but sometimes they arise by a single trunk. They cross the tuberosity of the maxilla and give off several twigs to the gums and neighboring parts of the mucous membrane of the cheek. They then enter the posterior alveolar canals on the infratemporal surface of the maxilla, and passing anteriorly in the substance of the bone, communicate with the middle superior alveolar nerve. They give off branches to the lining membrane of the maxillary sinus and three twigs to each molar tooth; these twigs enter the foramina at the apices of the roots of the teeth.

BRANCHES IN INFRAORBITAL CANAL

MIDDLE SUPERIOR ALVEOLAR BRANCH. The middle superior alveolar branch (*middle superior dental branch*) is given off from the nerve in the posterior part of the infraorbital canal, and runs downward and forward in a canal in the lateral wall of the maxillary sinus to supply the two premolar teeth. It forms a superior dental plexus with the anterior and posterior superior alveolar branches.

ANTERIOR SUPERIOR ALVEOLAR BRANCH. The large anterior superior alveolar branch (*anterior superior dental branch*) (Fig. 12-8) is given off from the nerve just before its exit from the infraorbital foramen; it courses in a canal in the anterior wall of the maxillary sinus, and divides into branches that supply the incisor and canine teeth. It communicates with the middle superior alveolar branch and gives off a nasal branch, which passes through a minute canal in the lateral wall of the inferior meatus and supplies the mucous membrane of the anterior part of the inferior meatus and the floor of the nasal cavity, communicating with the nasal branches from the pterygopalatine nerves.

The **infraorbital nerve** emerges through the infraorbital foramen and supplies the branches to the face (Fig. 12-10).

BRANCHES IN THE FACE

INFERIOR PALPEBRAL BRANCHES. The inferior palpebral branches (Fig. 12-8) pass superiorly, deep to the orbicularis oculi muscle. They supply the skin and conjunctiva of the lower eyelid, joining at the lateral angle of the orbit with the facial and zygomaticofacial nerves.

EXTERNAL NASAL BRANCHES. The external nasal branches (Fig. 12-8) supply the skin of the side of the nose and of the septum mobile nasi. They join with the terminal twigs of the nasociliary nerve.

SUPERIOR LABIAL BRANCHES. The superior labial branches (Fig. 12-8), the largest and most numerous, pass deep to the levator labii superioris muscle and are distributed to the skin of the upper lip, to the mucous membrane of the mouth, and to the labial glands. They communicate immediately inferior to the orbit with filaments from the facial nerve, forming with them the **infraorbital plexus.**

Mandibular Nerve

The **mandibular nerve** (*inferior maxillary nerve*) (Figs. 12-8; 12-11; 12-12), or third and largest division of the trigeminal, is a mixed nerve and has two roots: a *large sensory root* arising from the inferior angle of the trigeminal ganglion, and a *small motor root* (the entire motor root of the trigeminal nerve). The sensory fibers supply the skin of the temporal region, auricula, external meatus, cheek, lower lip, and lower part of the face; the mucous membrane of the cheek, tongue, and mastoid air cells; the lower teeth and gums; the mandible and temporomandibular joint; and part of the dura mater and skull. The motor fibers supply the muscles of mastication (masseter, temporalis, pterygoid), the mylohyoid and anterior belly of the digastric, and the tensores tympani and veli palatini. The two roots leave the middle cranial fossa through the foramen ovale, the motor part medial to the sensory, and unite just outside the skull. The main trunk thus formed is short, 2 or 3 mm, and divides into a smaller anterior and a larger posterior division. The otic ganglion lies close against the medial surface of the nerve, just outside the foramen ovale where the two roots fuse, and it surrounds the origin of the medial pterygoid nerve (Fig. 12-20).

A communication to the **otic ganglion**

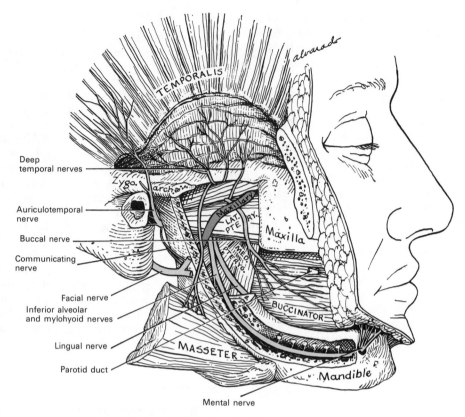

Fig. 12-11.   Mandibular division of the trigeminal nerve.

from the internal pterygoid nerve was formerly called a root of the ganglion, but the fibers pass through without synapses.

VARIATIONS.   The two divisions are separated by a fibrous band, the pterygospinous ligament, which may become ossified, and the anterior division then passes through a separate opening in the bone, the pterygospinous foramen (foramen of Civinini[1]).

The branches of the main trunk of the nerve are the meningeal branch, the medial pterygoid nerve, the anterior trunk, and the posterior trunk.

MENINGEAL BRANCH

The meningeal branch (*nervus spinosus, recurrent branch*) enters the skull through the foramen spinosum with the middle meningeal artery. It divides into two branches,

---

[1]Filippo Civinini (1805–1844): An Italian anatomist (Pisa).

which accompany the anterior and posterior divisions of the artery and supply the dura mater. The anterior branch communicates with the meningeal branch of the maxillary nerve; the posterior sends filaments to the mucous membrane of the mastoid air cells.

MEDIAL PTERYGOID NERVE

The **medial pterygoid nerve** (Fig. 12-20) is a slender branch that penetrates the otic ganglion and, after a short course, enters the deep surface of the muscle. It has two small branches that have a close association with the otic ganglion and have been described as branches of the ganglion, but their fibers pass through the ganglion without interruption. These are the **nerve to the tensor veli palatini** (Fig. 12-9), which enters the muscle near its origin, and the **nerve to the tensor tympani,** which lies close to and nearly parallel with the lesser superficial petrosal nerve and penetrates the cartilage of the auditory tube to supply the muscle.

ANTERIOR TRUNK

The anterior division of the mandibular nerve (Fig. 12-8) receives a small contribution of sensory fibers and all of the motor fibers from the motor root except those in the medial pterygoid and mylohyoid nerves. The following branches supply the muscles of mastication and the skin and mucous membrane of the cheek.

MASSETERIC NERVE. The masseteric nerve (Fig. 12-11) passes lateralward, above the lateral pterygoid, to the mandibular notch, which it crosses with the masseteric artery to enter the masseter muscle near its origin from the zygomatic arch. It gives a filament to the temporomandibular joint.

DEEP TEMPORAL NERVES. The deep temporal nerves (Fig. 12-11) are usually two, anterior and posterior, but a third or intermediate may be present. The anterior deep temporal frequently is given off from the buccal nerve; it emerges with the latter between the two heads of the lateral pterygoid muscle and turns superiorly into the anterior portion of the temporalis muscle. The posterior deep temporal nerve, and the intermediate nerve if present, pass over the superior border of the lateral pterygoid close to the bone of the temporal fossa and enter the deep surface of the muscle. The posterior branch sometimes arises in common with the masseteric nerve.

LATERAL PTERYGOID NERVE. The lateral pterygoid nerve enters the deep surface of the muscle. It frequently arises in conjunction with the buccal nerve.

BUCCAL NERVE. The buccal nerve (*buccinator nerve; long buccal nerve*) (Fig. 12-11) passes between the two heads of the lateral pterygoid muscle to reach its superficial surface; it follows or penetrates the inferior part of the temporalis muscle and emerges from under the anterior border of the masseter muscle. It ramifies on the surface of the buccinator muscle, forming a plexus of communications with the buccal branches of the facial nerve; it supplies the skin of the cheek over this muscle and sends penetrating branches to supply the mucous membrane of the mouth and part of the gums in the same area.

VARIATIONS. The buccal nerve may supply a branch to the lateral pterygoid as it passes through that muscle, and frequently it gives off the anterior deep temporal nerve. It may arise from the trigeminal ganglion, passing through its own foramen; it may be a branch of the inferior alveolar nerve; or it may be replaced by a branch of the maxillary nerve.

POSTERIOR TRUNK

The posterior division of the mandibular nerve is mainly sensory, but it has a small motor component. The following are its branches.

AURICULOTEMPORAL NERVE. The auriculotemporal nerve (Fig. 12-8) generally arises by two roots that join after encircling the middle meningeal artery close to the foramen spinosum. It runs posteriorly, deep to the lateral pterygoid muscle, along the medial side of the neck of the mandible, and it turns superiorly with the superficial temporal artery, between the auricula and the condyle of the mandible, under cover of the parotid gland. Escaping from beneath the gland, it passes over the root of the zygomatic arch and divides into superficial temporal branches (Fig. 12-18). The branches and communications of the auriculotemporal nerve are several.

The communications with the **facial nerve** (Fig. 12-11), usually two large communications, pass anteriorly from behind the neck of the mandible and join the facial nerve in the substance of the parotid gland at the posterior border of the masseter muscle. They carry sensory fibers that accompany the zygomatic, buccal, and mandibular branches of the facial nerve and supply the skin of these areas.

Communications with the **otic ganglion** (Fig. 12-20) join the roots of the auriculotemporal nerve close to their origin. They carry postganglionic parasympathetic fibers whose preganglionic fibers come from the glossopharyngeal nerve and supply the parotid gland with secretomotor fibers.

The **anterior auricular branches** (Fig. 12-11), usually two, supply the skin of the anterior superior part of the auricula, principally the helix and tragus.

The two branches to the **external acoustic meatus** (Fig. 12-11) enter the meatus between its bony and cartilaginous portions and supply the skin lining it; the upper one sends a filament to the tympanic membrane.

The **articular branches** consist of one or two twigs that enter the posterior part of the temporomandibular joint.

The **parotid branches** supply the parotid

gland, carrying the parasympathetic post-ganglionic fibers transmitted by the communication between the auriculotemporal nerve and the otic ganglion.

The **superficial temporal branches** accompany the superficial temporal artery to the vertex of the skull; they supply the skin of the temporal region and communicate with the facial and zygomaticotemporal nerves.

LINGUAL NERVE. The lingual nerve (Figs. 12-8; 12-12; 12-36) is at first deep to the lateral pterygoid muscle, running parallel with the inferior alveolar nerve and lying medial and anterior to it. It is frequently joined to the inferior alveolar nerve by a branch that may cross the maxillary artery. The chorda tympani nerve joins it here also. The lingual nerve runs between the medial pterygoid muscle and the mandible, and then crosses obliquely over the superior constrictor of the pharynx and the styloglossus to reach the side of the tongue. It passes between the hyoglossus and the deep part of the submandibular gland. Finally, crossing the lateral side of the submandibular duct, it runs along the undersurface of the tongue to its tip, lying immediately beneath the mucous membrane (Fig. 12-12).

The **chorda tympani** (Figs. 12-8; 12-12), a branch of the facial nerve, joins the lingual nerve posteriorly at an acute angle, 1 or 2 cm from the foramen ovale. It carries special sensory fibers for taste and parasympathetic preganglionic fibers for the submandibular ganglion.

The communications with the **submandibular ganglion** are usually two or more short nerves by which the ganglion seems to be suspended (Fig. 12-12). The proximal nerves carry the preganglionic parasympa-

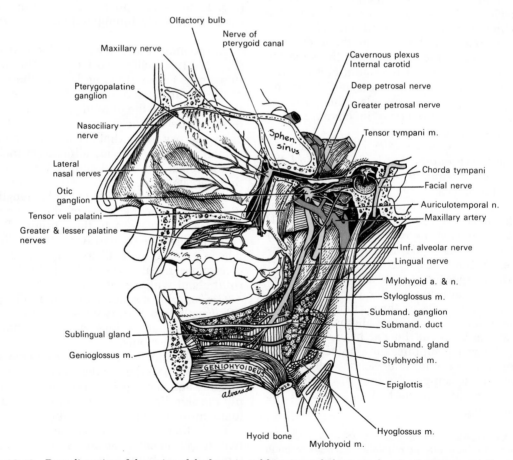

FIG. 12-12. Deep dissection of the region of the face viewed from its medial aspect, showing the pterygopalatine, otic, and submandibular ganglia and associated structures. (Redrawn from Töndury.)

thetic fibers communicated to the lingual by the chorda tympani. The distal communication contains postganglionic fibers for distribution to the sublingual gland.

Communications with the **hypoglossal nerve** (Fig. 12-36) form a plexus at the anterior margin of the hyoglossus muscle.

The **branches of distribution** supply the mucous membrane of the anterior two-thirds of the tongue, the adjacent mouth and gums, and the sublingual gland. The taste buds of the anterior two-thirds of the tongue are supplied by the fibers communicated through the chorda tympani.

INFERIOR ALVEOLAR NERVE. The inferior alveolar nerve (*inferior dental nerve*) (Fig. 12-11) accompanies the inferior alveolar artery, at first deep to the lateral pterygoid muscle and then between the sphenomandibular ligament and the ramus of the mandible, to the mandibular foramen. It enters the mandibular canal through the foramen, and passes anteriorly within the bone as far as the mental foramen, where it divides into two terminal branches.

The **mylohyoid nerve** (Fig. 12-8) leaves the inferior alveolar nerve just before it enters the mandibular foramen, and it continues inferiorly and anteriorly in a groove on the deep surface of the ramus of the mandible to reach the mylohyoid muscle. It supplies this muscle and crosses its superficial surface to reach the anterior belly of the digastric muscle (Fig. 12-8), which it also supplies.

The **dental branches** form a plexus within the bone and supply the molar and premolar teeth. Filaments enter the pulp canal of each root through the apical foramen and supply the pulp of the tooth.

The **incisive branch** is one of the terminal branches. It continues anteriorly within the bone, after the mental nerve separates from it, and forms a plexus that supplies the canine and incisor teeth.

The **mental nerve** (Fig. 12-11), the other terminal branch, emerges from the bone at the mental foramen and divides beneath the depressor anguli oris muscle into three branches: one is distributed to the skin of the chin; the other two, to the skin and mucous membrane of the lower lip. These branches communicate freely with branches of the facial nerve.

The otic and submandibular ganglia, although closely associated with the mandibular nerve, are not connected functionally, and are described, therefore, with the nerves that supply them with preganglionic fibers: the otic ganglion with the glossopharyngeal nerve, and the submandibular with the facial nerve.

VARIATIONS. The inferior alveolar and lingual nerves may form a single trunk or they may have communications of variable size; the chorda tympani may appear to join the inferior alveolar nerve, with a later communication carrying the fibers to the lingual nerve. The inferior alveolar nerve is occasionally perforated by the maxillary artery. It may have accessory roots from other branches of the mandibular nerve or a separate root from the trigeminal ganglion. The mylohyoid nerve may communicate with the lingual nerve, and it has been described as sending filaments to the depressor anguli oris, the platysma, the submandibular gland, and the integument below the chin.

TRIGEMINAL NERVE PAIN (TIC DOULOUREUX). The trigeminal nerve is more frequently the seat of severe neuritic or neuralgic pain than any other nerve in the body. The pain of a localized infection or irritation may be confined to that area, but commonly this is not the case. Involvement of an internal branch is likely to cause severe distress in a related cutaneous area by referred pain. As a general rule the diffusion of pain over the branches of the nerve is confined to one of the main divisions, although in severe cases it may radiate over the other main divisions.

The commonest example of this condition is the neuralgia that is often associated with dental caries. Although the tooth itself may not appear to be painful, the most distressing referred pains are experienced, which are relieved at once by treatment of the affected tooth. With the ophthalmic nerve, severe supraorbital pain is commonly associated with acute glaucoma or with frontal or ethmoidal sinusitis. Malignant growths or empyema of the maxillary sinus and diseased conditions in the nasal cavity, as well as dental caries, may cause neuralgia of the second division. Pain in the mandibular division is likely to occur in the ear or other distribution of the auriculotemporal nerve, although the actual disease may involve one of the lower teeth or the tongue.

When a focus of infection or irritation cannot be found, as is all too frequently the case, various measures may be taken to interrupt the pain fibers. Local injection of alcohol into the painful nerve may give temporary relief, and injection of a main division close to the ganglion has been performed with a certain measure of success. The main divisions have been incised surgically, and the entire ganglion has been removed in intractable cases. In the latter operation, bleeding is likely to be dangerous and the nerves to the muscles of mastication may be paralyzed. In other operations, the motor root is spared by cutting the sensory root inside the cranium before it reaches the ganglion.

VI. ABDUCENT NERVE

The **abducent nerve** (*sixth nerve*) (Fig. 12-6) supplies the rectus lateralis oculi muscle. Its superficial origin is in the furrow between the inferior border of the pons and the superior end of the pyramid of the medulla oblongata. It pierces the dura mater on the dorsum sellae of the sphenoid bone, runs through a notch below the posterior clinoid process, and traverses the cavernous sinus lateral to the internal carotid artery. It enters the orbit through the superior orbital fissure, superior to the ophthalmic vein, from which it is separated by a lamina of dura mater. After passing between the two heads of the rectus lateralis muscle, it enters its ocular surface.

Communication with the **sympathetic system** is by several filaments from the carotid and cavernous plexuses. Through a filament from the ophthalmic nerve, the abducent nerve communicates with the **trigeminal nerve.**

RELATION OF ORBITAL NERVES TO CAVERNOUS SINUS. Embedded in the lateral wall of the cavernous sinus, in order beginning with the most superior, are the oculomotor, trochlear, ophthalmic, and maxillary nerves (Fig. 12-13). The abducent nerve is suspended by connective tissue trabeculae within the sinus, lateral to the internal carotid artery and medial to the ophthalmic nerve. The maxillary nerve is related to the posterior portion of the sinus and it soon diverges from the other nerves. As the nerves approach the superior orbital fissure, the oculomotor and ophthalmic nerves divide into branches, and the abducent nerve approaches the others, so that their relative positions are considerably changed.

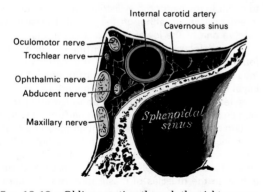

FIG. 12-13.   Oblique section through the right cavernous sinus.

VII. FACIAL NERVE

The **facial nerve** (*seventh nerve*) has two roots of unequal size (Fig. 12-14). The larger is the **motor root;** the smaller, lying between the motor root and the vestibulocochlear nerve, is called the **nervus intermedius** (*nerve of Wrisberg*[1]), and contains special sensory fibers for taste and parasympathetic fibers. The superficial origin of both roots is at the inferior border of the pons in the recess between the olive and the inferior cerebellar peduncle, the motor root being medial, the vestibulocochlear nerve lateral, and the nervus intermedius between.

The facial nerve is the motor nerve to the muscles of facial expression, to muscles in the scalp and external ear, and to the buccinator, platysma, stapedius, stylohyoid, and posterior belly of the digastric muscles. The sensory part supplies the anterior two-thirds of the tongue with taste, and parts of the external acoustic meatus, soft palate, and adjacent pharynx with general sensation. The parasympathetic part supplies secretomotor fibers for the submandibular, sublingual, lacrimal, nasal, and palatine glands.

From their superficial attachment, the two roots pass lateralward, with the vestibulocochlear nerve, into the internal acoustic meatus (Fig. 12-1). At the fundus of the meatus, the facial nerve separates from the vestibulocochlear nerve and enters the substance of the petrous portion of the temporal bone, through which it runs a serpentine course in the **facial canal** (*aqueductus Fallopii*[2]). At first it continues lateralward in the region between the cochlea and semicircular canals (Fig. 13-58); near the tympanic cavity it makes an abrupt bend posteriorly, runs in the medial wall of the cavity just above the oval window (Fig. 13-50), covered by a thin plate of bone that causes a slight prominence in the wall (Fig. 13-51), and then dips down beside the mastoid air cells to reach the stylomastoid foramen. Where it changes its course abruptly, there is an exaggeration of the bend into a U-shaped structure named the **geniculum;** at this point also, the two roots become fused and the nerve is swollen

[1]Heinrich August Wrisberg (1739–1808): A German anatomist (Göttingen).

[2]Gabriele Fallopius (1523–1552): An Italian anatomist (Padua).

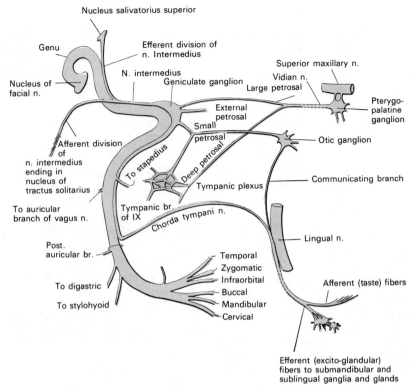

Fɪɢ. 12-14. Plan of the facial and intermediate nerves and their communication with other nerves.

by the presence of the geniculate ganglion (Fig. 12-15).

As it emerges from the stylomastoid foramen, the nerve runs anteriorly in the substance of the parotid gland, crosses the external carotid artery, and divides at the posterior border of the ramus of the mandible into two primary branches—a superior branch, the temporofacial, and an inferior branch, the cervicofacial. Numerous branches come off the cervicofacial branch in a plexiform arrangement to form the **parotid plexus;** these are distributed over the head, face, and upper part of the neck, supplying the superficial muscles in these regions.

Geniculate Ganglion

The **geniculate ganglion** (Fig. 12-15) is a small fusiform swelling of the geniculum, where the facial nerve bends abruptly backward at the hiatus of the facial canal. It is the sensory ganglion of the facial nerve. The central processes of its unipolar ganglion cells reach the brain stem through the **nervus intermedius;** the majority of the peripheral processes pass to the taste buds of the anterior two-thirds of the tongue through the chorda tympani and lingual nerves; many peripheral processes pass through the greater superficial petrosal and lesser palatine nerves to the soft palate; and a smaller number join the auricular branch of the vagus nerve to supply the skin of the external acoustic meatus and mastoid process (Foley et al., 1946).

The **glossopalatine nerve** is the name given by some authors (Hardesty, 1933) to that portion of the facial nerve contributed by the **nervus intermedius.** It comprises, therefore, the sensory part, including the geniculate ganglion, the chorda tympani, and greater petrosal nerve; and the parasympathetic part, including the submandibular and pterygopalatine ganglia and their branches. Although this separation is suggested by the similarity between this complex and the glossopharyngeal nerve, it has not been generally adopted.

Communications of the Facial Nerve

In the internal acoustic meatus, communications with the **vestibulocochlear nerve** probably contain fibers that leave the facial nerve, run with the vestibulocochlear nerve for a short distance, and return to the facial nerve.

At the geniculate ganglion, the facial nerve communicates with the **otic ganglion** by filaments that join the lesser petrosal nerve, and with the **sympathetic fibers** on the middle meningeal artery (*external superficial petrosal nerve*).

In the facial canal, just before it leaves the stylomastoid foramen, the facial nerve communicates with the **auricular branch of the vagus** as the latter runs across it in the substance of the bone.

After its exit from the stylomastoid foramen, the facial nerve communicates with the **glossopharyngeal** nerve, the **vagus** nerve, and the **great auricular** nerve from the cervical plexus. The **auriculotemporal** nerve from the mandibular division of the trigeminal usually sends two large communications that pass anteriorly from behind the neck of the mandible to join the facial nerve in the substance of the parotid gland (Fig. 12-11). They carry sensory fibers that accompany the terminal zygomatic, buccal, and mandibular branches to the skin of these areas.

Peripheral branches of the facial nerve communicate behind the ear, with the **lesser occipital nerve;** on the face, with **trigeminal branches;** and in the neck, with the **cervical cutaneous nerve.**

Branches of Facial Nerve from the Geniculate Ganglion

GREATER PETROSAL NERVE

The **greater petrosal nerve** (*greater superficial petrosal nerve*) arises from the geniculate ganglion (Fig. 12-15), and after a short course in the bone, it emerges through the hiatus of the facial canal into the middle cranial fossa. It runs rostralward beneath the dura mater and the trigeminal ganglion in a sulcus on the anterior surface of the petrous portion of the temporal bone, passes superior to the cartilage of the auditory tube (which fills the foramen lacerum), crosses the lateral side of the internal carotid artery, and unites with the deep petrosal nerve to

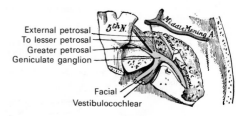

FIG. 12-15. The geniculate ganglion and facial nerve in the temporal bone.

form the nerve of the pterygoid canal (Vidian nerve) (Figs. 12-9; 12-12). The greater petrosal nerve is a mixed nerve, containing sensory and parasympathetic fibers. The parasympathetic fibers, from the nervus intermedius, become the motor root of the pterygopalatine ganglion. The bulk of the nerve consists of sensory fibers, which are the peripheral processes of cells in the geniculate ganglion and are distributed to the soft palate through the lesser palatine nerves, with a few filaments to the auditory tube.

Nerve of Pterygoid Canal

The **nerve of the pterygoid canal** (*Vidian nerve*) (Fig. 12-9), formed by the union of the greater petrosal and deep petrosal nerves at the foramen lacerum, enters the posterior opening of the pterygoid canal with the corresponding artery and is joined by a small *ascending sphenoidal branch* from the otic ganglion. The bony wall of the canal commonly causes a ridge in the floor of the sphenoidal sinus, and while the nerve is in the canal it gives off one or two filaments for the mucous membrane of the sinus. From the anterior opening of the canal, the nerve crosses the pterygopalatine fossa and enters the pterygopalatine ganglion.

Pterygopalatine Ganglion

The **pterygopalatine ganglion** (*sphenopalatine ganglion; Meckel's ganglion*) (Figs. 12-9; 12-12) is deeply placed in the pterygopalatine fossa, just inferior to the maxillary nerve as the latter crosses the fossa close to the pterygopalatine foramen. It is triangular or heartshaped and about 5 mm long. It is embedded in the fibrous tissue between the neighboring bones and is closely attached to the pterygopalatine branches of the maxillary division of the trigeminal nerve. It is a parasympathetic ganglion relaying chiefly

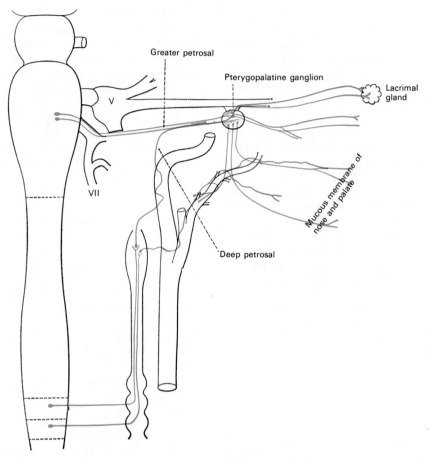

Greater petrosal

Pterygopalatine ganglion

Lacrimal gland

V

VII

Mucous membrane of nose and palate

Deep petrosal

Fig. 12-16. Autonomic connections of the pterygopalatine and superior cervical ganglia. Parasympathetic blue; sympathetic red.

secretomotor impulses from the facial nerve (Fig. 12-16).

The **parasympathetic root** (*visceral efferent*) of the pterygopalatine ganglion is the greater petrosal nerve, and its continuation, the nerve of the pterygoid canal. The fibers are preganglionic parasympathetic fibers, which leave the brain stem in the nervus intermedius.

COMMUNICATIONS OF THE PTERYGOPALATINE GANGLION. *Maxillary Nerve.* Two short trunks from the **maxillary nerve,** the pterygopalatine nerves, are commonly called the sensory root of the ganglion although they have no such functional connection with it. They contain mainly sensory fibers from the trigeminal ganglion that pass through or beside the pterygopalatine ganglion without synapses and continue on their way to the mucous membrane of the nasal cavity and palate. These trunks are important communications for the ganglion, however, since they are traversed by the postganglionic fibers of distribution on their way to the maxillary nerve, whence they reach the lacrimal gland and the small glands of the nasal cavity and palate.

Deep Petrosal Nerve The deep petrosal nerve is commonly called the sympathetic root of the pterygopalatine ganglion, although it is merely a communication between the ganglion and the sympathetic system. It contains postganglionic fibers from the superior cervical sympathetic ganglion, by way of the carotid plexus, and passes through the pterygopalatine ganglion without synapses and accompanies the branches of the pterygopalatine nerves (trigeminal branches) to their destination in the mucous membrane of the nasal cavity and palate.

The branches of distribution from
the pterygopalatine ganglion, containing the
postganglionic fibers of the cells within the
ganglion, are not independent nerves for the
most part, but find their way to their destina-
tions by accompanying other nerves, mainly
the branches of the maxillary nerve.

The fibers for the **lacrimal gland** pass back
to the main trunk of the maxillary nerve
through the pterygopalatine nerves. They
leave the maxillary through the zygomatic
and zygomaticotemporal nerves, pass
through a communication between the latter
and the lacrimal nerve in the orbit, and are
then distributed to the gland (Fig. 12-6).

Fibers for the **small glands** of the mucous
membrane of the **nasal cavity, pharynx,** and
palate join the greater and lesser palatine
nerves, the posterior superior nasal
branches, and the pharyngeal branch, and
are distributed with them.

*Branches of Facial Nerve
Within the Facial Canal*

NERVE TO STAPEDIUS MUSCLE

The **nerve to the stapedius muscle** arises
from the facial nerve as it passes downward
in the posterior wall of the tympanum. It
reaches the muscle through a minute opening in the base of the pyramid (see Middle
Ear, Chapter 13).

CHORDA TYMPANI NERVE

The **chorda tympani nerve** (Figs. 12-8; 12-
12) arises from the part of the facial nerve
that runs vertically downward in the posterior wall of the tympanum just before it
reaches the stylomastoid foramen. Entering
its own canal in the bone about 6 mm above
the stylomastoid foramen, the chorda passes
back cranialward almost parallel with the
facial nerve but diverges toward the lateral
wall of the tympanum. It emerges through an
aperture in the posterior wall of the tympanum (*iter chordae posterius*) between the
base of the pyramid and the attachment of
the tympanic membrane. It runs horizontally along the lateral wall of the tympanum
covered by the thin mucous membrane, and,
lying against the tympanic membrane, it
crosses the attached manubrium of the mal-

leus (Fig. 13-49). It leaves the tympanic cavity near the anterior border of the membrane
through the *iter chordae anterius,* traverses
a canal in the petrotympanic fissure (*canal
of Huguier*[1]), and emerges from the skull on
the medial surface of the spine of the sphenoid bone. After crossing the spine, usually
in a groove, and being joined by a small
communication from the otic ganglion, it
unites with the lingual nerve at an acute
angle between the medial and lateral pterygoid muscles.

The bulk of the fibers of the chorda tympani are special visceral afferents for taste,
which are distributed with the branches of
the lingual nerve to the anterior two-thirds
of the tongue. It also contains preganglionic
parasympathetic fibers (secretomotor) from
the nervus intermedius, which terminate in
synapses with cells in the submandibular
ganglion (Foley, 1945).

SUBMANDIBULAR GANGLION.   The **submandibular ganglion** (*submaxillary ganglion*)
(Figs. 12-8; 12-12) is a small mass, 2 to 5 mm
in diameter, situated above the deep portion
of the submandibular gland on the hyoglossus
muscle, near the posterior border of the mylohyoid muscle and suspended from the
lower border of the lingual nerve by two filaments approximately 5 mm long. The proximal filament is the **parasympathetic root,**
which conveys fibers originating in the
nervus intermedius and is communicated to
the lingual nerve by the chorda tympani.
These are preganglionic visceral efferent fibers (secretomotor) whose postganglionic
fibers innervate the submandibular, sublingual, lingual, and neighboring small salivary
glands (Fig. 12-17).

The **branches of distribution** are (*a*) five or
six filaments distributed to the submandibular gland and its duct, (*b*) to the small glands
about the floor of the mouth, and (*c*) the distal filament attaching the ganglion to the lingual nerve, which communicates the fibers
distributed to the sublingual and small lingual glands with the terminal branches of
the lingual nerve. Small groups of ganglion
cells are constantly found in the stroma of
the submandibular gland, usually near the
larger branches of the duct, and functionally
are considered a part of the submandibular
ganglion.

[1]Pierre Charles Huguier (1804–1873): A French surgeon
(Paris).

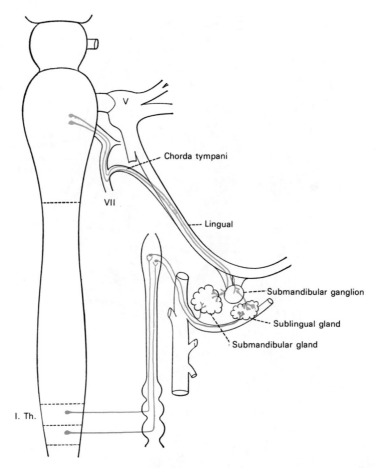

FIG. 12-17. Autonomic connections of the submandibular and superior cervical ganglia. Parasympathetic blue; sympathetic red.

A **communication** with the **sympathetic** bundles on the facial artery has been called the *sympathetic root of the ganglion,* but the fibers are postganglionic and have no synapses in the ganglion.

Visceral afferent fibers passing through the root to the lingual and thence to the chorda tympani have been called the *sensory root,* but they have no synapses in the ganglion and their cell bodies are in the geniculate ganglion.

*Branches of Facial Nerve
in the Face and Neck*

POSTERIOR AURICULAR NERVE. The posterior auricular nerve (Fig. 12-18) arises close to the stylomastoid foramen and runs cranialward anterior to the mastoid process; here it is joined by a filament from the auricular branch of the vagus, and communicates with the posterior branch of the great auricular nerve and with the lesser occipital nerve. Between the external acoustic meatus and the mastoid process it divides into auricular and occipital branches. The auricular branch supplies the auricularis posterior and the intrinsic muscles on the cranial surface of the auricula. The occipital branch, the larger of the two, passes backward along the superior nuchal line of the occipital bone and supplies the occipitalis muscle.

DIGASTRIC BRANCH. The digastric branch arises close to the stylomastoid foramen, and divides into several filaments, which supply the posterior belly of the digastric muscle.

STYLOHYOID BRANCH. The stylohyoid branch frequently arises in conjunction with the digastric branch; it is long and slender, and enters the stylohyoid muscle near its middle.

PAROTID PLEXUS. The terminal portion of

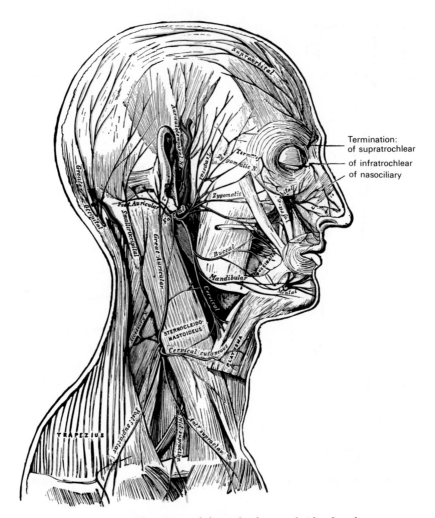

Termination:
of supratrochlear
of infratrochlear
of nasociliary

FIG. 12-18. The nerves of the scalp, face, and side of neck.

the facial nerve, within the substance of the parotid gland, divides into a temporofacial and a cervicofacial division which in turn, either within the gland or after leaving it, break up into a plexus and supply the facial muscles in different regions.

Temporal Branches. The temporal branches (Fig. 12-18) cross the zygomatic arch to the temporal region, supplying the auriculares anterior and superior. They communicate with the zygomaticotemporal branch of the maxillary nerve and with the auriculotemporal branch of the mandibular division of the trigeminal nerve. The more anterior branches supply the frontalis, the orbicularis oculi, and the corrugator muscles and join the supraorbital and lacrimal branches of the ophthalmic nerve.

Zygomatic Branches. The zygomatic branches (*malar branches*) run across the face in the region of the zygomatic arch to the lateral angle of the orbit, where they supply the orbicularis oculi, and communicate with filaments from the lacrimal nerve of the ophthalmic division and the zygomaticofacial branch of the maxillary division of the trigeminal nerve. The lower zygomatic branches commonly join the deep buccal branches and assist in forming the infraorbital plexus.

Buccal Branches. The buccal branches (*infraorbital branches*), larger than the rest, pass horizontally rostralward to be distributed inferior to the orbit and around the mouth. The superficial branches run beneath the skin and superficial to the muscles of the face, which they supply; some are distributed to the procerus muscle, communicating at the medial angle of the orbit with the infratrochlear and nasociliary branches of the ophthalmic nerve. The deep branches, commonly reinforced by zygomatic branches, pass deep to the zygomaticus and the levator labii superioris muscles, supplying them and forming an infraorbital plexus with the infraorbital branch of the maxillary division of the trigeminal nerve (Fig. 12-18). These branches also supply the small muscles of the nose. The more inferior deep branches supply the buccinator and orbicularis oris muscles, and communicate with filaments of the buccal branch of the mandibular division of the trigeminal nerve.

Mandibular Branch. The mandibular branch (r. marginalis mandibulae) passes rostralward deep to the platysma and depressor anguli oris supplying the muscles of the lower lip and chin and communicating with the mental branch of the inferior alveolar nerve.

Cervical Branch. The cervical branch (r. colli) runs rostralward deep to the platysma, which it supplies, and forms a series of arches across the side of the neck over the suprahyoid region. One branch joins the cervical cutaneous nerve from the cervical plexus.

VIII. VESTIBULOCOCHLEAR NERVE

The **vestibulocochlear nerve** (n. statoacusticus; n. octavius; eighth nerve) (Fig. 11-62) consists of two distinct sets of fibers, the **cochlear and vestibular nerves,** which differ in their peripheral endings, central connections, functions, and time of myelination. These two portions of the vestibulocochlear nerve are joined into a common trunk that enters the internal acoustic meatus with the facial nerve (Fig. 12-15). Centrally, the vestibulocochlear nerve divides into a lateral (cochlear) root and a medial (vestibular) root. As it passes distally in the internal auditory meatus, it divides into the various branches, which are distributed to the receptor areas in the membranous labyrinth (see Chapter 13). Both divisions of this nerve are sensory and the fibers arise from bipolar ganglion cells.

COCHLEAR NERVE. The cochlear nerve or root, the nerve of hearing, arises from bipolar cells in the spiral ganglion of the cochlea, situated near the inner edge of the osseous spiral lamina. The peripheral fibers pass to the organ of Corti.[1] The central fibers pass through the modiolus and then through the foramina of the tractus spiralis foraminosus or through the foramen centrale into the lateral or outer end of the internal acoustic meatus. The nerve passes along the internal acoustic meatus with the vestibular nerve and across the subarachnoid space, just above the flocculus, almost directly medialward toward the inferior peduncle to terminate in the cochlear nuclei.

The cochlear nerve is placed lateral to the vestibular root. Its fibers end in two nuclei: one, the ventral cochlear nucleus, lies immediately in front of the inferior cerebellar peduncle; the other, the dorsal cochlear nucleus, tuberculum acusticum, somewhat lateral to it (see Chapter 11).

VESTIBULAR NERVE. The vestibular nerve or root, the nerve of equilibration, arises from bipolar cells in the vestibular ganglion (ganglion of Scarpa[2]) which is situated in the superior part of the lateral end of the internal acoustic meatus. The peripheral fibers divide into three branches: the superior branch passes through the foramina in the area vestibularis superior and ends in the utricle and in the ampullae of the anterior and lateral semicircular ducts; the fibers of the inferior branch traverse the foramina in the area vestibularis inferior and end in the saccule; the posterior branch runs through the foramen singulare and supplies the ampulla of the posterior semicircular duct.

The fibers of the vestibular nerve enter the medulla oblongata, pass between the inferior cerebellar peduncle and the spinal tract of the trigeminal nerve, and bifurcate into ascending and descending branches. The descending branches form the spinal root of the vestibular nerve and terminate in the associated nucleus. The ascending branches pass to the medial, lateral, and superior vestibular nuclei, and to the nucleus fastigii and the vermis of the cerebellum.

[1]Alfonso Corti (1822–1888): An Italian anatomist (born in Sardinia, worked in Germany).
[2]Antonio Scarpa (1747–1832): An Italian anatomist and surgeon (Pavia).

IX. GLOSSOPHARYNGEAL NERVE

The **glossopharyngeal nerve** (*ninth nerve*) (Fig. 12-22), as its name implies, is distributed to the tongue and pharynx. It is a mixed nerve, its sensory fibers being both visceral and somatic, and its motor fibers both general and special visceral efferents. The somatic afferent fibers supply the mucous membrane of the pharynx, fauces, palatine tonsil, and posterior part of the tongue; special visceral afferents supply the taste buds of the posterior part of the tongue; general visceral afferents supply the blood pressure receptors of the carotid sinus. The special visceral efferent fibers supply the stylopharyngeus muscle; the general visceral efferents are mainly secretomotor for the parotid and small glands in the mucous membrane of the posterior part of the tongue and neighboring pharynx. The superficial origin is by three or four rootlets in series with those of the vagus nerve, attached to the superior part of the medulla oblongata in the groove between the olive and the inferior peduncle.

From its superficial origin, the nerve passes lateralward across the flocculus to the jugular foramen, through which it passes, lateral and anterior to the vagus and accessory nerves (Fig. 12-22), in a separate sheath of dura mater and lying in a groove on the lower border of the petrous portion of the temporal bone. After its exit from the skull, it runs anteriorly between the internal jugular vein and the internal carotid artery, superficial to the latter vessel and posterior to the styloid process and its muscles. It follows the posterior border of the stylopharyngeus muscle for 2 or 3 cm, then curves across its superficial surface to the posterior border of the hyoglossus muscle, where it penetrates more deeply to be distributed to the palatine tonsil, the mucous membrane of the fauces and base of the tongue, and the glands of that region. The portion of the nerve that lies in the jugular foramen has two enlargements, the superior and the inferior ganglia.

The **superior ganglion** (*jugular ganglion*) is situated in the upper part of the groove in which the nerve is lodged during its passage through the jugular foramen. It is very small, may be absent, and is usually regarded as a detached portion of the inferior ganglion.

The **inferior ganglion** (*petrous ganglion*) is situated in a depression in the lower border of the petrous portion of the temporal bone. These ganglia contain the cell bodies for the sensory fibers of the nerve (Fig. 12-21).

Communications of Glossopharyngeal Nerve

(1) The communications with the **vagus nerve** are two filaments, one joining the auricular branch, the other the superior ganglion. (2) The superior cervical **sympathetic** ganglion communicates with the inferior ganglion. (3) The communication with the **facial nerve** is between the trunk of the glossopharyngeal nerve below the inferior ganglion and the facial nerve after its exit from the stylomastoid foramen; it perforates the posterior belly of the digastric muscle.

Branches of Glossopharyngeal Nerve

TYMPANIC NERVE

The **tympanic nerve** (*nerve of Jacobson*) (Fig. 12-19) supplies parasympathetic fibers to the parotid gland through the otic ganglion and sensory fibers to the mucous membrane of the middle ear. It arises from the inferior ganglion and enters a small canal through an opening in the bony ridge that separates the carotid canal from the jugular fossa on the inferior surface of the petrous portion of the temporal bone. After a short cranialward course in the bone, it enters the tympanic cavity by an aperture in its floor near the medial wall. It continues upward in a groove on the surface of the promontory (Fig. 13-50), helps to form the tympanic plexus, reenters a canaliculus at the level of the processus cochleariformis, passes internal to the semicanal for the tensor tympani, and continues as the lesser petrosal nerve (Rosen, 1950).

TYMPANIC PLEXUS. The tympanic plexus lies in grooves on the surface of the promontory and is formed by the junction of the tympanic and caroticotympanic nerves. The **caroticotympanic nerves, superior** and **inferior,** are communications from the carotid plexus of the sympathetic, which enter the tympanic cavity by perforating the wall of the carotid canal. The plexus communicates

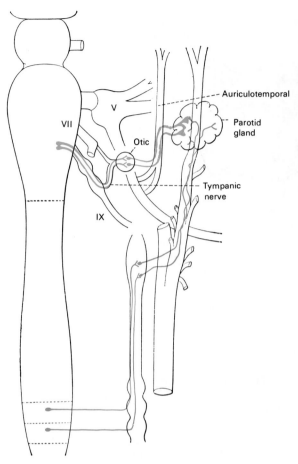

FIG. 12-19. Autonomic connections of the otic and superior cervical ganglia. Parasympathetic blue; sympathetic red.

with the greater petrosal nerve by a filament that passes through an opening on the labyrinthic wall, in front of the fenestra vestibuli.

Sensory branches are distributed through the plexus to the mucous membrane of the fenestra ovalis, fenestra rotunda, tympanic membrane, auditory tube, and mastoid air cells.

The **lesser petrosal nerve** (Figs. 12-9; 12-20) is the terminal branch or continuation of the tympanic nerve beyond the plexus. After penetrating the bone medial to the tensor tympani muscle, it emerges into the cranial cavity on the superior surface of the petrous portion of the temporal bone, immediately lateral to the hiatus of the facial canal. It leaves the cranial cavity again through the fissure between the petrous portion and the great wing of the sphenoid, or through a small opening in the latter bone, and terminates in the otic ganglion as its visceral motor or parasympathetic root. In the canal it is joined by a filament from the geniculate ganglion of the facial nerve.

OTIC GANGLION

The **otic ganglion** (Figs. 12-12; 12-20) is a flat, oval, or stellate ganglion, 2 to 4 mm in diameter, closely approximated to the medial surface of the mandibular division of the trigeminal, immediately outside of the foramen ovale, and it has the origin of the medial pterygoid nerve embedded in it. It is lateral to the cartilaginous portion of the auditory tube, anterior to the middle meningeal artery, and posterior to the origin of the tensor veli palatini muscle.

The **root** of the **otic ganglion,** which is **parasympathetic,** is the lesser petrosal nerve. It contains preganglionic fibers from the nucleus salivatorius inferior in the medulla oblongata, principally through the glossopharyngeal nerve but probably partly also through the facial nerve.

COMMUNICATIONS OF THE OTIC GANGLION. A communication with the **sympathetic** network on the middle meningeal artery has been called the sympathetic root of the ganglion, but these fibers are already postganglionic and pass through the ganglion without synapses.

A communication with the **medial pterygoid nerve** has been described as a motor root, and the continuation of the fibers to the tensor veli palatini and tensor tympani muscles has been described as branches of the ganglion, but they are trigeminal fibers that pass through the ganglion.

A communication with the **mandibular nerve** has been called a sensory root, but the fibers have no functional connection with the ganglion. A slender filament, the **sphenoidal branch,** connects with the nerve of the pterygoid canal, and a small branch communicates with the **chorda tympani.**

BRANCHES OF DISTRIBUTION OF THE OTIC GANGLION. The postganglionic fibers arising in the otic ganglion pass mainly through a communication with the auriculotemporal nerve and are distributed with its branches

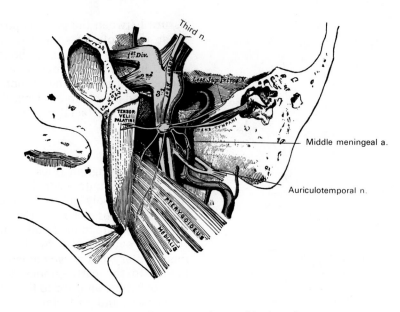

Fɪɢ. 12-20.   The otic ganglion and its branches.

to the **parotid gland.** Other filaments probably accompany other nerves to reach small glands in the mouth and pharynx.

CAROTID SINUS NERVE

The **carotid sinus nerve** (*nerve of Hering*[1]) (Fig. 12-21) arises from the main trunk of the glossopharyngeal nerve just beyond its emergence from the jugular foramen; it communicates with the nodose ganglion or the pharyngeal branch of the vagus nerve near its origin. Its continuation or principal branch runs down the anterior surface of the internal carotid artery to the carotid bifurcation, and it terminates in the wall of the dilated portion of the artery—the carotid sinus—supplying it with afferent fibers for its blood pressure receptors. It has a rather constant branch that joins the intercarotid plexus, formed principally by vagus and sympathetic branches, or it communicates with these nerves independently and reaches the carotid body. Glossopharyngeal fibers may traverse the plexus and its branches to the carotid body on their way to the carotid sinus (Sheehan et al., 1941). Its functional association with the carotid body is questionable.

[1]Heinrich Ewald Hering (1866–1948): A German physiologist (Cologne).

PHARYNGEAL BRANCHES

The **pharyngeal branches** are three or four filaments that join pharyngeal branches of the vagus and sympathetic nerves opposite the constrictor pharyngis medius muscle to form the **pharyngeal plexus** (Fig. 12-21). Branches from the plexus penetrate the muscular coat of the pharynx and supply its muscles and mucous membrane; the exact contribution of the glossopharyngeal nerve is uncertain.

BRANCH TO STYLOPHARYNGEUS

The **branch** to the **stylopharyngeus** is the only muscular branch of the glossopharyngeal nerve.

TONSILLAR BRANCHES

The **tonsillar branches** supply the palatine tonsil, forming around it a network from which filaments are distributed to the soft palate and fauces, where they communicate with the lesser palatine nerves.

LINGUAL BRANCHES

The **lingual branches** are two in number; one supplies the vallate papillae with afferent fibers for taste, and general afferents to the mucous membrane at the base of the

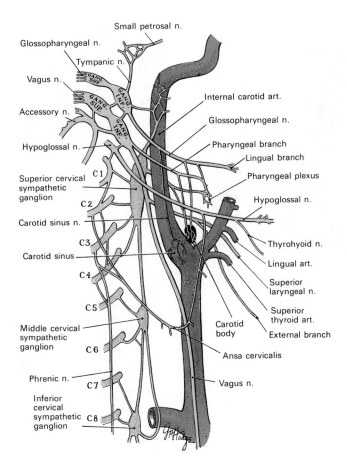

Fig. 12-21.   Carotid sinus nerve, carotid body, internal carotid artery, and neighboring cranial, spinal, and sympathetic nerve connections. Diagrammatic.

tongue; the other supplies the mucous membrane and glands of the posterior part of the tongue, and communicates with the lingual nerve.

CLINICAL COMMENT.   Neuralgic pain or "tic douloureux" of the glossopharyngeal nerve occurs in the ear, throat, base of the tongue, rim of the palate, and the lower lateral and posterior part of the pharynx. The most common trigger zone is the tonsillar fossa (Pastore and Meredith, 1949).

X. THE VAGUS NERVE

The **vagus nerve** (*tenth nerve; pneumogastric nerve*) (Figs. 12-22; 12-37), named from its wandering course, is the longest of the cranial nerves and has the most extensive distribution, passing through the neck and thorax into the abdomen. It has both somatic and visceral afferent fibers, and general and special visceral efferent fibers. The somatic sensory fibers supply the skin of the posterior surface of the external ear and the external acoustic meatus; the visceral afferent fibers supply the mucous membrane of the pharynx, larynx, bronchi, lungs, heart, esophagus, stomach, intestines, and kidney. General visceral efferent fibers (parasympathetic) are distributed to the heart, and supply the nonstriated muscle and glands of the esophagus, stomach, trachea, bronchi, biliary tract, and most of the intestine. Special visceral efferent fibers supply the voluntary muscles of the larynx, pharynx, and palate (except the tensor), but most of the latter fibers originate in the cranial part of the accessory nerve.

The superficial origin of the vagus is composed of eight or ten rootlets attached to the medulla oblongata in the groove between

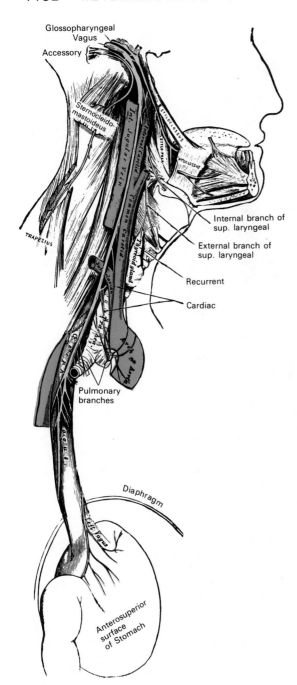

FIG. 12-22.   Course and distribution of the glossopharyngeal, vagus, and accessory nerves.

jugular foramen. The nerve leaves the cranial cavity through this opening, accompanied by the accessory nerve and contained in the same dural sheath, but separated by a septum from the glossopharyngeal nerve, which lies anteriorly. This portion of the vagus has two enlargements, the superior and inferior ganglia, which are the **sensory ganglia** of the nerve.

SUPERIOR GANGLION.   The **superior ganglion** (g. of the root; jugular ganglion) (Fig. 12-21) is a spherical swelling, about 4 mm in diameter, of the vagus nerve as it lies in the jugular foramen. The central processes of its unipolar (sensory) ganglion cells enter the medulla, usually in three or four large independent rootlets, slightly dorsal to the motor rootlets. Most of the peripheral processes of the ganglion cells enter the auricular branch of the vagus, but a few probably are distributed with the pharyngeal branches.

INFERIOR GANGLION.   The **inferior ganglion** (g. of the trunk; nodose ganglion) (Fig. 12-23) forms a fusiform swelling about 2.5 cm long on the vagus nerve after its exit from the jugular foramen and about 1 cm distal to the superior ganglion. The central processes of its unipolar (sensory) cells pass through the superior ganglion without traversing the region occupied by cells, and frequently accompany motor fibers in the rootlets for a short distance but enter the medulla slightly dorsal to them, in line with the superior rootlets. Some of the peripheral processes of the ganglion cells make up the internal ramus of the superior laryngeal nerve, and the rest are distributed with other branches of the vagus to the larynx, trachea, bronchi, esophagus, and other thoracic and abdominal viscera.

Communications of the Vagus Nerve

At the superior ganglion, several delicate filaments communicate (1) with the cranial portion of the **accessory nerve;** (2) with the inferior ganglion of the **glossopharyngeal nerve;** (3) with the **facial nerve** by means of the auricular branch; and (4) with the superior cervical ganglion of the **sympathetic system** (jugular nerve).

The cranial part of the **accessory nerve** joins the vagus just proximal to the inferior ganglion. It is the source of the greater part of the fibers in the motor branches of the vagus to the pharynx and larynx.

the olive and the inferior peduncle, inferior to those of the glossopharyngeal nerve and superior to those of the accessory nerve. The rootlets unite into a flat cord that passes beneath the flocculus of the cerebellum to the

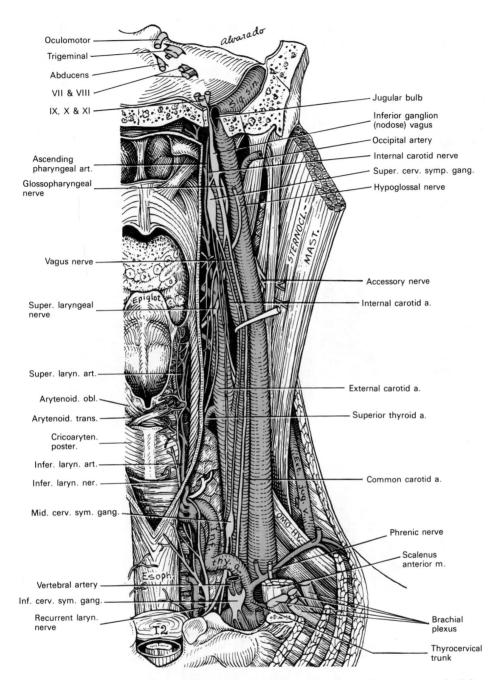

FIG. 12-23. Dorsal view of the pharynx and associated nerves and blood vessels after removal of the cervical vertebrae and part of the occipital bone. (Redrawn from Töndury.)

At the inferior ganglion, communication is (1) with the **hypoglossal nerve,** (2) with the superior cervical **sympathetic ganglion,** and (3) with the loop between the **first** and **second cervical** spinal nerves.

The resulting **vagus nerve trunk,** after it has been joined by the cranial part of the accessory nerve just distal to the inferior ganglion, passes vertically down the neck within the carotid sheath, deep to and between the internal jugular vein and the internal and common carotid arteries. Beyond the root of the neck the course of the nerve differs on the two sides of the body.

The **right vagus nerve** crosses the first part of the subclavian artery, lying superficial to it and between it and the brachiocephalic vein (Fig. 12-24). It continues along the side of the trachea to the dorsal aspect of the root of the lung, where it spreads out in the posterior pulmonary plexus (Fig. 12-25). Below this plexus, it splits into cords that enter into the formation of the plexus on the dorsal aspect of the esophagus. After sending communications to the left vagus, these cords unite with each other and with com-

munications from the left vagus to form a single trunk, the *posterior vagus nerve*, before passing through the esophageal hiatus in the diaphragm (Fig. 12-25). Below the diaphragm, the posterior vagus continues along the lesser curvature of the stomach on its posterior surface for a short distance and divides into a celiac and several gastric branches.

The **left vagus** enters the thorax between the left carotid and subclavian arteries, deep to the left brachiocephalic vein (Fig. 12-37). It crosses the left side of the arch of the aorta (Fig. 12-24), angling dorsally in its caudal-

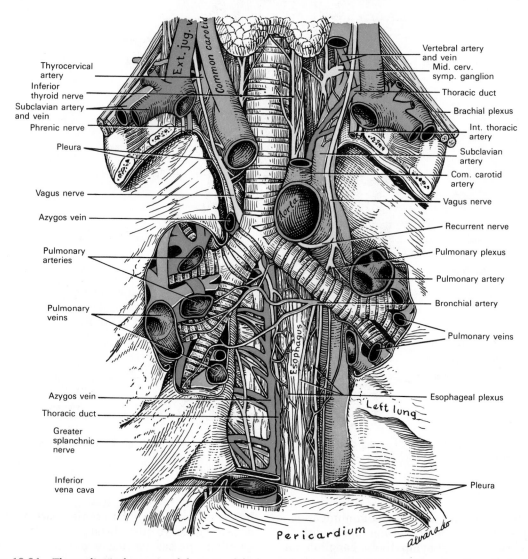

FIG. 12-24.   The mediastinal organs and the roots of the lungs after removal of the heart and pericardium. (Redrawn from Töndury.)

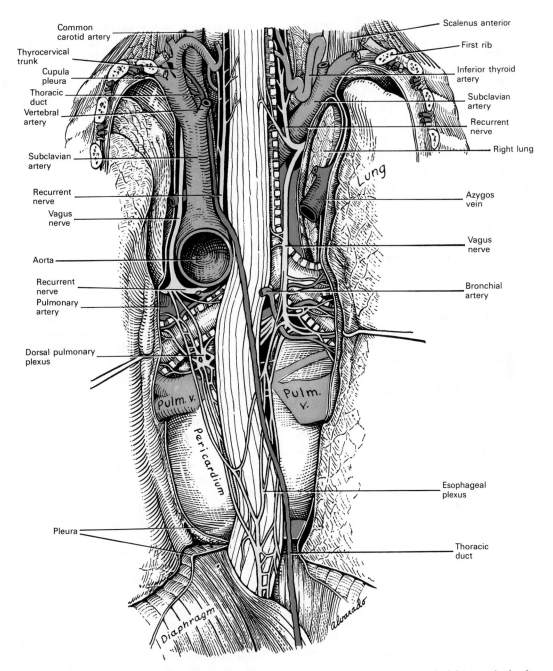

Common
carotid artery

Thyrocervical
trunk

Cupula
pleura

Thoracic
duct

Vertebral
artery

Subclavian
artery

Recurrent
nerve

Vagus
nerve

Aorta

Recurrent
nerve

Pulmonary
artery

Dorsal pulmonary
plexus

Pulm. v.

Pericardium

Pleura

Diaphragm

Scalenus anterior

First rib

Inferior thyroid
artery

Subclavian
artery

Recurrent
nerve

Right lung

Lung

Azygos
vein

Vagus
nerve

Bronchial
artery

Pulm.
v.

Esophageal
plexus

Thoracic
duct

Fig. 12-25.   Dorsal view of the mediastinal structures and roots of the lungs, after removal of the vertebral column, ribs, and thoracic aorta. (Redrawn from Töndury.)

ward course. It passes between the aorta and the left pulmonary artery just distal to the ligamentum arteriosum, and reaches the dorsal aspect of the root of the lung, where it flattens into the posterior pulmonary plexus (Fig. 12-25). It reaches the esophagus as a

variable number of strands; these follow the ventral aspect of the esophagus, send communications to the right vagus, and usually unite with each other and with substantial communications from the right vagus to form a single trunk, the *anterior vagus* nerve, before passing through the diaphragm. Below the diaphragm, the anterior vagus nerve, on the anterior aspect of the stomach, divides into an hepatic and several gastric branches (Jackson, 1949).

Branches in the Jugular Fossa

MENINGEAL BRANCH. The meningeal branch (dural branch) is a recurrent filament that arises at the superior ganglion and is distributed to the dura mater in the posterior cranial fossa.

AURICULAR BRANCH. The auricular branch (nerve of Arnold[1]) arises from the superior ganglion and soon communicates by a filament with the inferior ganglion of the glossopharyngeal nerve. It passes posterior to the internal jugular vein, and when it reaches the lateral wall of the jugular fossa, it enters the mastoid canaliculus, which crosses the facial canal in the bone about 4 mm superior to the stylomastoid foramen and communicates with the facial nerve. It is a somatic afferent nerve and it reaches the surface by passing through the tympanomastoid fissure. It divides into two branches: (a) one joins the posterior auricular nerve, and (b) the other is distributed to the skin of the back of the auricula and to the posterior part of the external acoustic meatus.

Branches of the Vagus Nerve in the Neck

PHARYNGEAL BRANCHES. The pharyngeal branches, usually two, arise at the upper part of the inferior ganglion; they contain sensory fibers from the ganglion and motor fibers from the communication with the accessory nerve. They pass across the internal carotid artery to the superior border of the constrictor pharyngis medius where they divide into several bundles that join branches of the glossopharyngeal, sympathetic, and external branch of the superior laryngeal nerves to form the **pharyngeal plexus** (Fig. 12-21).

Through the plexus, branches are distributed to the muscles and mucous membrane of the **pharynx,** and to the muscles of the **soft palate,** except the tensor veli palatini.

[1]Philipp Friedrich Arnold (1803–1890): A German anatomist (Zurich, Freiburg, Tübingen, and Heidelberg).

The nerves to the **carotid body** are filaments from the pharyngeal and possibly from the superior laryngeal branches, which join with similar filaments from the glossopharyngeal nerve and the superior cervical sympathetic ganglion to form the intercarotid plexus between the internal and external carotid arteries at the bifurcation. The vagus fibers are visceral afferents that terminate in the carotid body, which is a chemoreceptor sensitive to changes in oxygen tension of the blood. This is located at the carotid bifurcation (Sheehan et al., 1941).

SUPERIOR LARYNGEAL NERVE. The superior laryngeal nerve (Fig. 12-22) arises near the lower end of the inferior ganglion and passes caudalward and medialward deep to the internal carotid artery and along the pharynx toward the superior cornu of the thyroid cartilage. It has a communication with the superior cervical sympathetic ganglion and may contribute to the intercarotid plexus. It terminates by dividing into a smaller external and a larger internal branch.

The **external branch** (Fig. 12-22) continues caudalward beside the larynx, deep to the sternothyroid muscle and supplies motor fibers to the cricothyroid muscle and part of the constrictor pharyngis inferior. It contributes fibers to the pharyngeal plexus and communicates with the superior sympathetic cardiac nerve.

The **internal branch** swings anteriorly to reach the thyrohyoid membrane, which it pierces with the superior laryngeal artery. It supplies sensory fibers to the mucous membrane and parasympathetic secretomotor fibers to the associated glands through branches to the epiglottis, base of the tongue, aryepiglottic fold, and the larynx as far caudalward as the vocal folds. A filament passes caudalward beneath the mucous membrane on the inner surface of the thyroid cartilage and joins the recurrent nerve (Fig. 12-23).

SUPERIOR CARDIAC BRANCHES. Two or three superior cardiac branches arise from the vagus at the superior and inferior parts of the neck. The superior branches are small and communicate with the cardiac branches of the sympathetic. They can be traced to the deep part of the cardiac plexus. The inferior branch arises at the root of the neck just cranial to the first rib. On the right side it passes ventral or lateral to the brachiocephalic artery and joins the deep part of the cardiac plexus. On the left side it passes across the left side of the arch of the aorta and joins the superficial part of the cardiac plexus.

RECURRENT NERVE. The recurrent nerve (*inferior or recurrent laryngeal nerve*) (Fig. 12-25), as its name implies, arises far caudally and runs back cranialward in the neck to its destination, the muscles of the larynx. The origin and early part of its course are different on the two sides. **On the right side,** it arises in the root of the neck, as the vagus crosses superficial to the first part of the subclavian artery. It loops under the arch of this vessel and passes dorsal to it to the side of the trachea and esophagus (Fig. 12-74). **On the left side,** the recurrent nerve arises in the cranial part of the thorax, as the vagus crosses the left side of the arch of the aorta (Fig. 12-75). Just distal to the ligamentum arteriosum, it loops under the arch and passes around it to the side of the trachea. The further course on the two sides is similar; it passes deep to the common carotid artery and along the groove between the trachea and esophagus, medial to the overhanging deep surface of the thyroid lobe. Here it comes into close relationship with the terminal portion of the inferior thyroid artery. It runs under the caudal border of the constrictor pharyngis inferior muscle, enters the larynx through the cricothyroid membrane deep to the articulation of the inferior cornu of the thyroid with the cricoid cartilage, and is distributed to all the muscles of the larynx except the cricothyroid. Its branches are as follows:

Cardiac branches are given off as the nerve loops around the subclavian artery or the aorta. These are described below as the inferior cardiac branches of the vagus.

Tracheal and **esophageal branches,** more numerous on the left than on the right, are distributed to the mucous membranes and muscular coats (Fig. 12-23).

Pharyngeal branches are filaments to the constrictor pharyngis inferior.

Sensory and secretomotor filaments, which reach the recurrent through the communication with the internal branch of the superior laryngeal, supply the mucous membrane of the larynx below the vocal folds.

The **inferior laryngeal nerves** are the terminal branches that supply motor fibers to all the intrinsic muscles of the larynx except the cricothyroid.

VARIATIONS. When the right subclavian artery arises from the descending aorta, the recurrent nerve arises in the neck and passes directly to the larynx.

Branches of Vagus Nerve in the Thorax

INFERIOR CARDIAC BRANCHES. The inferior cardiac branches (*thoracic cardiac branches*) arise on the right side from the trunk of the vagus as it lies by the side of the trachea and from the recurrent nerve, and on the left side from the recurrent nerve only. They end in the deep part of the cardiac plexus.

The *visceral efferent fibers to the heart* in all the cardiac branches are preganglionic. After passage through the cardiac and coronary plexuses, these fibers form synapses with groups of ganglion cells in the heart wall, and the postganglionic fibers terminate about the conduction system and musculature of the heart.

The *visceral afferent fibers* from cells in the inferior ganglion traverse the cardiac plexus and cardiac nerves, supplying the heart and great vessels.

The *visceral afferent fibers* that supply the **aortic bodies** (*glomera aortica*) are carried mainly by the cardiac branches of the right vagus, and those that supply the **supracardial bodies** (*aortic paraganglia*) are carried mainly by those of the left vagus. These bodies are chemoreceptors similar to the carotid body; the cell bodies of the afferent fibers are in the inferior ganglion (Hollinshead; 1939, 1940).

DEPRESSOR NERVE OR NERVE OF CYON. In some animals, the afferent fibers of the vagus from the heart and great vessels are largely contained in a separate nerve whose stimulation causes depression of the activity of the heart. In man, these fibers are probably contained in the inferior cardiac branches (Mitchell, 1953).

ANTERIOR BRONCHIAL BRANCHES. The anterior bronchial branches (*anterior or ventral pulmonary branches*) are two or three small nerves on the ventral surface of the root of the lung that join with filaments from the sympathetic to form the **anterior pulmonary plexus.** From this plexus, filaments follow the ramifications of the bronchi and pulmonary vessels, or communicate with the cardiac or posterior pulmonary plexuses.

POSTERIOR BRONCHIAL BRANCHES. The posterior bronchial branches (*posterior or dorsal pulmonary branches*) (Fig. 12-25) are numerous offshoots from the main trunk of the vagus as it passes dorsal to the root of the lung. The vagus itself in this region is flat and spread out so that it combines with the bronchial branches and with the sympathetic communications to form what is called the **posterior pulmonary plexus.** The plexus has communications with the cardiac, aortic, and esophageal plexuses and its branches follow the ramifications of the bronchi and pulmonary vessels.

The *visceral efferent fibers* form synapses with small groups of ganglion cells in the walls of the bronchi, and the postganglionic fibers terminate in the nonstriated muscle and glands of the bronchi.

Afferent fibers supply the lungs and bronchi.

ESOPHAGEAL BRANCHES. The esophageal branches (Fig. 12-25) consist of superior filaments from the recurrent nerve and inferior branches from the trunk of the vagus and the esophageal plexus. They contain both visceral efferent (parasympathetic preganglionic) and visceral afferent fibers.

ESOPHAGEAL PLEXUS. The fibers of the vagus caudal to the root of the lung split into several bundles (Fig. 12-25), usually two to four larger bundles and a variable number of smaller parallel and communicating strands on each side, which spread out on the esophagus and become partly embedded in its adventitial coat. Filaments are given off for the innervation of the esophagus and there are communications with the splanchnic nerves and sympathetic trunk, forming as a whole the esophageal plexus. The bundles from the left vagus (Fig. 12-24) gradually swing around to the ventral surface of the esophagus; those from the right vagus (Fig. 12-25) swing to the dorsal surface. Just cranial to the diaphragm, the larger bundles from the left vagus usually combine with one or two strands from the right on the ventral surface of the esophagus to form a single

trunk, and this newly constituted vagus, which is not strictly the equivalent of the left vagus, passes through the esophageal hiatus of the diaphragm as the **anterior vagus.** A similar combination on the dorsal surface of the esophagus, mostly right vagus but with a communication from the left, passes through the esophageal hiatus as the **posterior vagus** (Jackson, 1949).

Branches of the Anterior and Posterior Vagi in the Abdomen

The branches in the abdomen (Figs. 12-71; 12-72) contain both *visceral efferent* (parasympathetic preganglionic) and *visceral afferent* fibers. The cells of the *postganglionic* fibers of the stomach and intestines are in the myenteric plexus (of Auerbach[1]) and the submucous plexus (of Meissner[2]); those of the glands are either in small local groups of ganglion cells or possibly in the celiac plexus.

GASTRIC BRANCHES. Usually four to six branches are given off by both anterior and posterior vagi at the cardiac end of the stomach. They fan out over their respective surfaces of the fundus and body and penetrate the wall to be distributed to the myenteric and submucous plexuses. On both anterior and posterior surfaces, one branch, longer than the others, follows along the lesser curvature and has been called the *principal nerve of the lesser curvature;* it is distributed to the pyloric vestibule rather than the pylorus itself.

HEPATIC BRANCHES. The hepatic branches from the anterior vagus are larger than those from the posterior vagus. They cross from the stomach to the liver in the lesser omentum and continue along the fissure for the ductus venosus to the porta hepatis, where they give off right and left branches to the liver. The large hepatic branch of the anterior vagus contributes to the plexus on the hepatic artery and has the following branches:

(1) Branches to the **gallbladder** and **bile ducts** come from the hepatic branches or the plexus on the artery.

[1]Leopold Auerbach (1828–1897): A German anatomist and neuropathologist (Breslau).
[2]Georg Meissner (1829–1905): A German physiologist (Basle and Göttingen).

(2) A **pancreatic branch** runs dorsalward to its destination.

(3) A branch along the right gastric artery is distributed to the **pylorus** and the first part of the **duodenum.**

(4) A branch accompanies the gastroduodenal artery and right gastroepiploic artery and is distributed to the **duodenum** and **stomach.**

CELIAC BRANCHES. A large terminal division of the posterior vagus follows the left gastric artery or runs along the crus of the diaphragm to the celiac plexus. The terminal branches cannot be followed once the nerve has entered the plexus, but the vagus fibers, through the secondary plexuses, reach the duodenum, pancreas, kidney, spleen, small intestine, and large intestine as far as the splenic flexure. Before the nerve enters the plexus it may give a branch to the superior mesenteric artery or to the aortic plexus (see Celiac Plexus).

XI. THE ACCESSORY NERVE

The **accessory nerve** (*eleventh nerve; spinal accessory nerve*) (Fig. 12-22) is a motor nerve consisting of two parts, a cranial and a spinal part.

CRANIAL PART. The cranial part (*r. internus; accessory portion*) arises by four or five delicate **cranial rootlets** from the side of the medulla oblongata, inferior to and in series with those of the vagus. From its deep origin and from its destination, it might well be considered a part of the vagus. It runs lateralward to the jugular foramen, where it interchanges fibers with the spinal part or becomes united with it for a short distance, and has one or two filaments of communication with the superior ganglion of the vagus. It then passes through the jugular foramen, separates from the spinal part and joins the vagus as the **ramus internus,** just proximal to the inferior ganglion. Its fibers are distributed through the pharyngeal branch of the vagus to the uvula, levator veli palatini, and the pharyngeal constrictor muscles, and through the superior and inferior laryngeal branches of the vagus to the muscles of the larynx and the esophagus.

SPINAL PART. The spinal part (*r. externus; spinal portion*) originates from motor cells in the lateral part of the ventral column of gray substance of the first five cervical segments of the spinal cord. The fibers pass through the lateral funiculus, emerge on the surface as the **spinal rootlets,** and join each other seriatim as they follow up the cord between the ligamentum denticulatum and the dorsal rootlets of the spinal nerves. The nerve passes through the foramen magnum into the cranial cavity, crosses the occipital bone to the jugular notch, and penetrates the dura mater over the jugular bulb as the **ramus externus** (Fig. 12-1). It passes through the jugular foramen in the same sheath of dura as the vagus, but is separated from it by a fold of the arachnoid. In the jugular foramen, it interchanges fibers with the cranial part or joins it for a short distance and separates from it again. At its exit from the foramen, it turns dorsalward, lying ventral to the internal jugular vein in two-thirds and dorsal to it in one-third of the bodies. It passes posterior to the stylohyoid and digastric muscles to the cranial part of the sternocleidomastoid muscle, which it pierces, and then courses obliquely caudalward across the posterior triangle of the neck to the ventral border of the trapezius muscle (Fig. 12-22). In the posterior triangle it is covered only by the outer investing layer of deep fascia, the subcutaneous fascia and the skin (Fig. 12-35).

COMMUNICATIONS AND BRANCHES. The accessory nerve communicates with the second, third, and fourth cervical nerves, and assuming a plexiform arrangement, it continues on the deep surface of the trapezius almost to its caudal border (Fig. 12-22). Experimental observations with monkeys indicate that the communications with the cervical nerves carry proprioceptive sensory fibers from cells in the dorsal root ganglia of the spinal nerves. The branches of the accessory nerve are as follows:

(1) **Sternocleidomastoid branches** are given off as the nerve penetrates this muscle.

(2) **Trapezius branches** are supplied from the part of the nerve lying deep to the muscle.

VARIATIONS. The lower limit of the origin of the spinal part may vary from C3 to C7. It may pass beneath the sternocleidomastoid muscle without piercing it, and in one instance it ended in that muscle, the trapezius being supplied by the third and fourth cervical nerves.

XII. HYPOGLOSSAL NERVE

The **hypoglossal nerve** (*twelfth nerve*) (Figs. 12-21; 12-36) is the motor nerve of the tongue. Its superficial origin from the medulla oblongata is by a series of rootlets in the ventrolateral sulcus between the pyramid and the olive.

The rootlets are collected into two bundles that perforate the dura mater separately, opposite the hypoglossal canal in the occipital bone, and unite after their passage through it; in some instances the canal is divided by a small bony spicule. As the nerve emerges from the skull, it is deeply placed beneath the internal carotid artery and the internal jugular vein and is closely bound to the vagus nerve. It runs caudalward and forward between the vein and artery, becomes superficial to them near the angle of the mandible, loops around the occipital artery, and passes anteriorly across the external carotid and lingual arteries caudal to the tendon of the digastric muscle (Fig. 12-36). It curves slightly cranialward above the hyoid bone, and passes deep to the tendon of the digastric and the stylohyoid muscle, between the mylohyoid and the hyoglossus muscles, and continues anteriorly among the fibers of the genioglossus as far as the tip of the tongue, distributing branches to the intrinsic muscles.

COMMUNICATIONS. The communications with the **vagus nerve** occur close to the skull; numerous filaments pass between the hypoglossal and the inferior ganglion of the vagus through the mass of connective tissue that binds the two nerves together. As the nerve winds around the occipital artery, it communicates by a filament with the **pharyngeal plexus.** The communication with the **sympathetic system** takes place opposite the atlas by branches of the superior cervical ganglion.

The **lingual nerve** communicates with the hypoglossal nerve near the anterior border of the hyoglossus muscle by numerous filaments that lie upon the muscle. The communication that occurs opposite the atlas, between the hypoglossal nerve and the loop connecting the anterior primary divisions of the **first** and **second cervical nerves** is especially significant because it contains the motor fibers for the nerves to the supra- and infrahyoid muscles. This communication probably also contains sensory fibers from the cranialmost cervical dorsal root ganglia.

BRANCHES. **Meningeal branches** are minute filaments that are given off in the hypoglossal canal and pass back to the dura mater of the posterior cranial fossa. They probably contain sensory fibers communicated to the hypoglossal nerve from the loop between the first and second cervical nerves.

The **superior root of the ansa cervicalis** (*descending hypoglossal*) (Figs. 12-26; 12-36) is a long slender branch that leaves the hypoglossal nerve as it loops around the occipital artery. It runs along the superficial surface of the carotid sheath to the middle of the neck, gives a branch to the superior belly of the omohyoid muscle, and becomes the medial arm of a loop, the **ansa cervicalis** (*ansa hypoglossi**). The lateral arm of the loop is the **inferior root of the ansa cervicalis** (*descending cervical*) from the second and third cervicals. The branches from the loop supply (1) the inferior belly of the omohyoid, (2) the sternohyoid, and (3) the sternothyroid muscles. The fibers in the descending hypoglossal nerve originate in the first cervical spinal segment, not in the hypoglossal nucleus, and pass through the communication between the first cervical nerve and the hypoglossal nerve described previously.

The **thyrohyoid branch** and the **geniohyoid branch** are also made up of fibers from the first cervical nerve. They leave the hypoglossal nerve near the posterior border of the hyoglossus muscle. The thyrohyoid branch runs obliquely across the greater cornu of the hyoid bone to reach the muscle it supplies.

The **muscular branches** that contain true hypoglossal fibers are distributed to the styloglossus, hyoglossus, genioglossus, and intrinsic muscles of the tongue. At the undersurface of the tongue, numerous slender

---

*The Paris Nomenclature has changed the name of this loop from ansa hypoglossi to ansa cervicalis. The names ansa hypoglossi, descending hypoglossal, and descending cervical are still important to mention because (1) there usually are several cervical loops and (2) the greatest importance of the name, ansa hypoglossi, is to the surgeon, who would identify it by its connection with the hypoglossal nerve rather than the cervical nerves. Knowledge of its cervical origin has been obtained largely through clinical observations and animal experimentation, not from dissection.

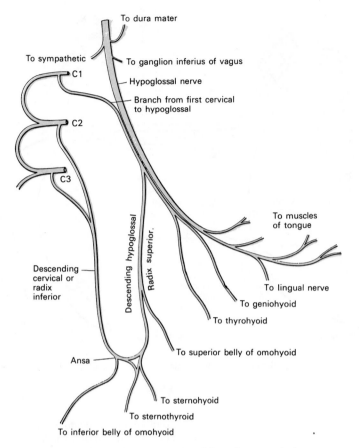

To dura mater

To sympathetic

To ganglion inferius of vagus

C1

Hypoglossal nerve

Branch from first cervical
to hypoglossal

C2

C3

To muscles
of tongue

Descending hypoglossal

Radix superior.

Descending
cervical or
radix
inferior

To lingual nerve

To geniohyoid

To thyrohyoid

To superior belly of omohyoid

Ansa

To sternohyoid

To sternothyroid

To inferior belly of omohyoid

FIG. 12-26. Plan of hypoglossal and first three cervical nerves.

branches pass cranialward into the substance of the organ to supply its intrinsic muscles. The branches of the hypoglossal nerve to the tongue probably contain proprioceptive sensory fibers with cells of origin in the first (if present) and second cervical dorsal root ganglia.

Spinal Nerves

The **spinal nerves** arise from the spinal cord within the spinal canal and pass out through the intervertebral foramina. The 31 pairs are grouped as follows: 8 cervical; 12 thoracic; 5 lumbar; 5 sacral; 1 coccygeal. The first cervical leaves the vertebral canal between the occipital bone and the atlas and is therefore called the **suboccipital nerve;** the eighth cervical nerve leaves between the seventh cervical and first thoracic vertebrae.

ROOTS OF THE SPINAL NERVES. Each spinal nerve is attached to the spinal cord by two roots, a ventral or motor root, and a dorsal or sensory root (Figs. 11-21; 12-27).

The **ventral root** (*anterior root; motor root*) emerges from the ventral surface of the spinal cord as a number of **rootlets** or filaments (*fila radicularia*), which usually combine to form two bundles near the intervertebral foramen.

The **dorsal root** (*posterior root; sensory root*) is larger than the ventral root because of the greater size and number of its rootlets; these are attached along the ventral lateral furrow of the spinal cord and unite to form two bundles, which enter the spinal ganglion.

The dorsal and ventral roots unite immediately beyond the spinal ganglion to form the spinal nerve, which then emerges through the intervertebral foramen. Both nerve roots receive a covering from the pia

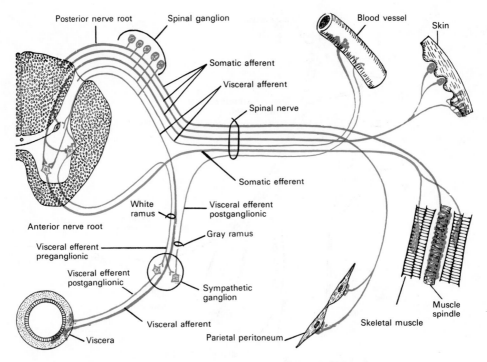

FIG. 12-27. Schema showing structure of a typical spinal nerve.

mater and are loosely invested by the arachnoid, the latter being prolonged as far as the points where the roots pierce the dura mater. The two roots pierce the dura separately, each receiving from this membrane a sheath which becomes continuous with the connective tissue of the epineurium after the roots join to form the spinal nerve.

The **spinal ganglion** (*ganglion spinale, dorsal root ganglion*) is a collection of nerve cells on the dorsal root of the spinal nerve. It is oval and proportional in size to the dorsal root on which it is situated; it is bifid medially where it is joined by the two bundles of rootlets. The ganglion is usually placed in the intervertebral foramen, immediately outside the dura mater, but there are exceptions to this rule; the ganglia of the first and second cervical nerves lie on the vertebral arches of the axis and atlas respectively, those of the sacral nerves are inside the vertebral canal, and that of the coccygeal nerve is within the sheath of the dura mater.

Size and Direction. In the cervical region, the roots of the first four nerves are small, the last four, large, and the dorsal roots, three times as large as the ventral, a larger proportion than in any other region; their individual filaments are also larger than in the ventral roots. The first cervical is an exception; its dorsal root is smaller than its ventral root. The roots of the first and second cervical nerves are short and run nearly horizontally to their exits from the vertebral canal. The roots of the third to the eighth nerves run obliquely caudalward, the obliquity and length successively increasing.

In the thoracic region the roots, with the exception of the first, are small, and the dorsal root is only slightly larger than the ventral root. They increase successively in length, and in the more caudal thoracic region they descend in contact with the spinal cord for a distance equal to the height of at least two vertebrae before they emerge from the vertebral canal.

The largest roots with the most numerous individual filaments are found in the lumbar and superior sacral regions. The roots of the coccygeal nerve are the smallest. The roots of the lumbar, sacral, and coccygeal nerves run vertically caudalward, and since the spinal cord ends near the lower border of the first lumbar vertebra, the roots of the successive segments are increasingly long. The name **cauda equina** is given to the resulting collection of nerve roots below the termination of the spinal cord. The largest nerve roots, and consequently the largest spinal nerves, are attached to the cervical and to the lumbar swellings of the spinal cord; these are the nerves largely distributed to the upper and lower limbs.

GRAY RAMI COMMUNICANTES. The gray rami communicantes (*postganglionic rami*) contain the postganglionic fibers from the adjacent sympathetic chain ganglia. These fibers are visceral efferents that run in the nerves and their branches toward the periphery, where they supply the smooth muscle in the blood vessel walls, the arrectores pilorum muscles, and the sweat glands. Since there are not as many sympathetic ganglia as there are spinal nerves, some ganglia supply rami to more than one nerve. Although variations are common, a simple plan may be given as follows: The first four cervical nerves receive their rami from the superior cervical ganglion; the fifth and sixth cervical nerves, from the middle ganglion; and the seventh and eighth cervical nerves, from the inferior cervical ganglion. The first ten thoracic nerves receive rami from corresponding ganglia, but the eleventh and twelfth receive rami from a single coalesced ganglion. The lumbar and sacral nerves receive their rami from a variable number of ganglia that correspond only approximately with the nerves.

Each spinal nerve usually receives two or three sympathetic gray rami communicantes. They join the nerve just distal to the union between the dorsal and ventral roots, and in the thoracic and upper lumbar region are regularly medial to the white rami communicantes. The latter are the branches of the spinal nerves that carry preganglionic fibers to the sympathetic chain. The ventral and dorsal primary divisions may each receive a ramus, but if the rami join the ventral division only, some of the fibers course back centrally until they can reach the dorsal division (Dass, 1952).

WHITE RAMUS COMMUNICANTES. The white ramus communicans (*preganglionic ramus*) is the branch of the spinal nerve through which the preganglionic fibers from the spinal cord reach the sympathetic chain and are thus the roots of the sympathetic ganglia. They arise from the twelve thoracic and first two lumbar nerves only and usually join the sympathetic chain at or near a ganglion. They leave the ventral primary division of the spinal nerve soon after it has emerged from the intervertebral foramen.

A small **meningeal branch** is given off from each spinal nerve immediately after it emerges from the intervertebral foramen. This branch reenters the vertebral canal through the foramen, supplies afferent fibers to the vertebrae and their ligaments, and carries sympathetic postganglionic fibers to the blood vessels of the spinal cord and its membranes.

PRIMARY DIVISIONS. The spinal nerve splits into its two primary divisions, ventral and dorsal, almost as soon as the two roots join, and both divisions receive fibers from both roots.

DORSAL PRIMARY DIVISIONS OF THE SPINAL NERVES

The **dorsal primary divisions** are smaller, as a rule, than the ventral divisions. As they arise from the spinal nerve, they are directed dorsalward, and with the exception of those of the first cervical, the fourth and fifth sacral, and the coccygeal, divide into **medial** and **lateral branches** for the supply of the muscles and skin of the dorsal part of the neck and trunk (Fig. 12-52).

Cervical Nerves

In the **first cervical** or **suboccipital nerve** the dorsal primary division is larger than the ventral. It emerges from the spinal canal superior to the posterior arch of the atlas and inferior to the vertebral artery to enter the suboccipital triangle. It supplies the muscles that bound this triangle, namely, the rectus .capitis posterior major and the obliqui superior and inferior, and it gives branches to the rectus capitis posterior minor and the semispinalis capitis (Fig. 12-28). A filament from the branch to the obliquus inferior joins the dorsal division of the second cervical nerve.

VARIATIONS. The first nerve occasionally has a cutaneous branch that accompanies the occipital artery to the scalp and communicates with the greater and lesser occipital nerves.

The dorsal division of the **second cervical nerve** is much larger than the ventral division and is the greatest of the cervical dorsal divisions. It emerges between the posterior arch of the atlas and the lamina of the axis, caudal to the obliquus inferior muscle, which it supplies. It communicates with the first cervical and then divides into a large medial branch and a small lateral branch.

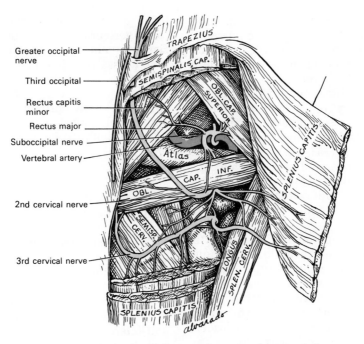

FIG. 12-28. Dorsal primary divisions of the upper three cervical nerves.

The **greater occipital nerve** (Figs. 12-28; 12-35) is the name given to the medial branch because of its size and distribution. It crosses obliquely between the obliquus inferior and the semispinalis capitis muscles, pierces the latter and the trapezius near their attachments to the occipital bone, and becomes subcutaneous (Fig. 12-28). It communicates with the third cervical nerve, runs upward on the back of the head with the occipital artery, and divides into branches that supply the scalp over the vertex and top of the head, communicating with the lesser occipital nerve. It gives muscular branches to the semispinalis capitis.

The **lesser occipital nerve** is described with the ventral primary divisions.

The **lateral branch** (*external branch*) supplies branches to the splenius and semispinalis capitis muscles and often communicates with the lateral branch of the third nerve.

The dorsal division of the **third cervical nerve** is intermediate in size between the second and fourth. Its medial branch runs between the semispinalis capitis and cervicis muscles, pierces the splenius and the trapezius, and gives branches to the skin.

The **third occipital nerve** (*least occipital nerve*) is the cutaneous part of the third nerve. It pierces the trapezius medial to the greater occipital nerve, with which it communicates, and is distributed to the skin of the lower part of the back of the head (Fig. 12-28). The lateral branch communicates with that of the second cervical nerve, supplies the same muscles, and gives a branch to the longissimus capitis muscle.

The dorsal primary divisions of the **fourth to eighth cervical nerves** divide into medial and lateral branches. The medial branches of the fourth and fifth cervical nerves run between the semispinalis capitis and cervicis, which they supply; near the spinous processes of the vertebrae they pierce the splenius and trapezius muscles to end in the skin (Fig. 12-31). Those of the lower three nerves are small and end in the semispinalis cervicis and capitis, the multifidus, and the interspinales. The lateral branches of the lower five nerves supply the splenius, iliocostalis cervicis, and the longissimus capitis and cervicis muscles.

VARIATIONS. The dorsal division of the first cervical nerve and the medial branches of the dorsal divisions of the second and third cervical nerves are sometimes joined by the communicating loops to form a **posterior cervical plexus** (of Cruveilhier).[1] The greater and lesser occipital nerves vary reciprocally with each other; the greater may communicate with the great auricular or posterior auricular nerves, and a branch to the auricula has been observed. The cutaneous branch of the fifth nerve may be lacking and the lower cervical nerves occasionally have cutaneous twigs.

Thoracic Nerves

The dorsal primary divisions of all the thoracic nerves have medial and lateral branches, but the cutaneous branches in the more cranial thorax are different from those in the more caudal thorax.

The **medial branch** (*internal branch*) (Fig. 12-52) from the dorsal divisions of the **upper six thoracic nerves** passes between and supplies the semispinalis and multifidus muscles, pierces the rhomboidei and trapezius, and approaching the surface close to the spinous process of the vertebra (Fig. 12-31), extends out laterally to the skin over the back. The medial branches of the lower six nerves end in the transversospinales and longissimus muscles, usually without cutaneous branches.

The **lateral branches** (*external branches*) run through or deep to the longissimus to the interval between it and the iliocostalis muscles, which they supply. They gradually increase in size from the first to the twelfth; the cranial six nerves end in the muscles, but the more caudal six have **cutaneous branches** that pierce the serratus posterior inferior and the latissimus dorsi along the junction between the fleshy and aponeurotic portions of the latter muscle (Fig. 12-31).

The cutaneous portions of both medial and lateral branches have a caudalward course that becomes more pronounced caudally, so that the twelfth nerve reaches down to the skin of the buttocks. The cutaneous part of the first thoracic nerve may be lacking. Both medial and lateral branches of some nerves may have cutaneous fibers, especially those of the sixth, seventh, and eighth thoracic nerves (Fig. 12-32).

Lumbar Nerves

The *medial branches* of the dorsal primary divisions of the lumbar nerves run close to the articular processes of the vertebrae and end in the multifidus muscle.

[1]Jean Cruveilhier (1791–1874): A French pathologist (Paris).

The **lateral branches** supply the sacrospinalis muscle. The upper three give off cutaneous nerves that pierce the aponeurosis of the latissimus dorsi at the lateral border of the sacrospinalis and cross the posterior part of the iliac crest to be distributed, as the **superior clunial nerves,** to the skin of the buttocks as far as the greater trochanter (Fig. 12-31).

Sacral and Coccygeal Nerves

The dorsal divisions of the sacral nerves are small and diminish in size distally; they emerge, except the last, through the posterior sacral foramina under cover of the multifidus muscle.

The **medial branches** of the first three are small and end in the multifidus.

The **lateral branches** of the first three nerves join with one another and with the last lumbar and fourth sacral nerves to form loops on the dorsal surface of the sacrum (Fig. 12-33). From these loops branches run to the dorsal surface of the sacrotuberous ligament and form a second series of loops under the gluteus maximus. From this second series two or three cutaneous branches pierce the gluteus maximus along a line from the posterior superior iliac spine to the tip of the coccyx, and they supply the skin over the medial part of the buttocks.

The dorsal divisions of the last two sacral nerves and the coccygeal nerves do not divide into medial and lateral branches, but unite with each other on the back of the sacrum, to form loops that supply the skin over the coccyx.

VENTRAL PRIMARY DIVISIONS OF THE SPINAL NERVES

The **ventral primary divisions** of the spinal nerves supply the ventral and lateral parts of the trunk and all parts of the limbs. They are, for the most part, larger than the dorsal divisions. In the thoracic region they remain independent of one another, but in the cervical, lumbar, and sacral regions they unite near their origins to form plexuses.

Cervical Nerves

The ventral division of the **first cervical** or **suboccipital nerve** issues from the vertebral canal cranial to the posterior arch of the

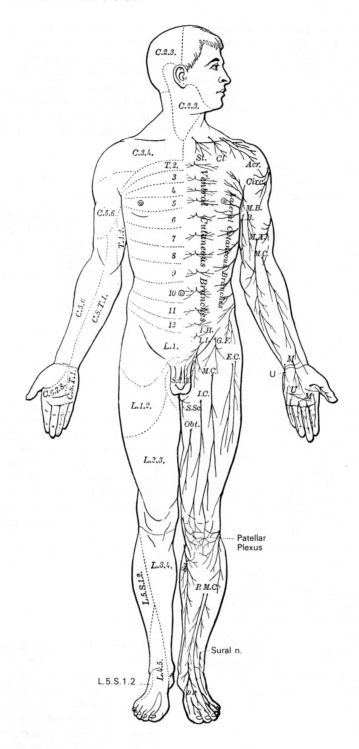

FIG. 12-29. Distribution of cutaneous nerves. Ventral aspect.

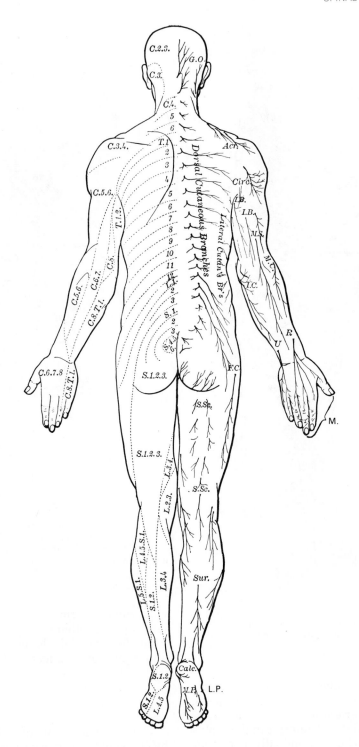

Fɪɢ. **12-30.** Distribution of cutaneous nerves. Dorsal aspect.

Fɪɢ. **12-29, 12-30.** These figures (and others like them) show areas of distribution on the skin supplied by each of the spinal nerves. The work of Sherrington has demonstrated that the "sensory root field" of a particular dorsal root ganglion overlaps that of the zones or "dermatomes" supplied by the ganglion above and below (Fig. 12-32). In fact, fibers carrying different modalities, i.e., pain and touch, vary in the amount of this overlap.

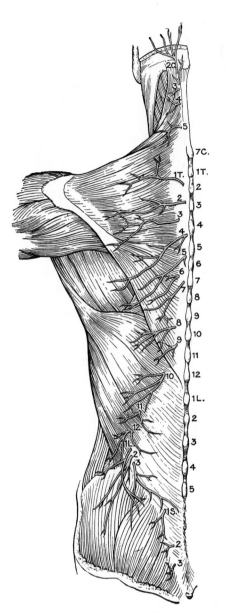

Fig. 12-31.   Diagram of the distribution of the cutaneous branches of the dorsal divisions of the spinal nerves.

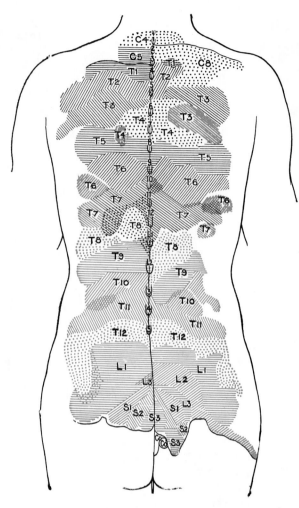

Fig. 12-32.   Areas of distribution of the cutaneous
branches of the dorsal divisions of the spinal nerves. The
areas of the medial branches are in black; those of the
lateral, in red. (H. M. Johnston.)

atlas and runs rostralward around the lateral
aspect of the superior articular process,
medial to the vertebral artery. In most instances it is medial and anterior to the rectus
capitis lateralis, but occasionally it pierces
the muscle.

The ventral divisions of the other **cervical
nerves** pass outward between the anterior
and posterior intertransversarii, lying on the
grooved cranial surfaces of the transverse
processes of the vertebrae. The first four cervical nerves form the cervical plexus; the
last four, together with the first thoracic,
form the brachial plexus. They all receive
gray rami communicantes from the sympathetic chain.

Cervical Plexus

The **cervical plexus** (Fig. 12-34) is formed
by the ventral primary divisions of the first
four cervical nerves; each nerve, except the
first, divides into a superior and inferior
branch, and the branches unite to form three
loops. The sympathetic rami may join the
nerves or the loops. The plexus is situated
opposite the cranial four cervical vertebrae,

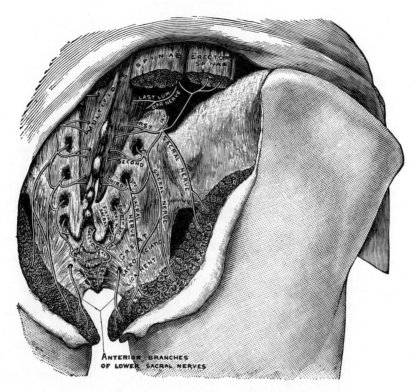

FIG. 12-33. The dorsal divisions of the sacral nerves.

ventrolateral to the levator scapulae and scalenus medius and deep to the sternocleidomastoid muscle.

The cervical plexus has communications with certain cranial nerves and muscular and cutaneous branches, which may be arranged in tabular form as follows; the numbers indicate the segmental components.

A. Communications of the Cervical Plexus

1. With vagus nerve	C 1, 2
2. With hypoglossal nerve	C 1, 2
3. With accessory nerve	C 2, 3, 4

B. Superficial or Cutaneous Branches

1. Lesser occipital	C 2
2. Great auricular	C 2, 3
3. Anterior cutaneous*	C 2, 3
4. Supraclavicular	C 3, 4

C. Deep or Muscular Branches

5. Rectus capitis anterior	C 1, 2
and lateralis	C 1
6. Longus capitus	C 1, 2, 3
and colli	C 2, 3, 4
7. Geniohyoid	C 1, (2)
Thyrohyoid	C 1, (2)
Omohyoid (superior)	C 1, (2)
8. Sternohyoid	C 2, 3
Sternothyroid	C 2, 3
Omohyoid (inferior)	C 2, 3
9. Phrenic	C 3, 4, 5
10. Sternocleidomastoid	C 2, 3
11. Trapezius	C 3, 4
12. Levator scapulae	C 3, 4
13. Scalenus medius	C 3, 4

*This name conforms with names in more caudal segments.

COMMUNICATIONS OF THE CERVICAL PLEXUS

VAGUS NERVE. The communication with the **vagus nerve** is between the loop connecting the first and second nerves and the inferior ganglion.

HYPOGLOSSAL NERVE. The communication with the **hypoglossal nerve** (Figs. 12-26; 12-34) is a short bundle that leaves the loop between the first and second nerves. The great bulk of its fibers are from the first nerve; it runs distally with hypoglossal nerve for 3 or 4 cm and leaves it again as the superior root (*descending hypoglossal*). It contains motor

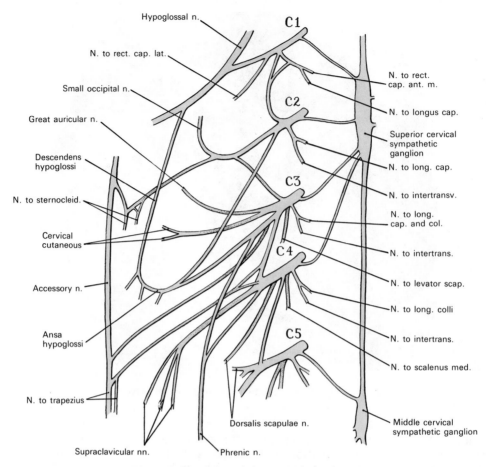

FIG. 12-34. Plan of the cervical plexus.

and proprioceptive fibers for certain hyoid muscles and is described later as part of the ansa cervicalis (*ansa hypoglossi*).

ACCESSORY NERVE. The communications with the **accessory nerve** (Fig. 12-34) leave the cervical plexus at several points: (*a*) one leaves the loop between the second and third nerves, frequently appearing to come from the lesser occipital nerve (C 2), and joins the fibers of the accessory, which supply the sternocleidomastoid; (*b*) a bundle leaves the third nerve, sometimes in association with the great auricular nerve, and joins the accessory fibers to the trapezius; (*c*) one or two bundles leave the fourth nerve and join the accessory nerve directly or enter into a network on the deep surface of the trapezius. These communications contain proprioceptive sensory fibers (Corbin and Harrison, 1939).

SUPERFICIAL OR CUTANEOUS BRANCHES OF THE CERVICAL PLEXUS

LESSER OCCIPITAL NERVE. The lesser occipital nerve (Fig. 12-35) arises from the second cervical nerve or the loop between the second and third nerves and ascends along the sternocleidomastoid muscle, curving around its posterior border. Near the insertion of the muscle on the cranium, it perforates the deep fascia and continues upward along the side of the head behind the ear, supplying the skin and communicating with the greater occipital nerve, the great auricular nerve, and the posterior auricular branch of the facial nerve. It has an auricular branch that supplies the upper and back part of the auricula, communicating with the mastoid branch of the great auricular nerve.

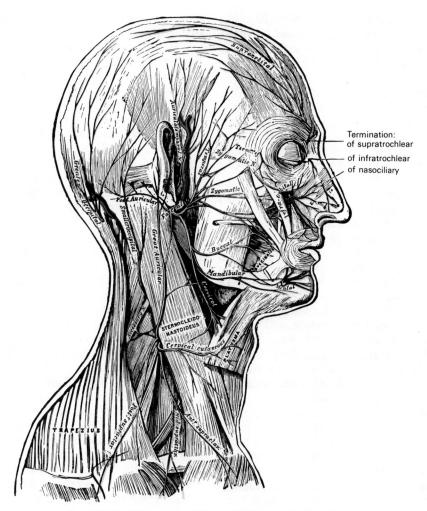

Termination:
of supratrochlear
of infratrochlear
of nasociliary

FIG. 12-35. The nerves of the scalp, face, and side of neck.

VARIATIONS. The lesser occipital nerve varies reciprocally with the greater occipital nerve; it is frequently duplicated or may be wanting.

GREAT AURICULAR NERVE. The great auricular nerve (Fig. 12-35), larger than the preceding, arises from the second and third nerves, winds around the posterior border of the sternocleidomastoid muscle, and after perforating the deep fascia, ascends on the surface of that muscle but deep to the platysma and divides into an anterior and a posterior branch.

The **anterior branch** (*facial branch*) is distributed to the skin of the face over the parotid gland. It communicates in the substance of the gland with the facial nerve.

The **posterior branch** (*mastoid branch*) supplies the skin over the mastoid process and back of the auricula except its upper part; a filament pierces the auricula to reach its lateral surface, where it is distributed to the lobule and lower part of the concha. It communicates with the smaller occipital nerve, the auricular branch of the vagus, and the posterior auricular branch of the facial nerve.

ANTERIOR CUTANEOUS NERVE. The anterior cutaneous nerve (Fig. 12-35) arises from the second and third cervical nerves and bends around the posterior border of the sternocleidomastoid at its middle. Crossing the surface of the muscle obliquely as it runs horizontally ventralward, it perforates the deep fascia, passing deep to the external jugular vein, pierces the platysma, and divides into ascending and descending branches.

The **ascending branches** (rr. *superiores*) pass cranialward to the submandibular region, pierce the platysma, and are distributed to the cranial, ventral, and lateral parts of the neck. One filament accompanies the external jugular vein toward the angle of the mandible and communicates with the cervical branch of the facial nerve under cover of the platysma.

The **descending branches** (rr. *inferiores*) pierce the platysma and are distributed to the skin of the ventral and lateral parts of the neck as far down as the sternum.

SUPRACLAVICULAR NERVES. The supraclavicular nerves (*descending branches*) (Fig. 12-35) arise from the third and fourth (mainly the fourth) cervical nerves. They emerge from under the posterior border of the sternocleidomastoid muscle and cross the posterior triangle of the neck under cover of the investing layer of deep fascia. Near the clavicle they perforate the fascia and platysma in three bundles or groups—medial, intermediate and lateral.

The **medial supraclavicular nerves** (*anterior supraclavicular nerves; suprasternal nerves*) cross the external jugular vein, the clavicular head of the sternocleidomastoid muscle, and the clavicle to supply the skin of the medial infraclavicular region as far as the midline. They furnish one or two filaments to the sternoclavicular joint.

The **intermediate supraclavicular nerves** (*middle supraclavicular nerves*) cross the clavicle and supply the skin over the pectoralis major and the deltoid and they communicate with the cutaneous branches of the more cranial intercostal nerves.

The **lateral supraclavicular nerves** (*posterior supraclavicular nerves; supracromial nerves*) pass obliquely across the outer surface of the trapezius and the acromion, and supply the skin of the cranial and dorsal parts of the shoulder.

VARIATIONS. One of the middle supraclavicular nerves may perforate the clavicle.

DEEP OR MUSCULAR BRANCHES OF THE CERVICAL PLEXUS

RECTUS CAPITIS BRANCHES. Branches to the rectus capitis anterior and rectus capitis

lateralis come from the loop between the first and second nerves.

LONGUS CAPITIS AND COLLI BRANCHES. Branches to the longus capitis and longus colli are given off separately; for the capitis from the first, second, and third; for the colli from the second, third, and fourth nerves.

HYOID MUSCULATURE BRANCHES. Branches to the hyoid musculature (Fig. 12-36) are described previously as branches of the hypoglossal nerve, but the fibers do not come from the hypoglossal nucleus in the brain; they originate instead from the cervical segments of the spinal cord. Fibers communicated to the hypoglossal nerve from the first cervical nerve or from the loop between the first and second nerves leave the hypoglossal nerve again to appear as (1) individual branches to the **geniohyoid, thyrohyoid** or **superior belly** of the **omohyoid,** or (2) as the **superior root** (*descendens hypoglossi*) to join the **inferior root** (*descendens cervicalis*) from the second and third cervical nerves to form the **ansa cervicalis** (*ansa hypoglossi*).

The nerve to the **geniohyoid** leaves the hypoglossal nerve distally and has a course similar to that of the thyrohyoid nerve.

The nerve to the **thyrohyoid** (Fig. 12-36) leaves the hypoglossal nerve near the posterior border of the hyoglossus and runs obliquely across the greater cornu of the hyoid bone and enters the muscle as a slender filament. It leaves the hypoglossal distal to the origin of the root superior of the ansa.

The fibers for the **superior belly** of the **omohyoid** leave the superior root before it forms the ansa and reach the muscle as a slender filament.

ANSA CERVICALIS. The ansa cervicalis (*ansa hypoglossi*) is a loop of slender nerves that may be somewhat plexiform, ventral and lateral to the common carotid artery and internal jugular vein. It is formed by fibers from the first cervical spinal segment that leave the hypoglossal nerve as the **superior root** (*descendens hypoglossi*), and the **inferior root** (*descendens cervicalis*) from the second and third cervical nerves. It usually lies superficial to the carotid sheath at about the level of the cricoid cartilage.

The **nerves** to the **sternohyoid** and **sternothyroid** muscles leave the convexity of the loop and run down the superficial surface of the carotid artery to enter the muscles at the root of the neck. They may be separate fila-

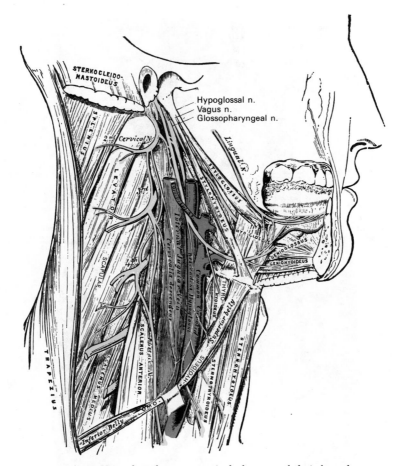

Fɪɢ. 12-36. Hypoglossal nerve, cervical plexus, and their branches.

ments or combined into a single nerve for a variable distance.

The nerve to the **inferior belly** of the **omohyoid** leaves the convexity of the loop and runs lateralward across the neck deep to the intermediate tendon of the muscle to reach the inferior belly.

VARIATIONS. The ansa is quite variable in position; it may occur at various levels. Frequently it is high near the bifurcation of the carotid, in which circumstance it may be within the carotid sheath. The superior root may appear to arise wholly or in part from the vagus. A branch to the sternocleidomastoid and filaments entering the thorax to join the vagus or sympathetic have been described.

PHRENIC NERVE. The **phrenic nerve** (*internal respiratory nerve of Bell*[1]) (Fig. 12-37) is generally known as the motor nerve to the

[1]Sir Charles Bell (1774–1842): A Scottish physiologist who worked in London.

diaphragm, but it contains about half as many sensory as motor fibers, and it should not be forgotten that the lower thoracic nerves also contribute to the innervation of the diaphragm. The phrenic nerve originates chiefly from the fourth cervical nerve but is augmented by fibers from the third and fifth nerves. It lies on the ventral surface of the scalenus anterior, gradually crossing from its lateral to its medial border (Fig. 12-36). Under cover of the sternocleidomastoid, it is crossed by the inferior belly of the omohyoid and the transverse cervical and suprascapular vessels. It continues with the scalenus anterior between the subclavian vein and artery, and as it enters the thorax, it crosses the origin of the internal thoracic artery and is joined by the pericardiacophrenic branch of this artery. It passes caudalward over the cupula of the pleura and ventral to the root of the lung, then along the lateral aspect of

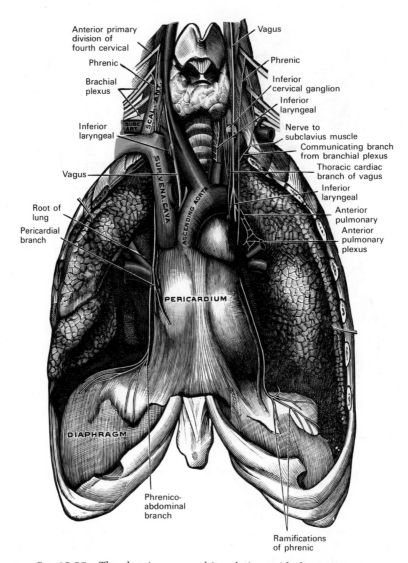

FIG. 12-37.   The phrenic nerve and its relations with the vagus nerve.

the pericardium, between it and the mediastinal pleura, until it reaches the diaphragm, where it divides into its terminal branches. At the root of the neck it is joined by a communication from the sympathetic trunk.

The right nerve is more deeply placed, is shorter, and runs more vertically caudalward than the left. In the upper part of the thorax it is lateral to the right brachiocephalic vein and the superior vena cava (Figs. 12-37; 12-78).

The left nerve is longer than the right because of the inclination of the heart toward the left and because of the more caudal position of the diaphragm on this side. At the

root of the neck it is crossed by the thoracic duct, and in the superior mediastinum it lies between the left common carotid and the subclavian arteries, and is lateral to the vagus as it crosses the left side of the arch of the aorta (Figs. 12-37; 12-79).

The **pleural branches** of the phrenic nerve are very fine filaments supplied to the costal and mediastinal pleura over the apex of the lung.

The **pericardial branches** are delicate filaments to the upper part of the pericardium.

The **terminal branches** pass through the diaphragm separately, and diverging from each other, are distributed on the abdominal surface to supply the diaphragm muscle and sensory fibers to the peritoneum. On the right side, a branch near the inferior vena cava communicates with the phrenic plexus, which accompanies the inferior phrenic artery from the celiac plexus, and where they join there is usually a small **phrenic ganglion.** On the left, there is a communication with the phrenic plexus also, but without a ganglion.

The **accessory phrenic nerve** is described later.

VARIATIONS. The phrenic nerve may receive fibers from the inferior root of the ansa cervicalis, or from the second or sixth nerves. At the root of the neck or in the thorax it may be joined by an accessory phrenic from the fifth nerve or from the nerve to the subclavius. It may arise from the nerve to the subclavius or give a branch to that muscle. It may pass ventral to the subclavian vein or perforate it.

OTHER BRANCHES. The branches to the **sternocleidomastoid** may be independent or partly communicate with the accessory nerve. They are proprioceptive sensory rather than motor nerves and are derived from the second and third nerves (Corbin and Harrison, 1938).

The branches to the **trapezius** are like those to the sternocleidomastoid and are derived from the third and fourth nerves.

Muscular branches to the **levator scapulae** are supplied by the third and fourth nerves.

Branches to the **scalenus medius** are from the third and fourth nerves.

Brachial Plexus

The **brachial plexus** (Figs. 12-38; 12-39), as its name implies, supplies the nerves to the upper limb. It is formed by the ventral primary divisions of the fifth to eighth cervical nerves and the first thoracic nerves. A communicating loop from the fourth to the fifth cervical nerve and one from the second to the first thoracic nerve also usually contribute to the plexus. It lies in the lateral part of the neck in the clavicular region, extending from the scalenus anterior to the axilla.

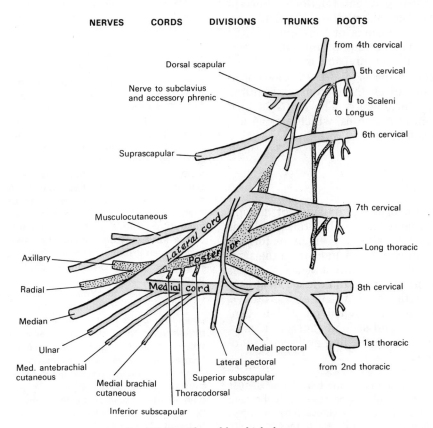

Fig. 12-38. Plan of brachial plexus.

COMPONENTS

The brachial plexus is composed of roots, trunks, divisions, cords, and terminal nerves. The **roots** of the brachial plexus are provided by the anterior primary divisions of the last four cervical nerves and the first thoracic nerve. The trunks are formed from these roots and are named according to their position relative to each other. The **superior trunk** is formed by the union of the fifth and sixth cervical nerves as they emerge between the scalenus medius and anterior. The **middle trunk** is formed by the seventh cervical nerve alone. The **inferior trunk** is formed by the union of the eighth cervical and first thoracic nerves.

The trunks, after a short course, split into **anterior** and **posterior divisions.** The anterior and posterior divisions of the superior and middle trunks are about equal in size, but the posterior division of the inferior trunk is much smaller than the anterior division because it receives a very small or no contribution from the first thoracic nerve. The **cords,** formed from these divisions, are named according to their relation to the axillary artery: **lateral, medial,** and **posterior.** The anterior divisions of the superior and middle trunks are united into the lateral cord. The anterior division of the inferior trunk becomes the medial cord. The posterior divisions of all three trunks are united into the posterior cord. The cords in turn break up into the nerves, which are the terminal branches.

SYMPATHETIC CONTRIBUTIONS
TO THE BRACHIAL PLEXUS

The ventral primary divisions of the spinal nerves that enter into the brachial plexus obtain their sympathetic fibers in the form of the gray rami communicantes from the sympathetic chain. The fifth and sixth nerves receive fibers from the middle cervical ganglion; the sixth, seventh, and eighth nerves, from the inferior cervical or stellate ganglion; the first and second thoracic nerves, from the stellate or the first and second thoracic ganglia.

RELATIONS. In the neck, the brachial plexus lies in the posterior triangle, being covered by the skin, platysma, and deep fascia; it is crossed by the supraclavicular nerves, the inferior belly of the omohyoid, the external jugular vein, and the transverse cervical artery. The roots emerge between the scaleni ante-

rior and medius, cranial to the third part of the subclavian artery, while the trunk formed by the union of the eighth cervical and first thoracic nerves is placed dorsal to the artery. The plexus next passes dorsal to the clavicle, the subclavius, and the transverse scapular vessels, and lies upon the first digitation of the serratus anterior and the subscapularis. In the axilla it is placed lateral to the first portion of the axillary artery; it surrounds the second part of the artery, one cord lying medial to it, one lateral to it, and one dorsal to it. In the axilla it gives off its terminal branches to the upper limb.

VARIATIONS. Variations of the brachial plexus are of several types: (a) variations in the contributions of the spinal nerves to the roots of the plexus, (b) variations in the formation of the trunks, divisions, or cords, (c) variations in the origin or combination of the branches, (d) variations in the relation to the artery.

(a) The fourth cervical nerve contributes to two-thirds of the plexuses, and T 2 contributes to more than one-third. When the contribution from C 4 is large and that from T 2 lacking, the plexus appears to have a more cranial position and has been termed **prefixed.** Similarly, when the contribution from T 2 is large and from C 4 lacking, the plexus appears to have a more caudal position and has been termed **postfixed.** It is doubtful whether this shifting of position is more common than one in which the plexus is spread out to include both C 4 and T 2, or contracted to exclude both. Variations in the contribution to the plexus may be correlated with the position of the limb bud at the time the nerves first grow into it in the embryo (Miller and Detwiler, 1936), and many variations are similar to the usual conditions found in the different primates.

(b) The trunks vary little in their formation from the cervical roots, but the superior and inferior trunks especially may appear to be absent because the nerves split into dorsal and ventral divisions before they combine into trunks. The cords in these instances are formed from the divisions of the nerves but the sources of the fibers can be readily traced and made to correspond with the usual pattern. Many of these instances may be the result of too vigorous a removal of the connective tissue sheaths of the nerves in dissection. The medial cord may receive a contribution from the middle trunk and the lateral cord may receive fibers from C 8 or the lower trunk.

(c)The median nerve may have small heads, either medial or lateral, in addition to the usual two. It appears to receive fibers from all segments entering the plexus in most instances. Many peculiarities involve combined origins of the median and musculocutaneous nerves, with separation into definitive nerves or branches farther down the arm; thus the musculocutaneous nerve gives a branch to the median in the arm in a fourth of the bodies, but a branch from the median to the ulnar is much less frequent. The musculocutaneous nerve frequently receives fibers from C 4 in addition to the usual C 5 and C 6, and appears to receive fibers from C 7 in only two-thirds. The nerve to the coracobrachialis muscle is a branch of the lateral cord or some part of the plexus (exclusive of C 8 and T 1) other than the musculocutaneous nerve in almost half of the bodies. The ulnar nerve may have a lateral head from the lateral cord, the lateral head of the median, or from C 7; in two-thirds of the plexuses the ulnar nerve may receive fibers from C 7 or possibly more cranial segments. The radial and axillary nerves may be formed from the trunks and divisions without the presence of a true posterior cord. The radial nerve appears to receive fibers from all segments contributing to the plexus in most instances, but participation of C 4 and T 1 is probably incidental because C 4 may not enter the plexus and in a few instances T 1 can be definitely excluded. The axillary nerve probably receives no fibers from C 8 and T 1 and the contribution from C 7 is undetermined.

(*d*) Among the variations in the relationship between the axillary artery and the brachial plexus, the most common is an artery superficial to the median nerve; also, the median nerve may be split by a branch of the artery. An aberrant axillary artery, i.e., one not derived from the seventh segmental artery, has a different relation to the roots of the plexus depending on whether it was derived from a segmental artery cranial or caudal to the seventh. The cords of the plexus may be split by arterial branches, and communicating loops of nerves may be formed around the artery or its branches.

	A. Branches from the Cervical Nerves	
1.	To the phrenic nerve	C 5
2.	To longus colli and the scaleni	C 5, 6, 7, 8
3.	Accessory phrenic	C 5

	B. Branches from the Roots	
4.	Dorsal scapular	C 5
5.	Long thoracic	C 5, 6, 7

	C. Branches from the Trunks	
6.	Nerve to the subclavius	C 5, 6
7.	Suprascapular	C 5, 6

	D. Branches from the Cords	
8.	Pectoral	C 5, 6, 7, 8, T 1
9.	Subscapular	C 5, 6
10.	Thoracodorsal	C 6, 7, 8
11.	Axillary	C 5, 6
12.	Medial brachial cutaneous	C 8, T 1
13.	Medial antebrachial cutaneous	C 8, T 1

	E. Terminal Nerves	
14.	Musculocutaneous	C 5, 6, 7
15.	Median	C 6, 7, 8, T 1
16.	Ulnar	C 8, T 1
17.	Radial	C 5, 6, 7, 8 (T 1)

BRANCHES OF THE BRACHIAL PLEXUS

The ventral primary divisions of the fifth to eighth cervical nerves give branches before they enter into the plexus, and the brachial plexus may be divided into the branches that arise from the roots, from the trunks, or from the cords and pass into the terminal nerves. From their topographic relation to the clavicle, the branches may be divided into supra- and infraclavicular branches.

Branches from the Cervical Nerves

Branches of the anterior primary divisions of the last four cervical and first thoracic nerves **before** they enter the **brachial plexus** include:

1. The fifth cervical nerve may contribute to the **phrenic nerve** at its origin.

2. **Muscular branches** from each of the four lower cervical nerves are supplied to the **longus colli** and the **scalenus anterior, medius,** and **posterior.**

3. **The accessory phrenic nerve** (Fig. 12-37) is an inconstant branch that may come from the nerve to the subclavius or from the fifth nerve. It passes ventral to the subclavian vein and joins the phrenic nerve at the root of the neck or in the thorax, forming a loop around the vein.

SURGICAL CONSIDERATIONS. Resection of the phrenic nerve for immobilization of the diaphragm may be only partially successful if the accessory is not resected also. In avulsion of the phrenic nerve, the subclavian vein is in danger of being torn by the loop between the accessory and the phrenic nerves.

Branches from the Roots (Supraclavicular Branches)

DORSAL SCAPULAR NERVES. The dorsal scapular nerve (*nerve to the rhomboidei; posterior scapular nerve*) (Fig. 12-38) arises from the fifth cervical nerve near the intervertebral foramen, frequently in common with a root of the long thoracic nerve. It pierces the scalenus medius and runs dorsally as well as caudalward on the deep surface of the levator scapulae to the vertebral border of the scapula. It supplies the rhomboideus major and minor, and along with the third and fourth nerves, gives a branch to the levator scapulae muscle.

LONG THORACIC NERVE. The long thoracic nerve (*external respiratory nerve of Bell; posterior thoracic nerve*) (Fig. 12-39) is the

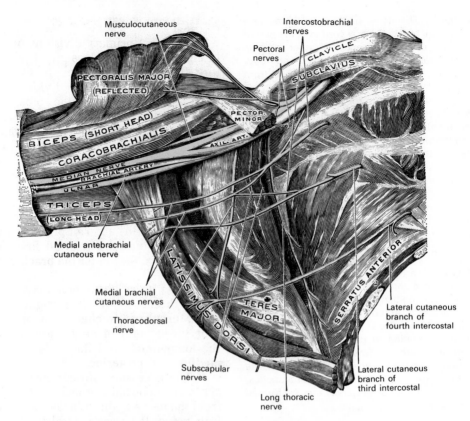

FIG. 12-39. The right brachial plexus (infraclavicular portion) in the axillary fossa; viewed from below and in front. The pectoralis major and minor muscles have been largely removed; their attachments have been reflected. (Spalteholz.)

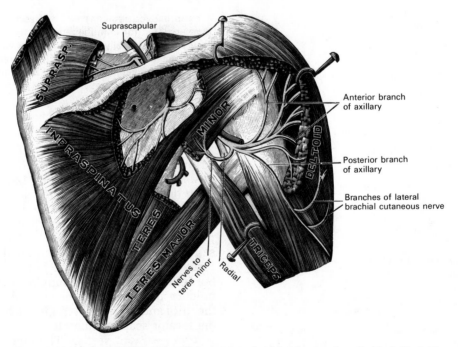

FIG. 12-40. Suprascapular and axillary nerves of right side, seen from behind. (Testut.)

nerve to the serratus anterior muscle. It arises by three roots: from the fifth, sixth, and seventh cervical nerves. Those from the fifth and sixth join just after they pierce the scalenus medius; the seventh joins them at the level of the first rib. The nerve runs caudalward dorsal to the brachial plexus and axillary vessels and continues along the lateral surfaces of the serratus anterior, under cover of the subscapularis muscle. It sends branches to all the digitations of the serratus; the fibers from the fifth nerve supply the upper part, the sixth the middle, and the seventh the lower part of the muscle.

VARIATIONS. The root from the fifth nerve may remain independent; the root from the seventh may be lacking.

Branches from the Trunks (Supraclavicular)

NERVE TO SUBCLAVIUS MUSCLE. The nerve to the subclavius is a small branch that arises from the superior trunk, although its fibers are mainly from the fifth nerve, and passes ventral to the distal part of the plexus, the subclavian artery, and the subclavian vein to reach the subclavius muscle (Fig. 12-37).

VARIATIONS. The accessory phrenic nerve may arise from the subclavius nerve, or the phrenic nerve may supply the branch to the subclavius muscle.

SUPRASCAPULAR NERVE. The suprascapular nerve (Fig. 12-40) arises from the superior trunk and takes a more or less direct course across the posterior triangle to the scapular notch, passing dorsal to the inferior belly of the omohyoid muscle and the anterior border of the trapezius. It passes through the notch, under the superior transverse ligament, runs deep to the supraspinatus muscle and around the lateral border of the spine of the scapula into the infraspinous fossa. In the supraspinous fossa it gives a branch to the supraspinatus and an articular filament to the shoulder joint. In the infraspinous fossa, it gives two branches to the infraspinatus and filaments to the shoulder joint and scapula.

Branches from the Cords (Infraclavicular Branches)

The infraclavicular branches arise from the three cords of the brachial plexus, but it should be emphasized that a particular branch of any cord need not contain fibers from all the cervical nerves contributing to that cord. For example, the axillary nerve, from the posterior cord, contains fibers from C 5 and 6 only, not from C 5, 6, 7, 8, and T 1. Likewise, a branch from one of the larger terminal nerves may not contain fibers from all the cervical segments contributing to that nerve; the branch of the radial nerve to the supinator muscle, for example, contains only fibers from C 6.

PECTORAL NERVES. The pectoral nerves (Fig. 12-46) are two nerves, one lateral and one medial to the axillary artery, that arise at the level of the clavicle and supply the pectoral muscles.

The **lateral (superior) pectoral nerve** is so named because it is lateral to the artery and arises from the lateral cord of the brachial plexus or from the anterior divisions of the superior and middle trunks just before they unite into the cord. It passes superficial to the first part of the axillary artery and vein, sends a communicating branch to the inferior pectoral branch, and then pierces the clavipectoral fascia to reach the deep surface of the clavicular and cranial sternocostal portions of the pectoralis major muscle.

The **medial (inferior) pectoral nerve** is so named because its origin is from the medial cord of the brachial plexus, medial to the artery. Its origin is more lateral in position with respect to the midline of the body than that of the superior branch. It passes between the axillary artery and vein, gives a branch that joins the communication from the superior branch to form a plexiform loop around the artery, and enters the deep surface of the pectoralis minor muscle. It supplies this muscle and two or three of its branches continue through the muscle to supply the more caudal part of the pectoralis major. The most distal branch may pass around the border of the minor. The loop gives off branches that supply both muscles.

SUBSCAPULAR NERVES. The subscapular nerves (Fig. 12-39), usually two in number, arise from the posterior cord of the brachial plexus, deep in the axilla.

The **superior subscapular nerve** (*short subscapular*), the smaller of the two, enters the superior part of the subscapularis muscle; it is frequently double.

The **inferior subscapular nerve** supplies the distal part of the subscapularis and ends in the teres major muscle.

VARIATIONS. The nerve to the teres major may be a separate branch of the posterior cord, or rarely, of the axillary nerve.

THORACODORSAL NERVES. The thoracodorsal nerve (*middle or long subscapular nerve*) (Fig. 12-39) is a branch of the posterior cord of the brachial plexus; it usually arises between the two subscapular nerves. It follows the course of the subscapular and thoracodorsal arteries along the posterior wall of the axilla, under cover of the ventral border of the latissimus dorsi, and terminates in branches that supply this muscle.

AXILLARY NERVE. The axillary nerve (*circumflex nerve*) (Fig. 12-40) is the last branch of the posterior cord of the brachial plexus before the latter becomes the radial nerve. It passes over the insertion of the subscapularis muscle dorsal to the axillary artery, crosses the teres minor, and leaves the axilla, accompanied by the posterior humeral circumflex artery, by passing through the quadrilateral space bounded by the surgical neck of the humerus, the teres major, the teres minor, and the long head of the triceps. It divides into two branches.

The **posterior branch** (*lower branch*) supplies the teres minor and the posterior part of the deltoid. It then pierces the deep fascia at the posterior border of the deltoid as the **lateral brachial cutaneous nerve,** to supply the skin over the distal two-thirds of the posterior part of this muscle and over the adjacent long head of the triceps brachii.

The **anterior branch** (*upper branch*) winds around the surgical neck of the humerus with the posterior humeral circumflex vessels, under cover of the deltoid as far as its anterior border. It supplies this muscle and sends a few small cutaneous filaments to the skin covering its distal part.

Articular filaments leave the nerve near its origin and in the quadrilateral space; these supply the anterior inferior part of the capsule of the shoulder joint.

MEDIAL BRACHIAL CUTANEOUS NERVE. The medial brachial cutaneous nerve (*nerve of Wrisberg*), a small nerve, arises from the medial cord of the brachial plexus and is distributed to the medial side of the arm. It passes through the axilla, at first lying dorsal and then medial to the axillary vein and brachial artery. It pierces the deep fascia in the middle of the arm and is distributed to the skin of the arm as far as the medial epi-

condyle and olecranon. A part of it forms a loop with the intercostobrachial nerve in the axilla, and there is a reciprocal relationship in size between these two nerves. It also communicates with the ulnar branch of the medial antebrachial cutaneous nerve or it may be a branch of the latter nerve.

MEDIAL ANTEBRACHIAL CUTANEOUS NERVE. The medial antebrachial cutaneous nerve (Figs. 12-41 to 12-44; 12-46) arises from the medial cord of the brachial plexus, medial to

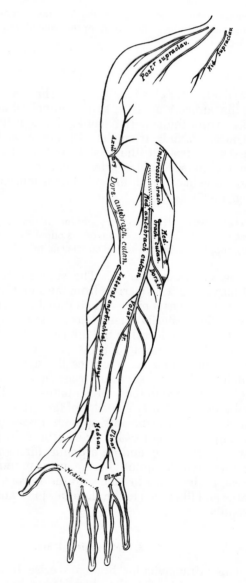

FIG. 12-41.   Cutaneous nerves of right upper limb. Anterior view.

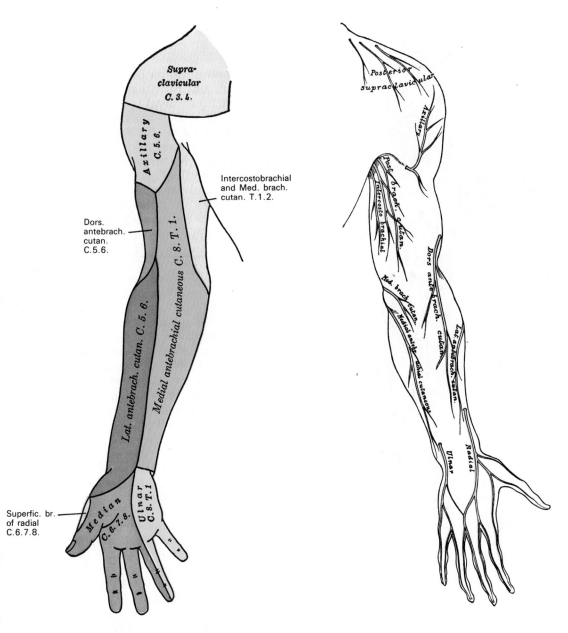

FIG. 12-42.   Diagram of segmental distribution of the cutaneous nerves of the right upper limb. Anterior view.

FIG. 12-43.   Cutaneous nerves of right upper limb. Posterior view.

the axillary artery. Near the axilla, it gives off a filament that pierces the fascia and supplies the skin over the biceps nearly as far as the elbow. The nerve runs down the ulnar side of the arm medial to the brachial artery, pierces the deep fascia with the basilic vein about the middle of the arm, and divides into an anterior and an ulnar branch.

The **ulnar branch** (*posterior branch*) passes obliquely distalward on the medial

side of the basilic vein, anterior to the medial epicondyle of the humerus to the dorsum of the forearm, and continues on its ulnar side as far as the wrist, supplying the skin. It communicates with the medial brachial cutaneous, the dorsal antebrachial cutaneous, and the dorsal branch of the ulnar nerve.

The **anterior branch** is larger and passes, usually superficial but occasionally deep, to

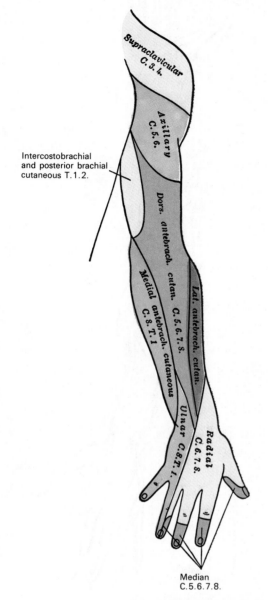

Intercostobrachial
and posterior brachial
cutaneous T.1.2.

FIG. 12-44.   Diagram of segmental distribution of the cutaneous nerves of the right upper limb. Posterior view.

MUSCULOCUTANEOUS NERVE.   The musculocutaneous nerve (Fig. 12-46) is formed by the splitting of the lateral cord of the brachial plexus at the inferior border of the pectoralis minor into two branches, the other branch being the lateral root of the median nerve. It pierces the coracobrachialis muscle, and lying between the brachialis and the biceps brachii, crosses to the lateral side of the arm. A short distance above the elbow it pierces the deep fascia lateral to the tendon of the biceps and continues into the forearm as the lateral antebrachial cutaneous nerve.

The branch to the **coracobrachialis** leaves the nerve close to its origin.

Muscular branches are supplied to the **biceps** and the greater part of the **brachialis.**

An articular filament given off from the nerve to the brachialis supplies the **elbow joint.**

A filament to the **humerus** enters the nutrient foramen with the artery.

The **lateral antebrachial cutaneous nerve** (Figs. 12-41 to 12-44) passes deep to the cephalic vein and divides opposite the elbow joint into an anterior and a dorsal branch.

The **anterior branch** (*volar branch*) follows along the radial border of the forearm to the wrist and supplies the skin over the radial half of its anterior surface. At the wrist it is superficial to the radial artery, and some of its filaments pierce the deep fascia to follow the vessel to the dorsal surface of the carpus. It terminates in cutaneous filaments at the thenar eminence after communicating with the superficial branch of the radial and the palmar cutaneous branch of the median nerve.

The **dorsal branch** (*posterior branch*) passes distally along the dorsal part of the radial surface of the forearm, supplying the skin almost to the wrist, and communicating with the dorsal antebrachial cutaneous nerve and the superficial branch of the radial nerve.

Variations.   The musculocutaneous and median nerves present frequent irregularities in their origins from the lateral cord of the plexus. The branch to the coracobrachialis muscle may be a separate nerve. In this condition the musculocutaneous nerve may continue with the median for a variable distance before it passes under the biceps. Some of the fibers of the median nerve may run for some distance in the musculocutaneous nerve before they join their proper trunk; less frequently the reverse is the case and fibers of the musculocutaneous nerve run with

the median basilic vein. It continues on the anterior part of the ulnar side of the forearm, distributing filaments to the skin as far as the wrist, and communicating with the palmar cutaneous branch of the ulnar nerve.

Terminal Branches

The **terminal branches** of the **brachial plexus** are the musculocutaneous, median, ulnar, and radial nerves.

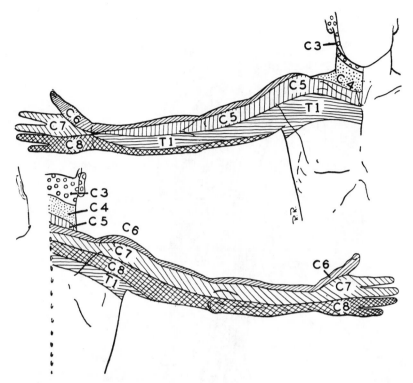

Fig. 12-45. Dermatome chart of the upper limb of man outlined by the pattern of hyposensitivity from loss of function of a single nerve root. (From Keegan and Garrett, Anat. Rec., *102*:415, 1948.)

the median. The musculocutaneous nerve may give a branch to the pronator teres or it may supply the dorsum of the thumb in the absence of the superficial branch of the radial nerve.

MEDIAN NERVE. The median nerve (Fig. 12-46) is the nerve to the radial side of the flexor portion of the forearm and hand. It takes its origin from the brachial plexus by two large roots, one from the lateral and one from the medial cord. The roots at first lie on each side of the third part of the axillary artery, then embracing it, they unite on its ventral surface to form the trunk of the nerve. In its course down the arm it accompanies the brachial artery, to which it is at first lateral, but gradually it crosses the ventral surface of the artery in the middle or distal part of the arm and lies medial to it at the bend of the elbow, where it is deep to the bicipital fascia and superficial to the brachialis muscle.

In the forearm, the median nerve passes between the two heads of the pronator teres, being separated from the ulnar artery by the deep head. It continues distally between the

flexores digitorum superficialis and profundus almost to the flexor retinaculum, where it becomes more superficial and lies between the tendons of the flexor digitorum superficialis and the flexor carpi radialis. It is deep to the tendon of the palmaris longus and slightly ulnarward of it, and it is the most superficial of the structures that pass through the tunnel under the flexor retinaculum.

In the palm of the hand, the median nerve is covered only by the skin and the palmar aponeurosis and rests on the tendons of the flexor muscles (Fig. 12-47). Immediately after emerging from under the retinaculum, it becomes enlarged and flattened and splits into muscular and digital branches.

Branches. The median nerve has no branches above the elbow joint unless, as occasionally happens, the nerve to the pronator teres arises there.

Articular branches to the **elbow joint** are one or two twigs given off as the nerve passes the joint.

Muscular branches leave the nerve near the elbow and supply all the superficial mus-

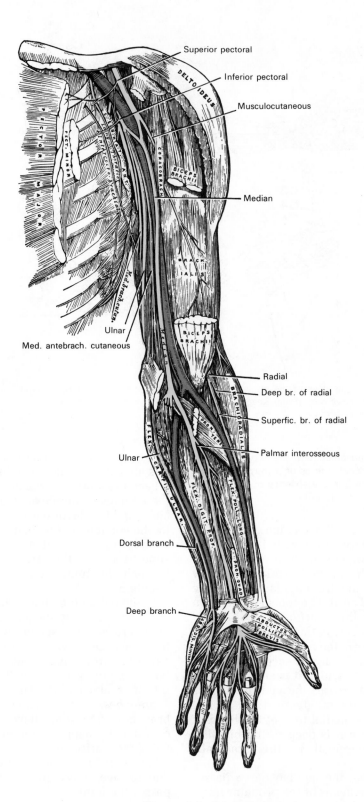

Fig. 12-46. Nerves of the left upper limb.

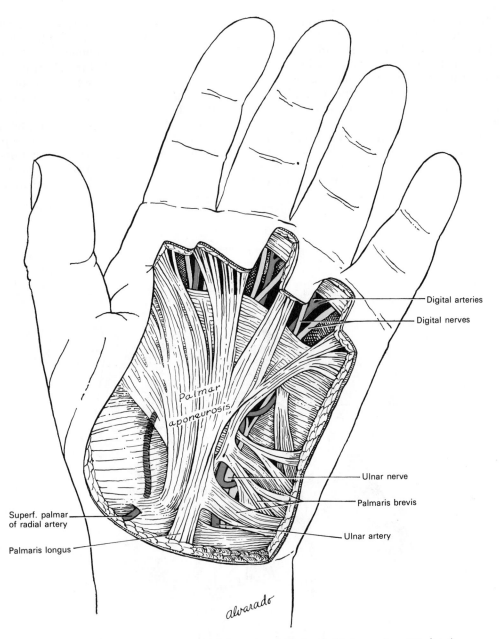

Digital arteries

Digital nerves

Ulnar nerve

Palmaris brevis

Ulnar artery

Superf. palmar
of radial artery

Palmaris longus

alvarado

FIG. 12-47. Superficial dissection of the palm of the hand. (Redrawn from Töndury.)

cles of the anterior part of the forearm except the flexor carpi ulnaris, i.e., the **pronator teres, flexor carpi radialis, palmaris longus,** and **flexor digitorum superficialis.**

The **anterior interosseous nerve** (Fig. 12-46) accompanies the anterior interosseous artery along the anterior surface of the interosseous membrane in the interval between the flexor pollicis longus and the flexor digi-

torum profundus, ending in the pronator quadratus and the wrist joint. It supplies all the deep anterior muscles of the forearm except the ulnar half of the profundus, i.e., the radial half of the flexor digitorum profundus, the flexor pollicis longus, and the pronator quadratus.

The **palmar branch of the median nerve** (Fig. 12-42) pierces the antebrachial fascia distal to the wrist and divides into a medial and a lateral branch. The medial branch supplies the skin of the palm and communicates with the palmar cutaneous branch of the ulnar nerve. The lateral branch supplies the skin over the thenar eminence and communicates with the lateral antebrachial cutaneous nerve.

The **muscular branch in the hand** (Fig. 12-48) is a short stout nerve that leaves the radial side of the median nerve, sometimes in company with the first common palmar digi-

tal nerve, just after the former passes under the flexor retinaculum. It supplies the muscles of the thenar eminence with the exception of the deep head of the short flexor, i.e., the abductor pollicis brevis, the opponens pollicis and the superficial head of the flexor pollicis brevis.

The **first common palmar digital nerve** (Figs. 12-48; 12-50) divides into three proper palmar digital nerves (*digital collaterals*), two of which supply the sides of the thumb; the third gives a twig to the first lumbrical

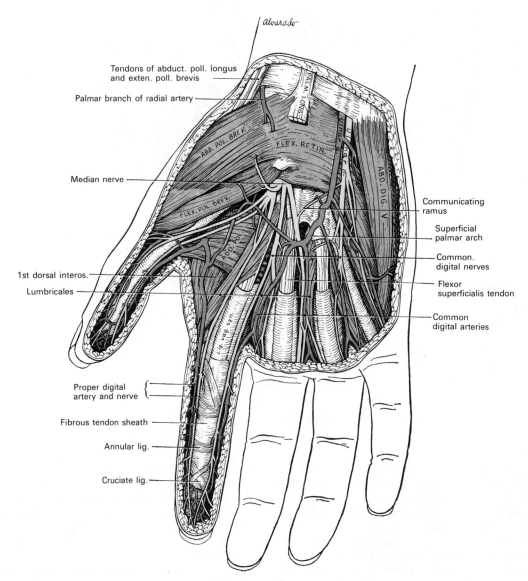

Fig. 12-48. Dissection of the palm of the hand, showing the superficial palmar arch, median and ulnar nerves, and synovial sheaths. (Redrawn from Töndury.)

and continues as the proper palmar digital nerve for the radial side of the index finger.

The **second common palmar digital nerve** gives a twig to the second lumbrical, and continuing to the web between the index and middle fingers, splits into proper digital nerves for the adjacent sides of these fingers.

The **third common palmar digital nerve** occasionally gives a twig to the third lumbrical, in which case it has a double innervation; it communicates with a branch of the ulnar nerve, and continues to the web between the middle and the ring fingers, where it splits into proper digital nerves for the adjacent sides of these digits.

The **proper digital nerves** (*digital collaterals*) (Figs. 12-47; 12-48; 12-50) supply the skin of the palmar surface and the dorsal surface over the terminal phalanx of their digits. At the end of the digit each nerve terminates in two branches, one of which ramifies in the skin of the ball, the other in the pulp under the nail. They communicate with the dorsal digital branches of the superficial radial nerve, and in the fingers they are superficial to the corresponding arteries.

Variations. In 16 per cent of bodies studied, the relation of the median nerve to the two heads of the pronator teres varies from that described; it may pass deep to the humeral head in the absence of an ulnar head, or deep to the ulnar head, or split the humeral head (Jamieson and Anson, 1952). There is overlapping of territory in the innervation of the flexor digitorum profundus by the median and ulnar nerves in 50 per cent of bodies; it is twice as common for the median to encroach on the ulnar. The portion of the profundus attached to the index finger is the only part of that muscle constantly supplied by one nerve, the median. In the majority of cases the profundus and the lumbrical of a particular digit are innervated by the same nerve. Encroachment of the median on the ulnar is less common for the lumbricals than the profundus. The median nerve may supply the first dorsal interosseus (Sunderland 1945, 1946).

ULNAR NERVE. The ulnar nerve (Fig. 12-46) occupies a superficial position along the medial side of the arm and is the nerve to the muscles and skin of the ulnar side of the forearm and hand. It is the terminal continuation of the medial cord of the brachial plexus, after the medial head of the median nerve has separated from it. It is medial, at first to the axillary artery and then to the brachial artery as far as the middle of the arm, and is parallel with and not far distant from the median and medial antebrachial cutaneous nerves. In the middle of the arm it angles dorsally, pierces the medial intermuscular septum, and follows along the medial head of the triceps muscle to the groove between the olecranon and the medial epicondyle of the humerus. In this position it is covered only by the skin and fascia and can readily be palpated as the "funny bone" of the elbow.

The ulnar nerve is accompanied in its course through the distal half of the arm by the superior ulnar collateral artery and the ulnar collateral branch of the radial nerve. It enters the forearm between the two heads of the flexor carpi ulnaris and continues between this muscle and the flexor digitorum profundus half way down the forearm. In the proximal part of the forearm it is separated from the ulnar artery by a considerable distance, but in the distal half it lies close to its ulnar side, radialward from the flexor carpi ulnaris, and covered only by the skin and fascia. Proximal to the wrist it gives off a large dorsal branch and continues into the hand, where it has muscular and digital branches.

Branches. The ulnar nerve usually has no branches proximal to the elbow. Distal to the elbow its branches are as follows:

The articular branches to the **elbow joint** are several small filaments that leave the nerve as it lies in the groove between the olecranon and the medial epicondyle of the humerus.

Two muscular branches arise near the elbow and supply the **flexor carpi ulnaris** and the ulnar half of the **flexor digitorum profundus.**

The **palmar cutaneous branch** of the ulnar nerve arises near the middle of the forearm and accompanies the ulnar artery into the hand. It gives filaments to the artery, perforates the flexor retinaculum, and ends in the skin of the palm, communicating with the palmar branch of the median nerve.

The **dorsal branch** (Figs. 12-43; 12-46) arises in the distal half of the forearm and reaches the dorsum of the wrist by passing between the flexor carpi ulnaris and the ulna. It pierces the deep fascia and divides into two **dorsal digital nerves** and a metacarpal communicating branch. The more medial digital nerve supplies the ulnar side of the little finger; the other digital branch, the adjacent sides of the little and ring fingers. The metacarpal branch supplies the skin of that area and continues toward the web between the ring and middle fingers, where it joins a similar branch of the superficial radial nerve to supply the adjacent sides of these two fingers. On the little finger the dorsal digital branches extend only as far as the base of the terminal phalanx, and on the ring finger as far as the base of the second phalanx; the more distal parts of these digits are supplied by the dorsal branches of the proper palmar digital nerves from the ulnar nerve.

The **palmar branch** or terminal portion of the ulnar nerve crosses the ulnar border of the wrist in company with the ulnar artery, superficial to the flexor retinaculum and under cover of the palmaris brevis, and divides into a superficial and deep branch.

The **superficial branch** (Figs. 12-48; 12-50) supplies the palmaris brevis and the skin of

the hypothenar eminence and divides into **digital branches.** A **proper palmar digital branch** goes to the ulnar side of the little finger; a **common palmar digital branch** divides into proper digital branches for the adjacent sides of the little and ring fingers and communicates with the branches of the median nerve. The proper digital branches are distributed to the fingers in the same manner as those of the median nerve.

The **deep branch** (Figs. 12-49; 12-50) passes between the abductor digiti minimi and flexor digiti minimi brevis accompanied by the deep branch of the ulnar artery; it then pierces the opponens digiti minimi and follows the course of the deep palmar arch across the interossei, deep to the midpalmar and thenar fascial clefts. Near its origin it gives branches to the three small muscles of the little finger, and as it crosses the hand it supplies the third and fourth lumbricals and all the interossei, both palmar and dor-

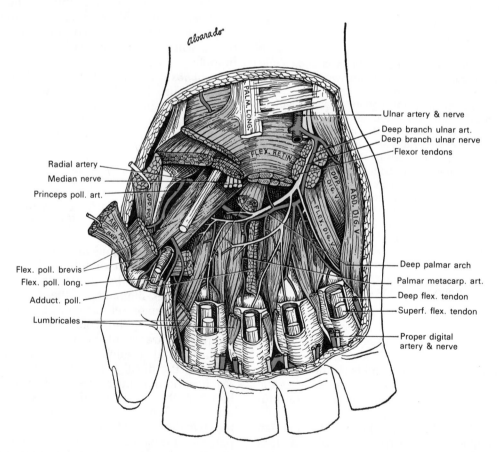

Fig. 12-49. Deep dissection of the palm of the hand, showing the deep palmar arch and the deep palmar branch of the ulnar nerve. (Redrawn from Töndury.)

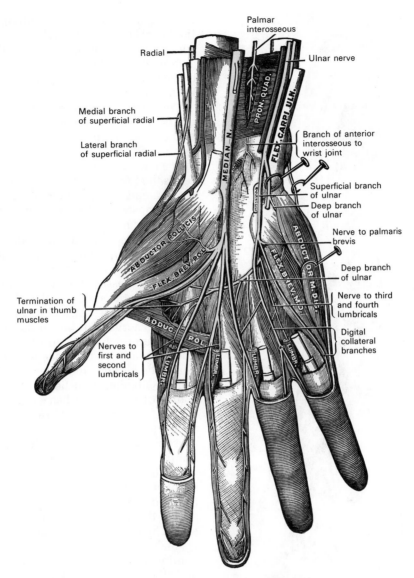

Fig. 12-50. Deep palmar nerves. (Testut.)

sal. It ends by supplying the adductores pollicis and the deep head of the flexor pollicis brevis. It also sends articular filaments to the wrist joint.

Variations. The ulnar nerve may pass in front of the medial epicondyle. It frequently has a communication with the median nerve in the forearm and rarely with the medial antebrachial cutaneous, median, or musculocutaneous nerves in the arm. It may send muscular branches to the medial head of the triceps, the flexor digitorum superficialis, the first and second lumbricals and the superficial head of the flexor pollicis brevis muscles. It may have defi-

ciencies on the dorsum of the hand that are supplied by the radial nerve or it may encroach upon the area usually supplied by that nerve. See variations of the median nerve and brachial plexus.

RADIAL NERVE.   The **radial nerve** (*musculospiral nerve*) (Fig. 12-51), the largest branch of the brachial plexus, is the continuation of the posterior cord; it supplies the extensor muscles of the arm and forearm as well as the skin covering them. It crosses the tendon of the latissimus dorsi, deep to the axillary artery, and after passing the inferior border

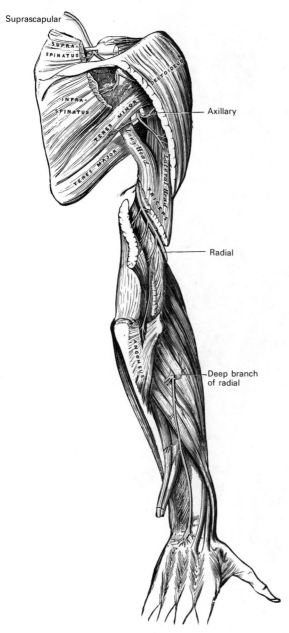

FIG. 12-51. The suprascapular, axillary, and radial nerves.

of the teres major, it winds around the medial side of the humerus and enters the substance of the triceps between the medial and long heads. It takes a spiral course down the arm close to the humerus in the groove that separates the origins of the medial and lateral heads of the triceps, accompanied by the deep brachial artery. Having reached the lateral side of the arm, it pierces the lateral intermuscular septum and runs between the brachialis and the brachioradialis anterior to the lateral epicondyle, where it divides into superficial and deep branches (Fig. 12-46).

Branches in the Arm. Both muscular and cutaneous branches of the radial nerve are found in the arm.

The **medial muscular branches** arise in the axilla and supply the medial and long heads of the triceps. The branch to the medial head is a long filament that accompanies the ulnar nerve and superior ulnar collateral artery as far as the lower third of the arm and is therefore named the **ulnar collateral nerve.**

The **posterior brachial cutaneous nerve** (*internal cutaneous branch of the musculospiral*) arises in the axilla to the medial side of the arm, supplying the skin on the dorsal surface nearly as far as the olecranon. In its course it crosses dorsal to, and communicates with, the intercostobrachial nerve.

The **posterior muscular branches** arise from the nerve as it lies in the spiral groove of the humerus; they supply the medial and lateral heads of the triceps and the anconeus. The nerve to the latter muscle is a long, slender filament that lies buried in the substance of the medial head of the triceps.

The **posterior antebrachial cutaneous nerve** (*external cutaneous branch of the musculospiral*) perforates the lateral head of the triceps at its attachment to the humerus and divides into proximal and distal branches. The smaller **proximal branch** lies close to the cephalic vein and supplies the skin of the dorsal part of the distal half of the arm. The **distal branch** pierces the deep fascia below the insertion of the deltoid and continues along the lateral side of the arm and elbow and the dorsal side of the forearm to the wrist. It supplies the skin in its course, and near its termination, it communicates with the dorsal branch of the lateral antebrachial cutaneous nerve.

The **lateral muscular branches** supply the brachioradialis, the extensor carpi radialis longus, and the lateral part of the brachialis.

Articular branches to the elbow come from the radial nerve between the brachialis and brachioradialis, from the ulnar collateral nerve, and from the nerve to the anconeus.

Branches in the Forearm. The **superficial branch of the radial nerve** (Fig. 12-46) runs along the lateral border of the forearm under cover of the brachioradialis muscle. In the proximal third of the forearm it gradually approaches the radial artery; in the middle third it lies just radialward to the artery; and in the lower third it quits the artery and angles dorsally under the tendon of the brachioradialis toward the dorsum of the wrist, where it pierces the deep fascia and divides into two branches.

The **lateral branch** is smaller, supplies the skin of the radial side and ball of the thumb, and communicates with the palmar branch of the lateral antebrachial cutaneous nerve.

The **medial branch** communicates above the wrist, with the dorsal branch of the lateral antebrachial cutaneous nerve, and on the dorsum of the hand, with the dorsal branch of the ulnar nerve. It then divides into four **dorsal digital nerves** (Fig. 12-43), which are distributed as follows: (1) to the ulnar side of the thumb, (2) to the radial side of the index finger, (3) to the adjacent sides of the index and middle fingers, and (4) to the dorsal branch of the ulnar nerve to form a communication, and to the adjacent sides of the middle and ring fingers.

The **deep branch of the radial nerve** (*posterior interosseous nerve*) (Fig. 12-51) winds to the dorsum of the forearm around the radial side of the radius between the planes of fibers of the supinator muscle and continues between the superficial and deep layers of muscles to the middle of the forearm. Considerably diminished in size, and named the **dorsal interosseous nerve,** it lies on the dorsal surface of the interosseous membrane, and under cover of the extensor pollicis longus, it continues to the dorsum of the carpus, where it ends.

Muscular branches of the deep branch of the radial nerve to the extensor carpi radialis brevis and the supinator are given off before the nerve turns dorsally. After passing through the supinator, branches are given to the extensor digitorum, extensor digiti minimi, extensor carpi ulnaris, the two extensores and abductor longus pollicis, and the extensor indicis.

Articular filaments from a terminal plexus of the radial nerve are distributed to the ligaments and articulations of the carpus and metacarpus.

VARIATIONS. The radial nerve may pass through the quadrilateral space with the axillary nerve, and when the profunda brachii supplies the nutrient artery of the humerus, it may be accompanied by a filament from the radial. The nerve to the brachialis

muscles is inconstant. The deep branch may pass superficial to the entire supinator. There is great variation in the distribution and overlapping of the radial and ulnar nerves on the back of the hand; the most frequent arrangement appears to be that the radial supplies three and a half digits, the ulnar, one and a half digits (Sunderland, 1945, 1946).

Thoracic Nerves

The ventral primary divisions (*anterior divisions*) (Fig. 12-52) of the thoracic nerves number 12 on each side. The first eleven, situated between the ribs, are termed **intercostals;** the twelfth lies below the last rib and is called the **subcostal nerve.** The intercostal nerves are distributed chiefly to the parietes of the thorax and abdomen, and they differ from the other spinal nerves in that each pursues an independent course, i.e., they do not enter into a plexus. The first two contribute to the upper limb as well as to the thorax; the next four are limited in their distribution to the thorax; the lower five supply the parietes of the thorax and the abdomen. The twelfth thoracic nerve is distributed to the abdominal wall and the skin of the buttock.

RAMI COMMUNICANTES

Each thoracic nerve contributes preganglionic sympathetic fibers to the sympathetic chain through a **white ramus communicans,** and receives postganglionic fibers from the chain ganglia through **gray rami communicantes.** Both rami are attached to the spinal nerves near their exit from the intervertebral foramina, the gray rami being more medial than the white. For more details see the section on the Sympathetic System.

FIRST THORACIC NERVE

The ventral primary division of the first thoracic nerve divides immediately into two parts: (1) the larger part leaves the thorax ventral to the neck of the first rib and becomes one of the roots of the brachial plexus (described previously); (2) the smaller part becomes the **first intercostal nerve.** The first intercostal nerve runs along the first intercostal space to the sternum, perforates the muscles and deep fascia, and ends as the first anterior cutaneous nerve of the thorax. It has no lateral cutaneous branch, as a rule, but it may have a communication with the intercostobrachial branch of the second

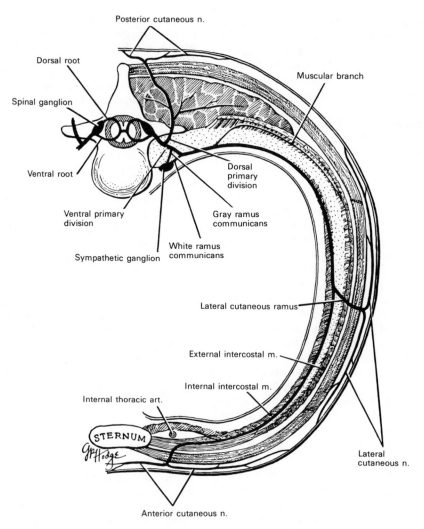

FIG. 12-52.   Plan of a typical intercostal nerve.

nerve. A communication between the first and second nerves inside the thorax occurs frequently and contains postganglionic sympathetic fibers from the second or even the third thoracic sympathetic ganglion.

SUPERIOR THORACIC NERVES

The ventral primary divisions of the second to the sixth thoracic nerves and the intercostal portion of the first thoracic are known as the **thoracic intercostal nerves** (Figs. 12-53; 12-54; 12-78). They pass ventralward in the intercostal spaces with the intercostal vessels; from the vertebral column to the angles of the ribs, they lie between the pleura and the posterior intercostal membrane; from the angle to the middle of the ribs, they pass between the intercostales interni and externi. They then enter the substance of the intercostales interni, where they remain concealed until they reach the costal cartilages; here they again emerge on the inner surface of the muscle and lie between it and the pleura. The nerves are inferior to the vessels and, like them, are at first close to the rib above, but as they proceed ventrally they may approach the middle of the intercostal space. Near the sternum they pass ventral to the transversus thoracis and the internal thoracic vessels, and near the sternum they pierce the intercostalis internus, the anterior intercostal membrane, the pectoralis major and the pec-

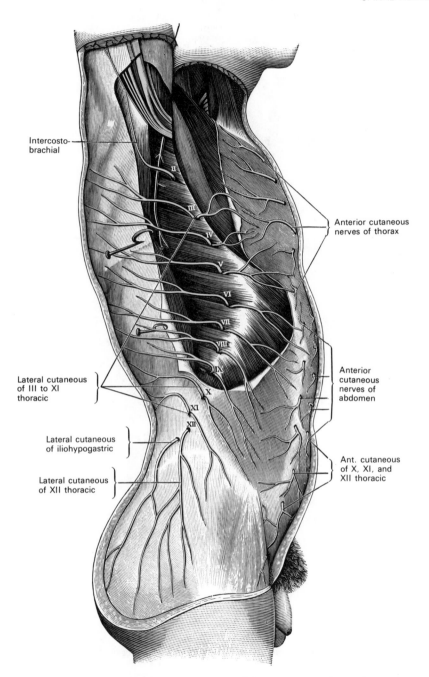

Intercosto-
brachial

Anterior cutaneous
nerves of thorax

Lateral cutaneous
of III to XI
thoracic

Anterior
cutaneous
nerves of
abdomen

Lateral cutaneous
of iliohypogastric

Ant. cutaneous
of X, XI, and
XII thoracic

Lateral cutaneous
of XII thoracic

FIG. 12-53. Cutaneous distribution of thoracic nerves. (Testut.)

toral fascia, terminating as the **anterior cutaneous nerves** of the thorax. They have short
medial and longer lateral branches that supply the skin and mamma.

MUSCULAR BRANCHES. **Muscular branches**
supply the intercostales interni and externi,
the subcostales, the levatores costarum, the

serratus posterior superior, and the transversus thoracis. At the ventral part of the thorax
some of these branches cross the costal cartilages from one intercostal space to another.

CUTANEOUS BRANCHES. The **lateral cutaneous nerves** (Figs. 12-53; 12-54) arise from
the intercostal nerves about midway be-

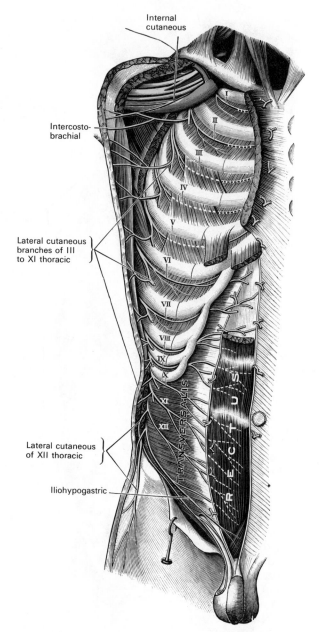

FIG. 12-54. Intercostal nerves, the superficial muscles having been removed. (Testut.)

tween the vertebral column and the sternum, pierce the intercostales externi and serratus anterior, and divide into anterior and posterior branches. The anterior branches supply the skin of the lateral and ventral part of the chest and mammae; those of the fifth and sixth nerves also supply the cranial digitations of the obliquus externus abdominis. The posterior branches supply the skin over the latissimus dorsi and the scapular region.

The **intercostobrachial nerve** (Figs. 12-39; 12-41; 12-46) arises from the second intercostal nerve as if it were a lateral cutaneous nerve, but it fails to divide into an anterior and a posterior branch. After piercing the intercostales and the serratus anterior, it crosses the axilla, embedded in the adipose tissue, to reach the medial side of the arm. It forms a loop of communication with the medial brachial cutaneous nerve of the brachial plexus and assists it in supplying the skin of the medial and posterior part of the arm. It also communicates with the posterior brachial cutaneous branch of the radial nerve.

An intercostobrachial branch frequently arises from the third intercostal nerve, supplying the axilla and medial side of the arm.

LOWER THORACIC NERVES

The ventral primary divisions of the seventh to eleventh thoracic nerves are continued beyond the intercostal spaces into the anterior abdominal wall and are named the **thoracoabdominal intercostal nerves** (Figs. 12-53; 12-54). They have the same arrangement as the upper nerves as far as the anterior ends of the intercostal spaces, where they pass dorsal to the costal cartilages. They run ventralward between the obliquus internus and transversus abdominis muscles, pierce the sheath of the rectus abdominis, which they supply, and terminate as the **anterior cutaneous branches.**

MUSCULAR BRANCHES. Muscular branches supply the intercostales, the obliqui, and transversus abdominis muscles; the last three also supply the serratus posterior inferior.

CUTANEOUS BRANCHES. The **lateral cutaneous branches** (Figs. 12-53; 12-54) arise midway along the nerves, pierce the intercostales externi and the obliquus externus in line with the lateral cutaneous branches of the upper intercostals, and divide into anterior and posterior branches. The anterior branches give branches to the digitations of the obliquus externus and extend caudalward and ventralward nearly as far as the margin of the rectus abdominis, supplying the skin. The posterior branches pass dorsally to supply the skin over the latissimus dorsi.

The **anterior cutaneous branches** penetrate the anterior layer of the sheath of the rectus abdominis and divide into medial and lateral branches that supply the skin of the anterior part of the abdominal wall.

TWELFTH THORACIC NERVE

The anterior primary division of the **twelfth thoracic** (Figs. 12-53; 12-54; 12-57) or **subcostal nerve** is larger than those above it. It runs along the inferior border of the twelfth rib, often communicates with the first lumbar nerve, and passes under the lateral lumbocostal arch. It crosses the ventral

surface of the quadratus lumborum, penetrates the transversus abdominis, and continues ventralward to be distributed in the same manner as the lower intercostal nerves. It communicates with the iliohypogastric nerve of the lumbar plexus and gives a branch to the pyramidalis. The lateral cutaneous branch is large and does not divide into anterior and posterior branches. It perforates the obliqui muscles, passes downward over the crest of the ilium ventral to a similar branch of the iliohypogastric nerve, and is distributed to the skin of the anterior part of the gluteal region, some filaments extending as far down as the greater trochanter.

Lumbar Nerves and Lumbosacral Plexus

The ventral primary divisions are increasingly large as they are placed more caudally in the lumbar region. Their course is lateralward and caudalward, either under cover of the psoas major muscle or between its fasciculi. The first three nerves and the larger part of the fourth nerve are connected by communicating loops, and they are frequently joined by a communication from the twelfth thoracic nerve, forming the lumbar plexus. The smaller part of the fourth nerve joins the fifth to form the lumbosacral trunk, which enters the sacral plexus. The fourth nerve, because it is divided between the two plexuses, is sometimes called the **nervus furcalis.**

Only the first two lumbar nerves contribute preganglionic sympathetic fibers to the sympathetic chain through **white rami communicantes.** All the lumbar nerves receive postganglionic fibers from the sympathetic chain through **gray rami communicantes.** For more details see the section on the Sympathetic Nervous System.

Muscular branches are supplied to the psoas major and the quadratus lumborum from the ventral primary divisions of the lumbar nerves before they enter the lumbar plexus.

The **lumbosacral plexus** is the name given to the combination of all the ventral primary divisions of the lumbar, sacral, and coccygeal nerves. The lumbar and sacral plexuses supply the lower limb, but in addition the sacral nerves supply the perineum through the pudendal plexus and the coccygeal region through the coccygeal plexus. For convenience in description, these plexuses will be considered separately.

LUMBAR PLEXUS

The **lumbar plexus** (Fig. 12-55) is formed by the ventral primary divisions of the first three and the greater part of the fourth lumbar nerves, with a communication from the twelfth thoracic usually joining the first lumbar nerve. It is situated on the inside of the posterior abdominal wall, either dorsal to the psoas major or among its fasciculi, and ventral to the transverse processes of the lumbar vertebrae. The lumbar plexus is not an intricate interlacement like the brachial plexus, but its branches usually arise from two or three nerves, so that the resulting junctions between adjacent nerves have the appearance of loops.

The manner in which the plexus is formed is the following (Figs. 12-55; 12-56): the first lumbar nerve, usually supplemented by a communication from the twelfth thoracic nerve, splits into a cranial and a caudal branch. The cranial branch forms the **iliohypogastric** and **ilioinguinal nerves;** the caudal and smaller branches unite with a branch from the second lumbar nerve to form the **genitofemoral nerve.** The remainder of the second nerve and the third and fourth nerves each split into a small ventral and a large dorsal portion; the ventral portions unite into one nerve, the **obturator.** The dorsal portions of the second and third nerves each divide again into unequal branches; the two smaller branches unite to form the **lateral femoral cutaneous nerve;** the two larger branches join the dorsal portion of the fourth nerve to form the **femoral nerve.** The **accessory obturator nerve,** which is present in about one out of five individuals, comes from the third and fourth nerves. A considerable part of the fourth nerve joins the fifth lumbar nerve in the **lumbosacral trunk.**

The **branches** of the lumbar plexus are as follows:

1. Iliohypogastric	L 1, (T 12)
2. Ilioinguinal	L 1
3. Genitofemoral	L 1, 2
4. Lateral femoral cutaneous	L 2, 3
5. Obturator	L 2, 3, 4
6. Accessory obturator	L 3, 4
7. Femoral	L 2, 3, 4

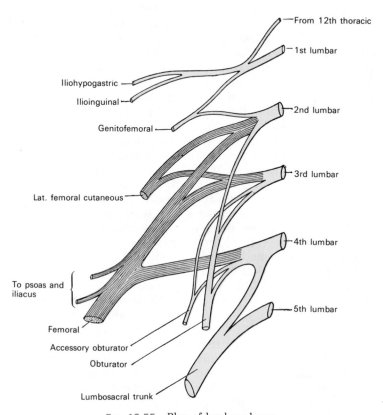

FIG. 12-55. Plan of lumbar plexus.

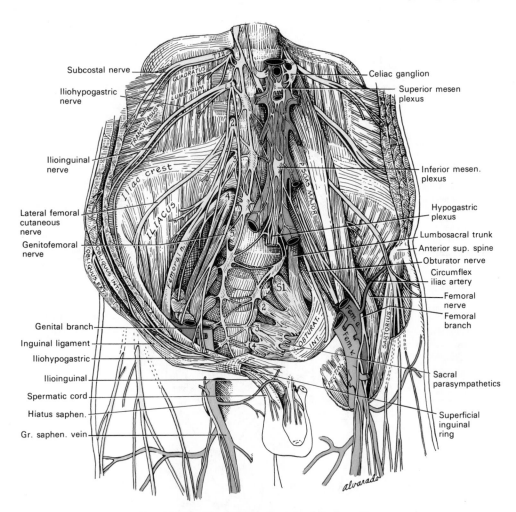

Fɪɢ. 12-56. The lumbar plexus and its branches.

These branches may be divided into two groups according to their distribution. The first three supply the caudal part of the parietes of the abdominal wall; the last four supply the anterior thigh and medial part of the leg.

ILIOHYPOGASTRIC NERVE. The iliohypogastric nerve (Figs. 12-56; 12-58) arises from the first lumbar nerve and from the communication with the twelfth thoracic nerve, when this is present. It emerges from the upper part of the lateral border of the psoas major, crosses the quadratus lumborum to the crest of the ilium, and penetrates the posterior part of the transversus abdominis near the crest of the ilium. Between the transversus and the obliquus internus it divides into a lateral and anterior cutaneous branch.

The **lateral cutaneous branch** (*iliac branch*) pierces the obliqui internus and externus immediately above the iliac crest, and is distributed to the skin of the gluteal region posterior to the lateral cutaneous branch of the twelfth thoracic nerve (Figs. 12-57; 12-64); these two nerves are inversely proportional in size.

The **anterior cutaneous branch** (*hypogastric branch*) (Figs. 12-54; 12-58) continues its course between the obliquus internus and tranversus, pierces the former, and becomes subcutaneous by passing through a perforation in the aponeurosis of the obliquus externus about 2 cm above the superficial inguinal ring. It is distributed to the skin of the hypogastric region (Fig. 12-27).

Muscular branches are supplied to the obliquus internus and transversus.

ILIOINGUINAL NERVE. The ilioinguinal

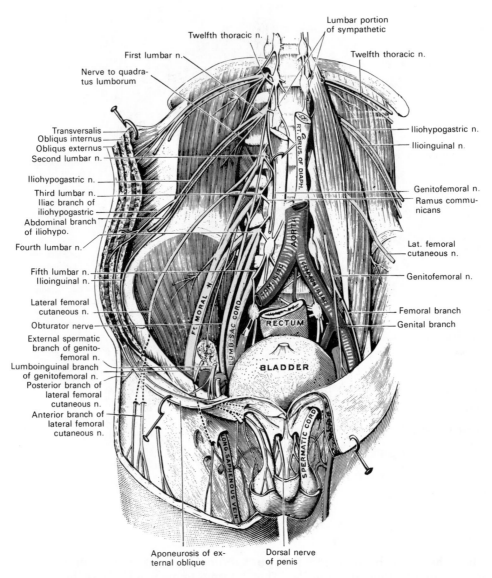

FIG. 12-57.   Deep and superficial dissection of the lumbar plexus. (Testut.)

nerve (Figs. 12-56; 12-57) arises from the lateral border of the psoas major just caudal to the iliohypogastric nerve and follows a similar course obliquely across the fibers of the quadratus lumborum to the crest of the ilium. It penetrates the transversus muscle near the anterior part of the crest, communicates with the iliohypogastric nerve, and pierces the obliquus internus, distributing filaments to it. It then accompanies the spermatic cord through the superficial inguinal ring, and is distributed to the skin over the proximal and medial part of the thigh, the

root of the penis, and the scrotum in the male (Fig. 12-57), or the mons pubis and the labium majus in the female.

Muscular twigs are supplied to the obliquus internus and transversus.

CLINICAL COMMENT. The optimal point for blocking the iliohypogastric and ilioinguinal nerves with local anesthesia is 4 to 6 cm posterior to the anterior superior spine of the ilium, along the lateral aspect of the external lip of the crest, where the nerves perforate the transversus abdominis muscle (Jamieson *et al.*, 1952).

GENITOFEMORAL NERVE. The genitofemoral nerve (**genitocrural nerve**) (Figs. 12-56; 12-57; 12-59) arises from the first and second lumbar nerves and passes caudalward through the substance of the psoas major until it emerges on its ventral surface opposite the third or fourth lumbar vertebra, where it is covered by transversalis fascia and peritoneum. On the surface of the muscle, or occasionally within its substance, it divides into a genital and a femoral branch.

The **genital branch** (*external spermatic nerve*) continues along the psoas major to the inguinal ligament, where it either pierces the transversalis and internal spermatic fascia or passes through the internal inguinal ring to reach the spermatic cord. It lies against the dorsal aspect of the cord, supplies the cremaster muscle, and is distributed to the skin of the scrotum and adjacent thigh. In the female it accompanies the round ligament of the uterus.

The **femoral branch** (*lumboinguinal nerve*) lies on the psoas major, lateral to the genital branch, and passes under the inguinal ligament with the external iliac artery. It enters the femoral sheath, superficial and lateral to the artery, then pierces the sheath and fascia lata to supply the skin of the proximal part of the anterior surface of the thigh. It gives a filament to the femoral artery and communicates with the anterior cutaneous branches of the femoral nerve (Figs. 12-58; 12-59).

VARIATIONS. The iliohypogastric or ilioinguinal nerves may arise from a common trunk, or the ilioinguinal nerve may join the iliohypogastric at the iliac crest, with the latter nerve then supplying the missing branches. The ilioinguinal nerve may be lacking and the genital branch supplies its branches, or the genital branch may be absent and the ilioinguinal substitutes for it. The femoral branch and lateral femoral cutaneous nerves or the anterior cutaneous branches of the femoral nerve may substitute for each other to a greater or lesser extent.

LATERAL FEMORAL CUTANEOUS NERVE. The **lateral femoral cutaneous nerve** (*external cutaneous nerve*) (Figs. 12-55; 12-58) arises from the dorsal portion of the ventral primary divisions of the second and third lumbar nerves. It emerges from the lateral border of the psoas major about its middle, and runs across the iliacus muscle obliquely toward the anterior superior iliac spine. It

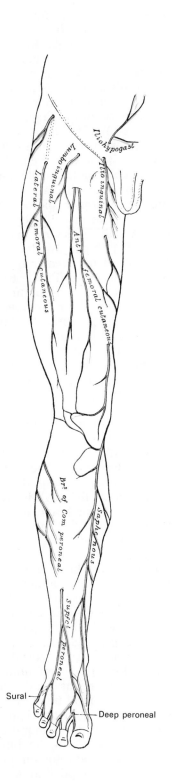

FIG. 12-58. Cutaneous nerves of right lower limb. Anterior view.

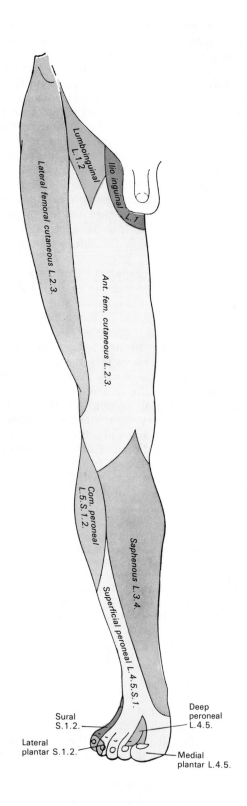

then passes under the inguinal ligament and over the sartorius muscle into the subcutaneous tissues of the thigh, dividing into an anterior and a posterior branch (Fig. 12-57).

The **anterior branch** becomes superficial about 10 cm distal to the inguinal ligament and is distributed to the skin of the lateral and anterior parts of the thigh as far as the knee (Fig. 12-58). The terminal filaments communicate with the anterior cutaneous branches of the femoral nerve and the infrapatellar branches of the saphenous nerve, forming the **patellar plexus.**

The **posterior branch** pierces the fascia lata and subdivides into filaments that pass posteriorly across the lateral and posterior surfaces of the thigh, supplying the skin from the level of the greater trochanter to the middle of the thigh (Fig. 12-64).

OBTURATOR NERVE.   The obturator nerve (Figs. 12-55; 12-56), the motor nerve to the adductores muscles, arises by three roots from the ventral portions of the second, third, and fourth lumbar nerves; the root from the third lumbar nerve is the largest, and that from the second is often very small. The obturator emerges from the medial border of psoas major near the brim of the pelvis, under cover of the common iliac vessels, and passes lateral to the hypogastric vessels and the ureter. It runs along the lateral wall of the lesser pelvis to enter the upper part of the obturator foramen with the obturator vessels. As it enters the thigh it divides into anterior and posterior branches, which are separated by some of the fibers of the obturator externus muscle and the adductor brevis muscle.

Anterior Branch.   The **anterior or superficial branch** (Fig. 12-60) communicates with the accessory obturator nerve, when this is present, and passes over the superior border of the obturator externus, deep to the pectineus and adductor longus and superficial to the adductor brevis. At the distal border of the adductor longus it communicates with the anterior cutaneous and saphenous branches of the femoral nerve, forming the **subsartorial plexus,** and terminates in filaments accompanying the femoral artery.

An **articular branch** to the hip joint is given off near the obturator foramen. **Muscular branches** are supplied to the adductor longus, gracilis, and usually to the adductor brevis; in rare instances it gives a branch to the pectineus muscle.

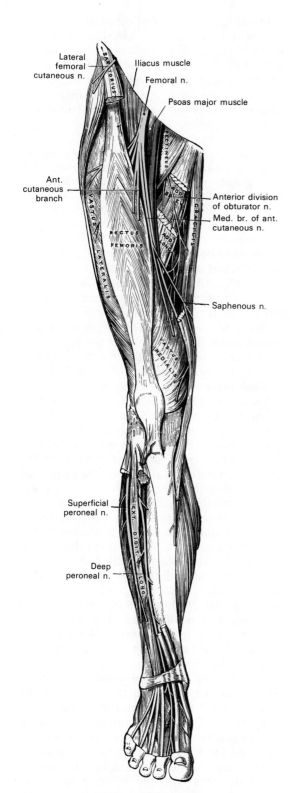

Lateral
femoral
cutaneous n.

Iliacus muscle

Femoral n.

Psoas major muscle

Ant.
cutaneous
branch

Anterior division
of obturator n.

Med. br. of ant.
cutaneous n.

Saphenous n.

Superficial
peroneal n.

Deep
peroneal n.

Fig. 12-60. Nerves of the right lower limb. Anterior view.

A **cutaneous branch** is occasionally found as a continuation of the communication with the anterior cutaneous and saphenous branches. It emerges from beneath the inferior border of the adductor longus and continues along the posterior margin of the sartorius to the medial side of the knee, where it pierces the deep fascia, communicates with the saphenous nerve, and is distributed to the skin of the medial side of the proximal half of the leg.

Posterior Branch. The **posterior branch** pierces the anterior part of the obturator externus, passes posterior to the adductor brevis and anterior to the adductor magnus, and divides into muscular and articular branches.

Muscular branches are given to the obturator externus as it passes between its fibers, to the adductor magnus, and to the adductor brevis when it is not supplied by the anterior branch.

The **articular branch** for the knee joint either perforates the adductor magnus or passes under the arch through which the femoral artery passes and enters the popliteal fossa. It accompanies the popliteal artery, supplying it with filaments, and reaches the back of the knee joint, where it perforates the oblique popliteal ligament and is distributed to the synovial membrane.

ACCESSORY OBTURATOR NERVE. The accessory obturator nerve (Fig. 12-56) is a small nerve present in about 29% of cases. It arises from the ventral part of the third and fourth lumbar nerves, follows along the medial border of the psoas major, and crosses over the superior ramus of the pubis instead of going through the obturator foramen with the obturator nerve. It passes deep to the pectineus muscle, supplying it with branches, communicates with the anterior branch of the obturator nerve, and sends a branch to the hip joint.

VARIATIONS. When the obturator nerve has a cutaneous branch, the medial cutaneous branch of the femoral nerve is correspondingly small. When the accessory obturator is lacking, the obturator nerve supplies two nerves to the hip joint; the accessory may be very small and be lost in the capsule of the hip joint.

FEMORAL NERVE. The femoral nerve (*anterior crural nerve*) (Figs. 12-55 to 12-60), the largest branch of the lumbar plexus and the principal nerve of the anterior part of the thigh, arises from the dorsal portions of the ventral primary divisions of the second, third, and fourth lumbar nerves. It emerges through the fibers of the psoas major at the distal part of its lateral border and passes down between it and the iliacus, being covered by the iliac portion of the transversalis fascia. The femoral nerve passes under the inguinal ligament, lateral to the femoral artery, and breaks up into branches soon after it enters the thigh. **Muscular branches** to the iliacus are given off within the abdomen. The **anterior cutaneous branches** are two large nerves: the intermediate and medial cutaneous nerves (Fig. 12-58).

Intermediate Cutaneous Nerve. The **intermediate cutaneous nerve** (*middle cutaneous nerve*) pierces the fascia lata (and generally the sartorius) about 7.5 cm distal to the inguinal ligament. It divides into two branches that descend in immediate proximity along the anterior thigh to supply the skin as far distally as the anterior knee. Here they communicate with the medial cutaneous nerve and the infrapatellar branch of the saphenous, thereby forming the **patellar plexus.** In the proximal part of the thigh the lateral branch of the intermediate cutaneous nerve communicates with the femoral branch of the genitofemoral nerve.

Medial Cutaneous Nerve. The **medial cutaneous nerve** (*internal cutaneous nerve*) passes obliquely across the proximal part of the sheath of the femoral artery, and divides into two branches, an anterior and a posterior, anterior to or at the medial side of that vessel. Before dividing, it gives off a few filaments that accompany the great saphenous vein and pierce the fascia lata to supply the integument of the medial side of the thigh. One of these filaments passes through the saphenous opening; a second becomes subcutaneous about the middle of the thigh; a third pierces the fascia at its lower third. The **anterior branch** runs distalward on the sartorius, perforates the fascia lata at the distal third of the thigh, and divides into two branches: one supplies the integument as far distally as the medial side of the knee; the other crosses to the lateral side of the patella, communicating in its course with the infrapatellar branch of the saphenous nerve. The posterior branch descends along the medial border of the sartorius muscle to the knee, where it pierces the fascia lata, communi-

cates with the saphenous nerve, and gives off several cutaneous branches. It then passes down to supply the integument of the medial side of the leg. Beneath the fascia lata, at the lower border of the adductor longus, it forms a plexiform network (**subsartorial plexus**) with branches of the saphenous and obturator nerves.

Variations. When the communicating branch from the obturator nerve is large and continues to the integument of the leg, the posterior branch of the medial cutaneous nerve is small and terminates in the plexus, occasionally giving off a few cutaneous filaments.

Nerve to Pectineus. The nerve to the pectineus arises immediately below the inguinal ligament, and passes deep to the femoral sheath to enter the anterior surface of the muscle; it is often double.

Nerve to Sartorius. The nerve to the sartorius arises in common with the intermediate cutaneous nerve and enters the deep surface of the proximal part of the muscle.

Saphenous Nerve. The saphenous nerve (Figs. 12-58; 12-60; 12-64) is the largest and longest branch of the femoral nerve; it supplies the skin of the medial side of the leg. It passes deep to the sartorius in company with the femoral artery, and lies anterior to the artery, crossing from its lateral to its medial side, within the fascial covering of the adductor canal (Fig. 8-61). At the tendinous arch in the adductor magnus, it quits the artery, penetrates the fascial covering of the adductor canal, continues along the medial side of the knee deep to the sartorius, and pierces the fascia lata between the tendons of the sartorius and the gracilis to become subcutaneous. It then accompanies the great saphenous vein along the tibial side of the leg, and at the medial border of the tibia in the distal third of the leg, it divides into two terminal branches.

One branch joins the medial cutaneous and obturator nerves in the middle of the thigh to form the **subsartorial plexus.**

A large **infrapatellar branch,** given off at the medial side of the knee, pierces the sartorius and fascia lata and is distributed to the skin in front of the patella. Above the knee it communicates with the anterior cutaneous branches of the femoral nerve; below the knee it communicates with other branches of the saphenous nerve; and on the lateral side of the joint, it communicates with branches of the lateral femoral cutaneous nerve to form the **patellar plexus.**

Branches below the knee are distributed to the skin of the anterior and medial side of the leg, communicating with the medial cutaneous nerve and the cutaneous branch of the obturator, if present. One branch continues along the margin of the tibia, and ends at the ankle. The other terminal branch passes anterior to the ankle and is distributed to the medial side of the foot, as far as the ball of the great toe, communicating with the medial branch of the superficial peroneal nerve.

Branches to the Quadriceps Femoris. The branch to the rectus femoris enters the proximal part of the deep surface of the muscle (Fig. 12-60). The large branch to the vastus lateralis accompanies the descending branch

of the lateral femoral circumflex artery to the distal part of the muscle. The branch to the vastus medialis muscle runs parallel with the saphenous nerve, lateral to the femoral vessels and outside of the adductor canal, and it enters the muscle near its middle. The two or three branches to the vastus intermedius, enter the anterior surface of the muscle about the middle of the thigh; a filament from one of these descends through the muscle to the articularis genu and the knee joint.

Articular Branches. The articular branch to the **hip joint** is derived from the nerve to the rectus femoris.

Articular branches to the **knee joint** are three in number. (1) A long slender filament derived from the nerve to the vastus lateralis penetrates the capsule of the joint on its anterior aspect. (2) A filament derived from the nerve to the vastus medialis can usually be

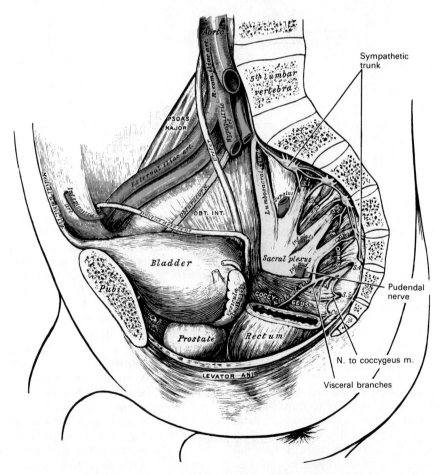

Fig. 12-61. Dissection of side wall of pelvis showing sacral and pudendal plexuses. (Testut.)

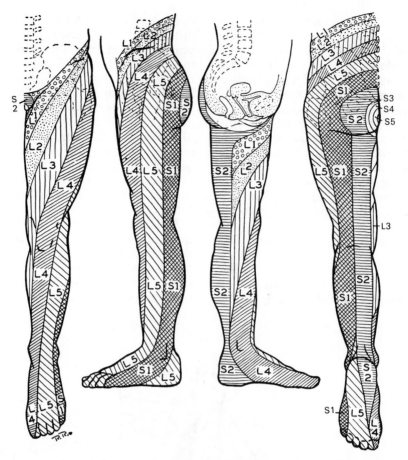

FIG. 12-62. Dermatome chart of the lower limb of man outlined by the pattern of hyposensitivity from loss of function of a single nerve root. (From Keegan and Garrett, Anat. Rec., *102*:417, 1948.)

traced downward on the surface of this muscle; it pierces the muscle and accompanies the articular branch of the descending genicular artery to the medial side of the articular capsule, which it penetrates, to supply the synovial membrane. (3) The branch to the vastus intermedius, which supplies the articularis genu, also is distributed to the knee joint.

Sacral and Coccygeal Nerves and Sacral Plexus

The ventral primary divisions of the sacral coccygeal nerves enter into the formation of the sacral and pudendal plexuses. Those from the first four sacral nerves enter the pelvis through the anterior sacral foramina; the fifth enters between the sacrum and coccyx, and the coccygeals enter below the first

piece of the coccyx. The first and second sacrals are large; the third, fourth, and fifth diminish progressively in size.

The **sacral plexus** (Figs. 12-61; 12-63; 12-70) is formed by the lumbosacral trunk from the fourth and fifth lumbar nerves and by the first, second, and third sacral nerves. They converge toward the caudal part of the greater sciatic foramen and unite into a large flat band, most of which is continued into the thigh as the sciatic nerve. The plexus lies against the posterior and lateral walls of the pelvis, between the piriformis and the internal iliac vessels, which are embedded in the pelvic subserous fascia. The nerves that enter the plexus, with the exception of the third sacral, split into ventral and dorsal portions, and the branches that arise from them are listed in the chart that appears at the top of page 1236.

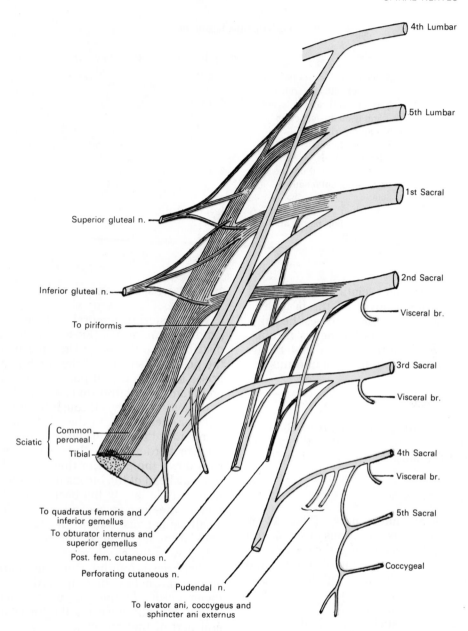

FIG. 12-63. Plan of sacral and coccygeal plexuses.

Labels in figure:
4th Lumbar
5th Lumbar
1st Sacral
Superior gluteal n.
Inferior gluteal n.
To piriformis
2nd Sacral
Visceral br.
3rd Sacral
Visceral br.
Sciatic { Common peroneal.
Tibial
4th Sacral
Visceral br.
5th Sacral
To quadratus femoris and inferior gemellus
To obturator internus and superior gemellus
Post. fem. cutaneous n.
Perforating cutaneous n.
Pudendal n.
To levator ani, coccygeus and sphincter ani externus
Coccygeal

NERVE TO QUADRATUS FEMORIS AND GEMELLUS INFERIOR

The **nerve to the quadratus femoris and gemellus inferior,** from the ventral portions of L 4, 5, and S 1, leaves the pelvis through the greater sciatic foramen, distal to the piriformis and ventral to the sciatic nerve. It runs ventral or deep to the tendon of the obturator internus and gemelli and enters the deep surface of the quadratus and gemellus inferior.

An **articular branch** is given to the hip joint.

NERVE TO OBTURATOR INTERNUS AND GEMELLUS SUPERIOR

The **nerve to the obturator internus and gemellus superior** (Fig. 12-66), from the ventral portions of L 5, and S 1, 2, leaves the pel-

Branches From the Sciatic Plexus

	Ventral Portions	Dorsal Portions
1. Nerve to quadratus femoris and gemellus inferior	L 4, 5, S 1	
2. Nerve to obturator internus and gemellus superior	L 5, S 1, 2	
3. Nerve to piriformis		S (1), 2
4. Superior gluteal		L 4, 5, S 1
5. Inferior gluteal		L 5, S 1, 2
6. Posterior femoral cutaneous	S 2, 3	S 1, 2
7. Perforating cutaneous		S 2, 3
8. Sciatic		
a. Tibial	L 4, 5, S 1, 2, 3	
b. Common peroneal		L 4, 5, S 1, 2
9. (Pudendal)	S 2, 3, 4	

vis through the greater sciatic foramen distal to the piriformis. It gives off the branch to the gemellus superior, then crosses the spine of the ischium, reenters the pelvis through the lesser sciatic foramen, and enters the pelvic surface of the obturator internus.

NERVE TO PIRIFORMIS

The **nerve to the piriformis,** from the dorsal portions of S 2 or S 1, 2, enters the ventral surface of the muscle; it may be double.

SUPERIOR GLUTEAL NERVE

The **superior gluteal nerve** (Figs. 12-66; 12-70), from the dorsal portions of L 4, 5, and S 1, leaves the pelvis through the greater sciatic foramen proximal to the piriformis, and accompanying the superior gluteal vessels and their branches, divides into a superior and an inferior branch.

The **superior branch** is distributed to the gluteus minimus muscle. The **inferior branch** crosses the gluteus minimus, gives filaments to the gluteus medius and minimus, and ends in a branch to the tensor fasciae latae.

INFERIOR GLUTEAL NERVE

The **inferior gluteal nerve,** from the dorsal portions of L 5, and S 1, 2, leaves the pelvis through the greater sciatic foramen distal to the piriformis and enters the deep surface of the gluteus maximus.

POSTERIOR FEMORAL CUTANEOUS NERVE

The **posterior femoral cutaneous nerve** (*small sciatic nerve*) (Figs. 12-64 to 12-66) is distributed to the skin of the perineum and the posterior surface of the thigh and leg. It arises from the dorsal portions of S 1 and 2 and from the ventral divisions of S 2 and 3, and leaves the pelvis through the greater sciatic foramen distal to the piriformis. It accompanies the inferior gluteal artery to the inferior border of the gluteus maximus and runs down the posterior thigh, superficial to the long head of the biceps femoris and deep to the fascia lata to the back of the knee. It pierces the deep fascia and accompanies the small saphenous vein to the middle of the back of the leg, its terminal twigs communicating with the sural nerve.

The three or four **inferior clunial branches** (*gluteal branches*) turn upward around the lower border of the gluteus maximus and supply the skin covering the lower and lateral part of that muscle.

The **perineal branches** (Fig. 12-66) arise at the lower border of the gluteus maximus, run medially over the origin of the hamstrings toward the groove between the thigh and perineum, and pierce the deep fascia to supply the skin of the external genitalia and adjacent proximal medial surface of the thigh. One long branch, the inferior pudendal (long scrotal nerve) (Fig. 12-70), runs ventrally in the fascia of the perineum to the skin of the scrotum and base of the penis in the male or to the skin of the labium majus in the female, communicating with the pos-

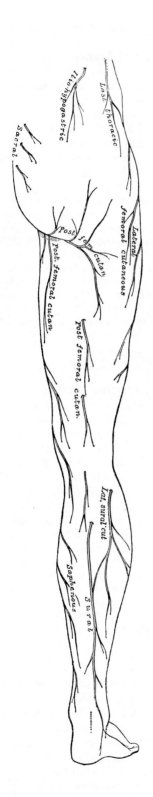

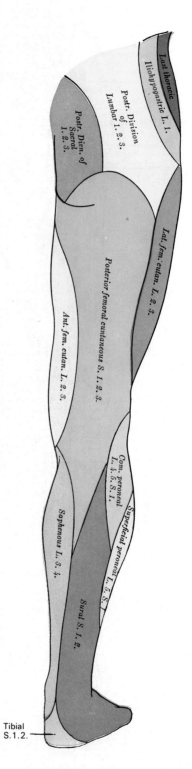

Fig. 12-64. Cutaneous nerves of right lower limb. Posterior view.

Fig. 12-65. Diagram of the segmental distribution of the cutaneous nerves of the right lower limb. Posterior view.

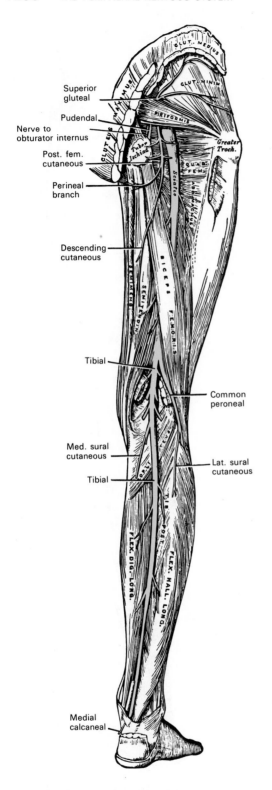

FIG. 12-66. Nerves of the right lower limb. Posterior view. (In this diagram the medial and lateral sural cutaneous are not in their normal position. They have been displaced by the removal of the superficial muscles.)

terior scrotal and inferior rectal branches of the pudendal nerve.

The **femoral branches** consist of numerous filaments from both sides of the nerve that are distributed to the skin of the posterior and medial sides of the thigh and the popliteal fossa (Fig. 12-64).

The **sural branches** are usually two terminal twigs that supply the skin of the posterior leg to a varying extent and communicate with the sural nerve.

VARIATIONS. When the tibial and peroneal nerves arise separately, the posterior femoral cutaneous nerve also arises from the sacral plexus in two parts. The ventral portion accompanies the tibial nerve and gives off the perineal and medial femoral branches; the dorsal portion passes through the piriformis with the peroneal nerve and supplies the gluteal and lateral femoral branches. The inferior pudendal branch may pierce the sacrotuberous ligament. The sural branches may be lacking or may extend down as far as the ankle.

PERFORATING CUTANEOUS NERVE

The **perforating cutaneous nerve** (*n. clunium inferior medialis*) arises from the posterior surface of the second and third sacral nerves. It pierces the caudal part of the sacrotuberous ligament, winds around the inferior border of the gluteus maximus, and is distributed to the skin over the medial and caudal parts of that muscle.

VARIATIONS. The perforating cutaneous nerve is lacking in one third of the bodies; its place may be taken by a branch of the posterior femoral cutaneous nerve, or by a branch from S 3 and 4, or S 4 and 5. It may pierce the gluteus maximus as well as the ligament, or instead of piercing the ligament it may accompany the pudendal nerve or go between the muscle and the ligament. It may arise in common with the pudendal nerve.

SCIATIC NERVE

The **sciatic nerve** (*n. ischiadicus; great sciatic nerve*) (Fig. 12-66) is the largest nerve in the body. It supplies the skin of the foot and most of the leg, the muscles of the posterior thigh, all the muscles of the leg and foot, and contributes filaments to all the joints of the lower limb. It is the continuation of the main part of the sacral plexus, arising from L 4, 5 and S 1, 2, 3. It passes out of the pelvis through the greater sciatic foramen and extends from the inferior border of the piriformis to the distal third of the thigh, where it splits into two large terminal divisions, the **tibial** and **common peroneal nerves.** In the proximal part of its course it rests upon the posterior surface of the ischium, between the ischial tuberosity and the greater trochanter of the femur, and crosses the obturator internus, gemelli, and quadratus femoris. It is accompanied by the posterior femoral cutaneous nerve and the inferior gluteal artery and is covered by the gluteus maximus. More distally it lies upon the adductor magnus and is crossed obliquely by the long head of the biceps femoris.

The tibial and common peroneal nerves represent two divisions within the sciatic nerve that are manifest at the origin of the nerve and preserve their identity throughout its length, although combined into one large nerve by a common connective tissue sheath. The tibial division takes its origin in the sacral plexus from the ventral divisions of L 4, 5, and S 1, 2, 3; the peroneal division originates from the dorsal divisions of L 4, 5 and S 1, 2. In some bodies these two divisions remain separate throughout their course, no true sciatic nerve being formed.

Articular Branches

Articular branches arise from the proximal part of the nerve and supply the hip joint, perforating the posterior part of the capsule. They may arise from the sacral plexus.

Muscular Branches

The **muscular branches** to the hamstrings, viz., the long head of the biceps femoris, the semitendinosus, and the semimembranosus muscles, and to the adductor magnus come from the tibial division; the branch to the short head of the biceps comes from the common peroneal division.

Tibial Nerve

The **tibial nerve** (*medial popliteal nerve*) (Fig. 12-66), the larger of the two terminal divisions of the sciatic nerve, is composed of fibers from the ventral portions of L 4, 5 and S 1, 2, 3. It continues in the same direction as the sciatic nerve, at first deep to the long head of the biceps, then through the middle of the popliteal fossa covered by adipose tis-

sue and fascia. After crossing the popliteus muscle, it passes between the heads of the gastrocnemius and under the soleus. It remains deep to these muscles down to the medial margin of the tendo calcaneus, along which it runs to the flexor retinaculum and there divides into the medial and lateral plantar nerves. At its origin it is some distance lateral to the popliteal artery and vein, but as it continues in a straight course it crosses superficial to the vessels in the popliteal fossa, and, lying medial to them, passes with them under the tendinous arch formed by the soleus. It accompanies the posterior tibial artery, at first being medial but soon crossing the artery and lying lateral to it down to the ankle.

ARTICULAR BRANCHES OF TIBIAL NERVE.   Articular branches supply the knee and ankle joints.

Three branches, accompanying the superior and inferior medial genicular and the middle genicular arteries, pierce the ligaments and supply the **knee joint.** The superior branch is inconstant. Just proximal to the bifurcation at the flexor retinaculum, an articular branch is given off to the **ankle joint.**

MUSCULAR BRANCHES OF TIBIAL NERVE. Muscular branches are supplied to the muscles of the posterior leg. Branches that arise as the nerve lies between the two heads of the gastrocnemius are distributed to both heads of the gastrocnemius, plantaris, soleus, and popliteus; the latter branch turns around the distal border and is distributed to the deep surface of the muscle. Arising more distally, either separately or by a common trunk, are branches to the soleus, tibialis posterior, flexor digitorum longus, and flexor hallucis longus; the branch to the last muscle accompanies the peroneal artery; that to the soleus enters the deep surface of the muscle.

MEDIAL SURAL CUTANEOUS NERVE.   The medial sural cutaneous nerve (*n. communicans tibialis*) remains superficial in the groove between the two heads of the gastrocnemius, accompanied by the small saphenous vein. Near the middle of the back of the leg, it pierces the deep fascia and is joined by the communicating ramus of the lateral sural cutaneous branch of the peroneal nerve to form the sural nerve.

SURAL NERVE.   The sural nerve (*short saphenous nerve*) (Fig. 12-64) is formed by the union of the medial sural cutaneous nerve and the communicating ramus of the lateral sural cutaneous nerve (peroneal anastomotic). Lying with the small saphenous vein near the lateral margin of the tendo calcaneus, it continues distally to the interval between the lateral malleolus and the calcaneus, supplying branches to the skin of the back of the leg and communicating with the posterior femoral cutaneous nerve. The nerve turns anteriorly below the lateral malleolus and is continued as the **lateral dorsal cutaneous nerve** along the lateral side of the foot and little toe (Fig. 12-58). It communicates on the dorsum of the foot with the intermediate dorsal cutaneous nerve, a branch of the superficial peroneal nerve.

MEDIAL CALCANEAL BRANCHES. The medial calcaneal branches (*internal calcaneal branches*) (Fig. 12-66) perforate the flexor retinaculum and are distributed to the skin of the heel and medial side of the sole of the foot.

MEDIAL PLANTAR NERVE. The medial plantar nerve (*internal plantar nerve*) (Fig. 12-67), the larger of the two terminal branches of the tibial nerve, accompanies the medial

plantar artery. From its origin under the flexor retinaculum, it passes deep to the abductor hallucis, and appearing between this muscle and the flexor digitorum brevis, it gives off the proper plantar digital nerve to the great toe. It finally divides opposite the bases of the metatarsal bones into three common digital nerves.

Branches of the Medial Plantar Nerve. The **plantar cutaneous branches** pierce the plantar aponeurosis between the abductor hallucis and the flexor digitorum brevis and are distributed to the skin of the sole of the foot (Fig. 12-68).

Muscular branches for the abductor hallucis and flexor digitorum brevis arise from the trunk of the nerve and enter the deep surfaces of the muscles. A branch for the flexor hallucis brevis springs from the proper digital nerve to the medial side of the great toe. A branch for the first lumbrical comes from the first common digital nerve.

The **articular branches** supply the joints of the tarsus and metatarsus.

The **proper digital nerve** of the great toe pierces the plantar aponeurosis posterior to the tarsometatarsal joint, sends a branch to the flexor hallucis brevis, and is distributed to the skin of the medial side of the great toe (Fig. 12-67).

The **three common digital nerves** pass between the divisions of the plantar aponeurosis; each splits into the two **proper digital nerves.** Those of the first common digital supply the adjacent sides of the great and

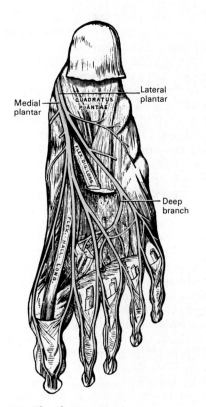

FIG. 12-67. The plantar nerves.

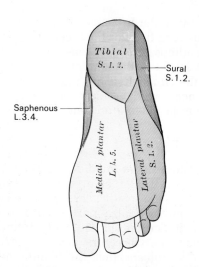

FIG. 12-68. Diagram of the segmental distribution of the cutaneous nerves of the sole of the foot.

second toes; those of the second, the adjacent sides of the second and third toes; those of the third, the adjacent sides of the third and fourth toes. The first common digital nerve gives a twig to the first lumbrical. The third communicates with the lateral plantar nerve. Each proper digital nerve gives off cutaneous and articular filaments along the digit, finally terminating in the ball of the toe, and opposite the distal phalanx sends a dorsal branch that is distributed to the structures around the nail.

LATERAL PLANTAR NERVE. The lateral plantar nerve (*external plantar nerve*) (Figs. 12-67; 12-68) supplies the skin of the fifth and lateral half of the fourth toes, as well as most of the deep muscles of the foot; its distribution is similar to that of the ulnar nerve in the hand. It passes distally with the lateral plantar artery to the lateral side of the foot, lying between the flexor digitorum brevis and the quadratus plantae, and in the interval between the former muscle and the abductor digiti minimi, it divides into a superficial and a deep branch.

Branches of the Lateral Plantar Nerve. **Muscular branches** to the quadratus plantae and the abductor digiti minimi are given off before the division into superficial and deep branches.

The **superficial branch** splits into a common digital nerve and a nerve that supplies the proper digital for the lateral side of the little toe, and muscular branches to the flexor digiti minimi brevis and the two interossei of the fourth intermetatarsal space. The common digital nerve has a communication with the third common digital branch of the medial plantar nerve and divides into two proper digital nerves that are distributed to the adjacent sides of the fourth and fifth toes.

The **deep branch** accompanies the lateral plantar artery on the deep surface of the tendons of the flexors longi and the adductor hallucis. It supplies all the interossei except those in the fourth intermetatarsal space, the second, third, and fourth lumbricals, and the adductor hallucis.

Common Peroneal Nerve

The **common peroneal nerve** (*lateral popliteal nerve; peroneal nerve*) (Fig. 12-66), the smaller of the two terminal divisions of the sciatic nerve, is composed of fibers from the dorsal portions of L 4, 5 and S 1, 2. It runs obliquely along the lateral side of the popliteal fossa, close to the medial border of the biceps femoris and between that muscle and the lateral head of the gastrocnemius, to the head of the fibula. It winds around the neck of the fibula and passes deep to the peroneus longus, where it divides into the superficial and deep peroneal nerves.

ARTICULAR BRANCHES. There are three articular branches to the knee: (*a*) one accompanies the superior lateral genicular artery to the knee, occasionally arising from the trunk of the sciatic nerve; (*b*) one accompanies the inferior lateral genicular artery to the joint; and (*c*) the third (recurrent) articular nerve is given off at the point of division of the common peroneal nerve and accompanies the anterior tibial recurrent artery through the substance of the tibialis anterior to the anterior part of the knee (Fig. 12-60, not labeled).

LATERAL SURAL CUTANEOUS NERVE. The **lateral sural cutaneous nerve** (*lateral cutaneous branch*) is distributed to the skin of the posterior and lateral surfaces of the leg (Fig. 12-65) and has an important communicating ramus. The **communicating ramus** (*peroneal anastomotic nerve*) (Figs. 12-64; 12-66) arises near the head of the fibula, crosses superficial to the lateral head of the gastrocnemius, and in the middle of the leg joins the medial sural cutaneous nerve to form the sural nerve.

DEEP PERONEAL NERVE. The deep peroneal nerve (*anterior tibial nerve*) (Figs. 12-60; 12-69), arising from the bifurcation of the common peroneal nerve between the fibula and the peroneus longus, continues deep to the extensor digitorum longus to the anterior surface of the interosseous membrane. It meets the anterior tibial artery in the proximal third of the leg and continues with it distally, passes under the extensor retinaculum, and terminates at the ankle in a medial and a lateral branch.

Muscular branches are supplied in the leg to the tibialis anterior, extensor digitorum longus, peroneus tertius, and extensor hallucis longus.

An **articular branch** is supplied to the ankle joint.

The **lateral terminal branch** (*external or tarsal branch*) passes across the tarsus deep to the extensor digitorum brevis, and having become enlarged like the dorsal interosseous nerve at the wrist, supplies the extensor digitorum brevis. Three minute interosseous

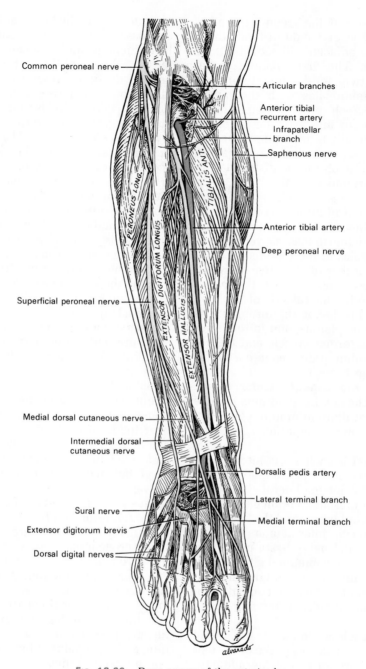

Common peroneal nerve

Articular branches

Anterior tibial
recurrent artery

Infrapatellar
branch

Saphenous nerve

Anterior tibial artery

Deep peroneal nerve

Superficial peroneal nerve

PERONEUS LONG.

EXTENSOR DIGITORUM LONGUS

TIBIALIS ANT.

EXTENSOR HALLUCIS

Medial dorsal cutaneous nerve

Intermedial dorsal
cutaneous nerve

Dorsalis pedis artery

Lateral terminal branch

Sural nerve

Medial terminal branch

Extensor digitorum brevis

Dorsal digital nerves

alvarado

FIG. 12-69.   Deep nerves of the anterior leg.

branches are given off from the enlargement; they supply the tarsal joints and the meta-tarsophalangeal joints of the second, third, and fourth toes. A muscular filament is sent to the second interosseus dorsalis from the first of these interosseous branches.

The **medial terminal branch** (*internal branch*) accompanies the dorsalis pedis ar-

tery along the dorsum of the foot, and at the first interosseous space, divides into two dorsal digital nerves (Fig. 12-70) which supply adjacent sides of the great and second toes, communicating with the medial dorsal cutaneous branch of the superficial peroneal nerve. An interosseous branch, given off before it divides, enters the first space, supplying the metatarsophalangeal joint of the great toe and sending a filament to the first interosseus dorsalis.

SUPERFICIAL PERONEAL NERVE. The superficial peroneal nerve (*musculocutaneous nerve*) (Figs. 12-69; 12-70) passes distally between the peronei and the extensor digitorum longus, pierces the deep fascia in the lower third of the leg, and divides into a medial and intermediate dorsal cutaneous nerve.

Muscular branches are given off in its course between the muscles to the peroneus longus and brevis.

Cutaneous filaments are supplied to the skin of the lower part of the leg.

The **medial dorsal cutaneous nerve** (*internal dorsal cutaneous branch*) (Fig. 12-69) passes in front of the ankle joint and divides into two dorsal digital branches. The medial one supplies the medial side of the great toe and communicates with the deep peroneal nerve. The lateral one supplies the adjacent sides of the second and third toes. It also supplies the skin of the medial side of the foot and ankle and communicates with the saphenous nerve.

The **intermediate dorsal cutaneous nerve** (*external dorsal cutaneous branch*) (Fig. 12-69) passes along the lateral part of the dorsum of the foot, supplying the skin of the lateral side of the foot and ankle and communicating with the sural nerve. It terminates by dividing into two dorsal digital branches, one of which supplies the adjacent sides of the third and fourth toes, the other the adjacent sides of the fourth and little toes. Frequently some of the lateral branches of the superficial peroneal nerve are absent and their places are then taken by branches of the sural nerve.

Pudendal Plexus

The **pudendal plexus** (Figs. 12-63; 12-70) is formed from the anterior branches of the second and third and all of the fourth sacral nerves. It is sometimes considered a part of the sacral plexus. It lies in the posterior hollow of the pelvis on the ventral surface of the piriformis muscle.

VISCERAL BRANCHES

The **visceral branches** arise from the second, third, and fourth sacral nerves. They contain (1) parasympathetic visceral efferent preganglionic fibers, which form synapses with the cells in the small scattered ganglia located in or near the walls of the pelvic viscera, and (2) visceral afferent fibers from the pelvic organs. These nerves join branches of the hypogastric plexus of the sympathetic and the sympathetic chain to form the pelvic plexus, which lies in the deeper portion of the pelvic subserous fascia (Ashley and Anson, 1946).

The various elements of the pelvic plexus cannot be followed by dissection, but the destination of the sacral parasympathetic fibers has been traced by clinical observation and animal experimentation.

The **branches to the bladder, prostate and seminal vesicles** approach these organs from their posterior and lateral sides and terminate in groups of ganglion cells in the pelvic plexus or in the walls of the organs. The postganglionic fibers are efferents to the muscles and glands, except the sphincters, to which they are inhibitory.

The **branches to the uterus** reach the ganglia in the uterine plexus. The postganglionic fibers are inhibitory except during pregnancy, when their function is said to be reversed.

The **branches to the external genitalia** leave the pelvic plexus to join the pudendal nerve. They are distributed through its branches to the corpora cavernosa, causing active dilatation of the cavernous blood sinuses.

The **branches to the alimentary tract** consist of (*a*) filaments that go directly to the rectum from the pelvic plexus, and (*b*) fibers that pass through the hypogastric plexus, the hypogastric nerves, and the inferior mesenteric plexus to reach the descending and sigmoid colon. These are preganglionic fibers that have synapses with the cells in the myenteric and submucous plexuses and are efferent to this part of the intestine, except for the sphincter ani internus, to which they are inhibitory.

MUSCULAR BRANCHES

The **muscular branches** (Fig. 12-61), derived mainly from the fourth (sometimes from the third and fifth) sacral, enter the pelvic surfaces of the levator ani and coccygeus. The nerve to the sphincter ani externus (perineal branch) reaches the ischiorectal fossa by piercing the coccygeus or by passing between it and the levator.

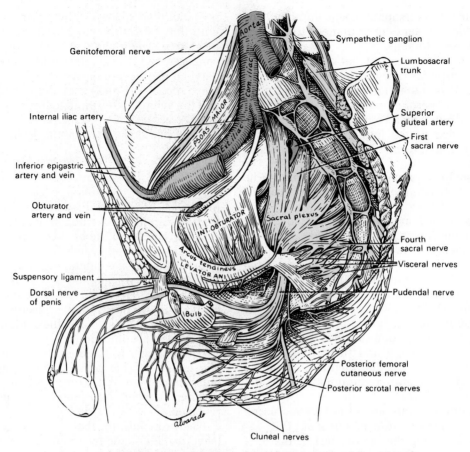

FIG. 12-70   Pudendal nerve, and sacral and pudendal plexus of the right side.

PUDENDAL NERVE

The **pudendal nerve** (*internal pudic nerve*) (Fig. 12-70) arises from S 2, 3 and 4, and passing between the piriformis and coccygeus, leaves the pelvis through the distal part of the greater sciatic foramen. It then crosses the spine of the ischium and reenters the pelvis through the lesser sciatic foramen. It accompanies the internal pudendal vessels along the lateral wall of the ischiorectal fossa in a tunnel formed by a splitting of the obturator fascia (*Alcock's canal*[1]), and as it approaches the urogenital diaphragm, it splits into two terminal branches.

INFERIOR RECTAL NERVE. The inferior rectal nerve (*inferior hemorrhoidal nerve*) arises from the pudendal nerve before its terminal division or occasionally directly

[1]Benjamin Alcock (1801-?): An Irish professor of anatomy (Cork, then went to America).

from the sacral plexus. Coursing medially across the ischiorectal fossa with the inferior rectal vessels, it breaks up into branches that are distributed to the sphincter ani externus and the integument around the anus. Branches of this nerve communicate with the perineal branch of the posterior femoral cutaneous nerve and with the posterior scrotal nerves.

PERINEAL NERVE. The perineal nerve (Fig. 12-70), the larger and more superficial of the two terminal branches of the pudendal, accompanies the perineal artery and at the urogenital diaphragm divides into superficial and deep branches.

The two **superficial branches** of the perineal nerve are the medial and the lateral posterior scrotal (or labial) nerves. They pierce the fascia of the urogenital diaphragm and run ventrally along the lateral part of the urogenital triangle in company with the posterior scrotal branches of the perineal artery. They are distributed to the skin of the scrotum in the male or to the skin of the labium majus in the female. The lateral branch communicates with the perineal branch of the posterior femoral cutaneous nerve.

The **deep branch** is mainly muscular, supplying branches to the transversus perinei superficialis, bulbocavernosus, ischiocavernosus, transversus perinei profundus, and sphincter urethrae membranacea. The nerve to the bulb is given off from the nerve to the bulbocavernosus; it pierces this muscle, and supplies the corpus cavernosum urethrae and the mucous membrane of the urethra. DORSAL NERVE OF PENIS. The **dorsal nerve** of the **penis** (Fig. 12-70), the deeper terminal branch of the pudendal nerve, accompanies the internal pudendal artery along the ramus of the ischium, and then runs ventrally along the margin of the inferior ramus of the pubis, lying between the superficial and deep layers of fascia of the urogenital diaphragm. Piercing the superficial layer, it gives a branch to the corpus cavernosum penis, and continuing forward in company with the dorsal artery of the penis between the layers of the suspensory ligament, it runs along the dorsum of the penis and is distributed to the skin of that organ, ending in the glans penis.

The **dorsal nerve** of the **clitoris** is smaller and has a distribution corresponding to that of the penis in the male.

Coccygeal Plexus

The **coccygeal plexus** (Fig. 12-63) is formed by the coccygeal nerve with communications from the fourth and fifth sacral nerves. From this delicate plexus, a few fine filaments, the **anococcygeal nerves,** pierce the sacrotuberous ligament and supply the skin in the region of the coccyx.

Visceral Nervous System

The **visceral nervous system** or visceral portion of the peripheral nervous system (*vegetative nervous system; involuntary nervous system; major sympathetic system; plexiform nervous system*) comprises the whole complex of fibers, nerves, ganglia, and plexuses by means of which impulses are conveyed from the central nervous system to the viscera and from the viscera to the central nervous system. It has the usual two groups of fibers necessary for reflex connections: (*a*) afferent fibers, receiving stimuli and carrying impulses toward the central nervous system, and (b) efferent fibers, carrying impulses from the appropriate centers to the active effector organs, which in this instance are the nonstriated muscle, cardiac muscle, and glands of the body.

Attention is called to the use of the term sympathetic system in this edition. In former editions it was used for the visceral nervous system and included both afferent and efferent, autonomic and parasympathetic components. Although the central nervous system is involved in all nervous or reflex control of the viscera, with the possible exception of the enteric plexus, it is more convenient for the purposes of description to separate the peripheral from the central portions. It must be emphasized, however, that the separation is artificial and should not be carried over into physiologic considerations. In the present edition the term sympathetic is used in the restricted meaning of the thoracolumbar division of the visceral efferent system. The terms sympathetic afferent and autonomic afferent are replaced by visceral afferent.

VISCERAL AFFERENT FIBERS

The **visceral afferent fibers** cannot be separated into a morphologically independent system because, like the somatic sensory fibers, their cell bodies are situated in the sensory ganglia of the cerebrospinal nerves. The distinction between somatic and visceral afferent fibers is one of peripheral distribution rather than one of fundamental anatomic and physiologic significance. The visceral afferents, however, commonly have modalities of sensation different from those of somatic afferents, and most of them are either vaguely localized or have no representation in consciousness. The visceral efferent fibers make reflex connections with visceral afferents. The number and extent of the visceral afferents are not clearly established, and the peripheral processes reach the ganglia by various routes. Many traverse the branches and plexuses of the autonomic system, most of them accompany blood vessels for at least a part of their course, and some run in the cerebrospinal nerves.

REFERRED PAIN. Although many, perhaps most, of the physiologic impulses carried by visceral afferent fibers fail to reach consciousness, pathologic conditions or excessive stimulation may activate those that carry pain. The central nervous system has a poorly developed power of localizing the source of such pain, and by some mechanism not clearly understood, the pain may be referred to the region supplied by the somatic afferent fibers whose central connections are the same as those of the visceral afferents. For example, the visceral afferents from the heart enter the upper thoracic nerves, and impulses traversing them may cause painful sensations in the axilla, down the ulnar surface of the arm, and in the precardial region. The study of clinical cases of referred pain has been useful in tracing the path of afferent fibers from the various viscera, and a knowledge of these paths may assist the diagnostician in locating a pathologic process.

The visceral afferent fibers are summarized below; for more complete descriptions of the individual nerves mentioned, consult the accounts given elsewhere.

HEAD. Visceral afferents from endings on the peripheral blood vessels of the face and scalp probably accompany the branches of the external carotid artery to the superior cervical ganglion and through communicating rami to the spinal nerves. Afferents on the blood vessels of the brain and meninges may accompany the branches of the internal carotid and vertebral arteries, passing through the upper cervical spinal nerves or possibly the ninth and tenth cranial nerves.

NOSE AND NASAL CAVITY. There is evidence that a few visceral afferent fibers from the nose are brought to the brain by the nervus terminalis (Larsell, 1950). Others traverse the branches of the sphenopalatine and palatine nerves to reach the facial nerve through the Vidian and greater superficial petrosal nerves.

MOUTH AND PHARYNX. The visceral afferents from the mouth, pharynx, and salivary glands pass through the pharyngeal plexus to the glossopharyngeal, vagus, and facial nerves.

NECK. Visceral afferents from the larynx, trachea, esophagus, and thyroid gland are carried by the vagus or reach the sympathetic trunk through the pharyngeal plexus and pass through rami communicantes to the cervical or upper thoracic nerves.

The **carotid sinus nerve** carries the visceral afferents from the pressoreceptor endings in the carotid sinus through the glossopharyngeal nerve. The chemoreceptor afferents from the **carotid body** reach the vagus nerve through branches of the pharyngeal or superior laryngeal nerves.

THORAX. The visceral afferents from the thoracic wall and parietal pleura join the intercostal nerves, after following the arteries for variable distances, and thus enter the spinal ganglia. Those from the parietal pericardium join either the phrenic nerve or the intercostal nerves.

Visceral afferents from the **heart** and the origins of the great vessels enter the cardiac plexus and either join the branches of the vagus or reach the upper thoracic spinal nerves and their ganglia by way of sympathetic branches, the sympathetic trunk, and rami communicantes.

The **aortic bodies** (*glomera aortica*) are chemoreceptors similar to the carotid bodies; their afferent fibers run in the right vagus and have cell bodies in the inferior ganglion. The supracardial bodies (aortic paraganglia) are also chemoreceptors with their afferent fibers in the left vagus and cell bodies in the inferior ganglion.

DEPRESSOR NERVE. In some animals, the afferent fibers of the vagus from the heart and great vessels are contained in a separate nerve whose stimulation depresses the activity of the heart. In man, these fibers are probably contained in the cardiac branches of the recurrent nerves.

Visceral afferents from the **lungs, bronchi,** and **pulmonary pleura** reach either the vagus nerve through the pulmonary plexuses or the spinal nerves and their ganglia through the sympathetic branches and rami communicantes.

ABDOMEN. Visceral afferent fibers from the **abdominal wall and parietal peritoneum** probably accompany the arteries in part of their course and finally reach the spinal ganglia through the spinal nerves. The myelinated fibers from the Pacinian[1] corpuscles in the mesentery and pancreas at its base run in the thoracic splanchnic nerves, then through the sympathetic trunk, and finally over the white rami communicantes to the spinal nerves and ganglia.

Visceral afferent fibers from the **stomach, small intestine, cecum, appendix, ascending and transverse colon, liver, gallbladder, bile ducts, pancreas,** and **suprarenals** traverse the celiac plexus and its secondary plexuses and branches. They accompany the arteries, pass through the splanchnic nerves, the

[1]Filippo Pacini (1812–1883): An Italian anatomist (Florence).

sympathetic trunk, and rami communicantes to reach the spinal nerves and ganglia. Some of these afferents may enter the vagus nerve.

The visceral afferents from the **kidney, ureter, testis, ductus deferens, ovary, and uterine tube** traverse the renal and celiac plexuses or parts of their secondary plexuses, pass through the lower thoracic and upper lumbar splanchnic nerves to the sympathetic trunk, and thence through white rami communicantes to the spinal nerves and ganglia.

PELVIS.   The visceral afferent fibers from the **descending colon, sigmoid,** and **rectum** traverse the pelvic plexus, hypogastric nerves and plexus, inferior mesenteric plexus, celiac plexus, and lumbar splanchnic nerves on their way to the sympathetic trunk, white rami communicantes, and spinal nerves and ganglia. Others from the rectum pass through the pelvic plexus into the visceral branches of the second, third, and fourth sacral nerves and their ganglia.

Visceral afferents from the **bladder, prostate, seminal vesicles,** and **urethra** pass through the pelvic plexus and through the hypogastric nerves and plexuses, splanchnic nerves, and sympathetic rami into the lumbar ganglia, or through the visceral branches of the sacral nerves into the sacral ganglia.

Visceral afferents from the **uterus** traverse the pelvic plexus, hypogastric nerves and plexus, lumbar splanchnic nerves, sympathetic trunk, rami communicantes, and lumbar spinal nerves and their ganglia.

Visceral afferent fibers from the **external genitalia** pass through either the pelvic plexus or the pudendal nerve and reach the sacral nerves and their ganglia.

UPPER EXTREMITY.   Visceral afferent fibers accompany the peripheral blood vessels for some distance, but may join the larger branches of the brachial plexus and reach the dorsal root ganglia through the spinal nerves, or they may follow the paths of the sympathetic fibers and reach the dorsal ganglia, especially the first two or three thoracic, through the white rami communicantes.

LOWER EXTREMITY.   Visceral afferents accompany the peripheral vessels and the femoral artery to the aortic plexus, then through the lumbar splanchnic nerves to the rami communicantes, spinal nerves and ganglia. Others may join the tibial or peroneal nerves and traverse the sacral and lumbar sympathetic trunk and rami communicantes to reach the lumbar nerves and their ganglia.

AUTONOMIC OR VISCERAL EFFERENT SYSTEM

The **visceral efferent** portions of the peripheral nervous system are combined into a morphologic and physiologic entity called the **autonomic nervous system** (*systema nervosum autonomicum*) (Fig. 12-71). The fundamental morphologic difference between the visceral and somatic motor systems is that two neurons are required to transmit an impulse from the central nervous system to the active effector organ in the viscera, whereas only a single neuron is required to carry an impulse from the central nervous system to a skeletal muscle fiber. As the name autonomic implies, this system has a certain amount of independence because, in most individuals, it is not under direct voluntary command. It is controlled by neurons within the central nervous system, nevertheless, and is connected with the latter at various levels. The enteric plexus is the only portion of the visceral system that seems to carry out reflex responses without involving the central nervous system (Kuntz, 1953).

The autonomic nervous system is composed of two divisions or systems that differ from each other morphologically and which for the most part are antagonistic to each other physiologically. The morphologic differences have to do (*a*) with the manner in which the two systems are connected with the central nervous system, and (*b*) with the location of their ganglia. The **sympathetic** or thoracolumbar system is connected with the central nervous system through the thoracic and upper lumbar segments of the spinal cord, and its ganglia tend to be placed near the spinal column rather than near the viscera innervated. The **parasympathetic** or craniosacral system is connected with the central nervous system through certain cranial nerves and through the middle three sacral segments of the spinal cord, and its ganglia tend to be placed peripherally near the organs innervated.

The sympathetic and parasympathetic systems both innervate many of the same organs, and in this double innervation the two systems are usually antagonistic to each other

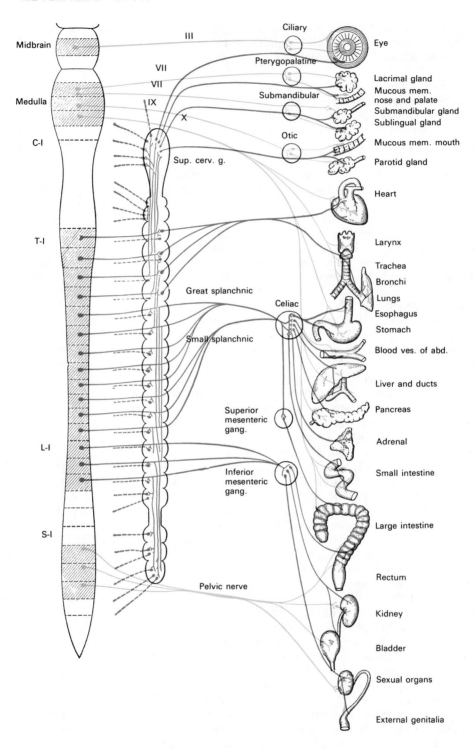

FIG. 12-71 Diagram of efferent autonomic nervous system. *Blue*, cranial and sacral outflow, parasympathetic. *Red*, thoracolumbar outflow, sympathetic. Interrupted lines designate postganglionic fibers to spinal and cranial nerves to supply vasomotor innervation to head, trunk and limbs, motor fibers to smooth muscles of skin and fibers to sweat glands. (Modified after Meyer and Gottlieb.) This is only a diagram and does not accurately portray all the details of distribution.

physiologically. No consistent rule can be given for the effect of each, but in general the sympathetic system mobilizes the energy for sudden activity such as that in rage or flight; for example, the pupils dilate, the heart beats faster, the peripheral blood vessels constrict, and the blood pressure rises. The parasympathetic system aims more toward restoring the reserves; for example, the pupils contract, the heart beats more slowly, and the alimentary tract and its glands become active.

The two systems frequently travel together, especially in the thorax, abdomen, and pelvis, forming extensive plexuses that contain the fibers of both. The arrangement of the bundles within these plexuses is very complicated and the identity of individual fibers cannot be determined with certainty. For purposes of description, therefore, a third subdivision of the autonomic system is recognized—the great autonomic plexuses.

Parasympathetic System

The parasympathetic system is the **craniosacral portion** of the **autonomic nervous system** and contains visceral efferent fibers that originate in certain cranial nerves and in the sacral portion of the spinal cord.

CRANIAL PORTION OF THE PARASYMPATHETIC SYSTEM

The **cranial outflow** includes fibers in the oculomotor, facial, glossopharyngeal, and vagus nerves. These nerves have been described previously and the details are repeated here only insofar as they apply to visceral efferent fibers.

OCULOMOTOR NERVE.   The oculomotor nerve contains efferent fibers for the nonstriated muscle making up the ciliaris and sphincter pupillae muscles of the eyeball (for diagram, see Fig. 12-5). The preganglionic fibers arise from the cells in the Edinger-Westphal nucleus[1,2] located in the anterior part of the oculomotor nucleus in the tegmentum of the midbrain. They run in the inferior division of the oculomotor nerve to the ciliary gan-

glion (Fig. 12-6) and there form synapses with the ganglion cells. The postganglionic fibers proceed in the short ciliary nerves to the eyeball, penetrate the sclera, and reach the muscles just named.

FACIAL NERVE.   The facial nerve contains efferent fibers for the lacrimal gland, the submandibular and sublingual glands, and the many small glands in the mucous membrane of the nasal cavity, palate, and tongue (for diagrams, see Figs. 12-16; 12-17). The preganglionic fibers arise from cells in the superior salivatory nucleus in the reticular formation, dorsomedial to the facial nucleus in the pons, and leave the brain in the nervus intermedius.

Pterygopalatine Ganglion.   Certain preganglionic fibers branch from the facial nerve at the geniculum via the greater petrosal nerve, course through the pterygoid canal, and terminate by forming synapses with the cells in the pterygopalatine ganglion (Fig. 12-12). Some of the postganglionic fibers reach the **lacrimal gland** via the maxillary, zygomatic, and lacrimal nerve route; others accompany the branches of the maxillary nerve to the **glands in the mucous membrane of the nasal cavity and nasopharynx;** and still others accompany the palatine nerves to the **glands of the soft palate, tonsils, uvula, roof of the mouth, and upper lip.**

Submandibular Ganglion.   Other preganglionic fibers leave the facial nerve in the chorda tympani and with it join the lingual nerve to reach the submandibular ganglion (Fig. 12-12). They form synapses in the ganglion or with groups of ganglion cells in the substance of the gland. The postganglionic fibers form the secretomotor supply to the **submandibular** and the **sublingual glands.**

Otic Ganglion.   Filaments communicating with the **otic ganglion** may contribute preganglionic fibers which join those from the glossopharyngeal nerve to supply the parotid gland.

GLOSSOPHARYNGEAL NERVE.   The glossopharyngeal nerve contains efferent fibers for the parotid gland and small glands in the mucous membrane of the tongue and floor of the mouth (for diagram, see Fig. 12-21) The preganglionic fibers arise in the inferior salivatory nucleus in the medulla oblongata, traverse the tympanic and lesser petrosal nerves, and form synapses in the **otic ganglion** (Fig. 12-12). Most of the post-

---

[1]Ludwig Edinger (1855–1918): A German neurologist and anatomist (Frankfurt-am-Main).
[2]Karl Friedrich Otto Westphal (1833–1890): A German psychiatrist and anatomist (Berlin).

ganglionic fibers join the auriculotemporal nerve and are distributed with its branches to the **parotid gland,** providing its secreto-motor fibers. Other postganglionic fibers are said to supply the glands of the mucous membrane of the tongue and floor of the mouth.

VAGUS NERVE. The vagus nerve contains efferent fibers for the nonstriated muscle and glands of the bronchial tree, of the alimentary tract as far as the transverse colon, of the gallbladder and bile ducts, the pancreas, and inhibitory fibers for the heart. The preganglionic fibers arise from cells in the dorsal motor nucleus of the vagus in the medulla oblongata; they run in the vagus nerve and its branches to ganglia situated in or near the organs innervated.

Heart. The preganglionic fibers for the heart reach the cardiac plexus by way of the superior and inferior cardiac nerves of the vagus, and are distributed by branches of the plexus to the ganglion cells in the heart wall (Fig. 12-77). The ganglion cells form numerous clusters in the connective tissue of the epicardium on the surface of the atrium, on the auricular appendages, and in the inter-atrial septum. The postganglionic fibers terminate in relation to specialized muscular elements in the sinoatrial and atrioventricular nodes, and atrioventricular bundle and its branches as far as the Purkinje[1] fibers (Stotler and McMahon, 1947).

Lungs. The preganglionic fibers from the bronchi leave the main trunk of the vagus nerve in the thorax as the anterior and posterior bronchial branches. They traverse the anterior and posterior pulmonary plexuses (Figs. 12-24; 12-25) and terminate in the clusters of ganglion cells scattered along the ramifications of the bronchial tree. The postganglionic fibers are distributed to the bronchial musculature and bronchial glands.

Alimentary Tract. The **alimentary tract** receives, through various branches, preganglionic fibers that end in the **myenteric plexus** of Auerbach and the **submucous plexus** of Meissner, forming synapses with the ganglion cells scattered in groups throughout these plexuses. The postganglionic fibers are the efferents from the muscular walls and the secreting cells of the tunica mucosa.

[1]Johannes Evangelista Purkinje (1787–1869): A Bohemian anatomist and physiologist (Breslau and Prague).

The preganglionic fibers reach the upper part of the **esophagus** through the recurrent nerve, and the lower part, below the hilum of the lung, through branches from the esophageal plexus (Fig. 12-76).

The **stomach** receives an average of four branches from the anterior vagus and six from the posterior vagus. The pylorus and duodenum receive fibers from the hepatic branch of the anterior vagus.

The **small intestine, cecum, appendix vermiformis, ascending and transverse colon** receive fibers from the posterior vagus that join the celiac plexus (Fig. 12-80) and accompany the branches of the superior mesenteric artery (Jackson, 1949).

The **gallbladder and bile ducts** receive preganglionic fibers through the vagus branches in the celiac plexus, which traverse the gastrohepatic ligament and terminate in the small clusters of ganglion cells in the wall of the gallbladder and in the region adjacent to the bile ducts. The postganglionic fibers are the efferents for the muscular walls and mucous membrane.

The **pancreas** receives fibers through the hepatic branches of the anterior and posterior vagi and, through the branches of the celiac plexus that accompany the arteries, supply this organ.

SACRAL PORTION OF THE PARASYMPATHETIC SYSTEM

The cells that give rise to the sacral outflow are in the second, third, and fourth sacral segments of the spinal cord and pass out with the corresponding sacral nerves. They leave the sacral nerves in the visceral branches and join the pelvic plexus (Fig. 12-81) in the deeper portions of the pelvic subserous fascia. Branches from this plexus contain preganglionic fibers for the scattered ganglia in or near the walls of the various pelvic viscera (Ashley and Anson, 1946).

The branches to the **bladder, prostate** and **seminal vesicles** supply efferent fibers to these organs, except for the fibers to the sphincter, which are inhibitory.

The branches to the **uterus** and **vagina** reach ganglia in the uterovaginal plexus; the postganglionic fibers are inhibitory except during pregnancy, when their function is said to be reversed.

The branches of the **pelvic plexus,** which join the pudendal nerve and are distributed to the **external genitalia,** cause active dilatation of the cavernous blood sinuses of the erectile corpora.

The visceral branches of the sacral nerves in laboratory animals are concentrated in a single trunk called the **"nervus erigens"** because of this function, but the term is not directly applicable to the conditions in man because there is no single nerve such as that used for experimentation in the animals.

Branches from the **pelvic plexus** containing preganglionic fibers join the **hypogastric nerve,** and through it and the inferior mesenteric plexus are distributed to the **descending** and **sigmoid colon** and **rectum.** They are efferent to this part of the intestine except for the sphincter ani internus, for which they are inhibitory. Small branches from the pelvic plexus go directly to the rectum.

Sympathetic System

The **sympathetic system** (Fig. 12-72) receives its fibers that connect with the central nervous system through the **thoracolumbar outflow** of visceral efferent fibers. These fibers are the axons of cells in the lateral column of gray matter in the thoracic and upper lumbar segments of the spinal cord. They leave the cord through the ventral roots of the spinal nerves and traverse short communications to reach the sympathetic trunk, where they may terminate in the chain ganglia of the trunk itself, or may continue into the collateral ganglia of the prevertebral plexuses. They are the preganglionic fibers and are mostly the small myelinated variety (3 μm or less in diameter). The postganglionic fibers are the axons of the cells in these chain and collateral ganglia and are generally unmyelinated. They are distributed to the heart, nonstriated muscle, and glands all over the body, which they reach by way of communications with the cerebrospinal nerves, by way of various plexuses, and by their own visceral branches of distribution.

VARIATIONS. Variability is a prominent characteristic of the sympathetic system, and although a description that will correspond with even the majority of individuals is impossible, certain general principles of organization can be recognized. These are incorporated in the account that follows and the general description is supplemented and made specific by giving a few of the common variations. Many of the details concerning the paths taken by the fibers are still unknown, and we must rely heavily upon information obtained from animal experimentation, although it may not have been confirmed by clinical observations with human patients.

SYMPATHETIC TRUNK. The **sympathetic trunk** consists of a series of ganglia called the central or chain ganglia (Fig. 12-76), which are connected by intervening cords, and extend along the lateral aspect of the vertebral column from the base of the skull to the coccyx. The cranial end of the trunk proper is formed by the superior cervical ganglion (Fig. 12-72), but there is a direct continuation into the head, the internal carotid nerve. The caudal ends of the two trunks converge at the coccyx and may merge into a single ganglion—the *ganglion impar.* Cross connections between the two cords in the sacral region are frequent, but rarely occur above the fifth lumbar.

The trunk contains, in addition to the ganglia, the preganglionic fibers, which are small (1 to 3 μm in diameter) and myelinated; the postganglionic fibers, which are mostly unmyelinated; and a smaller number of afferent fibers, which are both myelinated (medium, 5 μm, and large, 10 μm) and unmyelinated. All types of fibers may run up or down in the trunk for the distance between two or many segments. The cords intervening between the ganglia are usually single except in the lower cervical region, but doubling is frequent in any part of the trunk although it rarely extends farther than between two adjacent ganglia. Numerous small collections of ganglion cells occur outside of the major ganglia; these may be microscopic in size or grossly visible. They are called intermediary ganglia, and are found in the roots or branches of the trunk and even in the spinal nerves close to these communicating rami (Kuntz and Alexander, 1950).

GANGLIA. The **central** or **chain ganglia** of the sympathetic trunk are rounded, fusiform or irregular in shape, with diameters usually ranging from 1 to 10 mm, but neighboring ganglia may fuse into larger masses and the superior cervical ganglion is always larger. They contain multipolar neurons whose processes are the postganglionic fibers.

The **roots** of the ganglia of the sympathetic trunk are commonly called the **white rami communicantes** (*rami communicantes albi*) because of the whitish color imparted to

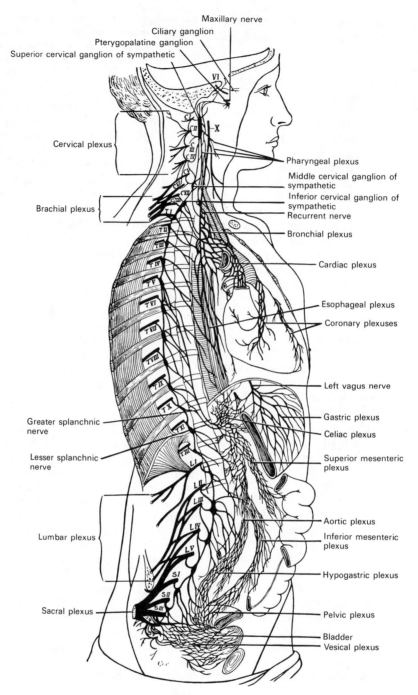

Maxillary nerve
Ciliary ganglion
Pterygopalatine ganglion
Superior cervical ganglion of sympathetic

Cervical plexus

Pharyngeal plexus
Middle cervical ganglion of sympathetic
Inferior cervical ganglion of sympathetic
Recurrent nerve
Bronchial plexus

Brachial plexus

Cardiac plexus

Esophageal plexus
Coronary plexuses

Left vagus nerve

Gastric plexus
Celiac plexus

Greater splanchnic nerve

Superior mesenteric plexus

Lesser splanchnic nerve

Aortic plexus
Inferior mesenteric plexus

Lumbar plexus

Hypogastric plexus

Sacral plexus

Pelvic plexus

Bladder
Vesical plexus

FIG. 12-72.   The right sympathetic chain and its connections with the thoracic, abdominal, and pelvic plexuses. (After Schwalbe.)

them by the preponderance of myelinated fibers they contain. These myelinated fibers are mainly the small (1 to 3 μm in diameter) preganglionic axons of the thoracolumbar outflow, whose cell bodies are in the lateral column of gray matter in the spinal cord. They emerge from the spinal cord with the somatic motor fibers in the ventral roots of the spinal nerves of the thoracic and the first and second lumbar segments. Many of the preganglionic fibers in each root fail to make synaptic connections in the ganglia at the level of entrance; some travel cranialward to the cervical ganglia, some caudalward to the lumbar and sacral ganglia, and many in the lower thoracic and upper lumbar levels pass out of the trunk and reach the celiac and related collateral ganglia through the splanchnic nerves. One preganglionic fiber may give collaterals to several of the chain ganglia and may terminate on as many as 15 to 20 ganglionic neurons.

The roots of the ganglia or white rami communicantes are to be distinguished from the branches to the spinal nerves or gray rami communicantes with which they are closely related. The white rami leave the anterior primary divisions of the spinal nerves close outside of the intervertebral foramina, but are regularly more distal than the gray rami. They contain medium and large myelinated fibers, probably afferents, and some unmyelinated fibers in addition to the small myelinated preganglionic fibers. In the lower thoracic and upper lumbar regions the white rami take an oblique course from the nerve of one segment to the ganglion of the segment below and have been called, therefore, the oblique rami, as opposed to the transverse gray rami. They are more likely to be attached to the intervening cords than the gray rami. The white rami, varying from 0.5 to 2 cm, are longer in the lower thoracic and lumbar region, where they are also usually double or triple.

The sympathetic trunk has branches of distribution, containing the postganglionic fibers that originate in the ganglia, and branches of communication, containing preganglionic fibers that pass through the trunk without synapses on their way to the collateral ganglia in the abdomen.

BRANCHES OF DISTRIBUTION. The branches of distribution are of several types: (1) branches to the spinal nerves, (2) branches to the cranial nerves, (3) branches accompanying arteries, (4) separate branches to individual organs, and (5) branches to the great autonomic plexuses.

1. The branches to the **spinal nerves** are commonly called the **gray rami communicantes** (Fig. 12-73) because the preponderance of unmyelinated fibers gives them a more grayish cast than the roots or white rami, which are adjacent to them in the thoracic area. These branches contribute postganglionic fibers, most of which accompany the cutaneous branches of the spinal nerves to supply the arrectores pilorum muscles, the sweat glands, and the vasoconstrictor fibers of the peripheral blood vessels.

Gray rami join all the spinal nerves, while the white rami arise only from the thoracic and first two lumbar segments. The gray ramus usually leaves the

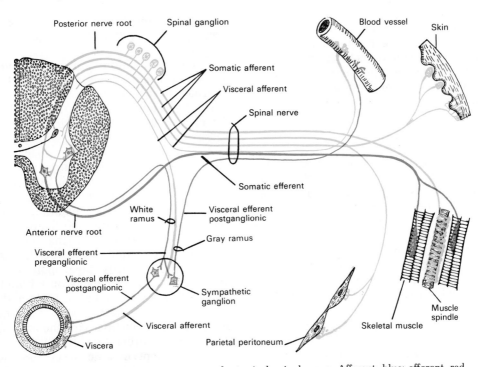

Fig. 12-73. Schema showing structure of a typical spinal nerve. Afferent, blue; efferent, red.

trunk at a ganglion near the same level as the spinal nerve, and in the thoracic and lumbar regions, where there are both gray and white rami, it is regularly proximal to the white.

Branches to the **cranial nerves** may go directly to the nerves or they may pass through plexuses on blood vessels. Direct communications are encountered to the superior and inferior ganglia of the vagus, the inferior ganglion of the glossopharyngeal, and the hypoglossal nerves.

Branches accompanying the **arteries** are too numerous to list here, but are given with the detailed descriptions of the different parts of the system in later pages. Prominent examples are the nerves on the internal and external carotid arteries and their branches.

Branches to **individual organs** may take an independent course but they commonly pass through plexuses (for example, the cardiac branches), or accompany blood vessels for some distance (for example, those to the bulb of the eye).

Branches to the **cardiac, pulmonary,** and **pelvic plexuses** probably contain postganglionic fibers, but most of the branches from the trunk to the abdominal plexuses probably contain preganglionic fibers.

BRANCHES OF COMMUNICATION. The principal branches of communication containing preganglionic fibers are the **splanchnic nerves.** They branch from the ganglia or the trunk of the thoracic and lumbar regions and supply the fibers that are the roots to the celiac, aorticorenal, and mesenteric ganglia. The postganglionic fibers from these ganglia supply the various abdominal and pelvic organs.

DIVISIONS OF THE SYMPATHETIC SYSTEM

The sympathetic nervous system is divided into portions according to topographic position as follows: (1) cephalic, (2) cervical, (3) thoracic, (4) abdominal or lumbar, and (5) pelvic. These parts are not independent and are chosen merely for convenience in description. In addition to these portions, concerned primarily with the sympathetic trunk and ganglia, there are autonomic plexuses, which are described separately.

Cephalic Portion of the Sympathetic System

The **cephalic portion** of the sympathetic system contains no part of the ganglionated cord, but is formed largely by a direct cephalic prolongation from the superior cervical ganglion, named the internal carotid nerve. In addition, there are nerves accompanying the vertebral artery and the various branches of the external carotid artery that supply fibers to many structures in the head, such as the pilomotor muscles, sweat glands, and peripheral arteries of the face, and the salivary and other glands.

INTERNAL CAROTID NERVE. The internal carotid nerve (Fig. 12-23), arising from the cephalic end of the superior cervical ganglion, accompanies the internal carotid artery, and after entering the carotid canal in the petrous portion of the temporal bone, divides into two branches that lie against the medial and lateral aspects of the artery. The lateral branch, the larger of the two, distributes filaments to the carotid artery and forms the internal carotid plexus. The medial branch also distributes filaments to the artery, and following the artery to the cavernous sinus, forms the cavernous plexus.

Internal Carotid Plexus. The internal carotid plexus (*carotid plexus*) (Fig. 12-21) is the continuation of the lateral branch of the internal carotid nerve. It surrounds the lateral aspect of the artery, occasionally containing a small carotid ganglion. In addition to filaments to the artery, it has the following branches:

A **communication** with the **trigeminal nerve** joins the latter at the trigeminal ganglion.

A **communication** joins the **abducent nerve** as it lies near the lateral aspect of the internal carotid artery.

The **deep petrosal nerve** (Fig. 12-9) leaves the plexus at the lateral side of the artery, passes through the cartilage of the auditory tube which fills the foramen lacerum, and joins the greater petrosal nerve to form the nerve of the pterygoid canal (Vidian nerve). In the pterygopalatine fossa, the Vidian nerve joins the pterygopalatine ganglion, and its contribution through the deep petrosal nerve has been called the sympathetic root of the ganglion. However the sympathetic fibers, already postganglionic, pass through or beside the ganglion without synapses and are distributed to the glands and blood vessels of the pharynx, nasal cavity, and palate by accompanying the branches of the maxillary nerve (Fig. 12-16).

The **caroticotympanic nerves** are two or three filaments that pass through foramina in the bony wall of the carotid canal and join the tympanic plexus on the promontory of the middle ear.

Cavernous Plexus. The cavernous plexus (Fig. 12-12) is the continuation of the medial branch of the internal carotid nerve. It lies inferior and medial to the part of the internal carotid artery enclosed by the cavernous sinus. It communicates with the adjacent cranial nerves and continues along the artery to its terminal branches.

The communication with the **oculomotor nerve** enters the orbit and joins the nerve at its point of division.

A communication joins the **trochlear nerve** as it lies in the lateral wall of the cavernous sinus.

The communication with the **ophthalmic** division of the **trigeminal** joins the latter nerve at its inferior surface.

Fibers to the **dilator pupillae muscle** of the iris traverse the communication with the ophthalmic nerve and accompany the nasociliary nerve and the long ciliary nerves to the posterior part of the bulb, where they penetrate the sclera and run forward to the iris (for diagram, see Fig. 12-4).

The communication with the **ciliary ganglion** leaves the anterior part of the cavernous plexus and enters the orbit through the superior orbital fissure. It may join the nasociliary nerve and reach the ganglion through the latter's branch to the ganglion, or it may take an independent course to the ganglion.

Filaments to the **hypophysis** accompany its blood vessels.

The **terminal filaments** from the internal carotid and cavernous plexuses continue along the anterior and middle cerebral arteries and the ophthalmic artery. Fibers on the cerebral arteries may be traced to the pia mater; those on the ophthalmic artery accompany all its branches in the orbit. The filaments on the anterior communicating artery may connect the sympathetic nerves of the right and left sides.

External Carotid Nerves. The external carotid nerves send filaments out along all the branches of the external carotid artery. The filaments on the facial artery join the **submandibular ganglion;** they have formerly been called the sympathetic root of the ganglion, but they pass through it without forming synapses and supply the submandibular and probably the sublingual glands (for diagram, see Fig. 12-17). The network of filaments on the middle meningeal artery gives

off the **small deep petrosal nerve,** which has been called the sympathetic root of the **otic ganglion,** but its fibers pass through the ganglion without synapses, some accompanying the auriculotemporal nerve to the parotid gland (for diagram, see Fig. 12-19), others forming the **external superficial petrosal nerve** which is a communication with the geniculate ganglion. Filaments on the facial, superficial temporal, and other arteries that are distributed to the skin supply the arrectores pilorum muscles and sweat glands, as well as the muscles constricting the arteries themselves.

Cervical Portion of the Sympathetic System

The **cervical portion** of the sympathetic trunk consists of three ganglia—superior, middle, and inferior—connected by intervening cords (Fig. 12-74). It is ventral to the transverse processes of the vertebrae, close to the carotid artery, being embedded in the fascia of the carotid sheath itself or in the connective tissue between the sheath and the longus colli and capitis muscles. It receives no roots or white rami communicantes from the cervical spinal nerves; its preganglionic fibers enter the trunk through the white rami from the upper five thoracic spinal nerves, mainly the second and third, and travel cranialward in the trunk to the three ganglia. The trunk also contains postganglionic fibers from various sources and visceral afferent fibers with their cell bodies in the dorsal root ganglia.

SUPERIOR CERVICAL GANGLION. The **superior cervical ganglion** (Fig. 12-22), much larger than the other cervical ganglia and usually the largest of all the trunk ganglia, is approximately 28 mm long and 8 mm wide, fusiform in shape, frequently broad and flat, and occasionally constricted into two or more parts. It is embedded in the connective tissue between the carotid sheath and the prevertebral fascia over the longus capitis, at the level of the second cervical vertebra. It is the cephalic end of the sympathetic chain and is connected with the middle ganglion caudally by a rather long interganglionic cord. It is believed to be formed by the coalescence of sympathetic primordia from the cranial four cervical segments of the body. Its branches follow.

Internal and External Carotid Nerves. The internal and external carotid nerves (described previously) leave the cephalic pole

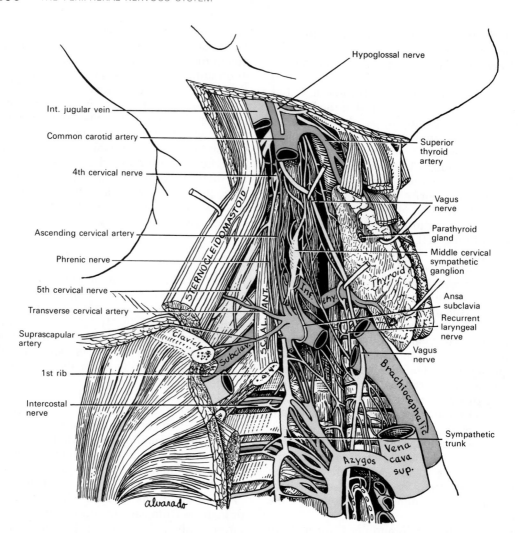

FIG. 12-74. Cervical portion of the sympathetic nervous system on the right side, with the common carotid artery and internal jugular vein removed and with the vagus nerve and thyroid gland drawn aside. (Redrawn from Töndury.)

of the ganglion, and serve as direct continuations of the sympathetic trunk into the head.

Cranial Nerves. Communications with the cranial nerves are delicate filaments that join (a) the inferior ganglion of the glossopharyngeal nerve, (b) the superior and inferior ganglia of the vagus, and (c) the hypoglossal nerve. The jugular nerve is a filament that passes cranialward to the base of the skull and divides to join the inferior ganglion of the glossopharyngeal and the superior ganglion of the vagus nerve.

Cervical Nerves. Branches to the upper two to four cervical spinal nerves are the **gray rami communicantes** of these nerves. They course lateralward and dorsalward,

and have been called the lateral or external branches of the ganglion. The branches to any one nerve are variable and may be multiple or absent.

Variations. The branches to the **first** and **second nerves** are constantly present. Since the spinal nerves in the neck are connected by loops, as parts of the cervical plexus, the branch to the first nerve may join the loop between the first and second nerves, and the branch to the second may join the loop between the second and third nerves. There may be two to four branches to each of these nerves.

The **third cervical nerve** receives a branch (ramus) from the ganglion in the majority of individuals, but in many instances it comes from the trunk below the superior ganglion. A branch for the third nerve often forms a loop with a lower branch, which then supplies roots to both the third and fourth nerves. Branches may join the third nerve itself or the loop between it and the fourth nerve.

The **fourth nerve** receives a branch from the ganglion only occasionally. Its rami frequently arise in common with those of either the third, fifth, or even the sixth nerves, and may come from the trunk or from nerves accompanying the vertebral artery.

Pharyngeal Branches. Pharyngeal branches, commonly four to six, leave the medial aspect of the ganglion, and in their course toward the pharynx, communicate with pharyngeal branches of the glossopharyngeal and vagus nerves opposite the constrictor pharyngis medius to form the **pharyngeal plexus.** Some of the filaments form a plexus on the lateral wall of the pharynx, others travel in the substance of the prevertebral fascia to the dorsum of the pharynx and form a posterior pharyngeal plexus. Filaments communicate with the superior laryngeal nerve.

Intercarotid Plexus. The intercarotid plexus receives one or two branches, either from the ganglion or from the external carotid nerves. They communicate with filaments from the pharyngeal branch of the vagus and the carotid branch of the glossopharyngeal nerve in the region of the carotid bifurcation, and are distributed in the plexus to the carotid sinus and the carotid body, where they probably serve a vasomotor function.

Superior Cardiac Nerve. The superior cardiac nerve (Figs. 12-74; 12-75) arises by two or three filaments from the ganglion, and occasionally also by a filament from the trunk between the superior and middle ganglia. It runs down the neck in the connective tissue of the posterior layers of the carotid sheath superficial to the longus colli, and crosses ventral or dorsal to the inferior thyroid artery and recurrent nerve. The course of the nerves on the two sides then differs. The *right nerve*, at the root of the neck, passes either ventral or dorsal to the subclavian artery, and along the brachiocephalic artery to the deep part of the cardiac plexus. The *left nerve* passes ventral to the common carotid and across the left side of the arch of the aorta to reach the superficial part of the cardiac plexus. The superior cardiac nerves may communicate with the middle and inferior cardiac sympathetic nerves, with the cardiac branches of the vagus, the external branch of the superior laryngeal nerve, the

recurrent nerve, the thyroid branch of the middle ganglion, the nerves on the inferior thyroid artery, and the tracheal and anterior pulmonary plexuses.

MIDDLE CERVICAL GANGLION. The middle cervical ganglion (Fig. 12-75), the smallest of the three cervical ganglia, varies in size, form, and position, and may be either absent or doubled. It probably represents a fusion of the two sympathetic primordia corresponding to the fifth and sixth cervical nerves. When single, the ganglion may have a high position, at the level of the transverse process of the sixth cervical vertebra (the **carotid** or *Chassaignac's tubercle*[1]), or a low position nearer the level of the seventh cervical vertebra. In the high position it lies on the longus colli above the cranial bend of the inferior thyroid artery. In the low position it lies in close association with the ventral or ventromedial aspect of the vertebral artery, 1 to 3 cm from the latter's origin. The middle cervical ganglion has no white ramus communicans; its preganglionic fibers probably leave the spinal cord through the white rami of the second and third thoracic nerves and reach the ganglion through the intervening sympathetic trunk.

Variations. The middle cervical ganglion was absent in 5, single in 10 and double in 10 of 25 dissections (Pick and Sheehan, 1946). It was present in 53 and double in two of 100 body halves (Jamieson et al., 1952), with 64% in the high position. A small ganglion in the low position may have no branches. Several small thickenings may occur along the trunk between the superior and inferior ganglia. The ganglion may be split, surrounding the inferior thyroid artery (Axford, 1928).

The **intermediate cervical sympathetic ganglion** (of Jonnesco, 1923) corresponds to a middle ganglion in the low position described above. According to Saccomanno (1943), it appears more constantly than a middle ganglion in the high position. A ganglion in the low position has been called the **thyroid ganglion** because of its close relationship to the inferior thyroid artery (Jamieson et al., 1952). The **vertebral ganglion** is a name given to a ganglion in the low position, on the deep part of the loop of the ansa subclavia, included as part of the stellate ganglionic configuration (Pick and Sheehan, 1946).

Cervical Nerves. The branches (gray rami communicantes) are constantly supplied to the fifth and sixth nerves. A ganglion in the

[1]Charles Marie Edouard Chassaignac (1805–1879): A French surgeon (Paris).

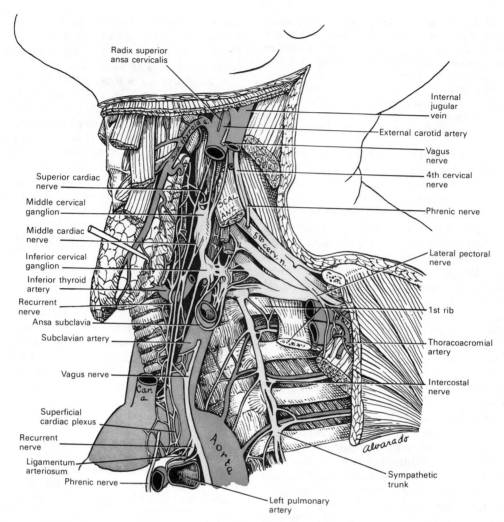

FIG. 12-75. Cervical portion of the sympathetic system on the left side, with the common carotid artery, internal jugular vein, vagus nerve, and subclavian artery partly removed, and the thyroid gland drawn forward. (Redrawn from Töndury.)

high position may send branches to the fourth nerve.

The gray rami of the **fifth cervical nerve** number from one to three. (1) The most constant arises from the middle ganglion or the trunk just above it, runs upward and laterally across the scalenus anterior, winds around the carotid tubercle and along a groove in the transverse process of the fifth cervical vertebra to the fifth nerve. It may pierce the scalenus anterior and it may divide, one branch going to the sixth nerve, or it may be prolonged to give a branch to the fourth or even the third nerve (Axford, 1928). (2) A ramus present in the majority of individuals leaves the trunk just above the ca-

rotid tubercle, pierces the longus colli either medial or lateral to the vertebral artery, and receives a communication from the nerve of the vertebral artery. (3) An inconstant ramus is a branch of that of the sixth nerve which accompanies the vertebral artery.

The rami of the **sixth cervical nerve** may be two to four in number, one from the middle and one from the inferior ganglion being constant. (1) A short ramus from the middle ganglion runs cranialward through the longus colli just above the carotid tubercle to join the sixth nerve as it lies in its groove, medial to the vertebral artery. (2) A long fine branch from the middle ganglion or the trunk just above it crosses the vertebral artery and scalenus anterior and joins the nerve lateral to the carotid tubercle, sometimes continuing on to the fifth nerve.

Middle Cardiac Nerve. The middle cardiac nerve (*great cardiac nerve*) (Fig. 12-75), the largest of the three cardiac nerves in the neck, arises from the middle cervical ganglion, from the trunk between the middle and inferior ganglia, or both. As it descends dorsal to the common carotid artery, it communicates with the superior cardiac nerve and the inferior laryngeal nerve. On the right side, at the root of the neck, it goes either deep or superficial to the subclavian artery, continues along the trachea, communicates with the recurrent nerve, and joins the right side of the deep part of the cardiac plexus. The left nerve enters the thorax between the common carotid and subclavian arteries and joins the left side of the deep part of the cardiac plexus.

Thyroid Nerves. Branches from the middle cervical ganglion form a plexus on the inferior thyroid artery, supply the thyroid gland, join the plexus on the common carotid artery, and may communicate with the inferior laryngeal and external branch of the superior laryngeal nerves (Braeucker, 1922).

The trunk between the middle and inferior cervical ganglia is constantly double, the two strands enclosing the subclavian artery. The superficial strand is usually much longer than the deep strand and forms a loop about the artery, supplying it with branches; it is called the **ansa subclavia** (of Vieusens[1]) (Fig. 12-75). Since it is a rather constant feature, it may be used to identify and distinguish the middle and inferior ganglia or the components representing them. Occasionally the loop is formed about the vertebral artery instead of the subclavian, or there may be individual loops about both.

INFERIOR CERVICAL GANGLION. The inferior cervical ganglion (Fig. 12-75) is situated between the base of the transverse process of the seventh cervical vertebra and the neck of the first rib, on the medial side of the costocervical artery. In most instances it is incompletely separated from or fused with the first thoracic ganglion, but it will be described as it appears when discrete, and the fused ganglion, called the stellate, also will be described below. It is larger than the middle ganglion, has an irregular shape, and probably represents the fusion of sympathetic primordia corresponding to the seventh and eighth cervical nerves. It has no white ramus, but receives its preganglionic fibers from the thoracic part of the trunk through its connection with the first thoracic ganglion.

Variations. The inferior cervical ganglion was independent in 5 out of 25 cases (Pick and Sheehan, 1946) and in 18 out of 100 cases (Jamieson et al., 1952). A white ramus has been described joining the eighth cervical nerve and the sympathetic (Pearson, 1952).

Cervical Nerves. Branches (gray rami communicantes) are constantly supplied to the **sixth, seventh,** and **eighth cervical nerves.**

The **sixth cervical nerve** commonly receives branches from the inferior cervical ganglion as well as from the middle cervical ganglion. (1) A constant, rather thick ramus from the deep part of the inferior cervical ganglion runs cranialward along the medial aspect of the vertebral artery, ventral to the vertebral veins and lateral to the longus colli, and enters the foramen in the transverse process of the sixth cervical vertebra with the vertebral artery. It communicates with the plexus on the vertebral artery and supplies rami for the sixth and seventh, sometimes the fifth or even more cephalic nerves. (2) An inconstant ramus from the inferior ganglion is similar to the last, but pierces the scalenus anterior instead of passing through the foramen (Axford, 1928).

The **seventh cervical nerve** receives two to five branches (gray rami) from the inferior cervical ganglion. (1) A constant, well-defined branch 15 to 25 mm long crosses ventral to the eighth cervical nerve, either deep to, superficial to, or piercing the scalenus anterior. It may be composed of two or three parallel filaments. (2) A constant branch accompanies the vertebral artery and is shared by the sixth nerve. (3) A frequent branch lies close to the vertebral vein, crosses the eighth cervical nerve, to which it may give filaments, and enters the foramen in the seventh cervical vertebra with the vein (Axford, 1928).

The **eighth cervical nerve** receives two to five rather short branches (averaging 10 mm in length) from the inferior cervical ganglion. (1) A constant, well-defined, thick branch runs cranialward and lateralward, often across the neck of the first rib, and joins the eighth cervical nerve deep to the scalenus anterior. It is dorsal to the first part of the subclavian artery and is intimately related to the superior intercostal artery. It may be represented by two to four parallel filaments. (2) A constant short thick branch from the upper pole of the inferior ganglion runs

---

[1]Raymond de Vieussens (1641–1715): A French anatomist (Montpellier).

vertically cranialward, medial and dorsal to the vertebral artery, a few millimeters lateral to the longus colli. It passes ventral to the transverse process of the first thoracic vertebra and medial to the first costocentral articulation, and joins the eighth cervical nerve as it emerges from its foramen. (3) A frequent ramus accompanies the vertebral vein with a similar branch to the seventh nerve.

Inferior Cardiac Nerve. The inferior cardiac nerve (Fig. 12-75) arises from either the inferior cervical ganglion, the first thoracic ganglion, the stellate ganglion, or the ansa subclavia. It passes deep to the subclavian artery and along the anterior surface of the trachea to the deep cardiac plexus. It communicates with the middle cardiac nerve and the recurrent laryngeal nerve, and supplies twigs to various cervical structures (Saccomanno, 1943).

Vertebral Nerve. The branches that accompany the vertebral artery through the vertebral foramina are large. They join similar branches from the first thoracic ganglion to form the vertebral nerve, which continues into the cranial cavity on the basilar, posterior cerebral, and cerebellar arteries. Communications between the vertebral nerve and the cervical spinal nerves frequently serve as rami of these nerves (Christensen et al., 1952).

Junction of Cervical and Thoracic Portions

The junction of the cervical with the thoracic portion of the sympathetic trunk requires special consideration, first, because the lowest cervical and highest thoracic ganglia are usually fused, and second, because the trunk makes an abrupt change of direction at this point. The cervical portion of the trunk lies upon the ventral aspect of the transverse processes, but is also in a plane ventral to the vertebral bodies because of the latter's small size and the presence of the longus colli muscle. The trunk drops back dorsalward as it enters the thorax, winding around the transverse process of the seventh cervical vertebra to reach the neck of the first rib.

STELLATE GANGLION. The stellate ganglion (*cervicothoracic ganglion*) (Fig. 12-75) is the name given to the ganglionic mass that results when, as is usually the case, the inferior cervical and first thoracic ganglia are fused. It varies in size and form, occasionally including the middle cervical or the second thoracic ganglia, and is located between the eighth cervical and first thoracic nerves. Its branches and communications are modifications of those that would be found if the component ganglia remained separate. The white ramus communicans, therefore, comes from the first thoracic nerve, or from the second also, if the mass includes the second ganglion.

The **branches** of the **stellate ganglion** that supply the gray rami to the spinal nerves include: a frequent branch to the sixth nerve, a constant branch to the seventh nerve, and constant double or multiple branches to the eighth cervical, first thoracic, and second thoracic nerves. The large branch to the vertebral artery leaves the superior border of the ganglion and forms the major portion of the vertebral nerve. Other branches are similar to those described for the independent ganglia (Kirgis and Kuntz, 1942).

Variations. In 25 bodies, the inferior cervical ganglion was independent in 5, fused with the first thoracic in 17, and fused with both the first and second thoracic in 3 (Pick and Sheehan, 1946). A stellate ganglion was present in 82 of 100 bodies (Jamieson et al., 1952). When the second thoracic is fused with the first, the inferior cervical is more likely to be independent.

Surgical Considerations. The branches (gray rami) from the stellate ganglion to the eighth cervical and first thoracic nerves carry the bulk of the sympathetic fibers to the upper limb (Kirgis and Kuntz, 1942). Other branches also carry fibers, however, and Woollard and Norish (1933) recommend that the middle cervical and second thoracic ganglia be included in the "stellate complex," and that the description of the independent ganglia be abandoned. The sympathetic rami of the brachial plexus are described in more detail in the discussion of the brachial plexus.

Comparative Anatomic Considerations. The stellate ganglion in cats includes the inferior cervical and first three thoracic ganglia (Saccomanno, 1943) and in rhesus monkeys the inferior cervical and first two thoracic ganglia (Sheehan and Pick, 1943). The function of the stellate ganglion in control of the heart, as it is revealed by animal experimentation with dogs, cats, and monkeys, does not agree entirely with that revealed by clinical observations. A partial explanation for this is provided by the fact that more thoracic ganglia are included in the stellate complex and a much greater bulk of the accelerator nerve arises from the ganglion in these animals than in man.

Thoracic Portion of the Sympathetic System

The **thoracic portion** of the sympathetic trunk (Figs. 12-76; 12-77) contains a series of ganglia that correspond approximately to the thoracic spinal nerves, but the coalescence of adjacent ganglia commonly reduces the number to fewer than 12. The ganglia are oval, fusiform, triangular, or irregular in shape. They lie against the necks of the ribs in the upper thorax, gradually become more ventrally placed in the lower thorax, and finally lie at the sides of the bodies of the lowest thoracic vertebrae. The trunk is covered by the costal portion of the parietal pleura, and the interganglionic cords are between the pleura and the intercostal vessels, which they cross.

The roots of the ganglia are the white rami communicantes supplied by each spinal nerve to the corresponding ganglion or the trunk nearby. They contain predominantly small myelinated fibers (1 to 3 μm in diameter), which leave the spinal cord through the ventral roots of the spinal nerves; the cell bodies are in the lateral column of gray matter in the spinal cord. Many of the preganglionic fibers fail to make synapses in their ganglia of entrance. From the upper five roots, these fibers take a cranial direction in the trunk, and for the most part, terminate in the cervical ganglia. Many of the fibers in the caudal six or seven rami traverse the trunk for a variable distance and then emerge into the splanchnic nerves, which are the roots of the celiac and related ganglia.

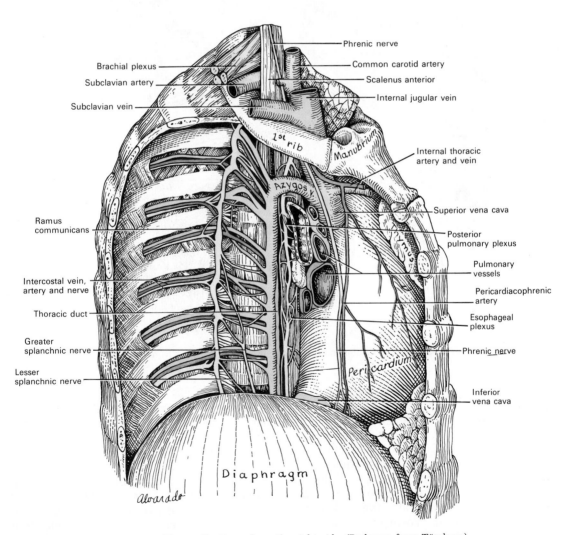

FIG. 12-76. The mediastinum from the right side. (Redrawn from Töndury.)

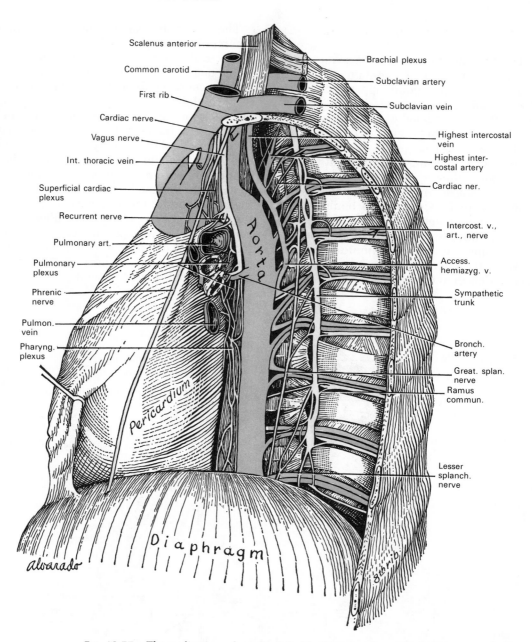

FIG. 12-77. The mediastinum from the left side. (Redrawn from Töndury.)

FIRST THORACIC GANGLION. The first thoracic ganglion, when independent, is larger than the rest. It has an elongated or crescentic shape. Because of the change in direction of the trunk as it passes from the neck into the thorax, the ganglion is elongated dorsoventrally. It lies at the medial end of the first intercostal space or ventral to the neck of the first rib, medial to the costocervical arterial trunk. It is usually combined with the inferior cervical ganglion into a stellate ganglion (as discussed previously), or it may coalesce with the second thoracic ganglion. The second to the tenth ganglia lie opposite the intervertebral disc or the cranial border of the next more caudal vertebra, slightly lower than the corresponding spinal nerve. In most individuals, the last thoracic ganglion, lying on the body of the twelfth vertebra, is larger, and by its connection to both the eleventh and twelfth nerves, takes the place of these two ganglia.

Variations. The first thoracic ganglion was independent of the stellate in 5 instances, and the second thoracic was independent in 22 out of 25 instances. Fusion occurred between the third and fourth three times; fourth and fifth five times; fifth and sixth once; sixth and seventh once; seventh and eighth four times; eighth and ninth twice; ninth and tenth twice (Pick and Sheehan, 1946). Small accessory ganglia occur at or near the junction of the communicating rami with the thoracic nerves, especially the upper four. Those at the white rami may provide sympathetic pathways to spinal nerves without traversing the sympathetic trunk (Ehrlich and Alexander, 1951).

The branches of the thoracic trunk are of three varieties: gray rami communicantes, visceral branches, and the splanchnic nerves.

GRAY RAMI COMMUNICANTES. The branches that form the gray rami communicantes are supplied to each spinal nerve. Usually two or three short branches are sent to each corresponding spinal nerve, and occasionally a slender branch reaches the next more caudal ganglion. When there are two or more branches to a single nerve, one branch may go to the anterior and one to the posterior primary division of the nerve. When the branches go only to the anterior primary division, the fibers turn back within the nerve to reach the posterior division (Dass, 1952). The gray rami are regularly proximal and transverse; the white rami, distal and oblique.

VISCERAL BRANCHES. Visceral branches are supplied to the cardiac, pulmonary, esophageal, and aortic plexuses.

Cardiac Plexus. Branches to the cardiac plexus from the first five thoracic ganglia vary in number both in different individuals and between the two sides in the same individual; there may be 15 or 20 in some cases. Larger ones usually come from ganglia; smaller ones, from intervening cords. They course medially close to the intercostal artery and vein, usually between the vessels and in the same connective tissue sheath, supplying filaments to them and communicating with the esophageal and pulmonary plexuses. From the right side they approach the deep part of the plexus, between the esophagus and the lateral aspect of the aorta. From the left side, they pass dorsal to the aorta and approach the deep part of the plexus from the right (Kuntz and Morehouse, 1930). Branches from the third, fourth, and fifth ganglia are more abundant than those from the first two; branches may

come from the sixth and seventh ganglia, but these usually enter the aortic network. The cross-sectional area of the thoracic cardiac nerves, as they enter the plexus, is twice as large as that of the cervical cardiac nerves (Saccomanno, 1943).

Esophageal Branches. Delicate esophageal branches from several of the thoracic ganglia, both upper and lower, may follow the intercostal vessels to the esophagus and join the plexus formed by the vagus, or filaments may be supplied by the cardiac and aortic plexuses, or by the splanchnic nerves.

Posterior Pulmonary Plexus. The posterior pulmonary plexus receives twigs from the second, third, and fourth ganglia that follow the intercostal arteries to the hilum of the lung.

Aortic Network. Branches to the aortic network come from the last five or six thoracic ganglia, from the cardiac plexus, and from the splanchnic nerves. Branches from these bundles accompany the branches of the artery, and probably supplement the splanchnic nerves.

SPLANCHNIC NERVES. The splanchnic nerves arise from the caudal six or seven thoracic and first lumbar ganglia. They are not, strictly speaking, true branches of the ganglia, since they contain only a small number of postganglionic fibers. They are composed principally of myelinated fibers, and accordingly have a whitish color and firm consistency similar to those of the somatic nerves. The small myelinated fibers (1 to 3 μm in diameter), which predominate, are the preganglionic fibers that pass through the chain ganglia without synapses to become the roots of the celiac and related ganglia. Also present are large myelinated fibers, which are probably visceral afferents with their cell bodies in the dorsal root ganglia of the spinal nerves. Many of the fibers of all types probably come from spinal cord segments higher than the ganglia from which the branches arise.

Greater Splanchnic Nerve. The greater splanchnic nerve (Figs. 12-76; 12-77) is formed by contributions from the fifth (or sixth) to the ninth (or tenth) thoracic ganglia, which leave the ganglia in a medial direction and angle across the vertebral bodies obliquely in their caudalward course. They are combined into a single nerve that pierces the crus of the diaphragm and, after making an abrupt bend or loop ventralward, ends in the celiac ganglion by entering the lateral border of its principal mass (Fig. 12-78). A small splanchnic ganglion occurs commonly in the nerve at the level of the eleventh or twelfth thoracic vertebra; it is considered part of the celiac ganglion formed by cells that failed to migrate as far as the large ganglion during embryonic development. Preganglionic fibers to the suprarenal glands are conveyed by the splanchnic nerves and pass through the celiac plexus without synapses in the ganglion.

Lesser Splanchnic Nerve. The lesser splanchnic nerve (Fig. 12-77) is formed by branches of the ninth and tenth thoracic ganglia or from the cord between them. It pierces the crus of the diaphragm with the greater splanchnic nerve and ends in the aorticorenal ganglion.

Lowest Splanchnic Nerve. The lowest splanchnic nerve (*least splanchnic nerve*), when present, is a branch of the last thoracic ganglion or of the lesser splanchnic nerve. It passes through the diaphragm with the sympathetic trunk and ends in the renal plexus.

Variations. The uppermost branch to the splanchnics in 25 dissections was the fourth ganglion once; fifth, twice; sixth, eleven times; seventh, seven times; eighth, four times (Pick and Sheehan, 1946). Filaments from the upper thoracic and stellate ganglia, from the cardiac nerves, or from the branches to the pulmonary and aortic plexuses sometimes continue downward to join the celiac plexus and have been considered a fourth splanchnic nerve. Lumbar splanchnic nerves are described below.

Abdominal Portion of the Sympathetic System

The abdominal portion of the sympathetic trunk (Fig. 12-79) is situated ventral to the bodies of the lumbar vertebrae, along the medial margin of the psoas major. The cord connecting the last thoracic and first lumbar ganglia bends ventrally as it passes under the medial lumbocostal arch of the diaphragm, bringing the trunk rather abruptly into its ventral relationship with the lumbar vertebrae. The left trunk is partly concealed by the aorta; the right, by the inferior vena cava.

LUMBAR GANGLIA. The lumbar ganglia have no fixed pattern. The number varies from two to six, with four or five occurring in three-fourths of the trunks, but massive fusions are frequent and two specimens with four ganglia may bear no resemblance to

each other. Although the five individual lumbar ganglia should not be expected in any particular instance, each one occurs with sufficient frequency to make an anatomic description possible. The numbering of the ganglia is based upon the spinal nerves with which they are connected as well as upon the relationship to the vertebrae (Pick and Sheehan, 1946).

The **roots of the lumbar ganglia** (*white rami communicantes*) are found only as far as the second lumbar spinal nerve, the caudal limit of the thoracolumbar outflow. The preganglionic fibers for the rest of the lumbar and for the sacral and coccygeal ganglia run caudally in the trunk, mainly from these first two lumbar roots. One or two roots (white rami) are supplied to each of the first three lumbar ganglia (or their representatives in fused ganglia) by the spinal nerve one segment above; the roots take an oblique course caudalward while the branches to the spinal nerves (gray rami) take a transverse course (Botar, 1932). Thus the twelfth thoracic nerve sends roots to the first lumbar ganglion; the first lumbar nerve, to the second ganglion; and the second lumbar nerve, to the third ganglion.

The ganglia of the lumbar trunk, when independently represented, lie on the bodies of the corresponding vertebrae or the intervertebral discs caudally. The first ganglion is close to or partly concealed by the medial lumbocostal arch. The ganglion on the second lumbar vertebra is the most constant, largest, and most easily palpated and identified by the surgeon. The fifth ganglion is relatively inaccessible to the surgeon because of the common iliac vessels.

Variations. In 25 bodies, the first lumbar ganglion (identified by its rami) was independent in 13, fused with other ganglia in 10, and separated into two parts in 2; the second ganglion was missing in 2, independent in 12, fused in 7, and split in 4; the third ganglion was independent in 2, fused in 17, split in 4, and connected only with L 3 nerve in 3; the fourth ganglion was independent in one, fused in 12, split in 12, and of these 11 connected with L 4 only; the fifth ganglion was independent in 4, fused in 3, split in 18, and of these 15 connected with L 5 only (Pick and Sheehan, 1946).

The **branches** of the **lumbar trunk** may be divided into three groups: (1) the gray rami communicantes, (2) the lumbar splanchnic

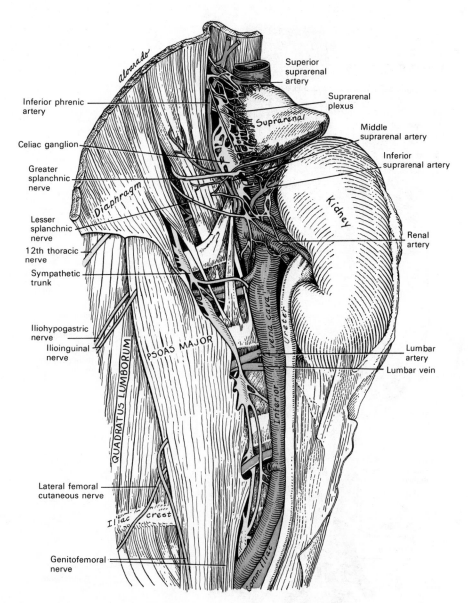

Fig. 12-78. The lumbar portion of the right sympathetic trunk, the celiac ganglion, splanchnic nerves, suprarenal gland, and kidney. (Redrawn from Töndury.)

nerves, and (3) the visceral branches through the celiac plexus.

GRAY RAMI COMMUNICANTES. The branches that are the gray rami communicantes are supplied to each of the lumbar nerves. They take a transverse path, in contrast to the oblique path of the white rami, and they are more proximal than the white rami in the segments where both are present. They are longer than those in the thoracic region because the lumbar trunk is more ventrally placed, at some distance from the spinal nerves, and they commonly accompany the lumbar arteries under the fibrous arches of the psoas, frequently splitting, doubling, and rejoining (Kuntz and Alexander, 1950).

LUMBAR SPLANCHNIC NERVES. The lumbar splanchnic nerves are two to four relatively short branches of the lumbar trunk at the level of the first, second, and third lumbar vertebrae, and are, therefore, caudal to the last root of the lesser splanchnic nerve. They

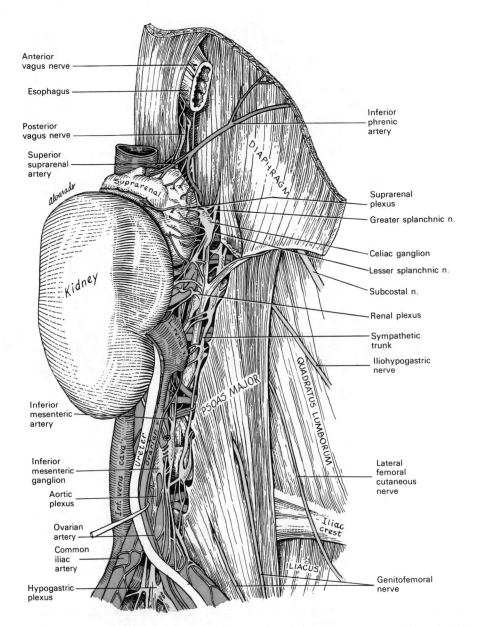

FIG. 12-79. The lumbar portion of the left sympathetic trunk, celiac and mesenteric ganglia, splanchnic nerves, hypogastric plexus, suprarenal gland, and kidney. (Redrawn from Töndury.)

pass either medialward, caudalward, and ventralward to join the aortic network, or, in the case of the most caudal lumbar splanchnic, caudalward and ventralward around the aorta to the inferior mesenteric ganglion or the hypogastric nerve. On the right side they pass between the aorta and the inferior vena cava (Trumble, 1934). Frequently they contain small groups of ganglion cells that are believed to be displaced from the sympathetic trunk (Harris, 1943).

CELIAC GANGLION. The celiac ganglion (*semilunar ganglion*) (Fig. 12-80) comprises two masses of ganglionic tissue approximately 2 cm in diameter, superficially resembling lymph nodes, which lie ventral and lateral to the abdominal aorta on each side at the level of the first lumbar vertebra. The ganglia are irregular in shape, are usually partly dispersed into several small ganglionic masses, and are connected with each other across the midline by a dense network

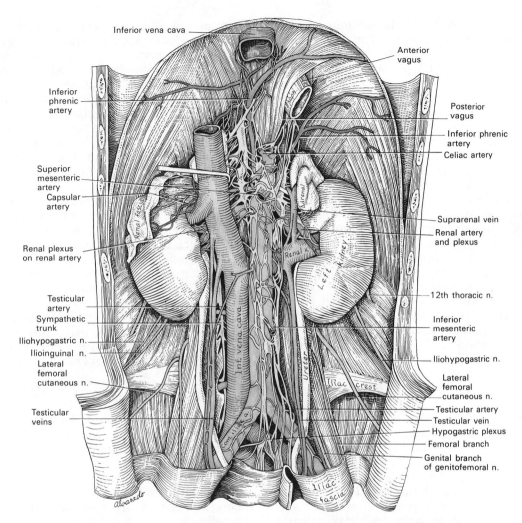

Fig. 12-80. View of the posterior abdominal wall, showing the celiac, aortic, and hypogastric plexuses of autonomic nerves. (Redrawn from Töndury.)

of bundles, especially caudal to the celiac artery. The ganglia lie on the ventral surface of the crura of the diaphragm, close to the medial border of the suprarenal glands. The right ganglion is covered by the inferior vena cava; the left is covered by the peritoneum of the lesser sac in close relation to the pancreas.

AORTICORENAL AND SUPERIOR MESENTERIC GANGLIA. The **aorticorenal** and **superior mesenteric ganglia** are masses that are more or less completely detached from the caudal portion of the celiac ganglion but represent portions of the larger celiac ganglionic complex. The aorticorenal ganglion lies at the origin of the renal artery; the superior mesenteric ganglion, at the origin of the corresponding artery. The **roots** of these ganglia

are the **splanchnic nerves** (discussed previously). Each greater splanchnic nerve enters the dorsal and lateral border of the main celiac ganglion (Fig. 12-80); the lesser splanchnic nerve enters the aorticorenal ganglion (Fig. 12-78), and the lowest splanchnic joins the renal plexus. They contain preganglionic fibers from the last six or seven thoracic spinal cord segments, which pass through the sympathetic trunk without synapses.

The **postganglionic fibers** that arise in the celiac ganglia form an extensive plexus (the celiac plexus) of nerve bundles and filaments, which branch off into subsidiary plexuses, in general following the branches of the abdominal aorta. The nerves from the celiac plexus pass down the aorta in the form of a network that is penetrated by the inferior mesenteric artery. The more cranial lumbar splanchnic nerves make a stout contribution to this network and join the caudal lumbar splanchnic nerves to form thick ganglionated nerve bundles on each side of the midline. These bundles converge and meet at the bifurcation of the aorta, with a free decussation of fibers, and then continue as the **right** and **left hypogastric nerves.** Small ganglionic nodes are present, especially if the network is compressed (Trumble, 1934).

INFERIOR MESENTERIC GANGLION.   The inferior mesenteric ganglion is more difficult to define in man than in many animals, but a considerable amount of ganglionic tissue is almost invariably present at the origin of the inferior mesenteric artery. The roots of the ganglion are provided by nerves from the celiac plexus, the celiac roots, and by the lumbar splanchnic nerves.

The inferior mesenteric ganglion in cats (Harris, 1943) is composed of three distinct masses arranged in a triangle about the origin of the inferior mesenteric artery. Ganglia are found in dogs, guinea pigs, rabbits, and monkeys also (Trumble, 1934).

Branches.   The branches of the inferior mesenteric ganglion are (*a*) nerves that accompany the inferior mesenteric artery and its branches to supply the **colon,** and (*b*) fibers that join each **hypogastric nerve** and continue from the bifurcation to join the pelvic plexus. The hypogastric nerve crosses the medial side of the ureter and contributes to the ureteric network of nerves. It contains mainly fine unmyelinated fibers but has many medium myelinated fibers (4 to 6 μm) and a few large ones, probably afferent. The hypogastric nerves fan out into an extensive network just under the parietal peritoneum in the subserous fascia. They supply the rectal, vesical, prostatic, ureteric, and ductus deferens nerves (Ashley and Anson, 1946).

Pelvic Portion of the Sympathetic System

The **pelvic portion** of the **sympathetic trunk** (Fig. 12-81) lies against the ventral surface of the sacrum, medial to the sacral foramina. It is the direct continuation of the lumbar trunk and contains four or five ganglia, smaller than those in other parts of the chain. Fusion of adjacent ganglia is common and cords connecting the trunks of the two sides across the midline occur regularly.

There are no white rami communicantes in the sacral region; small myelinated preganglionic fibers from the second, third, and fourth sacral nerves enter the pelvic plexus but they belong to the parasympathetic or craniosacral outflow rather than the sympathetic. The coccygeal ganglion is the most caudal ganglion of the sympathetic trunk; it is commonly a single ganglion, the **ganglion impar,** representing a fusion of the ganglia of the two sides, and usually lies in the midline but may be at one side.

The branches of the sacral and coccygeal ganglia which are the **gray rami communicantes** of the **sacral spinal nerves** are supplied to each of the sacral and the coccygeal nerves. In most instances, each ganglion, or its representative in a fused ganglion, supplies rami to two adjacent spinal nerves (Pick and Sheehan, 1946).

Visceral branches in variable numbers join the hypogastric and pelvic plexuses, and are supplied through them to the pelvic viscera and blood vessels (Trumble, 1934).

Great Autonomic Plexuses

The two subdivisions of the autonomic nervous system, the sympathetic and parasympathetic, are combined into extensive plexuses in the thorax, abdomen, and pelvis, named respectively the cardiac plexus, the celiac plexus, and the pelvic plexus. Experimental and clinical observations have made it possible to trace the sympathetic and parasympathetic components to some extent, but on the morphologic evidence of dissections, it is almost impossible to distinguish the ultimate paths of the fibers belonging to the two systems. These plexuses also contain visceral afferent fibers, described in earlier pages.

CARDIAC PLEXUS

The **cardiac plexus** (Fig. 7-19) is situated at the base of the heart, close to the arch of the aorta, and is traditionally subdivided into a superficial and a deep part for topographic reasons, although the functional associations do not justify this division. The sympathetic contribution is largely postganglionic; the parasympathetic, largely preganglionic, with scattered groups of ganglion cells.

SUPERFICIAL PART.   The superficial part of the cardiac plexus (Fig. 12-77) lies in the arch

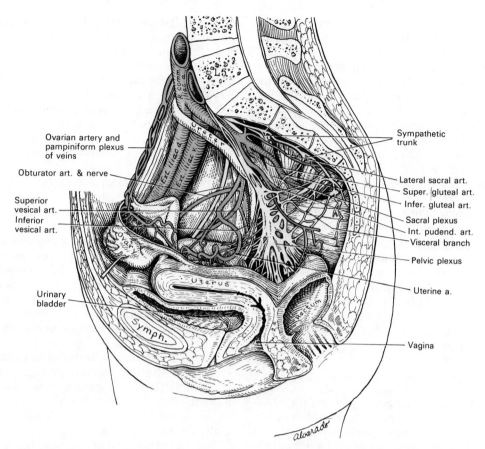

FIG. 12-81. Sagittal section of an adult female pelvis, with the peritoneum and subserous fascia partially dissected away to show the pelvic plexus of autonomic nerves. (Redrawn from Töndury.)

of the aorta somewhat on the left side between it and the bifurcation of the pulmonary artery. It is formed by the superior cervical cardiac branch of the left sympathetic and the lower of the two superior cardiac branches of the left vagus. A small ganglion, the cardiac ganglion (of Wrisberg[1]), is occasionally found in this plexus at the right side of the ligamentum arteriosum; it is probably a parasympathetic ganglion receiving preganglionic fibers from the vagus.

The branches of the superficial cardiac plexus are as follows: (a) to the **deep cardiac plexus,** (b) to the **anterior coronary plexus,** and (c) to the **left anterior pulmonary plexus.**

DEEP PART. The deep part of the cardiac plexus is situated deep to the arch of the aorta, between it and the bifurcation of the trachea and cranial to the pulmonary artery.

It is much more extensive than the superficial part and receives all the cardiac branches of both vagi and sympathetic trunks, except the two mentioned previously which enter the superficial part (left superior sympathetic and left lower superior of the vagus). It also receives the lower cardiac branches of the vagus and the recurrent nerve, visceral rami from the upper cranial thoracic sympathetic ganglia, and communications from the superficial part of the cardiac plexus.

The branches of the deep part of the cardiac plexus may be divided into right and left halves.

The right half of the deep cardiac plexus gives branches that follow the right pulmonary artery. Those ventral to the artery are more numerous, and after contributing a few filaments to the anterior pulmonary plexus, continue into the anterior coronary plexus; those dorsal to the artery distribute a few fil-

[1]Heinrich August Wrisberg (1739–1808): A German anatomist (Göttingen)

aments to the right atrium and are then continued onward to form part of the posterior coronary plexus.

The left half of the deep cardiac plexus communicates with the superficial plexus, gives filaments to the left atrium and to the anterior pulmonary plexus, and is then continued into the greater part of the posterior coronary plexus.

The **posterior coronary plexus** (*left coronary plexus*) is larger than the anterior, and accompanies the left coronary artery. It is formed chiefly by filaments from the left half and by a few filaments from the right half of the deep plexus. It is distributed to the left atrium and ventricle.

The **anterior coronary plexus** (*right coronary plexus*) is formed partly from the superficial and partly from the deep parts of the cardiac plexus. It accompanies the right coronary artery and is distributed to the right atrium and ventricle (Stotler and McMahon, 1947).

CELIAC PLEXUS

The **celiac plexus** (*solar plexus*) (Figs. 12-78; 12-80) is situated at the level of the upper part of the first lumbar vertebra. It contains two large ganglionic masses and a dense network of fibers surrounding the roots of the celiac and superior mesenteric arteries. The denser part of the plexus lies between the suprarenal glands, on the ventral surface of the crura of the diaphragm and abdominal aorta, and dorsal to the stomach and the omental bursa, but it has extensive prolongations caudalward on the aorta and out along its branches. The celiac ganglia are described with the abdominal portion of the sympathetic system. The preganglionic parasympathetic fibers reach the plexus through the anterior (left) and posterior (right) vagi on the stomach. The preganglionic sympathetic fibers reach the celiac, aorticorenal, and superior mesenteric ganglia through the greater and lesser splanchnic nerves. These nerves also supply preganglionic fibers to the cells in the medulla of the suprarenal glands, which correspond developmentally to postganglionic neurons.

SECONDARY PLEXUSES. The secondary plexuses and prolongations from the celiac plexus follow.

The **phrenic plexus** (Fig. 12-80) accompanies the inferior phrenic artery to the dia-phragm. It arises from the superior part of the celiac plexus and is larger on the right side than on the left. It communicates with the phrenic nerve; at the point of junction with the right phrenic nerve, near the vena caval foramen in the diaphragm, there may be a small ganglion, the **phrenic ganglion.** The phrenic plexus gives filaments to the inferior vena cava, the inferior phrenic arteries, and the suprarenal and hepatic plexuses.

The **hepatic plexus** accompanies the hepatic artery, ramifying upon its branches and upon those of the portal vein in the substance of the liver. A large network accompanies the gastroduodenal artery and is continued as the inferior gastric plexus on the right gastroepiploic artery along the greater curvature of the stomach, communicating with the splenic plexus. Extensions from the hepatic plexus supply the pancreas and gallbladder, and there is a communication with the phrenic plexus.

The **splenic plexus**, containing mainly fibers from the left celiac ganglion and the anterior (left) vagus, accompanies the splenic artery to the spleen and in its course sends filaments along its branches, especially to the pancreas.

The **superior gastric plexus** (*gastric or coronary plexus*) accompanies the left gastric artery along the lesser curvature of the stomach and communicates with the anterior (left) vagus.

The **suprarenal plexus** (Fig. 12-78) is composed principally of short, rather stout, branches from the celiac ganglion, with some contributions from the greater splanchnic nerve and the phrenic plexus. The fibers that it contains are predominantly preganglionic sympathetics, which pass through the celiac ganglion without synapses and are distributed to the medulla of the suprarenal gland, where the cells are homologous with postganglionic neurons. Postganglionic fibers to the blood vessels are also present (Swinyard, 1937).

The **renal plexus** (Fig. 12-78) is formed by filaments from the celiac plexus, the aorticorenal ganglion, the aortic plexus, and the smallest splanchnic nerve. It accompanies the renal artery into the kidney, giving some filaments to the spermatic plexus and to the inferior vena cava on the right side (Christensen et al., 1951).

The **spermatic plexus** receives filaments

from the renal and aortic plexuses and accompanies the testicular artery to the testis. In the female, the ovarian plexus arises from the renal plexus, accompanies the ovarian artery, and is distributed to the ovary, uterine tubes, and the fundus of the uterus.

The **superior mesenteric plexus** is essentially the lower part of the celiac plexus; it may appear more or less detached from the rest of the plexus and frequently it contains a separate ganglionic mass, the superior mesenteric ganglion. Its vagus fibers come principally from the posterior vagus. It surrounds and accompanies the superior mesenteric artery, being distributed with the latter's pancreatic, intestinal, ileocolic, right colic, and middle colic branches, which supply the corresponding organs.

The **abdominal aortic plexus** (*aortic plexus*) (Fig. 12-80) is formed from both right and left celiac plexuses and ganglia and from the lumbar splanchnic nerves. It lies upon the ventral and lateral surfaces of the aorta between the origins of the superior and inferior mesenteric arteries. From this plexus arise the spermatic, inferior mesenteric, external iliac, and hypogastric plexuses, and filaments to the inferior vena cava.

The **inferior mesenteric plexus** surrounds the origin of the inferior mesenteric artery and contains a ganglion (Fig. 12-80), the inferior mesenteric ganglion, or thickened bundles that contain ganglion cells. It is derived from the aortic plexus, through whose celiac and lumbar splanchnic contributions it receives preganglionic as well as postganglionic fibers. It surrounds the inferior mesenteric artery, and divides into a number of subsidiary plexuses that accompany its branches—the left colic, sigmoid, and superior rectal—to the corresponding organs (Harris, 1943).

The **superior hypogastric plexus** (*presacral nerve*) (Fig. 12-80) is the caudalward continuation of the aortic and inferior mesenteric plexuses. It extends from the level of the fourth lumbar to the first sacral vertebrae and lies in the subserous fascia just under the peritoneum. It is at first ventral to the aorta, then lies between the common iliac arteries; crosses the left common iliac vein, and enters the pelvis to lie against the middle sacral vessels and the vertebrae. At the first sacral vertebra it divides into two parts, the right and left hypogastric nerves.

The **hypogastric nerve** may be a single rather large nerve or several bundles forming a parallel network. It lies medial and dorsal to the common and internal iliac arteries, crosses the branches of the latter, and enters the inferior hypogastric plexus.

The **inferior hypogastric plexus** is a fanlike expansion from the hypogastric nerves at the proximal part of the rectum and bladder in the subserous fascia just above the sacrogenital fold. It receives filaments from the sacral portion of the sympathetic chain and from the deeper parts of the pelvic plexus.

PELVIC PLEXUS

The **pelvic plexus** (Fig. 12-81) of the autonomic system is formed by the hypogastric plexus, by rami from the sacral portion of the sympathetic chain, and by the visceral branches of the second, third, and fourth sacral nerves. Through its secondary plexuses it is distributed to all the pelvic viscera.

SECONDARY PLEXUSES. The **middle rectal plexus** (*middle hemorrhoidal plexus*) is contained in the tissue of the sacrogenital fold and is therefore the most superficial part of the pelvic plexus with relation to the peritoneum. It is usually independent of the middle rectal artery, and except for its terminal filaments to the caudal sigmoid colon and rectum, it is several centimeters from the bowel itself. From its superficial position, it appears to be the continuation of the inferior hypogastric plexus, but the latter has a number of other continuations and the rectal plexus supplies the lower bowel with parasympathetic fibers through contributions from the visceral branches of the sacral nerves. It communicates with the superior rectal branches from the inferior mesenteric plexus (Ashley and Anson, 1946).

The **vesical plexus** arises from the anterior part of the pelvic plexus, its fibers derived from the superficial or hypogastric network and the deeper bundles from the sacral nerves. The filaments are divisible into periureteric, prostatic, seminal vesicle, and lateral vesical groups.

The **prostatic plexus,** from the deeper ventral part of the pelvic plexus, is composed of large nerves that are distributed to the prostate, seminal vesicles, and corpora cavernosa. The nerves supplying the corpora consist of two sets, the greater and lesser cavernous nerves, which arise from the ventral part of the prostatic plexus, join with branches of the pudendal nerve, and pass beneath the pubic arch. Filaments at the base of the gland supply the prostatic and membranous urethra, the ejaculatory ducts, and the bulbourethral glands.

The **greater cavernous nerve** passes distalward along the dorsum of the penis, joins the dorsal nerve of the penis, and is distributed to the corpus cavernosum penis.

The **lesser cavernous nerves** perforate the fibrous covering of the penis near its root, and are distributed to the corpus spongiosum penis and the penile urethra.

The **vaginal plexus** arises from the caudal part of the pelvic plexus. It is distributed to the walls of the vagina and to the erectile tissue of the vestibule and clitoris.

The **uterine plexus** arises from the caudal portion of the pelvic plexus. It approaches the uterus from its caudal and lateral aspect in the base of the broad ligament, in the same region as the uterine artery. It is distributed to its musculature, supplies filaments to the uterine tube, and communicates with the ovarian plexus (Curtis et al., 1942).

References

(References are listed not only to those articles and books cited in the text, but to others as well which are considered to contain valuable resource information for the student who desires it.)

Cranial Nerves

I. Nervus Terminalis; Olfactory Nerve

Bojsen-Moller, F. 1975. Demonstration of terminalis, olfactory, trigeminal and perivascular nerves in the rat nasal septum. J. Comp. Neurol., 159:245–256.

Brookover, C. 1914. The nervus terminalis in man. J. Comp. Neurol., 24:131–136.

de Lorenzo, A. J. 1957. Electron microscopic observations of the olfactory mucosa and olfactory nerve. J. Biophys. Biochem. Cytol., 3:839–850.

Graziadei, G. A., M. S. Karlan, J. J. Bernstein, and P. P. C. Graziadei. 1980. Reinnervation of the olfactory bulb after section of the olfactory nerve in monkey (*Saimiri sciureus*). Brain Res., 189:343–354.

Larsell, O. 1950. The nervus terminalis. Ann. Otol., 59:414–438.

Loo, S. K. 1977. Fine structure of the olfactory epithelium in some primates. J. Anat. (Lond.), 123:135–146.

Pearson, A. A. 1941. The development of the nervus terminalis in man. J. Comp. Neurol., 75:39–66.

Rao, V. S. 1948. The lateral olfactory root in the human embryo. Anat. Rec., 101:617–620.

Schultz, E. W. 1941. Regeneration of olfactory cells. Proc. Soc. Exp. Biol. Med., 46:41–43.

Smith, C. G. 1942. Age incidence of atrophy of olfactory nerves in man: A contribution to the study of the process of aging. J. Comp. Neurol., 77:589–596.

Smith, C. G. 1951. Regeneration of sensory olfactory epithelium and nerves in adult frogs. Anat. Rec., 109:661–672.

II. Optic Nerve

Anderson, D. R., and W. Hoyt. 1969. Ultrastructure of intraorbital portion of human and monkey optic nerve. Arch. Ophthalmol., 82:506–530.

Arees, E. A. 1978. Growth patterns of axons in the optic nerve of chick during myelogenesis. J. Comp. Neurol., 180:73–84.

Arey, L. B. 1916. The function of the efferent fibers of the optic nerve of fishes. J. Comp. Neurol., 26:213–245.

Bruesch, S. R., and L. B. Arey. 1942. The number of myelinated and unmyelinated fibers in the optic nerve of vertebrates. J. Comp. Neurol., 77:631–665.

Donovan, A. 1967. The nerve fiber composition of the cat optic nerve. J. Anat. (Lond.), 101:1–11.

Francois, J., and A. Neetens. 1969. Physioanatomy of the axial vascularisation of the optic nerve. Documenta Ophthalmol., 26:38–49.

Goldberg, S., and M. Kotani. 1967. The projection of optic nerve fibers in the frog *Rana catesbiana* as studied by radioautography. Anat. Rec., 158:325–331.

Henkind, P., and M. Levitsky. 1969. Angioarchitecture of the optic nerve. I. The papilla. Amer. J. Ophthalmol., 68:979–986.

Hokoc, J. N., and E. Oswaldo-Cruz. 1978. Quantitative analysis of the opossum's optic nerve: An electron microscope study. J. Comp. Neurol., 178:773–782.

Hughes, A. 1977. The pigmented-rat optic nerve: Fibre count and fibre diameter spectrum. J. Comp. Neurol., 176:263–268.

Hughes, A., and H. Wassle. 1976. The cat optic nerve: Fibre total count and diameter spectrum. J. Comp. Neurol., 169:171–184.

Jayatilaka, A. D. P. 1967. A note on arachnoid villi in relation to human optic nerves. J. Anat. (Lond.), 101:171–173.

Kupfer, C., L. Chumbley, and J. de C. Downer. 1967. Quantitative histology of optic nerve, optic tract and lateral geniculate nucleus of man. J. Anat. (Lond.), 101:393–402.

Lieberman, M. F., A. E. Maumenee, and W. R. Green. 1976. Histologic studies of the vasculature of the anterior optic nerve. Amer. J. Ophthalmol., 82:405–423.

Rogers, K. T. 1957. Early development of the optic nerve in the chick. Anat. Rec., 129:97–107.

Steele, E. J., and M. J. Blunt. 1956. The blood supply of the optic nerve and chiasma in man. J. Anat. (Lond.), 90:486–493.

Stone, J., and J. E. Campion. 1978. Estimate of the number of myelinated axons in the cat's optic nerve. J. Comp. Neurol., 180:799–806.

Sturrock, R. R. 1975. A light and electron microscopic study of proliferation and maturation of fibrous astrocytes in the optic nerve of the human embryo. J. Anat. (Lond.), 119:223–234.

Tennekoon, G. I., S. R. Cohen, D. L. Price, and G. M. McKhann. 1977. Myelinogenesis in optic nerve. A morphological, autoradiographic and biochemical analysis. J. Cell Biol., 72:604–616.

Turner, J. W. A. 1943. Indirect injuries of the optic nerve. Brain, 66:140–151.

Vaney, D. I., and A. Hughes. 1976. The rabbit optic nerve. Fibre diameter spectrum, fibre count and comparison with retinal ganglion cell count. J. Comp. Neurol., 170:241–251.

Vaughn, J. E., P. L. Hinds, and R. P. Skoff. 1970. Electron microscopic studies of Wallerian degeneration in rat optic nerves. I. The multipotential glia. J. Comp. Neurol., 140:175–206.

Vaughn, J. E., and D. C. Pease. 1970. Electron microscopic studies of Wallerian degeneration in rat optic nerves. II. Astrocytes, oligodendrocytes and adventitial cells. J. Comp. Neurol., 140:207–226.

III. Oculomotor Nerve

Abd-El-Malek, S. 1938. On the localization of nerve centres of the extrinsic ocular muscles in the oculomotor nucleus. J. Anat. (Lond.), 72:518–523.

Abd-El-Malek, S. 1938. On the presence of sensory fibers in the ocular nerves. J. Anat. (Lond.), 72:524–530.

Bach y Rita, P., and F. Ito. 1966. Properties of stretch receptors in cat extra-ocular muscles. J. Physiol. (Lond.), 186:663–688.

Bortolami, R., A. Veggetti, E. Callegari, M. L. Lucchi, and G. Palmieri. 1977. Afferent fibers and sensory ganglion cells with the oculomotor nerve in some mammals and man. I. Anatomical investigations. Arch. Ital. Biol., 115:355–385.

Clark, W. E. Le Gros. 1926. The mammalian oculomotor nucleus. J. Anat. (Lond.), 60:426–448.

Corbin, K. B., and R. K. Oliver. 1942. The origin of fibers to the grape-like endings in the insertion third of the extra-ocular muscles. J. Comp. Neurol., 77:171–186.

Manni, E., R. Bortolami, V. E. Pettorossi, M. L. Lucchi, and E. Callegari. 1978. Afferent fibers and sensory ganglion cells within the oculomotor nerve in some mammals and man. II. Electrophysiological investigations. Arch. Ital. Biol., 116:16–24.

Pearson, A. A. 1944. The oculomotor nucleus in the human fetus. J. Comp. Neurol., 80:47–63.

Sunderland, S., and E. S. R. Hughes. 1946. The pupilloconstrictor pathway and the nerves to the ocular muscles in man. Brain, 69:301–309.

Warwick, R. 1954. The ocular parasympathetic nerve supply and its mesencephalic sources. J. Anat. (Lond.), 88:71–93.

IV. Trochlear Nerve

Batini, C., C. Buisseret-Delmas, and R. T. Kado. 1979. On the fibers of the III, IV and VI cranial nerves in the cat. Arch. Ital. Biol., 117:111–122.

Gacek, R. R. 1979. Location of trochlear vestibuloocular neurons in the cat. Exp. Neurol., 66:135–145.

Kerns, J. M., and L. A. Rothblat. 1981. The effects of monocular deprivation on the development of the rat trochlear nerve. Brain Res., 230:367–371.

Kerns, R. M. 1980. Postnatal differentiation of the rat trochlear nerve. J. Comp. Neurol., 189:291–306.

Mustafa, G. Y., and H. J. Gamble. 1979. Changes in axonal numbers in developing human trochlear nerve. J. Anat. (Lond.), 128:323–330.

Nathan, H., and Y. Goldhammer. 1973. The rootlets of the trochlear nerve. Anatomical observations in human bodies. Acta Anat., 84:590–596.

Pearson, A. A. 1943. The trochlear nerve in human fetuses. J. Comp. Neurol., 78:29–43.

Sherif, M. F., N. J. Papadopoulos, and E. N. Albert. 1981. Fiber contribution from the mesencephalic nucleus of the trigeminal nerve to the trochlear nerve in the cat: a histological quantitative study. Anat. Rec., 201:669–678.

Sohal, G. S. 1976. An experimental study of cell death in the developing trochlear nucleus. Exp. Neurol., 51:684–698.

Sohal, G. S., and R. K. Holt. 1978. Identification of trochlear motoneurons by retrograde transport of horseradish peroxidase. Exp. Neurol., 59:509–514.

Sohal, G. S., and T. A. Weidman. 1978. Development of the trochlear nerve: loss of axons during normal ontogenesis. Brain Res., 142:455–465.

Sohal, G. S., T. A. Weidman, and S. D. Stoney. 1978. Development of the trochlear nerve: effects of early removal of periphery. Exp. Neurol., 59:331–341.

Stoney, S. D., Jr., and G. S. Sohal. 1978. Development of the trochlear nerve: neuromuscular transmission and electrophysiologic properties. Exp. Neurol., 62:798–803.

Weidman, T. A., and G. S. Sohal. 1977. Cell and fiber composition of the trochlear nerve. Brain Res., 125:340–344.

Zaki, W. 1960. The trochlear nerve in man. Study relative to its origin, its intracerebral traject and its structure. Arch. Anat. Histol. Embryol., 43:105–120.

V. Trigeminal Nerve

Anderson, K. V., and H. S. Rosing. 1977. Location of feline trigeminal ganglion cells innervating maxillary canine teeth: a horseradish peroxidase analysis. Exp. Neurol., 57:302–306.

Arvidsson, J. 1975. Location of cat trigeminal ganglion cells innervating dental pulp of upper and lower canines studied by retrograde transport of horseradish peroxidase. Brain Res., .135–139.

Augustine, J. R., B. Vidić, . A. Young. 1971. An intermediate root of the trigeminal nerve in the dog. Anat. Rec., . 97–703.

Baumel, J. J., P. Vanderheiden, and J. E. McElenney. 1971. The auriculotemporal nerve of man. Amer. J. Anat., 130:431–440.

Beasley, W. L., and G. R. Holland. 1978. A quantitative analysis of the innervation of the pulp of the cat's canine tooth. J. Comp. Neurol., 179:487–494.

Bernick, S. 1948. Innervation of the human tooth. Anat. Rec., 101:81–107.

Burr, H. S., and G. B. Robinson. 1925. An anatomical study of the Gasserian ganglion, with particular reference to the nature and extent of Meckel's cave. Anat. Rec., 29:269–282.

Carter, R. B., and E. N. Keen. 1971. The intramandibular course of the inferior alveolar nerve. J. Anat. (Lond.), 108:433–440.

Christensen, K. 1934. The innervation of the nasal mucosa, with special reference to its afferent supply. Ann. Otol. Rhinol. Laryngol., 43:1066–1083.

Corbin, K. B. 1940. Observations on the peripheral distribution of fibers arising in the mesencephalic nucleus of the fifth cranial nerve. J. Comp. Neurol., 73:153–177.

Corbin, K. B., and F. Harrison. 1940. Function of mesencephalic root of fifth cranial nerve. J. Neurophysiol., 3:423–435.

Cushing, H. 1904. The sensory distribution of the fifth cranial nerve. Bull. Johns Hopkins Hosp., 15:213–232.

Fitzgerald, M. J. T. 1956. The occurrence of a middle superior alveolar nerve in man. J. Anat. (Lond.), 90:520–522.

Foley, J. O., H. R. Pepper, and W. H. Kessler. 1946. The ratio of nerve fibers to nerve cells in the geniculate ganglion. J. Comp. Neurol., 85:141–148.

Frazier, C. H., and E. Whitehead. 1925. The morphology of the gasserian ganglion. Brain, 48:458–475.

Gasser, R. F., and D. M. Wise. 1972. The trigeminal nerve in the baboon. Anat. Rec.,172:511–522.

Henderson, W. R. 1965. The anatomy of the gasserian ganglion and the distribution of pain in relation to injection and operation for trigeminal neuralgia. Ann. Roy. Coll. Surg. (Lond.), 37:346–373.

Holland, G. R. 1978. Fibre numbers and sizes in the inferior alveolar nerve of the cat. J. Anat. (Lond.), 127:343–352.

Jannetta, P. J., and R. W. Rand. 1966. Transtentorial retrogasserian rhizotomy in trigeminal neuralgia by microneurosurgical technique. Bull. Los Angeles Neurol. Soc., 31:93–99.

Jones, F. W. 1939. The anterior superior alveolar nerves and vessels. J. Anat. (Lond.), 73:583–591.

Mohuiddin, A. 1951. The postnatal development of the inferior dental nerve of the cat. J. Anat. (Lond.), 85:24–35.

Pearson, A. A. 1949. The development and connections of the mesencephalic root of the trigeminal nerve in man. J. Comp. Neurol., 90:1–46.

Pearson, A. A. 1977. The early innervation of the developing deciduous teeth. J. Anat. (Lond.), 123:563–578.

Roberts, W. H., and N. B. Jorgensen. 1962. A note on the distribution of the superior alveolar nerves in relation to the primary teeth. Anat. Rec., 141:81–84.

Starkie, C., and D. Stewart. 1931. The intra-mandibular course of the inferior dental nerve. J. Anat. (Lond.), 65:319–323.

Szentagothai, J. 1949. Functional representation in the motor trigeminal nucleus. J. Comp. Neurol., 90:111–120.

Vidić, B., and J. Stefanatos. 1969. The roots of the trigeminal nerve and their fiber components. Anat. Rec., 163:330.

Wedgwood, M. 1966. The peripheral course of the inferior dental nerve. J. Anat. (Lond.), 100:639–650.

Young, R. F., and R. Stevens. 1979. Unmyelinated axons in the trigeminal motor root of human and cat. J. Comp. Neurol., 183:205–214.

VI. Abducent Nerve

Bronson, R. T., and E. T. Hedley-Whyte. 1977. Morphometric analysis of the effects of exenteration and enucleation on the development of the third and sixth cranial nerves in the rat. J. Comp. Neurol., 176:315–330.

Cushing, H. 1911. Strangulation of the nervi abducentes by lateral branches of the basilar artery in cases of brain tumour. Brain, 33:204–235.

Gacek, R. R. 1979. Location of abducens afferent neurons in the cat. Exp. Neurol., 64:342–353.

Konigsmark, B. W., U. P. Kalyanarama, P. Corey, and E. A. Murphy. 1969. An evaluation of techniques in neuronal population estimates: the sixth nerve nucleus. Bull. Johns Hopkins Hosp., 125:146–158.

Sivanandasingham, P. 1978. Peripheral pathway of proprioceptive fibres from feline extra-ocular muscles. I. A histological study. Acta Anat., 100:173–184.

Spencer, R. F., and J. D. Porter. 1981. Innervation and structure of extraocular muscles in the monkey in comparison to those of the cat. J. Comp. Neurol., 198:649–665.

Spencer, R. F., and P. Sterling. 1977. Study of motoneurones and interneurones in the cat abducens nucleus identified by retrograde intraaxonal transport of horseradish peroxidase. J. Comp. Neurol., 176:65–86.

Vijayashanker, N., and H. Brody. 1977. A study of aging in the human abducens nucleus. J. Comp. Neurol., 173:433–438.

Wolff, E. 1928. A bend of the sixth cranial nerve, and its clinical significance. J. Anat. (Lond.), 63:150–151.

Wolff, E. 1928. A bend in the sixth nerve and its probable significance. Brit. J. Ophthalmol., 12:22–24.

VII. Facial Nerve

Blevins, C. E. 1964. Studies on the innervation of the stapedius muscle of the cat. Anat. Rec., 149:157–171.

Blunt, M. J. 1954. The blood supply of the facial nerve. J. Anat. (Lond.), 88:520–526.

Bosman, D. H. 1978. The distribution of the chorda tympani in the middle ear area in man and two other primates. J. Anat. (Lond.), 127:443–446.

Bowden, R. E. M., and Z. Y. Mahran. 1960. Experimental and histological studies of the extrapetrous portion of the facial nerve and its communications with the trigeminal nerve in the rabbit. J. Anat. (Lond.), 94:375–386.

Bruesch, S. R. 1944. The distribution of myelinated afferent fibers in the branches of the cat's facial nerve. J. Comp. Neurol., 81:169–191.

Bunnell, S. 1937. Surgical repair of the facial nerve. Arch. Otolaryngol., 25:235–259.

Coleman, C. C. 1944. Surgical lesions of the facial nerve: With comments on its anatomy. Ann. Surg., 119:641–655.

de Lorenzo, A. S. 1958. Electron microscopic observations on the taste buds of the rabbit. J. Biophys. Biochem. Cytol., 4:143–150.

Dobozi, M. 1975. Surgical anatomy of the geniculate ganglion. Acta Otolaryngol. (Stockholm), 80:116–119.

Durcan, D. J., J. J. Shea, and J. P. Sleeckx. 1967. Bifurcation of the facial nerve. Arch. Otolaryngol., 86:619–631.

Foley, J. O. 1945. The sensory and motor axons of the chorda tympani. Proc. Soc. Exp. Biol. Med., 60:262–267.

Foley, J. O., and F. S. Dubois. 1943. An experimental study of the facial nerve. J. Comp. Neurol., 79:79–105.

Gasser, R. F. 1970. The early development of the parotid gland around the facial nerve and its branches in man. Anat. Rec., 167:63–78.

Gomez, H. 1961. The innervation of lingual salivary glands. Anat. Rec., 139:69–76.

Higbee, D. 1949. Functional and anatomic relation of the sphenopalatine ganglion to the autonomic nervous system. Arch. Otolaryngol., 50:45–58.

Jannetta, P. J., M. Abbasy, J. C. Maroon, F. M. Ramos, and M. S. Alvin. 1977. Etiology and definitive microsurgical treatment of hemifacial spasm. Operative techniques and results in 47 patients. J. Neurosurg., 47:321–328.

Kubota, K., and T. Masegi. 1972. Proprioceptive afferents in facial nerves of some insectivores. Anat. Rec., 173:353–363.

Kudo, H., and S. Nori. 1974. Topography of the facial nerve in the human temporal bone. Acta Anat., *90*:467–480.

Laurenson, R. D. 1964. A rapid exposure of the facial nerve. Anat. Rec., *150*:317–318.

McCormack, L. J., E. W. Cauldwell, and B. J. Anson. 1945. The surgical anatomy of the facial nerve with special reference to the parotid gland. Surg. Gynecol. Obstet., *80*:620–630.

McKenzie, J. 1948. The parotid gland in relation to the facial nerve. J. Anat. (Lond.), *82*:183–186.

Miller, I. J., Jr. 1974. Branched chorda tympani neurons and interactions among taste receptors. J. Comp. Neurol., *158*:155–166.

Olmsted, J. M. D. 1922. Taste fibers and the chorda tympani nerve. J. Comp. Neurol., *34*:337–341.

Podvinec, M., and C. R. Pfaltz. 1976. Studies on the anatomy of the facial nerve. Acta Otolaryngol. (Stockholm), *81*:173–177.

Rhoton, A. L., Jr., S. Kobayashi, and W. H. Hollinshead. 1968. Nervus intermedius. J. Neurosurg., *29*:609–618.

Ruskell, G. L. 1970. An ocular parasympathetic nerve pathway of facial nerve origin and its influence on intraocular pressure. Exp. Eye Res., *10*:319–330.

Ruskell, G. L. 1971. The distribution of autonomic postganglionic nerve fibres in the lacrimal gland in the rat. J. Anat. (Lond.), *109*:229–242.

Shimazawa, A. 1971. Quantitative studies of the greater petrosal nerve of the mouse with the electron microscope. Anat. Rec., *170*:303–308.

Shimazawa, A. 1973. An electron microscopic analysis of the nerve of the pterygoid canal in the mouse. Anat. Rec., *175*:631–637.

Sunderland, S., and D. F. Cossar. 1953. The structure of the facial nerve. Anat. Rec., *116*:147–165.

Vidic, B. 1968. The origin and the course of the communicating branch of the facial nerve to the lesser petrosal nerve in man. Anat. Rec., *162*:511–515.

Vidic, B. 1970. The connections of the intra-osseous segment of the facial nerve in baboon (*papio* sp.). Anat. Rec., *168*:477–490.

Vidic, B., and P. A. Young. 1967. Gross and microscopic observations on the communicating branch of the facial nerve to the lesser petrosal nerve. Anat. Rec., *158*:257–261.

Wind, J. 1977. The facial nerve and human evolution. Adv. Otorhinolaryngol., *22*:215–219.

Engstrom, H., and H. Wersall. 1958. Structure and innervation of the ear sensory epithelia. Internat. Rev. Cytol., *6*:535–585.

Fernandez, C. 1951. The innervation of the cochlea (guinea pig). Laryngoscope, *61*:1152–1172.

Gacek, R., and G. L. Rasmussen. 1957. Fiber analysis of the acoustic nerve of the cat, monkey and guinea pig. Anat. Rec., *127*:417 (Abstract).

Gacek, R. R., and G. L. Rasmussen. 1961. Fiber analysis of the statoacoustic nerve of guinea pig, cat, and monkey. Anat. Rec., *139*:455–463.

Galambos, R., and H. Davis. 1943. The response of single auditory nerve fibers to acoustic stimulation. J. Neurophysiol., *6*:39–57.

Hardy, M. 1934. Observations on the innervation of the macula sacculi in man. Anat. Rec., *59*:403–418.

Osen, K. K. 1970. Course and termination of the primary afferents in the cochlear nuclei of the cat. An experimental anatomical study. Arch. Ital. Biol., *108*:21–51.

Petroff, A. E. 1955. An experimental investigation of the origin of efferent fiber projections to the vestibular neuroepithelium. Anat. Rec., *121*:352–353.

Rasmussen, A. T. 1940. Studies of the eighth cranial nerve of man. Laryngoscope, *50*:67–83.

Rasmussen, G. L., and R. Gacek. 1958. Concerning the question of an efferent component of the vestibular nerve of the cat. Anat. Rec., *130*:361–362.

Rasmussen, G. L., and W. F. Windle. 1960. *Neural Mechanisms of the Auditory and Vestibular Systems.* Charles C Thomas, Springfield, Ill., 422 pp.

Romand, R., A. Sans, M. R. Romand, and R. Marty. 1976. The structural maturation of the stato-acoustic nerve in the cat. J. Comp. Neurol., *170*:1–16.

Ross, M. D. 1969. The general visceral efferent component of the eighth cranial nerve. J. Comp. Neurol., *135*:453–478.

Stopp, P. E., and S. D. Comis. 1978. Afferent and efferent innervation of the guinea-pig cochlea: a light microscopic and histochemical study. Neuroscience, *3*:1197–1206.

Tasaki, I. 1954. Nerve impulses in individual auditory nerve fibers of the guinea pig. J. Neurophysiol., *17*:97–122.

Wersall, J. 1956. Studies on the structure and innervation of the sensory epithelium of the cristae ampullares in the guinea pig. Acta Otolaryngol. (Stockholm), Suppl., *126*:1–85.

VIII. Vestibulocochlear Nerve

Arnesen, A. R., and K. K. Osen. 1978. The cochlear nerve in the cat: Topography, cochleotopy and fiber spectrum. J. Comp. Neurol., *178*:661–678.

Davis, H. 1954. The excitation of nerve impulses in the cochlea. Ann. Otol. Rhinol. Laryngol., *63*:469–480.

Davis, H. 1958. Transmission and conduction in the cochlea. Laryngoscope, *68*:359–382.

Dohlman, G., J. Farkashidy, and F. Salonna. 1958. Centrifugal nerve fibers to the sensory epithelium of the vestibular labyrinth. J. Laryngol. Otol. (Lond.), *72*:984–991.

Dunn, R. F. 1978. Nerve fibers of the eighth nerve and their distribution to the sensory nerves of the inner ear in the bullfrog. J. Comp. Neurol., *182*:621–636.

Engstrom, H. 1958. On the double innervation of the sensory epithelia of the inner ear. Acta Otolaryngol. (Stockholm), *49*:109–118.

IX. Glossopharyngeal Nerve

Abbot, P. C., M. D. B. Daly, and A. Howe. 1972. Early ultrastructural changes in the carotid body after degenerative section of the carotid sinus nerve in the cat. Acta Anat., *83*:161–184.

Boyd, J. D. 1937. Observations on the human carotid sinus and its nerve supply. Anat. Anz., *84*:386–399.

Dixon, J. S. 1966. The fine structure of parasympathetic nerve cells in the otic ganglia of the rabbit. Anat. Rec., *156*:239–251.

Erickson, T. C. 1936. Paroxysmal neuralgia of the tympanic branch of the glossopharyngeal nerve: Report of a case in which relief was obtained by intracranial section of the glossopharyngeal nerve. Arch. Neurol. Psychiat., *35*:1070–1075.

Eyzaguirre, C., and J. Lewin. 1961. Effect of sympathetic stimulation on carotid nerve activity. J. Physiol. (Lond.), *159*:251–267.

Frenckner, P. 1951. Observations on the anatomy of the tympanic plexus and technique of tympanosympathectomy. Arch. Otolaryngol., 54:347–355.

Guth, L. 1957. The effects of glossopharyngeal nerve transection on the circumvallate papilla of the rat. Anat. Rec., 128:715–731.

Heymans, C., and J. J. Bouckaert. 1939. Les chémorecepteurs du sinus carotidien. Ergebn. der Physiol., 41:28–55.

Kienecker, E.-W., H. Knoche, and D. Bingmann. 1978. Functional properties of regenerating sinus nerve fibres in the rabbit. Neuroscience, 3:977–988.

Lawn, A. M. 1966. The localization, in the nucleus ambiguus of the rabbit, of the cells of origin of motor nerve fibers in the glossopharyngeal nerve and various branches of the vagus nerve by means of retrograde degeneration. J. Comp. Neurol., 127:293–305.

Mitchell, R. A., A. Sinha, and D. M. McDonald. 1972. Chemoreceptive properties of regenerated endings of the carotid sinus nerve. Brain Res., 28:681–685.

Nishio, J., T. Matsuya, K. Ibuki, and T. Miyazaki. 1976. Roles of the facial, glossopharyngeal and vagus nerves in velopharyngeal movement. Cleft Palate J., 13:201–214.

Partridge, E. J. 1918. The relations of the glossopharyngeal nerve at its exit from the cranial cavity. J. Anat. (Lond.), 52:332–334.

Pastore, P. N., and J. M. Meredith. 1949. Glossopharyngeal neuralgia. Arch. Otolaryngol., 50:789–794.

Reichert, F. L., and E. J. Poth. 1933. Recent knowledge regarding the physiology of the glossopharyngeal nerve in man with analysis of its sensory, motor, gustatory and secretory functions. Bull. Johns Hopkins Hosp., 53:131–139.

Ripley, R. C., J. W. Hollifield, and A. S. Nies. 1977. Sustained hypertension after section of the glossopharyngeal nerve. Amer. J. Med., 62:297–302.

Rosen, S. 1950. The tympanic plexus. Arch. Otolaryngol., 52:15–18.

Sheehan, D., J. H. Mulholland, and B. Shafiroff. 1941. Surgical anatomy of the carotid sinus nerve. Anat. Rec., 80:431–442.

Sprague, J. M. 1944. The innervation of the pharynx in the Rhesus monkey, and the formation of the pharyngeal plexus in primates. Anat. Rec., 90:197–208.

State, F. A., and R. E. M. Bowden. 1974. The effect of transection of the glossopharyngeal nerve upon the structure, cholinesterase activity and innervation of taste buds in rabbits. J. Anat. (Lond.), 118:77–100.

Tarlov, I. M. 1940. Sensory and motor roots of the glossopharyngeal nerve and the vagus-spinal accessory complex. Arch. Neurol. Psychiat., 44:1018–1021.

von Euler, U. S., and Y. Zotterman. 1942. Action potentials from the baroceptive and chemoceptive fibres in the carotid sinus nerve of the dog. Acta Physiol. Scandinav., 4:13–22.

Willis, A. G., and J. D. Tange. 1959. Studies on the innervation of the carotid sinus of man. Amer. J. Anat., 104:87–113.

Zapata, P., L. J. Stensaas, and C. Eyzaguirre. 1976. Axon regeneration following a lesion of the carotid nerve. Electrophysiological and ultrastructural observations. Brain Res., 113:235–253.

X. Vagus Nerve

Armstrong, W. G., and J. W. Hinton. 1951. Multiple divisions of the recurrent laryngeal nerve: an anatomic study. Arch. Surg., 62:532–539.

Berlin, D. D., and F. H. Lahey. 1929. Dissections of the recurrent and superior laryngeal nerves: The relation of the recurrent to the inferior thyroid artery and the relation of the superior to abductor paralysis. Surg. Gynecol. Obstet., 49:102–104.

Bowden, R. E. M. 1955. Surgical anatomy of the recurrent laryngeal nerve. Brit. J. Surg., 43:153–163.

Brocklehurst, R. J., and F. H. Edgeworth. 1940. The fibre components of the laryngeal nerves of Macaca mulatta. J. Anat. (Lond.), 74:386–389.

Chamberlin, J. A., and T. Winship. 1947. Anatomic variations of the vagus nerves—their significance in vagus neurectomy. Surgery, 22:1–19.

Chase, M. R., and S. W. Ranson. 1914. The structure of the roots, trunk and branches of the vagus nerve. J. Comp. Neurol., 24:31–60.

Dragstedt, L. R., H. J. Fournier, E. R. Woodward, E. B. Tovee, and P. V. Harper, Jr. 1947. Transabdominal gastric vagotomy: A study of the anatomy and surgery of the vagus nerves at the lower portion of the esophagus. Surg. Gynecol. Obstet., 85:461–466.

Dubois, F. S., and J. O. Foley. 1936. Experimental studies on the vagus and spinal accessory nerve in the cat. Anat. Rec., 64:285–307.

Evans, D. H. L., and J. G. Murray. 1954. Histological and functional studies on the fibre composition of the vagus nerve of the rabbit. J. Anat. (Lond.), 88:320–337.

Foley, J. O., and F. S. Dubois. 1937. Quantitative studies of the vagus nerve in the cat: I. The ratio of sensory to motor fibers. J. Comp. Neurol., 67:49–67.

Gordon, T., N. Niven-Jenkins, and G. Vrbová. 1980. Observations on neuromuscular connection between the vagus nerve and skeletal muscle. Neuroscience, 5:597–610.

Hoffman, H. H., and A. Kuntz. 1957. Vagus nerve components. Anat. Rec., 127:551–568.

Hoffman, H. H., and H. N. Schnitzlein. 1961. The numbers of nerve fibers in the vagus nerve of man. Anat. Rec., 139:429–435.

Hollinshead, W. H. 1939. The origin of the nerve fibers to the glomus aorticum of the cat. J. Comp. Neurol., 71:417–426.

Hollinshead, W. H. 1940. The innervation of the supracardial bodies in the cat. J. Comp. Neurol., 73:37–48.

Jackson, R. G. 1949. Anatomy of the vagus nerves in the region of the lower esophagus and the stomach. Anat. Rec., 103:1–18.

Johnson, F. E., and E. A. Boyden. 1952. The effect of double vagotomy on the motor activity of the human gallbladder. Surgery, 32:591–601.

Kyösola, K., L. Rechardt, L. Veijola, T. Waris, and O. Penttilä. 1980. Innervation of the human gastric wall. J. Anat. (Lond.), 131:453–470.

Legrand, M. C., G. H. Paff, and R. J. Boucek. 1966. Initiation of vagal control of heart rate in the embryonic chick. Anat. Rec., 155:163–166.

Lemere, F. 1932. Innervation of the larynx: I. Innervation of the laryngeal muscles. Amer. J. Anat., 51:417–437.

Lemere, F. 1932. Innervation of the larynx. II. Ramus anastomoticus and ganglion cells of the superior laryngeal nerve. Anat. Rec., 54:389–407.

Mitchell, G. A. G., and R. Warwick. 1955. The dorsal vagal nucleus. Acta Anat., 25:371–395.

Mohr, P. D. 1969. The blood supply of the vagus nerve. Acta Anat., 73:19–26.

Morrison, L. F. 1952. Recurrent laryngeal nerve paralysis. A revised conception based on the dissection of one hundred cadavers. Ann. Otol. Rhinol. Laryngol., *61*: 567–592.

Peden, J. K., C. F. Schneider, and R. D. Bikel. 1950. Anatomic relations of the vagus nerves to esophagus. Amer. J. Surg., *80*:32–34.

Ranson, S. W., and P. Mihalik. 1932. The structure of the vagus nerve. Anat. Rec., *54*:355–360.

Reed, A. F. 1943. The relations of the inferior laryngeal nerve to the inferior thyroid artery. Anat. Rec., *85*:17–23.

Réthi, A. 1951. Histological analysis of the experimentally degenerated vagus nerve. Acta Morphol., *1*:221–230.

Richardson, A. P., and J. C. Hinsey. 1933. Functional studies of the nodose ganglion of the vagus with degeneration methods. Proc. Soc. Exp. Biol. Med., *30*:1141–1143.

Schnitzlein, H. N., L. C. Rowe, and H. H. Hoffman. 1958. The myelinated component of the vagus nerves in man. Anat. Rec., *131*:649–667.

Sunderland, S., and W. E. Swaney. 1952. The intraneural topography of the recurrent laryngeal nerve in man. Anat. Rec., *114*:411–426.

Tarlov, I. M. 1942. Section of the cephalic third of the vagus-spinal accessory complex: Clinical and histologic results. Arch. Neurol. Psychiat., *47*:141–148.

Teitelbaum, H. A. 1933. The nature of the thoracic and abdominal distribution of the vagus nerves. Anat. Rec., *55*:297–317.

Todo, K., T. Yamamoto, H. Satomi, H. Ise, H. Takatama, and K. Takahashi. 1977. Origins of vagal preganglionic fibers to the sinoatrial and atrioventricular node regions in the cat heart as studied by the horseradish peroxidase method. Brain Res., *130*: 545–550.

XI. Accessory Nerve

Balagura, S., and R. G. Katz. 1980. Undecussated innervation to the sternocleidomastoid muscle: a reinstatement. Ann. Neurol., *7*:84–85.

Black, D. 1914. On the so-called "bulbar" portion of the accessory nerve. Anat. Rec., *8*:110–112.

Corbin, K. B., and F. Harrison. 1938. Proprioceptive components of cranial nerves. The spinal accessory nerve. J. Comp. Neurol., *69*:315–328.

Corbin, K. B., and F. Harrison. 1939. The sensory innervation of the spinal accessory and tongue musculature in the rhesus monkey. Brain, *62*:191–197.

Gordon, S. L., W. P. Graham, J. T. Black, and S. H. Miller. 1977. Accessory nerve function after surgical procedures in the posterior triangle. Arch. Surg., *112*:264–268.

Hill, J. H., and N. R. Olson. 1979. The surgical anatomy of the spinal accessory nerve and the internal branch of the superior laryngeal nerve. Laryngoscope, *89*: 1935–1942.

Pearson, A. A. 1938. The spinal accessory nerve in human embryos. J. Comp. Neurol., *68*:243–266.

Pearson, A. A., R. W. Sauter, and G. R. Herrin. 1964. The accessory nerve and its relation to the upper spinal nerves. Amer. J. Anat., *114*:371–391.

Romanes, G. J. 1940. The spinal accessory nerve in the sheep. J. Anat. (Lond.), *74*:336–347.

Straus, W. L., Jr., and A. B. Howell. 1936. The spinal accessory nerve and its musculature. Quart. Rev. Biol., *11*:387–402.

Windle, W. F. 1931. The sensory component of the spinal accessory nerve. J. Comp. Neurol., *53*:115–127.

Yee, J., F. Harrison, and K. B. Corbin. 1939. The sensory innervation of the spinal accessory and tongue musculature in the rabbit. J. Comp. Neurol., *70*:305–314.

XII. Hypoglossal Nerve

Allen, W. F. 1929. Effect of repeated traumatization of the central stump of the hypoglossal nerve on degeneration and regeneration of its fibers and cells. Anat. Rec., *43*:27–32.

Barnard, J. W. 1940. The hypoglossal complex of vertebrates. J. Comp. Neurol., *72*:489–524.

Downman, C. B. B. 1939. Afferent fibres of the hypoglossal nerve. J. Anat. (Lond.), *73*:387–395.

Pearson, A. A. 1939. The hypoglossal nerve in human embryos. J. Comp. Neurol., *71*:21–39.

Pearson, A. A. 1943. Sensory type neurons in the hypoglossal nerve. Anat. Rec., *85*:365–375.

Scotti, D., D. Melancon, and A. Oliver. 1978. Hypoglossal paralysis due to compression by a tortuous internal carotid artery in the neck. Neuroradiology, *14*:263–265.

Shaia, F. T., and M. D. Graham. 1977. The hypoglossal nerve. Its relationship to the temporal bone and jugular foramen. Laryngoscope, *87*:1137–1139.

Tarkhan, A. A., and I. Abou-el-naga. 1947. Sensory fibres in the hypoglossal nerve. J. Anat. (Lond.), *81*:23–32.

Tarkhan, A. A., and S. Abd El-Malek. 1950. On the presence of sensory nerve cells on the hypoglossal nerve. J. Comp. Neurol., *93*:219–228.

Vij, S., and R. Kanagasuntheram. 1972. Development of the nerve supply to the human tongue. Acta Anat., *81*:466–477.

Weddell, G., J. A. Harpman, D. Lambley, and L. Young. 1940. The innervation of the musculature of the tongue. J. Anat. (Lond.), *74*:255–267.

Zimny, R., T. Sobusiak, and Z. Mattosz. 1970/1971. The afferent components of the hypoglossal nerve. An experimental study with toluidine blue and silver impregnation methods. J. für Hirnforsch., *12*:83–100.

Spinal Nerves

A. General

Barry, A. 1956. A quantitative study of the prenatal changes in angulation of the spinal nerves. Anat. Rec., *126*:97–110.

Cave, A. J. E. 1929. The distribution of the first intercostal nerve and its relation to the first rib. J. Anat. (Lond.), *63*:367–379.

Collis, J. L., L. M. Satchwell, and L. D. Abrams. 1954. Nerve supply to the diaphragm. Thorax, *9*:22–25.

Corbin, K. B., and F. Harrison. 1939. The sensory innervation of the spinal accessory and tongue musculature in the rhesus monkey. Brain, *62*:191–197.

Corbin, K. B., and J. C. Hinsey. 1935. Intramedullary course of the dorsal root fibers of each of the first four cervical nerves. J. Comp. Neurol., *63*:119–126.

Davies, F., R. J. Gladstone, and E. P. Stibbe. 1932. The anatomy of the intercostal nerves. J. Anat. (Lond.), *66*:323–333.

Emery, D. G., H. Ito, and R. E. Coggeshall. 1977. Unmyelinated axons in thoracic ventral roots of the cat. J. Comp. Neurol., *172*:37–48.

Fraher, J. P. 1978. Quantitative studies on the maturation of central and peripheral parts of individual ventral motoneuron axons. I. Myelin sheath and axon calibre. J. Anat. (Lond.), *126*:509–534.

Fraher, J. P. 1978. Quantitative studies on the maturation

of central and peripheral parts of individual ventral motoneuron axons. II. Internodal length. J. Anat. (Lond.), 127:1–16.

Gershenbaum, M. R., and F. J. Roisen. 1978. A scanning electron microscopic study of peripheral nerve degeneration and regeneration. Neuroscience, 3:1241–1250.

Hidayet, M. A., H. A. Wahid, and A. S. Wilson. 1974. Investigations on the innervation of the human diaphragm. Anat. Rec., 179:507–516.

Hinsey, J. C. 1933. The functional components of the dorsal roots of spinal nerves. Quart. Rev. Biol., 8:457–464.

Hogg, I. D. 1941. Sensory nerves and associated structures in the skin of human fetuses of 8 to 14 weeks of menstrual age correlated with functional capability. J. Comp. Neurol., 75:371–410.

Horwitz, M. T., and L. M. Tocantins. 1938. An anatomical study of the long thoracic nerve and the related scapular bursae in the pathogenesis of local paralysis of the serratus anterior muscle. Anat. Rec., 71:375–385.

Keegan, J. J., and F. D. Garrett. 1948. The segmental distribution of the cutaneous nerves in the limbs of man. Anat. Rec., 102:409–437.

Kelley, W. O. 1950. Phrenic nerve paralysis: Special consideration of the accessory phrenic nerve. J. Thoracic Surg., 19:923–928.

Kiss, F., and H. C. Ballon. 1929. Contribution to the nerve supply of the diaphragm. Anat. Rec., 41:285–298.

Kozlov, V. I. 1969. Formation and structure of the spinal nerve. Acta Anat., 73:321–350.

Langford, L. A., and R. E. Coggeshall. 1979. Branching of sensory axons in the dorsal root and evidence for the absence of dorsal root efferent fibers. J. Comp. Neurol., 184:193–204.

McKinniss, M. E. 1936. The number of ganglion cells in the dorsal root ganglia of the second and third cervical nerves in human fetuses of various ages. Anat. Rec., 65:255–259.

Merendino, K. A., R. J. Johnson, H. H. Skinner, and R. X. Maguire. 1956. Intradiaphragmatic distribution of the phrenic nerve with particular reference to placement of diaphragmatic incisions and controlled segmental paralysis. Surg. Gynecol. Obstet., 39:189–198.

Mesulam, M.-M., and T. M. Brushart. 1979. Transganglionic and anterograde transport of horseradish peroxidase across dorsal root ganglia: a tetramethylbenzidine method for tracing central sensory connections of muscles and peripheral nerves. Neuroscience, 4:1107–1117.

Moyer, E. K., and B. F. Kaliszewski. 1958. The number of nerve fibers in motor spinal nerve roots of young, mature and aged cats. Anat. Rec., 131:681–699.

Moyer, E. K., and D. L. Kimmel. 1948. The repair of severed motor and sensory spinal nerve roots by the arterial sleeve method of anastomosis. J. Comp. Neurol., 88:285–317.

Pallie, W., and J. K. Manuel. 1968. Intersegmental anastomoses between dorsal spinal rootlets in some vertebrates. Acta Anat., 70:341–351.

Pearson, A. A., and J. J. Bass. 1961. Cutaneous branches of the posterior primary rami of the cervical nerves. Anat. Rec., 139:263 (Abstract).

Pearson, A. A., R. W. Sauter, and J. J. Bass. 1963. Cutaneous branches of the dorsal primary rami of the cervical nerves. Amer. J. Anat., 112:169–180.

Pearson, A. A., R. W. Sauter, and T. F. Buckley. 1966.

Further observations on the cutaneous branches of the dorsal primary rami of the spinal nerves. Amer. J. Anat., 118:891–903.

Perera, H., and F. R. Edwards. 1957. Intradiaphragmatic course of the left phrenic nerve in relation to diaphragmatic incisions. Lancet, 2:75–77.

Roofe, P. G. 1940. Innervation of the annulus fibrosus and posterior longitudinal ligament. Arch. Neurol. Psychiat., 44:100–103.

Schlaepfer, K. 1926. A further note on the motor innervation of the diaphragm. Anat. Rec., 32:143–150.

Sindou, M., C. Quoex, and C. Baleydier. 1974. Fiber organization at the posterior spinal cord-rootlet junction in man. J. Comp. Neurol., 153:15–26.

Stilwell, D. L., Jr. 1956. The nerve supply of the vertebral column and its associated structures in the monkey. Anat. Rec., 125:139–169.

Stilwell, D. L., Jr. 1957. Regional variations in the innervation of deep fasciae and aponeuroses. Anat. Rec., 127:635–653.

Sunderland, S., and K. C. Bradley. 1949. The cross-sectional area of peripheral nerve trunks devoted to nerve fibers. Brain, 72:428–449.

Tarlov, I. M. 1937. Structure of the nerve root: I. Nature of the junction between the central and the peripheral nervous system. Arch. Neurol. Psychiat., 37:555–583.

Tarlov, I. M. 1937. Structure of the nerve root: II. Differentiation of sensory from motor roots; observations on identification in roots of mixed cranial nerves. Arch. Neurol. Psychiat., 37:1338–1355.

Thornton, M. W., and M. R. Sweisthal. 1969. The phrenic nerve; its terminal divisions and supply to the crura of the diaphragm. Anat. Rec., 164:283–290.

Wilson, A. S. 1968. Investigations on the innervation of the diaphragm in cats and rodents. Anat. Rec., 162:425–432.

Wilson, A. S. 1970. Experimental studies on the innervation of the diaphragm in cats. Anat. Rec., 168:537–547.

Yee, J., and K. B. Corbin. 1939. The intramedullary course of the upper five cervical dorsal root fibers in the rabbit. J. Comp. Neurol., 70:297–304.

B. Brachial Plexus and Nerves of Upper Limb

Beaton, L. E., and B. J. Anson. 1939. The relation of the median nerve to the pronator teres muscle. Anat. Rec., 75:23–26.

Blunt, M. J. 1959. The vascular anatomy of the median nerve in the forearm and hand. J. Anat. (Lond.), 93:15–22.

Capener, N. 1966. The vulnerability of the posterior interosseous nerve of the forearm. A case report and an anatomical study. J. Bone Jt. Surg., 48B:770–773.

Clausen, E. G. 1942. Postoperative "anesthetic" paralysis of the brachial plexus: A review of the literature and report of nine cases. Surgery, 12:933–942.

Davis, L., J. Martin, and G. Perret. 1947. The treatment of injuries of the brachial plexus. Ann. Surg., 125:647–657.

De Jong, R. N. 1943. Syndrome of involvement of posterior cord of the brachial plexus. Arch. Neurol. Psychiat., 49:860–862.

Farber, J. S., and R. S. Bryan. 1968. The anterior interosseous nerve syndrome. J. Bone Jt. Surg., 50A:521–523.

Fullerton, P. M. 1963. The effect of ischaemia on nerve

conduction in the carpal tunnel syndrome. J. Neurol. Neurosurg. Psychiat., 26:385-397.

Gardner, E. 1948. The innervation of the elbow joint. Anat. Rec., 102:161-174.

Gardner, E. 1948. The innervation of the shoulder joint. Anat. Rec., 102:1-18.

Gitlin, G. 1957. Concerning the gangliform enlargement ('Pseudoganglion') on the nerve to the teres minor muscle. J. Anat. (Lond.), 91:466-470.

Gore, D. R. 1971. Carpometacarpal dislocation producing compression of the deep branch of the ulnar nerve. J. Bone Jt. Surg., 53A:1387-1390.

Gray, D. J., and E. Gardner. 1965. The innervation of the joints of the wrist and hand. Anat. Rec., 151:261-266.

Harness, D., and E. Sekeles. 1971. The double anastomotic innervation of thenar muscles. J. Anat. (Lond.), 109:461-466.

Harris, W. 1939. *The Morphology of the Brachial Plexus.* Oxford University Press, London, 117 pp.

Ip, M. C., and K. S. F. Chang. 1968. A study on the radial supply of the brachialis muscle. Anat. Rec., 162:363-371.

Jamieson, R. W., and B. J. Anson. 1952. The relation of the median nerve to the heads of origin of the pronator teres muscle. Quart. Bull. Northwestern Univ. Med. Sch., 26:34-35.

Johnson, R. K., and M. M. Shrewsbury. 1970. Anatomical course of the thenar branch of the median nerve—usually in a separate tunnel through the transverse carpal ligament. J. Bone Jt. Surg., 52A:269-273.

Kasai, T. 1963. About the N. cutaneous brachii lateralis inferior. Amer. J. Anat., 112:305-309.

Keegan, J. J., and F. D. Garrett. 1948. The segmental distribution of the cutaneous nerves in the limbs of man. Anat. Rec., 102:407-437.

Kerr, A. T. 1918. The brachial plexus of nerves in man, the variations in its formation and branches. Amer. J. Anat., 23:285-395.

Latarjet, M., J. H. Neidhart, A. Morrin, and J. M. Autissier. 1967. L'entrée du nerf musculo-cutané dans le muscle coraco-brachial. Comp. Rend. Assn. Anat., 138:755-765.

Linell, E. A. 1921. The distribution of nerves in the upper limb, with reference to variabilities and their clinical significance. J. Anat. (Lond.), 55:79-112.

Linscheid, R. L. 1965. Injuries to the radial nerve at the wrist. Arch. Surg., 91:942-946.

Marble, H. C., E. Hamlin, Jr., and A. L. Watkins. 1942. Regeneration in the ulnar, median and radial nerves. Amer. J. Surg., 55:274-294.

Miller, M. R., H. J. Ralston, III, and M. Kasahara. 1958. The pattern of cutaneous innervation of the human hand. Amer. J. Anat., 102:183-217.

Miller, R. A. 1939. Observations upon the arrangement of the axillary artery and brachial plexus. Amer. J. Anat., 64:143-163.

Miller, R. A., and S. R. Detwiler. 1936. Comparative studies upon the origin and development of the brachial plexus. Anat. Rec., 65:273-292.

Millesi, H., G. Meissl, and A. Berger. 1976. Further experience with interfascicular grafting of the median, ulnar and radial nerves. J. Bone Jt. Surg., 58A:209-218.

Ogden, J. A. 1972. An unusual branch of the median nerve. J. Bone Jt. Surg., 54A:1779-1781.

Olson, I. A. 1968. The origin of the lateral cutaneous nerve of forearm and its anaesthesia for modified brachial plexus block. J. Anat. (Lond.), 105:381-382.

Papathanassion, B. T. 1968. A variant of the motor branch of the median nerve in the hand. J. Bone Jt. Surg., 50B:156-157.

Rowntree, T. 1949. Anomalous innervation of the hand muscles. J. Bone Jt. Surg., 31B:505-510.

Singer, E. 1933. Human brachial plexus united into a single cord: Description and interpretation. Anat. Rec., 55:411-419.

Spinner, M. 1970. The anterior interosseous nerve syndrome with special attention to its variations. J. Bone Jt. Surg., 52A:84-94.

Stilwell, D. L., Jr. 1957. The innervation of deep structures of the hand. Amer. J. Anat., 101:75-99.

Sunderland, S. 1945. The innervation of the flexor digitorum profundus and lumbrical muscles. Anat. Rec., 93:317-321.

Sunderland, S. 1945. The intraneural topography of the radial, median and ulnar nerves. Brain, 68:243-299.

Sunderland, S. 1946. The innervation of the first dorsal interosseous muscle of the hand. Anat. Rec., 95:7-10.

Sunderland, S. 1948. The distribution of sympathetic fibres in the brachial plexus in man. Brain, 71:88-102.

Sunderland, S., and G. M. Bedbrook. 1949. The relative sympathetic contribution to individual roots of the brachial plexus in man. Brain, 72:297-301.

Taleisnik, J. 1973. The palmar cutaneous branch of the median nerve and the approach to the carpal tunnel. An anatomical study. J. Bone Jt. Surg., 55A:1212-1217.

Tracey, J. F., and E. W. Brannon. 1958. Management of brachial plexus injuries (traction type). J. Bone Jt. Surg., 40A:1031-1042.

Wallace, W. A., and R. E. Coupland. 1975. Variations in the nerves of the thumb and index finger. J. Bone Jt. Surg., 57B:491-494.

Wallace, W. A., and P. A. M. Weston. 1974. The arrangement of the digital nerves within the human thumb and index fingers. J. Anat. (Lond.), 118:381-382.

Wood, V. E., and G. K. Frykman. 1978. Unusual branching of the median nerve at the wrist. A case report. J. Bone Jt. Surg., 60A:267-268.

C. Lumbosacral Plexus and Nerves of Lower Limb

Beaton, L. E., and B. J. Anson. 1937. The relation of the sciatic nerve and of its subdivisions to the piriformis muscle. Anat. Rec., 70:1-5.

Beaton, L. E., and B. J. Anson. 1938. The sciatic nerve and the piriformis muscle: their interrelation a possible cause of coccygodynia. J. Bone Jt. Surg., 20:686-688.

Bordeen, C. R., and A. W. Elting. 1901. A statistical study of the variations in the formation and position of the lumbo-sacral plexus in man. Anat. Anz., 19:124-135 and 209-238.

Champetier, J., and C. Descours. 1968. The branches of the posterior tibial nerve in the tibiotarsal joint. Comp. Rend. Assn. Anat., 141:677-685.

Coggeshall, R. C., J. D. Coulter, and W. D. Willis, Jr. 1974. Unmyelinated axons in the ventral roots of the cat lumbosacral enlargement. J. Comp. Neurol., 153:39-58.

Davies, F. 1935. A note on the first lumbar nerve. J. Anat. (Lond.), 70:177-178.

Day, M. H. 1964. The blood supply of the lumbar and sacral plexuses in the human foetus. J. Anat. (Lond.), 98:105-116.

Dempsher, J., and J. M. Sprague. 1948. A study of the distribution of sensory and motor fibers in the lumbosacral plexus and in the thoracic nerves of the cat. Anat. Rec., *102*:195–204.

Gardner, E. 1944. The distribution and termination of nerves in the knee joint of the cat. J. Comp. Neurol., *80*:11–32.

Gardner, E. 1948. The innervation of the knee joint. Anat. Rec., *101*:109–130.

Gardner, E. 1948. The innervation of the hip joint. Anat. Rec., *101*:353–371.

Gardner, E., and D. Gray. 1968. The innervation of the joints of the foot. Anat. Rec., *161*:141–148.

Garnjobst, W. 1964. Injuries to the saphenous nerve following operations for varicose veins. Surg. Gynecol. Obstet., *119*:359–361.

Gutrecht, J. A., and P. J. Dyck. 1970. Quantitative teased-fiber and histologic studies of human sural nerve during postnatal development. J. Comp. Neurol., *138*:117–130.

Harrison, V. F., and O. A. Mortensen. 1962. Identification and voluntary control of single motor unit activity in the tibialis anterior muscle. Anat. Rec., *144*:109–116.

Horwitz, M. T. 1939. The anatomy of the lumbosacral nerve plexus—its relation to variations of vertebral segmentation and the posterior sacral nerve plexus. Anat. Rec., *74*:91–107.

Huelke, D. F. 1957. A study of the formation of the sural nerve in adult man. Amer. J. Phys. Anthrop., *15*:137–147.

Huelke, D. F. 1958. The origin of the peroneal communicating nerve in adult man. Anat. Rec., *132*:81–92.

Ip, M. C., G. Vrbová, and D. R. Westbury. 1977. The sensory reinnervation of hind limb muscles of the cat following denervation and de-efferentation. Neuroscience, *2*:423–434.

Jamieson, R. W., L. L. Swigart, and B. J. Anson. 1952. Points of parietal perforation of the ilioinguinal and iliohypogastric nerves in relation to optimal sites for local anesthesia. Quart. Bull. Northwest. Univ. Med. Sch., *26*:22–26.

Kaiser, R. A. 1949. Obturator neurectomy for coxalgia. An anatomical study of the obturator and the accessory obturator nerve. J. Bone Jt. Surg., *31A*:815–819.

Keegan, J. J. 1947. Relations of nerve roots to abnormalities of lumbar and cervical portions of the spine. Arch. Surg., *55*:246–270.

Miller, M. R., and M. Kasahara. 1959. The pattern of cutaneous innervation of the human foot. Amer. J. Anat., *105*:233–255.

Ochoa, J. 1971. The sural nerve of the human fetus: electron microscope observations and count of axons. J. Anat. (Lond.), *108*:231–245.

Ochoa, J., and W. E. P. Mair. 1969. The normal sural nerve in man. I. Ultrastructure and numbers of fibres and cells. Acta Neuropathol., *13*:197–216.

Oelrich, T. M., and D. A. Moosman. 1977. The aberrant course of the cutaneous component of the ilioinguinal nerve. Anat. Rec., *189*:233–236.

Roberts, W. H. B., and W. H. Taylor. 1973. Inferior rectal nerve variations as it relates to pudendal block. Anat. Rec., *177*:461–464.

Stilwell, D. L., Jr. 1957. The innervation of deep structures of the human foot. Amer. J. Anat., *101*:59–73.

Sunderland, S. 1948. The intraneural topography of the sciatic nerve and its popliteal divisions in man. Brain, *71*:242–273.

Sunderland, S., and E. S. R. Hughes. 1946. Metrical and non-metrical features of the muscular branches of

the sciatic nerve and its medial and lateral popliteal divisions. J. Comp. Neurol., *85*:205–222.

Trotter, M. 1932. The relation of the sciatic nerve to the piriformis muscle. Anat. Rec., *52*:321–323.

Ueyama, T. 1978. The topography of root fibres within the sciatic nerve trunk of the dog. J. Anat. (Lond.), *127*:277–290.

Wertheimer, L. G. 1952. The sensory nerves of the hip joint. J. Bone Jt. Surg., *34A*:477–487.

Wilson, A. S., P. G. Legg, and J. C. McNeur. 1969. Studies on the innervation of the medial meniscus in the human knee joint. Anat. Rec., *165*:485–491.

Woodburne, R. T. 1960. The accessory obturator nerve and the innervation of the pectineus muscle. Anat. Rec., *136*:367–369.

D. Visceral Afferent and Autonomic Nerves (Sympathetic and Sacral Parasympathetic)

Alm, P., J. Alumets, E. Brodin, R. Håkanson, G. Nilsson, N.-O. Sjöberg, and F. Sundler. 1978. Peptidergic (substance P) nerves in the genito-urinary tract. Neuroscience, *3*:419–426.

Alm, P., J. Alumets, R. Håkanson, and F. Sundler. 1977. Peptidergic (vasoactive intestinal peptide) nerves in the genito-urinary tract. Neuroscience, *2*:751–754.

Ashley, F. L., and B. J. Anson. 1946. The pelvic autonomic nerves in the male. Surg. Gynecol. Obstet., *82*:598–608.

Axford, M. 1928. Some observations on the cervical sympathetic ganglia. J. Anat. (Lond.), *62*:301–318.

Barlow, C. M., and W. S. Root. 1949. The ocular sympathetic path between the superior cervical ganglion and the orbit in the cat. J. Comp. Neurol., *91*:195–208.

Becker, R. F., and J. A. Grunt. 1957. The cervical sympathetic ganglia. Anat. Rec., *127*:1–14.

Bojsen-Miller, F., and J. Tranum-Jensen. 1969. On nerves and nerve endings in the conducting system of the moderator band (septomarginal trabecula). J. Anat. (Lond.), *108*:387–395.

Chiba, T. 1978. Monoamine fluorescence and electron microscopic studies on small intensely fluorescent (granule-containing) cells in human sympathetic ganglia. J. Comp. Neurol., *179*:153–168.

Christensen, K., E. Lewis, and A. Kuntz. 1951. Innervation of the renal blood vessels in the cat. J. Comp. Neurol., *95*:373–385.

Christensen, K., E. H. Polley, and E. Lewis. 1952. The nerves along the vertebral artery and innervation of the blood vessels of the hindbrain of the cat. J. Comp. Neurol., *96*:71–91.

Cleveland, D. A. 1932. Afferent fibers in the cervical sympathetic trunk, superior cervical ganglion and internal carotid nerve. J. Comp. Neurol., *54*:35–43.

Coggeshall, R. E., and S. L. Galbraith. 1978. Categories of axons in mammalian rami communicantes, Part II. J. Comp. Neurol., *181*:349–360.

Coggeshall, R. E., M. B. Hancock, and M. L. Applebaum. 1976. Categories of axons in mammalian rami communicantes. J. Comp. Neurol., *167*:105–124.

Curtis, A. H., B. J. Anson, F. L. Ashley, and T. Jones. 1942. The anatomy of the pelvic autonomic nerves in relation to gynecology. Surg. Gynecol. Obstet., *75*:743–750.

Dass, R. 1952. Sympathetic components of the dorsal primary divisions of human spinal nerves. Anat. Rec., *113*:493–501.

Ebbesson, S. O. E. 1963. A quantitative study of human

superior cervical sympathetic ganglia. Anat. Rec., 146:353–356.

Edwards, E. A. 1951. Operative anatomy of the lumbar sympathetic chain. Angiology, 2:184–198.

Edwards, L. F., and R. C. Baker. 1940. Variations in the formation of the splanchnic nerves in man. Anat. Rec., 77:335–342.

Ehrlich, E., Jr., and W. F. Alexander. 1951. Surgical implications of upper thoracic independent sympathetic pathways. Arch. Surg., 62:609–614.

Elftman, A. G. 1943. The afferent and parasympathetic innervation of the lungs and trachea of the dog. Amer. J. Anat., 72:1–27.

Ellison, J. P., and T. H. Williams. 1969. Sympathetic nerve pathways to the human heart and their variations. Amer. J. Anat., 124:149–162.

Foley, J. O., and H. N. Schnitzlein. 1957. The contribution of individual thoracic spinal nerves to the upper cervical sympathetic trunk. J. Comp. Neurol., 108:109–120.

Forbes, H. S., and S. Cobb. 1938. Vasomotor control of cerebral vessels. Brain, 61:221–233.

Gardner, E. 1943. Surgical anatomy of the external carotid plexus. Arch. Surg., 46:238–244.

Gardner, E., and R. O'Rahilly. 1976. The nerve supply and conducting system of the human heart at the end of the embryonic period proper. J. Anat. (Lond.), 121:571–587.

Gernandt, B., and Y. Zotterman. 1946. The splanchnic efferent outflow of impulses in the light of ergotamine action. Acta Physiol. Scand., 11:301–317.

Gray, D. J. 1947. The intrinsic nerves in the testis. Anat. Rec., 98:325–335.

Harris, A. J. 1943. An experimental analysis of the inferior mesenteric plexus. J. Comp. Neurol., 79:1–17.

Hinsey, J. C., K. Hare, and G. A. Wolf. 1942. Structure of the cervical sympathetic chain in the cat. Anat. Rec., 82:175–183.

Jamieson, R. W., D. B. Smith, and B. J. Anson. 1952. The cervical sympathetic ganglia. Quart. Bull. Northwest. Univ. Med. Sch., 26:219–227.

Jonnesco, T. 1923. Le Sympathique Cervico-thoracique. Masson, Paris, 91 pp.

Kirgis, H. D., and A. Kuntz. 1942. Inconstant sympathetic neural pathways; their relation to sympathetic denervation of upper extremity. Arch. Surg., 44:95–102.

Kisner, W. H., and H. Mahorner. 1952. An evaluation of lumbar sympathectomy. Amer. Surg., 18:30–35.

Kuntz, A. 1953. The Autonomic Nervous System. 4th ed. Lea & Febiger, Philadelphia, 605 pp.

Kuntz, A., and W. F. Alexander. 1950. Surgical implications of lower thoracic and lumbar independent sympathetic pathways. Arch. Surg., 61:1007–1018.

Kuntz, A., H. H. Hoffman, and E. M. Schaeffer. 1957. Fiber components of the splanchnic nerves. Anat. Rec., 128:139–146.

Kuntz, A., and A. Morehouse. 1930. Thoracic sympathetic cardiac nerves in man. Their relation to cervical sympathetic ganglionectomy. Arch. Surg., 20:607–613.

Kuntz, A., and R. E. Morris, Jr. 1946. Components and distribution of the spermatic nerves and the nerves of the vas deferens. J. Comp. Neurol., 85:33–44.

Langworthy, O. R. 1965. Commissures of autonomic fibers in the pelvic organs of rats. Anat. Rec., 151:583–587.

Lannon, J., and E. Weller. 1947. The parasympathetic supply of the distal colon. Brit. J. Surg., 34:373–378.

Larsell, O., and R. A. Fenton. 1936. Sympathetic innervation of the nose: Research report. Arch. Otolaryngol., 24:687–695.

Lever, J. D. 1976. Studies on sympathetic ganglia. J. Anat. (Lond.), 121:430–431.

Mitchell, G. A. G. 1938. The innervation of the ovary, uterine tube, testis and epididymis. J. Anat. (Lond.), 72:508–517.

Mitchell, G. A. G. 1951. The intrinsic renal nerves. Acta Anat., 13:1–15.

Mitchell, G. A. G. 1953. Anatomy of the Autonomic Nervous System. Williams & Wilkins Co., Baltimore, 356 pp.

Mizeres, N. J. 1963. The cardiac plexus in man. Amer. J. Anat., 112:141–151.

Mustalish, A. C., and J. Pick. 1964. On the innervation of the blood vessels in the human foot. Anat. Rec., 149:587–590.

Nadelhaft, I., W. C. Degroat, and C. Morgan. 1980. Location and morphology of parasympathetic preganglionic neurons in the sacral spinal cord of the cat revealed by retrograde axonal transport. J. Comp. Neurol., 193:265–282.

Nonidez, J. F. 1939. Studies on the innervation of the heart. I. Distribution of the cardiac nerves, with special reference to the identification of the sympathetic and parasympathetic postganglionics. Amer. J. Anat., 65:361–413.

Nonidez, J. F. 1941. Studies on the innervation of the heart. II. Afferent nerve endings in the large arteries and veins. Amer. J. Anat., 68:151–189.

Pearson, A. A. 1952. The connections of the sympathetic trunk in the cervical and upper thoracic levels in the human fetus. Anat. Rec., 106:231.

Pearson, A. A., and A. L. Eckhardt. 1960. Observations on the gray and white rami communicantes in human embryos. Anat. Rec., 138:115–127.

Pearson, A. A., and R. W. Sauter. 1970. Nerve contributions to the pelvic plexus and the umbilical cord. Amer. J. Anat., 128:485–498.

Perlow, S., and K. L. Vehe. 1935. Variations in the gross anatomy of the stellate and lumbar sympathetic ganglia. Amer. J. Surg., 30:454–458.

Pick, J., and D. Sheehan. 1946. Sympathetic rami in man. J. Anat. (Lond.), 80:12–20.

Randall, W. C., J. A. Armour, D. C. Randall, and O. A. Smith. 1971. Functional anatomy of the cardiac nerves in the baboon. Anat. Rec., 170:183–198.

Ray, B. S., J. C. Hinsey, and W. A. Geohegan. 1943. Observations on the distribution of the sympathetic nerves to the pupil and upper extremity as determined by stimulation of the anterior roots in man. Ann. Surg., 118:647–655.

Reed, A. F. 1951. The origins of the splanchnic nerves. Anat. Rec., 109:341.

Richardson, K. C. 1960. Studies on the structure of autonomic nerves in the small intestine. J. Anat. (Lond.), 94:457–472.

Richardson, K. C. 1962. The fine structure of autonomic nerve endings in smooth muscle of the rat vas deferens. J. Anat. (Lond.), 96:427–442.

Rubin, E., and D. Purves. 1980. Segmental organization of sympathetic preganglionic neurons in the mammalian spinal cord. J. Comp. Neurol., 192:163–174.

Saccomanno, G. 1943. The components of the upper thoracic sympathetic nerves. J. Comp. Neurol., 79:355–378.

Shashirina, M. I. 1964. Structural organization of superior cervical sympathetic ganglion of the cat. Fed. Proc., 23:T711–T714.

Shimazawa, A. 1972. Quantitative studies of the deep petrosal nerve of the mouse with the electron microscope. Anat. Rec., 172:483–488.

Smith, R. B. 1971. The occurrence and location of intrinsic cardiac ganglia and nerve plexuses in the human neonate. Anat. Rec., 169:33–40.

Southam, J. A. 1959. The inferior mesenteric ganglion. J. Anat. (Lond.), 93:304–308.

Stotler, W. A., and R. A. McMahon. 1947. Innervation and structure of conductive system of human heart. J. Comp. Neurol., 87:57–83.

Sulkin, N. M., and A. Kuntz. 1950. A histochemical study of the autonomic ganglia of the cat following prolonged preganglionic stimulation. Anat. Rec., 108:255–277.

Swinyard, C. A. 1937. The innervation of the suprarenal glands. Anat. Rec., 68:417–429.

Thompson, S. A., and J. A. Gosling. 1977. Morphological and histochemical observations on human fetal and postnatal pelvic autonomic ganglia. J. Anat. (Lond.), 124:259–260.

Trumble, H. C. 1934. The parasympathetic nerve supply to the distal colon. Med. J. Australia, II:149–151.

Webber, R. H. 1958. A contribution on the sympathetic nerves in the lumbar region. Anat. Rec., 130:581–604.

Woodburne, R. T. 1956. The sacral parasympathetic innervation of the colon. Anat. Rec., 124:67–76.

Woollard, H. H., and R. E. Norrish. 1933. Anatomy of the peripheral sympathetic nervous system. Brit. J. Surg., 21:83–103.

Wrete, M. 1959. The anatomy of the sympathetic trunks in man. J. Anat. (Lond.), 93:448–459.

13

Organs

of the

Special

Senses

Endings of the sensory nerves of the peripheral nervous system are found in end organs which may be divided into two large groups: (1) the organs of the special senses of smell, taste, sight, and hearing, and (2) the sensory endings of general sensation of heat, cold, touch, pain, pressure, etc. This chapter considers only the organs of the special senses.

Organ of Smell

The sensory endings for the sense of smell are located in the nose, and for this reason the entire nose has, by long tradition, been described in this chapter. In this edition, however, all but the olfactory area are described in Chapter 15, the Respiratory System, because the nose and nasal cavity are more important for the passage of air than for the sense of smell and they are customarily referred to by clinicians as the upper respiratory tract.

The **olfactory region** is located in the most superior part of both nasal fossae and occupies the mucous membrane covering the superior nasal concha and the septum opposite (Fig. 13-1). Thus it is confined to an area of the fossa whose walls are formed by the ethmoid bone.

The olfactory sensory endings are the least specialized of the special senses. They are modified epithelial cells liberally scattered within the columnar epithelium of the mucous membrane. The sensory cells are known as **olfactory cells** and the other epithelial cells, as **supporting cells.** Although the epithelium appears pseudostratified, the supporting cells are not ciliated and there are no goblet cells (Fig. 13-2).

The **olfactory cells** are bipolar, with a small amount of cytoplasm surrounding a large spherical nucleus. The slender peripheral or superficial process extends to the surface of the epithelial membrane and sends out a tuft of fine processes known as **olfactory hairs.** The central or deep process penetrates the basement membrane and in the underlying connective tissue joins neighboring processes to form the bundles of unmyelinated fibers of the **olfactory nerves** (Fig. 13-3). Mingled with the nerve bundles in the subepithelial tissue are numerous branched tubular glands of a serous secreting type (glands of Bowman[1]), which keep the membrane protected with moisture.

The bundles of nerve fibers form a plexus in the submucosa and finally collect into about 20 nerves that pass through the openings in the cribriform plate of the ethmoid bone as the **fila olfactoria.** The nerve fibers end by forming synapses with processes of the mitral cells in the glomeruli of the olfactory bulb (Fig. 13-3; also see page 1050).

Organ of Taste

The peripheral taste organs (*gustatory organs*) are the **taste buds** (*gustatory caliculi*) distributed over the tongue and occasionally on adjacent parts. They are spherical or ovoid nests of cells embedded in the strati-

---

[1]Sir William Bowman (1816–1892): An English anatomist, ophthalmologist, and physiologist (London).

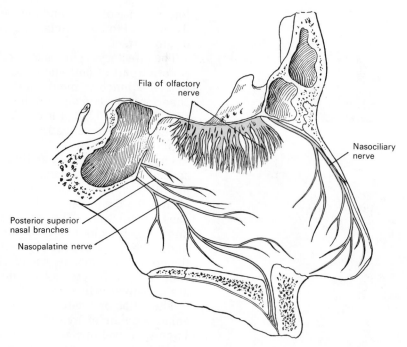

FIG. 13-1. The nerves of the right side of the septum of the nose.

fied squamous epithelium (Fig. 13-4), and are present in large numbers on the sides of the vallate papillae and to a lesser extent on the opposed walls of the fossae around them (Fig. 13-5). They are also found over the sides and back of the tongue, especially on the fungiform papillae (Fig. 16-36). They are plentiful over the lingual fimbria and are occasionally present on the oral surface of the soft palate and on the posterior surface of the epiglottis.

Each **taste bud** occupies an ovoid pocket that extends through the thickness of the epithelium. It has two openings—one at the surface and the other at the basement membrane. There are two types of cells within the bud: **gustatory cells** and **supporting cells** (Fig. 13-4). The supporting cells are elongated and extend between the basement membrane and the surface; they form an outer shell for the taste bud, arranged like the staves of a wooden cask. Supporting cells are also scattered through the bud between the gustatory cells. The gustatory cells occupy the interior portion of the bud; they are spindle-shaped and have a large round

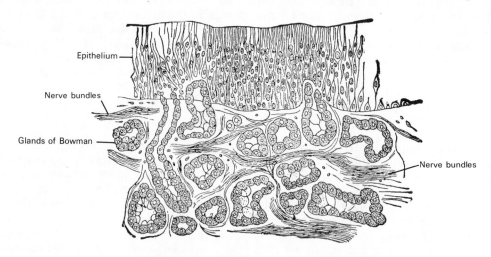

FIG. 13-2. Section of the olfactory mucous membrane (Cadiat).

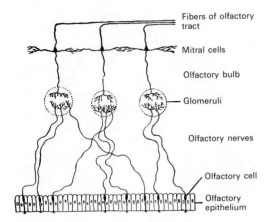

FIG. 13-3. Plan of olfactory neurons.

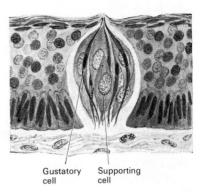

Gustatory cell Supporting cell

FIG. 13-4. Section through a taste bud. Semidiagrammatic 450 ×.

central nucleus. At the peripheral end of each gustatory cell protrudes a delicate hair-like process, the **gustatory hair,** through an opening at the surface, the **gustatory pore.** The central end of the gustatory cell does not end in an axon, as in the case of the olfactory cell, but remains within the taste bud, where it is in intimate contact with many fine terminations of nerves that pass into the taste bud through an opening in the basement membrane. The nerves are myelinated until they reach the taste buds but lose their sheaths as they enter the bud.

NERVES. The posterior third of the tongue, including the taste buds on the vallate papillae, is supplied by the glossopharyngeal nerve. The fibers for

taste to the anterior two-thirds leave the brain in the nervus intermedius, continue in the chorda tympani, and reach the tongue in the branches of the lingual nerve. It is believed that taste buds on the epiglottis are supplied by the vagus nerve.

Organ of Sight: The Eye

The **bulb of the eye,** or **organ of sight,** is contained in a bony cavity, the orbit, which is composed of parts of the frontal, maxillary, zygomatic, sphenoid, ethmoid, lacrimal, and palatine bones. The bulb is embedded in the orbital fat but separated from it by a thin membranous sac, the fascia bulbi.

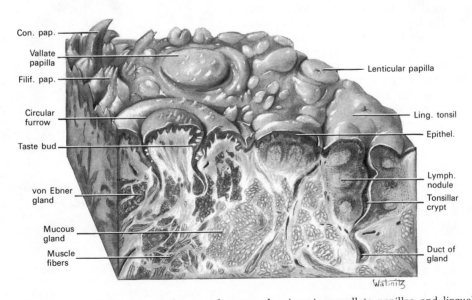

FIG. 13-5. Section of posterior part of dorsum of tongue showing circumvallate papillae and lingual tonsil. (Redrawn from Braus.)

Associated with it are certain accessory structures—the muscles, fasciae, eyebrows, eyelids, conjunctiva, and lacrimal apparatus.

DEVELOPMENT

The retina and optic nerve develop from the forebrain, the lens, from the overlying ectoderm, and the accessory structures, from the mesenchyme. The eyes begin to develop as a pair of diverticula from the lateral aspects of the forebrain. These diverticula make their appearance before the closure of the anterior end of the neural tube; after closure of the tube they are known as the **optic vesicles.** They project toward the sides of the head, and the peripheral part of each vesicle expands to form a hollow bulb, while the proximal part remains narrow and constitutes the **optic stalk.** When the peripheral part of the optic vesicle comes in contact with the overlying ectoderm, the latter thickens, invaginates, becomes severed from the ectoderm, and forms the **lens vesicle.** At the same time the optic vesicle partially encircles the lens vesicle, forming the **optic cup.** Its two layers are continuous with each other at the cup margin, which ultimately overlaps the front of the lens except for the future aperture of the pupil (Figs. 13-6 to 13-9). The process of invagination also causes an infolding of the posterior surface of the vesicle and the optic stalk, producing the **choroidal fissure** (Fig. 13-10). Mesenchyme and the retinal blood vessels (hyaloid artery) grow into the fissure. The fissure closes during the seventh week and the two-layered optic cup and optic stalk become complete. Sometimes the choroidal fissure persists, and when this occurs the choroid and iris in the region of the fissure remain undeveloped, giving rise to the condition known as *coloboma* of the choroid or iris.

RETINA. The retina is developed from the optic cup. The outer stratum of the cup persists as a single layer of cells which assume a columnar shape, acquire pigment, and form the pigmented layer of the retina; the pigment first appears in the cells near the edge of the cup. The cells of the inner stratum proliferate and form a layer of considerable thickness from which the nervous elements and the sustentacular fibers of the retina are developed. In the portion of the cup that overlaps the lens, the inner stratum is not differentiated into nervous elements, but

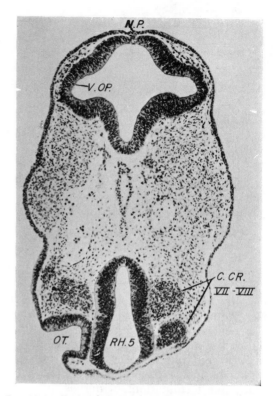

FIG. 13-6.   Brain of 3.9-mm human embryo, 22–23 somites. *C.CR.,* neural crest of nerves VII and VIII; *N.P.,* neuropore; *OT.,* otic vesicle; *RH,* rhombencephalon; *V.OP.,* optic vesicle. (Carnegie embryo No. 8943. Bartelmez and Dekeban, Contribution to Embryology, Vol. 37, 1962.

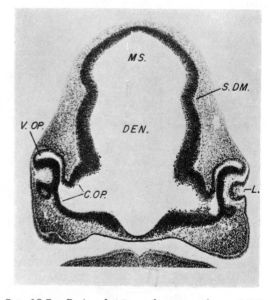

FIG. 13-7.   Brain of 6.7-mm human embryo. *C.OP.,* conus opticus; *DEN.,* diencephalon; *L.,* lens; *MS.,* mesencephalon; *S.DM.,* sulcus di-mesencephalicus; *V.OP.,* optic vesicle. (Carnegie Embryo No. 6502. Bartelmez and Dekaban, 1962.

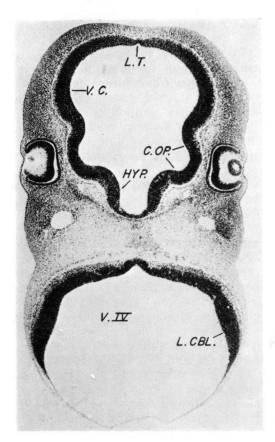

FIG. 13-8.   Brain of 7.5-mm human embryo. *C.OP.,* conus opticus; *HYP.,* hypophysis; *L.CBL.,* lamina cerebellaris; *L.T.,* lamina terminalis; *V.C.,* cerebral vesicle; *V.IV.,* fourth ventricle. (Carnegie Embryo No. 6506. Bartelmez and Dekeban, 1962.)

forms a layer of columnar cells that is applied to the pigmented layer; these two strata form the **pars ciliaris** and **pars iridica retinae.**

The cells of the inner or retinal layer of the optic cup become differentiated into spongioblasts and germinal cells, and the latter by their subdivisions give rise to neuroblasts. From the spongioblasts are formed the sustentacular fibers of Müller,[1] the outer and inner limiting membranes, and the groundwork of the molecular layers of the retina. The neuroblasts form the ganglionic and nuclear layers. The layer of rods and cones first develops in the central part of the optic cup and from there gradually extends toward the cup margin. All the layers of the retina are completed by the eighth month of fetal life.

OPTIC NERVE.   The optic stalk is converted into the optic nerve by the obliteration of its cavity and the growth of nerve fibers into it. Most of these fibers are centripetal and grow backward into the optic stalk from the nerve cells of the retina, but a few extend in the opposite direction and are derived from nerve cells in the brain. The fibers of the optic nerve receive their myelin sheaths about the tenth week after birth. The **optic chiasma** is formed by the meeting and partial decussation of the fibers of the two optic nerves. Behind the chiasma, the fibers grow

[1]Heinrich Müller (1820-1864): A German anatomist (Wurzburg).

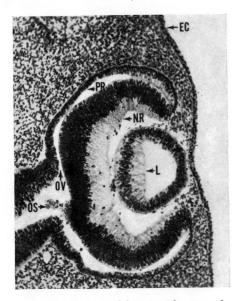

FIG. 13-9.   Optic cup and lens vesicle, 8-mm human embryo. *EC,* ectoderm; *L,* lens; *NR,* neural part of retina; *OS,* optic stalk; *OV,* optic vesicle; *PR,* pigment layer of retina. (Courtesy of Doctor Aeleta Barber.)

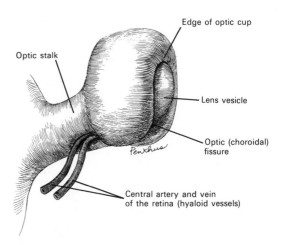

FIG. 13-10.   The optic cup and lens vesicle in a human embryo during the fifth week. Note the optic fissure and the developing central vessels of the retina (hyaloid vessels).

backward as the optic tracts to the thalamus and midbrain.

CRYSTALLINE LENS.   The crystalline lens is developed from the lens vesicle, which recedes within the margin of the cup and becomes separated by mesoderm from the overlying ectoderm. The cells forming the posterior wall of the vesicle lengthen and are converted into the lens fibers, which grow forward and fill the cavity of the vesicle (Fig. 13-11). The cells forming the anterior wall retain their cellular character and form the epithelium on the anterior surface of the adult lens.

HYALOID ARTERY.   A capillary net continuous with the primitive choroid net enters the choroid fissure. As the fissure closes, connections with the choroid net are cut off except at the edge of the cup. The vessel enclosed in the optic stalk is the hyaloid artery. Its branches surround the deep surface of the lens and drain into the choroid net at the margin of the cup. As the vitreous body increases, the hyaloid supplies branches to it. By the second month, the lens is invested by a vascular mesodermal capsule, the **capsula vasculosa lentis.** The blood vessels supplying the posterior part of this capsule are derived from the hyaloid artery; those for the

anterior part, from the anterior ciliary arteries. The portion of the capsule that covers the front of the lens is named the **pupillary membrane.** By the sixth month all the vessels of the capsule have atrophied except the hyaloid artery, which disappears during the ninth month; the position of this artery is indicated in the adult by the hyaloid canal, which extends from the optic disc to the posterior surface of the lens. With the loss of its blood vessels, the capsula vasculosa lentis disappears, but sometimes the pupillary membrane persists at birth, giving rise to the condition termed *congenital atresia of the pupil.*

CENTRAL ARTERY OF THE RETINA.   By the fourth month, branches of the hyaloid artery and veins that have developed during the third month begin to spread out in the retina and reach the ora serrata by the eighth month. After atrophy of the vitreous part of the hyaloid vessels, the proximal part in the optic nerve and retina becomes the central artery of the retina.

CHOROID.   The choroid is analogous to the leptomeninges of the brain and spinal cord. It develops from mesenchyme between the sclera (dura) and the optic cup (an extension of the brain wall). The mesenchyme is in-

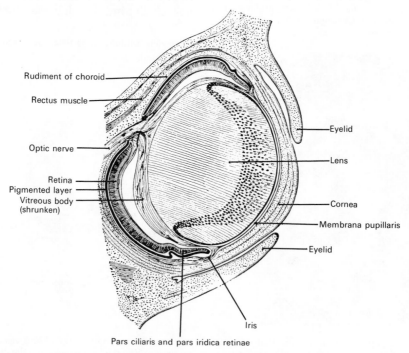

Rudiment of choroid
Rectus muscle
Optic nerve
Retina
Pigmented layer
Vitreous body (shrunken)
Pars ciliaris and pars iridica retinae
Iris
Eyelid
Lens
Cornea
Membrana pupillaris
Eyelid

FIG. 13-11.   Horizontal section through the eye of an 18-day-old rabbit embryo. 30 ×. (Kölliker.)

vaded by capillaries from the ciliary vessels. They form a rich plexus over the outer surface of the optic cup and connect with the hyaloid capillary plexus in the pupillary region. The mesenchyme forms a loose network of fibroblasts, pigmented cells, and collagenous fibrils, partially separated by mesothelial-lined, fluid-containing spaces.

CANAL OF SCHLEMM. The canal of Schlemm[1] (sinus venosus sclerae) and the circular vessels of the iris develop from the anterior extension of the plexus of the choroid. The former is analogous to a venous sinus of the dura mater; the anterior chamber is analogous to the subarachnoid spaces. The drainage of aqueous humor by way of the pectinate villi into the canal of Schlemm is analogous to the drainage of the subarachnoid fluid by way of the arachnoid granulations into the dural sinuses.

[1]Friedrich S. Schlemm (1795–1858): A German anatomist (Berlin).

VITREOUS BODY. The vitreous body (Fig. 13-12) develops between the lens and the optic cup as the two structures become separated. Some authors believe that initially the retina and perhaps the lens as well play roles in the formation of the vitreous body; others believe that the retina plays the sole role; and still others believe that both retina and invading mesenchyme are involved. Fixed and stained preparations show throughout the vitreous body a delicate network of fibrils continuous with the long processes of stellate mesenchymal cells and with the retina. Later the fibrils are limited to the ciliary region, where they are supposed to form the **zonula ciliaris.**

SCLERA. The sclera (Fig. 13-12) is derived from the mesenchyme surrounding the optic cup. Fibroblasts form a dense layer of collagenous fibers continuous with the sheath of the optic nerve. The sclera is analogous to the dura mater.

CORNEA. Most of the cornea (Fig. 13-12) is

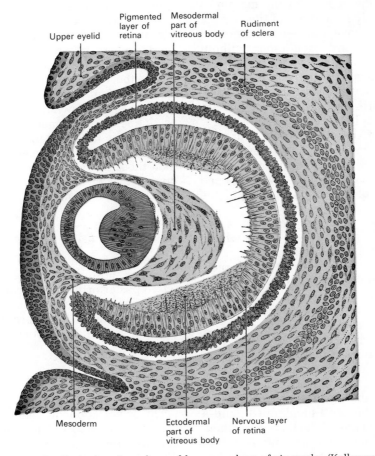

Upper eyelid — Pigmented layer of retina — Mesodermal part of vitreous body — Rudiment of sclera

Mesoderm — Ectodermal part of vitreous body — Nervous layer of retina

FIG. 13-12. Sagittal section of eye of human embryo of six weeks. (Kollmann.)

derived from mesenchyme that invades the region between the lens and ectoderm. The overlying ectoderm becomes the corneal epithelium. The endothelial (mesothelial) layer comes from mesenchymal cells that line the corneal side of the cleft (anterior chamber), which develops between the cornea and pupillary membrane. The factors responsible for the transparency of the cornea are unknown.

ANTERIOR CHAMBER.   The anterior chamber of the eye appears as a cleft in the mesoderm separating the lens from the overlying ectoderm. The layer of mesoderm anterior to the cleft forms the substantia propria of the cornea, that posterior to the cleft forms the stroma of the iris and the pupillary membrane.

PUPILLARY MEMBRANE.   In the fetus, the pupil is closed by a delicate vascular membrane, the pupillary membrane, which divides the space in which the iris is suspended into two distinct chambers. The vessels of this membrane are derived partly from those of the capsule of the margin of the iris and partly from those of the capsule of the lens; they have a looped arrangement and converge toward each other without anastomosing. About the sixth month, the membrane begins to disappear by absorption from the center toward the circumference, and at birth only a few fragments are present; in exceptional cases it persists.

The fibers of the **ciliary muscle** are derived from the mesoderm, but those of the sphincter and dilatator pupillary muscles are of ectodermal origin, being developed from the cells of the pupillary part of the optic cup.

EYELIDS.   The eyelids are formed as small cutaneous folds (Figs. 13-11; 13-12) which, about the middle of the third month, come together and unite in front of the cornea. They remain united until about the end of the sixth month.

LACRIMAL APPARATUS.   The **lacrimal sac** and **nasolacrimal duct** develop from a thickening of the ectoderm in the groove known as the **nasoöptic furrow,** between the lateral nasal and maxillary processes. This thickening forms a solid cord of cells that sinks into the mesoderm; during the third month the central cells of the cord break down, and a lumen, the nasolacrimal duct, is established. The lacrimal ducts arise as buds from the upper part of the cord of cells and secondarily establish openings (*puncta lacrimalia*) on the margins of the lids. The **epithelium** of the cornea and conjunctiva, and that which lines the ducts and alveoli of the lacrimal gland, is of ectodermal origin, as are also the **eyelashes** and the lining cells of the glands that open on the lid margins.

BULB OF THE EYE

The bulb of the eye is composed of segments of two spheres of different sizes. The transparent anterior segment, the cornea, is from a small sphere and forms about one-sixth of the bulb. The opaque posterior segment is from a larger sphere and forms about five-sixths of the bulb. The term **anterior pole** is applied to the central point of the anterior curvature of the bulb, and **posterior pole,** to the central point of its posterior curvature; a line joining the two poles forms the **optic axis.** The axes of the two bulbs are nearly parallel, and therefore do not correspond to the axes of the orbits, which diverge lateralward. The optic nerves follow the direction of the axes of the orbits and therefore are not parallel; each leaves its eyeball 3 mm to the nasal side and a little below the level of the posterior pole. The transverse and anteroposterior diameters of the bulb are somewhat greater than its vertical diameter; the former amounts to about 24 mm and the latter, to about 23.5 mm. In the female, all three diameters are rather less than those in the male. The anteroposterior diameter at birth is about 17.5 mm, and at puberty, 20 to 21 mm.

Tunics of the Eye

The bulb is composed of three tunics: (1) an outer fibrous tunic, consisting of the **sclera** posteriorly and the **cornea** anteriorly; (2) an intermediate vascular, pigmented tunic, comprising the **choroid, ciliary body, and iris;** and (3) an internal neural tunic, the **retina** (Fig. 13-13).

FIBROUS TUNIC

The sclera is opaque, and constitutes the posterior five-sixths of the tunic; the cornea is transparent, and forms the anterior sixth (Fig. 13-13).

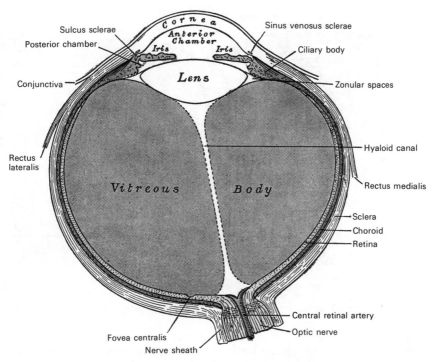

Fig. 13-13. Horizontal section of the eyeball.

SCLERA. The sclera received its name from its density and hardness; it is a tough, inelastic membrane that maintains the size and form of the bulb. Its thicker posterior part measures 1 mm. Its *external surface* is white and smooth, except at the points at which the recti and obliqui muscles are inserted. Its anterior part is covered by the conjunctival membrane; the rest of the surface is separated from the fascia bulbi (capsule of Tenon[1]) by the loose connective tissue of a fascial cleft (see p. 1307), which permits rotation of the eyeball within the fascia.

Its **inner surface** is brown and marked by grooves, in which the ciliary nerves and vessels are lodged; it is loosely attached to the pigmented suprachoroid lamina of the choroid. Posteriorly it is pierced by the optic nerve, and is continuous through the fibrous sheath of this nerve with the dura mater.

The optic nerve passes through the **lamina cribrosa sclerae** (Fig. 13-14), in which minute orifices transmit the neural filaments, and the fibrous septa dividing them from one another are continuous with those that separate the bundles of nerve fibers. One of these openings, larger than the rest and occupying the center of the lamina, transmits the central artery and vein of the retina. Around the exit of the optic nerve are numerous small apertures for the transmission of the ciliary vessels and nerves, and about midway between the cribrosa and the sclerocorneal junction are four or five large apertures from the transmission of the **vorticose veins.**

The sclera is directly continuous with the cornea anteriorly, the line of union being termed the **sclerocorneal junction.** Near the junction, the inner surface of the sclera projects into a circular ridge called the **scleral spur,** to which the ciliary muscle and the iris are attached. Anterior to this ridge is a circular depression, the scleral sulcus, which is crossed by trabecular tissue that separates the angle of the anterior chamber from the canal of Schlemm. The spaces of the trabecular tissue (spaces of Fontana[2]) connect on one side with the anterior chamber of the eye and on the other with the pectinate villi.

[1] Jacques René Tenon (1724–1816): A French surgeon and pathologist (Paris).

[2] Felice Fontana (1720–1805): An Italian physiologist and philosopher (Pisa and Florence).

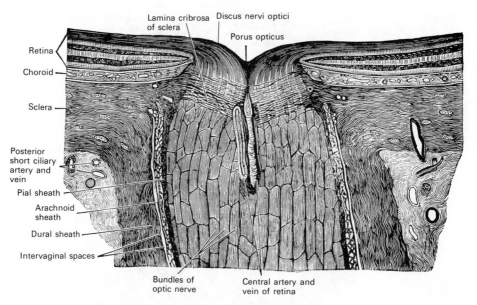

Retina

Choroid

Sclera

Posterior
short ciliary
artery and
vein

Pial sheath

Arachnoid
sheath

Dural sheath

Intervaginal spaces

Lamina cribrosa
of sclera

Discus nervi optici

Porus opticus

Bundles of
optic nerve

Central artery and
vein of retina

FIG. 13-14.   The terminal portion of the optic nerve as it traverses the lamina cribrosa of the sclera.

The aqueous humor filters through the walls of the villi into the canal of Schlemm.

CORNEA.   The cornea, the transparent part of the external tunic, is almost circular in outline, occasionally a little broader in the tranverse than in the vertical direction. It is convex anteriorly and projects like a dome beyond the sclera. Its degree of curvature varies in different individuals, and in the same individual at different periods of life, being more pronounced in youth than in advanced life. The cornea is dense and of uniform thickness throughout; its posterior circumference is perfectly circular in outline, and exceeds the anterior slightly in diameter. Immediately anterior to the sclerocorneal junction the cornea bulges inward as a thickened rim.

VASCULAR TUNIC

The vascular tunic (*uvea*) (Figs. 13-15; 13-16; 13-17) of the eye is composed of the choroid, the ciliary body, and the iris.

The choroid invests the posterior five-sixths of the bulb and extends as far anteriorly as the ora serrata of the retina. The ciliary body connects the choroid to the circumference of the iris. The iris is a circular diaphragm behind the cornea and presents near its center a rounded aperture, the **pupil.**

CHOROID.   The choroid is a thin, highly vascular membrane, dark brown or chocolate-colored, investing the posterior five-sixths of the bulb; it is pierced posteriorly by the optic nerve, and in this location is firmly adherent to the sclera. Its outer surface is loosely connected by the lamina suprachoroidea with the sclera; its inner surface is attached to the pigmented layer of the retina.

CILIARY BODY.   The ciliary body extends from the ora serrata of the retina to the outer edge of the iris and the sclerocorneal junction. It consists of the thickened vascular tunic of the eye and the ciliary muscle. Its

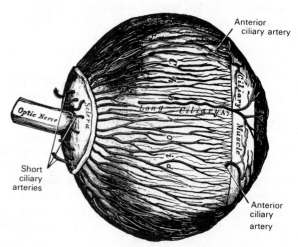

Anterior
ciliary artery

Short
ciliary
arteries

Anterior
ciliary
artery

FIG. 13-15.   The arteries of the choroid and iris. The greater part of the sclera has been removed. (Enlarged.)

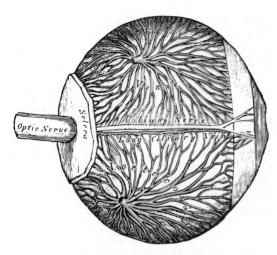

FIG. 13-16. The veins of the choroid. (Enlarged.)

inner surface is covered by the thin, pigmented **ciliary part of the retina.** The suspensory ligament of the lens is attached to the ciliary body. The ciliary body comprises two zones, the orbiculus ciliaris and the ciliary processes (Fig. 13-18).

The **orbiculus ciliaris** is 4 mm wide and extends from the ora serrata to the ciliary processes. Its thickness increases as it approaches the ciliary processes owing to the increase in thickness of the ciliary muscle. The choroid layer is thicker here than over the optical part of the retina. Its inner surface presents numerous small, radially arranged ridges, the ciliary processes.

The **ciliary processes** are formed by the inward folding of the various layers of the choroid, i.e., the choroid proper and the lamina basalis; the layers are received between corresponding foldings of the suspensory ligament of the lens. The ciliary processes are arranged radially on the posterior surface of the iris, forming a sort of frill around the margin of the lens (Fig. 13-18). Their numbers vary from 60 to 80; the larger ones are about 2.5 mm long, and the smaller ones, consisting of about one-third of the entire number, are situated in spaces between the larger processes, but with no regular arrangement. They are attached by their periphery to three or four ridges of the orbiculus ciliaris and are continuous with the layers of the choroid. Their other extremities are free and rounded, and are directed toward the circumference of the lens. Anteriorly, they are continuous with the periphery of the iris. Their posterior surfaces are connected with the suspensory ligament of the lens.

The **ciliaris muscle** (*ciliary muscle*) consists of unstriated fibers. It forms a grayish, semitransparent, circular band, about 3 mm broad, on the outer surface of the anterior part of the choroid. It is thickest anteriorly and consists of two sets of fibers, **meridional**

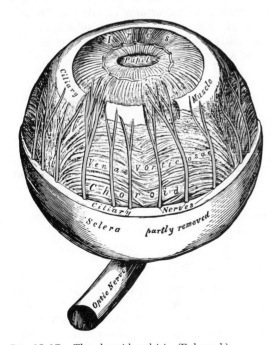

FIG. 13-17. The choroid and iris. (Enlarged.)

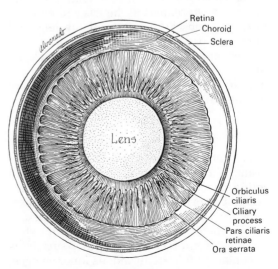

FIG. 13-18. Interior of anterior half of bulb of eye.

and **circular.** The meridional fibers, much more numerous than the circular fibers, arise from the posterior margin of the scleral spur; they run posteriorly and are attached to the ciliary processes and orbiculus ciliaris. The circular fibers are internal to the meridional ones, and in a meridional section they appear as a triangular zone behind the filtration angle and close to the circumference of the iris. They are well developed in hypermetropic eyes, but are rudimentary or absent in myopic eyes. The ciliaris muscle is the chief agent in accommodation, i.e., in adjusting the eye to the vision of near objects. When it contracts, it draws the ciliary processes centripetally, relaxes the suspensory ligament of the lens, and thus allows the lens to become more convex.

IRIS. The iris has received its name from its various colors in different individuals. It is a circular, contractile disc suspended in the aqueous humor between the cornea and lens; it is perforated a little to the nasal side of its center by a circular aperture, the **pupil.** At its periphery it is continuous with the ciliary body, and it is also connected with the posterior elastic lamina of the cornea by means of the pectinate ligament. Its flat surfaces are anterior, toward the cornea, and posterior, toward the ciliary processes and lens. The iris divides the space between the lens and the cornea into an anterior and a posterior chamber. The **anterior chamber** of the eye is bound anteriorly by the posterior surface of the cornea and posteriorly by the iris and the central part of the lens. The **posterior chamber** is a narrow cleft posterior to the peripheral part of the iris and anterior to the suspensory ligament of the lens and the ciliary processes (Fig. 13-13). In the adult, the two chambers communicate through the pupil, but in the fetus, until the seventh month, they are separated by the pupillary membrane (p. 1288).

INTERNAL TUNIC

The **internal tunic** is a continuous membrane, but has a posterior nervous part, the **retina** proper, and an anterior non-neural part, the **pars ciliaris** and **pars iridica retinae.**

RETINA. The retina is a delicate neural membrane upon which the images of external objects are received. Its outer surface is in contact with the choroid; its inner surface,

with the vitreous body. Posteriorly, it is continuous with the optic nerve. It gradually diminishes in thickness anteriorly and extends nearly as far as the ciliary body, where it appears to end in a jagged margin, the **ora serrata.** Here the neural tissue of the retina ends, but a thin prolongation of the membrane extends anteriorly over the back of the ciliary processes and iris, forming the **pars ciliaris retinae** and **pars iridica retinae** already referred to. This forward prolongation consists of the pigmented layer of the retina together with a stratum of columnar epithelium. The retina is soft and semitransparent. It has a purple tint in the fresh state, owing to the presence of a coloring material named **rhodopsin** or **visual purple,** but it soon becomes clouded, opaque, and bleached when exposed to sunlight. Exactly in the center of the posterior part of the retina, corresponding to the optical axis of the eye and at a point in which the sense of vision is most perfect, is an oval yellowish area, the **macula;** in the macula is a central depression, the **fovea centralis** (Fig. 13-18). At the fovea centralis the retina is exceedingly thin, and the dark color of the choroid is distinctly seen through it. About 3 mm to the nasal side of the macula is the **optic disc,** which marks the exit of the optic nerve. A depression in the center of the disc, the porus opticus, marks the point of entrance of the central retinal artery. The disc is the only part of the surface of the retina that is insensitive to light, and it is termed the **blind spot** (Fig. 13-19).

Structure of the Tunics of the Bulb

STRUCTURE OF THE SCLERA

The sclera is formed of bundles and laminae of collagenous fibers intermixed with fine elastic fibers, fibroblasts, and other connective tissue cells. Its *vessels* are not numerous, the capillaries being small and uniting at long and wide intervals. Its *nerves* are derived from the ciliary nerves.

STRUCTURE OF THE CORNEA

The cornea consists of five layers: (1) the **corneal epithelium,** continuous with that of the conjunctiva; (2) the **anterior lamina;** (3) the **substantia propria;** (4) the **posterior lam-**

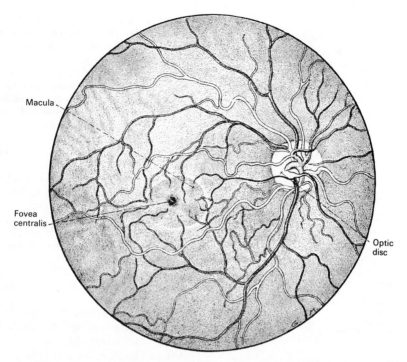

Fig. 13-19. Interior of posterior half of right eye as viewed through an ophthalmoscope. The distribution of central vessels of the retina is shown (veins darker than arteries) and their relation to the optic disc. The area of most acute vision, the macula, is shown. (Eycleshymer and Jones.)

ina; and (5) the **endothelium** (mesothelium) of the anterior chamber (Fig. 13-20).

The **corneal epithelium** (*anterior layer*) covers the front of the cornea and consists of several layers of cells. The cells of the deepest layer are columnar; then follow two or three layers of polyhedral cells, the majority of which are prickle cells similar to those found in the stratum mucosum of the cuticle. Lastly, there are three or four layers of squamous cells, with flat nuclei.

The **anterior lamina** (*anterior limiting layer; Bowman's membrane*) consists of closely interwoven fibrils, similar to those found in the substantia propria but containing no corneal corpuscles. It may be regarded as a condensed part of the substantia propria.

The **substantia propria corneae** is fibrous, tough, unyielding, and perfectly transparent. It is composed of about 60 flat superimposed lamellae made up of bundles of modified connective tissue, the fibers of which are directly continuous with those of the sclera. The fibers of each lamella are for the most part parallel with one another, but at right angles to those of adjacent lamellae, and frequently pass from one lamella to the next. The lamellae are held together by an interstitial cement substance, except for the **corneal spaces** which contain the **corneal corpuscles,** modified fibroblasts resembling the space in form, but not entirely filling them.

The **posterior elastic lamina** (*membrane of Descemet*[1]; *membrane of Demours*[2]) covers the posterior surface of the substantia propria. It is an elastic, transparent, homogeneous membrane of extreme thinness, which is not rendered opaque by either water, alcohol, or acids. When stripped from the substantia propria it curls up, or rolls upon itself with the attached surface innermost.

At the margin of the cornea, the posterior elastic lamina breaks up into fibers that form the trabecular tissue already described. Some of the fibers of this trabecular tissue are continued into the substance of the iris,

[1]Jean Descemet (1732–1810): A French anatomist and surgeon (Paris).
[2]Pierre Demours (1702–1795): A French ophthalmologist (Paris).

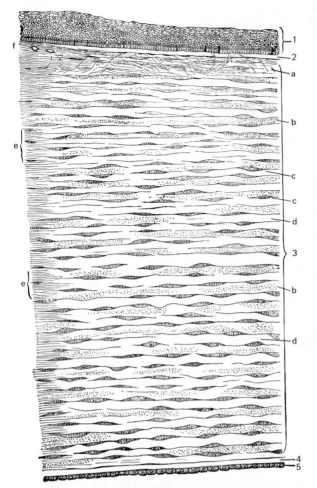

FIG. 13-20. Vertical section of human cornea from near the margin. (Waldeyer.) Magnified. 1. Epithelium. 2. Anterior lamina. 3. Substantia propria. 4. Posterior elastic lamina. 5. Endothelium of the anterior chamber. *a*, Oblique fibers in the anterior layer of the substantia propria. *b*, Lamellae, the fibers of which are cut across, producing a dotted appearance. *c*, Corneal corpuscles appearing fusiform in section. *d*, Lamellae, the fibers of which are cut longitudinally. *e*, Transition to the sclera, with more distinct fibrillation, and surmounted by a thicker epithelium. *f*, Small blood vessels cut across near the margin of the cornea.

forming the **pectinate ligament of the iris,** while others are connected with the forepart of the sclera and choroid.

The **endothelium** of the anterior chamber (*posterior layer; corneal endothelium*) is a mesothelial, rather than an endothelial, layer that covers the posterior surface of the elastic lamina. It is reflected onto the front of the iris, and also lines the spaces of the angle of the iris. It consists of a single stratum of polygonal, flattened, nucleated cells.

VESSELS AND NERVES. The cornea is a nonvascular structure. The capillary vessels, which end in loops at its circumference, are derived from the anterior ciliary arteries. Lymphatic vessels have not yet been demonstrated in it, but are represented by the channels in which the bundles of nerves run; these channels are lined by an endothelium.

The **nerves** are numerous and are derived from the ciliary nerves. Around the periphery of the cornea they form an *annular plexus,* from which fibers enter the substantia propria. They lose their myelin sheaths and ramify throughout its substance in a delicate network, and their terminal filaments form a firm and close plexus on the surface of the cornea proper, beneath the epithelium. This is termed the *subepithelial plexus,* and from it fibrils are given off that ramify between the epithelial cells, forming an *intraepithelial plexus.*

STRUCTURE OF THE CHOROID

The choroid consists mainly of a dense capillary plexus and of small arteries and veins that carry blood to and from this plexus. On its external surface is a thin membrane, the **lamina suprachoroidea,** composed of delicate nonvascular lamellae—each lamella consisting of a network of fine elastic fibers, among which are branched pigment cells. The potential spaces between the lamellae are lined by mesothelium and open freely into the **perichoroidal space.**

Internal to this lamina is the **choroid proper,** consisting of two layers: an outer layer, composed of small arteries and veins with pigment cells interspersed between them; and an inner layer, consisting of a capillary plexus.

The **outer layer** (*lamina vasculosa*) consists partly of the larger branches of the short ciliary arteries that run between the veins before they bend inward to end in the capillaries, but primarily this layer is formed of veins, named the **venae vorticosae** from their arrangement (Figs. 13-16; 13-17). The veins converge to four or five equidistant trunks, which pierce the sclera midway between the sclerocorneal junction and the exit of the optic nerve. Interspersed between the vessels are dark, star-shaped pigment cells, the processes of which, communicating with those of neighboring cells, form a delicate network or stroma, which loses its pigmented character toward the inner surface of the choroid.

The **inner layer** (*lamina choriocapillaris*) consists of an exceedingly fine capillary plexus formed by the short ciliary vessels; the network is closer and finer in the posterior than in the anterior part of the choroid. About 1.25 cm behind the cornea, the meshes of the plexus become larger and are continuous with those of the ciliary processes. These two laminae are connected by a **stratum intermedium,** which consists of fine elastic fibers. On the inner surface of the lamina choriocapillaris is a thin, structureless, or faintly fibrous membrane called the **lamina basalis;** it is closely connected with the stroma of the choroid and separates it from the pigmented layer of the retina.

One of the functions of the choroid is to provide nutrition for the retina and to convey vessels and nerves to the ciliary body and iris.

TAPETUM.   This name is applied to the outer and posterior part of the choroid, which in many animals has an iridescent appearance.

STRUCTURE OF THE CILIARY PROCESSES

The structure of the ciliary processes (Fig. 13-18) is similar to that of the choroid, but the vessels are larger and have chiefly a longitudinal direction. Their posterior surfaces are covered by a bilaminar layer of black pigment cells, which is continued anteriorly from the retina and is named the **pars ciliaris retinae.** In the stroma of the ciliary processes are also stellate pigment cells, but these are not as numerous as in the choroid itself.

STRUCTURE OF THE IRIS

The iris is composed of the following structures:

1. A **layer of flat mesothelial cells** is placed on a delicate hyaline basement membrane on the anterior surface. This layer is continuous with the mesothelium covering the posterior elastic lamina of the cornea, and in individuals with dark-colored irises, the cells contain pigment granules.

2. The **stroma** of the iris consists of fibers and cells. The former are made up of delicate collagenous bundles. A few fibers at the circumference of the iris have a circular direction, but the majority radiate toward the pupil, forming delicate meshes in which the vessels and nerves are contained. Interspersed between the bundles of connective tissue are numerous branched cells with fine

processes. In dark eyes, many of these cells contain pigment granules, but in blue eyes and the eyes of albinos, they are unpigmented.

3. The **muscular fibers** are involuntary and consist of circular and radiating fibers. The **circular fibers** form the **sphincter pupillae;** they are arranged in a narrow band about 1 mm wide that surrounds the margin of the pupil toward the posterior surface of the iris. The fibers near the free margin are closely aggregated; those near the periphery of the band are somewhat separated and form incomplete circles. The **radiating fibers** form the **dilatator pupillae;** they converge from the circumference toward the center and blend with the circular fibers near the margin of the pupil.

4. The posterior surface of the iris has a deep purple tint, being covered by two layers of pigmented columnar epithelium that are continuous at the periphery of the iris with the pars ciliaris retinae. This pigmented epithelium is named the **pars iridica retinae.**

The color of the iris is produced by the reflection of light from dark pigment cells underlying a translucent tissue, and it is determined, therefore, by the amount of the pigment and its distribution throughout the texture of the iris. The number and the situation of the pigment cells differ in different irises. In the albino, pigment is absent; in the various shades of blue eyes, the pigment cells are confined to the posterior surface of the iris, whereas in gray, brown, and black eyes, pigment is found also in the cells of the stroma and in those of the endothelium on the front of the iris.

CLINICAL NOTE.   The iris may be absent, either in part or altogether, as a congenital condition, and in some instances the pupillary membrane persists, although it is rarely complete. The iris may be the seat of a malformation termed a *coloboma,* which consists of a deficiency or cleft, due in a great number of cases to an arrest in development. In these cases the cleft is found at the lower aspect, extending directly downward from the pupil, and the gap frequently extends through the choroid to the porus opticus. In some rare cases the gap is found in other parts of the iris, and is not then associated with any deficiency of the choroid.

VESSELS AND NERVES.   The **arteries of the iris** are derived from the long and anterior ciliary arteries and from the vessels of the ciliary processes (see p. 1292). Each of the two long ciliary arteries, having reached the attached margin of the iris, divides into an upper and lower branch. These anastomose with corresponding branches from the opposite side and thus encircle the iris. Into this vascular circle, the **circulus arteriosus major,** the anterior ciliary arteries pour their blood, and from it vessels converge to the free margin of the iris, where they communicate to form a second circle, the **circulus arteriosus minor** (Fig. 13-21).

The **nerves of the choroid and iris** (Fig. 13-16) are the long and short ciliary nerves; the former are branches of the nasociliary nerve, and the latter are branches of the ciliary ganglion. They pierce the sclera around the entrance of the optic nerve, run forward in the perichoroidal space, and supply the blood vessels of the choroid. After reaching the iris, they form a plexus around its attached margin; from this are derived unmyelinated fibers that end in the sphincter and dilatator pupillae. Other fibers from the plexus end in a network on the anterior surface of the iris. The fibers, derived through the motor root of the ciliary ganglion from the oculomotor nerve, supply the sphincter pupillae, while those derived from the sympathetic root supply the dilatator pupillae.

STRUCTURE OF THE RETINA

The retina consists of an outer pigmented layer and an inner neural stratum or retina proper (Figs. 13-22; 13-23).

RETINA PROPER. The neural structures of the retina proper are supported by a series of nonneural or sustentacular fibers. In histologic sections made perpendicularly to the surface of the retina, ten layers are identified and named from within outward as follows:

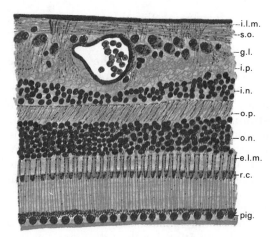

FIG. 13-22. Section of human retina. *e.l.m.*, external limiting membrane; *g.l.*, ganglionic layer; *i.l.m.*, internal limiting membrane; *i.n.*, inner nuclear layer; *i.p.*, inner plexiform layer; *o.n.*, outer nuclear layer; *o.p.*, outer plexiform layer; *pig.*, pigment cell layer; *r.c.*, layer of rods and cones; *s.o.*, stratum opticum. 500 ×. (Redrawn from Sobotta.)

1. Internal limiting membrane
2. Stratum opticum, layer of nerve fibers
3. Ganglion cell layer
4. Inner plexiform layer
5. Inner nuclear layer
6. Outer plexiform layer
7. Outer nuclear layer
8. External limiting membrane
9. Layer of rods and cones
10. Pigmented layer

1. The **internal limiting membrane** is a thin cribriform layer formed by the sustentacular fibers (p. 1301).

2. The **stratum opticum** or **layer of nerve fibers** is formed by the expansion of the fibers of the optic nerve; it is thickest near the **porus opticus,** gradually diminishing toward the **ora serrata.** As the nerve fibers pass through the lamina cribrosa sclerae, they lose their myelin sheaths and continue onward through the choroid and retina as simple axons. When they reach the internal surface of the retina, they radiate as a group of bundles from their point of entrance, and in many places they are arranged in plexuses. Most of the fibers are centripetal and are the direct continuations of the axons of the cells of the ganglionic layer, but a few of them are centrifugal and ramify in the inner plexiform and inner nuclear layers, where they end in enlarged extremities.

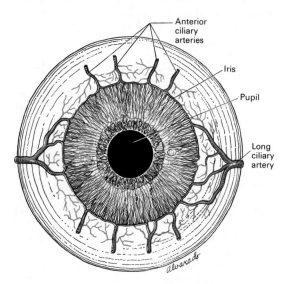

FIG. 13-21. Iris, anterior view.

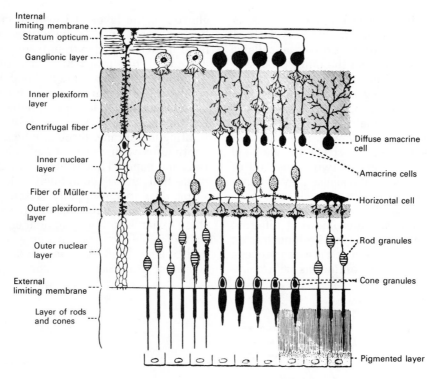

Internal
limiting membrane
Stratum opticum
Ganglionic layer
Inner plexiform
layer
Centrifugal fiber
Inner nuclear
layer
Fiber of Müller
Outer plexiform
layer
Outer nuclear
layer
External
limiting membrane
Layer of rods
and cones

Diffuse amacrine
cell
Amacrine cells
Horizontal cell
Rod granules
Cone granules
Pigmented layer

Fig. 13-23. Plan of retinal neurons. (After Cajal.)

3. The **ganglion cell layer** consists of a single layer of large ganglion cells, except in the macula, where there are several strata. The cells are somewhat flask-shaped; the rounded internal surface of each rests on the stratum opticum and sends an axon into the fiber layer. From the opposite end numerous dendrites extend into the inner plexiform layer, where they branch and form flat arborizations at different levels. The ganglion cells vary much in size, and the dendrites of the smaller ones usually arborize in the inner plexiform layer as soon as they enter it, while those of the larger cells ramify close to the inner nuclear layer.

4. The **inner plexiform layer** is made up of a dense reticulum of minute fibrils formed by the interlacement of the dendrites of the ganglion cells with those of the cells of the inner nuclear layer. Within this reticulum, a few branched spongioblasts are embedded.

5. The **inner nuclear layer** is made up of closely packed cells, of which there are three varieties: bipolar cells, horizontal cells and amacrine cells.

The **bipolar cells,** by far the most numerous, are round or oval, and each is prolonged into an inner and an outer process. They are divisible into rod bipolars and cone bipolars. The inner processes of the **rod bipolars** run through the inner plexiform layer and arborize around the bodies of the cells of the ganglionic layer; their outer processes end in the outer plexiform layer in tufts of fibrils around the button-like ends of the inner processes of the rod granules. The inner processes of the **cone bipolars** ramify in the inner plexiform layer in contact with the dendrites of the ganglionic cells.

The **horizontal cells** lie in the outer part of the inner nuclear layer and have somewhat flattened cell bodies. Their dendrites divide into numerous branches in the outer plexiform layer, while their axons run horizontally for some distance and finally ramify in the same layer.

The **amacrine cells** are in the inner part of the inner nuclear layer and are so named because it has not yet been shown that they possess axons. Their dendrites ramify extensively in the inner plexiform layer.

The cell bodies and nuclei of the sustentacular fibers are in this layer.

6. The **outer plexiform layer** is much thin-

ner than the inner layer but, like it, consists of a dense network of minute fibrils derived from the processes of the horizontal cells of the preceding layer and the outer processes of the rod and cone bipoplar cells, which ramify in it, forming arborizations around the enlarged ends of the rod fibers and with the branched foot plates of the cone fibers.

7. The **outer nuclear layer,** like the inner nuclear layer, contains strata of oval cell bodies. The latter are of two types: rod or cone nuclei, so named because they are connected respectively with the rods and cones of the next layer. The **rod nuclei** are much more numerous and are placed at different levels throughout the layer. Prolonged from either extremity of each cell is a fine process. The outer process is continuous with a sin-

gle rod of the layer of rods and cones; the inner process ends in the outer plexiform layer in an enlarged extremity and is embedded in the tuft into which the outer processes of the rod bipolar cells break up. In its course it has numerous varicosities.

The **cone nuclei,** fewer in number than the rod nuclei, are placed close to the external limiting membrane, through which they are continuous with the cones of the layer of rods and cones. They contain a piriform nucleus that almost completely fills the cell. From the inner extremity, a thick process passes into the outer plexiform layer, and there expands into a pyramidal enlargement or foot plate from which numerous fine fibrils are given off and come in contact with the outer processes of the cone bipolars.

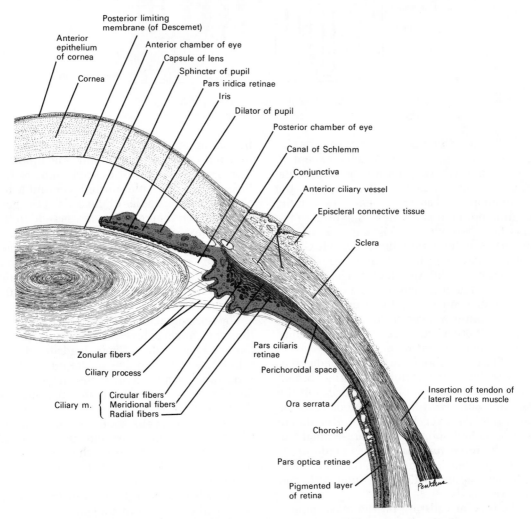

FIG. 13-24. A horizontal section through half of the anterior part of the eyeball.

8. The **external limiting membrane** is formed by the sustentacular cells (see below).

9. The **layer of rods and cones** (*Jacob's membrane*[1]).—The elements composing this layer are of two types, **rods** and **cones,** the former being much more numerous than the latter, except in the macula (Figs. 13-22; 13-23).

The **rods** are cylindrical, 40 to 60 μ long, 2 μ thick and arranged perpendicular to the surface. Each rod consists of an outer and an inner segment of about equal length. The outer segment is more slender than the inner segment, is strongly refractive and has transverse striae; it tends to break up into a number of thin discs superimposed on one another. It also exhibits faint longitudinal markings. The inner part of the inner segment is indistinctly granular and is called the myoid; its outer part presents a longitudinal striation and represents the ellipsoid, which is composed of fine, bright, highly refracting fibrils. The visual purple or **rhodopsin,** found in the outer segments, gives a purple or reddish color in a dark-adapted eye, and is the pigment for detection of low light intensity.

The **cones** are conical or flask-shaped; their broad ends rest upon the external limiting membrane, and the narrow, pointed extremity is turned to the choroid. Like the rods, each is made up of two segments, outer and inner; the outer segment is a short conical process, which, like the outer segment of the rod, exhibits transverse striae. The inner segment resembles the inner segment of the rods in structure, but it has a larger, thicker ellipsoid and a bulging deep granular part. The pigment in the outer segment of the cones is **iodopsin,** which is for color vision.

10. The **pigmented layer** consists of a single stratum of cells. When viewed from the outer surface, these cells are smooth and hexagonal; when seen in section, each cell consists of an outer nonpigmented part containing a large oval nucleus and an inner pigmented portion that extends as a series of straight thread-like processes between the rods, especially when the eye is exposed to light. In the eyes of albinos, the cells of this layer have no pigment.

[1]Arthur Jacob (1790–1874): Irish ophthalmologist and anatomist (Dublin).

SUPPORTING FRAMEWORK OF THE RETINA. The neural layers of the retina are connected by a supporting framework formed by the **sustentacular fibers of Müller;** these fibers pass through all the neural layers except that of the rods and cones. Each begins on the inner surface of the retina by an expanded, often forked base, which sometimes contains a spheroidal body that stains deeply with hematoxylin. The edges of the bases of adjoining fibers unite to form the **internal limiting membrane.** As the fibers pass through the nerve fiber and ganglionic layers, they give off a few lateral branches; in the inner nuclear layer they give off numerous lateral processes for the support of the bipolar cells, while in the outer nuclear layer they form a network around the rod and cone fibrils, and unite to form the **external limiting membrane** at the bases of the rods and cones. At the level of the inner nuclear layer, each sustentacular fiber contains a clear oval nucleus.

MACULA AND FOVEA CENTRALIS. In the macula, the nerve fibers are wanting as a continuous layer, the ganglionic layer consists of several strata of cells, and the rods are lacking. The cones are longer and narrower than in other parts, and in the outer nuclear layer there are only cone nuclei, the processes of which are very long and arranged in curved lines.

In the fovea centralis, the only parts present are (1) the cones; (2) the outer nuclear layer, the cone fibers of which are directed almost horizontally; (3) an exceedingly thin inner plexiform layer. The pigmented layer is thicker and its pigment more pronounced than elsewhere. The color of the macula seems to imbue all the layers except that of the cones; it is a rich yellow and deepest toward the center of the macula.

ORA SERRATA. At the **ora serrata** (Fig. 13-18), the neural layers of the retina end abruptly, and the retina continues onward as a single layer of columnar cells covered by the pigmented layer. This double layer is known as the **pars ciliaris retinae,** and can be traced forward from the ciliary processes on to the back of the iris, where it is termed the **pars iridica retinae.**

VESSELS. The **central retinal artery** (Fig. 13-19) and its accompanying vein pierce the optic nerve and enter the bulb of the eye through the porus opticus. The artery immediately bifurcates into an upper and a lower branch, and each of these again divides into a medial or nasal and a lateral or temporal branch. These branches first run between the hyaloid membrane and the neural layer, but they soon enter the neural layer and pass anteriorly, dividing dichotomously. From these branches a minute capillary plexus is given off that does not extend beyond the inner nuclear layer. The macula receives two small branches (superior and inferior macular arteries) from the temporal branches and small twigs directly from the central artery; these do not, however, reach as far as the fovea centralis, which has no blood vessels. The branches of the central retinal artery do not anastomose with each other—in other words, they are terminal arteries. In the fetus, a small vessel, the *hyaloid artery,* passes forward as a continuation of the central retinal artery through the vitreous humor to the posterior surface of the capsule of the lens.

REFRACTING MEDIA

The refracting media are five:

Cornea (see p. 1292)
Aqueous humor
Vitreous body
Zonula ciliaris
Crystalline lens

Aqueous Humor

The aqueous humor fills the anterior and posterior chambers. It is small in quantity, has an alkaline reaction, and consists mainly of water. The aqueous humor is secreted by the ciliary process. The fluid passes through the posterior chamber and the pupil into the anterior chamber. From the angle of the anterior chamber it passes into the spaces of Fontana to the pectinate villi, through which it is filtered into the venous canal of Schlemm.

Vitreous Body

The vitreous body fills the concavity of the pars optica retinae, to which it is firmly adherent, especially at the ora serrata. It is transparent, semigelatinous, and is hollowed anteriorly for the lens. Some indications of the hyaloid canal between the optic nerve and lens may persist.

No blood vessels penetrate the vitreous body, so that its nutrition must be carried on by vessels of the retina and ciliary processes, situated upon its exterior.

Zonula Ciliaris

The zonula ciliaris (*zonule of Zinn,*[1] *suspensory ligament of the lens*) consists of a series of straight fibrils that radiate from the ciliary body to the lens. It is attached to the capsule of the lens a short distance anterior to its equator. Scattered delicate fibers are also attached to the region of the equator itself. This ligament retains the lens in position, and is relaxed by the contraction of the meridional fibers of the ciliary muscle, so that the lens is allowed to become more convex. Posterior to the suspensory ligament there is a sacculated canal, the **spatia zonularia** (*canal of Petit*[2]), which encircles the equator of the lens; it can be easily inflated through a fine blowpipe inserted under the suspensory ligament.

Crystalline Lens

The crystalline lens, enclosed in its capsule, is situated immediately between the iris and the vitreous body and is encircled by the ciliary processes, which slightly overlap its margin.

The **capsule of the lens** (*capsula lentis*) is a transparent, structureless membrane that closely surrounds the lens and is thicker in front than behind. It is brittle but highly elastic, and when ruptured the edges roll up, with the outer surface innermost. It rests in the hyaloid fossa in the anterior part of the vitreous body posteriorly and is in contact with the free border of the iris anteriorly, but recedes from it at the circumference, thus forming the posterior chamber of the eye. It is retained in its position chiefly by the suspensory ligament of the lens, already described.

The **lens** is a transparent, biconvex body; the convexity of its anterior surface is less than that of its posterior surface. The central points of these surfaces are termed, respectively, the **anterior** and **posterior poles;** a line connecting the poles constitutes the **axis** of the lens, while the marginal circumference is termed the **equator.**

[1] Johann Gottfried Zinn (1727–1759): A German physician and botanist (Göttingen).
[2] François Pourfour du Petit (1664–1741): French surgeon and anatomist (Paris).

STRUCTURE OF THE LENS

The lens is made up of a soft cortical substance and a firm, central part, the **nucleus** (Fig. 13-25). Faint lines (*radii lentis*) radiate from the poles to the equator. In the adult there may be six or more of these lines, but in the fetus there are only three, and they diverge from each other at angles of 120° (Fig. 13-26). On the anterior surface, one line ascends vertically and the other two diverge downward; on the posterior surface, one ray descends vertically and the other two diverge upward. They correspond with the free edges of an equal number of septa that are composed of an amorphous substance. These septa dip into the substance of the lens and mark the interruptions of a series of concentrically arranged laminae (Fig. 13-25). Each lamina is built up of a number of hexagonal, ribbon-like lens fibers, the edges of which are more or less serrated. The serrations fit between those of neighboring fibers, while the ends of the fibers come into apposition at the septa. The fibers run in a curved manner from the septa on the anterior surface to those on the posterior surface. No fibers pass from pole to pole; they are arranged in such a way that those beginning near the pole on one surface of the lens end near the peripheral extremity of the plane on the other, and vice versa. The fibers of the outer layers of the lens are nucleated, and together they form a nuclear layer, most distinct toward the equator. The anterior surface of the lens is covered by a layer of transparent, columnar, nucleated epithelium. At the equator the cells become elongated, and their gradual transition into lens fibers can be traced (Fig. 13-27).

In the fetus, the lens is nearly spherical, and has a slightly reddish tint; it is soft and breaks down readily on the slightest pres-

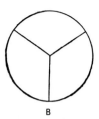

FIG. 13-26. Diagram showing the direction and arrangement of the radiating lines on the front and back of the fetal lens. *A,* From the front. *B,* From the back.

sure. A small branch from the central retinal artery runs forward, as already mentioned, through the vitreous body to the posterior part of the capsule of the lens, where its branches radiate and form a plexiform network that covers the posterior surface of the capsule. The branches are continuous around the margin of the capsule with the vessels of the pupillary membrane and with those of the iris. **In the adult,** the lens is colorless, transparent, firm, and devoid of vessels. **In old age** it becomes flattened on both surfaces and slightly opaque, has an amber tint, and increases in density (Fig. 13-28).

VESSELS AND NERVES OF THE BULB (Fig. 13-15). The **arteries** of the eye are the long, short, and anterior ciliary arteries and the central retinal artery. They have already been described (see p. 688).

The **ciliary veins** are seen on the outer surface of the choroid and are named, because of their arrangement, the *venae vorticosae* (Fig. 13-16). The veins converge to four or five equidistant trunks that pierce the sclera midway between the sclerocorneal junction and the porus opticus. Another set of veins accompanies the anterior ciliary arteries. All these veins open into the ophthalmic veins.

The **ciliary nerves** are derived from the nasociliary nerve and from the ciliary ganglion.

ACCESSORY ORGANS OF THE EYE

The **accessory organs of the eye** include the **ocular muscles,** the **fasciae,** the **eyebrows,** the **eyelids,** the **conjunctiva,** and the **lacrimal apparatus.**

Ocular Muscles

The ocular muscles are:

Levator palpebrae superioris
Rectus superior
Rectus inferior

FIG. 13-25. The crystalline lens, hardened and divided. (Enlarged.)

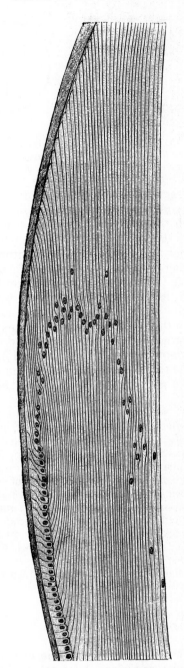

Fɪɢ. 13-27.   Section through the margin of the lens, showing the transition of the epithelium into the lens fibers. (Babuchin.)

Rectus medialis
Rectus lateralis
Obliquus superior
Obliquus inferior

The **levator palpebrae superioris** (Fig. 13-29) is a thin, flat muscle *arising* from the

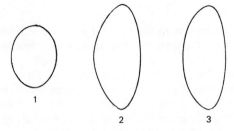

Fɪɢ. 13-28.   Profile views of the lens at different periods of life. 1. In the fetus. 2. In adult life. 3. In old age.

inferior surface of the small wing of the sphenoid, where it is separated from the optic foramen by the origin of the rectus superior. At its origin it is narrow and tendinous, but it becomes broad and fleshy and ends anteriorly in a wide aponeurosis that splits into three lamellae. The superficial lamella blends with the superior part of the orbital septum, and is prolonged over the superior tarsus to the palpebral part of the orbicularis oculi and to the deep surface of the skin of the upper eyelid. The middle lamella, largely made up of nonstriated muscular fibers, is inserted into the superior margin of the superior tarsus, while the deepest lamella blends with an expansion from the sheath of the rectus superior and with it is attached to the superior fornix of the conjunctiva.

The four **recti** muscles (Fig. 13-30) *arise* from a fibrous ring (*anulus tendineus communis*) that surrounds the superior, medial, and inferior margins of the optic foramen and encircles the optic nerve (Fig. 13-31). The ring is completed by a tendinous bridge prolonged over the inferior and medial part of the superior orbital fissure and attached to a tubercle on the margin of the great wing of the sphenoid, bounding the fissure. Two specialized parts of this fibrous ring may be made out: a lower, the *ligament* or *tendon of Zinn,* which gives origin to the **rectus inferior,** part of the **rectus medialis,** and the lower head of origin of the **rectus lateralis;** and an upper part which gives origin to the **rectus superior,** the rest of the rectus medialis, and the upper head of the rectus lateralis. This upper band is sometimes termed the *superior tendon of Lockwood.*[1] Each muscle passes anteriorly in the position implied by its name, to be *inserted* by a tendi-

[1]Charles Barrett Lockwood (1856–1914): An English surgeon (London).

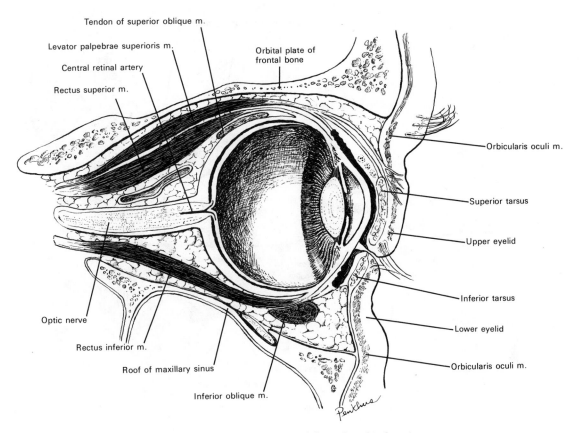

Fig. 13-29. Sagittal section of the right orbital cavity.

nous expansion into the sclera, about 6 mm from the margin of the cornea. Between the two heads of the rectus lateralis is a narrow interval through which pass the two divisions of the oculomotor nerve, the nasociliary nerve, the abducens nerve, and the ophthalmic vein. Although these muscles have a common origin and are inserted in a similar manner into the sclera, there are certain differences between their length and breadth. The rectus medialis is the broadest, the rectus lateralis the longest, and the rectus superior the thinnest and narrowest.

The **obliquus superior oculi** (*superior oblique*) is a fusiform muscle placed at the superior and medial side of the orbit. It *arises* immediately above the margin of the optic foramen, superior and medial to the origin of the rectus superior, and passing anteriorly, it ends in a rounded tendon that plays in a fibrocartilaginous ring or pulley attached to the trochlear fovea of the frontal bone. The contiguous surfaces of the tendon and ring are lined by a delicate synovial sheath and are enclosed in a thin fibrous investment. The tendon bends around the pulley at more than a right angle, passes beneath the rectus superior to the lateral part of the bulb of the eye, and is *inserted* into the sclera, posterior to the equator of the eyeball; the insertion of the muscle lies between the rectus superior and the rectus lateralis.

The **obliquus inferior oculi** (*inferior oblique*) is a thin, narrow muscle placed near the anterior margin of the floor of the orbit. It *arises* from the orbital surface of the maxilla, lateral to the lacrimal groove. Passing lateralward, at first between the rectus inferior and the floor of the orbit and then between the bulb and the rectus lateralis, it is *inserted* into the lateral part of the sclera between the rectus superior and the rectus lateralis, near but somewhat posterior to the insertion of the obliquus superior.

NERVES. The levator palpebrae superioris, the obliquus inferior, and the recti superior, inferior, and medialis are supplied by the oculomotor nerve; the obliquus superior is supplied by the trochlear nerve, and the rectus lateralis, by the abducens nerve (Fig. 13-31).

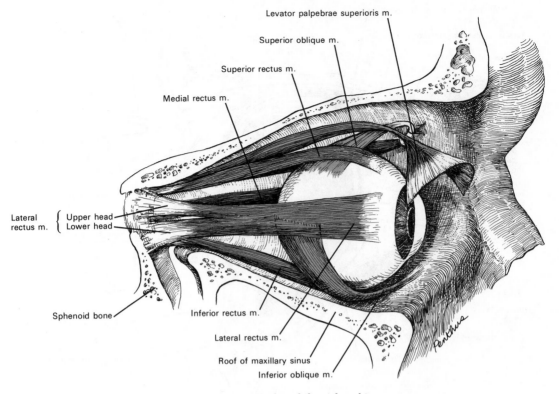

Levator palpebrae superioris m.

Superior oblique m.

Superior rectus m.

Medial rectus m.

Lateral rectus m. { Upper head
{ Lower head

Sphenoid bone

Inferior rectus m.

Lateral rectus m.

Roof of maxillary sinus

Inferior oblique m.

FIG. 13-30. Muscles of the right orbit.

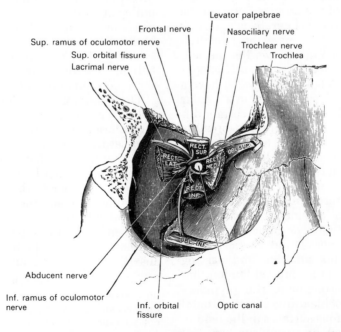

Levator palpebrae

Frontal nerve

Nasociliary nerve

Sup. ramus of oculomotor nerve

Sup. orbital fissure

Trochlear nerve

Trochlea

Lacrimal nerve

RECT. SUP.

OBL. SUP.

RECT. LAT.

RECT. MED.

RECT. INF.

OBL. INF.

Abducent nerve

Inf. ramus of oculomotor nerve

Inf. orbital fissure

Optic canal

FIG. 13-31. Dissection showing origins of right ocular muscles and the nerves entering the orbit through the superior orbital fissure.

ACTIONS. The levator palpebrae *raises* the upper eyelid and is the direct antagonist of the orbicularis oculi. The four recti are attached to the bulb of the eye in such a manner that, acting singly, they turn its corneal surface in the direction corresponding to their names. The movement produced by the rectus superior or rectus inferior is not a simple one because the axis of the orbit, and hence their direction of pull, diverges from the median plane, causing their pull to be accompanied by a certain deviation medialward, with a slight amount of rotation. These latter movements are corrected by the obliqui; the obliquus inferior corrects the medial deviation caused by the rectus superior and the obliquus superior corrects the deviation caused by the rectus inferior. The contraction of the rectus lateralis or rectus medialis, on the other hand, produces a purely horizontal movement. If any two neighboring recti of one eye act together, they carry the globe of the eye diagonally in these directions: upward and medialward, upward and lateralward, downward and medialward, or downward and lateralward. Sometimes the corresponding recti of the two eyes act in unison, and at other times the opposite recti act together. Thus, in turning the eyes to the right, the rectus lateralis of the right eye will act in unison with the rectus medialis of the left eye, but if both eyes are directed to an object in the midline at a short distance, the two recti mediales will act in unison. The movement of circumduction, as in looking around a room, is performed by the successive actions of the four recti. The obliqui rotate the eyeball on its anteroposterior axis, the superior muscle directing the cornea downward and lateralward, and the inferior muscle directing it upward and lateral-

ward. These movements are required for the correct viewing of an object when the head is moved laterally, as from shoulder to shoulder, in order that the picture may fall in all respects on the same part of the retina of either eye.

A layer of nonstriped muscle, the **orbitalis muscle** (of H. Müller), bridges the inferior orbital fissure.

Fasciae and Ligaments

The **fascia bulbi** (*vagina bulbi; capsule of Tenon*) (Fig. 13-32) is a thin membrane that envelops the eyeball from the optic nerve to the ciliary region, separating it from the orbital fat and forming a socket in which it plays. Its inner surface is smooth and is separated from the outer surface of the sclera by the **periscleral fascial cleft.** This space between the fascia and sclera (*spatium intervaginale*) is continuous with the subdural and subarachnoid cavities; it is traversed by delicate bands of connective tissue that extend between the fascia and the sclera. The fascia is perforated by the ciliary vessels and nerves, and it fuses with the sheath of the optic nerve and with the sclera around the lamina cribrosa. Anteriorly it blends with the ocular conjunctiva, with which it is attached to the ciliary region of the bulb.

The fascia bulbi is prolonged over tendons

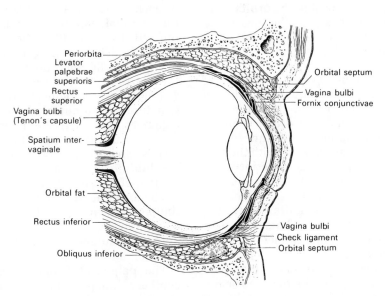

FIG. 13-32. Fascia bulbi (capsule of Tenon). Sagittal section, semidiagrammatic. (Redrawn from *Surgical Anatomy of the Human Body* by J. B. Deaver, P. Blakiston's Son & Co., 1926.)

of the individual ocular muscles as a tubular sheath (*fascia musculares*). The sheath of the tendon of the obliquus superior is carried as far as the fibrous pulley of that muscle; the sheath of the obliquus inferior reaches as far as the floor of the orbit, to which it gives off a slip. The sheaths of the recti are gradually lost in the perimysium, but they give off important expansions. The expansion from the rectus superior blends with the tendon of the levator palpebrae; that of the rectus inferior is attached to the inferior tarsus. The expansions from the sheaths of the recti medialis and especially the rectus lateralis are strong and are attached to the lacrimal and zygomatic bones respectively. As they probably check the actions of the two recti, these structures have been named the **medial** and **lateral check ligaments.** A thickening of the lower part of the fascia bulbi, named the **suspensory ligament of the eye,** is slung like a hammock below the eyeball, and is expanded in the center and narrow at the ends, and attached to the zygomatic and lacrimal bones respectively.

The **periorbita** is the name given to the periosteum of the orbit. It is loosely connected to the bones and can be readily separated from them. It is continuous with the dura mater by processes that pass through the optic foramen and superior orbital fissure and with the sheath of the optic nerve. At the margin of the orbit, a process from the periorbita assists in forming the **orbital septum.** The periorbita has two other processes: one to enclose the lacrimal gland, the other to hold the pulley of the obliquus superior in position.

The **orbital septum** (*palpebral ligament*) (Fig. 13-32) is a membranous sheet that is continuous with the part of the periorbita that is attached to the edge of the orbit. In the upper eyelid, the orbital septum blends by its peripheral circumference with the tendon of the levator palpebrae superioris and the superior tarsus; in the lower eyelid, the orbital septum blends with the inferior tarsus . Medially it is thin, and after separating from the medial palpebral ligament, it is fixed to the lacrimal bone immediately behind the lacrimal sac. The septum is perforated by the vessels and nerves that pass from the orbital cavity to the face and scalp. The eyelids are richly supplied with blood.

Eyebrows

The eyebrows (*supercilia*) are two arched eminences that surmount the superior margins of the orbits. The eyebrows consist of thickened integument that is connected beneath with the orbicularis oculi, corrugator, and frontalis muscles. They support numerous short, thick hairs.

Eyelids

The eyelids (*palpebrae*) are two thin, movable covers placed in front of the eye that protect it from injury by their closure. The upper eyelid is larger and more movable than the lower, and is furnished with an elevator muscle, the levator palpebrae superioris. When the eyelids are open, an elliptical space, the **palpebral fissure** (*rima palpebrarum*), is left between their margins, the angles of which correspond to the junctions of the upper and lower eyelids. These angles are called the **palpebral commissures** or **canthi.**

The **lateral palpebral commissure** (*external* or *lateral canthus*) is more acute than the medial commissure, and the eyelids here lie in close contact with the bulb of the eye. The **medial palpebral commissure** (*internal* or *medial canthus*) is prolonged for a short distance toward the nose, and the two eyelids are separated by a triangular space, the **lacus lacrimalis** (Fig. 13-37). At the basal angles of the lacus lacrimalis, on the margin of each eyelid, is a small conical elevation, the **lacrimal papilla,** the apex of which is pierced by a small orifice, the **punctum lacrimale.** This opening is the commencement of the lacrimal duct.

STRUCTURE OF THE EYELIDS. (Fig. 13-33). The eyelids are composed of the following structures, from outer to inner surfaces: integument, areolar tissue, fibers of the orbicularis oculi, tarsus, orbital septum, tarsal glands, and conjunctiva. The upper eyelid, in addition, has the aponeurosis of the levator palpebrae superioris (Fig. 13-34).

The **integument** is extremely thin and continuous at the margins of the eyelids with the conjunctiva.

The **subcutaneous areolar tissue** is lax, delicate, and seldom contains any fat.

The **palpebral fibers of the orbicularis oculi** are thin, pale, and have an involuntary as well as a voluntary action.

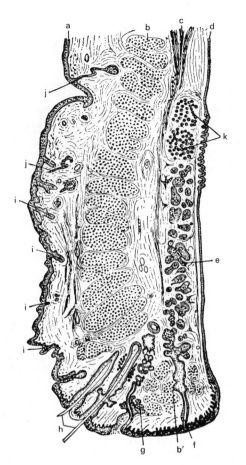

FIG. 13-33. Sagittal section through the upper eyelid. (After Waldeyer.) *a*, Skin. *b*, Orbicularis oculi. *b′*, Marginal fasciculus of orbicularis (ciliary bundle). *c*, Levator palpebrae. *d*, Conjunctiva. *e*, Tarsus. *f*, Tarsal gland. *g*, Ciliary gland. *h*, Eyelashes. *i*, Small hairs of skin. *j*, Sweat glands. *k*, Posterior tarsal glands.

The **tarsi** (*tarsal plates*) (Fig. 13-34) are two thin, elongated plates of dense connective tissue about 2.5 cm long; one is placed in each eyelid and contributes to its form and support. The **superior tarsus,** the larger, is semilunar, about 10 mm in breadth at the center, and gradually narrows toward its extremities. To the anterior surface of this plate the aponeurosis of the levator palpebrae superioris is attached. The **inferior tarsus,** the smaller plate, is thin, elliptical, and has a vertical diameter of about 5 mm. The free or ciliary margins of these plates are thick and straight. The attached or orbital margins are connected to the circumference of the orbit by the orbital septum. The lateral angles are attached to the zygomatic bone by the lateral palpebral raphe. The medial angles of the two plates end at the lacus lacrimalis and are attached to the frontal process of the maxilla by the medial palpebral ligament.

EYELASHES. The eyelashes (*cilia*) are attached to the free edges of the eyelids; they are short, thick, curved hairs, arranged in a double or triple row: those of the upper eyelid, more numerous and longer than those of the lower, curve upward; those of the lower eyelid curve downward, so that they do not interlace when the lids are closed. Near the attachment of the eyelashes are the openings of the **ciliary glands,** which are arranged in several rows close to the free margin of the lid. These are regarded as enlarged and modified sudoriferous glands.

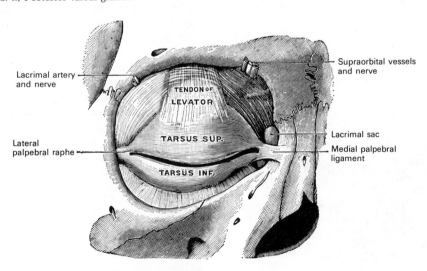

FIG. 13-34. The tarsi and their ligaments. Right eye; anterior view.

TARSAL GLANDS.   The tarsal glands (*Meibomian*[1] *glands*) (Figs. 13-35; 13-36) are situated on the inner surfaces of the eyelids between the tarsi and conjunctiva. They may be seen distinctly through the conjunctiva when the eyelids are everted, having an appearance similar to parallel strings of pearls. There are about 30 glands in the upper eyelid and somewhat fewer in the lower. They are embedded in grooves in the inner surfaces of the tarsi and correspond in length with the breadth of these plates; they are, consequently, longer in the upper than in the lower eyelid. Their ducts open on the free margins of the lids by minute apertures.

STRUCTURE OF THE TARSAL GLANDS.   The tarsal glands are modified sebaceous glands; each consists of a single straight tube or follicle with numerous small lateral diverticula. The tubes are supported by a basement membrane and are lined at their mouths by stratified epithelium; the deeper parts of the tubes and the lateral offshoots are lined by a layer of polyhedral cells.

Conjunctiva

The conjunctiva is the mucous membrane of the eye. It lines the inner surfaces of the eyelids or palpebrae and is reflected over the anterior part of the sclera.

The **palpebral portion** (*tunica conjunctiva palpebrarum*) is thick, opaque, highly vascular, and covered with numerous papillae; the deeper part has a considerable amount of

[1] Heinrich Meibom (1638–1700): A German physician and anatomist (Helmstadt).

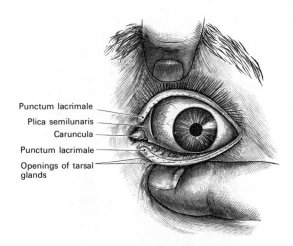

Punctum lacrimale
Plica semilunaris
Caruncula
Punctum lacrimale
Openings of tarsal glands

FIG. 13-36.   Anterior view of left eye with eyelids separated to show medial canthus.

lymphoid tissue. At the margins of the lids, the conjunctiva becomes continuous with the lining membrane of the ducts of the tarsal glands; through the lacrimal ducts, it is continuous with the lining membrane of the lacrimal sac and nasolacrimal duct. At the lateral angle of the upper eyelid, the ducts of the lacrimal gland open on the free surface of the conjunctiva, and at the medial angle it forms a semilunar fold, the **plica semilunaris** (Fig. 13-36). The line of reflection of the conjunctiva from the upper eyelid onto the bulb of the eye is named the **superior fornix,** and that from the lower lid is named the **inferior fornix.**

The **bulbar portion** (*tunica conjunctiva bulbi*) of the conjunctiva is loosely connected over the sclera; it is thin, transparent, destitute of papillae, and only slightly vascular. At the circumference of the *cornea,* the conjunctiva becomes the epithelium of the cornea, already described (see p. 1295). *Lymphatics* arise in the conjunctiva in a delicate zone around the cornea; these vessels run to the scleral conjunctiva. Lymphoid follicles are found in the conjunctiva, chiefly near the medial palpebral commissure. They were first described by Bruch[1], in his description of Peyer's patches[2] of the small intestine, as "identical structures existing in the under eyelid of the ox."

[1] Carl Wilhelm Ludwig Bruch (1819–1884): A German anatomist (Basel and Giessen).
[2] Johann Conrad Peyer (1653–1712): Swiss anatomist and logician (Schaffhausen).

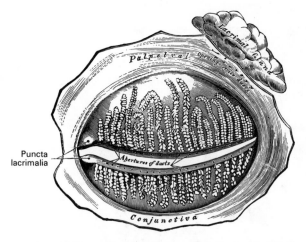

Puncta lacrimalia

FIG. 13-35.   The tarsal glands seen from the inner surface of the eyelids.

In and near the fornices, but more plentiful in the upper than in the lower eyelid, several convoluted tubular glands open on the surface of the conjunctiva.

CARUNCULA LACRIMALIS.   The caruncula lacrimalis is a small, reddish protuberance situated at the medial palpebral commissure and filling up the lacus lacrimalis. It consists of a small island of skin containing sebaceous and sudoriferous glands, and it is the source of the whitish secretion that constantly collects in this region. A few slender hairs are attached to its surface. Lateral to the caruncula is a slight semilunar fold of conjunctiva, the concavity of which is directed toward the cornea; this is called the **plica semilunaris.** Smooth muscular fibers have been found in this fold, and in some of the domesticated animals it contains a thin plate of cartilage.

NERVES.   The **nerves** in the conjunctiva are numerous and form rich plexuses. They appear to terminate in a peculiar form of tactile corpuscle called terminal bulbs.

Lacrimal Apparatus

The lacrimal apparatus (Fig. 13-37) consists of (a) the **lacrimal gland,** which secretes the tears, and its excretory ducts, which convey the fluid to the surface of the eye; and (b) the **lacrimal ducts,** the **lacrimal sac,** and the **nasolacrimal duct,** by which the fluid is conveyed into the cavity of the nose.

LACRIMAL GLAND.   The lacrimal gland (Figs. 13-34; 13-35) is superior and lateral to the bulb and is lodged in the lacrimal fossa on the medial side of the zygomatic process of the frontal bone. It is oval, about the size and shape of an almond, and consists of two portions. The **orbital part** (*pars orbitalis; superior lacrimal gland*) is connected to the periosteum of the orbit by a few fibrous bands and rests upon the tendons of the recti superioris and lateralis, which separate it from the bulb of the eye. The **palpebral part** (*pars palpebralis; inferior lacrimal gland*) is separated from the orbital part by a fibrous septum and projects into the back part of the upper eyelid, where its deep surface is related to the conjunctiva. The ducts of the

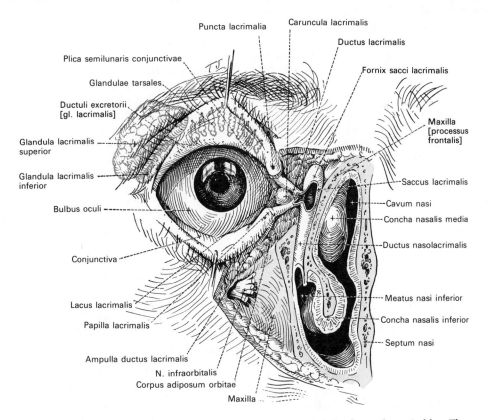

FIG. 13-37.   Topography of the lacrimal apparatus. The lacrimal and tarsal glands are shown in blue. The nose and a portion of the face have been cut away. (Eycleshymer and Jones.)

glands, numbering six to twelve, run obliquely beneath the conjunctiva for a short distance and open along the upper and lateral half of the superior conjunctival fornix.

LACRIMAL DUCTS. The lacrimal ducts (*lacrimal canals*), one in each eyelid, commence at minute orifices, termed **puncta lacrimalia,** on the summits of the **papillae lacrimales,** which are on the margins of the lids at the lateral extremity of the lacus lacrimalis. The **superior duct,** the smaller and shorter of the two, at first ascends and then bends at an acute angle and passes medialward and downward to the lacrimal sac. The **inferior duct** at first descends and then runs almost horizontally to the lacrimal sac. At the angles they are dilated into **ampullae;** their walls are dense, and their mucous lining is covered by stratified squamous epithelium placed on a basement membrane. Outside the basement membrane is a layer of striped muscle that is continuous with the lacrimal part of the orbicularis oculi; at the base of each lacrimal papilla the muscular fibers are circularly arranged and form a kind of sphincter.

LACRIMAL SAC. The lacrimal sac is the upper dilated end of the nasolacrimal duct. It is lodged in a deep groove formed by the lacrimal bone and the frontal process of the maxilla. It is oval and measures 12 to 15 mm long; its upper end is closed and round, and its lower end continues into the nasolacrimal duct. The superficial surface of the lacrimal sac is covered by a fibrous expansion derived from the medial palpebral ligament, and its deep surface is crossed by the lacrimal part of the orbicularis oculi (p. 441), which is attached to the crest on the lacrimal bone.

NASOLACRIMAL DUCT. The nasolacrimal duct (*nasal duct*) is a membranous canal, about 18 mm long, that extends from the lower part of the lacrimal sac to the inferior meatus of the nose, where it ends in a somewhat expanded orifice that is provided with an imperfect valve, the **plica lacrimalis** (*Hasneri*[1]), which is formed by a fold of the mucous membrane. The nasolacrimal duct is contained in an osseous canal that is formed by the maxilla, the lacrimal bone, and the inferior nasal concha. It is narrower in the middle than at either end, and is directed downward, backward, and a little lateralward. The mucous lining of the lacrimal sac and nasolacrimal duct is covered with columnar epithelium, which is ciliated in places.

STRUCTURE OF THE LACRIMAL GLAND. In structure and general appearance the lacrimal gland resembles the serous salivary glands.

STRUCTURE OF THE LACRIMAL SAC. The lacrimal sac consists of a fibrous elastic coat that is lined internally by mucous membrane. The mucous membrane is continuous through the lacrimal ducts with the conjunctiva, and through the nasolacrimal duct with the mucous membrane of the nasal cavity.

Organ of Hearing and Equilibration: The Ear

The **ear** (*organum vestibulocochleare; organum oticum; organum stato-acousticum*), or **organ of hearing and equilibration,** is divisible into three parts: the **external ear,** the **middle ear** or **tympanic cavity,** and the **internal ear** or **labyrinth.**

DEVELOPMENT

The first rudiment of the internal ear appears shortly after that of the eye in the form of a patch of thickened ectoderm, the **auditory plate,** over the region of the hindbrain. The auditory plate becomes depressed and converted into the **auditory pit** (Fig. 13-38). The mouth of the pit is then closed, and the **auditory vesicle** is formed (Fig. 13-39); from this the epithelial lining of the membranous labyrinth is derived. From the vesicle certain diverticula are given off that form the various parts of the membranous labyrinth (Figs.

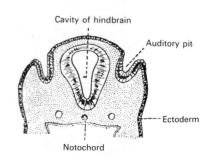

Cavity of hindbrain

Auditory pit

Ectoderm

Notochord

FIG. 13-38.   Section through the head of a human embryo, about 12 days old, in the region of the hindbrain. (Kollmann.)

[1]Joseph Ritter von Artha Hasner (1819–1892): A Bohemian ophthalmologist (Prague).

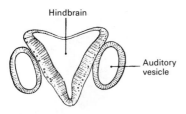

FIG. 13-39. Section through hindbrain and auditory vesicles of an embryo more advanced than that of Figure 13-38. (After His.)

13-40; 13-41). One from the middle part forms the ductus and saccus endolymphaticus; another from the anterior end gradually elongates, and forming a tube coiled on itself, becomes the cochlear duct, the vestibular extremity of which is subsequently constricted to form the canalis reuniens. Three other diverticula appear as disc-like evaginations on the surface of the vesicle. The central parts of the walls of the discs coalesce and disappear, while the peripheral portions persist to form the semicircular ducts; of these, the anterior is the first and the lateral is the last to be completed. The central part of the vesicle represents the membranous vestibule, and this is subdivided by a constriction into a smaller ventral part, the saccule, and a larger dorsal and posterior part, the utricle. This subdivision is effected by a fold that extends deeply into the proximal part of the ductus endolymphaticus, so that the utricle and saccule ultimately communicate with each other by means of a Y-shaped canal. The saccule opens into the cochlear duct through the canalis reuniens, and the semicircular ducts communicate with the utricle.

The mesodermal tissue surrounding the various parts of the epithelial labyrinth is converted into a cartilaginous capsule, and this is finally ossified to form the bony labyrinth. Between the cartilaginous capsule and the epithelial structures is a stratum of mesodermal tissue that is differentiated into three layers: an outer layer, which forms the periosteal lining of the bony labyrinth; an inner layer, which is in direct contact with the epithelial structures; and an intermediate layer, which consists of gelatinous tissue. The perilymphatic spaces are developed by the absorption of this gelatinous tissue. The modiolus and osseous spiral lamina of the cochlea are not preformed in cartilage but are ossified directly from connective tissue.

The **middle ear** and **auditory tube** are developed from the first pharyngeal pouch. The entodermal lining of the dorsal end of this pouch is in contact with the ectoderm of the corresponding pharyngeal groove; by the extension of the mesoderm between these two layers, the **tympanic membrane** forms. During the sixth or seventh month, the **tympanic antrum** appears as an upward and backward expansion of the tympanic cavity. There is some difference of opinion regarding the exact mode of development of the ossicles of the middle ear. The view generally held is that the **malleus** develops from the proximal end of the mandibular (Meckel's) cartilage (Fig. 2-48); the **incus** develops in the proximal end of the mandibular arch; and the **stapes** develops from the proximal end of the hyoid arch. The malleus, with the exception of its anterior process, is ossified from a single center that appears near the neck of the bone; the anterior process is ossified separately in membrane and joins the main part of the bone about the sixth month of fetal life. The incus is ossified from one center that appears in the upper part of its long crus and ultimately extends into its lenticular process. The stapes first appears as a ring (*anulus stapedius*) encircling a small vessel, the stapedial artery, which subsequently atrophies; it is ossified from a single center in its base.

The **external ear** and **external acoustic meatus** develop from the first branchial groove. The lower part of this groove extends inward as a funnel-shaped tube (primary meatus), from which the cartilaginous portion and a small part of the roof of the osseous portion of the meatus are developed. From the lower part of the funnel-shaped tube, an epithelial lamina extends downward and inward along the inferior wall of the primitive tympanic cavity; by the splitting of this lamina the inner part of the meatus (secondary meatus) is produced, while the inner portion of the lamina forms the cutaneous stratum of the tympanic membrane.

The **auricula** or **pinna** develops by the gradual differentiation or tubercles that appear around the margin of the first branchial groove (see Chapter 2 and Fig. 2-57).

The **acoustic nerve** rudiment appears about the end of the third week as a group of ganglion cells closely applied to the cranial edge of the auditory vesicle. Whether these cells are derived from the ectoderm adjoin-

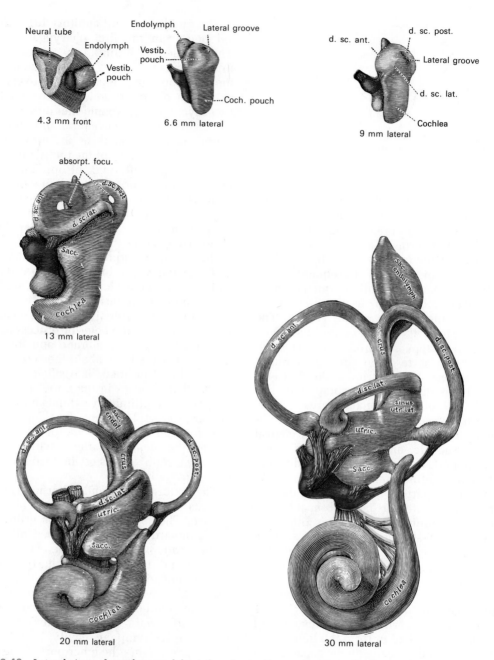

Fig. 13-40. Lateral views of membranous labyrinth and acoustic complex. 25 × dia. (Streeter.) *absorpt. focu,* Area of wall where absorption is complete; *amp.,* ampulla membranacea; *crus,* crus commune; *d. sc. lat.,* ductus semicircularis lateralis; *d. sc. post.,* ductus semicircularis posterior; *d. sc. ant.,* ductus semicircular anterior; *coch.* or *cochlea,* ductus cochlearis; *duct. endolymph,* ductus endolymphaticus; *d. reuniens,* ductus reuniens Henseni; *endol.* or *endolymphs,* appendix endolymphaticus; *rec. utr.,* recessus utriculi; *sacc.,* sacculus; *sac endol.,* saccus endolymphaticus; *sinu utr. lat.,* sinus utriculi lateralis; *utric.,* utriculus; *vestib. p.,* vestibular pouch.

ing the auditory vesicle or whether they have migrated from the wall of the neural tube is not yet certain. The ganglion gradually splits into two parts: the **vestibular ganglion** and the **spiral ganglion.** The peripheral branches of the vestibular ganglion pass in two divisions; the pars superior gives rami to

the superior ampulla of the anterior semicircular duct, to the lateral ampulla, and to the utricle; the pars inferior gives rami to the saccule and the posterior ampulla. The proximal fibers of the vestibular ganglion form the vestibular nerve; the proximal fibers of the spiral ganglion form the cochlear nerve.

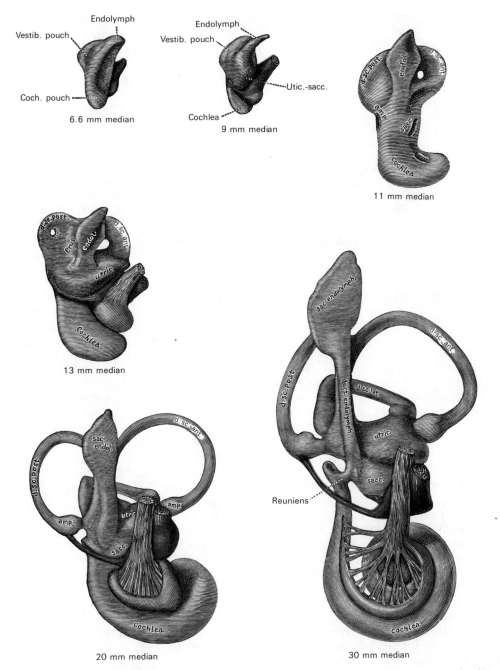

FIG. 13-41. Median views of membranous labyrinth and acoustic complex in human embryos. 25 × dia. (Streeter.)

EXTERNAL EAR

The **external ear** consists of the expanded portion named the **auricula** or **pinna** and the **external acoustic meatus.** The former projects from the side of the head and collects the vibrations of the air by which sound is produced; the latter leads inward from the bottom of the auricula and conducts the vibrations to the tympanic cavity.

Auricula

The **auricula** or **pinna** (Fig. 13-42) is ovoid, with its larger end directed upward. Its lateral surface is irregularly concave, directed slightly anteriorly, and has numerous eminences and depressions, to which names have been assigned. The prominent rim of the auricula is called the **helix.** Where the helix turns inferiorly, a small tubercle, the **auricu-**

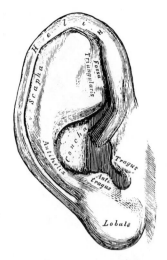

FIG. 13-42. The auricula, lateral surface.

lar tubercle of Darwin, is frequently seen; this tubercle is evident about the sixth month of fetal life, when the whole auricula closely resembles that of some adult monkeys. Another curved prominence, parallel with and anterior to the helix, is called the antihelix; this divides above into two crura, between which is a triangular depression, the fossa triangularis. The narrow curved depression between the helix and the antihelix is called the scapha. The antihelix describes a curve around a deep, capacious cavity, the concha, which is partially divided into two parts by the crus or commencement of the helix; the superior part is termed the cymba conchae, and the inferior part is called the cavum conchae. Anterior to the concha and projecting posteriorly over the meatus is a small pointed eminence, the tragus, so called because it is generally covered on its undersurface with a tuft of hair resembling a goat's beard. Opposite the tragus and separated from it by the intertragic notch is a small tubercle, the antitragus. Below this is the lobule, which is composed of areolar and adipose tissues and lacks the firmness and elasticity of the rest of the auricula.

The cranial surface of the auricula presents elevations that correspond to the depressions on its lateral surface, after which they are named, e.g., eminentia conchae, eminentia triangularis.

STRUCTURE OF THE AURICULA

The auricula is composed of a thin plate of yellow elastic cartilage covered with integument; it is connected to the surrounding parts by ligaments and muscles and to the commencement of the external acoustic meatus by fibrous tissue.

INTEGUMENT. The skin is thin, closely adherent to the cartilage, and covered with fine hairs furnished with sebaceous glands, which are most numerous in the concha and scaphoid fossa. On the tragus and antitragus, the hairs are strong and numerous. The skin of the auricula is continuous with that lining the external acoustic meatus.

CARTILAGE. The cartilage of the auricula (cartilago auriculae; cartilage of the pinna) (Figs. 13-42; 13-43) consists of a single piece; it gives form to this part of the ear, and upon its surface are found the eminences and depressions just described. It is absent from the lobule; it is also deficient between the tragus and beginning of the helix, the gap being filled by dense fibrous tissue. At the front of the auricula, where the helix bends upward, is a small projection of cartilage called the spina helicis; in the lower part of the helix, the cartilage is prolonged downward as a tail-like process, the cauda helicis; this is separated from the antihelix by a fissure, the fissura antitragohelicina. The cranial aspect of the cartilage exhibits a transverse furrow, the sulcus antihelicis transversus, which corresponds with the inferior crus of the antihelix and separates the eminentia conchae from the eminentia triangularis. The eminentia conchae is crossed by a vertical ridge (ponticulus), which gives attachment to the auricularis posterior muscle. In the cartilage of the auricula are two fissures, one behind the crus helicis and another in the tragus.

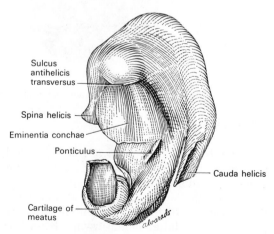

FIG. 13-43. Cranial surface of cartilage of right auricula.

LIGAMENTS. The ligaments of the auricula (*ligamenta auricularia* [*Valsalva*[1]]; *ligaments of the pinna*) consist of two sets: (1) **extrinsic,** connecting the auricula to the side of the head; (2) **intrinsic,** connecting various parts of the cartilage together.

The **extrinsic ligaments** are two, anterior and posterior. The *anterior ligament* extends from the tragus and spina helicis to the root of the zygomatic process of the temporal bone. The *posterior ligament* passes from the posterior surface of the concha to the outer surface of the mastoid process.

The chief **intrinsic ligaments** are: (1) a strong fibrous band that stretches from the tragus to the commencement of the helix, completing the meatus in front and partly encircling the boundary of the concha; and (2) a band between the antihelix and the cauda helicis. Other less important bands are found on the cranial surface of the pinna.

MUSCLES. The muscles of the auricula (Fig. 13-44) consist of two sets: (1) **extrinsic,** which connect the auricula with the skull and scalp and move it as a whole; and (2) **intrinsic,** which extend from one part of the auricle to another.

[1] Antonio Maria Valsalva (1666–1723): An Italian anatomist (Bologna).

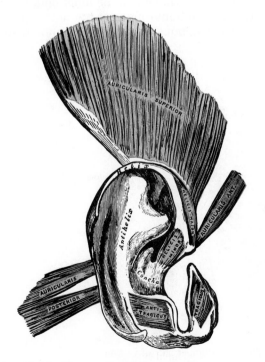

FIG. 13-44. The muscles of the auricula.

The **extrinsic muscles** are the auriculares anterior, superior, and posterior, described with the facial muscles in Chapter 6.

The **intrinsic muscles** are:

Helicis major
Helicis minor
Tragicus
Antitragicus
Transversus auriculae
Obliquus auriculae

The **helicis major** is a narrow vertical band situated upon the anterior margin of the helix. It *arises* from the spina helicis and is *inserted* into the anterior border of the helix, where the helix is about to curve backward.

The **helicis minor** is an oblique fasciculus that covers the crus helicis.

The **tragicus** is a short, flat vertical band on the lateral surface of the tragus.

The **antitragicus** *arises* from the outer part of the antitragus and is inserted into the cauda helicis and antihelix.

The **transversus auriculae** is placed on the cranial surface of the pinna. It consists of scattered fibers, partly tendinous and partly muscular, that extend from the eminentia conchae to the prominence corresponding with the scapha.

The **obliquus auriculae,** also on the cranial surface, consists of a few fibers that extend from the upper and back part of the concha to the convexity immediately above it.

VESSELS. The **arteries** of the auricula are the posterior auricular from the external carotid, the anterior auricular from the superficial temporal, and a branch from the occipital artery.

The **veins** accompany the corresponding arteries.

NERVES. The auriculares anterior and superior and the intrinsic muscles on the lateral surface are supplied by the temporal branch of the facial nerve. The auricularis posterior and the intrinsic muscles on the cranial surface are supplied by the posterior auricular branch of the same nerve.

The **sensory nerves** are: the great auricular, from the cervical plexus; the auricular branch of the vagus; the auriculotemporal branch of the mandibular nerve; and the lesser occipital from the cervical plexus.

External Acoustic Meatus

The **external acoustic meatus** (*external auditory canal or meatus*) extends from the bottom of the concha to the tympanic mem-

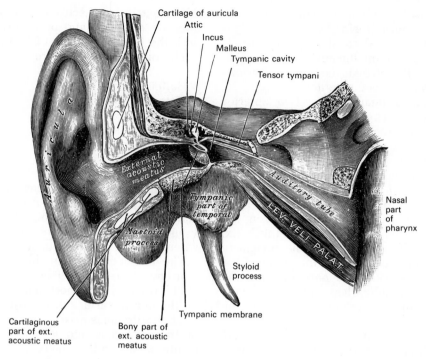

Fig. 13-45. External and middle ear, opened anteriorly; right side.

brane (Figs. 13-45; 13-46). It is about 4 cm long if measured from the tragus; from the bottom of the concha it is about 2.5 cm long. It forms an S-shaped curve, and is directed at first inward, forward, and slightly upward (*pars externa*); it then passes inward and backward (*pars media*), and lastly is carried inward, forward, and slightly downward (*pars interna*). It is an oval cylindrical canal, the greatest diameter being directed downward and backward at the external orifice, but nearly horizontal at the inner end. It has two constrictions, one near the inner end of the cartilaginous portion, and another, the

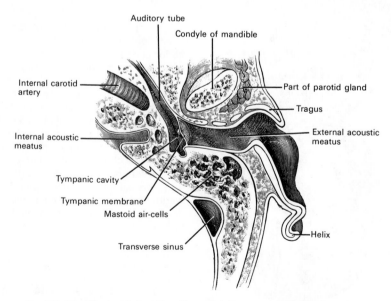

Fig. 13-46. Horizontal section through left ear; upper half of section.

isthmus, in the osseous portion about 2 cm from the bottom of the concha. The tympanic membrane, which closes the inner end of the meatus, is directed obliquely, so the floor and anterior wall of the meatus are longer than the roof and posterior wall.

The external acoustic meatus is formed partly by cartilage and membrane and partly by bone, and it is lined with skin.

The **cartilaginous portion** is about 8 mm long; it is continuous with the cartilage of the auricula and firmly attached to the circumference of the auditory process of the temporal bone. The cartilage is deficient at the cranial and posterior part of the meatus, where it is replaced by fibrous membrane; two or three deep fissures are present in the anterior part of the cartilage.

The **osseous portion** is about 16 mm long and is narrower than the cartilaginous portion. It is directed inward and a little forward, forming in its course a slight curve, the convexity of which is upward and backward. The inner end is smaller than the outer end and is sloped; the anterior wall projects beyond the posterior for about 4 mm. This end is marked, except at its cranial part, by a narrow groove, the **tympanic sulcus,** in which the circumference of the tympanic membrane is attached. Its outer end is dilated and rough in the greater part of its circumference for the attachment of the cartilage of the auricula. The anterior and inferior parts of the osseous portion are formed by a curved plate of bone—the tympanic part of the temporal bone—which in the fetus exists as a separate ring **(anulus tympanicus)** that is incomplete at its cranial part (Chapter 4).

The **skin** lining the meatus is very thin; it adheres closely to the cartilaginous and osseous portions of the tube and covers the outer surface of the tympanic membrane. After maceration, the thin pouch of epidermis, when withdrawn, preserves the form of the meatus. In the thick subcutaneous tissue of the cartilaginous part of the meatus are numerous ceruminous glands that secrete the ear wax; their structure resembles that of the sudoriferous glands.

RELATIONS OF THE MEATUS. Anterior to the osseous part is the condyle of the mandible, but the latter is frequently separated from the cartilaginous part by a portion of the parotid gland. The movements of the jaw influence to some extent the lumen of this latter portion. Posterior to the osseous part are the mastoid air cells, which are separated from the meatus by a thin layer of bone.

VESSELS. The **arteries** supplying the walls of the meatus are branches from the posterior auricular, maxillary, and superficial temporal arteries.

NERVES. The **nerves** are chiefly derived from the auriculotemporal branch of the mandibular nerve and the auricular branch of the vagus.

MIDDLE EAR OR TYMPANUM

The **middle ear** or **tympanic cavity** (*drum; tympanum*) is an irregular, laterally compressed space within the temporal bone. It is filled with air, which is conveyed to it from the nasal part of the pharynx through the auditory tube. The cavity contains a chain of tiny movable bones that bridge from its lateral to its medial wall and convey the vibrations communicated to the tympanic membrane across the cavity to the internal ear.

The tympanic cavity consists of two parts: the **tympanic cavity proper,** opposite the tympanic membrane, and the **attic** or **epitympanic recess,** cranial to the level of the membrane which contains the upper half of the malleus and the greater part of the incus. Including the attic, the vertical and anteroposterior diameters of the cavity are each about 15 mm. The transverse diameter measures about 6 mm superiorly and 4 mm inferiorly; opposite the center of the tympanic membrane it is only about 2 mm. The tympanic cavity is bounded laterally by the tympanic membrane; medially, by the lateral wall of the internal ear. It communicates posteriorly with the tympanic antrum and through it with the mastoid air cells; anteriorly, the tympanic cavity communicates with the auditory tube (Fig. 13-45).

Tegmental Wall or Roof

The **roof** is formed by a thin plate of bone, the **tegmen tympani,** which separates the tympanic from the cranial cavity. It is situated on the anterior surface of the petrous portion of the temporal bone close to its angle of junction with the squama temporalis. It is prolonged posteriorly to roof in the tympanic antrum, and anteriorly to cover the semicanal for the tensor tympani muscle. Its lateral edge corresponds with the remains of the petrosquamous suture.

Jugular Wall or Floor

The **floor** is narrow and consists of a thin plate of bone (**fundus tympani**), which separates the tympanic cavity from the jugular fossa. It has, near the labyrinthic wall, a small aperture for the passage of the tympanic branch of the glossopharyngeal nerve.

Membranous or Lateral Wall

The **lateral wall** (*outer wall*) is formed mainly by the tympanic membrane and partly by the ring of bone into which this membrane is inserted. The ring of bone is incomplete at its upper part, forming a notch (*of Rivinus[1]*), close to which are three small apertures: the **iter chordae posterius,** the **petrotympanic fissure,** and the **iter chordae anterius.**

The **iter chordae posterius** (*apertura tympanica canaliculi chordae*) is situated in the angle of junction between the mastoid and membranous wall of the tympanic cavity,

[1] Augustus Quirinus Rivinus (1652–1723): A German anatomist and botanist (Leipzig).

immediately posterior to the tympanic membrane and on a level with the superior end of the manubrium of the malleus. It leads into a minute canal in the mastoid bone that descends almost parallel with the facial nerve and meets the facial canal at an acute angle near the stylomastoid foramen. Through it the chorda tympani nerve enters the tympanic cavity (Fig. 13-47).

The **petrotympanic fissure** (*Glaserian fissure[2]*) (Fig. 13-47) opens just superior and anterior to the ring of bone into which the tympanic membrane is inserted; in this situation it is a mere slit about 2 mm long. It lodges the anterior process and anterior ligament of the malleus and gives passage to the anterior tympanic branch of the maxillary artery.

The **iter chordae anterius** (*canal of Huguier[3]*) is at the medial end of the petrotympanic fissure; through it the chorda tympani nerve leaves the tympanic cavity.

[2] Johann Heinrich Glaser (1629–1675): A Swiss anatomist (Basel).
[3] Pierre Charles Huguier (1804–1874): A French surgeon (Paris).

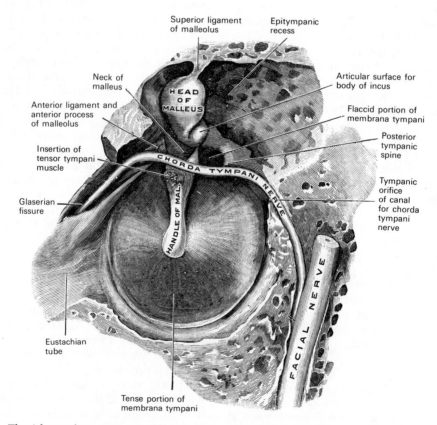

FIG. 13-47. The right membrana tympani with the malleus and the chorda tympani, viewed from within. (Spalteholz.)

Tympanic Membrane

The **tympanic membrane** (Figs. 13-47; 13-48) separates the tympanic cavity from the bottom of the external acoustic meatus. It is a thin, semitransparent membrane, nearly oval, directed very obliquely inferiorly and medially to form an angle of about 55° with the floor of the meatus. Its longest diameter, almost vertical, measures 9 to 10 mm; its shortest diameter measures 8 to 9 mm. The greater part of its circumference is thick and forms a **fibrocartilaginous ring,** which is fixed in the **tympanic sulcus** at the inner end of the meatus. This sulcus is deficient superiorly at the **notch of Rivinus,** and from the ends of this notch two bands, the **anterior** and **posterior malleolar folds,** are prolonged to the lateral process of the malleus. The small, somewhat triangular part of the membrane situated superior to these folds is lax and thin and is named the **pars flaccida,** as opposed to the rest of the membrane, which is called the **pars tensa.** The manubrium of the malleus is firmly attached to the medial surface of the membrane as far as its center, which it draws inward toward the tympanic cavity. The lateral surface of the membrane is thus concave, and the most depressed part of this concavity is named the **umbo.**

STRUCTURE OF THE TYMPANIC MEMBRANE

The tympanic membrane is composed of three strata: **lateral** (*cutaneous*), **intermediate** (*fibrous*), and **medial** (*mucous*). The **cutaneous stratum** is derived from the integument lining the meatus. The **fibrous stratum** consists of two layers: a **radiate stratum,** the fibers of which diverge from the manubrium of the malleus, and a **circular stratum,** the fibers of which are plentiful around the cir-

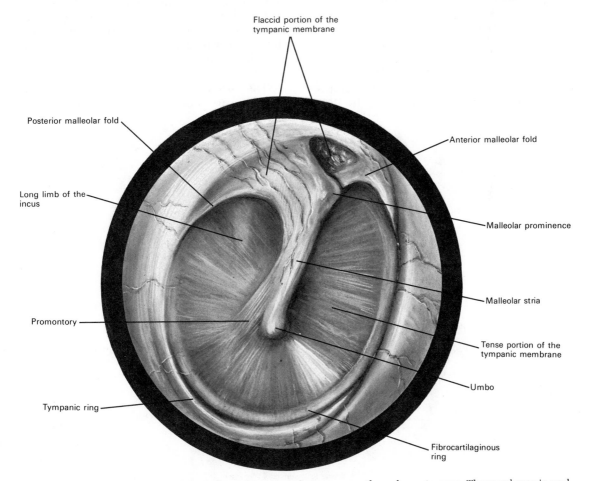

Flaccid portion of the tympanic membrane

Posterior malleolar fold

Long limb of the incus

Promontory

Tympanic ring

Anterior malleolar fold

Malleolar prominence

Malleolar stria

Tense portion of the tympanic membrane

Umbo

Fibrocartilaginous ring

Fig. **13-48.** The lateral aspect of the right tympanic membrane as seen through an otoscope. The membrane is oval and measures about 11 mm in its long diameter and about 9 mm across. (Redrawn after Sobotta.)

cumference but sparse and scattered near the center of the membrane. Branched or dendritic fibers are also present, especially in the posterior half of the membrane. The **mucous stratum** is the internal surface of the membrane derived from the lining of the cavity.

VESSELS AND NERVES. The **arteries** of the tympanic membrane are derived from the deep auricular branch of the maxillary artery, which ramifies beneath the cutaneous stratum, and from the stylomastoid branch of the posterior auricular artery and the tympanic branch of the maxillary artery, which are distributed on the mucous surface. The superficial **veins** open into the external jugular; those on the deep surface drain partly into the transverse sinus and veins of the dura mater, and partly into a plexus on the auditory tube.

The membrane receives its chief **nerve supply** from the auriculotemporal branch of the mandibular nerve; the auricular branch of the vagus and the tympanic branch of the glossopharyngeal nerve also supply it.

Labyrinthic or Medial Wall

The **medial wall** (*inner wall*) is directed vertically and presents for examination the **fenestrae vestibuli** and **cochleae,** the prom-

ontory, and the **prominence of the facial canal** (Fig. 13-49).

The **fenestra vestibuli** (*fenestra ovalis; oval window*) is a reniform opening leading from the tympanic cavity into the vestibule of the internal ear. Its long diameter is horizontal, and its convex border is upward. In the intact body it is occupied by the base of the stapes, the circumference of which is secured by the annular ligament to the margin of the foramen (Fig. 13-50).

The **fenestra cochleae** (*fenestra rotunda; round window*) is situated inferior and a little posterior to the fenestra vestibuli, from which it is separated by a rounded elevation, the **promontory.** It is placed at the bottom of a funnel-shaped depression and, in the macerated bone, leads into the cochlea of the internal ear; in the intact body it is closed by a membrane, the **secondary tympanic membrane,** which is concave toward the cochlea. This membrane consists of three layers: an external or mucous layer derived from the mucous lining of the tympanic cavity; an internal layer from the lining membrane of the cochlea; and an intermediate or fibrous layer.

The **promontory** (*promontorium*) is a rounded prominence formed by the outward

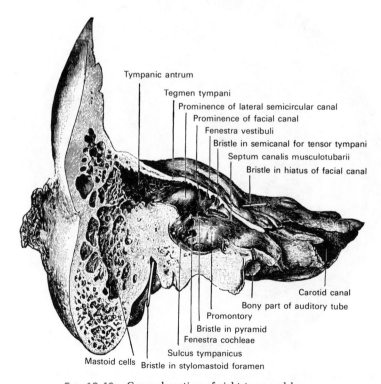

Tympanic antrum
Tegmen tympani
Prominence of lateral semicircular canal
Prominence of facial canal
Fenestra vestibuli
Bristle in semicanal for tensor tympani
Septum canalis musculotubarii
Bristle in hiatus of facial canal

Carotid canal
Bony part of auditory tube
Promontory
Bristle in pyramid
Fenestra cochleae
Sulcus tympanicus
Mastoid cells   Bristle in stylomastoid foramen

FIG. 13-49. Coronal section of right temporal bone.

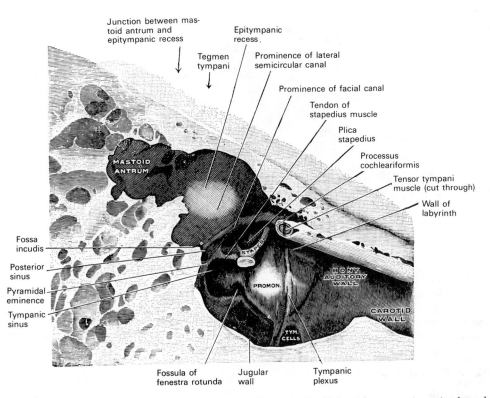

Junction between mastoid antrum and epitympanic recess

Epitympanic recess

Tegmen tympani

Prominence of lateral semicircular canal

Prominence of facial canal

Tendon of stapedius muscle

Plica stapedius

Processus cochleariformis

Tensor tympani muscle (cut through)

Wall of labyrinth

Fossa incudis

Posterior sinus

Pyramidal eminence

Tympanic sinus

Fossula of fenestra rotunda

Jugular wall

Tympanic plexus

MASTOID ANTRUM

STAPES

PROMON.

BONY AUDITORY WALL

CAROTID WALL

TYM. CELLS

FIG. 13-50. The medial wall and part of the posterior and anterior walls of the right tympanic cavity, lateral view. (Spalteholz.) (See Fig. 13-49.)

projection of the first turn of the cochlea. It is placed between the fenestrae, and its surface is furrowed by small grooves for the nerves of the tympanic plexus. A minute spicule of bone frequently connects the promontory to the pyramidal eminence.

The **prominence of the facial canal** (*prominentia canalis facialis; prominence of aqueduct of Fallopius*[1]) indicates the position of the bony canal in which the facial nerve is contained; this canal traverses the labyrinthic wall of the tympanic cavity superior to the fenestra vestibuli, and posterior to that opening, curves nearly vertically inferiorly along the mastoid wall (Fig. 13-51).

Mastoid or Posterior Wall

The **posterior wall** presents for examination the **entrance to the tympanic antrum,** the **pyramidal eminence,** and the **fossa incudis.**

The **entrance to the antrum** is a large, ir-

regular aperture that leads posteriorly from the epitympanic recess into a considerable air space, the **tympanic** or **mastoid antrum** (see Chapter 4). The antrum leads posteriorly and inferiorly into the **mastoid air cells,** which vary greatly in number, size, and form. The antrum and mastoid air cells are lined by mucous membrane that is continuous with that which lines the tympanic cavity. On the medial wall of the entrance to the antrum is a rounded eminence, situated superior and posterior to the prominence of the facial canal; it corresponds with the position of the ampullated ends of the anterior and lateral semicircular canals.

The **pyramidal eminence** (*pyramid*) is situated immediately posterior to the fenestra vestibuli and anterior to the vertical portion of the facial canal. It is hollow but contains the stapedius muscle. Its summit projects anteriorly toward the fenestra vestibuli and is pierced by a small aperture that transmits the tendon of the muscle. The cavity in the pyramidal eminence is prolonged posteriorly toward the facial canal and communicates with it by a minute aperture that trans-

[1]Gabriele Falloppio (1523–1563): An Italian anatomist and surgeon (Padua).

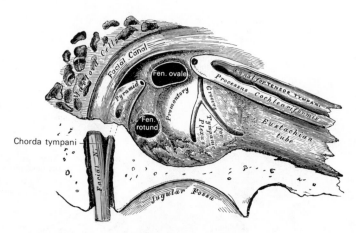

FIG. 13-51. View of the medial wall of the tympanum (enlarged). (See Fig. 13-49.)

mits a branch of the facial nerve to the stapedius muscle.

The **fossa incudis** is a small depression in the inferior and posterior part of the epitympanic recess; it lodges the short crus of the incus.

Carotid or Anterior Wall

The **anterior wall** corresponds with the carotid canal, from which it is separated by a thin plate of bone that is perforated by the tympanic branch of the internal carotid artery and the caroticotympanic nerve, which connects the sympathetic plexus on the internal carotid artery with the tympanic plexus on the promontory. At the superior part of the anterior wall are the orifice of the semicanal for the tensor tympani muscle and the tympanic orifice of the auditory tube, separated from each other by a thin horizontal plate of bone, the **septum canalis musculotubarii.** These canals run from the tympanic cavity medialward to the retiring angle between the squama and the petrous portion of the temporal bone.

The **semicanal for the tensor tympani** (*semicanalis m. tensoris tympani*) is the superior and the smaller of the two canals; it is cylindrical and lies beneath the tegmen tympani. It extends to the labyrinthic wall of the tympanic cavity and ends immediately above the fenestra vestibuli. The **septum canalis musculotubarii** (*processus cochleariformis*) passes posteriorly, forming its lateral wall and floor; it expands above the anterior end of the fenestra vestibuli and terminates there by curving laterally to form a pulley over which the tendon of the muscle passes.

Auditory Tube

The **auditory tube** (*tuba auditiva; Eustachian tube*[1]) is the channel through which the tympanic cavity communicates with the nasal part of the pharynx. Its length is about 36 mm, and in its medialward course it forms an angle of about 45° with the sagittal plane and an angle of 30 to 40° with the horizontal plane. Its walls are composed partly of bone and partly of cartilage and fibrous tissue (Fig. 13-45).

The **osseous portion** is about 12 mm long. It begins in the carotid wall of the tympanic cavity, inferior to the septum canalis musculotubarii, and gradually narrowing, ends at the angle of junction of the squama and the petrous portion of the temporal bone. Its extremity has a jagged margin that serves for the attachment of the cartilaginous portion.

The **cartilaginous portion,** about 24 mm long, is contained in a triangular plate of elastic fibrocartilage, the apex of which is attached to the margin of the medial end of the osseous portion of the tube, while the base lies directly under the mucous membrane of the nasal part of the pharynx, where it forms an elevation, the **torus tubarius** or cushion, posterior to the pharyngeal orifice of the tube. The superior edge of the cartilage is bent over laterally, forming a groove or furrow that is open inferiorly and laterally. The part of the canal lying in this furrow is completed by fibrous membrane. The cartilage lies in a groove between the petrous part of the temporal and the great

[1]Bartolommeo Eustachio (1520?–1574): An Italian anatomist (Rome).

wing of the sphenoid, and it ends opposite the middle of the medial pterygoid plate.

The cartilaginous and bony portions of the tube are not in the same plane, the former inclining inferiorly a little more than the latter. The diameter of the tube is not uniform throughout; it is greatest at the pharyngeal orifice, least at the junction of the bony and cartilaginous portions, and again increases toward the tympanic opening. The narrowest part of the tube is termed the **isthmus.** The position and relations of the **pharyngeal orifice** are described with the nasal part of the pharynx.

The mucous membrane of the tube is continuous with that of the nasal part of the pharynx and the tympanic cavity. It is covered with ciliated epithelium and is thin in the osseous portion, while in the cartilaginous portion it contains many mucous glands and, near the pharyngeal orifice, a considerable amount of adenoid tissue, the **tubal tonsil.** The tube is opened during deglutition by the salpingopharyngeus and dilatator tubae muscles. The latter arises from the hook of the cartilage and from the membranous part of the tube and blends below with the tensor veli palatini.

Auditory Ossicles

The tympanic cavity contains a chain of three movable ossicles, the **malleus, incus,** and **stapes.** The first is attached to the tympanic membrane, the last to the circumference of the fenestra vestibuli, the incus being placed between and connected to both by delicate articulations (Fig. 13-55).

MALLEUS. The malleus (Fig. 13-52), so named from its fancied resemblance to a hammer, consists of a **head, neck,** and three processes, viz., the **manubrium** and the **anterior** and **lateral processes.**

The **head** is the large superior extremity of the bone; it is oval and articulates posteriorly with the incus, being free in the rest of its extent. The facet for articulation with the incus is constricted near the middle and consists of a superior larger and an inferior smaller part, which are bent nearly at a right angle to each other. Opposite the constriction, the inferior margin of the facet projects in the form of a process, the **cog-tooth** or **spur of the malleus.**

The **neck** is the narrow contracted part joining the head with a prominence to which the anterior and posterior processes are attached.

The **manubrium mallei** is connected by its lateral margin with the tympanic membrane. It is directed slightly posteriorly as well as inferiorly; it decreases in size toward its free end, which is slightly curved and flattened transversely. On its medial side, near the head end, is a slight projection into which the tendon of the tensor tympani is inserted.

The **anterior process** is a delicate spicule directed anteriorly toward the petrotympanic fissure, to which it is connected by ligamentous fibers. In the fetus this is the longest process of the malleus, and it is directly continuous with the cartilage of Meckel.[1]

The **lateral process** is a slight conical projection that springs from the root of the manubrium. It is directed laterally and is attached to the superior part of the tympanic membrane and, by means of the anterior and posterior malleolar folds, to the extremities of the notch of Rivinus.

INCUS. The incus (Fig. 13-53) received its name from its resemblance to an anvil, but it looks more like a premolar tooth with two

[1]Johann Friedrich Meckel (the Younger) (1781–1833): A German anatomist and surgeon (Halle).

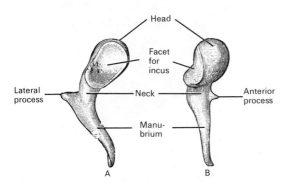

FIG. **13-52.** Left malleus. *A,* Posterior view. *B,* Medial view.

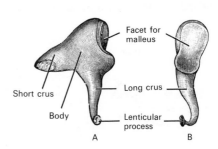

FIG. **13-53.** Left incus. *A,* Medial view. *B,* Anterior view.

roots that differ in length and are widely separated from each other. It consists of a **body**
and **two crura.**

The **body** is somewhat cubical but compressed transversely. On its anterior surface
is a deeply concavo-convex facet that articulates with the head of the malleus.

The two crura diverge from each other
nearly at right angles. The **short crus,** which
is somewhat conical, projects almost horizontally backward and is attached to the
fossa incudis in the posterior part of the
epitympanic recess. The **long crus** descends
nearly vertically parallel to the manubrium
of the malleus and, bending medialward,
ends in a rounded projection, the **lenticular
process,** which is tipped with cartilage and
articulates with the head of the stapes.

STAPES. The stapes (Fig. 13-54), so called
from its resemblance to a stirrup, consists of
a **head, neck, two crura,** and a **base.**

The **head** has a depression, which is covered by cartilage; it articulates with the lenticular process of the incus.

The **neck,** the constricted part of the bone
succeeding the head, has the insertion of the
tendon of the stapedius muscle attached
to it.

The **two crura** diverge from the neck and
are connected at their ends by a flattened
oval plate, the **base,** which forms the foot-
plate of the stirrup and is fixed to the margin
of the fenestra vestibuli by a ring of ligamentous fibers. Of the two crura, the anterior is
shorter and less curved than the posterior.

ARTICULATIONS OF THE AUDITORY OSSICLES.
The *incudomalleolar joint* is a saddleshaped
diarthrosis; it is surrounded by an articular
capsule, and the joint cavity is incompletely
divided into two by a wedgeshaped articular
disc or meniscus. The *incudostapedial joint*
is an enarthrosis surrounded by an articular
capsule. Some observers have described an
articular disc or meniscus in this joint; others regard the joint as a syndesmosis.

LIGAMENTS OF THE OSSICLES. The ossicles
are connected with the walls of the tympanic cavity by ligaments: three for the malleus and one each for the incus and stapes
(Fig. 13-55).

The **anterior ligament of the malleus** is attached by one end to the neck of the malleus,
just above the anterior process, and by the
other end to the anterior wall of the tympanic cavity, close to the petrotympanic
fissure. Some of its fibers are prolonged
through the fissure to reach the spine of the
sphenoid bone.

The **superior ligament of the malleus** is a
delicate, round bundle that descends from
the roof of the epitympanic recess to the
head of the malleus.

The **lateral ligament of the malleus** is a triangular band passing from the posterior part
of the notch of Rivinus to the head of the
malleus. Helmholtz described the anterior
ligament and the posterior part of the lateral
ligament as forming together the **axis ligament,** around which the malleus rotates.

The **posterior ligament of the incus** is a
short, thick band connecting the end of the
short crus of the incus to the fossa incudis.

A **superior ligament of the incus** has been
described, but it is little more than a fold of
mucous membrane.

The vestibular surface and the circumfer-

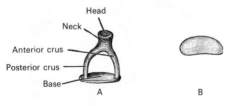

FIG. 13-54. *A,* Left stapes. *B,* Base of stapes, medial
surface.

Head

Neck

Anterior crus

Posterior crus

Base

A

B

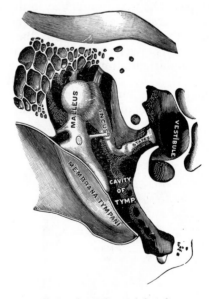

FIG. 13-55. Chain of ossicles and their ligaments, anterior view of a vertical transverse section of the tympanum. (Testut.)

ence of the base of the stapes are covered with hyaline cartilage. The cartilage that encircles the base is attached to the margin of the fenestra vestibuli by a fibrous ring, the **annular ligament of the base of the stapes.**

Muscles of the Tympanic Cavity

These are the tensor tympani and the stapedius (Figs. 13-45; 13-50).

The **tensor tympani** is contained in the bony canal above the osseous portion of the auditory tube, from which it is separated by the septum canalis musculotubarii. It *arises* from the cartilaginous portion of the auditory tube and the adjoining part of the great wing of the sphenoid, as well as from the osseous canal in which it is contained. Passing posteriorly through the canal, it ends in a slender tendon that enters the tympanic cavity, makes a sharp bend around the extremity of the septum, the **processus cochleariformis,** and is *inserted* into the manubrium of the malleus, near its root. It is supplied by a branch of the mandibular nerve that passes through the otic ganglion.

The **stapedius** *arises* from the wall of the conical cavity inside the pyramid. Its tendon emerges from the orifice at the apex of the eminence, and is *inserted* into the posterior surface of the neck of the stapes. It is supplied by a branch of the facial nerve.

ACTIONS. The tensor tympani draws the tympanic membrane medialward, and thus increases its tension. The stapedius pulls the head of the stapes posteriorly, tilting the base and possibly increasing the tension of the fluid within the internal ear. Both muscles have a snubbing action, reducing the oscillations of the ossicles and thereby protecting the inner ear from injury during a loud noise, such as that of a riveting machine.

Mucous Membrane of the Tympanic Cavity

This membrane is continuous with that of the pharynx through the auditory tube. It invests the auditory ossicles, the muscles, and the nerves contained in the tympanic cavity. It forms the medial layer of the tympanic membrane and the lateral layer of the secondary tympanic membrane, and is reflected into the tympanic antrum and mastoid cells, which it lines throughout. It forms several folds that extend from the walls of the tympanic cavity to the ossicles. Of these,

one descends from the roof of the cavity to the head of the malleus and superior margin of the body of the incus; a second invests the stapedius muscle. Other folds invest the chorda tympani nerve and the tensor tympani muscle. These folds separate pouch-like cavities and give the interior of the tympanum a somewhat honey-combed appearance. One of these pouches, the **pouch of Prussak**[1], is well marked and lies between the neck of the malleus and the membrana flaccida. Two other recesses may be mentioned: they are formed by the mucous membrane that envelops the chorda tympani nerve and are named the **anterior** and **posterior recesses** of **Tröltsch**[2]. In the tympanic cavity this membrane is pale, thin, slightly vascular, and covered for the most part with columnar ciliated epithelium, but over the pyramidal eminence, ossicles, and tympanic membrane it possesses a flat, nonciliated epithelium. In the tympanic antrum and mastoid cells, its epithelium is also nonciliated. In the osseous portion of the auditory tube the membrane is thin, but in the cartilaginous portion it is thick, highly vascular, and provided with numerous mucous glands. The epithelium that lines the tube is columnar and ciliated.

VESSELS AND NERVES. There are six **arteries,** two of which are larger than the others, viz., the tympanic branch of the maxillary artery, which supplies the tympanic membrane and the stylomastoid branch of the posterior auricular artery, which supplies the back part of the tympanic cavity and mastoid cells. The smaller arteries include the following: the petrosal branch of the middle meningeal artery, which enters through the hiatus of the facial canal; a branch from the ascending pharyngeal artery and another from the artery of the pterygoid canal, which accompany the auditory tube; and the tympanic branch from the internal carotid, which is given off in the carotid canal and perforates the thin anterior wall of the tympanic cavity. The **veins** terminate in the pterygoid plexus and the superior petrosal sinus.

The **nerves** constitute the tympanic plexus, which ramifies upon the surface of the promontory, and the chorda tympani, which merely passes through the cavity. The plexus is formed by (1) the tympanic branch of the glossopharyngeal nerve; (2) the caroticotympanic nerves; (3) the lesser petrosal nerve;

[1]Alexander Prussak (1839–1897): A Russian otologist (St. Petersburg).
[2]Anton Friedrich von Tröltsch (1829–1890): A German otologist (Würzburg).

and (4) a branch that joins the greater petrosal nerve.

The **tympanic branch of the glossopharyngeal** (*Jacobson's nerve*[1]) enters the tympanic cavity by an aperture in its floor close to the labyrinthic wall; it divides into branches that ramify on the promontory and enter into the formation of the tympanic plexus.

The **superior and inferior caroticotympanic nerves** from the carotid plexus of sympathetic fibers pass through the wall of the carotid canal and join the tympanic plexus. The branch to the greater petrosal passes through an opening on the labyrinthic wall, anterior to the fenestra vestibuli.

The **lesser petrosal nerve** is a continuation of the tympanic branch of the glossopharyngeal nerve beyond the tympanic plexus. It penetrates the bone near the geniculate ganglion of the facial nerve, with which it communicates by a filament. It continues through the bone and enters the middle cranial fossa through a small aperture situated lateral to the hiatus of the facial canal on the anterior surface of the petrous portion of the temporal bone. It leaves the middle fossa through a fissure or foramen of its own or through the foramen ovale, and ends in the otic ganglion, constituting its root and supplying it with preganglionic parasympathetic fibers.

The **branches of distribution** of the tympanic plexus supply the mucous membrane of the tympanic cavity, the fenestra vestibuli, the fenestra cochleae, and the auditory tube. The lesser petrosal may be looked upon as the continuation of the tympanic branch of the glossopharyngeal nerve through the plexus to the otic ganglion.

In addition to the tympanic plexus, there are nerves that supply the muscles. The tensor tympani is supplied by a branch from the mandibular nerve that passes through the otic ganglion, and the stapedius is supplied by a branch from the facial nerve.

The **chorda tympani nerve** crosses the tympanic cavity (Fig. 13-47). It leaves the facial canal about 6 mm before the facial nerve emerges from the stylomastoid foramen. It runs back upward in a canal of its own, almost parallel with the facial canal, and enters the tympanic cavity through the iter chordae posterius, where it becomes invested with mucous membrane. It traverses the tympanic cavity across the medial surface of the tympanic membrane and over the upper part of the manubrium of the malleus to the carotid wall, where it makes exit through the iter chordae anterius (*canal of Huguier*).

INTERNAL EAR OR LABYRINTH

The **internal ear** is the essential part of the organ of hearing, the ultimate distribution of the vestibulocochlear nerve. It is called the

labyrinth, from the complexity of its shape, and consists of two parts: the **osseous labyrinth,** a series of cavities within the petrous part of the temporal bone, and the **membranous labyrinth,** a series of communicating membranous sacs and ducts contained within the bony cavities.

Osseus Labyrinth

The **osseous labyrinth** (Figs. 13-56; 13-57) consists of three parts: **vestibule, semicircular canals,** and **cochlea.** These are cavities hollowed out of the substance of the bone and lined by periosteum; they contain a clear fluid, the **perilymph,** in which the membranous labyrinth is suspended.

VESTIBULE. The vestibule is the central part of the osseous labyrinth. It is situated medial to the tympanic cavity, posterior to the cochlea, and anterior to the semicircular canals. It is somewhat ovoid but flat transversely; it measures about 5 mm sagittally and vertically, and about 3 mm across. In its *lateral* or *tympanic wall* is the **fenestra vestibuli,** which in the intact body is closed by the base of the stapes and annular ligament. On its *medial wall* is a small circular depression, the **recessus sphericus,** which is perforated at its anterior and inferior part by several minute holes **(macula cribrosa media)** for the passage of filaments of the vestibulocochlear nerve to the saccule. Posterior to this depression is an oblique ridge, the **crista vestibuli,** the anterior end of which is named the **pyramid of the vestibule.** This ridge bifurcates to enclose a small depression, the

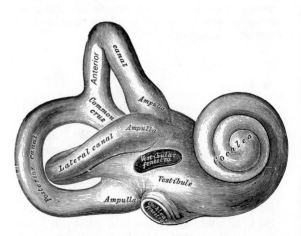

FIG. 13-56. Right osseous labyrinth with spongy bone removed. Lateral view.

fossa cochlearis, which is perforated by a number of holes for the passage of filaments of the vestibulocochlear nerve that supply the vestibular end of the ductus cochlearis. At the posterior part of the medial wall is the orifice of the **aquaeductus vestibuli,** which extends to the posterior surface of the petrous portion of the temporal bone. It transmits a small vein and contains a tubular prolongation of the membranous labyrinth, the **ductus endolymphaticus,** which ends in a cul-de-sac between the layers of the dura mater within the cranial cavity.

On the *superior wall* or *roof* is a transversely oval depression, the **recessus ellipticus,** which is separated from the recessus sphericus by the crista vestibuli already mentioned. The pyramid and adjoining part of the recessus ellipticus are perforated by a number of holes **(macula cribrosa superior).** The apertures in the pyramid transmit the nerves to the utricle; those in the recessus ellipticus transmit the nerves to the ampullae of the anterior and lateral semicircular

ducts. Posteriorly are the five orifices of the semicircular canals. Anteriorly is an elliptical opening that communicates with the scala vestibuli of the cochlea.

BONY SEMICIRCULAR CANALS.   There are three bony semicircular canals: **anterior, posterior,** and **lateral;** these are situated superior and posterior to the vestibule. They are unequal in length and compressed from side to side; each describes the greater part of a circle. Each measures about 0.8 mm in diameter, and presents a dilatation at one end, called the **ampulla,** which measures more than twice the diameter of the tube. They open into the vestibule by five orifices, with one of the apertures being common to two of the canals.

The **anterior semicircular canal** (*superior semicircular canal*); 15 to 20 mm long, is vertical in direction and is placed transversely to the long axis of the petrous portion of the temporal bone (Fig. 13-58), on the anterior surface of which its arch forms a round projection. It describes about two-thirds of a

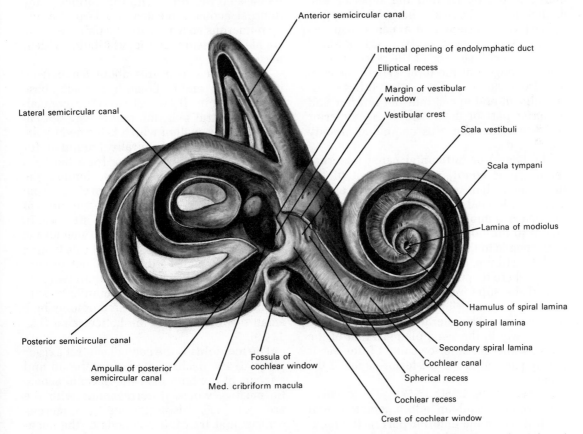

FIG. 13-57.   The right bony labyrinth with the semicircular canals and the cochlea opened. (Redrawn after Sobotta.)

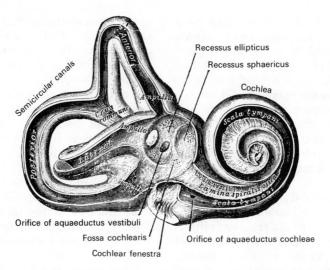

Recessus ellipticus

Recessus sphaericus

Cochlea

Semicircular canals

Orifice of aquaeductus vestibuli

Fossa cochlearis

Cochlear fenestra

Orifice of aquaeductus cochleae

FIG. 13-58.   Interior of right osseous labyrinth.

circle. Its lateral extremity is ampullated and opens into the superior part of the vestibule; the opposite end joins with the superior part of the posterior canal to form the **crus commune,** which opens into the superior and medial part of the vestibule.

The **posterior semicircular canal,** also vertical, is directed posteriorly, nearly parallel to the posterior surface of the petrous bone. It is the longest of the three canals, measuring from 18 to 22 mm. Its inferior or ampullated end opens into the inferior and posterior part of the vestibule; its superior end opens into the crus commune already mentioned.

The **lateral** or **horizontal canal** (*external semicircular canal*) is the shortest of the three canals. It measures 12 to 15 mm, and its arch is directed horizontally lateralward; thus, each semicircular canal stands at right angles to the other two. Its ampullated end corresponds to the superior and lateral angle of the vestibule, just above the fenestra vestibuli, where it opens close to the ampullated end of the superior canal. Its opposite end opens at the superior and posterior part of the vestibule. The lateral canal of one ear is very nearly in the same plane as that of the other, while the anterior canal of one ear is nearly parallel to the posterior canal of the other.

COCHLEA (Figs. 13-56; 13-57).   The cochlea bears some resemblance to a common snail shell. It forms the anterior part of the labyrinth, is conical, and is placed almost hori-

zontally in front of the vestibule. Its **apex** (*cupula*) is directed anteriorly and laterally, with a slight inclination inferiorly, toward the labyrinthic wall of the tympanic cavity. Its **base** corresponds with the bottom of the internal acoustic meatus and is perforated by numerous apertures for the passage of the cochlear division of the vestibulocochlear nerve.

The cochlea measures about 5 mm from base to apex, and its breadth across the base is about 9 mm. It consists of: (1) a conical-shaped central axis, the **modiolus;** (2) a canal, the inner wall of which is formed by the central axis, wound spirally around it for two and three-quarter turns, from the base to the apex; and (3) a delicate lamina, the **osseous spiral lamina,** which projects from the modiolus, and, following the windings of the canal, partially subdivides it into two. In the intact body, the **basilar membrane** stretches from the free border of this lamina to the outer wall of the bony cochlea and completely separates the canal into two passages, which, however, communicate with each other at the apex of the modiolus by a small opening, named the **helicotrema** (Fig. 13-59).

The **modiolus** is the conical central axis or pillar of the cochlea. Its base is broad and appears at the bottom of the internal acoustic meatus, where it corresponds with the area cochleae; it is perforated by numerous orifices that transmit filaments of the cochlear division of the vestibulocochlear nerve.

The nerves for the first turn and a half pass through the foramina of the tractus spiralis foraminosus; those for the apical turn pass through the foramen centrale. The canals of the tractus spiralis foraminosus pass through the modiolus and successively bend outward to reach the attached margin of the lamina spiralis ossea. Here they become enlarged and by their apposition form the **spiral canal of the modiolus,** which follows the course of the attached margin of the osseous spiral lamina and lodges the **spiral ganglion** (*ganglion of Corti*[1]). The foramen centrale is continued into a canal that runs through the middle of the modiolus to its apex. The modiolus diminishes rapidly in size in the second and succeeding coil.

The bony canal of the cochlea takes two and three-quarter turns around the modiolus. It is about 30 mm long and diminishes gradually in diameter from the base to the summit, where it terminates in the **cupula,** which forms the apex of the cochlea (Fig. 13-59). The beginning of this canal is about 3 mm in diameter; it diverges from the modiolus toward the tympanic cavity and vestibule and presents three openings. One open-

[1]Alfonso Corti (1822–1888): An Italian anatomist (Sardinia).

ing, the **fenestra cochleae,** communicates with the tympanic cavity; in the intact body this aperture is closed by the **secondary tympanic membrane.** Another opening has an elliptical form and opens into the vestibule. The third opening is the aperture of the aquaeductus cochleae, which leads to a minute, funnel-shaped canal that opens on the inferior surface of the petrous part of the temporal bone. It transmits a small vein, and also forms a communication between the subarachnoid cavity and the scala tympani.

The **osseous spiral lamina** is a bony shelf or ledge that projects from the modiolus into the interior of the canal, and like the canal, takes two and three-quarter turns around the modiolus. It reaches about half way toward the outer wall of the tube and partially divides its cavity into two passages or scalae, of which the superior is named the **scala vestibuli,** while the inferior is termed the **scala tympani** (Fig. 13-59). Near the summit of the cochlea, the lamina ends in a hook-shaped process, the **hamulus laminae spiralis;** this assists in forming the boundary of a small opening, the **helicotrema,** through which the two scalae communicate with each other. From the spiral canal of the modiolus numerous canals pass outward through the osseous spiral lamina as far as

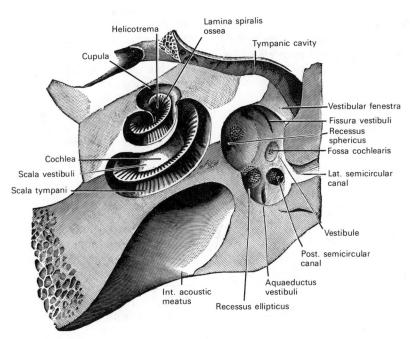

FIG. 13-59.   The cochlea and vestibule, viewed from above. All the hard parts that form the roof of the internal ear have been removed with a saw.

its free edge. In the inferior part of the first turn a second bony lamina, the **secondary spiral lamina,** projects inward from the outer wall of the bony tube; it does not, however, reach the primary osseous spiral lamina, so that if viewed from the vestibule a narrow fissure, the **vestibule fissure,** is seen between them.

The osseous labyrinth is lined by an exceedingly thin fibrous membrane. The attached surface is rough, fibrous, and closely adherent to the bone; its free surface is smooth, pale, covered with a layer of epithelium, and secretes a thin, limpid fluid, the **perilymph.** A delicate tubular process of this membrane is prolonged along the aqueduct of the cochlea to the inner surface of the dura mater.

Membranous Labyrinth

The **membranous labyrinth** (Figs. 13-60; 13-61; 13-62) is lodged within the bony cavities just described, and has the same general form as these; it is, however, considerably smaller, and is partly separated from the bony walls by a fluid, the **perilymph.** In certain places it is fixed to the walls of the cavity. The membranous labyrinth also contains fluid, the **endolymph,** and the ramifications of the vestibulocochlear nerve are distributed on its walls.

Within the osseous vestibule the membranous labyrinth does not quite preserve the form of the bony cavity, but consists of two membranous sacs, the **utricle** and the **saccule.**

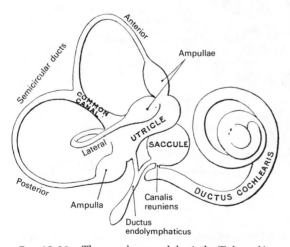

FIG. 13-60. The membranous labyrinth. (Enlarged.)

UTRICLE. The utricle, the larger of the two sacs, is oblong, compressed transversely, and occupies the superior and posterior part of the vestibule, lying in contact with the recessus ellipticus and the part inferior to it. That portion which is lodged in the recess forms a sort of pouch or cul-de-sac, the floor and anterior wall of which are thick and form the **macula of the utricle,** which receives the utricular filaments of the vestibulocochlear nerve. The cavity of the utricle communicates behind with the semicircular ducts by five orifices. From its anterior wall is given off the **ductus utriculosaccularis,** which opens into the ductus endolymphaticus.

SACCULE. The saccule is the smaller of the two vestibular sacs; it has a globular shape and lies in the recessus sphericus near the opening of the scala vestibuli of the cochlea. Its anterior part exhibits an oval thickening, the **macula of the saccule,** to which are distributed the saccular filaments of the vestibulocochlear nerve. Its cavity does not directly communicate with that of the utricle. From the posterior wall a canal, the **ductus endolymphaticus,** is given off; this duct is joined by the ductus utriculosaccularis, and then passes along the aquaeductus vestibuli and ends in a blind pouch **(saccus endolymphaticus)** on the posterior surface of the petrous portion of the temporal bone, where it is in contact with the dura mater. From the inferior part of the saccule a short tube, the **ductus reuniens** (*canalis reuniens of Hensen*[1]) passes inferiorly and opens into the ductus cochlearis near its vestibular extremity (Fig. 13-60).

SEMICIRCULAR DUCTS. The semicircular ducts (*membranous semicircular canals*) (Figs. 13-61; 13-62) are about one-fourth of the diameter of the osseous canals, but in number, shape, and general form they are precisely similar, and each has an ampulla at one end. They open by five orifices into the utricle, one opening being common to the medial end of the anterior duct and the superior end of the posterior duct. In each ampulla, the wall is thickened and projects into the cavity as a fiddle-shaped, transversely placed elevation, the **ampullary crest,** in which the nerves end.

The utricle, saccule, and semicircular ducts are held in position by numerous fi-

[1] Victor Hensen (1835–1924): German anatomist and physiologist (Kiel).

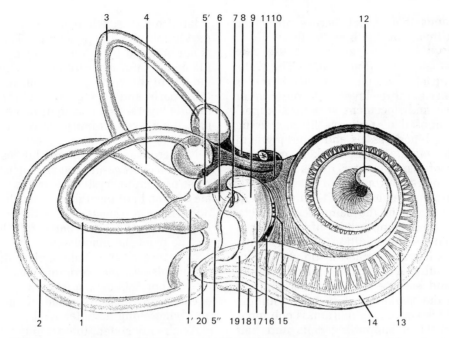

FIG. 13-61.   Right human membranous labyrinth, removed from its bony enclosure and viewed from the anterolateral aspect. (G. Retzius.)

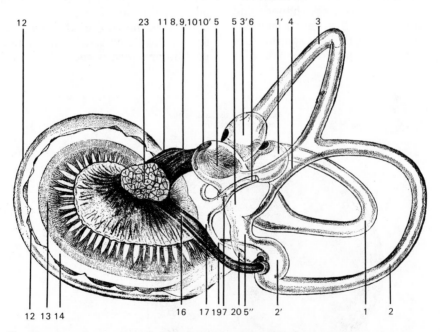

FIG. 13-62.   The same from the posteromedial aspect. 1. Lateral semicircular canal; 1′, its ampulla; 2. Posterior canal; 2′, its ampulla. 3. Anterior canal; 3′, its ampulla. 4. Conjoined limb of anterior and posterior canals (*sinus utriculi superior*). 5. Utricle; 5′, recessus utriculi; 5″, sinus utriculi posterior. 6. Ductus endolymphaticus. 7. Canalis utriculo-saccularis. 8. Nerve to ampulla of anterior canal. 9. Nerve to ampulla of lateral canal. 10. Nerve to recessus utriculi (in Fig. 13-61, the three branches appear conjoined); 10′, ending of nerve in recessus utriculi. 11. Facial nerve. 12. Lagena cochleae. 13. Nerve of cochlea within spiral lamina. 14. Basilar membrane. 15. Nerve fibers to macula of saccule. 16. Nerve to ampulla of posterior canal. 17. Saccule. 18. Secondary membrane of tympanum. 19. Canalis reuniens. 20. Vestibular end of ductus cochlearis. 23. Section of the facial and acoustic nerves within internal acoustic meatus. The separation between them is not apparent in this section. (G. Retzius.)

brous bands that stretch across the space between them and the bony walls.

STRUCTURE OF THE UTRICLE, SACCULE, AND SEMICIRCULAR DUCTS (Fig. 13-63). The walls of the utricle, saccule, and semicircular ducts consist of three layers. The *outer layer* is a loose and flocculent structure, apparently composed of ordinary fibrous tissue containing blood vessels and some pigment cells. The *middle layer,* thicker and more transparent than the outer, forms a homogeneous membrana propria and has numerous papilliform projections on its internal surface, especially in the semicircular ducts, which exhibit an appearance of longitudinal fibrillation on the addition of acetic acid. The *inner layer* is formed of polygonal nucleated epithelial cells. In the maculae of the utricle and saccule and in the ampullary crests of the semicircular ducts, the middle coat is thick and the epithelium is columnar and consists of **supporting cells** and **hair cells.** The supporting cells are fusiform; their deep ends are attached to the membrana propria while their free extremities are united to form a thin cuticle. The **neuroepithelium** or hair cells are flask-shaped, and their deep, rounded ends do not reach the membrana propria, but lie between the supporting cells. The deep part of each contains a large nucleus, while its more superficial part is granular and pigmented. The free end is surmounted by a long, tapering, hair-like filament that projects into the cavity. The filaments of the vestibulocochlear nerve enter these parts, and having pierced the outer and middle layers, they lose their myelin sheaths and their axons ramify between the hair cells.

Numerous small round bodies termed **otoconia** (*statoconia*), each consisting of a mass of minute crystalline grains of carbonate of lime held together in a mesh of gelatinous tissue, are suspended in the endolymph in contact with the free ends of the hairs projecting from the maculae.

COCHLEAR DUCT. The cochlear duct (*ductus cochlearis; membranous cochlea; scala media*) consists of a spirally arranged tube enclosed in the bony canal of the cochlea and lying along its outer wall.

As already stated, the osseous spiral lamina extends only part of the distance between the modiolus and the outer wall of the cochlea, while the **basilar membrane** (*membrana spiralis*) stretches from its free edge to the outer wall of the cochlea and completes the roof of the scala tympani. A second and more delicate membrane, the **vestibular membrane** (*Reissner's mem-*

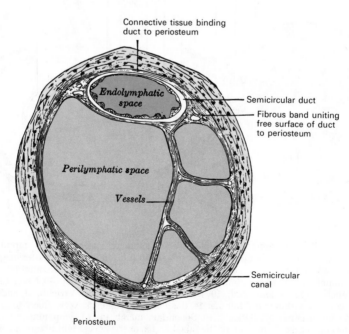

FIG. 13-63. Transverse section of a human semicircular canal and duct. (After Rüdinger.)

brane[1]), extends from the thickened perios-teum covering the osseous spiral lamina to the outer wall of the cochlea, where it is attached at some little distance above the outer edge of the basilar membrane. A canal is thus shut off between the scala tympani and the scala vestibuli; this is the **cochlear duct** or **scala media** (Figs. 13-64; 13-65). It is triangular on transverse section, its roof being formed by the vestibular membrane, its outer wall of the periosteum lining the bony canal, and its floor by the membrana spiralis and the outer part of the osseous spiral lamina. Its extremities are closed, the upper is termed the **lagena** and is attached to the cupula at the upper part of the helico-

trema; the inferior is lodged in the recessus cochlearis of the vestibule. Near the inferior end, the ductus cochlearis becomes continuous with the sacculus by means of a narrow, short canal, the **ductus reuniens** (Fig. 13-60).

The **spiral organ of Corti** rests upon the basilar membrane. The vestibular membrane is thin, homogeneous, and covered on its superior and inferior surfaces by a layer of epithelium. The periosteum, forming the outer wall of the cochlear duct, is greatly thickened and altered in character, and is called the **spiral ligament.** It projects inward below as a triangular prominence, the **basilar crest,** which gives attachment to the outer edge of the basilar membrane; immediately above the crest is a concavity, the **sulcus spiralis externus.** The upper portion of the spiral ligament contains numerous

[1]Ernst Reissner (1824–1878): A German anatomist (Dorpat, then Breslau).

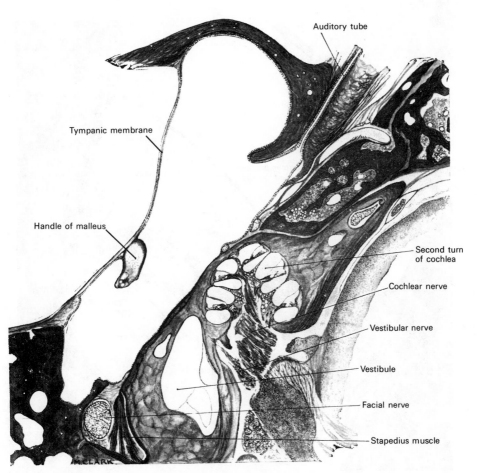

FIG. 13-64. A section through the left temporal bone. (Drawn from a section prepared at the Fehrens Institute, kindly lent by Prof. J. Kirk.)

capillary loops and small blood vessels, and it is termed the **stria vascularis.**

The osseous spiral lamina consists of two plates of bone, and between these are the canals for the transmission of the filaments of the vestibulocochlear nerve. On the upper plate of that part of the lamina which is outside the vestibular membrane, the periosteum is thickened to form the **limbus laminae spiralis.** This ends externally in a concavity. The **sulcus spiralis internus** (Fig. 13-65) represents, in section, the form of the letter **C;** the upper part, formed by the overhanging extremity of the limbus, is named the **vestibular lip;** the lower part, prolonged and tapering, is called the **tympanic lip,** and is perforated by numerous foramina for the passage of the cochlear nerve fibers. The upper surface of the vestibular lip is intersected at right angles by a number of furrows, between which are numerous eleva-tions; these present the appearance of teeth along the free surface and margin of the lip, and they have been named **dentes acoustici.** The limbus is covered by a layer of what appears to be squamous epithelium, but the deeper parts of the cells with their contained nuclei occupy the intervals between the elevations and between the auditory teeth. This layer of epithelium is continuous on the one hand with that lining the sulcus spiralis internus, and on the other with that covering the undersurface of the vestibular membrane.

BASILAR MEMBRANE. The basilar membrane stretches from the tympanic lip of the osseous spiral lamina to the basilar crest and consists of two parts, an inner and an outer. The inner part is thin and is named the **zona arcuata;** it supports the spiral organ of Corti. The outer is thicker and striated and is termed the **zona pectinata.** The undersurface

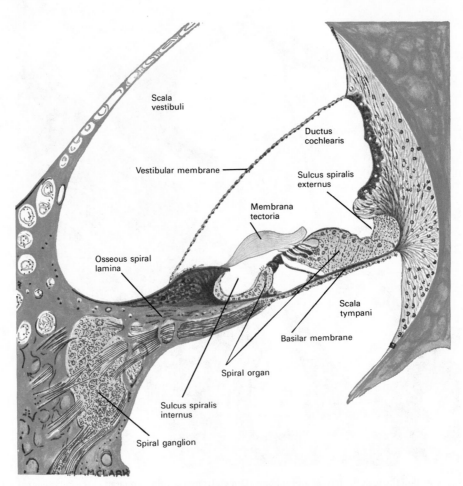

FIG. 13-65. A section through the second turn of the cochlea indicated in the previous figure. (Mallory's stain.)

of the membrane is covered by a layer of vascular connective tissue; one of the vessels in this tissue is somewhat larger than the rest, and is named the **vas spirale.**

STRUCTURE OF THE ORGAN OF CORTI. The **spiral organ** or **organ of Corti** (Fig. 13-66) is composed of a series of epithelial structures placed upon the inner part of the basilar membrane. The average length is 31.5 mm. The more central of these structures are two rows of rod-like bodies, the **inner and outer rods** or **pillars of Corti.** The bases of the rods are supported on the basilar membrane, those of the inner row at some distance from those of the outer row. The two rows incline toward each other, and coming into contact superiorly, enclose between them and the basilar membrane a triangular tunnel, the **tunnel of Corti.** On the inner side of the inner rods is a single row of hair cells, and on the outer side of the outer rods are three or four rows of similar cells, together with certain supporting cells termed the cells of Deiters and Hensen. The free ends of the outer hair cells occupy a series of apertures in a net-like membrane, the **reticular membrane,** and the entire organ is covered by the **tectorial membrane.**

Rods of Corti. Each consists of a base or foot-plate and elongated part or body, and an upper end or head. The body of each rod is finely striated, but in the head there is an oval, nonstriated portion that stains deeply with carmine. Occupying the angles between the rods and the basilar membrane are nucleated cells that partly envelop the

rods and extend onto the floor of Corti's tunnel; these may be looked upon as the undifferentiated parts of the cells from which the rods have been formed.

The **inner rods** number nearly 6000, and their bases rest on the basilar membrane close to the tympanic lip of the sulcus spiralis internus. The shaft or body of each is sinuously curved and forms an angle of about 60° with the basilar membrane. The head resembles the proximal end of the ulna and presents a deep concavity that accommodates a convexity on the head of the outer rod. The headplate, or portion overhanging the concavity, overlaps the headplate of the outer rod.

The **outer rods,** numbering nearly 4000, are longer and more obliquely set than the inner rods, forming with the basilar membrane an angle of about 40°. Their heads are convex internally; they fit into the concavities on the heads of the inner rods and are continued outward as thin flattened plates, termed **phalangeal processes,** which unite with the phalangeal processes of Deiters' cells to form the reticular membrane.

Hair Cells. The hair cells are short columnar cells; their free ends are on a level with the heads of Corti's rods, and each is surmounted by about 20 hair-like processes arranged in the form of a crescent with its concavity directed inward. The deep ends of the cells reach about half way along Corti's rods, and each contains a large nucleus; in contact with the deep ends of the hair cells are the terminal filaments of the cochlear division

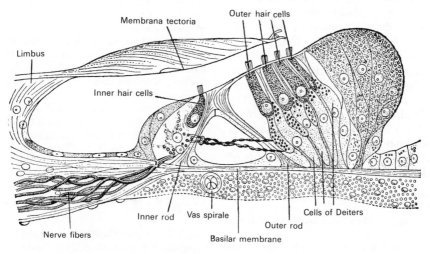

Fig. **13-66.** Section through the spiral organ of Corti. Magnified. (G. Retzius.)

of the vestibulocochlear nerve. The *inner hair cells*, about 3500 in number, are arranged in a single row on the medial side of the inner rods. Because their diameters are greater than those of the rods, each hair cell is supported by more than one rod. The free ends of the inner hair cells are encircled by a cuticular membrane that is fixed to the heads of the inner rods. Adjoining the inner hair cells are one or two rows of columnar supporting cells, which, in turn, are continuous with the cubical cells lining the sulcus spiralis internus. The *outer* hair cells number about 12,000 and are nearly twice as long as the inner. In the basal coil of the cochlea they are arranged in three regular rows; in the apical coil, in four somewhat irregular rows. The receptors for high tones are located in the basal turn of the cochlea.

Supporting Cells. Between the rows of the outer hair cells are rows of supporting cells, called the **cells of Deiters**[1]; their expanded bases are planted on the basilar membrane, while the opposite end of each presents a clubbed extremity or **phalangeal process.** Immediately to the outer side of Deiters' cells are five or six rows of columnar cells, the **supporting cells of Hensen.** Their bases are narrow, while their upper parts are expanded and form a rounded elevation on the floor of the ductus cochlearis. The columnar cells lying outside Hensen's cells are termed the **cells of Claudius.**[2] A space exists between the outer rods of Corti and the adjacent hair cells; this is called the **space of Nuel.**[3]

Reticular Lamina. The **reticular lamina** is a delicate framework perforated by rounded holes which are occupied by the free ends of the outer hair cells. It extends from the heads of the outer rods of Corti to the external row of the outer hair cells, and is formed by several rows of minute fiddle-shaped cuticular structures, called **phalanges,** between which are circular apertures containing the free ends of the hair cells. The innermost row of phalanges consists of the phalangeal processes of the outer rods of Corti; the outer rows are formed by the modified free ends of Deiters' cells.

[1] Otto Friedrich Carl Deiters, (1834–1863): A German anatomist (Bonn).
[2] Friedrich Matthias Claudius (1822–1869): An Austrian anatomist.
[3] Jean Pierre Nuel (1847–1920): A Belgian physiologist and otologist (Ghent and Liège).

Tectorial Membrane. Covering the sulcus spiralis internus and the spiral organ of Corti is the **tectorial membrane,** which is attached to the limbus laminae spiralis close to the inner edge of the vestibular membrane. Its inner part is thin and overlies the auditory teeth of Huschke;[4] its outer part is thick, and along its lower surface, opposite the inner hair cells, is a clear band, named **Hensen's stripe** owing to the intercrossing of its fibers. The lateral margin of the membrane is much thinner. It consists of fine, colorless fibers embedded in a transparent matrix of a soft collagenous, semisolid character with marked adhesiveness; the matrix may be a variety of soft keratin. The general transverse direction of the fibers inclines from the radius of the cochlea toward the apex.

Nerves. The **vestibulocochlear nerve** (*acoustic nerve; n. statoacusticus; nerve of hearing*) divides near the bottom of the internal acoustic meatus into an anterior or cochlear branch and a posterior or vestibular branch.

The **vestibular nerve** supplies the utricle, the saccule, and the ampullae of the semicircular ducts. On the trunk of the nerve, within the internal acoustic meatus, is the **vestibular ganglion** (*ganglion of Scarpa*[5]); the fibers of the nerve arise from the cells of this ganglion. On the distal side of the ganglion the nerve splits into a superior, an inferior, and a posterior branch. The filaments of the *superior branch* are transmitted through the foramina in the area vestibularis superior, and end in the macula of the utricle and in the ampullae of the anterior and lateral semicircular ducts. Those of the *inferior branch* traverse the foramina in the area vestibularis inferior and end in the macula of the saccule. The *posterior branch* runs through the foramen singulare at the posteroinferior part of the bottom of the meatus and divides into filaments to supply the ampulla of the posterior semicircular duct.

The **cochlear nerve** divides into numerous filaments at the base of the modiolus. Those for the basal and middle coils pass through the foramina in the tractus spiralis foraminosus and those for the apical coil pass through the canalis centralis; the nerves bend outward to pass between the lamellae of the osseous spiral lamina. Occupying the spiral canal of the modiolus is the **spiral ganglion of the cochlea** (*ganglion of Corti*) (Fig. 13-65), which consists of bipolar nerve cells that constitute the cells of origin of this nerve. Reaching the outer edge of the osseous spiral lamina, the fibers of the nerve pass through the foramina in the tympanic lip. Some end

[4] Emil Huschke (1797–1858): A German anatomist (Jena).
[5] Anthony Scarpa (1747–1832): Italian anatomist and surgeon (Pavia).

by arborizing around the bases of the inner hair cells, while others pass between Corti's rods and across the tunnel to end in a similar manner in relation to the outer hair cells. The cochlear nerve gives off a vestibular branch to supply the vestibular end of the ductus cochlearis; the filaments of this branch pass through the foramina in the fossa cochlearis.

Vessels. The **arteries of the labyrinth** are the internal auditory artery, which comes from the basilar artery, and the stylomastoid artery, which comes from the posterior auricular artery. The internal auditory artery divides at the bottom of the internal

acoustic meatus into two branches: cochlear and vestibular. The cochlear branch subdivides into 12 or 14 twigs that traverse the canals in the modiolus and are distributed, in the form of a capillary network, in the lamina spiralis and basilar membrane. The vestibular branches are distributed to the utricle, saccule, and semicircular ducts.

The **veins** of the vestibule and semicircular canals accompany the arteries and, receiving those of the cochlea at the base of the modiolus, unite to form the internal auditory veins, which end in the posterior part of the superior petrosal sinus or in the transverse sinus.

References

(References are listed not only to those articles and books cited in the text, but to others as well which are considered to contain valuable resource information for the student who desires it.)

Organ of Smell

Arstila, A., and J. Wersall. 1967. The ultrastructure of the olfactory epithelium of the guinea pig. Acta Otolaryngol., *64:* 187–204.

Bloom, G. 1954. Studies on the olfactory epithelium of the frog and the toad with the aid of light and electron microscopy. Z. Zellforsch., *41:*89–100.

deLorenzo, A. J. 1957. Electron microscopic observations of the olfactory mucosa and olfactory nerve. J. Biophys. Biochem. Cytol., 3:839–850.

Doty, R. L., ed. 1976. *Mammalian Olfaction, Reproductive Processes and Behavior.* Academic Press, New York, 344 pp.

Frisch, D. 1967. Ultrastructure of mouse olfactory mucosa. Amer. J. Anat., *121:*87–120.

Gasser, H. S. 1956. Olfactory nerve fibers. J. Gen. Physiol., *39:*473–498.

Graziadei, P. P. C. 1971. The olfactory mucosa of vertebrates. In *Handbook of Sensory Physiology,* Vol. 4, *Chemical Senses.* Edited by L. M. Beidler. Springer-Verlag, Berlin, pp. 27–58.

Gross, G. W., and L. M. Beidler. 1973. Fast axonal transport in the c-fibers of the garfish olfactory nerve. J. Neurobiol., 4:413–428.

Gross, G. W., and L. M. Beidler. 1975. A quantitative analysis of isotope concentration profiles and rapid transport velocities in the c-fibers of the garfish olfactory nerve. J. Neurobiol., 6:213–232.

Mathews, D. F. 1972. Response patterns of single neurons in the tortoise olfactory epithelium and olfactory bulb. J. Gen. Physiol., *60:*166–180.

Moulton, D. G., and L. M. Beidler. 1967. Structure and function of the peripheral olfactory system. Physiol. Rev., 47:1–52.

Moulton, D. G., G. Celebi, and R. P. Fink. 1970. Olfaction in mammals—two aspects: proliferation of cells in the olfactory epithelium and sensitivity to odours. In *Taste and Smell in Vertebrates.* Edited by G. E. W. Wolstenholme and J. Knight. Ciba Foundation Symposium, W. and A. Churchill, London, pp. 227–245.

Mozell, M. M. 1966. The spatiotemporal analysis of odorants at the level of the olfactory receptor sheet. J. Gen. Physiol., *50:*25–42.

Reese, T. 1965. Olfactory cilia in the frog. J. Cell Biol., 25:209–230.

Reese, T., and M. W. Brightman. 1970. Olfactory surface and central olfactory connections in some vertebrates. In *Taste and Smell in Vertebrates.* Edited by G. E. W. Wolstenholme and J. Knight. Ciba Foundation Symposium, W. and A. Churchill, London, pp. 115–143.

Smith, C. G. 1941. Incidence of atopy of olfactory nerves in man. Arch. Otolaryngol., *34:* 533–539.

Smith, C. G. 1951. Regeneration of sensory olfactory epithelium and nerves in adult frogs. Anat. Rec., *109:* 661–672.

Organ of Taste

Beidler, L. M. 1954. A theory of taste stimulation. J. Gen. Physiol., *38:*133–148.

Beidler, L. M., and R. L. S. Smallman. 1965. Renewal of cells within taste buds. J. Cell Biol., 27:263–272.

Benjamin, R. M., and H. Burton. 1968. Projection of taste nerve afferents to anterior opercular insular cortex in squirrel monkey (Saimiri sciureus). Brain Res., 7:221–231.

Benjamin, R. M., R. Emmers, and A. J. Blomquist. 1968. Projection of tongue nerve afferents to somatic sensory area I in squirrel monkey (Saimiri sciureus). Brain Res., 7:208–220.

Bradley, R. M., and I. B. Stern. 1967. The development of the human taste bud during the fetal period. J. Anat. (Lond.), *101:*743–752.

deLorenzo, A. J. 1958. Electron microscopic observations on the taste buds of the rabbit. J. Biophys. Biochem. Cytol., 4:143–150.

Farbman, A. I. 1965. Electron microscope study of the developing taste bud in rat fungiform papilla. Devel. Biol., *11:*110–135.

Farbman, A. I. 1965. Fine structure of the taste bud. J. Ultrastruct. Res., *12:*328–350.

Fishman, I. Y. 1957. Single fiber gustatory impulses in rat and hamster. J. Cell. Comp. Physiol., *49:*319–334.

Frank, M. 1973. An analysis of hamster afferent taste nerve response functions. J. Gen. Physiol., *61:* 588–618.

Fujimoto, S., and R. G. Murray. 1970. Fine structure of degeneration and regeneration in denervated rabbit vallate taste buds. Anat. Rec., 168:393–414.

Graziadei, P. P. C. 1969. The ultrastructure of vertebrate taste buds. In Olfaction and Taste III. Edited by C. Pfaffmann. The Rockefeller University Press, New York, pp. 315–330.

Guth, L. 1958. Taste buds on the cat's circumvallate papilla after reinnervation by glossopharyngeal, vagus, and hypoglossal nerves. Anat. Rec., 130:25–37.

Lalonde, E. R., and J. A. Eglitis. 1961. Number and distribution of taste buds on the epiglottis, pharynx, larynx, soft palate and uvula in a human newborn. Anat. Rec., 140:91–95.

Murray, R. G., and A. Murray. 1967. Fine structure of taste buds of rabbit foliate papillae. J. Ultrastruct. Res., 19:327–353.

Murray, R. G., A. Murray, and S. Fujimoto. 1969. Fine structure of gustatory cells in rabbit taste buds. J. Ultrastruct. Res., 27:444–461.

Oakley, B. 1970. Reformation of taste buds by crossed sensory nerves in the rat's tongue. Acta Physiol. Scand., 79:88–94.

Oakley, B., and R. M. Benjamin. 1966. Neural mechanisms of taste. Physiol. Rev., 46:173–211.

Pfaffmann, C. 1941. Gustatory afferent impulses. J. Cell. Comp. Physiol., 17:243–258.

Sato, T., and L. M. Beidler. 1975. Membrane resistance change of frog taste cells in response to water and NaCl. J. Gen. Physiol., 66:735–764.

Wolstenholme, G. E. W., and J. Knight, eds. 1970. Taste and Smell in Vertebrates. A Ciba Symposium. W. and A. Churchill, London, 402 pp.

The Eye

Adelmann, H. B. 1936. The problem of cyclopia. Q. Rev. Biol., 11:284–304.

Anderson, D. R., and W. Hoyt. 1966. Ultrastructure of intraorbital portion of human and monkey optic nerve. Arch. Ophthalmol., 82:506–530.

Anderson, R. L., and C. Beard. 1977. The levator aponeurosis. Attachments and their clinical significance. Arch. Ophthalmol., 95:1437–1441.

Ashton, N. 1951. Anatomical study of Schlemm's canal and aqueous veins by means of neoprene cats. Part I, Aqueous veins. Brit. J. Ophthalmol., 35:291–303.

Barber, A. 1955. Embryology of the Human Eye. C. V. Mosby, St. Louis, 236 pp.

Bartelmez, G. W., and A. S. Dekaban. 1962. The early development of the human brain. Carneg. Inst., Contr. Embryol., 37:13–32.

Baylor, D. A. 1974. Lateral interaction between vertebrate photoreceptors. Fed. Proc., 33:1074–1077.

Beatie, J. C., and D. L. Stillwell, Jr. 1961. Innervation of the eye. Anat. Rec., 141:45–61.

Bernhard, C. G., J. Boëthius, G. Gemne, and G. Struwe. 1970. Eye ultrastructure, colour reception and behaviour. Nature (Lond.), 226:865–866.

Bodemer, C. W. 1958. The origin and development of the extrinsic ocular muscles in the trout (Salmo trutta). J. Morphol., 102:119–155.

Boniuk, M. 1973. Eyelids, lacrimal apparatus, and conjunctiva. Arch. Ophthalmol., 90:239–250.

Boycott, B. B., and J. E. Dowling. 1969. Origin of the primate retina: Light microscopy. Philos. Trans. Roy. Soc. Lond. (Biol.), 255:109–184.

Breathnach, A. S., and L. M. A. Wylie. 1966. Ultrastructure of retinal pigment epithelium of the human fetus. J. Ultrastruct. Res., 16:584–597.

Charnwood, L. 1950. An Essay on Binocular Vision. Hafner Publishing Co., New York, 117 pp.

Cohen, A. I. 1965. New details of the ultrastructure of the outer segments and ciliary connections of the rods of human and macaque retinas. Anat. Rec., 152:63–80.

Cohen, A. I. 1965. The electron microscopy of the normal human lens. Invest. Ophthalmol., 4:433–446.

Collin, J. R., C. Beard, and I. Wood. 1978. Experimental and clinical data on the insertion of the levator palpebrae superioris muscle. Amer. J. Ophthalmol., 85:792–801.

Cooper, S., and P. M. Daniel. 1949. Muscle spindles in human extrinsic eye muscles. Brain, 72:1–24.

Cooper, S., P. M. Daniel, and D. Whitteridge. 1955. Muscle spindles and other sensory endings in the extrinsic eye muscles; the physiology and anatomy of these receptors and of their connections with the brain stem. Brain, 78:564–583.

Coulombre, A. J., and J. L. Coulombre. 1975. Mechanisms of ocular development. Internat. Ophthalmol. Clin., 15:7–18.

Coulombre, A. J., and E. S. Crelin. 1958. The role of the developing eye in the morphogenesis of the avian skull. Amer. J. Phys. Anthropol., 16:25–37.

Couly, G., J. Hureau, and P. Tessier. 1976. The anatomy of the external palpebrae ligament in man. J. Maxillofac. Surg., 4:195–197.

Davson, H. 1980. Physiology of the Eye. 4th ed. Academic Press, New York, 644 pp.

deRobertis, E. 1960. Some observations on the ultrastructure and morphogenesis of photoreceptors. J. Gen. Physiol., 43:1–13.

Detwiler, S. R. 1955. The eye and its structural adaptations. Proc. Amer. Philos. Soc., 99:224–238.

Dowling, J. E. 1965. Foveal receptors of the monkey retina: fine structure. Science, 147:57–59.

Dowling, J. E. 1970. Organization of vertebrate retinas. Invest. Ophthalmol., 9:655–680.

Dowling, J. E., and B. B. Boycott. 1966. Organisation of the primate retina: Electron microscopy. Proc. Roy. Soc. Lond. (Biol.), 166:80–111.

Ehinger, B., and B. Falck. 1966. Concomitant adrenergic and parasympathetic fibers in the rat iris. Acta Physiol. Scand., 67:201–207.

Fine, B. S., and A. J. Tousimis. 1961. The structure of the vitreous body and the suspensory ligaments of the lens. Arch. Ophthalmol., 65:95–110.

Fine, B. S., and L. E. Zimmerman. 1962. Muller's cells and the middle limiting membrane of the human retina. An electron microscopic study. Invest. Ophthalmol., 1:304–326.

Fisher, S. K., and K. A. Linberg. 1975. Intercellular junctions in the early human embryonic retina. J. Ultrastruct. Res., 51:69–78.

Friedenwald, J. S. 1949. The formation of the intraocular fluid. Amer. J. Ophthalmol., 32:9–27.

Génis-Gálvez, J. M. 1957. Innervation of the ciliary muscle. Anat. Rec., 127:219–230.

Gilbert, P. W. 1957. The origin and development of the human extrinsic ocular muscles. Carneg. Inst., Contr. Embryol., 36:59–78.

Gillilan, L. A. 1961. The collateral circulation of the human orbit. Arch. Ophthalmol., 65:684–694.

Greiner, J. V., H. I. Covington, and M. R. Allansmith.

1977. Surface morphology of the human upper tarsal conjunctiva. Amer. J. Ophthalmol., *83*:892–905.

Harayama, K., T. Amemiya, and H. Nishimura. 1980. Development of rectus muscles during fetal life: Insertion sites and width. Invest. Ophthalmol. Vis. Sci., *19*:468–474.

Hayreh, S. S. 1963. Arteries of the orbit in the human being. Brit. J. Surg., *50*:938–953.

Hayreh, S. S. 1969. Blood supply of the optic head and its role in optic atrophy, glaucoma and oedema of the optic disc. Brit. J. Ophthalmol., *53*:721–748.

Hogan, M. J. 1963. The vitreous, its structure and relation to the ciliary body and retina. Invest. Ophthalmol., *2*:418–445.

Hogan, M. J., J. A. Alvarado, and J. E. Weddell. 1971. *Histology of the Human Eye.* W. B. Saunders, Philadelphia, 687 pp.

Hollenberg, M. J., and M. H. Bernstein. 1966. Fine structure of the photoreceptor cells of the ground squirrel. Amer. J. Anat., *118*:359–374.

Hubbard, R., and A. Kropf. 1958. The action of light on rhodopsin. Proc. Nat. Acad. Sci. USA, 44:130–139.

Ishikawa, T. 1962. Fine structure of the human ciliary muscle. Invest. Ophthalmol., *1*:587–608.

Jack, R. L. 1972. Ultrastructure of the hyaloid vascular system. Arch. Ophthalmol., *87*:555–567.

Jakus, M. A. 1956. Studies on the cornea. II. The fine structure of Descemet's membrane. J. Biophys. Biochem. Cytol. (Suppl.), *2*:243–252.

Jakus, M. A. 1964. *Ocular Fine Structure, Selected Electron Micrographs.* Little, Brown and Co., Boston, 204 pp.

Jampel, R. S. 1970. The fundamental principle of the action of the oblique ocular muscles. Amer. J. Ophthalmol., *69*:623–638.

Kolb, B. 1970. Organization of the outer plexiform layer of the primate retina: electron microscopy of Golgi-impregnated cells. Philos. Trans. Roy. Soc. Lond. (Biol.), *258*:261–283.

Ladman, A. J. 1958. The fine structure of the rod-bipolar cell synapse in the retina of the albino rat. J. Biophys. Biochem. Cytol., *4*:459–466.

Leighton, D. A., and A. Tomlinson. 1972. Changes in axial length and other dimensions of the eyeball with increasing age. Acta Ophthalmol., *50*:815–826.

Limborgh, J., and I. Tonneyck-Müller. 1976. Experimental studies on the relationships between eye growth and skull growth. Ophthalmologica, *173*:317–325.

Mann, I. C. 1957. *Developmental Abnormalities of the Eye.* 2nd ed. British Medical Association, London, 419 pp.

Mann, I. C. 1964. *The Development of the Human Eye.* 3rd ed. British Medical Association, London, 316 pp.

Marshall, J., and P. L. Ansell. 1971. Membranous inclusions in the retinal pigmented epithelium: phagosomes and myeloid bodies. J. Anat. (Lond.), *110*: 91–104.

Oppenheimer, D. R., E. Palmer, and G. Weddell. 1958. Nerve endings in the conjunctiva. J. Anat. (Lond.), *92*:321–352.

O'Rahilly, R. 1961. A simple demonstration of the features of the bony orbit. Anat. Rec., *141*:315–316.

Pearson, A. A. 1980. The development of the eyelids: Part I. External features. J. Anat. (Lond.), *130*:33–42.

Polyak, S. L. 1941. *The Retina.* University of Chicago Press. Chicago, 607 pp.

Polyak, S. L. 1957. *The Vertebrate Visual System.* University of Chicago Press, Chicago, 1390 pp.

Prince, J. H. 1956. *Comparative Anatomy of the Eye.* Charles C Thomas, Springfield, Ill. 418 pp.

Raviola, E., and N. B. Gilula. 1973. Gap junctions between photoreceptor cells in the vertebrate retina. Proc. Nat. Acad. Sci., *70*:1677–1681.

Raviola, G. 1971. The fine structure of the ciliary zonule and ciliary epithelium. Invest. Ophthalmol., *10*: 851–869.

Raviola, G. 1974. Effects of paracentesis on the blood-aqueous barrier: an electron microscope study on Macaca mulatta using horseradish peroxidase as a tracer. Invest. Ophthalmol., *13*:828–858.

Raviola, G., and E. Raviola. 1967. Light and electron microscopic observations on the inner plexiform layer of the rabbit retina. Amer. J. Anat., *120*:403–425.

Renz, B. E., and C. M. Vygantas. 1977. Hyaloid vascular remnants in human neonates. Ann. Ophthalmol., *9*:179–184.

Reyer, R. W. 1954. Regeneration of the lens in the amphibian eye. Quart. Rev. Biol., *29*:1–46.

Richardson, K. C. 1964. The fine structure of the albino rabbit iris with special reference to the identification of adrenergic and cholinergic nerves and nerve endings in its intrinsic muscles. Amer. J. Anat., *114*:173–205.

Rushton, W. A. H. 1962. *Visual Pigments in Man.* Liverpool University Press, Liverpool, England, 38 pp.

Rushton, W. A. H., and G. H. Henry. 1968. Bleaching and regeneration of cone pigments in man. Vision Res., *8*:617–631.

Ruskell, G. L. 1970. An ocular parasympathetic nerve pathway of facial nerve origin and its influence on intraocular pressure. Exp. Eye Res., *10*:319–330.

Ruskell, G. L. 1971. The distribution of autonomic post-ganglionic nerve fibers in the lacrimal gland in the rat. J. Anat. (Lond.), *109*:229–242.

Sexton, R. R. 1970. Eyelids, lacrimal apparatus, and conjunctiva. Arch. Ophthalmol., *83*:361–377.

Sheldon, H. 1956. An electron microscope study of the epithelium in the normal mature and immature mouse cornea. J. Biophys. Biochem. Cytol., *2*:253–262.

Singh, S., and R. Dass. 1960. The central artery of the retina. I. Origin and course. Brit. J. Ophthalmol., *44*:193–212.

Singh, S., and R. Dass. 1960. The central artery of the retina. II. A study of its distribution and anastomoses. Brit. J. Ophthalmol., *44*:280–299.

Sjostrand, F. S. 1953. Ultrastructure of inner segments of retinal rods of the guinea pig as revealed by the electron microscope. J. Cell. Comp. Physiol., *42*:45–70.

Sjostrand, F. S. 1953. Ultrastructure of outer segments of rods and cones of the eye as revealed by the electron microscope. J. Cell. Comp. Physiol., *42*:15–44.

Sjostrand, F. S. 1958. Ultrastructure of retinal rod synapses of the guinea pig eye as revealed by three-dimensional reconstructions from serial sections. J. Ultrastruct. Res., *2*:122–170.

Sorsby, A., and M. Sheriden. 1960. The eye at birth: measurement of the principal diameters in forty-eight cadavers. J. Anat. (Lond.), *94*:192–197.

Stell, W. K., and D. O. Lightfoot. 1975. Color-specific interconnections of cones and horizontal cells in the retina of the goldfish. J. Comp. Neurol., *159*:473–502.

Stone, L. S. 1960. Regeneration of the lens, iris, and neural retina in a vertebrate eye. Yale J. Biol. Med., *32*:464–473.

Tripathi, R. C. 1970. Mechanism of the aqueous outflow across the trabecular wall of Schlemm's canal. Exp. Eye Res., *10*:111–116.

Villegas, G. M. 1964. Ultrastructure of the human retina. J. Anat. (Lond.), *98*:501–513.

Wald, G. 1955. The photoreceptor process in vision. Amer. J. Ophthalmol., *40*:18–41.

Walls, G. L. 1963. *The Vertebrate Eye and Its Adaptive Radiation.* Hafner, New York. 785 pp.

Wanko, T., and M. A. Gavin. 1959. Electron microscopic study of lens fibers. J. Biophys. Biochem. Cytol., *6*:97–102.

Wesley, R. E., C. D. McCord, Jr., and N. A. Jones. 1980. Height of the tarsus of the lower eyelid. Amer. J. Ophthalmol. *90*:102–105.

Westheimer, G., and S. M. Blair. 1973. The parasympathetic pathways to internal eye muscles. Invest. Ophthalmol., *12*:193–197.

Willis, N. R., M. J. Hollenberg, and C. R. Brackevelt. 1969. The fine structure of the lens of the fetal rat. Canad. J. Ophthalmol., *4*:307–318.

Wilson, D. B. 1980. Embryonic development of the head and neck: Part 4, Organs of special sense. Head Neck Surg., *2*:237–247.

Wislocki, G. B., and A. J. Ladman. 1955. The demonstration of a blood-ocular barrier in the albino rat by means of the intravitam deposition of silver. J. Biophys. Biochem. Cytol., *1*:501–510.

Wolff, E. 1968. *Anatomy of the Eye and Orbit.* 6th ed. Revised by R. J. Last. Lewis, London, 529 pp.

Wolter, J. R. 1959. Glia of the human retina. Amer. J. Ophthalmol., *48*:370–393.

Woolf, D. 1956. A comparative cytological study of the ciliary muscle. Anat. Rec., *124*:145–164.

Yamada, E. 1969. Some structural features of the fovea centralis in the human retina. Arch. Ophthalmol., *82*:151–159.

Young, R. W. 1967. The renewal of photoreceptor cell outer segments. J. Cell Biol., *33*:61–72.

Young, R. W. 1970. Visual cells. Sci. Amer., *223*:80–91.

Young, R. W. 1971. The renewal of rod and cone outer segments in the Rhesus monkey. J. Cell Biol., *49*:303–318.

Young, R. W. 1976. Visual cells and the concept of renewal. Invest. Ophthalmol., *15*:700–725.

Young, R. W., and D. Bok. 1969. Participation of the retinal pigment epithelium in the rod outer segment renewal process. J. Cell Biol., *42*:392–403.

Young, R. W., and B. Droz. 1968. The renewal of protein in retinal rods and cones. J. Cell Biol., *39*:169–184.

The Ear

Adrian, E. D. 1943. Discharges from vestibular receptors in the cat. J. Physiol. (Lond.), *101*:389–407.

Aimi, K. 1978. The tympanic isthmus: Its anatomy and clinical significance. Laryngoscope, *88*:1067–1081.

Allin, E. F. 1975. Evolution of the mammalian middle ear. J. Morphol., *147*:403–437.

Allison, G. R. 1978. Anatomy of the external ear. Clin. Plast. Surg., *5*:419–422.

Altmann, F. 1950. Normal development of the ear and its mechanics. Arch. Otolaryngol., *52*:725–766.

Altmann, F. 1951. Malformations of the Eustachian tube, the middle ear, and its appendages. Arch. Otolaryngol., *54*:241–266.

Altmann, F., and J. G. Waltner. 1947. The circulation of labyrinthine fluids. Experimental investigations in rabbits. Ann. Otol. Rhinol. Laryngol., *56*:684–708.

Amjad, A. H., A. A. Sheer, and J. Rosenthal. 1969. Human internal auditory canal. Arch. Otolaryngol., *89*:709–714.

Anson, B. J., and T. H. Bast. 1955. The ear and the temporal bone. Development and adult structure. Otolaryngology, *1*:1–111.

Anson, B. J., and T. H. Bast. 1956. Development and adult anatomy of the auditory ossicles in relation to the operation for mobilization of the stapes in otosclerotic deafness. Laryngoscope, *66*:785–795.

Anson, B. J., and E. W. Cauldwell. 1942. The developmental anatomy of the human stapes. Ann. Otol. Rhinol. Laryngol., *51*:891–904.

Anson, B. J., J. A. Donaldson, and B. B Shilling. 1972. Surgical anatomy of the chorda tympani. Ann. Otol. Rhinol. Laryngol., *81*:616–631.

Anson, B. J., J. S. Hanson, and S. F. Richany. 1960. Early embryology of the auditory ossicles and associated structures in relation to certain anomalies observed clinically. Ann. Otol. Rhinol. Laryngol., *69*:427–447.

Arslan, M. 1960. The innervation of the middle ear. Proc. Roy. Soc. Med. (Lond.), *53*:1068–1074.

Barlow, C. M., and W. S. Root. 1949. The ocular sympathetic path between the superior cervical ganglion and the orbit in the cat. J. Comp. Neurol., *91*:195–208.

Bast, T. H. 1930. Ossification of the otic capsule in human fetuses. Carneg. Inst., Contr. Embryol., *21*:53–82.

Belanger, L. F. 1953. Autoradiographic detection of S[35] in the membranes of the inner ear of the rat. Science, *118*:520–521.

Blevins, C. E. 1963. Innervation of the tensor tympani muscle of the cat. Amer. J. Anat., *113*:287–301.

Blevins, C. E. 1964. Studies on the innervation of the stapedius muscle of the cat. Anat. Rec., *149*:157–171.

Bolz, E. A., and D. J. Lim. 1972. Morphology of the stapediovestibular joint. Acta Otolaryngol., *73*:10–17.

Bosman, D. H. 1978. The distribution of the chorda tympani in the middle ear area in man and two other primates. J. Anat. (Lond.), *127*:443–445.

Bowden, R. E. 1977. Development of the middle and external ear in man. Proc. Roy. Soc. Med. (Lond.), *70*:807–815.

Bredberg, G. 1968. Cellular pattern and nerve supply of the human organ of Corti. Acta Otolaryngol., Suppl., *236*:1–135.

Cahill, D. R., and M. H. Snow. 1975. A quick, effective method for dissecting the middle ear. Anat. Rec., *181*:685–687.

Citron, L., D. Exley, and C. S. Hallpike. 1956. Formation, circulation and chemical properties of labyrinthine fluids. Brit. Med. Bull., *12*:101–104.

Costa, A. 1972. Embryogenesis of the ear and its central projection. Adv. Exp. Med. Biol., *30*:291–303.

Cullen, J. K., M. S. Ellis, C. I. Berlin, and R. J. Lousteau. 1972. Human acoustic nerve action potential recordings from the tympanic membrane without anesthesia. Acta Otolaryngol., *74*:15–22.

Dankbaar, W. A. 1970. The pattern of stapedial vibration. J. Acoust. Soc. Amer., *48*:1021–1022.

Davis, H. 1957. Biophysics and physiology of the inner ear. Physiol. Rev., *37*:1–49.

Dayal, V. S., J. Farkashidy, and A. Kokshanian. 1973. Embryology of the ear. Canad. J. Otolaryngol., *2*:136–142.

Djupesland, G., and E. Gronas. 1973. On the insertion of the human stapedial tendon. Canad. J. Otolaryngol., 2:119–123.

Donaldson, I. 1980. Surgical anatomy of the tympanic nerve. J. Laryngol. Otol., 94:163–168.

Duvall, A. J., A. Flock, and J. Wersall. 1966. The ultrastructure of the sensory hairs and associated organelles of the cochlear inner hair cell, with reference to directional sensitivity. J. Cell Biol., 29:497–505.

Engström, H., H. W. Ades, and A. Andersson. 1966. Structural Pattern of the Organ of Corti. Williams and Wilkins, Baltimore, 172 pp.

Engström, H., and J. Wersall. 1953a. Structure of the organ of Corti. 1. Outer hair cells. Acta Otolaryngol., 43:1–10.

Engström, H., and J. Wersall. 1953b. Structure of the organ of Corti. 2. Supporting structures and their relations to sensory cells and nerve endings. Acta Otolaryngol., 43:323–334.

Engström, H., and J. Wersall. 1958. Structure and innervation of the inner ear sensory epithelia. Internat. Rev. Cytol., 6:535–585.

Erulkar, S. D., M. L. Shelanski, B. L. Whitsel, and P. Ogle. 1964. Studies of muscle fibers of the tensor tympani of the cat. Anat. Rec., 149:279–297.

Fernandez, C. 1951. The innervation of the cochlea (guinea pig). Laryngoscope, 61:1152–1172.

Fernandez, C., J. M. Goldberg, and W. R. Abend. 1972. Response to static tilts of peripheral neurons innervating otolith organs of the squirrel monkey. J. Neurophysiol., 35:978–997.

Fleischer, G. 1978. Evolutionary principles of the mammalian middle ear. Adv. Anat. Embryol. Cell Biol., 55:3–70.

Flock, A. 1964. Structure of the macula utriculi with special reference to directional interplay of sensory responses as revealed by morphological polarization. J. Cell Biol., 22:413–432.

Galambos, R., and A. Rupert. 1959. Action of the middle ear muscles in normal cats. J. Acoust. Soc. Amer., 31:349–355.

Geisler, C. D., W. S. Rhode, and D. T. Kennedy. 1974. Responses to tonal stimuli of single auditory nerve fibers and their relationship to basilar membrane motion in the squirrel monkey. J. Neurophysiol., 37:1156–1172.

Gibbin, K. P. 1979. The histopathology of the incus after stapedectomy. Clin. Otolaryngol., 4:343–354.

Graham, M. D., C. Reams, and R. Perkins. 1978. Human tympanic membrane-malleus attachment. Preliminary study. Ann. Otol. Rhinol. Laryngol., 87:426–431.

Graves, G. O., and L. F. Edwards. 1944. The eustachian tube. A review of its descriptive, microscopic, topographic and clinical anatomy. Arch. Otolaryngol., 39:359–397.

Gussen, R. 1968. Articular and internal remodeling in the human otic capsule. Amer. J. Anat., 122:397–418.

Gussen, R. 1970. Pacinian corpuscles in the middle ear. J. Laryngol. Otol., 84:71–76.

Hamilton, D. W. 1968. The calyceal synapse of type I vestibular hair cells. J. Ultrastruct. Res., 23:98–114.

Hamilton, D. W. 1970. The cilium on mammalian vestibular hair cells. Anat. Rec., 164:253–258.

Hawkins, J. E., and L. G. Johnsson. 1968. Light microscopic observations of the inner ear in man and monkey. Ann. Otol. Rhinol. Laryngol., 77:608–628.

Hentzer, E. 1970. Histologic studies of the normal mucosa in the middle ear, mastoid cavities, and Eustachian tube. Ann. Otol. Rhinol. Laryngol., 79:825–833.

Iurato, S. 1967. Submicroscopic Structure of the Inner Ear. Pergamon Press, London, 367 pp.

Johnson, F. R., R. M. H. McMinn, and G. N. Atfield. 1968. Ultrastructural and biochemical observations on the tympanic membrane. J. Anat. (Lond.), 103:297–310.

Johnstone, B. M., and P. M. Sellick. 1972. The peripheral auditory apparatus. Q. Rev. Biophys., 5:1–57.

Johnstone, B. M., and K. J. Taylor. 1971. Physiology of the middle ear transmission system. J. Otolaryngol. Soc. Aust., 3:226–228.

Jordan, V. M. 1973. Anatomy of the internal auditory canal. Adv. Otorhinolaryngol., 20:288–295.

Kimura, R. S. 1966. Hairs of the cochlear sensory cells and their attachment to the tectorial membrane. Acta Otolaryngol., 61:55–72.

Kimura, R. S. 1969. Distribution, structure and function of dark cells in the vestibular labyrinth. Ann. Otol. Rhinol. Laryngol., 78:542–561.

Kimura, R. S., and H. F. Schuknecht. 1970. The ultrastructure of the human stria vascularis. Part I. Acta Otolaryngol., 69:415–427.

Kimura, R. S., and J. Wersall. 1962. Termination of the olivocochlear bundle in relation to the outer hair cells of the organ of Corti in guinea pig. Acta Otolaryngol., 55:11–32.

Lim, D. J. 1976. Functional morphology of the mucosa of the middle ear and Eustachian tube. Ann. Otol. Rhinol. Laryngol., 85:36–43.

Lindeman, H. H. 1973. Anatomy of the otolith organs. Adv. Otorhinolaryngol., 20:405–433.

Lindquist, P. G. 1964. Experiments in endolymph circulation. Acta Otolaryngol., Suppl., 188:198–201.

Lindquist, P. G., R. Kimura, and J. Wersall. 1963. Ultrastructural organization of the epithelial lining in the endolymphatic duct and sac in the guinea pig. Acta Otolaryngol., 57:65–80.

Lowenstein, O. 1956. Peripheral mechanisms of equilibrium. Brit. Med. Bull., 12:114–118.

MacPhee, R. D. 1979. Entotympanics, ontogeny and primates. Folia Primatol. (Basel), 31:23–47.

Manley, G. A., and B. M. Johnstone. 1974. Middle-ear function in the guinea-pig. J. Acoust. Soc. Amer., 46:571–576.

Marovitz, W. F., and E. S. Porubsky. 1971. The embryological development of the middle ear space—a new concept. Ann. Otol. Rhinol. Laryngol., 80:384–389.

Masuda, Y., R. Saito, Y. Endo, Y. Kondo, and Y. Ogura. 1978. Histological development of stapes footplate in human embryos. Acta Med. Okayama, 32:109–117.

McLellan, M. S., and C. H. Webb. 1961. Ear studies in the newborn infant. J. Pediatr., 58:523–527.

Möller, A. R., ed. 1973. Basic Mechanisms in Hearing. Academic Press, New York, 941 pp.

Montagna, W., C. R. Noback, and F. G. Zak. 1948. Pigment, lipids and other substances in the glands of the external auditory meatus of man. Amer. J. Anat., 83:409–436.

Papangelou, L. 1975. Study of the human internal auditory canal in relation to age and sex. J. Laryngol. Otol., 89:79–89.

Perry, E. T. 1957. The Human Ear Canal. Charles C Thomas, Springfield, Ill., 116 pp.

Potter, A. B. 1936. Function of the stapedius muscle. Ann. Otol. Rhinol. Laryngol., 45:638–643.

Proctor, B. 1971. Surgical anatomy of the posterior tympanum. Acta Otorhinolaryngol. Belg., 25:911–928.

Rasmussen, G. L., and W. F. Windle, eds. 1960. Neural

Mechanisms of the Auditory and Vestibular Systems. Charles C Thomas, Springfield, Ill., 422 pp.

Saito, R., M. Igarashi, B. R. Alford, and F. R. Guilford. 1971. Anatomical measurement of the sinus tympani. A study of horizontal serial sections of the human temporal bone. Arch. Otolaryngol., *94:* 418–425.

Sando, I. 1965. The anatomical interrelationships of the cochlear nerve fibers. Acta Otolaryngol., *59:*417–436.

Schuknecht, H. K. 1950. A clinical study of auditory damage following blows to the head. Ann. Otolaryngol., *59:*331–358.

Smith, C. A. 1968. Electron microscopy of the inner ear. Ann. Otol. Rhinol. Laryngol., *77:*629–643.

Smith, C. A. 1975. The inner ear: Its embryological development and microstructure. In *The Nervous System.* Edited by D. B. Tower. Vol. 3, *Human Communication and Its Disorders.* Raven Press, New York, pp. 1–18.

Smith, C. A., and E. W. Dempsey. 1957. Electron microscopy of the organ of Corti. Amer. J. Anat., *100:*337–368.

Spector, B. 1944. Storage of trypan blue in the internal ear of the rat. Anat. Rec., *88:*83–89.

Spoendlin, H. 1969. Innervation patterns in the organ of Corti of the cat. Acta Otolaryngol., *67:*239–254.

Spoendlin, H. 1972. Innervation densities of the cochlea. Acta Otolaryngol., *73:*235–248.

Sunderland, S. 1945. The arterial relations of the internal auditory meatus. Brain, *68:*23–27.

Tasaki, I. 1954. Nerve impulses in individual auditory nerve fibers of guinea pig. J. Neurophysiol., *17:*97–122.

Tonndorf, J., and S. M. Khanna. 1970. The role of the tympanic membrane in middle ear transmission. Ann. Otol. Rhinol. Laryngol., *79:*743–753.

von Bekesy, G. 1956. Current status of theories of hearing. Science, *123:*779–783.

Wersäll, J. 1956. Studies on the structure and innervation of the sensory epithelium of the cristae ampullares in the guinea pig. Acta Otolaryngol., Suppl., *126:*1–85.

Wersäll, J., and Å. Flock. 1964. Physiological aspects of the structure of vestibular end organs. Acta Otolaryngol., Suppl., *192:*85–89.

Wersäll, J., Å. Flock, and P. G. Lindquist. 1965. Structural basis for directional sensitivity in cochlear and vestibular sensory receptors. Cold Spring Harbor Symp. Quant. Biol., *30:*115–132.

Wiggers, H. C. 1937. The functions of the intra-aural muscles. Amer. J. Physiol., *120:*771–780.

Winerman, I., H. Nathan, and B. Arensburg. 1980. Posterior ligament of the incus: variations in its components. Ear Nose Throat J., *59:*227–231.

Wislocki, G. B., and A. J. Ladman. 1955. Selective histochemical staining of the otolithic membranes, cupulae and tectorial membrane of the inner ear. J. Anat. (Lond.), *89:*1–12.

Wolff, D., R. J. Bellucci, and A. A. Eggston. 1957. *Microscopic Anatomy of the Temporal Bone.* Williams and Wilkins, Baltimore, 414 pp.

Yntema, C. L. 1933. Experiments on the determination of the ear ectoderm in the embryo of Amblystoma punctatum. J. Exp. Zool., *65:*317–357.

Yoko, Y. 1971. Early formation of nerve fibers in the human otocyst. Acta Anat. (Basel), *80:*99–106.

Zorzetto, N., and O. J. Tamega. 1979. The anatomical relationship of the middle ear and the jugular bulb. Anat. Anz., *146:*470–480.

14

The Integument

Covering the surface of the body and sheltering it from injurious influences in the environment is the **skin** or **integument.** It protects the deeper tissues from injury, from drying, and from invasion by foreign organisms; it contains the peripheral endings of many of the sensory nerves. It plays an important part in the regulation of the body temperature and also has limited excretory and absorbing powers. It consists principally of a layer of dense connective tissue, named the **dermis** (*corium, cutis vera*), and an external covering of epithelium, termed the **epidermis** or **cuticle.** On the surface of the former layer are **sensitive** and **vascular papillae;** within or beneath it are certain organs with special functions, namely, the **sudoriferous** and **sebaceous glands** and the **hair follicles.**

Development

The epidermis and its appendages (hairs, nails, sebaceous and sweat glands) are developed from the ectoderm; the corium or true skin is of mesodermal origin. About the fifth week of fetal development, the epidermis consists of two layers of cells, the deeper one corresponding to the rete mucosum. The subcutaneous fat appears about the fourth month, and the papillae of the true skin, about the sixth month. A considerable desquamation of epidermis takes place during fetal life, and this desquamated epidermis, mixed with sebaceous secretion, constitutes the **vernix caseosa,** with which the skin is smeared during the last three months of fetal life. The nails are formed at the third month and begin to project from the epidermis about the sixth month. The hairs appear between the third and fourth months in the form of solid downgrowths of the deeper layer of the epidermis, the growing extremities of which become inverted by papillary projections from the corium. The central cells of the solid downgrowths undergo alteration to form the hair, while the peripheral cells are retained to form the lining cells of the hair follicle. About the fifth month the fetal hairs (**lanugo**) appear first on the head and then on the other parts; they drop after birth and give place to the permanent hairs. The cellular structures of the sudoriferous and sebaceous glands are formed from the ectoderm, whereas the connective tissue and blood vessels are derived from the mesoderm. All the sweat glands are fully formed at birth; they begin to develop as early as the fourth month.

Structure

EPIDERMIS. The **epidermis, cuticle,** or **scarf skin** is nonvascular, consists of stratified epithelium, and is accurately molded over the papillary layer of the dermis. It varies in thickness in different parts of the body. In some situations, as in the palms of the hands and soles of the feet, it is thick, hard, and horny in texture. The more superficial layers of cells, called the **horny layer** (*stratum corneum*), may be separated by maceration from a deeper stratum, which is called the **stratum mucosum** and which consists of several layers of differently shaped cells. The free surface of the epidermis is marked by a network of linear furrows of variable size, dividing the surface into polygonal or lozenge-shaped areas. Some of these furrows are large, such as those opposite the flexures of the joints, and they correspond to folds in the dermis. In other situations, as upon the back of the hand, they are exceedingly fine and intersect one another at various angles.

Upon the palmar surfaces of the hands and fingers and upon the soles of the feet, the epidermal ridges are distinct and disposed in curves; they depend upon the large size and peculiar arrangements of the papillae upon which the epidermis is placed. These ridges increase friction between contact surfaces to prevent slipping in walking or prehension. The direction of the ridges is at right angles to the force that tends to produce slipping or to the resultant of such forces when these forces vary in direction. In each individual the lines of the tips of the fingers and thumbs form distinct patterns unlike those of any other person. A method of determining the identity of a person is based on this fact by making impressions or **fingerprints** of these lines. The deep surface of the epidermis is accurately molded upon the papillary layer of the dermis, the papillae being covered by a basement membrane. When the epidermis is lifted off by maceration and inverted, it presents on its undersurface pits or depressions that correspond to the papillae, and ridges corresponding to the intervals between them. Fine tubular prolongations are continued from this layer into the ducts of the sudoriferous and sebaceous glands.

The **stratified squamous epithelium** of the epidermis is composed of several layers named according to various properties such as shape of cells, texture, composition and position. Beginning with the deepest, they are: (*a*) stratum basale, (*b*) stratum spinosum, (*c*) stratum granulosum, (*d*) stratum lucidum, and (*e*) stratum corneum.

The **stratum basale,** the deepest layer, is composed of columnar or cylindrical cells, giving it an alternate name **stratum cylindricum.** The ends of the cells in contact with the basement membrane have toothlike protoplasmic projections that fit into sockets in the membrane and appear to anchor the cells to the underlying dermis. The cells of this layer undergo division by mitosis, supplying new cells to make up for the continual loss of surface layers from abrasion. This layer has been appropriately named the **stratum germinativum** (Fig. 14-1).

The **stratum spinosum** is composed of several layers of polygonal cells, the number depending upon the area of the body from which the skin is taken. As a result of the slight shrinkage caused by technical procedures, these cells in ordinary histologic preparations appear to have *cytoplasmic bridges* connecting them with their neighbors. When they are pulled apart, they appear to have minute *spines* on their surface, giving them the name *prickle cells*. Electron microscopic studies have shown that each cell adheres to its neighbors at particular points called desmosomes, and that these points are drawn out into the spines by

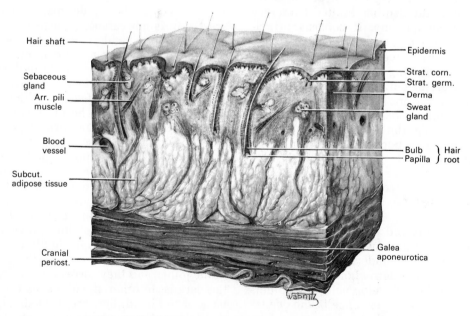

Hair shaft

Sebaceous gland

Arr. pili muscle

Blood vessel

Subcut. adipose tissue

Cranial periost.

Epidermis

Strat. corn.
Strat. germ.
Derma

Sweat gland

Bulb } Hair
Papilla } root

Galea aponeurotica

FIG. 14-1. Section through scalp of adult cadaver. 5 ×.

shrinkage. Ultrastructural studies have also shown that the cytoplasm of these cells contains minute fibrils, some of which are oriented toward the desmosomes. In light microscopic preparations these fibrils may be visible, and since they are associated with the desmosomes, they give the impression of running between cells and are called **tonofibrils.**

The **stratum granulosum** is composed of two or three rows of flat cells that lie parallel with the surface. They contain numerous large granules that stain deeply with hematoxylin. They are composed of **keratohyalin,** a substance that apparently is transformed into keratin in more superficial layers.

The **stratum lucidum** appears to be a homogenous translucent band, much thinner than the strata on either side of it. The cells contain droplets of **eleidin** and their nuclei and cell boundaries are not visible.

The **stratum corneum** is composed of squamous plates of scales fused together to make the outer horny layer. These plates are the remains of the cells and contain a fibrous protein, **keratin.** The most superficial layer sloughs off or desquamates. The thickness of this layer is correlated with the trauma to which an area is subjected, being very thick on the palms and soles but thin over protected areas.

PIGMENTATION. The black color of the skin in the Negro and the tawny color among some of the white races are due to the presence of pigment in the cells of the epidermis. This pigment is especially distinct in the cells of the stratum basale and is similar to that found in the cells of the pigmented layer of the retina. As the cells approach the surface and desiccate, the color becomes partially lost; the disappearance of the pigment from the superficial layers of the epidermis, however, is difficult to explain.

The pigment (**melanin**) consists of small dark brown or black granules, closely packed together within the cells, but not involving the nucleus.

The main purpose of the epidermis is protection. As the surface is worn away, new cells are supplied, and thus the true skin, the vessels, and the nerves that it contains are defended against damage.

DERMIS. The **dermis, corium, cutis vera** or **true skin** is tough, flexible, and elastic. Its thickness varies in different parts of the body. Thus it is very thick in the palms of the hands and soles of the feet. It is thicker on the dorsal aspect of the body than on the ventral aspects, and on the lateral sides of the limbs than on the medial sides. In the eyelids, scrotum, and penis, it is exceedingly thin and delicate.

The dermis consists of felted connective tissue, with a varying amount of elastic fibers and numerous blood vessels, lymphatics, and nerves. The connective tissue is arranged in two layers: a **deeper** or **reticular layer,** and a **superficial** or **papillary layer** (Fig. 14-2).

Smooth muscle cells are found in the dermis and the subcutaneous layers of the scrotum, penis, labia majora, and nipples. In the mammary papilla, the smooth muscle cells are disposed in circular bands and radiating bundles arranged in superimposed laminae. Where there are hairs, discrete bundles of smooth muscle called the **arrectores pilorum** are attached in the superficial layers of the corium and near the base of each hair follicle.

The **papillary layer** (*stratum papillare; superficial layer; corpus papillare of the corium*) consists of numerous small, highly sensitive, and vascular eminences, the **papillae,** which rise perpendicularly from its surface. The papillae are minute conical eminences with rounded or blunted extremities; occasionally they are divided into two or more parts and are received into corresponding pits on the undersurface of the cuticle. On the general surface of the body, especially in parts endowed with slight sensibility, they are few in number and exceedingly minute, but in some situations, as upon the palmar surfaces of the hands and fingers and the plantar surfaces of the feet and toes, the papillae are long, large, closely aggregated, and arranged in parallel curved lines, forming the elevated ridges seen on the free surface of the epidermis. Each ridge contains two rows of papillae, between which the ducts of the sudoriferous glands pass outward to open on the summit of the ridge. Each papilla consists of small and closely interlacing bundles of finely fibrillated tissue with a few elastic fibers; within this tissue is a capillary loop, and in some papillae, especially in the palms of the hand and the fingers, there are tactile corpuscles (Fig. 14-3).

The **reticular layer** (*stratum reticulare; deep layer*) consists of fibroelastic connec-

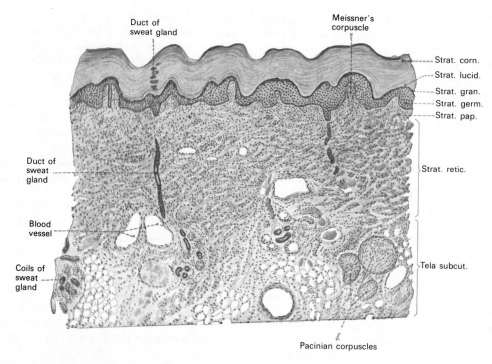

Duct of
sweat gland

Meissner's
corpuscle

Strat. corn.
Strat. lucid.
Strat. gran.
Strat. germ.
Strat. pap.

Strat. retic.

Duct of
sweat
gland

Blood
vessel

Coils of
sweat
gland

Tela subcut.

Pacinian corpuscles

FIG. 14-2. Section through the skin of the human foot, cut perpendicularly to the surface. (From Rauber-Kopsch, *Lehrbuch u. Atlas d. Anatomie d. Menschen,* 19th ed., Vol. II, Georg Thieme Verlag, Stuttgart, 1955.)

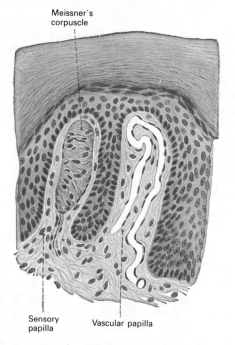

Meissner's
corpuscle

Sensory
papilla

Vascular papilla

FIG. 14-3. Vascular and sensory papillae in the skin of the human foot. (From Rauber-Kopsch, *Lehrbuch u. Atlas d. Anatomie d. Menschen,* 19th ed., Vol. II, Georg Thieme Verlag, Stuttgart, 1955.)

tive tissue, which is composed chiefly of collagenous bundles but contains yellow elastic fibers in varying number in different parts of the body. The cells are principally fibroblasts and histiocytes, but other types may be found. Near the papillary layer the collagenous bundles are small and compactly arranged; in the deeper layers they are larger and coarser, and between their meshes are sweat glands, sebaceous glands, hair shafts or follicles, and small collections of fat cells. The deep surface of the reticular layer merges with the adipose tissue of the subcutaneous fascia.

CLEAVAGE LINES OF THE SKIN (*Langer's lines*[1]). When a sharp conical instrument is used to make a penetrating wound, it does not leave a round hole in the skin, as might be expected, but a slit such as would be expected from a flat blade. Maps of the directions of these slits from puncture wounds over all parts of the body have been made from dissecting room material (Figs. 14-4;

[1] Carl Ritter von Eldenberg von Langer (1819–1887): Austrian anatomist (Vienna).

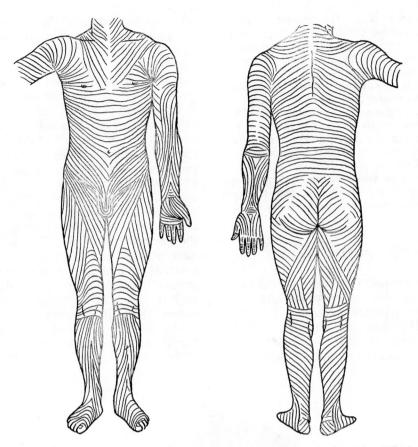

FIG. 14-4. Cleavage lines (Langer's lines) of the skin. Trunk and extremities. (Eller.)

14-5). These maps indicate that there are definite lines of **tension** or **cleavage lines** within the skin that are characteristic for each part of the body. In microscopic sections cut parallel with these lines, most of the collagenous bundles of the reticular layer are cut longitudinally, while in sections cut across the lines, the bundles are in cross section. The cleavage lines correspond closely with the crease lines on the surface of the skin in most parts of the body. The pattern of the cleavage lines, according to Cox (1941), varies with body configuration, but is constant for individuals of similar build regardless of age. There are limited areas of the body in which the orientation of the bundles is irregular and confused. The cleavage lines are of particular interest to the surgeon because an incision made parallel to the lines heals with a fine linear scar, while an incision across the lines may set up irregular tensions that result in an unsightly scar.

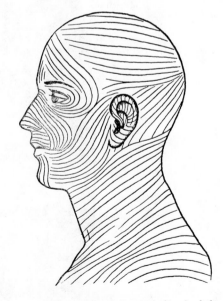

FIG. 14-5. Cleavage lines (Langer's lines) of the skin. Head and neck. (Eller.)

VESSELS AND NERVES. The **arteries** supplying the skin form a network in the subcutaneous tissue, branches of which supply the sudoriferous glands, the hair follicles, and the fat. Other branches unite in a plexus immediately beneath the dermis; from this plexus, fine capillary vessels pass into the papillae, forming, in the smaller ones, a single capillary loop, but in the larger, a more or less convoluted vessel. The **lymphatic vessels** of the skin form two networks, superficial and deep, which communicate with each other and with those of the subcutaneous tissue by oblique branches.

The **nerves** of the skin terminate partly in the epidermis and partly in the corium; at their terminals are found various types of nerve endings.

Appendages of the Skin

The appendages of the skin are the **nails,** the **hairs,** and the **sudoriferous** and **sebaceous glands** with their ducts.

NAILS. The nails (ungues) (Fig. 14-6) are flattened, elastic structures of a horny texture placed upon the dorsal surfaces of the terminal phalanges of the fingers and toes. Each nail is convex on its outer surface, concave within, and is implanted by a portion called the **root** into a groove in the skin. The exposed portion is called the **body,** and the distal extremity, the **free edge.** The nail is firmly adherent to the corium, being accurately molded upon its surface; the part beneath the body and root of the nail is called the **nail matrix,** because from it the nail is produced. Under the greater part of the body of the nail, the matrix is thick and raised into a series of longitudinal ridges that are highly vascular, and the color is seen through the transparent tissue. Near the root of the nail, the papillae are smaller, less vascular, and have no regular arrangement, and here the tissue of the nail is not firmly adherent to the connective tissue stratum but only in contact with it; hence this portion is whiter and is called the **lunula** because of its shape.

The cuticle as it passes forward on the dorsal surface of the finger or toe is attached to the surface of the nail a little in advance

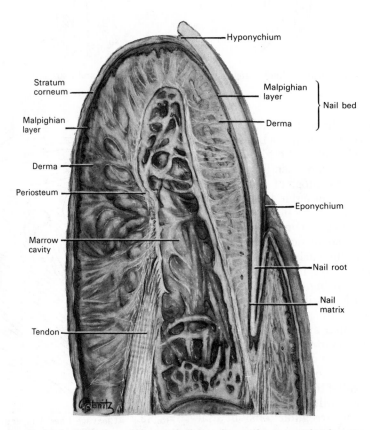

FIG. 14-6. Median longitudinal section through finger of adult cadaver. The marrow has been removed to show bony trabeculae. 5 ×.

of its root. At the extremity of the finger, it is connected with the undersurface of the nail; both epidermic structures are thus directly continuous with each other. The superficial, horny part of the nail consists of a greatly thickened stratum lucidum; the stratum corneum forms merely the thin cuticular fold (**eponychium**) that overlaps the lunula; and the deeper part consists of the stratum mucosum. The cells in contact with the papillae of the matrix are columnar and arranged perpendicularly to the surface; those that succeed them are round or polygonal, with the more superficial ones becoming broad, thin, flattened, and so closely packed as to make the limits of the cells indistinct. The nails grow in length by the proliferation of the cells of the stratum germinativum at the root of the nail, and they grow in thickness from the part of the stratum germinativum that underlies the lunula.

HAIR.   Hairs (*pili*) are found on nearly every part of the body's surface, but are absent from the palms of the hands, the soles of the feet, the dorsal surfaces of the terminal phalanges, the glans penis, the inner surface of the prepuce, and the inner surfaces of the labia. They vary much in length, thickness, and color in different parts of the body and among different races of mankind. In some parts, as in the skin of the eyelids, they are so short that they do not project beyond the follicles containing them; in others, as upon the scalp, they are of considerable length. Some hairs, such as the eyelashes, the pubic hairs, and the whiskers and beard, are remarkable for their thickness. Straight hairs are stronger than curly hairs, and present on transverse section a cylindrical or oval outline; curly hairs, on the other hand, are flat. A hair consists of a **root,** which is the part implanted in the skin, and a **shaft** or **scapus,** which is the portion projecting from the surface.

The **root of the hair** (*radix pili*) ends in an enlargement, the **hair bulb,** which is whiter and softer than the shaft, and is lodged in a follicular involution of the epidermis called the **hair follicle** (Fig. 14-7). The follicle of a very long hair extends into the subcutaneous cellular tissue. The hair follicle commences on the surface of the skin with a funnel-shaped opening and passes inward in an oblique direction for straight hairs or curved for curly hairs; it is dilated at its deep extremity, where it corresponds with the hair

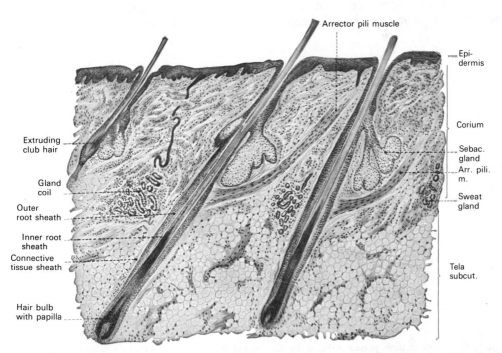

FIG. 14-7.   Section of human scalp with hairs cut longitudinally. (From Rauber-Kopsch, *Lehrbuch u. Atlas d. Anatomie d. Menschen,* 19th ed., Vol. II, Georg Thieme Verlag, Stuttgart, 1955.)

bulb. Opening into the follicle near its free extremity are the ducts of one or more sebaceous glands. At the bottom of each hair follicle is a small conical, vascular eminence or **papilla,** similar in every respect to those found upon the surface of the skin; it is continuous with the dermic layer of the follicle and supplied with nerve fibrils. The hair follicle consists of two coats—an **outer** or **dermic,** and an **inner** or **epidermic.**

The **outer** or **dermic coat** is formed mainly of fibrous tissue; it is continuous with the dermis, highly vascular, and supplied with numerous minute nervous filaments. It consists of three layers (Fig. 14-8). The most internal is a hyaline basement membrane, which is well marked in the larger hair follicles but not very distinct in the follicles of minute hairs. This membrane is limited to the deeper part of the follicle. Outside this is a compact layer of fibers and spindle-shaped cells arranged circularly around the follicle; this layer extends from the bottom of the follicle as high as the entrance of the ducts of the sebaceous glands. Externally a thick layer of connective tissue is arranged in longitudinal bundles, forming a more open texture and corresponding to the reticular part of the dermis; in this are contained the blood vessels and nerves.

The **inner** or **epidermic coat** adheres closely to the root of the hair and consists of two strata, named respectively the **outer** and **inner root sheaths.** The outer sheath corresponds with the stratum mucosum of the epidermis and resembles it in the rounded form and soft character of its cells; at the bottom of the hair follicle these cells become continuous with those of the root of the hair. The inner root sheath consists of (1) a delicate cuticle next to the hair, composed of a single layer of imbricated scales with atrophied nuclei; (2) one or two layers of horny, flat nucleated cells, known as **Huxley's layer**[1]; and (3) a single layer of cubical cells with clear, flat nuclei, called **Henle's layer.**[2]

The **hair bulb** is molded over the papilla and composed of polyhedral epithelial cells, which become elongated and spindle-shaped as they pass upward into the root of the hair. Some, however, remain polyhedral in the center. Some of these latter cells contain pigment granules, which are responsible for the color of the hair. It occasionally happens that these pigment granules completely fill the cells in the center of the bulb. This causes the dark tract of pigment, of greater or lesser length, that often is found in the axis of the hair.

[1] Thomas Henry Huxley (1825–1895): An English physiologist and naturalist (London).
[2] Friedrich Gustav Jakob Henle (1809–1885): A German anatomist and histologist (Göttingen).

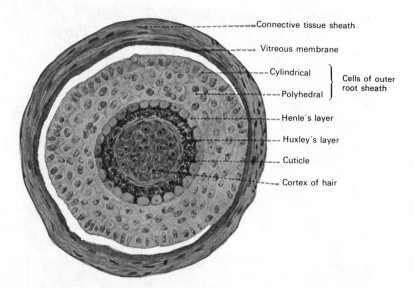

Connective tissue sheath

Vitreous membrane

Cylindrical

Polyhedral

Cells of outer root sheath

Henle's layer

Huxley's layer

Cuticle

Cortex of hair

FIG. 14-8. Cross section through hair and root sheath. The red granules in Huxley's layer are keratohyalin. The space between hair and sheath is an artifact. (From Rauber-Kopsch, *Lehrbuch u. Atlas d. Anatomie d. Menschen,* 19th ed., Vol. II, Georg Thieme Verlag, Stuttgart, 1955.)

The **shaft of the hair** (*scapus pili*) consists, from within outward, of three parts: the medulla, the cortex, and the cuticle. The **medulla** is usually wanting in the fine hairs covering the surface of the body, and commonly in those of the head. It is more opaque and deeply colored than the cortex when viewed by transmitted light, but when viewed by reflected light it is white. It is composed of rows of polyhedral cells that contain granules of eleidin and frequently air spaces. The **cortex** constitutes the chief part of the shaft; its cells are elongated and united to form flat fusiform fibers that contain pigment granules in dark hair and air in white hair. The **cuticle** consists of a single layer of flat scales that overlap one another from deep to superficial.

Connected with the hair follicles are minute bundles of involuntary muscular fibers, termed the **arrectores pilorum.** They *arise* from the superficial layer of the dermis and are *inserted* into the hair follicle, deep to the entrance of the duct of the sebaceous gland. They are placed on the side toward which the hair slopes, and by their action they diminish the obliquity of the follicle and make the hair stand erect (Fig. 14-7). The sebaceous gland is situated in the angle formed by the arrector muscle with the superficial portion of the hair follicle, and contraction of the muscle thus tends to squeeze the sebaceous secretion from the duct of the gland.

SEBACEOUS GLANDS. The sebaceous glands are small, sacculated, glandular organs lodged in the substance of the dermis. They are found in most parts of the skin, but are especially abundant in the scalp and face; they are also numerous around the apertures of the anus, nose, mouth, and external ear, but are wanting in the palms of the hands and soles of the feet. Each gland consists of a single duct, more or less capacious, which emerges from a cluster of oval or flask-shaped alveoli. The number of alveoli may vary from two to five in number, but in some instances there may be as many as 20. Each alveolus is composed of a transparent basement membrane enclosing epithelial cells. The outer or marginal cells are small, polyhedral, and continuous with the cells lining the duct. The remainder of the alveolus is filled with larger cells that contain lipid, except in the center, where the cells have disintegrated, leaving a cavity filled with their débris and a mass of fatty matter,

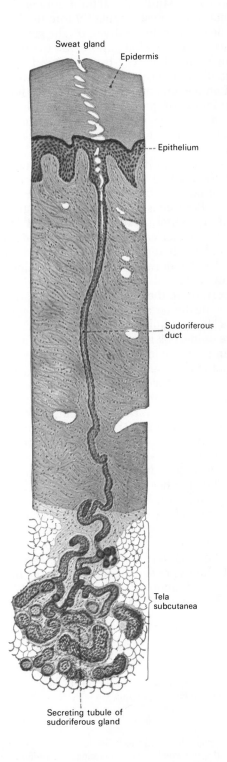

FIG. 14-9. Sweat gland from the sole of the human foot. (From Rauber-Kopsch, *Lehrbuch u. Atlas d. Anatomie d. Menschen,* 19th ed., Vol. II, Georg Thieme Verlag, Stuttgart, 1955.)

which constitute the **sebum cutaneum.** The ducts open most frequently into the hair follicles, but occasionally upon the general surface, as in the labia minora and the free margin of the lips. On the nose and face the glands are large, distinctly lobulated, and often become much enlarged from the accumulation of pent-up secretion. The tarsal glands of the eyelids are elongated sebaceous glands with numerous lateral diverticula.

SUDORIFEROUS GLANDS. The sudoriferous or sweat glands (*glandulae sudoriferae*) (Fig. 14-9) are found in almost every part of the skin. Each consists of a single tube, the deep part of which is irregularly coiled into an oval or spherical ball, named the **body** of the gland, while the superficial part, or **duct,** traverses the dermis and cuticle and opens on the surface of the skin by a funnel-shaped aperture. In the superficial layers of the dermis the duct is straight, but in the deeper layers it is convoluted or even twisted. Where the epidermis is thick, as in the palms of the hands and soles of the feet, the part of the duct that passes through it is spirally coiled. The size of the glands varies. They are especially large in those regions where the amount of perspiration is great, as in the axillae, where they form a thin, mammillated layer of a reddish color, which corresponds exactly to the situation of the hair in this region; the glands are also large in the groin.

The number of glands varies. They are plentiful on the palms of the hands and on the soles of the feet, where the orifices of the ducts are exceedingly regular and open on the curved ridges. They are least numerous in the neck and back. The palm has about 370 glands per square centimeter; the back of the hand, about 200; the forehead, 175; the breast, abdomen, and forearm, 155, and the leg and back, 60 to 80. It has been estimated that the total number is about 2,000,000. The following shows the average number of sweat glands per square centimeter of skin on the fingers in various races (Clark and Llamon, 1917):

American (white)	558.2
American (Negro)	597.2
Filipino	653.6
Moro	684.4
Negrito (adult)	709.2
Hindu	738.2
Negrito (youth)	950.0

Sweat glands are absent in the deeper portion of the external auditory meatus, the prepuce, and the glans penis.

The tube, both in the body of the gland and in the duct, consists of two layers—an outer layer of fine **areolar tissue** and an inner layer of **epithelium.** The outer layer is thin and continuous with the superficial stratum of the dermis. In the body of the gland, the epithelium consists of a single layer of cubical cells, between the deep ends of which the basement membrane is a layer of longitudinally or obliquely arranged nonstriped muscular fibers. The ducts are destitute of muscular fibers and are composed of a basement membrane lined by two or three layers of polyhedral cells; the lumen of the duct is coated by a thin cuticle. When the epidermis is carefully removed from the surface of the dermis, the ducts may be pulled out from the dermis in the form of short, thread-like processes. The ceruminous glands of the external acoustic meatus, the ciliary glands at the margins of the eyelids, the circumanal glands and probably the mammary glands are modified sudoriferous glands. The average quantity of sweat secreted in 24 hours varies from 700 to 900 gm.

References

(References are listed not only to those articles and books cited in the text, but to others as well which are considered to contain valuable resource information for the student who desires it.)

Skin

Billingham, R. E., and P. B. Medawar. 1953. A study of the branched cells of mammalian epidermis with special reference to the fate of their division products. Philos. Trans. Roy. Soc. Lond. (Biol.), 237:151–169.

Billingham, R. E., and W. K. Silvers. 1960. The melanocytes of mammals. Q. Rev. Biol., 35:1–40.

Breathnach, A. S. 1964. Observations on cytoplasmic organelles in Langerhans cells of human epidermis. J. Anat. (Lond.), 98:265–270.

Breathnach, A. S. 1965. The cell of Langerhans. Int. Rev. Cytol., 18:1–28.

Breathnach, A. S. 1968. The epidermal Langerhans cell. Brit. J. Dermatol., 80:688–689.

Breathnach, A. S. 1971. An Atlas of the Ultrastructure of Human Skin. Churchill, London, 406 pp.

Breathnach, A. S. 1981. Ultrastructure of embryonic skin. Curr. Probl. Dermatol., 9:1–28.

Brody, I. 1959. The keratinization of epidermal cells of normal guinea pig skin as revealed by electron microscopy. J. Ultrastruct. Res., 2:482–511.

Brody, I. 1960. The ultrastructure of the tonofibrils in the keratinization process of normal human epidermis. J. Ultrastruct. Res., 4:264–297.

Brody, I. 1977. Ultrastructure of stratum corneum. Int. J. Dermatol., 16:245–256.

Champion, R. H., T. Gillman, A. J. Rook, and R. T. Sims (eds.). 1970. *An Introduction to the Biology of the Skin*. Blackwell, Oxford, 450 pp.

Cox, H. T. 1941. The cleavage lines of the skin. Brit. J. Surg., 29:234–240.

Cummins, H., and C. Midlo. 1961. *Finger prints, Palms and Soles*. Dover Publications, New York, 319 pp.

Danielson, D. A. 1977. Wrinkling of the human skin. J. Biomech., 10:201–204.

Drochmans, P. 1963. Melanin granules: Their fine structure, formation, and degradation in normal and pathological tissues. Int. Rev. Exp. Pathol., 2:357–422.

Eady, R. A. J., et al. 1979. Mast cell population density, blood vessel density and histamine content in normal human skin. Brit. J. Dermatol., 100:623–633.

Enna, C. D., and R. F. Dyer. 1979. The histomorphology of the elastic tissue system in the skin of the human hand. Hand, 11:144–150.

Farquhar, M. G., and G. E. Palade. 1965. Cell junctions in amphibian skin. J. Cell Biol., 26:263–291.

Felsher, Z. 1947. Studies on the adherence of the epidermis to the corium. J. Invest. Dermatol., 8:35–47.

Fitzpatrick, T. B., and G. Szabo. 1959. The melanocyte: Cytology and cytochemistry. J. Invest. Dermatol., 32:197–209.

Forbes, G. 1938. Lymphatics of the skin, with a note on lymphatic watershed areas. J. Anat. (Lond.), 72:399–410.

Green, H. 1980. The keratinocyte as differentiated cell type. Harvey Lect., 74:101–139.

Haertsch, P. A. 1981. The blood supply to the skin of the leg; A post-mortem investigation. Brit. J. Plast. Surg., 34:470–477.

Hibbs, R. G., G. E. Burch, and J. H. Phillips. 1958. The fine structure of the small blood vessels of the normal human dermis and subcutis. Amer. Heart J., 56:662–670.

Hibbs, R. G., and W. H. Clark, Jr. 1959. Electron microscope studies of the human epidermis. The cell boundaries and topography of the stratum Malpighii. J. Biophys. Biochem. Cytol., 6:71–76.

Holbrook, K. A. 1979. Human epidermal embryogenesis. Int. J. Dermatol., 18:329–356.

Holbrook, K. A., and G. F. Odland. 1980. Regional development of the human epidermis in the first trimester embryo and the second trimester fetus (ages related to the timing of amniocentesis and fetal biopsy). J. Invest. Dermatol., 74:161–168.

Katzberg, A. A. 1958. The area of the dermo-epidermal junction in human skin. Anat. Rec., 131:717–726.

Lavker, R. M., and A. G. Matoltsy. 1970. Formation of horny cells. The fate of cell organelles and differentiation products in ruminal epithelium. J. Cell Biol., 44:501–512.

Leder, L. D. 1981. Intraepidermal mast cells and their origin. Amer. J. Dermatopathol., 3:247–250.

Marks, R. 1981. Measurement of biological ageing in human epidermis. Brit. J. Dermatol., 104:627–633.

Matoltsy, A. G. 1966. Membrane-coating granules of the epidermis. J. Ultrastruct. Res., 15:510–515.

Matoltsy, A. G., and S. J. Sinesi. 1957. A study of the mechanism of keratinization of human epidermal cells. Anat. Rec., 128:55–68.

Medawar, P. B. 1953. The micro-anatomy of the mammalian epidermis. Q. J. Micr. Sci., 94:481–506.

Meyer, W., K. Neurand, and B. Radke. 1982. Collagen fibre arrangement in the skin of the pig. J. Anat. (Lond.), 134:139–148.

Montagna, W. 1965. The skin. Sci. Amer., 212(2):56–66.

Montagna, W., and W. C. Lobitz, Jr. (eds.). 1964. *The Epidermis*. Academic Press, New York, 646 pp.

Montagna, W., and P. F. Parakkal. 1974. *The Structure and Function of the Skin*. 3rd ed. Academic Press, New York, 433 pp.

Munger, B. L. 1982. Multiple afferent innervation of primate facial hairs—Henry Head and Max von Frey revisited Brain Res., 257:1-43.

Odland, G. F. 1950. The morphology of the attachment between the dermis and the epidermis. Anat. Rec., 108:399–413.

Odland, G. F. 1958. The fine structure of the interrelationship of cells in the human epidermis. J. Biophys. Biochem. Cytol., 4:529–538.

Okajima, M. 1979. Dermal and epidermal structures of the volar skin. Birth Defects, 15:179–198.

Sagebiel, R. W. 1972. *In vivo* and *in vitro* uptake of ferritin by Langerhans cells of the epidermis. J. Invest. Dermatol., 58:47–54.

Sarkany, I., and G. A. Caron. 1965. Microtopography of the human skin. J. Anat. (Lond.), 99:359–364.

Selby, C. C. 1955. An electron microscope study of the epidermis of mammalian skin in thin sections. I. Dermo-epidermal junction and basal cell layer. J. Biophys. Biochem. Cytol., 1:429–444.

Selby, C. C. 1957. An electron microscopic study of thin sections of human skin. II. Superficial cell layers of footpad epidermis. J. Invest. Dermatol., 29:131–149.

Southwood, W. F. W. 1955. The thickness of the skin. Plast. Reconstr. Surg., 15:423–429.

Spearman, R. I. C. 1973. *The Integument: A Textbook of Skin Biology*. Cambridge University Press, London, 208 pp.

Taussig, J., and G. D. Williams. 1940. Skin color and skin cancer. Arch. Pathol., 30:721–730.

Tsuji, T. 1982. Scanning electron microscopy of dermal elastic fibres in transverse section. Brit. J. Dermatol., 106:545–550.

Zelickson, S. A. (ed.). 1967. *Ultrastructure of Normal and Abnormal Skin*. Lea & Febiger, Philadelphia, 431 pp.

Hair, Nails and Glands

Baker, B. L. 1951. The relationship of the adrenal, thyroid, and pituitary glands to the growth of hair. Ann. N. Y. Acad. Sci., 53:690–707.

Baran, R. 1981. Nail growth direction revisited. Why do nails grow out instead of up? J. Amer. Acad. Dermatol., 4:78–84.

Bean, W. B. 1980. Nail growth. Thirty-five years of observation. Arch. Intern. Med., 140:73–76.

Bell, M. 1971. A comparative study of sebaceous gland ultrastructure in subhuman primates. III. Macaques: Ultrastructure of sebaceous glands during fetal development of rhesus monkeys (*Macaca mulatta*). Anat. Rec., 170:331:342.

Birbeck, M. S. C., E. H. Mercer, and N. A. Barnicot. 1956.

The structure and formation of pigment granules in human hair. Exp. Cell Res., 10:505–514.

Birbeck, M. S. C., and E. H. Mercer. 1957. The electron microscopy of the human hair follicle. I. Introduction and the hair cortex. II. The hair cuticle. III. The inner root sheath and trichohyaline. J. Biophys. Biochem. Cytol., 3:203–230.

Butcher, E. O. 1946. Hair growth and sebaceous glands in skin transplanted under the skin and into the peritoneal cavity in the rat. Anat. Rec., 96:101–109.

Butcher, E. O. 1951. Development of the pilary system and the replacement of hair in mammals. Ann. N. Y. Acad. Sci., 53:508–516.

Chase, H. B. 1954. Growth of the hair. Physiol. Rev., 34:113–126.

Chase, H. B., W. Montagna, and J. D. Malone. 1953. Changes in the skin in relation to the hair growth cycle. Anat. Rec., 116:75–81.

Clark, E., and R. H. Lhamon. 1917. Observations on the sweat glands of tropical and northern races. Anat. Rec., 12:139–147.

Dawber, R. P. 1980. The ultrastructure and growth of human nails. Arch. Dermatol. Res., 269:197–204.

Dixon, A. D. 1961. The innervation of hair follicles in the mammalian lip. Anat. Rec., 140:147–158.

Duggins, O. H. 1954. Age changes in head hair from birth to maturity. IV. Refractive indices and birefringence of the cuticle of hair of children. Amer. J. Phys. Anthrop., N.S., 12:89–114.

Ellis, R. A. 1967. Eccrine, sebaceous and apocrine glands. In Ultrastructure of Normal and Abnormal Skin. Edited by A. S. Zelickson. Lea & Febiger, Philadelphia, pp. 132–162.

Hamilton, J. B. 1942. Male hormone stimulation is prerequisite and an incitant in common baldness. Amer. J. Anat., 71:451–480.

Hamilton, J. B. 1951. Patterned loss of hair in man: Types and incidence. Ann. N. Y. Acad. Sci., 53:708–728.

Hibbs, R. G. 1958. The fine structure of human eccrine sweat glands. Amer. J. Anat., 103:201–217.

Holbrook, K. A., and G. F. Odland. 1978. Structure of the human fetal hair canal and initial hair eruption. J. Invest. Dermatol., 71:385–390.

Inaba, M., C. T. McKinstry, and F. Umezawa. 1981. Clinical observations on the development and eventual character of hair in the axillae of human beings. J. Dermatol. Surg. Oncol., 7:340–342.

Ito, T., and S. Shibasaki. 1966. Electron microscopic study on human eccrine sweat glands. Arch. Histol. Jap., 27:81–115.

Johnson, P. L., and G. Bevelander. 1946. Glycogen and phosphatase in the developing hair. Anat. Rec., 95:193–199.

Mahler, F., et al. 1979. Blood pressure fluctuations in human nailfold capillaries. Amer. J. Physiol., 236:H888–H893.

Montagna, W., H. B. Chase, and W. C. Lobitz, Jr. 1953. Histology and cytochemistry of human skin. IV. The eccrine sweat glands. J. Invest. Dermatol., 20:415–423.

Munger, B. L. 1961. The ultrastructure and histophysiology of human eccrine sweat glands. J. Biophys. Biochem. Cytol., 11:385–402.

Munger, B. L., and S. W. Brusilow. 1961. An electron microscopic study of eccrine sweat glands of the cat foot and toe pads—Evidence for ductal reabsorption in the human. J. Biophys. Biochem. Cytol., 11:403–417.

Prakkal, P. F. 1966. The fine structure of the dermal papilla of the guinea pig hair follicle. J. Ultrastruct. Res., 14:133–142.

Riggott, J. M., and E. H. Wyatt. 1980. Scanning electron microscopy of hair from different regions of the body of the rat. J. Anat. (Lond.), 130:121–126.

Rogers, G. E. 1957. Electron microscope observations on the structure of sebaceous glands. Exp. Cell Res., 13:517–520.

Rogers, G. E. 1958. Some aspects of the structure of the inner root sheath of hair follicles revealed by light and electron microscopy. Exp. Cell Res., 14:378–387.

Runne, U., and C. E. Orfanos. 1981. The human nail: Structure, growth and pathological changes. Curr. Probl. Dermatol., 9:102–149.

Singh, J. D. 1982. Distribution of hair on the phalanges of the hand in Nigerians. Acta Anat., 112:31–35.

Sulzberger, M. B., F. Herrmann, R. Keller, and B. V. Pisha. 1950. Studies of sweating: III. Experimental factors in influencing the function of the sweat ducts. J. Invest. Dermatol., 14:91–112.

Sunderland, S., and L. J. Ray. 1952. The effect of denervation on nail growth. J. Neurol. Neurosurg. Psychiatry, 15:50–53.

Szabo, G. 1967. The regional anatomy of the human integument, with special reference to the distribution of hair follicles, sweat glands and melanocytes. Philos. Trans. Roy. Soc. Lond. (Biol.), 252:447–485.

Thomas, P. K., and D. G. Ferriman. 1957. Variation in facial and pubic hair growth in white women. Amer. J. Phys. Anthrop., N.S., 15:171–180.

Trotter, M. 1938. A review of the classifications of hair. Amer. J. Phys. Anthrop., 24:105–126.

Trotter, M. 1939. Classifications of hair color. Amer. J. Phys. Anthrop., 25:237–260.

Trotter, M., and O. H. Duggins. 1948. Age changes in head hair from birth to maturity. I. Index and size of hair of children. Amer. J. Phys. Anthrop., N.S., 6:489–506.

Trotter, M., and O. H. Duggins. 1950. Age changes in head hair from birth to maturity. III. Cuticular scale counts of hair of children. Amer. J. Phys. Anthrop., N.S., 8:467–484.

Uno, H., and W. Montagna. 1982. Reinnervation of hair follicle end organs and Meissner corpuscles in skin grafts of macaques. J. Invest. Dermatol., 78:210–214.

Weiner, J. S., and K. Hellmann. 1960. The sweat glands. Biol. Rev., 35:141–186.

Weinstein, G. D., and K. M. Mooney. 1980. Cell proliferation kinetics in the human hair root. J. Invest. Dermatol., 74:43–46.

Young, R. D. 1980. Morphological and ultrastructural aspects of the dermal papilla during the growth cycle of the vibrissal follicle in the rat. J. Anat. (Lond.), 131:355–365.

Zaias, N., and J. Alvarez. 1968. The formation of the primate nail plate. J. Invest. Dermatol., 51:120–136.

Zimmermann, A. A., and C. Cornbleet. 1948. The development of epidermal pigmentation in the negro fetus. J. Invest. Dermatol., 11:383–395.

15

The Respiratory System

Respiration, or exchange of gaseous substance between the air and the blood stream, is brought about in the respiratory system (*respiratory apparatus*): the nose, nasal passages, nasopharynx, larynx, trachea, bronchi, and lungs. The pleura, pleural cavities, and the topography of other structures in the thorax are also described in this chapter. The muscular actions associated with respiration are discussed with the description of the diaphragm in Chapter 6.

Development

The development of the nose is described in Chapter 2, Embryology.

The primordium of the principal respiratory organs appears as a median longitudinal groove in the ventral wall of the pharynx in an embryo of the third week (3 to 4 mm). The groove deepens and its lips fuse to form the **laryngotracheal tube** (Fig. 15-1), the cranial end of which opens into the pharynx by a slit-like aperture formed by the persistent ventral part of the groove. The tube is lined by endoderm, from which the epithelial lining of the respiratory tract is developed. The cranial part of the tube becomes the larynx, its next succeeding part the trachea, and from its caudal end two lateral outgrowths arise, the right and left **lung buds,** from which the bronchi and lungs develop.

The first rudiment of the larynx consists of two **arytenoid swellings,** which appear on either side of the cranial end of the laryngotracheal groove and are continuous ventrally with a transverse ridge (**furcula of His**[1]) that lies between the ventral ends of the third branchial arches and from which the epiglottis is subsequently developed (Figs. 16-4; 16-5). After the separation of the trachea from the esophagus the arytenoid swellings come into contact with each other and with the back of the epiglottis, and the entrance to the larynx assumes the form of a T-shaped cleft. The margins of the cleft adhere to each other and the laryngeal entrance is occluded for a time.

The mesodermal wall of the tube condenses to form the cartilages of the larynx and trachea. The arytenoid swellings differentiate into the arytenoid and corniculate cartilages, and the folds joining them to the epiglottis form the aryepiglottic folds, in which the cuneiform cartilages develop as derivatives of the epiglottis. The thyroid cartilage appears as two lateral plates, each chondrified from two centers and united in the ventral midline by membrane in which an additional center of chondrification develops. The cricoid cartilage arises from two cartilaginous centers that soon unite ventrally and gradually grow around to fuse on the dorsal aspect of the tube.

The right and left lung buds grow out dorsal to the common cardinal veins and are at first symmetrical, but their ends soon become lobulated, three lobules appearing on the right, and two on the left. These subdivisions are the early indications of the corresponding lobes of the lungs (Figs. 15-2; 15-3). The buds undergo further subdivision and ramification, and ultimately end in minute expanded extremities—the infundibula of the lung. After the sixth month the air sacs begin to appear on the infundibula in the form of minute pouches.

[1]Wilhelm His (1831–1904): A German anatomist and physiologist (Basel and Leipzig).

1357

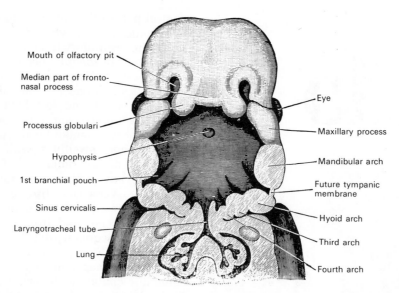

Mouth of olfactory pit

Median part of fronto-
nasal process

Eye

Processus globulari

Maxillary process

Hypophysis

Mandibular arch

1st branchial pouch

Future tympanic
membrane

Sinus cervicalis

Hyoid arch

Laryngotracheal tube

Third arch

Lung

Fourth arch

FIG. 15-1.   The head and neck of a 32-day-old human embryo seen from the ventral surface. The floor of the mouth and pharynx has been removed. (His.)

During the course of their development the lungs migrate in a caudal direction, so that by the time of birth the bifurcation of the trachea is opposite the fourth thoracic vertebra. As the lungs grow, they project into the part of the celom that will ultimately form the pleural cavities, and the superficial layer of the mesoderm enveloping the lung rudiment expands on the growing lung and is converted into the pulmonary pleura.

The pulmonary arteries are derived from the sixth aortic arches (Figs. 8-6 to 8-9).

Upper Respiratory Tract

NOSE

External Nose

The external nose is shaped like a pyramid, its free angle termed the **apex.** The two elliptical orifices, the **nares,** are separated from each other by a median septum, the **columna.** The interior surface of the **anterior**

nares or **nostrils** is provided with stiff hairs, or **vibrissae,** which arrest the passage of foreign substances carried with the current of air intended for respiration. The lateral surfaces of the nose form, by their union in the midline, the **dorsum nasi,** the direction of which varies considerably in different individuals. The upper part of the dorsum is supported by the nasal bones and is named the **bridge.** The lateral surface ends below in rounded eminences, the **alae nasi.**

STRUCTURE.   The framework of the external nose is composed of bones and cartilages; it is covered by the integument and is lined by mucous membrane.

Bones and Cartilage.   The **bony framework** occupies the upper part of the organ; it consists of the nasal bones and the frontal processes of the maxillae.

The **cartilaginous framework** consists of

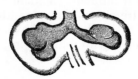

FIG. 15-2.   Lung buds from a human embryo of about four weeks, showing commencing lobulations. (His.)

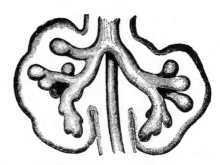

FIG. 15-3.   Lungs of a human embryo more advanced in development. (His.)

five large pieces: the **septal cartilage,** the **two lateral** and the **two greater alar cartilages,** and several smaller pieces, the **lesser alar cartilages** (Figs. 15-4 to 15-7). The various cartilages are connected to each other and to the bones by a tough fibrous membrane.

The **septal cartilage** is thicker at its margins than at its center and separates the nasal cavities. Its anterior margin is connected with the nasal bones, and is continuous with the anterior margins of the lateral cartilages. Inferiorly, it is connected to the medial crura of the greater alar cartilages by fibrous tissue. Its posterior margin is connected with the perpendicular plate of the ethmoid; its inferior margin, with the vomer and the palatine processes of the maxillae.

The septal cartilage may be prolonged posteriorly (especially in children) for some distance between the vomer and perpendicular plate of the ethmoid as a narrow process, the **sphenoidal process.** The septal cartilage does not reach as far as the lowest part of the nasal septum. This is formed by the medial crura of the greater alar cartilages.

The **lateral cartilage** (*upper lateral cartilage*) is situated below the inferior margin of the nasal bone; it is flat and triangular. Its anterior margin is thicker than the posterior, and is continuous above with the cartilage of the septum, but separated from it below by a narrow fissure. Its superior margin is attached to the nasal bone and the frontal process of the maxilla; its inferior margin is connected by fibrous tissue with the greater alar cartilage.

The **greater alar cartilage** (*lower lateral cartilage*) is a thin, flexible plate situated

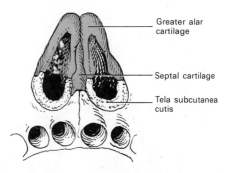

FIG. 15-5.   Cartilages of the nose. Inferior view.

immediately below the lateral cartilage and bent upon itself in such a manner that it forms the medial and lateral walls of the naris of its own side. The portion that forms the **medial wall** is loosely connected with the corresponding portion of the opposite cartilage, the two forming, together with the thickened integument and subjacent tissue, the **septum mobile nasi.** The part that forms the **lateral wall** is curved to correspond with the ala of the nose. It is oval, flat, and narrow posteriorly where it is connected with the frontal process of the maxilla by a tough fibrous membrane, in which are found three or four small cartilaginous plates, the **lesser alar cartilages** (*sesamoid cartilages*). Superiorly, the greater alar cartilage is connected by fibrous tissue to the lateral cartilage and anterior part of the cartilage of the septum; inferiorly, it falls short of the margin of the naris, the ala being completed by fatty and fibrous tissue covered by skin. Anteriorly, the greater alar cartilages are separated by a notch that corresponds with the apex of the nose.

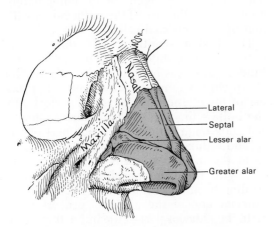

FIG. 15-4.   Cartilages of the nose. Side view.

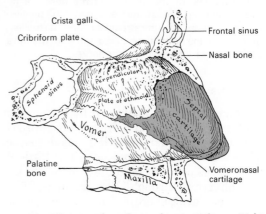

FIG. 15-6.   Bones and cartilages of septum of nose. Right side.

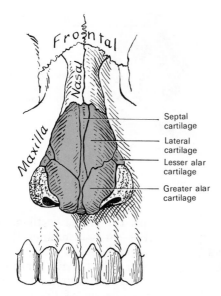

Septal cartilage

Lateral cartilage

Lesser alar cartilage

Greater alar cartilage

Fig. 15-7. Cartilages of the nose. Anterior view.

Muscles. The muscles acting on the external nose are described in Chapter 6.

Integument. The integument of the dorsum and sides of the nose is thin and loosely connected with the subjacent parts, but over the tip and alae it is thicker and more firmly adherent, and is furnished with a large number of sebaceous follicles, the orifices of which are usually quite distinct.

Vessels and Nerves. The arteries of the external nose are the alar and septal branches of the facial artery, which supply the alae and septum. The dorsum and sides are supplied from the dorsal nasal branch of the ophthalmic artery and the infraorbital branch of the maxillary artery. The veins end in the facial and ophthalmic veins.

The nerves for the muscles of the nose are derived from the facial nerve. The skin receives branches from the infratrochlear and nasociliary branches of the ophthalmic nerve and from the infraorbital branch of the maxillary nerve.

Nasal Cavity

The **nasal cavity** is divided by the median septum into two symmetrical and approximately equal chambers, the **nasal fossae.** They have their external openings through the nostrils or nares, and they open into the nasopharynx through the choanae. The nares are oval apertures measuring about 1.5 cm anteroposteriorly and 1 cm transversely. The choanae are two oval openings measuring approximately 2.5 cm vertically and 1.5 cm transversely.

The bony boundaries of the nasal cavity are described in Chapter 4, Osteology.

Inside the aperture of the nostril is a slight dilatation, the **vestibule,** which is bounded laterally by the ala and lateral crus of the greater alar cartilage and medially by the medial crus of the same cartilage. It is lined by skin containing hairs and sebaceous glands, and it extends as a small recess toward the apex of the nose. Each nasal fossa is divided into two parts: an **olfactory region,** consisting of the superior nasal concha and the opposed part of the septum, and a **respiratory region,** which comprises the rest of the cavity.

LATERAL WALL (Figs. 15-8; 15-9). On the lateral wall are the **superior, middle,** and **inferior nasal conchae,** overhanging the corresponding nasal passages or **meatuses.** Above the superior concha is a narrow recess, the **sphenoethmoidal recess,** into which the sphenoidal sinus opens. The **superior meatus** is a short oblique passage that extends about half way along the superior border of the middle concha; the posterior ethmoidal cells open into the anterior part of this meatus. The **middle meatus** is continued anteriorly into a shallow depression situated above the vestibule and named the **atrium** of the middle meatus. The lateral wall of this meatus is fully displayed only by raising or removing the middle concha. On it is a rounded elevation, the **bulla ethmoidalis,** and anterior and inferior to this is a curved cleft, the **hiatus semilunaris.**

The **bulla ethmoidalis** (Fig. 15-9) is caused by the bulging of the middle ethmoidal cells, which open on or immediately above it, and the size of the bulla varies with that of its contained cells.

The **hiatus semilunaris** is the narrow curved opening into a deep pocket, the **infundibulum.** It is bounded inferiorly by the sharp concave margin of the **uncinate process** of the ethmoid bone, and superiorly, by the bulla ethmoidalis. The anterior ethmoidal cells open into the anterior part of the infundibulum. In slightly over 50% of subjects, the infundibulum is directly continuous with the **frontonasal duct** or passage leading from the frontal air sinus. When the anterior end of the uncinate process fuses with the anterior part of the bulla, this continuity is interrupted, and the frontonasal

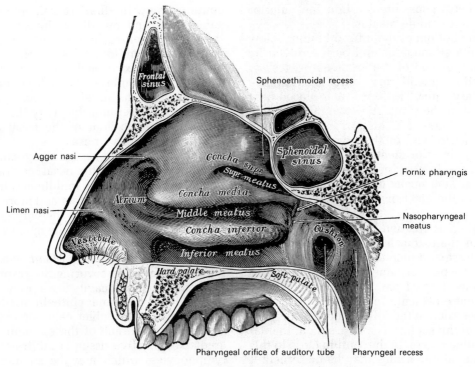

Fig. 15-8. Lateral wall of nasal cavity.

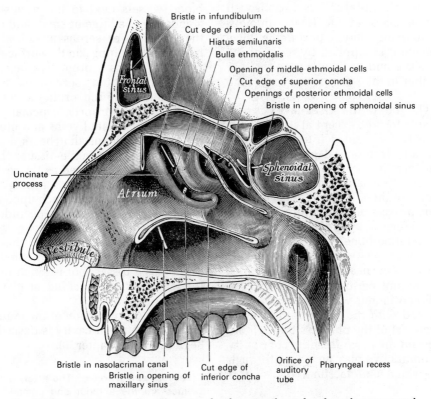

Fig. 15-9. Lateral wall of nasal cavity; the three nasal conchae have been removed.

duct then opens directly into the anterior end of the middle meatus.

In the bottom of the infundibulum, below the bulla ethmoidalis and partly hidden by the inferior end of the uncinate process, is the **ostium maxillare,** or opening from the maxillary sinus. This opening is placed near the roof of the sinus. An **accessory** opening from the sinus is frequently present below the posterior end of the middle nasal concha. The **inferior meatus** is inferior and lateral to the inferior nasal concha; the nasolacrimal duct opens into this meatus under cover of the anterior part of the concha.

MEDIAL WALL (Fig. 15-6). The medial wall or septum is frequently deflected from the median plane, thus lessening the size of one nasal fossa and increasing that of the other; ridges or spurs of bone growing into one or the other fossa from the septum are also sometimes present. Immediately over the incisive canal at the inferior edge of the cartilage of the septum is a depression, the **nasopalatine recess.** In the septum close to this recess a minute orifice may be discerned; it leads posteriorly into a blind pouch, the rudimentary **vomeronasal organ of Jacobson,**[1] which is supported by a strip of cartilage, the **vomeronasal cartilage.** This organ is well developed in many of the lower animals, where it apparently plays a part in the sense of smell, since it is supplied by twigs of the olfactory nerve and is lined by epithelium similar to that in the olfactory region of the nose.

ROOF. The **roof** of the nasal cavity is narrow from side to side, except at its posterior part, and may be divided into sphenoidal, ethmoidal, and frontonasal parts, after the bones that form it.

FLOOR. The **floor** is concave from side to side. Its anterior three-fourths are formed by the palatine process of the maxilla; its posterior fourth is formed by the horizontal process of the palatine bone.

MUCOUS MEMBRANE. The nasal mucous membrane lines the nasal cavity and adheres intimately to the periosteum or perichondrium. It is continuous with the skin through the nares, and with the mucous membrane of the nasal part of the pharynx through the choanae. From the nasal fossa its continuity with the conjunctiva may be traced through the nasolacrimal and lacrimal ducts; it is

continuous with the frontal, ethmoidal, sphenoidal, and maxillary sinuses through the several openings in the meatuses. Continuity with the middle ear is established with the nasopharynx through the auditory tube. The mucous membrane is thickest and most vascular over the nasal conchae. It is also thick over the septum, but very thin in the meatuses, on the floor of the nasal cavity, and in the various sinuses.

Owing to the thickness of the greater part of this membrane, the nasal cavities are much narrower and the middle and inferior nasal conchae appear larger and more prominent than in the skeleton; the various apertures communicating with the meatuses also are considerably narrowed.

Structure of the Mucous Membrane. The mucous membrane covering the **respiratory portion** of the nasal cavity has a pseudostratified, ciliated, columnar epithelium, liberally interspersed with goblet cells. Beneath the basement membrane of the epithelium, the areolar connective tissue is infiltrated with lymphocytes, which may be so numerous that they form a diffuse lymphoid tissue. The lamina propria is composed of a layer of glands toward the surface and a layer of blood vessels next to the periosteum. The glands may be large or small and contain either mucous or serous alveoli. They have individual openings on the surface, and their secretion forms a protective layer over the membrane which serves to warm and moisten the inspired air. In the deeper layers of the lamina propria, especially over the conchae, the dilated veins and blood spaces form a rich plexus that bears a superficial resemblance to erectile tissue; these areas easily become engorged as the result of irritation or inflammation, causing the membrane to swell and encroach upon the lumen of the meatuses, and even occlude the ostia of the sinuses.

The mucous membrane of the paranasal sinuses is columnar ciliated, similar to that of the nasal fossae, but is much thinner and for the most part lacking in glands in the lamina propria.

The **olfactory portion** of the mucous membrane of the nasal cavity is described at the beginning of Chapter 13.

Vessels and Nerves. The **arteries** of the nasal cavities are the anterior and posterior ethmoidal branches of the ophthalmic artery, which supply the ethmoidal cells, frontal sinuses, and roof of the

---

[1]Ludwig Levin Jacobson (1783–1843): A Danish anatomist and physician (Copenhagen).

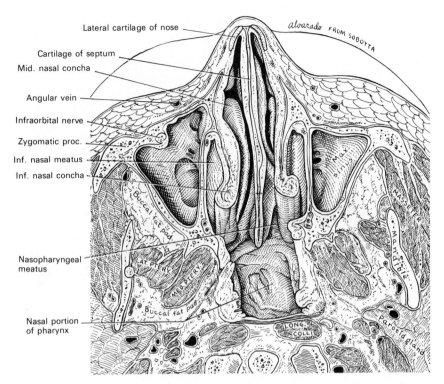

Lateral cartilage of nose

Cartilage of septum

Mid. nasal concha

Angular vein

Infraorbital nerve

Zygomatic proc.

Inf. nasal meatus

Inf. nasal concha

Nasopharyngeal
meatus

Nasal portion
of pharynx

FIG. 15-10. Transverse section through the anterior part of the head at a level just inferior to the apex of the dens (odontoid process). Inferior view.

nose; the sphenopalatine branch of the maxillary artery, which supplies the mucous membrane covering the conchae, the meatuses, and septum; the septal ramus of the superior labial branch of the facial artery; the infraorbital and alveolar branches of the maxillary artery, which supply the lining membrane of the maxillary sinus; and the pharyngeal branch of the maxillary artery, which is distributed to the sphenoidal sinus. The ramifications of these vessels form a close plexiform network beneath and in the substance of the mucous membrane.

Some **veins** open into the sphenopalatine vein; others join the facial vein. Some accompany the ethmoidal arteries and end in the ophthalmic veins, and a few communicate with the veins on the orbital surface of the frontal lobe of the brain through the foramina in the cribriform plate of the ethmoid bone. When the foramen cecum is patent, it transmits a vein to the superior sagittal sinus.

The **lymphatics** are described in Chapter 10.

The **nerves** of ordinary sensation are: the nasociliary branch of the ophthalmic nerve, filaments from the anterior alveolar branch of the maxillary nerve, the nerve of the pterygoid canal, the nasopalatine nerve, the anterior palatine nerve, and nasal branches of the pterygopalatine nerve.

The nasociliary branch of the ophthalmic nerve distributes filaments to the anterior part of the septum and lateral wall of the nasal cavity. Filaments from the anterior alveolar nerve supply the inferior meatus and inferior concha. The nerve of the ptery-

goid canal supplies the upper and back part of the septum and superior concha; the upper nasal branches from the pterygopalatine nerve have a similar distribution. The nasopalatine nerve supplies the middle of the septum. The anterior palatine nerve supplies the lower nasal branches to the middle and inferior conchae.

The **olfactory nerves,** the nerves of the special sense of smell, are a number of fine filaments distributed to the mucous membrane of the olfactory region. Their fibers arise from the bipolar olfactory cells and lack medullary sheaths. They unite into fasciculi that form a plexus beneath the mucous membrane and then ascend in grooves or canals in the ethmoid bone. They pass into the skull through the foramina in the cribriform plate of the ethmoid and enter the olfactory bulb, in which they ramify and form synapses with the dendrites of the mitral cells (see Chapter 13).

Accessory Sinuses of the Nose

The **accessory sinuses** or **air cells of the nose** (*paranasal sinuses*) (Figs. 15-9 to 15-12) are the **frontal, ethmoidal, sphenoidal,** and **maxillary.** They vary in size and form in different individuals and are lined by ciliated mucous membrane directly continuous with that of the nasal cavity.

FRONTAL SINUSES.   The frontal sinuses, sit-

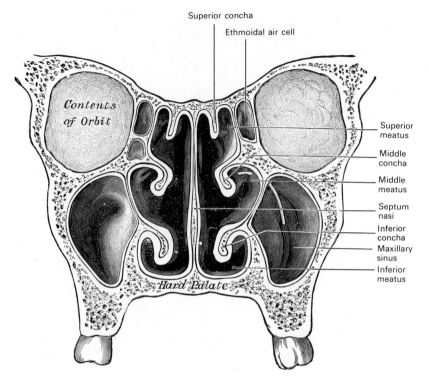

FIG. 15-11. Coronal section of nasal cavities.

uated behind the superciliary arches, are rarely symmetrical, and the septum between them frequently deviates to one side or the other of the midline. Their average measurements are: height, 3 cm; breadth, 2.5 cm; depth, 2.5 cm. A large frontal sinus may extend out over most of the orbit. Each opens into the anterior part of the corresponding middle meatus of the nose through the frontonasal duct, which enters the anterior part of the middle meatus. Absent at birth, the frontal sinuses are generally fairly well developed between the seventh and eighth years, but reach their full size only after puberty.

ETHMOIDAL AIR CELLS.   The ethmoidal air cells consist of numerous small, thin-walled cavities occupying the ethmoidal labyrinth and completed by the frontal, maxillary, lacrimal, sphenoidal, and palatine bones. They lie between the upper parts of the nasal cavities and the orbits, and are separated from these cavities by thin bony laminae. On either side they are arranged in three groups: **anterior, middle,** and **posterior.** The anterior and middle groups open into the middle meatus of the nose, the former by way of the infundibulum, the latter on or above the

bulla ethmoidalis. The posterior cells open into the superior meatus under cover of the superior nasal concha; sometimes one or more open into the sphenoidal sinus. The ethmoidal cells begin to develop after fetal life.

SPHENOIDAL SINUSES.   The sphenoidal sinuses (Fig. 15-9), contained within the body of the sphenoid, vary in size and shape, and owing to the lateral displacement of the intervening septum, they are rarely symmetrical. Their average measurements are: vertical height, 2.2 cm; transverse breadth, 2 cm; depth 2.2 cm. When exceptionally large, they may extend into the roots of the pterygoid processes or great wings, and may invade the basilar part of the occipital bone. Each sinus communicates with the sphenoethmoidal recess by means of an ostium in the upper part of its anterior wall. The sphenoidal sinuses are present as minute cavities at birth, but their main development takes place after puberty.

MAXILLARY SINUS.   The maxillary sinus (*antrum of Highmore*[1]), the largest of the

[1]Nathaniel Highmore (1613–1685): An English physician (Sherborne).

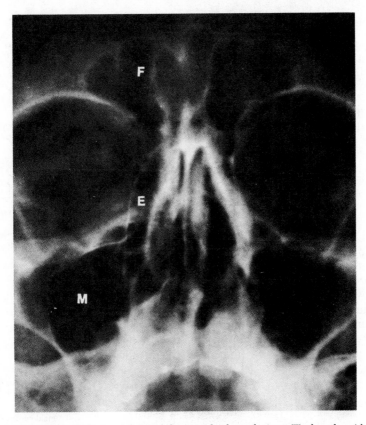

Fig. 15-12. Radiograph of adult skull, frontal view, showing the frontal sinus (F), the ethmoid air cells (E), and the maxillary sinus (M). (From Juhl, J.H., *Essentials of Roentgen Interpretation*, 4th ed., Harper and Row Publishers, Hagerstown, Md., 1981.)

accessory sinuses of the nose, is a pyramidal cavity in the body of the maxilla. Its base is formed by the lateral wall of the nasal cavity, and its apex extends into the zygomatic process. Its roof or orbital wall is frequently ridged by the bony wall of the infraorbital canal; its floor is formed by the alveolar process and is usually 1 to 10 mm below the level of the floor of the nose. Projecting into the floor are several conical elevations corresponding with the roots of the first and second molar teeth, and in some cases the floor is perforated by one or more of these roots. The size of the sinus varies in different skulls and even on the two sides of the same skull. The adult capacity varies from 9.5 cc to 20 cc (average about 14.75 cc). The measurements of an average-sized sinus are: vertical height opposite the first molar tooth, 3.75 cm; transverse breadth, 2.5 cm; anteroposterior depth, 3 cm. In the superior part of its medial wall is an ostium through which it communicates with the lower part of the infundibulum; a second or accessory orifice is frequently

present in the middle meatus posterior to the first. The maxillary sinus appears as a shallow groove on the medial surface of the bone about the fourth month of fetal life, but does not reach its full size until after the second dentition.

LARYNX

The **larynx** or **organ of voice** is the part of the air passage that connects the pharynx with the trachea. It produces a considerable projection in the midline of the neck called the Adam's apple. It forms the inferior part of the anterior wall of the pharynx, and is covered by the mucous lining of that cavity. Its vertical extent corresponds to the fourth, fifth, and sixth cervical vertebrae, but it is placed somewhat higher in the female and also during childhood; on either side of it lie the great vessels of the neck. The average measurements of the adult larynx in both males and females appear on page 1366.

	Males (mm)	Females (mm)
Length	44	36
Transverse diameter	43	41
Anteroposterior diameter	36	26
Circumference	136	112

Until puberty the larynx of the male differs little in size from that of the female. In the female its increase after puberty is only slight; in the male it undergoes considerable increase; all the cartilages are enlarged and the thyroid cartilage becomes prominent in the midline of the neck, while the length of the rima glottidis is nearly doubled.

The larynx is broad above, where it presents the form of a triangular box flattened dorsally and at the sides, and it is bounded ventrally by a prominent vertical ridge. Caudally, it is narrow and cylindrical. It is composed of cartilages, which are connected together by ligaments and moved by numerous muscles. It is lined by mucous membrane continuous with that of the pharynx and the trachea.

Cartilages of the Larynx

There are nine cartilages of the larynx (Fig. 15-13), three single and three paired, as follows:

Thyroid
Cricoid
Arytenoid (two)
Corniculate (two)
Cuneiform (two)
Epiglottis

THYROID CARTILAGE. The thyroid cartilage (Fig. 15-13) is the largest cartilage of the larynx. It consists of two laminae, the anterior borders of which are fused with each other at an acute angle in the midline of the neck; these form a subcutaneous projection named the **laryngeal prominence** (Adam's apple). This prominence is most distinct at its cranial part and is larger in the male than in the female. Immediately above it, the laminae are separated by a V-shaped notch, the **superior thyroid notch.** The laminae are irregularly quadrilateral in shape, and their posterior angles are prolonged into processes termed the **superior** and **inferior cornua.**

On the *outer surface* of each lamina an **oblique line** runs caudalward and ventral-

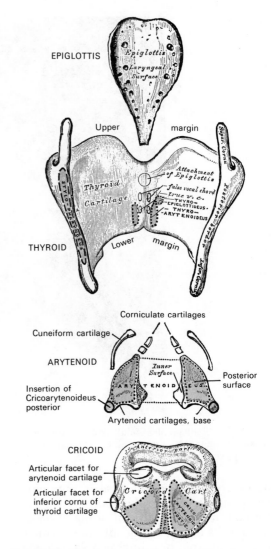

FIG. 15-13. The cartilages of the larynx. Posterior view.

ward from the superior thyroid tubercle, situated near the root of the superior cornu, to the inferior thyroid tubercle on the caudal border. This line gives attachment to the sternothyroid, thyrohyoid, and the inferior pharyngeal constrictor. The *inner surface* is smooth; cranially and dorsally it is slightly concave and covered by mucous membrane. Ventrally, in the angle formed by the junction of the laminae, are attached the stem of the epiglottis, the ventricular and vocal ligaments, the thyroarytenoid, thyroepiglotticus, and vocalis muscles, and the thyroepiglottic ligament.

The *cranial border* is concave dorsally and convex ventrally; it gives attachment to the corresponding half of the thyrohyoid membrane. The *caudal border* is concave dorsally and nearly straight ventrally, the two parts being separated by the inferior thyroid tubercle. A small part of it in and near the midline is connected to the cricoid cartilage by the middle cricothyroid ligament. The *dorsal border,* thick and rounded, receives the insertions of the stylopharyngeus and palatopharyngeus. It ends cranially, in the superior cornu, and caudally, in the inferior cornu. The **superior cornu** is long and narrow, directed cranialward, and ends in a conical extremity, which gives attachment to the lateral thyrohyoid ligament. The **inferior cornu** is short and thick; it is directed caudalward, with a slight inclination, and presents, on the medial side of its tip, a small oval articular facet for articulation with the side of the cricoid cartilage.

During infancy, the laminae of the thyroid cartilage are joined to each other by a narrow, lozenge-shaped strip named the **intrathyroid cartilage.** This strip extends from the cranial to the caudal border of the cartilage in the midline, and is distinguished from the laminae by being more transparent and more flexible.

CRICOID CARTILAGE.   The cricoid cartilage (Fig. 15-13) is smaller but thicker and stronger than the thyroid, and forms the caudal and dorsal parts of the wall of the larynx. It consists of two parts: a **posterior quadrate lamina** and a narrow **anterior arch,** one-fourth or one-fifth of the depth of the lamina.

The **lamina** of the cricoid cartilage is deep and broad and measures about 2 or 3 cm craniocaudally. On its dorsal surface, in the midline, is a vertical ridge, to the caudal part of which are attached the longitudinal fibers of the esophagus, and on either side of this ridge is a broad depression for the posterior cricoarytenoid.

The **arch** is narrow, convex, and measures vertically from 5 to 7 mm; it affords attachment externally, ventrally, and at the sides to the cricothyroid muscle, and dorsally, to part of the inferior constrictor of the pharynx.

On either side, at the junction of the lamina with the arch, is a small round articular surface for articulation with the inferior cornu of the thyroid cartilage.

The *caudal border* of the cricoid cartilage is horizontal and connected to the highest ring of the trachea by the cricotracheal ligament. The *cranial border* runs obliquely

cranialward and dorsalward, owing to the great depth of the lamina. It gives attachment, ventrally, to the middle cricothyroid ligament and at the side, to the conus elasticus and the lateral cricoarytenoids. Dorsally, it has a shallow notch in the middle, and on either side of this a smooth, oval convex surface for articulation with the base of an arytenoid cartilage. The inner surface of the cricoid cartilage is smooth and lined by mucous membrane.

ARYTENOID CARTILAGE.   The arytenoid cartilages (Fig. 15-13) are two in number, situated at the cranial border of the lamina of the cricoid cartilage, at the dorsum of the larynx. Each is pyramidal and has three surfaces, a base, and an apex.

The *dorsal surface* is triangular, smooth, concave, and gives attachment to the oblique and transverse arytenoid. The *ventrolateral surface* is somewhat convex and rough. On it, near the apex of the cartilage, is a rounded elevation (**colliculus**) from which a ridge (**crista arcuata**) curves at first dorsalward and then caudalward and ventralward to the vocal process. The caudal part of this crest intervenes between two depressions or **foveae:** an upper triangular depression and a lower oblong depression. The latter gives attachment to the vocalis muscle. The *medial surface* is narrow, smooth, and flat. It is covered by mucous membrane and forms the lateral boundary of the intercartilaginous part of the rima glottidis.

The **base** of each cartilage is broad, and has a concave smooth surface for articulation with the cricoid cartilage. Its lateral angle is short, round, and prominent; it projects dorsalward, and is termed the **muscular process.** It receives the insertion of the posterior cricoarytenoid dorsally and the lateral cricoarytenoid ventrally. Its anterior angle, also prominent but more pointed, projects horizontally ventralward; it gives attachment to the vocal ligament and is called the **vocal process** (Fig. 15-16).

The **apex** of each cartilage is pointed, curved dorsalward and medialward, and surmounted by a small conical, cartilaginous nodule, the **corniculate cartilage.**

CORNICULATE CARTILAGE.   The corniculate cartilages (*cartilages of Santorini*[1]) are two small conical nodules consisting of yellow

---

[1]Giovanni Domenico Santorini (1681–1737): An Italian anatomist (Venice).

elastic cartilage, which articulate with the summits of the arytenoid cartilages and serve to prolong them dorsalward and medialward. They are situated in the posterior parts of the aryepiglottic folds of mucous membrane, and they are sometimes fused with the arytenoid cartilages.

CUNEIFORM CARTILAGE.   The cuneiform cartilages (*cartilages of Wrisberg*[2]) are two small, elongated pieces of yellow elastic cartilage, placed one on either side, in the aryepiglottic fold, where they give rise to small

[2]Heinrich August Wrisberg (1739–1808): A German anatomist (Göttingen).

whitish elevations on the surface of the mucous membrane, just ventral to the arytenoid cartilages.

EPIGLOTTIS.   The epiglottis (Fig. 15-13) is a thin lamella of yellow elastic cartilage, shaped like a leaf. It projects obliquely upward behind the root of the tongue and ventral to the entrance of the larynx. The free extremity is broad and round; the attached part or stem is long, narrow, and connected by the **thyroepiglottic ligament** to the angle formed by the two laminae of the thyroid cartilage, a short distance caudal to the superior thyroid notch. The caudal part of its anterior surface is connected to the cranial

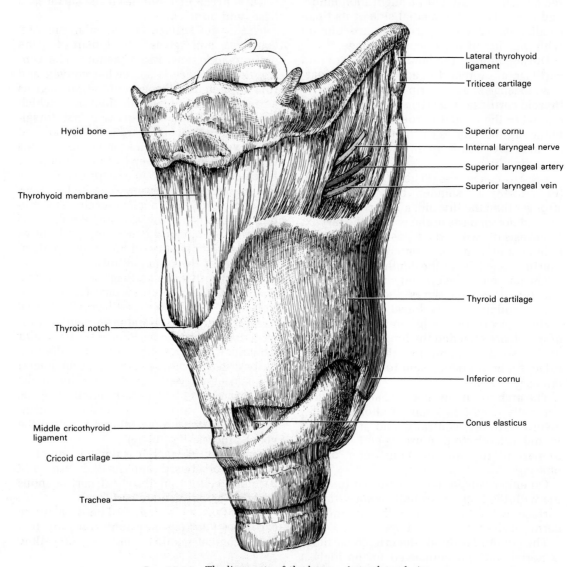

FIG. 15-14.   The ligaments of the larynx. Anterolateral view.

border of the body of the hyoid bone by an elastic ligamentous band, the **hyoepiglottic ligament.**

The *anterior or lingual surface* is curved ventralward and is covered on its cranial, free part by mucous membrane, which is reflected onto the sides and root of the tongue, forming a median and two lateral **glossoepiglottic folds.** The lateral folds are partly attached to the wall of the pharynx. The depressions between the epiglottis and the root of the tongue, on either side of the median fold, are named the **valleculae** (Fig. 15-18). The caudal part of the anterior surface lies dorsal to the hyoid bone, the hyothyroid membrane, and the upper part of the thyroid cartilage, but is separated from these structures by a mass of fatty tissue.

The *posterior or laryngeal surface* is smooth, concave from side to side, and concavoconvex vertically; its caudal part projects dorsalward as an elevation, the **tubercle** or **cushion.** The surface of the cartilage is indented by a number of small pits in which mucous glands are lodged. To its sides the aryepiglottic folds are attached.

STRUCTURE.   The corniculate and cuneiform cartilages, the epiglottis, and the apices of the arytenoids at first consist of hyaline cartilage, but later elastic fibers are deposited in the matrix, converting them into yellow elastic cartilage that shows little tendency to calcification. The thyroid, cricoid, and the greater part of the arytenoids consist of hyaline cartilage and become more or less ossified as age advances. Ossification commences about the twenty-fifth year in the thyroid cartilage and somewhat later in the cricoid and arytenoids; by the sixty-fifth year these cartilages may be completely converted into bone.

Ligaments

The ligaments of the larynx (Figs. 15-14; 15-15) are **extrinsic** (those that connect the thyroid cartilage and epiglottis with the hyoid bone, and the cricoid cartilage with the trachea) and **intrinsic** (those that connect the several cartilages of the larynx to each other).

EXTRINSIC LIGAMENTS.   The ligaments that connect the thyroid cartilage with the hyoid bone are the thyrohyoid membrane and a middle and two lateral thyrohyoid ligaments.

The **thyrohyoid membrane** (*hyothyroid membrane*) is a broad, fibroelastic layer attached to the cranial border of the thyroid cartilage and to the front of its superior cornu, and to the cranial margin of the dorsal surface of the body and greater cornua of the hyoid bone. It thus passes dorsal to the posterior surface of the body of the hyoid, and is separated from it by a bursa that facilitates the upward movement of the larynx during deglutition. Its middle thicker part is termed the **middle thyrohyoid ligament** (*middle hyothyroid ligament*); its lateral thinner portions are pierced by the superior laryngeal vessels and the internal branch of the superior laryngeal nerve. Its anterior surface is in relation with the thyrohyoid, sternohyoid, and omohyoid muscles, and with the body of the hyoid bone.

The **thyrohyoid ligament** (*lateral hyothyroid ligament*) is a round elastic cord that forms the posterior border of the thyrohyoid membrane and passes between the tip of the superior cornu of the thyroid cartilage and the extremity of the greater cornu of the hyoid bone. A small cartilaginous nodule (*cartilago triticea*), sometimes bony, is frequently found in it.

The epiglottis is connected with the hyoid bone by an elastic band, the **hyoepiglottic ligament,** which extends from the anterior surface of the epiglottis to the upper border of the body of the hyoid bone. The glossoepiglottic folds of mucous membrane may also be considered extrinsic ligaments of the epiglottis.

The **cricotracheal ligament** connects the cricoid cartilage with the first ring of the trachea. It resembles the fibrous membrane that connects the cartilaginous rings of the trachea to each other.

INTRINSIC LIGAMENTS.   Beneath the mucous membrane of the larnyx is a broad sheet of fibrous tissue that contains many elastic fibers and is termed the **elastic membrane of the larynx.** It is subdivided on either side by the interval between the ventricular and vocal ligaments. The cranial portion extends between the arytenoid cartilage and the epiglottis and is often poorly defined; the caudal part is a well-marked membrane that forms, with its fellow of the opposite side, the conus elasticus, which connects the thyroid, cricoid, and arytenoid cartilages to one another.

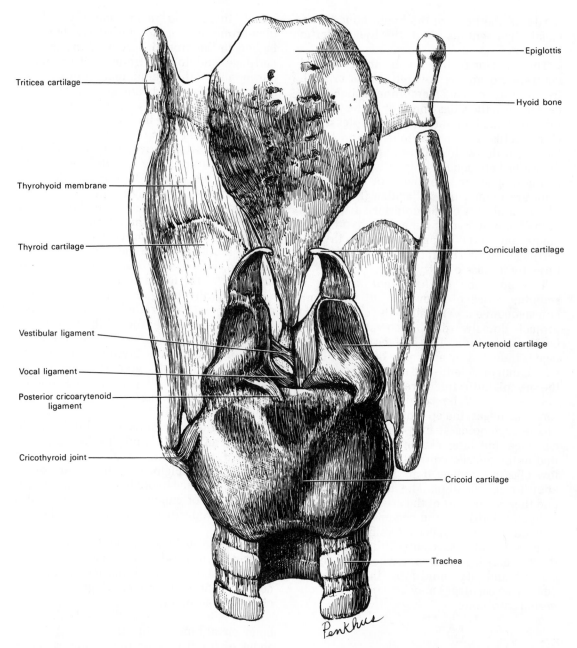

Triticea cartilage

Thyrohyoid membrane

Thyroid cartilage

Vestibular ligament

Vocal ligament

Posterior cricoarytenoid
ligament

Cricothyroid joint

Epiglottis

Hyoid bone

Corniculate cartilage

Arytenoid cartilage

Cricoid cartilage

Trachea

Penkhus

FIG. 15-15.   The ligaments of the larynx viewed from the posterior aspect. Note that the right half of the thyrohyoid membrane has been removed to show the contours of the thyroid cartilage, the epiglottis, and the hyoid bone.

The **conus elasticus** (*cricothyroid membrane*) is composed mainly of yellow elastic tissue. It consists of an anterior and two lateral portions. The **anterior part** or **middle cricothyroid ligament** (*ligamentum cricothyroideum medium; central part of cricothyroid membrane*) is thick, strong, and connects the ventral parts of the contiguous margins of the thyroid and cricoid cartilages. It is overlapped on either side by the crico-thyroid muscles, but between these it is subcutaneous; it is crossed horizontally by a small anastomotic arterial arch formed by the junction of the two cricothyroid arteries, branches of which pierce it. The **lateral portions** are thinner and lie close under the mucous membrane of the larynx; they extend from the superior border of the cricoid cartilage to the inferior margin of the vocal ligaments, with which they are continuous.

These ligaments may, therefore, be regarded as the free borders of the lateral portions of the conus elasticus, and they extend from the vocal processes of the arytenoid cartilages to the angle of the thyroid cartilage about midway between its cranial and caudal borders.

An **articular capsule,** strengthened posteriorly by a well-marked fibrous band, encloses the articulation of the inferior cornu of the thyroid with the cricoid cartilage on either side.

Each arytenoid cartilage is connected to the cricoid by a capsule and a posterior cricoarytenoid ligament. The **capsule** is thin and loose and is attached to the margins of the articular surfaces. The **posterior cricoarytenoid ligament** extends from the cricoid to the medial and dorsal part of the base of the arytenoid.

The **thyroepiglottic ligament** is a long, slender, elastic cord that connects the stem of the epiglottis with the angle of the thyroid cartilage immediately beneath the superior thyroid notch and cranial to the attachment of the ventricular ligaments.

MOVEMENTS. The articulation between the inferior cornu of the thyroid cartilage and the cricoid cartilage on either side is diarthrodial and permits rotatory and gliding movements. During the rotatory movement, the cricoid cartilage rotates upon the inferior cornua of the thyroid cartilage around an axis that passes transversely through both joints. The gliding movement consists of a limited shifting of the cricoid on the thyroid in different directions.

The articulation between the arytenoid cartilages and the cricoid is also diarthrodial and permits two varieties of movement: one is rotation of the arytenoid on a vertical axis, whereby the vocal process is moved lateralward or medialward and the rima glottidis is increased or diminished; the other is a gliding movement, which allows the arytenoid cartilages to approach or recede from each other. From the direction and slope of the articular surfaces, lateral gliding is accompanied by a forward and downward movement. The two movements of gliding and rotation are associated, the medial gliding being connected with the medialward rotation, and the lateral gliding with lateralward rotation. The posterior cricoarytenoid ligaments limit the forward movement of the arytenoid cartilages on the cricoid.

Interior of the Larynx

The **cavity of the larynx** extends from the laryngeal entrance to the caudal border of the cricoid cartilage, where it is continuous with that of the trachea. It is divided into two parts by the projection of the vocal folds, between which is a narrow triangular fissure or opening, the **rima glottidis.** The portion of the cavity of the larynx above the vocal folds is called the **vestibule.** It is wide and triangular; about the center of its base or anterior wall, however, is the backward projection of the tubercle of the epiglottis. It contains the ventricular folds, and between these and the vocal folds are the **ventricles of the larynx.** The portion caudal to the vocal folds is at first elliptical, but widens out, assumes a circular form, and is continuous with the tube of the trachea (Figs. 15-16; 15-17).

ENTRANCE OF THE LARYNX. The **entrance of the larynx** (Fig. 15-18) is bounded ventrally by the epiglottis; dorsally, by the apices of the arytenoid cartilages, the corniculate cartilages, and the interarytenoid notch; and on either side, by a fold of mucous membrane, enclosing ligamentous and muscular fibers, which is stretched between the side of the epiglottis and the apex of the arytenoid cartilage. This is the **aryepiglottic fold,** on the posterior part of the margin of which the cuneiform cartilage forms a more or less distinct whitish prominence, the **cuneiform tubercle.**

VESTIBULAR FOLDS. The vestibular folds (*plicae ventriculares; superior or false vocal cords*) are two thick folds of mucous membrane, each of which encloses a narrow band of fibrous tissue, the **vestibular ligament,** which is attached ventrally to the angle of the thyroid cartilage, immediately below the attachment of the epiglottis, and dorsally to the ventrolateral surface of the arytenoid cartilage, a short distance above the vocal process. The caudal border of this ligament, enclosed in mucous membrane, forms a free crescentic margin, which constitutes the upper boundary of the ventricle of the larynx.

VOCAL FOLDS. The vocal folds (*plicae vocales; inferior or true vocal cords*) enclose two strong bands named the **vocal ligaments** or vocal cords (*inferior thyroarytenoid*). Each ligament consists of a band of yellow elastic tissue attached ventrally to the angle of the thyroid cartilage and dorsally to the vocal process of the arytenoid. Its caudal border is continuous with the thin lateral part of the conus elasticus. Its cranial border forms the boundary of the ventricle of the larynx. Laterally, the vocalis muscle lies paral-

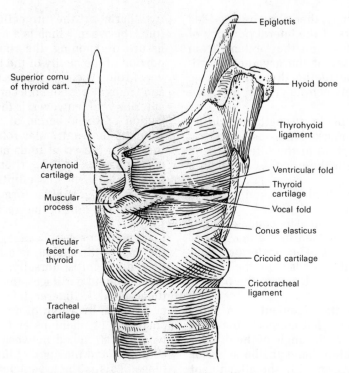

FIG. 15-16. A dissection to show the right half of the conus elasticus. The right lamina of the thyroid cartilage and the subjacent muscles have been removed.

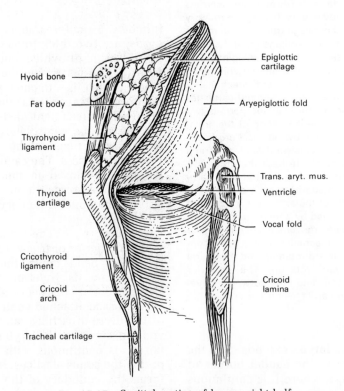

FIG. 15-17. Sagittal section of larynx, right half.

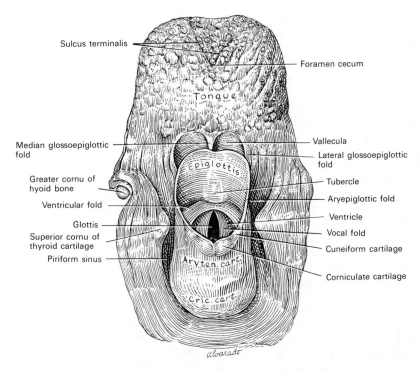

FIG. 15-18. The entrance to the larynx, posterior view.

lel with it. The ligament is covered medially by mucous membrane that is extremely thin and closely adheres to its surface.

VENTRICLE. The ventricle of the larynx (*ventriculus laryngis* [*Morgagnii*[1]]; *laryngeal sinus*) is a fusiform fossa, bounded by the free crescentic edge of the ventricular fold, by the straight margin of the vocal fold and by the mucous membrane covering the corresponding thyroarytenoid muscle. The anterior part of the ventricle leads by a narrow opening into a cecal pouch of mucous membrane of variable size called the **appendix.**

APPENDIX. The appendix of the laryngeal ventricle (*appendix ventriculi laryngis; laryngeal saccule*) is a membranous sac placed between the ventricular fold and the inner surface of the thyroid cartilage, occasionally extending as far as its cranial border or even higher. On the surface of its mucous membrane are the openings of 60 or 70 mucous glands, which are lodged in the submucous areolar tissue. This sac is enclosed in a fibrous capsule that is continuous caudally with the ventricular ligament. Its medial sur-

face is covered by a few delicate muscular fasciculi that *arise* from the apex of the arytenoid cartilage and become lost in the aryepiglottic fold of mucous membrane; laterally it is separated from the thyroid cartilage by the thyroepigloticus. These muscles compress the sac and express the secretion it contains upon the vocal folds to lubricate their surfaces.

RIMA GLOTTIDIS. The rima glottidis (Fig. 15-19) is the elongated fissure or opening

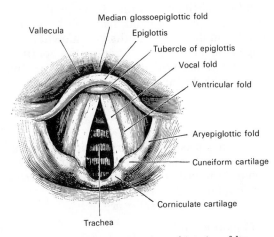

FIG. 15-19. Laryngoscopic view of interior of larynx.

---

[1]Giovanni Battista Morgagni (1682–1771): An Italian anatomist and pathologist (Padua).

between the vocal folds ventrally and the bases and vocal processes of the arytenoid cartilages dorsally. It is therefore subdivided into a larger anterior intramembranous part (*glottis vocalis*), which measures about three-fifths of the length of the entire aperture, and a posterior intercartilaginous part (*glottis respiratoria*). Posteriorly it is limited by the mucous membrane passing between the arytenoid cartilages. The rima glottidis is the narrowest part of the cavity of the larynx, and its level corresponds with the bases of the arytenoid cartilages. Its length, in the male, is about 23 mm; in the female, from 17 to 18 mm. The width and shape of the rima glottidis vary with the movements of the vocal cords and arytenoid cartilages during respiration and phonation. In the condition of rest, i.e., when these structures are uninfluenced by muscular action, as in quiet respiration, the intermembranous part is triangular, with its apex in front and its base behind (the latter is represented by a line, about 8 mm long, connecting the anterior ends of the vocal processes, while the medial surfaces of the arytenoids are parallel to each other, and hence the intercartilaginous part is rectangular). During extreme adduction of the vocal folds, as in the emission of a high note, the intermembranous part is reduced to a linear slit by the apposition of the vocal folds, while the intercartilaginous part is triangular, its apex corresponding to the anterior ends of the vocal processes of the arytenoids, which are approximated by the medial rotation of the cartilages. Conversely, in extreme abduction of the vocal folds, as in forced inspiration, the arytenoids and their vocal processes are rotated lateralward and the intercartilaginous part is triangular, but with its apex directed backward. In this condition the entire glottis is somewhat lozenge-shaped, the widest part of the aperture corresponding with the attachments of the vocal folds to the vocal processes.

Muscles

The muscles of the larynx are *extrinsic*, passing between the larynx and parts around (described in Chapter 6), and *intrinsic*, confined entirely to the larynx.

The intrinsic muscles are:

Cricothyroid
Posterior cricoarytenoid
Lateral cricoarytenoid
Arytenoid
Thyroarytenoid

CRICOTHYROID. The triangular cricothyroid muscle (Fig. 15-20) *arises* from the front and lateral part of the cricoid cartilage; its fibers are arranged in two groups. The caudal fibers constitute a **pars obliqua** and slant backward to the anterior border of the inferior cornu; the anterior fibers, forming a **pars recta,** run cranialward, backward, and lateralward to the posterior part of the lower border of the lamina of the thyroid cartilage. The medial borders of the muscles on the two sides are separated by a triangular interval, occupied by the middle cricothyroid ligament.

POSTERIOR CRICOARYTENOID. The posterior cricoarytenoid (Fig. 15-20) *arises* from the broad depression on the corresponding half of the posterior surface of the lamina of the cricoid cartilage; its fibers run cranialward and lateralward, and converge to be *inserted* into the back of the muscular process of the arytenoid cartilage. The uppermost fibers are nearly horizontal, the middle are oblique, and the lowest are almost vertical.

LATERAL CRICOARYTENOID. The lateral cricoarytenoid (Fig. 15-22) is oblong and

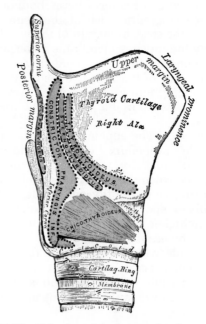

FIG. 15-20. Side view of the larynx, showing muscular attachments.

smaller than the posterior cricoarytenoid muscle. It *arises* from the cranial border of the arch of the cricoid cartilage, and passing obliquely upward and backward, is *inserted* into the front of the muscular process of the arytenoid cartilage.

ARYTENOID. The arytenoid (Fig. 15-21) is a single muscle that fills up the posterior concave surfaces of the arytenoid cartilages. It *arises* from the posterior surface and lateral border of one arytenoid cartilage, and is *inserted* into the corresponding parts of the opposite cartilage. It consists of oblique and transverse parts. The **oblique arytenoid,** the more superficial, forms two fasciculi that pass from the base of one cartilage to the apex of the opposite one, and therefore cross each other like the limbs of the letter X. A few fibers are continued around the lateral margin of the cartilage, and are prolonged into the aryepiglottic fold; they are sometimes described as a separate muscle, the **aryepiglotticus.** The **transverse arytenoid** crosses transversely between the two cartilages.

THYROARYTENOID. The thyroarytenoid (Figs. 15-22; 15-23) is a broad, thin, muscle that lies parallel with and lateral to the vocal fold and supports the wall of the ventricle and its appendix. It *arises* from the caudal

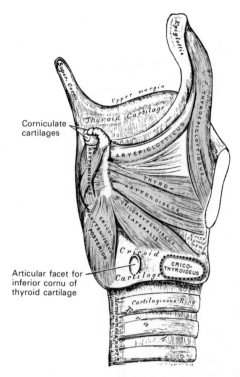

FIG. 15-22. Muscles of larynx, side view. Right lamina of thyroid cartilage removed.

half of the angle of the thyroid cartilage and from the middle cricothyroid ligament. Its fibers pass dorsalward and lateralward, to be *inserted* into the base and anterior surface of the arytenoid cartilage. The more medial fibers of the muscle can be differenti-

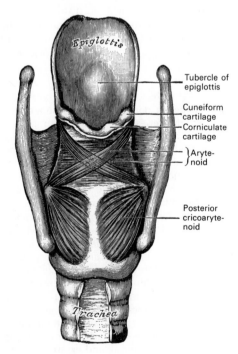

FIG. 15-21. Muscles of larynx, posterior view.

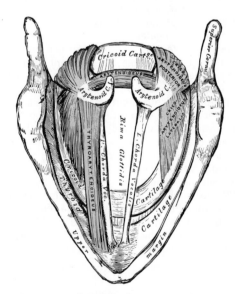

FIG. 15-23. Muscles of the larynx, seen from above. (Enlarged.)

ated as a band that is inserted into the vocal process of the arytenoid cartilage and into the adjacent portion of its anterior surface; it is termed the **vocalis,** and lies parallel with the vocal ligament, to which it adheres.

Many fibers of the thyroarytenoid are prolonged into the aryepiglottic fold, where some of them become lost, while others are continued to the margin of the epiglottis. They have received a distinctive name, **thyroepiglotticus,** and are sometimes described as a separate muscle. A few fibers extend along the wall of the ventricle from the lateral wall of the arytenoid cartilage to the side of the epiglottis and constitute the **ventricularis** muscle.

ACTIONS. The actions of the muscles of the larynx may be conveniently divided into two groups: (1) those that open and close the glottis; (2) those that regulate the degree of tension of the vocal folds.

The *posterior cricoarytenoids* separate the vocal folds and, consequently, **open** the glottis by rotating the arytenoid cartilages outward around a vertical axis passing through the cricoarytenoid joints, so that their vocal processes and the vocal folds attached to them become widely separated.

The *lateral cricoarytenoids* **close** the glottis by rotating the arytenoid cartilages inward, so as to approximate their vocal processes.

The *arytenoid* approximates the arytenoid cartilages, and thus **closes** the opening of the glottis, especially at its back part.

The *cricothyroids* produce **tension** and elongation of the vocal folds by drawing up the arch of the cricoid cartilage and tilting back the upper border of its lamina; the distance between the vocal processes and the angle of the thyroid is thus increased and the folds are elongated.

The *thyroarytenoids,* consisting of two parts having different attachments and different directions, have a rather complicated action. Their main action is to draw the arytenoid cartilages forward toward the thyroid and thus shorten and relax the vocal folds. Their lateral portions rotate the arytenoid cartilage inward and thus narrow the rima glottidis by bringing the two vocal folds together. Certain minute fibers of the vocalis division, inserting obliquely upon the vocal ligament and designated as the **aryvocalis muscle,** are considered by Strong (1935) to be chiefly responsible for the control of pitch, through their ability to regulate the length of the vibrating part of the vocal folds.

The manner in which the entrance of the larynx is closed during deglutition is discussed in Chapter 16.

Structure

MUCOUS MEMBRANE. The mucous membrane of the larynx is continuous with that lining the mouth, pharynx, trachea and bronchi, and is prolonged into the lungs. It lines the posterior surface and the upper part of the anterior surface of the epiglottis, to which it is closely adherent, and helps to form the aryepiglottic folds, which bound the entrance of the larynx. It lines the whole of the cavity of the larynx; by its reduplication, it forms the chief part of the vestibular fold, and from the ventricle, it is continued into the ventricular appendix. It is then reflected over the vocal ligament, where it is thin and intimately adherent, covers the inner surface of the conus elasticus and cricoid cartilage, and is ultimately continuous with the lining membrane of the trachea. The anterior surface and the upper half of the posterior surface of the epiglottis, the upper part of the aryepiglottic folds, and the vocal folds are covered by stratified squamous epithelium; all the rest of the laryngeal mucous membrane is covered by columnar ciliated cells, but patches of stratified squamous epithelium are found in the mucous membrane above the glottis.

GLANDS. The mucous membrane of the larynx is furnished with numerous mucus-secreting glands, the orifices of which are found in nearly every part; they are plentiful upon the epiglottis, being lodged in little pits in its substance; they are also found in large numbers along the margin of the aryepiglottic fold in front of the arytenoid cartilages, where they are termed the **arytenoid glands.** They exist also in large numbers in the ventricular appendages. None are found on the free edges of the vocal folds.

VESSELS. The chief **arteries** of the larynx are the laryngeal branches derived from the superior and inferior thyroid arteries. The **veins** accompany the arteries; those accompanying the superior laryngeal artery join the superior thyroid vein, which opens into the internal jugular vein, while those accompanying the inferior laryngeal artery join the inferior thyroid vein, which opens into the brachiocephalic vein. The **lymphatic vessels** consist of two sets, superior and inferior. The former accompany the superior laryngeal artery and pierce the thyrohyoid membrane, to end in the nodes situated near the bifurcation of the common carotid artery. Of the latter, some pass through the middle cricothyroid ligament and open into nodes lying in front of that ligament or in front of the upper part of the trachea, while others pass to the deep cervical nodes and to the nodes accompanying the inferior thyroid artery.

NERVES. The **nerves** are derived from the vagus nerve by way of the internal and external branches of the superior laryngeal nerve and from the recurrent nerve. The internal laryngeal branch is sensory.

It enters the larynx by piercing the posterior part of the thyrohyoid membrane above the superior laryngeal vessels, and divides into three branches: one is distributed to both surfaces of the epiglottis; a second, to the aryepiglottic fold; and a third, the largest, supplies the mucous membrane over the back of the larynx and communicates with the recurrent nerve. The external laryngeal branch supplies the cricothyroid muscle. The recurrent nerve passes cranialward beneath the caudal border of the constrictor pharyngis inferior immediately dorsal to the cricothyroid joint. It supplies all the muscles of the larynx except the cricothyroid. The sensory branches of the laryngeal nerves form subepithelial plexuses, from which fibers end between the cells covering the mucous membrane.

Over the posterior surface of the epiglottis, in the aryepiglottic folds, and less regularly in some other parts, taste buds similar to those in the tongue are found.

Fibers from the sympathetic nerves supply the blood vessels and glands.

TRACHEA AND BRONCHI

Trachea

The **trachea** or **windpipe** (Fig 15-24) is a cartilaginous and membranous tube, that extends from the larynx, on a level with the sixth cervical vertebra, to the cranial border of the fifth thoracic vertebra, where it divides into the two bronchi. The trachea is nearly but not quite cylindrical, being flat-

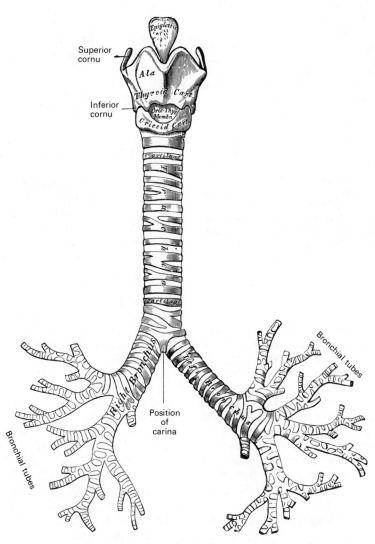

Fig. 15-24. Front view of cartilages of larynx, trachea, and bronchi.

THE RESPIRATORY SYSTEM

tened dorsally; its length measures about 11 cm; its diameter, from side to side, is from 2 to 2.5 cm, being always greater in the male than in the female. In the child the trachea is smaller, more deeply placed, and more movable than in the adult.

RELATIONS. The *ventral surface* of the trachea is covered, **in the neck,** by the isthmus of the thyroid gland, the inferior thyroid veins, the thyroid ima artery (when that vessel exists), the sternothyroid and sternohyoid muscles, the cervical fascia, and, more superficially, by the anastomosing branches between the anterior jugular veins. **In the thorax,** it is covered by the manubrium sterni, the remains of the thymus, the left brachiocephalic vein, the aortic arch, the brachiocephalic and left common carotid arteries, and the deep cardiac plexus. *Dorsally* it is in contact with the esophagus. *Laterally,* **in the neck,** it is in relation with the common carotid arteries, the right and left lobes of the thyroid gland, the inferior thyroid arteries, and the recurrent nerves; **in the thorax,** it lies in the superior mediastinum, and is in relation on the right side with the pleura and right vagus, and near the root of the neck with the brachiocephalic artery; on its left side are the left recurrent nerve, the aortic arch, and the left common carotid and subclavian arteries.

CLINICAL CONSIDERATIONS. The **operation for tracheotomy,** when performed as an emergency, is best carried out by piercing the neck 1 cm below the cricoid cartilage and opening the trachea between the second and third cartilaginous rings. The thyroid and cricoid cartilages are avoided because of subsequent scarring of the larynx. The isthmus of the thyroid gland is usually caudal to the first two tracheal cartilages and no blood vessels of appreciable size are present. Caudal to the thyroid gland, the trachea is situated much deeper than above it because of Burns's space,[1] and large veins are likely to be present.

Right Bronchus

The right bronchus is wider, shorter, and less abrupt in its divergence from the trachea than the left. It is about 2.5 cm long and enters the right lung nearly opposite the fifth thoracic vertebra. The azygos vein arches over it and the pulmonary artery lies at first inferior and then ventral to it. It gives rise to three subsidiary bronchi, one to each of the lobes. The superior lobe bronchus comes off above the pulmonary artery and has been called, therefore, the **eparterial bronchus.** The bronchi to the middle and inferior lobes

[1] Allen Burns (1781–1813): A Scottish anatomist (Glasgow).

separate below the pulmonary artery and are accordingly **hyparterial** in position.

The **right superior lobe bronchus** divides into three branches named, according to the bronchopulmonary segments that they enter, the bronchus for the apical segment, for the posterior segment, and for the anterior segment. The **right middle lobe bronchus** divides into two branches, the bronchus for the lateral and for the medial segments. The **right inferior lobe bronchus** first gives off the bronchus to the superior segment, and then divides into four bronchi for the basal segments: the medial basal, anterior basal, lateral basal, and posterior basal segments.

Left Bronchus

The left bronchus is smaller in caliber but about twice as long as the right (5 cm). It passes under the aortic arch and crosses ventral to the esophagus, thoracic duct, and descending aorta. It is superior to the pulmonary artery at first, then dorsal, and finally passes inferior to the artery before it divides into the bronchi for the superior and inferior lobes. Both lobar bronchi, therefore, are hyparterial in position.

The **left superior lobe bronchus** divides into two branches, one of which is distributed to a portion of the left lung corresponding to the right superior lobe; the other, to a portion corresponding to the right middle lobe. These two branches are called the superior division and inferior division bronchi to differentiate them from segmental bronchi. The *superior division bronchus* of the left superior lobe divides into branches for the apical posterior segment and the anterior segment. The *inferior division bronchus* divides into bronchi for the superior and inferior segments. The **left inferior lobe bronchus** gives off first the bronchus for the superior segment and then divides into three branches for basal segments: the anterior medial basal, lateral basal, and posterior basal segments.

The picture seen through a bronchoscope may be reproduced if a section is made across the trachea and a bird's-eye view taken of its interior (Fig. 15-25). At the bottom of the trachea, the septum that separates the two bronchi is visible as a spur (in bronchoscopic terminology) and is named the carina. The **carina** is placed to the left of the

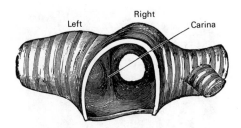

FIG. 15-25. Bifurcation of the trachea, viewed from above, with the interior showing the carina as it would be seen through a bronchoscope.

middle line and the right bronchus appears as a more direct continuation of the trachea than the left. Because of this asymmetry and because the right bronchus is larger in diameter than the left, foreign bodies that enter the trachea have a tendency to drop into the right bronchus rather than into the left.

Structure

The trachea and extrapulmonary bronchi are composed of imperfect rings of hyaline cartilage, fibrous tissue, muscular fibers, mucous membrane, and glands (Fig. 15-26).

CARTILAGES. The cartilages of the trachea number from 16 to 20; each forms an imperfect ring, which occupies the anterior two-thirds or so of the circumference of the tra-

chea, being deficient behind, where the tube is completed by fibrous tissue and non-striated muscular fibers. The cartilages are placed horizontally above each other, separated by narrow intervals. They measure about 4 mm deep and 1 mm thick. Their outer surfaces are flattened in a vertical direction, but internally, they are convex, the cartilages being thicker in the middle than at the margins. Two or more of the cartilages often unite, partially or completely, and they are sometimes bifurcated at their extremities. They are highly elastic, but may become calcified in advanced life. In the right bronchus the cartilages vary in number from six to eight; in the left, from nine to twelve. They are shorter and narrower than those of the trachea, but have the same shape and arrangement. The special tracheal cartilages are the first and the last (Fig. 15-24).

The **first cartilage** is broader than the rest and often divided at one end; it is connected by the cricotracheal ligament with the caudal border of the cricoid cartilage, with which or with the succeeding cartilage it is sometimes blended.

The **last cartilage** is thick and broad in the middle, because its lower border is prolonged into a triangular hook-shaped process that curves downward and backward between the origins of the two bronchi. It

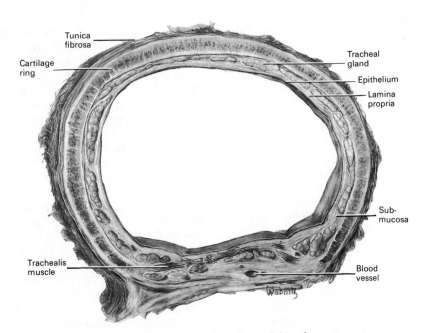

FIG. 15-26. Cross section of trachea, adult cadaver. 5 ×.

ends on each side in an imperfect ring, which encloses the commencement of each bronchus. The cartilage above the last is somewhat broader than the others at its center.

FIBROUS MEMBRANE. The cartilages are enclosed in an elastic fibrous membrane that consists of two layers; one, the thicker, passes over the outer surface of the ring, and the other passes over the inner surface. At the cranial and caudal margins of the cartilages the two layers blend together to form a single membrane that connects the rings with one another. In the space between the ends of the rings dorsally, the membrane forms a single layer also.

MUSCULAR TISSUE. The muscular tissue consists of two layers of nonstriated muscle, longitudinal and transverse, between the ends of the cartilages. The **longitudinal fibers** are external and consist of a few scattered bundles. The **transverse fibers** (trachealis muscle) are arranged internally in branching and anastomosing bands that extend more or less transversely (Fig. 15-26).

MUCOUS MEMBRANE. The mucous membrane is continuous above with that of the larynx and below with that of the bronchi. It consists of areolar and lymphoid tissue and has a well-marked basement membrane that supports a stratified epithelium, the surface layer of which is columnar and ciliated, while the deeper layers are composed of oval or rounded cells. Beneath the basement membrane is a distinct layer of longitudinal elastic fibers with a small amount of intervening areolar tissue. The submucous layer is composed of a loose meshwork of connective tissue that contains large blood vessels, nerves, and mucous glands; the ducts of the latter pierce the overlying layers and open on the surface (Fig. 15-27).

VESSELS AND NERVES. The trachea is supplied with blood by the inferior thyroid **arteries**. The **veins** end in the thyroid venous plexus. The **nerves** are derived from the vagus and the recurrent nerves and from the sympathetic trunk; they are distributed to the trachealis muscles and between the epithelial cells.

PLEURA

Each lung is invested by an exceedingly delicate serous membrane, the **pleura,** which is arranged in the form of a closed invagi-

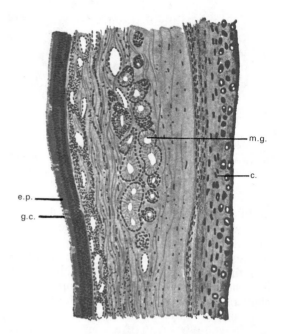

FIG. 15-27. Section through wall of human trachea. *c,* Cartilage; *e.p.,* pseudostratified ciliated epithelium; *g.c.,* goblet cell; *m.g.,* mucous gland. 8 ×. (Sobotta.)

nated sac. A portion of the serous membrane covers the surface of the lung and dips into the fissures between its lobes; it is called the **pulmonary pleura.** The rest of the membrane lines the inner surface of the chest wall, covers the diaphragm, and is reflected over the structures occupying the middle of the thorax; this portion is termed the **parietal pleura.** The two layers are continuous with each other around and below the root of the lung; in health they are in actual contact with each other, but the potential space between them is known as the **pleural cavity.** When the lung collapses or when air or fluid collects between the two layers, the cavity becomes apparent. The right and left pleural sacs are entirely separate from each other; between them are all the thoracic viscera except the lungs, and they touch each other only for a short distance ventrally, behind the upper part of the body of the sternum; the interval in the middle of the thorax between the two sacs is termed the mediastinum.

Different portions of the parietal pleura have received special names that indicate their position: thus, that portion which lines the inner surfaces of the ribs and intercostal muscles is the **costal pleura;** that clothing the convex surface of the diaphragm is the **dia-**

phragmatic pleura; that which rises into the neck, over the summit of the lung, is the **cupula of the pleura** (*cervical pleura*); and that which is applied to the other thoracic viscera is the **mediastinal pleura.**

Parietal Pleura

Commencing at the sternum, the pleura passes lateralward, lining the costal cartilages, ribs, and intercostal muscles (Figs. 15-28; 15-29). At the dorsal part of the thorax, it passes over the sympathetic trunk and its branches and is reflected upon the sides of the bodies of the vertebrae, where it is separated by a narrow interval, the posterior mediastinum, from the opposite pleura. From the vertebral column the pleura passes to the side of the pericardium, which it covers to a slight extent before it covers the dorsal part of the root of the lung, from the caudal border of which a triangular sheet descends vertically toward the diaphragm. This sheet is the posterior layer of a fold known as the **pulmonary ligament.**

From the dorsal portion of the lung root, the pleura may be traced over the costal surface of the lung, over its apex and base, and also over the sides of the fissures between the lobes, onto its mediastinal surface and the ventral part of its root. It is continued from the caudal margin of the root as the anterior layer of the pulmonary ligament, and from this it is reflected onto the pericardium (**pericardial pleura**), and then to the dorsal surface of the sternum. Cranial to the level of the root of the lung, however, the mediastinal pleura passes uninterruptedly from the vertebral column to the sternum over the structures in the superior mediastinum. It covers the cranial surface of the diaphragm and extends, ventrally, as low as the costal cartilage of the seventh rib; at the side of the chest, it extends to the caudal border of the tenth rib on the left side and to the cranial border of the same rib on the right side. Dorsally, it reaches as low as the twelfth rib, and sometimes even to the transverse process of the first lumbar vertebra.

The **cupula** projects through the superior

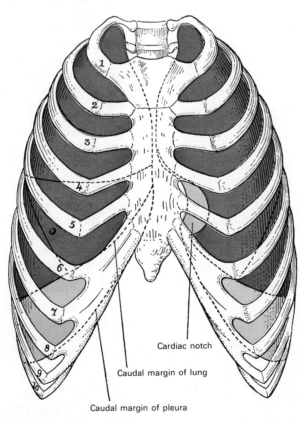

Cardiac notch

Caudal margin of lung

Caudal margin of pleura

FIG. 15-28. Front view of thorax, showing the relations of the pleurae and lungs to the chest wall. Phrenicocostal and costomediastinal sinus in blue; lungs in purple.

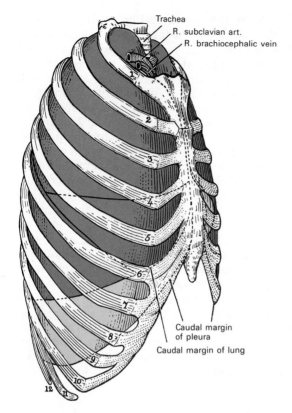

Trachea
R. subclavian art.
R. brachiocephalic vein

Caudal margin
of pleura

Caudal margin of lung

Fig. 15-29. Lateral view of thorax, showing the relations of the pleurae and lungs to the chest wall. Phrenicocostal and costomediastinal sinus in blue; lungs in purple.

opening of the thorax into the neck, extending from 2.5 to 5 cm above the sternal end of the first rib; this portion of the sac is strengthened by a dome-like expansion of fascia (**Sibson's**[1] **fascia**), which is attached ventrally to the inner border of the first rib and dorsally to the anterior border of the transverse process of the seventh cervical vertebra. This is covered and strengthened by a few spreading muscular fibers derived from the scalene muscles called the scalenus minimus.

In the ventral part of the chest, the two pleural sacs behind the manubrium are separated by an angular interval; the reflection is represented by a line drawn from the sternoclavicular articulation to the midpoint of the junction of the manubrium with the body of the sternum. From this point the two pleurae descend in close contact to the level of the fourth costal cartilages. The line of reflection on the right side is continued cau-

[1]Francis Sibson (1814–1876): An English physician (London).

dalward in nearly a straight line to the xiphoid process, and then turns lateralward, while on the left side the line of reflection diverges lateralward and is continued caudalward, close to the left border of the sternum, as far as the sixth costal cartilage. The caudal limit of the pleura is on a considerably lower level than the corresponding limit of the lung, but does not extend to the attachment of the diaphragm, so that caudal to the line of reflection from the chest wall on to the diaphragm, the pleura is in direct contact with the rib cartilages and the intercostales interni. Moreover, in ordinary inspiration the thin inferior margin of the lung does not extend as far as the line of the pleural reflection, so that the costal and diaphragmatic pleurae are in contact here, the intervening narrow slit being termed the **costodiaphragmatic recess** (phrenicocostal sinus). A similar condition exists behind the sternum and rib cartilages, where the anterior thin margin of the lung falls short of the line of pleural reflection, and where the slit-like cavity between the two layers of pleura forms the **costomediastinal recess** (*costomediastinal sinus*).

The line along which the right pleura is reflected from the chest wall to the diaphragm starts ventrally, immediately below the seventh sternocostal joint, and runs caudalward and dorsalward behind the seventh costal cartilage, thus crossing the tenth rib in the midaxillary line, from which it is prolonged to the level of the spinous process of the twelfth thoracic vertebra. The reflection of the left pleura follows at first the ascending part of the sixth costal cartilage, and in the rest of its course is slightly lower than that of the right side.

The free surface of the pleura is smooth, glistening, and moistened by a serous fluid; its attached surface adheres intimately to the lung and to the pulmonary vessels as they emerge from the pericardium; it also adheres to the cranial surface of the diaphragm. Throughout the rest of its extent it is easily separable from the adjacent parts.

The right pleural sac is shorter, wider, and reaches higher in the neck than the left.

Pulmonary Ligament

The root of the lung is covered by pleura, except at its caudal border where the investing layers come into contact. Here they form a sort of mesenteric fold, the **pulmonary ligament,** which extends between the mediastinal surface of the lung and the pericardium. Just above the diaphragm the ligament ends in a free falciform border. It retains the lower part of the lung in position.

Structure

Like other serous membranes, the pleura is composed of a single layer of flattened mesothelial cells resting upon a delicate connective tissue membrane, beneath which lies a stroma of collagenous tissue containing several prominent networks of yellow elastic fibers. Blood vessels, lymphatics, and nerves are distributed in the substance of the pleura.

VESSELS AND NERVES. The **arteries** of the pleura are derived from the intercostal, internal thoracic, musculophrenic, thymic, pericardiac, pulmonary, and bronchial vessels. The **veins** correspond to the arteries. The **lymphatics** are described in Chapter 10.

The **nerves** of the parietal pleura are derived from the phrenic, intercostal, vagus, and sympathetic nerves; those of the pulmonary pleura, from the vagus and sympathetic nerves through the pulmo-

nary plexuses at the hilum of the lung. It is believed that nerves accompany the ramifications of the bronchial arteries in the pulmonary pleura.

MEDIASTINUM

The **mediastinum** is interposed as a partition in the median portion of the thorax, separating the parietal pleural sacs of the two lungs (Fig. 15-31). It extends from the sternum ventrally to the vertebral column dorsally and comprises all the thoracic viscera, except the lungs and pleurae, embedded in a thickening and expansion of the subserous fascia of the thorax. It is divided arbitrarily, for the purposes of description, into cranial and caudal parts by a plane that extends from the sternal angle to the caudal border of the fourth thoracic vertebra. The cranial part is named the **superior mediastinum;** the caudal part is again subdivided into three parts: the **anterior mediastinum,** ventral to the pericardium; the **middle mediastinum,** containing the pericardium; and the **posterior mediastinum,** dorsal to the pericardium.

The **superior mediastinum** (Fig. 15-30) is bounded by the superior aperture of the thorax; by the plane of the superior limit of the pericardium; by the manubrium; by the

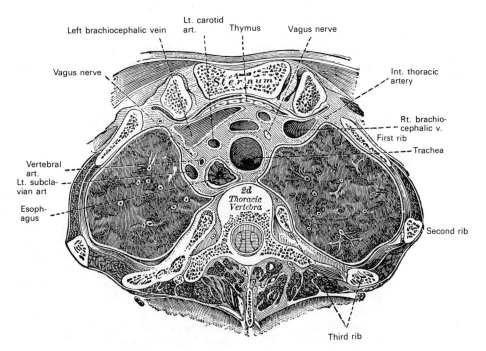

FIG. 15-30. Transverse section through the upper margin of the second thoracic vertebra, showing the superior mediastinum. (Braune.)

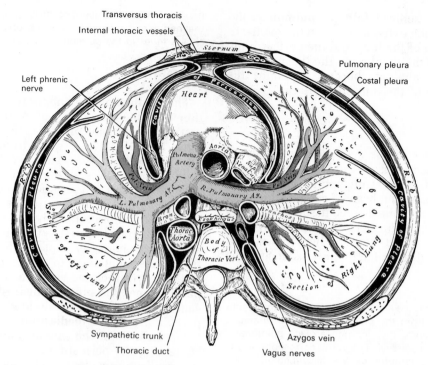

FIG. 15-31. A transverse section of the thorax showing the contents of the middle and the posterior mediastinum. The pleural and pericardial cavities are exaggerated since normally there is no space between parietal and visceral pleurae and between pericardium and heart.

upper four thoracic vertebrae; and laterally, by the mediastinal pleurae of the two lungs. It contains the origins of the sternohyoid and sternothyroid muscles and the lower ends of the longus colli muscles; the aortic arch; the brachiocephalic artery and the thoracic portions of the left common carotid and the left subclavian arteries; the brachiocephalic veins and the upper half of the superior vena cava; the left highest intercostal vein; the vagus, cardiac, phrenic, and left recurrent nerves; the trachea, esophagus, and thoracic duct; and the remains of the thymus, and some lymph nodes.

The **anterior mediastinum** (Fig. 15-31) is bounded ventrally by the body of the sternum and because of the position of the heart, the left transversus thoracis muscle and parts of the fourth, fifth, sixth, and seventh costal cartilages. It is bounded dorsally by the parietal pericardium and extends caudalward as far as the diaphragm. Besides a few lymph nodes and vessels, it contains only a thin layer of subserous fascia which is separated from the endothoracic or deep fascia superiorly by a fascial cleft, but there is a firm attachment in its caudal part that forms the pericardiosternal ligament.

The **middle mediastinum** (Fig. 15-31) is the broadest part of the interpleural septum. It contains the heart enclosed in the pericardium, the ascending aorta, the lower half of the superior vena cava with the azygos vein opening into it, the pulmonary trunk dividing into its two branches, the right and left pulmonary veins, and the phrenic nerves.

The **posterior mediastinum** (Fig. 15-31) is an irregularly shaped mass running parallel with the vertebral column and, because of the slope of the diaphragm, it extends caudally beyond the pericardium. It is bounded ventrally, by the pericardium and, more caudally, by the diaphragm; dorsally, by the vertebral column from the lower border of the fourth to the twelfth thoracic vertebra; and on either side, by the mediastinal pleurae. It contains the thoracic part of the descending aorta, the azygos and hemiazygos veins, the vagus and splanchnic nerves, the bifurcation of the trachea and the two bronchi, the esophagus, the thoracic duct, and many large lymph nodes.

Some authorities include the bifurcation of the trachea, the two bronchi, and the roots of the two lungs in the middle mediastinum.

Lungs (Pulmones)

The **lungs** are the essential organs of respiration; they are placed on either side within the thorax and are separated from each other by the heart and other contents of the mediastinum (Fig. 12-76). The substance of the lung has a light, porous, spongy texture; it floats in water and crepitates when handled owing to the presence of air in the alveoli. It is also highly elastic, hence the retracted state of these organs when they are removed from the closed cavity of the thorax. The surface is smooth, shiny, and marked out into numerous polyhedral areas that indicate the lobules of the organ: each of these areas is crossed by numerous lighter lines.

At birth the lungs are pinkish-white in color; in adult life the color is a slaty gray, mottled in patches, and as age advances, this mottling assumes a black color. The coloring matter consists of granules of carbon deposited in the areolar tissue near the surface of the organ. It increases in quantity as age advances, and is more abundant in males than in females. As a rule, the posterior border of the lung is darker than the anterior.

The right lung usually weighs about 625 gm, the left weighs 567 gm, but much variation is met according to the amount of blood or serous fluid they may contain. The lungs are heavier in males than in females, their proportion to the body being 1 to 37 in the former and 1 to 43 in the latter. The vital capacity, the quantity of air that can be exhaled by the deepest expiration after making the deepest inspiration, varies greatly with the individual; an average for an adult man is 3700 ml. The total volume of the fully expanded lungs is about 6500 ml; this includes both tissues and contained air. The tidal air, the amount of air breathed in or out during quiet respiration, is about 500 ml for the adult man. Various calculations indicate that the total epithelial area of the respiratory and nonrespiratory surfaces during ordinary deep inspiration of the adult is not greater than 70 square meters.

Each lung is conical and presents for examination an **apex,** a **base,** three **borders,** and two **surfaces.**

The **apex** is rounded and extends into the root of the neck, reaching 2.5 to 5 cm above the level of the sternal end of the first rib. A sulcus is produced by the subclavian artery as it curves ventral to the pleura and runs cranialward and lateralward immediately below the apex.

The **base** is broad, concave, and rests upon the convex surface of the diaphragm, which separates the right lung from the right lobe of the liver, and the left lung from the left lobe of the liver, the stomach, and the spleen. Since the diaphragm extends higher on the right than on the left side, the concavity on the base of the right lung is deeper than that on the left. Laterally and dorsally, the base is bounded by a thin, sharp margin that projects for some distance into the phrenicocostal sinus of the pleura, between the lower ribs and the costal attachment of the diaphragm. The base of the lung descends during inspiration and ascends during expiration.

SURFACES

The **costal surface** (*external or thoracic surface*) is smooth, convex, of considerable extent, and corresponds to the form of the cavity of the chest, being deeper dorsally than ventrally. It is in contact with the costal pleura and presents, in specimens that have been hardened in situ, slight grooves corresponding with the overlying ribs.

The **mediastinal surface** (*inner surface*) is divided into a mediastinal and a vertebral portion. The **mediastinal portion** is in contact with the mediastinal pleura. It has a deep concavity, the **cardiac impression,** which accommodates the pericardium. Dorsal and cranial to this concavity is a slight depression named the **hilum,** through which the structures that form the root of the lung enter and leave.

On the **right lung** (Fig. 15-33), immediately cranial to the hilum, is an arched furrow that accommodates the azygos vein. Running cranialward and then arching lateralward some little distance below the apex is a wide groove for the superior vena cava and right brachiocephalic vein. Dorsal to the hilum and the attachment of the pulmonary ligament is a vertical groove for the esophagus; this groove becomes less distinct caudally, owing to the inclination of the lower part of the esophagus to the left of the midline. Ventral and to the right of the lower part of esophageal groove is a deep concavity for the extrapericardiac portion of the thoracic part of the inferior vena cava.

On the **left lung** (Fig. 15-34), immediately

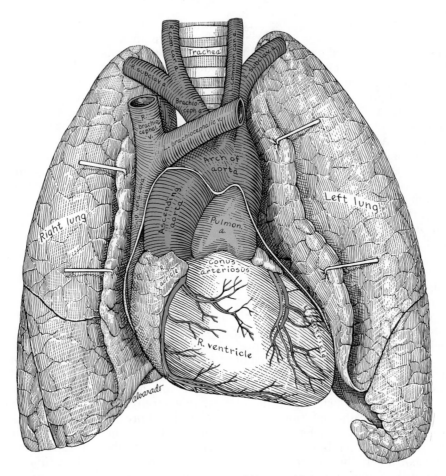

FIG. 15-32.   Front view of heart and lungs.

cranial to the hilum, is a well-marked curved furrow produced by the aortic arch, and running cranialward from this toward the apex is a groove that accommodates the left subclavian artery; a slight impression ventral to the latter and close to the margin of the lung lodges the left innominate vein. Dorsal to the hilum and pulmonary ligament is a vertical furrow produced by the descending aorta, and ventral to this, near the base of the lung, the lower part of the esophagus causes a shallow impression.

BORDERS

The **inferior border** is thin and sharp where it separates the base from the costal surface and extends into the phrenicocostal sinus; medially, where it divides the base from the mediastinal surface, it is blunt and rounded.

The **anterior border** is thin and sharp and overlaps the ventral surface of the heart and pericardium. The anterior border of the right lung is almost vertical and projects into the costomediastinal sinus; that of the left lung has an angular notch, the **cardiac notch,** in which the pericardium comes into contact with the sternum. Opposite this notch, the anterior margin of the left lung is situated a little distance lateral to the line of reflection of the corresponding part of the pleura.

FISSURES AND LOBES

RIGHT LUNG.   The right lung is divided into three lobes (superior, middle, and inferior) by two interlobar fissures.

The **oblique fissure** separates the inferior from the middle and superior lobes and corresponds with the oblique fissure in the left lung. Its direction is more vertical, however,

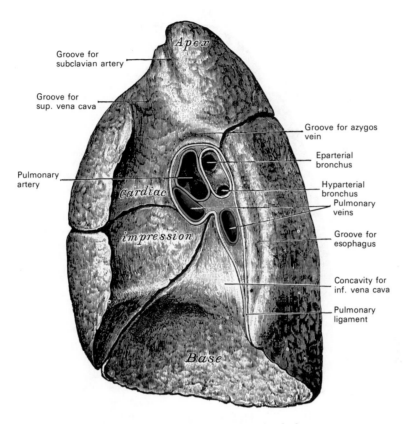

Groove for
subclavian artery

Groove for
sup. vena cava

Pulmonary
artery

Apex

Cardiac

impression

Base

Groove for azygos
vein

Eparterial
bronchus

Hyparterial
bronchus
Pulmonary
veins

Groove for
esophagus

Concavity for
inf. vena cava

Pulmonary
ligament

FIG. 15-33. Mediastinal surface of right lung.

and it cuts the lower border about 7.5 cm dorsal to its extremity.

The **horizontal fissure** separates the superior from the middle lobe. It begins in the previous fissure near the dorsal border of the lung, and running horizontally, cuts the ventral border on a level with the sternal end of the fourth costal cartilage; on the mediastinal surface it may be traced dorsal to the hilum. The middle lobe, the smallest lobe of the right lung, is wedge-shaped and includes the lower part of the ventral border and the ventral part of the base of the lung.

The right lung is shorter by 2.5 cm than the left because the diaphragm rises higher on the right side, and it is broader because of the inclination of the heart to the left side. Its total capacity is greater, however, and it weighs more than the left lung.

LEFT LUNG. The left lung is divided into two lobes, a superior and an inferior, by an **oblique fissure** that extends from the costal to the mediastinal surface of the lung both above and below the hilum. As seen on the surface, this fissure begins on the mediastinal surface of the lung at the hilum, and runs dorsalward and cranialward to the posterior border, which it crosses at a point about 6 cm below the apex. It then extends caudalward and ventralward over the costal surface, to reach the lower border near its extremity, and its further course can be followed across the mediastinal surface as far as the hilum. The superior lobe includes the apex, the anterior border, a considerable part of the costal surface, and the greater part of the mediastinal surface of the lung. The inferior lobe is larger than the superior and comprises almost the whole of the base, a large portion of the costal surface, and the greater part of the vertebral part of the medial surface.

ROOT

The root of the lung (*radix pulmonis*) (Figs. 15-33; 15-34) is formed by the bronchus, pulmonary artery, pulmonary veins, bronchial arteries and veins, pulmonary

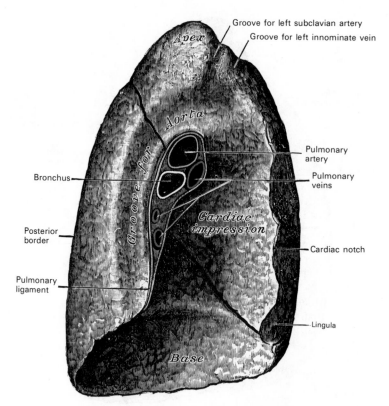

Groove for left subclavian artery
Groove for left innominate vein

Apex

Aorta

Pulmonary
artery

Bronchus

Groove for

Pulmonary
veins

Posterior
border

*Cardiac
impression*

Cardiac notch

Pulmonary
ligament

Lingula

Base

Fig. 15-34.   Mediastinal surface of left lung.

plexuses of nerves, lymphatic vessels, and bronchial lymph nodes. These structures are all embedded in mediastinal connective tissue and the entire mass is encircled by a reflection of the pleura. It corresponds to the hilum, which is near the center of the mediastinal surface of the lung, dorsal to the cardiac impression and closer to the posterior than the anterior border. The root of the right lung lies dorsal to the superior vena cava and the right atrium, and the azygos vein arches over it (Fig. 12-76). The root of the left lung is ventral to the descending aorta and inferior to the aortic arch (Fig. 12-77). The phrenic nerve, the pericardiophrenic artery and vein, and the anterior pulmonary plexus of nerves are ventral, while the vagus nerve and its posterior pulmonary plexus are dorsal to the root of the lung on both sides. Caudal to the root of each lung, the reflection of the pleura from mediastinum to lung is prolonged downward toward the diaphragm as the pulmonary ligament (p. 1382).

The chief structures of the roots of both lungs have a similar relation to each other in a dorsoventral direction, but there is a difference in their superior and inferior relations on the two sides. Thus the pulmonary

veins are ventral, the bronchi dorsal, and the pulmonary arteries between, on both sides. On the right side, the superior lobe bronchus is superior and the pulmonary artery is slightly lower; next are the bronchi to the middle and inferior lobes, and most inferior is the pulmonary vein (Fig. 15-33). On the left side, the pulmonary artery is superior, the pulmonary veins inferior, and the bronchus between (Fig. 15-34).

FURTHER SUBDIVISION OF THE LUNG

The importance of certain smaller units of structure of the lungs, called bronchopulmonary segments, has been emphasized by the thoracic surgeon, bronchoscopist, and radiologist. To interpret these units correctly, one should give particular attention to the concept that the lung is fundamentally the aggregate of all the branchings of the bronchus. According to this concept, a bronchopulmonary segment is that portion of the lung to which any particular bronchus is distributed, and the term might conceivably be applied to the lobule supplied by its lobular bronchus. In actual practice, however, it is customary to restrict the term bronchopulmonary segment to the portion of the lung supplied by the direct branches of the lobar bronchi (and of the division bronchi in the case of the left superior lobe). These segments are as definite as the lobes, and their relative extent and position can be demonstrated by introducing different colored gelatin into their bronchi. It is possible, in many cases, to follow the delicate connective tissue between the segments and dissect them apart. That the majority of extra fissures follow the planes of separation between the segments also demonstrates their fundamental nature.

BRONCHOPULMONARY SEGMENTS (Fig. 15-35). The bronchopulmonary segments are named according to their positions in the lobes, and the bronchus to each is named after its segment. The right superior lobe has three segments: apical, posterior, and anterior. The right middle lobe has a lateral and a medial segment. The right inferior lobe has a superior and four basal segments: the medial basal, anterior basal, lateral basal, and posterior basal. The left superior lobe is first separated into two divisions, the superior division corresponding to the right superior lobe, and an inferior division corresponding to the right middle lobe. The superior division of the left superior lobe has an apical posterior and an anterior segment. The inferior division of the left superior lobe has a superior and an inferior segment. The left inferior lobe has a superior and three basal segments: anteromedial basal, lateral basal, and posterior basal.

VARIATIONS. The branching of the lobar bronchi to form segmental bronchi is reasonably constant. The superior lobe bronchus is somewhat more constant than the inferior lobe bronchus. In approximately 95% of the specimens, it is possible to identify three segmental bronchi coming from the right superior lobe bronchus, although in some of these cases, two segmental bronchi may seem to arise from a short common stem. The size of the lung segment supplied by a particular bronchus may be larger or smaller than expected, even when the branching appears at first glance to follow the usual pattern, because it may have exchanged a smaller branch bronchus with an adjacent segmental bronchus. In the case of the lower lobe bronchus, the superior and the medial basal segmental bronchi are about as constant as the branches in the superior lobe, but there is more variation in the anterior basal, lateral basal, and posterior basal segments. In somewhat less than 50% of the specimens, branches come from the posterior aspect of the infe-

rior lobe bronchus between the superior segmental and the basal segmental bronchi, or from the stem below the anterior basal segmental bronchus.

The branchings of the bronchi are designated according to the parts of the lungs that they ventilate. The two main bronchi correspond to the right and left lungs. The right bronchus is subdivided into three lobar bronchi for the superior, middle, and inferior lobes; the left bronchus is subdivided into two, for the superior and inferior lobes. Each lobar bronchus is subdivided into branchings for the bronchopulmonary segments depicted in Figure 15-36.

For the purpose of relating the smaller branchings of the bronchi to the ramifications of the pulmonary arteries and veins, they have been named and numbered by Boyden (1955) (see Fig. 15-36). In the lists of segmental bronchi, arteries, and veins, corresponding structures of the right and left lungs are placed opposite each other.

BLOOD VESSELS OF THE LUNGS

The vessels that serve the special function of the lungs, that is, of aerating the blood, are the pulmonary arteries and veins and their connecting capillaries (Figs. 15-36 to 15-41). The blood supply for the purpose of nutrition to the tissues of the lungs is provided by the bronchial arteries and veins; these are discussed on page 1398.

PULMONARY VESSELS. The right and left **pulmonary arteries,** formed by the bifurcation of the pulmonary trunk, convey the deoxygenated blood to the right and left lungs. They divide into branches that accompany the bronchi, coursing chiefly along their dorsal surface. The bronchopulmonary segments are supplied by main intrasegmental branches of the pulmonary arteries, which are single for the most part, but which may arise as common trunks for adjacent segments (Boyden, 1955). The artery for one segment is likely to supply small branches to the neighboring segments. Distal to the alveolar duct, branches are distributed to each atrium, from which arise smaller radicles terminating in a dense capillary network in the walls of the alveoli.

The **pulmonary veins** are described in Chapter 9. The segmental veins are listed here according to the same classification used for the segmental arteries.

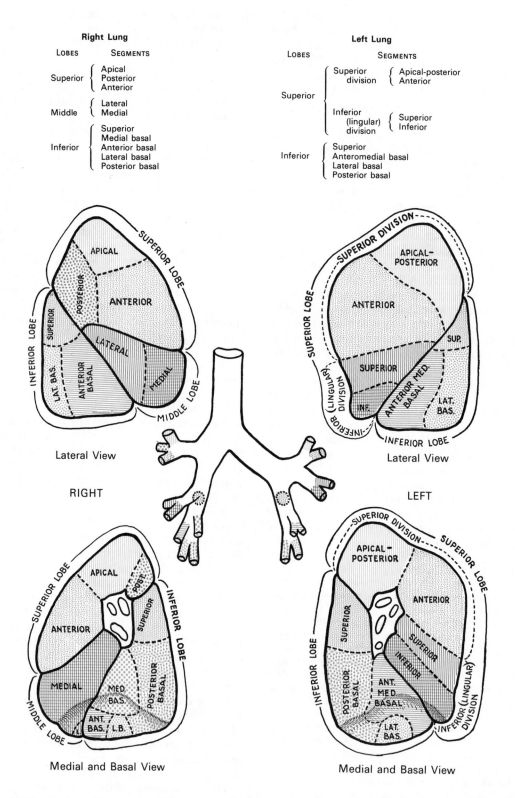

FIG. 15-35.   The bronchopulmonary segments. The segmental branches of the bronchi are shown in corresponding colors. (After J. F. Huber, 1947.)

SEGMENTAL BRONCHI

Right Superior Lobe

B^1–Apical bronchus
  B^1a–apical ramus
  B^1b–anterior ramus
B^2–Anterior bronchus
  B^2a–posterior ramus
    B^2a1–superior subramus
    B^2a2–inferior subramus
  B^2b–anterior ramus
B^3–Posterior bronchus
  B^3a–apical ramus
  B^3b–posterior ramus

Right Middle Lobe

B^4–Lateral bronchus
  B^4a–posterior ramus
  B^4b–anterior ramus
B^5–Medial bronchus
  B^5a–superior ramus
  B^5b–inferior ramus

Right Inferior Lobe

B^6–Superior bronchus
  B^6a–medial ramus
    B^6a1–paravertebral subramus
    B^6a2–posterior subramus
  B^6b–superior ramus
  B^6c–lateral ramus
B^*–Subsuperior bronchus
  BX^* (10)–accessory
    subsuperior bronchus
B^7–Medial basal bronchus
  B^7a–anterior ramus
  B^7b–medial ramus
B^8–Anterior basal bronchus
  B^8a–lateral ramus
  B^8b–basal ramus
B^9–Lateral basal bronchus
  B^9a–lateral ramus
  B^9b–basal ramus
B^{10}–Posterior basal bronchus
  BV^* (10)–accessory
    subsuperior ramus
  B^{10}a–laterobasal ramus
  B^{10}b–mediobasal ramus

Left Superior Lobe
Superior Divison

B^{1+3}–Apical posterior bronchus
  B^1a, B^3a–apical rami
  B^1b, B^3b–anterior and posterior rami
B^2–Anterior bronchus
  B^2a–posterior ramus

  B^2b–anterior ramus

Left Superior Lobe
Inferior (Lingular) Division

B^4–Superior lingular bronchus
  B^4a–posterior ramus
  B^4b–anterior ramus
B^5–Inferior lingular bronchus
  B^5a–superior ramus
  B^5b–inferior ramus

Left Inferior Lobe

B^6–Superior bronchus
  B^6a–medial ramus
    B^6a1–paravertebral subramus
    B^6a2–posterior subramus
  B^6b–superior ramus
  B^6c–lateral ramus
B^*–Subsuperior bronchus
  BX^* (9), BX^* (10)–accessory
    subsuperior bronchus
B^7–Medial basal bronchus
  B^7a–lateroanterior ramus
  B^7b–medioanterior ramus
B^8–Anterior basal bronchus
  B^8a–lateral ramus
  B^8b–basal ramus
B^9–Lateral basal bronchus
  BX^* (9)–accessory subsuperior ramus
  B^9b–basal ramus = B^9
B^{10}–Posterior basal bronchus
  BX^* (10)–accessory
    subsuperior ramus
  B^{10}a–laterobasal ramus
  B^{10}b–mediobasal ramus

SEGMENTAL ARTERIES

Right Superior Lobe

A^1–Apical segmental artery
$\quad$ A^1a–apical ramus
$\quad$ A^1b–anterior ramus
A^2–Anterior segmental artery
$\quad$ A^2a–posterior ramus
$\quad\quad$ A^2a1–superior subramus
$\quad\quad$ A^2a2–inferior subramus
$\quad$ A^2b–anterior ramus
A^3–Posterior segmental artery
$\quad$ A^3a–apical ramus
$\quad$ A^3b–posterior ramus

Middle Lobe

A^4–Lateral segmental artery

$\quad$ A^4a–posterior ramus
$\quad$ A^4b–anterior ramus
A^5–Medial segmental artery

A^5a–superior ramus
A^5b–inferior ramus

Right Inferior Lobe

A^6–Superior segmental artery
$\quad$ A^6a–medial ramus
$\quad$ A^6b–superior ramus
$\quad$ A^6c–lateral ramus
A^7–Medial basal artery
$\quad$ A^7a–anterior ramus
$\quad$ A^7b–medial ramus
A^8–Anterior basal artery
$\quad$ A^8a–lateral ramus
$\quad$ A^8b–basal ramus
A^9–Lateral basal artery
$\quad$ A^9a–lateral ramus
$\quad$ A^9b–basal ramus
A^{10}–Posterior basal artery
$\quad$ A^{10}a–laterobasal ramus
$\quad$ A^{10}b–mediobasal ramus

Left Superior Lobe
Superior Divison

A^{1+3}–Apical posterior segmental artery
$\quad$ A^1a–apical ramus
$\quad$ A^1b–anterior ramus
$\quad$ A^3a–apical ramus
$\quad$ A^3b–posterior ramus
A^2–Anterior segmental artery
$\quad$ A^2a–posterior ramus
$\quad$ A^2b–anterior ramus

Left Superior Lobe
Inferior or Lingular Division

A^4–Superior lingular artery
$\quad\quad$ (a. lingularis superior)
$\quad$ A^4a–posterior ramus
$\quad$ A^4b–anterior ramus
A^5–Inferior lingular artery
$\quad\quad$ (a. lingularis inferior)
$\quad$ A^5a–superior ramus
$\quad$ A^5b–inferior ramus

Left Inferior Lobe

A^6–Superior segmental artery
$\quad$ A^6a–medial ramus
$\quad$ A^6b–superior ramus
$\quad$ A^6c–lateral ramus
A^7–Medial basal artery
$\quad$ A^7a–anterior ramus
$\quad$ A^7b–medial ramus
A^8–Anterior basal artery
$\quad$ A^8a–lateral ramus
$\quad$ A^8b–basal ramus
A^9–Lateral basal artery
$\quad$ A^9a–lateral ramus
$\quad$ A^9b–basal ramus
A^{10}–Posterior basal artery
$\quad$ A^{10}a–laterobasal ramus
$\quad$ A^{10}b–mediobasal ramus

SEGMENTAL VEINS

Right Superior Lobe

V^1-Apical segmental vein
  V^1a-apical ramus
  V^1b-anterior ramus
V^3-Posterior segmental vein
  V^3a-apical ramus
  V^3b-posterior ramus
  V^3c-posterior intersegmental ramus
  V^3d-anterior ramus
V^2-Anterior segmental vein
  V^2a-superior ramus
  V^2b-Inferior ramus

Middle Lobe

V^4-Lateral segmental vein
  V^4a-posterior ramus
  V^4b-anterior ramus
V^5-Medial segmental vein
  V^5a-superior ramus
  V^5b-inferior ramus

Right Inferior Lobe

V^6-Superior segmental vein
  V^6a-medial ramus
  V^6b-superior ramus
  V^6c-lateral ramus
V^7-Medial basal segmental vein
  V^7a-anterior ramus
  V^7b-medial ramus
V^8-Anterior basal segmental vein
  V^8a-lateral ramus
  V^8b-basal ramus
V^9-Lateral basal segmental vein
  V^9a-lateral ramus
  V^9b-basal ramus
V^{10}-Posterior basal segmental vein
  V^{10}a-laterobasal ramus
  V^{10}b-mediobasal ramus

Left Superior Lobe
Superior Division

V^{1+3}-Apical posterior segmental vein
  V^1-Apical vein
    V^1a-apical ramus
    V^1b-anterior ramus
  V^3-Posterior vein
    V^3a-apical ramus
    V^3b-posterior ramus
    V^3c-posterior intersegmental ramus

V^2-Anterior segmental vein
  V^2a-superior ramus
  V^2b-inferior ramus
  V^2c-posterior ramus

Inferior Division

V^4-Superior lingular segmental vein
  V^4a-posterior ramus
  V^4b-anterior ramus
V^5-Inferior lingular segmental vein
  V^5a-superior ramus
  V^5b-inferior ramus

Left Inferior Lobe

V^6-Superior segmental vein
  V^6a-medial ramus
  V^6b-superior ramus
  V^6c-lateral ramus
V^7-Medial basal segmental vein
  V^7a-lateroanterior ramus
  V^7b-medioanterior ramus
V^8-Anterior basal segmental vein
  V^8a-lateral ramus
  V^8b-basal ramus
V^9-Lateral basal segmental vein
  V^9a-lateral ramus
  V^9b-basal ramus
V^{10}-Posterior basal segmental vein
  V^{10}a-laterobasal ramus
  V^{10}b-mediobasal ramus

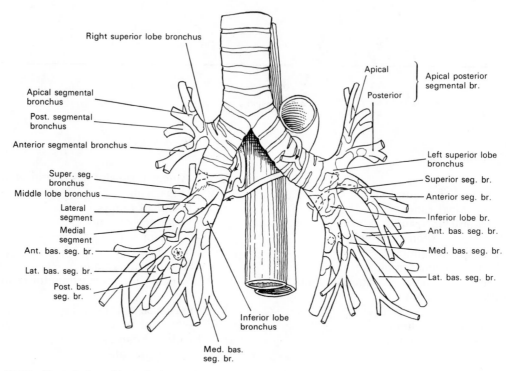

Right superior lobe bronchus

Apical

Posterior

Apical posterior segmental br.

Apical segmental bronchus

Post. segmental bronchus

Anterior segmental bronchus

Left superior lobe bronchus

Superior seg. br.

Anterior seg. br.

Inferior lobe br.

Ant. bas. seg. br.

Med. bas. seg. br.

Super. seg. bronchus

Middle lobe bronchus

Lateral segment

Medial segment

Ant. bas. seg. br.

Lat. bas. seg. br.

Post. bas. seg. br.

Lat. bas. seg. br.

Inferior lobe bronchus

Med. bas. seg. br.

FIG. 15-36. Ventral view of bronchial tree illustrating prevailing mode of branching. (Redrawn from E. A. Boyden, 1955, *Segmental Anatomy of the Lungs,* courtesy of Blakiston Division, McGraw-Hill Book Co., Inc.)

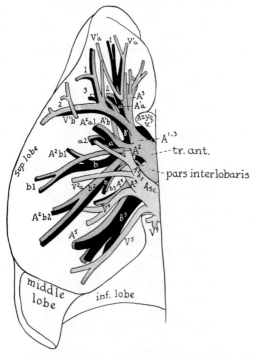

FIG. 15-37. Dissection of mediastinal surface of right superior lobe showing the prevailing pattern. The conventional colors are followed: red for arteries, blue for veins, and black for bronchi. (Redrawn from E. A. Boyden, 1955, *Segmental Anatomy of the Lungs,* courtesy of Blakiston Division, McGraw-Hill Book Co., Inc.)

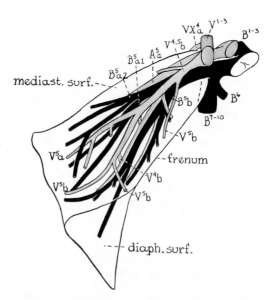

FIG. 15-38. Dissection of mediastinal surface of middle lobe. (Redrawn from E. A. Boyden, 1955, *Segmental Anatomy of the Lungs,* courtesy of Blakiston Division, McGraw-Hill Book Co., Inc.)

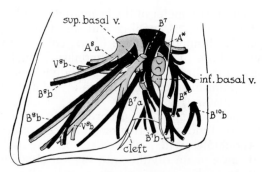

FIG. 15-39. Dissection of mediastinal surface of right inferior lobe; the mode of branching of the inferior pulmonary vein is atypical. (Redrawn from E. A. Boyden, 1955, *Segmental Anatomy of the Lungs,* courtesy of Blakiston Division, McGraw-Hill Book Co., Inc.)

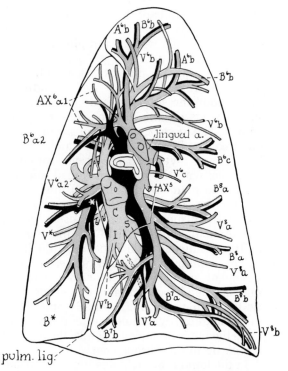

FIG. 15-41. Dissection of mediastinal surface of fairly typical left inferior lobe. (Redrawn from E. A. Boyden, 1955, *Segmental Anatomy of the Lungs,* courtesy of Blakiston Division, McGraw-Hill Book Co., Inc.)

Serous Coat

The serous coat is the pulmonary pleura (p. 1380). It is thin, transparent, and invests the entire organ as far as the root.

Subserous Alveolar Tissue

The subserous areolar tissue contains a large proportion of elastic fibers. It invests the entire surface of the lung and extends inward between the lobules.

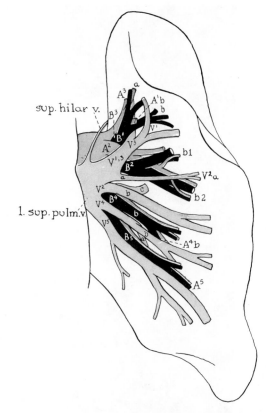

FIG. 15-40. Dissection of mediastinal surface of left superior lobe. (Redrawn from E. A. Boyden, 1955, *Segmental Anatomy of the Lungs,* courtesy of Blakiston Division, McGraw-Hill Book Co., Inc.)

Parenchyma

The parenchyma is composed of **secondary lobules** which, although closely connected together by an interlobular areolar tissue, are quite distinct from one another, and in the fetus may be teased asunder without much difficulty. The secondary lobules vary in size; those on the surface are large and pyramidal, with the base turned toward the surface; those in the interior are smaller and variously shaped. Each secondary lobule is composed of several **primary lobules,**

STRUCTURE

The lungs are composed of an external serous coat, a subserous areolar tissue, and the pulmonary substance or parenchyma.

the anatomic units of the lung. The primary lobule consists of an alveolar duct, the air spaces connected with it, and blood vessels, lymphatics, and nerves (Fig. 15-42).

INTRAPULMONARY BRONCHI. The intrapulmonary bronchi divide and subdivide throughout the entire organ, the smallest subdivisions constituting the lobular bronchioles. The *larger divisions* consist of: (1) an outer coat of fibrous tissue, in which irregular plates of hyaline cartilage are found at intervals, with the most developed appearing at the points of division; (2) internal to the fibrous coat, an interlacing network of circularly disposed smooth muscle fibers, the bronchial muscle; and (3) most internally, the mucous membrane, which is lined by columnar ciliated epithelium resting on a basement membrane. The corium of the mucous membrane contains numerous elastic fibers that run longitudinally and a certain amount of lymphoid tissue; it also contains the ducts of mucous glands, the acini of which lie in the fibrous coat.

In the **lobular bronchioles** (terminal bronchioles), the ciliated epithelial cells become cuboidal, and cartilage plates cease to exist

when the diameter of the bronchiole reaches about 1 mm. Branching and anastomosing bands of smooth muscle fibers, continuous with those of the intrapulmonary bronchi, invest the bronchiole and its subdivisions to the point of junction between alveolar duct and atrium (Fig. 15-43).

Each bronchiole, according to Miller (1947), divides into two or more **respiratory bronchioles,** with scattered alveoli, and each of these again divides into several **alveolar ducts,** with a greater number of alveoli connected with them (Fig. 15-44). Each alveolar duct is connected with a variable number of irregularly spherical spaces, which also possess alveoli, the **atria.** With each atrium, a

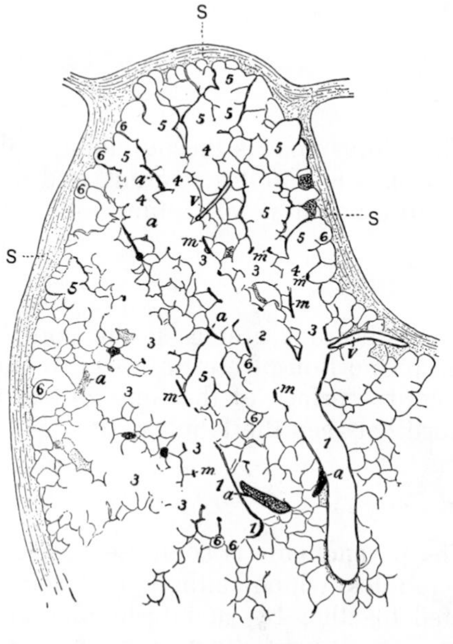

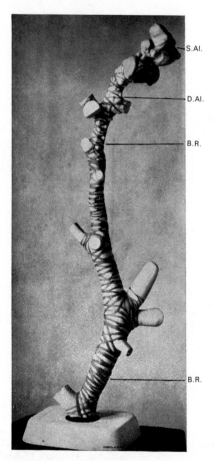

FIG. 15-43. View of a reconstruction, from a dog's lung, of the musculature of a noncartilaginous bronchiolus (lobular bronchiole), 0.565 mm in diameter, its branches, and its termination in a primary lobule of which only a single alveolar sac is shown completely reconstructed. B, bronchiolus; B.R. bronchiolus respiratorius; D.Al., ductulus alveolaris; S.Al., sacculus alveolaris. 125 × and reduced to 8. (Miller, The Lung, courtesy of Charles C Thomas.)

FIG. 15-42. Part of a secondary lobule from the depth of a human lung, showing parts of several primary lobules. 1, bronchiole; 2, respiratory bronchiole; 3, alveolar duct; 4, atria; 5, alveolar sac; 6, alveolus or air cell; m, smooth muscle; a, branch pulmonary artery; v, branch pulmonary vein; s, septum between secondary lobules. Camera drawing of one 50-μ section. 20 × diameters. (Miller, Jour. Morph.)

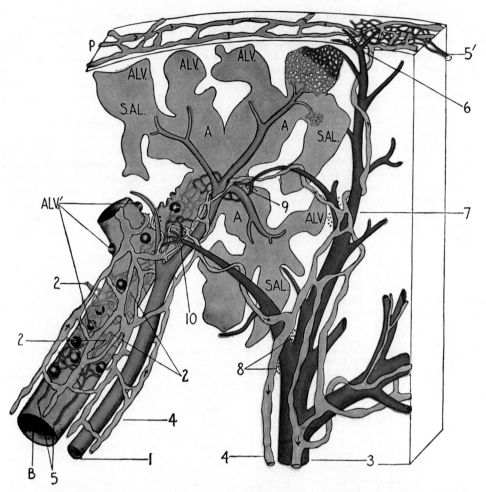

Fig. 15-44. General scheme of a primary lobule, showing the subdivisions of (B) a respiratory bronchiole into two alveolar ducts; and the atria (A), alveolar sacs (S.AL.) of one of these ducts. ALV', alveoli scattered along the bronchioles; P, pleura, 1, pulmonary artery, dividing into smaller radicles for each atrium, one of which terminates in a capillary plexus on the wall of an alveolus; 2, its branches to the respiratory bronchiole and alveolar duct; 3, pulmonary vein with its tributaries from the pleura 6, capillary plexus of alveolus, and wall of the atrium 9 and alveolar duct 10; 4, lymphatics; dotted areas at 7, 8, 9 and 10, indicating areas of lymphoid tissue; 5, bronchial artery terminating in a plexus on the wall of the bronchiole; 5', bronchial artery terminating in pleura. (Miller, The Lung, courtesy of Charles C Thomas.)

variable number (2 to 5) of **alveolar sacs** are connected which bear **alveoli** or air spaces on all parts of their circumference.

ALVEOLI. The alveoli are lined by a continuous layer of pulmonary alveolar epithelium. The nuclei of the epithelial cells protrude into the air spaces, and the perinuclear cytoplasm attenuates abruptly into thin sheets of cytoplasm. The cytoplasmic sheets, averaging about 0.2 μm in thickness, rest on a basement membrane and face on the alveolar air spaces. The alveolar walls contain blood capillaries and collagenous, reticular, and elastic connective tissue fibers. The bar-

rier between the capillary blood and alveolar air includes two thin layers of cytoplasm, alveolar epithelium and capillary endothelium with adherent basement membranes for each. Tissue space between these two membranes may be potential or real, but they are not adherent and variable amounts of interstitial elements may separate them.

The fetal lung resembles a gland in that the alveoli have a small lumen and are lined by cubical epithelium. After the first respiration the alveoli become distended, and the epithelium takes on the characteristics of the adult.

VESSELS. The **bronchial arteries** supply blood for the nutrition of the lung; the right lung usually receives a single artery and the left lung receives two. They are derived from the ventral side of the upper part of the thoracic aorta or from the upper aortic intercostal arteries. Some are distributed to the bronchial glands and to the walls of the bronchi and pulmonary vessels; those supplying the bronchi extend as far as the respiratory bronchioles, where they form capillary plexuses that unite with similar plexuses formed by the pulmonary artery, both of which give rise to small venous trunks forming one of the sources of the pulmonary vein. Others are distributed in the interlobular areolar tissue and end partly in the deep and partly in the superficial bronchial veins. Lastly, some ramify upon the surface of the lung, beneath the pleura, where they form a capillary network.

The **bronchial vein** is formed at the root of the lung. It receives superficial and deep veins from a limited area about the hilum; the larger part of the blood supplied by the bronchial arteries is returned by the pulmonary veins. The bronchial vein ends on the right side in the azygos vein, and on the left side in the highest intercostal vein or in the accessory hemiazygos vein.

The **lymphatics** are described on page 894.

NERVES. The lungs are supplied from the anterior and posterior pulmonary plexuses, formed chiefly by branches from the sympathetic and vagus nerves. The filaments from these plexuses accompany the bronchial tubes, supplying efferent fibers to the bronchial muscle and afferent fibers to the bronchial mucous membrane and probably to the alveoli of the lung. Small ganglia are found upon these nerves.

References

(References are listed not only to those articles and books cited in the text, but to others as well which are considered to contain valuable resource information for the student who desires it.)

Nose and Paranasal Sinuses

Ali, M. Y. 1965. Histology of the human nasopharyngeal mucosa, J. Anat. (Lond.), 99:657–672.

Allison, A. C. 1953. The morphology of the olfactory system in the vertebrates. Biol. Rev., 28:195–244.

Blanton, P. L., and N. L. Biggs. 1969. Eighteen hundred years of controversy: The paranasal sinuses. Amer. J. Anat., 124:135–147.

Brain, J. D. 1970. The uptake of inhaled gases by the nose. Ann. Otol. Rhinol. Laryngol., 79:529–539.

Bridger, M. W., and A. W. P. van Nostrand. 1978. The nose and paranasal sinuses—applied surgical anatomy. A histologic study of whole organ sections in three planes. J. Otolaryngol. (Suppl. 6), 7:1–33.

Burnham, H. H. 1935. An anatomical investigation of blood vessels of the lateral nasal wall and their relation to turbinates and sinuses. J. Laryngol. Otol., 50:569–593.

Cauna, N., and K. H. Hinderer. 1969. Fine structure of blood vessels of the human nasal respiratory mucosa. Ann. Otol. Rhinol. Laryngol., 78:865–879.

Cauna, N., K. H. Hinderer, and R. T. Wentges. 1969. Sensory receptor organs of the human nasal respiratory mucosa. Amer. J. Anat., 124:187–209.

Clark, G. M. 1971. The structural support of the nose. J. Otolaryngol. Soc. Austral., 3:235–240.

Dawes, J. D. K., and M. M. L. Prichard. 1953. Studies of the vascular arrangements of the nose. J. Anat. (Lond.), 87:311–322.

Dion, M. C., B. W. Jafek, and C. E. Tobin. 1978. The anatomy of the nose. External support. Arch. Otolaryngol., 104:145–150.

Drettner, B. 1979. The role of the nose in the functional unit of the respiratory system. Rhinology, 17:3–11.

Drumheller, G. W. 1973. Topology of the lateral nasal cartilages: The anatomical relationship of the lateral nasal to the greater alar cartilage, lateral crus. Anat. Rec., 176:321–327.

Fabricant, N. D., and G. Conklin. 1965. The Dangerous Cold—Its Cures and Complications. Macmillan Publishing Company, New York, 179 pp.

Jacobs, M. H. 1947. Anatomic study of the maxillary sinus from the standpoint of the oral surgeon. J. Oral Surg., 5:282–291.

Lucas, A. M. 1932. The nasal cavity and direction of fluid by ciliary movement in Macacus rhesus. Amer. J. Anat., 50:141–177.

Negus, V. E. 1958. The Comparative Anatomy and Physiology of the Nose and Paranasal Sinuses. Livingstone, Edinburgh, 402 pp.

Revskoi, Y. K. 1965. Variations in frontal sinus structure and their significance in selection of pilots. Fed. Proc., 24:T948–T950.

Ritter, F. N. 1970. The vasculature of the nose. Ann. Otol. Rhinol. Laryngol., 79:468–474.

Rose, J. M., C. M. Pomerat, and B. Danes. 1949. Tissue culture studies of ciliated nasal mucosa in man. Anat. Rec., 104:409–419.

Rosen, M. D., and B. G. Sarnat. 1954. A comparison of the volume of the left and right maxillary sinuses in dogs. Anat. Rec., 120:65–71.

Schaeffer, J. P. 1920. The Nose, Paranasal Sinuses, Nasolacrimal Passageways, and Olfactory Organ in Man. Blakiston Company, Philadelphia, 370 pp.

Sellers, L. M. 1949. The frontal sinus—A problem in diagnosis and treatment. The Mississippi Doctor, 27:317–320.

Stamm, W. K. 1981. Anatomy of the pterygopalatine foramen and the fontanella in the lateral nasal wall. Rhinology, 19:87–91.

Vidić, B. 1971. The morphogenesis of the lateral nasal wall in the early prenatal life of man. Amer. J. Anat., 130:121–139.

Warbrick, J. G. 1960. The early development of the nasal cavity and upper lip in the human embryo. J. Anat. (Lond.), 94:351–362.

Larynx, Trachea, and Bronchi

Baken, R. J., and C. R. Noback. 1971. Neuromuscular spindles in intrinsic muscles of a human larynx. J. Speech Hear. Res., 14:513–518.

Beck, C., and W. Mann. 1980. The inner laryngeal lymphatics. A lymphangioscopical and electron micro-scopical study. Acta Otolaryngol., 89:265–270.

Blanding, J. D., Jr., R. W. Ogilvie, C. L. Hoffman, and W. H. Knisely. 1964. The gross morphology of the arterial supply to the trachea, primary bronchi, and esophagus of the rabbit. Anat. Rec., 148:611–614.

Boyden, E. A. 1971. The structure of the pulmonary acinus in a child of six years and eight months. Amer. J. Anat., 132:275–299.

Boyden, E. A. 1974. The mode of origin of pulmonary acini and respiratory bronchioles in the fetal lung. Amer. J. Anat., 141:317–328.

Campbell, A. H., and A. G. Liddelow. 1967. Significant variations in the shape of the trachea and large bronchi. Med. J. Austral., 1:1017–1020.

Cauldwell, E. W., R. G. Siekert, R. E. Lininger, and B. J. Anson. 1948. The bronchial arteries. An anatomic study of 150 human cadavers. Surg. Gynecol. Obstet., 86:395–412.

English, D. T., and C. E. Blevins. 1969. Motor units of laryngeal muscles. Arch. Otolaryngol., 89:778–784.

Falk, D. 1975. Comparative anatomy of the larynx in man and the chimpanzee: Implications for language in Neanderthal. Amer. J. Phys. Anthropol., 43:123–132.

Fisher, A. W. F. 1964. The intrinsic innervation of the trachea. J. Anat. (Lond.), 98:117–124.

Gay, T., H. Hirose, M. Strome, and M. Sawashima. 1972. Electromyography of the intrinsic laryngeal muscles during phonation. Ann. Otol. Rhinol. Laryngol., 81:401–409.

Gray, G. W., and C. M. Wise. 1959. The Bases of Speech. 3rd Edition. Harper and Brothers, New York, 562 pp.

Greene, M. C. L. 1980. The Voice and Its Disorders. 4th Edition. J. B. Lippincott Co., Philadelphia, 484 pp.

Inoue, S., and G. P. Dionne. 1977. Tonofilaments in normal human bronchial epithelium and in squamous cell carcinoma. Amer. J. Pathol., 88:345–354.

Kahane, J. C. 1978. A morphological study of the human prepubertal and pubertal larynx. Amer. J. Anat., 151:11–19.

Keene, M. F. L. 1961. Muscle spindles in human laryngeal muscles. J. Anat. (Lond.), 95:25–29.

King, B. T., and R. L. Gregg. 1948. An anatomical reason for the various behaviors of paralyzed vocal cords. Ann. Otol. Rhinol. Laryngol., 57:925–944.

Kirchner, J. A., and B. D. Wyke. 1965. Articular reflex mechanisms in the larynx. Ann. Otol. Rhinol. Laryngol., 74:749–768.

Kotby, M. N., and L. K. Haugen. 1970. The mechanics of laryngeal function. Acta Otolaryngol., 70:203–211.

Latarje, M. 1954. La vascularisation sanguinea des bronches. Les bronches, 4:145.

von Leden, H., and P. Moore. 1961. Vibratory pattern of the vocal cords in unilateral laryngeal paralysis. Acta Otolaryngol., 53:493–506.

Lucier, G. E., J. Daynes, and B. J. Sessle. 1978. Laryngeal reflex regulation: Peripheral and central neural analyses. Exp. Neurol., 62:200–213.

MacKenzie, C. F., T. C. McAslan, B. Shin, D. Schellinger, and M. Helrich. 1978. The shape of the human adult trachea. Anesthesiology, 49:48–50.

Macklin, C. C. 1929. The musculature of the bronchi and lungs. Physiol. Rev., 9:1–60.

Marchand, P., J. C. Gilroy, and V. H. Wilson. 1950. An anatomical study of bronchial vascular system and its variations in disease. Thorax, 5:207–221.

Maue, W. M., and D. R. Dickson. 1971. Cartilages and ligaments of the adult human larynx. Arch. Otolaryngol., 94:432–439.

Miller, R. A. 1941. The laryngeal sacs of an infant and an adult gorilla. Amer. J. Anat., 69:1–17.

Miserocchi, G., J. Mortola, and G. Sant'Ambrogio. 1973. Localization of pulmonary stretch receptors in the airways of the dog. J. Physiol. (Lond.), 235:775–782.

Monkhouse, W. S., and W. F. Whimster. 1976. An account of the longitudinal mucosal corrugations of the human tracheo-bronchial tree, with observations on those of some animals. J. Anat. (Lond.), 122:681–695.

Mortola, J. P., and G. Sant'Ambrogio. 1979. Mechanics of the trachea and behaviour of its slowly adapting stretch receptors. J. Physiol. (Lond.), 286:577–590.

Negus, V. E. 1931. The Mechanism of the Larynx. C. V. Mosby, St. Louis, 528 pp.

Negus, V. E. 1949. The Comparative Anatomy and Physiology of the Larynx. Heinemann, London, 230 pp.

O'Donnell, S. R., and N. Saar. 1973. Histochemical localization of adrenergic nerves in the guinea-pig trachea. Brit. J. Pharmacol., 47:707–710.

Phalen, R. F., H. C. Yeh, G. M. Schum, and O. G. Raabe. 1978. Application of an idealized model to morphometry of the mammalian tracheobronchial tree. Anat. Rec., 190:167–176.

Pressman, J. J. 1942. Physiology of the vocal cords in phonation and respiration. Arch. Otolaryngol., 35:355–398.

Reid, L. 1979. Basic aspects of bronchial anatomy—and further knowledge of pulmonary neuro-anatomy. Scand. J. Respir. Dis. (Suppl.), 103:13–18.

Rhodin, J., and T. Dalhamm. 1956. Electron microscopy of the tracheal ciliated mucosa in rats. Z. Zellforsch., 44:345–412.

Salassa, J. R., B. W. Pearson, and W. S. Payne. 1977. Gross and microscopical blood supply of the trachea. Ann. Thorac. Surg., 24:100–117.

Schreider, J. P., and O. G. Raabe. 1981. Structure of the human respiratory acinus. Amer. J. Anat., 162:221–232.

Sellars, I. E., and E. N. Keen. 1978. The anatomy and movements of the cricoarytenoid joint. Laryngoscope, 88:667–674.

Smith, E. I. 1957. The early development of the trachea and esophagus in relation to atresia of the esophaghus and tracheoesophageal fistula. Carneg. Instn., Contr. Embryol., 36:41–56.

Sram, F., and J. Syka. 1977. Firing pattern of motor units in the vocal muscle during phonation. Acta Otolaryngol., 84:132–137.

Stell, P. M., I. Gregory, and J. Watt. 1980. Morphology of the human larynx. II. The subglottis. Clin. Otolaryngol., 5:389–395.

Stell, P. M., R. Gudrun, and J. Watt. 1981. Morphology of the human larynx. III. The supraglottis. Clin. Otolaryngol., 6:389–393.

Strong, L. H. 1935. The mechanism of laryngeal pitch. Anat. Rec., 63:13–28.

Tanabe, M., N. Isshiki, and K. Kitajima. 1972. Vibratory pattern of the vocal cord in unilateral paralysis of the cricothyroid muscle. An experimental study. Acta Otolaryngol., 74:339–345.

Turner, R. S. 1962. A note on the geometry of the tracheal bifurcation. Anat. Rec., 143:189–194.

Wessels, N. K. 1970. Mammalian lung development: Interactions in formation and morphogenesis of tracheal buds. J. Exp. Zool., 175:455–466.

Williams, A. F. 1951. The nerve supply of laryngeal muscles. J. Laryngol. Otol., 65:343–348.

Williams, A. F. 1954. The recurrent laryngeal nerve and the thyroid gland. J. Laryngol. Otol., 68:719–725.

Lungs, Bronchopulmonary Segments, and Pleura

Adams, F. H. 1966. Functional development of the fetal lung. J. Pediat., 68:794–801.

Alexander, H. L. 1933. The autonomic control of the heart, lungs, and bronchi. Ann. Intern. Med., 6:1033–1043.

Altman, P. L., J. F. Gibson, Jr., and C. C. Wang. 1958. *Handbook of Respiration.* Edited by D. S. Dittmer, and R. M. Grebe. W. B. Saunders, Philadelphia, 403 pp.

Appleton, A. B. 1944. Segments and blood-vessels of the lungs. Lancet, 2:592–594.

Avery, M. E. 1962. The alveolar lining layer. A review of studies on its role in pulmonary mechanics and in the pathogenesis of atelectasis. Pediatrics, 30:324–330.

Bloomer, W. E., A. A. Liebow, and M. R. Hales. 1960. *Surgical Anatomy of the Bronchovascular Segments.* Charles C Thomas, Springfield, Ill., 273 pp.

Boyden, E. A. 1953. A critique of the international nomenclature on bronchopulmonary segments. Dis. Chest, 23:266–269.

Boyden, E. A. 1955. *Segmental Anatomy of the Lungs.* The Blakiston Division, McGraw-Hill Book Co., New York, 276 pp.

Boyden, E. A. 1961. The nomenclature of the bronchopulmonary segments and their blood supply. Dis. Chest, 39:1–6.

Boyden, E. A. 1965. The terminal air sacs and their blood supply in a 37-day infant lung. Amer. J. Anat., 116:413–427.

Boyden, E. A. 1967. Notes on the development of the lung in infancy and early childhood. Amer. J. Anat., 121:749–762.

Boyden, E. A. 1969. The pattern of the terminal air spaces in a premature infant of 30–32 weeks that lived nineteen and a quarter hours. Amer. J. Anat., 126:31–40.

Boyden, E. A., and J. F. Hartman. 1946. An analysis of the variations in the bronchopulmonary segments of the left upper lobes of fifty lungs. Amer. J. Anat., 79:321–360.

Boyden, E. A., and D. H. Tompsett. 1965. The changing patterns in the developing lungs of infants. Acta Anat., 61:164–192.

Bradley, G. W. 1977. Control of the breathing pattern. Internat. Rev. Physiol., 14:185–217.

Brock, R. C. 1954. *The Anatomy of the Bronchial Tree.* 2nd Edition. Oxford University Press, London, 243 pp.

Čech, S. 1969. Adrenergic innervation of blood vessels in the lung of some mammals. Acta Anat., 74:169–182.

Clements, J. A. 1970. Pulmonary surfactant. Amer. Rev. Resp. Dis., 101:984–990.

Drinker, C. K. 1954. *The Clinical Physiology of the Lungs.* Charles C Thomas, Springfield, Ill., 84 pp.

Dunnill, M. S. 1962. Postnatal growth of the lung. Thorax, 17:329–333.

Elftman, A. G. 1943. The afferent and parasympathetic innervation of the lungs and trachea of the dog. Amer. J. Anat., 72:1–27.

Elliott, F. M., and L. Reid. 1965. Some new facts about the pulmonary artery and its branching pattern. Clin. Radiol., 16:193–198.

Empey, D. W. 1978. Diseases of the respiratory system. Introduction: Structure and function of the lungs. Brit. Med. J., 1:631–633.

Engel, S. 1962. *Lung Structure.* Charles C Thomas, Springfield, Ill., 300 pp.

Fenn, W. O., and H. Rahn. 1965. *Handbook of Physiology–Respiration.* Section 3, Vol. 2. American Physiological Society, Williams and Wilkins, Baltimore, pp. 927–1696.

Findlay, C. W., Jr., and H. C. Maier. 1951. Anomalies of the pulmonary vessels and their surgical significance. Surgery, 29:604–641.

Gaylor, J. B. 1934. The intrinsic nervous mechanism of the human lung. Brain, 57:143–160.

Harbord, R. P., and R. Woolner. 1959. *Symposium on Pulmonary Ventilation.* Williams and Wilkins, Baltimore, 109 pp.

Hasleton, P. S. 1972. The internal surface area of the adult human lung. J. Anat. (Lond.), 112:391–400.

von Hayek, H. 1960. *The Human Lung.* Translated by Vernon E. Krahl. Hafner Publishing Co., New York, 372 pp.

Heitzman, E. R., J. V. Scrivani, J. Martino, and J. Moro. 1971. The azygos vein and its pleural reflections. I. Normal roentgen anatomy. Radiology, 101:249–258.

Heuck, F. 1959. *Die Streifenatelektasen der Lunge.* Georg Thieme Verlag, Stuttgart, 108 pp.

Huber, J. F. 1949. Practical correlative anatomy of the bronchial tree and lungs. J. Nat. Med. Assoc., 41:49–60.

Hughes, G. M. 1974. *Comparative Physiology of Vertebrate Respiration.* 2nd Edition. Heinemann, London, 144 pp.

Jackson, C. L., and J. F. Huber. 1943. Correlated applied anatomy of the bronchial tree and lungs with a system of nomenclature. Dis. Chest, 9:319–326.

Kendall, M. W., and E. Eissmann. 1980. Scanning electron microscopic examination of human pulmonary capillaries using a latex replication method. Anat. Rec., 196:275–283.

Kikkawa, Y. 1970. Morphology of alveolar lining layer. Anat. Rec., 167:389–400.

Kirks, D. R., P. E. Kane, E. A. Free, and H. Taybi. 1976. Systemic arterial supply to normal basilar segments of the left lower lobe. Amer. J. Roentgenol., 126:817–821.

Kohn, K., and M. Richter. 1958. *Die Lungenarterienbahn bei angeborenen Herzfehlern.* Georg Thieme Verlag, Stuttgart, 112 pp.

Krahl, V. E. 1955. Current concept of the finer structure of the lung. Arch. Intern. Med., 96:342–356.

Krahl, V. E. 1964. Anatomy of the mammalian lung. In *Handbook of Physiology—Respiration.* Section 3, Vol. 1. Edited by W. O. Fenn and H. Rahn. American Physiological Society, Williams and Wilkins, Baltimore, pp. 213–284.

Lachman, E. 1942. A comparison of the posterior boundaries of lungs and pleura as demonstrated on the cadaver and on the roentgenogram of the living. Anat. Rec., 83:521–542.

Lachman, E. 1946. The dynamic concept of thoracic topography: A critical review of present day teaching of visceral anatomy. Amer. J. Roentgenol., 56:419–440.

Larsell, O. 1922. The ganglia, plexuses, and nerve-terminations of the mammalian lung and pleura pulmonalis. J. Comp. Neurol., 35:97–132.

Larsell, O., and R. S. Dow. 1933. The innervation of the human lung. Amer. J. Anat., 52:125–146.

Levin, D. L., A. M. Rudolph, M. A. Heymann, and R. H. Phibbs. 1976. Morphological development of the pulmonary vascular bed in fetal lambs. Circulation, 53:144–151.

Loosli, C. G., and R. F. Baker. 1962. The human lung: Microscopic structure and diffusion. In *Pulmonary*

Structure and Function. Edited by A. V. S. de Reuck and M. O'Conner. Ciba Foundation Symposium, J. & A. Churchill, London, pp. 194–204.

Loosli, C. G., and E. L. Potter. 1959. Pre and postnatal development of the respiratory portion of the human lung with special reference to the elastic fibers. Amer. Rev. Resp. Dis. (Suppl.), *80*:5–23.

Low, F. N. 1953. The pulmonary alveolar epithelium of laboratory mammals and man. Anat. Rec., *117*:241–263.

Low, F. N., and M. M. Sampaio. 1957. The pulmonary alveolar epithelium as an entodermal derivative. Anat. Rec., *127*:51–63.

Lunde, P. K., and B. A. Waaler. 1969. Transvascular fluid balance in the lung. J. Physiol. (Lond.), *205*:1–18.

McLaughlin, R. F., W. S. Tyler, and R. O. Canada. 1961. A study of the subgross pulmonary anatomy in various mammals. Amer. J. Anat., *108*:149–165.

Miller, W. S. 1907. The vascular supply of the pleura pulmonalis. Amer. J. Anat., *7*:389–407.

Miller, W. S. 1947. *The Lung.* 2nd Edition. Charles C Thomas, Springfield, Ill., 222 pp.

Negus, V. 1965. *The Biology of Respiration.* Williams and Wilkins, Baltimore, 228 pp.

Oliveros, L. G. 1959. *Veins of the Lungs.* Universidad de Salamanca, Salamanca, Spain, 280 pp.

Pattle, R. E. 1958. Properties, function, and origin of the alveolar lining layer. Proc. Roy. Soc. Lond. (Ser. B), *148*:217–240.

Pattle, R. E. 1965. Surface lining of lung alveoli. Physiol. Rev., *45*:48–79.

Pennell, T. C. 1966. Anatomical study of the peripheral pulmonary lymphatics. J. Thorac. Cardiovasc. Surg., *52*:629–634.

Peters, R. M. 1969. *The Mechanical Basis of Respiration.* Little, Brown, Boston, 393 pp.

de Reuck, A. V. S., and R. Porter (eds.). *Development of the Lung.* Ciba Foundation Symposium, J. and A. Churchill, London, 408 pp.

Reynolds, S. R. M. 1956. The fetal and neonatal pulmonary vasculature in the guinea pig in relation to hemodynamic changes at birth. Amer. J. Anat., *98*:97–127.

Sachis, P. N., D. L. Armstrong, L. E. Becker, and A. C. Bryan. 1982. Myelination of the human vagus nerve from 24 weeks postconceptual age to adolescence. J. Neuropathol. Exp. Neurol., *41*:466–472.

Simer, P. H. 1952. Drainage of pleural lymphatics. Anat. Rec., *113*:269–283.

Sorokin, S., H. A. Padykula, and E. Herman. 1959. Comparative histochemical patterns in developing mammalian lungs. Devel. Biol., *1*:125–151.

Spencer, H., and D. Leof. 1964. The innervation of the human lung. J. Anat. (Lond.), *98*:599–609.

Stransky, A., M. Szereda-Przestaszewska, and J. G. Widdicombe. 1973. The effects of lung reflexes on laryngeal resistance and motoneurone discharge. J. Physiol. (Lond.), *231*:417–438.

Thomas, L. B., and E. A. Boyden. 1952. Agenesis of the right lung. Surgery, *31*:429–435.

Tobin, C. E. 1952. The bronchial arteries and their connections with other vessels in the human lung. Surg. Gynec. Obstet., *95*:741–750.

Tobin, C. E. 1952. Methods of preparing and studying human lungs expanded and dried with compressed air. Anat. Rec., *114*:453–465.

Tobin, C. E. 1954. Lymphatics of the pulmonary alveoli. Anat. Rec., *120*:625–635.

Tobin, C. E. 1957. Human pulmonic lymphatics. Anat. Rec., *127*:611–633.

Tobin, C. E. 1966. Arteriovenous shunts in the peripheral pulmonary circulation in the human lung. Thorax, *21*:197–204.

Tobin, C. E., and M. O. Zariquiey. 1950. Bronchopulmonary segments and blood supply of the human lung. Med. Radiogr. Photogr., *26*:38–45.

Trapnell, D. H. 1970. The anatomy of the lymphatics of the lungs and chest wall. Thorax, *25*:255–256.

Waaler, B. A. 1971. Physiology of the pulmonary circulation. Angiologica, *8*:266–284.

Wang, K. P., and H. P. Tai. 1965. An analysis of variations of the segmental vessels of the right lower lobe in 50 Chinese lungs. Acta Anat. Sin., *8*:408–423.

Wells, L. J. 1954. Development of the human diaphragm and pleural sacs. Carneg. Instn., Contr. Embryol., *35*:107–134.

Wells, L. J., and E. A. Boyden. 1954. The development of the bronchopulmonary segments in human embryos of horizons XVII to XIX. Amer. J. Anat., *95*:163–201.

Woodburne, R. T. 1947. The costomediastinal border of the left pleura in the precordial area. Anat. Rec., *97*:197–210.

Wunder, C. C., et al. 1969. Nature of pleural space of dogs. J. Appl. Physiol., *27*:637–643.

Yernault, J. C., A. de Troyer, and D. Rodenstein. 1979. Sex and age differences in intrathoracic airways mechanics in normal man. J. Appl. Physiol., *46*:556–564.

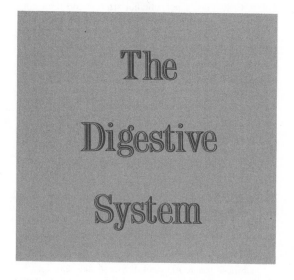

The Digestive System

16

lacteal vessels. Finally the small intestine ends in the **large intestine,** made up of **cecum, colon, rectum,** and **anal canal,** the last terminating on the surface of the body at the **anus.**

The accessory organs are the **teeth,** for purposes of mastication; the three pairs of **salivary glands**—the **parotid,** the **submandibular,** and the **sublingual**—the secretion from which mixes with the food in the mouth and converts it into a bolus and acts chemically on one of its constituents; the **liver** and **pancreas,** two large glands in the abdomen, the secretions of which, in addition to that of numerous minute glands in the walls of the alimentary canal, assist in the process of digestion.

Development of the Digestive System

The primitive digestive tube consists of two parts: (1) the **foregut,** within the cephalic flexure and dorsal to the heart; and (2) the **hindgut,** within the caudal flexure (Fig. 16-1).

The **digestive system,** which provides the apparatus for the digestion of food, consists of the digestive tube and certain accessory organs.

The **digestive tube** (*alimentary canal*) is a musculomembranous tube, about 9 meters long, that extends from the mouth to the anus; it is lined throughout its entire extent by mucous membrane. Different names have been applied to the various parts of its course: at its commencement is the **mouth,** where provision is made for the mechanical division of the food (*mastication*), and for its admixture with a fluid secreted by the salivary glands (*insalivation*). Beyond this are the **pharynx** and **esophagus,** which convey food into the **stomach,** where it is stored for a time and where the first stages of the digestive process take place. The stomach is followed by the **small intestine,** which is divided for purposes of description into three parts: the **duodenum,** the **jejunum,** and the **ileum.** In the small intestine the process of digestion is completed and the resulting products are absorbed into the blood and

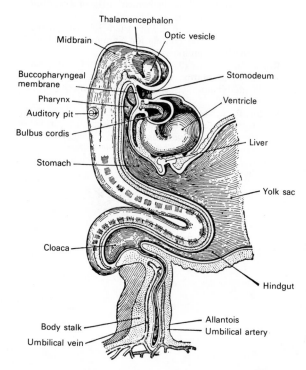

Fig. 16-1. Human embryo about 15 days old. Brain and heart represented from right side. Digestive tube and yolk sac in median section. (After His.)

Between these is the wide opening of the yolk sac, which is gradually narrowed and reduced to a small foramen leading into the vitelline duct. At first the foregut and hindgut end blindly. The anterior end of the foregut is separated from the stomodeum by the buccopharyngeal membrane (Fig. 16-1); the hindgut ends in the cloaca, which is closed by the cloacal membrane.

MOUTH AND ASSOCIATED STRUCTURES

MOUTH. The mouth is developed partly from the stomodeum and partly from the floor of the anterior portion of the foregut. By the growth of the head end of the embryo and the formation of the cephalic flexure, the pericardial area and the buccopharyngeal membrane come to lie on the ventral surface of the embryo. With the further expansion of the brain and the forward bulging of the pericardium, the buccopharyngeal membrane is depressed between these two prominences. This depression constitutes the **stomodeum** (Fig. 16-1), which is lined by ectoderm and is separated from the anterior end of the foregut by the buccopharyngeal membrane. This membrane is devoid of mesoderm, being formed by the apposition of the stomodal ectoderm with the foregut entoderm; at the end of the third week it disappears, and thus a communication is established between the mouth and the future pharynx. No trace of the membrane is found in the adult, and the communication just mentioned must not be confused with the permanent isthmus faucium. The lips, teeth, and gums are formed from the walls of the stomodeum, but the tongue develops in the floor of the pharynx.

The visceral arches extend ventrally between the stomodeum and the pericardium, and with the completion of the mandibular arch and the formation of the maxillary processes, the mouth assumes the appearance of a pentagonal orifice. The orifice is bounded cranially by the frontonasal process, caudally by the mandibular arch, and laterally by the maxillary processes (Fig. 16-2). With the inward growth and fusion of the palatine processes (Figs. 2-55; 2-56), the stomodeum is divided into an upper nasal and a lower buccal part. Along the free margins of the processes bounding the mouth cavity a shallow groove appears; this is

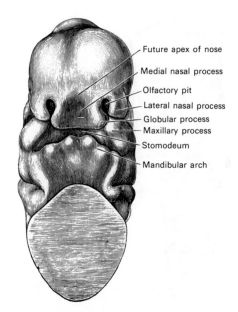

FIG. 16-2. Head end of human embryo about 30 to 31 days old. (From model by Peters.)

Labels: Future apex of nose, Medial nasal process, Olfactory pit, Lateral nasal process, Globular process, Maxillary process, Stomodeum, Mandibular arch

termed the **primary labial groove,** and from the bottom of it ectoderm grows downward into the underlying mesoderm. The central cells of the ectodermal downgrowth degenerate and a **secondary labial groove** forms, and as this deepens, the lips and cheeks separate from the alveolar processes of the maxillae and mandible.

SALIVARY GLANDS. The salivary glands arise as buds from the epithelial lining of the mouth. The parotid gland appears during the fourth week in the angle between the maxillary process and the mandibular arch. The submandibular gland appears in the sixth week, and the sublingual gland appears during the ninth week in the hollow between the tongue and the mandibular arch.

TONGUE. The tongue (Figs. 16-3 to 16-6) develops in the floor of the pharynx, and consists of an anterior or buccal part and a posterior or pharyngeal part, which are separated in the adult by the V-shaped sulcus terminalis. During the third week there appears, immediately dorsal to the ventral ends of the two halves of the mandibular arch, a rounded swelling named the **tuberculum impar,** which may help to form the buccal part of the tongue or may be purely a transitory structure. From the ventral ends of the fourth arch there arises a second and larger elevation, in the center of which is a median groove or furrow. This elevation was

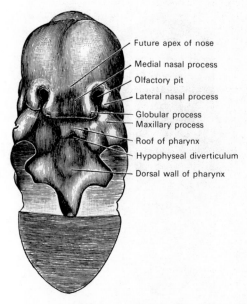

FIG. 16-3. Same embryo as shown in Figure 16-2, with front wall of pharynx removed.

named the **furcula** by His[1] and at first it is separated from the tuberculum impar by a depression, but later by a ridge, the **copula,** which is formed by the forward growth and fusion of the ventral ends of the second and third arches. The posterior or pharyngeal part of the tongue develops from the copula, which extends forward in the form of a V, to embrace between its two limbs the buccal part of the tongue. At the apex of the V, a pit-like invagination occurs to form the thyroid gland, and this depression is represented in the adult by the **foramen cecum** of the tongue. In the adult the union of the anterior and posterior parts of the tongue is

[1]Wilhelm His (1831–1904): A German anatomist and physiologist (Basel and Leipzig).

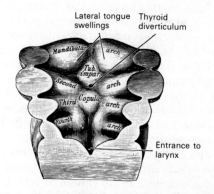

FIG. 16-4. Floor of pharynx of human embryo about 26 days old. (From model by Peters.)

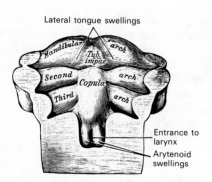

FIG. 16-5. Floor of pharynx of human embryo at the end of the fourth week. (From model by Peters.)

marked by the V-shaped **sulcus terminalis,** the apex of which is at the foramen cecum, while the two limbs run lateralward and forward, parallel to, but a little behind, the vallate papillae.

PALATINE TONSILS.   The palatine tonsils develop from the dorsal angles of the second branchial pouches. The entoderm that lines these pouches grows into the surrounding mesoderm in the form of solid buds. These buds become hollowed out by the degeneration and casting off of their central cells, forming the tonsillar crypts. Lymphoid cells accumulate around the crypts and become grouped to form the lymphoid follicles; the latter, however, are not well defined until after birth.

PHARYNX AND POSTPHARYNGEAL ALIMENTARY TUBE

The cranial part of the foregut becomes dilated to form the pharynx (Fig. 16-1), in relation to which the branchial arches are developed (see p. 45); the succeeding part re-

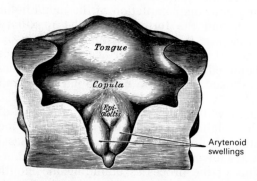

FIG. 16-6. Floor of pharynx of human embryo about 30 days old. (From model by Peters.)

mains tubular, and with the descent of the stomach is elongated to form the esophagus. About the fourth week a fusiform dilatation appears, which is the future stomach, and beyond this the gut opens freely into the yolk sac (Fig. 16-7, *A* and *B*). The opening is wide at first, but it gradually narrows into a tubular stalk, the **yolk stalk** or **vitelline duct.**

Between the stomach and the mouth of the yolk sac the liver diverticulum appears. From the stomach to the rectum the alimentary canal is attached to the notochord by a band of mesoderm, from which the common mesentery of the gut subsequently develops.

The stomach is also attached to the ventral abdominal wall as far as the umbilicus by

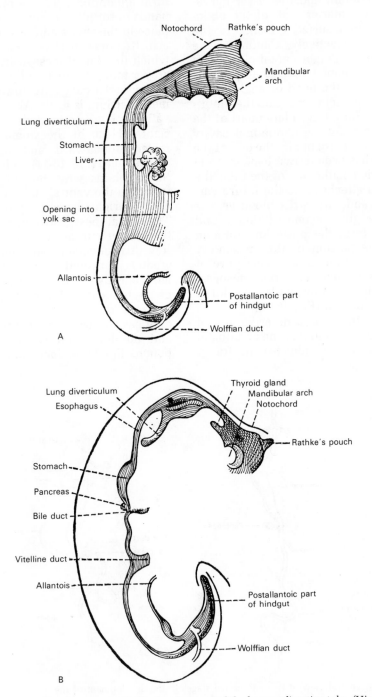

FIG. 16-7.   Sketches in profile of two stages in the development of the human digestive tube. (His.) *A*, 30 ×; *B*, 20 ×.

the septum transversum. The cranial portion of the septum takes part in the formation of the diaphragm, whereas the caudal portion, into which the liver grows, forms the **ventral mesogastrium** (Fig. 16-10). As the stomach undergoes further dilatation, its two curvatures become recognizable (Figs. 16-7, *B*; 16-8), the greater curvature oriented toward the vertebral column and the lesser curvature toward the anterior wall of the abdomen, with its two surfaces facing to the right and left respectively. Caudal to the stomach the gut undergoes great elongation and forms a V-shaped loop that projects ventralward; from the bend or angle of the loop the vitelline duct passes into the umbilicus (Fig. 16-9). For a time a large part of the loop extends beyond the abdominal cavity into the umbilical cord, but by the end of the third month it has been drawn back into the cavity. With the lengthening of the tube, the mesoderm that attaches it to the future vertebral column and carries the blood vessels to supply the gut, becomes thinner and drawn out to form the **posterior common mesentery.** The portion of this mesentery that is attached to the greater curvature of the stomach is named the **dorsal mesogastrium,** and the part that suspends the colon is termed the **mesocolon** (Fig. 16-10).

About the sixth week, a diverticulum of the gut appears just caudal to the opening of the vitelline duct and indicates the future cecum and vermiform appendix. The part of the loop on the distal side of the cecal diverticulum increases in diameter and forms the future ascending and transverse portions of the large intestine. Until the fifth month the cecal diverticulum has a uniform caliber, but from this time onward its distal part remains rudimentary and forms the vermiform appendix, while its proximal part expands to form the cecum. Changes also take place in the shape and position of the stomach. Its dorsal part or greater curvature, to which the dorsal mesogastrium is attached, grows much more rapidly than its ventral part or lesser curvature, to which the ventral mesogastrium is fixed. Also, the greater curvature swings toward the left, so that the right surface of the stomach becomes directed dorsalward and the left surface ventralward (Fig. 16-11), a change in position that explains why the left vagus nerve is found on the ventral surface of the stomach and the right vagus on the dorsal surface. The dorsal mesogastrium, being attached to the greater curvature, must necessarily follow this movement, and hence it becomes greatly elongated and drawn lateralward and ventralward from the vertebral column; as in the case of the stomach, the right surfaces of both the dorsal and ventral mesogastria are now directed dorsalward, and the left ventralward. In this way a pouch, the **lesser sac** (*omental bursa*), is

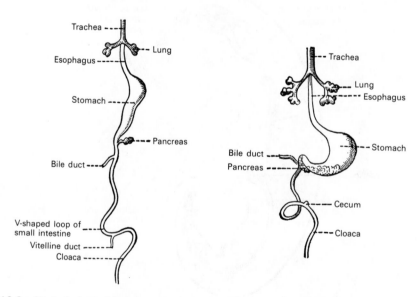

FIG. 16-8.   Ventral view of two successive stages in the development of the digestive tube. (His.)

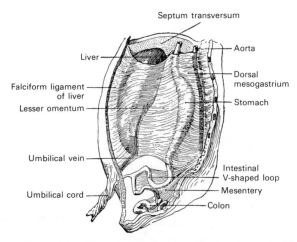

FIG. 16-9.   The primitive mesentery of a six-week-old human embryo, half schematic. (Kollmann.)

formed dorsal to the stomach. This increases in size as the digestive tube undergoes further development, and the entrance to the pouch becomes the future **epiploic foramen** or **foramen of Winslow.**[1]

The duodenum develops from the part of

[1]Jacob Benignus Winslow (1669–1760): A Danish anatomist who worked in Paris.

the tube that immediately succeeds the stomach; it undergoes little elongation, being more or less fixed in position by the liver and pancreas, which arise as diverticula from it. The duodenum is first suspended by a mesentery and projects ventralward in the form of a loop. The loop and its mesentery are subsequently displaced by the transverse colon, so that the right surface of the duodenal mesentery is directed dorsalward, and adhering to the parietal peritoneum, is lost. The remainder of the digestive tube becomes greatly elongated, and as a consequence the tube is coiled on itself. This elongation demands a corresponding increase in the width of the intestinal attachment of the mesentery, which becomes folded.

At this stage the small and large intestines are attached to the vertebral column by a continuous common mesentery, with the coils of the small intestine falling to the right of the midline and the large intestine falling to the left (Fig. 16-11).

Sometimes this condition persists throughout life, and it is then found that the duodenum does not cross from the right to the left side of the vertebral column, but lies entirely on the right side of the

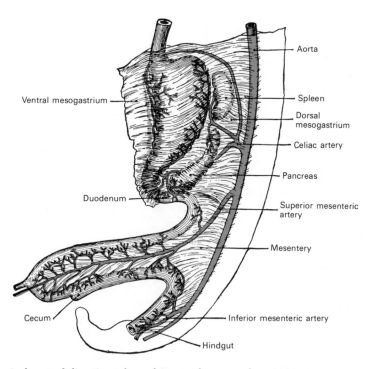

FIG. 16-10.   Abdominal part of digestive tube and its attachment to the primitive or common mesentery. Human embryo of six weeks. (After Toldt.)

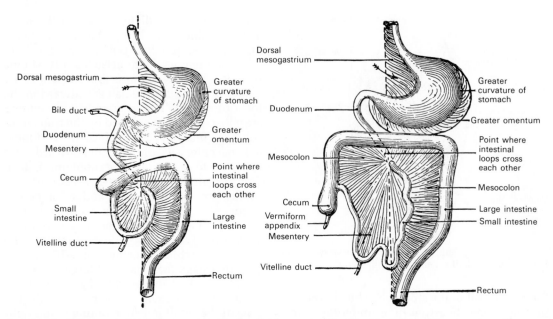

Fig. 16-11. Diagrams to illustrate two stages in the development of the digestive tube and its mesentery. The arrow indicates the entrance to the bursa omentalis. The ventral mesogastrium has been eliminated.

median plane, where it is continued into the jejunum; the arteries to the small intestine also arise from the right instead of the left side of the superior mesenteric artery.

The gut is now rotated counterclockwise upon itself, so that the large intestine is carried ventral to the small intestine, and the cecum is placed immediately caudal to the liver. About the sixth month the cecum descends into the right iliac fossa, and the large intestine forms an arch consisting of the ascending, transverse, and descending portions of the colon; the transverse portion crosses ventral to the duodenum and lies just caudal to the greater curvature of the stomach, the coils of the small intestine being disposed within this arch (Fig. 16-12). Sometimes the caudalward progress of the cecum is arrested, so that in the adult it may lie immediately caudal to the liver instead of in the right iliac region.

Further changes that take place in the lesser sac (omental bursa) and in the common mesentery give rise to the peritoneal relations seen in the adult. The omental bursa, which at first reaches only as far as the greater curvature of the stomach, grows caudalward to form the greater omentum; this extension lies ventral to the transverse colon and the coils of the small intestine (Fig. 16-13). Before the pleuroperitoneal opening is closed, the omental bursa sends a diverticulum cranialward on either side of the esophagus; the left diverticulum soon disappears, but the right diverticulum becomes constricted

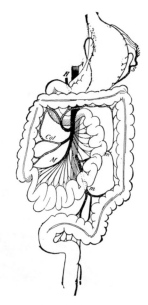

Fig. 16-12. Final disposition of the intestines and their vascular relations. (Jonnesco.) *A,* Aorta. *H,* Hepatic artery. *M, Col.,* Branches of superior mesenteric artery. *m, m',* Branches of inferior mesenteric artery. *S,* Splenic artery.

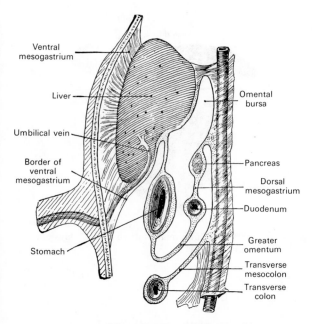

FIG. 16-13. Schematic figure of the omental bursa, etc. Human embryo of eight weeks. (Kollmann.)

and persists in most adults as a small sac lying within the thorax on the right side of the caudal end of the esophagus.

The ventral layer of the transverse mesocolon is at first distinct from the dorsal layer of the greater omentum, but ultimately the two blend, and hence the greater omentum appears to be attached to the transverse colon (Fig. 16-14). The mesenteries of the ascending and descending parts of the colon disappear in the majority of cases, while that of the small intestine assumes the oblique

attachment characteristic of its adult condition.

The lesser omentum is formed, as indicated previously, by the mesoderm or **ventral mesogastrium,** which attaches the stomach and duodenum to the anterior abdominal wall. The subsequent growth of the liver separates this leaf into two parts: the lesser omentum between the stomach and liver, and the falciform and coronary ligaments between the liver and the abdominal wall and diaphragm (Fig. 16-13).

RECTUM AND ANAL CANAL

The hindgut is at first prolonged caudalward into the body stalk as the tube of the allantois. With the growth and flexure of the caudal end of the embryo, the body stalk, with its allantoic tube, is carried cranialward to the ventral aspect of the body, and consequently a bend is formed at the junction of the hindgut and allantois. This bend becomes dilated into a pouch that constitutes the **entodermal cloaca;** the hindgut opens into its dorsal part and the allantois extends out ventrally from its ventral part. At a later stage the mesonephric (Wolffian)[1] and paramesonephric (Müllerian)[2] ducts open into its ventral portion. The cloaca is, for a time,

[1]Kaspar Friedrich Wolff (1733–1794): A Russian anatomist and physiologist (St. Petersburg).
[2]Johannes Peter Müller (1801–1858): A German anatomist and physiologist (Berlin).

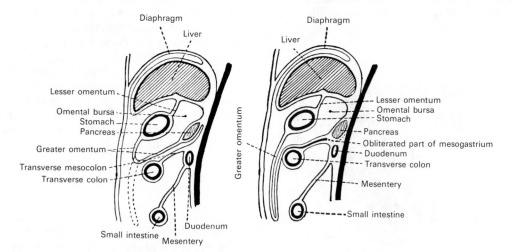

FIG. 16-14. Diagrams to illustrate the development of the greater omentum and transverse mesocolon.

shut off from the exterior by the **cloacal membrane,** which is formed by the apposition of the ectoderm and entoderm and reaches, at first, as far cranialward as the future umbilicus. The mesoderm subsequently progresses caudalward to form the lower part of the abdominal wall and symphysis pubis. By the growth of the surrounding tissues, the cloacal membrane comes to lie at the bottom of a depression, which is lined by ectoderm and is named the **ectodermal cloaca** (Fig. 16-15).

The entodermal cloaca is divided into a dorsal and a ventral part by means of a partition, the **urorectal septum** (Fig. 16-16), which grows caudalward from the ridge that separates the allantoic from the cloacal opening of the intestine; ultimately, this septum fuses with the cloacal membrane and divides it into an anal and a urogenital part. The dorsal part of the cloaca forms the rectum, and the ventral part forms the urogenital sinus and bladder. For a time a communication named the **cloacal duct** exists between the two parts of the cloaca below the urorectal septum; this duct occasionally persists as a passage between the rectum and urethra.

The anal canal is formed by an invagination of the ectoderm behind the urorectal septum. This invagination is termed the **proctodeum,** and it meets with the entoderm of the hindgut and forms the **anal membrane.** By the absorption of this membrane, the anal canal becomes continuous with the rectum (Fig. 17-12). A small part of the hindgut projects caudalward beyond the anal membrane; it is named the **postanal gut** (Fig. 16-15), and it usually becomes obliterated.

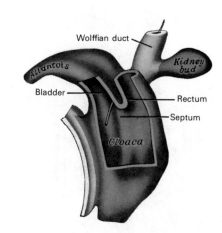

FIG. 16-16.   Cloaca of human embryo 25 to 27 days old. (From model by Keibel.)

The Mouth

The **cavity of the mouth** (*oral or buccal cavity*) (Fig. 16-17) is placed at the commencement of the digestive tube; it is a nearly oval shaped cavity that consists of two parts: an outer, smaller portion, the **vestibule,** and an inner, larger part, the **mouth cavity proper.**

VESTIBULE

The vestibule (*vestibulum oris*) is a slitlike space, bounded externally by the lips and cheeks, internally by the gums and teeth. It communicates with the surface of the body by the **rima** or **orifice of the mouth.** Superiorly and inferiorly, it is limited by the reflection of the mucous membrane from the lips and cheeks to the gums covering the upper and lower alveolar arches respectively. It receives the secretion from the parotid salivary glands and communicates, when the jaws are closed, with the mouth cavity proper by an aperture on either side behind the molar teeth and by narrow clefts between opposing teeth.

MOUTH CAVITY PROPER

The mouth cavity proper (Fig. 16-17) is bounded laterally and ventrally by the alveolar arches with their contained teeth; dorsally, it communicates with the pharynx by a constricted aperture termed the **isthmus**

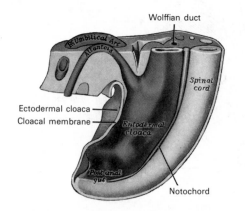

FIG. 16-15.   Tail end of human embryo 15 to 18 days old. (From model by Keibel.)

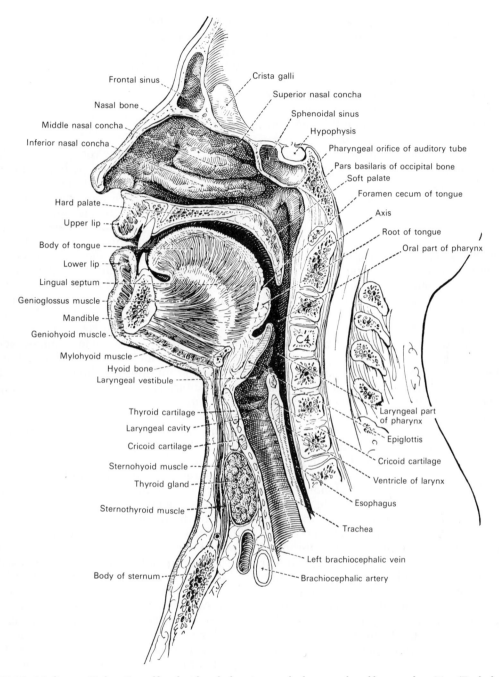

FIG. 16-17. Median sagittal section of head and neck showing nasal, pharyngeal, and laryngeal cavities. (Eycleshymer and Jones.)

faucium. It is roofed in by the hard and soft palates, while the greater part of the floor is formed by the tongue, the remainder by the reflection of the mucous membrane from the sides and under surface of the tongue to the gum lining the inner aspect of the mandible. It receives the secretions from the submandibular and sublingual salivary glands.

LIPS

The lips (*labia oris*) (Fig. 16-17), the two fleshy folds that surround the rima or orifice of the mouth, are covered externally by integument and internally by mucous membrane, between which are found the orbicularis oris muscle, the labial vessels, some nerves, areolar tissue, fat, and numerous small labial glands. The inner surface of each lip is connected in the midline to the corresponding gum by a median fold of mucous membrane, the **frenulum.**

The **labial glands** are situated between the mucous membrane and the orbicularis oris, around the orifice of the mouth. They are globular and about the size of small peas; their ducts open by minute orifices upon the mucous membrane. In structure they resemble the salivary glands.

CHEEKS

The cheeks (*buccae*) form the sides of the face and are continuous anteriorly with the lips. They are composed externally of integument and internally of mucous membrane; between the two are a muscular stratum, a large quantity of fat (*corpus adiposum buccae*), areolar tissue, vessels, nerves, and buccal glands.

The **buccal glands** are placed betweeen the mucous membrane and buccinator muscle; they are similar in structure to the labial glands but smaller. About five glands, larger than the rest, are placed between the masseter and buccinator muscles around the distal extremity of the parotid duct; their ducts open in the mouth opposite the last molar tooth. They are called **molar glands.**

STRUCTURE OF THE CHEEKS. The **mucous membrane** lining the cheek is reflected above and below upon the gums, and is continuous behind with the lining membrane of the soft palate. Opposite the second molar tooth of the maxilla is a papilla, on the summit of which is the aperture of the parotid duct (*papilla parotidea*). The principal muscle of the cheek is the buccinator, but other muscles enter into its formation, viz., the zygomaticus, risorius, and the platysma muscles.

GUMS

The gums (*gingivae*) are composed of dense fibrous tissue that is closely connected to the periosteum of the alveolar processes and surrounds the necks of the teeth. They are covered by a smooth and vascular mucous membrane, which is remarkable for its limited sensibility. Around the necks of the teeth this membrane presents numerous fine papillae and is reflected into the alveoli, where it is continuous with the periosteal membrane lining these cavities.

PALATE

The palate forms the roof of the mouth. It consists of two portions, the **hard palate** anteriorly and the **soft palate** posteriorly.

HARD PALATE. The hard palate (*palatum durum*) (Fig. 16-17) forms the roof of the mouth and separates the oral and nasal cavities. It is bounded anteriorly and at the sides by the alveolar arches and gums; posteriorly, it is continuous with the soft palate. Its bony support (*palatum osseum*), formed by the palatine process of the maxilla and the horizontal part of the palatine bone (Fig. 16-19), is covered by a dense structure formed by the periosteum and mucous membrane of the mouth, which are intimately adherent. Along the midline is a linear raphe, which ends anteriorly in a small papilla (*papilla incisiva*) that corresponds with the incisive canal. On either side and anterior to the raphe, the mucous membrane is thick, pale, and corrugated; posteriorly, it is thin, smooth, and of a deeper color. It is covered with stratified squamous epithelium and furnished with numerous palatal glands that lie between the mucous membrane and the surface of the bone.

SOFT PALATE. The soft palate (*palatum molle*) (Fig. 16-17) is suspended from the posterior border of the hard palate. It consists of a fold of mucous membrane that encloses muscular fibers, an aponeurosis, vessels, nerves, lymphoid tissue, and mucous glands. When elevated, as in swallowing and in sucking, it completely separates the nasal cavity and nasopharynx from the posterior part of the oral cavity and the oral portion of the pharynx (Fig. 16-17). When occupying its usual position, i.e., relaxed and pendent, its anterior surface is concave, continuous with the roof of the mouth, and marked by a median raphe. Its posterior surface is convex and continuous with the mucous membrane that covers the floor of the nasal cavities. It is attached to the posterior margin of the hard palate, and its sides are blended with the pharynx. Its posterior border, termed the

palatine velum (*velum palatinum*), is free and hangs like a curtain between the mouth and pharynx.

Hanging from the middle of its posterior border is a small, conical, pendulous process, the **palatine uvula.** Arching lateralward from the base of the uvula on either side are two curved folds of mucous membrane that contain muscular fibers and are called the **arches** or **pillars of the fauces** (Fig. 16-42).

The Teeth

16-24) appear at different periods of life. Those of the first set appear in infancy and are called the **deciduous** or **milk teeth.** Those of the second set appear in childhood, persist until old age, and are named **permanent teeth.**

The **deciduous teeth** number 20: four incisors, two canines, and four molars in each jaw. The **permanent teeth** number 32: four incisors, two canines, four premolars, and six molars in each jaw. The **dental formulae** may be represented as shown in Table 16-1.

TABLE 16-1. Dental Formulae

Deciduous Teeth

	mol.	can.	in.	in.	can.	mol.	
Upper jaw ...	2	1	2	2	1	2	} Total 20
Lower jaw ...	2	1	2	2	1	2	

Permanent Teeth

	pr.						pr.	
	mol.	mol.	can.	in.	in.	can.	mol.	mol.	
Upper jaw	3	2	1	2	2	1	2	3	} Total 32
Lower jaw	3	2	1	2	2	1	2	3	

GENERAL CHARACTERISTICS

Each tooth consists of three portions: the **crown,** projecting from the gum; the **root,** embedded in the alveolus; and the **neck,** the constricted portion between the crown and root (Fig. 16-24).

The **roots** of the teeth are firmly implanted in depressions within the alveoli; these depressions are lined with periosteum, which invests each tooth as far as the neck. At the margins of the alveoli, the periosteum is con-

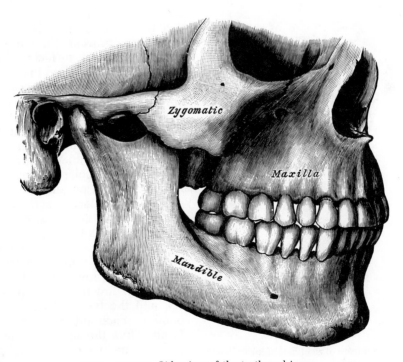

Fɪɢ. 16-18. Side view of the teeth and jaws.

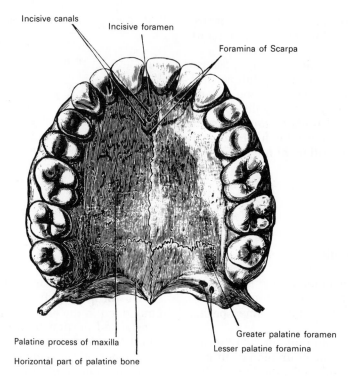

Incisive canals

Incisive foramen

Foramina of Scarpa

Palatine process of maxilla

Horizontal part of palatine bone

Greater palatine foramen

Lesser palatine foramina

Fig. 16-19. Permanent teeth of upper dental arch, seen from below.

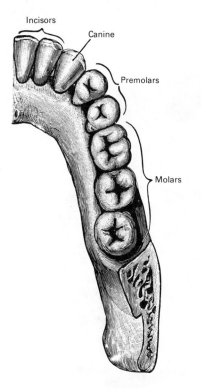

Incisors

Canine

Premolars

Molars

Fig. 16-20. Permanent teeth of right half of lower dental arch, seen from above.

tinuous with the fibrous structure of the gums.

Because of the curve of the dental arch, terms such as anterior and posterior, as applied to the teeth, are misleading and confusing. Special terms are therefore used to indicate the different surfaces of a tooth: the surface directed toward the lips or cheek is known as the **labial** or **buccal surface;** that directed toward the tongue is described as the **lingual surface;** those that touch neighboring teeth are termed **surfaces of contact.** In the case of the incisor and canine teeth, the surfaces of contact are medial and lateral; in the premolar and molar teeth they are anterior and posterior.

The superior dental arch is larger than the inferior, so that in the normal condition the teeth in the maxillae slightly overlap those of the mandible both anteriorly and at the sides. Since the upper central incisors are wider than the lower, the other teeth in the upper arch are thrown somewhat posteriorly, and the two sets do not quite correspond to each other when the mouth is closed. Thus the upper canine tooth rests partly on the lower canine and partly on the first premolar, and the cusps of the upper

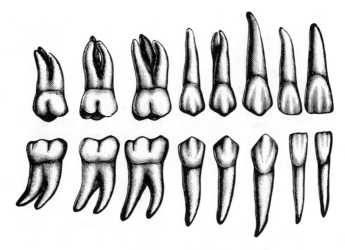

Fɪɢ. 16-21.   Permanent teeth. Right side. (Burchard.)

molar teeth lie behind the corresponding cusps of the lower molar teeth. The two series, however, end at nearly the same point posteriorly, mainly because the molars in the upper arch are smaller than those in the lower arch.

PERMANENT TEETH (FIGS. 16-20; 16-21)

INCISORS.   The incisors are so named for their sharp cutting edge, which is adapted for biting food. There are eight, and they form the four front teeth in each dental arch.

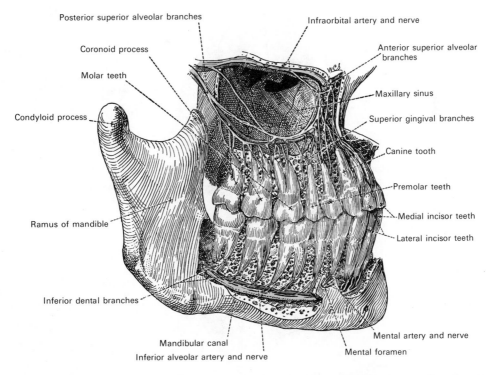

Fɪɢ. 16-22.   The permanent teeth, viewed from the right. The external layer of bone has been partly removed and the maxillary sinus has been opened to show the blood and nerve supply to the teeth. (Eycleshymer and Jones.)

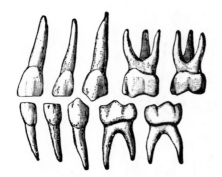

Fig. 16-23.   Deciduous teeth. Left side.

The **crown** of each incisor is directed vertically and is chisel-shaped, being bevelled at the expense of its lingual surface, so that it has a sharp horizontal cutting edge, which, before being subjected to attrition, presents three small prominent points separated by two slight notches. It is convex, smooth, and highly polished on its labial surface and concave on its lingual surface, where, in the teeth of the upper arch, it is frequently marked by an inverted V-shaped eminence situated near the gum. This is known as the **basal ridge** or **cingulum.** The **neck** is constricted. The **root** is long, single, conical, transversely flattened, thicker anteriorly, and slightly grooved on either side in the longitudinal direction.

The **upper incisors** are larger and stronger than the lower and are directed obliquely downward and forward. The central incisors are larger than the lateral incisors and their roots are more rounded.

The **lower incisors** are smaller than the upper; the central ones are smaller than the lateral and are the smallest of all the incisors. They are placed vertically and are somewhat bevelled anteriorly, where they

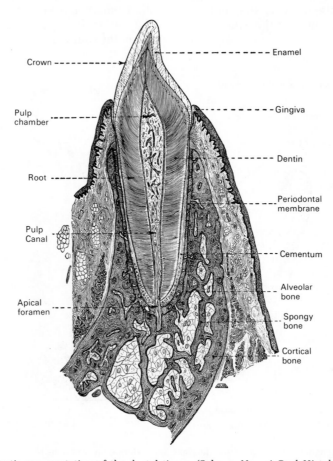

Fig. 16-24.   Diagrammatic representation of the dental tissue. (Schour, *Noyes' Oral Histology and Embryology.*)

have been worn down by contact with the overlapping edge of the upper teeth. The cingulum is absent.

CANINE TEETH. There are four canine teeth: two in the upper and two in the lower arch; one is placed lateral to each lateral incisor. They are larger and stronger than the incisors. Their roots sink deeply into the bones and cause well-marked prominences upon the surface of the alveolar arch.

The **crown** is large and conical, very convex on its labial surface, a little hollowed and uneven on its lingual surface, and tapering to a blunted point or cusp that projects beyond the level of the other teeth. The **root** is single but longer and thicker than that of the incisors. It is conical, compressed laterally, and marked by a slight groove on each side.

The **upper canine teeth** (popularly called *eye teeth*) are larger and longer than the lower, and usually present a distinct basal ridge.

The **lower canine teeth** (popularly called *stomach teeth*) are placed nearer the midline than the upper canines, so their summits correspond to the intervals between the upper canines and the lateral incisors.

PREMOLARS. There are eight premolar or **bicuspid** teeth: four in each arch. They are situated lateral and posterior to the canine teeth, and are smaller and shorter than the canine teeth.

The **crown** is compressed anteroposteriorly and is surmounted by two pyramidal eminences or cusps, a labial and a lingual, separated by a groove; hence their name **bicuspid.** Of the two cusps the labial is the larger and more prominent. The **neck** is oval. The **root** is generally single, compressed, and presents a deep groove anteriorly and posteriorly, which indicates a tendency in the root to become double. The apex is generally bifid.

The **upper premolars** are larger and have a greater tendency to form divided roots than the lower premolars; this is especially the case for the first upper premolar.

MOLARS. The molar teeth are the largest of the permanent set, and their broad crowns are adapted for grinding and crushing food. There are 12 molars: six in each arch, three being placed posterior to the second premolars on each side.

The **crown** of each is nearly cubical, convex on its buccal and lingual surfaces, and flat on its surfaces of contact; it is surmounted by four or five tubercles or cusps, which are separated from each other by a cruciate depression. Hence the molars are sometimes called **multicuspids.** The **neck** is distinct, large, and round.

As a rule, the first of the **upper molars** is the largest and the third is the smallest. The crown of the first has usually four tubercles; that of the second, three or four; that of the third, three. Each upper molar has three roots, and of these two are buccal and nearly parallel to each other; the third is lingual and diverges from the others. The roots of the third molar (*dens serotinus* or *wisdom tooth*) are more or less fused together.

The **lower molars** are larger than the upper. The crown of the first usually has five tubercles; those of the second and third have four or five. Each lower molar has two roots: an anterior, nearly vertical, and a posterior, directed obliquely backward. Both roots are grooved longitudinally, indicating a tendency to divide. The two roots of the third molar (*dens serotinus* or *wisdom tooth*) are more or less united.

DECIDUOUS TEETH

The deciduous teeth (*temporary* or *milk teeth*) (Fig. 16-23) are smaller than, but generally resemble in form, the permanent teeth that bear the same names. The posterior of the two molars is the largest of all the deciduous teeth, and is succeeded by the first molar. The first upper molar has only three cusps—two labial, one lingual; the second upper molar has four cusps. The first lower molar has four cusps; the second lower molar has five. The roots of the deciduous molars are smaller and more divergent than those of the permanent molars, but in other respects bear a strong resemblance to them.

STRUCTURE OF THE TEETH

In a vertical section of a tooth (Figs. 16-24 to 16-26), the **pulp cavity** is seen in the interior of the crown and the center of each root; it opens by a minute orifice at the extremity of the latter. It contains the **dental pulp,** a loose connective tissue richly supplied with

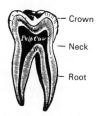

FIG. 16-25. Vertical section of a molar tooth.

vessels and nerves, which enter the cavity through the small aperture, the **apical foramen,** at the point of each root. Some of the cells of the pulp are arranged as a layer on the wall of the pulp cavity; these are named the **odontoblasts,** and during the development of the tooth, they are columnar, but later, after the dentin is fully formed, they become flattened and resemble osteoblasts. Each has two fine processes; the outer one passes into a dental canaliculus, the inner one is continuous with the processes of the connective tissue cells of the pulp matrix.

The solid portion of the tooth consists of (1) the **ivory** or **dentin,** which forms the bulk of the tooth; (2) the **enamel,** which covers the exposed part of the crown; and (3) a thin layer of bone, the **cement** or **crusta petrosa,** which is disposed on the surface of the root.

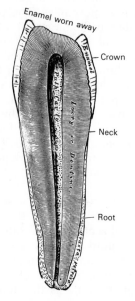

FIG. 16-26. Vertical section of a premolar tooth. (Magnified.)

Dentin

The **dentin** (*dentinum; substantia eburnea; ivory*) (Fig. 16-24) forms the principal mass of a tooth. It is a modification of osseous tissue, from which it differs, however, in structure. Microscopically it consists of a number of minute wavy and branching tubes, the **dental canaliculi,** embedded in a dense homogeneous substance, the **matrix.**

DENTAL CANALICULI. The dental canaliculi (*dentinal tubules*) (Fig. 16-27) are placed parallel with one another and open at their inner ends into the pulp cavity. In their course to the periphery they have two or three curves and twist spirally on themselves. These canaliculi vary in direction: thus in a tooth of the mandible they are vertical in the upper portion of the crown, becoming oblique and then horizontal in the neck and upper part of the root, while toward the lower part of the root they are inclined downward. In their course they divide and subdivide dichotomously, and especially in the root, give off minute branches that join together in loops in the matrix or end blindly. Near the periphery of the dentin, the finer ramifications of the canaliculi terminate imperceptibly by free ends. The dental canaliculi have definite walls, consisting of an elastic homogeneous membrane, the **dentinal sheath** (*of Neumann*[1]), which resists the action of acids; they contain slender cylindrical prolongations of the odontoblasts, the **dentinal fibers** (*Tomes' fibers*[2]).

MATRIX. The matrix (*intertubular dentin*) is translucent and contains the chief part of the inorganic matter of the dentin. In it are a number of fine fibrils, which are continuous with the fibrils of the dental pulp. After the organic matter has been removed by steeping a tooth in weak acid, the remaining organic matter may be torn into laminae that run parallel with the pulp cavity across the direction of the tubes.

A section of dry dentin often displays a series of somewhat parallel lines— the **incremental lines** (*of Salter*[3]). These lines are

[1]Ernst N. Neumann (1834–1918): A German pathologist (Königsberg).
[2]Sir John Tomes (1815–1895): An English dental surgeon (London).
[3]Sir Samuel A. Salter (1825–1897): An English dental surgeon and physician (London).

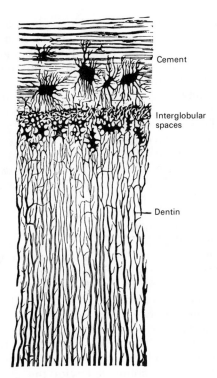

Cement

Interglobular spaces

Dentin

FIG. 16-27. Transverse section of a portion of the root of a canine tooth. 300 ×.

attrition) and becomes thinner toward the neck.

It consists of minute hexagonal rods or columns termed **enamel fibers** or **enamel prisms** (*prismata adamantina*). They lie parallel with one another, resting by one extremity upon the dentin, which has a number of minute depressions for their reception, and forming the free surface of the crown by the other extremity. The columns are directed vertically on the summit of the crown and horizontally at the sides; they are about 4 μ in diameter, and pursue a more or less wavy course. Each column is a six-sided prism and has numerous dark transverse shadings; these shadings are probably due to the manner in which the columns are developed in successive stages, producing shallow constrictions (as will be subsequently explained).

Another series of lines, which appear brown, the **parallel striae** or **colored lines** (*of Retzius*[2]), is seen on section. Some believe that these striae are produced by air in the interprismatic spaces; others believe that they are the result of true pigmentation. Numerous minute interstices appear among the enamel fibers near their dentinal ends, a provision calculated to allow the permeation of fluids from the dental canaliculi into the substance of the enamel.

composed of imperfectly calcified dentin arranged in layers. Because of the imperfection in the calcifying process, little irregular cavities are left, termed **interglobular spaces** (Fig. 16-27). Normally a series of these spaces is found toward the outer surface of the dentin, where they form a layer that is sometimes known as the **granular layer.** The name of these spaces was derived from the fact that they are surrounded by minute nodules or globules of dentin.

Other curved lines may be seen parallel to the surface. These are the *lines of Schreger*[1] and are caused by the optical effect of simultaneous curvature of the dentinal fibers.

Enamel

The enamel (*substantia adamantina*) is the hardest and most compact part of the tooth; it forms a thin crust over the exposed part of the crown as far as the commencement of the root. It is thickest on the grinding surface of the crown (until worn away by

Cement

The cement or **crusta petrosa** (*substantia ossea*) is disposed as a thin layer on the roots of the teeth from the termination of the enamel to the apex of each root, where it is usually very thick. In structure and chemical composition it resembles bone. It contains, sparingly, the lacunae and canaliculi that characterize true bone. The lacunae placed near the surface receive the canaliculi radiating from the sides of the lacunae toward the periodontal membrane, and those more deeply placed join with the adjacent dental canaliculi. In the thicker portions of the cement, the lamellae and Haversian canals[3] peculiar to bone are also found.

As age advances, the cement increases in thickness and gives rise to those bony

[1]Bernhard Nathaniel von Schreger (1766–1825): A German anatomist and surgeon (Altdorf).

[2]Magnus Gustav Retzius (1842–1919): A Swedish anatomist (Stockholm).

[3]Clopton Havers (1657–1702): An English physician (London).

growths or exostoses so common on the teeth of the aged; the pulp cavity also becomes partially filled with a hard substance that is intermediate in structure between dentin and bone (*osteodentin; secondary dentin*). It appears to be formed by a slow conversion of the dental pulp, which shrinks or even disappears.

DEVELOPMENT OF THE TEETH

In describing the development of the teeth, the mode of formation of the deciduous teeth must first be considered, and then that of the permanent series (Figs. 16-28 to 16-31).

Development of the Deciduous Teeth

The development of the deciduous teeth begins about the sixth week of fetal life as a thickening of the epithelium along the line of the future jaw, the thickening being due to a rapid multiplication of the more deeply situated epithelial cells. As the cells multiply they extend into the subjacent mesoderm, and thus form a ridge or strand of cells embedded in mesoderm.

About the seventh week a longitudinal splitting or cleavage of this strand of cells takes place, and it becomes divided into two strands; the separation begins anteriorly and extends laterally, the process occupying four or five weeks. Of the two strands thus formed, the **labial** forms the **labiodental lamina,** while the other, the **lingual,** is the ridge of cells in which the teeth, both deciduous

and permanent, develop. Hence it is known as the **dental lamina** or **common dental germ.** It forms a flat band of cells that grows into the substance of the embryonic jaw, at first horizontally inward, and then, as the teeth develop, vertically, i.e., upward in the upper jaw and downward in the lower jaw. While still maintaining a horizontal direction it has two edges—an *attached edge,* continuous with the epithelium lining the mouth, and a *free edge,* projecting inward, and embedded in the mesodermal tissue of the embryonic jaw. Along its line of attachment to the buccal epithelium is a shallow groove, the **dental furrow.**

About the ninth week, the dental lamina begins to develop ten enlargements along its free border in each jaw, and each corresponds to a future deciduous tooth. These enlargements consist of masses of epithelial cells, and the cells of the deeper part—that is, the part farthest from the margin of the jaw—increase rapidly and spread out in all directions. Each mass thus comes to assume a club shape, connected with the general epithelial lining of the mouth by a narrow neck and embraced by mesoderm. They are now known as **special dental germs** (Fig. 16-28). After a time the lower expanded portion inclines outward, forming an angle with the superficial constricted portion, which is sometimes known as the neck of the special dental germ.

About the tenth week the mesodermal tissue deep to these special dental germs becomes differentiated into papillae; these grow upward, and come in contact with the

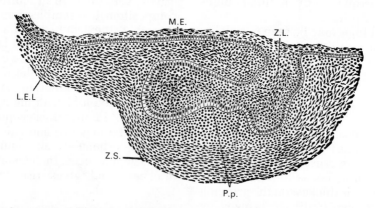

Fig. 16-28.   Sagittal section through the first lower deciduous molar of a human embryo 30-mm long. (Röse.) 100 ×.
L.E.L., Labiodental lamina, here separated from the dental lamina. *Z.L.,* Placed over the shallow dental furrow, points to the dental lamina, which is spread out below to form the enamel germ of the future tooth. *P.p.,* Bicuspid papilla, capped by the enamel germ. *Z.S.,* Condensed tissue forming dental sac. *M.E.,* Mouth epithelium.

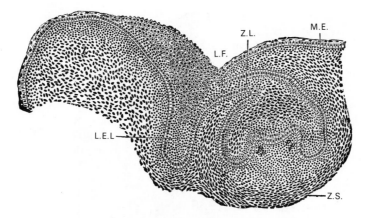

FIG. 16-29. Similar section through the canine tooth of an embryo 40-mm long. (Röse.) 100 ×. *L.F.*, Labiodental furrow. (Other abbreviations as in Fig. 16-28.)

epithelial cells of the special dental germs, which fold over them like a hood or cap. There is, then, at this stage a papilla (or papillae) that has already begun to assume somewhat the shape of the crown of the future tooth, and from which the dentin and pulp of the tooth are formed, surmounted by

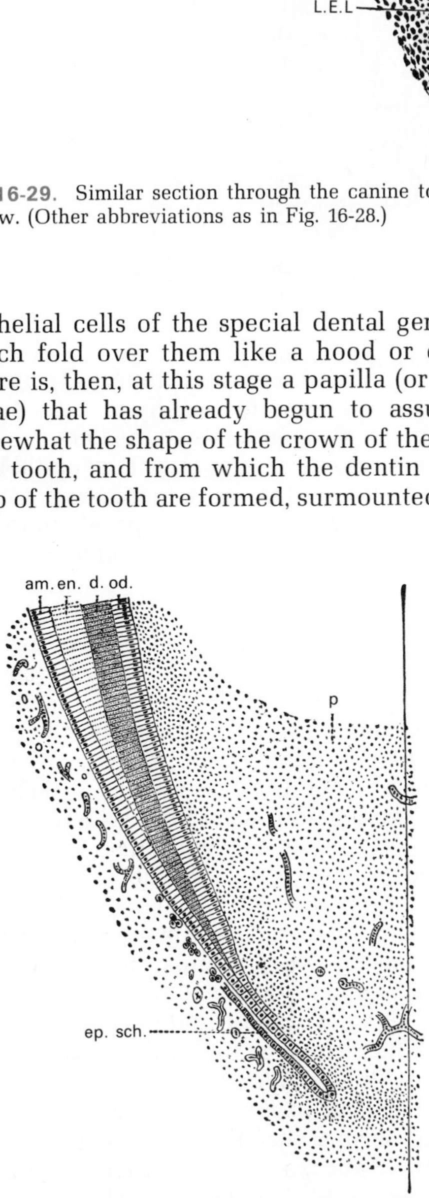

FIG. 16-30. Longitudinal section of the lower part of a growing tooth, showing the extension of the layer of ameloblasts beyond the crown to mark off the limit of formation of the dentin of the root. (Röse.) *am.*, Ameloblasts, continuous below with *ep. sch.*, the epithelial sheath of Hertwig. *d.*, Dentin. *en.*, Enamel. *od.*, Odontoblasts. *p.*, Pulp.

a dome or cap of epithelial cells from which the enamel is derived.

In the meantime, while these changes have been going on, the dental lamina has been extending posteriorly behind the special dental germ corresponding to the second deciduous molar tooth, and at about the seventeenth week it presents an enlargement, the special dental germ for the first permanent molar, soon followed by the formation of a papilla in the mesodermal tissue for the same tooth. This is followed, about the sixth month after birth, by a further extension posteriorly of the dental lamina, with the formation of another enlargement and its corresponding papilla for the second molar. And finally the process is repeated for the third molar, its papilla appearing about the fifth year of life.

After the formation of the special dental germs, the dental lamina undergoes atrophic changes and becomes cribriform, except on the lingual and lateral aspects of each of the special germs of the temporary teeth, where it undergoes a local thickening to form the special dental germ of each of the successional permanent teeth—i.e., the ten anterior ones in each jaw. Here the same process goes on as has been described in connection with those of the deciduous teeth: that is, the special dental germs recede into the substance of the gum behind the germs of the deciduous teeth. As they recede they become club-shaped, form expansions at their distal extremities, and finally meet papillae, which have been formed in the mesoderm, in the same manner as was the case in the decidu-

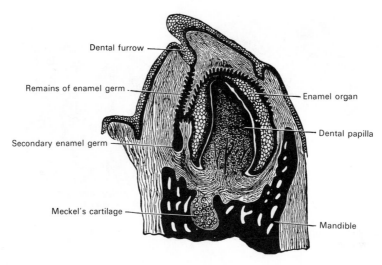

Dental furrow

Remains of enamel germ

Enamel organ

Dental papilla

Secondary enamel germ

Meckel's cartilage

Mandible

FIG. 16-31. Vertical section of the mandible of an early human fetus. 25 ×.

ous teeth. The apex of each papilla indents the dental germ, which encloses it, and forming a cap for it, becomes converted into the enamel, while the papilla forms the dentin and pulp of the permanent tooth.

The **special dental germs** consist at first of rounded or polyhedral epithelial cells; after the formation of the papillae, these cells undergo a differentiation into three layers. Those in immediate contact with the papilla become elongated and form a layer of well-marked columnar epithelium that coats the papilla. They are the cells that form the enamel fibers, and are therefore termed **enamel cells** or **ameloblasts** (Fig. 16-30). The cells of the outer layer of the special dental germ, which are in contact with the inner surface of the dental sac (presently to be described), are much shorter, cubical, and are named the **external enamel epithelium.** All the intermediate round cells of the dental germ between these two layers undergo a peculiar change. They become stellate and develop processes that unite to form a network into which fluid is secreted; this has the appearance of a jelly, and to it the name of enamel pulp is given. This transformed special dental germ is now known by the name of **enamel organ** (Fig. 16-31).

While these changes are occurring, a sac is formed around each enamel organ from the surrounding mesodermal tissue. This is known as the **dental sac,** and it is a vascular membrane of connective tissue. It grows up from below, and thus encloses the whole

tooth germ. As it grows it causes the neck of the enamel organ to atrophy and disappear, so that all communication between the enamel organ and the superficial epithelium is cut off. At this stage there are vascular papillae surmounted by caps of epithelial cells, the whole being surrounded by membranous sacs.

FORMATION OF THE ENAMEL (Fig. 16-30). The enamel is formed exclusively from the enamel cells or ameloblasts of the special dental germ, either by direct calcification of the columnar cells, which become elongated into the hexagonal rods of the enamel, or, as is more generally believed, as a secretion from the ameloblasts within which calcareous matter is subsequently deposited.

The process begins at the apex of each cusp, at the ends of the enamel cells in contact with the dental papilla. Here a fine globular deposit appears, apparently being shed from the end of the ameloblasts. It is known as the **enamel droplet,** and it resembles keratin in its resistance to the action of mineral acids. This droplet then becomes fibrous, calcifies, and forms the first layer of enamel; a second droplet now appears and calcifies, and so on; successive droplets of this keratin-like material are shed from the ameloblasts and form successive layers of enamel, the ameloblasts gradually receding as each layer is produced, until at the termination of the process they have almost disappeared.

The intermediate cells of the enamel pulp

atrophy and disappear, so that the newly formed calcified material and the external enamel epithelium come into apposition. This latter layer, however, soon disappears on the emergence of the tooth beyond the gum. After its disappearance, the crown of the tooth is still covered by a distinct membrane, which persists for some time. This is known as the **cuticula dentis, or Nasmyth's membrane,**[1] and is believed to be the last-formed layer of enamel derived from the ameloblasts, which has not become calcified. It forms a horny layer, which may be separated from the subjacent calcified mass by the action of strong acids. It is marked by the hexagonal impressions of the enamel prisms, and when stained by silver nitrate, shows the characteristic appearance of epithelium.

FORMATION OF THE DENTIN (Fig. 16-30). While these changes take place in the epithelium to form the enamel, contemporaneous changes in the differentiated mesoderm of the dental papillae result in the formation of the dentin. As stated before, the first germs of the dentin are the papillae, which correspond in number to the teeth and are formed from the soft mesodermal tissue that bounds the depressions containing the special enamel germs.

The papillae grow upward into the enamel germs and become covered by them; both are enclosed in a vascular connective tissue, the **dental sac,** in the manner described previously. Each papilla then constitutes the formative pulp from which the dentin and permanent pulp are developed; it consists of round cells and is highly vascular, and soon begins to assume the shape of the future tooth. The next step is the appearance of the **odontoblasts,** whose relation to the development of the teeth is similar to that of osteoblasts to the formation of bone. They are formed from the cells of the periphery of the papilla—that is, from the cells in immediate contact with the ameloblasts of the special dental germ. These cells become elongated; one end of the elongated cell rests against the epithelium of the special dental germs, and the other end is tapered and often branched.

By direct transformation of the peripheral ends of these cells, or by a secretion from them, a layer of uncalcified matrix (**prodentin**) is formed that caps the cusp or cusps of the papillae. This matrix becomes fibrillated, and in it islets of calcification appear and coalesce to form a continuous layer of calcified material that covers each cusp and constitutes the first layer of dentin. The odontoblasts, having thus formed the first layer, retire toward the center of the papilla, and as they do so, they produce successive layers of dentin from their peripheral extremities, i.e., they form the dentinal matrix in which calcification subsequently occurs.

As they thus recede from the periphery of the papilla, the odontoblasts leave behind filamentous processes of protoplasm, which are provided with finer side processes. These are surrounded by the calcified material and thus form the dental canaliculi, and, by their side branches, the anastomosing canaliculi. The processes of protoplasm contained within them constitute the **dentinal fibers** (*Tomes' fibers*).

In this way the entire thickness of the dentin is developed, each canaliculus being completed throughout its whole length by a single odontoblast. The central part of the papilla does not undergo calcification, but persists as the pulp of the tooth. In this process of dentin formation, an uncalcified matrix is first developed, and in this matrix islets of calcification appear, which subsequently blend together to form a cap to each cusp; in like manner, successive layers are produced, which ultimately become blended with each other. In certain places this blending is not complete, and portions of the matrix remain uncalcified between the successive layers; this gives rise to little spaces, which are the interglobular spaces alluded to previously.

FORMATION OF THE CEMENT. The root of the tooth begins to form shortly before the crown emerges through the gum but is not completed until some time afterward. It is produced by a downgrowth of the epithelium of the dental germ, which extends almost as far as the location of the apex of the future root and determines the form of this portion of the tooth. This fold of epithelium is known as the **epithelial sheath,** and on its papillary surface odontoblasts appear, which in turn form dentin, so that the process of dentin formation is identical in both the crown and the root. After the dentin of the root has developed, the vascular tissues

[1]Alexander Nasmyth (d. 1847): A Scottish dental surgeon (London).

of the dental sac break through the epithelial sheath and spread over the surface of the root as a layer of bone-forming material. Osteoblasts appear in this material, and the process of ossification is identical to the ordinary intramembranous ossification of bone. In this way the cement is formed, which consists of ordinary bone containing canaliculi and lacunae.

FORMATION OF THE ALVEOLI. About the fourteenth week of embryonic life, the dental lamina becomes enclosed in a trough or groove of mesodermal tissue, which at first is common to all the dental germs, but subsequently becomes divided by bony septa into loculi. Each loculus contains the special dental germ of a deciduous tooth and its corresponding permanent tooth. After birth, each cavity becomes subdivided, so as to form separate loculi (the future alveoli) for each deciduous tooth and for its corresponding permanent tooth. Although at one time the whole of the growing tooth is contained in the cavity of the alveolus, the latter never completely encloses it, since there is always an aperture over the top of the crown that is filled by soft tissue, by which the dental sac is connected with the surface of the gum, and which in the permanent teeth is called the **gubernaculum dentis.**

Development of the Permanent Teeth

The permanent teeth develop in two sets: (1) those that replace the deciduous teeth, and which, like them, number ten in each jaw: these are the **successional permanent teeth;** and (2) those that have no deciduous predecessors, but are added to the temporary dental series. There are three such teeth on either side in each jaw, and they are called **superadded permanent teeth.** These include the three molars of the permanent set, as the molars of the deciduous set are replaced by the premolars of the permanent set. The development of the successional permanent teeth—the ten anterior ones in either jaw—has already been indicated. During their development the permanent teeth, enclosed in their sacs, come to lie on the lingual side of the deciduous teeth, more distant from the margin of the future gum, and as already stated, are separated from the deciduous teeth by bony partitions. As the crown of the permanent tooth grows, these bony partitions and the root of the decidu-

ous tooth are absorbed, and finally nothing but the crown of the deciduous tooth remains. This is shed or removed and the permanent tooth takes its place.

The superadded permanent teeth develop in the manner already described, by extensions of the posterior part of the dental lamina in each jaw.

ERUPTION OF THE TEETH (FIGS. 16-32; 16-33)

When the calcification of the different tissues of the tooth is sufficiently advanced to enable it to bear the pressure to which it will be afterward subjected, eruption occurs with the tooth making its way through the gum. The gum is absorbed by the pressure of the crown of the tooth against it, which is itself pressed toward the surface by the increasing size of the root. At the same time the septa between the dental sacs ossify and constitute the alveoli; these firmly embrace the necks of the teeth and afford them a solid basis of support.

The eruption of the deciduous teeth commences about the seventh month after birth and is complete about the end of the second year. The teeth of the lower jaw erupt before those of the upper jaw.

The following are the most usual times of eruption:

Lower central incisors	6 to 9 months
Upper incisors	8 to 10 months
Lower lateral incisors and first molars	15 to 21 months
Canines	16 to 20 months
Second molars	20 to 24 months

There are, however, considerable variations in these times. Thus, according to Holt, at the age of 1 year, a child should have 6 teeth; at $1\frac{1}{2}$ years, 12 teeth; at 2 years, 16 teeth; and at $2\frac{1}{2}$ years, 20 teeth.

Calcification of the permanent teeth proceeds in the following order in the lower jaw (in the upper jaw it occurs a little later): the first molar, soon after birth; the central and lateral incisors and the canine, about six months after birth; the premolars, at the second year or a little later; the second molar, about the end of the second year; the third molar, about the twelfth year.

The eruption of the permanent teeth takes

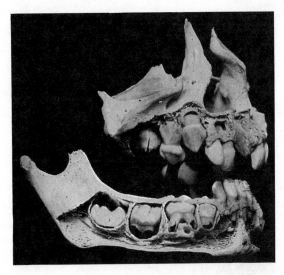

FIG. 16-32. Maxilla and mandible at about one year. (Noyes.)

place at the following periods, the teeth of the lower jaw preceding those of the upper by short intervals:

First molars . . . . .6th year
Two central incisors . . . . .7th year
Two lateral incisors . . . . .8th year
First premolars . . . . .9th year
Second premolars . . . . 10th year
Canines . . . . .11th to 12th year
Second molars . . . . .12th to 13th year
Third molars . . . . .17th to 25th year

Toward the sixth year, before the shedding of the deciduous teeth begins, there are 24 teeth in each jaw: the ten deciduous teeth and the crowns of all the permanent teeth except the third molars (Fig. 16-33). The third molars (*wisdom teeth; dentes serotini*) are irregular in their eruption and may be badly oriented or buried in bone to such an extent that they must be removed surgically. Not infrequently, one or all four may fail entirely to develop.

The Tongue

The **tongue** (*lingua*) is the principal organ of the sense of taste and an important organ of speech; it also assists in the mastication and deglutition of food. It is situated in the floor of the mouth, within the curve of the body of the mandible.

BORDERS AND SURFACES

ROOT. The root (*base*) is the posterior part of the tongue and it is connected with the hyoid bone by the hyoglossus and genioglossus muscles and the hyoglossal membrane; with the epiglottis by three folds (*glossoepiglottic folds*) of mucous mem-

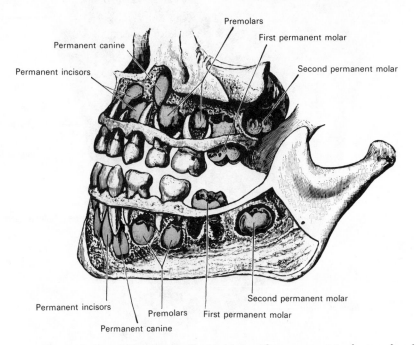

FIG. 16-33. The teeth of a child aged about seven years. The permanent teeth are colored *blue*.

brane; with the soft palate by the glossopalatine arches; and with the pharynx by the superior constrictor muscle of the pharynx and the mucous membrane.

APEX.   The apex (*tip*) is the somewhat attenuated anterior end, which rests against the lingual surfaces of the lower incisor teeth.

INFERIOR SURFACE.   The inferior surface (*undersurface*) (Fig. 16-39) is connected with the mandible by the mucous membrane, which is reflected over the floor of the mouth to the lingual surface of the gum. The mu-

cous membrane between the floor and the tongue in the midline is elevated into a distinct vertical fold, the **frenulum linguae.** On either side lateral to the frenulum is a slight fold of the mucous membrane, the **plica fimbriata,** the free edge of which occasionally exhibits a series of fringe-like processes.

The apex of the tongue, part of the inferior surface, the sides, and the dorsum are free.

DORSUM.   The dorsum of the tongue (Fig. 16-34) is convex and marked by a **median sulcus,** which divides it into symmetrical halves. This sulcus ends posteriorly about

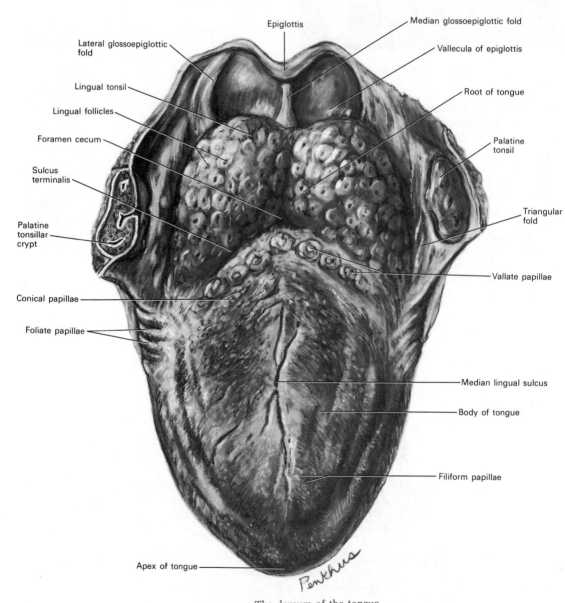

FIG. 16-34.   The dorsum of the tongue.

2.5 cm from the root of the organ in a depression called the **foramen cecum,** from which a shallow groove, the **sulcus terminalis,** runs lateralward and forward on either side to the margin of the tongue. The part of the dorsum of the tongue that is anterior to this groove, forming about two-thirds of its surface, is rough and covered with papillae; the posterior third is smoother, and contains numerous mucous glands and lymph follicles (**lingual tonsil**). The foramen cecum is the remains of the cranial part of the **thyroglossal duct** or diverticulum, from which the thyroid gland develops; the pyramidal lobe of the thyroid gland indicates the position of the lower part of the duct.

PAPILLAE

The papillae of the tongue (Fig. 16-35) are projections of the corium, thickly distributed over the anterior two-thirds of its dorsum, giving to this surface its characteristic roughness. The varieties of papillae are the **vallate papillae, fungiform papillae, filiform papillae,** and **papillae simplices.**

VALLATE PAPILLAE. The vallate papillae (*circumvallate papillae*) (Fig. 16-35) are large and vary in number from 8 to 12. They are situated on the dorsum of the tongue immediately anterior to the foramen cecum and sulcus terminalis, forming a row on either side like the limbs of the letter V. Each papilla consists of a projection of mucous membrane 1 to 2 mm wide, attached to the bottom of a circular depression of the mucous membrane, the margin of which is elevated to form a wall (*vallum*); between this and the papilla is a circular sulcus or furrow. The papilla is shaped like a truncated cone, of which the smaller end is attached to the tongue. The broader part of the cone shape is free, projecting a little above the surface of the tongue; it is studded with numerous small secondary papillae and covered by stratified squamous epithelium. The taste buds are especially numerous on the walls of the papilla within the circular furrow.

FUNGIFORM PAPILLAE. The fungiform papillae (Fig. 16-36), more numerous than the vallate, are found chiefly at the sides and apex of the tongue but are scattered irregularly and sparingly over the dorsum. They are easily recognized among the other papillae by their large size, round eminences, and deep red color. They are narrow at their attachment to the tongue but broad and round at their free extremities; they are covered with secondary papillae.

FILIFORM PAPILLAE. The filiform papillae (*conical papillae*) (Fig. 16-36) cover the anterior two-thirds of the dorsum. They are minute, filiform, and arranged in lines parallel with the two rows of the vallate papillae, except at the apex of the tongue, where their

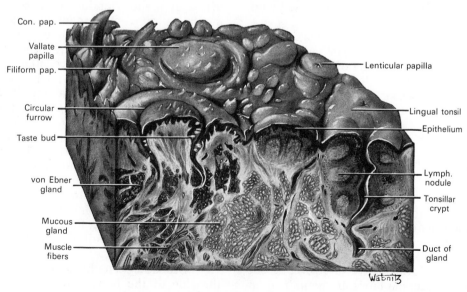

FIG. 16-35. Section of posterior part of dorsum of tongue showing circumvallate papillae and lingual tonsil. (Redrawn from Braus.)

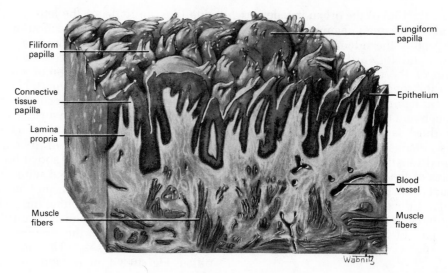

Filiform papilla

Connective tissue papilla

Lamina propria

Muscle fibers

Fungiform papilla

Epithelium

Blood vessel

Muscle fibers

Wabnitz

FIG. 16-36.   Section of anterior part of dorsum of tongue showing fungiform and filiform papillae. (Redrawn from Braus.)

direction is transverse. Projecting from their apices are numerous filamentous processes, or secondary papillae; these have a whitish tint owing to the thickness and density of the epithelium of which they are composed. This epithelium has undergone a peculiar modification: the cells are cornified and elongated into dense, imbricated, brushlike processes. The papillae also contain elastic fibers, which render them firmer and more elastic than the papillae of mucous membrane generally. The larger and longer papillae of this group are sometimes called **conical papillae.**

PAPILLAE SIMPLICES.   The papillae simplices are similar to those of the skin and cover the entire mucous membrane of the tongue, as well as the larger papillae. They consist of closely set microscopic projections into the layer of epithelium, each containing a capillary loop.

MUSCLES

The tongue is divided into lateral halves by a median fibrous septum, which extends throughout its entire length and is fixed inferiorly to the hyoid bone. In either half there are two sets of muscles, extrinsic and intrinsic; the former have their origins outside the tongue, the latter are contained entirely within it.

Extrinsic Muscles

The extrinsic muscles (Fig. 16-37) are:

Genioglossus
Hyoglossus
Chondroglossus
Styloglossus
Palatoglossus

GENIOGLOSSUS.   The genioglossus (*geniohyoglossus*) is a flat, fan-shaped muscle close to and parallel with the median plane. It *arises* by a short tendon from the superior mental spine on the inner surface of the symphysis menti, immediately above the geniohyoideus muscle. The inferior fibers extend downward, and are attached by a thin aponeurosis to the upper part of the body of the hyoid bone; a few fibers pass between the hyoglossus and chondroglossus to blend with the middle constrictor of the pharynx. The middle fibers pass posteriorly, and the superior ones upward, to enter the whole length of the undersurface of the tongue, from the root to the apex. The muscles of the two sides are separated at their insertions by the median fibrous septum of the tongue; anteriorly, they are more or less blended, the fasciculi decussating in the median plane.

HYOGLOSSUS.   The hyoglossus *arises* from the side of the body and from the whole length of the greater horn of the hyoid

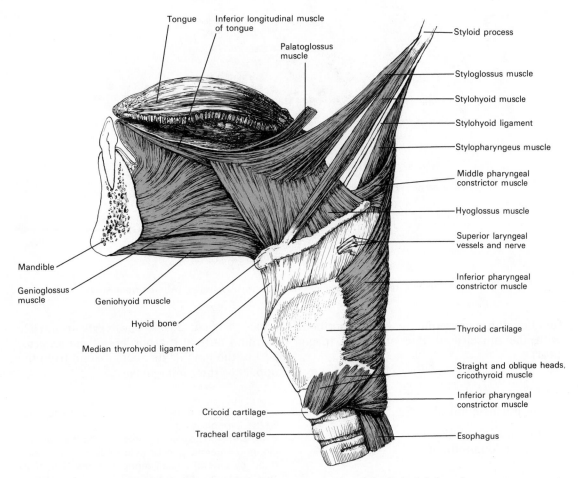

Labels (clockwise from top):
Tongue · Inferior longitudinal muscle of tongue · Palatoglossus muscle · Styloid process · Styloglossus muscle · Stylohyoid muscle · Stylohyoid ligament · Stylopharyngeus muscle · Middle pharyngeal constrictor muscle · Hyoglossus muscle · Superior laryngeal vessels and nerve · Inferior pharyngeal constrictor muscle · Thyroid cartilage · Straight and oblique heads, cricothyroid muscle · Inferior pharyngeal constrictor muscle · Esophagus · Tracheal cartilage · Cricoid cartilage · Median thyrohyoid ligament · Hyoid bone · Geniohyoid muscle · Genioglossus muscle · Mandible

FIG. 16-37.   The extrinsic muscles of the tongue as viewed from the left lateral aspect.

bone, and passes almost vertically upward to enter the side of the tongue, between the styloglossus and the inferior longitudinal muscle. The fibers arising from the body of the hyoid bone overlap those from the greater horn.

CHONDROGLOSSUS.   The chondroglossus is sometimes described as a part of the hyoglossus, but is separated from it by fibers of the genioglossus, which pass to the side of the pharynx. It is about 2 cm long, and *arises* from the medial side and base of the lesser horn and contiguous portion of the body of the hyoid bone, and passes directly upward to blend with the intrinsic muscular fibers of the tongue, between the hyoglossus and genioglossus.

A small slip of muscular fibers is occasionally found arising from the cartilago triticea in the lateral thyrohyoid ligament and enter-

ing the tongue with the most posterior fibers of the hyoglossus.

STYLOGLOSSUS.   The styloglossus, the shortest and smallest of the three styloid muscles, *arises* from the anterior and lateral surfaces of the styloid process and from the stylomandibular ligament. Curving down anteriorly between the internal and external carotid arteries, it divides upon the side of the tongue into two portions: one, longitudinal, enters the side of the tongue near its dorsal surface, blending with the fibers of the inferior longitudinal muscle anterior to the hyoglossus; the other, oblique, overlaps the hyoglossus and decussates with its fibers.

PALATOGLOSSUS.   The palatoglossus (*glossopalatinus*), although one of the muscles of the tongue, is more closely associated with the soft palate both in situation and

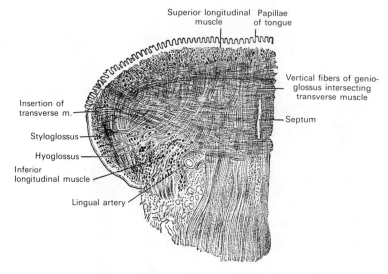

Fig. 16-38. Coronal section of tongue, showing intrinsic muscles. (Altered from Krause.)

function; it has consequently been described with the muscles of that structure (page 1441).

Intrinsic Muscles

The intrinsic muscles (Fig. 16-38) are:

Superior longitudinal
Inferior longitudinal
Transverse
Vertical

SUPERIOR LONGITUDINAL MUSCLE OF THE TONGUE. The superior longitudinal muscle is a thin stratum of oblique and longitudinal fibers immediately underlying the mucous membrane on the dorsum of the tongue. It *arises* from the submucous fibrous layer close to the epiglottis and from the median fibrous septum, and runs anteriorly to the edges of the tongue.

INFERIOR LONGITUDINAL MUSCLE OF THE TONGUE. The inferior longitudinal muscle is a narrow band situated on the inferior surface of the tongue between the genioglossus and the hyoglossus. It extends from the root to the apex of the tongue; some of its fibers are connected with the body of the hyoid bone, others blend with the fibers of the styloglossus.

TRANSVERSE MUSCLE. The transverse muscle consists of fibers that *arise* from the median fibrous septum and pass lateralward to be *inserted* into the submucous fibrous tissue at the sides of the tongue.

VERTICAL MUSCLE. The vertical muscle is found only at the borders of the anterior part of the tongue; its fibers extend from the upper- to the undersurface.

NERVES. The muscles of the tongue are supplied by the hypoglossal nerve.

ACTIONS. The movements of the tongue, although numerous and complicated, may be understood by carefully considering the direction of its muscle fibers. The genioglossi, by means of their posterior fibers, draw the root of the tongue anteriorly and protrude the apex from the mouth. The anterior fibers draw the tongue back into the mouth. The two parts acting in their entirety draw the tongue downward, making its superior surface concave from side to side, and forming a channel along which fluids may pass toward the pharynx, as in sucking. The hyoglossi depress the tongue and draw down its sides. The styloglossi draw the tongue upward and backward. The palatoglossi draw the root of the tongue upward.

The intrinsic muscles are mainly concerned with altering the shape of the tongue whereby it becomes shortened, narrowed, or curved in different directions; thus, the superior and inferior longitudinal muscles tend to shorten the tongue, but the former, in addition, turns the tip and sides upward to render the dorsum concave, while the latter pulls the tip downward and renders the dorsum convex. The transverse muscle narrows and elongates the tongue, and the vertical muscle flattens and broadens it. The complex arrangement of the muscular fibers of the tongue and the various directions in which they run give this organ the power to assume the forms necessary for the enunciation of different consonantal sounds.

STRUCTURE OF THE TONGUE

The tongue is invested by mucous membrane and a submucous fibrous layer.

MUCOUS MEMBRANE. The mucous membrane covering the inferior surface of the tongue is thin, smooth, and identical in structure with that lining the rest of the oral cavity. The mucous membrane of the dorsum of the tongue posterior to the foramen cecum and sulcus terminalis is thick and freely movable over the subjacent parts. It contains a large number of lymphoid follicles, which together constitute the **lingual tonsil** (Fig. 16-35). Each follicle forms a round eminence, the center of which is perforated by a minute orifice leading into a funnel-shaped cavity or recess; around this recess are grouped numerous oval or round nodules of lymphoid tissue, each enveloped by a capsule derived from the submucosa, and opening into the bottom of the recesses are the ducts of mucous glands. The mucous membrane on the anterior part of the dorsum of the tongue is thin, intimately adherent to the muscular tissue, and presents numerous minute surface eminences, the **papillae** of the tongue. It consists of a layer of connective tissue, the **corium,** covered with epithelium (Figs. 16-35; 16-36).

The epithelium is stratified squamous, similar to that of the skin, and each papilla has a separate investment from root to summit. The deepest cells may sometimes be detached as a separate layer, corresponding to the rete mucosum, but they never contain pigment.

CORIUM. The corium consists of a dense feltwork of fibrous connective tissue, with numerous elastic fibers, firmly connected with the fibrous tissue forming the septa between the muscular bundles of the tongue. It contains the ramifications of the numerous vessels and nerves from which the papillae are supplied, large plexuses of lymphatic vessels, and the glands of the tongue.

PAPILLAE. The papillae appear to resemble in structure those of the cutis, consisting of cone-shaped projections of connective tissue, covered with a thick layer of stratified squamous epithelium, and containing one or more capillary loops, among which nerves are distributed in great abundance. If the epithelium is removed, however, one sees that they are not simple elevations like the papillae of the skin, for the surface of each is studded with minute conical processes that form secondary papillae. In the vallate papillae, the nerves are numerous and large. They are also numerous in the fungiform papillae and end in a plexiform network from which brushlike branches proceed; their mode of termination is similar in the filiform papillae.

GLANDS OF THE TONGUE. The tongue has mucous and serous glands.

The **mucous glands** are similar in structure to the labial and buccal glands. They are found especially at the back part behind the vallate papillae, but are also present at the apex and marginal parts. In this connection the anterior lingual glands require special notice. They are situated on the undersurface of the apex of the tongue (Fig. 16-39), one on either side of the frenulum, where they are covered by a fasciculus of muscular fibers derived from the styloglossus and inferior longitudinal muscle. They are 12 to 25 mm long and about 8 mm broad, and each opens by three or four ducts on the undersurface of the apex.

The **serous glands** (v. Ebner's glands[1]) occur only at the back of the tongue in the neighborhood of the taste buds, their ducts opening for the most part into the fossae of the vallate papillae. These glands are racemose, the duct of each branching into several minute ducts, which end in alveoli lined by a single layer of more or less columnar epithelium. Their secretion is watery and probably assists in distributing the substance to be tasted over the taste area.

SEPTUM. The septum consists of a vertical layer of fibrous tissue that extends throughout the entire length of the median plane of the tongue, though it does not quite reach the dorsum. It is thicker behind than in front and occasionally contains a small fibrocartilage, about 6 mm long. It is well displayed by making a vertical section across the organ.

HYOGLOSSAL MEMBRANE. The hyoglossal membrane is a strong fibrous lamina that connects the undersurface of the root of the tongue to the body of the hyoid bone. This membrane receives some of the fibers of the genioglossi anteriorly.

TASTE BUDS. Taste buds, the end-organs of the gustatory sense, are scattered over the mucous membrane of the mouth and tongue

[1]Victor Ebner von Rofenstein (1842–1925): An Austrian histologist (Vienna).

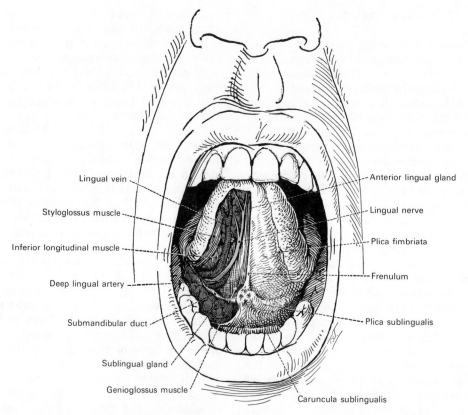

Lingual vein

Styloglossus muscle

Inferior longitudinal muscle

Deep lingual artery

Submandibular duct

Sublingual gland

Genioglossus muscle

Anterior lingual gland

Lingual nerve

Plica fimbriata

Frenulum

Plica sublingualis

Caruncula sublingualis

FIG. 16-39. The inferior surface of the tongue, with the right side dissected to show the blood vessels, nerve and salivary gland. (Eycleshymer and Jones.)

at irregular intervals. They occur especially in the sides of the vallate papillae. They are described in Chapter 13 under the Organs of the Senses.

VESSELS AND NERVES. The main **artery** of the tongue is the lingual branch of the external carotid artery but the facial and ascending pharyngeal arteries also give branches to it. The **veins** open into the internal jugular vein.

The **lymphatics of the tongue** have been described on page 882.

The **sensory nerves of the tongue** are: (1) the lingual branch of the mandibular nerve, which is distributed to the papillae at the anterior part and sides of the tongue, and forms the nerve of ordinary sensibility for its anterior two-thirds; (2) the chorda tympani branch of the facial nerve, which runs in the sheath of the lingual nerve and is generally regarded as the nerve of taste for the anterior two-thirds; this nerve is a continuation of the sensory root of the facial nerve (*nervus intermedius*); (3) the lingual branch of the glossopharyngeal nerve, which is distributed to the mucous membrane at the base and sides of the tongue and to the vallate papillae, and

which supplies both gustatory filaments and fibers of general sensation to this region; (4) the superior laryngeal nerve, which sends some fine branches to the root near the epiglottis.

Salivary Glands

Three pairs of large salivary glands pour their secretion into the mouth; they are the **parotid, submandibular,** and **sublingual** (Fig. 16-40).

PAROTID GLAND

The parotid gland, the largest of the three, varies in weight from 14 to 28 gm. It lies upon the side of the face, immediately below and in front of the external ear. The main portion of the gland is superficial, somewhat flat and quadrilateral, and is placed between the ramus of the mandible, the mastoid process, and the sternocleidmastoid muscle. Superiorly, it is broad and reaches nearly to the zygomatic arch; inferiorly, it tapers somewhat to about the level of a line joining the tip of the mastoid process to the angle of the mandible. The remainder of the gland is irregularly wedge-shaped, and extends deeply inward toward the pharyngeal wall.

The gland is enclosed within a capsule that is continuous with the deep cervical fascia; the layer covering the superficial surface is dense and closely adherent to the gland. A portion of the fascia that is attached to the styloid process and the angle of the mandible is thickened to form the *stylomandibular ligament,* which intervenes between the parotid and submandibular glands.

SURFACES. The **anterior surface** of the gland is molded on the posterior border of the ramus of the mandible, clothed by the medial pterygoid and masseter muscles. The inner lip of the groove dips, for a short distance, between the two pterygoid muscles, while the outer lip extends for some distance over the superficial surface of the masseter; a small portion of this lip immediately below the zygomatic arch is usually detached, and is named the **accessory part** of the gland.

The **posterior surface** is grooved longitudinally and abuts against the external acoustic meatus, the mastoid process, and the anterior border of the sternocleidomastoid muscle.

The **superficial surface,** slightly lobulated,

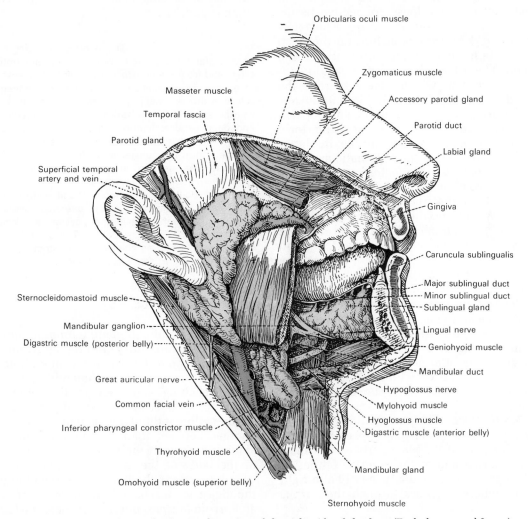

Orbicularis oculi muscle

Zygomaticus muscle

Accessory parotid gland

Parotid duct

Masseter muscle

Temporal fascia

Labial gland

Parotid gland

Superficial temporal artery and vein

Gingiva

Caruncula sublingualis

Major sublingual duct

Minor sublingual duct

Sublingual gland

Sternocleidomastoid muscle

Lingual nerve

Mandibular ganglion

Geniohyoid muscle

Digastric muscle (posterior belly)

Mandibular duct

Hypoglossus nerve

Great auricular nerve

Mylohyoid muscle

Common facial vein

Hyoglossus muscle

Digastric muscle (anterior belly)

Inferior pharyngeal constrictor muscle

Thyrohyoid muscle

Mandibular gland

Omohyoid muscle (superior belly)

Sternohyoid muscle

FIG. 16-40. The salivary glands in a dissection of the right side of the face. (Eycleshymer and Jones.)

is covered by the integument, the superficial fascia containing the facial branches of the great auricular nerve and some small lymph glands, and the connective tissue that forms the capsule of the gland.

The **deep surface** extends inward by means of two processes, one of which lies on the digastric muscle, the styloid process, and the styloid group of muscles, and projects under the mastoid process and the sternocleidomastoid muscle. The other process is situated anterior to the styloid process and sometimes passes into the posterior part of the mandibular fossa behind the temporomandibular joint. The deep surface is in contact with the internal and external carotid arteries, the internal jugular vein, and the vagus and glossopharyngeal nerves.

The gland is separated from the pharyngeal wall by some loose connective tissue.

STRUCTURES WITHIN THE GLAND. The *external carotid artery* lies at first on the deep surface and then in the substance of the gland. The artery gives off its *posterior auricular* branch, which emerges from the gland posteriorly. It then divides into its terminal branches, the *maxillary* and *superficial temporal*; the former runs anteriorly deep to the neck of the mandible; the latter runs upward across the zygomatic arch and gives off its *transverse facial* branch, which emerges from the anterior part of the gland. Superficial to the arteries are the *superficial temporal* and *maxillary veins*, which unite to form the *retromandibular vein*. A large anastomosing vein emerges from the gland and unites with the facial vein to form the *common facial vein*; the remainder of the vein unites in the gland with the posterior auricular vein to form the *external jugular vein*. On a still more superficial plane is the *facial nerve*, the branches of which emerge from the borders of the gland. Branches of the *great auricular nerve* pierce the gland to join the facial nerve, while the *auriculotemporal nerve* issues from the upper part of the gland.

Parotid Duct. The parotid duct (*Stensen's duct*[1]) is about 7 cm long. It begins at the anterior part of the gland, crosses the masseter muscle and at the anterior border of this muscle turns medially, nearly at a right angle, where it passes through the corpus

---

[1]Niels Stensen (1638–1686): A Danish physician, anatomist, and physiologist (Copenhagen and Florence).

adiposum of the cheek and pierces the buccinator muscle. It runs for a short distance obliquely forward between the buccinator muscle and mucous membrane of the mouth, and then opens upon the oral surface of the cheek by a small orifice opposite the second upper molar tooth. While crossing the masseter muscle, it receives the duct of the accessory gland. In this position it lies between the branches of the facial nerve; the accessory part of the gland and the transverse facial artery are superior to it.

The parotid duct has a wall of considerable thickness; its canal is about 3 to 4 mm in diameter, but at its orifice on the oral surface of the cheek its lumen is greatly reduced in size. It consists of a thick external fibrous coat that contains contractile fibers and an internal or mucous coat lined with low columnar epithelium.

VESSELS AND NERVES. The **arteries** supplying the parotid gland are derived from the external carotid artery and from the branches given off by that vessel in or near its substance. The **veins** empty themselves in the external jugular vein through some of its tributaries. The **lymphatics** end in the superficial and deep cervical lymph nodes, passing in their course through two or three nodes placed on the surface and in the substance of the parotid.

The **nerves** are derived from the plexus of the *sympathetic nerve* on the external carotid artery and from the auriculotemporal nerve. The fibers from the latter nerve are cranial *parasympathetics* derived from the glossopharyngeal and possibly the facial nerve, through the otic ganglion. The sympathetic fibers are regarded as chiefly vasoconstrictors; the parasympathetic fibers, as secretory.

SUBMANDIBULAR GLAND

The submandibular gland (*glandula mandibularis; submaxillary gland*) (Fig. 16-40) is round and about the size of a walnut. Much of it is situated in the submandibular triangle, reaching anteriorly to the anterior belly of the digastric muscle and posteriorly to the stylomandibular ligament, which intervenes between it and the parotid gland. It extends superiorly under the inferior border of the body of the mandible; inferiorly, it usually overlaps the intermediate tendon of the digastric muscle and the insertion of the stylohyoid muscle, and from its deep surface a tongue-like *deep process* extends anteriorly above the mylohyoid muscle.

SURFACES. The **superficial surface** of the submandibular gland consists of an upper and a lower part. The *upper part* is directed superficially; it lies partly against the submandibular depression on the inner surface of the body of the mandible and partly on the medial pterygoid muscle. The *lower part* is covered by the skin, superficial fascia, platysma, and deep cervical fascia; it is crossed by the facial vein and by filaments of the facial nerve. In contact with it, near the mandible, are the submandibular lymph nodes.

The **deep surface** is in relation with the mylohyoid, hyoglossus, styloglossus, stylohyoid, and posterior belly of the digastric muscles; in contact with it are the mylohyoid nerve and the mylohyoid and submental vessels. The facial artery is embedded in a groove in the posterior border of the gland.

The **deep process** of the gland extends anteriorly between the mylohyoid laterally and the hyoglossus and styloglossus muscles medially. Above it are the lingual nerve and submandibular ganglion; below it, the hypoglossal nerve and its accompanying vein.

SUBMANDIBULAR DUCT. The submandibular duct (*Wharton's duct*[1]; *submaxillary duct*) is about 5 cm long and its wall is much thinner than that of the parotid duct. It begins by numerous branches from the deep surface of the gland, and runs anteriorly between the mylohyoid, the hyoglossus, and the genioglossus muscles; then between the sublingual gland and the genioglossus, and opens by a narrow orifice on the summit of a small papilla, the **caruncula sublingualis,** at the side of the frenulum (Fig. 16-39). On the hyoglossus it lies between the lingual and hypoglossal nerves, but at the anterior border of the muscle it is crossed laterally by the lingual nerve; the terminal branches of the lingual nerve ascend on its medial side.

VESSELS AND NERVES. The **arteries** supplying the submandibular gland are branches of the facial and lingual arteries. Its **veins** follow the course of the arteries. The **secretomotor nerves** are from cranial parasympathetic fibers of the facial nerve, which pass by way of the chorda tympani and submandibular ganglion; the **sympathetic** (*vasomotor*) fibers come from the superior cervical ganglion by way of plexuses on the external carotid and facial arteries.

[1]Thomas Wharton (1616–1673): An English physician and anatomist (London).

SUBLINGUAL GLAND

The sublingual gland (Fig. 16-40) is the smallest of the three glands. It is situated beneath the mucous membrane of the floor of the mouth, at the side of the frenulum linguae, in contact with the sublingual depression on the inner surface of the mandible, close to the symphysis. It is narrow, flat, shaped somewhat like an almond, and weighs nearly 2 gm. It is in relation, *superiorly,* with the mucous membrane; *inferiorly,* with the mylohyoid muscle; *posteriorly,* with the deep part of the submandibular gland; *laterally,* with the mandible; and *medially,* with the genioglossus muscle, from which it is separated by the lingual nerve and the submandibular duct.

It has 8 to 20 excretory ducts. One or more join to form the **larger sublingual duct** (*duct of Bartholin*[2]), which opens into the submandibular duct. Of the **small sublingual ducts** (*ducts of Rivinus*[3]), some join the submandibular duct; others open separately into the mouth on the elevated crest of mucous membrane (*plica sublingualis*), caused by the projection of the gland, on either side of the frenulum linguae.

VESSELS AND NERVES. The sublingual gland is supplied with blood from the sublingual and submental arteries. Its nerves are derived in a manner similar to those of the submandibular gland.

STRUCTURE OF THE SALIVARY GLANDS

The salivary glands are compound racemose glands consisting of numerous lobes, which are made up of smaller lobules connected by dense areolar tissue, vessels, and ducts. Each lobule consists of the ramifications of a single duct; the branches terminate in dilated ends or alveoli, on which the capillaries are distributed. The alveoli are enclosed by a basement membrane, which is continuous with the membrana propria of the duct and consists of a network of branched and flattened nucleated cells.

The alveoli of the salivary glands are of two kinds, which differ in the appearance of

[2]Caspar Thoméson Bartholin, II (1655–1738): A Danish professor of medicine, anatomy, and physics (Copenhagen).
[3]Augustus Quirinus Rivinus (1652–1723): A German anatomist and botanist (Leipzig).

their secreting cells, in their size, and in the nature of their secretion: (1) the mucous variety secretes a viscid fluid that contains mucin; (2) the serous variety secretes a thinner and more watery fluid. The sublingual gland consists of mucous alveoli; the parotid, of serous alveoli. The submandibular contains both mucous and serous alveoli; the latter, however, are preponderant.

The cells in the **mucous alveoli** are cuboidal. In the fresh condition they contain large granules of mucinogen. In hardened preparations, a delicate protoplasmic network is seen, and the cells are clear and transparent. The flat nucleus is usually situated near the basement membrane.

Some alveoli have peculiar crescentic bodies lying between the cells and the membrana propria. They are termed the **crescents of Gianuzzi,**[1] or the **demilunes of Heidenhain**[2] (Fig. 16-41), and are composed of polyhedral granular cells. Fine canaliculi pass between the mucus-secreting cells to reach the demilunes.

In the **serous alveoli,** the cells almost completely fill the cavity, so that the lumen is

[1]Giuseppe Giannuzzi (1839–1876): An Italian physiologist and anatomist (Siena).
[2]Rudolf Peter Heidenhain (1834–1897): A German physiologist (Breslau).

hardly perceptible; they contain secretory granules embedded in a closely reticulated protoplasm (Fig. 16-41). The cells are more cubical than those of the mucous type; the nucleus of each is spherical and placed near the center of the cell, and the granules are smaller.

The appearance of both mucous and serous cells varies according to whether the gland is in a resting condition or has recently been active. In the former case the cells are large and contain many secretory granules; in the latter case they are shrunken and contain few granules, which are chiefly collected at the inner ends of the cells.

The **ducts** are lined at their origins by epithelium that differs little from the pavement form. As the ducts enlarge, the epithelial cells change to the columnar type, and the part of the cell next to the basement membrane is finely striated.

The lobules of the salivary glands are richly supplied with blood vessels, which form a dense network in the interalveolar spaces. Fine plexuses of nerves are also found in the interlobular tissue. The nerve fibrils pierce the basement membrane of the alveoli and end in branched varicose filaments between the secreting cells. In the hilum of the submandibular gland is a collection of nerve cells termed **Langley's ganglion.**[3]

ACCESSORY GLANDS

Besides the salivary glands proper, numerous other glands are found in the mouth. Many of these glands are found at the posterior part of the dorsum of the tongue behind the vallate papillae and also along its margins as far forward as the apex. Others lie around and in the palatine tonsil between its crypts, and large numbers are present in the soft palate, the lips, and the cheeks. These glands have the same structure as the larger salivary glands and are the mucous or mixed type.

The Fauces

The aperture by which the mouth communicates with the pharynx is called the fauces. The **isthmus faucium** (Fig. 16-42) is bounded,

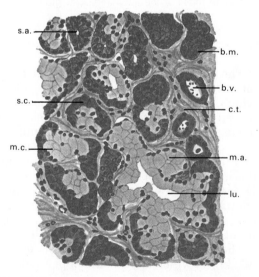

FIG. 16-41. Section of the human sublingual gland. *b.m.*, Basement membrane; *b.v.*, blood vessel; *c.t.*, connective tissue stroma; *lu.*, lumen; *m.a.*, mucous alveolus; *m.c.*, mucus-secreting cells; *s.a.*, serous alveolus; *s.c.*, serous crescent (crescent of Gianuzzi or demilune of Heidenhain). 250 ×. (Sobotta.)

[3]John Newport Langley (1852–1925): An English physiologist (Cambridge).

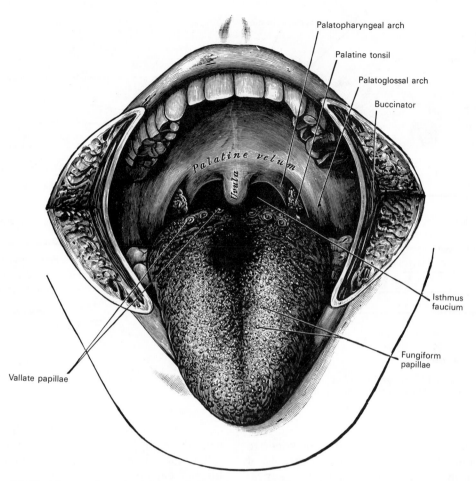

Palatopharyngeal arch

Palatine tonsil

Palatoglossal arch

Buccinator

Palatine velum

Uvula

Isthmus faucium

Fungiform papillae

Vallate papillae

FIG. 16-42.   The mouth cavity. The cheeks have been slit transversely and the tongue pulled forward.

superiorly, by the soft palate; inferiorly, by the dorsum of the tongue; and on either side, by the palatoglossal arch.

PALATOGLOSSAL ARCH

The palatoglossal arch (*arcus glossopalatinus; anterior pillar of fauces*) curves downward from the soft palate near the uvula to the side of the base of the tongue, and is formed by the projection of the palatoglossus muscle with its covering mucous membrane.

PALATOPHARYNGEAL ARCH

The palatopharyngeal arch (*arcus pharyngopalatinus; posterior pillar of fauces*) blends with the free border of the soft palate near the uvula and curves downward to the side of the pharynx; it is formed by the projection of the palatopharyngeus muscle and covered by mucous membrane. The two arches diverge inferiorly to form a triangular interval, in which the palatine tonsil is lodged.

PALATINE TONSIL

The palatine tonsil (*tonsil*) (Figs. 16-42; 16-43) is a prominent mass of lymphatic tissue situated at the side of the fauces between the palatoglossal and palatopharyngeal arches. Each tonsil consists of an aggregation of lymphatic nodules underlying the mucous membrane between the arches. The lymphatic tissue does not completely fill the interval between the two arches, however; a small depression, the **supratonsillar fossa,**

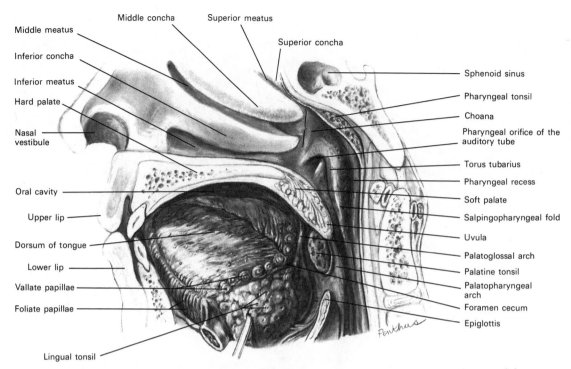

FIG. 16-43. The nasopharynx and oropharynx in midsagittal section, also showing the entire dorsum of the tongue and the right palatine tonsil. (Redrawn after Sobotta.)

occupies the upper part of the interval. The tonsil extends for a variable distance deep to the palatoglossal arch. The covering fold of mucous membrane reaches across the supra-tonsillar fossa between the two arches; this is sometimes termed the **plica semilunaris.** The remainder of the fold is called the **plica triangularis.** Between the plica triangularis and the surface of the tonsil is a space known as the **tonsillar sinus,** which may be obliterated by its walls becoming adherent. A large portion of the tonsil is below the level of the surrounding mucous membrane, i.e., is embedded, while the remainder projects as the visible tonsil (Fig. 16-43, A). In the child the tonsils are relatively (and frequently absolutely) larger than in the adult, and about one-third of the tonsil is embedded. After puberty, the embedded portion diminishes considerably in size and the tonsil assumes a disc-like form. The shape and size of the tonsil vary considerably in different individuals.

SURFACES. The **medial surface** of the tonsil is free except anteriorly, where it is covered by the plica triangularis; it has 12 to 15 orifices (*fossulae tonsillares*) leading into **crypts**

(*cryptae tonsillares*), which may branch and extend deeply into the tonsillar substance.

The **lateral** or **deep surface** adheres to a fibrous capsule which continues into the plica triangularis. It is separated from the inner surface of the superior pharyngeal constrictor usually by some loose connective tissue; this muscle intervenes between the tonsil and the facial artery with its tonsillar and ascending palatine branches. The internal carotid artery lies posterior and lateral to the tonsil at a distance of 20 to 25 mm.

WALDEYER'S RING. The palatine tonsils form part of a circular band of lymphatic tissue (Waldeyer's ring[1]), which guards the opening into the digestive and respiratory tubes (Fig. 16-43). The anterior part of the ring is formed by the submucous lymphatic collections (**lingual tonsil**) on the posterior part of the tongue; the lateral portions consist of the palatine tonsils and tubal tonsils at the openings of the auditory tubes. The ring is completed posteriorly by the pharyngeal tonsil on the posterior wall of the phar-

[1]Heinrich Wilhelm Gottfried Waldeyer (1836–1921): A German professor of pathologic anatomy (Berlin).

ynx. In the intervals between these main masses are smaller collections of lymphoid tissue.

STRUCTURE OF THE PALATINE TONSIL

(Fig. 16-44). Stratified squamous epithelium, like that of the palate and oral pharynx, covers the free surface of the tonsil and extends down into its substance to form the lining of the crypts. Each **crypt** is surrounded by a layer of lymphatic tissue containing numerous scattered lymphatic nodules, whose germinal centers are especially prominent in children and young adults. A thin connective tissue capsule, derived from the submucosa of the pharynx, encloses the whole tonsil and sends delicate septa between the lymphatic tissue layers surrounding the crypts. The epithelium of the crypts is so invaded by leukocytes in many places that it is scarcely distinguishable from the lymphatic tissue. Polymorphonuclear leukocytes from the blood as well as lymphocytes penetrate the epithelium, and when they are found as free swimming cells in the saliva, they are known as *salivary corpuscles*. Small mucous glands occur in the submucosa about the tonsil but their ducts, as a rule, do not open into the crypts.

VESSELS AND NERVES.

The **arteries** supplying the tonsil are the dorsalis linguae from the lingual artery, the ascending palatine and tonsillar from the facial artery, the ascending pharyngeal from the external carotid artery, the descending palatine branch of the maxillary artery, and a twig from the small meningeal artery.

The **veins**, which form the tonsillar plexus on the lateral side of the tonsil, drain into the pterygoid plexus or the facial vein.

The **lymphatic vessels**, beginning in the dense network of capillaries surrounding the lymphatic tissue, penetrate the pharyngeal wall and pass to the deep cervical nodes. The largest of these nodes, lying beside the posterior belly of the digastric muscle, is especially associated with the tonsil and is easily palpated when the latter is inflamed.

The **nerves** are derived from the middle and posterior palatine branches of the maxillary nerve and from the glossopharyngeal nerve.

PALATINE APONEUROSIS.

Attached to the posterior border of the hard palate is a thin, firm, fibrous lamella that supports the muscles and gives strength to the soft palate. Laterally it is continuous with the pharyngeal aponeurosis.

MUSCLES OF THE PALATE

The muscles of the palate (Fig. 16-45) are:

Levator veli palatini
Tensor veli palatini

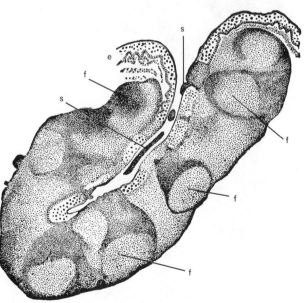

FIG. 16-44. Section through one of the crypts of the tonsil. (Stöhr.) Magnified. *e*, Stratified epithelium of general surface, continued into crypt. *f*, Nodules of lymphoid tissue—opposite each nodule, numbers of lymph cells are passing into or through the epithelium. *s*, Cells that have escaped to mix with the saliva as salivary corpuscles.

Musculus uvulae
Palatoglossus
Palatopharyngeus

LEVATOR VELI PALATINI. The levator veli palatini (*levator palati*) is situated lateral to the choanae and deep to the torus tubarius. It *arises* from the inferior surface of the apex of the petrous part of the temporal bone and from the medial lamina of the cartilage of the auditory tube. After passing above the superior concave margin of the superior constrictor, it spreads out in the palatine velum, its fibers extending obliquely downward and medialward to the midline, where they blend with those of the opposite side.

TENSOR VELI PALATINI. The tensor veli palatini (*tensor palati*) is a thin, ribbon-like muscle placed lateral and anterior to the levator. It *arises* by a flat lamella from the scaphoid fossa at the base of the medial pterygoid plate, from the spine of the sphenoid, and from the lateral wall of the cartilage of the auditory tube. Descending vertically between the medial pterygoid plate and the medial pterygoid muscle, it ends in a tendon that winds around the pterygoid hamulus, being retained in this situation by some of the fibers of origin of the medial pterygoid. Between the tendon and the hamulus is a small bursa. The tendon then passes medialward and is *inserted* into the palatine aponeurosis and into the surface behind the transverse ridge on the horizontal part of the palatine bone.

MUSCULUS UVULAE. The musculus uvulae (*azygos uvulae*) *arises* from the posterior

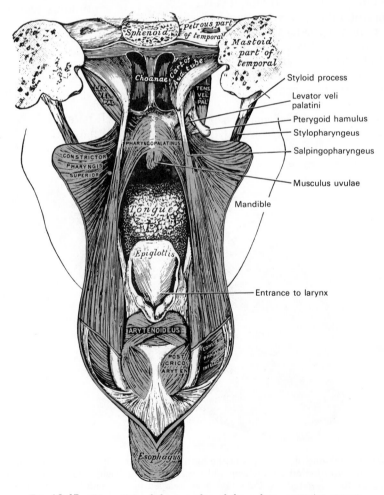

FIG. 16-45. Dissection of the muscles of the palate, posterior aspect.

nasal spine of the palatine bones and from the palatine aponeurosis. It descends to be *inserted* into the uvula.

PALATOGLOSSUS.   The palatoglossus (*glossopalatinus*) is a small fleshy fasciculus, narrower in the middle than at either end, which forms, with the mucous membrane covering its surface, the palatoglossal arch. It *arises* from the anterior surface of the soft palate, where it is continuous with the muscle of the opposite side. Passing down anterior to the palatine tonsil, it is *inserted* into the side of the tongue, some of its fibers spreading over the dorsum and others passing deeply to intermingle with the transverse muscle of the tongue.

PALATOPHARYNGEUS.   The palatopharyngeus (*pharyngopalatinus*) is a long, fleshy fasciculus narrower in the middle than at either end. With the mucous membrane covering its surface, it forms the palatopharyngeal arch. It is separated from the palatoglossus muscle by a triangular interval, in which the palatine tonsil is lodged. It *arises* from the soft palate, where it is divided into two fasciculi by the levator veli palatini and the musculus uvulae.

The **posterior fasciculus** lies in contact with the mucous membrane and joins with that of the opposite muscle in the midline; the **anterior fasciculus,** the thicker, lies in the soft palate between the levator and the tensor palatini and joins in the midline the corresponding part of the opposite muscle. Passing down posterior to the palatine tonsil, the palatopharyngeus joins the stylopharyngeus, and is *inserted* with that muscle into the posterior border of the thyroid cartilage. Some of its fibers are lost on the side of the pharynx and others pass across the midline posteriorly to decussate with the muscle of the opposite side.

NERVES.   The tensor veli palatini is supplied by a branch of the fifth cranial nerve. The remaining muscles of this group are supplied by the bulbar portion of the accessory nerve through the pharyngeal plexus (pharyngeal branch of the vagus nerve).

DEGLUTITION.   During the *first stage* of deglutition, the bolus of food is driven back into the fauces by the pressure of the tongue against the hard palate, the base of the tongue being, at the same time, retracted and the larynx raised with the pharynx. During the *second stage,* the entrance to the larynx is closed by drawing the arytenoid cartilages forward toward the cushion of the epiglottis—a movement produced by the contraction of the thyroarytenoid, the arytenoid, and the aryepiglotticus.

After leaving the tongue, the bolus passes on to the posterior or laryngeal surface of the epiglottis and glides along this for a distance; then the palatoglossi, the constrictors of the fauces, contract behind it. The soft palate is raised slightly by the levator veli palatini and made tense by the tensor veli palatini, and the palatopharyngei, by their contraction, pull the pharynx upward over the bolus and nearly come together; the uvula fills the slight interval between them. By these means the food is prevented from passing into the nasal part of the pharynx. At the same time, the palatopharyngei form an inclined plane directed obliquely downward and backward; along the undersurface of this plane, the bolus descends into the lower part of the pharynx. The salpingopharyngei raise the upper and lateral parts of the pharynx—i.e., those parts above the points where the stylopharyngei are attached to the pharynx.

STRUCTURE OF THE SOFT PALATE

MUCOUS MEMBRANE.   The mucous membrane of the soft palate is thin and covered with stratified squamous epithelium on both surfaces, except near the pharyngeal ostium of the auditory tube, where it is columnar and ciliated. The mucous membrane on the nasal surface of the soft palate in the fetus is covered throughout by columnar ciliated epithelium, which subsequently becomes squamous except at its free margin. Beneath the mucous membrane on the oral surface of the soft palate is a considerable amount of lymphoid tissue. The palatine glands form a continuous layer on its posterior surface and around the uvula.

VESSELS AND NERVES.   The **arteries** supplying the soft palate are the descending palatine branch of the maxillary artery, the ascending palatine branch of the facial artery, and the palatine branch of the ascending pharyngeal artery. The **veins** end chiefly in the pterygoid and tonsillar plexuses. The **lymphatic vessels** pass to the deep cervical nodes.

The **sensory nerves** are derived from the palatine, nasopalatine, and glossopharyngeal nerves.

The Pharynx

The **pharynx** is that part of the digestive tube placed posterior or dorsal to the nasal cavities, mouth, and larynx. It is a musculomembranous tube that extends from the inferior surface of the skull to the level of the cricoid cartilage ventrally and that of the sixth cervical vertebra dorsally.

The cavity of the pharynx (Fig. 16-43) is about 12.5 cm long and is broader in the transverse than in the anteroposterior diameter. Its greatest breadth is immediately below the base of the skull, where it projects on either side, behind the pharyngeal ostium of the auditory tube, as the **pharyngeal recess** (*fossa of Rosenmüller*[1]); its narrowest point is at its termination in the esophagus. It is limited, *superiorly*, by the body of the sphenoid and basilar part of the occipital bone; *inferiorly*, it is continuous with the esophagus. *Posteriorly*, it is separated by a fascial cleft from the cervical portion of the vertebral column, and the prevertebral fascia covering the longus colli and longus capitis muscles. *Anteriorly*, it is incomplete and is attached in succession to the medial pterygoid plate, pterygomandibular raphe, mandible, tongue, hyoid bone, and thyroid and cricoid cartilages. *Laterally*, it is connected to the styloid processes and their muscles, and is in contact with the common and internal carotid arteries, the internal jugular veins, the glossopharyngeal, vagus, and hypoglossal nerves, the sympathetic trunks, and small parts of the medial pterygoidei. Seven cavities communicate with it: the two nasal cavities, the two tympanic cavities, the mouth, the larynx, and the esophagus. The cavity of the pharynx may be subdivided into three parts: **nasal, oral,** and **laryngeal** (Fig. 16-17).

NASAL PART OF PHARYNX

The nasal part of the pharynx (*nasopharynx*) (Fig. 16-46) lies posterior to the nose and above the level of the soft palate: it differs from the oral and laryngeal parts of the pharynx in that its cavity always remains patent. It communicates through the choanae with the nasal cavities anteriorly. On its lateral wall is the **pharyngeal ostium of the auditory tube,** which is bounded posteriorly by the **torus,** a prominence caused by the protrusion of the medial end of the cartilage of the tube under the mucous membrane. A vertical fold of mucous membrane, the **salpingopharyngeal fold,** stretches down

[1] Johann Christian Rosenmüller (1771–1820): A German anatomist and surgeon (Leipzig).

from the torus and contains the salpingopharyngeus muscle. A second and smaller fold, the **salpingopalatine fold,** stretches from the upper part of the torus to the palate. Behind the ostium of the auditory tube is a deep recess, the **pharyngeal recess** (*fossa of Rosenmüller*) (Fig. 16-43).

On the posterior wall is a prominence produced by lymphatic tissue, which is known as the **pharyngeal tonsil;** during childhood this tonsil is likely to be hypertrophied into a considerable mass, at which time it is called **adenoids** (Fig. 16-43). Above the pharyngeal tonsil, in the midline, an irregular flask-shaped depression of the mucous membrane sometimes extends up as far as the basilar process of the occipital bone; this is known as the **pharyngeal bursa.**

ORAL PART OF PHARYNX

The oral part of the pharynx reaches from the soft palate to the level of the hyoid bone. It opens into the mouth anteriorly through the isthmus faucium. In its lateral wall, between the two palatine arches, is the **faucial** or **palatine tonsil.**

LARYNGEAL PART OF PHARYNX

The laryngeal part of the pharynx reaches from the hyoid bone to the lower border of the cricoid cartilage, where it is continuous with the esophagus. Anteriorly, the entrance of the larynx is formed by the epiglottis; laterally, its boundaries are the aryepiglottic folds. On either side of the laryngeal orifice is a recess, termed the **sinus piriformis,** which is bounded medially by the aryepiglottic fold and laterally by the thyroid cartilage and thyrohyoid membrane.

MUSCLES OF THE PHARYNX

The muscles of the pharynx (Fig. 16-47) are:

Inferior constrictor
Middle constrictor
Superior constrictor
Stylopharyngeus
Salpingopharyngeus
Palatopharyngeus

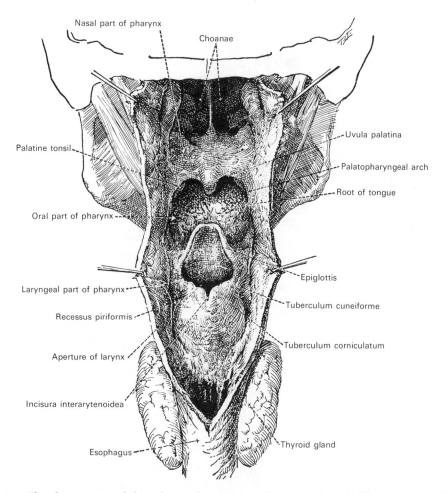

Nasal part of pharynx

Choanae

Uvula palatina

Palatine tonsil

Palatopharyngeal arch

Root of tongue

Oral part of pharynx

Epiglottis

Laryngeal part of pharynx

Tuberculum cuneiforme

Recessus piriformis

Tuberculum corniculatum

Aperture of larynx

Incisura interarytenoidea

Thyroid gland

Esophagus

FIG. 16-46. The pharynx viewed through a median incision of its posterior wall. (Eycleshymer and Jones.)

INFERIOR CONSTRICTOR. The inferior constrictor (Figs. 16-47; 16-48), the thickest of the three constrictors, *arises* from the sides of the cricoid cartilage in the interval between the cricothyroid muscle anteriorly and the articular facet for the inferior cornu of the thyroid cartilage posteriorly. It *arises* also from the oblique line on the side of the lamina of the thyroid cartilage, from the surface dorsal to this nearly as far as the posterior border, and from the inferior cornu. From these origins the fibers spread dorsalward and medialward to be *inserted* with the muscle of the opposite side into the fibrous raphe in the posterior median line of the pharynx. The inferior fibers are horizontal and continuous with the circular fibers of the esophagus; the rest ascend, increasing in obliquity, and overlap the middle constrictor.

MIDDLE CONSTRICTOR. The middle constrictor (Figs. 16-47; 16-48) is a fan-shaped

muscle, smaller than the inferior constrictor. It *arises* from the whole length of the superior border of the greater cornu of the hyoid bone, from the lesser cornu, and from the stylohyoid ligament. The fibers diverge from their origin: the lower ones descend beneath the inferior constrictor, the middle fibers pass transversely, and the upper fibers ascend and overlap the superior constrictor. It is *inserted* into the posterior median fibrous raphe, blending in the midline with the muscle of the opposite side.

SUPERIOR CONSTRICTOR. The superior constrictor (Figs. 16-47; 16-48) is a quadrilateral muscle, thinner and paler than the other two constrictors. It *arises* from the inferior third of the posterior margin of the medial pterygoid plate and its hamulus, from the pterygomandibular raphe, from the alveolar process of the mandible above the posterior end of the mylohyoid line, and by a few fibers from

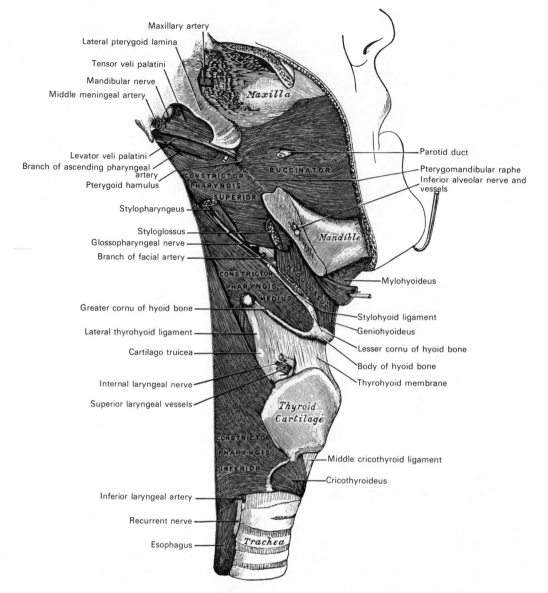

Maxillary artery

Lateral pterygoid lamina

Tensor veli palatini

Mandibular nerve

Middle meningeal artery

Maxilla

Levator veli palatini

Branch of ascending pharyngeal artery

Pterygoid hamulus

CONSTRICTOR PHARYNGIS SUPERIOR

BUCCINATOR

Parotid duct

Pterygomandibular raphe

Inferior alveolar nerve and vessels

Stylopharyngeus

Styloglossus

Glossopharyngeal nerve

Branch of facial artery

Mandible

CONSTRICTOR PHARYNGIS MEDIUS

HIOGLOSSUS

Mylohyoideus

Greater cornu of hyoid bone

Lateral thyrohyoid ligament

Cartilago truicea

Internal laryngeal nerve

Superior laryngeal vessels

Stylohyoid ligament

Geniohyoideus

Lesser cornu of hyoid bone

Body of hyoid bone

Thyrohyoid membrane

Thyroid Cartilage

CONSTRICTOR PHARYNGIS INFERIOR

Middle cricothyroid ligament

Cricothyroideus

Inferior laryngeal artery

Recurrent nerve

Esophagus

Trachea

Fig. 16-47.   The buccinator and muscles of the pharynx.

the side of the tongue. The fibers curve posteriorly to be *inserted* into the median raphe; they are also prolonged by means of an aponeurosis to the pharyngeal spine on the basilar part of the occipital bone. The superior fibers curve below the levator veli palatini and the auditory tube. The interval between the upper border of the muscle and the base of the skull is closed by the pharyngeal aponeurosis, and is known as the **sinus of Morgagni.**[1]

[1]Giovanni Battista Morgagni (1682–1771): An Italian anatomist and pathologist (Padua).

STYLOPHARYNGEUS.   The stylopharyngeus (Fig. 16-37), a long, slender muscle, *arises* from the medial side of the base of the styloid process, passes downward along the side of the pharynx between the superior and middle constrictors, and spreads out beneath the mucous membrane. Some of its fibers are lost in the constrictor muscles while others join with the palatopharyngeus and are *inserted* into the posterior border of the thyroid cartilage. The glossopharyngeal nerve runs on the lateral side of this muscle and crosses over it to reach the tongue.

SALPINGOPHARYNGEUS.   The salpingopha-

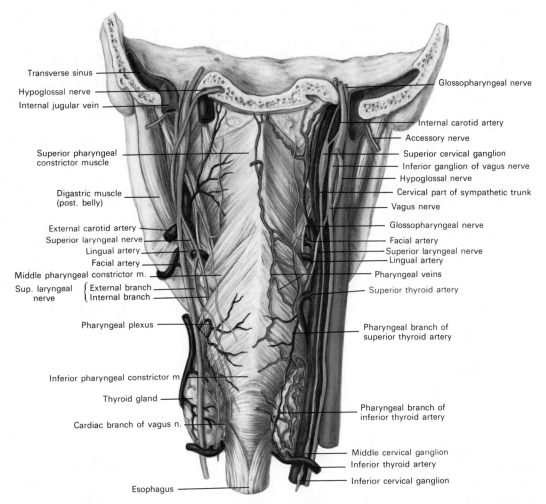

FIG. 16-48. The muscles of the pharynx and the vessels and nerves on its dorsal and lateral aspects; posterior view. (Redrawn after Sobotta.)

ryngeus (Fig. 16-45) *arises* from the inferior part of the auditory tube near its orifice; it passes downward and blends with the posterior fasciculus of the palatopharyngeus muscle.

PALATOPHARYNGEUS. The palatopharyngeus is described with the muscles of the palate.

NERVES. The constrictors and the salpingopharyngeus are supplied by branches from the pharyngeal plexus. The inferior constrictor is supplied by additional branches from the external laryngeal and recurrent nerves, and the stylopharyngeus is supplied by the glossopharyngeal nerve.

ACTIONS. When deglutition is about to be performed, the pharynx is drawn upward and dilated in different directions to receive the food propelled into it from the mouth. The stylopharyngei, which are much farther removed from one another at their ori-

gin than at their insertion, draw the sides of the pharynx upward and lateralward and so increase its transverse diameter; its breadth in the anteroposterior direction is increased by the larynx and tongue being carried forward in their ascent. As soon as the bolus of food is received in the pharynx, the elevator muscles relax, the pharynx descends, and the constrictors contract upon the bolus and convey it downward into the esophagus.

STRUCTURE OF THE PHARYNX

The pharynx is composed of three coats: **mucous, fibrous,** and **muscular.**

The **pharyngeal aponeurosis,** or **fibrous coat,** is situated between the mucous and muscular layers. It is thick superiorly where the muscular fibers are wanting, and is firmly connected to the basilar portion of the occipital and the petrous portions of the temporal bones. As it descends it diminishes in thickness and gradually blends with the fascia. A strong fibrous band attached to the pharyngeal spine on the inferior surface of the basilar portion of the occipital bone forms the median raphe, which serves as the insertion for all three of the pharyngeal constrictor muscles.

The **mucous coat** is continuous with that lining the nasal cavities, the mouth, the auditory tubes, and the larynx. In the nasal part of the pharynx it is covered by columnar ciliated epithelium; in the oral and laryngeal portions, the epithelium is stratified squamous. Beneath the mucous membrane are found racemose mucous glands; they are especially numerous at the cranial part of the pharynx around the orifices of the auditory tubes.

The **muscular coat** consists of the muscles of the pharynx.

The Esophagus

The **esophagus** or **gullet** (Fig. 16-49) is a muscular canal, about 23 to 25 cm long, which extends from the pharynx to the stomach. It begins in the neck at the inferior border of the cricoid cartilage, opposite the sixth cervical vertebra, and descends along the ventral aspect of the vertebral column. It passes through the superior and posterior mediastina and the diaphragm, and entering the abdomen, ends at the cardiac orifice of the stomach, opposite the eleventh thoracic vertebra. The general direction of the esophagus is vertical; but it has two slight curves in its course. At its commencement it is placed in the middle line; but it inclines to the left side as far as the root of the neck, gradually passes to the middle line again at the level of the fifth thoracic vertebra, and finally deviates to the left as it passes ventralward to the esophageal hiatus in the diaphragm. The esophagus also has anteroposterior flexures corresponding to the curvatures of the cervical and thoracic portions of the vertebral column. It is the narrowest part of the digestive tube, and is most contracted at its commencement and at the point where it passes through the diaphragm.

RELATIONS. *Cervical Portion.* The cervical portion of the esophagus is in relation, *ventrally,* with the trachea; and at the lower part of the neck,

where it projects to the left side, with the thyroid gland; *dorsally,* it rests upon the vertebral column and longus colli muscles; *on either side* it is in relation with the common carotid artery (especially the left, as it inclines to that side) and parts of the lobes of the thyroid gland. The recurrent nerves ascend between it and the trachea; to its left side is the thoracic duct.

Thoracic Portion. The thoracic portion of the esophagus is at first situated in the superior mediastinum between the trachea and the vertebral column, a little to the left of the midline. It then passes dorsal and to the right of the aortic arch and descends in the posterior mediastinum along the right side of the descending aorta. Then it runs ventral and a little to the left of the aorta and enters the abdomen through the diaphragm at the level of the tenth thoracic vertebra. Just before it perforates the diaphragm it has a distinct dilatation. It is in relation, *ventrally,* with the trachea, the left bronchus, the pericardium, and the diaphragm. *Dorsally,* it rests upon the vertebral column, the longus colli muscles, the right aortic intercostal arteries, the thoracic duct, and the hemiazygos veins; near the diaphragm, it rests upon the ventral surface of the aorta. On its *left* side in the superior mediastinum are the terminal part of the aortic arch, the left subclavian artery, the thoracic duct, and left pleura. Running upward in the angle between it and the trachea is the left recurrent nerve; caudally, it is in relation with the descending thoracic aorta. On its *right* side are the right pleura and the azygos vein, which it overlaps. Below the roots of the lungs the vagi descend in close contact with the esophagus; the right nerve passes downward dorsally and the left nerve is ventral to it, the two nerves forming a plexus around the tube.

In the caudal part of the posterior mediastinum, the thoracic duct lies to the right side of the esophagus. More cranially, it is placed dorsal to it, and, crossing about the level of the fourth thoracic vertebra, is continued upward on its left side.

Abdominal Portion. The abdominal portion of the esophagus lies in the esophageal groove on the posterior surface of the left lobe of the liver. It measures about 1.25 cm in length, and only its ventral and left aspects are covered by peritoneum. It is somewhat conical, with its base applied to the upper orifice of the stomach, and is known as the **cardiac antrum.**

STRUCTURE OF THE ESOPHAGUS

The esophagus has four coats: an **external** or **fibrous,** a **muscular,** a **submucous** or **areolar,** and an **internal** or **mucous coat** (Fig. 16-50).

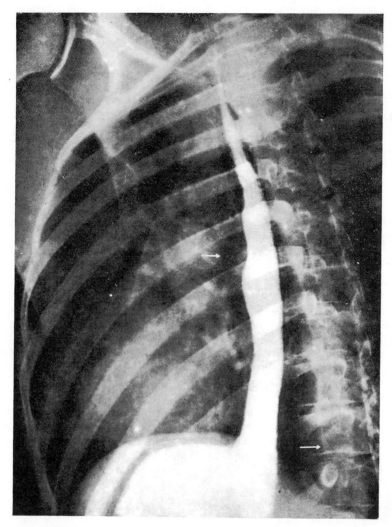

FIG. 16-49. Esophagus during the passage of a barium meal. Note that in the upper part of the esophagus, longitudinal folds in the mucous membrane can be identified. The upper arrow points to the shadow of the right bronchus; the lower arrow indicates the tenth thoracic vertebra. Note that the lower part of the esophagus inclines forward away from the vertebral column.

MUSCULAR COAT. The muscular coat is composed of two planes of considerable thickness: an external plane of longitudinal and an internal plane of circular fibers.

The *longitudinal fibers* are arranged at the commencement of the tube in three fasciculi: one ventral, which is attached to the vertical ridge on the posterior surface of the lamina of the cricoid cartilage by the *tendo cricoësophageus;* and one at either side, which is continuous with the muscular fibers of the pharynx. As they descend, these fibers blend together to form a uniform layer that covers the outer surface of the tube.

Accessory slips of muscular fibers pass between the esophagus and the left pleura, where the latter covers the thoracic aorta, the root of the left bronchus, or the back of the pericardium.

The *circular fibers* are continuous above with the inferior constrictor; their direction is transverse at the cranial and caudal parts of the tube, but oblique in the intermediate part.

The muscular fibers in the cranial part of the esophagus are red and consist chiefly of striated muscle; the intermediate part is mixed; and the lower part, with rare exceptions, contains only smooth muscle.

AREOLAR COAT. The areolar or submucous coat loosely connects the mucous and muscular coats. It contains blood vessels, nerves,

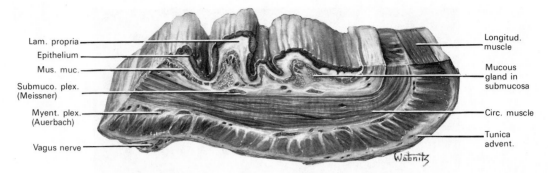

Lam. propria

Epithelium

Mus. muc.

Submuco. plex.
(Meissner)

Myent. plex.
(Auerbach)

Vagus nerve

Longitud.
muscle

Mucous
gland in
submucosa

Circ. muscle

Tunica
advent.

Wabnitz

FIG. 16-50.   Cross section through lower part of esophagus of adult cadaver. 5 ×.

and mucous glands. The **esophageal glands** are small, compound racemose glands of the mucous type; they are lodged in the submucous tissue, and each opens upon the surface by a long excretory duct.

MUCOUS COAT.   The mucous coat is thick and reddish cranially and pale caudally. It is disposed in longitudinal folds that disappear on distention of the tube. Its surface is covered throughout with a thick layer of stratified squamous epithelium. Beneath the mucous membrane, between it and the areolar coat, is a layer of longitudinally arranged, nonstriped muscular fibers; this is the **muscularis mucosae.** At the commencement of the esophagus it is absent or is represented by only a few scattered bundles; more caudally it forms a large stratum.

VESSELS AND NERVES.   The **arteries** supplying the esophagus are derived from the inferior thyroid branch of the thyrocervical trunk, from the descending thoracic aorta, from the bronchial arteries, from the left gastric branch of the celiac artery, and from the left inferior phrenic branch of the abdominal aorta. For the most part they have a longitudinal direction.

The **veins** end in the inferior thyroid, azygos, hemiazygos, and gastric veins, thereby forming an important anastomosis between the portal and systemic venous systems.

The **nerves** are derived from the recurrent vagus, which supplies the striated musculature of the organ, and from the vagus and sympathetic trunks, which supply fibers to the smooth musculature. These cranial parasympathetic and sympathetic fibers form plexuses between the two layers of the muscular coat and in the submucosa, as in the stomach and intestines.

The Abdomen

The **abdomen** contains the largest cavity in the body. The cranial boundary is formed by the diaphragm, which extends over it as a dome, so that the cavity extends high into the bony thorax; on the right side it reaches the cranial border of the fifth rib in the mammary line, while on the left side it falls below this level by about 2.5 cm. The caudal boundary is formed principally by the levator ani and coccygeus or diaphragm of the pelvis. To facilitate description, the abdomen is artificially divided into two parts: a cranial and larger part, the *abdomen proper* and a caudal and smaller part, the *pelvis.* The limit between these parts is marked by the superior aperture of the lesser pelvis.

The **abdomen proper** differs from the other great cavities of the body in being bounded for the most part by muscles and fasciae, so that it can vary in capacity and shape according to the condition of the viscera that it contains; in addition, the abdomen varies in form and extent with age and sex. In the adult male, with moderate distention of the viscera, it is oval but flattened dorsoventrally. In the adult female with a fully developed pelvis, it is ovoid, with the narrower pole cranialward; in young children it is also ovoid, but with the narrower pole caudalward.

BOUNDARIES

The abdomen is bounded *ventrally* and at the sides by the abdominal muscles and the iliacus; *dorsally* by the vertebral column and the psoas and quadratus lumborum muscles; *cranially* by the diaphragm; and *caudally* by the plane of the superior aperture of the lesser pelvis. The muscles forming the boundaries of the cavity are lined upon their inner surfaces by transversalis fascia.

The abdomen contains the greater part of the digestive tube, the liver and pancreas, the spleen, the kidneys, and the suprarenal glands. Most of these structures, as well as the wall of the cavity in which they are contained, are more or less covered by an extensive and complicated serous membrane, the peritoneum.

APERTURES

The apertures in the walls of the abdomen for the transmission of structures to or from it are, *cranially,* the **vena caval opening,** the **aortic hiatus,** and the **esophageal hiatus.** *Caudally,* there are two apertures on either side: one for the passage of the femoral vessels and lumboinguinal nerve, and the other for the transmission of the spermatic cord in the male and the round ligament of the uterus in the female. In the fetus, the **umbilicus** transmits the umbilical vessels, the allantois, and the vitelline duct through the ventral wall.

REGIONS

A simple but effective division of the abdomen commonly used by clinicians is established by two lines intersecting at right angles: the median sagittal line and a transverse line through the umbilicus divide the abdomen into **four quadrants,** an upper and lower right, and an upper and lower left. The more accurate and elaborate division described and depicted in the third chapter (p. 91) is used for reference in many descriptions of abdominal structures.

The following description is based upon the inspection of a cadaver. Always remember that the abdominal organs are highly movable during life, and the arbitrary placement in this account is used to provide a basic knowledge for adaptation to conditions that are seen in patients. Refer frequently to Figures 3-17, 3-19, 3-20, and 3-23 to relate the organs to the skeleton.

DISPOSITION OF THE ABDOMINAL VISCERA

When the anterior abdominal wall of the cadaver is removed, only a few of the undisturbed organs are plainly visible (Fig. 16-51). Cranially and to the right is the **liver,** with the falciform ligament marking its ventral surface into the right and left lobes. The narrow left lobe extends across the median line, partly concealing the lesser curvature of the stomach. The **stomach** occupies the left cranial region more or less completely, depending on the degree of its distention. The tip of the **gallbladder** may protrude from the caudal border of the right lobe of the liver, and between this and the stomach is the pylorus and the first part of the **duodenum.** Draped caudalward from the greater curvature of the stomach and the first part of the duodenum, a filmy apron, the **greater omentum** (or epiploon), extends caudalward, concealing most of the other intestines. Caudally and on the right, parts of the **descending** and **sigmoid colon** are usually visible and a variable number of coils of the small intestine appear in the caudal part of the cavity. Just above the pubis, the **urinary bladder** is visible if partly distended.

The **large** and **small intestines** are brought into view if the omentum is lifted and turned upward over the stomach and liver (Fig. 16-52). The large intestine forms an arch with the ascending colon on the right, the descending and sigmoid on the left, and the transverse colon, with the omentum attached, forming the top of the arch. The small intestine may be lifted and drawn to the left to display its attachment to the mesentery and the attachment of the colon to the transverse mesocolon. The blind end of the ascending colon, the **cecum,** lies in the hollow of the right ilium, and near the ileocecal junction, the **vermiform appendix** may or may not be visible.

Further inspection requires removal of some of the organs and structures, namely, the small intestine except the duodenum and terminal ileum, the greater omentum, the transverse colon with its mesocolon, and the left lobe of the liver (Fig. 16-53). It is then possible to see the terminal part of the esophagus as it passes through the diaphragm and joins the cardiac end of the stomach. The filmy **lesser omentum** stretches between the lesser curvature of the stomach and the liver and duodenum. The head of the **pancreas** lies in the loop formed by the other three parts of the duodenum. The **spleen** can be brought into view between the greater curvature of the stomach and the diaphragm if the stomach is pulled forcibly toward the liver. Since it is placed rather far dorsally, it may not be seen but it can easily be palpated. If the subject is not obese, a smooth rounded prominence just caudal to the right and left flexures of the

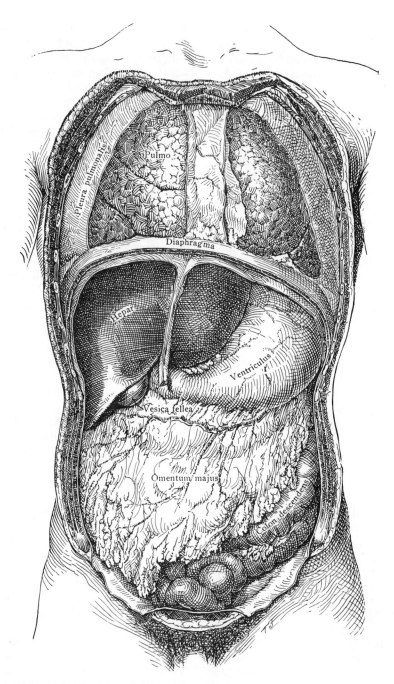

Fig. 16-51.   Ventral view of the thoracic and abdominal viscera in position after removal of the anterior thoracic and abdominal walls. (Eycleshymer and Jones.)

colon will represent the caudal poles of the two **kidneys.** The sigmoid colon is completely visible and its junction with the rectum may be detected. In the female, if the sigmoid colon is drawn to the left and caudalward, the **uterus** lying in the midline and the uterine tubes on each side may be brought into view. In the midline extending caudally, the prominence caused by the ab-

dominal **aorta** and its bifurcation into the iliac arteries should be discernible.

The Peritoneum

The **peritoneum** (*tunica serosa*) is the most extensive serous membrane in the body. It consists of a closed sac in the male,

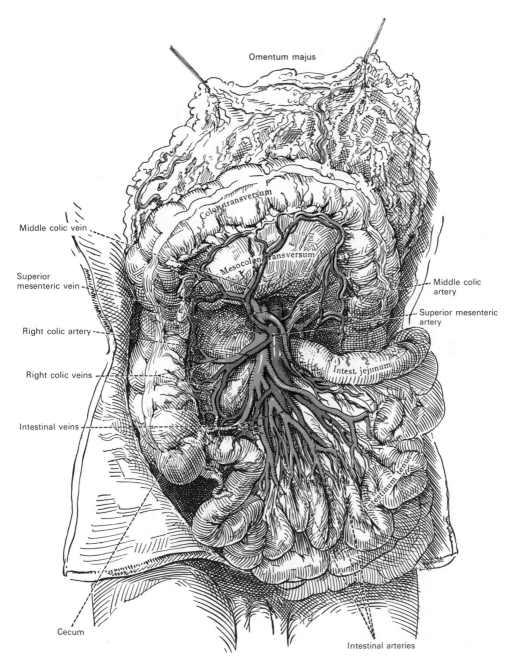

Fig. 16-52. Small and large intestines with their mesenteries and blood vessels viewed after the greater omentum has been drawn upward over the chest. (Eycleshymer and Jones.)

part of which is applied against the abdominal parietes, while the remainder is reflected over the contained viscera. In the female the peritoneum is not a closed sac, because the free ends of the uterine tubes open directly into the peritoneal cavity. The part that lines the abdominal wall is named the **parietal peritoneum;** that which is reflected over the contained viscera constitutes the **visceral peritoneum.**

The *free surface* of the membrane is a smooth layer of flat mesothelium lubricated by a small quantity of serous fluid, which allows the viscera to glide freely against the wall of the cavity or upon each other with the least possible friction. The *attached* surface is connected to the viscera and inner surface of the parietes by means of areolar tissue, which is termed the **subserous fascia.** The parietal portion is separated by a fascial

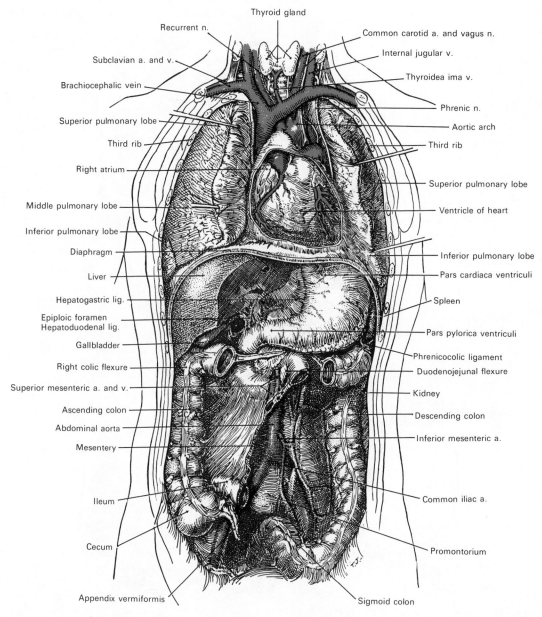

FIG. 16-53. Ventral view of the thoracic and abdominal viscera partially dissected. The anterior pleurae and pericardium have been removed; the structures at the root of the neck, dissected. The left lobe of the liver, the greater omentum, transverse colon, jejunum, and ileum have been removed. (Eycleshymer and Jones.)

cleft from the transversalis fascia that lines the abdomen and pelvis, but it adheres more closely to the undersurface of the diaphragm and in the midline of the anterior wall.

The space between the parietal and visceral layers of the peritoneum is named the **peritoneal cavity,** but under normal conditions this cavity is merely a potential one, since the parietal and visceral layers are in contact. The peritoneal cavity is divided by a narrow constriction into a greater sac and a lesser sac. The **greater sac,** commonly referred to simply as the peritoneal cavity, is related to the majority of the abdominal structures. The **lesser sac** is named the **omental bursa** and is related only to the dorsal surface of the stomach and closely surrounding structures. The constriction between the two sacs occurs between the liver and the duodenum and is named the **epi-**

ploic foramen or *foramen of Winslow*[1] (Figs. 16-53; 16-56).

Certain abdominal viscera are completely surrounded by peritoneum and are suspended from the wall by a thin sheet of peritoneum-covered connective tissue carrying the blood vessels. These sheets are given the general name of mesenteries. Other viscera are more closely attached to the abdominal wall and are only partly covered by peritoneum rather than suspended; these viscera are said to be **retroperitoneal.** No term corresponding to retroperitoneal has been adopted for the suspended viscera; "intraperitoneal" is unsatisfactory because this would imply that they are inside the peritoneal cavity. The term **mesentery,** when used specifically, refers only to the peritoneal suspension of the small intestine: "entery" refers to the intestine and "mes-" to the peritoneum. Thus, the names of the suspending folds of other organs use only the prefix "meso," for example, the mesocolon, mesoappendix, and mesovarium.

The complicated disposition and variations of the peritoneum in the adult body can be understood only through frequent reference to the changes in position and attachment of the viscera during embryonic and fetal development described in the first part of this chapter.

PARIETAL PERITONEUM

This discussion of the parietal peritoneum considers the abdominal and pelvic regions, including the omental bursa.

Abdominal Region

The peritoneum covering the internal surface of the ventral abdominal wall is smooth and relatively uninterrupted by folds and attachments. Certain vestiges of embryonic structures, however, have remained. Between the pubic bone and the umbilicus, a fibrous band in the midline, the middle umbilical ligament, represents the remains of the urachus, and 3 or 4 cm lateral to it are the medial umbilical ligaments, which are fibrous remnants of the obliterated umbilical arteries. The fold of peritoneum over the

urachal remnant is the **middle umbilical fold** (Fig. 16-54); the fold over the arterial remnants is the **medial umbilical fold.** Three or four centimeters lateral to the latter fold is the **lateral umbilical fold,** which is produced by a slight protrusion of the inferior epigastric artery and the interfoveolar ligament. The contracted bladder may have transverse folds extending laterally from its fundus, the **transverse vesical folds.**

Extending cranialward from the umbilicus is the **falciform ligament** of the liver. It is a thin fibrous sheet with a fibrous cord, the **ligamentum teres,** or remnant of the umbilical vein, in its free crescentic border. Its attachment to the wall is narrow from the umbilicus to the cranial surface of the liver, where its two leaves separate and the reflections extend laterally to become the anterior leaf of the **coronary ligament of the liver.**

The **anterior leaf** of the coronary ligament is the anterior boundary of an irregularly oval or triangular area of liver surface, approximately 15 cm in diameter, which has no peritoneal covering and is called the **bare area** of the liver. The posterior boundary of this area is the **posterior leaf** of the coronary ligament, where the peritoneum is reflected from the liver to the diaphragmatic portion of the dorsal abdominal wall. Between the anterior and posterior coronary leaves, the reflection extends laterally into a crescentic fold on each side, forming the **right** and **left triangular ligaments** of the liver (Fig. 16-55).

On the *left side,* beginning at the falciform and left triangular ligaments on the *anterior abdominal wall,* the peritoneum is uninterrupted by attachments as it passes around the lateral wall and over part of the posterior wall, until it reaches the **gastrophrenic** and **phrenicolienal ligaments.** The latter represent the adult position of the fetal dorsal mesogastrium. The phrenicolienal ligament angles laterally and its more caudal part is reflected from the cranial pole of the left kidney and then meets the splenic or left flexure of the colon. The peritoneum at the reflection of the flexure is prolonged laterally for several centimeters into a narrow fold, the **phrenicocolic ligament,** which represents the left lateral extent of the greater omentum.

The peritoneum of the *anterior and left lateral abdominal wall* from this ligament down to the pelvis and into the rectovesical fossa is uninterrupted by attachments. It is

---

[1]Jacob Benignus Winslow (1669-1760): A Danish anatomist who worked in Paris.

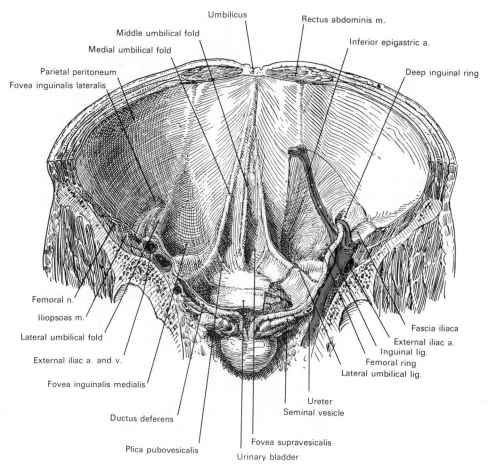

FIG. 16-54. The lower portion of the anterior abdominal wall, viewed from within. The peritoneum has been partially removed from the right side. (Eycleshymer and Jones.)

reflected over the descending colon without mesentery or mesocolon; then the **sigmoid mesocolon** swings laterally into a reflection over the rectum.

The peritoneum of the *right side* of the *anterior* and *lateral abdominal wall* extends caudally into the iliac region uninterrupted by attachments. Along the *right side* of the *posterior abdominal wall,* there is a reflection to the ascending colon and cecum. Between the right triangular ligament and the attachment of the right or hepatic flexure of the colon, an unattached peritoneal surface 3 or 4 cm wide extends medialward and cranialward between the anterior leaf of the coronary ligament of the liver cranially and the **transverse mesocolon** and reflection over the superior part of the duodenum caudally. Medial to the duodenum, the peritoneum of the posterior wall covers the inferior vena cava, and this structure marks the position of the **epiploic foramen,** which

leads toward the left into the **lesser sac** or **omental bursa.**

The reflections of peritoneum from the posterior wall that form the *boundaries of the lesser sac,* beginning at the vena cava, are as follows: the coronary ligament of the liver, along the caudate lobe to the diaphragmatic end of the lesser omentum, the esophagus and the gastrophrenic ligament, the phrenicolienal ligament, containing the tail of the pancreas, the transverse mesocolon, and the first part of the duodenum (Fig. 16–57).

The reflection of peritoneum to the ascending colon extends caudalward along the *right dorsolateral wall* to the cecum in the right iliac fossa. After rounding the blind end of the cecum, the peritoneum from the dorsal wall is reflected to the mesentery of the ileum and the jejunum. The attachment of the mesentery extends medialward as well as cranialward until it reaches the junc-

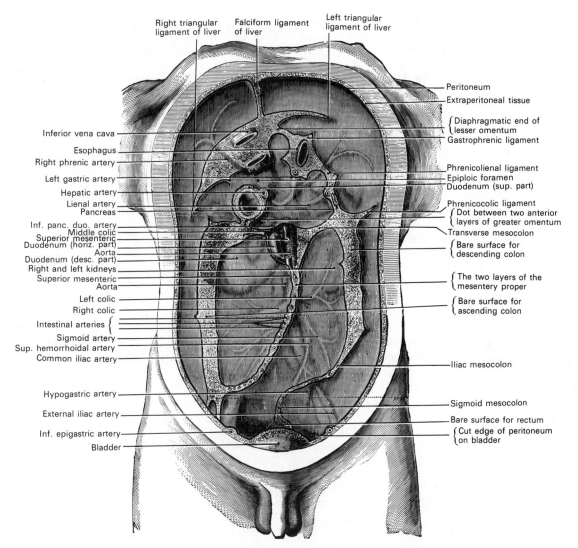

Right triangular ligament of liver
Falciform ligament of liver
Left triangular ligament of liver

Peritoneum
Extraperitoneal tissue
{ Diaphragmatic end of lesser omentum
Gastrophrenic ligament
Phrenicolienal ligament
Epiploic foramen
Duodenum (sup. part)
Phrenicocolic ligament
{ Dot between two anterior layers of greater omentum
Transverse mesocolon
{ Bare surface for descending colon
{ The two layers of the mesentery proper
{ Bare surface for ascending colon

Iliac mesocolon

Sigmoid mesocolon
Bare surface for rectum
{ Cut edge of peritoneum on bladder

Inferior vena cava
Esophagus
Right phrenic artery
Left gastric artery
Hepatic artery
Lienal artery
Pancreas
Inf. panc. duo. artery
Middle colic
Superior mesenteric
Duodenum (horiz. part)
Aorta
Duodenum (desc. part)
Right and left kidneys
Superior mesenteric
Aorta
Left colic
Right colic
Intestinal arteries {
Sigmoid artery
Sup. hemorrhoidal artery
Common iliac artery

Hypogastric artery
External iliac artery
Inf. epigastric artery
Bladder

FIG. 16-55. Diagram devised by Delepine to show the lines along which the peritoneum leaves the wall of the abdomen to invest the viscera.

tion of the duodenum with the jejunum at the caudal leaf of the transverse mesocolon. A triangular area of dorsal wall to the right of the mesentery is thus completely surrounded by peritoneal reflections: the ascending colon laterally, the transverse mesocolon cranially, and the mesentery medially. The peritoneum in this triangular area covers part of the right kidney and duodenum.

Pelvic Region

The peritoneum of the dorsal wall between the mesentery, transverse mesocolon, and descending and sigmoid colons extends into the right side of the pelvis between the rectum and the bladder. Here it follows closely the surfaces of the pelvic viscera and the inequalities of the pelvic walls, and presents important differences in the two sexes.

MALE. In the male (Fig. 16-55), the peritoneum encircles the sigmoid colon, from which it is reflected to the posterior wall of the pelvis as a fold, the **sigmoid mesocolon.** It then leaves the sides and finally the ventral surface of the rectum, and is continued onto the cranial ends of the seminal vesicles and the bladder. On either side of the rectum it forms the **pararectal fossa,** which varies in size with the distention of the rectum. Ventral to the rectum, the peritoneum forms the **rectovesical excavation,** which is limited laterally by peritoneal folds extending from the sides of the bladder to the rectum and sacrum. Because of their positions, these folds are known as the **rectovesical** or **sacrogenital folds.**

The peritoneum of the ventral pelvic wall covers the superior surface of the bladder and on either side of this viscus forms a depression, termed the **paravesical fossa,** which is limited laterally by the fold of peritoneum covering the ductus deferens. The size of this fossa depends on the state of distention of the bladder; when the bladder is empty, a variable fold of peritoneum, the **plica vesicalis transversa,** divides the fossa into two portions. Between the paravesical and pararectal fossae, the only elevations of the peritoneum are those produced by the ureters and the internal iliac vessels (Fig. 17–15).

FEMALE. In the female, pararectal and paravesical fossae similar to those in the male are present: the lateral limit of the paravesical fossa is the peritoneum investing the round ligament of the uterus. The rectovesical excavation, however, is divided by the uterus and vagina into a small anterior vesicouterine excavation and a large, deep, posterior rectouterine excavation. The sacrogenital folds form the margins of the latter, and are continued onto the dorsum of the uterus to form a transverse fold, the **torus uterinus.** The **broad ligaments** extend from the sides of the uterus to the lateral walls of the pelvis; they contain, in their free margins, the uterine tubes, and the ovaries are suspended from their posterior layers. On the lateral pelvic wall dorsal to the attachment of the broad ligament, in the angle between the elevations produced by the diverging internal iliac and external iliac vessels, is the slight **ovarian fossa,** in which the ovary normally lies (Fig. 17–59).

Omental Bursa

The omental bursa or lesser peritoneal sac is so named because a part of its wall is formed by the two omenta (to be described later). On the dorsal abdominal wall, the peritoneum of the greater sac is continuous with that of the lesser sac as it crosses the ventral surface of the inferior vena cava a short distance to the right of the midline (Fig. 16-57). This passage of continuity between the two sacs is the **epiploic foramen** or *foramen of Winslow.*

EPIPLOIC FORAMEN. The epiploic foramen (Fig 16-56) is a peritoneal-covered passage that usually admits two fingers. It can easily be located with a probing finger in the abdominal cavity by pushing a way between the inferior vena cava and the free edge of the hepatoduodenal ligament in the region between the neck of the gallbladder and the first part of the duodenum (Fig. 16-56). The epiploic foramen is bounded *ventrally* by the free border of the lesser omentum, with the common bile duct, the hepatic artery, and the portal vein between its two layers; *dorsally* by the peritoneum covering the inferior vena cava; *cranially* by the peritoneum on the caudate process of the liver; and *caudally* by the peritoneum covering the commencement of the duodenum and the hepatic artery before it passes between the two layers of the lesser omentum.

BOUNDARIES OF OMENTAL BURSA. The boundaries of the omental bursa will now be evident. It is bounded *ventrally* by the caudate lobe of the liver, the lesser omentum, the stomach, and the greater omentum. *Dorsally,* it is limited by the greater omentum, the transverse colon, the transverse mesocolon, the ventral surface of the pancreas, the left suprarenal gland, and the cranial end of the left kidney. To the right of the esophageal opening of the stomach it is formed by the part of the diaphragm that supports the caudate lobe of the liver. *Laterally,* the bursa extends from the epiploic foramen to the hilum of the spleen, where it is limited by the phrenicolienal and gastrolienal ligaments.

The omental bursa, therefore, consists of a series of pouches or **recesses** to which the following terms are applied: (1) the **vestibule,** a narrow channel that continues from the epiploic foramen over the head of the pancreas to the gastropancreatic fold; this fold extends from the omental tuberosity of the pancreas to the right side of the fundus of the stomach and contains the left gastric artery and the coronary vein; (2) the **superior omental recess,** which is found between the caudate lobe of the liver and the diaphragm; (3) the **lienal recess,** located between the spleen and the stomach; (4) the **inferior omental recess,** which comprises the remainder of the bursa (Fig. 16-57).

In the fetus the bursa reaches as low as the free margin of the greater omentum, but in the adult its vertical extent is usually more

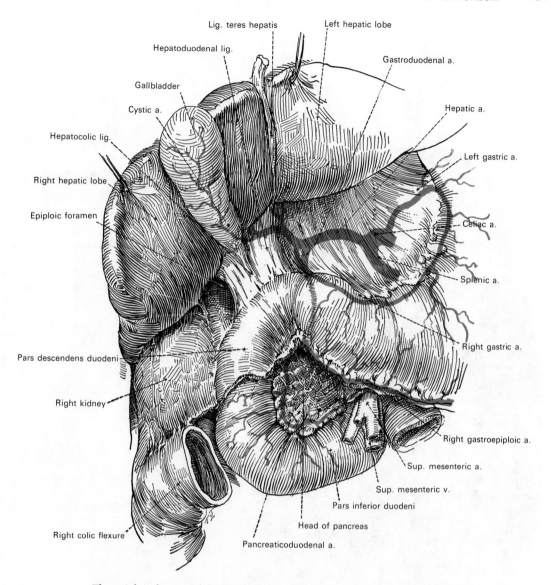

Lig. teres hepatis

Left hepatic lobe

Hepatoduodenal lig.

Gastroduodenal a.

Gallbladder

Cystic a.

Hepatic a.

Hepatocolic lig.

Left gastric a.

Right hepatic lobe

Celiac a.

Epiploic foramen

Splenic a.

Pars descendens duodeni

Right gastric a.

Right kidney

Right gastroepiploic a.

Sup. mesenteric a.

Sup. mesenteric v.

Pars inferior duodeni

Head of pancreas

Right colic flexure

Pancreaticoduodenal a.

Fɪɢ. 16-56.   The epiploic foramen (of Winslow) and neighboring structures. (Eycleshymer and Jones.)

limited owing to adhesions between the layers of the omentum. During a considerable part of fetal life the transverse colon is suspended from the posterior abdominal wall by a mesentery of its own, the greater omentum passing ventral to the colon (Fig. 16-13). This condition occasionally persists throughout life, but as a rule adhesion occurs between the mesentery of the transverse colon and the posterior layer of the greater omentum, so that the colon appears to receive its peritoneal covering by the splitting of the two posterior layers of the latter fold (Fig. 16-14). In the adult the omental bursa inter-

venes between the stomach and the structures on which the viscus lies, and therefore performs the functions of a serous bursa for the stomach.

VISCERAL PERITONEUM

The visceral peritoneum covers all or part of each abdominal organ and forms the various mesenteries, omenta, and ligaments associated with them. Numerous peritoneal folds extend between the various organs or connect them to the parietes; they serve to

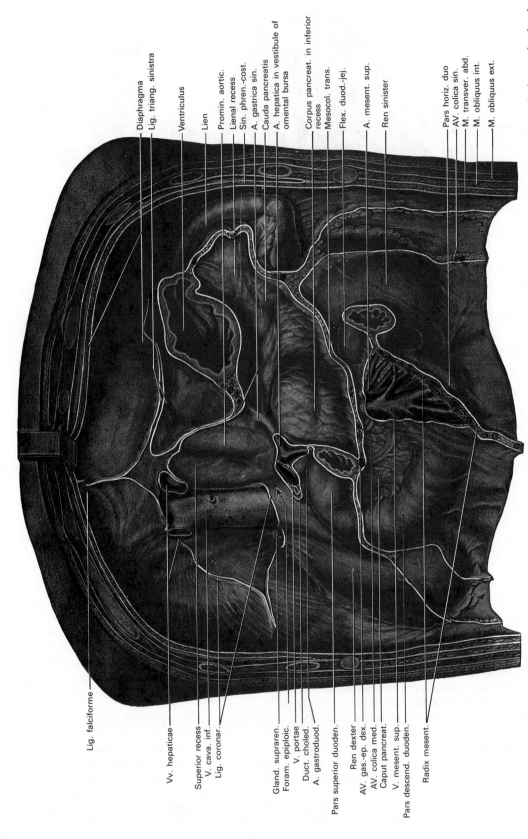

Diaphragma
Lig. triang. sinistra
Ventriculus
Lien
Promin. aortic.
Lienal recess
Sin. phren.-cost.
A. gastrica sin.
Cauda pancreatis
A. hepatica in vestibule of omental bursa
Corpus pancreat. in inferior recess
Mesocol. trans.
Flex. duod.-jej.
A. mesent. sup.
Ren sinister
Pars horiz. duo
AV. colica sin.
M. transver. abd.
M. obliquus int.
M. obliquus ext.

Lig. falciforme

Vv. hepaticae

Superior recess
V. cava. inf.
Lig. coronar.

Gland. supraren.
Foram. epiploic.
V. portae
Duct. choled.
A. gastroduod.

Pars superior duoden.

Ren dexter
AV. gas.-ep. dex.
AV. colica med.
Caput pancreat.
V. mesent. sup.
Pars descend. duoden.

Radix mesent.

FIG. 16-57. View of the posterior wall of the lesser sac of the peritoneum (omental bursa), showing the attachment of the ligaments of the liver, the hepatoduodenal ligament, the root of the mesentery and transverse mesocolon, and reflection of peritoneum from the stomach. The aorta, pancreas, kidneys, spleen, and duodenum are covered by periotneum. (Töndury, *Angewandte und topographische Anatomie*, courtesy of Georg Thieme Verlag.)

hold the viscera in position and at the same time enclose the vessels and nerves proceeding to them.

Mesenteries

The mesenteries are: the mesentery proper, the transverse mesocolon, and the sigmoid mesocolon. In addition, an ascending and a descending mesocolon are sometimes present.

MESENTERY PROPER. The mesentery proper is the broad, fan-shaped fold of peritoneum that connects the convolutions of the jejunum and ileum with the dorsal wall of the abdomen. Its *root*—the part connected with the structures ventral to the vertebral column—is about 15 cm long, extending obliquely from the duodenojejunal flexure at the left side of the second lumbar vertebra to the right sacroiliac articulation (Fig. 16–55). Its *intestinal border* is about 6 meters long, and here the two layers separate to enclose the intestine and form its peritoneal coat. The cranial end is narrow but widens rapidly to about 20 cm, and it is thrown into numerous plaits or folds. This part of the mesentery suspends the small intestine and contains between its layers the intestinal branches of the superior mesenteric artery, with the accompanying tributaries of the portal vein, the plexuses of nerves, the lacteal vessels, and the mesenteric lymph nodes.

TRANSVERSE MESOCOLON. The transverse mesocolon is a broad fold that connects the transverse colon to the dorsal wall of the abdomen. It is continuous with the greater omentum along the ventral surface of the transverse colon. Its two peritoneal layers diverge along the anterior border of the pancreas. It contains between its layers the vessels that supply the transverse colon.

SIGMOID MESOCOLON. The sigmoid mesocolon is the fold of peritoneum that retains the sigmoid colon in connection with the pelvic wall. Its line of attachment forms a U-shaped curve, with the apex of the curve placed about the point of division of the left common iliac artery. It is continuous with the **iliac mesocolon** on the left and ends in the median plane at the level of the third sacral vertebra over the rectum. The sigmoid and superior rectal vessels run between the two layers of this fold.

OTHER FEATURES. In most cases the peritoneum covers only the ventral surface and sides of the ascending and descending parts of the colon. Sometimes, however, these parts of the colon are surrounded by the serous membrane and are attached to the posterior abdominal wall by an ascending and a descending mesocolon respectively. A fold of peritoneum, the **phrenicocolic ligament,** is continued from the left colic flexure to the diaphragm opposite the tenth and eleventh ribs; it forms a pocket that supports the spleen and therefore has received the name, **sustentaculum lienis.**

The **appendices epiploicae** are small pouches of peritoneum that are filled with fat and situated along the colon and upper part of the rectum. They are appended chiefly to the transverse and sigmoid parts of the colon.

Omenta

There are two omenta, the lesser and the greater.

LESSER OMENTUM. The lesser omentum (*omentum minus; small omentum; gastrohepatic omentum*) extends to the liver from the lesser curvature of the stomach and the commencement of the duodenum. It is continuous with the two layers of peritoneum that cover respectively the ventral and dorsal surfaces of the stomach and the first part of the duodenum. The two peritoneal layers leave the stomach and duodenum as a thin membrane and ascend to the porta of the liver. To the left of the porta, the omentum is attached to the bottom of the fossa for the ductus venosus, along which it extends to the diaphragm, where the two layers separate to embrace the end of the esophagus. At the right, the omentum ends in a free margin, which constitutes the ventral boundary of the epiploic foramen. The portion of the lesser omentum located between the liver and stomach is termed the **hepatogastric ligament;** that between the liver and duodenum is the **hepatoduodenal ligament** (Fig. 16-56). Between the two layers of the hepatoduodenal ligament, close to the right free margin, are the hepatic artery, the common bile duct, the portal vein, lymphatics, and the hepatic plexus of nerves; all of these structures are enclosed in a fibrous capsule **(Glisson's capsule[1]).** The right and left gastric vessels run

[1]Francis Glisson (1597–1677): An English physician and anatomist (Cambridge).

between the layers of the lesser omentum, where they are attached to the stomach.

GREATER OMENTUM. The greater omentum (*omentum majus; gastrocolic omentum*) (Fig. 16-51) is a filmy apron draped over the transverse colon and coils of the small intestine. It is attached along the greater curvature of the stomach and the first part of the duodenum; its left border is continuous with the gastrolienal ligament. If it is lifted and turned back cranialward over the stomach and liver, one sees that it adheres to the transverse colon along the latter's whole length across the abdomen (Fig. 16-52). The greater omentum may be composed of a single membrane covered with peritoneum on both surfaces from the colon to its free caudal border, but in some people a variable amount of omentum near the colon is often composed of two peritoneum-covered membranes with a pocket between them. If the greater omentum contains this pocket, only the dorsal of the two membranes is attached to the colon, and the pocket opens into the omental bursa. This is best understood by referring to the conditions in the fetus depicted in the diagrams of Figure 16-14.

The membrane forming the greater omentum is thin, transparent, and fenestrated, except where there are blood vessels and accumulations of fat. The membrane between the stomach and the colon is the **gastrocolic ligament.** The right and left gastroepiploic blood vessels run between the two peritoneal layers of this ligament near their attachment to the greater curvature of the stomach, and they must be avoided when the ligament is cut to gain access to the omental bursa.

The greater omentum is a remarkable structure. It is greatly movable in the living individual and seems to have the ability to spread itself in areas where its presence is useful for the bodily economy. For example, in a patient with a ruptured appendix, the omentum may be found covering the area, walling off the infection to form an abscess and preventing generalized peritonitis. Occasionally the omentum finds its way into the sac of a hernia, where it may form adhesions, close up the sac, and produce a "spontaneous repair" of the hernia. In a cadaver it may be free and spread over the intestines or bunched up near the colon or in a recess. Adhesions, except along the stomach and colon, are abnormal and the result of inflammatory processes.

Ligaments

The term "ligament" has two meanings. When applied to structures related to a joint, it is a strong fibrous cord or sheet. When applied to the peritoneum it is merely a layer of serous membrane and has little or no tensile strength. The **ligaments of the peritoneum** are the parts of the membrane that extend between two structures and usually derive their names from these two structures—the gastrolienal ligament, for example. The ligaments are described in detail along with the organs to which they are related.

PERITONEAL RECESSES OR FOSSAE

In certain parts of the abdominal cavity, recesses of peritoneum form cul-de-sacs or pouches **(retroperitoneal fossae),** which have surgical importance in connection with the possible occurrence of *intraabdominal* or *retroperitoneal hernias* (Mayo et al., 1941). The largest of these is the omental bursa (already described), but several other smaller fossae require mention. These may be divided into three groups: duodenal, cecal, and intersigmoidal fossae.

Duodenal Fossae

Three appear fairly constantly (Figs. 16-58; 16-59): the inferior duodenal fossa, the superior duodenal fossa, and the duodenojejunal fossa. Two appear occasionally or rarely: the paraduodenal fossa and the retroduodenal fossa.

INFERIOR DUODENAL FOSSA. The inferior duodenal fossa, present in about 75% of bodies, is situated opposite the third lumbar vertebra on the left side of the ascending portion of the duodenum. It is bounded by a thin sharp fold of peritoneum, and if the tip of the index finger is introduced into the fossa, it passes a little distance caudalward behind the ascending portion of the duodenum.

SUPERIOR DUODENAL FOSSA. The superior duodenal fossa, present in about 50% of bodies, often coexists with the inferior fossa. It lies on the left of the ascending portions of the duodenum, at the level of the second lumbar vertebra. It extends cranialward behind the sickle-shaped duodenojejunal fold, and has a depth of about 2 cm.

DUODENOJEJUNAL FOSSA. The duodenojejunal fossa exists in about 20% of bodies. It

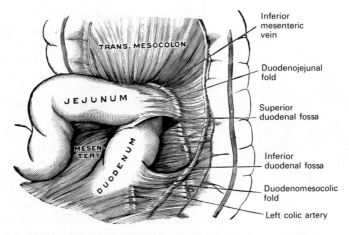

FIG. 16-58. Superior and inferior duodenal fossae. (Poirier and Charpy.)

lies between the right and left duodeno-mesocolic folds, extending cranialward behind the duodenojejunal junction toward the pancreas between the aorta on the right and the kidney on the left. It has a depth of 2 to 3 cm; its orifice is nearly circular and will admit the tip of the finger.

PARADUODENAL FOSSA. The paraduodenal fossa, rarely found, lies a short distance to the left of the ascending portion of the duodenum behind a peritoneal fold that contains the inferior mesenteric vein. The presence of a large fossa, containing most of the small intestine, has been explained as an abnormal rotation of the gut in embryonic

development (Mayo et al., 1941). A similar condition is usual in squirrel monkeys.

RETRODUODENAL FOSSA. The retroduodenal fossa, only occasionally present, lies dorsal to the horizontal and ascending parts of the duodenum and ventral to the aorta.

Cecal Fossae

The cecal fossae (*pericecal folds*) are three principal pouches or recesses in the neighborhood of the cecum (Figs. 16-60 to 16-62):

Superior ileocecal fossa
Inferior ileocecal fossa
Cecal fossa

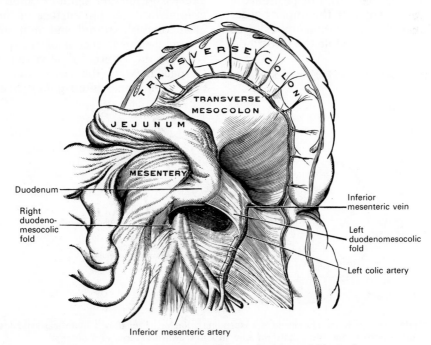

FIG. 16-59. Duodenojejunal fossa. (Poirier and Charpy.)

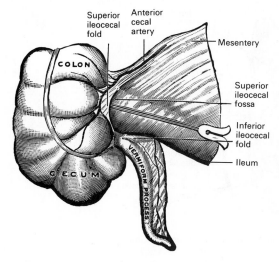

FIG. 16-60.   Superior ileocecal fossa. (Poirier and Charpy.)

SUPERIOR ILEOCECAL FOSSA.   The superior ileocecal fossa is formed by a fold of peritoneum that arches over the branch of the ileocolic artery that supplies the ileocolic junction. The fossa is a narrow chink situated between the mesentery of the small intestine, the ileum, and the small portion of the cecum behind.

INFERIOR ILEOCECAL FOSSA.   The inferior ileocecal fossa is situated behind the angle of junction of the ileum and the cecum. It is formed by the ileocecal fold of peritoneum (*bloodless fold of Treves*[1]), the upper border of which is fixed to the ileum opposite its mesenteric attachment, while the lower bor-

[1]Frederick Treves (1853–1923): An English surgeon (London).

der, passing over the ileocecal junction, joins the **mesenteriole of the vermiform appendix** and sometimes the appendix itself. Between this fold and the mesenteriole of the appendix is the inferior ileocecal fossa. It is bounded cranially by the dorsal surface of the ileum and the mesentery; ventrally and caudally by the ileocecal fold, and dorsally by the upper part of the mesenteriole of the vermiform appendix.

CECAL FOSSA.   The cecal fossa is situated immediately behind the cecum, which has to be raised to bring it into view. It varies much in size and extent. In some cases it is sufficiently large to admit the index finger, and extends upward behind the ascending colon in the direction of the kidney; in others it is merely a shallow depression. It is bounded on the right by the **cecal fold,** which is attached by one edge to the abdominal wall from the caudal border of the kidney to the iliac fossa, and by the other edge to the posterolateral aspect of the colon. In some instances additional fossae, the **retrocecal fossae,** are present.

Intersigmoid Fossa

The intersigmoid fossa (*recessus intersigmoideus*) (Fig. 16-84) is constant in the fetus and during infancy, but disappears in a certain percentage of individuals as age advances. When the sigmoid colon is drawn cranialward, the left surface of the sigmoid mesocolon is exposed, and on it will be seen a funnel-shaped recess of the peritoneum lying on the external iliac vessels in the interspace between the psoas and iliacus muscles. This is the intersigmoid fossa, which

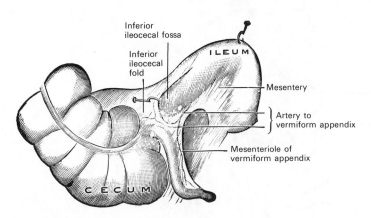

FIG. 16-61.   Inferior ileocecal fossa. The cecum and ascending colon have been drawn lateralward and downward, the ileum upward and backward, and the vermiform appendix downward. (Poirier and Charpy.)

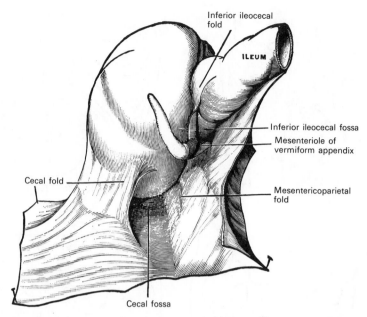

Inferior ileocecal
fold

ILEUM

Inferior ileocecal fossa

Mesenteriole of
vermiform appendix

Cecal fold

Mesentericoparietal
fold

Cecal fossa

FIG. 16-62.   The cecal fossa. The ileum and cecum are drawn dorsalward and cranialward. (Souligoux.)

lies dorsal to the sigmoid mesocolon and ventral to the parietal peritoneum. The fossa varies in size; in some instances it is a mere dimple, whereas in others it will admit the whole of the index finger.

The Stomach

The **stomach** (*ventriculus; gaster*) is situated in the right upper quadrant of the abdomen, partly covered by the ribs. It lies in a recess in the epigastric and left hypochondriac regions bounded by the anterior abdominal wall and the diaphragm, between the liver and the spleen.

SHAPE AND POSITION

The shape and position of the stomach are so greatly modified by changes within itself and in the surrounding viscera that no single form can be described as typical. The chief modifications are determined by (1) the amount of the stomach contents, (2) the stage that the digestive process has reached, (3) the degree of development of the gastric musculature, and (4) the condition of the adjacent intestines. It is possible, however, by comparing a series of stomachs, to de-

termine certain points more or less common to all (Figs. 16-63; 16-64).

According to Moody (1926), radiographs of the normal erect living body show the ordinary range of variation of the most caudal part of the *greater curvature* to be 7.3 cm above to 13.5 cm below the interiliac line in males and 6.5 cm above to 13.7 cm below the line in females. It is below the interiliac line in 74.4% of males and 87% of females. With the body horizontal, the most caudal part of the greater curvature in males is 16.5 cm above to 7.3 cm below the interiliac line, and in females, 15.5 cm above to 8.4 cm below the line. The most common position in the erect male (26%) is 2.6 to 5 cm below the interiliac line, and in the horizontal male (22.4%), 2.5 to 5 cm above the interiliac line. The most common position in the erect female (22.4%) is 5 to 7.5 cm below the interiliac line and in the horizontal female (24%), 2.5 to 5 cm above the interiliac line (Fig. 16-63).

The position of the *pylorus* in the erect living body of the male varies from 14.5 cm above to 8 cm below the iliac line, and in the female, 15 cm above to 2.5 cm below the interiliac line. The range of position of the pylorus in regard to the sagittal axis of the erect body varies in males from 8.8 cm to the right to 2 cm to the left of the axis. In 84% it is to the right of the axis. In females the posi-

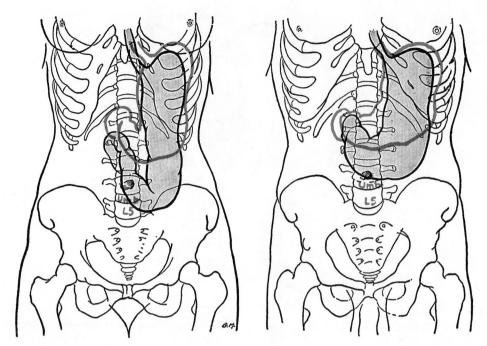

FIG. 16-63. Average position of the stomach based on x-ray studies. Standing position in black; reclining position in red. (Eycleshymer and Jones.)

tion ranges from 6 cm to the right to 2.6 cm to the left of the sagittal axis. In 18.5% it is to the right. The most common position in both males and females is 2.5 to 5 cm to the right.

OPENINGS

The opening from the esophagus into the stomach is known as the **cardiac orifice,** so named because of its close relationship with the part of the diaphragm upon which the heart rests. The short abdominal portion of the esophagus is situated to the left of the midline at the level of the tenth thoracic vertebra. It curves sharply to the left and dilates into the **cardiac antrum.** The right margin of the esophagus is continuous with the lesser curvature of the stomach; the left margin joins the greater curvature at an acute angle, forming the **incisura cardiaca** (*cardiac notch*).

The opening of the stomach into the duodenum is the **pyloric orifice,** which lies to the right of the midline at the level of the cranial border of the first lumbar vertebra. Its position is usually indicated on the surface of the stomach by a circular groove, the duodeno-pyloric constriction.

CURVATURES

The **lesser curvature** (Fig. 16-67) forms the right or concave border of the stomach. Nearer to its pyloric than its cardiac end is a well-marked notch, the **angular incisure** (Fig. 16-64), which varies somewhat in position with the state of distention; it serves to separate the stomach into a right and a left portion. The hepatogastric ligament (*lesser omentum*), which contains the left gastric artery and the right gastric branch of the

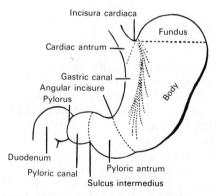

FIG. 16-64. Diagram showing the subdivisions of the human stomach. (F. T. Lewis.)

hepatic artery, is attached to the lesser curvature.

The **greater curvature,** directed to the left and ventralward, is four or five times as long as the lesser curvature. Starting from the cardiac orifice at the incisura cardiaca, it arches cranialward and to the left; the highest point of the convexity is level with the sixth left costal cartilage. From this level it curves more gradually caudalward, with a slight convexity to the left as low as the cartilage of the ninth rib; it then continues to the right, to end at the pylorus. Directly opposite the angular incisure of the lesser curvature, the greater curvature presents a dilatation, which is the **pyloric antrum.** This dilatation is limited on the right by a slight groove, the **sulcus intermedius,** which is about 2.5 cm from the pyloric ostium. The portion between the sulcus intermedius and the pyloric ostium is termed the **pyloric canal.** Attached along the greater curvature are certain mesenteric membranes derived from the embryonic dorsal mesogastrium. From the fundus, the gastrophrenic ligament extends to the diaphragm, and from the cranial part of the body, the gastrolienal ligament runs to the spleen. The greater omentum is suspended from the remainder of the curvature.

SURFACES

When the stomach is contracted, the two surfaces between the curvatures face somewhat cranialward and caudalward, but when it is partly distended, they face ventralward and dorsalward (Fig. 16-53).

VENTRAL SURFACE. The ventral surface (*anterosuperior surface*) (Fig. 16-53) is covered with peritoneum of the greater omentum. At the left and cranially it is against the diaphragm, which separates it from the base of the left lung, the heart, and the seventh, eighth, and ninth ribs and the corresponding intercostal spaces. The right portion is in relation with the left and quadrate lobes of the liver and the anterior abdominal wall. When the stomach is empty, the transverse colon may lie on the caudal part of this surface.

DORSAL SURFACE. The dorsal surface (*posteroinferior surface*) (Fig. 16-57) is covered with peritoneum of the lesser sac or omental bursa. It is in relation with the diaphragm, the spleen, the left suprarenal gland, the cranial part of the left kidney, the ventral surface of the pancreas, the left colic flexure, and the transverse mesocolon. These structures form a shallow **stomach bed** in which the organ rests (Fig. 16-57). The transverse mesocolon separates the stomach from the duodenojejunal junction and the rest of the small intestine. When the stomach is distended, especially if the patient is in the upright posture, the pyloric half of the stomach slides caudalward over the mesocolon and colon and may reach into the pelvis (Figs. 16-65; 16-66).

COMPONENT PARTS

A plane passing through the angular incisure on the lesser curvature divides the stomach into a left portion or **body** and a right or **pyloric portion** (Fig. 16-64). The cranial portion of the body is the **fundus,** which is marked off from the remainder of the body by a plane passing horizontally through the cardiac orifice. To the right of a plane through the sulcus intermedius at right angles to the long axis of this portion is the pyloric canal (Fig. 16-64), which ends in the thickened muscular ring, the **pyloric sphincter.**

A stomach examined during the process of digestion will be divided by a muscular constriction into a large, dilated left portion and a narrow, contracted, tubular right portion. The constriction appears in the body of the stomach and does not follow any of the anatomic landmarks; indeed, it shifts gradually toward the left as digestion progresses, i.e., more of the body is gradually absorbed into the tubular part.

INTERIOR OF THE STOMACH

When examined after death, the stomach is usually fixed at some stage of the digestive process. A common form is that shown in Figure 16-67. If the viscus were laid open by a section through the plane of its two curvatures, it would consist of two segments: a large globular portion on the left and a narrow tubular part on the right. These correspond to the clinical subdivisions of fundus and pyloric portions already described, and are separated by a constriction that indents the body and the greater curvature but does

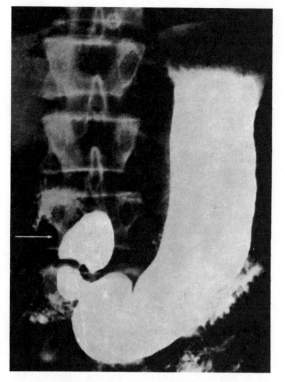

vature and at the pylorus, but elsewhere resemble a honeycomb. The folds involve both the mucosa and the submucosa, but they are transient and movable, and are gradually obliterated as the stomach distends.

The surface of the membrane does not appear smooth when examined with a lens, because closely scattered everywhere are the openings of the **gastric pits** or **foveolae.** On the cut surface of the stomach wall, the mucous membrane appears quite thick (5 mm) because tubular gastric glands extend down into it from the foveolae (Fig. 16-69).

The **lining epithelium** that covers the surface and extends down into the foveolae is composed of tall columnar cells of characteristic and rather uniform appearance; these are called **theca cells.** They secrete mucus, and within the part of the cell toward the surface, the precursor of the mucus can be seen in a shallow pocket or theca, much smaller and more difficult to identify

FIG. 16-65. Normal stomach after a barium meal. The tone of the muscular wall is good and supports the weight of the column in the body of the organ. The arrow points to the duodenal cap, below which a gap in the barium indicates the position of the pylorus.

not involve the lesser curvature. To the left of the cardiac orifice is the incisura cardiaca: the projection of this notch into the cavity of the stomach increases as the organ distends, and it supposedly acts as a valve preventing regurgitation into the esophagus. In the pyloric portion are seen the elevation corresponding to the angular incisure and the circular projection from the duodenopyloric constriction, which forms the pyloric valve; the separation of the pyloric canal from the rest of the pyloric part is scarcely indicated.

STRUCTURE OF THE STOMACH

The wall of the stomach has four coats: **mucous, submucous, muscular,** and **serous.**

MUCOUS MEMBRANE. The mucous membrane (Fig. 16-68) that lines the stomach has a soft, velvety appearance and pinkish color in the fresh state. It is thrown into thick folds, known as **rugae,** which tend to have a longitudinal direction along the lesser cur-

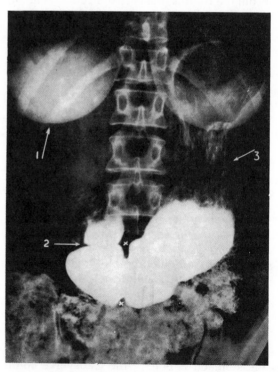

FIG. 16-66. Atonic stomach after a barium meal. This stomach contains the same amount of barium as the stomach in Fig. 16-65. Arrow 1 points to the shadow of the right breast; arrow 2, to the pylorus; arrow 3, to the upper part of the body of the stomach, where longitudinal folds can be seen in the mucous membrane. × × marks a wave of peristalsis.

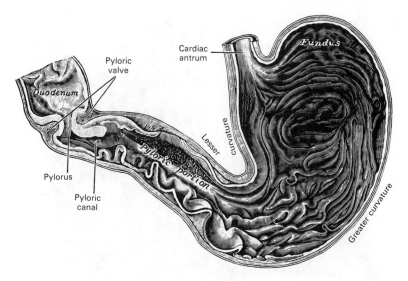

Interior of the stomach.
FIG. 16-67.

than the pocket of a goblet cell. They do not have a cuticular border like that seen in the cells of the intestine, and there are no goblet cells.

Gastric Glands. The gastric glands are simple tubular glands that open in groups of two or three into the bottoms of the gastric pits. Three kinds may be distinguished: (*a*) fundic glands, (*b*) cardiac glands, and (*c*) pyloric glands (the last two, being local modifications, will be described with the appropriate parts). Frequently, the **fundic glands** are simply called gastric glands, because they

are the most characteristic glands of the stomach and they occur throughout the fundus and body. The epithelial cells are of two types: chief cells and parietal cells.

Chief Cells. The **chief cells** are again subdivided into neck chief cells and body chief cells. The **neck chief cells** provide a transition between the lining epithelium of the foveolae and the secreting part of the glands. They are cuboidal or columnar, have a basophilic cytoplasm and, since they appear to be the source of new epithelial cells, they frequently contain mitotic nuclei. The **body**

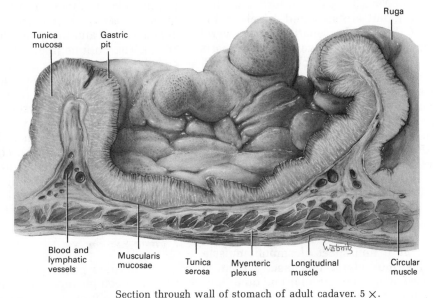

Section through wall of stomach of adult cadaver. 5 ×.
FIG. 16-68.

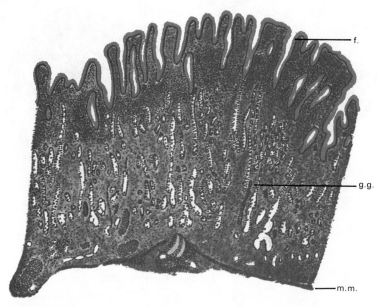

Fig. 16-69. Section through the body of the stomach of a 22-year-old man. *f*, Gastric fovea; *g.g.,* gastric gland; *m.m.,* muscularis mucosae. 50 ×. (Sobotta.)

chief cells extend down to the bottom of the glands. They resemble the neck chief cells but they are slightly more basophilic and have a prominent basal striation. They contain many small secretion granules, which can be made clearly visible with special stains; these cells secrete pepsin.

Parietal Cells. The **parietal cells** are the most characteristic cells in the gastric glands. Their name originated from the fact that they are pushed back against the basement membrane. They do not form a continuous layer, but are scattered all along the walls of the glands, separated by several chief cells and usually overlapped by parts of the neighboring body chief cells, which intervene between them and the internal surface of the gland. They are four or five times as large as chief cells, have a granular, intensely acidophilic cytoplasm, and are the source of the hydrochloric acid in the gastric juice (Fig. 16-70). The tissue surrounding and intervening between the gastric pits and the glands is composed of areolar connective tissue, blood and lymphatic capillaries with scattered lymphocytes, and occasionally even lymphatic nodules. The boundary between the mucosa and the submucosa is marked by a thin sheet of smooth muscle cells, the **muscularis mucosae,** which is composed of an inner circular and outer longitudinal layer.

SUBMUCOUS COAT. The submucous coat is composed of areolar connective tissue and blood and lymphatic vessels. It extends up into the rugae.

MUSCULAR COAT. In addition to the two layers of smooth muscle fibers characteristic of the digestive tube—inner circular and outer longitudinal—the muscular coat of the stomach has a layer of oblique fibers. The inner circular layer is well represented over the entire organ (Fig. 16-68). The outer longitudinal layer is not as uniform as the circular layer and is more concentrated along the lesser curvature and the greater curvature (Fig. 16-71). The oblique fibers are internal to the circular fibers, occur chiefly at the cardiac end of the stomach, and spread over the ventral and dorsal surfaces (Figs. 16-71; 16-72).

SEROUS COAT. The serous coat is composed of a small amount of areolar tissue that connects the mesothelial layer of the peritoneum to the muscular coat. It contains some of the larger blood vessels and lymphatics. A narrow strip along the lesser and greater curvatures, where the two omenta are attached, is not covered by peritoneum.

Cardiac Region

The cardiac portion of the stomach has certain peculiarities. The stratified squamous epithelium of the esophagus is ab-

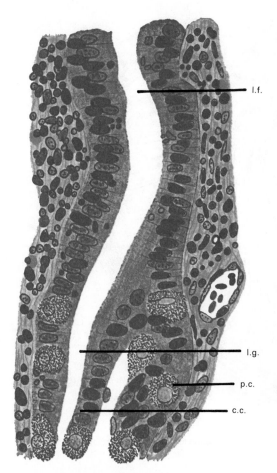

l.f.

l.g.

p.c.

c.c.

FIG. 16-70. Higher magnification of transition from fovea to gastric gland shown in Figure 16-69. *c.c.*, Chief cell; *l.f.*, lumen of fovea; *l.g.*, lumen of gland; *p.c.*, parietal cell. 5000 ×. (Sobotta.)

ruptly replaced by the columnar epithelium of the stomach. The **cardiac glands** are longer, more twisted, and contain no parietal cells; they resemble the esophageal glands and secrete mucus. Aggregations of lymph nodes or lymphatic nodules are common. The muscular layers are continuous with those of the esophagus.

Pylorus

The pylorus is distinctively marked by the thickening of the circular layer into the **pyloric sphincter,** which acts as a valve to close the lumen. The lining epithelium makes an abrupt transition from the gastric type with its theca cells to the intestinal epithelium type with striated cuticular border and interspaced goblet cells. The **pyloric glands** are devoid of parietal cells; they are longer

and more tortuous than the gastric glands and have the appearance of mucus-secreting cells. At the transition from stomach to duodenum, the gastric glands can be distinguished from the Brunner[1] glands of the duodenum because the gastric glands lie in the tunica mucosa, that is, inside the muscularis mucosae, whereas the Brunner glands lie in the submucosa.

VESSELS AND NERVES. The arteries supplying the stomach are: the left gastric, the right gastric, and the right gastroepiploic branches of the hepatic artery, and the left gastroepiploic and short gastric branches of the lienal artery. They supply the muscular coat, ramify in the submucous coat, and are finally distributed to the mucous membrane. The arrangement of the vessels in the mucous membrane is somewhat peculiar. The arteries break up at the base of the gastric tubules into a plexus of fine capillaries that run upward between the tubules, anastomosing with each other and ending in a plexus of larger capillaries that surround the mouths of the tubes and also form hexagonal meshes around the ducts.

From these the **veins** arise, and pursue a straight course downward, between the tubules, to the submucous tissue; they end either in the lienal and superior mesenteric veins, or directly in the portal vein.

The **lymphatics** are numerous: they consist of a superficial and a deep set, and they pass to lymph nodes along the two curvatures of the organ.

The **nerves** are the terminal branches of the right and left vagi, the former usually being distributed upon the dorsal and the latter upon the ventral part of the organ. Numerous sympathetic fibers that arise chiefly from the various subdivisions of the celiac plexus accompany the different blood vessels to the organ. Small sympathetic filaments may also arise directly from the phrenic and splanchnic trunks. Nerve plexuses are found in the submucous coat and between the layers of the muscular coat, as in the intestine. From these plexuses, fibrils are distributed to the muscular tissue and the mucous membrane.

Small Intestine

The **small intestine** extends from the pylorus to the ileocecal junction; it is about 7 meters long and gradually diminishes in diameter from its commencement to its termination. It is contained in the central and caudal part of the abdominal cavity and is surrounded cranially and at the sides by the

[1]Johann Konrad Brunner (1653-1727): A German anatomist (Heidelberg, then Strassburg).

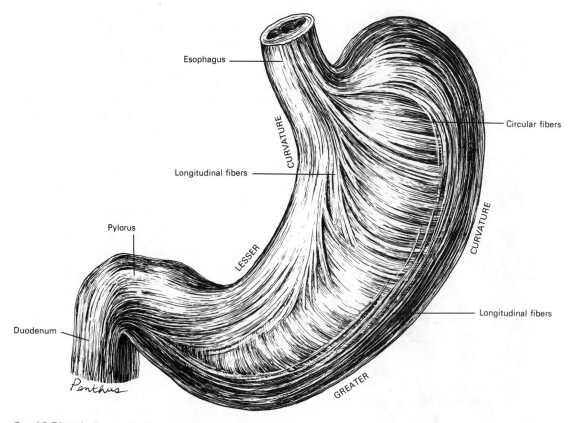

Esophagus

Circular fibers

CURVATURE

Longitudinal fibers

Pylorus

LESSER

CURVATURE

Longitudinal fibers

Duodenum

GREATER

Penthus

FIG. 16-71. The longitudinal and circular muscle fibers of the stomach; anterior surface. (Redrawn after Spalteholz.)

large intestine; a portion of it extends below the superior aperture of the pelvis and lies ventral to the rectum. It is in relation, ventrally, with the greater omentum and abdominal parietes, and is connected to the vertebral column by a reduplication of peritoneum, the **mesentery.** The small intestine is divisible into three portions: the **duodenum,** the **jejunum,** and the **ileum.**

DUODENUM

The duodenum (Fig. 16-57) received its name from being about equal in length to the breadth of 12 fingers (25 cm). It is the shortest, the widest, and the most fixed part of the small intestine, and it has no mesentery, being only partially covered by peritoneum. It has almost a circular course, so that its termination is not far removed from its starting point. For purposes of description the duodenum may be divided into four portions: **superior, descending, horizontal,** and **ascending.**

SUPERIOR PORTION. The superior portion is about 5 cm long, beginning at the pylorus and extending as far as the neck of the gallbladder. It is the most movable of the four portions. It is almost completely covered by peritoneum, but a small part of its posterior surface near the neck of the gallbladder and the inferior vena cava has no covering. The cranial border of its first half has the hepatoduodenal ligament attached to it; the caudal border is attached to the greater omentum. It is in such close relation with the gallbladder that it is usually stained by bile after death, especially on its ventral surface. It is in relation cranially with the quadrate lobe of the liver and the gallbladder; dorsally with the gastroduodenal artery, the common bile duct, and the portal vein; and caudally with the head and neck of the pancreas.

DESCENDING PORTION. The descending portion is 7 to 10 cm long and extends from the neck of the gallbladder, on a level with the first lumbar vertebra, along the right side of the vertebral column to the cranial border of the body of the fourth lumbar vertebra. It

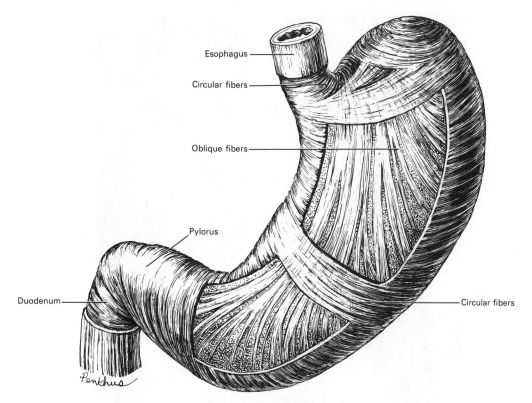

Esophagus

Circular fibers

Oblique fibers

Pylorus

Duodenum

Circular fibers

Penthus

FIG. 16-72. The oblique fibers of the stomach; anterior surface. (Redrawn after Spalteholz.)

is crossed in its middle third by the transverse colon, the posterior surface of which is not covered by peritoneum and is connected to the duodenum by a small quantity of connective tissue. The supra- and infracolic portions are covered ventrally by peritoneum (the infracolic part by the right leaf of the mesentery), but dorsally it is not covered by peritoneum. The descending portion is in relation ventrally with the gallbladder, the duodenal impression on the right lobe of the liver, and the transverse colon; dorsally, it has a variable relation to the right kidney in the neighborhood of the hilum and is connected to it by loose areolar tissue. The renal vessels, the inferior vena cava, and the psoas muscle are also dorsal to it. At its medial side are the head of the pancreas and the common bile duct; to its lateral side is the right colic flexure. The common bile duct and the pancreatic duct together perforate the medial side of this portion of the intestine 7 to 10 cm from the pylorus (Fig. 16-108); the accessory pancreatic duct sometimes pierces it about 2 cm proximal to this.

HORIZONTAL PORTION. The horizontal portion (*pars horizontalis; third or preaortic or transverse portion*) is 5 to 7.5 cm long. It begins at the right side of the fourth lumbar vertebra and passes from right to left, with a slight inclination cranialward, ventral to the great vessels and crura of the diaphragm, and joins the ascending portion ventral to the abdominal aorta. It is crossed by the superior mesenteric vessels and the mesentery. Its ventral surface is covered by peritoneum, except near the midline, where it is crossed by the superior mesenteric vessels. Its posterior surface is not covered by peritoneum, except toward its left extremity, where the posterior layer of the mesentery may sometimes cover it to variable extents. This surface rests upon the right crus of the diaphragm, the inferior vena cava, and the aorta. The cranial surface is in relation with the head of the pancreas.

ASCENDING PORTION. The ascending portion (*pars ascendens; fourth portion*) of the duodenum is about 2.5 cm long. It ascends on the left side of the aorta as far as the level of the cranial border of the second lumbar vertebra, where it turns abruptly ventralward to become the jejunum and forms the **duodenojejunal flexure.** It lies ventral to the

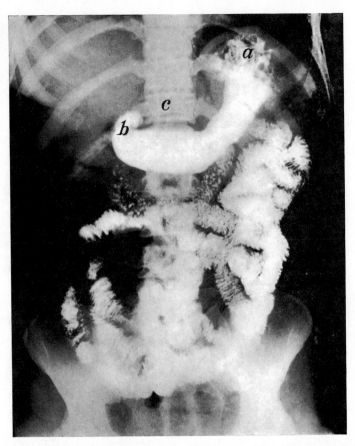

Fig. 16-73. Stomach and small intestines after a barium meal. a, Barium has settled out of the fundus of the stomach; b, pylorus; c, eighth thoracic vertebra. (Department of Radiology, University of Pennsylvania.)

left psoas major muscle and left renal vessels, and it is covered ventrally and partly at the sides by peritoneum that is continuous with the left portion of the mesentery.

The superior part of the duodenum, as stated previously, is somewhat movable, but the rest is practically fixed and bound to neighboring viscera and the posterior abdominal wall by the peritoneum. In addition, the duodenojejunal flexure is held in place by a fibrous muscular band, the **musculus suspensorius duodeni** or **ligament of Treitz**.[1] This structure commences in the connective tissue around the celiac artery and right crus of the diaphragm, and passes caudalward to insert into the superior border of the duodenojejunal curve and a part of the ascending duodenum. It possesses, according to Treitz, smooth muscular fibers

[1] Wenzel Treitz (1819–1872): An Austrian physician and professor of anatomy and pathology (Krakow and Prague).

mixed with the fibrous tissue of which it is principally composed (Haley and Peden, 1943). It has little importance as a muscle, but acts as a suspensory ligament.

VESSELS AND NERVES. The **arteries** supplying the duodenum are the right gastric and superior pancreaticoduodenal branches of the hepatic artery, and the inferior pancreaticoduodenal branch of the superior mesenteric artery.

The **veins** end in the lienal and superior mesenteric veins.

The **nerves** are derived from the celiac plexus.

JEJUNUM AND ILEUM

The remainder of the small intestine is named **jejunum** and **ileum** (Fig. 16-52), the former term being given to the proximal two-fifths and the latter to the distal three-fifths. There is no morphologic line of distinction between the two and the division is arbitrary; the character of the intestine gradually undergoes a change so that a portion of the bowel taken from the first part of the jejunum would present characteristic and marked differences from the last part of the ileum.

JEJUNUM. The jejunum (*intestinum jejunum*) is wider, its diameter being about 4 cm, and its wall is thicker, more vascular, and of a deeper color than the ileum. The circular folds (*plicae circulares; valvulae conniventes*) of its mucous membrane are large and thickly set, and its villi are larger than those in the ileum. The aggregated lymph nodules are usually absent in the proximal part of the jejunum, and in the distal part they are found less frequently than in the ileum; they are also smaller and tend to assume a circular form. By grasping the jejunum between the finger and thumb, one can feel the circular folds through the walls, but this is not true of the lower part of the ileum; in this way it is possible to distinguish the upper from the lower part of the small intestine.

ILEUM. The ileum is narrower, its diameter being 3.75 cm, and its coats are thinner and less vascular than those of the jejunum. It possesses few circular folds, and they are small and disappear entirely toward its caudal end, but **aggregated lymph nodules** (*Peyer's patches*[1]) are larger and more numerous than in the jejunum. The jejunum for the most part occupies the umbilical and left iliac regions, whereas the ileum occupies chiefly the umbilical, hypogastric, right iliac, and pelvic regions. The terminal part of the ileum usually lies in the pelvis, from which it ascends over the right psoas muscle and right iliac vessels; it ends in the right iliac fossa by opening into the medial side of the commencement of the large intestine.

The jejunum and ileum are attached to the posterior abdominal wall by an extensive fold of peritoneum, the **mesentery,** which allows free motion so that each coil can accommodate itself to changes in form and position. The mesentery is fan-shaped; its posterior border, or root is about 15 cm long. It is attached to the posterior abdominal wall from the left side of the body of the second lumbar vertebra to the right sacroiliac articulation, crossing successively the horizontal part of the duodenum, the aorta, the inferior vena cava, the ureter, and the right psoas

muscle (Fig. 16-53). Its breadth between its vertebral and intestinal borders averages about 20 cm and is greater in the middle than at its ends. Between its two layers are blood vessels, nerves, lacteals, and lymph glands, together with a variable amount of fat.

VARIATIONS. **Meckel's diverticulum**[2] (*diverticulum ilei*) is the name given to a pouch that projects from the lower part of the ileum in about 2% of subjects. Its average position is about 1 meter proximal to the ileocecal valve, and its average length is about 5 cm. Its caliber is generally similar to that of the ileum, and its blind extremity may be free or connected with the abdominal wall or some other portion of the intestine by a fibrous band. It represents the remains of the proximal part of the vitelline duct, the duct of communication between the yolk sac and the primitive digestive tube in early fetal life.

STRUCTURE OF THE SMALL INTESTINE

The internal surface has two types of irregularities or projections that are characteristic of the small intestine. They are the large **circular folds** and the minute **villi** (Figs. 16-74; 16-75).

The **circular folds** (*plicae circulares; valvulae conniventes; valves of Kerkring*[3]) (Fig. 16-74) are valvelike folds that project 3 to 10 mm into the lumen. The majority extend transversely around the inside of the cylinder of the intestine for about one-half to two-thirds of its circumference, but others complete the circle or form a spiral extending more than once around, even making two or three turns. The folds also have different heights, with high and low ones tending to alternate. The size and frequency of the folds are different in the three parts of the small intestine. The folds are formed by both the tela submucosa and the tunica mucosa. The core of submucosal connective tissue is firm, making these folds permanent structures that are not obliterated by distention, as are the transient rugae of the stomach.

The **villi** (Fig. 16-74) are tiny fingerlike projections, of a size just at the borderline of visibility with the naked eye, which are crowded together over the entire mucous

---

[1]Johann Conrad Peyer (1653–1712): A Swiss anatomist (Schaffhausen).

[2]Johann Friedrich Meckel (1781–1833): A German anatomist (Halle).
[3]Theodorus Kerckring (1640–1693): A German anatomist who worked in Amsterdam.

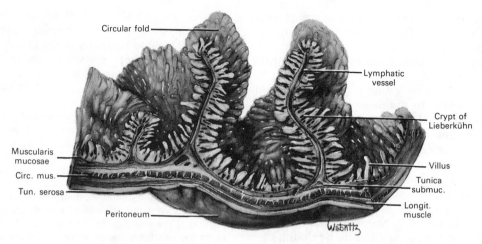

Fig. 16-74. Section through wall of small intestine (jejunum) of adult cadaver. 5 ×.

surface, giving it a velvety appearance. They are irregular in size and shape, are larger in some parts of the intestine than in others, and become considerably flattened by distention of the intestine. The villi are composed entirely of tissue belonging to the tunica mucosa.

The wall of the small intestine is composed of four coats; **mucous, submucous, muscular,** and **serous.**

MUCOUS MEMBRANE. The mucous membrane is composed of the villi, the intestinal glands, a connective tissue framework, and muscularis mucosae. The surface epithelium covering the villi is a simple columnar type in which the majority of cells have a characteristic striated free border. Recent observations with the electron microscope have demonstrated that the striated appearance is due to innumerable closely set projections of

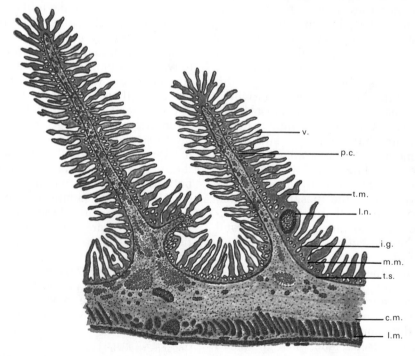

Fig. 16-75. A longitudinal section through the jejunum of a 24-year-old man, including section through two plicae circulares. *c.m.,* circular muscle; *i.g.,* intestinal gland; *l.m.,* longitudinal muscle; *l.n.,* lymphatic nodule; *m.m.,* muscularis mucosae; *p.c.,* plica circularis; *t.m.,* tunica mucosa; *v.,* villus. 15 ×. (Sobotta.)

cytoplasm that are too small (diameter 0.08 μm) to be distinguished with the best light microscopes. Scattered liberally among the striated epithelial cells are numerous mucus-secreting goblet cells.

Villi. Each villus has a core of delicate areolar and reticular connective tissue that provides a basement membrane for the epithelium and supports the rich network of capillary blood vessels and the usually single lymphatic vessel (Fig. 16-76). The lymphatic capillary begins blindly near the tip of the villus, occupies a more or less central position, and opens into the lymphatic vessels in the submucosa. This central lymphatic capillary is called a **lacteal,** the name having been given because it is filled with a white milky fluid, known as chyle, during the digestion of a meal rich in fat. Scattered single strands of smooth muscle run parallel with the lacteal and appear to be extensions of the muscularis mucosae into the villus.

Intestinal Glands. The intestinal glands (*crypts of Lieberkühn*[1]) are simple tubular glands that open into the depressions between the villi and form a rather uniform layer of glandular tissue between the bases of the villi and the muscularis mucosae (Fig. 16-77). The striated surface epithelium and

[1]Johann Nathaniel Lieberkühn (1711–1756): A German anatomist (Berlin).

goblet cells extend quite far down into the crypts, but at the bottom or fundus is a group of glandular secreting cells known as the **cells of Paneth.**[2] These cells contain large secretion granules that stain a bright red with eosin, and it is probable that they secrete the digestive enzymes of the small intestine. Mitotic divisions are frequently observed in the cells of the wall of the crypt and it is believed that proliferation of these cells replaces surface cells lost from natural attrition.

Muscularis Mucosae. The muscularis mucosae is a thin sheet of nonstriated muscle cells, composed of inner circular and outer longitudinal layers at the boundary between the mucosa and submucosa.

SUBMUCOUS COAT. The submucous coat is composed of fibroelastic and areolar connective tissue. It is a strong layer, forming the core of the circular folds. It contains the blood vessels and lymphatics that supply the mucous membrane and, near the muscularis, the **submucous nerve plexus of Meissner**[3] (Fig. 16-78). Small collections of lymphocytes and solitary lymphatic nodules may occur in any part of the small intestine;

[2]Josef Paneth (1857–1890): An Austrian physician (Vienna).

[3]George Meissner (1829–1905): A German professor of anatomy and physiology (Basel and Göttingen).

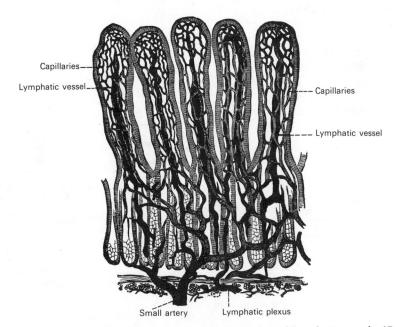

Capillaries—

Lymphatic vessel––

––– Capillaries

––– Lymphatic vessel

Small artery   Lymphatic plexus

FIG. 16-76. Villi of small intestine, showing blood vessels and lymphatic vessels. (Cadiat.)

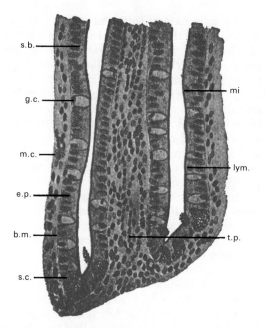

s.b.

g.c.

m.c.

e.p.

b.m.

s.c.

mi

lym.

t.p.

FIG. 16-77. Intestinal glands (crypts of Lieberkühn) from human duodenum. *b.m.*, Basement membrane; *e.p.*, epithelium; *g.c.*, goblet cell; *lym.*, lymphocyte; *m.c.*, mast cell; *mi.*, mitosis; *s.b.*, striated border; *s.c.*, glandular secreting cell (Paneth cell); *t.p.*, tunica propria. 300 ×. (Sobotta.)

plexus.[1] The muscularis is somewhat thicker in the proximal than in the distal part of the intestine.

SEROUS COAT. The serous coat is composed of the peritoneum and the areolar connective tissue connecting it to the muscular coat. The small intestine is covered by peritoneum except along the narrow strip or border attached to the mesentery and the retroperitoneal parts of the duodenum.

SPECIAL FEATURES. *Duodenum.* The circular folds are not found in the first 2.5 to 5 cm beyond the pylorus, but in the descending part, distal to the openings of the bile and pancreatic ducts, they are especially large and numerous. The villi are also especially large and numerous in the duodenum. The common bile duct and the pancreatic duct pierce the left side of the descending portion about 7 to 10 cm from the pylorus (Fig. 16-108). At the opening of the ducts there is a thickening of their muscle coats commonly called the **sphincter of Oddi**[2] (*sphincter ampullae hepatopancreaticae*) which usually causes a protrusion, the **papilla of Vater**,[3] at the proximal end of a longitudinal fold. There is a slight depression

[1]Leopold Auerbach (1828–1897): A German professor of neuropathology (Breslau).
[2]Ruggero Oddi (19th century): An Italian physiologist (Perugia).
[3]Abraham Vater (1684–1751): A German professor of anatomy (Wittenberg).

groups of nodules, known as Peyer's patches, occur in the ileum. The lymphatic nodules usually occupy the submucosa, infiltrate the muscularis mucosae, and extend out to the free surface, often appearing to obliterate some of the villi (Fig. 16-81).

MUSCULAR COAT. The muscular coat is composed of the usual two layers in the alimentary tube: outer longitudinal and inner circular. Between these two layers is a net of nervous tissue containing nonmyelinated nerve fibers and ganglion cells (Fig. 16-79) known as the **myenteric** or **Auerbach's**

FIG. 16-78. The nerve plexus of the submucosa from the rabbit. 50 ×.

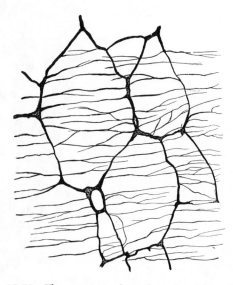

FIG. 16-79. The myenteric plexus from the rabbit. 50 ×.

and absence of the circular folds around the papilla (Fig. 16-80).

The **duodenal glands** (*Brunner's glands*[1]) differ from the other intestinal glands because they occupy the submucosa. The usual intestinal glands or crypts are found in the duodenum, and the ducts from the duodenal glands, after penetrating the muscularis mucosae, open into the bottoms of occasional crypts. The duodenal glands are compound tubuloalveolar glands; they resemble the pyloric glands of the stomach but are larger and, as mentioned, lie in the submucosa. The cells stain lightly in histologic preparations, but they give some of the reactions of mucus. The duodenal glands are largest and most numerous near the pylorus, forming there a rather thick complete layer, but they diminish in the horizontal portion and disappear near the duodenojejunal junction.

Jejunum. The circular folds and villi are almost as large and numerous in the proximal part of the jejunum as in the duodenum, but they gradually decrease in size and number toward the ileum.

Ileum. The circular folds and villi are smaller and less numerous in the ileum than in the jejunum, and toward the terminal part the folds may be widely scattered or even lacking.

Aggregated lymphatic nodules or **Peyer's patches** (*Peyer's glands; tonsillae intestinales*) are groups of lymphatic nodules

[1] Johann Konrad Brunner (1653–1727): A German professor of anatomy (Heidelberg and Strassburg).

spread out as a single layer in the mucous membrane of the wall of the ileum opposite the mesenteric attachment (Fig. 16-81). The patches are circular or oval, approximately 1 cm wide, and may extend along the intestine for 3 to 5 cm. They are largest and most frequent in the distal ileum but are occasionally seen even in the jejunum. They can be recognized in gross specimens at autopsy as thickened whitish patches in which circular folds are absent and the villi are sparse or lacking. In the dissecting room, where the subjects are usually of advanced age, the patches are difficult to identify because of the atrophy of the lymphatic tissue that occurs in older individuals.

VESSELS AND NERVES The jejunum and ileum are supplied by the **superior mesenteric artery,** the intestinal branches of which, having reached the attached border of the bowel, run between the serous and muscular coats with frequent inosculations to the free border, where they also anastomose with other branches running around the opposite surface of the gut. From these vessels numerous branches are given off that pierce the muscular coat, supplying it and forming an intricate plexus in the submucous tissue. From this plexus minute vessels pass to the glands and villi of the mucous membrane. The **veins** have a course and arrangement similar to that of the arteries.

The **lymphatics** of the small intestine (lacteals) are arranged in two sets, those of the mucous membrane and those of the muscular coat. The lymphatics of the villi commence in these structures in the manner described above. They form an intricate plexus in the mucous and submucous tissue and are joined by the lymphatics from the lymph spaces at the bases of the solitary nodules; from this they pass

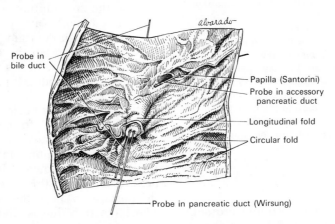

FIG. 16-80. Mucous membrane of the descending portion of the duodenum, showing the plica longitudinalis and the duodenal papilla (of Vater). (Redrawn from Spalteholz.)

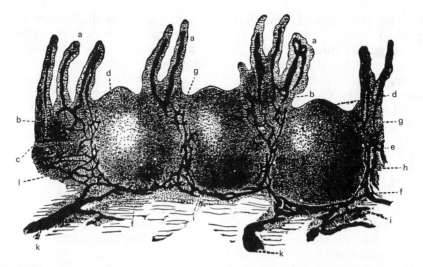

Fig. 16-81. Vertical section of a human aggregated lymphatic nodule, injected through its lymphatic canals. *a*, Villi with their chyle passages. *b*, Intestinal glands. *c*, Muscularis mucosae. *d*, Cupola or apex of solitary nodule. *e*, Mesial zone of nodule. *f*, Base of nodule. *g*, Points of exit of the lacteals from the villi and entrance into the true mucous membrane. *h*, Retiform arrangement of the lymphatics in the mesial zone. *i*, Course of the latter at the base of the nodule. *k*, Confluence of the lymphatics opening into the vessels of the submucous tissue. *l*, Follicular tissue of the latter.

to larger vessels at the mesenteric border of the gut. The lymphatics of the muscular coat are situated to a great extent between the two layers of muscular fibers, where they form a close plexus; throughout their course they communicate freely with the lymphatics from the mucous membrane, and empty in the same manner as these into the origins of the lacteal vessels at the attached border of the gut.

The **nerves** of the small intestine are derived from the plexuses of autonomic nerves around the superior mesenteric artery. They represent cranial parasympathetic fibers of the vagus and postganglionic sympathetic fibers from the celiac plexus. From this source they run to the **myenteric plexus** (*Auerbach's plexus*) (Fig. 16-79) of nerves and ganglia situated between the circular and longitudinal muscular fibers, from which the nervous branches are distributed to the muscular coats of the intestine. From this a secondary plexus, the **plexus of the submucosa** (*Meissner's plexus*) (Fig. 16-78) is derived, formed by branches that have perforated the circular muscular fibers. This plexus lies in the submucous coat of the intestine; it also contains ganglia from which nerve fibers pass to the muscularis mucosae and to the mucous membrane. The nerve bundles of the submucous plexus are finer than those of the myenteric plexus.

Large Intestine

The **large intestine** (*intestinum crassum*) (Figs. 16-52; 16-53; 16-82; 16-83) extends from the ileum to the anus. It is about 1.5 meters long, being one-fifth of the whole extent of the intestinal canal. Its caliber, when not unduly distended with feces, is largest at its commencement at the cecum and gradually diminishes as far as the rectum, where there is a dilatation of considerable size just above the anal canal. It differs from the small intestine in several ways: it has a sacculated form; it possesses certain appendages on its external coat, the **appendices epiploicae;** and its longitudinal muscular fibers are arranged in three **longitudinal bands or taeniae.**

The large intestine describes an arch around the convolutions of the small intestine. It commences in the right iliac region, ascends through the right lumbar and hypochondriac regions to the caudal surface of the liver, and there bends to the left to form the **right colic flexure.** It passes transversely across the epigastric and umbilical regions to the left hypochondriac region, where it bends caudalward, forming the **left colic flexure,** and descends through the left lumbar and iliac regions toward the pelvis. In the left iliac region it forms the **sigmoid flexure,** and from this it is continued along the posterior wall of the pelvis to the anus. The large intestine is divided into the **cecum, appendix, colon, rectum,** and **anal canal.**

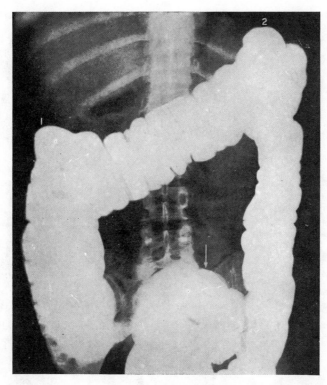

Fig. 16-82. Large intestine after a barium enema. *1*, Right colic flexure; *2*, left colic flexure. The arrow points to the pelvic colon. Note the sacculations of the gut, and the different levels of the two flexures.

CECUM

The **cecum** (*intestinum caecum*) (Fig. 16-84), the commencement of the large intestine, is the large blind pouch extending caudalward beyond the ileocecal valve. Its size is estimated variously by different authors, but on an average it is 6.25 cm long and 7.5 cm wide. It is situated in the right iliac fossa, cranial to the lateral half of the inguinal ligament. It rests on the iliacus muscle and usually lies in contact with the ventral abdominal wall, but the greater omentum and some coils of small intestine, if the cecum is empty, may lie ventral to it.

The common position of the cecum in the erect living body is not in the right iliac fossa but in the cavity of the true pelvis. As a rule, it is enveloped entirely by peritoneum, but in 5% of cases, part of the posterior surface is connected to the iliac fascia by connective tissue. The cecum lies quite free in the abdominal cavity and enjoys a considerable amount of movement, so that it may become herniated into the right inguinal canal, and

occasionally it has been found in an inguinal hernia on the left side.

The cecum varies in shape, but according to Treves, in man it may be classified as one of four types. In the *first* of his four types (about 2%), the cecum is conical and the appendix rises from its apex. The three longitudinal bands start from the appendix and are equidistant from each other. In the *second* type, the conical cecum has become quadrate because a saccule has grown out on either side of the anterior longitudinal band. These saccules are of equal size, and the appendix arises from between them, instead of from the apex of a cone (about 3%).

The *third* type is the normal type in man. In this case, the two saccules, which in the second type are uniform, have grown at unequal rates: the right with greater rapidity than the left. Consequently, an apparently new apex is formed by the caudalward growth of the right saccule, and the original apex, with the appendix attached, is pushed over to the left toward the ileocecal junction. The three longitudinal bands still start from

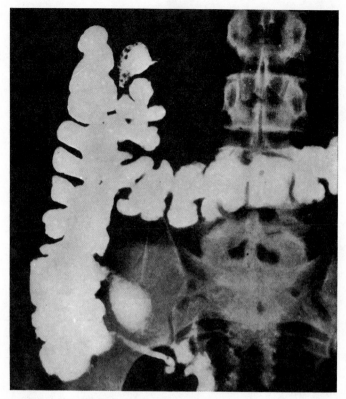

FIG. 16-83. Part of the large intestine after a barium meal. Note the vermiform appendix, which passes from the medial side of the cecum medially and slightly downward into the true pelvis. At a slightly higher level the terminal part of the ileum can be recognized. The first part of the transverse colon runs downward in front of, and slightly medial to, the ascending colon, before it turns to the left.

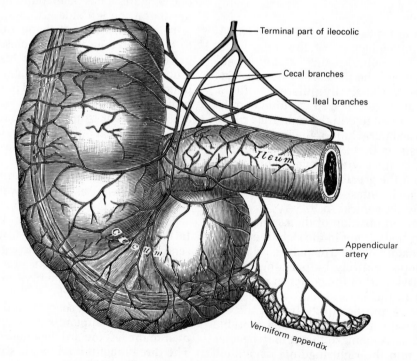

FIG. 16-84. The cecum and appendix, with their arteries.

the base of the vermiform appendix, but they are no longer equidistant from each other because the right saccule has grown between the anterior and posterior bands, pushing them over to the left. This type occurs in about 90% of persons. The *fourth* type is merely an exaggerated condition of the third; the right saccule is still larger and at the same time the left saccule has become atrophied, so that the original apex of the cecum, with the vermiform appendix, is close to the ileocecal junction, and the anterior band courses medialward to the same location (about 4%).

VERMIFORM APPENDIX. The vermiform appendix (*processus vermiformis*) (Fig. 16-85) is a long, worm-shaped tube, 5 to 10 mm in diameter, which starts from what was originally the apex of the cecum. It may project in one of several directions: cranialward behind the cecum; to the left behind the ileum and mesentery; or caudalward into the lesser pelvis. It varies from 2 to 20 cm in

length, its average being about 8 cm. It is suspended by a peritoneal mesenteriole derived from the left leaf of the mesentery, which is more or less triangular in shape and usually extends along the entire length of the tube. Between its two layers and close to its free margin lies the appendicular artery (Fig. 16-84).

The lumen of the vermiform appendix is small, extends throughout the whole length of the tube, and communicates with the cecum by an orifice distal to the ileocecal opening. It is sometimes guarded by a semilunar valve formed by a fold of mucous membrane, but this is by no means constant.

ILEOCECAL VALVE. The ileum ends by opening into the medial part of the large intestine, at the junction of the cecum with the colon. The opening is guarded by a valve, the ileocecal valve (Fig. 16-85), which consists of two segments or lips that project into the lumen of the large intestine. If the intestine has been inflated and dried, the lips

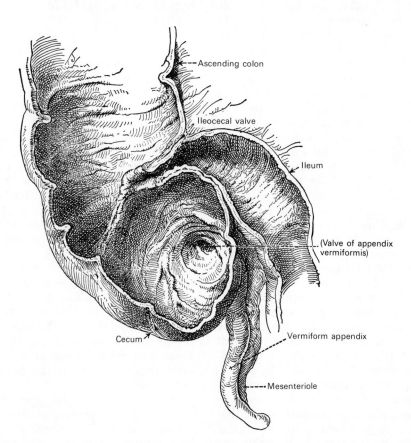

FIG. 16-85. The cecum, colic valve, and vermiform appendix, with anterior wall of terminal ileum and cecum removed. (Eycleshymer and Jones.)

have a semilunar shape, one toward the colon, the other toward the cecum. At the ends of the aperture the two segments of the valve coalesce and continue as narrow membranous ridges around the canal for a short distance, forming the **frenula of the valve.** In the fresh condition or in specimens that have been hardened in situ, the lips project as thick cushion-like folds into the lumen of the large gut; the opening between them may have the appearance of a slit or may be somewhat oval.

COLON

The colon is divided into four parts: the **ascending, transverse, descending,** and **sigmoid.**

ASCENDING COLON. The ascending colon passes cranialward from its commencement at the cecum to the caudal surface of the right lobe of the liver, where it is lodged in a shallow depression, the **colic impression,** to the right of the gallbladder; here it bends abruptly to the left, forming the **right colic** (*hepatic*) **flexure** (Fig. 16-53). It does not have a mesentery but is kept in contact with the posterior wall of the abdomen by the peritoneum, which covers its ventral surface and sides. Its dorsal surface is connected by loose areolar tissue with the iliacus and quadratus lumborum muscles, with the aponeurotic origin of transverse abdominal muscle, and with the ventral and lateral part of the right kidney. It is in relation, ventrally, with the convolutions of the ileum and the abdominal wall.

TRANSVERSE COLON. The transverse colon, the longest and most movable part of the colon, passes from the right colic flexure in the right hypochondriac region across the abdomen into the left hypochondriac region, where it curves sharply on itself beneath the caudal end of the spleen, forming the **left colic** (*splenic*) **flexure.** The transverse colon is festooned caudalward 7.5 to 10 cm below the interiliac line in most males in the erect posture, and 10 to 12.5 cm below in most females. It is completely invested by peritoneum, except for the **transverse mesocolon,** which is connected to the inferior border of the pancreas by a large duplicature of that membrane. It is in relation, by its cranial surface, with the liver and gallbladder, the greater curvature of the stomach, and the

spleen; by its caudal surface, with the small intestine; and by its ventral surface, with the anterior layers of the greater omentum and the abdominal wall. Its dorsal surface is in relation from right to left with the descending portion of the duodenum, the head of the pancreas, and some of the convolutions of the jejunum and ileum.

The **left colic** or **splenic flexure** (Fig. 16-53) occurs at the junction of the transverse and descending parts of the colon, and is in relation with the caudal end of the spleen and the tail of the pancreas. The flexure is so acute that the end of the transverse colon usually lies in contact with the front of the descending colon. Its position is more cranial and posterior to that of the right colic flexure. It is attached to the diaphragm opposite the tenth and eleventh ribs by a peritoneal fold, named the **phrenicocolic ligament,** which also assists in supporting the spleen.

DESCENDING COLON. The descending colon passes caudalward through the left hypochondriac and lumbar regions along the lateral border of the left kidney. At the caudal end of the kidney it turns medialward toward the lateral border of the psoas muscle and then descends, in the angle between the psoas and the quadratus lumborum muscles, to the crest of the ilium, where it becomes the iliac colon. The peritoneum covers its ventral surface and sides, while its dorsal surface is connected by areolar tissue with the caudal and lateral part of the left kidney, with the aponeurotic origin of the transversus abdominis, and with the quadratus lumborum muscle (Fig. 16-55). It is smaller in caliber and placed more deeply than the ascending colon and its posterior surface is more covered with peritoneum than is the ascending colon.

The part of the descending colon in the left iliac fossa may be called the **iliac colon** (Fig. 16-86). It is about 12 to 15 cm long. It begins at the level of the iliac crest, and ends in the sigmoid colon at the superior aperture of the lesser pelvis. It curves medialward ventral to the iliopsoas muscle and is covered by peritoneum on its sides and anterior surface only.

SIGMOID COLON. The sigmoid colon (*pelvic colon; sigmoid flexure*) (Fig. 16-86) forms a loop that averages about 40 cm in length and normally lies within the pelvis, but because of its freedom of movement, it may be

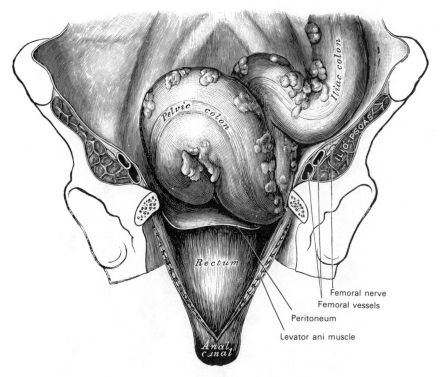

FIG. 16-86. Iliac colon, sigmoid or pelvic colon, and rectum seen from the front, after removal of pubic bones and bladder.

displaced into the abdominal cavity. It begins at the superior aperture of the lesser pelvis, where it is continuous with the iliac colon and forms one or two loops before it reaches the level of the third piece of the sacrum, where it bends downward and ends in the rectum. It is completely surrounded by peritoneum and is attached to the pelvic wall by an extensive mesentery, the **sigmoid mesocolon,** which gives it a considerable range of movement in its central portion. Dorsal to the sigmoid colon are the external iliac vessels, the left piriformis muscle, and the left sacral plexus of nerves; ventrally, some coils of the small intestine separate it from the bladder in the male and from the uterus in the female.

RECTUM

The rectum (Fig. 16-87) is continuous with the sigmoid colon at the level of the third sacral vertebra; it passes caudalward, lying in the sacrococcygeal curve, and ends in the anal canal. It has two dorsoventral curves: a cranial, with its convexity dorsalward, and a caudal, with its convexity ventralward. Two lateral curves are also described, one to the right, opposite the junction of the third and fourth sacral vertebrae, and the other to the left, opposite the left sacrococcygeal articulation; these, however, are of little importance. The rectum is about 12 cm long, and at its commencement its caliber is similar to that of the sigmoid colon, but near its termination it is dilated to form the **rectal ampulla.**

The rectum has no sacculations comparable to the haustrae of the colon, but when the lower part is contracted, its mucous membrane is thrown into longitudinal folds. There are certain permanent transverse folds that have a semilunar shape, **plicae transversales recti** (*Houston's valves*[1]) (Fig. 16-88). Usually there are three of these folds, but sometimes a fourth and occasionally only two are present. One is situated near the commencement of the rectum, on the right side; a second, about 3 cm beyond the first, extends inward from the left side of the tube; a third, the largest and most constant,

[1]John Houston (1802–1845): An Irish surgeon (Dublin).

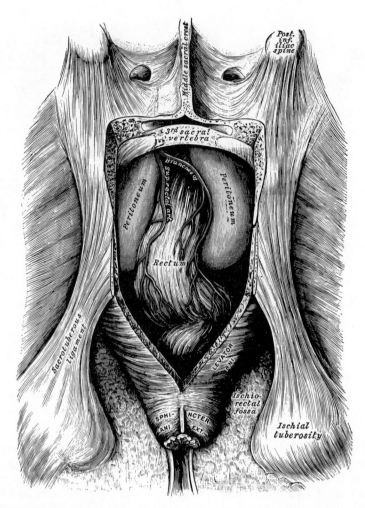

FIG. 16-87.   The posterior aspect of the rectum exposed by removing the lower part of the sacrum and the coccyx.

projects dorsalward, opposite the fundus of the urinary bladder. When a fourth is present, it is situated nearly 2.5 cm above the anus on the left and posterior wall of the tube. These folds are about 12 mm wide and contain some of the circular fibers of the gut. In the empty state of the intestine they overlap each other, as Houston remarks, "so effectually as to require considerable maneuvering to conduct a bougie or finger along the canal." Their function seems to be "to support the weight of fecal matter, and prevent its urging toward the anus, where its presence always excites a sensation demanding its discharge."

The peritoneum covers the ventral and lateral surface of the proximal two-thirds of the rectum, but distally it covers the ventral surface only, from which it is reflected onto the seminal vesicles in the male and the posterior vaginal wall in the female.

The level at which the peritoneum is reflected from the rectum to the viscus ventral to it has considerable surgical importance in connection with the removal of the distal part of the rectum. In the male, the distance of the rectovesical excavation from the anus is about 7.5 cm (i.e, the height to which an ordinary index finger can reach). In the female, the rectouterine excavation is about 5.5 cm from the anal orifice. The rectum is surrounded by dense fascia loosely attached to the rectal wall by areolar tissue, that is, there is a fascial cleft that allows distention.

RELATIONS OF THE RECTUM.   The proximal part of the rectum is in relation, dorsally, with the superior rectal vessels, the left piriformis muscle, and the left

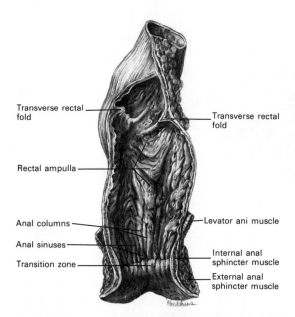

Transverse rectal fold

Transverse rectal fold

Rectal ampulla

Anal columns

Anal sinuses

Transition zone

Levator ani muscle

Internal anal sphincter muscle

External anal sphincter muscle

FIG. 16-88.   Coronal section of the rectum and anal canal showing the internal surface. (Redrawn after Sobotta.)

sacral plexus of nerves, which separate it from the pelvic surfaces of the sacral vertebrae; in its distal part it lies directly on the sacrum, coccyx, and levatores ani muscle, a dense fascia alone intervening. Ventrally, it is separated, in the male, from the fundus of the bladder and in the female, from the intestinal surface of the uterus and its appendages by some convolutions of the small intestine and frequently by the sigmoid colon. In the male, it is in relation with the triangular portion of the fundus of the bladder, the vesiculae seminales, the ductus deferentes, and more anteriorly the posterior surface of the prostate; in the female, it is in relation with the posterior wall of the vagina.

ANAL CANAL

The anal canal (Figs. 16-87; 16-91; 16-92), or terminal portion of the large intestine, begins where the ampulla of the rectum narrows (at the level of the apex of the prostate in the male) and ends at the anus. It measures 2.5 to 4 cm long and forms an angle with the lower part of the rectum. It has no peritoneal covering, but is invested by the sphincter ani internus, supported by the levatores ani, and surrounded at its termination by the sphincter ani externus. Dorsal to it is a mass of muscular and fibrous tissue, the **anococcygeal body;** ventral to it, in the male, but separated by the perineal center, are the membranous portion of the urethra and bulb

of the penis and the fascia of the urogenital diaphragm. In the female it is separated from the lower end of the vagina by a mass of muscular and fibrous tissue named the **perineal body.**

The proximal half of the anal canal has vertical folds produced by an infolding of the mucous membrane around a plexus of veins, known as the **rectal columns** (of Morgagni[1]; Fig. 16-88). They are separated from each other by furrows (**rectal sinuses**), which end distally in small valve-like folds, termed **anal valves,** which join together the distal ends of the rectal columns. When these columns are swollen and inflamed, they are known as hemorrhoids.

STRUCTURE OF THE LARGE INTESTINE

The wall of the cecum and colon has certain characteristic folds and irregularities that show up most prominently on the internal surface. Unlike those in the small intestine, these folds include all four layers and may be seen on the external surface. The longitudinal bands of the muscular coat, which are described below, cause a puckering of the wall, so that between them it bulges into sacculations called **haustrae.** The wall between the haustrae is thrown into folds that have a crescentic form on the interior of the colon and are called the **semilunar folds,** in contrast with the circular folds of the small intestine.

The large intestine has four coats: **mucous, submucous, muscular,** and **serous.**

MUCOUS COAT.   The mucous coat is smooth, that is, devoid of villi, and covers the inner surface of haustrae and semilunar folds in a coat of uniform thickness. The surface is covered with simple columnar epithelium containing large numbers of goblet cells. The glands of the large intestine are simple, straight, tubular glands containing the same type of epithelium as the surface; they are packed closely together and they open on the surface in tiny round holes that can readily be seen with a hand lens (Fig. 16-89). There is a delicate **muscularis mucosae** composed of inner circular and outer longitudinal fibers. Collections of lymphocytes and solitary lymphatic nodules occur

[1]Giovanni Battista Morgagni (1682–1771): An Italian professor of anatomy (Padua).

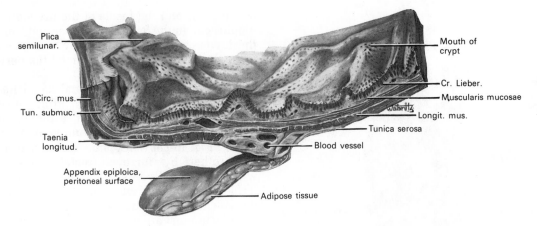

Plica semilunar.

Mouth of crypt

Cr. Lieber.

Muscularis mucosae

Circ. mus.

Tun. submuc.

Longit. mus.

Tunica serosa

Taenia longitud.

Blood vessel

Appendix epiploica, peritoneal surface

Adipose tissue

FIG. 16-89. Section through colon of adult cadaver. 5 ×.

frequently, especially near the colic valve and in the rectum.

SUBMUCOUS COAT. The submucous coat is a rather uniform layer of areolar tissue containing blood and lymphatic vessels and connecting the mucosa with the muscularis.

MUSCULAR COAT. The muscular coat is composed of the usual inner circular and outer longitudinal layers of nonstriated muscle.

The circular fibers form a thin layer over the cecum and colon, which is somewhat thickened in the semilunar folds between the haustrae, uniformly thickened in the rectum, and in the anal canal, constitutes the strong circular nonstriated muscle, the **sphincter ani internus.**

The longitudinal muscle fibers are concentrated into three longitudinal bands that are equally spaced and about 12 mm wide. They are easily seen in a gross specimen (Fig. 16-53) and are called **taeniae coli.** They have specific positions in relation to the position of the colon itself: (1) the posterior taenia is placed along the attached border; (2) the anterior taenia is easily visible on the exposed surface of the ascending and descending colon, but is covered by the attachment of the greater omentum on the transverse colon; and (3) the lateral taenia is found on the medial side of the ascending and descending colon and on the dorsal aspect of the transverse colon. The anterior taenia is a useful guide for locating the position of the appendix vermiformis, because the latter is a direct extension from it (Fig. 16-84).

The taeniae are shorter than the other coats of colon and cecum, causing the intervening wall to bulge into the sacculations known as **haustrae** which are typical of this part of the intestine (Fig. 16-52) and are responsible for the deep depressions in its outline in roentgenograms (Fig. 16-83). Between adjacent haustrae are the crescentic folds or plicae semilunares, which encroach on the lumen of the intestine.

SEROUS COAT. The serous coat, derived from the peritoneum, is complete over the cecum, appendix, transverse colon, and sigmoid colon except for their mesenteric attachments. It is incomplete on the rectum and on the ascending and descending colons where they are attached to the posterior abdominal wall.

SPECIAL FEATURES. *Ileocecal Valve.* Each lip of the valve is formed by a reduplication of the mucous membrane and the circular muscular fibers of the intestine. The longitudinal fibers and peritoneum are continued uninterruptedly from the small to the large intestine.

The surfaces of the valve directed toward the ileum are covered with villi and have the characteristic structure of the mucous membrane of the small intestine, while those turned toward the large intestine are destitute of villi and are marked with the orifices of the numerous tubular glands peculiar to the mucous membrane of the large intestine. These differences in structure continue as far as the free margins of the valve. It is generally maintained that this valve prevents reflux from the cecum into the ileum, but in all probability it acts as a sphincter around the end of the ileum and prevents the contents of the ileum from passing too quickly into the cecum.

Appendices Epiploicae.   The appendices epiploicae are characteristic features of the large intestine and may be used for its identification. They are small, round, irregular masses of fat, that average 0.5 to 1.0 cm in diameter, are almost completely covered by peritoneum, and are suspended from the surface of the colon and cecum by slender stalks (Fig. 16-89). They are usually attached along the taeniae, and are most numerous on the transverse colon.

Vermiform Appendix or Process.   The appendix has the same four coats as the colon. The epithelial lining and the glands are similar, but the glands are much fewer and the mucosa and submucosa are much thickened and almost entirely occupied by lymphatic nodules and lymphocytes (Fig. 16-90). The longitudinal fibers of the tunica muscularis are evenly distributed, not arranged into taeniae as in the colon, and the circular muscle is more prominent than the longitudinal.

Rectum.   The mucous membrane in the rectum is thicker and more vascular than that in the colon and is more loosely attached to the muscularis, as in the esophagus. The longitudinal fibers of the tunica muscularis are spread out into a layer that completely surrounds the rectum but is thicker on the anterior and posterior walls.

Anal Canal.   The mucous membrane in the anal canal is thick and vascular. Beneath the longitudinal folds or **rectal columns** of Morgagni are dilated veins, often knotted and tortuous, where the tributaries of the superior and inferior rectal veins anastomose. The epithelium changes abruptly at about 1.5 to 2 cm above the anal opening; the transition is marked by a white line, below which the epithelium is stratified squamous that is continuous with the skin. In the region of this line are the openings of the **anal glands,** which are greatly enlarged, modified skin glands (Fig. 16-91).

The proper circular muscle layer, continuous with that of the rectum, is greatly thickened in the anal canal, forming the nonstriated **internal anal sphincter** (*sphincter ani internus*). In the tissue surrounding the anus is a circular ring of striated muscle, the **external anal sphincter,** and diverging from this to the walls of the pelvis is the **levator ani.** Just beneath the integument at the anal orifice is a more delicate striated muscle, the **corrugator cutis ani;** its fibers are closely associated with the tributaries of the inferior rectal veins draining the plexuses in the rectal columns, and spasm of its fibers may seriously retard the venous drainage.

VESSELS AND NERVES   (Fig. 16-92). The **arteries** supplying the colon are derived from the colic and sigmoid branches of the mesenteric arteries. They give off large branches that ramify between and sup-

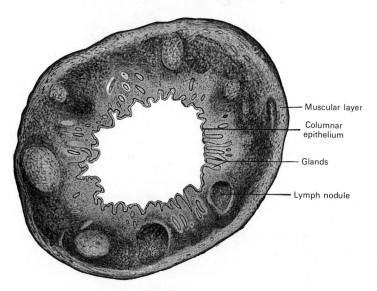

FIG. 16-90.   Transverse section of human appendix. 20 ×.

Muscular layer

Columnar epithelium

Glands

Lymph nodule

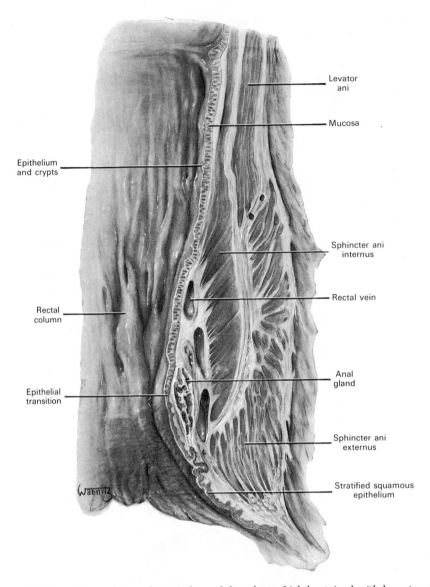

Levator
ani

Mucosa

Epithelium
and crypts

Sphincter ani
internus

Rectal vein

Rectal
column

Anal
gland

Epithelial
transition

Sphincter ani
externus

Stratified squamous
epithelium

Wabritz

FIG. 16-91. Section of the rectum and anus of an adult cadaver. Lightly stained with hematoxylin. 5 ×.

ply the muscular coats and, after dividing into small vessels in the submucous tissue, pass to the mucous membrane. The rectum is supplied by the superior rectal branch of the inferior mesenteric artery, and the anal canal is supplied by the middle rectal branch from the internal iliac artery and the inferior rectal branch from the internal pudendal artery. The superior rectal artery, the continuation of the inferior mesenteric, divides into two branches that run down either side of the rectum to within about 12.5 cm of the anus; they here split up into about six branches, which pierce the muscular coat and descend between it and the mucous membrane in the longitudinal direction, parallel with each other as far as the

sphincter ani internus, where they anastomose with the other rectal arteries and form a series of loops around the anus.

The **veins** of the rectum commence in a plexus of vessels that surrounds the anal canal. In the vessels forming this plexus are smaller saccular dilatations just within the margin of the anus; from the plexus about six large vessels are given off. These ascend between the muscular and mucous coats for about 12.5 cm, running parallel to each other; they then pierce the muscular coat (Fig. 16-92) and unite to form a single trunk, the superior rectal vein. This arrangement is termed the **rectal** or **hemorrhoidal plexus**; it communicates with the tributaries of the

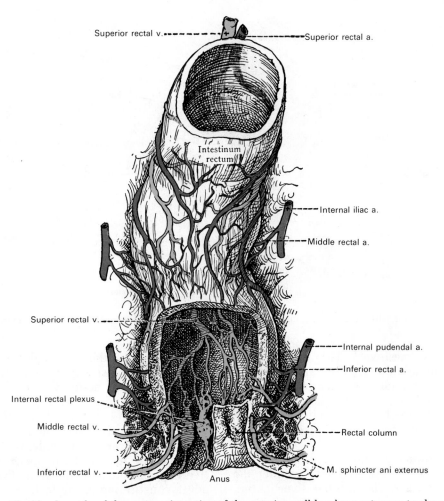

Superior rectal v.

Superior rectal a.

Intestinum rectum

Internal iliac a.

Middle rectal a.

Superior rectal v.

Internal pudendal a.

Inferior rectal a.

Internal rectal plexus

Middle rectal v.

Rectal column

M. sphincter ani externus

Inferior rectal v.

Anus

Fig. 16-92.   The blood supply of the rectum. A portion of the anterior wall has been cut away to show the rectal columns and the internal rectal (hemorrhoidal) plexus. (Eycleshymer and Jones.)

middle and inferior rectal veins at its commencement, thus establishing a communication between the systemic and portal circulations.

The **lymphatics** of the large intestine are described in Chapter 10.

The **nerves** to the region of the colon supplied by the superior mesenteric artery are derived in the same manner as those for the small intestine; nerves to the more distal portions of the colon and to the rectum are derived from sympathetic and sacral parasympathetic fibers through the inferior mesenteric and hypogastric plexuses. They are distributed in a similar way to those found in the small intestine.

The Liver

The **liver** (*hepar*), the largest gland in the body, is situated in the cranial and right parts of the abdominal cavity, occupying almost the whole of the right hypochondrium, the greater part of the epigastrium, and not uncommonly extending into the left hypochondrium as far as the mammary line.

DEVELOPMENT OF THE LIVER. The liver arises in the form of a diverticulum or hollow outgrowth from the ventral surface of that portion of the primitive gut which afterward becomes the descending part of the duodenum (Fig. 16-93). This diverticulum is lined by entoderm; it grows cranialward and ventralward into the septum transversum, which is a mass of mesoderm between the vitelline duct and the pericardial cavity. Two solid buds of cells that extend into the tisues represent the right and the left lobes of the liver. The solid buds of cells grow into columns or cylinders, termed the **hepatic cylinders,** which branch and anastomose to

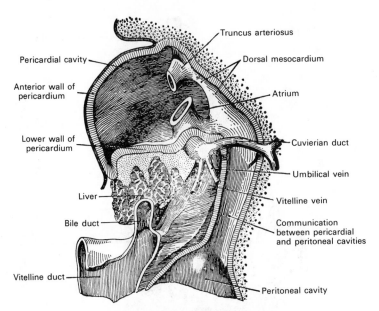

FIG. 16-93.   Liver with the septum transversum. Human embryo 3 mm long. (After model and figure by His.)

form a close meshwork. This network proceeds to invade the vitelline and umbilical veins, and eventually breaks up these vessels into a series of capillary-like vessels that ramify in the meshes of the cellular network and ultimately form the **sinusoids** of the liver.

The continued growth and ramification of the hepatic cylinders gradually produce the mass of the liver. The original diverticulum from the duodenum becomes the common bile duct, and from this the cystic duct and gallbladder arise as a solid outgrowth that later acquires a lumen. The opening of the common duct is at first in the ventral wall of the duodenum; later, owing to the rotation of the gut, the opening is carried to the left and then dorsalward to the position it occupies in the adult.

As the liver undergoes enlargement, both it and the ventral mesogastrium of the foregut are gradually differentiated from the septum transversum, and the liver projects caudalward into the abdominal cavity from the caudal surface of the septum transversum. By the growth of the liver, the ventral mesogastrium is divided into two parts, of which the ventral part forms the falciform and coronary ligaments and the dorsal part forms the lesser omentum. About the third month the liver almost fills the abdominal cavity, and its left lobe is nearly as large as its right.

From this period the relative development of the liver is less active, especially that of the left lobe, which actually undergoes some degeneration and becomes smaller than the right, but up to the end of fetal life the liver remains relatively larger than in the adult.

The **adult liver,** in the male, weighs 1.4 to 1.6 kg; in the female, 1.2 to 1.4 kg. It is relatively much larger in the fetus than in the adult, constituting in the former about one-eighteenth, and in the latter about one thirty-sixth of the entire body weight. Its greatest transverse measurement is 20 to 22.5 cm. Vertically, near its lateral or right surface, it measures about 15 to 17.5 cm; its greatest dorsoventral diameter, of 10 to 12.5 cm, is on a level with the cranial end of the right kidney. Opposite the vertebral column this measurement is reduced to about 7.5 cm. Its consistency is that of a soft solid; it is friable, easily lacerated and highly vascular; its color is dark reddish brown, and its specific gravity is 1.05.

The liver is irregularly hemispherical in shape, with an extensive, relatively smooth, convex diaphragmatic surface and a more irregular concave visceral surface. The diaphragmatic surface has four parts: ventral, superior, dorsal, and right. The human liver has four lobes: a large right lobe, a smaller left lobe, and much smaller caudate and quadrate lobes.

SURFACES AND BORDERS

DIAPHRAGMATIC SURFACE. The diaphragmatic surface has the following portions: ventral, superior, dorsal, and right.

The **ventral** or **anterior portion** (Fig. 16-94) is separated by the diaphragm from the sixth to tenth ribs and their costal cartilages on the right side and from the seventh and eighth cartilages on the left. In the median region it lies dorsal to the xyphoid process and that part of the muscular anterior abdominal wall between the diverging costal margins. It is completely covered by peritoneum except along the line of attachment of the falciform ligament.

The **superior portion** (Fig. 16-95) is separated by the dome of the diaphragm from the pleura and lungs on the right and from the pericardium and heart on the left. The area near the heart is marked by a shallow concavity, the **cardiac fossa** or **impression.** The surface is mostly covered by peritoneum, but along its dorsal part it is attached to the diaphragm by the superior reflection of the coronary ligament which separates the part covered with peritoneum from the so-called bare area.

The **dorsal** or **posterior portion** (Fig. 16-95) is broad and rounded on the right but narrow on the left. The central part presents a deep concavity that is molded to fit against the vertebral column and crura of the diaphragm; close to the right of this concavity the **inferior vena cava** lies almost buried in its **fossa.** Two or three centimeters to the left of the vena cava is the narrow **fossa for the ductus venosus.** The caudate lobe lies between these two fossae. To the right of the vena cava and partly on the visceral surface is a small triangular depressed area, the **suprarenal impression,** for the right suprarenal gland. To the left of the fossa for the ductus venosus is the **esophageal groove** for the antrum cardiacum of the esophagus.

A large part of the dorsal portion of the diaphragmatic surface is not covered by peritoneum; it is attached to the diaphragm by

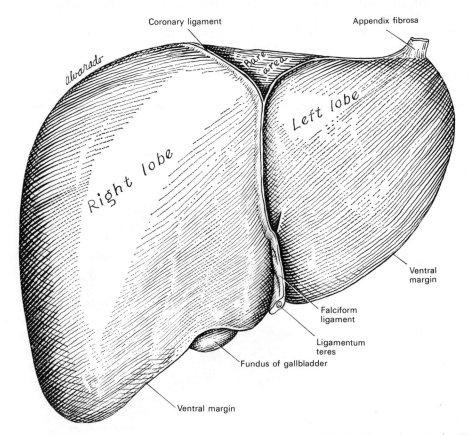

FIG. 16-94. Ventral or anterior portion of diaphragmatic surface of the liver. (Redrawn from Rauber-Kopsch.)

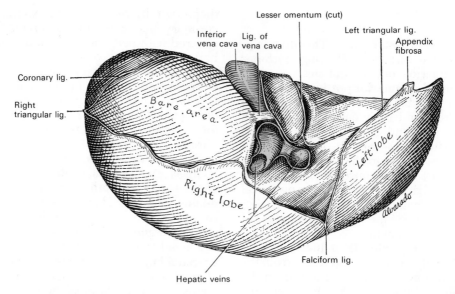

FIG. 16-95. Superior portion of diaphragmatic surface of liver. (Redrawn from Rauber-Kopsch.)

loose connective tissue. The uncovered area, frequently called the **bare area,** is bounded by the superior and inferior reflections of the coronary ligament.

The **right portion** merges with the other three parts of the diaphragmatic surface and continues down to the right margin, which separates it from the visceral surface.

VISCERAL SURFACE. The visceral surface is concave, facing dorsalward, caudalward,

and to the left (Fig. 16-96). It contains several fossae and impressions for neighboring viscera. A prominent marking of the left central part is the **porta hepatis,** a fissure for the passage of the blood vessels and bile duct. The visceral surface is covered by peritoneum except where the gallbladder is attached to it and at the porta.

The right lobe, lying to the right of the gallbladder, has three impressions. Farthest

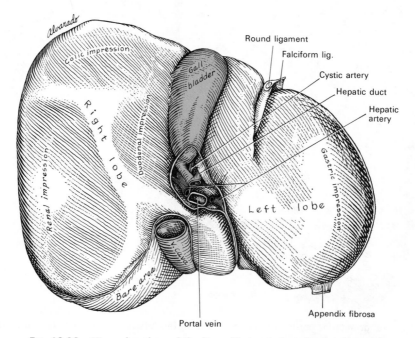

FIG. 16-96. Visceral surface of the liver. (Redrawn from Rauber-Kopsch.)

to the right is the **colic impression,** a flattened or shallow area for the right colic flexure; more dorsally a larger and deeper hollow is the **renal impression** for the right kidney, and the **duodenal impression** is a narrow and poorly marked area lying along the neck of the gallbladder. Between the gallbladder and **fossa for the umbilical vein** is the quadrate lobe. It is in relation with the pyloric end of the stomach, the superior portion of the duodenum, and the transverse colon.

The left lobe, lying to the left of the umbilical vein fossa, has two prominent markings. A large hollow extending out to the margin is the **gastric impression** for the ventral surface of the stomach. Toward the right it merges into a round eminence, the **tuber omentale,** which fits into the lesser curvature of the stomach and lies over the ventral surface of the lesser omentum. Just ventral to the inferior vena cava is a narrow strip of liver tissue, the **caudate process,** which connects the right inferior angle of the caudate lobe to the right lobe. Its peritoneal covering forms the ventral boundary of the epiploic foramen.

INFERIOR BORDER. The inferior border is thin and sharp, and marked opposite the attachment of the falciform ligament by a deep notch, the **umbilical notch,** and opposite the cartilage of the ninth rib by a second notch for the fundus of the gallbladder. In adult males this border generally corresponds with the lower margin of the thorax in the right mammary line, but in women and children it usually projects below the ribs. In the erect position it often extends below the interiliac line.

The **left extremity of the liver** is thin and flat from above downward.

FISSURES AND FOSSAE

The **left sagittal fossa** (*longitudinal fissure*) is a deep groove in the visceral surface that extends from the notch on the inferior margin of the liver to the cranial border of the organ. It is not named in the *Nomina Anatomica* but is worthy of mention because it separates the right and left lobes. The porta joins it, at right angles, and divides it into two parts. The ventral part is the **fissure for the ligamentum teres,** which lodges the umbilical vein in the fetus and its re-

mains (the ligamentum teres) in the adult; it lies between the quadrate lobe and the left lobe of the liver, and is often partially bridged over by a prolongation of the hepatic substance, the **pons hepatis.** The dorsal part, or **fossa for the ductus venosus,** lies between the left lobe and the caudate lobe; it lodges the ductus venosus in the fetus, and in the adult, the **ligamentum venosum,** a slender fibrous cord that represents the obliterated remains of that vessel.

The **porta** or **transverse fissure** (Fig. 16-97) is a short but deep fissure, about 5 cm long, extending transversely across the visceral surface of the left portion of the right lobe, nearer its dorsal surface than its ventral border. It joins nearly at right angles with the left sagittal fossa and separates the quadrate lobe ventrally from the caudate lobe and process dorsally. It transmits the portal vein and the hepatic artery, nerves, duct, and lymphatics. The hepatic duct lies ventral and to the right; the hepatic artery, to the left; and the portal vein, dorsal to and between the duct and the artery.

The **fossa for the gallbladder** is a shallow, oblong fossa placed on the visceral surface of the right lobe, parallel with the left sagittal fossa. It extends from the inferior free margin of the liver, which is notched by it, to the right extremity of the porta.

The **fossa for the inferior vena cava** is a short, deep depression and occasionally a complete canal if the substance of the liver surrounds the vena cava. It lies on the posterior surface between the caudate lobe and the bare area of the liver, and is separated from the porta by the caudate process. The orifices of the hepatic veins perforate the floor of this fossa to enter the inferior vena cava.

LOBES

RIGHT LOBE. The right lobe is six times as large as the left. It occupies the right hypochondrium, and is separated from the left lobe on its diaphragmatic surface by the falciform ligament, and on its visceral surface, by the left sagittal fossa. It has a somewhat quadrilateral form, its visceral and posterior surfaces being marked by three fossae: the porta and the fossae for the gallbladder and inferior vena cava, which separate its left part into two smaller lobes, the

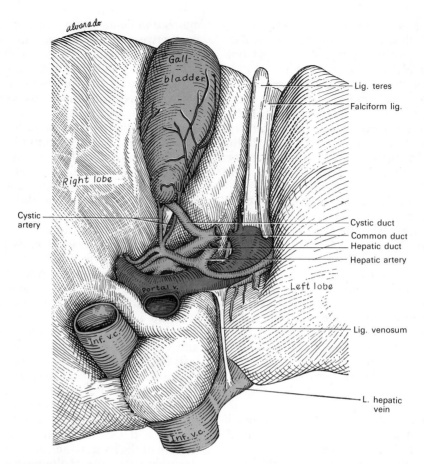

FIG. 16-97. The gallbladder and portal area of the visceral surface of the liver with blood vessels and ducts exposed. (Redrawn from Rauber-Kopsch.)

quadrate and **caudate lobes.** The impressions on the right lobe have already been described.

QUADRATE LOBE. The quadrate lobe is situated on the visceral surface of the right lobe. It is bounded ventrally by the inferior margin of the liver; dorsally, by the porta; on the right, by the fossa for the gallbladder; and on the left, by the fossa for the umbilical vein. It has an oblong shape, with the dorsoventral diameter greater than the transverse diameter.

CAUDATE LOBE. The caudate lobe (*Spigelian lobe*[1]) is situated upon the dorsal surface of the right lobe of the liver, opposite the tenth and eleventh thoracic vertebrae. It is bounded inferiorly, by the porta; on the right, by the fossa for the inferior vena cava; and on the left, by the fossa for the ductus venosus. It is nearly vertical in position and

somewhat concave in the transverse direction. The **caudate process** is a small elevation of the hepatic substance that extends obliquely lateralward from the lower extremity of the caudate lobe to the visceral surface of the right lobe. It is situated dorsal to the porta and separates the fossa for the gallbladder from the commencement of the fossa for the inferior vena cava.

LEFT LOBE. The left lobe is smaller and flatter than the right. It is situated in the epigastric and left hypochondriac regions. Its cranial surface is slightly convex and is molded to the diaphragm; its caudal surface presents the gastric impression and omental tuberosity already mentioned.

INTERNAL LOBES AND SEGMENTS

The liver substance is divisible internally, on the basis of function and surgical importance, into lobes and segments that do not

[1] Adriaan van der Spieghel (1578–1625): A Flemish anatomist who worked in Padua.

coincide completely with the divisions established by external markings. These subdivisions must take into account the two great functions of the liver: (1) that of a gland, with the bile passages serving as its ducts, and (2) that of a vascular and storage organ. As a vascular organ it is necessary to devise units or segments including both inflow and outflow of blood, i.e., the branchings of both portal and hepatic veins. This is complicated by two facts: the portal veins have peripheral and the hepatic veins central positions in the lobules and greater subdivisions, and the portal vein has a bilateral type of distribution, whereas the hepatic veins have three main stems. It has been customary, however, to describe the internal lobes and segments on the basis of the portal venous radicles along with the bile ducts and to superimpose upon this a description based upon the hepatic venous radicles.

HEPATIC TRIAD. The three structures of the triad—bile duct, portal vein, and hepatic artery (Fig. 16-98)—gather in the hepatoduodenal ligament ventral to the epiploic fora-men (of Winslow) at the porta hepatis. The hepatic duct is placed ventrally and to the right; the hepatic artery, to the left; and the portal vein, dorsally between the artery and duct. After their primary branching into right and left at the porta, they continue in a similar relationship throughout the organ. They are contained within a fibrous sheath, with prolongations to enclose their smallest branchings, called the **perivascular fibrous capsule** (*Glisson's capsule*) (Fig. 16-102).

The primary branching of the portal vein and hepatic artery at the porta establishes a right internal lobe and a left internal lobe. The **right internal lobe** comprises most of the right lobe of external marking described previously. It is subdivided into two segments; the **posterior segment** is more superior and somewhat larger than the **anterior segment.** Both segments are divided into superior and inferior portions.

The **left internal lobe** includes the left, the caudate, and the quadrate lobes of external marking. The division of the left lobe into a **medial segment** and a **lateral segment** is

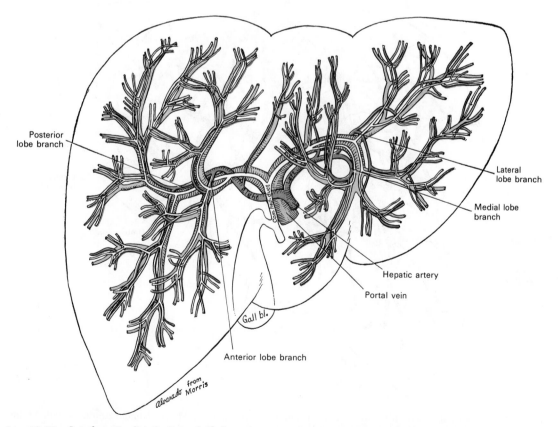

Posterior lobe branch

Lateral lobe branch

Medial lobe branch

Hepatic artery

Portal vein

Gall bl.

Anterior lobe branch

from Morris

alvarado

FIG. 16-98. Intrahepatic distribution of the hepatic artery, portal vein, and biliary ducts. (From *Morris' Human Anatomy,* 12th edition, 1966. Courtesy of Blakiston Division of McGraw-Hill Book Company.)

marked on the surface by the attachment of the falciform ligament, the ligamentum teres, and the ligamentum venosum. The two segments have superior and inferior portions.

VASCULAR SEGMENTS. The three principal hepatic veins—right, middle, and left—provide the basis for classifying parts of the internal segments described above into **true vascular segments:** a right, a left, and two middle segments. A variable number of minor hepatic veins assist in the drainage of these vascular segments (Fig. 16-99).

The **right dorsal** or **dorsocaudal vascular segment** corresponds with the dorsal segment of the internal right lobe; it is drained by the right hepatic vein.

The **left lateral vascular segment** corresponds with the lateral segment of the left internal lobe; it is drained by the left hepatic vein.

The **ventral middle vascular segment** contains both the anterior segment of the right lobe and the medial segment of the internal left lobe. It is drained by the middle hepatic vein.

The **dorsal middle vascular segment** contains the caudate lobe and caudate process, parts of the left internal lobe. It is drained usually by two minor hepatic veins, the cranial caudate hepatic vein and the caudal caudate hepatic vein.

LIGAMENTS

The liver is connected to the undersurface of the diaphragm and to the ventral wall of the abdomen by five ligaments. Four of these—the **falciform,** the **coronary,** and the two **lateral ligaments**—are peritoneal folds; the fifth, the **round ligament,** is a fibrous cord that represents the obliterated umbilical vein. The liver is also attached to the lesser curvature of the stomach by the hepatogastric ligament, and to the duodenum by the hepatoduodenal ligament.

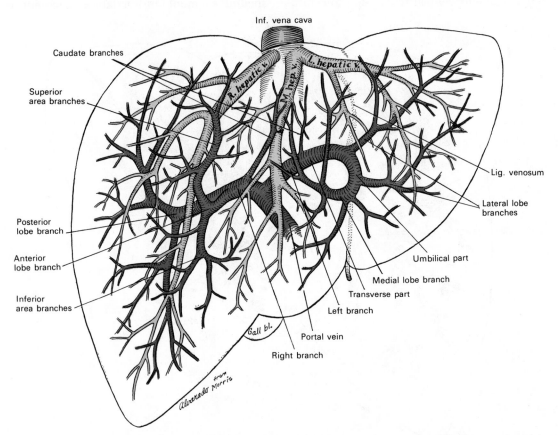

FIG. 16-99.   Intrahepatic distribution of the hepatic and portal veins. (From *Morris' Human Anatomy,* 12th edition, 1966. Courtesy of Blakiston Division of McGraw-Hill Book Company.)

FALCIFORM LIGAMENT. The falciform ligament is situated in a parasagittal plane, but lies obliquely so that one surface faces ventralward and is in contact with the peritoneum dorsal to the right rectus muscle and the diaphragm, while the other is directed dorsalward and is in contact with the left lobe of the liver. It is attached by its left margin to the abdominal surface of the diaphragm, and the dorsal surface of the sheath of the right rectus as far caudalward as the umbilicus; by its right margin it extends from the notch on the inferior margin of the liver, as far dorsalward as the bare area. It is composed of two layers of peritoneum closely united together. Its base or free edge contains between its layers the round ligament and the parumbilical veins.

CORONARY LIGAMENT. The coronary ligament (*ligamentum coronarium hepatis*) consists of an anterior and a posterior layer. The *anterior layer* is formed by the reflection of the peritoneum from the cranial margin of the bare area of the liver to the undersurface of the diaphragm, and is continuous with the right layer of the falciform ligament. The *posterior layer* is reflected from the caudal margin of the bare area to the right kidney and suprarenal gland, and it is termed the **hepatorenal ligament.**

TRIANGULAR LIGAMENTS. There are two triangular ligaments (*lateral ligaments*): right and left.

The **right triangular ligament** is situated at the right extremity of the bare area, and is a small fold that passes to the diaphragm, being formed by the apposition of the anterior and posterior layers of the coronary ligament.

The **left triangular ligament** is a large fold that connects the posterior part of the upper surface of the left lobe to the diaphragm; its anterior layer is continuous with the left layer of the falciform ligament. It terminates on the left in a strong fibrous band, the **appendix fibrosa hepatis.**

ROUND LIGAMENT. The round ligament (*ligamentum teres hepatis*) is a fibrous cord resulting from the obliteration of the umbilical vein. It ascends from the umbilicus in the free margin of the falciform ligament to the umbilical notch of the liver, from which it may be traced in its proper fossa on the inferior surface of the liver to the porta, where it becomes continuous with the *ligamentum venosum.*

FIXATION OF THE LIVER

Several factors contribute to maintaining the liver in place. The attachments of the liver to the diaphragm by the coronary and triangular ligaments and the intervening connective tissue of the bare area, together with the intimate connection of the inferior vena cava by the connective tissue and hepatic veins, support the posterior part of the liver. The lax falciform ligament certainly gives no support, though it probably limits lateral displacement. During descent of the diaphragm with deep breathing, the liver rolls ventralward, shifting the inferior border caudalward so that it can be palpated.

STRUCTURE OF THE LIVER

VESSELS AND NERVES. The vessels connected with the liver are the **hepatic artery**, the **portal vein**, and the **hepatic veins**.

The **hepatic artery** and **portal vein**, accompanied by numerous nerves, ascend to the porta between the layers of the lesser omentum. The **bile duct** and the lymphatic vessels descend from the porta between the layers of the same omentum. The relative positions of the three structures are as follows: the bile duct lies to the right, the hepatic artery to the left, and the portal vein dorsal to and between the other two (Fig. 16-108). They are enveloped in a loose areolar tissue, the **fibrous capsule of Glisson**, which accompanies the vessels in their course through the portal canals in the interior of the organ (Fig. 16-102).

The **hepatic veins** (Fig. 16-99) convey the blood from the liver and are described on page 836. They have little cellular investment, and what there is binds their walls closely to the walls of the canals through which they run, so that on section of the organ, they remain widely open and are solitary, and may be easily distinguished from the branches of the portal vein, which are more or less collapsed, and always accompanied by an artery and duct.

The **lymphatic vessels** of the liver are described under the lymphatic system (Chap. 10).

The **nerves** of the liver are derived from the right and left vagus and the celiac plexus of the sympathetic nerves. The fibers form plexuses along the hepatic artery and portal vein, enter the porta, and accompany the vessels and ducts to the interlobular spaces. The hepatic vessels are said to receive only sympathetic fibers, while both sympathetic and parasympathetic fibers are distributed to the walls of the bile ducts and gallbladder, where they form plexuses similar to the enteric plexuses of the intestinal wall.

Parenchyma

The substance of the liver or parenchyma is composed of lobules held together by an extremely fine areolar tissue, in which ramify the portal vein, hepatic artery, hepatic veins, lymphatics, and nerves, the whole being invested by a serous and a fibrous coat.

SEROUS COAT. The serous coat is derived from the peritoneum and invests the greater part of the surface of the organ. It adheres intimately to the fibrous coat.

FIBROUS COAT. The fibrous coat lies beneath the serous investment and covers the entire surface of the organ. It is difficult to demonstrate, except where the serous coat is deficient. At the porta it is continuous with the fibrous capsule of Glisson, and on the surface of the organ, with the areolar tissue separating the lobules.

LOBULES. The lobules form the principal mass of the parenchyma. Their outlines, about 2 mm in diameter, give a mottled appearance to the surface of the organ. They are roughly hexagonal in shape, with their columns of cells clustered around an intralobular vein, the smallest radicle of the hepatic vein. The adjacent faces of these neighboring hexagonal (or more irregularly polygonal) lobules are fitted together with a minimum of delicate connective tissue. In the pig, the individual lobules have complete connective tissue capsules and the hexagonal shape is more evident than in the human liver.

PORTAL CANAL. Portal canal is the name given to the channels through the parenchyma by which the smallest radicles of the portal vein, hepatic artery, and bile duct are distributed. These three are bound together by delicate connective tissue, the **capsula fibrosa perivascularis** or **Glisson's capsule.** They are situated between the lobules, and in a microscopic section cut perpendicularly through an intralobular vein, there will be three portal canals at the periphery of the lobule, about equally distant from each other at three angles of the hexagon. Polygonal shapes such as the hexagon in Figure 16-100 are common, however.

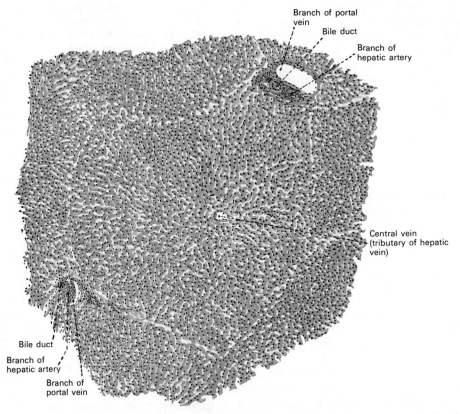

Branch of portal vein

Bile duct

Branch of hepatic artery

Central vein (tributary of hepatic vein)

Bile duct

Branch of hepatic artery

Branch of portal vein

FIG. 16-100. Human liver lobule surrounded by parts of six neighboring lobules. (From Rauber-Kopsch, *Lehrbuch u. Atlas d. Anatomie d. Menschen,* 19th ed., Vol. II, courtesy Georg Thieme Verlag, Stuttgart, 1955.)

Microscopic Appearance

Each lobule consists of a mass of **hepatic cells** arranged in irregular radiating columns and plates, between which are the blood channels (*sinusoids*) (Fig. 16-100). Between the cells are also minute bile capillaries. Therefore, each lobule has all the essentials of a secreting gland: (1) **cells** by which the secretion is formed; (2) **blood vessels** in close relation with the cells, containing the blood from which the secretion is derived; (3) **ducts,** by which the secretion, when formed, is carried away.

HEPATIC CELLS. The hepatic cells are polyhedral. They vary in size from 12 to 25 μm in diameter. They contain one or sometimes two distinct nuclei. The nucleus exhibits an intranuclear network and one or two refractile nucleoli. The cells usually contain granules, some of which are protoplasmic, while others consist of glycogen, fat, or an iron compound. In lower vertebrates such as the frog, the cells are arranged in tubes with the bile duct forming the lumen and the blood vessels located externally.

BLOOD VESSELS. The blood in the capillary plexuses around the liver cells is brought to the liver principally by the portal vein, but also to a certain extent by the hepatic artery.

The **hepatic artery,** which enters the liver at the porta with the portal vein and hepatic duct, ramifies with these vessels through the portal canals. It gives off **vaginal branches,** which ramify in the fibrous capsule of Glisson and appear destined chiefly for the nutrition of the coats of the vessels and ducts. It also gives off **capsular branches,** which reach the surface of the organ, ending in its fibrous coat in stellate plexuses. Finally, it gives off **interlobular branches,** which form a plexus outside each lobule to supply the walls of the interlobular veins and the accompanying bile ducts. From these plexuses, capillaries join directly with the sinusoids of the liver lobule at its periphery.

The **portal vein** also enters at the porta and runs through the portal canals (Fig. 16-99) enclosed in Glisson's capsule. In its course it divides into branches, which finally break up into the **interlobular plexuses** in the interlobular spaces. These interlobular plexuses give off portal venules, which divide into small branches and twigs as they pass to the surfaces of the lobules to join the hepatic sinusoids (Fig. 16-101).

Hepatic sinusoids are large, richly anastomosing, modified capillary channels lying between the cords of liver cells. They traverse the liver lobule from its periphery to the intralobular or central vein. At the periphery of the lobule they connect with interlobular branches of the portal vein and hepatic artery. Thus all the blood that enters the liver passes through the sinusoids to the central veins. The sinusoids are lined by

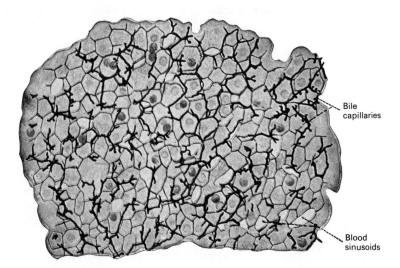

FIG. 16-101. Bile capillaries in the liver of a rabbit demonstrated by silver chromate, counterstained with alum carmine. (From Rauber-Kopsch, *Lehrbuch u. Atlas d. Anatomie d. Menschen,* 19th ed., Vol. II, courtesy Georg Thieme Verlag, Stuttgart, 1955.)

modified endothelium and contain many macrophages (v. Kupffer cells)[1] attached to their walls.

HEPATIC VEINS. At the center of the lobule, the sinusoids empty into one large vein that runs down the center of the lobule from apex to base; this is called the **intralobular** or **central vein.** At the base of the lobule this vein opens directly into the **sublobular vein,** with which the lobule is connected. The sublobular veins unite to form larger and larger trunks, and end at last in the hepatic veins; these converge to form three large trunks that open into the inferior vena cava while that vessel is situated in its fossa on the posterior surface of the liver.

BILE DUCTS. The bile ducts commence by little passages in the liver cells that communicate with canaliculi termed **intercellular biliary passages** (*bile capillaries*). These passages are merely little channels or spaces left between the contiguous surfaces of two cells or in the angle where three or more liver cells meet (Fig. 16-101). They are always separated from the blood capillaries by at least half the width of a liver cell. The channels thus formed radiate to the circumference of the lobule and open into the interlobular bile ducts, which run in Glisson's capsule, accompanying the portal vein and hepatic artery (Fig. 16-102). These join with other ducts to form two main trunks that leave the liver at the porta, and by their union form the **hepatic duct.**

[1] Karl Wilhelm von Kupffer (1829–1902): A German anatomist (Kiel, Königsberg and Munich).

Structure of the Ducts. The walls of the biliary ducts consist of a connective tissue coat that contains muscle cells arranged both circularly and longitudinally, and an epithelial layer that consists of short columnar cells resting on a distinct basement membrane.

EXCRETORY APPARATUS OF THE LIVER

The excretory apparatus of the liver consists of (1) the **hepatic duct,** formed by the junction of the two main ducts, which pass out of the liver at the porta; (2) the **gallbladder,** which serves as a reservoir for the bile; (3) the **cystic duct,** or the duct of the gallbladder; and (4) the **common bile duct,** which is formed by the junction of the hepatic and cystic ducts.

Hepatic Duct

Two main trunks of nearly equal size, one from the right, the other from the left lobe, unite to form the hepatic duct (Fig. 16-98), which passes to the right for about 4 cm between the layers of the lesser omentum, where it is joined at an acute angle by the cystic duct to form the common bile duct. The hepatic and part of the common duct are accompanied by the hepatic artery and portal vein.

Gallbladder

The gallbladder (*vesica fellea*) (Fig. 16-103) is a conical or pear-shaped musculo-membranous sac lodged in a fossa on the vis-

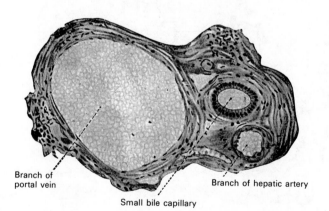

Branch of portal vein

Branch of hepatic artery

Small bile capillary

FIG. 16-102. Cross section of a portal canal with Glisson's capsule of interlobular connective tissue surrounding branches of the portal vein, hepatic artery, and hepatic duct. (From Rauber-Kopsch, *Lehrbuch u. Atlas d. Anatomie d. Menschen,* 19th ed., Vol. II, courtesy Georg Thieme Verlag, Stuttgart, 1955.)

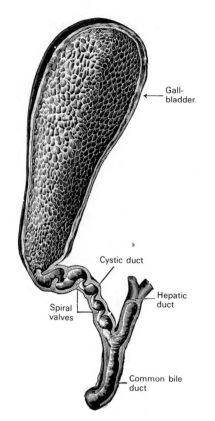

Fig. 16-103. The gallbladder and bile ducts laid open. (Spalteholz.)

ceral surface of the right lobe of the liver and extending from near the right extremity of the porta to the inferior border of the organ. It is 7 to 10 cm in length, 2.5 cm in breadth at its widest part, and holds from 30 to 35 ml of bile. It is divided into a fundus, body, and neck. The **fundus,** or broad extremity, is directed caudalward and projects beyond the inferior border of the liver; the **body** and **neck** are directed cranialward and dorsalward to the left. The surface of the gallbladder is attached to the liver by connective tissue and vessels. The caudal surface is covered by peritoneum, which is reflected on to it from the surface of the liver. Occasionally the whole of the organ is invested by the serous membrane and is then connected to the liver by a kind of mesentery.

RELATIONS. The **body** is in relation with the commencement of the transverse colon and farther dorsally usually with the descending portion of the duodenum, but sometimes with the superior portion of the duodenum or pyloric end of the stomach. The **fundus** is in relation, ventrally, with the abdominal parietes immediately below the ninth costal cartilage, and dorsally with the transverse colon. The **neck** is narrow and curves upon itself like the letter S; at its point of connection with the cystic duct it presents a well-marked constriction.

STRUCTURE OF THE GALLBLADDER. The gallbladder consists of three coats: **serous, fibromuscular,** and **mucous** (Fig. 16-104).

Serous Coat. The external or serous coat is derived from the peritoneum; it completely invests the fundus, but covers the body and neck only on their caudal surfaces.

Fibromuscular Coat. The fibromuscular coat, a thin but strong layer forming the framework of the sac, consists of dense fibrous tissue that interlaces in all directions and is mixed with smooth muscular fibers, which are disposed chiefly in a longitudinal direction with a few running transversely.

Mucous Coat. The internal or mucous coat is loosely connected with the fibrous layer. It is generally yellowish-brown and is elevated into minute rugae. Opposite the neck of the gallbladder the mucous membrane projects inward in oblique ridges or folds, forming a sort of spiral valve.

The mucous membrane is continuous through the hepatic duct with the mucous membrane that lines the ducts of the liver, and through the common bile duct with the mucous membrane of the duodenum. It is covered with high columnar epithelium and secretes mucin; in some animals it secretes a nucleoprotein instead of mucin.

VESSELS AND NERVES. The **arteries** to the gallbladder are derived from the cystic artery; its **veins** drain into liver capillaries and into the portal vein. The **lymphatics** are described under the lymphatic system. The **nerves** are described on page 1270.

Cystic Duct

The cystic duct, about 4 cm long, runs dorsalward, caudalward, and to the left from the neck of the gallbladder; it joins the hepatic duct to form the common bile duct. The mucous membrane lining its interior is thrown into a series of 5 to 12 crescentic folds similar to those found in the neck of the gallbladder. They project into the duct in regular succession, and are directed obliquely around the tube, presenting the ap-

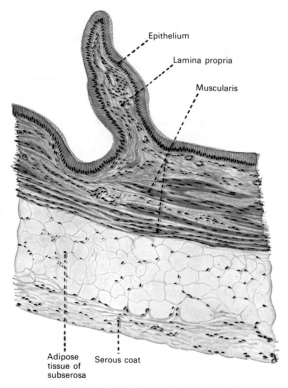

Epithelium

Lamina propria

Muscularis

Adipose tissue of subserosa

Serous coat

FIG. 16-104. Cross section through the wall of the human gallbladder. (From Rauber-Kopsch, *Lehrbuch u. Atlas d. Anatomie d. Menschen,* 19th ed., Vol. II, courtesy Georg Thieme Verlag, Stuttgart, 1955.)

pearance of a continuous spiral valve. They constitute the **spiral valve** (*of Heister*[1]), which is found only in primates and represents a device to prevent distention or collapse of the cystic duct due to changing pressures in the gallbladder or common duct associated with the assumption of an erect posture. When the duct is distended, the spaces between the folds are dilated, giving its exterior a twisted appearance.

Common Bile Duct

The common bile duct (*ductus choledochus*) (Fig. 16-108) is formed by the junction of the cystic and hepatic ducts; it is about 7.5 cm long and has the diameter of a soda fountain straw.

It descends along the right border of the lesser omentum dorsal to the superior portion of the duodenum, ventral to the portal vein and to the right of the hepatic artery. It

[1]Lorenz Heister (1683–1758): A German anatomist and surgeon (Helmstadt).

then runs in a groove near the right border of the posterior surface of the head of the pancreas; here it is situated ventral to the inferior vena cava and is occasionally completely embedded in the pancreatic substance. At its termination it is closely associated with the terminal portion of the pancreatic duct as it passes obliquely through the muscular and mucous coats of the duodenum. The walls of the terminal portions of both ducts are thickened by the presence of a sphincter muscle, the sphincter of Oddi, which usually causes a protrusion into the lumen of the duodenum, the **duodenal papilla** or papilla of Vater. A common orifice for the two ducts is present in about 60% of persons, and separate openings in about 40%. The ducts become narrower rather than wider as they traverse the papilla, and the length of common channel shared by bile and pancreatic ducts is less than one half the papilla in 75% of persons. The term "ampulla of Vater" should therefore be discarded in favor of the "papilla of Vater" (Sterling, 1954).

STRUCTURE. The coats of the large biliary ducts are **external** or **fibrous** and **internal** or **mucous.** The **fibrous coat** is composed of strong fibroareolar tissue, with a certain amount of muscular tissue arranged, for the most part, in a circular manner around the duct. The **mucous coat** is continuous with the lining membrane of the hepatic ducts and gallbladder and also with that of the duodenum; like the mucous membrane of these structures, its epithelium is columnar. It is provided with numerous mucous glands that are lobulated and open by minute orifices scattered irregularly in the larger ducts.

The Pancreas

The **pancreas** (Fig. 16-57) is situated transversely across the posterior wall of the abdomen, in the epigastric and left hypochondriac regions. Its length varies from 12.5 to 15 cm; its weight in the female is 84.88 ± 14.95 gm and in the male, 90.41 ± 16.08 gm (Schaefer, 1926).

DEVELOPMENT

The pancreas is developed in two parts, dorsal and ventral (Figs. 16-105; 16-106). The former arises as a diverticulum from the dor-

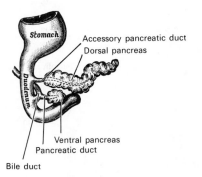

FIG. 16-105. Pancreas of a human embryo of five weeks. (Kollmann.)

sal aspect of the duodenum a short distance above the hepatic diverticulum, and growing cranialward and dorsalward into the dorsal mesogastrium, it forms a part of the head and uncinate process and the whole of the body and tail of the pancreas. The ventral part appears as a diverticulum from the primitive bile duct and forms the remainder of the head and uncinate process of the pancreas. The duct of the dorsal part **(accessory pancreatic duct)** therefore opens independently into the duodenum, while that of the ventral part **(pancreatic duct)** opens with the common bile duct. About the sixth week the two parts of the pancreas meet and fuse and a communication is established between their ducts. After this has occurred, the terminal part of the accessory duct, i.e., the part between the duodenum and the point of meeting of the two ducts, undergoes little or no enlargement, while the pancreatic duct increases in size and forms the main duct of the gland. The opening of the accessory duct into the duodenum is sometimes obliterated, and even when it remains patent it is proba-

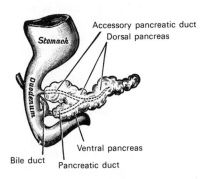

FIG. 16-106. Pancreas of a human embryo at end of sixth week. (Kollmann.)

ble that the whole of the pancreatic secretion is conveyed through the pancreatic duct.

At first the pancreas lies between the two layers of the dorsal mesogastrium, which give it a complete peritoneal investment, and its surfaces look to the right and left. With the change in the position of the stomach, the dorsal mesogastrium is drawn caudalward and to the left, and the right side of the pancreas is directed dorsalward and the left side, ventralward. The right surface becomes applied to the posterior abdominal wall, and the peritoneum that covered it undergoes absorption; thus, in the adult, the gland appears to lie behind the peritoneal cavity.

The **adult pancreas** is a compound racemose gland, analogous in its structures to the salivary glands, though softer and less compactly arranged than those organs. Its shape is long and irregularly prismatic; its right extremity, being broad, is called the **head,** which is connected to the main portion of the organ, or **body,** by a slight constriction, the **neck,** while its left extremity gradually tapers to form the **tail.**

HEAD

The head of the pancreas is lodged within the curve of the duodenum (Fig. 16-107). Its cranial border is overlapped by the superior part of the duodenum and its caudal border overlaps the horizontal part of the duodenum; its right and left borders overlap and insinuate themselves around the descending and ascending parts of the duodenum respectively. The angle of junction of the caudal and left lateral borders forms a prolongation, termed the **uncinate process.** In the groove between the duodenum and the right lateral and caudal borders are the anastomosing superior and inferior pancreaticoduodenal arteries (Fig. 8-58); the common bile duct descends to its termination in the descending part of the duodenum close to the right border.

The greater part of the right half of the **anterior surface** is in contact with the transverse mesocolon, with only areolar tissue intervening. From its cranial part the **neck** springs, its right limit being marked by a groove for the gastroduodenal artery. The caudal part of the right half, below the transverse colon, is covered by peritoneum that is continuous with the inferior layer of the

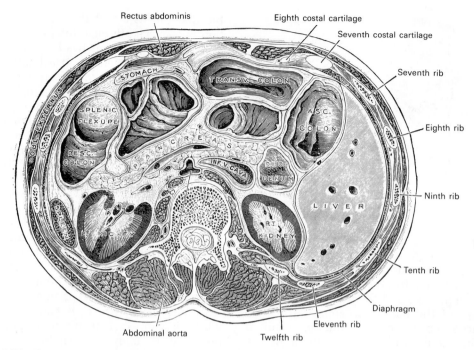

Rectus abdominis   Eighth costal cartilage
Seventh costal cartilage
Seventh rib
STOMACH   TRANSV. COLON
SPLENIC   ASC.
FLEXURE   COLON
Eighth rib
OBL. EXTERNUS   DESC.   PANCREAS   INF. V. CAVA   DUO-   Ninth rib
COLON   DENUM
LIVER
SPLENIC   LT. KIDNEY   RT.   Tenth rib
KIDNEY
Diaphragm
Abdominal aorta   Eleventh rib
Twelfth rib

Fig. 16-107. Transverse section through the middle of the first lumbar vertebra, showing the relations of the pancreas. (Braune.)

transverse mesocolon, and it is in contact with the coils of the small intestine. The superior mesenteric artery passes down ventral to the left half across the uncinate process; the superior mesenteric vein runs cranialward on the right side of the artery and, dorsal to the neck, joins with the lienal vein to form the portal vein (Fig. 9-33).

The **posterior surface** is in relation with the inferior vena cava, the common bile duct, the renal veins, the right crus of the diaphragm, and the aorta.

NECK

The neck is about 2.5 cm long, and is directed at first ventralward and then to the left to join the body. Its ventral surface supports the pylorus; its dorsal surface is in relation with the commencement of the portal vein. On the right it is grooved by the gastroduodenal artery.

BODY

The body has a prismatic shape and three surfaces: **anterior, posterior,** and **inferior;** and three borders: **superior, anterior,** and **inferior.**

The **anterior surface** is somewhat concave and is covered by the dorsal surface of the stomach, which rests upon it; the two organs are separated by the omental bursa. Where it joins the neck there is a well-marked prominence, the **tuber omentale,** which abuts against the posterior surface of the lesser omentum.

The **posterior surface** is devoid of peritoneum and is in contact with the aorta, the lienal vein, the left kidney and its vessels, the left suprarenal gland, the origin of the superior mesenteric artery, and the crura of the diaphragm.

The **inferior surface** is narrow on the right but broader on the left and is covered by peritoneum. It is in contact with the duodenojejunal flexure and some coils of the jejunum. Its left extremity rests on the left colic flexure.

The **superior border** is blunt and flat to the right and narrow and sharp to the left near the tail. It commences on the right in the omental tuberosity and is in relation with the celiac artery, from which the hepatic artery courses to the right just above the gland, while the lienal artery runs toward the left in a groove along this border.

The **anterior border** separates the anterior surface from the inferior surface, and along this border the two layers of the transverse mesocolon diverge from each other, one passing over the anterior surface, the other over the inferior surface.

The **inferior border** separates the posterior surface from the inferior surface. The superior mesenteric vessels emerge under its right extremity.

TAIL

The tail is the narrow part extending to the left as far as the caudal part of the gastric surface of the spleen. It lies in the phrenico-lienal ligament and is in contact with the left colic flexure.

PANCREATIC DUCT

The pancreatic duct (*duct of Wirsung*[1]) extends transversely through the substance of the pancreas (Fig. 16-108). It commences by the junction of the small ducts of the lobules situated in the tail of the pancreas, and running from left to right through the body, it receives the ducts of the various lobules composing the gland. It reaches the neck considerably augmented in size, and turning caudalward, dorsalward, and to the right, it comes into relation with the common bile duct, which lies to its right side. Leaving the

[1]Johann Georg Wirsung (1600–1643): A German anatomist who worked in Padua.

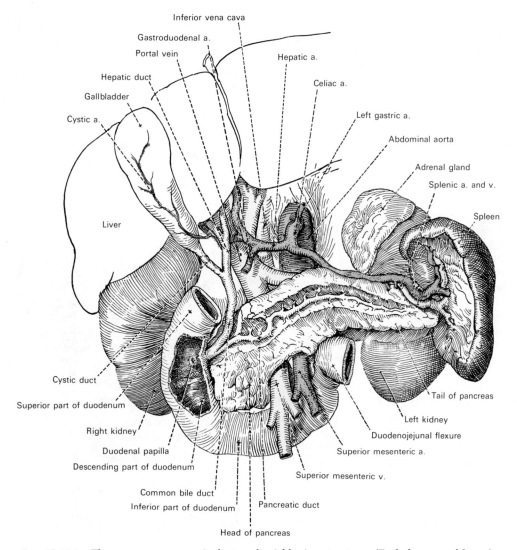

Fig. 16-108. The pancreas, pancreatic duct, and neighboring structures. (Eycleshymer and Jones.)

head of the gland, it passes very obliquely through the muscular and mucous coats of the duodenum and ends by an orifice common to it and the common bile duct upon the summit of the duodenal papilla, which is situated at the medial side of the descending portion of the duodenum, 7.5 to 10 cm below the pylorus. In 40% of bodies the pancreatic duct and the common bile duct open separately into the duodenum. Frequently an additional duct is given off from the pancreatic duct in the neck of the pancreas; this opens into the duodenum about 2.5 cm above the duodenal papilla. It receives the ducts from the lower part of the head and is known as the **accessory pancreatic duct** (*duct of Santorini*[1]).

FUNCTION. The pancreas is a gland of both external (exocrine) and internal (endocrine) secretion. The greater part of its bulk is formed by the exocrine gland. Its secretion, the **pancreatic juice,** is conveyed by the pancreatic duct to the duodenum, where its several enzymes aid in the digestion of proteins, carbohydrates, and fats. The endocrine gland is formed by small clumps of cells known as islets of Langerhans[2] scattered throughout the pancreas. The secretion of these cells, called **insulin,** is taken up by the bloodstream and is an important factor in the control of sugar metabolism in the body.

STRUCTURE OF THE PANCREAS

EXOCRINE PANCREAS. The exocrine pancreas is a compound tubulo-acinar or racemose gland, resembling the parotid gland in microscopic structure. It has a connective tissue covering but no distinct capsule. It is made up of lobes and lobules that are identifiable but indistinctly marked with connective tissue in the human pancreas.

The main **pancreatic duct** extends throughout the length of the gland, receiving smaller ducts from the lobes and lobules along its course (Fig. 16-108). Although two ducts are formed in the embryo, they are usually combined into a single system in the adult. The main duct is lined with cylindrical epithelium containing occasional goblet cells; the smaller intermediate and intercalated ducts have lower cuboidal cells.

The **secreting acini** are almost filled by the

[1]Giovanni Domenico Santorini (1681–1737): An Italian anatomist (Venice).
[2]Paul Langerhans (1847–1888): A German anatomist and pathologist (Freiburg).

protruding apices of the glandular cells, which are the serous or zymogenic type. The basal portion of these cells, resting on a basement membrane, contains the nucleus and a *basophilic substance,* which has a striated appearance and is identified as the cytoplasmic reticulum by the electron microscope. The apical portion of these cells contains *zymogen granules,* easily seen in the usual eosin-stained preparations because of their red color. During digestion the granules become reduced in number. A characteristic feature of the pancreas is the presence of **centroacinar cells.** They represent a continuation of the terminal duct into the secreting acini. They are smaller than the secreting cells and have a clear cytoplasm (Munger, 1958; Oram, 1955).

ENDOCRINE PANCREAS. The **islets of Langerhans** can be identified in the usual histological preparations as clusters of less deeply stained cells. They are scattered throughout the gland but usually appear near the center of a lobule (Fig. 16-109). The cells are arranged in cords or plates one or two cells thick between which are abundant capillaries. There are three types of cells, which can be demonstrated by the use of special stains: **alpha cells, beta cells,** and **delta cells.** The beta cells are most abundant and are the source of insulin (Ferner, 1952; Lacy, 1957).

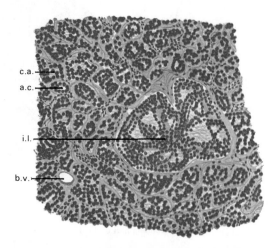

FIG. 16-109. Section of portion of lobule from human pancreas. *ac.,* Acinus of exocrine secreting gland; *b.v.,* blood vessel; *c.a.,* centroacinar cells; *i.l.,* islet of endocrine-secreting cells (islet of Langerhans). 200 ×. (Redrawn from Sobotta.)

VESSELS. The **arteries** of the pancreas are mainly branches of the splenic artery, which form several arcades with the pancreatic branches of the gastro-duodenal and superior mesenteric arteries (Michels, 1955). The **veins** are tributaries of the splenic and superior mesenteric portions of the portal vein. The **lymphatics** drain into the regional nodes associated with the major arteries to the gland.

NERVES. The autonomic innervation is both para-sympathetic and sympathetic through the splenic subdivision of the celiac plexus. The **sympathetic nerves** are postganglionic; the **parasympathetic nerves** are preganglionic nerves from the vagus, which synapse with ganglion cells scattered throughout the gland. **Sensory nerves** to the pancreas are conveyed mainly by way of the splanchnic nerves (Alvarado, 1955).

References

(References are listed not only to those articles and books cited in the text, but to others as well which are considered to contain valuable resource information for the student who desires it.)

General and Development of the Digestive Tube

Anson, B. J., R. Y. Lyman, and H. H. Lander. 1936. The abdominal viscera in situ: A study of 125 consecutive cadavers. Anat. Rec., 67:17–21.

Arey, L. B., M. J. Tremaine, and F. L. Monzingo. 1935. The numerical and topographical relations of taste buds to human circumvallate papillae throughout the life span. Anat. Rec., 64:9–25.

Bates, M. N. 1948. The early development of the hypoglossal musculature in the cat. Amer. J. Anat., 83:329–355.

Beattie, J. 1924. The early stages of the development of the ileo-colic sphincter. J. Anat. (Lond.), 59:56–59.

Cantino, D. 1970. An histochemical study of the nerve supply to the developing alimentary tract. Experientia, 26:766–767.

Cash, J. R. 1921. On the development of the lymphatics in the stomach of the embryo pig. Carneg. Inst., Contr. Embryol., 13:1–16.

Dixon, A. D. 1958. The development of the jaws. Dent. Prac., 9:10–20.

De Garis, C. F. 1941. Topography and development of the cecum-appendix. Ann. Surg., 113:540–548.

Harbin, R. M. 1930. Meckel's diverticulum. Surg. Gynecol. Obstet., 51:863–868.

Hawkes, R. C., G. N. Holland, W. S. Moore, E. J. Roebuck, and B. S. Worthington. 1981. Nuclear magnetic resonance (NMR) tomography of the normal abdomen. J. Comput. Assist. Tomogr., 5:613–618.

Huntington, G. S. 1903. *The Anatomy of the Human Peritoneum and Abdominal Cavity*. Lea Brothers and Co., Philadelphia, 292 pp.

Jacobson, E. D. 1977. Control of the splanchnic circulation. Yale J. Biol. Med., 50:301–306.

Jit, I. 1952. The development and the structure of the suspensory muscle of the duodenum. Anat. Rec., 113:395–407.

Johnson, F. P. 1910. The development of the mucous membrane of the oesophagus, stomach and small intestine in the human embryo. Amer. J. Anat., 10:521–561.

Johnson, F. P. 1914. The development of the rectum in the human embryo. Amer. J. Anat., 16:1–57.

Klueber, K. M., H. L. Langdon, and Y. Barnwell. 1979. The morphology of the vertical and transverse intrinsic musculature of the tongue in the 15-week human fetus. Acta Morphol. Neer. Scand., 17:301–310.

Kraus, B. S., and R. E. Jordan. 1965. *The Human Dentition Before Birth*. Lea & Febiger, Philadelphia, 218 pp.

Lawson, T. L., L. L. Berland, and W. D. Foley. 1981. Coronal upper abdominal anatomy: technique and gastrointestinal applications. Gastrointest. Radiol., 6:115–128.

Lewis, F. T. 1912. The form of the stomach in human embryos with notes upon the nomenclature of the stomach. Amer. J. Anat., 13:477–503.

Michels, N. A. 1955. *Blood Supply and Anatomy of the Upper Abdominal Organs, with a Descriptive Atlas*. J. B. Lippincott Co., Philadelphia, 581 pp.

Moody, R. O., W. E. Chamberlain, and R. G. Van Nuys. 1926. Visceral anatomy of healthy adults. Amer. J. Anat., 37:273–288.

Moody, R. O., and R. G. Van Nuys. 1928. Some results of a study of roentgenograms of the abdominal viscera. Amer. J. Roentgenol., 20:348–358.

Odgers, P. N. B. 1930. Some observations on the development of the ventral pancreas in man. J. Anat. (Lond.), 65:1–7.

Provenza, D. V. 1964. *Oral Histology; Inheritance and Development*. J. B. Lippincott Co., Philadelphia, 548 pp.

Read, J. B., and G. Burnstock. 1970. Development of the adrenergic innervation and chromaffin cells in the human fetal gut. Devel. Biol., 22:513–534.

Ruzicka, F. F., Jr., and P. Rossio. 1970. Normal vascular anatomy of the abdominal viscera. Radiol. Clin. North Amer., 8:3–29.

Sbarbati, R. 1979. Quantitative aspects of the embryonic growth of the intestine and stomach. J. Anat. (Lond.), 129:795–803.

Schwegler, R. A., Jr., and E. A. Boyden. 1937. The development of the pars intestinalis of the common bile duct in the human fetus, with special reference to the origin of the ampulla of Vater and the sphincter of Oddi. I. The involution of the ampulla. Anat. Rec., 67:441–467. II. The early development of the musculus proprius. Anat. Rec., 68:17–41. III. The composition of the musculus proprius. Anat. Rec., 68:193–219.

Smith, E. I. 1957. The early development of the trachea and esophagus in relation to atresia of the esophagus and tracheoesophageal fistula. Carneg. Inst., Contr. Embryol., 36:41–58.

Tench, E. M. 1936. Development of the anus in the human embryo. Amer. J. Anat., 59:333–345.

Oral Cavity; Teeth, Tongue and Salivary Glands

Abd-El-Malek, S. A. 1938. A contribution to the study of the movements of the tongue in animals, with special reference to the cat. J. Anat. (Lond.), 73:15–30.

Anderson, D. J., A. G. Hannum, and B. Matthews. 1970. Sensory mechanisms in mammalian teeth and their supporting structures. Physiol. Rev., 50:171–195.

Arvidson, K. 1979. Location and variation in number of taste buds in human fungiform papillae. Scand. J. Dent. Res., 87:435–442.

Barron, D. H. 1936. A note on the course of the proprioceptor fibres from the tongue. Anat. Rec., 66:11–15.

Bennett, G. A., and R. C. Hutchinson. 1946. Experimental studies on the movements of the mammalian tongue. II. The protrusion mechanism of the tongue (dog). Anat. Rec., 94:57–83.

Bernick, S., and R. Patek. 1969. Lymphatic vessels of the dental pulp in dogs. J. Dent. Res., 48:959–964.

Bjorndal, A. M., W. G. Henderson, A. E. Skidmore, and F. H. Kellner. 1974. Anatomic measurements of human teeth extracted from males between the ages of 17 and 21 years. Oral Surg., Oral Med., Oral Pathol., 38:791–803.

Carleton, A. 1938. Observations on the problem of proprioceptive innervation of the tongue. J. Anat. (Lond.), 72:502–507.

Carter, R. B., and E. N. Keen. 1971. The intramandibular course of the inferior alveolar nerve. J. Anat. (Lond.), 108:433–440.

Castelli, W. A. 1963. Vascular architecture of the human adult mandible. J. Dent. Res., 42:786–792.

Cohen, L. 1959. Venous drainage of the mandible. Oral Surg., Oral Med., Oral Pathol., 12:1447–1449.

Corbin, K. B., and F. Harrison. 1939. The sensory innervation of the spinal accessory and tongue musculature in the Rhesus monkey. Brain, 62:191–197.

Doran, G. A. 1975. Review of the evolution and phylogeny of the mammalian tongue. Acta Anat., 91:118–129.

Fireman, S. M., and A. M. Noyek. 1976. Dental anatomy and radiology and the maxillary sinus. Otolaryngol. Clin. North Amer., 9:83–91.

Fitzgerald, M. J. T., and J. H. Scott. 1958. Observations on the anatomy of the superior dental nerves. Brit. Dent. J., 104:205–208.

Freitag, P., and M. B. Engel. 1970. Autonomic innervation in rabbit salivary glands. Anat. Rec., 167:87–106.

Froeschels, E. 1949. Uvula and tonsils. Arch. Otolaryngol., 50:216–219.

Furseth, R. 1969. The fine structure of the cellular cementum of young human teeth. Arch. Oral Biol., 14:1147–1158.

Garn, S. M., K. Koski, and A. B. Lewis. 1957. Problems in determining the tooth eruption sequence in fossil and modern man. Amer. J. Phys. Anthrop., 15:313–331.

Garrett, J. R. 1966. The innervation of salivary glands. I. Cholinesterase-positive nerves in normal glands of the cat. J. Roy. Microscop. Soc., 85:135–148. II. The ultrastructure of nerves in normal glands of the cat. J. Roy. Microscop. Soc., 85:149–162.

Genis-Galvez, J. M., L. Santos-Gutierrez, and M. Martin-Lopez. 1966. On the double innervation of the parotid gland. An experimental study. Acta Anat., 63:398–403.

Hayes, E. R., and R. Elliott. 1942. Distribution of the taste buds on the tongue of the kitten, with particular reference to those innervated by the chorda tympani branch of the facial nerve. J. Comp. Neurol., 76:227–238.

Howkins, C. H. 1935. Blood supply of the lower jaw. Proc. Roy. Soc. Med., 29:506–507.

Hunt, J. R. 1915. The sensory field of the facial nerve. A further contribution to the symptomatology of the geniculate ganglion. Brain, 38:418–446.

Jamieson, J. K., and J. F. Dobson. 1920. The lymphatics of the tongue: with particular reference to the removal of lymphatic glands in cancer of the tongue. Brit. J. Surg., 8:80–87.

Jande, S. S., and L. F. Bélanger. 1970. Fine structural study of rat molar cementum. Anat. Rec., 167:439–463.

Jones, F. W. 1939. The anterior superior alveolar nerve and vessels. J. Anat. (Lond.), 73:583–591.

Kramer, I. R. H. 1960. The vascular architecture of the human dental pulp. Arch. Oral Biol. Med., 2:177–189.

Kubota, K., T. Negishi, and T. Masegi. 1975. Topological distribution of muscle spindles in the human tongue and its significance in proprioception. Bull. Tokyo Med. Dent. Univ., 22:235–242.

Kuehn, D. P., and N. A. Azzam. 1978. Anatomical characteristics of palatoglossus and the anterior faucial pillar. Cleft Palate J., 15:349–359.

Kuntz, A., and C. A. Richins. 1946. Components and distribution of the nerves of the parotid and submandibular glands. J. Comp. Neurol., 85:21–32.

Kurahashi, Y., and S. Yoshiki. 1972. Electron microscopic localization of alkaline phosphatase in the enamel organ of the young rat. Arch. Oral Biol. Med., 17:155–163.

Langworthy, O. R. 1924. A study of the innervation of the tongue musculature with particular reference to the proprioceptive mechanism. J. Comp. Neurol., 36:273–297.

Lavelle, C. L. 1978. Correlations between tooth dimensions of man and apes. Acta Anat., 102:358–364.

Leppi, T. J., and S. S. Spicer. 1966. The histochemistry of mucins in certain primate salivary glands. Amer. J. Anat., 118:833–860.

Lewinsky, W., and D. Stewart. 1936. The innervation of the periodontal membrane. J. Anat. (Lond.), 71:98–102.

Lewinsky, W., and D. Stewart. 1937. The innervation of the periodontal membrane of the cat, with some observations on the function of the end-organs found in that structure. J. Anat. (Lond.), 71:222–235.

MacGregor, A. 1936. An experimental investigation of the lymphatic system of the teeth and jaws. Proc. Roy. Soc. Med., 29:1237–1272.

McKenzie, J. 1948. The parotid gland in relation to the facial nerve. J. Anat. (Lond.), 82:183–186.

Mikhail, Y., S. Abd El-Rahman, and L. Morris. 1980. Observations on the structure of the subepithelial nerve plexus in the tongue. Acta Anat., 107:311–317.

Miller, I. J., Jr., and A. J. Preslar. 1975. Spatial distribution of rat fungiform papillae. Anat. Rec., 181:679–684.

Munger, B. L. 1964. Histochemical studies on seromucous and mucous secreting cells of human salivary glands. Amer. J. Anat., 115:411–429.

Olmsted, J. M. D. 1922. Taste fibers and the chorda tympani nerve. J. Comp. Neurol., 34:337–341.

Provenza, D. V., and R. F. Sisca. 1971. Electron microscopic study of human dental primordia. Arch. Oral Biol. Med., 16:121–133.

Reith, E. J. 1967. The early stage of amelogenesis as observed in molar teeth of young rats. J. Ultrastruct. Res., 17:503–526.

Reith, E. J. 1970. The stages of amelogenesis as observed in molar teeth of young rats. J. Ultrastruct. Res., 30:111–151.

Rich, A. R. 1920. The innervation of the tensor veli

palatini and levator veli palatini muscles. Bull. Johns Hopkins Hosp., 31:305-310.

Richardson, G. S., and E. M. Pullen. 1948. The uvula: Its structure and function and its importance. Arch. Otolaryngol., 47:379-394.

Robinson, C., H. D. Briggs, P. J. Atkinson, and J. A. Weatherell. 1982. Chemical changes during formation and maturation of human deciduous enamel. Arch. Oral Biol. Med., 26:1027-1033.

Robinson, P. P. 1979. The course, relations and distribution of the inferior alveolar nerve and its branches in the cat. Anat. Rec., 195:265-271.

Rohan, R. F., and L. Turner. 1956. The levator palati muscle. J. Anat. (Lond.), 90:153-154.

Rönnholm, E. 1962. The amelogenesis of human teeth as revealed by electron microscopy. II. The development of the enamel crystallites. J. Ultrastruct. Res., 6:249-303.

Scheinin, A., and E. I. Light. 1969. Innervation of the dental pulp. II. A fluorescence microscopy study in the rat incisor. Acta Odont. Scand., 27:313-319.

Schneyer, L. H., and C. A. Schneyer (eds.). 1967. Secretory Mechanisms of Salivary Glands. Academic Press, New York, 389 pp.

Schwartz, H. G., and G. Weddell. 1938. Observations on the pathways transmitting the sensation of taste. Brain, 61:99-115.

Scott, J. H. 1957. The shape of the dental arches. J. Dent. Res., 36:996-1003.

Selvig, K. A. 1965. The fine structure of human cementum. Acta Odont. Scand., 23:423-441.

Shackleford, J. M., and C. E. Klapper. 1962. A sexual dimorphism of hamster submaxillary mucin. Anat. Rec., 142:495-503.

Shackleford, J. M., and L. H. Schneyer. 1971. Ultrastructural aspects of the main excretory duct of rat submandibular gland. Anat. Rec., 169:679-698.

Shapiro, H. H. 1954. Maxillofacial Anatomy with Practical Applications. J. B. Lippincott Co., Philadelphia, 392 pp.

Sicher, H., and E. L. Du Brul. 1975. Oral Anatomy. 6th ed. C. V. Mosby Co., St. Louis, 554 pp.

Silva, D. G., and D. G. Kailis. 1972. Ultrastructural studies on the cervical loop and the development of the amelo-dentinal junction in the cat. Arch. Oral Biol. Med., 17:279-289.

Starkie, C., and D. Stewart. 1931. The intramandibular course of the inferior dental nerve. J. Anat. (Lond.), 65:319-323.

Strong, L. H. 1956. Muscle fibers of the tongue functional in constant production. Anat. Rec., 126:61-79.

Suarez, F. R. 1980. The clinical anatomy of the tonsillar (Waldeyer's) ring. Ear Nose Throat J., 59:447-453.

Tamarin, A., and L. M. Sreebny. 1965. The rat submaxillary salivary gland. A correlative study by light and electron microscopy. J. Morphol., 117:295-352.

Tandler, B. 1962. Ultrastructure of the human submaxillary gland. I. Architecture and histological relationships of the secretory cells. Amer. J. Anat., 111:287-307.

Teleford, I. R. 1946. Pigment studies on the incisor teeth of vitamin E deficient rats of the Long-Evans strain. Proc. Soc. Exp. Biol. Med., 63:89-91.

Tiegs, O. W. 1938. Further remarks on the terminations of nerves in human teeth. J. Anat. (Lond.), 72:234-246.

Tonge, C. H. 1953. The early development of teeth. Proc. Roy. Soc. Med., 46:313-318.

Van der Sprenkel, H. B. 1936. Microscopical investigation of the innervation of the tooth and its surroundings. J. Anat. (Lond.), 70:233-241.

Weddell, G., J. A. Harpman, D. G. Lambley, and L. Young. 1940. The innervation of the musculature of the tongue. J. Anat. (Lond.), 74:255-267.

Wedgwood, M. 1966. The peripheral course of the inferior dental nerve. J. Anat. (Lond.), 100:639-650.

Whillis, J. 1946. Movements of the tongue in swallowing. J. Anat. (Lond.), 80:115-116.

Whiteside, B. 1927. Nerve overlap in the gustatory apparatus of the rat. J. Comp. Neurol., 44:363-377.

Wood, B. A. 1981. Tooth size and shape and their relevance to studies in hominid evolution. Phil. Trans. Roy. Soc. Lond. (Biol.), 292:65-76.

Yee, J., F. Harrison, and K. B. Corbin. 1939. The sensory innervation of the spinal accessory and tongue musculature in the rabbit. J. Comp. Neurol., 70:305-314.

Zalewski, A. A. 1981. Regeneration of taste buds after reinnervation of a denervated tongue papilla by a normally nongustatory nerve. J. Comp. Neurol., 200:309-314.

Pharynx and Esophagus

Arey, L. B., and M. J. Tremaine. 1933. The muscle content of the lower oesophagus of man. Anat. Rec., 56:315-320.

Batson, O. V. 1942. Veins of the pharynx. Arch. Otolaryngol., 36:212-219.

Bosma, J. F. 1957. Deglutition: pharyngeal stage. Physiol. Rev., 37:275-300.

Browne, D. 1928. The surgical anatomy of the tonsil. J. Anat. (Lond.), 63:82-86.

Butler, H. 1951. The veins of the oesophagus. Thorax, 6:276-296.

Chu, C. H. U. 1968. Solitary neurons in human tongue. Anat. Rec., 162:505-510.

Doubilet, H., B. G. P. Shafiroff, and J. H. Mulholland. 1948. Anatomy of the peri-esophageal vagi. Ann. Surg., 127:128-135.

Fuller, A. P., J. A. Fozzard, and G. H. Wright. 1959. Sphincteric action of crico-pharyngeus: radiographic demonstration. Brit. J. Radiol., 32:32-35.

Hett, G. S., and H. G. Butterfield. 1909. The anatomy of the palatine tonsils. J. Anat. Physiol., 44:35-55.

Jackson, R. G. 1949. Anatomy of the vagus nerves in the region of the lower esophagus and the stomach. Anat. Rec., 103:1-18.

Jones, F. W. 1940. The nature of the soft palate. J. Anat. (Lond.), 74:147-170.

Kelemen, G. 1943. The palatine tonsil in the sixth decade. Ann. Otol. Rhinol. Laryngol., 52:419-443.

Lasjaunias, P., F. Guibert-Tranier, and J. P. Braun. 1981. The pharyngo-cerebellar artery or ascending pharyngeal artery origin of the posterior inferior cerebellar artery. J. Neuroradiol., 8:317-325.

Lerche, W. 1950. The Esophagus and Pharynx in Action: A Study of Structure in Relation to Function. Charles C Thomas, Springfield, Ill., 222 pp.

MacManus, J. E., J. T. Dameron, and J. R. Paine. 1950. The extent to which one may interfere with the blood supply of the esophagus and obtain healing on anastomosis. Surgery, 28:11-23.

Minear, W. L., L. B. Arey, and J. T. Milton. 1937. Prenatal and postnatal development and form of crypts of human palatine tonsil. Arch. Otolaryngol., 25:487-519.

Mitchell, G. A. G. 1938. The nerve-supply of the gastro-oesophageal junction. Brit. J. Surg., 26:333-345.

Morley, J. 1945. Pharyngeal diverticula. Brit. J. Surg., 33:101–106.

Nishimura, T., and T. Takasu. 1969. The adrenergic innervation in the esophagus and respiratory tract of the rabbit. Acta Otolaryngol., 67:444–452.

Peden, J. K., C. F. Schneider, and R. D. Bickel. 1950. Anatomic relations of the vagus nerves to the esophagus. Amer. J. Surg., 80:32–34.

Potter, S. E., and E. A. Holyoke. 1950. Observations on the intrinsic blood supply of the esophagus. Arch. Surg., 61:944–948.

Quisling, R. G., and J. F. Seeger. 1979. Ascending pharyngeal artery collateral circulation simulating internal carotid artery hypoplasia. Neuroradiology, 18:277–280.

Sauer, M. E. 1951. The cricoesophageal tendon. A recommendation for its inclusion in official anatomical nomenclature. Anat. Rec., 109:691–697.

Sengupta, B. N., and S. Sengupta. 1978. Muscle spindles in the inferior constrictor pharyngis muscle of the crab-eating monkey (Macaca irus). Acta Anat., 100:132–135.

Shapiro, A. L., and G. L. Robillard. 1950. The esophageal arteries: their configurational anatomy and variations in relation to surgery. Ann. Surg., 131:171–185.

Shek, J. L., C. A. Prietto, W. M. Tuttle, and E. J. O'Brien. 1950. An experimental study of the blood supply of the esophagus and its relation to esophageal resection and anastomoses. J. Thoracic Surg., 19: 523–533.

Sprague, J. M. 1944. The innervation of the pharynx in the rhesus monkey, and the formation of the pharyngeal plexus in primates. Anat. Rec., 90:197–208.

Swenson, O., K. Merrill, Jr., E. C. Pierce, II, and H. F. Rheinlander. 1950. Blood and nerve supply to the esophagus: An experimental study. J. Thoracic Surg., 19:462–476.

Swigart, L. L., R. G. Siekert, W. C. Hambley, and B. J. Anson. 1950. The esophageal arteries—an anatomic study of 150 specimens. Surg. Gynecol. Obstet., 90:234–243.

Todd, T. W., and R. H. Fowler. 1927. The muscular relations of the tonsil. Amer. J. Anat., 40:355–371.

Uddmann, R., J. Alumets, R. Håkason, F. Sundler, and B. Walles. 1980. Peptidergic (enkephalin) innervation of the mammalian esophagus. Gastroenterology, 78:732–737.

Wardill, W. E. M., and J. Whillis. 1936. Movements of the soft palate: With special reference to the function of the tensor palati muscle. Surg. Gynecol. Obstet., 62:836–839.

Whillis, J. 1930. A note on the muscles of the palate and the superior constrictor. J. Anat. (Lond.), 65:92–95.

Williams, D. B., and W. S. Payne. 1982. Observations on esophageal blood supply. Mayo Clin. Proc., 57:448–453.

Wong, Y. K., and G. M. Novotny. 1978. Retropharyngeal space—a review of anatomy, pathology, and clinical presentation. J. Otolaryngol., 7:528–536.

Wood, G. B. 1934. The peritonsillar spaces: An anatomic study. Arch. Otolaryngol., 20:837–841.

Yoshida, Y., T. Miyazaki, M. Hirano, T. Shin, T. Totoki, and T. Kanaseki. 1981. Localization of efferent neurons innervating the pharyngeal constrictor muscles and the cervical esophagus muscle in the cat by means of the horseradish peroxidase method. Neurosci. Lett., 22:91–95.

Zhu, T. L., J. S. Tan, Z. X. Zhang, B. K. Zhang, and Y. P. Ma. 1980. Vagus nerve anatomy at the lower esophagus and stomach. A study of 100 cadavers. Chinese Med. J., 93:629–636.

Peritoneum and Hernia

Anson, B. J., E. H. Morgan, and C. B. McVay. 1949. The anatomy of the hernial regions: I. Inguinal hernia. Surg. Gynecol. Obstet., 89:417–424.

Ball, C. F. 1935. Left paraduodenal hernia: Two cases, one with rupture through the wall of the hernial sac. Amer. J. Surg., 29:481–484.

Baron, M. A. 1941. Structure of the intestinal peritoneum in man. Amer. J. Anat., 69:439–497.

Baumeister, C., and McM. Hanchett. 1938. Right paraduodenal hernia: A case favoring the theory of Treitz. Arch. Surg., 37:327–332.

Bennett, R. J., Jr. 1936. Intra-abdominal hernia: Report of 2 cases. Amer. J. Surg., 34:374–375.

Bryan, R. C. 1935. Right paraduodenal hernia. Amer. J. Surg., 28:703–730.

Bryant, A. L. 1947. Spigelian hernia. Amer. J. Surg., 73:396–397.

Cogswell, H. D., and C. A. Thomas. 1941. Right paraduodenal hernia. Ann. Surg., 114:1035–1041.

Creizel, A., and J. Gárdonyi. 1979. A family study of congenital inguinal hernia. Amer. J. Med. Genet., 4:247–254.

Edwards, H. 1943. Inguinal hernia. Brit. J. Surg., 31:172–185.

Gardner, C. E., Jr. 1950. The surgical significance of anomalies of intestinal rotation. Ann. Surg., 131:879–898.

Gardner, J. H., E. A. Holyoke, and R. P. Giovacchini. 1957. Cleavage lines of the visceral and parietal peritoneum. Anat. Rec., 127:241–256.

Hakami, M., S. Habibzadeh, and S. H. Mosavy. 1975. Hernia through the foramen of Winslow. Amer. Surg., 41:355–357.

Halpert, B. 1938. Right retromesocolic hernia. Surgery, 3:579–584.

Hurwitz, A. 1981. Measuring gastric volume by dye dilution. Gut, 22:85–93.

Jones, F. W. 1913. The function of the coelom and the diaphragm. J. Anat. (Lond.), 47:282–318.

Madden, J. L. 1956. Anatomic and technical considerations in the treatment of esophageal hiatal hernia. Surg. Gynecol. Obstet., 102:187–194.

Mahorner, H. 1940. Umbilical and midline ventral herniae. Ann. Surg., 111:979–997.

Mayo, C. W., L. K. Stalker, and J. M. Miller. 1941. Intra-abdominal hernia. Review of 39 cases in which treatment was surgical. Ann. Surg., 114:875–885.

Moody, R. O. 1927. The position of the abdominal viscera in healthy, young British and American adults. J. Anat. (Lond.), 61:223–231.

Raftery, A. T. 1979. Regeneration of peritoneum: a fibrinolytic study. J. Anat. (Lond.), 129:659–664.

Sheehan, D. 1933. The afferent nerve supply of the mesentery and its significance in the causation of abdominal pain. J. Anat. (Lond.), 67:233–249.

Watson, L. F. 1938. Embryologic and anatomic considerations in etiology of inguinal and femoral hernias. Amer. J. Surg., 42:695–703.

Stomach

Baker, B. L., and G. D. Abrams. 1954. Effect of hypophysectomy on the cytology of the fundic glands of the stomach and on the secretion of pepsin. Amer. J. Physiol., 177:409–412.

Barlow, T. E., F. H. Bentley, and D. N. Walder. 1951. Arteries, veins, and arteriovenous anastomoses in the human stomach. Surg. Gynecol. Obstet., *93:*657-671.

Benjamin, H. B. 1951. The neurovascular mechanism of the stomach and duodenum. Surg. Gynecol Obstet., *92:*314-320.

Bowie, D. J. 1940. The distribution of the chief or pepsin-forming cells in the gastric mucosa of the cat. Anat. Rec., *78:*9-17.

Carey, J. M., and W. H. Hollinshead. 1955. An anatomic study of the esophageal hiatus. Surg. Gynecol. Obstet., *100:*196-200.

Cauldwell, E. W., and B. J. Anson. 1943. The visceral branches of the abdominal aorta: Topographical relationships. Amer. J. Anat., *73:*27-57.

Cobb, J. L. S., and T. Bennett. 1970. An ultrastructural study of mitotic division in differentiated gastric smooth muscle cells. Z. Zellforsch., *108:*177-189.

Edlich, R. F., J. W. Borner, J. Kuphal, and O. H. Wangensteen. 1970. Gastric blood flow. I. Its distribution during gastric distention. Amer. J. Surg., *120:*35-37.

Goldsmith, H. S., and H. Akiyama. 1979. A comparative study of Japanese and American gastric dimensions. Ann. Surg., *190:*690-693.

Holly, A. D. 1959. The fine structure of the gastric parietal cell in the mouse. J. Anat. (Lond.), *93:*217-225.

Horton, B. T. 1931. Pyloric musculature, with special reference to pyloric block. Amer. J. Anat., *41:*197-225.

Hunt, T. E. 1957. Mitotic activity in the gastric mucosa of the rat after fasting and refeeding. Anat. Rec., *127:*539-550.

Hunt, T. E., and E. A. Hunt. 1962. Radioautographic study of proliferation in the stomach of the rat using thymidine-H^3 and compound 48/80. Anat. Rec., *142:*505-517.

Ito, S. 1967. Anatomic structure of the gastric mucosa. In *Handbook of Physiology,* Vol. 2, Section 6, *Alimentary Canal.* Edited by D. F. Code. American Physiological Society, Washington, pp. 705-742.

Jackson, R. G. 1949. Anatomy of the vagus nerves in the region of the lower esophagus and the stomach. Anat. Rec., *103:*1-18.

Jansson, G. 1969. Extrinsic nervous control of gastric motility. An experimental study in the cat. Acta Physiol. Scand., Suppl. *326:*1-42.

Lachman, E. 1957. Roentgenologic manifestations of emotional disturbances in the stomach. Amer. J. Roentgenol., *77:*162-166.

Leela, K., and R. Kanagasuntheram. 1968. A microscopic study of the human pyloro-duodenal junction and proximal duodenum. Acta Anat., *71:*1-12.

Lendrum, F. C. 1937. Anatomic features of the cardiac orifice of the stomach: with special reference to cardiospasm. Arch. Intern. Med., *59:*474-511.

Lipkin, M., P. Sherlock, and B. Bell. 1963. Cell proliferation kinetics in the gastrointestinal tract of man. II. Cell renewal in stomach, ileum, colon and rectum. Gastroenterology, *45:*721-729.

MacDonald, W. C., J. S. Trier, and N. B. Everett. 1964. Cell proliferation and migration in the stomach, duodenum and rectum of man; radioautographic studies. Gastroenterology, *46:*405-417.

Mackintosh, C. E., and L. Kreel. 1977. Anatomy and radiology of the areae gastricae. Gut, *18:*855-864.

McCrea, E. D. 1926. The nerves of the stomach and their relation to surgery. Brit. J. Surg., *13:*621-648.

Mitchell, G. A. G. 1940. A macroscopic study of the nerve supply of the stomach. J. Anat. (Lond.), *75:*50-63.

Moody, R. O., R. G. Van Nuys, and C. H. Kidder. 1929. The form and position of the empty stomach in healthy young adults as shown in roentgenograms. Anat. Rec., *43:*359-379.

Rubin, W. 1972. An unusual intimate relationship between endocrine cells and other types of epithelial cells in the human stomach. J. Cell Biol., *52:*219-227.

Simer, P. H. 1935. Omental lymphatics in man. Anat. Rec., *63:*253-262.

Skandalakis, J. E., S. W. Gray, R. E. Soria, J. L. Sorg, and J. S. Rowe. 1980. Distribution of the vagus nerve to the stomach. Amer. Surg., *46:*130-139.

Spira, J.-J. 1957. Comparison of cardiac and pyloric sphincters. Lancet, *273:*1008.

Weber, J. 1958. The basophilic substance of the gastric chief cells and its relation to the process of secretion. Acta Anat., *33:*Suppl. 31, 1-79.

Williams, T. B. 1928. Vascular studies of the pylorus. Anat. Rec., *38:*273-291.

Small and Large Intestine

Andrew, A. 1971. The origin of intramural ganglia. IV. The origin of enteric ganglia: a critical review. J. Anat. (Lond.), *108:*169-184.

Baumgarten, H. G., A.-F. Holstein, and Ch. Owman. 1970. Auerbach's plexus of mammals and man. Z. Zellforsch., *106:*376-397.

Blair, J. B., E. A. Holyoke, and R. R. Best. 1950. A note on the lymphatics of the middle and lower rectum and anus. Anat. Rec., *108:*635-644.

Brown, J. O., and R. J. Echenberg. 1964. Mucosal reduplications associated with the ampullary portion of the major duodenal papilla in humans. Anat. Rec., *150:*293-301.

Buirge, R. E. 1943. Gross variations in the ileocecal valve. A study of the factors underlying incompetency. Anat. Rec., *86:*373-385.

Buirge, R. E. 1944. Experimental observations on the human ileocecal valve. Surgery, *16:*356-369.

Carey, E. J. 1921. Studies on the structure and function of the small intestine. Anat. Rec., *21:*189-215.

Clementi, F., and G. E. Palade. 1969. Intestinal capillaries. I. Permeability to peroxidase and ferritin. J. Cell Biol., *41:*33-58.

Cokkinis, A. J. 1930. Observations on the mesenteric circulation. J. Anat. (Lond.), *64:*200-205.

Cornes, J. S. 1965. Number, size, and distribution of Peyer's patches in the human small intestine. I. The development of Peyer's patches. Gut, *6:*225-229. II. The effect of age on Peyer's patches. Gut, *6:*230-233.

Drummond, H. 1914. The arterial supply of the rectum and pelvic colon. Brit. J. Surg., *1:*677-685.

Edwards, L. F. 1941. The retroduodenal artery. Anat. Rec., *81:*351-355.

Elsen, J., and L. B. Arey. 1966. On spirality in the intestinal wall. Amer. J. Anat., *118:*11-20.

Fehér, E. 1982. Electron microscopic study of retrograde axonal transport of horseradish peroxidase in the wall of the small intestine in the cat. Acta Anat., *112:*69-78.

Ferraz de Carvalho, C. A., and J. J. Faintuch. 1974. Functional value of the elastic fiber changes at the terminal segment of the human ileum. Acta Anat., *89:*461-472.

Fleischner, F. G., and C. Bernstein. 1950. Roentgen-anatomical studies of the normal ileocecal valve. Radiology, *54:*43-58.

Friedman, S. M. 1946. The position and mobility of the duodenum in the living subject. Amer. J. Anat., *79:*147-165.

Gabella, G. 1981. Ultrastructure of the nerve plexuses of

the mammalian intestine: the enteric glial cells. Neuroscience, 6:425–436.

Gershon, M. D., D. Sherman, and A. R. Gintzler. 1981. An ultrastructural analysis of the developing enteric nervous system of the guinea-pig small intestine. J. Neurocytol., 10:271–296.

Gladstone, R. J., and C. P. G. Wakeley. 1924. The relative frequency of the various positions of the vermiform appendix: As ascertained by an analysis of 3000 cases; with an account of its development. Brit. J. Surg., 11:503–520.

Goligher, J. C. 1949. The blood-supply to the sigmoid colon and rectum: With reference to the technique of rectal resection with restoration of continuity. Brit. J. Surg., 37:157–162.

Greenberg, M. W. 1950. Blood supply of the rectosigmoid and rectum. Ann. Surg., 131:100–108.

Haber, J. J. 1947. Meckel's diverticulum. Amer. J. Surg., 73:468–485.

Haley, J. C., and J. K. Peden. 1943. The suspensory muscle of the duodenum. Amer. J. Surg., 59:546–550.

Hardcastle, J. D., and C. V. Mann. 1969. A study of large bowel peristalsis in man. J. Roy. Coll. Surg. (Edinb.), 14:286–287.

Hunter, R. H. 1934. The ileo-caecal junction. J. Anat. (Lond.), 68:264–269.

Jacobson, L. F., and R. J. Noer. 1952. The vascular pattern of the intestinal villi in various laboratory animals and man. Anat. Rec., 114:85–101.

Kellogg, E. L. 1931. Abnormalities in the shape and position of the duodenum. Amer. J. Surg., 12:462–465.

Krause, W. J., and C. R. Leeson. 1967. The origin, development and differentiation of Brunner's glands in the rat. J. Anat. (Lond.), 101:309–320.

Lannon, J., and E. Weller. 1947. The parasympathetic supply of the distal colon. Brit. J. Surg., 34:373–378.

Leeson, T. S., and C. R. Leeson. 1968. The fine structure of Brunner's glands in man. J. Anat. (Lond.), 103:263–276.

Low, A. 1907. A note on the crura of the diaphragm and the muscle of Treitz. J. Anat. Physiol., 42:93–96.

Marsh, M. N., and J. A. Swift. 1969. A study of the small intestinal mucosa using the scanning electron microscope. Gut, 10:940–949.

Marshak, R. H., A. E. Lindner, and D. Maklansky. 1976. Ischemia of the small intestine. Amer. J. Gastroenterol., 66:390–400.

McMinn, R. M. H., and J. E. Mitchell. 1954. The formation of villi following artificial lesions of the mucosa in the small intestine of the cat. J. Anat. (Lond.), 88:99–107.

Milligan, E. T. C., and C. N. Morgan. 1934. Surgical anatomy of the anal canal: With special reference to anorectal fistulae. Lancet, 2:1150–1156.

Nesselrod, J. P. 1936. An anatomic restudy of the pelvic lymphatics. Ann. Surg., 104:905–916.

Noer, R. J. 1943. The blood vessels of the jejunum and ileum: A comparative study of man and certain laboratory animals. Amer. J. Anat., 73:293–334.

Ogilvie, H. 1952. The first part of the duodenum. Lancet, 1:1077–1081.

Schofield, G. C., and A. M. Atkins. 1970. Secretory immunoglobulin in columnar epithelial cells of the large intestine. J. Anat. (Lond.), 107:491–504.

Schofield, G. C., and D. G. Silva. 1968. The fine structure of enterochromaffin cells in the mouse colon. J. Anat. (Lond.), 103:1–13.

Shah, M. A., and M. Shah. 1946. The arterial supply of the vermiform appendix. Anat. Rec., 95:457–460.

Stephens, F. D. 1963. Congenital Malformations of the Rectum, Anus and Genito-Urinary Tracts. Williams and Wilkins Co., Baltimore, 380 pp.

Sunderland, S. 1942. Blood supply of the distal colon. Austral. New Zealand J. Surg., 11:253–263.

Underhill, B. M. L. 1955. Intestinal length in man. Brit. Med. J., 2:1243–1246.

Wakeley, C. P. G. 1933. The position of the vermiform appendix as ascertained by an analysis of 10,000 cases. J. Anat. (Lond.), 67:277–283.

Wakeley, C. P. G., and R. J. Gladstone. 1928. The relative frequency of the various positions of the appendix vermiformis as ascertained by an analysis of 5000 cases. J. Anat. (Lond.), 63:157–158.

Wilkie, D. P. D. 1911. The blood supply of the duodenum: With special reference to the supraduodenal artery. Surg. Gynecol. Obstet., 13:399–405.

Wilmer, H. A. 1941. The blood supply of the first part of the duodenum: With description of the gastroduodenal plexus. Surgery, 9:679–687.

Wood, J. D. 1981. Intrinsic neural control of intestinal motility. Ann. Rev. Physiol., 43:33–51.

Wotton, R. 1963. Lipid absorption. In International Review of Cytology. Vol. 15. Edited by G. H. Bourne. Academic Press, New York, pp. 399–420.

Yule, E. 1927. The arterial supply to the duodenum. J. Anat. (Lond.), 61:344–345.

Liver, Bile Ducts and Gallbladder

Alexander, W. F. 1940. The innervation of the biliary system. J. Comp. Neurol., 72:357–370.

Bergh, G. S. 1942. The effect of food upon the sphincter of Oddi in human subjects. Amer. J. Digest. Dis., 9:40–43.

Boyden, E. A. 1937. The sphincter of Oddi in man and certain representative mammals. Surgery, 1:25–37.

Boyden, E. A., and C. Van Buskirk. 1943. Rate of emptying of biliary tract following section of vagi or of all extrinsic nerves. Proc. Soc. Exp. Biol. Med., 53:174–175.

Browne, E. Z. 1940. Variations in origin and course of the hepatic artery and its branches: Importance from surgical viewpoint. Surgery, 8:424–445.

Burcharth, F., and S. N. Rasmussen. 1974. Localization of the porta hepatis by ultrasonic scanning prior to percutaneous transhepatic portography. Brit. J. Radiol., 47:598–600.

Cain, J. C., J. H. Grinlay, J. L. Bollman, E. V. Flock, and F. C. Mann. 1947. Lymph from the liver and thoracic duct: An experimental study. Surg. Gynecol. Obstet., 85:559–562.

Camparini, L. 1969. Lymph vessels of the liver in man. Microscopic morphology and histotopography. Angiologica, 6:262–274.

Daniel, P. M., and M. M. L. Prichard. 1951. Variations in the circulation of the portal venous blood within the liver. J. Physiol., 114:521–537.

Daseler, E. H., B. J. Anson, W. C. Hambley, and A. F. Reimann. 1947. The cystic artery and constituents of the hepatic pedicle: A study of 500 specimens. Surg. Gynecol. Obstet., 85:47–63.

Di Dio, L. J. A., and E. A. Boyden. 1962. The choledochoduodenal junction in the horse—a study of the musculature around the ends of the bile and pancreatic ducts in a species without a gall bladder. Anat. Rec., 143:61–69.

Dixon, C. F., and A. L. Lichtman. 1945. Congenital absence of the gall bladder. Surgery, 17:11–21.

Douglass, B. E., A. H. Baggenstoss, and W. H. Hollinshead. 1950. The anatomy of the portal vein and its tributaries. Surg. Gynecol. Obstet., 91:562–576.

Elias, H., and D. Petty. 1952. Gross anatomy of the blood vessels and ducts within the human liver. Amer. J. Anat., 90:59–111.

Franksson, C. 1947. The innervation at the common bile duct-duodenal junction from a surgical point of view. Acta Chir. Scand. 96:163–177.

Gelfand, D. W. 1980. Anatomy of the liver. Radiol. Clin. North Amer., 18:187–194.

Gilfillan, R. S. 1950. Anatomic study of the portal vein and its main branches. Arch. Surg., 61:449–461.

Glazer, H. B., R. J. Stanley, and R. E. Koehler. 1980. Cystic artery hemorrhage: a complication of penetrating duodenal ulcer. Radiology, 136:623–625.

Goor, D. A., and P. A. Ebert. 1972. Anomalies of the biliary tree: Report of a repair of an accessory bile duct and review of literature. Arch. Surg., 104:302–309.

Gupta, S. C., and C. D. Gupta. 1979. The hepatic veins—a radiographic and corrosion cast study. Indian J. Med. Res., 70:333–344.

Gupta, S. C., C. D. Gupta, and A. K. Arora. 1977. Subsegmentation of the human liver. J. Anat. (Lond.), 124:413–423.

Hammond, W. S. 1939. On the origin of the cells lining the liver sinusoids in the cat and the rat. Amer. J. Anat., 65:199–227.

Hard, W. L., and R. K. Hawkins. 1950. The role of the bile capillaries in the secretion of phosphatase by the rabbit liver. Anat. Rec., 106:395–411.

Hardy, K. J., I. C. Wheatley, A. I. E. Anderson, and R. J. Bond. 1976. The lymph nodes of the porta hepatis. Surg. Gynecol. Obstet., 143:225–228.

Hatfield, P. M., and R. E. Wise. 1976. Anatomic variation in the gallbladder and bile ducts. Semin. Roentgenol., 11:157–164.

Healey, J. E., Jr., and P. C. Schroy. 1953. Anatomy of the biliary ducts within the human liver; analysis of the prevailing pattern of branchings and the major variations of the biliary ducts. Arch. Surg., 66:599–616.

Healey, J. E., Jr., P. C. Schroy, and R. J. Sorensen. 1953. The intrahepatic distribution of the hepatic artery in man. J. Internat. Coll. Surg., 20:133–148.

Heath, T., and B. House. 1970. Origin and distribution of portal blood in the cat and the rabbit. Amer. J. Anat., 127:71–80.

Hillman, B. J., C. J. D'Orsi, E. H. Smith, and R. J. Bartrum. 1979. Ultrasonic appearance of the falciform ligament. Amer. J. Roentgenol., 132:205–206.

Hjortsjö, C.-H. 1951. The topography of the intrahepatic duct systems. Acta Anat., 11:599–615.

Jędrzejewski, K. 1979. On the anatomical nomenclature of hepatic segments and biliary ductules in man. Folia Morphol., 38:529–535.

Johnson, F. E., and E. A. Boyden. 1952. The effect of double vagotomy on the motor activity of the human gall bladder. Surgery, 32:591–601.

Kennedy, P. A., and G. F. Madding. 1977. Surgical anatomy of the liver. Surg. Clin. North Amer., 57:233–244.

Lichtenstein, M. E., and A. C. Ivy. 1937. The function of the "valves" of Heister. Surgery, 1:38–52.

Michels, N. A. 1951. The hepatic, cystic and retroduodenal arteries and their relations to the biliary ducts: With samples of the entire celiacal blood supply. Ann. Surg., 133:503–524.

Michels, N. A. 1953. Variational anatomy of the hepatic, cystic and retroduodenal arteries. Arch. Surg., 66:20–34.

Moosman, D. A. 1975. Where and how to find the cystic artery during cholecystectomy. Surg. Gynecol. Obstet., 141:769–772.

Nawar, N. N. Y., Y. Mikhail, and I. A. Bahi el Din. 1980. Fetal hepatic vessels and subsegmentation with evidence of further subdivision. Acta Anat., 108:389–393.

Popper, H. P., and F. Schaffner. 1957. Liver: Structure and Function. Blakiston Division, McGraw-Hill Book Co., New York, 777 pp.

Rappaport, A. M. 1958. The structural and functional unit in the human liver (liver acinus). Anat. Rec., 130:673–689.

Rienhoff, W. F., Jr., and K. L. Pickrell. 1945. Pancreatitis: An anatomic study of the pancreatic and extrahepatic biliary systems. Arch. Surg., 51:205–219.

Severn, C. B. 1972. A morphological study of the development of the human liver. II. Establishment of liver parenchyma, extrahepatic ducts, and associated venous channels. Amer. J. Anat., 133:85–107.

Shapiro, A. L., and G. L. Robillard. 1948. The arterial blood supply of the common and hepatic bile ducts with reference to the problems of common duct injury and repair: Based on a series of twenty-three dissections. Surgery, 23:1–11.

Sleight, D. R., and N. R. Thomford. 1970. Gross anatomy of the blood supply and biliary drainage of the canine liver. Anat. Rec., 166:153–160.

Sterling, J. A. 1954. The common channel for bile and pancreatic ducts. Surg. Gynecol. Obstet., 98:420–424.

Sutherland, S. D. 1966. The intrinsic innervation of the gallbladder in Macaca rhesus and Cavia procellus. J. Anat. (Lond.), 100:261–268.

Vandamme, J. P., J. Bonte, and G. Van der Schueren. 1969. A revaluation of hepatic and cystic arteries. The importance of the aberrant hepatic branches. Acta Anat., 73:192–209.

Williams, W. L. 1948. Vital staining of damaged liver cells. I. Reactions to acid azo dyes following acute chemical injury. Anat. Rec., 101:133–147.

Zientarski, B. 1976. Blood vessels of the caudate lobe of the human liver. Folia Morphol., 35:95–103.

Pancreas and Pancreatic Ducts

Acosta, J. M., J. C. Buceta, J. E. Pons, L. E. Prin de Figari, O. A. M. Rubio Galli, E. E. Weinschelbaum, and J. C. Soloaga. 1969. Distribution and volume of the islets of Langerhans in the canine pancreas. Acta Physiol. Lat. Amer., 19:175–180.

Alvarado, F. 1955. Distribution of nerves within the pancreas; experimental investigation. J. Internat. Coll. Surg., 23:675–699.

Baradi, A. F., and D. J. Brandic. 1969. Observations on the morphology of pancreatic secretory capillaries. Z. Zellforsch., 101:568–580.

Bencosme, S. A. 1955. The histogenesis and cytology of the pancreatic islets in the rabbit. Amer. J. Anat., 96:103–151.

Clark, E. 1913. The number of islands of Langerhans in the human pancreas. Anat. Anz., 43:81–94.

Conklin, J. L. 1962. Cytogenesis of the human fetal pancreas. Amer. J. Anat., 111:181–193.

Dawson, W., and J. Langman. 1961. An anatomical-radiological study on the pancreatic duct pattern in man. Anat. Rec., 139:59–68.

Di Magno, E. P., R. G. Shorter, W. F. Taylor, and V. L. W. Go. 1982. Relationships between pancreaticobiliary ductal anatomy and pancreatic ductal and parenchymal histology. Cancer, *49*:361–368.

Falconer, C. W. A., and E. Griffiths. 1950. The anatomy of the blood vessels in the region of the pancreas. Brit. J. Surg., *37*:334–344.

Ferner, H. 1952. *Das Inselsystem des Pankreas.* Georg Thieme Verlag, Stuttgart, 186 pp.

Filly, R. A., and E. N. Carlsen. 1976. Newer ultrasonographic anatomy in the upper abdomen: II. The major systemic veins and arteries with a special note on localization of the pancreas. J. Clin. Ultrasound, *4*:91–96.

Goldstein, M. B., and E. A. Davis, Jr. 1968. The three dimensional architecture of the islets of Langerhans. Acta Anat., *71*:161–171.

Hard, W. L. 1944. The origin and differentiation of the alpha and beta cells in the pancreatic islets of the rat. Amer. J. Anat., *75*:369–403.

Howard, J., and R. Jones. 1947. The anatomy of the pancreatic ducts: The etiology of acute pancreatitis. Amer. J. Med. Sci., *214*:617–622.

Kuntz, A., and C. A. Richins. 1949. Effects of direct and reflex nerve stimulation of the exocrine secretory activity of pancreas. J. Neurophysiol., *12*:29–35.

Lacy, P. E. 1957. Electron microscopic identification of different cell types in the islets of Langerhans of the guinea pig, rat, rabbit and dog. Anat. Rec., *128*:255–267.

Latta, J. S., and H. T. Harvey. 1942. Changes in the islets of Langerhans of the albino rat induced by insulin administration. Anat. Rec., *82*:281–295.

McCuskey, R. S., and T. M. Chapman. 1969. Microscopy of the living pancreas in situ. Amer. J. Anat., *126*:395–407.

Munger, B. L. 1958. A phase and electron microscopic study of cellular differentiation in pancreatic acinar cells of the mouse. Amer. J. Anat., *103*:1–33.

Oram, V. 1955. Cytoplasmic basophilic substance of the exocrine pancreatic cells; experimental, histological and densitometric study. Acta Anat., Suppl. *23*:1–114.

Pierson, J. M. 1943. The arterial blood supply of the pancreas. Surg. Gynecol. Obstet., *77*:426–432.

Reichardt, W., and R. Cameron. 1980. Anatomy of the pancreatic veins. A postmortem and clinical phlebographic investigation. Acta Radiol. Diag., *21*:33–41.

Reynolds, B. M. 1970. Observations of subcapsular lymphatics in normal and diseased human pancreas. Ann. Surg., *171*:559–566.

Richins, C. A. 1945. The innervation of the pancreas. J. Comp. Neurol., *83*:223–236.

Schaefer, J. H. 1926. The normal weight of the pancreas in the adult human being: a biometric study. Anat. Rec., *32*:119–132.

Singh, I. 1963. The terminal part of the accessory pancreatic duct and its musculature in the Rhesus monkey. J. Anat. (Lond.), *97*:107–110.

Wharton, G. K. 1932. The blood supply of the pancreas, with special reference to that of the islands of Langerhans. Anat. Rec., *53*:55–81.

Winborn, W. B. 1963. Light and electron microscopy of the islets of Langerhans of the Saimiri monkey pancreas. Anat. Rec., *147*:65–93.

Wong, K. C., and J. Lister. 1981. Human fetal development of the hepato-pancreatic duct junction—a possible explanation of congenital dilatation of the biliary tract. J. Ped. Surg., *16*:139–145.

Woodburne, R. T., and L. L. Olsen. 1951. The arteries of the pancreas. Anat. Rec., *111*:255–270.

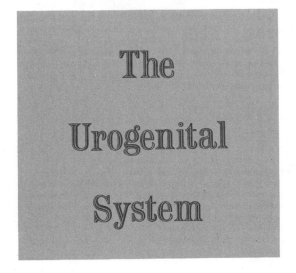

The Urogenital System

Urogenital or genitourinary is the term used for the system of **urinary organs,** which provide for the formation and discharge of the urine, and for the **genital organs,** which are concerned with the process of reproduction.

Development of Urinary Organs

The urogenital glands and ducts develop from the intermediate cell mass situated between the primitive segments and the lateral plates of mesoderm. The permanent organs of the adult are preceded by structures which, with the exception of the ducts, disappear almost entirely before the end of fetal life. These paired structures are: the **pronephros,** the **mesonephros,** and the **mesonephric** and **paramesonephric ducts.** The pronephros disappears very early. The structural elements of the mesonephros degenerate almost entirely, but in their place the genital gland develops. The mesonephric duct remains as the duct of the male genital gland; the paramesonephric, as that of the female. The final kidney is a new organ, the **metanephros.**

PRONEPHROS AND MESONEPHRIC DUCT

In the lateral part of the intermediate cell mass, immediately under the ectoderm, in the region from the fifth cervical to the third thoracic segments, a series of short evaginations from each segment grows dorsolaterally and caudally, fusing successively caudalward to form the **pronephric duct** (Fig. 17-1). This continues to grow caudally until it opens into the ventral part of the cloaca; beyond the pronephros it is termed the **mesonephric** or **Wolffian duct.**[1]

The original evaginations form a series of transverse tubules, each of which communicates by means of a funnel-shaped ciliated opening, the **nephrostome,** with the coelomic cavity, and in the course of each duct a **glomerulus** also is developed. A secondary glomerulus is formed ventral to each of these, and the complete group constitutes the **pronephros.** The pronephros undergoes rapid atrophy and, in 4-mm embryos, disappears except for the pronephric ducts, which persist as the excretory ducts of the succeeding kidneys, the mesonephroi.

MESONEPHROS

On the medial side of the mesonephric duct, from the sixth cervical to the third lumbar segments, a series of **mesonephric tubules** develops; at a later stage in the development they increase in number by outgrowths from the original tubules (Fig. 17-1). These tubules first appear as solid masses of cells that later develop a lumen; one end grows toward and finally opens into the mesonephric duct, the other dilates and is invaginated by a tuft of capillary blood vessels to form a glomerulus. The tubules collectively constitute the **mesonephros** or **Wolffian body.** By the fifth or sixth week this body produces an elongated spindle-shaped eminence termed the **urogenital fold,** which projects into the coelomic cavity at the side of the dorsal mesentery, reaching from the

[1] Kaspar Friedrich Wolff (1733–1794): A German anatomist and embryologist (St. Petersburg).

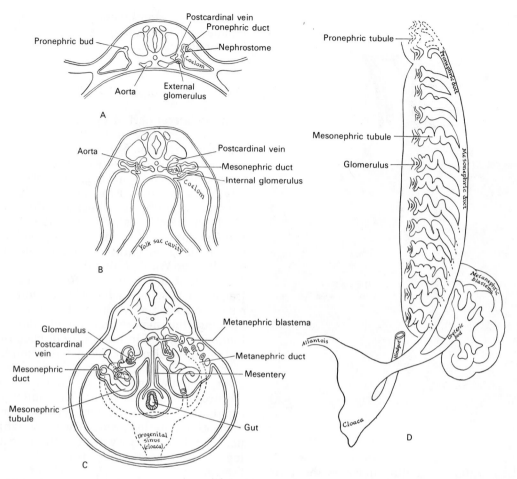

Fig. 17-1. Pronephros, mesonephros and metanephros. *A,* Diagrammatic cross section of embryo through somites 7 to 14. Earlier stages of development on left. *B,* Typical cross section through somites containing mesonephros. Earlier stage of development on left. *C,* Typical cross section through caudal portion of the mesonephros and including part of the metanephros. Excretory ducts and cloaca outlined in dash line. *D,* Ventrolateral view of embryonic excretory system (schematic). (Redrawn from Allan, F. D., *Essentials of Embryology,* Oxford University Press, 1960.)

septum transversum cranially to the fifth lumbar segement caudally. The reproductive glands develop in the urogenital folds.

The mesonephric bodies are the permanent kidneys in fish and amphibians, but in reptiles, birds, and mammals, they atrophy and for the most part disappear synchronously with the development of the permanent kidneys. The atrophy begins during the sixth or seventh week and rapidly proceeds, so that by the beginning of the fourth month only the ducts and a few of the tubules remain.

METANEPHROS AND PERMANENT KIDNEY

The rudiments of the permanent kidneys make their appearance about the end of the first or the beginning of the second month.

Each kidney has a two-fold origin, part arising from the metanephros, and part as a diverticulum from the caudal end of the mesonephric duct, close to where the latter opens into the cloaca (Figs. 17-2; 17-3).

The metanephros arises in the intermediate cell mass caudal to the mesonephros, which it resembles in structure. The diverticulum from the mesonephric duct grows dorsalward and cranialward along the dorsal abdominal wall, where its blind extremity expands and subsequently divides into several buds that form the rudiments of the pelvis and calyces of the kidney; by continued growth and subdivision it gives rise to the collecting tubules of the kidney. The caudal portion of the diverticulum becomes the ureter. The secretory tubules are developed from the metanephros, which is

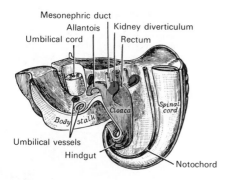

FIG. 17-2. Tail end of human embryo 25 to 29 days old. (From model by Keibel.)

molded over the growing end of the diverticulum from the mesonephric duct.

The tubules of the metanephros, unlike those of the pronephros and mesonephros, do not open into the mesonephric duct. One end expands to form a glomerulus, while the rest of the tubule rapidly elongates to form the convoluted and straight tubules, the loops of Henle,[1] and the connecting tubules; these last establish communication with the collecting tubules derived from the ultimate ramifications of the ureteric diverticulum. The mesoderm around the tubules condenses to form the connective tissue of the kidney. The ureter opens at first into the caudal end of the mesonephric duct; after the sixth week it is separated from the mesonephric duct and opens independently into the part of the cloaca that ultimately becomes the bladder (Figs. 17-2; 17-3).

The secretory tubules of the kidney be-

[1]Friedrich Gustav Jacob Henle (1809–1885): A German anatomist (Zurich, Heidelberg, and Göttingen).

come arranged into pyramidal masses or lobules (Fig. 17-4). The lobulated condition of the kidneys exists for some time after birth, and traces of it may be found even in the adult. The kidney of the ox and many other animals remains lobulated throughout life.

URINARY BLADDER

The bladder is formed partly from the endodermal cloaca and partly from the ends of the mesonephric ducts; the allantois has no share in its formation. After the rectum has separated from the dorsal part of the cloaca, the ventral part becomes subdivided into three portions: (1) an anterior **vesicourethral portion** continuous with the allantois, into which the mesonephric ducts open; (2) an intermediate narrow channel, the **pelvic portion;** and (3) a posterior **phallic portion,** closed externally by the urogenital membrane. The second and third parts together constitute the **urogenital sinus** (Figs. 17-5; 17-6). The vesicourethral portion absorbs the ends of the mesonephric ducts and the associated ends of the renal diverticula, and these give rise to the trigone of the bladder and part of the prostatic urethra. The remainder of the vesicourethral portion forms the body of the bladder and part of the prostatic urethra; its apex is prolonged to the umbilicus as a narrow canal (the **urachus**), which later is obliterated and becomes the middle umbilical ligament.

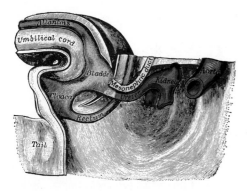

FIG. 17-3. Tail end of human embryo 32 to 33 days old. (From model by Keibel.)

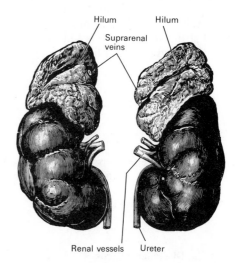

FIG. 17-4. The kidneys and suprarenal glands of a newborn child. Anterior aspect.

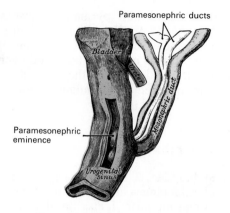

FIG. 17-5.   Urogenital sinus of female human embryo $8\frac{1}{2}$ to 9 weeks old. (From model by Keibel.)

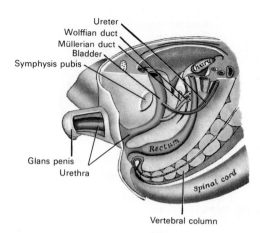

FIG. 17-6.   Tail end of male human embryo, from $8\frac{1}{2}$ to 9 weeks old. (From model by Keibel.)

Development of Generative Organs

GENITAL GLANDS

The first appearance of the genital gland is essentially the same in the two sexes; it consists of a thickening of the epithelial layer that lines the peritoneal cavity on the medial side of the urogenital fold. The thick plate of epithelium pushes the mesoderm before it and forms a distinct projection, the **genital ridge** (Fig. 17-7). From it the testis in the male and the ovary in the female are developed. At first the mesonephros and genital ridge are suspended by a common mesentery, but as the embryo grows the genital ridge gradually becomes pinched off from the mesonephros, with which it is at first continuous, though it still remains connected to the remnant of this body by a fold of peritoneum,

the **mesorchium** or **mesovarium.** About the seventh week the distinction of sex in the genital ridge begins to be perceptible.

TESTIS

At first the testis is a collection of cells derived from the coelomic epithelium with a surface covering and a central mass. As sexual differentiation begins, a series of cords appears in the central mass (Fig. 17-8), surrounded by a concentration of cells that later is converted into the tunica albuginea. The tunica separates the cords from the surface epithelium, excluding it from any part in forming the parenchyma of the testis. The more peripheral parts of the cords develop into the seminiferous tubules; the central parts run together toward the future mediastinum testis, forming a network that becomes the rete testis.

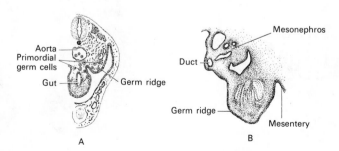

FIG. 17-7.   A, Diagrammatic cross section to show sites and migration of primordial germ cells from mesentery to gonad. B, Undifferentiated gonad. (Redrawn from Allan, F. D., *Essentials of Embryology,* Oxford University Press, 1960.)

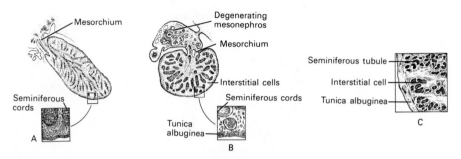

FIG. 17-8. Differentiation of testis; *A, B,* and *C* represent progressively later stages. (Redrawn from Allan, F. D., *Essentials of Embryology,* Oxford University Press, 1960.)

MALE MESONEPHRIC DUCT

In the male, the mesonephric duct (*Wolffian duct*) (see page 1515) persists and forms the epididymis, the ductus deferens, and the ejaculatory duct; the seminal vesicle arises during the third month as a lateral diverticulum from its caudal end (Fig. 17-9). A large part of the cranial portion of the mesonephros atrophies and disappears, but a few tubules may persist as the appendix of the epididymis; this vestigial structure ends blindly. From the remainder of the cranial tubules the efferent ducts of the testis form.

The caudal tubules are represented by the ductuli aberrantes and the paradidymis.

DESCENT OF THE TESTES

At an early period of fetal life, the testes are placed at the dorsal part of the abdominal cavity; they are covered by the peritoneum and each is attached to the mesonephros by a peritoneal fold, the **mesorchium.**

From the ventral part of the mesonephros, a fold of peritoneum termed the **inguinal fold** grows ventralward to meet and fuse with the **inguinal crest,** a peritoneal fold that grows dorsalward from the anterolateral abdominal wall. The testis thus acquires an indirect connection with the anterior abdominal wall, and at the same time a portion of the peritoneal cavity lateral to these fused folds is marked off as the future saccus vaginalis.

In the inguinal crest a peculiar structure, the **gubernaculum testis,** makes its appearance. This is at first a slender band that extends from the part of the skin of the groin that afterward forms the scrotum through the inguinal canal to the body and epididy-mis of the testis. As development advances, the peritoneum covering the gubernaculum forms two folds, one above the testis and the other below it. The fold cranial to the testis is the **plica vascularis,** which ultimately contains the testicular vessels; the caudal fold, the **plica gubernatrix,** contains the lower part of the gubernaculum, which has now grown into a thick cord. It ends caudally at the abdominal inguinal ring beside a tube of peritoneum, the **saccus** or **processus vaginalis,** which protrudes down the inguinal canal. By the fifth month the cranial part of the gubernaculum has disappeared, but the caudal part now consists of a central core of unstriped muscle fiber and an outer layer of striped elements connected with the abdominal wall. The main portion of the gubernaculum is attached to the skin at the point where the scrotum develops, and as the pouch forms, most of the caudal end of the gubernaculum is carried with it; other bands extend to the medial side of the thigh and to the perineum.

The tube of peritoneum constituting the **saccus** or **processus vaginalis** projects itself downward into the inguinal canal and emerges at the superficial inguinal ring, pushing before it a part of the obliquus internus and the fascia of the obliquus externus, which form respectively the cremaster muscle and the external spermatic fascia. It forms a gradually elongating sac that eventually reaches the bottom of the scrotum, and behind this sac the testis descends. Since the growth of the gubernaculum is not commensurate with the growth of the body of the fetus, the latter increasing more rapidly in relative length, it has been assumed that this prevents cranialward displacement of the testis. In addition, the gubernaculum may actually shorten, in which case it might

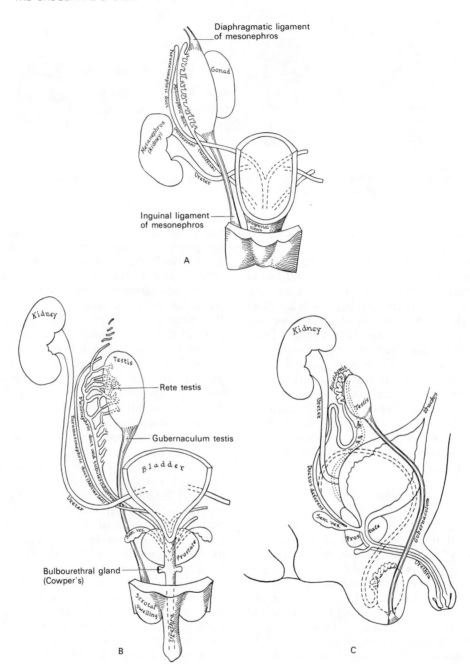

FIG. 17-9. Diagrammatic representation of the differentiation of the male genital ducts. *A*, Undifferentiated stage. *B*, Early differentiation. *C*, The condition just prior to birth when testis is undescended. Dash lines indicate position of testis after descent and dotted lines the position of the obliterated paramesonephric duct. Vestigial structures retained from the latter are labeled. (Redrawn from Allan, F. D., *Essentials of Embryology*, Oxford University Press, 1960.)

exert traction on the testis and tend to displace it toward the scrotum. By the end of the eighth month the testis has reached the scrotum, preceded by the processus vaginalis, which communicates by its upper extremity with the peritoneal cavity. Just before birth the upper part of the saccus

vaginalis normally closes, and this obliteration extends gradually downward to within a short distance of the testis. The peritoneum surrounding the testis is then cut off entirely from the general peritoneal cavity and becomes the **tunica vaginalis** (Fig. 17-10).

Occasionally the testis fails to descend or

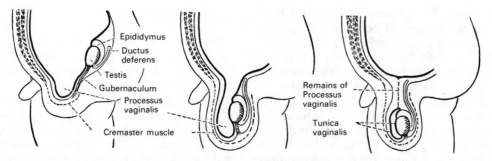

FIG. 17-10.   Diagram illustrating descent of the testis: *A*, Before descent. The processus vaginalis is present before descent begins, the testis lying behind the peritoneum. *B*, Descent nearly complete but processus vaginalis not obliterated. *C*, Processus vaginalis obliterated except for the terminal portion, which persists as the tunica vaginalis of the adult.

the descent is incomplete, a condition known as *cryptorchidism*. In such cases administration of an extract of the pituitary gland may cause the testis to occupy its normal position. This would seem to indicate that the descent may to some extent be under hormonal control.

PROSTATE

The prostate arises between the third and fourth months as a series of solid diverticula from the epithelium lining the urogenital sinus and vesicourethral part of the cloaca. These buds arise in five distinct groups, grow rapidly in length, and soon acquire a lumen. Eventually the prostatic urethra and ejaculatory ducts are embedded in a five-lobed gland, the parts of which are called the median, anterior, posterior, and lateral lobes. The lateral lobes are the largest. There are no distinct dividing lines between the parts of the formed gland, and the divisions are important only because of their individual peculiarities in disease processes.

Skene's ducts[1] in the female urethra are regarded as the homologues of the prostatic glands.

The **bulbourethral glands of Cowper**[2] in the male and the **greater vestibular glands of Bartholin**[3] in the female also arise as diverticula from the epithelial lining of the urogenital sinus.

OVARY

The ovary (Fig. 17-11), formed from the genital ridge, is at first a collection of cells derived from the coelomic epithelium; later the mass is differentiated into a central part of medulla covered by a surface layer, the **germinal epithelium.** Between the cells of the germinal epithelium a number of larger cells, the **primitive ova,** are found. These are

[1] Alexander Johnston Chalmers Skene (1837–1900): An American gynecologist (New York).
[2] William Cowper (1666–1709): An English surgeon (London).
[3] Caspar Bartholin (1655–1738): A Danish anatomist (Copenhagen).

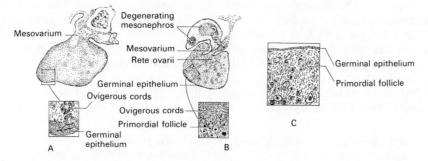

FIG. 17-11.   Differentiation of ovary; *A*, *B*, and *C* represent progressively later stages. (Redrawn from Allan, F. D., *Essentials of Embryology*, Oxford University Press, 1960.)

carried into the subjacent stroma by bud-like ingrowths **(genital cords)** of the germinal epithelium. The surface epithelium ultimately forms the permanent epithelial covering of this organ; it soon loses its connection with the central mass and a tunica albuginea develops between them. The ova are derived chiefly from the cells of the central mass; these are separated from one another by the growth of connective tissue in an irregular manner. Each ovum acquires a covering of connective tissue (follicle) cells, and in this way the rudiments of the ovarian follicles are formed.

Primordial germ cells (Fig. 17-3) are, according to some authors, set aside at a very early age from the somatic cells. They are first recognized in the yolk sac. Later they migrate through the mesentery into the primitive germinal epithelium, to be carried into the gonad with the sex cords, and later still they develop either into ova in the female or sperm cells in the male. Some authors deny their existence while others claim they all degenerate and take no part in the formation of the adult sex cells.

FEMALE MESONEPHROS AND MESONEPHRIC DUCT

In the female, these structures atrophy (see page 1516). The remains of the mesonephric tubules may be divided into three groups. One group of tubules from the cranial portion of the mesonephros persists and produces one or more **vesicular appendices** in the fringes of the uterine tube. The middle and largest group, together with a segment of the mesonephric duct, persist as the **epoöphoron** (*organ of Rosenmüller*[1]). Persistent portions of the mesonephric duct are known as **Gartner's ducts**[2] (Fig. 17-55). These may exist as part of the epoöphoron or as isolated segments as far as the hymen. The third and most caudal group of remaining mesonephric ducts constitutes the **paroöphoron,** which usually disappears completely before the adult stage. Any one of these vestigial tubules that persists in the adult as a stalked vesicle is called an **hydatid.**

PARAMESONEPHRIC DUCTS

Shortly after the formation of the mesonephric ducts, the paramesonephric ducts (*Müllerian ducts*[3]) develop (Fig. 17-12). Each arises on the lateral aspect of the corresponding mesonephric duct as a tubular invagination of the cells lining the coelom (Fig. 17-12). The orifice of the invagination remains patent and undergoes enlargement and modification to form the abdominal ostium of the uterine tube. The ducts pass caudalward lateral to the mesonephric ducts, but toward the caudal end of the embryo they cross to the medial side of these ducts, and thus come to lie side by side between and caudal to the latter—the four ducts forming what is termed the **genital cord.** The paramesonephric ducts end in an epithelial elevation, the **paramesonephric eminence,** on the ventral part of the cloaca between the orifices of the mesonephric ducts; at a later date they open into the cloaca in this situation.

In the male the paramesonephric (*Müllerian*) ducts atrophy, but traces of their cranial ends are represented by the **appendices testis** (*hydatids of Morgagni*[4]), while their caudal portions fuse to form the utriculus in the floor of the prostatic portion of the urethra.

In the female the paramesonephric (*Müllerian*) ducts persist and undergo further development. The portions that lie in the genital ridge fuse to form the uterus and vagina; the parts cranial to this ridge remain separate, and each forms the corresponding uterine tube—the abdominal ostium of which is developed from the cranial extremity of the original tubular invagination from the coelom. The fusion of the paramesonephric ducts begins in the third month, and the septum formed by their fused medial walls disappears. Entodermal epithelium of the urogenital sinus invades the region where the vagina forms, replacing the paramesonephric epithelium almost entirely, and for a time the vagina is represented by a solid rod of epithelial cells, but in fetuses of five months the lumen reappears. About the fifth month an annular constriction marks the position of the neck of the uterus, and

---

[1] Johann Christian Rosenmüller (1771–1820): A German anatomist and surgeon (Leipzig).
[2] Hermann Treschow Gartner (1785–1827): A Danish surgeon and anatomist (Copenhagen).

[3] Johannes Peter Müller (1801–1858): A German professor of anatomy and physiology (Berlin).
[4] Giovanni Battista Morgagni (1682–1771): An Italian anatomist and pathologist (Padua).

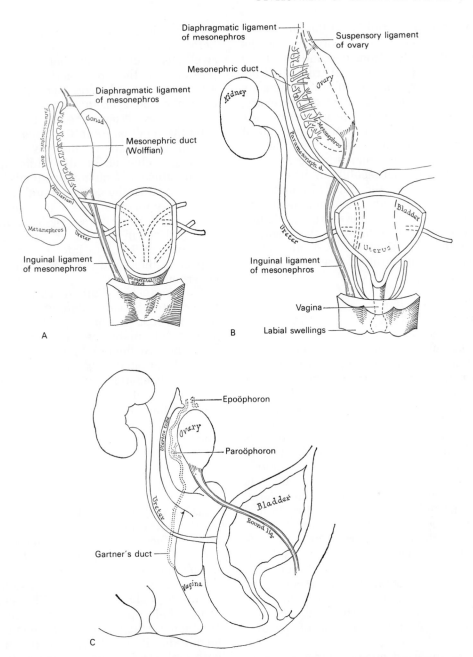

FIG. 17-12. Diagrammatic representation of the differentiation of the female genital ducts. *A,* Undifferentiated stage. *B,* Early differentiation. *C,* The condition existing at birth. In the latter, the dotted lines indicate the position of the mesonephric duct and tubules. Vestigial structures derived from these are labeled. (Redrawn from Allan, F. D., *Essentials of Embryology,* Oxford University Press, 1960.)

after the sixth month the walls of the uterus begin to thicken. A ring-like outgrowth of epithelium occurs at the caudal end of the uterus and marks the future vaginal fornices. The hymen arises at the site of the paramesonephric eminence. It represents the separation between vagina and urogenital sinus.

DESCENT OF THE OVARIES

In the female there is also a gubernaculum, which effects a considerable change in the position of the ovary (Fig. 17-12), though not so extensive a change as that affecting the testis. The gubernaculum in the female lies in contact with the fundus of the uterus and acquires adhesions to this organ; thus the ovary is prevented from descending below this level. The part of the gubernaculum between the ovary and the uterus ultimately becomes the proper **ligament of the ovary,** while the part between the uterus and the labium majus forms the **round ligament of the uterus.** A pouch of peritoneum analogous to the saccus vaginalis in the male accompanies the gubernaculum in the inguinal canal; this is called the **canal of Nuck.**[1]

In rare cases the gubernaculum may fail to develop adhesions to the uterus, and then the ovary descends through the inguinal canal into the labium majus. Under these circumstances its position resembles that of the testis.

EXTERNAL ORGANS OF GENERATION

As is shown in Figure 17-2, the cloacal membrane, composed of ectoderm and endoderm, initially extends from the umbilicus to the tail bud. The growing mesoderm extends to the midventral line for some distance caudal to the umbilicus and forms the lower part of the abdominal wall; it ends caudally in a prominent swelling, the **genital tubercle** (Fig. 17-13). Dorsal to this tubercle the urogenital part of the cloacal membrane separates the ingrowing sheets of mesoderm.

The first rudiment of the penis (or clitoris) is a structure termed the **phallus;** it is derived from the phallic portion of the cloaca which has extended on the end and sides of the undersurface of the genital tubercle. The terminal part of the phallus, representing the future glans, becomes solid; the remainder, which is hollow, is converted into a longitudinal groove by the absorption of the urogenital membrane.

In the female a deep groove forms around the phallus and separates it from the rest of the genital tubercle. The tissue at the sides of the genital tubercle grows caudalward as the **genital swellings,** which ultimately form the labia majora; the tubercle itself becomes the mons pubis. The labia minora arise by the continued growth of the lips of the groove on the undersurface of the phallus; the remainder of the phallus forms the clitoris.

In the male the early changes are similar, but the pelvic portion of the cloaca undergoes much greater development, pushing before it the phallic portion. The genital swellings extend around between the pelvic portion and the anus, and form a scrotal area; during the changes associated with the descent of the testes this area is drawn out to form the scrotal sacs. The penis is developed from the phallus. As in the female, the urogenital membrane undergoes absorption, forming a channel on the undersurface of the phallus; this channel extends only as far forward as the corona glandis.

The **corpora cavernosa** of the penis (or clitoris) and of the urethra arise from the mesodermal tissue in the phallus. They are dense structures at first, but later vascular spaces appear in them and they gradually become cavernous.

The **prepuce** in both sexes is formed by the growth of a solid plate of ectoderm into the superficial part of the phallus; on coronal section this plate presents the shape of a horseshoe. By the breaking down of its more centrally situated cells, the plate is split into two lamellae, and a cutaneous fold, the prepuce, is liberated and forms a hood over the glans. Adherent prepuce is not a secondary adhesion but a hindered central desquamation.

URETHRA

As already described, in both sexes the phallic portion of the cloaca extends to the undersurface of the genital tubercle as far forward as the apex. At the apex the walls of the phallic portion come together and fuse, the lumen is obliterated, and a solid plate, the **urethral plate,** is formed. The remainder of the phallic portion is tubular for a time, and then, by the absorption of the urogenital membrane, it establishes a communication with the exterior; this opening is the **primitive urogenital ostium,** which extends forward to the corona glandis.

In the female this condition is largely retained; the portion of the groove on the clitoris broadens out while the body of the clitoris enlarges, and thus the adult urethral opening is situated behind the base of the clitoris.

In the male, by the greater growth of the pelvic portion of the cloaca, a longer urethra is formed, and the primitive ostium is carried distalward with the phallus, but it still ends at the corona glandis. Later it closes

[1]Anton Nuck (1650–1692): A Dutch anatomist (Leiden).

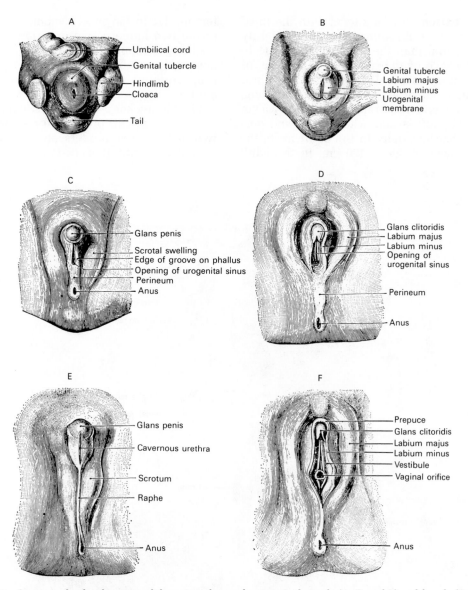

A
- Umbilical cord
- Genital tubercle
- Hindlimb
- Cloaca
- Tail

B
- Genital tubercle
- Labium majus
- Labium minus
- Urogenital membrane

C
- Glans penis
- Scrotal swelling
- Edge of groove on phallus
- Opening of urogenital sinus
- Perineum
- Anus

D
- Glans clitoridis
- Labium majus
- Labium minus
- Opening of urogenital sinus
- Perineum
- Anus

E
- Glans penis
- Cavernous urethra
- Scrotum
- Raphe
- Anus

F
- Prepuce
- Glans clitoridis
- Labium majus
- Labium minus
- Vestibule
- Vaginal orifice
- Anus

FIG. 17-13. Stages in the development of the external sexual organs in the male (A, C, and E) and female (B, D, and F). (Drawn from the Ecker-Ziegler models.)

from proximally to distally. Meanwhile the urethral plate of the glans breaks down centrally to form a median groove continuous with the primitive ostium. This groove also closes proximally, so that the external urethral opening is shifted distalward to the end of the glans (Fig. 17-6).

Urinary Organs

The urinary organs comprise the **kidneys,** which produce the urine; the **ureters,** or ducts, which convey urine to the **urinary bladder,** where it is retained for a time; and the **urethra,** through which it is discharged from the body.

KIDNEYS

The **kidneys** (*renes*) are situated in the dorsal part of the abdomen, one on either side of the vertebral column, covered by the peritoneum and surrounded by a mass of fat and loose areolar tissue. Their cranial extremities are on a level with the cranial border of the twelfth thoracic vertebra, their caudal extremities on a level with the third lumbar. The right kidney is usually slightly more caudal than the left, probably due to the presence of the liver. The long axis of each kidney is parallel with the vertebral column. Each kidney is about 11.25 cm long, 5 to 7.5 cm wide, and rather more than 2.5 cm thick. The left is somewhat longer and narrower than the right. In the adult male, the kidney weighs 125 to 170 gm; in the adult female, 115 to 155 gm. The combined weight of the two kidneys in proportion to that of the body is about 1 to 240. The kidneys in the newborn are about three times as large in proportion to the body weight as in the adult.

The kidney has a characteristic form similar to a bean, and presents for examination two surfaces, two borders, and a cranial and caudal extremity (Fig. 17-14).

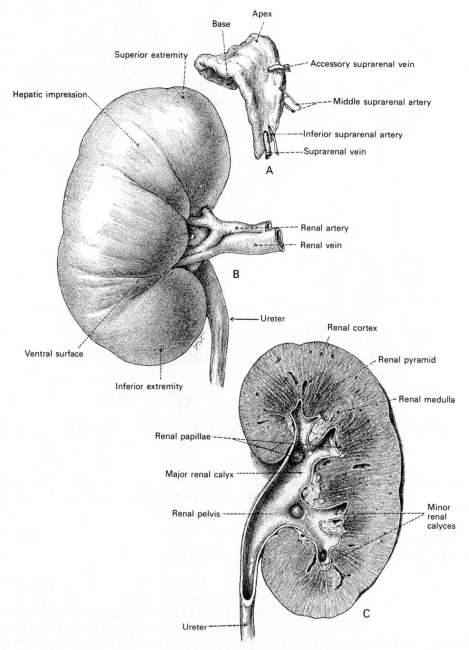

FIG. 17-14.   The right kidney and suprarenal gland. *A*, Suprarenal gland. *B*, Kidney, surface view. *C*, Kidney, longitudinal section showing pelvis. (Eycleshymer and Jones.)

Surfaces

The **ventral surface** (Fig. 17-15) of each kidney is convex and faces ventralward and slightly lateralward. Its relations to adjacent viscera differ so completely on the two sides that separate descriptions are necessary.

VENTRAL SURFACE OF RIGHT KIDNEY. A narrow portion at the cranial extremity is in relation with the right suprarenal gland (Fig. 17-17). A large area just caudal to this and involving about three-fourths of the surface lies in the renal impression on the visceral surface of the liver, and a narrow but somewhat variable area near the medial border is in contact with the descending part of the duodenum. The caudal part of the ventral surface is in contact laterally with the right colic flexure, and medially, as a rule, with the small intestine. The areas in relation with the liver and small intestine are covered by peritoneum; the suprarenal, duodenal, and colic areas are devoid of peritoneum.

VENTRAL SURFACE OF LEFT KIDNEY. A small area along the cranial part of the medial border is in relation with the left suprarenal gland, and close to the lateral border is a long strip in contact with the renal impression on the spleen (Fig. 17-17). A somewhat quadrilateral field, about the middle of the ventral surface, marks the site of contact with the body of the pancreas, on the deep surface of which are the lienal vessels. Above this is a small triangular portion, between the suprarenal and splenic areas, in contact with the posterior surface of the stomach. Caudal to the pancreatic area, the lateral part is in relation with the left colic flexure, the medial part, with the small intes-

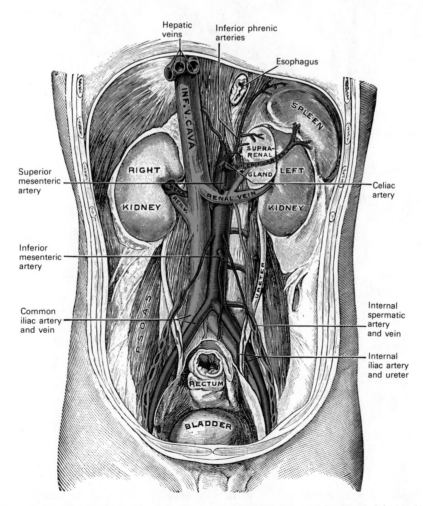

FIG. 17-15. Ventral view of abdominal viscera after removal of the peritoneum of the dorsal abdominal wall, showing kidneys, suprarenal glands, and great vessels. (Corning.)

tine. The areas in contact with the stomach and spleen are covered by the peritoneum of the omental bursa, while that in relation to the small intestine is covered by the peritoneum of the greater sac; dorsal to the latter are some branches of the left colic vessels. The suprarenal, pancreatic, and colic areas are devoid of peritoneum.

DORSAL SURFACE.   The dorsal surface (Figs. 17-16; 17-18) of each kidney is directed dorsalward and medialward. It is embedded in areolar and fatty tissue and entirely devoid of peritoneal covering. It lies upon the diaphragm; the medial and lateral lumbocostal arches; the psoas major, the quadratus

lumborum, and the tendon of the transversus abdominis muscles; the subcostal and one or two of the upper lumbar arteries; and the last thoracic, iliohypogastric, and ilioinguinal nerves. The cranial extremity of the right kidney rests upon the twelfth rib; the left, usually on the eleventh and twelfth ribs. The diaphragm separates the kidney from the pleura, which dips down to form the phrenicocostal sinus, but frequently the muscular fibers of the diaphragm are defective or absent over a triangular area immediately above the lateral lumbocostal arch, in which case the perinephric areolar tissue is in contact with the diaphragmatic pleura.

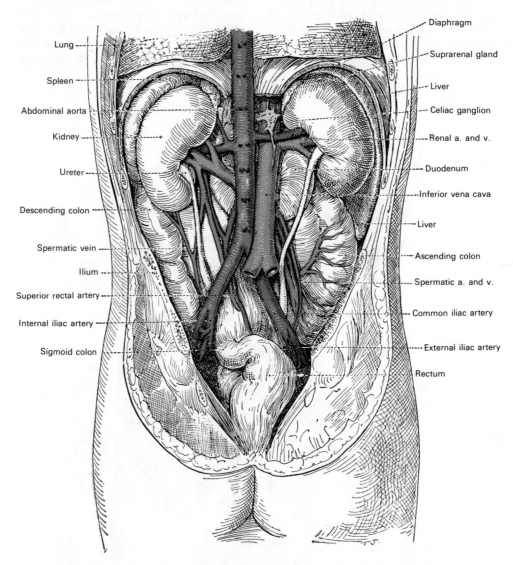

FIG. 17-16.   Dissection of abdominal viscera, dorsal view showing relation of the kidneys. (After Corning's *Topographischen Anatomie* in Eycleshymer and Jones.)

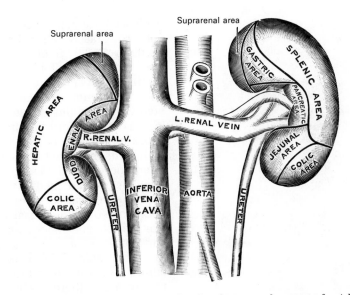

FIG. 17-17.   The ventral surfaces of the kidneys, showing the areas of contact of neighboring viscera.

Borders

The **lateral border** (*external border*) is convex and directed toward the posterolateral wall of the abdomen. On the left side it is in contact, at its cranial part, with the spleen.

The **medial border** (*internal border*) is concave in the center and convex toward either extremity; it is directed ventralward and a little caudalward. Its central part has a deep longitudinal fissure, bounded by prominent overhanging ventral and dorsal lips. This fissure, named the **hilum,** transmits the vessels, nerves, and ureter. Above the hilum the medial border is in relation with the suprarenal gland; below the hilum, with the ureter.

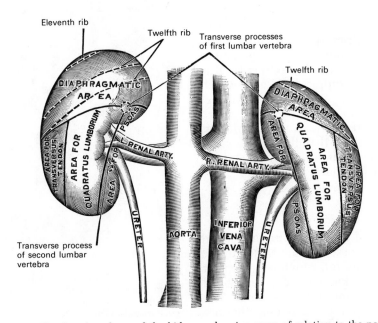

FIG. 17-18.   The dorsal surfaces of the kidneys, showing areas of relation to the parietes.

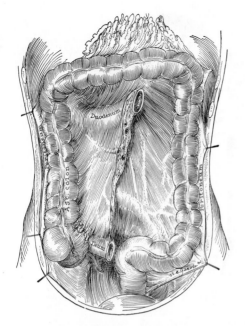

FIG. 17-19.   Visceral and parietal peritoneum associated with large intestine. Small intestine and mesentery removed. (Tobin, 1944.)

The relative positions of the main structures in the hilum are as follows: the vein is ventral, the artery is in the middle, and the ureter is dorsal and directed caudalward. Frequently, however, branches of both artery and vein are placed dorsal to the ureter.

Extremities

The **cranial extremity** (*extremitas superior*) is thick and round and is nearer the median line than the caudal extremity. It is surmounted by the suprarenal gland, which also covers a small portion of the ventral surface.

The **caudal extremity** (*extremitas inferior*) is smaller and thinner than the superior and farther from the median line. It extends to within 5 cm of the iliac crest.

Renal Fascia

The kidney and its vessels are embedded in a mass of fatty tissue, termed the adipose capsule or *perirenal fat,* which is thickest at the margins of the kidney and is prolonged through the hilum into the renal sinus. The kidney and the adipose capsule together are

enclosed in a specialized lamination of the subserous fascia called the renal fascia (Figs. 17-19 to 17-22). It occupies a position between the internal investing layer of deep fascia (transversalis, endoabdominal fascia) and the stratum of subserous fascia associated with the intestine and its blood vessels (Fig. 17-20). In forming the renal fascia, the subserous fascia of the lateral abdominal wall splits into two fibrous lamellae near the lateral border of the kidney (Fig. 17-22). Both lamellae extend medially, the ventral one over the ventral surface of the kidney, the dorsal one over the dorsal surface. The anterior lamella continues over the renal vessels and aorta to join the similar membrane of the other side. The posterior lamella also continues across the midline but lies deep to the aorta, and there it is more adherent to the underlying deep fascia than in the region of the kidney. The renal fascia is connected to the fibrous tunic of the kidney by numerous trabeculae that traverse the adipose capsule and are strongest near the lower end of the organ. Dorsal to the fascia renalis is a large quantity of fat, which constitutes the *paranephric body (pararenal fat).*

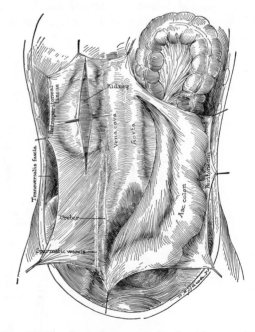

FIG. 17-20.   Peritoneum associated with ascending colon dissected free and displaced to expose the deeper stratum of subserous fascia associated with kidney and great vessels. (Tobin, 1944.)

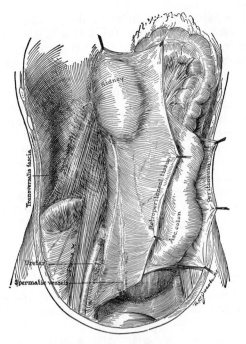

<figure>FIG. 17-21. Deeper stratum of subserous fascia dissected free and displaced to expose the transversalis fascia. (Tobin, 1944.)</figure>

Fixation of the Kidney

The kidneys are not rigidly fixed to the abdominal wall, and since they are in contact with the diaphragm, they move with it during respiration. They are held in position by the renal fascia just described and by the large renal arteries and veins. That the adipose capsule and the paranephric fat body play an important part in holding the kidney in position is indicated by the occurrence of a condition called movable kidney in emaciated individuals.

General Structure of the Kidney

The kidney is invested by a fibrous tunic or capsule that forms a firm, smooth covering to the organ. The tunic can be stripped off easily, but in doing so numerous fine processes of connective tissue and small blood vessels are torn. When the capsule is stripped off, the surface of the kidney is smooth and deep red. In infants, fissures extending for some depth may be seen on the surface of the organ, a remnant of the lobular contruction of the gland.

If a vertical section of the kidney were made from its convex to its concave border, it would be seen that the hilum expands into a central cavity, the **renal sinus;** this contains the cranial part of the renal pelvis and the calyces, surrounded by some fat in which the branches of the renal vessels and nerves are embedded. The renal sinus is lined by a prolongation of the fibrous tunic, which is continuous with the covering of the pelvis of the kidney around the lips of the hilum.

The **minor renal calyces,** numbering 4 to 13, are cup-shaped tubes, each of which embraces usually one but occasionally two or more of the renal papillae. They unite to form two or three short tubes, the major calyces, and these in turn join to form a

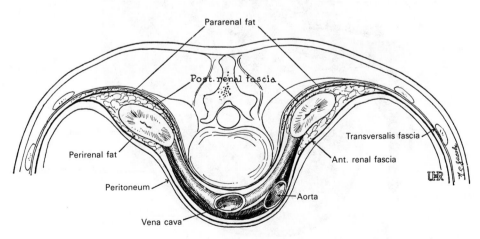

<figure>FIG. 17-22. Transverse section, showing relations of renal fascia. (Tobin, 1944.)</figure>

funnel-shaped sac, the **renal pelvis.** Spirally arranged muscles surrounding the calyces may have a milking action on these tubes, thereby aiding the flow of urine into the renal pelvis. As the pelvis leaves the renal sinus, it diminishes rapidly in caliber and merges insensibly into the **ureter,** the excretory duct of the kidney.

The kidney is composed of an internal **medullary** and an external **cortical substance** (Fig. 17-23).

The **medullary substance** consists of a series of striated conical masses, termed the **renal pyramids,** of which there are 8 to 18. Their bases are directed toward the circumference of the kidney, while their apices converge toward the renal sinus, where they form prominent papillae projecting into the lumen of the minor calyces.

The **cortical substance** is reddish brown and soft and granular. It lies immediately beneath the fibrous tunic, arches over the bases of the pyramids, and dips in between adjacent pyramids toward the renal sinus.

The parts dipping between the pyramids are named **renal columns** (of Bertini[1]), while the portions that connect the renal columns to each other and intervene between the bases of the pyramids and the fibrous tunic are called the **cortical arches.** If the cortex is examined with a lens, it is seen to consist of a series of lighter colored, conical areas, termed the *radiate part,* and a darker colored intervening substance, which from the complexity of its structure is named the *convoluted part.* The rays gradually taper toward the circumference of the kidney and consist of a series of outward prolongations from the base of each renal pyramid.

Microscopic Anatomy

The **renal tubules** (Fig. 17-23), of which the kidney is for the most part composed, commence in the cortical substance. After

[1]Exupère Joseph Bertin (1712–1781): A French anatomist (Rheims).

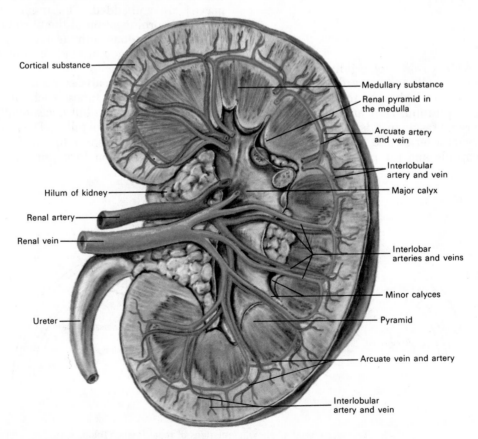

Longitudinal section through the kidney near the hilum showing the renal cortex and medulla and the blood vessels.

pursuing a circuitous course through the cortical and medullary substances, they finally end at the apices of the renal pyramids by open mouths, so that the fluid they contain is emptied through the calyces into the pelvis of the kidney. If the surface of one of the papillae is examined with a lens, it is seen to be studded over with 16 to 20 minute openings, the orifices of the renal tubules. Pressure applied to a fresh kidney will cause urine to exude from these orifices.

The tubules commence in the cortex and renal columns as the **renal corpuscles** or **Malpighian**[1] **bodies.** They are small, round, deep red masses, varying in size but averaging about 0.2 mm in diameter. Each of these bodies is composed of two parts: a central **glomerulus** of vessels, and a double-walled membranous envelope, the **glomerular capsule** (*capsule of Bowman*[2]), which is the invaginated pouch-like commencement of a renal tubule.

GLOMERULUS. The glomerulus is a tuft of nonanastomosing capillaries, among which there is a scanty amount of connective tissue. This capillary tuft is derived from an arteriole, the *afferent vessel,* which enters the capsule, generally at a point opposite to that at which the capsule joins the tubule (Fig. 17-24). Upon entering the capsule, the afferent arteriole divides into 2 to 10 primary branches, which in turn subdivide into about 50 capillary loops that generally do not anastomose. These loops are 300 to 500 μm long. The capillaries join to form the *efferent arteriole,* which leaves Bowman's capsule adjacent to the afferent vessel, the latter generally being the larger of the two. The total surface area of the capillaries of all glomeruli is about 1 square meter.

For a variable distance before the afferent arteriole enters the glomerulus, the muscle cells of the media adjacent to the distal convolution of its own nephron are modified, appearing as relatively large afibrillar cells. This structure presents evidence of glandular activity. At times it appears partially to invest the artery in a nest-like group of cells embedded in a delicate fibrillar network and is known as the **juxtaglomerular apparatus.**

GLOMERULAR CAPSULE. The **glomerular** or **Bowman's capsule,** which surrounds the glo-

merulus, consists of a double-walled sac. The outer wall (parietal layer) is continuous with the inner wall (visceral layer) at the points of entrance and exit of the afferent and efferent vessels respectively. The cavity between the two layers is continuous with the lumen of the proximal convoluted tubule. The parietal layer is smooth. The visceral layer covers the glomerulus and dips in between the capillary loops, almost completely surrounding each one. Both layers of the capsule consist of flat epithelial cells that have a basement membrane. This covers the outer surface of the parietal layer and is continuous with that of the tubule cells. The basement membrane of the visceral layer is in contact with the glomerular capillaries. Microdissection experiments of living glomeruli show that they lie in a gelatinous matrix. The walls of the capillaries, the overlying cells of the visceral layer of the capsule, and the gelatinous matrix constitute a filter mechanism through which nonprotein constituents of the blood plasma can enter the tubule.

RENAL TUBULE. A renal tubule, beginning with the capsule of Bowman as it surrounds the glomerulus and ending where the tubule joins the excretory duct or collecting tubule, constitutes a **nephron**—the structural and functional unit of the kidney. There are about 1,250,000 of these units in each kidney.

During its course, a tubule has many changes in shape and direction. It is contained partly in the cortical and partly in the medullary substance. At its junction with the glomerular capsule, it exhibits a somewhat constricted portion, which is termed the **neck.** Beyond this the tubule becomes convoluted, and pursues a long course in the cortical substance constituting the **proximal convoluted tube.** The convolutions disappear as the tube approaches the medullary substance in a more or less spiral manner. Throughout this portion of its course the renal tubule is contained entirely in the cortical substance and has a fairly uniform caliber. It now enters the medullary substance, suddenly becomes much smaller, quite straight, and dips down for a variable depth into the pyramid, constituting the thin or **descending limb of Henle's loop.** Bending on itself, it forms what is termed the **loop of Henle** and ascending, it becomes suddenly enlarged, forming the thick or **ascending limb of Henle's loop,** which enters the cortical substance where it again becomes dila-

[1]Marcello Malpighi (1628-1694): An Italian anatomist (Bologna).

[2]Sir William Bowman (1816–1892): An English physician (London).

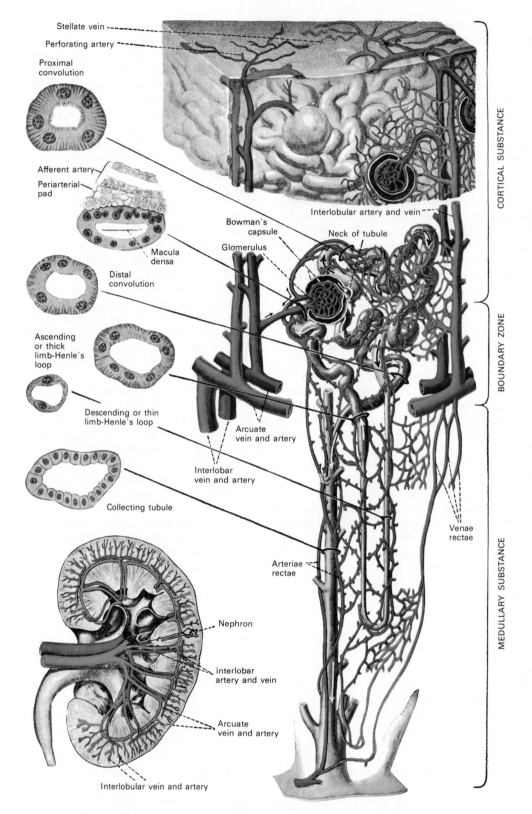

FIG. 17-24.   Diagram of a portion of kidney lobule illustrating a nephron, typical histologic sections of the various divisions of a nephron, and the disposition of the renal vessels. The section of the collecting tubule is reproduced at a lower magnification than the divisions of the nephron. (Courtesy of R. G. Williams.)

ted and tortuous. It is now called the **distal convoluted tubule.** The terminal part of the ascending limb of Henle's loop crosses in contact with, or sometimes lies parallel to, the afferent arteriole of its own glomerulus. The turns of the distal convoluted tubule into which it merges lie among the coils of the proximal portion of the nephron and terminate in a narrow part that enters a collecting tubule.

The **straight** or **collecting tubes** commence in the radiate part of the cortex, where they receive the curved ends of the distal convoluted tubules. They unite at short intervals with one another, the resulting tubes presenting a considerable increase in caliber, so that a series of comparatively large tubes passes from the bases of the rays into the renal pyramids. In the medulla the tubes of each pyramid converge to join a central tube (*duct of Bellini*[1]), which finally opens on the summit of one of the papillae; the contents of the tube are therefore discharged into one of the minor calyces.

Structure of the Renal Tubules. The various parts of the nephron present quite different cellular appearances and these appearances vary depending upon the functional state of the cells. The **proximal convoluted tubule** is about 14 mm long and 59 μm in diameter. It is composed of one layer of large cuboidal cells with central spherical nuclei. The cells dovetail laterally with one another and the lateral cell limits are rarely seen. The distal ends of the cells bulge into the lumen and are covered with a brush border. The cytoplasm is abundant and coarsely granular. Parallel striations in the cytoplasm perpendicular to the basement membrane are due to mitochondria.

The transition between the epithelium of the proximal convoluted tubule and the thin segment is abrupt. The **thin segment** may be absent in nephrons beginning near the surface of the kidney. It is composed of squamous cells with pale-staining cytoplasm and flat nuclei. The epithelial change from the descending limb to the **ascending** or **thick limb** is also quite abrupt. The cells become cuboidal and deeper-staining, with perpendicular striations in the basal parts but without distinct cell boundaries or brush borders.

At about the junction of the ascending limb with the distal convoluted tubule, the nephron comes into contact with the juxtaglomerular cells of its own afferent arteriole. Here the epithelium of the tubule is greatly modified, having high cells and crowded nuclei. This constitutes the **macula densa** or epithelial plaque.

Transition to the tortuous **distal convoluted tubule** is gradual. This segment of the nephron is about 5 mm long and 35 μm in diameter. The cells are lower and the lumen larger than in the proximal tubule. They do not have a brush border and the boundaries are fairly distinct. The distal convoluted tubule merges into a short connecting segment that joins the collecting or excretory tubule.

The collecting tubules have a typical epithelium which is quite different from that in the various portions of the nephron. In the smallest tubes, the cells are cuboidal and distinctly outlined with round nuclei and clear cytoplasm. As the tubules become larger, the cells are higher, finally becoming tall columnar cells in the ducts of Bellini. The columnar epithelium becomes continuous with the cells covering the surface of the papillae.

The length of the nephron varies from 30 to 38 mm, while the length of the collecting tubules is estimated at 20 to 22 mm.

RENAL BLOOD VESSELS. The kidney is plentifully supplied with blood by the renal artery, a large branch of the abdominal aorta (Fig. 17-24). Before entering the kidney substance the number and disposition of the branches of the renal artery exhibit great variation. In most cases, the renal artery divides into two primary branches, a larger anterior and a smaller posterior branch. The anterior branch supplies exclusively the anterior or ventral half of the organ and the posterior supplies the posterior or dorsal half. Therefore, there is a line (Brödel's line[2]) in the long axis of the lateral border of the kidney that passes between the two main arterial divisions in which there are no large vessels, a feature that is utilized to minimize hemorrhage when nephrotomy is done. The primary branches subdivide and diverge until they come to lie on the anterior and posterior aspects, respectively, of the calyces. Further subdivisions occur that enter the kidney substance and run between the pyramids. These are known as **interlobar arteries.** When these vessels reach the corticomedullary zone, they make more or less well-defined arches over the bases of the pyramids and are then called **arcuate arteries.** These vessels give off a series of branches called **interlobular arteries.**

[1]Lorenzo Bellini (1643–1704): An Italian anatomist (Pisa).

[2]Max Brödel (1870–1941): An American physician and medical artist (Baltimore).

The interlobular arteries and the terminal parts of the arcuate vessels run vertically and nearly parallel toward the cortex and periphery of the kidney. The interlobular arteries may terminate: (1) as an afferent glomerular artery to one or more glomeruli; (2) in a capillary plexus around the convoluted tubules in the cortices without relation to the glomerulus, therefore a nutrient artery; and (3) as a perforating capsular vessel. Divisions classified under (2) must be regarded as exceptional.

The most important and numerous branches of the interlobular arteries are the **afferent glomerular vessels.** These break up into capillary loops, the glomerulus, within Bowman's capsule. The loops unite to form the efferent arteriole. The **efferent glomerular vessel** forms a plexus about the convoluted tubule and part of Henle's loop and sends one or more branches toward the pelvis, the **arteria recta,** which supplies the collecting tubules and loops of Henle.

The arteriae rectae are derived chiefly from the efferent arterioles of glomeruli located in the boundary zone. They are frequently known as **arteriae rectae spuriae** to distinguish them from a few straight vessels arising directly from the arcuate or interlobular arteries without relation to glomeruli and hence called **arteriae rectae verae.** It is likely that the so-called arteriae rectae verae were once derived from an efferent glomerular vessel but that the glomerulus atrophied, thus giving them the appearance of true nutrient branches from the arcuate or interlobular arteries. No arteriae rectae are derived from the vessels of the cortical zone. It has not been determined whether anastomoses occur between the capillaries of adjacent nephrons. However, there is free anastomosis between the branches of the arteriae rectae. The arteriae rectae, or straight arteries, surround the limbs of Henle and pass down between the straight collecting tubules, where they form terminal plexuses around the tubes.

This plexus drains into the **venae rectae,** which in turn carry the blood to the interlobular veins, thence into the arcuate veins, then into the interlobar veins and finally into the renal veins, which discharge into the inferior vena cava. Note that the chief renal arteries have counterparts in the venous system. All veins from the dorsal half of the kidney cross to the ventral half between the minor calyces to join the ventral collecting veins before leaving the kidney. Whatever arterial anastomoses there are within the kidney occur beyond the glomeruli; the renal arteries and their branches, therefore, are terminal as far as the arteriae rectae. An exception to this condition is found in cases in which arteriovenous anastomoses have been described. These connections have been found in three locations: between the arteries and veins of the sinus renalis, between the subcapsular vessels, and between the interlobular arteries and veins. In the latter position particularly, arterial blood could reach the tubules of the nephron by retrograde flow through the veins. The constancy with which this type of anastomosis occurs or its role in the circulation of the kidney has not been accu-

rately determined. Anastomosis between veins is rich. The perforating capsular vessels, the terminations of interlobular arteries, form connections with nonrenal vessels in the fat that surrounds the kidney. They frequently drain into the subcapsular or **stellate veins,** which in turn drain into the interlobular veins.

The **lymphatics** of the kidney are described in Chapter 10.

NERVES. The nerves of the kidney, although small, number about 15. They have small ganglia and are derived from the renal plexus, which is formed by branches from the celiac plexus, the lower and outer part of the celiac ganglion and aortic plexus, and from the lesser and lowest splanchnic nerves. They communicate with the testicular plexus, a circumstance that may explain the occurrence of pain in the testis in disorders of the kidney. They accompany the renal artery and its branches and are distributed to the blood vessels and to the cells of the urinary tubules.

CONNECTIVE TISSUES. Although the tubules and vessels are closely packed, a small amount of connective tissue (*intertubular stroma*), continuous with the fibrous tunic, binds them firmly together and supports the blood vessels, lymphatics, and nerves.

VARIATIONS. Malformations of the kidney are common. One kidney may be absent entirely, but the number of these cases is small. One kidney may be congenitally atrophied, in which case the kidney is very small but usually healthy in structure. These cases are important and must be taken into account when nephrectomy is contemplated. A more common malformation is the fusion of the two kidneys together. They may be joined together only at their lower ends by means of a thick mass of renal tissue, forming a horseshoe-shaped body, or they may be completely united, forming a disc-like kidney, from which two ureters descend into the bladder. These fused kidneys are generally situated in the midline of the abdomen, but may be displaced as well. In some mammals, *e.g.*, ox and bear, the kidney consists of a number of distinct lobules; this lobulated condition is characteristic of the kidney of the human fetus, and traces of it may persist in the adult. Sometimes the pelvis is duplicated, while a double ureter is not uncommon. In some rare instances a third kidney may be present.

One or both kidneys may be misplaced as a congenital condition and remain fixed in this abnormal position. They are then often misshapen. They may be situated higher, though this is uncommon, or lower than normal, or removed farther from the vertebral column than usual. They may be displaced into the iliac fossa, over the sacroiliac joint, onto the promontory of the sacrum, or into the pelvis between the rectum and bladder or by the side of the uterus. In these latter cases they may cause serious trouble.

The kidney may also be displaced as a congenital condition but not fixed; it is then known as a *floating kidney.* This disorder is believed to be due to the fact that the kidney is completely enveloped by peritoneum, which then passes backward to the vertebral column as a double layer, forming a mesonephron that permits movement. The kidney may also be misplaced as an acquired condition; in these cases the kidney is mobile in the tissues that surround it, moving with the capsule in the perinephric tissues. This condition is known as *movable kidney,* and is more common in females than in males. It occurs in badly nourished people or in those who have become emaciated from any cause. It must not be confused with the *floating kidney,* which is a congenital condition due to the development of a mesonephron. The two conditions cannot, however, be distinguished until the abdomen is opened or the kidney explored from the loin. Accessory renal arteries that enter one or both poles of the kidney instead of at the hilum are fairly common.

URETERS

The **ureters** (Figs. 17-15; 17-16) are the two tubes that convey the urine from the kidneys to the urinary bladder. The renal portion of each tube commences within the sinus of the corresponding kidney as a number of short cup-shaped tubes, termed **calyces,** which encircle the renal papillae (Fig. 17-25). Since a single calyx may enclose more than one papilla, the calyces are generally fewer in number than the pyramids—the former varying from 7 to 13, the latter from 8 to 18. The calyces join to form two or three short tubes, and these unite to form a funnel-shaped dilatation named the **renal pelvis.** The pelvis is situated partly inside and partly outside the renal sinus, where it becomes continuous with the ureter, usually on a level with the spinous process of the first lumbar vertebra.

The **ureter proper** varies in length from 28 to 34 cm, the right being about 1 cm shorter than the left. It is a thick-walled narrow tube, not of uniform caliber, varying from 1 mm to 1 cm in diameter. It runs caudalward and medialward on the psoas major muscle, and entering the pelvic cavity, finally opens into the fundus of the bladder.

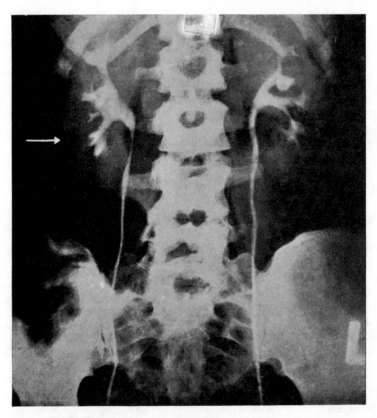

FIG. 17-25. Ureters, pelves, and minor calyces after intravenous injection of a radiopaque medium (Uroselectan). Note cupping of minor calyces; the relation of the ureter to the transverse processes of the lumbar vertebrae, and the psoas major. The arrow points to the shadow of the right kidney. Anterior view.

Abdominal Part

The abdominal part lies behind the peritoneum on the medial part of the psoas major embedded in the subserous fascia, and is crossed obliquely by the testicular vessels. It enters the pelvic cavity by crossing either the termination of the common iliac vessels or the commencement of the external iliac vessels (Fig. 17-15).

At its origin, the *right ureter* is usually covered by the descending part of the duodenum. In its course, it lies to the right of the inferior vena cava, and is crossed by the right colic and ileocolic vessels, while near the superior aperture of the pelvis, it passes dorsal to the caudal part of the mesentery and the terminal part of the ileum. The *left ureter* is crossed by the left colic vessels, and near the superior aperture of the pelvis, it passes behind the sigmoid colon and its mesentery.

Pelvic Part

This part runs at first caudalward on the lateral wall of the pelvic cavity, along the anterior border of the greater sciatic notch and under cover of the peritoneum. It lies ventral to the internal iliac artery medial to the obturator nerve and the obturator, inferior vesical, and middle rectal arteries. Opposite the lower part of the greater sciatic foramen it inclines medialward and reaches the lateral angle of the bladder, where it is situated ventral to the upper end of the seminal vesicle; here the ductus deferens crosses to its medial side, and the vesical veins surround it. Finally, the ureters run obliquely for about 2 cm through the wall of the bladder and open by slit-like apertures into the cavity of the viscus at the lateral angles of the trigone. When the bladder is distended, the openings of the ureters are about 5 cm apart, but when it is empty and contracted, the distance between them is diminished by one-half. Owing to their oblique course through the coats of the bladder, the upper and lower walls of the terminal portions of the ureters become closely applied to each other when the viscus is distended, and acting as valves, prevent regurgitation of urine from the bladder. The ureter normally undergoes constriction at three points in its course: (1) at the ureteropelvic junction (average diameter, 2 mm); (2) where it crosses the iliac vessels (average diameter, 4 mm); (3) where it joins the bladder (average diameter, 1 to 5 mm). Between these points the abdominal ureter averages 10 mm in diameter, and the pelvic ureter, 5 mm.

In the **female,** the ureter forms, as it lies in relation to the wall of the pelvis, the posterior boundary of a shallow depression named the **ovarian fossa,** in which the ovary is situated. It then runs medialward and ventralward on the lateral aspect of the cervix uteri and upper part of the vagina to reach the fundus of the bladder. In this part of its course it is accompanied for about 2.5 cm by the uterine artery, which then crosses over the ureter and ascends between the two layers of the broad ligament. The ureter is about 2 cm distant from the side of the uterine cervix.

Structure

The ureter is composed of three coats: **fibrous, muscular,** and **mucous** (Fig. 17-26).

The **fibrous coat** is continuous at one end with the fibrous tunic of the kidney on the floor of the sinus, while at the other it is lost in the fibrous structure of the bladder.

In the renal pelvis the **muscular coat** consists of two layers, longitudinal and circular: the longitudinal fibers become lost upon the sides of the papillae at the extremities of the calyces; the circular fibers may be traced

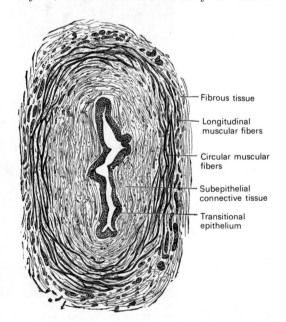

Fibrous tissue

Longitudinal muscular fibers

Circular muscular fibers

Subepithelial connective tissue

Transitional epithelium

Fɪɢ. 17-26. Transverse section of ureter.

surrounding the medullary substance in the same situation. In the ureter proper the muscular fibers are distinct and arranged in three layers: an external longitudinal layer, a middle circular layer, and an internal layer, which is less distinct than the other two, but has a general longitudinal direction. This internal layer is found only in the neighborhood of the bladder.

The **mucous coat** is smooth, having only a few longitudinal folds that become effaced by distention. It is continuous with the mucous membrane of the bladder below, while it is prolonged over the papillae of the kidney above. Its epithelium is transitional and resembles that found in the bladder (see Fig. 17-33). It consists of several layers of cells, of which the innermost—that is to say, the cells in contact with the urine—are somewhat flattened, with concavities on their deep surfaces into which the rounded ends of the cells of the second layer fit. These intermediate cells more or less resemble columnar epithelium; they are pear-shaped, with rounded internal extremities that fit into the concavities of the cells of the first layer, and narrow external extremities that are wedged between the cells of the third layer. The external or third layer consists of conical or oval cells that vary in number in different parts and have processes that extend into the basement membrane. Beneath the epithelium and separating it from the muscular coats is a dense layer of fibrous tissue containing many elastic fibers.

VESSELS AND NERVES. The **arteries** supplying the ureter are branches from the renal, testicular, internal iliac, and inferior vesical arteries.

The **nerves** are derived from the inferior mesenteric, testicular, and pelvic plexuses. The lower one-third of the ureter contains nerve cells that are probably incorporated in vagus efferent chains. The afferent supply of the ureter is contained in the eleventh and twelfth thoracic and first lumbar nerves. The vagus supply to the ureter probably also has afferent components.

VARIATIONS. The upper portion of the ureter is sometimes double; rarely it is double the greater part of its extent, or even completely so. In such cases there are two openings into the bladder. Asymmetry in these variations is common.

URINARY BLADDER

The **urinary bladder** (Fig. 17-27) is a musculomembranous sac that acts as a reservoir for the urine. As its size, position, and rela-

tions vary according to the amount of fluid it contains, it is necessary to study it as it appears (a) when *empty*, and (b) when *distended* (Fig. 17-28). In both conditions, the position of the bladder varies with the condition of the rectum, being pushed upward and forward when the rectum is distended.

Empty Bladder

When hardened in situ, the empty bladder has the form of a flattened tetrahedron, with its vertex tilted ventralward. It has a fundus, a vertex, a superior and an inferior surface. The **fundus** (Fig. 17-44) is triangular and is directed caudalward and dorsalward toward the rectum, from which it is separated by the rectovesical fascia, the seminal vesicles, and the terminal portions of the ductus deferentes. The **vertex** is directed ventralward toward the upper part of the symphysis pubis, and from it the median umbilical ligament is continued cranialward on the anterior abdominal wall to the umbilicus. The peritoneum covering the ligament is the median umbilical fold.

The **superior surface** is triangular, bounded on either side by a lateral border that separates it from the inferior surface, and by a posterior border, represented by a line joining the two ureters, which intervenes between it and the fundus. The lateral borders extend from the ureters to the vertex, and from them the peritoneum is carried to the walls of the pelvis. On either side of the bladder the peritoneum shows a depression, named the **paravesical fossa.** The superior surface is covered by peritoneum and is in relation with the sigmoid colon and some of the coils of the small intestine. When the bladder is empty and firmly contracted, this surface is convex and the lateral and posterior borders are round, whereas if the bladder is relaxed it is concave, and the interior of the viscus, as seen in a median sagittal section, has the appearance of a V-shaped slit, with a shorter posterior and a longer anterior limb— the apex of the V corresponding with the internal orifice of the urethra.

The **inferior surface** is uncovered by peritoneum. It may be divided into a posterior or prostatic area and two lateral surfaces. The prostatic area rests upon and is in direct continuity with the base of the prostate; and from it the urethra emerges. The lateral portion of the inferior surface is directed cau-

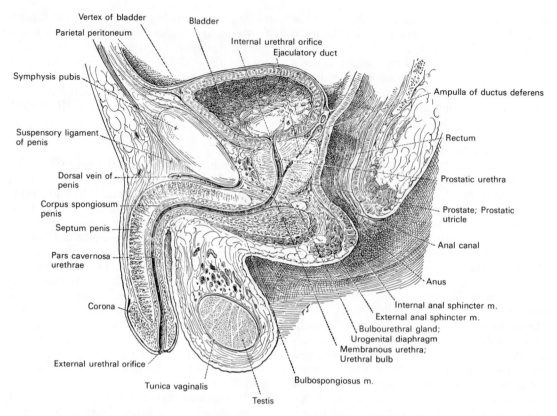

FIG. 17-27. Median sagittal section through male pelvis, viewed from left side. (Eycleshymer and Jones.)

dalward and lateralward and is separated from the symphysis pubis by the prevesical fascial cleft (*cavum Retzii*)[1].

When the bladder is empty, it is placed entirely within the pelvis, below the level of the obliterated hypogastric arteries, and below the level of those portions of the ductus deferentes that are in contact with the lateral wall of the pelvis; after they cross the ureters, the ductus deferentes come into contact with the fundus of the bladder. As the viscus fills, its fundus, being more or less fixed, is only slightly depressed, while its superior surface gradually rises into the abdominal cavity, carrying with it its peritoneal covering and at the same time rounding off the posterior and lateral borders.

Distended Bladder

When the bladder is moderately full, it contains about 0.5 L and assumes an oval form; the long diameter of the oval measures

[1] Anders Adolf Retzius (1796–1860): A Swedish anatomist (Stockholm).

about 12 cm and is directed cranialward and ventralward. In this condition it presents a posterosuperior and an anteroinferior surface and two lateral surfaces—a fundus and a summit.

The **posterosuperior surface** is covered by peritoneum. Dorsally, it is separated from the rectum by the rectovesical excavation, while its anterior part is in contact with the coils of the small intestine.

The **anteroinferior surface** is devoid of peritoneum; it rests against the pubic bones, above which it is in contact with the back of the anterior abdominal wall. The caudal parts of the lateral surfaces are destitute of peritoneum and are in contact with the lateral walls of the pelvis. The line of peritoneal reflection from the lateral surface is raised to the level of the obliterated hypogastric artery.

The **fundus** undergoes little alteration in position, being only slightly lowered. It exhibits, however, a narrow triangular area, which is separated from the rectum merely by the rectovesical fascia. This area is

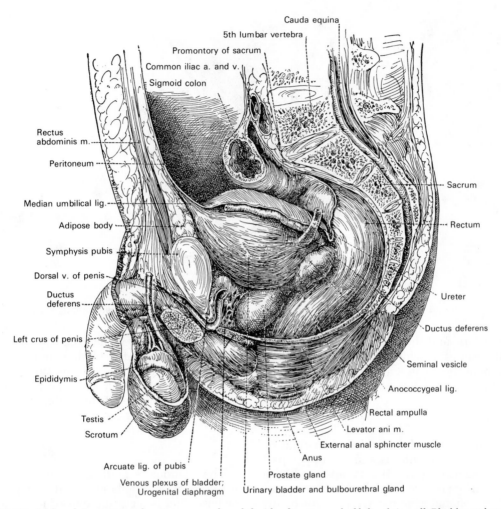

Cauda equina

5th lumbar vertebra

Promontory of sacrum

Common iliac a. and v.

Sigmoid colon

Rectus abdominis m.

Peritoneum

Median umbilical lig.

Adipose body

Symphysis pubis

Dorsal v. of penis

Ductus deferens

Left crus of penis

Epididymis

Testis

Scrotum

Arcuate lig. of pubis

Venous plexus of bladder; Urogenital diaphragm

Prostate gland

Urinary bladder and bulbourethral gland

Anus

External anal sphincter muscle

Levator ani m.

Rectal ampulla

Anococcygeal lig.

Seminal vesicle

Ductus deferens

Ureter

Rectum

Sacrum

FIG. 17-28. Male pelvic organs and perineum seen from left side after removal of left pelvic wall. Bladder and rectum moderately distended. (Eycleshymer and Jones.)

bounded below by the prostate, above by the rectovesical fold of peritoneum, and laterally by the ductus deferentes. The ductus deferentes frequently come in contact with each other above the prostate, and under such circumstances the lower part of the triangular area is obliterated. The line of reflection of the peritoneum from the rectum to the bladder appears to undergo little or no change when the bladder is distended; it is situated about 10 cm from the anus.

The **summit** is directed cranialward and ventralward above the point of attachment of the median umbilical ligament, and hence the peritoneum that follows the ligament forms a pouch of varying depth between the summit of the bladder and the anterior abdominal wall.

Bladder in the Child

In the newborn child the internal urethral orifice is at the level of the upper border of the symphysis pubis (Figs. 17-29; 17-30); the bladder therefore lies relatively at a much higher level in the infant than in the adult. Its anterior surface is in contact with about the lower two-thirds of that part of the abdominal wall which lies between the symphysis pubis and the umbilicus. Its fundus is clothed with peritoneum as far as the level of the internal orifice of the urethra. Although the bladder of the infant is usually described as an abdominal organ, only about one-half of it lies above the plane of the superior aperture of the pelvis. The internal urethral orifice sinks rapidly during the first

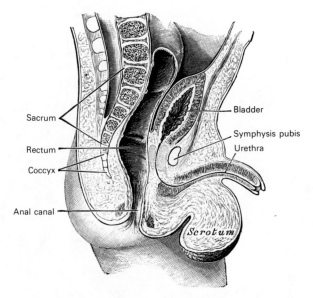

Sacrum

Rectum

Coccyx

Anal canal

Bladder

Symphysis pubis

Urethra

Scrotum

FIG. 17-29. Sagittal section through the pelvis of a newly born male child.

three years, and then more slowly until the ninth year, after which it remains stationary until puberty, when it again slowly descends and reaches its adult position.

Female Bladder

In the female, the bladder is in relation dorsally with the uterus and the upper part of the vagina (Fig. 17-31). It is separated from the anterior surface of the body of the uterus by the vesicouterine excavation, but below the level of this excavation it is connected to the front of the cervix uteri and the upper part of the anterior wall of the vagina by areolar tissue. When the bladder is empty the uterus rests upon its superior surface. The female bladder is said by some to be more capacious than that of the male, but probably the opposite is the case.

Ligaments

The bladder is held in position by ligamentous attachments at its inferior portion or base, that is, near the exit of the urethra, and at the vertex. The remainder of the wall, enclosed in subserous fascia, is free to move

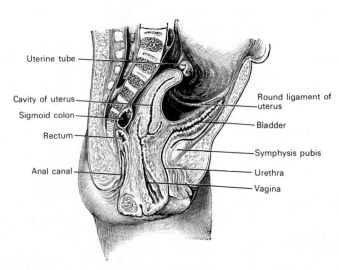

Uterine tube

Cavity of uterus

Sigmoid colon

Rectum

Anal canal

Round ligament of uterus

Bladder

Symphysis pubis

Urethra

Vagina

FIG. 17-30. Sagittal section through the pelvis of a newly born female child.

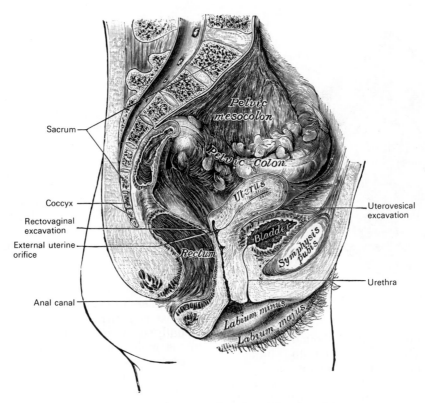

Sacrum

Coccyx

Rectovaginal
excavation

External uterine
orifice

Anal canal

Pelvic
mesocolon

Pelvic Colon

Uterus

Bladder

Rectum

Symphysis
pubis

Labium minus

Labium majus

Uterovesical
excavation

Urethra

FIG. 17-31. Median sagittal section of female pelvis.

during the expansion and contraction of filling and emptying.

The base of the bladder is attached to the internal investing layer of deep fascia on the pubic bone by strong fibrous bands that may contain muscle fibers, the pubovesicales. In the male, because the prostate is firmly bound to the bladder in this region, these attachments are between the prostate and pubic bone rather than directly to the bladder, and they are named the **medial** and **lateral puboprostatic ligaments.** In the female the attachments are directly between bladder and pubis and are therefore called the **pubovesical ligaments.**

The base of the bladder is secured posteriorly to the side of the rectum and the sacrum by condensations of the subserous fascia underlying the sacrogenital folds. They are called the **rectovesical ligaments,** or, since they may contain smooth muscle bundles, the **rectovesical** muscles.

The **median umbilical ligament** is a fibrous or fibromuscular cord, the remains of the urachus, which extends from the vertex of the bladder to the umbilicus. It is broad at its attachment to the bladder and becomes narrow as it nears the umbilicus.

In addition to these fibrous or true ligaments, there is a series of folds where the peritoneum is reflected from the bladder to the abdominal wall; these are called **false ligaments of the bladder.** Anteriorly there are three folds: the **median umbilical fold** on the median umbilical ligament, and two **lateral umbilical folds** on the obliterated umbilical arteries. The reflections of the peritoneum onto the side wall of the pelvis form the **lateral false ligaments,** while the **sacrogenital folds** constitute **posterior false ligaments.**

Interior of the Bladder

The mucous membrane lining the greater part of the bladder is loosely attached to the muscular coat, and appears wrinkled or folded when the bladder is contracted (Fig. 17-32); in the distended condition the folds are effaced. Over a small triangular area, termed the **trigone,** immediately above and behind the internal orifice of the urethra, the

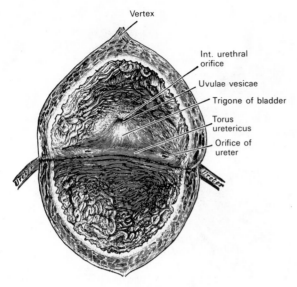

FIG. 17-32.   The interior of bladder.

Labels on figure: Vertex; Int. urethral orifice; Uvulae vesicae; Trigone of bladder; Torus uretericus; Orifice of ureter; Ureter

mucous membrane is firmly bound to the muscular coat and is always smooth. The anterior angle of the trigone is formed by the internal orifice of the urethra: its posterolateral angles are formed by the orifices of the ureters. Stretching dorsal to the latter openings is a slightly curved ridge, the **torus uretericus,** which forms the base of the trigone and is produced by an underlying bundle of nonstriped muscular fibers. The lateral parts of this ridge extend beyond the openings of the ureters and are named the **plicae uretericae;** they are produced by the terminal portions of the ureters as they traverse the bladder wall obliquely. When the bladder is illuminated, the torus uretericus appears as a pale band and forms an important guide during the operation of introducing a catheter into the ureter.

The **orifices of the ureters** are placed at the posterolateral angles of the trigone and are usually slit-like in form. In the contracted bladder, they are about 2.5 cm apart and about the same distance from the internal urethral orifice; in the distended viscus, these measurements may be increased to about 5 cm.

The **internal urethral orifice** is placed at the apex of the trigone in the most dependent part of the bladder; it usually has a somewhat crescentic form. The mucous membrane immediately behind it presents a slight elevation, the **uvula vesicae,** caused by the middle lobe of the prostate.

Structure

The bladder is composed of the four coats: **serous, muscular, submucous,** and **mucous** (Fig. 17-33).

SEROUS COAT.   The serous coat is a partial one and is derived from the peritoneum. It invests the superior surface and the upper parts of the lateral surfaces, and is reflected from these onto the abdominal and pelvic walls.

MUSCULAR COAT.   The muscular coat consists of three layers of smooth muscular fibers: an external layer, composed of fibers having for the most part a longitudinal arrangement; a middle layer, in which the fibers are arranged more or less in a circular manner; and an internal layer, in which the fibers have a general longitudinal arrangement (Fig. 17-34).

The *fibers of the external layer* arise from the posterior surface of the body of the pubis in both sexes (*musculi pubovesicales*), and in the male from the adjacent part of the prostate and its capsule. They pass, in a more or less longitudinal manner, up the inferior surface of the bladder, over its vertex, and then descend along its fundus, to become attached to the prostate in the male and to the front of the vagina in the female. At the sides of the bladder, the fibers are arranged obliquely and intersect one another.

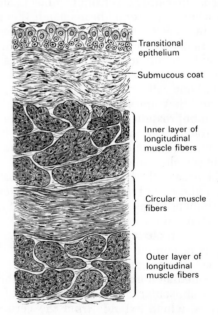

Labels on figure: Transitional epithelium; Submucous coat; Inner layer of longitudinal muscle fibers; Circular muscle fibers; Outer layer of longitudinal muscle fibers

FIG. 17-33.   Vertical section of bladder wall.

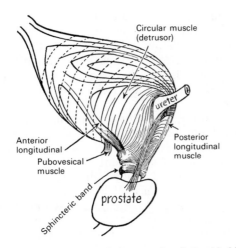

Circular muscle
(detrusor)

ureter

Anterior
longitudinal

Posterior
longitudinal
muscle

Pubovesical
muscle

Sphincteric band

prostate

FIG. 17-34. Diagram of the muscle of the bladder.
(After McCrea.)

This layer has been named the **detrusor urinae muscle.**

The *fibers of the middle circular layer* are thinly and irregularly scattered on the body of the organ, and although to some extent they are placed transversely to the long axis of the bladder, they are largely arranged obliquely. Toward the lower part of the bladder, around the internal urethral orifice, they are disposed in a thick circular layer, forming the **sphincter vesicae,** which is continuous with the muscular fibers of the prostate.

The *internal longitudinal layer* is thin. Its fasciculi have a reticular arrangement but with a tendency to assume for the most part a longitudinal direction. Two bands of oblique fibers, originating behind the orifices of the ureters, converge to the back part of the prostate and are inserted by means of a fibrous process into the middle lobe of that organ. They are the **muscles of the ureters,** described in the early 19th century by Sir C. Bell, who supposed that during the contraction of the bladder they serve to retain the oblique direction of the ureters and so prevent the reflux of the urine into them.

SUBMUCOUS COAT. The submucous coat consists of a layer of areolar tissue connecting the muscular and mucous coats and intimately united to the latter.

MUCOUS COAT. The mucous coat is thin, smooth, and a pale rose color. It is continuous above through the ureters with the lining membrane of the renal tubules and below with that of the urethra. The loose texture of the submucous layer allows the mucous coat to be thrown into folds or *rugae* when the

bladder is empty. Over the trigone the mucous membrane is closely attached to the muscular coat, and is not thrown into folds, but is smooth and flat. The epithelium covering it is transitional, consisting of a superficial layer of polyhedral flattened cells, each with one, two, or three nuclei; beneath these is a stratum of large club-shaped cells, with their narrow extremities directed downward and wedged between smaller spindle-shaped cells, containing oval nuclei (Fig. 17-33). The epithelium varies accordingly as the bladder is distended or contracted. In the former condition the superficial cells are flattened and those of the other layers are shortened; in the latter they present the appearance described above. There are no true glands in the mucous membrane of the bladder, though certain mucous follicles that occur especially near the neck of the bladder have been regarded as such.

VESSELS AND NERVES. The **arteries** supplying the bladder are the superior, middle, and inferior vesical arteries, derived from the internal iliac artery. The obturator and inferior gluteal arteries also supply small visceral branches to the bladder, and in the female additional branches are derived from the uterine and vaginal arteries.

The **veins** form a complicated plexus on the inferior surface and the fundus near the prostate; they end in the internal iliac veins.

The **lymphatics** are described in Chapter 10.

The **nerves** of the bladder are (1) fine medullated fibers from the third and fourth sacral nerves, and (2) nonmedullated fibers from the hypogastric plexus. They are connected with ganglia in the outer and submucous coats and are finally distributed, all as nonmedullated fibers, to the muscular layer and epithelial lining of the viscus.

VARIATIONS. A defect of development is known as *extroversion of the bladder.* In this condition the lower part of the abdominal wall and the anterior wall of the bladder are wanting, so that the fundus of the bladder appears on the abdominal surface and is pushed forward by the pressure of the viscera within the abdomen, forming a red vascular tumor on which the openings of the ureters are visible. The penis, except the glans, is rudimentary and is cleft on its dorsal surface, exposing the floor of the urethra, a condition known as *epispadias.* The pelvic bones are also arrested in development.

MALE URETHRA

The **male urethra** (Fig. 17-35) extends from the internal urethral orifice in the urinary bladder to the external urethral orifice at the

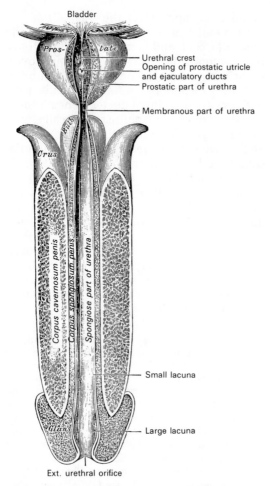

Fig. 17-35. The male urethra laid open on its anterior (upper) surface.

end of the penis. It has a double curve in the ordinary relaxed state of the penis (Fig. 17-27). Its length varies from 17.5 to 20 cm, and it is divided into three portions, **prostatic, membranous,** and **spongiose,** the structure and relations of which are essentially different. Except during the passage of the urine or semen, the greater part of the urethral canal is a mere transverse cleft or slit, with its surfaces in contact. At the external orifice the slit is vertical, in the membranous portion it is irregular or stellate, and in the prostatic portion it is somewhat arched.

Prostatic Portion

This portion, the widest and most dilatable part of the canal, is about 3 cm long. It runs almost vertically through the prostate from its base to its apex, lying nearer its anterior than its posterior surface. The form of the canal is spindle-shaped, being wider in the middle than at either extremity and narrowest below, where it joins the membranous portion. A transverse section of the canal as it lies in the prostate is horseshoe-shaped, with the convexity directed ventralward.

Upon the posterior wall or floor is a narrow longitudinal ridge, the **urethral crest,** formed by an elevation of the mucous membrane and its subjacent tissue. It is 15 to 17 mm long and about 3 mm high. On either side of the crest is a slightly depressed fossa, the **prostatic sinus,** the floor of which is perforated by numerous apertures, the **orifices of the prostatic ducts** from the lateral lobes of the prostate; the ducts of the middle lobe open behind the crest. At the distal part of the urethral crest, below its summit, is a median elevation, the **colliculus seminalis** or **verumontanum,** upon or within the margins of which are the orifices of the prostatic utricle and the slit-like openings of the ejaculatory ducts.

The **prostatic utricle** forms a cul-de-sac about 6 mm long, which runs upward and backward in the substance of the prostate behind the middle lobe. Its walls are composed of fibrous tissue, muscular fibers, and mucous membrane, and numerous small glands open on its inner surface. Some authors call it the **uterus masculinus,** because it is developed from the united lower ends of the atrophied paramesonephric or Müllerian ducts, and therefore is homologous with the uterus and vagina in the female.

Membranous Portion

This portion is the shortest, least dilatable, and with the exception of the external orifice, the narrowest part of the canal. It extends between the apex of the prostate and the bulb of the penis, perforating the urogenital diaphragm about 2.5 cm from the pubic symphysis. The dorsal part of the urethral bulb lies in apposition with the superficial layer of the urogenital diaphragm, but its upper portion diverges somewhat from this fascia. The anterior wall of the membranous urethra is thus prolonged for a short distance ventral to the urogenital diaphragm; it measures about 2 cm in length, while the posterior wall, which is between the two fasciae of the diaphragm, is only 1.25 cm long.

The membranous portion of the urethra is completely surrounded by the fibers of the sphincter urethrae. Ventral to it, the deep dorsal vein of the penis enters the pelvis between the transverse ligament of the pelvis and the arcuate pubic ligament; on either side near its termination are the bulbourethral glands.

Spongiose Portion

The spongiose portion (*penile portion*) is the longest part of the urethra and is contained in the corpus spongiosum. It is about 15 cm long, and extends from the termination of the membranous portion to the external urethral orifice. Commencing at the superficial layer of the urogenital diaphragm, it passes ventralward and upward to the front of the symphysis pubis, and then, in the flaccid penis, it bends downward. It is narrow and has a uniform size in the body of the penis, measuring about 6 mm in diameter; it is dilated proximally, within the bulb, and again anteriorly within the glans penis, where it forms the **fossa navicularis urethrae.**

External Urethral Orifice

This is the most contracted part of the urethra; it is a vertical slit, about 6 mm long, bounded on each side by a small labium.

The lining membrane of the urethra, especially that on the floor of the spongiose portion, contains the orifices of numerous mucous glands and follicles situated in the submucous tissue; these are named the **urethral glands** (of *Littré*)[1]. There are also small pit-like recesses, or **lacunae,** of varying sizes. These orifices are directed distalward, so that they may easily intercept the point of a catheter in its passage along the canal. One of these lacunae, larger than the rest, is situated on the upper surface of the fossa navicularis; it is called the **lacuna magna.** The bulbourethral glands open into the cavernous portion about 2.5 cm distal to the inferior fascia of the urogenital diaphragm.

Structure

The urethra is composed of mucous membrane supported by a submucous tissue that connects it with the various structures through which it passes.

---

[1]Alexis Littré (1658–1725): A French surgeon (Rheims and Paris).

MUCOUS COAT. The mucous coat is continuous with the mucous membrane of the bladder, ureters, and kidneys; externally, it is continuous with the integument covering the glans penis. It is prolonged into the ducts of the glands that open into the urethra, viz., the bulbourethral glands and the prostate, and into the ductus deferentes and seminal vesicles through the ejaculatory ducts. In the spongiose and membranous portions, the mucous membrane is arranged in longitudinal folds when the tube is empty. Small papillae are found upon it, near the external urethral orifice. Its epithelial lining is columnar except near the external orifice, where it is squamous and stratified.

SUBMUCOSA. The submucosa has a characteristic structure. It is composed of a thick stroma of connective tissue rich in elastic fibers. These fibers connect freely with the spongy tissue of the penis, which prevents ready removal of the mucosa in this region. However, in the membranous and prostatic portions, which change but little during erection, the urethra is quite free and may on dissection be stripped readily.

VESSELS AND NERVES. The **urethral artery,** a branch of the internal pudendal artery in the perineum, supplies the membranous and penile urethra. The **veins** of the urethra and corpus spongiosum drain into the deep vein of the penis and the pudendal plexus. The **lymphatics** are described in Chapter 10. The **nerves** come from the pudendal nerve.

CONGENITAL DEFECTS. The defect most frequently met is one in which there is a cleft of the floor of the urethra owing to an arrest of union in the midline. This is known as *hypospadias,* and the cleft may vary in extent. In the simplest and most common form, the deficiency is confined to the glans penis. The urethra ends at the point where the extremity of the prepuce joins the body of the penis, in a small valve-like opening. The prepuce is also cleft on its undersurface and forms a sort of hood over the glans. There is a depression on the glans in the position of the normal meatus. This condition produces no disability and requires no treatment.

In more severe cases the spongiose portion of the urethra is cleft throughout its entire length, and the opening of the urethra occurs at the junction of the penis and the scrotum. The undersurface of the penis in the midline has a furrow lined by a moist mucous membrane, on either side of which is often more or less dense fibrous tissue stretching from the glans to the opening of the urethra, which prevents complete erection from taking place. Great discomfort is induced during micturition and sexual intercourse is impossible. The condition may be remedied by a series of plastic operations.

The worst form of this condition occurs when the

urethra is deficient as far back as the perineum, and the scrotum is cleft. The penis is small and bound down between the two halves of the scrotum, resembling a hypertrophied clitoris. The testes are often undescended. The condition of parts, therefore, greatly resembles the external genitalia of the female, and many children, the victims of this malformation, have been raised as girls. The halves of the scrotum, deficient of testes, resemble the labia, the cleft between them looks like the orifice of the vagina, and the diminutive penis is mistaken for an enlarged clitoris.

A more uncommon form of malformation involves an apparent deficiency of the upper wall of the urethra; this is named *epispadias*. The deficiency may vary in extent; complete deficiency is associated with extroversion of the bladder. In less extensive cases without extroversion, there is an infundibuliform opening into the bladder. The penis is usually dwarfed and turned upward, so that the glans lies over the opening. Congenital stricture also occurs occasionally, and in such cases multiple strictures may be present throughout the whole length of the spongiose portion.

FEMALE URETHRA

The **female urethra** (Fig. 17-31) is a narrow membranous canal, about 4 cm long, extending from the bladder to the external orifice in the vestibule. It is dorsal to the symphysis pubis and embedded in the anterior wall of the vagina. Its direction is obliquely downward and forward; it is slightly curved, with the concavity directed forward. Its diameter when undilated is about 6 mm. It perforates the fasciae of the urogenital diaphragm, and its external orifice is situated directly ventral to the vaginal opening and about 2.5 cm dorsal to the glans clitoridis. The lining membrane is thrown into longitudinal folds, one of which, placed along the floor of the canal, is termed the **urethral crest.** Many small urethral glands open into the urethra. The largest of these are the paraurethral glands (*of Skene*), the ducts of which open just within the urethral orifice.

Structure

The female urethra consists of three coats: **muscular, erectile,** and **mucous.**

COATS. The **muscular coat** is continuous with that of the bladder; it extends the whole length of the tube, and consists of circular fibers. In addition, between the superior and inferior fasciae of the urogenital diaphragm,

the female urethra is surrounded by the sphincter urethrae, as in the male.

A thin layer of **spongy erectile tissue,** containing a plexus of large veins intermixed with bundles of unstriped muscular fibers, lies immediately beneath the mucous coat.

The **mucous coat** is continuous externally with that of the vulva, and internally with that of the bladder. It is lined by stratified squamous epithelium, which becomes transitional near the bladder. Its external orifice is surrounded by a few mucous follicles.

VESSELS AND NERVES. The **arteries** are derived from the inferior vesical and internal pudendal arteries. The **veins** drain into the vesical and vaginal veins. The **lymphatics** are described in Chapter 10. The **nerves** are from the pelvic plexus and pudendal nerves.

Male Genital Organs

The **male genitals** include the **testes,** the **ductus deferentes,** the **vesiculae seminales,** the **ejaculatory ducts,** and the **penis,** together with the following accessory structures, viz., the **prostate** and the **bulbourethral glands.**

TESTES AND THEIR COVERINGS

The **testes** (Figs. 17-36; 17-37) are two parenchymatous organs that produce the semen; they are suspended in the scrotum by the spermatic cords. At an early period of fetal life the testes are contained in the abdominal cavity, behind the peritoneum. Before birth they descend to the inguinal canal, along which they pass, and emerging at the superficial inguinal ring, descend into the scrotum, becoming invested in their course by coverings derived from the serous, muscular, and fibrous layers of the abdominal parietes, as well as by the scrotum.

Coverings

The coverings of the testes are:

Skin ⎫
Dartos tunic ⎭ Scrotum
External spermatic fascia
Cremasteric layer
Internal spermatic fascia
Tunica vaginalis

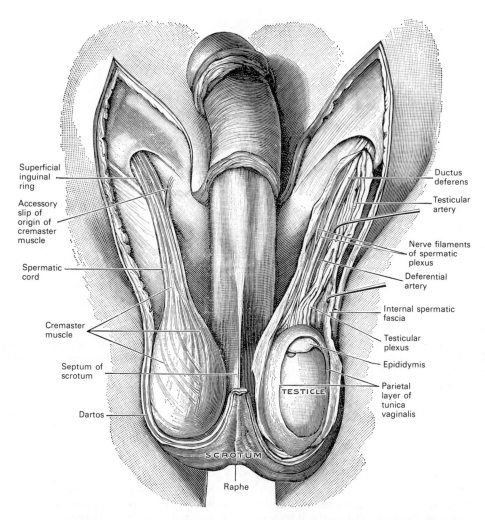

Superficial
inguinal
ring

Accessory
slip of
origin of
cremaster
muscle

Spermatic
cord

Cremaster
muscle

Septum of
scrotum

Dartos

Ductus
deferens

Testicular
artery

Nerve filaments
of spermatic
plexus

Deferential
artery

Internal spermatic
fascia

Testicular
plexus

Epididymis

Parietal
layer of
tunica
vaginalis

TESTICLE

SCROTUM

Raphe

FIG. 17-36. The scrotum. The penis has been turned upward and the anterior wall of the scrotum has been removed.
On the right side, the spermatic cord, the internal spermatic fascia, and the cremaster muscles are displayed; on the left
side the internal spermatic fascia has been divided by a longitudinal incision passing along the front of the cord and
the testicle, and a portion of the parietal layer of the tunica vaginalis has been removed to display the testicle and a
portion of the head of the epididymis covered by the visceral layer of the tunica vaginalis. (Toldt.)

SCROTUM.   The scrotum is a cutaneous
pouch containing the testes and parts of the
spermatic cords. It is divided on its surface
into two lateral portions by a ridge or **raphe,**
which is continued ventralward to the undersurface of the penis, and dorsalward
along the midline of the perineum to the
anus. The external aspect of the scrotum
varies under different circumstances. Under
the influence of warmth and in old and de-
bilitated persons, it becomes elongated and
flaccid, but under the influence of cold, and
in the young and robust, it is short, corrugated, and closely applied to the testes. Of
the two lateral portions, the left hangs lower

than the right, to correspond with the greater
length of the left spermatic cord.

The **scrotum** consists of two layers, the
skin or **integument** and the **dartos tunic.**

Integument.   The integument is thin,
brownish, and generally thrown into folds or
rugae. It is provided with sebaceous follicles, the secretion of which has a characteristic odor, and is beset with thinly scattered,
crisp kinky hairs, the roots of which are visible through the skin.

Dartos Tunic.   The dartos tunic contains a
thin layer of nonstriped muscular fibers that
are continuous around the base of the scrotum with the two layers of the superficial

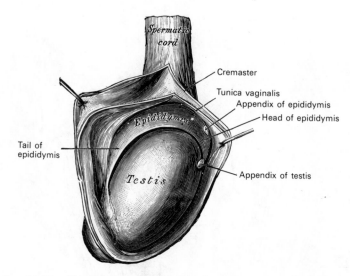

FIG. 17-37.   The right testis, exposed by laying open the tunica vaginalis.

fascia of the groin and the perineum. It sends inward a septum, which divides the scrotal pouch into two cavities for the testes, and extends between the raphe and the undersurface of the penis as far as its root.

The dartos tunic is closely united to the skin externally, but is separated from the subjacent parts by a distinct fascial cleft, upon which it glides with the greatest facility. It contains no fat and is highly vascular.

EXTERNAL SPERMATIC FASCIA.   The external spermatic fascia (*intercrural or intercolumnar fascia*) is a thin membrane prolonged distalward over the cord and testis. It is continuous at the superficial inguinal ring with the aponeurosis of the obliquus externus abdominis and is separated from the enclosing dartos by a fascial cleft.

CREMASTERIC LAYER.   The cremasteric layer consists of the scattered bundles of the cremaster muscle connected into a continuous membrane by the cremasteric fascia. It forms the *middle spermatic layer* and corresponds to the obliquus internus abdominis and its fasciae.

INTERNAL SPERMATIC LAYER.   The internal spermatic fascia (*infundibuliform fascia; tunica vaginalis communis*) is a thin membrane that is often difficult to separate from the cremasteric layer but more easily separates from the cord and testis, which it encloses. It is continuous at the deep inguinal ring with the transversalis fascia.

The **tunica vaginalis** is described with the testes.

VESSELS AND NERVES.   The **arteries** that supply the coverings of the testes are the superficial and deep external pudendal branches of the femoral artery, the superficial perineal branch of the internal pudendal artery, and the cremasteric branch from the inferior epigastric artery. The **veins** follow the course of the corresponding arteries. The **lymphatics** end in the inguinal lymph nodes. The **nerves** are the ilioinguinal and genitofemoral branches of the lumbar plexus, the two superficial perineal branches of the pudendal nerve, and the pudendal branch of the posterior femoral cutaneous nerve.

The scrotum forms an admirable covering for the protection of the testes. These bodies, lying suspended and loose in the cavity of the scrotum and surrounded by serous membrane, are capable of great mobility, and can therefore easily slip about within the scrotum and thus avoid injuries from blows or squeezes. The skin of the scrotum is elastic and capable of great distention, and on account of the looseness and amount of subcutaneous tissue, the scrotum becomes greatly enlarged in cases of edema, to which this part is especially liable as a result of its dependent position.

The **inguinal canal** is described in Chapter 6.

Spermatic Cord

The spermatic cord (Fig. 17-38) extends from the deep inguinal ring, where the structures of which it is composed converge, to the testis. In the abdominal wall, the cord passes obliquely along the inguinal canal, lying at first inferior to the obliquus internus and superior to the fascia transversalis, but

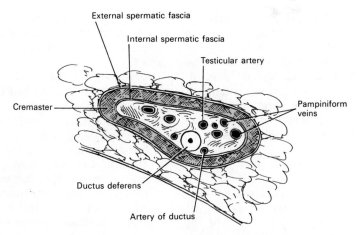

External spermatic fascia

Internal spermatic fascia

Testicular artery

Cremaster

Pampiniform veins

Ductus deferens

Artery of ductus

Fɪɢ. 17-38.   Section across the spermatic cord as it lies in the scrotum canal.

nearer the pubis, it rests upon the inguinal and lacunar ligaments, having the aponeurosis of the obliquus externus ventral to it and the inguinal falx dorsal to it. It then escapes at the ring and descends nearly vertically into the scrotum. The left cord is longer than the right; consequently the left testis hangs somewhat lower than the right.

STRUCTURE OF THE SPERMATIC CORD. The spermatic cord is composed of arteries, veins, lymphatics, nerves, and the excretory duct of the testis. These structures are connected together by the **internal spermatic fascia,** which is continuous with the subserous fascia of the abdomen at the deep inguinal ring; they are invested by the layers brought down by the testis in its descent.

Arteries. The arteries of the cord are the testicular and external spermatic arteries and the artery to the ductus deferens.

The *testicular artery,* a branch of the abdominal aorta, escapes from the abdomen at the deep inguinal ring, and accompanies the other constituents of the spermatic cord along the inguinal canal and through the superficial inguinal ring into the scrotum. It then descends to the testis, and becoming tortuous, divides into several branches, two or three of which accompany the ductus deferens and supply the epididymis, anastomosing with the artery of the ductus deferens. The others supply the substance of the testis.

The *external spermatic artery* is a branch of the inferior epigastric artery. It accompanies the spermatic cord and supplies the coverings of the cord, anastomosing with the testicular artery.

The *artery of the ductus deferens,* a branch of the superior vesical artery is a long, slender vessel that accompanies the ductus deferens, ramifying upon

its coats and anastomosing with the testicular artery near the testis.

Veins. The testicular veins (Fig. 9-32) emerge from the back of the testis and receive tributaries from the epididymis. They unite and form a convoluted plexus, the **plexus pampiniformis,** which forms the chief mass of the cord. The vessels composing this plexus are numerous and ascend along the cord in front of the ductus deferens. Below the superficial inguinal ring they unite to form three or four veins, which pass along the inguinal canal, and entering the abdomen through the abdominal inguinal ring, coalesce to form two veins. These again unite to form a single vein that opens on the right side into the inferior vena cava, at an acute angle, and on the left side into the left renal vein, at a right angle.

The **lymphatic vessels** are described in Chapter 10.

The **nerves** are the spermatic plexus from the sympathetic, joined by filaments from the pelvic plexus, which accompany the artery of the ductus deferens.

Testes

The testes are oval (Fig. 17-37), compressed laterally and suspended in the scrotum by the spermatic cords. They average 4 to 5 cm in length, 2.5 cm in breadth, and 3 cm in the anteroposterior diameter. The weight of one gland varies from 10.5 to 14 gm. Each organ has an oblique position in the scrotum. The cranial extremity is directed ventralward and a little lateralward; the caudal, dorsalward and a little medialward. The anterior convex border faces ventralward and caudalward; the posterior or straight border, to which the cord is attached, dorsalward and cranialward.

The anterior border and lateral surfaces, as well as both extremities of the organ, are convex, free, smooth, and invested by the visceral layer of the tunica vaginalis. The posterior border, to which the cord is attached, receives only a partial investment from that membrane. Lying upon the lateral edge of this posterior border is a long, narrow, flat body named the **epididymis.**

EPIDIDYMIS

The epididymis consists of a central portion or **body;** an upper enlarged extremity, the **head;** and a lower pointed extremity, the **tail,** which is continuous with the **ductus deferens,** the duct of the testis. The head is intimately connected with the upper end of the testis by means of the efferent ductules of the gland; the tail is connected with the lower end by cellular tissue, and a reflection of the tunica vaginalis. The lateral surface, head, and tail of the epididymis are free and covered by the serous membrane; the body is also completely invested by it, except along its posterior border, while between the body and the testis is a pouch, named the **sinus of the epididymis** (*digital fossa*). The epididymis is connected to the back of the testis by a fold of the serous membrane.

APPENDAGES OF THE TESTIS AND EPIDIDYMIS

On the proximal extremity of the testis, just beneath the head of the epididymis, is a minute oval, sessile body, the **appendix of the testis;** it is the remnant of the upper end of the Müllerian duct. On the head of the epididymis is a second small stalked appendage (sometimes duplicated); it is named the **appendix of the epididymis** and is usually regarded as a detached efferent duct.

TUNICS

The testis is invested by three tunics: the **tunica vaginalis, tunica albuginea,** and **tunica vasculosa.**

TUNICA VAGINALIS. The tunica vaginalis is the serous covering of the testis. It is a pouch of serous membrane derived from the saccus vaginalis of the peritoneum, which in the fetus preceded the descent of the testis from the abdomen into the scrotum. After its descent, the portion of the pouch that extends from the deep inguinal ring to near the upper part of the gland becomes obliterated; the lower portion remains as a closed sac that invests the testis, and may be described as consisting of a **visceral** and a **parietal lamina** (Fig. 17-37).

Visceral Lamina. The visceral lamina covers the greater part of the testis and epididymis, connecting the latter to the testis by means of a distinct fold. From the posterior border of the gland it is reflected onto the internal surface of the scrotal coverings.

Parietal Lamina. The parietal lamina is more extensive than the visceral, extending cranialward for some distance ventral and medial to the cord, and reaching below the testis. The inner surface of the tunica vaginalis is smooth and covered by a layer of mesothelial cells. The interval between the visceral and parietal laminae constitutes the cavity of the tunica vaginalis.

The obliterated portion of the saccus vaginalis may generally be seen as a fibrous thread lying in the loose areolar tissue around the spermatic cord. Sometimes this may be traced as a distinct band from the upper end of the inguinal canal, where it is connected with the peritoneum, down to the tunica vaginalis; sometimes it gradually becomes lost on the spermatic cord. Occasionally no trace of it can be detected. In some cases it happens that the pouch of peritoneum does not become obliterated, but the sac of the peritoneum communicates with the tunica vaginalis. This may give rise to one of the varieties of oblique inguinal hernia. In other cases the pouch may contract, but not become entirely obliterated; it then forms a minute canal leading from the peritoneum to the tunica vaginalis.

TUNICA ALBUGINEA. The tunica albuginea is the fibrous covering of the testis. It is a dense membrane with a bluish-white color, composed of bundles of white fibrous tissue that interlace in every direction. It is covered by the tunica vaginalis, except at the points of attachment of the epididymis to the testis and along its posterior border, where the testicular vessels enter the gland. It is applied to the tunica vasculosa over the glandular substance of the testis, and at its posterior border, is reflected into the interior of the gland, forming an incomplete vertical septum called the **mediastinum testis** (*corpus Highmori*[1]).

Mediastinum Testis. The mediastinum

---

[1]Nathaniel Highmore (1613–1685): An English surgeon (Sherborne).

testis extends from the upper to near the lower extremity of the gland, and is wider above than below. From its front and sides numerous imperfect septa (*trabeculae*) are given off, which radiate toward the surface of the organ and are attached to the tunica albuginea. They divide the interior of the organ into a number of incomplete lobules that are somewhat cone-shaped, being broad at their bases at the surface of the gland and becoming narrower as they converge to the mediastinum. The mediastinum supports the vessels and ducts of the testis in their passage to and from the substance of the gland.

TUNICA VASCULOSA. The tunica vasculosa is the vascular layer of the testis, consisting of a plexus of blood vessels held together by delicate areolar tissue. It clothes the inner surface of the tunica albuginea and the different septa in the interior of the gland and therefore forms an internal investment to all the spaces of which the gland is composed.

STRUCTURE. The glandular structure of the testis consists of numerous lobules. Their number, in a single testis, is estimated to be between 250 and 400. They differ in size according to their position, those in the middle of the gland being larger and longer. The lobules (Fig. 17-39) are conical in shape, the base being directed toward the circumference of the organ, the apex toward the mediastinum. Each lobule is contained in one of the intervals between the fibrous septa that extend between the mediastinum

testis and the tunica albuginea, and consists of one to three or more minute convoluted tubes, the **seminiferous tubules.**

Seminiferous Tubules. The tubules may be separately unraveled by careful dissection under water, and may be seen to commence either by free cecal ends or by anastomotic loops (Fig. 17-40). They are supported by loose connective tissue that contains here and there groups of interstitial cells containing yellow pigment granules. The total number of tubules is estimated at 840, and the average length of each is 70 to 80 cm. Their diameter varies from 0.12 to 0.3 mm. The tubules are pale in early life, but in old age they acquire a deep yellow tinge from containing much fatty matter. Each tubule consists of a basement layer formed of laminated connective tissue containing numerous elastic fibers with flattened cells between the layers and covered externally by a layer of flattened epithelioid cells.

Basement Membrane. Within the basement membrane are epithelial cells arranged in several irregular layers, which are not

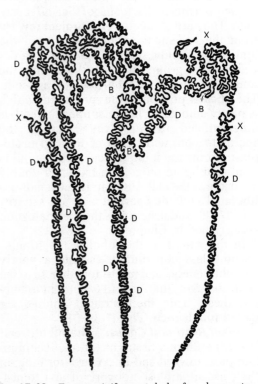

FIG. 17-40. Four seminiferous tubules from human testis showing anastomosing loops. *B*, branching or fork; *D*, diverticulum; *X*, broken end. (Johnson, Anat. Rec., 1934; courtesy of Wistar Institute.)

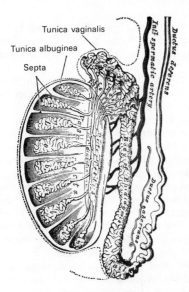

FIG. 17-39. Vertical section of the testis, to show the arrangement of the ducts.

always clearly separated, but which may be arranged in three different groups. Among these cells may be seen the **spermatozoa** in different stages of development.

(1) Lining the basement membrane and forming the outer zone is a layer of cubical cells, with small nuclei; some of these enlarge to become **spermatogonia.** The nuclei of some of the spermatogonia may be in process of mitotic division (*karyokinesis*). As a consequence, daughter cells are formed, which constitute the second zone (Fig. 17-41).

(2) Within this layer, a number of larger polyhedral cells with clear nuclei are arranged in two or three layers; these are the **intermediate cells** or **spermatocytes.** Most of these cells are in a stage of karyokinetic division, and the cells that result from this division form those of the next layer, the **spermatids.**

(3) The third layer of cells consists of the spermatids, and each of these, without further subdivision, becomes a **spermatozoön.** The spermatids are small polyhedral cells, the nucleus of each of which contains half the usual number of chromosomes. In addition to these three layers, other cells, termed the **supporting cells** (*cells of Sertoli*[1]), are seen. They are elongated and project inward from the basement membrane toward the lumen of the tube. As the spermatozoa develop, they group themselves around the inner extremities of the supporting cells. The nuclear portion of the spermatid, which is partly embedded in the supporting cell, is differentiated to form the head of the spermatozoön, while part of the cell protoplasm forms the middle piece and the tail is produced by an outgrowth from the double centriole of the cell. Ultimately the heads are liberated and the spermatozoa are set free (Fig. 17-42). The structure of the spermatozoön is described in Chapter 2.

In the apices of the lobules, the tubules become less convoluted, assume a nearly straight course, and unite together to form 20 to 30 larger ducts, about 0.5 mm in diameter; these, from their straight course, are called **tubuli recti** (Fig. 17-39).

Tubuli Recti and Cones. The **tubuli recti** enter the fibrous tissue of the mediastinum and pass upward and backward, forming, in their ascent, a close network of anastomosing tubes that are merely channels in the fibrous stroma, lined by flattened epithelium and having no proper walls; this constitutes the **rete testis.** At the upper end of the mediastinum, the vessels of the rete testis terminate in 12 to 20 ducts, the **ductuli efferentes;** they perforate the tunica albuginea and carry the seminal fluid from the testis to the epididymis. Their course is at first straight; they then become enlarged and exceedingly convoluted, and form a series of conical masses, the **coni vasculosi,** which together constitute the head of the epididymis. Each cone consists of a single convoluted duct, 15 to 20 cm long, the diameter of which gradually decreases from the testis to the epididymis. Opposite the bases of the cones the efferent vessels open at narrow intervals into a single duct, which constitutes by its complex convolutions the body and tail of the epididymis. When the convolutions of this tube are unraveled, it measures upward of 6 meters in length; it increases in diameter and thickness as it approaches the ductus deferens. The convolutions are held together by fine areolar tissue and by bands of fibrous tissue.

The tubuli recti have very thin walls; like the channels of the rete testis, they are lined by a single layer of flattened epithelium. The walls of the ductuli efferentes and the tube of the epididymis are much thicker owing to the presence of muscular tissue, which is principally arranged in a circular manner. These tubes are lined by columnar ciliated epithelium.

VESSELS AND NERVES.   The **arteries** to the testis are described in Chapter 8; the **veins,** in Chapter 9; the **lymphatics,** in Chapter 10; and the **nerves,** in Chapter 12.

VARIATIONS.   The testis, developed in the lumbar region, may be arrested or delayed in its transit to the scrotum (*cryptorchism*). It may be retained in the abdomen or it may be arrested at the deep inguinal ring or in the inguinal canal. It may just pass out of the superficial inguinal ring without finding its way to the bottom of the scrotum. When retained in the abdomen, it gives rise to no symptoms or signs other than the absence of the testis from the scrotum, but when it is retained in the inguinal canal it is subjected to pressure and may become inflamed and painful. The retained testis is probably functionally useless, so that a man in whom both testes are retained (*anorchism*) is sterile, though he may not be impotent. The absence of one testis is termed *monorchism.* When a testis is retained in the inguinal canal it is often complicated with a congenital

---

[1]Enrico Sertoli (1842–1910): An Italian histologist (Milan).

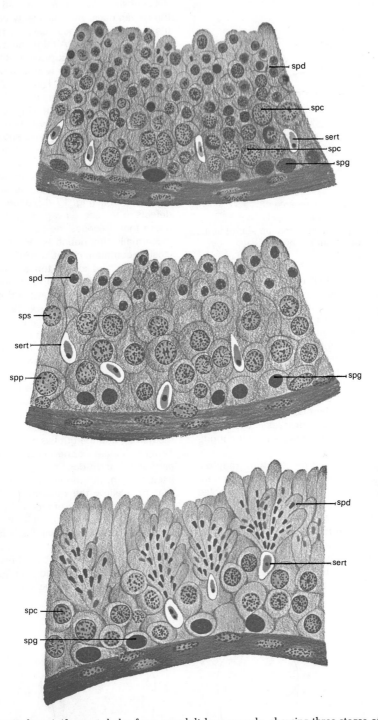

Fig. 17-41. Sections of seminiferous tubules from an adult human male, showing three stages of development. *1.* Spermatogonia and spermatids predominate. *2.* Spermatocytes predominate. *3.* Maturing spermatozoa or spermato- blast predominate. sert., supporting cells (*Sertoli cells*); spc., spermatocytes; spd., spermatids; spg., spermatogonia; spp., primary spermatocytes; sps., secondary spermatocytes. 400 ×. (Redrawn from Sobotta.)

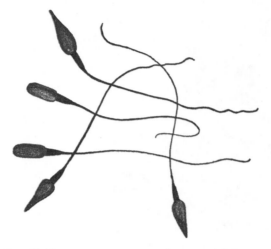

Fig. 17-42. Mature spermatozoa from an adult human male. 1000 ×. (Redrawn from Sobotta.)

hernia, because the funicular process of the peritoneum was not obliterated.

The testis may also descend through the inguinal canal but miss the scrotum and assume some abnormal position. The most common form occurs when the testis, emerging at the subcutaneous inguinal ring, slips down between the scrotum and thigh and comes to rest in the perineum. This is known as *perineal ectopia testis.* With each variety of abnormality in the position of the testis, it is common to find concurrently a congenital hernia, or if a hernia is not actually present, the funicular process is usually patent, almost invariably so if the testis is in the inguinal canal.

The testis, finally reaching the scrotum, may occupy an abnormal position in it. It may be inverted, so that its posterior or attached border is directed forward and the tunica vaginalis is situated behind.

Fluid collections of a serous character are frequently found in the scrotum. To these the term *hydrocele* is applied. The most common form is the ordinary *vaginal hydrocele,* in which the fluid is contained in the sac of the tunica vaginalis, which is normally separated from the peritoneal cavity by the

whole extent of the inguinal canal. In another form, the *congenital hydrocele,* the fluid is in the sac of the tunica vaginalis, but this cavity communicates with the general peritoneal cavity, its tubular process remaining pervious. A third variety, known as an *infantile hydrocele,* occurs when the tubular process becomes obliterated only at its upper part, at or near the deep inguinal ring. It resembles the vaginal hydrocele except in shape, the collection of fluid extending up the cord into the inguinal canal. Fourthly, the funicular process may become obliterated both at the deep inguinal ring and above the epididymis, leaving a central unobliterated portion, which may become distended with fluid, giving rise to a condition known as *encysted hydrocele of the cord.*

CONGENITAL HERNIA. Some varieties of oblique inguinal hernia (Fig. 17-43) depend upon congenital defects in the saccus vaginalis, the pouch of peritoneum that precedes the descent of the testis.

Normally this pouch is closed before birth, with closure commencing at two points: at the deep inguinal ring and at the top of the epididymis, and gradually extending until the whole of the intervening portion is converted into a fibrous cord. Failure in the completion of this process produces variations in the relation of the hernial protrusion to the testis and tunica vaginalis. These constitute distinct varieties of inguinal hernia, viz., the hernia of the funicular process and the complete congenital variety.

When the saccus vaginalis remains patent throughout, the cavity of the tunica vaginalis communicates directly with that of the peritoneum. The intestine descends along this pouch into the cavity of the tunica vaginalis, which constitutes the sac of the hernia, and the gut lies in contact with the testis. Though this form of hernia is termed *complete congenital,* the term does not imply that the hernia existed at birth, but merely that a condition is present that may allow the descent of the hernia at any moment. As a matter of fact, congenital hernias frequently do not appear until adult life.

When the processus vaginalis is occluded at the lower point only, i.e., just above the testis, the intestine descends into the pouch of peritoneum as far as the testis, but is prevented from entering the sac of the tunica vaginalis by the septum that has formed

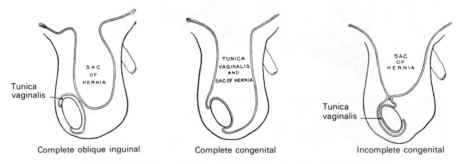

Fig. 17-43. Varieties of oblique inguinal hernia. See also Figure 17-10.

between it and the pouch. This is known as *hernia into the funicular process* or *incomplete congenital hernia;* it differs from the former in that instead of enveloping the testis, it lies above it.

DUCTUS DEFERENS

The **ductus deferens** (*vas deferens; seminal duct*), the excretory duct of the testis, is the continuation of the canal of the epididymis. Commencing at the lower part of the tail of the epididymis, it is at first tortuous, but gradually becomes less twisted as it ascends along the posterior border of the testis and medial side of the epididymis, and traverses the inguinal canal to the deep inguinal ring as a constituent of the spermatic cord (Fig. 17-38). Here it separates from the other structures of the cord, curves around the lateral side of the inferior epigastric artery, and ascends for about 2.5 cm ventral to the external iliac artery. It is next directed dorsalward and slightly caudalward, and crossing the external iliac vessels obliquely, enters the pelvic cavity, where it lies between the peritoneal membrane and the lateral wall of the pelvis. There it descends on the medial side of the obliterated umbilical artery and the obturator nerve and vessels. It then crosses ventral to the ureter, and reaching the medial side of this tube, bends to form an acute angle, and runs medialward and slightly ventralward between the fundus of the bladder and the upper end of the seminal vesicle (Fig. 17-44). Reaching the medial side of the seminal vesicle, it is di-

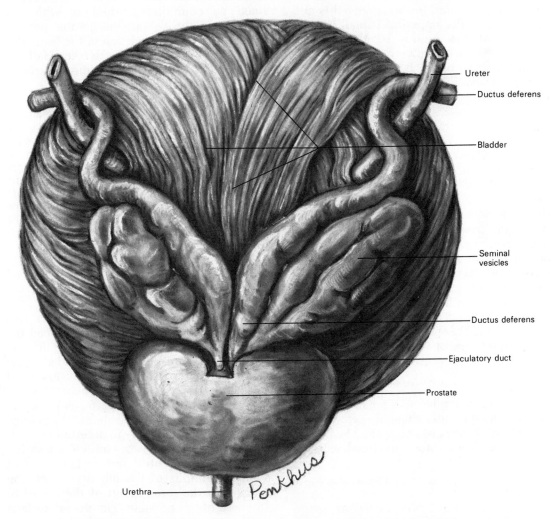

Ureter

Ductus deferens

Bladder

Seminal vesicles

Ductus deferens

Ejaculatory duct

Prostate

Urethra

FIG. 17-44.   The fundus of the bladder showing the seminal vesicles, ductus deferens, ejaculatory ducts, and the prostate.

rected caudalward and medialward in contact with it, gradually approaching the opposite ductus. Here it lies between the fundus of the bladder and the rectum, where it is enclosed, together with the seminal vesicle, in a sheath derived from the rectovesical portion of the subserous fascia. Lastly, it is directed downward to the base of the prostate, where it becomes greatly narrowed, and is joined at an acute angle by the duct of the seminal vesicle to form the ejaculatory duct, which traverses the prostate behind its middle lobe and opens into the prostatic portion of the urethra, close to the orifice of the prostatic utricle.

The ductus deferens presents a hard and cord-like sensation to the palpating fingers; its walls are dense and its canal is extremely small. At the fundus of the bladder it becomes enlarged and tortuous, and this portion is termed the **ampulla.** A small triangular area of the fundus of the bladder, between the ductus deferentes laterally and the peritoneum of the bottom of the rectovesical excavation, is in contact with the rectum.

Ductuli Aberrantes

A long narrow tube, the **ductulus aberrans inferior** (*vas aberrans of Haller[1]*), is occasionally found connected with the lower part of the canal of the epididymis or with the commencement of the ductus deferens. Its length varies from 3.5 to 35 cm, and it may become dilated toward its extremity; more commonly it retains the same diameter throughout. Its structure is similar to that of the ductus deferens. Occasionally it is found unconnected with the epididymis. A second tube, the **ductulus aberrans superior,** occurs in the head of the epididymis; it is connected with the rete testis.

Paradidymis (organ of Giraldés[2])

This term is applied to a small collection of convoluted tubules situated in front of the lower part of the cord above the head of the epididymis. These tubes are lined with co-

[1]Albrecht von Haller (1708–1777): A Swiss physiologist (Bern and Göttingen).
[2]Joachim Albin Cardozo Cazado Giraldés (1808–1875): A Portuguese surgeon (Paris).

lumnar ciliated epithelium and probably represent the remains of a part of the Wolffian body.

Structure

The ductus deferens consists of three coats: an **external** or **areolar coat,** a **muscular coat,** and an **internal** or **mucous coat.**

In the greater part of the tube, the muscular coat consists of two layers of unstriped muscular fiber: an outer longitudinal layer, and an inner circular layer. In addition, at the commencement of the ductus is a third layer, consisting of longitudinal fibers placed internal to the circular stratum, between it and the mucous membrane. The internal or mucous coat is pale and arranged in longitudinal folds. The mucous coat is lined by columnar epithelium, which is nonciliated throughout the greater part of the tube. A variable portion of the testicular end of the tube is lined by two strata of columnar cells, and the cells of the superficial layer are ciliated.

VESSELS AND NERVES.   The **artery of the ductus deferens,** a branch of the superior vesical artery, is a long slender vessel that accompanies the ductus, ramifying upon its coats and anastomosing with the testicular artery near the testis. The ampulla is supplied from the middle and inferior vesical and middle rectal arteries. The **veins** drain into the pampiniform plexus, the vesical veins, and prostatic plexus. The **lymphatics** are described in Chapter 10. The **nerves** are from the pelvic plexus accompanying the artery.

SEMINAL VESICLES

The **seminal vesicles** (Fig. 17-44) are two lobulated membranous pouches, placed between the fundus of the bladder and the rectum, which secrete a fluid to be added to the secretion of the testes. Carefully executed necropsies have shown that in man the seminal vesicles do not store sperm, as the name implies. They are usually about 7.5 cm long, but vary in size, not only in different individuals but also in the same individual on the two sides. The **ventral surface** is in contact with the fundus of the bladder, extending from near the termination of the ureter to the base of the prostate. The **dorsal surface** rests upon the rectum, from which it is separated by the rectovesical fascia. The **upper extremities** of the two vesicles diverge from each other and are in relation with the ductus deferentes and the terminations of the ureters; these are partly covered by peritoneum. The **lower extremities** are pointed and converge toward the base of the prostate. The ampullae of the ductus deferentes lie along their medial margin.

Each vesicle consists of a single tube coiled upon itself and giving off several irregular blind diverticula; the separate coils, as well as the diverticula, are bound together by fibrous tissue. When uncoiled, the tube is about the diameter of a quill (5 mm) and varies in length from 10 to 15 cm. It ends superiorly in a cul-de-sac. The inferior extremity becomes constricted into a narrow straight duct that joins with the corresponding ductus deferens to form the ejaculatory duct.

Structure

The seminal vesicles are composed of three coats: an **external** or **areolar coat;** a **middle** or **muscular coat,** thinner than that in the ductus deferens and arranged in an outer longitudinal and an inner circular layer; and an **internal** or **mucous coat,** which is pale, whitish brown, and has a delicate reticular structure. The epithelium is columnar, and in the diverticula goblet cells are present, the secretion of which increases the bulk of the seminal fluid.

VESSELS AND NERVES. The **arteries** that supply the seminal vesicles are derived from the middle and inferior vesical and middle rectal arteries. The **veins** and **lymphatics** accompany the arteries. The **nerves** are derived from the pelvic plexuses.

EJACULATORY DUCT

The **ejaculatory duct** (Figs. 17-27; 17-45), one on either side of the midline, is formed by the union of the duct from the seminal vesicle with the ductus deferens; it is about 2 cm long. It commences at the base of the prostate, and runs ventralward and caudalward between its middle and lateral lobes and along the sides of the prostatic utricle; it ends in a slit-like orifice close to or just within the margins of the utricle. The ducts diminish in size and also converge toward their terminations.

Structure

The coats of the ejaculatory ducts are extremely thin. They are: an **outer fibrous layer,** which is almost entirely lost after the entrance of the ducts into the prostate; a **layer of muscular fibers** consisting of a thin outer circular layer and an inner longitudinal layer; and **mucous membrane.**

THE PENIS

The **penis** is attached to the front and sides of the pubic arch. In the flaccid condition it is cylindrical, but when erect it assumes the form of a triangular prism with rounded angles, one side of the prism forming the dorsum. It is composed of three cylindrical masses of cavernous tissue bound together by fibrous tissue and covered with skin. The two lateral masses are known as the **corpora cavernosa penis;** the third mass is median and is termed the **corpus spongiosum penis,** which contains the greater part of the urethra (Figs. 17-47; 17-48).

The integument covering the penis is remarkable for its thinness, its dark color, its looseness of connection with the deeper parts of the organ, and its absence of adipose tissue. At the root of the penis it is continuous with that over the pubes, scrotum, and perineum. At the neck it leaves the surface and becomes folded upon itself to form the **prepuce** or foreskin. The internal layer of the prepuce is directly continuous along the line of the neck with the integument over the glans. Immediately behind the external urethral orifice it forms a small secondary reduplication, attached along the bottom of a depressed median raphe, which extends from the meatus to the neck; this fold is termed the **frenulum** of the prepuce.

The integument covering the glans is continuous with the urethral mucous membrane at the orifice; it is devoid of hairs, but projecting from its free surface are a number of small, sensitive papillae. Scattered **preputial glands** (of Tyson[1]) are present on the neck of the penis and inner layer of the prepuce. They secrete a sebaceous material with a peculiar odor that readily undergoes decomposition; when mixed with discarded epithe-

[1]Edward Tyson (1650–1708): An English physician and anatomist (London).

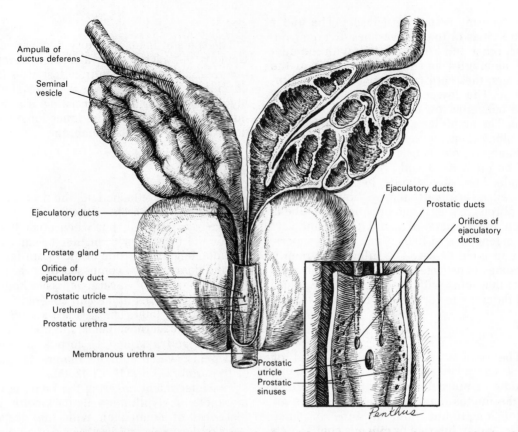

Fig. 17-45. Anterior view of the seminal vesicles, the ampullae of the ductus deferentes, the prostate gland and the prostatic urethra. The anterior walls of the left ampulla, left seminal vesicle and the prostatic urethra have been cut away. The lower right figure shows a more detailed view of the prostatic urethra and the orifices of the prostatic and ejaculatory ducts.

lial cells it is called **smegma.** The prepuce covers a variable amount of the glans and is separated from it by a potential space—the **preputial space**—which has two shallow fossae, one on either side of the frenulum.

Fascia

The **subcutaneous fascia of the penis** is directly continuous with that of the scrotum, and like it, contains a **dartos** tunic with its layer of scattered smooth muscle cells. It is not divisible into a superficial and a deep layer and it contains no adipose tissue. A fascial cleft between the superficial and deep fasciae gives the skin great movability.

The **deep fascia of the penis** (*Buck's fascia*[1]) (Fig. 17-46) forms a tubular investment for the shaft of the penis as far distally as the

[1] Gurdon Buck (1807–1877): An American surgeon (New York).

corona glandis. Proximally, it invests the crura and bulb and is firmly attached with them to the ischiopubic rami and the superficial layer of the urogenital diaphragm. At the anterior or distal extremities of the bulbospongiosus and ischiocavernosi muscles, it splits into a superficial and deep lamina. The superficial lamina covers the superficial surface of these muscles as the external perineal fascia of the perineum; the deep lamina is the continuation of the proper deep fascia of the penis (Buck's fascia). A septum of fascia extends inward between the corpora cavernosa and corpus spongiosum penis, providing separate tubular investments for these columns of erectile tissue.

CLINICAL CONSIDERATIONS. The deep fascia of the penis (Buck's fascia) encloses the organ in a strong capsule. An abscess, a hematoma, or an extravasation of urine from rupture of the penile urethra would be confined to the penis by this envelope. A rupture of the urethra in its membranous portion,

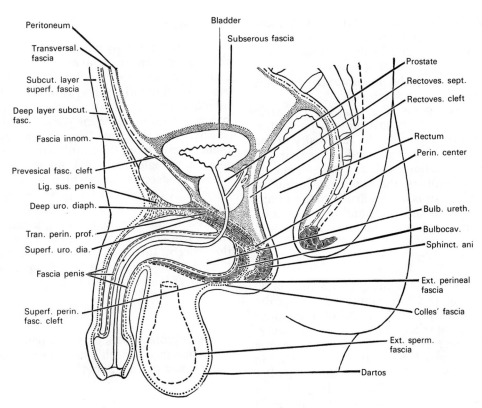

FIG. 17-46. Fasciae of pelvis and perineum in median sagittal section. Diagram.

however, would allow urine to enter the fascial cleft that is between the deep and superficial fasciae and is continuous with the cleft in the scrotum, under Colles' fascia, and under Scarpa's fascia.

Corpora Cavernosa Penis

The distal three-fourths of these two cylindrical masses of erectile tissue are intimately bound together and make up the greater part of the shaft of the penis (Fig. 17-47). At the pubic symphysis, however, their posterior portions diverge from each other as two gradually tapering structures called the **crura.**

The corpora retain a uniform diameter in the shaft and terminate anteriorly in a bluntly rounded extremity approximately 1 cm from the end of the penis, being embedded in a cap formed by the glans penis.

The corpora cavernosa penis are surrounded by a strong fibrous envelope consisting of superficial and deep fibers. The superficial fibers have a longitudinal direction and form a single tube that encloses both corpora. The deep fibers are arranged circularly around each corpus, and form the

septum of the penis by their junction in the median plane. This is thick and complete behind but imperfect in front, where it consists of a series of vertical bands arranged like the teeth of a comb; it is therefore

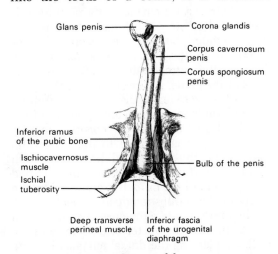

FIG. 17-47. The ventral aspect of the corpora cavernosa penis and the corpus spongiosum penis. These cylindrical masses of tissue are shown in the erect position with the glans and distal part of the corpus spongiosum detached from the corpora cavernosa and turned to the right.

named the **septum pectiniforme.** A shallow groove that marks their junction on the upper surface lodges the deep dorsal vein of the penis, while a deeper and wider groove between them on the undersurface contains the corpus spongiosum penis.

Each **crus penis,** the tapering posterior portion of a corpus cavernosum penis, terminates just ventral to the tuberosity of the ischium in a bluntly pointed process, and as it meets its fellow, it has a slight enlargement, the bulb of the corpus cavernosum penis. The crus is bound firmly to the ramus of the ischium and pubis and is enclosed by the fibers of the ischiocavernosus muscle.

Corpus Spongiosum Penis

This part of the penis contains the penile urethra (Fig. 17-47). Its middle portion, in the shaft of the penis, is a uniform cylinder somewhat smaller than a corpus cavernosum penis. At each end it expands, the distal extremity forming the glans penis, the proximal forming the **bulb of the penis.** Between the expansions, it lies in the groove on the undersurface of the corpora cavernosa penis.

The **glans penis** is the anterior end of the corpus spongiosum expanded into an obtuse cone similar to the cap of a mushroom. It is molded over and securely attached to the blunt extremity of the corpora cavernosa penis, and extends farther over their dorsal than their ventral surfaces. Its periphery is larger in diameter than the shaft, projecting in a rounded border, the **corona glandis.** Proximal to the corona is a constriction forming the **retroglandular sulcus** and the **neck of the penis.** At the summit of the glans is the slit-like external orifice of the urethra.

The **bulb** is the conical enlargement of the proximal 4 or 5 cm of the corpus spongiosum. It is just superficial to the urogenital diaphragm, the superficial layer of which blends with its fibrous capsule and is called the ligament of the bulb. It is enclosed by the fibers of the bulbospongiosus muscle.

The urethra enters the corpus spongiosum 1 or 2 cm from the posterior extremity of the bulb by piercing the dorsal surface, i.e., the surface that blends with the urogenital diaphragm. The posterior most expanded portion of the bulb, accordingly, projects backward toward the anus beyond the entrance of the urethra.

Ligaments

The **fundiform ligament** is an extensive thickening of the deep layer of subcutaneous fascia (Scarpa's[1]) of the anterior abdominal wall just above the pubis where it is firmly attached to the rectus sheath. The fibrous bands extend down to the dorsum and sides of the root of the penis. The **suspensory ligament,** shorter than the fundiform, is a strong fibrous triangle derived from the external investing deep fascia, which attaches the dorsum of the root of the penis to the inferior end of the linea alba, the symphysis pubis, and the arcuate pubic ligament. Serving as ligaments also are the attachments of the crura to the ischiopubic rami and of the bulb to the urogenital diaphragm, which have been described previously.

Muscles

The voluntary muscles of the penis are the bulbospongiosus, ischiocavernosus, and the transversus perinei superficialis.

Structure of the Penis

From the internal surface of the fibrous envelope, tunica albuginea, of the corpora cavernosa penis, as well as from the sides of the septum, numerous bands or cords are given off that cross the interior of the corpora cavernosa in all directions, subdividing them into separate compartments and giving the entire structure a spongy appearance (Fig. 17-48). These bands and cords are called **trabeculae;** they consist of white fibrous tissue, elastic fibers, and plain muscular fibers. In them are contained numerous arteries and nerves.

The component fibers that form the trabeculae are larger and stronger around the circumference than at the centers of the corpora cavernosa; they are also thicker proximally than distally. The interspaces or cavernous spaces (blood sinuses), on the contrary, are larger at the center than at the circumference, their long diameters being directed transversely. They contain blood and are lined by a layer of flat cells similar to those in the endothelial lining of veins.

The fibrous envelope of the corpus spongiosum is thinner, whiter, and more elastic

[1]Antonio Scarpa (1747–1832): An Italian anatomist (Pavia).

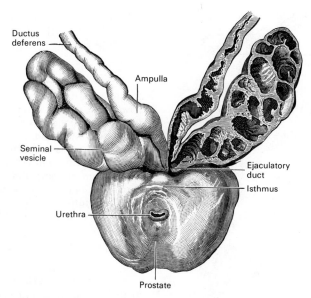

Ductus deferens

Ampulla

Seminal vesicle

Ejaculatory duct

Isthmus

Urethra

Prostate

FIG. 17-52.   Prostate with seminal vesicles and seminal ducts, viewed ventrally and cranially. (Spalteholz.)

sponds in origin and fate to the processus vaginalis in the inguinal region. It is about 4 cm distant from the anus.

Near its cranial border is a depression through which the two ejaculatory ducts enter the prostate. This depression divides the posterior surface into a lower larger and an upper smaller part. The cranial smaller part constitutes the **middle lobe** of the prostate and intervenes between the ejaculatory ducts and the urethra; it varies greatly in size, and in some cases is destitute of glandular tissue. The caudal larger portion sometimes has a shallow median furrow, which imperfectly separates it into a **right** and a **left lateral lobe.** These form the main mass of the gland and are directly continuous with each other dorsal to the urethra. Cranial to the urethra the lobes are connected by a band named the **isthmus,** which consists of the same tissues as the capsule and is devoid of glandular substance.

ANTERIOR SURFACE. The anterior surface (*ventral surface; pubic surface*) measures about 2.5 cm craniocaudally but it is narrow and convex from side to side. It is placed about 2 cm dorsal to the pubic symphysis, from which it is separated by a plexus of veins and a quantity of loose fat. It is connected to the pubic bone on either side by the **puboprostatic ligaments.** The urethra emerges from this surface a little cranial and ventral to the apex of the gland.

LATERAL SURFACES. The lateral surfaces are prominent and are covered by the ventral portions of the levatores ani which, however, are separated from the gland by a plexus of veins.

Structure

The prostate is immediately enveloped by a thin but firm fibrous capsule distinct from that derived from the subserous fascia and separated from it by a plexus of veins. This capsule adheres firmly to the prostate and is structurally continuous with the stroma of the gland, being composed of the same tissues: nonstriped muscle and fibrous tissue. The substance of the prostate is dense, a pale reddish-gray, and not easily torn. It consists of glandular substance and muscular tissue (Fig. 17-53).

The **muscular tissue** constitutes the proper stroma of the prostate; the connective tissue being scanty and simply forming, between the muscular fibers, thin trabeculae in which the vessels and nerves of the gland ramify. The muscular tissue is arranged as follows: immediately beneath the fibrous capsule is a dense layer that forms an investing sheath for the gland; around the urethra, as it lies in the prostate, is another dense layer of circular fibers that is continuous above with the internal layer of the muscular coat of the bladder and blends below

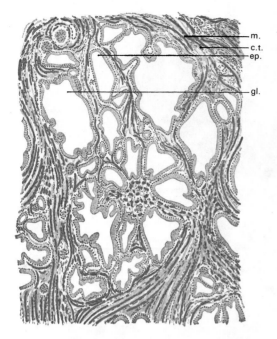

FIG. 17-53. Section through prostate of a 22-year-old man. *c.t.*, Connective tissue stroma; *ep.*, glandular epithelium; *gl.*, lumen of gland; *m.*, smooth muscle fiber in stroma. 50 ×. (Sobotta.)

with the fibers surrounding the membranous portion of the urethra. Between these two layers, strong bands of muscular tissue, which decussate freely, form meshes in which the glandular structure of the organ is embedded. In the part of the gland that is situated in front of the urethra the muscular tissue is especially dense, and there is little or no gland tissue, while in the part that is behind the urethra, the muscular tissue presents a wide-meshed structure, which is densest at the base of the gland (i.e., near the bladder) and becomes looser and more sponge-like toward the apex of the organ.

The **glandular substance** is composed of numerous follicular pouches, the lining of which frequently shows papillary elevations. The follicles open into elongated canals that join to form 12 to 20 small excretory ducts. They are connected by areolar tissue, supported by prolongations from the fibrous capsule and muscular stroma, and enclosed in a delicate capillary plexus. The epithelium that lines the canals and the terminal vesicles is columnar. The prostatic ducts open into the floor of the prostatic portion of the urethra and are lined with two layers of epithelium, the inner layer consist-

ing of columnar and the outer of small cubical cells. Small colloid masses, known as **amyloid bodies,** are often found in the gland tubes.

VESSELS AND NERVES. The **arteries** supplying the prostate are derived from the internal pudendal, inferior vesical, and middle rectal arteries. Its **veins** form the prostatic plexus, which surrounds the sides and base of the gland. The plexus receives the dorsal vein of the penis and empties into the internal iliac veins. It has anastomotic connections with the vertebral system of veins, through which metastases of carcinoma may reach the bones or even the brain. The **lymphatics** are described in Chapter 10.

The **nerves** are derived from the pelvic plexus.

BULBOURETHRAL GLANDS

The **bulbourethral glands** (*Cowper's glands*) (Fig. 17-27) are two small, round, and somewhat lobulated yellow bodies, about the size of peas, placed dorsal and lateral to the membranous portion of the urethra between the two layers of the fascia of the urogenital diaphragm. They lie close to the bulb and are enclosed by the transverse fibers of the sphincter urethrae.

The excretory duct of each gland, nearly 2.5 cm long, passes obliquely forward beneath the mucous membrane and opens by a minute orifice on the floor of the cavernous portion of the urethra about 2.5 cm in front of the urogenital diaphragm.

Structure

Each gland is made up of several lobules held together by a fibrous investment. Each lobule consists of acini, lined by columnar epithelial cells, opening into one duct, which joins with the ducts of other lobules outside the gland to form the single excretory duct.

Female Genital Organs

The **female genital organs** consist of an internal and an external group. The **internal organs** are situated within the pelvis and consist of the **ovaries,** the **uterine tubes,** the **uterus,** and the **vagina.** The **external organs** are superficial to the urogenital diaphragm and below the pubic arch. They comprise the **mons pubis,** the **labia majora** and **minora pudendi,** the **clitoris,** the **vestibule,** the **bulb of the vestibule,** and the **greater vestibular glands.**

OVARIES

The **ovaries** are homologous with the testes in the male. They are two nodular bodies, situated one on either side of the uterus in relation to the lateral wall of the pelvis and attached to the broad ligament of the uterus, dorsal and caudal to the uterine tubes (Fig. 17-54). The ovaries are grayish-pink and have either a smooth or a puckered, uneven surface. They are each about 4 cm long, 2 cm wide, and about 8 mm thick, and weigh 2 to 3.5 gm. Each ovary has a lateral and a medial surface, a cranial or tubal and a caudal or uterine extremity, and a ventral or mesovarian and a dorsal free border. It lies in a shallow depression, named the **ovarian fossa,** on the lateral wall of the pelvis; this fossa is bounded by the external iliac vessels, the obliterated umbilical artery, and the ureter. The exact position of the ovary has been the subject of considerable differences of opinion, and the description given here refers to the ovary of the nulliparous woman. The ovary becomes displaced during the first pregnancy and probably never again returns to its original position. In the erect posture the long axis of the ovary is vertical.

The *tubal extremity* is near the external iliac vein. Attached to it are the ovarian fimbriae of the uterine tube and a fold of peritoneum, the **suspensory ligament of the ovary,** which is directed cranialward over the iliac vessels and contains the ovarian vessels. The *uterine end* is directed caudalward toward the pelvic floor; it is usually narrower than the tubal end and is attached to the lateral angle of the uterus, immediately behind the uterine tube, by a round cord termed the **ligament of the ovary,** which lies within the broad ligament and contains some nonstriped muscular fibers.

The *lateral surface* is in contact with the parietal peritoneum, which lines the ovarian fossa; the *medial surface* is covered largely by the fimbriated extremity of the uterine tube. The *mesovarian border* is straight and directed toward the obliterated umbilical artery; it is attached to the dorsal surface of the broad ligament by a short fold named the **mesovarium.** Between the two layers of this fold, the blood vessels and nerves pass to the hilum of the ovary. The *free border* is convex and directed toward the ureter. The uterine tube arches over the ovary, runs along its mesovarian border, then curves over its tubal pole and finally passes along its free border and medial surface.

Epoöphoron

The epoöphoron (*parovarium; organ of Rosenmüller*) (Fig. 17-55) lies in the mesosalpinx between the ovary and the uterine tube, and consists of a few short tubules **(ductuli**

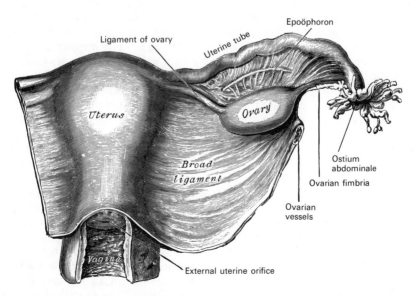

FIG. 17-54. Uterus and right broad ligament, seen dorsally. The broad ligament has been spread out and the ovary drawn caudalward.

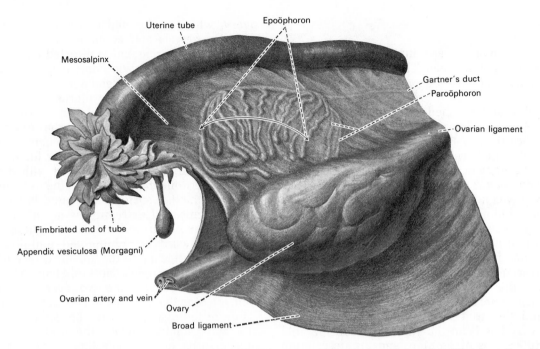

FIG. 17-55. Broad ligaments of adult, showing remains of the Wolffian body and duct. (Modified from Farre.)

transversi), which converge toward the ovary while their opposite ends open into a rudimentary duct, the **ductus longitudinalis epoöphori** (*duct of Gartner*).

Paroöphoron

The paroöphoron consists of a few scattered rudimentary tubules, best seen in the child, situated in the broad ligament between the epoöphoron and the uterus. The ductuli transversi of the epoöphoron and the tubules of the paroöphoron are remnants of the tubules of the Wolffian body or mesonephros; the ductus longitudinalis epoöphori is a persistent portion of the mesonephric or Wolffian duct. In the fetus the ovaries are situated, like the testes, in the lumbar region, near the kidneys, but they gradually descend into the pelvis.

Structure

The surface of the ovary is covered by a layer of columnar cells, which constitutes the **germinal epithelium** (*of Waldeyer*[1]). This

[1]Heinrich Wilhelm Gottfried Waldeyer (1836–1921): A German anatomist (Breslau and Berlin).

epithelium, which is in linear continuity with the peritoneum, gives to the ovary a dull gray color compared with the shining smoothness of the peritoneum, and the transistion between the squamous epithelium of the peritoneum and the columnar cells that cover the ovary is marked by a line near the attached border of the ovary. The ovary consists of a number of vesicular ovarian follicles embedded in the meshes of a stroma or framework (Fig. 17-56).

STROMA. The stroma is a peculiar soft tissue consisting mostly of spindle-shaped cells with a small amount of ordinary connective tissue abundantly supplied with blood vessels. At the surface of the organ, this tissue is much condensed and forms a layer **(tunica albuginea)** composed of short connective tissue fibers, with fusiform cells between them.

OVARIAN FOLLICLES. The stroma immediately beneath the tunica albuginea, containing a large number of minute vesicles about 35 μm in diameter, constitutes the **cortex** of the ovary. These vesicles are the **primary follicles** (Fig. 17-56). They contain an ovum about 20 μm in diameter surrounded by a single layer of follicular cells. When these follicles begin to mature, the ovum swells and the follicular cells become larger and

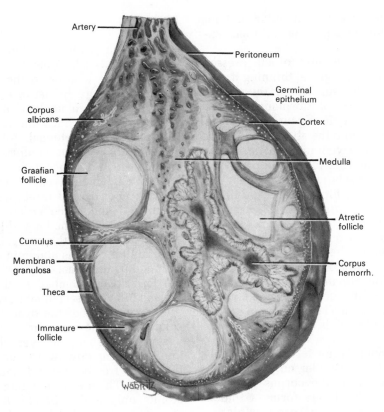

Fig. 17-56. Section through ovary of 23-year-old woman. Lightly stained with hematoxylin. 5 ×.

proliferate by mitosis. When the follicular cells form many layers, a cavity filled with fluid forms at one side. This is the beginning of **vesicular** or **ripening follicles.** These recede from the surface toward the highly vascular stroma in the center of the organ, termed the **medullary substance** (*zona vasculosa of Waldeyer*). This stroma forms the tissue of the hilum by which the ovary is attached and through which the blood vessels enter; it does not contain follicles.

When the follicle is mature, it forms a clear vesicle 10 to 12 mm in diameter and is known as a **Graafian[1] follicle.** The follicular cells are arranged in a layer three or four cells thick around the large accumulation of follicular fluid. At one side of the follicle they form a mass of cells protruding into the cavity, constituting the **cumulus oophorus.** The ovum, now about 100 μm in diameter, is in the center of the cumulus, with the layer of follicular cells immediately surrounding

arranged in a regular row known as the **corona radiata.**

THECA FOLLICULI. During the stages of growth of the follicle, the cells of the stroma around it become flattened into a sheath or theca (Fig. 17-57). When the follicle is ma-

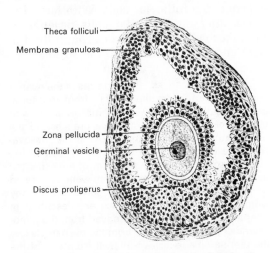

Fig. 17-57. Section of a vesicular ovarian follicle. 50 ×.

[1]Reijnier de Graaf (1641–1673): A Dutch physician and anatomist (Delft).

ture, the theca develops an outer and inner stratum, the **theca interna** and **theca externa.**

OVULATION. The mature follicle presses against the tunica albuginea, thinning it and bulging out on the surface of the ovary. Under the influence of hormones of the hypophysis, the follicle ruptures, allowing the fluid to escape into the abdominal cavity and the ovum to be swept into the fimbriated opening of the uterine tube. The cavity of the follicle collapses but contains some fluid tinged with blood, which gives it the name **corpus hemorrhagicum** (Fig. 17-56). The remaining follicular cells become greatly enlarged and take on a yellow color, forming the **corpus luteum.** The cells of the theca interna also enlarge, forming *theca lutein cells* in contrast to the *follicular lutein cells.* If the ovum is fertilized, the corpus luteum continues to grow, attaining a size of 30 mm by the end of nine months. This is a **corpus luteum graviditatis.**

When the ovum is not fertilized, the follicle forms a much smaller **corpus luteum menstruationis.** After the corpus degenerates, the connective tissue forms a folded whitish scar known as a **corpus albicans** (Fig. 17-56).

The periodic changes in the ovary and uterus during the menstrual cycle are discussed later.

ATRESIA OF FOLLICLES. Many of the follicles do not reach full maturity and rupture; they undergo instead a degenerative process known as atresia. In smaller follicles, the ovum disintegrates and is absorbed with the degenerating follicular cells. With large follicles, after the follicular cells degenerate, the theca cells form a layer around the follicle and are absorbed into the stroma.

VESSELS AND NERVES. The **arteries** of the ovaries and uterine tubes are the ovarian from the aorta. Each anastomoses freely in the mesosalpinx, with the uterine artery giving off some branches to the uterine tube and others that traverse the mesovarium and enter the hilum of the ovary. The **veins** emerge from the hilum in the form of a plexus, the **pampiniform plexus;** the ovarian vein is formed from this plexus, and leaves the pelvis in company with the artery. The **lymphatics** are described in Chapter 10. The **nerves** are derived from the hypogastric or pelvic plexus and from the ovarian plexus; the uterine tube receives a branch from one of the uterine nerves.

UTERINE TUBE

The uterine tubes (*Fallopian tube*[1]; *oviduct*) (Figs. 17-58; 17-59; 17-61) convey the ova from the ovaries to the cavity of the uterus. They are bilateral, extending from the superior lateral angle of the uterus to the side of the pelvis. Each one is suspended by a mesenteric peritoneal fold called the **mesosalpinx,** which comprises the upper free margin and adjacent movable portion of the broad ligament. Each tube is about 10 cm long and consists of three portions: (1) the **isthmus,** or medial constricted third; (2) the **ampulla,** or intermediate dilated portion, which curves over the ovary; and (3) the **infundibulum** with its **abdominal ostium,** surrounded by **fimbriae,** one of which, the **ovarian fimbria,** is attached to the ovary. The uterine tube is directed lateralward as far as the uterine pole of the ovary, and then ascends along the mesovarian border of the ovary to the tubal pole, over which it arches; finally it turns downward and ends in relation to the free border and medial surface of the ovary. The uterine opening is minute and admits only a fine bristle; the abdominal opening is somewhat larger. Frequently one or more small pedunculated vesicles are connected with the fimbriae of the uterine tube or with the broad ligament close to them. These are termed the **appendices vesiculosae** (*hydatids of Morgagni*).

Structure

The uterine tube consists of three coats: **serous, muscular,** and **mucous.** The **external** or **serous coat** is peritoneal. The **middle** or **muscular coat** consists of an external longitudinal and an internal circular layer of nonstriped muscular fibers continuous with those of the uterus. The **internal** or **mucous coat** is continuous with the mucous lining of the uterus, and at the abdominal ostium of the tube, with the peritoneum. It is thrown into longitudinal folds, **plicae tubariae,** which are much more extensive in the ampulla than in the isthmus. The lining epithelium is columnar and ciliated. This form of epithelium is also found on the inner surface of the fimbriae, while on the outer or serous

[1] Gabriele Fallopio (1523–1562): An Italian anatomist and student of Vesalius (Padua).

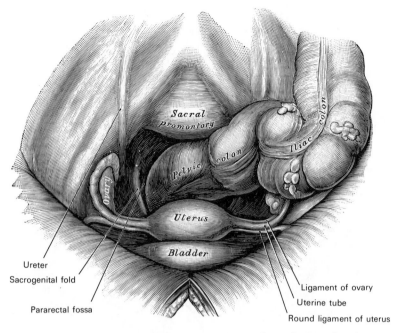

Fɪɢ. 17-58. Female pelvis and its contents, seen from above and in front.

surfaces of these processes the epithelium merges into the mesothelium of the peritoneum.

Fertilization of the ovum is believed to occur in the tube (see Chapter 2), and the fertilized ovum then normally passes into the uterus. The ovum, however, may adhere to and develop in the uterine tube, giving rise to the commonest variety of *ectopic gestation*. In such cases the amnion and chorion are formed, but a true decidua is never present. The gestation usually ends by extrusion of the ovum through the abdominal ostium, although the tube may rupture into the peritoneal cavity; this is accompanied by severe hemorrhage and requires surgical intervention.

VESSELS AND NERVES. The **arteries** that supply the uterine tube are branches of the uterine and ovarian arteries. One branch from the uterine artery, much stronger than the others, has special surgical importance (Sampson's[1] artery). It arises just before the uterine artery anastomoses with the ovarian artery, and continues along the tube to its fimbriated extremity, giving off branches along its course. The **veins** drain into the uterine plexus. The **lymphatics**

[1]John Albertson Sampson (1873–1943): An American gynecologist (Albany).

drain partly with the uterine and partly with the ovarian vessels.

The **nerves** are derived from the pelvic plexus of sympathetic and parasympathetic sacral nerves.

THE UTERUS

The **uterus** (*womb*) (Figs. 17-54; 17-58 to 17-60) is a thick-walled, hollow, muscular organ situated in the pelvis between the bladder and the rectum. The uterine tubes open into its upper abdominal part, and at its perineal end its cavity communicates with that of the vagina. It changes remarkably in size, structure, and position during pregnancy, and although it returns again almost to its former condition, the following description is based on the uterus of an adult woman who has never been pregnant.

The uterus is pear-shaped, somewhat flattened dorsoventrally, and elongated, with its long axis parallel with the median plane but curved to correspond with the axis of the pelvis. It measures about 7.5 cm in length, 5 cm in breadth at its cranial part, and nearly 2.75 cm in thickness; it weighs 30 to 40 gm.

The uterus is divided structurally and functionally into two parts, the body and the cervix. On the surface, about midway be-

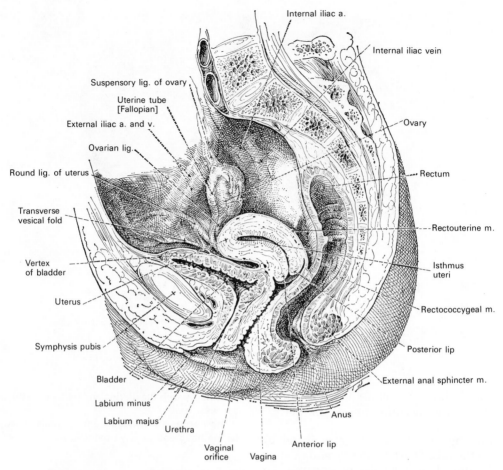

Internal iliac a.

Internal iliac vein

Suspensory lig. of ovary

Uterine tube
[Fallopian]

External iliac a. and v.

Ovarian lig.

Round lig. of uterus

Transverse
vesical fold

Vertex
of bladder

Uterus

Symphysis pubis

Bladder

Labium minus

Labium majus

Urethra

Vaginal
orifice

Vagina

Ovary

Rectum

Rectouterine m.

Isthmus
uteri

Rectococcygeal m.

Posterior lip

External anal sphincter m.

Anus

Anterior lip

FIG. 17-59. Median sagittal section of female pelvis. The bladder is empty; the uterus and vagina are slightly dilated. (Eycleshymer and Jones.)

tween the two ends, is a slight constriction known as the **isthmus.** Corresponding to this on the inside is a narrowing of the cavity, formerly known as the **internal orifice** of the uterus. The part cranial to the isthmus is the body; caudal to it, the cervix.

Body

The part of the body (*corpus uteri*) that extends toward the abdomen as a free, rounded extremity above the entrance of the uterine tubes is called the **fundus.** The body gradually narrows from the fundus to the isthmus (Fig. 17-61).

The ventral or **vesical surface** faces toward the bladder, and the peritoneum covering its surface is reflected, at the junction with the cervix, into the vesicouterine excavation and thence to the bladder.

The dorsal or **intestinal surface** is more

convex and the peritoneum covering its surface is more extensive than that of the vesical surface. It extends over the cervix and over the cranial portion of the vagina. The peritoneal excavation between it and the rectum is the rectouterine fossa and is usually occupied by coils of small intestine.

The **fundus** is convex in all directions and its free surface, covered by peritoneum, is usually in contact with coils of small intestine or a distended sigmoid colon.

The **lateral margins** are slightly convex. At the upper end of each, the uterine tube joins the uterine wall. Ventral and caudal to this point the **round ligament of the uterus** is fixed, while dorsal to it is the attachment of the **ligament of the ovary.** These three structures lie within a fold of peritoneum that is reflected from the margin of the uterus to the wall of the pelvis, and is named the **broad ligament.**

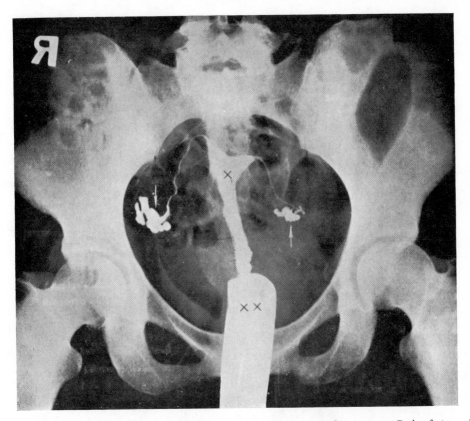

FIG. 17-60. Genital tract in the female after an injection of barium sulfate into the uterus. ⨯, Body of uterus. Note the two cornua leading to the uterine tubes. ⨯⨯, Speculum in vagina. The arrows indicate the infundibula of the uterine tubes. Some of the barium has passed through the pelvic opening of the tube into the general peritoneal cavity.

Cervix

The cervix (*neck*) is the portion of the uterus between the isthmus and the vagina. It is about 2 cm long, and the end toward the perineum protrudes into the cavity of the vagina as a free extremity. The attachment of the vagina around its periphery divides it into a vaginal and a supravaginal portion.

The **supravaginal portion** is separated ventrally from the bladder by fibrous tissue **(parametrium)**, which extends also to its sides and lateralward between the layers of the broad ligaments. The uterine arteries reach the margins of the cervix in this fibrous tissue, where they cross over the ureters. The ureters, on either side, run ventralward in the parametrium about 2 cm from the cervix. Dorsally, the supravaginal cervix is covered by peritoneum, which is prolonged caudalward on the posterior vaginal wall before it is reflected over to the rectum.

The **vaginal portion** of the cervix projects into the cavity of the vagina, the free end resting against the dorsal wall. On its rounded extremity is a round or oval depression that marks the **uterine orifice** (*external os*) leading into the cavity of the cervix. The orifice is bounded by two lips, the **ventral lip** and the **dorsal lip.** Between the dorsal lip and the vaginal wall is the **posterior fornix,** the most cranial part of the vaginal cavity, and between the ventral lip and the vaginal wall is the **anterior fornix** (Fig. 17-59).

The **cavity of the uterus** is compressed dorsoventrally into a mere slit. The orifices of the uterine tubes and the internal surface of the fundus form the base of a triangle whose apex is the isthmus, leading into the canal of the cervix. The total length of the uterine cavity from the external orifice to the fundus is about 6.25 cm.

Canal of Cervix

The canal of the cervix (Fig. 17-61) is somewhat flattened and wider in the middle than at either extremity. It communicates

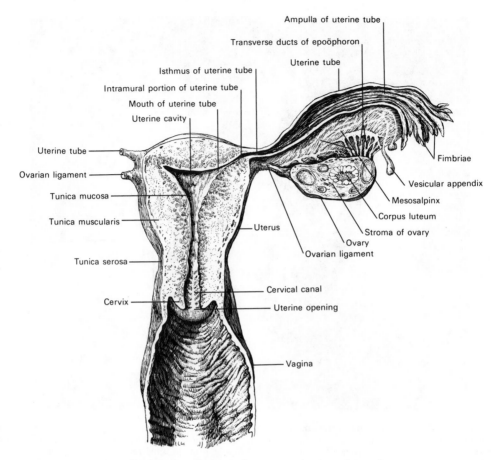

Ampulla of uterine tube

Transverse ducts of epoöphoron

Uterine tube

Isthmus of uterine tube

Intramural portion of uterine tube

Mouth of uterine tube

Uterine cavity

Uterine tube

Ovarian ligament

Tunica mucosa

Tunica muscularis

Tunica serosa

Cervix

Fimbriae

Vesicular appendix

Mesosalpinx

Corpus luteum

Stroma of ovary

Ovary

Ovarian ligament

Uterus

Cervical canal

Uterine opening

Vagina

FIG. 17-61.   Section through the vagina, uterus, uterine tube, and ovary.

through the isthmus with the cavity of the uterine body, and through the external orifice with the vaginal cavity. The wall of the canal has an anterior and a posterior longitudinal ridge, from each of which proceed a number of small oblique columns, the **palmate folds,** which give the appearance of branches from the stem of a tree. To this arrangement, the name **arbor vitae uterina** is applied. The folds on the two walls are not opposed exactly, but fit between one another to close the cervical canal.

Ligaments

The principal ligaments of the uterus are the **broad ligaments,** the **round ligaments,** the **uterosacral ligaments,** and the **cardinal ligaments.** In addition, certain peritoneal folds are called ligaments: the **vesicouterine, rectouterine,** and **sacrogenital ligaments.**

BROAD LIGAMENTS. The broad ligaments (Figs. 17-54; 17-55) are two fibrous sheets, cov-

ered on both surfaces with peritoneum, that extend from each side of the uterus to the lateral wall and bottom of the pelvis. The uterus and these lateral extensions together form a septum across the cavity of the pelvis, dividing it into a **vesicouterine** and a **rectouterine fossa** (Figs. 17-58; 17-59, not labeled). The broad ligament is thicker at its inferior pelvic attachment than at its free border. Between the two peritoneal sheets or *leaves of the ligament* are:

1. Parametrium
2. Uterine artery
3. Uterine tube
4. Round ligament
5. Epoöphoron and paroöphoron
6. (Ovary)
7. (Ureter)

1. The **parametrium** is the extension of the subserous connective tissue of the uterus laterally into the broad ligament. The name has been applied also to the whole broad lig-

ament below the attachment of the ovary, but sometimes the latter is referred to as the **mesometrium.** It contains scattered smooth muscle bundles, is anchored to the lateral pelvic wall, and is continuous with the cardinal ligament of the intrapelvic fascia.

2. The **uterine artery** of each side enters the base of the broad ligament at the latter's attachment to the lateral wall of the pelvis and traverses the pelvis between the two leaves close to their reflection to the pelvic floor. It crosses the ureter just before it reaches the cervical portion of the uterus, and in the parametrium it follows the lateral border of the uterus up to the isthmus of the uterine tube. It follows the tube laterally and anastomoses with the ovarian artery. The **ovarian artery** crosses the external iliac vessels in a vertical direction, enters the most superior lateral portion of the broad ligament enclosed in a fibrous cord, the **suspensory ligament of the ovary** (*infundibulopelvic ligament*), and follows the attached border of the ovary to the anastomoses with the uterine artery.

3. The superior free border of the broad ligament is occupied by the **uterine tube,** except at its lateral extremity, where it forms the suspensory ligament of the ovary which also attaches the infundibulum of the tube to the lateral wall of the pelvis. The free portion of the broad ligament extending down as far as the attachment of the ovary is called the **mesosalpinx.** It is less fixed than the rest of the ligament and affords a movable mesenteric support for the tube.

4. The **round ligaments** are described below with the other ligaments of the uterus.

5. The **epoöphoron** and **paroöphoron** lie in the mesosalpinx.

6. The **ovary** is secured to the posterior surface of the broad ligament by a mesenteric attachment, the **mesovarium,** derived from the posterior leaf. It does not lie, therefore, between the leaves of the broad ligament. Folds of the posterior leaf at each end of the ovary cover the suspensory ligament of the ovary and the **ligamentum ovarii proprium,** a musculofibrous cord joining the ovary to the uterus.

7. The **ureter** crosses the attached inferior border of the broad ligament obliquely, as it courses along the pelvic floor toward the base of the bladder. It comes to within 1 or 2 cm of the uterus at this point and lies close to the uterine artery, between the latter and the pelvic diaphragm.

ROUND LIGAMENT. The round ligament is a flat band attached to the superior part of the lateral border of the uterus just caudal and ventral to the isthmus of the uterine tube. It traverses the pelvis between the leaves of the broad ligament but causes a prominent protrusion of the anterior leaf only. It reaches the pelvic wall lateral to the lateral vesicoumbilical fold, ascends over the external iliac vessels and inguinal ligament, and penetrates the abdominal wall through the deep inguinal ring. It passes through the inguinal canal, and its constituent fibers spread out to help form the substance of the labia majora. In the fetus, the peritoneum is prolonged in the form of a tubular process for a short distance into the inguinal canal beside the ligament. This process is called the **canal of Nuck** (Fig. 17-62). It is generally obliterated in the adult, but sometimes remains pervious even in advanced life. It is analogous to the saccus vaginalis, which precedes the descent of the testis.

CARDINAL LIGAMENT. The cardinal ligament (*of Mackenrodt*[1]) is a fibrous sheet of the subserous fascia embedded in the adipose tissue on each side of the lower cervix uteri and vagina. To form it, the fasciae over the ventral and dorsal walls of the vagina and cervix come together at the lateral border of these organs, and the resulting sheet extends across the pelvic floor as a deeper

[1] Alwin Mackenrodt (1859–1925): A German gynecologist and pathologist (Berlin).

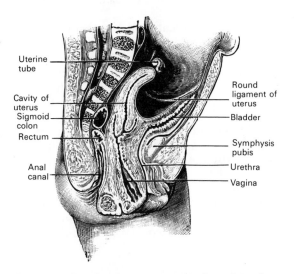

Uterine tube

Cavity of uterus

Sigmoid colon

Rectum

Anal canal

Round ligament of uterus

Bladder

Symphysis pubis

Urethra

Vagina

FIG. 17-62. Sagittal section through the pelvis of a newly born female child.

continuation of the broad ligament. As the sheet reaches the lateral portion of the pelvic diaphragm, it forms ventral and dorsal extensions that are attached to the internal investing layer of deep fascia (supra-anal fascia) on the inner surface of the levator ani, coccygeus and piriformis muscles. This attachment is commonly visible as a white line 2 or 3 cm below the arcus tendineus of the levator ani, and is called the **arcus tendineus of the pelvic fascia.** The ventral extension is continuous with the tissue supporting the bladder; the dorsal extension blends with the uterosacral ligaments. The vaginal arteries cross the pelvis in close association with this ligament, giving it additional substance and support, and bundles of smooth muscle may be embedded in it.

UTEROSACRAL LIGAMENT.   The uterosacral ligament is a prominent fibrous band of subserous fascia that takes a curved course along the lateral wall of the pelvis from the cervix uteri to the sacrum. It is a posterior continuation of the tissue that forms the cardinal ligament. It is attached to the deep fascia and periosteum of the sacrum and contains a bundle of smooth muscle named the **rectouterinus.** The ligaments on the two sides project out from the wall as crescentic shelves, which narrow the diameter of the cavity ventral to the lower rectum and mark it off as the *cul-de-sac of Douglas.*[1]

OTHER LIGAMENTS AND FOLDS.   The **vesicouterine fold** or **anterior ligament** is the reflection of peritoneum from the anterior surface of the uterus, at the junction of the cervix and body, to the posterior surface of the bladder.

The **rectovaginal fold** or **posterior ligament** is the peritoneum reflected from the wall of the posterior fornix of the vagina onto the anterior surface of the rectum.

The **sacrogenital** or **rectouterine folds** (*of Douglas*) are two crescentic folds that cover the uterosacral ligaments.

The **rectouterine excavation** (*pouch or cul-de-sac of Douglas*) is a deep pouch formed by the most inferior or caudal portion of the parietal peritoneum. Its ventral boundary is the supravaginal cervix and posterior fornix of the vagina; dorsal boundary, the rectum; and lateral boundary, the sacrogenital folds covering the uterosacral ligaments.

[1]James Douglas (1675–1742): An English physician and anatomist (London).

Support of the Uterus

The principal support of the uterus is the pelvic diaphragm, especially the levator ani and its investing layers of fascia, and unless it is intact the other structures are unable to carry out their supporting function. The uterus is held in its proper position within the pelvis by its attachment to the vagina and by the cardinal, broad, and uterosacral ligaments. The blood vessels reinforce these ligaments. The round ligaments and the peritoneal folds have relatively slight importance as mechanical supports. The padding of adipose tissue about the ligaments and organs, in well-nourished individuals, is an important element of support. There is a great variation in the size and development of the supporting structures in different individuals, and they may be thickened or strengthened in response to physiologic and pathologic changes.

Position of the Uterus

The form, size, and situation of the uterus vary at different periods of life and under different circumstances.

In the fetus and infant, the uterus is contained in the abdominal cavity, projecting beyond the superior aperture of the pelvis (Fig. 17-62). The cervix is considerably larger than the body.

At puberty, the uterus is piriform and weighs 14 to 17 gm. It has descended into the pelvis, the fundus being just below the level of the superior aperture of this cavity. The palmate folds are distinct and extend to the upper part of the cavity of the organ.

The position of the uterus *in the adult* is liable to considerable variation. With the bladder and rectum empty, the body of the uterus is nearly horizontal when the individual is standing. The fundus is about 2 cm behind the symphysis pubis and slightly cranial to it. The uterus and vagina are at an angle of about 90° with each other. The external os is half way between the spines of the ischia. As the bladder fills, the uterus is bent back toward the sacrum.

During menstruation, the organ is enlarged, more vascular, and its surfaces rounder; the external orifice is rounded, its labia swollen, and the lining membrane of the body thickened, softer, and darker.

During pregnancy, the uterus becomes enormously enlarged, and in the eighth month reaches the epigastric region. The increase in size is partly due to growth of preexisting muscle and partly to the development of new fibers.

After parturition, the uterus nearly regains its usual size, weighing about 42 gm, but its cavity is larger than in the virgin state; its vessels are tortuous, and its muscular layers are more defined; the external orifice is more marked, and its edges present one or more fissures.

In old age, the uterus becomes atrophied, paler, and denser in texture; a more distinct constriction separates the body and cervix. The isthmus is frequently, and the ostium occasionally, obliterated, while the lips almost entirely disappear.

Structure

The uterus is composed of three coats: an **external** or **serous,** a **middle** or **muscular,** and an **internal** or **mucous coat.**

SEROUS COAT. The serous coat (*perimetrium*) is derived from the peritoneum; it invests the fundus and the whole of the intestinal surface of the uterus, but covers the vesical surface only as far as the junction of the body and cervix. In the lower fourth of the intestinal surface, the peritoneum, though covering the uterus, is not closely connected with it, being separated from it by a layer of loose cellular tissue and some large veins.

MUSCULAR COAT. The muscular coat (*myometrium*) forms the chief bulk of the substance of the uterus. In the non-gravid young woman it is dense, firm, and grayish, and cuts almost like cartilage. It is thick opposite the middle of the body and fundus, and thin at the orifices of the uterine tubes. It consists of bundles of unstriped muscular fibers disposed in a thick, felt-like structure, intermixed with areolar tissue, blood vessels, lymphatic vessels, and nerves. Muscle fibers are continued on to the uterine tube, the round ligament, and the ligament of the ovary, some passing at each side into the broad ligament, and others running backward from the cervix into the uterosacral ligaments. During pregnancy the muscular tissue becomes more prominently developed, the fibers being greatly enlarged.

MUCOUS MEMBRANE. The mucous membrane (*endometrium*) (Fig. 17-63) is smooth and closely adherent to the subjacent muscular tissue. It is continuous through the fim-

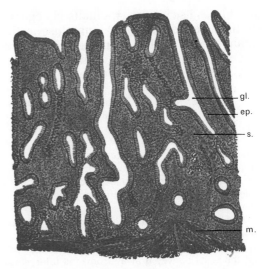

FIG. 17-63. Vertical section of mucous membrane of human uterus. *ep.,* epithelium; *gl.,* lumen of gland; *m.,* muscularis; *s.,* connective tissue stroma. (Sobotta.)

briated extremity of the uterine tubes with the peritoneum, and through the uterine ostium with the lining of the vagina.

In the *body of the uterus,* the mucous membrane is smooth, soft, pale red, lined by a single layer of high columnar ciliated epithelium, and presents, when viewed with a lens, the orifices of numerous tubular follicles arranged perpendicularly to the surface. The structure of the corium differs from that of ordinary mucous membranes; it consists of an embryronic nucleated and highly cellular form of connective tissue containing numerous large lymphatics. In it are the tube-like **uterine glands.**

In the *cervix,* the mucous membrane is sharply differentiated from that of the uterine cavity. It is thrown into numerous oblique ridges that diverge from an anterior and posterior longitudinal raphe. In the upper two-thirds of the canal, the mucous membrane is provided with numerous deep glandular follicles that secrete a clear viscid alkaline mucus; in addition, extending through the whole length of the canal is a variable number of little cysts that have become occluded and distended with retained secretion. These are called the **ovula Nabothi.**[1] The mucous membrane covering the lower half of the cervical canal has numerous papillae. The epithelium of the upper two-thirds is cylindrical and ciliated, but below

[1] Martin Naboth (1675–1721): A German anatomist (Leipzig).

this it loses its cilia and gradually changes to stratified squamous epithelium close to the ostium uteri. On the vaginal surface of the cervix the epithelium is stratified squamous similar to that lining the vagina.

VESSELS AND NERVES. The **arteries** of the uterus are the uterine, from the internal iliac artery, and the ovarian, from the abdominal aorta. They are remarkable for their tortuous course in the substance of the organ and for their frequent anastomoses. The termination of the ovarian artery meets that of the uterine artery and forms an anastomotic trunk from which branches are given off to supply the uterus in a circular disposition.

The **veins** are large and correspond with the arteries. They end in the uterine plexuses. In the impregnated uterus, the arteries carry the blood to, and the veins convey it away from, the intervillous space of the placenta. The **lymphatics** are described in Chapter 10.

The **nerves** are derived from the hypogastric and ovarian plexuses and from the third and fourth sacral nerve. Afferent fibers from the uterus enter the spinal cord solely through the eleventh and twelfth thoracic nerves.

Menstrual Cycle

After puberty and throughout the child-bearing period the endometrium undergoes periodic changes called the **menses.** The manifestation of this cycle is bleeding from the uterus, and the beginning of the bleeding is used as the point of time from which the cycle is measured. The common length of a cycle is 28 days, but this may vary markedly from individual to individual and between cycles of the same individual.

The changes in the uterus are closely correlated with cyclic changes in the ovary, and the two organs must be considered together. Although the changes are a continuous process, the cycle is usually divided into four phases: (a) proliferative, (b) secretory or progravid, (c) premenstrual, and (d) menstrual.

PROLIFERATIVE PHASE. The portion of the endometrium adjacent to the myometrium, known as the *basalis,* remains after the menstrual flow, allowing the glands and lining epithelium to be restored to an inactive condition such as that shown in Figure 17-63. After the cessation of the menstrual flow, a hormone from the adenohypophysis, the **follicle-stimulating hormone,** causes the growth of one of the ovarian follicles. As the follicle enlarges it produces the hormone **es-trogen,** which stimulates the proliferation of the endometrium, greatly enlarging the uterine glands.

SECRETORY OR PROGRAVID PHASE. When the ovarian follicle has ripened into a mature Graafian follicle, it ruptures, frequently at the middle of the cycle, liberating the ovum. The ruptured follicle, under the influence of the pituitary **luteinizing hormone,** develops into a corpus luteum, whose proliferating cells produce the hormone **progesterone.** The progesterone stimulates the enlarged endometrial glands to secrete and prepare the uterus for the reception of the fertilized ovum.

PREMENSTRUAL PHASE. If the ovum is fertilized, the corpus luteum continues to grow into the corpus luteum of pregnancy, but if the ovum is not fertilized, the corpus luteum begins to degenerate and it no longer produces hormones. The endometrium responds with changes in the blood supply, deterioration of the tissues, and fragmentation of the glands and epithelium.

MENSTRUAL PHASE. That portion of the endometrium containing the enlarged glands and abundant capillaries is known as the *functionalis,* in distinction from the depths of the glands and tissue adjacent to the myometrium, known as the *basalis.* During the menstrual discharge, the functionalis sloughs away and the ruptured vessels produce the *bleeding* of the menstrual flow. The basalis remains to establish the new endometrium for the next succeeding cycle.

THE VAGINA

The **vagina** (Figs. 17-61; 17-64) extends from the vestibule to the uterus and is situated dorsal to the bladder and ventral to the rectum. Its axis forms an angle of over 90° with that of the uterus. Its walls are ordinarily in contact, and the usual shape of its lower part on transverse section is that of an H, the transverse limb being slightly curved forward or backward, while the lateral limbs are somewhat convex toward the median line; its middle part has the appearance of a transverse slit. The length of the vagina is 6 to 7.5 cm along the ventral wall and 9 cm along the dorsal wall. It is constricted at its commencement, dilated in the middle, and narrowed near its uterine extremity; it surrounds the vaginal portion of the cervix uteri, its attachment extending higher up on the dorsal than on the ventral wall of the uterus. The recess dorsal to the cervix is the **posterior fornix;** the smaller ventral and lateral recesses are called the **anterior** and **lateral fornices.**

RELATIONS.   The **ventral surface** of the vagina is in relation with the fundus of the bladder and with the urethra. Its **dorsal surface** is separated from the rectum by the rectouterine excavation in its upper fourth and by the rectovaginal fascia in its middle two-fourths; the lower fourth is separated from the anal canal by the perineal body. As the terminal portions of the ureters pass ventralward and medialward to reach the fundus of the bladder, they run close to the lateral fornices of the vagina, and as they enter the bladder they are slightly in front of the ventral fornix.

Structure

The vagina consists of an **internal mucous lining** and a **muscular coat** separated by a layer of erectile tissue.

MUCOUS MEMBRANE.   The mucous membrane is continuous above with that lining the uterus. Its inner surface has two longitudinal ridges, one on its anterior and one on its posterior wall. These ridges are called the **columns of the vagina** and from them numerous transverse ridges or rugae extend outward on either side. These rugae are divided by furrows of variable depth, giving to the mucous membrane the appearance of being studded over with conical projections or papillae; they are most numerous near the orifice of the vagina, especially before parturition. The epithelium covering the mucous membrane is of the stratified squamous variety. The submucous tissue is loose and contains a plexus of large veins, together with smooth muscular fibers derived from the muscular coat. It contains mucous crypts, but no true glands.

MUSCULAR COAT.   The muscular coat consists of two layers: an external longitudinal layer, which is by far the stronger, and an internal circular layer. The longitudinal fibers are continuous with the superficial muscular fibers of the uterus. The strongest fasciculi are those attached to the rectovesical fascia on either side. The two layers are not distinctly separable from each other, but are connected by oblique decussating fasciculi, which pass from the one layer to the other. In addition to this, the vagina at its lower end is surrounded by the erectile tissue of the bulb of the vestibule and a band of striped muscular fibers, the **bulbospongiosus** muscle.

External to the muscular coat is a layer of connective tissue containing a large plexus of blood vessels.

VESSELS AND NERVES.   The **vaginal artery** arises from the uterine artery or from the adjacent internal iliac artery, supplies the mucous membrane, and anastomoses with the uterine, inferior vesicle, and middle rectal arteries. The branch of the uterine artery to the cervix descends on the dorsal or ventral wall, forming the azygos artery of the vagina, which anastomoses with the vaginal artery. The **veins** run along both sides, forming plexuses with the uterine, vesical, and rectal veins, and ending in a vein that opens into the internal iliac vein. The **lymphatics** are described in Chapter 10.

The **nerves** are derived from the vaginal plexus and from the pudendal nerve.

EXTERNAL GENITAL ORGANS

The **external genital organs** of the female (Fig. 17-64) are: (1) the **mons pubis,** (2) the **labia majora,** (3) the **labia minora,** (4) the **clitoris,** (5) the **vestibule of the vagina,** (6) the **bulb of the vestibule,** and (7) the **greater vestibular glands.** The term **pudendum** or **vulva,** as generally applied, includes all these parts.

Mons Pubis

The **mons pubis** (*mons Veneris*), the rounded eminence in front of the pubic symphysis, is formed by a collection of fatty tissue beneath the integument. It becomes covered with hair at puberty.

Labia Majora

The **labia majora** are two prominent longitudinal cutaneous folds that extend caudalward and dorsalward from the mons pubis and form the lateral boundaries of a fissure or cleft, the **pudendal cleft,** into which the vagina and urethra open. Each labium has two surfaces: an outer surface, pigmented and covered with strong, crisp hairs; and an inner surface, smooth and beset with large sebaceous follicles. Between the two is a considerable quantity of areolar tissue, fat, and a tissue resembling the dartos tunic of

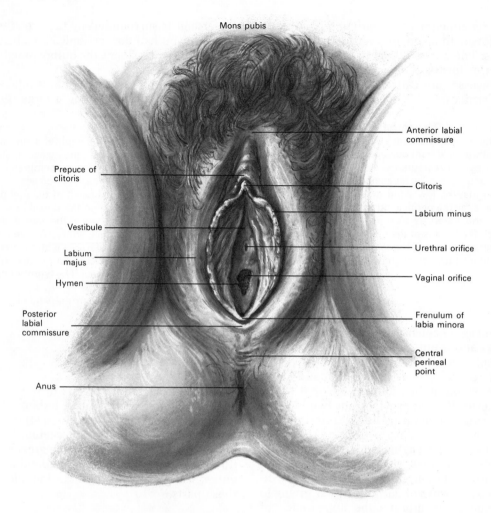

Mons pubis

Anterior labial
commissure

Prepuce of
clitoris

Clitoris

Labium minus

Vestibule

Urethral orifice

Labium
majus

Vaginal orifice

Hymen

Posterior
labial
commissure

Frenulum of
labia minora

Central
perineal
point

Anus

FIG. 17-64.   The female external genitalia. The labia majora and labia minora have been drawn apart.

the scrotum, besides vessels, nerves, and glands. The labia are thicker in front, where they join to form the **anterior labial commissure.** Posteriorly they are not really joined, but appear to become lost in the neighboring integument, ending close to, and nearly parallel with each other. Together with the connecting skin between them, they form the **posterior labial commissure** or posterior boundary of the pudendum. The labia majora correspond to the scrotum in the male.

Labia Minora

The **labia minora** (*nymphae*) are two small folds situated between the labia majora and extending from the clitoris obliquely back-

ward for about 4 cm on either side of the orifice of the vagina; they end between the labia majora and the vaginal orifice. In the virgin, the posterior ends of the labia minora are usually joined across the midline by a fold of skin, named the **frenulum of the labia** or **fourchette.** Anteriorly, each labium minus divides into two portions: the upper division passes above the clitoris to meet its fellow of the opposite side, forming a fold that overhangs the glans clitoridis and is named the **preputium clitoridis;** the lower division passes beneath the clitoris and becomes united to its undersurface, forming, with the corresponding structure of the opposite side, the **frenulum of the clitoris.** On the opposed surfaces of the labia minora are numerous sebaceous follicles.

Clitoris

The **clitoris** is an erectile structure, homologous with the penis. It is situated beneath the anterior labial commissure, partially hidden between the anterior ends of the labia minora. The **free extremity** (*glans clitoridis*) is a small rounded tubercle. It consists of two corpora cavernosa, composed of erectile tissue enclosed in a dense layer of fibrous membrane; each corpus is connected to the rami of the pubis and ischium by a crus similar to that of the penis. The clitoris, like the penis, is provided with a suspensory ligament and with two small muscles, the ischiocavernosi, which are inserted into the crura of the clitoris.

Vestibule

The cleft between the labia minora and behind the glans clitoridis is named the **vestibule of the vagina:** in it are seen the urethral and vaginal orifices and the openings of the ducts of the greater vestibular glands.

The **external urethral orifice** (*urinary meatus*) is placed about 2.5 cm dorsal to the glans clitoridis and immediately ventral to that of the vagina; it usually assumes the form of a short, sagittal cleft with slightly raised margins.

The **vaginal orifice** is a median slit caudal and dorsal to the opening of the urethra; its size varies inversely with that of the **hymen.**

The **hymen** is a thin fold of mucous membrane situated at the orifice of the vagina; the inner edges of the fold are normally in contact with each other, and the vaginal orifice appears as a cleft between them. The hymen varies much in shape. When stretched, its commonest form is that of a ring, generally broadest posteriorly; sometimes it is represented by a semilunar fold, with its concave margin turned toward the pubes. Occasionally it is cribriform or its free margin forms a membranous fringe. It may be entirely absent or may form a complete septum across the lower end of the vagina; the latter condition is known as **imperforate hymen.** It may persist after copulation, so its presence cannot be considered a sign of virginity. When the hymen has been ruptured, small rounded elevations known as the **carunculae hymenales** are found as its

remains. Between the hymen and the frenulum of the labia is a shallow depression, named the **navicular fossa.**

Bulb of Vestibule

The **bulb of the vestibule** (*vaginal bulb*) is the homologue of the bulb and adjoining part of the corpus spongiosum penis of the male. It consists of two elongated masses of erectile tissue, placed one on either side of the vaginal orifice and united to each other in front by a narrow median band termed the **pars intermedia.** Each lateral mass measures a little over 2.5 cm long. Their posterior ends are expanded and are in contact with the greater vestibular glands; their anterior ends are tapered and joined to each other by the pars intermedia; their deep surfaces are in contact with the superficial layer of the urogenital diaphragm. Superficially, they are covered by the bulbospongiosus muscle.

Greater Vestibular Glands

The **greater vestibular glands** (*Bartholin's glands*) are the homologues of the bulbourethral glands in the male. They consist of two small, roundish bodies situated one on either side of the vaginal orifice in contact with the posterior end of each lateral mass of the bulb of the vestibule. Each gland opens by means of a duct, about 2 cm long, immediately lateral to the hymen, in the groove between it and the labium minus.

VESSELS AND NERVES. The **arteries** are derived from the internal pudendal artery through the perineal artery, the artery of the bulb, and deep and dorsal arteries of the clitoris. The **veins** drain into the pudendal plexus, which empties into the vaginal and inferior vesical veins. The **lymphatics** follow the external pudendal vessels and drain into the inguinal or external iliac nodes. The **nerves** are derived from the pudendal nerve and the pelvic plexus.

MAMMARY GLAND

The mammary gland (*breast*) is an accessory of the reproductive system functionally, since it secretes milk for nourishment of the infant, but structurally and developmentally it is closely related to the integument. It reaches its typical development in women during the early childbearing period, but is present only in a rudimentary form in infants, children, and men.

In the adult nullipara, each mamma forms a discoidal, hemispherical, or conical eminence on the anterior chest wall, extending from the second to the sixth or seventh rib, and from the lateral border of the sternum into the axilla. It protrudes 3 to 5 cm from the chest wall, and its craniocaudal diameter, approximately 10 to 12 cm, is somewhat less than its transverse diameter. Its average weight is 150 to 200 gm, increasing to 400 or 500 gm during lactation. The left mamma is generally slightly larger than the right.

The **glandular tissue** forms 15 or 20 lobes arranged radially about the nipple, each lobe having its own individual excretory duct. The glandular tissue does not occupy the entire eminence called the breast; a variable but considerable amount of adipose tissue fills out the stroma between and around the lobes. The central portion is predominantly glandular, the peripheral predominantly fat. The connective tissue stroma in many places is concentrated into fibrous bands that course vertically through the substance of the breast, attaching the deep layer of the subcutaneous fascia to the dermis of the skin. These bands are known as **suspensory ligaments of the breast** or **Cooper's ligaments**[1]. The entire breast is contained within the subcutaneous fascia. The deep surface is separated from the underlying external investing layer of deep fascia by a fascial cleft that allows considerable mobility. The deep surface of the breast is concave, molded over the ventral chest wall mostly in contact with the pectoral fascia, but laterally with the axillary and serratus anterior fascia, and inferiorly it may reach the obliquus externus and rectus abdominis muscles.

Papilla

The **mammary papilla** or **nipple** projects as a small cylindrical or conical body, a little below the center of each breast at about the level of the fourth intercostal space. It is perforated at the tip by 15 or 20 minute openings, the apertures of the lactiferous ducts. The characteristic skin of the nipple, pigmented, wrinkled, and roughened by papil-

lae, extends outward on the surface of the breast for 1 or 2 cm to form the **areola.** The color of the nipple and areola in nulliparae varies from rosy pink to brown, depending on the complexion of the individual. During the second month and progressing through pregnancy, the skin becomes darker and the areola becomes larger. Following lactation the pigmentation diminishes but is never entirely lost and may be used to differentiate nulliparous from parous individuals.

Areola

The **areola** is made rough by the presence of numerous large sebaceous glands, which produce small elevations of its surface. These **areolar glands** (*glands of Montgomery*[2]) secrete a lipoid material that lubricates and protects the nipple during nursing. The subcutaneous tissue of the areola contains circular and radiating smooth muscle bundles, which cause the nipple to become erect in response to stimulation.

Development

The primordium of the mamma is first recognizable during the sixth week of intrauterine life as a bandlike thickening of the ectoderm of the anterolateral body wall. It extends from the axilla to the inguinal region and is called the **milk line.** The thickening of the ectoderm pushes into the subcutaneous mesoderm as it enlarges. After the eighth week, only the portion of the ridge destined to become mamma is identifiable. During the remainder of fetal life the epithelial cells proliferate, gradually forming buds and cords of cells projecting into the subcutaneous tissue, and by birth, little more than the main ducts have formed in both sexes. The glands remain in this infantile condition in the male.

ADOLESCENT HYPERTROPHY. The female gland changes only slightly from the infantile condition until the approach of puberty. At this time the mamma enlarges due to an increase in glandular tissue, particularly the ducts, and to a deposit of adipose tissue. The mammary papilla and areola enlarge, increase slightly in pigmentation, acquire smooth muscle, and become sensitive. After

---

[1] Sir Astley Paston Cooper (1768–1841): An English surgeon (London).

[2] William Fetherstone Montgomery (1797–1859): An Irish obstetrician (Dublin).

the onset of the menses, the mamma change with each period. In the premenstrual phase there is a vascular engorgement, increase in the glands, and enlargement of their lumens. During the postmenstrual phase the mamma regresses and then remains in an inactive stage until the next premenstrual phase.

HYPERTROPHY OF PREGNANCY. Visible enlargement of the breast begins after the second month of pregnancy and is accompanied by an increased pigmentation and enlargement of the papilla, areola, and areolar glands. The duct system develops first, reaching its completion during the first six months; the acini and secreting portion follow during the last three months. The adipose tissue is almost completely replaced by parenchyma.

The secretion from the mammary gland during the first two or three days after parturition is thin and yellowish and is called **colostrum.** The secretion of true milk begins on the third or fourth day and continues through the nursing period.

INVOLUTION AFTER LACTATION. At the termination of nursing, the gland gradually regresses by loss of the glandular tissue; the ducts and acini return to their former size and number, and the interstices are filled with adipose tissue. There is a slight decrease in size of the breast as a whole and it tends to become more flabby and pendulous than the nulliparous breast. The pigmentation of the nipple and the areola decreases but does not entirely disappear.

MENOPAUSAL INVOLUTION. At the end of the childbearing period, the mammae regress and the glandular tissue reverts toward the infantile condition. The adipose tissue disappears more slowly, especially in obese individuals, but eventually a senile atrophy occurs that leaves the mamma a shriveled pendulous fold of skin.

HORMONAL RELATIONSHIPS. The hypertrophy of puberty and the cyclic engorgements accompanying menstruation are responses to variations in the concentration of the ovarian sex hormones from the follicles and corpora lutea. The hypertrophy of pregnancy occurs in response to an increase in the corpus luteum hormone. The presence of the ovary is necessary only during the first part of pregnancy; later, the hormones are supplied by the placenta. Lactation is a response to the lactogenic hormone of the hypophysis, but is influenced by the nervous system through the stimulus of suckling. Suppression of the ovarian hormones after the menopause results in the involution. Frequently the mammary glands in the newborn of both sexes secrete a fluid called "witches' milk," under the influence of the hormones passed through the placenta from the maternal circulation.

Structure

The mamma consists of gland tissue, of fibrous tissue connecting its lobes, and of fatty tissue in the intervals between the lobes (Figs. 17-65; 17-66). The gland tissue, when freed from fibrous tissue and fat, is pale red, firm, flat, and thicker in the center than at the circumference.

The subcutaneous surface of the mamma has numerous irregular processes that project toward the skin and are joined to it by bands of connective tissue. It consists of numerous lobes, and these are composed of lobules, connected together by areolar tissue, blood vessels, and ducts. The smallest lobules consist of a cluster of rounded alveoli, which open into the smallest branches of the lactiferous ducts; these ducts unite to form larger ducts, and these end in a single canal, corresponding to one of the chief subdivisions of the gland. The number of excretory ducts varies from 15 to 20; they are termed the lactiferous tubules. They converge toward the areola, beneath which they form dilatations or **ampullae,** which serve as reservoirs for the milk. At the base of the papillae they become contracted and pursue a straight course to its summit, perforating it by separate orifices considerably narrower than the ducts themselves. The ducts are composed of areolar tissue containing longitudinal and transverse elastic fibers; muscular fibers are entirely absent. The ducts are lined by columnar epithelium resting on a basement membrane.

The epithelium of the mamma differs according to the state of activity of the organ. In the gland of a woman who is not pregnant or suckling, the alveoli are very small and solid, being filled with a mass of granular polyhedral cells. During pregnancy the alveoli enlarge, and the cells undergo rapid multiplication. At the commencement of lactation, the cells in the center of the alveolus undergo fatty degeneration, and are eliminated in the first milk as **colostrum corpus-**

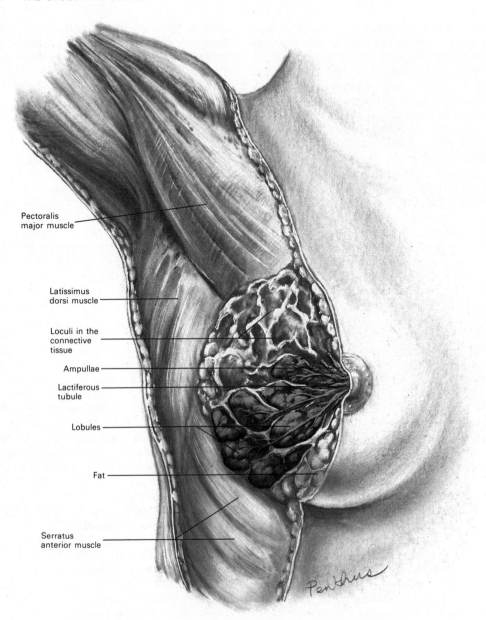

Pectoralis
major muscle

Latissimus
dorsi muscle

Loculi in the
connective
tissue

Ampullae

Lactiferous
tubule

Lobules

Fat

Serratus
anterior muscle

FIG. 17-65.   Dissection of the lateral aspect of the right mammary gland during the period of lactation.

cles. The peripheral cells of the alveolus remain, and form a single layer of granular, short columnar cells with spherical nuclei, lining the basement membrane. The cells during the state of activity of the gland are capable of forming oil globules in their interior, which are then ejected into the lumen of the alveolus and constitute the milk globules. When the acini are distended by the accumulation of the secretion, the lining epithelium becomes flattened.

The **fibrous tissue** invests the entire surface of the mamma. Bands of fibrous tissue traverse the gland and connect the overlying skin to the underlying deep layer of subcutaneous fascia. These constitute the ligaments of Cooper.

The **fatty tissue** covers the surface of the gland and occupies the interval between its lobes. It usually exists in considerable abundance and determines the form and size of the areola and papilla (Fig. 17-67).

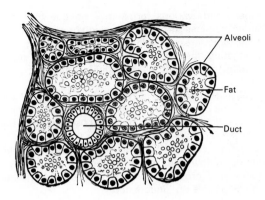

FIG. 17-66.   Section of portion of breast.

VESSELS AND NERVES.   The **arteries** supplying the mammae are derived from the thoracic branches of the axillary, the intercostal, and the internal thoracic arteries. The **veins** describe an anastomotic circle around the base of the papilla, called the **circulus venosus.** From this, large branches transmit the blood to the circumference of the gland and end in the axillary and internal thoracic veins. The **lymphatics** are described in Chapter 10. The **nerves** are derived from the anterior and lateral cutaneous branches of the fourth, fifth, and sixth thoracic nerves.

VARIATIONS.   Independent of the physiologic variations, conditions of underdeveloped breasts (*hypomastia*), hypertrophy, and inequality on the two sides are common. Variations in position, cranially or caudally, are not infrequent, as might be

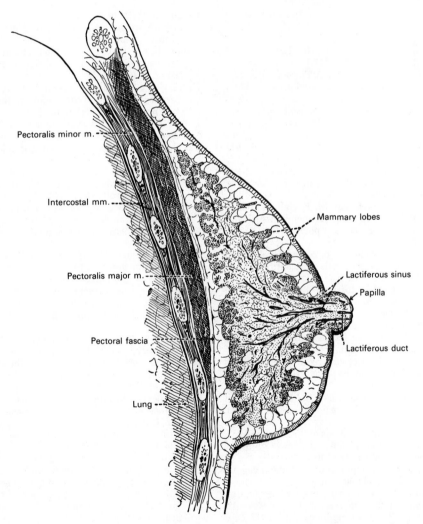

FIG. 17-67.   Sagittal section through lactating mammary gland. (Eycleshymer and Jones.)

expected from the method of embryologic development. Absence of the breast, *amastia,* is rare; an increase in the number of mammae, **polymastia,** is not rare. The supernumerary mammae occur somewhere along the milk line in most instances, but other locations have been reported. Not infrequently they secrete milk during normal periods of lactation. When only the nipples of the supernumerary mammae are present the condition is **polythelia;** the extra nipples occur along the milk line in the majority of cases and may be found in males, though less frequently than in females.

Racial variations may be due to genetic influences, which cause discoidal, hemispherical, pear-shaped, or conical forms to predominate. On the other hand, various forms may be due to intentional practices such as suppressing development by tight brassiéres at adolescence or tremendously elongating the breasts by manipulation to make nursing convenient for an infant strapped on the back.

Gynecomastia is a condition in which the mammae of a male are enlarged. In pseudogynecomastia, the increase is due to adipose tissue. In the true gynecomastia, to some extent the epithelial tissue, but especially the firmer connective tissue elements are involved. A background of endocrine dysfunction is probably involved in this disorder, but this is not clearly understood.

References

(References are listed not only to those articles and books cited in the text, but to others as well which are considered to contain valuable resource information for the student who desires it.)

Development of the Urinary and Generative Organs

Alexander, D. P., and D. A. Nixon. 1961. The foetal kidney. Brit. Med. Bull., *17:*112–117.

Alexander, J. C., K. B. King, and C. S. Fromm. 1950. Congenital solitary kidney with crossed ureter. J. Urol., 64:230–234.

Allan, F. D. 1969. *Essentials of Human Embryology.* 2nd ed. Oxford University Press, New York. 344 pp.

Altschule, M. D. 1930. The changes in the mesonephric tubules of human embryos ten to twelve weeks old. Anat. Rec., 46:81–91.

Aoki, A. 1966. Development of the human renal glomerulus. I. Differentiation of the filtering membrane. Anat. Rec., 155:339–351.

Asaad, E. I., Z. H. Badawi, and M. F. Gaballah. 1975. Nature and fate of the ureteric membrane closing the primitive ureterovesical opening. Acta Anat., 92:524–536.

Backhouse, K. M., and M. Butler. 1956. The development of the coverings of the testis and cord. J. Anat. (Lond.), 92:645.

Barnstein, J. N., and H. W. Mossman. 1938. The origin of the penile urethra and bulbo-urethral glands with particular reference to the red squirrel (*Tamiasciurus hudsonicus*). Anat. Rec., 72:67–85.

Begg, R. C. 1930. The urachus: Its anatomy, histology and development. J. Anat. (Lond.), 64:170–183.

Black, J. L., and B. H. Erickson. 1968. Oogenesis and ovarian development in the prenatal pig. Anat. Rec., 161:45–56.

Bloomfield, A., and J. E. Frazer. 1927. The development of the lower end of the vagina. J. Anat. (Lond.), 62:9–32.

Bremer, J. L. 1916. The interrelations of the mesonephros, kidney and placenta in different classes of mammals. Amer. J. Anat., 19:179–210.

Bulmer, D. 1957. The development of the human vagina. J. Anat. (Lond.), 91:490–509.

Bulmer, D. 1959. Histochemical observations on the foetal vaginal epithelium. J. Anat. (Lond.), 93:36–42.

Calabrisi, P. 1956. The nerve supply of the erectile cavernous tissue of the genitalia in the human embryo and fetus. Anat. Rec., 125:713–723.

Chiquoine, A. D. 1954. The identification, origin and migration of the primordial germ cells in the mouse embryo. Anat. Rec., 118:135–146.

Davies, J., and D. V. Davies. 1950. The development of the mesonephros of the sheep. Proc. Zool. Soc. Lond., 120:73–93.

Dröes, J. T. 1974. Observations on the musculature of the urinary bladder and the urethra in the human foetus. Brit. J. Urol., 46:179–185.

Everett, N. B. 1943. Observational and experimental evidences relating to the origin and differentiation of the definitive germ cells in mice. J. Exp. Zool., 92:49–91.

Faulconer, R. J. 1951. Observations on the origin of the Müllerian groove in human embryos. Carneg. Inst., Contr. Embryol., 34:159–164.

Forsberg, J.-G. 1965. Mitotic rate and radioautographic studies on the derivation and differentiation of the epithelium in the mouse vaginal anlage. Acta Anat., 62:266–282.

Fraenkel, L., and G. N. Papanicolaou. 1938. Growth, desquamation and involution of the vaginal epithelium of fetuses and children with a consideration of the related hormonal factors. Amer. J. Anat., 62:427–452.

Fraser, E. A. 1920. The pronephros and early development of the mesonephros in the cat. J. Anat. (Lond.), 54:287–304.

Fraser, E. A. 1950. The development of the vertebrate excretory system. Biol. Rev., 25:159–187.

Frazer, J. E. 1935. The terminal part of the Wolffian duct. J. Anat. (Lond.), 69:455–468.

Gersh, I. 1937. The correlation of structure and function in developing mesonephros and metanephros. Carneg. Inst., Contr. Embryol., 26:33–58.

Gillman, J. 1948. The development of the gonads in man, with a consideration of the role of fetal endocrines and the histogenesis of ovarian tumors. Carneg. Inst., Contr. Embryol., 32:81–131.

Glenister, T. W. 1954. The origin and fate of the urethral plate in man. J. Anat. (Lond.), 88:413–425.

Glenister, T. W. 1956. A consideration of the processes involved in the development of the prepuce in man. Brit. J. Urol., 28:243–249.

Glenister, T. W. 1958. A correlation of the normal and abnormal development of the penile urethra and of the infra-umbilical abdominal wall. Brit. J. Urol., 30:117–126.

Glenister, T. W. 1962. The development of the utricle and of the so-called 'middle' or 'median' lobe of the human prostate. J. Anat. (Lond.), 96:443–455.

Gordon-Taylor, G. 1936. On horseshoes and on horseshoe kidney, concave downwards. Brit. J. Urol., 8:112–118.

Gruenwald, P. 1941. The relation of the growing Müllerian duct to the Wolffian duct and its importance for the genesis of malformations. Anat. Rec., 81:1–19.

Gruenwald, P. 1942. The development of the sex cords in the gonads of man and mammals. Amer. J. Anat., 70:359–398.

Gruenwald, P. 1943. The normal changes in the position of the embryonic kidney. Anat. Rec., 85:163–176.

Gyllensten, L. 1949. Contributions to the embryology of the urinary bladder; Part I. Development of the definitive relations between the openings of the Wolffian ducts and the ureters. Acta Anat., 7:305–344.

Hammond, G., L. Yglesias, and J. E. Davis. 1941. The urachus, its anatomy and associated fasciae. Anat. Rec., 80:271–287.

Holyoke, E. A. 1949. The differentiation of embryonic gonads transplanted to the adult omentum in the albino rat. Anat. Rec., 103:675–699.

Hooker, C. W., and L. C. Strong. 1944. Hermaphroditism in rodents, with a description of a case in the mouse. Yale J. Biol. Med., 16:341–352.

Hunter, R. H. 1930. Observations on the development of the human female genital tract. Carneg. Inst., Contr. Embryol., 22:91–107.

Hunter, R. H. 1935. Notes on the development of the prepuce. J. Anat. (Lond.), 70:68–75.

Johnson, F. P. 1920. The later development of the urethra in the male. J. Urol., 4:447–501.

Kampmeier, O. F. 1926. The metanephros or so-called permanent kidney in part provisional and vestigial. Anat. Rec., 33:115–120.

Koff, A. K. 1933. Development of the vagina in the human fetus. Carneg. Inst., Contr. Embryol., 24:59–90.

Leeson, T. S. 1957. The fine structure of the mesonephros of the 17-day rabbit embryo. Exp. Cell Res., 12:670–672.

Lemeh, C. N. 1960. A study of the development and structural relationships of the testis and gubernaculum. Surg. Gynecol. Obstet., 110:164–172.

Lewis, O. J. 1958. The development of the blood vessels of the metanephros. J. Anat. (Lond.), 92:84–97.

Lewis, O. J. 1958. The vascular arrangement of the mammalian renal glomerulus as revealed by a study of its development. J. Anat. (Lond.), 92:433–440.

Lowsley, O. S. 1912. The development of the human prostate gland with reference to the development of other structures at the neck of the urinary bladder. Amer. J. Anat., 13:299–349.

MacDonald, M. S., and J. L. Emery. 1959. The late intrauterine and postnatal development of human renal glomeruli. J. Anat. (Lond.), 93:331–340.

Mahoney, P. J., and D. Ennis. 1936. Congenital patent urachus. N. Engl. J. Med., 215:193–195.

McKelvey, J. L., and J. S. Baxter. 1935. Abnormal development of the vagina and genitourinary tract. Amer. J. Obstet. Gynecol., 29:267–271.

Meyer, R. 1946. Normal and abnormal development of the ureter in the human embryo—a mechanistic consideration. Anat. Rec., 96:355–371.

Mintz, B. 1960. Embryological phases of mammalian gametogenesis. J. Cell. Comp. Physiol., 56(Suppl.): 31–48.

Monie, I. W. 1949. Double ureter in two human embryos. Anat. Rec., 103:195–204.

Monroe, C. W., and B. Spector. 1962. The epithelium of the hydatid of Morgagni in the human adult female. Anat. Rec., 142:189–193.

O'Connor, R. J. 1938. Experiments on the development of the pronephric duct. J. Anat. (Lond.), 73:145–154.

Patton, W. H. G. 1948. A case of true hermaphrodism in an adult African. Anat. Rec., 101:479–485.

Pinkerton, J. H. M., D. G. McKay, E. C. Adams, and A. T. Hertig. 1961. Development of the human ovary: A study using histochemical technic. Obstet. Gynecol., 18:152–181.

Price, D. 1936. Normal development of the prostate and seminal vesicles of the rat with a study of experimental post-natal modifications. Amer. J. Anat., 60:79–128.

Price, D. 1947. An analysis of the factors influencing growth and development of the mammalian reproductive tract. Physiol. Zool., 20:213–247.

Roosen-Runge, E. C., and J. Lund. 1972. Abnormal sex cord formation and an intratesticular adrenal cortical nodule in a human fetus. Anat. Rec., 173:57–68.

Shikinami, J. 1926. Detailed form of the wolffian body in human embryos of the first eight weeks. Carneg. Inst., Contr. Embryol., 18:49–61.

Spaulding, M. H. 1921. The development of the external genitalia in the human embryo. Carneg. Inst., Contr. Embryol., 13:67–88.

Torrey, T. W. 1954. The early development of the human nephros. Carneg. Inst., Contr. Embryol., 35:175–198.

Wachstein, M., and M. Bradshaw. 1965. Histochemical localization of enzyme activity in the kidneys of three mammalian species during their postnatal development. J. Histochem. Cytochem., 13:44–56.

van Wagenen, G., and M. E. Simpson. 1965. Embryology of the Ovary and Testis—Homo Sapiens and Macaca Mulatta. Yale Univ. Press, New Haven. 256 pp.

Weaver, M. E. 1966. Persistent urachus—an observation in miniature swine. Anat. Rec., 154:701–703.

Wharton, L. R., Jr. 1949. Double ureters and associated renal anomalies in early human embryos. Carneg. Inst., Contr. Embryol., 33:103–112.

Williams, D. I. 1951. The development of the trigone of the bladder. Brit. J. Urol., 23:123–128.

Wilson, K. W. 1926. Origin and development of the rete ovarii and the rete testis in the human embryo. Carneg. Inst., Contr. Embryol., 17:69–88.

Witschi, E. 1948. Migration of the germ cells of human embryos from the yolk sac to the primitive gonadal folds. Carneg. Inst., Contr. Embryol., 32:67–80.

Zamboni, L., and C. de Martino. 1968. Embryogenesis of the human renal glomerulus. I. A histologic study. Arch. Pathol., 86:279–291.

Urinary Organs

Kidney

Anson, B. J., E. W. Cauldwell, J. W. Pick, and L. E. Beaton. 1947. The blood supply of the kidney, suprarenal gland, and associated structures. Surg. Gynecol. Obstet., 84:313–320.

Anson, B. J., and E. H. Daseler. 1961. Common variations in renal anatomy, affecting blood supply, form, and topography. Surg. Gynecol. Obstet., 112:439–449.

Anson, B. J., J. W. Pick, and E. W. Cauldwell. 1942. The

anatomy of commoner renal anomalies: Ectopic and horseshoe kidneys. J. Urol., 47:112–132.

Barajas, L. 1966. The development and ultrastructure of the juxtaglomerular cell granule. J. Ultrastruct. Res., 15:400–413.

Benjamin, J. A., and C. E. Tobin. 1951. Abnormalities of the kidneys, ureters, and perinephric fascia: Anatomic and clinical study. J. Urol., 65:715–731.

Boyer, C. C. 1956. The vascular pattern of the renal glomerulus as revealed by plastic reconstruction from serial sections. Anat. Rec., 125:433–441.

Bulger, R. E. 1965. The shape of rat kidney tubular cells. Amer. J. Anat., 116:237–255.

Bulger, R. E. 1971. Ultrastructure of the junctional complexes from the descending thin limbs of the loop of Henle from rats. Anat. Rec., 171:471–475.

Bulger, R. E. 1973. Rat renal capsule: Presence of layers of unique squamous cells. Anat. Rec., 177:393–407.

Bulger, R. E., R. E. Cronin, and D. C. Dobyan. 1979. Survey of the morphology of the dog kidney. Anat. Rec., 194:41–65.

Bulger, R. E., and D. C. Dobyan. 1983. Recent structure-function relationships in normal and injured mammalian kidneys. Anat. Rec., 205:1–11.

Cameron, G., and R. Chambers. 1938. Direct evidence of function in kidney of an early human fetus. Amer. J. Physiol., 123:482–485.

Carroll, N., G. W. Crock, C. C. Funder, C. R. Green, K. N. Ham, and J. D. Tange. 1974. Scanning electron microscopy of the rat renal papilla. J. Anat. (Lond.), 117:447–452.

Creasey, M., and D. B. Moffat. 1971. The effect of changes in conditions of water balance on the vascular bundles of the rat kidney. J. Anat. (Lond.), 109:437–442.

Davies, J. 1954. Cytological evidence of protein absorption in fetal and adult mammalian kidneys. Amer. J. Anat., 94:45–71.

Doležel, S. 1975. The connective tissue skeleton in the mammalian kidney and its innervation. Acta Anat., 93:194–209.

Dunihue, F. W., and E. B. Whorton. 1972. Juxtaglomerular granular cells and nephron autoregulation. Anat. Rec., 173:479–483.

Edelman, R., and P. M. Hartcroft. 1961. Localization of renin in juxtaglomerular cells of rabbit and dog through the use of the fluorescent-antibody technique. Circ. Res., 9:1069–1077.

Evan, A. P., P. G. Simone, S. Solomon, and E. F. Loker. 1972. Structural changes in the proximal tubule of kidneys from hypophysectomized rats. Anat. Rec., 174:265–277.

Farquhar, M. G., S. L. Wissig, and G. E. Palade. 1961. Glomerular permeability. I. Ferritin transfer across the normal glomerular capillary wall. J. Exp. Med., 113:47–66.

Fetterman, G. H., N. A. Shuplock, F. Philipp, and H. S. Gregg. 1965. The growth and maturation of human glomeruli and proximal convolutions from term to adulthood. Studies by microdissection. Pediatrics, 35:601–619.

Fine, H., and E. N. Keen. 1966. The arteries of the human kidney. J. Anat. (Lond.), 100:881–894.

Fine, H., and E. N. Keen. 1976. Some observations on the medulla of the kidney. Brit. J. Urol., 48:161–169.

Glenn, J. F. 1959. Analysis of fifty-one patients with horseshoe kidney. N. Engl. J. Med., 261:684–687.

Gosling, J. A., and J. S. Dixon. 1969. The fine structure of the vasa recta and associated nerves in the rabbit kidney. Anat. Rec., 165:503–514.

Graham, R. C., Jr., and M. J. Karnovsky. 1966. Glomerular permeability. Ultrastructural and cytochemical studies using peroxidases as protein tracers. J. Exp. Med., 124:1123–1134.

Graham, R. C., Jr., and M. J. Karnovsky. 1966. The early stages of absorption of injected horseradish peroxidase in the proximal tubules of mouse kidney: ultrastructural cytochemistry by a new technique. J. Histochem. Cytochem., 14:291–302.

Graves, F. T. 1954. The anatomy of the intrarenal arteries and its application to segmental resection of the kidney. Brit. J. Surg., 42:132–139.

Graves, F. T. 1956. The aberrant renal artery. J. Anat. (Lond.), 90:553–558.

Grossman, J. 1954. A note on the radiological demonstration of perirenal space. J. Anat. (Lond.), 88:407–409.

Groth, C. G. 1972. Landmarks in clinical renal transplantation. Surg. Gynecol. Obstet., 134:323–328.

Hammersen, F., and J. Staubesand. 1961. Arteries and capillaries of the human renal pelvis with special reference to the so-called spiral arteries. I. Angioarchitectural studies on the kidneys. Z. Anat. Entwicklungsgesch., 122:314–347.

Harman, P. J., and H. Davies. 1948. Intrinsic nerves in the mammalian kidney. I. Anatomy in mouse, rat, cat, and macaque. J. Comp. Neurol., 89:225–243.

Hodson, C. J. 1978. The renal parenchyma and its blood supply. Curr. Probl. Diagn. Radiol., 7:1–32.

Holmes, M. J., P. J. O'Morchoe, and C. C. C. O'Morchoe. 1977. Morphology of the intrarenal lymphatic system. Capsular and hilar communications. Amer. J. Anat., 149:333–351.

Inke, G. 1981. Gross internal structure of the human kidney. Prog. Clin. Biol. Res., 59:71–78.

Kaye, K. W., and M. E. Goldberg. 1982. Applied anatomy of the kidney and ureter. Urol. Clin. North Amer., 9:3–13.

Kurtz, S. M. 1958. The electron microscopy of the developing human renal glomerulus. Exp. Cell Res., 14:355–367.

Kurtz, S. M., and J. D. Feldman. 1962. Experimental studies on the formation of the glomerular basement membrane. J. Ultrastruct. Res., 6:19–27.

Lewis, O. J. 1958. The vascular arrangement of the mammalian renal glomerulus as revealed by a study of its development. J. Anat. (Lond.), 92:433–440.

Ljungqvist, A., and C. Lagergren. 1962. Normal intrarenal arterial pattern in adult and ageing human kidney. J. Anat. (Lond.), 96:285–300.

Martin, C. P. 1942. A note on the renal fascia. J. Anat. (Lond.), 77:101–103.

Merklin, R. J., and N. A. Michels. 1958. The variant renal and suprarenal blood supply with data on the inferior phrenic, ureteral and gonadal arteries. J. Internat. Coll. Surg., 29:41–76.

Mitchell, G. A. G. 1935. The innervation of the kidney, ureter, testicle and epididymis. J. Anat. (Lond.), 70:10–32.

Mitchell, G. A. G. 1950. The nerve supply of the kidneys. Acta Anat., 10:1–37.

Mitchell, G. A. G. 1950. The renal fascia. Brit. J. Surg., 37:257–266.

Moffat, D. B. 1981. New ideas on the anatomy of the kidney. J. Clin. Pathol., 34:1197–1206.

Moffat, D. B., and J. Fourman. 1964. Ectopic glomeruli in the human and animal kidney. Anat. Rec., 149:1–11.

Moody, R. O., and R. G. van Nuys. 1940. The position and mobility of the kidneys in healthy young men and women. Anat. Rec., 76:111–133.

Pfeiffer, E. W. 1968. Comparative anatomical observations of the mammalian renal pelvis and medulla. J. Anat. (Lond.), 102:321–331.

Pick, J. W., and B. J. Anson. 1940. The renal vascular pedicle. J. Urol., 44:411–434.

Priman, J. 1929. A consideration of normal and abnormal positions of the hilum of the kidney. Anat. Rec., 42:355–363.

Reis, R. H., and G. Esenther. 1959. Variations in the pattern of renal vessels and their relation to the type of posterior vena cava in man. Amer. J. Anat., 104:295–318.

Smith, G. T. 1963. The renal vascular patterns in man. J. Urol., 89:275–288.

Smith, H. W. 1956. *Principles of Renal Physiology.* Oxford University Press, New York. 237 pp.

Sulkin, N. M. 1949. Cytologic studies of the remaining kidney following unilateral nephrectomy in the rat. Anat. Rec., 105:95–111.

Sykes, D. 1963. The arterial supply of the human kidney with special reference to accessory renal arteries. Brit. J. Surg., 50:368–374.

Sykes, D. 1964. The correlation between renal vascularisation and lobulation of the kidney. Brit. J. Urol., 36:549–555.

Tobin, C. E. 1944. The renal fascia and its relation to the transversalis fascia. Anat. Rec., 89:295–311.

Vordermark, J. S., II. 1981. Segmental anatomy of the kidney. Urology, 17:521–531.

Zimny, M. L., and E. Rigamer. 1966. Glomerular ultrastructure in the kidney of a hibernating animal. Anat. Rec., 154:87–94.

Zwemer, R. L., and R. M. Wotton. 1944. Fat excretion in the guinea pig kidney. Anat. Rec., 90:107–114.

Ureters, Bladder and Urethra

Ambrose, S. S., and W. P. Nicolson, III. 1962. The causes of vesicoureteral reflux in children. J. Urol., 87:688–694.

Ambrose, S. S., and W. P. Nicolson. 1964. Ureteral reflex in duplicated ureters. J. Urol., 92:439–443.

Barrington, F. J. F. 1933. The localization of the paths subserving micturition in the spinal cord of the cat. Brain, 56:126–148.

Barrington, F. J. F. 1941. The component reflexes of micturition in the cat: Part III. Brain, 64:239–243.

Begg, R. C. 1927. The urachus and umbilical fistulae. Surg. Gynecol. Obstet., 45:165–178.

Begg, R. C. 1930. The urachus: Its anatomy, histology and development. J. Anat. (Lond.), 64:170–183.

Beneventi, F. A. 1943. A study of the posterior urethra in the newborn female. Surg. Gynecol. Obstet., 76:64–76.

Bödeker, J., K. Bandhauer, C.-P. Kölln, and R. Nagel. 1977. Analysis of bladder configuration and pressure from pressoreceptor stimulation in rabbits. Invest. Urol., 14:407–410.

Bradley, W. E., G. L. Rockswold, G. W. Timm, and F. B. Scott. 1976. Neurology of micturition. J. Urol., 115:481–486.

Bumpus, H. C., Jr., and W. Antopol. 1934. Distribution of blood to the prostatic urethra: A demonstration. J. Urol., 32:354–358.

Daniel, O., and R. Shackman. 1952. The blood supply of the human ureter in relation to uterocolic anastomosis. Brit. J. Urol., 24:334–343.

Dees, J. E. 1940. Anomalous relationship between ureter and external iliac artery. J. Urol., 44:207–215.

Denny-Brown, D., and E. G. Robertson. 1933. On the physiology of micturition. Brain, 56:149–190.

Denny-Brown, D., and E. G. Robertson. 1933. The state of the bladder and its sphincters in complete transverse lesions on the spinal cord and cauda equina. Brain, 56:397–463.

Deter, R. L., G. T. Caldwell, and A. I. Folsom. 1946. A clinical and pathological study of the posterior female urethra. J. Urol., 55:651–662.

Dixon, J. S., and J. A. Gosling. 1971. Histochemical and electron microscopic observations on the innervation of the upper segment of the mammalian ureter. J. Anat. (Lond.), 110:57–66.

Duff, P. A. 1950. Retrocaval ureter: Case report. J. Urol., 63:496–499.

van Duzen, R. E., and W. W. Looney. 1932. Further studies on the trigone muscle: The anatomy and practical considerations. J. Urol., 27:129–144.

Emmett, J. L., R. V. Daut, and J. H. Dunn. 1948. Role of the external urethral sphincter in the normal bladder and cord bladder. J. Urol., 59:439–454.

Evans, J. P. 1936. Observations on the nerves of supply to the bladder and urethra of the cat, with a study of their action potentials. J. Physiol., 86:396–414.

Fearnsides, E. G. 1917. The innervation of the bladder and urethra: A review. Brain, 40:149–187.

Finestone, E. O. 1941. Urinary extravasation (periurethral phlegmon): Pathogenesis and experimental study. Surg. Gynecol. Obstet., 73:218–227.

Firth, J. A., and R. M. Hicks. 1973. Interspecies variation in the fine structure and enzyme cytochemistry of mammalian transitional epithelium. J. Anat. (Lond.), 116:31–43.

Fletcher, T. F., and W. E. Bradley. 1978. Neuroanatomy of the bladder-urethra. J. Urol., 119:153–160.

Gill, W. B., and G. A. Curtis. 1977. The influence of bladder fullness on upper urinary tract dimensions and renal excretory function. J. Urol., 117:573–576.

Glenister, T. W. 1954. The origin and fate of the urethral plate in man. J. Anat. (Lond.), 88:413–425.

Glenn, J. F. 1959. Agenesis of the bladder. JAMA, 169:2016–2018.

Gosling, J. A. 1970. The musculature of the upper urinary tract. Acta Anat., 75:408–422.

Gosling, J. A. 1979. The structure of the bladder and urethra in relation to function. J. Urol. Clin. North Amer., 6:31–38.

Gosling, J. A., and J. S. Dixon. 1974. Sensory nerves in the mammalian urinary tract. An evaluation using light and electron microscopy. J. Anat. (Lond.), 117:133–144.

Gosling, J. A., and J. S. Dixon. 1975. The structure and innervation of smooth muscle in the wall of the bladder neck and proximal urethra. Brit. J. Urol., 47:549–558.

Gosling, J. A., J. S. Dixon, and M. Dunn. 1977. The structure of the rabbit urinary bladder after experimental distension. Invest. Urol., 14:386–389.

deGroat, W. C., and A. M. Booth. 1980. Physiology of the urinary bladder and urethra. Ann. Intern. Med., 92:312–315.

Gross, R. E., and T. C. Moore. 1950. Duplication of the urethra. Arch. Surg., 60:749–761.

Gruenwald, P., and S. N. Surks. 1943. Pre-ureteric vena cava and its embryological explanation. J. Urol., 49:195–201.

Hanna, M. K., R. D. Jeffs, J. M. Sturgess, and M. Barkin. 1976. Ureteral structure and ultrastructure. Part I. The normal human ureter. J. Urol., 116:718–724.

Heimburger, R. F. 1949. The sacral innervation of the human bladder. Arch. Neurol. Psychiat., 62:686–687.

Hoyes, A. D., and P. Barber. 1976. Parameters of fixation of the putative pain afferents in the ureter: Preservation of the dense cores of the large vesicles in the axonal terminals. J. Anat. (Lond.), 122:113–120.

Hoyes, A. D., and P. Barber. 1978. On the calibre of the ureteric lumen. J. Anat. (Lond.), 126:203–207.

Hoyes, A. D., R. Bourne, and B. G. H. Martin. 1975. Ultrastructure of the submucous nerves of the rat ureter. J. Anat. (Lond.), 119:123–132.

Hoyes, A. D., N. I. Ramus, and B. G. H. Martin. 1972. Fine structure of the epithelium of the human fetal bladder. J. Anat. (Lond.), 111:415–425.

Hunter, De W. T., Jr. 1954. A new concept of urinary bladder musculature. J. Urol., 71:695–704.

Kaye, K. W., and M. E. Goldberg. 1982. Applied anatomy of the kidney and ureter. Urol. Clin. North Amer., 9:3–13.

Kleyntjens, F., and O. R. Langworthy. 1937. Sensory nerve endings on the smooth muscle of the urinary bladder. J. Comp. Neurol., 67:367–380.

Klück, P. 1980. The autonomic innervation of the human urinary bladder, bladder neck and urethra: A histochemical study. Anat. Rec., 198:439–447.

Kuntz, A., and G. Saccomanno. 1944. Sympathetic innervation of the detrusor muscle. J. Urol., 51:535–542.

Langworthy, O. R., and E. L. Murphy. 1939. Nerve endings in the urinary bladder. J. Comp. Neurol., 71:487–509.

McCallum, R. W. 1979. The adult male urethra: Normal anatomy, pathology, and method of urethrography. Radiol. Clin. North Amer., 17:227–244.

Michaels, J. P. 1948. Study of ureteral blood supply and its bearing on necrosis of the ureter following the Wertheim operation. Surg. Gynecol. Obstet., 86:36–44.

Milroy, E. J., and A. T. K. Cockett. 1973. Lymphatic system of canine bladder. An anatomic study. Urology, 2:375–377.

Newell, Q. U. 1939. Injury to ureters during pelvic operations. Ann. Surg., 109:981–986.

Notley, R. G. 1969. The innervation of the upper ureter in man and in the rat: An ultrastructural study. J. Anat. (Lond.), 105:393–402.

Oelrich, T. M. 1980. The urethral sphincter muscle in the male. Amer. J. Anat., 158:229–246.

Parker, A. E. 1936. The lymph collectors from the urinary bladder and their connections with the main posterior lymph channels of the abdomen. Anat. Rec., 65:443–460.

Parker, A. E. 1936. The lymph vessels from the posterior urethra, their regional lymph nodes and relationships to the main posterior abdominal lymph channels. J. Urol., 36:538–557.

Powell, T. O. 1944. Studies in the lymphatics of the female urinary bladder. Surg. Gynecol. Obstet., 78:605–609.

Purinton, P. T., and J. E. Oliver, Jr. 1979. Spinal cord origin of innervation to the bladder and urethra of the dog. Exp. Neurol., 65:422–434.

Ricci, J. V., J. R. Lisa, and C. H. Thom. 1950. The female urethra: A histologic study as an aid in urethral surgery. Amer. J. Surg., 79:499–505.

Robson, M. C., and E. B. Ruth. 1962. Bilocular bladder: An anatomical study of a case: With a consideration of urinary tract anomalies. Anat. Rec., 142:63–71.

Rockswold, G. L., W. E. Bradley, and S. N. Chou. 1980. Innervation of the external urethral and external anal sphincters in higher primates. J. Comp. Neurol., 193:521–528.

Rolnick, H. C., and F. K. Arnheim. 1949. An anatomic study of the external urethral sphincter in relation to prostatic surgery. J. Urol., 61:591–603.

Shehata, R. 1976. The arterial supply of the urinary bladder. Acta Anat., 96:128–134.

Shehata, R. 1977. A comparative study of the urinary bladder and the intramural portion of the ureter. Acta Anat., 98:380–395.

Shehata, R. 1979. Venous drainage of the urinary bladder. Acta Anat., 105:61–64.

Simon, H. E., and N. A. Brandeberry. 1946. Anomalies of the urachus: Persistent fetal bladder. J. Urol., 55:401–408.

Tanagho, E. A., F. H. Meyers, and D. R. Smith. 1968. The trigone: Anatomical and physiological considerations. I. In relation to the uterovesical junction. J. Urol., 100:623–632.

Tanagho, E. A., and E. R. Miller. 1970. Initiation of voiding. Brit. J. Urol., 42:175–183.

Tanagho, E. A., and R. C. B. Pugh. 1963. The anatomy and function of the uterovesical junction. Brit. J. Urol., 35:151–165.

Tobin, C. E., and J. A. Benjamin. 1944. Anatomical study and clinical consideration of the fasciae limiting urinary extravasation from the penile urethra. Surg. Gynecol. Obstet., 79:195–204.

Trotter, M., and J. C. Finerty. 1948. An anomalous urinary bladder. Anat. Rec., 100:259–269.

Tyler, D. E. 1962. Stratified squamous epithelium in the vesical trigone and urethra: Findings correlated with the menstrual cycle and age. Amer. J. Anat., 111:319–335.

Uehling, D. T. 1978. The normal caliber of the adult female urethra. J. Urol., 120:176–177.

Wesson, M. B. 1920. Anatomical, embryological and physiological studies of the trigone and neck of the bladder. J. Urol., 4:279–315.

Wesson, M. B. 1950. Anatomy, physiology, embryology, and congenital abnormalities of the bladder. In Cyclopedia of Medicine, Surgery, and Specialties. 3rd ed, vol. 2. F. A. Davis Company, Philadelphia. pp. 211–229.

Wharton, L. R. 1932. Innervation of the ureter, with respect to denervation. J. Urol., 28:639–673.

Woodburne, R. T. 1961. The sphincter mechanism of the urinary bladder and the urethra. Anat. Rec., 141:11–20.

Woodburne, R. T. 1965. The ureter, ureterovesical juction, and vesical trigone. Anat. Rec., 151:243–249.

Woodburne, R. T., and J. Lapides. 1972. The ureteral lumen during peristalsis. Amer. J. Anat., 133:255–258.

Male Genital Organs

Testis and Coverings; Inguinal Canal, Spermatic Cord and Hernias

Backhouse, K. M., and H. Butler. 1960. The gubernaculum testis of the pig (*Sus scropha*). J. Anat. (Lond.), 94:107–120.

Bedford, J. M., M. Berrios, and G. L. Dryden. 1982. Biology of the scrotum. IV. Testis location and temperature sensitivity. J. Exp. Zool., 224:379–388.

Campbell, M. 1951. Undescended testicle and hypospadias. Amer. J. Surg., 82:8–17.

Carr, I., E. J. Clegg, and G. A. Meek. 1968. Sertoli cells as phagocytes: An electron microscopic study. J. Anat. (Lond.), 102:501–509.

Chang, K. S. F., F. K. Hsu, S. T. Chan, and Y. B. Chan. 1960. Scrotal asymmetry and handedness. J. Anat. (Lond.), 94:543–548.

Clark, G., and J. A. Gavan. 1962. Skeletal effects of prepubertal castration in the male chimpanzee. Anat. Rec., 143:179–181.

Clermont, Y. 1963. The cycle of the seminiferous epithelium in man. Amer. J. Anat., 112:35–51.

Clermont, Y. 1969. Two classes of spermatogonial stem cells in the monkey (Cercopithecus aethiops). Amer. J. Anat., 126:57–71.

Clermont, Y., and C. P. Leblond. 1955. Spermiogenesis of man, monkey, ram and other mammals as shown by "periodic acid-Schiff" technique. Amer. J. Anat., 96:229–253.

Comhaire, F., M. Kunnen, and C. Nahoum. 1981. Radiological anatomy of the internal spermatic vein(s) in 200 retrograde venograms. Internat. J. Androl., 4:379–387.

Crelin, E. S., and D. K. Blood. 1961. The influence of the testes on the shaping of the bony pelvis in mice. Anat. Rec., 140:375–379.

Dias, P. L. 1983. The effects of vasectomy on testicular volume. Brit. J. Urol., 55:83–84.

Dym, M. 1973. The fine structure of the monkey (Macaca) Sertoli cell and its role in maintaining the blood-testis barrier. Anat. Rec., 175:639–656.

Farris, E. J. 1950. *Human Fertility and Problems of the Male.* Author's Press, Inc., White Plains, New York. 211 pp.

Fawcett, D. W., and M. H. Burgos. 1960. Studies on the fine structure of the mammalian testis. II. The human interstitial tissue. Amer. J. Anat., 107:245–269.

Fawcett, D. W., and S. Ito. 1965. The fine structure of bat spermatozoa. Amer. J. Anat., 116:567–609.

Fawcett, D. W., W. B. Neaves, and M. N. Flores. 1973. Comparative observations on intertubular lymphatics and the organization of the interstitial tissue of the mammalian testis. Biol. Reprod., 9:500–532.

Flickinger, C., and D. W. Fawcett. 1967. The junctional specializations of Sertoli cells in the seminiferous epithelium. Anat. Rec., 158:207–221.

Flickinger, C. J. 1972. Ultrastructure of the rat testis after vasectomy. Anat. Rec., 174:477–493.

Gray, D. J. 1947. The intrinsic nerves of the testis. Anat. Rec., 98:325–335.

Hamilton, J. B. 1942. Male hormone stimulation is prerequisite and an incitant in common baldness. Amer. J. Anat., 71:451–480.

Hamilton, J. B. 1946. A secondary sexual character that develops in men but not in women upon ageing of an organ present in both sexes. Anat. Rec., 94:466–467. Abstract.

Hamilton, J. B. 1948. The role of testicular secretions as indicated by the effects of castration in man and by studies of pathological conditions and the short lifespan associated with maleness. Recent Prog. Hor. Res., 3:257–322.

Hill, E. C. 1909. The vascularization of the human testis. Amer. J. Anat., 9:463–474.

Hoffer, A. P., and J. Greenberg. 1978. The structure of the epididymis, efferent ductules and ductus deferens of the guinea pig: a light microscope study. Anat. Rec., 190:659–677.

Kormano, M., and H. Suoranta. 1971. Microvascular organization of the adult human testis. Anat. Rec., 170:31–40.

Kuntz, A., and R. E. Morris, Jr. 1946. Components and distribution of the spermatic nerves and the nerves of the vas deferens. J. Comp. Neurol., 85:33–44.

Leblond, C. P., and Y. Clermont. 1952. Definition of the stages of the cycle of the seminiferous epithelium in the rat. Ann. N. Y. Acad. Sci., 55:548–573.

Morehead, J. R., and C. F. Morgan. 1963. Renewal of spermatogenesis following 28-day cryptorchidism in the rat as affected by injections of testosterone-propionate. Anat. Rec., 145:262. Abstract.

Narbaitz, R. 1962. The primordial germ cells in the male human embryo. Carneg. Inst., Contr. Embryol., 37:117–119.

Nathan, H., P. V. Tobias, and M. D. Wellsted. 1976. An unusual case of right and left testicular arteries arching over the left renal vein. Brit. J. Urol., 48:135–138.

Nicander, L. 1967. An electron microscopical study of cell contacts in the seminiferous tubules of some mammals. Z. Zellforsch. mikrosk. Anat., 83:375–397.

Niemi, M., and M. Kormano. 1965. Cyclical changes in the significance of lipids and acid phosphatase activity in the seminiferous tubules of the rat testis. Anat. Rec., 151:159–170.

Pfeiffer, C. A., and A. Kirschbaum. 1943. Relation of interstitial cell hyperplasia to secretion of male hormone in the sparrow. Anat. Rec., 85:211–227.

Roosen-Runge, E. C., and L. O. Giesel, Jr. 1950. Quantitative studies on spermatogenesis in the albino rat. Amer. J. Anat., 87:1–30.

Roosen-Runge, E. C., and A. F. Holstein. 1978. The human rete testis. Cell Tissue Res., 189:409–433.

Ross, M. H. 1974. The organization of the seminiferous epithelium in the mouse testis following ligation of the efferent ductules. A light microscopic study. Anat. Rec., 180:565–580.

Rothwell, B., and M. D. Tingari. 1973. The ultrastructure of the boundary tissue of the seminiferous tubule in the testis of the domestic fowl (Gallus domesticus). J. Anat. (Lond.), 114:321–328.

Scorer, C. G. 1964. The descent of the testis. Arch. Dis. Child., 39:605–609.

Smith, R. D. 1949. The permeability of Colles' fascia for urine. J. Urol., 62:535–541.

Vitale, R., D. W. Fawcett, and M. Dym. 1973. The normal development of the blood-testis barrier and the effects of clomiphene and estrogen treatment. Anat. Rec., 176:333–344.

Wells, L. J. 1943. Descent of the testis; anatomical and hormonal considerations. Surgery, 14:436–472.

Wells, L. J., and D. State. 1947. Misconception of the gubernaculum testis. Surgery, 22:502–508.

Williams, R. G. 1949. Some responses of living blood vessels and connective tissue to testicular grafts in rabbits. Anat. Rec., 104:147–161.

Wislocki, G. B. 1933. Observations on the descent of the testes in the macaque and in the chimpanzee. Anat. Rec., 57:133–148.

Woolard, H. H., and E. A. Carmichael. 1933. The testis and referred pain. Brain, 56:293–303.

Wyndham, N. R. 1943. A morphological study of testicular descent. J. Anat. (Lond.), 77:179–188.

Seminal Vesicles; Prostate Gland; Ductus Deferens

Aboul-Azm, T. E. 1979. Anatomy of the human seminal vesicles and ejaculatory ducts. Arch. Androl., 3:287–292.

Aitken, R. N. C. 1959. Observations on the development of the seminal vesicles, prostate and bulbourethral glands in the ram. J. Anat. (Lond.), 93:43–51.

Batson, O. V. 1940. The function of the vertebral veins and their role in the spread of metastases. Ann. Surg., 112:138–149.

Clegg, E. J. 1955. The arterial supply of the human prostate and seminal vesicles. J. Anat. (Lond.), 89:209–216.

Clegg, E. J. 1956. The vascular arrangements within the human prostate gland. Brit. J. Urol., 28:428–435.

LeDuc, I. E. 1939. Anatomy of the prostate and pathology of early benign hypertrophy. J. Urol., 42:1217–1241.

Fisher, E. R., and W. Jeffrey. 1965. Ultrastructure of human normal and neoplastic prostate; with comments related to prostatic effects of hormonal stimulation in the rabbit. Amer. J. Clin. Pathol., 44:119–134.

Flickinger, C. J. 1974. Synthesis, intracellular transport, and release of secretory protein in the seminal vesicle of the rat, as studied by electron microscope radioautography. Anat. Rec., 180:407–426.

Flocks, R. H. 1937. The arterial distribution within the prostate gland: Its rôle in transurethral prostatic resection. J. Urol., 37:524–548.

Furness, J. B., and T. Iwayama. 1972. The arrangement and identification of axons innervating the vas deferens of the guinea-pig. J. Anat. (Lond.), 113:179–196.

Glenister, T. W. 1962. The development of the utricle and of the so-called "middle" or "median" lobe of the human prostate. J. Anat. (Lond.), 96:443–455.

Gosling, J. A., and J. S. Dixon. 1972. Differences in the manner of autonomic innervation of the muscle layers of the guinea-pig ductus deferens. J. Anat. (Lond.), 112:81–91.

Hutch, J. A., and O. N. Rambo, Jr. 1970. A study of the anatomy of the prostate, prostatic urethra and urinary sphincter system. J. Urol., 104:443–452.

Kuntz, A., and R. E. Morris, Jr. 1946. Components and distribution of the spermatic nerves and the nerves of the vas deferens. J. Comp. Neurol., 85:33–44.

Lowsley, O. S. 1930. Embryology, anatomy and surgery of the prostate gland: With a report of operative results. Amer. J. Surg., 8:526–541.

McMahon, S. 1938. An anatomical study by injection technique of the ejaculatory ducts and their relations. J. Anat. (Lond.), 72:556–574.

Richardson, K. C. 1962. The fine structure of autonomic nerve endings in smooth muscle of the rat vas deferens. J. Anat. (Lond.), 96:427–442.

Riva, A. 1967. Fine structure of human seminal vesicular epithelium. J. Anat. (Lond.), 102:71–86.

Riva, A., and R. A. Stockwell. 1969. A histochemical study of human seminal vesicle epithelium. J. Anat. (Lond.), 104:253–262.

Swyer, G. I. M. 1944. Post-natal growth changes in the human prostate. J. Anat. (Lond.), 78:130–145.

Tanahashi, Y., H. Watanabe, D. Igari, K. Harada, and M. Saitoh. 1975. Volume estimation of the seminal vesicles by means of transrectal ultrasonotomography: A preliminary report. Brit. J. Urol., 47:695–702.

Tobin, C. E., and J. A. Benjamin. 1945. Anatomical and surgical restudy of Denonvilliers' fascia. Surg. Gynecol. Obstet., 80:373–388.

Yamauchi, A., and G. Burnstock. 1969. Post-natal development of smooth muscle cells in the mouse vas deferens: A fine structural study. J. Anat. (Lond.), 104:1–15.

Bulbourethral Glands and Penis

Alvarez-Morujo, A. 1967. Terminal arteries of the penis. Acta Anat., 67:387–398.

Ashdown, R. R., and H. Gilanpour. 1974. Venous drainage of the corpus cavernosum penis in impotent and normal bulls. J. Anat. (Lond.), 117:159–170.

Ashdown, R. R., S. W. Ricketts, and R. C. Wardley. 1968. The fibrous architecture of the integumentary coverings of the bovine penis. J. Anat. (Lond.), 103:567–572.

Aykroyd, O. E., and S. Zuckerman. 1938. The effect of sex-hormones on the bulbourethral glands of rhesus monkeys. J. Anat. (Lond.), 73:135–144.

Carter, J. P., N. N. Isa, N. Hashem, and F. O. Raasch, Jr. 1968. Congenital absence of the penis: A case report. J. Urol., 99:766–768.

Casey, W. C., and R. W. Woods. 1982. Anatomy and histology of penile deep dorsal vein: Venous cushions and proximal "sphincter." Urology, 19:284–286.

Christensen, G. C. 1954. Angioarchitecture of the canine penis and the process of erection. Amer. J. Anat., 95:227–261.

Cohen, S. J. 1968. Diphallus with duplication of colon and bladder. Proc. Roy. Soc. Med., 61:305.

Congdon, E. D., and J. M. Essenberg. 1955. Subcutaneous attachments of the human penis and scrotum. A study of 55 series of gross sections. Amer. J. Anat., 97:331–357.

Dail, W. G., Jr., and A. P. Evan, Jr. 1974. Experimental evidence indicating that the penis of the rat is innervated by short adrenergic neurons. Amer. J. Anat., 141:203–217.

Deysach, L. J. 1939. The comparative morphology of the erectile tissue of the penis with especial emphasis on the probable mechanism of erection. Amer. J. Anat., 64:111–131.

Feldman, K. W., and D. W. Smith. 1975. Fetal phallic growth and penile standards for newborn male infants. J. Pediatr., 86:395–398.

Fitzpatrick, T. 1975. The corpus cavernosum intercommunicating venous drainage system. J. Urol., 113:494–496.

Fitzpatrick, T. J. 1974. Venography of the deep dorsal venous and valvular systems. J. Urol., 111:518–520.

Fitzpatrick, T. J. 1982. The penile intercommunicating venous valvular system. J. Urol., 127:1099–1100.

Fitzpatrick, T. J., and J. F. Cooper. 1975. A cavernosogram study on the valvular competence of the human deep dorsal vein. J. Urol., 113:497–499.

Forshall, I., and P. P. Rickham. 1956. Transposition of the penis and scrotum. Brit. J. Urol., 28:250–252.

Fujimoto, S., and Y. Takeshige. 1975. The wall structure of the arteries in the corpora cavernosa penis of rabbits: Light and electron microscopy. Anat. Rec., 181:641–657.

Goldstein, A. M., J. P. Meehan, R. Zakhary, P. A. Buckley, and F. A. Rogers. 1982. New observations on microarchitecture of corpora cavernosa in man and possible relationship to mechanism of erection. Urology, 20:259–266.

Hart, B. L. 1972. The action of extrinsic penile muscles during copulation in the male dog. Anat. Rec., 173:1–5.

Hart, B. L., and R. L. Kitchell. 1965. External morphology

of the erect glans penis of the dog. Anat. Rec., 152:193–198.

Huguet, J. F., J. Clerissi, and C. Juhan. 1981. Radiologic anatomy of pudendal artery. Europ. J. Radiol., 1:278–284.

Leeson, T. S., and C. R. Leeson. 1965. The fine structure of cavernous tissue in the adult rat penis. Invest. Urol., 3:144–154.

Lierse, W. 1982. Blood vessels and nerves of the human penis. Urol. Int., 37:145–151.

Rukstinat, G. J., and R. J. Hasterlik. 1939. Congenital absence of the penis. Arch. Pathol., 27:984–993.

Shira, M., K. Sasaki, and A. Rikimaru. 1972. Histochemical investigation on the distribution of adrenergic and cholinergic nerves in human penis. Tohoku J. Exp. Med., 107:403–404.

Tobin, C. E., and J. A. Benjamin. 1944. Anatomical study and clinical consideration of the fasciae limiting urinary extravasation from the penile urethra. Surg. Gynecol. Obstet., 79:195–204.

Tobin, C. E., and J. A. Benjamin. 1949. Anatomic and clinical re-evaluation of Camper's, Scarpa's and Colles' fasciae. Surg. Gynecol. Obstet., 88:545–559.

Uhlenhuth, E., R. D. Smith, E. C. Day, and E. B. Middleton. 1949. A re-investigation of Colles' and Buck's fasciae in the male J. Urol., 62:542–563.

Wesson, M. B. 1945. The value of Buck's and Colles' fasciae. J. Urol., 53:365–372.

Female Genital Organs

Ovaries and Ovulation

Baker, T. G., and P. Neal. 1974. Oogenesis in human fetal ovaries maintained in organ culture. J. Anat. (Lond.), 117:591–604.

Baker, T. G., and P. Neal. 1974. Organ culture of cortical fragments and Graafian follicles from human ovaries. J. Anat. (Lond.), 117:361–371.

Bassett, D. L. 1943. The changes in the vascular pattern of the ovary of the albino rat during the estrous cycle. Amer. J. Anat., 73:251–291.

Byskov, A. G., and S. Lintern-Moore. 1973. Follicle formation in the immature mouse ovary: The role of the rete ovarii. J. Anat. (Lond.), 116:207–217.

Curry, T. E., Jr., C. A. Hodson, M. L. Capps, P. D. Mozley, D. E. D. Jones, and H. W. Burden. 1982. A light and electron microscopic study of the human ovarian ligament. Acta Anat., 112:178–184.

Dahl, E. 1971. Studies of the fine structure of ovarian interstitial tissue. I. A comparative study of the fine structure of the ovarian interstitial tissue in the rat and the domestic fowl. J. Anat. (Lond.), 108:275–290.

Deanesly, R. 1970. Oögenesis and the development of the ovarian interstitial tissue in the ferret. J. Anat. (Lond.), 107:165–178.

Deanesly, R. 1972. Origins and development of interstitial tissue in ovaries of rabbit and guinea-pig. J. Anat. (Lond.), 113:251–260.

Duke, K. L. 1947. The fibrous connective tissue of the rabbit ovary from sex differentiation to maturity. Anat. Rec., 98:507–525.

Everett, J. W. 1945. The microscopically demonstrable lipids of cyclic corpora lutea in the rat. Amer. J. Anat., 77:293–323.

Everett, J. W. 1956. Functional corpora lutea maintained for months by autografts of rat hypophyses. Endocrinology, 58:786–796.

Everett, N. B. 1942. The origin of ova in the adult opossum. Anat. Rec., 82:77–91.

Farris, E. J. 1946. The time of ovulation in the monkey. Anat. Rec., 95:337–345.

Ginther, O. J., D. J. Dierschke, S. W. Walsh, and C. H. Del Campo. 1974. Anatomy of arteries and veins of uterus and ovaries in rhesus monkeys. Biol. Reprod., 11:205–219.

Green, J. A. 1957. Some effects of advancing age on the histology and reactivity of the mouse ovary. Anat. Rec., 129:333–347.

Greenwald, G. S. 1964. Ovarian follicular development in the pregnant hamster. Anat. Rec., 148:605–609.

Hill, R. T., E. Allen, and T. C. Kramer. 1935. Cinemicrographic studies of rabbit ovulation. Anat. Rec., 63:239–245.

Kent, H. A., Jr. 1962. Polyovular follicles and multinucleate ova in the ovaries of young hamsters. Anat. Rec., 143:345–349.

Koering, M. J. 1974. Comparative morphology of the primate ovary. Contrib. Primatol., 3:38–81.

Krehbiel, R., and J. C. Plagge. 1962. Distribution of ova in the rat uterus. Anat. Rec., 143:239–241.

Langman, L., and H. S. Burr. 1942. Electrometric timing of human ovulation. Amer. J. Obstet. Gynecol., 44:223–230.

Mangoushi, M. A. 1974. The fate of grafted fetal ovaries in the rat. J. Anat. (Lond.), 118:601–610.

Mangoushi, M. A. 1975. Scrotal allografts of fetal ovaries. J. Anat. (Lond.), 120:595–599.

Marvin, H. N. 1947. Diestrus and the formation of corpora lutea in rats with persistent estrus, treated with desoxycorticosterone acetate. Anat. Rec., 98:383–391.

Mitchell, G. A. G. 1938. The innervation of the ovary, uterine tube, testis and epididymis. J. Anat. (Lond.), 72:508–517.

Mori, H., and K. Matsumoto. 1973. Development of the secondary interstitial gland in the rabbit ovary. J. Anat. (Lond.), 116:417–430.

Mossman, H. W., M. J. Koering, and D. Ferry, Jr. 1964. Cyclic changes of interstitial gland tissue of the human ovary. Amer. J. Anat., 115:235–255.

Ohlson, L. 1978. Topography of ovarian veins in pregnancy. Acta Radiol. Diagn. (Stockholm), 19:669–674.

Pankratz, D. S. 1938. Some observations on the graafian follicles in an adult human ovary. Anat. Rec., 71:211–219.

Pincus, G. 1939. The comparative behavior of mammalian eggs *in vivo* and *in vitro*. IV. The development of fertilized and artificially activated rabbit eggs. J. Exp. Zool., 82:85–129.

Rennie, P., G. R. Davenport, and W. Welborn. 1965. Cellular changes, protein and glycogen content in developing rabbit corpora lutea. Anat. Rec., 153:289–293.

Skalko, R. G., J. R. Ruby, and R. F. Dyer. 1968. Demonstration of mast cells in the postnatal mouse ovary. Anat. Rec., 161:459–463.

Watzka, M. 1957. Weibliche Genitalorgan: Das Ovarium. In *Handbuch der mikroskopischen Anatomie des Menschen*. Vol. 7, Pt. 3. Springer-Verlag, Berlin. 178 pp.

Uterus and Menstruation; Uterine Tubes; Female Pelvis

Alden, R. H. 1942. The oviduct and egg transport in the albino rat. Anat. Rec., 84:137–169.

Alden, R. H. 1945. Implantation of the rat egg. I. Experimental alteration of uterine polarity. J. Exp. Zool., 100:229–235.

Asdell, S. A. 1964. *Patterns of Mammalian Reproduction*. 2nd ed. Comstock Publ. Assoc., Ithaca, New York. 670 pp.

Bartelmez, G. W. 1953. Factors in the variability of the menstrual cycle. Anat. Rec., 115:101–120.

Beck, A. J., and F. Beck. 1967. The origin of intraarterial cells in the pregnant uterus of the macaque (*Macaca mulatta*). Anat. Rec., 158:111–113.

Blandau, R. J., and D. L. Odor. 1949. The total number of spermatozoa reaching various segments of the reproductive tract in the female albino rat at intervals after insemination. Anat. Rec., 103:93–109.

Bo, W. J. 1956. The relationship between vitamin A deficiency and estrogen in producing uterine metaplasia in the rat. Anat. Rec., 124:619–627.

Bo, W. J., D. L. Odor, and M. L. Rothrock. 1969. Ultrastructure of uterine smooth muscle following progesterone or progesterone-estrogen treatment. Anat. Rec., 163:121–132.

Bradley, C. F., and C. E. Graham. 1972. Effects of estrogen on the transition zone of the mouse uterine cervix. Anat. Rec., 173:235–248.

Bulmer, D., and S. Peel. 1974. An autoradiographic study of cellular proliferation in the uterus and placenta of the pregnant rat. J. Anat. (Lond.), 117:433–441.

Burrows, H. 1949. *Biological Actions of Sex Hormones*. 2nd ed. Cambridge University Press, London. 616 pp.

Curtis, A. H., and B. J. Anson. 1938. Bilateral double infundibulum of the uterine tube. Anat. Rec., 71:177–179.

Curtis, A. H., B. J. Anson, F. L. Ashley, and T. Jones. 1942. The blood vessels of the female pelvis in relation to gynecological surgery. Surg. Gynecol. Obstet., 75:421–423.

El-Banna, A. A., and E. S. E. Hafez. 1970. Profile analysis of the oviductal wall in rabbits and cattle. Anat. Rec., 166:469–478.

Ford, D. H. 1956. A study of the changes in vaginal alkaline phosphatase activity during the estrous cycle in adult and in young "first-estrous" rats. Anat. Rec., 125:261–277.

Gardner, W. U. 1955. Localization of strain differences in vaginal sensitivity to estrogens. Anat. Rec., 121:297–298.

Goss, C. M. 1962. On the anatomy of the uterus. Anat. Rec., 144:77–83.

Greulich, W. W., and H. Thoms. 1944. The growth and development of the pelvis of individual girls before, during, and after puberty. Yale J. Biol. Med., 17:91–98.

Hirsch, E. F., and M. E. Martin. 1943. The distribution of nerves in the adult human myometrium. Surg. Gynecol. Obstet., 76:697–702.

Jarcho, J. 1946. Malformations of the uterus; review of the subject, including embryology, comparative anatomy, diagnosis and report of cases. Amer. J. Surg., 71:106–166.

Knudsen, J. F., A. Costoff, and V. B. Mahesh. 1974. Correlation of serum gonadotropins, ovarian and uterine histology in immature and prepubertal rats. Anat. Rec., 180:497–508.

Krueger, W. A., and H. C. Maibenco. 1972. DNA replication and cell division in the hamster uterus. Anat. Rec., 173:229–234.

Lisa, J. R., J. D. Gioia, and I. C. Rubin. 1954. Observations on the interstitial portion of the fallopian tube. Surg. Gynecol. Obstet., 99:159–169.

Markee, J. E. 1940. Menstruation in intraocular endometrial transplants in the rhesus monkey. Carneg. Inst.,

Contr. Embryol., 28:221–308.

Markee, J. E. 1951. Physiology of reproduction. Ann. Rev. Physiol., 13:367–396.

Mastroianni, L., Jr. 1962. The structure and function of the fallopian tube. A correlative review. Clin. Obstet. Gynecol., 5:781–790.

McLean, J. M., and R. J. Scothorne. 1970. The lymphatics of the endometrium in the rabbit. J. Anat. (Lond.), 107:39–48.

McLean, J. M., and R. J. Scothorne. 1972. The fate of skin allografts in the rabbit uterus. J. Anat. (Lond.), 112:423–432.

Nakanishi, H., H. Wansbrough, and C. Wood. 1967. Postganglionic sympathetic nerve innervating human fallopian tube. Amer. J. Physiol., 213:613–619.

Nilsson, O. 1962. Electron microscopy of the glandular epithelium in the human uterus. I. Follicular phase. J. Ultrastruct. Res., 6:413–421.

Nilsson, O. 1962. Electron microscopy of the glandular epithelium in the human uterus. II. Early and late leuteal phase. J. Ultrastruct. Res., 6:422–431.

O'Leary, J. L., and J. A. O'Leary. 1966. Uterine artery ligation in the control of intractable postpartum hemorrhage. Amer. J. Obstet. Gynecol., 94:920–924.

Owman, C., E. Rosengren, and N.-O. Sjöberg. 1967. Adrenergic innervation of the human female reproductive organs: A histochemical and chemical investigation. Obstet. Gynecol., 30:763–773.

Papanicolaou, G. N. 1933. The sexual cycle in the human female as revealed by vaginal smears. Amer. J. Anat., 52:519–637.

Papanicolaou, G. N., H. F. Traut, and A. A. Marchetti. 1948. *The Epithelia of Woman's Reproductive Organs*. The Commonwealth Fund, New York. 53 pp.

Peppler, R. D. 1976. Effect of uterine artery ligation on ovulation in the rat. Anat. Rec., 184:183–185.

Power, R. M. H. 1944. The exact anatomy and development of the ligaments attached to the cervix uteri. Surg. Gynecol. Obstet., 79:390–396.

Ramsey, E. M. 1955. Vascular patterns in the endometrium and the placenta. Angiology, 6:321–338.

Reynolds, S. R. M. 1965. *Physiology of the Uterus*. Hafner Publishing Company, New York. 619 pp.

Sampson, J. A. 1937. The lymphatics of the mucosa of the fimbriae of the fallopian tube. Amer. J. Obstet. Gynecol., 33:911–930.

Schlegel, J. U. 1945/46. Arteriovenous anastomoses in the endometrium in man. Acta Anat., 1:284–325.

Song, J. 1964. *The Human Uterus, Morphogenesis and Embryological Basis for Cancer*. Charles C Thomas, Springfield, Ill. 196 pp.

Sorger, T., R. Pittman, and A. L. Soderwall. 1983. Principal features of the nerve supply to the ovary, oviduct and tubal third of the uterus in the golden hamster. Biol. Reprod., 28:461–482.

Sweeney, W. J., III. 1962. The interstitial portion of the uterine tube—Its gross anatomy, course and length. Obstet. Gynecol., 19:1–8.

Thoms, H., and W. W. Greulich. 1940. A comparative study of male and female pelves. Amer. J. Obstet. Gynecol., 39:56–62.

Toth, S., and A. Toth. 1974. Undescribed muscle bundle of the human uterus: Fasciculus cervicoangularis. Amer. J. Obstet. Gynecol., 118:979–984.

Uhlenhuth, E. 1953. *Problems in the Anatomy of the Pelvis*. J. B. Lippincott Company, Philadelphia. 206 pp.

Wermuth, E. G. 1939. Anastomoses between the rectal and uterine veins forming a connexion between the somatic and portal venous system in the recto-uterine pouch. J. Anat. (Lond.), 74:116–126.

Wislocki, G. B., and E. W. Dempsey. 1939. Remarks on the lymphatics of the reproductive tract of the female rhesus monkey (*Macaca mulatta*). Anat. Rec., *75*:341-363.

Woodruff, J. D., and C. J. Pauerstein. 1969. *The Fallopian Tube.* Williams and Wilkins Company, Baltimore. 361 pp.

Vagina and Female External Genital Organs

Abrams, R. M., P. S. Kalra, and C. J. Wilcox. 1978. Vaginal blood flow during the menstrual cycle. Amer. J. Obstet. Gynecol., *132*:396-400.

Barker, T. E., and B. E. Walker. 1966. Initiation of irreversible differentiation in vaginal epithelium. Anat. Rec., *154*:149-160.

Bryan, A. L., J. A. Nigro, and V. S. Counseller. 1949. One hundred cases of congenital absence of the vagina. Surg. Gynecol. Obstet., *88*:79-86.

Bulmer, D. 1957. The development of the human vagina. J. Anat. (Lond.), *91*:490-509.

Bulmer, D. 1959. Histochemical observations on the foetal vaginal epithelium. J. Anat. (Lond.), *93*:36-42.

Capraro, V. J., and M. B. Gallego. 1976. Vaginal agenesis. Amer. J. Obstet. Gynecol., *124*:98-107.

Hisano, N. 1977. Postnatal development of vagina, clitoris and urethral glands of the golden hamster. Acta Anat., *97*:371-378.

Huffman, J. W. 1943. The development of the periurethral glands in the human female. Amer. J. Obstet. Gynecol., *46*:773-785.

Huffman, J. W. 1951. Clinical significance of the paraurethral ducts and glands. Arch. Surg., *62*:615-626.

Jarosz, S. J., T. J. Kuehl, and W. R. Dukelow. 1977. Vaginal cytology, induced ovulation and gestation in the squirrel monkey (*Saimiri sciureus*). Biol. Reprod., *16*:97-103.

Kuhn, R. J., and V. E. Hollyock. 1982. Observations on the anatomy of the rectovaginal pouch and septum. Obstet. Gynecol., *59*:445-447.

Mahran, M., and A. M. Saleh. 1964. The microscopic anatomy of the hymen. Anat. Rec., *149*:313-318.

Ohta, Y., and T. Iguchi. 1976. Development of the vaginal epithelium showing estrogen-independent proliferation and cornification in neonatally androgenized mice. Endocrinol. Jpn., *23*:333-340.

O'Rahilly, R. 1977. The development of the vagina in the human. Birth Defects, *13*:123-136.

Plapinger, L. 1982. Surface morphology of uterine and vaginal epithelia in mice during normal postnatal development. Biol. Reprod., *26*:961-972.

Ricci, J. V., J. R. Lisa, C. H. Thom, and W. L. Kron. 1947. The relationship of the vagina to adjacent organs in reconstructive surgery: A histologic study. Amer. J. Surg., *74*:387-410.

Shaw, W. 1947. A study of the surgical anatomy of the vagina, with special reference to vaginal operations. Brit. Med. J., *1*:477-482.

Uhlenhuth, E., W. M. Wolfe, E. M. Smith, and E. B. Middleton. 1948. The rectogenital septum. Surg. Gynecol. Obstet., *86*:148-163.

Wagner, G., and B. Ottesen. 1980. Vaginal blood flow during sexual stimulation. Obstet. Gynecol., *56*:621-624.

Mammary Gland

Agate, F. J., Jr. 1952. The growth and secretory activity of the mammary glands of the pregnant rhesus monkey (*Macaca mulatta*) following hypophysectomy. Amer. J. Anat., *90*:257-283.

Brown, P. J., F. J. Bourne, and H. R. Denny. 1975.

Immunoglobulin-containing cells in pig mammary gland. J. Anat. (Lond.), *120*:329-335.

Cholnoky, T. de 1939. Supernumerary breast. Arch. Surg., *39*:926-941.

Cunningham, L. 1977. The anatomy of the arteries and veins of the breast. J. Surg. Oncol., *9*:71-85.

Gaffney, E. V., F. P. Polanowski, S. E. Blackburn, and J. P. Lambiase. 1976. Origin, concentration and structural features of human mammary gland cells cultured from breast secretions. Cell Tissue Res., *172*:269-279.

Giacometti, L., and W. Montagna. 1962. The nipple and areola of the human female breast. Anat. Rec., *144*:191-197.

Gladstone, R. J. 1930. Axillary mamma in a man. J. Anat. (Lond.), *64*:239-246.

Karsner, H. T. 1946. Gynecomastia. Amer. J. Pathol., *22*:235-315.

Maliniac, J. W. 1943. Arterial blood supply of the breast. Arch. Surg., *47*:329-343.

McKiernan, J. F., and D. Hull. 1981. Prolactin, maternal oestrogens, and breast development in the newborn. Arch. Dis. Child., *56*:770-774.

Morehead, J. R. 1982. Anatomy and embryology of the breast. Clin. Obstet. Gynecol., *25*:353-357.

Murad, T. M. 1970. Ultrastructural study of rat mammary gland during pregnancy. Anat. Rec., *167*:17-36.

Radnor, C. J. P. 1972. Myoepithelial cell differentiation in rat mammary glands. J. Anat. (Lond.), *111*:381-398.

Radnor, C. J. P. 1972. Myoepithelium in the prelactating and lactating mammary glands of the rat. J. Anat. (Lond.), *112*:337-353.

Radnor, C. J. P. 1972. Myoepithelium in involuting mammary glands of the rat. J. Anat. (Lond.), *112*:355-365.

Richardson, K. C. 1949. Contractile tissues in the mammary gland, with special reference to myoepithelium in the goat. Proc. Roy. Soc. Lond. Series B, *136*:30-45.

Romano, S. A., and E. M. McFetridge. 1938. The limitations and dangers of mammography by contrast mediums. JAMA, *110*:1905-1910.

Sakki, S. 1974. Angiography of the female breast. Ann. Clin. Res., *6*:Suppl., 12, 1-47.

Sala, N. L., E. C. Luther, J. C. Arballo, and J. Cordero Funes. 1974. Roles of temperature, pressure, and touch in reflex milk ejection in lactating women. J. Appl. Physiol., *37*:840-843.

Spratt, J. S. 1979. Anatomy of the breast. Major Probl. Clin. Surg., *5*:1-13.

Sykes, P. A. 1969. The nerve supply of the human nipple. J. Anat. (Lond.), *105*:201.

Thorek, M. 1942. *Plastic Surgery of the Breast and Abdominal Wall.* Charles C Thomas, Springfield, Ill. 446 pp.

Traurig, H. H., and C. F. Morgan. 1962. Autoradiographic investigation of mouse mammary gland growth by the incorporation of tritiated thymidine into its epithelial components. Anat. Rec., *142*:286-287.

Wellings, S. R., and J. R. Philip. 1964. The function of the Golgi apparatus in lactating cells of the BALB/cCrgl mouse. An electron microscopic and autoradiographic study. Z. Zellforsch. mikrosk. Anat., *61*:871-882.

Williams, W. L. 1942. Normal and experimental mammary involution in the mouse as related to the inception and cessation of lactation. Amer. J. Anat., *71*:1-41.

Williams, W. L. 1945. The effects of lactogenic hormone on post-parturient unsuckled mammary glands of the mouse. Anat. Rec., *93*:171-183.

18

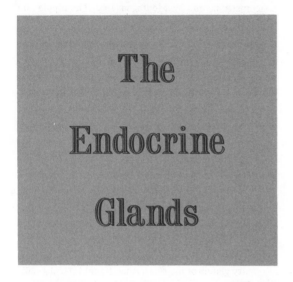

The

Endocrine

Glands

Endocrine glands or **ductless glands** are grouped together because of their common characteristics of not having ducts and of discharging their specific secretions, called hormones, into the blood stream.

Hormones are chemical messengers, and although they are carried by the blood to all parts of the body, only certain organs or types of cells are able to respond to their stimulation. The specific organ that does respond to a particular hormone is called its **target organ.** The body produces several hormones other than those secreted by the endocrine glands described in this chapter, such as insulin, from the islands of Langerhans[1] in the pancreas; androgens and estrogens, from the gonads; and secretin, from the mucous membrane of the gastrointestinal tract. The structures secreting these hormones are described in other chapters.

The endocrine glands included in this chapter are the thyroid and parathyroid glands, the hypophysis or pituitary gland,

[1]Paul Langerhans (1847–1888): A German anatomist and pathologist (Berlin and Freiburg).

the suprarenal glands, and the paraganglia. The pineal gland is discussed in Chapter 11, with the epithalamus. The thymus was once classified with the ductless glands but is now included in the lymphatic system.

Thyroid Gland

DEVELOPMENT

The thyroid gland is developed from a median diverticulum of the ventral wall of the pharynx (Fig. 18-1) that appears about the fourth week on the summit of the tuberculum impar, but later is found in the furrow immediately caudal to the tuberculum. It grows distalward and caudalward as a tubular duct, which bifurcates and subsequently subdivides into a series of cellular cords, from which the isthmus and lateral lobes of the thyroid gland are developed. The connection of the median diverticulum with the pharynx is termed the **thyroglossal duct;** its continuity is subsequently interrupted and it degenerates, its cranial end being represented by the foramen cecum of the tongue, and its caudal end, by the pyramidal lobe of the thyroid gland.

ANATOMY

The thyroid gland (Fig. 18-2) is a highly vascular organ situated at the front of the neck. It consists of right and left lobes con-

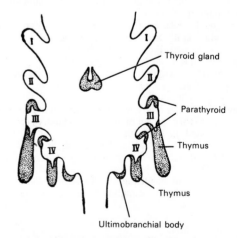

FIG. 18-1. Scheme showing development of brachial epithelial bodies. *I, II, III, IV,* Branchial pouches. (Modified from Kohn.)

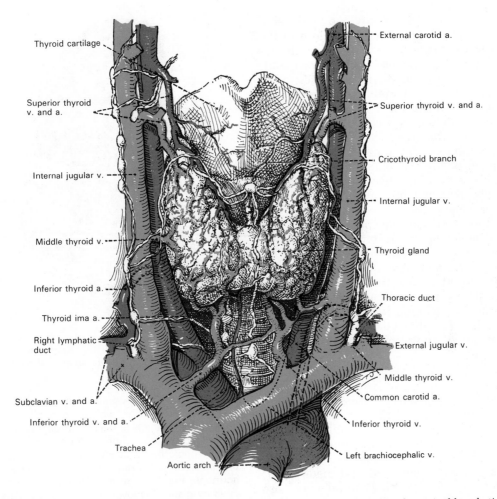

FIG. 18-2. Blood supply of the thyroid gland, viewed from in front. The thoracic duct and principal lymphatics are also shown. (Ecyleshymer and Jones.)

nected across the midline by a narrow portion, the isthmus. Its weight is somewhat variable, but usually about 30 gm. It is slightly heavier in women and becomes enlarged during pregnancy.

The **lobes** have a conical shape, the apex of each being directed cranialward and lateralward as far as the junction of the middle with the caudal third of the thyroid cartilage; the base looks caudalward and is on a level with the fifth or sixth tracheal ring. Each lobe is about 5 cm long; its greatest width is about 3 cm, and its thickness about 2 cm. The **lateral** or **superficial surface** is convex and covered by the skin, the subcutaneous and deep fasciae, the sternocleidomastoid, the superior belly of the omohyoid, the sternohyoid, and the sternothyroid. Beneath the sternothyroid muscle, the vis-

ceral layer of the deep fascia forms a capsule for the gland. The **deep** or **medial surface** is molded over the underlying structures: the trachea, the inferior pharyngeal constrictor and the posterior part of the cricothyroid muscle, the esophagus (particularly on the left side of the neck), the superior and inferior thyroid arteries, and the recurrent nerves. The **ventral border** is thin and inclines obliquely from above downward toward the midline of the neck, while the **dorsal border** is thick and overlaps the common carotid artery and, as a rule, the parathyroids.

The **isthmus** connects the lower thirds of the lobes. It measures about 1.25 cm in breadth and in depth, and usually covers the second and third rings of the trachea. Its situation and size have many variations. In the midline of the neck it is covered by the skin and fascia, and on either side, close to the midline, by the sternothyroid. Across its cranial border runs an anastomotic branch uniting the two superior thyroid arteries; at its caudal border are the inferior thyroid veins. Sometimes the isthmus is altogether wanting.

A third conical lobe, called the **pyramidal lobe,** frequently arises from the cranial part of the isthmus or from the adjacent portions of either lobe (most commonly the left) and ascends as far as the hyoid bone. It is occasionally detached or may be divided into two or more parts.

A fibrous or muscular band is sometimes found attached to the body of the hyoid bone and to the isthmus of the gland or its pyramidal lobe. When muscular, it is termed the **levator glandulae thyroideae.**

Small detached portions of thyroid tissue are sometimes found in the vicinity of the lateral lobes or above the isthmus. They are called **accessory thyroid glands.**

STRUCTURE

The thyroid gland is invested by a thin capsule of connective tissue, which projects into its substance and imperfectly divides it into masses of irregular form and size. When the fresh organ is cut open, its interior appears to be brownish-red and made up of closed vesicles containing colloid, which are separated from each other by intermediate connective tissue (Fig. 18-3).

The thyroid follicles (acini, vesicles) of the adult are closed sacs of inconstant shape and size. They are normally of microscopic dimensions, although macroscopically visible in colloid goiter. The follicles located peripherally may be larger than those centrally placed. The cuboidal epithelium of the normal thyroid rests directly on the delicate connective tissue surrounding the follicle. No basement membrane can be seen. The capillaries and lymphatics are thus in close contact with the secretory epithelium of the gland. The follicle normally is filled with colloid which contains the active principle of the gland, thyroxin. The thyroid epithelium may secrete this hormone directly into the colloid-filled lumen of the follicle where it is stored, or the hormone may be secreted directly into the capillaries. The stored colloid may be absorbed and liberated into the capillaries.

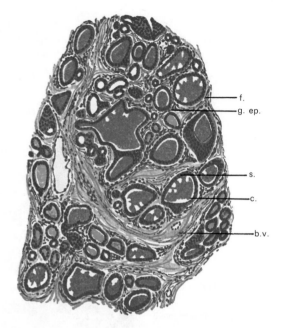

FIG. 18-3. Section of thyroid gland of a 21-year-old man. b.v., blood vessel; c., colloid; f., follicle; g.ep., glandular epithelium; s., connective tissue stroma. 120 ×. (Sobotta.)

FUNCTION. The hormone, thyroxin, requires iodine for its elaboration, which is normally obtained in adequate amounts in the diet. Proper functioning of the thyroid gland and normal histologic structure depend upon the adequacy of available iodine. The thyroid itself is activated or regulated by another hormone, the thyrotropic hormone of the anterior pituitary gland. Removal of the thyroid results in a marked reduction of the oxidative processes of the body. This lowered metabolic rate is characteristic of hypothyroidism. In infancy and childhood, the thyroid gland is essential to normal growth of the body.

VESSELS AND NERVES. (Fig. 18-4). The **arteries** supplying the thyroid gland are the superior and inferior thyroid arteries and sometimes an additional branch (thyroidea ima), from the brachiocephalic artery or the arch of the aorta, which ascends upon the front of the trachea. The arteries are remarkable for their large size and frequent anastomoses. The **veins** (Fig. 18-2) form a plexus on the surface of the gland and on the front of the trachea; from this plexus the superior, middle, and inferior thyroid veins arise; the superior and middle veins end in the internal jugular vein; the inferior, in the brachiocephalic vein. The capillary blood vessels form a dense plexus in the connective tissue around the vesicles, lying between the epithelium of the vesicles and the endothelium of the lymphatics, which surround a greater or smaller part of the circumference of the vesicle. The **lymphatic vessels** run in the interlobular connective tissue, not uncommonly surrounding the arteries they accompany, and communicate with a network in the capsule of the gland; they may con-

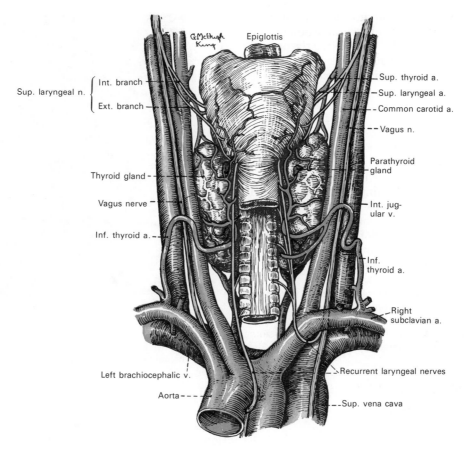

Sup. laryngeal n. { Int. branch
Ext. branch

Epiglottis

Sup. thyroid a.
Sup. laryngeal a.
Common carotid a.
Vagus n.
Parathyroid gland
Int. jugular v.
Inf. thyroid a.
Right subclavian a.

Thyroid gland
Vagus nerve
Inf. thyroid a.

Left brachiocephalic v.
Aorta

Recurrent laryngeal nerves
Sup. vena cava

FIG. 18-4.   Posterior view of larynx, trachea, thyroid, and parathyroids. Relations of thyroid arteries and laryngeal nerves are shown. (Nordland, 1930.)

tain colloid material. They end in the thoracic and right lymphatic trunks. The **nerves** are derived from the middle and inferior cervical ganglia of the sympathetic system.

Parathyroid Glands

DEVELOPMENT

The parathyroid bodies are developed as outgrowths from the third and fourth branchial pouches (Fig. 18-1).

ANATOMY

The **parathyroid glands** (Fig. 18-4) are small brownish-red bodies, situated as a rule between the dorsal borders of the lateral lobes of the thyroid gland and its capsule. They differ from the thyroid gland in struc-

ture, being composed of masses of cells arranged in a more or less columnar fashion with numerous intervening capillaries. They measure on an average about 6 mm long and 3 to 4 mm wide, and usually have the appearance of flattened oval discs. The parathyroid glands are divided, according to their situation, into **superior** and **inferior** glands. The superior, usually numbering two, are the more constant in position, situated, one on either side at the level of the caudal border of the cricoid cartilage, beside the junction of the pharynx and esophagus. The inferior glands, also usually two, may be applied to the caudal edge of the lateral lobes; placed at some little distance caudal to the thyroid gland; or found in relation to one of the inferior thyroid veins.

Four parathyroid glands are usually found in the human body; fewer than four were found in less than 1% of over a thousand persons (Pepere), but more than four (five or six) were found in over 33% of 122 bodies examined by Civalleri. In addition, numerous minute islands of parathyroid tissue may be found scattered in the connective tissue and fat of the neck around the parathyroid glands proper, quite distinct from them.

STRUCTURE

Microscopically the parathyroids consist of intercommunicating columns of cells (*principal* or *chief cells*) supported by connective tissue containing a rich supply of blood capillaries (Fig. 18-5). Most of the cells are clear, but some larger cells contain oxyphil granules (*eosinophil cells*). Vesicles containing colloid have been described as occurring in the parathyroid.

FUNCTION.   The parathyroids secrete a hormone, parathyrin or parathormone, necessary for calcium metabolism. The tetany that follows parathyroidectomy can be relieved by feeding or injecting calcium salts or parathyroid extracts.

Hypophysis Cerebri or Pituitary Gland

DEVELOPMENT

The hypophysis has a dual origin that corresponds to the two distinct parts of the adult gland. At an early embryonic stage, about the fourth week, when the neural tube and the primitive digestive tube are still in close proximity, outgrowths or diverticula from both tubes come in contact with each other to form the primordium of the hypophysis. The part from the neural tube is in the diencephalic floor of the hypothalamus, and is called the **infundibular process.** The part from the alimentary tube is from the portion of the future pharynx developed from ectoderm, the stomodaeum, and is known as **Rathke's pouch.**[1] The infundibular process retains a connection with the hypothalamus as the stalk, preserves some resemblance to neural tissue, and becomes the *neural lobe.* The cells of Rathke's pouch adhere to the infundibular process and grow

[1]Martin Heinrich Rathke (1793–1860): A German embryologist and anatomist (Danzig and Königsberg).

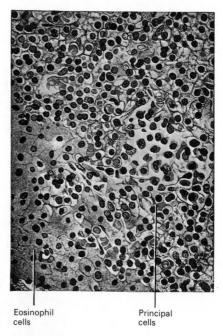

FIG. 18-5.   Section of a human parathyroid gland to show principal and eosinophil cells. 250 ×.

partly around it, forming the *pars tuberalis* (Fig. 18-6). The remainder of the pouch forms a double layered cup and the connection with the pharynx gradually disappears. The rostral portion of the cup thickens greatly and becomes the glandular anterior lobe or *pars distalis.* The other layer of the cup remains adherent to the neural lobe, thickens but little, and develops into the *pars intermedia.* The cavity of the original diverticulum eventually disappears except

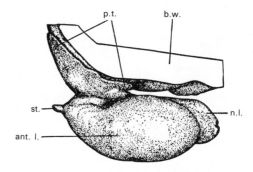

FIG. 18-6.   Model of hypophysis and adjacent brain wall from a thirty-day embryo, viewed from the left side. 25 ×. b. w., brain wall; p. t., pars tuberalis; st., stalk; ant. l., anterior lobe; n. l., neural lobe. (Atwell, Amer. J. Anat.; courtesy of Wistar Institute.)

for the vestigial lumina in the form of one or more narrow vesicles of variable length in the pars intermedia.

ANATOMY

The **hypophysis cerebri** or **pituitary gland** is couched in the sella turcica or hypophyseal fossa of the body of the sphenoid bone. It is attached to the hypothalamus by the **stalk** or **infundibulum.** After the brain has been removed from the cranial cavity, the cut end of the stalk is visible protruding through a hole in the diaphragma sellae. The latter is a shelf of dura mater stretching between the clinoid processes and covering over the sella turcica. The two principal portions of the gland, the anterior and posterior lobes, are not easily distinguished in the whole gland, but are visible in a median sagittal section because of a difference in color.

The hypophysis is a small gland, 1.2 to 1.5 cm in its greatest diameter, which is from side to side; its rostrocaudal diameter is approximately 1 cm, and its thickness is 0.5 cm. Its weight in an adult male is 0.5 to 0.6 gm, but varies according to the stature rather than the weight of an individual. It is larger in women, and since a slight increase occurs during pregnancy, in a multipara it may weigh as much as 1 gm.

RELATIONS.   Many important structures lie in close proximity to the hypophysis (Fig. 18-7). The internal carotid artery emerges from the dural covering of the cavernous sinus immediately lateral to it. The intercavernous and circular sinuses are enclosed in the diaphragma sellae above it. The optic nerves, chiasma, and optic tracts lie between it and the bulk of the brain. In skulls with large sphenoid sinuses, the hypophysis may be separated from this air space only by a thin plate of bone and the dural lining of the sella. The practical importance of these relations is emphasized in patients with tumors of the hypophysis. These tumors may compress the carotid arteries or the optic nerves, causing changes in the retina known as choked discs, which are visible with the ophthalmoscope.

LOBES

The **hypophysis** is divided into two parts on the basis of embryologic development, adult morphology, and function. The nam-

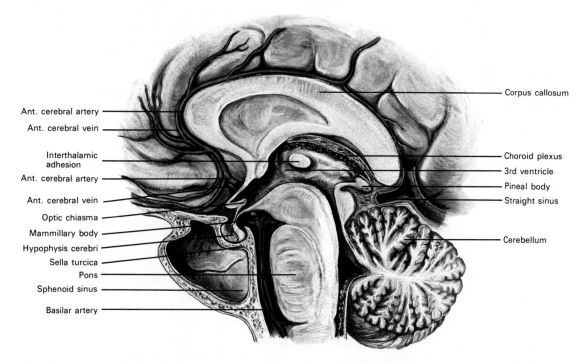

Ant. cerebral artery
Ant. cerebral vein
Interthalamic adhesion
Ant. cerebral artery
Ant. cerebral vein
Optic chiasma
Mammillary body
Hypophysis cerebri
Sella turcica
Pons
Sphenoid sinus
Basilar artery

Corpus callosum
Choroid plexus
3rd ventricle
Pineal body
Straight sinus
Cerebellum

Fɪɢ. 18-7.   A midsagittal section through the cerebrum showing the corpus callosum, anterior cerebral vessels, cerebellum, brainstem and the hypophysis cerebri located within the sella turcica.

ing of these main parts and their subdivisions results in some overlapping, which is best shown in Table 18-1 and Figure 18-8.

Anterior Lobe

The **anterior lobe** or **adenohypophysis** is larger than the posterior lobe, occupying about three-fourths of the gland. It appears darker in a sagittal section because of its vascularity and the composition of glandular tissue.

STRUCTURE. The **anterior lobe** or **pars distalis** is composed of cords of epithelial cells richly supplied with sinusoidal capillaries. The epithelial cells are of two main types: those that are colored by various stains, the **chromophils,** and those that remain pale or unstained, the **chromophobes.** The chromophils in turn are two kinds: those whose granules take up acid dyes, named **acidophils,** and those that take up basic dyes, the **basophils.** The basophils can again be subdivided into beta and delta types by their coloration with certain dyes. There is no definite pattern of distribution of the two chromophils, either in different parts of the gland or within individual cords. The chromophobes are smaller than the chromophils, less discrete, and tend to accumulate in the center of cords and clumps. The lack of mitotic division in the hypophysis and experimental observations indicate that the chromophils undergo cycles of secretory activity in which they discharge their specific granules and pass through a resting stage in the form of chromophobes.

Posterior Lobe

The **posterior lobe** or **neurohypophysis** has a pale color, corresponding to the overlying brain with which it is continuous by means of the infundibulum and median eminence of the tuber cinereum.

STRUCTURE. The specific cell is the pituicyte, which resembles neuroglia in shape due to its long processes. The outlines of the cells and their processes are distinguishable only in special preparations. They contain a variable number of granules and lipid globules. The inclusions, blackened by osmic acid, correspond with the neurosecretion, which is stored by them but is produced by the nerve cells of the supraoptic and paraventricular nuclei of the hypothalamus.

The **pars tuberalis** is a layer of cuboidal cells covering the stalk and neighboring area. It is richly supplied with arterial blood.

The **pars intermedia** is composed of a thin layer of epithelial cells between the neural lobe and the pars distalis. It encloses the remnants of the vestigial lumen of Rathke's pouch, which appear as vesicles containing a "colloid" resembling that of the thyroid gland in histologic preparations but having no relation to it chemically. The lining cells also may be ciliated.

VESSELS. The **arteries** to the hypophysis are the superior hypophyseal arteries, from the internal carotid or posterior communicating arteries, and the inferior hypophyseal arteries, which are also branches of the internal carotid but traverse the cavernous sinus. The branches of the superior arteries supply the stalk and adjacent parts of the anterior lobe (Fig. 18-9). The branches of the inferior arteries supply the posterior lobe. The blood supply of the pars distalis is mainly through a **portal system** of veins. The blood from the capillaries of the pars tuberalis and adjacent stalk collects into veins that pass along the stalk and break up into the numerous sinusoidal capillaries of the pars distalis.

The **veins** of the hypophysis are the lateral hypophyseal veins, which drain into the cavernous and intercavernous sinuses.

TABLE 18-1.   Major Divisions and Subdivisions of the Mammalian Hypophysis

Anterior lobe (*lobus anterior*)	Pars distalis (Pars anterior) / Pars tuberalis	Adenohypophysis (glandular portion)
Posterior lobe (*lobus posterior*)	Pars intermedia	
	Pars nervosa (Neural lobe) (Infundibular process)	Neurohypophysis (neural portion)
	Infundibulum and Median eminence of Tuber cinereum	

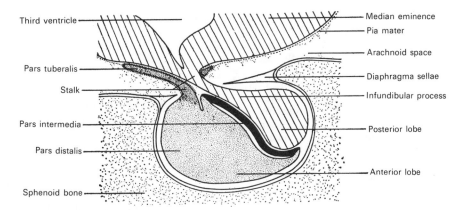

FIG. 18-8.   Diagrammatic median sagittal section of the hypophysis cerebri or pituitary gland. (Atwell.)

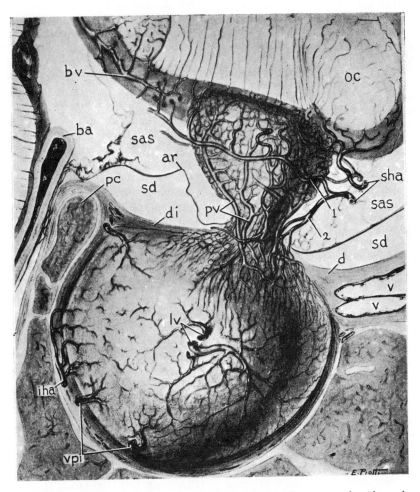

FIG. 18-9.   Schematic drawing of the hypophysis of the adult rhesus monkey. *ar*, arachnoid membrane; *ba*, basilar artery; *bv*, basilar vein; *d*, dura; *di*, sellar diaphragm; *iha*, inferior hypophyseal artery; *lv*, lateral hypophyseal veins; *oc*, optic chiasma; *pc*, posterior clinoid process; *pv*, portal venules; *sas*, subarachnoid space; *sd*, subdural space; *sha*, superior hypophyseal arteries (1, branches to hypophyseal stalk; 2, branches to anterior lobe); *v*, dural veins; *vpi*, veins of infundibular process. (Wislocki, *Pituitary Gland*, Williams & Wilkins, Baltimore, 1938.)

NERVES. The pars distalis has no specific innervation. Fibers from the superior cervical ganglion of the sympathetic system have been traced along the blood vessels but have not been conclusively associated with the glandular cells.

The neurohypophysis is supplied by fibers from the supraoptic and paraventricular nuclei of the hypothalamus. Similar osmiophilic neurosecretory granules are found in the cells of these nuclei and in their processes, which extend down to the posterior lobe in the **hypothalamico-hypophyseal** system in the infundibulum. Experimental evidence indicates that the hormones extracted from the neurohypophysis originate from these neurosecretions and are stored in the gland.

FUNCTION. The hypophysis supplies several hormones. Among those of the anterior lobe are a growth hormone, *somatotropin,* which affects general body growth; a thyrotropic hormone, which acts on the thyroid gland; an adrenocorticotropic hormone (ACTH), which acts on the suprarenal cortex; two gonadotropic hormones, one that stimulates ovarian follicles (FSH) and another that stimulates the lutein cells (LH); and *prolactin,* which promotes milk secretion by the mammary gland. Attempts to associate the specific types of cells with the elaboration of these hormones have been partly successful. The growth hormone is produced by acidophils. Tumors of the gland composed of acidophils are found in patients with tremendous overgrowth of various parts of the body in a condition called acromegaly, and the absence of acidophils is notable in pituitary dwarfs. Acidophils also appear to be responsible for prolactin. Experimental evidence points to the basophils as the source of the follicle-stimulating, thyroid-stimulating, and luteinizing hormones. The source of adrenotropic hormone is controversial but may be a specific cell developed from chromophobes.

From the neurohypophysis, two hormones have been extracted: oxytocin (Pituitrin), which stimulates the contraction of smooth muscle and is sometimes used in obstetric practice to make the uterus contract; and the antidiuretic principle or vasopressin, which inhibits diuresis by the kidneys and also raises the blood pressure. As mentioned previously, it is doubtful that the neural lobe elaborates these hormones. A lack of the antidiuretic principle in patients produces a condition called diabetes insipidus.

Suprarenal or Adrenal Gland

The suprarenal gland (*adrenal gland*) in man and other mammals is a combination of two distinct glands that remain independent in fishes and other more primitive vertebrates. The two organs in the human gland are fused together but remain distinct and identifiable as the cortex and medulla of the adult gland.

DEVELOPMENT

The **cortex** of the suprarenal gland is first recognizable in an embryo of the sixth week as a groove in the *coelom* at the base of the mesentery near the cranial end of the mesonephros. The cells at the bottom of the groove proliferate rapidly to form a mass in the mesenchyme extending toward the aorta. During the seventh and eighth weeks the cells become arranged into cords with dilated blood spaces between; the connection with the coelomic mesothelium is lost, and a capsule of connective tissue encloses the gland. During the remainder of fetal development the cortical tissue is composed of two zones: an outer zone of more undifferentiated cells and an inner zone of cell cords that appear to be differentiated and active in secretion. After birth the inner zone atrophies and the outer zone differentiates into the three zones of the adult gland.

The **medulla** of the suprarenal gland is developed from cells of the *neural crest,* which migrate ventrally along with the cells that form the sympathetic ganglia. These cells later detach themselves from the ganglia and become small knots of glandular cells scattered along the vertebral column. During the seventh and eighth weeks, a large group of the neural cells, migrating along the suprarenal vein, invades the cortex, thus establishing the primordium of the suprarenal medulla. These glandular cells of sympathetic origin contain a substance, probably the precursor of the secretion, epinephrine, which is colored brown by chromic acid. This has given them their name of **chromaffin** or **pheochrome cells.** Many of the small masses of chromaffin cells persist throughout life and are given the name **paraganglia.**

ANATOMY

The suprarenal gland, as the name suggests, is located at the cranial pole of each of the two kidneys. They are quite different in shape on the two sides (Figs. 18-10; 18-11). The right gland resembles a pyramid, or the cocked hat of colonial days; the left is semilunar and tends to be slightly larger. The

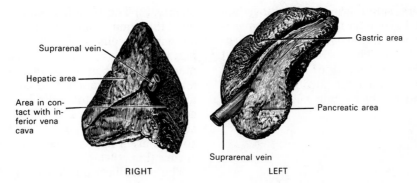

Fɪɢ. 18-10. Suprarenal glands, ventral view.

length and width vary from 3 to 5 cm and the thickness 4 to 6 mm. The average weight is 3.5 to 5 gm.

RELATIONS. The **right suprarenal** is situated dorsal to the inferior vena cava and right lobe of the liver, and ventral to the diaphragm and cranial end of the right kidney. The *ventral surface* has two areas: a medial, narrow, and nonperitoneal surface, which lies cranial to the inferior vena cava; and a lateral, somewhat triangular, surface in contact with the liver. The cranial part of the latter surface is devoid of peritoneum, and is in relation with the bare area of the liver near its caudal and medial angle, while its caudal portion is covered by peritoneum, reflected onto it from the inferior layer of the coronary ligament; occasionally the duodenum overlaps the inferior portion. A little below the apex and near the ventral border of the gland is a short furrow termed the **hilum,** from which the suprarenal vein emerges to join the inferior vena cava. The *posterior surface* is divided into cranial and caudal parts by a curved ridge: the cranial, slightly convex, rests upon the diaphragm; the caudal, concave, is in contact with the cranial end and the adjacent part of the ventral surface of the kidney.

The **left suprarenal** has a crescentic shape, its concavity being adapted to the medial border of the cranial part of the left kidney. Its *ventral surface* has two areas: a cranial one, covered by the peritoneum of the omental bursa, which separates it from the cardiac end of the stomach and sometimes from the cranial extremity of the spleen; and a caudal one, which is in contact with the pancreas and splenic artery and therefore is not covered by the peritoneum. On the ventral surface, near its caudal end, is a furrow or **hilum,** from which the suprarenal vein emerges. Its *posterior surface* has a vertical ridge, which divides it into two areas; the lateral area rests on the kidney, and the medial and smaller area rests on the left crus of the diaphragm.

The surface of the suprarenal gland is surrounded by areolar tissue containing much fat. It is closely invested by a thin fibrous capsule, which is difficult to remove because of the numerous fibrous processes and vessels entering the organ through the furrows on its ventral surface and base.

ACCESSORY GLANDS

Accessory suprarenals occur frequently in the connective tissue near the main gland, but they may occur near the testis or the

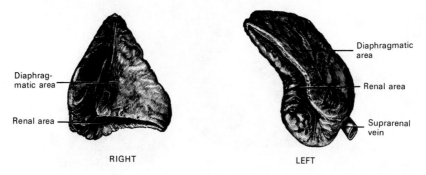

Fɪɢ. 18-11. Suprarenal glands, dorsal view.

ovary, where small groups of embryonic cells might have migrated with the mesonephros, especially in certain hermaphroditic conditions. They are usually small round bodies composed of cortical cells. Larger ones may have the cortical zones represented and rarely contain central medullary tissue.

STRUCTURE

The suprarenal gland consists of two portions, an outer **cortex** that takes up the greater part of the organ and an inner **medulla** (Fig. 18-12). In the fresh condition the outer cortex is a deep yellow, the inner part a dark red or brown, and the medulla a pale pink. At the transition between the cortex and medulla, there is no connective tissue barrier, and the cords of the two types of cells are intermingled for a short distance.

The **cortical portion** consists of a fine connective-tissue network, in which is embedded the glandular epithelium. The epithelial cells are polyhedral and possess rounded nuclei; many of the cells contain coarse granules, and others have lipoid globules. Owing to differences in the arrangement of the cells, three distinct zones can be made out: (1) The **zona glomerulosa,** situated beneath the capsule, consists of cells arranged in rounded groups, with indications here and there of an alveolar structure; the cells of this zone are highly granular and stain deeply. (2) The **zona fasciculata,** continuous with the zona glomerulosa, is composed of columns of cells arranged in a radial manner; these cells contain finer granules and in many instances globules of lipoid material. (3) The **zona reticularis,** in contact with the medulla, consists of cylindrical masses of cells irregularly arranged; these cells often contain pigment granules, which give this zone a darker appearance than the rest of the cortex (Fig. 18-13).

The **medullary portion** is highly vascular and consists of large chromaffin cells arranged in a network. The irregular polyhedral cells have a finely granular cytoplasm that is probably concerned with the secretion of epinephrine. In the meshes of the cellular network are large anastomosing venous sinuses (sinusoids), which are in close relationship with the chromaffin or medullary cells. In many places the endothelial lining of the blood sinuses is in direct contact with the medullary cells. Some authors believe the endothelium is absent in places and here the medullary cells are directly bathed by the blood. This intimate relationship between the chromaffin cells and the blood stream undoubtedly facilitates the discharge of the internal secretion into the blood. There is a loose meshwork of supporting connective tissue containing nonstriped muscle fibers. This portion of the gland is richly supplied with nonmedullated nerve fibers, and here and there sympathetic ganglia are found.

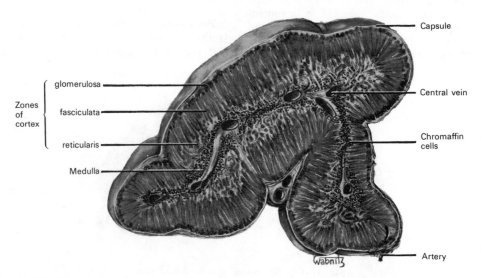

FIG. 18-12. Section through suprarenal gland of adult cadaver. Lightly stained with hematoxylin. 5 ×.

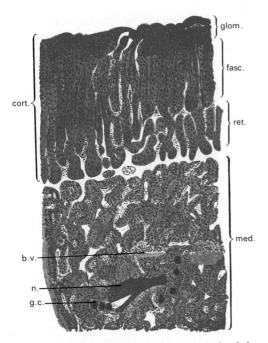

Labels on figure: glom., fasc., cort., ret., med., b.v., n., g.c.

FIG. 18-13. Section of human suprarenal gland. b.v., central vein; cort., cortex; fasc., zona fasciculata; glom., zona glomerulosa; ret., zona reticularis; med., medulla; g.c., ganglion cell; n., nerve fibers. 100 ×. (Redrawn from Sobotta.)

VESSELS AND NERVES. The suprarenals are highly vascular organs. The arteries are numerous and comparatively large. These arteries are derived from the aorta and the inferior phrenic and the renal arteries. Three sets of branches penetrate the capsule, one of which breaks up in capillaries that supply the capsule. The second set breaks up into capillaries that supply the cell cords of the cortex and empty into veins in the medulla. The third set of arteries traverses the cortex to supply the medulla only and breaks up into the sinusoids of the medulla. Venous blood from the capsule is collected into veins of the capsule. Blood from the two other sets of arteries is gathered into the central vein in the medulla, which emerges at the hilum as the suprarenal vein. On the right side this vein opens into the inferior vena cava; on the left, into the left renal vein.

Lymphatics accompany the large blood vessels and end in the lumbar nodes.

The **nerves** are numerous and are derived from the celiac and renal plexuses. They enter the lower and medial part of the capsule, traverse the cortex, and end around the cells of the medulla. They have numerous small ganglia in the medullary portion of the gland.

FUNCTION. The adrenal medulla elaborates an internal secretion, epinephrine (Adrenalin), which has definite sympathomimetic actions such as causing constriction of arterioles, acceleration of the heart rate, and contraction of the radial muscle of the iris. In man an injection of the drug epinephrine

hydrochloride increases systolic blood pressure, pulse rate, minute volume of the heart, and volume of respiration. These effects are transitory and disappear after one or two hours. A rise in blood sugar and basal metabolic rate follows injection of this drug due to an increased enzymatic breakdown of glycogen in liver and muscle. The duration of epinephrine action is brief, the hormone being rapidly inactivated in the body.

The adrenal cortex elaborates one or more hormones essential to maintenance of life. Various fractions serve in the maintenance of physiologic "steady states," the regulation of the distribution of water and electrolytes, and many aspects of carbohydrate metabolism and muscular efficiency.

Paraganglia, Glomera, and "Glands" or "Bodies"

The various structures that have been given these names are for the most part small, difficult to demonstrate, and frequently misnamed. The following is an attempt to sort them out and give their appropriate names and synonyms.

PARAGANGLIA

The **paraganglia** (*chromaffin bodies; pheochrome bodies*) are small groups of chromaffin cells connected with the ganglia of the sympathetic trunk and the ganglia of the celiac, renal, suprarenal, aortic, and hypogastric plexuses. They are sometimes found in connection with the ganglia of other sympathetic plexuses.

AORTIC BODIES OR LUMBAR PARAGANGLIA

The **aortic glands** or **bodies** (*corpora para-aortica, glands or organs of Zuckerkandl*[1]) are the largest of these groups of chromaffin cells, about 1 cm long in the newborn. They lie one on either side of the aorta in the region of the inferior mesenteric artery. They decrease in size with age and after puberty are only visible with the microscope. Other groups of chromaffin cells have been found associated with the sympathetic plexuses of the abdomen independently of the ganglia.

[1]Emil Zuckerkandl (1849–1910): An Austrian anatomist (Graz and Vienna).

GLOMUS COCCYGEUM

The **glomus coccygeum** (*coccygeal gland or body; Luschka's gland*[2]) is placed ventral to, or immediately distal to, the tip of the coccyx. It is about 2.5 mm in diameter and is irregularly oval in shape; several smaller nodules are found around or near the main mass.

It consists of irregular masses of round or polyhedral cells, with each mass grouped

[2]Hubert von L. Luschka (1820–1875): A German anatomist (Tübingen).

around a dilated sinusoidal capillary vessel. Each cell contains a large round or oval nucleus, the protoplasm surrounding which is clear and not stained by chromic salts. It is not a chromaffin paraganglion, and since its structure is not like that of the chemoreceptors (carotid bodies), its function is unknown.

The **glomus caroticum** or **carotid body** is not a paraganglion but a chemoreceptor.

The **glomus aorticum** is better named the **cardioaortic body** or glomus aorticum supracardiale. It is a chemoreceptor similar to the carotid body.

References

(References are listed not only to those articles and books cited in the text, but to others as well which are considered to contain valuable resource information for the student who desires it.)

Endocrinology, General

Fawcett, D. W., J. A. Long, and A. L. Jones. 1969. The ultrastructure of endocrine glands. Recent Progr. Hormone Res., 25:315–380.

Frankel, H. H., R. R. Patek, and S. Bernick. 1962. Long term studies of the rat reticuloendothelial system and endocrine gland responses to foreign particles. Anat. Rec., 142:359–373.

Gorbman, A., and H. A. Bern. 1962. A Textbook of Comparative Endocrinology. John Wiley & Sons, Inc., New York, 468 pp.

Hall, P. F. 1959. The Functions of the Endocrine Glands. W. B. Saunders Co., Philadelphia, 290 pp.

Hollinshead, W. H. 1952. Anatomy of the endocrine glands. Surg. Clin. North Amer., 32:1115–1140.

Martini, L., and W. F. Ganong (eds.). 1969. Frontiers in Neuroendocrinology. Raven Press, New York. 8 vols.

Talbot, N. B., E. H. Sobel, J. W. McArthur, and J. D. Crawford. 1952. Functional Endocrinology from Birth through Adolescence. Harvard University Press, Cambridge, 638 pp.

Wilkins, L. 1965. The Diagnosis and Treatment of Endocrine Disorders in Childhood and Adolescence. 3rd ed. Charles C Thomas, Springfield, Ill., 619 pp.

Young, W. C. 1961. Sex and Internal Secretions. Williams & Wilkins Co., Baltimore. 2 vols., 1609 pp.

Thyroid Gland

Anast, C. S. 1966. Thyrocalcitonin—a review. Clin. Orthop., 47:179–187.

Andros, G., and S. H. Wollman. 1964. Autoradiographic localization of iodide[125] in the thyroid epithelial cell. Proc. Soc. Exp. Biol. Med., 115:775–777.

Axelrad, A. A., C. P. Leblond, and H. Isler. 1955. Role of iodine deficiency in the production of goiter by the Remington diet. Endocrinology, 56:387–403.

Bachhuber, C. A. 1943. Complications of thyroid surgery: Anatomy of the recurrent laryngeal nerve, middle thyroid vein and inferior thyroid artery. Amer. J. Surg., 60:96–100.

Bailey, H. 1925. Thyroglossal cysts and fistulae. Brit. J. Surg., 12:579–589.

Berlin, D. D. 1935. The recurrent laryngeal nerves in total ablation of the normal thyroid gland: An anatomical and surgical study. Surg. Gynecol. Obstet., 60:19–26.

Biddulph, D. M., and H. C. Maibenco. 1972. Response of hamster thyroid light cells to plasma calcium. Anat. Rec., 173:25–43.

Brookhaven Symposia in Biology No. 7. 1955. The Thyroid. Brookhaven National Laboratory, Upton, New York, 271 pp.

Bussolati, G., and A. G. E. Pearse. 1967. Immunofluorescent localization of calcitonin in the 'C' cells of the pig and dog thyroid. J. Endocrinol., 37:205–209.

Calvert, R. 1973. Ultrastructural localization of alkaline phosphatase activity in the developing thyroid gland of the rat. Anat. Rec., 177:359–375.

Carpenter, G. R., and J. L. Emery. 1976. Inclusions in the human thyroid. J. Anat. (Lond.), 122:77–89.

Copp, D. H., A. G. F. Davidson, and B. Cheney. 1961. Evidence for a new parathyroid hormone which lowers blood calcium. Proc. Canad. Fed. Biol. Soc., 4:17.

Dozois, R. R., and O. H. Beahrs. 1977. Surgical anatomy and technique of thyroid and parathyroid surgery. Surg. Clin. North Amer., 57:647–661.

Follis, R. H., Jr. 1959. Experimental colloid goiter in the hamster. Proc. Soc. Exp. Biol. Med., 100:203–206.

Fowler, C. H., and W. A. Hanson. 1929. Surgical anatomy of the thyroid gland with special reference to the relations of the recurrent laryngeal nerve. Surg. Gynecol. Obstet., 49:59–66.

Friedman, J., and L. G. Raisz. 1965. Thyrocalcitonin: Inhibitor of bone resorption in tissue culture. Science, 150:1465–1467.

Gross, J. 1954. Thyroid hormones. Brit. Med. Bull., 10:218–223.

Gross, J. 1957. The dynamic cytology of the thyroid gland. Internat. Rev. Cytol., 6:265–288.

Harries, D. J. 1955. Thyroid enlargement and the cricothyroid muscle. Brit. Med. J., 1:1012–1013.

Heimann, P. 1966. Ultrastructure of hyman thyroid: A study of normal thyroid, untreated and treated diffuse toxic goiter. Acta Endocrinol., 53 (Suppl. 110), 102 pp.

Hirsch, P. F., and P. L. Munson. 1969. Thyrocalcitonin. Physiol. Rev., 49:548–622.

Hirsch, P. F., E. F. Voelkel, and P. L. Munson. 1964. Thyrocalcitonin: Hypocalcemic hypophosphatemic principle of the thyroid gland. Science, 146:412–413.

Holmgren, H., and B. Naumann. 1949. A study of the nerves of the thyroid gland and their relationship to glandular function. Acta Endocrinol., 3:215–235.

Hubert, L. 1947. Thyroglossal cysts and sinuses: Analysis of forty-three cases. Arch. Otolaryngol., 45:105–111.

Hunt, T. E. 1944. Mitotic activity in the thyroid gland of female rats. Anat. Rec., 90:133–138.

Isler, H., S. K. Sarkar, B. Thompson, and B. Tonkin. 1968. The architecture of the thyroid gland: A 3-dimensional investigation. Anat. Rec., 161:325–336.

Kameda, Y. 1982. Immunohistochemical study of C cell follicles in dog thyroid glands. Anat. Rec., 204:55–60.

Keyes, E. L. 1940. Demonstration of the nerve to the levator glandulae thyreoideae muscle. Anat. Rec., 77:293–295.

King, W. L. M., and J. de J. Pemberton. 1942. So called lateral aberrant thyroid tumors. Surg. Gynecol. Obstet., 74:991–1001.

Kingsbury, B. F. 1939. The question of a lateral thyroid in mammals with special reference to man. Amer. J. Anat., 65:333–359.

Krafka, J., Jr. Intratracheal thyroid occurring in a seven months human fetus. Ann. Surg., 106:457–459.

Krausen, A. S. 1976. The inferior thyroid veins—the ultimate guardians of the trachea. Laryngoscope, 86:1849–1855.

Lahey, F. H., and N. W. Swinton. 1934. Intrathoracic goiter. Surg. Gynecol. Obstet., 59:627–637.

Leblond, C. P., and J. Gross. 1948. Thyroglobulin formation in the thyroid follicle visualized by "coated autograph" technique. Endocrinology, 43:306–324.

Lie, T. A., and A. A. W. Op de Coul. 1977. Collateral circulation through thyroid arteries after occlusion of the common carotid artery. Clin. Neurol. Neurosurg., 80:83–85.

Loré, J. M., Jr. 1983. Practical anatomical considerations in thyroid tumor surgery. Arch. Otolaryngol., 109:568–574.

MacIntyre, I. 1967. Calcitonin: A general review. Calc. Tiss. Res., 1:173–182.

Mahorner, H. R., H. D. Caylor, C. F. Schlotthauer, and J. de J. Pemberton. 1927. Observations on the lymphatic connections of the thyroid gland in man. Anat. Rec., 36:341–348.

Marks, S. C. 1969. The parafollicular cell of the thyroid gland as the source of an osteoblast-stimulating factor: Evidence from experimentally osteopetrotic mice. J. Bone Joint Surg., 51A:875–890.

Marshall, C. F. 1895. Variations in the form of the thyroid gland in man. J. Anat. Physiol., 29:234–239.

Marshall, S. F., and W. F. Becker. 1949. Thyroglossal cysts and sinuses. Ann. Surg., 129:642–651.

Masson, J. C., and S. C. Mueller. 1933. Ovarian tumors of thyroid tissue. Surg. Gynecol. Obstet., 56:931–938.

Mojab, K., and B. C. Ghosh. 1976. Thyroid angiography. Amer. J. Surg., 132:620–622.

Nadler, N. J., B. A. Young, C. P. Leblond, and B. Mitmaker. 1964. Elaboration of thyroglobulin in the thyroid follicle. Endocrinology, 74:333–354.

Nonidez, J. F. 1931. Innervation of the thyroid gland. II. Origin and course of the thyroid nerves in the dog. Amer. J. Anat., 48:299–329.

Nonidez, J. F. 1932. The origin of the "parafollicular" cell, a second epithelial component of the thyroid gland of the dog. Amer. J. Anat., 49:479–505.

Nonidez, J. F. 1935. Innervation of the thyroid gland: III.

Distribution and termination of the nerve fibers in the dog. Amer. J. Anat., 57:135–169.

Nordland, M. 1930. The larynx as related to surgery of the thyroid based on an anatomical study. Surg. Gynecol. Obstet., 51:449–459.

Pearse, A. G. E. 1966. The cytochemistry of the thyroid C cells and their relationship to calcitonin. Proc. Roy. Soc. London, Series B, 164:478–487.

Pearse, A. G. E., and A. F. Carvalheira. 1967. Cytochemical evidence for an ultimobranchial origin of rodent thyroid C cells. Nature, 214:929–930.

Plagge, J. C. 1943. Effects of hypotonic solutions upon the living thyroid gland. Anat. Rec., 87:345–353.

Pratt, G. W. 1916. The thyroidea ima artery. J. Anat. Physiol., 1:239–242.

Ramsay, A. J., and G. A. Bennett. 1943. Studies on the thyroid gland. I. The structure, extent and drainage of the "lymph-sac" of the thyroid gland (Felis domestica). Anat. Rec., 87:321–339

Rasmussen, H., and M. M. Pechet. 1970. Calcitonin. Scientific Amer., 223:42–50.

Reed, A. F. 1943. The relations of the inferior laryngeal nerve to the inferior thyroid artery. Anat. Rec., 85:17–23.

Rhinehart, D. A. 1912. The nerves of the thyroid and parathyroid bodies. Amer. J. Anat., 13:91–102.

Richter, K. M. 1944. Some new observations bearing on the effect of hyperthyroidism on genital structure and function. J. Morphol., 74:375–393.

Roeder, C. A. 1932. Operations on the superior pole of the thyroid. Arch Surg., 24:426–439.

Rogers, L. 1929. The thyroid arteries considered in relation to their surgical importance. J. Anat. (Lond.), 64:50–61.

Sharer, R. F. 1936. Substernal thyroid. Amer. J. Surg., 32:56–62.

Shepard, T. H., H. J. Andersen, and H. Andersen. 1964. The human fetal thyroid. I. Its weight in relation to body weight, crown-rump length, foot length and estimated gestation age. Anat. Rec., 148:123–128.

Shepard, T. H., H. Andersen, and H. J. Andersen. 1964. Histochemical studies of the human fetal thyroid during the first half of fetal life. Anat. Rec., 149:363–379.

Smith, I. H., and H. C. Moloy. 1930. The effect of nerve stimulation and nerve degeneration on the mitochondria and histology of the thyroid gland. Anat. Rec., 45:393–406.

Smith, S. D., and R. S. Benton. 1978. A rare origin of the superior thyroid artery. Acta Anat., 101:91–93.

Stewart, J. D. 1932. Circulation of the hyman thyroid. Arch. Surg., 25:1157–1165.

Taylor, S. (ed.). 1968. Calcitonin. Heinemann, London, 402 pp.

Teitelbaum, S. L., K. E. Moore, and W. Shieber. 1970. C cell follicles in the dog thyroid: Demonstrated by in vivo perfusion. Anat. Rec., 168:69–77.

Tzinas, S., C. Droulias, N. Harlaftis, J. T. Atkin, Jr., S. W. Gray, and J. E. Skandalakis. 1976. Vascular patterns of the thyroid gland. Amer. J. Surg., 42:639–644.

Warren, S., and J. D. Feldman. 1949. The nature of lateral "aberrant" thyroid tumors. Surg. Gynecol. Obstet., 88: 31–44.

Weller, G. L., Jr. 1933. Development of the thyroid, parathyroid and thymus glands in man. Carneg. Inst., Contr. Embryol., 24:93–139.

Werner, S. C., and S. H. Ingbar (eds.). 1978. The Thyroid. 4th ed. Harper and Row, New York, 1047 pp.

Williams, R. G. 1944. Some properties of living thyroid

cells and follicles. Amer. J. Anat., 75:95–119.

Young, B. A., A. D. Care, and T. Duncan. 1968. Some observations on the light cells of the thyroid gland of the pig in relation to thyrocalcitonin production. J. Anat. (Lond.), 102:275–288.

Young, B. A., and C. P. Leblond. 1963. The light cell as compared to the follicular cell in the thyroid gland of the rat. Endocrinology, 73:669–686.

Ziegelman, E. F. 1933. Laryngeal nerves: Surgical importance in relation to the thyroid arteries, thyroid gland and larynx. Arch. Otolaryngol., 18:793–808.

Parathyroid Glands

Albright, F., and E. C. Reifenstein. 1948. *The Parathyroid Glands and Metabolic Bone Disease.* Williams and Wilkins Co., Baltimore, 393 pp.

Baker, B. L. 1942. A study of the parathyroid glands of the normal and hypophysectomized monkey (*Macaca mulatta*). Anat. Rec., 83:47–73.

Barnicot, N. A. 1948. The local action of the parathyroid and other tissues on bone in intracerebral grafts. J. Anat. (Lond.), 82:233–248.

Bhaskar, S. N., I. Schour, R. O. Greep, and J. P. Weinmann. 1952. The corrective effect of parathyroid hormone on genetic anomalies in the dentition and the tibia of the *ia* rat. J. Dent. Res., 31:257–270.

Black, B. M. 1948. Surgical aspects of hyperparathyroidism: Review of sixty-three cases. Surg. Gynecol. Obstet., 87:172–182.

Capen, C. C., and G. N. Rowland. 1968. The ultrastructure of the parathyroid glands of young cats. Anat. Rec., 162:327–339.

Chang, H. 1951. Grafts of parathyroid and other tissues to bone. Anat. Rec., 111:23–47.

Curtis, G. M. 1930. The blood supply of the human parathyroids. Surg. Gynecol. Obstet., 51:805–809.

DeRobertis, E. 1940. The cytology of the parathyroid gland of rats injected with parathyroid extract. Anat. Rec., 78:473–495.

Drake, T. G., F. Albright, and B. Castleman. 1937. Parathyroid hyperplasia in rabbits produced by parenteral phosphate administration. J. Clin. Invest., 16:203–206.

Dufour, D. R., and S. Y. Wilkerson. 1983. Factors related to parathyroid weight in normal persons. Arch. Pathol. Lab. Med., 107:167–172.

Fleischmann, W. 1951. *Comparative Physiology of the Thyroid and Parathyroid Glands.* Charles C Thomas, Springfield, Ill., 78 pp.

Gaillard, P. J., R. V. Talmage, and A. M. Budy (eds.). 1965. *The Parathyroid Glands: Ultrastructure, Secretion and Function.* University of Chicago Press, Chicago, 353 pp.

Gilmour, J. R. 1938. The gross anatomy of the parathyroid glands. J. Pathol. Bacteriol., 46:133–149.

Gilmour, J. R., and W. J. Martin. 1937. The weight of the parathyroid glands. J. Pathol. Bacteriol., 44:431–462.

Godwin, M. C. 1937. The development of the parathyroids in the dog with emphasis upon the origin of accessory glands. Anat. Rec., 68:305–325.

Gray, S. W., J. E. Skandalakis, J. T. Atkin, Jr., C. Droulias, and M. D. Vohman. 1976. Parathyroid glands. Amer. Surg., 42:653–656.

Greep, R. O., and R. V. Talmadge (eds.). 1961. *The Parathyroids.* Charles C Thomas, Springfield, Ill., 473 pp.

Grollman, A. 1954. The role of the kidney in the parathyroid control of the blood calcium as determined by studies on the nephrectomized dog. Endocrinology, 55:166–172.

Ham, A. W., N. Littner, T. G. H. Drake, E. C. Robertson, and F. F. Tisdall. 1940. Physiological hypertrophy of the parathyroids, its cause and its relation to rickets. Amer. J. Pathol., 16:277–286.

Heinbach, W. F., Jr. 1933. A study of the number and location of the parathyroid in man. Anat. Rec., 57:251–261.

Howard, J. E. 1971. The biological mechanisms of transport and storage of calcium. Canad. Med. Assoc. J., 104:699–703.

Jordan, R. K., B. MacFarlane, and R. J. Scothorne. 1975. An electron microscopic study of the histogenesis of the parathyroid gland in the sheep. J. Anat. (Lond.), 119:235–254.

Lahey, F. H., and G. E. Haggart. 1935. Hyperparathyroidism: Clinical diagnosis and the operative technique of parathyroidectomy. Surg. Gynecol. Obstet., 60:1033–1051.

Lever, J. D. 1958. Cytological appearances in the normal and activated parathyroid of the rat. A combined study by electron and light microscopy with certain quantitative assessments. J. Endocrinol., 17:210–217.

MacCallum, W. G. 1906. The surgical relations of the parathyroid glands. Brit. Med. J., 2:1282–1286.

Millzner, R. J. 1931. The normal variations in the position of the human parathyroid glands. Anat. Rec., 48:399–405.

Munger, B. L., and S. I. Roth. 1963. The cytology of the normal parathyroid glands of man and Virginia deer; a light and electron microscopic study with morphologic evidence of secretory activity. J. Cell Biol., 16:379–400.

Murley, R. S., and P. M. Peters. 1961. Inadvertent parathyroidectomy. Proc. Roy. Soc. Med., 54:487–489.

Norris, E. H. 1937. The parathyroid glands and the lateral thyroid in man: Their morphogenesis, histogenesis, topographic anatomy and prenatal growth. Carneg. Inst., Contr. Embryol., 26:247–294.

Norris, E. H. 1946. Anatomical evidence of prenatal function of the human parathyroid glands. Anat. Rec., 96:129–141.

Norris, E. H. 1947. The parathyroid adenoma: A study of 322 cases. Internat. Abstr. Surg., 84:1–41.

Pyrtek, L. J., and R. L. Painter. 1964. An anatomic study of the relationship of the parathyroid glands to the recurrent laryngeal nerve. Surg. Gynecol. Obstet., 119:509–512.

Raybuck, H. E. 1952. The innervation of the parathyroid glands. Anat. Rec., 112:117–123.

Rienhoff, W. F., Jr. 1950. The surgical treatment of hyperparathyroidism: With a report of 27 cases. Ann. Surg., 131:917–944.

Walton, A. J. 1931. The surgical treatment of parathyroid tumours. Brit. J. Surg., 19:285–291.

Weller, G. L., Jr. 1933. Development of the thyroid, parathyroid and thymus gland in man. Carneg. Inst., Contr. Embryol., 24:93–139.

Weymouth, R. J., and H. R. Seibel. 1969. An electron microscopic study of the parathyroid glands in man: Evidence of secretory material. Acta Endocrinol., 61:334–342.

Yeghiayan, E., J. M. Rojo-Ortega, and J. Genest. 1972. Parathyroid vessel innervation: An ultrastructural study. J. Anat. (Lond.), 112:137–142.

Hypophysis Cerebri

Adams, J. H., P. M. Daniel, and M. M. L. Prichard. 1969. The blood supply of the pituitary gland of the ferret with special reference to infarction after stalk section. J. Anat. (Lond.), 104:209–225.

Anand Kumar, T. C., and D. S. Vincent. 1974. Fine structure of the pars intermedia in the rhesus monkey, *Macaca mulatta*. J. Anat. (Lond.), 118:155–169.

Aplington, H. W., Jr. 1962. Cellular changes in the pituitary of Necturus following thyroidectomy. Anat. Rec., 143:133–145.

Arey, L. B. 1950. The craniopharyngeal canal reviewed and reinterpreted. Anat. Rec., 106:1–16.

Atwell, W. J. 1926. The development of the hypophysis cerebri in man, with special reference to the pars tuberalis. Amer. J. Anat., 37:159–193.

Bain, J., and C. Ezrin. 1970. Immunofluorescent localization of the LH cell of the human adenohypophysis. J. Clin. Endocrinol. Metabol., 30:181–184.

Baker, B. L. 1970. Studies on hormone localization with emphasis on the hypophysis. J. Histochem. Cytochem., 18:1–8.

Baker, B. L., S. Pek, A. R. Midgley, Jr., and B. E. Gersten. 1970. Identification of the corticotropin cell in rat hypophyses with peroxidase-labeled antibody. Anat. Rec., 166:557–567.

Bargmann, W. 1966. Neurosecretion. Internat. Rev. Cytol., 19:183–201.

Bargmann, W., and E. Scharrer. 1951. The site of origin of the hormones of the posterior pituitary. Amer. Sci., 39:255–259.

Beckmann, J. W., and L. S. Kubie. 1929. A clinical study of twenty-one cases of tumour of the hypophyseal stalk. Brain, 52:127–170.

Bodian, D. 1951. Nerve endings, neurosecretory substance and lobular organization of the neurohypophysis. Bull. Johns Hopkins Hosp., 89:354–376.

Bodian, D. 1966. Herring bodies and neuro-apocrine secretion in the monkey. An electron microscopic study of the fate of the neurosecretory product. Bull. Johns Hopkins Hosp., 118:282–326.

Bogdanove, E. M. 1967. Analysis of histophysiologic responses of the rat hypophysis to androgen treatment. Anat. Rec., 157:117–135.

Boyd, W. H. 1980. Intrinsic neuronal elements of the hypophyseal intermediate lobe. Anat. Anz., 148:392–408.

Brooks, C. M. 1938. A study of the mechanism whereby coitus excites the ovulation-producing activity of the rabbit's pituitary. Amer. J. Physiol., 121:157–177.

Carmel, P. W., J. Lobo Antunes, and M. Ferin. 1979. Collection of blood from the pituitary stalk and portal veins in monkeys, and from the pituitary sinusoidal system of monkey and man. J. Neurosurg., 50:75–80.

Chatterjee, P. 1976. Development and cytodifferentiation of the rabbit pars intermedia. II. Neonatal to adult. J. Anat. (Lond.), 122:415–433.

Conklin, J. L. 1968. The development of the human fetal adenohypophysis. Anat. Rec., 160:79–91.

Crafts, R. C., and B. S. Walker. 1947. The effects of hypophysectomy on serum and storage iron in adult female rats. Endocrinology, 41:340–346.

Dandy, W. E. 1913. The nerve supply to the pituitary body. Amer. J. Anat., 15:333–343.

Dandy, W. R., and E. Goetsch. 1911. The blood supply of the pituitary body. Amer. J. Anat., 11:137–150.

Daniel, P. M. 1976. Anatomy of the hypothalamus and pituitary gland. J. Clin. Pathol., Suppl., 7:1–7.

Dearden, N. M., and R. L. Holmes. 1976. Cyto-differentiation and portal vascular development in the mouse adenohypophysis. J. Anat. (Lond.), 121:551–569.

Dey, F. L. 1943. Evidence of hypothalamic control of hypophyseal gonadotrophic functions in the female guinea pig. Endocrinology, 33:75–82.

Duncan, D. 1956. An electron microscope study of the neurohypophysis of a bird, *Gallus domesticus*. Anat. Rec., 125:457–471.

Etkin, W. 1943. The developmental control of pars intermedia by brain. J. Exp. Zool., 92:31–47.

Farquhar, M. G. 1961. Fine structure and function in capillaries of the anterior pituitary gland. Angiology, 12:270–292.

Fisher, C., W. R. Ingram, W. K. Hare, and S. W. Ranson. 1935. The degeneration of the supraoptico-hypophyseal system in diabetes insipidus. Anat. Rec., 63:29–52.

Green, H. T. 1957. The venous drainage of the human hypophysis cerebri. Amer. J. Anat., 100:435–469.

Green, J. D. 1948. The histology of the hypophysial stalk and median eminence in man with special reference to blood vessels, nerve fibers and a peculiar neurovascular zone in this region. Anat. Rec., 100:273–295.

Green, J. D. 1951. The comparative anatomy of the hypophysis, with special reference to its blood supply and innervation. Amer. J. Anat., 88:225–311.

Green, J. D., and G. W. Harris. 1949. Observation of the hypophysio-portal vessels of the living rat. J. Physiol., 108:359–361.

Guillemin, R., and R. Burgus. 1972. The hormones of the hypothalamus. Sci. Amer., 227:24–33.

Hair, G. W. 1938. The nerve supply of the hypophysis of the cat. Anat. Rec., 71:141–160.

Halmi, N. S. 1952. Two types of basophils in the rat pituitary: "Thyrotrophs" and "gonadotrophs" vs. beta and delta cells. Endocrinology, 50:140–142.

Harris, G. W. 1955. *Neural Control of the Pituitary Gland*. Williams & Wilkins Co. Baltimore, 298 pp.

Harris, G. W., and B. T. Donovan (eds.). 1966. *The Pituitary Gland*. Butterworth, London. 3 vols.

Hegre, E. S. 1946. The developmental stage at which the intermediate lobe of the hypophysis becomes determined. J. Exp. Zool., 103:321–333.

Herbert, D. C., and T. Hayashida. 1970. Prolactin localization in the primate pituitary by immunofluorescence. Science, 169:378–379.

Herlant, M. 1964. The cells of the adenohypophysis and their functional significance. Internat. Rev. Cytol., 17:299–382.

Holmes, R. L., and S. Zuckerman. 1959. The blood supply of the hypophysis in *Macaca mulatta*. J. Anat. (Lond.), 93:1–8.

Hunt, T. E. 1949. Mitotic activity in the hypophysis of the rat during pregnancy and lactation. Anat. Rec., 105:361–373.

Hunt, T. E. 1951. The effect of hypophyseal extract on mitotic activity of the rat hypophysis. Anat. Rec., 111:713–725.

Hymer, W. C., and W. H. McShan. 1963. Isolation of rat pituitary granules and the study of their biochemical properties and hormonal activities. J. Cell Biol., 17:67–86.

Ifft, J. D. 1953. The effect of superior cervical ganglionectomy on the cell population of the rat adenohypophysis and on the estrous cycle. Anat. Rec., *117*:395–404.

Ingram, W. R., and C. Fisher. 1936. The relation of the posterior pituitary to water exchange in the cat. Anat. Rec., *66*:271–293.

King, J. C., T. H. Williams, and A. A. Arimura. 1975. Localization of luteinizing hormone-releasing hormone in rat hypothalamus using radioimmunoassay. J. Anat. (Lond.), *120*:275–288.

von Lawzewitsch, I., G. H. Dickmann, L. Amezúa, and C. Pardal. 1972. Cytological and ultrastructural characterization of the human pituitary. Acta Anat., *81*:286–316.

von Lawzewitsch, I., and R. Sarrat. 1972. Comparative anatomy and the evolution of the neurosecretory hypothalamic-hypophyseal system. Acta Anat., *81*:13–22.

Leclercq, T. A., and F. Grisoli. 1983. Arterial blood supply of the normal human pituitary gland. An anatomical study. J. Neurosurg., *58*:678–681.

Leininger, C. R., and S. W. Ranson. 1943. The effect of hypophysial stalk transection upon gonadotrophic function in the guinea pig. Anat. Rec., *87*:77–83.

Markee, J. E., C. H. Sawyer, and W. H. Hollinshead. 1946. Activation of the anterior hypophysis by electrical stimulation in the rabbit. Endocrinology, *38*:345–357.

Markee, J. E., C. H. Sawyer, and W. H. Hollinshead. 1948. Adrenergic control of the release of luteinizing hormone from the hypophysis of the rabbit. Recent Prog. Hormone Res., *2*:117–131.

McConnell, E. M. 1953. The arterial blood supply of the human hypophysis cerebri. Anat. Rec., *115*:175–203.

McGrath, P. 1971. The volume of the human pharyngeal hypophysis in relation to age and sex. J. Anat. (Lond.), *110*:275–282.

McGrath, P. 1976. Further observations on the pharyngeal hypophysis and the postsphenoid in the mature male rat. J. Anat. (Lond.), *121*:193–201.

McNary, W. F., Jr. 1957. Progressive cytological changes in the hypophysis associated with endocrine interaction following exposure to cold. Anat. Rec., *128*:233–253.

Melchionna, R. H., and R. A. Moore. 1938. The pharyngeal pituitary gland. Amer. J. Pathol., *14*:763–773.

Messier, B. 1965. Number and distribution of thyrotropic cells in the mouse pituitary gland. Anat. Rec., *153*:343–348.

Nakane, P. K. 1970. Classifications of anterior pituitary cell types with immunoenzyme histochemistry. J. Histochem. Cytochem., *18*:9–20.

Oboussier, H. 1979. Problems of body structure and size of the hypophysis. Acta Anat., *104*:374–381.

Palay, S. L. 1953. Neurosecretory phenomena in the hypothalamo–hypophysial system of man and monkey. Amer. J. Anat., *93*:107–141.

Pearse, A. G. E. 1952. Observations on the localisation, nature and chemical constitution of some components of the anterior hypophysis. J. Pathol. Bacteriol., *64*:791–809.

Phifer, R. F., S. S. Spicer, and D. N. Orth. 1970. Specific demonstration of the human hypophyseal cells which produce adrenocorticotropic hormone. J. Clin. Endocrinol., *31*:347–361.

Popa, G., and U. Fielding. 1930. A portal circulation from the pituitary to the hypothalamic region. J. Anat. (Lond.), *65*:88–91.

Popa, G. T., and U. Fielding. 1930. The vascular link between the pituitary and the hypothalamus. Lancet, *2*:238–240.

Purves, H. D., and W. E. Griesbach. 1951. The significance of the Gomori staining of the basophils of the rat pituitary. Endocrinology, *49*:652–662.

Rasmussen, A. T. 1936. The proportions of the various subdivisions of the normal adult human hypophysis cerebri, etc. In *The Pituitary Gland,* W. Timme, ed. Res. Publ. Assoc. Nerv. Ment. Dis., Vol. 17. Williams & Wilkins Co., Baltimore, pp. 118–150.

Rinehart, J. F., and M. G. Farquhar. 1955. The fine vascular organization of the anterior pituitary gland. Anat. Rec., *121*:207–239.

Salazar, H., and R. R. Peterson. 1964. Morphologic observations concerning the release and transport of secretory products in the adenohypophysis. Amer. J. Anat., *115*:199–215.

Sawyer, C. H. 1947. Cholinergic stimulation of the release of melanophore hormone by the hypophysis in salamander larvae. J. Exp. Zool., *106*:145–179.

Sawyer, C. H., J. E. Markee, and W. H. Hollinshead. 1947. Inhibition of ovulation in the rabbit by the adrenergic-blocking agent dibenamine. Endocrinology, *41*:395–402.

Scharrer, E., and B. Scharrer. 1945. Neurosecretion. Physiol. Rev., *25*:171–181.

Scharrer, E. A., and G. J. Wittenstein. 1952. The effect of the interruption of the hypothalamo-hypophyseal neurosecretory pathway in the dog. Anat. Rec., *112*:387. Abstract.

Shanklin, W. M. 1951. The histogenesis and histology of an integumentary type of epithelium in the human hypophysis. Anat. Rec., *109*:217–231.

Stanfield, J. P. 1960. The blood supply of the human pituitary gland. J. Anat. (Lond.), *94*:257–273.

Szentágothai, J., B. Flerkó, B. Mess, and B. Halász. 1968. *Hypothalamic Control of the Anterior Pituitary.* 3rd ed. Akadémiai Kiadó, Budapest, 398 pp.

Wislocki, G. B. 1937. The meningeal relations of the hypophysis cerebri. II. An embryological study of the meninges and blood vessels of the human hypophysis. Amer. J. Anat., *61*:95–129.

Wislocki, G. B. 1938. The vascular supply of the hypophysis cerebri of the rhesus monkey and man. In *The Pituitary Gland,* W. Timme, ed. Res. Publ. Assoc. Nerv. Ment. Dis., Vol. 17. Williams & Wilkins Co., Baltimore, pp. 48–68.

Xuereb, G. P., M. M. L. Prichard, and P. M. Daniel. 1954. The arterial supply and venous drainage of the human hypophysis cerebri. Quart. J. Exp. Physiol., *39*:199–218.

Xuereb, G. P., M. M. L. Prichard, and P. M. Daniel. 1954. The hypophysial portal system of vessels in man. Quart. J. Exp. Physiol., *39*:219–227.

Zeitlin, H., and E. Oldberg. 1940. Craniopharyngioma in the third ventricle of the brain: Partial surgical removal and pathologic study. Arch. Neurol. Psychiat., *43*:1195–1204.

Suprarenal Glands

Alden, R. H., and J. S. Davis. 1962. Role of adrenals in uterine lipid metabolism. Anat. Rec., *142*:53–56.

Al-Lami, F. 1969. Light and electron microscopy of the adrenal medulla of *Macaca mulatta* monkey. Anat. Rec., *164*:317–332.

Al-Lami, F. 1970. Follicular arrangements in hamster adrenomedullary cells: Light and electron microscopic study. Anat. Rec., *168*:161–177.

Anson, B. J., E. W. Cauldwell, J. W. Pick, and L. E. Beaton. 1947. The blood supply of the kidney, suprarenal gland, and associated structures. Surg. Gyne-col. Obstet., *84*:313–320.

Baker, D. D., and R. N. Baillif. 1939. Role of capsule in suprarenal regeneration studied with aid of colchicine. Proc. Soc. Exp. Biol. Med., *40*:117–121.

Bartlett, C. J. 1915. Direct union between adrenals and kidneys (subcapsular location of adrenals). Anat. Rec., *10*:67–77.

Belt, A. E., and T. O. Powell. 1934. Clinical manifestations of the chromaffin cell tumors arising from the suprarenal medulla: Suprarenal sympathetic syndrome. Surg. Gynecol. Obstet., *59*:9–24.

Bennett, H. S. 1940. The life history and secretion of the cells of the adrenal cortex of the cat. Amer. J. Anat., *67*:151–227.

Bennett, H. S. 1941. Cytological manifestations of secretion in the adrenal medulla of the cat. Amer. J. Anat., *69*:333–381.

Bloodworth, J. M. B., and K. L. Powers. 1968. The ultrastructure of the normal dog adrenal. J. Anat. (Lond.), *102*:457–476.

Bourne, G. H. 1949. *The Mammalian Adrenal Gland.* Clarendon Press, Oxford, 239 pp.

Brown, W. J., L. Barajas, and H. Latta. 1971. The ultrastructure of the human adrenal medulla: With comparative studies of white rat. Anat. Rec., *169*:173–184.

Cahill, G. F. 1935. Air injections to demonstrate the adrenals by X-ray. J. Urol., *34*:238–243.

Cahill, G. F., M. M. Melicow, and H. H. Darby. 1942. Adrenal cortical tumors. The types of nonhormonal and hormonal tumors. Surg. Gynecol. Obstet., *74*:281–305.

Carmichael, S. W., D. E. Haines, and C. A. Pinkstaff. 1975. The suprarenal glands of a prosimian primate, the lesser bushbaby (*Galago senegalensis*). J. Morphol., *145*:239–249.

Cerny, J. C. 1977. Anatomy of the adrenal gland. Urol. Clin. North Amer., *4*:169–177.

Cori, C. F., and A. D. Welch. 1942. The adrenal medulla. In *Glandular Physiology and Therapy.* American Medical Association, Chicago, pp. 307–326.

Coupland, R. E. 1965. Electron microscopic observations on the structure of the rat adrenal medulla. I. The ultrastructure and organization of chromaffin cells in the normal adrenal medulla. J. Anat. (Lond.), *99*:231–254.

Coupland, R. E. 1965. Electron microscopic observations on the structure of the rat adrenal medulla. II. Normal innervation. J. Anat. (Lond.), *99*:255–272.

Crowder, R. E. 1957. The development of the adrenal gland in man, with special reference to origin and ultimate location of cell types, and evidence in favor of the "cell migration" theory. Carneg. Inst., Contr. Embryol., *36*:193–210.

Culp, O. S. 1939. Adrenal heterotopia: A survey of the literature and report of a case. J. Urol., *41*:303–309.

Currie, A. R., T. Symington, and J. K. Grant (eds.). 1962. *The Human Adrenal Cortex.* Williams & Wilkins Co., Baltimore, 643 pp.

Deane, H. W., and R. O. Greep. 1946. A morphological and histochemical study of the rat's adrenal cortex after hypophysectomy, with comments on the liver. Amer. J. Anat., *79*:117–145.

Denber, H. C. B. 1947. The question of regeneration of nerve fibers to the human adrenal gland after bilateral sympathectomy. Ann. Surg., *126*:332–339.

Dolishnii, N. V. 1965. Plastic properties of the arteries of the adrenals. Fed. Proc., *24*:T941–T944.

Dougherty, T. F., and A. White. 1945. Functional alterations in lymphoid tissue induced by adrenal cortical secretion. Amer. J. Anat., *77*:81–116.

Eisenstein, A. B. (ed.). 1967. *The Adrenal Cortex.* Little, Brown & Co., Boston, 685 pp.

Elfvin, L.-G. 1965. The fine structure of the cell surface of chromaffin cells in the rat adrenal medulla J. Ultrastruct. Res., *12*:263–286.

von Euler, U. S. 1951. Hormones of the sympathetic nervous system and the adrenal medulla. Brit. Med. J., *1*:105–108.

Everett, N. B. 1949. Autoplastic and homoplastic transplants of the rat adrenal cortex and medulla to the kidney. Anat. Rec., *103*:335–347.

Feldman, J. D. 1950. Histochemical reactions of adrenal cortical cells. Anat. Rec., *107*:347–358.

Feldman, J. D. 1951. Endocrine control of the adrenal gland. Anat. Rec., *109*:41–69.

Freeman, N. E., R. H. Smithwick, and J. C. White. 1934. Adrenal secretion in man. Amer. J. Physiol., *107*:529–534.

Gagnon, R. 1956. The venous drainage of the human adrenal gland. Rev. Canad. Biol., *14*:350–359.

Harrison, R. G., and M. J. Hoey. 1960. *The Adrenal Circulation.* Blackwell Scientific Publications, Oxford, 77 pp.

Hench, P. S., E. C. Kendall, C. H. Slocumb, and H. F. Polley. 1949. The effect of a hormone of the adrenal cortex (17 hydroxy-11-dehydrocorticosterone: Compound E) and of pituitary adrenocorticotropic hormone on rheumatoid arthritis. Ann. Rheum. Dis., *8*:97–104.

Hoar, R. M., and A. J. Salem. 1962. The production of congenital malformations in guinea pigs by adrenalectomy. Anat. Rec., *143*:157–167.

Hoerr, N. L. 1936. Histological studies on lipins. II. A cytological analysis of the liposomes in the adrenal cortex of the guinea pig. Anat. Rec., *66*:317–342.

Hollinshead, W. H. 1936. The innervation of the adrenal glands. J. Comp. Neurol., *64*:449–467.

Hollinshead, W. H., and H. Finkelstein. 1937. Regeneration of nerves to the adrenal gland. J. Comp. Neurol., *67*:215–220.

Holmes, W. N. 1955. Histological variations in the adrenal cortex of the golden hamster with special reference to the x zone. Anat. Rec., *122*:271–293.

Houser, R. G., F. A. Hartman, R. A. Knouff, and F. W. McCoy. 1962. Adrenals in some Panama monkeys. Anat. Rec., *142*:41–51.

Hunt, T. E., and E. A. Hunt. 1964. The proliferative activity of the adrenal cortex using a radioautographic technic with Thymidine-H[3]. Anat. Rec., *149*:387–395.

Idelman, S. 1970. Ultrastructure of the mammalian adrenal cortex. Internat. Rev. Cytol., *27*:181–281.

Jayne, E. P. 1953. Cytology of the adrenal gland of the rat at different ages. Anat. Rec., *115*:459–483.

Johannisson, E. 1968. The foetal adrenal cortex in the human. Its ultrastructure at different stages of development and in different functional states. Acta Endocrinol., *58* (Suppl. 130), 107 pp.

Jones, I. C. 1957. *The Adrenal Cortex.* Cambridge University Press, Cambridge, 315 pp.

Keene, M. F. L., and E. E. Hewer. 1927. Observations on the development of the human suprarenal gland. J. Anat. (Lond.), *61*:302–324.

Kitchell, R. L., and L. J. Wells. 1952. Functioning of the hypophysis and adrenals in fetal rats: Effects of hypophysectomy, adrenalectomy, castration, injected ACTH and implanted sex hormones. Anat. Rec., *112*:561–591.

Knouff, R. A., J. B. Brown, and B. M. Schneider. 1941. Correlated chemical and histological studies of the adrenal lipids. I. The effect of extreme muscular activity on the adrenal lipids of the guinea pig. Anat. Rec., *79*:17–38.

Lever, J. D. 1955. Electron microscopic observations on the adrenal cortex. Amer. J. Anat., *97*:409–429.

Levy, S. E., and A. Blalock. 1939. A method for transplanting the adrenal gland of the dog with reestablishment of its blood supply: Report of observations. Ann. Surg., *109*:84–98.

Long, C. N. H. 1956. Pituitary-adrenal relationships. Ann. Rev. Physiol., *18*:409–432.

Long, J. A., and A. L. Jones. 1967. Observations on the fine structure of the adrenal cortex of man. Lab. Invest., *17*:355–370.

Long, J. A., and A. L. Jones. 1970. Alterations in the fine structure of the oppossum adrenal cortex following sodium deprivation. Anat. Rec., *166*:1–26.

MacFarland, W. E., and H. A. Davenport. 1941. Adrenal innervation. J. Comp. Neurol., *75*:219–233.

MacKinnon, I. L., and P. C. B. MacKinnon. 1958. Seasonal rhythm in the morphology of the suprarenal cortex in women of child-bearing age. J. Endocrinol., *17*:456–462.

MacKinnon, P. C. B., and I. L. MacKinnon. 1960. Morphologic features of the human suprarenal cortex in men aged 20–86 years. J. Anat. (Lond.), *94*:183–191.

Merklin, R. J. 1962. Arterial supply of the suprarenal gland. Anat. Rec., *144*:359–371.

Mikhail, Y., and Z. Mahran. 1965. Innervation of the cortical and medullary portions of the adrenal gland of the rat during postnatal life. Anat. Rec., *152*:431–437.

Moon, H. D. (ed.). 1961. *The Adrenal Cortex*. Paul B. Hoeber, New York, 315 pp.

Orda, R., and Z. Rudberg. 1976. The adreno-renal-ureteral sheath: Surgical-anatomical study. Urol. Internat., *31*:179–182.

Pohorecky, L. A., and R. J. Wurtman. 1971. Adrenocortical control of epinephrine synthesis. Pharmacol. Rev., *23*:1–35.

Priestley, J. T. 1952. Lesions of the adrenal glands. Surg. Clin. North Amer., *32*:1053–1064.

Prunty, F. T. G. 1962. The adrenal cortex: Introduction. Brit. Med. Bull., *18*:89–91.

Reiter, R. J., and R. A. Hoffman. 1967. Adrenocortical cytogenesis in the adult male golden hamster. A radioautographic study using tritiated-thymidine. J. Anat. (Lond.), *101*:723–729.

Salmon, T. N., and R. L. Zwemer. 1941. A study of the life history of cortico-adrenal gland cells of the rat by means of trypan blue injections. Anat. Rec., *80*:421–429.

Schaberg, A. 1955. Regeneration of the adrenal cortex in vitro. Anat. Rec., *122*:205–221.

Shelton, J. H., and A. L. Jones. 1971. The fine structure of the mouse adrenal cortex and the ultrastructural changes in the zona glomerulosa with low and high sodium diets. Anat. Rec., *170*:147–182.

Sheridan, M. N., and W. D. Belt. 1964. Fine structure of the guinea pig adrenal cortex. Anat. Rec., *149*:73–97.

Swinyard, C. A. 1937. The innervation of the suprarenal glands. Anat. Rec., *68*:417–429.

Swinyard, C. A. 1940. Volume and cortico-medullary ratio of the adult human suprarenal gland. Anat. Rec., *76*:69–79.

Swinyard, C. A. 1943. Growth of the human suprarenal glands. Anat. Rec., *87*:141–150.

Uotila, U. U. 1940. The early embryological development of the fetal and permanent adrenal cortex in man. Anat. Rec., *76*:183–203.

Whiteley, H. J., and H. B. Stoner. 1957. The effect of pregnancy on the human adrenal cortex. J. Endocrinol., *14*:325–334.

Williams, R. G. 1947. Studies of adrenal cortex: Regeneration of the transplanted gland and the vital quality of autogenous grafts. Amer. J. Anat., *81*:199–231.

Wislocki, G. B., and S. J. Crowe. 1924. Experimental observations on the adrenals and the chromaffin system. Bull. Johns Hopkins Hosp., *35*:187–193.

Wotton, R. M., and R. L. Zwemer. 1943. A study of the cytogenesis of cortico-adrenal cells in the cat. Anat. Rec., *86*:409–416.

Yates, R. D. 1964. A light and electron microscopic study correlating the chromaffin reaction and granule ultrastructure in the adrenal medulla of the Syrian hamster. Anat. Rec., *149*:237–249.

Yates, R. D., J. G. Wood, and D. Duncan. 1962. Phase and electron microscopic observations on two cell types in the adrenal medulla of the Syrian hamster. Tex. Rep. Biol. Med., *20*:494–502.

Young, J. Z. 1939. Partial degeneration of the nerve supply of the adrenal. A study in autonomic innervation. J. Anat. (Lond.), *73*:540–550.

Zwemer, R. L., and R. Truszkowski. 1937. The importance of corticoadrenal regulation of potassium metabolism. Endocrinology, *21*:40–49.

Paraganglia, Glomera and Bodies

Adams, W. E. 1958. *The Comparative Morphology of the Carotid Body and Carotid Sinus*. Charles C Thomas, Springfield, Ill., 272 pp.

Al-Lami, F., and R. G. Murray. 1968. Fine structure of the carotid body of normal and anoxic cats. Anat. Rec., *160*:697–718.

Biscoe, T. J. 1971. Carotid body: Structure and function. Physiol. Rev., *51*:437–495.

Biscoe, T. J., and W. E. Stehbens. 1965. Electron microscopic observations on the carotid body. Nature, *208*:708–709.

Boyd, J. D. 1937. The development of the human carotid body. Carneg. Inst., Contr. Embryol., *26*:1–31.

Comroe, J. H., Jr. 1939. The location and function of the chemoreceptors of the aorta. Amer. J. Physiol., *127*:176–191.

Coupland, R. E. 1952. The prenatal development of the abdominal para-aortic bodies in man. J. Anat. (Lond.), *86*:357–372.

Coupland, R. E. 1954. Post-natal fate of the abdominal para-aortic bodies in man. J. Anat. (Lond.), *88*:455–464.

Coupland, R. E. 1965. *The Natural History of the Chromaffin Cell*. Longmans, London, 279 pp.

Deane, B. M., A. Howe, and M. Morgan. 1975. Abdominal vagal paraganglia: Distribution and comparison with carotid body, in the rat. Acta Anat., *93*:19–28.

Hervonen, A., S. Partanen, A. Vaalasti, M. Partanen, L. Kanerva, and H. Alho. 1978. The distribution and endocrine nature of the abdominal paraganglia of adult man. Amer. J. Anat., *153*:563–572.

Hollinshead, W. H. 1937. The innervation of the abdominal chromaffin tissue. J. Comp. Neurol., *67*:133–143.

Hollinshead, W. H. 1940. Chromaffin tissue and paraganglia. Quart. Rev. Biol., *15:*156–171.

Hollinshead, W. H. 1940. The innervation of the supracardial bodies in the cat. J. Comp. Neurol., *73:*37–48.

Hollinshead, W. H. 1943. A cytological study of the carotid body of the cat. Amer. J. Anat., *73:*185–213.

Hollinshead, W. H. 1945. Effects of anoxia upon carotid body morphology. Anat. Rec., *92:*255–261.

Lever, J. D., P. R. Lewis, and J. D. Boyd. 1959. Observations on the fine structure and histochemistry of the carotid body in the cat and rabbit. J. Anat. (Lond.), *93:*478–490.

Nonidez, J. F. 1936. Observations on the blood supply and the innervation of the aortic paraganglion of the cat. J. Anat. (Lond.), *70:*215–224.

Partanen, M., C. C. Chiueh, and S. I. Rapoport. 1981. Age related increase in catecholamine-containing paraganglia in male Fischer-344 rats. Anat. Rec., *201:*563–566.

Phillips, J. R. 1947. Paroxysmal hypertension due to paraganglioma. Amer. J. Surg., *73:*111–115.

Schmidt, C. F., and J. H. Comroe, Jr. 1940. Functions of the carotid and aortic bodies. Physiol. Rev., *20:*115–157.

Smith, C. A. 1948. Paraganglioma (pheochromocytoma): Case report. J. Urol., *60:*697–701.

Thompson, S. A., and J. A. Gosling. 1976. Histochemical light microscopic study of catecholamine containing paraganglia in the human pelvis. Cell Tissue Res., *170:*539–548.

Willis, A. G., and J. D. Tange. 1959. Studies on the innervation of the carotid sinus of man. Amer. J. Anat., *104:*87–113.

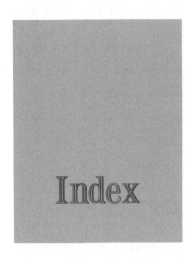

Index